DISCARD
Otterbein University
Courtright Memorial Library
Courtright Memorial Library
Otterbein College
Westerville, Ohio 43081

Springer Handbook
of Acoustics

Springer Handbooks provide a concise compilation of approved key information on methods of research, general principles, and functional relationships in physical sciences and engineering. The world's leading experts in the fields of physics and engineering will be assigned by one or several renowned editors to write the chapters comprising each volume. The content is selected by these experts from Springer sources (books, journals, online content) and other systematic and approved recent publications of physical and technical information.

The volumes are designed to be useful as readable desk reference books to give a fast and comprehensive overview and easy retrieval of essential reliable key information, including tables, graphs, and bibliographies. References to extensive sources are provided.

Springer Handbook
of Acoustics

Thomas D. Rossing (Ed.)

With CD-ROM, 962 Figures and 91 Tables

Editor:

Thomas D. Rossing
Stanford University
Center for Computer Research in Music and Acoustics
Stanford, CA 94305, USA

Editorial Board:

Manfred R. Schroeder, University of Göttingen, Germany
William M. Hartmann, Michigan State University, USA
Neville H. Fletcher, Australian National University, Australia
Floyd Dunn, University of Illinois, USA
D. Murray Campbell, The University of Edinburgh, UK

Library of Congress Control Number: 2006927050

ISBN: 978-0-387-30446-5   e-ISBN: 0-387-30425-0
Printed on acid free paper

© 2007, Springer Science+Business Media, LLC New York
All rights reserved. This work may not be translated or copied in whole or in part without the written permission of the publisher (Springer Science+Business Media, LLC New York, 233 Spring Street, New York, NY 10013, USA), except for brief excerpts in connection with reviews or scholarly analysis. Use in connection with any form of information storage and retrieval, electronic adaptation, computer software, or by similar or dissimilar methodology now known or hereafter developed is forbidden. The use in this publication of trade names, trademarks, service marks, and similar terms, even if they are not identified as such, is not to be taken as an expression of opinion as to whether or not they are subject to proprietary rights.

The use of designations, trademarks, etc. in this publication does not imply, even in the absence of a specific statement, that such names are exempt from the relevant protective laws and regulations and therefore free for general use.

Product liability: The publisher cannot guarantee the accuracy of any information about dosage and application contained in this book. In every individual case the user must check such information by consulting the relevant literature.

Production and typesetting: LE-TeX GbR, Leipzig
Handbook Manager: Dr. W. Skolaut, Heidelberg
Typography and layout: schreiberVIS, Seeheim
Illustrations:
schreiberVIS, Seeheim & Hippmann GbR, Schwarzenbruck
Cover design: eStudio Calamar Steinen, Barcelona
Cover production: WMXDesign GmbH, Heidelberg
Printing and binding: Stürtz AG, Würzburg

SPIN 11309031   100/3100/YL   5 4 3 2 1 0

Foreword

The present handbook covers a very wide field. Its 28 chapters range from the history of acoustics to sound propagation in the atmosphere; from nonlinear and underwater acoustics to thermoacoustics and concert hall acoustics. Also covered are musical acoustics, including computer and electronic music; speech and singing; animal (including whales) communication as well as bioacoustics in general, psychoacoustics and medical acoustics. In addition, there are chapters on structural acoustics, vibration and noise, including optical methods for their measurement; microphones, their calibration, and microphone and hydrophone arrays; acoustic holography; model analysis and much else needed by the professional engineer and scientist.

Among the authors we find many illustrious names: Yoichi Ando, Mack Breazeale, Babrina Dunmire, Neville Fletcher, Anders Gade, William Hartmann, William Kuperman, Werner Lauterborn, George Maling, Brian Moore, Allan Pierce, Thomas Rossing, Johan Sundberg, Eric Young, and many more. They hail from countries around the world: Australia, Canada, Denmark, France, Germany, Japan, Korea, Sweden, the United Kingdom, and the USA. There is no doubt that this handbook will fill many needs, nay be irreplaceable in the art of exercising today's many interdisciplinary tasks devolving on acoustics. No reader could wish for a wider and more expert coverage. I wish the present tome the wide acceptance and success it surely deserves.

Prof. Dr. M. R. Schroeder
University Professor
Speach and Acoustics Laboratory
University of Göttingen,
Germany

Göttingen, March 2007   Manfred R. Schroeder

Preface

"A handbook," according to the dictionary, "is a book capable of being conveniently carried as a ready reference." Springer has created the Springer Handbook series on important scientific and technical subjects, and we feel fortunate that they have included acoustics in this category.

Acoustics, the science of sound, is a rather broad subject to be covered in a single handbook. It embodies many different academic disciplines, such as physics, mechanical and electrical engineering, mathematics, speech and hearing sciences, music, and architecture. There are many technical areas in acoustics; the Acoustical Society of America, for example, includes 14 technical committees representing different areas of acoustics. It is impossible to cover all of these areas in a single handbook. We have tried to include as many as possible of the "hot" topics in this interdisciplinary field, including basic science and technological applications. We apologize to the reader whose favorite topics are not included.

Prof. em. T. D. Rossing
Northern Illinois University
Presently visiting Professor
of Music at Stanford University

We have grouped the 28 chapters in the book into eight parts: Propagation of Sound; Physical and Nonlinear Acoustics; Architectural Acoustics; Hearing and Signal Processing; Music, Speech, and Electroacoustics; Biological and Medical Acoustics; Structural Acoustics and Noise; and Engineering Acoustics. The chapters are of varying length. They also reflect the individual writing styles of the various authors, all of whom are authorities in their fields. Although an attempt was made to keep the mathematical level of the chapters as even as possible, readers will note that some chapters are more mathematical than others; this is unavoidable and in fact lends some degree of richness to the book.

We are indebted to many persons, especially Werner Skolaut, the manager of the Springer Handbooks, and to the editorial board, consisting of Neville Fletcher, Floyd Dunn, William Hartmann, and Murray Campbell, and for their advice. Each chapter was reviewed by two authoritative reviewers, and we are grateful to them for their services. But most of all we thank the authors, all of whom are busy people but devoted much time to carefully preparing their chapters.

Stanford, April 2007   Thomas D. Rossing

List of Authors

Iskander Akhatov
North Dakota State University
Center for Nanoscale Science and Engineering,
Department of Mechanical Engineering
111 Dolve Hall
Fargo, ND 58105-5285, USA
e-mail: iskander.akhatov@ndsu.edu

Yoichi Ando
3917-680 Takachiho
899-6603 Makizono, Kirishima, Japan
e-mail: andoy@cameo.plala.or.jp

Keith Attenborough
The University of Hull
Department of Engineering
Cottingham Road
Hull, HU6 7RX, UK
e-mail: k.attenborough@hull.ac.uk

Whitlow W. L. Au
Hawaii Institute of Marine Biology
P.O. Box 1106
Kailua, HI 96734, USA
e-mail: wau@hawaii.edu

Kirk W. Beach
University of Washington
Department of Surgery
Seattle, WA 98195, USA
e-mail: kwbeach@u.washington.edu

Mack A. Breazeale
University of Mississippi
National Center for Physical Acoustics
027 NCPA Bldg.
University, MS 38677, USA
e-mail: breazeal@olemiss.edu

Antoine Chaigne
Unité de Mécanique (UME)
Ecole Nationale Supérieure de Techniques
Avancées (ENSTA)
Chemin de la Hunière
91761 Palaiseau, France
e-mail: antoine.chaigne@ensta.fr

Perry R. Cook
Princeton University
Department of Computer Science
35 Olden Street
Princeton, NJ 08540, USA
e-mail: prc@cs.princeton.edu

James Cowan
Resource Systems Group Inc.
White River Junction, VT 85001, USA
e-mail: jcowan@rsginc.com

Mark F. Davis
Dolby Laboratories
100 Potrero Ave,
San Francisco, CA 94103, USA
e-mail: mfd@dolby.com

Barbrina Dunmire
University of Washington
Applied Physics Laboratory
1013 NE 40th Str.
Seattle, WA 98105, USA
e-mail: mrbean@u.washington.edu

Neville H. Fletcher
Australian National University
Research School of Physical Sciences and
Engineering
Canberra, ACT 0200, Australia
e-mail: neville.fletcher@anu.edu.au

Anders Christian Gade
Technical University of Denmark
Acoustic Technology, Oersted.DTU
Building 352
2800 Lyngby, Denmark
e-mail: acg@oersted.dtu.dk,
anders.gade@get2net.dk

Colin Gough
University of Birmingham
School of Physics and Astronomy
Birmingham, B15 2TT, UK
e-mail: c.gough@bham.ac.uk

William M. Hartmann
Michigan State University
1226 BPS Building
East Lansing, MI 48824, USA
e-mail: *hartmann@pa.msu.edu*

Finn Jacobsen
Ørsted DTU, Technical University of Denmark
Acoustic Technology
Ørsteds Plads, Building 352
2800 Lyngby, Denmark
e-mail: *fja@oersted.dtu.dk*

Yang-Hann Kim
Korea Advanced Institute of Science
and Technology (KAIST)
Department of Mechanical Engineering
Center for Noise and Vibration Control (NOVIC)
Acoustics and Vibration Laboratory
373-1 Kusong-dong, Yusong-gu
Daejeon, 305-701, Korea
e-mail: *yanghannkim@kaist.ac.kr*

William A. Kuperman
University of California at San Diego
Scripps Institution of Oceanography
9500 Gilman Drive
La Jolla, CA 92093-0701, USA
e-mail: *wkuperman@ucsd.edu*

Thomas Kurz
Universität Göttingen
Friedrich-Hund-Platz 1
37077 Göttingen, Germany
e-mail: *t.kurz@dpi.physik.uni-goettingen.de*

Marc O. Lammers
Hawaii Institute of Marine Biology
P.O. Box 1106
Kailua, HI 96734, USA
e-mail: *lammers@hawaii.edu*

Werner Lauterborn
Universität Göttingen
Drittes Physikalisches Institut
Friedrich-Hund-Platz 1
37077 Göttingen, Germany
e-mail:
w.lauterborn@dpi.physik.uni-goettingen.de

Björn Lindblom
Stockholm University
Department of Linguistics
10691 Stockholm, Sweden
e-mail: *lindblom@ling.su.se,
blindblom@mail.utexas.edu*

George C. Maling, Jr.
Institute of Noise Control Engineering of the USA
60 High Head Road
Harpswell, ME 04079, USA
e-mail: *maling@alum.mit.edu*

Michael McPherson
The University of Mississippi
Department of Physics and Astronomy
123 Lewis Hall
P. O. Box 1848
University, MS 38677, USA
e-mail: *mcph@phy.olemiss.edu*

Nils-Erik Molin
Luleå University of Technology
Experimental Mechanics
SE-971 87 Luleå, Sweden
e-mail: *nem@ltu.se*

Brian C. J. Moore
University of Cambridge
Department of Experimental Psychology
Downing Street
Cambridge, CB2 3EB, UK
e-mail: *bcjm@cam.ac.uk*

Alan D. Pierce
Boston University
College of Engineering
Boston, MA , USA
e-mail: *adp@bu.edu*

Thomas D. Rossing
Stanford University
Center for Computer Research in Music and
Acoustics (CCRMA)
Department of Music
Stanford, CA 94305, USA
e-mail: *rossing@ccrma.stanford.edu*

Philippe Roux
Université Joseph Fourier
Laboratoire de Geophysique Interne et
Tectonophysique
38041 Grenoble, France
e-mail: *philippe.roux@obs.ujf-grenoble.fr*

Johan Sundberg
KTH–Royal Institute of Technology
Department of Speech, Music, and Hearing
SE-10044 Stockholm, Sweden
e-mail: *pjohan@speech.kth.se*

Gregory W. Swift
Los Alamos National Laboratory
Condensed Matter and Thermal Physics Group
Los Alamos, NM 87545, USA
e-mail: *swift@lanl.gov*

George S. K. Wong
Institute for National Measurement Standards
(INMS)
National Research Council Canada (NRC)
1200 Montreal Road
Ottawa, ON K1A 0R6, Canada
e-mail: *George.Wong@nrc-cnrc.gc.ca*

Eric D. Young
Johns Hopkins University
Baltimore, MD 21205, USA
e-mail: *eyoung@jhu.edu*

Contents

List of Abbreviations XXI

1 Introduction to Acoustics
Thomas D. Rossing 1
1.1 Acoustics: The Science of Sound 1
1.2 Sounds We Hear 1
1.3 Sounds We Cannot Hear: Ultrasound and Infrasound 2
1.4 Sounds We Would Rather Not Hear: Environmental Noise Control 2
1.5 Aesthetic Sound: Music 3
1.6 Sound of the Human Voice: Speech and Singing 3
1.7 How We Hear: Physiological and Psychological Acoustics 4
1.8 Architectural Acoustics 4
1.9 Harnessing Sound: Physical and Engineering Acoustics 5
1.10 Medical Acoustics 5
1.11 Sounds of the Sea 6
References 6

Part A Propagation of Sound

2 A Brief History of Acoustics
Thomas D. Rossing 9
2.1 Acoustics in Ancient Times 9
2.2 Early Experiments on Vibrating Strings, Membranes and Plates .. 10
2.3 Speed of Sound in Air 10
2.4 Speed of Sound in Liquids and Solids 11
2.5 Determining Frequency 11
2.6 Acoustics in the 19th Century 12
2.7 The 20th Century 15
2.8 Conclusion 23
References 23

3 Basic Linear Acoustics
Alan D. Pierce 25
3.1 Introduction 27
3.2 Equations of Continuum Mechanics 28
3.3 Equations of Linear Acoustics 35
3.4 Variational Formulations 40
3.5 Waves of Constant Frequency 45
3.6 Plane Waves 47

	3.7	Attenuation of Sound	49
	3.8	Acoustic Intensity and Power	58
	3.9	Impedance	60
	3.10	Reflection and Transmission	61
	3.11	Spherical Waves	65
	3.12	Cylindrical Waves	75
	3.13	Simple Sources of Sound	82
	3.14	Integral Equations in Acoustics	87
	3.15	Waveguides, Ducts, and Resonators	89
	3.16	Ray Acoustics	94
	3.17	Diffraction	98
	3.18	Parabolic Equation Methods	107
	References		108
4	**Sound Propagation in the Atmosphere**		
	Keith Attenborough		113
	4.1	A Short History of Outdoor Acoustics	113
	4.2	Applications of Outdoor Acoustics	114
	4.3	Spreading Losses	115
	4.4	Atmospheric Absorption	116
	4.5	Diffraction and Barriers	116
	4.6	Ground Effects	120
	4.7	Attenuation Through Trees and Foliage	129
	4.8	Wind and Temperature Gradient Effects on Outdoor Sound	131
	4.9	Concluding Remarks	142
	References		143
5	**Underwater Acoustics**		
	William A. Kuperman, Philippe Roux		149
	5.1	Ocean Acoustic Environment	151
	5.2	Physical Mechanisms	155
	5.3	SONAR and the SONAR Equation	165
	5.4	Sound Propagation Models	167
	5.5	Quantitative Description of Propagation	177
	5.6	SONAR Array Processing	179
	5.7	Active SONAR Processing	185
	5.8	Acoustics and Marine Animals	195
	5.A	Appendix: Units	201
	References		201

Part B Physical and Nonlinear Acoustics

6	**Physical Acoustics**		
	Mack A. Breazeale, Michael McPherson		207
	6.1	Theoretical Overview	209
	6.2	Applications of Physical Acoustics	219

	6.3	Apparatus	226
	6.4	Surface Acoustic Waves	231
	6.5	Nonlinear Acoustics	234
	References		237

7 Thermoacoustics
Gregory W. Swift 239

	7.1	History	239
	7.2	Shared Concepts	240
	7.3	Engines	244
	7.4	Dissipation	249
	7.5	Refrigeration	250
	7.6	Mixture Separation	253
	References		254

8 Nonlinear Acoustics in Fluids
Werner Lauterborn, Thomas Kurz, Iskander Akhatov 257

	8.1	Origin of Nonlinearity	258
	8.2	Equation of State	259
	8.3	The Nonlinearity Parameter B/A	260
	8.4	The Coefficient of Nonlinearity β	262
	8.5	Simple Nonlinear Waves	263
	8.6	Lossless Finite-Amplitude Acoustic Waves	264
	8.7	Thermoviscous Finite-Amplitude Acoustic Waves	268
	8.8	Shock Waves	271
	8.9	Interaction of Nonlinear Waves	273
	8.10	Bubbly Liquids	275
	8.11	Sonoluminescence	286
	8.12	Acoustic Chaos	289
	References		293

Part C Architectural Acoustics

9 Acoustics in Halls for Speech and Music
Anders Christian Gade 301

	9.1	Room Acoustic Concepts	302
	9.2	Subjective Room Acoustics	303
	9.3	Subjective and Objective Room Acoustic Parameters	306
	9.4	Measurement of Objective Parameters	314
	9.5	Prediction of Room Acoustic Parameters	316
	9.6	Geometric Design Considerations	323
	9.7	Room Acoustic Design of Auditoria for Specific Purposes	334
	9.8	Sound Systems for Auditoria	346
	References		349

10 Concert Hall Acoustics Based on Subjective Preference Theory
Yoichi Ando 351
- 10.1 Theory of Subjective Preference for the Sound Field 353
- 10.2 Design Studies 361
- 10.3 Individual Preferences of a Listener and a Performer 370
- 10.4 Acoustical Measurements of the Sound Fields in Rooms 377
- **References** 384

11 Building Acoustics
James Cowan 387
- 11.1 Room Acoustics 387
- 11.2 General Noise Reduction Methods 400
- 11.3 Noise Ratings for Steady Background Sound Levels 403
- 11.4 Noise Sources in Buildings 405
- 11.5 Noise Control Methods for Building Systems 407
- 11.6 Acoustical Privacy in Buildings 419
- 11.7 Relevant Standards 424
- **References** 425

Part D Hearing and Signal Processing

12 Physiological Acoustics
Eric D. Young 429
- 12.1 The External and Middle Ear 429
- 12.2 Cochlea 434
- 12.3 Auditory Nerve and Central Nervous System 449
- 12.4 Summary 452
- **References** 453

13 Psychoacoustics
Brian C. J. Moore 459
- 13.1 Absolute Thresholds 460
- 13.2 Frequency Selectivity and Masking 461
- 13.3 Loudness 468
- 13.4 Temporal Processing in the Auditory System 473
- 13.5 Pitch Perception 477
- 13.6 Timbre Perception 483
- 13.7 The Localization of Sounds 484
- 13.8 Auditory Scene Analysis 485
- 13.9 Further Reading and Supplementary Materials 494
- **References** 495

14 Acoustic Signal Processing
William M. Hartmann 503
- 14.1 Definitions 504
- 14.2 Fourier Series 505

14.3	Fourier Transform	507
14.4	Power, Energy, and Power Spectrum	510
14.5	Statistics	511
14.6	Hilbert Transform and the Envelope	514
14.7	Filters	515
14.8	The Cepstrum	517
14.9	Noise	518
14.10	Sampled data	520
14.11	Discrete Fourier Transform	522
14.12	The z-Transform	524
14.13	Maximum Length Sequences	526
14.14	Information Theory	528
References		530

Part E Music, Speech, Electroacoustics

15 Musical Acoustics
Colin Gough 533

15.1	Vibrational Modes of Instruments	535
15.2	Stringed Instruments	554
15.3	Wind Instruments	601
15.4	Percussion Instruments	641
References		661

16 The Human Voice in Speech and Singing
Björn Lindblom, Johan Sundberg 669

16.1	Breathing	669
16.2	The Glottal Sound Source	676
16.3	The Vocal Tract Filter	682
16.4	Articulatory Processes, Vowels and Consonants	687
16.5	The Syllable	695
16.6	Rhythm and Timing	699
16.7	Prosody and Speech Dynamics	701
16.8	Control of Sound in Speech and Singing	703
16.9	The Expressive Power of the Human Voice	706
References		706

17 Computer Music
Perry R. Cook 713

17.1	Computer Audio Basics	714
17.2	Pulse Code Modulation Synthesis	717
17.3	Additive (Fourier, Sinusoidal) Synthesis	719
17.4	Modal (Damped Sinusoidal) Synthesis	722
17.5	Subtractive (Source-Filter) Synthesis	724
17.6	Frequency Modulation (FM) Synthesis	727
17.7	FOFs, Wavelets, and Grains	728

17.8	Physical Modeling (The Wave Equation)	730
17.9	Music Description and Control	735
17.10	Composition	737
17.11	Controllers and Performance Systems	737
17.12	Music Understanding and Modeling by Computer	738
17.13	Conclusions, and the Future	740
References		740

18 Audio and Electroacoustics
Mark F. Davis 743
18.1	Historical Review	744
18.2	The Psychoacoustics of Audio and Electroacoustics	747
18.3	Audio Specifications	751
18.4	Audio Components	757
18.5	Digital Audio	768
18.6	Complete Audio Systems	775
18.7	Appraisal and Speculation	778
References		778

Part F Biological and Medical Acoustics

19 Animal Bioacoustics
Neville H. Fletcher 785
19.1	Optimized Communication	785
19.2	Hearing and Sound Production	787
19.3	Vibrational Communication	788
19.4	Insects	788
19.5	Land Vertebrates	790
19.6	Birds	795
19.7	Bats	796
19.8	Aquatic Animals	797
19.9	Generalities	799
19.10	Quantitative System Analysis	799
References		802

20 Cetacean Acoustics
Whitlow W. L. Au, Marc O. Lammers 805
20.1	Hearing in Cetaceans	806
20.2	Echolocation Signals	813
20.3	Odontocete Acoustic Communication	821
20.4	Acoustic Signals of Mysticetes	827
20.5	Discussion	830
References		831

21 Medical Acoustics
Kirk W. Beach, Barbrina Dunmire 839
| 21.1 | Introduction to Medical Acoustics | 841 |

21.2	Medical Diagnosis; Physical Examination	842
21.3	Basic Physics of Ultrasound Propagation in Tissue	848
21.4	Methods of Medical Ultrasound Examination	857
21.5	Medical Contrast Agents	882
21.6	Ultrasound Hyperthermia in Physical Therapy	889
21.7	High-Intensity Focused Ultrasound (HIFU) in Surgery	890
21.8	Lithotripsy of Kidney Stones	891
21.9	Thrombolysis	892
21.10	Lower-Frequency Therapies	892
21.11	Ultrasound Safety	892
	References	895

Part G Structural Acoustics and Noise

22 Structural Acoustics and Vibrations
Antoine Chaigne 901

22.1	Dynamics of the Linear Single-Degree-of-Freedom (1-DOF) Oscillator	903
22.2	Discrete Systems	907
22.3	Strings and Membranes	913
22.4	Bars, Plates and Shells	920
22.5	Structural–Acoustic Coupling	926
22.6	Damping	940
22.7	Nonlinear Vibrations	947
22.8	Conclusion. Advanced Topics	957
	References	958

23 Noise
George C. Maling, Jr. 961

23.1	Instruments for Noise Measurements	965
23.2	Noise Sources	970
23.3	Propagation Paths	991
23.4	Noise and the Receiver	999
23.5	Regulations and Policy for Noise Control	1006
23.6	Other Information Resources	1010
	References	1010

Part H Engineering Acoustics

24 Microphones and Their Calibration
George S. K. Wong 1021

24.1	Historic References on Condenser Microphones and Calibration	1024
24.2	Theory	1024
24.3	Reciprocity Pressure Calibration	1026
24.4	Corrections	1029

	24.5	Free-Field Microphone Calibration	1039
	24.6	Comparison Methods for Microphone Calibration	1039
	24.7	Frequency Response Measurement with Electrostatic Actuators	1043
	24.8	Overall View on Microphone Calibration	1043
	24.A	Acoustic Transfer Impedance Evaluation	1045
	24.B	Physical Properties of Air	1045
	References		1048

25 Sound Intensity
Finn Jacobsen 1053
25.1	Conservation of Sound Energy	1054
25.2	Active and Reactive Sound Fields	1055
25.3	Measurement of Sound Intensity	1058
25.4	Applications of Sound Intensity	1068
References		1072

26 Acoustic Holography
Yang-Hann Kim 1077
26.1	The Methodology of Acoustic Source Identification	1077
26.2	Acoustic Holography: Measurement, Prediction and Analysis	1079
26.3	Summary	1092
26.A	Mathematical Derivations of Three Acoustic Holography Methods and Their Discrete Forms	1092
References		1095

27 Optical Methods for Acoustics and Vibration Measurements
Nils-Erik Molin 1101
27.1	Introduction	1101
27.2	Measurement Principles and Some Applications	1105
27.3	Summary	1122
References		1123

28 Modal Analysis
Thomas D. Rossing 1127
28.1	Modes of Vibration	1127
28.2	Experimental Modal Testing	1128
28.3	Mathematical Modal Analysis	1133
28.4	Sound-Field Analysis	1136
28.5	Holographic Modal Analysis	1137
References		1138

Acknowledgements 1139
About the Authors 1141
Detailed Contents 1147
Subject Index 1167

List of Abbreviations

A

ABR	auditory brainstem responses
AC	articulation class
ACF	autocorrelation function
ADC	analog-to-digital converter
ADCP	acoustic Doppler current profiler
ADP	ammonium dihydrogen phosphate
AM	amplitude modulated
AMD	air moving device
AN	auditory nerve
ANSI	American National Standards Institute
AR	assisted resonance
ASW	apparent source width
AUV	automated underwater vehicle

B

BB	bite block
BEM	boundary-element method
BER	bit error rate
BF	best frequency
BR	bass ratio

C

CAATI	computed angle-of-arrival transient imaging
CAC	ceiling attenuation class
CCD	charge-coupled device
CDF	cumulative distribution function
CMU	concrete masonry unit
CN	cochlear nucleus
CND	cumulative normal distribution
CSDM	cross-spectral-density matrix

D

DAC	digital-to-analog converter
DL	difference limen
DOF	degree of freedom
DRS	directed reflection sequence
DSL	deep scattering layer
DSP	digital speckle photography
DSP	digital signal processing
DSPI	digital speckle-pattern interferometry

E

EARP	equal-amplitude random-phase
EDT	early decay time
EDV	end diastolic velocity
EEG	electroencephalography
EOF	empirical orthogonal function
EOH	electro-optic holography
ERB	equivalent rectangular bandwidth
ESPI	electronic speckle-pattern interferometry

F

FCC	Federal Communications Commission
FEA	finite-element analysis
FEM	finite-element method
FERC	Federal Energy Regulatory Commission
FFP	fast field program
FFT	fast Fourier transform
FIR	finite impulse response
FM	frequency modulated
FMDL	frequency modulation detection limen
FOM	figure of merit
FRF	frequency response function
FSK	frequency shift keying

G

GA	genetic algorithm

H

HVAC	heating, ventilating and air conditioning

I

IACC	interaural cross-correlation coefficient
IACF	interaural cross-correlation function
IAD	interaural amplitude difference
ICAO	International Civil Aircraft Organization
IF	intermediate frequency
IFFT	inverse fast Fourier transform
IHC	inner hair cells
IIR	infinite impulse response
IM	intermodulation
IRF	impulse response function
ISI	intersymbol interference
ITD	interaural time difference
ITDG	initial time delay gap

J

JND	just noticeable difference

K

KDP	potassium dihydrogen phosphate

L

LDA	laser Doppler anemometry
LDV	laser Doppler vibrometry
LEF	lateral energy fraction
LEV	listener envelopment
LL	listening level
LOC	lateral olivocochlear system
LP	long-play vinyl record
LTAS	long-term-average spectra

M

MAA	minimum audible angle
MAF	minimum audible field
MAP	minimum audible pressure
MCR	multichannel reverberation
MDOF	multiple degree of freedom
MEG	magnetoencephalogram
MEMS	microelectromechanical system
MFDR	maximum flow declination rate
MFP	matched field processing
MIMO	multiple-input multiple-output
MLM	maximum-likelihood method
MLS	maximum length sequence
MOC	medial olivocochlear system
MRA	main response axis
MRI	magnetic resonance imaging
MTF	modulation transfer function
MTS	multichannel television sound
MV	minimum variance

N

NDT	nondestructive testing
NMI	National Metrology Institute
NRC	noise reduction coefficient

O

OAE	otoacoustic emission
ODS	operating deflexion shape
OHC	outer hair cells
OITC	outdoor–indoor transmission class
OR	or operation
OSHA	Occupational Safety and Health Administration

P

PC	phase conjugation
PCM	pulse code modulation
PD	probability of detection
PDF	probability density function
PE	parabolic equation
PFA	probability of false alarm
PIV	particle image velocimetry
PL	propagation loss
PLIF	planar laser-induced fluorescent
PM	phase modulation
PMF	probability mass function
PS	phase stepping
PS	peak systolic
PSD	power spectral density
PSK	phase shift keying
PTC	psychophysical tuning curve
PVDF	polyvinylidene fluoride
PZT	lead zirconate titanate

Q

QAM	quadrature amplitude modulation

R

RASTI	rapid speech transmission index
REL	resting expiratory level
RF	radio frequency
RIAA	Recording Industry Association of America
RMS	root-mean-square
ROC	receiving operating characteristic
RUS	resonant ultrasound spectroscopy

S

s.c.	supporting cells
S/N	signal-to-noise
SAA	sound absorption average
SAC	spatial audio coding
SAW	surface acoustic wave
SBSL	single-bubble sonoluminescence
SDOF	single degree of freedom
SE	signal excess
SEA	statistical energy analysis
SG	spiral ganglion
SI	speckle interferometry
SIL	speech interference level
SIL	sound intensity level
SISO	single-input single-output
SL	sensation level
SM	scala media
SNR	signal-to-noise ratio
SOC	superior olivary complex
SP	speckle photography
SPL	sound pressure level
SR	spontaneous discharge rate
ST	scala tympani

STC	sound transmission class	**U**	
STI	speech transmission index	UMM	unit modal mass
SV	scala vestibuli	**V**	
SVR	slow vertex response		
T		VBR	variable bitrate
TDAC	time-domain alias cancellation	VC	vital capacity
TDGF	time-domain Green's function	**W**	
THD	total harmonic distortion		
TL	transmission loss	WS	working standard
TLC	total lung capacity	**X**	
TMTF	temporal modulation transfer function		
TNM	traffic noise model	XOR	exclusive or
TR	treble ratio		
TR	time reversal		
TTS	temporary threshold shift		
TVG	time-varied gain		

1. Introduction to Acoustics

This brief introduction may help to persuade the reader that acoustics covers a wide range of interesting topics. It is impossible to cover all these topics in a single handbook, but we have attempted to include a sampling of hot topics that represent current acoustical research, both fundamental and applied.

Acoustics is the science of sound. It deals with the production of sound, the propagation of sound from the source to the receiver, and the detection and perception of sound. The word *sound* is often used to describe two different things: an auditory sensation in the ear, and the disturbance in a medium that can cause this sensation. By making this distinction, the age-old question "If a tree falls in a forest and no one is there to hear it, does it make a sound?" can be answered.

1.1	Acoustics: The Science of Sound	1
1.2	Sounds We Hear	1
1.3	Sounds We Cannot Hear: Ultrasound and Infrasound	2
1.4	Sounds We Would Rather Not Hear: Environmental Noise Control	2
1.5	Aesthetic Sound: Music	3
1.6	Sound of the Human Voice: Speech and Singing	3
1.7	How We Hear: Physiological and Psychological Acoustics	4
1.8	Architectural Acoustics	4
1.9	Harnessing Sound: Physical and Engineering Acoustics	5
1.10	Medical Acoustics	5
1.11	Sounds of the Sea	6
References		6

1.1 Acoustics: The Science of Sound

Acoustics has become a broad interdisciplinary field encompassing the academic disciplines of physics, engineering, psychology, speech, audiology, music, architecture, physiology, neuroscience, and others. Among the branches of acoustics are architectural acoustics, physical acoustics, musical acoustics, psychoacoustics, electroacoustics, noise control, shock and vibration, underwater acoustics, speech, physiological acoustics, etc.

Sound can be produced by a number of different processes, which include the following.

Vibrating bodies: when a drumhead or a noisy machine vibrates, it displaces air and causes the local air pressure to fluctuate.

Changing airflow: when we speak or sing, our vocal folds open and close to let through puffs of air. In a siren, holes on a rapidly rotating plate alternately pass and block air, resulting in a loud sound.

Time-dependent heat sources: an electrical spark produces a crackle; an explosion produces a bang due to the expansion of air caused by rapid heating. Thunder results from rapid heating by a bolt of lightning.

Supersonic flow: shock waves result when a supersonic airplane or a speeding bullet forces air to flow faster than the speed of sound.

1.2 Sounds We Hear

The range of sound intensity and the range of frequency to which the human auditory system responds is quite remarkable. The intensity ratio between the sounds that bring pain to our ears and the weakest sounds we can hear is more than 10^{12}. The frequency ratio between the highest and lowest frequencies we can hear is nearly 10^3, or more than nine octaves (each octave representing a doubling of frequency). Human vision is also quite remarkable, but the frequency range does not begin to compare to that of human hearing. The frequency range of vision is a little less than one octave (about $4 \times 10^{14} - 7 \times 10^{14}$ Hz). Within this one octave range we can identify more than 7 million colors. Given that the frequency range of the ear is nine times greater, one can imagine how many sound *colors* might be possible.

Humans and other animals use sound to communicate, and so it is not surprising that human hearing is most sensitive over the frequency range covered by human speech. This is no doubt a logical outcome of natural selection. This same match is found throughout much of the animal kingdom. Simple observations show that small animals generally use high frequencies for communication while large animals use low frequencies. In Chap. 19, it is shown that song frequency f scales with animal mass M roughly as $f \propto M^{-1/3}$.

The least amount of sound energy we can hear is of the order of 10^{-20} J (cf. sensitivity of the eye: about one quantum of light in the middle of the visible spectrum $\approx 4 \times 10^{-19}$ J). The upper limit of the sound pressure that can be generated is set approximately by atmospheric pressure. Such an ultimate sound wave would have a sound pressure level of about 191 dB. In practice, of course, nonlinear effects set in well below this level and limit the maximum pressure. A large-amplitude sound wave will change waveform and finally break into a shock, approaching a sawtooth waveform. Nonlinear effects are discussed in Chap. 8.

1.3 Sounds We Cannot Hear: Ultrasound and Infrasound

Sound waves below the frequency of human hearing are called *infrasound*, while sound waves with frequency above the range of human hearing are called *ultrasound*. These sounds have many interesting properties, and are being widely studied. Ultrasound is very important in medical and industrial imaging. It also forms the basis of a growing number of medical procedures, both diagnostic and therapeutic (see Chap. 21). Ultrasound has many applications in scientific research, especially in the study of solids and fluids (see Chap. 6).

Frequencies as high as 500 MHz have been generated, with a wavelength of about 0.6 μm in air. This is on the order of the wavelength of light and within an order of magnitude of the mean free path of air molecules. A gas ceases to behave like a continuum when the wavelength of sound becomes of the order of the mean free path, and this sets an upper limit on the frequency of sound that can propagate. In solids the assumption of continuum extends down to the intermolecular spacing of approximately 0.1 nm, with a limiting frequency of about 10^{12} Hz. The ultimate limit is actually reached when the wavelength is twice the spacing of the unit cell of a crystal, where the propagation of multiply scattered sound resembles the diffusion of heat [1.1].

Natural phenomena are prodigious generators of infrasound. When Krakatoa exploded, windows were shattered hundreds of miles away by the infrasonic wave. The ringing of both the Earth and the atmosphere continued for hours. The sudden shock wave of an explosion propels a complex infrasonic signal far beyond the shattered perimeter. Earthquakes generate intense infrasonic waves. The faster moving P (primary) waves arrive at distant locations tens of seconds before the destructive S (secondary) waves. (The P waves carry information; the S waves carry energy.) Certain animals and fish can sense these infrasonic precursors and react with fear and anxiety.

A growing amount of astronomical evidence indicates that primordial sound waves at exceedingly low frequency propagated in the universe during its first 380 000 years while it was a plasma of charged particles and thus opaque to electromagnetic radiation. Sound is therefore older than light.

1.4 Sounds We Would Rather Not Hear: Environmental Noise Control

Noise has been receiving increasing recognition as one of our critical environmental pollution problems. Like air and water pollution, noise pollution increases with population density; in our urban areas, it is a serious threat to our quality of life. Noise-induced hearing loss is a major health problem for millions of people employed in noisy environments. Besides actual hearing loss, humans are affected in many other ways by high levels of noise. Interference with speech, interruption of sleep, and other physiological and psychological effects of noise have been the subject of considerable study. Noise control is discussed in Chap. 23. The propagation of sound in air in Chap. 4, and building acoustics is the subject of Chap. 11.

Fortunately for the environment, even the noisiest machines convert only a small part of their total energy into sound. A jet aircraft, for example, may produce a kilowatt of acoustic power, but this is less than 0.02% of its mechanical output. Automobiles emit approximately 0.001% of their power as sound. Nevertheless, the shear number of machines operating in our society makes it crucial that we minimize their sound output and take measures to prevent the sound from propagating throughout our environment. Although reducing the emitted noise is best done at the source, it is possible, to some extent, to block the transmission of this noise from the source to the receiver. Reduction of classroom noise, which impedes learning in so many schools, is receiving increased attention from government officials as well as from acousticians [1.2].

1.5 Aesthetic Sound: Music

Music may be defined as an art form using sequences and clusters of sounds. Music is carried to the listener by sound waves. The science of musical sound is often called musical acoustics and is discussed in Chap. 15.

Musical acoustics deals with the production of sound by musical instruments, the transmission of music from the performer to the listener, and the perception and cognition of sound by the listener. Understanding the production of sound by musical instruments requires understanding how they vibrate and how they radiate sound. Transmission of sound from the performer to the listener involves a study of concert hall acoustics (covered in Chaps. 9 and 10) and the recording and reproduction of musical sound (covered in Chap. 15). Perception of musical sound is based on psychoacoustics, which is discussed in Chap. 13.

Electronic musical instruments have become increasingly important in contemporary music. Computers have made possible artificial musical intelligence, the synthesis of new musical sounds and the accurate and flexible re-creation of traditional musical sounds by artificial means. Not only do computers talk and sing and play music, they listen to us doing the same, and our interactions with computers are becoming more like our interactions with each other. Electronic and computer music is discussed in Chap. 17.

1.6 Sound of the Human Voice: Speech and Singing

It is difficult to overstate the importance of the human voice. Of all the members of the animal kingdom, we alone have the power of articulate speech. Speech is our chief means of communication. In addition, the human voice is our oldest musical instrument. Speech and singing, the closely related functions of the human voice, are discussed in a unified way in Chap. 16.

In the simplest model of speech production, the vocal folds act as the source and the vocal tract as a filter of the source sound. According to this model, the spectrum envelope of speech sound can be thought of as the product of two components:

Speech sound = source spectrum × filter function.

The nearly triangular waveform of the air flow from the glottis has a spectrum of harmonics that diminish in amplitude roughly as $1/n^2$ (i.e., at a rate of −12 dB/octave). The formants or resonances of the vocal tract create the various vowel sounds. The vocal tract can be shaped by movements of the tongue, the lips, and the soft palate to tune the formants and articulate the various speech sounds.

Sung vowels are fundamentally the same as spoken vowels, although singers do make vowel modifications in order to improve the musical tone, especially in their high range. In order to produce tones over a wide range of pitch, singers use muscular action in the larynx, which leads to different registers.

Much research has been directed at computer recognition and synthesis of speech. Goals of such research include voice-controlled word processors, voice control of computers and other machines, data entry by voice, etc.In general it is more difficult for a computer to understand language than to speak it.

1.7 How We Hear: Physiological and Psychological Acoustics

The human auditory system is complex in structure and remarkable in function. Not only does it respond to a wide range of stimuli, but it precisely identifies the pitch, timbre, and direction of a sound. Some of the hearing function is done in the organ we call the ear; some of it is done in the central nervous system as well.

Physiological acoustics, which is discussed in Chap. 12, focuses its attention mainly on the peripheral auditory system, especially the cochlea. The dynamic behavior of the cochlea is a subject of great interest. It is now known that the maximum response along the basilar membrane of the cochlea has a sharper peak in a living ear than in a dead one.

Resting on the basilar membrane is the delicate and complex organ of Corti, which contains several rows of hair cells to which are attached auditory nerve fibers. The inner hair cells are mainly responsible for transmitting signals to the auditory nerve fibers, while the more-numerous outer hair cells act as biological amplifiers. It is estimated that the outer hair cells add about 40 dB of amplification to very weak signals, so that hearing sensitivity decreases by a considerable amount when these delicate cells are destroyed by overexposure to noise.

Our knowledge of the cochlea has now progressed to a point where it is possible to construct and implant electronic devices in the cochlea that stimulate the auditory nerve. A cochlear implant is an electronic device that restores partial hearing in many deaf people [1.3]. It is surgically implanted in the inner ear and activated by a device worn outside the ear. An implant has four basic parts: a microphone, a speech processor and transmitter, a receiver inside the ear, and electrodes that transmit impulses to the auditory nerve and thence to the brain.

Psychoacoustics (psychological acoustics), the subject of Chap. 13, deals with the relationships between the physical characteristics of sounds and their perceptual attributes, such as loudness, pitch, and timbre.

The threshold of hearing depends upon frequency, the lowest being around 3–4 kHz, where the ear canal has a resonance, and rising considerably at low frequency. Temporal resolution, such as the ability to detect brief gaps between stimuli or to detect modulation of a sound, is a subject of considerable interest, as is the ability to localize the sound source. Sound localization depends upon detecting differences in arrival time and differences in intensity at our two ears, as well as spectral cues that help us to localize a source in the median plane.

Most sound that reaches our ears comes from several different sources. The extent to which we can perceive each source separately is sometimes called segregation. One important cue for perceptual separation of nearly simultaneous sounds is onset and offset disparity. Another is spectrum change with time. When we listen to rapid sequence of sounds, they may be grouped together (fusion) or they may be perceived as different streams (fission). It is difficult to judge the temporal order of sounds that are perceived in different streams.

1.8 Architectural Acoustics

To many lay people, an acoustician is a person who designs concert halls. That is an important part of architectural acoustics, to be sure, but this field incorporates much more. Architectural acousticians seek to understand and to optimize the sound environment in rooms and buildings of all types, including those used for work, residential living, education, and leisure. In fact, some of the earliest attempts to optimize sound transmission were practised in the design of ancient amphitheaters, and the acoustical design of outdoor spaces for concerts and drama still challenge architects.

In a room, most of the sound waves that reach the listener's ear have been reflected by one or more surfaces of the room or by objects in the room. In a typical room, sound waves undergo dozens of reflections before they become inaudible. It is not surprising, therefore, that the acoustical properties of rooms play an important role in determining the nature of the sound heard by a listener. Minimizing extraneous noise is an important part of the acoustical design of rooms and buildings of all kinds. Chapter 9 presents the principles of room acoustics and applies them to performance and assembly halls, including theaters and lecture halls, opera halls, concert halls, worship halls, and auditoria.

The subject of concert hall acoustics is almost certain to provoke a lively discussion by both performers and serious listeners. Musicians recognize the importance of the concert hall in communication between performer and listener. Opinions of new halls tend to polarize toward extremes of very good or very bad. In considering concert and opera halls, it is important to seek a common language for musicians and acousticians in order to understand how objective measurements relate to subjective qualities [1.4, 5]. Chapter 10 discusses subjective preference theory and how it relates to concert hall design.

Two acoustical concerns in buildings are providing the occupants with privacy and with a quiet environment, which means dealing with noise sources within the building as well as noise transmitted from outside. The most common noise sources in buildings, other than the inhabitants, are related to heating, ventilating, and air conditioning (HVAC) systems, plumbing systems, and electrical systems. Quieting can best be done at the source, but transmission of noise throughout the building must also be prevented. The most common external noise sources that affect buildings are those associated with transportation, such as motor vehicles, trains, and airplanes. There is no substitute for massive walls, although doors and windows must receive attention as well. Building acoustics is discussed in Chap. 11.

1.9 Harnessing Sound: Physical and Engineering Acoustics

It is sometimes said that physicists study nature, engineers attempt to improve it. Physical acoustics and engineering acoustics are two very important areas of acoustics. Physical acousticians investigate a wide range of scientific phenomena, including the propagation of sound in solids, liquids, and gases, and the way sound interacts with the media through which it propagates. The study of ultrasound and infrasound are especially interesting. Physical acoustics is discussed in Chap. 6.

Acoustic techniques have been widely used to study the structural and thermodynamic properties of materials at very low temperatures. Studying the propagation of ultrasound in metals, dielectric crystals, amorphous solids, and magnetic materials has yielded valuable information about their elastic, structural and other properties. Especially interesting has been the propagation of sound in superfluid helium. Second sound, an unusual type of temperature wave, was discovered in 1944, and since that time so-called third sound, fourth sound, and fifth sound have been described [1.6].

Nonlinear effects in sound are an important part of physical acoustics. Nonlinear effects of interest include waveform distortion, shock-wave formation, interactions of sound with sound, acoustic streaming, cavitation, and acoustic levitation. Nonlinearity leads to distortion of the sinusoidal waveform of a sound wave so that it becomes nearly triangular as the shock wave forms. On the other hand, local disturbances, called *solitons*, retain their shape over large distances.

The study of the interaction of sound and light, called acoustooptics, is an interesting field in physical acoustics that has led to several practical devices. In an acoustooptic modulator, for example, sound waves form a sort of moving optical diffraction grating that diffracts and modulates a laser beam.

Sonoluminescence is the name given to a process by which intense sound waves can generate light. The light is emitted by bubbles in a liquid excited by sound. The observed spectra of emitted light seem to indicate temperatures hotter than the surface of the sun. Some experimental evidence indicates that nuclear fusion may take place in bubbles in deuterated acetone irradiated with intense ultrasound.

Topics of interest in engineering acoustics cover a wide range and include: transducers and arrays, underwater acoustic systems, acoustical instrumentation, audio engineering, acoustical holography and acoustical imaging, ultrasound, and infrasound. Several of these topics are covered in Chaps. 5, 18, 24, 25, 26, 27, and 28. Much effort has been directed into engineering increasingly small transducers to produce and detect sound. Microphones are being fabricated on silicon chips as parts of integrated circuits.

The interaction of sound and heat, called thermoacoustics, is an interesting field that applies principles of physical acoustics to engineering systems. The thermoacoustic effect is the conversion of sound energy to heat or visa versa. In thermoacoustic processes, acoustic power can pump heat from a region of low temperature to a region of higher temperature. This can be used to construct heat engines or refrigerators with no moving parts. Thermoacoustics is discussed in Chap. 7.

1.10 Medical Acoustics

Two uses of sound that physicians have employed for many years are *auscultation*, listening to the body with a stethoscope, and *percussion*, sound generation by the striking the chest or abdomen to assess transmission or resonance. The most exciting new developments in medical acoustics, however, involve the use of ultrasound, both diagnostic imaging and therapeutic applications.

There has been a steady improvement in the quality of diagnostic ultrasound imaging. Two important commercial developments have been the advent of real-time three-dimensional (3-D) imaging and the development of hand-held scanners. Surgeons can now carry out procedures without requiring optical access. Although measurements on isolated tissue samples show that acoustic attenuation and backscatter correlate with pathology, implementing algorithms to obtain this information on a clinical scanner is challenging at the present time.

The therapeutic use of ultrasound has blossomed in recent years. Shock-wave lithotripsy is the predominant surgical operation for the treatment of kidney stones. Shock waves also appear to be effective at helping heal broken bones. High-intensity focused ultrasound is used to heat tissue selectivity so that cells can be destroyed in a local region. Ultrasonic devices appear to hold promise for treating glaucoma, fighting cancer, and controlling internal bleeding. Advanced therapies, such as puncturing holes in the heart, promoting localized drug delivery, and even carrying out brain surgery through an intact skull appear to be feasible with ultrasound [1.7].

Other applications of medical ultrasound are included in Chap. 21.

1.11 Sounds of the Sea

Oceans cover more than 70% of the Earth's surface. Sound waves are widely used to explore the oceans, because they travel much better in sea water than light waves. Likewise, sound waves are used, by humans and dolphins alike, to communicate under water, because they travel much better than radio waves. Acoustical oceanography has many military, as well as commercial applications. Much of our understanding of underwater sound propagation is a result of research conducted during and following World War II. Underwater acoustics is discussed in Chap. 5.

The speed of sound in water, which is about 1500 m/s, increases with increasing static pressure by about 1 part per million per kilopascal, or about 1% per 1000 m of depth, assuming temperature remains constant. The variation with temperature is an increase of about 2% per °C temperature rise. Refraction of sound, due to these changes in speed, along with reflection at the surface and the bottom, lead to waveguides at various ocean depths. During World War II, a *deep channel* was discovered in which sound waves could travel distances in excess of 3000 km. This phenomenon gave rise to the deep channel or sound fixing and ranging (SOFAR) channel, which could be used to locate, by acoustic means, airmen downed at sea.

One of the most important applications of underwater acoustics is sound navigation and ranging (SONAR). The purpose of most sonar systems is to detect and localize a target, such as submarines, mines, fish, or surface ships. Other SONARs are designed to measure some quantity, such as the ocean depth or the speed of ocean currents.

An interesting phenomenon called cavitation occurs when sound waves of high intensity propagate through water. When the rarefaction tension phase of the sound wave is great enough, the medium ruptures and cavitation bubbles appear. Cavitation bubbles can be produced by the tips of high-speed propellers. Bubbles affect the speed of sound as well as its attenuation [1.7, 8].

References

1.1 U. Ingard: Acoustics. In: *Handbook of Physics*, 2nd edn., ed. by E.U. Condon, H. Odishaw (McGraw-Hill, New York 1967)

1.2 B. Seep, R. Glosemeyer, E. Hulce, M. Linn, P. Aytar, R. Coffeen: *Classroom Acoustics* (Acoustical Society of America, Melville 2000, 2003)

1.3 M.F. Dorman, B.S. Wilson: The design and function of cochlear implants, Am. Scientist **19**, 436–445 (2004)

1.4 L.L. Beranek: *Music, Acoustics and Architecture* (Wiley, New York 1962)

1.5 L. Beranek: *Concert Halls and Opera Houses*, 2nd edn. (Springer, Berlin, Heidelberg, New York 2004)

1.6 G. Williams: Low-temperature acoustics. In: *McGraw-Hill Encyclopedia of Physics*, 2nd edn., ed. by S. Parker (McGraw-Hill, New York 1993)

1.7 R.O. Cleveland: Biomedical ultrasound/bioresponse to vibration. In: *ASA at 75*, ed. by H.E. Bass, W.J. Cavanaugh (Acoustical Society of America, Melville 2004)

1.8 H. Medwin, C.S. Clay: *Fundamentals of Acoustical Oceanography* (Academic, Boston 1998)

Part A Propagation of Sound

2 A Brief History of Acoustics
Thomas D. Rossing, Stanford, USA

3 Basic Linear Acoustics
Alan D. Pierce, Boston, USA

4 Sound Propagation in the Atmosphere
Keith Attenborough, Hull, UK

5 Underwater Acoustics
William A. Kuperman, La Jolla, USA
Philippe Roux, Grenoble, France

2. A Brief History of Acoustics

Although there are certainly some good historical treatments of acoustics in the literature, it still seems appropriate to begin a handbook of acoustics with a brief history of the subject. We begin by mentioning some important experiments that took place before the 19th century. Acoustics in the 19th century is characterized by describing the work of seven outstanding acousticians: Tyndall, von Helmholtz, Rayleigh, Stokes, Bell, Edison, and Koenig. Of course this sampling omits the mention of many other outstanding investigators.

To represent acoustics during the 20th century, we have selected eight areas of acoustics, again not trying to be all-inclusive. We select the eight areas represented by the first eight technical areas in the Acoustical Society of America. These are architectural acoustics, physical acoustics, engineering acoustics, structural acoustics, underwater acoustics, physiological and psychological acoustics, speech, and musical acoustics. We apologize to readers whose main interest is in another area of acoustics. It is, after all, a broad interdisciplinary field.

2.1	Acoustics in Ancient Times	9
2.2	Early Experiments on Vibrating Strings, Membranes and Plates	10
2.3	Speed of Sound in Air	10
2.4	Speed of Sound in Liquids and Solids	11
2.5	Determining Frequency	11
2.6	Acoustics in the 19th Century	12
	2.6.1 Tyndall	12
	2.6.2 Helmholtz	12
	2.6.3 Rayleigh	13
	2.6.4 George Stokes	13
	2.6.5 Alexander Graham Bell	14
	2.6.6 Thomas Edison	14
	2.6.7 Rudolph Koenig	14
2.7	The 20th Century	15
	2.7.1 Architectural Acoustics	15
	2.7.2 Physical Acoustics	16
	2.7.3 Engineering Acoustics	18
	2.7.4 Structural Acoustics	19
	2.7.5 Underwater Acoustics	19
	2.7.6 Physiological and Psychological Acoustics	20
	2.7.7 Speech	21
	2.7.8 Musical Acoustics	21
2.8	Conclusion	23
References		23

2.1 Acoustics in Ancient Times

Acoustics is the science of sound. Although sound waves are nearly as old as the universe, the scientific study of sound is generally considered to have its origin in ancient Greece. The word acoustics is derived from the Greek word *akouein*, to hear, although Sauveur appears to have been the first person to apply the term acoustics to the science of sound in 1701 [2.1].

Pythagoras, who established mathematics in Greek culture during the sixth century BC, studied vibrating strings and musical sounds. He apparently discovered that dividing the length of a vibrating string into simple ratios produced consonant musical intervals. According to legend, he also observed how the pitch of the string changed with tension and the tones generated by striking musical glasses, but these are probably just legends [2.2].

Although the Greeks were certainly aware of the importance of good acoustical design in their many fine theaters, the Roman architect Vitruvius was the first to write about it in his monumental *De Architectura*, which includes a remarkable understanding and analysis of theater acoustics: "We must choose a site in which the voice may fall smoothly, and not be returned by reflection so as to convey an indistinct meaning to the ear."

2.2 Early Experiments on Vibrating Strings, Membranes and Plates

Much of early acoustical investigations were closely tied to musical acoustics. Galileo reviewed the relationship of the pitch of a string to its vibrating length, and he related the number of vibrations per unit time to pitch. Joseph Sauveur made more-thorough studies of frequency in relation to pitch. The English mathematician Brook Taylor provided a dynamical solution for the frequency of a vibrating string based on the assumed curve for the shape of the string when vibrating in its fundamental mode. Daniel Bernoulli set up a partial differential equation for the vibrating string and obtained solutions which d'Alembert interpreted as waves traveling in both directions along the string [2.3].

The first solution of the problem of vibrating membranes was apparently the work of S. D. Poisson, and the circular membrane was handled by R. F. A. Clebsch. Vibrating plates are somewhat more complex than vibrating membranes. In 1787 *E. F. F. Chladni* described his method of using sand sprinkled on vibrating plates to show nodal lines [2.4]. He observed that the addition of one nodal circle raised the frequency of a circular plate by about the same amount as adding two nodal diameters, a relationship that Lord Rayleigh called Chladni's law. Sophie Germain wrote a fourth-order equation to describe plate vibrations, and thus won a prize provided by the French emperor Napoleon, although Kirchhoff later gave a more accurate treatment of the boundary conditions. *Rayleigh*, of course, treated both membranes and plates in his celebrated book *Theory of Sound* [2.5].

Chladni generated his vibration patterns by "strewing sand" on the plate, which then collected along the nodal lines. Later he noticed that fine shavings from the hair of his violin bow did not follow the sand to the nodes, but instead collected at the antinodes. *Savart* noted the same behavior for fine lycopodium powder [2.6]. Michael *Faraday* explained this as being due to acoustic streaming [2.7]. Mary *Waller* published several papers and a book on Chladni patterns, in which she noted that particle diameter should exceed $100\,\mu m$ in order to collect at the nodes [2.8]. Chladni figures of some of the many vibrational modes of a circular plate are shown in Fig. 2.1.

Fig. 2.1 Chladni patterns on a circular plate. The first four have two, three, four, and five nodal lines but no nodal circles; the second four have one or two nodal circles

2.3 Speed of Sound in Air

From earliest times, there was agreement that sound is propagated from one place to another by some activity of the air. Aristotle understood that there is actual motion of air, and apparently deduced that air is compressed. The Jesuit priest Athanasius Kircher was one of the first to observe the sound in a vacuum chamber, and since he could hear the bell he concluded that air was not necessary for the propagation of sound. Robert Boyle, however, repeated the experiment with a much improved pump and noted the much-observed decrease in sound intensity as the air is pumped out. We now know that sound propagates quite well in rarified air, and that the decrease in intensity at low pressure is mainly due to the impedance mismatch between the source and the medium as well as the impedance mismatch at the walls of the container.

As early as 1635, Gassendi measured the speed of sound using firearms and assuming that the light of the flash is transmitted instantaneously. His value came out to be $478\,m/s$. Gassendi noted that the speed of sound did not depend on the pitch of the sound, contrary to the view of Aristotle, who had taught that high notes are transmitted faster than low notes. In a more careful experiment, Mersenne determined the speed of sound to be $450\,m/s$ [2.9]. In 1650, G. A. Borelli and V. Viviani of the Accademia del Cimento of Florence obtained a value of 350 m/s for the speed of sound [2.10]. Another Italian, G. L. Bianconi, showed that the speed of sound in air increases with temperature [2.11].

The first attempt to calculate the speed of sound through air was apparently made by Sir Isaac Newton. He assumed that, when a pulse is propagated through a fluid, the particles of the fluid move in simple harmonic motion, and that if this is true for one particle, it must be true for all adjacent ones. The result is that the speed of sound is equal to the square root of the ratio of the atmospheric pressure to the density of the air. This leads to values that are considerably less than those measured by Newton (at Trinity College in Cambridge) and others.

In 1816, Pierre Simon Laplace suggested that in Newton's and Lagrange's calculations an error had been made in using for the volume elasticity of the air the pressure itself, which is equivalent to assuming the elastic motions of the air particles take place at constant temperature. In view of the rapidity of the motions, it seemed more reasonable to assume that the compressions and rarefactions follow the adiabatic law. The adiabatic elasticity is greater than the isothermal elasticity by a factor γ, which is the ratio of the specific heat at constant pressure to that at constant volume. The speed of sound should thus be given by $c = (\gamma p/\rho)^{1/2}$, where p is the pressure and ρ is the density. This gives much better agreement with experimental values [2.3].

2.4 Speed of Sound in Liquids and Solids

The first serious attempt to measure the speed of sound in liquid was probably that of the Swiss physicist Daniel Colladon, who in 1826 conducted studies in Lake Geneva. In 1825, the Academy of Sciences in Paris had announced as the prize competition for 1826 the measurement of the compressibility of the principal liquids. Colladon measured the static compressibility of several liquids, and he decided to check the accuracy of his measurements by measuring the speed of sound, which depends on the compressibility. The compressibility of water computed from the speed of sound turned out to be very close to the statically measured values [2.12]. Oh yes, he won the prize from the Academy.

In 1808, the French physicist *J. B. Biot* measured the speed of sound in a 1000 m long iron water pipe in Paris by direct timing of the sound travel [2.13]. He compared the arrival times of the sound through the metal and through the air and determined that the speed is much greater in the metal. Chladni had earlier studied the speed of sound in solids by noting the pitch emanating from a struck solid bar, just as we do today. He deduced that the speed of sound in tin is about 7.5 times greater than in air, while in copper it was about 12 times greater. Biot's values for the speed in metals agreed well with Chladni's.

2.5 Determining Frequency

Much of the early research on sound was tied to musical sound. Vibrating strings, membranes, plates, and air columns were the bases of various musical instruments. Music emphasized the importance of ratios for the different tones. A string could be divided into halves or thirds or fourths to give harmonious pitches. It was also known that pitch is related to frequency. Marin Mersenne (1588–1648) was apparently the first to determine the frequency corresponding to a given pitch. By working with a long rope, he was able determine the frequency of a standing wave on the length, mass, and tension of the rope. He then used a short wire under tension and from his rope formula he was able to compute the frequency of oscillation [2.14]. The relationship between pitch and frequency was later improved by Joseph Sauveur, who counted beats between two low-pitched organ pipes differing in pitch by a semitone. Sauveur deduced that "the relation between sounds of low and high pitch is exemplified in the ratio of the numbers of vibrations which they both make in the same time." [2.1]. He recognized that two sounds differing a musical fifth have frequencies in the ratio of 3:2. We have already commented that Sauveur was the first to apply the term *acoustics* to the science of sound. "I have come then to the opinion that there is a science superior to music, and I call it *acoustics*; it has for its object sound in general, whereas music has for its objects sounds agreeable to the ear." [2.1]

Tuning forks were widely used for determining pitch by the 19th century. Johann Scheibler (1777–1837) developed a tuning-fork *tonometer* which consisted of some 56 tuning forks. One was adjusted to the pitch of A above middle C, and another was adjusted by ear to be one octave lower. The others were then adjusted to differ successively by four vibrations per second above the lower A. Thus, he divided the octave into 55 intervals, each of about four vibrations per second. He then measured the number of beats in each interval, the sum total of such beats giving him the absolute frequency. He determined the frequency of the lower A to be 220 vibrations per second and the upper A to be 440 vibrations per second [2.15].

Felix Savart (1791–1841) used a rapidly rotating toothed wheel with 600 teeth to produce sounds of high frequency. He estimated the upper frequency threshold of hearing to be 24 000 vibrations per second. Charles Wheatstone (1802–1875) pioneered the use of rapidly rotating mirrors to study periodic events. This technique was later used by Rudolph Koenig and others to study speech sounds.

2.6 Acoustics in the 19th Century

Acoustics really blossomed in the 19th century. It is impossible to describe but a fraction of the significant work in acoustics during this century. We will try to provide a skeleton, at least, by mentioning the work of a few scientists. Especially noteworthy is the work of Tyndall, von Helmholtz, and Rayleigh, so we begin with them.

2.6.1 Tyndall

John Tyndall was born in County Carlow, Ireland in 1820. His parents were unable to finance any advanced education. After working at various jobs, he traveled to Marburg, Germany where he obtained a doctorate. He was appointed Professor of Natural Philosophy at the Royal Institution in London, where he displayed his skills in popular lecturing. In 1872 he made a lecture tour in the United States, which was a great success. His first lectures were on heat, and in 1863 these lectures were published under the title *Heat as a Mode of Motion*.

In 1867 he published his book *On Sound* with seven chapters. Later he added chapters on the transmission of sound through the atmosphere and on combinations of musical tones. In two chapters on vibrations of rods, plates, and bells, he notes that longitudinal vibrations produced by rubbing a rod lengthwise with a cloth or leather treated with rosin excited vibrations of higher frequency than the transverse vibrations. He discusses the determination of the waveform of musical sounds. By shining an intense beam of light on a mirror attached to a tuning fork and then to a slowly rotating mirror, as Lissajous had done, he spread out the waveform of the oscillations.

Tyndall is well remembered for his work on the effect of fog on transmission of sound through the atmosphere. He had succeeded Faraday as scientific advisor to the Elder Brethren of Trinity House, which supervised lighthouses and pilots in England. When fog obscures the lights of lighthouses, ships depend on whistles, bells, sirens, and even gunfire for navigation warnings. In 1873 Tyndall began a systematic study of sound propagation over water in various weather conditions in the straits of Dover. He noted great inconsistencies in the propagation.

2.6.2 Helmholtz

Hermann von Helmholtz was educated in medicine. He had wanted to study physics, but his father could not afford to support him, and the Prussian government offered financial support for students of medicine who would sign up for an extended period of service with the military. He was assigned a post in Potsdam, where he was able to set up his own laboratory in physics and physiology. The brilliance of his work led to cancelation of his remaining years of army duty and to his appointment as Professor of Physiology at Königsberg. He gave up the practice of medicine and wrote papers on physiology, color perception, and electricity. His first important paper in acoustics appears to have been his *On Combination Tones*, published in 1856 [2.16].

His book *On Sensations of Tone* (1862) combines his knowledge of both physiology and physics and music as well. He worked with little more than a stringed instrument, tuning forks, his siren, and his famous resonators to show that pitch is due to the fundamental frequency but the quality of a musical sound is due to the presence of upper partials. He showed how the ear can separate out the various components of a complex tone. He concluded that the quality of a tone depends solely on the number and relative strength of its partial tone and not on their relative phase.

In order to study vibrations of violin stings and speech sounds, von Helmholtz invented a vibration microscope, which displayed Lissajous patterns of vibration. One lens of the microscope is attached to the prong of a tuning fork, so a fixed spot appears to move up and down. A spot of luminous paint is then applied to the string, and a bow is drawn horizontally across the vertical string. The point on the horizontally vibrating violin string forms a Lissajous pattern as it moves. By viewing patterns for a bowed violin string, von Helmholtz was able to determine the actual motion of the string, and such motion is still referred to as Helmholtz motion.

Much of Helmholtz's book is devoted to discussion of hearing. Using a double siren, he studied difference tones and combination tones. He determined that beyond about 30 beats per second, the listener no longer hears individual beats but the tone becomes jarring or rough. He postulated that individual nerve fibers acted as vibrating strings, each resonating at a different frequency. Noting that skilled musicians "can distinguish with certainty a difference in pitch arising from half a vibration in a second in the doubly accented octave", he concluded that some 1000 different pitches might be distinguished in the octave between 50 and 100 cycles per second, and since there are 4500 nerve fibers in the cochlea, this represented about one fiber for each two cents of musical interval. He admitted, however "that we cannot precisely ascertain what parts of the ear actually vibrate sympathetically with individual tones."

2.6.3 Rayleigh

Rayleigh was a giant. He contributed to so many areas of physics, and his contributions to acoustics were monumental. His book *Theory of Sound* still has an honored place on the desk of every acoustician (alongside von Helmholtz's book, perhaps). In addition to his book, he published some 128 papers on acoustics. He anticipated so many interesting things. I have sometimes made the statement that every time I have a good idea about sound Rayleigh steals it and puts it into his book.

John William Strutt, who was to become the third Baron Rayleigh, was born at the family estate in Terling England in 1842. (Milk from the Rayleigh estate has supplied many families in London to this day.) He enrolled at Eton, but illness caused him to drop out, and he completed his schooling at a small academy in Torquay before entering Trinity College, Cambridge. His ill health may have been a blessing for the rest of the world. After nearly dying of rheumatic fever, he took a long cruise up the Nile river, during which he concentrated on writing his *Science of Sound*.

Soon after he returned to England, his father died and he became the third Baron Rayleigh and inherited title to the estate at Terling, where he set up a laboratory. When James Clerk Maxwell died in 1879, Rayleigh was offered the position as Cavendish Professor of Physics at Cambridge. He accepted it, in large measure because there was an agricultural depression at the time and his farm tenants were having difficulties in making rent payments [2.15].

Rayleigh's book and his papers cover such a wide range of topics in acoustics that it would be impractical to attempt to describe them here. His brilliant use of mathematics set the standard for subsequent writings on acoustics. The first volume of his book develops the theory of vibrations and its applications to strings, bars, membranes, and plates, while the second volume begins with aerial vibrations and the propagation of waves in fluids.

Rayleigh combined experimental work with theory in a very skillful way. Needing a way to determine the intensity of a sound source, he noted that a light disk suspended in a beam of sound tended to line up with its plane perpendicular to the direction of the fluid motion. The torque on the disk is proportional to the sound intensity. By suspending a light mirror in a sound field, the sound intensity could be determined by means of a sensitive optical lever. The arrangement, known as a Rayleigh disk, is still used to measure sound intensity.

Another acoustical phenomenon that bears his name is the propagation of Rayleigh waves on the plane surface of an elastic solid. Rayleigh waves are observed on both large and small scales. Most of the shaking felt from an earthquake is due to Rayleigh waves, which can be much larger than the seismic waves. Surface acoustic wave (SAW) filters and sensors make use of Rayleigh waves.

2.6.4 George Stokes

George Gabriel Stokes was born in County Sligo, Ireland in 1819. His father was a Protestant minister, and all of his brothers became priests. He was educated at Bristol College and Pembroke College, Cambridge. In 1841 he graduated as senior wrangler (the top First Class degree) in the mathematical tripos and he was the first Smith's prize man. He was awarded a Fellowship at Pembroke College and later appointed Lucasian professor of mathematics at Cambridge. The position paid rather poorly, however, so he accepted an additional position as professor of physics at the Government School of Mines in London.

William Hopkins, his Cambridge tutor, advised him to undertake research into hydrodynamics, and in 1842 he published a paper *On the steady motion of incompressible fluids*. In 1845 he published his classic paper *On the theories of the internal friction of fluids in motion*, which presents a three-dimensional equation of motion of a viscous fluid that has come to be known as the Stokes–Navier equation. Although he discovered that Navier, Poisson, and Saint-Venant had also considered the problem, he felt that his results were obtained with sufficiently different assumptions to justify publication. The Stokes–Navier equation of motion of a viscous, compressible fluid is still the starting point for much of the theory of sound propagation in fluids.

2.6.5 Alexander Graham Bell

Alexander Graham Bell was born in Edinburgh, Scotland in 1847. He taught music and elocution in Scotland before moving to Canada with his parents in 1868, and in 1871 he moved to Boston as a teacher of the deaf. In his spare time he worked on the *harmonic telegraph*, a device that would allow two or more electrical signals to be transmitted on the same wire. Throughout his life, Bell had been interested in the education of deaf people, which interest lead him to invent the microphone and, in 1876, his *electrical speech machine*, which we now call a telephone. He was encouraged to work steadily on this invention by Joseph Henry, secretary of the Smithsonian Institution and a highly respected physicist and inventor.

Bell's telephone was a great financial, as well as technical success. Bell set up a laboratory on his estate near Braddock, Nova Scotia and continued to improve the telephone as well as to work on other inventions. The magnetic transmitter was replaced by Thomas Edison's carbon microphone, the rights to which he obtained as a result of mergers and patent lawsuits [2.15].

2.6.6 Thomas Edison

The same year that Bell was born in Scotland (1847), Thomas A. Edison, the great inventor, was born in Milan, Ohio. At the age of 14 he published his own small newspaper, probably the first newspaper to be sold on trains. Also aged 14 he contracted scarlet fever which destroyed most of his hearing. His first invention was an improved stock-ticker for which he was paid $40 000.

Shortly after setting up a laboratory in Menlo Park, New Jersey, he invented (in 1877) the first phonograph. This was followed (in 1879) by the incandescent electric light bulb and a few years later by the Vitascope, which led to the first silent motion pictures. Other inventions included the dictaphone, mimeograph and storage battery.

The first published article on the phonograph appeared in *Scientific American* in 1877 after Edition visited the New York offices of the journal and demonstrated his machine. Later he demonstrated his machine in Washington for President Hayes, members of Congress and other notables. Many others made improvements to Edison's talking machine, but the credit still goes to Edison for first showing that the human voice could be recorded for posterity.

In its founding year (1929), the Acoustical Society of America (ASA) made Thomas Edison an honorary fellow, an honor which was not again bestowed during the 20 years that followed.

2.6.7 Rudolph Koenig

Rudolph Koenig was born in Koenigsberg (now Kaliningrad), Russia in 1832 and attended the university there at a time when von Helmholtz was a Professor of Physiology there. A few years after taking his degree, Koenig moved to Paris where he studied violin making under Villaume. He started his own business making acoustical apparatus, which he did with great care and talent. He devoted more than 40 years to making the best acoustical equipment of his day, many items of which are still in working order in museums and acoustics laboratories. Koenig, who never married, lived in the small front room of his Paris apartment, which was also his office and stock room, while the building and testing of instruments was done in the back rooms by Koenig and a few assistants. We will attempt to describe but a few of his acoustical instruments, but they have been well documented by *Greenslade* [2.17], *Beyer* [2.18], and others. The two largest collections of Koenig apparatus in North America are at the Smithsonian Institution and the University of Toronto.

Koenig made tuning forks of all sizes. A large 64 Hz fork formed the basis for a tuning-fork clock. A set of forks covering a range of frequencies in small steps was called a *tonometer* by Johann Scheibler. For his own use, Koenig made a tonometer consisting of 154 forks ranging in frequency from 16 to 21 845 Hz. Many tuning forks were mounted on hollow wooden resonators. He made both cylindrical and spherical Helmholtz resonators of all sizes.

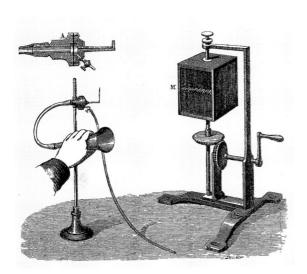

Fig. 2.2 Koenig's manometric flame apparatus. The image of the oscillating flame is seen in the rotating mirror (after [2.17])

is divided into two parts by a thin flexible membrane. Sounds waves are collected by a funnel, pass down the rubber tube, and cause the membrane to vibrate. Vibrations of the membrane cause a periodic change in the supply of gas to the burner, so the flame oscillates up and down at the frequency of the sound. The oscillating flame is viewed in the rotating mirror.

Koenig made apparatus for both the Fourier analysis and the synthesis of sound. At the 1876 exhibition, the instrument was used to show eight harmonics of a sung vowel. The Fourier analyzer included eight Helmholtz resonators, tuned to eight harmonics, which fed eight manometric flames. The coefficients of the various sinusoidal terms related to the heights of the eight flame images. The Helmholtz resonators could be tuned to different frequencies. The Fourier synthesizer had 10 electromagnetically-driven tuning forks and 10 Helmholtz resonators. A hole in each resonator could be opened or closed by means of keys [2.17].

To his contemporaries, Koenig was probably best known for his invention (1862) of the manometric flame apparatus, shown in Fig. 2.2, which allowed the visualization of acoustic signals. The manometric capsule

2.7 The 20th Century

The history of acoustics in the 20th century could be presented in several ways. In his definitive history, *Beyer* [2.15] devotes one chapter to each quarter century, perhaps the most sensible way to organize the subject. One could divide the century at the year 1929, the year the Acoustical Society of America was founded. One of the events in connection with the 75th anniversary of this society was the publication of a snapshot history of the Society written by representatives from the 15 technical committees and edited by Henry *Bass* and William *Cavanaugh* [2.19]. Since we make no pretense of covering all areas of acoustics nor of reporting all acoustical developments in the 20th century, we will merely select a few significant areas of acoustics and try to discuss briefly some significant developments in these. For want of other criteria, we have selected the nine areas of acoustics that correspond to the first eight technical committees in the Acoustical Society of America.

2.7.1 Architectural Acoustics

Wallace Clement Sabine (1868–1919) is generally considered to be the father of architectural acoustics. He was the first to make quantitative measurements on the acoustics of rooms. His discovery that the product of total absorption and the duration of residual sound is a constant still forms the basis of sound control in rooms. His pioneering work was not done entirely by choice, however. As a 27-year old professor at Harvard University, he was assigned by the President to determine corrective measures for the lecture room at Harvard's Fogg Art Museum. As he begins his famous paper on reverberation [2.20] "The following investigation was not undertaken at first by choice but devolved on the writer in 1895 through instructions from the Corporation of Harvard University to propose changes for remedying the acoustical difficulties in the lecture-room of the Fogg Art Museum, a building that had just been completed."

Sabine determined the reverberation time in the Fogg lecture room by using an organ pipe and a chronograph. He found the reverberation time in the empty room to be 5.62 seconds. Then he started adding seat cushions from the Sanders Theater and measuring the resulting reverberation times. He developed an empirical formula $T = 0.164 V/A$, where T is reverberation time, V is volume (in cubic feet) and A is the average absorption

Fig. 2.3 Interior of Symphony Hall in Boston whose acoustical design by Wallace Clement Sabine set a standard for concert halls

coefficient times the total area (in square feet) of the walls, ceiling and floor. This formula is still called the Sabine reverberation formula.

Following his success with the Fogg lecture room, Sabine was asked to come up with acoustical specifications for the New Boston Music Hall, now known as Symphony Hall, which would ensure hearing superb music from every seat. Sabine answered with a shoebox shape for the building to keep out street noise. Then, using his mathematical formula for reverberation time, Sabine carefully adjusted the spacing between the rows of seats, the slant of the walls, the shape of the stage, and materials used in the walls to produce the exquisite sound heard today at Symphony Hall (see Fig. 2.3).

Vern Knudsen (1893–1974), physicist at University of California Los Angeles (UCLA) and third president of the Acoustical Society of America, was one of many persons who contributed to architectural acoustics in the first half of the 20th century. His collaboration with Hans Kneser of Germany led to an understanding of molecular relaxation phenomena in gases and liquids. In 1932 he published a book on *Architectural Acoustics* [2.21], and in 1950, a book *Architectural Designing in Acoustics* with *Cyril Harris* [2.22], which summarized most of what was known about the subject by the middle of the century.

In the mid 1940s Richard Bolt, a physicist at the Massachusetts Institute of Technology (MIT), was asked by the United Nations (UN) to design the acoustics for one of the UN's new buildings. Realizing the work that was ahead of him, he asked Leo Beranek to join him. At the same time they hired another MIT professor, Robert Newman, to help with the work with the United Nations; together they formed the firm of Bolt, Beranek, and Newman (BBN), which was to become one of the foremost architectural consulting firms in the world. This firm has provided acoustical consultation for a number of notable concert halls, including Avery Fisher Hall in New York, the Koussevitzky Music Shed at Tanglewood, Davies Symphony Hall in San Francisco, Roy Thompson Hall in Toronto, and the Center for the Performing Arts in Tokyo [2.23]. They are also well known for their efforts in pioneering the Arpanet, forerunner of the Internet.

Recipients of the Wallace Clement Sabine award for accomplishments in architectural acoustics, include Vern Knudsen, Floyd Watson, Leo Beranek, Erwin Meyer, Hale Sabine, Lothar Cremer, Cyril Harris, Thomas Northwood, Richard Waterhouse, Harold Marshall, Russell Johnson, and Alfred Warnock. The work of each of these distinguished acousticians could be a chapter in the history of acoustics but space does not allow it.

2.7.2 Physical Acoustics

Although all of acoustics, the science of sound, incorporates the laws of physics, we usually think of physical acoustics as being concerned with fundamental acoustic wave propagation phenomena, including transmission, reflection, refraction, interference, diffraction, scattering, absorption, dispersion of sound and the use of acoustics to study physical properties of matter and to produce changes in these properties. The foundations for physical acoustics were laid by such 19th century giants as von Helmholtz, Rayleigh, Tyndall, Stokes, Kirchhoff, and others.

Ultrasonic waves, sound waves with frequencies above the range of human hearing, have attracted the attention of many physicists and engineers in the 20th century. An early source of ultrasound was the Galton whistle, used by Francis Galton to study the upper threshold of hearing in animals. More powerful sources of ultrasound followed the discovery of the piezoelectric effect in crystals by Jacques and Pierre Curie. They found that applying an electric field to the plates of certain natural crystals such as quartz produced changes in thickness. Later in the century, highly efficient ceramic piezoelectric transducers were used to produce high-intensity ultrasound in solids, liquids, and gases.

Probably the most important use of ultrasound nowadays is in ultrasonic imaging, in medicine (sonograms) as well as in the ocean (sonar). Ultrasonic waves are used in many medical diagnostic procedures. They are directed toward a patient's body and reflected when they reach boundaries between tissues of different densities. These reflected waves are detected and displayed on a monitor. Ultrasound can also be used to detect malignancies and hemorrhaging in various organs. It is also used to monitor real-time movement of heart valves and large blood vessels. Air, bone, and other calcified tissues absorb most of the ultrasound beam; therefore this technique cannot be used to examine the bones or the lungs.

The father of sonar (sound navigation and ranging) was Paul Langevin, who used active echo ranging sonar at about 45 kHz to detect mines during World War I. Sonar is used to explore the ocean and study marine life in addition to its many military applications [2.24]. New types of sonar include synthetic aperture sonar for high-resolution imaging using a moving hydrophone array and computed angle-of-arrival transient imaging (CAATI).

Infrasonic waves, which have frequencies below the range of human hearing, have been less frequently studied than ultrasonic waves. Natural phenomena are prodigious generators of infrasound. When the volcano Krakatoa exploded, windows were shattered hundreds of miles away by the infrasonic wave. The *ringing* of both earth and atmosphere continued for hours. It is believed that infrasound actually formed the upper pitch of this natural volcanic explosion, tones unmeasurably deep forming the actual central harmonic of the event. Infrasound from large meteoroids that enter our atmosphere have very large amplitudes, even great enough to break glass windows [2.25]. Ultralow-pitch earthquake sounds are keenly felt by animals and sensitive humans. Quakes occur in distinct stages. Long before the final breaking release of built-up earth tensions, there are numerous and succinct precursory shocks. Deep shocks produce strong infrasonic impulses up to the surface, the result of massive heaving ground strata. Certain animals (fish) can actually hear infrasonic precursors.

Aeroacoustics, a branch of physical acoustics, is the study of sound generated by (or in) flowing fluids. The mechanism for sound or noise generation may be due to turbulence in flows, resonant effects in cavities or waveguides, vibration of boundaries of structures etc. A flow may alter the propagation of sound and boundaries can lead to scattering; both features play a significant part in altering the noise received at a particular observation point. A notable pioneer in aeroacoustics was Sir James Lighthill (1924–1998), whose analyses of the sounds generated in a fluid by turbulence have had appreciable importance in the study of nonlinear acoustics. He identified quadrupole sound sources in the inhomogeneities of turbulence as a major source of the noise from jet aircraft engines, for example [2.26].

There are several sources of nonlinearity when sound propagates through gases, liquids, or solids. At least since the time of Stokes, it has been known that in fluids compressions propagate slightly faster than rarefactions, which leads to distortion of the wave front and even to the formation of shock waves. Richard Fay (1891–1964) noted that the waveform takes on the shape of a sawtooth. In 1935 Eugene Fubini–Ghiron demonstrated that the pressure amplitude in a nondissipative fluid is proportional to an infinite series in the harmonics of the original signal [2.15]. Several books treat nonlinear sound propagation, including those by *Beyer* [2.27] and by *Hamilton* and *Blackstock* [2.28].

Measurements of sound propagation in liquid helium have led to our basic understanding of cryogenics and also to several surprises. The attenuation of sound shows a sharp peak near the so-called lambda point at which helium takes on superfluid properties. This behavior was explained by Lev Landau (1908–1968) and others. Second sound, the propagation of waves consisting of periodic oscillations of temperature and entropy, was discovered in 1944 by V. O. Peshkov. Third sound, a surface wave of the superfluid component was reported in 1958, whereas fourth sound was discovered in 1962 by K. A. Shapiro and Isadore Rudnick. Fifth sound, a thermal wave, has also been reported, as has *zero* sound [2.15].

While there a number of ways in which light can interact with sound, the term *optoacoustics* typically refers to sound produced by high-intensity light from a laser. The optoacoustic (or photoacoustic) effect is characterized by the generation of sound through interaction of electromagnetic radiation with matter. Absorption of single laser pulses in a sample can effectively generate optoacoustic waves through the thermoelastic effect. After absorption of a short pulse the heated region thermally expands, creating a mechanical disturbance that propagates into the surrounding medium as a sound wave. The waves are recorded at the surface of the sample with broadband ultrasound transducers

Sonoluminescence uses sound to produce light. Sonoluminescence, the emission of light by bubbles in a liquid excited by sound, was discovered by H. Frenzel and H. Schultes in 1934, but was not considered very interesting at the time. A major breakthrough occurred when Felipe Gaitan and his colleagues were able to produce single-bubble sonoluminescence, in which a single bubble, trapped in a standing acoustic wave, emits light with each pulsation [2.29].

The wavelength of the emitted light is very short, with the spectrum extending well into the ultraviolet. The observed spectrum of emitted light seems to indicate a temperature in the bubble of at least $10\,000\,°C$, and possibly a temperature in excess of one million degrees C. Such a high temperature makes the study of sonoluminescence especially interesting for the possibility that it might be a means to achieve thermonuclear fusion. If the bubble is hot enough, and the pressures in it high enough, fusion reactions like those that occur in the Sun could be produced within these tiny bubbles.

When sound travels in small channels, oscillating heat also flows to and from the channel walls, leading to a rich variety of *thermoacoustic* effects. In 1980, Nicholas *Rott* developed the mathematics describing acoustic oscillations in a gas in a channel with an axial temperature gradient, a problem investigated by Rayleigh and Kirchhoff without much success [2.30]. Applying Rott's mathematics, *Hofler*, et al. invented a standing-wave thermoacoustic refrigerator in which the coupled oscillations of gas motion, temperature, and heat transfer in the sound wave are phased so that heat is absorbed at low temperature and waste heat is rejected at higher temperature [2.31].

Recipients of the ASA silver medal in physical acoustics since it was first awarded in 1975 have included Isadore Rudnick, Martin Greenspan, Herbert McSkimin, David Blackstock, Mack Breazeale, Allan Pierce, Julian Maynard, Robert Apfel, Gregory Swift, and Philip Marston.

2.7.3 Engineering Acoustics

It is virtually impossible to amplify sound waves. Electrical signals, on the other hand, are relatively easy to amplify. Thus a practical system for amplifying sound includes input and output *transducers*, together with the electronic amplifier. Transducers have occupied a central role in engineering acoustics during the 20th century.

The transducers in a sound amplifying system are microphones and loudspeakers. The first microphones were Bell's magnetic transmitter and the loosely packed carbon microphones of Edison and Berliner. A great step forward in 1917 was the invention of the condenser microphone by Edward Wente (1889–1972). In 1962, James West and Gerhard Sessler invented the foil electret or electret condenser microphone, which has become the most ubiquitous microphone in use. It can be found in everything from telephones to children's toys to medical devices. Nearly 90% of the approximately one billion microphones manufactured annually are electret designs.

Ernst W. Siemens was the first to describe the *dynamic* or moving-coil loudspeaker, with a circular coil of wire in a magnetic field and supported so that it could move axially. John Stroh first described the conical paper diaphragm that terminated at the rim of the speaker in a section that was flat except for corrugations. In 1925, Chester W. Rice and Edward W. Kellogg at General Electric established the basic principle of the direct-radiator loudspeaker with a small coil-driven mass-controlled diaphragm in a baffle with a broad mid-frequency range of uniform response. In 1926, Radio Corporation of America (RCA) used this design in the *Radiola* line of alternating current (AC)-powered radios. In 1943 James Lansing introduced the Altec-Lansing 604 duplex radiator which combined an efficient 15 inch woofer with a high-frequency compression driver and horn [2.32].

In 1946, Paul Klipsch introduced the Klipschorn, a corner-folded horn that made use of the room boundaries themselves to achieve efficient radiation at low frequency. In the early 1940s, the Jensen company popularized the vented box or bass reflex loudspeaker enclosure. In 1951, specific loudspeaker driver parameters and appropriate enclosure alignments were described by Neville Thiele and later refined by Richard Small. Thiele–Small parameters are now routinely published by loudspeaker manufacturers and used by professionals and amateurs alike to design vented enclosures [2.33].

The Audio Engineering Society was formed in 1948, the same year the microgroove $33\,1/3$ rpm long-play vinyl record (LP) was introduced by Columbia Records. The founding of this new society had the unfortunate effect of distancing engineers primarily interested in audio from the rest of the acoustics engineering community.

Natural piezoelectric crystals were used to generate sound waves for underwater signaling and for ultrasonic research. In 1917 Paul Langevin obtained a large crystal of natural quartz from which $10 \times 10 \times 1.6$ cm slices could be cut. He constructed a transmitter that sent out a beam powerful enough to kill fish in its near field [2.15]. After World Wart II, materials, such as potassium dihydrogen phosphate (KDP), ammonium dihydrogen phosphate (ADP) and barium titanate replaced natural quartz in transducers. There are several piezoelectric ceramic compositions in common use today: barium titanate, lead zirconate titanate (PZT) and modified iterations such as lead lanthanum zirconate titanate (PLZT), lead metaniobate and lead magnesium niobate (PMN, including electrostrictive formulations). The PZT compositions are the most widely used in applications involving light shutters, micro-positioning devices, speakers and medical array transducers

Recipients of the ASA silver medal in engineering acoustics have included Harry Olson, Hugh Knowles, Benjamin Bauer, Per Bruel, Vincent Salmon, Albert Bodine, Joshua Greenspon, Alan Powell, James West, Richard Lyon, and Ilene Busch-Vishniac. Interdisciplinary medals have gone to Victor Anderson, Steven Garrett, and Gerhard Sessler.

2.7.4 Structural Acoustics

The vibrations of solid structures was discussed at some length by Rayleigh, Love, Timoshenko, Clebsch, Airey, Lamb, and others during the 19th and early 20th centuries. Nonlinear vibrations were considered by G. Duffing in 1918. R. N. Arnold and G. B. Warburton solved the complete boundary-value problem of the free vibration of a finite cylindrical shell. Significant advances have been made in our understanding of the radiation, scattering, and response of fluid-loaded elastic plates by G. Maidanik, E. Kerwin, M. Junger, and D. Feit.

Statistical energy analysis (SEA), championed by Richard Lyon and Gideon Maidanik, had its beginnings in the early 1960s. In the 1980s, Christian Soize developed the fuzzy structure theory to predict the mid-frequency dynamic response of a master structure coupled with a large number of complex secondary subsystems. The structural and geometric details of the latter are not well defined and therefore labeled as *fuzzy*.

A number of good books have been written on the vibrations of simple and complex structures. Especially noteworthy, in my opinion, are books by *Cremer* et al. [2.34], *Junger* and *Feit* [2.35], *Leissa* [2.36], [2.37], and *Skudrzyk* [2.38]. Statistical energy analysis is described by *Lyon* [2.39]. Near-field acoustic holography, developed by Jay Maynard and Earl Williams, use pressure measurements in the near field of a vibrating object to determine its source distribution on the vibrating surface [2.40]. A near-field acoustic hologram of a rectangular plate driven at a point is shown in Fig. 2.4.

The ASA has awarded its Trent–Crede medal, which recognizes accomplishment in shock and vibration, to Carl Vigness, Raymond Mindlin, Elias Klein, J. P. Den

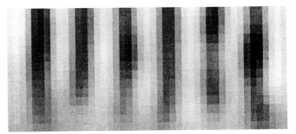

Fig. 2.4 Near-field hologram of pressure near a rectangular plate driven at 1858 Hz at a point (courtesy of Earl Williams)

Hartog, Stephen Crandall, John Snowdon, Eric Ungar, Miguel Junger, Gideon Maidanik, Preston Smith, David Feit, and Sabih Hayek.

2.7.5 Underwater Acoustics

The science of underwater technology in the 20th century is based on the remarkable tools of transduction that the 19th century gave us. It was partly motivated by the two world wars and the cold war and the threats raised by submarines and underwater mines. Two nonmilitary commercial fields that have been important driving forces in underwater acoustics are geophysical prospecting and fishing. The extraction of oil from the seafloor now supplies 25% of our total supply [2.41].

Essential to understanding underwater sound propagation is detailed knowledge about the speed of sound in the sea. In 1924, *Heck* and *Service* published tables on the dependence of sound speed on temperature, salinity, and pressure [2.42]. Summer conditions, with strong solar heating and a warm atmosphere, give rise to sound speeds that are higher near the surface and decrease with depth, while winter conditions, with cooling of the surface, reverses the temperature gradient. Thus, sound waves will bend downward under summer conditions and upward in the winter.

Submarine detection can be either passive (listening to the sounds made by the submarine) or active (transmitting a signal and listen for the echo). Well into the 1950s, both the United States and United Kingdom chose active high-frequency systems since passive systems at that time were limited by the ship's radiated noise and the self noise of the arrays. During World War II, underwater acoustics research results were secret, but at the end of the war, the National Defense Research Council (NDRC) published the results. The Sub-Surface Warfare Division alone produced 22 volumes [2.41]. Later the NDRC was disbanded and projects were transferred to the Navy (some reports have been published by IEEE).

The absorption in seawater was found to be much higher than predicted by classical theory. O. B. Wilson and R. W. Leonard concluded that this was due to the relaxation frequency of magnesium sulfate, which is present in low concentration in the sea [2.43]. Ernest Yeager and Fred Fisher found that boric acid in small concentrations exhibits a relaxation frequency near 1 kHz. In 1950 the papers of Tolstoy and Clay discussed propagation in shallow water. At the Scripps Institution of Oceanography Fred Fisher and Vernon Simmons made resonance measurements of seawater in a 200 l glass sphere over a wide range of frequencies and temperature, confirming the earlier results and improving the empirical absorption equation [2.41].

Ambient noise in the sea is due to a variety of causes, such as ships, marine mammals, snapping shrimp, and dynamic processes in the sea itself. Early measurements of ambient noise, made under Vern Knudsen, came to be known as the Knudsen curves. Wittenborn made measurements with two hydrophones, one in the sound channel and the other below it. A comparison of the noise levels showed about a 20 dB difference over the low-frequency band but little difference at high frequency. It has been suggested that a source of low-frequency noise is the collective oscillation of bubble clouds.

The ASA pioneers medal in underwater acoustics has been awarded to Harvey Hayes, Albert Wood, Warren Horton, Frederick Hunt, Harold Saxton, Carl Eckart, Claude Horton, Arthur Williams, Fred Spiess, Robert Urick, Ivan Tolstoy, Homer Bucker, William Kuperman, Darrell Jackson, and Frederick Tappert.

2.7.6 Physiological and Psychological Acoustics

Physiological acoustics deals with the peripheral auditory system, including the cochlear mechanism, stimulus encoding in the auditory nerve, and models of auditory discrimination.

This field of acoustics probably owes more to Georg von Békésy (1899–1972) than any other person. Born in Budapest, he worked for the Hungarian Telephone Co., the University of Budapest, the Karolinska Institute in Stockholm, Harvard University, and the University of Hawaii. In 1962 he was awarded the Nobel prize in physiology and medicine for his research on the ear. He determined the static and dynamic properties of the basilar membrane, and he built a mechanical model of the cochlea. He was probably the first person to observe eddy currents in the fluid of the cochlea. Josef Zwislocki (1922–) reasoned that the existence of such fluid motions would inevitably lead to nonlinearities, although Helmholtz had pretty much assumed that the inner ear was a linear system [2.44].

In 1971 William Rhode succeeded in making measurements on a live cochlea for the first time. Using the Mössbauer effect to measure the velocity of the basilar membrane, he made a significant discovery. The frequency tuning was far sharper than that reported for dead cochleae. Moreover, the response was highly nonlinear, with the gain increasing by orders of magnitude at low sound levels. There is an active amplifier in the cochlea that boosts faint sounds, leading to a strongly compressive response of the basilar membrane. The work of Peter Dallos, Bill Brownell, and others identified the outer hair cells as the cochlear amplifiers [2.45].

It is possible, by inserting a tiny electrode into the auditory nerve, to pick up the electrical signals traveling in a single fiber of the auditory nerve from the cochlea to the brain. Each auditory nerve fiber responds over a certain range of frequency and pressure. Nelson Kiang and others have determined that tuning curves of each fiber show a maximum in sensitivity. Within several hours after death, the basilar membrane response decreases, the frequency of maximum response shifts down, and the response curve broadens.

Psychological acoustics or *psychoacoustics* deals with subjective attributes of sound, such as loudness, pitch, and timbre and how they relate to physically measurable quantities such as the sound level, frequency, and spectrum of the stimulus.

At the Bell Telephone laboratories, Harvey Fletcher, first president of the Acoustical Society of America, and W. A. Munson determined contours of equal loudness by having listeners compare a large number of tones to pure tones of 1000 Hz. These contours of equal loudness came to be labeled by an appropriate number of *phon*s. S. S. Stevens is responsible for the loudness scale of *sone*s and for ways to calculate the loudness in *sone*s. His proposal to express pitch in *mels* did not become as widely adopted, however, probably because musicians and others prefer to express pitch in terms of the musical scale.

The threshold for detecting pure tones is mostly determined by the sound transmission through the outer and middle ear; to a first approximation the inner ear (the cochlea) is equally sensitive to all frequencies, except the very highest and lowest. In 1951, J. C. R. *Licklider* (1915–1990), who is well known for his work on developing the Internet, put the results of several hearing surveys together [2.46]. Masking of one tone by another was discussed in a classic paper by *Wegel* and *Lane* who showed that low-frequency tones can mask higher-frequency tones better than the reverse [2.47].

Two major theories of pitch perception gradually developed on the basis of experiments in many laboratories. They are usually referred to as the place (or frequency) theory and the periodicity (or time) theory. By observing wavelike motions of the basilar membrane caused by sound stimulation, Békésy provided support for the place theory. In the late 1930s, however, J. F. Schouten and his colleagues performed pitch-shift experiments that provided support for the periodicity theory of pitch. Modern theories of pitch perception often combine elements of both of these [2.48].

Recipients of the ASA von Békésy medal have been Jozef Zwislocki, Peter Dallos, and Murray Sachs, while the silver medal in psychological and physiological acoustics has been awarded to Lloyd Jeffress, Ernest Wever, Eberhard Zwicker, David Green, Nathaniel Durlach, Neal Viemeister, and Brian Moore.

2.7.7 Speech

The production, transmission, and perception of speech have always played an important role in acoustics. Harvey Fletcher published his book *Speech and Hearing* in 1929, the same year as the first meeting of the Acoustical Society of America. The first issue of the *Journal of the Acoustical Society of America* included papers on speech by G. Oscar Russell, Vern Knudsen, Norman French and Walter Koenig, Jr.

In 1939, Homer Dudley invented the vocoder, a system in which speech was analyzed into component parts consisting of the pitch fundamental frequency of the voice, the noise, and the intensities of the speech in a series of band-pass filters. This machine, which was demonstrated at New York World's Fair, could speak simple phrases.

An instrument that is particularly useful for speech analysis is the *sound spectrograph*, originally developed at the Bell Telephone laboratories around 1945. This instrument records a sound-level frequency–time plot for a brief sample of speech on which the sound level is represent by the degree of blackness in a two-dimensional time–frequency graph, as shown in Fig. 2.5. Digital versions of the sound spectrograph are used these days, but the display format is similar to the original machine.

Phonetic aspects of speech research blossomed at the Bell Telephone laboratories and elsewhere in the 1950s. Gordon Peterson and his colleagues produced

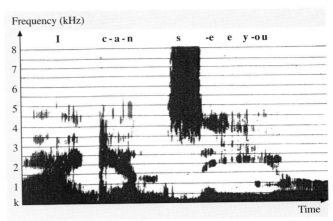

Fig. 2.5 Speech spectrogram of a simple sentence ("I can see you") recorded on a sound spectrograph

several studies of vowels. Gunnar Fant published a complete survey of the field in *Acoustic Theory of Speech Production* [2.49]. The pattern playback, developed at Haskins Laboratories, dominated early research using synthetic speech in the United States. James Flanagan (1925–) demonstrated the significance of the use of our understanding of fluid dynamics in analyzing the behavior of the glottis. Kenneth Stevens and Arthur House noted that the bursts of air from the glottis had a triangular waveform that led to a rich spectrum of harmonics.

Speech synthesis and automatic recognition of speech have been important topics in speech research. Dennis Klatt (1938–1988) developed a system for synthesizing speech, and shortly before his death he gave the first completely intelligible synthesized speech paper presented to the ASA [2.50]. *Fry* and *Denes* constructed a system in which speech was fed into an acoustic recognizer that compares "the changing spectrum of the speech wave with certain reference patterns and indicates the occurrence of the phoneme whose reference pattern best matches that of the incoming wave" [2.51].

Recipients of the ASA silver medal in speech communication have included Franklin Cooper, Gunnar Fant, Kenneth Stevens, Dennis Klatt, Arthur House, Peter Ladefoged, and Patricia Kuhl.

2.7.8 Musical Acoustics

Musical acoustics deals with the production of musical sound, its transmission to the listener, and its perception. Thus this interdisciplinary field overlaps architectural acoustics, engineering acoustics, and psychoacoustics. The study of the singing voice also overlaps the study of speech. In recent years, the scientific study of musical performance has also been included in musical acoustics.

Because the transmission and perception of sound have already been discussed, we will concentrate on the production of musical sound by musical instruments, including the human voice. It is convenient to classify musical instruments into families in accordance with the way they produce sound: string, wind, percussion, and electronic.

Bowed string instruments were probably the first to attract the attention of scientific researchers. The modern violin was developed largely in Italy in the 16th century by Gaspara da Salo and the Amati family. In the 18th century, Antonio Stradivari, a pupil of Nicolo Amati, and Guiseppi Guarneri created instruments with great brilliance that have set the standard for violin makers since that time. Outstanding contributions to our understanding of violin acoustics have been made by Felix Savart, Hermann von Helmholtz, Lord Rayleigh, C. V. Raman, Frederick Saunders, and Lothar Cremer, all of whom also distinguished themselves in fields other than violin acoustics. In more recent times, the work of Professor Saunders has been continued by members of the Catgut Acoustical Society, led by Carleen Hutchins. This work has made good use of modern tools such as computers, holographic interferometers, and fast Fourier transform (FFT) analyzers. One noteworthy product of modern violin research has been the development of an octet of scaled violins, covering the full range of musical performance.

The piano, invented by Bartolomeo Cristofori in 1709, is one of the most versatile of all musical instruments. One of the foremost piano researcher of our time is Harold Conklin. After he retired from the Baldwin Piano Co, he published a series of three papers in *JASA* (J. Acoustical Society of America) that could serve as a textbook for piano researchers [2.52]. Gabriel Weinreich explained the behavior of coupled piano strings and the aftersound which results from this coupling. Others who have contributed substantially to our understanding of piano acoustics are Anders Askenfelt, Eric Jansson, Juergen Meyer, Klaus Wogram, Ingolf Bork, Donald Hall, Isao Nakamura, Hideo Suzuki, and Nicholas Giordano. Many other string instruments have been studied scientifically, but space does not allow a discussion of their history here.

Pioneers in the study of wind instruments included Arthur Benade (1925–1987), John Backus (1911–1988), and John Coltman (1915–). Backus, a research physicist, studied both brass and woodwind instruments, especially the nonlinear flow control properties of woodwind reeds. He improved the capillary method for measuring input impedance of air columns, and he developed synthetic reeds for woodwind instruments. Benade's extensive work led to greater understanding of mode conversion in flared horns, a model of woodwind instrument bores based on the acoustics of a lattice of tone holes, characterization of wind instruments in terms of cutoff frequencies, and radiation from brass and woodwind instruments. His two books *Horns, Strings and Harmony* and *Fundamentals of Musical Acoustics* have both been reprinted by Dover Books. Coltman, a physicist and executive at the Westinghouse Electric Corporation, devoted much of his spare time to the study of the musical, historical, and acoustical aspects of the flute and organ pipes. He collected more than 200 instruments of the flute family, which he used in his studies. More recently, flutes, organ pipes, and other wind instruments have been studied by Neville Fletcher and his colleagues in Australia.

The human voice is our oldest musical instrument, and its acoustics has been extensively studied by Johan Sundberg and colleagues in Stockholm. A unified discussion of speech and the singing voice appears in this handbook.

The acoustics of percussion instruments from many different countries have been studied by Thomas *Rossing* and his students, and many of them are described in his book *Science of Percussion Instruments* [2.53] as well as in his published papers.

Electronic music technology was made possible with the invention of the vacuum tube early in the 20th century. In 1919 Leon Theremin invented the aetherophone (later called the Theremin), an instrument whose vacuum-tube oscillators can be controlled by the proximity of the player's hands to two antennae. In 1928, Maurice Martenot built the Ondes Martenot. In 1935 Laurens Hammond used magnetic tone-wheel generators as the basis for his electromechanical organ, which became a very popular instrument. Analog music synthesizers became popular around the middle of the 20th century. In the mid 1960s, Robert Moog and Donald Buchla built successful voltage-controlled music synthesizers which gave way to a revolution in the way composers could easily synthesize new sounds. Gradually, however, analog music synthesizers gave way to digital techniques making use of digital computers. Although many people contributed to the development of computer music, Max Mathews is often called the father of computer music, since he developed the MUSIC I program that begat many successful music synthesis programs and blossomed into a rich resource for musical expression [2.54].

The ASA has awarded its silver medal in musical acoustics to Carleen Hutchins, Arthur Benade, John Backus, Max Mathews, Thomas Rossing, Neville Fletcher, and Johan Sundberg.

2.8 Conclusion

This brief summary of acoustics history has only scratched the surface. Many fine books on the subject appear in the list of references, and readers are urged to explore the subject further. The science of sound is a fascinating subject that draws from many different disciplines.

References

2.1 R.B. Lindsay: *Acoustics: Historical and Philosophical Development* (Dowden, Hutchinson & Ross, Stroudsburg, PA 1973) p. 88, Translation of Sauveur's paper
2.2 F.V. Hunt: *Origins in Acoustics* (Acoustical Society of America, Woodbury, NY 1992)
2.3 R.B. Lindsay: The story of acoustics, J. Acoust. Soc. Am. **39**, 629–644 (1966)
2.4 E.F.F. Chladni: *Entdeckungen über die Theorie des Klanges* (Breitkopf und Härtel, Leipzig 1787)
2.5 Lord Rayleigh J.W. Strutt: *The Theory of Sound*, Vol. 1, 2nd edn. (Macmillan, London 1894), reprinted by Dover, 1945
2.6 M. Savart: Recherches sur les vibrations normales, Ann. Chim. **36**, 187–208 (1827)
2.7 M. Faraday: On a peculiar class of acoustical figures; and on certain forms assumed by groups of particles upon vibrating elastic surfaces, Philos. Trans. R. Soc. **121**, 299–318 (1831)
2.8 M.D. Waller: *Chladni Figures: A Study in Symmetry* (Bell, London 1961)
2.9 L.M.A. Lenihan: Mersenne and Gassendi. An early chapter in the history of shound, Acustica **2**, 96–99 (1951)
2.10 D.C. Miller: *Anecdotal History of the Science of Sound* (Macmillan, New York 1935) p. 20
2.11 L.M.A. Lenihan: The velocity of sound in air, Acustica **2**, 205–212 (1952)
2.12 J.-D. Colladon, J.K.F. Sturm: Mémoire sur la compression des liquides et la vitesse du son dans l'eau, Ann. Chim. Phys. **36**, 113 (1827)
2.13 J.B. Biot: Ann. Chim. Phys. **13**, 5 (1808)
2.14 M. Mersenne: *Harmonie Universelle* (Crmoisy, Paris 1636), Translated into English by J. Hawkins, 1853
2.15 R.T. Beyer: *Sounds of Our Times* (Springer, New York 1999)
2.16 H. von Helmholtz: On sensations of tone, Ann. Phys. Chem **99**, 497–540 (1856)
2.17 T.B. Greenslade jr: The Acoustical Apparatus of Rudolph Koenig, The Phys. Teacher **30**, 518–524 (1992)
2.18 R.T. Beyer: Rudolph Koenig, 1832–1902, ECHOES **9**(1), 6 (1999)
2.19 H.E. Bass, W.J. Cavanaugh (Eds.): *ASA at 75* (Acoustical Society of America, Melville 2004)
2.20 W.C. Sabine: *Reverberation* (The American Architect, 1900), Reprinted in Collected Papers on Acoustics by Wallace Clement Sabine, Dover, New York, 1964
2.21 V.O. Knudsen: *Architectural Acoustics* (Wiley, New York 1932)
2.22 V.O. Knudsen, C. Harris: *Acoustical Designing in Architecture* (Wiley, New York 1950), Revised edition published in 1978 by the Acoustical Society of America
2.23 L. Beranek: *Concert and Opera Halls, How They Sound* (Acoustical Society of America, Woodbury 1996)
2.24 C.M. McKinney: The early history of high frequency, short range, high resolution, active sonar, ECHOES **12**(2), 4–7 (2002)
2.25 D.O. ReVelle: Gobal infrasonic monitoring of large meteoroids, ECHOES **11**(1), 5 (2001)
2.26 J. Lighthill: *Waves in Fluids* (Cambridge Univ. Press, Cambridge 1978)
2.27 R. Beyer: *Nonlinear Acoustics* (US Dept. of the Navy, Providence 1974), Revised and reissued by ASA, NY 1997
2.28 M.F. Hamilton, D.T. Blackstock (Eds.): *Nonlinear Acoustics* (Academic, San Diego 1998)
2.29 D.F. Gaitan, L.A. Crum, C.C. Church, R.A. Roy: Sonoluminescence and bubble dynamics for a single, stable, cavitation bubble, J. Acoust. Soc. Am. **91**, 3166–3183 (1992)
2.30 N. Rott: Thermoacoustics, Adv. Appl. Mech. **20**, 135–175 (1980)
2.31 T. Hofler, J.C. Wheatley, G. W. Swift, and A. Migliori: Acoustic cooling engine:, 1988 US Patent No. 4,722,201.

2.32 G.L. Augspurger: Theory, ingenuity, and wishful wizardry in loudspeaker design–A half-century of progress?, J. Acoust. Soc. Am. **77**, 1303–1308 (1985)

2.33 A.N. Thiele: Loudspeakers in vented boxes, Proc. IRE Aust. **22**, 487–505 (1961), Reprinted in J. Aud. Eng. Soc. 19, 352–392, 471–483 (1971)

2.34 L. Cremer, M. Heckl, E.E. Ungar: *Structure Borne Sound: Structural Vibrations and Sound Radiation at Audio Frequencies*, 2nd edn. (Springer, New York 1990)

2.35 M.C. Junger, D. Feit: *Sound, Structures, and Their Interaction*, 2nd edn. (MIT Press, 1986)

2.36 A.W. Leissa: *Vibrations of Plates* (Acoustical Society of America, Melville, NY 1993)

2.37 A.W. Leissa: *Vibrations of Shells* (Acoustical Society of America, Melville, NY 1993)

2.38 E. Skudrzyk: *Simple and Complex Vibratory Systems* (Univ. Pennsylvania Press, Philadelphia 1968)

2.39 R.H. Lyon: *Statistical Energy Analysis of Dynamical System, Theory and Applicationss* (MIT Press, Cambridge 1975)

2.40 E.G. Williams: *Fourier Acoustics: Sound Radiation and Nearfield Acoustic Holography* (Academic, San Diego 1999)

2.41 R.R. Goodman: A brief history of underwater acoustics. In: *ASA at 75*, ed. by H.E. Bass, W.J. Cavanaugh (Acoustical Society of America, Melville 2004)

2.42 N.H. Heck, J.H. Service: *Velocity of Sound in Seawater* (SUSC&GS Special Publication 108, 1924)

2.43 O.B. Wilson, R.W. Leonard: Measurements of sound absorption in aqueous salt solutions by a resonator method, J. Acoust. Soc. Am. **26**, 223 (1954)

2.44 H. von Helmholtz: *Die Lehre von den Tonempfindungen* (Longmans, New York 1862), Translated by Alexander Ellis as *On the Sensations of Tone* and reprinted by Dover, 1954

2.45 M.B. Sachs: The History of Physiological Acoustics. In: *ASA at 75*, ed. by H.E. Bass, W.J. Cavanaugh (Acoustical Society of America, Melville 2004)

2.46 J.C.R. Licklider: Basic correlates of the auditory stimulus. In: *Handbook of Experimental Psychology*, ed. by S.S. Stevens (J. Wiley, New York 1951)

2.47 R.L. Wegel, C.E. Lane: The auditory masking of one pure tone by another and its probable relation to the dynamics of the inner ear, Phys. Rev. **23**, 266–285 (1924)

2.48 B.C.J. Moore: Frequency analysis and pitch perception. In: *Human Psychophysics*, ed. by W.A. Yost, A.N. Popper, R.R. Fay (Springer, New York 1993)

2.49 G. Fant: *Acoustical Thoery of Speech Production* (Mouton, The Hague 1960)

2.50 P. Ladefoged: The study of speech communication in the acoustical society of america. In: *ASA at 75*, ed. by H.E. Bass, W.J. Cavanaugh (Acoustical Society of America, Melville 2004)

2.51 D.B. Fry, P. Denes: Mechanical speech recognition. In: *Communication Theory*, ed. by W. Jackson. (Butterworth, London 1953)

2.52 H.A. Conklin Jr: Design and tone in the mechanoacoustic piano, Parts I, II, and III, J. Acoust. Soc. Am. **99**, 3286–3296 (1996), **100**, 695–708 (1996), and **100**, 1286–1298 (1996)

2.53 T.D. Rossing: *The Science of Percussion Instruments* (World Scientific, Singapore 2000)

2.54 T.D. Rossing, F.R. Moore, P.A. Wheeler: *Science of Sound*, 3rd edn. (Addison-Wesley, San Francisco 2002),

3. Basic Linear Acoustics

This chapter deals with the physical and mathematical aspects of sound when the disturbances are, in some sense, small. Acoustics is usually concerned with small-amplitude phenomena, and consequently a linear description is usually applicable. Disturbances are governed by the properties of the medium in which they occur, and the governing equations are the equations of continuum mechanics, which apply equally to gases, liquids, and solids. These include the mass, momentum, and energy equations, as well as thermodynamic principles. The viscosity and thermal conduction enter into the versions of these equations that apply to fluids. Fluids of typical great interest are air and sea water, and consequently this chapter includes a summary of their relevant acoustic properties. The foundation is also laid for the consideration of acoustic waves in elastic solids, suspensions, bubbly liquids, and porous media.

This is a long chapter, and a great number of what one might term classical acoustics topics are included, especially topics that one might encounter in an introductory course in acoustics: the wave theory of sound, the wave equation, reflection of sound, transmission from one media to another, propagation through ducts, radiation from various types of sources, and the diffraction of sound.

3.1	Introduction	27
3.2	**Equations of Continuum Mechanics**	28
	3.2.1 Mass, Momentum, and Energy Equations	28
	3.2.2 Newtonian Fluids and the Shear Viscosity	30
	3.2.3 Equilibrium Thermodynamics	30
	3.2.4 Bulk Viscosity and Thermal Conductivity	30
	3.2.5 Navier–Stokes–Fourier Equations	31
	3.2.6 Thermodynamic Coefficients	31
	3.2.7 Ideal Compressible Fluids	32
	3.2.8 Suspensions and Bubbly Liquids	32
	3.2.9 Elastic Solids	33
3.3	**Equations of Linear Acoustics**	35
	3.3.1 The Linearization Process	35
	3.3.2 Linearized Equations for an Ideal Fluid	36
	3.3.3 The Wave Equation	36
	3.3.4 Wave Equations for Isotropic Elastic Solids	36
	3.3.5 Linearized Equations for a Viscous Fluid	37
	3.3.6 Acoustic, Entropy, and Vorticity Modes	37
	3.3.7 Boundary Conditions at Interfaces	39
3.4	**Variational Formulations**	40
	3.4.1 Hamilton's Principle	40
	3.4.2 Biot's Formulation for Porous Media	42
	3.4.3 Disturbance Modes in a Biot Medium	43
3.5	**Waves of Constant Frequency**	45
	3.5.1 Spectral Density	45
	3.5.2 Fourier Transforms	45
	3.5.3 Complex Number Representation	46
	3.5.4 Time Averages of Products	47
3.6	**Plane Waves**	47
	3.6.1 Plane Waves in Fluids	47
	3.6.2 Plane Waves in Solids	48
3.7	**Attenuation of Sound**	49
	3.7.1 Classical Absorption	49
	3.7.2 Relaxation Processes	50
	3.7.3 Continuously Distributed Relaxations	52
	3.7.4 Kramers–Krönig Relations	52
	3.7.5 Attenuation of Sound in Air	55
	3.7.6 Attenuation of Sound in Sea Water	57
3.8	**Acoustic Intensity and Power**	58
	3.8.1 Energy Conservation Interpretation	58
	3.8.2 Acoustic Energy Density and Intensity	58
	3.8.3 Acoustic Power	59
	3.8.4 Rate of Energy Dissipation	59
	3.8.5 Energy Corollary for Elastic Waves	60

3.9	**Impedance**		60	3.13.7	Multipole Series	85
	3.9.1	Mechanical Impedance	60	3.13.8	Acoustically Compact Sources	86
	3.9.2	Specific Acoustic Impedance	60	3.13.9	Spherical Harmonics	86
	3.9.3	Characteristic Impedance	60	3.14	**Integral Equations in Acoustics**	87
	3.9.4	Radiation Impedance	61		3.14.1 The Helmholtz–Kirchhoff Integral	87
	3.9.5	Acoustic Impedance	61		3.14.2 Integral Equations for Surface Fields	88
3.10	**Reflection and Transmission**		61	3.15	**Waveguides, Ducts, and Resonators**	89
	3.10.1	Reflection at a Plane Surface	61		3.15.1 Guided Modes in a Duct	89
	3.10.2	Reflection at an Interface	62		3.15.2 Cylindrical Ducts	90
	3.10.3	Theory of the Impedance Tube	62		3.15.3 Low-Frequency Model for Ducts	90
	3.10.4	Transmission through Walls and Slabs	63		3.15.4 Sound Attenuation in Ducts	91
	3.10.5	Transmission through Limp Plates	64		3.15.5 Mufflers and Acoustic Filters	92
	3.10.6	Transmission through Porous Blankets	64		3.15.6 Non-Reflecting Dissipative Mufflers	93
	3.10.7	Transmission through Elastic Plates	64		3.15.7 Expansion Chamber Muffler	93
					3.15.8 Helmholtz Resonators	93
3.11	**Spherical Waves**		65	3.16	**Ray Acoustics**	94
	3.11.1	Spherically Symmetric Outgoing Waves	65		3.16.1 Wavefront Propagation	94
	3.11.2	Radially Oscillating Sphere	66		3.16.2 Reflected and Diffracted Rays	95
	3.11.3	Transversely Oscillating Sphere	67		3.16.3 Inhomogeneous Moving Media	96
	3.11.4	Axially Symmetric Solutions	68		3.16.4 The Eikonal Approximation	96
	3.11.5	Scattering by a Rigid Sphere	73		3.16.5 Rectilinear Propagation of Amplitudes	97
3.12	**Cylindrical Waves**		75	3.17	**Diffraction**	98
	3.12.1	Cylindrically Symmetric Outgoing Waves	75		3.17.1 Posing of the Diffraction Problem	98
	3.12.2	Bessel and Hankel Functions	77		3.17.2 Rays and Spatial Regions	98
	3.12.3	Radially Oscillating Cylinder	81		3.17.3 Residual Diffracted Wave	99
	3.12.4	Transversely Oscillating Cylinder	81		3.17.4 Solution for Diffracted Waves	102
3.13	**Simple Sources of Sound**		82		3.17.5 Impulse Solution	102
	3.13.1	Volume Sources	82		3.17.6 Constant-Frequency Diffraction	103
	3.13.2	Small Piston in a Rigid Baffle	82		3.17.7 Uniform Asymptotic Solution	103
	3.13.3	Multiple and Distributed Sources	82		3.17.8 Special Functions for Diffraction	104
	3.13.4	Piston of Finite Size in a Rigid Baffle	83		3.17.9 Plane Wave Diffraction	105
	3.13.5	Thermoacoustic Sources	84		3.17.10 Small-Angle Diffraction	106
	3.13.6	Green's Functions	85		3.17.11 Thin-Screen Diffraction	107
				3.18	**Parabolic Equation Methods**	107
				References		108

List of symbols

B_V	bulk modulus at constant entropy
c	speed of sound
c_p	specific heat at constant pressure
c_v	specific heat at constant volume
D/Dt	convective time derivative
$\mathfrak{D}$	rate of energy dissipation per unit volume
e_i	unit vector in the direction of increasing x_i
f	frequency, cycles per second
f	force per unit volume
f_1, f_2	volume fractions
g	acceleration associated with gravity
h	absolute humidity, fraction of molecules that are water
$H(\omega)$	transfer function
I	acoustic intensity (energy flux)
k	wavenumber
$\mathfrak{L}$	Lagrangian density

M	average molecular weight	∂	partial differentiation operator
$\boldsymbol{n}$	unit vector normal to a surface	ϵ	expansion parameter
p	pressure	ϵ_{ij}	component of the strain tensor
$\hat{p}(\omega)$	Fourier transform of $p(t)$, complex amplitude	ζ	location of pole in the complex plane
$\boldsymbol{q}$	heat flux vector	η	loss factor
Q	quality factor	θ_I	angle of incidence
R_0	universal gas constant	κ	coefficient of thermal conduction
Re	real part	λ	wavelength
s	entropy per unit mass	λ_L	Lamè constant for a solid
S	surface, surface area	μ	(shear) viscosity
$\boldsymbol{t}$	traction vector, force per unit area	μ_B	bulk viscosity
T	absolute temperature	μ_L	Lamè constant for a solid
$\mathfrak{T}$	kinetic energy per unit volume	ν	Poisson's ratio
u	internal energy per unit mass	ξ	internal variable in irreversible thermodynamics
$\boldsymbol{u}$	locally averaged displacement field of solid matter	ξ_j	component of the displacement field
$\boldsymbol{U}$	locally averaged displacement field of fluid matter	π	ratio of circle circumference to diameter
		ρ	density, mass per unit volume
$\mathfrak{U}$	strain energy per unit volume	σ_{ij}	component of the stress tensor
$\boldsymbol{v}$	fluid velocity	τ_B	characteristic time in the Biot model
V	volume	τ_ν	relaxation time associated with ν-th process
w	acoustic energy density	ϕ	phase constant
α	attenuation coefficient	Φ	scalar potential
β	coefficient of thermal expansion	$\chi(x)$	nominal phase change in propagation distance x
γ	specific heat ratio		
δ_{ij}	Kronecker delta	$\boldsymbol{\Psi}$	vector potential
$\delta(t)$	delta function	ω	angular frequency, radians per second

3.1 Introduction

Small-amplitude phenomena can be described to a good approximation in terms of linear algebraic and linear differential equations. Because acoustic phenomenon are typically of small amplitude, the analysis that accompanies most applications of acoustics consequently draws upon a linearized theory, briefly referred to as linear acoustics.

One reason why one chooses to view acoustics as a linear phenomenon, unless there are strong indications of the importance of nonlinear behavior, is the intrinsic conceptual simplicity brought about by the principle of superposition, which can be loosely described by the relation

$$\mathfrak{L}(af_1 + bf_2) = a\mathfrak{L}(f_1) + b\mathfrak{L}(f_2), \tag{3.1}$$

where f_1 and f_2 are two *causes* and $\mathfrak{L}(f)$ is the set of possible mathematical or computational steps that predicts the *effect* of the cause f, such steps being describable so that they are the same for any such cause. The quantities a and b are arbitrary numbers. Doubling a cause doubles the effect, and the effect of a sum of two causes is the sum of the effects of each of the separate causes.

Thus, for example, a complicated sound field caused by several sources can be regarded as a sum of fields, each of which is caused by an individual source. Moreover, if there is assurance that the governing equations have time-independent coefficients, then the concept of frequency has great utility. Waves of each frequency propagate independently, so one can analyze the generation and propagation of each frequency separately, and then combine the results.

Nevertheless, linear acoustics should always be regarded as an approximation, and the understanding of nonlinear aspects of acoustics is of increasing importance in modern applications. Consequently, part of the task of the present chapter is to explain in what sense linear acoustics is an approximation. An extensive discussion of the extent to which the linearization approximation is valid is not attempted, but a discussion is given of the manner in which the linear equations result from nonlinear equations that are regarded as more nearly exact.

The historical origins [3.1] of linear acoustics reach back to antiquity, to *Pythagoras* and *Aristotle* [3.2], both of whom are associated with ancient Greece, and also to persons associated with other ancient civilizations. The mathematical theory began with Mersenne, Galileo, and Newton, and developed into its more familiar form during the time of *Euler* and *Lagrange* [3.3]. Prominent contributors during the 19-th century include Poisson, Laplace, Cauchy, Green, Stokes, Helmholtz, Kirchhoff, and Rayleigh. The latter's book [3.4], *The Theory of Sound*, is still widely read and quoted today.

3.2 Equations of Continuum Mechanics

Sound can propagate through liquids, gases, and solids. It can also propagate through composite media such as suspensions, mixtures, and porous media. In any portion of a medium that is of uniform composition, the general equations that are assumed to apply are those that are associated with the theory of continuum mechanics.

3.2.1 Mass, Momentum, and Energy Equations

The primary equations governing sound are those that account for the conservation of mass and energy, and for changes in momentum. These may be written in the form of either partial differential equations or integral equations. The former is the customary starting point for the derivation of approximate equations for linear acoustics. Extensive discussions of the equations of continuum mechanics can be found in texts by *Thompson* [3.5], *Fung* [3.6], *Shapiro* [3.7], *Batchelor* [3.8], *Truesdell* [3.9]), and *Landau* and *Lifshitz* [3.10], and in the encyclopedia article by *Truesdell* and *Toupin* [3.11].

The conservation of mass is described by the partial differential equation,

$$\frac{\partial \rho}{\partial t} + \nabla \cdot (\rho \mathbf{v}) = 0 \,, \tag{3.2}$$

where ρ is the (possibly position- and time-dependent) mass density (mass per unit volume of the material), and $\mathbf{v}$ is the local and instantaneous particle velocity, defined so that $\rho \mathbf{v} \cdot \mathbf{n}$ is the net mass flowing per unit time per unit area across an arbitrary stationary surface within the material whose local unit outward normal vector is $\mathbf{n}$.

The generalization of Newton's second law to a continuum is described by Cauchy's equation of motion, which is written in Cartesian coordinates as

$$\rho \frac{\mathrm{D} \mathbf{v}}{\mathrm{D} t} = \sum_{ij} \mathbf{e}_i \frac{\partial \sigma_{ij}}{\partial x_j} + \mathbf{g} \rho \,. \tag{3.3}$$

Here the Eulerian description is used, with each field variable regarded as a function of actual spatial position coordinates and time. (The alternate description is the Lagrangian description, where the field variables are regarded as functions of the coordinates that the material being described would have in some reference configuration.) The σ_{ij} are the Cartesian components of the stress tensor. The quantities $\mathbf{e}_i$ are the unit vectors in a Cartesian coordinate system. These stress tensor components are such that

$$\mathbf{t} = \sum_{i,j} \mathbf{e}_i \sigma_{ij} n_j \tag{3.4}$$

is the traction vector, the *surface* force per unit area on any given surface with local unit outward normal $\mathbf{n}$.

The second term on the right of (3.3) is a body-force term associated with gravity, with $\mathbf{g}$ representing the vector acceleration due to gravity. In some instances, it may be appropriate (as in the case of analyses of transduction) to include body-force terms associated with external electromagnetic fields, but such are excluded from consideration in the present chapter.

The time derivative operator on the left side of (3.3) is Stokes' total time derivative operator [3.12],

$$\frac{\mathrm{D}}{\mathrm{D}t} = \frac{\partial}{\partial t} + \mathbf{v} \cdot \nabla \,, \tag{3.5}$$

with the two terms corresponding to: (i) the time derivative as would be seen by an observer at rest, and (ii) the convective time derivative. This total time derivative applied to the particle velocity field yields the particle acceleration field, so the right side of (3.3) is the apparent force per unit volume on an element of the continuum.

The stress tensor's Cartesian component σ_{ij} is, in accordance with (3.4), the i-th component of the surface force per unit area on a segment of the (internal and hypothetical) surface of a small element of the continuum, when the unit outward normal to the surface is in

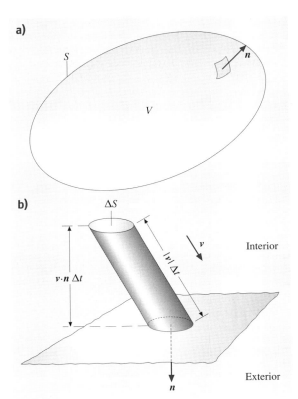

Fig. 3.1 Sketches supporting the identification of $\rho \mathbf{v}\cdot\mathbf{n}$ as mass flowing per unit area and per unit time through a surface whose unit outward normal is $\mathbf{n}$. All the matter in the slanted cylinder's volume flows through the area ΔS in time Δt

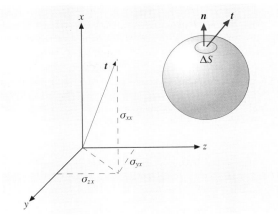

Fig. 3.2 Traction vector on a surface whose unit normal $\mathbf{n}$ is in the $+x$-direction

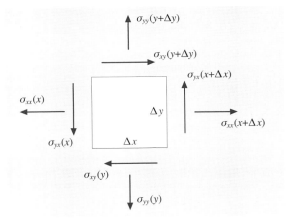

Fig. 3.3 Two-dimensional sketch supporting the identification of the force per unit volume associated with stress in terms of partial derivatives of the stress components

the j-th direction. This surface force is caused by interactions with the neighboring particles just outside the surface or by the transfer of momentum due to diffusion of molecules across the surface.

The stress-force term on the right side of (3.3) results from the surface-force definition of the stress components and from a version of the divergence theorem, so that the stress-related force per unit volume is

$$f_{\text{stress}} \to \frac{1}{V}\int_S \mathbf{t}(\mathbf{n})\,\mathrm{d}S \to \frac{1}{V}\int_S \sum_{ij} \mathbf{e}_i \sigma_{ij} n_j \, \mathrm{d}S$$

$$\to \frac{1}{V}\int_V \sum_{ij} \mathbf{e}_i \frac{\partial \sigma_{ij}}{\partial x_j}\,\mathrm{d}V \to \sum_{ij} \mathbf{e}_i \frac{\partial \sigma_{ij}}{\partial x_j}. \quad (3.6)$$

Here V is an arbitrary but small volume enclosing the point of interest, and S is the surface enclosing that volume; the quantity $\mathbf{n}$ is the unit outward normal vector to the surface.

Considerations of the net torque that such surface forces would exert on a small element, and the requirement that the angular acceleration of the element be finite in the limit of very small element dimensions, leads to the conclusion,

$$\sigma_{ij} = \sigma_{ji}\,, \quad (3.7)$$

so the stress tensor is symmetric.

The equations described above are supplemented by an energy conservation relation,

$$\rho\frac{\mathrm{D}}{\mathrm{D}t}\left(\tfrac{1}{2}v^2 + u\right) = \sum_{ij}\frac{\partial}{\partial x_j}\left(\sigma_{ij}v_i\right) + \rho\mathbf{g}\cdot\mathbf{v} - \nabla\cdot\mathbf{q}\,,$$

$$(3.8)$$

which involves the internal energy u per unit mass and the heat flux vector $\mathbf{q}$. The latter has the interpretation that its dot product with any unit normal vector represents the heat energy flowing per unit time and per unit area across any internal surface with the corresponding unit normal. An alternate version,

$$\rho \frac{Du}{Dt} = \sum_{ij} \sigma_{ij} \frac{\partial v_i}{\partial x_j} - \nabla \cdot \mathbf{q}, \quad (3.9)$$

of the energy equation results when the vector dot product of $\mathbf{v}$ with (3.3) is subtracted from (3.8). [In carrying through the derivation, one recognizes that $\mathbf{v}\cdot(\rho D\mathbf{v}/Dt)$ is $(1/2)\rho Dv^2/Dt$.]

3.2.2 Newtonian Fluids and the Shear Viscosity

The above equations apply to both fluids and solids. Additional equations needed to complete the set must take into account the properties of the material that make up the continuum. Air and water, for example, are fluids, and their viscous behavior corresponds to that for newtonian fluids, so that the stress tensor is given by

$$\sigma_{ij} = \sigma_n \delta_{ij} + \mu \phi_{ij}, \quad (3.10)$$

which involves the rate-of-shear tensor,

$$\phi_{ij} = \frac{\partial v_i}{\partial x_j} + \frac{\partial v_j}{\partial x_i} - \frac{2}{3}\nabla \cdot \mathbf{v} \delta_{ij}, \quad (3.11)$$

which is defined so that the sum of its diagonal elements is zero. The quantity σ_n that appears in (3.10) is the average normal stress component, and μ is the shear viscosity, while δ_{ij} is the Kronecker delta, equal to unity if the two indices are equal and otherwise equal to zero.

3.2.3 Equilibrium Thermodynamics

Further specification of the equations governing sound requires some assumptions concerning the extent to which equilibrium or near-equilibrium thermodynamics applies to acoustic disturbances. In the simplest idealization, which neglects relaxation processes, one assumes that the thermodynamic variables describing the perturbations are related just as for quasi-equilibrium processes, so one can use an equation of state of the form

$$s = s(u, \rho^{-1}), \quad (3.12)$$

which has the corresponding differential relation [3.13],

$$T\,ds = du + p\,d\rho^{-1}, \quad (3.13)$$

in accordance with the second law of thermodynamics. Here s is the entropy per unit mass, T is the absolute temperature, and p is the absolute pressure. The reciprocal of the density, which appears in the above relations, is recognized as the specific volume. (Here the adjective specific means per unit mass.)

For the case of an ideal gas with temperature-independent specific heats, which is a common idealization for air, the function in (3.12) is given by

$$s = \frac{R_0}{M}\ln\left(u^{1/(\gamma-1)}\rho^{-1}\right) + s_0. \quad (3.14)$$

Here s_0 is independent of u and ρ^{-1}, while M is the average molecular weight (average molecular mass in atomic mass units), and R_0 is the universal gas constant (equal to Boltzmann's constant divided by the mass in an atomic mass unit), equal to 8314 J/kgK. The quantity γ is the specific heat ratio, equal to approximately 7/5 for diatomic gases, to 5/3 for monatomic gases, and to 9/7 for polyatomic gases whose molecules are not collinear. For air (which is primarily a mixture of diatomic oxygen, diatomic nitrogen, with approximately 1% monatomic argon), γ is 1.4 and M is 29.0. (The expression given here for entropy neglects the contribution of internal vibrations of diatomic molecules, which cause the specific heats and γ to depend slightly on temperature. When one seeks to explain the absorption of sound in air, a nonequilibrium entropy [3.14] is used which depends on the fractions of O_2 and N_2 molecules that are in their first excited vibrational states, in addition to the quantities u and ρ^{-1}.)

For other substances, a knowledge of the first and second derivatives of s, evaluated at some representative thermodynamic state, is sufficient for most linear acoustics applications.

3.2.4 Bulk Viscosity and Thermal Conductivity

The average normal stress must equal the negative of the thermodynamic pressure that enters into (3.13), when the fluid is in equilibrium. Consideration of the requirement that the equations be the same regardless of the choice of directions of the coordinate axes, plus the assumption of a newtonian fluid, lead to the relation [3.12],

$$\sigma_n = -p + \mu_B \nabla \cdot \mathbf{v}, \quad (3.15)$$

which involves the bulk viscosity μ_B. Also, the Fourier model for heat conduction [3.15] requires that the heat flux vector $\mathbf{q}$ be proportional to the gradient of the temperature,

$$q = -\kappa \nabla T,\qquad(3.16)$$

where κ is the coefficient of thermal conduction.

Transport Properties of Air
Regarding the values of the viscosities and the thermal conductivity that enter into the above expressions, the values for air are approximated [3.16–18] by

$$\mu_S = \mu_0 \left(\frac{T}{T_0}\right)^{3/2} \frac{T_0 + T_S}{T + T_S},\qquad(3.17)$$

$$\mu_B = 0.6\mu_S,\qquad(3.18)$$

$$\kappa = \kappa_0 \left(\frac{T}{T_0}\right)^{3/2} \frac{T_0 + T_A e^{-T_B/T_0}}{T + T_A e^{-T_B/T}}.\qquad(3.19)$$

Here T_S is 110.4 K, T_A is 245.4 K, T_B is 27.6 K, T_0 is 300 K, μ_0 is 1.846×10^{-5} kg/(ms), and κ_0 is 2.624×10^{-2} W/(mK).

Transport Properties of Water
The corresponding values of the viscosities and the thermal conductivity of water depend only on temperature and are given by [3.19, 20]

$$\mu_S = 1.002 \times 10^{-3} e^{-0.0248 \Delta T},\qquad(3.20)$$

$$\mu_B = 3\mu_S,\qquad(3.21)$$

$$\kappa = 0.597 + 0.0017 \Delta T - 7.5 \times 10^{-6}(\Delta T)^2.\qquad(3.22)$$

The quantities that appear in these equations are understood to be in MKS units, and ΔT is the temperature relative to 283.16 K (10 °C).

3.2.5 Navier–Stokes–Fourier Equations

The assumptions represented by (3.12, 13, 15), and (3.16) cause the governing continuum mechanics equations for a fluid to reduce to

$$\frac{\partial \rho}{\partial t} + \nabla \cdot (\rho \mathbf{v}) = 0,\qquad(3.23)$$

$$\rho \frac{D\mathbf{v}}{Dt} = -\nabla p + \nabla(\mu_B \nabla \cdot \mathbf{v}) + \sum_{ij} \mathbf{e}_i \frac{\partial}{\partial x_j}(\mu \phi_{ij}) + \mathbf{g}\rho,\qquad(3.24)$$

$$\rho T \frac{Ds}{Dt} = \tfrac{1}{2}\mu \sum_{ij} \phi_{ij}^2 + \mu_B (\nabla \cdot \mathbf{v})^2 + \nabla \cdot (\kappa \nabla T),\qquad(3.25)$$

which are known as the Navier–Stokes–Fourier equations for compressible flow.

3.2.6 Thermodynamic Coefficients

An implication of the existence of an equation of state of the form of (3.12) is that any thermodynamic variable can be regarded as a function of any other two (independent) thermodynamic variables. The pressure p, for example, can be regarded as a function of ρ and T, or of s and ρ. In the expression of differential relations, such as that which gives dp as a linear combination of ds and $d\rho$, it is helpful to express the coefficients in terms of a relatively small number of commonly tabulated quantities. A standard set of such includes:

1. the square of the sound speed,

$$c^2 = \left(\frac{\partial p}{\partial \rho}\right)_s,\qquad(3.26)$$

2. the bulk modulus at constant entropy,

$$B_V = \rho c^2 = \rho \left(\frac{\partial p}{\partial \rho}\right)_s,\qquad(3.27)$$

3. the specific heat at constant pressure,

$$c_p = T \left(\frac{\partial s}{\partial T}\right)_p,\qquad(3.28)$$

4. the specific heat at constant volume,

$$c_v = T \left(\frac{\partial s}{\partial T}\right)_\rho,\qquad(3.29)$$

5. the coefficient of thermal expansion,

$$\beta = \rho \left(\frac{\partial (1/\rho)}{\partial T}\right)_p.\qquad(3.30)$$

(The subscripts on the partial derivatives indicate the independent thermodynamic quantity that is kept fixed during the differentiation.) The subscript V on B_V is included here to remind one that the modulus is associated with changes in volume. For a fixed mass of fluid the decrease in volume per unit volume per unit increase in pressure is the bulk compressibility, and the reciprocal of this is the bulk modulus.

The coefficients that are given here are related by the thermodynamic identity,

$$\gamma - 1 = T\beta^2 c^2 / c_p,\qquad(3.31)$$

where γ is the specific heat ratio,

$$\gamma = \frac{c_p}{c_v}.\qquad(3.32)$$

In terms of the thermodynamic coefficients defined above, the differential relations of principal interest in acoustics are

$$d\rho = \frac{1}{c^2} dp - \left(\frac{\rho \beta T}{c_p}\right) ds, \quad (3.33)$$

$$dT = \left(\frac{T\beta}{\rho c_p}\right) dp + \left(\frac{T}{c_p}\right) ds. \quad (3.34)$$

Thermodynamic Coefficients for an Ideal Gas

For an ideal gas, which is the common idealization for air, it follows from (3.12) and (3.13) (with the abbreviation R for R_0/M) that the thermodynamic coefficients are given by

$$c^2 = \gamma R T = \frac{\gamma p}{\rho}, \quad (3.35)$$

$$c_p = \frac{\gamma R}{\gamma - 1}, \quad (3.36)$$

$$c_v = \frac{R}{\gamma - 1}, \quad (3.37)$$

$$\beta = \frac{1}{T}. \quad (3.38)$$

For air, R is 287 J/kgK and γ is 1.4, so c_p and c_v are 1005 J/kgK and 718 J/kgK. A temperature of 293.16 K and a pressure of 10^5 Pa yield a sound speed of 343 m/s and a density ρ of 1.19 kg/m^3.

Thermodynamic Properties of Water

For pure water [3.21], the sound speed is approximately given in MKS units by

$$c = 1447 + 4.0\Delta T + 1.6 \times 10^{-6} p. \quad (3.39)$$

Here c is in meters per second, ΔT is temperature relative to 283.16 K (10 °C), and p is absolute pressure in pascals. The pressure and temperature dependence of the density [3.19] is approximately given by

$$\rho \approx 999.7 + 0.048 \times 10^{-5} p$$
$$- 0.088 \Delta T - 0.007(\Delta T)^2. \quad (3.40)$$

The coefficient of thermal expansion is given by

$$\beta \approx (8.8 + 0.022 \times 10^{-5} p + 1.4 \Delta T) \times 10^{-5}, \quad (3.41)$$

and the coefficient of specific heat at constant pressure is approximately given by

$$c_p \approx 4192 - 0.40 \times 10^{-5} p - 1.6 \Delta T. \quad (3.42)$$

The specific heat ratio is very close to unity, the deviation being described by

$$\frac{\gamma - 1}{\gamma} \approx 0.0011\left(1 + \frac{\Delta T}{6} + 0.0024 \times 10^{-5} p\right)^2. \quad (3.43)$$

The values for sea water are somewhat different, because of the presence of dissolved salts. An approximate expression for the speed of sound in sea water [3.22] is given by

$$c \approx 1490 + 3.6\Delta T + 1.6 \times 10^{-6} p + 1.3 \Delta S. \quad (3.44)$$

Here ΔS is the deviation of the salinity in parts per thousand from a nominal value of 35.

3.2.7 Ideal Compressible Fluids

If viscosity and thermal conductivity are neglected at the outset, then the Navier–Stokes equation (3.24) reduces to

$$\rho \frac{D\mathbf{v}}{Dt} = -\nabla p + \mathbf{g}\rho, \quad (3.45)$$

which is known as the Euler equation. The energy equation (3.25) reduces to the isentropic flow condition

$$\frac{Ds}{Dt} = 0. \quad (3.46)$$

Moreover, the latter in conjunction with the differential equation of state (3.33) yields

$$\frac{Dp}{Dt} = c^2 \frac{D\rho}{Dt}. \quad (3.47)$$

The latter combines with the conservation of mass relation (3.2) to yield

$$\frac{Dp}{Dt} + \rho c^2 \nabla \cdot \mathbf{v} = 0, \quad (3.48)$$

where ρc^2 is recognized as the bulk modulus B_V.

In many applications of acoustics, it is these equations that are used as a starting point.

3.2.8 Suspensions and Bubbly Liquids

Approximate equations of the same general form as those for an ideal fluid result for fluids that have small particles of a different material suspended in them. Let the fluid itself have ambient density ρ_1 and let the material of the suspended particles have ambient density ρ_2.

The fractions of the overall volume that are occupied by the two materials are denoted by f_1 and f_2, so that $f_1 + f_2 = 1$. Thus, a volume V contains a mass

$$M = \rho_1 f_1 V + \rho_2 f_2 V. \quad (3.49)$$

The equivalent density M/V is consequently

$$\rho_{eq} = \rho_1 f_1 + \rho_2 f_2 . \quad (3.50)$$

The equivalent bulk modulus is deduced by considering the decrease $-\Delta V$ of such a volume due to an increment Δp of pressure with the assumption that the pressure is uniform throughout the volume, so that

$$-\Delta V = V_1 \frac{\Delta p}{B_{V,1}} + V_2 \frac{\Delta p}{B_{V,2}} , \quad (3.51)$$

where $V_1 = f_1 V$ and $V_2 = f_2 V$ are the volumes occupied by the two materials. The equivalent bulk modulus consequently satisfies the relation

$$\frac{1}{B_{V,eq}} = \frac{f_1}{B_{V,1}} + \frac{f_2}{B_{V,2}} . \quad (3.52)$$

The bulk modulus for any material is the density times the sound speed squared, so the effective sound speed for the mixture satisfies the relation

$$\frac{1}{\rho_{eq} c_{eq}^2} = \frac{f_1}{\rho_1 c_1^2} + \frac{f_2}{\rho_2 c_2^2} , \quad (3.53)$$

whereby

$$\frac{1}{c_{eq}^2} = \left(\frac{f_1}{\rho_1 c_1^2} + \frac{f_2}{\rho_2 c_2^2} \right) (\rho_1 f_1 + \rho_2 f_2) . \quad (3.54)$$

The latter relation, rewritten in an equivalent form, is sometimes referred to as *Wood's* equation [3.23], and can be traced back to a paper by *Mallock* [3.24].

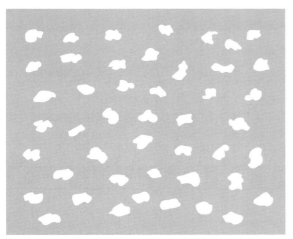

Fig. 3.4 Sketch of a fluid with embedded particles suspended within it

3.2.9 Elastic Solids

For analyses of acoustic wave propagation in solids (including crystals), the idealization of a linear elastic solid [3.6, 25–28] is often used. The solid particle displacements are regarded as sufficiently small that convective time derivatives can be neglected, so the Cauchy equation, given previously by (3.3), reduces, with the omission of the invariably negligible gravity term, to an equation, whose i-th Cartesian component is

$$\rho \frac{\partial^2 \xi_i}{\partial t^2} = \sum_{j=1}^{3} \frac{\partial \sigma_{ij}}{\partial x_j} . \quad (3.55)$$

Here ρ is the density; $\xi_i(\mathbf{x}, t)$ is the i-th Cartesian component of the displacement of the particle nominally at $\mathbf{x}$. This displacement is measured relative to the particle's ambient position. Also, the ambient stress is presumed sufficiently small or slowly varying with position that it is sufficient to take the stress components that appear in Cauchy's equation as being those associated with the disturbance. These should vanish when there is no strain associated with the disturbance.

The stress tensor components σ_{ij} are ordinarily taken to be linearly dependent on the (linear) strain components,

$$\epsilon_{ij} = \frac{1}{2} \left(\frac{\partial \xi_i}{\partial x_j} + \frac{\partial \xi_j}{\partial x_i} \right) , \quad (3.56)$$

so that

$$\sigma_{ij} = \sum_{kl} K_{ijkl} \epsilon_{kl} . \quad (3.57)$$

The number of required distinct elastic coefficients is limited for various reasons. One is that the stress tensor is symmetric and another is that the strain tensor is symmetric. Also, if the quasistatic work required to achieve a given state of deformation is independent of the detailed history of the deformation, then there must be a strain-energy density function $\mathfrak{U}$ that, in the limit of small deformations, is a bilinear function of the strains. There are only six different strain components and the number of their distinct products is $6+5+4+3+2+1 = 21$, and such is the maximum number of coefficients that one might need. Any intrinsic symmetry, such as exists in various types of crystals, will reduce this number further [3.29].

Strain Energy

The increment of work done during a deformation by stress components on a small element of a solid that is initially a small box (Δx by Δy by Δz) is

$$\delta[\text{Work}] = \sum_i \sigma_{ii}\left(\frac{\partial}{\partial x_i}\delta\xi_i\right)\Delta x \Delta y \Delta z$$
$$+ \sum_{i \neq j} \sigma_{ij}\left(\frac{\partial}{\partial x_j}\delta\xi_i\right)\Delta x \Delta y \Delta z, \quad (3.58)$$

where here σ_{ij} is interpreted as the i-th component of the force per unit area on a surface whose outward normal is in the direction of increasing x_j. With use of the symmetry of the stress tensor and of the definition of the strain tensor components, one obtains the incremental strain energy per unit volume as

$$\delta\mathfrak{U} = \sum_{ij} \sigma_{ij}\delta\epsilon_{ij}. \quad (3.59)$$

If the reference state, where $\mathfrak{U} = 0$, is taken as that in which the strain is zero, and given that each stress component is a linear combination of the strain components, then the above integrates to

$$\mathfrak{U} = \sum_{ij} \frac{1}{2}\sigma_{ij}\epsilon_{ij}. \quad (3.60)$$

Here the stresses are understood to be given in terms of the strains by an appropriate stress–strain relation of the general form of (3.57).

Because of the symmetry of both the stress tensor and the strain tensor, the internal energy $\mathfrak{U}$ per unit can be regarded as a function of only the six distinct strain components: $\epsilon_{xx}, \epsilon_{yy}, \epsilon_{zz}, \epsilon_{xy}, \epsilon_{yz}$, and ϵ_{xz}. Consequently, with such an understanding, it follows from the differential relation (3.59) that

$$\frac{\partial \mathfrak{U}}{\partial \epsilon_{xx}} = \sigma_{xx}; \quad \frac{\partial \mathfrak{U}}{\partial \epsilon_{xy}} = 2\sigma_{xy}, \quad (3.61)$$

with analogous relations for derivatives with respect to the other distinct strain components.

Isotropic Solids

If the properties of the solid are such that they are independent of the orientation of the axes of the Cartesian coordinate system, then the solid is said to be isotropic. This idealization is usually regarded as good if the solid is made up of a random assemblage of many small grains. The tiny grains are possibly crystalline with directional properties, but the orientation of the grains is random, so that an element composed of a large number of grains has no directional orientation. For such an isotropic solid the number of

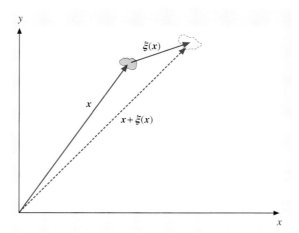

Fig. 3.5 Displacement field vector $\boldsymbol{\xi}$ in a solid. A material point normally at $\boldsymbol{x}$ is displaced to $\boldsymbol{x} + \boldsymbol{\xi}(\boldsymbol{x})$

different coefficients in the stress–strain relation (3.57) is only two, and the relation can be taken in the general form

$$\sigma_{ij} = 2\mu_\text{L}\epsilon_{ij} + \lambda_\text{L}\delta_{ij}\sum_{k=1}^{3}\epsilon_{kk}. \quad (3.62)$$

This involves two material constants, termed the Lamè constants, and here denoted by λ_L and μ_L (the latter being the same as the shear modulus G).

Equivalently, one can express the strain components in terms of the stress components by the inverse of the above set of relations. The generic form is sometimes written

$$\epsilon_{ij} = \frac{1+\nu}{E}\sigma_{ij} - \frac{\nu}{E}\sum_k \sigma_{kk}\delta_{ij}, \quad (3.63)$$

where E is the elastic modulus and ν is Poisson's ratio. The relation of these two constants to the Lamè constants is such that

$$\lambda_\text{L} = \frac{\nu E}{(1+\nu)(1-2\nu)}, \quad (3.64)$$

$$\mu_\text{L} = G = \frac{E}{2(1+\nu)}. \quad (3.65)$$

In undergraduate texts on the mechanics of materials, the relations (3.63) are often written out separately for the diagonal and off-diagonal strain elements, so

Table 3.1 Representative numbers for the elastic properties of various materials. (A comprehensive data set, giving values appropriate to specific details concerning the circumstances and the nature of the specimens, can be found at the web site for MatWeb, Material Property Data)

Material	E (Pa)	ν	λ_L (Pa)	μ_L (Pa)	ρ (kg/m^3)
Aluminum	7.0×10^{10}	0.33	5.1×10^{10}	2.6×10^{10}	2.8×10^3
Brass	10.5×10^{10}	0.34	8.3×10^{10}	3.9×10^{10}	8.4×10^3
Copper	12.0×10^{10}	0.34	9.5×10^{10}	4.5×10^{10}	8.9×10^3
Iron	12.0×10^{10}	0.25	4.8×10^{10}	4.8×10^{10}	7.2×10^3
Lead	1.4×10^{10}	0.43	3.2×10^{10}	0.5×10^{10}	11.3×10^3
Steel	20.0×10^{10}	0.28	9.9×10^{10}	7.8×10^{10}	7.8×10^3
Titanium	11.0×10^{10}	0.34	8.7×10^{10}	4.1×10^{10}	4.5×10^3

that,

$$\epsilon_{xx} = \frac{1}{E}[\sigma_{xx} - \nu(\sigma_{yy} + \sigma_{zz})] \,, \qquad (3.66)$$

$$\gamma_{xy} = 2\epsilon_{xy} = \frac{1}{G}\sigma_{xy} \,, \qquad (3.67)$$

with analogous relations for the other strain components. The nomenclature γ_{xy} is used for what is often termed the shear strain, this being twice as large as that shear strain that is defined as a component of a strain tensor.

The strain energy density $\mathfrak{U}$ of an isotropic solid results from (3.60) and (3.62), the result being

$$\mathfrak{U} = \mu_L \sum_{ij} \epsilon_{ij}^2 + \frac{1}{2}\lambda_L \left(\sum_k \epsilon_{kk}\right)^2 . \qquad (3.68)$$

In the idealized and simplified model of acoustic propagation in solids, the conservation of mass and energy is not explicitly invoked. The density that appears on the right side of (3.55) is regarded as time independent, and as that appropriate for the undisturbed state of the solid.

3.3 Equations of Linear Acoustics

The equations that one ordinarily deals with in acoustics are linear in the field amplitudes, where the fields of interest are quantities that depend on position and time. The basic governing equations are partial differential equations.

3.3.1 The Linearization Process

Sound results from a time-varying perturbation of the dynamic and thermodynamic variables that describe the medium. For sound in fluids (liquids and gases), the quantities appropriate to the ambient medium (i. e., the medium in the absence of a disturbance) are customarily represented [3.30] by the subscript 0, and the perturbations are represented by a prime on the corresponding symbol. Thus one expresses the total pressure as

$$p = p_0 + p' \,, \qquad (3.69)$$

with corresponding expressions for fluctuations in specific entropy, fluid velocity, and density. The linear equations that govern acoustical disturbances are then determined by the first-order terms in the expansion of the governing nonlinear equations in the primed variables. The zeroth-order terms cancel out completely because the ambient variables should themselves correspond to a valid state of motion of the medium. Thus, for example, the linearized version of the conservation of mass relation in (3.2) is

$$\frac{\partial \rho'}{\partial t} + \nabla \cdot (\boldsymbol{v}_0 \rho' + \boldsymbol{v}' \rho_0) = 0 \,. \qquad (3.70)$$

The possible forms of the linear equations differ in complexity according to what is assumed about the medium's ambient state and according to what dissipative terms are included in the original governing equations. In the establishment of a rationale for using simplified models, it is helpful to think in terms of the order of magnitudes and characteristic scales of the coefficients in more nearly comprehensive models. If the spatial region of interest has bounding dimensions that are smaller than any scale length over which the ambient variables vary by a respectable fraction, then it may be appropriate to idealize the coefficients in the governing equations as if the ambient medium were spatially uniform or homogeneous. Similarly, if the characteristic wave periods or propagation times are much less than characteristic time scales for the ambient medium, then it may be appropriate to idealize the coefficients as being time independent. Examination of orders of magnitudes of terms suggests that the ambient velocity may be neglected if it is much less than the sound speed c. The first-order perturbation to the gravitational force term can ordinarily be neglected [3.31] if the previously stated conditions are met and if the quantity g/c is sufficiently smaller than any characteristic frequency of the disturbance. Analogous inferences can be made about the thermal conductivity and viscosity. However, the latter may be important [3.32] in the interaction of acoustic waves in fluids with adjacent solid bodies.

A discussion of the restrictions on using linear equations and of neglecting second- and higher-order terms in the primed variables is outside the scope of the present chapter, but it should be noted that one regards p' as small [3.30] if it is substantially less than $\rho_0 c^2$, and $|v'|$ as small if it is much less than c, where c is the sound speed defined via (3.26). It is not necessary that p' be much less than p_0, and it is certainly not necessary that $|v'|$ be less than $|v_0|$.

3.3.2 Linearized Equations for an Ideal Fluid

The customary equations for linear acoustics neglect dissipation processes and consequently can be derived from the equations for flow of a compressible ideal fluid, given in (3.45) through (3.48). If one neglects gravity at the outset, and assumes the ambient fluid velocity is zero, then the ambient pressure is constant. In such a case, (3.48) leads after linearization to

$$\frac{\partial p}{\partial t} + \rho c^2 \nabla \cdot \mathbf{v} = 0, \tag{3.71}$$

and the Euler equation (3.45) leads to

$$\rho \frac{\partial \mathbf{v}}{\partial t} = -\nabla p. \tag{3.72}$$

Here a common notational convention is used to delete primes and subscripts. The density ρ here is understood to be the ambient density ρ_0, while p and $\mathbf{v}$ are understood to be the acoustically induced perturbations to the pressure and fluid velocity.

These two coupled equations for p and $\mathbf{v}$ remain applicable when the ambient density and sound speed vary with position.

3.3.3 The Wave Equation

A single partial differential equation for the acoustic part of the pressure results when one takes the time derivative of (3.71) and then uses (3.72) to reexpress the time derivative of the fluid velocity in terms of pressure. The resulting equation, as derived by Bergmann [3.31] for the case when the density varies with position, is

$$\nabla \cdot \left(\frac{1}{\rho}\nabla p\right) - \frac{1}{\rho c^2}\frac{\partial^2 p}{\partial t^2} = 0. \tag{3.73}$$

If the ambient density is independent of position, then this reduces to

$$\nabla^2 p - \frac{1}{c^2}\frac{\partial^2 p}{\partial t^2} = 0, \tag{3.74}$$

which dates back to Euler and which is what is ordinarily termed the wave equation of linear acoustics. Often this is written as

$$\Box^2 p = 0, \tag{3.75}$$

in terms of the d'Alembertian operator defined by

$$\Box^2 = \nabla^2 - c^{-2}\partial^2/\partial t^2. \tag{3.76}$$

3.3.4 Wave Equations for Isotropic Elastic Solids

When the ambient density and the Lamé constants are independent of position, the Cauchy equation and the stress–strain relations described by (3.55, 56), and (3.62) can be combined to the single vector equation,

$$\frac{\partial^2 \boldsymbol{\xi}}{\partial t^2} = (c_1^2 - c_2^2)\nabla(\nabla \cdot \boldsymbol{\xi}) + c_2^2 \nabla^2 \boldsymbol{\xi}, \tag{3.77}$$

which is equivalently written as

$$\frac{\partial^2 \boldsymbol{\xi}}{\partial t^2} = c_1^2 \nabla(\nabla \cdot \boldsymbol{\xi}) - c_2^2 \nabla \times (\nabla \times \boldsymbol{\xi}). \tag{3.78}$$

Here the quantities c_1 and c_2 are defined by

$$c_1^2 = \frac{\lambda_L + 2\mu_L}{\rho}, \tag{3.79}$$

$$c_2^2 = \frac{\mu_L}{\rho}. \tag{3.80}$$

Because any vector field may be decomposed into a sum of two fields, one with zero curl and the other with zero divergence, one can set the displacement vector $\boldsymbol{\xi}$ to

$$\boldsymbol{\xi} = \nabla \Phi + \nabla \times \boldsymbol{\Psi} \tag{3.81}$$

in terms of a scalar and a vector potential. This expression will satisfy (3.77) identically, provided the two potentials separately satisfy

$$\nabla^2 \Phi - \frac{1}{c_1^2} \frac{\partial^2 \Phi}{\partial t^2} = 0, \tag{3.82}$$

$$\nabla^2 \boldsymbol{\Psi} - \frac{1}{c_2^2} \frac{\partial^2 \boldsymbol{\Psi}}{\partial t^2} = 0. \tag{3.83}$$

Both of these equations are of the form of the simple wave equation of (3.74). The first corresponds to longitudinal wave propagation, and the second corresponds to shear wave propagation. The quantities c_1 and c_2 are referred to as the dilatational and shear wave speeds, respectively.

If the displacement field is irrotational, so that the curl of the displacement vector is zero (as is so for sound in fluids), then each component of the displacement field satisfies (3.82). If the displacement field is solenoidal, so that its divergence is zero, then each component of the displacement satisfies (3.83).

3.3.5 Linearized Equations for a Viscous Fluid

For the restricted but relatively widely applicable case when the Navier–Stokes–Fourier equations are presumed to hold and the characteristic scales of the ambient medium and the disturbance are such that the ambient flow is negligible, and the coefficients are idealizable as constants, the linear equations have the form appropriate for a disturbance in a homogeneous, time-independent, non-moving medium, these equations being

$$\frac{\partial \rho'}{\partial t} + \rho_0 \nabla \cdot \boldsymbol{v}' = 0, \tag{3.84}$$

$$\rho_0 \frac{\partial \boldsymbol{v}'}{\partial t} = -\nabla p' + \left(\frac{1}{3}\mu + \mu_B\right) \nabla (\nabla \cdot \boldsymbol{v}') + \mu \nabla^2 \boldsymbol{v}', \tag{3.85}$$

$$\rho_0 T_0 \frac{\partial s'}{\partial t} = \kappa \nabla^2 T', \tag{3.86}$$

$$\rho' = \frac{1}{c^2} p' - \left(\frac{\rho \beta T}{c_p}\right)_0 s', \tag{3.87}$$

$$T' = \left(\frac{T\beta}{\rho c_p}\right)_0 p' + \left(\frac{T}{c_p}\right)_0 s'. \tag{3.88}$$

The primes on the perturbation field variables are needed here to distinguish them from the corresponding ambient quantities.

3.3.6 Acoustic, Entropy, and Vorticity Modes

In general, any solution of the linearized equations for a fluid with viscosity and thermal conductivity can be regarded [3.30, 32–34] as a sum of three basic types of solutions. The common terminology for these is: (i) the acoustic mode, (ii) the entropy mode, and (iii) the vorticity mode. Thus, with appropriate subscripts denoting the various fundamental mode types, one writes the fluid velocity perturbation in the form,

$$\boldsymbol{v}' = \boldsymbol{v}_{\text{ac}} + \boldsymbol{v}_{\text{ent}} + \boldsymbol{v}_{\text{vor}}, \tag{3.89}$$

with similar decompositions for the other field variables. The decomposition is intended to be such that, for waves of constant frequency, each field variable's contribution from any given mode satisfies a partial differential equation that is second order [rather than of some higher order as might be derived from (3.84)–(3.88)] in the spatial derivatives. That such a decomposition is possible follows from a theorem [3.35, 36] that any solution of a partial differential equation of, for example, the form,

$$(\nabla^2 + \lambda_1)(\nabla^2 + \lambda_2)(\nabla^2 + \lambda_3)\psi = 0, \tag{3.90}$$

can be written as a sum,

$$\psi = \psi_1 + \psi_2 + \psi_3, \tag{3.91}$$

where the individual terms each satisfy second-order differential equations of the form,

$$(\nabla^2 + \lambda_i)\psi_i = 0. \tag{3.92}$$

This decomposition is possible providing no two of the λ_i are equal. [In regard to (3.90), one should note that each of the three operator factors is a second-order differential operator, so the overall equation is a sixth-order partial differential equation. One significance of the theorem is that one has replaced the seemingly formidable problem of solving a sixth-order partial differential equation by that of solving three second-order partial differential equations.]

Linear acoustics is primarily concerned with solutions of the equations that govern the acoustic mode, while conduction heat transfer is primarily concerned with solutions of the equations that govern the entropy mode. However, the simultaneous consideration of all three solutions is often necessary when one seeks to satisfy boundary conditions at solid surfaces. The decomposition becomes relatively simple when the characteristic angular frequency is sufficiently small that

$$\omega \ll \frac{\rho c^2}{(4/3)\mu + \mu_B}, \quad (3.93)$$

$$\omega \ll \frac{\rho c^2 c_p}{\kappa}. \quad (3.94)$$

These conditions are invariably well met in all applications of interest. For air the right sides of the two inequalities correspond to frequencies of the order of 10^9 Hz, while for water they correspond to frequencies of the order of 10^{12} Hz. If the inequalities are not satisfied, it is also possible that the Navier–Stokes–Fourier model is not appropriate, since the latter presumes that the variations are sufficiently slow that the medium can be regarded as being in quasi-thermodynamic equilibrium.

The acoustic mode and the entropy mode are both characterized by the vorticity (curl of the fluid velocity) being identically zero. (This characterization is consistent with the theorem that any vector field can be decomposed into a sum of a field with zero divergence and a field with zero curl.) In the limit of zero viscosity and zero thermal conductivity [which is a limit consistent with (3.93) and (3.94)], the acoustic mode is characterized by zero entropy perturbations, while the entropy mode is a time-independent entropy perturbation, with zero pressure perturbation. These identifications allow the approximate equations governing the various modes to be developed in the limit for which the inequalities are valid. In carrying out the development, it is convenient to use an expansion parameter defined by,

$$\epsilon = \frac{\omega \mu}{\rho_0 c^2}, \quad (3.95)$$

where ω is a characteristic angular frequency, or equivalently, the reciprocal of a characteristic time for the perturbation. In the ordering, ϵ is regarded as small, μ_B/μ and $\kappa/\mu c_p$ are regarded as of order unity, and all time derivative operators are regarded as of order ω.

Acoustic Mode

To zeroth order in ϵ, the acoustic mode is what results when the thermal conductivity and the viscosities are identically zero and the entropy perturbation is also zero. In such a case, the relations (3.71) and (3.72) are valid and one obtains the wave equation of (3.74). All of the field variables in the acoustic mode satisfy this wave equation to zeroth order in ϵ. This is so because the governing equations have constant coefficients.

When one wishes to take into account the corrections that result to first order in ϵ, the entropy perturbation is no longer regarded as identically zero, but its magnitude can be deduced by replacing the T' that appears in (3.86) by only the first term in the sum of (3.88), the substitution being represented by

$$T_{ac} = \frac{T_0 \beta}{\rho_0 c_p} p_{ac}. \quad (3.96)$$

To the same order of approximation one may use the zeroth-order wave equation (3.74) to express the right side of (3.86) in terms of a second-order time derivative of the acoustic-mode pressure. The resulting equation can then be integrated once in time with the constant of integration set to zero because the acoustic mode's entropy perturbation must vanish when the corresponding pressure perturbation is identically zero. Thus one obtains the acoustic-mode entropy as

$$s_{ac} \approx \frac{\kappa \beta}{\rho_0^2 c^2 c_p} \frac{\partial p_{ac}}{\partial t}, \quad (3.97)$$

correct to first order in ϵ.

This expression for the entropy perturbation is substituted into (3.87) and the resulting expression for ρ_{ac} is substituted into the conservation of mass relation, represented by (3.84), so that one obtains

$$\frac{\partial p_{ac}}{\partial t} - \frac{(\gamma-1)\kappa}{\rho_0 c^2 c_p} \frac{\partial^2 p_{ac}}{\partial t^2} + \rho_0 c^2 \nabla \cdot \mathbf{v}_{ac} = 0. \quad (3.98)$$

The vorticity in the acoustic mode is zero,

$$\nabla \times \mathbf{v}_{ac} = 0, \quad (3.99)$$

so the momentum balance equation (3.85) simplifies to one where the right side is a gradient. Then with an appropriate substitution for the divergence of the fluid velocity from (3.84), one finds that the momentum balance equation to first order in ϵ takes the form

$$\rho_0 \frac{\partial \mathbf{v}_{ac}}{\partial t} = -\nabla \left\{ p_{ac} + \frac{1}{\rho_0 c^2}[(4/3)\mu + \mu_B] \frac{\partial p_{ac}}{\partial t} \right\}. \quad (3.100)$$

The latter and (3.84), in a manner similar to that in which the wave equation of (3.73) is derived, yield the dissipative wave equation represented by

$$\nabla^2 p_{ac} - \frac{1}{c^2}\frac{\partial^2 p_{ac}}{\partial t^2} + \frac{2\delta_{cl}}{c^4}\frac{\partial^3 p_{ac}}{\partial t^3} = 0. \quad (3.101)$$

The quantity δ_{cl}, defined by

$$\delta_{cl} = \frac{1}{2\rho_0}\left[(4/3)\mu + \mu_B + (\gamma-1)(\kappa/c_p)\right], \quad (3.102)$$

is a constant that characterizes the classical dissipation in the acoustic mode.

Entropy Mode

For the entropy mode, the pressure perturbation vanishes to zeroth order in ϵ, while the entropy perturbation does not. Thus for this mode, (3.87) and (3.88) approximate in zeroth order to

$$\rho_{ent} = -\left(\frac{\rho\beta T}{c_p}\right)_0 s_{ent}, \quad (3.103)$$

$$T_{ent} = \left(\frac{T}{c_p}\right)_0 s_{ent}. \quad (3.104)$$

The latter when substituted into the energy equation (3.86) yields the diffusion equation of conduction heat transfer,

$$\frac{\partial s_{ent}}{\partial t} = \frac{\kappa}{\rho_0 c_p}\nabla^2 s_{ent}. \quad (3.105)$$

All of the entropy-mode field variables, including T_{ent}, satisfy this diffusion equation to lowest order in ϵ.

The mass conservation equation, with an analogous substitution from (3.103) and another substitution from (3.88), integrates to

$$\boldsymbol{v}_{ent} = \frac{\beta T_0 \kappa}{\rho_0 c_p^2}\nabla s_{ent}, \quad (3.106)$$

so there is a weak fluid flow caused by entropy or temperature gradients in the entropy mode. The substitution of this expression for the fluid velocity into the linearized Navier–Stokes equation (3.85) subsequently yields a lowest-order expression for the entropy-mode contribution to the pressure perturbation, this being

$$p_{ent} = \frac{\beta T_0}{c_p}\left[(4/3)\mu + \mu_B - (\kappa/c_p)\right]\frac{\partial s_{ent}}{\partial t}. \quad (3.107)$$

Vorticity Mode

For the vorticity mode, the divergence of the fluid velocity is identically zero,

$$\nabla \cdot \boldsymbol{v}_{vor} = 0. \quad (3.108)$$

The mass conservation equation (3.84) and the thermodynamic relations, (3.87) and (3.88), consequently require that all of the thermodynamic field quantities with subscript 'vor' are zero. The Navier–Stokes equation (3.85) accordingly requires that each component of the fluid velocity associated with this mode satisfy the vorticity diffusion equation

$$\rho_0 \frac{\partial \boldsymbol{v}_{vor}}{\partial t} = \mu \nabla^2 \boldsymbol{v}_{vor}. \quad (3.109)$$

3.3.7 Boundary Conditions at Interfaces

For the model of a fluid without viscosity or thermal conduction, such as is governed by (3.71)–(3.75), the appropriate boundary conditions at an interface are that the normal component of the fluid velocity be continuous and that the pressure be continuous;

$$\boldsymbol{v}_1 \cdot \boldsymbol{n}_{12} = \boldsymbol{v}_2 \cdot \boldsymbol{n}_{12}; \qquad p_1 = p_2. \quad (3.110)$$

At a rigid nonmoving surface, the normal component must consequently vanish,

$$\boldsymbol{v} \cdot \boldsymbol{n} = 0, \quad (3.111)$$

but no restrictions are placed on the tangential component of the velocity. The model also places no requirements on the value of the temperature at a solid surface.

When viscosity is taken into account, no matter how small it may be, the model demands additional boundary conditions. An additional condition invariably imposed is that the tangential components of the velocity also be continuous, the rationale [3.37] being that a fluid should not slip any more freely with respect to an interface than it does with itself; this lack of slip is invariably observed [3.38] when the motion is examined sufficiently close to an interface. Similarly, when thermal conduction is taken into account, the rational interpretation of the model requires that the temperature be continuous at an interface. Newton's third law requires that the shear stresses exerted across an interface be continuous, and the conservation of energy requires that the normal component of the heat flux be continuous.

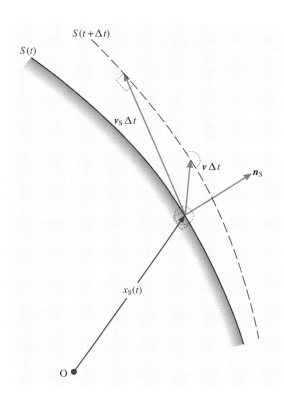

Fig. 3.6 Graphical proof that the normal component of fluid velocity should be continuous at any interface. The interface surface is denoted by $S(t)$, and $x_S(t)$ is the location of any material particle on either side of the interface, but located just at the interface. Both it and the particle on the other side of the interface that neighbors it at time t will continue to stay on the interface, although they may be displaced tangentially

The apparent paradox, that tiny but nonzero values of the viscosity and thermal conductivity should drastically alter the basic nature of the boundary conditions at an interface, is resolved by acoustic boundary-layer theory. Under normal circumstances, disturbances associated with sound are almost entirely in the acoustic mode except in oscillating acoustic boundary layers near interfaces. In these boundary layers, a portion of the disturbance is in the entropy and vorticity modes. These modes make insignificant contributions to the total perturbation pressure and normal velocity component, but have temperature and tangential velocity components, respectively, that are comparable to those of the acoustic mode. The boundary and continuity conditions can be satisfied by the sum of the three mode fields, but not by the acoustic mode field alone. However, the entropy and vorticity mode fields typically die out rapidly with distance from an interface. The characteristic decay lengths [3.30] when the disturbance has an angular frequency ω are

$$\ell_{\text{vor}} = \left(\frac{2\mu}{\omega\rho}\right)^{1/2}, \quad (3.112)$$

$$\ell_{\text{ent}} = \left(\frac{2\kappa}{\omega\rho c_p}\right)^{1/2}. \quad (3.113)$$

Given that the parameter ϵ in (3.95) is much less than unity, these lengths are much shorter than a characteristic wavelength. If they are, in addition, much shorter than the dimensions of the space to which the acoustic field is confined, then the acoustic boundary layer has a minor effect on the acoustic field outside the boundary layer. However, because dissipation of energy can occur within such boundary layers, the existence of a boundary layer can have a long-term accumulative effect. This can be manifested by the larger attenuation of sound waves propagating in pipes [3.39] and in the finite Q (quality factors) of resonators.

3.4 Variational Formulations

Occasionally, the starting point for acoustical analysis is not a set of partial differential equations, such as those described in the previous section, but instead a variational principle. The attraction of such an alternate approach is that it is a convenient departure point for making approximations. The most fundamental of the variational principles is the version of Hamilton's principle [3.40, 41] that applies to a continuum.

3.4.1 Hamilton's Principle

The standard version is that where the displacement field $\xi(x, t)$ is taken as the field variable. The medium is assumed to have no ambient motion, and viscosity and thermal conduction are ignored.

At any given instant and near any given point there is a kinetic energy density taken as

$$\mathfrak{T} = \frac{1}{2}\rho \sum_i \left(\frac{\partial \xi_i}{\partial t}\right)^2 . \tag{3.114}$$

This is the kinetic energy per unit volume; the quantity ρ is understood to be the ambient density, this being an appropriate approximation given that the principle should be consistent with a linear formulation.

The second density of interest is the strain energy density, $\mathfrak{U}$, which is presumed to be a bilinear function of the displacement field components and of their spatial derivatives. For an isotropic linear elastic solid, this is given by (3.68), which can be equivalently written, making use of the definition of the strain components, as

$$\mathfrak{U} = \frac{1}{4}\mu_L \sum_{ij} \left(\frac{\partial \xi_i}{\partial x_j} + \frac{\partial \xi_j}{\partial x_i}\right)^2 + \frac{1}{2}\lambda_L (\nabla \cdot \boldsymbol{\xi})^2 . \tag{3.115}$$

For the case of a compressible fluid, the stress tensor is given by

$$\sigma_{ij} = -p\delta_{ij} = \rho c^2 \nabla \cdot \boldsymbol{\xi}\, \delta_{ij} . \tag{3.116}$$

Here p is the pressure associated with the disturbance, and the second version follows from the time integration of (3.71). The general expression (3.60) consequently yields

$$\mathfrak{U} = \frac{1}{2}\rho c^2 (\nabla \cdot \boldsymbol{\xi})^2 \tag{3.117}$$

for the strain energy density in a fluid.

There are other examples that can be considered for which there are different expressions for the strain energy density.

Given the kinetic and strain energy densities, one constructs a Lagrangian density,

$$\mathfrak{L} = \mathfrak{T} - \mathfrak{U} . \tag{3.118}$$

Hamilton's principle then takes the form

$$\delta \iiiint \mathfrak{L}\, dx\, dy\, dz\, dt = 0 . \tag{3.119}$$

Here the integration is over an arbitrary volume and arbitrary time interval. The symbol δ implies a variation from the exact solution of the problem of interest. In such a variation, the local and instantaneous value $\boldsymbol{\xi}(\boldsymbol{x}, t)$ is considered to deviate by an infinitesimal amount $\delta \boldsymbol{\xi}(\boldsymbol{x}, t)$.

The variation of the Lagrangian is calculated in the same manner as one determines differentials using the chain rule of differentiation. One subsequently uses the fact that a variation of a derivative is the derivative of a variation. One then integrates by parts, whenever necessary, so that one eventually obtains, an equation of the form

$$\iiiint \sum_i [\text{expression}]_i\, \delta\xi_i\, dx\, dy\, dz\, dt$$
$$+ [\text{boundary terms}] = 0 . \tag{3.120}$$

Here the boundary terms result from the integrations by parts, and they involve values of the variations at the boundaries (integration limits) of the fourfold integration domain. If one takes the variations to be zero at these boundaries, but otherwise arbitrary within the domain, one concludes that the coefficient expressions of the $\delta\xi_i$ within the integrand must be identically zero when the displacement field is the correct solution. Consequently, one infers that the displacement field components must satisfy

$$[\text{expression}]_i = 0 . \tag{3.121}$$

The detailed derivation yields the above with the form

$$\frac{\partial \mathfrak{L}}{\partial \xi_i} - \frac{\partial}{\partial t}\left(\frac{\partial \mathfrak{L}}{\partial(\partial \xi_i/\partial t)}\right) - \sum_j \frac{\partial}{\partial x_j}\left(\frac{\partial \mathfrak{L}}{\partial(\partial \xi_i/\partial x_j)}\right)$$
$$+ \sum_{jk} \frac{\partial^2}{\partial x_j \partial x_k}\left(\frac{\partial \mathfrak{L}}{\partial(\partial^2 \xi_i/\partial x_j \partial x_k)}\right) - \ldots = 0 . \tag{3.122}$$

This, along with its counterparts for the other displacement components, is sometimes referred to as the Lagrange–Euler equation(s). The terms that are required correspond to those spatial derivatives on which the strain energy density $\mathfrak{U}$ depends.

For the case of an isotropic solid, the Lagrange–Euler equations correspond to those of (3.77), which is equivalently written, given that the Lamé coefficients are independent of position, as

$$\rho \frac{\partial^2 \boldsymbol{\xi}}{\partial t^2} = \rho(c_1^2 - c_2^2)\nabla(\nabla \cdot \boldsymbol{\xi}) + \rho c_2^2 \nabla^2 \boldsymbol{\xi} . \tag{3.123}$$

For a fluid, the analogous result is

$$\rho \frac{\partial^2 \boldsymbol{\xi}}{\partial t^2} = \nabla(\rho c^2 \nabla \cdot \boldsymbol{\xi}) . \tag{3.124}$$

The more familiar Bergmann version (3.73) of the wave equation results from this if one divides both sides by ρ, then takes the divergence, and subsequently sets

$$p = -\rho c^2 \nabla \cdot \boldsymbol{\xi} \,. \tag{3.125}$$

Here the variational principle automatically yields a wave equation that is suitable for inhomogeneous media, where ρ and c vary with position.

3.4.2 Biot's Formulation for Porous Media

A commonly used model for waves in porous media follows from a formulation developed by *Biot* [3.42], with a version of Hamilton's principle used in the formulation. The medium consists of a solid matrix that is connected, so that it can resist shear stresses, and within which are small pores filled with a compressible fluid.

Two displacement fields are envisioned, one associated with the local average displacement $\boldsymbol{U}$ of the fluid matter, the other associated with the average displacement $\boldsymbol{u}$ of the solid matter. The average here is understood to be an average over scales large compared to representative pore sizes or grain sizes, but yet small compared to whatever lengths are of dominant interest. Because nonlinear effects are believed minor, the kinetic and strain energies per unit volume, again averaged over such length scales, are taken to be quadratic in the displacement fields and their derivatives. The overall medium is presumed to be isotropic, so the two energy densities must be unchanged under coordinate rotations. These innocuous assumptions lead to the general expressions

$$\mathfrak{T} = \frac{1}{2}\rho_{11}\frac{\partial \boldsymbol{u}}{\partial t}\cdot\frac{\partial \boldsymbol{u}}{\partial t} + \rho_{12}\frac{\partial \boldsymbol{U}}{\partial t}\cdot\frac{\partial \boldsymbol{u}}{\partial t} + \frac{1}{2}\rho_{22}\frac{\partial \boldsymbol{U}}{\partial t}\cdot\frac{\partial \boldsymbol{U}}{\partial t}\,, \tag{3.126}$$

$$\mathfrak{U} = \frac{1}{4}N\sum_{ij}\left(\frac{\partial u_i}{\partial x_j} + \frac{\partial u_j}{\partial x_i}\right)^2 + \frac{1}{2}A(\nabla\cdot\boldsymbol{u})^2 + Q(\nabla\cdot\boldsymbol{u})(\nabla\cdot\boldsymbol{U}) + \frac{1}{2}R(\nabla\cdot\boldsymbol{U})^2\,. \tag{3.127}$$

where there are seven, possibly position-dependent, constants. The above form assumes that there is no potential energy associated with shear deformation in the fluid and that there is no potential energy associated with the relative displacements of the solid and the fluid. The quantities $\boldsymbol{U}$ and $\boldsymbol{u}$ here, which are displacements, should not be confused with velocities, although these symbols are used for velocities in much of the literature. They are used here for displacements, because they were used as such by Biot in his original paper.

Because the two energy densities must be nonnegative, the constants that appear here are constrained so that

$$\rho_{11} \geq 0\,; \quad \rho_{22} \geq 0\,; \quad \rho_{11}\rho_{22} - \rho_{12}^2 \geq 0\,, \tag{3.128}$$

$$N \geq 0\,; \quad A+2N \geq 0\,; \quad R \geq 0\,;$$
$$(A+2N)R - Q^2 \geq 0\,. \tag{3.129}$$

The coupling constants, ρ_{12} and Q, can in principle be either positive or negative, but their magnitudes are constrained by the equations above.

It is presumed that this idealization for the energy densities will apply best at low frequencies, and one can conceive of a possible low-frequency disturbance where the solid and the fluid move locally together, with the locking being caused by the viscosity in the fluid and by the boundary condition that displacements at interfaces must be continuous. In such circumstances the considerations that lead to the composite density of (3.50) should apply, so one should have

$$\rho_{11} + 2\rho_{12} + \rho_{22} = \rho_{\text{eq}} = f_s\rho_s + f_f\rho_f\,. \tag{3.130}$$

Here f_s and f_f are the volume fractions of the material that are solid and fluid, respectively. The latter is referred to as the porosity.

In general, the various constants have to be inferred from experiment, and it is difficult to deduce them from first principles. An extensive discussion of these constants with some representative values can be found in the monograph by *Stoll* [3.43].

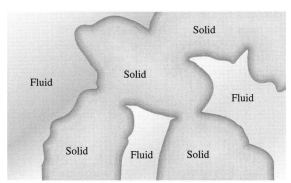

Fig. 3.7 Sketch of a portion of a porous medium in which a compressible fluid fills the pores within a solid matrix. Grains that touch each other are in welded contact, insofar as small-amplitude disturbances are concerned

Because the friction associated with the motion of the fluid with respect to the solid matrix is inherently nonconservative (i.e., energy is lost), the original version (3.119) of Hamilton's principle does not apply. The applicable extension includes a term that represents the time integral of the virtual work done by nonconservative forces during a variation. If such nonconservative forces are taken to be distributed over the volume, with the forces per unit volume on the fluid denoted by $\mathfrak{F}$ and those on the solid denoted by $\mathfrak{f}$, then the modified version of Hamilton's principle is

$$\iiiint (\delta\mathcal{L} + \mathfrak{F}\cdot\delta U + \mathfrak{f}\cdot\delta u)\, dx\, dy\, dz\, dt = 0. \tag{3.131}$$

Moreover, one infers that

$$\mathfrak{F} = -\mathfrak{f}, \tag{3.132}$$

since the virtual work associated with friction between the solid matrix and the fluid must vanish if the two are moved together.

The Lagrange–Euler equations that result from the above variational principle are

$$\frac{\partial}{\partial t}\left(\frac{\partial\mathcal{L}}{\partial(\partial U_i/\partial t)}\right) + \sum_j \frac{\partial}{\partial x_j}\left(\frac{\partial\mathcal{L}}{\partial(\partial U_i/\partial x_j)}\right) = \mathfrak{F}_i, \tag{3.133}$$

$$\frac{\partial}{\partial t}\left(\frac{\partial\mathcal{L}}{\partial(\partial u_i/\partial t)}\right) + \sum_j \frac{\partial}{\partial x_j}\left(\frac{\partial\mathcal{L}}{\partial(\partial u_i/\partial x_j)}\right) = \mathfrak{f}_i. \tag{3.134}$$

Biot, in his original exposition, took the internal distributed forces to be proportional to the relative velocities, reminiscent of dashpots, so that

$$\mathfrak{F}_i = -\mathfrak{f}_i = -b\left(\frac{\partial U_i}{\partial t} - \frac{\partial u_i}{\partial t}\right). \tag{3.135}$$

Here the quantity b can be regarded as the apparent dashpot constant per unit volume. This form is such that the derived equations, in the limit of vanishingly small frequencies, are consistent with Darcy's law [3.44] for steady fluid flow through a porous medium.

The equations that result from this formulation, when written out explicitly, are

$$\frac{\partial^2}{\partial t^2}(\rho_{11}u + \rho_{12}U) - \nabla(A'\nabla\cdot u + Q\nabla\cdot U)$$
$$+ \nabla\times[N(\nabla\times u)] = b\frac{\partial}{\partial t}(U - u), \tag{3.136}$$

$$\frac{\partial^2}{\partial t^2}(\rho_{12}u + \rho_{22}U) - \nabla(Q\nabla\cdot u + R\nabla\cdot U)$$
$$= -b\frac{\partial}{\partial t}(U - u), \tag{3.137}$$

with the abbreviation $A' = A + 2N$. Here, and in what follows, it is assumed that the various material constants are independent of position, although some of the derived equations may be valid to a good approximation even when this is not the case.

3.4.3 Disturbance Modes in a Biot Medium

Disturbances that satisfy the equations derived in the previous section can be represented as a superposition of three basic modal disturbances. These are here denoted as the acoustic mode, the Darcy mode, and the shear mode, and one writes

$$u = u_{\text{ac}} + u_{\text{D}} + u_{\text{sh}}, \tag{3.138}$$
$$U = U_{\text{ac}} + U_{\text{D}} + U_{\text{sh}}. \tag{3.139}$$

At low frequencies, the motion in the acoustic mode and in the shear mode is nearly such that the fluid and solid displacements are the same,

$$U_{\text{ac}} \approx u_{\text{ac}}; \qquad U_{\text{sh}} \approx u_{\text{sh}}. \tag{3.140}$$

The lowest-order (in frequency divided by the dashpot parameter b) approximation results from taking the sum of (3.136) and (3.137) and then setting $u = U$, yielding

$$\rho_{\text{eq}}\frac{\partial^2}{\partial t^2}U - \nabla(B_{\text{V,B}}\nabla\cdot U) + \nabla\times[G_{\text{B}}(\nabla\times U)] = 0, \tag{3.141}$$

with the abbreviations

$$\rho_{\text{eq}} = \rho_{11} + 2\rho_{12} + \rho_{22}, \tag{3.142}$$
$$B_{\text{V,B}} = A + 2N + 2Q + R, \tag{3.143}$$
$$G_{\text{B}} = N, \tag{3.144}$$

for the apparent density, bulk modulus, and shear modulus of the Biot medium. The same equation results for the solid displacement field u in this approximation.

Acoustic Mode

For the acoustic mode, the curl of each displacement field is zero, so

$$\rho_{\text{eq}}\frac{\partial^2}{\partial t^2}U - \nabla(B_{\text{V,B}}\nabla\cdot U) = 0. \tag{3.145}$$

One can identify an apparent pressure disturbance associated with this mode, so that

$$p_{\text{ac}} \approx -B_{\text{V,B}}\nabla\cdot U, \tag{3.146}$$

and this will satisfy the wave equation

$$\nabla^2 p_{ac} - \frac{1}{c_{ac}^2} \frac{\partial^2 p_{ac}}{\partial t^2} = 0, \quad (3.147)$$

where

$$c_{ac}^2 = \frac{B_{V,B}}{\rho_{eq}} \quad (3.148)$$

is the square of the apparent sound speed for disturbances carried by the acoustic mode.

The equations just derived can be improved by an iteration procedure to take into account that the dashpot constant has a finite (rather than an infinite) value. To the next order of approximation, one obtains

$$\nabla^2 \boldsymbol{U}_{ac} - \frac{1}{c_{ac}^2} \frac{\partial^2}{\partial t^2} \boldsymbol{U}_{ac} = -\frac{\tau_B}{c_{ac}^2} \frac{\partial^3}{\partial t^3} \boldsymbol{U}_{ac}. \quad (3.149)$$

This wave equation with an added (dissipative) term holds for each component of $\boldsymbol{U}_{ac}$ and of $\boldsymbol{u}_{ac}$, as well as for the pressure p_{ac}. The time constant τ_B which appears here is given by

$$\tau_B = \frac{D^2}{bB_{V,B}^2 \rho_{eq}}, \quad (3.150)$$

where

$$D = (\rho_{11} + \rho_{12})(R+Q) \\ - (\rho_{22} + \rho_{12})(A + 2N + Q) \quad (3.151)$$

is a parameter that characterizes the mismatch between the material properties of the fluid and the solid. The appearance of the dissipative term in the wave equation is associated with the imperfect locking of the two displacement fields, so that

$$\boldsymbol{U}_{ac} - \boldsymbol{u}_{ac} \approx \frac{D}{bB_{V,B}} \frac{\partial \boldsymbol{u}_{ac}}{\partial t}. \quad (3.152)$$

One should note that (3.149) is of the same general mathematical form as the dissipative wave equation (3.101).

Shear mode

For the shear mode, the divergence of each displacement field is zero, so (3.141) reduces to

$$\rho_{eq} \frac{\partial^2}{\partial t^2} \boldsymbol{U} + \nabla \times [G_B (\nabla \times \boldsymbol{U})] = 0, \quad (3.153)$$

where this holds to lowest order for both $\boldsymbol{U}_{sh}$ and $\boldsymbol{u}_{sh}$, these being equal in this approximation. A mathematical identity for the curl of a curl reduces this to

$$\nabla^2 \boldsymbol{u}_{sh} - \frac{1}{c_{sh}^2} \frac{\partial^2 \boldsymbol{u}_{sh}}{\partial t^2} = 0, \quad (3.154)$$

where

$$c_{sh}^2 = \frac{G_B}{\rho_{eq}}. \quad (3.155)$$

An iteration process that takes into account that the two displacement fields are not exactly the same yields

$$\boldsymbol{u}_{sh} - \boldsymbol{U}_{sh} \approx \frac{\rho_{22} + \rho_{12}}{b} \frac{\partial \boldsymbol{U}_{sh}}{\partial t}, \quad (3.156)$$

with the result that the wave equation above is replaced by a dissipative wave equation

$$\nabla^2 \boldsymbol{u}_{sh} - \frac{1}{c_{sh}^2} \frac{\partial^2 \boldsymbol{u}_{sh}}{\partial t^2} = -\frac{(\rho_{22} + \rho_{12})^2}{G_B b} \frac{\partial^3 \boldsymbol{u}_{sh}}{\partial t^3}. \quad (3.157)$$

Darcy Mode

For the Darcy mode, the curl of both displacement fields is zero,

$$\nabla \times \boldsymbol{U}_D = 0; \qquad \nabla \times \boldsymbol{u}_D = 0. \quad (3.158)$$

The inertia term in both (3.136) and (3.137) is negligible to a lowest approximation, and the compatibility of the two equations requires

$$(A' + Q)\boldsymbol{u}_D = -(R+Q)\boldsymbol{U}_D, \quad (3.159)$$

so that the fluid and the solid move in opposite directions.

Given the above observation and the neglect of the inertia terms, either of (3.136) and (3.137) reduces to the diffusion equation

$$\nabla^2 \boldsymbol{U}_D = \kappa_D \frac{\partial \boldsymbol{U}_D}{\partial t}, \quad (3.160)$$

with the same equation also being satisfied by $\boldsymbol{u}_D$. Here

$$\kappa_D = \frac{bB_{V,B}}{A'R - Q^2} \quad (3.161)$$

is a constant whose reciprocal characterizes the tendency of the medium to allow diffusion of fluid through the solid matrix. This is independent of any of the inertial densities, and it is always positive.

3.5 Waves of Constant Frequency

One of the principal terms used in describing sound is that of frequency. Although certainly not all acoustic disturbances are purely sinusoidal oscillations of a quantity (such as pressure or displacement) that oscillates with constant frequency about a mean value, it is usually possible to characterize (at least to an order of magnitude) any such disturbance by one or a limited number of representative frequencies, these being the reciprocals of the characteristic time scales. Such a description is highly useful, because it gives one some insight into what detailed physical effects are relevant and of what instrumentation is applicable. The simplest sort of acoustic signal would be one where a quantity such as the fluctuating part p of the pressure is oscillating with time t as

$$p = A\cos(\omega t - \phi). \tag{3.162}$$

Here the amplitude A and phase constant ϕ are independent of t (but possibly dependent on position). The quantity ω is called the angular frequency and has the units of radians divided by time (for example, rad/s or simply s^{-1}, when the unit of time is the second). The number f of repetitions per unit time is what one normally refers to as the frequency (without a modifier), such that $\omega = 2\pi f$. The value of f in hertz (abbreviated to Hz) is the frequency in cycles (repetitions) per second. The period $T = 1/f$ is the time interval between repetitions.

The human ear responds [3.45] almost exclusively to frequencies between roughly 20 Hz and 20 kHz. Consequently, sounds composed of frequencies below 20 Hz are said to be infrasonic; those composed of frequencies above 20 kHz are said to be ultrasonic. The scope of acoustics, given its general definition as a physical phenomenon, is not limited to audible frequencies.

3.5.1 Spectral Density

The term frequency spectrum is often used in relation to sound. Often the use is somewhat loose, but there are circumstances for which the terminology can be made precise. If the fluctuating physical quantity p associated with the acoustic disturbance is a sum of sinusoidal disturbances, the n-th being $p_n = A_n \cos(2\pi f_n t - \phi_n)$ and having frequency f_n, no two frequencies being the same, then the set of mean squared amplitudes can be taken as a description of the spectrum of the signal. Also, if the sound is made up of many different frequencies, then one can use the time-averaged sum of the p_n^2 that correspond to those frequencies within a given specified frequency band as a measure of the strength of the signal within that frequency band. This sum divided by the width of the frequency band often approaches a quasi-limit as the bandwidth becomes small, but with the number of included terms still being moderately large, and with this quasi-limit being a definite smooth function of the center frequency of the band. This quasi-limit is called the spectral density p_f^2 of the signal. Although an idealized quantity, it can often be repetitively measured to relatively high accuracy; instrumentation for the measurement of spectral densities or of integrals of spectral densities over specified (such as octaves) frequency bands is widely available commercially.

The utility of the spectral density concept rests on the principle of superposition and on a version of Parseval's theorem, which states that, when the signal is a sum of discrete frequency components, then if averages are taken over a sufficiently long time interval,

$$(p^2)_{\text{av}} = \sum_n (p_n^2)_{\text{av}}. \tag{3.163}$$

Consequently, in the quasi-limit corresponding to the spectral density description, one has the mean squared pressure expressed as an integral over spectral density,

$$(p^2)_{\text{av}} = \int_0^\infty p_f^2(f)\,\mathrm{d}f. \tag{3.164}$$

The spectral density at a given frequency times a narrow bandwidth centered at that frequency is interpreted as the contribution to the mean squared acoustic pressure from that frequency band.

If the signal is a linear response to a source (or excitation) characterized by some time variation $s(t)$ with spectral density $s_f^2(f)$, then the spectral density of the response is

$$p_f^2(f) = |H_{ps}(2\pi f)|^2 s_f^2(f), \tag{3.165}$$

where the proportionality factor $|H_{ps}(2\pi f)|^2$ is the square of the absolute magnitude of a transfer function that is independent of the excitation, but which does depend on frequency. For any given frequency, this transfer function can be determined from the response to a known sinusoidal excitation with the same frequency. Thus the analysis for constant-frequency excitation is fundamental, even when the situation of interest may involve broadband excitation.

3.5.2 Fourier Transforms

Analogous arguments can be made for transient excitation and transient signals, where time-dependent quantities are represented by integrals over Fourier transforms, so that

$$p(t) = \int_{-\infty}^{\infty} \hat{p}(\omega) e^{-i\omega t} \, d\omega, \quad (3.166)$$

with

$$\hat{p}(\omega) = \frac{1}{2\pi} \int_{-\infty}^{\infty} p(t) e^{i\omega t} \, dt. \quad (3.167)$$

Here $\hat{p}(\omega)$ is termed the Fourier transform of $p(t)$, and $p(t)$ is conversely termed the inverse Fourier transform of $\hat{p}(\omega)$. Definitions of the Fourier transform vary in the acoustics literature. What is used here is what is commonly used for analyses of wave propagation problems. Whatever definition is used must be accompanied by a definition of the inverse Fourier transform, so that the inverse Fourier transform produces the original function. In all such cases, given that $p(t)$ and ω are real, one has

$$\hat{p}(-\omega) = \hat{p}(\omega)^*, \quad (3.168)$$

where the asterisk denotes the complex conjugate. Thus, the magnitude of the Fourier transform need only be known for positive frequencies. The phase changes sign when one changes the sign of the frequency.

Parseval's theorem for Fourier transforms is

$$\int_{-\infty}^{\infty} p^2(t) \, dt = 4\pi \int_{0}^{\infty} |\hat{p}(\omega)|^2 \, d\omega, \quad (3.169)$$

so that $4\pi |\hat{p}(\omega)|^2 \Delta\omega$ can be regarded as to the contribution to the time integral of $p^2(t)$ from a narrow band of (positive) frequencies of width $\Delta\omega$.

3.5.3 Complex Number Representation

Insofar as the governing equations are linear with coefficients independent of time, disturbances that vary sinusoidally with time can propagate without change of frequency. Such sinusoidally varying disturbances of constant frequency have the same repetition period (the reciprocal of the frequency) at every point, but the phase will in general vary from point to point.

When one considers a disturbance of fixed frequency or one frequency component of a multifrequency disturbance, it is convenient to use a complex number representation, such that each field amplitude is written [3.4, 46]

$$p = \mathrm{Re}(\hat{p} e^{-i\omega t}). \quad (3.170)$$

Here $\hat{p}$ is called the complex amplitude of the acoustic pressure and in general varies with position. (The use of the $e^{-i\omega t}$ time dependence is traditional among acoustics professionals who are primarily concerned with the wave aspects of acoustics. In the vibrations literature, one often finds instead a postulated $e^{j\omega t}$ time dependence. The use of j as the square root of -1 instead of i is a carryover from the analysis of oscillations in electrical circuits, where i is reserved for electrical current.)

If one inserts expressions such as (3.170) into a homogeneous linear ordinary or partial differential equation with real time-independent coefficients, the result can always be written in the form

$$\mathrm{Re}\left(\Phi e^{-i\omega t}\right) = 0, \quad (3.171)$$

where the quantity Φ is an expression depending on the complex amplitudes and their spatial derivatives, but not depending on time. The requirement that (3.171) should hold for all values of time consequently can be satisfied if and only if

$$\Phi = 0. \quad (3.172)$$

Moreover, the form of the expression Φ can be readily obtained from the original equation with a simple prescription: replace all field variables by their complex amplitudes and replace all time derivatives using the substitution

$$\frac{\partial}{\partial t} \to -i\omega. \quad (3.173)$$

Thus, for example, the linear acoustic equations given by (3.71) and (3.72) reduce to

$$-i\omega \hat{p} + \rho c^2 \nabla \cdot \hat{\boldsymbol{v}} = 0, \quad (3.174)$$
$$-i\omega \rho \hat{\boldsymbol{v}} = -\nabla \hat{p}. \quad (3.175)$$

The wave equation in (3.74) reduces (with $k = \omega/c$ denoting the wavenumber) to

$$\nabla^2 \hat{p} + k^2 \hat{p} = 0, \quad (3.176)$$

which is the Helmholtz equation [3.47] for the complex pressure amplitude.

3.5.4 Time Averages of Products

When dealing with waves of constant frequency, one often makes use of averages over a wave period of the product of two quantities that oscillate with the same frequency. Let

$$p = \text{Re}\left(\hat{p}e^{-i\omega t}\right); \qquad v = \text{Re}\left(\hat{v}e^{-i\omega t}\right) \quad (3.177)$$

be two such quantities. Then the time average of their product is

$$[p(t)v(t)]_{\text{av}} = |\hat{p}||\hat{v}|[\cos(\omega t - \phi_p)\cos(\omega t - \phi_v)]_{\text{av}}, \quad (3.178)$$

where ϕ_p and ϕ_v are the phases of $\hat{p}$ and $\hat{v}$.

The trigonometric identity

$$\cos(A)\cos(B) = \frac{1}{2}\left[\cos(A+B) + \cos(A-B)\right], \quad (3.179)$$

with appropriate identifications for A and B, yields a term which averages out to zero and a term which is independent of time. Thus one has

$$[p(t)v(t)]_{\text{av}} = \frac{1}{2}|\hat{p}||\hat{v}|\cos(\phi_p - \phi_v), \quad (3.180)$$

or, equivalently

$$[p(t)v(t)]_{\text{av}} = \frac{1}{2}|\hat{p}||\hat{v}|\text{Re}(e^{i(\phi_p - \phi_v)}), \quad (3.181)$$

which in turn can be written

$$[p(t)v(t)]_{\text{av}} = \frac{1}{2}\text{Re}(\hat{p}\hat{v}^*). \quad (3.182)$$

The asterisk here denotes the complex conjugate. Because the real part of a complex conjugate is the same as the real part of the original complex number, it is immaterial whether one takes the complex conjugate of $\hat{p}$ or $\hat{v}$, but one takes the complex conjugate of only one.

3.6 Plane Waves

A solution of the wave equation that plays a central role in many acoustical concepts is that of a plane traveling wave.

3.6.1 Plane Waves in Fluids

The mathematical representation of a plane wave is such that all acoustic field quantities vary with time and with one Cartesian coordinate. That coordinate is taken here as x; consequently, all the acoustic quantities are independent of y and z. The Laplacian reduces to a second partial derivative with respect to x and the d'Alembertian can be expressed as the product of two first-order operators, so that the wave equation takes the form

$$\left(\frac{\partial}{\partial x} - \frac{1}{c}\frac{\partial}{\partial t}\right)\left(\frac{\partial}{\partial x} + \frac{1}{c}\frac{\partial}{\partial t}\right)p = 0. \quad (3.183)$$

The solution of this equation is a sum of two expressions, each of which is such that operation by one or the other of the two factors in (3.183) yields zero. Such a solution is represented by

$$p(x,t) = f(x - ct) + g(x + ct), \quad (3.184)$$

where f and g are two arbitrary functions. The argument combination $x - ct$ of the first term is such that the function f with that argument represents a plane wave traveling forward in the $+x$-direction at a velocity c. The second term in (3.184) similarly represents a plane wave traveling backwards in the $-x$-direction, also at a velocity c. For a traveling plane wave, not only the shape, but also the amplitude is conserved during propagation. A typical situation for which wave propagation can be adequately described by traveling plane waves is that of low-frequency sound propagation in a duct. Diverging or converging (focused) waves can often be approximately regarded as planar within regions of restricted extent.

The fluid velocity that corresponds to the plane wave solution above has y and z components that are identically zero, but the x-component, in accordance with (3.71) and (3.72), is

$$v_x = \frac{1}{\rho c}f(x - ct) - \frac{1}{\rho c}g(x + ct). \quad (3.185)$$

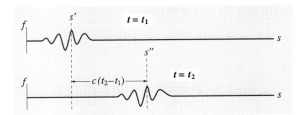

Fig. 3.8 Waveform propagating in the $+s$-direction with constant speed c. Shown are plots of the function $f(s - ct)$ versus s at two successive values of the time t

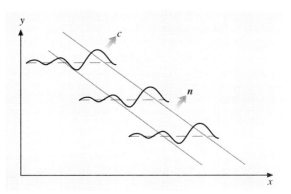

Fig. 3.9 Plane wave propagating in the direction of the unit vector n

The general rule that emerges from this is that, for a plane wave propagating in the direction corresponding to unit vector n, the acoustic part of the pressure is

$$p = f(n \cdot x - ct) \tag{3.186}$$

for some generic function f of the indicated argument, while the acoustically induced fluid velocity is

$$v = \frac{n}{\rho c} p . \tag{3.187}$$

Because the fluid velocity is in the same direction as that of the wave propagation, such waves are said to be longitudinal. (Electromagnetic plane waves in free space, in contrast, are transverse. Both the electric field vector and the magnetic field vector are perpendicular to the direction of propagation.)

For a plane wave traveling along the $+x$-axis at a velocity c, the form of (3.186) implies that one can represent a constant-frequency acoustic pressure disturbance by the expression

$$p = |P| \cos [k(x - ct) + \phi_0] = \mathrm{Re}(P \mathrm{e}^{\mathrm{i}k(x-ct)}) , \tag{3.188}$$

with the angular frequency identified as

$$\omega = ck . \tag{3.189}$$

Alternately, if the complex amplitude representation of (3.170) is used, one writes the complex amplitude of the acoustic pressure as

$$\hat{p} = P \mathrm{e}^{\mathrm{i}kx} , \tag{3.190}$$

where P is a complex number related to the constants that appear in (3.188) as

$$P = |P| \mathrm{e}^{\mathrm{i}\phi_0} . \tag{3.191}$$

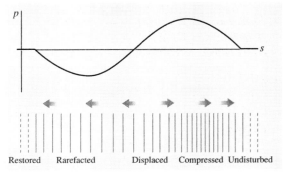

Fig. 3.10 Fluid velocity and pressure in a one-cycle sinusoidal pulse propagating in the $+x$-direction

Here $|P|$ is the amplitude of the disturbance, ϕ_0 is a phase constant, and k is a constant termed the wavenumber. The wavelength λ is the increment in propagation distance x required to change the argument of the cosine by 2π radians, so

$$k = \frac{2\pi}{\lambda} . \tag{3.192}$$

Also, the increment in t required to change the argument by 2π is the period T, which is the reciprocal of the frequency f, so one has

$$\lambda = \frac{c}{f} . \tag{3.193}$$

3.6.2 Plane Waves in Solids

Plane acoustic waves in isotropic elastic solids [3.26] have properties similar to those of waves in fluids. Dilatational (or longitudinal) plane waves are such that the curl of the displacement field vanishes, so the displacement vector must be parallel to the direction of propagation. (Dilation means expansion and is the antonym of compression. Dilatational plane waves could equally well be termed compressional plane waves. Seismologists refer to dilational waves as P-waves, where the letter P stands for primary. They refer to shear waves as S-waves, where the letter S stands for secondary.) A comparison of (3.82) with the wave equation (3.74) indicates that such a wave must propagate with a speed c_l determined by (3.79). Thus, a wave propagating in the $+x$-direction has no y and z components of displacement, and has an x-component described by

$$\xi_x = F(x - c_1 t) , \tag{3.194}$$

where F is an arbitrary function. The stress components can be deduced from (3.56, 57, 79, 80). These equations, as well as symmetry considerations, require for a dilatational wave propagating in the x-direction, that the off-diagonal elements of the stress tensor vanish. The diagonal elements are given by

$$\sigma_{xx} = \rho c_1^2 F'(x - c_1 t), \tag{3.195}$$

$$\sigma_{yy} = \sigma_{zz} = \rho(c_1^2 - 2c_2^2) F'(x - c_1 t). \tag{3.196}$$

Here the primes denote derivatives with respect to the total argument.

The divergence of the displacement field in a shear wave is zero, so a plane shear wave must cause a displacement perpendicular to the direction of propagation.

Shear waves are therefore transverse waves. Equation (3.83), when considered in a manner similar to that described above for the wave equation for waves in fluids, leads to the conclusion that plane shear waves must propagate with a speed c_2. A plane shear wave polarized in the y-direction and propagating in the x-direction will have only a y-component of displacement, given by

$$\xi_y = F(x - c_2 t). \tag{3.197}$$

The only nonzero stress components are the shear stresses

$$\sigma_{yx} = \rho c_2^2 F'(x - c_2 t) = \sigma_{xy}. \tag{3.198}$$

3.7 Attenuation of Sound

Plane waves of constant frequency propagating through bulk materials have amplitudes that typically decrease exponentially with increasing propagation distance, such that the magnitude of the complex pressure amplitude varies as

$$|\hat{p}(x)| = |\hat{p}(0)| e^{-\alpha x}. \tag{3.199}$$

The quantity α is the plane wave attenuation coefficient and has units of nepers per meter (Np/m); it is an intrinsic frequency-dependent property of the material. This exponential decrease of amplitude is called attenuation or absorption of sound and is associated with the transfer of acoustic energy to the internal energy of the material. (If $|\hat{p}|^2$ decreases to a tenth of its original value, it is said to have decreased by 10 decibels (dB), so an attenuation constant of α nepers per meter is equivalent to an attenuation constant of $[20/(\ln 10)]\alpha$ decibels per meter, or 8.6859α decibels per meter.)

3.7.1 Classical Absorption

The attenuation of sound due to the classical processes of viscous energy absorption and thermal conduction is derivable [3.33] from the dissipative wave equation (3.101) given previously for the acoustics mode. Dissipative processes enter into this equation through a parameter δ_{cl}, which is defined by (3.102). To determine the attenuation of waves governed by such a dissipative wave equation, one sets the perturbation pressure equal to

$$p_{ac} = P e^{-i\omega t} e^{ikx}, \tag{3.200}$$

where P is independent of position and k is a complex number. Such a substitution, with reasoning such as that which leads from (3.171) to (3.172), yields an algebraic equation that can be nontrivially (amplitude not identically zero) satisfied only if k satisfies the relation

$$k^2 = \frac{\omega^2}{c^2} + i\frac{2\delta_{cl}\omega^3}{c^3}. \tag{3.201}$$

The root that corresponds to waves propagating in the $+x$-direction is that which evolves to (3.189) in the limit of no absorption, and for which the real part of k is positive. Thus to first order in δ_{cl}, one has the complex wavenumber

$$k = \frac{\omega}{c} + i\frac{\delta_{cl}\omega^2}{c^3}. \tag{3.202}$$

The attenuation coefficient is the imaginary part of this, in accordance with (3.199), so

$$\alpha_{cl} = \frac{\delta_{cl}\omega^2}{c^3}. \tag{3.203}$$

This is termed the classical attenuation coefficient for acoustic waves in fluids and is designated by the subscript 'cl'. The distinguishing characteristic of this classical attenuation coefficient is its quadratic increase with increasing frequency.

The same type of frequency dependence is obeyed by the acoustic and shear wave modes for the Biot model of porous media in the limit of low frequencies. From (3.149) one derives

$$\alpha_{ac} = \frac{\tau_B}{2c_{ac}}\omega^2, \tag{3.204}$$

and, from (3.157), one derives

$$\alpha_{\text{sh}} = \frac{c_{\text{sh}}(\rho_{22}+\rho_{12})^2}{2G_B b}\omega^2 \ . \tag{3.205}$$

3.7.2 Relaxation Processes

For many substances, including air, sea water, biological tissues, marine sediments, and rocks, the variation of the absorption coefficient with frequency is not quadratic [3.48–50], and the classical model is insufficient to predict the magnitude of the absorption coefficient. The substitution of a different value of the bulk viscosity is insufficient to remove the discrepancy, because this would still yield the quadratic frequency dependence. The successful theory to account for such discrepancies in air and sea water and other fluids is in terms of relaxation processes. The physical nature of the relaxation processes vary from fluid to fluid, but a general theory in terms of irreversible thermodynamics [3.51–53] yields appropriate equations.

The equation of state for the instantaneous (rather than equilibrium) entropy is written as

$$s = s(u, \rho^{-1}, \xi) \ , \tag{3.206}$$

where ξ represents one or more internal variables. The differential relation of (3.13) is replaced by

$$T\,\mathrm{d}s = \mathrm{d}u + p\,\mathrm{d}\rho^{-1} + \sum_\nu A_\nu \mathrm{d}\xi_\nu \ , \tag{3.207}$$

where the affinities A_ν are defined by this equation. These vanish when the fluid is in equilibrium with a given specified internal energy and density. The pressure p here is the same as enters into the expression (3.15) for the average normal stress, and the T is the same as enters into the Fourier law (3.16) of heat conduction. The mass conservation law (3.2) and the Navier–Stokes equation (3.24) remain unchanged, but the energy equation, expressed in (3.25) in terms of entropy, is replaced by the entropy balance equation

$$\rho\frac{\mathrm{D}s}{\mathrm{D}t} + \nabla\cdot\frac{\boldsymbol{q}}{T} = \sigma_s \ . \tag{3.208}$$

Here the quantity σ_s, which indicates the rate of irreversible entropy production per unit volume, is given by

$$T\sigma_s = \mu_B(\nabla\cdot\boldsymbol{v})^2 + \tfrac{1}{2}\mu\sum_{ij}\phi_{ij}^2 + \frac{\kappa}{T}(\nabla T)^2$$

$$+ \rho\sum_\nu A_\nu \frac{\mathrm{D}\xi_\nu}{\mathrm{D}t} \ . \tag{3.209}$$

One needs in addition relations that specify how the internal variables ξ_ν relax to their equilibrium values. The simplest assumption, and one which is substantiated for air and sea water, is that these relax independently according to the rule [3.54]

$$\frac{\mathrm{D}\xi_\nu}{\mathrm{D}t} = -\frac{1}{\tau_\nu}\left(\xi_\nu - \xi_{\nu,\text{eq}}\right) \ . \tag{3.210}$$

The relaxation times τ_ν that appear here are positive and independent of the internal variables.

When the linearization process is applied to the nonlinear equations for the model just described of a fluid with internal relaxation, one obtains the set of equations

$$\frac{\partial \rho'}{\partial t} + \rho_0 \nabla\cdot\boldsymbol{v}' = 0 \ , \tag{3.211}$$

$$\rho_0\frac{\partial \boldsymbol{v}'}{\partial t} = -\nabla p' + (1/3\mu + \mu_B)$$
$$\times \nabla(\nabla\cdot\boldsymbol{v}') + \mu\nabla^2\boldsymbol{v}' \ , \tag{3.212}$$

$$\rho_0 T_0 \frac{\partial s'}{\partial t} = \kappa\nabla^2 T' \ , \tag{3.213}$$

$$\rho' = \frac{1}{c^2}p' - \left(\frac{\rho\beta T}{c_p}\right)_0 s'$$
$$+ \sum_\nu \left(\xi_\nu' - \xi_{\nu,\text{eq}}'\right)a_\nu \ , \tag{3.214}$$

$$T' = \left(\frac{T\beta}{\rho c_p}\right)_0 p' + \left(\frac{T}{c_p}\right)_0 s'$$
$$+ \sum_\nu \left(\xi_\nu' - \xi_{\nu,\text{eq}}'\right)b_\nu \ , \tag{3.215}$$

$$\frac{\partial \xi_\nu'}{\partial t} = -\frac{1}{\tau_\nu}\left(\xi_\nu' - \xi_{\nu,\text{eq}}'\right) \ , \tag{3.216}$$

$$\xi_{\nu,\text{eq}}' = m_\nu s' + n_\nu p' \ . \tag{3.217}$$

Here a_ν, b_ν, m_ν, and n_ν are constants whose values depend on the ambient equilibrium state.

For absorption of sound, the interest is in the acoustic mode, and so approximations corresponding to those discussed in the context of (3.93) through (3.101) can also be made here. The quantities a_ν and b_ν are treated as small, so that one obtains, to first order in ϵ and in these quantities, a wave equation of the form

$$\nabla^2 p_{\text{ac}} - \frac{1}{c^2}\frac{\partial^2}{\partial t^2}\left(p_{\text{ac}} - \frac{2\delta_{\text{cl}}}{c^2}\frac{\partial p_{\text{ac}}}{\partial t}\right.$$
$$\left. - 2\sum_\nu \frac{(\Delta c)_\nu}{c}\tau_\nu \frac{\partial p_\nu}{\partial t}\right) = 0 \ . \tag{3.218}$$

The auxiliary internal variables that enter here satisfy the supplemental relaxation equations

$$\frac{\partial p_\nu}{\partial t} = -\frac{1}{\tau_\nu}(p_\nu - p_{\mathrm{ac}}) \ . \tag{3.219}$$

Here the notation is such that the p_ν are given in terms of previously introduced quantities by

$$p_\nu = \xi'_\nu / n_\nu \ . \tag{3.220}$$

The sound speed increments that appear in (3.218) represent the combinations

$$(\Delta c)_\nu = 1/2 a_\nu n_\nu c^2 \ . \tag{3.221}$$

Equations (3.218) and (3.219) are independent of the explicit physical interpretation of the relaxation processes. Insofar as acoustic propagation is concerned, any relaxation process is characterized by two parameters, the relaxation time τ_ν and the sound speed increment $(\Delta c)_\nu$. The various parameters that enter into the irreversible thermodynamics formulation of (3.214) through (3.216) affect the propagation of sound only as they enter into the values of the relaxation times and of the sound speed increments. The replacement of internal variables by quantities p_ν with the units of pressure implies no assumption as to the precise nature of the relaxation process. [An alternate formulation for a parallel class of relaxation processes concerns structural relaxation [3.55]. The substance can locally have more than one state, each of which has a different compressibility. The internal variables are associated with the deviations of the probabilities of the system being in each of the states from the probabilities that would exist were the system in quasistatic equilibrium. The equations that result are mathematically the same as (3.218) and (3.219).]

The attenuation of plane waves governed by (3.218) and (3.219) is determined when one inserts substitutions of the form of (3.200) for p_{ac} and the p_ν. The relaxation equations yield the relations

$$\hat{p}_\nu = \frac{1}{1 - i\omega\tau_\nu} \hat{p}_{\mathrm{ac}} \ . \tag{3.222}$$

These, when inserted into the complex amplitude version of (3.218), yield the dispersion relation

$$k^2 = \frac{\omega^2}{c^2}\left(1 + i\frac{2\omega\delta_{\mathrm{cl}}}{c^2} + \sum_\nu \frac{2(\Delta c)_\nu}{c}\frac{i\omega\tau_\nu}{1 - i\omega\tau_\nu}\right) . \tag{3.223}$$

To first order in the small parameters, δ_{cl} and $(\Delta c)_\nu$, this yields the complex wavenumber

$$k = \frac{\omega}{c}\left(1 + i\frac{\omega\delta_{\mathrm{cl}}}{c^2} + \sum_\nu \frac{(\Delta c)_\nu}{c}\frac{i\omega\tau_\nu}{1 - i\omega\tau_\nu}\right), \tag{3.224}$$

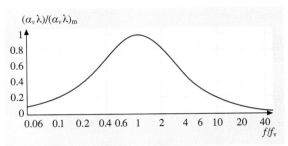

Fig. 3.11 Attenuation per wavelength (in terms of characteristic parameters) for propagation in a medium with a single relaxation process. Here the wavelength λ is $2\pi c/\omega$, and the ratio f/f_ν of frequency to relaxation frequency is $\omega\tau_\nu$. The curve as constructed is independent of c and $(\Delta c)_\nu$

which corresponds to waves propagating in the $+x$-direction. The attenuation coefficient, determined by the imaginary part of (3.224), can be written

$$\alpha = \alpha_{\mathrm{cl}} + \sum_\nu \alpha_\nu \ . \tag{3.225}$$

The first term is the classical attenuation determined by (3.203). The remaining terms correspond to the incremental attenuations produced by the separate relaxation processes, these being

$$\alpha_\nu = \frac{(\Delta c)_\nu}{c^2} \frac{\omega^2 \tau_\nu}{1 + (\omega\tau_\nu)^2} \ . \tag{3.226}$$

Any such term increases quadratically with frequency at low frequencies, as would be the case for classical absorption (with an increased bulk viscosity), but it approaches a constant value at high frequencies.

The labeling of the quantities $(\Delta c)_\nu$ as sound speed increments follows from an examination of the real part of the complex wavenumber, given by

$$k_{\mathrm{R}} = \frac{\omega}{c}\left(1 - \sum_\nu \frac{(\Delta c)_\nu}{c}\frac{(\omega\tau_\nu)^2}{1 + (\omega\tau_\nu)^2}\right) . \tag{3.227}$$

The ratio, of ω to k_{R}, is identified as the phase velocity v_{ph} of the wave. In the limit of low frequencies, the phase velocity predicted by (3.227) is the quantity c, which is the sound speed for a quasi-equilibrium process. In the limit of high frequencies, however, to first order in the $(\Delta c)_\nu$, the phase velocity approaches the limit

$$v_{\mathrm{ph}} \to c + \sum_\nu (\Delta c)_\nu \ . \tag{3.228}$$

Consequently, each $(\Delta c)_\nu$ corresponds to the net increase in phase velocity of the sound wave that occurs

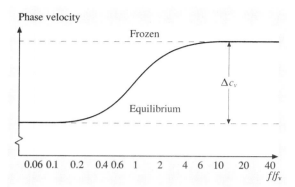

Fig. 3.12 Change in phase velocity as a function of frequency for propagation in a medium with a single relaxation process. The asymptote at zero frequency is the equilibrium sound speed c, and that in the limit of infinite frequency is the frozen sound speed $c + \Delta c_v$

when the frequency is increased from a value small compared to the relaxation frequency to one large compared to the relaxation frequency. Here the term relaxation frequency is used to denote the reciprocal of the product of 2π with the relaxation time.

3.7.3 Continuously Distributed Relaxations

It is sometimes argued that, for heterogeneous media such as biological tissue and rocks, the attenuation is due to a statistical distribution of relaxation times, so that the sum in (3.224) should be replaced by an integral, yielding

$$k = \frac{\omega}{c}\left(1 + i\frac{\omega \delta_{cl}}{c^2} + \frac{1}{c}\int_0^\infty \frac{d(\Delta c)}{d\tau}\frac{i\omega\tau}{1 - i\omega\tau}d\tau\right), \tag{3.229}$$

and so that the attenuation constant due to the relaxation processes becomes

$$\sum_\nu \alpha_\nu \to \frac{1}{c^2}\int_0^\infty \frac{d(\Delta c)}{d\tau}\frac{\omega^2\tau}{1+(\omega\tau)^2}d\tau. \tag{3.230}$$

Here, the quantity

$$\frac{d(\Delta c)}{d\tau}\Delta\tau \tag{3.231}$$

is interpreted as the additional increment in phase velocity at high frequencies due to all the relaxation processes whose relaxation times are in the interval $\Delta\tau$.

Such a continuous distribution of relaxation processes is intrinsically capable of explaining a variety of experimentally observed frequency dependencies of $\alpha(\omega)$. A simple example is that where

$$\frac{d(\Delta c)}{d\tau} = \frac{K}{\tau^q}, \tag{3.232}$$

with K being a constant independent of τ, and with q being a specified exponent, with $0 < q < 2$. In such a case the integral in (3.230) becomes

$$\frac{1}{c^2}\int_0^\infty \frac{d(\Delta c)}{d\tau}\frac{\omega^2\tau}{1+(\omega\tau)^2}d\tau = \frac{K}{c^2}\omega^q\int_0^\infty \frac{u^{1-q}}{1+u^2}du \tag{3.233}$$

giving a power-law dependence on ω that varies as ω^q. Although the hypothesis of (3.232) is probably unrealistic over all ranges of relaxation time τ, its being nearly satisfied over an extensive range of such times could yield predictions that would explain a power-law dependence. (The integral that appears here is just a numerical constant, which is finite provided $0 < q < 2$. The integral emerges when one changes the integration variable from τ to $u = \omega\tau$.)

The corresponding increment in the reciprocal $1/v_{ph} = k_R/\omega$ of the phase velocity v_{ph} is

$$\Delta\left(\frac{1}{v_{ph}}\right) = -\frac{1}{c^2}\int_0^\infty \frac{d(\Delta c)}{d\tau}\frac{\omega^2\tau^2}{1+(\omega\tau)^2}d\tau. \tag{3.234}$$

Insertion of (3.232) into this yields

$$\Delta\left(\frac{1}{v_{ph}}\right) = -\frac{K}{c^2}\omega^{q-1}\int_0^\infty \frac{u^{2-q}}{1+u^2}du, \tag{3.235}$$

which varies with ω as ω^{q-1}. In this case, however, the integral exists only if $1 < q < 3$, so the overall analysis has credibility only if $1 < q < 2$.

3.7.4 Kramers–Krönig Relations

In attempts to explain or predict the frequency dependence of attenuation and phase velocity in materials, some help is found from the principle of causality [3.56]. Suppose one has a plane wave traveling in the $+x$-direction and that, at $x = 0$, the acoustic pressure is a transient given by

$$p(0,t) = \int_{-\infty}^\infty \hat{p}(\omega)e^{-i\omega t}d\omega = 2\mathrm{Re}\int_0^\infty \hat{p}(\omega)e^{-i\omega t}d\omega. \tag{3.236}$$

Then the transient at a distant positive value of x is given by

$$p(x,t) = 2\mathrm{Re}\int_0^\infty \hat{p}(\omega)\mathrm{e}^{-\mathrm{i}\omega t}\mathrm{e}^{\mathrm{i}kx}\,\mathrm{d}\omega\,, \qquad (3.237)$$

where $k = k(\omega)$ is the complex wavenumber, the imaginary part of which is the attenuation coefficient.

The causality argument is now made that, if $p(0, t)$ vanishes at all times before some time t_0 in the past, then so should $p(x, t)$. One defines $k(\omega)$ for negative values of frequency so that

$$k(-\omega) = -k^*(\omega)\,. \qquad (3.238)$$

This requires that the real and imaginary parts of the complex wavenumber be such that

$$k_\mathrm{R}(-\omega) = -k_\mathrm{R}(\omega)\,; \qquad k_\mathrm{I}(-\omega) = k_\mathrm{I}(\omega)\,, \qquad (3.239)$$

so that the real part is odd, and the imaginary part is even in ω. This extension to negative frequencies allows one to write (3.237) in the form

$$p(x,t) = \int_{-\infty}^{\infty} \hat{p}(\omega)\mathrm{e}^{-\mathrm{i}\omega t}\mathrm{e}^{\mathrm{i}kx}\,\mathrm{d}\omega\,. \qquad (3.240)$$

The theory of complex variables assures one that this integral will indeed vanish for sufficiently early times if: (i) one can regard the integrand as being defined for complex values of ω, and (ii) one can regard the integrand as having certain general properties in the upper half of the complex ω plane. In particular, it must be analytic (no poles, no branch cuts, no points where a power series does not exist) in this upper half of the comp;ex plane. Another condition is that, when t has a sufficiently negative value, the integrand goes to zero as $|\omega| \to \infty$ in the upper half-plane. The possibility of the latter can be tested with the question of whether

$$\mathrm{Re}\,[-\mathrm{i}\omega t + \mathrm{i}kx)] \to -\infty \qquad \text{as} \qquad \omega \to \mathrm{i}\infty\,, \qquad (3.241)$$

with $t < 0$ and $x > 0$. In any event, for all these conditions to be met, the complex wavenumber $k(\omega)$, considered as a complex function of the complex variable ω, has to be analytic in the upper half-plane. One cannot say at the outset, without having a well-defined causal mathematical model, just how it behaves at infinity, but various experimental data suggests it approaches a polynomial with a leading exponent of not more than two. Also, analysis on specific cases suggests it is always such that k/ω is analytic and, moreover, finite at $\omega = 0$.

Contour Integral Identity

Given these general properties of the complex wavenumber, one considers the contour integral

$$\begin{aligned}&I_\mathrm{A}(\zeta_1, \zeta_2, \zeta_3, \zeta_4)\\&= \oint \frac{k/\omega}{(\omega - \zeta_1)(\omega - \zeta_2)(\omega - \zeta_3)(\omega - \zeta_4)}\,\mathrm{d}\omega\,,\end{aligned} \qquad (3.242)$$

where the contour is closed, the integration proceeds in the counterclockwise sense, the contour is entirely in the upper half plane, and where the contour encloses the four poles at ζ_1, ζ_2, etc., all of which are in the upper half-plane.

(The number of poles that one includes is somewhat arbitrary, and four is a judicious compromise for the present chapter. The original [3.57, 58] analysis directed toward acoustical applications included only two, but this led to divergent integrals when the result was applied to certain experimental laws that had been extrapolated to high frequencies. The more poles one includes, the less dependent is the prediction on details of extrapolations to frequencies outside the range of experimental data.)

The integral can be deformed to one which proceeds along the real axis from $-\infty$ to ∞ and which is then completed by a semicircle of infinite radius that encloses the upper half-plane. Because of the supposed behavior of k at ∞ in the upper half-plane, the latter integral is zero and one has

$$\begin{aligned}&I_\mathrm{A}(\zeta_1, \zeta_2, \zeta_3, \zeta_4)\\&= \int_{-\infty}^{\infty} \frac{k/\omega}{(\omega - \zeta_1)(\omega - \zeta_2)(\omega - \zeta_3)(\omega - \zeta_4)}\,\mathrm{d}\omega\,.\end{aligned} \qquad (3.243)$$

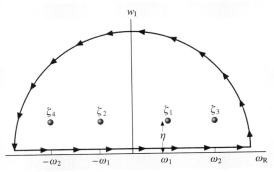

Fig. 3.13 Contour integral in the complex frequency plane used in the derivation of one version of the Kramers–Krönig relations

Furthermore, the residue theorem yields

$$I_A(\zeta_1, \zeta_2, \zeta_3, \zeta_4)$$
$$= \frac{2\pi i (k/\omega)_1}{(\zeta_1 - \zeta_2)(\zeta_1 - \zeta_3)(\zeta_1 - \zeta_4)}$$
$$+ \frac{2\pi i (k/\omega)_2}{(\zeta_2 - \zeta_1)(\zeta_2 - \zeta_3)(\zeta_2 - \zeta_4)}$$
$$+ \frac{2\pi i (k/\omega)_3}{(\zeta_3 - \zeta_1)(\zeta_3 - \zeta_2)(\zeta_3 - \zeta_4)}$$
$$+ \frac{2\pi i (k/\omega)_4}{(\zeta_4 - \zeta_1)(\zeta_4 - \zeta_2)(\zeta_4 - \zeta_3)}, \quad (3.244)$$

where $(k/\omega)_1$ is k/ω evaluated at $\omega = \zeta_1$, etc.

To take advantage of the symmetry properties (3.239), one sets

$$\zeta_1 = \omega_1 + i\eta, \quad \zeta_3 = -\omega_1 + i\eta;$$
$$\zeta_2 = \omega_2 + i\eta; \quad \zeta_4 = -\omega_2 + i\eta, \quad (3.245)$$

so that the poles occur in pairs, symmetric about the imaginary axis. One next lets the small positive parameter η go to zero, and recognizes that

$$\lim_{\epsilon \to 0} \lim_{\eta \to 0} \int_{\omega_1 - \epsilon}^{\omega_1 + \epsilon} \frac{f(\omega)}{\omega - \omega_1 - i\eta} \, d\omega = \pi i f(\omega_1). \quad (3.246)$$

Thus in this limit, the integral (3.243) becomes represented as a principal value plus a set of four *narrow gap* terms that are recognized as one half of the right side of (3.244). The resulting mathematical identity is

$$\text{Pr} \int_{-\infty}^{\infty} \frac{k/\omega}{(\omega^2 - \omega_1^2)(\omega^2 - \omega_2^2)} \, d\omega = \frac{i\pi}{\omega_1^2 - \omega_2^2}$$
$$\times \left(\frac{k(\omega_1) + k(-\omega_1)}{2\omega_1^2} - \frac{k(\omega_2) + k(-\omega_2)}{2\omega_2^2} \right). \quad (3.247)$$

Real Part Within the Integrand

When one uses the symmetry properties (3.239), the above reduces to

$$2\,\text{Pr} \int_0^{\infty} \frac{k_R(\omega)}{\omega(\omega^2 - \omega_1^2)(\omega^2 - \omega_2^2)} \, d\omega$$
$$= -\frac{\pi}{\omega_1^2 - \omega_2^2} \left(\frac{k_I(\omega_1)}{\omega_1^2} - \frac{k_I(\omega_2)}{\omega_2^2} \right). \quad (3.248)$$

The significant thing about this result and the key to its potential usefulness is that the left side involves only the real part of $k(\omega)$ and the right side only the imaginary. Thus if one knew the real part completely and knew the imaginary part for only one frequency, one could find the imaginary part for any other frequency with the relation

$$\frac{k_I(\omega)}{\omega^2} = \frac{k_I(\omega_1)}{\omega_1^2} + \frac{2(\omega_1^2 - \omega^2)}{\pi}$$
$$\times \text{Pr} \int_0^{\infty} \frac{k_R(\omega')}{\omega'(\omega'^2 - \omega_1^2)(\omega'^2 - \omega^2)} \, d\omega'. \quad (3.249)$$

Here the notation is slightly altered: ω' is the dummy variable of integration, ω_1 is the value of the angular frequency at which one presumably already knows k_I, and ω is that angular frequency for which the value is predicted by the right side. Relations such as this which allow predictions of one part of $k(\omega)$ from another part are known as Kramers–Krönig relations.

Other Kramers–Krönig relations can be obtained from (3.247) with a suitable replacement for k/ω in the integrand and/or taking limits with ω_1 or ω_2 approaching either 0 or ∞. One can also add or subtract simple functions to the integrand where the resulting extra integrals are known. For example, if one sets k/ω to unity in (3.247) one obtains the identity

$$\text{Pr} \int_0^{\infty} \frac{d\omega'}{(\omega'^2 - \omega_1^2)(\omega'^2 - \omega^2)} = 0. \quad (3.250)$$

Thus, (3.249) can be equivalently written

$$\frac{k_I(\omega)}{\omega^2} = \frac{k_I(\omega_1)}{\omega_1^2} + \frac{2(\omega_1^2 - \omega^2)}{\pi}$$
$$\times \text{Pr} \int_0^{\infty} \frac{[k_R(\omega')/\omega'] - [k_R(\omega')/\omega']_0}{(\omega'^2 - \omega_1^2)(\omega'^2 - \omega^2)} \, d\omega', \quad (3.251)$$

where the subscript 0 indicates that the quantity is evaluated in the limit of zero frequency. A further step is to take the limit as $\omega_1 \to 0$, so that one obtains

$$\frac{k_I(\omega)}{\omega^2} = \left[\frac{k_I(\omega)}{\omega^2}\right]_0 - \frac{2\omega^2}{\pi}$$
$$\times \text{Pr} \int_0^{\infty} \frac{[k_R(\omega')/\omega'] - [k_R(\omega')/\omega']_0}{\omega'^2(\omega'^2 - \omega^2)} \, d\omega'. \quad (3.252)$$

The numerator in the integrand is recognized as the difference in the reciprocals of the phase velocities, at $\omega = \omega'$ and at $\omega = 0$. The validity of the equality requires the integrand to be sufficiently well behaved near

$\omega' = 0$, and this is consistent with k_R being odd in ω and representable locally as a power series. Validity also requires that k_I/ω^2 be finite at $\omega = 0$, which is consistent with the relaxation models mentioned in the section below. The existence of the integral also places a restriction on the asymptotic dependence of k_R as $\omega \to \infty$.

A possible consequence of the last relation is that

$$\lim_{\omega \to \infty} \left(\frac{k_I(\omega)}{\omega^2} \right)$$

$$= \left(\frac{k_I(\omega)}{\omega^2} \right)_0$$

$$+ \frac{2}{\pi} \int_0^\infty \frac{[k_R(\omega')/\omega'] - [k_R(\omega')/\omega']_0}{\omega'^2} d\omega' . \quad (3.253)$$

The validity of this requires that the integral exists, which is so if the phase velocity approaches a constant at high frequency. An implication is that the attenuation must approach a constant multiplied by ω^2 at high frequency, where the constant is smaller than that at low frequency if the phase velocity at high frequency is higher than that at low frequency.

Imaginary Part Within the Integrand

Analogous results, only with the integration over the imaginary part of k, result when k/ω is replaced by k in (3.247). Doing so yields

$$2 \Pr \int_0^\infty \frac{k_I(\omega)}{(\omega^2 - \omega_1^2)(\omega^2 - \omega_2^2)} d\omega$$

$$= \frac{\pi}{\omega_1^2 - \omega_2^2} \left(\frac{k_R(\omega_1)}{\omega_1} - \frac{k_R(\omega_2)}{\omega_2} \right) , \quad (3.254)$$

which in turn yields

$$\frac{k_R(\omega)}{\omega}$$

$$= \frac{k_R(\omega_1)}{\omega_1} - \frac{2(\omega_1^2 - \omega^2)}{\pi}$$

$$\times \Pr \int_0^\infty \frac{k_I(\omega')}{(\omega'^2 - \omega_1^2)(\omega'^2 - \omega^2)} d\omega' . \quad (3.255)$$

Then, taking of the limit $\omega_1 \to 0$, one obtains

$$\frac{k_R(\omega)}{\omega}$$

$$= \left(\frac{k_R(\omega)}{\omega} \right)_0 + \frac{2\omega^2}{\pi} \Pr \int_0^\infty \frac{k_I(\omega')}{\omega'^2(\omega'^2 - \omega^2)} d\omega'.$$

$$(3.256)$$

This latter expression requires that k_I/ω^2 be integrable near $\omega = 0$.

Attenuation Proportional to Frequency

Some experimental data for various materials suggest that, for those materials, k_I is directly proportional to ω over a wide range of frequencies. The relation (3.256) is inapplicable if one seeks the corresponding expression for phase velocity. Instead, one uses (3.255), treating ω_1 as a parameter. If one inserts

$$k_I(\omega) = K\omega \quad (3.257)$$

into (3.255), the resulting integral can be performed analytically, with the result

$$\frac{k_R(\omega)}{\omega} = \frac{k_R(\omega_1)}{\omega_1} - \frac{2}{\pi} K \ln \left(\frac{\omega}{\omega_1} \right) . \quad (3.258)$$

(The simplest procedure for deriving this is to replace the infinite upper limit by a large finite number and then separate the integrand using the method of partial fractions. After evaluation of the individual terms, one takes the limit as the upper limit goes to infinity, and discovers appropriate cancelations.) The properties of the logarithm are such that the above indicates that the quantity

$$\frac{k_R(\omega)}{\omega} + \frac{2}{\pi} K \ln(\omega) = \text{constant} \quad (3.259)$$

is independent of ω. This deduction is independent of the choice of ω_1, but the analysis does not tell one what the constant should be. A concise restating of the result is that there is some positive number ω_0, such that

$$\frac{k_R(\omega)}{\omega} = \frac{2}{\pi} K \ln \left(\frac{\omega_0}{\omega} \right) . \quad (3.260)$$

The result is presumably valid at best only over the range of frequencies for which (3.257) is valid. Since negative phase velocities are unlikely, it must also be such that the parameter ω_0 is above this range. This approximate result also predicts a zero phase velocity in the limit of zero frequency, and this is also likely to be unrealistic. But there may nevertheless be some range of frequencies for which both (3.257) and (3.260) would give a good fit to experimental data.

3.7.5 Attenuation of Sound in Air

In air, the relaxation processes that affect sound attenuation are those associated with the (quantized) internal vibrations of the diatomic molecules O_2 and N_2. The ratio of the numbers of molecules in the ground and first excited vibrational states is a function of temperature when the gas is in thermodynamic equilibrium, but during an acoustic disturbance the redistribution of molecules to what is appropriate to the concurrent gas temperature is not instantaneous. The knowledge that the relaxation processes are vibrational relaxation processes and that only the ground and first excited states are appreciably involved allows the sound speed increments to be determined from first principles. One need only determine the difference in sound speeds resulting from the two assumptions: (i) that the distribution of vibrational energy is frozen, and (ii) that the vibrational energy is always distributed as for a gas in total thermodynamic equilibrium. The resulting sound speed increments [3.14] are

$$\frac{(\Delta c)_v}{c} = \frac{(\gamma-1)^2}{2\gamma} \frac{n_v}{n} \left(\frac{T_v^*}{T}\right)^2 e^{-T_v^*/T}, \quad (3.261)$$

where n is the total number of molecules per unit volume and n_v is the number of molecules of the type corresponding to the designation parameter v. The quantity T is the absolute temperature, and T_v^* is a characteristic temperature, equal to the energy jump ΔE between the two vibrational states divided by Boltzmann's constant. The value of T_1^* (corresponding to O_2 relaxation) is 2239 K, and the value of T_2^* (corresponding to N_2 relaxation) is 3352 K. For air the fraction n_1/n of molecules that are O_2 is 0.21, while the fraction n_2/n of molecules that are N_2 is 0.78. At a representative temperature of 20 °C, the calculated value of $(\Delta c)_1$ is 0.11 m/s, while that for $(\Delta c)_2$ is 0.023 m/s.

Because relaxation in a gas is caused by two-body collisions, the relaxation times at a given absolute temperature must vary inversely with the absolute pressure. The relatively small (and highly variable) number of water-vapor molecules in the air has a significant effect on the relaxation times because collisions of diatomic molecules with H_2O molecules are much more likely to cause a transition between one internal vibrational quantum state and another. Semi-empirical expressions for the two relaxation times for air are given [3.59, 60] by

$$\frac{p_{\text{ref}}}{p} \frac{1}{2\pi\tau_1} = 24 + 4.04 \times 10^6 h \left(\frac{0.02 + 100h}{0.391 + 100h}\right), \quad (3.262)$$

$$\frac{p_{\text{ref}}}{p} \frac{1}{2\pi\tau_2} = \left(\frac{T_{\text{ref}}}{T}\right)^{1/2} (9 + 2.8 \times 10^4 h\, e^{-F}) \quad (3.263)$$

$$F = 4.17 \left[\left(\frac{T_{\text{ref}}}{T}\right)^{1/3} - 1\right]. \quad (3.264)$$

The subscript 1 corresponds to O_2 relaxation, and the subscript 2 corresponds to N_2 relaxation. The quantity h here is the fraction of the air molecules that are H_2O molecules; the reference temperature is 293.16 K; and the reference pressure is 10^5 Pa. The value of h can be determined from the commonly reported relative humidity (RH, expressed as a percentage) and the vapor pressure of water at the local temperature according to the defining relation

$$h = 10^{-2}(\text{RH}) \frac{p_{\text{vp}}(T)}{p}. \quad (3.265)$$

However, as indicated in (3.262) and (3.263), the physics of the relaxation process depends on the absolute humidity and has no direct involvement with the value of the vapor pressure of water. A table of the vapor pressure of water may be found in various references; some representative values in pascals are 872, 1228, 1705, 2338,

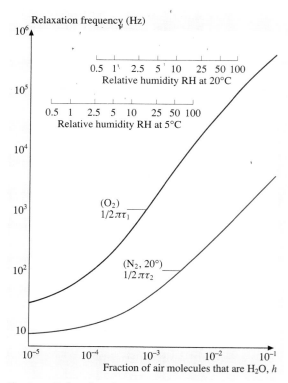

Fig. 3.14 Relaxation frequencies of air as a function of absolute humidity. The pressure is 1.0 atmosphere

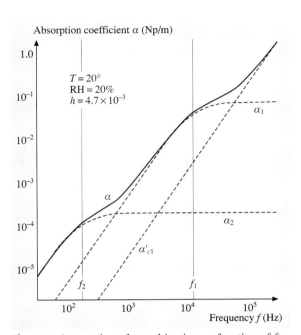

Fig. 3.15 Attenuation of sound in air as a function of frequency (log–log plot). The slopes of the initial portions of the dashed lines correspond to a quadratic dependence on frequency

the attenuation being the same as that corresponding to classical processes, and with the intrinsic bulk viscosity associated with molecular rotation. Nevertheless, even through the coefficient of the square of the frequency drops over two intervals, the overall trend is that the attenuation constant always increases with increasing frequency.

3.7.6 Attenuation of Sound in Sea Water

The relaxation processes contributing to the attenuation of sound are associated with dissolved boric acid $B(OH)_3$ (subscript 1) and magnesium sulfate $MgSO_4$ (subscript 2). From approximate formulas derived by *Fisher* and *Simmons* [3.61] from a combination of experiment and theory one extracts the identifications

$$\frac{\delta_{cl}}{c^3} = 1.42 \times 10^{-15} F(T_C) G(P_{atm}), \quad (3.266)$$

with

$$F(T_C) = 1 - 4.24 \times 10^{-2} T_C + 8.53 \times 10^{-4} T_C^2 \\ - 6.23 \times 10^{-6} T_C^3, \quad (3.267)$$

$$G(P_{atm}) = 1 - 3.84 \times 10^{-4} P_{atm} + 7.57 \times 10^{-8} P_{atm}^2, \quad (3.268)$$

$$\frac{(\Delta c)_1}{c^2} = 1.64 \times 10^{-9} (1 + 2.29 \times 10^{-2} T_C \\ - 5.07 \times 10^{-4} T_C^2) \frac{S}{35}, \quad (3.269)$$

$$\frac{(\Delta c)_2}{c^2} = 8.94 \times 10^{-9} (1 + 0.0134 T_C) \\ \times (1 - 10.3 \times 10^{-4} P_{atm} \\ + 3.7 \times 10^{-7} P_{atm}^2) \frac{S}{35}, \quad (3.270)$$

$$\frac{1}{2\pi\tau_1} = 1320\, T\, e^{-1700/T}, \quad (3.271)$$

$$\frac{1}{2\pi\tau_2} = 15.5 \times 10^6\, T\, e^{-3052/T}. \quad (3.272)$$

Here T_C is temperature in °C, T is absolute temperature, while P_{atm} is the absolute pressure in atmospheres; S is the salinity in parts per thousand (which is typically of the order of 35 for sea water). All of the quantities on the left sides of these equations are in MKS units.

4243, and 7376 Pa at temperatures of 5, 10, 15, 20, 30, and 40°C, respectively.

Because of the two relaxation frequencies, the frequency dependence of the attenuation coefficient for sound in air has three distinct regions. At very low frequencies, where the frequency is much lower than that associated with molecular nitrogen, the attenuation associated with vibrational relaxation of nitrogen molecules dominates. The dependence is nearly quadratic in frequency, with an apparent bulk viscosity that is associated with the nitrogen relaxation. In an intermediate region, where the frequency is substantially larger than that associated with nitrogen relaxation, but still substantially less than that associated with oxygen relaxation, the dependence is again quadratic in frequency, but the coefficient is smaller, and the apparent bulk viscosity is that associated with oxygen relaxation. Then in the higher-frequency range, substantially above both relaxation frequencies, the quadratic frequency dependence is again evident, but with an even smaller coefficient,

3.8 Acoustic Intensity and Power

A complete set of linear acoustic equations, regardless of the idealization incorporated into its formulation, usually yields a corollary [3.30, 62]

$$\frac{\partial w}{\partial t} + \nabla \cdot \mathbf{I} = -\mathfrak{D}, \tag{3.273}$$

where the terms are quadratic in the acoustic field amplitudes, the quantity w contains one term that is identifiable as a kinetic energy of fluid motion per unit volume, and the quantity $\mathfrak{D}$ is either zero or positive.

3.8.1 Energy Conservation Interpretation

The relation (3.273) is interpreted as a statement of energy conservation. The quantity w is the energy density, or energy per unit volume associated with the wave disturbance, while the vector quantity $\mathbf{I}$ is an intensity vector or energy flux vector. Its interpretation is such that its dot product with any unit vector represents the energy flowing per unit area and time across a surface whose normal is in that designated direction. The quantity $\mathfrak{D}$ is interpreted as the energy that is dissipated per unit time and volume.

This interpretation of an equation such as (3.273) as a conservation law follows when one integrates both sides over an arbitrary fixed volume V within the fluid and reexpresses the volume integral of the divergence of $\mathbf{I}$ by a surface integral by means of the divergence theorem (alternately referred to as Gauss's theorem). Doing this yields

$$\frac{\partial}{\partial t} \iiint_V w \, dV + \iint_S \mathbf{I} \cdot \mathbf{n} \, dS = -\iiint_V \mathfrak{D} \, dV, \tag{3.274}$$

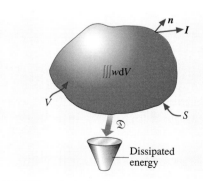

Fig. 3.16 Hypothetical volume inside a fluid, within which the acoustic energy is being dissipated and out of which acoustic energy is flowing

where $\mathbf{n}$ is the unit normal vector pointing out of the surface S enclosing V. This relation states that the net rate of increase of *acoustical energy* within the volume must equal the *acoustic power* flowing into the volume across its confining surface minus the energy that is being dissipated per unit time within the volume.

3.8.2 Acoustic Energy Density and Intensity

For the ideal case, when there is no ambient velocity and when viscosity and thermal conduction are neglected, the energy corollary results from (3.71) and (3.72). The derivation begins with one's taking the dot product of the fluid velocity $\mathbf{v}$ with (3.72), and then using vector identities and (3.71) to reexpress the right side as the sum of a divergence and a time derivative. The result yields the identification of the energy density as

$$w = \frac{1}{2}\rho v^2 + \frac{1}{2}\frac{1}{\rho c^2} p^2, \tag{3.275}$$

and yields the identification of the acoustic intensity as

$$\mathbf{I} = p\mathbf{v}. \tag{3.276}$$

For this ideal case, there is no dissipative term on the right side. The corollary of (3.273) remains valid even if the ambient density and sound speed vary from point to point.

The first term in the expression for w is recognized as the acoustic kinetic energy per unit volume, and the second term is identified as the potential energy per unit volume due to compression of the fluid.

Intensity Carried by Plane Waves

For a plane wave, it follows from (3.186) and (3.187) that the kinetic and potential energies are the same [3.63], and that the energy density is given by

$$w = \frac{1}{\rho c^2} p^2. \tag{3.277}$$

The intensity becomes

$$\mathbf{I} = \mathbf{n}\frac{p^2}{\rho c}. \tag{3.278}$$

For such a case, the intensity and the energy density are related by

$$\mathbf{I} = c\mathbf{n}w. \tag{3.279}$$

This yields the interpretation that the energy in a sound wave is moving in the direction of propagation with the sound speed. Consequently, the sound speed can be regarded as an energy propagation velocity. (This is in accord with the fact that sound waves in this idealization are nondispersive, so the group and phase velocities are the same.)

3.8.3 Acoustic Power

Many sound fields can be idealized as being steady, such that long-time averages are insensitive to the duration and the center time of the averaging interval. Constant-frequency sounds and continuous noises fall into this category.

In the special case of constant-frequency sounds, when complex amplitudes are used to described the acoustic field, the general theorem (3.182) for averaging over products of quantities oscillating with the same frequency applies, and one finds

$$w_{av} = \frac{1}{4}\rho \hat{\boldsymbol{v}} \cdot \hat{\boldsymbol{v}}^* + \frac{1}{4}\frac{1}{\rho c^2}|\hat{p}|^2 , \qquad (3.280)$$

$$\boldsymbol{I}_{av} = \frac{1}{2}\mathrm{Re}\left\{\hat{p}^* \hat{\boldsymbol{v}}\right\} . \qquad (3.281)$$

For steady sounds, the time derivative of the acoustic energy density will average to zero over a sufficiently long time period, so the acoustic energy corollary of (3.273), in the absence of dissipation, yields the time-averaged relation

$$\nabla \cdot \boldsymbol{I}_{av} = 0 . \qquad (3.282)$$

This implies that the time-averaged vector intensity field is solenoidal (zero divergence) in regions that do not contain acoustic sources. This same relation holds for any frequency component of the acoustic field or for the net contribution to the field from any given frequency band. In the following discussion, the intensity $\boldsymbol{I}_{av}$ is understood to refer to such a time average for some specified frequency band.

The relation (3.282) yields the integral relation

$$\int_S \boldsymbol{I}_{av} \cdot \boldsymbol{n} \, \mathrm{d}S = 0 , \qquad (3.283)$$

which is interpreted as a statement that the net acoustic power flowing out of any region not containing sources must be zero when averaged over time and for any given frequency band.

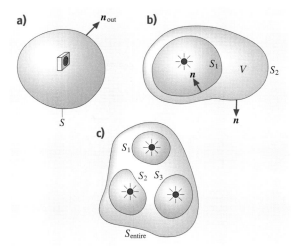

Fig. 3.17 Surfaces enclosing one or more sources. The integration of the time-averaged acoustic intensity component in the direction of the unit outward normal over any such surface yields the time average of the total power generated with the volume enclosed by the surface

For a closed surface that encloses one or more sources, such that the governing linear acoustic equations do not apply at every point within the volume, the reasoning above allows one to define the time-averaged net acoustic power of these sources as

$$\mathfrak{P}_{av} = \int_S \boldsymbol{I}_{av} \cdot \boldsymbol{n} \, \mathrm{d}S , \qquad (3.284)$$

where the surface S encloses the sources. It follows from (3.283) that the acoustic power of a source computed in such a manner will be the same for any two choices of the surface S, provided that both surfaces enclose the same source and no other sources. The value of the integral is independent of the size and of the shape of S. This result is of fundamental importance for the measurement of source power. Instrumentation to measure the time-averaged intensity directly has become widely available in recent years and is often used in determining the principal noise sources in complicated environments.

3.8.4 Rate of Energy Dissipation

Under circumstances in which viscosity, thermal conduction, and internal relaxation are to be taken into account, the linear acoustic equations for the acoustic mode, which yield the dissipative wave equation of (3.218) and the relaxation equations described by

(3.219), yield [3.14] an energy corollary of the form of (3.273). To a satisfactory approximation the energy density and intensity remain as given by (3.275) and (3.276). The energy dissipation per unit volume is no longer zero, but is instead

$$\mathfrak{D}_{av} = 2\frac{\delta_{cl}}{\rho c^4}\left(\frac{\partial p}{\partial t}\right)^2 + 2\sum_{\nu}\frac{(\Delta c)_\nu}{\rho c^3}\tau_\nu\left(\frac{\partial p_\nu}{\partial t}\right)^2. \tag{3.285}$$

For constant-frequency plane waves propagating in the x-direction, the time average of the time derivative of the energy density is zero, and the time averages of I and $\mathfrak{D}$ will both be quadratic in the wave amplitude $|\hat{p}|$. The identification of the attenuation coefficient α must then be such that

$$\mathfrak{D}_{av} = 2\alpha I_{av}. \tag{3.286}$$

The magnitude of the acoustic intensity decreases with propagation distance x as

$$I_{av}(x) = I_{av}(0)\mathrm{e}^{-2\alpha x}. \tag{3.287}$$

This identification of the attenuation coefficient is consistent with that given in (3.225).

3.8.5 Energy Corollary for Elastic Waves

An energy conservation corollary of the form of (3.273) also holds for sound in solids. The appropriate identifications for the energy density w and the components I_i of the intensity are

$$w = \frac{1}{2}\rho\sum_i\left(\frac{\partial \xi_i}{\partial t}\right)^2 + \frac{1}{2}\sum_{i,j}\epsilon_{ij}\sigma_{ij} \tag{3.288}$$

$$I_i = -\sum_j\sigma_{ij}\frac{\partial \xi_j}{\partial t}. \tag{3.289}$$

3.9 Impedance

Complex ratios of acoustic variables are often intrinsic quantities independent of the detailed nature of the acoustic disturbance.

3.9.1 Mechanical Impedance

The ratio of the complex amplitude of a sinusoidally varying force to the complex amplitude of the resulting velocity at a point on a vibrating object is called the mechanical impedance at that point. It is a complex number and usually a function of frequency. Other definitions [3.64] of impedance are also in widespread use in acoustics.

3.9.2 Specific Acoustic Impedance

The specific acoustic impedance or unit-area acoustic impedance $Z_S(\omega)$ for a surface is defined as

$$Z_S(\omega) = \frac{\hat{p}}{\hat{v}_{in}}, \tag{3.290}$$

where $\hat{v}_{in}$ is the component of the fluid velocity directed into the surface under consideration. Typically, the specific acoustic impedance, often referred to briefly as the impedance without any adjective, is used to describe the acoustic properties of materials. In many cases, surfaces of materials abutting fluids can be characterized as locally reacting, so that Z_S is independent of the detailed nature of the acoustic pressure field. In particular, the locally reacting hypothesis implies that the velocity of the material at the surface is unaffected by pressures other than in the immediate vicinity of the point of interest. At a nominally motionless and passively responding surface, and when the hypothesis is valid, the appropriate boundary condition on the complex amplitude $\hat{p}$ that satisfies the Helmholtz equation is given by

$$i\omega\rho\hat{p} = -Z_S\nabla\hat{p}\cdot\boldsymbol{n}, \tag{3.291}$$

where $\boldsymbol{n}$ is the unit normal vector pointing out of the material into the fluid. A surface that is perfectly rigid has $|Z_S| = \infty$. The other extreme, where $Z_S = 0$, corresponds to the ideal case of a pressure release surface. This is, for example, what is normally assumed for the upper surface of the ocean in underwater sound. Since a passive surface absorbs energy from the sound field, the time-averaged intensity component into the surface should be positive or zero. This observation leads to the requirement that the real part (specific acoustic resistance) of the impedance should always be nonnegative. The imaginary part (specific acoustic reactance) may be either positive or negative.

3.9.3 Characteristic Impedance

For extended substances, a related definition is that of characteristic impedance Z_{char}, defined as the ratio of $\hat{p}$ to the complex amplitude $\hat{v}$ of the fluid velocity in the direction of propagation when a plane wave is propagating through the substance. As indicated by (3.187), this characteristic impedance, when the fluid is lossless, is ρc, regardless of frequency and position in the field. In the general case when the propagation is dispersive and there is a plane wave attenuation, one has

$$Z_{\text{char}} = \frac{\rho \omega}{k}, \qquad (3.292)$$

where $k(\omega)$ is the complex wavenumber.

The MKS units $[(\text{kg/m}^{-3})(\text{m/s})]$ of specific acoustic impedance are referred to as MKS rayl (after Rayleigh). The characteristic impedance of air under standard conditions is approximately 400 MKS rayls, and that of water is approximately 1.5×10^6 MKS rayl.

3.9.4 Radiation Impedance

The radiation impedance Z_{rad} is defined as $\hat{p}/\hat{v}_n$, where $\hat{v}_n$ corresponds to the outward normal component of velocity at a vibrating surface. (Specific examples involving spherical waves are given in a subsequent section of this chapter.)

Given the definition of the radiation impedance, the time-averaged power flow per unit area of surface out of the surface is

$$\begin{aligned} I_{\text{rad}} &= \frac{1}{2}\text{Re}(\hat{p}\hat{v}_n^*) = \frac{1}{2}|\hat{v}_n|^2 \text{Re}(Z_{\text{rad}}) \\ &= \frac{1}{2}|\hat{p}|^2 \text{Re}\left(\frac{1}{Z_{\text{rad}}}\right). \end{aligned} \qquad (3.293)$$

The total power radiated is the integral of this over the surface of the body.

3.9.5 Acoustic Impedance

The term acoustic impedance Z_A is reserved [3.64] for the ratio of $\hat{p}$ to the volume velocity complex amplitude. Here volume velocity is the net volume of fluid flowing past a specified surface element per unit time in a specified directional sense. One may speak, for example, of the acoustic impedance of an orifice in a wall, of the acoustic impedance at the mouth of a Helmholtz resonator, and of the acoustic impedance at the end of a pipe.

3.10 Reflection and Transmission

When sound impinges on a surface, some sound is reflected and some is transmitted and possibly absorbed within or on the the other side of the surface. To understand the processes that occur, it is often an appropriate idealization to take the incident wave as a plane wave and to consider the surface as flat.

3.10.1 Reflection at a Plane Surface

When a plane wave reflects at a surface with finite specific acoustic impedance Z_S, a reflected wave is formed such that the angle of incidence θ_I equals the angle of reflection (law of mirrors). Here both angles are reckoned from the line normal to the surface and correspond to the directions of the two waves. If one takes the y-axis as pointing out of the surface and the surface as coinciding with the $y = 0$ plane, then an incident plane wave propagating obliquely in the $+x$-direction will have a complex pressure amplitude

$$\hat{p}_{\text{in}} = \hat{f} e^{ik_x x} e^{-ik_y y}, \qquad (3.294)$$

where $\hat{f}$ is a constant. (For transient reflection, the quantity $\hat{f}$ can be taken as the Fourier transform of the incident pressure pulse at the origin.) The two indicated wavenumber components are $k_x = k \sin \theta_I$ and $k_y = k \cos \theta_I$.

The reflected wave has a complex pressure amplitude given by

$$\hat{p}_{\text{refl}} = \mathfrak{R}(\theta_I, \omega) \hat{f} e^{ik_x x} e^{ik_y y}, \qquad (3.295)$$

where the quantity $\mathfrak{R}(\theta_I, \omega)$ is the pressure amplitude reflection coefficient.

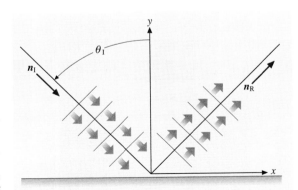

Fig. 3.18 Reflection of a plane wave at a planar surface

Analysis that makes use of the boundary condition (3.291) leads to the identification

$$\mathfrak{R}(\theta_\mathrm{I}, \omega) = \frac{\xi(\omega)\cos\theta_\mathrm{I} - 1}{\xi(\omega)\cos\theta_\mathrm{I} + 1} \qquad (3.296)$$

for the reflection coefficient, with the abbreviation

$$\xi(\omega) = \frac{Z_\mathrm{S}}{\rho c}, \qquad (3.297)$$

which represents the ratio of the specific acoustic impedance of the surface to the characteristic impedance of the medium.

3.10.2 Reflection at an Interface

The above relations also apply, with an appropriate identification of the quantity Z_S, to sound reflection [3.65] at an interface between two fluids with different sound speeds and densities. Translational symmetry requires that the disturbance in the second fluid have the same apparent phase velocity (ω/k_x) (trace velocity) along the x-axis as does the disturbance in the first fluid. This requirement is known as the trace velocity matching principle [3.4, 30] and leads to the observation that k_x is the same in both fluids. One distinguishes two possibilities: the trace velocity is higher than the sound speed c_2 or lower than c_2.

For the first possibility, one has the inequality

$$c_2 < \frac{c_1}{\sin\theta_\mathrm{I}}, \qquad (3.298)$$

and a propagating plane wave (transmitted wave) is excited in the second fluid, with complex pressure amplitude

$$\hat{p}_\mathrm{trans} = \mathfrak{T}(\omega, \theta_\mathrm{I})\hat{f}\,\mathrm{e}^{\mathrm{i}k_x x}\,\mathrm{e}^{\mathrm{i}k_2 y \cos\theta_\mathrm{II}}, \qquad (3.299)$$

where $k_2 = \omega/c_2$ is the wavenumber in the second fluid and θ_II (angle of refraction) is the angle at which the transmitted wave is propagating. The trace velocity matching principle leads to Snell's law,

$$\frac{\sin\theta_\mathrm{I}}{c_1} = \frac{\sin\theta_\mathrm{II}}{c_2}. \qquad (3.300)$$

The change in propagation direction from θ_I to θ_II is the phenomenon of refraction.

The requirement that the pressure be continuous across the interface yields the relation

$$1 + \mathfrak{R} = \mathfrak{T}, \qquad (3.301)$$

while the continuity of the normal component of the fluid velocity yields

$$\frac{\cos\theta_\mathrm{I}}{\rho_1 c_1}(1 - \mathfrak{R}) = \frac{\cos\theta_\mathrm{II}}{\rho_2 c_2}\mathfrak{T}. \qquad (3.302)$$

From these one derives the reflection coefficient

$$\mathfrak{R} = \frac{Z_\mathrm{II} - Z_\mathrm{I}}{Z_\mathrm{II} + Z_\mathrm{I}}, \qquad (3.303)$$

which involves the two impedances defined by

$$Z_\mathrm{I} = \frac{\rho_1 c_1}{\cos\theta_\mathrm{I}}, \qquad (3.304)$$

$$Z_\mathrm{II} = \frac{\rho_2 c_2}{\cos\theta_\mathrm{II}}. \qquad (3.305)$$

The other possibility, which is the opposite of that in (3.298), can only occur when $c_2 > c_1$ and, moreover, only if θ_I is greater than the critical angle

$$\theta_\mathrm{cr} = \arcsin(c_1/c_2). \qquad (3.306)$$

In this circumstance, an inhomogeneous plane wave, propagating in the x-direction, but dying out exponentially in the $+y$-direction, is excited in the second medium. Instead of (3.299), one has the transmitted pressure given by

$$\hat{p}_\mathrm{trans} = \mathfrak{T}(\omega, \theta_\mathrm{I})\hat{f}\,\mathrm{e}^{\mathrm{i}k_x x}\,\mathrm{e}^{-\beta k_2 y}, \qquad (3.307)$$

with

$$\beta = [(c_2/c_1)^2 \sin^2\theta_\mathrm{I} - 1]^{1/2}. \qquad (3.308)$$

The previously stated equations governing the reflection and transmission coefficients are still applicable, provided one replaces $\cos\theta_\mathrm{II}$ by $\mathrm{i}\beta$. This causes the magnitude of the reflection coefficient $\mathfrak{R}$ to become unity,

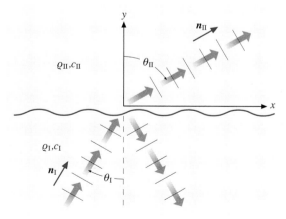

Fig. 3.19 Reflection of a plane wave at an interface between two fluids

so the time-averaged incident energy is totally reflected. Acoustic energy is present in the second fluid, but its time average over a wave period stays constant once the steady state is reached.

3.10.3 Theory of the Impedance Tube

Impedance tubes are commonly used in the measurement of specific acoustic impedances; the underlying theory [3.66, 67] is based for the most part on (3.295), (3.296), and (3.297) above. The incident and the reflected waves propagate along the axis of a cylindrical tube with the sample surface at one end. A loudspeaker at the other end creates a sinusoidal pressure disturbance that propagates down the tube. Reflections from the end covered with the test material create an incomplete standing-wave pattern inside the tube.

The wavelength of the sound emitted by the source can be adjusted, but it should be kept substantially larger than the pipe diameter, so that the plane wave assumption holds. With k_x identified as being 0, the complex amplitude that corresponds to the sum of the incident and reflected waves has an absolute magnitude given by

$$|\hat{p}| = |\hat{f}||1 + \mathfrak{R}e^{2iky}|, \quad (3.309)$$

where y is now the distance in front of the sample. The second factor varies with y and repeats at intervals of a half-wavelength, and varies from a minimum value of $1 - |\mathfrak{R}|$ to a maximum value of $1 + |\mathfrak{R}|$. Consequently,

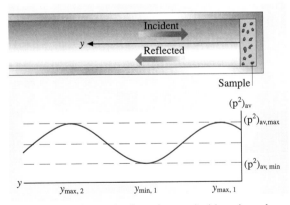

Fig. 3.20 Incident and reflected waves inside an impedance tube. The time-averaged pressure within the tube has minimum and maximum values whose ratio depends on the impedance of the sample at the end. Another measured parameter is the distance back from the sample at which the first maximum occurs

the ratio of the peak acoustic pressure amplitude $|\hat{p}|_{\max}$ (which occurs at one y-position) to the minimum acoustic pressure amplitude $|\hat{p}|_{\min}$ (which occurs at a position a quarter-wavelength away) determines the magnitude of the reflection coefficient via the relation

$$\frac{|\hat{p}|_{\min}}{|\hat{p}|_{\max}} = \frac{1 - |\mathfrak{R}|}{1 + |\mathfrak{R}|}. \quad (3.310)$$

The phase δ of the reflection coefficient can be determined with use of the observation that the peak amplitudes occur at y-values where $\delta + 2ky$ is an integer multiple of 2π, while the minimum amplitudes occur where it is π plus an integer multiple of 2π. Once the magnitude and phase of the reflection coefficient are determined, the specific acoustic impedance can be found from (3.296) and (3.297).

3.10.4 Transmission through Walls and Slabs

The analysis of transmission of sound through a wall or a partition [3.68] is often based on the idealization that the wall is of unlimited extent.

If the incoming plane wave has an angle of incidence θ_I and if the fluid on the opposite side of the wall has the same sound speed, then the trace velocity matching principle requires that the transmitted wave be propagating in the same direction.

A common assumption when the fluid is air is that the compression in the wall is negligible, so the wall is treated as a slab that has a uniform velocity v_{sl} throughout its thickness. The slab moves under the influence of the incident, reflected, and transmitted sound fields according to the relation (corresponding to Newton's

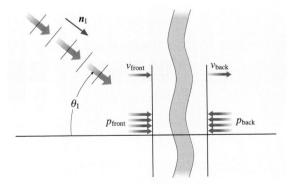

Fig. 3.21 Transmission of an incident plane wave through a thin flexible slab

second law) given by

$$m_{sl}\frac{\partial v_{sl}}{\partial t} = p_{\text{front}} - p_{\text{back}} + \text{bending term}, \quad (3.311)$$

where m_{sl} is the mass per unit surface area of the slab.

The front side is here taken as the side facing the incident wave; the transmitted wave propagates away from the back side. The bending term (discussed further below) accounts for any tendency of the slab to resist bending. If the slab is regarded as nonporous then the normal component of the fluid velocity both at the front and the back is regarded the same as the slab velocity itself. If it is taken as porous [3.69, 70] then these continuity equations are replaced by the relations

$$v_{\text{front}} - v_{sl} = v_{\text{back}} - v_{sl} = \frac{1}{R_f}(p_{\text{front}} - p_{\text{back}}), \quad (3.312)$$

where R_f is the specific flow resistance. The latter can be measured in steady flow for specific materials. For a homogeneous material, it is given by the product of the slab thickness h and the flow resistivity, the latter being a commonly tabulated property of porous materials. [The law represented by (3.312) is the thin slab, or blanket, counterpart of the Darcy's law mentioned in a preceding section, where Biot's model of porous media is discussed.]

In general, when one considers the reflection at and transmission through a slab, one can define a slab specific impedance Z_{sl} such that, with regard to complex amplitudes,

$$\hat{p}_{\text{front}} - \hat{p}_{\text{back}} = Z_{sl}\hat{v}_{\text{front}} = Z_{sl}\hat{v}_{\text{back}}, \quad (3.313)$$

where Z_{sl} depends on the angular frequency ω and the trace velocity $v_{tr} = c/\sin\theta_I$ of the incident wave over the surface of the slab. The value of the slab specific impedance can be derived using considerations such as those that correspond to (3.312) and (3.313). In terms of the slab specific impedance, the transmission coefficient $\mathfrak{T}$ is

$$\mathfrak{T} = \left(1 + \frac{1}{2}\frac{Z_{sl}}{\rho c}\cos\theta_I\right)^{-1}. \quad (3.314)$$

The fraction τ of incident power that is transmitted is $|\mathfrak{T}|^2$.

3.10.5 Transmission through Limp Plates

If the slab can be idealized as a limp plate (no resistance to bending) and not porous, the slab specific impedance Z_{sl} is $-i\omega m_{sl}$ and one obtains

$$\tau = \left[1 + \left(\frac{\omega m_{sl}}{2\rho c}\right)^2\cos^2\theta_I\right]^{-1} \approx \left(\frac{2\rho c}{\omega m_{sl}\cos\theta_I}\right)^2 \quad (3.315)$$

for the fraction of incident power that is transmitted. The latter version, which typically holds at moderate to high audible frequencies, predicts that τ decreases by a factor of 4 when the slab mass m_{sl} per unit area is increased by a factor of 2. In the noise control literature, this behavior is sometimes referred to as the mass law.

3.10.6 Transmission through Porous Blankets

For a porous blanket that has a specific flow resistance R_f the specific slab impedance becomes

$$Z_{sl} = \left(\frac{1}{R_f} - \frac{1}{i\omega m_{sl}}\right)^{-1}, \quad (3.316)$$

and the resulting fraction of incident power that is transmitted can be found with a substitution into (3.314), with τ equated to $|\mathfrak{T}|^2$.

3.10.7 Transmission through Elastic Plates

If the slab is idealized as a Bernoulli–Euler plate with elastic modulus E, Poisson's ratio ν, and thickness h, the bending term in (3.311) has a complex amplitude given by

$$\text{bending term} = -\frac{B_{pl}k_x^4\hat{v}_{sl}}{(-i\omega)}, \quad (3.317)$$

where

$$B_{pl} = \frac{1}{12}\frac{Eh^3}{(1-\nu^2)} \quad (3.318)$$

is the plate bending modulus. The slab specific impedance is consequently given by

$$Z_{sl} = -i\omega m_{sl}\left[1 - \left(\frac{f}{f_c}\right)^2\sin^4\theta_I\right], \quad (3.319)$$

where

$$f_c = \frac{c^2}{2\pi}\left(\frac{m_{sl}}{B_{pl}}\right)^{1/2} \quad (3.320)$$

gives the coincidence frequency, the frequency at which the phase velocity of freely propagating bending waves in the plate equals the speed of sound in the fluid.

Although the simple result of (3.319) predicts that the fraction of incident power transmitted is unity at a frequency of $f_c/\sin^2\theta_I$, the presence of damping processes in the plate causes the fraction of incident power transmitted always to be less than unity. A simple way of taking this into account makes use of a loss factor η (assumed to be much less than unity) for the plate, which corresponds to the fraction of stored elastic energy that is dissipated through damping processes during one radian (a cycle period divided by 2π). Because twice the loss factor times the natural frequency is the time coefficient for exponential time decay of the amplitude when the system is vibrating in any natural mode, the former can be regarded as the negative of the imaginary part of a complex frequency. Then, because the natural frequency squared is always proportional to the elastic modulus, and because the loss factor is invariably small, the loss factor can be formally introduced into the mathematical model by the replacement of the real elastic modulus by the complex number $(1-i\eta)E$. When this is done, one finds

$$Z_{sl} = \omega\eta m_{pl}\left(\frac{f}{f_c}\right)^2 \sin^4\theta_I - i\omega m_{sl}$$
$$\times\left[1 - \left(\frac{f}{f_c}\right)^2 \sin^4\theta_I\right] \quad (3.321)$$

for the slab impedance that is to be inserted into (3.314). The extra term ordinarily has very little effect on the fraction of incident power that is transmitted except when the (normally dominant) imaginary term is close to zero. When the frequency f is $f_c/\sin^2\theta_I$, one finds the value of τ to be

$$\tau = \left(1 + \frac{1}{2}\frac{\omega\eta m_{pl}}{\rho c}\cos\theta_I\right)^{-2}, \quad (3.322)$$

rather than identically unity.

3.11 Spherical Waves

In many circumstances of interest, applicable idealizations are waves that locally resemble waves spreading out radially from sources or from scatterers. The mathematical description of such waves has some similarities to plane waves, but important distinctions arise. The present section is concerned with a number of important situations where the appropriate coordinates are spherical coordinates.

3.11.1 Spherically Symmetric Outgoing Waves

For a spherically symmetric wave spreading out radially from a source in an unbounded medium, the symmetry implies that the acoustic field variables are a function of only the radial coordinate r and of time t. The Laplacian reduces then to

$$\nabla^2 p = \frac{\partial^2 p}{\partial r^2} + \frac{2}{r}\frac{\partial p}{\partial r} = \frac{1}{r}\frac{\partial^2(rp)}{\partial r^2}, \quad (3.323)$$

so the wave equation of (3.74) takes the form

$$\frac{\partial^2(rp)}{\partial r^2} - \frac{1}{c^2}\frac{\partial^2(rp)}{\partial t^2} = 0. \quad (3.324)$$

The solution of this is

$$p(r,t) = \frac{f(r-ct)}{r} + \frac{g(r+ct)}{r}. \quad (3.325)$$

Causality considerations (no sound before the source is turned on) lead to the conclusion that the second term

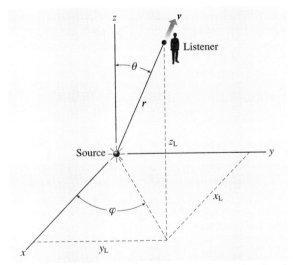

Fig. 3.22 Spherical coordinates. The common situation is when the source is at the origin and the listener (sound receiver) has coordinates (r, θ, ϕ)

on the right side of (3.325) is not an appropriate solution of the wave equation when the source is concentrated near the origin. The expression

$$p(r,t) = \frac{f(r-ct)}{r}, \quad (3.326)$$

which describes the acoustic pressure in an outgoing spherically symmetric wave, has the property that listeners at different radii will receive (with a time shift corresponding to the propagation time) waveforms of the same shape, but of different amplitudes. The factor of $1/r$ is characteristic of spherical spreading and implies that the peak waveform amplitudes in a spherical wave decrease with radial distance as $1/r$.

Spherical waves of constant frequency have complex amplitudes governed by the Helmholtz equation (3.176), with the Laplacian given as stated in (3.323). The complex pressure amplitude corresponding to (3.326) has the form

$$p = A \frac{e^{ikr}}{r}. \tag{3.327}$$

The fluid velocity associated with an outgoing spherical wave is purely radial and has the form

$$v_r = \frac{1}{\rho c}[-r^{-2}F(r-ct) + r^{-1}f(r-ct)]. \tag{3.328}$$

Here the function F is such that its derivative is the function f that appears in (3.326). Because the first term (a near-field term) decreases as the square rather than the first power of the reciprocal of the radial distance, the fluid velocity asymptotically approaches

$$v_r \to \frac{p}{\rho c}, \tag{3.329}$$

which is the same as the plane wave relation of (3.187).

For outgoing spherical waves of constant frequency, the complex amplitude of the fluid velocity is

$$\hat{v}_r = \frac{1}{\rho c}\left(1 - \frac{1}{ikr}\right)\hat{p}. \tag{3.330}$$

In this expression, there is a term that is in phase with the pressure and another term that is 90° ($\pi/2$) out of phase with it.

The time-averaged intensity for spherical waves, in accord with (3.281), is

$$I_{\mathrm{av}} = \frac{1}{2}\mathrm{Re}(\hat{p}\hat{v}_r^*). \tag{3.331}$$

The expression for v_r given above allows this to be simplified to

$$I_{\mathrm{av}} = \frac{1}{2}\frac{1}{\rho c}|\hat{p}|^2. \tag{3.332}$$

Then, with the expression (3.327) inserted for $\hat{p}$, one obtains

$$I_{\mathrm{av}} = \frac{1}{2}\frac{1}{\rho c}\frac{|A|^2}{r^2}. \tag{3.333}$$

The result confirms that the time-averaged intensity falls off as the square of the radial distance. This behavior is what is termed spherical spreading.

The spherical spreading law also follows from energy conservation considerations. The time-averaged power flowing through a spherical surface of radius r is the area $4\pi r^2$ of the surface times the time-averaged intensity. This power should be independent of the radius r since there is no external energy input or attenuation that is included in the considered model, so the intensity must fall off as $1/r^2$.

3.11.2 Radially Oscillating Sphere

The classic example of a source that generates outgoing spherical waves is an impenetrable sphere whose radius $r_{\mathrm{sp}}(t)$ oscillates with time with some given velocity amplitude v_{o}, so that

$$r_{\mathrm{sp}}(t) = a + \frac{v_{\mathrm{o}}}{\omega}\sin(\omega t). \tag{3.334}$$

Here a is the nominal radius of the sphere, and v_{o}/ω is the amplitude of the deviations of the actual radius from that value. For the linear acoustics idealization to be valid, it is required that this deviation be substantially less than a, so that

$$v_{\mathrm{o}} \ll \omega a. \tag{3.335}$$

The boundary condition on the fluid dynamic equations should ideally be

$$v_r = v_{\mathrm{o}}\cos(\omega t) \quad \text{at} \quad r = r_{\mathrm{sp}}(t), \tag{3.336}$$

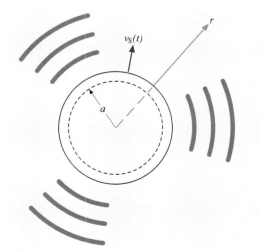

Fig. 3.23 Parameters used for the discussion of constant frequency sound radiated by a radially oscillating sphere

but, also in keeping with the linear acoustics idealization, it is replaced by

$$v_r = v_0 \cos(\omega t) \quad \text{at} \quad r = a. \tag{3.337}$$

The corresponding boundary condition on the complex amplitude is

$$\hat{v}_r = v_0 \quad \text{at} \quad r = a. \tag{3.338}$$

If the complex amplitude of the acoustic part of the pressure is taken of the general form of (3.327) then the radial fluid velocity, in accord with (3.330), has the complex amplitude

$$\hat{v}_r = \frac{1}{\rho c}\left(1 - \frac{1}{ikr}\right) A \frac{e^{ikr}}{r}. \tag{3.339}$$

The approximate boundary condition (3.338) consequently allows one to identify the constant A as

$$A = v_0 \left(\frac{ika^2 \rho c}{ika - 1}\right) e^{-ika}, \tag{3.340}$$

so that the acoustical part of the pressure has the complex amplitude

$$\hat{p} = v_0 \left(\frac{ika\rho c}{ika - 1}\right)\left(\frac{a}{r}\right) e^{-ik[r-a]}. \tag{3.341}$$

Radiation Impedance

The ratio of the complex amplitude of the pressure to that of the fluid velocity in the outward direction at a point on a vibrating surface is termed the specific radiation impedance (specific here meaning per unit area), so that

$$Z_{\text{rad}} = \frac{\hat{p}}{\hat{v}_n}, \tag{3.342}$$

where $\hat{v}_n$ is the component of the complex amplitude of the fluid velocity in the outward normal direction. For the case of the radially oscillating sphere of nominal radius a, the analysis above yields

$$Z_{\text{rad}} = \rho c \left(\frac{ika}{ika-1}\right). \tag{3.343}$$

Usually, this is referred to simply as the radiation impedance, without the qualifying adjective *specific*.

The time-averaged power radiated by an oscillating body, in accord with (3.281), is

$$\mathfrak{P}_{\text{av}} = \frac{1}{2}\int \mathrm{Re}\left(\hat{p}^* \hat{v}_n\right) \mathrm{d}S, \tag{3.344}$$

where the integral extends over the surface. Given the definition of the radiation impedance, this can be written in either of the equivalent forms

$$\mathfrak{P}_{\text{av}} = \frac{1}{2}\int |\hat{p}|^2 \mathrm{Re}\left(\frac{1}{Z_{\text{rad}}}\right) \mathrm{d}S$$
$$= \frac{1}{2}\int |\hat{v}_n|^2 \mathrm{Re}\left(Z_{\text{rad}}\right) \mathrm{d}S. \tag{3.345}$$

For the case of the radially oscillating sphere, the quantity $\hat{v}_n$ is v_0, and

$$\mathrm{Re}\left(Z_{\text{rad}}\right) = \rho c \left(\frac{k^2 a^2}{1 + k^2 a^2}\right). \tag{3.346}$$

The latter increases monotonically from 0 to 1 as the frequency increases from 0 to ∞. The surface area of the sphere is $4\pi a^2$, so the time-averaged acoustic power is

$$\mathfrak{P}_{\text{av}} = (2\pi a^2)(\rho c)v_0^2 \left(\frac{k^2 a^2}{1 + k^2 a^2}\right). \tag{3.347}$$

3.11.3 Transversely Oscillating Sphere

Another basic example for which the natural description is in terms of spherical coordinates is that of a rigid sphere oscillating back and forth in the z-direction about the origin.

The velocity of an arbitrary point on the surface of the sphere can be taken as $v_c \mathbf{e}_z \cos(\omega t)$, where v_c is the velocity amplitude of the oscillation and $\mathbf{e}_z$ is the unit vector in the direction of increasing z. Consistent with

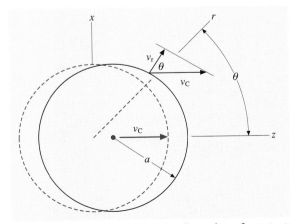

Fig. 3.24 Parameters used for the discussion of constant-frequency sound radiated by a transversely oscillating sphere

the desire to use a linear approximation, one approximates the unit normal vector for a given point on the surface of the sphere to be the same as when the sphere is centered at the origin, so that $\mathbf{n} \approx \mathbf{e}_r$, where the latter is the unit vector in the radial direction. The normal component of the fluid velocity is then approximately

$$v_n = v_c (\mathbf{e}_z \cdot \mathbf{e}_r) \cos(\omega t) . \tag{3.348}$$

The dot product is $\cos\theta$, where θ is the polar angle.

One also makes the approximation that the boundary condition is to be imposed, not at the actual (moving) location of the point on the surface, but at the place in space where that point is when the sphere is centered at the origin. All these considerations lead to the linear acoustics boundary condition

$$\hat{v}_r = v_c \cos\theta \quad \text{at} \quad r = a \tag{3.349}$$

for the complex amplitude of the fluid velocity.

The feature distinguishing this boundary condition from that for the radially oscillating sphere is the factor $\cos\theta$. The plausible conjecture that both $\hat{v}_r$ and $\hat{p}$ continue to have the same θ dependence for all values of r is correct in this case, and one can look for a solution of the Helmholtz equation that has such a dependence, such as

$$\hat{p} = B \frac{\partial}{\partial z}\left(\frac{e^{ikr}}{r}\right) = B \cos\theta \frac{d}{dr}\left(\frac{e^{ikr}}{r}\right) . \tag{3.350}$$

The first part of this relation follows because a derivative of a solution with respect to any Cartesian coordinate is also a solution and because, as demonstrated in a previous part of this section, e^{ikr}/r is a solution. The second part follows because $r^2 = z^2 + x^2 + y^2$, so $\partial r/\partial z = z/r = \cos\theta$. The quantity B is a complex numerical constant that remains to be determined.

The radial component of Euler's equation (3.175) for the constant-frequency case requires that

$$-i\omega\rho\hat{v}_r = -\frac{\partial \hat{p}}{\partial r} , \tag{3.351}$$

where on the right side the differentiation is to be carried out at constant θ. Given the expression (3.350), the corresponding relation for the radial component of the fluid velocity is consequently

$$\hat{v}_r = \frac{B}{i\omega\rho} \cos\theta \frac{d^2}{dr^2}\left(\frac{e^{ikr}}{r}\right) . \tag{3.352}$$

The boundary condition at $r = a$ is satisfied if one takes B to have a value such that

$$v_c = \frac{B}{i\omega\rho}\left[\frac{d^2}{dr^2}\left(\frac{e^{ikr}}{r}\right)\right]_{r=a} . \tag{3.353}$$

The indicated algebra yields

$$B = -\left(\frac{i\omega\rho a^3 v_c}{2 + 2ika + k^2 a^2}\right) e^{-ika} , \tag{3.354}$$

so the complex amplitude of the acoustic part of the pressure becomes

$$\hat{p} = \rho c v_c \frac{k^2 a^2}{2 + 2ika + k^2 a^2}\left(1 - \frac{1}{ikr}\right)\left(\frac{a}{r}\right)$$
$$\times e^{-ik[r-a]} \cos\theta , \tag{3.355}$$

while the radial component of the fluid velocity is

$$\hat{v}_r = v_c \left(\frac{2 + 2ikr + k^2 r^2}{2 + 2ika + k^2 a^2}\right)\left(\frac{a^3}{r^3}\right) e^{-ik[r-a]} \cos\theta . \tag{3.356}$$

The radiation impedance is

$$Z_{\text{rad}} = \rho c \left(\frac{k^2 a^2 + ika}{2 + 2ika + k^2 a^2}\right) . \tag{3.357}$$

Various simplifications result when considers limiting cases for the values of kr and ka. A case of common interest is when $ka \ll 1$ (small sphere) and $kr \gg 1$ (far field), so that

$$\hat{p} = \left(\frac{\omega^2 \rho v_c a^3}{c}\right) \frac{e^{ikr}}{r} \cos\theta . \tag{3.358}$$

3.11.4 Axially Symmetric Solutions

The example discussed in the previous subsection of radiation from a transversely oscillating sphere is one of a class of solutions of the linear acoustic equations where the field quantities depend on the spherical coordinates r and θ but not on the azimuthal angle ϕ. The Helmholtz equation for such circumstances has the form

$$\frac{1}{r^2}\frac{\partial}{\partial r}\left(r^2 \frac{\partial \hat{p}}{\partial r}\right) + \frac{1}{r^2 \sin\theta}\frac{\partial}{\partial \theta}\left(\sin\theta \frac{\partial \hat{p}}{\partial \theta}\right) + k^2 \hat{p} = 0 . \tag{3.359}$$

A common technique is to build up solutions of this equation using the principle of superposition, with the individual terms being factored solutions of the form

$$\hat{p}_\ell = P_\ell(\cos\theta)\Re_\ell(kr) , \tag{3.360}$$

where ℓ is an integer that distinguishes the various particular separated solutions. Insertion of this product into the Helmholtz equation leads to the conclusion that each factor must satisfy an appropriate ordinary differential equation, the two differential equations being (with η replacing kr)

$$\frac{1}{\sin\theta}\frac{d}{d\theta}\left(\sin\theta \frac{dP_\ell}{d\theta}\right) + \lambda_\ell P_\ell = 0, \tag{3.361}$$

$$\frac{d}{d\eta}\left(\eta^2 \frac{d\mathcal{R}_\ell}{d\eta}\right) - \lambda_\ell \mathcal{R}_\ell + \eta^2 \mathcal{R}_\ell = 0. \tag{3.362}$$

Here λ_ℓ is a constant, termed the separation constant. Equivalently, with P_ℓ regarded as a function of $\xi = \cos\theta$, the first of these two differential equations can be written

$$\frac{d}{d\xi}\left[(1-\xi^2)\frac{dP_\ell}{d\xi}\right] + \lambda_\ell P_\ell = 0. \tag{3.363}$$

Legendre Polynomials

Usually, one desires solutions that are finite at both $\theta = 0$ and $\theta = \pi$, or at $\xi = 1$ and $\xi = -1$, but the solutions for the θ-dependent factor are usually singular at one of the other of these two end points. However, for special values (eigenvalues) of the separation constant λ_ℓ, there exist particular solutions (eigenfunctions) that are finite at both points. To determine these functions, one postulates a series solution of the form

$$P_\ell(\xi) = \sum_{n=0}^{\infty} a_{\ell,n} \xi^n, \tag{3.364}$$

and derives the recursion relation

$$n(n-1)a_{\ell,n} = [(n-1)(n-2) - \lambda_\ell] a_{\ell,n-2}. \tag{3.365}$$

The series diverges as $\xi \to \pm 1$ unless it only has a finite number of terms, and such may be so if for some n the quantity in brackets on the right side is zero. The general choice of the separation constant that allows this is

$$\lambda_\ell = \ell(\ell+1), \tag{3.366}$$

where ℓ is an integer, so that the recursion relation becomes

$$n(n-1)a_{\ell,n} = [(n-\ell-2)(n+\ell-1)]a_{\ell,n-2}. \tag{3.367}$$

However, $a_{\ell,0}$ and $a_{\ell,1}$ can be chosen independently and the recursion relation can only terminate one of the two possible infinite series. Consequently, one must choose

$$\begin{aligned} a_{\ell,1} &= 0 \quad \text{if } \ell \text{ even}; \\ a_{\ell,0} &= 0 \quad \text{if } \ell \text{ odd}. \end{aligned} \tag{3.368}$$

If ℓ is even the terms correspond to $n=0$, $n=2$, $n=4$, up to $n=\ell$, while if ℓ is odd the terms correspond to $n=1$, $n=3$, up to $n=\ell$. The customary normalization is that $P_\ell(1) = 1$, and the polynomials that are derived are termed the Legendre polynomials. A general expression that results from examination of the recursion relation for the coefficients is

$$P_\ell(\xi) = a_{\ell,\ell}\left(\xi^\ell - \frac{\ell(\ell-1)}{2(2\ell-1)}\xi^{\ell-2} \right. \\ \left. + \frac{\ell(\ell-2)(\ell-1)(\ell-3)}{(2)(4)(2\ell-1)(2\ell-3)}\xi^{\ell-4} + \cdots \right), \tag{3.369}$$

where the last term has ξ raised to either the power of 1 or 0, depending on whether ℓ is odd or even. Equivalently, if one sets,

$$a_{\ell,\ell} = K_\ell \frac{(2\ell)!}{2^\ell (\ell!)^2}, \tag{3.370}$$

where K_ℓ is to be selected, the series has the relatively simple form

$$P_\ell(\xi) = K_\ell \sum_{m=0}^{M(\ell)} (-1)^m \frac{(2\ell-2m)!}{2^\ell m!(\ell-m)!(\ell-2m)!}\xi^{\ell-2m} \\ = K_\ell \sum_{m=0}^{M(\ell)} b_{\ell,m}\xi^{\ell-2m}. \tag{3.371}$$

Here $M(\ell) = \ell/2$ if ℓ is even, and $M(\ell) = (\ell-1)/2$ if ℓ is odd, so $M(0) = 0$, $M(1) = 0$, $M(2) = 1$, $M(3) = 1$, $M(4) = 2$, etc.

The coefficients $b_{\ell,m}$ as defined here satisfy the relation

$$(\ell+1)b_{\ell+1,m} = (2\ell+1)b_{\ell,m} - \ell b_{\ell-1,m-1}, \tag{3.372}$$

as can be verified by algebraic manipulation. A consequence of this relation is

$$(\ell+1)\frac{P_{\ell+1}(\xi)}{K_{\ell+1}} = (2\ell+1)\xi\frac{P_\ell(\xi)}{K_\ell} - \ell\frac{P_{\ell-1}(\xi)}{K_{\ell-1}} \tag{3.373}$$

when $\ell \geq 1$.

The customary normalization is to take $P_\ell(1) = 1$. The series for $\ell = 0$ and $\ell = 1$ are each of only one term, and the normalization requirement leads to $K_0 = 1$ and $K_1 = 1$. The relation (3.373) indicates that the normalization requirement will result, via induction, for all successive ℓ if one takes $K_\ell = 1$ for all ℓ. With this definition, the relation (3.373) yields the recursion relation among polynomials of different orders

$$(\ell+1)P_{\ell+1}(\xi) = (2\ell+1)\xi P_\ell(\xi) - \ell P_{\ell-1}(\xi). \tag{3.374}$$

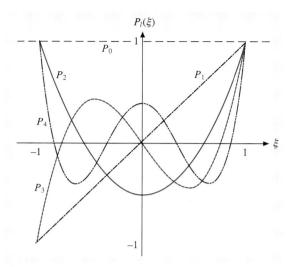

Fig. 3.25 Legendre polynomials for various orders

The latter holds for $\ell \geq 1$. With this, for example, given that $P_0(\xi) = 1$ and that $P_1(\xi) = \xi$, one derives

$$(2)P_2(\xi) = (3)\xi\xi - (1)(1) \,. \tag{3.375}$$

The first few of these polynomials are

$$P_0(\xi) = 1 \,, \tag{3.376}$$

$$P_1(\xi) = \xi \,, \tag{3.377}$$

$$P_2(\xi) = \frac{1}{2}(3\xi^2 - 1) \,, \tag{3.378}$$

$$P_3(\xi) = \frac{1}{2}(5\xi^3 - 3\xi) \,, \tag{3.379}$$

$$P_4(\xi) = \frac{1}{8}(35\xi^4 - 30\xi^2 + 3) \,, \tag{3.380}$$

$$P_5(\xi) = \frac{1}{8}(63\xi^5 - 70\xi^3 + 15\xi) \,, \tag{3.381}$$

with the customary identification of $\xi = \cos\theta$.

An alternate statement for the series expression (3.371), given $K_\ell = 1$, is the Rodrigues relation,

$$P_\ell(\xi) = \frac{1}{2^\ell \ell!} \frac{\mathrm{d}^\ell}{\mathrm{d}\xi^\ell}(\xi^2 - 1)^\ell \,. \tag{3.382}$$

This can be verified by using the binomial expansion

$$(\xi^2 - 1)^\ell = (-1)^\ell \sum_{n=1}^{\ell} (-1)^n \frac{\ell!}{n!(\ell-n)!} \xi^{2n} \,, \tag{3.383}$$

so that

$$\frac{\mathrm{d}^\ell}{\mathrm{d}\xi^\ell}(\xi^2 - 1)^\ell = (-1)^\ell \sum_{n=\ell-M}^{\ell} (-1)^n \frac{\ell!}{n!(\ell-n)!}$$
$$\times \frac{(2n)!}{(2n-\ell)!} \xi^{2n-\ell} \,, \tag{3.384}$$

or, with the change of summation index to $m = \ell - n$,

$$\frac{\mathrm{d}^\ell}{\mathrm{d}\xi^\ell}(\xi^2 - 1)^\ell = \sum_{m=0}^{M(\ell)} (-1)^m \frac{\ell!}{m!(\ell-m)!}$$
$$\times \frac{(2\ell - 2m)!}{(\ell - 2m)!} \xi^{\ell - 2m} = \ell! 2^\ell P_\ell(\xi) \,. \tag{3.385}$$

Another derivable property of these functions is that they are orthogonal in the sense that

$$\int_0^\pi P_\ell(\cos\theta) P_{\ell'}(\cos\theta) \sin\theta \, \mathrm{d}\theta = 0 \quad \text{if } \ell \neq \ell' \,. \tag{3.386}$$

This is demonstrated by taking the differential equations (3.361) satisfied by P_ℓ and $P_{\ell'}$, multiplying the first by $P_{\ell'} \sin\theta$, multiplying the second by $P_\ell(\theta) \sin\theta$, then subtracting the second from the first, with a subsequent integration over θ from 0 to π. Given that $\lambda_\ell \neq \lambda_{\ell'}$ and that the two polynomials are finite at the integration limits, the conclusion is as stated above.

If the two indices are equal, the chosen normalization, whereby $P_\ell(1) = 1$, leads to

$$\int_0^\pi [P_\ell(\cos\theta)]^2 \sin\theta \, \mathrm{d}\theta = \frac{2}{2\ell + 1} \,. \tag{3.387}$$

The general derivation of this makes use of the Rodrigues relation (3.382) and of multiple integrations by parts. To carry through the derivation, one must first verify, for arbitrary nonnegative integers s and t, and with $s < t$, that

$$\frac{\mathrm{d}}{\mathrm{d}\xi^s}(\xi^2 - 1)^t = 0 \quad \text{at } \xi = \pm 1 \,, \tag{3.388}$$

which is accomplished by use of the chain rule of differential calculus. With the use of this relation and of the

Rodrigues relation, one has

$$\int_{-1}^{1} [P_\ell(\xi)]^2 \, d\xi$$

$$= \left(\frac{1}{2^\ell \ell!}\right)^2 (-1)^\ell \int_{-1}^{1} (\xi^2 - 1)^\ell$$

$$\times \frac{d^{2\ell}}{d\xi^{2\ell}} (\xi^2 - 1)^\ell \, d\xi \qquad (3.389)$$

$$= \left(\frac{1}{2^\ell \ell!}\right)^2 (2\ell)! \int_{-1}^{1} (1 - \xi^2)^\ell \, d\xi$$

$$= \left(\frac{1}{2^\ell \ell!}\right)^2 (2\ell)! 2 \int_{0}^{\pi/2} \sin\theta^{2\ell+1} \, d\theta . \qquad (3.390)$$

The trigonometric integral I_ℓ in the latter expression is evaluated using the trigonometric identity

$$\frac{d}{d\theta}(\sin^{2\ell}\theta \cos\theta) = -\sin^{2\ell+1}\theta$$

$$+ 2\ell(\sin^{2\ell-2}\theta)(1 - \sin^2\theta) , \qquad (3.391)$$

the integral of which yields the recursion relation

$$I_\ell = \frac{2\ell}{(2\ell+1)} I_{\ell-1} , \qquad (3.392)$$

and from this one infers

$$I_\ell = \frac{(2^\ell \ell!)^2}{(2\ell+1)!} I_0 . \qquad (3.393)$$

The integral for $\ell = 0$ is unity, so one has

$$\int_{-1}^{1} [P_\ell(\xi)]^2 \, d\xi = \left(\frac{1}{2^\ell \ell!}\right)^2 (2\ell)! 2 \left(\frac{(2^\ell \ell!)^2}{(2\ell+1)!}\right)$$

$$= \frac{2}{2\ell+1} . \qquad (3.394)$$

Spherical Bessel Functions

The ordinary differential equation (3.362) for the factor $\mathfrak{R}_\ell(\eta)$ takes the form

$$\frac{d}{d\eta}\left(\eta^2 \frac{d\mathfrak{R}_\ell}{d\eta}\right) - \ell(\ell+1)\mathfrak{R}_\ell + \eta^2 \mathfrak{R}_\ell = 0 , \qquad (3.395)$$

with the identification for the separation constant that results from the requirement that the θ-dependent factor

be finite at $\theta = 0$ and $\theta = \pi$. For $\ell = 0$, a possible solution is

$$\mathfrak{R}_0 = A_0 \frac{e^{i\eta}}{\eta} , \qquad (3.396)$$

as can be verified by direct substitution, with A_0 being any constant. Since there is no corresponding θ dependence this is the same as the solution (3.327) for an outgoing spherical wave.

For arbitrary positive integer ℓ, a possible solution is

$$h_\ell^{(1)}(\eta) = -i\eta^\ell \left(-\frac{1}{\eta}\frac{d}{d\eta}\right)^\ell \frac{e^{i\eta}}{\eta} , \qquad (3.397)$$

so that, in particular,

$$h_0^{(1)}(\eta) = -i\frac{e^{i\eta}}{\eta} ;$$

$$h_1^{(1)}(\eta) = i\frac{d}{d\eta}\frac{e^{i\eta}}{\eta} = -\left(1 + \frac{i}{\eta}\right)\frac{e^{i\eta}}{\eta} . \qquad (3.398)$$

Alternately, both the real and imaginary parts should be solutions, so if one writes

$$h_\ell^{(1)}(\eta) = j_\ell(\eta) + iy_\ell(\eta) , \qquad (3.399)$$

then

$$j_\ell(\eta) = \eta^\ell \left(-\frac{1}{\eta}\frac{d}{d\eta}\right)^\ell \frac{\sin\eta}{\eta} , \qquad (3.400)$$

$$y_\ell(\eta) = -\eta^\ell \left(-\frac{1}{\eta}\frac{d}{d\eta}\right)^\ell \frac{\cos\eta}{\eta} , \qquad (3.401)$$

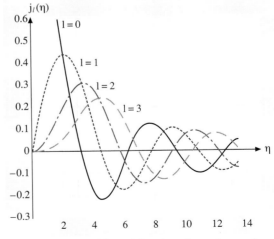

Fig. 3.26 Spherical Bessel functions for various orders

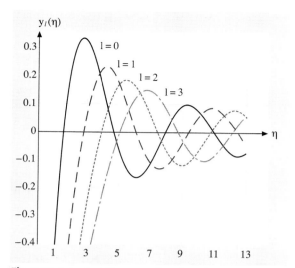

Fig. 3.27 Spherical Neumann functions for various orders

are each solutions. The function $h_\ell^{(1)}(\eta)$ is referred to as the spherical Hankel function of the ℓ-th order and first kind (the second kind has i replaced by $-$i), while $j_\ell(\eta)$ is the spherical Bessel function of the ℓ-th order, and $y_\ell(\eta)$ is the spherical Neumann function of the ℓ-th order.

A proof that $h_\ell^{(1)}(\eta)$, as defined above, satisfies the ordinary differential equation for the corresponding value of ℓ proceeds by induction. The assertion is true for $\ell = 0$ and it is easily demonstrated also to be true for $\ell = 1$, and the definition (3.397) yields the recursion relation

$$h_{\ell+1}^{(1)}(\eta) = \frac{\ell}{\eta} h_\ell^{(1)} - \frac{d}{d\eta} h_\ell^{(1)}. \tag{3.402}$$

The differential equation for $h_\ell^{(1)}$ and what results from taking the derivative of that equation gives one the relations,

$$\frac{d^2}{d\eta^2} h_\ell^{(1)} + \frac{2}{\eta} \frac{d}{d\eta} h_\ell^{(1)} + \left(1 - \frac{\ell(\ell+1)}{\eta^2}\right) h_\ell^{(1)} = 0, \tag{3.403}$$

$$\frac{d^2}{d\eta^2} \frac{h_\ell^{(1)}}{\eta} + \frac{2}{\eta} \frac{d}{d\eta} \frac{h_\ell^{(1)}}{\eta} + \frac{2}{\eta^2} \frac{d}{d\eta} h_\ell^{(1)} + \left(1 - \frac{\ell(\ell+1)}{\eta^2}\right) \frac{h_\ell^{(1)}}{\eta} = 0, \tag{3.404}$$

$$\frac{d^2}{d\eta^2} \frac{dh_\ell^{(1)}}{d\eta} + \frac{2}{\eta} \frac{d}{d\eta} \frac{dh_\ell^{(1)}}{d\eta} - \frac{2}{\eta^2} \frac{d}{d\eta} h_\ell^{(1)} + \frac{2}{\eta^3} \ell(\ell+1) h_\ell^{(1)} + \left(1 - \frac{\ell(\ell+1)}{\eta^2}\right) \frac{dh_\ell^{(1)}}{d\eta} = 0. \tag{3.405}$$

Multiplication of the second of these by ℓ, then subtracting the third, subsequently making use of the recursion relation, yields

$$\frac{d^2}{d\eta^2} h_{\ell+1}^{(1)} + \frac{2}{\eta} \frac{d}{d\eta} h_{\ell+1}^{(1)} + \frac{2(\ell+1)}{\eta^2} \frac{d}{d\eta} h_\ell^{(1)} - \frac{2}{\eta^3} \ell(\ell+1) h_\ell^{(1)} + \left(1 - \frac{\ell(\ell+1)}{\eta^2}\right) h_{\ell+1}^{(1)} = 0. \tag{3.406}$$

A further substitution that makes use of the recursion relation yields

$$\frac{d^2}{d\eta^2} h_{\ell+1}^{(1)} + \frac{2}{\eta} \frac{d}{d\eta} h_{\ell+1}^{(1)} - \frac{2(\ell+1)}{\eta^2} h_{\ell+1}^{(1)} + \left(1 - \frac{\ell(\ell+1)}{\eta^2}\right) h_{\ell+1}^{(1)} = 0, \tag{3.407}$$

or, equivalently,

$$\frac{d^2}{d\eta^2} h_{\ell+1}^{(1)} + \frac{2}{\eta} \frac{d}{d\eta} h_{\ell+1}^{(1)} + \left(1 - \frac{(\ell+1)(\ell+2)}{\eta^2}\right) h_{\ell+1}^{(1)} = 0, \tag{3.408}$$

which is the differential equation that one desires $h_{\ell+1}^{(1)}$ to satisfy; it is the $\ell + 1$ counterpart of (3.403).

Another recursion relation derivable from (3.397) is

$$(\ell+1) h_{\ell+1}^{(1)} = -(2\ell+1) \frac{d}{d\eta} h_\ell^{(1)} + \ell h_{\ell-1}^{(1)}. \tag{3.409}$$

Limiting approximate expressions for the spherical Hankel function and the spherical Bessel function are derivable from the definition (3.397). For $\eta \ll 1$, one has

$$h_0^{(1)}(\eta) \approx -\frac{i}{\eta}; \quad h_\ell^{(1)}(\eta) \approx -i \frac{1 \cdot 3 \cdot 5 \cdots (2\ell-1)}{\eta^{\ell+1}}, \tag{3.410}$$

$$j_0(\eta) \approx 1 - \frac{1}{6} \eta^2; \quad j_\ell(\eta) \approx \frac{\eta^\ell}{1 \cdot 3 \cdot 5 \cdots (2\ell+1)}. \tag{3.411}$$

(The second term in the expansion for j_0 is needed in the event that one desires a nonzero first approximation for the derivative.) In the asymptotic limit, when $\eta \gg 1$, one has

$$h_\ell^{(1)}(\eta) \to (-\mathrm{i})^{\ell+1}\frac{\mathrm{e}^{\mathrm{i}\eta}}{\eta}. \tag{3.412}$$

As is evident from the $\mathrm{e}^{\mathrm{i}\eta} = \mathrm{e}^{\mathrm{i}kr}$ factor that appears here, the spherical Hankel function of the first kind corresponds to a radially outgoing wave, given that one is using the $\mathrm{e}^{-\mathrm{i}\omega t}$ time dependence. Its complex conjugate, the spherical Hankel function of the second kind, corresponds to a radially incoming wave.

Plane Wave Expansion

A mathematical identity concerning the expansion of a plane wave in terms of Legendre polynomials and spherical Bessel functions results from consideration of

$$\mathrm{e}^{\mathrm{i}kr\cos\theta} = \sum_{\ell=0}^{\infty} \mathcal{J}_\ell(kr) P_\ell(\cos\theta). \tag{3.413}$$

Because the Legendre polynomials are a complete set, such an expansion is possible and expected to converge. The coefficients $\mathcal{J}_\ell(kr)$ are determined with the use of the orthogonality conditions, (3.386) and (3.387), to be given by

$$\mathcal{J}_\ell(\eta) = \frac{2\ell+1}{2}\int_0^\pi \mathrm{e}^{\mathrm{i}\eta\cos\theta} P_\ell(\cos\theta) \sin\theta\,\mathrm{d}\theta$$

$$= \frac{2\ell+1}{2}\int_{-1}^{1} \mathrm{e}^{\mathrm{i}\eta\xi} P_\ell(\xi)\,\mathrm{d}\xi. \tag{3.414}$$

The integral above evaluates, for $\ell = 0$, to

$$\mathcal{J}_0(\eta) = \frac{1}{2}\frac{1}{\mathrm{i}\eta}(\mathrm{e}^{\mathrm{i}\eta} - \mathrm{e}^{-\mathrm{i}\eta}) = j_\ell(\eta), \tag{3.415}$$

and, for $\ell = 1$, to

$$\mathcal{J}_1(\eta) = \frac{3}{2}\frac{\mathrm{d}}{\mathrm{d}(\mathrm{i}\eta)}\left(\frac{1}{\mathrm{i}\eta}(\mathrm{e}^{\mathrm{i}\eta}-\mathrm{e}^{-\mathrm{i}\eta})\right)$$

$$= 3\mathrm{i}\left(\frac{\sin\eta}{\eta^2} - \frac{\cos\eta}{\eta}\right) = 3\mathrm{i} j_1(\eta). \tag{3.416}$$

To proceed to larger values of ℓ, one makes use of the recursion relations, (3.374) and (3.409), and infers the general relation

$$\mathcal{J}_\ell(\eta) = (2\ell+1)\mathrm{i}^\ell j_\ell(\eta). \tag{3.417}$$

This relation certainly holds for $\ell = 0$ and $\ell = 1$, and a general proof for arbitrary integer ℓ follows by induction. One assumes the relation is true for $\ell - 1$ and ℓ, and considers

$$\mathcal{J}_{\ell+1}(\eta)$$

$$= \frac{2\ell+3}{2}\int_{-1}^{1} \mathrm{e}^{\mathrm{i}\eta\xi} P_{\ell+1}(\xi)\,\mathrm{d}\xi \tag{3.418}$$

$$= \frac{(2\ell+3)(2\ell+1)}{2(\ell+1)}\int_{-1}^{1} \mathrm{e}^{\mathrm{i}\eta\xi}\xi P_\ell(\xi)\,\mathrm{d}\xi$$

$$- \frac{(2\ell+3)\ell}{2(\ell+1)}\int_{-1}^{1} \mathrm{e}^{\mathrm{i}\eta\xi} P_{\ell-1}(\xi)\,\mathrm{d}\xi \tag{3.419}$$

$$= \frac{(2\ell+3)}{(\ell+1)}(-\mathrm{i})\frac{\mathrm{d}}{\mathrm{d}\eta}\mathcal{J}_\ell$$

$$- \frac{(2\ell+3)\ell}{(\ell+1)(2\ell-1)}\mathcal{J}_{\ell-1} \tag{3.420}$$

where the second version results from the recursion relation (3.374). Then, with the appropriate insertions from (3.417), one has

$$\mathcal{J}_{\ell+1}(\eta) = (2\ell+3)\mathrm{i}^{\ell+1}$$

$$\times\left[-\frac{(2\ell+1)}{(\ell+1)}\frac{\mathrm{d}}{\mathrm{d}\eta}j_\ell(\eta) \right.$$

$$\left. + \frac{\ell}{(\ell+1)}j_{\ell-1}(\eta)\right]. \tag{3.421}$$

The recursion relation (3.409) replaces the quantity in brackets by $j_{\ell+1}$, so one obtains

$$\mathcal{J}_{\ell+1}(\eta) = (2\ell+3)\mathrm{i}^{\ell+1} j_{\ell+1}(\xi), \tag{3.422}$$

which is the $\ell+1$ counterpart of (3.417). Thus the appropriate plane wave expansion is identified as

$$\mathrm{e}^{\mathrm{i}kr\cos\theta} = \sum_{\ell=0}^{\infty}(2\ell+1)\mathrm{i}^\ell j_\ell(kr) P_\ell(\cos\theta). \tag{3.423}$$

3.11.5 Scattering by a Rigid Sphere

An application of the functions introduced above is the scattering of an incident plane wave by a rigid sphere (radius a). One sets the complex amplitude of the acoustic part of the pressure to

$$\hat{p} = \hat{P}_{\mathrm{inc}}\mathrm{e}^{\mathrm{i}kz} + \hat{p}_{\mathrm{sc}}. \tag{3.424}$$

The first term represents the incident wave, which has an amplitude $\hat{P}_{\mathrm{inc}}$, constant frequency ω and which is proceeding in the $+z$-direction. The second term represents the scattered wave. This term is required to be made up of waves that propagate out from the sphere, which is centered at the origin. With some generality, one can use the principle of superposition and set

$$\hat{p}_{sc} = \hat{P}_{inc} \sum_{\ell=0}^{\infty} A_\ell P_\ell(\cos\theta) h_\ell^{(1)}(kr), \quad (3.425)$$

with the factors in the terms in the sum representing the Legendre polynomials and spherical Hankel functions. The individual terms in the sum are sometimes referred to as partial waves. The analytical task is to determine the coefficients A_ℓ. (The expression above applies to scattering from any axisymmetric body, including spheroids and penetrable objects, where the properties might vary with r and θ. The discussion here, however, is limited to that of the rigid sphere.)

With the plane wave expansion (3.423) inserted for the direct wave, the sum of incident and scattered waves takes the form

$$\hat{p} = \hat{P}_{inc} \sum_{\ell=0}^{\infty} \left[(2\ell+1)\mathrm{i}^\ell j_\ell(kr) + A_\ell h_\ell^{(1)}(kr)\right] \\ \times P_\ell(\cos\theta). \quad (3.426)$$

the boundary condition imposed by the rigidity of the sphere is that $\hat{v}_r = 0$ at $r = a$, or equivalently that $\partial \hat{p}/\partial r = 0$ at $r = a$. With the aid of the radial component of the Euler equation (3.175) and of the linear independence of the various Legendre polynomials, this yields

$$\frac{\mathrm{d}}{\mathrm{d}r}\left[(2\ell+1)\mathrm{i}^\ell j_\ell(kr) + A_\ell h_\ell^{(1)}(kr)\right] = 0$$

$$\text{at} \quad r = a. \quad (3.427)$$

The desired coefficients are consequently

$$A_\ell = -(2\ell+1)\mathrm{i}^\ell \frac{\left[\frac{\mathrm{d}}{\mathrm{d}r} j_\ell(kr)\right]_{r=a}}{\left[\frac{\mathrm{d}}{\mathrm{d}r} h_\ell^{(1)}(kr)\right]_{r=a}}. \quad (3.428)$$

Far-Field Scattering

In the limit of large kr, the asymptotic expression (3.412) for the spherical Hankel function can be used, and the scattered wave takes the asymptotic form

$$\hat{p}_{sc} \to \hat{P}_{inc} f(\theta) \frac{e^{ikr}}{kr}, \quad (3.429)$$

where

$$f(\theta) = \sum_{\ell=0}^{\infty} (-\mathrm{i})^{\ell+1} A_\ell P_\ell(\cos\theta) \quad (3.430)$$

is a complex dimensionless function of only the angle θ.

This asymptotic form holds for scattering from any axisymmetric object of bounded extent. In the more general case when the object is not axisymmetric, one should regard the function f as being also a function of the azimuthal angle ϕ, so that

$$f(\theta) \to f(\theta, \phi). \quad (3.431)$$

The prediction in all cases is that the far-field scattered wave resembles an outgoing spherical wave, but with an amplitude that depends on the direction of propagation.

The far-field intensity associated with the scattered wave is asymptotically entirely in the radial direction, and in accord with (3.281), its time average is given by

$$I_{sc} = \frac{1}{2} \frac{1}{\rho c} |\hat{p}_{sc}|^2, \quad (3.432)$$

or,

$$I_{sc} = \frac{1}{2} \frac{|\hat{P}_{inc}|^2}{\rho c} |f(\theta,\phi)|^2 \frac{1}{k^2 r^2}. \quad (3.433)$$

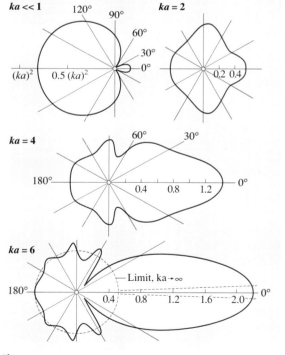

Fig. 3.28 Angular distribution of plane wave scattering by a rigid sphere of radius a. The quantity $(\mathrm{d}\sigma/\mathrm{d}\Omega)^{1/2}/a$ is plotted versus the polar angle θ

The differential scattering cross section is the power scattered per unit solid angle and per unit incident intensity, and is consequently given by

$$\frac{d\sigma}{d\Omega} = \frac{r^2 I_{sc}}{I_{inc}} = \frac{1}{k^2}|f(\theta,\phi)|^2 . \quad (3.434)$$

(Here the notation is such that σ corresponds to the cross-sectional area, and Ω corresponds to the solid angle in steradians, with the total solid angle in a sphere being 4π steradians.)

The total scattering cross section is the total scattered power divided by the incident intensity, so that

$$\sigma = \frac{\mathfrak{P}_{sc}}{I_{inc}} = \int \frac{d\sigma}{d\Omega} d\Omega = \frac{\mathfrak{P}_{sc}}{I_{inc}}$$

$$= \int_0^{2\pi}\int_0^{\pi} \frac{1}{k^2}|f(\theta,\phi)|^2 \sin\theta \, d\theta \, d\phi . \quad (3.435)$$

For the case of an axially symmetric scatterer, this reduces to

$$\sigma = \frac{2\pi}{k^2}\int_0^\pi |f(\theta,\phi)|^2 \sin\theta \, d\theta$$

$$= \frac{2\pi}{k^2}\sum_\ell \sum_{\ell'}(-i)^{\ell-\ell'} A_\ell A_{\ell'} \int_0^\pi P_\ell P_{\ell'} \sin\theta \, d\theta .$$

$$(3.436)$$

The orthogonality of the Legendre polynomials reduces this to

$$\sigma = \frac{4\pi}{k^2}\sum_\ell \frac{1}{(2\ell+1)}|A_\ell|^2 , \quad (3.437)$$

so the powers scattered by the individual partial waves are additive, even though there may be an intricate interference pattern in the angular directionality.

Rayleigh Scattering

When the object causing the scattering is much smaller than a wavelength, considerable simplification results, and the characteristic features that result are associated with the term *Rayleigh scattering*. In the case of scattering by a rigid sphere, only the $\ell = 0$ and $\ell = 1$ terms are significant, and both have comparable influence. The small-ka approximations for the spherical Bessel function and the spherical Hankel function yield

$$A_0 \approx i\frac{1}{3}(ka)^3 ; \qquad A_1 \approx i(ka)^3 . \quad (3.438)$$

Then with the appropriate identifications of the spherical Hankel function and the Legendre polynomial for $\ell = 0$ and $\ell = 1$, one obtains

$$\hat{p}_{sc} = -\hat{P}_{inc}\frac{k^2 a^3}{3}\left[1 - \frac{3}{2}\cos\theta\left(1 + \frac{i}{kr}\right)\right]\frac{e^{ikr}}{r} .$$

$$(3.439)$$

The corresponding expressions for the differential scattering cross section and the total cross section are

$$\frac{d\sigma}{d\Omega} = \frac{k^4 a^6}{9}\left(1 - \frac{3}{2}\cos\theta\right)^2 , \quad (3.440)$$

$$\sigma = \frac{7}{9}(ka)^4 \pi a^2 . \quad (3.441)$$

A feature of this expression is the characteristically strong dependence on frequency. The amplitude of the scattered wave varies as the square of the frequency, and the far-field intensity and scattering cross sections vary as the fourth power of the frequency. This type of frequency dependence is invariably true for all types of small scatterers (a notable exception being when the scatterer has an internal resonance at the incident frequency) and is the distinguishing feature of Rayleigh scattering.

3.12 Cylindrical Waves

The present section is concerned with a number of important situations where the appropriate coordinates are the cylindrical coordinates (w, ϕ, z).

3.12.1 Cylindrically Symmetric Outgoing Waves

For cylindrically symmetric waves, there is no dependence on the azimuthal angle ϕ or on the axial coordinate z, so the Laplacian in cylindrical coordinates reduces to

$$\nabla^2 \psi = \frac{1}{w}\frac{\partial}{\partial w}\left(w\frac{\partial \psi}{\partial w}\right) = \frac{\partial^2 \psi}{\partial w^2} + \frac{1}{w}\frac{\partial \psi}{\partial w}$$

$$= \frac{1}{\sqrt{w}}\frac{\partial^2(\sqrt{w}\psi)}{\partial w^2} + \frac{\sqrt{w}\psi}{4w^{5/2}} , \quad (3.442)$$

where w is here the radial distance from the z-axis. Consequently, the wave equation of (3.74) takes the

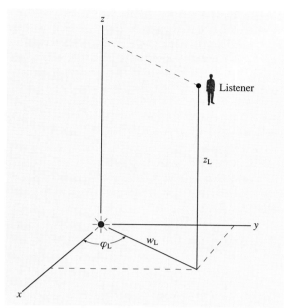

Fig. 3.29 Cylindrical coordinates. A hypothetical listener point has cylindrical coordinates w_L, ϕ_L, and z_L

form

$$\frac{\partial^2(\sqrt{w}\,p)}{\partial w^2} - \frac{1}{c^2}\frac{\partial^2(\sqrt{w}\,p)}{\partial t^2} + \frac{\sqrt{w}\,p}{4w^2} = 0, \quad (3.443)$$

and the Helmholtz equation of (3.176) takes the form

$$\frac{d^2\hat{p}}{dw^2} + \frac{1}{w}\frac{d\hat{p}}{dw} + k^2\hat{p} = 0. \quad (3.444)$$

General Transient Solution

A heuristic construction of an outgoing wave solution of (3.443) begins with an infinite line array of identical and synchronous point sources, equally spaced along the z-axis, and each giving rise to an identical spherical wave. Such waves differ, however, at any given distant point in that the radial distance R from the source depends on the z-coordinate z_0 of the source, so that

$$R = [w^2 + (z - z_0)^2]^{1/2}. \quad (3.445)$$

With the taking of an appropriate limit as the source separation interval goes to zero, one finds the general solution for an outgoing cylindrical wave to be

$$p(r, t) = \int_{-\infty}^{\infty} \frac{f[t - (R/c)]}{R}\,dz_0. \quad (3.446)$$

Here the function $f(t)$ is arbitrary. The argument $t - (R/c)$ of this function in the integrand is the retarded time, the time at which the incremental contribution was generated. With some effort, one can show that the integral in (3.446) is independent of z, and that it satisfies the cylindrical wave equation (3.443). A consequence of the above result is that outgoing cylindrical waves must have a tail. Even if the function $f(t)$ were to be zero except within a narrow time interval, the integral will give a nonzero value for times arbitrarily later than the time when the signal is first received. This is in contrast to the case for outgoing spherical waves, where there may be no tail.

If the distance w is sufficiently great, the resulting waveform will be such that it will be frozen in shape, and will decrease in amplitude as the inverse of the square root of w. The last term in (3.443) becomes insignificant and the appropriate limiting approximation is of the form

$$p(w, t) = \frac{1}{\sqrt{w}} F[t - (w/c)]. \quad (3.447)$$

The function $F(t)$ that appears here is related to the $f(t)$ that appears in (3.446). The relationship (derived further

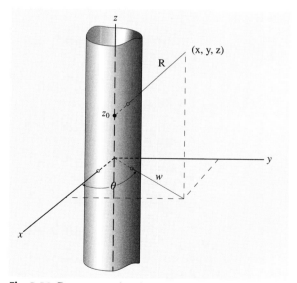

Fig. 3.30 Parameters involved in the construction of a cylindrical wave as a linear superposition of spherical waves from a continuous smear of simultaneous spherical sources spaced along a line. The quantity z_0 represents a source point along the z-axis, and R is the distance from that source point to a given listener point

below) is

$$F(t) = (2c)^{1/2} \int_{-\infty}^{t} \frac{f(\tau)}{(t-\tau)^{1/2}} \,d\tau \,. \tag{3.448}$$

Since $f(t)$ is arbitrary, one can equally well regard $F(t)$ as arbitrary. The expression (3.447) is readily seen to be an appropriate approximate solution of (3.442), since that wave equation implies that $\sqrt{wp}$ should satisfy the equation for plane waves when w is large.

Constant-Frequency Solution

The corresponding cylindrical wave solution for the Helmholtz equation is obtained by setting

$$f(t) \to A\, e^{-i\omega t}\,, \tag{3.449}$$

so that

$$\hat{p}(r) = A \int_{-\infty}^{\infty} \frac{e^{ikR}}{R} \,dz_0\,, \tag{3.450}$$

or, equivalently,

$$\hat{p}(r) = i\pi A H_0^{(1)}(kw)\,. \tag{3.451}$$

Here A is an arbitrary constant and the quantity $H_0^{(1)}(\eta)$ is the Hankel function of the first kind and zeroth order,

$$H_0^{(1)}(\eta) = \frac{1}{i\pi} \int_{-\infty}^{\infty} \frac{e^{i[\eta^2+\zeta^2]^{1/2}}}{[\eta^2+\zeta^2]^{1/2}} \,d\zeta\,. \tag{3.452}$$

3.12.2 Bessel and Hankel Functions

The Hankel function that appears above is discussed extensively in the literature, and corresponds to an outgoing wave. The differential equation that it satisfies is obtained from the Helmholtz equation for cylindrical waves by setting $kw = \eta$, so that

$$\left(\frac{d^2}{d\eta^2} + \frac{1}{\eta}\frac{d}{d\eta} + 1\right) H_0^{(1)}(\eta) = 0\,. \tag{3.453}$$

This is a special case (with $\nu = 0$) of the second-order differential equation

$$\left[\frac{d^2}{d\eta^2} + \frac{1}{\eta}\frac{d}{d\eta} + \left(1 - \frac{\nu^2}{\eta^2}\right)\right] \mathcal{J}_\nu(\eta) = 0\,, \tag{3.454}$$

whose general solution is

$$\mathcal{J}_\nu(\eta) = a\, J_\nu(\eta) + b Y_\nu(\eta)\,. \tag{3.455}$$

Here $J_\nu(\eta)$ and $Y_\nu(\eta)$ are the Bessel function and the Neumann function. The second, frequently denoted by $N_\nu(\eta)$, is sometimes referred to as Weber's function. The Hankel functions of the first and second kinds are given by

$$\begin{aligned} H_\nu^{(1)}(\eta) &= J_\nu(\eta) + i Y_\nu(\eta)\,;\\ H_\nu^{(2)}(\eta) &= J_\nu(\eta) - i Y_\nu(\eta)\,. \end{aligned} \tag{3.456}$$

Both $J_\nu(\eta)$ and $Y_\nu(\eta)$ are real when ν and η are real and positive, so the two Hankel functions are a complex conjugate pair under such circumstances.

An alternate mathematical representation for the Hankel function $H_0^{(1)}$ that is more amenable to numerical computation is

$$H_0^{(1)}(\eta) = \frac{2}{\pi}\int_0^{\pi/2} e^{i\eta\cos\phi}\,d\phi + \frac{2}{i\pi}\int_0^{\infty} e^{-\eta\sinh s}\,ds\,, \tag{3.457}$$

and from this one identifies expressions for the Bessel and Neumann functions as

$$J_0(\eta) = \frac{2}{\pi}\int_0^{\pi/2} \cos(\eta\cos\phi)\,d\phi\,, \tag{3.458}$$

$$Y_0(\eta) = \frac{2}{\pi}\int_0^{\pi/2} \sin(\eta\cos\phi)\,d\phi - \frac{2}{\pi}\int_0^{\infty} e^{-\eta\sinh s}\,ds\,. \tag{3.459}$$

In the limit of small η, the function $H_0^{(1)}$ has a logarithmic singularity, and is approximately given by

$$H_0^{(1)}(\eta) \approx 1 - \frac{1}{4}\eta^2 - \frac{2i}{\pi}[\ln(2/\eta) - \gamma]\,, \tag{3.460}$$

where

$$\gamma = \int_0^1 \frac{1 - e^{-s}}{s}\,ds - \int_1^{\infty} \frac{e^{-s}}{s}\,ds = 0.5772157\cdots \tag{3.461}$$

is the Euler–Mascheroni constant. In the limit of large η, the corresponding limiting expression is

$$\lim_{\eta\to\infty} H_0^{(1)}(\eta) = \left(\frac{2}{\pi\eta}\right)^{1/2} e^{-i\pi/4} e^{i\eta}\,. \tag{3.462}$$

An expression valid for all positive ν and positive η is

$$H_\nu^{(1)}(\eta) = \frac{1}{\pi i}\int_{-\infty}^{\infty+i\pi} e^{\eta\sinh\alpha - \nu\alpha}\,d\alpha \tag{3.463}$$

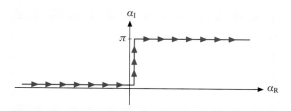

Fig. 3.31 Contour integral used for the definition of the Hankel function of the first kind

The latter is a contour integral, where a possible contour is from $-\infty$ along the negative real axis to the origin, then up the imaginary axis to $i\pi$, then parallel to the positive real axis with the imaginary part held fixed to $i\pi$ out to $\infty + i\pi$. The convergence along the latter leg is guaranteed because

$$\sinh(s + i\pi) = -\sinh(s) \, . \tag{3.464}$$

Thus, when written out explicitly for the contour just described, it becomes

$$H_\nu^{(1)}(\eta) = \frac{1}{\pi i} \int_0^\infty e^{-\eta \sinh s + \nu s} \, ds$$
$$+ \frac{1}{\pi} \int_0^\pi e^{i\eta \sin\theta - i\nu\theta} \, d\theta$$
$$+ \frac{e^{-i\nu\pi}}{\pi i} \int_0^\infty e^{-\eta \sinh s - \nu s} \, ds \, . \tag{3.465}$$

For $\nu = 0$, this reduces to the expression (3.457). That the above is indeed a solution of the ordinary differential equation (3.454) follows from substitution of the contour integral expression into the differential equation multiplied by η^2, followed by differentiation under the integral sign, so that one has

$$\frac{1}{\pi i} \int_{-\infty}^{\infty + i\pi} (\eta^2 \sinh^2\alpha + \eta \sinh\alpha + \eta^2 - \nu^2)$$
$$\times e^{\eta \sinh\alpha - \nu\alpha} \, d\alpha$$
$$= \frac{1}{\pi i} \int_{-\infty}^{\infty + i\pi} \frac{d}{d\alpha}[(\eta \cosh\alpha + \nu) e^{\eta \sinh\alpha - \nu\alpha}] \, d\alpha = 0 \, .$$
$$\tag{3.466}$$

The latter equality follows because the quantity being differentiated in the integrand vanishes at both ends of the integration contour.

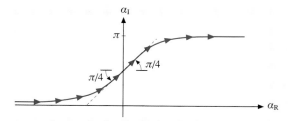

Fig. 3.32 Contour integral along the path of steepest descents, used in the derivation of the asymptotic expression for the Hankel function

For arbitrary positive ν, the corresponding limiting expression at large η for the Hankel function can be derived using the method of steepest descents. The saddle point α_{sp} for the integrand in (3.463) occurs on the imaginary axis where $\cosh\alpha = \nu/\eta$, which is close to $\alpha = i\pi/2$, so that the exponent near the saddle point is approximately

$$\eta \sinh\alpha - \nu\alpha \approx i(\eta - \nu\pi/2) + i\frac{1}{2}\eta(\alpha - \alpha_{sp})^2 \, .$$
$$\tag{3.467}$$

The path of steepest descent crosses the imaginary axis proceeding obliquely upwards at an angle of $\pi/4$, so the resulting asymptotic expression is

$$\lim_{\eta \to \infty} H_\nu^{(1)}(\eta) = \left(\frac{2}{\pi\eta}\right)^{1/2} e^{-i[(\nu/2)+(1/4)]\pi} e^{i\eta} \, .$$
$$\tag{3.468}$$

The corresponding limiting expressions for $J_\nu(\eta)$ and $Y_\nu(\eta)$ result from taking the real and imaginary parts of this. The above reduces to (3.462) when $\nu = 0$. This asymptotic behavior, plus the requirement that the function satisfy the ordinary differential equation (3.454), is

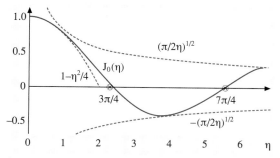

Fig. 3.33 Graph of the Bessel function of the zeroth order, shown along with common approximations used for small and large values of the argument

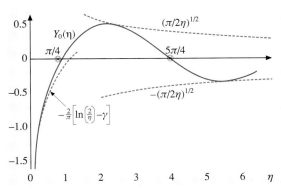

Fig. 3.34 Graph of the Neumann function of the zeroth order, shown along with common approximations used for small and large values of the argument

sufficient to define the Hankel function uniquely. The definition (3.463) is also such that it is consistent with the requirement that the real part (the Bessel function) be finite for real positive ν.

For small values of the argument η and for nonzero integer n these functions have the limiting expressions

$$J_n(\eta) \to \frac{1}{n!}\left(\frac{\eta}{2}\right)^n , \qquad (3.469)$$

$$Y_n(\eta) \to -\frac{(n-1)!}{\pi}\left(\frac{2}{\eta}\right)^n . \qquad (3.470)$$

These expressions also apply for noninteger positive index ν providing that one replaces $n! \to \Gamma(\nu+1)$ and $(n-1)! \to \Gamma(\nu)$, with the use of the Gamma function. The expression for the Neumann function obtained in this way is not appropriate in the limit $\nu \to 0$, as the Gamma function has a singularity when its argument is zero. Instead, one must use the logarithmic expression given by the imaginary part of (3.460).

The definition given above of the Hankel functions for arbitrary index ν yields the recursion relations

$$\frac{d}{d\eta}H_0^{(1)}(\eta) = -H_1^{(1)}(\eta) , \qquad (3.471)$$

$$\frac{d}{d\eta}H_\nu^{(1)}(\eta) = H_{\nu-1}^{(1)}(\eta) - \frac{\nu}{\eta}H_\nu^{(1)}(\eta) . \qquad (3.472)$$

Analogous recursion relations hold for the Bessel and Neumann functions.

Factored Solutions

One principal use of the Hankel functions and the Bessel functions of arbitrary order is in the synthesis of solutions of the Helmholtz equation in cylindrical coordinates,

$$\frac{\partial^2 \hat{p}}{\partial w^2} + \frac{1}{w}\frac{\partial \hat{p}}{\partial w} + \frac{1}{w^2}\frac{\partial^2 \hat{p}}{\partial \phi^2} + \frac{\partial^2 \hat{p}}{\partial z^2} + k^2 \hat{p} = 0 . \qquad (3.473)$$

One uses factored solutions of the form

$$\hat{p} = \mathfrak{W}(w)\Phi(\phi)\mathcal{Z}(z) . \qquad (3.474)$$

If one takes the latter two factors to satisfy the ordinary differential equations

$$\frac{d^2}{d\phi^2}\Phi + \nu^2\Phi = 0 , \qquad (3.475)$$

$$\frac{d^2}{dz^2}\mathcal{Z} + \alpha^2 \mathcal{Z} = 0 , \qquad (3.476)$$

then the ordinary differential equation that results for the radial-coordinate-dependent factor is

$$\frac{d^2\mathfrak{W}}{dw^2} + \frac{1}{w}\frac{d\mathfrak{W}}{dw} - \frac{\nu^2}{w^2}\mathfrak{W} + (k^2 - \alpha^2)\mathfrak{W} = 0 . \qquad (3.477)$$

The latter is recognized as the ordinary differential equation satisfied by Bessel functions and Hankel functions of the ν-th order, so possible solutions are

$$\begin{aligned}\mathfrak{W} &= H_\nu^{(1)}[(k^2-\alpha^2)^{1/2}w] \; ; \\ &\quad J_\nu[(k^2-\alpha^2)^{1/2}w] \; ; \\ &\quad Y_\nu[(k^2-\alpha^2)^{1/2}w] \; .\end{aligned} \qquad (3.478)$$

One can use combinations of such factored solutions for the synthesis of solutions of basic problems such as the scattering of a plane wave by a rigid cylinder. The solution is analogous to that discussed in a preceding section for scattering by a rigid sphere. For such problems in cylindrical coordinates, the applicable plane wave expansion theorem is

$$e^{ikw\cos\phi} = \sum_{n=0}^{\infty} 2\epsilon_n \cos(n\phi) J_n(kw) , \qquad (3.479)$$

where

$$\epsilon_n = \left[\frac{1}{\pi}\int_0^{2\pi}\cos^2(n\phi)\,d\phi\right]^{-1} \qquad (3.480)$$

is $1/2$ if $n = 0$ and 1 if $n \geq 0$. This expansion formula results because of the orthogonality of the trigonometric functions and because

$$J_n(kw) = \frac{1}{2\pi}\int_0^{2\pi} e^{ikw\cos\phi}\cos(n\phi)\,d\phi . \qquad (3.481)$$

Cylindrical Waves at Large Distances

The asymptotic expression (3.462) indicates that, at sufficiently large w, the solution of the Helmholtz equation for outgoing cylindrical waves can be represented by

$$\hat{p} = \frac{B}{\sqrt{w}} e^{ikw}, \qquad (3.482)$$

where B is a constant which is related to the A introduced previously through

$$B = i\pi \left(\frac{2}{\pi k}\right)^{1/2} e^{-i\pi/4} A. \qquad (3.483)$$

Consequently, the relation between the functions $F(t)$ and $f(t)$ that appear in (3.446) and (3.447) must be such that

$$F(t) = 2(2\pi c)^{1/2} \mathrm{Re} \int_0^\infty \frac{1}{\omega^{1/2}} e^{i\pi/4} \hat{f}(\omega) e^{-i\omega t}\, d\omega, \qquad (3.484)$$

where $\hat{f}(\omega)$ is the Fourier transform of $f(t)$. To do the requisite integration, one inserts the counterpart of (3.167) for $\hat{f}(\omega)$ and then interchanges the order of integration. The result,

$$F(t) = (2c)^{1/2} \int_{-\infty}^t \frac{f(\tau)}{(t-\tau)^{1/2}}\, d\tau, \qquad (3.485)$$

is what was previously given in (3.448).

Fluid Velocity for Cylindrical Waves

The fluid velocity induced by outgoing cylindrical waves is not as simply related to the corresponding acoustic pressure as that induced by a plane wave, although symmetry directs that the velocity must be in the appropriate radial direction when the propagation is cylindrically symmetric. For the constant-frequency case, when $\hat{p}$ is given by

$$\hat{p} = C H_0^{(1)}(kw), \qquad (3.486)$$

the radial component in cylindrical coordinates of the linearized Euler equation (3.175) requires

$$\hat{v}_w = \frac{C}{i\omega\rho} \frac{d}{dw} H_0^{(1)}(kw). \qquad (3.487)$$

At small distances, when (3.460) applies, one has

$$\hat{v}_w \approx \frac{C}{i\omega\rho} \frac{2i}{\pi} \frac{1}{w} = \frac{2C}{\pi\omega\rho}\frac{1}{w}. \qquad (3.488)$$

This dependence on $1/w$ is characteristic of cylindrically spreading flow from a line source. The complex amplitude, $-i\omega d\hat{m}/d\ell$, of the mass efflux rate, the mass of fluid flowing out per unit axial length per unit time of the source, is $2\pi w \rho$ times the above, so that

$$-i\omega \frac{d\hat{m}}{d\ell} = \frac{4C}{\omega}. \qquad (3.489)$$

This gives one a physical interpretation of the constant C, so that $-4iC$ is the complex amplitude of the second derivative with respect to time of the mass that has been expelled from a small cylinder surrounding the source per unit length of cylinder.

In the limit of large w, the asymptotic relation for the Hankel function is applicable, so that

$$\frac{d}{dw} H_0^{(1)} \approx ik H_0^{(1)}(kw), \qquad (3.490)$$

One consequently recovers the plane wave relation

$$\hat{v}_w = \frac{1}{\rho c} \hat{p}. \qquad (3.491)$$

Since this holds for Fourier transforms and since ρc is independent of frequency, it should hold for transient cylindrical waves. (Here large w implies large compared to a characteristic wavelength, or compared to c divided by a characteristic angular frequency.)

The previous discussion concerning acoustic intensity implies that the intensity of an outgoing cylindrical wave should decrease with w as $1/w$ so that the power flow through any cylindrical surface around the z-axis should stay constant. This has to remain so, even when one expresses the field in terms of Hankel functions, so one would expect those functions to have some mathematical property that guarantees this. To show that this is indeed the case, one expresses the intensity for the constant-frequency case as

$$I_{w,\mathrm{av}} = [\mathrm{Re}(\hat{p}e^{-i\omega t}) \mathrm{Re}(\hat{v}_w e^{-i\omega t})]_\mathrm{av}$$
$$= \frac{1}{2} \mathrm{Re}(\hat{p}^* \hat{v}_w), \qquad (3.492)$$

in accordance with (3.182).

Then, with the explicit substitution of the expressions involving Bessel functions one obtains

$$I_{r,\mathrm{av}} = \frac{|C|^2}{2\rho c} \mathrm{Re}[(J_0 - iY_0)(Y_0' - iJ_0')]$$
$$= \frac{|C|^2}{2\rho c} W(J_0, Y_0), \qquad (3.493)$$

where

$$W(J_0, Y_0) = J_0 Y_0' - Y_0 J_0' \qquad (3.494)$$

is the Wronskian for the Bessel and Neumann functions. Here the primes denote differentiation with respect to the argument $\eta = kw$ of the indicated function. One can derive from the differential equation (3.453) that these two functions independently satisfy the Wronskian relation

$$\eta W(J_0, Y_0) = \text{constant}, \quad (3.495)$$

and this is so regardless of the specialized definitions of the two functions. The constant can be evaluated from values and derivatives at any given point. In particular, one can use the asymptotic expressions, and the derived constant is $2/\pi$. Thus one has

$$I_{w,\text{av}} = \frac{1}{\pi} \frac{|C|^2}{\rho cw} = \frac{1}{\rho cw} \frac{1}{16\pi} \left| \omega^2 \frac{d\hat{m}}{d\ell} \right|^2. \quad (3.496)$$

3.12.3 Radially Oscillating Cylinder

A classic example for the radiation of cylindrical waves is that where a cylinder has nominal radius a, and an instantaneous radius

$$w_{\text{cyl}}(t) = a + \frac{v_0}{\omega} \sin(\omega t). \quad (3.497)$$

Here v_0/ω is the amplitude of the deviations of the actual radius from the nominal value a. For the linear acoustics idealization to be valid, it is required that this deviation be substantially less than a, so that

$$v_0 \ll \omega a. \quad (3.498)$$

The boundary condition on the fluid dynamic equations should ideally be

$$v_w = v_0 \cos(\omega t) \quad \text{at} \quad w = w_{\text{cyl}}(t), \quad (3.499)$$

but (also in keeping with the linear acoustics idealization) it is replaced by

$$v_w = v_0 \cos(\omega t) \quad \text{at} \quad w = a. \quad (3.500)$$

The corresponding boundary condition on the complex amplitude is

$$\hat{v}_w = v_0 \quad \text{at} \quad w = a. \quad (3.501)$$

If one takes the complex amplitude of the acoustic pressure to be of the form

$$\hat{p} = A\, H_0^{(1)}(kw), \quad (3.502)$$

where A is a complex number to be determined, then the radial component of the fluid velocity is

$$\hat{v}_w = \frac{A}{i\omega\rho} \frac{d}{dw} H_0^{(1)}(kw). \quad (3.503)$$

The imposing of the boundary condition determines the value of A and one obtains

$$\hat{p} = i\omega\rho v_0 \frac{H_0^{(1)}(kw)}{\left[\frac{d}{dw} H_0^{(1)}(kw)\right]_{w=a}}. \quad (3.504)$$

3.12.4 Transversely Oscillating Cylinder

If the cylinder is rigid and oscillating back and forth in the x-direction, the analysis is similar to that for the transversely oscillating sphere. The complex amplitude of the acoustic part of the pressure should have the general form

$$\hat{p} = B \frac{d}{dx} H_0^{(1)}(kw) = B \cos\phi \frac{d}{dw} H_0^{(1)}(kw), \quad (3.505)$$

where B is to be determined from the boundary condition. The latter can also be written

$$\hat{p} = -kB \cos\phi H_1^{(1)}(kw), \quad (3.506)$$

in accord with the recursion relation (3.471). The latter version is of the standard form for a factored solution of the Helmholtz equation in cylindrical coordinates.

Euler's equation gives the radial component of the fluid velocity as

$$\hat{v}_w = \frac{1}{i\omega\rho} B \cos\phi \frac{d^2}{dw^2} H_0^{(1)}(kw). \quad (3.507)$$

The appropriate approximate boundary condition is that

$$\hat{v}_w = v_c \cos\phi \quad \text{at} \quad w = a. \quad (3.508)$$

This allows one to identify

$$B = \frac{i\omega\rho v_c}{\left[\frac{d^2}{dw^2} H_0^{(1)}(kw)\right]_{w=a}}. \quad (3.509)$$

3.13 Simple Sources of Sound

Whatever generates an acoustic wave is termed a source. In acoustical analysis, sources are incorporated into the governing equations through either boundary conditions or source terms.

3.13.1 Volume Sources

Sources that are some distance from bounding surfaces and that are small compared to a wavelength can frequently be described by source terms. The simplest such source would be one that causes a net amount of mass of fluid to flow out of or into a fixed surface that encases it. The example of a radially oscillating sphere, discussed in a preceding section, is a classic example of such a source, and the surface through which the mass is flowing can be taken as a fixed spherical surface just outside the actual moving surface of the sphere. This mass passing out per unit time divided by the ambient density ρ is a quantity $Q_S(t)$ termed the source strength function or the source volume velocity. If such a source is concentrated at a point x_0, then the appropriate inhomogeneous wave equation which would replace (3.74) would be

$$\nabla^2 p - \frac{1}{c^2}\frac{\partial^2 p}{\partial t^2} = -\rho \dot{Q}_S(t)\delta(\mathbf{x}-\mathbf{x}_0)\,, \quad (3.510)$$

where $\delta(\mathbf{x})$ is the Dirac delta function, which has a volume integral of unity and that is concentrated at the point where its argument vanishes.

The solution of the above inhomogeneous wave equation for an isolated source at the origin ($\mathbf{r}_0 = 0$) in an unbounded region is

$$p = \frac{\rho}{4\pi r}\dot{Q}_S[t-(r/c)] = \frac{S[t-(r/c)]}{r}\,, \quad (3.511)$$

where

$$S(t) = \frac{\rho}{4\pi}\frac{\mathrm{d}}{\mathrm{d}t}Q_S(t) \quad (3.512)$$

is called the monopole strength. The replacement of the argument by the retarded time $t-(r/c)$ accounts for the transit time lag r/c for the sound to propagate from the source to the listener.

For a radially oscillating small sphere, $Q_S(t)$ can be taken as the time derivative of the instantaneous volume within the sphere. Thus the radiated acoustic pressure is proportional to the volume acceleration, the second derivative with respect to time of the sphere's volume, so that

$$p = \frac{\rho}{4\pi r}\left(\frac{\mathrm{d}^2 V}{\mathrm{d}t^2}\right)_{t\to t-(r/c)}. \quad (3.513)$$

This rule holds for bodies that are not necessarily spherically shaped, provided the largest dimensions are small compared to a characteristic wavelength and provided the acoustic pressure is measured at a sufficiently large distance from the source.

3.13.2 Small Piston in a Rigid Baffle

Another example of a volume source is that of a very small (relative to a wavelength) piston mounted in an infinite rigid baffle. If the piston has area A and outward normal velocity $v_n(t)$, then symmetry (or, equivalently, the inclusion of an image source) requires that the radiated sound be the same as from an isolated source with twice the volume velocity of the piston, so one has

$$p = \frac{\rho}{2\pi r}A\dot{v}_n[t-(r/c)]\,. \quad (3.514)$$

3.13.3 Multiple and Distributed Sources

For assemblies of sources, each concentrated at a point, the generalization of (3.510) is to replace the right side by a sum of individual source terms; the solution to this inhomogeneous wave equation is given by

$$p = \sum_n \frac{S[t-(R_n/c)]}{R_n}\,, \quad (3.515)$$

where R_n is the distance of the listener from the n-th source.

When the source is continuously distributed in space, the point-source term is replaced by a smoothly varying function, so the inhomogeneous wave equation is now of the form

$$\nabla^2 p - \frac{1}{c^2}\frac{\partial^2 p}{\partial t^2} = -\rho \dot{q}_S(\mathbf{x},t) = -4\pi s(\mathbf{x},t)\,. \quad (3.516)$$

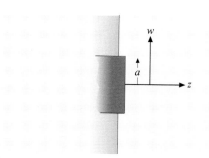

Fig. 3.35 Small piston source in a rigid baffle

The function q_S is termed the source strength density (source strength per unit volume). For an unbounded medium, the solution of this latter equation is given by the definite integral

$$p(\mathbf{x}, t) = \iiint \frac{1}{R} s(\mathbf{x}_0, t - (R/c)) \, dV_0, \quad (3.517)$$

where the volume integration ranges over source position $\mathbf{x}_0$, and where $R = |\mathbf{x} - \mathbf{x}_0|$ is the distance between the listener and source positions.

3.13.4 Piston of Finite Size in a Rigid Baffle

The concept of multiple sources and the mathematical description above allows an extension of (3.514) to the case when a finite piston is oscillating in a rigid baffle. An area element ΔA located on the $z = 0$ plane at (x_0, y_0) acts as a volume source where the time rate of change of the volume is $2v_n(x_0, y_0, t)\Delta A$. Here, as in (3.514), the factor of two results from the requirement that the contribution from this source element should give no increment of normal velocity on the surface outside of the area element. The normal velocity v_n is here the outward velocity of the piston, which is the velocity in the $+z$-direction, and is that appropriate to the surface point (x_0, y_0).

Superposition of the contribution from all the area elements subsequently yields, for the acoustic pressure in the radiated wave,

$$p = \frac{\rho}{2\pi} \iint \frac{v_n[x_0, y_0, t - (R/c)]}{R} \, dx_0 \, dy_0, \quad (3.518)$$

where

$$R = \left[(x - x_0)^2 + (y - y_0)^2 + z^2\right]^{1/2} \quad (3.519)$$

is the distance from the surface point $(x_0, y_0, 0)$ to the listener point (x, y, z). The integration extends over the portion of the surface that is moving.

If the surface is moving at a constant angular frequency ω, so that

$$v_n(x_0, y_0, t) = \text{Re}\left[\hat{v}_n(x_0, y_0) e^{-i\omega t}\right], \quad (3.520)$$

then (3.518) yields the complex amplitude of the radiated pressure field as

$$\hat{p} = \frac{-i\omega\rho}{2\pi} \iint \hat{v}_n(x_0, y_0) \frac{e^{ikR}}{R} \, dx_0 \, dy_0, \quad (3.521)$$

with $k = \omega/c$.

Appropriate formulas for the far field when the piston is of limited extent and centered at the origin result when the quantity R in the denominator is replaced by r and when it is replaced by (with the use of the binomial expansion)

$$R \approx r - \frac{x}{r} x_0 - \frac{y}{r} y_0 \quad (3.522)$$

in the exponent. Here r is the radial distance in spherical coordinates, and one can take

$$\frac{x}{r} = \sin\theta\cos\phi; \quad \frac{y}{r} = \sin\theta\sin\phi, \quad (3.523)$$

with θ as the polar angle and ϕ as the azimuthal solution. In this far-field limit, the acoustic pressure takes the form

$$\hat{p} \to F(\theta, \phi) \frac{e^{ikr}}{r}, \quad (3.524)$$

where the directivity function is given by

$$F(\theta, \phi) = \frac{-i\omega\rho}{2\pi} \iint \hat{v}_n(x_0, y_0) e^{-ikx_0 \sin\theta\cos\phi}$$
$$\times e^{-iky_0 \sin\theta\sin\phi} \, dx_0 \, dy_0. \quad (3.525)$$

The integrals (3.521) and (3.525) can be evaluated in terms of standard special functions when the piston is rigid and of a simple shape.

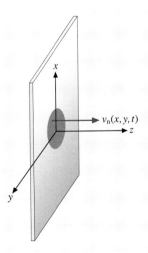

Fig. 3.36 Quantities involved in the integration to determine sound radiation from a planar surface in a state of nonuniform vibration

Rectangular Piston

If the piston is rectangular, with dimensions a and b, and centered at the origin, the far-field pattern is given by

$$F(\theta, \phi) = \frac{-i\omega\rho v_0}{2\pi} ab \frac{\sin[(1/2)a \sin\theta \cos\phi]}{(1/2)a \sin\theta \cos\phi} \times \frac{\sin[(1/2)b \sin\theta \sin\phi]}{(1/2)b \sin\theta \sin\phi}. \quad (3.526)$$

Circular Piston

If the piston is circular of radius a, the corresponding expression is independent of the azimuthal angle and given by

$$F(\theta) = \frac{-i\omega\rho v_0}{2\pi} \int_0^a \int_0^{2\pi} e^{-ikw_0 \sin\theta \cos\phi_0} d\phi_0 \, w_0 \, dw_0. \quad (3.527)$$

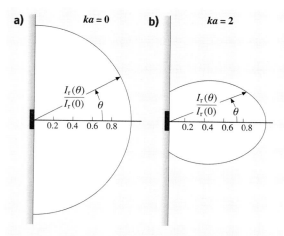

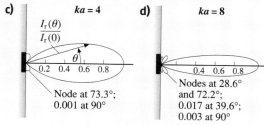

Fig. 3.37 Far-field radiation patterns of a vibrating circular piston in an otherwise rigid baffle for various values of ka. The quantity plotted is the intensity at a large radial distance r relative to that when the polar angle θ is zero, which is $[2J_1(\xi)/\xi]^2$, with $\xi = ka \sin\theta$

The integral over the azimuthal angle on the piston yields a Bessel function, so one has

$$F(\theta) = -i\omega\rho v_0 \int_0^a J_0(kw_0 \sin\theta) w_0 \, dw_0. \quad (3.528)$$

The recursion relation (3.472) implies that

$$\frac{d}{d\eta} J_1(\eta) = J_0(\eta) - \frac{1}{\eta} J_1(\eta), \quad (3.529)$$

so

$$\eta J_0(\eta) = \eta \frac{d}{d\eta} J_1(\eta) + J_1(\eta) = \frac{d}{d\eta}(\eta J_1(\eta)). \quad (3.530)$$

Consequently the integration in (3.528) can be performed in terms of the Bessel function of the first order, with the result

$$F(\theta) = -i\omega\rho v_0 a^2 \left(\frac{J_1(ka \sin\theta)}{ka \sin\theta}\right). \quad (3.531)$$

3.13.5 Thermoacoustic Sources

The differential equation (3.516) arises when one suddenly adds heat to a fluid [3.71, 72], as with a laser or by combustion, so that the entropy s per unit mass changes according to the thermodynamic relation

$$\rho T Ds/Dt = h, \quad (3.532)$$

so that, to first order, the equation of state (3.87) yields

$$\frac{\partial p'}{\partial t} = c^2 \frac{\partial \rho'}{\partial t} + \frac{c^2 \beta}{c_p} h. \quad (3.533)$$

Here h is the heat added per unit time and unit volume, β is the volume expansion coefficient defined by (3.30), and c_p is the specific heat at constant pressure.

Equation (3.533) changes the basic linear acoustic equations (3.71) and (3.72) to

$$\frac{\partial p}{\partial t} + \rho c^2 \nabla \cdot \mathbf{v} = \frac{c^2 \beta}{c_p} h, \quad (3.534)$$

$$\rho_0 \frac{\partial \mathbf{v}}{\partial t} + \nabla p = 0, \quad (3.535)$$

so the energy conservation corollary of (3.273) becomes

$$\frac{\partial w}{\partial t} + \nabla \cdot \mathbf{I} = \frac{\beta}{\rho c_p} ph, \quad (3.536)$$

and the wave equation becomes

$$\nabla^2 p - \frac{1}{c^2} \frac{\partial^2 p}{\partial t^2} = -\frac{\beta}{c_p} \frac{\partial h}{\partial t}. \quad (3.537)$$

The appropriate identification for the monopole strength density function $s(\mathbf{x}, t)$ in the integral expression of (3.517) is consequently given by

$$s(\mathbf{x}, t) = \frac{1}{4\pi} \frac{\beta}{c_p} \frac{\partial h}{\partial t} . \tag{3.538}$$

3.13.6 Green's Functions

The idealization of a point source of constant frequency is of basic importance in formulating solutions to complicated acoustic radiation problems. The resulting complex pressure amplitude of the field resulting from such a single source can be expressed as a constant times a Green's function, where the Green's function is a solution of the inhomogeneous Helmholtz equation

$$(\nabla^2 + k^2) G(\mathbf{x}|\mathbf{x}_0) = -4\pi \delta(\mathbf{x} - \mathbf{x}_0) . \tag{3.539}$$

(The inclusion of the factor 4π on the right is done in much of the literature, but not universally; its objective here is that the mathematical form of the Green's function be simpler.) If the medium surrounding the source is unbounded, then the Green's function is the free-space (no external boundaries) Green's function, which can be identified from (3.511) to have the form

$$G(\mathbf{x}|\mathbf{x}_0) = \frac{1}{R} e^{ikR} , \tag{3.540}$$

where R is the distance between the source and observation (listener) point. When external boundary conditions are imposed on the Green's function that satisfies (3.539), then the Green's function will have a form different from that of the free-space Green's function. However, in all such cases, the Green's function will approach $1/R$ plus a bounded function when the listener position approaches the source point.

An example of a Green's function that is not the free-space Green's function is that which corresponds to a point source on one side of an infinitely-extended rigid plane. If the source point is at (x_0, y_0, z_0) where $z_0 > 0$, and if the rigid plane is the $z = 0$ plane, then

$$G(\mathbf{x}|\mathbf{x}_0) = \frac{1}{R} e^{ikR} + \frac{1}{R_i} e^{ikR_i} , \tag{3.541}$$

where

$$R_i = [(x - x_0)^2 + (y - y_0)^2 + (z + z_0)^2]^{1/2} . \tag{3.542}$$

Because this Green's function is even in y, it automatically satisfies the boundary conditions that its normal derivative vanish at the rigid boundary.

3.13.7 Multipole Series

Radiation fields from sources of limited spatial extent in unbounded environments can be described either in terms of multipoles or spherical harmonics. That such descriptions are feasible can be demonstrated with the aid of the constant-frequency version of (3.517):

$$\hat{p}(\mathbf{x}) = \int \hat{s}(\mathbf{x}_0) \frac{e^{ikR}}{R} \, dV_0 , \tag{3.543}$$

which is the solution of the inhomogeneous Helmholtz equation with a continuous distribution of monopole sources taken into account.

The multipole series results from this when one expands $R^{-1} e^{ikR}$ in a power series in the coordinates of the source position $\mathbf{x}_0$ and then integrates term by term. Up to second order one obtains

$$\hat{p} = \hat{S} \frac{e^{ikr}}{r} - \sum_{\nu=1}^{3} \hat{D}_\nu \frac{\partial}{\partial x_\nu} \frac{e^{ikr}}{r}$$
$$+ \sum_{\mu,\nu=1}^{3} \hat{Q}_{\mu\nu} \frac{\partial^2}{\partial x_\mu \partial x_\nu} \frac{e^{ikr}}{r} , \tag{3.544}$$

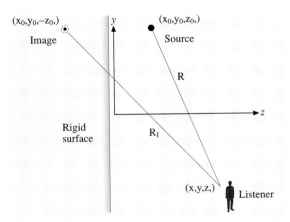

Fig. 3.38 Parameters involved in the construction of the Green's function corresponding to a point source outside a perfectly reflecting (rigid) wall. The quantity R is the direct distance from the source, and R_I is the distance from the image source

where the coefficients are expressed by

$$\hat{S} = \int \hat{s}(\boldsymbol{x}) \mathrm{d}V \,, \tag{3.545}$$

$$\hat{D}_\nu = -\int x_\nu \hat{s}(\boldsymbol{x}) \mathrm{d}V \,, \tag{3.546}$$

$$\hat{Q}_{\mu\nu} = \frac{1}{2!} \int x_\mu x_\nu \hat{s}(\boldsymbol{x}) \mathrm{d}V \,. \tag{3.547}$$

The three terms in (3.544) are said to be the monopole, dipole, and quadrupole terms, respectively. The coefficients $\hat{S}$, $\hat{D}_\nu$, and $\hat{Q}_{\mu\nu}$ are similarly labeled. The D_ν are the components of a dipole moment vector, while the $Q_{\mu\nu}$ are the components of a quadrupole moment tensor. The general validity [3.73, 74] of such a description extends beyond the manner of derivation and is not restricted to sound generated by a continuous source distribution embedded in the fluid. It applies in particular to the sound radiated by a vibrating body of arbitrary shape.

3.13.8 Acoustically Compact Sources

If the source is acoustically compact, so that its largest dimension is much shorter than a wavelength, the multipole series converges rapidly, so one typically only need retain the first nonzero term. Sources exist whose net monopole strength is zero, and sources also exist whose dipole moment vector components are all zero as well. Consequently, compact sources are frequently classed as monopole, dipole, and quadrupole sources. The prototype of a monopole source is a body of oscillating volume. One for a dipole source is a rigid solid undergoing translational oscillations; another would be a vibrating plate or shell whose thickness changes negligibly. In the former case, the detailed theory shows that in the limit of sufficiently low frequency, the dipole moment vector is given by

$$\hat{D}_\nu = \hat{F}_\nu + m_\mathrm{d} \hat{a}_{C,\nu} \,, \tag{3.548}$$

where $\hat{F}_\nu$ is associated with the force which the moving body exerts on the surrounding fluid and where $\hat{a}_{C,\nu}$ is associated with the acceleration of the geometric center of the body. The quantity m_d is the mass of fluid displaced by the body.

The simplest example of a dipole source is that of a rigid sphere transversely oscillating along the z-axis about the origin. (This is discussed in general terms in a preceding section of this chapter.) If the radius of the sphere is a and if $ka \ll 1$, then the force and acceleration have only a z-component, and the force amplitude is given by

$$\hat{F}_z = \frac{1}{2} m_\mathrm{d} \hat{a}_{C,z} \,. \tag{3.549}$$

The dipole moment when the center velocity has amplitude $\hat{v}_C$ is consequently of the form

$$\hat{D}_z = -\frac{3}{2} \mathrm{i}\omega[(4/3)\rho\pi a^3]\hat{v}_C \,. \tag{3.550}$$

Taking into account that the derivative of r with respect to z is $\cos\theta$, one finds the acoustic field from (3.544) to be given by

$$\hat{p} = -\hat{D}_z \cos\theta \frac{\mathrm{d}}{\mathrm{d}r} \frac{\mathrm{e}^{\mathrm{i}kr}}{r} \,. \tag{3.551}$$

When $kr \gg 1$, this approaches the limiting form

$$\hat{p} \to -\mathrm{i}k D_z \cos\theta \frac{\mathrm{e}^{\mathrm{i}kr}}{r} \,. \tag{3.552}$$

The far-field intensity, in accord with (3.276), has a time average of $|\hat{p}|^2/(2\rho c)$ and is directed in the radial direction in this asymptotic limit. The drop of intensity as $1/r^2$ with increasing radial distance is the same as for spherical spreading, but the intensity varies with direction as $\cos^2\theta$.

Vibrating bodies that radiate as quadrupole sources (and which therefore have no dipole radiation) usually do so because of symmetry. Vibrating bells [3.75] and tuning forks are typically quadrupole radiators.

A dipole source can be represented by two similar monopole sources, 180° out of phase with each other, and very close together. Since they are radiating out of phase, there is no total mass flow input into the medium. Such a dipole source will have a net acoustic power output substantially lower than that of either of the component monopoles when radiating individually. Similarly, a quadrupole can be formed by two identical but oppositely directed dipoles brought very close together. If the two dipoles have a common axis, then a longitudinal quadrupole results; when they are side by side, a lateral quadrupole results. In either case, the quadrupole radiation is much weaker than would be that from either dipole when radiating separately.

3.13.9 Spherical Harmonics

The closely related description of source radiation in terms of spherical harmonics results from (3.543) when one inserts the expansion [3.76–78]

$$\frac{\mathrm{e}^{\mathrm{i}kR}}{R} = \sum_{\ell=0}^{\infty} (2\ell+1) j_\ell(kr_0) h_\ell^{(1)}(kr) P_\ell(\cos\Theta) \,, \tag{3.553}$$

where j_ℓ is the spherical Bessel function and $h_\ell^{(1)}$ is the spherical Hankel function of order ℓ and of the first kind. (The expansion here assumes $r > r_0$; otherwise, one interchanges r and r_0 in the above.) The quantity $P_\ell(\cos \Theta)$ is the Legendre function of order ℓ, while Θ is the angle between the directions of x and x_0. Alternately, one uses the expansion

$$P_\ell(\cos \Theta) = \sum_{m=-\ell}^{\ell} \frac{(\ell-|m|)!}{(\ell+|m|)!} Y_\ell^m(\theta, \phi) Y_\ell^{-m}(\theta_0, \phi_0), \tag{3.554}$$

where the spherical harmonics are defined by

$$Y_\ell^m(\theta, \phi) = e^{im\phi} P_\ell^{|m|}(\cos \theta). \tag{3.555}$$

Here the functions $P_\ell^{|m|}(\cos \theta)$ are the associated Legendre functions. [The value of $P_0(\cos \Theta)$ is identically 1.]

If such an expansion is inserted into (3.543) and if r is understood to be sufficiently large that there are no sources beyond that radius, one has the relation

$$\hat{p}(r, \theta, \phi) = a_{00} h_0^{(1)}(kr)$$
$$+ \sum_{\ell=1}^{\infty} \sum_{m=-\ell}^{\ell} a_{\ell m} h_\ell^{(1)}(kr) Y_\ell^m(\theta, \phi), \tag{3.556}$$

with coefficients given by

$$a_{\ell m} = ik(2\ell+1) \frac{(\ell-|m|)!}{(\ell+|m|)!}$$
$$\times \int \hat{s}(r_0, \theta_0, \phi_0) j_\ell(kr_0) Y_\ell^{-m}(\theta_0, \phi_0) \, dV_0. \tag{3.557}$$

These volume integrations are to be carried out in spherical coordinates. The general result of (3.557) holds for any source of limited extent; any such wave field in an unbounded medium must have such an expansion in terms of spherical Hankel functions and spherical harmonics.

The spherical Hankel functions have the asymptotic (large-r) form

$$h_\ell^{(1)}(kr) \to (-i)^{(\ell+1)} \frac{e^{ikr}}{kr}, \tag{3.558}$$

so the acoustic radiation field must asymptotically approach

$$\hat{p} \to \hat{F}(\theta, \phi) \frac{e^{ikr}}{r}, \tag{3.559}$$

where the function $\hat{F}(\theta, \phi)$ is a function of θ and ϕ that has an expansion in terms of spherical harmonics. In this asymptotic limit, the acoustic intensity is in the radial direction and given by

$$I_{r,\mathrm{av}} = \frac{1}{2} \frac{|\hat{F}|^2}{\rho c r^2}. \tag{3.560}$$

For fixed θ and ϕ, the time-averaged intensity must asymptotically decrease as $1/r^2$. The coefficient of $1/r^2$ in the above describes the far-field radiation pattern of the source, having the units of watts per steradian.

Although the two types of expansions, multipoles and spherical harmonics, are related, the relationship is not trivial. The quadrupole term in (3.544), for example, cannot be equated to the sum of the $\ell = 2$ terms in (3.556). It is possible to have spherically symmetric quadrupole radiation, so an $\ell = 0$ term would have to be included.

3.14 Integral Equations in Acoustics

There are many common circumstances where the determination of acoustic fields and their properties is approached via the solution of integral equations rather than of partial differential equations.

3.14.1 The Helmholtz–Kirchhoff Integral

For the analysis of radiation of sound from a vibrating body of limited extent in an unbounded region, there is an applicable integral corollary of the Helmholtz equation of (3.176) which dates back to 19th century works of *Helmholtz* [3.47] and *Kirchhoff* [3.79].

One considers a closed surface S where the outward normal component of the particle velocity has complex amplitude $\hat{v}_n(x_S)$ and complex pressure amplitude $\hat{p}_S(x_S)$ at a point x_S on the surface. For notational convenience, one introduces a quantity defined by

$$\hat{f}_S = -i\omega \rho \hat{v}_n, \tag{3.561}$$

where $\hat{v}_n(x_S)$ is the normal component $n(x_S) \cdot \hat{v}(x_S)$ of the complex fluid velocity vector amplitude $\hat{v}(x)$ at the surface. One can regard $\hat{f}_S$ as a convenient grouping of symbols, either as a constant multiplied by the normal velocity, or as a constant multiplied by the normal ac-

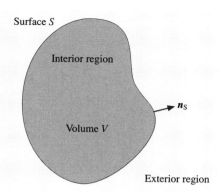

Fig. 3.39 Arbitrary surface in an arbitrary state of vibration at constant frequency, with unit normal n_S

celeration, or as the normal component of the apparent body force, per unit volume, exerted on the fluid at the surface point x_S. Because of the latter identification, the use of the symbol f is appropriate. The subscript S is used to denote values appropriate to the surface.

Then, given that there are no sources outside the surface, a mathematical derivation, involving the differential equation of (3.539) for the free-space Green's function of (3.540), involving the Helmholtz equation, and involving the divergence theorem, yields

$$\hat{p}(x) = \mathfrak{M}(x, \hat{p}_S, \hat{f}_S) \qquad (3.562)$$

for the complex pressure amplitude $\hat{p}$ at a point x outside the surface. Here the right side is given by the expression

$$\mathfrak{M} = \frac{1}{4\pi} \int \{ \hat{f}_S(x'_S) G(x|x'_S) \\ + \hat{p}_S(x'_S) n(x'_S) \cdot [\nabla' G(x|x')]_{x'=x'_S} \} \, dS' \, . \qquad (3.563)$$

In the integrand of this expression, the point x'_S (after the evaluation of any requisite normal derivatives) is understood to range over the surface S, with the point x held fixed during the integration. The unit outward normal vector $n(x'_S)$ points out of the enclosed volume V at the surface point x'_S.

In (3.563), the integral $\mathfrak{M}$ is a function of the point x, but a functional (function of a function) of the function arguments $\hat{p}_S$ and $\hat{f}_S$.

3.14.2 Integral Equations for Surface Fields

The functions $\hat{p}_S$ and $\hat{f}_S$ cannot be independently prescribed on the surface S. Specifying either one is a sufficient inner boundary condition on the Helmholtz

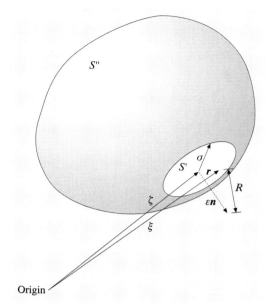

Fig. 3.40 Sketch illustrating the methodology for the derivation of an integral equation from the Helmholtz–Kelvin relation. The exterior point at $\xi = \zeta + n\epsilon$ is gradually allowed to approach an arbitrary point ζ on the surface, and the integration is represented as a sum of integrals over S' and S''. One considers the parameters ϵ and σ as small, but with $\sigma \gg \epsilon$, first takes the limit as $\epsilon \to 0$, then takes the limit as $\sigma \to 0$

equation. The corollary of (3.562) applies only if both functions correspond to a physically realizable radiation field outside the surface S. If this is so and if the point x is formally set to a point inside the enclosed volume, then analogous mathematics involving the properties of the Green's function leads to the deduction [3.80, 81]

$$\mathfrak{M}(x_{\text{inside}}, \hat{p}_S, \hat{f}_S) = 0 \, , \qquad (3.564)$$

which is a general relation between the surface value functions $\hat{p}_S$ and $\hat{f}_S$, holding for any choice of the point x inside the enclosed volume.

Equation (3.562) allows one to derive two additional relations (distinguished by the subscripts I and II) between the surface values of $\hat{p}_S$ and $\hat{f}_S$. One results when the off-surface point x is allowed to approach an arbitrary but fixed surface point x_S. For one of the terms in the integral defining the quantity $\mathfrak{M}$, the limit as x approaches x_S of the integral is not the same as the integral over the limit of the integrand as x approaches x_S. With this subtlety taken into account, one obtains

$$\hat{p}_S(x_S) - \mathfrak{L}_I(x_S, \hat{p}_S) = \mathfrak{H}_I(x_S, \hat{f}_S), \quad (3.565)$$

where the two linear operators $\mathfrak{L}_I$ and $\mathfrak{H}_I$ are

$$\mathfrak{L}_I(x_S, \hat{p}_S)$$
$$= \frac{1}{2\pi} \int \hat{p}(x'_S) n(x'_S) [\nabla' G(x_S | x')]_{x' = x'_S} \, dS', \quad (3.566)$$

$$\mathfrak{H}_I(x_S, \hat{f}_S)$$
$$= \frac{1}{2\pi} \int \hat{f}(x'_S) G(x'_S | x_S) \, dS'. \quad (3.567)$$

These linear operators operate on the surface values of $\hat{p}_S$ and $\hat{f}_S$, respectively, with the result in each case being a function of the position of the surface point x_S.

The second type of surface relationship [3.82] is obtained by taking the gradient of both sides of (3.562), subsequently setting x to $x_S + \epsilon n(x_S)$, where x_S is an arbitrary point on the surface, taking the dot product with $n(x_S)$, then taking the limit as ϵ goes to zero. The order of the processes, doing the integration and taking the limit, cannot be blindly interchanged, and some mathematical manipulations making use of the properties of the Green's function are necessary before one can obtain a relation in which all integrations are performed after all necessary limits are taken. The result is

$$-\mathfrak{L}_{II}(x_S, \hat{p}_S) = \hat{f}_S(x_S) + \mathfrak{H}_{II}(x_S, \hat{f}_S), \quad (3.568)$$

where the relevant operators are as given by

$$\mathfrak{L}_{II}(x, \hat{p}_S) = [n(x_S) \times \nabla] \cdot \frac{1}{2\pi} \int [n(x'_S) \times \nabla' \hat{p}_S(x'_S)]$$
$$\times G(x_S | x'_S) \, dS' + \frac{k^2}{2\pi}$$
$$\times \int n(x_S) \cdot n(x'_S) \hat{p}_S(x'_S) G(x_S | x'_S) \, dS', \quad (3.569)$$

$$\mathfrak{H}_{II}(x_S, \hat{f}_S) = \frac{1}{2\pi} \int \hat{f}_S(x'_S) n(x_S)$$
$$\times [\nabla G(x | x'_S)]_{x = x_S} \, dS'. \quad (3.570)$$

With regard to (3.569), one should note that the operator $n(x_S) \times \nabla$ involves only derivatives tangential to the surface, so that the integral on which it acts needs only be evaluated at surface points x_S.

A variety of numerical techniques have been used and discussed in recent literature to solve either (3.564), (3.565), or (3.568), or some combination [3.83] of these for the surface pressure $\hat{p}_S$, given the surface force function $\hat{f}_S$. Once this is done, the radiation field at any external point x is found by numerical integration of the corollary integral relation of (3.562).

3.15 Waveguides, Ducts, and Resonators

External boundaries can channel sound propagation, and in some cases can create a buildup of acoustic energy within a confined space.

3.15.1 Guided Modes in a Duct

Pipes or ducts act as guides of acoustic waves, and the net flow of energy, other than that associated with wall dissipation, is along the direction of the duct. The general theory of guided waves applies and leads to a representation in terms of guided modes.

If the duct axis is the x-axis and the duct cross section is independent of x, the guided mode series [3.4, 84] has the form

$$\hat{p} = \sum_n X_n(x) \Psi_n(y, z), \quad (3.571)$$

where the $\Psi_n(y, z)$ are eigenfunctions of the equation

$$\left(\frac{\partial^2}{\partial y^2} + \frac{\partial^2}{\partial z^2} \right) \Psi_n + \alpha_n^2 \Psi_n = 0, \quad (3.572)$$

with the α_n^2 being the corresponding eigenvalues. The appropriate boundary condition, if the duct walls are idealized as being perfectly rigid, is that the normal component of the gradient of Ψ_n vanishes at the walls. Typically, the Ψ_n are required to conform to some normalization condition, such as

$$\int \Psi_n^2 \, dA = A, \quad (3.573)$$

where A is the duct cross-sectional area.

The general theory leads to the conclusion that one can always find a complete set of Ψ_n, which with the

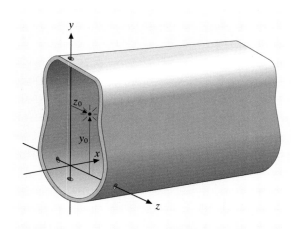

Fig. 3.41 Generic sketch of a duct that carries guided sound waves

rigid wall boundary condition imposed, are such that the cross-sectional eigenfunctions are orthogonal, so that

$$\int \Psi_n \Psi_m \, dA = 0 \quad (3.574)$$

if n and m correspond to different guided modes. The eigenvalues α_n^2, moreover, are all real and nonnegative. However, for cross sections that have some type of symmetry, it may be that more than one linearly independent eigenfunction Ψ_n (modes characterized by different values of the index n) correspond to the same numerical value of α_n^2. In such cases the eigenvalue is said to be degenerate.

The variation of guided mode amplitudes with source excitation is ordinarily incorporated into the axial wave functions $X_n(x)$, which satisfy the one-dimensional Helmholtz equation

$$\frac{d^2 X_n}{dx^2} + (k^2 - \alpha^2) X_n = 0 \, . \quad (3.575)$$

Here $k = \omega/s$ is the free-space wavenumber. The form of the solution depends on whether α_n^2 is greater or less than k^2. If $\alpha_n^2 < k^2$, the mode is said to be a propagating mode and the solution for X_n is given by

$$X_n = A_n e^{ik_n x} + B_n e^{-ik_n x} \, , \quad (3.576)$$

where the k_n, defined by

$$k_n = (k^2 - \alpha_n^2)^{1/2} \, , \quad (3.577)$$

are the modal wavenumbers. However, if the value of α_n^2 is greater than k^2, the mode is evanescent (not propagating), and one has

$$X_n = A_n e^{-\beta_n x} + B_n e^{\beta_n x} \, , \quad (3.578)$$

where β_n is given by

$$\beta_n = (\alpha_n^2 - k^2)^{1/2} \, . \quad (3.579)$$

Unless the termination of the duct is relatively close to the source, waves that grow exponentially with distance from the source are not meaningful, so only the term that corresponds to exponentially dying waves is ordinarily kept in the description of sound fields in ducts.

3.15.2 Cylindrical Ducts

For a duct with circular cross section and radius a, the index n is replaced by an index set (q, m, s), and the eigenfunctions Ψ_n are described by

$$\Psi_n = K_{qm} J_m \left(\frac{\eta_{qm} w}{a} \right) \begin{Bmatrix} \cos m\phi \\ \sin m\phi \end{Bmatrix} \, . \quad (3.580)$$

Here either the cosine ($s = 1$) or the sine ($s = -1$) corresponds to an eigenfunction. The quantities K_{qm} are normalization constants, and the J_m are Bessel functions of order m. The corresponding eigenvalues are given by

$$\alpha_n = \eta_{qm}/a \, , \quad (3.581)$$

where the η_{qm} are the zeros of $\eta J'_m(\eta)$, arranged in ascending order with the index q ranging upwards from 1. The smaller roots [3.85] for the axisymmetric modes are $\eta_{1,0} = 0.00$, $\eta_{2,0} = 3.83171$, and $\eta_{3,0} = 7.01559$, while those corresponding to $m = 1$ are $\eta_{1,1} = 1.84118$, $\eta_{2,1} = 5.33144$, and $\eta_{3,1} = 8.53632$.

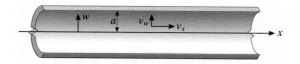

Fig. 3.42 Cylindrical duct

3.15.3 Low-Frequency Model for Ducts

In many situations of interest, the frequency of the acoustic disturbance is so low that only one guided mode can propagate, and all other modes are evanescent. Given that the walls can be idealized as rigid, there is always one mode that can propagate, this being the plane wave

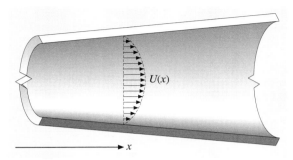

Fig. 3.43 Volume velocity in a duct

Fig. 3.44 Two dissimilar ducts with a junction modeled as a continuous volume velocity two-port. The difference in pressures is related to the volume velocity by the junction impedance

mode for which the eigenvalue α_0 is identically zero. The other modes will all be evanescent if the value of k is less than the corresponding α_n for each such mode. This would be so if the frequency is less than the lowest cutoff frequency for any of the nonplanar modes. For the circular duct case discussed above, for example, this would require that the inequality,

$$f < \frac{1.84118}{2\pi} \frac{c}{a}, \tag{3.582}$$

be satisfied.

When the single-guided-mode assumption is valid, and even if the duct cross-sectional area should vary with distance along the duct, the acoustic field equations can be replaced to a good approximation by the acoustic transmission-line equations,

$$\frac{\partial p}{\partial t} + \frac{\rho c^2}{A} \frac{\partial U}{\partial x} = 0, \tag{3.583}$$

$$\rho \frac{\partial U}{\partial t} = -A \frac{\partial p}{\partial x}. \tag{3.584}$$

Here U, equal to $A v_x$, is the volume velocity, the volume of fluid passing through the duct per unit time.

One of the implications of the low-frequency model described by (3.583) and (3.584) is that the volume velocity and the pressure are both continuous, even when the duct has a sudden change in cross-sectional area. The pressure continuity assumption is not necessarily a good approximation, but becomes better with decreasing frequency. An improved model (continuous volume-velocity two-port) takes the volume velocities on both sides of the junction as being equal, and sets the difference of the complex amplitudes of the upstream and downstream pressures just ahead and after the junction to

$$\hat{p}_{\text{ahead}} - \hat{p}_{\text{after}} = Z_J \hat{U}_{\text{junction}}, \tag{3.585}$$

where the junction's acoustic impedance Z_J is taken in the simplest approximation as $-i\omega M_{A,J}$, and where $M_{A,J}$ is a real number independent of frequency that is called the acoustic inertance of the junction. Approximate expressions for this acoustic inertance can be found in the literature [3.78, 86–88]; a simple rule is that it is ordinarily less than $8\rho/3A_{\min}$, where $A_{\min}$ is the smaller of the two cross-sectional areas. One may note, moreover, that Z_J goes to zero when the frequency goes to zero.

When an incident wave is incident at a junction, reflected and transmitted waves are created in the two ducts. The pressure amplitude reflection and transmission coefficients are given by

$$\mathfrak{R} = \frac{Z_J + \rho c/A_2 - \rho c/A_1}{Z_J + \rho c/A_2 + \rho c/A_1}, \tag{3.586}$$

$$\mathfrak{T} = \frac{2\rho c/A_2}{Z_J + \rho c/A_2 + \rho c/A_1}. \tag{3.587}$$

At sufficiently low frequencies, these are further approximated by replacing Z_J by zero.

3.15.4 Sound Attenuation in Ducts

The presence of the duct walls affects the attenuation of sound and usually causes the attenuation coefficient to be much higher than for plane waves in open space.

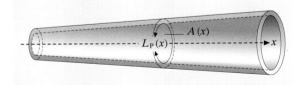

Fig. 3.45 Geometrical parameters pertaining to acoustic attenuation in a duct. The total length L_P around the perimeter gives the transverse extent of the dissipative boundary layer at the duct wall

A simple theory for predicting this attenuation uses the concepts of acoustic, entropy, and vorticity modes, as is indicated by the decomposition of (3.89) in a previous section of this chapter. The entropy and vorticity mode fields exist in an acoustic boundary layer near the duct walls and are such that they combine with the acoustic mode field to satisfy the appropriate boundary conditions at the walls.

With viscosity and thermal conduction thus taken into account in the acoustic boundary layer, and with the assumption that the duct walls are much better heat conductors than the fluid itself, the approximate attenuation coefficient for the plane wave guided mode is given by [3.89]

$$\alpha = \alpha_{\text{walls}}$$
$$= \left(\frac{\omega}{8\rho c^2}\right)^{1/2} [\mu^{1/2} + (\gamma - 1)(\kappa/c_p)^{1/2}] \frac{L_P}{A} , \quad (3.588)$$

where L_P is the length of the perimeter of the duct cross section.

A related effect [3.90] is that sound travels slightly slower in ducts than in an open environment. The complex wavenumber corresponding to propagation down the axis of the tube is approximately

$$k \approx \frac{\omega}{c} + (1+\mathrm{i})\alpha_{\text{walls}} , \quad (3.589)$$

and the phase velocity is consequently approximately

$$v_{\text{ph}} = \frac{\omega}{k_R} \approx c - \frac{c^2 \alpha_{\text{walls}}}{\omega} . \quad (3.590)$$

For a weakly attenuated and weakly dispersive wave, the group velocity [3.74], the speed at which energy travels, is approximately

$$v_{\text{gr}} \approx \left(\frac{\mathrm{d}k_R}{\mathrm{d}\omega}\right)^{-1} , \quad (3.591)$$

so, for the present situation,

$$v_{\text{gr}} \approx c - \frac{c^2 \alpha_{\text{walls}}}{2\omega} . \quad (3.592)$$

Thus both the phase velocity and the group velocity are less than the nominal speed of sound.

3.15.5 Mufflers and Acoustic Filters

The analysis of mufflers [3.91, 92] is often based on the idealization that their acoustic transmission characteristics are independent of sound amplitudes, so they act as linear devices. The muffler is regarded as an insertion into a duct, which reflects waves back toward the source and which alters the transmission of sound beyond the muffler. The properties of the muffler vary with frequency, so the theory analyzes the muffler's effects on individual frequency components. The frequency is assumed to be sufficiently low that only the plane wave mode propagates in the inlet and outlet ducts. Because of the assumed linear behavior of the muffler, and the single-mode assumption, the muffler conforms to the model of a linear two-port, the ports being the inlet and outlet. The model leads to the prediction that one may characterize the muffler at any given frequency by a matrix, such that

$$\begin{pmatrix} \hat{p}_{\text{in}} \\ \hat{p}_{\text{out}} \end{pmatrix} = \begin{pmatrix} K_{11} & K_{12} \\ K_{21} & K_{22} \end{pmatrix} \begin{pmatrix} \hat{U}_{\text{in}} \\ \hat{U}_{\text{out}} \end{pmatrix} , \quad (3.593)$$

where the coefficients K_{ij} represent the acoustical properties of the muffler. Reciprocity [3.30] requires that the determinant of the matrix be unity. Also, for a symmetric muffler, K_{12} and K_{21} must be identical.

It is ordinarily a good approximation that the waves reflected at the entrance of the muffler back to the source are significantly attenuated, so that they have negligible amplitude when they return to the muffler. Similarly, the assumption that the waves transmitted beyond the muffler do not return to the muffler (anechoic termination)

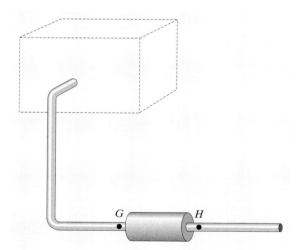

Fig. 3.46 Sketch of a prototype muffler configuration. Point G is a point on the input side of the muffler, and point H is on the output side. The noise source is upstream on the input side, and the tailpipe is downstream on the output side

is ordinarily valid. With these assumptions, the insertion of the muffler causes the mean squared pressure downstream of the muffler to drop by a factor

$$\tau = \left(\frac{1}{4}|K_{11} + K_{22} + \frac{\rho c}{A}K_{21} + \frac{A}{\rho c}K_{12}|^2\right)^{-1}, \quad (3.594)$$

where A is the cross section of the duct ahead of and behind the muffler.

Acoustic mufflers can be divided into two broad categories: reactive mufflers and dissipative mufflers. In a reactive muffler, the basic property of the muffler is that it reflects a substantial fraction of the incident acoustic energy back toward the source. The dissipation of energy within the muffler itself plays a minor role; the reflection is caused primarily by the geometrical characteristics of the muffler. In a dissipative muffler, however, a low transmission of sound is achieved by internal dissipation of acoustic energy within the muffler. Absorbing material along the walls is ordinarily used to achieve this dissipation.

3.15.6 Non-Reflecting Dissipative Mufflers

When a segment of pipe of cross section A and length L is covered with a duct lining material that attenuates the amplitude of traveling plane waves by a factor of $e^{-\alpha L}$,

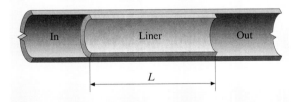

Fig. 3.47 Sketch of a nonreflective dissipative muffler

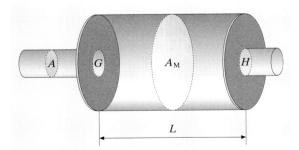

Fig. 3.48 Sketch of an expansion chamber muffler

the muffler's K-matrix is given by

$$K = \begin{pmatrix} \cos(kL + i\alpha L) & -i\frac{\rho c}{A}\sin(kL + i\alpha L) \\ -i\frac{A}{\rho c}\sin(kL + i\alpha L) & \cos(kL + i\alpha L) \end{pmatrix}, \quad (3.595)$$

so the fractional drop in mean squared pressure reduces to

$$\tau = e^{-2\alpha L}. \quad (3.596)$$

3.15.7 Expansion Chamber Muffler

The simplest reactive muffler is the expansion chamber, which consists of a duct of length L and cross section A_M connected at both ends to a pipe of smaller cross section A_P.

The K-matrix for such a muffler is found directly from (3.595) by setting α to zero (no dissipation in the chamber), but replacing A by A_M. The corresponding result for the fractional drop in mean squared pressure is

$$\tau = \left[\cos^2 kL + \frac{1}{4}(m + m^{-1})^2 \sin^2 kl\right]^{-1}, \quad (3.597)$$

where m is the expansion ratio A_M/A_P. The relative reduction in mean square pressure is thus periodic with frequency. A maximum reduction occurs when the length L is an odd multiple of quarter-wavelengths. The greatest reduction in mean squared pressure is when τ has its minimum value, which is that given by

$$\tau = \frac{4}{(m + m^{-1})^2}. \quad (3.598)$$

3.15.8 Helmholtz Resonators

An idealized acoustical system [3.47, 93] that is the prototype of commonly encountered resonance phenomena

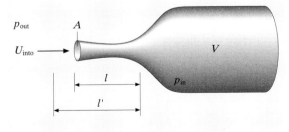

Fig. 3.49 Sketch of a Helmholtz resonator, showing the characteristic parameters involved in the analysis

consists of a cavity of volume V with a neck of length ℓ and cross-sectional area A.

In the limit where the acoustic frequency is sufficiently low that the wavelength is much larger than any dimension of the resonator, the compressible fluid in the resonator acts as a spring with spring constant

$$k_{sp} = \frac{\rho c^2 A^2}{V}, \quad (3.599)$$

and the fluid in the neck behaves as a lumped mass of magnitude

$$m = \rho A \ell' . \quad (3.600)$$

Here ℓ' is ℓ plus the end corrections for the two ends of the neck. If ℓ is somewhat larger than the neck radius a, and if both ends are terminated by a flange, then the two end corrections are each $0.82a$. (The determination of end corrections has an extensive history; a discussion can be found in the text by the author [3.30].) The resonance frequency ω_r, in radians per second, of the resonator is given by

$$\omega_r = 2\pi f_r = (k_{sp}/m)^{1/2} = (M_A C_A)^{-1/2} , \quad (3.601)$$

where

$$M_A = \rho \ell'/S \quad (3.602)$$

gives the acoustic inertance of the neck and

$$C_A = V/\rho c^2 \quad (3.603)$$

gives the acoustic compliance of the cavity. The ratio of the complex pressure amplitude just outside the mouth of the neck to the complex volume velocity amplitude of flow into the neck is the acoustic impedance Z_{HR} of the Helmholtz resonator and given, with the neglect of damping, by

$$Z_{HR} = -i\omega M_A + \frac{1}{-i\omega C_A} , \quad (3.604)$$

which vanishes at the resonance frequency.

If a Helmholtz resonator is inserted as a side branch into the wall of a duct, it acts as a reactive muffler that has a high insertion loss near the resonance frequency of the resonator. The analysis assumes that the acoustic

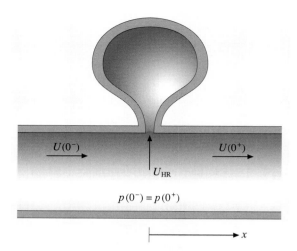

Fig. 3.50 Helmholtz resonator as a side-branch in a duct

pressures in the duct just before and just after the resonator are the same as the pressure at the mouth of the resonator. Also, the acoustical analog of Kirchhoff's circuit law for currents applies, so that the volume velocity flowing in the duct ahead of the resonator equals the sum of the volume velocities flowing into the resonator and through the duct just after the resonator. These relations, with (3.604), allow one to work out expressions for the amplitude reflection and transmission coefficients at the resonator, the latter being given by

$$\mathfrak{T} = \frac{2Z_{HR}}{2Z_{HR} + \rho c / A_D} . \quad (3.605)$$

The fraction of incident power that is transmitted is consequently given by

$$\tau = \left(1 + \frac{1}{4\beta^2 (f/f_r - f_r/f)^2}\right)^{-1} , \quad (3.606)$$

where β is determined by

$$\beta^2 = (M_A/C_A)(A_D/\rho c)^3 . \quad (3.607)$$

The fraction transmitted is formally zero at the resonance frequency, but if β is large compared with unity, the bandwidth over which the drop in transmitted power is small is narrow compared with f_r.

3.16 Ray Acoustics

When the medium is slowly varying over distances comparable to a wavelength and if the propagation distances are substantially greater than a wavelength, it is often convenient to regard acoustic fields as being carried along rays. These can be regarded as lines or paths in space.

3.16.1 Wavefront Propagation

A wavefront is a hypothetical surface in space over which distinct waveform features are simultaneously received. The theory of plane wave propagation predicts that wavefronts move locally with speed c when viewed in a coordinate system in which the ambient medium appears at rest. If the ambient medium is moving with velocity v, the wave velocity cn seen by someone moving with the fluid becomes $v + cn$ in a coordinate system at rest. Here n is the unit vector normal to the wavefront; it coincides with the direction of propagation if the coordinate system is moving with the local ambient fluid velocity v. A ray can be defined [3.94] as the time trajectory of a point that is always on the same wavefront, and for which the velocity is

$$v_{\text{ray}} = v + nc. \tag{3.608}$$

To determine ray paths without explicit knowledge of wavefronts, it is appropriate to consider a function $\tau(x)$, which gives the time at which the wavefront of interest passes the point x. Its gradient s is termed the wave slowness vector and is related to the local wavefront

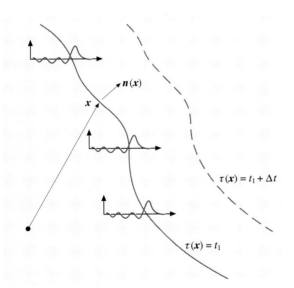

Fig. 3.51 Sketch illustrating the concept of a wavefront as a surface along which characteristic waveform features are simultaneously received. The time a given waveform passes a point x is $\tau(x)$ and the unit normal in the direction of propagation is $n(x)$

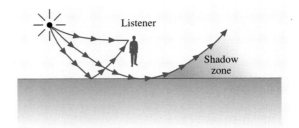

Fig. 3.52 Multipaths and shadow zones. The situation depicted is when the sound speed decreases with distance above the ground so that rays are refracted upward

normal by the equations

$$s = \frac{n}{c + v \cdot n}, \tag{3.609}$$

$$n = \frac{cs}{\Omega}, \tag{3.610}$$

where the quantity Ω is defined by

$$\Omega = 1 - v \cdot s. \tag{3.611}$$

Given the concepts of ray position and the slowness vector, a purely kinematic derivation leads to the ray-tracing equations

$$\frac{d x}{d t} = \frac{c^2 s}{\Omega} + v, \tag{3.612}$$

$$\frac{d s}{d t} = -\frac{\Omega}{c}\nabla c - (s \cdot \nabla)v - s \times (\nabla \times v). \tag{3.613}$$

Simpler versions result when there is no ambient flow, or when the ambient variables vary with only one coordinate. If the ambient variables are independent of position, then the ray paths are straight lines.

The above equations are often used for analysis of propagation through inhomogeneous media (moving or nonmoving) when the ambient variables do not vary significantly over distances of the order of a wavelength, even though they may do so over the total propagation distance. The rays that connect the source and listener locations are termed the eigenrays for that situation. If there is more than one eigenray, one has multipath reception. If there is no eigenray, then the listener is in a shadow zone.

3.16.2 Reflected and Diffracted Rays

This theory is readily extended to take into account solid boundaries and interfaces. When an incident ray strikes a solid boundary, a reflected ray is generated whose direction is determined by the trace velocity matching principle, such that angle of incidence equals angle of reflection. At an interface a transmitted ray is also generated, whose direction is predicted by Snell's law.

An extension of the theory known as the geometrical theory of diffraction and due primarily to *Keller* [3.95] allows for the possibility that diffracting edges can be a source of diffracted rays which have a multiplicity of directions. The allowable diffracted rays must have the same trace velocity along the diffracting edge as does the incident ray.

The ray paths connecting the source and listener satisfy *Fermat*'s principle [3.96], that the travel time along an actual eigenray is stationary relative to other geometrically admissible paths. *Keller*'s geometrical theory of diffraction [3.95] extends this principle to include paths which have discontinuous directions and which may have portions that lie on solid surfaces.

3.16.3 Inhomogeneous Moving Media

Determination of the amplitude variation along ray paths requires a more explicit use of the equations of linear acoustics. If there is no ambient fluid velocity then an appropriate wave equation is (3.73). When there is also an ambient flow present then an appropriate generalization [3.97, 98] for the acoustic mode is

$$\frac{1}{\rho} \nabla \cdot (\rho \nabla \Phi) - D_t \left(\frac{1}{c^2} D_t \Phi \right) = 0 \,. \tag{3.614}$$

Here D_t, defined by

$$D_t = \frac{\partial}{\partial t} + \boldsymbol{v} \cdot \nabla \tag{3.615}$$

is the time derivative following the ambient flow. The dependent variable Φ is related to the acoustic pressure perturbation by

$$p = -\rho D_t \Phi \,. \tag{3.616}$$

This wave equation is derived from the linear acoustics equations for circumstances when the ambient variables vary slowly with position and time and neglects terms of second order in $1/\omega T$ and $c/\omega L$, where ω is a characteristic frequency of the disturbance, T is a characteristic time scale for temporal variations of the ambient quantities, and L is a characteristic length scale for spatial variations of these approximations. The nature of the entailed approximations is consistent with the notion that geometrical acoustics is a high-frequency approximation and yields results that are the same in the high-frequency limit as would be derived from the complete set of linear acoustic equations.

3.16.4 The Eikonal Approximation

The equations of geometrical acoustics follow [3.99] from (3.614), if one sets the potential function Φ equal to the expression

$$\Phi(\boldsymbol{x}, t) = \Psi(\boldsymbol{x}) F(t - \tau) \,. \tag{3.617}$$

Here the function F is assumed to be a rapidly varying function of its argument. The quantities Ψ and τ vary only with position. (It is assumed here that the ambient variables do not vary with time.)

When such a substitution is made, one obtains terms that involve the second, first, and zeroth derivatives of the function F. The geometrical acoustics or eikonal approximation results when one requires that the coefficients of F'' and F' both vanish identically, and thus assumes that the terms involving undifferentiated F are of lesser importance in the high-frequency limit. The setting to zero of the coefficient of the second derivative yields

$$(1 - \boldsymbol{v} \cdot \nabla \tau)^2 = c^2 (\nabla \tau)^2 \,, \tag{3.618}$$

which is termed the eikonal equation [3.100, 101]. The setting to zero of the coefficient of the first derivative yields

$$\nabla \cdot \left\{ \frac{\rho \Psi^2}{c^2} [c^2 \nabla \tau + (1 - \boldsymbol{v} \cdot \nabla \tau) \boldsymbol{v}] \right\} = 0 \,, \tag{3.619}$$

which is termed the transport equation. The convenient property of these equations is that they are independent of the nature of the waveform function F.

The eikonal equation, which is a nonlinear first-order partial differential equation for the eikonal $\tau(\boldsymbol{x})$ can be solved by Cauchy's method of characteristics, with $\boldsymbol{s} = \nabla \tau$. What results are the ray-tracing equations (3.612) and (3.613). However, the time variable t is replaced by the eikonal τ. The manner in which these equations solve the eikonal equation is such that if τ and its gradient $\boldsymbol{s}$ are known at any given point then the ray-tracing equations determine these same quantities at all other points on the ray that passes through this initial point. The initial value of $\boldsymbol{s}$ must conform to the eikonal equation itself, as is stated in (3.618). If this is so, then the ray-tracing equations insure that it conforms to the same equation, with the appropriate ambient variables, all along the ray.

Solution of the transport equation, given by (3.619), is facilitated by the concept of a ray tube. One conceives

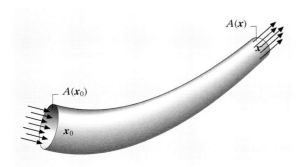

Fig. 3.53 Sketch of a ray tube. Here x_0 and x are two points on the central ray

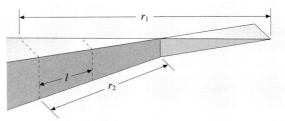

Fig. 3.54 Rectilinear propagation of sound. The wavefront at any point has two principal radii of curvature, which change with the propagation distance. In the case shown, the wavefront at the left is such that it is convex, with two different focal lengths, r_1 and r_2, these being the two radii of curvature. Over a short propagation distance ℓ, these reduce to $r_1 - \ell$ and $r_2 - \ell$

of a family of adjacent rays which comprise a narrow tube of cross-sectional area A, this area varying with distance ℓ along the path of the central ray of the tube. Then the transport equation reduces to

$$\frac{d}{d\ell}\left(A\frac{\rho\Psi^2}{c^2}\Omega v_{\text{ray}}\right) = 0, \quad (3.620)$$

so the quantity in brackets is constant along the ray tube.

Alternately, the relation of (3.616) implies that the acoustic pressure is locally given to an equivalent approximation by

$$p = P(\mathbf{x}) f(t - \tau), \quad (3.621)$$

where f is the derivative of F, and P is related to Ψ by

$$P = -\rho\Omega\Psi. \quad (3.622)$$

Then the transport equation (3.619) requires that

$$B = \frac{A P^2 v_{\text{ray}}}{\rho c^2 \Omega} \quad (3.623)$$

be constant along a ray tube, so that

$$\frac{dB}{d\ell} = 0. \quad (3.624)$$

This quantity B is often referred to as the Blokhintzev invariant.

If there is no ambient flow then the constancy of B along a ray tube can be regarded as a statement that the average acoustic power flowing along the ray tube is constant, for a constant-frequency disturbance, or in general that the power appears constant along a trajectory moving with the ray velocity. The interpretation when there is an ambient flow is that the wave action is conserved [3.102].

Geometrical acoustics (insofar as amplitude prediction is concerned) breaks down at caustics, which are surfaces along which the ray tube areas vanish. Modified versions [3.103, 104] involving Airy functions and higher transcendental functions, of the geometrical acoustics theory have been derived which overcome this limitation.

3.16.5 Rectilinear Propagation of Amplitudes

A simple but important limiting case of geometrical acoustics is for propagation in homogeneous nonmoving media. The rays are then straight lines and perpendicular to wavefronts. A given wavefront has two principal radii of curvature, R_{I} and R_{II}. With the convention that these are positive when the wavefront is concave in the corresponding direction of principal curvature, and negative when it is convex, the curvature radii increase by a distance ℓ when the wave propagates through that distance. The ray tube area is proportional to the product of the two curvature radii, so the acoustic pressure varies with distance ℓ along the ray as

$$p = \left(\frac{R_{\text{I}} R_{\text{II}}}{(R_{\text{I}}+\ell)(R_{\text{II}}+\ell)}\right)^{1/2} f[t - (\ell/c)]. \quad (3.625)$$

Here R_{I} and R_{II} are the two principal radii of curvature when ℓ is zero. The relations (3.326) and (3.447) for spherical and cylindrical spreading are both limiting cases of this expression.

3.17 Diffraction

The bending of sound around corners or around objects into shadow zones is typically referred to as diffraction. The prototype for problems of diffraction around an edge is that of a plane wave impinging on a rigid half-plane (knife edge). The problem was solved exactly by *Sommerfeld* [3.105], and the solution has subsequently been rederived by a variety of other methods. The present chapter considers a generalization of this problem where a point source is near a rigid wedge. The Sommerfeld problem emerges as a limiting case.

3.17.1 Posing of the Diffraction Problem

The geometry adopted here is such that the right surface of the wedge occupies the region where $y = 0$ and $x > 0$; the diffracting edge is the z-axis. A cylindrical coordinate system is used where w is the radial distance from the z axis and where ϕ is the polar angle measured counterclockwise from the x-axis. The other surface of the wedge is at $\phi = \beta$, where $\beta > \pi$. The region occupied by the wedge has an interior angle of $2\pi - \beta$.

One seeks solutions of the inhomogeneous wave equation

$$\nabla^2 p - \frac{1}{c^2}\frac{\partial^2 p}{\partial t^2} = -4\pi S(t)\delta(\mathbf{x} - \mathbf{x_0}) \tag{3.626}$$

for the region

$$0 < \phi < \beta; \quad w \geq 0; \quad -\infty < z < \infty. \tag{3.627}$$

The point source is located at $w = w_S$, $\phi = \phi_S$, and $z = 0$. It is assumed that the source is on the back side, so that $\phi_S > \pi$. (A person on the front face of the wedge should not be able to see the source.)

The quantity $S(t)$ is the time-dependent source strength, so that the transient acoustic pressure field near the source has a singular part given by

$$p \approx \frac{S\left[t - (R_D/c)\right]}{R_D}, \tag{3.628}$$

where R_D is the direct distance from the source,

$$R_D = \left[(w\cos\phi - w_S\cos\phi_S)^2 \right.$$
$$\left. + (w\sin\phi - w_S\sin\phi_S)^2 + z^2\right]^{1/2}, \tag{3.629}$$

or

$$R_D = \left[w^2 + w_S^2 - 2ww_S\cos(\phi - \phi_S) + z^2\right]^{1/2}. \tag{3.630}$$

Boundary conditions imposed by the rigid surfaces of the wedge are

$$\frac{\partial p}{\partial \phi} = 0 \quad \text{at} \quad \phi = 0 \quad \text{and at} \quad \phi = \beta. \tag{3.631}$$

There is also the requirement that the field at large radial distance w should represent an outgoing wave.

3.17.2 Rays and Spatial Regions

For the problem just posed, one can distinguish three regions, one where both a direct wave and a reflected wave are possible, one where only the direct wave is

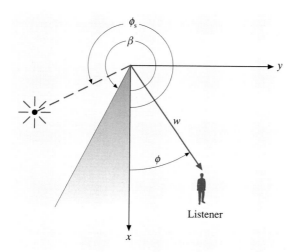

Fig. 3.55 Coordinate system and angles associated with the diffraction of sound by a wedge

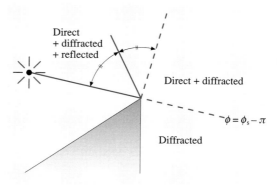

Fig. 3.56 Spatial regions associated with sound diffraction by a wedge

possible, and one where neither the direct wave nor the reflected wave is possible. The borders of these two regions are the angles ϕ_A and ϕ_B, where

$$\phi_A = \phi_S - \pi \, ; \qquad \phi_B = 2\beta - \phi_S - \pi \, . \qquad (3.632)$$

Reflected Wave Possible

In this region, characterized by the largest values of the azimuthal angle ϕ, the restriction is

$$\beta > \phi > \phi_B \, . \qquad (3.633)$$

For such azimuthal angles, one can have an incident ray that hits the back side ($\phi = \beta$) of the wedge, reflects according to the law of mirrors, and then passes through the listener point. This reflected wave come from the image source location at $\phi = 2\beta - \phi_S$, $w = w_S$, $z = 0$. The path distance for this reflected wave arrival is

$$R_R = \left[w^2 + w_S^2 - 2ww_S \cos(\phi + \phi_S - 2\beta) + z^2 \right]^{1/2} . \qquad (3.634)$$

The requirement that defines the border of this region is that any hypothetical straight line from the image source to the listener must pass through the back side of the wedge. Equivalently stated, there must be a reflection point on the back face of the wedge itself.

In this region, the geometrical acoustics (GA) portion of the acoustic pressure field is given by

$$p_{GA} = \frac{S[t - (R_D/c)]}{R_D} + \frac{S[t - (R_R/c)]}{R_R} \, . \qquad (3.635)$$

Along the plane where $\phi = \phi_B$, the reflected path distance becomes

$$R_R \to L \quad \text{as} \quad \phi \to \phi_B \, , \qquad (3.636)$$

where

$$L = \left[(w + w_0)^2 + z^2 \right]^{1/2} . \qquad (3.637)$$

The discontinuity in the geometrical acoustics portion caused by the dropping of the reflected wave is consequently

$$p_{GA}(\phi_B + \epsilon) - p_{GA}(\phi_B - \epsilon) = \frac{S[t - (L/c)]}{L} \, . \qquad (3.638)$$

Here the dependence of interest is that on the angular coordinate ϕ; the quantity ϵ is an arbitrarily small positive number.

Direct Wave, No Reflected Wave

This is the region of intermediate values of the azimuthal angle ϕ, where

$$\phi_B > \phi > \phi_A \, , \qquad (3.639)$$

and the listener can still see the source directly, but the listener cannot see the reflection of the source on the backside of the wedge. There is a direct wave, but no reflected wave.

Here the geometrical acoustics (GA) portion of the acoustic pressure field is given by

$$p_{GA} = \frac{S[t - (R_D/c)]}{R_D} \, . \qquad (3.640)$$

At the lower angular boundary ϕ_A of this region the direct wave propagation distance becomes

$$R_D \to L \quad \text{as} \quad \phi \to \phi_A \, , \qquad (3.641)$$

where L is the same length as defined above in (3.637).

The discontinuity in the geometrical acoustics portion caused by the dropping of the incident wave is consequently

$$p_{GA}(\phi_A + \epsilon) - p_{GA}(\phi_A - \epsilon) = \frac{S[t - (L/c)]}{L} \, , \qquad (3.642)$$

which is the same expression as for the discontinuity at ϕ_B.

No Direct Wave, No Reflected Wave

In this region, where

$$\phi_A > \phi > 0 \, , \qquad (3.643)$$

the listener is in the true shadow zone and there is no geometrical acoustics contribution, so

$$p_{GA} = 0 \, . \qquad (3.644)$$

3.17.3 Residual Diffracted Wave

The complete solution to the problem as posed above, for a point source outside a rigid wedge, is expressed with some generality as

$$p = p_{GA} + p_{\text{diffr}} \, . \qquad (3.645)$$

The geometrical acoustics portion is as defined above; in one region it is a direct wave plus a reflected wave; in a second region it is only a direct wave; in a third region (shadow region) it is identically zero. The diffracted wave is what is needed so that the sum will be a solution of the wave equation that satisfies the appropriate boundary conditions.

Whatever the diffracted wave is, it must have a discontinuous nature at the radial planes $\phi = \phi_A$ and $\phi = \phi_B$. This is because p_{GA} is discontinuous at these two values of ϕ and because the total solution should be continuous. Thus, one must require

$$p_{\text{diffr}}(\phi_B + \epsilon) - p_{\text{diffr}}(\phi_B - \epsilon) = -\frac{S[t-(L/c)]}{L}, \quad (3.646)$$

$$p_{\text{diffr}}(\phi_A + \epsilon) - p_{\text{diffr}}(\phi_A - \epsilon) = -\frac{S[t-(L/c)]}{L}. \quad (3.647)$$

Also, except in close vicinity to these radial planes, the diffracted wave should spread out at large w as an outgoing wave. One may anticipate that, at large values of w, the amplitude decreases with w as $1/\sqrt{w}$, as in cylindrical spreading.

Diffracted Path Length

In the actual solution given further below for the diffracted wave, a predominant role is played by the parameter L given by (3.637). This can be interpreted as the net distance from the source to the listener along the shortest path that touches the diffracting edge. This path can alternately be called the diffracted ray.

Such a path must touch the edge at a point where each of the two segments make the same angle with the edge. This is equivalent to a statement of Keller's law of geometrical diffraction [3.95]. Here, when the listener is at $z = z_L$, the source is at $z = 0$, and the edge is the z-axis, the z-coordinate where the diffracted path leaves the edge is $[w_S/(w_S + w_L)]z_L$. Thus one has

$$L = \frac{[w_S^2(w_S + w_L)^2 + w_0^2 z_L^2]^{1/2}}{w_S + w_L} + \frac{[w_L^2(w_S + w_L)^2 + w_L^2 z_L^2]^{1/2}}{w_S + w_L}. \quad (3.648)$$

The first term corresponds to the segment that goes from the source to the edge, while the second corresponds to the segment that goes from the edge to the listener. All this simplifies to

$$L = [(w_S + w)^2 + z^2]^{1/2}. \quad (3.649)$$

Here the listener z-coordinate is taken simply to be z, and its radial coordinate is taken simply to be w. The time interval L/c is the shortest time that it takes for a signal to go from the source to the listener via the edge.

As is evident from the discussion above of the discontinuities in the geometrical acoustics portion of the field, this diffracted wave path is the same as the reflected wave path on the radial plane $\phi = \phi_B$ and is the same as the direct wave path on the radial plane $\phi = \phi_A$.

There is another characteristic length that enters into the solution, this being

$$Q = [(w_S - w)^2 + z^2]^{1/2}. \quad (3.650)$$

It is a shorter length than L and is the distance from the source to a hypothetical point that has the same coordinates as that of the listener, except that the ϕ coordinate is the same as that of the source. These two quantities are related, so that

$$L^2 - Q^2 = 4w_S w, \quad (3.651)$$

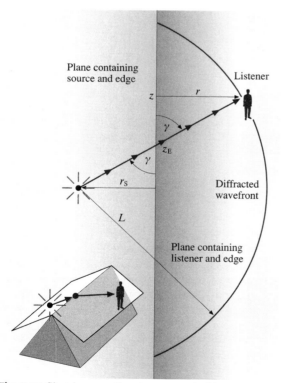

Fig. 3.57 Sketch supporting the geometrical interpretation of the parameter L as the diffracted ray length. The segment from source to edge makes the same angle γ with the edge as does the segment from edge to listener, and the sum of the two segment lengths is L

which is independent of the z-coordinate separation of the source and listener.

Satisfaction of Boundary Conditions

A function that satisfies the appropriate boundary conditions (3.631) can be built up from the functions

$$\sin(\nu x_q) \quad \text{and} \quad \cos(\nu x_q); \quad q = 1, 2, 3, 4,$$
(3.652)

where

$$\nu = \frac{\pi}{\beta},$$
(3.653)

$$x_1 = \pi + \phi + \phi_S = \phi - \phi_B + 2\beta,$$
$$x_2 = \pi - \phi - \phi_S = -\phi - \phi_A,$$
$$x_3 = \pi + \phi - \phi_S = \phi - \phi_A,$$
$$x_4 = \pi - \phi + \phi_S = 2\beta - \phi_B - \phi.$$
(3.654)

The trigonometric functions that appear above are not periodic with period 2π in the angles ϕ and ϕ_S, so the definition must be restricted to ranges of these quantities between 0 and β.

The satisfaction of the boundary conditions at $\phi = 0$ and $\phi = \beta$ is achieved by any function of the general form

$$\sum_{q=1}^{4} \Psi(\cos x_q, \sin x_q),$$
(3.655)

where the function Ψ is an arbitrary function of two arguments. To demonstrate that the boundary condition at $\phi = 0$ is satisfied, it is sufficient to regard the function $\Psi(\cos x_q, \sin x_q)$ as only a function of x_q. One derives

$$\left(\frac{d}{d\phi} \sum_{q=1}^{4} \Psi(x_q)\right)_{\phi=0}$$
$$= \{\nu \Psi'[\nu(\pi + \phi_S)] - \Psi'[\nu(\pi - \phi_S)] + \Psi'[\nu(\pi - \phi_S)] - \Psi'[\nu(\pi + \phi_S)]\} = 0.$$
(3.656)

The first and fourth terms cancel, and the second and third terms cancel.

The boundary condition at $\phi = \beta$, however, requires explicit recognition that the postulated function Ψ depends on x_q only through the two trigonometric functions. Both are periodic in ϕ with period 2β, because $\nu\beta = \pi$. Moreover, the sum of the first and fourth terms is periodic with period β, and the same is so for the sum of the second and third terms. The entire sum has this periodicity, so if the derivative with respect to ϕ vanishes at $\phi = 0$, it also has to vanish at $\phi = \beta$.

Discontinuities in the Diffraction Term

Two of the individual x_q defined above are such that $\cos(\nu x_q)$ is unity at one or the other of the two radial planes $\phi = \phi_A$ and $\phi = \phi_B$. One notes that

$$\cos(\nu x_3) = 1 \quad \text{at} \quad \phi = \phi_A,$$
(3.657)

$$\cos(\nu x_1) = 1 \quad \text{at} \quad \phi = \phi_B.$$
(3.658)

Near the radial plane $\phi = \phi_A = \phi_S - \pi$, one has

$$x_3 = \phi - \phi_A; \quad \cos(\nu x_3) \approx 1 - \frac{1}{2}(\nu^2)(\phi - \phi_A)^2;$$
$$\sin(\nu x_3) \approx (\nu)(\phi - \phi_A).$$
(3.659)

Similarly, near the radial plane $\phi = \phi_B = 2\beta - \phi_S - \pi$,

$$x_1 = 2\beta + \phi - \phi_B; \quad \cos(\nu x_1) \approx 1 - \frac{1}{2}(\nu^2)(\phi - \phi_B)^2;$$
$$\sin(\nu x_1) \approx (\nu)(\phi - \phi_B).$$
(3.660)

This behavior allows one to construct a discontinuous function of the generic form

$$U(\phi) = \sum_{q=1}^{4} \int_{L}^{\infty} \frac{M(\xi) \sin(\nu x_q)}{F(\xi) - \cos(\nu x_q)} d\xi,$$
(3.661)

where $M(\xi)$ and $F(\xi)$ are some specified functions of the dummy integration variable, such that $F \geq 1$, and $F = 1$ at only one value of ξ. This allows one to compute the discontinuity at ϕ_A, for example, as

$$U(\phi_A + \epsilon) - U(\phi_A - \epsilon)$$
$$= 2\nu\epsilon \int_{L}^{\infty} \frac{M(\xi)}{F(\xi) - 1 + \frac{1}{2}\nu^2 \epsilon^2} d\xi$$
(3.662)

where it is understood that one is to take the limit of $\epsilon \to 0$. The dominant contribution to the integration should come from the value of ξ at which $F = 1$. There are various possibilities where this limit is finite and nonzero. The one of immediate interest is where $F(L) = 1$, where the derivative is positive at $\xi = L$, and where M is singular as $1/(\xi - L)^{1/2}$ near $\xi = L$. For these circumstances the limit is

$$\lim_{\epsilon \to 0} [U(\phi_A + \epsilon) - U(\phi_A - \epsilon)]$$
$$= 4\pi \left(\frac{1}{2F'}\right)_L^{1/2} [(\xi - L)^{1/2} M]_L$$
(3.663)

where the subscript L implies that the indicated quantity is to be evaluated at $\xi = L$.

3.17.4 Solution for Diffracted Waves

The details [3.30, 106–111] that lead to the solution vary somewhat in the literature and are invariably intricate. The procedure typically involves extensive application of the theory of the functions of a complex variable. The following takes some intermediate results from the derivation in the present author's text [3.30] and uses the results derived above to establish the plausibility of the solution.

The diffracted field at any point depends linearly on the time history of the source strength, and the shortest time for source to travel as a diffracted wave from source to listener is L/c, so the principles of causality and linearity allow one to write, with all generality, the diffracted wave as

$$p_{\text{diffr}} = -\frac{1}{\beta} \int_L^\infty S\left(t - \frac{\xi}{c}\right) K_\nu(\xi) \, d\xi , \qquad (3.664)$$

where the function $K_\nu(\xi)$ depends on the positions of the source and listener, as well as on the angular width of the wedge. The factor $-1/\beta$ in front of the integral is for mathematical convenience.

The boundary conditions at the rigid wedge are satisfied if one takes the dependence on ϕ and ϕ_S to be of the form of (3.655). The proper jumps will be possible if one takes the general dependence on the trigonometric functions to be as suggested in (3.661). The appropriate identification (substantiated further below) is

$$K_\nu = \frac{1}{(\xi^2 - Q^2)^{1/2}} \frac{1}{(\xi^2 - L^2)^{1/2}}$$
$$\times \sum_{q=1}^4 \frac{\sin(\nu x_q)}{\cosh \nu s - \cos(\nu x_q)} . \qquad (3.665)$$

The quantity s is a function of ξ, defined as

$$s = 2 \tanh^{-1} \left(\frac{\xi^2 - L^2}{\xi^2 - Q^2}\right)^{1/2} , \qquad (3.666)$$

so that

$$\xi^2 = L^2 + (L^2 - Q^2) \sinh^2(s/2) , \qquad (3.667)$$

$\cosh \nu s$

$$= \frac{1}{2} \left[\left(\frac{(\xi^2 - Q^2)^{1/2} - (\xi^2 - L^2)^{1/2}}{(\xi^2 - Q^2)^{1/2} + (\xi^2 - L^2)^{1/2}}\right)^\nu \right.$$
$$\left. + \left(\frac{(\xi^2 - Q^2)^{1/2} + (\xi^2 - L^2)^{1/2}}{(\xi^2 - Q^2)^{1/2} - (\xi^2 - L^2)^{1/2}}\right)^\nu \right] . \qquad (3.668)$$

That this solution has the proper jump behavior at ϕ_A and ϕ_B follows because, for ξ near L,

$$-\frac{1}{\beta} \frac{1}{(\xi^2 - Q^2)^{1/2}} \frac{1}{(\xi^2 - L^2)^{1/2}}$$
$$\to -\left(\frac{1}{\beta (L^2 - Q^2)^{1/2} (2L)^{1/2}}\right) \frac{1}{(\xi - L)^{1/2}} , \qquad (3.669)$$

$$\cosh \nu s \to 1 + \frac{4\nu^2 L}{L^2 - Q^2}(\xi - L) . \qquad (3.670)$$

The result in (3.663) consequently yields

$$p_{\text{diffr}}(\phi_A + \epsilon) - p_{\text{diffr}}(\phi_A - \epsilon)$$
$$= -\left(\frac{4\pi}{\beta(L^2 - Q^2)^{1/2}(2L)^{1/2}}\right)\left(\frac{(L^2 - Q^2)^{1/2}}{2\nu(2L)^{1/2}}\right)$$
$$\times S[t - (L/c)] = -\frac{S[t - (L/c)]}{L} , \qquad (3.671)$$

which is the same as required in (3.647).

(The solution has here been shown to satisfy the boundary conditions and the jump conditions. An explicit proof that the total solution satisfies the wave equation can be found in the text by the author [3.30].)

3.17.5 Impulse Solution

An elegant feature of the above transient solution is that no integration is required for the case when the source function is a delta function (impulse source). If

$$S(t) \to A\delta(t) , \qquad (3.672)$$

then the integration properties of the delta function yield

$$p_{\text{diffr}} \to -\frac{Ac}{\beta} H(ct - L) \frac{1}{(c^2 t^2 - Q^2)^{1/2}}$$
$$\times \frac{1}{(c^2 t^2 - L^2)^{1/2}} \sum_{q=1}^4 \frac{\sin(\nu x_q)}{\cosh \nu s - \cos(\nu x_q)} , \qquad (3.673)$$

where $\cosh \nu s$ is as given by (3.668), but with ξ replaced by ct. The function $H(ct - L)$ is the unit step function, equal to zero when its argument is negative and equal to unity when its argument is positive.

(In recent literature, an equivalent version of this solution is occasionally referred to as the *Biot–Tolstoy* solution [3.111], although its first derivation was by *Friedlander* [3.110]. In retrospect, it is a simple deduction from the constant-frequency solution given by *MacDonald* [3.106], *Whipple* [3.107], *Bromwich*

[3.108], and *Carslaw* [3.109]. In the modern era with digital computers, where the numerical calculation of Fourier transforms is virtually instantaneous and routine, an explicit analytical solution for the impulse response is often a good starting point for determining diffraction of waves of constant frequency.)

3.17.6 Constant-Frequency Diffraction

The transient solution (3.664) above yields the solution for when the source is of constant frequency. One makes the substitution

$$S(t) \to \hat{S} e^{-i\omega t}, \tag{3.674}$$

and obtains

$$\hat{p}_{\text{diffr}} = -\frac{\hat{S}}{\beta} \int_L^\infty e^{ik\xi} K_\nu(\xi)\, d\xi. \tag{3.675}$$

The integral that appears here can be reduced to standard functions in various limits. A convenient first step is to change the integration variable to the parameter s, so that

$$\xi^2 = L^2 + (L^2 - Q^2)\sinh^2(s/2), \tag{3.676}$$

$$\frac{d\xi}{[\xi^2 - Q^2]^{1/2}[\xi^2 - L^2]^{1/2}} = \frac{ds}{2\xi}, \tag{3.677}$$

and so that the range of integration on s is from 0 to ∞. One notes that the integrand is even in s, so that it is convenient to extend the integration from $-\infty$ to ∞, and then divide by two. Also, one can use trigonometric identities to combine the x_1 term with the x_2 term and to combine the x_3 term with the x_4 term. All this yields the result

$$\hat{p}_{\text{diffr}} = \frac{\hat{S}\sin\nu\pi}{2\beta} \sum_{+,-} \int_{-\infty}^\infty \frac{e^{ik\xi}}{\xi} F_\nu(s, \phi \pm \phi_S)\, ds, \tag{3.678}$$

where

$$F_\nu(s, \phi) = \frac{U_\nu(\phi) - J(\nu s)\cos\nu\phi}{J^2(\nu s) + 2J(\nu s)V_\nu^2(\phi) + U_\nu^2(\phi)}, \tag{3.679}$$

with the abbreviations

$$J(\nu s) = \cosh\nu s - 1, \tag{3.680}$$

$$U_\nu(\phi) = \cos\nu\pi - \cos\nu\phi, \tag{3.681}$$

$$V_\nu(\phi) = (1 - \cos\nu\phi\cos\nu\pi)^{1/2}. \tag{3.682}$$

The requisite integrals exist, except when either $U_\nu(\phi + \phi_S)$ or $U_\nu(\phi - \phi_S)$ should be zero. The angles at which one or the other occurs correspond to boundaries between different regions of the diffracted field.

3.17.7 Uniform Asymptotic Solution

Although the integration in (3.678) can readily be completed numerically, considerable insight and useful formulas arise when one considers the limit when both the source and the listener are many wavelengths from the edge. One argues that the dominant contribution to the integration comes from values of s that are close to 0 (where the phase of the exponential is stationary), so one approximates

$$\xi \to L + \frac{(L^2 - Q^2)}{8L}s^2 = L + \frac{\pi}{2k}\Gamma^2 s^2, \tag{3.683}$$

$$\frac{e^{ik\xi}}{\xi} \to \frac{e^{ikL}}{L} e^{i(\pi/2)\Gamma^2 s^2}, \tag{3.684}$$

$$J(\nu s) = \cosh\nu s - 1 \to \frac{\nu^2}{2}s^2, \tag{3.685}$$

$$F_\nu(s, \phi) \to \frac{U_\nu(\phi)}{U_\nu^2(\phi) + \nu^2 V_\nu^2(\phi)s^2},$$

$$= \frac{1}{2\nu V_\nu(\phi)}\left(\frac{1}{M_\nu(\phi) + is} + \frac{1}{M_\nu(\phi) - is}\right). \tag{3.686}$$

The expression (3.686) uses the abbreviation

$$M_\nu(\phi) = \frac{U_\nu(\phi)}{\nu V_\nu(\phi)} = \frac{\cos\nu\pi - \cos\nu\phi}{\nu(1 - \cos\nu\phi\cos\nu\pi)^{1/2}}, \tag{3.687}$$

while (3.684) uses the abbreviation

$$\Gamma^2 = k\frac{(L^2 - Q^2)}{4\pi L} = \frac{kww_S}{\pi L}, \tag{3.688}$$

where the latter version follows from the definitions (3.649) and (3.650).

One also notes that the symmetry in the exponential factor allows one to identify

$$\int_{-\infty}^\infty \frac{e^{i(\pi/2)\Gamma^2 s^2}}{M_\nu + is}\, ds = \int_{-\infty}^\infty \frac{e^{i(\pi/2)\Gamma^2 s^2}}{M_\nu - is}\, ds, \tag{3.689}$$

so the number of required integral terms is halved. A further step is to change the integration variable to

$$u = (\pi/2)^{1/2}\Gamma e^{-i\pi/4} s. \tag{3.690}$$

All this results in the uniform asymptotic expression

$$\hat{p}_{\text{diffr}} = \hat{S} \frac{e^{ikL}}{L} \frac{e^{i\pi/4}}{\sqrt{2}} \sum_{+,-} \frac{\sin \nu\pi}{V_\nu(\phi \pm \phi_S)}$$
$$\times A_D[\Gamma M_\nu(\phi \pm \phi_S)] \qquad (3.691)$$

for the diffracted wave. Here one makes use of the definition

$$A_D(X) = \frac{1}{\pi 2^{1/2}} \int_{-\infty}^{\infty} \frac{e^{-u^2} \, du}{(\pi/2)^{1/2} X - e^{-i\pi/4} u}, \qquad (3.692)$$

of a complex-valued function of a single real variable. The properties of this function are discussed in detail below.

3.17.8 Special Functions for Diffraction

The diffraction integral (3.692) is a ubiquitous feature in diffraction theories. It changes sign,

$$A_D(-X) = -A_D(X), \qquad (3.693)$$

when its argument X changes sign (as can be demonstrated by changing the integration variable to $-u$ after changing the sign of X). The function is complex, so one can write in general

$$A_D(X) = \text{sign}(X)[f(|X|) - ig(|X|)], \qquad (3.694)$$

where the functions f and g (referred to as the auxiliary Fresnel functions) represent the real and negative imaginary parts of $A_D(|X|)$.

To relate $A_D(X)$ to the often encountered Fresnel integrals, one makes use of the identity [with $\zeta = (\pi/2)^{1/2}|X|$ being positive]

$$\frac{1}{\zeta - e^{-i\pi/4} u} = e^{-i\pi/4} \int_0^\infty \exp[i(\zeta e^{i\pi/4} - u)q] \, dq. \qquad (3.695)$$

Then, one has

$$\int_{-\infty}^{\infty} \frac{e^{-u^2} \, du}{\zeta - e^{-i\pi/4} u}$$
$$= e^{-i\pi/4} e^{-i\zeta^2} \int_0^\infty e^{-y^2} \left(\int_{-\infty}^{\infty} e^{-(u+iq/2)^2} \, du \right) dq, \qquad (3.696)$$

with the abbreviation $y = q/2 + e^{-i\pi/4}\zeta$. The inner integral over u yields $\sqrt{\pi}$, and one sets

$$\int_0^\infty e^{-y^2} \, dq = 2 \int_{\zeta e^{-i\pi/4}}^{\infty} e^{-y^2} \, dy$$
$$= 2 \int_{\zeta e^{-i\pi/4}}^{0} e^{-y^2} \, dy + 2 \int_0^\infty e^{-y^2} \, dy. \qquad (3.697)$$

The second term in the latter expression is $\sqrt{\pi}$, and in the first term one changes the variable of integration to $t = (2/\pi)^{1/2} y e^{i\pi/4}$, so that

$$\int_{\zeta e^{-i\pi/4}}^{0} e^{-y^2} \, dy = -(\pi/2)^{1/2} e^{-i\pi/4}$$
$$\times \int_0^{(2/\pi)^{1/2}\zeta} e^{i(\pi/2)t^2} \, dt. \qquad (3.698)$$

The analytical steps outlined above lead to the result

$$A_D(X) = \frac{(1-i)}{2} e^{-i[\pi/2]X^2}$$
$$\times \{\text{sign}(X) - (1-i)[C(X) + iS(X)]\}, \qquad (3.699)$$

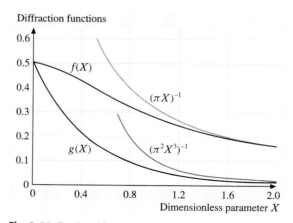

Fig. 3.58 Real and imaginary parts of the diffraction integral. Also shown are the appropriate asymptotic forms that are applicable at large arguments

where $C(X)$ and $S(X)$ are the Fresnel integrals

$$C(X) = \int_0^X \cos\left(\frac{\pi}{2}t^2\right) dt; \quad (3.700)$$

$$S(X) = \int_0^X \sin\left(\frac{\pi}{2}t^2\right) dt. \quad (3.701)$$

In accordance with (3.693), the above representation of $A_D(X)$ is an odd function of X, the two Fresnel integrals themselves being odd in X. The discontinuity in $A_D(X)$ is readily identified from this representation as

$$A_D(0+\epsilon) - A_D(0-\epsilon) = 1 - i; \quad (3.702)$$

The behavior of the two auxiliary Fresnel functions at small to moderate values of $|X|$ can be determined by expanding the terms in (3.699) in a power series in X, so that one identifies

$$f(|X|) \approx \frac{1}{2} - \frac{\pi}{4}X^2 + \frac{\pi}{3}|X|^3 - \ldots, \quad (3.703)$$

$$g(|X|) \approx \frac{1}{2} - |X| + \frac{\pi}{2}X^2 - \ldots. \quad (3.704)$$

To determine the behavior at large values of $|X|$, it is expedient to return to the expression (3.692) and expand the integrand in powers of $1/X$ and then integrate term by term. Doing this yields

$$f(|X|) \to \frac{1}{\pi|X|} - \frac{3}{\pi^3|X|^5} + \ldots, \quad (3.705)$$

$$g(|X|) \to \frac{1}{\pi^2|X|^3} - \frac{15}{\pi^4|X|^7} + \ldots. \quad (3.706)$$

These are asymptotic series, so one should retain at most only those terms which are of decreasing value. The leading terms are a good approximation for $|X| > 2$.

3.17.9 Plane Wave Diffraction

The preceding analysis, for when the source is a point source at a finite distance from the edge, can be adapted to the case when one idealizes the incident wave as a plane wave, propagating in the direction of the unit vector

$$\mathbf{n}_{\text{inc}} = n_x \mathbf{e}_x + n_y \mathbf{e}_y + n_z \mathbf{e}_z, \quad (3.707)$$

which points from a distant source (cylindrical coordinates w_S, ϕ_S, z_S) toward the coordinate origin, so that

$$n_x = \frac{-w_S \cos\phi_S}{(w_S^2 + z_S^2)^{1/2}}; \quad n_y = \frac{-w_S \sin\phi_S}{(w_S^2 + z_S^2)^{1/2}};$$

$$n_z = \frac{-z_S}{(w_S^2 + z_S^2)^{1/2}}. \quad (3.708)$$

Without any loss of generality, one can consider z_S to be negative, so that n_z is positive. Then at the point on the edge where $z = 0$, the direct wave makes an angle

$$\gamma = \cos^{-1}\left(\frac{-z_S}{(w_S^2 + z_S^2)^{1/2}}\right) \quad (3.709)$$

with the edge of the wedge, and this definition yields

$$\sin\gamma = \frac{w_S}{(w_S^2 + z_S^2)^{1/2}}; \quad \cos\gamma = n_z. \quad (3.710)$$

One lets $\hat{p}_{\text{inc}}$ be the complex amplitude of the incident wave at the origin (the point on the edge where $z = 0$), so that

$$\hat{S}\frac{1}{(w_S^2 + z_S^2)^{1/2}} e^{i(w_S^2 + z_S^2)^{1/2}} \to \hat{p}_{\text{inc}}, \quad (3.711)$$

and one holds this quantity constant while letting $(w_S^2 + z_S^2)^{1/2}$ become arbitrarily large. Then, with appropriate use of Taylor series (or binomial series) expansions, one has

$$R_D \to (w_S^2 + z_S^2)^{1/2} - w\sin\gamma\cos(\phi - \phi_S) + z\cos\gamma, \quad (3.712)$$

$$R_R \to (w_S^2 + z_S^2)^{1/2} - w\sin\gamma\cos(\phi + \phi_S - 2\beta) + z\cos\gamma, \quad (3.713)$$

$$L \to (w_S^2 + z_S^2)^{1/2} + w\sin\gamma + z\cos\gamma, \quad (3.714)$$

$$\Gamma \to \left(\frac{kw\sin\gamma}{\pi}\right)^{1/2}. \quad (3.715)$$

In this limit, the geometrical acoustics portion of the solution, for waves of constant frequency, is given by one of the following three expressions. In the region $\beta > \phi > \phi_B$, one has

$$\hat{p}_{\text{GA}} = \hat{p}_{\text{inc}} e^{ikn_z z} (e^{ikn_x x} e^{ikn_y y} + e^{-ikw\sin\gamma\cos(\phi + \phi_S - 2\beta)}), \quad (3.716)$$

where the two terms correspond to the incident wave and the reflected wave. In the region $\phi_B > \phi > \phi_A$, one has

$$\hat{p}_{\text{GA}} = \hat{p}_{\text{inc}} e^{ikn_z z} e^{ikn_x x} e^{ikn_y y}, \quad (3.717)$$

which is the incident wave only. Then, in the shadow region, where $\phi_A > \phi > 0$, one has

$$\hat{p}_{GA} = 0, \qquad (3.718)$$

and there is neither an incident wave nor a reflected wave.

The diffracted wave, in this plane wave limit, becomes

$$\hat{p}_{\text{diffr}} = \hat{p}_{\text{inc}} e^{ikz\cos\gamma} e^{ikw\sin\gamma} \frac{e^{i\pi/4}}{\sqrt{2}}$$
$$\times \sum_{+,-} \frac{\sin\nu\pi}{V_\nu(\phi \pm \phi_S)} A_D [\Gamma M_\nu(\phi \pm \phi_S)], \qquad (3.719)$$

where Γ is now understood to be given by (3.715). This result is, as before, for the case when the listener is many wavelengths from the edge, so that the parameter Γ is large compared with unity.

3.17.10 Small-Angle Diffraction

Simple approximate formulas emerge from the above general results when one limits one's attention to the diffracted field near the edge of the shadow zone boundary.

The incident wave is taken as having its direction lying in the (x, y) plane and one introduces rotated coordinates (x', y'), so that the y'-axis is in the direction of the incident sound, with the coordinate origin remaining at the edge of the wedge. One regards y' as

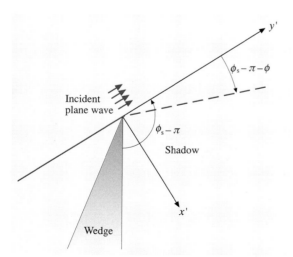

Fig. 3.59 Geometry and parameters used in discussion of small-angle diffraction of a plane wave by a rigid wedge

being large compared to a wavelength. The magnitude $|x'|$ is regarded as substantially smaller than y', but not necessarily small compared to a wavelength.

In the plane wave diffraction expression (3.719) the angle γ is $\pi/2$, and the only term of possible significance is that corresponding to the minus sign, so one has

$$\hat{p}_{\text{diffr}} = \hat{p}_{\text{inc}} e^{ikw} \frac{e^{i\pi/4}}{\sqrt{2}} \frac{\sin\nu\pi}{V_\nu(\phi - \phi_S)}$$
$$\times A_D [\Gamma M_\nu(\phi - \phi_S)]. \qquad (3.720)$$

Also, because $|x'|$ is small compared with y', one can assume that $|\phi_S - \pi - \phi|$ is small compared with unity, so that

$$\cos\nu(\phi - \phi_S) \approx \cos\nu\pi + (\nu\sin\nu\pi)(\phi - \phi_S + \pi), \qquad (3.721)$$

$$V_\nu(\phi - \phi_S) \approx \sin\nu\pi, \qquad (3.722)$$

$$M_\nu(\phi - \phi_S) \approx \phi_S - \pi - \phi \approx \frac{x'}{y'}. \qquad (3.723)$$

Further approximations that are consistent with this small-angle diffraction model are to set $w \to y'$ in the expression for Γ, but to set

$$w \to y' + \frac{1}{2}\frac{(x')^2}{y'} \qquad (3.724)$$

in the exponent. The second term is needed if one needs to take into account any phase shift relative to that of the incident wave.

The approximations just described lead to the expression

$$\hat{p}_{\text{diffr}} = \hat{p}_{\text{inc}} e^{iky'} \frac{1+i}{2} e^{(\pi/2)X^2} A_D(X), \qquad (3.725)$$

with

$$X = \left(\frac{k}{\pi y'}\right) x'. \qquad (3.726)$$

A remarkable feature of this result is that it is independent of the wedge angle β. It applies in the same approximate sense equally for diffraction by a thin screen and by a right-angled corner.

The total acoustic field in this region just above and just where the shadow zone begins can be found by adding the incident wave for $x' < 0$. In the shadow zone there is no incident wave, and one accounts for this by using a step function $H(-X)$. Thus the total field is approximated by

$$\hat{p}_{\text{GA}} + \hat{p}_{\text{diffr}} \to p_{\text{inc}} e^{iky'}$$
$$\times \left[H(-X) + \frac{(1+i)}{2} e^{i(\pi/2)X^2} A_D(X) \right]. \qquad (3.727)$$

Fig. 3.60 Characteristic diffraction pattern as a function of the diffraction parameter X. The function is the absolute magnitude of the complex amplitude of the received acoustic pressure, incident plus diffracted, relative to that of the incident wave

One can now substitute into this the expression (3.699), with the result

$$\hat{p}_{GA} + \hat{p}_{diffr} = \hat{p}_{inc}\,e^{iky}\left(\frac{1-i}{2}\right)$$
$$\times\left\{\left[\frac{1}{2}-C(X)\right]+i\left[\frac{1}{2}-S(X)\right]\right\}, \quad (3.728)$$

or, equivalently,

$$\hat{p}_{GA} + \hat{p}_{diffr} = \hat{p}_{inc}\,e^{iky}\left(\frac{1-i}{2}\right)\int_X^\infty e^{i(\pi/2)u^2}\,du\,. \quad (3.729)$$

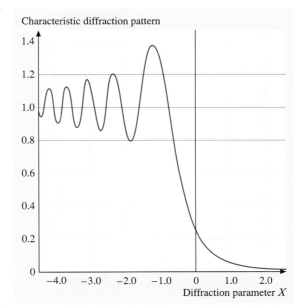

Characteristic diffraction pattern

The definitions of the diffraction integral and of the Fresnel integrals are such that, in the latter expressions, one does not need a step function.

The expressions just given, or their equivalents, are what are most commonly used in assessments of diffraction. The plot of the magnitude squared, relative to that of the incident wave,

$$\left|\frac{\hat{p}_{GA}+\hat{p}_{diffr}}{\hat{p}_{inc}}\right|^2 = \frac{1}{2}\left|\int_X^\infty e^{i(\pi/2)u^2}\,du\right|^2, \quad (3.730)$$

shows a monotonic decay in the shadow region ($X>0$) and a series of diminishing ripples about unity in the illuminated region ($X<0$). The peaks are interpreted as resulting from constructive interference of the incident and diffracted waves. The valleys result from destructive interference.

3.17.11 Thin-Screen Diffraction

For the general case of plane wave diffraction, not necessarily at small angles, a relatively simple limiting case is when the wedge is a thin screen, so that $\beta = 2\pi$ and the wedge index ν is $1/2$. Additional simplicity results for the case where the incident wave's direction $\mathbf{n}_{inc}$ is perpendicular to the edge, so that $\gamma = \pi/2$. In this limiting case, $\sin \nu\pi = 1$ and $\cos \nu\pi = 0$, so, with reference to (3.682), one finds $V_\nu(\phi \pm \phi_S) = 1$, and with reference to (3.687), one finds

$$M_\nu(\phi\pm\phi_S) = -2\cos\frac{1}{2}(\phi\pm\phi_S)\,. \quad (3.731)$$

Then, with use of the fact that $A_D(X)$ is odd in X, one obtains

$$\hat{p}_{diffr} = \frac{-\hat{p}_{inc}\,e^{i(kw+\pi/4)}}{\sqrt{2}}$$
$$\times \sum_{+,-} A_D\left[\left(\frac{4kw}{\pi}\right)^{1/2}\cos\frac{1}{2}(\phi\pm\phi_S)\right]. \quad (3.732)$$

Although derived here for the asymptotic limit when kw is large, this result is actually exact and holds for all values of w. It was derived originally by *Sommerfeld* in 1896 [3.105].

3.18 Parabolic Equation Methods

It is often useful in propagation analyses to replace the Helmholtz equation or its counterpart for inhomogeneous media by a partial differential equation that is first order, rather than second order, in the coordinate that corresponds to the primary direction of propagation. Approximations that achieve this are known generally as parabolic equation approximations and stem from early work by *Fock* [3.112] and *Tappert* [3.113]. Their chief advantage is that the resulting equation is no longer elliptic, but parabolic, so it is no longer necessary to check whether a radiation condition is satisfied at large propagation distances. This is especially convenient for numerical computation, for the resulting algorithms march out systematically in propagation distance, without having to look far ahead for the determination of the field at a subsequent point.

A simple example of a parabolic equation approximation is that of two-dimensional propagation in a medium where the sound speed and density vary primarily with the coordinate y, but weakly with x. A wave of constant angular frequency ω is propagating primarily in the x-direction but is affected by the y variations of the ambient medium, and by the presence of interfaces that are nearly parallel to the x-direction. The propagation is consequently governed by the two-dimensional reduced wave equation,

$$\frac{\partial}{\partial x}\left(\frac{1}{\rho}\frac{\partial \hat{p}}{\partial x}\right) + \frac{\partial}{\partial y}\left(\frac{1}{\rho}\frac{\partial \hat{p}}{\partial y}\right) + \frac{\omega^2}{\rho c^2}\hat{p} = 0 \quad (3.733)$$

which can be obtained from (3.73) by use of the complex amplitude convention.

In the parabolic approximation, the exact solution $\hat{p}$ of (3.733), with a radiation condition imposed, is represented by

$$\hat{p}(x,y) = e^{i\chi(x)} F(x,y), \quad (3.734)$$

where

$$\chi(x) = \int_0^x k_0 \, dx. \quad (3.735)$$

Here $k_0(x)$ is a judiciously chosen reference wavenumber (possibly dependent on the primary coordinate x). The crux of the method is that the function $F(x,y)$ is taken to satisfy an approximate partial differential equation [parabolic equation (PE)] that differs from what would result were (3.734) inserted into (3.733) with no discarded terms.

The underlying assumption is that the x-dependence of the complex pressure amplitude is mostly accounted for by the exponential factor, so the characteristic scale for variation of the amplitude factor F with the coordinate x is much greater than the reciprocal of k_0. Consequently second derivatives of F with respect to x can be neglected in the derived differential equation. The y derivatives of the ambient variables are also neglected. The resulting approximate equation is given by

$$\frac{\partial}{\partial x}\left(\frac{k_0}{\rho}F\right) + \frac{k_0}{\rho}\frac{\partial F}{\partial x} = i\frac{\partial}{\partial y}\left(\frac{1}{\rho}\frac{\partial F}{\partial y}\right)$$
$$+ i\frac{k_0^2}{\rho}(n^2 - 1)F, \quad (3.736)$$

where n, the apparent index of refraction, is an abbreviation for $k_0^{-1}\omega/c$. If the reference wavenumber k_0 is selected to be independent of x, then (3.736) reduces to

$$2k_0 \frac{\partial F}{\partial x} = i\rho \frac{\partial}{\partial y}\left(\frac{1}{\rho}\frac{\partial F}{\partial y}\right) + ik_0^2(n^2 - 1)F. \quad (3.737)$$

Although the computational results must depend on the selection of the reference wavenumber k_0, results for various specific cases tend to indicate that the sensitivity to such a selection is not great.

References

3.1 F.V. Hunt: *Origins of Acoustics* (Yale Univ. Press, New Haven 1978)
3.2 R.B. Lindsay: *Acoustics: Historical and Philosophical Development* (Dowden, Hutchinson, Ross, Stroudsburg 1972)
3.3 C.A. Truesdell III: The theory of aerial sound, Leonhardi Euleri Opera Omnia, Ser. 2 **13**, XIX–XXII (1955)
3.4 J.W.S. Rayleigh: *The Theory of Sound*, Vol. 1 and 2 (Dover, New York 1945), The 1st ed. was published in 1878; the 2nd ed. in 1896
3.5 P.A. Thompson: *Compressible-Fluid Dynamics* (McGraw-Hill, New York 1972)
3.6 Y.C. Fung: *Foundations of Solid Mechanics* (Prentice-Hall, Englewood Cliffs 1965)
3.7 A.H. Shapiro: *The Dynamics and Thermodynamics of Compressible Fluid Flow* (Ronald, New York 1953)
3.8 G.K. Batchelor: *An Introduction to Fluid Dynamics* (Cambridge Univ. Press, Cambridge 1967)
3.9 C. Truesdell: *The Mechanical Foundations of Elasticity and Fluid Mechanics* (Gordon Breach, New York 1966)
3.10 L.D. Landau, E.M. Lifshitz: *Fluid Mechanics* (Pergamon, New York 1959)
3.11 C. Truesdell, R. Toupin: The classical field theories. In: *Principles of Classical Mechanics and Field Theory*, Handbook of Physics, Vol. 3, ed. by S. Flügge (Springer, Berlin, Heidelberg 1960) pp. 226–858
3.12 G.G. Stokes: On the theories of the internal friction of fluids in motion, and of the equilibrium and motion

3.12 of elastic solids, Trans. Camb. Philos. Soc. **8**, 75–102 (1845)
3.13 L.D. Landau, E.M. Lifshitz, L.P. Pitaevskii: *Statistical Physics, Part 1* (Pergamon, New York 1980)
3.14 A.D. Pierce: Aeroacoustic fluid dynamic equations and their energy corollary with O_2 and N_2 relaxation effects included, J. Sound Vibr. **58**, 189–200 (1978)
3.15 J. Fourier: *Analytical Theory of Heat* (Dover, New York 1955), The 1st ed. was published in 1822
3.16 J. Hilsenrath: *Tables of Thermodynamic and Transport Properties of Air* (Pergamon, Oxford 1960)
3.17 W. Sutherland: The viscosity of gases and molecular force, Phil. Mag. **5**(36), 507–531 (1893)
3.18 M. Greenspan: Rotational relaxation in nitrogen, oxygen, and air, J. Acoust. Soc. Am. **31**, 155–160 (1959)
3.19 R.A. Horne: *Marine Chemistry* (Wiley-Interscience, New York 1969)
3.20 J.M.M. Pinkerton: A pulse method for the measurement of ultrasonic absorption in liquids: Results for water, Nature **160**, 128–129 (1947)
3.21 W.D. Wilson: Speed of sound in distilled water as a function of temperature and pressure, J. Acoust. Soc. Am. **31**, 1067–1072 (1959)
3.22 W.D. Wilson: Speed of sound in sea water as a function of temperature, pressure, and salinity, J. Acoust. Soc. Am. **32**, 641–644 (1960)
3.23 A.B. Wood: *A Textbook of Sound*, 2nd edn. (Macmillan, New York 1941)
3.24 A. Mallock: The damping of sound by frothy liquids, Proc. Roy. Soc. Lond. A **84**, 391–395 (1910)
3.25 S.H. Crandall, N.C. Dahl, T.J. Lardner: *An Introduction to the Mechanics of Solids* (McGraw-Hill, New York 1978)
3.26 L.D. Landau, E.M. Lifshitz: *Theory of Elasticity* (Pergamon, New York 1970)
3.27 A.E.H. Love: *The Mathematical Theory of Elasticity* (Dover, New York 1944)
3.28 J.D. Achenbach: *Wave Propagation in Elastic Solids* (North-Holland, Amsterdam 1975)
3.29 M.J.P. Musgrave: *Crystal Acoustics* (Acoust. Soc. Am., Melville 2003)
3.30 A.D. Pierce: *Acoustics: An Introduction to Its Physical Principles and Applications* (Acoust. Soc. Am., Melville 1989)
3.31 P.G. Bergmann: The wave equation in a medium with a variable index of refraction, J. Acoust. Soc. Am. **17**, 329–333 (1946)
3.32 L. Cremer: Über die akustische Grenzschicht von starren Wänden, Arch. Elekt. Uebertrag. **2**, 136–139 (1948), in German
3.33 G.R. Kirchhoff: Über den Einfluß der Wärmeleitung in einem Gase auf die Schallbewegung, Ann. Phys. Chem. Ser. 5 **134**, 177–193 (1878)
3.34 L.S.G. Kovasznay: Turbulence in supersonic flow, J. Aeronaut. Sci. **20**, 657–674 (1953)
3.35 T.Y. Wu: Small perturbations in the unsteady flow of a compressible, viscous, and heat-conducting fluid, J. Math. Phys. **35**, 13–27 (1956)
3.36 R. Courant: *Differential and Integral Calculus*, Vol. 1, 2 (Wiley-Interscience, New York 1937)
3.37 H. Lamb: *Hydrodynamics* (Dover, New York 1945)
3.38 A.H. Shapiro: *Shape and Flow: The Fluid Dynamics of Drag* (Doubleday, Garden City 1961) pp. 59–63
3.39 R.F. Lambert: Wall viscosity and heat conduction losses in rigid tubes, J. Acoust. Soc. Am. **23**, 480–481 (1951)
3.40 S.H. Crandall, D.C. Karnopp, E.F. Kurtz Jr., D.C. Pridmore-Brown: *Dynamics of Mechanical and Electromechanical Systems* (McGraw-Hill, New York 1968)
3.41 A.D. Pierce: Variational principles in acoustic radiation and scattering. In: *Physical Acoustics*, Vol. 22, ed. by A.D. Pierce (Academic, San Diego 1993) pp. 205–217
3.42 M.A. Biot: Theory of propagation of elastic waves in a fluid saturated porous solid. I. Low frequency range, J. Acoust. Soc. Am. **28**, 168–178 (1956)
3.43 R.D. Stoll: *Sediment Acoustics* (Springer, Berlin, Heidelberg 1989)
3.44 J. Bear: *Dynamics of Fluids in Porous Media* (Dover, New York 1988)
3.45 H. Fletcher: Auditory patterns, Rev. Mod. Phys. **12**, 47–65 (1940)
3.46 C.J. Bouwkamp: A contribution to the theory of acoustic radiation, Phillips Res. Rep. **1**, 251–277 (1946)
3.47 H. Helmholtz: Theorie der Luftschwingunden in Röhren mit offenen Enden, J. Reine Angew. Math. **57**, 1–72 (1860)
3.48 R.T. Beyer, S.V. Letcher: *Physical Ultrasonics* (Academic, New York 1969)
3.49 J.J. Markham, R.T. Beyer, R.B. Lindsay: Absorption of sound in fluids, Rev. Mod. Phys. **23**, 353–411 (1951)
3.50 R.B. Lindsay: *Physical Acoustics* (Dowden, Hutchinson, Ross, Stroudsburg 1974)
3.51 J. Meixner: Absorption and dispersion of sound in gases with chemically reacting and excitable components, Ann. Phys. **5**(43), 470–487 (1943)
3.52 J. Meixner: Allgemeine Theorie der Schallabsorption in Gasen un Flüssigkeiten ujnter Berücksichtigung der Transporterscheinungen, Acustica **2**, 101–109 (1952)
3.53 J. Meixner: Flows of fluid media with internal transformations and bulk viscosity, Z. Phys. **131**, 456–469 (1952)
3.54 K.F. Herzfeld, F.O. Rice: Dispersion and absorption of high frequency sound waves, Phys. Rev. **31**, 691–695 (1928)
3.55 L. Hall: The origin of ultrasonic absorption in water, Phys. Rev. **73**, 775–781 (1948)
3.56 J.S. Toll: Causality and the dispersion relations: Logical foundations, Phys. Rev. **104**, 1760–1770 (1956)

3.57 V.L. Ginzberg: Concerning the general relationship between absorption and dispersion of sound waves, Soviet Phys. Acoust. **1**, 32–41 (1955)

3.58 C.W. Horton, Sr.: Dispersion relationships in sediments and sea water, J. Acoust. Soc. Am. **55**, 547–549 (1974)

3.59 L.B. Evans, H.E. Bass, L.C. Sutherland: Atmospheric absorption of sound: Analytical expressions, J. Acoust. Soc. Am. **51**, 1565–1575 (1972)

3.60 H.E. Bass, L.C. Sutherland, A.J. Zuckerwar: Atmospheric absorption of sound: Update, J. Acoust. Soc. Am. **87**, 2019–2021 (1990)

3.61 F.H. Fisher, V.P. Simmons: Sound absorption in sea water, J. Acoust. Soc. Am. **62**, 558–564 (1977)

3.62 G.R. Kirchhoff: *Vorlesungen über mathematische Physik: Mechanik*, 2nd edn. (Teubner, Leipzig 1877)

3.63 J.W.S. Rayleigh: On progressive waves, Proc. Lond. Math. Soc. **9**, 21 (1877)

3.64 L.L. Beranek: Acoustical definitions. In: *American Institute of Physics Handbook*, ed. by D.E. Gray (McGraw-Hill, New York 1972), Sec. 3app. 3-2 to 3-30

3.65 G. Green: On the reflexion and refraction of sound, Trans. Camb. Philos. Soc. **6**, 403–412 (1838)

3.66 E.T. Paris: On the stationary wave method of measuring sound-absorption at normal incidence, Proc. Phys. Soc. Lond. **39**, 269–295 (1927)

3.67 W.M. Hall: An acoustic transmission line for impedance measurements, J. Acoust. Soc. Am. **11**, 140–146 (1939)

3.68 L. Cremer: Theory of the sound blockage of thin walls in the case of oblique incidence, Akust. Z. **7**, 81–104 (1942)

3.69 L.L. Beranek: Acoustical properties of homogeneous, isotropic rigid tiles and flexible blankets, J. Acoust. Soc. Am. **19**, 556–568 (1947)

3.70 L.L. Beranek: *Noise and Vibration Control* (McGraw-Hill, New York 1971)

3.71 B.T. Chu: *Pressure waves generated by addition of heat in a gaseous medium (1955), NACA Technical Note 3411* (National Advisory Committee for Aeronautics, Washington 1955)

3.72 P.J. Westervelt, R.S. Larson: Laser-excited broadside array, J. Acoust. Soc. Am. **54**, 121–122 (1973)

3.73 H. Lamb: *Dynamical Theory of Sound* (Dover, New York 1960)

3.74 M.J. Lighthill: *Waves in Fluids* (Cambridge Univ. Press, Cambridge 1978)

3.75 G.G. Stokes: On the communication of vibration from a vibrating body to a surrounding gas, Philos. Trans. R. Soc. Lond. **158**, 447–463 (1868)

3.76 P.M. Morse: *Vibration and Sound* (Acoust. Soc. Am., Melville 1976)

3.77 P.M. Morse, H. Feshbach: *Methods of Theoretical Physics*, Vol. 1, 2 (McGraw-Hill, New York 1953)

3.78 P.M. Morse, K.U. Ingard: *Theoretical Acoustics* (McGraw-Hill, New York 1968)

3.79 G.R. Kirchhoff: Zur Theorie der Lichtstrahlen, Ann. Phys. Chem. **18**, 663–695 (1883), in German

3.80 L.G. Copley: Fundamental results concerning integral representations in acoustic radiation, J. Acoust. Soc. Am. **44**, 28–32 (1968)

3.81 H.A. Schenck: Improved integral formulation for acoustic radiation problems, J. Acoust. Soc. Am. **44**, 41–58 (1968)

3.82 A.-W. Maue: Zur Formulierung eines allgemeinen Beugungsproblems durch eine Integralgleichung, Z. Phys. **126**, 601–618 (1949), in German

3.83 A.J. Burton, G.F. Miller: The application of integral equation methods to the numerical solution of some exterior boundary value problems, Proc. Roy. Soc. Lond. A **323**, 201–210 (1971)

3.84 H.E. Hartig, C.E. Swanson: Transverse acoustic waves in rigid tubes, Phys. Rev. **54**, 618–626 (1938)

3.85 M. Abramowitz, I.A. Stegun: *Handbook of Mathematical Functions* (Dover, New York 1965)

3.86 F. Karal: The analogous acoustical impedance for discontinuities and constrictions of circular cross-section, J. Acoust. Soc. Am. **25**, 327–334 (1953)

3.87 J.W. Miles: The reflection of sound due to a change of cross section of a circular tube, J. Acoust. Soc. Am. **16**, 14–19 (1944)

3.88 J.W. Miles: The analysis of plane discontinuities in cylindrical tubes, II, J. Acoust. Soc. Am. **17**, 272–284 (1946)

3.89 W.P. Mason: The propagation characteristics of sound tubes and acoustic filters, Phys. Rev. **31**, 283–295 (1928)

3.90 P.S.H. Henry: The tube effect in sound-velocity measurements, Proc. Phys. Soc. Lond. **43**, 340–361 (1931)

3.91 P.O.A.L. Davies: The design of silencers for internal combustion engines, J. Sound Vibr. **1**, 185–201 (1964)

3.92 T.F.W. Embleton: Mufflers. In: *Noise and Vibration Control*, ed. by L.L. Beranek (McGraw-Hill, New York 1971) pp. 362–405

3.93 J.W.S. Rayleigh: On the theory of resonance, Philos. Trans. R. Soc. Lond. **161**, 77–118 (1870)

3.94 E.A. Milne: Sound waves in the atmosphere, Philos. Mag. **6**(42), 96–114 (1921)

3.95 J.B. Keller: Geometrical theory of diffraction. In: *Calculus of Variations and its Applications*, ed. by L.M. Graves (McGraw-Hill, New York 1958) pp. 27–52

3.96 P. Ugincius: Ray acoustics and Fermat's principle in a moving inhomogeneous medium, J. Acoust. Soc. Am. **51**, 1759–1763 (1972)

3.97 D.I. Blokhintzev: The propagation of sound in an inhomogeneous and moving medium, J. Acoust. Soc. Am. **18**, 322–328 (1946)

3.98 A.D. Pierce: Wave equation for sound in fluids with unsteady inhomogeneous flow, J. Acoust. Soc. Am. **87**, 2292–2299 (1990)

3.99 A. Sommerfeld, J. Runge: Anwendung der Vektorrechnung geometrische Optik, Ann. Phys. **4**(35), 277–298 (1911)

3.100 G.S. Heller: Propagation of acoustic discontinuities in an inhomogeneous moving liquid medium, J. Acoust. Soc. Am. **25**, 950–951 (1953)

3.101 J.B. Keller: Geometrical acoustics I: The theory of weak shock waves, J. Appl. Phys. **25**, 938–947 (1954)

3.102 W.D. Hayes: Energy invariant for geometric acoustics in a moving medium, Phys. Fluids **11**, 1654–1656 (1968)

3.103 R.B. Buchal, J.B. Keller: Boundary layer problems in diffraction theory, Commun. Pure Appl. Math. **13**, 85–144 (1960)

3.104 D. Ludwig: Uniform asymptotic expansions at a caustic, Commun. Pure Appl. Math. **19**, 215–250 (1966)

3.105 A. Sommerfeld: *Mathematical Theory of Diffraction* (Springer, Berlin 2004), (R. Nagem, M. Zampolli, G. Sandri, translators)

3.106 H.M. MacDonald: A class of diffraction problems, Proc. Lond. Math. Soc. **14**, 410–427 (1915)

3.107 F.J.W. Whipple: Diffraction by a wedge and kindred problems, Proc. Lond. Math. Soc. **16**, 481–500 (1919)

3.108 T.J.A. Bromwich: Diffraction of waves by a wedge, Proc. Lond. Math. Soc. **14**, 450–463 (1915)

3.109 H.S. Carslaw: Diffraction of waves by a wedge of any angle, Proc. Lond. Math. Soc. **18**, 291–306 (1919)

3.110 F.G. Friedlander: *Sound Pulses* (Cambridge Univ. Press, Cambridge 1958)

3.111 M.A. Biot, I. Tolstoy: Formulation of wave propagation in infinite media by normal coordinates, with an application to diffraction, J. Acoust. Soc. Am. **29**, 381–391 (1957)

3.112 V.A. Fock: *Electromagnetic Diffraction and Propagation Problems* (Pergamon, London 1965)

3.113 F.D. Tappert: The parabolic propagation method, Chap. 5. In: *Wave Propagation and Underwater Acoust.*, ed. by J.B. Keller, J.S. Papadakis (Springer, Berlin, Heidelberg 1977) pp. 224–287

4. Sound Propagation in the Atmosphere

Propagation of sound close to the ground outdoors involves geometric spreading, air absorption, interaction with the ground, barriers, vegetation and refraction associated with wind and temperature gradients. After a brief survey of historical aspects of the study of outdoor sound and its applications, this chapter details the physical principles associated with various propagation effects, reviews data that demonstrate them and methods for predicting them. The discussion is concerned primarily with the relatively short ranges and spectra of interest when predicting and assessing community noise rather than the frequencies and long ranges of concern, for example, in infrasonic global monitoring or used for remote sensing of the atmosphere. Specific phenomena that are discussed include spreading losses, atmospheric absorption, diffraction by barriers and buildings, interaction of sound with the ground (ground waves, surface waves, ground impedance associated with porosity and roughness, and elasticity effects), propagation through shrubs and trees, wind and temperature gradient effects, shadow zones and incoherence due to atmospheric turbulence. The chapter concludes by suggesting a few areas that require further research.

4.1	**A Short History of Outdoor Acoustics**	113
4.2	**Applications of Outdoor Acoustics**	114
4.3	**Spreading Losses**	115
4.4	**Atmospheric Absorption**	116
4.5	**Diffraction and Barriers**	116
	4.5.1 Single-Edge Diffraction	116
	4.5.2 Effects of the Ground on Barrier Performance	118
	4.5.3 Diffraction by Finite-Length Barriers and Buildings	119
4.6	**Ground Effects**	120
	4.6.1 Boundary Conditions at the Ground	120
	4.6.2 Attenuation of Spherical Acoustic Waves over the Ground	120
	4.6.3 Surface Waves	122
	4.6.4 Acoustic Impedance of Ground Surfaces	122
	4.6.5 Effects of Small-Scale Roughness	123
	4.6.6 Examples of Ground Attenuation under Weakly Refracting Conditions	124
	4.6.7 Effects of Ground Elasticity	125
4.7	**Attenuation Through Trees and Foliage**	129
4.8	**Wind and Temperature Gradient Effects on Outdoor Sound**	131
	4.8.1 Inversions and Shadow Zones	131
	4.8.2 Meteorological Classes for Outdoor Sound Propagation	133
	4.8.3 Typical Speed of Sound Profiles	135
	4.8.4 Atmospheric Turbulence Effects	138
4.9	**Concluding Remarks**	142
	4.9.1 Modeling Meteorological and Topographical Effects	142
	4.9.2 Effects of Trees and Tall Vegetation	142
	4.9.3 Low-Frequency Interaction with the Ground	143
	4.9.4 Rough-Sea Effects	143
	4.9.5 Predicting Outdoor Noise	143
References		143

4.1 A Short History of Outdoor Acoustics

Early experiments on outdoor sound were concerned with the speed of sound [4.1]. Sound from explosions and the firing of large guns contains substantial low-frequency content (< 100 Hz) and is able to travel for considerable distances outdoors. The Franciscan friar, Marin Mersenne (1588–1648), suggested timing the interval between seeing the flash and hearing the report of guns fired at a known distance. William Derham (1657–1735), the rector of a small church near London, observed and recorded the influence of wind and temperature on sound speed. Derham noted the difference in the sound of church bells at the same distance over newly fallen snow and over a hard frozen surface.

Before enough was known for the military exploitation of outdoor acoustics, there were many unwitting influences of propagation conditions on the course of battle [4.2]. In June 1666, Samuel Pepys noted that the sounds of a naval engagement between the British and Dutch fleets were heard clearly at some spots but not at others a similar distance away or closer [4.3].

During the First World War, acoustic shadow zones, similar to those observed by Pepys, were observed during the battle of Antwerp. Observers also noted that battle sounds from France only reached England during the summer months whereas they were best heard in Germany during the winter. After the war there was great interest in these observations among the scientific community. Large amounts of ammunition were detonated throughout England and the public was asked to listen for sounds of explosions. Despite the continuing interest in atmospheric acoustics after World War 1, the advent of the submarine encouraged greater efforts in underwater acoustics research during and after World War 2.

4.2 Applications of Outdoor Acoustics

Although much current interest in sound propagation in the atmosphere relates to the prediction and control of noise from land and air transport and from industrial sources, outdoor acoustics has continued to have extensive military applications in source acquisition, ranging and identification [4.4]. Acoustic disturbances in the atmosphere give rise to solid particle motion in porous ground, induced by local pressure variations as well as air-particle motion in the pores. There is a distinction between the seismic disturbances associated with direct seismic excitation of the ground and solid-particle motion in the ground induced by airborne sounds. This has enabled the design of systems that distinguish between airborne and ground-borne sources and the application of acoustical techniques to the detection of buried land mines [4.5]. The many other applications of studies of outdoor sound propagation include aspects of animal bioacoustics [4.6] and acoustic remote sounding of the atmosphere [4.7].

Atmospheric sound propagation close to the ground is sensitive to the acoustical properties of the ground surface as well as to meteorological conditions. Most natural ground surfaces are porous. The surface porosity allows sound to penetrate and hence to be absorbed and delayed through friction and thermal exchanges. There is interference between sound traveling directly between source and receiver and sound reflected from the ground. This interference, known as the ground effect [4.8, 9], is similar to the Lloyd's mirror effect in optics but is not directly analogous. Usually, the propagation of light may be described by rays. At the lower end of the audible frequency range (20–500 Hz) and near grazing incidence to the ground, the consequences of the curvature of the expanding sound waves from a finite source are significant. Consequently, ray-based modeling is not appropriate and it is necessary to use full-wave techniques. Moreover, there are few outdoor surfaces that are mirror-like to incident sound waves. Most ground surfaces cause changes in phase as well as amplitude during reflection. Apart from the relevance to outdoor noise prediction, the sensitivity of outdoor sound propagation to ground surface properties has suggested acoustical techniques for determining soil physical properties such as porosity and air permeability [4.10–12]. These are relatively noninvasive compared with other methods.

The last decade has seen considerable advances in numerical and analytical methods for outdoor sound prediction [4.13]. Details of these are beyond the scope of this work but many are described in the excellent text by *Salomons* [4.14]; a review of recent progress may be found in *Berengier* et al. [4.15]. Among the numerical methods borrowed and adapted from underwater acoustics are the fast field program (FFP) and that based on the parabolic equation (PE). The more important advances are based predominantly on the parabolic equation method since it enables predictions that allow for changes in atmospheric and ground conditions with range whereas the FFP intrinsically does not. At present methods for predicting outdoor noise are undergoing considerable assessment and change in Europe as a result of a recent European Commission (EC) directive [4.16] and the associated requirements for noise mapping. Perhaps the most sophisticated scheme for this purpose is NORD2000 [4.17]. A European project HARMONOISE [4.18] is developing a comprehensive source-independent scheme for outdoor sound prediction. As in NORD2000 various relatively simple formulae, predicting the effect of topography for example, are being derived and tested against numerical predictions.

4.3 Spreading Losses

Distance alone will result in wavefront spreading. In the simplest case of a sound source radiating equally in all directions, the intensity I [W^{-2}] at a distance rm from the source of power P [W], is given by

$$I = \frac{P}{4\pi r^2}, \quad (4.1)$$

This represents the power per unit area on a spherical wavefront of radius r. In logarithmic form the relationship between sound pressure level L_p and sound power L_W, may be written

$$L_p = L_W - 20 \log r - 11 \, \text{dB}. \quad (4.2)$$

From a point sound source, this means a reduction of $20 \log 2$ dB, i.e., 6 dB, per distance doubling in all directions (a point source is omnidirectional). Most sources appear to be point sources when the receiver is at a sufficient distance from them. If the source is directional then (4.2) is modified by inclusion of the directivity index DI.

$$L_p = L_W + DI - 20 \log r - 11 \, \text{dB}. \quad (4.3)$$

The directivity index is $10 \log DF$ dB where DF is the directivity factor, given by the ratio of the actual intensity in a given direction to the intensity of an omnidirectional source of the same power output. Such directivity is either inherent or location-induced. A simple case of location-induced directivity arises if the point source, which would usually create spherical wavefronts of sound, is placed on a perfectly reflecting flat plane. Radiation from the source is thereby restricted to a hemisphere. The directivity factor for a point source on a perfectly reflecting plane is 2 and the directivity index is 3 dB. For a point source at the junction of a vertical perfectly reflecting wall with a horizontal perfectly reflecting plane, the directivity factor is 4 and the directivity index is 6 dB. It should be noted that these adjustments ignore phase effects and assume incoherent reflection [4.19].

From an infinite line source, the wavefronts are cylindrical, so wavefront spreading means a reduction of 3 dB per distance doubling. Highway traffic may be approximated by a line of incoherent point sources on an acoustically hard surface. If a line source of length l consists of contiguous omnidirectional incoherent elements of length dx and source strength $P \, dx$, the intensity at a location halfway along the line and at a perpendicular distance d from it, so that $dx = r d\theta / \cos \theta$, where r is the distance from any element at angle θ from the perpendicular, is given by

$$I = \int_{-l/2}^{l/2} \frac{P}{2\pi r^2} \, dx = \frac{P}{2\pi d} \left[2 \tan^{-1} \left(\frac{l}{2d} \right) \right],$$

This results in

$$L_p = L_W - 10 \log d - 8 + 10 \log \left[2 \tan^{-1} \left(\frac{l}{2d} \right) \right] \text{dB}. \quad (4.4)$$

Figure 4.1 shows that the attenuation due to wavefront spreading from the finite line source behaves as that from an infinite line at distances much less than the length of the source and as that from a point source at distances greater than the length of the source.

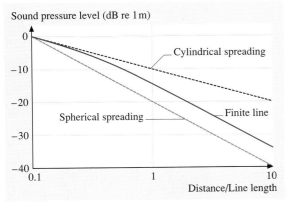

Fig. 4.1 Comparison of attenuation due to geometrical spreading from point, infinite line and finite line sources

4.4 Atmospheric Absorption

A proportion of sound energy is converted to heat as it travels through the air. There are heat conduction, shear viscosity and molecular relaxation losses [4.20]. The resulting air absorption becomes significant at high frequencies and at long range so air acts as a low-pass filter at long range. For a plane wave, the pressure p at distance x from a position where the pressure is p_0 is given by

$$p = p_0 e^{-\alpha x/2} \quad (4.5)$$

The attenuation coefficient α for air absorption depends on frequency, humidity, temperature and pressure and may be calculated using (4.6) through (4.8) [4.21].

$$\alpha = f^2 \left[\left(\frac{1.84 \times 10^{-11}}{\left(\frac{T_0}{T}\right)^{\frac{1}{2}} \frac{p_s}{p_0}} \right) \right.$$
$$+ \left(\frac{T_0}{T}\right)^{2.5} \left(\frac{0.10680 \, e^{-3352/T} f_{r,N}}{f^2 + f_{r,N}^2} \right.$$
$$\left. \left. + \frac{0.01278 \, e^{-2239.1/T} f_{r,O}}{f^2 + f_{r,O}^2} \right) \right] \frac{\text{Np}}{\text{m} \cdot \text{atm}}, \quad (4.6)$$

where $f_{r,N}$ and $f_{r,O}$ are relaxation frequencies associated with the vibration of nitrogen and oxygen molecules respectively and are given by:

$$f_{r,N} = \frac{p_s}{p_{s0}} \left(\frac{T_0}{T}\right)^{\frac{1}{2}}$$

$$\times \left(9 + 280 H e^{-4.17\left[(T_0/T)^{1/3} - 1\right]} \right), \quad (4.7)$$

$$f_{r,O} = \frac{p_s}{p_{s0}} \left(24.0 + 4.04 \times 10^4 H \frac{0.02 + H}{0.391 + H} \right), \quad (4.8)$$

where f is the frequency. T is the absolute temperature of the atmosphere in degrees Kelvin, $T_0 = 293.15$ K is the reference value of T (20 °C), H is the percentage molar concentration of water vapor in the atmosphere $= \rho_{\text{sat}} r_h p_0 / p_s$, r_h is the relative humidity (%); p_s is the local atmospheric pressure and p_0 is the reference atmospheric pressure (1 atm $= 1.01325 \times 10^5$ Pa); $\rho_{\text{sat}} = 10^{C_{\text{sat}}}$, where $C_{\text{sat}} = -6.8346(T_0/T)^{1.261} + 4.6151$. These formulae give estimates of the absorption of pure tones to an accuracy of $\pm 10\%$ for $0.05 < H < 5$, $253 < T < 323$, $p_0 < 200$ kPa.

Outdoor air absorption varies through the day and the year [4.22, 23].

Absolute humidity H is an important factor in the diurnal variation and usually peaks in the afternoon. Usually the diurnal variations are greatest in the summer. It should be noted that the use of (arithmetic) mean values of atmospheric absorption may lead to overestimates of attenuation when attempting to establish worst-case exposures for the purposes of environmental noise impact assessment. Investigations of local climate statistics, say hourly means over one year, should lead to more accurate estimates of the lowest absorption values.

4.5 Diffraction and Barriers

Purpose-built noise barriers have become a very common feature of the urban landscape of Europe, the Far East and America. In the USA, over 1200 miles of noise barriers were constructed in 2001 alone. The majority of barriers are installed in the vicinity of transportation and industrial noise sources to shield nearby residential properties. Noise barriers are cost effective only for the protection of large areas including several buildings and are rarely used for the protection of individual properties. Noise barriers of usual height are generally ineffective in protecting the upper levels of multi-storey dwellings. In the past two decades environmental noise barriers have become the subject of extensive studies, the results of which have been consolidated in the form of national and international standards and prediction models [4.24–26]. Extensive guides to the acoustic and visual design of noise barriers are available [4.27, 28]. Some issues remain to be resolved relating to the degradation of the noise-barrier performance in the presence of wind and temperature gradients, the influence of localized atmospheric turbulence, temporal effects from moving traffic, the influence of local vegetation, the aesthetic quality of barriers and their environmental impact.

4.5.1 Single-Edge Diffraction

A noise barrier works by blocking the direct path from the noise source to the receiver. The noise then reaches the receiver only via diffraction around the barrier edges. The calculation of barrier attenuation is therefore mainly dependent on the solution of the diffraction problem. Exact integral solutions of the diffraction problem were available as early as the late 19th [4.29] and early 20th century [4.30]. For practical calculations however it is necessary to use approximations to the exact solutions. Usually this involves assuming that the source and receiver are more than a wavelength from the barrier and the receiver is in the shadow zone, which is valid in almost all applications of noise barriers. The Kirchhoff–Fresnel approximation [4.31], in terms of the Fresnel numbers for thin rigid screens (4.10), and the geometrical theory of diffraction [4.32] for wedges and thick barriers have been used for deriving practical formulae for barrier calculations. For a rigid wedge barrier, the solution provided by *Hadden* and *Pierce* [4.33] is relatively easy to calculate and highly accurate. A line integral solution, based on the Kirchhoff–Fresnel approximation [4.34] describes the diffracted pressure from a distribution of sources along the barrier edge and has been extended to deal with barriers with jagged edges [4.35]. There is also a time-domain model [4.36].

As long as the transmission loss through the barrier material is sufficiently high, the performance of a barrier is dictated by the geometry (Fig. 4.2).

The total sound field in the vicinity of a semi-infinite half plane depends on the relative position of source, receiver, and the thin plane. The total sound field p_T in each of three regions shown in Fig. 4.2 is given as follows:

In front of the barrier: $p_T = p_i + p_r + p_d$, (4.9a)

Above the barrier: $p_T = p_i + p_d$, (4.9b)

In the shadow zone: $p_T = p_d$. (4.9c)

The Fresnel numbers of the source and image source are denoted, respectively, by N_1 and N_2, and are defined as follows:

$$N_1 = \frac{R' - R_1}{\lambda/2} = \frac{k}{\pi}(R' - R_1),$$ (4.10a)

and $$N_2 = \frac{R' - R_2}{\lambda/2} = \frac{k}{\pi}(R' - R_2).$$ (4.10b)

where $R' = r_s + r_r$ is the shortest source–edge–receiver path.

The attenuation (Att) of the screen, sometimes known as the insertion loss IL, is often used to assess the acoustics performance of the barrier. It is defined as follows,

$$\text{Att} = \text{IL} = 20 \lg \left(\left| \frac{p_w}{p_{w/o}} \right| \right) \text{dB},$$ (4.11)

where p_w and $p_{w/o}$ is the total sound field with or without the presence of the barrier. Note that the barrier attenuation is equal to the insertion loss only in the absence of ground effect.

Maekawa [4.37] has provided a chart that expresses the attenuation of a thin rigid barrier based on the Fresnel number N_1 associated with the source. The chart was derived empirically from extensive laboratory experimental data but use of the Fresnel number was suggested by the Kirchhoff–Fresnel diffraction theory. Maekawa's chart extends into the illuminated zone where N_1 is taken to be negative. Maekawa's chart has been proved to be very successful and has become the de facto standard empirical method for barrier calculations. Many of the barrier calculation methods embodied in national and international standards [4.24, 38] stem from this chart. There have been many attempts to fit the chart with simple formulae. One of the simplest formulae [4.39] is

$$\text{Att} = 10 \lg(3 + 20N) \text{dB}.$$ (4.12)

The *Maekawa* curve can be represented mathematically by [4.40]

$$\text{Att} = 5 + 20 \lg \frac{\sqrt{2\pi N_1}}{\tanh \sqrt{2\pi N_1}}.$$ (4.13)

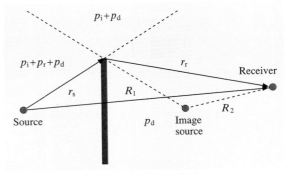

Fig. 4.2 Diffraction of sound by a thin barrier

An improved version of this result, using both Fresnel Numbers (4.10), is [4.41]

$$\text{Att} = \text{Att}_s + \text{Att}_b + \text{Att}_{sb} + \text{Att}_{sp}, \quad (4.14a)$$

where
$$\text{Att}_s = 20\lg\frac{\sqrt{2\pi N_1}}{\tanh\sqrt{2\pi N_1}} - 1, \quad (4.14b)$$

$$\text{Att}_b = 20\lg\left[1+\tanh\left(0.6\lg\frac{N_2}{N_1}\right)\right], \quad (4.14c)$$

$$\text{Att}_{sb} = \left(6\tanh\sqrt{N_2} - 2 - \text{Att}_b\right) \times \left(1 - \tanh\sqrt{10N_1}\right), \quad (4.14d)$$

$$\text{Att}_{sp} = -10\lg\frac{1}{(R'/R_1)^2 + (R'/R_1)}. \quad (4.14e)$$

The term Att_s is a function of N_1, which is a measure of the relative position of the receiver from the source. The second term depends on the ratio of N_2/N_1, which depends on the proximity of either the source or the receiver to the half plane. The third term is only significant when N_1 is small and depends on the proximity of the receiver to the shadow boundary. The last term, a function of the ratio R'/R_1, accounts for the diffraction effect due to spherical incident waves.

4.5.2 Effects of the Ground on Barrier Performance

Equations (4.12–4.14) only predict the amplitude of sound and do not include wave interference effects. Such interference effects result from the contributions from different diffracted wave paths in the presence of ground (see also Sect. 4.6).

Consider a source S_g located at the left side of the barrier, a receiver R_g at the right side of the barrier and E is the diffraction point on the barrier edge (see Fig. 4.3).

The sound reflected from the ground surface can be described by an image of the source S_i. On the receiver side, sound waves will also be reflected from the ground. This effect can be considered in terms of an image of the receiver R_i. The pressure at the receiver is the sum of four terms that correspond to the sound paths $S_g E R_g$, $S_i E R_g$, $S_g E R_i$ and $S_i E R_i$. If the surface is a perfectly reflecting ground, the total sound field is the sum of the diffracted fields of these four paths,

$$P_T = P_1 + P_2 + P_3 + P_4, \quad (4.15)$$

where

$$P_1 = P(S_g, R_g, E),$$
$$P_2 = P(S_i, R_g, E),$$
$$P_3 = P(S_g, R_i, E),$$
$$P_4 = P(S_i, R_i, E).$$

$P(S, R, E)$ is the diffracted sound field due to a thin barrier for given positions of source S, receiver R and the point of diffraction at the barrier edge E. If the ground has finite impedance (such as grass or a porous road surface) then the pressure corresponding to rays reflected from these surfaces should be multiplied by the appropriate spherical wave reflection coefficient(s) to allow for the change in phase and amplitude of the wave on reflection as follows,

$$P_T = P_1 + Q_s P_2 + Q_R P_3 + Q_s Q_R P_4, \quad (4.16)$$

where Q_s and Q_R are the spherical wave reflection coefficients for the source and receiver side respectively. The spherical wave reflection coefficients can be calculated for different types of ground surfaces and source/receiver geometries (Sect. 4.6).

Usually, for a given source and receiver position, the acoustic performance of the barrier on the ground is assessed by use of either the excess attenuation (EA) or the insertion loss (IL). They are defined as follows,

$$\text{EA} = SPL_f - SPL_b, \quad (4.17)$$
$$\text{IL} = SPL_g - SPL_b, \quad (4.18)$$

where SPL_f is the free field noise level, SPL_g is the noise level with the ground present and SPL_b is the noise level with the barrier and ground present. Note that, in the absence of a reflecting ground, the numerical value of EA (which was called Att previously) is the same as IL. If the calculation is carried out in terms of amplitude only, then the attenuation Att_n for each sound path can be directly determined from the appropriate Fresnel number F_n for that path. The excess attenuation

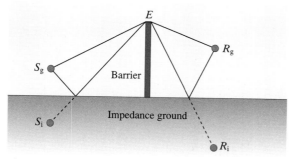

Fig. 4.3 Diffraction by a barrier on impedance ground

of the barrier on a rigid ground is then given by

$$A_T = 10 \lg \left(10^{-\left|\frac{Att_1}{10}\right|} + 10^{-\left|\frac{Att_2}{10}\right|} \right.$$
$$\left. + 10^{-\left|\frac{Att_3}{10}\right|} + 10^{-\left|\frac{Att_4}{10}\right|} \right). \quad (4.19)$$

The attenuation for each path can either be calculated by empirical or analytical formulae depending on the complexity of the model and the required accuracy.

A modified form of the empirical formula for the calculation of barrier attenuation is [4.24]

$$\text{IL} = 10 \log_{10}\left[3 + \left(C_2 \frac{\delta_1}{\lambda}\right) C_3 K_{\text{met}}\right], \quad (4.20)$$

where $C_2 = 20$ and includes the effect of ground reflections; $C_2 = 40$ if ground reflections are taken into account elsewhere; C_3 is a factor to take into account a double diffraction or finite barrier effect, $C_3 = 1$ for a single diffraction and $\delta_1 = (r_s + r_r) - R_1$ (Fig. 4.2). The C_3 expression for double diffraction is given later.

The term K_{met} in (4.15) is a correction factor for average downwind meteorological effects, and is given by

$$K_{\text{met}} = e^{-\frac{1}{2000}\sqrt{\frac{r_s r_r r_o}{2\delta_1}}}$$

for $\delta_1 > 0$ and $K_{\text{met}} = 1$ for $\delta_1 \leq 0$.

The formula reduces to the simple formula (4.12) when the barrier is thin, there is no ground and if meteorological effects are ignored.

There is a simple approach capable of modeling wave effects in which the phase of the wave at the receiver is calculated from the path length via the top of the screen, assuming a phase change in the diffracted wave of $\pi/4$ [4.42]. This phase change is assumed to be constant for all source–barrier–receiver geometries. The diffracted wave, for example, for the path $S_g E R_g$ would thus be given by

$$P_1 = \text{Att}_1 \, e^{-i\left[k(r_0 + r_r) + \pi/4\right]}. \quad (4.21)$$

This approach provides a reasonable approximation for the many situations of interest where source and receiver are many wavelengths from the barrier and the receiver is in the shadow zone.

For a thick barrier of width w, the International Standards Organization (ISO) standard ISO 9613-2 [4.24] provides the following form of correction factor C_3 for use in (4.20):

$$C_3 = \frac{\left[1 + \left(\frac{5\lambda}{w}\right)^2\right]}{\left[\frac{1}{3} + \left(\frac{5\lambda}{w}\right)^2\right]},$$

where for double diffraction, $\delta_1 = (r_s + r_r + w) - R_1$.

Note that this empirical method is for rigid barriers of finite thickness and does not take absorptive surfaces into account.

4.5.3 Diffraction by Finite-Length Barriers and Buildings

All noise barriers have finite length and for certain conditions sound diffracting around the vertical ends of the barrier may be significant. This will also be the case for sound diffracting around buildings. Figure 4.4 shows eight diffracted ray paths contributing to the total field behind a finite-length barrier situated on finite-impedance ground. In addition to the four normal ray paths diffracted at the top edge of the barrier (Fig. 4.3), four more diffracted ray paths result from the vertical edges – two ray paths from each one. The two rays at either side are, respectively, the direct diffracted ray and the diffracted–reflected ray. Strictly, there are two further ray paths at each side which involve two reflections at the ground as well as diffraction at the vertical edge but usually these are neglected.

The reflection angles of the two diffracted–reflected rays are independent of the barrier position. They will either reflect at the source side or on the receiver side of the barrier, which are dependent on the relative positions of the source, receiver and barrier. The total field is given by

$$P_T = P_1 + Q_s P_2 + Q_R P_3 + Q_s Q_R P_4 + P_5$$
$$+ Q_R P_6 + P_7 + Q_R P_8, \quad (4.22)$$

where P_1–P_4 are those given earlier for the diffraction at the top edge of the barrier.

Although accurate diffraction formulas may be used to compute P_i ($i = 1, \ldots, 8$), a simpler approach is to assume that each diffracted ray has a constant phase shift of $\pi/4$ regardless of the position of source, receiver and diffraction point.

To predict the attenuation due to a single building, the double diffraction calculations mentioned earlier could be used. For a source or receiver situated in a built-up area, ISO 9613-2 [4.24] proposes an empirical method for calculating the combined effects of screening and multiple reflections. The net attenuation A_{build} dB (< 10 dB) is given by

$$A_{build} = A_{build,1} + A_{build,2} \qquad (4.23)$$
$$A_{build,1} = 0.1 B d_0,$$
$$A_{build,2} = -10 \log \left[1 - (p/100)\right],$$

where B is the area density ratio of buildings (total plan area/total ground area) and d_0 is the length of the refracted path from source to receiver that passes through buildings. $A_{build,2}$ is intended to be used only where there are well defined but discontinuous rows of buildings near to a road or railway and, p is the percentage of the length of facades relative to the total length of the road or railway. As with barrier attenuation, the attenuation due to buildings is to be included only when it is predicted to be greater than that due to ground effect. The ISO scheme offers also a frequency dependent attenuation coefficient (dB/m) for propagation of industrial noise through an array of buildings on an industrial site. It should be

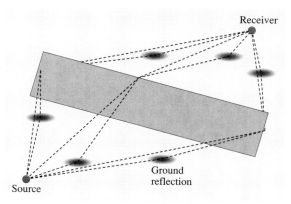

Fig. 4.4 Ray paths around a finite-length barrier or building on the ground

noted that there are considerable current efforts devoted to studying sound propagation through buildings and along streets but this work is not surveyed here.

4.6 Ground Effects

Ground effects (for elevated source and receiver) are the result of interference between sound traveling directly from source to receiver and sound reflected from the ground when both source and receiver are close to the ground. Sometimes the phenomenon is called ground absorption but, since the interaction of outdoor sound with the ground involves interference, there can be enhancement associated with constructive interference as well as attenuation resulting from destructive interference. Close to ground such as nonporous concrete or asphalt, the sound pressure is doubled more or less over a wide range of audible frequencies. Such ground surfaces are described as *acoustically hard*. Over porous surfaces, such as soil, sand and snow, enhancement tends to occur at low frequencies since the longer the sound wave the less able it is to penetrate the pores. However, at higher frequencies, sound is able to penetrate porous ground so the surface reflection is changed in amplitude and phase. The presence of vegetation tends to make the surface layer of ground including the root zone more porous. Snow is significantly more porous than soil and sand. The layer of partly decayed matter on the floor of a forest is also highly porous. Porous ground surfaces are sometimes called *acoustically soft*.

4.6.1 Boundary Conditions at the Ground

For most environmental noise predictions porous ground may be considered to be rigid, so only one wave type, i.e., the wave penetrating the pores, need be considered. With this assumption, the speed of sound in the ground (c_1) is typically much smaller than that (c) in the air, i.e., $c \gg c_1$. The propagation of sound in the air gaps between the solid particles of which the ground is composed, is impeded by viscous friction. This in turn means that the index of refraction in the ground $n_1 = c/c_1 \gg 1$ and any incoming sound ray is refracted towards the normal as it propagates from air into the ground. This type of ground surface is called *locally reacting* because the air–ground interaction is independent of the angle of incidence of the incoming waves. The acoustical properties of locally reacting ground may be represented simply by its relative normal-incidence surface impedance (Z), or its inverse (the relative admittance β) and the ground is said to form a finite-impedance boundary. A perfectly hard ground has infinite impedance (zero admittance). A perfectly soft ground has zero impedance (infinite admittance). If the ground is not locally reacting, i.e., it is externally reacting, the impedance condition is replaced by two separate conditions governing the continuity of pressure and the continuity of the normal component of air particle velocity.

4.6.2 Attenuation of Spherical Acoustic Waves over the Ground

The idealized case of a point (omnidirectional) source of sound at height z_s and a receiver at height z, separated by a horizontal distance r above a finite-impedance plane (admittance β) is shown in Fig. 4.5.

Between source and receiver, a direct sound path of length R_1 and a ground-reflected path of length R_2 are identified. With the restrictions of long range ($r \approx R_2$), high frequency ($kr \gg 1$, $k(z+z_s) \gg 1$, where $k = \omega/c$ and $\omega = 2\pi f$, f being frequency) and with both the source and receiver located close ($r \gg z + z_s$) to a relatively hard ground surface ($|\beta|^{big} \ll 1$), the total sound field at (x, y, z) can be determined from

$$p(x, y, z) = \frac{e^{-ikR_1}}{4\pi R_1} + \frac{e^{-ikR_2}}{4\pi R_2} + \Phi_p + \phi_s, \quad (4.24)$$

where

$$\Phi_p \approx 2i\sqrt{\pi} \left(\tfrac{1}{2}kR_2\right)^{1/2} \beta e^{-w^2} \text{erfc}(iw) \frac{e^{-ikR_2}}{4\pi R_2}. \quad (4.25)$$

and w, sometimes called the numerical distance, is given by

$$w \approx \tfrac{1}{2}(1-i)\sqrt{kR_2}(\cos\theta + \beta). \quad (4.26)$$

ϕ_s represents a surface wave and is small compared with Φ_p under most circumstances. It is included in careful computations of the complementary error function erfc(x) [4.43]. In all of the above a time dependence of $e^{i\omega t}$ is understood.

After rearrangement, the sound field due to a point monopole source above a locally reacting ground becomes

$$p(x, y, z) = \frac{e^{-ikR_1}}{4\pi R_1} + \left[R_p + (1-R_p)F(w)\right] \times \frac{e^{-ikR_2}}{4\pi R_2}, \quad (4.27)$$

where $F(w)$, sometimes called the boundary loss factor, is given by

$$F(w) = 1 - i\sqrt{\pi}\, w \exp(-w^2)\text{erfc}(iw) \quad (4.28)$$

and describes the interaction of a spherical wavefront with a ground of finite impedance [4.44]. The term in the square bracket of (4.27) may be interpreted as the spherical wave reflection coefficient

$$Q = R_p + (1 - R_p)F(w), \quad (4.29)$$

which can be seen to involve the plane wave reflection coefficient R_p and a correction. The second term of Q allows for the fact that the wavefronts are spherical rather than plane. Its contribution to the total sound field has been called the *ground wave*, in analogy with the corresponding term in the theory of amplitude-modulated (AM) radio reception [4.45]. It represents a contribution from the vicinity of the image of the source in the ground plane. If the wavefront is plane ($R_2 \to \infty$) then $|w| \to \infty$ and $F \to 0$. If the surface is acoustically hard, then $|\beta| \to 0$, which implies $|w| \to 0$ and $F \to 1$. If $\beta = 0$, corresponding to a perfect reflector, the sound field consists of two terms: a direct-wave contribution and a wave from the image source corresponding to specular reflection and the total sound field may be written

$$p(x, y, z) = \frac{e^{-ikR_1}}{4\pi R_1} + \frac{e^{-ikR_2}}{4\pi R_2}.$$

This has a first minimum corresponding to destructive interference between the direct and ground-reflected components when $k(R_2 - R_1) = \pi$, or $f = c/2(R_2 - R_1)$. Normally, for source and receiver close to the ground, this destructive interference is at too high a frequency to be of importance in outdoor sound prediction. The higher the frequency of the first minimum in the ground effect, the more likely that it will be destroyed by turbulence (Sect. 4.8).

For $|\beta| \ll 1$ but at grazing incidence ($\theta = \pi/2$), so that $R_p = -1$ and

$$p(x, y, z) = 2F(w)e^{-ikr}/r, \quad (4.30)$$

the numerical distance w is given by

$$w = \tfrac{1}{2}(1-i)\beta\sqrt{kr}. \quad (4.31)$$

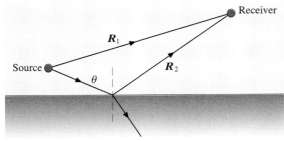

Fig. 4.5 Sound propagation from a point source to a receiver above a ground surface

If the plane wave reflection coefficient had been used instead of the spherical wave reflection coefficient for grazing incidence, it would have led to the prediction of zero sound field when both source and receiver are on the ground. Equation (4.22) is the most widely used analytical solution for predicting sound field above a locally reacting ground in a homogeneous atmosphere. There are many other accurate asymptotic and numerical solutions available but no significant numerical differences between various predictions have been revealed for practical geometries and typical outdoor ground surfaces.

4.6.3 Surface Waves

Although, numerically, it is part of the calculation of the complementary error function, physically the surface wave is a separate contribution propagating close to and parallel to the porous ground surface. It produces elliptical motion of air particles as a result of combining motion parallel to the surface with that normal to the surface in and out of the pores. The surface wave decays with the inverse square root of range rather than inversely with range as is true for other components. At grazing incidence on an impedance plane, with normalized admittance $\beta = \beta_r + i\beta_x$, the condition for the existence of the surface wave is

$$\frac{1}{\beta_x^2} > \frac{1}{\beta_r^2} + 1. \tag{4.32}$$

For a surface with large impedance, i.e., where $|\beta| \to 0$, the condition is simply that the imaginary part of the ground impedance (the reactance) is greater than the real part (the resistance). This type of surface impedance is possessed by cellular or lattice layers placed on smooth, hard surfaces. Surface waves due to a point source have been generated and studied extensively over such surfaces in laboratory experiments [4.46–48]. The outdoor ground type most likely to produce a measurable surface wave is a thin layer of snow over a frozen ground. By using blank pistol shots in experiments over snow, Albert [4.49] has confirmed the existence of the predicted type of surface wave outdoors.

There are some cases where it is not possible to model the ground surface as an impedance plane, i.e., n_1 is not sufficiently high to warrant the assumption that $n_1 \gg 1$. In this case, the refraction of the sound wave depends on the angle of incidence as sound enters the porous medium. This means that the apparent impedance depends not only on the physical properties of the ground surface but also, critically, on the angle of incidence. It is possible to define an *effective* admittance, β_e defined by:

$$\beta_e = \varsigma_1 \sqrt{n_1^2 - \sin^2 \theta}, \tag{4.33}$$

where $\varsigma_1 = \rho/\rho_1$ is the ratio of air density to the (complex) density of the rigid, porous ground.

This allows use of the same results as before but with the admittance replaced by the effective admittance for a semi-infinite non-locally reacting ground [4.50]. There are some situations where there is a highly porous surface layer above a relatively nonporous substrate. This is the case with forest floors consisting of partly decomposed litter layers above soil with a relatively high flow resistivity, with freshly fallen snow on a hard ground or with porous asphalt laid on a nonporous substrate. The minimum depth d_m for such a multiple layer ground to be treated as a semi-infinite externally reacting ground to satisfy the above condition depends on the acoustical properties of the ground and the angle of incidence. We can consider two limiting cases. If the propagation constant within the surface layer is denoted by $k_1 = k_r - ik_x$, and for normal incidence, where $\theta = 0$, the required condition is simply

$$d_m > 6/k_x. \tag{4.34}$$

For grazing incidence where $\theta = \pi/2$, the required condition is

$$d_m > 6 \left(\sqrt{\frac{(k_r^2 - k_x^2 - 1)^2}{4} + k_r^2 k_x^2} - \frac{k_r^2 - k_x^2 - 1}{2} \right)^{\frac{1}{2}}. \tag{4.35}$$

It is possible to derive an expression for the effective admittance of ground with an arbitrary number of layers. However, sound waves can seldom penetrate more than a few centimeters in most outdoor ground surfaces. Lower layers contribute little to the total sound field above the ground and, normally, consideration of ground structures consisting of more than two layers is not required for predicting outdoor sound. Nevertheless, the assumption of a double-layer structure [4.50] has been found to enable improved agreement with data obtained over snow [4.51]. It has been shown rigorously that, in cases where the surface impedance depends on angle, replacing the normal surface impedance by the grazing incidence value is sufficiently accurate for predicting outdoor sound [4.52].

4.6.4 Acoustic Impedance of Ground Surfaces

For most applications of outdoor acoustics, porous ground may be considered to have a rigid, rather than elastic, frame. The most important characteristic of a porous ground surface that affects its acoustical character is its flow resistivity. Soil scientists tend to refer to air permeability, which is proportional to the inverse of flow resistivity. Flow resistivity is a measure of the ease with which air can move into and out of the ground. It represents the ratio of the applied pressure gradient to the induced volume flow rate per unit thickness of material and has units of $\mathrm{Pa\,s\,m^{-2}}$. If the ground surface has a high flow resistivity, it means that it is difficult for air to flow through the surface. Flow resistivity increases as porosity decreases. For example, conventional hot-rolled asphalt has a very high flow resistivity ($10\,000\,000\,\mathrm{Pa\,s\,m^{-2}}$) and negligible porosity, whereas drainage asphalt has a volume porosity of up to 0.25 and a relatively low flow resistivity ($<30\,000\,\mathrm{Pa\,s\,m^{-2}}$). Soils have volume porosities of between 10% and 40%. A wet compacted silt may have a porosity as low as 0.1 and a rather high flow resistivity ($4\,000\,000\,\mathrm{Pa\,s\,m^{-2}}$). Newly fallen snow has a porosity of around 60% and a fairly low flow resistivity ($<10\,000\,\mathrm{Pa\,s\,m^{-2}}$). The thickness of the surface porous layer and whether or not it has acoustically hard substrate are also important factors.

A widely used model [4.53] for the acoustical properties of outdoor surfaces involves a single parameter, the *effective* flow resistivity σ_e, to characterize the ground. According to this single-parameter model, the propagation constant k and normalized characteristic impedance Z are given by

$$\frac{k}{k_1} = \left[1 + 0.0978(f/\sigma_e)^{-0.700} - \mathrm{i}0.189(f/\sigma_e)^{-0.595}\right], \quad (4.36a)$$

$$Z = \frac{\rho_1 c_1}{\rho c} = 1 + 0.0571(f/\sigma_e)^{-0.754} - \mathrm{i}0.087(f/\sigma_e)^{-0.732}. \quad (4.36b)$$

This model may be used for a locally reacting ground as well as for an extended reaction surface. On the other hand, there is considerable evidence that (4.36a) tends to overestimate the attenuation within a porous material with high flow resistivity. On occasion better agreement with grassland data has been obtained by assuming that the ground surface is that of a hard-backed porous layer of thickness d [4.54] such that the surface impedance Z_S is given by

$$Z_S = Z \coth(\mathrm{i}kd). \quad (4.36c)$$

A model based on an exponential change in porosity with depth has been suggested [4.55, 56]. Although this model is suitable only for high flow resistivity, i.e., a locally reacting ground, it has enabled better agreement with measured data for the acoustical properties of many outdoor ground surfaces than equation (4.36b). The two adjustable parameters are the effective flow resistivity (σ_e) and the effective rate of change of porosity with depth (α_e). The impedance of the ground surface is predicted by

$$Z = 0.436(1-\mathrm{i})\sqrt{\frac{\sigma_e}{f}} - 19.74\mathrm{i}\frac{\alpha_e}{f}. \quad (4.37)$$

More-sophisticated theoretical models for the acoustical properties of rigid, porous materials introduce porosity, the tortuosity (or *twistiness*) of the pores and factors related to pore shape [4.57] and multiple layering. Models that introduce viscous and thermal characteristic dimensions of pores [4.58] are based on a formulation by *Johnson* et al. [4.59]. Recently, it has been shown possible to obtain explicit relationships between the characteristic dimensions and grain size by assuming a specific microstructure of identical stacked spheres [4.60]. Other developments allowing for a log-normal pore-size distribution, while assuming pores of identical shape [4.57, 61], are based on the work of *Yamamoto* and *Turgut* [4.62]. As mentioned previously, sometimes it is important to include multiple layering as well. A standard method for obtaining ground impedance is the template method based on short-range measurements of excess attenuation [4.63]. Some values of parameters deduced from data according to (4.36) and (4.37) show that there can be a wide range of parameter values for grassland.

4.6.5 Effects of Small-Scale Roughness

Some surface impedance spectra derived directly from measurements of complex excess attenuation over uncultivated grassland [4.64] indicate that the surface impedance tends to zero above 3 kHz [4.65]. The effects of incoherent scatter from a randomly rough porous surface may be used to explain these measurements. Using a boss approach, an approximate effective admittance for grazing incidence on a hard surface containing randomly distributed two-dimensional (2-D) roughness normal to the roughness axes, may be written [4.66]

$$\beta \approx \left(\frac{3V^2 k^3 b}{2}\right)\left(1 + \frac{\delta^2}{2}\right) + \mathrm{i} V k (\delta - 1), \quad (4.38)$$

where V is the roughness volume per unit area of surface (equivalent to mean roughness height), b is the mean center-to-center spacing, δ is an interaction factor depending on the roughness shape and packing density and k is the wave number. An interesting consequence of (4.38) is that a surface that would be acoustically hard if smooth has, effectively, a finite impedance at grazing incidence when rough. The real part of the admittance allows for incoherent scatter from the surface and varies with the cube of the frequency and the square of the roughness volume per unit area. The same approach can be extended to give the effective normalized surface admittance of a porous surface containing 2-D roughness [4.65, 69]. For a randomly rough porous surface near grazing incidence [4.70] it is possible to obtain the following approximation:

$$Z_r \approx Z_s - \left(\frac{\langle H \rangle R_s}{\gamma \rho_0 c_0}\right)\left(\frac{2}{\nu} - 1\right), \quad [\mathrm{Re}(Z_r) \geq 0], \quad (4.39)$$

where $\nu = 1 + \frac{2}{3}\pi \langle H \rangle$, $\langle H \rangle$ is the root mean square roughness height and Z_s is the impedance of the porous surface if it were smooth. This can be used with an impedance model or measured smooth surface impedance to predict the effect of surface roughness for long wavelengths.

Potentially, cultivation practices have important influences on ground effect since they change the surface properties. *Aylor* [4.71] noted a significant change in the excess attenuation at a range of 50 m over a soil after disking without any noticeable change in the meteorological conditions. Another cultivation practice is sub-soiling, which is intended to break up soil compaction 300 mm or more beneath the ground surface caused, for example, by the repeated passage of heavy vehicles. It is achieved by creating cracks in the compacted layer by means of a single- or double-bladed tine with sharpened leading edges. Sub-soiling only has a small effect on the surface profile of the ground. Plowing turns the soil surface over to a depth of about 0.15 m. Measurements taken over cultivated surfaces before and after sub-soiling and plowing have been shown to be consistent with the predicted effects of the resulting changes in surface roughness and flow resistivity [4.65, 72].

4.6.6 Examples of Ground Attenuation under Weakly Refracting Conditions

Pioneering studies of the combined influences of ground surface and meteorological conditions [4.67, 68] were carried out using a fixed Rolls Royce Avon jet engine as a source at two airfields. The wind speeds and temperatures were monitored at two heights and therefore it was possible to deduce something about the wind and temperature gradients during the measurements. However, perhaps because the role of turbulence was not appreciated (Sect. 4.8.3), the magnitude of turbulence was not monitored. This was the first research to note and quantify the change in ground effect with type of surface. Examples of the resulting data, quoted as the difference between sound pressure levels at 19 m (the reference location) and more distant locations corrected for the decrease expected from spherical spreading and air absorption, are shown in Fig. 4.6. During slightly downwind conditions with low wind speed ($< 2\,\mathrm{m\,s}^{-1}$) and small temperature gradients ($< 0.01°/\mathrm{m}$), the ground attenuation over grass-covered ground at Hatfield, although still a major propagation factor of more than 15 dB near 400 Hz, was less than that over the other grass-covered ground at Radlett and its maximum value occurred at a higher frequency. Snowfall during the pe-

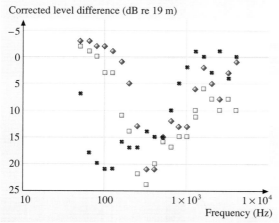

Fig. 4.6 Parkin and Scholes data for the level difference between microphones at a height of 1.5 m and at distances of 19 m and 347 m from a fixed jet engine source (nozzle-center height 1.82 m) corrected for wavefront spreading and air absorption. □ and ◇ represent data over airfields (grass covered) at Radlett and Hatfield respectively with a positive vector wind between source and receiver of 1.27 m/s (5 ft/s); × represent data obtained over approximately 0.15 m thick (6–9 inch) snow at Hatfield and a positive vector wind of 1.52 m/s (6 ft/s). (After [4.67, 68])

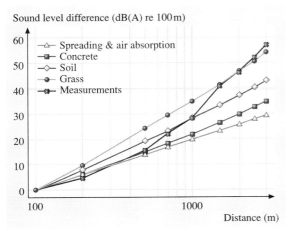

Fig. 4.7 Measured differences between the A-weighted sound level at 100 m and those measured at ranges up to 3 km during an Il-86 aircraft's engine test in the direction of maximum jet noise generation ($\approx 40°$ from exhaust axis) and predictions for levels due to a point source at the engine center height assuming spherical spreading plus air absorption and various types of ground

riod of the measurements also enabled observation of the large change resulting from the presence of snow at low frequencies, i.e., over 20 dB attenuation in the 63 Hz and 125 Hz third-octave bands.

Noise measurements have been made to distances of 3 km during aircraft engine run-ups with the aim of defining noise contours in the vicinity of airports [4.73].

Measurements were made for a range of power settings during several summer days under weakly refracting weather conditions (wind speed < 5 m/s, temperature 20–25 °C). 7–10 measurements were made at every measurement station (in accordance with International Civil Aviation Organization (ICAO) annex 16 requirements) and the results have been averaged. Example results are shown in Fig. 4.7. It can be shown that these data are consistent with nearly acoustically neutral, i.e., zero gradient of the sound of speed, conditions. Note that at 3 km, the measured levels are more than 30 dB less than would be expected from wavefront spreading and air absorption only.

Up to distances of 500–700 m from the engine, the data suggest attenuation rates near to the concrete or spherical spreading plus air absorption predictions. Beyond 700 m the measured attenuation rate is nearer to the soil prediction or between the soil and grass predictions. These results are consistent with the fact that the run-ups took place over the concrete surface of an apron

and further away (i.e., between 500–700 m in various directions) the ground surface was soil and/or grass.

4.6.7 Effects of Ground Elasticity

Noise sources such as blasts and heavy weapons shooting create low-frequency impulsive sound waves that propagate over long distances, and can create a major environmental problem for military training fields. Such impulsive sound sources tend to disturb neighbors more through the vibration and rattling induced in buildings, than by the direct audible sound itself [4.74, 75]. Human perception of whole-body vibration includes frequencies down to 1 Hz [4.76]. Moreover, the fundamental natural frequencies of buildings are in the range 1–10 Hz. Planning tools to predict and control such activities must therefore be able to handle sound propagation down to these low frequencies.

Despite their valuable role in many sound-propagation predictions, locally reacting ground impedance models have the intrinsic theoretical shortcoming that they fail to account for air-to-ground coupling through interaction between the atmospheric sound wave and elastic waves in the ground. This occurs particularly at low frequencies. Indeed air–ground coupled surface waves at low frequencies have been of considerable interest in geophysics, both because of the measured ground-roll caused by intense acoustic sources and the possible use of air sources in ground layering studies. Theoretical analyses have been carried out for spherical wave incidence on a ground consisting either of a fluid or solid layer above a fluid or solid half-space [4.77]. However, to describe the phenomenon of acoustic-to-seismic coupling accurately it has been found that the ground must be modeled as an elastic porous material [4.78, 79]. The resulting theory is relevant not only to predicting the ground vibration induced by low-frequency acoustic sources but also, as we shall see, in predicting the propagation of low-frequency sound above the ground.

The classical theory for a porous and elastic medium predicts the existence of three wave types in the porous medium: two dilatational waves and one rotational wave. In a material consisting of a dense solid frame with a low-density fluid saturating the pores, the first kind of dilatational wave has a velocity very similar to the dilatational wave (or geophysical P wave) traveling in the drained frame. The attenuation of the first dilatational wave type is however, higher than that of the P wave in the drained frame. The extra attenuation comes from the viscous forces in the pore fluid acting on the pore walls. This wave has negligible dispersion and the attenuation is proportional to the square of the frequency, as is the case for the rotational wave. The viscous coupling leads to some of the energy in this propagating wave being carried into the pore fluid as the second type of dilatational wave.

In air-saturated soils, the second dilatational wave has a much lower velocity than the first and is often called the slow wave, while the dilatational wave of the first kind being called the fast wave. The attenuation of the slow wave stems from viscous forces acting on the pore walls and from thermal exchange with the pore walls. Its rigid-frame limit is very similar to the wave that travels through the fluid in the pores of a rigid, porous solid [4.80]. It should be remarked that the slow wave is the only wave type considered in the previous discussions of ground effect. When the slow wave is excited, most of the energy in this wave type is carried in the pore fluid. However, the viscous coupling at the pore walls leads to some propagation within the solid frame. At low frequencies, it has the nature of a diffusion process rather than a wave, being analogous to heat conduction. The attenuation for the slow wave is higher than that of the first in most materials and, at low frequencies, the real and imaginary parts of the propagation constant are nearly equal.

The rotational wave has a very similar velocity to the rotational wave carried in the drained frame (or the S wave of geophysics). Again there is some extra attenuation due to the viscous forces associated with differential movement between solid and pore fluid. The fluid is unable to support rotational waves, but is driven by the solid.

The propagation of each wave type is determined by many parameters relating to the elastic properties of the solid and fluid components. Considerable efforts have been made to identify these parameters and determine appropriate forms for specific materials.

In the formulation described here, only equations describing the two dilatational waves are introduced. The coupled equations governing the propagation of dilatational waves can be written as [4.81]

$$\nabla^2(He - C\xi) = \frac{\partial^2}{\partial t^2}(\rho e - \rho_f \xi), \quad (4.40)$$

$$\nabla^2(Ce - M\xi) = \frac{\partial^2}{\partial t^2}(\rho_f e - m\xi) - \frac{\eta}{k}\frac{\delta\xi}{\delta t}F(\lambda), \quad (4.41)$$

where $e = \nabla \cdot u$ is the dilatation or volumetric strain vector of the skeletal frame; $\xi = \Omega \nabla \cdot (u - U)$ is the relative dilatation vector between the frame and the fluid; u is the displacement of the frame, U is the displacement of the fluid, $F(\lambda)$ is the viscosity correction function, ρ is the total density of the medium, ρ_f is the fluid density, μ is the dynamic fluid viscosity and k is the permeability.

The second term on the right-hand side of (4.41), $\frac{\mu}{k}\frac{\partial \xi}{\partial t}F(\lambda)$, allows for damping through viscous drag as the fluid and matrix move relative to one another; m is a dimensionless parameter that accounts for the fact that not all the fluid moves in the direction of macroscopic pressure gradient as not all the pores run normal to the surface and is given by $m = \frac{\tau \rho_f}{\Omega}$, where τ is the tortuosity and Ω is the porosity.

H, C and M are elastic constants that can be expressed in terms of the bulk moduli K_s, K_f and K_b of the grains, fluid and frame, respectively and the shear modulus μ of the frame [4.58].

Assuming that e and ξ vary as $e^{i\omega t}$, $\partial/\partial t$ can be replaced by $i\omega$ and (4.40) can be written [4.80] as

$$\nabla^2(Ce - M\xi) = -\omega[\rho_f e - \rho(\omega)\xi], \quad (4.42)$$

where $\rho(\omega) = \frac{\tau \rho_f}{\Omega} - \frac{i\mu}{\omega k}F(\lambda)$ is the dynamic fluid density. The original formulation of $F(\lambda)$ [and hence of $\rho(\omega)$] was a generalization from the specific forms corresponding to slit-like and cylindrical forms but assuming pores with identical shape. Expressions are also available for triangular and rectangular pore shapes and for more arbitrary microstructures [4.57, 58, 61]. If plane waves of the form $e = A \exp[-i(lx - \omega t)]$ and $\xi = B \exp[-i(lx - \omega t)]$ are assumed, then the dispersion equations for the propagation constants may be derived. These are:

$$A(l^2 H - \omega^2 \rho) + B(\omega^2 \rho_f - l^2 C) = 0, \quad (4.43)$$

and

$$A(l^2 C - \omega^2 \rho_f) + B[m\omega^2 - l^2 M - i\omega F(\lambda)\eta/k] = 0. \quad (4.44)$$

A nontrivial solution of (4.43) and (4.44) exists only if the determinant of the coefficient vanishes, giving

$$\begin{vmatrix} l^2 H - \omega^2 \rho & \omega^2 \rho_f - l^2 C \\ l^2 C - \omega^2 \rho_f & m\omega^2 - l^2 M - i\omega F(\lambda)\frac{\eta}{k} \end{vmatrix} = 0. \quad (4.45)$$

There are two complex roots of this equation from which both the attenuation and phase velocities of the two dilatational waves are calculated.

At the interface between different porous elastic media there are six boundary conditions that may be

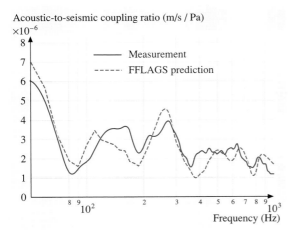

Fig. 4.8 Measured and predicted acoustic-to-seismic coupling ratio for a layered soil (range 3.0 m, source height 0.45 m)

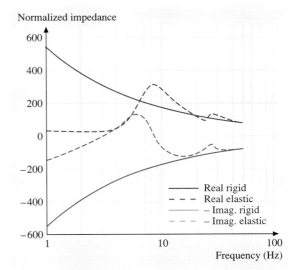

Fig. 4.9 Predicted surface impedance at a grazing angle of 0.018° for poro-elastic and rigid porous ground (four-layer system, Table 4.1)

applied. These are

1. Continuity of total normal stress
2. Continuity of normal displacement
3. Continuity of fluid pressure
4. Continuity of tangential stress
5. Continuity of normal fluid displacement
6. Continuity of tangential frame displacement

At an interface between a fluid and the poro-elastic layer (such as the surface of the ground) the first four boundary conditions apply.

The resulting equations and those for a greater number of porous and elastic layers are solved numerically [4.58].

The spectrum of the ratio between the normal component of the soil particle velocity at the surface of the ground and the incident sound pressure, the acoustic-to-seismic coupling ratio or transfer function, is strongly influenced by discontinuities in the elastic wave properties within the ground. At frequencies corresponding to peaks in the transfer function, there are local maxima in the transfer of sound energy into the soil [4.78]. These are associated with constructive interference between down- and up-going waves within each soil layer. Consequently there is a relationship between near-surface layering in soil and the peaks or *layer resonances* that appear in the measured acoustic-to-seismic transfer function spectrum: the lower the frequency of the peak in the spectrum, the deeper the associated layer. Figure 4.8 shows example measurements and predictions of the acoustic-to-seismic transfer function spectrum at the soil surface [4.12]. The measurements were made using a loudspeaker sound source and a microphone positioned close to the surface, vertically above a geophone buried just below the surface of a soil that had a loose surface layer. Seismic refraction survey measurements at the same site were used to determine the wave speeds. The predictions have been made by using a computer

Table 4.1 Ground profile and parameters used in the calculations for Figs. 4.9 and 4.10

Layer	Flow resistivity ($kPa\,s\,m^{-2}$)	Porosity	Thickness (m)	P-wave speed (m/s)	S-wave speed (m/s)	Damping
1	1 740	0.3	0.5	560	230	0.04
2	1 740	0.3	1.0	220	98	0.02
3	1 740 000	0.01	150	1500	850	0.001
4	1 740 000	0.01	150	1500	354	0.001
5	1 740 000	0.01	Half-space	1500	450	0.001

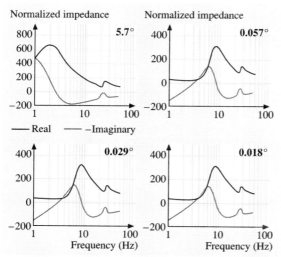

Fig. 4.10 Normalized surface impedance predicted for the four-layer structure, speed of sound in air = 329 m/s (corresponding to an air temperature of −4 °C) for grazing angles between 0.018° and 5.7°

code known as the fast field program for air–ground systems (FFLAGS) that models sound propagation in a system of fluid layers and porous elastic layers [4.79].

This numerical theory (FFLAGS) may be used also to predict the ground impedance at low frequencies. In Fig. 4.9, the predictions for the surface impedance at a grazing angle of 0.018° are shown as a function of frequency for the layered porous and elastic system described by Table 4.1 and compared with those for a rigid porous ground with the same surface flow resistivity and porosity.

The influence of ground elasticity is to reduce the magnitude of the impedance considerably below 50 Hz. Potentially this is very significant for predictions of low-frequency noise, e.g., blast noise, at long range.

Figure 4.10 shows that the surface impedance of this four-layer poro-elastic system varies between grazing angles of 5.7° and 0.57° but remains more or less constant for smaller grazing angles. The predictions show two resonances. The lowest-frequency resonance is the most angle dependent. The peak in the real part changes from 2 Hz at 5.7° to 8 Hz at 0.057°. On the other hand the higher-frequency resonance peak near 25 Hz remains relatively unchanged with range.

The peak at the lower frequency may be associated with the predicted coincidence between the Rayleigh wave speed in the ground and the speed of sound in air (Fig. 4.11). Compared with the near pressure dou-

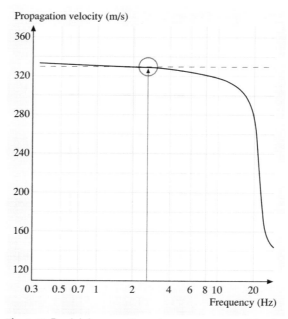

Fig. 4.11 Rayleigh-wave dispersion curve predicted for the system described by Table 4.1

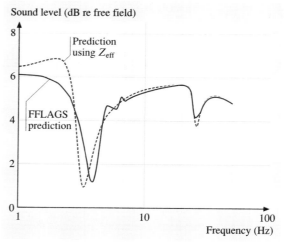

Fig. 4.12 Excess attenuation spectra predicted for source height 2 m, receiver height 0.1 m and horizontal range of 6.3 km by FFLAGS (assumed speed of sound in air of 329 m/s) and by classical theory using impedance calculated for 0.018° grazing angle

bling predicted by classical ground impedance models, the predicted reduction of ground impedance at low frequencies above layered elastic ground can be interpreted as the result of coupling of a significant fraction of the incident sound energy into ground-borne Rayleigh waves.

Numerical theory has also been used to explore the consequences of this coupling for the excess attenuation of low-frequency sound above ground [4.82].

Figure 4.12 shows the excess attenuation spectra predicted for source height 2 m, receiver height 0.1 m and horizontal range of 6.3 km over a layered ground profile corresponding to Table 4.1 (assumed a speed of sound in air of 329 m/s) and by classical theory for a point source above an impedance (locally reacting) plane (4.22) using the impedance calculated for a 0.018° grazing angle (Z_{eff}, broken lines in Fig. 4.9).

The predictions show a significant extra attenuation for 2–10 Hz. The predictions also indicate that, for an assumed speed of sound in air of 329 m/s, and, apart from an enhancement near 2 Hz, the excess attenuation spectrum might be predicted tolerably well by using modified classical theory instead of a full poro-elastic layer calculation.

It is difficult to measure the surface impedance of the ground at low frequencies [4.83, 84]. Consequently the predictions of significant ground elasticity effects have been validated only by using data for acoustic-to-seismic coupling, i. e., by measurements of the ratio of ground surface particle velocity relative to the incident sound pressure [4.82].

4.7 Attenuation Through Trees and Foliage

A mature forest or woodland may have three types of influence on sound. First is the ground effect. This is particularly significant if there is a thick litter layer of partially decomposing vegetation on the forest floor. In such a situation the ground surface consists of a thick highly porous layer with rather low flow resistivity, thus giving a primary excess attenuation maximum at lower frequencies than observed over typical grassland. This is similar to the effect, mentioned earlier, over snow. Secondly the trunks and branches scatter the sound out of the path between source and receiver. Thirdly the foliage attenuates the sound by viscous friction. To predict the total attenuation through woodland, *Price* et al. [4.85] simply added the predicted contributions to attenuation for large cylinders (representing trunks), small cylinders (representing foliage), and the ground. The predictions are in qualitative agreement with their measurements, but it is necessary to adjust several parameters to obtain quantitative agreement. Price et al. found that foliage has the greatest effect above 1 kHz and the foliage attenuation increased in approximately a linear fashion with frequency. Figure 4.13 shows a typical variation of at-

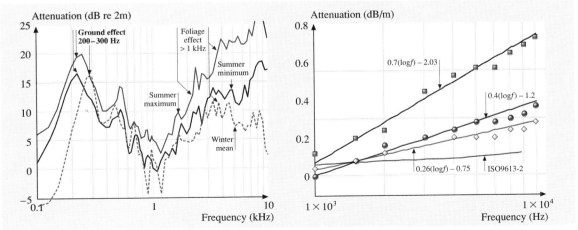

Fig. 4.13 (*Left*) Measured attenuation through alternate bands of Norway spruce and oak (planted in 1946) with hawthorn, roses and honeysuckle undergrowth; visibility less than 24 m. (*Right*) Linear fits to attenuation above 1 kHz in mixed conifers (*squares*), mixed deciduous summer (*circles*) and spruce monoculture (*diamonds*). Also shown is the foliage attenuation predicted according to ISO 9613-2. (After [4.24])

tenuation with frequency and linear fits to the foliage attenuation.

Often the insertion loss of tree belts alongside highways is considered relative to that over open grassland. An unfortunate consequence of the lower-frequency ground effect observed in mature tree stands is that the low-frequency destructive interference resulting from the relatively soft ground between the trees is associated with a constructive interference maximum at important frequencies (near 1 kHz) for traffic noise. Consequently many narrow tree belts alongside roads do not offer much additional attenuation of traffic noise compared with the same distances over open grassland. A Danish study found relative attenuation of 3 dB in the A-weighted L_{eq} due to traffic noise for tree belts 15–41 m wide [4.86]. Data obtained in the UK [4.87] indicates a maximum reduction of the A-weighted L_{10} level due to traffic noise of 6 dB through 30 m of dense spruce compared with the same depth of grassland. This study also found that the effectiveness of the vegetation was greatest closest to the road. A relative reduction of 5 dB in the A-weighted L_{10} level was found after 10 m of vegetation. For a narrow tree belt to be effective against traffic noise it is important that (a) the ground effect is similar to that for grassland, (b) there is substantial reduction of coherence between ground-reflected and direct sound at frequencies of 1 kHz and above, and that the attenuation through scattering is significant.

If the belt is sufficiently wide then the resulting greater extent of the ground-effect dip can compensate for its low frequency. Through 100 m of red pine forest, *Heisler* et al. [4.88] have found 8 dB reduction in the A-weighted L_{eq} due to road traffic compared with open grassland. The edge of the forest was 10 m from the edge of the highway and the trees occupied a gradual downward slope from the roadway extending about 325 m in each direction along the highway from the study site. Compared with open grassland *Huisman* [4.89] has predicted an extra 10 dB(A) attenuation of road traffic noise through 100 m of pine forest. He has remarked also that, whereas downward-refracting conditions lead to higher sound levels over grassland, the levels in woodland are comparatively unaffected. This suggests that extra attenuation obtained through the use of trees should be relatively robust to changing meteorology.

Defrance et al. [4.90] have compared results from both numerical calculations and outdoor measurements for different meteorological situations. A numerical parabolic equation code has been developed [4.91] and adapted to road traffic noise situations [4.92] where road line sources are modeled as series of equivalent point sources of height 0.5 m. The data showed a reduction in A-weighted L_{eq} due to the trees of 3 dB during downward-refracting conditions, 2 dB during homogeneous conditions and 1 dB during upward-refracting conditions. The numerical predictions suggest that in downward-refracting conditions the extra attenuation due to the forest is 2–6 dB(A) with the receiver at least 100 m away from the road. In upward-refracting conditions, the numerical model predicts that the forest may increase the received sound levels somewhat at large distances but this is of less importance since levels at larger distances tend to be relatively low anyway. In homogeneous conditions, it is predicted that sound propagation through the forest is affected only by scattering by trunks and foliage. *Defrance* et al. [4.90] have concluded that a forest strip of at least 100 m wide appears to be a useful natural acoustical barrier. Nevertheless both the data and numerical simulations were compared to sound levels without the trees present, i.e., over ground from which the trees had simply been removed. This means that the ground effect both with and without trees would have been similar. This is rarely likely to be the case.

A similar numerical model has been developed recently [4.93] including allowance for ground effect, wind speed gradient through foliage and assuming effective wave numbers deduced from multiple scattering theory for the scattering effects of trunks, branches and foliage. Again, the model predicts that the large wind speed gradient in the foliage tends to refract sound towards the ground and has an important effect particularly during upwind conditions. However, neither of the PE models [4.91, 93] include back-scatter or turbulence effects. The neglect of the back-scatter is inherent to the PE, which is a *one-way* prediction method. While this is not a serious problem for propagation over flat ground because the back-scatter is small, nor over an impermeable barrier because the back-scattered field, though strong, does not propagate through the barrier, back scatter is likely to be significant for a forest. Indeed acoustic reflections from the edges of forests are readily detectable.

4.8 Wind and Temperature Gradient Effects on Outdoor Sound

The atmosphere is constantly in motion as a consequence of wind shear and uneven heating of the Earth's surface (Fig. 4.14).

Any turbulent flow of a fluid across a rough solid surface generates a boundary layer. Most interest, from the point of view of community noise prediction, focuses on the lower part of the meteorological boundary layer called the surface layer. In the surface layer, turbulent fluxes vary by less than 10% of their magnitude but the wind speed and temperature gradients are largest. In typical daytime conditions the surface layer extends over 50–100 m. Usually, it is thinner at night. Turbulence may be modeled in terms of a series of moving eddies or *turbules* with a distribution of sizes.

In most meteorological conditions, the speed of sound changes with height above the ground. Usually, temperature decreases with height (the adiabatic lapse condition). In the absence of wind, this causes sound waves to bend, or refract, upwards. Wind speed adds or subtracts from sound speed. When the source is downwind of the receiver the sound has to propagate upwind. As height increases, the wind speed increases and the amount being subtracted from the speed of sound increases, leading to a negative gradient in the speed of sound. Downwind, sound refracts downwards. Wind effects tend to dominate over temperature effects when both are present. Temperature inversions, in which air temperature increases up to the inversion height, cause sound waves to be refracted downwards below that height. Under inversion conditions, or downwind, sound levels decrease less rapidly than would be expected from wavefront spreading alone.

In general, the relationship between the speed of sound profile $c(z)$, temperature profile $T(z)$ and wind speed profile $u(z)$ in the direction of sound propagation is given by

$$c(z) = c(0)\sqrt{\frac{T(z) + 273.15}{273.15}} + u(z), \quad (4.46)$$

where T is in °C and u is in m/s.

4.8.1 Inversions and Shadow Zones

If the air temperature first increases up to some height before resuming its usual decrease with height, then there is an inversion. Sound from sources beneath the inversion height will tend to be refracted towards the ground. This is a favorable condition for sound propagation and may lead to higher levels than would be the case under acoustically neutral conditions. This will also be true for receivers downwind of a source. In terms of rays between source and receiver it is necessary to take into account any ground reflections. However, rather than use plane wave reflection coefficients to describe these ground reflections, a better approximation is to use spherical wave reflection coefficients (Sect. 4.6.2).

There are distinct advantages in assuming a linear effective speed of sound profile in ray tracing and ignoring the vector wind since this assumption leads to circular ray paths and relatively tractable analytical solutions. With this assumption, the effective speed of sound c can be written,

$$c(z) = c_0(1 + \zeta z), \quad (4.47)$$

where ζ is the normalized sound velocity gradient $[(dc/dz)/c_0]$ and z is the height above ground. If it also assumed that the source–receiver distance and the effective speed of sound gradient are sufficiently small that there is only a single *ray bounce*, i.e., a single ground reflection between the source and receiver, it is possible to use a simple adaptation of the formula (4.22), replacing the geometrical ray paths defining the direct and reflected path lengths by curved ones. Consequently, the sound field is approximated by

$$p = \left[\exp(-ik_0\xi_1) + Q\exp(-ik_0\xi_2)\right]/4\pi d, \quad (4.48a)$$

where Q is the appropriate spherical wave reflection coefficient, d is the horizontal separation between the source and receiver, and ξ_1 and ξ_2 are, respectively,

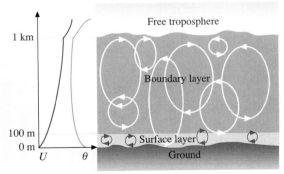

Fig. 4.14 Schematic representation of the daytime atmospheric boundary layer and turbulent eddy structures. The curve on the *left* shows the mean wind speed (U) and the potential temperature profiles ($\theta = T + \gamma_d z$, where $\gamma_d = 0.098$ °C/km is the dry adiabatic lapse rate, T is the temperature and z is the height)

the acoustical path lengths of the direct and reflected waves. These acoustical path lengths can be determined by [4.94, 95]

$$\xi_1 = \int_{\phi_<}^{\phi_>} \frac{d\phi}{\varsigma \sin\phi} = \varsigma^{-1} \log_e \left[\frac{\tan(\phi_>/2)}{\tan(\phi_</2)}\right] \quad (4.48b)$$

and

$$\xi_2 = \int_{\theta_<}^{\theta_>} \frac{d\theta}{\varsigma \sin\theta}$$

$$= \varsigma^{-1} \log_e \left[\tan(\theta_>/2)\tan^2(\theta_0/2)/\tan(\theta_</2)\right], \quad (4.48c)$$

where $\phi(z)$ and $\theta(z)$ are the polar angles (measured from the positive z-axis) of the direct and reflected waves.

The subscripts $>$ and $<$ denote the corresponding parameters evaluated at $z_>$ and $z_<$ respectively, $z_> \equiv \max(z_s, z_r)$ and $z_< \equiv \min(z_s, z_r)$.

The computation of $\phi(z)$ and $\theta(z)$ requires the corresponding polar angles (ϕ_0 and θ_0) at $z = 0$ [4.96]. Once the polar angles are determined at $z = 0$, $\phi(z)$ and $\theta(z)$ at other heights can be found by using Snell's law:

$$\sin\vartheta = (1 + \varsigma z)\sin\vartheta_0,$$

where $\vartheta = \phi$ or θ. Substitution of these angles into (4.48b) and (4.48c) and, in turn, into (4.48a) makes it possible to calculate the sound field in the presence of a linear sound-velocity gradient.

For downward refraction, additional rays will cause a discontinuity in the predicted sound level because of the inherent approximation used in ray tracing. It is possible to determine the critical range r_c at which there are two additional ray arrivals. For $\varsigma > 0$, this critical range is given by:

$$r_c = \frac{\left\{\left[\sqrt{(\varsigma z_>)^2 + 2\varsigma z_>} + \sqrt{(\varsigma z_<)^2 + 2\varsigma z_<}\right]^{2/3}\right\}^{3/2}}{\varsigma}$$
$$+ \frac{\left\{\left[\sqrt{(\varsigma z_>)^2 + 2\varsigma z_>} - \sqrt{(\varsigma z_<)^2 + 2\varsigma z_<}\right]^{2/3}\right\}^{3/2}}{\varsigma}. \quad (4.49)$$

Figure 4.15 shows that, for source and receiver at 1 m height, if predictions are confined to a horizontal separation of less than 1 km and a normalized speed of sound gradient of less than 0.0001 m^{-1} (corresponding, for example, to a wind speed gradient of less than

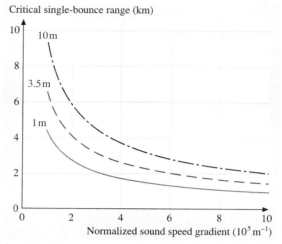

Fig. 4.15 Maximum ranges for which the single-bounce assumption is valid for a linear speed of sound gradient based on (4.35) and assuming equal source and receiver heights: 1 m (*solid line*); 3.5 m (*broken line*) and 10 m (*dot–dash line*)

0.1 s^{-1}) then, it is reasonable to assume a single ground bounce in ray tracing. The critical range for the single-bounce assumption increases as the source and receiver heights increase.

A negative sound gradient means upward refraction and the creation of a sound shadow at a distance from the source that depends on the gradient. The presence of a shadow zone means that the sound level decreases faster than would be expected from distance alone. A combination of a slightly negative temperature gradient, strong upwind propagation and air absorption has been observed, in carefully monitored experiments, to reduce sound levels, 640 m from a 6 m-high source over relatively hard ground, by up to 20 dB more than expected from spherical spreading [4.97]. Since shadow zones can be areas in which there is significant excess attenuation, it is important to be able to locate their boundaries approximately.

For upward-refracting conditions, ray tracing is incorrect when the receiver is in the shadow and penumbra zones. The shadow boundary can be determined from geometrical considerations. For a given source and receiver heights, the critical range r'_c is determined as:

$$r'_c = \frac{\sqrt{(\varsigma' z_>)^2 + 2\varsigma' z_>} + \sqrt{\varsigma'^2 z_<(2z_> - z_<) + 2\varsigma' z_<}}{\varsigma'}, \quad (4.50)$$

where

$$\varsigma' = \frac{|\varsigma|}{1-|\varsigma|z_>}.$$

Figure 4.16 shows that, for source and receiver heights of 1 m and a normalized speed of sound gradient of $0.0001\,\mathrm{m}^{-1}$, the distance to the shadow zone boundary is about 300 m. As expected the distance to the shadow-zone boundary is predicted to increase as the source and receiver heights are increased.

A good approximation of (4.36), for the distance to the shadow zone, when the source is close to the ground and ζ is small, is

$$r_\mathrm{c} = \left[\frac{2c_0}{-\frac{\mathrm{d}c}{\mathrm{d}z}}\right]^{1/2}\left(\sqrt{h_\mathrm{s}}+\sqrt{h_\mathrm{r}}\right),\qquad(4.51)$$

where h_s and h_r are the heights of source and receiver respectively and $\mathrm{d}c/\mathrm{d}z$ must be negative for a temperature-induced shadow zone.

Conditions of weak refraction may be said to exist where, under downward-refracting conditions, the ground-reflected ray undergoes only a single bounce and, under upward-refracting conditions, the receiver is within the illuminated zone.

When wind is present, the combined effects of temperature lapse and wind will tend to enhance the shadow zone upwind of the source, since wind speed tends to increase with height. Downwind of the source, however, the wind will counteract the effect of temperature lapse, and the shadow zone will be destroyed. In any case, an acoustic shadow zone is never as complete as an optical one would be, as a result of diffraction and turbulence. In the presence of wind with a wind speed gradient of $\mathrm{d}u/\mathrm{d}z$, the formula for the distance to the shadow-zone boundary is given by

$$r_\mathrm{c} = \left[\frac{2c_0}{\frac{\mathrm{d}u}{\mathrm{d}z}\cos\beta - \frac{\mathrm{d}c}{\mathrm{d}z}}\right]^{1/2}\left(\sqrt{h_\mathrm{s}}+\sqrt{h_\mathrm{r}}\right),\qquad(4.52)$$

where β is the angle between the direction of the wind and the line between source and receiver.

Note that there will be a value of the angle β, (say β_c), given by

$$\frac{\mathrm{d}u}{\mathrm{d}z}\cos\beta_c = \frac{\mathrm{d}c}{\mathrm{d}z}$$

or

$$\beta_\mathrm{c} = \cos^{-1}\left(\frac{\mathrm{d}c}{\mathrm{d}z}\bigg/\frac{\mathrm{d}u}{\mathrm{d}z}\right)\qquad(4.53)$$

at and beyond which there will not be a shadow zone. This represents the critical angle at which the effect of wind counteracts that of the temperature gradient.

4.8.2 Meteorological Classes for Outdoor Sound Propagation

There is a considerable body of knowledge about meteorological influences on air quality in general and the dispersion of plumes from stacks in particular. Plume behavior depends on vertical temperature gradients and hence on the degree of mixing in the atmosphere. Vertical temperature gradients decrease with increasing wind. The stability of the atmosphere in respect to plume dispersion is described in terms of Pasquill classes. This classification is based on incoming solar radiation, time of day and wind speed. There are six Pasquill classes (A–F) defined in Table 4.2.

Data are recorded in this form by meteorological stations and so, at first sight, it is a convenient classification system for noise prediction.

Class A represents a very unstable atmosphere with strong vertical air transport, i.e., mixing. Class F represents a very stable atmosphere with weak vertical transport. Class D represents a meteorologically neutral atmosphere. Such an atmosphere has a logarithmic wind speed profile and a temperature gradient corresponding to the normal decrease with height (adiabatic lapse rate). A meteorologically neutral atmosphere occurs for high wind speeds and large values of cloud cover. This means that a meteorologically neutral atmosphere may be far

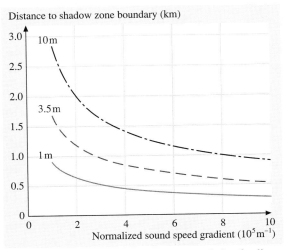

Fig. 4.16 Distances to shadow-zone boundaries for linear speed of sound gradient based on (4.36) assuming equal source and receiver heights: 1 m (*solid line*); 3.5 m (*broken line*) and 10 m (*dot–dash line*)

Table 4.2 Pasquill (meteorological) stability categories

Wind speed[a] (m/s)	Daytime Incoming solar radiation (mW/cm²)				1 h before sunset or after sunrise	Nighttime cloud cover (octas)		
	> 60	30–60	< 30	Overcast		0–3	4–7	8
≤ 1.5	A	A–B	B	C	D	F or G[b]	F	D
2.0–2.5	A–B	B	C	C	D	F	E	D
3.0–4.5	B	B–C	C	C	D	E	D	D
5.0–6.0	C	C–D	D	D	D	D	D	D
> 6.0	D	D	D	D	D	D	D	D

[a] measured to the nearest 0.5 m/s at 11 m height
[b] Category G is an additional category restricted to nighttime with less than 1 octa of cloud and a wind speed of less than 0.5 m/s

Table 4.3 CONCAWE meteorological classes for noise prediction

Meteorological category	Pasquill stability category and wind speed (m/s) Positive is towards receiver		
	A, B	C, D, E	F, G
1	$v < -3.0$	–	–
2	$-3.0 < v < -0.5$	$v < -3.0$	–
3	$-0.5 < v < +0.5$	$-3.0 < v < -0.5$	$v < -3.0$
4[a]	$+0.5 < v < +3.0$	$-0.5 < v < +0.5$	$-3.0 < v < -0.5$
5	$v > +3.0$	$+0.5 < v < +3.0$	$-0.5 < v < +0.5$
6	–	$v > +3.0$	$+0.5 < v < +3.0$

[a] Category with assumed zero meteorological influence

from *acoustically* neutral. Typically, the atmosphere is unstable by day and stable by night. This means that classes A–D might be appropriate classes by day and D–F by night. With practice, it is possible to estimate Pasquill stability categories in the field, for a particular time and season, from a visual estimate of the degree of cloud cover.

The Pasquill classification of meteorological conditions has been adopted as a classification system for noise-prediction schemes [4.38, 98]. However, it is clear from Table 4.2, that the meteorologically neutral category (C), while being fairly common in a temperate climate, includes a wide range of wind speeds and is therefore not very suitable as a category for noise prediction. In the CONservation of Clean Air and Water in Europe (CONCAWE) scheme [4.98], this problem is addressed by defining six noise-prediction categories based on Pasquill categories (representing the temperature gradient) and wind speed. There are 18 subcategories depending on wind speed. These are defined in Table 4.3.

CONCAWE category 4 is specified as that in which there is zero meteorological influence. So CONCAWE category 4 is equivalent to acoustically neutral conditions.

The CONCAWE scheme requires octave band analysis. Meteorological corrections in this scheme are based primarily on analysis of data from *Parkin* and *Scholes* [4.67, 68] together with measurements made at several industrial sites. The excess attenuation in each octave band for each category tends to approach asymptotic limits with increasing distance. Values at 2 km for CONCAWE categories 1 (strong wind from receiver to source, hence upward refraction) and 6 (strong downward refraction) are listed in Table 4.4.

Wind speed and temperature gradients are not independent. For example, very large temperature and

Table 4.4 Values of the meteorological corrections for CONCAWE categories 1 and 6

Octave band centre frequency (Hz)	63	125	250	500	1000	2000	4000
Category 1	8.9	6.7	4.9	10.0	12.2	7.3	8.8
Category 6	−2.3	−4.2	−6.5	−7.2	−4.9	−4.3	−7.4

Table 4.5 Estimated probability of occurrence of various combinations of wind and temperature gradient

Temperature gradient	Zero wind	Strong wind	Very strong wind
Very large negative temperature gradient	Frequent	Occasional	Rare or never
Large negative temperature gradient	Frequent	Occasional	Occasional
Zero temperature gradient	Occasional	Frequent	Frequent
Large positive temperature gradient	Frequent	Occasional	Occasional
Very large positive temperature gradient	Frequent	Occasional	Rare or never

Table 4.6 Meteorological classes for noise prediction based on qualitative descriptions

W1	Strong wind (> 3–5 m/s) from receiver to source
W2	Moderate wind (≈ 1–3 m/s) from receiver to source, or strong wind at 45°
W3	No wind, or any cross wind
W4	Moderate wind (≈ 1–3 m/s) from source to receiver, or strong wind at 45°
W5	Strong wind (> 3–5 m/s) from source to receiver
TG1	Strong negative: daytime with strong radiation (high sun, little cloud cover), dry surface and little wind
TG2	Moderate negative: as T1 but one condition missing
TG3	Near isothermal: early morning or late afternoon (e.g., one hour after sunrise or before sunset)
TG5	Moderate positive: nighttime with overcast sky or substantial wind
TG6	Strong positive: nighttime with clear sky and little or no wind

wind speed gradients cannot coexist. Strong turbulence associated with high wind speeds does not allow the development of marked thermal stratification.

Table 4.5 shows a rough estimate of the probability of occurrence of various combinations of wind and temperature gradients (TG) [4.97].

With regard to sound propagation, the component of the wind vector in the direction between source and receiver is most important. So the wind categories (W) must take this into account. Moreover, it is possible to give more detailed but qualitative descriptions of each of the meteorological categories (W and TG, see Table 4.6).

In Table 4.7, the revised categories are identified with qualitative predictions of their effects on noise levels. The classes are not symmetrical around zero meteorological influence. Typically there are more meteorological condition combinations that lead to attenuation than lead to enhancement. Moreover, the increases in noise level (say 1–5 dB) are smaller than the decreases (say 5–20 dB).

Using the values at 500 Hz as a rough guide for the likely corrections on overall A-weighted broadband levels it is noticeable that the CONCAWE meteorological corrections are not symmetrical around zero. The CONCAWE scheme suggests meteorological variations of between 10 dB less than the acoustically neutral level for strong upward refraction between source and receiver and 7 dB more than the acoustically neutral level for strong downward refraction between the source and receiver.

Table 4.7 Qualitative estimates of the impact of meteorological condition on noise levels

	W1	W2	W3	W4	W5
TG1	–	Large attenuation	Small attenuation	Small attenuation	–
TG2	Large attenuation	Small attenuation	Small attenuation	Zero meteorological influence	Small enhancement
TG3	Small attenuation	Small attenuation	Zero meteorological influence	Small enhancement	Small enhancement
TG4	Small attenuation	Zero meteorological influence	Small enhancement	Small enhancement	Large enhancement
TG5	–	Small enhancement	Small enhancement	Large enhancement	–

4.8.3 Typical Speed of Sound Profiles

Outdoor sound prediction requires information on wind speed, direction and temperature as a function of height near to the propagation path. These determine the speed of sound profile. Ideally, the heights at which the meteorological data are collected should reflect the application. If this information is not available, then there are alternative procedures. It is possible, for example, to generate an approximate speed of sound profile from temperature and wind speed at a given height using the Monin–Obukhov similarity theory [4.99] and to input this directly into a prediction scheme.

According to this theory, the wind speed component (m/s) in the source–receiver direction and temperature (°C) at height z are calculated from the values at ground level and other parameters as follows:

$$u(z) = \frac{u_*}{k}\left[\ln\left(\frac{z+z_M}{z_M}\right) + \psi_M\left(\frac{z}{L}\right)\right], \quad (4.54)$$

$$T(z) = T_0 + \frac{T_*}{k}\left[\ln\left(\frac{z+z_H}{z_H}\right) + \psi_H\left(\frac{z}{L}\right)\right] + \Gamma z, \quad (4.55)$$

where the parameters are defined in Table 4.8.

For a neutral atmosphere, $1/L = 0$ and $\psi_M = \psi_H = 0$.

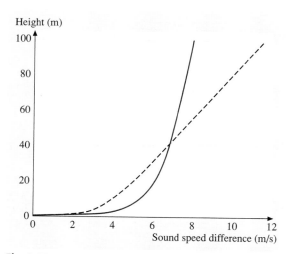

Fig. 4.17 Two downward-refracting speed of speed of sound profiles relative to the sound speed at the ground obtained from similarity theory. The *continuous curve* is approximately logarithmic corresponding to a large Obukhov length and to a cloudy, windy night. The *broken curve* corresponds to a small Obukhov length as on a calm clear night and is predominantly linear away from the ground

The associated speed of sound profile, $c(z)$, is calculated from (4.33).

Note that the resulting profiles are valid in the surface or boundary layer only but not at zero height. In fact, the profiles given by the above equations, sometimes called Businger–Dyer profiles [4.100], have been found to give good agreement with measured profiles up to 100 m. This height range is relevant to sound propagation over distances up to 10 km [4.101]. However improved profiles are available that are valid to greater heights. For example [4.102],

$$\psi_M = \psi_H = -7\ln(z/L) - 4.25/(z/L) \\ + 0.5/(z/L)^2 - 0.852 \quad \text{for } z > 0.5L. \quad (4.56)$$

Often z_M and z_H are taken to be equal. The roughness length varies, for example, between 0.0002 (still water) and 0.1 (grass). More generally, the roughness length can be estimated from the Davenport classification [4.103].

Figure 4.17 gives examples of speed of sound (difference) profiles, $[c(z) - c(0)]$, generated from (4.53) through (4.56) using

1. $z_M = z_H = 0.02$, $u_* = 0.34$, $T_* = 0.0212$, $T_{av} = 10$, $T_0 = 6$, (giving $L = 390.64$),
2. $z_M = z_H = 0.02$, $u_* = 0.15$, $T_* = 0.1371$, $T_{av} = 10$, $T_0 = 6$, (giving $L = 11.76$),

and $\Gamma = -0.01$. These parameters are intended to correspond to a cloudy windy night and a calm, clear night respectively [4.89].

Salomons et al. [4.104] have suggested a method to obtain the remaining unknown parameters, u_*, T_*, and L from the relationship

$$L = \frac{u_*^2}{kgT_*} \quad (4.57)$$

and the Pasquill category (P).

From empirical meteorological tables, approximate relationships between the Pasquill class P the wind speed u_{10} at a reference height of 10 m and the fractional cloud cover N_c have been obtained. The latter determines the incoming solar radiation and therefore the heating of the ground. The former is a guide to the degree of mixing. The approximate relationship is

$$P(u_{10}, N_c) = 1 + 3\left[1 + \exp(3.5 - 0.5u_{10} - 0.5N_c)\right]^{-1}$$
during the day
$$= 6 - 2\left[1 + \exp(12 - 2u_{10} - 2N_c)\right]^{-1}$$
during the night $\quad (4.58)$

Table 4.8 Definitions of parameters used in equations

$u*$	Friction velocity (m/s)	(depends on surface roughness)
z_M	Momentum roughness length	(depends on surface roughness)
z_H	Heat roughness length	(depends on surface roughness)
$T*$	Scaling temperature K	The precise value of this is not important for sound propagation. A convenient value is 283 K
k	Von Karman constant	$(= 0.41)$
T_0	Temperature (°C) at zero height	Again it is convenient to use 283 K
Γ	Adiabatic correction factor	$= -0.01\,°C/m$ for dry air. Moisture affects this value but the difference is small
L	Obukhov length (m) $> 0 \rightarrow$ stable, $< 0 \rightarrow$ unstable	$= \pm \frac{u_*^2}{kgT_*}(T_{av} + 273.15)$, the thickness of the surface or boundary layer is given by $2Lm$
T_{av}	Average temperature (°C)	It is convenient to use $T_{av} = 10$ so that $(T_{av} + 273.15) = \theta_0$
ψ_M	Diabatic momentum profile correction (mixing) function	$= -2\ln\left(\frac{(1+\chi_M)}{2}\right) - \ln\left(\frac{(1+\chi_M^2)}{2}\right) + 2\arctan(\chi_M) - \pi/2$ if $L<0$ $= 5(z/L)$ if $L > 0$
ψ_H	Diabatic heat profile correction (mixing) function	$= -2\ln\left(\frac{(1+\chi_H)}{2}\right)$ if $L<0$ $= 5(z/L)$ if $L > 0$ or for $z \leq 0.5L$
χ_M	Inverse diabatic influence or function for momentum	$= \left(1 - \frac{16z}{L}\right)^{0.25}$
χ_H	Inverse diabatic influence function for momentum	$= \left(1 - \frac{16z}{L}\right)^{0.5}$

A proposed relationship between the Obukhov length L [m] as a function of P and roughness length $z_0 < 0.5$ m is:

$$\frac{1}{L(P,z_0)} = B_1(P)\log(z_0) + B_2(P), \quad (4.59a)$$

where

$$B_1(P) = 0.0436 - 0.0017P - 0.0023P^2 \quad (4.59b)$$

and

$$B_2(P) = \min(0, 0.045P - 0.125) \text{ for } 1 \leq P \leq 4$$
$$\max(0, 0.025P - 0.125) \text{ for } 4 \leq P \leq 6. \quad (4.59c)$$

Alternatively, values of B_1 and B_2 may be obtained from Table 4.9.

Equations (4.59) give

$$L = L(u_{10}, N_c, z_0). \quad (4.60)$$

Also u_{10} is given by (4.39) with $z = 10$ m, i.e.,

$$u(z) = \frac{u_*}{k}\left[\ln\left(\frac{10+z_M}{z_M}\right) + \psi_M\left(\frac{10}{L}\right)\right]. \quad (4.61)$$

Equations (4.40), (4.41) and (4.61) may be solved for u_*, T_*, and L. Hence it is possible to calculate ψ_M, ψ_H, $u(z)$ and $T(z)$.

Figure 4.18 shows the results of this procedure for a ground with a roughness length of 0.1 m and two upwind and downwind daytime classes defined by the parameters listed in the caption.

As a consequence of atmospheric turbulence, instantaneous profiles of temperature and wind speed show considerable variations with both time and position. These variations are eliminated considerably by averaging over a period of the order of 10 minutes. The Monin–Obukhov or Businger–Dyer models give good descriptions of the averaged profiles over longer periods.

The Pasquill category C profiles shown in Fig. 4.18 are approximated closely by logarithmic curves of the form

$$c(z) = c(0) + b\ln\left[\frac{z}{z_0} + 1\right], \quad (4.62)$$

Table 4.9 Values of the constants B1 and B2 for the six Pasquill classes

Pasquill class	A	B	C	D	E	F
B_1	0.04	0.03	0.02	0	−0.02	−0.05
B_2	−0.08	−0.035	0	0	0	0.025

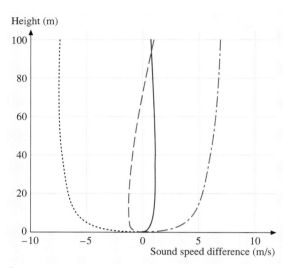

Fig. 4.18 Two daytime speed of sound profiles (upwind – *dashed* and *dotted*; downwind – *solid* and *dash-dot*) determined from the parameters listed in Table 4.9

where the parameter b (> 0 for downward refraction and < 0 for upward refraction) is a measure of the strength of the atmospheric refraction. Such logarithmic speed of sound profiles are realistic for open ground areas without obstacles particularly in the daytime.

A better fit to nighttime profiles is obtained with power laws of the form

$$c(z) = c(0) + b\left(\frac{z}{z_0}\right)^\alpha, \tag{4.63}$$

where $\alpha = 0.4(P-4)^{\frac{1}{4}}$.

The temperature term in the effective speed of sound profile given by (4.33) can be approximated by truncating a Taylor expansion after the first term to give

$$c(z) = c[T_0] + \frac{1}{2}\sqrt{\frac{\kappa R}{T_0}}[T(z) - T_0] + u(z). \tag{4.64}$$

When combined with (4.39) this leads to a linear dependence on temperature and a logarithmic dependence on wind speed with height. By comparing with 12 months of meteorological data obtained at a 50 m-high meteorological tower in Germany, *Heimann* and *Salomons* [4.105] found that (4.44) is a reasonably accurate approximation to vertical profiles of effective speed of sound even in unstable conditions and in situations where Monin–Obukhov theory is not valid. By making a series of sound level predictions (using the parabolic equation method) for different meteorological conditions it was found that a minimum of 25 meteorological classes is necessary to ensure 2 dB or less deviation in the estimated annual average sound level from the reference case with 121 categories.

There are simpler, linear-segment profiles deduced from a wide range of meteorological data that may be used to represent worst-case noise conditions, i.e., best conditions for propagation. The first of these profiles may be calculated from a temperature gradient of $+15\,°\mathrm{C/km}$ from the surface to 300 m and $8\,°\mathrm{C/km}$ above that, assuming a surface temperature of $20\,°\mathrm{C}$. This type of profile can occur during the daytime or at night downwind due to wind shear in the atmosphere or a very high temperature inversion. If this is considered too extreme, or too rare a condition, then a second possibility is a shallow inversion, which occurs regularly at night. A typical depth is 200 m. The profile may be calculated from a temperature gradient of $+20\,°\mathrm{C/km}$ from the surface to 200 m and $-8\,°\mathrm{C/km}$ above that assuming a surface temperature of $20\,°\mathrm{C}$.

The prediction of outdoor sound propagation also requires information about turbulence.

4.8.4 Atmospheric Turbulence Effects

Shadow zones due to atmospheric refraction are penetrated by sound scattered by turbulence and this sets a limit of the order of 20–25 dB to the reduction of sound levels within the sound shadow [4.106, 107].

Sound propagating through a turbulent atmosphere will fluctuate both in amplitude and phase as a result of fluctuations in the refractive index caused by fluctuations in temperature and wind velocity. When predicting outdoor sound, it is usual to refer to these fluctuations in wind velocity and temperature rather than the cause of the turbulence. The amplitude of fluctuations in sound level caused by turbulence initially increase with increasing distance of propagation, sound frequency and strength of turbulence but reach a limiting value fairly quickly. This means that the fluctuation in overall sound levels from distant sources (e.g., line of sight from an aircraft at a few km) may have a standard deviation of no more than about 6 dB [4.107].

There are two types of atmospheric instability responsible for the generation of turbulent kinetic energy: shear and buoyancy. Shear instabilities are associated with mechanical turbulence. High-wind conditions and a small temperature difference between the air and ground are the primary causes of mechanical turbulence. Buoyancy or convective turbulence is associated with thermal instabilities. Such turbulence prevails when the ground is much warmer than the overlying air, such as, for example, on a sunny day. The irregularities in the temperature and wind fields are directly related to the scattering of sound waves in the atmosphere.

Fluid particles in turbulent flow often move in *loops* (Fig. 4.9) corresponding to swirls or eddies. Turbulence can be visualized as a continuous distribution of eddies in time and space. The largest eddies can extend to the height of the boundary layer, i.e., up to 1–2 km on a sunny afternoon. However, the outer scale of usual interest in community noise prediction is of the order of meters. In the size range of interest, sometimes called the inertial subrange, the kinetic energy in the larger eddies is transferred continuously to smaller ones. As the eddy size becomes smaller, virtually all of the energy is dissipated into heat. The length scale at which at which viscous dissipation processes begin to dominate for atmospheric turbulence is about 1.4 mm.

The size of eddies of most importance to sound propagation, for example in the shadow zone, may be estimated by considering Bragg diffraction [4.108]. For a sound with wavelength λ being scattered through angle θ, (Fig. 4.19), the important scattering structures have a spatial periodicity D satisfying

$$\lambda = 2D \sin(\theta/2). \tag{4.65}$$

At a frequency of 500 Hz and a scattering angle of 10°, this predicts a size of 4 m.

When acoustic waves propagate nearly horizontally, the (overall) variance in the effective index of refraction $\langle \mu^2 \rangle$ is related approximately to those in velocity and temperature by [4.109]

$$\langle \mu^2 \rangle = \frac{\langle u'^2 \rangle}{c_0^2} \cos^2 \phi + \frac{\langle v'^2 \rangle}{c_0^2} \sin^2 \phi$$
$$+ \frac{\langle u'T \rangle}{c_0} \cos \phi + \frac{\langle T^2 \rangle}{4T_0^2}, \tag{4.66}$$

where T, u' and v' are the fluctuations in temperature, horizontal wind speed parallel to the mean wind and horizontal wind speed perpendicular to the mean wind, respectively. ϕ is the angle between the wind and the wavefront normal. Use of similarity theory gives [4.110]

$$\langle \mu^2 \rangle = \frac{5u_*^2}{c_0^2} + \frac{2.5u_*T_*}{c_0 T_0} \cos \phi + \frac{T_*^2}{T_0^2}, \tag{4.67}$$

where u_* and T_* are the friction velocity and scaling temperature ($= -Q/u_*$, Q being the surface temperature flux), respectively.

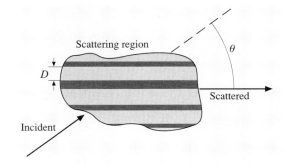

Fig. 4.19 Bragg reflection condition for acoustic scattering from turbulence

Typically, during the daytime, the velocity term in the effective index of refraction variance always dominates over the temperature term. This is true, even on sunny days, when turbulence is generated by buoyancy rather than shear. Strong buoyant instabilities produce vigorous motion of the air. Situations where temperature fluctuations have a more significant effect on acoustic scattering than velocity fluctuations occur most often during clear, still nights.

Although the second term (the covariance term) in (4.47) may be at least as important as the temperature term [4.110], it is often ignored for the purpose of predicting acoustic propagation. Estimations of the fluctuations in terms of u_* and T_* and the Monin–Obukhov length L are given by [4.111], for $L > 0$ (stable conditions, e.g., at night)

$$\sqrt{\langle u'^2 \rangle} = \sigma_u = 2.4u_*, \quad \sqrt{\langle v'^2 \rangle} = \sigma_v = 1.9u_*,$$
$$\sqrt{\langle T^2 \rangle} = \sigma_T = 1.5T_*$$

for $L < 0$ (unstable conditions, e.g., daytime)

$$\sigma_u = \left(12 - 0.5\frac{z}{L}\right)^{1/3} u_*,$$
$$\sigma_v = 0.8\left(12 - 0.5\frac{z}{L}\right)^{1/3} u_*,$$
$$\sigma_T = 2\left(1 - 18\frac{z}{L}\right)^{-1/2} T_*.$$

For line-of-sight propagation, the mean squared fluctuation in the phase of plane sound waves (sometimes called the strength parameter) is given by [4.112]

$$\Phi^2 = 2\langle \mu^2 \rangle k_0^2 XL,$$

where X is the range and L is the inertial length scale of the turbulence. Alternatively, the variance in the log-amplitude fluctuations in a plane sound wave propagating through turbulence is given by [4.113]

$$\langle \chi^2 \rangle = \frac{k_0^2 X}{4} \left(L_T \frac{\sigma_T^2}{T_0^2} + L_v \frac{\sigma_v^2}{c_0^2} \right),$$

where L_T, σ_T^2 and L_v, σ_v^2 are integral length scales and variances of temperature and velocity fluctuations respectively.

There are several models for the size distribution of turbulent eddies. In the Gaussian model of turbulence statistics, the energy spectrum $\phi_n(K)$ of the index of refraction is given by

$$\phi_n(K) = \langle \mu^2 \rangle \frac{L^2}{4\pi} \exp\left(-\frac{K^2 L^2}{4}\right), \qquad (4.68)$$

where L is a single length scale (integral or outer length scale) proportional to the correlation length (inner length scale) ℓ_G, i. e.,

$$L = \ell_G \sqrt{\pi}/2.$$

The Gaussian model has some utility in theoretical models of sound propagation through turbulence since it allows many results to be obtained in simple analytical form. However, as shown below, it provides a poor overall description of the spectrum of atmospheric turbulence [4.112].

In the von Karman spectrum, known to work reasonably well for turbulence with high Reynolds number, the spectrum of the variance in index of refraction is given by

$$\phi_n(K) = \langle \mu^2 \rangle \frac{L}{\pi(1 + K^2 \ell_K^2)} \qquad (4.69)$$

where $L = \ell_K \sqrt{\pi} \Gamma(5/6)/\Gamma(1/3)$.

Figure 4.20 compares the spectral function $[K\phi(K)/\langle\mu^2\rangle]$ given by the von Karman spectrum for $\langle\mu^2\rangle = 10^{-2}$ and $\ell_K = 1$ m with two spectral functions calculated assuming a Gaussian turbulence spectrum, respectively for $\langle\mu^2\rangle = 0.818 \times 10^{-2}$ and $\ell_G = 0.93$ m and $\langle\mu^2\rangle = 0.2 \times 10^{-2}$ and $\ell_G = 0.1$ m. The variance and inner length scale for the first Gaussian spectrum have been chosen to match the von Karman spectrum exactly for the low wave numbers (larger eddy sizes). It also offers a reasonable representation near to the spectral peak. Past the spectral peak and at high wave numbers, the first Gaussian spectrum decays far too rapidly. The second Gaussian spectrum clearly matches the von Karman spectrum over a narrow range of smaller eddy sizes. If this happens to be the wave number range of interest in

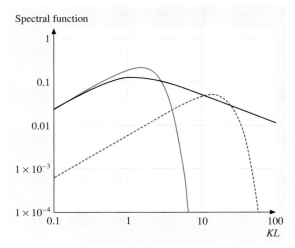

Fig. 4.20 A von Karman spectrum of turbulence and two Gaussian spectra chosen to match it at low wave numbers and over a narrow range of high wave numbers, respectively

scattering from turbulence, then the Gaussian spectrum may be satisfactory.

Most recent calculations of turbulence effects on outdoor sound have relied on estimated or best-fit values rather than measured values of turbulence parameters. Under these circumstances, there is no reason to assume spectral models other than the Gaussian one.

Typically, the high-wave-number part of the spectrum is the main contributor to turbulence effects on sound propagation. This explains why the assumption of a Gaussian spectrum results in best-fit parameter values that are rather less than those that are measured.

Turbulence destroys the coherence between direct and ground-reflected sound and consequently reduces the destructive interference in the ground effect. Equation (4.22) may be modified [4.114] to obtain the mean-squared pressure at a receiver in a turbulent but *acoustically neutral* (no refraction) atmosphere

$$\langle p^2 \rangle = \frac{1}{R_1^2} + \frac{|Q|^2}{R_2^2} + \frac{2|Q|}{R_1 R_2} \cos\left[k(R_2 - R_1) + \theta\right] T, \qquad (4.70)$$

where θ is the phase of the reflection coefficient, ($Q = |Q|\mathrm{e}^{-\mathrm{i}\theta}$), and T is the coherence factor determined by the turbulence effect.

Hence the sound pressure level P is given by;

$$P = 10\log_{10}\langle p^2 \rangle. \qquad (4.71)$$

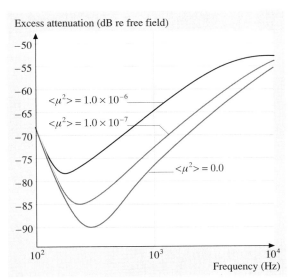

Fig. 4.21 Excess attenuation versus frequency for a source and receiver above an impedance ground in an acoustically neutral atmosphere predicted by (4.46)–(4.52) for three values of $\langle \mu^2 \rangle$ between 0 and 10^{-6}. The assumed source and receiver heights are 1.8 and 1.5 m, respectively, and the assumed separation is 600 m. A two-parameter impedance model (4.31) has been used with values of 300 000 N s m^{-4} and 0.0 m^{-1}

For a Gaussian turbulence spectrum, the coherence factor T is given by [4.114]

$$T = e^{-\sigma^2(1-\rho)}, \quad (4.72)$$

where σ^2 is the variance of the phase fluctuation along a path and ρ is the phase covariance between adjacent paths (e.g. direct and reflected)

$$\sigma^2 = A\sqrt{\pi}\langle \mu^2 \rangle k^2 R L_0, \quad (4.73)$$

of the index of refraction, and L_0 is the outer (inertial) scale of turbulence.

The coefficient A is given by:

$$A = 0.5, \quad R > kL_0^2, \quad (4.74a)$$
$$A = 1.0, \quad R < kL_0^2. \quad (4.74b)$$

The parameters $\langle \mu^2 \rangle$ and L_0 may be determined from field measurements or estimated.
The phase covariance is given by

$$\rho = \frac{\sqrt{\pi}}{2} \frac{L_0}{h} \mathrm{erf}\left(\frac{h}{L_0}\right), \quad (4.75)$$

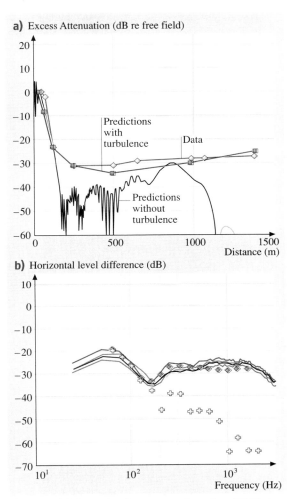

Fig. 4.22 (a) Data for the propagation of a 424 Hz tone as a function of range out to 1.5 km [4.115] compared to FFP predictions with and without turbulence [4.116] and **(b)** Broadband data to 2500 Hz for four 26 s averages (*lines*) of the spectra of the horizontal level differences between sound level measurements at 6.4 m high receivers, 152.4 m and 762 m from a fixed jet engine source (nozzle exit centered at 2.16 m height) compared to FFP predictions with (*diamonds*) and without (*crosses*) turbulence [4.116]

where h is the maximum transverse path separation and $\mathrm{erf}(x)$ is the error function defined by

$$\mathrm{erf}(x) = \frac{2}{\sqrt{\pi}} \int_0^x e^{-t^2} \, dt, \quad (4.76)$$

For a sound field consisting *only* of direct and reflected paths (which will be true at short ranges) in the absence of refraction, the parameter h is given by:

$$\frac{1}{h} = \frac{1}{2}\left(\frac{1}{h_s} + \frac{1}{h_r}\right), \qquad (4.77)$$

where h_s and h_r are the source and receiver heights respectively. *Daigle* [4.117] uses half this value to obtain better agreement with data.

When $h \to 0$, then $\rho \to 1$ and $T \to 1$. This is the case near grazing incidence.

For $h \to$ large, then $T \to$ maximum. This will be the case for a greatly elevated source and/or receiver. The mean-squared refractive index may be calculated from the measured instantaneous variation of wind speed and temperature with time at the receiver. Specifically

$$\langle \mu^2 \rangle = \frac{\sigma_w^2 \cos^2 \alpha}{C_0^2} + \frac{\sigma_T^2}{4T_0^2},$$

where σ_w^2 is the variance of the wind velocity, σ_T^2 is the variance of the temperature fluctuations, α is the wind vector direction, and C_0 and T_0 are the ambient sound speed and temperature, respectively. Typical values of best-fit mean-squared refractive index are between 10^{-6} for calm conditions and 10^{-4} for strong not perfect, represents a large improvement over the results of calculations without turbulence and, in the case turbulence. A typical value of L_0 is 1 m but in general a value equal to the source height should be used.

Figure 4.21 shows example results of computations of excess attenuation spectra using (4.50) through (4.55). Note that increasing turbulence reduces the depth of the main ground-effect dip.

Figures 4.22a, b show comparisons between measured data and theoretical predictions using a full-wave numerical solution (FFP) with and without turbulence [4.116]. The data in Fig. 4.22a were obtained with a loudspeaker source [4.118] in strong upwind conditions modeled by the logarithmic speed of sound gradient [4.115]

$$c(z) = 340.0 - 2.0 \ln\left(z/6 \times 10^{-3}\right), \qquad (4.78)$$

where z is the height in *m*.

Those in Fig. 4.22b were obtained with a fixed jet engine source during conditions with a negative temperature gradient (modeled by two linear segments) and a wind speed that was more or less constant with height at 4 ms^{-1} in the direction between receivers and source.

The agreement obtained between the predictions from the FFP including turbulence and the data, while of Fig. 4.22a, is similar to that obtained with predictions given by the parabolic equation method [4.115].

4.9 Concluding Remarks

There are several areas of outdoor acoustics in which further research is needed. This concluding section discusses current trends and lists some objectives that are yet to be fulfilled.

4.9.1 Modeling Meteorological and Topographical Effects

In Sect. 4.5 it was mentioned that the degradation of noise barrier performance in the presence of wind and temperature gradients has still to be established. Much current research is concerned with atmospheric and topographical effects and with the interaction between meteorology and topography [4.119–121]. Recent trends include combining mesoscale meteorological models with acoustic propagation models. The steady improvement in readily available computational power has encouraged the development of computationally intensive finite-difference time-domain (FDTD) models, which are based on numerical solutions of Euler's equations [4.122, 123] and models combining computational fluid dynamics (CFD) and FDTD [4.124]. Work remains to be done on the acoustical effects of barriers and hills.

4.9.2 Effects of Trees and Tall Vegetation

There is a need for more data on the effects of trees on sound propagation, if only to counteract the conventional wisdom that they have no practical part to play in reducing noise along the propagation path between source and receiver. This belief stems from the idea that there are so many holes in a belt of trees that a belt of practical width does not offer a significant barrier. Yet, as discussed in Sect. 4.7 there is increasing evidence that dense plantings with foliage to ground level can offer significant reductions. Moreover dense periodic plantings could exploit *sonic crystal* effects [4.125], whereby three dimensional (3-D) diffraction gratings produce stop bands or, after deliberate introduction of defects, can be used to direct sound away from receivers.

There is little data on the propagation of sound through crops and shrubs [4.126]. Indeed attenuation rates over tall crops and other vegetation remain uncertain both from the lack of measurements and from the lack of sufficiently sophisticated modeling. Improvement in multiple-scattering models of attenuation through leaves or needles might result from more sophisticated treatment of the boundary conditions. Corn and wheat plants have a large number of long ribbon-like leaves that scatter sound. Such leaves may be modeled either as elliptic cylinders or oblate spheroids, with a minor axis of length zero.

4.9.3 Low-Frequency Interaction with the Ground

Predictions of significant ground elasticity effects have been validated only by data for acoustic-to-seismic coupling, i.e., by measurements of the ratio of ground surface particle velocity relative to the incident sound pressure. Nevertheless given the significant predicted effects of layered ground elasticity on low-frequency propagation above ground it remains of interest to make further efforts to measure them.

4.9.4 Rough-Sea Effects

Coastal communities are regularly exposed to aircraft noise. To reduce the noise impact of sonic booms from civil supersonic flights, it is likely that, wherever possible, the aircraft will pass through the sound barrier over the sea. This means that prediction of aircraft noise in general, and sonic boom characteristics in particular, in coastal areas will be important and will involve propagation over the sea as well as the land. Given that the specific impedance of seawater is greater than that of air by four orders of magnitude, the sea surface may be considered to be acoustically hard. However, it is likely that sound propagation is modified during near-grazing propagation above a rough sea surface. Such propagation is likely to be of interest also when predicting sound propagation from near-ground explosions. Although the sea surface is continuously in motion associated with winds and currents, so that the scattered field is not constant, a sonic boom or blast waveform is sufficiently short compared with the period of the sea surface motion that the roughness may be considered to be static. As long as the incident acoustic wavelengths are large compared with the water wave heights and periods, the effect of the diffraction of sound waves by roughness may be modeled by an effective impedance. This is a convenient way to incorporate the acoustical properties of a rough sea surface into sonic boom and blast sound propagation models. Some modeling work has been carried out on this basis [4.127]. However it remains to be validated by data.

4.9.5 Predicting Outdoor Noise

Most current prediction schemes for outdoor sources such as transportation and industrial plant are empirical. To some extent, this is a reflection of the complexity arising from the variety of source properties, path configurations and receiver locations. On the other hand, apart from distinctions between source types and properties, the propagation factors are common. As discussed in this chapter, there has been considerable progress in modeling outdoor sound. The increasing computational power that is available for routine use makes several of the semi-analytical results and recent numerical codes of more practical value. Consequently, it can be expected that there will be a trend towards schemes that implement a common propagation package but employ differing source descriptors [4.18].

References

4.1 F.V. Hunt: *Origins in Acoustics* (Acoustical Society of America, AIP, New York 1992)
4.2 C.D. Ross: Outdoor sound propagation in the US civil war, Appl. Acoust. **59**, 101–115 (2000)
4.3 M.V. Naramoto: A concise history of acoustics in warfare, Appl. Acoust. **59**, 137–147 (2000)
4.4 G. Becker, A. Gudesen: Passive sensing with acoustics on the battlefield, Appl. Acoust. **59**, 149–178 (2000)
4.5 N. Xiang, J.M. Sabatier: Land mine detection measurements using acoustic-to-seismic coupling. In: *Detection and Remediation Technologies for Mines and Minelike Targets*, Proc. SPIE, Vol. 4038, ed. by V. Abinath, C. Dubey, J.F. Harvey (The Internatonal Society for Optical Engineering, Bellingham, WA 2000) pp. 645–655
4.6 A. Michelson: Sound reception in different environments. In: *Perspectives in Sensory Ecology*, ed. by M.A. Ali (Plenum, New York 1978) pp. 345–373
4.7 V.E. Ostashev: *Acoustics in Moving Inhomogeneous Media* (E & FN Spon, London 1999)
4.8 J.E. Piercy, T.F.W. Embleton, L.C. Sutherland: Review of noise propagation in the atmosphere, J. Acoust. Soc. Am. **61**, 1403–1418 (1977)

4.9 K. Attenborough: Review of ground effects on outdoor sound propagation from continuous broadband sources, Appl. Acoust. **24**, 289–319 (1988)

4.10 J.M. Sabatier, H.M. Hess, P.A. Arnott, K. Attenborough, E. Grissinger, M. Romkens: In situ measurements of soil physical properties by acoustical techniques, Soil Sci. Am. J. **54**, 658–672 (1990)

4.11 H.M. Moore, K. Attenborough: Acoustic determination of air-filled porosity and relative air permeability of soils, J. Soil Sci. **43**, 211–228 (1992)

4.12 N. D. Harrop: The exploitation of acoustic-to-seismic coupling for the determination of soil properties, Ph. D. Thesis, The Open University, 2000

4.13 K. Attenborough, H.E. Bass, X. Di, R. Raspet, G.R. Becker, A. Güdesen, A. Chrestman, G.A. Daigle, A. L'Espérance, Y. Gabillet, K.E. Gilbert, Y.L. Li, M.J. White, P. Naz, J.M. Noble, H.J.A.M. van Hoof: Benchmark cases for outdoor sound propagation models, J. Acoust. Soc. Am. **97**, 173–191 (1995)

4.14 E.M. Salomons: *Computational Atmospheric Acoustics* (Kluwer, Dordrecht 2002)

4.15 M.C. Bérengier, B. Gavreau, P. Blanc-Benon, D. Juvé: A short review of recent progress in modelling outdoor sound propagation, Acta Acustica united with Acustica **89**, 980–991 (2003)

4.16 Directive of the European parliament and of the Council relating to the assessment and management of noise 2002/EC/49: 25 June 2002, Official Journal of the European Communities **L189**, 12–25 (2002)

4.17 J. Kragh, B. Plovsing: *Nord2000. Comprehensive Outdoor Sound Propagation Model. Part I-II. DELTA Acoustics & Vibration Report, 1849-1851/00, 2000* (Danish Electronics, Light and Acoustics, Lyngby, Denmark 2001)

4.18 HARMONOISE contract funded by the European Commission IST-2000-28419, (http://www.harmonoise.org, 2000)

4.19 K.M. Li, S.H. Tang: The predicted barrier effects in the proximity of tall buildings, J. Acoust. Soc. Am. **114**, 821–832 (2003)

4.20 H.E. Bass, L.C. Sutherland, A.J. Zuckewar: Atmospheric absorption of sound: Further developments, J. Acoust. Soc. Am. **97**, 680–683 (1995)

4.21 D.T. Blackstock: *Fundamentals of Physical Acoustics, University of Texas Austin, Texas* (Wiley, New York 2000)

4.22 C. Larsson: Atmospheric Absorption Conditions for Horizontal Sound Propagation, Appl. Acoust. **50**, 231–245 (1997)

4.23 C. Larsson: Weather Effects on Outdoor Sound Propagation, Int. J. Acoust. Vibration **5**, 33–36 (2000)

4.24 ISO 9613-2: *Acoustics: Attenuation of Sound During Propagation Outdoors. Part 2: General Method of Calculation* (International Standard Organisation, Geneva, Switzerland 1996)

4.25 ISO 10847: *Acoustics – In-situ determination of insertion loss of outdoor noise barriers of all types* (International Standards Organisation, Geneva, Switzerland 1997)

4.26 ANSI S12.8: *Methods for Determination of Insertion Loss of Outdoor Noise Barriers* (American National Standard Institute, Washington, DC 1998)

4.27 G.A. Daigle: *Report by the International Institute of Noise Control Engineering Working Party on the Effectiveness of Noise Walls* (Noise/News International I-INCE, Poughkeepsie, NY 1999)

4.28 B. Kotzen, C. English: *Environmental Noise Barriers: A Guide to their Acoustic and Visual Design* (E&Fn Spon, London 1999)

4.29 A. Sommerfeld: Mathematische Theorie der Diffraction, Math. Ann. **47**, 317–374 (1896)

4.30 H.M. MacDonald: A class of diffraction problems, Proc. Lond. Math. Soc. **14**, 410–427 (1915)

4.31 S.W. Redfearn: Some acoustical source–observer problems, Philos. Mag. Ser. **7**, 223–236 (1940)

4.32 J.B. Keller: The geometrical theory of diffraction, J. Opt. Soc. **52**, 116–130 (1962)

4.33 W.J. Hadden, A.D. Pierce: Sound diffraction around screens and wedges for arbitrary point source locations, J. Acoust. Soc. Am. **69**, 1266–1276 (1981)

4.34 T.F.W. Embleton: Line integral theory of barrier attenuation in the presence of ground, J. Acoust. Soc. Am. **67**, 42–45 (1980)

4.35 P. Menounou, I.J. Busch-Vishniac, D.T. Blackstock: Directive line source model: A new model for sound diffracted by half planes and wedges, J. Acoust. Soc. Am. **107**, 2973–2986 (2000)

4.36 H. Medwin: Shadowing by finite noise barriers, J. Acoust. Soc. Am. **69**, 1060–1064 (1981)

4.37 Z. Maekawa: Noise reduction by screens, Appl. Acoust. **1**, 157–173 (1968)

4.38 K.J. Marsh: The CONCAWE Model for Calculating the Propagation of Noise from Open-Air Industrial Plants, Appl. Acoust. **15**, 411–428 (1982)

4.39 R.B. Tatge: Barrier-wall attenuation with a finite sized source, J. Acoust. Soc. Am. **53**, 1317–1319 (1973)

4.40 U.J. Kurze, G.S. Anderson: Sound attenuation by barriers, Appl. Acoust. **4**, 35–53 (1971)

4.41 P. Menounou: A correction to Maekawa's curve for the insertion loss behind barriers, J. Acoust. Soc. Am. **110**, 1828–1838 (2001)

4.42 Y.W. Lam, S.C. Roberts: A simple method for accurate prediction of finite barrier insertion loss, J. Acoust. Soc. Am. **93**, 1445–1452 (1993)

4.43 M. Abramowitz, I.A. Stegun: *Handbook of Mathematical Functions with Formulas, Graphs, and Mathematical Tables* (Dover, New York 1972)

4.44 K. Attenborough, S.I. Hayek, J.M. Lawther: Propagation of sound above a porous half-space, J. Acoust. Soc. Am. **68**, 1493–1501 (1980)

4.45 A. Banos: *Dipole radiation in the presence of conducting half-space.* (Pergamon, New York 1966), Chap. 2–4

4.46 R.J. Donato: Model experiments on surface waves, J. Acoust. Soc. Am. **63**, 700–703 (1978)

4.47 G.A. Daigle, M.R. Stinson, D.I. Havelock: Experiments on surface waves over a model impedance using acoustical pulses, J. Acoust. Soc. Am. **99**, 1993–2005 (1996)

4.48 Q. Wang, K.M. Li: Surface waves over a convex impedance surface, J. Acoust. Soc. Am. **106**, 2345–2357 (1999)

4.49 D.G. Albert: Observation of acoustic surface waves in outdoor sound propagation, J. Acoust. Soc. Am. **113**, 2495–2500 (2003)

4.50 K.M. Li, T. Waters-Fuller, K. Attenborough: Sound propagation from a point source over extended-reaction ground, J. Acoust. Soc. Am. **104**, 679–685 (1998)

4.51 J. Nicolas, J.L. Berry, G.A. Daigle: Propagation of sound above a finite layer of snow, J. Acoust. Soc. Am. **77**, 67–73 (1985)

4.52 J.F. Allard, G. Jansens, W. Lauriks: Reflection of spherical waves by a non-locally reacting porous medium, Wave Motion **36**, 143–155 (2002)

4.53 M.E. Delany, E.N. Bazley: Acoustical properties of fibrous absorbent materials, Appl. Acoust. **3**, 105–116 (1970)

4.54 K.B. Rasmussen: Sound propagation over grass covered ground, J. Sound Vib. **78**, 247–255 (1981)

4.55 K. Attenborough: Ground parameter information for propagation modeling, J. Acoust. Soc. Am. **92**, 418–427 (1992), see also R. Raspet, K. Attenborough: Erratum: Ground parameter information for propagation modeling. J. Acoust. Soc. Am. 92, 3007 (1992)

4.56 R. Raspet, J.M. Sabatier: The surface impedance of grounds with exponential porosity profiles, J. Acoust. Soc. Am. **99**, 147–152 (1996)

4.57 K. Attenborough: Models for the acoustical properties of air-saturated granular materials, Acta Acust. **1**, 213–226 (1993)

4.58 J.F. Allard: *Propagation of Sound in Porous Media: Modelling Sound Absorbing Material* (Elsevier, New York 1993)

4.59 D.L. Johnson, T.J. Plona, R. Dashen: Theory of dynamic permeability and tortuosity in fluid-saturated porous media, J. Fluid Mech. **176**, 379–401 (1987)

4.60 O. Umnova, K. Attenborough, K.M. Li: Cell model calculations of dynamic drag parameters in packings of spheres, J. Acoust. Soc. Am. **107**, 3113–3119 (2000)

4.61 K.V. Horoshenkov, K. Attenborough, S.N. Chandler-Wilde: Padé approximants for the acoustical properties of rigid frame porous media with pore size distribution, J. Acoust. Soc. Am. **104**, 1198–1209 (1998)

4.62 T. Yamamoto, A. Turgut: Acoustic wave propagation through porous media with arbitrary pore size distributions, J. Acoust. Soc. Am. **83**, 1744–1751 (1988)

4.63 ANSI S1.18-1999: *Template method for Ground Impedance, Standards Secretariat* (Acoustical Society of America, New York 1999)

4.64 S. Taherzadeh, K. Attenborough: Deduction of ground impedance from measurements of excess attenuation spectra, J. Acoust. Soc. Am. **105**, 2039–2042 (1999)

4.65 K. Attenborough, T. Waters-Fuller: Effective impedance of rough porous ground surfaces, J. Acoust. Soc. Am. **108**, 949–956 (2000)

4.66 P. Boulanger, K. Attenborough, S. Taherzadeh, T.F. Waters-Fuller, K.M. Li: Ground effect over hard rough surfaces, J. Acoust. Soc. Am. **104**, 1474–1482 (1998)

4.67 P.H. Parkin, W.E. Scholes: The horizontal propagation of sound from a jet close to the ground at Radlett, J. Sound Vib. **1**, 1–13 (1965)

4.68 P.H. Parkin, W.E. Scholes: The horizontal propagation of sound from a jet close to the ground at Hatfield, J. Sound Vib. **2**, 353–374 (1965)

4.69 K. Attenborough, S. Taherzadeh: Propagation from a point source over a rough finite impedance boundary, J. Acoust. Soc. Am. **98**, 1717–1722 (1995)

4.70 Airbus France SA (AM-B), Final Technical Report on Project n° GRD1-2000-25189 SOBER (2004)

4.71 D.E. Aylor: Noise reduction by vegetation and ground, J. Acoust. Soc. Am. **51**, 197–205 (1972)

4.72 K. Attenborough, T. Waters-Fuller, K.M. Li, J.A. Lines: Acoustical properties of Farmland, J. Agric. Eng. Res. **76**, 183–195 (2000)

4.73 O. Zaporozhets, V. Tokarev, K. Attenborough: Predicting Noise from Aircraft Operated on the Ground, Appl. Acoust. **64**, 941–953 (2003)

4.74 C. Madshus, N.I. Nilsen: Low frequency vibration and noise from military blast activity –prediction and evaluation of annoyance, Proc. InterNoise 2000, Nice, France (2000)

4.75 Øhrstrøm E.: (1996) Community reaction to noise and vibrations from railway traffic. Proc. Internoise 1996, Liverpool U.K.

4.76 ISO 2631-2 (1998), Mechanical vibration and shock – Evaluation of human exposure to whole body vibration – Part 2: Vibration in buildings

4.77 W.M. Ewing, W.S. Jardetzky, F. Press: *Elastic Waves in Layered Media* (McGraw-Hill, New York 1957)

4.78 J.M. Sabatier, H.E. Bass, L.M. Bolen, K. Attenborough: Acoustically induced seismic waves, J. Acoust. Soc. Am. **80**, 646–649 (1986)

4.79 S. Tooms, S. Taherzadeh, K. Attenborough: Sound propagating in a refracting fluid above a layered fluid saturated porous elastic material, J. Acoust. Soc. Am. **93**, 173–181 (1993)

4.80 K. Attenborough: On the acoustic slow wave in air filled granular media, J. Acoust. Soc. Am. **81**, 93–102 (1987)

4.81 R.D. Stoll: Theoretical aspects of sound transmission in sediments, J. Acoust. Soc. Am. **68**(5), 1341–1350 (1980)

4.82 C. Madshus, F. Lovholt, A. Kaynia, L.R. Hole, K. Attenborough, S. Taherzadeh: Air-ground interaction in long range propagation of low frequency sound and vibration – field tests and model verification, Appl. Acoust. **66**, 553–578 (2005)

4.83 G.A. Daigle, M.R. Stinson: Impedance of grass covered ground at low frequencies using a phase difference technique, J. Acoust. Soc. Am. **81**, 62–68 (1987)

4.84 M.W. Sprague, R. Raspet, H.E. Bass, J.M. Sabatier: Low frequency acoustic ground impedance measurement techniques, Appl. Acoust. **39**, 307–325 (1993)

4.85 M.A. Price, K. Attenborough, N.W. Heap: Sound attenuation through trees: Measurements and Models, J. Acoust. Soc. Am. **84**, 1836–1844 (1988)

4.86 J. Kragh: Road traffic noise attenuation by belts of trees and bushes, Danish Acoustical Laboratory Report no.31 (1982)

4.87 L. R. Huddart: The use of vegetation for traffic noise screening, TRRL Research Report 238 (1990)

4.88 G. M. Heisler, O. H. McDaniel, K. K. Hodgdon, J. J. Portelli, S. B. Glesson: Highway Noise Abatement in Two Forests, Proc. NOISE-CON 87, PSU, USA (1987)

4.89 W.H.T. Huisman: *Sound propagation over vegetation-covered ground. Ph.D. Thesis* (University of Nijmegen, The Netherlands 1990)

4.90 J. Defrance, N. Barriere, E. Premat: Forest as a meteorological screen for traffic noise. Proc. 9th ICSV, Orlando (2002)

4.91 N. Barrière, Y. Gabillet: Sound propagation over a barrier with realistic wind gradients. Comparison of wind tunnel experiments with GFPE computations, Acustica-Acta Acustica **85**, 325–334 (1999)

4.92 N. Barrière: *Etude théorique et expérimentale de la propagation du bruit de trafic en forêt. Ph.D. Thesis* (Ecole Centrale de Lyon, France 1999)

4.93 M. E. Swearingen, M. White: Sound propagation through a forest: a predictive model. Proc. 11th LRSPS, Vermont (2004)

4.94 K.M. Li, K. Attenborough, N.W. Heap: Source height determination by ground effect inversion in the presence of a sound velocity gradient, J. Sound Vib. **145**, 111–128 (1991)

4.95 I. Rudnick: Propagation of sound in open air. In: *Handbook of Noise Control, Chap. 3*, ed. by C.M. Harris (McGraw–Hill, New York 1957) pp. 3:1–3:17

4.96 K.M. Li: On the validity of the heuristic ray-trace based modification to the Weyl Van der Pol formula, J. Acoust. Soc. Am. **93**, 1727–1735 (1993)

4.97 V. Zouboff, Y. Brunet, M. Berengier, E. Sechet: Proc. 6th International Symposium on Long range Sound propagation, ed. D.I.Havelock and M.Stinson, Ottawa, 251-269 (1994)

4.98 The propagation of noise from petroleum and petrochemical complexes to neighbouring communities, CONCAWE Report no.4/81, Den Haag (1981)

4.99 A.S. Monin, A.M. Yaglom: *Statistical Fluid Mechanics: mechanics of turbulence*, Vol. 1 (MIT press, Cambridge 1979)

4.100 R.B. Stull: *An Introduction to Boundary Layer Meteorology* (Kluwer, Dordrecht 1991) pp. 347–347

4.101 E.M. Salomons: Downwind propagation of sound in an atmosphere with a realistic sound speed profile: A semi-analytical ray model, J. Acoust. Soc. Am. **95**, 2425–2436 (1994)

4.102 A.A.M. Holtslag: Estimates of diabatic wind speed profiles from near surface weather observations, Boundary-Layer Meteorol. **29**, 225–250 (1984)

4.103 A.G. Davenport: Rationale for determining design wind velocities, J. Am. Soc. Civ. Eng. **ST-86**, 39–68 (1960)

4.104 E. M. Salomons, F. H. van den Berg, H. E. A. Brackenhoff: Long-term average sound transfer through the atmosphere based on meteorological statistics and numerical computations of sound propagation. Proc. 6th International Symposium on Long range Sound propagation, ed. D.I.Havelock and M.Stinson, NRCC, Ottawa, 209-228 (1994)

4.105 D. Heimann, E. Salomons: Testing meteorological classifications for the prediction of long-term average sound levels, J. Am. Soc. Civ. Eng. **65**, 925–950 (2004)

4.106 T.F.W. Embleton: Tutorial on sound propagation outdoors, J. Acoust. Soc. Am. **100**, 31–48 (1996)

4.107 L.C. Sutherland, G.A. Daigle: Atmospheric sound propagation. In: *Encylclopedia of Acoustics*, ed. by M.J. Crocker (Wiley, New York 1998) pp. 305–329

4.108 M. R. Stinson, D. J. Havelock, G. A. Daigle: Simulation of scattering by turbulence into a shadow zone region using the GF-PE method, 283 – 307 Proc. 6th LRSPS Ottawa, NRCC, 1996

4.109 D. K. Wilson: A brief tutorial on atmospheric boundary-layer turbulence for acousticians. 111 – 122, Proc. 7th LRSPS, Ecole Centrale, Lyon, 1996

4.110 D.K. Wilson, J.G. Brasseur, K.E. Gilbert: Acoustic scattering and the spectrum of atmospheric turbulence, J. Acoust. Soc. Am. **105**, 30–34 (1999)

4.111 A. L'Esperance, G. A. Daigle, Y. Gabillet: Estimation of linear sound speed gradients associated to general meteorological conditions. Proc. 6th LRSPS Ottawa, NRCC, (1996)

4.112 D. K. Wilson: On the application of turbulence spectral/correlation models to sound propagation in the atmosphere, Proc. 8th LRSPS, Penn State (1988)

4.113 V.E. Ostashev, D.K. Wilson: Relative contributions from temperature and wind velocity fluctuations to the statistical moments of a sound field in a turbulent atmosphere, Acustica – Acta Acustica **86**, 260–268 (2000)

4.114 S.F. Clifford, R.T. Lataitis: Turbulence effects on acoustic wave propagation over a smooth surface, J. Acoust. Soc. Am. **73**, 1545–1550 (1983)

4.115 K.E. Gilbert, R. Raspet, X. Di: Calculation of turbulence effects in an upward refracting atmosphere, J. Acoust. Soc. Am. **87**(6), 2428–2437 (1990)

4.116 S. Taherzadeh, K. Attenborough: Using the Fast Field Program in a refracting, turbulent medium,

4.117 Report 3 for ESDU International (unpublished) (1999).

4.117 G.A. Daigle: Effects of atmospheric turbulence on the interference of sound waves above a finite impedance boundary, J. Acoust. Soc. Am. **65**, 45–49 (1979)

4.118 F.M. Weiner, D.N. Keast: Experimental study of the propagation of sound over ground, J. Acoust. Soc. Am. **31**(6), 724–733 (1959)

4.119 E.M. Salomons: Reduction of the performance of a noise screen due to screen induced wind-speed gradients. Numerical computations and wind-tunnel experiments, J. Acoust. Soc. Am. **105**, 2287–2293 (1999)

4.120 E.M. Salomons, K.B. Rasmussen: Numerical computation of sound propagation over a noise screen based on an analytic approximation of the wind speed field, Appl. Acoust. **60**, 327–341 (2000)

4.121 M. West, Y.W. Lam: Prediction of sound fields in the presence of terrain features which produce a range dependent meteorology using the generalised terrain parabolic equation (GT-PE) model. Proc. InterNoise 2000, Nice, France 2, 943 (2000)

4.122 E.M. Salomons, R. Blumrich, D. Heimann: Eulerian time-domain model for sound propagation over a finite impedance ground surface. Comparison with frequency-domain models, Acust. Acta Acust. **88**, 483–492 (2002)

4.123 R. Blumrich, D. Heimann: A linearized Eulerian sound propagation model for studies of complex meteorological effects, J. Acoust. Soc. Am. **112**, 446–455 (2002)

4.124 T. Van Renterghem, D. Botteldooren: Numerical simulation of the effect of trees on down-wind noise barrier performance, Acust. Acta Acust. **89**, 764–778 (2003)

4.125 J.V. Sanchez-Perez, C. Rubio, R. Martinez-Sala, R. Sanchez-Grandia, V. Gomez: Acoustic barriers based on periodic arrays of scatterers, Appl. Phys. Lett. **81**, 5240–5242 (2002)

4.126 D.E. Aylor: Sound transmission through vegetation in relation to leaf area density, leaf width and breadth of canopy, J. Acoust. Soc. Am. **51**, 411–414 (1972)

4.127 P. Boulanger, K. Attenborough: Effective impedance spectra for predicting rough sea effects on atmospheric impulsive sounds, J. Acoust. Soc. Am. **117**, 751–762 (2005)

5. Underwater Acoustics

It is well established that sound waves, compared to electromagnetic waves, propagate long distances in the ocean. Hence, in the ocean as opposed to air or a vacuum, one uses sound navigation and ranging (SONAR) instead of radar, acoustic communication instead of radio, and acoustic imaging and tomography instead of microwave or optical imaging or X-ray tomography. Underwater acoustics is the science of sound in water (most commonly in the ocean) and encompasses not only the study of sound propagation, but also the masking of sound signals by interfering phenomenon and signal processing for extracting these signals from interference. This chapter we will present the basics physics of ocean acoustics and then discuss applications.

5.1	**Ocean Acoustic Environment**		151
	5.1.1	Ocean Environment	151
	5.1.2	Basic Acoustic Propagation Paths	152
	5.1.3	Geometric Spreading Loss	154
5.2	**Physical Mechanisms**		155
	5.2.1	Transducers	155
	5.2.2	Volume Attenuation	157
	5.2.3	Bottom Loss	158
	5.2.4	Scattering and Reverberation	159
	5.2.5	Ambient Noise	160
	5.2.6	Bubbles and Bubbly Media	162
5.3	**SONAR and the SONAR Equation**		165
	5.3.1	Detection Threshold and Receiver Operating Characteristics Curves	165
	5.3.2	Passive SONAR Equation	166
	5.3.3	Active SONAR Equation	167
5.4	**Sound Propagation Models**		167
	5.4.1	The Wave Equation and Boundary Conditions	168
	5.4.2	Ray Theory	168
	5.4.3	Wavenumber Representation or Spectral Solution	169
	5.4.4	Normal-Mode Model	169
	5.4.5	Parabolic Equation (PE) Model	172
	5.4.6	Propagation and Transmission Loss	174
	5.4.7	Fourier Synthesis of Frequency-Domain Solutions	175
5.5	**Quantitative Description of Propagation**		177
5.6	**SONAR Array Processing**		179
	5.6.1	Linear Plane-Wave Beam-Forming and Spatio-Temporal Sampling	179
	5.6.2	Some Beam-Former Properties	181
	5.6.3	Adaptive Processing	182
	5.6.4	Matched Field Processing, Phase Conjugation and Time Reversal	182
5.7	**Active SONAR Processing**		185
	5.7.1	Active SONAR Signal Processing	185
	5.7.2	Underwater Acoustic Imaging	187
	5.7.3	Acoustic Telemetry	191
	5.7.4	Travel-Time Tomography	192
5.8	**Acoustics and Marine Animals**		195
	5.8.1	Fisheries Acoustics	195
	5.8.2	Marine Mammal Acoustics	198
5.A	**Appendix: Units**		201
References			201

During the two World Wars, both shallow and deep-water acoustics studies were pursued, but during the Cold War, emphasis shifted sharply to deep water. The post-World War II (WWII) history of antisubmarine warfare (ASW) actually started in 1943 with Ewing and Worzel discovering the deep sound channel (DSC) caused by a minimum in the temperature-dependent sound speed. (Brekhovskikh of the Soviet Union also discovered it independently, but later.) This minimum has been mapped (dotted line in Fig. 5.1), and typically varies from the cold surface at the poles to a depth of about 1300 m at the equator. Since sound refracts toward lower sound speeds, the DSC produces a refraction-generated waveguide (gray lines) contained within the ocean, such that sound paths oscillate about the sound speed minimum and can propagate thousands of kilometers.

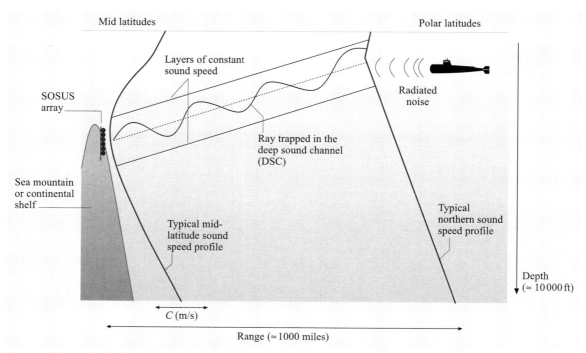

Fig. 5.1 Schematic of a long-range passive detection of a submarine in polar waters by a surveillance system in a temperate region [5.1]

Exploiting the DSC, the US Navy created the multi-billion dollar sound ocean surveillance system (SOSUS) network to monitor Soviet ballistic-missile nuclear submarines. Acoustic antennas were placed on ocean mountains or continental rises whose height extended into the DSC. These antennas were hard-wired to land stations using reliable undersea telephone cable technology. Submarines typically go down to depths of a few hundred meters. With many submarines loitering in polar waters, they were coupling into the DSC at shallower depths. Detections were made on very narrow-band radiation caused by imperfect, rotating machinery such as propellers. The advantage of detecting a set of narrow-band lines is that most of the broadband ocean noise can be filtered out. Though it was a Cold War, the multi-decade success of SOSUS was, in effect, a major Naval victory. The system was compromised by a spy episode, when the nature of the system was revealed. The result was a Soviet submarine quietening program and over the years, the Soviet fleet became quieter, reducing the long-range capability of the SOSUS system. The end of the Cold War led to an emphasis on the issue of detecting very quiet diesel–electric submarines in the noisy, shallow water that encompasses about 5% of the World's oceans on the continental shelves, roughly the region from the beach to the shelfbreak at about $\approx 200\,\mathrm{m}$ depth. However, there are also signs of a rekindling of interest in deep-water problems.

Parallel to these military developments, the field of ocean acoustics also grew for commercial, environmental and other purposes. Since the ocean environment has a large effect on acoustic propagation and therefore SONAR performance, acoustic methods to map and otherwise study the ocean were developed. As active SONARs were being put forward as a solution for the detection of quiet submarines, there was a growing need to study the effects of sound on marine mammals. Commercially, acoustic methods for fish finding and counting were developed as well as bottom-mapping techniques, the latter being having both commercial and military applications. All in all, ocean acoustics research and development has blossomed in the last half-century and many standard monographs and textbooks are now available (e.g. [5.2–10]).

5.1 Ocean Acoustic Environment

The acoustic properties of the ocean, such as the paths along which sound from a localized source travel, are mainly dependent on the ocean sound speed structure, which in turn is dependent on the oceanographic environment. The combination of water column and bottom properties leads to a set of generic sound-propagation paths descriptive of most propagation phenomena in the ocean.

5.1.1 Ocean Environment

Sound speed in the ocean water column is a function of temperature, salinity and ambient pressure. Since the ambient pressure is a function of depth, it is customary to express the sound speed (c) in m/s as an empirical function of temperature (T) in degrees centigrade, salinity (S) in parts per thousand and depth (z) in meters, for example [5.7, 11, 12]

$$c = 1449.2 + 4.6T - 0.055T^2 + 0.00029T^3 \\ + (1.34 - 0.01T)(S - 35) + 0.016z \ . \quad (5.1)$$

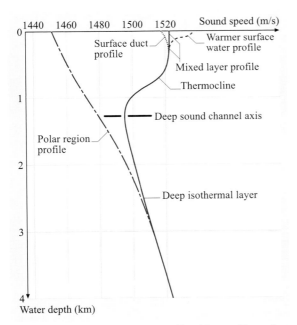

Fig. 5.2 Generic sound-speed profiles. The profiles reflect the tendencies that sound speed varies directly with temperature and hydrostatic pressure. Near-surface mixing can lead to almost isovelocity in that region. In polar waters, the coldest region is at the surface

Figure 5.2 shows a typical set of sound speed profiles indicating the greatest variability near the surface. In a warmer season (or warmer part of the day, sometimes referred to as the *afternoon effect*), the temperature increases near the surface and hence the sound speed increases toward the sea surface. In nonpolar regions where mixing near the surface due to wind and wave activity is important, a *mixed layer* of almost constant temperature is often created. In this isothermal layer the sound speed increases with depth because of the increasing ambient pressure, the last term in (5.1). This is the *surface duct* region. Below the mixed layer is the thermocline where the temperature and hence the sound speed decreases with depth. Below the thermocline, the temperature is constant and the sound speed increases because of increasing ambient pressure. Therefore, between the deep isothermal region and the mixed layer, there is a depth at minimum sound speed referred to as the axis of the *deep sound channel*. However, in polar regions, the water is coldest near the surface so that the minimum sound speed is at the surface. Figure 5.3 is a contour display of the sound speed structure of the North and South Atlantic with the deep sound channel axis indicated by the heavy dashed line. Note that the deep sound channel becomes shallower toward the poles. Aside from sound speed effects, the ocean volume is absorptive and causes attenuation that increases with acoustic frequency.

Shallower water such as that in continental shelf and slope regions is not deep enough for the depth-pressure term in (5.1) to be significant. Thus the winter profile tends to isovelocity, simply because of mixing, whereas the summer profile has a higher sound speed near the surface due to heating; both are schematically represented in Fig. 5.4.

The sound speed structure regulates the interaction of sound with the boundaries. The ocean is bounded above by air, which is a nearly perfect reflector; however, the sea surface is often rough, causing sound to scatter in directions away from the specular reflecting angle. The ocean bottom is typically a complicated, rough, layered structure supporting elastic waves. Its geoacoustic properties are summarized by density, compressional and shear speed, and attenuation profiles. The two basic interfaces, air/sea and sea/bottom, can be thought of as the boundaries of an acoustic waveguide whose internal index of refraction is determined by the fun-

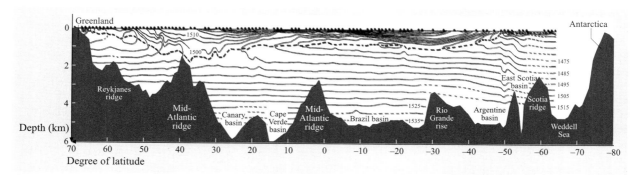

Fig. 5.3 Sound-speed contours at 5 m/s intervals taken from the North and South Atlantic along 30.50° W. The *dashed line* indicates the axis of the deep sound channel (after [5.13]). The sound channel is deepest near the equator and comes to the surface at the poles

damental oceanographic parameters represented in the sound speed equation (5.2).

Due to the density stratification of the water column, the interior of the ocean supports a variety of waves, just as the ocean surface does. One particularly important type of wave in both shallow water and deep water is the internal gravity wave (IW) [5.14]. This wave type is bounded in frequency between the inertial frequency, $f = 2\Omega \sin\theta$, where Ω is the rotation frequency of the earth and θ is the latitude, and the highest buoyancy frequency (or Brunt–Vaisala frequency) $N_{max}(z)$, where $N^2(z) = -(g/\rho)\,d\rho/dz$, and $\rho(z)$ is the density of the fluid as a function of depth z. The inertial frequency varies from two cycles per day at the poles to zero cycles per day at the equator, and the maximum buoyancy frequency is usually on the order of 5–10 cycles per hour.

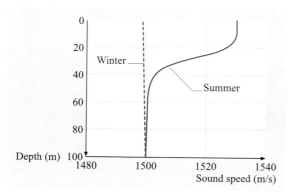

Fig. 5.4 Typical summer and winter shallow-water sound-speed profiles. Warming causes the high-speed region near the surface in the summer. Without strong heating, mixing tends to make the shallow-water region isovelocity in the winter

Two categories of IWs are found in stratified coastal waters: linear and nonlinear waves. The linear waves, found virtually everywhere, obey a standard linear wave equation for the displacement of the surfaces of constant density (*isopycnal* surfaces). The nonlinear IWs, which are generated under somewhat more specialized circumstances than the linear waves (and thus are not always present), can obey a family of nonlinear wave equations. The most useful and illustrative of them is the familiar Korteweg–deVries equation (KdV), which governs the horizontal components of the nonlinear internal waves. The vertical component of the nonlinear internal waves obeys a normal-mode equation.

5.1.2 Basic Acoustic Propagation Paths

Sound propagation in the ocean can be qualitatively broken down into three classes: very-short-range, deep-water and shallow-water propagation.

Very-Short-Range Propagation

The pressure amplitude from a point source in free space falls off with range r as r^{-1}; this geometric loss is called *spherical spreading*. Most sources of interest in the deep ocean are closer to the surface than to the bottom. Hence, the two main short-range paths are the direct path and the surface reflected path. When these two paths *interfere*, they produce a spatial distribution of sound often referred to as a Lloyd mirror pattern, as shown in insert of Fig. 5.5. Also, with reference to Fig. 5.5, note that the *transmission loss* is a decibel measure of the decay with distance of acoustic intensity from a source relative to its value at unit distance (see Appendix), the latter being proportional to the square of the acoustic amplitude.

Long-Range Propagation Paths

Figure 5.6 is a schematic of propagation paths in the ocean resulting from the sound-speed profiles (indicated by the dashed line) described above in Fig. 5.2. These paths can be understood from *Snell's law*,

$$\frac{\cos\theta(z)}{c(z)} = \text{constant}, \quad (5.2)$$

which relates the ray angle $\theta(z)$ with respect to the horizontal, to the local sound speed $c(z)$ at depth z. The equation requires that, the higher the sound speed, the smaller the angle with the horizontal, meaning that sound bends away from regions of high sound speed; or said another way, sound bends toward regions of low sound speed. Therefore, paths 1, 2, and 3 are the simplest to explain since they are paths that oscillate about the local sound speed minima. For example, path 3 depicted by a ray leaving a source near the deep sound channel axis at a small horizontal angle propagates in the deep sound channel. This path, in temperate latitudes where the sound speed minimum is far from the surface, permits propagation over distances of thousands of kilometers.

The upper turning point of this path typically interacts with the thermocline, which is a region of strong internal wave activity. Path 4, which is at slightly steeper angles and is usually excited by a near-surface source, is *convergence zone* propagation, a spatially periodic (35–65 km) refocusing phenomenon producing zones of high intensity near the surface due to the upward refracting nature of the deep sound-speed profile. Regions between these zones are referred to as shadow regions. Referring back to Fig. 5.2, there may be a depth in the deep isothermal layer at which the sound speed is the same as it is at the surface. This depth is called the *critical depth* and is the lower limit of the deep sound channel. If the critical depth is in the water column, the environment supports long-distance propagation without bottom interaction whereas if there is no critical depth in the water column, the ocean bottom is the lower boundary of the deep sound channel. The *bottom bounce* path 5 is also a periodic phenomenon but with a shorter cycle distance and shorter propagation distance because of losses when sound is reflected from the ocean bottom. Finally, note that when bottom baths are described in the general context of the spectral properties of waveguide propagation, they are described in terms of the continuous horizontal wavenumber region as explained in the discussion associated with Fig. 5.32a.

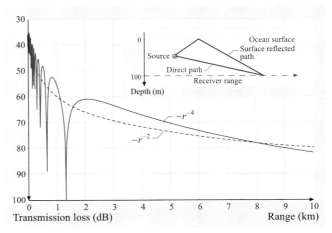

Fig. 5.5 The insert shows the geometry of the Lloyd's mirror effect. The plots show a comparison of Lloyd's mirror (*full line*) to spherical spreading (*dashed line*). Transmission losses are plotted in decibels corresponding to losses of $10\log r^2$ and $10\log r^4$, respectively, as explained in Sect. 5.1.3

Shallow Water and Waveguide Propagation

In general the ocean can be thought of as an acoustic waveguide [5.1]; this waveguide physics is particularly evident in shallow water (inshore out to the continental slope, typically to depths of a few hundred meters). Snell's law applied to the summer profile in Fig. 5.4 produces rays which bend more toward the bottom than winter profiles in which the rays tend to be straight. This implies two effects with respect to the ocean bottom: (1) for a given range, there are more bounces off

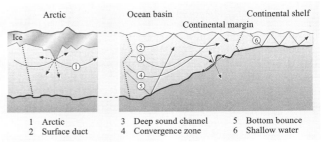

1 Arctic 3 Deep sound channel 5 Bottom bounce
2 Surface duct 4 Convergence zone 6 Shallow water

Fig. 5.6 Schematic representation of various types of sound propagation in the ocean. An intuitive guide is that Snell's law has sound turning toward lower-speed regions. That alone explains all the refractive paths: 1, 2, 3 and 4. It will also explain any curvature associated with paths 5 and 6. Thus, the summer profile would have path 6 curving downward (this curvature is not shown in the figure) while the deep sound-speed profile below the minimum curves upward (4)

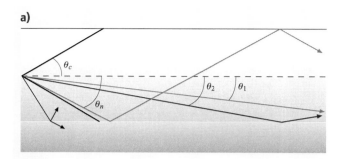

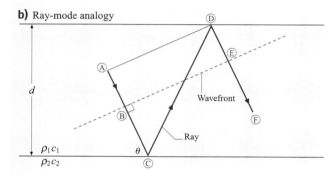

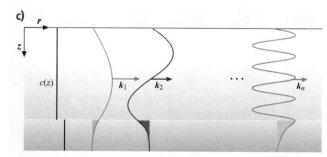

Fig. 5.7a–c Ocean waveguide propagation. (**a**) Long-distance propagation occurs within the critical angle cone of $2\theta_c$. (**b**) For the example shown, the condition for constructive interference is that the phase change along BCDE be a multiple of 2π. (**c**) The constructive interference can be interpreted as discrete modes traveling in the waveguide, each with their own horizontal wavenumber

than winter paths, summer shallow-water propagation is lossier than winter. This result is tempered by rough winter surface conditions that generate large scattering losses at the higher frequencies.

For simplicity we consider an isovelocity waveguide bounded above by the air–water interface and below by a two-fluid interface that is classically defined as a Pekeris waveguide. From Sect. 5.2.3, we have perfect reflection with a 180° phase change at the surface and for grazing angles lower than the bottom critical angle, there will also be perfect bottom reflection. Therefore, as schematically indicated in Fig. 5.7a, ray paths within a cone of $2\theta_c$ will propagate unattenuated down the waveguide. Because the up- and down-going rays have equal amplitudes, preferred angles will exist for which constructive interference occurs. These particular angles are associated with the normal modes of the waveguide, as formally derived from the wave equation in the Sect. 5.4. However, it is instructive to understand the geometric origin of the waveguide modal structure. Figure 5.7b is a schematic of a ray reflected from the bottom and then the surface of a Pekeris waveguide. Consider a ray along the path ACDF and its wavefront, which is perpendicular to the ray. The two down-going rays of equal amplitude, AC and DF, will constructively interfere if points B and E have a phase difference of a multiple of 2π (and similarly for up-going rays). The phase change at the two boundaries must be included. There is a discrete set of angles up to the critical angle for which this constructive interference takes place and, hence, for which sound propagates. This discrete set, in terms of wave physics, are called the normal modes of the waveguide, illustrated in Fig. 5.7c. They correspond to the ray schematic of Fig. 5.7a. Mode propagation is further discussed in the Sect. 5.4.4.

5.1.3 Geometric Spreading Loss

The energy per unit time emitted by a sound source is flowing through a larger area with increasing range. Intensity is the power flux through a unit area, which translates to the energy flow per unit time through a unit area. The simplest example of geometric loss is spherical spreading for a point source in free space where the area increases as r^2, where r is the range from the point source. So spherical spreading results in an intensity decay proportional to r^{-2}. Since intensity is proportional to the square of the pressure amplitude, the fluctuations in pressure induced by the sound p, decay as r^{-1}. For range-independent ducted propagation, that is, where rays are refracted or reflected back towards the horizon-

the ocean bottom in summer than in the winter, and (2) the ray angles intercepting the bottom are steeper in the summer than in the winter. A qualitative understanding of the reflection properties of the ocean bottom should therefore be very revealing of sound propagation in summer versus winter. Basically, near-grazing incidence is much less lossy than larger, more-vertical angles of incidence. Since summer propagation paths have more bounces, each of which are at steeper angles

tal direction, there is no loss associated with the vertical dimension. In this case, the spreading surface is the area of the cylinder whose axis is in the vertical direction passing through the source $2\pi rH$, where H is the depth of the duct (waveguide), and is constant. Geometric loss in the near-field Lloyd-mirror regime requires consideration of interfering beams from direct and surface reflected paths. To summarize, the geometric spreading laws for the pressure field (recall that intensity is proportional to the square of the pressure.) are:

- Spherical spreading loss: $p \propto r^{-1}$;
- Cylindrical spreading loss: $p \propto r^{-1/2}$;
- Lloyd-mirror loss: $p \propto r^{-2}$.

5.2 Physical Mechanisms

The physical mechanisms associated with the generation, reception, attenuation and scattering of sound in the ocean are discussed in this section.

5.2.1 Transducers

A transducer converts some sort of energy to sound (source) or converts sound energy (receiver) to an electrical signal [5.15]. In underwater acoustics piezoelectric and magnetostrictive transducers are commonly used; the former connects electric polarization to mechanical strain and the latter connects the magnetization of a ferromagnetic material to mechanical strain. Piezoelectric transducers represent more than 90% of the sound sources used in the ocean. Magnetostrictive transducers are more expensive, have poor efficiency and a narrow frequency bandwidth. However, they allow large vibration amplitudes and are relevant to low-frequency high-power applications. In addition there are: electrodynamic transducers in which sound pressure oscillations move a current-carrying coil through a magnetic field causing a back emf, and electrostatic transducers in which charged electrodes moving in a sound field change the capacitance of the system. Explosives, air guns, electric discharges, and lasers are also used as wide-band sources.

Hydrophones, underwater acoustic receivers, are commonly piezoelectric devices with good sensitivity and low internal noise levels. Hydrophones usually work on large frequency bandwidths since they do not need to be adjusted to a resonant frequency. They are associated with low-level electronics such as preamplifiers and filters.

Because the field of transducers is large by itself, we concentrate in this section on some very practical issues that are immediately necessary to either convert received voltage levels to pressure levels or transmitter excitation to pressure levels. Practical issues about transducers and hydrophones deal with the understanding of specification sheets given by the manufacturer. Among those, we will describe, based on a practical example, the definition and the use of the following quantities:

- Transmitting voltage response
- Open-circuit receiving response
- Transmitting and receiving beam patterns at specific frequencies
- Impedance and/or admittance versus frequency
- Resonant frequency, maximum voltage and maximum source level (for a transducer)

Figure 5.8 is a specification sheet provided by the ITC (Internationsal Transducer Corp.) for a deep-water omnidirectional transducer. Figure 5.8a corresponds to the transmitting sensitivity versus frequency. The units are in dB re μPa/V@1m, which means that, at the resonant frequency 11.5 kHz for example, the transducer excited with a 1 V amplitude transmits at one meter a pressure p_t such that $20 \log_{10} \left(\frac{p_t}{1 \times 10^{-6}} \right) = 149$ dB, i.e. $p_t \approx 28.2$ Pa. Similarly, Fig. 5.8b shows the receiving sensitivity versus frequency. The units are now in dB re 1V/μPa which means that, at 11.5 kHz for example, the transducer converts a 1 μPa amplitude field into a voltage V_r such that $20 \log_{10} \left(\frac{V_r}{1} \right) = -186$ dB, i.e. $V_r \approx 5 \times 10^{-10}$ V. Figure 5.8c shows the admittance versus frequency. The complex admittance Y is the inverse of the complex impedance Z. The real part of the admittance is called the conductance G; the imaginary part is the susceptance B. Those curves directly yield the calculation of the electrical impedance of the transducer. For example, the impedance of ITC-1007 at the resonant frequency is $|Z| = 1/(\sqrt{G^2 + B^2}) \approx 115\,\Omega$. When used as a source, the transducer electrical impedance has to match the output impedance of the power amplifier to allow for a good power transfer through the transducer. In the case where the impedances do not match, a customized matching box will be necessary. Knowing the electrical impedance $|Z|$ and the input power $I = 10\,000$ W, the maximum voltage can be determined as $U_{max} = \sqrt{|Z|I} \approx 1072$ V. According to the trans-

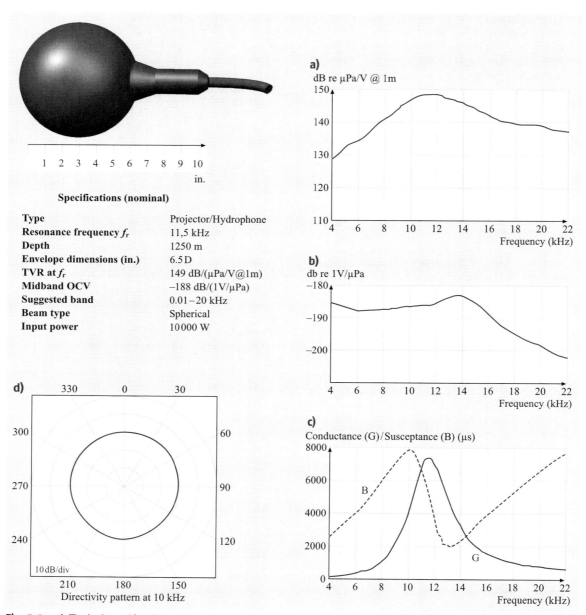

Fig. 5.8a–d Typical specification sheet of a powerful underwater acoustic transponder (*top left*). (**a**) Transmitting voltage response. (**b**) Receiving voltage response. (**c**) Real (*full line*) and imaginary part (*dashed line*) of the water admittance Y. (**d**) Directionality pattern at one frequency (Courtesy of International Transducer Corp.)

mitting voltage response, this corresponds to a source level of nearly 210 dB re μPa at the resonant frequency. Finally, Fig. 5.8d represents the directionality pattern at a given frequency. It shows that the ITC-1007 is omnidirectional at 10 kHz.

When transducers have to be coupled to a power amplifier or another electronic device, it may be useful to model the transducer as an electronic circuit (Fig. 5.9a). The frequency dependence of the conductance G and susceptance B (Fig. 5.8c) yield the components of the equivalent circuit, as shown in Fig. 5.9b. Similarly, an important parameter is the quality factor Q, which measures the ratio between the mechanical energy transmitted by the transducer and the energy dissipated (Fig. 5.10). Finally, the equivalent circuit leads to the measure of the electroacoustic power efficiency k^2 that corresponds to the ratio of the output acoustic power and the input electric power.

Hydrophones are usually described with the same characteristics as transducers but they are only designed to work in reception. To this goal, hydrophones are usually connected to a preamplifier with high input impedance to avoid any loss in the signal reception. A typical hydrophone exhibits a flat receiving response on a large bandwidth far away from its resonance frequency (Fig. 5.11a). As expected, the sensitivity of a hydrophone is much higher than the sensitivity of a transducer. Low electronic noise below the ocean ambient-noise level is also an important characteristic for hydrophones (Fig. 5.11b). Finally, hydrophones are typically designed to be omnidirectional (Fig. 5.11c).

5.2.2 Volume Attenuation

Attenuation is characterized by an exponential decay of the sound field. If A_0 is the root-mean-square (rms) amplitude of the sound field at unit distance from the source, then the attenuation of the sound field causes the amplitude to decay with distance along the path, r:

$$A = A_0 \exp(-\alpha r), \qquad (5.3)$$

where the unit of α is Nepers/distance. The attenuation coefficient can be expressed in decibels per unit distance by the conversion $\alpha' = 8.686\alpha$. Volume attenuation increases with frequency and the frequency dependence of attenuation can be roughly divided into four regimes as displayed in Fig. 5.12. In region I, leakage out of the sound channel is believed to be the main cause of attenuation. The main mechanisms associated with regions II and III are boric acid and magnesium sulfate chemical relaxation. Region IV is dominated by the shear and bulk viscosity associated with fresh water. A summary of the approximate frequency dependence (f in kHz) of attenuation (in units of dB/km) is given by

$$\alpha'(\text{dB/km}) = 3.3 \times 10^{-3} + \frac{0.11 f^2}{1 + f^2}$$
$$+ \frac{43 f^2}{4100 + f^2} + 2.98 \times 10^{-4} f^2, \qquad (5.4)$$

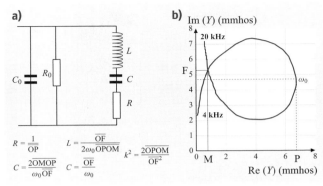

Fig. 5.9 (a) Representation of the transducer as an electronic circuit around the resonant frequency. The resistor R_0 corresponds to the dielectric loss in the transducer and is commonly supposed infinite. C_0 is the transducer capacity, L and C are the mass and rigidity of the material, respectively. R includes both the mechanic loss and the energy mechanically transmitted by the transducer. (b) The values of C_0, L, C and R are obtained from the positions of the points F, M and P in the real–imaginary admittance curve given in the specification sheet

with the terms sequentially associated with regions I–IV in Fig. 5.12.

In Fig. 5.6, the losses associated with path 3 only include volume attenuation and scattering because this path does not involve boundary interactions. The volume scattering can be biological in origin or arise from interaction with internal wave activity in the vicinity

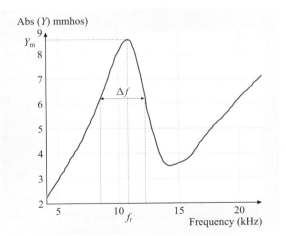

Fig. 5.10 Frequency dependence of the admittance curve that allows the calculation of the quality factor $Q = f_r/\Delta f$ of the transducer at the resonant frequency f_r. Y_m is the maximum of the admittance

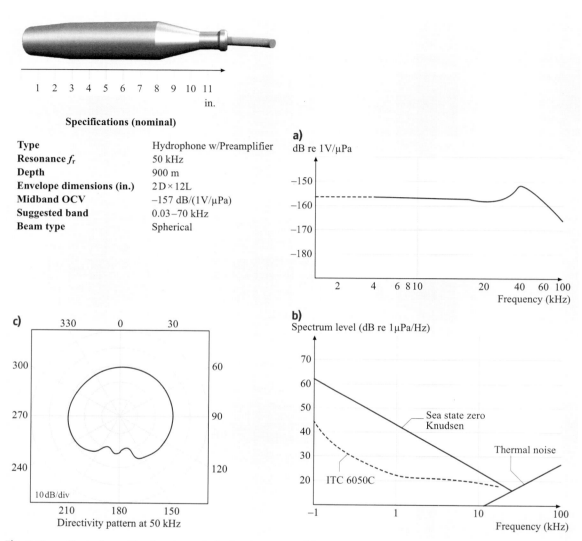

Fig. 5.11a–c Typical specification sheet of a hydrophone (top). (**a**) Receiving frequency response. (**b**) Spectral noise level of the hydrophone to be compared to the ocean ambient noise level. (**c**) Directionality pattern at one frequency (Courtesy of International Transducer Corp.)

of the upper part of the deep sound channel where paths are refracted before they interact with the surface. Both of these effects are small at low frequencies. This same internal wave region is also on the lower boundary of the surface duct, allowing scattering out of the surface duct, thereby also constituting a loss mechanism for the surface duct. This mechanism also leaks sound into the deep sound channel, a region that without scattering would be a shadow zone for a surface duct source. This type of scattering from internal waves is also a source of fluctuation of the sound field.

5.2.3 Bottom Loss

The structure of the ocean bottom affects those acoustic paths which interact with it. This bottom interaction is summarized by *bottom reflectivity*, the amplitude ratio of reflected and incident plane waves at the ocean–bottom interface as a function of grazing angle, θ (Fig. 5.13a).

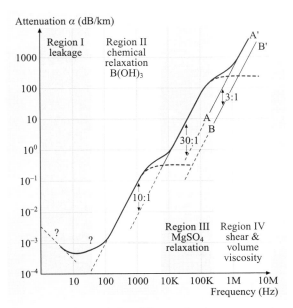

Fig. 5.12 Regions of the different dominant attenuation processes for sound propagating in seawater (after [5.16]). The attenuation is given in dB per kilometer

For a simple bottom, which can be represented by a semi-infinite half-space with constant sound speed c_2 and density ρ_2, the reflectivity is given by

$$R(\theta) = \frac{\rho_2 k_{1z} - \rho_1 k_{2z}}{\rho_2 k_{1z} + \rho_1 k_{2z}}, \tag{5.5}$$

with the subscripts 1 and 2 denoting the water and ocean bottom, respectively; the wavenumbers are given by

$$k_{iz} = (\omega/c_i) \sin \theta_i = k \sin \theta_i; \; i = 1, 2. \tag{5.6}$$

The incident and transmitted grazing angles are related by Snell's law,

$$c_2 \cos \theta_1 = c_1 \cos \theta_2, \tag{5.7}$$

and the incident grazing angle θ_1 is also equal to the angle of the reflected plane wave.

For this simple water–bottom interface for which we take $c_2 > c_1$, there exists a critical grazing angle θ_c below which there is perfect reflection,

$$\cos \theta_c = \frac{c_1}{c_2}. \tag{5.8}$$

For a lossy bottom, there is no perfect reflection, as also indicated in a typical reflection curve in Fig. 5.13b. These results are approximately frequency independent.

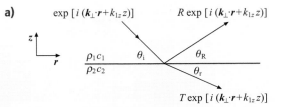

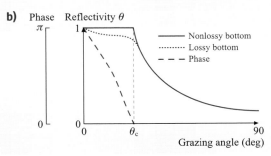

Fig. 5.13a,b The reflection and transmission process. Grazing angles are defined relative to the horizontal. (a) A plane wave is incident on an interface separating two media with densities and sound speeds ρ, c. $R(\theta)$ and $T(\theta)$ are reflection and transmission coefficients. Snell's law is a statement that $k_\perp$, the horizontal component of the wave vector, is the same for all three waves. (b) Rayleigh reflection curve $R(\theta)$ as a function of the grazing angle indicating critical angle θ_c. The dashed curve shows that, if the second medium is lossy, there is no perfect reflection below the critical angle. Note that for the non-lossy bottom, there is complete reflection below the critical angle, but with a phase change

However, for a layered bottom, the reflectivity has a complicated frequency dependence. It should be pointed out that, if the density of the second medium vanishes, the reflectivity reduces to the pressure release case of $R(\theta) = -1$.

5.2.4 Scattering and Reverberation

Scattering caused by rough boundaries or volume heterogeneities is a mechanism for loss (attenuation), reverberant interference and fluctuation. Attenuation from volume scattering is addressed in Sect. 5.2.2. In most cases, it is the mean or coherent (or specular) part of the acoustic field which is of interest for a SONAR or communications application and scattering causes part of the acoustic field to be randomized. Rough surface scattering out of the *specular direction* can be thought of as an attenuation of the mean acoustic field and typically increases with increasing frequency. A formula often used to describe reflectivity from a rough boundary is

$$R'(\theta) = R(\theta) \exp\left(-\frac{\Gamma^2}{2}\right), \quad (5.9)$$

where $R(\theta)$ is the reflection coefficient of the smooth interface and Γ is the Rayleigh roughness parameter defined as $\Gamma \equiv 2k\sigma \sin\theta$ where $k = 2\pi/\lambda$, λ is the acoustic wavelength, and σ is the rms roughness height [5.18–20].

The scattered field is often referred to as reverberation. Surface, bottom or volume scattering strength, $S_{S,B,V}$ is a simple parameterization of the production of reverberation and is defined as the ratio in decibels of the sound scattered by a unit surface area or volume referenced to a unit distance I_{scat} to the incident plane wave intensity I_{inc}

$$S_{S,B,V} = 10 \log \frac{I_{\text{scat}}}{I_{\text{inc}}}. \quad (5.10)$$

The Chapman–Harris [5.21] curves predicts the ocean surface scattering strength in the 400–6400 Hz region,

$$S_S = 3.3\beta \log \frac{\theta}{30} - 42.4 \log \beta + 2.6;$$
$$\beta = 107(wf^{1/3})^{-0.58}, \quad (5.11)$$

where θ is the grazing angle in degrees, w the wind speed in m/s and f the frequency in Hz. Ocean surface scattering is further discussed in [5.22].

The simple characterization of bottom backscattering strength utilizes Lambert's rule for diffuse scattering,

$$S_B = A + 10 \log \sin^2 \theta, \quad (5.12)$$

where the first term is determined empirically. Under the assumption that all incident energy is scattered into the water column with no transmission into the bottom, A is -5 dB. Typical realistic values for A [5.23] which have been measured are -17 dB for the large basalt mid-Atlantic ridge cliffs and -27 dB for sediment ponds.

The volume scattering strength is typically reduced to a surface scattering strength by taking S_V as an average volume scattering strength within some layer at a particular depth; then the corresponding surface scattering strength is

$$S_S = S_V + 10 \log H, \quad (5.13)$$

where H is the layer thickness. The column or integrated scattering strength is defined as the case for which H is the total water depth.

Volume scattering usually decreases with depth (about 5 dB per 300 m) with the exception of the deep scattering layer. For frequencies less than 10 kHz, fish with air-filled swim bladders are the main scatterers. Above 20 kHz, zooplankton or smaller animals that feed upon phytoplankton and the associated biological chain are the scatterers. The deep scattering layer (DSL) is deeper in the day than in the night, changing most rapidly during sunset and sunrise. This layer produces a strong scattering increase of 5–15 dB within 100 m of the surface at night and virtually no scattering in the daytime at the surface since it migrates down to hundreds of meters. Since higher pressure compresses the fish swim bladder, the backscattering acoustic resonance (Sect. 5.2.6) tends to be at a higher frequency during the day when the DSL migrates to greater depths. Examples of day and night scattering strengths are shown in Fig. 5.14.

Finally, as explained in Sect. 5.2.6, near-surface bubbles and bubble clouds can be thought of as either volume or surface scattering mechanisms acting in concert with the rough surface. Bubbles have resonances (typically greater than 10 kHz) and at these resonances, scattering is strongly enhanced. Bubble clouds have collective properties; among these properties is that a bubbly mixture, as specified by its void fraction (total bubble gas volume divided by water volume) has a considerable lower sound speed than water.

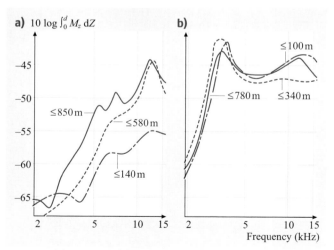

Fig. 5.14 (a) Day and (b) night scattering strength measurements using an explosive source as a function of frequency (after [5.17]). The spectra measured at various times after the explosion are labeled with the depth of the nearest scatterer that could have contributed to the reverberation. The ordinate corresponds to S_V in (5.13)

5.2.5 Ambient Noise

There are essentially two types of ocean acoustic noise: manmade and natural. Generally, shipping is the most important source of manmade noise, though noise from offshore oil rigs is becoming increasingly prevalent. See also Table 5.2 in the Marine Mammal section for specific examples of manmade noise. Typically, natural noise dominates at low frequencies (below 10 Hz) and high frequencies (above a few hundred Hz). Shipping fills in the region between ten and a few hundred Hz and this component is increasing over time [5.25, 26]. A summary of the spectrum of noise is shown in Fig. 5.15. The higher-frequency noise is usually parameterized according to the sea state (also Beaufort number) and/or wind. Table 5.1 summarizes the description of the sea state.

The sound-speed profile affects the vertical and angular distribution of noise in the deep ocean. When there is a critical depth (Sect. 5.1.2), sound from surface sources travels long distances without interacting with the ocean bottom, but a receiver below this critical depth should sense less surface noise because propagation involves interaction with lossy boundaries, the surface and/or bottom. This is illustrated in Fig. 5.16a,b which shows a deep-water environment with measured ambient noise. Figure 5.16c is an example of vertical directionality of noise which also follows the propagation physics discussed above. The shallower depth is at the axis of the deep sound channel while the other is at the critical depth. The pattern is narrower at the critical depth where the sound paths tend to be horizontal since the rays are turning around at the lower boundary of the deep sound channel.

In a range-independent ocean, Snell's law predicts a horizontal noise notch at depths where the speed of sound is less than the near-surface sound speed. Returning to (5.2) and reading off the sound speeds from Fig. 5.16 at the surface ($c = 1530$ m/s) and say, 300 m

Table 5.1 Descriptor of the ocean sea surface [5.24]

Sea criteria	Beaufort scale	Wind speed range knots (m/s)	mean knots (m/s)	12 h wind Wave height[a,b] ft (m)	Fully arisen sea Wave height[a,b] ft (m)	Duration[b,c] h	Fetch[b,c] naut. miles (km)	Sea-state scale
Mirror–like	0	< 1 (< 0.5)						0
Ripples	1	1–3 (0.5–1.7)	2 (1.1)					0.5
Small wavelets	2	4–6 (1.8–3.3)	5 (2.5)	< 1 (< 0.30)	< 1 (< 0.30)			1
Large wavelets, scattered whitecaps	3	7–10 (3.4–5.4)	8.5 (4.4)	1–2 (0.30–0.61)	1–2 (0.30–0.61)	< 2.5	< 10 (< 19)	2
Small waves, frequent whitecaps	4	11–16 (5.5–8.4)	13.5 (6.9)	2–5 (0.61–1.5)	2–6 (0.61–1.8)	2.5–6.5	10–40 (19–74)	3
Moderate waves, many whitecaps	5	17–21 (8.5–11.1)	19 (9.8)	5–8 (1.5–2.4)	6–10 (1.8–3.0)	6.5–11	40–100 (74–185)	4
Large waves, whitecapes every– where, spray	6	22–27 (11.2–14.1)	24.5 (12.6)	8–12 (2.4–3.7)	10–17 (3.0–5.2)	11–18	100–200 (185–370)	5
Heaped–up sea, blown spray, streaks	7	28–33 (14.2–17.2)	30.5 (15.7)	12–17 (3.7–5.2)	17–26 (5.2–7.9)	18–29	200–400 (370–740)	6
Moderately high, long waves, spindrift	8	34–40 (17.3–20.8)	37 (19.0)	17–24 (5.2–7.3)	26–39 (7.9–11.9)	29–42	400–700 (740–1300)	7

Notes:
a The average height of the highest one-third of the waves (significant wave height)
b Estimated from data given in US Navy Hydrographic Office (Washington, D.C.) publications HO 604 (1951) and HO 603 (1955)
c The minimum fetch and duration of the wind needed to generate a fully arisen sea

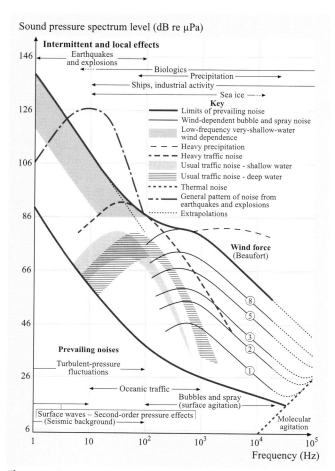

Fig. 5.15 Composite of ambient noise spectra (after [5.24])

bottom properties. Thus, if the bottom is not very lossy, surface sources exciting low-order modes can come from large distances (and hence large areas). These paths are close to the horizontal and noise will then tend to have a strong horizontal component. On the other hand, a very lossy bottom will prevent long-range propagating paths from contributing to the noise field and the noise will tend to be local and subsequently vertical.

5.2.6 Bubbles and Bubbly Media

Bubbles not only occur naturally in the ocean, but the swim bladders of fish can also be thought of as bubbles. The physics of bubbles is a large area of activity in acoustics [5.7, 28]. Here we will confine ourselves to some aspects relevant to ocean acoustics. First we discuss some properties of bubbles in terms of resonators, scatterers and then go on to some aspects of bubbly media and scattering from bubbles.

Bubble scattering follows two regimes depending on the magnitude of the bubble radius oscillations in response to the incident fluctuating pressure field [5.29]:

1. For small pressure amplitudes, the response is linear. The first step in any linear analysis is the identification of the resonance frequency of an oscillating bubble and the measurement of the bubble scattering cross section.
2. Due to nonlinear terms in the governing equations, the response of a bubble will be affected by nonlinearities as the amplitude of the pressure field is increased. In this case, the bubble may continue to oscillate stably (stable acoustic cavitation) generating (sub)harmonics in the scattered field. Under other circumstances, the change in bubble size during a single cycle of oscillation becomes so large that the bubble undergoes a cycle of explosive cavitation growth and violent collapse. Such a response is termed *transient acoustic cavitation* and is distinguished from stable acoustic cavitation by the fact that the bubble radius changes by several orders of magnitude during each cycle.

The Bubble as a Scatterer

The calculation of the natural acoustic resonance of an oscillating bubble in the linear regime requires considerable algebra combining: (1) the equation of motion, (2) mass conservation, and (3) continuity relations at the bubble surface. These developments go beyond the scope of this chapter. In the following, we simply summarize the final results, which are the expression of the

(1500 m/s), a horizontal ray ($\theta = 0$) launched from the ocean surface would have an angle with respect to the horizontal of about $11°$ at 300 m depth. All other rays would arrive with greater vertical angles. Hence we expect this horizontal notch. However, the horizontal notch is often not seen at shipping noise frequencies. That is because shipping tends to be concentrated in continental-shelf regions and propagation down a continental slope converts high-angle rays to lower angles at each bounce. There are also deep-sound-channel shoaling effects that result in the same trend in angle conversion.

The vertical directionality of noise in shallow water has a simple environmental dependence [5.27]. For example, in the summer with a downward refracting profile, the same discussion as above leads to a horizontal noise notch. However the vertical directionality of noise from the surface in the winter tends to be driven by bubble natural acoustic resonance ω_0^2, and the expression for the scattered field from an acoustic bubble under an incident pressure field at frequency ω.

The acoustic resonance in the linear regime of a single bubble of radius a is:

$$\omega_0^2 = \left[\frac{3\gamma p_0}{\rho_w a^2} + (3\gamma - 1)\frac{2T}{\rho_w a^3}\right], \quad (5.14)$$

where p_0 is the ambient pressure outside the bubble, T is surface tension (tensile force/length in units N/m), ρ_w is the density of water and γ is the ratio of specific heats. Neglecting surface tension in (5.14) for bubble sizes larger than 1 μm and considering the acoustic expansion/compression process to be adiabatic ($c_{air}^2 = \gamma p_0/\rho_{air}$), we obtain the approximate expression

$$\omega_0 = \frac{1}{a}\sqrt{\frac{3c_{air}^2 \rho_{air}}{\rho_w}}, \quad (5.15)$$

and for $c_{air} \simeq 340$ m/s, we get $f_0 \simeq \frac{3}{a}$ with a in m and f_0 in Hz.

We now consider an incident plane wave at frequency ω in the regime $ka = \omega a/c = 2\pi a/\lambda \ll 1$. The far-field expression for the spatial part of the radiated acoustic field p_r is

$$p_r(r) = -\frac{a}{r} p_i \exp\left[-\frac{i\omega}{c}(r-a)\right]$$
$$\times \left[1 - \frac{\omega_0^2}{\omega^2}\left(1 - \frac{i\omega a}{c}\right)\right]^{-1}. \quad (5.16)$$

First, we note that in the high-frequency limit, we recover $p_r(r,t) = -\frac{a}{r} p_i\left(t - \frac{r-a}{c}\right)$ that was given by the boundary conditions at the bubble surface.

To understand the effect of the resonance frequency, consider two cases. The first case is $\omega \gg \omega_0$ for which we obtain:

$$p_r(r) = -\frac{a}{r} p_i e^{-\frac{i\omega}{c}(r-a)}, \quad (5.17)$$

whereas for the case $\omega = \omega_0$ we get:

$$p_r(r) = -\frac{ic}{\omega r} p_i e^{-\frac{i\omega}{c}(r-a)} = -\frac{i\lambda}{2\pi r} p_i e^{-\frac{i\omega}{c}(r-a)}. \quad (5.18)$$

Comparing the two equations, (5.18) appears to be the field radiating from a sphere of radius $\lambda/2\pi$ which is much larger than a. For example, neglecting surface tension, at 1 atm, $\omega_0 a \approx 20$ so that $\lambda \approx 500a$. This resonance effect is also quite apparent when considering the scattering cross section of the bubble.

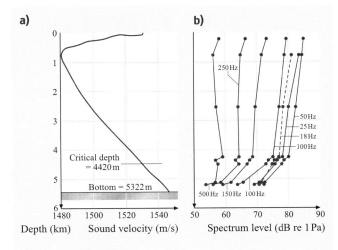

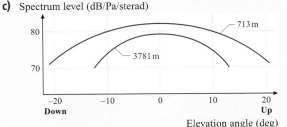

Fig. 5.16a–c Noise in the deep ocean. (**a**) Sound-speed profile and (**b**) noise level as a function of depth in the Pacific (after [5.30]). (**c**) The vertical directionality of noise at the axis of the deep sound channel and at the critical depth in the Pacific (after [5.31])

The scattering cross section σ_s is the ratio of the total scattered power (intensity × area = pressure × velocity × enclosing area) to the incident plane-wave intensity (given by $p_i^2/2\rho_w c$, with the factor of 1/2 coming from averaging over a cycle) and therefore has the units of area. We perform this calculation in the far field using (5.16) to obtain

$$\sigma_s = \frac{4\pi a^2}{\left[1 - \frac{\omega_0^2}{\omega^2}\left(1 - \frac{i\omega a}{c}\right)\right]^2} \xrightarrow{\omega=\omega_0} \frac{\lambda^2}{\pi}, \quad (5.19)$$

which is consistent with a surface area associated with the discussion below (5.18). The resonance makes the bubble appear larger in surface area than its dimension. On the other hand, for the case $\omega \ll \omega_0$, we have

$$\sigma_s = 4\pi a^2 \left(\frac{\omega}{\omega_0}\right)^4, \quad (5.20)$$

Fig. 5.17 (a) Wave breaking on the shore; **(b)** magnified view of individual bubbles within the plumes taken a second or so after wave breaking. The void fraction of air in these plumes is a few percent and bubbles range in size from less than 50 μm to a few mm radius. Bubble plumes found beneath breaking waves in 15–20 m/s winds in the open ocean have a similar size distribution (Courtesy of Grant Deane, Scripps Institution of Oceanography)

which is Rayleigh scattering. The analogous mechanism in electromagnetics explains why the sky is blue: blue has a higher frequency than red so it is scattered more. The above derivations only included radiation damping and we have not included lossy effects caused by thermal conductivity and shear viscosity. Looking at, for example (5.16), we can think of the radiation damping constant to be $\delta_r = ka$ or, in other words, at resonance $\sigma_s = 4\pi a^2/\delta_r^2$. Finally, we mention that the extinction cross section is the sum of the scattering cross section and the absorption cross section. The damping coefficients for thermal conductivity and shear viscosity are typically experimentally determined.

Bubbly Media

The region immediately below the surface of the ocean is a bubbly medium (Fig. 5.17). The existence of bubbles changes the effective compressibility of the water. We define a volume fraction of bubble (also called the void fraction) μ so that the density of the mixture is simply $\rho_m = \mu \rho_b + (1-\mu)\rho_w$ where the subscripts b and w refer to a bubble and water, respectively. We consider low frequencies with respect to resonance and we use the compressibility, the inverse of the bulk modulus $B = \rho(\delta p/\delta \rho)$, since it is additive and permits us to write down the compressibility of the mixture,

$$K_m = \mu K_b + (1-\mu)K_w \rightarrow \frac{1}{B_m}$$
$$= \mu \frac{1}{B_b} + (1-\mu)\frac{1}{B_w} \,. \quad (5.21)$$

Using the above discussion relating sound speed and the adiabatic bulk modulus for the air bubble with $\gamma = 1.4$ (the bulk modulus of water is 2.3×10^9 Pa) we then obtain:

$$\frac{1}{\rho_m c_m^2} = \frac{\mu}{1.4 p_b} + \frac{(1-\mu)}{\rho_w c_w^2} \rightarrow c_m^2$$
$$= \left(\frac{1.4 p_b}{\mu \rho_w c_w^2 + 1.4 p_b (1-\mu)}\right) c_w^2 \,, \quad (5.22)$$

where we have taken $\rho_m \approx \rho_w$. Substituting some typical numbers, we use atmospheric pressure, $p_b \approx 10^5$ Pa, $\rho_w \approx 10^3$ kg/m^3, $c_w = 1500$ m/s and we consider two void fractions: μ of 0.0001 and 0.001 (large). The corresponding sound speeds in the bubbly mixture are about 930 m/s and 370 m/s, respectively. A small amount a bubbles significantly changes the compressibility of the medium and therefore drastically changes the sound speeds. In reality, the speed of sound through the bubbly medium varies with frequency since compressibility is the ratio of the fractional change in volume to the incident pressure. The volume change is related to the bubble surface displacement and hence to velocity or radiated pressure as per (5.16). Therefore, the bubble compressibility is actually frequency dependent and a more rigorous treatment of propagation in bubbly media would show the dispersion of the speed through the bubbly medium.

When sound propagates through a bubbly medium, the scattering will also cause attenuation due to scattering and absorption. As stated above, the extinction cross section, σ_e, is a measure of this phenomenon. For a bubbly medium (and for the simple case of single-sized bubbles), an acoustic beam will be altered by the absorption and scattering out of the beam. For an incident plane wave of intensity I_0, the power removed by each bubble is $I_0\sigma_e$ so that the rate of change of intensity as the beam travels through a bubble medium of N bubbles per unit volume is

$$\frac{dI}{dx} = -I_0 \sigma_e N \rightarrow I = I_0 \exp(-\sigma_e N x) \ . \quad (5.23)$$

Therefore, a bubbly medium changes the sound speed, absorbs sound, and is dispersive.

5.3 SONAR and the SONAR Equation

A major application of underwater acoustics is SONAR system technology. The performance of SONAR is often approximately described by the SONAR equation. The methodology of the SONAR equation is analogous to an accounting procedure involving acoustic signal, interference and system characteristics. Figure 5.18 provides a schematic of passive and active SONARs.

5.3.1 Detection Threshold and Receiver Operating Characteristics Curves

The detection threshold (DT) [5.16] is a decibel number that essentially incorporates the SONAR system's (which includes the operator) ability to decide that a detection is made or not made. The detection process includes the following probabilities:

- the probability of detection (PD): the probability that a signal is detected if it is present;
- 1-PD: the probability the signal will not be detected if it is present;
- the probability of false alarm (PFA): the probability that a signal is detected when it is not present;
- 1-PFA: the probability that the signal will not be detected when it is not present.

In practical terms, since the signal and noise are fluctuating, the detection is made (over a time interval) when the fluctuating sum of the signal and noise exceeds a threshold that is determined from empirically derived probability density functions (PDFs) of noise and signal plus noise. For example, the case that the noise alone rises above the threshold contributes to the PFA. Therefore, the process for determining a detection threshold level will depend on the PD and PFA. Typically numbers might be a PD of 0.5 and PFA of 0.0001. The probabilities will themselves be a function of the relation between the signal and noise statistics, as represented by their mean and variance. The detection index d succinctly characterizes this relation in that it indicates how easy it is to observe a signal in noise,

$$d = \frac{(M_{sn} - M_n)^2}{\sigma_n^2} \ , \quad (5.24)$$

where M_{sn} is the mean of the signal plus noise, M_n is the noise mean and σ_n^2 is the noise variance. Figure 5.19 shows schematically the implications of the detection index where the relative proximity of the two probability density functions (PDFs) determine the detection

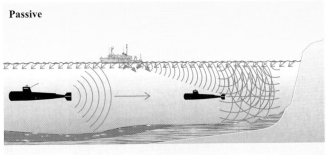

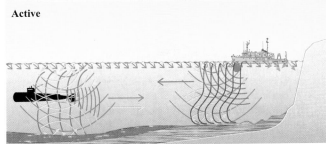

Fig. 5.18 Passive and active SONAR for submarine detection. *Passive*: the submarine on the right tries to detect sounds (blue) from the other submarine using a towed array (antenna). These sounds are distorted by the shallow-water environment and are embedded in ocean surface noise (green) and surface shipping noise (red). *Active*: the ship on the right sends out a pulse (red) and an echo (blue), distorted by the shallow-water environment, is returned to the ship SONAR which tries to distinguish it from backscattered reverberation (yellow) and ocean noise (green) [5.1]

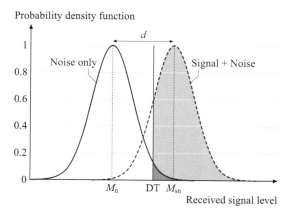

Fig. 5.19 Probability density functions (PDF) of noise and signal+noise. Various probabilities as explained in the text are related to specific regions under the curves. Thus, the probability of detection (PD) is the light-shaded area under the signal+noise PDF curve that is to the right of DT, the detection threshold (*thick vertical line*). Similarly, the probability of false alarm is the dark shaded area under the noise PDF curve that is to the right of DT

statistics. We first note that the PD is the area under the signal-plus-noise curve to the right of the DT and the PFA is the area under the noise curve to the right of the DT. Then, for example, if the mean of the signal-plus-noise PDF was further to the right [higher signal-to-noise ratio (SNR)], the detection index and PD would be larger.

We could then move the DT to the right, changing the PD, but the PFA would be smaller. Receiver operating characteristics (ROC) curves are plots of PD versus PFA parameterized by d (Fig. 5.19). Figure 5.20 gives a typical example of using this methodology and refer the reader to the SONAR literature [5.16] for a more complete treatment. A square-law detector is commonly used for a passive system dealing with unknown signal. In that case, it has been shown that the detection index for a small-SNR, narrow-band signal in Gaussian noise is given by

$$d = \omega t \left(\frac{S}{N}\right)^2, \tag{5.25}$$

where ω, t, S and N are the bandwidth (taken to be larger than the width of the spectral line of the signal), integration time, signal power and noise power in the bandwidth, respectively. If 1 Hz is the noise bandwidth reference, the detection index is $d = \omega t (S/N_0)$, where N_0 is the noise in a 1 Hz band. This gives a relation between the signal-to-noise ratio at the output of an energy detector of a narrow-band signal in Gaussian noise to the detection index:

$$\text{SNR} = 5 \log \left(\frac{d}{\omega t}\right) \equiv \text{DT}, \tag{5.26}$$

where the equivalence symbol is for a detection criteria of a specified PD, PFA.

In reality, there are correction factors for the detection threshold related to the length of observations used to make a decision, human factors, and others that we omit from this discussion. Since the criteria are specified through the ROC curves, we can now estimate the DT for specific cases. We simply go to the the ROC curves for a selected PD versus PFA and read off the detection index, from which we can compute the detection threshold. For example, from Fig. 5.20, for a PD of 50% and a PFA of 0.01%, the detection index is $d = 16$. Using unit bandwidth and integration time, the detection threshold is DT$= 5 \log 16 = 6$ dB. This methodology is an example of the meaning of the DT term in the SONAR equation below, though the relationship between DT and d is different depending on the type of receiver and SONAR.

5.3.2 Passive SONAR Equation

A passive SONAR system uses the radiated sound from a target to detect and locate the target. A radiating object of source level (SL) (all units are in decibels) is received at a hydrophone of a SONAR system at a lower signal level S because of the transmission loss (TL) it suffers (e.g., cylindrical spreading plus attenuation or a TL computed from one of the propagation models of Sect. 5.4),

$$S = \text{SL} - \text{TL}. \tag{5.27}$$

The noise, N at a single hydrophone is subtracted from (5.27) to obtain the signal-to-noise ratio at a single hydrophone,

$$\text{SNR} = \text{SL} - \text{TL} - \text{N}. \tag{5.28}$$

Typically a SONAR system consists of an array or antenna of hydrophones which provides signal-to-noise enhancement through a beam-forming process (Sect. 5.6). This process is quantified in decibels by the array gain (AG) (see Sect. 5.6.2) that is added to the single-hydrophone SNR to give the SNR at the output of the beam-former,

$$\text{SNR}_{\text{BF}} = \text{SL} - \text{TL} - \text{N} + \text{AG}. \tag{5.29}$$

As discussed, because detection involves additional factors including SONAR operator ability, it is necessary to specify a detection threshold, DT level above the SNR_{BF} at which there is a 50% (by convention) probability of detection. The difference between these two quantities is called the signal excess (SE),

$$SE = SL - TL - N + AG - DT. \quad (5.30)$$

This decibel bookkeeping leads to an important SONAR engineering descriptor called the figure of merit (FOM) which is the transmission loss that gives a zero signal excess,

$$FOM = SL - N + AG - DT. \quad (5.31)$$

The FOM encompasses the various parameters a SONAR engineer must deal with: expected source level, the noise environment, array gain and the detection threshold. Conversely since the FOM is a transmission loss, one can use the output of a propagation model (or if appropriate, a simple geometric loss plus attenuation) to estimate the minimum range at which a 50% probability of detection can be expected. This range changes with oceanographic conditions and is often referred to as the range of the day in navy SONAR applications. Finally, we mention that formal detection theory involves a criterion involving the probability of detection PD versus probability of false alarm PFA. Plots of PD versus PFA as parameterized by, say SNR, are called receiver operating characteristic (ROC) curves [5.16]. Clearly, for a PFA of 1, the probability of detection goes to unity.

5.3.3 Active SONAR Equation

A monostatic active SONAR transmits a pulse to a target and its echo is detected at a receiver co-located with the transmitter. A bistatic active SONAR has the receiver in a different location than the transmitter. The main differences between the passive and active cases are the addition of a target strength (TS) term; reverberation (reverberation level, RL) is usually the dominant source of interference, as opposed to noise; and the transmission loss is over two paths: transmitter to target and target to

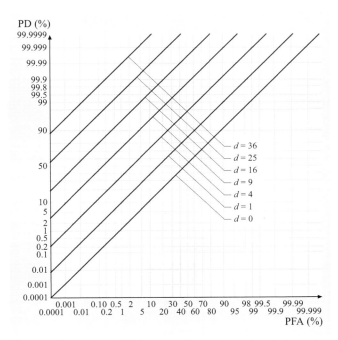

Fig. 5.20 Example of ROC curves. For a given signal plus noise, different threshold settings correspond to different PD and PFA. The ROC curves summarizes the relation between PD and PFA for different thresholds as parameterized through the detection index. (After [5.16])

receiver. In the monostatic case, the transmission loss is 2TL, where TL is the one-way transmission loss, and in the bistatic case, the transmission loss is the sum (in dB) over paths from the transmitter to the target and the target to the receiver, $TL_1 + TL_2$. The concept of the detection threshold is useful for both passive and active SONARs. Hence, for signal excess, we have

$$SE = SL - TL_1 + TS - TL_2$$
$$- (RL + N) + AG - DT. \quad (5.32)$$

The corresponding FOM for an active system is defined for the maximum allowable two-way transmission loss with $TS = 0\,\text{dB}$.

5.4 Sound Propagation Models

The wave equation describing sound propagation is derived from the equations of hydrodynamics and its coefficients and boundary conditions are descriptive of the ocean environment. There are essentially four types of models (computer solutions to the wave equation [5.4]) to describe sound propagation in the sea: ray theory, the spectral method or fast field program (FFP), normal mode (NM) and parabolic equation (PE). All of these models allow for the fact that the ocean environment varies with depth. A model that also takes into account horizontal variations in the environment (i.e., sloping bottom or spatially variable oceanography) is termed range-dependent. For high frequencies (a few kilohertz or above), ray theory is the most practical. The other three model types are more applicable and useable at lower frequencies (below 1 kHz). Between these frequencies, one can choose either, noting that the wave solution is the most accurate and should probably be used in all cases where the the calculation is still feasible and/or practical. The models discussed here are essentially two-dimensional models since the index of refraction has much stronger dependence on depth than on horizontal distance. Nevertheless, bottom topography and strong ocean features can cause horizontal refraction (out of the range–depth plane). Ray models are most easily extendable to include this added complexity. Full three-dimensional wave models are extremely computationally intensive. A compromise that often works for weakly three-dimensional problems is the $N \times 2D$ approximation that combines two-dimensional solutions along radials to produce a three-dimensional solution.

5.4.1 The Wave Equation and Boundary Conditions

The wave equation for the pressure p in cylindrical coordinates with the range coordinates denoted by $\boldsymbol{r} = (x, y)$ and the depth coordinate denoted by z (taken positive downward) for a source-free region is

$$\nabla^2 p(\boldsymbol{r}, z, t) - \frac{1}{c^2(\boldsymbol{r}, z)} \frac{\partial^2 p(\boldsymbol{r}, z, t)}{\partial t^2} = 0, \quad (5.33)$$

where $c(\boldsymbol{r}, z)$ is the sound speed in the wave-propagating medium.

It is convenient to solve (5.33) in the frequency domain by assuming a solution with a frequency dependence of $\exp(-i\omega t)$ to obtain the Helmholtz equation ($K \equiv \omega/c$),

$$\nabla^2 p(\boldsymbol{r}, z) + K^2 p(\boldsymbol{r}, z) = 0 \quad (5.34)$$

with

$$K^2(\boldsymbol{r}, z) = \frac{\omega^2}{c^2(\boldsymbol{r}, z)}. \quad (5.35)$$

The range-dependent environment manifests itself as the coefficient $K^2(\boldsymbol{r}, z)$ of the partial differential equation for the appropriate sound speed profile. The range-dependent bottom type and topography appears as boundary conditions. In underwater acoustics both fluid and elastic (shear-supporting sediments and bottom strata) media are of interest. For simplicity we only consider fluid media below. This is most often a good approximation to describing the bottom near the bottom interface since the material typically has a low shear speed which can be inserted into the fluid equation solution by a perturbation procedure [5.32].

The most common plane-interface boundary conditions encountered in underwater acoustics are the pressure release condition at the ocean surface,

$$p = 0, \quad (5.36)$$

and at the ocean–bottom interface, continuity of pressure

$$p_1 = p_2, \quad (5.37)$$

and vertical particle velocity

$$\frac{1}{\rho_1} \frac{\partial p_1}{\partial z} = \frac{1}{\rho_2} \frac{\partial p_2}{\partial z}, \quad (5.38)$$

where the ρ_i are the densities of the two media. These latter boundary conditions applied to the plane-wave fields in Fig. 5.13a yield the Rayleigh reflection coefficient given by (5.5).

The Helmholtz equation for an acoustic field from a point source is

$$\nabla^2 G(\boldsymbol{r}, z) + K^2(\boldsymbol{r}, z) G(\boldsymbol{r}, z) = -\delta^2(\boldsymbol{r} - \boldsymbol{r}_s)\delta(z - z_s), \quad (5.39)$$

where the subscript 's' denotes the source coordinates. The acoustic field from a point source, $G(\boldsymbol{r})$ is either obtained by solving the boundary-value problem of (5.39) (spectral method or normal modes) or by approximating (5.39) using an initial-value problem (ray theory, parabolic equation).

5.4.2 Ray Theory

Ray theory is a geometrical, high-frequency approximate solution to (5.39) of the form

$$G(\boldsymbol{R}) = A(\boldsymbol{R}) \exp[iS(\boldsymbol{R})], \quad (5.40)$$

where the exponential term allows for rapid variations as a function of range and $A(\boldsymbol{R})$ is a more slowly varying *envelope* that incorporates both geometrical spreading and loss mechanisms. The geometrical approximation is that the amplitude varies slowly with range (i.e.,

$(1/A)\nabla^2 A \ll K^2)$ so that (5.34) yields the eikonal equation

$$(\nabla S)^2 = K^2 \,. \tag{5.41}$$

The ray trajectories are perpendicular to surfaces of constant phase (wavefronts), S, and may be expressed mathematically as follows:

$$\frac{d}{dl}\left(K\frac{d\boldsymbol{R}}{dl}\right) = \nabla K \,, \tag{5.42}$$

where l is the arc length along the direction of the ray and $\boldsymbol{R}$ is the displacement vector to a point on the ray. The direction of average flux (energy) follows that of the trajectories and the amplitude of the field at any point can be obtained from the density of rays.

The ray theory method is computationally rapid and extends to range-dependent problems. Furthermore, the ray traces give a physical picture of the acoustic paths. It is helpful in describing how sound redistributes itself when propagating long distances over paths that include shallow and deep environments and/or mid-latitude to polar regions. The disadvantage of conventional ray theory is that it does not include diffraction, including effects that describe the low-frequency dependence (degree of trapping) of ducted propagation.

5.4.3 Wavenumber Representation or Spectral Solution

The wave equation can be solved efficiently with spectral methods when the ocean environment does not vary with range. The term fast field program (FFP) had been used because the spectral methods became practical with the advent of the fast Fourier transform (FFT). Assume a solution of (5.39) of the form

$$G(\boldsymbol{r}, z) = \frac{1}{2\pi}\int d^2 \boldsymbol{k}\, g(\boldsymbol{k}, z, z_s) \exp[i\boldsymbol{k}\cdot(\boldsymbol{r}-\boldsymbol{r}_s)] \,, \tag{5.43}$$

which then leads to the equation for the depth-dependent Green's function $g(\boldsymbol{k}, z, z_s)$,

$$\frac{d^2 g}{dz^2} + \left[K^2(z) - k^2\right] g = -\frac{1}{2\pi}\delta(z-z_s) \,. \tag{5.44}$$

Furthermore, we assume azimuthal symmetry, $kr > 2\pi$ and $\boldsymbol{r}_s = 0$ so that (5.43) reduces to

$$G(r, z) = \frac{\exp(-i\pi/4)}{(2\pi r)^{1/2}} \times \int_{-\infty}^{+\infty} dk\, k^{1/2} g(k, z, z_s) \exp(ikr) \,. \tag{5.45}$$

This integral is then evaluated using the fast Fourier transform algorithm. Although the method was initially labeled *fast field* it is fairly slow because of the time required to calculate the Green's functions [solve (5.44)]. However, it has advantages when one wishes to calculate the near-field region or to include shear-wave effects in elastic media; it is also often used as a benchmark for other less exact techniques. With a great deal of additional computational effort, this method is extendable to range-dependent environments.

5.4.4 Normal-Mode Model

Rather than solve (5.44) for each g for the complete set of ks (typically thousands of times), one can utilize a normal-mode expansion of the form

$$g(\boldsymbol{k}, z) = \sum a_n(\boldsymbol{k}) u_n(z) \,, \tag{5.46}$$

where the quantities u_n are eigenfunctions of the following eigenvalue problem:

$$\frac{d^2 u_n}{dz^2} + \left[K^2(z) - k_n^2\right] u_n(z) = 0 \,. \tag{5.47}$$

The eigenfunctions u_n are zero at $z = 0$, satisfy the local boundary conditions descriptive of the ocean bottom properties and satisfy a radiation condition for $z \to \infty$. For pressure, they form an orthonormal set in a Hilbert space with weighting function $\rho^{-1}(z)$, the local density. The range of discrete eigenvalues corresponding to the poles in the integrand of (5.45) is given by the condition

$$\min[K(z)] < k_n < \max[K(z)] \,. \tag{5.48}$$

These discrete eigenvalues correspond to discrete angles within the critical angle cone in Fig. 5.7a as discussed in Sect. 5.1.2. The eigenvalues k_n typically have a small imaginary part α_n, which serves as the modal attenuation representative of all the losses in the ocean environment. Solving (5.39) using the normal-mode expansion given by (5.46) yields (for the source at the origin).

$$G(r, z) = \frac{i}{4\rho(z_s)} \sum_n u_n(z_s) u_n(z) H_0^1(k_n r) \,. \tag{5.49}$$

The asymptotic form of the Hankel function can be used in the above equation to obtain the well-known normal-mode representation of a waveguide in cylindrical coordinates:

$$G(r, z) = \frac{i}{(8\pi r)^{1/2}\rho(z_s)} \exp(-i\pi/4) \times \sum_n \frac{u_n(z_s) u_n(z)}{k_n^{1/2}} \exp(ik_n r) \,. \tag{5.50}$$

Equation (5.50) is a far-field solution of the wave equation and neglects the continuous spectrum of modes $[k_n < \min[K(z)]$ of (5.48)]. For the purposes of illustrating the various portions of the acoustic field, we note that k_n is a horizontal wavenumber so that a ray angle associated with a mode with respect to the horizontal can be taken to be $\theta = \cos^{-1}[k_n/K(z)]$. For a simple waveguide the maximum sound speed is the bottom sound speed corresponding to $\min[K(z)]$. At this value of $K(z)$, we have from Snell's law $\theta = \theta_c$, the bottom critical angle. In effect, if we look at a ray picture of the modes, the continuous portion of the mode spectrum corresponds to rays with grazing angles greater than the bottom critical angle of Fig. 5.13b and therefore outside the cone of Fig. 5.7a. This portion undergoes severe loss. Hence, we note that the continuous spectrum is the near (vertical) field and the discrete spectrum is the far (more-horizontal, profile-dependent) field falling within the cone in Fig. 5.7a.

The advantages of the normal-mode procedure are that: the solution is available for all source and receiver configurations once the eigenvalue problem is solved; it is easily extended in moderately range-dependent conditions using the adiabatic approximation; it can be applied (with more effort) to extremely range-dependent environments using coupled-mode theory. However, it does not include a full representation of the near field.

Adiabatic Mode Theory

All of the range-independent normal-mode calculation method developed for environmental ocean acoustic modeling applications can be adapted to mildly range-dependent conditions using adiabatic mode theory. The underlying assumption is that individual propagating normal modes adapt to the local environment, but do not scatter or couple into each other. The coefficients of the mode expansion, a_n in (5.46), now become mild functions of range, i.e., $a_n(k) \to a_n(k, r)$. This modifies (5.45) as follows:

$$G(r, z) = \frac{i\rho(z_s)}{(8\pi r)^{1/2}} \exp(-i\pi/4)$$
$$\times \sum_n \frac{u_n(z_s) v_n(z)}{\overline{k_n}^{1/2}} \exp(i\overline{k_n} r), \quad (5.51)$$

where the range-averaged wavenumber (eigenvalue) is

$$\overline{k_n} = \frac{1}{r} \int_0^r k_n(r') \, dr', \quad (5.52)$$

and the $k_n(r')$ are obtained at each range segment from the eigenvalue problem (5.47) evaluated for the environment at that particular range along the path. The quantities u_n and v_n are the sets of modes at the source and the field positions, respectively.

Simply stated, the adiabatic mode theory leads to a description of sound propagation such that the acoustic field is a function of the modal structure at both the source and the receiver and some average propagation conditions between the two. Thus, for example, when sound emanates from a shallow region where only two discrete modes exist and propagates into a deeper region with the same bottom (same critical angle), the two modes from the shallow region adapt to the form of the first two modes in the deep region. However, the deep region can support many more modes; intuitively, we therefore expect the resulting two modes in the deep region will take up a smaller, more-horizontal part of the cone of Fig. 5.7a than they take up in the shallow region. This means that sound rays going from shallow to deep regions tend to become more horizontal, which is consistent with a ray picture of down-slope propagation. Finally, fully coupled mode theory for range-dependent environments has been developed but requires intensive computation.

Mode Dispersion in a Waveguide

Acoustical propagation in simple free space is nondispersive. That is, all plane waves travel with speeds independent of frequency. Further, their group and phase speeds are the same as the medium sound speed. However, geometric dispersion is a property of waveguide propagation.

We consider a shallow-water waveguide of the Pekeris type, i.e., a homogeneous water column overlying a homogeneous and denser fluid bottom. A waveguide of this type supports modal propagation, where each modes is characterized by a depth-dependent modal amplitude $u_n(z)$ and a horizontally projected propagating wavenumber k_n. Each mode is characterized by a group velocity v_{gn} and a phase velocity $v_{\varphi n}$, which are related through the formula:

$$\frac{1}{v_{gn}} = v_{\varphi n} \int_0^\infty \frac{u_n^2(z)}{\rho(z) c^2(z)} \, dz. \quad (5.53)$$

In the case of a perfect waveguide with a constant sound speed c and no penetration in the bottom, (5.53) takes the simplest form $v_{gn} v_{pn} = c^2$. The difference in group velocities between modes results in modal dispersion in a waveguide (Fig. 5.21).

In shallow water, lower-order modes usually travel faster than higher-order modes (Figs. 5.22, 23). In deep water, where the deep sound channel behaves as a waveguide, the fastest modes are the higher-order modes (Fig. 5.24), a property of refraction-dominated modal propagation. Since modal wavenumbers k_n are frequency dependent, the modes can be plotted in a frequency–wavenumber space that is the Fourier transform of the time-range representation of the dispersed field in Fig. 5.25. As already seen in Fig. 5.21, Fig. 5.26 shows that modes have a cut-off frequency and that there exist a finite number of propagating modes at a given frequency.

Dispersion and The Waveguide Invariant [5.3]

There is actually a fairly robust parameter, the waveguide invariant, that describes dispersion over a (sometimes large) interval of a group of modes. The waveguide invariant β has two important interpretations; first, it is related the local change in the modal group velocity with respect to the change in phase velocity,

$$\frac{1}{\beta} = -\frac{\partial S_g}{\partial S_p}, \qquad (5.54)$$

where S_g and S_p and group and phase slowness, where slowness is the inverse of speed.

It turns out that β often has a rather robust value for certain circumstances. For example, it is unity for many shallow-water situations that are dominated by bottom reflection; on the other hand it is negative for refraction-dominated propagation.

Figure 5.27 shows a calculation for a Pacific deep-water case. Note from the definition that β is negative up to a phase speed of about 1540; this region is one of refraction such as deep sound channel and convergence zone propagation. Beyond 1540, β is positive; this is the bottom bounce region dominated by reflections rather than refraction.

The invariant also relates the change in range in the locations of the interference peaks of a transmission loss curve to a change in frequency

$$\frac{\Delta r}{r} = \frac{1}{\beta}\frac{\Delta\omega}{\omega}. \qquad (5.55)$$

In Fig. 5.28, we show a set of TL curves for different frequencies; the interference peaks are shifted according to the invariant formula. Another way of representing this shift is through the striations, where TL is the third dimension of a frequency–range plot as shown in Fig. 5.29, which was derived from shallow-water transmission loss data. If one represents range as the product of the velocity of the radiator and time, then, the range axis can be replaced by time and one

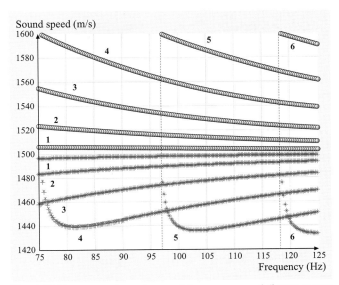

Fig. 5.21 Frequency dependence of the group speed (lower curves 1–6) and phase speed (upper curves 1–6) of modes 1–6 in a Pekeris waveguide with a 100 m water depth, a bottom sound speed c_b and density of 1600 m/s and 1800 kg/m^3. Sound speed in water c_w is constant and equal to 1500 m/s. The bold vertical lines show the cut-off frequencies of modes 4–6, respectively

has a frequency–time plot, often called a spectrogram. The TL curve is a slice through the spectrogram for a given frequency and time converted to the appropriate range.

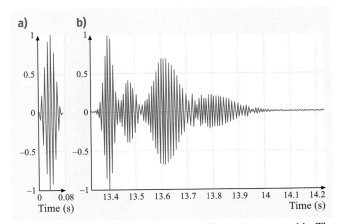

Fig. 5.22a,b Acoustic dispersion in a shallow water waveguide. The *right panel* (**a**) corresponds to the waveguide response at $R = 20$ km to the emitted signal (central frequency $= 100$ Hz, 50 Hz bandwidth) displayed in the *left panel*. (**b**) Source depth is at 40 m and the receiver depth at 60 m. Waveguide parameters are the same as in Fig. 5.21

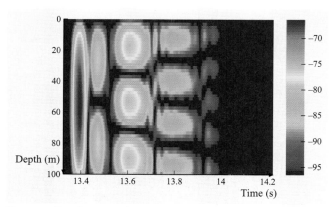

Fig. 5.23 Depth-versus-time representation of the field intensity after a $R = 20$ km propagation in a shallow-water waveguide. The waveguide characteristics are the same as in Fig. 5.21. Source depth is 40 m. The color scale is in dB with a 0 dB source level amplitude at the source

Doppler Shift in a Waveguide

The theory of Doppler shifts involving either a moving source and/or receiver is well known in acoustics, particularly in free space. However, in a waveguide, even if we limit ourselves to horizontal motion, the results are slightly more complex. The individual paths associated with modal propagation, as per Fig. 5.7a, all have finite and different angles with respect to the horizontal. Thus, each mode has a different Doppler shift. The waveguide

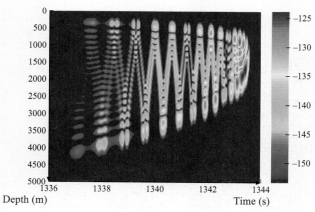

Fig. 5.24 Depth-versus-time representation of the field intensity after a $R = 2000$ km propagation in a deep-water waveguide. We used the Munk profile as a depth-dependent sound-speed profile. Source depth is 900 m, source frequency is 22.5 Hz with a 15 Hz bandwidth. The color scale is in dB with a 0 dB source level amplitude at the source

theory for both source and/or receiver has been worked out (see [5.33], in which there is also a review of the pertinent literature). We return to (5.50) for a harmonic source of angular frequency ω, which therefore results in an additional, identical (when there is no motion) factor for each term of $e^{-i\omega t}$ in the time domain. We now consider a source with a frequency spectrum $S(\omega)$. For constant, horizontal velocities, the normal-mode field results in the *receiver reference frame* are still valid

$$\psi(\mathbf{r}_0 + \mathbf{v}_r t, z, \omega) = \frac{i}{4\rho(z_s)}$$
$$\times \sum_n S(\Omega_n) u_n(z_s) u_n(z) H_0^1(k_n r), \quad (5.56)$$

but with Doppler shifted modal wavenumbers

$$k_n \to k_n \left(1 + \frac{v_r}{v_{gn}} \cos \theta_r \right), \quad (5.57)$$

and Doppler-shifted frequencies (now for each modal term in the summation) of

$$\Omega_n = \omega - k_n (v_s \cos \theta_s - v_r \cos \theta_r), \quad (5.58)$$

where v_{gn} is the group velocity of the n-th mode, $v_s \cos \theta_s$ is the radial speed of the source, and $v_r \cos \theta_r$ is the radial speed of the receiver. Here, radial refers to the projection of the velocities onto the horizontal line between the source and receiver. Note that this latter expression, when multiplied through by the wavenumber, shows that the frequency shift is proportional to the ratio of speeds to the modal phase speed, as opposed to the wavenumber shift which involves the group speed.

5.4.5 Parabolic Equation (PE) Model

The PE method was introduced into ocean acoustics and made viable with the development of the *Tappert split-step algorithm*, which utilized fast Fourier transforms at each range step [5.6]. Subsequent numerical developments greatly expanded the applicability and accuracy of the parabolic equation method.

Standard PE Split-Step Algorithm

The PE method is presently the most practical and all-encompassing wave-theoretic range-dependent propagation model. In its simplest form, it is a far-field narrow-angle ($\approx \pm 20°$) with respect to the horizontal – adequate for many underwater propagation problems – approximations to the wave equation. Assuming azimuthal symmetry about a source, we express the solution of (5.34) in cylindrical coordinates in a source-free region in the form

$$p(r,z) = \psi(r,z)H(r), \quad (5.59)$$

and we define $K^2(r,z) \equiv K_0^2 n^2$, n therefore being an *index of refraction* c_0/c, where c_0 is a reference sound speed. Substituting (5.59) into (5.34) and taking K_0^2 as the separation constant we end up with a Bessel equation for H that has a Hankel function as the outgoing solution. If we use the asymptotic form of the Hankel function, $H_0^1(K_0 r)$, and invoke the *paraxial* (narrow-angle) approximation,

$$\frac{\partial^2 \psi}{\partial r^2} \ll 2K_0 \frac{\partial \psi}{\partial r}, \quad (5.60)$$

we obtain the parabolic equation (in r),

$$\frac{\partial^2 \psi}{\partial z^2} + 2iK_0 \frac{\partial \psi}{\partial r} + K_0^2(n^2 - 1)\psi = 0, \quad (5.61)$$

where we note that n is a function of range and depth. We use a marching solution to solve the parabolic equation. There has been an assortment of numerical solutions but the one that still remains a standard is the so-called *split-step* range-marching algorithm,

$$\psi(r+\Delta r, z) = \exp\left[\frac{iK_0}{2}(n^2-1)\Delta r\right] F^{-1} \\ \times \left\{\left[\exp\left(-\frac{i\Delta r}{2K_0}s^2\right)\right] F[\psi(r,z)]\right\}. \quad (5.62)$$

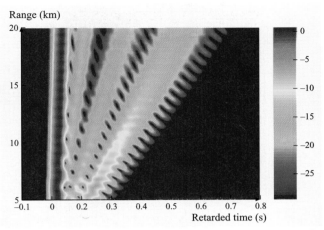

Fig. 5.25 Range-versus-time representation of the field intensity for the same waveguide and source–receiver depth as in Fig. 5.21. The figure clearly shows that modes travel at different speed. The retarded time t is $t - R/c_\omega$. The color scale is in dB with a 0 dB reference at range 5 km

The Fourier transforms F, are performed using FFTs. Equation (5.62) is the solution for n constant, but the error introduced when n (profile or bathymetry) varies with range and depth can be made arbitrarily small by increasing the transform size and decreasing the range-step size. It is possible to modify the split-step algorithm to increase its accuracy with respect to higher-angle propagation.

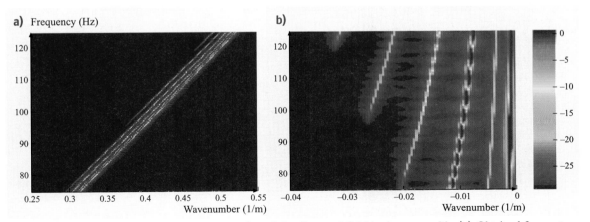

Fig. 5.26a,b Frequency–wavenumber representation of the dispersed field in the waveguide. (**a**) Obtained from a two-dimensional FFT of the range–time plot in Fig. 5.25. The wavenumbers of propagating modes are bounded by the upper and lower sound speeds in waveguide, ω/c_ω and ω/c_b. (**b**) A rotated version of (**a**), so that the sound speed in water appears infinite. This representation is easier to see the mode separation and the mode cut-off frequency. The color scale is in dB

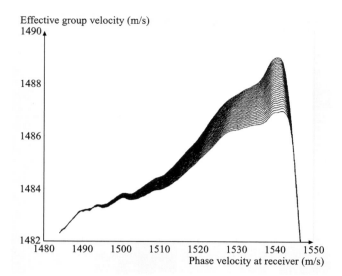

Fig. 5.27 Group speed versus phase speed in the Pacific Deep water case. The *curves* are for different frequencies starting with the lowest frequency 60 Hz being on *top*. Low phase velocity corresponds to low mode number

when the environment supports acoustic paths that become more vertical such as when the bottom has a very high speed and hence, large critical angle with respect to the horizontal. In addition, for elastic propagation, the compressional and shear waves span a wide angle interval. Finally, Fourier synthesis for pulse modeling requires high accurate in phase and the high-angle PEs are more accurate in phase, even at low angles.

Equation (5.61) with the second-order range derivative which was neglected because of (5.60) can be written in operator notation as:

$$[P^2 + 2iK_0 P + K_0^2(Q^2 - 1)]\psi = 0 , \qquad (5.63)$$

where

$$P \equiv \frac{\partial}{\partial r}, \quad Q \equiv \sqrt{n^2 + \frac{1}{K_0^2}\frac{\partial^2}{\partial z^2}} . \qquad (5.64)$$

Factoring (5.64) assuming weak range dependence and retaining only the factor associated with outgoing propagation yields a one-way equation

$$P\psi = iK_0(Q - 1)\psi , \qquad (5.65)$$

which is a generalization of the parabolic equation beyond the narrow-angle approximation associated with (5.60). If we define $Q = \sqrt{1+q}$ and expand Q in a Taylor series as a function of q, the standard PE method is recovered by $Q \approx 1 + 0.5q$. The wide-angle PE to arbitrary accuracy in angle, phase, etc. can be obtained from a Padé series representation of the Q operator,

$$Q \equiv \sqrt{1+q} = 1 + \sum_{j=1}^{n} \frac{a_{j,n} q}{1 + b_{j,n} q} + O(q^{2n+1}) , \qquad (5.66)$$

where n is the number of terms in the Padé expansion and

$$a_{j,n} = \frac{2}{2n+1} \sin^2\left(\frac{j\pi}{2n+1}\right), \quad b_{j,n} = \cos^2\left(\frac{j\pi}{2n+1}\right) . \qquad (5.67)$$

The solution of (5.65) using Eqs. (5.66) and (5.67) has been implemented using finite-difference techniques for fluid and elastic media.

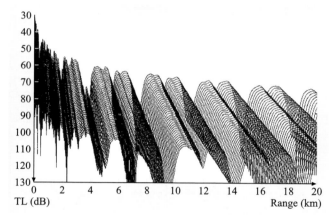

Fig. 5.28 Transmission loss versus range from 130 to 170 Hz calculated by OASES (Ocean Acoustic Seismic Exploration Synthesis) for an environment composed of a 200 m homogeneous fluid layer overlying a homogeneous, absorbing fluid half-space. The fluid layer has a 1450 m/s sound speed, whereas the bottom half-space has a sound speed of 1500 m/s and an attenuation of 10 dB/λ. The curves for increasing frequencies are progressively offset in 1 dB steps. The source and receiver depths are 20 and 100 m, respectively

Generalized or Higher-Order PE Methods

Methods of solving the parabolic equation, including extensions to higher-angle propagation, elastic media, and direct time-domain solutions including nonlinear effects have recently appeared (for example, see [5.34]). In particular, accurate high-angle solutions are important

5.4.6 Propagation and Transmission Loss

Propagation loss (PL) and transmission loss (TL) are decibel (see Appendix) quantities that are either measured or derived from propagation models. They represent the loss in intensity of the acoustic field as a function of range as referenced to some location. For modeling, we typically use the intensity at one meter range from the source using the assumption of spherical spreading over that one meter. Hence, if P is the output of a propagation code, the propagation loss is

$$\text{PL} = 20 \log \left| \frac{P}{p_0(r=1)} \right| \equiv -\text{TL}, \qquad (5.68)$$

where p_0 is the pressure of the source in free space. Transmission loss is a positive quantity.

Often, one sees references to coherent and incoherent propagation loss, which can be confusing. Deterministic physics has all acoustic propagation as coherent. If the medium has some randomness, than phase information is lost by some sort of averaging process. This leads to the idea of incoherent propagation. In normal-mode theory, the results is that the cross terms involving differences of modal wavenumbers do not contribute and the propagation curve is therefore smooth, without the interference pattern (Fig. 5.30c). Clearly such a calculation is easy with modes, but there is no equivalent simple way for the other wave models other than actually to introduce randomness. Figure 5.30 shows an application of this incoherent sum of modes. In shallow water an incoherent sum of modes is more or less equivalent to frequency smoothing, over, for example a third of an octave. The environment shown in Fig. 5.30a corresponds to experimental data contours of third-octave transmission loss as a function of frequency and range shown in Fig. 5.30b. Normal-mode model runs are shown in Fig. 5.30c, in which the modes are summed incoherently. Figure 5.30d is a model result of the same type of mode computations, but over the whole frequency spectrum; note the excellent agreement. This kind of analysis can be iteratively used to estimate the environment. Figures 5.30b and 5.30d are referred to an optimum frequency curves in the sense that 300–400 Hz appears to be the optimum frequency of propagation. For example, a vertical slice at 60 km indicates that the minimum loss is in that frequency interval. The curves in Fig. 5.30c are horizontal slices of the contour plot in Fig. 5.30d. The optimum frequency comes from the combination of high loss into the bottom at low frequency because

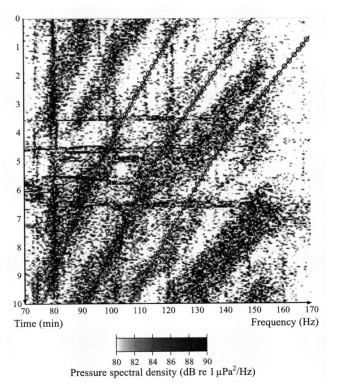

Fig. 5.29 A data example of spectrogram of a broadband moving source. The striations are the peaks in the interference pattern of the transmission loss. The time axis is converted to range if one knows the path and speed of the source [5.35]

of increased penetration into the lossy bottom and high losses at high frequency caused by scattering and other water volume attenuation effects that tend to have a frequency-squared dependence. Hence, the existence of an intermediate frequency at which bottom loss is small and frequency-squared dependence is not yet dominant.

Finally we note that there is also an incoherent sum of rays in which ray intensities and not amplitudes with phases are summed. This is not the equivalent of an incoherent sum of mode. For example, an incoherent sum of modes does not produce convergence zones while incoherent rays show convergence-zone properties.

5.4.7 Fourier Synthesis of Frequency-Domain Solutions

Using a Fourier transform, the frequency-domain solution $p(r, z, \omega)$ of the wave equation can be transformed

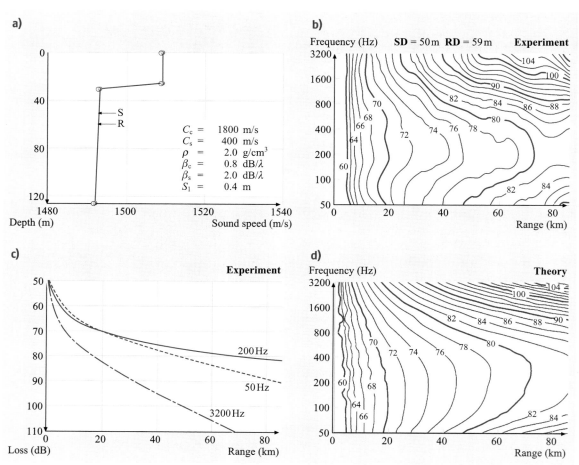

Fig. 5.30a–d Optimum frequency curves for a shallow-water environment. (**a**) The shallow-water environment. Source and receiver depths are marked by the arrows. (**b**) Contour of third-octave transmission loss data over frequency and range. (**c**) Incoherent normal-mode transmission loss. (**d**) Contour of incoherent transmission loss looking very much like the data in (**b**)

into the time-dependent solution $p(r, z, t)$. We have

$$p(r, z, t) = \frac{1}{2\pi} \int_{-\infty}^{+\infty} S(\omega) p(r, z, \omega) \, e^{-i\omega t} \, d\omega \,. \tag{5.69}$$

$S(\omega)$ is the source spectrum defined as

$$S(\omega) = \int_{-\infty}^{+\infty} s(t) e^{i\omega t} \, dt \,, \tag{5.70}$$

where $s(t)$ is the pulse source for which the time-domain solution $p(r, z, t)$ is investigated. Before (5.69) can be used, the transfer function $p(r, z, \omega)$ has to be determined from one of the frequency-domain [or continuous-wave (CW)] propagation models described above at a number of discrete frequencies within the frequency band of interest. This means the ocean response to a pulse $s(t)$ is obtained by combining the convenient and computationally efficient CW codes (spectral integrals, normal-mode or parabolic equations) with a Fourier synthesis approach. Even if this last step is conceptually simple, there are several numerical issues that have to be addressed like frequency and time windowing, and aliasing.

The first step in evaluating the frequency integral by means of a (fast) Fourier transform is to reduce the integration interval. The source pulse $s(t)$ being known, a frequency interval $[\omega_{\min}, \omega_{\max}]$ is determined, outside which the source does not emit any significant energy. Moreover, the final time-domain solution $p(r, z, t)$ being real, we know that $p(r, z, -\omega) = \overline{p(r, z, \omega)}$. This reduces the integration performed in (5.69) to the interval $[\omega_{\min}, \omega_{\max}]$

$$p(r, z, t) = \mathrm{Re}\left[\frac{1}{\pi}\int_{\omega_{\min}}^{\omega_{\max}} S(\omega) p(r, z, \omega)\, e^{-i\omega t}\, d\omega\right]. \tag{5.71}$$

Now, the set of discrete frequencies on which the integration has to be performed on the interval $[\omega_{\min}, \omega_{\max}]$ depends on both the choice of the sampling frequency F_s of the signal and some a priori knowledge of the time spread T of the final solution $p(r, z, t)$. First, the sampling frequency F_s has to satisfy the Nyquist criterium $F_s > 2f_{\max} = (\omega_{\max}/\pi)$. Typically, to allow for a reasonably looking graphical signal display, an appropriate value is $F_s = 8f_{\max}$. Second, the time window T of the signal after propagation through the ocean should be taken as short as possible. However, this time window must be chosen in accordance with the waveguide physics. A conservative way to estimate the time spread is to consider that T is bounded by the slowest and fastest available speeds for the particular environment:

$$T = t_{\max} - t_{\min} \geq R\left(\frac{1}{c_{\min}} - \frac{1}{c_{\max}}\right), \tag{5.72}$$

where R is the propagation range. With the proper choice of T and F_s, we construct an N-point time vector $\mathbf{t} = [0 : \frac{1}{F_s} : T]$, with $N = F_s T + 1$ and a frequency vector $\frac{\omega}{2\pi} = [0 : \frac{F_s}{N} : F_s(1 - \frac{1}{N})]$. The time vector $\mathbf{t}$ corresponds to the time axis on which the time-dependent solution $p(r, z, t)$ will be computed from the frequency bins of the frequency vector $\boldsymbol{\omega}$ inside the interval $[\omega_{\min}, \omega_{\max}]$. Before computing the Fourier transform, it is necessary to add a retarded time to the integral $\exp(-i t_{\min}\omega)$ so that the time vector $\mathbf{t}$ starts at the earliest possible arrival $t = t_{\min} + [0 : \frac{1}{F_s} : T]$. From (5.80), a reasonable choice for $t_{\min}$ is $t_{\min} = (R/c_{\max})$. It follows:

$$p(r, z, t) = Re\left\{\mathrm{IFFT}\left[S(\omega) p(r, z, \omega) e^{-i\omega t_{\min}}\right]_{\omega_{\min}}^{\omega_{\max}}\right\}, \tag{5.73}$$

where the inverse fast Fourier transform (IFFT) is performed on the frequency vector $\boldsymbol{\omega}$ in the interval $[\omega_{\min}, \omega_{\max}]$. We do not discuss here the numerous (fast) Fourier transform algorithms, which can be found in signal processing or numerical methods books. Typical results of this broadband modeling are presented in Figs. 5.22–5.25.

5.5 Quantitative Description of Propagation

All of the models described above attempt to describe reality and to solve in one way or another the Helmholtz equation. They therefore should be consistent and there is much insight to be gained from understanding this consistency. The models ultimately compute propagation loss which is taken as the decibel ratio (see Appendix) of the pressure at the field point to a reference pressure, typically one meter from the source.

Figure 5.31 shows convergence-zone-type propagation for a simplified profile. The ray trace in Fig. 5.31b shows the cyclic focusing discussed in Sect. 5.1.2. The same profile is used to calculate normal modes, shown in Fig. 5.31c which, when summed according to (5.50) referenced to 1 m, exhibits the same cyclic pattern as the ray picture. Figure 5.31d shows both the normal-mode (wave theory) and ray theory result. Ray theory exhibits sharply bounded shadow regions as expected whereas the normal-mode theory, which includes diffraction, shows that the acoustic field does exist in the shadow regions and the convergence zones have structure.

Normal-mode models sum the discrete modes that roughly correspond to angles of propagation within the cone of Fig. 5.7a. The spectral method can include the full field, discrete plus continuous, the latter corresponding to larger angles. The discussion below (5.50) defines these angles in terms of horizontal wavenumbers, and the eigenvalues of the normal-mode problem are a discrete set of horizontal wavenumbers. Hence the integrand (Green's function) of the spectral method has peaks at the eigenvalues associated with the normal modes. These peeks are shown on the right of Fig. 5.32a. The smoother portion of the spectrum is the continuous part, corresponding to the larger angles. Therefore, the consistency we expect between the normal-mode and the spectral method and the physics of Fig. 5.7 is that the continuous portion of the spectral solution decays rapidly with range so that there should be complete agreement at

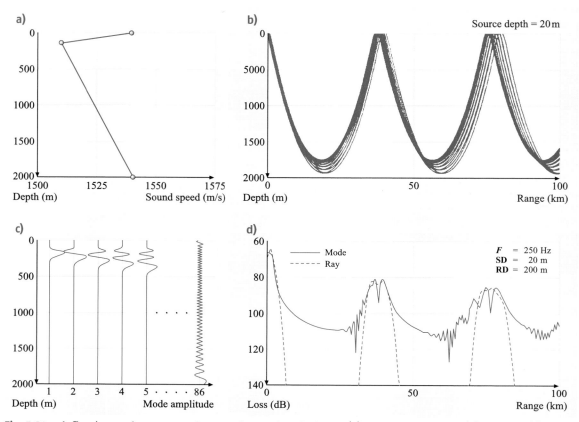

Fig. 5.31a–d Consistency between ray theory and normal mode theory. (**a**) Sound-speed profile, (**b**) ray trace, (**c**) normal modes, (**d**) propagation calculations. The shadow zones in (**d**) are much sharper for ray theory which does not include diffraction effects

long range between the normal-mode and spectral solutions. The Lloyd's mirror effect, a near-field effect, should also be exhibited in the spectral solution but not the normal-mode solution. These aspects are apparent in Fig. 5.32b. The PE solution is in good agreement with the other solutions but with some phase error associated with the average wavenumber that must be chosen in the split-step method. The PE solution, which contains part of the continuous spectrum including the Lloyd's mirror beams, is more accurate than the normal-mode solution at short range; however, the generalized PE can be made arbitrarily accurate at short range by including more expansion terms in (5.66).

Range-dependent results are shown in Fig. 5.33. A ray trace, a ray trace field result, a PE result and data are plotted together for a range-dependent sound-speed profile environment. The models agree with the data in general, with the exception that the ray results predict too sharp a leading edge for the convergence zone.

Up-slope propagation is modeled with the PE in Fig. 5.34. As the field propagates up-slope, sound is dumped into the bottom in what appears to be discrete beams. The flat region has three modes and each is cut off successively as sound propagates into the shallower water. The ray picture also has a consistent explanation for this phenomenon. The rays for each mode become steeper as they propagate up-slope. When the ray angle exceeds the critical angle the sound is significantly transmitted into the bottom.

5.6 SONAR Array Processing

Signal processing is common to many fields [5.36–38]. In this section we emphasize applications to underwater acoustics, mainly concentrating on spatial processing of pressure fields. We note, though, that there is also growing interest in processing vector fields such as acoustic displacement, velocity or acceleration [5.39]. Further, the array processing discussed below for passive SONARs is also applicable to active SONAR signal processing. Spatial sampling of a sound field is usually done by an array of transducers, although the synthetic aperture array, in which a sensor is moved through space to obtain measurements in both the time and space domains, is an important exception. Spatial sampling is analogous to temporal sampling with the sampling interval replaced by the sensor spacing. The Nyquist criterion requires that the sensor spacing be at least twice the spatial wavenumber of the measured sound field. Finally, we note that recently work of processing ambient noise fields have proven to yield information about the ocean environment [5.40, 41].

5.6.1 Linear Plane-Wave Beam-Forming and Spatio-Temporal Sampling

The simplest example of array processing is phase shading in the frequency domain (or time delay in the time domain) to search for the bearing of a plane-wave signal. This procedure is referred to as plane wave beam-forming, or delay and sum beam-forming in the time domain. For simplicity we consider a liner array and we take θ as the bearing angle associated with the plane-wave signal as shown in Fig. 5.35a.

Frequency-Domain Processing
A plane wave can be represented as

$$s(\theta) = \exp(\mathrm{i}\boldsymbol{k}\cdot\boldsymbol{r})\,, \tag{5.74}$$

where we have suppressed the time dependence $\exp(-\mathrm{i}\omega t)$ and $k = |\boldsymbol{k}| = (\omega/c)$. The field is summed in phase if the receiving element (hydrophone or microphone) inputs at position $\boldsymbol{d}$ is multiplied by the complex conjugate of the plane-wave phase factor,

$$\omega_i^* = \exp(-\mathrm{i}\boldsymbol{k}_\mathrm{s}\cdot\boldsymbol{d_i}) = \exp[-\mathrm{i}d(k\sin\theta_\mathrm{s})]\,, \tag{5.75}$$

where θ_s is a scanning angle. This process will have a maximum when the scanning angle equals the incident angle of the signal.

The output of this beam-forming process is denoted $B(\theta_\mathrm{s})$, but often it is the power output of the beam-former that is of interest:

$$\begin{aligned}B(\theta_\mathrm{s}) &= \left|\sum_{i=1}^{m}\omega_i^*(\theta_\mathrm{s})\left[s_i(\theta)+n_i\right]\right|^2 \\ &= \sum_{i,j=1}^{m}\omega_i^*(\theta_\mathrm{s})\left(s_{ij}+n_{ij}\right)\omega_j(\theta_\mathrm{s})\,,\end{aligned} \tag{5.76}$$

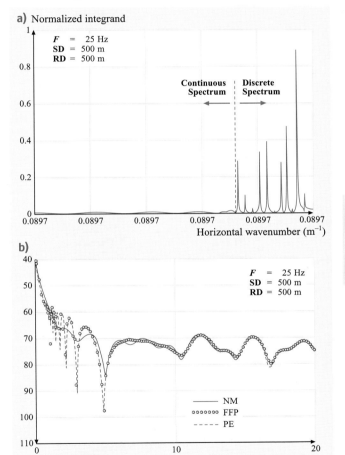

Fig. 5.32a,b Relationship between spectral (FFP), normal-mode (NM) and split-step parabolic-equation model (PE) computations. (a) FFP Green's function from (5.44). (b) Normal-mode, spectral (FFP) and PE propagation results showing some agreement in the near field and complete agreement in the far field. Higher-order PE methods will give identical results to the FFP

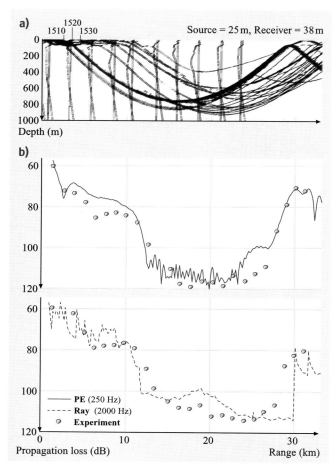

Fig. 5.33a,b Model and data comparison for a range-dependent case. (**a**) Profiles and ray trace for a case of a surface duct disappearing. (**b**) 250 Hz PE and 2 kHz ray-trace comparisons with data. Again, the diffraction-less ray theory shows more abrupt change in field

the signal and noise with elements $K_{ij} = s_{ij} + n_{ij}$ since the signal and noise are assumed to be independent. Equation (5.76) can be rewritten as

$$B(\theta_s) = \boldsymbol{w}^\dagger(\theta_s) \boldsymbol{K}(\theta_{\text{true}}) \boldsymbol{w}(\theta_s) \equiv \boldsymbol{w}^\dagger \boldsymbol{K} \boldsymbol{w}, \quad (5.77)$$

where † denotes the complex transpose. The CSDM or the covariance of the field is composed of uncorrelated signal and noise covariances,

$$\boldsymbol{K} = \boldsymbol{K}_s + \boldsymbol{K}_n. \quad (5.78)$$

The data across the array as represented in the matrix $\boldsymbol{K}$ contains the information that the source is in the direction θ_{true}. Sometimes $\boldsymbol{w}(\theta_s)$ is referred to as a replica and this beam-forming process is viewed as matching the received data across the array with a replica. The type of beam-former represented by (5.77) is called a linear or Bartlett beam-former.

For the sample covariance estimation we assume we have an array with N sensors located at $\boldsymbol{d}_i$, $i = 1, 2, \cdots, N$ and a narrow-band model as illustrated in Fig. 5.35b. These covariances are estimated by segmenting the received data, $r_i(t)$ into *snapshots* using a sampling window, $W(t)$, that is unity in the interval $[0, T_w]$,

$$R_i^l(f) = \int_{T_l}^{T_l + T_w} r_i(t) W(t - T_l) e^{-i2\pi f t} dt, \quad (5.79)$$

where here the notation uses frequency f, rather than angular frequency ω. In most beam-forming algorithms the data vectors are averaged to form the sample covariance matrix

$$\hat{K}(f) = \frac{1}{L} \sum_{l=1}^{L} R^l(f) R^l(f)^H, \quad (5.80)$$

where L is the number of snapshots.

Time–Domain Processing

Time delaying a signal is the time-domain analog to phase shading of the signal in the frequency domain. In general, the time domain is a viable alternative to frequency analysis when processing coherent broadband signals such as temporal impulse. Time-delay beam-forming is then the frequency equivalent of the phase beam-forming described in Sect. 5.6.1. Indeed, it can be derived formally by taking the Fourier transform of the beam-forming process with the result that the beam-

where s_i and n_i are the signal and noise at the i-th receiving element and where $s_{ij} + n_{ij}$ are elements of a cross-spectral density matrix which, when obtained from data, would involve Fourier transforms and ensemble averages as mentioned in the discussion below (5.78) augmented by Fig. 5.35b. In writing down the right-hand side of (5.76), the signal and noise fields were assumed to be mutually incoherent.

We can write the above expression in matrix notation where the boldface lower-case letters denote vectors and boldface upper-case letters denote matrices. Define a steering column vector $\boldsymbol{w}$ whose i-th element is w_i and a cross-spectral density matrix (CSDM) $\boldsymbol{K}$ of former output is:

$$b(\theta, t) = \int_{-\infty}^{\infty} B(\theta, \omega) \exp(i\omega t)\,d\omega$$

$$= \int_{-\infty}^{\infty} \sum_i R_i(\omega) \exp\left(-i\frac{\omega}{c_i} d_i \sin\theta\right)$$
$$\times \exp(i\omega t)\,d\omega$$

$$= \sum_i R_i\left(t - \frac{d_i}{c_i}\sin\theta\right), \quad (5.81)$$

where $R_i(t)$ is the time domain data, d_i the location and c_i the sound speed at the i-th phone. Physically speaking, the delay $\tau_i = (d_i/c_i)\sin\theta$ is, to a first approximation, the time delay associated with the phase shift in the frequency-domain array processing. A more rigorous approach in the case of a strong depth dependence of the sound-speed profile leads to:

$$\tau_i = \int_{d_0}^{d_i} \sqrt{\frac{1}{c^2(z)} - \frac{\cos^2(\theta)}{c^2(d_0)}}\, dz, \quad (5.82)$$

where d_0 is the depth of a reference hydrophone on the array. Note that, in the case of a uniform sound speed profile over the array, the above processes are plane-wave beam-forming.

Examples of the practical use of time-domain versus frequency-domain beam-forming are discussed below. Consider an incident field on a 20-element vertical array made of seven pulses arriving at different angles in a noisy waveguide environment (Fig. 5.36). The SNR ratio does not allow an accurate detection and identification of the echoes. Time-delay beam-forming results applied on these broadband signals is presented in Fig. 5.37. The SNR of the time-domain processing is a combination of array gain and frequency-coherent processing. In comparison, phase beam-forming performed on the same data and averaged incoherently over frequencies show a degraded detection of the incident echoes (Fig. 5.38).

5.6.2 Some Beam-Former Properties

Figure 5.39 shows the output results of some frequency-domain plane-wave beam-formers for the cases of one and two incident signals. To be noted are the side-lobes of the Bartlett processor and the high-resolution performance of the adaptive processors (discussed in the next section). Some of the general attributes of an array beam-former are:

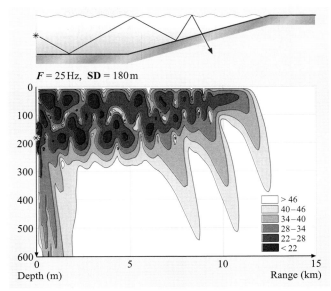

Fig. 5.34 Relation between up-slope propagation (from PE calculation) showing individual mode cut-off and energy dumping in the bottom, and a corresponding ray schematic. In this case, three modes were excited in the flat region. The ray picture shows that a ray becomes steeper as it bounces up slope and when its grazing angle exceeds the critical angle, it is partially transmitted into the bottom (and subsequently with more and more transmission with each successive, higher angle bounce)

- The main response axis (MRA): generally, one normalizes the beam pattern to have 0 dB, or unity gain in the steered direction.
- Beam width: an array with finite extent, or aperture, must have a finite beam width centered about the MRA, which is termed the *main lobe*.
- Side-lobes: angular or wavenumber regions where the array has a relatively strong response. Sometimes they can be comparable to the MRA, but in a well-designed array, they are usually -20 dB or lower, i.e., the response of the array is less than 0.1 of a signal in the direction of a side-lobe.
- Wavenumber processing: rather than scan through incident angles, one can scan through wavenumbers, $k\sin\theta_s \equiv \kappa_s$; scanning through all possible values of κ_s results in nonphysical angles that correspond to waves not propagating at the acoustic medium speed. Such waves can exist, such as those associated with array vibrations. The beams associated with these wavenumbers are sometimes referred to as virtual beams. An important aspect of these beams is that their

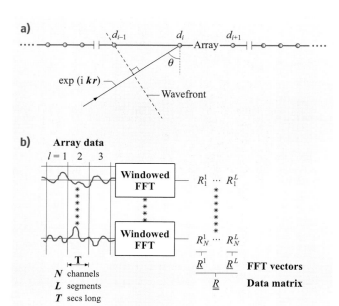

Fig. 5.35a,b Array processing. (**a**) Geometry for plane-wave beam-forming. (**b**) The data is transformed to the frequency domain in which the plane-wave processing takes place. The cross-spectral-density matrix (CSDM) is an outer product of the data vector for a given frequency

side-lobes can be in the physical angle region, thereby interfering with acoustic propagating signals.
- Array gain: defined as the decibel ratio of the signal-to-noise ratios of the array output to a single phone

output. If the noise field is isotropic, the array gain is also termed the directionality index.

5.6.3 Adaptive Processing

There are high-resolution methods to suppress side-lobes, usually referred to as adaptive methods since the signal processing procedure constructs weight vectors that depend on the received data itself. We briefly describe one of these procedures: the minimum-variance distortionless processor (MV), sometimes also called the maximum-likelihood method (MLM) directional spectrum-estimation procedure.

We seek a weight vector w_{MV} applied to the matrix K such that its effect will be to minimize the output of the beam-former (5.77) except in the look direction, where we want the signal to pass undistorted. The weight vector is therefore chosen to minimize the functional [5.38]. From (5.72)

$$F = w_{MV} K w_{MV} + \alpha (w_{MV} w - 1) \, . \tag{5.83}$$

The first term is the mean-square output of the array and the second term incorporates the constraint of unity gain by means of the Lagrangian multiplier α. Following the method of Lagrange multipliers, we obtain the MV weight vector,

$$w_{MV} = \frac{K^{-1} w}{w K^{-1} w} \, . \tag{5.84}$$

This new weight vector depends on the received data as represented by the cross-spectral density matrix; hence, the method is adaptive. Substituting back into (5.77) gives the output of our MV processor,

$$B_{MV}(\theta_s) = [w(\theta_s) K^{-1}(\theta_{true}) w(\theta_s)]^{-1} \, . \tag{5.85}$$

The MV beam-former should have the same peak value at θ_{true} as the Bartlett beam-former, (5.77) but with side-lobes suppressed and a narrower main beam, indicating that it is a high-resolution beam-former. Examples are shown in Fig. 5.39. Of particular practical interest for this type of processor is the estimation procedure associated with Fig. 5.35b and (5.80). One must take sufficient snapshots to allow the stable inversion of the CSDM. This requirement may conflict with source motion when the source moves through multiple beams along the time interval needed to construct the CSDM.

5.6.4 Matched Field Processing, Phase Conjugation and Time Reversal

Matched field processing (MFP) [5.42] is a generalization of plane-wave beam-forming in which data

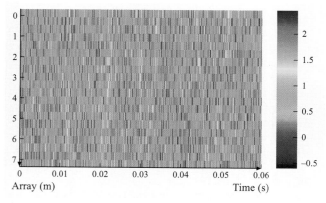

Fig. 5.36 Depth-versus-time representation of simulated broadband coherent data received on a vertical array in the presence of ambient noise. The wavefronts corresponding to different sources are nearly undistinguishable

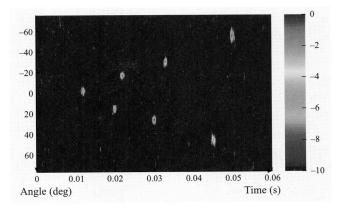

Fig. 5.37 Angle-versus-time representation of the simulated data in Fig. 5.36 after time-delay beam-forming. All sources appear clearly above the noise with their corresponding arrival angle

on an array is correlated with replicas from a (waveguide) propagation model for candidate locations $\hat{r}, \hat{z}$ (Fig. 5.40). Localization of a source is accomplished with a resolution consistent with the modal structure and SNR. The central difficulty with MFP is specifying the coefficients and boundary conditions of the acoustic wave equation for shallow water propagation, *i.e.*, knowing the ocean environment in order to generate the replicas. An alternative to performing this model-based processing is phase conjugation (PC), in the frequency domain or, time reversal (TR) in the time domain, in which the conjugate or time-reversed data is used as source excitations on a transmit array co-located with the receive array (Fig. 5.41a) [5.43]. The PC/TR process is equivalent to correlating the data with the *actual* transfer function from the array to the original source location. In other words, both MFP and PC are signal processing analogs to the mechanical lens adjustment feedback technique used in adaptive optics: MFP uses data together with a model (note the feedback arrow in Fig. 5.40) whereas PC/TR is an active form of adaptive optics simply retransmitting phase-conjugate/time-reversed signal through the same medium (*e.g.*, see result of Fig. 5.41). Though time reversal is thought of as as active process, it is presented in this section because of its relation to passive MFP.

Ocean Time–Reversal Acoustics

Phase conjugation, first demonstrated in nonlinear optics and its Fourier conjugate version, time reversal is a process that has recently implemented in ultrasonic laboratory acoustic experiments [5.44]. Implementation of time reversal in the ocean for single elements [5.45] and using a finite spatial aperture of sources, referred to as a *time-reversal mirror* (TRM) [5.46], is now well established.

The geometry of a time-reversal experiment is shown in Fig. 5.41. Using the well-established theory of PC and TRM in a waveguide, we just write down the result of the phase-conjugation and time-reversal process, respectively, propagating toward the focal position

$$P_{\text{pc}}(r, z, \omega) = \sum_{j=1}^{J} G_\omega(r, z, z_j) G_\omega^*(R, z, z_{\text{ps}}) S^*(\omega)$$

(5.86)

and

$$P_{\text{trm}}(r, z, t) = \frac{1}{(2\pi)^2} \sum_{j=1}^{J} \iint G(r, z, t''; 0, z_j, t') \\ \times G(R, z_j, t'; 0, z_{\text{ps}}, 0) \\ \times S(t'' - t + T) \, \mathrm{d}t \, \mathrm{d}t'' ,$$

(5.87)

where S is the source function, $G_\omega^*(R, z, z_{\text{ps}})$ is the frequency-domain Green's function and $G(R, z_j, t'; 0, z_{\text{ps}}, 0)$ is the time-domain Green's function (TDGF) from the probe source at depth z_{ps} to each element of the SRA at range R and depth z_j. Emphasizing the time-domain process, $G(r, z, t''; 0, z_j, t')$ is the

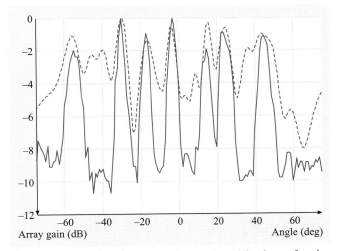

Fig. 5.38 Comparison between coherent time-delay beam-forming (*in red*) and incoherent frequency-domain beam-forming (*in blue*) for the simulated data shown in Fig. 5.35. When data come from coherent broadband sources, time-domain bema-forming show better performance than frequency analysis

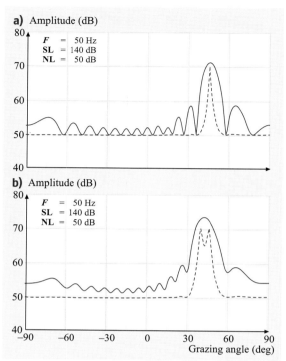

Fig. 5.39a,b Simulated beam-former outputs. (**a**) Single sources at a bearing of 45°. (**b**) Two sources with 6.3° angular separation. *Solid line*: linear processor (Bartlett). *Dashed line*: minimum variance distortion-less processor (MV) showing that the side-lobes are suppressed

TDGF from each element of the SRA back to range r and depth z. The focused field at the probe source position is $P_{\text{trm}}(R, z_{\text{ps}}, t)$. The summation is performed on the J elements of the TRM. The signal $S(t'' - t + T)$ of duration τ is the time-reversed version of the original probe source signal and the derivation of (5.87) uses the causality requirement, $T > 2\tau$, i.e., the time reversal interval T must contain the total time of reception and transmission at the SRA, 2τ.

Figure 5.41a,b,c shows the result of a TRM experiment. The size of the focus is consistent with the spatial structure of the highest-order mode surviving the two-way propagation and can also be mathematically described by a virtual array of sources using image theory in a waveguide.

Matched Field Processing

Linear matched field processing (MFP) can be thought of as the passive signal-processing implementation of phase conjugation. Referring to the phase-conjugation process described by (5.86), rename S^*G^* as the data at each array element and call the data vector on the array $R(a_{\text{true}})$, where a_{true} represents the (unknown) location of the source (Fig. 5.35b). Now in phase conjugation, G represents an actual propagation from the source array. In MPF, we do the propagation numerically using one of the acoustic models, but rather than use the actual Green's function, we use a normalized version of it called a replica: $\omega(a) = G(a)/|G(a)|$, where $G(a)$ is a vector of Green's functions of dimension of the number of array elements that represents the propagation from a candidate source position to the array elements and is the magnitude of the vector over the array elements. Taking the square of the PC process with replica's replacing the Green's functions yields the beam-former of the matched field processor

$$B_{\text{mf}}(a) = \omega^H(a) K(a_{\text{true}}) \omega(a), \quad (5.88)$$

where a realization of the CSDM on the array is then $K(a_{\text{true}}) = R(a_{\text{true}}) R^H(a_{\text{true}})$ and a sample CSDM is built up as per (5.80).

MFP works because the unique spatial structure of the field permits localization in range, depth and azimuth depending on the array geometry and complexity of the ocean environment. The process is shown schematically in Fig. 5.40. MFP is usually implemented by systematically placing a test point source at each point of a search grid, computing the acoustic field (replicas) at all the elements of the array and then correlating this modeled field with the data from the real point source, $K(a_{\text{true}}) = RR^H$, whose location is unknown. When the test point source is co-located with the true point source, the correlation will be a maximum. The scalar function $B_{\text{mf}}(a)$ is also referred to as the ambiguity function (or surface) of the matched field processor because it also contains ambiguous peaks which are analogous to the side-lobes of a conventional plane-wave beam-former. Side-lobe suppression can often be accomplished by using the adaptive beam-forming methods discussed in the plane-wave section.

Adaptive processors are very sensitive to the accuracy of the replica functions which, in turn, require almost impossible knowledge of the environment. Hence, much work has been done on developing robust forms of adaptive processing such as the white-noise constraint method and others [5.47, 48]. An example of matched field processing performed incoherently on eight tones between 50 Hz and 200 Hz is shown in Fig. 5.42.

5.7 Active SONAR Processing

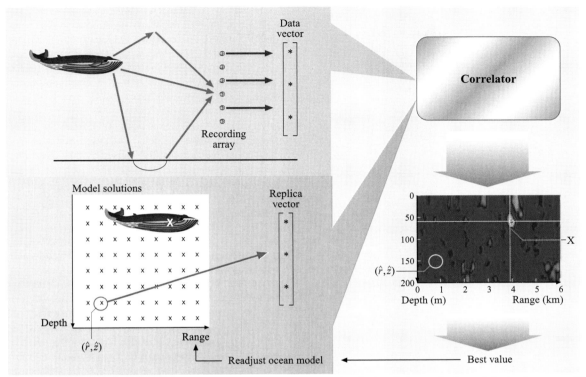

Fig. 5.40 Matched field processing (MFP). Here, the example consists in localizing a singing whale in the ocean. If your model of waveguide propagation is sufficiently accurate for this environment, then comparing the recorded sounds – the whale's data vector – one frequency at a time, for example, with replica data based on best guesses of the location $(\hat{r}, \hat{z})$ that the model provides, will eventually yield its location. The red peak in the data indicates the location of highest correlation. The small, circled × represents a bad guess, which thus does not compare well with the measured data. The feedback loop suggests a way to optimize the model: by fine-tuning the focus – the peak resolution in the plot – one can readjust the model's bases (for example, the sound-speed profile). That feedback describes a signal-processing version of adaptive optics. Matched field processing can then be used to perform acoustic tomography in the ocean [5.49]

An active SONAR system transmits a pulse and extracts information from the echo it receives as opposed to a passive SONAR system, which extracts information from signals received from radiating sources. An active SONAR system and its associated waveform is designed to detect targets and estimate their range, Doppler (speed) and bearing or to determine some properties of the medium such as ocean bottom bathymetry, ocean currents, winds, particulate concentration, etc. The spatial processing methods already discussed are applicable to the active problem so that in this section we emphasize the temporal aspects of active signal processing.

5.7.1 Active SONAR Signal Processing

The basic elements of an active SONAR are: the (waveform) transmitter, the channel through which the signal, echo and interference propagates, and the receiver [5.50]. The receiver consists of some sort of matched filter, a square-law device, and possibly a threshold device for the detector and range, Doppler and bearing scanners for the estimator.

The matched filter maximizes the ratio of the peak output signal power to the variance of the noise and is implemented by correlating the received signal with

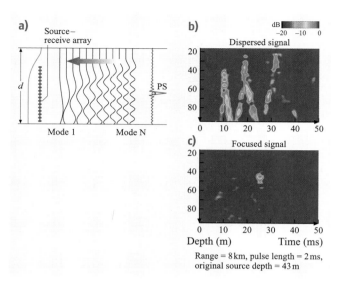

Fig. 5.41a–c Ocean acoustic time-reversal mirror (**a**) The acoustic field from a probe source (PS) is received on a source–receive array (SRA). (**b**) The signal received on the SRA with the first mode arriving first. At the SRA the signal is digitized, time-reversed and retransmitted. (**c**) The time-reversed signal received on an array at the same range as PS. The signal has been refocused at the probe source position with a resolution determined by the highest-order mode surviving the two-way process

the transmitted signal. A simple description of the received signal, $r(t)$, is that it is an attenuated, delayed, and Doppler-shifted version of the transmitted signal $s_t(t)$,

$$r(t) \to \text{Re}\left[\alpha e^{i\tilde{\theta}} s_t(t-T) e^{2\pi i f_c t} e^{2\pi i f_d t} + n(t)\right], \quad (5.89)$$

where α is the attenuation transmission loss and target cross section, θ is a random phase from the range uncertainty compared to a wavelength, T is the range delay time, f_c is the center frequency of the transmitted signal and f_d is the Doppler shift caused by the target. The correlation process will have an output related to the following process,

$$C(a) = \left| \int \tilde{r}(t)\tilde{s}(t;a)\,dt \right|^2, \quad (5.90)$$

where $\tilde{s}(t;a)$ is a replica of the transmitted signal modified by a parameter set a which include the propagation–reflection process, e.g., range delay and Doppler rate. For the detection problem, the correlation receiver is used to generate a sufficient statistic which is the basis for a threshold comparison in making a decision if a target is present. The performance of the detector is described by receiving operating characteristic (ROC) curves which plot the detection of probability versus false-alarm probability, as parameterized by a statistics of the signal and noise levels. The parameters a set the range and Doppler value in the particular resolution cell of concern. To estimate these parameters, the correlation is done as a function of a.

For a matched filter operating in a background of white noise detecting a point target in a given range–Doppler resolution cell, the detection signal-to-noise ratio depends on the average energy-to-noise ratio and not on the shape of the signal. The waveform becomes

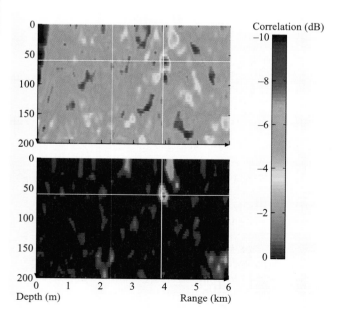

Fig. 5.42a,b Matched field processing example for a vertical array in a shallow-water environment. Specific information regarding the experiment can be found on the web at http://www.mpl.ucsd.edu/swellex96/. (**a**) Bartlett result with significant side-lobes only 3 dB down. (**b**) Adaptive processor results shows considerable side-lobe suppression. The processor is actually a *white-noise constrained* MVDP for which the diagonal of the CSDM is deliberately loaded by a specific algorithm to stabilize the processor to some uncertainty in the environment and/or array configuration. The ambiguity surfaces in (**a**) and (**b**) are an incoherent average over eight tones at 53, 85, 101, 117, 133, 149, 165, 181, and 197 Hz

a factor when there is a reverberant environment and when one is concerned with estimating target range and Doppler. A waveform's potential for range and Doppler resolution can be ascertained from the ambiguity function of the transmitted signal. This ambiguity function is related to the correlation process of (5.90) for a transmitted signal scanned as a function of range and Doppler,

$$\Theta(\hat{T}, T_t, \hat{f}_d, f_{d_t})$$
$$\propto \left| \int \tilde{s}(t-T)\tilde{s}_t(t-\hat{T}) e^{2\pi i (f_{d_t} - \hat{f}_d)t} \, dt \right|^2, \quad (5.91)$$

where T_t and f_{d_t} are the true target range (time) and Doppler, and $\hat{T}$ and $\hat{f}_d$ are the scanning estimates of the range and Doppler. Figure 5.43 are sketches of ambiguities functions of some typical waveforms. The range resolution is determined by the reciprocal of the bandwidth and the Doppler resolution by the reciprocal of the duration. A coded or pseudo-random (PR) sequence can attain good resolution of both by appearing as long-duration noise with a wide bandwidth. Ambiguity functions can be used to design desirable waveforms for particular situations. However, one must also consider the randomizing effect of the real ocean. The *scattering function* describes how a transmitted signal statistically redistributes its energy in the reverberant ocean environment which causes multipath and Doppler spread. In particular, in a reverberation-limited environment, only increasing the transmitted power does not change the signal-to-reverberation level. Signal design should minimize the overlap of the ambiguity function of the target displaced to its range and Doppler and the scattering function.

5.7.2 Underwater Acoustic Imaging

Imaging can be divided into categories concerned with water column and bottom properties. Here we describe several applications of active SONAR to imaging the ocean.

Water Column Imaging. Backscatter from particulate objects that move along with the water motion (such as biological matter, bubbles sediments) contains velocity information because of the Doppler shift. An acoustic Doppler current profiler (ADCP) might typically consist of three or four source–receivers pointed in slightly different directions but generally up (from the bottom) or down (from a ship). The multiple directions are for resolving motion in different directions. The Doppler shift

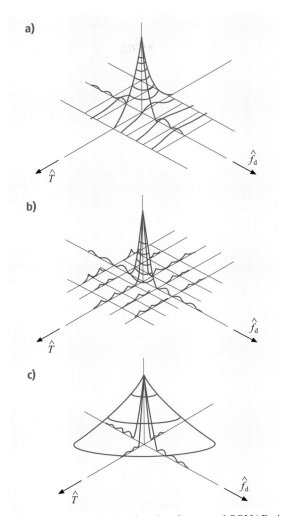

Fig. 5.43a–c Ambiguity function for several SONAR signals: (**a**) rectangular pulse; (**b**) coded pulses; (**c**) chirped FM pulse

of the returning scattered field is simply $-2f(v/c)$ (as opposed to the more-complicated long-range waveguide Doppler shift discussed in Sect. 5.4.4), where f is the acoustic frequency, v is the radial velocity of the scatterer (water motion), and c is the sound speed. With three or four narrow-beam transducers, the current vector can be ascertained as a function of distance from an ADCP by gating the received signal and associating a distance with the time gated segment of the signal. Water-column motion associated with internal waves can be determined by this process also where essentially one uses a kind of ADCP points in the horizontal direction. For elab-

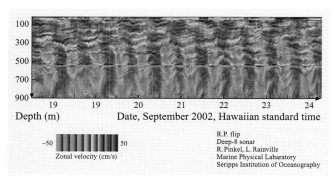

Fig. 5.44 Depth–time representation of the East–West component of water velocity over the Kaena ridge west of Oahu. The dominant signal is the 12.4 h tide, which has downward propagating crests of long vertical wavelength. The SONAR is at ≈390 m, where there is a fine scar in the record. The scar at ≈ 550 m is the echo of the sea floor at 1100 m, aliased back into the record (Courtesy Rob. Pinkel, Scripps Institution of Oceanography)

to two complete courses on submarine acoustic imaging methods available: (1) from the Ocean Mapping Group (University of New Brunswick, Canada) at http://www.omg.unb.ca/GGE/; (2) under a PDF format (*Multibeam Sonar: Theory of Operation*) from L3 Communication SeaBeam Instruments at http://www.mbari.org/data/mbsystem/formatdoc/. The scope of the paragraph below is to describe the basics and keywords associated with bottom mapping in underwater acoustics.

Active SONARs are devices that emit sound with specific waveforms and listen for the echoes from remote objects in the water. Among many applications, SONAR systems are used as echo-sounders for measuring water depths by transmitting acoustic pulses from the ocean surface and listening for their reflection (or echo) from the sea floor. The time between transmission of a pulse and the return of its echo is the time it takes the sound to travel to the bottom and back. Knowing this time and the speed of sound in water allows one to calculate the range to the bottom. This technique has been widely used to map much of the world's water-covered areas and has permitted ships to navigate safely through the world's oceans. The purpose of a large-scale bathymetric survey is to produce accurate depth measurements for many neighboring points on the sea floor such that an accurate picture of the geography of the bottom can be established. To do this efficiently, two things are required of the SONAR used: it must produce accurate

orate images of currents and the internal wave motion (Fig. 5.44), the Doppler measurements of backscattering off zooplankton are combined with array-processing techniques used in bottom mapping as discussed below.

Bottom Mapping. Because of many industrial applications, there exists a huge literature on multibeam systems for bottom mapping. Among the material available on the web, we would like to refer

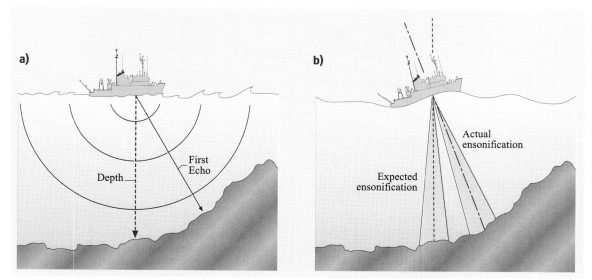

Fig. 5.45a,b Surveying an irregular sea floor (**a**) with a wide-beam SONAR and (**b**) with a narrow-beam SONAR (Courtesy L-3 Communications SeaBeam Instruments)

depth measurements that correspond to well-defined locations on the sea floor (that is, specific latitudes and longitudes); and it must be able to make large numbers of these measurements in a reasonable amount of time. In addition, information derived from echo sounding has aided in laying transoceanic telephone cables, exploring and drilling for offshore oil, locating important underwater mineral deposits, and improving our understanding of the Earth's geological processes.

The earliest, most basic and still most widely used echo-sounding devices are *single-beam echo sounders*. The purpose of these instruments is to make serial measurements of the ocean depth at many locations. Recorded depths can be combined with their physical locations to build a three-dimensional map of the ocean floor. In general, single-beam depth sounders are set up to make measurements from a vessel in motion. Until the early 1960s most depth sounding used single-beam echo sounders. These devices make a single depth measurement with each acoustic pulse (or ping) and include both wide- and narrow-beam systems (Fig. 5.45a,b). Relatively inexpensive wide-beam sounders detect echoes within a large solid angle under a vessel and are useful for finding potential hazards for safe navigation. However, these devices are unable to provide much detailed information about the sea bottom. On the other hand, more-expensive narrow-beam sounders are capable of providing high spatial resolution with the small solid angle encompassed by their beam, but can cover only a limited survey area with each ping. Neither system provides a method for creating detailed maps of the sea floor that minimizes ship time and is thus cost-effective.

A *multibeam SONAR* is an instrument that can map more than one location on the ocean floor with a single ping and with higher resolution than those of conventional echo sounders. Effectively, the function of a narrow single-beam echo sounder is performed at several different locations on the bottom at once. These bottom locations are arranged such that they map a contiguous area of the bottom – usually a strip of points in a direction perpendicular to the path of the survey vessel (Fig. 5.46). Clearly, this is highly advantageous. Multibeam SONARs can map complete scans of the bottom in roughly the time it takes for the echo to return from the farthest angle. Because they are far more complex, the cost of a multibeam SONAR can be many times that of a single-beam SONAR. However, this cost is more than compensated by the savings associated with reduced ship operating time. As a consequence, multibeam SONARs are the survey instrument of choice in most mapping applications, particularly in

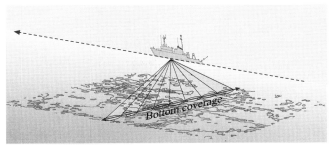

Fig. 5.46 Bottom mapping with a multibeam SONAR system (Courtesy L-3 Communications SeaBeam Instruments)

deep ocean environments where ship operating time is expensive [5.51]. Multibeam SONARs often utilize the *Mills Cross* technique which takes advantage of the high resolution obtained from two intersecting focusing regions from a perpendicular linear source and receive array.

Instead of measuring the depth to the ocean bottom, a *side-scan SONAR* reveals information about the sea-floor composition by taking advantage of the different sound absorbing and reflecting characteristics of different materials. Some types of material, such as metals or recently extruded volcanic rock, are good reflectors. Clay and silt, on the other hand, do not reflect sound well. Strong reflectors create strong echoes, while weak reflectors create weaker echoes. Knowing these characteristics, you can use the strength of acoustic returns to examine the composition of the sea floor. Reporting the strength of echoes is essentially what a side-scan SONAR is designed to do. Combining bottom-composition information provided by a side-scan SONAR with the depth information from a range-finding SONRA can be a powerful tool for examining the characteristics of an ocean bottom.

The name *side scan* is used for historical reasons – because these SONARs were originally built to be sensitive to echo returns from bottom locations on either side of a survey ship, instead of directly below, as was the case for a traditional single-beam depth sounder [5.52].

Side-scan SONAR employs much of the same hardware and processes as conventional depth-sounding SONAR. Pulses are transmitted by a projector (or array of projectors), and hydrophones receive echoes of those pulses from the ocean floor and pass them to a receiver system. Where side-scan SONAR differs from a depth-sounding system is in the way it processes these returns. While a single-beam echo sounder is only concerned with the time between transmission and the earliest return echo, this first returned echo only marks when

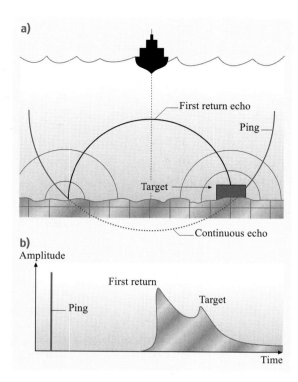

Fig. 5.47 (a) Schematic of the spherical wavefronts scattered by a detailed bottom. (b) Amplitude-versus-time representation of the backscattered signal in a side-scan SONAR system (Courtesy L-3 Communications SeaBeam Instruments)

things start to get interesting to a side-scan SONAR (Fig. 5.47a). As it continues its spherical propagation, the transmitted pulse still interacts with the bottom, thus creating at the receiver a continuous series of weakening echoes in time (Fig. 5.47b).

In the example presented in Fig. 5.47, the side-scan SONAR detects a bottom feature (the box). From the amplitude-versus-time plot, an observer can tell there is a highly reflective feature on the bottom. From the time difference between the first echo (which is presumed to be due to the bottom directly below the SONAR system) and the echo of the reflective feature, the observer can compute the range to the feature from the SONAR.

As a practical instrument, the simplistic side-scan SONAR described above is not very useful. While it provides the times of echoes, it does not provide their direction. Most side-scan SONARs deal with this problem by introducing some directionality into their projected pulses, and, to some degree, their receivers. This is done by using a line array of projectors to send pulses. The long axis of the line array is oriented parallel to the direction of travel of the SONAR survey vessel (often the arrays are towed behind the ship). In practice, side-scan SONARs tend to mount both the line array and hydrophones on a towfish, a device that is towed in the water below the surface behind a survey vessel (Fig. 5.48). As the survey vessel travels, the towfish transmits and listens to the echoes of a series of pulses. The echoes of each pulse are used to build up amplitude-versus-time plots for each side of the vessel. To adjust for the decline in the strength of echoes due to attenuation, a time-varying gain is applied to the amplitude values so that sea-floor features with similar reflectivities have similar amplitudes. Eventually the noise inherent in the system (which remains constant) becomes comparable to the amplitude of the echo, and the amplified trace becomes dominated by noise. Recording of each trace is usually cut off before this occurs so that the next ping can be transmitted.

Figure 5.49 shows an example of ship-wreck discovery using side-scan SONAR data performed from an automated underwater vehicle (AUV) in a 1500 m-deep ocean.

Often one is interested in sub-bottom profiling that requires high spatial and therefore temporal resolution to image closely spaced interfaces. Frequency-modulated (FM) sweeps provide such high-resolution high-intensity signals after matched filtering. Thus, for example, a matched filter output of a 1 s FM sweep from 2–12 kHz would compresses the energy of the 1 s pulse into one that has the temporal resolution of a 0.1 ms. With such high resolution, reflection coefficients from chirp SONARs can be related to sedimentary characteristics [5.53, 54]. Figure 5.50 is an example of a chirp SONAR output indicating very finely layered interfaces. Figure 5.50a shows the range dependency of the seabed

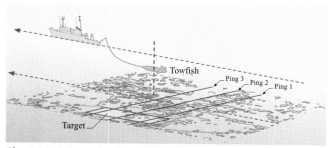

Fig. 5.48 Schematic of a side-scan SONAR imaging the ocean floor with successive pings (Courtesy L-3 Communications SeaBeam Instruments)

along the cross-shelf track taken by a chirp SONAR. Sand ridges with less acoustic penetration occupy most of the mid-shelf area with a few km spacing. In the outer-shelf area, dipping layers over the distinct *R reflector* are detected. The spikes in the water column at the mid-shelf area are schools of fish near the bottom, which were mostly seen during the surveys conducted in daylight. Figure 5.50b shows the sub-bottom profile of the along-shelf track with acoustic penetration as deep as 40 m. Along-shelf track is relatively less range-dependent. However, several scour marks (≈ 100 m wide, a few m deep) are detected on the sea floor. These scour marks are attributed to gouging by iceberg keels and the resultant deformations of deeper sublayers is also displayed [A. Turgut, personal communications].

5.7.3 Acoustic Telemetry

Because electromagnetic waves do not propagate in the ocean, underwater acoustic data transmission have many applications, including:

- Communication between two submarines or a submarine and a support vessel
- Communication between a ship and an automated underwater vehicle (AUV) either to get data available without recovering the instruments or to control the instruments onboard the AUV or the AUV itself
- Data transmission to an automated system

Underwater acoustic communications [5.55] are typically achieved using digital signals. We usually distinguish between coherent and incoherent transmissions. For example, incoherent transmissions might consist of transmitting the symbols "0" and "1" at different frequencies (frequency shift keying (FSK)), whereas coherent transmissions might encode using different phases of a given sinusoid (phase shift keying (PSK)) [Fig. 5.51].

Acoustic telemetry can be performed from one source to one receiver in a single-input single-output mode (SISO). To improve performance in terms of data rate and error rate, acoustic networks are now commonly used. In particular, recent works in underwater acoustic communications deal with multiple-input multiple-output (MIMO) configurations (Fig. 5.52) in shallow-water oceans [5.56].

The trend toward MIMO is justified by the fact that the amount of information – known as the information capacity, I – that can be sent between arrays of transmitters and receivers is larger than in a SISO case. Indeed, the SISO information capacity is given by Shannon's famous formula [5.57]:

Fig. 5.49 Side-scan SONAR data obtained in 2001 from the HUGIN 3000 AUVs of a ship sunk in the Gulf of Mexico during World War II. C&C Technology Inc. (Courtesy of the National D-Day Museum, New Orleans)

$$I = \log_2\left(1 + \frac{S}{N}\right) \text{ bits s}^{-1}\text{Hz}^{-1}, \quad (5.92)$$

where S is the received signal power, N the noise power and the Shannon capacity I is measured in bits per second per Hertz of bandwidth available for transmission. Equation (5.92) states that the channel capacity increases with the signal-to-noise ratio. In a MIMO configuration with M_t transmitters and M_r receivers, (5.92) is changed into [5.58]:

$$I \sim M_i \log_2\left(1 + \frac{SM_r/M_t}{N}\right) \text{ bits s}^{-1}\text{Hz}^{-1}, \quad (5.93)$$

with $M_r \geq M_t$ to be able to decode the M_t separate transmitted signals. Sending M_t different bitstreams is advantageous since it gives a factor of M_t outside the log, linearly increasing the channel capacity I compared to a logarithmic increase when playing on the output power S.

Beside the optimized allocation of power between the M_r and M_t receivers and transmitters (the so-called water-filling approach [5.59]), other particular issues in underwater acoustic telemetry deal with Doppler tracking, channel estimation and signal-to-noise ratios. Those combined parameters often result in a tradeoff between data rate and error rate. For example, low frequencies propagate further with better signal-to-noise ratios and hence lower error rates but the data rate is poor. High frequencies provide high data rates but suffer from strong loss in the acoustic channel, potentially resulting in large error rates.

Another difficulty has to do with multipath propagation and/or reverberation in the ocean, causing intersymbol interference (ISI) and scattering, also called

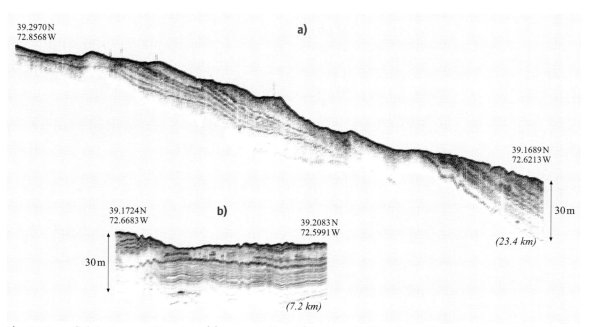

Fig. 5.50a,b Sub-bottom profiles along (**a**) cross-shelf and (**b**) along-shelf tracks on the New Jersey shelf collected by a hull-mounted chirp SONAR (2–12 kHz) during the shallow-water acoustic technology experiment (SWAT2000) (Courtesy A. Turgut, Naval Research Laboratory)

the fading effect (Fig. 5.53). As a consequence, the performance of a given system depends strongly on the acoustic environment.

Figure 5.54 shows examples of experimental channel impulse responses recorded at different frequencies in shallow-water waveguides. The difference in temporal dispersion (relative to the acoustic period) between Fig. 5.54a and Fig. 5.54c shows that high-frequency transmissions suffer from stronger ISI and shorter coherence time.

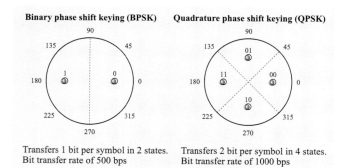

Fig. 5.51 Example of signal space diagrams for coherent communications

In the presence of ISI, Eqs. (5.92, 93) can be generalized so that the channel capacity I still depends on the signal-to-noise ratio S/N where the noise N now includes ISI [5.60]. However, there exist many ways to reduce propagation effects on the quality of the acoustic transmission. One technological solution is to use directional arrays/antenna. Another one is to further code digital signals (CDMA or turbo codes [5.61]) in order to detect and potentially correct transmission errors. But, most importantly, there are many powerful signal-processing techniques derived from telecommunications to take into account at the receiver of the channel impulse response. An efficient one is adaptive equalization that uses a time-dependent channel estimate in order to decode the next symbols from the previously decoded one [5.62]. Other methods to deal with multipath complexities use time-reversal methods, either alone or with equalization methods [5.63–65].

5.7.4 Travel-Time Tomography

Tomography [5.9, 66] generally refers to applying some form of inverse theory to observations in order to infer properties of the propagation medium. The received field from a source emitting a pulse will be spread in time as a result of multiple paths, in which different paths have different arrival times (Fig. 5.55). Hence the arrival times are related to the acoustic sampling of the medium. In contrast, standard medical X-ray tomography utilizes the different attenuation of the paths rather than arrival time for the inversion process. Since the ocean sound speed is a function of temperature and other oceanographic parameters, the arrival structure is ultimately related to a map of these oceanographic parameters. Thus, measuring the fluctuations of the arrival times through the experiments theoretically leads to the knowledge of the spatial-temporal scale of the ocean fluctuations.

Tomographic inversion in the ocean has typically relied on these three points. First, only the arrival time (and not the amplitude) of the multipath structure is used as an observable (Table 5.2) [5.67]. Enhanced time-of-arrival resolution is typically obtained using pulse-compression techniques [5.68] as mentioned in the bottom-mapping section above. Depending on the experimental configuration, a choice of compression is to be made between M-sequences [5.69], which are strongly Doppler sensitive but have low temporal side-lobes and frequency-modulated chirps that are Doppler insensitive with higher side-lobes (Fig. 5.43). Second, the inversion is performed by comparing the experimental arrival times to those given by a model (Fig. 5.56). Last, tomographic inversion algorithm classically deals with a linearized problem. This means that the model has to match the experimental data so that the inversion only deals with small perturbations.

Thus, ocean tomography starts from a sound-speed profile $c(\mathbf{r})$ on which small perturbations are added $\delta c(\mathbf{r}, t) \ll c$. The ocean model $c(\mathbf{r})$ has to be accurate enough to relate without ambiguity an experimentally measured travel time T_i to a model-deduced ray path Γ_i. Typically, some baseline oceanographic information is known so that one searches for departures from this baseline information. The perturbation infers a change of travel time δT_i along the ray such that, in a first linear approximation

$$\delta T_i \approx \int_{\Gamma_i} \frac{-\delta c}{c^2} \, \mathrm{d}s \,, \tag{5.94}$$

where Γ_i correspond to the Fermat path of the unperturbed ray. An efficient implementation of the inversion procedure utilizes a projection of the local sound-speed fluctuations $\delta c(\mathbf{r}, t)$ on a set of chosen functions $\Psi_k(\mathbf{r})$ that constitutes a basis of the ocean structure. We have

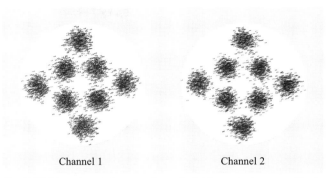

Fig. 5.52 Experimental examples of eight quadrature amplitude modulation (QAM) transmissions in a multiple-input multiple-output configuration (MIMO) at 3.5 kHz in a 9 km-long, 120 m-deep shallow-water ocean. The SNR is 30 dB on channel 1 and 2, the symbol duration is 1 ms (data rate is 8 kB/s per channel) and bit error rate (BER) is 1×10^{-4} (Courtesy H.C. Song, Scripps Institution of Oceanography)

then

$$\delta c(\mathbf{r}, t) = \sum_{k=1}^{N} p_k(t) \psi_k(\mathbf{r}) \,, \tag{5.95}$$

where $p_k(t)$ is a set of unknown parameters. In its most primitive form, the ocean can be discretized into elementary cells, each cell being characterized by an unknown sound-speed perturbation p_k. Combining the two above

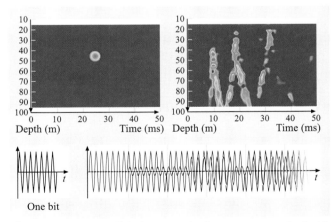

Fig. 5.53 A coherent digital communication system must deal with the intersymbol interference caused by dispersive multipath environment of the ocean waveguide (*top right*). When a sequence of phase-shifted symbols (*in black*) are sent, the resulting transmission (*in brown*) is fading out because of symbol interference

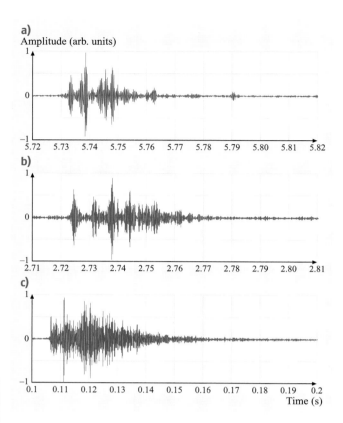

Fig. 5.54a–c Examples of transfer functions recorded at sea in different shallow-water environments at various frequencies. (a) Central frequency = 3.5 kHz with a 1 kHz bandwidth, 10 km range in a 120 m-deep waveguide. (b) Central frequency = 6.5 kHz with a 2 kHz bandwidth, 4 km range in a 50 m-deep waveguide. (c) Central frequency = 15 kHz with a 10 kHz bandwidth, 160 m range in a 12 m-deep waveguide

Allowing for some noise in the measurement and assuming a set of arrival times δT_i, $i \in [1, M]$, (5.94) can be rewritten in an algebraic form [5.71]:

$$\delta \boldsymbol{T} = \boldsymbol{G}\boldsymbol{p} + \boldsymbol{n} \ . \tag{5.97}$$

There exists many algorithms to obtain an estimate of the parameters $\tilde{\boldsymbol{p}}$ of the parameters $\boldsymbol{p}$ from the data $\delta \boldsymbol{T}$ knowing the matrix $\boldsymbol{G}$. For example, when $N > M$, a least-mean-square estimator gives:

$$\tilde{\boldsymbol{p}} = G^T (G G^T)^{-1} \delta \boldsymbol{T} \ . \tag{5.98}$$

Considerations about pertinent functions $\Psi_k(\mathbf{r})$ such as empirical orthogonal functions (EOFs) and the optimal inversion procedure can be found in the literature [5.72, 73].

Tomographic experiments have been performed to greater than megameter ranges. For example, two major experiments in the 1990s were performed by the Thetis 2 group in the western Mediterranean over a seasonal cycle [5.74] and by the North Pacific Acoustic Laboratory (NPAL, http://npal.ucsd.edu) in the North Pacific basin. The NPAL experiment was directed at using travel-time data obtained from a few acoustic sources and receivers located throughout the North Pacific Ocean to study temperature variations at large scale [5.75, 76]. The idea behind the project, known as acoustic thermometry of ocean climate (ATOC), is that sound travels slightly faster in warm water than in cold water. Thus, precisely measuring the time it takes for a sound signal to travel between two points reveals the average temperature along the path. Sound travels at about 1500 m/s in water, while typical changes in the sound speed of the ocean as a result of temperature changes are only 5–10 m/s. The tomography

equations, it follows:

$$\delta T_i = \int_{\Gamma_i} \frac{-1}{c^2} \sum_{k=1}^{N} p_k \psi_k(\boldsymbol{r}) \, \mathrm{d}s$$

$$= \sum_{k=1}^{N} p_k \int_{\Gamma_i} \frac{-\Psi_k(\boldsymbol{r})}{c^2} \, \mathrm{d}s = \sum_{k=1}^{N} p_k G_{ik} \ . \tag{5.96}$$

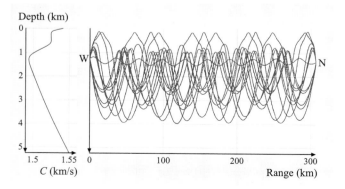

Fig. 5.55 Left: reference sound-speed profile $C_0(z)$. Right: corresponding ray traces. Table 5.2 identifies all plotted rays. Note the shallow and deep turning groups of rays as well as the axial ray (after [5.70]) (Courtesy Peter Worcester, Scripps Institution of Oceanography)

technique is very sensitive, and experiments so far have shown that temperatures in the ocean can be measured to a precision of $0.01\,°C$, which is needed to detect subtle variations and trends of the ocean basin.

New trends in ocean tomography deal with full-wave inversion involving the use of both travel times and amplitudes of echoes for a better accuracy in the inversion algorithm. In this case, a full-wave kernel for acoustic propagation has to be used [5.77], which includes the sensitivity of the whole Green's functions (both travel times and amplitudes) to sound speed variations. In the case of high frequencies, the application of the travel-time-sensitivity kernel to an ocean acoustic waveguide gives a picture close to the ray-theoretic one. However, in the low-frequency case of interest in ocean acoustic tomography, there are significant deviations. Low-frequency travel times are sensitive to sound-speed changes in Fresnel-zone-scale areas surrounding the eigen-rays, but not on the eigen-rays themselves, where the sensitivity is zero. This diffraction phenomenon known as the *banana–doughnut* [5.78] debate is still actively discussed in the field of ocean and seismic tomography [5.79].

Table 5.2 Identification of rays (after [5.70]). The identifier is $\pm n(\theta_s, \theta_R, \hat{z}^+, \hat{z}^-)$, where positive (negative) rays depart upward (downward) from the source, n is the total number of upper and lower turning points, θ_s is the departure angle at the source, θ_R is the arrival angle at the receiver, and $\hat{z}^+$ and $\hat{z}^-$ are the upper and lower turning depths, respectively (Courtesy Peter Worcester, Scripps Institution of Oceanography)

	$\pm n$	θ_s (deg)	θ_R (deg)	$\hat{z}^+$ (m)	$\hat{z}^-$ (m)
1	8	11.6	11.6	126	3801
2	-8	-11.6	-11.7	125	3803
3	9	12.0	-12.0	99	3932
4	11	11.1	-11.1	617	3624
5	-11	-10.8	10.2	737	3303
6	12	9.7	9.7	776	3170
7	-12	-9.6	-9.7	780	3156
8	13	9.3	-9.3	809	3046
9	-13	-8.2	8.3	881	2746
10	14	7.9	8.0	901	2653
11	-14	-7.8	-7.9	905	2638
12	15	7.4	-7.5	925	2546
13	19	3.5	-3.8	1118	1790
	20	1.2	1.7	1221	1507

5.8 Acoustics and Marine Animals

In the context of contemporary acoustics, marine animal life is typically divided into the categories of marine mammals and non-marine mammals which include fish and others sea animals. The acoustics dealing with fish relates to either finding, counting and catching them. The acoustics concerned with marine mammals is for either studying their behavior or determining to what extent manmade sounds are harmful to them.

5.8.1 Fisheries Acoustics

An extensive introduction to fisheries acoustics is [5.80] and can be found online at http://www.fao.org/docrep/X5818E/x5818e00.htm#Contents. For more-detailed information, we recommend two special issues in [5.81, 82].

Following the development of SONAR systems, acoustic technology has had a major impact on fishing and on fisheries research. SONARs and echo sounders are now used as standard tools to search for concentrations of fish or to assess for biomass stock. With SONAR, it is possible to sample the water column much faster than trawl fishing. Moreover, SONARs have helped in our understanding of how fish are distributed in the ocean and how they behave over time. Depending on the fish density in the water column, echo-counting or echo-integration [5.83] are used to evaluate the fish biomass in the water column. Research on the signature of the echo return for a specific fish [5.84] or for a fish school [5.85, 86] is still an active area of research.

Fisheries acoustics experiments are performed with SONAR in the same way as bottom profiling [5.87–89]. The ship covers the area of interest by transect lines [5.90] while the SONAR sends and receives acoustic pulses (pings) as often as allowed by the SONAR system and the water depth (the signal from the previous ping has to die out before the next ping is transmitted). Typical SONAR frequencies are 38 kHz, 70 kHz, and 120 kHz. An echogram is a display of the instantaneous received intensity along the ship track. The echograms in Fig. 5.57a,b reveal individual fish echoes as well as fish school backscattered signals.

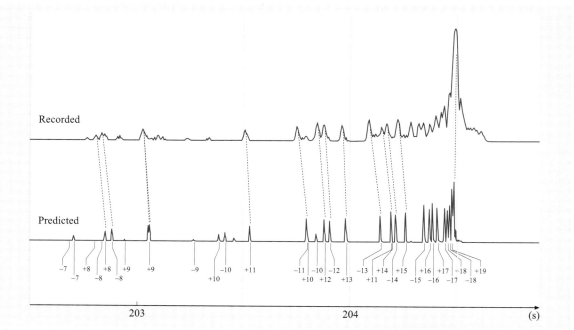

Fig. 5.56 Comparison of the predicted and measured arrival patterns (after [5.70]). The predicted arrival pattern was calculated from $C_0(z)$ as shown in Fig. 5.55. Geometric arrivals are labeled $\pm n$ as in Table 5.2. The peaks connected by *dashed lines* are the ray arrivals actually used (Courtesy Peter Worcester, Scripps Institution of Oceanography)

From a practical point of view, the relationship between an acoustic target and its received echo can be understood from the SONAR equation. With the procedure of Sect. 5.3.2 and 5.3.3, the echo level (in decibels) of a given target is

$$\text{EL} = \text{SL} + \text{RG} + \text{TS} - 2\text{TL}, \tag{5.99}$$

where SL is the source level and TL is the geometrical spreading loss in the medium. The target strength TS is defined as $\text{TS} = 10\log_{10}(\sigma/4\pi)$, where σ is the backscattering cross section of the target. The receiver gain RG is often set up as a time-varied gain (TVG), which compensates for the geometrical spreading loss.

In (5.99), the target is supposed to be on the acoustic axis of the transducer. In general, a single-beam SONAR cannot distinguish between a small target on the transducer axis and a big target off-axis. The two echoes may have the same amplitude because the lower TS for the small target will be compensated by the off-axis power loss for the big target. One way to remove this ambiguity is to use a dual-beam SONAR [5.91] (or split-beam SONAR) that provides the position of the target in the beam pattern.

Echo-integration consists of integrating in time (as changed into depth from $R = ct/2$) the received intensity coming from the fish along the ship transects [5.92]. This value is then converted to a biomass via the backscattering cross section σ. In the case of a single target (Fig. 5.57a), the intensity of the echo E_1 is

$$E_1 = \overline{\sigma}\phi^2(r)\left[\frac{\exp(-2\beta r)}{r^4}\right], \tag{5.100}$$

where $\phi^2(R)$ is the depth-dependent (or time-varying) gain; the term in bracket describes the geometrical spreading $(1/r^4)$ and loss $[\exp(-2\beta r)]$ of the echo during its round trip between the source and the receiver and $\overline{\sigma}$ is the scattering cross section of the target averaged over the bandwidth of the transmitted signal. In (5.100) we did not include the effect of the beam pattern and the sensitivity of the system in emission-reception. These parameters depend on the type of SONAR used and are often measured in situ from a calibration experiment [5.93, 94]. In the case of a distributed targets in the water column [5.95, 96] (Fig. 5.57b), the received intensity E_h is integrated over a layer h at a depth r, which

Fig. 5.57a,b Typical echogram obtained during an echo-integration campaign using a single-beam SONAR on the Gambie river. (**a**) Biomass made of individual fish. (**b**) Biomass made of fish school (Courtesy Jean Guillard, INRA)

gives:

$$E_h = h\overline{\sigma} n \phi^2(r) \left[\frac{\exp(-2\beta r)}{r^2} \right], \qquad (5.101)$$

where n corresponds to the density of fish per unit volume in the layer h. Equation (5.101) is based on the linearity principle assumption, which states that, on the average over many pings, the intensity E_h is equal to the sum of the intensity produced by each fish individually [5.97].

In the case of either a single or multiple targets, the idea is to relate directly the echo-integrated result E_1 or E_h to the fish scattering amplitudes $\overline{\sigma}$ or $n\overline{\sigma}$. To that goal, the time-varying gain $\phi^2(r)$ has to compensate appropriately for the geometric and attenuation loss. For a volume integration, $\phi^2(r) = R^2 \exp(2\beta r)$, the so-called $20 \log r$ gain is used while a $40 \log r$ gain is applied in the case of individual echoes.

As a matter of fact, acoustic instruments, such as echo sounders and SONAR, are unique as underwater sampling tools since they detect objects at ranges of many hundreds of meters, independent of water clarity and depth. However, until recently, these instruments could only operate in two dimensions, providing observational slices through the water column. New trends in fisheries acoustics incorporate the use of multibeam SONAR – typically used in bottom mapping, see Sect. 5.7.2 – which provides detailed data describing the internal and external three-dimensional (3-D) structure of underwater objects such as fish schools [5.98] (Fig. 5.58). Multibeam SONARs are now used in a wide variety of fisheries research applications including: (1) three-dimensional descriptions of school structure and position in the water column [5.99]; knowledge of schooling is vital for understanding many aspects of

Fig. 5.58a–d Images of a fish school of *Sardinella aurita* from a multibeam SONAR. *Arrows* indicate vessel route. (**a**) 3-D reconstruction of the school. Multibeam SONAR receiving beams are shown at the front of the vessel. Remaining panels show cross sections of density in fish from: (**b**) the horizontal plane, (**c**) the vertical plane along-ships, (**d**) the vertical plane athwart ships. Red cross-hairs indicate location of the other two cross sections (Courtesy Francois Gerlotto, IRD)

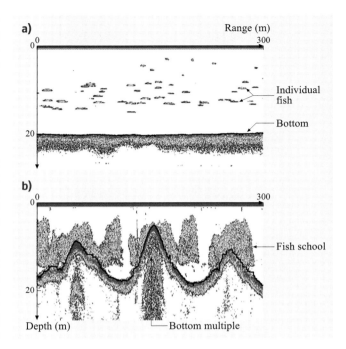

fish (and fisheries) ecology [5.100]; (2) detailed internal images of fish schools, providing insights into the organization of fish within a school, for example, indicating the presence of large gaps or vacuoles and areas of higher densities or nuclei (Fig. 5.58b, d) [5.101].

In general, one tries to convert the echo-integration result into a biomass or fish density. This requires the knowledge of the target strength TS or the equivalent backscattering cross section σ of the target under

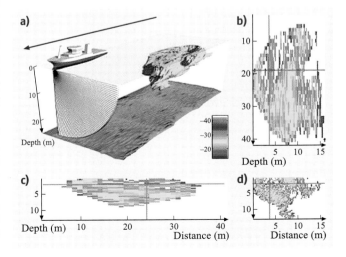

investigation [5.102]. Several models attempt to develop equations expressing σ/λ^2 as a function of L/λ, where L is the fish length. However, physiological and behavioral differences between species make a more-empirical approach necessary, which often is of the form:

$$TS = 20 \log L + b, \quad (5.102)$$

where b depends on the acoustic frequency and the fish species. For example, we have $b = -71.3$ dB for herring and $b = -67.4$ dB for cod at 38 kHz, which makes the cod scattering cross section twice that of herring for the same fish length. In situ measurements of TS with fish at sea [5.103, 104] or in cages [5.105] have also been attempted to classify fish species acoustically.

Similar works on size distribution assessment have been performed at higher frequency (larger than 500 kHz) on small fish or zooplankton using multi-element arrays [5.106] and wide-band or multi-frequency methods [5.107, 108]. The advantage of combining the acoustic signature of fish at various frequencies is to provide an accurate target sizing. When multi-elements are used, then the position of each individual inside the acoustic beam is known. Using simultaneously multi-element arrays and broadband approaches may also be the key for the unsolved problem of species identification in fisheries acoustics.

In conclusion, SONAR systems and echo integration are now well-established techniques for the measurement of fish abundance. They provide quick results and accurate information about the pelagic fish distribution in the area covered by the ship during the survey. The recent development of multibeam SONAR has improved the reliability of the acoustic results and now provides 3-D information about fish-school size and shape. However, some important sources of error remain when performing echo-integration, among which are: (1) the discrimination between fish echoes and unwanted target echoes, (2) the difficulty of adequately sampling a large area in a limited time, (3) the problems related to the fish behavior (escaping from the transect line, for example) and (4) the physiological parameters that determine the fish target strength.

5.8.2 Marine Mammal Acoustics

In order to put into perspective both the sound levels that mammals emit and hear, we mention an assortment of sounds and noise found in the ocean. Note that the underwater acoustic decibel scale used is, as per the Appendix, relative to 1 μPa. For source levels, one is also referencing the sound level at 1 m. Lightening can be as high has 260 dB and a seafloor volcanic eruption can be as high as 255 dB. Heavy rain increases the background noise by as much as 35 dB in a band from a few hundred Hz to 20 kHz Snapping shrimp individually can have broadband source levels greater than 185 dB while fish choruses can raise ambient noise levels 20 dB in the range of 50–5000 Hz. Of course, the Wenz curves in Fig. 5.15 show a distribution of natural and manmade noise levels whereas specific examples of manmade noise are given in Table 5.3.

Table 5.3 Examples of manmade noise

Ships underway	Broadband source level (dB re 1 μPa at 1 m)
Tug and barge (18 km/hour)	171
Supply ship (example: Kigoriak)	181
Large tanker	186
Icebreaking	193
Seismic survey	**Broadband source level (dB re 1 μPa at 1 m)**
Air-gun array (32 guns)	259 (peak)
Military SONARS	**Broadband source level (dB re 1 μPa at 1 m)**
AN/SQS-53C (US Navy tactical mid-frequency SONAR, center frequencies 2.6 and 3.3 kHz)	235
AN/SQS-56 (US Navy tactical mid-frequency sonar, center frequencies 6.8 to 8.2 kHz)	223
Surveillance Towed Array Sensor System Low Frequency Active (SURTASS-LFA) (100–500 Hz)	215 dB per projector, with up to 18 projectors in a vertical array operating simultaneously
Ocean acoustic studies	**Broadband source level (dB re 1 μPa at 1 m)**
Heard island feasibility test (HIFT) (center frequency 57 Hz)	206 dB for a single projector, with up to 5 projectors in a vertical array operating simultaneously
Acoustic thermometry of ocean climate (ATOC)/North Pacific acoustic laboratory (NPAL) (center frequency 75 Hz)	195

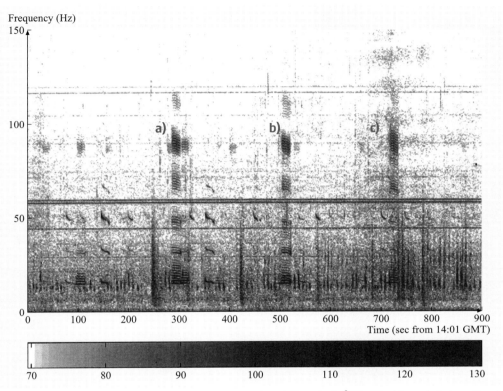

Fig. 5.59 Whale spectrogram power spectral density are in units of dB re 1 μPa²/Hz. The blue-whale broadband signals denoted by (*a*), (*b*) and (*c*) are designated "type A calls". The FM sweeps are "type B calls". The multiple vertical energy bands between 20 and 30 Hz have the appearance of fin-whale vocalizations [5.109]

Marine mammal sounds span the spectrum from 10–200 000 kHz. Examples are: blue (see spectrogram in Fig. 5.59) and fin whales in the 20 Hz region with source levels as high as 190 dB, Wedell seals in the 1–10 kHz region producing 193 dB levels; bottlenose dolphin, 228 dB in a noisy background, sperm whale clicks are the loudest recorded levels at 232 dB. A list of typical levels is shown in Table 5.4. Most of the levels listed are substantial and strongly suggest acoustics as a modality for monitoring marine mammals. Thus, for example, Fig. 5.60 shows the acoustically derived track of a blue whale over 43 days and thousands of kilometers as determined from SOSUS arrays (see Introduction to this chapter) in the Atlantic Ocean.

The issues mostly dealt with in marine mammal acoustics are: understanding the physiology and behavior associated with the production and reception of sounds, and the effects that manmade sounds have on marine mammals from actual physical harm to causing changes in behavior. Physical harm includes actual

Table 5.4 Marine-mammal sound levels

Source	Broadband source level (dB re 1 μPa at 1 m)
Sperm whale clicks	163–223
Beluga whale echo-location click	206–225 (peak to peak)
White-beaked dolphin echo-location clicks	194–219 (peak to peak)
Spinner dolphin pulse bursts	108–115
Bottlenose dolphin whistles	125–173
Fin whale moans	155–186
Blue whate moans	155–188
Gray whale moans	142–185
Bowhead whale tonals, moans and song	128–189
Humpback whale song	144–174
Humpback whale fluke and flipper slap	183–192
Southern right whale pulsive call	172–187
Snapping shrimp	183–189 (peak to peak)

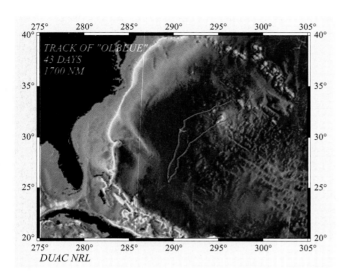

Fig. 5.60 Track of a blue whale in the Atlantic Ocean determined by US Navy personnel operating SOSUS stations. (Courtesy Clyde Nishimura, Naval Research Laboratory and Chris Clark, Cornell University)

damage (acoustic trauma) and permanent and temporary threshold shifts in hearing. A major ongoing effort is to determine the safe levels of manmade sounds. Acoustics has now become an important research tool in the marine mammal arena (e.g. [5.110–115]).

Tables 5.3 and 5.4 have been taken form University of Rhode web site: http://www.dosits.org/science/ssea/2.htm with references to [5.24, 25, 116].

Advances in the bandwidth and data-storage capabilities of sea-floor autonomous acoustic recording packages (ARPs) have enabled the study of odontocete (toothed-whale) long-term acoustic behavior. Figure 5.61 illustrates one day of broadband (10 Hz–100 kHz) acoustic data collected in the Santa Barbara Channel. The passage of vocalizing dolphins is recorded by the aggregate spectra from their echolocation clicks (>20 kHz) and whistles (5–15 kHz). Varying proportions of clicks and whistles are seen for each calling

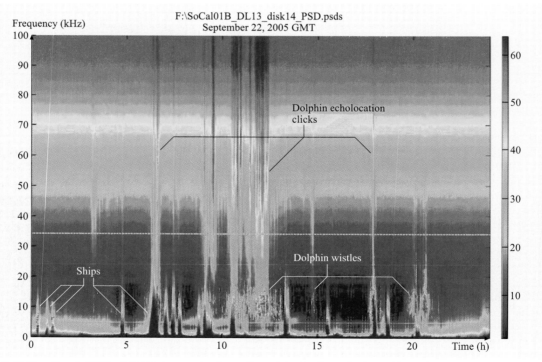

Fig. 5.61 Broadband (10 Hz–100 kHz) acoustic data collected in the Santa Barbara Channel, illustrating spectra from dolphin echolocation clicks (> 20 kHz) and whistles (5–15 kHz) as seen in a daily sonogram (Sept 22, 2005). Passages of individual commercial ships are seen at mid and low frequencies (< 15 kHz) (Courtesy John Hildebrand, Scripps Institution of Oceanography)

bout. These data allow for study of the acoustic behavior under varying conditions (day–night) and for the determination of the seasonal presence of calling animals.

5.A Appendix: Units

The decibel (dB) is the dominant unit in underwater acoustics and denotes a ratio of intensities (not pressures) expressed in terms of a logarithmic (base 10) scale. Two intensities, I_1 and I_2 have a ratio I_1/I_2 in decibels of $10 \log I_1/I_2$ dB. Absolute intensities can therefore be expressed by using a reference intensity. The presently accepted reference intensity in underwater acoustics is based on a reference pressure of one micropascal. Therefore, taking I_2 as the intensity of a plane wave of pressure $1\,\mu$Pa, a sound wave having an intensity, of, say, one million times that of a plane wave of rms pressure $1\,\mu$Pa has a level of $10 \log(10^6/1) \equiv 60$ dB re $1\,\mu$Pa. Pressure (p) ratios are expressed in dB re $1\,\mu$Pa by taking $20 \log p_1/p_2$, where it is understood that the reference originates from the intensity of a plane wave of pressure equal to $1\,\mu$Pa.

The average intensity, I, of a plane wave with rms pressure p in a medium of density ρ and sound speed c is $I = p^2/\rho c$. In seawater, $(\rho c)_{\text{water}}$ is 1.5×10^6 Pa s m^{-1} so that a plane wave of rms pressure $1\,\mu$Pa has an intensity of 6.76×10^{-19} W/m^2.

For reference, we also mention the relevant units in air, where the reference pressure is related, more or less, to the minimum level of sound we can hear is. This is a pressure level 26 dB higher than the water reference. Further, since the intensity level associated with this reference is 10^{-12} W/m^2. Therefore, one should be careful when relating units between water and air, as the latter's reference intensity is higher.

References

5.1 W.A. Kuperman, J.F. Lynch: Shallow-water acoustics, Phys. Today **57**, 55–61 (2004)
5.2 L.M. Brekhovskikh: *Waves in Layered Media*, 2nd edn. (Academic, New York 1980)
5.3 L.M. Brekhovskikh, Y.P. Lysanov: *Fundamentals of Ocean Acoustics* (Springer, Berlin, Heidelberg 1991)
5.4 F.B. Jensen, W.A. Kuperman, M.B. Porter, H. Schmidt: *Computational Ocean Acoustics* (Springer, Berlin, New York 2000)
5.5 B.G. Katsnelson, V.G. Petnikov: *Shallow Water Acoustics* (Springer, Berlin, Heidelberg 2001)
5.6 J.B. Keller, J.S. Papadakis (Eds.): *Wave Propagation in Underwater Acoustics* (Springer Berlin, New York 1977)
5.7 H. Medwin, C.S. Clay: *Fundamentals of Acoustical Oceanography* (Academic, San Diego 1997)
5.8 H. Medwin: *Sounds in the Sea: From Ocean Acoustics to Acoustical Oceanography* (Cambridge Univ. Press, Cambridge 2005)
5.9 W. Munk, P. Worcester, C. Wunsch: *Ocean Acoustic Tomography* (Cambridge Univ. Press, Cambridge 1995)
5.10 D. Ross: *Mechanics of Underwater Noise* (Pergamon, New York 1976)
5.11 B.D. Dushaw, P.F. Worcester, B.D. Cornuelle, B.M. Howe: On equations for the speed of sound in seawater, J. Acoust. Soc. Am. **93**, 255–275 (1993)
5.12 J.L. Spiesberger, K. Metzger: A new algorithm for sound speed in seawater, J. Acoust. Soc. Am. **89**, 2677–2688 (1991)
5.13 J. Northrup, J.G. Colborn: Sofar channel axial sound speed and depth in the Atlantic Ocean, J. Geophys. Res. **79**, 5633–5641 (1974)
5.14 W.H. Munk: Internal waves and small scale processes. In: *Evolution of Physical Oceanography*, ed. by B. Warren, C. Wunsch (MIT, Cambridge 1981) pp. 264–291
5.15 O.B. Wilson: *An Introduction to the Theory and Design of Sonar Transducers* (Government Printing Office, Washington 1985)
5.16 R.J. Urick: *Principles of Underwater Sound*, 3rd edn. (McGraw-Hill, New York 1983)
5.17 R.P. Chapman, J.R. Marchall: Reverberation from deep scattering layers in the Western North Atlantic, J. Acoust. Soc. Am. **40**, 405–411 (1966)
5.18 J.A. Ogilvy: Wave scattering from rough surfaces, Rep. Prog. Phys. **50**, 1553–1608 (1987)
5.19 E.I. Thorsos: Acoustic scattering from a "Pierson-Moskowitz" sea surface, J. Acoust. Soc. Am. **89**, 335–349 (1990)
5.20 P.H. Dahl: On bistatic sea surface scattering: Field measurements and modeling, J. Acoust. Soc. Am. **105**, 2155–2169 (1999)
5.21 R.P. Chapman, H.H. Harris: Surface backscatterring strengths measured with explosive sound sources, J. Acoust. Soc. Am. **34**, 1592–1597 (1962)
5.22 M. Nicholas, P.M. Ogden, F.T. Erskine: Improved empirical descriptions for acoustic surface backscatter in the ocean, IEEE J. Ocean. Eng. **23**, 81–95 (1998)

5.23 N.C. Makris, S.C. Chia, L.T. Fialkowski: The bi-azimuthal scattering distribution of an abyssal hill, J. Acoust. Soc. Am. **106**, 2491–2512 (1999)

5.24 G.M. Wenz: Acoustics ambient noise in the ocean: spectra and sources, J. Acoust. Soc. Am **34**, 1936–1956 (1962)

5.25 R.K. Andrew, B.M. Howe, J.A. Mercer, M.A. Dzieciuch: Ocean ambient sound: Comparing the 1960s with the 1990s for a receiver off the California coast, Acoust. Res. Lett. Online **3**(2), 65–70 (2002)

5.26 M.A. McDonald, J.A. Hildebrand, S.W. Wiggins: Increases in deep ocean ambient noise in the Northeast Pacific west of San Nicolas Island, California, J. Acoust. Soc. Am. **120**, 711–718 (2006)

5.27 W.A. Kuperman, F. Ingenito: Spatial correlation of surface generated noise in a stratified ocean, J. Acoust. Soc. Am. **67**, 1988–1996 (1980)

5.28 T.G. Leighton: *The Acoustic Bubble* (Academic, London 1994)

5.29 C.E. Brennen: *Cavitation and bubble dynamics* (Oxford Univ. Press, Oxford 1995)

5.30 G.B. Morris: Depth dependence of ambient noise in the Northeastern Pacific Ocean, J. Acoust. Soc. Am. **64**, 581–590 (1978)

5.31 V.C. Anderson: Variations of the vertical directivity of noise with depth in the North Pacific, J. Acoust. Soc. Am. **66**, 1446–1452 (1979)

5.32 F. Ingenito, S.N. Wolf: Acoustic propagation in shallow water overlying a consolidated bottom, J. Acous. Soc. Am. **60**(6), 611–617 (1976)

5.33 H. Schmidt, W.A. Kuperman: Spectral and modal representation of the Doppler-shifted field in ocean waveguides, J. Acoust. Soc. Am. **96**, 386–395 (1994)

5.34 M.D. Collins, D.K. Dacol: A mapping approach for handling sloping interfaces, J. Acoust. Soc. Am. **107**, 1937–1942 (2000)

5.35 W.A. Kuperman, G.L. D'Spain (Eds.): *Ocean Acoustics Interference Phenomena and Signal Processing* (American Institution of Physics, Melville 2002)

5.36 H.L. Van Trees: *Detection Estimation and Modulation Theory* (Wiley, New York 1971)

5.37 H.L. Van Trees: *Optimum Array Processing* (Wiley, New York 2002)

5.38 D.H. Johnson, D.E. Dudgeon: *Array Signal Processing: Concepts and Techniques* (Prentice Hall, Englewood Cliffs 1993)

5.39 G.L. D'Spain, J.C. Luby, G.R. Wilson, R.A. Gramann: Vector sensors and vector line arrays: Comments on optiman array gain and detection, J. Acoust. Soc. Am. **120**, 171–185 (2006)

5.40 P. Roux, W.A. Kuperman, NPAL Group: Extracting coherent wavefronts from acoustic ambient noise in the ocean, J. Acoust. Soc. Am. **116**, 1995–2003 (2004)

5.41 M. Siderius, C.H. Harrison, M.B. Porter: A passive fathometer for determining bottom depth and imaging seabed layering using ambient noise, J. Acoust. Soc. Am. **12**, 1315–1323 (2006)

5.42 A.B. Baggeroer, W.A. Kuperman, P.N. Mikhalevsky: An Overview of matched field methods in ocean acoustics, IEEE J. Ocean. Eng. **18**(4), 401–424 (1993)

5.43 W.A. Kuperman, D.R. Jackson: Ocean acoustics, matched-field processing and phase conjugation. In: *Imaging of complex media with acoustic and seismic waves*, Topics Appl. Phys. **84**, 43–97 (2002)

5.44 M. Fink: Time Reversed Physics, Phys. Today **50**, 34–40 (1997)

5.45 A. Parvelescu, C.S. Clay: Reproducibility of signal transmissions in the ocean, Radio Electron. Eng. **29**, 223–228 (1965)

5.46 W.A. Kuperman, W.S. Hodgkiss, H.C. Song, T. Akal, C. Ferla, D. Jackson: Phase conjugation in the ocean: Experimental Demonstration of an acoustical time-reversal mirror in the ocean, J. Acoust. Soc. Am. **102**, 25–40 (1998)

5.47 H. Cox, R.M. Zeskind, M.O. Owen: Robust Adaptive Beamforming, IEEE Trans. Acoust. Speech Signal Proc. **35**, 10 (1987)

5.48 J.L. Krolik: Matched-field minimum variance beamforming in a random ocean channel, J. Acoust. Soc. Am. **92**, 1408–1419 (1992)

5.49 A. Tolstoy, O. Diachok, L.N. Frazer: Acoustic tomography via matched field processing, J. Acoust. Soc. Am. **89**, 1119–1127 (1991)

5.50 A.B. Baggeroer: *Applications of Digital Signal Processing*, ed. by A.V. Oppenheim (Prentice Hall, Englewood Cliffs 1978)

5.51 C. De Moustier: Beyond bathymetry: mapping acoustic backscattering from the deep seafloor with Sea Beam, J. Acoust. Soc. Am. **79**, 316–331 (1986)

5.52 P. Cervenka, C. De Moustier: Sidescan sonar image processing techniques, IEEE J. Ocean. Eng. **18**, 108–122 (1993)

5.53 C. De Moustier, H. Matsumoto: Seafloor acoustic remote sensing with multibeam echo-sounders and bathymetric sidescan sonar systems, Marine Geophys. Res. **15**, 27–42 (1993)

5.54 A. Turgut, M. McCord, J. Newcomb, R. Fisher: *Chirp sonar sediment characterization at the northern Gulf of Mexico Littoral Acoustic Demonstration Center experimental site*, Proc. OCEANS 2002 MTS/IEEE Conf. (IEEE, New York 2002)

5.55 D.B. Kilfoyle, A.B. Baggeroer: The state of the art in underwater acoustic telemetry, IEEE J. Ocean. Eng. **25**(1), 4–27 (2000)

5.56 D.B. Kilfoyle, J.C. Preissig, A.B. Baggeroer: Spatial modulation of partially coherent multiple-input/multiple-output channels, IEEE Trans. Signal Proc. **51**, 794–804 (2003)

5.57 C.E. Shannon: A mathematical theory of communication, Bell Syst. Tech. J. **27**, 379 (1948)

5.58 S.H. Simon, A.L. Moustakas, M. Stoychev, H. Safar: Communication in a disordered world, Phys. Today **54**(9) (2001)

5.59 G. Raleigh, J.M. Cioffi: Spatio-temporal coding for wireless communication, IEEE Trans. Commun. **46**, 357–366 (1998)

5.60 G.J. Foschini, M.J. Gans: On limits of wireless communications in a fading environment whe using multiple antennas, Wireless Personnal Communications **6**, 311–335 (1998)

5.61 C. Berrou, A. Glavieux: Near optimum error correcting coding and decoding: turbo-codes, IEEE Trans. Commun. **2**, 1064–1070 (1996)

5.62 J.G. Proakis: *Digital Communications* (McGraw Hill, New York 1989)

5.63 M. Stojanovich: Retrofocusing techniques for high rate acoustic communications, J. Acoust. Soc. Am. **117**, 1173–1185 (2005)

5.64 H.C. Song, W.S. Hodgkiss, W.A. Kuperman, M. Stevenson, T. Akal: Improvement of time-reversal communications using adaptive channel equalizers, IEEE J Ocean. Eng. **31**, 487–496 (2006)

5.65 T.C. Yang: Temporal resolution of time-reversal and passive-phase onjugation for underwater acoustic communications, IEEE J. Ocean. Eng. **28**, 229–245 (2003)

5.66 C. Wunsch: *The ocean circulation inverse problem* (Cambridge Univ. Press, Cambridge 1996)

5.67 E.K. Skarsoulis, B.D. Cornuelle: Travel-time sensitivity kernels in ocean acoustic tomography, J. Acoust. Soc. Am. **116**, 227 (2004)

5.68 T.G. Birdsall, K. Metzger, M.A. Dzieciuch: Signals, signal processing, and general results, J. Acoust. Soc. Am. **96**, 2343 (1994)

5.69 P.N. Mikhalevsky, A.B. Baggeroer, A. Gavrilov, M. Slavinsky: Continuous wave and M-sequence transmissions across the Arctic, J. Acoust. Soc. Am. **96**, 3235 (1994)

5.70 B.M. Howe, P.F. Worcester, R.C. Spindel: Ocean acoustic tomography: mesoscale velocity, J. Geophys. Res. **92**, 3785–3805 (1987)

5.71 R.A. Knox: Ocean acoustic tomography: A primer. In: *Ocean Circulation Models: Combining Data and Dynamics*, ed. by D.L.T. Anderson, J. Willebrand (Kluwer Academic, Dordrecht 1989)

5.72 M.I. Taroudakis: Ocean Acoustic Tomography, Lect. Notes ECUA, 77–95 (2002)

5.73 A.J.W. Duijndam, M.A. Schonewille, C.O.H. Hindriks: Reconstruction of band-limited signals, irregularly sampled along one spatial direction, Geophysics **64**(2), 524–538 (1999)

5.74 U. Send, The THETIS-2 Group: Acoustic observations of heat content across the Mediterranean Sea, Nature **38**, 615–617 (1997)

5.75 J.A. Colosi, E.K. Scheer, S.M. Flatté, B.D. Cornuelle, M.A. Dzieciuch, W.H. Munk, P.F. Worcester, B.M. Howe, J.A. Mercer, R.C. Spindel, K. Metzger, T.G. Birdsall, A.B. Baggeroer: Comparisons of measured and predicted acoustic fluctuations for a 3250 km propagation experiment in the eastern North Pacific Ocean, J. Acoust. Soc. Am. **105**, 3202–3218 (1999)

5.76 J.A. Colosi, A.B. Baggeroer, B.D. Cornuelle, M.A. Dzieciuch, W.H. Munk, P.F. Worcester, B.D. Dushaw, B.M. Howe, J.A. Mercer, R.C. Spindel, T.G. Birdsall, K. Metzger, A.M.G. Forbes: Analysis of multipath acoustic field variability and coherence in the finale of broadband basin-scale transmissions in the North Pacific Ocean, J. Acoust. Soc. Am. **117**, 1538–1564 (2005)

5.77 M. De Hoop, R. van der Hilst: On sensitivity kernels for wave equation transmission tomography, Geophys. J. Int. **160**, 621–633 (2005)

5.78 R. van der Hilst, M. de Hoop: Banana-doughnut kernels and mantle tomography, Geophys. J. Int. **163**, 956–961 (2005)

5.79 F. Dahlen, G. Nolet: Comment on the paper on sensitivity kernels for wave equation transmission tomography by de Hoop and van der Hilst, Geophys. J. Int. **163**, 949–951 (2005)

5.80 K.A. Johannesson, R.B. Mitson: *Fisheries acoustics – A practical manual for aquatic biomass estimation*, FAO Fish. Tech. Paper 240 (FAO, Rome 1983)

5.81 G.L. Thomas: *Fisheries Acoustics, Special Issue*, Fish. Res. 14(2/3) (Elsevier, Amsterdam 1992)

5.82 A. Duncan: *Shallow Water Fisheries Acoustics* 35(1/2) (Elsevier, Amsterdam 1998)

5.83 B.J. Robinson: Statistics of single fish echoes observed at sea, ICES C. M. B **16**, 1–4 (1976)

5.84 D. B.Reeder, J.M. Jech, T.K. Stanton: Broadband acoustic backscatter and high-resolution morphology of fish: Measurement and modeling, J. Acoust. Soc. Am. **116**, 2489 (2004)

5.85 A. Alvarez, Z. Ye: Effects of fish school structures on acoustic scattering, ICES J. Marine Sci. **56**(3), 361–369 (1999)

5.86 D. Reid, C. Scalabrin, P. Petitgas, J. Masse, R. Aukland, P. Carrera, S. Georgakarakos: Standard protocols for the analysis of school based data from echo sounder surveys, Fish. Res. **47**(2/3), 125–136 (2000)

5.87 D.V. Holliday: Resonance Structure in Echoes from Schooled Pelagic Fish, J. Acoust. Soc. Am. **51**, 1322–1332 (1972)

5.88 D.N. MacLennan: Acoustical measurement of fish abundance, J. Acoust. Soc. Am. **87**(1), 1–15 (1990)

5.89 D.N. Mac Lennan, E.J. Simmonds: *Fisheries Acoustics* (Chapman Hall, London 1992)

5.90 P. Petitgas: A method for the identification and characterization of clusters of schools along the transect lines of fisheries-acoustic surveys, ICES J. Marine Sci. **60**(4), 872–884 (2003)

5.91 Y. Takao, M. Furusawa: Dual-beam echo integration method for precise acoustic surveys, ICES J. Marine Sci. **53**(2), 351–358 (1996)

5.92 E. Marchal, P. Petitgas: Precision of acoustic fish stock estimates: separating the number of schools

5.93 T.K. Stanton: Effects of transducer motion on echo-integration techniques, J. Acoust. Soc. Am. **72**, 947–949 (1982)

5.94 K.G. Foote, H.P. Knudsen, G. Vestnes, D.N. MacLennan, E.J. Simmonds: Calibration of acoustic instruments for fish density estimation: a practical guide, ICES Coop. Res. Rep. **144**, 69 (1987)

5.95 K.G. Foote: Energy in acoustic echoes from fish aggregations, Fish. Res. **1**, 129–140 (1981)

5.96 P. Petitgas, D. Reid, P. Carrera, M. Iglesias, S. Georgakarakos, B. Liorzou, J. Massé: On the relation between schools, clusters of schools, and abundance in pelagic fish stocks, ICES J. Marine Sci. **58**(6), 1150–1160 (2001)

5.97 K.G. Foote: Linearity of fisheries acoustics, with addition theorems, J. Acoust. Soc. Am. **73**(6), 1932–1940 (1983)

5.98 F. Gerlotto, M. Soria, P. Freon: From two dimensions to three: the use of multibeam sonar for a new approach in fisheries acoustics, Can. J. Fish. Aquat. Sci. **56**, 6–12 (1999)

5.99 L. Mayer, Y. Li, G. Melvin: 3D visualization for pelagic fisheries research and assessment, ICES J. Marine Sci. **59**(1), 216–225 (2002)

5.100 C. Scalabrin, N. Diner, A. Weill, A. Hillion, M.-C. Mouchot: Narrowband acoustic identification of monospecific fish shoals, ICES J. Marine Sci. **53**, 181–188 (1996)

5.101 J. Guillard, P. Brehmer, M. Colon, Y. Guennegan: 3-D characteristics of young-of-year pelagic fish schools in lake, Aqu. Liv. Res. **19**, 115–122 (2006)

5.102 K.G. Foote: Fish target strengths for use in echo integrator surveys, J. Acoust. Soc. Am. **82**, 981 (1987)

5.103 E. Josse, A. Bertrand: In situ acoustic target strength measurements of tuna associated with a fish aggregating device, ICES J. Marine Sci. **57**(4), 911–918 (2000)

5.104 B.J. Robinson: *In situ measurements of the target strengths of pelagic fishes*, ICES/FAO Symp. Fish. Acoust (FAO, Rome 2000) pp. 1–7

5.105 S. Gauthier, G.A. Rose: Target strength of encaged Atlantic redfish, ICES J. Marine Sci. **58**, 562–568 (2001)

5.106 J.S. Jaffe: Target localization for a three-dimensional multibeam sonar imaging system, J. Acoust. Soc. Am. **105**(6), 3168–3175 (1999)

5.107 D.V. Holliday, R.E. Pieper: Volume scattering strengths and zooplankton distributions at acoustic frequencies between 0.5 and 3 MHz, J. Acoust. Soc. Am. **67**, 135 (1980)

5.108 C.F. Greenlaw, D.V. Holliday, D.E. McGehee: Multistatic, multifrequency scattering from zooplankton, J. Acoust. Soc. Am. **103**, 3069 (1998)

5.109 A.M. Thode, G.L. D'Spain, W.A. Kuperman: Mathed-field processing, geoacoustic inversion, and source signature recovery of blue whale vocalizations, J. Acoust. Soc. Am. **107**, 1286–1300 (2000)

5.110 M. Johnson, P. Tyack, D. Nowacek: A digital recording tag for measuring the response of marine mammals to sound, J. Acoust. Soc. Am. **108**, 2582–2583 (2000)

5.111 D.K. Mellinger, K.M. Stafford, G.C. Fox: Seasonal occurrence of sperm whale (Physeter macrocephalus) sounds in the Gulf of Alaska, 1999-2001, Marine Mammal Sci. **20**(1), 48–62 (2004)

5.112 W.J. Richardson, C.R. Greene, C.I. Malme, D.H. Thomson: *Marine Mammals and Noise* (Academic, San Diego 1995)

5.113 A. Sirovic, J.A. Hildebrand, S.M. Wiggins, M.A. McDonald, S.E. Moore, D. Thiele: Seasonality of blue and fin whale calls and the influence of sea ice in the western Antarctic penninsula, Deep Sea Res. **51**(II), 2327–2344 (2004)

5.114 A.M. Thode, D.K. Mellinger, S. Stienessen, A. Martinez, K. Mullin: Depth-dependent acoustic features of diving sperm whales (Physeter macrocephalus) in the Gulf of Mexico, J. Acoust. Soc. Am. **112**, 308–321 (2002)

5.115 P. Tyack: Acoustics communications under the sea. In: *Animal Acoustic Communication: Recent Advances*, ed. by S.L. Hopp, M.J. Owren, C.S. Evans (Springer, Berlin, Heidelberg 1988) pp. 163–220

5.116 National Research Council: *Ocean Noise and Marine Mammals* (National Academy, Washington 2003) p. 192

Part B Physical and Nonlinear Acoustics

6 Physical Acoustics
Mack A. Breazeale, University, USA
Michael McPherson, University, USA

7 Thermoacoustics
Gregory W. Swift, Los Alamos, USA

8 Nonlinear Acoustics in Fluids
Werner Lauterborn, Göttingen, Germany
Thomas Kurz, Göttingen, Germany
Iskander Akhatov, Fargo, USA

6. Physical Acoustics

An overview of the fundamental concepts needed for an understanding of physical acoustics is provided. Basic derivations of the acoustic wave equation are presented for both fluids and solids. Fundamental wave concepts are discussed with an emphasis on the acoustic case. Discussions of different experiments and apparatus provide examples of how physical acoustics can be applied and of its diversity. Nonlinear acoustics is also described.

- 6.1 Theoretical Overview 209
  - 6.1.1 Basic Wave Concepts 209
  - 6.1.2 Properties of Waves 210
  - 6.1.3 Wave Propagation in Fluids 215
  - 6.1.4 Wave Propagation in Solids 217
  - 6.1.5 Attenuation 218
- 6.2 Applications of Physical Acoustics 219
  - 6.2.1 Crystalline Elastic Constants 219
  - 6.2.2 Resonant Ultrasound Spectroscopy (RUS) 220
  - 6.2.3 Measurement Of Attenuation (Classical Approach) 221
  - 6.2.4 Acoustic Levitation 222
  - 6.2.5 Sonoluminescence 222
  - 6.2.6 Thermoacoustic Engines (Refrigerators and Prime Movers) 223
  - 6.2.7 Acoustic Detection of Land Mines 224
  - 6.2.8 Medical Ultrasonography 224
- 6.3 Apparatus 226
  - 6.3.1 Examples of Apparatus 226
  - 6.3.2 Piezoelectricity and Transduction 226
  - 6.3.3 Schlieren Imaging 228
  - 6.3.4 Goniometer System 230
  - 6.3.5 Capacitive Receiver 231
- 6.4 Surface Acoustic Waves 231
- 6.5 Nonlinear Acoustics 234
  - 6.5.1 Nonlinearity of Fluids 234
  - 6.5.2 Nonlinearity of Solids 235
  - 6.5.3 Comparison of Fluids and Solids.. 236
- References 237

Physical acoustics involves the use of acoustic techniques in the study of physical phenomena as well as the use of other experimental techniques (optical, electronic, etc.) to study acoustic phenomena (including the study of mechanical vibration and wave propagation in solids, liquids, and gasses). The subject is so broad that a single chapter cannot cover the entire subject. For example, recently the 25th volume of a series of books entitled *Physical Acoustics* was published [6.1]. *Mason* [6.2] began the series in 1964. The intermediate volumes are not repetitious, but deal with different aspects of physical acoustics. Even though all of physical acoustics cannot be covered in this chapter, some examples will illustrate the role played by physical acoustics in the development of physics.

Since much of physics involves the use and study of waves, it is useful to begin by mentioning some different types of waves and their properties. The most basic definition of a wave is a disturbance that propagates through a medium. A simple analogy can be made with a stack of dominoes that are lined up and knocked over. As the first domino falls into the second, it is knocked over into the third, which is knocked over into the next one, and so on. In this way, the disturbance travels down the entire chain of dominoes (which we may think of as particles in a medium) even though no particular domino has moved very far. Thus, we may consider the motion of an individual domino, or the motion of the disturbance which is traveling down the entire chain of dominoes. This suggests that we define two concepts, the average particle velocity of the individual dominoes and the wave velocity (of the disturbance) down the chain of dominoes. Acoustic waves behave in a similar manner. In physical acoustics it is necessary to distinguish between particle velocity and wave (or phase) velocity.

There are two basic types of waves: longitudinal waves, and transverse waves. These waves are defined according to the direction of the particle motion in the medium relative to the direction in which the wave travels. Longitudinal waves are waves in which the particle motion in the medium is in the same direction as the wave is traveling. Transverse waves are those in which the particle motion in the medium is at a right angle to the direction of wave propagation. Figure 6.1 is a depiction of longitudinal and transverse waves. Another less common type of wave, known as a torsional wave, can also propagate in a medium. Torsional waves are waves in which particles move in a circle in a plane perpendicular to the direction of the wave propagation. Figure 6.2 shows a commonly used apparatus for the demonstration of torsional waves.

There also are more-complicated types of waves that exist in acoustics. For example, surface waves (Rayleigh waves, Scholte–Stonley waves, etc.) can propagate along the boundary between two media. Another example of a more complicated type of wave propagation is that of Lamb waves, which can propagate along thin plates. A familiar example of Lamb waves are the waves that propagate along a flag blowing in a wind.

In acoustics, waves are generally described by the pressure variations that occur in the medium (solid or fluid) due to the wave. As an acoustic wave passes through the medium, it causes the pressure to vary as the acoustic energy causes the distance between the molecules or atoms of the fluid or solid to change periodically. The total pressure is given by

$$p_T(x,t) = p_0(x,t) + p_1(x,t) \,. \tag{6.1}$$

Here p_0 represents the ambient pressure of the fluid and p_1 represents the pressure fluctuation caused by the acoustic field. Since pressure is defined as the force per unit area, it has units of newtons per square meter (N/m^2). The official SI designation for pressure is the pascal (1 Pa = 1 N/m^2). Atmospheric pressure at sea level is 1 atmosphere (atm) = 1.013×10^5 Pa. The types of sounds we encounter cause pressure fluctuations in the range from 10^{-3}–10 Pa.

One can also describe the strength of the sound wave in terms of the energy that it carries. Experimentally, one can measure the power in the acoustic wave, or the amount of energy carried by the wave per unit time. Rather than trying to measure the power at every point in space, it is usual to measure the power only at the location of the detector. So, a more convenient measurement is the power density, also referred to as the acoustic intensity I. In order to make a definition which does not depend on the geometry of the detector, one considers the power density only over an infinitesimal area of size dA:

$$I = \frac{dP}{dA} \,, \tag{6.2}$$

where dP is the portion of the acoustic power that interacts with the area dA *of the detector* oriented perpendicular to the direction of the oncoming acoustic wave. The units of acoustic intensity are watts per square meter (W/m^2).

The human ear can generally perceive sound pressures over the range from about 20 μPa up to about 200 Pa (a very large dynamic range). Because the range of typical acoustic pressures is so large, it is convenient to work with a relative measurement scale rather than an absolute measurement scale. These scales are expressed using logarithms to compress the dynamic range. In acoustics, the scale is defined so that every factor of ten increase in the amount of energy carried by the wave is represented as a change of 1 bel (named after Alexander Graham Bell). However, the bel is often too large to be useful. For this reason, one uses the decibel scale (1/10 of a bel). Therefore, one can write the sound intensity level (SIL) as the logarithm of two intensities:

$$\text{SIL(dB)} = 10 \log \left(\frac{I}{I_{\text{ref}}} \right) \,, \tag{6.3}$$

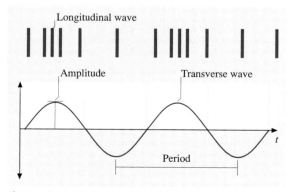

Fig. 6.1 Longitudinal and transverse waves

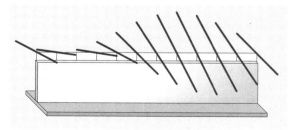

Fig. 6.2 Apparatus for the demonstration of torsional waves

where I is the intensity of the sound wave and I_{ref} is a reference intensity. (One should note that the bel, or decibel, is not a unit in the typical sense; rather, it is simply an indication of the relative sound level).

In order for scientists and engineers to communicate meaningfully, certain standard reference values have been defined. For the intensity of a sound wave in air, the reference intensity is defined to be $I_{ref} = 10^{-12}$ W/m^2.

In addition to measuring sound intensity levels, it is also common to measure sound pressure levels (SPL). The sound pressure level is defined as

$$\text{SPL(dB)} = 20 \log \left(\frac{p}{p_{ref}} \right), \tag{6.4}$$

where p is the acoustic pressure and p_{ref} is a reference pressure. (The factor of 20 comes from the fact that I is proportional to p^2.) For sound in air, the reference pressure is defined as 20 µPa (2×10^{-5} Pa).

For sound in water, the reference is 1 µPa [historically other reference pressures, for example, 20 µPa and 0.1 Pa, have been defined). It is important to note that sound pressure levels are meaningful only if the reference value is defined. It should also be noted that this logarithmic method of defining the sound pressure level makes it easy to compare two sound levels. It can be shown that $\text{SPL}_2 - \text{SPL}_1 = 20 \log \left(\frac{p_2}{p_1} \right)$; hence, an SPL difference depends only on the two pressures and not on the choice of reference pressure used.

Since both optical and acoustic phenomena involve wave propagation, it is illustrative to contrast them. Optical waves propagate as transverse waves. Acoustic waves in a fluid are longitudinal; those in a solid can be transverse or longitudinal. Under some conditions, waves may propagate along interfaces between media; such waves are generally referred to as surface waves. Sometimes acoustic surface waves correspond with an optical analogue. However, since the acoustic wavelength is much larger than the optical wavelength, the phenomenon may be much more noticeable in physical acoustics experiments. Many physical processes produce acoustic disturbances directly. For this reason, the study of the acoustic disturbance often gives information about a physical process. The type of acoustic wave should be examined to determine whether an optical model is appropriate.

6.1 Theoretical Overview

6.1.1 Basic Wave Concepts

Although the domino analogy is useful for conveying the idea of how a disturbance can travel through a medium, real waves in physical systems are generally more complicated. Consider a spring or slinky that is stretched along its length. By rapidly compressing the end of the spring, one can send a pulse of energy down the length of the spring. This pulse would essentially be a longitudinal wave pulse traveling down the length of the spring. As the pulse traveled down the length, the material of the spring would compress or *bunch up* in the region of the pulse and *stretch out* on either side of it. The compressed regions are known as condensations and the stretched regions are known as rarefactions.

It is this compression that one could actually witness traveling down the length of the spring. No part of the spring itself would move very far (just as no domino moved very far), but the disturbance would travel down the entire length of the spring. One could also repeatedly drive the end of the spring back and forth (along its length). This would cause several pulses (each creating compressions with stretched regions around them) to propagate along the length of the spring, with the motion of the spring material being along the direction of the propagating disturbance. This would be a multipulse longitudinal wave.

Now, let us consider an example of a transverse wave, with particle motion perpendicular to the direction of propagation. Probably the simplest example is a string with one end fastened to a wall and the opposite end driven at right angles to the direction along which the string lies. This drive sends pulses down the length of the string. The motion of the particles in the string is at right angles to the motion of the disturbance, but the disturbance itself (whether one pulse or several) travels down the length of the string. Thus, one sees a transverse wave traveling down the length of the string.

Any such periodic pulsing of disturbances (whether longitudinal or transverse) can be represented mathematically as a combination of sine and/or cosine waves through a process known as Fourier decomposition. Thus, without loss of generality, one can illustrate additional wave concepts by considering a wave whose shape is described mathematically by a sine wave.

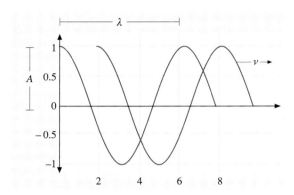

Fig. 6.3 One cycle of a sinusoidal wave traveling to the right

Figure 6.3 shows one full cycle of a sinusoidal wave which is moving to the right (as a sine-shaped wave would propagate down a string, for example). The initial wave at $t = 0$ and beginning at $x = 0$ can be described mathematically. Let the wave be traveling along the x-axis direction and let the particle displacement be occurring along the y-axis direction. In general, the profile or shape of the wave is described mathematically as

$$y(x) = A \sin(kx + \varphi), \quad (6.5)$$

where A represents the maximum displacement of the string (i.e., the particle displacement) in the y-direction and k represents a scaling factor, the wave number. The argument of the sine function in (6.5) is known as the phase. For each value of x, the function $y(x)$ has a unique value, which leads to a specific y value (sometimes called a point of constant phase). The term φ is known as the phase shift because it causes a shifting of the wave profile along the x-axis (forward for a positive phase shift and backward for a negative phase shift). Such a sine function varies between $+A$ and $-A$, and one full cycle of the wave has a length of $\lambda = \frac{2\pi}{k}$. The length λ is known as the wavelength, and the maximum displacement A is known as the wave amplitude.

As this disturbance shape moves toward the right, its position moves some distance $\Delta x = x_f - x_o$ during some time interval Δt, which means the disturbance is traveling with some velocity $c = \frac{\Delta x}{\Delta t}$. Thus, the distance the wave has traveled shows this profile both before and after it has traveled the distance Δx (Fig. 6.3). The traveling wave can be expressed as a function of both position (which determines its profile) and time (which determines the distance it has traveled). The equation for a traveling wave, then, is given by

$$y(x, t) = A \sin[k(x - ct) + \varphi], \quad (6.6)$$

where $t = 0$ gives the shape of the wave at $t = 0$ (which here is assumed constant), and $y(x, t)$ gives the shape and position of the wave disturbance as it travels. Again, A represents the amplitude, k represents the wave number, and φ represents the phase shift. Equation (6.6) is applicable for all types of waves (longitudinal, transverse, etc.) traveling in any type of medium (a spring, a string, a fluid, a solid, etc.).

Thus far, we have introduced several important basic wave concepts including wave profile, phase, phase shift, amplitude, wavelength, wave number, and wave velocity. There is one additional basic concept of great importance, the wave frequency. The frequency is defined as the rate at which (the number of times per second) a point of constant phase passes a point in space. The most obvious points of constant phase to consider are the maximum value (crest) or the minimum value (trough) of the wave. One can think of the concept of frequency less rigorously as the number of pulses generated per second by the source causing the wave. The velocity of the wave is the product of the wavelength and the frequency.

One can note that, rather than consisting of just one or a few pulses, (6.6) represents a continually varying wave propagating down some medium. Such waves are known as continuous waves. There is also a type of wave that is a bit between a single pulse and an infinitely continuous wave. A wave that consists of a finite number of cycles is known as a wave packet or a tone burst. When dealing with a tone burst, the concepts of phase velocity and group velocity are much more evident. Generally speaking, the center of the tone burst travels at the phase velocity – the ends travel close to the group velocity.

6.1.2 Properties of Waves

All waves can exhibit the following phenomena: reflection, refraction, interference and diffraction. (Transverse waves can also exhibit a phenomenon known as polarization, which allows oscillation in only one plane.)

Reflection

The easiest way to understand reflection is to consider the simple model of a transverse wave pulse traveling down a taut string that is affixed at the opposite end (as seen in Fig. 6.4. (A pulse is described here for purposes of clarity, but the results described apply to continu-

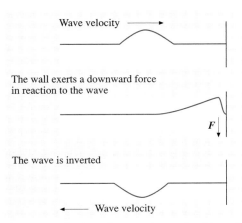

Fig. 6.4 Reflection of a wave from a fixed boundary

ous waves as well.) As the pulse reaches the end of the string, the particles start to move upward, but they cannot because the string is fastened to the pole. (This is known as a fixed or rigid boundary.) The pole exerts a force on the particles in the string, which causes the pulse to rebound and travel in the opposite direction. Since the force of the pole on the string in the y-direction must be downward (to counteract the upward motion of the particles), there is a 180° phase shift in the wave. This is seen in the figure where the reflected pulse has flipped upside down relative to the incident pulse.

Figure 6.5 shows a different type of boundary from which a wave (or pulse) can reflect; in this case the end of the string is on a massless ring that can slide freely up and down the pole. (This situation is known as a free boundary.) As the wave reaches the ring, it

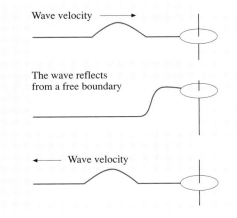

Fig. 6.5 Reflection of a wave from a free boundary

drives the ring upwards. As the ring moves back down, a reflected wave, which travels in a direction opposite to that of the incoming wave, is also generated. However, there is no 180° phase shift upon reflection from a free boundary.

Acoustic Impedance

Another important wave concept is that of wave impedance, which is usually denoted by the variable Z. When the reflection of the wave is not total, part of the energy in the wave can be reflected and part transmitted. For this reason, it is necessary to consider acoustic waves at an interface between two media and to be able to calculate how much of the energy is reflected and how much is transmitted. The definition of media impedance facilitates this. For acoustic waves, the impedance Z is defined as the ratio of sound pressure to particle velocity. The units for impedance are the Rayl, so named in honor of Lord Rayleigh. 1 Rayl = 1 Pa s/m.

Often one speaks of the *characteristic impedance* of a medium (a fluid or solid); in this case one is referring to the medium in the "open space" condition where there are no obstructions to the wave which would cause the wave to reflect or scatter. The characteristic impedance of a material is usually denoted by Z_0, and it can be determined by the product of the mean density of the medium ρ with the speed of sound in the medium. In air, the characteristic impedance near room temperature is about, 410 Rayl.

The acoustic impedance concept is particularly useful. Consider a sound wave that passes from an initial medium with one impedance into a second medium with a different impedance. The efficiency of the energy transfer from one medium into the next is given by the ratio of the two impedances. If the impedances (ρc) are identical, their ratio will be 1; and all of the acoustic energy will pass from the first medium into the second across the interface between them. If the impedances of the two media are different, some of the energy will be reflected back into the initial medium when the sound field interacts with the interface between the two media. Thus the impedance enables one to characterize the acoustic transmission and reflection at the boundary of the two materials. The difference in Z, which leads to some of the energy being reflected back into the initial medium, is often referred to as the impedance mismatch. When the (usual) acoustic boundary conditions apply and require that the particle velocity and pressure be continuous across the interface between the two media, one can calculate the percentage of the energy that is reflected back into the medium. This is given in fractional form by the reflection coefficient given by

$$R = \left(\frac{Z_2 - Z_1}{Z_2 + Z_1}\right)^2, \quad (6.7)$$

where Z_1 and Z_2 are the impedances of the two media. Since both values of Z must be positive, R must be less than one. The fraction of the energy transmitted into the second medium is given by $T = 1 - R$ because 100% of the energy must be divided between T and R.

Refraction

Refraction is a change of the direction of wave propagation as the wave passes from one medium into another across an interface. Bending occurs when the wave speed is different in the two media. If there is an angle between the normal to the plane of the boundary and the incident wave, there is a brief time interval when part of the wave is in the original medium (traveling at one velocity) and part of the wave is in the second medium (traveling at a different velocity). This causes the bending of the waves as they pass from the first medium to the second. (There is no bending at normal incidence.)

Reflection and refraction can occur simultaneously when a wave impinges on a boundary between two media with different wave propagation speeds. Some of the energy of the wave is reflected back into the original medium, and some of the energy is transmitted and refracted into the second medium. This means that a wave incident on a boundary can generate two waves: a reflected wave and a transmitted wave whose direction of propagation is determined by Snell's law.

All waves obey Snell's law. For optical waves the proper form of Snell's law is:

$$n_1 \sin\theta_1 = n_2 \sin\theta_2, \quad (6.8)$$

where n_1 and n_2 are the refractive indices and θ_1 and θ_2 are propagation directions. For acoustic waves the proper form of Snell's law is:

$$\frac{\sin\theta_1}{v_1} = \frac{\sin\theta_2}{v_2}, \quad (6.9)$$

where v_1 is the wave velocity in medium 1 and v_2 is the wave velocity in medium 2. These two forms are very similar since the refractive index is $n = c/C_m$, where c is the velocity of light in a vacuum and C_m is the velocity of light in the medium under consideration.

Interference

Spatial Interference. Interference is a phenomenon that occurs when two (or more) waves add together. Consider two identical transverse waves traveling to the right (one after the other) down a string towards a boundary at the end. When the first wave encounters the boundary, it reflects and travels in the leftward direction. When it encounters the second, rightward moving wave the two waves add together linearly (in accordance with the principle of superposition). The displacement amplitude at the point in space where two waves combine is either greater than or less than the displacement amplitude of each wave. If the resultant wave has an amplitude that is smaller than that of either of the original two waves, the two waves are said to have

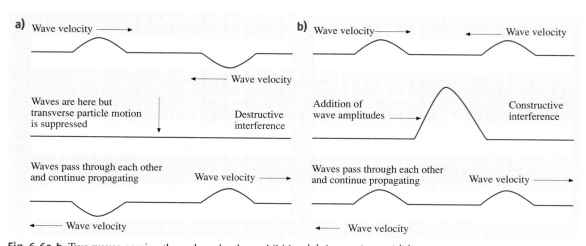

Fig. 6.6a,b Two waves passing through each other exhibiting (a) destructive and (b) constructive interference

destructively interfered with one another. If the combined wave has an amplitude that is greater than either of its two constituent waves, then the two waves are said to have constructively interfered with each other. The maximum possible displacement of the combination is the sum of the maximum possible displacements of the two waves (complete constructive interference); the minimum possible displacement is zero (complete destructive interference for waves of equal amplitude). It is important to note that the waves interfere only as they pass through one another. Figure 6.6 shows the two special cases of complete destructive and complete constructive interference (for clarity only a portion of the wave is drawn).

If a periodic wave is sent down the string and meets a returning periodic wave traveling in the opposite direction, the two waves interfere. This results in wave superposition, i.e., the resulting amplitude at any point and time is the sum of the amplitudes of the two waves. If the returning wave is inverted (due to a fixed boundary reflection) and if the length of the string is an integral multiple of the half-wavelength corresponding to the drive frequency, conditions for resonance are satisfied.

This resonance produces a standing wave. An example of a standing wave is shown in Fig. 6.7. The points of maximum displacement are known as the standing-wave antinodes. The points of zero displacement are known as the standing-wave nodes.

Resonance Behavior. Every vibrating system has some characteristic frequency that allows the vibration amplitude to reach a maximum. This characteristic frequency is determined by the physical parameters (such as the geometry) of the system. The frequency that causes maximum amplitude of vibration is known as the resonant frequency, and a system driven at its resonant frequency is said to be in resonance. Standing waves are simply one type of resonance behavior.

Longitudinal acoustic waves can also exhibit resonance behavior. When the distance between a sound emitter and a reflector is an integer number of half wavelengths, the waves interfere and produce standing waves. This interference can be observed optically, acousticly or electronically. By observing a large number of standing waves one can obtain an accurate value of the wavelength, and hence an accurate value of the wave velocity.

One of the simplest techniques for observing acoustic resonances in water, or any other transparent liquid, is to illuminate the resonance chamber, then to focus a microscope on it. The microscope field of view images the nodal lines that are spaced half an acoustic wavelength apart. A screw that moves the microscope perpendicular to the lines allows one to make very accurate wavelength measurements, and hence accurate sound velocity measurements.

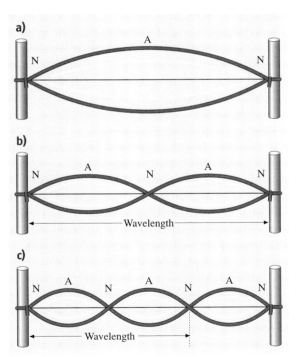

Fig. 6.7a–c Standing waves in a string with nodes and antinodes indicated. (**a**) The fundamental; (**b**) the second harmonic; (**c**) the third harmonic

Temporal Interference. So far we have considered the interference of two waves that are combining in space (this is referred to as spatial interference). It is also possible for two waves to interfere because of a difference in frequency (which is referred to as temporal interference). One interesting example of this is the phenomenon of wave beating.

Consider two sinusoidal acoustic waves with slightly different frequencies that arrive at the same point in space. Without loss of generality, we can assume that these two waves have the same amplitude. The superposition principle informs us that the resultant pressure caused by the two waves is the sum of the pressure caused by each wave individually. Thus, we have for the total pressure

$$p_T(t) = A\left[\cos(\omega_1 t) + \cos(\omega_2 t)\right] . \tag{6.10}$$

By making use of a standard trigonometric identity, this can be rewritten as

$$p_T(t) = 2A \cos\left(\frac{(\omega_1 - \omega_2)}{2}t\right) \cos\left(\frac{(\omega_1 + \omega_2)}{2}t\right).$$
(6.11)

Since the difference in frequencies is small, the two waves can be in phase, causing constructive interference and reinforcing one another. Over some period of time, the frequency difference causes the two waves to go out of phase, causing destructive interference (when $\omega_1 t$ eventually leads $\omega_2 t$ by 180°). Eventually, the waves are again in phase and, they constructively interfere again. The amplitude of the combination will rise and fall in a periodic fashion. This phenomenon is known as the beating of the two waves. This beating phenomenon can be described as a separate wave with an amplitude that is slowly varying according to

$$p(t) = A_0(t) \cos(\omega_{\text{avg}} t)$$
(6.12)

where

$$A_0(t) = 2A \cos\left(\frac{\omega_1 - \omega_2}{2}t\right)$$
(6.13)

and

$$\omega_{\text{avg}} = \frac{(\omega_1 + \omega_2)}{2}.$$
(6.14)

The cosine in the expression for $A_0(t)$ varies between positive and negative 1, giving the largest amplitude in each case. The period of oscillation of this amplitude variation is given by

$$T_b = \frac{2\pi}{\omega_1 - \omega_2} = \frac{2\pi}{\omega_b} = \frac{1}{f_b}.$$
(6.15)

The frequency f_b of the amplitude variation is known as the beat frequency. Figure 6.8 shows the superposition of two waves that have two frequencies that are different but close together. The beat frequency corresponds to the difference between the two frequencies that are beating.

The phenomenon of beating is often exploited by musicians in tuning their instruments. By using a reference (such as a 440 Hz tuning fork that corresponds to the A above middle C) one can check the tuning of the instrument. If the A above middle C on the instrument is out of tune by 2 Hz, the sounds from the tuning fork and the note generate a 2 Hz beating sound when they are played together. The instrument is then tuned until the beating sound vanishes. Then the frequency of the instrument is the same as that of the tuning fork. Once the A is in tune, the other notes can be tuned relative to

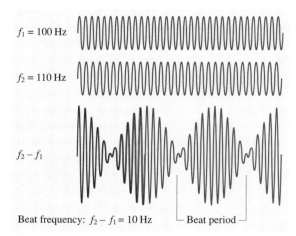

Fig. 6.8 The beating of two waves with slightly different frequencies

the A by counting the beats per unit time when different notes are played in various combinations.

Multi-frequency Sound

When sound consists of many frequencies (not necessarily all close together), one needs a means of characterizing the sound level. One may use a weighted average of the sound over all the frequencies present, or one may use information about how much energy is distributed over a particular range of frequencies. A selected range of frequencies is known as a frequency band. By means of filters (either acoustic or electrical, depending on the application) one can isolate the various frequency bands across the entire frequency spectrum. One can then talk of the acoustic pressure due to a particular frequency band. The *band pressure level* is given by

$$\text{PL}_{\text{band}} = 20 \log_{10}\left(\frac{p_{\text{band}}}{p_{\text{ref}}}\right)$$
(6.16)

where p_{band} is the room-mean-square (rms) average pressure of the sound in the frequency band range and p_{ref} is the standard reference for sound in air, 20 µPa. The average of the pressures of the frequency bands over the complete spectrum is the average acoustic signal. However, the presence of multiple frequencies complicates the situation: one does not simply add the frequency band pressures or the band pressure levels. Instead, it is the p^2 values which must be summed:

$$p_{\text{rms}}^2 = \sum p_{\text{band}}^2$$
(6.17)

or:

$$\text{SPL} = 10 \log_{10}\left(\frac{p_{\text{rms}}^2}{p_{\text{ref}}^2}\right) = 10 \log_{10} \sum \left(\frac{p_{\text{band}}}{p_{\text{ref}}}\right)^2 \quad (6.18)$$

or, simplifying further

$$\text{SPL} = 10 \log_{10} \sum (10)^{(\text{PL}_{\text{band}}/10)}. \quad (6.19)$$

The octave is a common choice for the width of a frequency band. With a one-octave filter, only frequencies up to twice the lowest frequency of the filter are allowed to pass. One should also note that, when an octave band filter is labeled with its center frequency, this is determined by the geometric mean (not the arithmetic mean), i.e.,

$$f_{\text{center}} = \sqrt{f_{\text{low}} f_{\text{high}}} = \sqrt{2 f_{\text{low}}^2} = \sqrt{2} f_{\text{low}}. \quad (6.20)$$

Coherent Signals

Two signals are coherent if there is a fixed relative phase relation between them. Two loudspeakers driven by the same source would be coherent. However, two loudspeakers being driven by two compact-disc (CD) players (even if each player was playing a copy of the same CD) would not be coherent because there is no connection causing a constant phase relationship. For two incoherent sources, the total pressure is

$$p_{\text{tot}}^2 = (p_{1(\text{rms})} + p_{2(\text{rms})})^2 = p_{1(\text{rms})}^2 + p_{2(\text{rms})}^2 \quad (6.21)$$

(where the $2 p_{1(\text{rms})} p_{2(\text{rms})}$ term has averaged out to zero). For coherent sources, however, the fixed phase relationship allows for the possibility of destructive or constructive interference. Therefore, the signal can vary in amplitude between $(p_{1(\text{rms})} + p_{2(\text{rms})})^2$ and $(p_{1(\text{rms})} - p_{2(\text{rms})})^2$.

Diffraction

In optics it is usual to begin a discussion of diffraction by pointing out grating effects. In acoustics one seldom encounters grating effects. Instead, one encounters changes in wave direction of propagation resulting from diffraction. Thus, in acoustics it is necessary to begin on a more fundamental level.

The phenomenon of diffraction is the bending of a wave around an edge. The amount of bending that occurs depends on the relative size of the wavelength compared to the size of the edge (or aperture) with which it interacts. When considering refraction or reflection it is often convenient to model the waves by drawing a single ray in the direction of the wave's propagation. However, the ray approach does not provide a means to model the bending caused by diffraction. The bending of a wave by diffraction is the result of wave interference. The more accurate approach, then, is to determine the magnitude of each wave that is contributing to the diffraction and to determine how it is interfering with other waves to cause the diffraction effect observed.

It is useful to note that diffraction effects can depend on the shape of the wavefronts that encounter the edge around which the diffraction is occurring. Near the source the waves can have a very strong curvature relative to the wavelength. Far enough from the source the curvature diminishes significantly (creating essentially plane waves). Fresnel diffraction occurs when curved wavefronts interact. Fraunhofer diffraction occurs when planar wavefronts interact. In acoustics these two regions are known as the near field (the Fresnel zone) and the far field (the Fraunhofer zone), respectively. In optics these two zones are distinguished by regions in which two different integral approximations are valid.

6.1.3 Wave Propagation in Fluids

The propagation of an acoustic wave is described mathematically by the acoustic wave equation. One can use the approach of continuum mechanics to derive equations appropriate to physical acoustics [6.3]. In the continuum approach one postulates fields of density, stress, velocity, etc., all of which must satisfy basic conservation laws. In addition, there are constitutive relations which characterize the medium. For illustration, acoustic propagation through a compressible fluid medium is considered first. As the acoustic disturbance passes through a small area of the medium, the density of the medium at that location fluctuates. As the crest of the acoustic pressure wave passes through the region, the density in that region increases; this compression is known as the acoustic condensation. Conversely, when the trough of the acoustic wave passes through the region, the density of the medium at that location decreases; this expansion is known as the acoustic rarefaction.

In a gas, the constitutive relationship needed to characterize the pressure fluctuations is the ideal gas equation of state, $PV = nRT$, where P is the pressure of the gas, V is the volume of the gas, n is the number of moles of the gas, R is the universal gas constant ($R = 8.3145 \, \text{J/mol K}$), and T is the temperature of the gas. In a given, small volume of the gas, there is a variation of the density ρ from its equilibrium value ρ_0 caused by the change in pressure $\Delta P = P - P_0$ as the disturbance passes through that volume. (Here P is the pressure at any instant in time and P_0 is the equilibrium pressure.)

In many situations, there are further constraints on the system which simplify the constitutive relationship. A gas may act as a heat reservoir. If the processes occur on a time scale that allows heat to be exchanged, the gas is maintained at a constant temperature. In this case, the constitutive relationship can be simplified and expressed as $P_0 V_0 = PV = $ a constant. Since the number of gas molecules (and hence the mass) is constant, we can express this as the isothermal condition

$$\frac{P}{P_0} = \frac{\rho}{\rho_0}, \qquad (6.22)$$

which relates the instantaneous pressure to the equilibrium pressure.

Most acoustic processes occur with no exchange of heat energy between adjacent volumes of the gas. Such processes are known as adiabatic or isentropic processes. Under such conditions, the constitutive relation is modified according to the adiabatic condition. For the adiabatic compression of an ideal gas, it has been found that the relationship

$$PV^\gamma = P_0 V_0^\gamma \qquad (6.23)$$

holds, where γ is the ratio of the specific heat of the gas at constant pressure to the specific heat at constant volume. This leads to an adiabatic constraint for the acoustic process given by

$$\frac{P}{P_0} = \left(\frac{\rho}{\rho_0}\right)^\gamma. \qquad (6.24)$$

When dealing with a real gas, one can make use of a Taylor expansion of the pressure variations caused by the fluctuations in density:

$$P = P_0 + \left[\frac{\partial P}{\partial \rho}\right]_{\rho_0} (\Delta \rho)$$
$$+ \frac{1}{2}\left[\frac{\partial^2 P}{\partial \rho^2}\right]_{\rho_0} (\Delta \rho)^2 + \ldots, \qquad (6.25)$$

where

$$\Delta \rho = (\rho - \rho_0). \qquad (6.26)$$

When the fluctuations in density are small, only the first-order terms in $\Delta \rho$ are nonnegligible. In this case, one can rearrange the above equation as

$$\Delta P = P - P_0 = \left[\frac{\partial p}{\partial \rho}\right]_{\rho_0} \Delta \rho = B \frac{\Delta \rho}{\rho_0}, \qquad (6.27)$$

where $B = \rho_0 \left[\frac{\partial p}{\partial \rho}\right]_{\rho_0}$ is the adiabatic bulk modulus of the gas. This equation describes the relationship between the pressure of the gas and its density during an expansion or contraction. [When the fluctuations are not small one has finite-amplitude (or nonlinear) effects which are considered later.]

Let us now consider the physical motion of a fluid as the acoustic wave passes through it. We begin with an infinitesimal volume element dV that is fixed in space, and we consider the motion of the particles as they pass through this region. Since mass is conserved, the net flux of mass entering or leaving this fixed volume must correspond to a change in the density of the fluid contained within that volume. This is expressed through the continuity equation,

$$\frac{\partial \rho}{\partial t} = -\nabla \cdot (\rho \boldsymbol{u}). \qquad (6.28)$$

The rate of mass change in the region is

$$\frac{\partial \rho}{\partial t} dV, \qquad (6.29)$$

and the net influx of mass into this region is given by

$$-\nabla \cdot (\rho \boldsymbol{u}) \, dV. \qquad (6.30)$$

Consider a volume element of fluid as it moves with the fluid. This dV of fluid contains some infinitesimal amount of mass, dm. The net force acting on this small mass of fluid is given by Newton's second law, $d\boldsymbol{F} = \boldsymbol{a} \, dm$. It can be shown that Newton's second law leads to a relationship between the particle velocity and the acoustic pressure. This relationship, given by

$$\rho_0 \frac{\partial \boldsymbol{u}}{\partial t} = -\nabla p, \qquad (6.31)$$

is known as the linear Euler equation.

By combining our adiabatic pressure condition with the continuity equation and the linear Euler equation, one can derive the acoustic wave equation. This equation takes the form

$$\nabla^2 p = \frac{1}{c^2} \frac{\partial^2 p}{\partial t^2}, \qquad (6.32)$$

where c is the speed of the sound wave which is given by

$$c = \sqrt{B/\rho_0}. \qquad (6.33)$$

6.1.4 Wave Propagation in Solids

Similarly, one can consider the transmission of an acoustic wave through a solid medium. As an example, consider the one-dimensional case for propagation of an acoustic wave through a long bar of length L and cross-sectional area S. In place of the pressure, we consider the stress applied to the medium, which is given by the relationship

$$\sigma = F/S \qquad (6.34)$$

where σ is the stress, F is the force applied along the length (L) of the bar, and S is the cross sectional area of the bar.

A stress applied to a material causes a resultant compression or expansion of the material. This response is the strain ζ. Strain is defined by the relationship,

$$\zeta = \Delta L / L_o , \qquad (6.35)$$

where ΔL is the change in length of the bar, and L_o is the original length of the bar.

Let us consider the actual motion of particles as an acoustic wave passes through some small length of the bar dx. This acoustic wave causes both a stress and a strain. The Hooke's law approximation, which can be used for most materials and vibration amplitudes, provides the constitutive relationship that relates the stress applied to the material with the resulting strain. Hooke's law states that stress is proportional to strain, or

$$\sigma = -Y\zeta . \qquad (6.36)$$

If we consider the strain over our small length dx, we can write this as

$$\frac{F}{S} = -Y\left(\frac{dL}{dx}\right) \qquad (6.37)$$

or

$$F = -YS\left(\frac{dL}{dx}\right) . \qquad (6.38)$$

The net force acting on our segment dx is given by

$$dF = -\left(\frac{\partial F}{\partial x}\right) dx = YS \left(\frac{\partial^2 L}{\partial x^2}\right) dx . \qquad (6.39)$$

Again, we can make use of Newton's second law, $F = ma$, and express this force in terms of the mass and acceleration of our segment. The mass of the segment of length dx is simply the density times the volume, or $dm = \rho dV = \rho S \, dx$. Thus,

$$dF = \left(\frac{\partial^2 L}{dt^2}\right) dm = \rho S \left(\frac{\partial^2 L}{dt^2}\right) dx , \qquad (6.40)$$

where $\frac{\partial^2 L}{dt^2}$ is the acceleration of the particles along the length dx as the acoustic wave stresses it. Equating our two expressions for the net force acting on our segment, dF,

$$\frac{\partial^2 L}{\partial x^2} = \frac{1}{c^2}\left(\frac{\partial^2 L}{\partial t^2}\right) , \qquad (6.41)$$

where

$$c = \sqrt{Y/\rho} \qquad (6.42)$$

is the speed at which the acoustic wave is traveling through the bar. The form of the equation for the propagation of an acoustic wave through a solid medium is very similar to that for the propagation of an acoustic wave through a fluid.

Both of the wave equations developed so far have implicitly considered longitudinal compressions; for example, the derivation of the wave equation for an acoustic wave traveling down a thin bar assumed no transverse components to the motion. However, if we consider transverse motion, the resulting wave equation is of the same form as that for longitudinal waves. For longitudinal waves, the solution to the wave equation is given by

$$L(x, t) = A\cos(\omega t \pm kx + \phi) \text{ (longitudinal)} , \qquad (6.43)$$

where $L(x, t)$ represents the amount of compression or rarefaction at some position x and time t. For transverse waves, the solution to the wave equation is given by

$$y(x, t) = A\cos(\omega t \pm kx + \phi) \text{ (transverse)} , \qquad (6.44)$$

where $y(x, t)$ represents the vibration orthogonal to the direction of wave motion as a function of x and t. In both cases, A is the vibration amplitude, $k = 2\pi/\lambda$ is the wave number, and ϕ is the phase shift (which depends on the initial conditions of the system).

One can also consider an acoustic disturbance traveling in two dimensions; let us first consider a thin,

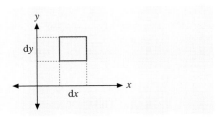

Fig. 6.9 Definitions necessary for the representation of a two-dimensional wave

stretched membrane as seen in Fig. 6.9. Let σ_s be the areal surface density (with units of kg/m^2), and let Γ be the tension per unit length. The transverse displacement of an infinitesimally small area of the membrane dS will now be a function of the spatial coordinates x and y along the two dimensions and the time t. Let us define the infinitesimal area d$S = $ dx dy and the displacement of dS as the acoustic disturbance passes through it to be $u(x, y, t)$. Newton's second law can now be applied to our areal element dS

$$\left[(\Gamma \partial u/\partial x)_{x+\mathrm{d}x, y} - (\Gamma \partial u/\partial y)_{x, y}\right] \mathrm{d}y$$
$$+ \left[(\Gamma \partial u/\partial y)_{x, y+\mathrm{d}y} - (\Gamma \partial u/\partial x)_{x, y}\right] \mathrm{d}x$$
$$= \Gamma \left(\frac{\partial^2 u}{\partial x^2} + \frac{\partial^2 u}{\partial y^2}\right) \mathrm{d}x \, \mathrm{d}y \qquad (6.45)$$

and

$$\Gamma \left(\frac{\partial^2 u}{\partial x^2} + \frac{\partial^2 u}{\partial y^2}\right) \mathrm{d}x \, \mathrm{d}y = \sigma_s \frac{\partial^2 u}{\partial t^2} \mathrm{d}x \, \mathrm{d}y \qquad (6.46)$$

or

$$\nabla^2 u = \left(\frac{\partial^2 u}{\partial x^2} + \frac{\partial^2 u}{\partial y^2}\right) = \frac{1}{c^2} \frac{\partial^2 u}{\partial t^2}, \qquad (6.47)$$

where $c = \sqrt{\Gamma/\sigma_s}$ is the speed of the acoustic wave in the membrane. Equation (6.47) now describes a two-dimensional wave propagating along a membrane.

The extension of this idea to three dimensions is fairly straightforward. As an acoustic wave passes through a three-dimensional medium, it can cause pressure fluctuations (leading to a volume change) in all three dimensions. The displacement u is now a function four variables $u = u(x, y, z, t)$. The volume change of a cube is given by

$$\Delta V = \Delta x \Delta y \Delta z \left(1 + \frac{\partial u}{\partial x}\right)\left(1 + \frac{\partial u}{\partial y}\right)\left(1 + \frac{\partial u}{\partial z}\right), \qquad (6.48)$$

$$\Delta V \approx \Delta x \Delta y \Delta z \left(1 + \frac{\partial u}{\partial x} + \frac{\partial u}{\partial y} + \frac{\partial u}{\partial z}\right), \qquad (6.49)$$

where the higher-order cross terms are negligibly small and have been dropped. This can be rewritten as

$$\Delta V = \Delta x \Delta y \Delta z \left(1 + \nabla \cdot \boldsymbol{u}\right). \qquad (6.50)$$

If the mass is held constant, and only the volume changes, the density change is given by

$$\rho_0 + \rho_1 = \frac{\rho_0}{1 + \nabla \cdot \boldsymbol{u}}. \qquad (6.51)$$

But since the denominator is very close to unity, we may rewrite this using the binomial expansion. Solving for ρ_1 we have

$$\rho_1 \approx -\rho_0 \nabla \cdot \boldsymbol{u}. \qquad (6.52)$$

Again we can consider Newton's second law; without loss of generality, consider the force exerted along the x-direction from the pressure due to the acoustic wave, which is given by

$$F_x = [p(x, t) - p(x + \Delta x, t)] \Delta y \Delta z, \qquad (6.53)$$

where $\Delta y \Delta z$ is the infinitesimal area upon which the pressure exerts a force. We can rewrite this as

$$F_x = -\Delta y \Delta z \left[\left(\frac{\partial p}{\partial x}\right) \Delta x\right], \qquad (6.54)$$

where the results for F_y and F_z are similar. Combining these, we can express the total force vector as

$$\boldsymbol{F} = -\nabla p \Delta x \Delta y \Delta z. \qquad (6.55)$$

Using the fact that the mass can be expressed in terms of the density, we can rewrite this as

$$-\nabla p = \rho_0 \frac{\partial^2 \boldsymbol{u}}{\partial t^2}. \qquad (6.56)$$

As in the case with fluids, we need a constitutive relationship to finalize the expression. For most situations in solids, the adiabatic conditions apply and the pressure fluctuation is a function of the density only. Making use of a Taylor expansion, we have

$$p \approx p(\rho_0) + (\rho - \rho_0) \frac{\mathrm{d}p}{\mathrm{d}\rho} \qquad (6.57)$$

but since $p = p_0 + p_1$, we can note that

$$p_1 = \rho_1 \frac{\partial p}{\partial \rho}. \qquad (6.58)$$

Using (6.52), we can eliminate ρ_1 from (6.58) to yield

$$p_1 = -\rho_0 \left(\frac{\mathrm{d}p}{\mathrm{d}\rho}\right) \nabla \cdot \boldsymbol{u}. \qquad (6.59)$$

We can eliminate the divergence of the displacement vector from this equation by taking the divergence of (6.56) and the second time derivative of (6.59). This gives two expressions that both equal $-\rho_0 \partial^2 (\nabla \cdot \boldsymbol{u})/\mathrm{d}t^2$ and thus are equal to each other. From this we can determine the full form of our wave equation

$$\frac{\partial^2 p_1}{\partial t^2} = c^2 \nabla^2 p_1, \qquad (6.60)$$

where c is again the wave speed and is given by

$$c^2 = \frac{dp}{d\rho}. \tag{6.61}$$

It should be noted that the above considerations were for an isotropic solid. In a crystalline medium, other complications can arise. These will be noted in a subsequent section.

6.1.5 Attenuation

In a real physical system, there are mechanisms by which energy is dissipated. In a gas, the dissipation comes from thermal and viscous effects. In a solid, dissipation comes from interactions with dislocations in the solid (holes, displaced atoms, interstitial atoms of other materials, etc.) and grain boundaries between adjacent parts of the solid, In practice, loss of energy over one acoustic cycle is negligible. However, as sound travels over a longer path, one expects these energy losses to cause a significant decrease in amplitude. In some situations, these dissipation effects must be accounted for in the solution of the wave equation.

The solution of the wave equation can be written in exponential form,

$$A = A_o e^{i(k'x - \omega t)}. \tag{6.62}$$

If one wishes to account for dissipation effects, one can assume that the wave number k' has an imaginary component, i.e.

$$k' = k + i\alpha, \tag{6.63}$$

where k and α are both real and $i = \sqrt{-1}$. Using this value of k' for the new wave number we have

$$A = A_o e^{i(k'x - \omega t)} = A_o e^{i[(k + i\alpha)x - \omega t]}. \tag{6.64}$$

Simplifying, one has

$$A = A_o e^{i(kx - \omega t) + i^2 \alpha x} = A_o e^{i(kx - \omega t) - \alpha x}$$
$$= A_o e^{-\alpha x} e^{i(kx - \omega t)}, \tag{6.65}$$

where α is known as the absorption coefficient. The resulting equation is modulated by a decreasing exponential function; i.e., an undriven acoustic wave passing through a lossy medium is damped to zero amplitude as $x \to \infty$.

6.2 Applications of Physical Acoustics

There are several interesting phenomena associated with the application of physical acoustics. The first is the wave velocity itself. Table 6.1 shows the wave velocity of sound in various fluids (both gases and liquids). Table 6.2 shows the wave velocity of sound in various solids (both metals and nonmetals). The velocity increases as one goes from gases to liquids to solids. The velocity variation from gases to liquids comes from the fact that gas molecules must travel farther before striking another gas molecule. In a liquid, molecules are closer together, which means that sound travels faster.

The change from liquids to solids is associated with increase of binding strength as one goes from liquid to solid. The rigidity of a solid leads to higher sound velocity than is found in liquids.

6.2.1 Crystalline Elastic Constants

Another application of physical acoustics involves the measurement of the crystalline elastic constants in a lattice. In Sect. 6.1.2, we considered the propagation of an acoustic field along a one-dimensional solid (in which

Table 6.1 Typical values of the sound velocity in fluids (25 °C)

Gas	Velocity (m/s)	Liquid	Velocity (m/s)
Air	331	Carbon tetrachloride (CCl_4)	929
Carbon dioxide (CO_2)	259	Ethanol (C_2H_6O)	1207
Hydrogen (H_2)	1284	Ethylene glycol ($C_2H_6O_2$)	1658
Methane (CH_4)	430	Glycerol ($C_3H_8O_3$)	1904
Oxygen (O_2)	316	Mercury (Hg)	1450
Sulfur dioxide (SO_2)	213	Water (distilled)	1498
Helium (H_2)	1016	Water (sea)	1531

Table 6.2 Typical values of the sound velocity in solids (25 °C)

Metals	Longitudinal velocity (m/s)	Shear (transverse) velocity (m/s)
Aluminum (rolled)	6420	3040
Beryllium	12890	8880
Brass (0.70 Cu, 0.30 Zn)	4700	2110
Copper (rolled)	5010	2270
Iron	5960	3240
Tin (rolled)	3320	1670
Zinc (rolled)	4210	2440
Lead (rolled)	1960	610
Nonmetals	**Longitudinal velocity (m/s)**	**Shear (transverse) velocity (m/s)**
Fused silica	5968	3764
Glass (Pyrex)	5640	3280
Lucite	2680	1100
Rubber (gum)	1550	–
Nylon	2620	1070

the internal structure of the solid played no role in the propagation of the wave). Real solids exist in three dimensions, and the acoustic field propagation depends on the internal structure of the material. The nature of the forces between the atoms (or molecules) that make up the lattice cause the speed of sound to be different along different directions of propagation (since the elastic force constants are different along the different directions). The measurement of crystal elastic constants depends on the ability to make an accurate determination of the wave velocity in different directions in a crystalline lattice. This can be done by cutting (or lapping) crystals in such a manner that parallel faces are in the directions to be measured.

For an isotropic solid one can determine the compressional modulus and the shear modulus from a single sample since both compressional and shear waves can be excited. For cubic crystals at least two orientations are required since there are three elastic constants (and still only two waves to be excited). For other crystalline symmetries a greater number of measurements is required.

6.2.2 Resonant Ultrasound Spectroscopy (RUS)

Recently a new technique for measuring crystalline elastic constants, known as resonant ultrasound spectroscopy (RUS), has been developed [6.4]. Typically, one uses a very small sample with a shape that has known acoustic resonant modes (usually a small parallelepiped, though sometimes other geometries such as cylinders are used). The sample is placed so that the driving transducer makes a minimal contact with the surface of the sample [the boundaries of the sample must be pressure-free and (shearing) traction-free for the technique to work]. Figure 6.10 shows a photograph of a small parallelepiped sample mounted in an RUS apparatus.

After the sample is mounted, the transducer is swept through a range of frequencies (usually from a few hertz to a few kilohertz) and the response of the material is

Fig. 6.10 Sample mounted for an RUS measurement. Sample dimensions are 2.0 mm × 2.5 mm × 3.0 mm

measured. Some resonances are caused by the geometry of the sample (just as a string of fixed length has certain resonant frequencies determined by the length of the string). In the RUS technique, some fairly sophisticated software eliminates the geometrical resonances; the remaining resonances are resonances of the internal lattice structure of the material. These resonances are determined by the elastic constants. The RUS technique, then, is used to evaluate all elastic constants from a single sample from the spectrum of resonant frequencies produced by the various internal resonances.

Measurement of Attenuation with the RUS Technique

The RUS technique is also useful for measuring the attenuation coefficients of solid materials. The resonance curves generated by the RUS experiment are plots of the response amplitude of the solid as a function of input frequency (for a constant input amplitude). Every resonance curve has a parameter known as the Q of the system. The Q value can be related to the maximum amplitude $1/e$ value, which in turn can be related to the attenuation coefficient. Thus, the resonance curves generated by the RUS experiment can be used to determine the attenuation in the material at various frequencies.

6.2.3 Measurement Of Attenuation (Classical Approach)

Measurement of attenuation at audible frequencies and below is very difficult. Attenuation is usually measured at ultrasonic frequencies since the plane-wave approximation can be satisfied. The traditional means of measuring the acoustic attenuation requires the measurement of the echo train of an acoustic tone burst as it travels through the medium. Sound travels down the length of the sample, reflects from the opposite boundary and returns to its origin. During each round trip, it travels a distance twice the length of the sample. The transducer used to emit the sound now acts as a receiver and measures the amplitude as it strikes the initial surface. The sound then continues to reflect back and forth through the sample. On each subsequent round trip, the sound amplitude is diminished.

Measured amplitude values are then fit to an exponential curve, and the value of the absorption coefficient is determined from this fit. Actually, this experimental arrangement measures the insertion loss of the system, the losses associated with the transducer and the adhesive used to bond the transducer to the sample as well as the attenuation of sound in the sample. However, the values of the insertion loss of the system and the attenuation inside the sample are usually very close to each other. If one needs the true attenuation in the sample, one can

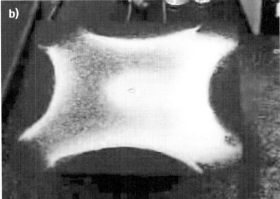

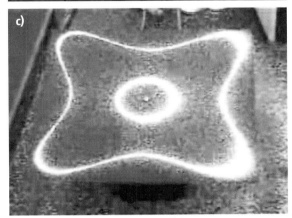

Fig. 6.11 (a) Fine rice on a plate; (b) as the plate is excited acoustically the rice begins to migrate to the nodes; (c) the Chladni pattern has formed

use various combinations of transducers to account for the other losses in the system [6.5].

Losses in a sample can come from a number of sources: viscosity, thermal conductivity, and molecular relaxation. Viscosity and thermal conductivity are usually referred to as classical losses. They can be calculated readily. Both are linearly dependent on frequency. Relaxation is a frequency-dependent phenomenon. The maximum value occurs when the sound frequency is the same as the relaxation frequency, which is determined by the characteristics of the medium. Because of this complication, determination of the attenuation to be expected can be difficult.

6.2.4 Acoustic Levitation

Acoustic levitation involves the use of acoustic vibrations to move objects from one place to the other, or to keep them fixed in space. Chladni produced an early form of acoustic levitation to locate the nodal planes in a vibrating plate. Chladni discovered that small particles on a plate were moved to the nodal planes by plate vibrations. An example of Chladni figures is shown in Fig. 6.11. Plate vibrations have caused powder to migrate toward nodal planes, making them much more obvious.

The use of radiation pressure to counterbalance gravity (or buoyancy) has recently led to a number of situations in which levitation is in evidence.

1. The force exerted on a small object by radiation pressure can be used to counterbalance the pull of gravity [6.6].
2. The radiation force exerted on a bubble in water by a stationary ultrasonic wave has been used to counteract the hydrostatic or buoyancy force on the bubble. This balance of forces makes it possible for the bubble to remain at the same point indefinitely. Single-bubble sonoluminescence studies are now possible [6.7].
3. Latex particles having a diameter of 270 μm or clusters of frog eggs can be trapped in a potential well generated by oppositely directed focused ultrasonic beams. This makes it possible to move the trapped objects at will. Such a system has been called *acoustic tweezers* by *Wu* [6.8].

6.2.5 Sonoluminescence

Sonoluminescence is the conversion of high-intensity acoustic energy into light. Sonoluminescence was first discovered in water in the early 1930s [6.9, 10]. However, interest in the phenomenon languished for several decades.

In the late seventies, a new type of sonoluminescence was found to occur in solids [6.11, 12]. This can occur when high-intensity Lamb waves are generated along a thin plate of ferroelectric material which is driven at the frequency of mechanical resonance in a partial vacuum (the phenomenon occurs at about 0.1 atm). The acoustic fields interact with dislocations and defects in the solid which leads to the generation of visible light in ferroelectric materials. Figure 6.12 shows a diagram of the experimental setup and a photograph of the light emitted during the excitation of solid-state sonoluminescence.

In the early 1990s, sonoluminescence emissions from the oscillations of a single bubble in water were discovered [6.7]. With single-bubble sonoluminescence, a single bubble is placed in a container of degassed water (often injected by a syringe). Sound is used to push the bubble to the center of the container and to set the bubble into high-amplitude oscillation. The dynamic range for the bubble radius through an oscillation can be as great

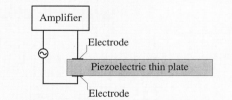

a) Lamb waves generated by the mechanical response of the plate to the driving voltage at the electrodes

Fig. 6.12 (a) Block diagram of apparatus for solid-state sonoluminescence and **(b)** photograph of light emitted from a small, thin plate of LiNbO$_3$

as 50 μm to 5 μm through one oscillation [6.13–15]. The light is emitted as the bubble goes through its minimum radius. Typically, one requires high-amplitude sound corresponding to a sound pressure level in excess of 110 dB. The frequency of sound needed to drive the bubble into sonoluminescence is in excess of 30 kHz, which is just beyond the range of human hearing.

Another peculiar feature of the sonoluminescence phenomenon is the regularity of the light emissions. Rather than shining continuously, the light is emitted as a series of extremely regular periodic flashes. This was not realized in the initial sonoluminescence experiments, because the rate at which the flashes appear requires a frequency resolution available only in the best detectors. The duration of the pulses is less than 50 ps. The interval between pulses is roughly 35 μs. The time between flashes varies by less than 40 ps.

The discovery of single-bubble sonoluminescence has caused a resurgence of interest in the phenomenon. The light emitted appears to be centered in the near-ultraviolet and is apparently black body in nature (unfortunately water absorbs much of the higher-frequency light, so a complete characterization of the spectrum is difficult to achieve). Adiabatic compression of the bubble through its oscillation would suggest temperatures of about 10 000 K (with pressures of about 10 000 atm). The temperatures corresponding to the observed spectra are in excess of 70 000 K [6.13]. In fact, they may even be much higher.

Since the measured spectrum suggests that the actual temperatures and pressures within the bubble may be quite high, simple compression does not seem to be an adequate model for the phenomenon. It is possible that the collapsing bubble induces a spherically symmetric shockwave that is driven inward towards the center of the bubble. These shocks could possibly drive the temperatures and pressures in the interior of the bubble high enough to generate the light. (Indeed, some physicists have suggested that sonoluminescence might enable the ignition of fusion reactions, though as of this writing that remains speculation).

6.2.6 Thermoacoustic Engines (Refrigerators and Prime Movers)

Another interesting application of physical acoustics is that of thermoacoustics. Thermoacoustics involves the conversion of acoustic energy into thermal energy or the reverse process of converting thermal energy into sound [6.16]. Figure 6.13 shows a photograph of a thermoacoustic engine. To understand the processes

Fig. 6.13 A thermoacoustic engine

involved in thermoacoustics, let us consider a small packet of gas in a tube which has a sound wave traveling through it (thermoacoustic effects can occur with either standing or progressive waves). As the compression of the wave passes through the region containing the packet of gas, three effects occur:

1. The gas compresses adiabatically, its temperature increases in accordance with Boyle's law (due to the compression), and the packet is displaced some distance down the tube.
2. As the rarefaction phase of the wave passes through the gas, this process reverses.
3. The wall of the tube acts as a heat reservoir. As the packet of gas goes through the acoustic process, it deposits heat at the wall during the compression phase (the wall literally conducts the heat away from the compressed packet of gas).

This process is happening down the entire length of the tube; thus a temperature gradient is established down the tube.

To create a useful thermoacoustic device, one must increase the surface area of wall that the gas is in contact with so that more heat is deposited down the tube. This is accomplished by inserting a *stack* into the tube. In the inside of the tube there are several equally spaced plates which are also in contact with the exterior walls of the tube. Each plate provides additional surface area for the deposition of thermal energy and increases the overall thermoacoustic effect. (The stack must not impede the wave traveling down the tube, or the thermoacoustic effect is minimized.) Modern stacks use much more complicated geometries to improve the efficiency of the thermoacoustic device (indeed the term stack now is a misnomer, but the principle remains the same). Figure 6.14 shows a photograph of a stack used in a modern thermoacoustic engine. The stack shown in Fig. 6.14 is made of a ceramic material.

Fig. 6.14 A stack used in a thermoacoustic engine. Pores allow the gas in the tube to move under the influence of the acoustic wave while increasing the surface area for the gas to deposit heat

A thermoacoustic device used in this way is a thermoacoustic refrigerator. The sound generated down the tube (by a speaker or some other device) is literally pumping heat energy from one end of the tube to the other. Thus, one end of the tube gets hot and the other cools down (the cool end is used for refrigeration).

The reverse of this thermoacoustic refrigeration process can also occur. In this case, the tube ends are fitted with heat exchangers (one exchanger is hot, one is cold). The heat delivered to the tube by the heat exchangers does work on the system and generates sound in the tube. The frequency and amplitudes of the generated waves depend on the geometry of the tube and the stacks. When a thermoacoustic engine is driven in this way (converting heat into sound), it is known as a prime mover.

Though much research to improve the efficiency of thermoacoustic engines is ongoing, the currently obtainable efficiencies are quite low compared to standard refrigeration systems. Thermoacoustic engines, however, offer several advantages: they have no moving parts to wear out, they are inexpensive to manufacture, and they are highly reliable, which is useful if refrigeration is needed in inaccessible places.

One practical application for thermoacoustic engines is often cited: the liquefaction of natural gas for transport. This is accomplished by having two thermoacoustic engines, one working as a prime mover and the other working as a refrigerator. A small portion of the gas is ignited and burned to produce heat. The heat is applied to the prime mover to generate sound. The sound is directed into a second thermoacoustic engine that acts as a refrigerator. The sound from the prime mover pumps enough heat down the refrigerator to cool the gas enough to liquefy it for storage. The topic of thermoacoustics is discussed in greater detail in Chap. 7.

6.2.7 Acoustic Detection of Land Mines

Often, during wars and other armed conflicts, mine fields are set up and later abandoned. Even today it is not uncommon for mines originally planted during World War II to be discovered still buried and active. According to the humanitarian organization CARE, 70 people are killed each day by land mines, with the vast majority being civilians. Since most antipersonnel mines manufactured today contain no metal parts, electromagnetic-field-based metal detectors cannot locate them for removal. Acoustic detection of land mines offers a potential solution for this problem.

The approach used in acoustic land-mine detection is conceptually simple but technically challenging. Among the first efforts made at acoustic land-mine detection were those of *House* and *Pape* [6.17]. They sent sounds into the ground and examined the reflections from buried objects. *Don* and *Rogers* [6.18] and *Caulfield* [6.19] improved the technique by including a reference beam that provided information for comparison with the reflected signals. Unfortunately, this technique yielded too many false positives to be practical because one could not distinguish between a mine and some other buried object such as a rock or the root of a tree.

Newer techniques involving the coupling of an acoustic signal with a seismic vibration have been developed with much more success [6.20–24]. They make use of remote sensing and analysis by computer. Remote measurement of the acoustic field is done with a laser Doppler vibrometer (LDV), which is a an optical device used to measure velocities and displacements of vibrating bodies without physical contact. With this technique, the ground is excited acousticly with lower frequencies (usually on the order of a few hundred Hz). The LDV is used to measure the vibration of the soil as it responds to this driving force. If an object is buried under the soil in the region of excitation, it alters the resonance characteristics of the soil and introduces nonlinear effects. With appropriate digital signal processing and analysis, one can develop a system capable of recognizing different types of structures buried beneath the soil. This reduces the number of false positives and makes the system more efficient.

6.2.8 Medical Ultrasonography

For the general public, one of the most familiar applications of physical acoustics is that of medical ultrasonography. Medical ultrasonography is a medical diagnostic technique which can use sound information to construct images for the visualization of the size, structure and lesions of internal organs and other bodily tissues. These images can be used for both diagnostic and treatment purposes (for example enabling a surgeon to visualize an area with a tumor during a biopsy). The most familiar application of this technique is obstetric ultrasonography, which uses the technique to image and monitor the fetus during a pregnancy. An ultrasonograph of a fetus is shown in Fig. 6.15.

This technique relies on the fact that in different materials the speed of sound and acoustic impedance are different. A collimated beam of high-frequency sound is projected into the body of the person being examined. The frequencies chosen will depend on the application. For example, if the tissue is deeper within the body, the sound must travel over a longer path and attenuation affects can present difficulties. Using a lower ultrasonic frequency reduces attenuation. Alternatively, if a higher resolution is needed, a higher frequency is used. In each position where the density of the tissue changes, there is an acoustic impedance mismatch. Therefore, at each interface between various types of tissues some of the sound is reflected. By measuring the time between echoes, one can determine the spatial position of the various tissues.

If a single, stationary transducer is used; one gets spatial information that lies along a straight line. Typically, the probe contains a phased array of transducers that are used to generate image information from different directions around the area of interest. The different transducers in the probe send out acoustic pulses that are reflected from the various tissues. As the acoustic signals return, the transducers receive them and convert the information into a digital, pictorial representation. One must also match the impedances between the surface of the probe and the body. The head of the probe is usually soft rubber, and the contact between the probe and the body is impedance-matched by a water-based gel.

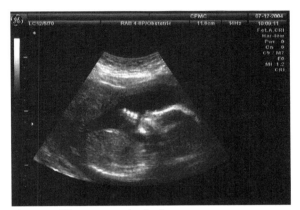

Fig. 6.15 Ultrasonograph of a fetus

To construct the image, each transducer (which themselves are separated spatially somewhat) measures the strength of the echo. This measurement indicates how much sound has been lost due to attenuation (different tissues have different attenuation values). Additionally, each transducer measures the delay time between echoes, which indicates the distance the sound has traveled before encountering the interface causing the echo (actually since the sound has made a round trip during the echo the actual displacement between the tissues is $\frac{1}{2}$ of the total distance traveled). With this information a two-dimensional image can be created. In some versions of the technique, computers can be used to generate a three-dimensional image from the information as well.

A more esoteric form of ultrasonograpy, Doppler ultrasonography, is also used. This technique requires separate arrays of transducers, one for broadcasting a continuous-wave acoustic signal and another for receiving it. By measuring a frequency shift caused by the Doppler effect, the probe can detect structures moving towards or away from the probe. For example, as a given volume of blood passes through the heart or some other organ, its velocity and direction can be determined and visualized. More-recent versions of this technique make use of pulses rather than continuous waves and can therefore use a single probe for both broadcast and reception of the signal. This version of the technique requires more-advanced analysis to determine the frequency shift. This technique presents advantages because the timing of the pulses and their echoes can be measured to provide distance information as well.

6.3 Apparatus

Given the diverse nature of physical acoustics, any two laboratories conducting physical acoustics experiments might have considerably different types of equipment and apparatus (for example, experiments dealing with acoustic phenomena in water usually require tanks to hold the water). As is the case in any physics laboratory, it is highly useful to have access to a functioning machine shop and electronics shop for the design, construction, and repair of equipment needed for the physical acoustics experiment that is being conducted. Additionally, a wide range of commercial equipment is available for purchase and use in the lab setting.

6.3.1 Examples of Apparatus

Some typical equipment in an acoustic physics lab might include some of the following:

- Loudspeakers;
- Transducers (microphones/hydrophones);
- Acoustic absorbers;
- Function generators, for generating a variety of acoustic signals including single-frequency sinusoids, swept-frequency sinusoids, pulses, tone bursts, white or pink noise, etc.;
- Electronics equipment such as multimeters, impedance-matching networks, signal-gating equipment, etc.;
- Amplifiers (acoustic, broadband, Intermediate Frequency (IF), lock-in);
- Oscilloscopes. Today's digital oscilloscopes (including virtual oscilloscopes implemented on personal computers) can perform many functions that previously required several different pieces of equipment; for example, fast Fourier transform (FFT) analysis previously required a separate spectrum analyzer; waveform averaging previously required a separate boxcar integrator, etc.);
- Computers (both for control of apparatus and analysis of data).

For audible acoustic waves in air the frequency range is typically 20–20 000 Hz. This corresponds to a wavelength range of 16.5 m–16.5 mm. Since these wavelengths often present difficulties in the laboratory, and since the physical principles apply at all frequencies, the laboratory apparatus is often adapted to a higher frequency range. The propagating medium is often water since a convenient ultrasonic range of 1–100 MHz gives a convenient wavelength range of 0.014–1.4 mm. The experimental arrangements to be described below are for some specialized applications and cover this wavelength range, or somewhat lower. The results, however, are useful in the audio range as well.

6.3.2 Piezoelectricity and Transduction

In physical acoustics transducers play an important role. A transducer is a device that can convert a mechanical vibration into a current or vice versa. Transducers can be used to generate sound or to detect sound. Audible frequencies can be produced by loudspeakers and received by microphones, which often are driven electromagnetically. (For example, a magnet interacting with a current-carrying coil experiences a magnetic force that can accelerate it, or if the magnet is moved it can induce a corresponding current in the coil.) The magnet could be used to drive the cone of a loudspeaker to convert a current into an audible tone (or speech, or music, etc.). Similarly, one can use the fact that the capacitance between two parallel-plate capacitors varies as a function of the separation distance between two plates. A capacitor with one plate fixed and the other allowed to vibrate in response to some form of mechanical forcing (say the force caused by the pressure of an acoustic wave) is a transducer. With a voltage across the two plates, an electrical current is generated as the plates move with respect to one another under the force of the vibration of an acoustic wave impinging upon it. These types of transducers are described in Chap. 24.

The most common type of transducers used in the laboratory for higher-frequency work are piezoelectric-element-based transducers. In order to understand how these transducers work (and are used) we must first examine the phenomenon of piezoelectricity.

Piezoelectricity is characterized by the both direct piezoelectric effect, in which a mechanical stress applied to the material causes a potential difference across the surface of the faces to which the stress is applied, and the secondary piezoelectric effect, in which a potential difference applied across the faces of the material causes a deformation (expansion or contraction). The deformation caused by the direct piezoelectric effect is on the order of nanometers, but leads to many uses in acoustics such as the production and detection of sound, microbalance applications (where very small masses are measured by determining the change in the resonance frequency of a piezoelectric crystal when it is loaded with the small mass), and frequency generation/control for oscillators.

Piezoelectricity arises in a crystal when the crystal's unit cell lacks a center of symmetry. In a piezoelectric crystal, the positive and negative charges are separated by distance. This causes the formation of electric dipoles, even though the crystal is electrically neutral overall. The dipoles near one another tend to orient themselves along the same direction. These small regions of aligned dipoles are known as domains because of the similarity to the magnetic analog. Usually, these domains are oriented randomly, but can be aligned by the application of a strong electric field in a process known as poling (typically the sample is poled at a high temperature and cooled while the electric field is maintained). Of the 32 different crystal classifications, 20 exhibit piezoelectric properties and 10 of those are polar (i. e. spontaneously polarized). If the dipole can be reversed by an applied external electric field, the material is additionally known as a ferroelectric (in analogy to ferromagnetism).

When a piezoelectric material undergoes a deformation induced by an external mechanical stress, the symmetry of the charge distribution in the domains is disturbed. This gives rise to a potential difference across the surfaces of the crystal. A 2 kN force ($\approx$ 500 lbs) applied across a 1 cm cube of quartz can generate up to 12 500 V of potential difference.

Several crystals are known to exhibit piezoelectric properties, including tourmaline, topaz, rochelle salt, and quartz (which is most commonly used in acoustic applications). In addition, some ceramic materials with perovskite or tungsten bronze structures, including barium titanate ($BaTiO_3$), lithium niobate ($LiNbO_3$), PZT [$Pb(ZrTi)O_3$], $Ba_2NaNb_5O_5$, and $Pb_2KNb_5O_{15}$, also exhibit piezoelectric properties. Historically, quartz was the first piezoelectric material widely used for acoustics applications. Quartz has a very sharp frequency response, and some unfavorable electrical properties such as a very high electrical impedance which requires impedance matching for the acoustic experiment. Many of the ceramic materials such as PZT or lithium niobate have a much broader frequency response and a relatively low impedance which usually does not require impedance matching. For these reasons, the use of quartz has largely been supplanted in the acoustics lab by transducers fashioned from these other materials. Although in some situations (if a very sharp frequency response is desired), quartz is still the best choice.

In addition to these materials, some polymer materials behave as electrets (materials possessing a quasi-permanent electric dipole polarization). In most piezoelectric crystals, the orientation of the polarization is limited by the symmetry of the crystal. However, in an electret this is not the case. The electret material polyvinylidene fluoride (PVDF) exhibits piezoelectricity several times that of quartz. It can also be fashioned more easily into larger shapes.

Since these materials can be used to convert a sinusoidal electrical current into a corresponding sinusoidal mechanical vibration as well as convert a mechanical vibration into a corresponding electrical current, they provide a connection between the electrical system and the acoustic system. In physical acoustics their ability to produce a single frequency is especially important. The transducers to be described here are those which are usually used to study sound propaga-

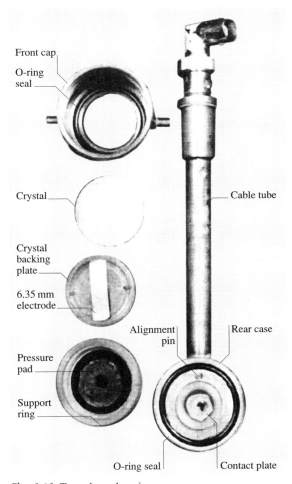

Fig. 6.16 Transducer housing

tion in liquids or in solids. The frequencies used are ultrasonic.

The first transducers were made from single crystals. Single crystals of x-cut quartz were preferred for longitudinal waves, and y-cut for transverse waves, their thickness determined by the frequency desired. Single crystals are still used for certain applications; however, the high-impedance problems were solved by introduction of polarized ceramics containing barium titanate or strontium titanate. These transducers have an electrical impedance which matches the 50 Ω impedance found on most electrical apparatus. Low-impedance transducers are currently commercially available.

Such transducers can be used to generate or receive sound. When used as a receiver in liquids such transducers are called hydrophones. For the generation of ultrasonic waves in liquids, one surface of the transducer material should be in contact with the liquid and the other surface in contact with air (or other gas). With this arrangement, most of the acoustic energy enters the liquid. In the laboratory it is preferable to have one surface at the ground potential and electrically drive the other surface, which is insulated. Commercial transducers which accomplish these objectives are available. They are designed for operation at a specific frequency. A transducer housing is shown in Fig. 6.16. The transducer crystal and the (insulated) support ring can be changed to accommodate the frequency desired. In the figure a strip back electrode is shown. For generating acoustic vibration over the entire surface, the back of the transducer can be coated with an electrode. The width which produces a Gaussian output in one dimension is described in [6.25]. The use of a circular electrode that can produce a Gaussian amplitude distribution in two dimensions is described in [6.26].

For experiments with solids the transducer is often bonded directly to the solid without the need for an external housing. For single-transducer pulse-echo operation the opposite end of the sample must be flat because it must reflect the sound. For two-transducer operation one transducer is the acoustic transmitter, the other the receiver.

With solids it is convenient to use coaxial transducers because both transducer surfaces must be electrically accessible. The grounded surface in a coaxial transducer can be made to wrap around the edges so it can be accessed from the top as well. An example is shown in Fig. 6.17. The center conductor is the high-voltage terminal. The outer conductor is grounded; it is in electrical contact with the conductor that covers the other side of the transducer.

6.3.3 Schlieren Imaging

Often it is useful to be able to see the existence of an acoustic field. The Schlieren arrangement shown in Fig. 6.18 facilitates the visualization of sound fields in water. Light from a laser is brought to focus on a circular aperture by lens 1. The circular aperture is located at a focus of lens 2, so that the light emerges parallel. The water tank is located in this parallel light. Lens 3 forms a real image of the contents of the water tank on the screen, which is a photographic negative for photographs. By using a wire or an ink spot on an optical flat at the focus of lens 3, one produces conditions for dark-field illumination. The image on the screen, then, is an image of the ultrasonic wave propagating through the water. The fact that the ultrasonic wave is propagating through the water means that the individual wavefronts are not seen. To see the individual wavefronts, one must add a stroboscopic light source synchronized with the ultrasonic wave. In this way the light can be on at the frequency of the sound and produce a photograph which shows the individual wavefronts.

On some occasions it may be useful to obtain color images of the sound field; for example, one can show an incident beam in one color and its reflection from an interface in a second color for clarity. The resultant photographs can be beautiful; however, to a certain extent their beauty is controlled by the operator. The merit of color Schlieren photography may be more from its aesthetic or pedagogical value, rather than its practical application. This is the reason that color Schlieren

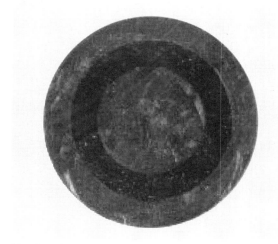

Fig. 6.17 Coaxial transducer

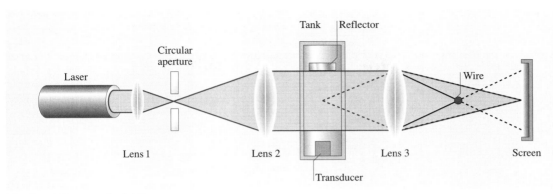

Fig. 6.18 Diagram of a Schlieren system

photographs are seldom encountered in a physical acoustic laboratory. For completeness, however, it may be worthwhile to describe the process.

The apparatus used is analogous to that given in Fig. 6.18. The difference is that a white-light source is

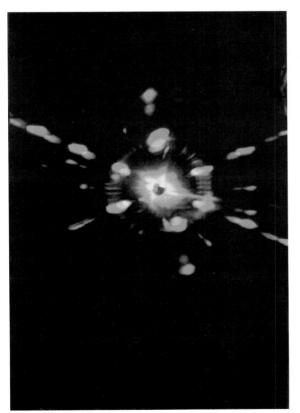

Fig. 6.19 Spectra formed by the diffraction of light through a sound field

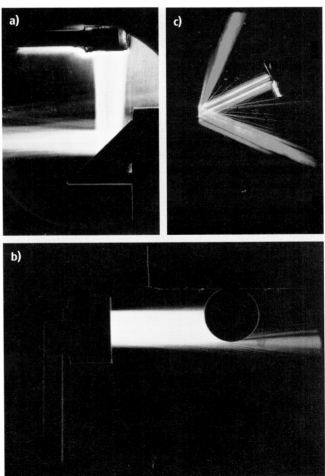

Fig. 6.20 Schlieren photographs showing reflection and diffraction of ultrasonic waves by a solid immersed in water

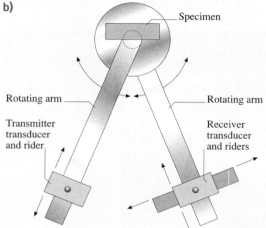

Fig. 6.21 (a) Photograph of a goniometer system; (b) diagram of a goniometer system

used in the place of the monochromatic light from the laser. As the light diffracts, the diffraction pattern formed at the focus of lens 3 is made up of complete spectra (with each color of the spectra containing a complete image of the acoustic field). A photograph showing the spectra produced is shown in Fig. 6.19. If the spectra are allowed to diverge beyond this point of focus, they will combine into a white-light image of the acoustic field. However, one can use a slit (rather than using a wire or an ink spot at this focus to produce dark-field illumination) to pass only the desired colors. The position of the slit selects the color of the light producing the image of the sound field. If one selects the blue-colored incident beam from one of the spectra, and the red-colored reflected beam from another spectrum (blocking out all the other colors), a dual-colored image results. Since each diffraction order contains enough information to produce a complete sound field image, the operator has control over its color. Spatial filtering, then, becomes a means of controlling the color of the various parts of the image. Three image examples are given in Fig. 6.20. Figure 6.20a is reflection of a sound beam from a surface. Figure 6.20b shows diffraction of sound around a cylinder. Figure 6.20c shows backward displacement at a periodic interface. The incident beam is shown in a different color from that of the reflected beam and the diffraction order.

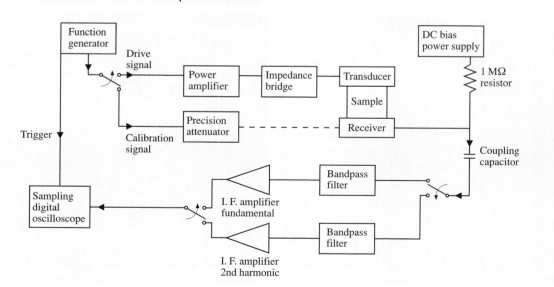

Fig. 6.22 Block Diagram of apparatus for absolute amplitude measurements

water–solid interfaces. By immersing the goniometer in other liquids, the type of liquid can also be changed.

6.3.5 Capacitive Receiver

In many experiments, one measures acoustic amplitude relative to some reference amplitude, which is usually determined by the parameters of the experiment. However, in some studies it is necessary to make a measurement of the absolute amplitude of acoustic vibration. This is especially true of the measurement of the nonlinearity of solids. For the measurement of the absolute acoustic amplitude in a solid, a capacitive system can be used. If the end of the sample is coated with a conductive material, it can act as one face of a parallel-plate capacitor. A bias voltage is put across that capacitance, which enables it to work as a capacitive microphone. As the acoustic wave causes the end of the sample to vibrate, the capacitor produces an electrical signal. One can relate the measured electrical amplitude to the acoustic amplitude because all quantities relating to them can be determined.

The parallel-plate approximation, which is very well satisfied for plate diameters much larger than the plate separation, is the only approximation necessary. The electrical apparatus necessary for absolute amplitude measurements in solids is shown in the block diagram of Fig. 6.22. A calibration signal is used in such a manner that the same oscilloscope can be used for the calibration and the measurements. The mounting system for room-temperature measurements of a sample is shown in Fig. 6.23. Since stray capacitance affects the impedance of the resistor, this impedance must be measured at the frequencies used. The voltage drop in the resistor can be measured with either the calibration signal or the signal from the capacitive receiver. A comparison of the two completes the calibration. With this system acoustic amplitudes as small as 10^{-14} m (which is approximately the limit set by thermal noise) have been measured in copper single crystals [6.27].

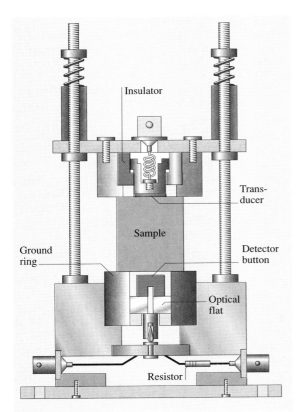

Fig. 6.23 Mounting system for room-temperature measurement of absolute amplitudes of ultrasonic waves in solids

6.3.4 Goniometer System

For studies of the propagation of an ultrasonic pulse in water one can mount the ultrasonic transducers on a goniometer such as that shown in Fig. 6.21. This is a modification of the pulse-echo system. The advantage of this system is that the arms of the goniometer allow for the adjustments indicated in Fig. 6.21b. By use of this goniometer it is possible to make detailed studies of the reflection of ultrasonic waves from a variety of

6.4 Surface Acoustic Waves

It has been discovered that surface acoustic waves are useful in industrial situations because they are relatively slow compared with bulk waves or electromagnetic waves. Many surface acoustic wave devices are made by coating a solid with an interdigitated conducting layer. In this case, the surface acoustic wave produces the desired delay time and depends for its generation on a fringing field (or the substrate may be piezoelectric). The inverse

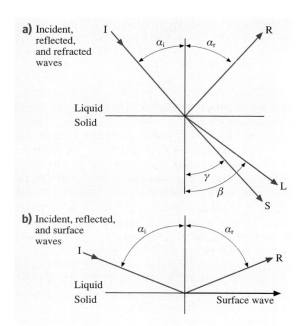

Fig. 6.24 Sound waves at a liquid–solid interface

process can be used to receive surface acoustic waves and convert them into an electrical signal, which can then be amplified.

Another type of surface wave is possibly more illustrative of the connection between surface acoustic waves and physical acoustics. This is the surface acoustic wave generated when the trace velocity of an incident wave is equal to the velocity of the surface acoustic wave. This occurs when a longitudinal wave is incident on a solid from a liquid. This is analogous to the optical case of total internal reflection [6.28], but new information comes from the acoustic investigation.

The interface between a liquid and a solid is shown in Fig. 6.24, in which the various waves and their angles are indicated. The directions in which the various waves propagate at a liquid–solid interface can be calculated from Snell's law, which for this situation can be written

$$\frac{\sin\theta_i}{v} = \frac{\sin\theta_r}{v} = \frac{\sin\theta_L}{v_L} = \frac{\sin\theta_S}{v_S}, \quad (6.66)$$

where the velocity of the longitudinal wave in the liquid, that in the solid, and the velocity of the shear wave in the solid are, respectively, v, v_L, and v_S. The propagation directions of the various waves are indicated in Fig. 6.24.

Since much of the theory has been developed in connection with geology, the theoretical development of *Ergin* [6.29] can be used directly. Ergin has shown that the energy reflected at an interface is proportional to the square of the amplitude reflection coefficient, which can be calculated directly [6.29]. The energy reflection coefficient is given by

$$R_E = \left(\frac{\cos\beta - A\cos\alpha\,(1-B)}{\cos\beta + A\cos\alpha\,(1-B)}\right)^2, \quad (6.67)$$

where

$$A = \frac{\rho_1 V_L}{\rho V} \quad \text{and}$$

$$B = 2\sin\gamma\sin 2\gamma\left(\cos\gamma - \frac{v_S}{v_L}\cos\beta\right). \quad (6.68)$$

The relationship among the angles α, β and γ can be determined from Snell's law as given in (6.66). The book by *Brekhovskikh* [6.30] is also a good source of information on this subject.

Typical plots of the energy reflection coefficient as a function of incident angle are given in Fig. 6.25 in

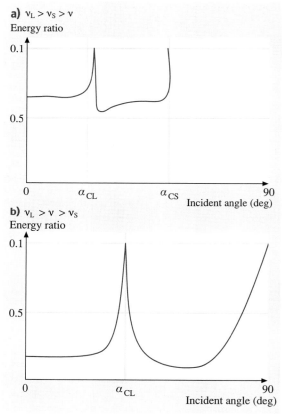

Fig. 6.25 Behavior of energy reflected at a liquid–solid interface

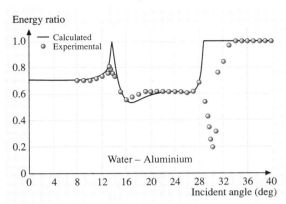

Fig. 6.26 A 2 MHz ultrasonic wave reflected at a liquid–solid interface

which the critical angles are indicated. Usually, $v_L > v_S > v$, so the curve in Fig. 6.25a is observed. It will be noticed immediately that there is a critical angle for both the longitudinal and transverse waves in the solid. In optics there is no longitudinal wave; therefore the curve has only one critical angle.

If one uses a pulse-echo system to verify the behavior of an ultrasonic pulse at an interface between a liquid and a solid, one gets results that can be graphed as shown in Fig. 6.26. At an angle somewhat greater than the critical angle for a transverse wave in the solid, one finds a dip in the data. This dip is associated with the generation of a surface wave. The surface wave is excited when the projection of the wavelength of the incident wave onto the interface matches the wavelength of the surface wave. The effect of the surface wave can be seen in the Schlieren photographs in Fig. 6.27.

Figure 6.27 shows the reflection at a water–aluminum interface at an angle less than that for excitation of a surface wave (a Rayleigh surface wave), at the angle at which a surface wave is excited, and an angle greater. When a surface wave is excited the reflected beam contains two (or more) components: the specular beam (reflected in a normal manner) and a beam displaced down the interface. Since most of the energy is contained in the displaced beam, the minimum in the data shown in Fig. 6.24 is caused by the excitation of the displaced beam by the surface wave. This has been shown to be the case by displacing the receiver to follow the displaced beam with a goniometer system, as shown in Fig. 6.21. This minimizes the dip in data shown in Fig. 6.24. *Neubauer* has shown that the ultrasonic beam excited by the surface wave is 180° out of phase with the specularly reflected

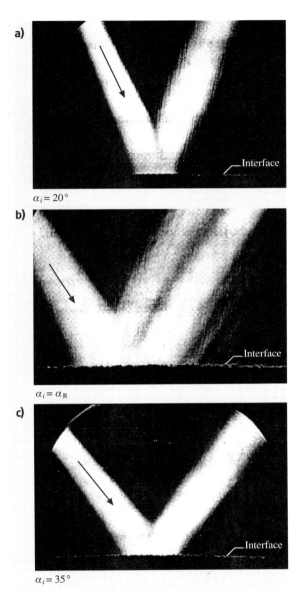

Fig. 6.27 Schlieren photographs showing the behavior of a 4 MHz ultrasonic beam reflected at a water–aluminium interface

beam [6.31]. Destructive interference resulting from phase cancelation causes these beams to be separated by a null strip. Although a water–aluminum interface has been used in these examples, the phenomenon occurs at all liquid–solid interfaces. It is less noticeable at higher ultrasonic frequencies since the wavelength is smaller.

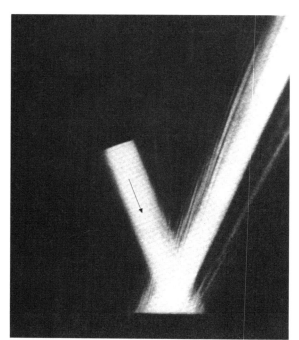

Fig. 6.28 Schlieren photograph showing backward displacement of a 6 MHz ultrasonic beam at a corrugated water–brass interface

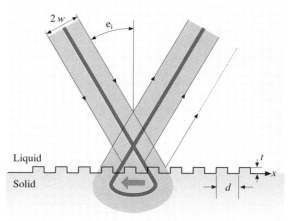

Fig. 6.29 Diagram of incident beam coupling to a backward-directed leaky wave to produce backward displacement of the reflected beam

At a corrugated interface it is possible that the incident beam couples to a negatively directed surface wave so that the reflected beam is displaced in the negative direction. This phenomenon was predicted for optical waves by *Tamir* and *Bertoni* [6.32]. They determined that the optimum angle for this to occur is given by

$$\sin\theta_i = V_{\text{liq}} \left(\frac{1}{fd} - \frac{1}{V_R} \right) \qquad (6.69)$$

where d is the period, f is the frequency, V_{liq} is the wave propagation velocity in the liquid, and V_R is the propagation velocity of the leaky surface wave. Figure 6.28 is a Schlieren photograph which shows what happens in this case [6.33]. Figure 6.29 is a diagram of the phenomenon [6.33]. The incident beam couples to a backward-directed surface wave to produce backward displacement of the reflected beam.

6.5 Nonlinear Acoustics

There are several sources of nonlinearity whether the propagating medium be a gas, liquid, or solid. They are described in more detail in Chap. 8. Even in an ideal medium in which one considers only interatomic forces, there is still a source of nonlinear behavior since compression requires a slightly different force from dilatation. With Hooke's law (strain is proportional to stress) one assumes that they are equal. This is seldom true. The subject of nonlinear acoustics has been developed to the point that it is now possible to go beyond the linear approximation with many substances.

6.5.1 Nonlinearity of Fluids

If one assumes an ideal gas and keeps the first set of nonlinear terms, *Beyer* has shown that the equation of motion in one dimension becomes [6.34]

$$\frac{\partial^2 \xi}{\partial t^2} = \frac{c_o^2}{\left(1 + \frac{\partial \xi}{\partial a}\right)^{\gamma+1}} \frac{\partial^2 \xi}{\partial a^2}. \qquad (6.70)$$

This form of the nondissipative wave equation in one dimension in Lagrangian coordinates includes nonlinear terms. In this equation γ is the ratio of specific heats and

$$c_o^2 = \frac{\gamma p_o}{\rho_o} \frac{1}{\left(1+\frac{\partial \xi}{\partial a}\right)^{\gamma-1}}.\qquad(6.71)$$

One can also generalize this equation in a form which applies to all fluids. By expanding the equation of state $p = p_o \left(\frac{\rho}{\rho_o}\right)^\gamma$ in powers of the condensation $s = \frac{\rho-\rho_o}{\rho_o}$, one obtains

$$p = p_o + As + \frac{B}{2!}s^2 + \ldots \quad \text{and}$$
$$c^2 = c_o^2 \left[1 + \left(\frac{B}{A}\right)s + \ldots\right].\qquad(6.72)$$

This makes it possible to obtain the nonlinear wave equation in the form [6.24]:

$$\frac{\partial^2 \xi}{\partial t^2} = \frac{c_o^2}{\left(1+\frac{\partial \xi}{\partial a}\right)^{2+\frac{B}{A}}} \frac{\partial^2 \xi}{\partial a^2}.\qquad(6.73)$$

In this form one can recognize that the quantity $2 + B/A$ for fluids plays the same role as $\gamma + 1$ for ideal gases. Values of B/A for fluids given in Table 6.3 indicate that nonlinearity of fluids, even in the absence of bubbles of air, cannot always be ignored. The nonlinearity of fluids is discussed in greater detail in Chap. 8.

6.5.2 Nonlinearity of Solids

The propagation of a wave in a nonlinear solid is described by first introducing third-order elastic constants. When extending the stress–strain relationship (which essentially is a force-based approach) it becomes difficult to keep a consistent approximation among the various nonlinear terms. However, If one instead uses an energy approach, a consistent approximation is automatically maintained for all the terms of higher order.

Beginning with the elastic potential energy, one can define both the second-order constants (those determining the wave velocity in the linear approximation) and the third-order elastic constants simultaneously. The elastic potential energy is

$$\phi(\eta) = \frac{1}{2!} \sum_{ijkl} C_{ijkl} \eta_{ij} \eta_{kl}$$
$$+ \frac{1}{3!} \sum_{ijklmn} C_{ijklmn} \eta_{ij} \eta_{kl} \eta_{mn} + \ldots,\qquad(6.74)$$

Table 6.3 Values of B/A

Substance	T (°C)	B/A
Distilled water	0	4.2
	20	5.0
	40	5.4
	60	5.7
	80	6.1
	100	6.1
Sea water (3.5%)	20	5.25
Methanol	20	9.6
Ethanol	0	10.4
	20	10.5
	40	10.6
N-propanol	20	10.7
N-butanol	20	10.7
Acetone	20	9.2
Beneze	20	9/0
Chlorobenzene	30	9.3
Liquid nitrogen	b.p.	6.6
Benzyl alcohol	30	10.2
Diethylamine	30	10.3
Ethylene glycol	30	9.7
Ethyl formate	30	9.8
Heptane	30	10.0
Hexane	30	9.9
Methyl acetate	30	9.7
Mercury	30	7.8
Sodium	110	2.7
Potassium	100	2.9
Tin	240	4.4
Monatomic gas	20	0.67
Diatomic gas	20	0.40

Table 6.4 K_2 and K_3 for the principal directions in a cubic crystal

Direction	K_2	K_3
[100]	C_{11}	C_{111}
[110]	$\frac{C_{11}+C_{12}+2C_{44}}{2}$	$\frac{C_{111}+3C_{112}+12C_{166}}{4}$
[111]	$\frac{C_{11}+2C_{12}+4C_{44}}{3}$	$\frac{1}{9}(C_{111}+6C_{112}+12C_{144}+24C_{166}+2C_{123}+16C_{456})$

Table 6.5 Comparison of room-temperature values of the ultrasonic nonlinearity parameters of solids. BCC = body-centered cubic; FCC = face-centered cubic

Material or structure	Bonding	β_{avg}
Zincblende	Covalent	2.2
Flourite	Ionic	3.8
FCC	Metallic	5.6
FCC (inert gas)	van der Waals	6.4
BCC	Metallic	8.2
NaCl	Ionic	14.6
Fused silica	Isotropic	−3.4
$YBa_2Cu_3O_{7-\delta}$ (ceramic)	Isotropic	14.3

Table 6.6 Parameters entering into the description of finite-amplitude waves in gases, liquids and solids

Parameter	Ideal gas	Liquid	Solid
c_0^2	$\frac{\gamma P_0}{\rho_0}$	$\frac{A}{\rho_0}$	$\frac{K_2}{\rho_0}$
Nonlinearity parameter β	$\gamma + 1$	$\frac{B}{A} + 2$	$-\left(\frac{K_3}{K_2} + 3\right)$

where the η are the strains, C_{ijkl} are the second-order elastic constants, and C_{ijklmn} are the third-order elastic constants. For cubic crystals there are only three second-order elastic constants: C_{11}, C_{12} and C_{44}, and only six third-order elastic constants: C_{111}, C_{112}, C_{144}, C_{166}, C_{123} and C_{456}. This makes the investigation of cubic crystals relatively straightforward [6.35]. By using the appropriate form of Lagrange's equations, specializing to a specific orientation of the coordinates with respect to the ultrasonic wave propagation direction, and neglecting attenuation and higher-order terms, one can write the nonlinear wave equation for propagation in the directions that allow the propagation of purely longitudinal waves (with no excitation of transverse waves). In a cubic crystalline lattice there are three of these *pure mode* directions for longitudinal waves and the nonlinear wave equation has the form [6.35]

$$\rho_o \frac{\partial^2 u}{\partial t^2} = K_2 \frac{\partial^2 u}{\partial a^2} + (3K_2 + K_3) \frac{\partial u}{\partial a} \frac{\partial^2 u}{\partial a^2} + \dots, \quad (6.75)$$

where both K_2 and K_3 depend on the orientation considered. The quantity K_2 determines the wave velocity: $K_2 = c_o^2 \rho_o$. The quantity K_3 contains only third-order elastic constants. The quantities K_2 and K_3 are given for the three pure-mode directions in a cubic lattice in Table 6.4. The ratio of the coefficient of the nonlinear term to that of the linear term has a special significance. It is often called the *nonlinearity parameter* β and its magnitude is $\beta = 3 + \frac{K_3}{K_2}$. Since K_3 is an inherently negative quantity and is usually greater in magnitude than $3K_2$, a minus sign is often included in the definition:

$$\beta = -\left(3 + \frac{K_3}{K_2}\right). \quad (6.76)$$

The nonlinearity parameters of many cubic solids have been measured. As might be expected, there is a difference between the quantities measured in the three pure-mode directions (usually labeled as the [100], [110] and [111] directions). These differences, however, are not great. If one averages them, one gets the results shown in Table 6.5. The nonlinearity parameters cover the range 2–15. This means that for cubic crystals the coefficient of the nonlinear term in the nonlinear wave equation is 2–15 times as large as the coefficient of the linear term. This gives an impression of the approximation involved when one ignores nonlinear acoustics.

There is also a source of nonlinearity of solids that appears to come from the presence of domains in lithium niobate; this has been called *acoustic memory* [6.36].

It is possible to measure all six third-order elastic constants of cubic crystals. To do so, however, it is necessary to make additional measurements. The procedure that minimizes errors in the evaluation of third-order elastic constants from combination of nonlinearity parameters with the results of hydrostatic-pressure measurements has been considered by *Breazeale* et al. [6.37] and applied to the evaluation of the third-order elastic constants of two perovskite crystals.

6.5.3 Comparison of Fluids and Solids

To facilitate comparison between fluids and solids, it is necessary to use a binomial expansion of the denominator of (6.73)

$$\left(1 + \frac{\partial \xi}{\partial a}\right)^{-\left(\frac{B}{A} + 2\right)} = 1 + \left(\frac{B}{A} + 2\right) \frac{\partial \xi}{\partial a} + \dots \quad (6.77)$$

Using this expression, (6.73) becomes

$$\frac{\partial^2 \xi}{\partial t^2} = c_o^2 \frac{\partial^2 \xi}{\partial a^2} + c_o^2 \left(\frac{B}{A} + 2\right) \frac{\partial \xi}{\partial a} \frac{\partial^2 \xi}{\partial a^2} + \dots \quad (6.78)$$

This form of the equation can be compared directly with (6.74) for solids. The ratio of the coefficient of the nonlinear term to that of the linear term can be evaluated directly. The nonlinearity parameters of the various substances are listed in Table 6.6. Use of Table 6.6 allows one to make a comparison between the nonlinearity of fluids as listed in Table 6.3 and the nonlinearity parameters of solids listed in Table 6.5. Nominally, they are of the same order of magnitude. This means that solids exhibit intrinsic nonlinearity that is comparable to that exhibited by fluids. Thus, the approximation made by assuming that Hooke's law (strain is proportional to stress) is valid for solids is comparable to the approximation made in the derivation of the linear wave equation for fluids.

References

6.1 R.N. Thurston, A.D. Pierce (Eds.): *Physical Acoustics*, Vol. XXV (Academic, San Diego 1999)
6.2 W.P. Mason (Ed.): *Physical Acoustics*, Vol. I A,B (Academic, New York 1964)
6.3 R.N. Thurston: Wave propagation in fluids and normal solids. In: *In: Physical Acoustics*, Vol. I A, ed. by P.W. Mason (Academic, New York 1964)
6.4 A. Migliori, T.W. Darling, J.P. Baiardo, F. Freibert: Resonant ultrasound spectroscopy (RUS). In: *Handbook of Elastic Properties of Solids, Liquids, and Gases*, Vol. I, ed. by M. Levy, H. Bass, R. Stern (Academic, New York 2001) p.10, Cha
6.5 R. Truell, C. Elbaum, B. Chick: *Ultrasonic Methods in Solid State Physics* (Academic, New York 1969)
6.6 C. Allen, I. Rudnick: A powerful high-frequency siren, J. Acoust. Soc. Am. **19**, 857–865 (1947)
6.7 D.F. Gaitan, L.A. Crum, C.C. Church, R.A. Roy: Sonoluminescence and bubble dynamics for a single, stable, cavitation bubble, J. Acoust. Soc. Am. **91**, 3166–3183 (1992)
6.8 J. Wu: Acoustical tweezers, J. Acoust. Soc. Am. **89**, 2140–2143 (1991)
6.9 N. Marinesco, J.J. Trillat: Action des ultrasons sur les plaques photographiques, C. R. Acad. Sci. Paris **196**, 858 (1933)
6.10 H. Frenzel, H. Schultes: Luminiszenz im ultraschallbeschickten Wasse, Z. Phys. Chem. **27**, 421 (1934)
6.11 I.V. Ostrovskii, P. Das: Observation of a new class of crystal sonoluminescence at piezoelectric crystal surface, Appl. Phys. Lett. **70**, 167–169 (1979)
6.12 I. Miyake, H. Futama: Sonoluminescence in X-rayed KCl crystals, J. Phys. Soc. Jpn. **51**, 3985–3989 (1982)
6.13 B.P. Barber, R. Hiller, K. Arisaka, H. Fetterman, S.J. Putterman: Resolving the picosecond characteristics of synchronous sonoluminescence, J. Acoust. Soc. Am. **91**, 3061 (1992)
6.14 J.T. Carlson, S.D. Lewia, A.A. Atchley, D.F. Gaitan, X.K. Maruyama, M.E. Lowry, M.J. Moran, D.R. Sweider: Spectra of picosecond sonoluminescence. In: *Advances in Nonlinear Acoustics*, ed. by H. Hobaek (World Scientific, Singapore 1993)
6.15 R. Hiller, S.J. Putterman, B.P. Barber: Spectrum of synchronous picosecond sonoluminescence, Phys. Rev. Lett. **69**, 1182–1184 (1992)
6.16 G.W. Swift: Thermoacoustic engines, J. Acoust. Soc. Am. **84**, 1145–1180 (1988)
6.17 L. J. House, D. B. Pape: Method and Apparatus for Acoustic Energy – Identification of Objects Buried in Soil, US Patent 5357063 (1993)
6.18 C.G. Don, A.J. Rogers: Using acoustic impulses to identify a buried non-metallic objects, J. Acoust. Soc. Am. **95**, 2837–2838 (1994)
6.19 D. D. Caulfield: Acoustic Detection Apparatus, US Patent 4922467 (1989)
6.20 D.M. Donskoy: Nonlinear vibro-acoustic technique for land mine detection, SPIE Proc. **3392**, 211–217 (1998)
6.21 D.M. Donskoy: Detection and discrimination of non-metallic mines, SPIE Proc. **3710**, 239–246 (1999)
6.22 D.M. Donskoy, N. Sedunov, A. Ekimov, M. Tsionskiy: Optimization of seismo-acoustic land mine detection using dynamic impedances of mines and soil, SPIE Proc. **4394**, 575–582 (2001)
6.23 J.M. Sabatier, N. Xiang: Laser-doppler based acoustic-to-seismic detection of buried mines, Proc. SPIE **3710**, 215–222 (1999)
6.24 N. Xiang, J.M. Sabatier: Detection and Remediation Technologies for Mines and Minelike Targets VI, Proc. SPIE **4394**, 535–541 (2001)
6.25 F.D. Martin, M.A. Breazeale: A simple way to eliminate diffraction lobes emitted by ultrasonic transducers, J. Acoust. Soc. Am. **49**, 1668–1669 (1971)
6.26 G. Du, M.A. Breazeale: The ultrasonic field of a Gaussian transducer, J. Acoust. Soc. Am. **78**, 2083–2086 (1985)
6.27 R.D. Peters, M.A. Breazeale: Third harmonic of an initially sinusoidal ultrasonic wave in copper, Appl. Phys. Lett. **12**, 106–108 (1968)
6.28 F. Goos, H. Hänchen: Ein neuer und fundamentaler Versuch zur Totalreflexion, Ann. Phys. **1**, 333–346 (1947), 6. Folge
6.29 K. Ergin: Energy ratio of seismic waves reflected and refracted at a rock-water boundary, Bull. Seismol. Soc. Am. **42**, 349–372 (1952)
6.30 L.M. Brekhovskikh: *Waves in Layered Media* (Academic, New York 1960)
6.31 W.G. Neubauer: Ultrasonic reflection of a bounded beam at Rayleigh and critical angles for a plane liquid-solid interface, J. Appl. Phys. **44**, 48–55 (1973)

6.32 T. Tamir, H.L. Bertoni: Lateral displacement of optical beams at multilayered and periodic structures, J. Opt. Soc. Am. **61**, 1397–1413 (1971)

6.33 M.A. Breazeale, M.A. Torbett: Backward displacement of waves reflected from an interface having superimposed periodicity, Appl. Phys. Lett. **29**, 456–458 (1976)

6.34 R.T. Beyer: *Nonlinear Acoustics* (Naval Ships Systems Command, Washington 1974)

6.35 M.A. Breazeale: Third-order elastic constants of cubic crystals. In: *Handbook of Elastic Properties of Solids, Liquids and Gases*, Vol. I, ed. by M. Levy, H. Bass, R. Stern (Academic, New York 2001) pp. 489–510, Chap. 21

6.36 M.S. McPherson, I. Ostrovskii, M.A. Breazeale: Observation of acoustical memory in $LiNbO_3$, Phys. Rev. Lett. **89**, 115506 (2002)

6.37 M.A. Breazeale, J. Philip, A. Zarembowitch, M. Fischer, Y. Gesland: Acoustical measurement of solid state nonlinearity: Aplication to $CsCdF_3$ and $KZnF_3$, J. Sound Vib. **88**, 138–140 (1983)

7. Thermoacoustics

Thermodynamic and fluid-dynamic processes in sound waves in gases can convert energy from one form to another. In these *thermoacoustic* processes [7.1, 2], high-temperature heat or chemical energy can be partially converted to acoustic power, acoustic power can produce heat, acoustic power can pump heat from a low temperature or to a high temperature, and acoustic power can be partially converted to chemical potential in the separation of gas mixtures. In some cases, the thermoacoustic perspective brings new insights to decades-old technologies. Well-engineered thermoacoustic devices using extremely intense sound approach the power conversion per unit volume and the efficiency of mature energy-conversion equipment such as internal combustion engines, and the simplicity of few or no moving parts drives the development of practical applications.

This chapter surveys thermoacoustic energy conversion, so the reader can understand how thermoacoustic devices work and can estimate some relevant numbers. After a brief history, an initial section defines vocabulary and establishes preliminary concepts, and subsequent sections

7.1	History	239
7.2	**Shared Concepts**	240
	7.2.1 Pressure and Velocity	240
	7.2.2 Power	243
7.3	**Engines**	244
	7.3.1 Standing-Wave Engines	244
	7.3.2 Traveling-Wave Engines	246
	7.3.3 Combustion	248
7.4	**Dissipation**	249
7.5	**Refrigeration**	250
	7.5.1 Standing-Wave Refrigeration ..	250
	7.5.2 Traveling-Wave Refrigeration ..	251
7.6	**Mixture Separation**	253
References		254

explain engines, dissipation, refrigeration, and mixture separation. Combustion thermoacoustics is mentioned only briefly. Transduction and measurement systems that use heat-generated surface and bulk acoustic waves in solids are not discussed.

7.1 History

The history of thermoacoustic energy conversion has many interwoven roots, branches, and trunks. It is a complicated history because invention and technology development outside of the discipline of acoustics have sometimes preceded fundamental understanding; at other times, fundamental science has come first.

Rott [7.3, 4] developed the mathematics describing acoustic oscillations in a gas in a channel with an axial temperature gradient, with lateral channel dimensions of the order of the gas thermal penetration depth (typically ≈ 1 mm), this being much shorter than the wavelength (typically ≈ 1 m). The problem had been investigated by Rayleigh and by Kirchhoff, but without quantitative success. In Rott's time, the motivation to understand the problem arose largely from the cryogenic phenomenon known as Taconis oscillations – when a gas-filled tube is cooled from ambient temperature to a cryogenic temperature, the gas sometimes oscillates spontaneously, with large heat transport from ambient to the cryogenic environment. *Yazaki* [7.5] demonstrated convincingly that Rott's analysis of Taconis oscillations was quantitatively accurate.

A century earlier, *Rayleigh* [7.6] understood the qualitative features of such heat-driven oscillations: "If heat be given to the air at the moment of greatest condensation (i.e., greatest density) or be taken from it at the moment of greatest rarefaction, the vibration is encouraged." He had studied Sondhauss oscillations [7.7], the glassblowers' precursor to Taconis oscillations.

Applying Rott's mathematics to a situation where the temperature gradient along the channel was too weak to satisfy Rayleigh's criterion for spontaneous oscillations, *Hofler* [7.8] invented a standing-wave thermoacoustic refrigerator, and demonstrated [7.9] again that Rott's approach to acoustics in small channels was quantitatively accurate. In this type of refrigerator, the coupled oscillations of gas motion, temperature, and heat transfer in the sound wave are phased in time so that heat is absorbed from a load at a low temperature and waste heat is rejected to a sink at a higher temperature.

Meanwhile, completely independently, pulse-tube refrigeration was becoming the most actively investigated area of cryogenic refrigeration. This development began with *Gifford*'s [7.10] accidental discovery and subsequent investigation of the cooling associated with square-wave pulses of pressure applied to one end of a pipe that was closed at the other end. Although the relationship was not recognized at the time, this phenomenon shared much physics with Hofler's refrigerator. *Mikulin*'s [7.11] attempt at improvement in heat transfer in one part of this basic pulse-tube refrigerator led unexpectedly to a dramatic improvement of performance, and *Radebaugh* [7.12] realized that the resulting *orifice* pulse-tube refrigerator was in fact a variant of the Stirling cryocooler.

Development of Stirling engines and refrigerators started in the 19th century, the engines at first as an alternative to steam engines [7.13]. Crankshafts, multiple pistons, and other moving parts seemed at first to be essential. An important modern chapter in their development began in the 1970s with the invention of *free-piston* Stirling engines and refrigerators, in which each piston's motion is determined by interactions between the piston's dynamics and the gas's dynamics rather than by a crankshaft and connecting rod. Analysis of such complex, coupled phenomena is complicated, because the oscillating motion causes oscillating pressure differences while simultaneously the oscillating pressure differences cause oscillating motion. *Ceperley* [7.14, 15] added an explicitly acoustic perspective to Stirling engines and refrigerators when he realized that the time phasing between pressure and motion oscillations in the heart of their regenerators is that of a traveling acoustic wave. Many years later, acoustic versions of such engines were demonstrated by *Yazaki* [7.16], *deBlok* [7.17], and *Backhaus* [7.18], the last achieving a heat-to-acoustic energy efficiency comparable to that of other mature energy-conversion technologies.

7.2 Shared Concepts

7.2.1 Pressure and Velocity

For a monofrequency wave, oscillating variables can be represented with complex notation, such as

$$p(x,t) = p_m + \text{Re}\left[p_1(x)e^{i\omega t}\right] \tag{7.1}$$

for the pressure p, where p_m is the mean pressure, Re (z) indicates the real part of z, $\omega = 2\pi f$ is the angular frequency, f is the frequency, and the complex number p_1 specifies both the amplitude and the time phase of the oscillating part of the pressure. For propagation in the x direction through a cross-sectional area A in a duct, p_1 is a function of x. In the simple lossless, uniform-area situation the sinusoidal x dependence can be found from the wave equation, which can be written with $i\omega$ substituted for time derivatives as

$$\omega^2 p_1 + c^2 \frac{d^2 p_1}{dx^2} = 0, \tag{7.2}$$

where $c^2 = (\partial p/\partial \rho)_s$ is the square of the adiabatic sound speed, with ρ the density and s the entropy. In thermoacoustics, intuition is well served by thinking of (7.2) as two first-order differential equations coupling two variables, pressure p_1 and the x component of volume velocity, U_1:

$$\frac{dp_1}{dx} = -\frac{i\omega \rho_m}{A} U_1, \tag{7.3}$$

$$\frac{dU_1}{dx} = -\frac{i\omega A}{\rho_m c^2} p_1. \tag{7.4}$$

For a reasonably short length of duct Δx, these can be written approximately as

$$\Delta p_1 = -i\omega L\, U_1, \tag{7.5}$$

$$p_1 = -\frac{1}{i\omega C}\Delta U_1, \tag{7.6}$$

where

$$L = \frac{\rho_m \Delta x}{A}, \tag{7.7}$$

$$C = \frac{A \Delta x}{\rho_m c^2}. \tag{7.8}$$

Table 7.1 The acoustic–electric analog. The basic building blocks of thermoacoustic energy-conversion devices are inertances L, compliances C, viscous and thermal resistances R_{visc} and R_{therm}, gains (or losses) G, and power transducers

Acoustic variables	Electric variables
Pressure p_1	Voltage
Volume velocity U_1	Current
Inertance L	Series inductance
Viscous resistance R_{visc}	Series resistance
Compliance C	Capacitance to ground
Thermal-hysteresis resistance R_{therm}	Resistance to ground
Gain/loss along temperature gradient, G	Proportionally controlled current injector
Stroke-controlled transducer	Current source
Force-controlled transducer	Voltage source

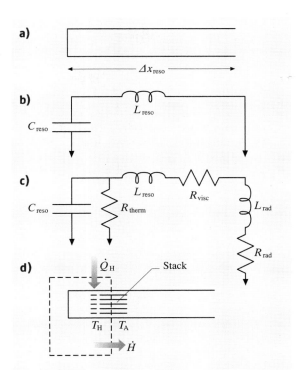

Fig. 7.1a–d Quarter-wavelength resonator. (**a**) A quarter-wavelength resonator: a tube closed at one end and open at the other. (**b**) A simple lossless lumped-element model of the resonator as a compliance C_{reso} in series with an inertance L_{reso}. (**c**) A more complicated lumped-element model, including thermal-hysteresis resistance R_{therm} in the compliance, viscous resistance R_{visc} in the inertance, and radiation impedance $R_{rad} + i\omega L_{rad}$. (**d**) An illustration of the first law of thermodynamics for a control volume, shown enclosed by the dashed line, which intersects the stack of a well-insulated standing-wave engine. In steady state, the thermal power $\dot{Q}_H$ that flows into the system at the hot heat exchanger must equal the total power $\dot{H}$ that flows along the stack

As introduced in Chap. 6 (Physical Acoustics) (7.3) is the acoustic Euler equation, the inviscid form of the momentum equation. It describes the force that the pressure gradient must exert to overcome the gas's inertia. In the lumped-element form (7.5), the geometrical and gas-property aspects of the inertia of the gas in the duct are represented by the duct's inertance L. Similarly introduced in Chap. 6, (7.4) and (7.6) are, respectively, the differential and lumped-element versions of the continuity equation combined with the gas's adiabatic equation of state. These describe the compressibility of the gas in response to being squeezed along the x-direction. In the lumped-element form (7.6), the geometrical and gas-property aspects of the compressibility of the gas in the duct are represented by the duct's compliance C.

Accurate calculation of the wave in a long duct or a long series of ducts requires analytical or numerical integration of differential equations [7.1], but for the purposes of this chapter we will consider lumped-element approximations, because they are satisfactorily accurate for many estimates, they allow a *circuit model* representation of devices, and they facilitate intuition based on experience with alternating-current (AC) electrical circuits. Table 7.1 shows the analogy between acoustic devices and electrical circuits.

For example, the closed–open resonator shown in Fig. 7.1a has a resonance frequency that can be calculated by setting its length Δx_{reso} equal to a quarter wavelength of sound; the result is $f_{reso} = c/4\Delta x_{reso}$. The simplest lumped-element approximation of the quarter-wavelength resonator is shown in Fig. 7.1b. We assign compliance C_{reso} to the left half of the resonator, because the compressibility of the gas is more important than its inertia in the left half, where $|p_1|$ is high and $|U_1|$ is low. Similarly, we assign inertance L_{reso} to the right half of the resonator, because the inertia is more important in the right half where $|U_1|$ is high. Setting $\Delta x = \Delta x_{reso}/2$ in (7.7) and (7.8), and recalling that the resonance frequency of an electrical LC circuit is given by $\omega^2 = 1/LC$, we find $f_{reso} = c/\pi \Delta x_{reso}$, differing from the exact result only by a factor of $4/\pi$. Such accuracy will be acceptable for estimation and understanding in this chapter.

Circuit models can include as much detail as is necessary to convey the essential physics [7.1]. When the viscosity is included in the momentum equation, the lumped-element form becomes

$$\Delta p_1 = -(i\omega L + R_{\text{visc}})U_1 , \qquad (7.9)$$

and when the thermal conductivity and a temperature gradient in the x direction are included in the continuity equation, the lumped-element form becomes

$$\Delta U_1 = -\left(i\omega C + \frac{1}{R_{\text{therm}}}\right)p_1 + GU_{\text{in},1} . \qquad (7.10)$$

Figure 7.1c is a better model than Fig. 7.1b for the closed–open resonator of Fig. 7.1a, because it includes thermal-hysteresis resistance in the compliance, viscous resistance in the inertance, and the inertial and resistive radiation impedance at the open end. This model would yield reasonably accurate estimates for the free-decay time and quality factor of the resonator as well as its resonance frequency.

Ducts filled with porous media and having temperature gradients in the x-direction play a central role in thermoacoustic energy conversion [7.1]. The pore size is characterized by the hydraulic radius r_h, defined as the ratio of gas volume to gas–solid contact area. In a circular pore, r_h is half the circle's radius; in the gap between parallel plates, r_h is half the gap. The term *stack* is used for a porous medium whose r_h is comparable to the gas thermal penetration depth

$$\delta_{\text{therm}} = \sqrt{2k/\omega\rho_m c_p} , \qquad (7.11)$$

where k is the gas thermal conductivity and c_p is its isobaric heat capacity per unit mass, while a porous medium with $r_h \ll \delta_{\text{therm}}$ is known as a *regenerator*. The viscous penetration depth

$$\delta_{\text{visc}} = \sqrt{2\mu/\omega\rho_m} , \qquad (7.12)$$

where μ is the viscosity, is typically comparable to, but smaller than, the thermal penetration depth. If the distance between a particular mass of gas and the nearest solid wall is much greater than the penetration depths, thermal and viscous interactions with the wall do not affect that gas.

In such porous media with axial temperature gradients, the gain/loss term $GU_{\text{in},1}$ in (7.10) is responsible for the creation of acoustic power by engines and for the thermodynamically required consumption of acoustic power by refrigerators. This term represents cyclic thermal expansion and contraction of the gas as it moves along and experiences the temperature gradient in the solid. In regenerators, the gain G is nearly real, and ΔU_1 caused by the motion along the temperature gradient is in phase with U_1 itself, because of the excellent thermal contact between the gas in small pores and the

Table 7.2 Expressions for the lumped-element building blocks L, R_{visc}, C, R_{therm}, and GU_1, and for the total power $\dot{H}$, in the boundary-layer limit $r_h \gg \delta$ and in the small-pore limit $r_h \ll \delta$. The symbol "$\sim$" in the small-pore limit indicates that the numerical prefactor depends on the shape of the pore

	Boundary-layer limit	**Small-pore limit**
L	$\dfrac{\rho_m \Delta x}{A}$	$\sim \dfrac{\rho_m \Delta x}{A}$
R_{visc}	$\dfrac{\mu \Delta x}{A r_h \delta_{\text{visc}}}$	$\sim \dfrac{2\mu \Delta x}{A r_h^2}$
C	$\dfrac{A\Delta x}{\rho_m c^2} = \dfrac{A\Delta x}{\gamma p_m}$	$\dfrac{\gamma A \Delta x}{\rho_m c^2} = \dfrac{A\Delta x}{p_m}$
R_{therm}	$\dfrac{\rho_m^2 c_p^2 T_m r_h \delta_{\text{therm}}}{k A \Delta x}$	$\sim \dfrac{3 k T_m}{4\omega^2 r_h^2 A \Delta x}$
GU_1	$\dfrac{1-i}{2}\dfrac{1}{1+\sqrt{\sigma}}\dfrac{\delta_{\text{therm}}}{r_h}\dfrac{\Delta T_m}{T_m}U_1$	$\dfrac{\Delta T_m}{T_{\text{in},m}}U_{\text{in},1}$
$\dot{H}$	$\dfrac{\delta_{\text{therm}}}{4 r_h (1+\sigma)} \text{Re}\left\{\tilde{p}_1 U_1 \left[i(1+\sqrt{\sigma}) + (1-\sqrt{\sigma})\right]\right\}$ $-\dfrac{\delta_{\text{therm}}\rho_m c_p (1-\sigma\sqrt{\sigma})|U_1|^2}{4 r_h A \omega (1-\sigma^2)}\dfrac{\Delta T}{\Delta x}$ $+\dot{E} - (Ak + A_{\text{solid}} k_{\text{solid}})\dfrac{\Delta T}{\Delta x}$	$-A_{\text{solid}} k_{\text{solid}} \dfrac{\Delta T}{\Delta x}$

walls of the pores. (Positive G indicates gain; negative G indicates loss.) In stacks, a nonzero imaginary part of G reflects imperfect thermal contact in the pores and the resultant time delay between the gas's cyclic motion along the solid's temperature gradient and the gas's thermal expansion and contraction.

Table 7.2 gives expressions for the impedances in the boundary-layer limit, $r_h \gg \delta_{therm}$ and $r_h \gg \delta_{visc}$, which is appropriate for large ducts, and in the small-pore limit appropriate for regenerators. Boundary-layer-limit entries are usefully accurate in stacks, in which $r_h \sim \delta$. General expressions for arbitrary r_h and many common pore shapes are given in [7.1].

The lumped-element approach summarized in Table 7.1 and the limiting expressions given in Table 7.2 are sufficient for most of this overview, but quantitatively accurate thermoacoustic analysis requires slightly more sophisticated techniques and includes more phenomena [7.1]. Every differential length dx of duct has dL, dC, dR_{visc}, and $d(1/R_{therm})$, and if $dT_m/dx \neq 0$ it also has dG, so a finite-length element is more analogous to an electrical transmission line than to a few lumped impedances. In addition to smoothly varying x dependencies for all variables, important phenomena include turbulence, which increases R_{visc} above the values given in Table 7.2; pore sizes which are in neither of the limits given in Table 7.2; nonlinear terms in the momentum and continuity equations, which cause frequency doubling, tripling, etc., so that the steady-state wave is a superposition of waves of many frequencies; and streaming flows caused by the wave itself. Many of these subjects are introduced in Chap. 8 (Nonlinear Acoustics). Thermoacoustics software that includes most or all of these phenomena and has the properties of several commonly used gases is available [7.19, 20].

For estimating the behavior of thermoacoustic devices, it is useful to remember some properties of common ideal gases [7.21]. The equation of state is

$$p = \frac{\rho R_{univ} T}{M}, \qquad (7.13)$$

where $R_{univ} = 8.3 \, \text{J}/(\text{mole K})$ is the universal gas constant and M is the molar mass. The ratio of isobaric to isochoric specific heats, γ, is 5/3 for monatomic gases such as helium and 7/5 for diatomic gases such as nitrogen and air near ambient temperature, and appears in both the adiabatic sound speed

$$c = \sqrt{\frac{\gamma R_{univ} T}{M}} \qquad (7.14)$$

and the isobaric heat capacity per unit mass

$$c_p = \frac{\gamma R_{univ}}{(\gamma - 1) M}. \qquad (7.15)$$

The viscosity of many common gases (e.g., air, helium, and argon) is about

$$\mu \simeq (2 \times 10^{-5} \, \text{kg/m s}) \left(\frac{T}{300 \, \text{K}}\right)^{0.7}, \qquad (7.16)$$

and the thermal conductivity k can be estimated by remembering that the Prandtl number

$$\sigma = \frac{\mu c_p}{k} \qquad (7.17)$$

is about 2/3 for pure gases, but somewhat lower for gas mixtures [7.22].

7.2.2 Power

In addition to ordinary acoustic power $\dot{E}$, the time-averaged thermal power $\dot{Q}$, total power $\dot{H}$, and exergy flux $\dot{X}$ are important in thermoacoustic energy conversion. These thermoacoustic powers are related to the simpler concepts of work, heat, enthalpy, and exergy that are encountered in introductory [7.21] and advanced [7.23] thermodynamics.

Just as acoustic intensity is the time-averaged product of pressure and velocity, acoustic power $\dot{E}$ is the time-averaged product of pressure and volume velocity. In complex notation,

$$\dot{E} = \frac{1}{2} \text{Re} \left(\tilde{p}_1 U_1\right), \qquad (7.18)$$

where the tilde denotes complex conjugation. At transducers, it is apparent that acoustic power is closely related to thermodynamic work, because a moving piston working against gas in an open space with volume V transforms mechanical power to acoustic power (or vice versa) at a rate $f \oint p \, dV$, which is equal to (7.18) for sinusoidal pressure and motion. Resistances R dissipate acoustic power; the gain/loss term GU_1 in components with temperature gradients can either consume or produce acoustic power, and inertances L and compliances C neither dissipate nor produce acoustic power, but simply pass it along while changing p_1 or U_1.

Time-averaged thermal power $\dot{Q}$ (i.e., time-averaged heat per unit time) is added to or removed from the gas at heat exchangers, which are typically arrays of tubes, high-conductivity fins, or both, spanning a duct. Thermal power can be supplied to an engine by high-temperature combustion products flowing through such tubes or by circulation of a high-temperature heat-transfer fluid through such tubes and a source of heat elsewhere. Thermal power is often removed from engines and refrigerators by ambient-temperature water flowing through such tubes.

Of greatest importance is the total time-averaged power

$$\dot{H} = \int \left[\frac{1}{2} \rho_m \mathrm{Re}\left(\tilde{h}_1 u_1\right) - k \frac{dT_m}{dx} \right] dA , \quad (7.19)$$

based on the x component u of velocity and the enthalpy h per unit mass, which is the energy of most utility in fluid mechanics. Fig. 7.1d uses a simple standing-wave engine (discussed more fully below) to illustrate the centrality of $\dot{H}$ to the first law of thermodynamics in thermoacoustics. The figure shows a heat exchanger and stack in a quarter-wavelength resonator. When heat is applied to the hot heat exchanger, the resonance is driven by processes (described below) in the stack. The dashed line in Fig. 7.1d encloses a volume whose energy must be constant when the engine is running in steady state. If the side walls of the engine are well insulated and rigid within that volume, then the only energy flows per unit time into and out of the volume are $\dot{Q}_H$ and whatever power flows to the right through the stack. We define the total power flow through the stack to be $\dot{H}$, and the first law of thermodynamics ensures that $\dot{H} = \dot{Q}_H$ in this simple engine.

As shown in (7.19), the total power $\dot{H}$ is the sum of ordinary steady-state heat conduction (most importantly in the solid parts of the stack or regenerator and the surrounding duct walls) and the total power carried convectively by the gas itself. Analysis of the gas contribution requires spatial and temporal averaging of the enthalpy transport in the gas [7.4], and shows that the most important contributions are acoustic power flowing through the pores of the stack and a shuttling of energy by the gas that occurs because entropy oscillations in the gas are nonzero and have a component in phase with velocity. Remarkably, these two phenomena nearly cancel in the small pores of a regenerator.

Finally, the exergy flux $\dot{X}$ represents the rate at which thermodynamic work can be done, in principle, with unrestricted access to a thermal bath at ambient temperature [7.1, 23]. Exergy flux is sometimes used in complex systems to analyze sources of inefficiency according to location or process.

7.3 Engines

Implicit in Rayleigh's criterion [7.6] for spontaneous oscillations, "If heat be given to the air at the moment of greatest (density) or be taken from it at the moment of greatest rarefaction, the vibration is encouraged," is the existence of a vibration in need of encouragement, typically a resonance with a high quality factor. Today, we would express Rayleigh's criterion in either of two ways, depending on whether we adopt a Lagrangian perspective, focusing on discrete masses of gas as they move, or an Eulerian perspective, focusing on fixed locations in space as the gas moves past them. In the Lagrangian perspective, some of the gas in a thermoacoustic engine must experience $\oint p \, dV > 0$, where V is the volume of an identified mass of gas. In the Eulerian perspective, part of the device must have $d\dot{E}/dx > 0$, arising from $\mathrm{Re}\,(\tilde{p}_1 \, dU_1/dx) > 0$ and the G term in (7.10).

By *engine* we mean a prime mover, i.e., something that converts heat to work or acoustic power. We describe three varieties: standing-wave engines, traveling-wave engines, and pulse combustors.

7.3.1 Standing-Wave Engines

Continuing the quarter-wavelength example introduced in Fig. 7.1, Fig. 7.2 shows a simple standing-wave engine, of a type that is available as a classroom demonstration [7.24]. A stack and hot heat exchanger are near the left end of the resonator, and its right end is open to the air. When heat $\dot{Q}_H$ is first applied to the hot heat exchanger, the heat exchanger and the adjacent end of the stack warm up, establishing an axial temperature gradient from hot to ambient in the stack. When the gradient is steep enough (as explained below), the acoustic oscillations start spontaneously, and grow in intensity as more heat is added and a steady state is approached. In the steady state, total power $\dot{H}$ equal to $\dot{Q}_H$ (minus any heat leak to the room) flows through the stack, creating acoustic power, some of which is dissipated elsewhere in the resonator and some of which is radiated into the room. To the right of the stack, where an ambient-temperature heat exchanger would often be located, the heat is carried rightward and out of this resonator by streaming-driven

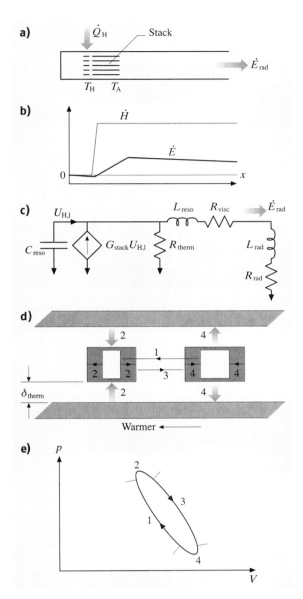

Fig. 7.2a–e A standing-wave engine. (**a**) A hot heat exchanger and a stack in a quarter-wavelength resonator. Heat $\dot{Q}_H$ is injected at the hot heat exchanger, and the device radiates acoustic power $\dot{E}_{rad}$ into the surroundings. (**b**) Total power flow $\dot{H}$ and acoustic power $\dot{E}$ as functions of position x in the device. Positive power flows in the positive-x direction. (**c**) Lumped-element model of the device. (**d**) Close-up view of part of one pore in the stack of (**a**), showing a small mass of gas going through one full cycle, imagined as four discrete steps. *Thin arrows* represent motion, and *wide, open arrows* represent heat flow. (**e**) Plot showing how the pressure p and volume V of that small mass of gas evolve with time in a clockwise elliptical trajectory. *Tick marks* show approximate boundaries between the four numbered steps of the cycle shown in (**d**)

convection in the resonator, while fresh air at ambient temperature T_A streams inwards.

The most important process in the standing-wave engine is illustrated in Fig. 7.2d and Fig. 7.2e from a Lagrangian perspective. Figure 7.2d shows a greatly magnified view of a mass of gas inside one pore of the stack. The sinusoidal thermodynamic cycle of that mass of gas in pressure p and volume V is shown in Fig. 7.2e; the mass's temperature, entropy, density, and other properties also vary sinusoidally in time. However, for qualitative understanding of the processes, we describe the cycle as if it were a time series of four discrete steps, numbered 1–4 in Fig. 7.2d and Fig. 7.2e. In step 1, the gas is simultaneously compressed to a smaller volume and moved leftward by the wave a distance $2|\xi_1|$, which is much smaller than the length Δx of the stack. While the gas is moving leftwards, the pressure changes by $2|p_1|$ and the temperature would change by $2|T_1| = 2T_m(\gamma-1)|p_1|/\gamma p_m$ if the process were adiabatic. This suggests the definition of a critical temperature gradient

$$\nabla T_{crit} = |T_1|/|\xi_1|. \tag{7.20}$$

Thermal contact in the stack's large pores is imperfect, so step 1 is actually not too far from adiabatic. However, the temperature gradient imposed on the stack is greater than ∇T_{crit}, so the gas arrives at its new location less warm than the adjacent pore walls. Thus, in step 2, heat flows from the solid into the gas, warming the gas and causing thermal expansion of the gas to a larger volume. In step 3, the gas is simultaneously expanded to a larger volume and moved rightward by the wave. It arrives at its new location warmer than the adjacent solid, so in step 4 heat flows from the gas to the solid, cooling the gas and causing thermal contraction of the gas to a smaller volume. This brings it back to the start of the cycle, ready to repeat step 1.

Although the mass of gas under consideration returns to its starting conditions each cycle, its net effect on its surroundings is nonzero. First, its thermal expansion occurs at a higher pressure than its thermal contraction, so $\oint p\,dV > 0$: the gas does work on its surroundings, satisfying Rayleigh's criterion. This work is responsible for the positive slope of $\dot{E}$ versus x in the stack in Fig. 7.2b and is represented by the gain element $G_{stack}U_{H,1}$ in

Fig. 7.2c. All masses of gas within the stack contribute to this production of acoustic power; one can think of the steady-state operation as due to all masses of gas within the stack adding energy to the oscillation every cycle, to make up for energy lost from the oscillation elsewhere, e.g., in viscous drag in the resonator and acoustic radiation to the surroundings. Second, the gas absorbs heat from the solid at the left extreme of its motion, at a relatively warm temperature, and delivers heat to the solid farther to the right at a lower temperature. In this way, all masses of gas within the stack pass heat along the solid, down the temperature gradient from left to right; within a single pore, the gas masses are like members of a bucket brigade (a line of people fighting a fire by passing buckets of water from a source of water to the fire while passing empty buckets back to the source). This transport of heat is responsible for most of $\dot{H}$ inside the stack, shown in Fig. 7.2b.

This style of engine is called *standing wave* because the time phasing between pressure and motion is close to that of a standing wave. (If it were *exactly* that of a standing wave, $\dot{E}$ would have to be exactly zero at all x.) To achieve the time phasing between pressure and volume changes that is necessary for $\oint p \, dV > 0$, imperfect thermal contact between the gas and the solid in the stack is required, so that the gas can be somewhat thermally isolated from the solid during the motion in steps 1 and 3 but still exchange significant heat with the solid during steps 2 and 4. This imperfect thermal contact occurs because the distance between the gas and the nearest solid surface is of the order of δ_{therm}, and it causes R_{therm} to be significant, so standing-wave engines are inherently inefficient. Nevertheless, standing-wave engines are exceptionally simple. They include milliwatt classroom demonstrations like the illustration in Fig. 7.2, similar demonstrations with the addition of a water- or air-cooled heat exchanger at the ambient end of the stack, research engines [7.25, 26] up to several kW, and the Taconis and Sondhauss oscillations [7.5, 7]. Variants based on the same physics of intrinsically irreversible heat transfer include the *no-stack* standing-wave engine [7.27], which has two heat exchangers but no stack, and the Rijke tube [7.28], which has only a single, hot heat exchanger and uses a superposition of steady and oscillating flow of air through that heat exchanger to create $\oint p \, dV > 0$.

7.3.2 Traveling-Wave Engines

One variety of what acousticians call traveling-wave engines has been known for almost two centuries as a Stirling engine [7.13, 29, 30], and is illustrated in Fig. 7.3a, Fig. 7.3b, Fig. 7.3f, and Fig. 7.3g. A regenerator bridges the gap between two heat exchangers, one at ambient temperature T_A and the other at hot temperature T_H; this assembly lies between two pistons, whose oscillations take the gas through a sinusoidal cycle that can be approximated as four discrete steps: compression, displacement rightward toward higher temperature, expansion, and displacement leftward toward lower temperature. For a small mass of gas in a single pore in the heart of the regenerator, the four steps of the cycle are illustrated in Fig. 7.3f. In step 1, the gas is compressed by rising pressure, rejecting heat to the nearby solid. In step 2, it is moved to the right, toward higher temperature, absorbing heat from the solid and experiencing thermal expansion as it moves. In step 3, the gas is expanded by falling pressure, and absorbs heat from the nearby solid. In step 4, the gas is moved leftward, toward lower temperature, rejecting heat to the solid and experiencing thermal contraction as it moves. The Stirling engine accomplishes $\oint p \, dV > 0$ in Fig. 7.3g for each mass of gas in the regenerator, and this work production allows the hot piston to extract more work from the gas in each cycle than the ambient piston delivers to the gas.

The similarities and differences between this process and the standing-wave process of Fig. 7.2 are instructive. Here, the pore size is $\ll \delta_{therm}$, so the thermal contact between the gas and the solid in the regenerator is excellent and the gas is always at the temperature of the part of the solid to which it is adjacent. Thus, the thermal expansion and contraction occur during the motion parts of the cycle, instead of during the stationary parts of the cycle in the standing-wave engine, and the pressure must be high during the rightward motion and low during the leftward motion to accomplish $\oint p \, dV > 0$. This is the time phasing between motion and pressure that occurs in a traveling wave [7.14, 15], here traveling from left to right. The small pore size maintains thermodynamically reversible heat transfer, so R_{therm} is negligible, and traveling-wave engines have inherently higher efficiency than standing-wave engines. One remaining source of inefficiency is the viscous resistance R_{visc} in the regenerator, which can be significant because the small pores necessary for thermal efficiency cause R_{visc} to be large. To minimize the impact of R_{visc}, traveling-wave engines have $|p_1| > \rho_m c |U_1|/A$, so the magnitude of the specific acoustic impedance $|z_{ac}| = |p_1| A/|U_1|$ is greater than that of a traveling wave.

The gain G listed in Table 7.2 takes on a particularly simple form in the tight-pore limit: $G_{reg} = \Delta T_m / T_{in,m}$.

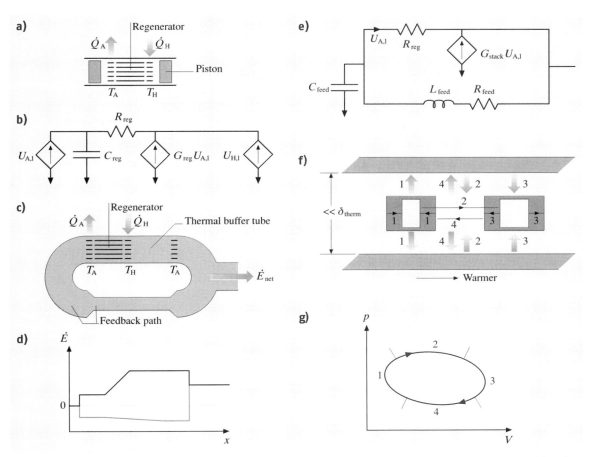

Fig. 7.3a–g Traveling-wave engines. (**a**) A Stirling engine. From *left to right*, the ambient piston, the ambient heat exchanger, the regenerator, the hot heat exchanger, and the hot piston. Time-averaged thermal power $\dot{Q}_H$ is injected into the gas at hot temperature T_H, waste thermal power $\dot{Q}_A$ is removed at ambient temperature T_A, and net mechanical power is extracted by the two pistons. (**b**) Lumped-element model of the engine in (**a**). (**c**) Acoustic–Stirling hybrid engine, with the same processes as (**a**) in the regenerator and its two adjacent heat exchangers, but with additional acoustic components replacing the two pistons. (**d**) Acoustic power $\dot{E}$ as a function of position x in the device shown in (**c**). Positive power flows in the positive-x direction, so the bottom branch of the curve represents power flowing leftward through the feedback path. Total power $\dot{H}$ is not shown, because it is essentially zero in the regenerator and essentially identical to $\dot{E}$ in the open parts of the device. (**e**) Lumped-element model of the engine in (**c**). (**f**) Close-up view of part of one pore in the regenerator of (**a**) or (**c**), showing a small mass of gas going through one full cycle, imagined as four discrete steps. (**g**) Plot showing how the pressure p and volume V of that small mass of gas evolve with time in a clockwise elliptical trajectory. *Tick marks* show approximate boundaries between the four steps of the cycle shown in (**f**)

In the engine of Fig. 7.3a, the initial temperature is T_A, and $\Delta T_m = T_H - T_A$, so the extra volume velocity that the lumped-element model injects at the right end of the regenerator is $U_{A,1}(T_H - T_A)/T_A$ and the total volume velocity at the right end of the regenerator is $U_{A,1} T_H/T_A$. Thus, the regenerator acts like an amplifier of volume velocity, with amplification T_H/T_A. If R_{reg} is small so that p_1 is nearly the same on both sides of the regenerator, then the regenerator amplifies $\dot{E}$ by nearly T_H/T_A.

Figure 7.3c-e illustrates a thermoacoustic–Stirling hybrid engine [7.16–18], in which the processes in the regenerator and its adjacent heat exchangers are the same

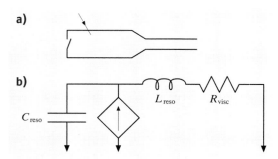

Fig. 7.4a,b Pulsed combustor. (**a**) From *left to right*, check valve for admitting fresh air, fuel injection, combustion cavity, neck. (**b**) Lumped-element model of the combustor

as in the Stirling engine and in Fig. 7.3f and Fig. 7.3g, but with acoustic elements replacing the Stirling engine's pistons. The toroidal topology of the thermoacoustic–Stirling hybrid allows some of the acoustic power that leaves the hot end of the regenerator to be fed back to the ambient end of the regenerator, eliminating the need for the Stirling engine's ambient piston. The feedback path has an inertance L_{feed} (with an unavoidable but small viscous resistance R_{feed}) and a compliance C_{feed}, with $1/L_{feed}C_{feed} \gg \omega^2$ and with ωL_{feed} significantly smaller than the regenerator viscous resistance R_{reg}. These choices let the $L_{feed}C_{feed}$ feedback path boost p_1 as acoustic power flows through it, providing the extra p_1 needed to drive U_1 into the regenerator at its ambient end.

The Stirling engine in Fig. 7.3a has a hot piston, which extracts acoustic power from the gas. From the hot piston's face, mechanical power flows from T_H to T_A along a temperature gradient in a moving part, either the piston itself or a connecting rod. In the thermoacoustic–Stirling hybrid engine of Fig. 7.3c, this thermal isolation function is accomplished by the thermal buffer tube, a thermally stratified column of moving gas. A well-designed thermal buffer tube passes acoustic power with little attenuation and has $\dot{H} \simeq \dot{E}$, so almost no thermal power flows from hot to ambient along the tube and little thermal power need be removed at the auxiliary heat exchanger at the thermal buffer tube's ambient end. Minimizing streaming and attendant heat convection in thermal buffer tubes, which otherwise causes $\dot{H} \neq \dot{E}$, is a topic of current research in thermoacoustics.

Traditional Stirling engines are used for propulsion in some submarines [7.31] and for auxiliary power in boats [7.32], and are under development for residential cogeneration of electricity and space heating [7.33], concentrated solar electricity generation, and nuclear generation of electricity for spacecraft [7.34]. In most of these applications, piston motion is converted to electricity via relative motion of wires and permanent magnets, either with a rotary alternator for crankshaft-coupled pistons or a linear alternator (reminiscent of a loudspeaker) for resonant *free-piston* configurations. Thermoacoustic-Stirling hybrid engines are under consideration for small-scale natural-gas liquefaction [7.25, 35] and for spacecraft power [7.36]. In the former case, acoustic power is fed directly from the engine to cryogenic acoustic refrigerators, without transduction to electrical power.

7.3.3 Combustion

In the standing-wave and traveling-wave engines, Rayleigh's criterion $\oint p \, dV > 0$ is met with volume changes that arise from temperature changes; those temperature changes, in turn, arise from thermal contact between the gas and nearby solid surfaces. In pulsed combustion, the volume changes needed to meet Rayleigh's criterion arise from both temperature and mole-number changes, which in turn are due to time-dependent chemical reactions whose rate is controlled by the time-dependent pressure or time-dependent velocity [7.37, 38].

Figure 7.4 illustrates one configuration in which pulsed combustion can occur. At the closed end of a closed–open resonator, a check valve periodically lets fresh air into the resonator and a fuel injector adds fuel, either steadily or periodically. If the rate of the exothermic chemical reaction increases with pressure (e.g., via the temperature's adiabatic dependence on pressure), positive dV occurs when p is high, meeting Rayleigh's criterion. A four-step diagram of the process, resembling Fig. 7.2d and Fig. 7.3f, is not included in Fig. 7.4 because the process is fundamentally not cyclic: a given mass of gas does not return to its starting conditions, but rather each mass of fresh-air–fuel mixture burns and expands only once.

Combustion instabilities can occur in rockets, jet engines, and gas turbines, with potentially devastating consequences if the pressure oscillations are high enough to cause structural damage. Much of the literature on thermoacoustic combustion is devoted to understanding such oscillations and using active or passive means to prevent them. However, some devices such as high-efficiency residential gas-fired furnaces deliberately use pulsed combustion as illustrated in Fig. 7.4 to pump fresh air in and exhaust gases out of the combustor. This eliminates the need to leave the exhaust gases hot enough for strong chimney convection, so a larger fraction of the heat of combustion can be delivered to the home.

7.4 Dissipation

The dissipative processes represented above by $R_{\rm visc}$ and $R_{\rm therm}$ occur whenever gas-borne sound interacts with solid surfaces. Figure 7.5 illustrates this in the case of a short length dx of a large-radius duct with no axial temperature gradient. The origin of the viscous dissipation of acoustic power is viscous shear within the viscous penetration depth $\delta_{\rm visc}$, as shown in Fig. 7.5c. Most people find viscous dissipation intuitively plausible, imagining the frictional dissipation of mechanical energy when one surface rubs on another. More subtle is the thermal relaxation of acoustic power, illustrated in Fig. 7.5d and Fig. 7.5e. Gas is pressurized nearly adiabatically in step 1, then shrinks during thermal equilibration with the surface in step 2. It is depressurized nearly adiabatically in step 3, and then thermally expands during thermal equilibration with the surface during step 4. As shown in Fig. 7.5e, the net effect is $\oint p \, dV < 0$: the gas absorbs acoustic power from the wave, because the contraction occurs at high pressure and the expansion at low pressure. To avoid a hopelessly cluttered illustration, Fig. 7.5d shows the thermal-hysteresis process superimposed on the left–right oscillating motion in steps 1 and 3, but the thermal-hysteresis process occurs even in the absence of such motion.

Differentiating (7.18) with respect to x shows that the dissipation of acoustic power in the duct in length

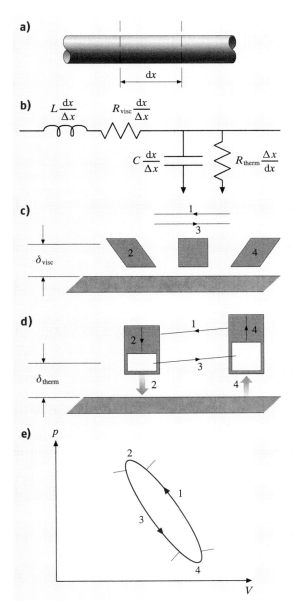

Fig. 7.5a–e Boundary dissipation in acoustic waves. (a) A duct with no temperature gradient, with one short length dx identified. (b) Each length dx has inertance, viscous resistance, compliance, and thermal hysteresis resistance. (c) The dissipation of acoustic power by viscous resistance is due to shear in the gas within roughly $\delta_{\rm visc}$ of the boundary, here occurring during steps 1 and 3 of the cycle. (d) and (e) The dissipation of acoustic power by thermal relaxation hysteresis occurs within roughly $\delta_{\rm therm}$ of the boundary. Gas is pressurized nearly adiabatically in step 1, then shrinks during thermal equilibration with the surface in step 2. It is depressurized nearly adiabatically in step 3, and then thermally expands during thermal equilibration with the surface during step 4. The net effect is that the gas absorbs acoustic power from the wave

dx is given by

$$d\dot{E} = \frac{1}{2}\,\text{Re}\left(\frac{d\tilde{p}_1}{dx}U_1 + \tilde{p}_1\frac{dU_1}{dx}\right)dx\,, \qquad (7.21)$$

and examination of (7.10) and (7.9) associates R_{visc} with the first term and R_{therm} with the second term. Expressions for R_{visc} and R_{therm} in the boundary-layer approximation are given in Table 7.2, and allow the expression of (7.21) in terms of the dissipation of acoustic power per unit of surface area S:

$$\begin{aligned}d\dot{E} &= \frac{1}{4}\rho_m|u_1|^2\omega\delta_{\text{visc}}\,dS\\ &+ \frac{1}{4}(\gamma-1)\frac{|p_1|^2}{\rho_m c^2}\omega\delta_{\text{therm}}\,dS\,,\end{aligned}\qquad (7.22)$$

where $u_1 = U_1/A$ is the velocity outside the boundary layer, parallel to the surface. Each term in this result expresses a dissipation as the product of a stored energy per unit volume $\rho_m|u_1|^2/4$ or $(\gamma-1)|p_1|^2/4\rho_m c^2$, a volume $\delta_{\text{visc}}\,dS$ or $\delta_{\text{therm}}\,dS$, and a rate ω.

7.5 Refrigeration

7.5.1 Standing-Wave Refrigeration

The thermal-hysteresis process described in Sect. 7.4 consumes acoustic power without doing anything thermodynamically useful. Standing-wave refrigeration consumes acoustic power via a similar process, but achieves a thermodynamic purpose: pumping heat up a temperature gradient, either to remove heat from a temperature below ambient or (less commonly) to deliver heat to a temperature above ambient. Figure 7.6a–c shows a standing-wave refrigerator of the style pioneered by *Hofler* [7.8, 9] and recently studied by *Tijani* [7.39]. At the left end, a driver such as a loudspeaker injects acoustic power $\dot{E}$, which flows rightward through the stack, causing a leftward flow of total energy $\dot{H}$.

The most important process in a standing-wave refrigerator is illustrated in Fig. 7.6d and Fig. 7.6e from a Lagrangian perspective. Fig. 7.6d shows a greatly magnified view of a small mass of gas inside one pore of the stack. The sinusoidal thermodynamic cycle of that mass of gas in pressure p and volume V is shown in Fig. 7.6e; the mass's temperature, entropy, density, and other properties also vary sinusoidally in time. However, for qualitative understanding of the processes, we describe them as if they are a time series of four discrete steps, numbered 1–4 in Fig. 7.6d and Fig. 7.6e. In step 1, the gas is simultaneously compressed to a smaller volume and moved leftward by the wave. Thermal contact is imperfect in the pores of a stack, so during step 1 the gas experiences a nearly adiabatic temperature increase due to the pressure increase that causes the compression. It arrives at its new location warmer than the adjacent solid because the temperature gradient in the solid is less than the critical temperature gradient ∇T_{crit} defined in (7.20). Thus, in step 2, heat flows from the gas into the solid, cooling the gas and causing thermal contraction of the gas to a smaller volume. In step 3, the gas is simultaneously expanded to a larger volume and moved rightward by the wave. It arrives at its new location cooler than the adjacent solid, so in step 4 heat flows from the solid to the gas, warming the gas and causing thermal expansion of the gas to a larger volume. This brings it back to the start of the cycle, ready to repeat step 1.

The mass of gas shown in Fig. 7.6d has two time-averaged effects on its surroundings. First, the gas absorbs heat from the solid at the right extreme of its motion, at a relatively cool temperature, and delivers heat to the solid farther to the left at a higher temperature. In this way, all masses of gas within the stack pass heat along the solid, up the temperature gradient from right to left—within a single pore, the gas masses are like members of a bucket brigade passing water. This provides the desired refrigeration or heat-pumping effect. If the left heat exchanger is held at ambient temperature, as shown in Fig. 7.6a, then the system is a refrigerator, absorbing thermal power $\dot{Q}_C$ from an external load at the right heat exchanger at one end of the bucket brigade at T_C, as waste thermal power is rejected to an external ambient heat sink at the other end of the bucket brigade at T_A. (The system functions as a heat pump if the right heat exchanger is held at ambient temperature; then the left heat exchanger is above ambient temperature.) Second, the gas's thermal expansion occurs at a lower pressure than its thermal contraction, so $\oint p\,dV < 0$: the gas absorbs work from its surroundings. This work is responsible for the negative slope of $\dot{E}$ versus x in the stack in Fig. 7.6b and is represented by the gain element $G_{\text{stack}}U_{A,1}$ in Fig. 7.6c. All masses of gas within the stack contribute to this consumption of acoustic power, which must be supplied by the driver.

As in the standing-wave engine, the time phasing between pressure and motion in the standing-wave refrigerator is close to that of a standing wave. Imperfect thermal contact between the gas and the solid in the stack is required to keep the gas rather isolated from the solid during the motion in steps 1 and 3 but still able to exchange significant heat with the solid during steps 2 and 4. This imperfect thermal contact occurs because the distance between the gas and the nearest solid surface is of the order of δ_{therm}, and it causes R_{therm} to be significant, so standing-wave refrigerators are inherently inefficient.

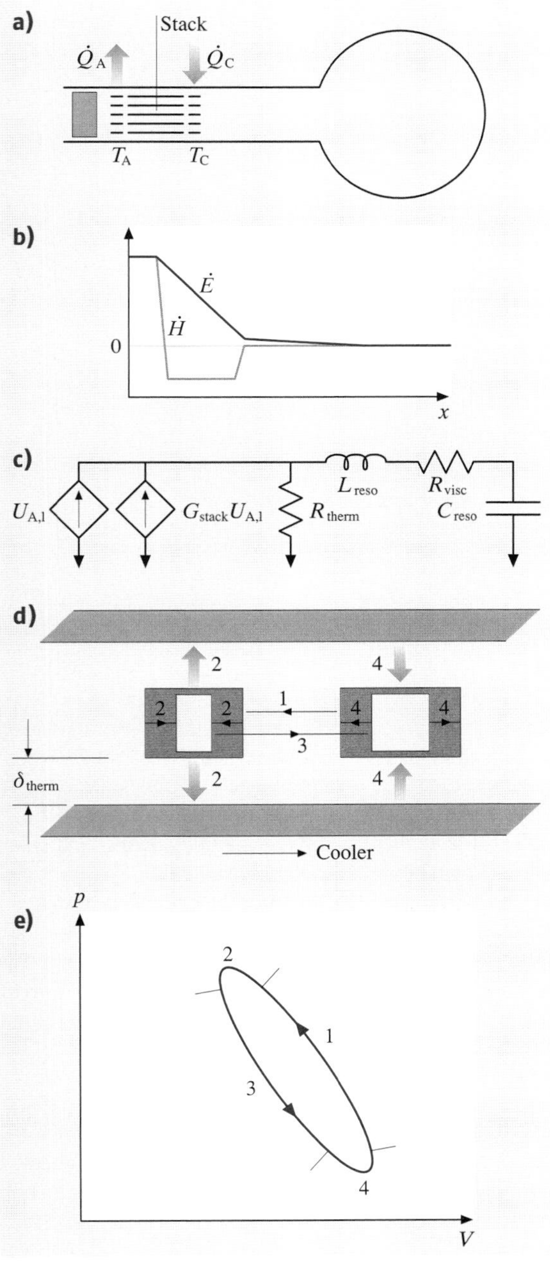

7.5.2 Traveling-Wave Refrigeration

Several varieties of traveling-wave refrigerator are commercially available or under development. At their core is a regenerator, in which the process shown in Fig. 7.7a,b operates. In step 1 of the process, the gas is compressed by rising pressure, rejecting heat to the nearby solid. In step 2, it is moved to the right, toward lower temperature, rejecting heat to the solid and experiencing thermal contraction as it moves. In step 3, the gas is expanded by falling pressure, and absorbs heat from the nearby solid. In step 4, the gas is moved leftward, toward higher temperature, absorbing heat from the solid and experiencing thermal expansion as it moves. The heat transfers between gas and solid in steps 2 and 4 are equal and opposite, so the net thermal effect of each mass of gas on the solid is due to steps 1 and 3, and is to move heat from right to left, up the temperature gradient. As before, the motion of any particular mass of gas is less than the length of the regenerator, so the heat is passed bucket-brigade fashion from the cold end of the regenerator to the ambient end. Each mass of gas absorbs $\oint p\, dV$ of acoustic power from the wave as shown in Fig. 7.7b, because the thermal contraction in step 2 occurs at high pres-

Fig. 7.6a–e A standing-wave refrigerator. (**a**) From *left to right*, a piston, ambient heat exchanger, stack, cold heat exchanger, tube, and tank. Acoustic power is supplied to the gas by the piston to maintain the standing wave, and results in thermal power $\dot{Q}_C$ being absorbed by the gas from a load at cold temperature T_C while waste thermal power $\dot{Q}_A$ is rejected by the gas to an external heat sink at ambient temperature T_A. (**b**) Total power flow $\dot{H}$ and acoustic power $\dot{E}$ as functions of position x in the device. Positive power flows in the positive-x direction. (**c**) Lumped-element model of the device. (**d**) Close-up view of part of one pore in the stack of (**a**), showing a small mass of gas going through one full cycle, imagined as four discrete steps. (**e**) Plot showing how the pressure p and volume V of that small mass of gas evolve with time in a counter-clockwise elliptical trajectory. Tick marks show approximate boundaries between the four steps of the cycle shown in (**d**)

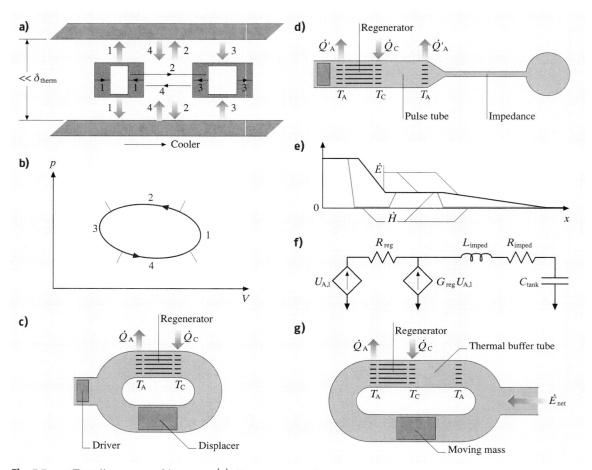

Fig. 7.7a–g Traveling-wave refrigerators. (**a**) Close-up view of part of one pore in the regenerator of a traveling-wave refrigerator, showing a small mass of gas going through one full cycle, imagined as four discrete steps. (**b**) Plot showing how the pressure p and volume V of that small mass of gas evolve with time in a counterclockwise elliptical trajectory. Tick marks show approximate boundaries between the four steps of the cycle shown in (**a**). (**c**) A traditional Stirling refrigerator. From *left to right*, the ambient driver piston; the ambient heat exchanger, the regenerator, and the cold heat exchanger are across the top; the displacer is below. Time-averaged thermal power $\dot{Q}_C$ is removed by the gas from an external heat load at cold temperature T_C, while waste thermal power $\dot{Q}_A$ is rejected by the gas to an external heat sink at ambient temperature T_A and net acoustic power is supplied to the gas by the driver. (**d**) Pulse-tube refrigerator, with the same processes as (**c**) in the regenerator and its two adjacent heat exchangers, but with the displacer replaced by acoustic components. (**e**) Acoustic power $\dot{E}$ and total power $\dot{H}$ as functions of position x in the device shown in (**d**). Positive power flows in the positive-x direction. (**f**) Lumped-element model of the pulse-tube refrigerator in (**d**). (**g**) A refrigerator in which the thermal-buffer function and moving-mass function of (**c**)'s displacer are in two separate components

sure and the thermal expansion in step 4 occurs at low pressure. The small pore size, $r_h \ll \delta_{\mathrm{therm}}$, maintains thermodynamically reversible heat transfer, so R_{therm} is negligible, and traveling-wave refrigerators have an inherently high efficiency. Acoustically, the process shown in Fig. 7.7a represents acoustic power traveling from left to right through the regenerator, and being partly consumed as it goes. Different varieties of traveling-wave refrigerator use different methods to create the necessary amplitudes and relative time phasing of motion and pressure to achieve this acoustic power flow through the regenerator.

The traditional Stirling refrigerator [7.40] uses two moving pistons, which can be either crankshaft-coupled in a configuration like that of the engine in Fig. 7.3a or of the *free-displacer* variety shown in Fig. 7.7c. In the free-displacer Stirling refrigerator (often called *free piston*), the solid displacer moves in response to the gas-pressure forces on it, without linkage to any external motor. Its area and mass are selected to give its motion the desired amplitude and time phase [7.29]. Acoustic power is transmitted through it, from right to left, so that the acoustic power that flows out of the right end of the regenerator is fed through the displacer to the left, added to the acoustic power supplied by the driver, and injected into the left end of the regenerator. The displacer must be a thermal insulator, because its right end is at T_C and its left end at T_A. Crankshaft-coupled Stirling refrigerators were used in the 19th century to keep beef cold on the long sea voyage from South America to Britain. Free-piston Stirling cryocoolers are in common use today for cooling infrared sensors in military night-vision goggles and surveillance satellites, and a free-piston Stirling refrigerator built into a small, portable *ice chest* is commercially available at low cost for picnics [7.41].

The pulse-tube refrigerator [7.12], illustrated in Fig. 7.7d–f, uses only one piston. The acoustic power flowing out of the right end of the regenerator is absorbed in an acoustic impedance, instead of being fed back to the left end of the regenerator. This gives the pulse-tube refrigerator a lower efficiency than the Stirling refrigerator, but for cryogenic applications the reduced efficiency is a small price to pay for the elimination of the Stirling's cold moving part. The lumped-element model of the impedance and adjacent tank shown in Fig. 7.7f correctly suggests that proper design of L_{imped}, R_{imped}, and C_{tank} can create almost any desired ratio of $|U_1|$ to $|p_1|$ and time phasing between U_1 and p_1 at the right end of the regenerator. The desired impedance is often achieved by choosing $R_{\text{visc}} \sim \omega L_{\text{imped}} \gg 1/\omega C_{\text{imped}}$. The so-called pulse tube, and the ambient heat exchanger to its right, thermally isolate the cold heat exchanger from the dissipation of acoustic power in the impedance, just as the thermal buffer tube performs that function in Fig. 7.3c. Pulse-tube refrigerators are in common use today in satellites [7.42] and are under development for many other applications such as small-scale oxygen liquefaction and cooling superconducting equipment.

Another variation of the theme is shown in Fig. 7.7g, where the thermal buffer tube and the moving mass perform the thermal-insulation and inertial functions of the free displacer of Fig. 7.7c in two separate locations [7.43, 44]. This variety is under development for commercial food refrigeration [7.45]. Yet another variety [7.46] replaces the inertial moving mass of Fig. 7.7g with inertial moving gas, similar to the thermoacoustic–Stirling hybrid engine of Fig. 7.3c.

7.6 Mixture Separation

In thermoacoustic mixture separation, acoustic power causes the components of a gas mixture to separate [7.47, 48]. The process is loosely analogous to the pumping of heat through the stack in a standing-wave refrigerator. The expenditure of acoustic power results in an increase in the Gibbs free energy of the mixture's components, and the efficiency of the process is comparable to that of some other practical separation processes [7.49].

Figure 7.8 illustrates the process for a binary gas mixture of heavy and light molecules, whose motions are indicated by filled and open arrows, respectively. The motion of the molecules in steps 1 and 3 of the process is bulk motion of the gas, and the motion during steps 2 and 4 is thermal diffusion, in which light molecules diffuse toward higher temperature and heavy molecules diffuse toward lower temperature. In step 1, the gas in the center of the pore moves leftward and its pressure rises, so its temperature rises nearly adiabatically. Trapped by the viscous boundary layer near the wall of the tube is some other gas that does not move and whose thermal contact with the pore wall keeps its temperature from rising. During step 2, the temperature difference between the central gas and the peripheral gas causes light atoms to diffuse into the center and heavy atoms to diffuse out of the center, enriching the center in light atoms. Bulk motion rightward in step 3 carries this central light-enriched gas to the right. Simultaneously, the pressure drops in step 3, and hence the central temperature drops nearly adiabatically. Thus thermal diffusion in step 4 pulls heavy molecules into the center and drives light molecules out of the center, leaving the center enriched in heavy molecules so that the leftward motion during step 1 carries heavy-enriched gas leftward. The net effect of steps 1 and 3 is to move heavy molecules leftward and light molecules rightward. The process

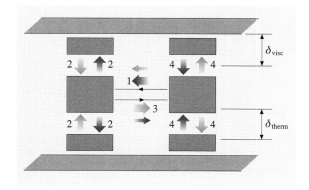

Fig. 7.8 Close-up view of part of one pore in a thermoacoustic mixture separator, showing a small body of gas in the center of the pore going through one full cycle, imagined as four discrete steps, and exchanging mass with neighboring, immobile gas. The *gray* and *brown arrows* in steps 2 and 4 signify thermally driven diffusion of light and heavy molecules between the central gas and the peripheral gas close to the wall of the pore. In steps 1 and 3, the *gray* and *brown arrows* signify bulk motion of the gas, without diffusion

takes place when the mole-fraction gradient is less than a critical gradient, analogous to the critical temperature gradient below which standing-wave refrigeration occurs.

As is evident from Fig. 7.8, the process works best for tubes with r_h somewhat larger than δ_{therm} and δ_{visc}. However, unlike the description above, the process actually works best near traveling-wave phasing, because two 45° phase shifts were ignored in that description. High separation purities require long tubes, and high mole fluxes will require many passages in parallel, perhaps in structures similar to the stacks of standing-wave engines and refrigerators.

In a 2 m-long tube, a 50–50 helium–argon mixture has been separated to yield 70% helium, 30% argon at one end and 30% helium, 70% argon at the other end; and a measurable enrichment of ^{22}Ne from natural neon has been achieved [7.50].

References

7.1 G.W. Swift: *Thermoacoustics: A Unifying Perspective for some Engines and Refrigerators* (Acoustical Society of America Publications, Sewickley PA 2002)

7.2 S.L. Garrett: Resource letter TA-1, Thermoacoustic engines and refrigerators, Am. J. Phys. **72**, 11–17 (2004)

7.3 N. Rott: Damped and thermally driven acoustic oscillations in wide and narrow tubes, Z. Angew. Math. Phys. **20**, 230–243 (1969)

7.4 N. Rott: Thermally driven acoustic oscillations, Part III: Second-order heat flux, Z. Angew. Math. Phys. **26**, 43–49 (1975)

7.5 T. Yazaki, A. Tominaga, Y. Narahara: Experiments on thermally driven acoustic oscillations of gaseous helium, J. Low Temp. Phys. **41**, 45–60 (1980)

7.6 J.W. Strutt (Baron Rayleigh): The explanation of certain acoustical phenomena, Nature **18**, 319–321 (1878)

7.7 K.T. Feldman: Review of the literature on Sondhauss thermoacoustic phenomena, J. Sound Vibrat. **7**, 71–82 (1968)

7.8 T. Hofler, J.C. Wheatley, G.W. Swift, A. Migliori: Acoustic cooling engine, US Patent No. 4,722,201. (1988)

7.9 T.J. Hofler: *Thermoacoustic refrigerator design and performance*. Ph.D. thesis, Physics Department (University of California, San Diego 1986)

7.10 W.E. Gifford, R.C. Longsworth: Pulse tube refrigeration progress, Adv. Cryogenic Eng. B **10**, 69–79 (1965)

7.11 E.L. Mikulin, A.A. Tarasov, M.P. Shkrebyonock: Low-temperature expansion pulse tubes, Adv. Cryogenic Eng. **29**, 629–637 (1984)

7.12 R. Radebaugh: *Development of the pulse tube refrigerator as an efficient and reliable cryocooler* (Proc. Inst. Refrigeration, London 2000) pp. 11–29

7.13 G. Walker: *Stirling Engines* (Clarendon, Oxford 1960)

7.14 P.H. Ceperley: A pistonless Stirling engine – The traveling wave heat engine, J. Acoust. Soc. Am. **66**, 1508–1513 (1979)

7.15 P.H. Ceperley: Gain and efficiency of a short traveling wave heat engine, J. Acoust. Soc. Am. **77**, 1239–1244 (1985)

7.16 T. Yazaki, A. Iwata, T. Maekawa, A. Tominaga: Traveling wave thermoacoustic engine in a looped tube, Phys. Rev. Lett. **81**, 3128–3131 (1998)

7.17 C.M. de Blok: Thermoacoustic system, 1998. Dutch Patent: International Application Number PCT/NL98/00515. US Patent 6,314,740, November 13 (2001)

7.18 S. Backhaus, G.W. Swift: A thermoacoustic-Stirling heat engine, Nature **399**, 335–338 (1999)

7.19 D. Gedeon: *A globally implicit Stirling cycle simulation*, Proceedings of the 21st Intersociety Energy Conversion Engineering Conference (Am. Chem. Soc.,

7.20 W.C. Ward, G.W. Swift: Design environment for low amplitude thermoacoustic engines (DeltaE), J. Acoust. Soc. Am. **95**, 3671–3672 (1994), Software and user's guide available either from the Los Alamos thermoacoustics web site www.lanl.gov/thermoacoustics/ or from the Energy Science and Technology Software Center, US Department of Energy, Oak Ridge, Tennessee

Washington 1986) pp. 550–554, Software available from Gedeon Associates, Athens, Ohio

7.21 R.E. Sonntag, C. Borgnakke, G.J. Van Wylen: *Fundamentals of Thermodynamics* (Wiley, New York 2003)

7.22 F.W. Giacobbe: Estimation of Prandtl numbers in binary mixtures of helium and other noble gases, J. Acoust. Soc. Am. **96**, 3568–3580 (1994)

7.23 A. Bejan: *Advanced Engineering Thermodynamics*, 2nd edn. (Wiley, New York 1997)

7.24 Acoustic Laser Kit, Graduate Program in Acoustics, P. O. Box 30, State College, PA 16804-0030 (www.acs.psu.edu/thermoacoustics/refrigeration/laserdemo.htm)

7.25 G.W. Swift, J.J. Wollan: Thermoacoustics for liquefaction of natural gas, GasTIPS **8**(4), 21–26 (2002), Also available at www.lanl.gov/thermoacoustics/Pubs/GasTIPS.pdf

7.26 D.L. Gardner, G.W. Swift: A cascade thermoacoustic engine, J. Acoust. Soc. Am. **114**, 1905–1919 (2003)

7.27 R.S. Wakeland, R.M. Keolian: Thermoacoustics with idealized heat exchangers and no stack, J. Acoust. Soc. Am. **111**, 2654–2664 (2002)

7.28 K.T. Feldman: Review of the literature on Rijke thermoacoustic phenomena, J. Sound Vibrat. **7**, 83–89 (1968)

7.29 I. Urieli, D.M. Berchowitz: *Stirling Cycle Engine Analysis* (Adam Hilger, Bristol 1984)

7.30 A.J. Organ: *Thermodynamics and Gas Dynamics of the Stirling Cycle Machine* (Cambridge Univ. Press, Cambridge 1992)

7.31 H. Nilsson, C. Bratt: Test results from a 15 kW air-independent Stirling power generator. In: *Proc. 6th International Symposium on Unmanned Untethered Submersible Technology* (IEEE, 1989) pp. 123–128

7.32 WhisperGen Limited, Christchurch, New Zealand

7.33 S. Qiu, D.L. Redinger, J.E. Augenblick: The next generation infinia free-piston Stirling engine for micro-CHP applications. In: *Proc. 12th International Stirling Engine Conference* (Durham University, 2005) pp. 156–165

7.34 M.A. White: A new paradigm for high-power Stirling applications, Proc. Space Nuclear Conference, Stirling Technology Company, Kennewick WA (2005)

7.35 B. Arman, J. Wollan, V. Kotsubo, S. Backhaus, G. Swift: Operation of thermoacoustic Stirling heat engine driven large multiple pulse tube refrigerators. In: *Cryocoolers 13*, ed. by R.G. Ross (Springer, Berlin, New York 2005) pp. 181–188

7.36 S. Backhaus, E. Tward, M. Petach: Traveling-wave thermoacoustic electric generator, Appl. Phys. Lett. **85**, 1085–1087 (2004)

7.37 B. Zinn: Pulsating combustion. In: *Advanced Combustion Methods*, ed. by F.J. Weinberg (Academic, London 1986) pp. 113–181

7.38 F.E.C. Culick: Combustion instabilities and Rayleigh's criterion. In: *Modern Research Topics in Aerospace Propulsion*, ed. by C. Casci, G. Angelino, L. DeLuca, W.A. Sirignano (Springer, Berlin, New York 1991) pp. 508–517, (in honor of Corrado Casci)

7.39 M.E.H. Tijani, J.C.H. Zeegers, A.T.A.M. deWaele: The optimal stack spacing for thermoacoustic refrigeration, J. Acoust. Soc. Am. **112**, 128–133 (2002)

7.40 G. Walker: *Cryocoolers* (Plenum, New York 1983)

7.41 N.W. Lane: Commercialization status of free-piston Stirling machines. In: *Proc. 12th International Stirling Engine Conference* (Durham University, 2005) pp. 30–37

7.42 E. Tward, C.K. Chan, C. Jaco, J. Godden, J. Chapsky, P. Clancy: Miniature space pulse tube cryocoolers, Cryogenics **39**, 717–720 (1999)

7.43 R.W.M. Smith, M.E. Poese, S.L. Garrett, R.S. Wakeland: Thermoacoustic device, US Patent No. 6,725,670.(2004)

7.44 M. E. Poese, R. W. M. Smith, R. S. Wakeland, S. L. Garrett: Bellows bounce thermoacoustic device, US Patent No. 6,792,764.(2004)

7.45 S.L. Garret: Pennsylvania State University, private communication

7.46 G.W. Swift, D.L. Gardner, S. Backhaus: Acoustic recovery of lost power in pulse tube refrigerators, J. Acoust. Soc. Am. **105**, 711–724 (1999)

7.47 D.A. Geller, G.W. Swift: Saturation of thermoacoustic mixture separation, J. Acoust. Soc. Am. **111**, 1675–1684 (2002)

7.48 P.S. Spoor, G.W. Swift: Thermoacoustic separation of a He-Ar mixture, Phys. Rev. Lett. **85**, 1646–1649 (2000)

7.49 D.A. Geller, G.W. Swift: Thermodynamic efficiency of thermoacoustic mixture separation, J. Acoust. Soc. Am. **112**, 504–510 (2002)

7.50 D.A. Geller, G.W. Swift: Thermoacoustic enrichment of the isotopes of neon, J. Acoust. Soc. Am. **115**, 2059–2070 (2004),

8. Nonlinear Acoustics in Fluids

At high sound intensities or long propagation distances at sufficiently low damping acoustic phenomena become nonlinear. This chapter focuses on nonlinear acoustic wave properties in gases and liquids. The origin of nonlinearity, equations of state, simple nonlinear waves, nonlinear acoustic wave equations, shock-wave formation, and interaction of waves are presented and discussed. Tables are given for the nonlinearity parameter *B/A* for water and a range of organic liquids, liquid metals and gases. Acoustic cavitation with its nonlinear bubble oscillations, pattern formation and sonoluminescence (light from sound) are modern examples of nonlinear acoustics. The language of nonlinear dynamics needed for understanding chaotic dynamics and acoustic chaotic systems is introduced.

8.1 **Origin of Nonlinearity** 258
8.2 **Equation of State** 259
8.3 **The Nonlinearity Parameter** *B/A* 260
8.4 **The Coefficient of Nonlinearity** β 262
8.5 **Simple Nonlinear Waves** 263
8.6 **Lossless Finite-Amplitude Acoustic Waves** 264
8.7 **Thermoviscous Finite-Amplitude Acoustic Waves** 268
8.8 **Shock Waves** 271
8.9 **Interaction of Nonlinear Waves** 273
8.10 **Bubbly Liquids** 275
    8.10.1 Incompressible Liquids 276
    8.10.2 Compressible Liquids 278
    8.10.3 Low-Frequency Waves:
           The Korteweg–de Vries Equation .. 279
    8.10.4 Envelopes of Wave Trains: The
           Nonlinear Schrödinger Equation .. 282
    8.10.5 Interaction of Nonlinear Waves.
           Sound–Ultrasound Interaction 284
8.11 **Sonoluminescence** 286
8.12 **Acoustic Chaos** 289
    8.12.1 Methods of Chaos Physics 289
    8.12.2 Chaotic Sound Waves 291
References 293

Acoustics belongs to those areas of physics, where nonlinear phenomena were observed first [8.1–4]. All musical instruments including the human voice belong to nonlinear acoustics as they produce sound waves and these, moreover, are not simple sinusoidal waves. Generation processes are necessarily nonlinear as new phenomena appear that are not proportional to already existing ones. Propagation, on the other hand, is normally linear for propagation distances that are not too long and only starts to show nonlinear effects at high intensities. As mankind started to speak early in its history and also is supposed to have made music early on, nonlinear acoustics belongs to the oldest parts of science. Sorge in 1745 and Tartini in 1754 report that, when two tones of frequency f_1 and $f_2 > f_1$ are presented to the ear at high intensities, the difference frequency $f_2 - f_1$ is additionally heard. For a long time there was a debate of whether these tones were generated in the ear or were already present in the propagating medium. Today we know that in this case the ear generates the difference frequency. But we also know that a difference frequency can be generated in a nonlinear medium during the propagation of a sound wave [8.5]. This fact is used in the parametric acoustic array to transmit and receive highly directed low-frequency sound beams. Under suitable circumstances generally a series of combination frequencies $f_{mn} = mf_1 + nf_2$ with $m, n = 0, \pm 1, \pm 2, \ldots$ can be generated.

A special case is a single harmonic wave of frequency f propagating in a nonlinear medium. It generates the higher harmonics $f_m = mf$. This leads to a steepening of the wave and often to the formation of shock waves. Further nonlinear effects are subsumed under the heading of self-action, as changes in the medium are produced by the sound wave that act back on the wave. Examples are self-focusing and self-transparency.

Also forces may be transmitted to the medium setting it into motion, as exemplified by the phenomenon of acoustic streaming. Acoustic radiation forces may be used to levitate objects and keep them in traps. In other areas of science this is accomplished by electrical or electromagnetical forces (ion traps, optical tweezers). Nonlinear acoustics is the subject of many books, congress proceedings and survey articles [8.1–4, 6–19].

In liquids a special phenomenon occurs, the rupture of the medium by sound waves, called acoustic cavitation [8.20–27]. This gives rise to a plethora of special effects bringing acoustics in contact with almost any other of the natural sciences. Via the dynamics of the cavities, or the cavitation bubbles produced, dirt can be removed from surfaces (ultrasonic cleaning), light emission may occur upon bubble collapse (sonoluminescence [8.28]) and chemical reactions are initiated (sonochemistry [8.29, 30]).

Recently, the general theory of nonlinear dynamics lead to the interesting finding that nonlinear dynamical systems may not just show this simple scenario of combination tones and self-actions, but complicated dynamics resembling stochastic behavior. This is known as deterministic chaos or, in the context of acoustics, *acoustic chaos* [8.31, 32].

Nonlinear acoustics also appears in wave propagation in solids. In this case, further, entirely different, nonlinear phenomena appear, because not only longitudinal but also transverse waves are supported. A separate chapter of this Handbook is devoted to this topic.

8.1 Origin of Nonlinearity

All acoustic phenomena necessarily become nonlinear at high intensities. This can be demonstrated when looking at the propagation of a harmonic sound wave in air. In Fig. 8.1 a harmonic sound wave is represented graphically. In the upper diagram the sound pressure p is plotted versus location x. The static pressure p_{stat} serves as a reference line around which the sound pressure oscillates. This pressure p_{stat} is normally about 1 bar. In the lower diagram the density distribution is given schematically in a grey scale from black to white for the pressure range of the sound wave.

Assuming that the sound pressure amplitude is increased steadily, a point is reached where the sound pressure amplitude attains p_{stat} giving a pressure of zero in the sound pressure minimum. It is impossible to go below this point as there cannot be less than zero air molecules in the minimum. However, air can be compressed above the pressure p_{stat} just by increasing the force. Therefore, beyond a sound pressure amplitude $p = p_{stat}$ no harmonic wave can exist in a gas; it must become nonlinear and contain harmonics. The obvious reason for the nonlinearity in this case (a gas)

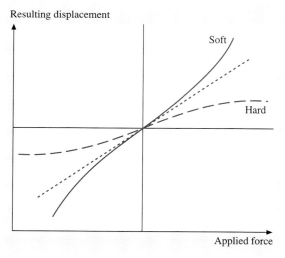

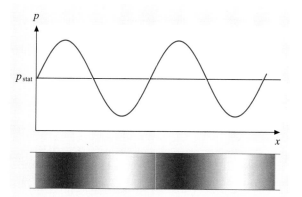

Fig. 8.1 Graphical representation of a harmonic sound wave

Fig. 8.2 Symmetric nonlinear expansion and compression laws compared to a linear law (*straight dotted line*) with soft spring behavior (*solid line*) and hard spring behavior (*dashed line*)

is the asymmetry in density between compression and expansion.

However, symmetric expansion and compression also lead to nonlinear effects when they are not proportional to an applied force or stress. Figure 8.2 shows two types of symmetric nonlinearities: hard and soft spring behavior upon expansion and compression with reference to a linear law represented by a straight (dotted) line. An example of hard spring behavior is a string on a violin, because the average string tension increases with increasing wave amplitude. It is intrinsically symmetric, as positive and negative elongation are equivalent. An example of soft spring behavior is the pendulum, also having a symmetric force–displacement law in a homogeneous gravitational field. An example of an asymmetric mixed type, but with overall soft spring behavior, is a bubble in water oscillating in a sound field. Upon compression a bubble shows hard spring behavior, upon expansion soft spring behavior in such a way that soft spring behavior dominates.

Acoustic waves also show nonlinear behavior in propagation, even without any nonlinearity of the medium. This is due to the cumulative effect of distortion of the wave profile by convection, introduced by the particle velocity that itself constitutes the sound wave. Larger particle velocities propagate faster than slower ones, leading to distortion of an acoustic wave upon propagation. This property can be considered an intrinsic self-action of the wave, leading to an intrinsic nonlinearity in acoustics. This aspect of nonlinearity is discussed below in the context of the coefficient of nonlinearity β.

8.2 Equation of State

The compression and expansion of a medium is described by the equation of state, for which an interrelation between three quantities is needed, usually the pressure p, density ϱ and specific entropy s (entropy per unit mass). Often the variation of entropy can be neglected in acoustic phenomena and the calculations can be carried out at constant entropy; the equations are then called isentropic. A basic quantity in acoustics, the velocity of sound or sound speed c of a medium, is related to these quantities:

$$c^2 = \left(\frac{\partial p}{\partial \varrho}\right)_s . \tag{8.1}$$

The subscript index s indicates that the entropy is to be held constant to give the isentropic velocity of sound.

Even an ideal gas is nonlinear, because it obeys the isentropic equation of state

$$p = p_0 \left(\frac{\varrho}{\varrho_0}\right)^\gamma , \tag{8.2}$$

where p_0 and ϱ_0 are the reference (ambient) pressure and density, respectively, and

$$\gamma = \frac{c_p}{c_v} , \tag{8.3}$$

where γ is the quotient of the specific heat c_p at constant pressure and c_v that at constant volume.

An empirical formula, the Tait equation, is often used for liquids, obviously constructed similarly to the equation of state of the ideal gas:

$$p = P \left(\frac{\varrho}{\varrho_0}\right)^{\gamma_L} - Q \tag{8.4}$$

or

$$\frac{p+Q}{p_\infty + Q} = \left(\frac{\varrho}{\varrho_0}\right)^{\gamma_L} . \tag{8.5}$$

The two quantities Q and γ_L have to be fitted to the experimental pressure–density curve. For water $\gamma_L = 7$, $P = p_\infty + Q = 3001$ bar and $Q = 3000$ bar are used. (Note that γ_L in the Tait equation is not the quotient of the specific heats.)

There is another way of writing the equation of state where the pressure is developed as a Taylor series as a function of density and entropy [8.17]. In the isentropic case it reads [8.33–35]

$$p - p_0 = \left(\frac{\partial p}{\partial \varrho}\right)_{s,\varrho=\varrho_0} (\varrho - \varrho_0)$$
$$+ \frac{1}{2}\left(\frac{\partial^2 p}{\partial \varrho^2}\right)_{s,\varrho=\varrho_0} (\varrho - \varrho_0)^2 + \ldots \tag{8.6}$$

or

$$p - p_0 = A\frac{\varrho - \varrho_0}{\varrho_0} + \frac{B}{2}\left(\frac{\varrho - \varrho_0}{\varrho_0}\right)^2$$
$$+ \frac{C}{6}\left(\frac{\varrho - \varrho_0}{\varrho_0}\right)^3 + \ldots \tag{8.7}$$

with

$$A = \varrho_0 \left(\frac{\partial p}{\partial \varrho}\right)_{s,\varrho=\varrho_0} = \varrho_0 c_0^2 , \quad (8.8)$$

$$B = \varrho_0^2 \left(\frac{\partial^2 p}{\partial \varrho^2}\right)_{s,\varrho=\varrho_0} , \quad (8.9)$$

$$C = \varrho_0^3 \left(\frac{\partial^3 p}{\partial \varrho^3}\right)_{s,\varrho=\varrho_0} . \quad (8.10)$$

Here, c_0 is the velocity of sound under the reference conditions. The higher-order terms of the Taylor series can normally be neglected.

8.3 The Nonlinearity Parameter B/A

To characterize the strength of the nonlinearity of a medium properly the relation of B to the linear coefficient A is important [8.43, 44]. Therefore A and B are combined as B/A, the *nonlinearity parameter*, a pure number.

Table 8.1 B/A values for pure water at atmospheric pressure

T(°C)	B/A	Year	Ref.
0	4.2	1974	[8.36]
20	5	1974	[8.36]
20	4.985±0.063	1989	[8.37]
25	5.11±0.20	1983	[8.38]
26	5.1	1989	[8.39]
30	5.31	1985	[8.40]
30	5.18±0.033	1991	[8.41]
30	5.280±0.021	1989	[8.37]
30	5.38±0.12	2001	[8.42]
40	5.4	1974	[8.36]
40	5.54±0.12	2001	[8.42]
50	5.69±0.13	2001	[8.42]
60	5.7	1974	[8.36]
60	5.82±0.13	2001	[8.42]
70	5.98±0.13	2001	[8.42]
80	6.1	1974	[8.36]
80	6.06±0.13	2001	[8.42]
100	6.1	1974	[8.36]

It is easy to show that for an ideal gas we get from (8.2) together with (8.8) and (8.9)

$$\frac{B}{A} = \gamma - 1 , \quad (8.11)$$

so that $B/A = 0.67$ for a monatomic gas (for example noble gases) and $B/A = 0.40$ for a diatomic gas (for example air). Moreover, it can be shown that

$$\frac{C}{A} = (\gamma - 1)(\gamma - 2) \quad (8.12)$$

for an ideal gas, leading to negative values.

When there is no analytic expression for the equation of state recourse must be made to the direct definitions of A and B and appropriate measurements. From the expressions for A (8.8) and B (8.9) a formula for the nonlinearity parameter B/A is readily found:

$$\frac{B}{A} = \frac{\varrho_0}{c_0^2} \left(\frac{\partial^2 p}{\partial \varrho^2}\right)_{s,\varrho=\varrho_0} . \quad (8.13)$$

To determine B/A from this formula, as well as the density and sound velocity the variation in pressure effected by an isentropic (adiabatic) variation in density has to be measured. Due to the small density variations with pressure in liquids and the error-increasing second derivative this approach is not feasible.

Fortunately, equivalent expressions for B/A have been found in terms of more easily measurable

Table 8.2 Pressure and temperature dependence of the B/A values for water [8.42]

T(K) P(MPa)	303.15	313.15	323.15	333.15	343.15	353.15	363.15	373.15
0.1	5.38±0.12	5.54±0.12	5.69±0.13	5.82±0.13	5.98±0.13	6.06±0.13	–	–
5	5.46±0.12	5.59±0.12	5.76±0.13	5.87±0.13	6.04±0.13	6.07±0.13	6.03±0.13	6.05±0.13
10	5.55±0.12	5.62±0.12	5.78±0.13	5.94±0.13	6.03±0.13	6.11±0.13	6.06±0.13	6.01±0.13
15	5.57±0.12	5.66±0.12	5.83±0.13	5.96±0.13	6.07±0.13	6.09±0.13	6.11±0.13	6.08±0.13
20	5.61±0.12	5.68±0.13	5.81±0.13	5.98±0.13	6.10±0.13	6.14±0.14	6.12±0.13	6.06±0.13
30	5.63±0.12	5.70±0.13	5.84±0.13	5.95±0.13	6.07±0.13	6.16±0.14	6.09±0.13	6.08±0.13
40	5.73±0.13	5.77±0.13	5.86±0.13	6.02±0.13	6.11±0.13	6.14±0.14	6.16±0.14	6.14±0.14
50	5.82±0.13	5.84±0.13	5.93±0.13	6.04±0.13	6.13±0.13	6.16±0.14	6.12±0.13	6.09±0.13

Table 8.3 B/A values for organic liquids at atmospheric pressure

Substance	$T(°C)$	B/A	Ref.
1,2-DHCP	30	11.8	[8.36]
1-Propanol	20	9.5	[8.45]
1-Butanol	20	9.8	[8.45]
1-Pentanol	20	10	[8.45]
1-Pentanol	20	10	[8.45]
1-Hexanol	20	10.2	[8.45]
1-Heptanol	20	10.6	[8.45]
1-Octanol	20	10.7	[8.45]
1-Nonanol	20	10.8	[8.45]
1-Decanol	20	10.7	[8.45]
Acetone	20	9.23	[8.46]
	20	8.0	[8.45]
	40	9.51	[8.46]
Benzene	20	9	[8.36]
	20	8.4	[8.47]
	25	6.5	[8.48]
	40	8.5	[8.48]
Benzyl alcohol	30	10.19	[8.46]
	50	9.97	[8.46]
Carbon bisulfide	10	6.4	[8.48]
	25	6.2	[8.48]
	40	6.1	[8.48]
Carbon tetrachloride	10	8.1	[8.48]
	25	8.7	[8.48]
	25	7.85 ± 0.31	[8.38]
	40	9.3	[8.48]
Chlorobenzene	30	9.33	[8.46]
Chloroform	25	8.2	[8.48]
Cyclohexane	30	10.1	[8.36]
Diethylamine	30	10.3	[8.46]
Ethanol	0	10.42	[8.46]
	20	10.52	[8.46]
	20	9.3	[8.45]
	40	10.6	[8.46]
Ethylene glycol	25	9.88 ± 0.4	[8.38]
	26	9.6	[8.39]
	30	9.7	[8.36]
	30	9.93	[8.40]
	30	9.88 ± 0.035	[8.41]
Ethyl formate	30	9.8	[8.36]
Heptane	30	10	[8.36]
	40	10.05	[8.49]
Hexane	25	9.81 ± 0.39	[8.38]
	30	9.9	[8.36]
	40	10.39	[8.49]

Table 8.3 (cont.)

Substance	$T(°C)$	B/A	Ref.
Methanol	20	8.6	[8.45]
	20	9.42	[8.46]
	30	9.64	[8.46]
Methyl acetate	30	9.7	[8.36]
Methyl iodide	30	8.2	[8.36]
Nitrobenzene	30	9.9	[8.36]
n-Butanol	0	10.71	[8.46]
	20	10.69	[8.46]
	40	10.75	[8.46]
n-Propanol	0	10.47	[8.46]
	20	10.69	[8.46]
	40	10.73	[8.46]
Octane	40	9.75	[8.49]
Pentane	30	9.87	[8.49]
Toluene	20	5.6	[8.48]
	25	7.9	[8.48]
	30	8.929	[8.50]

Table 8.4 B/A values for liquid metals and gases at atmospheric pressure

Substance	$T(°C)$	B/A	Ref.
Liquid metals			
Bismuth	318	7.1	[8.36]
Indium	160	4.6	[8.36]
Mercury	30	7.8	[8.36]
Potassium	100	2.9	[8.36]
Sodium	110	2.7	[8.36]
Tin	240	4.4	[8.36]
Liquid gases			
Argon	-187.15	5.01	[8.51]
	-183.15	5.67	[8.51]
Helium	-271.38	4.5	[8.52]
Hydrogen	-259.15	5.59	[8.51]
	-257.15	6.87	[8.51]
	-255.15	7.64	[8.51]
	-253.15	7.79	[8.51]
Methane	-163.15	17.95	[8.51]
	-153.15	10.31	[8.51]
	-143.15	6.54	[8.51]
	-138.15	5.41	[8.51]
Nitrogen	-203.15	7.7	[8.51]
	-195.76	6.6	[8.52]
	-193.15	8.03	[8.51]
	-183.15	9.00	[8.51]
Other substances			
Sea water (3.5% NaCl)	20	5.25	[8.36]
Sulfur	121	9.5	[8.36]

quantities, namely the variation of sound velocity with pressure and temperature. Introducing the definition of the sound velocity the following formulation chain holds: $(\partial^2 p/\partial \varrho^2)_{s,\varrho=\varrho_0} = (\partial c^2/\partial \varrho)_{s,\varrho=\varrho_0} = 2c_0(\partial c/\partial \varrho)_{s,\varrho=\varrho_0} = 2c_0(\partial c/\partial p)_{s,p=p_0}(\partial p/\partial \varrho)_{s,\varrho=\varrho_0} = 2c_0^3(\partial c/\partial p)_{s,p=p_0}$. Insertion of this result into (8.9) yields

$$B = \varrho_0^2 \left(\frac{\partial^2 p}{\partial \varrho^2}\right)_{s,\varrho=\varrho_0} = 2\varrho_0^2 c_0^3 \left(\frac{\partial c}{\partial p}\right)_{s,p=p_0} \quad (8.14)$$

and into (8.13) results in

$$\frac{B}{A} = 2\varrho_0 c_0 \left(\frac{\partial c}{\partial p}\right)_{s,p=p_0} . \quad (8.15)$$

Here B/A is essentially given by the variation of sound velocity c with pressure p at constant entropy s. This quantity can be measured with sound waves when the pressure is varied sufficiently rapidly but still smoothly (no shocks) to maintain isentropic conditions.

Equation (8.15) can be transformed further [8.44, 53] using standard thermodynamic manipulations and definitions. Starting from $c = c(p, t, s = \text{const.})$ it follows that $dc = (\partial c/\partial p)_{T,p=p_0} dp + (\partial c/\partial p)_{p,T=T_0} dT$ or

$$\left(\frac{\partial c}{\partial p}\right)_{s,p=p_0} = \left(\frac{\partial c}{\partial p}\right)_{T,p=p_0}$$
$$+ \left(\frac{\partial c}{\partial T}\right)_{p,T=T_0} \left(\frac{\partial T}{\partial p}\right)_{s,p=p_0} . \quad (8.16)$$

From the general thermodynamic relation $T ds = c_p dT - (\alpha_T/\varrho)T dp = 0$ for the isentropic case the relation

$$\left(\frac{\partial T}{\partial p}\right)_{s,p=p_0} = \frac{T_0 \alpha_T}{\varrho_0 c_p} . \quad (8.17)$$

follows, where

$$\alpha_T = \frac{1}{V}\left(\frac{\partial V}{\partial T}\right)_{p,T=T_0} = -\frac{1}{\varrho_0}\left(\frac{\partial \varrho}{\partial T}\right)_{p,T=T_0} \quad (8.18)$$

is the isobaric volume coefficient of the thermal expansion and c_p is the specific heat at constant pressure of the liquid. Insertion of (8.17) into (8.16) together with (8.15) yields

$$\frac{B}{A} = 2\varrho_0 c_0 \left(\frac{\partial c}{\partial p}\right)_{T,p=p_0}$$
$$+ 2\frac{c_0 T_0 \alpha_T}{\varrho_0 c_p}\left(\frac{\partial c}{\partial T}\right)_{p,T=T_0} . \quad (8.19)$$

This form of B/A divides its value into an isothermal (first) and isobaric (second) part. It has been found that the isothermal part dominates.

For liquids, B/A mostly varies between 2 and 12. For water under normal conditions it is about 5. Gases (with B/A smaller than one, see before) are much less nonlinear than liquids. Water with gas bubbles, however, may have a very large value of B/A, strongly depending on the volume fraction, bubble sizes and frequency of the sound wave. Extremely high values on the order of 1000 to 10 000 have been reported [8.54, 55]. Tables with values of B/A for a large number of materials can be found in [8.43, 44].

As water is the most common fluid two tables (Table 8.1, 2) of B/A for water as a function of temperature and pressure are given (see also [8.56]). Two more tables (Table 8.3, 4) list B/A values for organic liquids and for liquid metals and gases, both at atmospheric pressure.

8.4 The Coefficient of Nonlinearity β

In a sound wave the propagation velocity dx/dt of a quantity (pressure, density) as observed by an outside stationary observer changes along the wave. As shown by *Riemann* [8.57] and *Earnshaw* [8.58], for a forward-traveling plane wave it is given by

$$\frac{dx}{dt} = c + u , \quad (8.20)$$

c being the sound velocity in the medium without the particle velocity u introduced by the sound wave. The sound velocity c is given by (8.1), $c^2 = (\partial p/\partial \varrho)_s$, and contains the nonlinear properties of the medium. It is customary to incorporate this nonlinearity in (8.20) in a second-order approximation as [8.59, 60]

$$\frac{dx}{dt} = c_0 + \beta u , \quad (8.21)$$

introducing a *coefficient of nonlinearity* β. Here c_0 is the sound velocity in the limit of vanishing sound pressure amplitude. The coefficient of nonlinearity β is related to the parameter of nonlinearity B/A as derived from the Taylor expansion of the isentropic equation of state [8.7] via

$$\beta = 1 + \frac{B}{2A} . \quad (8.22)$$

The number 1 in this equation comes from the u in [8.22]. For an ideal gas [8.2] $B/A = \gamma - 1$ (8.11) and thus

$$\beta = 1 + \frac{\gamma - 1}{2} = \frac{\gamma + 1}{2}. \tag{8.23}$$

The quantity β is made up of two terms. They can be interpreted as coming from the nonlinearity of the medium (second term) and from convection (first term). This convective part is inevitably connected with the sound wave and is an inherent (kinematic) nonlinearity that is also present when there is no nonlinearity in the medium (thermodynamic nonlinearity) [8.60].

The introduction of β besides B/A is justified because it not only incorporates the nonlinearity of the medium, as does B/A, but also the inherent nonlinearity of acoustic propagation. The deformation of a wave as it propagates is described by β.

8.5 Simple Nonlinear Waves

In the linear case the wave equation is obtained for the quantities pressure $p - p_0$, density $\varrho - \varrho_0$ and each component of the particle velocity $\boldsymbol{u} - \boldsymbol{u}_0$ (p_0 ambient pressure, ϱ_0 ambient density, $\boldsymbol{u}_0$ constant (mean) velocity, often $\boldsymbol{u}_0 = 0$) from the equations for a compressible fluid. In the spatially one-dimensional case, when φ denotes one of the perturbational quantities, one gets

$$\varphi_{tt} - c_0^2 \varphi_{xx} = 0, \tag{8.24}$$

where c_0 is the propagation velocity of the perturbation, given by $c_0^2 = (\partial p / \partial \varrho)_{s,\varrho=\varrho_0}$, and the subscripts t and x denote partial differentiation with respect to time and space, respectively. A simple way to incorporate the influence of nonlinearities of the medium without much mathematical effort consists in considering the propagation velocity no longer as constant. One proceeds as follows. As nonlinear waves do not superpose interaction-free, and because waves running in opposite directions (*compound waves*) would cause problems, the wave equation above is written as

$$\left(\frac{\partial}{\partial t} - c_0 \frac{\partial}{\partial x}\right)\left(\frac{\partial}{\partial t} + c_0 \frac{\partial}{\partial x}\right)\varphi = 0 \tag{8.25}$$

and only one part, i.e. a wave running in one direction only, called a *progressive* or *traveling wave*, is taken, for instance:

$$\varphi_t + c_0 \varphi_x = 0, \tag{8.26}$$

a forward, i.e. in the $+x$-direction, traveling wave. The general solution of this equation is

$$\varphi(x,t) = f(x - c_0 t). \tag{8.27}$$

The function f can be a quite general function of the argument $x - c_0 t$. A nonlinear extension can then be written as

$$\varphi_t + v(\varphi)\varphi_x = 0, \tag{8.28}$$

whereby now the propagation velocity $v(\varphi)$ is a function of the perturbation φ. In this way the simplest propagation equation for nonlinear waves is obtained; it already leads to the problem of breaking waves and the occurrence of shock waves.

A solution to the nonlinear propagation equation (8.28) can be given in an implicit way:

$$\varphi(x,t) = f[x - v(\varphi)t], \tag{8.29}$$

as can be proven immediately by insertion. This seems of little use for real calculations of the propagation of the perturbation profile. However, the equation allows for a simple physical interpretation, that is, that the quantity φ propagates with the velocity $v(\varphi)$. This leads to a simple geometric construction for the propagation of a perturbation (Fig. 8.3).

To this end the *initial-value problem* (restricted to progressive waves in one direction only and cutting out one wavelength traveling undisturbed) is considered:

$$\varphi(x, t=0) = f(\xi). \tag{8.30}$$

To each ξ value belongs a value $f(\xi)$ that propagates with the velocity $v[f(\xi)]$:

$$\left.\frac{dx}{dt}\right|_f = v[f(\xi)] \tag{8.31}$$

or

$$x = \xi + v[f(\xi)]t. \tag{8.32}$$

This is the equation of a straight line in the (x,t)-plane that crosses $x(t=0) = \xi$ with the derivative $v(f(\xi))$. Along this straight line φ stays constant. These lines are called *characteristics* and were introduced by *Riemann* [8.57]. In this way the solution of (hyperbolic) partial differential equations can be transformed to the solution of ordinary differential equations, which is of great advantage numerically.

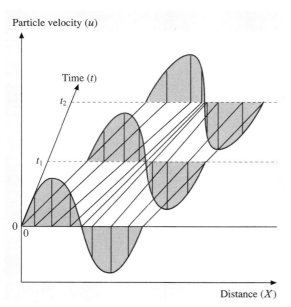

Fig. 8.3 Characteristics and the construction of the waveform for nonlinear progressive wave propagation (after Beyer [8.3])

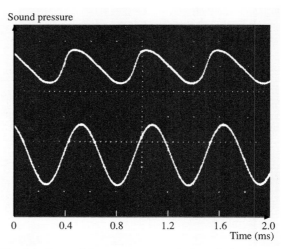

Fig. 8.4 Form of the sound wave at the launch site (lower) and after three meters of propagation (upper) in a tube filled with air. Measurement done at the Drittes Physikalisches Institut, Göttingen

The initial-value problem (8.30) is difficult to realize experimentally. Normally a sound wave is launched by a loudspeaker or transducer vibrating at some location. For one-dimensional problems a piston in a tube is a good approximation to a *boundary condition* $\varphi(x = 0, t) = g(t)$, when the finite amplitude of the piston can be neglected. (The piston cannot vibrate and stay stationary at $x = 0$.) Also, simple waves are produced from the outset without the complications of compound waves and their mixed flow being produced by an initial condition in which the perturbation values are given in space at a fixed time ($t = 0$) [8.61].

The steepening of a wavefront has been measured this way by sending a strong sound wave through a nonlinear medium and monitoring the wave with a microphone. Figure 8.4 shows the steepening of a sound wave in air after three meters of propagation. The wave at a frequency of 2 kHz was guided inside a plastic tube. As the horizontal axis denotes time, the steepening is on the left flank, the flank of the propagation direction that passes the microphone first.

8.6 Lossless Finite-Amplitude Acoustic Waves

How elastic waves propagate through a medium is a main topic in acoustics and has been treated to various degrees of accuracy. The starting point is a set of equations from fluid mechanics. For plane acoustic waves in nondissipative fluids this set is given by three equations: the equation of continuity, the Euler equation and the equation of state:

$$\frac{\partial \varrho}{\partial t} + u \frac{\partial \varrho}{\partial x} + \varrho \frac{\partial u}{\partial x} = 0, \tag{8.33}$$

$$\frac{\partial u}{\partial t} + u \frac{\partial u}{\partial x} + \frac{1}{\varrho} \frac{\partial p}{\partial x} = 0, \tag{8.34}$$

$$p = p(\varrho). \tag{8.35}$$

The quantities ϱ, u and p are the density of the fluid, the particle velocity and the pressure, respectively. The three equations can be condensed into two in view of the unique relation between p and ϱ. From (8.35) it follows that $\partial p/\partial x = c^2 \partial \varrho/\partial x$ and with (8.34) we obtain

$$\frac{\partial u}{\partial t} + u \frac{\partial u}{\partial x} + \frac{c^2}{\varrho} \frac{\partial \varrho}{\partial x} = 0, \tag{8.36}$$

where c is the sound velocity of the medium. The following relation between u and ϱ holds for forward-traveling waves [8.57, 58] (see also [8.1]):

$$\frac{\partial u}{\partial x} = \frac{c}{\varrho}\frac{\partial \varrho}{\partial x} \quad (8.37)$$

giving the propagation equation

$$\frac{\partial u}{\partial t} + (u+c)\frac{\partial u}{\partial x} = 0 \quad (8.38)$$

for a plane progressive wave without dissipation. The propagation velocity c (here for the particle velocity we use u) due to the nonlinearity of the medium to a second-order approximation follows from $u + c = c_0 + \beta u$ (see (8.21) and (8.22)) as

$$c = c_0 + \frac{B}{2A}u \ . \quad (8.39)$$

The corresponding relation for an ideal gas,

$$c = c_0 + \frac{\gamma - 1}{2}u \ , \quad (8.40)$$

is exact.

Comparing the propagation equation (8.38) with (8.28) where a general function $v(\varphi)$ was introduced for the nonlinear propagation velocity with φ a disturbance of the medium, for instance u, the function can now be specified as

$$v(u) = u + c(u) \ . \quad (8.41)$$

This finding gives rise to the following degrees of approximation.

The linear approximation

$$v(u) = c_0 \ , \quad (8.42)$$

the kinematic approximation, where the medium is still treated as linear,

$$v(u) = u + c_0 \ , \quad (8.43)$$

the quadratic approximation

$$v(u) = u + c_0 + \frac{B}{2A}u = c_0 + \beta u \ , \quad (8.44)$$

and so forth, as further approximations are included. From these equations it follows that $v(u=0) = c_0$ regardless of the degree of approximation. This means that the wave as a whole propagates with the linear velocity c_0; its form, however, changes. This holds true as long as the form of the wave stays continuous. Also, there is no truly linear case. Even if the medium is considered linear the kinematic approximation reveals that there is a distortion of the wave. Only in the limit of infinitely small amplitude of the particle velocity u (implying also an infinitely small amplitude of acoustic pressure $p - p_0$) does a disturbance propagate linearly. There is no finite amplitude that propagates linearly. This is different from the transverse waves that appear in solids, for instance, and in electromagnetic wave propagation. In these cases, linear waves of finite amplitude exist, because they do not experience distortion due to a kinematic term u.

A solution to the propagation equation (8.38) to a second-order approximation,

$$\frac{\partial u}{\partial t} + (c_0 + \beta u)\frac{\partial u}{\partial x} = 0 \ , \quad (8.45)$$

can be given in implicit form as for (8.28):

$$u(x, t) = f[x - (c_0 + \beta u)t] \ . \quad (8.46)$$

For the boundary condition (source or signaling problem)

$$u(x = 0, t) = u_a \sin \omega t \quad (8.47)$$

the implicit solution reads with the function $f = u_a \sin(\omega t - kx)$ and $k = \omega/v(u)$:

$$u(x, t) = u_a \sin\left(\omega t - \frac{\omega}{c_0 + \beta u}x\right) \ , \quad (8.48)$$

or, when expanding and truncating the denominator in the argument of the sine wave:

$$u(x, t) = u_a \sin\left[\omega t - \frac{\omega}{c_0}\left(1 - \beta\frac{u}{c_0}\right)x\right] \ . \quad (8.49)$$

With the wave number

$$k_0 = \frac{\omega}{c_0} \ , \quad (8.50)$$

$\omega = 2\pi f$, where f is the frequency of the sound wave, and

$$M_a = \frac{u_a}{c_0} \ , \quad (8.51)$$

is the initial (peak) acoustic Mach number, the solution reads:

$$\frac{u(x, t)}{u_a} = \sin\left[\omega t - k_0 x\left(1 + \beta M_a \frac{u(x, t)}{u_a}\right)\right] \ . \quad (8.52)$$

This implicit solution can be turned into an explicit one, as shown by Fubini in 1935 [8.62] (see also [8.59, 63]), when the solution is expanded in a Fourier series ($\varphi = \omega t - k_0 x$):

$$\frac{u}{u_a} = \sum_{n=1}^{\infty} B_n \sin(n\varphi) \quad (8.53)$$

with

$$B_n = \frac{1}{\pi}\int_0^{2\pi} \frac{u}{u_a}\sin(n\varphi)\mathrm{d}\varphi \ . \qquad (8.54)$$

After insertion of (8.52) into (8.54) and some mathematical operations the Bessel functions J_n of the first kind of order n yield

$$B_n = \frac{2}{\beta M_a k_0 n x} J_n(\beta M_a k_0 n x) \ . \qquad (8.55)$$

The explicit solution then reads

$$\frac{u(x,t)}{u_a} = \frac{2}{\beta M_a k_0} \sum_{n=1}^{\infty} \frac{J_n(\beta M_a k_0 n x)}{n x} \sin n(\omega t - k_0 x) \ . \qquad (8.56)$$

The region of application of this solution is determined by $\beta M_a k_0$. The inverse has the dimensions of a length:

$$x_\perp = \frac{1}{\beta M_a k_0} \ , \qquad (8.57)$$

and is called the *shock distance*, because at this distance the wave develops a vertical tangent, the beginning of becoming a shock wave. The solution is valid only up to this distance $x = x_\perp$.

To simplify the notation the dimensionless normalized distance σ may be introduced:

$$\sigma = \frac{x}{x_\perp} \ . \qquad (8.58)$$

The Fubini solution then reads

$$\frac{u}{u_a} = 2\sum_{n=1}^{\infty} \frac{J_n(n\sigma)}{n\sigma} \sin n(\omega t - k_0 x) \ . \qquad (8.59)$$

When a pure spatial sinusoidal wave is taken as the initial condition [8.64]:

$$u(x,t=0) = u_a \sin k_0 x \ , \qquad (8.60)$$

the implicit solution reads

$$u(x,t) = u_a \sin k_0 [x - (c_0 + \beta u)t] \ , \qquad (8.61)$$

or, with the acoustic Mach number $M_a = u_a/c_0$ as before:

$$\frac{u(x,t)}{u_a} = \sin k_0 \left[x - c_0 t \left(1 + \beta M_a \frac{u(x,t)}{u_a}\right)\right] \ . \qquad (8.62)$$

When again doing a Fourier expansion the explicit solution emerges:

$$\frac{u(x,t)}{u_a} = \frac{2}{\beta M_a k_0 c_0} \sum_{n=1}^{\infty} (-1)^{n+1}$$
$$\times \frac{J_n(\beta M_a k_0 c_0 n t)}{n t} \sin n k_0 (x - c_0 t) \ . \qquad (8.63)$$

The region of application is determined by $\beta M_a k_0 c_0$. The inverse has the dimension of time:

$$t_\perp = \frac{1}{\beta M_a k_0 c_0} \ . \qquad (8.64)$$

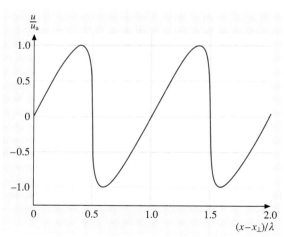

Fig. 8.5 Waveform of the Fubini solution (8.66) at the shock formation time $t_\perp$ ($\sigma_t = 1$) for the initial-value problem. λ is the wavelength of the sound wave

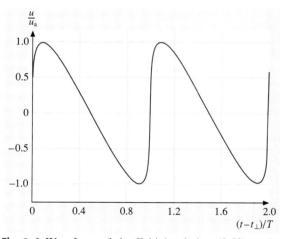

Fig. 8.6 Waveform of the Fubini solution (8.59) at the shock distance $x_\perp$ ($\sigma = 1$) for the source problem. T is the period of the sound wave. Compare the experiment in Fig. 8.4

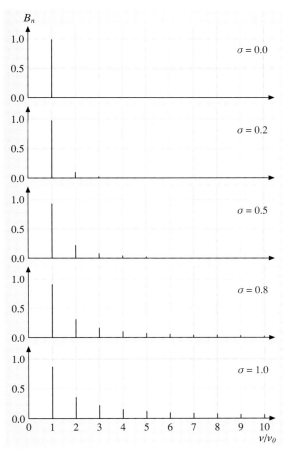

Fig. 8.7 Growth of the harmonics upon propagation for the Fubini solution (8.59) (source problem) at different normalized distances σ up to the shock distance. [8.38]

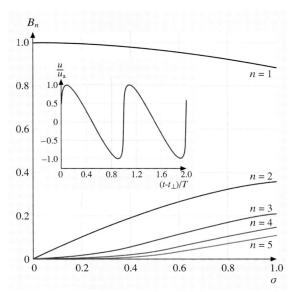

Fig. 8.8 Growth of the first five spectral components B_n of a plane wave as a function of the normalized distance σ for the Fubini solution. The *inset* gives the waveform at $\sigma = 1$ for the source problem, T being the period of the sound wave. [8.64]

The quantity $t_\perp$ is called the *shock formation time*, because at this time the wave develops a vertical tangent. The solution is valid only up to this time $t = t_\perp$. To simplify the notation the dimensionless normalized time σ_t may be introduced

$$\sigma_t = \frac{t}{t_\perp}. \tag{8.65}$$

The Fubini solution then reads:

$$\frac{u(x,t)}{u_a} = 2\sum_{n=1}^{\infty} (-1)^{n+1} \frac{J_n(n\sigma_t)}{n\sigma_t} \sin nk_0(x - c_0 t). \tag{8.66}$$

When comparing $t_\perp$ with $x_\perp$, the relation

$$c_0 t_\perp = x_\perp \tag{8.67}$$

is noted. This means that the shock distance is reached in the shock formation time when the wave travels at the linear sound speed c_0. This is in agreement with the earlier observation that the wave travels at the linear sound speed regardless of the nonlinearity as long as the wave stays continuous. In the case of the quadratic approximation the range in space and time for which this property holds can be quantified explicitly.

To give an impression of what the wave looks like when having propagated the shock distance $x_\perp$ in the shock formation time $t_\perp$, Fig. 8.5 shows two wavelengths at the shock formation time $t_\perp$ and Fig. 8.6 shows two periods at the shock distance $x_\perp$.

A set of spectra of the waveform for different normalized distances σ is plotted in Fig. 8.7 for the source problem, where all harmonics have positive value. The growth of the harmonics B_n (8.55) in the spectrum on the way to the shock distance is visualized. Similar plots have been given by *Fubini Ghiron* [8.65]. A plot of the first five Fourier coefficients B_n as a function of σ is given in Fig. 8.8. In the inset the waveform at the shock distance $x_\perp$ is plotted for two periods of the wave.

The solutions (8.59) and (8.66) are given for the particle velocity u. This quantity is difficult to measure. Instead, in experiments, pressure is the variable of choice. When combining the equation of state for an ideal gas with the definition of sound velocity (see (8.2) and (8.1)) a relation between pressure and sound velocity is obtained:

$$\frac{p}{p_0} = \left(\frac{c}{c_0}\right)^{2\gamma/(\gamma-1)}. \tag{8.68}$$

Insertion of the expression $c = c_0 + [(\gamma-1)/2]u$ (8.40) yields:

$$p = p_0 \left(1 + \frac{\gamma-1}{2}\frac{u}{c_0}\right)^{2\gamma/(\gamma-1)}. \tag{8.69}$$

Inserting the solution for the particle velocity u into this equation will give the pressure p and the acoustic pressure $p - p_0$. This operation is easy to perform numerically but not analytically. However, the approximation scheme for moderate-strength finite-amplitude waves, $|u/c_0| \ll 1$, is only consistent when expanding (8.69) and taking only the linear part. This leads to the *linear* relation between acoustic pressure $p - p_0$ and acoustic particle velocity u (local linearity property for weak nonlinearity) [8.61, 66]:

$$p - p_0 = \varrho_0 c_0 u, \tag{8.70}$$

where the expression $\varrho_0 c_0$ is known as the impedance. When the boundary condition (8.47) is rewritten as

$$(p - p_0)(x = 0, t) = \varrho_0 c_0 u_a \sin \omega t$$
$$= (p_a - p_0) \sin \omega t \tag{8.71}$$

with $p_a - p_0 = \varrho_0 c_0 u_a$, the relation

$$\frac{p - p_0}{p_a - p_0} = \frac{u}{u_a} \tag{8.72}$$

is obtained. The normalized Fubini solution then reads in terms of the acoustic pressure for the boundary-value or source problem

$$\frac{p - p_0}{p_a - p_0} = 2 \sum_{n=1}^{\infty} \frac{J_n(n\sigma)}{n\sigma} \sin n(\omega t - k_0 x), \tag{8.73}$$

and for the initial-value problem:

$$\frac{p - p_0}{p_a - p_0} = 2 \sum_{n=1}^{\infty} (-1)^{n+1} \frac{J_n(n\sigma_t)}{n\sigma_t} \sin nk_0(x - c_0 t). \tag{8.74}$$

The steepening of acoustic waves upon propagation may occur in musical wind instruments, e.g., the trombone [8.67]. It has been shown that depending on the playing level the emitted sound waves steepen when going from piano to fortissimo, even up to shock waves. Indeed, the brightness (the *metallic* sound) of loudly played brass instruments (in particular the trumpet and the trombone as opposed to the saxhorns) is attributed to the increasing harmonic content connected with wave steepening, as exemplified in Fig. 8.7.

8.7 Thermoviscous Finite-Amplitude Acoustic Waves

Because of the inherently nonlinear nature of acoustic wave propagation steep gradients of physical quantities (pressure, density, temperature, ...) inevitably appear after some time of traveling, after which losses can no longer be neglected. In the steep gradients brought about by nonlinearity, linear phenomena that could be neglected before become important [8.68]. These are losses by *diffusive* mechanisms, in particular viscosity and heat conduction, and spreading phenomena by *dispersive* mechanisms as in frequency dispersion. When losses balance nonlinearity, the characteristic waveforms of *shock waves* appear, when frequency dispersion balances nonlinearity the characteristic form of *solitons* appear. Both are given about equal space in this treatment. The inclusion of thermoviscous losses is treated in this chapter, the inclusion of frequency dispersion, small in pure air and water, but large in water with bubbles, is treated in Chap. 8.10 on liquids containing bubbles.

The extension of (8.45) when thermoviscous losses are included as a small effect leads to the Burgers equation

$$\frac{\partial u}{\partial t} + (c_0 + \beta u)\frac{\partial u}{\partial x} = \frac{1}{2}\delta \frac{\partial^2 u}{\partial x^2}. \tag{8.75}$$

Here δ, comprising the losses, has been called the diffusivity of sound by Lighthill [8.69]:

$$\delta = \frac{1}{\rho_0}\left(\frac{4}{3}\mu + \mu_B\right) + \frac{\kappa}{\rho_0}\left(\frac{1}{c_v} - \frac{1}{c_p}\right)$$
$$= \nu\left(\frac{4}{3} + \frac{\mu_B}{\mu} + \frac{\gamma-1}{Pr}\right) \tag{8.76}$$

where μ is the shear viscosity, μ_B the bulk viscosity, $\nu = \mu/\varrho_0$ the kinematic viscosity, κ the thermal conductivity, c_v and c_p the specific heats at constant volume and constant pressure, respectively, and Pr being the Prandtl number, $Pr = \mu c_\mathrm{p}/\kappa$. The equation is an approximation to a more exact second-order equation that, however, does not lend itself to an exact solution as does the Burgers equation above. In the context of acoustics this relation was first derived by *Mendousse*, although for viscous losses only [8.70]. The derivations make use of a careful comparison of the order of magnitude of derivatives retaining the leading terms. The above form of the Burgers equation is best suited to initial-value problems. Source problems are best described when transforming (8.75) to new coordinates (x, τ) with the retarded time $\tau = t - x/c_0$:

$$\frac{\partial u}{\partial x} - \frac{\beta}{c_0^2} u \frac{\partial u}{\partial \tau} = \frac{\delta}{2c_0^3} \frac{\partial^2 u}{\partial \tau^2} . \tag{8.77}$$

The equation can be normalized to a form with only one parameter:

$$\frac{\partial W}{\partial \sigma} - W \frac{\partial W}{\partial \varphi} = \frac{1}{\Gamma} \frac{\partial^2 W}{\partial \varphi^2} , \tag{8.78}$$

where $W = u/u_0$, $\sigma = x/x_\perp$, $\varphi = \omega \tau = \omega t - k_0 x$, $\Gamma = \beta M_0 k_0/\alpha = 2\pi \beta M_0/\alpha\lambda$, with α being the damping constant for linear waves:

$$\alpha = \frac{\delta k_0^2}{2c_0} . \tag{8.79}$$

Γ is called the Gol'dberg number after *Gol'dberg* [8.71] who introduced this normalization (*Blackstock* [8.72]). It can be written as

$$\Gamma = \frac{1/x_\perp}{\alpha} \tag{8.80}$$

where $1/x_\perp = \beta M_0 k_0$ is the strength of the nonlinearity and α is the strength of the damping. The Gol'dberg number is therefore a measure of the importance of nonlinearity in relation to damping. For $\Gamma > 1$ nonlinearity takes over and for $\Gamma < 1$ damping takes over. For $\Gamma \gg 1$ nonlinearity has time to accumulate its effects, for $\Gamma \ll 1$ damping does not allow nonlinear effects to develop.

The Burgers equation is exactly integrable by the Hopf–Cole transformation [8.73, 74]:

$$W = \frac{2}{\Gamma} \frac{1}{\zeta} \frac{\partial \zeta}{\partial \varphi} = \frac{2}{\Gamma} \frac{\partial \ln \zeta}{\partial \varphi} \tag{8.81}$$

that is best done in two steps [8.10]

$$W = \frac{\partial \psi}{\partial \varphi} , \tag{8.82}$$

$$\psi = \frac{2}{\Gamma} \ln \zeta . \tag{8.83}$$

By this transformation the nonlinear Burgers equation is reduced to the linear heat conduction or diffusion equation

$$\frac{\partial \zeta}{\partial \sigma} = \frac{1}{\Gamma} \frac{\partial^2 \zeta}{\partial \varphi^2} . \tag{8.84}$$

For this equation a general explicit solution is available:

$$\zeta(\sigma, \varphi)$$
$$= \sqrt{\frac{\Gamma}{4\pi\sigma}} \int_{-\infty}^{+\infty} \zeta(0, \varphi') \exp\left(-\Gamma \frac{(\varphi' - \varphi)^2}{4\sigma}\right) \mathrm{d}\varphi' . \tag{8.85}$$

For a specific solution the initial or boundary conditions must be specified. A common problem is the piston that starts to vibrate sinusoidally at time $t = 0$. This problem has been treated by *Blackstock* [8.72] whose derivation is closely followed here. The boundary condition is given by

$$\begin{aligned} u(0, t) &= 0 & \text{for} \quad t \leq 0 , \\ u(0, t) &= u_\mathrm{a} \sin \omega t & \text{for} \quad t > 0 , \end{aligned} \tag{8.86}$$

or in terms of the variable $W(\sigma, \varphi)$

$$\begin{aligned} W(0, \varphi) &= 0 & \text{for} \quad \varphi \leq 0 , \\ W(0, \varphi) &= \sin \varphi & \text{for} \quad \varphi > 0 . \end{aligned} \tag{8.87}$$

To solve the heat conduction equation the boundary condition is needed for $\zeta(\sigma, \varphi)$. To this end the Hopf–Cole transformation (8.81) is reversed:

$$\zeta(\sigma, \varphi) = \exp\left(\frac{\Gamma}{2} \int_{-\infty}^{\varphi} W(\sigma, \varphi') \mathrm{d}\varphi'\right) . \tag{8.88}$$

Insertion of $W(0, \varphi)$ yields as the boundary condition for ζ:

$$\begin{aligned} \zeta(0, \varphi) &= 1 & \text{for} \quad \varphi \leq 0 \\ \zeta(0, \varphi) &= \mathrm{e}^{\frac{1}{2}\Gamma(1 - \cos \varphi)} & \text{for} \quad \varphi > 0 \end{aligned} \tag{8.89}$$

and insertion into (8.85) yields the solution in terms of ζ. Using $\bar{\sigma} = \sqrt{4\sigma/\Gamma}$ and $q = (\varphi' - \varphi)/\bar{\sigma}$ the solution for the vibrating piston in terms of ζ reads

$$\zeta(\sigma, \varphi) = \frac{1}{\sqrt{\pi}} \int_{-\infty}^{-\varphi/\bar{\sigma}} \mathrm{e}^{-q^2} \mathrm{d}q$$
$$+ \frac{1}{\sqrt{\pi}} \mathrm{e}^{\frac{1}{2}\Gamma} \int_{-\varphi/\bar{\sigma}}^{\infty} \mathrm{e}^{-\frac{1}{2}\Gamma \cos(\bar{\sigma}q + \varphi)} \mathrm{e}^{-q^2} \mathrm{d}q . \tag{8.90}$$

This solution is involved and also contains the transients after starting the oscillation. When $\varphi \to \infty$ these transients decay and the steady-state solution is obtained:

$$\zeta(\sigma, \varphi | \varphi \to \infty)$$
$$= \frac{1}{\sqrt{\pi}} e^{\frac{1}{2}\Gamma} \int_{-\infty}^{\infty} e^{-\frac{1}{2}\Gamma \cos(\bar{\sigma}q+\varphi)} e^{-q^2} dq. \quad (8.91)$$

With the help of the modified Bessel functions of order n, $I_n(z) = i^{-n} J_n(iz)$, and the relation [8.75]

$$e^{z\cos\theta} = I_0(z) + 2\sum_{n=1}^{\infty} I_n(z) \cos n\theta \quad (8.92)$$

the expression

$$e^{-\frac{1}{2}\Gamma \cos(\bar{\sigma}q+\varphi)}$$
$$= I_0\left(-\frac{1}{2}\Gamma\right) + 2\sum_{n=1}^{\infty} I_n\left(-\frac{1}{2}\Gamma\right) \cos n(\bar{\sigma}q+\varphi)$$
$$= I_0\left(\frac{1}{2}\Gamma\right) + 2\sum_{n=1}^{\infty} (-1)^n I_n\left(\frac{1}{2}\Gamma\right) \cos n(\bar{\sigma}q+\varphi) \quad (8.93)$$

is valid. Inserting this into (8.91) and integrating yields, observing that $\int_{-\infty}^{+\infty} \exp(-q^2 x^2) \cos[p(x+\lambda)]dx = (\sqrt{\pi}/q)\exp[-p^2/(4q^2)]\cos p\lambda$ [8.76]:

$$\zeta(\sigma, \varphi | \varphi \to \infty)$$
$$= e^{\frac{1}{2}\Gamma}\left[I_0\left(\frac{1}{2}\Gamma\right) + 2\sum_{n=1}^{\infty}(-1)^n I_n\left(\frac{1}{2}\Gamma\right)e^{-n^2\sigma/\Gamma}\cos n\varphi\right]. \quad (8.94)$$

This is the exact steady-state solution for the oscillating piston problem given for ζ as a Fourier series. Transforming to $W(\sigma, \varphi)$ via (8.81) gives

$$W(\sigma, \varphi)$$
$$= \frac{4\Gamma^{-1}\sum_{n=1}^{\infty}(-1)^{n+1}n I_n\left(\frac{1}{2}\Gamma\right)e^{-n^2\sigma/\Gamma}\sin n\varphi}{I_0\left(\frac{1}{2}\Gamma\right) + 2\sum_{n=1}^{\infty}(-1)^n I_n\left(\frac{1}{2}\Gamma\right)e^{-n^2\sigma/\Gamma}\cos n\varphi}. \quad (8.95)$$

Finally, for $u(x, t)$ the solution reads

$$\frac{u(x,t)}{u_a}$$
$$= \frac{4\Gamma^{-1}\sum_{n=1}^{\infty}(-1)^{n+1}n I_n\left(\frac{1}{2}\Gamma\right)e^{-n^2\alpha x}\sin n(\omega t - k_0 x)}{I_0\left(\frac{1}{2}\Gamma\right) + 2\sum_{n=1}^{\infty}(-1)^n I_n\left(\frac{1}{2}\Gamma\right)e^{-n^2\alpha x}\cos n(\omega t - k_0 x)}. \quad (8.96)$$

The equation for the acoustic pressure $p - p_0$ again can be obtained via the approximation (8.72) as before in the lossless case: $(p - p_0)/(p_a - p_0) = u/u_a$ gives the identical equation.

There are regions of the parameter Γ and the independent variables x and t where the solution is difficult to calculate numerically. However, in these cases approximations can often be formulated.

For $\Gamma \to \infty$ and $\sigma = x/x_\perp \ll 1$ the Fubini solution is recovered.

For $\sigma \gg \Gamma$, i.e. far away from the source, the first terms in the numerator and the denominator in (8.96) dominate, leading to

$$\frac{u(x,t)}{u_a} = \frac{4}{\Gamma}\frac{I_1(\Gamma/2)}{I_0(\Gamma/2)} e^{-\alpha x} \sin(\omega t - k_0 x). \quad (8.97)$$

When additionally $\Gamma \gg 1$, i.e. nonlinearity dominates over attenuation, $I_0(\Gamma/2) \approx I_1(\Gamma/2)$ and therefore

$$u(x,t) = u_a \frac{4}{\Gamma} e^{-\alpha x} \sin(\omega t - k_0 x)$$
$$= 4u_a \alpha x_\perp e^{-\alpha x} \sin(\omega t - k_0 x)$$
$$= \frac{4\alpha c_0^2}{\beta\omega} e^{-\alpha x} \sin(\omega t - k_0 x) \quad (8.98)$$
$$= \frac{2\delta\omega}{\beta c_0} e^{-\alpha x} \sin(\omega t - k_0 x). \quad (8.99)$$

This series of equations for the amplitude of the sinusoidal wave radiated from a piston in the far field lends itself to several interpretations. As it is known that for $\Gamma \gg 1$ harmonics grow fast at first, these must later decay leaving the fundamental to a first approximation. The amplitude of the fundamental in the far field is independent of the amplitude u_a of the source, as can be seen from the third row (8.98). This means that there is a saturation effect. Nonlinearity together with attenuation does not allow the amplitude in the far field to increase in proportion to u_a because of the extra damping introduced by the generation of harmonics that are damped more strongly than the fundamental. This even works asymptotically because otherwise u_a would finally appear. The damping constant is frequency dependent (8.79), leading to the last row (8.99). It is seen that the asymptotic amplitude of the fundamental grows with the frequency $f = \omega/2\pi$. This continues until other effects come into play, for instance relaxation effects in the medium. Equations (8.98) and (8.99) for u are different from the equations for the acoustic pressure, as they cannot be normalized with u_a or $p_a - p_0$, respectively.

They are related via the impedance $\varrho_0 c_0$ (8.70):

$$(p-p_0)(x,t) = \frac{4\alpha\varrho_0 c_0^3}{\beta\omega} e^{-\alpha x} \sin(\omega t - k_0 x) \quad (8.100)$$

$$= \frac{2\delta\varrho_0\omega}{\beta} e^{-\alpha x} \sin(\omega t - k_0 x) . \quad (8.101)$$

Good approximations for $\Gamma \gg 1$ have been given by Fay [8.78]:

$$\frac{u(x,t)}{u_a} = \sum_{n=1}^{\infty} \frac{2/\Gamma}{\sinh[n(1+x/x_\perp)/\Gamma]} \sin n(\omega t - k_0 x) , \quad (8.102)$$

and Blackstock [8.72]:

$$\frac{u(x,t)}{u_a} = \frac{2}{\Gamma} \sum_{n=1}^{\infty} \frac{1-(n/\Gamma^2)\coth[n(1+x/x_\perp)/\Gamma]}{\sinh[n(1+x/x_\perp)/\Gamma]}$$
$$\times \sin n(\omega t - k_0 x) . \quad (8.103)$$

With an error of less than 1% at $\Gamma = 50$ Fay's solution is valid for $\sigma > 3.3$ and Blackstock's solution for $\sigma > 2.8$, rapidly improving with σ. The gap in σ from about one to three between the Fubini and Fay solution has been closed by Blackstock. He connected both solutions using weak shock theory giving the Fubini–Blackstock–Fay solution ([8.77], see also [8.1]). Figure 8.9 shows the first three harmonic components of the Fubini–Blackstock–Fay solution from [8.79]. Similar curves have been given by Cook [8.80]. He developed a numerical scheme for calculating the harmonic content of a wave as it propagates by including the losses in small spatial steps linearly for each harmonic component. As there are occurring only harmonic waves that do not break the growth in each small step is given by the Fubini solution.

In the limit $\Gamma \to \infty$ the Fay solution reduces to

$$\frac{u(x,t)}{u_a} = \sum_{n=1}^{\infty} \frac{2}{1+x/x_\perp} \sin n(\omega t - k_0 x) , \quad (8.104)$$

a solution that is also obtained by weak shock theory.

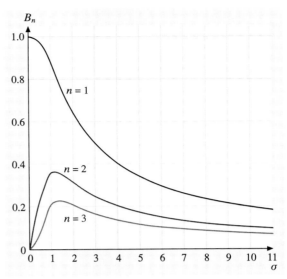

Fig. 8.9 Growth and decay of the first three harmonic components B_n of a plane wave as a function of the normalized distance σ according to the Fubini–Blackstock–Fay solution (after Blackstock [8.77])

When Fay's solution (8.102) is taken and $\sigma \gg \Gamma$ (farfield), the old-age region of the wave is reached. Then $\sinh[n(1+x/x_\perp)/\Gamma] \simeq \frac{1}{2}(e^{nx/x_\perp \Gamma} - e^{-nx/x_\perp \Gamma}) \simeq \frac{1}{2} e^{nx/x_\perp \Gamma} = \frac{1}{2} e^{n\alpha x}$ and

$$u(x,t) = \frac{4\alpha c_0^2}{\beta\omega} \sum_{n=1}^{\infty} e^{-n\alpha x} \sin n(\omega t - k_0 x) \quad (8.105)$$

is obtained similarly as for the fundamental (8.98). Additionally all harmonics that behave like the fundamental, i.e. that are not dependent on the initial peak particle velocity u_a, are obtained. Moreover, they do not decay (as linear waves do) proportionally to $e^{-n^2\alpha x}$ but only proportionally to $e^{-n\alpha x}$.

In the range $\sigma > 1$ shock waves may develop. These are discussed in the next section.

8.8 Shock Waves

The characteristics of Fig. 8.3 must cross. According to the geometrical construction the profile of the wave then becomes multivalued. This is easily envisaged with water surface waves. With pressure or density waves, however, there is only one pressure or one density at one place. The theory is therefore oversimplified and must be expanded with the help of physical arguments and the corresponding mathematical formulation. Shortly before overturning, the gradients of pressure and density become very large, and it is known that damping effects brought about by viscosity and heat conduction can no longer be neglected. When these effects are taken into account, the waves no longer turn over. Instead, a thin zone is formed, a *shock wave*, in which the values of pressure, density, temperature, etc. vary rapidly. A theoretical formulation has been given in the preceding section for thermoviscous sound waves based on the Burgers equation.

It has been found that damping effects do not necessarily have to be included in the theory, but that an extended damping free theory can be developed by introducing certain shock conditions that connect the variables across a discontinuity. A description of this theory can be found in the book by *Whitham* [8.10]. The result can be described in a simple construction. In Fig. 8.10 the profile of the wave has been determined according to the methods of characteristics for a time where the characteristics have already crossed. The shock wave then has to be inserted in such a way that the areas between the shock wave and the wave profile to the left and the right are equal (the equal area rule). Moreover, it can be shown that the velocity of the shock wave is approximately given by

$$v_s = \frac{1}{2}(v_1 + v_2), \tag{8.106}$$

where v_1 and v_2 are the propagation velocities of the wave before and behind the shock, respectively.

Shock waves are inherently difficult to describe by Fourier analysis as a large number of harmonics are needed to approximate a jump-like behavior. A time-domain description is often more favorable here. Such solutions have been developed for the propagation of sound waves in the shock regime. For $\Gamma > \sigma > 3$, *Khokhlov* and *Soluyan* [8.80] have given the following

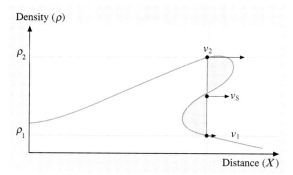

Fig. 8.10 Condition for the location of the shock wave for an overturning wave, the equal area rule

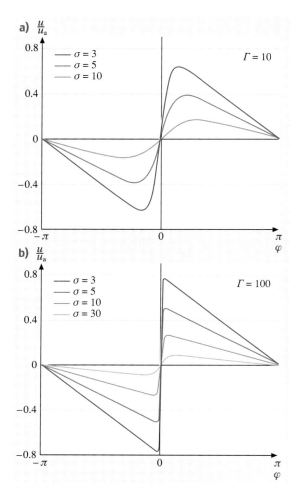

Fig. 8.11a,b Waveforms of the Khokhlov–Soluyan solution for Gol'dberg numbers $\Gamma = 10$ (a) and $\Gamma = 100$ (b) for different σ (the *age* of the wave)

solution

$$\frac{u(x,t)}{u_a} = \frac{1}{1+x/x_\perp}\left(-\varphi + \pi \tanh \frac{\pi \Gamma \varphi}{2(1+x/x_\perp)}\right) \tag{8.107}$$

for $-\pi < \varphi = \omega t - k_0 x < \pi$, covering a full cycle of the wave.

Figure 8.11 shows the waveforms for $\Gamma = 10$ and $\Gamma = 100$ for different $\sigma = x/x_\perp$. When the wave is followed at fixed Γ for increasing σ (i.e. increasing x), the decay and change of the waveform upon propagation can be observed.

The solution and the waveforms are remarkable as they are exact solutions of the Burgers equation, but

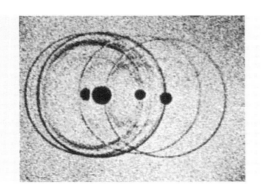

Fig. 8.12 Shock waves (*dark circles*) and bubbles (*dark spheres*) from laser-induced breakdown in water. Reconstructed image from a hologram 1.75 μs after breakdown. The size of the picture is 1.1×1.6 mm

approximate solutions with respect to the boundary condition.

When the wave gets very steep, i. e., in the case $\Gamma \to \infty$, it attains the shape of a sawtooth ($\sigma > 3$):

$$\frac{u(x,t)}{u_a} = \begin{cases} \dfrac{1}{1+x/x_\perp}[-\varphi+\pi] & 0 < \varphi \le \pi \,, \\ \dfrac{1}{1+x/x_\perp}[-\varphi-\pi] & -\pi \le \varphi < 0 \,, \end{cases} \quad (8.108)$$

where $\varphi = \omega t - k_0 x$. The amplitude of the jump is

$$\frac{u(x,t)}{u_a} = \frac{\pi}{1+x/x_\perp} \,, \quad (8.109)$$

the actual jump height from peak to peak being two times this value and falling off rapidly with x. This solution, too, is an exact solution of the Burgers equation, albeit for $1/\Gamma \equiv 0$, i. e., with the diffusion term missing in the Burgers equation.

The Fourier form solution of the sawtooth wave (8.108) is given by

$$\frac{u(x,t)}{u_a} = \frac{2}{1+x/x_\perp} \sum_{n=1}^{\infty} \frac{1}{n} \sin n(\omega t - k_0 x) \,. \quad (8.110)$$

This form can also be derived as approximation from the Fay solution (see (8.104)). This can be considered a consistency proof in the approximation approaches, one proceeding in the time domain, the other in the frequency domain.

An example of the occurrence of shock waves, albeit spherical ones, is given in Fig. 8.12. A laser pulse has been focused into water, leading to four spots of breakdown, the emission of four shock waves and the growth of four bubbles. Laser-produced bubbles are used in cavitation research to investigate bubble dynamics in liquids and their effects [8.81]. Focused shock waves are used in lithotripsy to fragment kidney stones to small enough pieces to fit through the urinary tract [8.82].

8.9 Interaction of Nonlinear Waves

Linear waves do not interact, in the sense that upon superposition they propagate independently, each according to their own parameters. This makes Fourier theory a powerful technique for linear systems. In nonlinear systems, however, the harmonic components of which an initial perturbation is composed interact and produce new components that again interact, etc. Even a single harmonic oscillation creates higher harmonics, as set out in the preceding sections.

For the description of strong nonlinear acoustic wave interaction it is very fortunate that the Burgers equation (8.78) allows for a linearizing transformation. The superposition can then be evaluated in the linear variables, giving the result of the nonlinear interaction when transforming back.

Of special interest is the interaction of two (primary) waves in the course of which the sum and, in particular, the difference frequency can be generated. A highly directional high-frequency source can be transformed into a highly directed low-frequency source (the parametric array). As the low-frequency beam generated has lower damping than the high-frequency beams it propagates to longer distances. Applications are sound navigation and ranging (SONAR) and surface and subsurface scanning of the ocean bottom.

Taking two waves with amplitudes $u_{1,2}$ and frequencies $\omega_{1,2}$, starting at $t = 0$, the boundary condition becomes

$$u(x=0,t) = \begin{cases} 0 & t < 0 \,, \\ u_1 \sin(\omega_1 t) + u_2 \sin(\omega_2 t) & t \ge 0 \,, \end{cases} \quad (8.111)$$

or with the normalized variables $W = u/u_0$, $\varphi = \omega \tau = \omega(t - x/c_0)$, $\sigma = x/x_\perp$,

$$W(\sigma=0,\varphi)$$
$$=\begin{cases}0 & \varphi<0,\\ W_1\sin(\Omega_1\varphi)+W_2\sin(\Omega_2\varphi), & \varphi\geq 0.\end{cases} \quad (8.112)$$

Here, $\Omega_{1,2}=\omega_{1,2}/\omega$ denote the input frequencies normalized to the reference frequency ω and $W_{1,2}=u_{1,2}/u_0$ the amplitudes normalized to the reference amplitude u_0. The inverse Hopf–Cole transformation (8.88) gives the initial condition for the linearized problem in the variable ζ. For $\varphi<0$ we obtain $\zeta(\sigma=0,\varphi)=1$, while for $\varphi\geq 0$:

$$\zeta(\sigma=0,\varphi)$$
$$=\exp\left[\frac{\Gamma}{2}\int_0^\varphi W(\sigma=0,\varphi')\,\mathrm{d}\varphi'\right]$$
$$=\exp\left\{-\frac{\Gamma}{2}\left[\frac{W_1}{\Omega_1}\cos(\Omega_1\varphi')+\frac{W_2}{\Omega_2}\cos(\Omega_2\varphi')\right]_0^\varphi\right\}. \quad (8.113)$$

To simplify notation, introduce the Gol'dberg numbers $\Gamma_{1,2}=\Gamma W_{1,2}/\Omega_{1,2}$ which agree with the definition (8.80) applied to the two input waves, respectively. Then

$$\zeta(\sigma=0,\varphi\geq 0)=\underbrace{\exp\left(\frac{\Gamma_1}{2}+\frac{\Gamma_2}{2}\right)}_{=:C}$$
$$\times\exp\left[-\frac{\Gamma_1}{2}\cos(\Omega_1\varphi)-\frac{\Gamma_2}{2}\cos(\Omega_2\varphi)\right]. \quad (8.114)$$

With this initial condition, the solution (8.85) of the diffusion equation (8.84) reads:

$$\zeta(\sigma,\varphi)=\sqrt{\frac{\Gamma}{4\pi\sigma}}\int_{-\infty}^0 \exp\left(-\Gamma\frac{(\varphi'-\varphi)^2}{4\sigma}\right)\mathrm{d}\varphi'$$
$$+C\sqrt{\frac{\Gamma}{4\pi\sigma}}\int_0^{+\infty}\exp\left[-\frac{\Gamma_1}{2}\cos(\Omega_1\varphi')\right.$$
$$\left.-\frac{\Gamma_2}{2}\cos(\Omega_2\varphi')-\Gamma\frac{(\varphi'-\varphi)^2}{4\sigma}\right]\mathrm{d}\varphi'$$
$$=\sqrt{\frac{\Gamma}{4\pi\sigma}}\int_{-\infty}^{-\varphi}\exp\left(-\Gamma\frac{\varphi'^2}{4\sigma}\right)\mathrm{d}\varphi'$$
$$+C\sqrt{\frac{\Gamma}{4\pi\sigma}}\int_{-\varphi}^{+\infty}\exp\left\{-\frac{\Gamma_1}{2}\cos[\Omega_1(\varphi'+\varphi)]\right.$$
$$\left.-\frac{\Gamma_2}{2}\cos[\Omega_2(\varphi'+\varphi)]-\Gamma\frac{\varphi'^2}{4\sigma}\right\}\mathrm{d}\varphi'.$$

This general solution is quite complicated as the integration limits depend on the phase φ. To dismiss the transients caused by the starting of the waves at $\varphi=0$, again the limit $\varphi\to\infty$ is considered. This means that at a certain position $x=\sigma x_\perp$ in the medium the wave is examined a long time after the passage of the initial perturbation that started at $x=0$, $t=0$. Then, the first integral of (8.115) vanishes, and the second integral can be evaluated from $-\infty$ to $+\infty$ by using the expansion (8.92), written in the following form:

$$\exp(z\cos\theta)=I_0(z)+2\sum_{n=1}^\infty I_n(z)\cos(n\theta)$$
$$=:\sum_{n=0}^\infty b_n I_n(z)\cos(n\theta), \quad (8.115)$$

where the numerical factors $b_0=1$, $b_n=2$ for $n>0$ have been introduced for convenient notation. Substituting this series in (8.115) and noting that I_n is an even (odd) function for even (odd) n, we get

$$\zeta(\sigma,\varphi)$$
$$=C\sqrt{\frac{\Gamma}{4\pi\sigma}}\sum_{m=0}^\infty\sum_{n=0}^\infty(-1)^{m+n}b_m b_n I_m\left(\frac{\Gamma_1}{2}\right)I_n\left(\frac{\Gamma_2}{2}\right)$$
$$\times\int_{-\infty}^{+\infty}\cos[m\Omega_1(\varphi'+\varphi)]\cos[n\Omega_2(\varphi'+\varphi)]$$
$$\times\exp\left(-\Gamma\frac{\varphi'^2}{4\sigma}\right)\mathrm{d}\varphi'. \quad (8.116)$$

Using $\cos(\alpha)\cos(\beta)=(1/2)[\cos(\alpha+\beta)+\cos(\alpha-\beta)]$, and proceeding as in the derivation of (8.94) we get

$$\zeta(\sigma,\varphi)$$
$$=\frac{C}{2}\sum_{m=0}^\infty\sum_{n=0}^\infty(-1)^{m+n}b_m b_n I_m\left(\frac{\Gamma_1}{2}\right)I_n\left(\frac{\Gamma_2}{2}\right)$$
$$\times\left[\exp\left(-\frac{(\Omega_{mn}^+)^2}{\Gamma}\sigma\right)\cos(\Omega_{mn}^+\varphi)\right.$$
$$\left.+\exp\left(-\frac{(\Omega_{mn}^-)^2}{\Gamma}\sigma\right)\cos(\Omega_{mn}^-\varphi)\right].$$

Here, a short-hand notation for the combination frequencies has been introduced: $\Omega_{mn}^\pm=m\Omega_1\pm n\Omega_2$. Transforming back to the original variable yields

$$W(\sigma,\varphi)$$

$$= \frac{2}{\Gamma} \frac{1}{\zeta(\sigma,\varphi)} \frac{\partial \zeta(\sigma,\varphi)}{\partial \varphi}$$

$$= \frac{\sum_{m=0}^{\infty}\sum_{n=0}^{\infty}(-w_{mn})[\Omega_{mn}^{+}e_{mn}^{+}\sin(\Omega_{mn}^{+}\varphi)+\Omega_{mn}^{-}e_{mn}^{-}\sin(\Omega_{mn}^{-}\varphi)]}{(\Gamma/2)\sum_{m=0}^{\infty}\sum_{n=0}^{\infty}w_{mn}[e_{mn}^{+}\cos(\Omega_{mn}^{+}\varphi)+e_{mn}^{-}\cos(\Omega_{mn}^{-}\varphi)]},$$
(8.117)

where

$$w_{mn} = (-1)^{m+n} b_m b_n I_m\left(\frac{\Gamma_1}{2}\right) I_n\left(\frac{\Gamma_2}{2}\right),$$

$$e_{mn}^{+} = \exp\left(-\frac{(\Omega_{mn}^{+})^2}{\Gamma}\sigma\right),$$

$$e_{mn}^{-} = \exp\left(-\frac{(\Omega_{mn}^{-})^2}{\Gamma}\sigma\right).$$

It is seen that for $\sigma > 0$ the solution in φ contains all combination frequencies $m\Omega_1 \pm n\Omega_2$ of the two input frequencies due to the nonlinear interaction. Note also that for $\Omega_1 = 1$, $\Gamma_1 = \Gamma$, $\Omega_2 = 0$, $\Gamma_2 = 0$ the solution (8.96) is recovered. In its full generality, the expression is rather cumbersome to analyze. However, as higher frequencies are more strongly damped for $\sigma \to \infty$ only a few frequencies remain with sufficient amplitude. In particular, if Ω_1 and $\Omega_2 < \Omega_1$ are close, the difference frequency $\Delta\Omega = \Omega_1 - \Omega_2$ ($m = n = 1$) will be small and give a strong component with

$$W_-(\sigma,\varphi) = -\frac{4\Delta\Omega}{\Gamma} \frac{I_1\left(\frac{\Gamma_1}{2}\right) I_1\left(\frac{\Gamma_2}{2}\right)}{I_0\left(\frac{\Gamma_1}{2}\right) I_0\left(\frac{\Gamma_2}{2}\right)}$$

$$\times \exp\left(-\frac{(\Delta\Omega)^2}{\Gamma}\sigma\right) \sin(\Delta\Omega\varphi), \quad (8.118)$$

or, returning to physical coordinates and constants,

$$u_-(x,t)$$

$$= -\frac{2\Delta\omega\,\delta}{c_0\beta} \frac{I_1\left(\frac{\Gamma_1}{2}\right) I_1\left(\frac{\Gamma_2}{2}\right)}{I_0\left(\frac{\Gamma_1}{2}\right) I_0\left(\frac{\Gamma_2}{2}\right)}$$

$$\times \exp\left(-\frac{\delta(\Delta\omega)^2}{2c_0^3}x\right) \sin[\Delta\omega(t-x/c_0)]$$

$$= u_-^{(0)} \exp\left(-\frac{\delta(\Delta\omega)^2}{2c_0^3}x\right) \sin[\Delta\omega(t-x/c_0)].$$
(8.119)

The Gol'dberg numbers of the interacting waves determine the quantity $u_-^{(0)}$ of the resulting difference-frequency wave. For $\Gamma_{1,2} \ll 1$, i.e., for low-power waves, $I_0(\Gamma/2) \approx 1$ and $I_1(\Gamma/2) \approx \Gamma/4$, thus the value

$$u_-^{(0)} = -\frac{2\Delta\omega\,\delta}{c_0\beta} \frac{\Gamma_1\Gamma_2}{16} = -\frac{\Delta\omega\beta c_0}{2\delta} \frac{u_{01}u_{02}}{\omega_1\omega_2} \quad (8.120)$$

is proportional to the product of the amplitudes of the interacting waves. For very intense waves, $\Gamma_{1,2} \gg 1$, as $\lim_{\xi\to\infty} I_1(\xi)/I_0(\xi) = 1$, the quantity $u_-^{(0)}$ of the difference-frequency wave becomes independent of u_{01} and u_{02},

$$u_-^{(0)} = -\frac{2\Delta\omega\,\delta}{c_0\beta}. \quad (8.121)$$

The difference-frequency wave (8.121) has been given for the far field where the two incoming waves have essentially ceased to interact. There, the wave propagates linearly and is exponentially damped by thermoviscous dissipation.

8.10 Bubbly Liquids

Liquids with bubbles have strongly pronounced nonlinear acoustic properties mainly due to the nonlinear oscillations of the bubbles and the high compressibility of the gas inside. Within recent decades theoretical and experimental investigations have detected many kinds of nonlinear wave phenomena in bubbly fluids. To mention just a few of these: ultrasound self-focusing [8.83, 84], acoustic chaos [8.31], sound self-transparency [8.85], wavefront conjugation [8.85], the acoustic phase echo [8.86] intensification of sound waves in nonuniform bubbly fluids [8.87, 88], subharmonic wave generation [8.89], structure formation in acoustic cavitation [8.90, 91], difference-frequency sound generation [8.92, 93]. These phenomena are discussed in several books and review papers [8.2, 9, 81, 94–97].

In this section we are going to discuss some phenomena related to nonlinear acoustic wave propagation in liquids with bubbles. First, the mathematical model for pressure-wave propagation in bubbly liquids will be presented. Second, this model will be used to investigate long nonlinear pressure waves, short pressure wave trains, and some nonlinear interactions between them.

Let α_l and α_g be the volume fractions and ρ_l and ρ_g the densities of the liquid and the gas (vapor), respectively. Then the density of the two-phase mixture, ρ, in the bubbly liquid mixture is given by

$$\rho = \alpha_l \rho_l + \alpha_g \rho_g, \qquad (8.122)$$

where $\alpha_l + \alpha_g = 1$. Note that the gas (vapor) volume fraction depends on the instantaneous bubble radius R,

$$\alpha_g = \frac{4}{3}\pi R^3 n, \qquad (8.123)$$

where n is the number density of bubbles in the mixture. Equation (8.123) is true in general and applicable to the case in which bubbles oscillate in an acoustic field, as considered in the following.

8.10.1 Incompressible Liquids

First, we consider the case when the liquid can be treated as incompressible. The bubbly fluid model in this case is based on the *bubble-in-cell* technique [8.94], which divides the bubbly liquid mixture into cells, with each cell consisting of a liquid sphere of radius R_c with a bubble of radius R at its center. A similar approach has been used recently to model wet foam drop dynamics in an acoustic field [8.98] and the acoustics of bubble clusters in a spherical resonator [8.80].

It should be mentioned here that the bubble-in-cell technique may be treated as a first-order correction to a bubbly fluid dynamics model due to a small bubble volume fraction. This technique is not capable of capturing the entire bubbly fluid dynamics for high concentration of bubbles when bubbles lose their spherical shape.

According to the bubble-in-cell technique, the radius, R_c, of the liquid sphere comprising each cell and the embedded bubble, are related to the local void fraction at any instant by

$$\frac{R}{R_c} = \alpha_g^{1/3}. \qquad (8.124)$$

We note that R_c, R, and α_g are variable in time.

The liquid conservation of mass equation and the assumed liquid incompressibility imply that the radial velocity around a single bubble depends on the radial coordinate:

$$v' = \frac{R^2 \dot{R}}{r'^2}, \quad R \le r' \le R_c, \qquad (8.125)$$

where $R(t)$ is the instantaneous radius of the bubble, primed quantities denote local (single-cell) variables, namely, r' is the radial coordinate with origin at the bubble's center, v' is the radial velocity of the liquid at radial location r', and the dot denotes a time derivative (i.e., $\dot{R} = dR/dt$).

The dynamics of the surrounding liquid is analyzed by writing the momentum equation for an incompressible, inviscid liquid as

$$\frac{\partial v'}{\partial t} + v' \frac{\partial v'}{\partial r'} + \frac{1}{\rho_l} \frac{\partial p'}{\partial r'} = 0. \qquad (8.126)$$

Integrating this equation over $r' \le R_c$ from R to some arbitrary r' and using (8.125) yields

$$p' = p_R - \rho_l \left[R\ddot{R} + \frac{3}{2}\dot{R}^2 \right.$$
$$\left. - \left(R\ddot{R} + 2\dot{R}^2\right)\frac{R}{r'} + \frac{1}{2}\dot{R}^2\left(\frac{R}{r'}\right)^4 \right], \qquad (8.127)$$

where p' is the liquid pressure at location r' around a bubble, and p_R is the liquid pressure at the bubble interface. Evaluating (8.127) at the outer radius of the cell, $r' = R_c$, and using (8.124), results in

$$\frac{p_R - p_c}{\rho_l} = \left(1 - \alpha_g^{1/3}\right) R\ddot{R}$$
$$+ \frac{3}{2}\left(1 - \frac{4}{3}\alpha_g^{1/3} + \frac{1}{3}\alpha_g^{4/3}\right) \dot{R}^2. \qquad (8.128)$$

This equation bears a resemblance to the well-known Rayleigh–Plesset equation that governs the motion of a single bubble in an infinite liquid. Indeed, in the limit of vanishing void fraction, $\alpha_g \to 0$, the pressure at the outer radius of the cell (p_c) becomes the liquid pressure far from the bubble ($p_c \to p_\infty$), and (8.128) reduces to the Rayleigh–Plesset equation

$$\frac{p_R - p_\infty}{\rho_l} = R\ddot{R} + \frac{3}{2}\dot{R}^2. \qquad (8.129)$$

Let us assume that the gas pressure inside the bubble p_g is spatially uniform. The momentum jump condition requires that the gas pressure differs from the liquid pressure at the bubble wall due to surface tension and viscous terms according to

$$p_R = p_g - \frac{2\sigma}{R} - \frac{4\mu_l \dot{R}}{R}, \qquad (8.130)$$

where σ is the surface tension and μ_l is the liquid viscosity. Furthermore, the gas pressure may be assumed to be governed by a polytropic gas law of the form

$$p_g = \left(p_0 + \frac{2\sigma}{R_0}\right)\left(\frac{R}{R_0}\right)^{-3\kappa}, \qquad (8.131)$$

where R_0 is the equilibrium radius of the bubble (i.e., the bubble radius at ambient pressure) and κ is a polytropic exponent. Combining (8.128)–(8.131) gives the following relationship between the liquid pressure at the edge of the cell (p_c) and the radius of the bubble (R):

$$p_c = \left(p_0 + \frac{2\sigma}{R}\right)\left(\frac{R}{R_0}\right)^{-3\kappa} - \frac{2\sigma}{R} - \frac{4\mu_l \dot{R}}{R}$$
$$- \rho_l \left[\left(1 - \alpha_g^{1/3}\right) R\ddot{R} + \frac{3}{2}\left(1 - \frac{4}{3}\alpha_g^{1/3} + \frac{1}{3}\alpha_g^{4/3}\right) \dot{R}^2\right]. \quad (8.132)$$

Equation (8.132) can be integrated to find the bubble radius, $R(t)$, given the time-dependent pressure at $r' = R_c$ and the initial conditions.

Next, we connect the various cells at points on their outer radii and replace the configuration with an equivalent fluid whose dynamics approximate those of the bubbly liquid. We assume that the velocity of the translational motion of the bubbles in such a fluid is equal to the velocity of the bubbly fluid $\mathbf{v}$. The fluid is required to conserve mass, thus

$$\frac{\partial \rho}{\partial t} + \nabla \cdot (\rho \mathbf{v}) = 0, \quad (8.133)$$

where ρ is given by (8.122). The fluid also satisfies Euler's equation,

$$\rho \frac{d\mathbf{v}}{dt} + \nabla p = 0, \quad (8.134)$$

where $d/dt = \partial/\partial t + \mathbf{v} \cdot \nabla$ is the material derivative, and p is the mean fluid pressure, which is approximately the pressure at the outer radius of a cell, p_c.

The mass and bubble densities are related by requiring that the total mass of each cell does not change in time. The initial volume of one cell is $V_0 = 1/n_0$, so the initial mass of the cell is $M_0 = \rho_0/n_0$, where $\rho_0 = \alpha_{l0}\rho_{l0} + \alpha_{g0}\rho_{g0}$. Requiring that the mass of each cell remains constant gives

$$\frac{\rho}{n} = \frac{\rho_0}{n_0}. \quad (8.135)$$

This set of equations describes the nonlinear dynamics of a bubbly liquid mixture over a wide range of pressures and temperatures.

We can now linearize these equations by assuming that the time-dependent quantities only vary slightly from their equilibrium values. Specifically, we write $\rho = \rho_0 + \tilde{\rho}$, $n = n_0 + \tilde{n}$, $p = p_0 + \tilde{p}$, $\mathbf{v} = \tilde{\mathbf{v}}$, $\alpha_g = \alpha_{g0} + \tilde{\alpha}_g$, and $R = R_0 + \tilde{R}$. The perturbed quantities are assumed to be small, so that any product of the perturbations may be neglected. When these assumptions are introduced into (8.132), we arrive at the following linearized equation:

$$\tilde{p} = -\left[\left(p_0 + \frac{2\sigma}{R_0}\right)\frac{3\kappa}{R_0} - \frac{2\sigma}{R_0^2}\right]\tilde{R}$$
$$- \frac{4\mu_l}{R_0}\frac{\partial \tilde{R}}{\partial t} - \rho_l\left(1 - \alpha_{g0}^{1/3}\right)R_0\frac{\partial^2 \tilde{R}}{\partial t^2}. \quad (8.136)$$

Linearizing (8.135) gives

$$\tilde{n} = \frac{n_0}{\rho_0}\tilde{\rho}. \quad (8.137)$$

Similarly, linearizing the density in (8.122) taking into account that $\rho_g/\rho_l \ll 1$ and using (8.137) yields

$$\tilde{\rho} = -4\pi R_0^2 \rho_0 n_0 \tilde{R}, \quad (8.138)$$

where $\rho_0 \approx \rho_l(1 - \alpha_{g0})$. By combining (8.136) and (8.138) we obtain a pressure–density relationship for the bubbly fluid:

$$\tilde{p} = \frac{C_b^2}{\omega_b^2}\left(\omega_b^2 \tilde{\rho} + 2\delta_b \frac{\partial \tilde{\rho}}{\partial t} + \frac{\partial^2 \tilde{\rho}}{\partial t^2}\right), \quad (8.139)$$

where C_b represents the low-frequency sound speed in the bubbly liquid

$$C_b^2 = \frac{3\kappa p_0 + (3\kappa - 1)\frac{2\sigma}{R_0}}{3\alpha_{g0}(1 - \alpha_{g0})\rho_l}, \quad (8.140)$$

and ω_b is the natural frequency of a bubble in the cell

$$\omega_b^2 = \frac{\omega_M^2}{1 - \alpha_{g0}^{1/3}}. \quad (8.141)$$

We note that ω_M is the natural frequency of a single bubble in an infinite liquid (i.e. the so-called Minnaert frequency),

$$\omega_M^2 = \frac{3\kappa p_0 + (3\kappa - 1)\frac{2\sigma}{R_0}}{\rho_l R_0^2}. \quad (8.142)$$

In (8.139) the parameter δ_b represents the dissipation coefficient due to the liquid viscosity,

$$\delta_b = \frac{2\mu_l}{\rho_l R_0^2 \left(1 - \alpha_{g0}^{1/3}\right)}. \quad (8.143)$$

Actually acoustic wave damping in bubbly liquids occurs due to several different physical mechanisms:

liquid viscosity, gas/vapor thermal diffusivity, acoustic radiation, etc. The contribution of each of these dissipation mechanisms to the total dissipation during bubble oscillations depends on frequency, bubble size, the type of gas in a bubble and liquid compressibility [8.94]. For convenience one can use an effective dissipation coefficient in the form of a viscous dissipation:

$$\delta_{\text{eff}} = \frac{2\mu_{\text{eff}}}{\rho_1 R_0^2 \left(1 - \alpha_{g0}^{1/3}\right)}, \quad (8.144)$$

where μ_{eff} denotes an effective viscosity (instead of just the liquid viscosity), which implicitly includes all the aforementioned dissipation mechanisms in a bubbly liquid. The value for μ_{eff} should be chosen to fit experimental observations and the theoretical data of amplitude versus frequency for a single bubble.

We can also linearize (8.133) and (8.134) to obtain

$$\frac{\partial \tilde{\rho}}{\partial t} + \rho_0 \nabla \cdot \mathbf{v} = 0, \quad (8.145)$$

$$\rho_0 \frac{\partial \mathbf{v}}{\partial t} + \nabla \tilde{p} = 0. \quad (8.146)$$

Combining the time derivative of (8.145) with the divergence of (8.146) results in

$$\frac{\partial^2 \tilde{\rho}}{\partial t^2} = \nabla^2 \tilde{p}. \quad (8.147)$$

Combining this result with (8.139) gives a wave equation for the bubbly liquid of the form

$$\frac{\partial^2 \tilde{p}}{\partial t^2} - C_b^2 \left(1 + \frac{2\delta_{\text{eff}}}{\omega_b^2} \frac{\partial}{\partial t} + \frac{1}{\omega_b^2} \frac{\partial^2}{\partial t^2}\right) \nabla^2 \tilde{p} = 0. \quad (8.148)$$

Equation (8.148) describes the propagation of a linear acoustical pressure perturbations in a liquid with bubbles when the liquid can be treated as incompressible. It also accounts for the effect of a small but finite void fraction on wave propagation [8.9, 94, 95].

8.10.2 Compressible Liquids

When the pressure waves are very intense, and the void fraction is small ($\alpha_g \ll 1$), one should take into account liquid compressibility, which may lead to acoustic radiation by the oscillating bubbles. In this case, correction terms of $\alpha_g^{1/3}$, $\alpha_g^{4/3}$ in (8.128) may be neglected, and in order to account for acoustic radiation (8.128) should be rewritten as follows [8.99, 100]:

$$R\ddot{R} + \frac{3}{2}\dot{R}^2 = \frac{p_R - p}{\rho_{10}} + \frac{R}{\rho_{10}C_1} \frac{d}{dt}(p_R - p), \quad (8.149)$$

where C_1 is the speed of sound in the liquid. In (8.149) the liquid density, ρ_1, is taken as a constant ρ_{10} although the equation of state of the liquid can be approximated as

$$p = p_0 + C_1^2 (\rho_1 - \rho_{10}). \quad (8.150)$$

After linearization (8.149) becomes [compare with (8.136)]:

$$\tilde{p} = -\left[\left(p_0 + \frac{2\sigma}{R_0}\right)\frac{3\kappa}{R_0} - \frac{2\sigma}{R_0^2}\right]\tilde{R}$$
$$-\frac{4\mu_1}{R_0}\frac{\partial \tilde{R}}{\partial t} - \rho_{10}R_0\left(1 + \frac{R_0}{C_1}\frac{\partial}{\partial t}\right)^{-1}\frac{\partial^2 \tilde{R}}{\partial t^2}. \quad (8.151)$$

In order to evaluate the last term in (8.151) and to incorporate acoustic radiation losses into an *effective* viscosity scheme we use the following approximation for the differential operator:

$$\left(1 + \frac{R_0}{C_1}\frac{\partial}{\partial t}\right)^{-1} \approx 1 - \frac{R_0}{C_1}\frac{\partial}{\partial t}. \quad (8.152)$$

Then (8.151) becomes

$$\tilde{p} \approx -\left[\left(p_0 + \frac{2\sigma}{R_0}\right)\frac{3\kappa}{R_0} - \frac{2\sigma}{R_0^2}\right]\tilde{R} - \frac{4\mu_1}{R_0}\frac{\partial \tilde{R}}{\partial t}$$
$$-\rho_{10}R_0\frac{\partial^2 \tilde{R}}{\partial t^2} + \frac{\rho_{10}R_0^2}{C_1}\frac{\partial^3 \tilde{R}}{\partial t^3}. \quad (8.153)$$

The third derivative in (8.153) can be estimated using the approximation of a freely oscillating bubble

$$\frac{\partial^2 \tilde{R}}{\partial t^2} \approx -\frac{\left(p_0 + \frac{2\sigma}{R_0}\right)\frac{3\kappa}{R_0} - \frac{2\sigma}{R_0^2}}{\rho_{10}R_0}\tilde{R}. \quad (8.154)$$

Substituting (8.154) in the third-derivative term of (8.153) we get

$$\tilde{p} = -\left[\left(p_0 + \frac{2\sigma}{R_0}\right)\frac{3\kappa}{R_0} - \frac{2\sigma}{R_0^2}\right]\tilde{R} - \frac{4\mu_{\text{eff}}}{R_0}\frac{\partial \tilde{R}}{\partial t}$$
$$-\rho_{10}R_0\frac{\partial^2 \tilde{R}}{\partial t^2}, \quad (8.155)$$

where

$$\mu_{\text{eff}} = \mu_1 + \mu_r, \quad \mu_r = \frac{(3\kappa - 1)\sigma}{2C_1} + \frac{3\kappa p_0}{4C_1}R_0. \quad (8.156)$$

It is easy to see from (8.156) that acoustic radiation may lead to a very large dissipation. For example, for

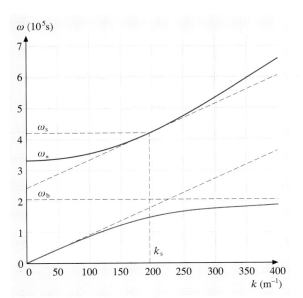

Fig. 8.13 The dispersion relation for water with air bubbles (the liquid is treated as compressible, dissipation is ignored): $C_1 = 1500$ m/s, $p_0 = 10^5$ Pa, $\kappa = 1.4$, $\sigma = 0.0725$ N/m, $\rho_{10} = 10^3$ kg/m, $\alpha_{g0} = 10^{-4}$, $R_0 = 10^{-4}$ m

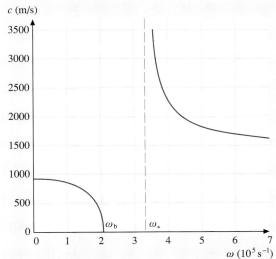

Fig. 8.14 The phase velocity for water with air bubbles (the liquid is treated as compressible, dissipation is ignored): $C_1 = 1500$ m/s, $p_0 = 10^5$ Pa, $\kappa = 1.4$, $\sigma = 0.0725$ N/m, $\rho_{10} = 10^3$ kg/m, $\alpha_{g0} = 10^{-4}$, $R_0 = 10^{-4}$ m

air bubbles of $R_0 = 10^{-4}$ m in water, effective viscosity due to acoustic radiation μ_r may be about seven times larger than the viscosity of water, μ_l.

Equation (8.135), and eventually its linearized form (8.137), are valid in the case of a compressible liquid. However, the linearized form of (8.122) changes. Now instead of (8.138) we have

$$\tilde{\rho} = \tilde{\rho}_1 - 4\pi R_0^2 \rho_0 n_0 \tilde{R} \tag{8.157}$$

or, accounting for the liquid equation of state (8.150),

$$\tilde{\rho} = C_1^{-2} \tilde{p} - 4\pi R_0^2 \rho_0 n_0 \tilde{R} . \tag{8.158}$$

Substituting (8.158) into (8.147) we have

$$C_1^{-2} \frac{\partial^2 \tilde{p}}{\partial t^2} - \nabla^2 \tilde{p} = 4\pi R_0^2 n_0 \rho_0 \frac{\partial^2 \tilde{R}}{\partial t^2} . \tag{8.159}$$

Equation (8.159) together with (8.155) leads to the following wave equation for a bubbly fluid in the case of a compressible liquid:

$$\frac{\partial^2 \tilde{p}}{\partial t^2} - C_b^2 \left(1 + \frac{2\delta_{\mathrm{eff}}}{\omega_b^2} \frac{\partial}{\partial t} + \frac{1}{\omega_b^2} \frac{\partial^2}{\partial t^2}\right) \times$$
$$\left(\nabla^2 - C_1^{-2} \frac{\partial^2}{\partial t^2}\right) \tilde{p} = 0 . \tag{8.160}$$

Here ω_b and δ_{eff} are calculated according to (8.141) and (8.144) in which α_{g0} is taken to be zero, and C_b is calculated according to (8.140). It is easy to see that (8.160) reduces to (8.148) in the case when $C_1 \to \infty$.

The linear wave equation (8.160) admits the harmonic wave-train solution

$$\tilde{p} = p_a \exp[\mathrm{i}(kx - \omega t)] , \tag{8.161}$$

in which the frequency ω and wave number k are related through the following dispersion relation

$$k^2 = \omega^2 \left(\frac{1}{C_1^2} + \frac{1}{C_b^2 \left(1 + \mathrm{i} \dfrac{2\omega \delta_{\mathrm{eff}}}{\omega_b^2} - \dfrac{\omega^2}{\omega_b^2}\right)} \right) . \tag{8.162}$$

The graph of this dispersion relation is shown in Fig. 8.13. The corresponding phase velocity is given in Fig. 8.14.

8.10.3 Low-Frequency Waves: The Korteweg–de Vries Equation

In order to analyze the low-frequency nonlinear acoustics of a bubbly liquid the low-frequency limit of the dispersion relation (8.162) is first considered:

$$\omega = Ck + i\alpha_1 k^2 - \alpha_2 k^3, \quad (8.163)$$

where

$$\frac{1}{C^2} = \frac{1}{C_1^2} + \frac{1}{C_b^2}, \quad \alpha_1 = \frac{\delta_{\text{eff}} C^4}{\omega_b^2 C_b^2}, \quad \alpha_2 = \frac{C^5}{2 C_b^2 \omega_b^2}. \quad (8.164)$$

Noting that

$$\omega = -i\frac{\partial}{\partial t}, \quad k = i\frac{\partial}{\partial x}, \quad (8.165)$$

(8.163) can be treated as the Fourier-space equivalent of an operator equation which when operating on $\tilde{p}$, yields

$$\frac{\partial \tilde{p}}{\partial t} + C\frac{\partial \tilde{p}}{\partial x} - \alpha_1 \frac{\partial^2 \tilde{p}}{\partial x^2} + \alpha_2 \frac{\partial^3 \tilde{p}}{\partial x^3} = 0. \quad (8.166)$$

Equation (8.166) should be corrected to account for weak nonlinearity. A systematic derivation is normally based on the *multiple-scales* technique [8.101]. Here we show how this derivation can be done less rigorously, but more simply. We assume that the nonlinearity in bubbly liquids comes only from bubble dynamics. Then (8.155) for weakly nonlinear bubble oscillations has an additional nonlinear term and becomes:

$$\tilde{p} = -B_1 \tilde{R} + B_2 \tilde{R}^2 - \frac{4\mu_{\text{eff}}}{R_0}\frac{\partial \tilde{R}}{\partial t} - \rho_{10} R_0 \frac{\partial^2 \tilde{R}}{\partial t^2}. \quad (8.167)$$

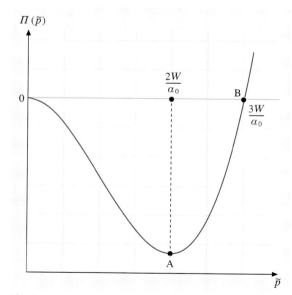

Fig. 8.15 The potential well used to illustrate the "soliton" solution for the KdV equation

$$B_1 = \left(p_0 + \frac{2\sigma}{R_0}\right)\frac{3\kappa}{R_0} - \frac{2\sigma}{R_0^2},$$

$$B_2 = \left(p_0 + \frac{2\sigma}{R_0}\right)\frac{3\kappa(3\kappa+1)}{2R_0^2} - \frac{2\sigma}{R_0^3}. \quad (8.168)$$

Equation (8.167) has to be combined with the linear wave equation (8.159), which in the case of plane waves can be written:

$$C_1^{-2}\frac{\partial^2 \tilde{p}}{\partial t^2} - \frac{\partial^2 \tilde{p}}{\partial x^2} = 4\pi R_0^2 n_0 \rho_0 \frac{\partial^2 \tilde{R}}{\partial t^2}. \quad (8.169)$$

Taking into account that all the terms except the first one on the right-hand side of (8.167) are small, i.e., of the second order of magnitude, one can derive the following nonlinear wave equation for pressure perturbations in a bubbly liquid:

$$C^{-2}\frac{\partial^2 \tilde{p}}{\partial t^2} - \frac{\partial^2 \tilde{p}}{\partial x^2}$$
$$= 4\pi R_0^2 n_0 \rho_0$$
$$\times \left(\frac{B_2}{B_1^3}\frac{\partial^2 \tilde{p}^2}{\partial t^2} + \frac{4\mu_{\text{eff}}}{R_0 B_1^2}\frac{\partial^3 \tilde{p}}{\partial t^3} + \frac{\rho_{10} R_0}{B_1^2}\frac{\partial^4 \tilde{p}}{\partial t^4}\right). \quad (8.170)$$

Equation (8.170) is derived for plane weakly nonlinear pressure waves traveling in a bubbly liquid in both directions. Namely, the left part of this equation contains a classic wave operator when applied to a pressure perturbation function describes waves traveling left to right and right to left. The right part of this equation contain terms of second order of smallness and is therefore responsible for a slight change of these waves. Thus, (8.170) may be structured as follows:

$$\left(C^{-1}\frac{\partial}{\partial t} + \frac{\partial}{\partial x}\right)\left(C^{-1}\frac{\partial}{\partial t} - \frac{\partial}{\partial x}\right)\tilde{p} = O\left(\varepsilon^2\right). \quad (8.171)$$

If one considers only waves traveling left to right then

$$C^{-1}\frac{\partial}{\partial t} + \frac{\partial}{\partial x} = O(\varepsilon), \quad (8.172)$$

and we can use the following derivative substitution (see [8.10])

$$\frac{\partial}{\partial t} \approx -C\frac{\partial}{\partial x}. \quad (8.173)$$

Then, after one time integration over space, and translation to the frame moving left to right with a speed of sound C, (8.170) leads to the Burgers–Korteweg–de Vries (BKdV) equation for a pressure wave [8.94–96, 102]:

$$\frac{\partial \tilde{p}}{\partial t} + \alpha_0 \tilde{p} \frac{\partial \tilde{p}}{\partial \xi} - \alpha_1 \frac{\partial^2 \tilde{p}}{\partial \xi^2} + \alpha_2 \frac{\partial^3 \tilde{p}}{\partial \xi^3} = 0,$$

$$\xi = x - Ct, \quad (8.174)$$

where

$$\alpha_0 = \frac{C^3}{C_b^2} \cdot \frac{B_2}{B_1^2} \quad (8.175)$$

and the parameters α_1, α_2, and C are presented in (8.164).

Let us first discuss a nondissipative case, $\alpha_1 = 0$. Then (8.174) is called the Korteweg–de Vries (KdV) equation. The low-frequency dispersion relation (8.163) shows that every Fourier component of initial arbitrary pressure perturbation propagates with its own speed, and the shape of the acoustic signal changes. Since $\alpha_2 > 0$, the long-wavelength waves propagate with higher speed than lower-wavelength waves. This means that dispersion may begin to compete with nonlinearity, which stops shock-wave formation. The competition between dispersion and nonlinearity leads to the formation of a so-called *soliton*.

In order to derive the shape of soliton, here we consider a steady solution of (8.174) in the moving frame

$$\eta = \xi - Wt. \quad (8.176)$$

Then, assuming that there is no pressure perturbation at infinity, (8.174) can be reduced as follows:

$$\alpha_2 \frac{d^2 \tilde{p}}{d\eta^2} = -\frac{\partial \Pi}{\partial \tilde{p}}, \quad \Pi(\tilde{p}) = -\frac{W \tilde{p}^2}{2} + \frac{\alpha_0 \tilde{p}^3}{6}. \quad (8.177)$$

It is instructive to use the analogy that (8.177) is similar to the equation of a point-like particle in a potential well shown schematically in Fig. 8.15. The structure of the pressure *solitary* wave can be viewed as follows: initially the particle is at point O, which corresponds to the fact that far away the pressure perturbation is equal to zero; then the particle slides down to the bottom of the potential well, point A, and due to a conservation law climbs up to point B; at point B it stops and moves all the way back to point O. The solution of (8.177) represents the shape of the soliton, which in the immovable frame is

$$\tilde{p} = \frac{3W}{\alpha_0} \cosh^{-2} \left\{ \sqrt{\frac{W}{4\alpha_2}} [x - (C+W)t] \right\}. \quad (8.178)$$

It was mentioned above that a soliton may be interpreted as a result of the balance between two competitors: nonlinearity and dispersion. To consider this competition in more detail the KdV equation is considered in the frame moving with the speed of sound in a bubbly liquid

$$\frac{\partial \tilde{p}}{\partial t} + \alpha_0 \tilde{p} \frac{\partial \tilde{p}}{\partial \xi} + \alpha_2 \frac{\partial^3 \tilde{p}}{\partial \xi^3} = 0, \quad \xi = x - Ct. \quad (8.179)$$

Then the solitary wave has the following shape:

$$\tilde{p} = \frac{3W}{\alpha_0} \cosh^{-2} \left\{ \sqrt{\frac{W}{4\alpha_2}} (\xi - Wt) \right\} \quad (8.180)$$

with a pressure amplitude equal to $3W/\alpha_0$ and a thickness of about $\sqrt{\alpha_2/W}$.

In order to evaluate the effect of nonlinearity let us consider the simple nonlinear equation

$$\frac{\partial \tilde{p}}{\partial t} + \alpha_0 \tilde{p} \frac{\partial \tilde{p}}{\partial \xi} = 0. \quad (8.181)$$

The factor $\alpha_0 \tilde{p}$ stands for the velocity of transportation of pressure perturbations. Thus, part of the pressure profile which has higher pressure will be translated faster than the part which has lower pressure. According to the solitary solution (8.180) the maximum translation velocity is equal to $3W$. The time needed for the pressure peak to be translated the distance of about the thickness of the pressure wave can be estimated as $t_{\rm nl} \sim \alpha_2^{1/2} W^{-3/2}$. This amount of time would be needed to transform the smooth solitary shape to a shock wave.

In order to evaluate the effect of dispersion let us consider the simple dispersive equation

$$\frac{\partial \tilde{p}}{\partial t} + \alpha_2 \frac{\partial^3 \tilde{p}}{\partial \xi^3} = 0. \quad (8.182)$$

This equation admits harmonic wave-train solutions of the form $\exp(ik\xi + i\alpha_2 k^3 t)$. The characteristic wave numbers contributing to the wave shape of thickness $\sqrt{\alpha_2/W}$ are $k \sim \sqrt{W/\alpha_2}$. Such wave numbers will gradually change the wave phase on π and thus lead to a substantial deformation of the wave shape. The time interval needed for this, $t_{\rm d}$, is calculated as follows: $\alpha_2 \left(\sqrt{W/\alpha_2}\right)^3 t_{\rm d} = \pi$, thus, $t_{\rm d} \sim \alpha_2^{1/2} W^{-3/2}$, which is $\sim t_{\rm nl}$. So, the time that is needed for nonlinearity to transform the solitary wave into a shock wave is equal to the time taken for dispersion to smooth out this wave.

One of the most important results obtained from the BKdV equation is the prediction of an oscillatory shock wave. In order to derive the shape of an oscillatory shock wave we again consider the steady-state solution of (8.174) in the moving frame (8.176). Then, assuming that there is no pressure perturbation at infinity, (8.174) can be reduced to

$$\alpha_2 \frac{d^2 \tilde{p}}{d\eta^2} - \alpha_1 \frac{d\tilde{p}}{d\eta} = -\frac{\partial \Pi}{\partial \tilde{p}},$$

$$\Pi(\tilde{p}) = -\frac{W\tilde{p}^2}{2} + \frac{\alpha_0 \tilde{p}^3}{6}. \quad (8.183)$$

Equation (8.183) represents the equation of a point-like particle in a potential well shown schematically in Fig. 8.15 with friction. Here $\tilde{p}$ stands for the coordinate of the particle, minus ξ stands for the time (time passes from $-\infty$ to $+\infty$), α_2 is the mass of the particle, and α_1 is the friction coefficient. An oscillatory shock wave corresponds to motion of the particle from point O to the bottom of potential well (point A) with some decaying oscillations around the bottom. Thus, the amplitude of the shock wave is equal to $2W/\alpha_0$. Eventually oscillations take place only for relatively small values of the friction coefficient α_1. To derive a criterion for an oscillatory shock wave one should investigate a solution of (8.183) in close vicinity to point A. Assuming that

$$\tilde{p} = \frac{2W}{\alpha_0} + \psi(\xi), \quad (8.184)$$

where ψ is small, and linearizing (8.183) we get

$$\alpha_2 \frac{d^2 \psi}{d\xi^2} - \alpha_1 \frac{d\psi}{d\xi} + W\psi = 0. \quad (8.185)$$

Equation (8.185) has a solution

$$\psi = A_1 \exp(-\lambda \xi) + A_2 \exp(\lambda \xi),$$

$$\lambda = \frac{\alpha_1}{2\alpha_2} \pm \sqrt{\frac{\alpha_1^2}{4\alpha_2^2} - \frac{W}{\alpha_1}}. \quad (8.186)$$

It is easy to see that the shock wave has an oscillatory structure if

$$\alpha_1 < \sqrt{4\alpha_2 W}. \quad (8.187)$$

In terms of the bubbly fluid parameters, (8.187) gives a critical effective viscosity

$$\mu_{\text{eff}} < \mu_{\text{crit}} = \rho_l R_0^2 \omega_b \frac{C_b}{C} \sqrt{\frac{1}{2} \frac{W}{C}}. \quad (8.188)$$

The KdV equation for describing the evolution of long-wavelength disturbances in an ideal liquid with adiabatic gas bubbles was first proposed in [8.102]. The solitons and shock waves in bubbly liquids were systematically investigated in [8.95] in the framework of the BKdV equation. Good correlation with experimental data was obtained. In [8.94] wave-propagation phenomena in bubbly liquids were analyzed using a more advanced and complete mathematical model that is valid for higher intensities of pressure waves. In [8.103] the effect of polydispersivity on long small-amplitude nonlinear waves in bubbly liquids was considered. It was shown that evolution equations of BKdV type can be applied for modeling of such waves. In particular, it was shown that for polydisperse bubbly liquids an effective monodisperse liquid model can be used with the bubble sizes found as ratios of some bubble size-distribution moments and corrections to the equation coefficients.

8.10.4 Envelopes of Wave Trains: The Nonlinear Schrödinger Equation

When the amplitude of an acoustic wave in a fluid is small enough the respective linearized system of equations will admit the harmonic wave-train solution

$$u = a \exp\{i[kx - \omega(k)t]\}. \quad (8.189)$$

If the amplitude is not small enough the nonlinearity cannot be neglected. The nonlinear terms produce higher harmonics, which react back on the original wave. Thus, the effect of nonlinearity on this sinusoidal oscillation is to cause a variation in its amplitude and phase in both space and time. This variation can be considered as small (in space) and slow (in time) for off-resonant situations. To account for a variation of amplitude and phase it is convenient to consider the amplitude as a complex function of space and time A.

Here we follow [8.104] to show what kind of equation arises as the evolution equation for a carrier-wave envelope in a general dispersive system. A linear system has a dispersion relation which is independent of the wave amplitude

$$\omega = \omega(k). \quad (8.190)$$

However, it is instructive to assume that the development of a harmonic wave in a weakly nonlinear system can be represented by a dispersion relation which is amplitude dependent

$$\omega = \omega(k, |A|^2). \quad (8.191)$$

Such a situation was initially uncovered in nonlinear optics and plasma where the refractive index or dielectric constant of a medium may be dependent on the electric field. That same situation occurs in bubbly liquids as well.

Let us consider a carrier wave of wave number k_0 and frequency $\omega_0 = \omega(k_0)$. A Taylor expansion of the dispersion relation (8.191) around k_0 gives

$$\Omega = \left(\frac{\partial \omega}{\partial k}\right)_0 K + \frac{1}{2}\left(\frac{\partial^2 \omega}{\partial k^2}\right)_0 K^2 + \left(\frac{\partial \omega}{\partial (|A|^2)}\right)_0 |A|^2 , \tag{8.192}$$

where $\Omega = \omega - \omega_0$ and $K = k - k_0$ are the frequency and wave number of the variation of the wave-train amplitude. Equation (8.192) represents a dispersion relation for the complex amplitude modulation. Noting that

$$\Omega = -\mathrm{i}\frac{\partial}{\partial t}, \quad K = \mathrm{i}\frac{\partial}{\partial x} , \tag{8.193}$$

(8.192) can be treated as the Fourier-space equivalent of an operator equation that when operating on A yields

$$\mathrm{i}\left[\frac{\partial}{\partial t} + \left(\frac{\partial \omega}{\partial k}\right)_0 \frac{\partial}{\partial x}\right] A + \frac{1}{2}\left(\frac{\partial^2 \omega}{\partial k^2}\right)_0 \frac{\partial^2 A}{\partial x^2}$$
$$- \left(\frac{\partial \omega}{\partial (|A|^2)}\right)_0 |A|^2 A = 0 . \tag{8.194}$$

In the frame moving with group velocity

$$\xi = x - C_\mathrm{g} t, \quad C_\mathrm{g} = (\partial \omega/\partial k)_0 , \tag{8.195}$$

(8.194) represent the classical nonlinear Schrödinger (NLS) equation

$$\mathrm{i}\frac{\partial A}{\partial t} = \beta_\mathrm{d}\frac{\partial^2 A}{\partial \xi^2} + \gamma_\mathrm{n}|A|^2 A , \tag{8.196}$$

$$\beta_\mathrm{d} = -\frac{1}{2}\left(\frac{\partial^2 \omega}{\partial k^2}\right)_0 , \quad \gamma_\mathrm{n} = \left(\frac{\partial \omega}{\partial (|A|^2)}\right)_0 , \tag{8.197}$$

which describes the evolution of wave-train envelopes.

Here A is a complex amplitude that can be represented as follows:

$$A(t,\xi) = a(t,\xi)\exp[\mathrm{i}\varphi(t,\xi)] , \tag{8.198}$$

where a and φ are the (real) amplitude and phase of the wave train. A spatially uniform solution of (8.196) corresponds to an unperturbed wave train. To analyze the stability of this uniform solution let us represent (8.196) as a set of two scalar equations:

$$-a\frac{\partial \varphi}{\partial t} = \beta_\mathrm{d}\left[\frac{\partial^2 a}{\partial \xi^2} - a\left(\frac{\partial \varphi}{\partial \xi}\right)^2\right] + \gamma_\mathrm{n} a^3 , \tag{8.199}$$

$$\frac{\partial a}{\partial t} = \beta_\mathrm{d}\left(2\frac{\partial a}{\partial \xi}\frac{\partial \varphi}{\partial \xi} + a\frac{\partial^2 \varphi}{\partial \xi^2}\right) . \tag{8.200}$$

It is easy to verify that a spatially uniform solution $(\partial/\partial \xi \equiv 0)$ of (8.199) and (8.200) is:

$$a = a_0, \quad \varphi = -\gamma_\mathrm{n} a_0^2 t . \tag{8.201}$$

The evolution of a small perturbation of this uniform solution

$$a = a_0 + \tilde{a}, \quad \varphi = -\gamma_\mathrm{n} a_0^2 t + \tilde{\varphi} , \tag{8.202}$$

is given by the following linearized equations:

$$a_0\frac{\partial \tilde{\varphi}}{\partial t} + \beta_\mathrm{d}\frac{\partial^2 \tilde{a}}{\partial \xi^2} + 2\gamma_\mathrm{n} a_0^2 \tilde{a} = 0 . \tag{8.203}$$

$$\frac{\partial \tilde{a}}{\partial t} - \beta_\mathrm{d} a_0 \frac{\partial^2 \tilde{\varphi}}{\partial \xi^2} = 0 . \tag{8.204}$$

Now let us consider the evolution of a periodic perturbation with wavelength $L = 2\pi/K$ that can be written as

$$\begin{pmatrix}\tilde{a}\\ \tilde{\varphi}\end{pmatrix} = \begin{pmatrix}a_1\\ \varphi_1\end{pmatrix}\exp(\sigma t + \mathrm{i}K\xi) . \tag{8.205}$$

The stability of the uniform solution depends on the sign of the real part of the growth-rate coefficient σ. To compute σ we substitute the perturbation (8.205) into the linearized equations (8.203) and (8.204) and obtain the following formula for σ

$$\sigma^2 = \beta_\mathrm{d}^2 K^2\left(\frac{2\gamma_\mathrm{n}}{\beta_\mathrm{d}}a_0^2 - K^2\right) . \tag{8.206}$$

This shows that the sign of the $\beta_\mathrm{d}\gamma_\mathrm{n}$ product is crucial for wave-train stability. If $\beta_\mathrm{d}\gamma_\mathrm{n} < 0$ then σ is always an imaginary number, and a uniform wave train is stable to small perturbations of any wavelength; if $\beta_\mathrm{d}\gamma_\mathrm{n} > 0$ then in the case that

$$K < K_\mathrm{cr} = \sqrt{\frac{2\gamma_\mathrm{n}}{\beta_\mathrm{d}}} a_0 \tag{8.207}$$

σ is real, and a long-wavelength instability occurs.

This heuristic derivation shows how the equation for the evolution of the wave-train envelope arises. The NLS equation has two parameters: β_d, γ_n. Eventually, the parameter β_d can be calculated from the linear dispersion relation discussed in detail earlier. However, the parameter γ_n has to be calculated from more-systematic, nonlinear arguments.

The general method of derivation is often given the name of the *method of multiple scales* [8.101]. A specific multiscale technique for weakly nonlinear oscillations of bubbles was developed in [8.105, 106]. This technique has been applied to analyze pressure-wave propagation in bubbly liquids.

In [8.107] the NLS equation describing the propagation of weakly nonlinear modulation waves in bubbly liquids was obtained for the first time. It was derived using the multiple asymptotic technique from the governing continuum equations for bubbly liquids presented in detail in this chapter. The NLS equation was applied for determination of the regions of modulation stability/instability in the parametric plane for mono- and polydisperse bubbly liquids.

To understand the behavior of the coefficients of the NLS equation one can consider the simplified model of a monodisperse system presented before. For this we note that the dispersion relation for bubbly liquids with compressible carrier liquid has two branches (Fig. 8.13). The low-frequency branch is responsible for bubble compressibility, and the high-frequency branch is due to liquid compressibility.

The first (low-frequency) branch corresponds to frequencies from zero to ω_b, which is the resonance frequency for bubbles of given size, the second (high-frequency) branch corresponds to frequencies from some $\omega_\star$ to infinity. There are no solutions in the form of spatially oscillating waves for frequencies between ω_b and $\omega_\star$; in this range there are exponentially decaying solutions only. This region is known as the *window of acoustic non-transparency*.

The coefficient β_d is calculated as shown in (8.197). For bubbly liquids this quantity is always positive for the low-frequency branch and always negative for the high-frequency branch.

The coefficient γ_n (see (8.197)) represents the nonlinearity of the system and is a complicated function of frequency ω. If it is positive, then long-wavelength modulation instability occurs. An important region of instability appears near the frequency ω_s, where $C_g(\omega_s) = C_e$, with C_e being the equilibrium speed of sound ($C_e = \omega/k$ for $\omega \to 0$).

For such frequencies the coefficient γ_n has a singularity and changes sign when the frequency is changing around this point. Physically this corresponds to the Benjamin–Feir instability, and can be explained by transfer of energy from mode 0 (DC mode, or mode of zero frequency, for which perturbations propagate with velocity C_e) to the principal mode of the wave train, which moves with the group velocity C_g.

Accounting for a small dissipation in bubbly liquids leads to the Landau–Ginzburg equation, which describes how the wave amplitude should decay due to dissipation. This part includes the dissipation effect due to viscosity and due to internal bubble heat transfer. The difference between the viscous and thermal effects for bubbles is that for an oscillating bubble there is a $\pi/2$ shift in phase between the bubble volume and the heat flux, while the loss of energy due to viscous dissipation occurs in phase with bubble oscillation. This phase shift results in the effect that the pressure in the mixture has some additional shift in phase relative to the mixture density (determined mainly by the bubble volume). This, in fact, can be treated as a weak dispersive effect and included to the dispersion relation in the third-order approximation.

It is interesting to note that the system of governing equations presented can be also obtained as Euler equations for some Lagrangian [8.108, 109]. The variation formulation might be very useful for the analysis of nonlinear acoustics phenomena in bubbly liquids.

8.10.5 Interaction of Nonlinear Waves. Sound–Ultrasound Interaction

The nature of the interactions among individual wave components is quite simple and can be explained by considering the nature of dispersive waves in general.

Suppose u represents some property of the fluid motion, for example the fluid velocity. Then infinitesimal (linear) plane one-dimensional sound waves in a pure liquid are governed by the classical linear wave equation

$$\frac{\partial^2 u}{\partial t^2} - c^2 \frac{\partial^2 u}{\partial x^2} = 0, \qquad (8.208)$$

where c is the speed of sound. This equation has elementary solutions of the type

$$u = A \exp[i(kx \pm \omega t)], \quad \omega = ck, \qquad (8.209)$$

describing the propagation of a sinusoidal wave with a definite speed c independent of the wavelength ($\lambda = 2\pi/k$, k is the wavenumber). These waves are called nondispersive.

However, many of the waves encountered in fluids do not obey the classical wave equation, and their phase velocity may not be independent of the wavenumber. An example is wave propagation in bubbly liquids.

More generally, for an infinitesimal disturbance, the governing equation is of the type

$$L(u) = 0, \qquad (8.210)$$

where L is a linear operator involving derivatives with respect to position and time. The form of this operator depends upon the fluid system and the particular type of wave motion considered. Again, this linear equation admits solutions of the form

$$u = A \exp[i(kx \pm \omega t)], \qquad (8.211)$$

provided that ω and k satisfy a so-called dispersion relation

$$D(\omega, k) = 0 \qquad (8.212)$$

that is characteristic for the fluid system and of the type of wave motion considered.

These infinitesimal wave solutions are useful since in many circumstances the nonlinear effects are weak. Thus, the infinitesimal amplitude solution in many cases represents a very useful first approximation to the motion. The interaction effects can be accounted for by considering an equation of the form

$$L(u) = \epsilon N(u), \qquad (8.213)$$

where N is some nonlinear operator whose precise nature again depends upon the particular fluid system considered, and ϵ is a small parameter characterizing the amplitude of the motion. Solutions to this equation can be built by successive approximation.

Let us consider two interacting wave trains, given to a first approximation by solutions to the linear equation (8.210):

$$u_1 = A_1 \exp[i(k_1 x - \omega_1 t)],$$
$$u_2 = A_2 \exp[i(k_2 x - \omega_2 t)], \qquad (8.214)$$

where the individual wave numbers and frequencies are related by the dispersion relation (8.212): $D(\omega_i, k_i) = 0$; $i = 1, 2$. These expressions are to be substituted into the small nonlinear term $\epsilon N(u)$ of (8.213).

If the lowest-order nonlinearity is quadratic, this substitution leads to expressions of the type

$$\sim \exp i[(k_1 \pm k_2)x - (\omega_1 \pm \omega_2)t]. \qquad (8.215)$$

These terms act as a small-amplitude forcing function to the linear system and provide an excitation at the wave numbers ($k_3 = k_1 \pm k_2$) and frequencies ($\omega_3 = \omega_1 \pm \omega_2$). In general, the response of the linear system to this forcing is expected to be small ($\sim \epsilon$). However, if the wavelength and frequency of the forcing are related by the dispersion relation (8.212) as well: $D(\omega_3, k_3) = 0$, then we have a case which normally is referred to as nonlinear resonance wave interaction. Similarly to a well-known resonance behavior of a linear oscillator, the amplitude of the third component

$$u_3 = A_3 \exp[i(k_3 x - \omega_3 t)] \qquad (8.216)$$

grows linearly at the beginning. Later the energy drain from the first two components begins to reduce their amplitudes and consequently the amplitude of the forcing function.

Let us now consider a set of three wave trains undergoing resonant interactions. The solutions representing these wave trains would be expected to look like the solutions of the linear problem, although the amplitudes $A_i (i = 1, 2, 3)$ may possibly vary slowly with time:

$$u = \sum_{i=1}^{3} \{A_i(\epsilon t) \exp[i(k_i x - \omega_i t)]$$
$$+ A_i^*(\epsilon t) \exp[-i(k_i x - \omega_i t)]\}. \qquad (8.217)$$

Here it is used that u is necessarily real. Substitution of this set into (8.213) with subsequent averaging over a few wavelengths leads to a system of equations for the amplitudes of the interacting wave trains. This system basically describes the energy balance between these wave components.

Thus, the existence of energy transfer of this kind clearly depends upon the existence of these resonance conditions. The classic example of dispersion relations which allow for such resonance conditions are capillary gravity waves on the surface of deep water [8.110]. This example, however, is far from the scope of the present section.

In this section the nonlinear acoustics of liquids with bubbles is discussed as an example of a multiphase fluids with complicated and pronounced nonlinearity and dispersion. The dispersion encountered there may also lead to a special nonlinear wave interaction called *long-wave–short-wave resonance* in this context.

In [8.111] it was suggested a new form of triad resonance between three waves of wave numbers k_1, k_2, and k_3 and frequencies ω_1, ω_2, and ω_3, respectively, such that

$$k_1 = k_2 + k_3, \quad \omega_1 = \omega_2 + \omega_3. \qquad (8.218)$$

It was suggested to consider k_1 and k_2 to be very close: $k_1 = k + \varepsilon$, $k_2 = k - \varepsilon$ and $k_3 = 2\varepsilon$ ($\varepsilon \ll k$). The resonance conditions (8.218) are automatically achieved if

$$\omega(k + \varepsilon) - \omega(k - \varepsilon) = \omega_3 \qquad (8.219)$$

or

$$2\varepsilon \frac{d\omega}{dk} = \omega_3. \qquad (8.220)$$

That means the group velocity of the short wave (wave number k) equals the phase velocity of the long wave (wave number 2ε). It is called the long-wave–short-wave resonance. At first sight this resonance would appear hard to achieve but it is clearly possible from Fig. 8.13.

Let us identify the lower (or low-frequency) and upper (or high-frequency) branches of the dispersion relation with subscripts "−" and "+", respectively. The long-wave asymptotics of the low-frequency branch, $\omega_l(k_l) = \omega_- |_{k \to 0}$, and the short-wave asymptotics of the high-frequency branch, $\omega_s(k_s) = \omega_+ |_{k \to \infty}$, can be written as follows:

$$\omega_l = c_e k_l + O(k_l^3), \quad \omega_s = c_f k_s + O(k_s^{-1}), \quad (8.221)$$

where

$$c_e = \frac{\omega_l}{k_l} \Big|_{k_l \to 0}, \quad c_f = \frac{d\omega_s}{dk_s} \Big|_{k_s \to \infty}, \quad (8.222)$$

are the *equilibrium* and the *frozen* speeds of sound in the bubbly mixture.

It is easy to see that dispersion relation (8.162) allows the existence of the long-wave–short-wave resonance. Since $c_f > c_e$, there is always some wave number k_s and frequency ω_s (Fig. 8.13) such that its group velocity is equal to the equilibrium velocity of the mixture.

It should be noted that the long-wave–short-wave resonance in bubbly liquids has nothing to do with the resonance of bubble oscillations. Suppose we consider water with air bubbles of radius $R_0 = 0.1$ mm under normal conditions ($p_0 = 0.1$ MPa, $\rho_{l0} = 10^3$ kg/m^3) and with a volume gas content $\alpha_{g0} = 2.22 \times 10^{-4}$. The wavelengths of the short and long waves are then $\lambda_s \approx 0.67$ cm and $\lambda_l \approx 6.7$ m, respectively. The equilibrium speed of sound is $c_e \approx 730$ m/s. Thus, the frequency of the long wave is $f_l = c_e / \lambda_l \approx 110$ Hz (audible sound). Obviously, the short-wave frequency $f_s \geq 35.6$ kHz lies in the ultrasound region. Moreover, a decrease in the bubble radius R_0 leads to increasing the short-wave frequency f_s. Hence, the long-wave–short-wave interaction in a bubbly fluid can be considered as the interaction of ultrasound and audible sound propagating in this medium.

The long-wave–short-wave resonance interaction for pressure waves in bubbly liquids was investigated in [8.112–115] using the method of multiple scales. In particular, it was shown that in a nondissipative bubbly medium this type of resonance interaction is described by the following equations [8.116]

$$\frac{\partial L}{\partial \tau} + \frac{\alpha}{2c_e} \frac{\partial |S|^2}{\partial \xi} = 0, \quad i\frac{\partial S}{\partial \tau} + \beta \frac{\partial^2 S}{\partial \xi^2} = \delta L S, \quad (8.223)$$

where L and S are normalized amplitudes of the long- and short-wave pressure perturbations, τ is the (slow) time and ξ is the space coordinate in the frame moving with the group velocity; α and β are parameters of interaction.

It turns out that in bubbly fluids, parameters of interaction can vanish simultaneously at some specific ultrasound frequency [8.115]. The interaction between sound and ultrasound is then *degenerate*, i.e., the equations for interaction are separated. However, such a degeneracy does not mean the absence of interaction. The quasi-monochromatic ultrasonic signal will still generate sound but of much smaller intensity than in the nondegenerate case.

8.11 Sonoluminescence

Sound waves not only give rise to self-steepening and shock waves but may even become the driving agent to the emission of light, a phenomenon called sonoluminescence that is mediated by bubbles in the liquid that are driven to strong oscillations and collapse [8.20–30]. The detection of the conditions where a single bubble can be trapped in the pressure antinode of a standing sound wave [8.117] has furthered this field considerably. In Fig. 8.16 a basic arrangement for single-bubble sonoluminescence (SBSL) can be seen where a bubble is trapped in a rectangular container filled with water.

Fig. 8.16 Photograph of a rectangular container with light-emitting bubble. There is only one transducer glued to the bottom plate of the container (dimensions: 50 mm × 50 mm × 60 mm). The piezoelectric transducer (dimensions: height 12 mm, diameter 47 mm) sets up a sound field of high pressure (approx. 1.3 bar) at a frequency of approx. 21 kHz. (Courtesy of R. Geisler)

The sound field is produced by a circular disc of piezoelectric material glued to the bottom of the container. The bright spot in the upper part of the container is the light from the bubble. It is of blueish white color. The needle sticking out into the container in the upper left part of the figure is a platinum wire for producing a bubble via a current pulse. The bubble is then driven via acoustic radiation forces into its position where it is trapped almost forever. The bubble exerts large oscillations that can be photographed [8.118]. Figure 8.17 shows a sequence of the bubble oscillation of a light-emitting bubble. A full cycle is shown after which the oscillation repeats. Indeed, this repetition has been used to sample the series of photographs by shifting the instant of exposure by 500 ns from frame to frame with respect to the phase of the sound wave. The camera is looking directly into the illuminating light source (back illumination). The bubble then appears dark on a bright background because the light is deflected off its surface. The white spot in the middle of the bubble results from the light passing the spherical bubble undeflected.

Upon the first large collapse a shock wave is radiated (Fig. 8.18). At the 30 ns interframe time the 12 frames given cover 330 ns, less than one interframe time Fig. 8.17. The shock wave becomes visible via deflection of the illuminating light. The bubble expands on a much slower time scale than the speed of the shock wave and is seen as a tiny black spot growing in the middle of the ring of the shock wave. Measurements of the shock velocity near the bubble at collapse (average from 6 to 73 μm) leads to about 2000 m/s, giving a shock pressure of about 5500 bar [8.119]. Higher-resolution measurements have increased these values to about 4000 m/s and 60 kbar [8.120].

The bubble oscillation as measured photographically can be compared with theoretical models for the mo-

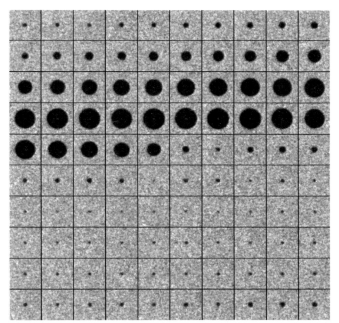

Fig. 8.17 Photographic series of a trapped, sonoluminescing bubble driven at 21.4 kHz and a pressure amplitude of 132 kPa. The interframe time is 500 ns. Picture size is 160 × 160 μm. (Courtesy of R. Geisler)

tion of a bubble in a liquid. A comparison is shown in Fig. 8.19. The overall curve is reproduced well, although the steep collapse is not resolved experimentally. For details the reader is referred to [8.81].

There is the question whether shock waves are also emitted into the interior of the bubble [8.121–124]. Theoretically, shock waves occur in certain models, when the collapse is of sufficient intensity. Recently, the problem has been approached by molecular dynamics calculations [8.125–128]. As the number of molecules

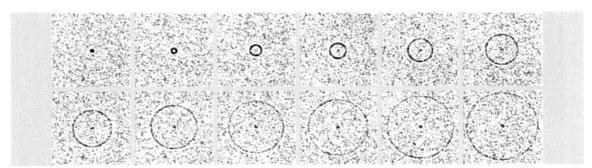

Fig. 8.18 Shock wave from a trapped, sonoluminescing bubble driven at 21.4 kHz and a pressure amplitude of 132 kPa. The interframe time is 30 ns. Picture size is 160 × 160 μm. (Courtesy of R. Geisler)

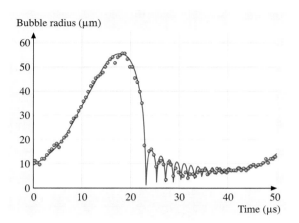

Fig. 8.19 Radius-time curve of a trapped bubble in a water-glycerine mixture derived from photographic observations. A numerically calculated curve (Gilmore model) is superimposed on the experimental data points (*open circles*). The calculation is based on the following parameters: driving frequency $f_0 = 21.4\,\text{kHz}$, ambient pressure $p_0 = 100\,\text{kPa}$, driving pressure $p_0 = 132\,\text{kPa}$, vapor pressure $p_v = 0$, equilibrium radius $R_n = 8.1\,\mu\text{m}$, density of the liquid $\rho = 1000\,\text{kg/m}^3$, viscosity $\mu = 0.0018\,\text{Ns/m}^2$ (measured) and surface tension $\sigma = 0.0725\,\text{N/m}$. The gas within the bubble is assumed to obey the adiabatic equation of state for an ideal gas with $\kappa = 1.2$. (Measured points courtesy of R. Geisler)

in a small bubble is relatively small, around 10^9 to 10^{10} molecules, this approach promises progress for the question of internal shock waves. At present molecular dynamics simulations are feasible with several million particles inside the bubble. Figure 8.20 gives a graphical view on the internal temperature distribution inside a collapsing sonoluminescence bubble with near-adiabatic conditions (reflection of the molecules at the inner boundary of the bubble) for six different times around maximum compression. The liquid

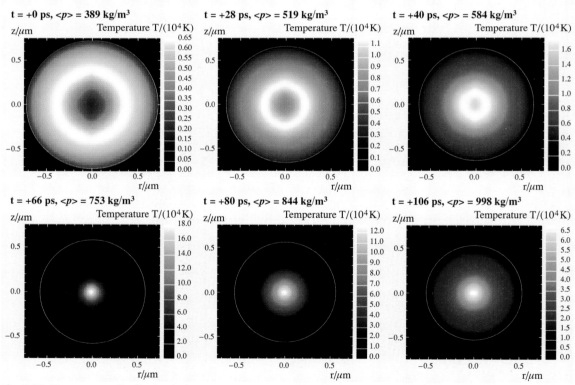

Fig. 8.20 Temperature distribution inside a collapsing sonoluminescence bubble filled with argon, one million particles, radius of the bubble at rest 4.5 μm, sound pressure amplitude 130 kPa, sound field frequency 26.5 kHz, total time covered 106 ps. (Courtesy of B. Metten [8.126])

Fig. 8.21 Distribution of temperature inside an acoustically driven bubble in dependence on time. Molecular dynamics calculation with argon and water vapor including chemical reactions. (after [8.128])

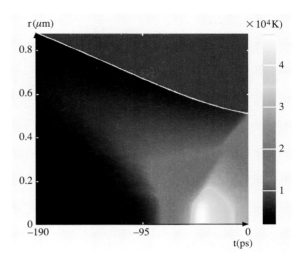

motion and the inner molecular dynamics calculations are coupled via the gas pressure (at the interface), which is determined from the molecular dynamics of the 10^6 particles and inserted into the Rayleigh–Plesset equation giving the motion of the bubble wall. A strong focusing of energy inside the bubble is observed under these conditions. An inward-traveling compression wave focuses at the center and yields a core temperature of more than 10^5 K. This temperature is an overestimate as no energy-consuming effects are included.

The temperature variation inside the bubble with time can be visualized for spherical bubbles in a single picture in the following way (Fig. 8.21). The vertical axis represents the radial direction from the bubble center, the horizontal axis represents time and the color gives the temperature in Kelvin according to the color-coded scale (bar aside the figure). The upper solid line in the figure depicts the bubble wall; the blue part is water. In this case an argon bubble with water vapor is taken and chemical reactions are included. Then the temperature is lowered significantly by the endothermal processes of chemical species formation, for instance of OH^- radicals. The temperature drops to about 40 000 K in the center [8.128].

8.12 Acoustic Chaos

Nonlinearity in acoustics not only leads to shock waves and, as just demonstrated, to light emission, but to even deeper questions of nonlinear physics in general [8.4]. In his curiosity, man wants to know the future. Knowledge of the future may also be of help to master one's life by taking proper provisions. The question therefore arises: how far can we look into the future and where are the difficulties in doing so? This is the question of predictability or unpredictability. In the attempt to solve this question we find deterministic systems, stochastic systems and, nowadays, chaotic systems. The latter are special in that they combine deterministic laws that nevertheless give unpredictable output. They thus form a link between deterministic and stochastic systems.

8.12.1 Methods of Chaos Physics

A description of the methods developed to handle chaotic systems, in particular in acoustics, has been given in [8.32]. The main question in the context

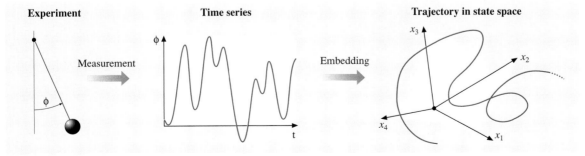

Fig. 8.22 Visualization of the embedding procedure

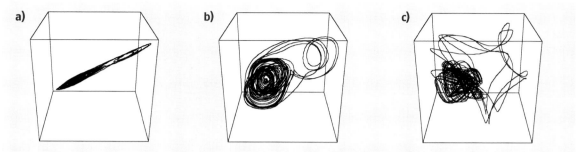

Fig. 8.23 Three embeddings with different delay times t_l for experimental data obtained from a chaotically oscillating, periodically driven pendulum. (Courtesy of U. Parlitz)

of experiments is how to connect the typically one-dimensional measurements (a time series) with the high-dimensional state space of theory (given by the number of variables describing the system). This is done by what is called *embedding* Fig. 8.22 [8.129, 130] and has led to a new field called *nonlinear time-series analysis* [8.131–133].

The embedding procedure runs as follows. Let $\{p(kt_s), k = 1, 2, \ldots, N\}$ be the measured time series, then an embedding into a n-dimensional state space is achieved by grouping n elements each of the time series to vectors $\boldsymbol{p}_k^{(n)} = [p(kt_s), p(kt_s + t_l), \ldots, p(kt_s + (n-1)t_l)]$, $k = 1, 2, \ldots, N - n$, in the n-dimensional state space. There t_s is the sampling interval at which samples of the variable p are taken. The delay time t_l is usually taken as a multiple of t_s, $t_l = lt_s$, $l = 1, 2, \ldots$, to avoid interpolation. The sampling time t_s, the delay time t_l and the embedding dimension n must all be chosen properly according to the problem under investigation. Figure 8.23 demonstrates the effect of different delay

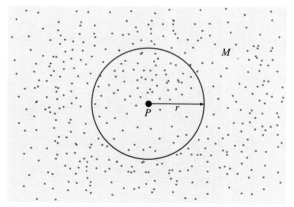

Fig. 8.24 Notations for the definition of the pointwise dimension

Fig. 8.25 Example of pointwise dimensions for one-dimensional (a curve) and two-dimensional (a surface) point sets

times with experimental data from a chaotically oscillating periodically driven pendulum. When the delay time t_l is too small then from one embedding vector to the next there is little difference and the embedded points all lie along the diagonal in the embedding space Fig. 8.23a. On the other hand, when the delay time t_l is too large, the reconstructed trajectory obtained by connecting successive vectors in the embedding space becomes folded and the reconstruction becomes fuzzy Fig. 8.23c. The embedding dimension n is usually not known beforehand when an unknown system is investigated for its properties. Therefore the embedding is done for increasing n until some criterion is fulfilled. The basic idea for a criterion is that a structure in the point set obtained by the embedding should appear. Then some law must be at work that generates it. The task after embedding is therefore to find and characterize the structure and to find the laws behind it. The characterization of the point set obtained can be done by determining its static structure (e.g., its dimension) and its dynamic structure (e.g., its expansion properties via the dynamics of nearby points).

The dimension easiest to understand is the pointwise dimension. It is defined at a point P of a point set M by $N(r) \sim r^D$, $r \to 0$, where $N(r)$ is the number of

Fig. 8.26 The point set of a strange attractor of an experimental, driven pendulum. (Courtesy of M. Kaufmann)

points of the set M inside the ball of radius r around P (Fig. 8.24). Examples for a line and a surface give the right integer dimensions of one and two (Fig. 8.25), because for a line we have $N(r) \sim r^1$ for $r \to 0$ and thus $D = 1$, and for a surface we have $N(r) \sim r^2$ for $r \to 0$ and thus $D = 2$.

Chaotic or strange attractors usually give a fractal (noninteger) dimension. The pointwise dimension may not be the same in this case for each point P. The distribution of dimensions then gives a measure of the inhomogeneity of the point set. Figure 8.26 gives an example of a fractal point set. It has been obtained by sampling the angle and the angular velocity of a pendulum at a given phase of the periodic driving for parameters in a chaotic regime.

To investigate the dynamic properties of a (strange or chaotic) set the succession of the points must be retained. As chaotic dynamics is brought about by a stretching and folding mechanism and shows sensitive dependence on initial conditions (see, e.g., [8.32]) the behavior of two nearby points is of interest: do they separate or approach under the dynamics? This

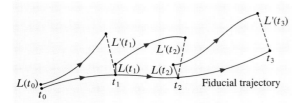

Fig. 8.27 Notations for the definition of the largest Lyapunov exponent

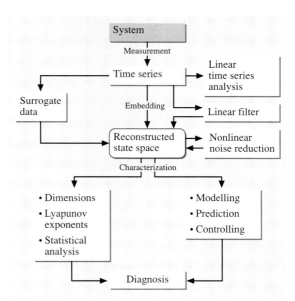

Fig. 8.28 Operations in nonlinear time-series analysis. (Courtesy of U. Parlitz)

behavior is quantified by the notion of the Lyapunov exponent. In Fig. 8.27 the calculation scheme [8.134] is depicted. It gives the maximum Lyapunov exponent via the formula

$$\lambda_{\max} = \frac{1}{t_m - t_0} \sum_{k=1}^{m} \log_2 \frac{L'(t_k)}{L(t_{k-1})} \quad \left[\frac{\text{bits}}{\text{s}}\right]. \quad (8.224)$$

As space does not permit us to dig deeper into the operations possible with embedded data, only a graph depicting the possible operations in nonlinear time-series analysis is given in Fig. 8.28. Starting from the time series obtained by some measurement, the usual approach is to do a linear time-series analysis, for instance by calculating the Fourier transform or some correlation. The new way for chaotic systems proceeds via embedding to a reconstructed state space. There may be some filtering involved in between, but this is a dangerous operation as the results are often difficult to predict. For the surrogate data operation and nonlinear noise reduction the reader is referred to [8.131–133]. From the characterization operations we have mentioned here the dimension estimation and the largest Lyapunov exponent. Various statistical analyses can be done and the data can also be used for modeling, prediction and controlling the system. Overall, nonlinear time series analysis is a new diagnostic tool for describing (nonlinear) systems.

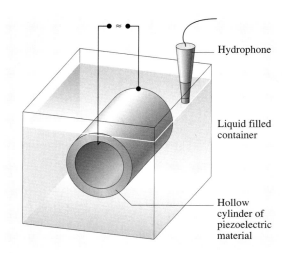

Fig. 8.29 Cylindrical transducer submerged in water for bubble oscillation, cavitation noise, and sonoluminescence studies

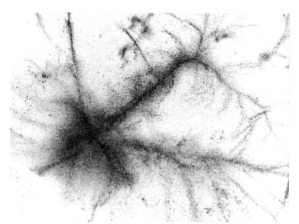

Fig. 8.30 Example of the filamentary bubble pattern (acoustic Lichtenberg figure) inside the cylinder in the projection along the axis of the transducer. (Courtesy of A. Billo)

Fig. 8.31 Power spectra of cavitation noise taken at a driving frequency of $f_0 = 23$ kHz, marked in the figures by small triangle. The driving pressure was increased from one graph to the next, giving first a periodic, anharmonic signal (*top*), then subharmonic motion with lines at multiples of $f_0/2$ (period-2 motion, *top next*), then period-4 motion with lines at multiples of $f_0/4$ (*third plot*) and finally a broadband spectrum indicative of chaotic behavior (*bottom*)▶

8.12.2 Chaotic Sound Waves

One of the first examples of chaotic dynamics and where these methods were first applied was acoustic cavitation [8.4, 135, 136]. To investigate this phenomenon, the rupture of liquids by sound waves and the phenomena associated with it, an experimental arrangement as depicted in Fig. 8.29 was used. A cylindrical transducer of piezoelectric material is used to generate a strong sound field in water. A typical transducer was of 76 mm

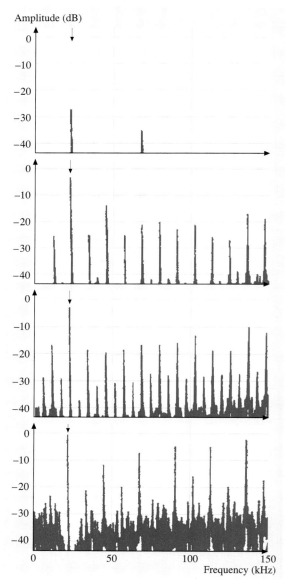

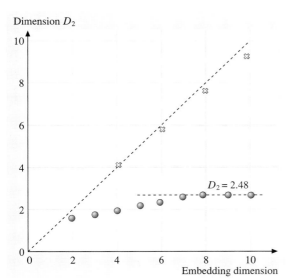

Fig. 8.32 Dimension estimation by embedding experimental acoustic cavitation sound pressure data into spaces of increasing dimension (*dots*). The *crosses* (×) are from an experiment, where noise from a noise generator has been embedded

length, 76 mm inner diameter, and 5 mm wall thickness submerged in a water-filled glass container. Cylinders of different size were available driven at different frequencies between 7 kHz and 23 kHz. A standing acoustic wave is set up inside the cylinder having its maximum pressure amplitude along the axis of the transducer. A hydrophone monitors the sound output of the liquid. Some electronics and a computer surround the experiment to drive the cylinder and to store the sampled pressure data from the hydrophone. When cavitation is fully developed, a dentritic filamentary bubble pattern is set up in the cylinder (Fig. 8.30).

When the sound field amplitude is raised the whole bubble field undergoes a period-doubling cascade to chaos [8.81, 137], a strong indication that chaotic dynamics really has been reached and that the broadband spectrum then encountered (Fig. 8.31) is not of stochastic origin. The cascade of period-doubled sound in the liquid has its origin in the cascade of period-doubled oscillations of the nonlinearly oscillating bubbles [8.81], and thus shows chaotic sound.

Embedding of the sound data into spaces of increasing dimension yields a structure with low fractal dimensions between two and three (Fig. 8.32). This is a quite surprising result in view of the thousands of oscillating bubbles of different sizes. Indeed, new results on the superposition of the sound output of a population of driven bubbles have indicated that the low dimensionality may be a result of the (weak) synchronization of their motion from the periodic driving and the scale of observation [8.138].

Chaotic sound waves also appear in other contexts [8.139]. In musical acoustics several instruments are susceptible to chaotic sound emission. Examples are the bowed string [8.140, 141], the clarinet [8.142] and gongs and cymbals [8.143].

The human speech production process is intrinsically nonlinear. To produce voice sounds the vocal cords are set into vibration through the air flow between them. The cords are a coupled system of nonlinear oscillators giving rise to bifurcations and chaos [8.144].

Sound can be generated by heat. When a gas-filled tube, closed at one end and open at the other is heated at the closed end and/or cooled at the open end, sound waves are emitted at a sufficiently high temperature difference (Sondhauss and Taconis oscillation). This phenomenon has been used to study its route to chaos [8.145]. It has been used to build a new type of refrigerator and heat pump [8.146].

References

8.1 M.F. Hamilton, D.T. Blackstock (Eds.): *Nonlinear Acoustics* (Academic, San Diego 1998)
8.2 K. Naugolnykh, L. Ostrovsky: *Nonlinear Wave Processes in Acoustics* (Cambridge Univ. Press, Cambridge 1998)
8.3 R.T. Beyer: *Nonlinear Acoustics* (Acoust. Soc. Am., Woodbury 1997)
8.4 W. Lauterborn, T. Kurz, U. Parlitz: Experimental nonlinear physics, Int. J. Bifurcation Chaos **7**, 2003–2033 (1997)
8.5 A.L. Thuras, R.T. Jenkins, H.T. O'Neil: Extraneous frequencies generated in air carrying intense sound waves, J. Acoust. Soc. Am. **6**, 173–180 (1935)
8.6 D.T. Blackstock, M.J. Crocker, D.G. Crighton, E.C. Everbach, M.A. Breazeale, A.B. Coppens, A.A. Atchley, M.F. Hamilton, W.E. Zorumski, W. Lauterborn, K.S. Suslick, L.A. Crum: Part II, Nonlinear Acoustics and Cavitation. In: *Encyclopedia of Acoustics*, Vol. 281, ed. by M.J. Crocker (Wiley, New York 1997) pp. 191–281

8.7 N.S. Bakhvalov, Y.M. Zhileikin, E.A. Zabolotskaya: *Nonlinear Theory of Sound Beams* (American Institute of Physics, Melville 1987)

8.8 B.K. Novikov, O.V. Rudenko, V.I. Timoshenko: *Nonlinear Underwater Acoustics* (American Institute of Physics, New York 1987)

8.9 O.V. Rudenko, S.I. Soluyan: *Theoretical Foundations of Nonlinear Acoustics* (Consultants Bureau, London 1977)

8.10 G.B. Whitham: *Linear and Nonlinear Waves* (Wiley, London 1974)

8.11 L.D. Rozenberg (Ed.): *High-Intensity Ultrasonic Fields* (Plenum, New York 1971)

8.12 M.F. Hamilton, D.T. Blackstock (Eds.): *Frontiers of Nonlinear Acoustics* (Elsevier, London 1990)

8.13 H. Hobaek (Ed.): *Advances in Nonlinear Acoustics* (World Scientific, Singapore 1993)

8.14 R.J. Wei (Ed.): *Nonlinear Acoustics in Perspective* (Nanjing Univ. Press, Nanjing 1996)

8.15 W. Lauterborn, T. Kurz (Eds.): *Nonlinear Acoustics at the Turn of the Millennium* (American Institute of Physics, Melville 2000)

8.16 O.V. Rudenko, O.A. Sapozhnikov (Eds.): *Nonlinear Acoustics at the Beginning of the 21st Century* (Faculty of Physics, Moscow State University, Moscow 2002)

8.17 S. Makarov, M. Ochmann: Nonlinear and thermoviscous phenomena in acoustics, Part I, Acustica **82**, 579–606 (1996)

8.18 S. Makarov, M. Ochmann: Nonlinear and thermoviscous phenomena in acoustics, Part II, Acustica **83**, 197–222 (1997)

8.19 S. Makarov, M. Ochmann: Nonlinear and thermoviscous phenomena in acoustics, Part III, Acustica **83**, 827–864 (1997)

8.20 C.E. Brennen: *Cavitation and Bubble Dynamics* (Oxford Univ. Press, Oxford 1995)

8.21 T.G. Leighton: *The Acoustic Bubble* (Academic, London 1994)

8.22 J.R. Blake, J.M. Boulton-Stone, N.H. Thomas (Eds.): *Bubble Dynamics and Interface Phenomena* (Kluwer, Dordrecht 1994)

8.23 F.R. Young: *Cavitation* (McGraw-Hill, London 1989)

8.24 L. van Wijngaarden (Ed.): *Mechanics and Physics of Bubbles in Liquids* (Martinus Nijhoff, The Hague 1982)

8.25 W. Lauterborn (Ed.): *Cavitation and Inhomogeneities in Underwater Acoustics* (Springer, Berlin, Heidelberg 1980)

8.26 E.A. Neppiras: Acoustic cavitation, Phys. Rep. **61**, 159–251 (1980)

8.27 H.G. Flynn: Physics of acoustic cavitation. In: *Physical Acoustics*, ed. by W.P. Mason (Academic, New York 1964) pp. 57–172

8.28 F.R. Young: *Sonoluminescence* (CRC, Boca Raton 2005)

8.29 T.J. Mason, J.P. Lorimer: *Applied Sonochemistry* (Wiley-VCH, Weinheim 2002)

8.30 K.S. Suslick (Ed.): *Ultrasound: Its Chemical, Physical and Biological Effects* (VCH, New York 1988)

8.31 W. Lauterborn: Acoustic chaos. In: *Encyclopedia of Physical Science and Technology*, 3rd edn., ed. by R.A. Meyers (Academic, San Diego 2002) pp. 117–127

8.32 W. Lauterborn, U. Parlitz: Methods of chaos physics and their application to acoustics, J. Acoust. Soc. Am. **84**, 1975–1993 (1988)

8.33 F.E. Fox, W.A. Wallace: Absorption of finite amplitude sound waves, J. Acoust. Soc. Am. **26**, 994–1006 (1954)

8.34 R.T. Beyer: Parameter of nonlinearity in fluids, J. Acoust. Soc. Am. **32**, 719–721 (1960)

8.35 A.B. Coppens, R.T. Beyer, M.B. Seiden, J. Donohue, F. Guepin, R.H. Hodson, C. Townsend: Parameter of nonlinearity. II, J. Acoust. Soc. Am. **38**, 797–804 (1965)

8.36 R.T. Beyer: *Nonlinear Acoustics* (Naval Ship Systems Command, Washington 1974), Table 3-1

8.37 E.C. Everbach: *Tissue Composition Determination via Measurement of the Acoustic Nonlinearity Parameter B/A, Ph.D. Dissertation* (Yale Univ., New Haven 1989) p. 66

8.38 Z. Zhu, M.S. Roos, W.N. Cobb, K. Jensen: Determination of the acoustic nonlinearity parameter B/A from phase measurements, J. Acoust. Soc. Am. **74**, 1518–1521 (1983)

8.39 X. Gong, Z. Zhu, T. Shi, J. Huang: Determination of the acoustic nonlinearity parameter in biological media using FAIS and ITD methods, J. Acoust. Soc. Am. **86**, 1–5 (1989)

8.40 W.K. Law, L.A. Frizell, F. Dunn: Determination of the nonlinearity parameter B/A of biological media, Ultrasound Med. Biol. **11**, 307–318 (1985)

8.41 J. Zhang, F. Dunn: A small volume thermodynamic system for B/A measurement, J. Acoust. Soc. Am. **89**, 73–79 (1991)

8.42 F. Plantier, J.L. Daridon, B. Lagourette: Measurement of the B/A nonlinearity parameter under high pressure: Application to water, J. Acoust. Soc. Am. **111**, 707–715 (2002)

8.43 E.C. Everbach: Parameters of nonlinearity of acoustic media. In: *Encyclopedia of Acoustics*, ed. by M.J. Crocker (Wiley, New York 1997) pp. 219–226

8.44 I. Rudnick: On the attenuation of finite amplitude waves in a liquid, J. Acoust. Soc. Am. **30**, 564–567 (1958)

8.45 J. Banchet, J.D.N. Cheeke: Measurements of the acoustic nonlinearity parameter B/A in solvents: Dependence on chain length and sound velocity, J. Acoust. Soc. Am. **108**, 2754–2758 (2000)

8.46 A.B. Coppens, R.T. Beyer, M.B. Seiden, J. Donohue, F. Guepin, R.H. Hodson, C. Townsend: Parameter of nonlinearity in fluids II, J. Acoust. Soc. Am. **38**, 797–804 (1965)

8.47 O. Nomoto: Nonlinear parameter of the 'Rao Liquid', J. Phys. Soc. Jpn. **21**, 569–571 (1966)

8.48 R.T. Beyer: Parameter of nonlinearity in fluids, J. Acoust. Soc. Am. **32**, 719–721 (1960)

8.49 K.L. Narayana, K.M. Swamy: Acoustic nonlinear parameter (B/A) in n-pentane, Acustica **49**, 336–339 (1981)

8.50 S.K. Kor, U.S. Tandon: Scattering of sound by sound from Beyers (B/A) parameters, Acustica **28**, 129–130 (1973)

8.51 K.M. Swamy, K.L. Narayana, P.S. Swamy: A study of (B/A) in liquified gases as a function of temperature and pressure from ultrasonic velocity measurements, Acustica **32**, 339–341 (1975)

8.52 H.A. Kashkooli, P.J. Dolan Jr., C.W. Smith: Measurement of the acoustic nonlinearity parameter in water, methanol, liquid nitrogen, and liquid helium-II by two different methods: A comparison, J. Acoust. Soc. Am. **82**, 2086–2089 (1987)

8.53 R.T. Beyer: The parameter B/A. In: *Nonlinear Acoustics*, ed. by M.F. Hamilton, D.T. Blackstock (Academic, San Diego 1998) pp. 25–39

8.54 L. Bjørnø: Acoustic nonlinearity of bubbly liquids, Appl. Sci. Res. **38**, 291–296 (1982)

8.55 J. Wu, Z. Zhu: Measurement of the effective nonlinearity parameter B/A of water containing trapped cylindrical bubbles, J. Acoust. Soc. Am. **89**, 2634–2639 (1991)

8.56 M.P. Hagelberg, G. Holton, S. Kao: Calculation of B/A for water from measurements of ultrasonic velocity versus temperature and pressure to 10 000 kg/cm^2, J. Acoust. Soc. Am. **41**, 564–567 (1967)

8.57 B. Riemann: Ueber die Fortpflanzung ebener Luftwellen von endlicher Schwingungsweite (On the propagation of plane waves of finite amplitude), Abhandl. Ges. Wiss. Göttingen **8**, 43–65 (1860), in German

8.58 S. Earnshaw: On the mathematical theory of sound, Philos. Trans. R. Soc. London **150**, 133–148 (1860)

8.59 D.T. Blackstock: Propagation of plane sound waves of finite amplitude in nondissipative fluids, J. Acoust. Soc. Am. **34**, 9–30 (1962)

8.60 M.F. Hamilton, D.T. Blackstock: On the coefficient of nonlinearity β in nonlinear acoustics, J. Acoust. Soc. Am. **83**, 74–77 (1988), Erratum, ibid, p. 1976

8.61 D.G. Crighton: Propagation of finite-amplitude waves in fluids. In: *Encyclopedia of Acoustics*, ed. by M.J. Crocker (Wiley, New York 1997) pp. 203–218

8.62 B.D. Cook: New procedure for computing finite-amplitude distortion, J. Acoust. Soc. Am. **34**, 941–946 (1962), see also the footnote on page 312 of 8.1

8.63 W. Keck, R.T. Beyer: Frequency spectrum of finite amplitude ultrasonic waves in liquids, Phys. Fluids **3**, 346–352 (1960)

8.64 L.E. Hargrove: Fourier series for the finite amplitude sound waveform in a dissipationless medium, J. Acoust. Soc. Am. **32**, 511–512 (1960)

8.65 E. Fubini Ghiron: Anomalie nella propagazione di onde acustiche di grande ampiezza (Anomalies in the propagation of acoustic waves of large amplitude), Alta Frequenza **4**, 530–581 (1935), in Italian

8.66 D.T. Blackstock: Nonlinear Acoustics (Theoretical). In: *American Institute of Physics Handbook*, 3rd edn., ed. by D.E. Gray (McGraw Hill, New York 1972) pp. 3/183–3/205

8.67 A. Hirschberg, J. Gilbert, R. Msallam, A.P.J. Wijnands: Shock waves in trombones, J. Acoust. Soc. Am. **99**, 1754–1758 (1996)

8.68 M.J. Lighthill: Viscosity effects in sound waves of finite amplitude. In: *Surveys in Mechanics*, ed. by G.K. Batchelor, R.M. Davies (Cambridge Univ. Press, Cambridge 1956) pp. 249–350

8.69 J. Lighthill: *Waves in Fluids* (Cambridge Univ. Press, Cambridge 1980)

8.70 J.S. Mendousse: Nonlinear dissipative distortion of progressive sound waves at moderate amplitudes, J. Acoust. Soc. Am. **25**, 51–54 (1953)

8.71 Z.A. Gol'dberg: Second approximation acoustic equations and the propagation of plane waves of finite amplitude, Sov. Phys.-Acoust. **2**, 346–350 (1956)

8.72 D.T. Blackstock: Thermoviscous attenuation of plane, periodic, finite amplitude sound waves, J. Acoust. Soc. Am. **36**, 534–542 (1964)

8.73 E. Hopf: The partial differential equation $u_t + uu_x = \mu u_{xx}$, Comm. Pure Appl. Math. **3**, 201–230 (1950)

8.74 J.D. Cole: On a quasi-linear parabolic equation occurring in aerodynamics, Quart. Appl. Math. **9**, 225–236 (1951)

8.75 M. Abramowitz, I.A. Stegun (Eds.): *Handbook of Mathematical Functions* (Dover, New York 1970)

8.76 L.S. Gradsteyn, I.M. Ryshik: *Table of Integrals, Series and Products* (Academic, Orlando 1980)

8.77 D.T. Blackstock: Connection between the Fay and Fubini solutions for plane sound waves of finite amplitude, J. Acoust. Soc. Am. **39**, 1019–1026 (1966)

8.78 R.D. Fay: Plane sound waves of finite amplitude, J. Acoust. Soc. Am. **3**, 222–241 (1931)

8.79 R.V. Khokhlov, S.I. Soluyan: Propagation of acoustic waves of moderate amplitude through dissipative and relaxing media, Acustica **14**, 241–246 (1964)

8.80 I.S. Akhatov, R.I. Nigmatulin, R.T. Lahey Jr.: The analysis of linear and nonlinear bubble cluster dynamics, Multiphase Sci. Technol. **17**, 225–256 (2005)

8.81 W. Lauterborn, T. Kurz, R. Mettin, C.D. Ohl: Experimental and theoretical bubble dynamics, Adv. Chem. Phys. **110**, 295–380 (1999)

8.82 C. Chaussy (Ed.): *Extracorporeal Shock Wave Lithotripsy* (Karger, New York 1986)

8.83 G.A. Askar'yan: Self-focusing of powerful sound during the production of bubbles, J. Exp. Theor. Phys. Lett. **13**, 283–284 (1971)

8.84 P. Ciuti, G. Jernetti, M.S. Sagoo: Optical visualization of non-linear acoustic propagation in cavitating liquids, Ultrasonics **18**, 111–114 (1980)

8.85 Yu.A. Kobelev, L.A. Ostrovsky: Nonlinear acoustic phenomena due to bubble drift in a gas-liquid mixture, J. Acoust. Soc. Am. **85**, 621–627 (1989)

8.86 S.L. Lopatnikov: Acoustic phase echo in liquid with gas bubbles, Sov. Tech. Phys. Lett. **6**, 270–271 (1980)

8.87 I.Sh. Akhatov, V.A. Baikov: Propagation of sound perturbations in heterogeneous gas-liquid systems, J. Eng. Phys. **9**, 276–280 (1986)

8.88 I.Sh. Akhatov, V.A. Baikov, R.A. Baikov: Propagation of nonlinear waves in gas-liquid media with a gas content variable in space, Fluid Dyn. **7**, 161–164 (1986)

8.89 E. Zabolotskaya: Nonlinear waves in liquid with gas bubbles, Trudi IOFAN (Institute of General Physics of the Academy of Sciences, Moscow) **18**, 121–155 (1989), in Russian

8.90 I. Akhatov, U. Parlitz, W. Lauterborn: Pattern formation in acoustic cavitation, J. Acoust. Soc. Am. **96**, 3627–3635 (1994)

8.91 I. Akhatov, U. Parlitz, W. Lauterborn: Towards a theory of self-organization phenomena in bubble-liquid mixtures, Phys. Rev. E **54**, 4990–5003 (1996)

8.92 O.A. Druzhinin, L.A. Ostrovsky, A. Prosperetti: Low-frequency acoustic wave generation in a resonant bubble-layer, J. Acoust. Soc. Am. **100**, 3570–3580 (1996)

8.93 L.A. Ostrovsky, A.M. Sutin, I.A. Soustova, A.I. Matveyev, A.I. Potapov: Nonlinear, low-frequency sound generation in a bubble layer: Theory and laboratory experiment, J. Acoust. Soc. Am. **104**, 722–726 (1998)

8.94 R.I. Nigmatulin: *Dynamics of Multiphase Systems* (Hemisphere, New York 1991)

8.95 V.E. Nakoryakov, V.G. Pokusaev, I.R. Shreiber: *Propagation of Waves in Gas- and Vapour-liquid Media* (Institute of Thermophysics, Novosibirsk 1983)

8.96 L. van Wijngaarden: One-dimensional flow of liquids containing small gas bubbles, Ann. Rev. Fluid Mech. **4**, 369–396 (1972)

8.97 U. Parlitz, R. Mettin, S. Luther, I. Akhatov, M. Voss, W. Lauterborn: Spatiotemporal dynamics of acoustic cavitation bubble clouds, Philos. Trans. R. Soc. London Ser. A **357**, 313–334 (1999)

8.98 J.G. McDaniel, I. Akhatov, R.G. Holt: Inviscid dynamics of a wet foam drop with monodisperse bubble size distribution, Phys. Fluids **14**, 1886–1984 (2002)

8.99 A. Prosperetti, A. Lezzi: Bubble dynamics in a compressible liquid. Part 1. First order theory, J. Fluid Mech. **168**, 457–478 (1986)

8.100 R.I. Nigmatulin, I.S. Akhatov, N.K. Vakhitova, R.T. Lahey Jr.: On the forced oscillations of a small gas bubble in a spherical liquid-filled flask, J. Fluid Mech. **414**, 47–73 (2000)

8.101 A. Jeffrey, T. Kawahara: *Asymptotic Methods in Nonlinear Wave Theory* (Pitman, London 1982)

8.102 L. van Wijngaarden: On the equations of motion for mixtures of fluid and gas bubbles, J. Fluid Mech. **33**, 465–474 (1968)

8.103 N.A. Gumerov: Propagation of long waves of finite amplitude in a liquid with polydispersed gas bubbles, J. Appl. Mech. Tech. Phys. **1**, 79–85 (1992)

8.104 A. Hasegawa: *Plasma Instabilities and Nonlinear Effects* (Springer, Berlin, Heidelberg 1975)

8.105 N.A. Gumerov: The weak non-linear fluctuations in the radius of a condensed drop in an acoustic field, J. Appl. Math. Mech. (PMM U.S.S.R.) **53**, 203–211 (1989)

8.106 N.A. Gumerov: Weakly non-linear oscillations of the radius of a vapour bubble in an acoustic field, J. Appl. Math. Mech. **55**, 205–211 (1991)

8.107 N.A. Gumerov: On quasi-monochromatic weakly nonlinear waves in a low-dissipative bubbly liquid, J. Appl. Math. Mech. **56**, 50–59 (1992)

8.108 S. Gavrilyuk: Large amplitude oscillations and their "thermodynamics" for continua with "memory", Euro. J. Mech. B/Fluids **13**, 753–764 (1994)

8.109 S.L. Gavrilyuk, V.M. Teshukov: Generalized vorticity for bubbly liquid and dispersive shallow water equations, Continuum Mech. Thermodyn. **13**, 365–382 (2001)

8.110 L.F. McGoldrick: Resonant interactions among capillary-gravity waves, J. Fluid Mech. **21**, 305–331 (1965)

8.111 D.J. Benney: A general theory for interactions between long and short waves, Studies Appl. Math. **56**, 81–94 (1977)

8.112 I.Sh. Akhatov, D.B. Khismatulin: Effect of dissipation on the interaction between long and short waves in bubbly liquids, Fluid Dyn. **35**, 573–583 (2000)

8.113 I.Sh. Akhatov, D.B. Khismatulin: Two-dimensional mechanisms of interaction between ultrasound and sound in bubbly liquids: Interaction equations, Acoust. Phys. **47**, 10–15 (2001)

8.114 I.Sh. Akhatov, D.B. Khismatulin: Mechanisms of interaction between ultrasound and sound in liquids with bubbles: Singular focusing, Acoust. Phys. **47**, 140–144 (2001)

8.115 D.B. Khismatulin, I.Sh. Akhatov: Sound – ultrasound interaction in bubbly fluids: Theory and possible applications, Phys. Fluids **13**, 3582–3598 (2001)

8.116 V.E. Zakharov: Collapse of Langmuir waves, Sov. Phys. J. Exp. Theor. Phys. **72**, 908–914 (1972)

8.117 D.F. Gaitan, L.A. Crum, C.C. Church, R.A. Roy: An experimental investigation of acoustic cavitation and sonoluminescence from a single bubble, J. Acoust. Soc. Am. **91**, 3166–3183 (1992)

8.118 Y. Tian, J.A. Ketterling, R.E. Apfel: Direct observations of microbubble oscillations, J. Acoust. Soc. Am. **100**, 3976–3978 (1996)

8.119 J. Holzfuss, M. Rüggeberg, A. Billo: Shock wave emission of a sonoluminescing bubble, Phys. Rev. Lett. **81**, 5434–5437 (1998)

8.120 R. Pecha, B. Gompf: Microimplosions: Cavitation collapse and shock wave emission on a nanosecond time scale, Phys. Rev. Lett. **84**, 1328–1330 (2000)

8.121 E. Heim: Über das Zustandekommen der Sonolumineszenz (On the origin of sonoluminescence). In: *Proceedings of the Third International Congress on Acoustics, Stuttgart, 1959*, ed. by L. Cremer (Elsevier, Amsterdam 1961) pp. 343–346, in German

8.122 C.C. Wu, P.H. Roberts: A model of sonoluminescence, Proc. R. Soc. Lond. A **445**, 323–349 (1994)

8.123 W.C. Moss, D.B. Clarke, D.A. Young: Calculated pulse widths and spectra of a single sonoluminescing bubble, Science **276**, 1398–1401 (1997)

8.124 V.Q. Vuong, A.J. Szeri, D.A. Young: Shock formation within sonoluminescence bubbles, Phys. Fluids **11**, 10–17 (1999)

8.125 B. Metten, W. Lauterborn: Molecular dynamics approach to single-bubble sonoluminescence. In: *Nonlinear Acoustics at the Turn of the Millennium*, ed. by W. Lauterborn, T. Kurz (Am. Inst. Physics, Melville 2000) pp. 429–432

8.126 B. Metten: *Molekulardynamik-Simulationen zur Sonolumineszenz*, Dissertation (Georg-August Universität, Göttingen 2000)

8.127 S.J. Ruuth, S. Putterman, B. Merriman: Molecular dynamics simulation of the response of a gas to a spherical piston: Implication for sonoluminescence, Phys. Rev. E **66**, 1–14 (2002), 036310

8.128 W. Lauterborn, T. Kurz, B. Metten, R. Geisler, D. Schanz: Molecular dynamics approach to sonoluminescent bubbles. In: *Theoretical and Computational Acoustics 2003*, ed. by A. Tolstoy, Yu-Chiung Teng, E.C. Shang (World Scientific, New Jersey 2004) pp. 233–243

8.129 N.H. Packard, J.P. Crutchfield, J.D. Farmer, R.S. Shaw: Geometry from a time series, Phys. Rev. Lett. **45**, 712–716 (1980)

8.130 F. Takens: Detecting strange attractors in turbulence. In: *Dynamical Systems and Turbulence*, ed. by D.A. Rand, L.S. Young (Springer, Berlin, Heidelberg 1981) pp. 366–381

8.131 H.D.I. Abarbanel: *Analysis of Observed Chaotic Data* (Springer, Berlin, New York 1996)

8.132 H. Kantz, T. Schreiber: *Nonlinear Time Series Analysis* (Cambridge Univ. Press, Cambridge 1997)

8.133 U. Parlitz: Nonlinear time serie analysis. In: *Nonlinear Modeling – Advanced Black Box Techniques*, ed. by J.A.K. Suykens, J. Vandewalle (Kluwer Academic, Dordrecht 1998) pp. 209–239

8.134 A. Wolf, J.B. Swift, H.L. Swinney, J.A. Vastano: Determining Lyapunov exponents from time series, Physica D **16**, 285–317 (1985)

8.135 W. Lauterborn, E. Cramer: Subharmonic route to chaos observed in acoustics, Phys. Rev. Lett. **47**, 1445–1448 (1981)

8.136 W. Lauterborn, J. Holzfuss: Acoustic chaos, Int. J. Bifurcation Chaos **1**, 13–26 (1991)

8.137 W. Lauterborn, A. Koch: Holographic observation of period-doubled and chaotic bubble oscillations in acoustic cavitation, Phys. Rev. A **35**, 1774–1976 (1987)

8.138 S. Luther, M. Sushchik, U. Parlitz, I. Akhatov, W. Lauterborn: Is cavitation noise governed by a low-dimensional chaotic attractor?. In: *Nonlinear Acoustics at the Turn of the Millennium*, ed. by W. Lauterborn, T. Kurz (Am. Inst. Physics, Melville 2000) pp. 355–358

8.139 W. Lauterborn: Nonlinear dynamics in acoustics, Acustica acta acustica Suppl. 1 **82**, S46–S55 (1996)

8.140 M.E. McIntyre, R.T. Schumacher, J. Woodhouse: On the oscillation of musical instruments, J. Acoust. Soc. Am. **74**, 1325–1345 (1983)

8.141 N.B. Tuffilaro: Nonlinear and chaotic string vibrations, Am. J. Phys. **57**, 408–414 (1989)

8.142 C. Maganza, R. Caussé, Laloë: Bifurcations, period doublings and chaos in clarinet-like systems, Europhys. Lett. **1**, 295–302 (1986)

8.143 K.A. Legge, N.H. Fletcher: Nonlinearity, chaos, and the sound of shallow gongs, J. Acoust. Soc. Am. **86**, 2439–2443 (1989)

8.144 H. Herzel: Bifurcation and chaos in voice signals, Appl. Mech. Rev. **46**, 399–413 (1993)

8.145 T. Yazaki: Experimental observation of thermoacoustic turbulence and universal properties at the quasiperiodic transition to chaos, Phys. Rev. **E48**, 1806–1818 (1993)

8.146 G.W. Swift: Thermoacoustic engines, J. Acoust. Soc. Am. **84**, 1145–1180 (1988)

Part C Architectural Acoustics

9 Acoustics in Halls for Speech and Music
Anders Christian Gade, Lyngby, Denmark

10 Concert Hall Acoustics Based on Subjective Preference Theory
Yoichi Ando, Makizono, Kirishima, Japan

11 Building Acoustics
James Cowan, White River Junction, USA

300

9. Acoustics in Halls for Speech and Music

This chapter deals specifically with concepts, tools, and architectural variables of importance when designing auditoria for speech and music. The focus will be on cultivating the useful components of the sound in the room rather than on avoiding noise from outside or from installations, which is dealt with in Chap. 11. The chapter starts by presenting the subjective aspects of the room acoustic experience according to consensus at the time of writing. Then follows a description of their objective counterparts, the objective room acoustic parameters, among which the classical *reverberation time* measure is only one of many, but still of fundamental value. After explanations on how these parameters can be measured and predicted during the design phase, the remainder of the chapter deals with how the acoustic properties can be controlled by the architectural design of auditoria. This is done by presenting the influence of individual design elements as well as brief descriptions of halls designed for specific purposes, such as drama, opera, and symphonic concerts. Finally, some important aspects of loudspeaker installations in auditoria are briefly touched upon.

9.1	Room Acoustic Concepts	302
9.2	Subjective Room Acoustics	303
	9.2.1 The Impulse Response	303
	9.2.2 Subjective Room Acoustic Experiment Techniques	303
	9.2.3 Subjective Effects of Audible Reflections	305
9.3	Subjective and Objective Room Acoustic Parameters	306
	9.3.1 Reverberation Time	306
	9.3.2 Clarity	308
	9.3.3 Sound Strength	308
	9.3.4 Measures of Spaciousness	309
	9.3.5 Parameters Relating to Timbre or Tonal Color	310
	9.3.6 Measures of Conditions for Performers	310
	9.3.7 Speech Intelligibility	311
	9.3.8 Isn't One Objective Parameter Enough?	312
	9.3.9 Recommended Values of Objective Parameters	313
9.4	Measurement of Objective Parameters	314
	9.4.1 The Schroeder Method for the Measurement of Decay Curves	314
	9.4.2 Frequency Range of Measurements	314
	9.4.3 Sound Sources	315
	9.4.4 Microphones	315
	9.4.5 Signal Storage and Processing	315
9.5	Prediction of Room Acoustic Parameters	316
	9.5.1 Prediction of Reverberation Time by Means of Classical Reverberation Theory	316
	9.5.2 Prediction of Reverberation in Coupled Rooms	318
	9.5.3 Absorption Data for Seats and Audiences	319
	9.5.4 Prediction by Computer Simulations	320
	9.5.5 Scale Model Predictions	321
	9.5.6 Prediction from Empirical Data	322
9.6	Geometric Design Considerations	323
	9.6.1 General Room Shape and Seating Layout	323
	9.6.2 Seating Arrangement in Section	326
	9.6.3 Balcony Design	327
	9.6.4 Volume and Ceiling Height	328
	9.6.5 Main Dimensions and Risks of Echoes	329
	9.6.6 Room Shape Details Causing Risks of Focusing and Flutter	329
	9.6.7 Cultivating Early Reflections	330
	9.6.8 Suspended Reflectors	331
	9.6.9 Sound-Diffusing Surfaces	333
9.7	Room Acoustic Design of Auditoria for Specific Purposes	334
	9.7.1 Speech Auditoria, Drama Theaters and Lecture Halls	334

9.7.2	Opera Halls	335	9.8 **Sound Systems for Auditoria**	346
9.7.3	Concert Halls for Classical Music	338	9.8.1 PA Systems	346
9.7.4	Multipurpose Halls	342	9.8.2 Reverberation-Enhancement Systems	348
9.7.5	Halls for Rhythmic Music	344	**References**	349
9.7.6	Worship Spaces/Churches	346		

Current knowledge about room acoustic design is based on several centuries of practical (trial-and-error) experience plus a single century of scientific research. Over the last couple of decades scientific knowledge has matured to a level where it can explain to a usable degree:

1. the subjective aspects related to how we perceive the acoustics of rooms
2. how these subjective aspects are governed by objective properties in the sound fields
3. how these objective properties of the sound field are governed by the physical variables in the architectural design

This chapter is organized in the same manner. The various subjective acoustic aspects will form the basis for our discussion and be a guide through most aspects of relevance in room acoustic design. However, some trends in current design practice based on the experience and intuition of individual acoustical designers will also be commented on. It is hoped that this approach will also have the advantage of stimulating the reader's ability to judge a room from her/his own listening experience, an ability which is important not only for designers of concert halls, but also for those responsible for creating and enjoying the sound in these halls: musicians, sound-system designers, recording engineers and concert enthusiasts.

9.1 Room Acoustic Concepts

In order to help the reader maintain a clear perspective all the way through this chapter, Fig. 9.1 illustrates the universe of architectural acoustics.

In the upper half of the figure, we have the phenomena experienced in the real world. Going from left to right, we have the auditoria, in which we can experience objective sound fields causing subjective impressions of the acoustic conditions. In all of these three domains, we find a huge number of degrees of freedom: halls can differ in a myriad of ways (from overall dimensions to detailed shaping of small details like door handles), the sound fields which we describe by the impulse response concept – as explained in a moment – contain a wealth of individual sound components (reflections), each being a function of time, level, direction and frequency. Also, every individual may have his/her own way of expressing what the room does to the sound heard.

In other words, like in many other aspects of life, the real world is so complex that we need to simplify the problem – reduce the degrees of freedom – through definition of abstract, well-formulated and meaningful concepts and parameters in all three domains (the lower row of boxes). First, we must try to define a vocabulary to describe the subjective acoustic impression: we will isolate a set of subjective acoustic parameters that are valid to most people's listening experience (as indicated by the box in the lower-right corner), then we will try to deduce those properties from the sound fields that are responsible for our experience of each of the subjective aspects: we will define a set of objective, measurable parameters that correlate well with the subjective parameters, and finally – in order to be able to guide the architect – we must find out which aspects of the design govern the important objective and in turn the subjective

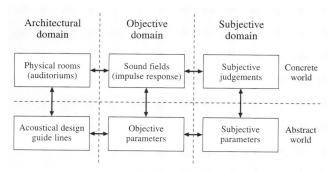

Fig. 9.1 Overview of concepts related to subjective room acoustics (after [9.1])

parameters, so we can assist in building halls meeting the specific needs of their users.

A rough historical overview: systematic search for objective acoustic parameters peaked in the 1960s and 1970s, while the relationships between the objective parameters and design variables were the topic of many research efforts in the 1980s and 1990s. However, we cannot yet claim to have the answers to all questions regarding the acoustic design of auditoria. Hopefully, there will always be room for intuition and individual judgment, not only in room acoustic design; but also as a source of inspiration for continued research efforts. Still it should be clear that room acoustic design today is based on solid research; it is not a snake-oil business.

9.2 Subjective Room Acoustics

9.2.1 The Impulse Response

The basic source of information regarding the audible properties of the sound field in a room is the impulse response signal. Actually, this signal – when recorded with a multichannel technique preserving the information about direction of incidence – can be shown to contain all information about the acoustics of a room between two specific source and receiver positions.

Consider a short sound impulse being emitted by the source on the stage, as shown in Fig. 9.2. A spherical wave propagates away from the source in all directions and the sound first heard in the listener position originates from that part of the wave that has propagated directly from the source to the receiver, called *the direct sound*. This is shown on the left in the lower part of Fig. 9.2, which shows the received signal versus time at a given position in the room.

This component is soon followed by other parts of the wave that have been reflected one or more times by the room boundaries or by objects in the room before reaching the receiver, called *the early reflections*. Besides arriving later than the direct sound, normally these reflections are also weaker because the intensity is reduced as the area of the spherical wavefront increases with time (spherical distance attenuation) and because a fraction of the energy is being absorbed each time the wave hits a more or less sound-absorbing room boundary or object in the room.

The sound wave will continue to be reflected and to pass the receiver position until all the energy has been absorbed by the boundaries/objects or by the air. The density of these later reflections increases with time (proportional to t^2), but the attenuation due to absorption at the room boundaries ensures that eventually all sound dies out. This decay is often heard as reverberation in the room, as Sabine did, when he carried out his famous experiment in the Fogg Art Museum more than 100 years ago [9.2]. This event marked the start of subjective room acoustics as a science.

9.2.2 Subjective Room Acoustic Experiment Techniques

With each of the reflections specified in time, level, spectrum and direction of incidence, the impulse response contains a huge amount of information. Now, the question is how we can reduce this to what is necessary for an explanation of why we perceive the acoustics in different rooms as being different.

Experiments trying to answer this question have mainly been carried out in the second half of the 20th century when electro-acoustic means for recording, measurement and simulation of sound fields had become available. Such experiments take place in the four rightmost boxes in Fig. 9.1: a number of impulse responses – or rather, music or speech convolved (filtered) through these responses – are presented to a number of subjects. The subjective evaluations are collected either as simple preferences between the sound fields presented in pairs or as scalings along specific subjective parameters suggested by the experimenter. The experimenter will also choose a number of different objective parameters, and calculate their values for each of the impulse responses presented. The level of correlation between these objective values and the objective scalings will then indicate which of the parameters and therefore which properties of the impulse responses are responsible for the subjective evaluations.

As can be imagined, the results of such experiments will be strongly dependent on the range of variation and degree of realism of the impulse responses presented – as well as by the sound quality of the equipment used for the presentation. Besides, the results will always be limited by the imagination of the experimenter regarding his/her:

1. suggestions of proper semantic descriptors for the subjects' evaluation (except in cases where the subjects are only asked to give preferences) and

2. suggestions of calculation of parameters from the impulse response.

In other words, before the experiments, the experimenter must have some good hypotheses regarding the contents of the lower-mid and lower-right boxes in Fig. 9.1.

Over the years, a variety of different techniques have been applied for presenting room sound fields to test subjects:

1. Collection of opinions about existing halls recalled from the memory of qualified listeners
2. Listening tests in different halls or in a hall with variable acoustics
3. Listening tests with presentation of recordings from existing halls or of *dry* speech/music convolved with impulse responses recorded in existing halls
4. Listening tests with sound fields synthesized in anechoic rooms (via multiple loudspeakers in an anechoic room) or impulse responses generated/modified by computer convolved with *dry* speech/music and presented over headphones.

It should be mentioned here that our acoustic memory is very short. Therefore, unless the subjects are highly experienced in evaluating concert hall acoustics, comparison of acoustic experiences separated by days,

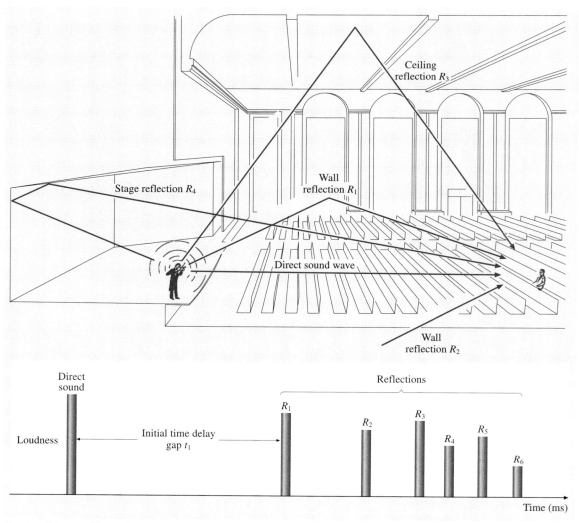

Fig. 9.2 Diagram illustrating the generation of an impulse response in a room (after [9.3])

weeks or even years is less reliable than comparison in the lab, where different recordings can be presented immediately one after the other. On the other hand, it is often difficult to get really qualified people to come to the lab to participate in scientific tests. Therefore, results from interview surveys can still be highly relevant.

These different techniques for presenting the sound fields all have different limitations. Experiments in the lab involving electro-acoustic reproduction or the generation of sound fields may lack fidelity, while listening in a real hall to a real orchestra will normally lack flexibility and control of the independent variables. Another important difference between these two methods is the absence of realistic visual cues in the lab. Enjoyment of a performance is a holistic experience, a fact which is now given increased scientific attention. In all cases elaborate statistical analysis of the results are needed in order to separate the significant facts from the error variance present in any experiment involving subjective judgements. Since the 1960s, multidimensional statistical tool such as factor analysis and multidimensional scaling have been very useful in distinguishing between the different subjective parameters which are present – consciously or not – in our evaluation. In any case, published results are more likely to be close to the truth, if they have been verified by several experiments – and experimenters – using different experimental approaches.

9.2.3 Subjective Effects of Audible Reflections

As the impulse response consists of the direct sound followed by a series of reflections, the simplest case possible would be to have only the direct sound plus a single reflection (and perhaps some late, artificial reverberation). Many experiments in simulated sound fields have been conducted using such a simple setup, and in spite of the obvious lack of realism, we can still learn a lot about the subjective effects of reflections from these experiments.

First of all, due to the psycho-acoustic phenomenon called forward masking, the reflection will be inaudible if it arrives very soon after the direct sound and/or its level is very low relative to the direct sound. Thus, there exists a level threshold of audibility depending on delay and direction of incidence relative to the direct sound. Only if the level of the reflection is above this threshold will the reflection have an audible effect, which again depends on its level, delay and direction of incidence. The possible effects are (at least):

- Increased level (energy addition)
- Increased clarity, if it arrives within 50–80 ms after the direct sound
- Increased spaciousness, if the reflection direction in the horizontal plane is different from that of the direct as this causes the signals received at the two ears to be different (If the angle between the direct and reflected sounds differs in the vertical plane only, the effect is rather a change in *timbre* or coloration of the sound.)
- Echo, typically observed for delays beyond 50 ms and at high reflection levels. If the delay is very long, say 200 ms, then the echo may even be detected at a much lower level;
- Coloration. If the delay is short, say below 30 ms, and the level is high, phase addition of the direct sound and the reflection will create a *comb filter* effect, which severely deteriorates the original frequency spectrum
- Change in localization direction, in cases where the reflection is louder than the direct sound. This may happen either due to amplification of the reflection via a concave surface or due to excess attenuation of the direct sound, for instance by an orchestra pit rail.

Audibility thresholds, echo risk and other effects also depend on spectral and temporal properties of the signal itself. Thus, with speech and fast, staccato played music (not to speak of artificial, impulsive *click* sounds), echoes are much easier to detect than when the signal

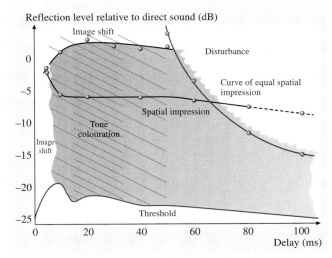

Fig. 9.3 Various audible effects of a single reflection arriving from the side. The signal used was music (after [9.4])

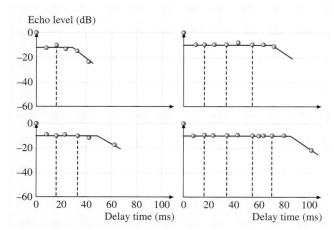

Fig. 9.4 Thresholds of perception of a new reflection (*circles* and *horizontal lines*) being added to impulse responses already containing one, two, three, or four other reflections with delay and relative levels, as indicated by the *vertical lines* (after [9.5])

is slow, legato music. On the other hand, these slower signals are more sensitive for the detection of coloration effects.

The temporal properties of sounds are sometimes described by the autocorrelation function. According to Ando (see Chap. 10 in this book) not only the thresholds of audibility but also the preferred delay of early reflections and the preferred reverberation time depend strongly on the autocorrelation of the signal. However, such a strong effect on concert hall preference – derived from listening experiments in rather simplified simulated sound fields in the lab – is not in accordance with everyday listening experiences in concert halls, in which we gladly accept listening to different music pieces of different tempo from different seats (different reflection delays) in the same hall (fixed reverberation time).

Figure 9.3 illustrates the various audible effects of a single reflection arriving from a 40° angle relative to a frontal direct sound. In this experiment music was used as the stimulus signal. Below the lower threshold (reflection level below about −20 dB and almost independent of the delay), the reflection is inaudible. In the upper-right region it causes a distinct echo, while in the shaded area we experience *spatial impression* or a broadening of the area from which the sound seem to originate. For delays lower than about 30 ms, this single reflection also causes an unpleasant change in *timbre* or *tonal color*, which is due to the combination of the direct and reflected signal forming a comb filter giving systematic changes in the spectrum. If the level of the reflection is increased to be louder than the direct sound (above 0 dB) while the delay is still below say 50 ms, we experience an *image shift* or a change in localization from the direction of the direct sound towards the direction from which the reflection is coming.

In practice, of course, the impulse response from a room contains many reflections, and we therefore need to know how the threshold of perceptibility of the incoming reflection changes in cases where the impulse response contains other reflections already. As seen from Fig. 9.4, the threshold level increases substantially in the delay range already occupied by other reflections. This phenomenon is primarily due to masking.

From this we can conclude that many of the details in a complex impulse response will be masked and only some of the dominant components or some of its overall characteristics seriously influence our perception of the acoustics. This is the reason why it is possible to create rather simple objective room acoustic parameters, as listed in the following section, which still describe the main features of the room acoustic experience. Thus, we seem to have some success describing the complex real world (the upper-middle and right boxes in Fig. 9.1) by simpler, abstract concepts. Also, it will be seen that the subjective effects caused by a single reflection remain important in the evaluation of more-realistic and complicated impulse responses.

9.3 Subjective and Objective Room Acoustic Parameters

From a consensus of numerous subjective experiments in real rooms and in simulated sound fields (representing all of the previously mentioned subjective research techniques), we now have a number of subjective parameters and corresponding objective measures available. Most of these are generally recognized as relevant descriptors of major aspects in our experience of the acoustics in rooms. This situation has promoted standardization of measurement methods and many of the objective parameters are now described in an appendix to the International Organization for Standardization (ISO) standard [9.6]. In order to maintain a clear distinction between the subjective and objective parameters in the following, the names for the subjective parameters will be printed in italics.

9.3.1 Reverberation Time

Reverberance is probably the best known of all subjective room acoustic aspects. When a room creates too much *reverberance*, speech loses intelligibility because important details (consonants) are masked by louder, lingering speech sounds (the vowels). For many forms of music, however, *reverberance* can add an attractive fullness to the sound by bonding adjacent notes together and blending the sounds from the different instruments/voices in an ensemble.

The reverberation time T which is the traditional objective measure of this quality, was invented 100 years ago by W. C. Sabine. T is defined as the time it takes for the sound level in the room to decrease by 60 dB after a continuous sound source has been shut off. In practice, the evaluation is limited to a smaller interval of the decay curve, from -5 dB to -35 dB (or -5 dB to -25 dB) below the start value; but still relating to a 60 dB decay (Fig. 9.5), i.e.:

$$T = 60\,\text{dB}\frac{(t_{-35}) - (t_{-5})}{(-5\,\text{dB}) - (-35\,\text{dB})}. \quad (9.1)$$

In this equation, t_{-x} denotes the time when the decay has decreased to X dB below its start value, or, if we let $R(t)$ represent the squared value of the decaying sound pressure and shut off the sound source at time $t = 0$:

$$10\log_{10}\left(\frac{R(t_{-X})}{R(0)}\right) = -X\,\text{dB}. \quad (9.2)$$

With the fluctuations always present in decay curves, T should rather be determined from the decay rate, A dB/s, as found from a least-squares regression line (determined from the relevant interval of the decay curve). Hereby we get for T:

$$T = \frac{60\,\text{dB}}{A\frac{\text{dB}}{\text{s}}} = \frac{60}{A}\,\text{s}. \quad (9.3)$$

Ways to obtain the decay curve from the impulse response will be further explained in the section on measurement techniques.

Due to masking, the entire decay process is only perceivable during breaks in the speech or music. Besides, the rate of decay is often different in the beginning and further down the decay curve. During running music or speech, the later, weaker part of the reverberation will be masked by the next syllable or musical note. Therefore an alternative measure, early decay time (EDT) has turned out to be better correlated with the *reverberance* perceived during running speech and music. This pa-

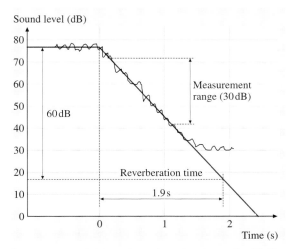

Fig. 9.5 The definition of reverberation time (after [9.4])

rameter, like T, also measures the rate of the decay; but now evaluated from the initial part, the interval between 0 and -10 dB, only. Thus,

$$\text{EDT} = 6(t_{-10}) \quad \text{or} \quad \text{EDT} = \frac{60}{A_{(0\,\text{dB} \to -10\,\text{dB})}}\,\text{s}. \quad (9.4)$$

The detailed behavior of the early part of the reverberation curve is influenced by the relative levels and distribution in time of the early reflections, which in turn vary depending on the positions of the source and receiver in the room. Likewise, the value of EDT is often found to vary throughout a hall, which is seldom the case with T.

In spite of the fact that EDT is a better descriptor of *reverberance* than T, T is still regarded the basic and most important objective parameter. This is mainly due to the general relationship between T and many of the other room acoustic parameters and because a lot of room acoustic theory relates to this concept, not least diffuse field theory, which is the basis for measurements of sound power, sound absorption, and sound insulation. T is also important by being referred to in legislation regarding room acoustic conditions in buildings.

Talking about diffuse field theory, it is often of relevance to compare the measured values of certain objective parameters with their expected values according to diffuse field theory and the measured or calculated reverberation time. As diffuse field theory predicts the decay to be purely exponential, the distribution in time of the impulse response squared should follow the function:

$$h^2(t) = A \exp\left(\frac{-13.8}{T}t\right), \quad (9.5)$$

in which the constant -13.8 is determined by the requirement that for $t = T$:

$$10 \log_{10}\left(\exp\left(\frac{-13.8}{T}t\right)\right) = 60\,\text{dB}. \quad (9.6)$$

With an exponential decay, the decay curve in dB becomes a straight line. Consequently, the expected value of EDT, EDT_{exp}, equals T.

When evaluating measurement results, it is also relevant to compare differences with the smallest change that can be detected subjectively. For EDT, this so-called subjective difference limen is about 5% [9.7].

9.3.2 Clarity

Clarity describes the degree to which every detail of the performance can be perceived as opposed to everything being blurred together by later-arriving, reverberant sound components. Thus, *clarity* is to a large extent a property complementary to *reverberance*.

When reflections are delayed by no more than 50–80 ms relative to the direct sound, the ear will integrate these contributions and the direct sound together, which means that we mainly perceive the effect as if the clear, original sound has been amplified relative to the later, reverberant energy. Thus, an objective parameter that compares the ratio between energy in the impulse response before and after 80 ms has been found to be a reasonably good descriptor of *clarity*

$$C = 10 \log_{10}\left[\int_0^{80\,\text{ms}} h^2(t)\,dt \bigg/ \int_{80\,\text{ms}}^{\infty} h^2(t)\,dt\right]. \quad (9.7)$$

The higher the value of C, the more the early sound dominates, and the higher the impression of *clarity*.

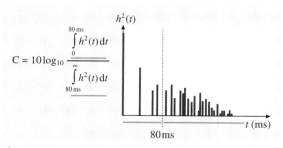

Fig. 9.6 The definition of C: the ratio between early and late energy in the impulse response

With exponential decay, the expected value of C becomes a function of T alone:

$$C_{\text{exp}} = 10 \log_{10}\left[\exp\left(\frac{1.104}{T}\right) - 1\right]\,\text{dB}. \quad (9.8)$$

The subjective difference limen for C (the equal best difference perceivable) is about 0.5 dB.

The definition of C is illustrated in Fig. 9.6.

Another parameter, which is also used to describe the balance between early and late sound or the balance between *clarity* and *reverberance*, is the center time t_s, which describes the center of gravity of the squared impulse response:

$$t_s = \int_0^{\infty} t h^2(t)\,dt \bigg/ \int_0^{\infty} h^2(t)\,dt. \quad (9.9)$$

A low value of t_s corresponds to a clear sound, whereas higher values indicate dominance of the late, reverberant energy. The main advantage of t_s is that it does not contain a sharp time limit between early and late energy, since this sharp distinction is not justified by our knowledge about the functioning of our hearing system. The subjective difference limen for t_s is about 10 ms, and the expected diffuse field value is simply given by:

$$t_{s,\text{exp}} = \frac{T}{13.8} \quad (9.10)$$

9.3.3 Sound Strength

The influence of the room on the perceived *loudness* is another important aspect of room acoustics. A relevant measurement of this property is simply the difference in dB between the level of a continuous, calibrated sound source measured in the room and the level the same source generates at 10 m distance in anechoic surroundings. This objective measure called the (relative) strength G can also be obtained from impulse response recordings from the ratio between the total energy of the impulse response and the energy of the direct sound with the latter being recorded at a fixed distance (10 m) from the impulsive sound source:

$$G = 10 \log_{10} \frac{\int_0^{\infty} h^2(t)\,dt}{\int_0^{t_{\text{dir}}} h_{10\,\text{m}}^2(t)\,dt}. \quad (9.11)$$

Here the upper integration limit in the denominator t_{dir} should be limited to the duration of the direct sound pulse (which in practice will depend on the bandwidth selected). A distance different from 10 m can be used, if a correction for the distance attenuation is applied as well.

The expected value of G according to diffuse field theory becomes a function of T as well as of the room volume, V:

$$G_{\mathrm{exp}} = 10\log_{10}\left(\frac{T}{V}\right) + 45\,\mathrm{dB}\,. \tag{9.12}$$

The subjective difference limen for G is about 1.0 dB. The definition of G is illustrated in Fig. 9.7.

9.3.4 Measures of Spaciousness

Spaciousness is the feeling that the sound is arriving from many different directions in contrast to a monophonic impression of all sound reaching the listener through a narrow opening. It is now clear that there are two aspects of *spaciousness*, both of which are attractive, particularly when listening to music:

- *apparent source width* (ASW): the impression that the sound image is wider than the visual, physical extent of the source(s) on stage. This should not be confused with localization errors, which of course should be avoided.

and

- *listener envelopment* (LEV): the impression of being inside and surrounded by the reverberant sound field in the room.

Both aspects have been found to be dependent on the direction of incidence of the impulse response reflections. When a larger portion of the early reflection energy (up to about 80 ms) arrives from lateral directions (from the sides) the ASW increases. When the level of the late, lateral reflections is high, strong LEV results.

The lateral components of the impulse response energy can be recorded using a figure-of-eight microphone with the sensitive axis held horizontal and perpendicular to the direction towards the sound source (so that the source lies in the deaf plane of the microphone). For measurement of the lateral energy fraction (LEF), the early part (up to 80 ms) of this lateral sound energy is compared with the energy of the direct sound plus all early reflections picked up by an ordinary omnidirectional microphone:

$$\mathrm{LEF} = \int_{t=5\,\mathrm{ms}}^{t=80\,\mathrm{ms}} h_1^2(t)\,\mathrm{d}t \Bigg/ \int_{t=0\,\mathrm{ms}}^{t=80\,\mathrm{ms}} h^2(t)\,\mathrm{d}t\,, \tag{9.13}$$

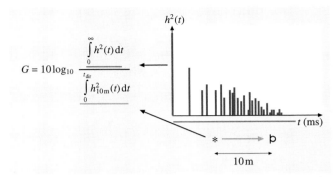

Fig. 9.7 The definition of strength G: the total energy in the impulse response measured relative to the direct sound level at 10 m distance from the source

where h_1 is the impulse response pressure recorded with a figure-of-eight microphone, whereas h is captured through the (usual) omnidirectional microphone.

It is mainly the energy at low and mid frequencies that contribute to the *spaciousness*. Consequently, LEF is normally averaged over the four octaves 125–1000 Hz. The higher the value of LEF, the wider the ASW. The literature contains many data on LEF values in different, existing concert halls.

LEF does not have an expected value related to T. In a completely diffuse field, LEF would be constant with a value of 0.33, which is higher than that normally found in real halls. The subjective difference limen for LEF is about 5%.

The definition of LEF is illustrated in Fig. 9.8.

The ASW aspect of *spaciousness* is not only dependent on the ratio between early lateral and total early

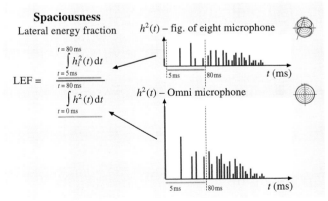

Fig. 9.8 The definition of lateral energy fraction LEF: the ratio between early reflection energy arriving from the sides and total early energy

sound; but also on the total level of the sound. The higher the G value (and the louder the sound source), the broader the acoustic image of the source. However, at the time of writing, there is no solid way of incorporating the influence of level into the objective measure.

Listener envelopment seems to be determined mainly by the spatial distribution and the level of the late reflections (arriving after 80 ms). A parameter called *late lateral strength* (LG) relating the late lateral energy to the direct sound at 10 m distance from the source has been proposed to measure this quality [9.8]:

$$\text{LG} = \int_{t=80\,\text{ms}}^{t=\infty} h_l^2(t)\,\mathrm{d}t \bigg/ \int_{t=0\,\text{ms}}^{t=t_{\text{dir}}} h_{10\,\text{m}}^2(t)\,\mathrm{d}t \,. \quad (9.14)$$

LG is likely to be included in the ISO 3382 standard (from 2007); but so far LG has only been found correlating with subjective responses in laboratory experiments using synthetic sound fields. Therefore, it may be wise not to put too much emphasis on this parameter until its value has also been confirmed by experiments in real halls.

In contrast to reflections arriving in the median plane, lateral reflections will cause the instantaneous pressure at the two ears of the listener to be different. Such a dissimilarity of the signals at the two ears can also be measured by means of the interaural cross-correlation coefficient IACC:

$$\text{IACC}_{t_1, t_2} = \max \left| \frac{\int_{t_1}^{t_2} h_L(t) h_R(t+\tau)\,\mathrm{d}t}{\sqrt{\int_{t_1}^{t_2} h_L^2(t)\,\mathrm{d}t \int_{t_1}^{t_2} h_R^2(t)\,\mathrm{d}t}} \right| \quad (9.15)$$

in which t_1 and t_2 define the time interval of the impulse response within which the correlation is calculated, h_L and h_R represent the impulse response measured at the entrance of the left and right ear, respectively, and τ is the interval within which we search for the maximum of the correlation. Normally, the range of τ is chosen between -1 and $+1$ ms, covering roughly the time it takes for sound to travel the distance from one ear to the other. As the correlation is normalized by the product of the energy of h_L and h_R, the IACC will take a value between 0 and 1. Results are often reported as $1 - \text{IACC}$ in order to obtain a value increasing with increasing dissimilarity, corresponding to an increasing impression of *spaciousness*. If the time interval for the calculation (t_1, t_2) is (0 ms, 100 ms), then the IACC will measure the ASW, while a later interval $(t_1, t_2) = (100\,\text{ms}, 1000\,\text{ms})$ may be used to describe the LEV. The literature on measured values in halls mainly contain data on IACC related to the 0–100 ms time interval.

Although the LEF and IACC parameters relate to the same subjective quality, they are not highly correlated in practice. Another puzzling fact is that LEF and IACC emphasize different frequency regions being of importance. LEF is primarily measured in the four lowest octaves, 125 Hz, 250 Hz, 500 Hz and 1000 Hz while IACC should rather be measured in the octave bands above 500 Hz. IACC values would always be high in the lower octaves, because the distance between the ears (< 30 cm) is small compared to 1/4 of the wave length (≈ 70 cm at 125 Hz).

9.3.5 Parameters Relating to Timbre or Tonal Color

Timbre describes the degree to which the room influences the frequency balance between high, middle and low frequencies, i.e. whether the sound is *harsh*, *bright*, *hollow*, *warm*, or whatever other adjective one would use to describe *tonal color*. Traditionally, a graph of the frequency variation of T (per 1/1 or 1/3 octave) has been used to indicate this quality; but a convenient single-number parameter intended to measure the warmth of the hall has been suggested [9.9]: the bass ratio (BR) given by:

$$\text{BR} = \frac{T_{125\,\text{Hz}} + T_{250\,\text{Hz}}}{T_{500\,\text{Hz}} + T_{1000\,\text{Hz}}} \,. \quad (9.16)$$

Likewise, a *treble ratio* (TR) can be formed as:

$$\text{TR} = \frac{T_{2000\,\text{Hz}} + T_{4000\,\text{Hz}}}{T_{500\,\text{Hz}} + T_{1000\,\text{Hz}}} \,. \quad (9.17)$$

However, in some halls, a lack of bass sound is experienced in spite of high T values at low frequencies. Therefore EDT or perhaps G versus frequency would be a better – and intuitively more logical – parameter for measurement of *timbre*. Likewise, BR or TR could be based on G rather than T values.

Besides the subjective parameters mentioned above, a quality called *intimacy* is also regarded as being important when listening in auditoria. Beranek [9.9] originally suggested this quality to be related to the initial time delay gap ITDG; but this has not been experimentally verified (At the time of writing, Beranek no longer advocates this idea but mentions the observation that, if the ITDG exceeds 45 ms, the hall has no intimacy [9.10]). It is likely that *intimacy* is related to a combination of some of the other parameters already mentioned, such as a high sound level combined with a clear and enveloping sound, as will naturally be experienced fairly close to the source or in small rooms.

9.3.6 Measures of Conditions for Performers

In rooms for performance of music it is also relevant to consider the acoustic conditions for the musicians, partly because it is important to ensure that the musicians are given the best working conditions possible and partly because the product as heard by the audience will also be better if the conditions are optimal for the performers.

Musicians will be concerned about *reverberance* and *timbre* as mentioned above; but also about at least two aspects which are unique for their situation: *ease of ensemble* and *support* [9.11] and [9.12].

Ease of ensemble relate to how well musicians can hear – and play together with – their colleagues. If ensemble is difficult to achieve, the result – as perceived by musicians and listeners alike – might be less precision in rhythm and intonation and a lack of balance between the various instruments. *Ease of ensemble* has been found to be related to the amount of early reflection energy being distributed on the stage. So far, the only widely recognized objective parameter suggested for objective measurement of this quality is *early support* ST_{early}.

ST_{early} is calculated from the ratio between the early reflection energy and the direct sound in an impulse response recorded on the stage with a distance of only one meter between the source and receiver:

$$ST_{early} = 10 \log_{10} \frac{\int_{20\,ms}^{100\,ms} h_{1m}^2(t)\,dt}{\int_{0\,ms}^{t_{dir}} h_{1m}^2(t)\,dt} \quad (9.18)$$

Support relates to the degree to which the room supports the musicians' efforts to create the tone on their own instruments, whether they find it easy to play or whether they have to force their instruments to fill the room. Having to force the instrument leads to playing fatigue and inferior quality of the sound.

Support is also related to the amount of reflected energy on the stage measured using only a one meter source/microphone distance. For measurement of *support*, however, one needs to consider also the later reflections. This is because for many types of instruments, (especially strings), the early reflections will be masked by the strong direct sound from the instrument itself. Consequently, it seems relevant to define a measure

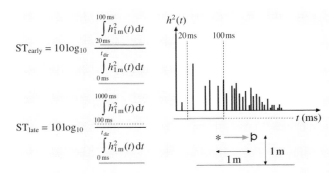

Fig. 9.9 The definition of early support ST_{early} and late support ST_{late}, the ratio between early and late reflection energy respectively and the direct sound in the impulse response. Both are measured at 1 m distance from the source

such as *late support*, ST_{late}:

$$ST_{late} = 10 \log_{10} \frac{\int_{100\,ms}^{1000\,ms} h_{1m}^2(t)\,dt}{\int_{0\,ms}^{t_{dir}} h_{1m}^2(t)\,dt} \quad (9.19)$$

which relates to late response (after 100 ms) from the room to the sound emitted.

ST_{early} and ST_{late} are typically measured in the four octaves 150 Hz, 500 Hz, 1000 Hz, 2000 Hz. The lowest octave is omitted primarily because it is not possible to isolate the direct sound from the early reflections in a narrow band measurement. The definitions of early and late support are illustrated in Fig. 9.9.

Also the support parameters are planned to be included in the ISO 3382 standard (from 2007); but like LG the experiences and amount of data available from existing halls are still rather limited.

The amount of *reverberance* on the stage may be measured by means of EDT, (with a source–receiver distance of, say, 5 m or more).

9.3.7 Speech Intelligibility

All the objective parameters mentioned above (except the basic T parameter), are mainly relevant in larger auditoria intended for performance of music. In auditoria used for speech, such as lecture halls or theaters, the influence of the acoustics on intelligibility is a major issue.

Currently, the most common way to assess objectively *speech intelligibility* in rooms is by measurement of the speech transmission index STI.

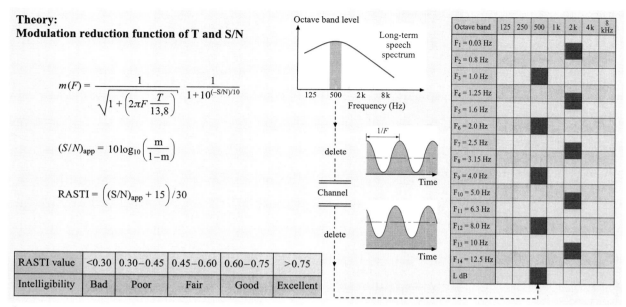

Fig. 9.10 Illustration of the theory and principle in the measurement of STI or RASTI. The scale for the evaluation of RASTI values is shown at the bottom (after [9.13])

As illustrated in Fig. 9.10, this measure is based on the idea that speech can be regarded as an amplitude-modulated signal in which the degree of modulation carries the speech information. If the transmission path adds noise or reverberation to the signal, the degree of modulation in the signal will be reduced, resulting in reduced intelligibility.

The modulation transfer is tested by emitting noise in seven octave bands, each modulated with 14 different modulation frequencies as listed in the table in Fig. 9.10 and then calculating the ratio between the original and the received degree of modulation, the modulation reduction factor, in each of these 98 combinations. A weighted average of the modulation reduction factor then results in a number between 0 and 1, corresponding to very poor and excellent conditions respectively.

A faster measurement method using only two carrier bands of noises and four plus five modulation frequencies (indicated by the dark squares in Fig. 9.10) is called rapid STI (RASTI). The STI/RASTI method is described in an International Electrotechnical Commission (IEC) standard: IEC 286-16.

Although the original method of STI or RASTI measurement employs modulated noise signals, it is also possible to calculate the modulation reduction factor from the impulse response. Thus, the modulation reduction factor versus modulation frequency F, which is called the modulation transfer function (MTF), can be found as the Fourier transform of the squared impulse response normalized by the total impulse response energy.

According to *Bradley* [9.14], a simpler parameter, called the *useful-to-detrimental sound ratio* U_{80} is equally suitable for measurement of speech intelligibility. U_{80} is simply a modified version of the clarity parameter defined in Sect. 9.3.2, in which a correction is made according to the ratio between the levels of speech and background noise.

9.3.8 Isn't One Objective Parameter Enough?

It is important to notice that all the parameters mentioned above – apart from T – may show large differences between different seats within the same hall. Actually, differences within a single hall are sometimes as large (both objectively and subjectively) as the differences between two different halls.

It should also be mentioned that the objective parameters related to the three subjectively different aspects: *reverberance/clarity*, *loudness* and *spaciousness* show low mutual correlation when measured in different seats or in different halls. In other words, they behave as orthogonal factors and do not monitor the same properties of the sound fields. Consequently, it is obviously necessary to apply one objective parameter for each of these subjective aspects.

It is also clear that different listeners will judge the importance of the various subjective parameters differently. While some may base their judgment primarily on *loudness* or *reverberance* others may be more concerned about *spaciousness* or *timbre*.

For these reasons, one should be very sceptical when people attempt to rank concert halls along a single, universal, one-dimensional quality scale or speak about the best hall in the world. Fortunately, the diversity of room acoustic design and the complexity of our perception does not justify such simplifications.

9.3.9 Recommended Values of Objective Parameters

In view of the remarks above, it may be regarded risky to recommend values for the various objective acoustic parameters. On the other hand, there is a long tradition of suggesting optimal values for reverberation time T as a function of hall volume and type of performance. Besides, most of the other parameters will seldom deviate drastically from the expected value determined by T. As we shall see later, this deviation is primarily influenced by how we shape the room to control early reflections. Designing for strong early reflections can increase *clarity*, *sound strength* and *spaciousness*. Current taste regarding concert hall acoustics for classical music seems to be in favor of high values for these factors in addition to strong *reverberance*. This could lead to suggested ranges for the various parameters in small and large halls, as shown in Table 9.1. These values relate to empty halls with well-upholstered seating, assuming that, when fully occupied with musicians and audience, the T values will drop by no more than say 0.2 s. The reason for relating the values to the unoccupied condition is that it is seldom possible to make acoustic measurements in the occupied state and almost all reference data existing regarding parameters other than T are from unoccupied halls. On the other hand, suggesting well-upholstered seats is a sound recommendation in most cases, which will ensure only small changes depending on the degree of occupancy. This will also justify extrapolation of the unoccupied values of the other parameters to the occupied condition, as mentioned later in this section.

The values listed in the table were arrived at by first choosing values for T ensuring high *reverberance*, while the correlative values for EDT, G and C stem from the diffuse field expected values; but these have been slightly changed to promote high *clarity* and high *levels*. As this is particularly important in larger halls, it has been suggested to use an EDT value 0.1 s lower than T ($=\text{EDT}_{\text{exp}}$) and C values 1 dB higher than C_{exp} in a 2500 m^3 hall, but 0.2 s less than T and 2 dB higher than C_{exp}, respectively, in the 25 000 m^2 hall. Likewise, it is suggested to use G values 2 dB less than G_{exp} in the small halls; but only 1 dB less than or even equal to G_{exp} in large halls. For halls of size between the 2500 and 25 000 m^3 listed in the table, one may of course interpolate to taste. Please note that, in general, G is found to be 2–3 dB lower that G_{exp} (Sect. 9.5.6) so the G values suggested here are actually 1–2 dB higher than those found in "normal" halls.

In the occupied condition, one may expect EDT, C and G to be reduced by amounts equal to the difference between the expected values calculated using T for the empty and occupied conditions respectively. In any case,

Table 9.1 Suggested position-averaged values of objective room acoustic parameters in unoccupied concert halls for classical music. It is assumed that the seats are highly absorptive, so that T will be reduced by no more than 0.2 s when the hall is fully occupied. 2.4 s should be regarded as an upper limit mainly suitable for terraced arena or DRS halls with large volumes per seat (Sect. 9.7.3), whereas 2.0 s will be more suitable for shoe-box-shaped halls, which are often found to be equally reverberant with lower T values and lower volumes per seat

Parameter	Symbol	Chamber music	Symphony
Hall size	V/N	2500 m^3/300 seats	25 000 m^3/2000 seats
Reverberation time	T	1.5 s	2.0–2.4 s
Early decay time	EDT	1.4 s	2.2 s
Strength	G	10 dB	3 dB
Clarity	C	3 dB	−1 dB
Lateral energy fraction	LEF	0.15–0.20	0.20–0.25
Interaural cross correlation	1 − IACC	0.6	0.7
Early support	ST_{early}	−10 dB	−14 dB

if the empty–occupied difference in T is limited to 0.2 s, as suggested here, then the difference in these parameters will be fairly limited, with changes in both C and G being less than one dB - given that the audience does not introduce additional grazing incidence attenuation of direct sound and early reflections (Sect. 9.6.2).

For opera, one may aim towards the same relationships between the diffuse field expected and the desired values of EDT, C and G, as mentioned above for symphonic halls; but the goal for the reverberation time will normally be set somewhat lower, 1.4–1.8 s in order to obtain a certain intelligibility of the libretto and perhaps to make breaks in the music sound more dramatic (Sect. 9.7.2).

For rhythmic music, T values of 0.8–1.5 s, depending on room size, are appropriate, and certainly the value should not increase towards low frequencies in this case.

In very large rock venues such as sports arenas, it may actually be very difficult to get T below 3–5 s; but even then the conditions can be satisfactory with a properly designed sound system. The reason is that in such large room volumes the level of the reverberant sound can often be controlled, if highly directive loudspeakers are being used. In these spaces, the biggest challenge is often to avoid highly delayed echoes, which are very annoying. For amplified music, only recommendations for T are relevant, because the acoustic aspects related to the other parameters will be determined primarily by the sound system (Sect. 9.7.5).

In order to ensure adequate speech intelligibility in lecture halls and theaters, values for STI/RASTI should be at least 0.6. In reverberant houses of worship values higher than 0.55 are hard to achieve even with a well-designed loudspeaker system.

9.4 Measurement of Objective Parameters

Whenever the acoustic specifications of a building are of importance we need to be able to predict the acoustic conditions during the design process as well as document the conditions after completion. Therefore, in this and the following sections, techniques for measurement and prediction of room acoustic conditions will be presented.

In Sect. 9.2.1 it was claimed that the impulse response contains all relevant information about the acoustic conditions in a room. If this is correct, it must also be true that all relevant acoustic parameters can be derived from impulse response measurements. Fortunately, this has also turned out to be the truth.

9.4.1 The Schroeder Method for the Measurement of Decay Curves

Originally, reverberation decays were recorded from the decay of the sound level following the termination of a stationary sound source. However, *Schroeder* [9.15] has shown that the reverberation curve can be measured with increased precision by *backwards* integration of the impulse response, $h(t)$, as follows:

$$R(t) = \int_t^\infty h^2(t)\,dt = \int_0^\infty h^2(t)\,dt - \int_0^t h^2(t)\,dt \quad (9.20)$$

in which $R(t)$ is equivalent to the decay of the squared pressure decay. The method is called backwards integration because the fixed upper integration limit, *infinite time*, is not known when we start the recording.

Therefore, in the days of magnetic tape recorders, the integration was done by playing the tape backwards. In this context *infinite time* means the duration of the recording, which should be comparable with the reverberation time. Traditional noise decays contain substantial, random fluctuations because of interference by the different eigenmodes present within the frequency range of the measurement (normally 1/3 or 1/1 octaves). These fluctuations will be different each time the measurement is carried out because the modes will be excited with random phase by the random noise (and the random time of the noise interruption). When, instead, the room is exited by a repeatable impulse, the response will be repeatable as well without such random fluctuations. In fact, Schroeder showed that the ensemble average of interrupted noise decays (recorded in the same source and receiver positions) will converge towards $R(t)$ as the number of averages goes towards infinity.

In (9.20), the right-hand side indicates that the impulse response decay can be derived by subtracting the running integration (from time zero to t) from the total energy. Contrary to the registration of noise decays, this means that the entire impulse response must be stored before the decay function is calculated. However, with modern digital measurement equipment, this is not a problem. From the $R(t)$ data, T is calculated by linear regression as explained in Sect. 9.3.1. As the EDT is evaluated from a smaller interval of the decay curve than T, the EDT is more sensitive to random decay fluctuations than is T. Therefore, EDT should never be derived from interrupted noise decays – only from integrated impulse responses.

9.4.2 Frequency Range of Measurements

T may be measured per 1/3 octave; but regarding the other objective parameters mentioned in this chapter, it makes no sense to use a frequency resolution higher than 1/1 octave. The reason is that the combination of narrow frequency intervals and narrow time windows in the other parameters will lead to extreme fluctuations with position of the measured values due to the acoustic relation of uncertainty. Such fluctuations have no parallel in our subjective experience. *Bradley* [9.16] has shown that, with 1/1 octave resolution, a change in microphone position of only 30 cm can already cause fluctuations in the clarity C and strength G of about 0.5–1 dB, which is equal to the subjective difference limen. Higher fluctuations would not make sense, as normally we do not experience an audible change when moving our head 30 cm in a concert hall.

The frequency range in which room acoustic measurements are made is normally the six octaves from 125 to 4000 Hz, but it may be extended to 63–8000 Hz when possible. Particularly for reverberation measurements in halls for amplified music, the 63 Hz octave is important.

9.4.3 Sound Sources

With the possibility of deriving all of the objective parameters described above from impulse response measurements, techniques for creating impulses are described briefly in the following.

Basically, the measurements require an omnidirectional sound source emitting short sound pulses (or other signals that can be processed to become impulses) covering the frequency interval of interest, an omnidirectional microphone and a medium for the storage of the electrical signal generated by the microphone.

Impulsive sources such as pistols firing blank cartridges, bursting balloons, or paper bags and electrical spark sources may be used; but these sources are primarily used for noncritical measurements, as their directivity and acoustic power are not easy to control.

For measurements requiring higher precision and repeatability, omnidirectional loudspeakers are preferable, as the power and directivity characteristics of loudspeakers are more stable and well defined. Omnidirectional loudspeakers are often built as six units placed into the sides of a cube, 12 units placed into a dodecahedron or 20 units in an icosahedron, see Fig. 9.11. The requirements regarding omnidirectivity of the source are described in the ISO 3382 standard.

Loudspeakers, however, cannot always emit sufficient acoustic power if an impulsive signal is applied directly. Therefore, special signals of longer duration such as maximum-length sequences or tone sweeps have been developed, which have the interesting property that their autocorrelation functions are band-limited delta pulses. This means that the loudspeaker can emit a large amount of energy without challenging its limited peak-power capability while impulse responses with high time resolution can still be obtained afterwards through post-processing (a correlation/convolution process) of the recorded noise or sweep responses.

In certain cases, other source directivity patterns than omnidirectional can be of interest. For example, a directivity like that of a human talker may be relevant if the measurement is related to the intelligibility of speech or to measurements on the stage in an opera house. A suitable talker/singer directivity can be obtained by using a small loudspeaker unit (membrane diameter about three to four inches) mounted in a closed box of size comparable to that of the human head. Using only one of the loudspeakers in a cube or dodecahedron speaker of limited size is another possibility.

9.4.4 Microphones

For most of the parameters mentioned, the microphone should be omnidirectional, while omni plus figure-of-eight directivity and dummy heads are used when the impulse responses are to be used for the calculation of LEF and IACC.

Three-dimensional intensity probes, sound field microphones or microphone arrays have also been suggested in attempts to obtain more-complete information about the amplitude and direction of incidence of individual reflections [9.17] or for wave field analysis [9.18]. However, such measurements are mainly for research or diagnostic purposes. As such, they do not fulfill the goal set up in Sect. 9.1 of reducing the information to what is known to be subjectively relevant. A selection of relevant microphones are shown in Fig. 9.12.

9.4.5 Signal Storage and Processing

For the storage of the impulse responses the following choices appear.

- The hard disk on a personal computer (PC) equipped with a sound card or special analog-to-digital (A/D) converter card

- Real-time (1/3 and 1/1 octave) analyzers or sound level meters with sufficient memory for storage of a series of short time-averaged root-mean-square (RMS) values
- Fast Fourier transform (FFT) analysers capable of recording adequately long records
- Tape recorders, preferably a digital audio tape (DAT) recorder, or even MP3 recorders given that the data reduction does not deteriorate the signal.

Of these, real-time analyzers and sound level meters will process the signal through an RMS detector before storage, which means that not the full signal, but only energy with a limited time resolution will be recorded.

Today, calculation of the objective parameters will always be carried out by a PC or by a computer built into a dedicated instrument. A large number of PC-based systems and software packages particularly suited for room acoustic measurements are available. Some systems come with a specific analog-to-digital (A/D), digital-to-analog (D/A) and signal-processing card, which must be installed in or connected to the PC, while others can use the sound card already available in most PCs as well as external sound cards for hobby music use.

9.5 Prediction of Room Acoustic Parameters

Beyond theoretical calculation of the reverberation time, the prediction techniques presented in this section range from computer simulations and scale model measurements to simple linear models based on empirical data collected from existing halls.

In all cases, the main objective is the prediction of the acoustic conditions described in terms of the acoustic parameters presented in Sect. 9.3. Just as we can record the sound in a hall, both scale models and numerical computer simulations can also provide *auralizations*, which are synthesized audible signals that allow the client and architect to listen to and participate in the acoustic evaluation of proposed design alternatives. Such audible simulations will often have much greater impact and be more convincing than the acoustician's verbal interpretation of the numerical results. Therefore, the acoustician must be very careful to judge the fidelity and the degree of realism before presenting auralizations.

9.5.1 Prediction of Reverberation Time by Means of Classical Reverberation Theory

Reverberation time as defined in Sect. 9.3.1 is still the most important objective parameter for the characterization of the acoustics of a room. Consequently, the prediction of T is a basic tool in room acoustic design. Calculation of reverberation time according to the Sabine equation was the first attempt in this direction and today is still the most commonly used. However, during the 100 years since its invention, several other reverberation time formulae have been suggested (by Eyring, Fitzroy, Millerton and Sette, Metha and Mulholland Kut-truff and others), all with the purpose of correcting some obvious shortcomings of the Sabine method.

The Sabine equation has been described in Sect. 11.1.4. In larger rooms, the absorption in the air plays a major role at high frequencies. If we include this factor, the Sabine equation reads:

$$T = \frac{0.161 V}{S\alpha^* + 4mV}, \quad (9.21)$$

where V is the room volume, S is the total area of the room surfaces, α^* is the area-weighted average absorption coefficient of the room surfaces

$$\alpha^* = \overline{\alpha} = \frac{1}{S} \sum_i S_i \alpha_i \quad (9.22)$$

and m accounts for the air absorption, which is a function of the relative humidity and increases rapidly with fre-

Fig. 9.11 Two types of omnidirectional sound sources for room acoustic measurements: blank pistol (in the *raised hand of the seated person*) and an omnidirectional loudspeaker (icosahedron with 20 units)

Table 9.2 Air absorption coefficient m for different frequencies and relative humidity; valid for an air temperature of 20 °C (after [9.19])

Relative humidity (%)	0.5 kHz	1 kHz	2 kHz	4 kHz	8 kHz	Air absorption
40	0.4	1.1	2.6	7.2	23.7	10^{-3} m^{-1}
50	0.4	1.0	2.4	6.1	19.2	10^{-3} m^{-1}
60	0.4	0.9	2.3	5.6	16.2	10^{-3} m^{-1}
70	0.4	0.9	2.1	5.3	14.3	10^{-3} m^{-1}
80	0.3	0.8	2.0	5.1	13.3	10^{-3} m^{-1}

quency [9.19]. Values of m versus frequency and relative humidity are listed in Table 9.2.

The Sabine equation assumes the absorption to be evenly distributed on all surfaces, and that the sound field is *diffuse*, which means that in each point of the room:

1. the energy density is the same, and
2. there is equal probability of the sound propagating in all directions.

Among the alternative reverberation formulae, only the Eyring method will be described here, as it may give more-accurate predictions in cases where the room is highly absorptive.

In contrast to Sabine, the Eyring theory assumes that the sound field is composed of plane sound waves that lose a fraction of their energy equal to the absorption coefficient of the surface whenever the wave hits a surface in the room. Thus, after n reflections, the energy in all wavefronts and the average energy density in the room has been reduced to $(1-\overline{\alpha})^n$ times the original value. The average distance of propagation during this process is $l_\mathrm{m} n = ct$, in which l_m is the *mean free path*,

Fig. 9.12 A selection of microphones types used for recording impulse responses in rooms. *From left*: artificial head (with both internal and external microphones), *sound field microphone* (with four capsules for 3-D recordings of reflections) and a two-channel (stereo) microphone with omnidirectional and figure-of-eight capsules

equal to the average distance traveled by the wave between two reflections, and t is the time elapsed since the wave propagation started. *Kosten* [9.20] has shown, that regardless of the room shape, l_m equals $4V/S$, if all directions of propagation are equally probable. These assumptions lead to the Eyring formula, which can be expressed by substituting α^* in (9.21) by

$$\alpha^* = \alpha_\mathrm{e} = \ln\left(\frac{1}{1-\overline{\alpha}}\right) \qquad (9.23)$$

For low values of $\overline{\alpha}$, $\overline{\alpha}$ and α_e are almost identical, causing the Sabine and the Eyring formulae to give nearly identical results; but the difference becomes noticeable when $\overline{\alpha}$ exceeds about 0.3. In general, α_e will always be larger than $\overline{\alpha}$, leading to $T_\mathrm{Sabine} > T_\mathrm{Eyring}$. In the extreme case of the mean absorption approaching 1, α_e converges towards ∞, whereby $T_\mathrm{Eyring} \to 0$ as one would expect for a totally absorbing room. However, T_Sabine converges towards the finite positive value $0.16 V/S$, which of course is unrealistic.

Both the Sabine and the Eyring theories suffer from at least two shortcomings. Neither of them consider the actual distribution of the free paths (see the comments on the *mean free path* above), which is highly dependent on the room shape, nor the often very uneven distribution of the absorption in the room, which causes different directions of sound propagation not to be equally probable. Typical situations where these factors become important are rooms with highly absorbing acoustic ceilings and hard walls and floors – or the reverse situation of an auditorium, where almost all absorption is found in the seating. In these situations the energy is certainly not evenly distributed, there is a considerable distribution of the path lengths, and there is a much higher probability for sound energy to travel towards the absorbing surface than away from it.

A common example in which the Sabine and Eyring theories fall short is the simple case of a rectangular room with plane walls, absorbing ceiling and low height compared to length and width. In such a room the main part of the late reverberant energy will be stored in

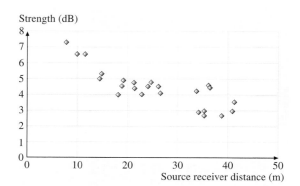

Fig. 9.13 Strength versus distance measured in the Boston Symphony Hall; values averaged over the four octaves 250–2000 Hz. The reverberation distance in this hall is about 5 m

one- and two-dimensional horizontal room modes with long free paths, whereas the three-dimensional modes with shorter free paths will soon have lost their energy by interacting with the absorbing ceiling. Hereby, different modes end up having very different decay rates, leading to bent or double-sloped decay curves and to measured T values often longer than calculated. In fact, only the unnatural situation with an even absorption distribution and all free paths of equal length will result in a straight-line decay and a well-defined T value.

From practical measurements of the distribution of G values, it is also clear that the assumption of an even distribution of the energy density is not met in large auditoria. Contrary to the Sabine theory, which predicts a constant level beyond the reverberation distance (Fig. 11.6), we observe that the level continues to decrease with distance from the source – even far beyond the reverberation distance, where the contribution of the direct sound is negligible. An example of this is seen in Fig. 9.13. (Based on empirical data, simple equations predicting this attenuation of G with distance as a function of room geometry have been derived as described in Sect. 9.5.6.)

Still, rather than relying on another of the reverberation formulas to provide a more correct result, today it may be wiser just to apply the Sabine and perhaps the Eyring equations, keeping in mind their limitations. Thus, *Beranek* [9.21] has suggested that, for Sabine calculations in concert halls, one should apply different sets of seat absorption values depending on the room shape. These values should be measured in nearly the same type of room as the hall in question.

When higher prediction accuracy is required, it is recommended to carry out simulations in a computer model as described in Sect. 9.5.4. However, also with this calculation approach, the question about which absorption values to use is relevant.

9.5.2 Prediction of Reverberation in Coupled Rooms

Auditoria with coupled spaces form another situation in which the decay curve may end up being nonlinear and the energy density will be different in different areas of the room. Typical of such cases are open-plan office floors coupled to an atrium covering several storeys, seating areas in auditoria subdivided by (deep) balconies, theaters with proscenium openings between the auditorium and the stage house, churches subdivided into main and side naves and concert halls with reverberation chambers. Diffuse field theory can be used to illuminate important properties of the reverberation conditions in coupled rooms.

We consider a sound source placed in a room, room 1, with volume V_1 and physical absorption area A_{10}. This room is coupled by an opening with area S to another room, room 2, with volume V_2 and absorption area A_{20}. From diffuse field theory (the energy balance equations) we find that the apparent absorption area of room 1, A_1, depends on the ratio between the absorption area of the attached room A_{20} and the area of the opening S:

$$A_1 = A_{10} + \frac{A_{20} S}{A_{20} + S} . \tag{9.24}$$

It is seen that for small openings, $S \ll A_{20}$, $A_1 \approx A_{10} + S$, and for large openings, $S \gg A_{20}$, $A_1 \approx A_{10} + A_{20}$, as we would expect.

As for the reverberation curve, a double slope may be observed, particularly if both source and receiver are placed in the less reverberant of the two rooms. The two reverberation times T_I and T_{II} defining the double slope can be described in terms of the reverberation times for each of the two rooms separately if we consider the opening to be totally absorbing:

$$T_1 = \frac{0.161 V_1}{A_{10} + S} \quad \text{and} \quad T_2 = \frac{0.161 V_2}{A_{20} + S} . \tag{9.25}$$

However, for our purpose the math gets simpler if we apply damping coefficients: $\delta_1 = 6.9/T_1$ and $\delta_2 = 6.9/T_2$ instead of the reverberation times. Then, with coupling, the two damping coefficients, δ_I and δ_{II}, corresponding to the slope in the first and second part of the decay, respectively, can be calculated from:

$$\delta_{I,II} = \frac{\delta_1 + \delta_2}{2} \pm \sqrt{\frac{(\delta_1 - \delta_2)^2}{4} + \frac{S^2 \delta_1 \delta_2}{(A_{10} + S)(A_{20} + S)}} \quad (9.26)$$

upon which

$$T_I = 6.9/\delta_I \quad \text{and} \quad T_{II} = 6.9/\delta_{II}. \quad (9.27)$$

Another important feature of the double-slope decay is how far down the knee point appears. If the opening area is large and/or the source or receiver is placed close to the opening, then a substantial exchange of energy can occur and the knee point may appear so early that even the EDT can be influenced by the decay from the coupled room with longer T. If the opening area is small, then the knee point appears late and far down on the curve, in which case the influence of the coupled room can only be perceived after final cords. This is the case with some of the existing concert halls equipped with coupled reverberation chambers. Further descriptions of coupled spaces can be found in [9.5].

9.5.3 Absorption Data for Seats and Audiences

Regardless of which reverberation formula is used, the availability of reliable absorption values is fundamental for the accurate prediction of T as well as of most other room acoustic parameters. Beyond the absorption values of various building materials (as listed in Table 11.1), realistic figures for the absorption of the chairs and seated audience are of special concern, because these elements normally account for most of the absorption in auditoria. The absorption of chairs – both with and without seated people – depends strongly on the degree of upholstery. Average values of absorption for three different degrees of chair upholstery have been published by *Beranek* [9.22] and quoted in Table 9.3. As mentioned earlier, chair absorption values for use specifically in Sabine calculations and for different types of hall geometries can be found in [9.21].

Sometimes audience absorption data are quoted as absorption area per chair or per person. However, as demonstrated by *Beranek* [9.3], the use of absorption coefficients multiplied by the area covered by the seats plus an exposed perimeter correction is more representative than absorption area per person, because audience absorption tends to vary proportionally with the area covered, while the absorption coefficient does not change much if the density of chairs is changed within that area.

Minimizing the total sound absorption area is often attempted in auditoria for classical music in order to obtain a strong, reverberant sound field. One way to achieve this is not to make the row-to-row distance and the width of the chairs larger than absolutely necessary. Another factor is the seat design itself. Here one should aim at a compromise between minimum absorption and still a minimum difference between the absorption of the occupied and empty seat. This is facilitated by back rests not being higher than the shoulders of the seated person and only the surfaces covered by the person being upholstered. The rear side of the back rest should be made from a hard material like varnished wood. On the other hand, a minimum difference between the occupied and empty chair implies that the upholstery on the seat and back rest should be fairly thick (e.g., 80 mm on the

Table 9.3 Absorption coefficients of seating areas in concert halls for three different degrees of upholstery; both empty and occupied (after [9.22])

Octave centre frequency (Hz)	125	250	500	1000	2000	4000
Heavily upholstered chairs Unoccupied	0.70	0.76	0.81	0.84	0.84	0.81
Heavily upholstered chairs Occupied	0.72	0.80	0.86	0.89	0.90	0.90
Medium upholstered chairs Unoccupied	0.54	0.62	0.68	0.70	0.68	0.66
Medium upholstered chairs Occupied	0.62	0.72	0.80	0.83	0.84	0.85
Lightly upholstered chairs Unoccupied	0.36	0.47	0.57	0.62	0.62	0.60
Lightly upholstered chairs Occupied	0.51	0.64	0.75	0.80	0.82	0.83

seat and 50 mm on the back rest), the bottom of the tip-up seats should be perforated or otherwise absorptive, and when the seat is in its vertical position, there should be a wide opening between the seat and the back rest so that the sound can still access the upholstered areas.

9.5.4 Prediction by Computer Simulations

Computer simulations take into account the geometry and actual distribution of the absorption materials in a room as well as the actual source and receiver positions. The results provided are values for all relevant acoustic parameters as well as the generation of audio samples for subjective judgments. Computer simulation is useful not only for the design of concert halls and theaters, but also for large spaces such as factories, open-plan offices, atria, traffic terminals, in which both reverberation, intelligibility and noise-mapping predictions are of interest. Besides, they can be used in *archaeological acoustics* for virtual reconstruction of the acoustics of ancient buildings such as Roman theaters (http://server.oersted.dtu.dk/www/oldat/erato/).

The first computer models appeared in the 1960s, and today a number of commercially available software packages are in regular use by acoustic consultants as well as by researchers all over the world.

In computer simulations the room geometry is represented by a three-dimensional (3-D) computer-aided design (CAD) model. Thus, the geometry information can often be imported from a 3-D model file already cre-

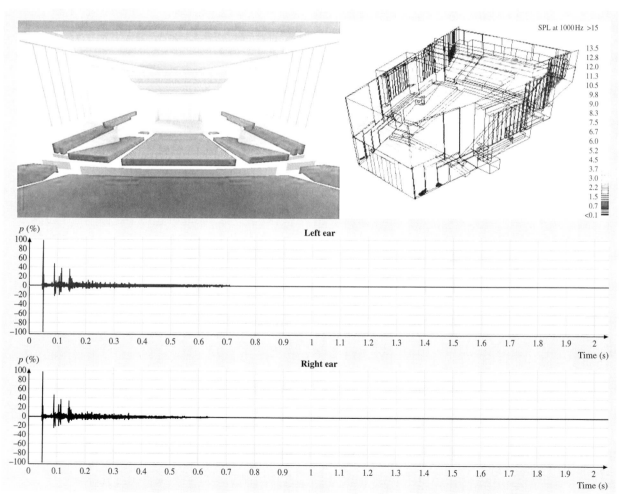

Fig. 9.14 Examples of output from a room acoustic simulation software package (ODEON); see text for explanation

ated by the architects. However, one needs to break up eventual curved surfaces into plane elements and often it is also advisable to simplify details in the geometry provided by the architect. Alternatively, the acoustician can build the room model by writing a file of coordinate points or mathematical expressions for all surface corners. Of course this will be more time consuming in cases of complex room geometry, but some programs offer intelligent programming routines for the modeling of geometric elements.

When the geometry has been completed, the absorption values for each octave band, the scatter, and the eventual degree of acoustic transparency have to be assigned to each surface. In addition, the source and receiver characteristics (power and directivity versus frequency, position and orientation) must also be entered.

In the calculation of sound propagation, most models disregard phase properties and use an energy approximation. The sound propagation is studied by means of rays (up to millions depending on the room complexity) being traced from the source position through reflections until their energy has been reduced below the dynamic range of interest (often determined by T or the signal-to-noise ratio in auralizations). Reflection directions are chosen so that angle of reflection equals angle of incidence or with more or less random angles depending on the scatter value attributed to the surface. For the calculation of early reflections, some models also apply image source theory, which actually makes phase considerations possible if the complex impedances of the various surface materials and not just their energy absorption coefficients are known. However, the late reverberation part is normally calculated by some kind of ray method, as image source calculations become impractical after a few orders of reflection. Recently, however, attempts to carry out complete calculations inspired by finite-element (FEM) or boundary-element (BEM) methods have been reported even for rather complex rooms [9.23], whereby phase and diffraction phenomena can be handled.

As a result of tracing the history of all rays emitted in the room model, source and receiver specific impulse responses are produced with detailed information about the direction of incidence, delay and level versus frequency of each reflection component. From these, all the parameters mentioned in Sect. 9.3 can be calculated. Most programs can generate a grid of receiver point over selected audience areas and code the numeric results using a color scale to facilitate a fast overview of the detailed position results for any of the parameters.

Also, automatic plots of relevant statistics regarding the distribution with frequency and position are provided.

Impulse responses suitable for use in *auralization* can be generated by applying available head-related transfer functions, which have been recorded as a function of direction of sound incidence on an artificial head. Such transfer functions not only represent the directivity of the human receiver, but are also important for correct perception of direction of incidence of sound field components when listening through headphones. Alternatively, the directional information of the individual reflections making up the impulse response can be coded into a surround-sound format for listening via a multiple-loudspeaker setup.

Figure 9.14 shows a view of a room model as well as two binaural impulse responses and a color mapping of a calculated parameter (generated by the ODEON software).

9.5.5 Scale Model Predictions

Acoustic scale models have been used since the 1930s [9.24] to study the behavior of sound in rooms. With proper care in detailing the geometry and the choice of materials for building the model, this technique can provide quite accurate predictions of impulse responses and acoustic parameters. Actually, scale modeling is still regarded as more reliable than computer models in cases where the room contains a lot of irregular surfaces or objects that diffuse or diffract sound waves. However, the use of physical scale models is limited to larger, acoustically demanding projects, as building and testing the model is far more time consuming and costly than computer modeling.

The diffusion and diffraction caused by an object or surface in a room depends on the ratio between the linear dimensions and the wavelength of the sound: l/λ. Therefore, in a $1:M$ scale model, the diffraction/diffusion conditions will be modeled correctly if the wavelengths of the measurement signals in the model are chosen such that:

$$\frac{l}{\lambda} = \frac{l_M}{\lambda_M} = \frac{l/M}{\lambda/M} \quad \Rightarrow \quad f_M = fM \,. \tag{9.28}$$

This means that we should increase the frequency used in the model measurements by the factor M. Fortunately, sound sources and microphones exist that are capable of handling frequencies up to around 100 kHz, whereby the frequency range up to 10 kHz can be handled in a 1 : 10 scale model or 2 kHz in a 1 : 50 scale model. The sources may be small loudspeaker units (piezoelectric or electro magnetic) and the microphones 1/4" or even 1/8" condenser types.

Ideally, the absorption coefficient α_M of the materials chosen for building each surface in the hall model should fulfil the equation:

$$\alpha_M(fM) = \alpha(f). \tag{9.29}$$

In practice, however, we need only to distinguish between mainly reflecting and mainly absorptive materials. The reflecting surfaces are quite easy, as these can be made of any hard material (plywood, plaster) with a couple of layers of varnish if necessary. It is far more important to build the seats/audience so that both head diffraction and absorption is correct for the scale chosen. In the end it is often necessary to fine-tune the reverberation time in the model by applying absorption to some secondary surfaces (surfaces that do not generate primary early reflections).

Another important issue is to compensate for the excessive air absorption at high frequencies in the model. This can be achieved either by drying the air in the model, by exchanging the air for nitrogen, or simply as a calculated compensation since the attenuation is a simple linear function of time once the, frequency, humidity and temperature have been specified. However, in the case of compensation the signal-to-noise ratio at high frequencies will be severely limited, which reduces the decay range for the estimation of T and the sound quality if auralization is attempted. Figures 9.15 and 9.16 show an example of a 1:10 scale model and its audience respectively.

9.5.6 Prediction from Empirical Data

In the 1980s, when most of the current objective acoustic parameters had been defined, several researchers started making systematic measurement surveys of concert, opera and multipurpose halls. Many of these measured data were published in [9.3]. Combining acoustic data with data on the geometry of the halls has made it possible to derive some simple rules of thumb regarding how, and how much, acoustic parameters are affected by the choice of gross room shape and dimensions.

Through statistical multiple linear-regression analysis of data from more than 50 halls, a number of relationships between acoustical and geometrical properties have been found [9.25]. A number of the linear regression models derived are shown in Table 9.4. In these models, the variations in the acoustic parameters are described as functions of the expected value according to diffuse field theory (G_{exp} and C_{exp}, Sect. 9.3) plus various geometrical variables such as: average room height H, average width W, room volume V, number of seats N, number of rear balcony levels 'no. rear balc.' etc. When the models were derived, the measured reverberation time (T) was used to calculate the expected values of strength G_{exp} and clarity C_{exp}. When using these models for prediction of the values in a hall not yet built, the expected values could be based on a Sabine calculation of T instead. It should be mentioned that, apart from $\Delta G(10\,\text{m})$, the attenuation in strength per 10 m increase in source–receiver distance, the predicted values should be considered as representing the position average of values to be found in the hall.

The three rightmost columns in Table 9.4 contain information about how well each prediction formula matched the measured data. When comparing the amounts of variance explained by the different models, it is seen that the diffuse field prediction is responsible for the largest portion of the variance. This may be interpreted as hall volume and total absorption area being the most important factors governing the behavior

Table 9.4 Regression models for the relationships between room acoustic parameters and room design variables as derived from a database containing data from more than 50 halls (after [9.25])

Room acoustic parameter	Regression models: $f(\text{PAR}_{expected}, \text{geometry})$	Correlation coefficient	% of variance	STD residuals
C (dB)	$-0.1 + 1.0 C_{exp}$	0.76	58	1.0 dB
	$-1.4 + 0.95 C_{exp} + 0.47 W/H + 0.031$ floor slope	0.83	68	0.9 dB
	$-1.2 + 1.03 C_{exp} + 0.43 W/H + 0.013$ wall angle	0.84	70	0.9 dB
	$-1.77 + 1.10 C_{exp} + 0.055 W + 0.027$ stage ceil. angle	0.86	74	0.8 dB
G (dB)	$-2.0 + 0.94 G_{exp}$	0.91	83	0.9 dB
	$-5.61 + 1.06 G_{exp} + 0.17 V/N + 0.04$ distance	0.94	89	0.9 dB
$\Delta G(10\,\text{m})$ (dB)	$-1.85 + 0.42$ no. rear balc.	0.50	25	0.7 dB
	$-1.41 + 0.35$ no. rear balc. -3.93 distance$/(HW)$	0.55	31	0.6 dB
LEF (–)	$0.39 - 0.0061$ width	0.70	49	0.05
	$0.37 - 0.0051$ width $- 0.00069$ wall angle	0.72	53	0.05

of C and G. However, as all the independent variables listed gave a significant improvement of the model accuracy, this also demonstrated that consideration of the geometrical factors can improve the prediction accuracy. Equally importantly, the acoustic effects of changes in certain design variables can be quantified quickly.

Prediction of Clarity

The first C-model illustrates that absorption and volume, as reflected in T, are responsible for the main part of the variation in C. The other three models all illustrate the positive, but sometimes unwanted, effect of average hall width on clarity. Moreover, it is seen that introducing a moderate 15° slope of the main audience floor (without changing the other variables: average width-to-height ratio, volume and absorption area) causes C to increase by about 0.5 dB on average. Similarly, changing the basic design from rectangular to a 70° fan shape makes C increase by about 1 dB. The last C-model illustrates the effect of tilting the angle of the stage ceiling towards the audience. Changing the slope from horizontal to 20° results in about 0.5 dB higher C values.

Predictions of Strength

The G model only considering G_{exp} is rather accurate. The constants in both G-models illustrate the fact that G is always a few dB lower than G_{exp}, as also predicted by *Barron's revised theory* [9.26].

The second G-model contains an independent variable: the ratio between the volume and the number of seats. The positive influence of this ratio is not immediately evident because V/N, which is strongly correlated with T, is expected to be incorporated into the variable G_{exp} already (although this result has also been confirmed by other researchers [9.27]).

As mentioned earlier, G shows a steady and significant decrease with distance in most halls, as was also found by *Barron* [9.26]. This phenomenon is described quantitatively by the two models for estimation of the rate of decrease in G per 10 m source–receiver distance $\Delta G(10\,\text{m})$.

Both listed $\Delta G(10\,\text{m})$ models indicate a reduced distance attenuation when the number of rear balconies is increased. This may be related to the fact that the level is increased in the more-distant seats when these are placed on a balcony, whereby they are closer to the stage and to the reflecting ceiling.

In the second $\Delta G(10\,\text{m})$ model, distance/HW appears as an independent variable. This variable equals the distance from the stage to the rearmost seat divided by the product of the average room height H and average hall width W. As the coefficient to this variable is negative, the natural result appears to be that attenuation with distance will increase if the hall is long, narrow or has a low ceiling.

Predictions of Lateral Energy Fraction

At the bottom of Table 9.4 are listed two models for LEF as a function of hall geometry only, as diffuse field theory has no effect on LEF variation. The effects of the width and angle between the side walls are understandable.

Simple prediction formulae as listed in Table 9.4 are particularly useful in the very early phases of the design process, in which it is natural for the architect to produce and test many different sketches in short succession, leaving no time to carry out computer simulation of each proposal. The importance of the knowledge embedded in these rule-of-thumb equations is highlighted by the fact that many aspects of the acoustics of a new hall are settled when one of these sketches is selected for the further development of the project.

9.6 Geometric Design Considerations

With the connections between room acoustic parameters and hall geometry described in the previous section, we have made the first approach to the third question set up in the introduction to this chapter, which is of real interest to the acoustic designer: how do we control the acoustic parameters by the architectural design variables? Referring again to Fig. 9.1, we are beginning to fill the lower-left box with architectural parameters and establish the relationship between these and the objective parameters. The major geometric factors of importance in auditorium design will be dealt with in the present section: seating layout in plan (determining the gross plan shape of the room) and in section (determining sight lines), use of balconies, choice of wall structure, room height, ceiling shape and use of free-hanging reflectors.

9.6.1 General Room Shape and Seating Layout

When people stop to watch or listen to a spontaneous performance, for instance in an open square in the city, the way they arrange themselves depends on the type

Fig. 9.15 A 1:10 scale model of the Danish Broadcasting concert hall in Ørestad, Copenhagen. The model is fitted with audience and orchestra (acoustic design by Nagata Acoustics, Japan)

Fig. 9.16 Model audience used in a 1:10 scale model of the Danish Broadcasting concert hall shown in Fig. 9.15. *Upper-left corner*: Close view of model orchestra musicians and spark sound source. *Lower left*: close-up view of model chairs and polystyrene audience with hollow chest. Dress and hair is made from wool felt. *Right*: model audience and chairs placed in model reverberation chamber for absorption testing (reverberation room designed by Nagata Acoustics)

Fig. 9.17 *Organic* formations of audience depending on the type of performance

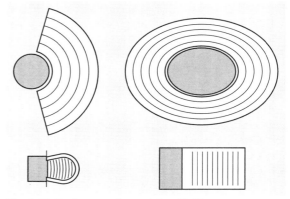

Fig. 9.18 Basic room shapes (after [9.4])

of performance. It will be governed by the fact that any newcomer will look for the best position available relative to his/her need to see or hear properly. The choice will be a compromise between choosing a position close to the performer(s) and next to other members of the audience or farther away but close to an eventual main center line for vision and sound radiation. In Fig. 9.17, such organic audience arrangements are shown for two different types of performances:

1. action or dialogue theater such as dancing, fighting, debating, circus performance, for which the visual and acoustic emission are more or less omnidirectional and
2. monologue or proscenium stage theater performance and concerts with acoustic instruments – all of which have a more-limited visual and/or acoustic directivity.

When we set up walls around the gathered people, we arrive at two classical plan shapes of auditoria developed early during our cultural history. These are shown at the top of Fig. 9.18: the fan-shaped Greek/Roman theater (left), and the amphitheater, circus or arena (right). Both were originally open theaters; but the shapes were maintained in roofed buildings. In the bottom of Fig. 9.18, later, basic forms are shown: the horseshoe, the Italian opera plan and the rectangular concert hall. The latter form was originally a result of traditional building forms and limitations in roof span with wooden beams.

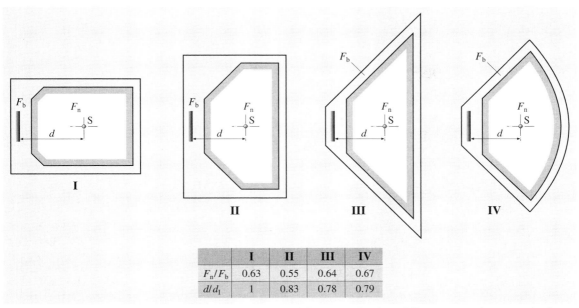

Fig. 9.19 Mean stage–listener distance and net efficiency of floor area for four different room shapes. F_n is the net area occupied by the audience, F_b is the total floor area, d is the average source–listener distance; $d_1 = d$ for case I (after [9.28])

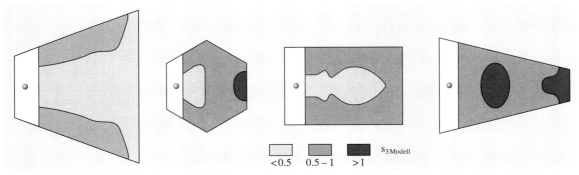

Fig. 9.20 Calculated distributions of early lateral reflection levels in rooms of different plan shape: fan, hexagon, rectangular and reverse fan. *Darker areas* represent higher LEF levels (after [9.5])

When building an auditorium, efficient use of the available floor space and the average proximity of the audience to the performers are very important parameters. Depending on the plan shape of the room different values appear as shown in Fig. 9.19.

Low average distance is the main reason for the frequent use of the fan shape (III and IV in Fig. 9.19), although the directivity of sound sources (like the human voice) as well as the quality of lines of sight sets limits on the opening angle of the fan.

Another limitation of the fan shape is that it does not generate the strong lateral reflections so important in halls for classical music. This was already seen in the empirical equations in Sect. 9.5.6 and confirmed in Fig. 9.20, showing the calculated distributions of early lateral reflection levels in rooms of different plan shapes.

The reason is that sound waves from the source hitting the side walls will produce reflections that will run almost parallel to the direct sound and so hit the listener from an almost frontal direction, therefore not contribut-

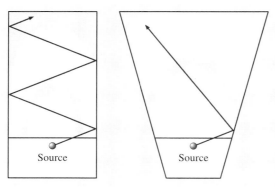

Fig. 9.21 Directions of side-wall reflections depending on room shape (after [9.4])

ing to the LEF or causing dissimilarity between the two ear signals. This is also illustrated in Fig. 9.21. In addition, many of the side-wall reflections will not reach the center of the hall at all.

9.6.2 Seating Arrangement in Section

When several rows of seated audience are placed on a flat floor, the direct sound from a source on the stage and the reflection off the audience surface will hit a receiver in this audience area after having traveled almost identical distances, and the reflection will hit the audience surface at a very small angle relative to the surface plane. At this grazing angle of incidence, all the energy will be reflected regardless of the diffuse absorption value of the surface, but the phase of the pressure in the reflected wave will also be shifted 180°. Hereby the direct sound and the grazing incidence reflection will almost cancel each other over a wide frequency range. Typical attenuation values found at seats beyond the tenth row can amount to 10–20 dB in the range 100–800 Hz relative to an unobstructed direct sound. Moreover, the higher frequencies above 1000 Hz will also be attenuated by 10 dB or more due to scattering of the sound by the heads of people sitting in the rows in front.

The same values of attenuation mentioned above for the direct sound are likely also to be valid for first-order reflections off vertical side walls. The result will be weaker direct sound and early reflections in the horizontal stalls area and subjective impressions of reduced *clarity*, *intimacy* and *warmth* in the stalls seats.

To avoid this grazing incidence attenuation, the seating needs to be sloped relative to the direction towards the source. The vertical angle of the audience floor can be designed to be constant or to vary gradually with

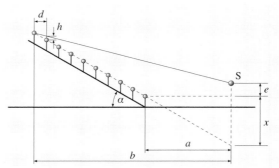

Fig. 9.22 Relevant parameters for the calculation of the necessary angle α for a constantly sloping floor section. a: distance from the source to the first seating row, b: distance from the source to the last seating row, d: distance between the seat rows, e: source height above the head of listeners in the first row (determined by the height of the stage floor over the stalls floor), h: line-of-sight clearance at the last row. Rows closer to the source will have clearance values larger than h (after [9.28])

distance. If a constant slope is planned, the angle α necessary to obtain a certain clearance h at the last row (equal to the vertical distance between sight lines from this and the preceding row) can be calculated from the variables indicated in Fig. 9.22:

$$\tan \alpha = \frac{hb}{da} - \frac{e}{a} . \tag{9.30}$$

In particular, it is of interest to use this formula to see how far away from the source it is possible to maintain a certain slope (one obvious example being $\alpha = 0$, corresponding to a horizontal floor):

$$b = \frac{d}{h}(e + a \tan \alpha) . \tag{9.31}$$

Normally, a clearance value of minimum 8 cm will be sufficient to avoid grazing-incidence attenuation; but sight lines will still be unsatisfactory unless staggered seating is applied so that each listener is looking towards the stage between two other heads and not over a person sitting directly in front. However, if the clearance is increased to about 12 cm, then both the acoustic and visual conditions will be satisfactory, but this implies a steeper floor slope.

In rows closer to the sound source a linear slope will cause the clearance to be higher than the design goal – and so higher than necessary. Often this is not desirable as it may cause the last row to be elevated high above the source and the steeply sloped floor will reduce

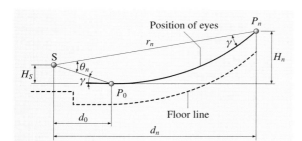

Fig. 9.23 Relevant parameters for the calculation of a floor slope with constant clearance. The variables are explained in the text (after [9.29])

the effective volume of the room. Therefore, it can be relevant to consider instead a curved floor slope with constant clearance, as illustrated in Fig. 9.23.

In Fig. 9.23, the height of a person's head in the n-th row H_n relative to the height of the head in the first row is calculated according to the formula (derived from a logarithmic spiral)

$$H_n = d_0 \gamma + d(\theta - \gamma)$$
$$= \gamma \left[d_n \log_e \left(\frac{d_n}{d_0} \right) - (d_n - d_0) \right]. \qquad (9.32)$$

In this expression d_0 is the distance from the source to the first row, d_n the distance to the n-th row and γ the desired angle between the tangent to the seating plane and the line of sight to the source. For a one meter row-to-row distance, an angle γ of 8° will correspond to a clearance of about 12 cm.

A curved slope can also be obtained by simply applying the linear-slope formula to each of smaller sections of seat rows successively (e.g. for every five rows). In this case the variables a and e in (9.30) should refer to the first row in the relevant section.

As mentioned earlier, a steep slope will reduce the volume and so the reverberation time T. Besides, geometric factors may reduce T beyond what a Sabine calculation may predict. The reason is that the elevated seating – and its mirror image in the ceiling – will cover a larger solid angle as seen from the source. Hereby a larger part of the emitted sound energy will quickly reach the absorbing seating and leave less for fueling a reverberant sound field in the room. The result is higher *clarity* and less *reverberance* in halls with steeply raked seating. This is in line with the empirical equation for clarity as a function of T (C_{exp}) and floor slope listed in Table 9.4, Sect. 9.5.6. Consequently, it is a good idea to limit the clearance/angle in

halls for classical music to about 8 cm or 6°, whereas for speech auditoriums and drama theaters, in which both intelligibility and good visual sight lines have priority, values closer to 12–15 cm or 8°–10° may be preferable.

9.6.3 Balcony Design

The introduction of balconies allows a hall with a larger seat count to be built where a limited footprint area is available; but often balconies are introduced simply in order to avoid longer distances between stage and listeners. The average stage–listener distance can be reduced significantly by elevating the rearmost seat rows and moving them forward on rear or side balconies above parts of the audience in the stalls. In this way not only is the direct sound increased by shortening the direct sound path, but the elevated seats also become closer to a possibly reflecting ceiling. The result is higher strength as well as higher clarity in the balcony seats compared to the alternative placement of seats at the back of an even deeper hall.

Since the line-of-sight and clearance criteria still have to be fulfilled for balcony seats, the slope often becomes quite steep on balconies, as illustrated in Fig. 9.24 showing a long section in a hall with constant clearance on main floor and balcony seats. For safety reasons the slope must be limited to no more than about 35°.

When balconies are introduced, the acoustics in the seats below the balconies need special attention. It is important to ensure sufficient opening height below balcony fronts relative to the depth of the seating under the balcony (Fig. 9.25). Otherwise, particularly the level of the reverberant energy in the overhung seats will be reduced, causing the overall sound level to be weak and lacking in fullness. Rules of thumb in design are $H \geq 2D$ for theaters (in which reverberant sound is less important) and $H \geq D$ for concert halls (in which fullness is

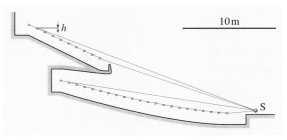

Fig. 9.24 Section of a hall with a rear balcony and constant clearance. The result is an increased slope for the elevated balcony (after [9.30])

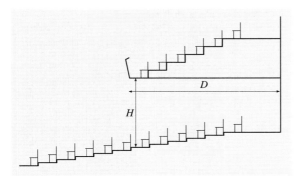

Fig. 9.25 Important parameters for maintaining proper sound in seats overhung by balconies (after [9.9])

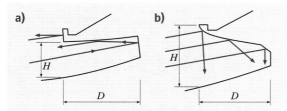

Fig. 9.26a,b Poor (**a**) and good (**b**) design of balcony profile (after [9.29])

Sometimes, the requirement for limited depth of balcony overhangs is also specified as a percentage of the ceiling area needed to be visible from the last row of seating or as a vertical opening angle β between the balcony soffit and the head of listener below the balcony front as seen from the last row. In the latter case, Barron has suggested 25° as suitable for drama halls and 35° for concert halls.

As illustrated in Fig. 9.26, it is often advantageous to let the soffit be sloped, so that it will help distribute reflected sound to the overhung seats. If a diffusing profile is added to the balcony soffit, the reflection off this surface may even gain a lateral component, so that *spaciousness* will not be reduced. The drawing also illustrates how the otherwise vertical balcony front and the right angle between the rear wall and balcony soffit have been modified to avoid sound being reflected back to the stage where it could generate an echo. However, in a rectangular hall with balconies along the side walls, vertical soffits in combination with the side-wall surface can increase the early and lateral reflection energy in the stalls area as shown in the half cross section in Fig. 9.27.

In general, as large-scale modulations in the hall geometry, balconies also provide low-frequency diffusion, which is considered an advantage.

9.6.4 Volume and Ceiling Height

As the people present are often the main source of sound absorption in an auditorium, the ratio between the volume and area covered by the audience and performers is an important factor for the reverberation time achievable. The general relationship valid for concert halls is shown in Fig. 9.28.

Because the variation in area per person does not vary significantly between halls, the volume per seat is also used as a rule of thumb for the calculation of the required volume. For speech theaters, a volume per seat of $5-7\,\mathrm{m}^3$ will result in a reverberation time around 1 s, whereas at least $10\,\mathrm{m}^3$ per seat is needed to obtain a T value of 2 s.

Since the audience normally occupies most of the floor area, the volume criterion can also be translated to a ceiling height criterion. In such cases, the abscissa in Fig. 9.28 can be interpreted as roughly equal to the required room height. A ceiling height of about 15 m is required if a T value of 2 s is the target, whereas a height of just 5–6 m will result in an auditorium with a reverberation time of about 1 s. It should be emphasized that, in the discussion above, we have assumed the other room surfaces to be reflective.

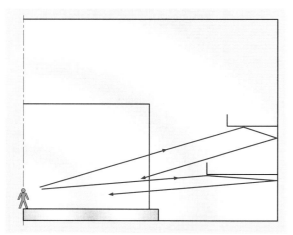

Fig. 9.27 A half cross section of a hall with side balconies that, in combination with the side wall, return early reflection energy towards the stalls area from a lateral angle (after [9.4])

more important) [9.9]. The criteria may also be stated in terms of the vertical viewing angle from the seat to the main hall volume. Recently, *Beranek* [9.3] has suggested this be a minimum of 45°.

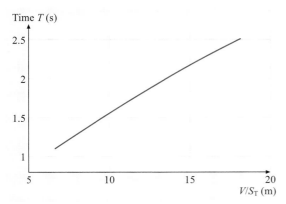

Fig. 9.28 Reverberation time T in occupied concert halls versus the ratio between the volume V and area occupied by the audience and orchestra S_T (including aisle space up to 1 m width). Derived from empirical data (after [9.9])

These guidelines on volume should be regarded as rough rules of thumb only, as room shape, floor slope, the presence of balconies and the distribution of absorption on other surfaces can cause significant variations in the objective values obtained for T and other parameters. In smaller halls (say below 1500 seats) for symphonic concerts, it is better to start with a volume slightly larger than required rather than the opposite, as one can always add a little absorption – but not more volume – to a hall near completion. On the other hand, in larger halls, where maintaining a suitably high G value is also of concern, a better strategy is to limit the volume and minimize the absorption instead. One way of minimizing absorption is to place some of the seats under balcony overhangs, where they are less exposed to the reverberant sound field. However, the other side of the coin is the poor conditions in these seats, as explained in Sect. 9.6.3.

9.6.5 Main Dimensions and Risks of Echoes

In order for early reflections to contribute to the *intelligibility* of speech or to the *clarity* of music, they must not be delayed more than about 30 ms and 80 ms, respectively, relative to the direct sound. If the first reflection arrives later than 50 ms, it is likely to be perceived as a disturbing echo when impulsive sounds are emitted (Sect. 9.2).

In large rooms it is a challenge to shape the surfaces so that reflections from the main reflecting areas arrive within 50 ms at all seats. It is possible to identify which surfaces are able to generate echoes at the receiver point

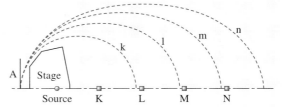

Fig. 9.29 Echo ellipses drawn over the plan of an auditorium (after [9.4])

in question by means of simple geometrical studies of plan and sectional drawings as seen in Fig. 9.29. Ellipses are drawn so that the source and relevant receiver positions are placed at the focal points and so that the sum of distances from the focal points to any point on the ellipse equals the distance between the focal points plus 17 m (times the scale of the drawing). Then, if a surface outside the ellipse marked "m" in the figure directs sound towards seats near point M, this surface must be made absorbing, diffusing or be reoriented so that the reflection is directed towards areas farther away from the source than M.

In particular, it is important to check the risk of echoes being generated by the rear wall behind the audience and from a high ceiling.

9.6.6 Room Shape Details Causing Risks of Focusing and Flutter

Concave surfaces can cause problems as they may focus the sound in certain areas while leaving others with too little sound. Thus, vaulted ceilings as seen in Fig. 9.30 are only acceptable if the radius of curvature is less than half the height of the room (or rather half the distance from peoples' heads to the ceiling) so that the focus center is placed high above the listeners. Situations with

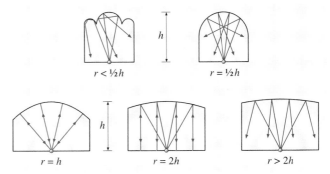

Fig. 9.30 Focusing by concave ceilings (after [9.31])

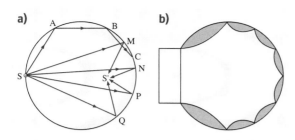

Fig. 9.31a,b Circular room with focused and creeping waves (**a**), and with the surface shape modified (**b**) so that these phenomena are avoided

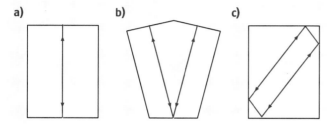

Fig. 9.32 Room shapes causing risk of flutter echoes

a slight curvature, $r \geq 2h$, may also be acceptable if the floor is covered with absorptive seating so that multiple reflections between floor and ceiling will not arise.

Circular rooms are also unfavorable both due to the risk of focusing and the risk of a whispering-gallery effect (boundary waves creeping along the wall). Modifications of the concave wall within the overall circular shape as shown in Fig. 9.31 can solve the problem, at least at mid and high frequencies.

Another frequent problem is regular, repeated reflections, so-called flutter echoes, which arise between parallel, smooth reflecting walls or along other simple reflection paths that can repeat themselves as shown in Fig. 9.32. Particularly in small rooms such reflections may cause strong coloration of the sound (through comb filtering as explained in Sect. 9.2.3) while in larger rooms they may be perceived as a long series of distinct echoes evenly spaced in time.

In any case, the flutter can be avoided by simple absorptive or diffusive treatment of at least one of the opposite reflecting surfaces, as shown in Fig. 9.33. Besides, if it is possible to change the angle between the opposite walls just by a few degrees (3−5°), the problem will also disappear.

9.6.7 Cultivating Early Reflections

In most rooms accommodating more than say 25 listeners, it is of relevance to consider how early reflections can help distribute the sound evenly from the source(s) to the audience. This will increase the early reflection energy and so improve *clarity/intelligibility* and perhaps even reduce the reverberation time by directing more sound energy towards the absorbing audience.

In rooms for speech, the ceiling height should be moderate (Sect. 9.6.4) so that this surface becomes the main surface for the distribution of early reflection energy to the audience. Figure 9.34 shows examples of how this can be accomplished in rooms with both low and high ceilings.

In this figure the two situations at the top need improvement. In the low-ceiling room to the left the echo from the rear wall/ceiling combination can be removed by introducing a slanted reflector covering the corner. In this way the sound is redirected to benefit the last rows, which may need this to compensate for the weak direct sound at this large distance. Alternatively, the energy from this rear corner may simply be attenuated by one or both surfaces near the corner being supplied with absorption.

If the ceiling is so high that echoes can be generated in the front part of the room, as shown to the right, the front part may likewise be treated with slanted reflectors that redirect the sound to the remote seat rows, or this ceiling area may simply be made sound absorbing. The lower-right section in the figure also illustrates how the

Fig. 9.33a–c Room with flutter echoes (**a**), and measures against this by means of diffusion (**b**) or absorption (**c**) (after [9.28])

floor can be tilted to reduce volume and allow the rear seats benefit from being closer to the reflecting ceiling.

If for architectural reasons the ceiling is higher than necessary in a room for speech or rhythmic music, substantial areas of the room surfaces must be treated with sound-absorbing material to control the reverberation time. However, more sound absorption will reduce the overall sound level in the room so this is only recommended if the sound can be amplified.

If the ceiling is to be partly absorbing, the ceiling area that should be kept reflecting in order to distribute useful first-order reflections to all listeners can be generated geometrically by drawing the image of the source in the ceiling surface and connecting lines from this image point to the boundaries of the seating area. Where these lines cross the actual ceiling we find the boundaries for the useful reflecting area, as shown in Fig. 9.35.

In larger auditoria, a more-detailed shaping of the ceiling profile may be needed to ensure even distribution of the reflected sound to the listeners. Notice in Fig. 9.36 that the concave overall shape of the ceiling can be maintained if just the size of the ceiling panel segments is large enough to redirect the reflections at suitably low frequencies.

Local reshaping of surfaces while maintaining the overall form is also illustrated in Fig. 9.37, in which the weak lateral reflections in a fan-shaped hall are improved by giving the walls a zigzag shape with almost parallel sets of smaller areas along the splayed side walls. In the example shown in the photo, the panels are even separated from the wall surface and tilted downward.

9.6.8 Suspended Reflectors

In many cases it is advantageous to reduce the delay and increase the level of a reflection without changing the room volume. In this case it is obvious to suggest individual reflecting surfaces suspended within the room boundaries, as shown to the right in Fig. 9.38. If the reflector is placed over an orchestra pit, it can improve mutual hearing among the musicians in the pit as well as increase the early energy from the singing on the stage to the seating primarily in the stalls area.

Suspended reflectors are also suitable for acoustic renovation of existing, listed buildings, because the main building structure is not seriously affected (as in the case of the Royal Albert Hall in London [9.3]).

For a given source position, a small reflector will cover only a limited area of receiver positions. There-

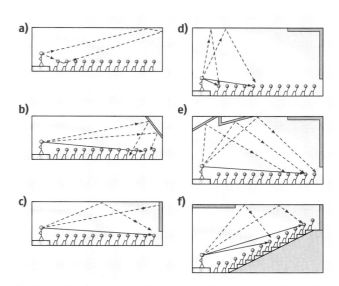

Fig. 9.34 Means of controlling early reflections in auditoria (after [9.28])

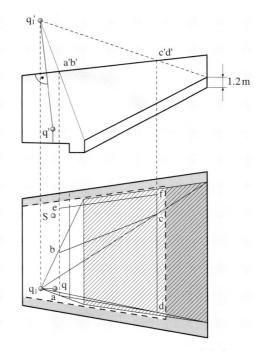

Fig. 9.35 Sketch identifying the area responsible for covering the audience area with a first-order ceiling reflection (after [9.32])

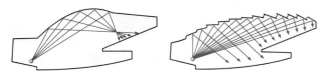

Fig. 9.36 Reshaping of ceiling profile to avoid focusing and provide more-even sound distribution

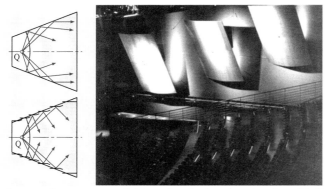

Fig. 9.37 Improvement of side-wall reflections in a fan-shaped hall by local reshaping or adding of panels parallel to the long axis of the room (drawing after [9.32])

the ratio between panel size and the wavelength, and focusing/diffusion due to panel curvature.

The attenuation with distance is as for a spherical wave in free field: 6 dB per doubling of the total distance from the source to the reflector a_1 plus the distance from the reflector to the receiver a_2. Normally, the attenuation due to absorption is negligible, as reflectors should obviously be made from a reflective material. The only exception could be when a thin, lightweight material is used, such as plastic foil or coated canvas. In this case the attenuation ΔL_{abs} can be calculated from the mass law

$$\Delta L_{abs} = -10 \log_{10} \left[1 + \left(\frac{\rho c}{\pi f m \cos \theta} \right)^2 \right] \quad (9.33)$$

where ρc equals the specific impedance of air ($\approx 414\,\mathrm{kg\,m^{-2}\,s^{-1}}$), f is the frequency in Hz, m is the mass per square meter of the material and θ is the angle of incidence relative to the normal direction.

As a simple design guide, diffraction can be said to cause the specular reflection to be attenuated by 6 dB per octave below a limiting frequency f_g given by:

$$f_g = \frac{c a^*}{2 S \cos \theta} \quad (9.34)$$

where a^* is the weighted *characteristic distance* of the reflector from the source and receiver:

$$a^* = \frac{2 a_1 a_2}{a_1 + a_2}, \quad (9.35)$$

S is the area of the reflector, and θ is the angle of incidence as before. Above the frequency specified in (9.34), the influence of diffraction can be neglected.

If the reflector is curved, the attenuation ΔL_{curv} can be calculated as:

$$\Delta L_{curv} = -10 \log_{10} \left| 1 + \frac{a^*}{R \cos \theta} \right|. \quad (9.36)$$

fore, how the reflector(s) influence the balance heard between different sources such as the different sections in a symphony orchestra must be thoroughly considered. With small reflectors, the balance between sections may be perceived as very different in different parts of the audience. Often it will be advantageous to give the reflector(s) a slightly convex shape to extend the area covered by the reflection and to soften its level.

The nature of the reflection off a panel of limited size is strongly frequency dependent and may be influenced by several factors: distance to source and receiver, absorption by the panel material, diffraction governed by

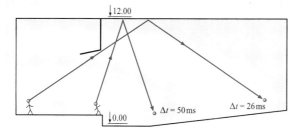
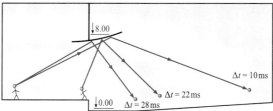

Fig. 9.38 The influence on reflection paths of a reflecting panel suspended in front of a stage (after [9.33])

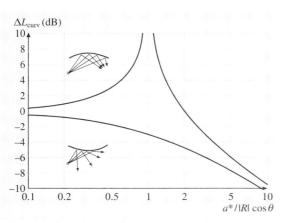

Fig. 9.39 Attenuation of a reflection caused by concave and convex reflectors (after [9.34])

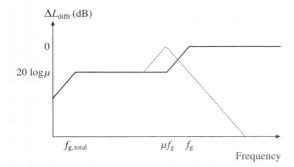

Fig. 9.40 Approximate attenuation due to diffraction from a reflector array. All non-horizontal lines have a slope of 6 dB per octave. See text for further explanation (after [9.34])

Here, the only new variable R is the radius of curvature. R should be entered positive for convex surfaces and negative for concave surfaces. When the reflector is concave, strong focussing occurs when $R = -a^*/\cos\theta$. In Fig. 9.39, graphs of ΔL_{curv} are shown for both the concave and convex cases. It is noted that the values can be both positive and negative. In the case where the surface is curved in both the x- and y-directions of the reflecting plane, the ΔL_{curv} correction should be applied twice using the appropriate radii of curvature in the two directions.

In certain cases an array of (identical) reflectors are suspended as an alternative to having one large reflector. Such an array can be characterized by the area of each element, S, the total area covered by the array S_{total} and the degree of coverage, μ, equal to the ratio between the total area of the n reflector elements nS and S_{total}. In this case, the attenuation due to diffraction depends on frequency as shown in Fig. 9.40.

Here, f_g is the limit frequency of the individual elements calculated according to (9.34), while $f_{g,\text{total}}$ is the lower-limit frequency related to the entire array area S_{total}

$$f_{g,\text{total}} = \frac{ca^*}{2S_{\text{total}}\cos\theta}. \tag{9.37}$$

First we look at the situation above the frequency f_g in Fig. 9.40. As in the case of the single reflector, the attenuation is zero as long as the specular reflection point falls on one of the reflector elements. If the specular point falls between two elements, the attenuation can be approximated by the dotted line, according to which we have an attenuation of 6 dB per octave as frequency increases. In the frequency range between $f_{g,\text{total}}$ and μf_g, the attenuation is roughly frequency independent and governed by the array density μ, and below $f_{g,\text{total}}$ the level again rolls off by 6 dB per octave. It is seen that, in order to obtain a wide range of frequency-independent reflection, one should aim to use many small reflectors rather than fewer, larger elements. Fortunately, the strong position-dependent variation in reflection level at high frequencies can be reduced by making the individual elements slightly convex.

It should be remembered that whenever surfaces are shaped with the purpose of efficiently directing sound energy from the source towards the absorbing audience, this energy is no longer available for the creation of the reverberant sound field. In other words, the more we try to improve the *clarity* and *intelligibility*, the more we reduce reverberation in terms of T and EDT (regardless of the physical absorption area), and the more the sound field will deviate from the diffuse field predicted by any simple reverberation formula. If substantial reverberation is needed as well, the solution may be to increase the volume per seat beyond the 10 m³ as already suggested for large halls in Table 9.1. As seen in Sect. 9.7.3 this is in line with the current trend in concert hall design.

9.6.9 Sound-Diffusing Surfaces

It is often stated that diffusion of sound from irregular surfaces adds a pleasant, smooth character to the sound in a concert hall. It is a fact that reflections from large smooth surfaces can produce an unpleasant harsh sound. It is also a fact that most of the surfaces in famous old and highly cherished halls such as the Musikvereinsaal break up and scatter the reflected sound in many directions due to rich ornamentation, many window and door niches and the coffered ceiling (Fig. 9.45). Even the large chandeliers in this hall break up the sound waves.

Many present-day architects are influenced by modernism and are less enthusiastic about filling their designs with rich ornamentation. Therefore, close cooperation is often needed between architect and acoustician in order to reach a surface structure that will scatter the sound sufficiently, although this is difficult to define exactly as good guidelines do not exist for what percentage of the wall/ceiling area one should make diffusive.

Schroeder [9.35] has done pioneering work on creating algorithms for the design of surface profiles that will ensure uniform distribution of scattered sound within a well-defined frequency range. These algorithms are based on number theory using the maximum length, quadratic residue or primitive root number sequences. However, in most cases of practical auditorium design, perfectly uniform scattering is not needed, and good results can be obtained even after a relaxed architectural adaptation of the ideas, as shown in Fig. 9.41.

The primary requirement to achieve diffusion is that the profile depth of the relief is comparable with a quarter of the wavelength and the linear extension of the diffusing element (before it is eventually repeated) is comparable with the wavelength at the lowest frequency of interest to be diffused. Given this, many different solutions can be imagined such as soft, curved forms cast in plaster, broken stone surfaces, wooden sticks, irregular brick walls etc.

Fig. 9.41 Diffusion treatment by means of wooden sticks screwed on an otherwise plane, wooden wall surface

Finally, it should be mentioned that strict periodic structures with sharp edges equally spaced along the surface can cause problems of harsh sound at frequencies above the design frequency range. Actually, it is safer to apply more-organic, chaotic structures, or to ask the workers not to be too careful with the ruler.

9.7 Room Acoustic Design of Auditoria for Specific Purposes

This section will present some examples showing how the principles outlined in Sect. 9.5 are implemented in the current design of different types of auditoria. The main focus will be on concert halls for classical music as well as drama and opera theaters. It will also be shown that design choices are not only compromises between acoustic and other concerns, but sometimes even between different room acoustic aspects. This is often the case with multipurpose auditoria. In any case, the compromises are more serious the larger the seating capacity.

9.7.1 Speech Auditoria, Drama Theaters and Lecture Halls

In the Western world, rooms for spoken drama as well as for opera have their roots in the Greek and Roman theaters in which the audience is arranged in sloping concentric rows, the *cavea*. In Roman times, this audience area covered a 180° angle around the orchestra or apron stage. The advantage of such an arrangement is of course the possibility to accommodate a large audience within a limited maximum distance from the stage. Many modern theaters and speech auditoria have adopted this layout – perhaps extended with one or more sets of balconies.

An example of such a modern speech auditorium (the lecture hall at Herlev Hospital, Denmark) is seen in Fig. 9.42 along with a Roman theater for comparison.

In a drama theater, a moderate reverberation time around 1 s will provide a good compromise between high intelligibility and a high strength value, as both are needed for unamplified speech. The T value may be reduced to about 0.5 s in small rooms for 50–100 people and also increased slightly for audiences above 1000. To obtain a high strength value in a large theater, it is important to choose a modest volume per seat (about 5 m^3) rather than to control the reverberation in an unneces-

Fig. 9.42 Photos of Roman theater (in Aspendos, Turkey) and of a modern speech auditorium (Herlev Hospital, Denmark)

sarily large volume by applying additional absorption beyond that provided by the audience chairs. Thus, the height of the ceiling should be chosen fairly low – which is also advantageous for its ability to distribute early reflections to the audience. According to diffuse field theory, the steady-state energy density equivalent to G_{exp} is inversely proportional to the total absorption area A in the room.

High *clarity* and sound *level* are also promoted by careful shaping of the reflecting ceiling and a substantial sloping of the audience floor. Due to the directivity of the human voice, a reflecting rear wall (or stage scenery in a theatre) is of less importance, unless an actor should turn his back to the audience. Also the wall surfaces close to the stage opening are important, because they will send sound across the room so that sound also reaches people behind the speaker when he turns the head towards one side. The rear wall/balcony soffits behind/over the rear part of the audience should diffuse or redirect the sound to benefit nearby listeners.

As *spaciousness* is not a quality of particular importance in a speech auditorium, reflections from the side walls farther from the stage are often given lower priority than a low average stage-to-audience distance, which is promoted by a fan-shaped floor plan. However, as indicated by the problems encountered in the Olivier Theater in London [9.4], there should be limits to the opening angle of large fan-shaped theaters.

In a modern theater, the floor slope will often be more modest than seen in Fig. 9.42. In contrast to the lecture hall, a modern drama theater also needs to be highly flexible regarding the layout of stage and audience areas. The classical proscenium frame inherited from the Italian Baroque theater is often considered to limit intimacy as well as the creativity of the stage-set designers and directors. If the classical proscenium stage is not needed, intimacy and audience involvement can be maximized by an arena arrangement, with audience all around the stage. In other cases, the stage can be extended into the audience chamber, for instance by removing the first rows of seats and raising the floor below to stage level. If such a flexible forestage is considered, the line-of-sight design should be adjusted accordingly, particularly if the hall features balconies.

An example of a flexible modern theater (without balcony) is seen in Fig. 9.43. In this room, the circular side/rear wall was made highly sound absorbing, leaving most of the distribution of early reflection energy to the ceiling.

Dedicated drama theaters are seldom built with a seat count higher than 500–1000, because visual as well as acoustic intimacy, including close view of facial expressions and freedom from artificial amplification, are given high priority. Thus, the maximum distance from the stage front to any seat should not exceed about 20 m for drama performances.

From this discussion it can be concluded that a reflecting ceiling of modest height is the main source of early reflection energy in theaters. However, lighting staff will often want to penetrate this surface with slots for stage lights and rigging or place light bridges across the room, which will act as possible barriers for the sound paths going via the ceiling. This challenge must always be considered in the acoustic design of drama theaters.

9.7.2 Opera Halls

There is a 350 year tradition of performing opera in halls of the Italian Baroque theater style. These are halls with a horse-shoe plan shape with several balcony levels running continuously from one side of the proscenium arch

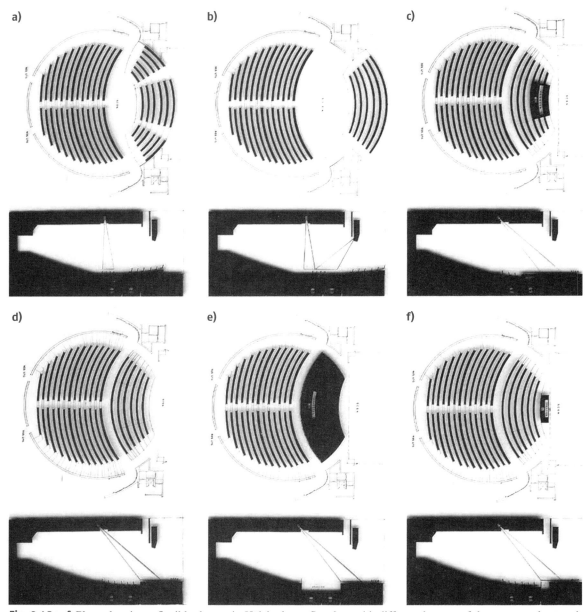

Fig. 9.43a–f Plans showing a flexible theater in Hëlsingborg, Sweden, with different layouts of the stage, orchestra pit and audience seating. (**a**) and (**b**): arena with small and large stage respectively, (**c**), (**e**), and (**f**): proscenium theater with three different sizes of orchestra pits. (**d**): proscenium theater without pit

to the other along the curved side and rear walls. The balconies are either divided into boxes separated by thin walls or are open with long seat rows. The orchestra is placed in front of the proscenium opening that connects the auditorium with the stage house.

This tradition has undergone few changes during the past 350 years. Since the 19th century, the orchestra has been placed in a lowered pit separated from the stalls by a closed barrier. In this arrangement, seats in the stalls and lower rear balcony experience an attenuation of the direct sound from the orchestra, because the pit rail blocks the free line of sight to the orchestra. The result may be an improved singer-to-orchestra balance in these seating areas. The pit rail will also reflect sound back and forth between the stage and pit, giving improved contact between the singers and the lowered orchestra.

Another development is that the separate boxes have been exchanged for open balconies, which improves the possibility for the now exposed side walls to help distribute the sound to seating farther backwards in the auditorium. This development is the result of advances in building technology as well as of changes in social attitudes.

The multiple levels of horseshoe-shaped balconies facilitates a short average distance between the audience and the singers, promoting both the visual and the acoustic intimacy, which is important for both parties. However, balcony seats close to the proscenium and on the upper balcony levels often have a very restricted view of the stage.

The short audience–singer distance obtained by the horseshoe form gives high levels of both strength and clarity. On the other hand, the horseshoe form and the proscenium arch also cause some acoustic problems:

- with all wall surfaces covered by balconies, only small free wall areas are left to generate the early reflections that assist an even distribution of the sound, clarity and the generation of reverberation in the room;
- often focusing effects occur from the curved wall surfaces and a cupola-shaped ceiling which, along with the placement of the orchestra out of sight in the pit, give a high risk of false localization of certain instruments;
- only part of the energy generated by the singer will reach the auditorium, because the singers are placed in a large stage house coupled to the auditorium via a proscenium opening of limited size, and this energy even depends on the acoustic properties of the ever-changing stage sets.

Recent developments in the acoustic design of opera halls have mainly aimed at avoiding these problems. Only a few attempts have been tried to radically change the overall shape (e.g. the Deutsche Oper in Berlin, and the Opéra Bastille in Paris) but without much success. Therefore, we are left with tradition and acoustic properties of horseshoe halls to shape the current acoustic ideals for opera, which are

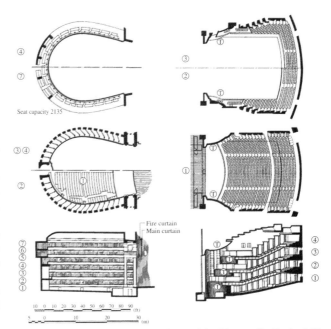

Fig. 9.44 Plan and section drawings of the Teatro alla Scala, Milan, Italy, from 1778 (*left*) and the New National Theatre Opera, Tokyo, Japan, from 1997 (*right*) (after [9.3])

- sufficient level of the singers' voices relative to the orchestra,
- some emphasis on *clarity* and *intelligibility* (although not as much as in speech auditoria),
- a certain *fullness of tone* and *reverberance*; but less than in a classical concert hall.

In recent years the trend has moved towards higher T values than found in the old halls, like Opéra Garnier, Paris (2131 seats, 1.1 s), Teatro Alla Scala, Milano (2289 seats, 1.2 s). This might well be due to the tendency to perform operas in the original language and display the translated text on a text board above the proscenium. Thus, for many recent opera halls we find T values of 1.4–1.8 s. Examples are: Göteborg (1390 seats, $T = 1.7$ s), Glyndebourne (1200 seats, $T = 1.4$ s), Tokyo (1810 seats, $T = 1.5$ s), and Semper Oper, Dresden (reconstruction, 1290 seats, $T = 1.6$ s).

As the width of the traditional horseshoe shape is about equal to its length, the lateral sound is not very pronounced and so *spaciousness* is seldom regarded a factor of particular importance in opera hall acoustics.

In order to ensure a sufficient level and *clarity* of the singers' sound, efficient sound reflections off the proscenium frame, ceiling, balcony fronts and from the corners between balcony soffits and rear/side walls are important. Of course the scenery may also be designed to help in this respect, but generally this is out of the hands of the acoustician.

If a seat count larger than, say, 1500 is demanded it is better to extend the top balcony backwards, as seen in Covent Garden in London and the Metropolitan in New York rather than increase the overall dimensions of the horseshoe, which would give poorer acoustics for all seats. As this extended balcony is close to the ceiling surface, the acoustics here can often be very intimate, clear and loud, in spite of the physical and visual distance to the stage being very large. In the Metropolitan Opera in New York the farthest seat is about 60 m from the stage.

The culmination of the old Baroque theater tradition is represented by the Teatro alla Scala in Milan, Italy, which dates from 1778. Drawings of this theater are shown in Fig. 9.44 along with a modern opera with almost the same number of seats, the New National Opera in Tokyo from 1997.

The horseshoe-shaped Milan opera contains six levels of balconies subdivided into boxes with the openings covered with curtains so that the audience could decide whether to concentrate on the play or on socializing with their guests or family in the box. The audience in the boxes receive a rather weak sound unless they are very close to the opening.

This aspect has been improved in the modern opera with open balconies. Steeper slopes on the stalls floor and even on the side balconies have also improved the direct sound propagation to all seats. Fewer balcony levels have created higher openings at the balcony fronts, allowing the reverberant sound to reach the seats in the last rows below the balconies, providing more early reflection energy from the walls to all seats.

In the Tokyo opera, much emphasis has been put on retracting the balconies away from the proscenium so that extensive side-wall areas close to the proscenium are available for projecting the sound from the stage into the auditorium. Along with the ceiling above the pit, these wall areas almost form a trumpet flare. In the section, it is also seen that the side balconies are tilted downwards towards the stage in order to improve lines of sight. Finally, the side-wall areas and the side-balcony fronts have been made straight to avoid focusing and improve lateral reflection energy.

9.7.3 Concert Halls for Classical Music

Public concerts with large orchestras started in Europe about 250 years ago in halls that resembled those in which the composers normally found audience for their nonreligious music such as the salons and ballrooms of the nobility and courts. It is likely that this tradition as well as the limited span of wooden beam ceilings are the main reasons why many of the old concert rooms were built in what we now call the classical shoe-box shape: a narrow rectangle with a flat floor and ceiling and with the orchestra placed on a platform at one end.

The most famous of the shoe-box-shaped halls is the Musikvereinsaal in Vienna, Austria, shown in Fig. 9.45. This hall, dating from 1870, has a volume of 15 000 m³ and seats 1670 people. Today, we understand many of the factors responsible for its highly praised acoustics:

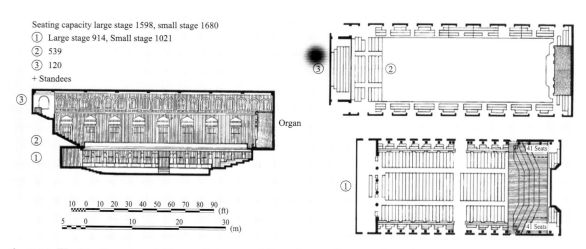

Fig. 9.45 The Musikvereinsaal, Vienna. Balcony and floor plans and long section (after [9.3])

- The volume per seat is only $9\,\mathrm{m}^3$; but T is still 2.0 s in the fully occupied hall because the seating area is relatively small (the seats are narrow and the row-to-row distance is small) and because of the flat main floor. Thus, the sound becomes both loud and reverberant.
- The reverberation also benefits from the fact that the absorbing audience (apart from a few on the top rear balcony) is placed in the lower half of the volume, whereby a *reservoir* of reverberant energy can exist in the upper part of the room.
- The close side walls create strong lateral reflections giving high *clarity* and *envelopment*, and all surfaces are modulated with rich ornamentation, which prevents echoes and causes the room reflections to sound soft and pleasant.

There are however elements of the acoustics, and lack of comfort, that we would not accept in a modern concert hall. The sound is somewhat unclear and remote in the rear part of the flat main floor due to grazing-incidence attenuation. Also, in the side balconies only people sitting close to the balcony front can see well. Other old and famous shoe-box halls (such as the Boston Symphony Hall) have too-shallow balcony overhangs, which results in weak reverberant energy below the balconies.

Because of the general and proven success of the shoe-box design, many new halls are built with this shape. However, in most of the modern shoe-box halls, the drawbacks mentioned above are avoided by a slightly sloping stalls floor and a different layout of the side balconies.

One of the modern shoe-box halls in which the above modifications have been introduced is the Kyoto Concert Hall, Japan, shown in Fig. 9.46. Comparing Fig. 9.45 and Fig. 9.46, the increased floor slope and the subdivided side balconies turned towards the stage are clearly seen.

Recalling the influence of floor slope on T (and other parameters, Sect. 9.5.6) the increased slope will result in a reduced T and higher C. In modern designs T will also tend to be reduced by more-spacious seating and less-shallow balcony overhangs, which result in a larger absorptive audience area being exposed to the sound field. All these factors call for a larger volume per seat in modern shoe-box halls than found in the classical ones, if a high T value is to be maintained. On the other hand, this might well be the reason why the newer halls seldom provide the same intensity and warmth as experienced when listening from one of the sparsely upholstered wooden chairs in Vienna.

Early in the 20th century the development of concrete building technology allowed architects to shape the halls more freely. At the same time it became popular to treat sound as light, by designing the surfaces by means of geometric acoustics and by trying to transmit the sound as efficiently as possible from the source on stage to the (absorbing) audience. Another desire was to put the audience as close as possible to the stage.

The result was the fan-shaped concert hall, which dominated modernist architecture between about 1920 and 1980. An example is seen in Fig. 9.47: Kleinhans Music Hall in Buffalo, USA. Comparison with Fig. 9.45 reveals a very wide hall with a relatively low ceiling and the stage surrounded by concave, almost parabolic, walls and with the ceiling projecting the sound into the absorbing audience area after just a single reflection. This phenomenon, as well as the low volume per seat (about $6.4\,\mathrm{m}^3$), are the reasons for the resulting low reverberation time and the weak sense of *reverberance*. The audience receives most of the sound from frontal directions or from the ceiling, so the sound is clear but lacks envelopment. Other examples of modernist fan-shaped halls are, the Salle Pleyel in Paris (1927) and the old Danish Broadcasting Concert Hall (Copenhagen, 1946). In conclusion, few concert halls from the modernist era have been acoustically successful.

Fortunately, mainly as a result of advances in room acoustic research, two alternative and more-promising designs appeared in the 1960s and 1970s: the *vineyard*

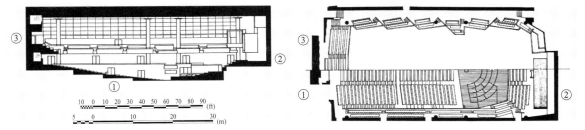

Fig. 9.46 Kyoto Concert Hall, Japan. Built 1995, volume $20\,000\,\mathrm{m}^3$, 1840 seats, $RT = 2.0\,\mathrm{s}$ (after [9.3])

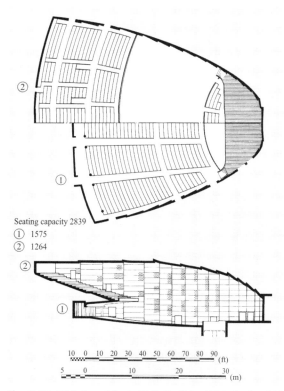

Fig. 9.47 Kleinhans Music Hall, Buffalo, USA. Built 1940, volume $18\,240\,\text{m}^3$, 2839 seats, RT $= 1.5$ s. Plans and long section (after [9.3])

and the *directed reflection sequence* (DRS). Both are arena-shaped with the possibility of the audience being placed all around the orchestra platform. Even more than the fan shape, this causes the audience to come closer to the orchestra for the sake of intimacy. However, both designs are primarily relevant for larger halls, say more than 1500 seats. For halls seating more than about 2000, these designs may even be more successful than the classical rectangular shape, which, with so many seats, possess a risk of some listeners being placed too far from the stage.

The first vineyard concert hall was the Philharmonie Berlin, shown in Fig. 9.48, which opened in 1963. This hall has no balconies. Instead, the seating area is subdivided into terraces elevated relative to each other, whereby the terrace fronts and sides can act as *local* reflectors for specific seating areas. By careful design of these terraces, it is possible to provide plenty of early, and even lateral, reflections to most of the seats in an arena-shaped hall. With no seats being overhung by balconies, a large absorptive area is exposed to the sound, whereby a generous volume per seat is recommended. Thus, in the Sapporo Concert Hall in Japan (1997) as well as in the new Danish Radio Concert Hall in Copenhagen, the volume per seat is about $14\,\text{m}^3$. This is a quite high value for concert halls, as a 2 s T value should normally be obtained with just $10\,\text{m}^3$ per seat.

In *directed reflection sequence* halls such as the Christchurch Town Hall shown in Fig. 9.49, most of the early reflections are provided by huge, suspended and tilted reflectors. These are distinctly separate from

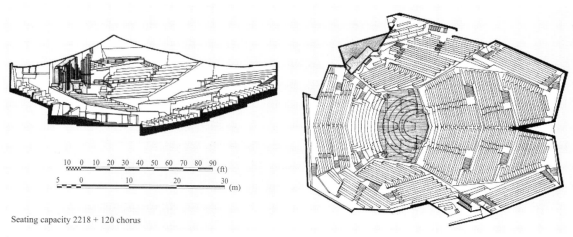

Fig. 9.48 Plan and section of Philharmonie Berlin, Germany. (Built 1963, 2335 seats, $V = 21\,000\,\text{m}^3$, $T = 2.1$ s) Acoustician: Lothar Cremer (after [9.3])

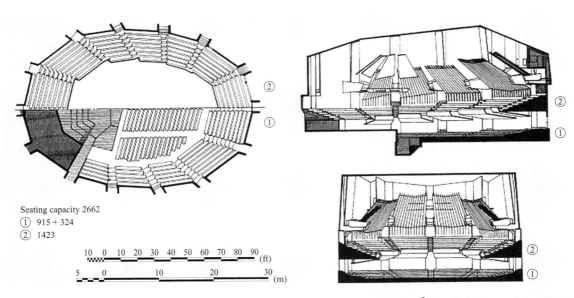

Fig. 9.49 Christchurch Town Hall, New Zealand. (Built 1972, 2662 seats, $V = 20\,500\,\text{m}^3$, $T = 2.4\,\text{s}$) Acoustician: Harold Marshall (after [9.3])

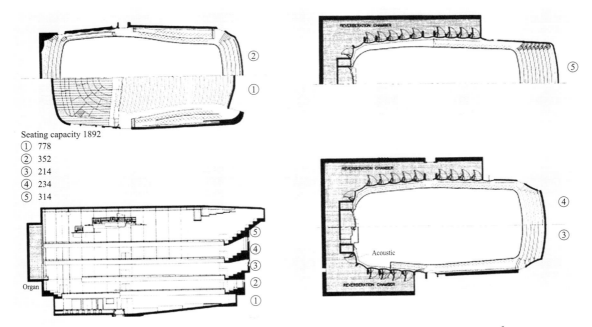

Fig. 9.50 The KKL Concert hall in Luzern, Switzerland. (Built 1999, 1892 seats, $V = 17\,800 + 6200\,\text{m}^3$, $T = 1.8\text{--}2.2\,\text{s}$) Acoustician: Russell Johnson (after [9.3])

the boundaries, which define the reverberant volume. The fronts and soffits of the sectioned balconies surrounding the main floor and stage in the elliptical plan provide additional early reflections. Because of the arena layout combined with the extensive use of balconies (containing more than half of the seats) the hall is very

 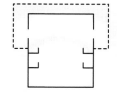 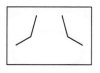

Fig. 9.51 Three concert hall concepts with the potential of combining high clarity with long reverberation. *From left to right*: vineyard (plan), shoe box with coupled reverberation chambers (cross section), and directed reflection sequence (cross section)

intimate, and the large reflecting surfaces provide a stunningly clear, dynamic and almost too loud sound. Still, the reverberation is long and rich because of the significant volume without absorption above/behind the reflectors.

This aim towards a combination of high *clarity* and high *reverberance*, probably developed through our extensive listening to recorded music, requires close reflecting surfaces as well as a large volume. The natural consequence of this demand is to separate the early reflecting surfaces from the room boundaries. Both the terrace fronts in the vineyard halls and the suspended reflectors in the DRS halls offer this possibility. A third way to achieve high *clarity* as well as high *reverberance* is a narrow shoe box with added surrounding volume, as found in a number of halls since about 1990. In these halls the volume of the narrow tall shoe-box hall providing the early reflections is quite moderate; but an extra volume is coupled to the main hall through openings that can be closed by heavy doors, so that the total volume can be varied. However, in such halls one should be careful to make the coupling area large enough for the added volume to have any significant effect (unless one sits close to one of the open doors), otherwise it is hard to justify the enormous costs of the extra volume and door system. With weak coupling, a double-sloped decay curve with a knee point perhaps 20 dB down is created, whereby the added, longer reverberation becomes barely audible except during breaks in the music.

A recent design of such a rectangular hall with a coupled volume is shown in Fig. 9.50.

The sketches in Fig. 9.51 summarize the basic design of the three types of halls mentioned. It is seen that they all possess the possibility of separating the surfaces that generate the early reflections from the volume boundaries that generate the reverberation.

9.7.4 Multipurpose Halls

Many modern halls built for cultural purposes often have to accommodate a variety of different types of events, from classical to pop/rock concerts, drama and musicals, opera, conferences, banquets, exhibitions, cinema and perhaps even sports events. From an acoustical point of view, the first concern in these cases is whether a variable reverberation time will be necessary. The answer is yes in most cases where some functions primarily require intelligibility of speech while others require a substantial reverberation, such as for classical music.

A hall in which these demands have been met by means of variable absorption is Dronningesalen, at the Royal Library in Copenhagen (Fig. 9.52). For this hall both chamber-music concerts and conferences with amplified speech were given high priority. The variable absorption is provided by means of moveable panels on the side walls as well as by folded curtains and roller blinds on the end walls. Combining these measures in different ways, T values in the range from 1.1 to above 1.8 s can be obtained (in the empty hall) as seen in Fig. 9.52.

In many cases, stage performances with extensive scenery are also required. The most common way to accomplish this is to design a stage house, mount an orchestra shell on the stage and raise the pit to stage level for concerts. If a hall is also to be used for banquets and exhibitions requiring a flat floor, it is common to place the stalls seats on a telescopic riser system on a flat floor instead of having a fixed, sloped seating. When not in use, this riser system is stored along the rear wall or under a balcony. If the reverberation time suits classical music concerts with the sloped seating in place, a substantial area of variable absorption is needed to reduce the T value when the chairs are absent. Another problem may be that the telescopic systems offer only a rather steep and linear slope, which is not optimal for classical concerts. If the chairs are fixed on the riser steps, the minimum step height is about 25 cm, corresponding to a rake of about 25%.

If a change in T is accomplished by means of variable absorption, G will be reduced along with lowering T. However, if instead T is lowered by reducing the volume, for instance by moving a wall or lowering the ceiling, G will remain approximately constant or perhaps even increase. This can be advantageous when the low-T setting is to be used, for instance for unamplified drama or for chamber music in a larger symphony concert hall.

An example of a hall with variable volume is found in Umeå, Sweden (Fig. 9.53). In this hall the auditorium volume can be adjusted by moving the proscenium wall between three positions. When the proscenium is stored against the rear wall in the *concert* format, the entire volume is available for concerts with large symphonic orchestras on an open stage. In this situation, the stage tower can be closed off by horizontal panels at ceiling level. But when the proscenium is moved forward to the *opera* setting, the auditorium becomes smaller and a proscenium stage area is created, while the ceiling panels are removed for access to the fly tower above. In the third position, *drama*, the auditorium volume is further reduced to create an intimate theater. The variable elements also include a hinged side-wall section to improve the shape of the room for theater as well as moveable reflectors and variable absorption curtains above the grid ceiling level.

It should be mentioned that many of the problems related to the successful design of speech and music auditoria become more severe with increased size of the room. In other words, it is much easier to design a hall for less than 1000 people, like the two examples presented above, than for 2000 plus. The problems become even more complicated if multipurpose function is requested.

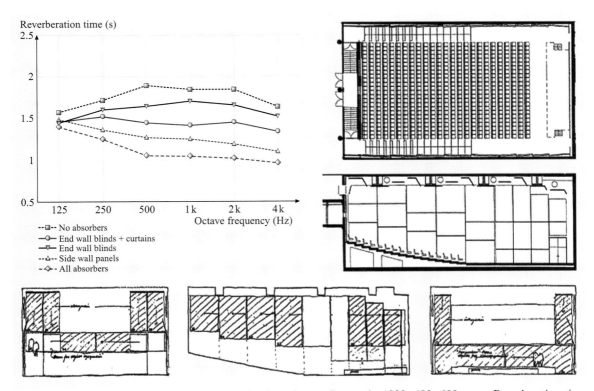

Fig. 9.52 Dronningesalen in the Royal Library in Copenhagen, Denmark; 1999, 400–600 seats. Reverberation time curves, plan, section and wall elevations with *hatched areas* indicating variable absorption

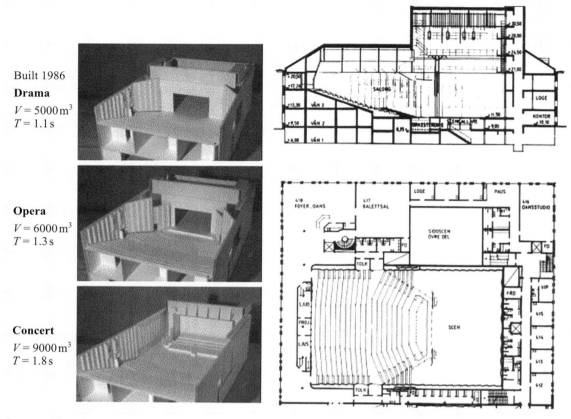

Fig. 9.53 The Idun multipurpose hall in Umeå, Sweden, with variable volume and stage configurations. The hall seats up to 860 people. Brief data (*left*), photos from cardboard model, long section and plan (*right*)

Still, a few examples of such successful designs exist. The most innovative is probably the Segerstrom Hall in Orange County, California, USA, which is shown in Fig. 9.54. This hall features a special layout of the almost 3000 audience seats, which are distributed on four levels. Each of these forms an almost rectangular, or even reverse fan, shape. However, the orientation of each level is shifted so that the overall plan becomes a wide fan. In this way the virtues of the fan shape for audience proximity to the stage are combined with the advantages of the rectangular shape for strong, early reflections from lateral directions. Thus, although the total width is about 50 m, the lateral energy level in this hall resembles that found in rectangular, 25 m-wide concert halls with typical seating capacity up to, say, 2000 people [9.36].

9.7.5 Halls for Rhythmic Music

Halls intended for rhythmic, amplified music concerts range in size from less than 100 to perhaps more than 50 000 listeners (in roofed sports arenas). The reason for this broad range of audience capacity is that the size and number of the sound amplification loudspeakers as well as of event video screens can be chosen at will to ensure good vision and adequate (often too high) sound levels for any size of audience. In any case, the reverberation time should be close to the optimal for speech, i.e., not more than 1 s if possible.

In huge arenas, of course, the reverberation time will be higher; but if efficient absorption is applied to the critical surfaces, a higher reverberation time need not compromise clarity. This is because the reverber-

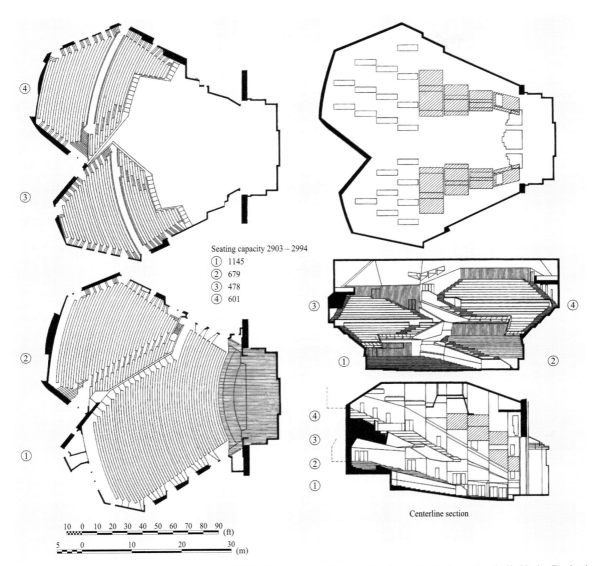

Fig. 9.54 Segerstrom Hall, Orange County, California, seating 2900 people. Acoustic design: Marshall, Hyde, Paoletti (after [9.3])

ation level will be low compared to the direct sound from the loudspeakers. In large spaces, designed to have a moderate T value, one must be careful to avoid echoes from distant, hard surfaces facing the often highly directional loudspeakers. Even small surfaces such as doors can cause echo problems if most other surfaces in the room are efficiently absorbing.

It is also very important to maintain modest low-frequency T values. A bass ratio larger than unity can create serious problems with control of the low-frequency levels. This is largely because most loudspeakers are almost omnidirectional at low frequencies, at which they willingly excite the room *reverberance*. In contrast, with high loudspeaker directivity in mid- and high-frequency range, it is easy to control the level in the audience area at these frequencies without spilling additional sound energy into the room.

Unfortunately, low-frequency reverberation control can be difficult to achieve, because low-frequency absorbers are normally less efficient (have lower α values) than mid-/high-frequency absorbers, and in a room in which these already occupy most of the available surfaces, it can be difficult to also find spaces for dedicated low-frequency absorbers. In particular, this poses a problem in buildings made from heavy concrete or masonry (which on the other hand is favorable for insulation of low-frequency sound towards the neighbors). When the low-frequency reverberation is too high the string bass and pedal drum will sound muddy. The audience area layout should be shaped to allow an even coverage by a sensible loudspeaker set up, and in large halls, a sloping floor is not suitable if a standing audience is expected. Fortunately, a sloping floor is not always mandatory, since in larger venues, the loudspeakers can be elevated to avoid grazing-incidence attenuation and elevated video projection screens improve the visual experience. It is advantageous to design the hall as well as the loudspeaker system so that zones with lower levels are also created, whereby people can have a chance to rest their ears, for instance while picking up refreshments in the bar. This may be accomplished by placing the bar in a smaller, partly decoupled, space with a shallower room height and an efficiently absorbing ceiling.

9.7.6 Worship Spaces/Churches

The liturgies in old Christian and Muslim worship were developed in highly reverberant cathedrals and mosques, both of which have their roots in Byzantine architecture. When new churches are built for the traditional Catholic or Protestant Christian denominations, it is customary to aim at a rich reverberation, which suit the pipe organ literature and congregational singing, while the intelligibility of the words by the priests is taken care of by a distributed public address (PA) sound system.

The same is true for new large mosques, in which the speech is amplified while the long reverberation supports traditional singing by the imams. Apart form these songs music is not performed in mosques.

In contrast to this, several Christian denominations, particularly in the US, have built new churches or temples (often accommodating several thousand people) and brought in newer art forms making use of large rock-music sound systems. In these spaces, therefore, the solution is often for the acoustician to create fairly dry, natural acoustics and to install both a heavy PA sound system as well as an artificial, electronic reverberation-enhancement system. Both types of sound systems are briefly described in Sect. 9.8.

9.8 Sound Systems for Auditoria

It is important to realize that sound systems will always be installed in both new and existing halls. As these systems are often used to amplify natural sound sources appearing on the stage in the hall, they may be regarded as a means for creating variable acoustics of the space. In this regard, the following will focus on how these systems interact with the natural acoustics, rather than talking about specific details or trends in the design of the loudspeakers themselves.

In principle, one may consider two different types of loudspeaker to be installed in an auditorium: traditional PA systems intended for the modification of the early part of the impulse response of the hall, and reverberation-enhancement systems intended for modification of the later part. Below, important aspects of both types will be briefly explored.

9.8.1 PA Systems

Normally, PA sound systems are installed with the purpose of increasing the sound level and/or the intelligibility of the performance. Use of the sound system for the reproduction of prerecorded sounds is of less interest in this context, as in this case it just acts like any other sound source working under the acoustic conditions provided by the natural acoustics of the room. In Sect. 9.2 we learned that, in order to improve *clarity/intelligibility* as well as the *level*, we need to increase the clarity C by adding sound components to the early part of the impulse response. (The alternative, to increase C by reducing the late part, can so far only be accomplished by adding physical absorption; but some day in the future, active systems with multiple microphones and loudspeakers integrated into *acoustic wallpaper* could be imagined as well.) Doing this requires that the main part of the sound energy leaving the loudspeakers hits the absorbing audience area. Alternatively, if substantial parts of the energy hit other, reflecting room surfaces, it may be reflected randomly and feed the reverberant field, which will reduce the clarity. For this reason, it is very important to apply loudspeakers with well-controlled directivity for this purpose, especially in auditoria with T values larger than optimal for speech.

For the acoustician, the most important technical specification for PA loudspeakers relates to their directivity, either in terms of vertical and horizontal radiation angles or in terms of the Q factor. Sometimes, the term *directivity index* $\mathrm{DI} = 10\log(Q)$ is used instead. Q is defined as the ratio between

1. the intensity emitted in the axis direction of the directive loudspeaker, and
2. the averaged intensity emitted by the same loudspeaker integrated over all directions.

The higher the Q value, the more the sound emitted will be concentrated in the forward direction (or rather, within the radiation angles specified).

When the loudspeaker is placed in a room, the sound field close to the loudspeaker will be dominated by the direct sound, whereas at longer distances, this component is negligible compared to the (diffuse) reverberant sound, as illustrated in Fig. 11.6. The (squared) sound pressure as a function of distance from the source in a room can be written as follows:

$$p^2(r) = \rho c\, P_{\mathrm{SP}} \left(\frac{Q_{\mathrm{SP}}}{4\pi r^2} + \frac{4}{A} \right). \tag{9.38}$$

In (9.38) the first term in the parenthesis is the direct sound component and the second one, $4/A$, the diffuse, reverberant component, P_{SP} is the emitted sound power, Q_{SP} is the Q factor of the loudspeaker, and A is the total sound absorption area of the room. Notice that in (9.38), for simplification, we assume a diffuse field intensity that is constant throughout the room, although in Sect. 9.5.1 and Fig. 9.13 we demonstrated that this is seldom the case.

Equation (9.38) shows that the direct component will dominate as long as the direct term is larger than the diffuse term. This will be the case for distances below the *critical distance* or the *reverberation distance*, $r_{\mathrm{cr,SP}}$. At this distance, the two components have equal magnitude, whereby

$$r_{\mathrm{cr,SP}} = \sqrt{\frac{Q_{\mathrm{SP}} A}{16\pi}} = \sqrt{\frac{Q_{\mathrm{SP}} V}{100\pi T}}. \tag{9.39}$$

As seen from (9.39), we maximize the range within which we are able to provide the listeners with a clear sound from the speakers by increasing Q (as well as A).

One should be careful not to apply a larger number of loudspeakers than needed for coverage of the audience area, as each new loudspeaker will contribute to the generation of the reverberant diffuse field, while only one at a time will send direct sound towards a given audience position. Likewise, the number of open microphones should be minimized (which is often done automatically by an intelligent microphone mixer), as all open microphones will pick up and amplify the unclear reverberant field in the room.

The number of active loudspeakers and microphones should also be minimized in order to reduce the risk of feedback, which is a consequence of the microphone(s) picking up the sound from the loudspeaker(s) and the amplifier gain being set to high. If the amplification through the entire closed loop (microphone–amplifier–loudspeaker–room and back to the microphone) exceeds unity, the system will start to oscillate at a random frequency, which is emitted at maximum power level into the room, a most annoying experience.

The risk of feedback can be reduced by minimizing the loudspeaker–listener distance and the source–microphone distance, whereby a suitable listening level can be obtained with a moderate amplifier gain setting. Use of head-borne microphones represents the ultimate reduction of the source–microphone distance. Consequently, with such microphones the risk of feedback is substantially reduced. Also, the microphone–loudspeaker distance should be maximized and directional microphones and loudspeakers should be used and positioned so that the transmission from the loudspeaker back into the microphone is minimized. Thus, the loudspeakers and the microphones should be placed pointing away from each other with the microphone pointing towards the stage and the loudspeaker pointing toward the audience.

Considering the factors mentioned above, we still need to choose a configuration of the loudspeakers for a given room. Often the choice is between

1. a central system, consisting of a single or a limited number of highly directional units arranged close together in a cluster over the stage, or
2. a distributed system of several smaller units placed closer to the listeners.

Fig. 9.55 Alternative solutions for loudspeaker coverage in an auditorium: a central (cluster) system H1, the same with a delayed secondary unit H2, or a highly distributed (pew back) system with small loudspeakers placed in each row of seats

These options are illustrated schematically in Fig. 9.55 as H1 and H3, respectively. H2 illustrates a compromise with a secondary unit or cluster assisting the main cluster, H1.

In meeting rooms or churches, highly distributed systems with small loudspeakers built into fixed meeting tables or pews can be very successful. With each loudspeaker being within a distance of say 1 m from the listener, each unit can be set to a very moderate level. Therefore, it is often possible to serve the listeners with a clear direct sound from such speakers without the total number of speakers exciting too much of the room reverberation. If the system is subdivided into independent sections, it is even possible to install automatic switches that only activate loudspeakers in zones where people are sitting. In this way, the reverberant field can be even further reduced in cases of fewer people attending the meeting or service. This is actually very fortunate as, especially in churches, fewer people also means less absorption and louder reverberation. The performance regarding intelligibility of pew-back systems with perhaps one hundred, moderately directive speakers is often equal to or better than when a highly directive central-cluster system is used. This is simply due to the moderate sound level needed from each of the closely placed loudspeakers compared to the much higher level needed for the cluster to reach the distant listeners. As can be imagined, the sound level distribution produced by the pew-back system will also be more uniform.

Of course, the highly distributed systems are not suitable for amplification of stage performances, where a high sound level of music is needed. For this purpose, a main cluster over the stage or a stereo system with loudspeakers on each side of the stage is preferable. For such systems, it is also more important to maintain a correct localization of the sound as coming from the performers on the stage.

In cases where some additional, distributed loudspeakers are needed to assist coverage in certain parts of the room, for instance on and under deep balconies, these are equipped with an electronic delay. If this delay is set so that the sound from the distributed assisting loudspeakers arrives 5–25 ms after the sound from the main loudspeakers, the listener will still perceive the sound as coming from the direction of the source. This is even true when the assisting loudspeakers are up to 5 dB louder than the main cluster at the listener's position. This very convenient phenomenon, called the precedence effect or Hass effect, is generally used as a guideline for the choice of level and delay settings in distributed loudspeaker systems.

9.8.2 Reverberation-Enhancement Systems

For auditoria where a long reverberation time is only needed occasionally or when the available room volume is simply inadequate, one may consider creating more reverberation by means of an electronic reverberation-enhancement system. Like the systems described in the previous section, these also consist of a number of loudspeakers and microphones connected by amplifiers; but there are many important differences.

The basic difference is that enhancement systems primarily attempt to add energy to the late part of the impulse response. Therefore, delays must be built into the signal processing between microphones and loudspeakers. In the first systems of this kind, *assisted*

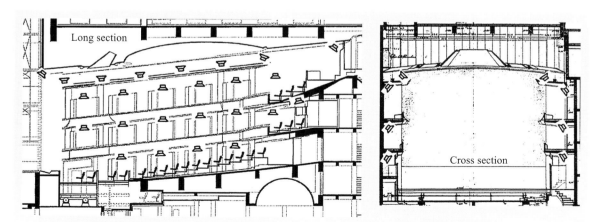

Fig. 9.56 Example of distribution of loudspeakers for reverberation enhancement in a theatre

resonance (AR) and *multichannel reverberation* (MCR), which appeared in the 1960s, this delay was provided by letting the loop gain of a large number of microphone–loudspeaker channels be tuned to just below the feedback limit. By spacing the microphones and loudspeakers far apart in the room (normally close to the ceiling), the sound would spend sufficient time traveling several times between loudspeakers and microphones for an audible prolongation of the reverberation to be perceived. In more-modern systems, electronic delays and reverberators are normally used.

In order to create the illusion of diffuse reverberation coming from the room itself, a large number of loudspeakers with low directivity must be distributed over the main surfaces in the room, as illustrated in Fig. 9.56. Since many speakers are needed, each can be fairly small and with limited acoustic power capability. The density and placement of these speakers as well as the balancing of their levels is essential to prevent localization of individual loudspeakers from any seat.

Regarding the microphones, these may be highly directive, but they must be placed quite far from the sound sources (and preferably out of sight) in order to cover the stage without the reverberant level being strongly dependent on the position of the sound source.

In view of what was explained in Sect. 9.8.1 about the measures needed to reduce the risk of feedback in loudspeaker systems, it is no wonder that most reverberation-enhancement manufacturers have to implement a solid strategy against uncontrolled feedback. One very efficient approach is to introduce time-varying delays in the sound-processing equipment, as this seems to destroy sharp room resonances when they start to build up to form a hauling frequency. Still, most enhancement systems work with a loop gain close to the feedback limit (which is actually participating actively in increasing the reverberation time), which is acceptable as long as it is under control and does not cause audible coloration of the sound.

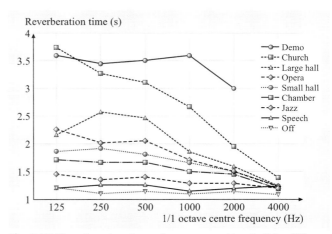

Fig. 9.57 Reverberation time values measured in a hall for different settings of the reverberation enhancement system (LARES)

It is essential for the acceptance of enhancement systems that neither audience nor performers experience the sound as coming from a loudspeaker system. Likewise, artefacts such as coloration due to feedback or the poor frequency characteristics of speakers are unacceptable.

The maximum increase in reverberation time possible with feedback-based systems like AR and MCR was typically about 50%. With modern digital reverberators, a much larger range can be achieved, as shown in Fig. 9.57. Increasing the reverberation time by a factor of three or more is not impossible. However, the illusion, when you still see the physical, much-dryer room around you, tends to break down when the value increases much above the 50%, which ironically is what the old systems could produce.

From a purist standpoint, reverberation enhancement can be regarded as an acoustic prosthesis, but in many cases a pragmatic approach to the use of these systems can increase the range of programs that a cultural venue can host.

References

9.1   A.C. Gade: *Subjective Room Acoustic Experiments with Musicians*. PhD thesis, Rep. No. 32 (Technical University of Denmark, Copenhagen 1982)

9.2   W.C. Sabine: *Collected Papers on Acoustics* (Peninsula, Los Altos 1994), Reissued

9.3   L.L. Beranek: *Concert Halls and Opera Houses, Music Acoustics and Architecture*, 2nd edn. (Springer, Berlin, New York 2004)

9.4   M.F.E. Barron: *Auditorium Acoustics and Architectural Design* (Spun, London 1993)

9.5   H. Kuttruff: *Room Acoustics* (Applied Science, London 1976)

9.6   ISO: *ISO 3382*, Acoustics – Measurement of the Reverberation Time of Rooms with Reference to Other Acoustical Parameters (ISO, Geneva 1997)

9.7 M. Vorländer: *International Round Robin on Room Acoustical Computer Simulations, Proc. 15th ICA* (Trondheim, Norway 1995) pp. 689–692

9.8 J.S. Bradley, G.A. Soloudre: The influence of late arriving energy on spetial impression, J. Acoust. Soc. Am. **97**(4), 2263–2271 (1995)

9.9 L.L. Beranek: *Music, Acoustics and Architecture* (Wiley, New York 1962)

9.10 Personal communication (summer 2006)

9.11 A.C. Gade: Investigations of musicians' room acoustic conditions in concert halls I, methods and laboratory experiments, Acustica **69**, 193–203 (1989)

9.12 A.C. Gade: Investigations of musicians' room acoustic conditions in concert halls II, Field experiments and synthesis of results, Acustica **69**, 249–262 (1989)

9.13 T. Houtgast, H.J.M. Steeneken: A tool for evaluating auditoria, Brüel Kjær Tech. Rev. **3**, 3–12 (1985)

9.14 J.S. Bradley: Predictors of speech intelligibility in rooms, J. Acoust. Soc. Am. **80**, 837–845 (1986)

9.15 M.R. Schroeder: New method of measuring reverberation time, J. Acoust. Soc. Am. **37**, 409 (1965)

9.16 J.S. Bradley: Accuracy and reproducibility of auditorium acoustics measures, Proc. IOA **10**, 399–406 (1988), Part 2

9.17 K. Sekigouchi, S. Kimura, T. Hanyuu: Analysisi of sound field on spatial information using a four-channel microphone system based on regular tetrahedron peak point method, Appl. Acoust. **37**, 305–323 (1992)

9.18 D. de Vries, A.J. Berkhout, J.J. Sonke: *Array Technology for Measurement and Analysis of Sound Fields in Enclosures, AES 100th Convention 1996 Copenhagen, Preprint 4266 (R-4)* (AES, San Francisco 1996)

9.19 C.M. Harris: Absorption of air versus humidity and temperature, J. Acoust. Soc. Am. **40**, 148–159 (1966)

9.20 C.W. Kosten: The mean free path in room acoustics, Acustica **10**, 245 (1960)

9.21 L.L. Beranek: Analysis of Sabine and Eyring equations and their applications to concert hall audience and chair absorption, J. Acoust. Soc. Am. **120** (2006)

9.22 L.L. Beranek, T. Hidaka: Sound absorption in concert halls by seats, occupied and unoccupied, and by the hall's interior surfaces, J. Acoust. Soc. Am. **104**, 3169–3177 (1998)

9.23 S. Sakamoto, T. Yokota, H. Tachibana: Numerical sound field analysis in halls using the finite difference time domain method. In: *Proc. Int. Symp. Room Acoustics: Design and Science (RADS)*, Awaji Island (ICA, Kyoto 2004)

9.24 F. Spandöck: Raumakustische Modellversuche, Ann. Phys. **20**, 345 (1934)

9.25 A.C. Gade: Room acoustic properties of concert halls: quantifying the influence of size, shape and absorption area, J. Acoust. Soc. Am. **100**, 2802 (1996)

9.26 M. Barron, L.-J. Lee: Energy relations in Concert Auditoria, J. Acoust. Soc. Am. **84**, 618–628 (1988)

9.27 W.-H. Chiang: *Effects of Various Architectural Parameters on Six Room Acoustical Measures in Auditoria, PhD thesis* (University of Florida, Gainesville 1994)

9.28 W. Fasold, H. Winkler: *Bauphysikalische Entwurfslehre, Band 5; Raumakustik* (Verlag Bauwesen, Berlin 1976)

9.29 Z. Maekawa, P. Lord: *Environmental and Architectural Acoustics* (Spon, London 1994)

9.30 W. Furrer, A. Lauber: *Raum- und Bauakustik Lärmabwehr* (Birkhaüser, Basel 1976)

9.31 J.H. Rindel: *Anvendt Rumakustik, Note 4201, Oersted DTU* (Technical University of Denmark, Copenhagen 1984), (In Danish)

9.32 L.I. Makrinenko: *Acoustics of Auditoriums in Public Buildings* (ASA, Melville 1994)

9.33 J. Sadowski: *Akustyka architektoniczna* (PWN, Warszawa 1976)

9.34 J.H. Rindel: *Acoustic design of reflectors in Auditoria, Note 4212 Oersted DTU* (Technical University of Denmark, Copenhagen 1999)

9.35 M.R. Schroeder: Binaural dissimilarity and optimum ceilings for concert halls: More lateral sound diffusion, J. Acoust. Soc. Am. **65**, 958–963 (1979)

9.36 J.R. Hyde: *Segerstrom Hall in Orange County – Design, Measurements and Results after years of operation*, Proc. IOA, Vol. 10 (IOA, St. Albans 1988), Part 2

10. Concert Hall Acoustics Based on Subjective Preference Theory

This chapter describes the theory of subjective preference for the sound field applied to designing concert halls. Special attention is paid to the process of obtaining scientific results, rather than only describing a final design method. Attention has also been given to enhancing satisfaction in the selection of the most preferred seat for each individual in a given hall. We begin with a brief historical review of concert hall acoustics and related fields since 1900.

A neurally grounded theory of subjective preference for the sound field in a concert hall, based on a model of the human auditory–brain system, is described [10.1]. Most generally, subjective preference itself is regarded as a primitive response of a living creature and entails judgments that steer an organism in the direction of maintaining its life. Brain activities relating to the scale value of subjective preference, obtained by paired-comparison tests, have been determined. The model represents *relativity*, relating the autocorrelation function (ACF) mechanism and the interaural cross-correlation function (IACF) mechanism for signals arriving at the two ear entrances. The representations of ACF have a firm neural basis in the temporal patterning signal at each of the two ears, while the IACF describes the correlations between the signals arriving at the two ear entrances. Since Helmholtz, it has been well appreciated that the cochlea carries out a rough spectral analysis of sound signals. However, by the use the of the spectrum of an acoustic signal, it was hard to obtain factors or cues to describe subjective responses directly. The auditory representations from the cochlea to the cortex that have been found to be related to subjective preference in a deep way involve these temporal response patterns, which have a very different character from those related to power spectrum analyses. The scale value of subjective preference of the sound field is well described by four orthogonal factors. Two are temporal factors (the initial delay time between the direct sound and the first reflection, Δt_1, and the reverberation time, T_{sub}) associated with the left cerebral hemisphere, and two are spatial factors [the binaural listening level (LL) and the magnitude of the IACF, the IACC] associated with the right hemisphere. The theory of subjective preference enables us to calculate the acoustical quality at any seat in a proposed concert hall, which leads to a seat selection system.

The temporal treatment enables musicians to choose the music program and/or performing style most suited to a performance in a particular concert hall. Also, for designing the stage enclosure for music performers, a temporal factor is proposed. Acoustical quality at each seating position examined in a real hall is confirmed by both temporal and spatial factors.

10.1	**Theory of Subjective Preference for the Sound Field**................	353
	10.1.1 Sound Fields with a Single Reflection	353
	10.1.2 Optimal Conditions Maximizing Subjective Preference	356
	10.1.3 Theory of Subjective Preference for the Sound Field....................	357
	10.1.4 Auditory Temporal Window for ACF and IACF Processing	360
	10.1.5 Specialization of Cerebral Hemispheres for Temporal and Spatial Factors of the Sound Field	360
10.2	**Design Studies**	361
	10.2.1 Study of a Space-Form Design by Genetic Algorithms (GA)	361
	10.2.2 Actual Design Studies.................	365
10.3	**Individual Preferences of a Listener and a Performer**...............	370
	10.3.1 Individual Subjective Preference of Each Listener	370
	10.3.2 Individual Subjective Preference of Each Cellist	374
10.4	**Acoustical Measurements of the Sound Fields in Rooms**	377
	10.4.1 Acoustic Test Techniques	377
	10.4.2 Subjective Preference Test in an Existing Hall......................	380
	10.4.3 Conclusions	383
References		384

In the first half of the 20th century, studies were mostly concentrated on temporal aspects of the sound field. In 1900 *Sabine* [10.2] initiated the science of architectural acoustics, developing a formula to quantify reverberation time. In 1949, *Haas* [10.3] investigated the echo disturbance effect from adjustment of the delay time of the early reflection by moving head positions on a magnetic tape recorder. He showed the disturbance of speech echo to be a function of the delay time, with amplitude as a parameter.

After investigations with a number of existing concert halls throughout the world, *Beranek* in 1962 [10.4] proposed a rating scale with eight factors applied to sound fields from data obtained by questionnaires on existing halls given to experienced listeners. Too much attention, however, has been given to monaural temporal factors of the sound field since Sabine's formulation of reverberation theory. For example, the Philharmonic Hall of Lincoln Center in New York, opened in 1962, was not satisfactory to listeners even after many improvements.

On the spatial aspect of the sound field, *Damaske* in 1967 [10.5] investigated subjective diffuseness by arranging a number of loudspeakers around the listener. In 1968, *Keet* [10.6] reported the variation of apparent source width (ASW) in relation to the short-term cross-correlation coefficient between two signals fed into the stereo loudspeakers as well as the sound pressure level. *Marshall* in 1968 [10.7] stressed the importance of early lateral reflections of just 90 degrees, and *Barron* in 1971 [10.8] reported *spatial impressions* or *envelopment* of sound fields in relation to the short-term interaural cross-correlation coefficient.

As a typical spatial factor of the sound field, *Damaske* and *Ando* in 1972 [10.9] defined the IACC as the maximum absolute value of the interaural cross-correlation function (IACF) within the possible maximum interaural delay range for the human head, such that

$$\text{IACC} = |\phi_{\text{lr}}(\tau)|_{\max} , \text{ for } |\tau| < 1 \text{ ms} \quad (10.1)$$

and proposed a method of calculating the IACF for a sound field. In 1974, *Schroeder* et al. in [10.10] reported results of paired-comparison tests asking listeners which of two music sound fields was preferred. In an anechoic chamber, sound fields in existing concert halls were reproduced at the ears of listeners through dummy head recordings and two loudspeaker systems, with filters reproducing spatial information. They found that two significant factors, the reverberation time and the IACC, had a strong influence on subjective preference.

In 1977, *Ando* discussed subjective preference in relation to the temporal and spatial factors of the sound field simulated with a single reflection [10.11]. In 1983, he described a theory of subjective preference in relation to four orthogonal factors consisting of the temporal and spatial factors for the sound field, which enable us to calculate the scale value of the subjective preference at each seat [10.12, 13]. *Cocchi* et al. reconfirmed this theory in an existing hall in 1990 [10.14]. In 1997, *Sato* et al. [10.15] reconfirmed this clearly by use of the paired-comparison judgment in an existing hall, switching identical loudspeakers located at different positions on the stage, instead of changing the seats of each subject. This method may avoid effects of other physical factors than the acoustic factors. In addition to the orthogonal factors, they found the interaural delay in the IACF, τ_{IACC}, as a measure of image shift of the sound source that is to be kept at zero realizes good balance in the sound field. Using this method, the dissimilarity distance has also been described by the temporal and spatial factors of the sound field in 2002 [10.16].

Thus far, the theory has been based on the global subjective attributes for a number of subjects. The theory may be applied for enhancing each individual's satisfaction by adjusting the weighting coefficient of each orthogonal factor [10.1], even though a certain amount of individual differences exist [10.17]. The seat selection system [10.18] was introduced in 1994 after construction of the Kirishima International Concert Hall.

For the purpose of identifying the model of the auditory–brain system, experiments point toward the possibility of developing the correlation between brain activities, measurable with electroencephalography (EEG) [10.19, 20]. Correspondences between subjective preference and brain activity have been found from EEG and magnetoencephalography (MEG) studies in human subjects [10.21, 22], but details of these results are not included in this chapter due to limited space. Results show that orthogonal factors comprise two temporal factors: the initial time delay gap between the direct sound and the first reflection (Δt_1) and the subsequent reverberation time (T_{sub}) are associated with the left hemisphere, and two spatial factors: the listening level (LL) and the magnitude of the IACF (IACC) are associated with the right hemisphere. The information corresponding to subjective preference of the sound field can be found in the effective duration of the autocorrelation function (ACF) of the alpha (α) waves of both EEG and MEG. A repetitive feature in the α-wave, as measured in its ACF at the preferred condition, has been found. The evidence ensures that the basic theory of subjective preference may also be applied to each individual preference [10.21].

Since individual differences of subjective preference in relation to the IACC in the spatial factor are small enough, at the first stage of acoustic design we can determine the architectural space form of the room. The temporal factors are closely related to the dimensions of a specific concert hall, which can be altered to exhibit specific types of music, such as organ music, chamber music or choral works.

On the neural mechanism in the auditory pathway, a possible mechanism for the interaural time difference and correlation processors in the time domain was proposed by *Jeffress* in 1948 [10.23], and by *Licklider* in 1951 [10.24]. By recording left and right auditory brainstem responses (ABR), Ando et al. in 1991 [10.25] revealed that the maximum neural activity (wave V at the inferior colliculus) corresponds well to the magnitude of the interaural cross-correlation function. Also, the left and right waves $IV_{l,r}$ are close to the sound energies at the right- and left-ear entrances. In fact, the time-domain analysis of the firing rate of the auditory nerve of a cat reveals a pattern of ACF rather than the frequency-domain analysis as reported by *Secker-Walker* and *Searle* in 1990 [10.26]. Cariani and Delgutte in 1996 showed that pooled inter-spike interval distributions resemble the short-time or the running autocorrelation function (ACF) for the low-frequency component. In addition, pooled interval distributions for sound stimuli consisting of the high-frequency component resemble the envelope for running ACF [10.27, 28].

Remarkably, primary sensations such as the pitch of the missing fundamental [10.29], loudness [10.30], and duration sensation [10.31] can be well described by the temporal factors extracted from the ACF [10.32, 33]. *Timbre* investigated by the dissimilarity judgment [10.16] of the sound field has been described by both the temporal and spatial factors. The typical spatial attributes of the sound field, such as subjective diffuseness [10.34] and apparent source width (ASW), as well as subjective preference, are well described by the spatial factor [10.32–36].

Besides the design of concert halls, other acoustical applications such as speech identification [10.36], environmental noise measurement [10.37], and sound localization in the median plane [10.38] should benefit from guidelines derived from this model.

10.1 Theory of Subjective Preference for the Sound Field

Subjective preference judgment is the most primitive response in any subjective attribute and entails judgments that steer an organism in the direction of maintaining and/or animating *life*. Subjective preference, therefore, may relate to an aesthetic issue. It is known that judgment in an absolute manner presents a problem in reliability; rather, data is judged in a relative manner such as by paired-comparison tests. This is the simplest method, in that any person may participate, and the resulting scale value may be utilized in the wide range of applications. From the results of subjective preference studies in relation to the temporal factor and the spatial factor of the sound field, the theory of subjective preference is derived. Examples of calculating subjective preference at each listener's position are demonstrated in Sect. 10.2 for the global listener and Sect. 10.3 for the individual listener. The relationship between the resulting scale value of subjective preference in an existing hall and the physical factors obtained by calculation using architectural plan drawings, has been examined by factor analysis in Sect. 10.4.

10.1.1 Sound Fields with a Single Reflection

Preferred Delay Time of a Single Reflection
First of all, the simplest sound field, which consists of the direct sound with the horizontal angle to a listener: $\xi_0 = 0°$ (the elevation angle $\eta_0 = 0°$), and a single reflection from a fixed direction $\xi_1 = 36°$ ($\eta_1 = 9°$), was investigated. These angles were selected since they are typical in a concert hall. The delay time Δt_1 of the reflection was adjusted in the range of 6–256 ms. The paired-comparison test was performed for all pairs in an anechoic chamber using normal hearing subjects with two different music motifs A and B (Table 10.1). The effective duration of the ACF of the sound source defined by τ_e may be obtained by the delay at which the envelope of the normalized autocorrelation function (ACF) becomes 0.1 (10-percentile delay) as shown in Fig. 10.1. The value of $(\tau_e)_{min}$ indicated in Table 10.1 is obtained from the minimum value of the running ACF, $2T = 2$ s, with an interval of 100 ms. The recommended $2T$ is given by (10.14). As far as these source signals are concerned, values of (τ_e) extracted from the long-term ACF

Table 10.1 Music and speech source signals and their minimum effective duration of the running ACF, $(\tau_e)_{min}$

Sound source	Title	Composer or writer	$(\tau_e)_{min}$ (ms)
Music motif A	Royal Pavane	Orlando Gibbons	125
Music motif B	Sinfonietta, opus 48; IV movement	Malcolm Arnold	40
Speech S	Poem read by a female	Doppo Kunikida	10

[1] The value of $(\tau_e)_{min}$ is the minimum value extracted from the running ACF, $2T = 2$ s, with a running interval of 100 ms. The recommended value of $2T$ is given by (10.14)

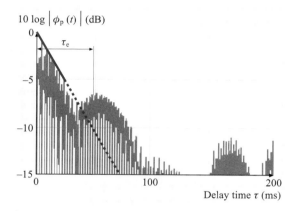

Fig. 10.1 Determination of the effective duration of the running ACF, τ_e. The value of τ_e may be obtained by the delay for which the envelope of the normalized ACF becomes 0.1 or -10 dB (10-percentile delay)

were similar to the minimum values from the running ACF, $(\tau_e)_{min}$.

For simplicity, the score was simply obtained, in this section, by accumulating a score giving $+1$ and -1 corresponding to positive and negative judgments, respectively, and the total score is divided by $S(F-1)$ to get the normalized score, where S is the number of subjects and F is the number of sound fields tested. The normalized scores for two motifs and the percentage of preference for speech signal as a function of the delay time are shown in Fig. 10.2.

Obviously, the most preferred delay time with the maximum score differs greatly between the two motifs. When the amplitude of reflection $A_1 = 1$, the most preferred delays are around 130 ms for music motif A, 35 ms for music motif B (Fig. 10.2a), and 16 ms for

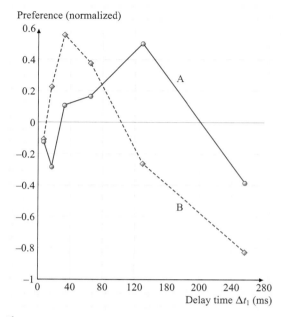

Fig. 10.2 Preference scores of the sound fields as a function of the delay time $A_1 = 1.0$ (six sound fields and 13 subjects [10.11]). Preference scores are directly obtained for each pair of sound fields by assigning the values $+1$ and -1, corresponding to positive and negative judgments, respectively. The normalized score is obtained by accumulating the scores for all sound fields (F) tested and all subjects (S), and then dividing by the factor $S(F-1)$ for 13 subjects. A: music motif A, Royal Pavane by Gibbons, $(\tau_e)_{min} = 125$ ms; B: music motif B, Sinfonietta, opus 48, III movement by Malcolm Arnold, $(\tau_e)_{min} = 40$ ms

speech [10.36]. Later, it was found that this corresponds well to the minimum values of the effective durations of the running ACF [10.39] of the source signals, so that the most preferred delay time were 125 ms (motif A), 40 ms (motif B) and 10 ms (speech). After inspection, the preferred delay is found roughly at certain durations of the ACF, defined by τ_p, such that the envelope of the ACF becomes $0.1A_1$. Thus, $[\Delta t_1]_p \approx \tau_e$ only when $A_1 = 1$. As shown in Fig. 10.3, changing the amplitude A_1, collected data of $[\Delta t_1]_p$ are expressed approximately by

$$[\Delta t_1]_p = \tau_p \approx (1 - \log_{10} A_1)(\tau_e)_{min} . \qquad (10.2)$$

Note that the amplitude of reflection relative to that of the direct sound should be measured by the most accurate method (ex. the square-root value of the ACF at the origin of the delay time).

Two reasons can be given for why the preference decreases for the short delay range of reflection, $0 < \Delta t_1 < [\Delta t_1]_p$:

1. Tone coloration effects occur because of the interference phenomenon in the coherent time region; and
2. The IACC increases when Δt_1 is near 0. The definition of the IACC, which may be extracted from the IACF, as given by [10.1], is shown in Fig. 10.4.

On the other hand, echo disturbance effects can be observed when Δt_1 is greater than $[\Delta t_1]_p$.

Preferred Horizontal Direction of a Single Reflection to a Listener

The direction was specified by loudspeakers located at $\xi_1 = 0°$ ($\eta_1 = 27°$), and $\xi_1 = 18°, 36°, \ldots, 90°$ ($\eta_1 = 9°$), where the delay time of the reflection was fixed at 32 ms [10.11]. Results of the preference tests for the two motifs are shown in Fig. 10.5. No fundamental differences are observed between the curves of the sound source in spite of the large differences in the value of $(\tau_e)_{\min}$. The preferred score increases roughly with decreasing IACC, the typical spatial factor. The correla-

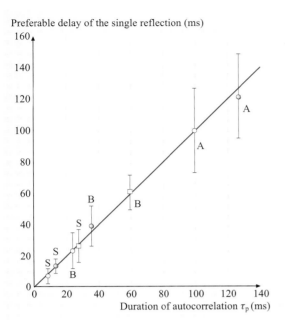

Fig. 10.3 The relationship between the preferred delay time of the single reflection and the duration of the ACF such that its envelope becomes 0.1 A_1 (for $k = 0.1$ and $c = 1.0$). The ranges of the preferred delays are graphically obtained at 0.1 below the maximum score. A, B and S refer to motif A, motif B, and speech, respectively. Different symbols indicate the center values obtained at the reflection amplitudes of $+6$ dB (○), 0 dB (●), and -6 dB (□), respectively (13–19 subjects). (After [10.11])

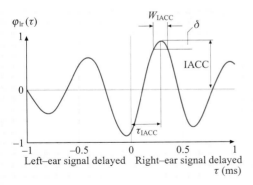

Fig. 10.4 Definition of the three spatial factors IACC, τ_{IACC} and W_{IACC}, extracted from the interaural cross-correlation function (IACF)

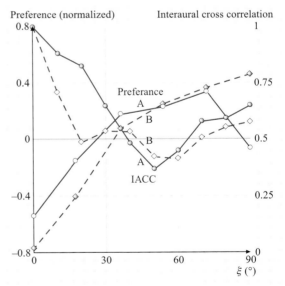

Fig. 10.5 Preference scores and the IACC of the sound fields with extreme music motifs A and B, as a function of the horizontal angle of a single reflection, $A_1 = 0$ dB (six sound fields and 13 subjects)

tion coefficient between the score and the IACC is -0.8 ($p < 0.01$). The score with motif A at $\xi_1 = 90°$ drops to a negative value, indicating that the lateral reflections coming only from around $\xi_1 = 90°$, thus, are not always preferred. The figure shows that there is a preference for angles less than $\xi_1 = 90°$, and on average there may be an optimum range centered on about $\xi_1 = 55°$. Similar results can be seen in the data from speech signals [10.40]. These results are due to the spatial factor independent of the temporal factor, which consists of the source signal.

10.1.2 Optimal Conditions Maximizing Subjective Preference

According to a systematic investigation of simulating the sound field with multiple reflections and reverberation by the aid of a computer and the listening test, the optimum design objectives and the linear scale value of subjective preference may be derived. The optimum design objectives can be described in terms of the subjectively preferred sound qualities, which are related to the four orthogonal factors describing the sound signals arriving at the two ears. They clearly lead to comprehensive criteria for achieving the optimal design of concert halls as summarized below [10.11–13].

Listening Level (LL)
The listening level is, of course, the primary criterion for listening to sound in a concert hall. The preferred listening level depends upon the music and the particular passage being performed. The preferred levels obtained with 16 subjects were similar for two extreme music motifs: 77–79 dBA in peak ranges for music motif A (Royal Pavane by Gibbons) with a slow tempo, and 79–80 dBA for music motif B (Sinfonietta by Arnold) with a fast tempo Fig. 10.7a.

Early Reflection after the Direct Sound (Δt_1)
An approximate relationship for the most preferred delay time has been discovered in terms of the ACF envelope of source signals and the total amplitude of reflections A. Generally, it is expressed by $[\Delta t_1]_p = \tau_p$

$$|\phi_p(\tau)|_{\text{envelope}} \approx kA^c, \text{ at } \tau = \tau_p, \quad (10.3)$$

where k and c are constants that depend on the subjective attributes [10.13, Fig. 42]. If the envelope of ACF is exponential, then

$$[\Delta t_1]_p = \tau_p \approx \left[\log_{10}(1/k) - c \log_{10} A\right] (\tau_e)_{\min}, \quad (10.4)$$

where the total pressure amplitude of reflection is given by

$$A = \left(A_1^2 + A_2^2 + A_3^2 + \ldots\right)^{1/2}. \quad (10.5)$$

The relationship given by (10.2) for a single reflection may be obtained by putting $A = A_1$, $k = 0.1$ and $c = 1$.

The value of $(\tau_e)_{\min}$ is observed at the most active part of a piece of music containing artistic information such as a vibrato, a quick passage in the music flow, and/or a very sharp sound signal. Echo disturbance, therefore, may be perceived at $(\tau_e)_{\min}$. Even for a long musical composition, the minimum part of $(\tau_e)_{\min}$ of the running ACF in the whole music, which determines the preferred temporal condition, may be taken into consideration for the choice of music program to be performed in a given concert hall. A method of controlling the minimum value $(\tau_e)_{\min}$ in performance, which determines the preferred temporal condition for vocal music has been discussed for blending the sound source and a given concert hall [10.41, 42]. If vibrato is introduced during singing, for example, it decreases $(\tau_e)_{\min}$, blending the sound field with a short reverberation time.

Reverberation Time after the Early Reflection (T_{sub})
It has been observed that the most preferred frequency response to the reverberation time is a flat curve [10.39]. The preferred reverberation time, which is equivalent to that defined by *Sabine* [10.2], is expressed approxi-

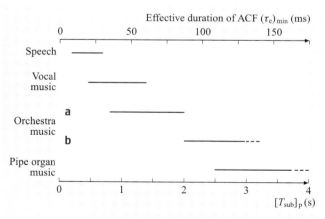

Fig. 10.6 Recommended reverberation time for several sound programs

mately by

$$[T_{sub}]_p \approx 23(\tau_e)_{min} . \qquad (10.6)$$

The total amplitudes of reflections A tested were 1.1 and 4.1, which cover the usual conditions of sound fields in a room. Recommended reverberation times for several sound sources are shown in Fig. 10.6. A lecture and conference room must be designed for speech, and an opera house mainly for vocal music but also for orchestra music. For orchestral music, there may be two or three types of concert hall designs according to the effective duration of the ACF. For example, symphony no. 41 by Mozart, *Le Sacre du Printemps* by Stravinsky, and Arnold's Sinfonietta have short ACFs and fit orchestra music of type (A). On the other hand, symphony no. 4 by Brahms and symphony no. 7 by Bruckner are typical of orchestra music (B). Much longer ACFs are typical for pipe-organ music, for example, by Bach.

The most preferred reverberation times for each sound source given by (10.6) might play important roles for the selection of music motifs to be performed. Of interest is that the most preferred reverberation time expressed by (10.6) implies about four times the *reverberation time* containing the source signal itself.

Magnitude of the Interaural Cross-Correlation Function (IACC)

All individual data indicated a negative correlation between the magnitude of the IACC and subjective preference, i.e., dissimilarity of signals arriving at the two ears is preferred. This holds only under the condition that the maximum value of the IACF is maintained at the origin of the time delay, keeping a balance of the sound field at the two ears. If not, then an image shift of the source may occur (Sect. 10.4.2). To obtain a small magnitude of the IACC in the most effective manner, the directions from which the early reflections arrive at the listener should be kept within a certain range of angles from the median plane centered on $\pm 55°$. It is obvious that the sound arriving from the median plane $\pm 0°$ makes the IACC greater. Sound arriving from $\pm 90°$ in the horizontal plane is not always advantageous, because the similar detour paths around the head to both ears cannot decrease the IACC effectively, particularly for frequency ranges higher than 500 Hz. For example, the most effective angles for the frequency ranges of 1 kHz and 2 kHz are centered on $\pm 55°$ and $\pm 36°$, respectively. To realize this condition simultaneously, a geometrical uneven surface has been proposed [10.43].

10.1.3 Theory of Subjective Preference for the Sound Field

Theory
Since the number of orthogonal acoustic factors of the sound field, which are included in the sound signals

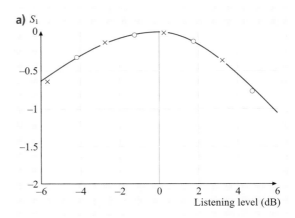

Fig. 10.7a–d Scale values of subjective preference obtained by the paired-comparison test for simulated sound fields in an anechoic chamber. Different symbols indicate scale values obtained from different source signals [10.12]. Even if different signals are used, a consistency of scale values as a function of each factor is observed, fitting a single curve. (**a**) As a function of the listening level, LL. The most preferred listening level, $[LL]_p = 0$ dB;

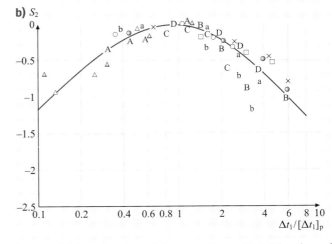

Fig. 10.7 (b) As a function of the normalized initial delay time of the first reflection by the most preferred delay time calculated by (10.4), $\Delta t_1/[\Delta t_1]_p$;

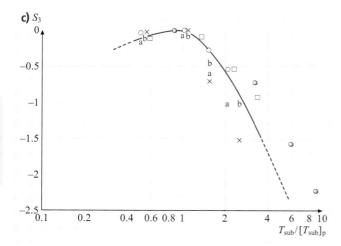

Fig. 10.7 (c) As a function of the normalized reverberation time for the most preferred calculated by (10.6), $T_{sub}/[T_{sub}]_p$;

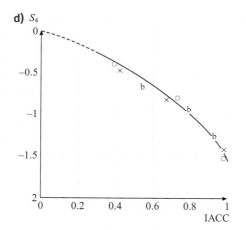

Fig. 10.7 (d) As a function of the IACC (After [10.12])

at both ears, is limited [10.12]; the scale value of any one-dimensional subjective response may be expressed by:

$$S = g(x_1, x_2, \ldots x_i) \,. \tag{10.7}$$

It has been verified by a series of experiments that four objective factors act independently on the scale value when changing two of the four factors simultaneously [10.13]. Results indicate that the units of the scale value of subjective preference derived by a series of experiments with different sound sources and different subjects have appeared to be constant [10.12], so that we may add scale values to obtain the total scale value

such as

$$\begin{aligned} S &= g(x_1) + g(x_2) + g(x_3) + g(x_4) \\ &= S_1 + S_2 + S_3 + S_4 \,, \end{aligned} \tag{10.8}$$

where S_i ($i = 1, 2, 3, 4$) is the scale value obtained relative to each objective parameter. Equation (10.8) indicates a four-dimensional continuity.

A Common Formula for the Four Normalized Orthogonal Factors

The dependence of the scale value on each objective parameter is shown graphically in Fig. 10.7. From the nature of the scale value, it is convenient to put a zero value at the most preferred conditions, as shown in this figure. These results of the scale value of subjective preference obtained from the different test series, using different music programs, yield the following common formula

$$S_i \approx -\alpha_i \, |x_i|^{3/2} \,, \quad i = 1, 2, 3, 4 \tag{10.9}$$

where values of α_i are weighting coefficients as listed in Table 10.2, which were obtained with a number of subjects. These coefficients depend on the individual. If α_i is close to zero, then a lesser contribution of the factor x_i on subjective preference is signified.

The factor x_1 is given by the sound pressure level (SPL) difference, measured by the A-weighted network, so that

$$x_1 = 20 \log P - 20 \log [P]_p \,, \tag{10.10}$$

P and $[P]_p$ being, respectively, the sound pressure at a specific seat and the most preferred sound pressure that may be assumed at a particular seat position in the

Table 10.2 Four orthogonal factors of the sound field and their weighting coefficients α_i obtained by a paired-comparison test of subjective preference with a number of subjects in conditions without any image shift of the source sound ($\tau_{IACC} = 0$)

i	x_i	α_i	
		$x_i > 0$	$x_i < 0$
1	$20 \log P - 20 \log[P]_p$ (dB)	0.07	0.04
2	$\log(\Delta t_1/[\Delta t_1]_p)$	1.42	1.11
3	$\log(T_{sub}/[T_{sub}]_p)$	0.45 + 0.75A	2.36 − 0.42 A
4	IACC	1.45	−

room under investigation:

$$x_2 = \log\left(\Delta t_1 / [\Delta t_1]_p\right), \quad (10.11)$$

$$x_3 = \log\left(T_{\text{sub}} / [T_{\text{sub}}]_p\right), \quad (10.12)$$

$$x_4 = \text{IACC}. \quad (10.13)$$

Scale values of preference have been formulated approximately in terms of the 3/2 powers of the normalized objective parameters, expressed in the logarithm for the parameters, x_1, x_2 and x_3. Thus, scale values are not greatly changed in the neighborhood of the most preferred conditions, but decrease rapidly outside of this range. The remarkable fact is that the spatial binaural parameter x_4 is expressed in terms of the 3/2 powers of its real values, indicating a greater contribution than those of the temporal parameters.

Limitation of Theory

Since experiments were conducted to find the optimal conditions, this theory holds in the range of preferred conditions obtained by the test. In order to demonstrate the independence of the four orthogonal factors, under the conditions of fixed Δt_1 and T_{sub} around the preferred conditions, scale values of subjective preference calculated by (10.8) for the LL with (10.10) and the IACC with (10.13) with constants listed in Table 10.2 are shown in Fig. 10.8. Agreement between the calcu-

Table 10.3 Hemispheric specializations determined by analyses of AEP (SVR), EEG and MEG[1]

Factors changed	AEP (SVR) $A(P_1 - N_1)$	EEG, ratio of ACF τ_e values of α-waves	MEG, ACF τ_e value of α-wave
Temporal			
Δt_1	L>R (speech)[2]	L>R (music)	L>R (speech)
T_{sub}	–	L>R (music)	–
Spatial			
LL	R>L (speech)	–	–
IACC	R>L (vowel /a/) R>L (band noise)	R>L (music)[3]	–

[1] See also [10.44] for a review of these investigations
[2] The sound source used in experiments is indicated in the bracket
[3] Flow of α-wave (EEG) from the right hemisphere to the left hemisphere for music stimuli when changing the IACC was observed by $|\phi(\tau)|_{\max}$ between α-waves recorded at different electrodes [10.45]

lated and observed values are satisfactory [10.13]. Even though both LL and the IACC are spatial factors, which are associated with the right cerebral hemisphere (Table 10.3), these are quite independent of each other. The same is true for the temporal factors of Δt_1 and T_{sub} associated with the left hemisphere. Of course, the spatial factor and the temporal factor are highly independent.

Example of Calculating the Sound Quality at Each Seat

As a typical example, we shall discuss the quality of the sound field at each seating position in a concert hall with a shape similar to that of Symphony Hall in Boston. Suppose that a single source is located at the center, 1.2 m above the stage floor. Receiving points at a height of 1.1 m above the floor level correspond to the ear positions. Reflections with their amplitudes, delay times, and directions of arrival at the listeners are taken into account using the image method. Contour lines of the total scale value of preference calculated for music motif B are shown in Fig. 10.9. Results shown in Fig. 10.9b demonstrate the effects of the reflection from the sidewalls adjusted to the stage, which produce decreasing values of the IACC for the audience area. Thus the preference value at each seat is increased compared with that

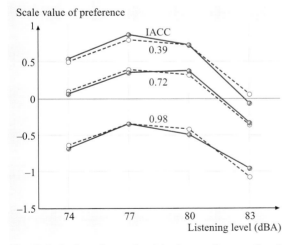

Fig. 10.8 Scale values of subjective preference for the sound field with music motif A as a function of the listening level and as a parameter of the IACC [10.13]. *Solid line*: calculated values based on (10.8) together with (10.9) taking the two factors (10.10) and (10.13) into consideration; *Dashed line*: measured values

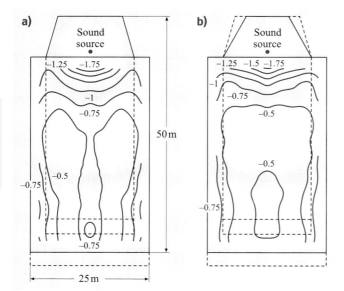

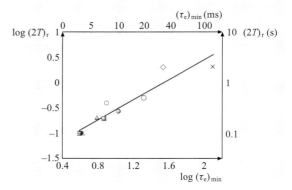

Fig. 10.9a,b An example of calculating the scale value with the four orthogonal factors using (10.8) through (10.13) with weighting coefficients (Table 10.2). (**a**) Contour lines of the total scale value for the Boston Symphony Hall, with original side reflectors on the stage. (**b**) Contour lines of the total scale values for the side reflectors optimized on the stage

Fig. 10.10 Recommended temporal window $(2T)_r$ for the ACF proceeding as a function of the minimum value of effective duration of the ACF $(\tau_e)_{min}$. Different symbols represent experimental results using different sound sources

in Fig. 10.9a. In this calculation, the reverberation time is assumed to be 1.8 s throughout the hall and the most preferred listening level, $[LL]_p = 20\log[P]_p$ in (10.10), is set for a point on the center line 20 m from the source position.

10.1.4 Auditory Temporal Window for ACF and IACF Processing

Auditory Temporal Window for ACF Processing
In analyzing the running ACF, the so-called *auditory–temporal window* $2T$ must be carefully determined. The initial part of the ACF within the effective duration of the ACF contains important information about the source signal. In order to determine the auditory–temporal window, successive loudness judgments in pursuit of the running SPL have been conducted. Results shown in Fig. 10.10 indicate that the recommended signal duration $(2T)_r$ to be analyzed is approximately expressed by [10.46]

$$(2T)_r \approx 30(\tau_e)_{min}, \quad (10.14)$$

where $(\tau_e)_{min}$ is the minimum effective duration, extracted from the running ACF [10.47]. This signifies an adaptive temporal window depending on the temporal characteristics of the sound signal in the auditory system. For example, the temporal window may differ according to music pieces $[(2T)_r = 0.5–5\,\text{s}]$ and to the vowels $[(2T)_r = 50–100\,\text{ms}]$ and consonants $[(2T)_r = 5–10\,\text{ms}]$ in continuous speech signals. It is worth noticing that the time constant represented by "fast" or "slow" of the sound level meter should be replaced by the temporal window, which depends on the effective duration of the ACF of the source signal. The running step (R_s), which signifies a degree of overlap of signal to be analyzed, is not critical. It may be selected as $K_2(2T)_r$; K_2 is chosen, say, in the range 0.25–0.5.

Auditory Temporal Window for IACF Processing
For the sound source fixed on the stage in a concert hall as usual, the value of $2T$ can be selected longer than 1.0 s for the measurement of the spatial factor at a fixed audience seat. But, when a sound signal is moving in the horizontal direction on the stage, we must know a suitable temporal window for $2T$ in analyzing the running IACF, which describes the moving image of sound localization. For a sound source moving sinusoidally in the horizontal plane with less than 0.2 Hz, $2T$ may be selected in a wide range from 30 to 1000 ms. If a sound source is moving below and/or at 4.0 Hz, $2T = 30–100\,\text{ms}$ is acceptable. In order to obtain a reliable result, it is recommended that such a temporal window for the IACF covering a wide range of movement velocity in the horizontal localization be fixed at about 30 ms.

10.1.5 Specialization of Cerebral Hemispheres for Temporal and Spatial Factors of the Sound Field

The independent influence of the aforementioned temporal and spatial factors on subjective preference judgments has been achieved by the specialization of the human cerebral hemispheres [10.44]. Recording over the left and right hemispheres of the slow vertex response (SVR) with latency of less than 500 ms, electroencephalograms (EEG) and magnetoencephalograms (MEG) have revealed various pieces of evidence (Table 10.3), with the most significant results being:

1. The left and right amplitudes of the evoked SVR, $A(P_1 - N_1)$ indicate that the left and right hemispheric dominance are due to temporal factors (Δt_1) and spatial factors (LL and IACC), respectively [10.48, 49].
2. Both left and right latencies of N_2 of SVR correspond well to the IACC [10.49].
3. Results of EEGs for the cerebral hemispheric specialization of the temporal factors, i.e., Δt_1 and T_{sub} indicated left-hemisphere dominance [10.50, 51], while the IACC indicated right-hemisphere dominance [10.45]. Thus, a high degree of independence between temporal and spatial factors was indicated.
4. The scale value of subjective preference is well described in relation to the value of τ_e extracted from the ACF of α-wave signals over the left hemisphere and the right hemisphere according to changes in temporal and spatial factors of the sound field, respectively [10.45, 50, 51].
5. Amplitudes of MEGs recorded when Δt_1 was changed reconfirm the left-hemisphere specialization [10.52].
6. The scale values of individual subjective preference relate directly to the value of τ_e extracted from the ACF of the α-wave of the MEG [10.52]. It is worth noting that the amplitudes of the α-wave in both the EEG and the MEG do not correspond well to the scale value of subjective preference.

In addition to the aforementioned activities in the time domain, in both the left and right hemispheres, spatial activity waves were analyzed by the cross-correlation function of alpha waves from the EEGs and MEGs. The results showed that a larger area of the brain is activated when the preferred sound field is presented [10.53] than when a less preferred one. This implies that the brain repeats a similar temporal rhythm in the α-wave range over a wider area of the scalp under the preferred conditions.

It has been reported that the left hemisphere is mainly associated with speech and time-sequential identifications, and the right is concerned with nonverbal and spatial identification [10.54, 55]. However, when the IACC was changed using speech and music signals, right-hemisphere dominance was observed, as indicated in Table 10.3. Therefore, hemispheric dominance is a relative response depending on which factor is changed in the comparison pair, and no absolute behavior could be observed.

To date, it has been discovered that the LL and the IACC are dominantly associated with the right cerebral hemisphere, and the temporal factors, Δt_1 and T_{sub} are associated with the left. This implies that such specialization of the human cerebral hemisphere may relate to the highly independent influence of spatial and temporal criteria on any subjective attribute. It is remarkable, for example, that *cocktail party effects* might well be explained by such specialization of the human brain, because speech is processed in the left hemisphere, and spatial information is processed in the right hemisphere independently.

10.2 Design Studies

Using the scale values in the four orthogonal factors of the sound field obtained by a number of listeners, the *principle of superposition* expressed by (10.8) together with (10.9) through (10.13) can be applied to calculate the scale value of preference for each seat. Comparison of the total preference values for different configurations of concert halls allows a designer to choose the best scheme. Temporal factors relating to its dimensions and the absorbing material on its walls are carefully determined according to the purpose of the hall in terms of a range of specific music programs (Fig. 10.6). In this section, we discuss mainly the spatial form of the hall, the magnitude of the interaural correlation function and the binaural listening level.

10.2.1 Study of a Space-Form Design by Genetic Algorithms (GA)

A large number of concert halls have been built since the time of the ancient Greeks, but only the halls with good sound quality are well liked. In order to increase the measure of success, a genetic algorithm (GA) system [10.56], a form of evolutionary computing, can be applied to the acoustic design of concert halls [10.57,58]. The GA system is applied here to generate the alternative scheme on the left-hand side of Fig. 10.11 for listeners.

Procedure

In this calculation, linear scale values of subjective preference S_1 and S_4 given by (10.9) are employed as fitting functions due to the LL and IACC, because the geometrical shape of a hall is directly affected by these spatial factors. The spatial factor for a source on the stage was calculated at a number of seating positions. For the sake of convenience, the single omnidirectional source was assumed to be at the center of the stage, 1.5 m above the stage floor. The receiving points that correspond to the ear positions were 1.1 m above the floor of the hall. The image method was used to determine the amplitudes, delay times, and directions of arrival of reflections at these receiving points. Reflections were calculated up to the second order. In fact, there was no change in the relative relationship among the factors obtained from calculations performed up to the second, third, and fourth order of reflection. The averaged values of the IACC for five music motifs (motifs A through E [10.13]) were used for the calculation.

Those hall shapes that produced greater scale values are selected as parent chromosomes. An example of the encoding of the chromosome is given in Fig. 10.12. The first bit indicated the direction of motion for the vertex.

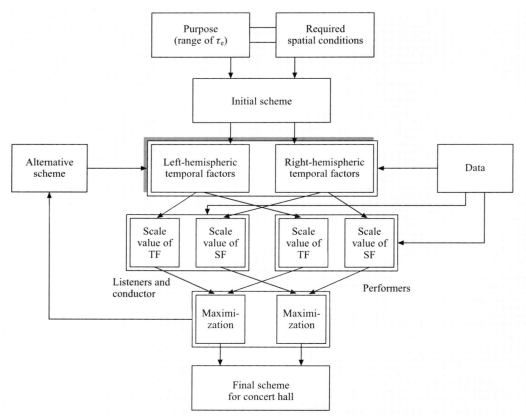

Fig. 10.11 The procedure for designing the sound field in a concert hall maximizing the scale values of subjective preference for a number of listeners (including a conductor) and performers. Data for global values of subjective preference may be utilized when designing a public hall

The other $(n-1)$ bits indicated the range over which the vertex moved. To create a new generation, the room shapes are modified and the corresponding movement of the vertices of the walls is encoded in chromosomes, i.e., binary strings. After GA operations that include crossover and mutation, new offspring are created. The fitness of the offspring is then evaluated in terms of the scale value of subjective preference. This process is repeated until the end condition of about 2000 generations is satisfied.

Shoe-Box Optimized

First of all, the proportions of the shoe-box hall were optimized (model 1). The initial geometry is shown in Fig. 10.13. In this form, the hall was 20 m wide, the stage was 12 m deep, the room was 30 m long, and the ceiling was 15 m above the floor. The point source was located at the center of the stage and 4.0 m from the front of the stage; 72 listening positions were selected. The range of motion for each sidewall and the ceiling was ±5 m from the respective initial positions, and the distance through which each was moved was coded on the chromosome of the GA. Scale values at the listening positions other than those within 1 m of the sidewalls were included in the averages ($\overline{S_1}$ and $\overline{S_4}$). These values were employed as the measure of fitness. In this calculation, the most preferred listening level $[LL]_p$ was chosen at the frontal seat near the stage. Results of optimization of the hall for $\overline{S_1}$ and $\overline{S_4}$ are shown in Fig. 10.14a and Fig. 10.14b, respectively. The width and length were almost the same in the two results, but the indicated heights were quite different. The height of the ceiling that maximizes $\overline{S_1}$ was as low as possible within the allowed range of motion to obtain a constant LL (Fig. 10.14a). The height that maximizes $\overline{S_4}$, on the other hand, was at the upper limit of the allowed range of motion to obtain small values of the IACC (Fig. 10.14b). Table 10.4 shows the comparison of the proportions obtained here and those of the Grosser Musikvereinssaal, which is a typical example of an excellent concert hall. The length/width ratios are almost the same. The height/width ratio of

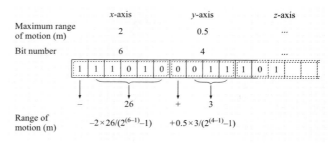

Fig. 10.12 An example of the binary strings used in encoding of the chromosome to represent modifications to the room shape

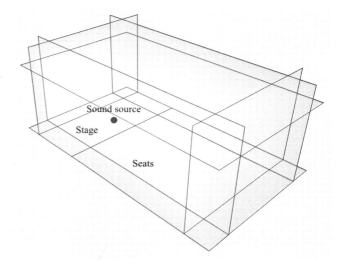

Fig. 10.13 The initial scheme of a concert hall (model 1). The range of sidewall and ceiling variation was ±5 m from the initial scheme

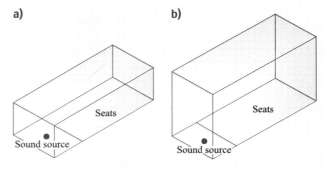

Fig. 10.14a,b Results for the model 1. (a) Geometry optimized for $\overline{S_1}$. (b) Geometry optimized for $\overline{S_4}$. (After [10.57])

Table 10.4 Comparison of proportions for the optimized spatial form of *shoe-box* type and the *Grosser Musikvereinssaal*

	Length/width	Height/width
Optimized for $\overline{S_1}$ for the listening level	2.50	0.71
Optimized for $\overline{S_4}$ for the IACC	2.57	1.43
Grosser Musikvereinssaal	2.55	0.93

the Grosser Musikvereinsaal is intermediate between our results for the two factors. For the ceiling of the hall, the height that maximized $\overline{S_1}$ was the lowest within the allowed range of motion (Fig. 10.14a). This is due to the fact that more energy can be provided from the ceiling to the listening position throughout the seats. To maximize $\overline{S_4}$, on the other hand, the ceiling took on the maximum height in the possible range of motion (Fig. 10.14b). Reflections from the flat ceiling did not decrease the IACC, but those from the sidewalls did.

Modification from the Shoe-Box

Next, to obtain even better sound fields, a little more complicated form (model 2), as shown in Fig. 10.15, was examined. The floor plan optimized according to the above results was applied as a starting point. The hall in its initial form was 14 m wide, the stage was 9 m deep, the room was 27 m long, and the ceiling was 15 m above the stage floor. The sound source was again 4.0 m from the front of the stage, but was 0.5 m to one side of the centerline and 1.5 m above the stage floor. The front and rear walls were vertically bisected to obtain two faces, and each stretch of wall along the side of the seating area was divided into four faces. The walls were kept vertical (i. e., tilting was not allowed) to examine only the plan of the hall in terms of maximizing $\overline{S_1}$ and $\overline{S_4}$. Each wall was moved independently of the other walls. The openings between the walls, in the acoustical simulation using the image method, were assumed not to reflect the sound. Forty-nine listening positions distributing throughout the seating area on a 2 m×4 m grid were selected. In the GA operation, the sidewalls were moved, so that none of these 49 listening positions were excluded. The moving range of each vertex was ±2 m in the direction of the line normal to the surface. The coordinates of the two bottom vertices of each surface were encoded on the chromosomes for the GA. In this calculation, the most preferred listening level was set for a point on the hall's long axis (central line), 10 m from the source position.

Leaf-Shape Concert Hall

The result of optimizing the hall for $\overline{S_1}$ is shown in Fig. 10.16 and the contour lines of equal $\overline{S_1}$ values are shown in Fig. 10.17. To maximize $\overline{S_1}$, the rear wall of the stage and the rear wall of the audience area took on concave shapes. The result of optimizing for $\overline{S_4}$ is shown in Fig. 10.18 and contour lines of equal $\overline{S_4}$ values are shown in Fig. 10.19. To maximize $\overline{S_4}$, on the other hand, the rear wall of the stage and the rear wall of the audience area took on convex shapes. With regard to the sidewalls, both $\overline{S_1}$ and $\overline{S_4}$ were maximized by the *leaf-shaped* plan, which is discussed in the following section.

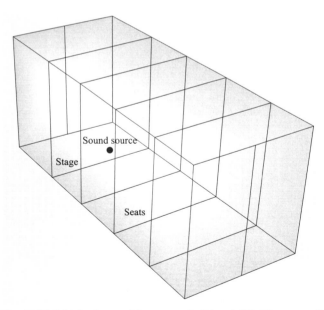

Fig. 10.15 Initial scheme of the concert hall (model 2). The rear wall of the stage and the rear wall of the audience area were divided into two. Sidewalls were divided into four

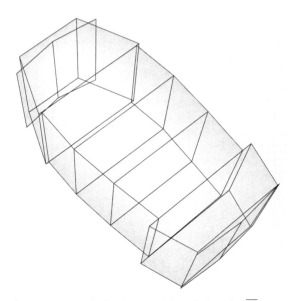

Fig. 10.16 A result for the model 2 optimized for $\overline{S_1}$

As for the conflicting requirements for S_1 and S_4, the maximization of S_4 may take priority over that of S_1, because the preference increases with decreasing IACC without exception [10.1], while there is a large individual difference in the preferred LL [10.59]. It is worth noting that listeners themselves can usually choose the best seat with respect to the preferred LL in a real concert hall.

A conductor and/or music director must be aware of the sound field characteristics of a concert hall. One is then able to select a program of music so that the sound is best in that hall in terms of the temporal factors (Fig. 10.6).

10.2.2 Actual Design Studies

After testing more than 200 listeners, a small value of the IACC, which corresponds to different sound signals arriving at two ears, was demonstrated to be the preferred condition for individuals without exception. A practical application of this design theory was done in the Kirishima International Concert Hall (Miyama Conceru), which was characterized by a *leaf shape* (Fig. 10.20).

Temporal Factors of the Sound Field for Listeners

When the space is designed for pipe-organ performance, the range of $(\tau_e)_{\min}$, which may be selected to be centered on 200 ms, determines the typical temporal factor of the hall: $[T_{\text{sub}}]_p \approx 4.6\,\text{s}$ (10.6). When designed for the performance of chamber music, the range is selected to be near the value of 65 ms ($[T_{\text{sub}}]_p \approx 1.5\,\text{s}$). The conductor and/or the sound coordinator select suitable musical motifs with a satisfactory range of effective duration of the ACF to achieve a music performance that blends the music and the sound field in a hall (Fig. 10.6). To adjust the preferred condition of Δt_1, on the other hand, since the value of $(\tau_e)_{\min}$ for violins is usually shorter than that of contrabasses in the low-frequency range, the position of the violins is shifted closer to the left wall on the stage, and the position of the contrabasses is shifted closer to the center, as viewed from the audience.

Spatial Factors of the Sound Field for Listeners

The IACC should be kept as small as possible, maintaining $\tau_{\text{IACC}} = 0$. This is realized by suppressing the strong reflection from the ceiling, and by appropriate reflections from the sidewall at certain angles. When the source signal mainly contains frequency compo-

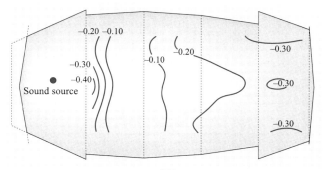

Fig. 10.17 Contour lines of equal $\overline{S_1}$ values calculated for the geometry shown in Fig. 10.16

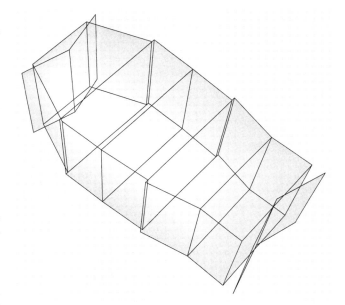

Fig. 10.18 A result for model 2 optimized for $\overline{S_4}$

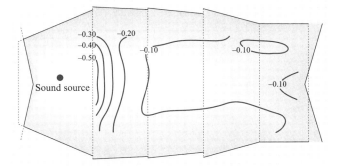

Fig. 10.19 Contour lines of equal $\overline{S_4}$ values calculated for the geometry shown in Fig. 10.18

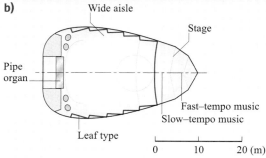

Fig. 10.20a,b A leaf shape for the plan proposed for the Kirishima International Concert Hall. (**a**) Original leaf shape. (**b**) Proposed shape for the plan. As usual, the sound field in *circled seating area* close to the stage must be carefully designed to obtain reflections from the walls on the stage and tilted sidewalls

threshold level. This may be one of the reasons why a more diffuse sound field can be perceived with increasing power of the sound source. When the source is weak enough, that only the direct sound is heard, the actual IACC being processed in the auditory–brain system approaches unity, resulting in no diffuse sound impression. Thus, the IACC should be small enough with only strong early reflections.

Sound Field for Musicians

For music performers, the temporal factor is considered to be much more critical than the spatial factor (Sect. 10.3.2). Since musicians perform over a sequence of time, reflections with a suitable delay in terms of the value of $(\tau_e)_{min}$ of the source signals are of particular importance. Without any spatial subjective diffuseness, the preferred directions of reflections are in the median plane of music performers, resulting in IACC $\approx$ 1.0 [10.62, 63]. In order to satisfy these acoustic conditions, some design iterations are required, maximizing scale values for both musicians and listeners and leading to the final scheme of the concert hall as shown in Fig. 10.11.

Sound Field for the Conductor

It is recommended that the sound field for the conductor on the stage should be designed as that of a listener with appropriate reflections from the sidewalls on the stage [10.64].

Acoustic Design with Architects

From the historical viewpoint, architects have been more concerned with spatial criteria from the visual standpoint, and less so from the point of view of temporal criteria for blending human experience and the environment with design. On the other hand, acousticians

nents around 1 kHz, the reflection from the side walls is adjusted to be centered roughly 55° to each listener, measured from the median plane. Under actual hearing conditions, the perceived IACC depends on whether or not the amplitudes of reflection exceed the hearing

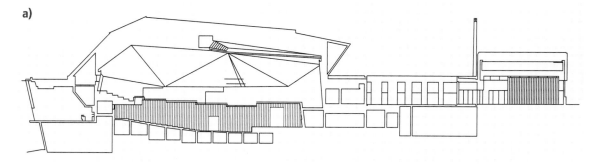

Fig. 10.21a–f Scheme of the Kirishima International Concert Hall, Kagoshima, Japan designed by the architect Maki and associates (1997) [10.60, 61]. (**a**) Longitudinal section;

have mainly been concerned with temporal criteria, represented primarily by the reverberation time, from the time of *Sabine* [10.2] onward. No comprehensive theory of design including the spatial criterion as represented by the IACC existed before 1977, so that discussions between acousticians and architects were rarely on the same subject. As a matter of fact, both temporal and spatial factors are deeply interrelated with both acoustic design and architectural design [10.65, 66].

As an initial design sketch of the Kirishima International Concert Hall, a plan shape like a leaf (Fig. 10.20a) was presented at the first meeting for discussion with the architect Fumihiko Maki and associates with the explanation of the temporal and spatial factors of sound field.

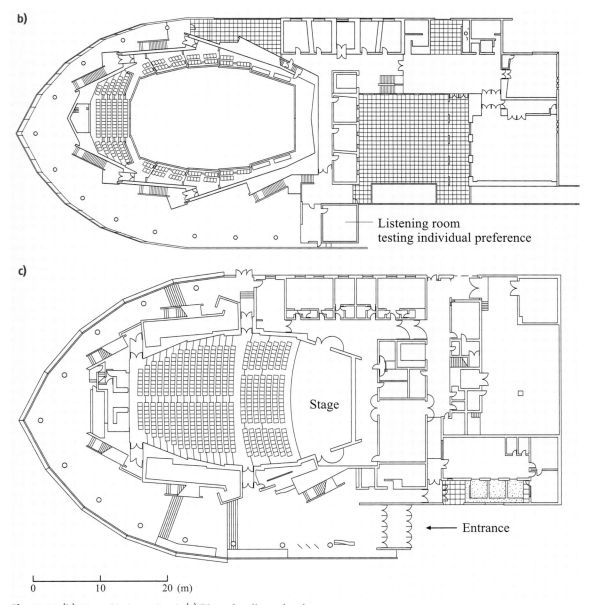

Fig. 10.21 (b) Plan of balcony level; **(c)** Plan of audience level

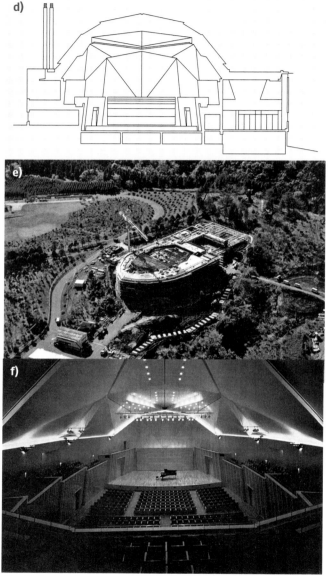

Fig. 10.21 (d) Cross section; **(e)** The Kirishima International Concert Hall, Kagoshima under construction. The leaf shape can be seen; **(f)** Tilt sidewalls and triangular ceilings after construction of the hall

The final architectural schemes, together with the special listening room for testing individual preference of sound field and selecting the appropriate seats for maximizing individual preference of the sound field, are shown in Fig. 10.21b. In these figures, the concert courtyard, the small concert hall, several rehearsal rooms and dressing rooms are also shown. The concert hall under construction is shown in Fig. 10.21e, in which the leaf shape may be seen; it was opened in 1994 (Fig. 10.21f).

Details of Acoustic Design

For listeners on the main floor. In order to obtain a small value of the IACC for most of the listeners, ceilings were designed using a number of triangular plates with adjusted angles, and the side walls were given a 10% tilt with respect to the main audience floor, as are shown in Fig. 10.21d and Fig. 10.23. In addition, diffusing elements were designed on the sidewalls to avoid the image shift of sound sources on the stage caused by the strong reflection in the high-frequency range above 2 kHz. These diffusers on the sidewalls were designed as a modification of the Schroeder diffuser [10.69] without the wells, as shown by the detail of Fig. 10.24.

For music performers on stage. In order to provide reflections from places near the median plane of each of the performers on the stage, the back wall on the stage is carefully designed as shown in the lower left in Fig. 10.23. The tilted back wall consists of six sub-walls with angles adjusted to provide appropriate reflections within the median plane of the performer. It is worth noting that the tilted sidewalls on the stage provide good reflections to the audience sitting close to the stage, at the same time resulting in a decrease of the IACC. Also, the sidewall on the stage of the left-hand side looking from the audience may provide useful reflections arriving from the back for a piano soloist.

Stage floor structure. For the purpose of suppressing the vibration [10.70] of the stage floor and anomalous sound radiation from the stage floor during a performance, the joists form triangles without any neighboring parallel structure, as shown in Fig. 10.25. The thickness of the floor is designed to be relatively thin (27 mm) in order to radiate sound effectively from the vibration of instruments such as the cello and contrabass. During rehearsal, music performers may control the radiation power somewhat by adjusting their position or by the use of a rubber pad between the floor and the instrument.

After some weeks, Maki and Ikeda indicated a scheme of the concert hall as shown in Fig. 10.21 [10.60, 61]. Without any change of plan and cross sections, the calculated results indicated the excellent sound field as shown in Fig. 10.22 [10.67, 68].

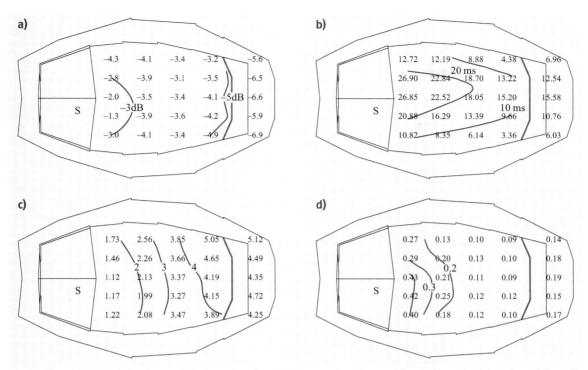

Fig. 10.22a–d Calculated orthogonal factors at each seat with a performing position S for an initial design of the hall. In the final design of the hall, the width was enlarged by about 1 m to increase the number of seats. The designed reverberation time was about 1.7 s for the 500 Hz band. (**a**) Relative listening level; (**b**) initial time delay gap between the direct sound and the first reflection Δt_1 [ms]; (**c**) A-value, the total amplitude of reflections; (**d**) IACC for white noise

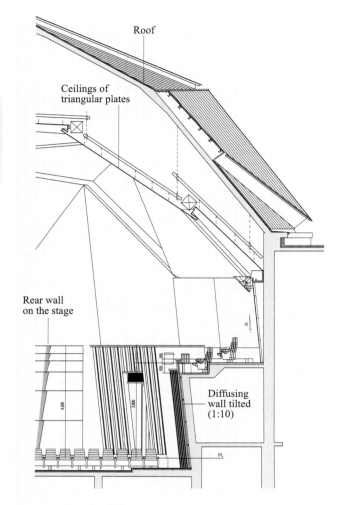

Fig. 10.23 Details of the cross section, including a sectional detail of the rear wall on the stage at the lower left of the figure

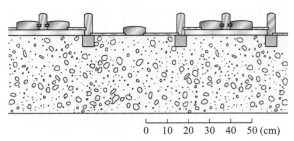

Fig. 10.24 Detail of the diffusing sidewalls effective for the higher-frequency range above 1.5 kHz, avoiding image shift of the sound source on the stage. The surface is deformed from the Schroeder diffuser by removal of the well partitions. (After [10.69])

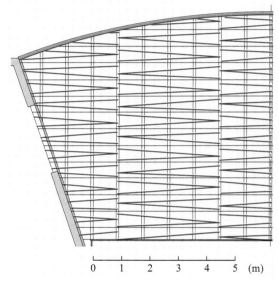

Fig. 10.25 Detail of the triangular joist arrangement for the stage floor, avoiding anomalous radiation due to the normal modes of vibration from certain music instruments that touch the floor

10.3 Individual Preferences of a Listener and a Performer

The minimum unit of society to be satisfied by the environment is one individual, which leads to a unique personal existence. Here, we demonstrate that the individual subjective preferences of each listener and each cellist may be described by the theory in Sect. 10.1, which resulted from observing a number of subjects.

10.3.1 Individual Subjective Preference of Each Listener

In order to enhance individual satisfaction for each listener, a special facility for seat selection, testing each listener's own subjective preference [10.59, 71], was first introduced at the Kirishima International Concert Hall in 1994. The sound simulation system employed multiple loudspeakers. It used arrows for testing the subjective preference of four listeners at the same time. Since the four orthogonal factors of the sound field influence the preference judgment almost independently [10.1], each single factor is varied, while the other three are fixed at the preferred condition for the average listener. Results of testing acousticians who participated in the International Symposium on Music and Concert Hall Acoustics (MCHA95), which was held in Kirishima in May 1995, are presented here [10.1].

Individual Preference and Seat Selection

The music source was orchestral, the *Water Music* by Händel; the effective duration of the ACF was 62 ms [10.11]. The total number of listeners participating was 106. Typical examples of the test results for listener BL as a function of each factor are shown in Fig. 10.26. Scale values of this listener were rather close to the averages for subjects previously collected: the most preferred $[LL]_p$ was 83 dBA, the value

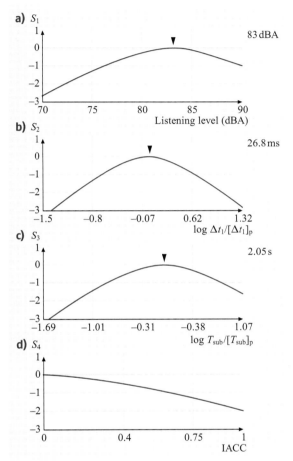

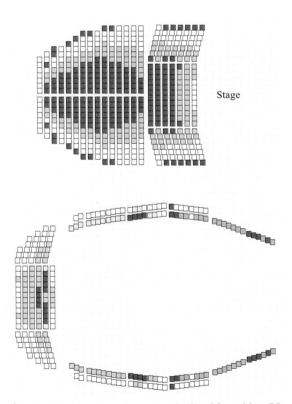

Fig. 10.26a–d Scale values of preference obtained by paired-comparison tests for the four orthogonal factors, subject BL. (**a**) The most preferred listening level was 83 dBA, the individual weighting coefficient in (10.9): $\alpha_1 = 0.06$; (**b**) the preferred initial time delay gap between the direct sound and first reflection was 26.8 ms, the individual weighting coefficient in (10.9): $\alpha_2 = 1.86$, where $[\Delta t_1]_p$ calculated by (10.4) with $(\tau_e)_{min} = 62$ ms for the music used ($A = 4$) is 24.8 ms; (**c**) the preferred subsequent reverberation time is 2.05 s, the individual weighting coefficient in (10.9): $\alpha_3 = 1.46$, where $[T_{sub}]_p$ calculated by (10.6) with $(\tau_e)_{min} = 62$ ms for the music used, is 1.43 s; (**d**) individual weighting coefficient in (10.9) for IACC: $\alpha_4 = 1.96$.

Fig. 10.27 Preferred seating area calculated for subject BL. The seats are classified in three parts according to the scale value of subjective preference calculated by the summation of S_1–S_4 (10.8) together with (10.9). Black indicates preferred seating areas, about one third of all seats in this concert hall, for subject BL

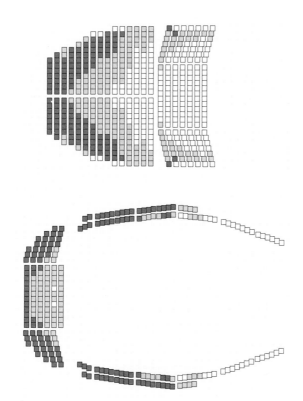

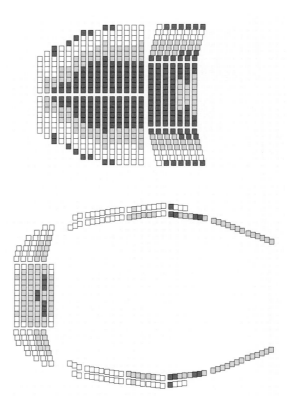

Fig. 10.28 Preferred seating area calculated for subject KH

Fig. 10.29 Preferred seating area calculated for subject KK

$[\Delta t_1]_p \approx [(1 - \log_{10} A)(\tau_e)_{\min}]$ was 26.8 ms (the global preferred value calculated by (10.4) with the total sound pressure as was simulated $A = 4.0$ is 24.8 ms. And the most preferred reverberation time is 2.05 s [the global preferred value calculated by (10.6) is 1.43 s]. Thus, as

was designed, the center area of seats was preferred for listener BL (Fig. 10.27). With regard to the IACC, the result for all listeners was that the scale value of prefer-

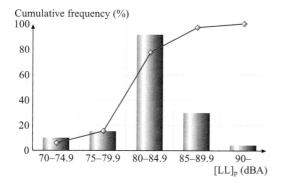

Fig. 10.30 Cumulative frequency of preferred listening level $[LL]_p$ (106 subjects). About 60% of subjects preferred the range of 80–84.9 dBA

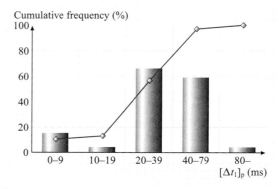

Fig. 10.31 Cumulative frequency of the preferred initial time delay gap between the direct sound and the first reflection $[\Delta t_1]_p$ (106 subjects). About 45% of subjects preferred the range of 20–39 ms. The value of $[\Delta t_1]_p$ calculated using (10.4) is 24.8 ms

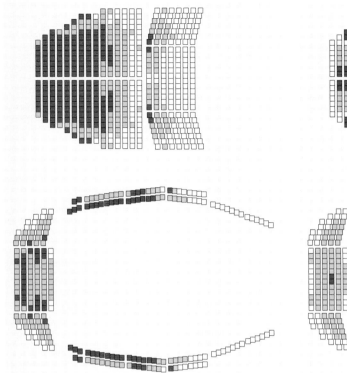

Fig. 10.32 Preferred seating area calculated for subject DP

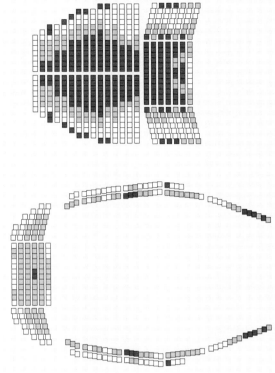

Fig. 10.33 Preferred seating area calculated for subject CA

ence increased with decreasing the IACC value. Since listener KH preferred a very short delay time of Δt_1, his preferred seats were located close to the boundary wall, as shown in Fig. 10.28. Listener KK indicated a preferred listening level exceeding 90 dBA. For this listener, the front seating area close to the stage was preferable, as shown in Fig. 10.29. For listener DP, whose preferred listening level was rather weak (76.0 dBA) and the preferred initial delay time was short (15.0 ms), the preferred seat was in the rear part of hall, as shown in Fig. 10.32. The preferred initial time delay gap for listener CA exceeds 100.0 ms, but was not critical. Thus, any initial delay times were acceptable, but the IACC was critical. Therefore, the preferred areas of seats were located as shown in Fig. 10.33.

Cumulative Frequency of Preferred Values

Cumulative frequencies of the preferred values with 106 listeners are shown in Fig. 10.30 through Fig. 10.34 for three factors. As indicated in Fig. 10.30, about 60% of listeners preferred the range 80–84.9 dBA when listening to music, but some listeners indicated that the most preferred LL was above 90 dBA, and the total range of the preferred LL was scattered, exceeding a 20 dB range. As shown in Fig. 10.31, about 45% of listeners preferred initial delay times of 20–39 ms, which were

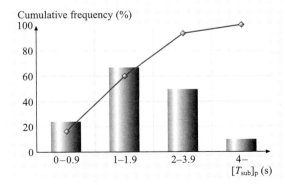

Fig. 10.34 Cumulative frequency of the preferred subsequent reverberation time $[T_{sub}]_p$ (106 subjects). About 45% of subjects preferred the range 1.0–1.9 s. The value of $[T_{sub}]_p$ calculated using (10.6) is 1.43 s

around the calculated preference of 24.8 ms (10.4) with $k = 0.1$, $c = 1$ and $A = 4.0$; however, some listeners indicated 0–9 ms and others more than 80 ms. With regard to the reverberation time, as shown in Fig. 10.34, about 45% of listeners preferred 1.0–1.9 s, which is centered on the preferred value of 1.43 s calculated by (10.6), but some listeners indicated preferences lower than 0.9 s or more than 4.0 s.

Independence of the Preferred Conditions

It was thought that both the initial delay time and the subsequent reverberation time appear to be mutually related, due to a kind of *liveness* of the sound field. Also, it was assumed that there is a strong interdependence between these factors for each individual. However, as shown in Fig. 10.35, there was little correlation between preference values of $[\Delta t_1]_p$ and $[T_{sub}]_p$ (the correlation is 0.06). The same is true for the correlation between values of $[T_{sub}]_p$ and $[LL]_p$ and for that between values of $[LL]_p$ and $[\Delta t_1]_p$, a correlation of less than 0.11. Figure 10.36 shows the three-dimensional plots of the preferred values of $[LL]_p$, $[\Delta t_1]_p$ and $[T_{sub}]_p$ for the 106 listeners. Looking at a continuous distribution in preferred values, no specific groupings of individuals can be classified from the data.

Another important fact is that there was no correlation between the weighting coefficients α_i and α_j, $i \neq j$ (i, $j = 1, 2, 3, 4$) in (10.9) for each individual listener [10.1]. A listener indicating a relatively small value of one factor will not always indicate a relatively small value for another factor. Thus, a listener can be critical about a preferred condition as a function of a certain factor, while insensitive to other factors, resulting in characteristic differences from other listeners.

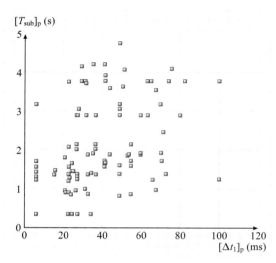

Fig. 10.35 The relationship between the preferred values of $[\Delta t_1]_p$ and $[T_{sub}]_p$ for each subject. No significant correlation between values was achieved

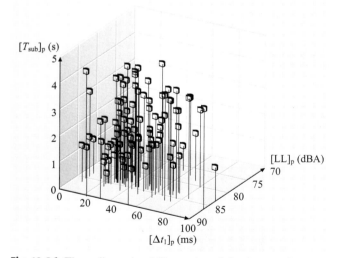

Fig. 10.36 Three-dimensional illustration of the three preferred orthogonal factors for the sound field for each individual subject (*squares*). All 106 listeners tested indicated that a smaller value of the IACC was preferred, and these data are not included in this figure. Preferred conditions are distributed in a certain range of each factor, so that subjects could not be classified into any specific groups

10.3.2 Individual Subjective Preference of Each Cellist

To realize an excellent concert, we need to know the optimal conditions not only in the stage enclosure design for performers, but also in the style of the performance. The primary issue is that the stage enclosure is designed to provide a sound field in which performers can play easily. *Marshall* et al. [10.72] investigated the effects of stage size on the playing of an ensemble. The parameters related to stage size in their study were the delay time and the amplitude of reflections. *Gade* [10.73] performed a laboratory experiment to investigate the preferred conditions for the total amplitude of the reflections of performers. *Nakayama* [10.62] showed a relationship between the preferred delay time and the effective duration of the long-time ACF of the source signal for alto-recorder soloists (data was rearranged [10.1]). When we listen to a wide range of music signals containing a large fluctuation, it is more accurately expressed by the minimum value of the effective duration $(\tau_e)_{min}$ of the running ACF of the source signals [10.39]. For individual singers, *Noson* et al. [10.74, 75] reported that the most preferred condition of the single reflection for an individual singer may be described by $(\tau_e)_{min}$ and a modified amplitude of reflection according to the overestimate and bone conduction effects (for control of $(\tau_e)_{min}$ see [10.41, 42]).

Preferred Delay Time of the Single Reflection for Cellists

As a good example, the preferred delay time of the single reflection for individual cello soloist is described by the minimum value of the effective duration of the running ACF of the music motifs played by that cellist [10.76]. The same music motifs (motifs I and II) used in the experiment by Nakayama were applied here [10.62]. The tempo of motif I was faster than that of motif II, as shown in Fig. 10.37. A microphone in front of the cellist picked up the music signals performed three times by each of five cellists.

Figure 10.38 shows an example of the regression curve for the scale value of preference fitted by (10.9). The peak of this curve denotes the most preferred delay time $[\Delta t_1]_p$. The values for individual cellists are listed in Table 10.5. Global and individual results (except for that of subject E) for music motif II were longer than those for music motif I.

The most preferred delay time of a single reflection is approximately expressed by the duration τ'_p of the ACF as similar to that of listeners (10.4), so that

$$[\Delta t_1]_p = \tau'_p \approx \left[\log_{10}(1/k') - c' \log_{10} A'\right](\tau_e)_{min}, \tag{10.15}$$

where the values k' and c' are constants that depend on the individual performer and musical instrument used. A substantial difference from (10.4) of listeners is that the amplitude of the reflection A' is defined by $A' = 1$ relative to -10 dB of the direct sound as measured at the ear's entrance. This is due to the phenomenon of *missing reflection* (i.e., a performer overestimating the reflection) [10.1].

Individual Subjective Preference

Using the quasi-Newton method, the resulting constants on average were $k' \approx 1/2$ and $c' \approx 1$ for the five cellists. (It is worth noting that the coefficients k' and c' for alto-recorder soloists were 2/3 and 1/4, respectively [10.1].) After setting $k' = 1/2$, the coefficient c' for each individual was figured out as listed in Table 10.6. The average value of the coefficient c' for the five cellists

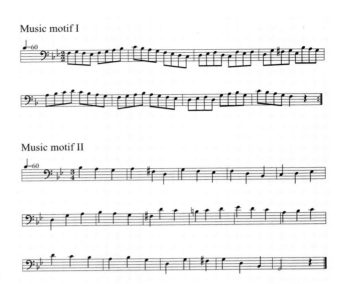

Fig. 10.37 Music motifs II and I composed by Tsuneko Okamoto for the investigation [10.62]

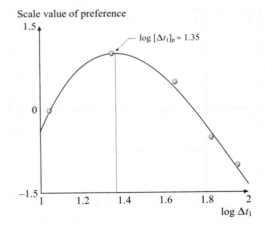

Fig. 10.38 An example of the regression curve for the preferred delay time (subject D, music motif I, -15 dB), $\log_{10}[\Delta t_1]_p \approx 1.35$, thus $[\Delta t_1]_p \approx 22.6$ ms

obtained was 1.03. The relationship between the most preferred delay time $[\Delta t_1]_p$ obtained by the judgment and the duration τ'_p ($= [\Delta t_1]_p$) of the ACF calculated by (10.15) is shown in Fig. 10.40. Different symbols indicate values obtained in different test series with two music motifs. The correlation coefficient between calculated values of $[\Delta t_1]_p$ and measured values is 0.91 ($p < 0.01$). The scale values of preference for each of the five cellists as a function of the delay time of the sin-

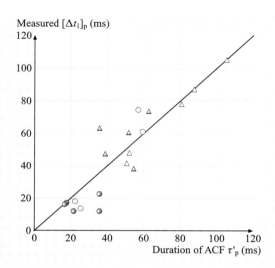

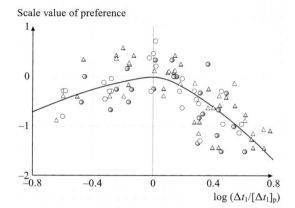

gle reflection normalized by the calculated $[\Delta t_1]_p$ are shown in Fig. 10.40.

Fig. 10.39 The relationship between the most preferred delay time $[\Delta t_1]_p$ measured and the duration $[\Delta t_1]_p = \tau'_p$ calculated using (10.15). Correlation coefficient, $r = 0.91$ ($p < 0.01$). ●: music motif I, -15 dB; ○: music motif I, -21 dB; ▲: music motif II, -15 dB; △: music motif II, -21 dB

Fig. 10.40 Scale values of preference for each of five cellists as a function of the delay time of the single reflection normalized by its most preferred delay time calculated by (10.15). ●: music motif I, -15 dB; ○: music motif I, -21 dB; ▲: music motif II, -15 dB; △: music motif II, -21 dB. The regression curve is expressed by (10.9), $i = 2$

Table 10.5 Judged and calculated preferred delay times of a single reflection for each cello soloist. Calculated values of $[\Delta t_1]_p$ are obtained by (10.15) using the amplitude of the reflection A'_1 and $(\tau_e)_{min}$ for music signals performed by each cellist

A(dB)	A'(dB) ($= A + 10$)	A'	Cellist	Judged $[\Delta t_1]_p$(ms)		Calculated $[\Delta t_1]_p$(ms)	
				Motif I	Motif II	Motif I	Motif II
-15	-5	0.56	A	16.2	47.9	16.3	38.5
			B	<12.0	73.8	35.2	62.7
			C	<12.0	60.8	211.3	51.3
			D	22.6	38.2	35.1	53.9
			E	17.6	63.6	17.3	35.2
			Global	18.0	48.3	24.3	47.5
-21	-11	0.28	A	18.1	48.4	21.8	51.5
			B	61.2	105.0	59.3	105.6
			C	–	77.9	–	80.6
			D	74.6	86.8	56.9	87.4
			E	<14.0	42.2	24.8	50.2
			Global	30.4	71.8	37.6	73.4

Table 10.6 The coefficient c' for each cellist in (10.15), calculating the preferred delay time of reflection for individual results and the global result, with $k' = 1/2$ (fixed)

	Cellist					Averaged
	A	B	C	D	E	(global)
Coefficient c'	0.47	1.61	1.10	1.30	0.67	≈ 1.03

Subjective Responses as a Function of $[\Delta t_1]_p / (\tau_e)_{min}$

Figure 10.41 shows the relative amplitude of the single reflection to that of the direct sound for the preference of cello soloists as a function of the delay time of the single reflection normalized by the minimum value of the effective duration $(\tau_e)_{min}$ of the running ACF of the source signal. Several other subjective responses in terms of the amplitude are shown together as a function of the delay time of the single reflection normalized by the value of the effective duration τ_e of the long-time ACF. All these values can generally be expressed by (10.4) in relation to the effective duration of the ACF with the constants k and c, which depend on different subjective responses. An alto-recorder soloist's preference is also plotted in this figure [10.1]. The values for performers are below or close to the threshold of perception of listeners [10.77]. These reconfirm the phenomenon of *missing reflection* for performers.

In order to blend the source music under performance and the sound field in a given concert hall, a performer, to some extent, can control the value of $(\tau_e)_{min}$ by introducing vibrato. Such an introduction of vibrato may decease the value of $(\tau_e)_{min}$ to obtain a more preferred condition for listeners as well, even though there is a short reverberation time in a given concert hall [10.41, 42].

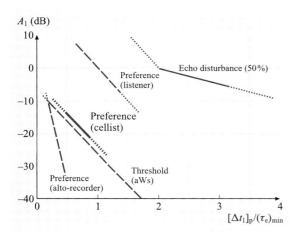

Fig. 10.41 Relative amplitude of the single reflection for the subjective preference of cello soloists as a function of the delay time of a single reflection normalized by the value of $(\tau_e)_{min}$. Also, the amplitudes of several subjective responses as a function of the delay time of the single reflection normalized by the value of τ_e of the source signal. Note that threshold has been rearranged using the typical ACF of the speech signal [10.1]

10.4 Acoustical Measurements of the Sound Fields in Rooms

Acoustical measurements were made in an existing hall for the purpose of testing acoustic factors that were calculated using the architectural scheme at the design stage. Also, subjective preference judgments for different source locations on the stage were performed by the paired-comparison test at each set of seats. The relationship between the resulting scale values of subjective preference and the physical factors obtained by simulation using architectural plan drawings was examined by factor analysis. The accumulation and understanding of the field data, in turn, may improve details of future methods for calculating acoustic factors.

Fig. 10.42 A system for measuring the four orthogonal factors of sound fields and evaluating subjective qualities at each seat in a room. TS: test signal (maximum-length sequence signal); IPR: impulse response analyzer; RH: right-hemispheric factors (listening level and the IACC); LH: left-hemispheric factors (Δt_1, T_{sub}, and the A value); CP: comparators with the most preferred condition based on the effective duration of ACF, $(\tau_e)_{min}$; ACF: autocorrelation function; SIG: source signals; $g_r(x)$: scale values from the right-hemispheric factors; $g_l(x)$: scale values from the left-hemispheric factors; Σ: total scale value of subjective preference

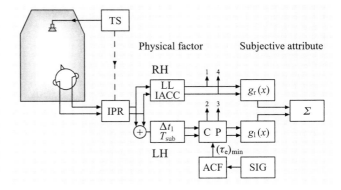

10.4.1 Acoustic Test Techniques

Binaural Impulse Response

A diagnostic system for measuring the impulse response at the two ear entrances determining the four orthogonal factors, and for further evaluation of the subjective attributes of sound field at each seat in a hall is shown in Fig. 10.42. A test signal is radiated from the loudspeaker to measure impulse responses using two small microphones placed at the ear entrances of a real head (1.1 m above the floor). Then spatial factors associated with the right-hemisphere specialization (LL and IACC) and the temporal factors of left-hemisphere specialization (Δt_1 and T_{sub}) are analyzed. When the effective duration of the ACF of the source signal (τ_e) is calculated, the total scale value may be obtained by adding the scale values of the orthogonal factors referred to the most preferred conditions. The value of $(\tau_e)_{min}$ is used to determine the most preferred temporal values for $[\Delta t_1]_p$ and $[T_{sub}]_p$ (Sect. 10.1.2). If the source signal is fed into the ACF processor, then outputs (numbered 1–4) may be used to control the sound field simultaneously with an electroacoustic system, without any manual adjustment, preserving the preferred conditions of the four factors.

Examples of measuring binaural impulse responses at a seat close to the stage (seat a, left ear) and at a rear seat b (right ear) in the Kirishima International Concert Hall are demonstrated in Fig. 10.43. In this measurement, an omnidirectional dodecahedron loudspeaker with 12 full-range drivers was placed on the stage 1.5 m above the floor for a sound source. The total amplitude of reflections A at a seat close to the stage is usually smaller than that at seat far from the stage.

Another powerful signal to be radiated from the loudspeaker to measure the impulse binaural responses is the pulse signal generated by inverse Fourier transformation [10.78].

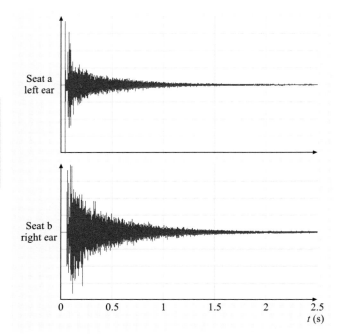

Fig. 10.43a,b Examples of the impulse response measured at seats: (**a**) (near to the stage) and (**b**) (far from the stage) in the Kirishima International Concert Hall. The amplitude of the impulse response measured at seat (**a**) is attenuated because of a strong direct sound.

Reverberation Time

After the impulse response is obtained, the reverberation time is measured by *Schroeder's* method [10.79, 80]. The integrated decay curve as a function of time may be obtained by squaring and integrating the impulse response of the sound field in a room, such that

$$\langle s^2(t) \rangle = K \int_{t+T}^{t} h^2(x)\,dx \qquad (10.16)$$

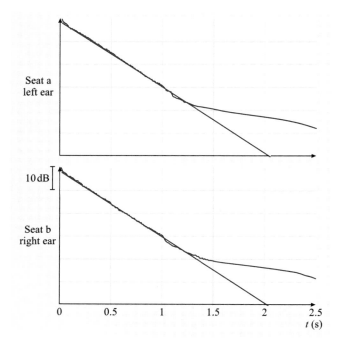

Fig. 10.44 Integrated decay curves obtained from the impulse responses at the two ear entrances, at seat (**a**) in the Kirishima International Concert Hall

where the time T should be chosen sufficiently longer than the reverberation time.

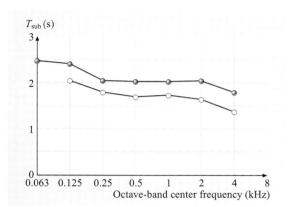

Fig. 10.45 Reverberation time measured in the Kirishima International Concert Hall. ●: measured values without audience; ○: estimated with full audience

For the 500 Hz octave band, examples of the measured decay curve and the decay rate of both left and right ears at seating position a are shown in Fig. 10.47. The reverberation times measured are both 2.07 s. The measured reverberation times with octave band filters in the Kirishima International Concert Hall (without audience) are plotted as filled circles in Fig. 10.45. The empty circles are estimated values of the reverberation time for a full audience. It is worth noticing that *Jordan* [10.81] showed that the values of the early decay time (EDT) measured over the first 10 dB of decay are close to values of the reverberation time averaged with the interval of -5 to -35 dB.

The total amplitude of reflection A, defined by (10.5), is obtained as its square

$$A^2 = \frac{\int_\infty^\varepsilon h^2(x)\,\mathrm{d}x}{\int_\varepsilon^0 h^2(x)\,\mathrm{d}x}, \tag{10.17}$$

where ε signifies a small delay time just large enough to cover the duration of the direct sound.

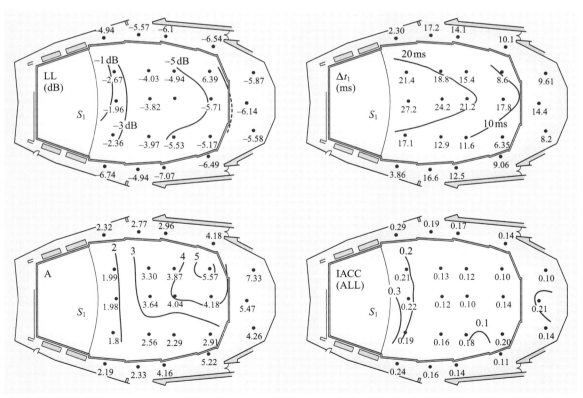

Fig. 10.46a–d Orthogonal factors measured at each seat in the Kirishima International Concert Hall, other than the reverberation time, which is almost constant throughout the hall: **(a)** listening level; **(b)** Δt_1; **(c)** A value, the total amplitude of reflections; **(d)** IACC

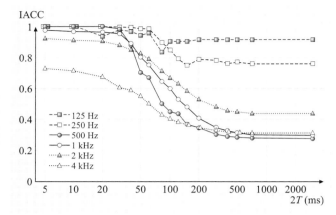

Fig. 10.47 Measured IACC as a function of the integration interval $2T$ of the impulse responses for each octave band range. The value of IACC converged for $2T > 200$ ms (After [10.82])

Measurement of Acoustic Factors at Each Seat in a Concert Hall

In Sect. 10.3.1 we discussed the seat selection system, designed for the purpose of enhancing individual satisfaction. To begin with, four orthogonal factors are measured at each seat in a concert hall [10.67, 68]. Measured values of the listening level (LL), the total amplitude of reflection (A), the initial time delay gap (Δt_1) between the direct sound and the first reflection excluding the reflection from the floor, and the IACC at each seat in the Kirishima International Concert Hall are shown in Fig. 10.46. The reverberation times at all the seats had almost the same value, about 2.05 s for the 500 Hz band.

Even though the final scheme of the concert hall was changed in terms of the width of the hall (one meter wider) from the scheme at the design stage, values of each physical factor measured as shown in Fig. 10.46 are not very different from the values calculated (Fig. 10.22).

Recommended Method for IACC Measurement

There are two purposes for measuring IACC, as needed for subjective evaluations and acoustic comparison of existing halls:

1. In order to evaluate the subjective quality of the sound field in an existing hall, the IACF (with values of IACC, τ_{IACC} and W_{IACC} defined in Fig. 10.4 as well as LL) together with the other three factors Δt_1, T_{sub} and A are measured. Without any octave band filtering, measurements must be performed after passage of the music or the speech signal through an A-weighting network, under identical conditions with subjective judgment.
2. In order to compare values of the IACC as well as T_{sub} for the sound field in existing halls, measurements with octave band filtering are performed. With a fixed sound source on the stage, the IACC is defined by a long integration interval, which includes the effects of the direct sound and all reflections, including reverberation, without any temporal subdivisions.

A typical example of measuring the IACC as a function of the integration interval, which was performed in Symphony Hall, Boston, is shown in Fig. 10.47 [10.82]. It is remarkable that the measured values of IACC almost converged for $2T \approx 200$ ms, and the values are not so different for longer intervals.

If a room is used for performing dance, ice-skating or a party, then the listeners face in various directions. In this case, the values of IACC and τ_{IACC} are measured as a function of the direction of the head. The measured results with the 500 Hz octave band noise in an oblong atrium of a hotel at the distance 10 m from the source position are demonstrated in Fig. 10.48 [10.83]. When the listener is facing the sound source, then IACC = 0.41, and $\tau_{IACC} = 0$, and thus no image shift occurs. These values are nearly unchanged for head directional angles less than 30° when the listener is facing the lateral side at 90°, then the IACC is greater than 0.50, and τ_{IACC} is about 600 μs, due to the interaural delay time.

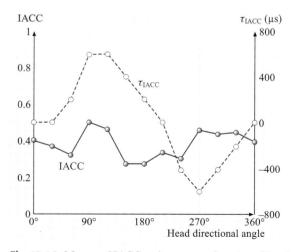

Fig. 10.48 Measured IACC and τ_{IACC} as a function of head direction relative to the sound source in a narrow hotel atrium (After [10.83])

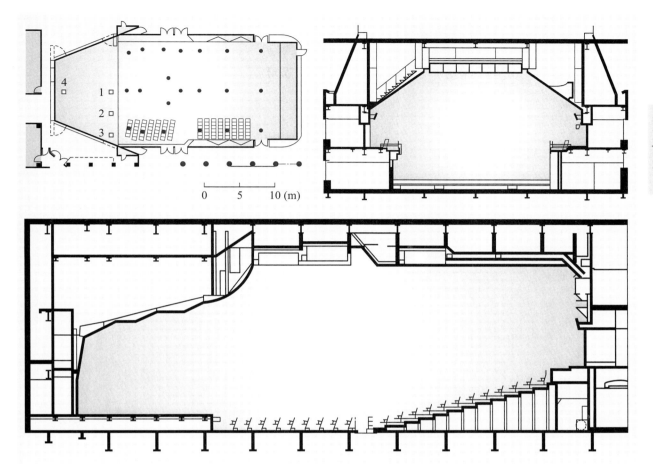

Fig. 10.49 *Left panel* plan; *right* and *bottom panel* cross sections of the Uhara Hall, Kobe. Four source positions, 1, 2, 3 and 4 on the stage, which were switched in the paired-comparison test of subjective preference without moving subjects from seat to seat. There were 21 listening positions including neighboring seats

10.4.2 Subjective Preference Test in an Existing Hall

The subjective preference judgments for different source locations on the stage at Uhara Hall, Kobe, were performed by a paired-comparison test at each set of seats. The relationship between the resulting scale value of subjective preference and the physical factors obtained by calculation, using architectural plan drawings, was examined by factor analysis [10.84]. Calculated scale values of subjective preference were reconfirmed for the Uhara Hall (Fig. 10.49). The physical factors at each set of seats for four source locations on the stage were calculated. In the simulation, the directional characteristics of the four loudspeakers used in preference tests were not taken into consideration for the sake of convenience. The simulation calculation was performed up to the third order of reflection. Due to a floor structure with a fair amount of acoustic transparency, the floor reflection was not taken into account for the calculation, and part of the diffuser ceiling was regarded as a nonreflective plane for the sake of convenience. In the calculation of the IACC, the listeners faced toward the center of the stage, so that the IACC was not always a maximum at the interaural time delay $\tau = 0$.

The hall contains 650 seats with a volume of $4870 \, \text{m}^3$. Four identical loudspeakers with the same characteristics were placed 0.8 m above the stage floor,

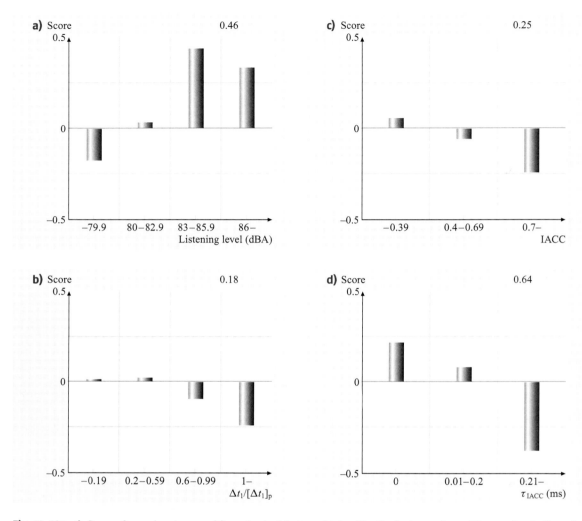

Fig. 10.50a–d Scores for each category of four physical factors obtained by the factor analyses. The number indicated at the upper left part of each figure signifies the partial correlation coefficient between the score and each factor. (**a**) Listening level; (**b**) normalized initial time delay gap between the direct sound and the first reflection; (**c**) IACC; (**d**) interaural time delay of the IACC, τ_{IACC}, found as the most significant factor in this investigation with loudspeaker reproductions on the stage. Tendencies obtained here are similar to those of the scale value shown in Fig. 10.7, which were obtained from the simulated sound field

and sixty-four listeners, divided into 21 groups, were seated in the specified set of seats. Without moving from seat to seat and excluding the effects of other physical factors such as visual and tactual senses on judgments, subjective preference tests by the paired-comparison method were conducted, switching only the loudspeakers on the stage. As a source signal, music motif B was selected in the tests. Scale values of preference were obtained by applying the law of comparative judgment and were reconfirmed by the goodness of the fit [10.85, 86]. The session was repeated five times, exchanging seats, and thus data for 14–16 subjects in total were obtained for each set of seats.

Results of Multiple-Dimensional Analyses

In order to examine the relationship between scale values of subjective preference and physical factors obtained by simulation of an architectural scheme, the data were analyzed by factor analysis [10.79, 80, 87].

Of the four orthogonal factors, the reverberation time was almost constant for the source location and the seat location throughout the hall, and thus was not involved in the analysis. As previously discussed, as a condition for calculating the scale value of preference, the maximum value of the interaural cross-correlation function must be maintained at $\tau_{IACC} = 0$ to ensure frontal localization of the sound source. However, the IACC was not always maintained at $\tau = 0$ due to the loudspeaker locations, because the subjects were facing the center of stage. In this analysis, therefore, the effect of the interaural time delay of the IACC was added as an additional factor. Thus, the outside variable to be predicted with factor analysis was the scale value obtained by subjective judgments, and the explanatory factors were: (1) the listening level, (2) the initial time delay gap, (3) the IACC, and (4) the interaural time delay (τ_{IACC}).

Scores for Each Factor

The scores for each category of the factors obtained from the factor analysis are shown in Fig. 10.50. As shown in Fig. 10.50a, the scores for the listening level indicate a peak at the subcategory of 83–85.9 dB, with decreasing scores moving away from the preferred listening level. For the IACC, the preference score increases with a decrease in the IACC (Fig. 10.50c). It is worth noting that the scores for the aforementioned two factors are in good agreement with preference scale values obtained from preference judgments for a simulated sound field (Fig. 10.7a and 10.7d). The scores of the initial time delay gap normalized to the optimum value ($\Delta t_1/[\Delta t_1]_p$) peaked at smaller values (Fig. 10.50b) than the most preferred value of the initial time delay gap obtained from the simulated sound fields (Fig. 10.7b). It is considered that, due to the limited range of the Δt_1 in the existing concert hall and the limited data for the short range of the Δt_1, the effects of the Δt_1 of the sound fields was rather minor in this investigation. Concerning τ_{IACC}, as shown in Fig. 10.50d, the score decreases monotonically as the delay is increased. This may be caused by an image shift without balancing of the sound field.

Measured and Calculated Subjective Preference Values

The relationship between the scale value obtained by subjective judgments and the total score at each center of three or four seats is shown in Fig. 10.51. The scale values of preference are well predicted with the total score for four loudspeaker locations ($r = 0.70$, $p < 0.01$). In some cases there is a certain degree of

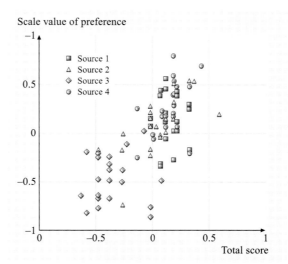

Fig. 10.51 Relationship between the scale value of subjective preference obtained by the paired-comparison test in the existing hall and the total score calculated by using the scores shown in Fig. 10.50. The correlation coefficient was $r = 0.70$ ($p < 0.01$)

apparent coherence between physical factors, for example, the calculated listening level and the IACC for sound fields in existing concert halls. However, these factors are theoretically orthogonal, and therefore the preference scores obtained were in good agreement with the calculated preference scale values obtained by the simulation.

So far the subjective preference of source locations on the stage has been examined at each set of seats. The rear source position (#4) on the stage is more preferred than that of the other source locations. The side source position (#3) indicates a low preference, due to the interaural time delay. The initial time delay gap resulted in a small influence on the total score because of its limited range.

10.4.3 Conclusions

Results of the analysis demonstrate that the theory of calculating subjective preference by the use of orthogonal parameters obtained in the laboratory is supported by experiments in a real hall. This may hold only when the maximum value of the interaural cross-correlation is maintained at $\tau = 0$. However, this condition is usually realized by introducing certain diffusing elements on the sidewalls in a real concert hall when the listeners face a performer.

References

10.1 Y. Ando: *Architectural Acoustics, Blending Sound Sources, Sound Fields, and Listeners* (AIP/Springer, New York 1998)

10.2 W.C. Sabine: *Reverberation (The American Architect and the Engineering Record, 1900), Prefaced by L.L. Beranek: Collected papers on acoustics* (Peninsula, Los Altos 1992)

10.3 H. Haas: Über den Einfluss eines Einfachechos auf die Hörsamkeit von Sprache, Acustica **1**, 49–58 (1951)

10.4 L.L. Beranek: *Music, Acoustics and Architecture* (Wiley, New York 1962)

10.5 P. Damaske: Subjektive Untersuchungen von Schallfeldern, Acustica **19**, 199–213 (1967/68)

10.6 M.V. Keet: *The influence of early lateral reflections on the spatial impression* (Proc. 6th Int. Congress of Acoustics, Tokyo 1968), Tech. Dig., paper E-2-4

10.7 A.H. Marshall: Acoustical determinants for the architectural design of concert halls, Architect. Sci. Rev. (Australia) **11**, 81–87 (1968)

10.8 M. Barron: The subjective effects of first reflections in concert halls – The need for lateral reflections, J. Sound Vibr. **15**, 475–494 (1971)

10.9 P. Damaske, Y. Ando: Interaural crosscorrelation for multichannel loudspeaker reproduction, Acustica **27**, 232–238 (1972)

10.10 M.R. Schroeder, D. Gottlob, K.F. Siebrasse: Comparative study of European concert halls: Correlation of subjective preference with geometric and acoustic parameters, J. Acoust. Soc. Am. **65**, 958–963 (1974)

10.11 Y. Ando: Subjective preference in relation to objective parameters of music sound fields with a single echo, J. Acoust. Soc. Am. **62**, 1436–1441 (1977)

10.12 Y. Ando: Calculation of subjective preference at each seat in a concert hall, J. Acoust. Soc. Am. **74**, 873–887 (1983)

10.13 Y. Ando: *Concert Hall Acoustics* (Springer, Berlin, Heidelberg 1985)

10.14 A. Cocchi, A. Farina, L. Rocco: Reliability of scale-model research: A concert hall case, Appl. Acoust. **30**, 1–13 (1990)

10.15 S. Sato, Y. Mori, Y. Ando: The subjective evaluation of source locations on the stage by listeners, music and concert hall acoustics. In: *Conf. Proc. MCHA 1995*, ed. by Y. Ando, D. Noson (Academic, London 1997) pp. 117–123, Chap. 12

10.16 T. Hotehama, S. Sato, Y. Ando: Dissimilarity judgments in relation to temporal and spatial factors for the sound fields in an existing hall, J. Sound Vibr. **258**, 429–441 (2002)

10.17 H. Sakai, P.K. Singh, Y. Ando: Inter-individual differences in subjective preference judgments of sound fields. In: *Music and Concert Hall Acoustics, Conf. Proc. MCHA 1995*, ed. by Y. Ando, D. Noson (Academic, London 1997) pp. 125–130, Chap 13

10.18 M. Sakurai, Y. Korenaga, Y. Ando: A sound simulation system for seat selection. In: *Music and Concert Hall Acoustics, Conf. Proc. MCHA 1995*, ed. by Y. Ando, D. Noson (Academic, London 1997) pp. 51–59, Chap. 6

10.19 Y. Ando: Evoked potentials relating to the subjective preference of sound fields, Acustica **76**, 292–296 (1992)

10.20 S. Sato, K. Nishio, Y. Ando: Propagation of alpha waves corresponding to subjective preference from the right hemisphere to the left with change in the IACC of a sound field, J. Temporal Des. Architect. Environ. **3**, 60–69 (2003)

10.21 Y. Soeta, S. Nakagawa, M. Tonoike, Y. Ando: Magnetoencephalographic responses corresponding to individual subjective preference of sound fields, J. Sound Vibr. **258**, 419–428 (2002)

10.22 Y. Soeta, S. Nakagawa, M. Tonoike, Y. Ando: Spatial analysis of magnetoencephalographic alpha waves in relation to subjective preference of a sound field, J. Temporal Des. Architect. Environ. **3**, 28–35 (2003)

10.23 L.A. Jeffress: A place theory of sound localization, J. Comp. Physiol. Psychol. **41**, 35–39 (1948)

10.24 J.C.R. Licklider: A duplex theory of pitch perception, Experientia **7**, 128–134 (1951)

10.25 Y. Ando, K. Yamamoto, H. Nagamastu, S.H. Kang: Auditory brainstem response (ABR) in relation to the horizontal angle of sound incidence, Acoust. Lett. **15**, 57–64 (1991)

10.26 H.E. Secker-Walker, C.L. Searle: Time domain analysis of auditory-nerve-fiber firing rates, J. Acoust. Soc. Am. **88**, 1427–1436 (1990)

10.27 P.A. Cariani, B. Delgutte: Neural correlates of the pitch of complex tones. I. Pitch and pitch salience, J. Neurophysiol. **76**, 1698–1716 (1996)

10.28 P.A. Cariani, B. Delgutte: Neural correlates of the pitch of complex tones. II. Pitch shift, pitch ambiguity, phase-invariance, pitch circularity, and the dominance region for pitch, J. Neurophysiol. **76**, 1717–1734 (1996)

10.29 M. Inoue, Y. Ando, T. Taguti: The frequency range applicable to pitch identification based upon the auto-correlation function model, J. Sound Vibr. **241**, 105–116 (2001)

10.30 S. Sato, T. Kitamura, H. Sakai, Y. Ando: The loudness of "complex noise" in relation to the factors extracted from the autocorrelation function, J. Sound Vibr. **241**, 97–103 (2001)

10.31 K. Saifuddin, T. Matsushima, Y. Ando: Duration sensation when listening to pure tone and complex tone, J. Temporal Des. Architect. Environ. **2**, 42–47 (2002)

10.32 Y. Ando, H. Sakai, S. Sato: Formulae describing subjective attributes for sound fields based on

10.33 Y. Ando: A theory of primary sensations measuring environmental noise, J. Sound Vibr. **241**, 3–18 (2001)

10.34 Y. Ando, Y. Kurihara: Nonlinear response in evaluating the subjective diffuseness of sound field, J. Acoust. Soc. Am. **80**, 833–836 (1986)

10.35 Y. Ando, S.H. Kang, H. Nagamatsu: On the auditory-evoked potentials in relation to the IACC of sound field, J. Acoust. Soc. Jpn. (E) **8**, 183–190 (1987)

10.36 Y. Ando, S. Sato, H. Sakai: Fundamental subjective attributes of sound fields based on the model of auditory – brain system. In: *Computational Acoustics in Architecture*, ed. by J.J. Sendra (WIT, Southampton 1999)

10.37 Y. Ando, R. Pompoli: Factors to be measured of environmental noise and its subjective responses based on the model of auditory-brain system, J. Temporal Des. Architect. Environ. **2**, 2–12 (2002)

10.38 S. Sato, Y. Ando, V. Mellert: Cues for localization in the median plane extracted from the autocorrelation function, J. Sound Vibr. **241**, 53–56 (2001)

10.39 Y. Ando, T. Okano, Y. Takezoe: The running autocorrelation function of different music signals relating to preferred temporal parameters of sound fields, J. Acoust. Soc. Am. **86**, 644–649 (1989)

10.40 Y. Ando, K. Kageyama: Subjective preference of sound with a single early reflection, Acustica **37**, 111–117 (1977)

10.41 K. Kato, Y. Ando: A study of the blending of vocal music with the sound field by different singing styles, J. Sound Vibr. **258**, 463–472 (2002)

10.42 K. Kato, K. Fujii, K. Kawai, Y. Ando, T. Yano: Blending vocal music with the sound field – the effective duration of autocorrelation function of Western professional singing voices with different vowels and pitches, Proc. Int. Symp. Musical Acoust., ISMA (Acoustical Society of Japan, Kyoto 2004) pp. 37–40

10.43 Y. Ando, M. Sakamoto: Superposition of geometries of surface for desired directional reflections in a concert hall, J. Acoust. Soc. Am. **84**, 1734–1740 (1988)

10.44 Y. Ando: Investigations on cerebral hemisphere activities related to subjective preference of the sound field, published for 1983 – 2003, J. Temporal Design Architect. Environ. **3**, 2–27 (2003)

10.45 S. Sato, K. Nishio, Y. Ando: Propagation of alpha waves corresponding to subjective preference from the right hemisphere to the left with change in the IACC of a sound field, J. Temporal Des. Architect. Environ. **3**, 60–69 (2003)

10.46 K. Mouri, K. Akiyama, Y. Ando: Preliminary study on recommended time duration of source signals to be analyzed, in relation to its effective duration of autocorrelation function, J. Sound Vibr. **241**, 87–95 (2001)

10.47 T. Taguti, Y. Ando: Characteristics of the short-term autocorrelation function of sound signals in piano performances. In: *Music and Concert Hall Acoustics, Conf. Proc. MCHA 1995*, ed. by Y. Ando, D. Noson (Academic, London 1997) pp. 233–238, Chap. 23

10.48 Y. Ando, S.H. Kang, Morita: On the relationship between auditory-evoked potentials and subjective preference for sound field, J. Acoust. Soc. Jpn. (E) **8**, 197–204 (1987)

10.49 Y. Ando, S.H. Kang, H. Nagamatsu: On the auditory-evoked potential in relation to the IACC of sound field, J. Acoust. Soc. Jpn. (E) **8**, 183–190 (1987)

10.50 Y. Ando, C. Chen: On the analysis of autocorrelation function of α-waves on the left and right cerebral hemispheres in relation to the delay time of single sound reflection, J. Architect. Planning Environ. Eng., Architectural Institute of Japan (AIJ) **488**, 67–73 (1996), (in Japanese)

10.51 C. Chen, Y. Ando: On the relationship between the autocorrelation function of the α-waves on the left and right cerebral hemispheres and subjective preference for the reverberation time of music sound field, J. Architect. Planning Environ. Eng., Architectural Institute of Japan (AIJ) **489**, 73–80 (1996), (in Japanese)

10.52 Y. Soeta, S. Nakagawa, M. Tonoike, Y. Ando: Magnetoencephalographic responses corresponding to individual subjective preference of sound fields, J. Sound Vibr. **258**, 419–428 (2002)

10.53 Y. Soeta, S. Nakagawa, M. Tonoike, Y. Ando: Spatial analysis of magnetoencephalographic alpha waves in relation to subjective preference of a sound field, J. Temporal Des. Architect. Environ. **3**, 28–35 (2003)

10.54 D. Kimura: The asymmetry of the human brain, Sci. Am. **228**, 70–78 (1973)

10.55 R.W. Sperry: Lateral specialization in the surgically separated hemispheres. In: *The Neurosciences: Third study program*, ed. by F.O. Schmitt, F.C. Worden (MIT, Cambridge 1974), Chap. 1

10.56 J.H. Holland: *Adaptation in Natural and Artificial Systems* (Univ. Michigan Press, Ann Arbor 1975)

10.57 S. Sato, K. Otori, A. Takizawa, H. Sakai, Y. Ando, H. Kawamura: Applying genetic algorithms to the optimum design of a concert hall, J. Sound Vibr. **258**, 517–526 (2002)

10.58 S. Sato, T. Hayashi, A. Takizawa, A. Tani, H. Kawamura, Y. Ando: Acoustic design of theatres applying genetic algorithms, J. Temporal Des. Architect. Environ. **4**, 41–51 (2004)

10.59 H. Sakai, P. K. Singh, Y. Ando: Inter-individual differences in subjective preference judgments of sound fields. In: *Music and Concert Hall Acoustics, Conf. Proc. MCHA 1995*, ed. by Y. Ando, D. Noson (Academic, London 1997), Chap. 13

10.60 F. Maki: Sound and figure: concert hall design. In: *Music and Concert Hall Acoustics, Conf. Proc. MCHA 1995*, ed. by Y. Ando, D. Noson (Academic, London 1997), Chap. 1

10.61 Y. Ikeda: Designing a contemporary classic concert hall using computer graphics. In: *Music and Concert Hall Acoustics, Conf. Proc. MCHA 1995*, ed. by Y. Ando, D. Noson (Academic, London 1997), Chap. 2

10.62 I. Nakayama: Preferred time delay of a single reflection for performers, Acustica **54**, 217–221 (1984)

10.63 I. Nakayama, T. Uehara: Preferred direction of a single reflection for a performer, Acustica **65**, 205–208 (1988)

10.64 J. Meyer: *Influence of Communication on Stage on the Musical Quality* (Proc. 15th Int. Congress Acoustics, Trondheim 1995) pp. 573–576

10.65 Y. Ando, B. P. Johnson, T. Bosworth: Theory of planning physical environments incorporating spatial and temporal values, Memoir of Graduate School Science and Technology **14-A**, 67–92 (1996)

10.66 Y. Ando: On the temporal design of environments, J. Temporal Des. Architect. Environ. **4**, 2–14 (2004)

10.67 Y. Ando, S. Sato, T. Nakajima, M. Sakurai: Acoustic design of a concert hall applying the theory of subjective preference, and the acoustic measurement after construction, Acustica Acta Acustica. **83**, 635–643 (1997)

10.68 T. Nakajima, Y. Ando: Calculation and measurement of acoustic factors at each seat in the Kirishima International Concert Hall. In: *Music and Concert Hall Acoustics, Conf. Proc. MCHA 1995*, ed. by Y. Ando, D. Noson (Academic, London 1997), Chap. 5

10.69 M.R. Schroeder: *Number Theory in Science and Communication* (Springer, Berlin, Heidelberg 1984)

10.70 P.M. Morse: *Vibration and Sound* (McGraw-Hill, New York 1948)

10.71 Y. Ando, P.K. Singh: Global subjective evaluations for design of sound fields and individual subjective preference for seat selection. In: *Music and Concert Hall Acoustics, Conf. Proc. MCHA 1995*, ed. by Y. Ando, D. Noson (Academic, London 1997), Chap. 4

10.72 A.H. Marshall, D. Gottlob, H. Alrutz: Acoustical conditions preferred for ensemble, J. Acoust. Soc. Am. **64**, 1437–1442 (1978)

10.73 A.C. Gade: Investigations of musicians' room acoustic conditions in concert halls. Part I: Methods and laboratory experiments, Acustica **69**, 193–203 (1989)

10.74 D. Noson, S. Sato, H. Sakai, Y. Ando: Singer responses to sound fields with a simulated reflection, J. Sound Vibr. **232**, 39–51 (2000)

10.75 D. Noson, S. Sato, H. Sakai, Y. Ando: Melisma singing and preferred stage acoustics for singers, J. Sound Vibr. **258**, 473–485 (2002)

10.76 S. Sato, S. Ohta, Y. Ando: Subjective preference of cellists for the delay time of a single reflection in a performance, J. Sound Vibr. **232**, 27–37 (2000)

10.77 H.P. Seraphim: Über die Wahrnehmbarkeit mehrerer Rückwürfe von Sprachschall, Acustica **11**, 80–91 (1961)

10.78 N. Aoshima: Computer-generated pulse signal applied for sound measurement, J. Acoust. Soc. Am. **69**, 1484–1488 (1981)

10.79 C. Hayashi: Multidimensional quantification I., Proc. Jpn. Acad. **30**, 61–65 (1954)

10.80 C. Hayashi: Multidimensional quantification II., Proc. Jpn. Acad. **30**, 165–169 (1954)

10.81 V.L. Jordan: Acoustical criteria for auditoriums and their relation to model techniques, J. Acoust. Soc. Am. **47**, 408–412 (1969)

10.82 T. Hidaka: Personal communication (1996)

10.83 Y. Kobayasi, H. Tokuhiro, M. Owaki, K. Okuno, S. Yamada, Y. Ando: Acoustical design and characteristics of the atrium for music performance in a hotel. In: *Music and Concert Hall Acoustics, Conf. Proc. MCHA 1995*, ed. by Y. Ando, D. Noson (Academic, London 1997), Chap. 27

10.84 S. Sato, Y. Mori, Y. Ando: The subjective evaluation of source locations on the stage by listeners. In: *Music and Concert Hall Acoustics, Conf. Proc. MCHA 1995*, ed. by Y. Ando, D. Noson (Academic, London 1997), Chap. 12

10.85 L.L. Thurstone: A law of comparative judgment, Psychol. Rev **34**, 273–289 (1951)

10.86 F. Mosteller: Remarks on the method of paired comparison, III. Psychometrika **16**, 207–218 (1951)

10.87 C. Hayashi: On the prediction of phenomena from qualitative data and the quantification of qualitative data from the mathematico-statistical point of view, Ann. Inst. Statist. Math., 69–98 (1952)

11. Building Acoustics

This chapter summarizes and explains key concepts of building acoustics. These issues include the behavior of sound waves in rooms, the most commonly used rating systems for sound and sound control in buildings, the most common noise sources found in buildings, practical noise control methods for these sources, and the specific topic of office acoustics. Common noise issues for multi-dwelling units can be derived from most of the sections of this chapter. Books can be and have been written on each of these topics, so the purpose of this chapter is to summarize this information and provide appropriate resources for further exploration of each topic.

11.1	**Room Acoustics**	387
	11.1.1 Room Modes	388
	11.1.2 Sound Fields in Rooms	389
	11.1.3 Sound Absorption	390
	11.1.4 Reverberation	394
	11.1.5 Effects of Room Shapes	394
	11.1.6 Sound Insulation	395
11.2	**General Noise Reduction Methods**	400
	11.2.1 Space Planning	400
	11.2.2 Enclosures	400
	11.2.3 Barriers	402
	11.2.4 Mufflers	402
	11.2.5 Absorptive Treatment	402
	11.2.6 Direct Impact and Vibration Isolation	402
	11.2.7 Active Noise Control	402
	11.2.8 Masking	403
11.3	**Noise Ratings for Steady Background Sound Levels**	403
11.4	**Noise Sources in Buildings**	405
	11.4.1 HVAC Systems	405
	11.4.2 Plumbing Systems	406
	11.4.3 Electrical Systems	406
	11.4.4 Exterior Sources	406
11.5	**Noise Control Methods for Building Systems**	407
	11.5.1 Walls, Floor/Ceilings, Window and Door Assemblies	407
	11.5.2 HVAC Systems	412
	11.5.3 Plumbing Systems	415
	11.5.4 Electrical Systems	416
	11.5.5 Exterior Sources	417
11.6	**Acoustical Privacy in Buildings**	419
	11.6.1 Office Acoustics Concerns	419
	11.6.2 Metrics for Speech Privacy	419
	11.6.3 Fully Enclosed Offices	422
	11.6.4 Open-Plan Offices	422
11.7	**Relevant Standards**	424
	References	425

11.1 Room Acoustics

When a sound wave in a room encounters a boundary surface, some of its energy is reflected back into the room, as illustrated in Fig. 11.1. The physical laws regarding reflected sound energy are analogous to those for the laws of optics. Just as light bounces off a mirror at the same angle as its angle of incidence, sound waves have equal angles of incidence and reflection. Acoustically reflective surfaces are typically smooth and hard.

Some common room acoustics problems caused by reflection are echoes and room resonance. Echoes result from the limitations of our hearing mechanisms to process sounds. When the difference in arrival times between two sounds is less than 60 ms, we hear the combination of the two sounds as a single sound. However, when this difference exceeds 60 ms, we hear the two distinct sounds (Fig. 11.2). When these two sounds are generated by the same source, this effect (which is called an echo) can cause difficulty in understanding speech, especially when arrival times differ by more than 100 ms. These kinds of delays occur when a person hears a sound wave coming directly from a source and another coming from a reflecting surface. Given that the speed of sound in air is roughly 300 m per second, the 100 ms delay translates to a distance of roughly 30 m. There-

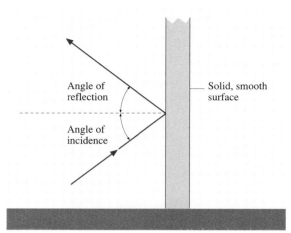

Fig. 11.1 Reflection of a sound wave off a hard, smooth surface

fore, a difference in sound travel path of more than 30 m between a sound wave traveling directly from a source to a listener and a sound wave traveling from a source to a reflecting surface and then to a listener causes an echo.

When parallel reflective surfaces are tall and more than 10 m from each other, a rapid succession of mid-frequency echoes can occur, known as flutter echo, resulting in a flapping sound similar to that generated by birds or bats. When these surfaces are fairly close to each other, as in a narrow hallway or stairwell, flutter echo sounds more like a high-pitched buzzing since the smaller spacing accommodates higher frequencies (shorter wavelengths).

Echoes are normally perceived as discrete reflections that can be clearly heard and identified, but many reflections off all surfaces within a room can combine to produce the phenomenon known as reverberation. Reverberation can raise the sound level in a room and can also reduce speech intelligibility, but it is desirable for certain types of music. More discussion on reverberation and its control can be found later in this chapter.

11.1.1 Room Modes

Room modes occur at specific frequencies when two reflective walls are parallel to each other. Since the surfaces reflect the sound, their mirror images reflect off each wall to set up a stationary pressure pattern in a room. This phenomenon is called a single-dimensional (or axial) standing wave, as shown in Fig. 11.3, and it is the simplest form of room modes.

Axial standing waves occur at frequencies described by the equation below:

$$f_n = nc/2d , \qquad (11.1)$$

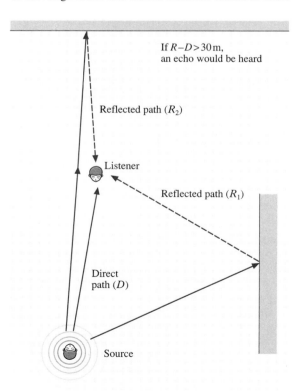

Fig. 11.2 Generation of an echo

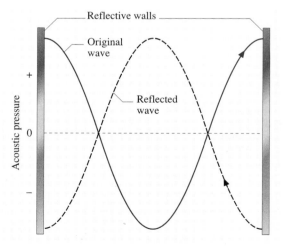

Fig. 11.3 Generation of a standing wave between parallel, reflective surfaces

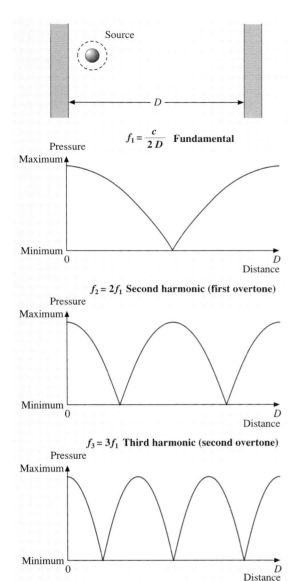

Fig. 11.4 Pressure patterns of the lowest axial standing wave modes in a room with parallel reflective surfaces

where $n = 1, 2, 3, \ldots$, c is the speed of sound, and d is the distance between the parallel reflective surfaces. This means that axial standing wave room modes occur at integer multiples of specific 1/2-wavelengths of sound, as is shown in Fig. 11.4. Standing waves can become more complex in two and three dimensions, where they are known as tangential and oblique modes, respectively.

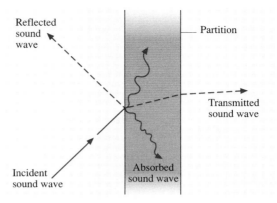

Fig. 11.5 The interaction of a sound wave with a partition

A design problem with standing waves is that they generate uneven sound distribution patterns. Some areas will have higher levels of sound (because a standing wave is reinforcing the pressure at those locations) and some areas will have lower levels of sound (because the standing wave is canceling much of the pressure at those locations), but only at specific frequencies.

The topic of building acoustics stresses the importance of controlling sound. When one talks about controlling sound, it is often assumed that one is referring to the reduction of sound. However, there are cases (e.g., auditorium or concert hall design) in which we want to preserve the sound energy but we would like to control its spatial spreading characteristics. The primary ways to reduce sound are through absorption and insulation. Using absorption on an auditorium's side walls may eliminate unwanted reflections but may also eliminate the possibility of some people hearing sound coming from the stage. We therefore must clarify how we plan to control the sound. Redirection and diffusion can have favorable acoustic results for even sound distribution in large rooms.

Discussions about noise control usually refer to the reduction of sound (since noise is defined as unwanted sound). Figure 11.5 shows what happens to a sound wave that interacts with a room's surface. Part is absorbed, part is redirected, and the rest is transmitted through the surface.

11.1.2 Sound Fields in Rooms

Several different fields, or acoustic environments, exist in all rooms. The size of these fields depends on the dimensions of the room, the reflective qualities of the room's surfaces, and the frequency of interest. The

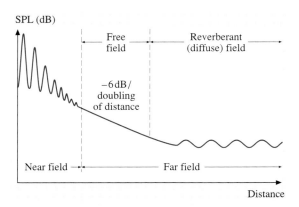

Fig. 11.6 Sound fields in rooms

near field is the region within 1/4-wavelength (of the lowest frequency of interest) of a sound source or large reflective surface. Sound pressure levels can fluctuate dramatically in the near field of a source and sound pressures cancel and enhance each other near large reflective surfaces, so sound pressure level measurements should be avoided in near fields.

The far field is the region beyond the near field. It is contrasted with the near field to designate the region that is appropriate for recording sound pressure level measurements from a sound source. As one moves out of the near field and into the far field, sound pressure levels drop off at a rate of 6 dB per doubling of distance, for a point source, in accordance with the inverse square law that is simplified in the following equation:

$$SPL_2 = SPL_1 - [20 \times \log(d_2/d_1)], \quad (11.2)$$

where SPL_1 is the sound pressure level at the location closer to the sound source, SPL_2 is the sound pressure level at the location farther from the sound source, d_1 is the distance from the source at which SPL_1 is measured, and d_2 is the distance from the source at which SPL_2 is measured.

The region in which (11.2) is valid is known as the free field. Free-field conditions exist in large open outdoor spaces or in rooms having highly absorptive surfaces, in which there are no obstructions in the sound travel path between the source and listener. The free field is sometimes referred to as the direct field when source measurements are taking place that only consider the sound wave traveling from the source to the listener with no influence by reflected sound coming from room surfaces.

As sound pressure levels decay from sources within a room, they eventually drop to a relatively constant level which is determined by the amount of reflected sound within a room. This lower limit is known as the reverberant or diffuse field. The size of the reverberant field depends on the size of the room and the size and characteristics of its reflective surfaces. Within the reverberant field, sound pressures are similar independent of location. Figure 11.6 puts all of the fields described above into perspective.

11.1.3 Sound Absorption

Absorption converts sound energy into heat energy. It is useful for reducing sound levels *within* rooms but not *between* rooms. Each material with which a sound wave interacts absorbs some sound. The most common measurement of that is the absorption coefficient, typically denoted by the Greek letter α. The absorption coefficient is a ratio of absorbed to incident sound energy. If a material does not absorb any sound incident upon it, its absorption coefficient is 0. In other words, a material with an absorption coefficient of 0 reflects all sound incident upon it. In practice, all materials absorb some sound, so this is a theoretical limit. If a material absorbs all sound incident upon it, its absorption coefficient is 1. As with the lower limit for absorption coefficients, all materials reflect some sound, so this is also a theoretical limit. Therefore, absorption coefficients range between 0 and 1.

Absorption coefficients vary with frequency. Typical absorptive materials are porous and characterized by absorption coefficients that increase with frequency. They therefore have limited effectiveness for lower frequencies, especially below 250 Hz. There are absorbers that have been designed to absorb these lower frequencies, and these will be discussed shortly. However, for typical cases, it is convenient to use a single number (incorporating multiple frequency components) to describe the absorption characteristics of a material. This value has been defined by the American Society for Testing and Materials (ASTM) in Standard C 423 as the noise reduction coefficient (NRC). The NRC is the arithmetic (as opposed to the logarithmic) average of a material's absorption coefficients at 250, 500, 1000, and 2000 Hz, rounded to the closest 0.05.

Table 11.1 lists absorption coefficients and NRC values for common materials. Note that the values listed in Table 11.1 are for general reference purposes only, and specific values should be based on manufacturers specifications. Also note that absorption coefficients and NRC values have no units associated with them. In general, materials with NRC values less than 0.20 are considered

Table 11.1 Absorption coefficients and NRC values for common materials

Material	125 Hz	250 Hz	500 Hz	1000 Hz	2000 Hz	4000 Hz	NRC
Painted drywall	0.10	0.08	0.05	0.03	0.03	0.03	0.05
Plaster	0.02	0.03	0.04	0.05	0.04	0.03	0.05
Smooth concrete	0.10	0.05	0.06	0.07	0.09	0.08	0.05
Coarse concrete	0.36	0.44	0.31	0.29	0.39	0.25	0.35
Smooth brick	0.03	0.03	0.03	0.04	0.05	0.07	0.05
Glass	0.05	0.03	0.02	0.02	0.03	0.02	0.05
Plywood	0.58	0.22	0.07	0.04	0.03	0.07	0.10
Metal blinds	0.06	0.05	0.07	0.15	0.13	0.17	0.10
Thick panel	0.25	0.47	0.71	0.79	0.81	0.78	0.70
Light drapery	0.03	0.04	0.11	0.17	0.24	0.35	0.15
Heavy drapery	0.14	0.35	0.55	0.72	0.70	0.65	0.60
Helmholtz resonator	0.20	0.95	0.85	0.49	0.53	0.50	0.70
Ceramic tile	0.01	0.01	0.01	0.01	0.02	0.02	0.00
Linoleum	0.02	0.03	0.03	0.03	0.03	0.02	0.05
Carpet	0.05	0.05	0.10	0.20	0.30	0.40	0.15
Carpet on concrete	0.05	0.10	0.15	0.30	0.50	0.55	0.25
Carpet on rubber	0.05	0.15	0.13	0.40	0.50	0.60	0.30
Upholstered seats	0.19	0.37	0.56	0.67	0.61	0.59	0.55
Occupied seats	0.39	0.57	0.80	0.94	0.92	0.87	0.80
Water surface	0.01	0.01	0.01	0.01	0.02	0.03	0.00
Soil	0.15	0.25	0.40	0.55	0.60	0.60	0.45
Grass	0.11	0.26	0.60	0.69	0.92	0.99	0.60
Cellulose spray (1″)	0.08	0.29	0.75	0.98	0.93	0.76	0.75

to be reflective while those with NRC values greater than 0.40 are considered to be absorptive. When significant sound energy must be absorbed, as may be the case for eliminating echoes or standing waves, materials having higher absorption coefficients are usually recommended.

A few cautionary notes apply. NRC values are convenient to use for rating the absorption characteristics of a material. However, they should only be used when the sound sources of interest are within the 200–2000 Hz range. For sources outside of this range, and especially below this range, materials effective for the specific frequency of interest must be used. Also note that some manufacturers specify NRC and absorption coefficient values that are greater than 1.0. Methods used to measure absorption coefficients can artificially raise their values above 1.0; yet such values inaccurately imply that more energy is absorbed by a material than is incident upon it, which is a physical impossibility. Therefore, any published absorption coefficient or NRC values greater than 1.0 should be not be considered as greater than 1.0.

Another absorption metric gaining increasing application is the sound absorption average (SAA). The SAA is a single number rating that is the average, rounded off to the nearest 0.01, of the sound absorption coefficients of a material for the twelve one-third octave bands from 200 through 2500 Hz. Although the SAA replaces the NRC rating, as directed by the ASTM C 423 since the year 2000, most product literature still uses NRC values.

The method by which these materials are mounted affects their sound absorption effectiveness, especially for low frequencies (below 500 Hz). Figure 11.7 shows

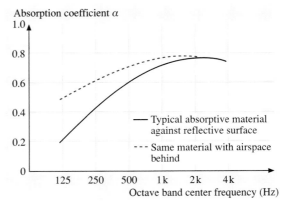

Fig. 11.7 The general effect of an air space between absorptive material and its mounting surface

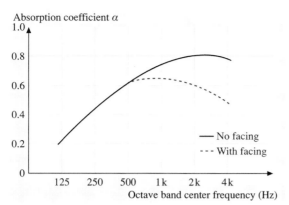

Fig. 11.8 The general effect of non-acoustically transparent facings on absorption

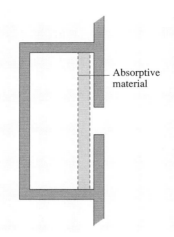

Fig. 11.10 Cross section of a Helmholtz resonator concrete masonry unit. The absorptive material helps to spread the absorptive properties over a larger frequency range than would otherwise occur

the effect of air spaces between an absorptive material and a mounting surface.

Facings on absorptive materials can degrade their absorptive properties, especially for higher frequencies (above 2000 Hz). Acoustically transparent facings, such as grill cloth and open-weave fabrics, will have little effect on absorption properties, but facing materials that are not acoustically transparent, such as perforated metal and wood slats, can produce the kind of effects shown in Fig. 11.8.

Of critical importance with these facings is the open area ratio. Since there are so many perforated metal designs, it is best to check with the manufacturer for the most appropriate opening ratio and design for the situation. Figure 11.9, however, offers some guidelines for wood-slat ceiling designs.

As Table 11.1 shows, the absorption coefficients of most porous materials increase with increasing frequency. This means that they are not as effective at low frequencies as at higher ones. If absorption is required for frequencies below 250 Hz, special materials must be used. Each of these materials has an air space behind its light or open surface to provide the extra absorption. Two common materials used for this purpose are Helmholtz resonators and diaphragmatic absorbers. Helmholtz resonators have narrow openings that lead to the outside on one end and into a larger air cavity on the other, as is shown in Fig. 11.10.

As viewed by an incoming sound wave, the air in the narrow neck functions as a mass and the air in the larger cavity functions as a spring. A mass on a spring will resonate at a frequency appropriate to that mass

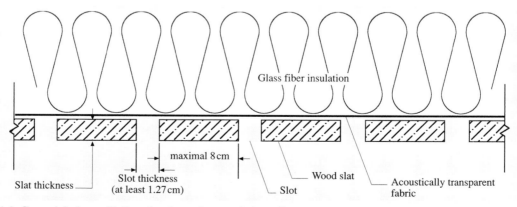

Fig. 11.9 General design guidelines for absorptive wood-slat ceilings

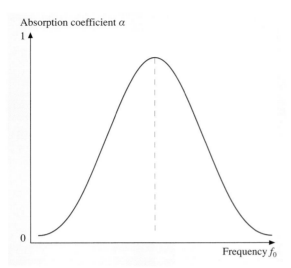

Fig. 11.11 General absorption characteristics of Helmholtz resonators not filled with absorptive material

and spring stiffness. Thus, at and near the resonant frequency of such an acoustic chamber, sound will be absorbed from an incoming sound wave, as is shown in Fig. 11.11.

Such devices incorporated in wall constructions are usually resonant below 250 Hz, depending on the dimensions of the neck and size of the cavity. There are commercially available products that incorporate this design into concrete masonry units. When viewing these products installed as partitions, the wall surfaces have slots in them, as are shown in Fig. 11.12. Table 11.1 has a listing for these products which shows their su-

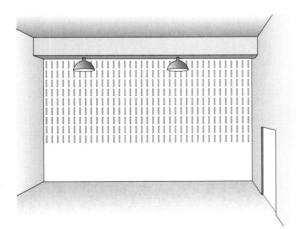

Fig. 11.12 Wall built from Helmholtz resonator concrete masonry units

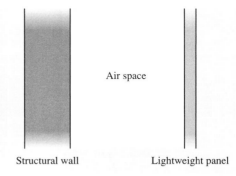

Fig. 11.13 Cross section of a generalized diaphragmatic absorber

perior absorption in a narrow low-frequency range. The porous and coarse nature of the surface provides modest absorption at higher frequencies also.

Diaphragmatic absorbers work according to similar principles to Helmholtz resonators, except they are comprised of an air space behind a lightweight wall, as shown in Fig. 11.13.

Reverberant Field Noise Reduction

Adding absorptive materials to a room's surfaces, in addition to reducing or eliminating echoes and standing waves, can reduce the overall reverberant field noise level within a room. The mathematical basis for this is to determine the total absorption for all of a room's surfaces using the following equation:

$$A = \sum_{i=1}^{n} \alpha_i S_i = \alpha_1 S_1 + \alpha_2 S_2 + \ldots + \alpha_n S_n \, , \quad (11.3)$$

where α_1, α_2, etc., are the absorption coefficients of each surface material in the room while S_1, S_2, etc. are the surface areas corresponding to the surfaces having the same subscript value as those for the absorption coefficients. The total absorption (A) is in units of sabins, for which one sabin is defined as the total absorption provided by a one square foot piece of material having an absorption coefficient of 1.

The reverberant field noise reduction is defined as

$$\mathrm{NR} = 10 \times \log(A_2/A_1) \, , \quad (11.4)$$

where A_1 is the total absorption within a room before absorptive treatment is applied and A_2 is the total absorption within that same room after absorptive treatment is applied. The practical limits of NR are 12–15 dBA with typical cases yielding 6–8 dBA of noise reduction within a room from the proper use of absorptive materials. Bear in mind that this will not reduce sound pressure levels close to a source but will only reduce reflected sound pressure levels, which limits the effectiveness of the noise reduction to locations in a room's reverberant field.

11.1.4 Reverberation

Absorption is useful in reducing or eliminating unwanted reflections off surfaces. The standing waves discussed in Sect. 11.1.1 can be eliminated by covering one of the parallel surfaces with absorptive material. Absorption can also be used to eliminate echoes. The rear walls of auditoriums are prime candidates for absorptive materials since rear walls have the greatest potential to cause echoes. The most common use of absorption, however, is to control reverberation.

Reverberation is the build up of sound within a room resulting from repeated sound wave reflections off all of its surfaces. Reverberation can increase sound levels within a room by up to 15 dBA, as well as distort speech intelligibility. Reverberation is desirable for rooms in which music is being played, especially classical and cathedral-style music, to add a pleasant persistence to the sounds. Therefore, there are different reverberation characteristics that would be appropriate for different room uses.

Reverberation is described by a parameter known as the reverberation time (denoted RT_{60}). RT_{60} can be defined in two ways – physically and mathematically. Physically, RT_{60} is the time (in seconds) that it takes for a sound source to reduce in sound pressure level (within a room) by a factor of 60 dB after that sound source has been silenced. Mathematically, in what is known as the sabine equation, RT_{60} is directly proportional to the volume of a room and inversely proportional to the absorption of the materials in the room, as follows:

$$RT_{60} = 0.161 \cdot V/A , \qquad (11.5)$$

where V is the room volume in cubic meters and A is the total absorption of the room's surfaces in metric sabins.

This means that RT_{60} increases as the size of a room increases and as the absorption of the room's surfaces decreases. Conversely, RT_{60} decreases as the size of a room decreases and as the absorption of the room's surfaces increases. There are then two principal ways to control RT_{60}: (1) by changing a room's size and (2) by changing the amount of absorption on its surfaces. Although it is possible to reduce a large room's volume by dividing the space with walls, it is more practical to adjust RT_{60} by adding sound absorptive materials.

Table 11.2 Optimum mid-frequency RT_{60} values for various occupied facilities

Type of facility	Optimum mid-frequency RT_{60} (s)
Broadcast studio	0.5
Classroom	1.0
Lecture/conference room	1.0
Movie/drama theater	1.0
Multipurpose auditorium	1.3 to 1.5
Contemporary church	1.4 to 1.6
Rock concert hall	1.5
Opera house	1.4 to 1.6
Symphony hall	1.8 to 2.0
Cathedral	3.0 or higher

Since the absorption performance of materials varies with frequency, so does RT_{60}. Table 11.2 offers generally accepted ranges of RT_{60} for different uses in the mid-frequency (500–1000 Hz) range. As you can see from the table, lower RT_{60} values are desirable for rooms used mainly for human speech and higher RT_{60} values are desirable for rooms used mainly for music. The optimum mid-frequency RT_{60} for a fully occupied room is different for various types of music. Because contemporary orchestral repertoires emphasize late classical and romantic music, the optimum mid-frequency RT_{60} for a fully occupied concert hall is usually stated as 1.8–2.0 s.

Optimum RT_{60} values generally increase by 10% of the values in Table 11.2 for each halving of frequency below 500 Hz and decrease by 10% of the values in Table 11.2 for each doubling of frequency above 1000 Hz. A room with a low (less than 0.8 s) RT_{60} is called a *dead* room and a room with a high (greater than 1.7 s) RT_{60} is called a *live* room. Multipurpose facilities should have RT_{60} values between the live and dead range limits.

Note that the sabine equation is based on the assumption that the absorptive properties of the surface materials are spread evenly throughout the room. When this is clearly not the case, other (more complex) RT_{60} calculation methods would be more appropriate.

11.1.5 Effects of Room Shapes

Although absorption is necessary in many circumstances, eliminating reflections is not always useful. This is of key importance in rooms where an audience is listening to a performance or lecture. In this type of room, it is desirable that all audience members hear the sound not only clearly, but without preference to seating location. Without an electronic sound system, this can only be accomplished by reflections off side walls and ceilings. Discrete echoes can be eliminated by avoiding smooth, flat reflective surfaces and by having irregular and convex surfaces to diffuse the sound evenly throughout the audience. For smaller rooms such as recording studios that require diffusion, commercially available sound-diffusing panels called quadratic residue diffusers are available. These panels can also be used for larger spaces, as well as irregularly shaped surfaces.

Concave reflective surfaces focus sound in certain areas and defocus sound from others, causing hot spots where sound is concentrated and dead spots where sound cannot be heard. Concave reflective surfaces should be avoided for this reason. If aesthetics dictate the need for a concave surface, it would be best to install an absorptive or diffusive surface (as needed) and cover it with acoustically transparent fabric in the concave shape.

Reflective rear walls in auditoriums are notorious for generating echoes because of their associated large sound travel path differences. For this reason, reflective surfaces should be avoided for rear walls. Reflective surfaces are beneficial, especially for concert halls, when they are close to the stage and along side walls. Reflective surfaces close to the stage assist in several ways, by sending sound into the audience rather than allowing it to be lost behind the stage and by enhancing the sound through lateral reflections off side walls to spread the sound more evenly throughout an audience. Another benefit of reflective surfaces near the stage is that they permit the performers to hear each other, something that is critical to concert performances. These so-called early reflections are usually generated by shells on the stage or by hanging reflective panels.

11.1.6 Sound Insulation

The description of the insulation of sound is similar in many ways to the description of the absorption of sound. As with absorption, there is a transmission coefficient that ranges from the ideal limits of 0 to 1. The transmission coefficient, denoted by the Greek letter τ, is the unitless ratio of transmitted to incident sound energy. Unlike the absorption coefficient, however, the limit of $\tau = 1$ is possible in practice since a transmission coefficient of 1 implies that all of the sound energy is transmitted through a partition. This would be the case

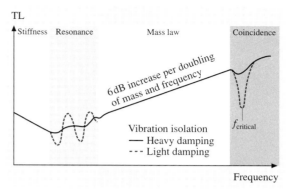

Fig. 11.14 General characteristics of the transmission loss spectrum

for an open window or door, where the sound energy has no obstruction to its path. The other extreme of $\tau = 0$ (implying no sound transmission), however, is not a practical value since some sound will always transmit through a partition.

Unlike absorption, the principal descriptor for sound insulation is a decibel level based on the transmission coefficient. This value is known as the transmission loss (TL), and is based on the following equation:

$$\mathrm{TL} = 10 \cdot \log(1/\tau) \, . \tag{11.6}$$

The transmission loss can be loosely defined as the amount of sound reduced by a partition between a sound source and a listener. The complete sound reduction of a partition between two rooms also takes into account the absorptive characteristics of the listener's room, as follows:

$$\mathrm{SPL_S} - \mathrm{SPL_L} = \mathrm{TL} + 10 \cdot \log(A_\mathrm{L}/S) \, , \tag{11.7}$$

where $\mathrm{SPL_S}$ is the average sound pressure level in the room enclosing the sound source, $\mathrm{SPL_L}$ is the average sound pressure level in the adjacent listener's room, A_L is the total absorption in the listener's room, TL is the transmission loss of the partition between the two rooms, and S is the surface area of the partition between the two rooms. Note that TL is the quantity that is typically reported in manufacturers' literature since it is measured in a laboratory independent of the installation.

Since the logarithm of 1 is 0, the condition in which the transmission coefficient is 1 translates to a TL of 0 dB. This concurs with the notion that an open air space in a wall allows the free passage of sound. Although an open air space itself would have a TL value of 0 dB associated with it, a wall with an open air space in it

Table 11.3 Critical frequencies for common building materials

Material	Thickness (cm)	Critical frequency (Hz)
Concrete	8	100
Plywood	1.2	1700
Gypsum wall board	1.2	3100
Steel or aluminum	0.3	4100
Lead	1.2	4400
Glass	0.3	4900
Plexiglass	0.3	9800

would have a TL value greater than 0 (up to 10) dB, depending on the size of the opening and the location with respect to the wall. The practical upper limit of TL is roughly 70 dB.

As for absorption, TL is frequency dependent. Typical partitions have TL values that increase with increasing frequency, as is shown in Fig. 11.14, with the exception of a dip in TL around a bending wave resonance frequency, known as the critical frequency. This effect is known as the coincidence dip. The extent of this dip depends on the vibrational damping of the partition and the frequency at which it occurs depends on the density and thickness of the material.

Table 11.3 lists critical frequencies for common building materials. Critical frequencies generally decrease at the same rate as the material thickness increases. Therefore, a doubling in material thickness translates to cutting the critical frequency in half for the same homogeneous material. Coincidence effects can be minimized by using multilayered partitions with different materials and different material thicknesses combined into a single partition.

There is a single-number rating for TL that takes the entire frequency spectrum into account, established by ASTM standard E413. This value, known as the sound transmission class (STC), is not derived by the simple averaging method used for NRC values. Instead, the TL frequency spectrum is matched to a standard curve within the limits imposed by the ASTM standard. Figure 11.15 shows how STC is calculated from the TL spectrum. Note that STC values have no units associated with them, but they are based on decibels. Similar to NRC, STC is useful to describe the sound insulation efficiency of a partition over the human speech frequency range of roughly 500–2000 Hz. If sound insulation outside of that frequency range (especially for frequencies below 250 Hz) is required, the TL values relevant to the frequency range of interest must be used.

Table 11.4 lists TL and STC values for common partitions. As with Table 11.1, these values are for general reference purposes only, and specific values should be based on manufacturers' specifications. Table 11.5 gives some meaning to the numbers by listing speech privacy ratings for different ranges of STC values.

As with any construction system, the actual in-field performance seldom matches the theoretical or laboratory STC rating for the construction. This is most often caused by flanking paths, poor construction practices, and/or lack of sealing cracks, gaps and openings. The noise isolation class (NIC) is the metric that considers these conditions. NIC is determined using the same method and template as is used for determining STC (shown on Fig. 11.15), using measured field data instead of using controlled laboratory data. The NIC value is typically 5–10 points less than the STC rating of a given partition design.

STC does not effectively consider the transmission of impact noise on floors from foot traffic as it may affect floors below. For this purpose the impact insulation class (IIC) was established by ASTM standard

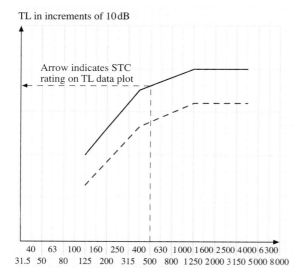

Fig. 11.15 The standard STC calculation curve, from ASTM E 413. STC values are derived by plotting the TL spectrum on this curve and determining the highest curve shape that meets the criteria of: (1) the sum of the differences between individual TL values and the *solid line* is less than 32 dB and (2) no individual 1/3-octave band TL value is more than 8 dB below the *solid line*. The TL value at 500 Hz corresponding to this spectrum is the STC value

Table 11.4 Transmission loss and STC values for common partitions

Partition	125 Hz	250 Hz	500 Hz	1000 Hz	2000 Hz	4000 Hz	STC
1/2 inch drywall on both sides of wooden studs	17	31	33	40	38	36	33
1/2 inch drywall on wooden studs with 2 inches of insulation	15	30	34	44	46	41	37
Double layer of 1/2 inch drywall on wooden studs	25	34	41	51	48	50	41
1/2 inch drywall on staggered wooden studs	23	28	39	46	54	44	39
1/2 inch drywall on staggered wooden studs with 2 inches of insulation	29	38	45	52	58	50	48
1/2 inch drywall on metal studs	22	27	43	47	37	46	39
1/2 inch drywall on metal studs with 2 inches of insulation	26	41	52	54	45	51	45
8 inch thick concrete masonry units	36	44	50	54	58	56	53
Open-plane office partition	10	12	12	12	12	11	12
4 inch thick brick wall	32	34	40	47	55	61	45
1/2 inch drywall inside/1 inch stucco outside on wooden studs	21	33	41	46	47	51	42
Single-paned 1/8 inch thick glass	18	21	26	31	33	22	26
1/2 inch thick laminated glass	31	34	38	40	37	46	40
Double-paned 1/8 inch thick glass with 2 inch air gap	13	25	35	44	49	43	37
Hollow wooden door, 1 3/4 inch thick	14	19	23	18	17	21	19
Solid wooden door, 1 3/4 inch thick	29	31	31	31	39	43	34
Hollow metal door, 1 3/4 inch thick	24	23	29	31	24	40	28
Filled metal door, 1 3/4 inch thick	26	34	40	48	44	52	43
Wood joist floor/ceiling with 1/2 inch plywood subfloor and 1/2 inch drywall	23	32	36	45	49	56	37
8 inch thick concrete slab floor	32	38	47	52	57	63	50
Wood plank shingled roof	29	33	37	44	55	63	43
Wood plank shingled roof with 1/2 inch drywall ceiling, 4 inches of insulation	35	42	49	62	67	79	53
Corrugated steel roof with 1 inch of sprayed cellulose	17	22	26	30	35	41	30

E 989. The IIC stresses the low-frequency range, as is shown in Fig. 11.16, but note that IIC does not address very low (below 100 Hz) frequencies, which may be associated with resonances of a building's structural components.

One other single-number rating worth noting here is the outdoor–indoor transmission class (OITC), established in ASTM E 1332. The OITC is a single-number rating for the effectiveness of a building façade in reducing noise to interior spaces. The calculation of OITC is based on an A-weighted source spectrum of typical transportation (aircraft, rail, and traffic) sources in the 80–4000 Hz range and the façade's transmission loss spectrum in the same frequency range.

Non-Homogeneous Partitions

Up to this point we have been discussing homogeneous partitions. Actual designs, however, include walls composed of different materials, such as those with windows and doors. Each of these wall components has different associated TL characteristics. Wall components having lower TL characteristics than the rest of the wall can significantly degrade a wall's sound reduction effectiveness. This can be calculated using the composite

Table 11.5 Speech privacy associated with STC ratings*

STC range	Sound privacy
0 to 20	No privacy (voices heard clearly between rooms)
20 to 40	Some privacy (voices heard in low** background noise)
40 to 55	Adequate privacy (only raised voices heard in low background noise)
55 to 65	Complete privacy (only high level noise heard in low background noise)
70	Practical limit

* assuming no significant flanking paths or openings in walls
** in the 35 dBA range

Table 11.6 Transmission loss reduction as a function of air opening*

Wall area having air opening (%)	Resultant wall TL (dB)	Resultant reduction in TL (dB)
0.01	39	6
0.1	30	15
0.5	23	22
1	20	25
5	13	32
10	10	35
20	7	38
50	3	42
75	1	44
100	0	45

* based on original wall TL of 45 dB

transmission coefficient, which is defined as:

$$\tau_{\text{comp}} = \left(\sum_{i=1}^{n} \tau_i S_i\right)/S$$
$$= (\tau_1 S_1 + \tau_2 S_2 + \ldots + \tau_n S_n)/S, \quad (11.8)$$

where τ_1, τ_2, etc., are the transmission coefficients of each wall component while S_1, S_2, etc. are the surface areas corresponding to the surfaces having the same subscript value as those for the transmission coefficients, and S is the surface area of the entire partition. The composite transmission coefficient that is calculated using (11.8) can then be replaced in (11.6) to yield the composite transmission loss for the partition.

Air gaps (e.g., around doors, windows or penetrations) are notorious for compromising the sound reduction effectiveness of walls. Table 11.6 shows the TL degradation caused by varying sizes of air gaps in a wall originally rated at a TL of 45 dB. As the table shows, an air gap just one tenth of one percent the size of a wall can lower the TL rating from 45 to 30 dB. Figure 11.17 offers a graphical method for making this determination, not only for openings but also for doors and windows that have lower TL values than the walls in which they are installed. This emphasizes the importance of sealing walls, avoiding air gaps, and placing airtight seals around doors.

A rule that governs most of the TL spectrum of homogeneous partitions, known as the mass law, states that TL increases at a rate of 6 dB with each doubling of mass and with each doubling of frequency (shown in Fig. 11.14). This rule can make it impractical to solve sound privacy issues with homogeneous partitions. For example, if a 1/2 meter-thick concrete wall does not provide enough sound insulation for a specific situation,

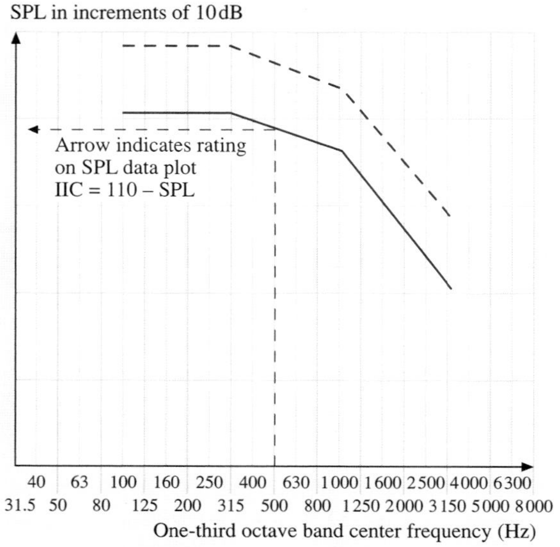

Fig. 11.16 The standard IIC calculation curve, from ASTM E 989. IIC values are derived by plotting the SPL spectrum (from a tapping machine specified in ASTM E 492) on this curve and determining the lowest curve shape that meets the criteria of: (1) the sum of the differences between individual SPL values and the *solid line* is less than 32 dB and (2) no individual 1/3-octave band TL value is more than 8 dB above the *solid line*. The SPL value at 500 Hz corresponding to this spectrum, subtracted from 110, is the IIC value

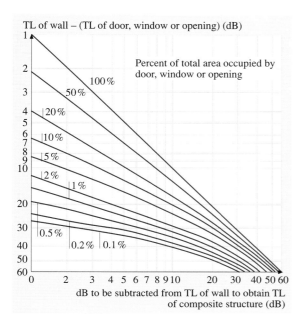

Fig. 11.17 Chart for determining the reduction in TL wall rating by the installation of a door, window, or opening

doubling the thickness of that wall to 1 meter would only offer an additional 6 dB of sound reduction. A more practical way of avoiding this kind of issue is by using multilayered partitions.

Multilayered partitions are comprised of layers of different materials. Each time sound passes through a different material, its level is reduced. Therefore, this method can be used to reduce costs, weight and space restrictions while providing adequate sound reduction. A sharp change in density of material is most effective in raising TL. Air spaces between wall sections and materials are effective by setting up such environments and also breaking any rigid connections between sides of a partition. A rigid connection can provide a vibration channel for sound to pass through with little reduction. For example, the noise reduction effectiveness of a studded wall filled with fiber insulation between studs can be short-circuited because sound may travel unimpeded

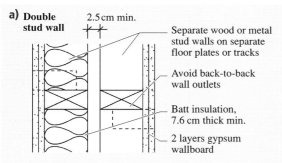

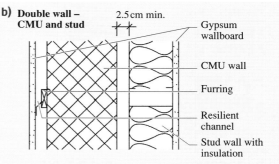

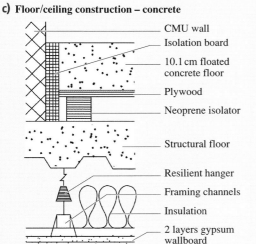

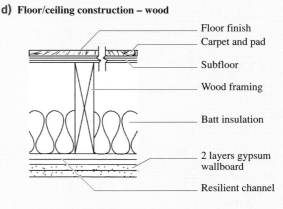

Fig. 11.18 Examples of multilayered partition and floor/ceiling designs that result in high TL

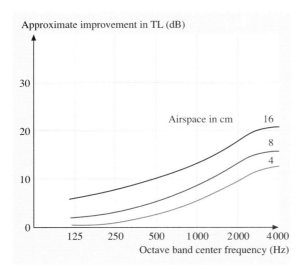

Fig. 11.19 General TL improvement that can be expected from air spaces in partitions

through the studs to the other side of the wall. Figure 11.18 shows examples of multilayered partitions that are effective for sound insulation.

Figure 11.19 shows the TL benefit of different air spaces in partitions. In general, the TL of a partition increases at a rate of 5 dB with each doubling of air space (within a partition) beginning with an air space of 5 cm.

As is mentioned above, transmission loss generally increases with frequency. If significant sound reduction is required for frequencies below 250 Hz, it is most practical to use multilayered partitions with air spaces and staggered studs or resilient channels between components. Resilient channels are shaped metal bars (the S shape is common) or resilient clips that connect two wall components. Only one side of their shape touches each side of the wall, and they effectively reduce vibrational energy from traveling through them. Resilient clips resemble isolation hangers, which are discussed later in this chapter. Another option for greater transmission loss at low frequencies is massive concrete slabs.

11.2 General Noise Reduction Methods

Noise can be controlled at its source, in the path between the source and the listener, or at the listener. Table 11.7 summarizes the general options available. If the noise can be controlled sufficiently well at its source, it is unnecessary to consider the path or listener locations. Likewise, if the noise can be controlled sufficiently well in the path between the source and listener, it is unnecessary to consider the listener's location for noise control measures.

The options for noise control at the source are generally self-explanatory. Most often, due to economic or logistic issues, noise control options are limited to the path between the source and the listener. Given many misconceptions about these options, it is useful to discuss some of them further.

11.2.1 Space Planning

The most effective noise control method for buildings is space planning. This is true both for room layout within a building and for the placement of the building itself with respect to exterior noise sources. Inherently noisy spaces adjacent to spaces inherently requiring quiet should be avoided as much as possible. The noisiest indoor spaces tend to be mechanical rooms, restrooms, and elevator shafts.

For placement with respect to loud exterior sources, buildings should be as far as possible from noisy streets or rail lines, taking advantage of shielding from other buildings and topographical features. If building placement near noisy outdoor sources is unavoidable, special attention should be paid to the design of the façade(s) facing the noisy sources, avoiding operable windows, doors, and penetrations.

Another aspect to consider for outdoor sound is the potential effect of a building's noise sources on the surrounding community. Rooftop mechanical units are the most common noise sources that may need to be insulated from community intrusions, especially when municipal codes are involved. The community noise aspect of a building is often considered in the building's permitting process.

11.2.2 Enclosures

Enclosures can be effective at reducing noise levels as long as they are properly designed. The following points should be considered when designing noise enclosures (also illustrated in Fig. 11.20a–d).

1. The enclosure must completely surround the noise source, with no air gaps. As mentioned earlier in

Table 11.7 Generalized noise reduction options

Control at the source	Control in the path	Control at the listener
Maintenance	Enclosure(s)	Relocate listener
Avoid resonance	Barrier(s)	Enclose listener
Relocate source/ space planning	Muffler(s)	Hearing protection
Remove unnecessary sources	Absorptive treatment	Masking
Use quieter model(s)	Vibration isolation	
Redesign source to be quieter	Active noise control	

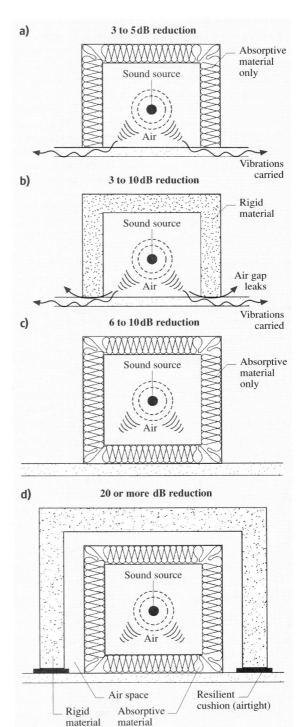

Fig. 11.20a–d Noise reduction effectiveness of enclosures sitting on a solid surface (**a**) enclosing the source on five sides with absorptive material only, (**b**) enclosing the source on five sides with one layer of rigid, nonporous material, (**c**) completely enclosing the source with absorptive material only, and (**d**) surrounding the source in an isolated, multilayered enclosure

this chapter, air gaps can significantly compromise the noise reduction effectiveness of partitions. An enclosure with any side open is not an enclosure but a barrier, and the noise reduction effectiveness of barriers is limited by diffraction to 15 dBA, independent of the barrier material (as long as the material would provide at least 20 dBA of noise reduction on its own). Enclosures, on the other hand, can provide up to 70 dBA of reduction.

2. The enclosure must be isolated from floors or any structural members of a building. An enclosure covering the sides and top of a noise source but not the bottom (since the source is sitting on the floor or ground) can compromise its effectiveness for several reasons. First, the chances of the sides of the enclosure perfectly sealing to the ground are slim, and therefore, air gaps would result. Second, vibrations will be carried along the ground or floor since the source is in direct contact with it. The only way to reduce these vibrations is to vibrationally isolate the source from the ground or floor using tuned springs (appropriate for the source), pads, or a floating floor.

3. The enclosure should be constructed using nonporous materials. Sound absorptive material can be effective in reducing noise when it is used as part of a multilayered enclosure (on the inside); however, absorptive material by itself is not effective in reducing noise.

4. The enclosure designer must consider that some sources require ventilation. This cannot translate to leaving an opening in the enclosure without compromising the noise control effectiveness of the enclosure. Ventilation systems must be developed that minimize noise transmission.
5. The enclosure should be constructed using multi-layered construction for maximum efficiency. As is mentioned earlier in this chapter, doubling the mass of an enclosure would add 6 dB to its noise reduction effectiveness. This can easily lead to excessive weight for an effective homogeneous enclosure. As for single partitions, multilayered enclosures can add more than 20 dB of effectiveness (under similar space requirements) to massive enclosures with a fraction of the weight.

11.2.3 Barriers

A barrier is contrasted from an enclosure by being open to the air on at least one side. Because of diffraction, noise barriers are limited to 15 dBA of noise reduction capability, independent of the material. This limited effectiveness is compromised even more if there are reflective ceilings above the barrier because sound reflected off the ceiling minimizes the barrier's effectiveness. Therefore, wherever noise barriers are used indoors, an absorptive ceiling should be installed above them. It is also important to have no air spaces within or under the barriers, since this will compromise their already limited effectiveness.

The noise reduction effectiveness of barriers is typically rated by the insertion loss (denoted by IL). The IL is the reduction in sound pressure level, at a specific location, resulting from the installation of a barrier. The IL is then the difference between conditions with and without a barrier.

To provide any insertion loss, a barrier must break the line of sight between the sound source and listener. In other words, if you can see a sound source on the other side of a barrier, that barrier is providing no sound reduction (from that source) for you. Breaking this line of sight typically provides a minimum of 3–5 dBA of insertion loss, with insertion loss increasing as one goes further into the shadow zone of the barrier.

Bear in mind that a row of trees is ineffective for outdoor noise control. Only solid walls or earth berms breaking the line of sight between the noise source and building occupants will produce any meaningful noise reduction.

11.2.4 Mufflers

Mufflers are devices that are inserted in the path of ductwork or piping with the specific intention of reducing sound traveling through that conduit. The effectiveness of mufflers is typically rated using insertion loss. Mufflers must be designed for each purpose to preserve the required static or back pressure characteristics of the equipment and to reduce noise in the appropriate frequency range(s). For that reason, each muffler is unique to its installation.

11.2.5 Absorptive Treatment

Absorptive treatment within a room can reduce reverberation and, in this process, reduce noise levels by up to 10 dBA. Bear in mind, though, that absorptive treatment is only effective for reducing reverberation within a room and not for reducing transmission of sound between rooms.

11.2.6 Direct Impact and Vibration Isolation

Mechanical equipment can generate vibration that can travel through a building's structural members to affect remote locations within a building. It is therefore prudent to isolate any heavy equipment from any structural members of buildings. This can be accomplished by mounting the equipment on springs, pads, and/or inertia blocks; however, the selection of specific isolating devices (especially springs) should be performed by a specialist trained in vibration analysis. The main reason for this is that each vibration isolation device is tuned to a specific frequency range. If this is not matched properly with the treated equipment, the devices can amplify the vibration and cause more of a problem than would have occurred without any treatment.

Mechanical equipment and, to a lesser degree, footfalls generate direct impacts that should also be addressed by properly isolating floors from building structures.

11.2.7 Active Noise Control

Passive noise control involves all of the noise control methods discussed above, in which the sound field is not directly altered. Active noise control involves electronically altering the character of the sound wave to reduce its level. In this case, a microphone measures the noise and a processor generates a mirror image of

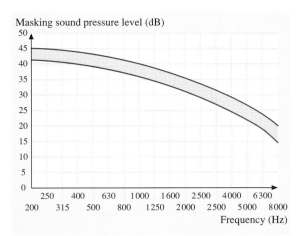

Fig. 11.21 Typical characteristics of effective masking systems (frequency response range in *shaded area*)

(180° out of phase from) that source. This mirror image is then reproduced by a loudspeaker in the path of the original sound. This new sound cancels enough of the original signal to reduce levels by up to 40 dB under the appropriate conditions. Although this is a very powerful noise control tool, active noise control is only practical in local environments and for tones below 500 Hz. Ventilation ducts provide ideal environments for active noise control systems because they are enclosed environments and their noise is often dominated by low frequency tones (associated with the fan characteristics).

11.2.8 Masking

As long as background sound levels are low in a building, one way of easing a noise problem is to add a more pleasing sound to the environment to make the noise less noticeable. This is especially desirable in open-plan offices where people can clearly hear activities in other offices and areas (since their offices are not completely enclosed). Any desirable sound can provide masking, but most often electronic masking systems are used, comprised of loudspeakers placed between

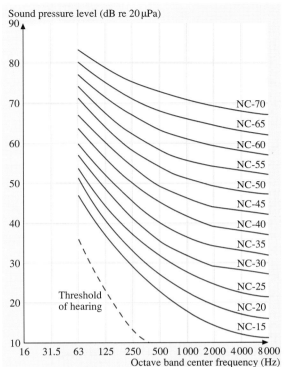

Fig. 11.22 NC curves

dropped ceilings and structural ceilings. These loudspeakers are connected to signal processors that are set to generate sounds similar to those generated by typical heating, ventilating and air conditioning (HVAC) systems. Although many people think of masking system sounds as white noise, which has an equal amount of energy in all audible frequencies, a typical masking system frequency response is more like that shown in Fig. 11.21, where less emphasis is placed on higher frequencies. Whatever the response, it is advisable to have an electronic masking system installed by a contractor with experience in this area. If the system is set at too high a level or with a harsh frequency response, the resulting environment may be more unpleasant with the masking system than without.

11.3 Noise Ratings for Steady Background Sound Levels

There are two standardized methods for rating the background sound levels inside rooms. Each of these methods is referenced in ANSI standard S12.2 and the *American Society of Heating, Refrigerating and Air-Conditioning Engineers (ASHRAE) Handbook*. These are the balanced noise criterion (NCB) and the room criterion (RC)

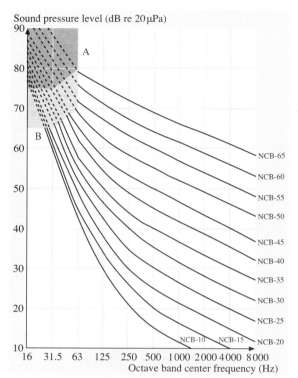

Fig. 11.23 NCB curves. Sound in the *region labeled A* can cause perceptible noise-induced vibration such as rattling of doors, windows, or fixtures. Sound in the *region labeled B* may generate lower levels of these noise-induced vibrations curves. The NCB curves are updated versions of the widely used noise criterion (NC) curves, different in that they are extended down to the 16 and 31.5 Hz octave

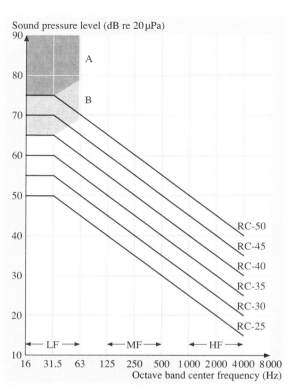

Fig. 11.24 RC curves. The *A and B regions* have the same meanings as the NCB curves in Fig. 11.23. The LF, MF, and HF regions are designated as low-frequency (rumble), medium-frequency (roar), and high-frequency (hiss) ranges, respectively

bands. Figure 11.22 shows the NC curves and Fig. 11.23 shows the NCB curves.

Table 11.8 Recommended background noise criteria for common spaces

Category of space	Specific uses	NC, NCB or RC(N) range	dBA limit
Sensitive listening spaces	Broadcast and recording studios, concert halls	15 to 20	25 dBA
Performance spaces	Theaters, churches, video and teleconferencing	20 to 25	30 dBA
Presentation spaces	Large conference rooms, small auditoriums, movie theaters, courtrooms, meeting and banquet rooms, executive offices	25 to 30	35 dBA
Private spaces	Offices, small conference rooms, classrooms, private residences, hospitals, hotels, libraries	30 to 35	40 dBA
Public spaces	Restaurants, lobbies, open-plan offices, clinics	35 to 40	45 dBA
Service and support spaces	Computer equipment rooms, public circulation areas, arenas, convention centers	40 to 45	50 dBA

As for the room criterion (RC) method, the NCB method was developed to better consider low-frequency noise generated by HVAC systems. In each case, standardized curves are used in conjunction with measured noise levels in a room to yield a single-number rating for HVAC noise. The NC and NCB rating values result from plotting the measured octave band sound pressure level spectrum on each chart and noting the lowest standard curve that is not exceeded by the measured values. This single number is then rated against design guidelines established in the two references above. Determining the RC rating is more complicated, based on the 1000 Hz value of the standard RC curve (in Fig. 11.24) that matches the arithmetic average of the 500, 1000, and 2000 Hz values of the SPL room spectrum. The quality of the sound is also rated by the RC method, based on the deviations of the measured SPL spectrum from the standard RC curve. A *neutral* spectrum, designated by N and used in most RC specifications, implies that levels below 500 Hz do not exceed an RC curve by more than 5 dB and levels above 1000 Hz do not exceed an RC curve by more than 3 dB. When deviations greater than these exist, the RC rating is given the designation of the frequency range (from Fig. 11.24) in which the greatest deviations occur. The levels in the octave bands below 500 Hz are much lower for RC curves than those for the NCB curves. The purpose of this lower limit is to consider fluctuations or surges at the ventilation outlets better. These fluctuations are not considered for NCB ratings.

Table 11.8 lists generally recommended NC, NCB and RC(N) ranges for common spaces.

11.4 Noise Sources in Buildings

The most common noise sources in buildings, other than the inhabitants, are related to HVAC systems, plumbing systems, electrical systems, and exterior sources. Figure 11.25 shows the most common paths for intrusive noise in buildings.

Although there is legislation that often deals with noise disturbances directly related to people as sources, these other sources are typically unregulated and their associated noise levels are governed by the desires of developers, builders, and occupants.

11.4.1 HVAC Systems

Noise sources associated with HVAC systems can be separated into two general categories – mechanical equipment and duct-borne/airflow noise.

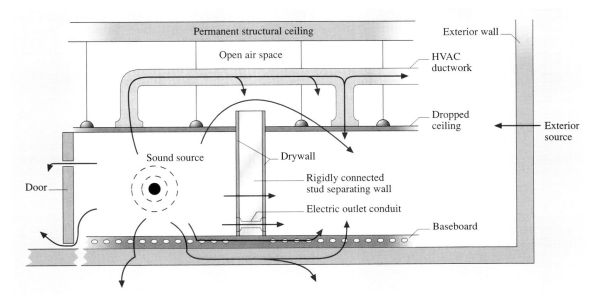

Fig. 11.25 Common noise leaks occur through these nine paths

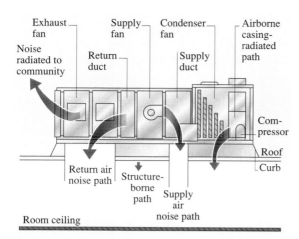

Fig. 11.26 Common noise paths for a typical rooftop HVAC unit

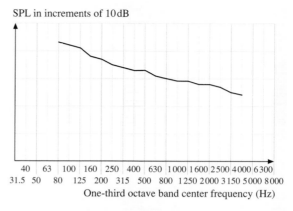

Fig. 11.27 Average transportation noise spectrum shape (adapted from data in ASTM E 1332)

Mechanical Equipment

Common mechanical equipment associated with HVAC systems includes pumps, compressors, chillers, generators, and air handlers. Rotating components of these pieces of equipment, such as gears and fans, generate most of the noise that causes concerns in buildings. Rotating mechanical equipment typically generate tones, their frequencies being associated with their rotational speeds. When mechanical equipment is housed in rooms within buildings, its associated noise can affect rooms throughout a building. When mechanical equipment is placed on rooftops or slabs outside buildings, its associated noise can also affect the surrounding communities. Figure 11.26 shows noise sources and paths for a typical rooftop HVAC unit.

Mechanical equipment not only generates noise in buildings, but also generates vibrations that can propagate throughout a building's structural members if not properly isolated. These vibrations can excite building members far from the sources and cause remote building components to rattle and generate their own noise.

Duct-Borne/Airflow Noise

Air is typically carried throughout a building using a system of ductwork. Ducts carry the tones generated by fans and they cause additional noise by inducing turbulence in the airflow. Some of this noise is carried through the ductwork to rooms in buildings (out of grilles in each room) and some of it is radiated directly from the duct walls (known as breakout noise).

11.4.2 Plumbing Systems

The most common plumbing system noise sources are water flowing through pipes and noise radiating from the walls of pipes.

11.4.3 Electrical Systems

The most common electrical system noise sources are transformers and noise radiated from associated conduit (by carrying vibrations that excite walls).

11.4.4 Exterior Sources

The most common exterior noise sources that affect building inhabitants are those associated with transportation (such as vehicular traffic, rail, and aircraft), industrial operations, and mechanical equipment from nearby buildings or mounted outside the same building. Figure 11.27 shows an average spectrum shape for transportation noise sources used by ASTM E 1332. This shows the concentration of sound in the low frequency (below 500 Hz) range, emphasizing the need for exterior-wall noise control design measures that effectively address low frequencies.

11.5 Noise Control Methods for Building Systems

The principles discussed in Sect. 11.2 can be put into practice using the general design concepts listed in this section. Specific dimensions and sound ratings have intentionally been omitted from the drawings in this section because these drawings stress the design principles rather than specific installations. These design concepts can be used to develop specific designs for each project.

11.5.1 Walls, Floor/Ceilings, Window and Door Assemblies

A key issue with all of these designs is to stop the path of sound and vibration through rigid connections between building components. Generous use of air spaces and resilient materials is the most effective method for this to occur. Another key issue is avoiding openings in and around partitions (for the reasons stated in Sect. 11.1.6). All openings must therefore be completely sealed. These two themes will be repeated in most of the designs listed in this section.

Walls

The most effective wall system for noise reduction is a double-wall system, in which wall sections (including studs) are separated by an air space. The larger the depth of the air space, the more effective the wall will be in controlling sound. Double walls can be especially effective for controlling low-frequency sound, as long as the walls are properly sealed and isolated from the building structure.

Figure 11.28 shows the general concept of a double-studded wall that is effective for noise control. Although metal studs are shown in the drawing, wooden studs would provide comparable results in this design because of the air gap. Figure 11.29 shows a full floor-to-structural-ceiling double-wall construction concept.

Note that the wall must be sealed with a non-hardening material at its perimeter to avoid any sound leakage through air gaps (since a hardening material may crack and leave gaps). Figure 11.30 shows a double-wall concept using a concrete masonry unit (CMU) and studded component.

All of these double-wall designs will provide STC values in excess of 55; however, if low-frequency noise

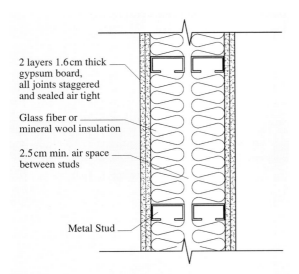

Fig. 11.28 Cross section of a double-wall design

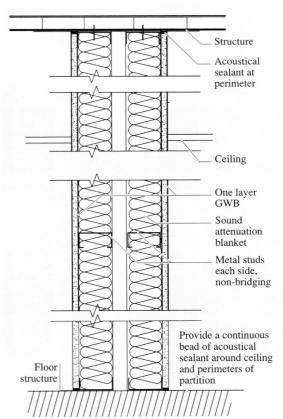

Fig. 11.29 Cross section of a double-wall design extending from floor to structural ceiling

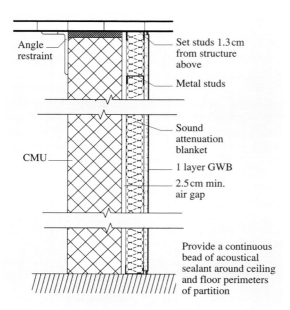

Fig. 11.30 Cross section of a floor to structural ceiling double-wall design with a CMU wall

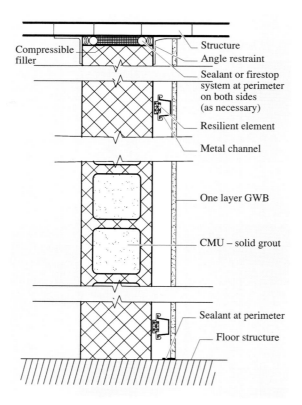

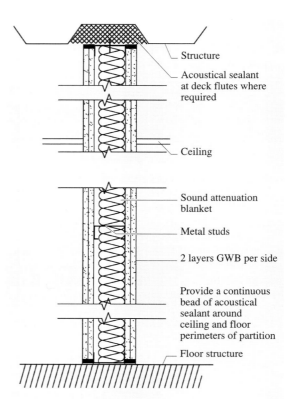

Fig. 11.32 Cross section of a single-studded sound insulating wall meeting a fluted metal structural deck

control is an issue, the STC rating does not address that and larger air spaces between wall components will increase the low-frequency noise control effectiveness.

Double-wall constructions are often impractical because of space restrictions. In these cases, an alternative to the CMU construction in Fig. 11.30 would be that shown in Fig. 11.31, in which the air space is replaced by a resilient element.

When working with gypsum board, Fig. 11.32 shows a design concept that will provide an STC of at least 50 as long as it is sealed properly. Note that a fluted structural deck may provide air gaps at the top of the wall that would need to be sealed to preserve the noise reduction effectiveness of the system.

When a room needs to be isolated from the building structure (which would be the case when structure-borne vibrations cause unwanted noise and vibration in certain rooms in which sensitive activities are tak-

Fig. 11.31 Cross section of alternate construction to a double-wall design in Fig. 11.30

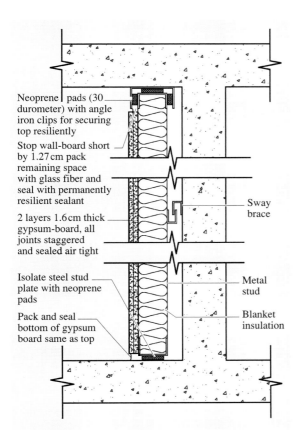

Fig. 11.33 Cross section of a resiliently mounted wall on a concrete building structure

ing place), Fig. 11.33 shows the conceptual design for a resiliently mounted wall. Figure 11.34 shows appropriate designs for intersections of double-studded walls.

A common path for sound leaks between rooms in buildings with large windows as facades is the connection between walls and those windows. Figures 11.35 and 11.36 show designs that effectively minimize sound transmission between rooms having this situation, Fig. 11.35 with a permanent installation and Fig. 11.36 with a curtain wall design.

Floor/Ceilings

Noise control design for floor/ceiling assemblies typically involves a significant amount of vibration isolation to minimize the transmission of footfall and other impact noise between floors. The design concepts that minimize vibration between floors can also be used to minimize sound transmission between floors.

The most common noise control design for floors is the floating floor system, in which a layer of resilient material is placed between the floor structure and the subflooring. More complicated measures are required for effective control of the high vibration and noise levels that can be associated with mechanical equipment in buildings. Figures 11.37–11.43 offer design concepts for resiliently mounted ceilings for mechanical rooms or other noisy spaces. Figure 11.44 shows a sound isolating design for a metal roof deck and Fig. 11.45 shows a sound isolating design for a concrete roof deck.

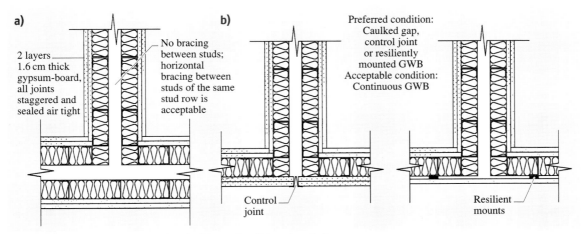

Fig. 11.34 Cross section of effective double-stud wall junction designs

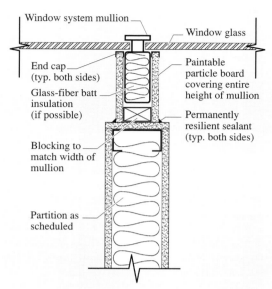

Fig. 11.35 Cross section of an effective mullion/partition junction

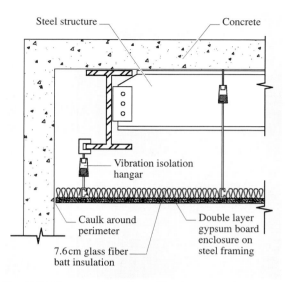

Fig. 11.37 Cross section of a resiliently mounted gypsum-board ceiling

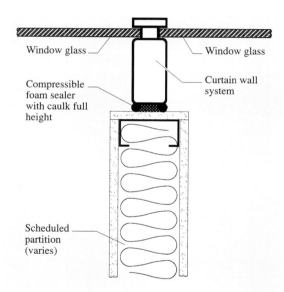

Fig. 11.36 Cross section of an effective mullion/partition junction with a curtain wall system

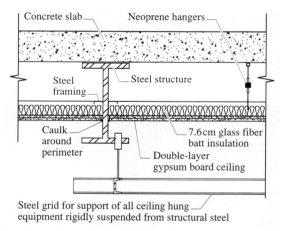

Fig. 11.38 Cross section of a sound insulating gypsum board ceiling framed from I-beams

Windows and Doors

Windows and doors often have the lowest TL ratings of a wall system and are therefore often the sources of noise leaks through partitions. In addition to their lower TL ratings than the main wall, installation with air gaps around them can significantly compromise their noise control effectiveness.

Many window manufacturers provide models that offer significant sound insulation and each of these manufacturers offers its own acoustical test data and design parameters. As for walls, larger air spaces between double-paned designs will provide better sound insulation than single-paned windows of the same thickness. For those wishing to manufacture custom sound-insulating windows, Fig. 11.46 shows some design

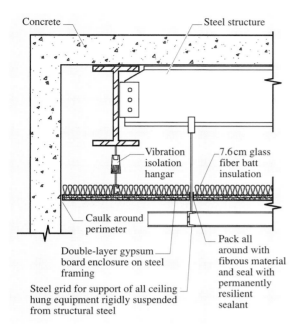

Fig. 11.39 Cross section of an effective sound insulating ceiling for mechanical rooms

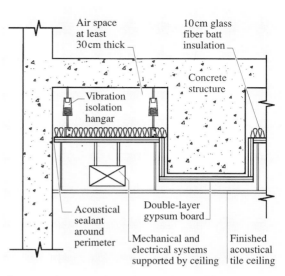

Fig. 11.41 Cross section of a resilient ceiling supporting mechanical and electrical system equipment

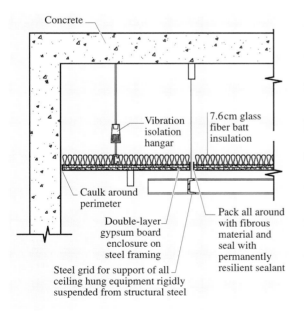

Fig. 11.40 Cross section of a ceiling resiliently mounted from a concrete slab

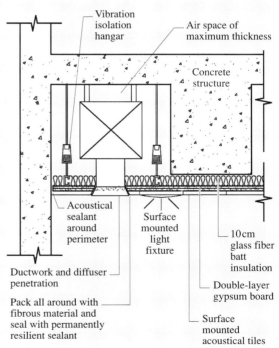

Fig. 11.42 Cross section of a resilient ceiling with HVAC equipment

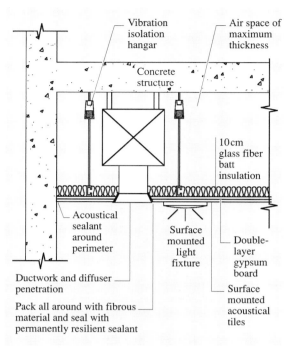

Fig. 11.43 Cross section of an alternative resilient ceiling design

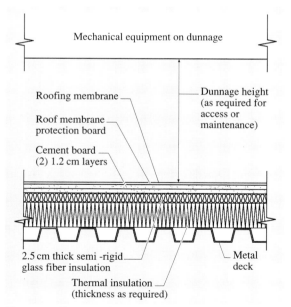

Fig. 11.44 Cross section of a sound isolating sandwich design for a metal roof deck

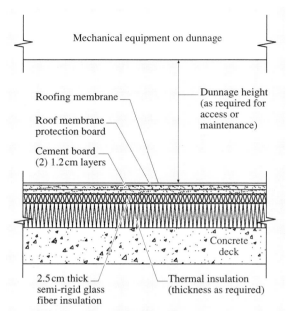

Fig. 11.45 Cross section of a sound isolating sandwich design for a concrete roof deck

guidelines for high-TL interior windows and Fig. 11.47 shows design guidelines for high-TL exterior windows.

As for windows there are many door manufacturers that provide sound insulation data for their prefabricated products. In general, solid-core doors are more effective than hollow-core doors for noise reduction. To maximize the effectiveness of these doors, they should be gasketed around their perimeters and they should either have drop seals or rubber skirts on their undersides, which should contact solid floor thresholds.

11.5.2 HVAC Systems

ASHRAE publishes thorough discussions on HVAC noise sources and their control in their *Fundamentals and Applications Handbooks*. The basic principles to bear in mind for minimizing HVAC noise are minimizing airflow turbulence and isolating vibrations, each of which is able to carry noise far from the source to remote locations in buildings.

Airflow Turbulence

Airflow turbulence noise can be minimized by

- lowering flow rates as much as possible,

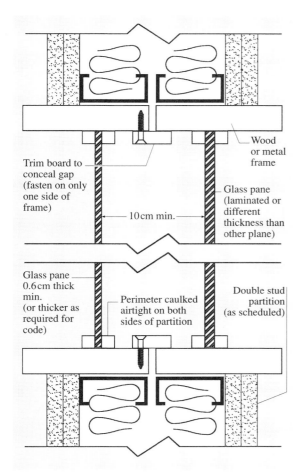

Fig. 11.46 Cross section of a high-TL interior window in a double-wall design

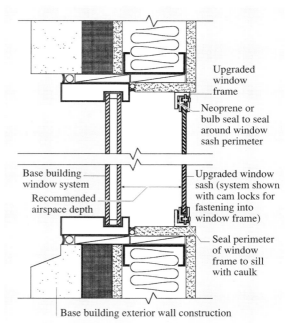

Fig. 11.47 Cross section of an upgraded exterior window for high-TL design

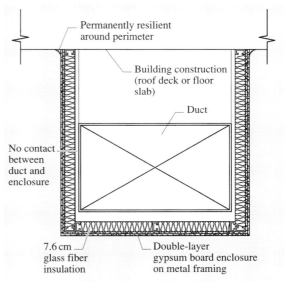

Fig. 11.48 Cross section of a double-layer gypsum-board duct enclosure

- distributing the flow as evenly as possible throughout the ducted system,
- providing long ducted passages and transitions that avoid sharp changes in direction or discontinuities in cross-sectional area,
- installing properly designed silencers and absorptive lining in the flow path that do not unduly compromise the efficiency of the HVAC system,
- wrapping ductwork with lagging materials, and
- enclosing ductwork and HVAC equipment.

Figures 11.48 through 11.50 offer effective noise control designs for duct and hanging HVAC equipment enclosures.

Another issue to consider with ductwork is sound leaks through wall and ceiling duct penetrations. If these penetrations are not sealed properly, they will diminish the noise reduction effectiveness of the wall or ceiling

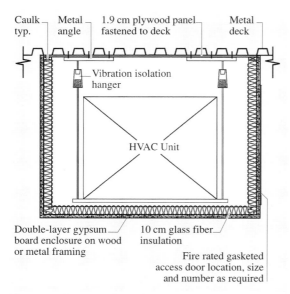

Fig. 11.49 Cross section of a duct enclosure resiliently mounted to a metal deck

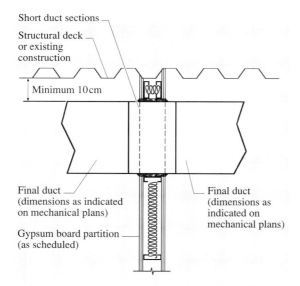

Fig. 11.51 Cross section of a duct penetrating a gypsum-board partition near a deck

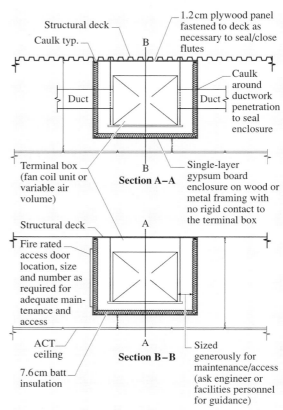

Fig. 11.50 Sections of a ducted terminal box enclosure mounted to a structural deck

that they are penetrating. Figure 11.51 shows an effective design for a single duct penetrating a wall and Fig. 11.52 shows an effective design for multiple duct penetrations. Figure 11.53 shows design considerations for the placement of a duct silencer near a wall penetration.

Vibration Isolation

Springs and pads are the most common tools used to minimize the transfer of vibration from HVAC units to the building structure. The key here is to eliminate any rigid connections between the units and the structure. Figure 11.54 shows a typical spring isolator properly mounted to reduce this transfer of vibrations effectively.

These benefits, however, will become short-circuited if improper or misaligned isolators are installed. Figure 11.55 shows improperly loaded spring isolators supporting a piece of equipment, one being overcompressed and the other being undercompressed.

Figure 11.56 shows an example of a spring isolating pipe hanger that is misaligned. The rigid connections established by this misalignment compromise the effectiveness of the system.

Figure 11.57 shows a common design in which pipe supports are rigidly attached to a floor. Figure 11.58

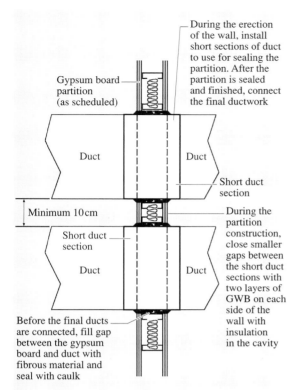

Fig. 11.52 Cross section of sealed multiple duct penetrations in a partition

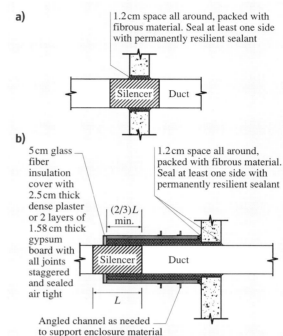

Fig. 11.53a,b Cross section of a duct penetration with appropriate silencer locations. (**a**) preferred location; (**b**) alternate location

Fig. 11.54 A properly mounted spring isolator

shows an effective design to reduce the transfer of vibrations between the piping and the floor using neoprene pads and spring hangers. Note that any spring hangers must be appropriately sized for each situation. When lateral restraints are required for large equipment, Fig. 11.59 offers a general design for effective vibration isolation.

Enclosures for Floor-Based Equipment

When barriers and source treatments do not provide enough noise control, the only practical noise control option for many floor-based units may be a full enclosure. Key issues for enclosures of HVAC equipment are access (for maintenance) and ventilation. A practical design for indoor environments is shown in Fig. 11.60, in which a heavy curtain system (which can be drawn along tracks for access) surrounds the equipment. There are hoods built into the sliding curtains to provide the necessary ventilation for the equipment.

Exterior enclosures are discussed in Sect. 11.5.5 below.

11.5.3 Plumbing Systems

As for HVAC systems, the most common plumbing-system noise issues are related to reducing flow turbulence and vibration isolation. For plumbing sys-

Fig. 11.55 Examples of misaligned and misloaded spring isolators

Fig. 11.57 Pipe supports rigidly attached to a structural floor. This can channel vibrations through the building structure

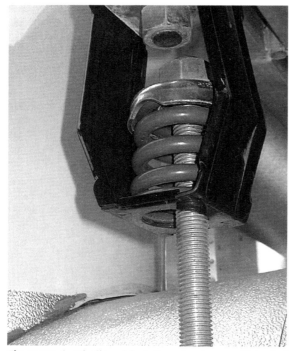

Fig. 11.56 A misaligned isolating pipe hanger. Note that the post is touching the support, thus short-circuiting the isolating effect of the hanger

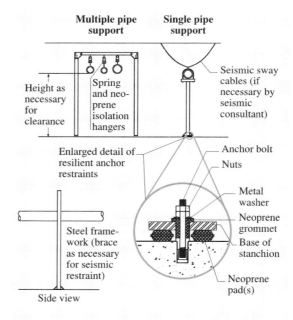

Fig. 11.58 Pipe floor supports with resilient anchors

tems we are dealing with liquid flow rather than air flow and pipe noise and vibration isolation rather than duct noise and vibration isolation. As for ducted systems, pipes can be wrapped with lagging materials to reduce breakout noise radiated from the pipe walls. Figures 11.61 and 11.62 show the general design principles to bear in mind for pipe penetrations through common

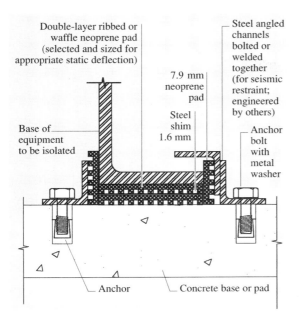

Fig. 11.59 Cross section of a lateral restraint for an equipment base on neoprene pads

Fig. 11.60 A heavy curtain enclosure system for floor-based equipment

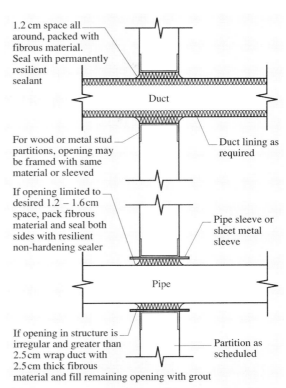

Fig. 11.61 Cross sections of sealed wall penetrations with a duct and a pipe

walls and Fig. 11.63 shows these principles in practice in an installation.

11.5.4 Electrical Systems

Electrical system noise control issues typically deal with transformer noise control and the vibration channel set up by rigid conduit between noise and vibration sources and noise-sensitive spaces. When transformers cause noise issues, they are most commonly addressed by enclosing the transformers. Conduit issues can be addressed by using a flexible conduit that is not stretched. Figure 11.64 shows an example of stretched electrical conduit, which provides a clear vibration channel from the source to remote locations.

An air space between an electrical box and other equipment will break any vibration channel that would otherwise have been established, as is shown in Fig. 11.65.

11.5.5 Exterior Sources

Environmental noise sources, such as transportation or industrial sources, can only be controlled in buildings by using exterior-wall designs that effectively reduce noise in the frequency ranges of those sources. Exterior noise sources, such as rooftop mechanical equipment or mechanical equipment in lots beside buildings (associated with the operations of the building in question) can be controlled by appropriately designed enclosure systems.

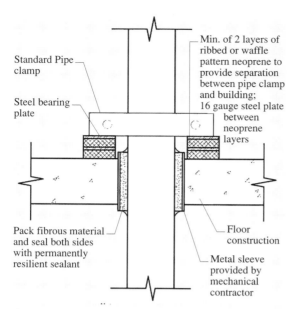

Fig. 11.62 General considerations for sealing pipe penetrations

Fig. 11.64 An example of stretched flexible conduit

Fig. 11.63 Examples of sealed pipe and duct penetrations

Fig. 11.65 An example of separating an electrical box from mechanical equipment to isolate vibrations

Figure 11.66 shows an enclosure for a rooftop HVAC unit and Fig. 11.67 shows an enclosure for a rooftop chiller unit.

Note in Fig. 11.66 the relief hoods that provide ventilation while directing the sound away from the community. When designing these types of enclosures, it would be prudent to face any ventilation openings away from noise-sensitive locations, including skylights on the roof of the building supporting the unit. Figure 11.68 shows a chiller enclosure at ground level next to a building near a residential community.

11.6 Acoustical Privacy in Buildings

Privacy issues can be a concern for spaces in which speech is a predominant source. A partial list of these areas includes offices, dwelling units, classrooms, health care facilities, libraries, hotels, and motels. The office environment provides a good example of privacy concerns and how they are handled. These solutions can be used for sound privacy in most situations.

11.6.1 Office Acoustics Concerns

The most common concerns related to office acoustics are privacy and distractions. Distractions can affect productivity. The concern for privacy has recently received much broader attention due to the passage of numerous privacy-protection laws by the US federal government and the European Parliament. Current legislation in the USA requires that the transfer of health records, in whatever format, conforms with requirements for privacy (through the Health Insurance Portability and Accounting Act of 1996). This includes privacy for digital transfer of health records between health care providers and presumably aural privacy as well. Similar concerns are also entering the realm of potential legislation for the financial world (through the Gramm Leach Bliley Financial Services Modernization Act of 1999, relating to privacy of financial transactions).

11.6.2 Metrics for Speech Privacy

Many metrics are available for rating office environments, particularly those related to speech privacy. Listed below are the most popular.

Articulation Index

Speech intelligibility and acoustics in open-plan offices have been rated in terms of the articulation index (AI) until recently (through ANSI S3.5), but AI is still being used by many consultants. AI is a measure of the ratio between a voice level and steady background noise (background noise from mechanical equipment, traffic, or electronic sound masking). AI was originally developed to evaluate communication systems, and has been widely used to assess conditions for speech intelligibility in rooms. AI values range from near 0 (low speech level and relatively high background noise resulting in poor intelligibility and good speech privacy) to 1.0 (high voice level and low background noise resulting in excellent communication and no speech privacy).

Fig. 11.66 An example of a rooftop HVAC unit enclosure

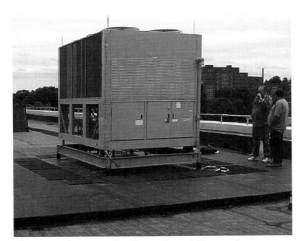

Fig. 11.67 An example of a rooftop chiller enclosure

Fig. 11.68 An example of a ground-level chiller enclosure

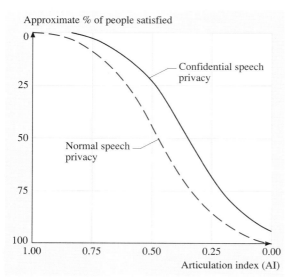

Fig. 11.69 Correlation between articulation index and speech privacy (adapted from [11.1])

When privacy is desired, it is necessary to have a low AI. When communication is desired, it is necessary to have a high AI so people can clearly understand speech.

Figure 11.69 shows how AI relates to speech privacy. This graph provides general guidelines from experience. Speech privacy goals are divided into three categories – minimal distraction, normal speech privacy, and confidential speech privacy. Minimal distraction corresponds to an AI of 0.35 or less. Normal speech privacy, in which the average person can work without distraction although occasional parts of outside conversations can be heard, corresponds to an AI of 0.20 or less. Confidential privacy, where the average person can carry on discussions with assurance that he or she will not be understood by neighbors, corresponds to an AI of 0.05 or less. There is no privacy when AI exceeds 0.40.

Speech Intelligibility Index

AI has been replaced by the speech intelligibility index (SII) in recent versions of ANSI S3.5. SII is still, like AI, a weighted speech-to-noise ratio. However, it is somewhat more complex to calculate than AI and includes revised frequency weightings and the masking effect of one frequency band on nearby frequency bands. Like AI, it has values that range between 0 and 1, but for the same conditions SII values are slightly larger than AI values.

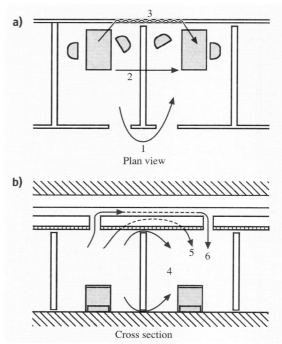

Fig. 11.70a,b The most common paths for sound leakage between closed offices. (**a**) Plan view; (**b**) cross section (1-diffraction between openings, 2-transmission through partitions, 3-structure-borne transmission, 4-transmission through air gaps in wall seals, 5-diffraction through plenum when partition does not seal to structural deck, and 6-transmission through ductwork)

Privacy Index

Low AI values indicate a higher degree of privacy. Since this may be confusing to laypersons who want to focus on better privacy conditions, a metric call privacy index (PI) has been published in ASTM E 1130. The Privacy Index (PI) is defined as:

$$PI = (1 - AI) \times 100, \qquad (11.9)$$

presented as a percentage. Higher PI values indicate more privacy; lower PI values indicate less privacy. For example, an AI of 0.10 corresponds to a PI of 90%.

Generally accepted practice for the design of open-plan offices refers to two levels of speech privacy. *Confidential privacy* is defined as a condition in which speech may be detected but not understood. *Normal privacy* allows modest amounts of intelligibility, but normal work patterns are usually not interrupted. These terms are standardized in ASTM E 1374.

Table 11.9 AI, SII and PI for open plan offices

AI	SII	PI	Privacy condition	Office environment
> 0.65	> 0.75	< 35%	Good communication	Necessary when communication is desirable (conference rooms, classrooms, auditoriums, etc.)
> 0.40	> 0.45	< 60%	No privacy	Clear intelligibility of conversations and distraction
0.35	0.45	65%	Freedom from distraction	Reasonable work conditions not requiring heavy concentration or speech privacy; can hear and understand neighboring conversations
0.20	0.27	80%	Normal speech privacy	Only occasional intelligibility from a neighbor's conversation; work patterns not interrupted
< 0.05	< 0.10	> 95%	Confidential speech privacy	Aware of neighbor's conversation but it is not intelligible

Table 11.9 compares AI, SII, and PI for the same types of communication and privacy conditions.

Articulation Class

In an open-plan office, the effectiveness of the ceiling in determining noise reduction is partially related to the height of the barrier that separates two workstations, the distance between a talker and listener, and the absorption characteristics. Therefore, another metric has been developed to describe a ceiling's contribution to noise reduction between typical work areas. This is the articulation class (AC), measured in an office mock-up environment in accordance with ASTM E 1110 and E 1111. The AC is the sum of the weighted sound attenuation numbers in a series of 15 test bands for a carefully specified office layout, barrier height, and ceiling height. AC values usually exceed 100 and typically range from about 150 to 220, with higher values indicating more privacy and less speech intelligibility. AC only considers the effects of noise reduction while AI, SII, and PI also consider the effects of speech level and spectrum characteristics, and background level and spectrum characteristics. Subjective ratings of AC for levels of privacy (like those listed in Table 11.9) have not been analyzed at this time.

Speech Interference Level

The international community uses the speech interference level (SIL), as defined in ISO 9921-1, to rate speech communication environments. SIL is defined as the arithmetic average of sound pressure levels in the ambient environment at the 500, 1000, 2000, and 4000 Hz octave-band frequencies. SIL can also be approximated by subtracting 8 dB from the overall dBA ambient level. Speech intelligibility is rated by the difference between the SIL and the A-weighted sound pressure level of speech at the listener's location. If this signal-to-noise ratio exceeds 10 dB, that indicates satisfactory speech communication. Table 11.10 shows general characteristics of assessing speech intelligibility against this signal-to-noise ratio.

Speech Transmission Index

While the metrics above are typically used for office environments to rate privacy between spaces, there is another metric that is widely used to describe speech intelligibility within performance and lecture spaces. This metric, known as the speech transmission index (STI), can be measured with commercially available instruments, often in terms of the metric known as the rapid speech transmission index (RASTI) to streamline the monitoring effort. STI measurements consider octave-

Table 11.10 ISO 9921-1 speech intelligibility ratings based on SIL

Signal-to-noise ratio at listener's position (dBA-SIL)	Speech intelligibility rating
< −6	Insufficient
−6 to −3	Unsatisfactory
−3 to 0	Sufficient
0 to 6	Satisfactory
6 to 12	Good
12 to 18	Very good
> 18	Excellent

band levels between 125 and 8000 Hz while RASTI measurements only consider octave-band levels at 500 and 2000 Hz. STI measurement procedures are standardized in the International Electrotechnical Commission (IEC) standard IEC 60268-16. STI, as AI and SII, ranges from 0 to 1 and, unlike AI and SII, STI takes room reverberation into account in its calculations. The meanings of STI ratings are generally similar to those for SII, with higher values being good for speech intelligibility and bad for privacy, and low values representing the opposite. While STI can be measured directly with instruments, AI and SII cannot.

Ceiling Attenuation Class

With current lightweight construction, walls often do not extend to the structure. In these cases, the path for sound through the ceiling plenum is the weakest sound path between offices. This sound path is rated with a ceiling attenuation class (CAC) that is analogous to an STC rating. The CAC value is measured in accordance with ASTM E 1414 and measures the sound transfer from one standard-sized office through an acoustic tile ceiling, then into a standard plenum, and then back into the neighboring office again through an acoustic tile ceiling. Mineral fiber acoustic tiles typically have a rating of CAC 35–39. Fiberglass tiles have significantly lower CAC values. In cases where there is a return air grille in the ceiling so air can be exhausted through the plenum, the field performance will also be significantly degraded.

11.6.3 Fully Enclosed Offices

If confidential privacy is required, employees must be located in closed offices with sealed doors and walls that extend to the structural ceiling. This point needs to be stressed because many closed offices are designed with unsealed doors and walls extending to suspended, nonstructural ceilings, allowing space for HVAC and electrical equipment. Typical suspended ceilings provide acoustic absorption for reflected sound but provide little in the way of transmission loss. This means that sound will travel through suspended ceiling panels and over walls as if the walls are barriers, thus limiting the acoustic privacy of these offices almost to the open-plan condition. Figure 11.70 shows the most common sound leakage paths between closed offices.

Speech privacy between two rooms, in general, is a function of two key parameters – noise reduction and background noise level. One must not only reduce the sound level of the unwanted source, but one must also generate sufficient background noise to mask the unwanted source, making it less intelligible and less distracting. Confidential speech privacy can be achieved in closed offices by using walls having a high transmission loss. If these types of walls are used, the background sound levels can remain low. However, for normal speech privacy in an open-plan office, one must strike a balance between a lower transmission loss (because of the limited effectiveness of barriers) and a higher background noise level (that can often be set using an electronic masking system).

For typical contemporary offices, the range of construction options is often limited to just the number of layers of gypsum board on the wall, the type of stud framing, and whether or not there will be insulation in the cavity of the wall. A single metal stud with a single layer of gypsum board on each side of the stud and no insulation in the cavity will have a performance in the range of STC 40. If the layers of gypsum board are doubled (that is, two layers on each side), then the overall STC value will increase by about 5 STC points. There is negligible difference between the performance of the wall with gypsum board that is 1.2 cm thick or 1.6 cm thick. Insulation in the cavity of a wall with a single stud may provide an additional 3–5 STC points performance. Wood studs instead of metal studs can provide a more rigid path for the transfer of sound (in a single-wall design), and may degrade the performance by 3–5 STC points. Further improvements above STC 50 require the use of resilient channels or staggered stud framing, as discussed in Sect. 11.5.1.

11.6.4 Open-Plan Offices

The open-plan office design has become very popular through the latter half of the 20th century and into the 21st. This design scheme saves money, promotes teamwork, and improves flexibility for future renovations. Many employees, however, view this design as a series of compromises in terms of space, prestige, and (most of all) privacy. As employees consider changing from the traditional closed office to open-plan cubicles, they often have concerns about their abilities to work productively in what they anticipate to be a noisier, more distracting workplace. The overwhelming complaint about open-plan office design is the lack of acoustic privacy.

The first step in the acoustic design of an office is to determine the needs of the employees in each area. Employees in some companies may need to communicate freely as part of their jobs. These workers would not need any acoustic privacy. Some may need visual privacy (using barriers that block their line of sight with others) but

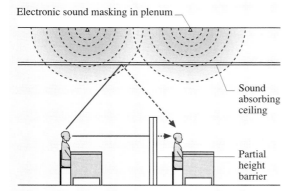

Fig. 11.71 Acoustic considerations for open-plan office design in cross section

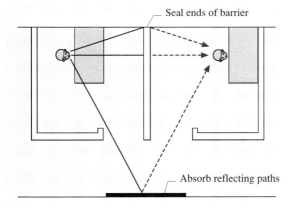

Fig. 11.72 Acoustic considerations for open-plan office design in plan view

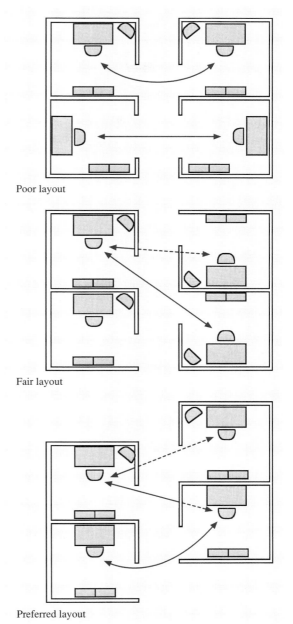

Fig. 11.73 Acoustical considerations for open-plan office layout

need only a minimal amount of speech privacy. Some may require an environment free from distractions for the performance of detailed work. It is often this last category of employees that feels most compromised by being moved into cubicles. This category of employee needs normal speech privacy, in which conversations in adjacent areas can be understood but are not distracting to concentration on tasks. Normal speech privacy is attainable for open-plan offices with the proper acoustic design.

Confidential speech privacy, in which no part of a conversation can be understood from an adjacent space, cannot be expected from open-plan office designs.

Assuming a goal of normal speech privacy, there are three acoustic design principles that must be addressed to achieve that goal in an open-plan office. These three principles are to absorb, block and cover speech in nearby workstations. These three principles are briefly discussed below.

Absorption

It is critical to have as much absorption as possible in rooms designed for open-plan offices. This will stop sound from reflecting off room surfaces to travel appreciable distances and from causing remote distractions. If one has to choose surfaces for absorptive treatment, the ceiling is the most critical surface to start with. This is because reflections off ceilings significantly compromise the already limited noise reduction capabilities of barriers. Ceiling finishes should have NRC ratings of at least 0.90 or an AC of at least 180. Ceiling treatments can be part of suspended ceilings or surface-applied. Hard, sound-reflecting lighting fixtures can degrade the performance of the ceiling finish, so it is ideal to use indirect floor lighting to keep the ceiling as absorbing as possible. If ceiling fixtures must be used, parabolic fixtures are more appropriate than others because they diffuse the sound more evenly. Figure 11.71 shows ceiling considerations. Absorptive finishes are also important on the interior of workstation panels and on any walls that may reflect sound from one cubicle to another. Figure 11.72 shows how to place absorptive materials strategically on walls to eliminate reflected paths between cubicles. Carpeting is important for floors to control impact noise (such as footfalls, dropped objects and furniture movement) as well as to absorb reflected sound energy.

Blocking

Sound is blocked in open-plan offices by cubicle barriers. To be most effective, these barriers must be designed to minimize the compromises caused by diffraction. In this regard, barriers should be at least 1.5 m tall and should have STC ratings of at least 25. The amount by which STC ratings exceed 25 will not make a difference in barrier performance (because of diffraction). They should be arranged to block the line of sight between workers (as is shown in Fig. 11.73) and should be flush with floors, window sills, and side walls.

Masking

A typical contemporary office, with sealed windows and properly maintained HVAC systems, often has too low of a background noise level to allow for normal speech privacy. The most practical way to improve privacy under these conditions is to increase the background noise level by adding an electronic sound masking system. These systems can include loudspeakers distributed throughout ceiling plenums, hanging unobtrusively from ceilings, or in each cubicle. These systems, when set up properly, sound like typical HVAC noise to employees. Figure 11.21 shows a typical spectrum for sound masking systems.

11.7 Relevant Standards

The following standards are referenced in this chapter:

US Standards

ANSI (American National Standards Institute)

S3.5	Standard Methods for Calculation of the Speech Intelligibility Index
S12.2	Standard Criteria for Evaluating Room Noise

ASTM (American Society of Testing and Materials)

C 423	Standard Test Method for Sound Absorption and Sound Absorption Coefficients by the Reverberation Room Method
E 90	Standard Test Method for Laboratory Measurement of Airborne Sound Transmission Loss of Building Partitions and Elements
E 413	Classification for Rating Sound Insulation
E 492	Standard Test Method for Laboratory Measurement of Impact Sound Transmission Through Floor-Ceiling Assemblies Using the Tapping Machine
E 989	Standard Classification for Determination of Impact Insulation Class (IIC)
E 1041	Standard Guide for Measurement of Masking Sound in Open Offices
E 1110	Standard Classification for Determination of Articulation Class
E 1111	Standard Test Method for Measuring the Interzone Attenuation of Ceiling Systems
E 1130	Standard Test Method for Objective Measurement of Speech Privacy in Open Offices Using Articulation Index

E 1179	Standard Specification for Sound Sources Used for Testing Open Office Components and Systems
E 1332	Standard Classification for Determination of Outdoor–Indoor Transmission Class
E 1374	Standard Guide for Open Office Acoustics and Applicable ASTM Standards
E 1375	Standard Test Method for Measuring the Interzone Attenuation of Furniture Panels Used as Acoustical Barriers
E 1376	Standard Test Method for Measuring the Interzone Attenuation of Sound Reflected by Wall Finishes and Furniture Panels
E 1414	Standard Test Method for Airborne Sound Attenuation Between Rooms Sharing a Common Ceiling Plenum

International Standards

International Electrotechnical Commission (IEC) and International Organization for Standardization (ISO)

IEC 60268-16	Objective Rating Of Speech Intelligibility By Speech Transmission Index
ISO 9921-1	Speech Interference Level And Communication Distances For Persons With Normal Hearing Capacity In Direct Communication (SIL method)

References

11.1   W.J. Cavanaugh, W.R. Farrell, P.W. Hirtle, B.G. Watters: Speech privacy in buildings, J. Acoust. Soc. Am. **34**(4), 475–492 (1962)

Part D Hearing and Signal Processing

12 Physiological Acoustics
Eric D. Young, Baltimore, USA

13 Psychoacoustics
Brian C. J. Moore, Cambridge, UK

14 Acoustic Signal Processing
William M. Hartmann, East Lansing, USA

12. Physiological Acoustics

The analysis of sound in the peripheral auditory system solves three important problems. First, sound energy impinging on the head must be captured and presented to the transduction apparatus in the ear as a suitable mechanical signal; second, this mechanical signal needs to be transduced into a neural representation that can be used by the brain; third, the resulting neural representation needs to be analyzed by central neurons to extract information useful to the animal. This chapter provides an overview of some aspects of the first two of these processes. The description is entirely focused on the mammalian auditory system, primarily on human hearing and on the hearing of a few commonly used laboratory animals (mainly rodents and carnivores). Useful summaries of non-mammalian hearing are available [12.1]. Because of the large size of the literature, review papers are referenced wherever possible.

12.1	**The External and Middle Ear**	429
	12.1.1 External Ear	429
	12.1.2 Middle Ear	432
12.2	**Cochlea**	434
	12.2.1 Anatomy of the Cochlea	434
	12.2.2 Basilar-Membrane Vibration and Frequency Analysis in the Cochlea	436
	12.2.3 Representation of Sound in the Auditory Nerve	441
	12.2.4 Hair Cells	443
12.3	**Auditory Nerve and Central Nervous System**	449
	12.3.1 AN Responses to Complex Stimuli	449
	12.3.2 Tasks of the Central Auditory System	451
12.4	**Summary**	452
	References	453

12.1 The External and Middle Ear

The external and middle ears capture sound energy and couple it efficiently to the cochlea. The problem that must be solved here is that sound in air is not efficiently absorbed by fluid-filled structures such as the inner ear. Most sound energy is reflected from an interface between air and water, and terrestrial animals gain up to 30 dB in auditory sensitivity by virtue of their external and middle ears.

Comprehensive reviews of the external and middle ears are available [12.2, 3]. Figure 12.1 shows an anatomical picture of the ear (Fig. 12.1a) and a schematic showing the important signals in this part of the system (Fig. 12.1b). The external ear collects sound from the ambient field (p_F) and conducts it to the eardrum where sound pressure fluctuations (p_T) are converted to mechanical motion of the middle-ear bones, the malleus (with velocity v_M), the incus, and the stapes; the velocity of the stapes (v_S) is the input signal to the cochlea, producing a pressure p_V in the scala vestibuli. The following sections describe the properties of the transfer function from p_F to p_V.

12.1.1 External Ear

The acoustical properties of the external ear transform the sound pressure (p_F) in the free field to the sound pressure at the eardrum (p_T). Two main aspects of this transformation important for hearing are:

1. efficient capture of sound impinging on the head; and
2. the directional sensitivity of the external ear, which provides a cue for localization of sound sources.

Figures 12.2a and 12.2b show external-ear transfer functions for the human and cat ear. These are the dB ratio of the sound pressure near the eardrum p_T (see the caption) to the sound in free field (p_F). The free-field sound p_F is the sound pressure that would exist at the approximate location of the eardrum if the subject were removed from the sound field completely. Thus the transformation from p_F to p_T contains all the effects of putting the subject in the sound field, including those due to refraction and reflection of sound from the

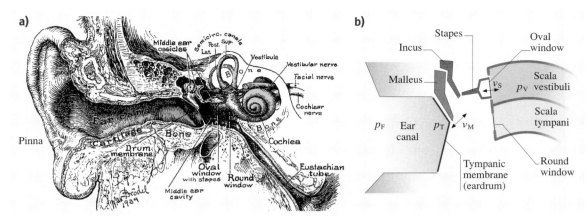

Fig. 12.1 (a) Drawing of a cross section of the human ear, showing the pinna, the ear canal, eardrum (drum membrane), middle-ear bones (ossicles), and the labyrinth. The latter contains the vestibular organs of balance and the cochlea, or inner ear. (After [12.8]) **(b)** Schematic of the mammalian external and middle ears showing the signals referred to in the text. The middle-ear bones are the malleus, incus, and stapes. The stapes terminates on the cochlea, on a flexible membrane called the oval window. The velocity of the stapes (v_S) is the input signal to the cochlea; it produces a pressure variation p_V in the fluids of the scala vestibuli, one of the chambers making up the cochlea. The structure of the cochlea and definitions of the scala vestibuli and scala tympani are shown in a later figure. The function of the external and middle ears is to capture the energy in the external sound field p_F and transfer it to stapes motion v_S or equivalently, sound pressure p_V in the scala vestibuli

surfaces of the body, the head, and the ear and the effects of propagation down the ear canal. In each case several transfer functions are shown. The dotted lines show p_T/p_F for sound originating directly in front of the subject [12.4, 5]; these curves are averaged across a number of ears. The solid lines show transfer functions measured in one ear from different locations in space [12.6, 7]. These have a fixed azimuth (i. e., position along the horizon, with 0° being directly in front of the subject) and vary in elevation (again 0° is directly in front) in the median plane. Elevations are given in the legends. External-ear transfer functions differ in detail from animal to animal. In particular, the large fluctuations at high frequencies, above 3 kHz in the human ear and 5 kHz in the cat ear, differ from subject to subject. These fluctuations are not seen in the averaged ears because they are substantially reduced by averaging across subjects, although some aspects of them remain [12.7]. However, the general features illustrated here are typical.

Capture of Sound Energy by the External Ear

Over most of the range of frequencies, the sound pressure is higher at the eardrum than in the free field in both the human (Fig. 12.2a) and the cat (Fig. 12.2b) subjects. The amplification results from resonances in the external ear canal and in the cavities of the pinna; the resonances produce the broad peak between 2 and 5 kHz [12.2]. Although this seems to be an amplification of the sound field, the pressure gain is not by itself sufficient to specify fully the sound-collecting function of the external ear, because it is really the power collection that is important and sound power is the product of pressure and velocity. A useful summary measure of power collection is the effective cross-sectional area of the ear a_{ETM} [12.2, 9]. a_{ETM} measures the ear's ability to collect sound from a diffuse pressure field incident on the head, in the sense that the sound power into the middle ear is equal to a_{ETM} multiplied by the power density in the sound field. If a_{ETM} is larger (smaller) than the anatomical area of the eardrum, for example, it means that the external ear is more (less) efficient at collecting sound than a simple eardrum-sized membrane located on the surface of the head.

Figure 12.2c shows calculations of the magnitude of a_{ETM} for the cat and human ear [12.2]. For comparison, the horizontal line shows the cross-sectional area of the external opening of the cat's pinna. Around 2–3 kHz, a_{ETM} equals the pinna area for the cat. The dotted line marked "ideal" is the maximum possible value of a_{ETM}, based on the power incident on a sphere in a diffuse field [12.2, 9]. At frequencies above 2 kHz, the cat ear is close to this maximum and the human ear is some-

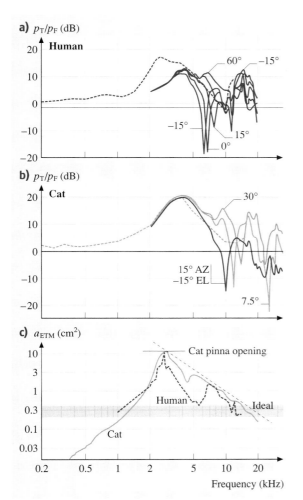

Fig. 12.2a–c Transformation of sound pressure by the external ear, considered as the ratio p_T/p_F for the human ear (**a**) and the cat ear (**b**). p_T is the pressure measured by a probe microphone in the ear canal, within 2 mm of the eardrum for all cases except the *solid curves* in (**a**), which show the pressure at the external end of the ear canal, with the ear canal blocked. The *dashed curves* are averaged across many ears, with the sound originating straight ahead, and the *solid curves* are data from one ear at various sound-source directions. The *numbers* next to the curves or in the legend identify the elevation of the sound source relative to the head. In both cases, the azimuth of the source is fixed. (**c**) Capture of sound energy by the external ear, considered as the effective cross-sectional area of the ear, mapped to the eardrum and plotted versus frequency. Curves for the cat and human ear are shown. The *shaded bar* shows the range of cross-sectional areas of the ear canal, for comparison. (Fig. 12.2a after [12.4, 7]; Fig. 12.2b after [12.5, 6]; Fig. 12.2c after [12.2])

what lower. At these frequencies, the external ear is collecting the sound power in a diffuse field nearly as efficiently as is possible. At lower frequencies, the efficiency drops rapidly; *Shaw* [12.9] attributes the drop off to three effects:

1. loss of pressure gain below the canal resonance
2. increase in the wavelength of sound compared to the size of the ear
3. increase in the impedance of the middle ear at low frequencies, decreasing the ability of the ear to absorb sound.

Figure 12.2c shows that the external ear can be considered as an acoustical horn that captures sound and couples it to the eardrum. A feeling for the effectiveness of the external ear's sound collection can be gotten by comparing the values of a_{ETM} to the cross-sectional area of the ear canal, which is 0.2–0.4 cm^2 in cat, shown by the shaded bar [12.10].

Directional Sensitivity

Directional sensitivity means that the transfer function of the external ear depends on the direction to the sound source. As a result, the spectra of broadband sounds (like noise) will be different, depending on the direction in space from which they originate. Listeners, both humans and non-human mammals, are sensitive to these differences and are able to extract sound localization information from them. For example, human observers use these so-called *spectral cues* to localize sounds in elevation and to disambiguate binaural difference cues, helping to distinguish sounds in front and behind the listener [12.11]. Cats depend on spectral cues to localize sounds in the frontal plane, both in azimuth and elevation [12.12].

The directional sensitivity of the external ear is generated by sound reflecting off the structures of the pinna. In humans the pinna is the cartilaginous structure attached to the external ear canal on the side of the head, and includes the cavities that lead to the entrance of the ear canal [12.13, 14]. In the cat, the pinna is a collection of similar cartilaginous shapes and cavities leading to the ear canal, most of which are located on the surface of the head at the ear canal opening. The external auricle, which forms the movable part of the cat's ear, serves as an additional sound collector and a way for the cat to change the directionality of its ear under voluntary control [12.15]. The nature of spectral cues in human and cat are shown in Figs. 12.2a,b. The most prominent spectral cue is the notch at mid frequencies (3–10 kHz in human and 7–20 kHz in cat). The notch is created as an interference pattern when sound propagates through the pinna. Essentially, there are multiple sound paths through the pinna due to reflections. These echoed sounds reach the ear canal after variable delays, depending on the path taken. At the ear canal, sounds can cancel at frequencies where the delays are approximately a half-cycle. The interference patterns so generated produce the notches shown in Fig. 12.2. The notches occur at relatively high frequencies, where the wavelength of sound is comparable to the lengths of the reflection paths in the pinna. In cats, the notch moves toward higher frequencies as the elevation (or the azimuth, not shown) increases [12.6, 16]. At higher frequencies, above 10 kHz in humans and 20 kHz in cat, there are additional complex changes in the transfer function of the external ear that are not easily summarized. Cats appear to be sensitive to spectral cues at all frequencies, but use the notch for localizing sounds and the higher-frequency characteristics for discriminating the locations of two sounds [12.17].

12.1.2 Middle Ear

The function of the middle ear is to transfer sound from the air to the fluids of the cochlea (e.g., from p_T to p_V in Fig. 12.1b). The process can be considered as an impedance transformation [12.18]. The specific acoustic impedance of a medium is the ratio of the sound pressure to the particle velocity of a plane wave propagating through the medium and is a property of the medium itself. When sound impinges on an interface between two media with different impedances, such as an air–water interface, energy is reflected from the boundary. In the case of air and water, the air is a low-impedance medium (low pressure, large velocity) and the water is high impedance medium (high pressure, low velocity). At the boundary, the pressure and velocity must be continuous, i.e. equal across the boundary. If the pressure is equal across the boundary, then the velocities must be different, larger in air than in water because of the differing impedances. A reflection occurs to allow both boundary conditions to be satisfied, i.e., the boundary conditions can only be satisfied by creating a third reflected wave at the boundary. The reflected wave, of course, takes energy away from the wave that propagates through the boundary. Maximum energy transfer across the boundary occurs when the impedance of the source medium equals that of the second medium (no reflection). In the middle ear, the challenge is to transform the low impedance of air to the high impedance of the cochlea, so as to couple as much energy as possible into the cochlea.

The function of the middle ear as a impedance transformer is schematized in Fig. 12.3a which shows a drawing of the eardrum and the ossicles in approximately the anatomically correct position. In this model, the eardrum and oval window are assumed to act as pistons and the ossicles are assumed to act as a lever system, rotating approximately in the plane of the figure around the axis shown by the white dot in the head of the malleus [12.3]. The area of the eardrum (A_{TM}) is larger than the area of the oval window (A_{OW}), which increases the pressure by the ratio $p_V/p_T = A_{TM}/A_{OW}$. The difference in the lengths of the malleus and incus further increases the pressure by the lever ratio L_M/L_I, and decreases the velocity by the same amount. Thus the net pressure ratio is $A_{TM}L_M/A_{OW}L_I$. The area ratio amounts to a factor of 10–40 and the lever ratio is about 1.2–2.5 in mammalian ears [12.3]. For the cat, the pressure ratio should be about 36 dB. The impedance change is the ratio of the pressure and velocity ratios, $A_{TM}/A_{OW}(L_M/L_I)^2$, which is about 29 for the human ear [12.18]. This is less than the ideal value of 3500 for an air–water interface or 135 estimated for an air–cochlea interface, but is still a significant improvement (about 15 dB).

In reality, the motion of middle-ear components is more complex than is assumed in this model [12.2]. First, the eardrum does not displace as a simple piston; second the malleus and incus undergo a more complex motion than the one-dimensional rotation assumed in Fig. 12.3a; and third, there are losses in the motion of the ossicles that are not included in the model. Figure 12.3b shows a comparison of the actual pressure transformation in the middle ear of a cat, given as p_V/p_T. The dashed line is the prediction of the transformer model (36 dB). The cat ear gives a smaller pressure p_V than the transformer model for the reasons listed above, along with the fact that additional impedances in the middle ear must be considered when doing this calculation, such as the middle-ear cavity behind the eardrum, whose air is compressed and expanded as the eardrum moves [12.2]. Figure 12.3c shows the transfer function as it is usually displayed, as a transfer admittance from the eardrum pressure p_T to the velocity in the cochlea v_S.

The overall function of the external and middle ears in collecting sound from a diffuse field in the environment and delivering it to the cochlea is shown in

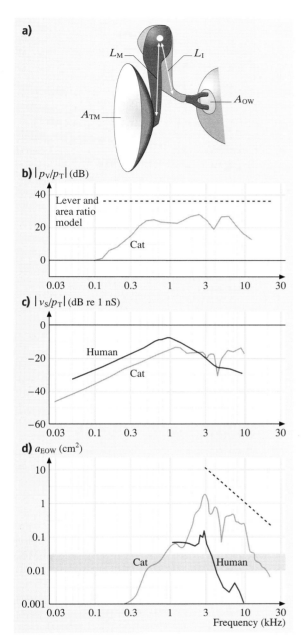

Fig. 12.3 (a) Schematic drawing of the mammalian middle ear from the eardrum (TM on the *left*) to the oval window (OW on the *right*). The malleus, incus, and stapes are shown in their approximate anatomical arrangement. The areas of the TM and OW are shown along with the lever arms of the malleus (L_M) and incus (L_I). These lever arms are drawn as if the malleus and incus rotate in the plane of the paper around an axis indicated by the *white dot*. In reality, the motion is more complex. (b) Ratio of the sound pressure in scala vestibuli (p_V) to the sound pressure at the eardrum (p_T) as a function of frequency, from measurements by Décory (unpublished doctoral thesis, 1989 [12.2]). The *dashed line* is the prediction of the model in (a) for typical dimensions of the cat middle ear. (c) Transfer admittance of the middle ear in human and cat, given as the velocity at the stapes (v_S) divided by the pressure at the eardrum (p_T). (d) Performance of the external and middle ears in sound collection plotted as the effective area of the ear, referenced to the oval window. This is the cross-sectional area across which the ear collects sound power in a diffuse sound field, plotted against frequency; the *shaded bar* shows the range of cross sectional areas of the oval window for comparison. Comparing with Fig. 12.2c shows the effect of the middle ear. (After [12.2])

Fig. 12.3d [12.2]. This figure plots the effective area of the ear as a sound collector as in Fig. 12.2c, except now it refers to the sound power delivered to the oval window. Again, the dashed line shows the performance of an ideal spherical receiver and is the same line as in Fig. 12.2c. The effective area has a bandpass shape, as in Fig. 12.2c, with a maximum at 3 kHz. The sharp drop off in performance below 3 kHz was seen in the external ear analysis and occurs because energy is not absorbed at the eardrum at low frequencies. At higher frequencies the effective area tracks the ideal receiver, but is about 10–15 dB smaller than the performance at the eardrum, which approximates the ideal for the cat. This decrease reflects the losses in the middle ear discussed above. Although the external and middle ear do not approach ideal performance, they do serve to couple sound into the cochlea. As a comparison to the effective area in Fig. 12.3d, the cross-sectional area of the oval window is about 0.01–0.03 cm^2 in the cat and human (shaded bar). Thus the effective area is larger than the area of the oval window over the mid frequencies. Moreover, if there were no middle ear, the collecting cross section of the oval window would be smaller by about 15–30 dB because of the impedance mismatch between the air and cochlear fluids [12.19].

12.2 Cochlea

The cochlea contains the ear's transduction apparatus, by which sound energy is converted into electrical activity in nerve cells. The conversion occurs in transduction cells, called inner hair cells, and is transferred to neurons in the auditory nerve (AN) that connect to the brain. However the function of the cochlea is more than a transducer. The cochlea also does a frequency analysis, in which a complex sound such as speech or music is separated into its component frequencies, a process not unlike a Fourier or wavelet transform. Auditory perception is largely based on the frequency content of sounds; such features as the identity of sounds (which speech sound?), their pitches (which musical note?), and the extent to which they interact in the ear (e.g., to make it hard to hear one sound because of the presence of another) are determined by the mixture of frequencies making them up (Chap. 13).

In this section, the steps in the transduction process in the inner ear or cochlea are described. The problems solved in this process include frequency analysis, mentioned above, but also regulating the sensitivity of process. Transduction must be very sensitive at low sound levels, so that sounds can be detected near the limit imposed by Brownian motion of the components of the cochlea [12.20]. At the same time, the cochlea must function over the wide dynamic range of sound intensities (up to 100 dB or more) that we encounter in the world. Responses to this wide dynamic range must be maintained in neurons with much more limited dynamic ranges, 20–40 dB [12.21]. In part this is accomplished by compressing the sound in the transduction process, so that acoustic sound intensities varying over a range of 60–100 dB are mapped into neural excitation over a much more limited dynamic range. Thus cochlear sensitivity must be high at low sound levels to allow soft sounds to be heard and must be reduced at high sound levels to maintain responsiveness without saturation. Both frequency tuning and dynamic-range adjustment depend on the same cochlear element, the second set of transducer cells called outer hair cells. Whereas inner hair cells convey the representation of sound to the AN, the outer hair cells participate in cochlear transduction itself, making the cochlea more sensitive, increasing its frequency selectivity, and compressing its dynamic range. These processes are described in more detail below.

The remainder of this chapter assumes some knowledge of neural excitation and synaptic transmission. These subjects are too extensive to review here, but can be quickly grasped from an introductory text on neurophysiology or neuroscience.

12.2.1 Anatomy of the Cochlea

The inner ear consists of the cochlea and the AN. The cochlea contains the transduction apparatus; it is a coiled structure, as shown in Fig. 12.1a. A cross-sectional view of a cochlea is shown in Fig. 12.4a. This figure is a low-resolution sketch that is provided to show the locations of the major parts of the structure. The section is cut through the center of the cochlear coil and shows approximately 3.5 turns of the coil. The cochlea actually consists of three fluid-filled chambers (or scalae) that coil together. A cross section of one turn of the spiral showing the three chambers is in Fig. 12.4b. The scala tympani (ST) and scala media (SM) are separated by the basilar membrane, a complex structure consisting of connective tissue and several layers of epithelial cells. The scala vestibuli (SV) is separated from the SM by Reissner's membrane, a thin epithelial sheet consisting of two cell layers. Reissner's membrane is not important mechanically in the cochlea and is usually ignored when discussing cochlear function. In the following, the mechanical properties of the basilar membrane are discussed without reference to the Reissner's membrane and it is usual to speak of the basilar membrane as if it separated the SV and ST.

What Reissner's membrane does do is to separate the fluids of the SV from the SM. The fluids in the SV and ST are typical of the fluids filling extracellular spaces in the body; this fluid, called *perilymph* is basically a dilute NaCl solution, with a variety of other constituents but with low K^+ concentration. In contrast, the fluid in SM, called *endolymph*, has a high K^+ concentration with low Na^+ and Ca^{2+} concentrations. Such a solution is rare in extracellular spaces, but is commonly found bathing the apical surfaces of hair cells in sensory organs from insects to mammals, suggesting that the high potassium concentration is important for hair-cell function. The endolymph is generated in the stria vascularis, a specialized epithelium in the lateral wall of the SM by a complex multicellular active-transport system [12.22].

Mounted on the basilar membrane is the organ of Corti, which contains the actual cochlear transduction apparatus, shown in more detail in Fig. 12.4c. The organ of Corti consists of supporting cells and hair cells. There are two types of hair cells, called inner (IHC) and outer

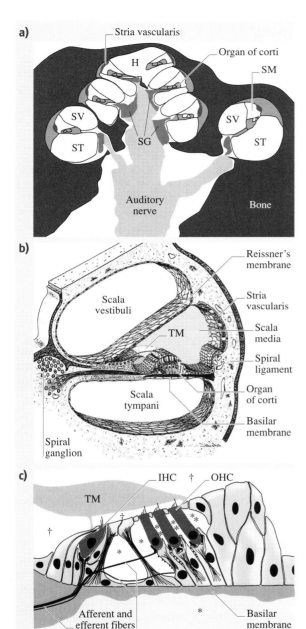

Fig. 12.4 (a) Sketch of a cross section of the cochlea, cut through the center of the coil. The cochlear spirals are cut in cross section, shown at higher resolution in (b). The cochlea consists of three fluid-filled chambers that spiral together: ST – scala tympani, SV – scala vestibuli, and SM – scala media. H is the helicotrema, a connection between the SV and ST at the apex of the cochlea. The cochlear transducer has three essential functional parts:
1. the basilar membrane which separates the SM and ST;
2. the organ of Corti, which contains the hair cells, and
3. the stria vascularis which secretes the fluid in the SM and provides an energy source to the transducer.

The AN fibers have their cell bodies in the spiral ganglion (SG); their axons project to the brain in the AN. (b) Cross section of the cochlear spiral showing the three scalae. The scalae are fluid-filled; the SV and ST contain perilymph. The SM contains endolymph. (c) Schematic drawing of the organ of Corti showing the inner (IHC) and outer hair cells (OHC) and several kinds of supporting cells. The spaces containing perilymph are marked with *asterisks* and the spaces containing endolymph with *daggers* (†). TM is the tectorial membrane, which lies on the organ of Corti and is important in stimulating the hair cells. Nerve fibers enter the organ from the SG (off to the left). Both afferent and efferent fibers are present. Afferents are the SG neurons that are connected synaptically to hair cells, mainly IHCs, and carry information to the brain. Efferents are the axons of neurons located in the brain which connect to OHCs and to the afferents under the IHCs (Fig. 12.5). (After [12.23])

(OHC), because of their positions. The AN fibers and their cell bodies in the spiral ganglion (*SG* in Fig. 12.4a) occupy the center of the cochlear coil. Spiral ganglion cells have two processes: a distal process that invades the organ of Corti and innervates one or more hair cells, usually an IHC, and a central process that travels in the AN to the brain.

The arrangement of the nerve fibers in the cochlea is reviewed in detail by *Ryugo* [12.24] and by *Warr* [12.25] and is summarized in Fig. 12.5. Both afferent and efferent fibers are present. The afferents are the neurons mentioned above with their cell bodies in the SG which carry information from the hair cells to the brain. The efferents are the axons of neurons in the brain that terminate in the cochlea and allow central control of cochlear transduction.

Considering the afferents firsts, there are two groups of SG neurons, called type I and type II. Type I neurons make up about 90–95% the population; their distal processes travel directly to the IHC in the organ of Corti. Each type I neuron innervates one IHC, and each hair cell receives a number of type I fibers; the exact number varies with species and location in the cochlea, but is typically 10–20. The connection between the IHC and type I hair cells is a standard chemical synapse, by which the hair cell excites the AN fiber. Type I AN fibers project into the core areas of the cochlear nucleus, the

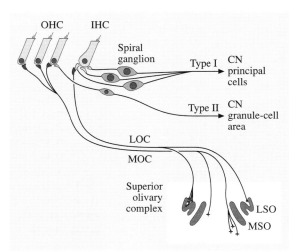

Fig. 12.5 A schematic of the wiring diagram of the mammalian cochlea. The hair cells are on the left. Afferent terminals on hair cells, those in which the hair cell excites an AN fiber, are shown *lightly shaded*; efferent terminals, in which an axon from a cell body in the central nervous system contacts a hair cell or another terminal in the organ of Corti, are shown *heavily shaded*. All the synapses in the cochlea are chemical. Inner hair cells (IHC) are contacted by the distal processes of AN fibers. These fibers have bipolar cell bodies in the spiral ganglion; their axons are myelinated and travel centrally (type I fibers) to innervate principal neurons in the cochlear nucleus (CN). Outer hair cells (OHC) are innervated by a small population of type II spiral ganglion cells whose axons terminate in granule cell regions in the CN. The efferent fibers to the cochlea are called olivocochlear neurons; they originate in the superior olivary complex (SOC, box at lower right), another part of the central auditory system. There are two kinds of efferents: the medial olivocochlear system (MOC), originates near the medial nucleus of the SOC and innervates OHCs. The lateral olivocochlear system (LOC) originates near the lateral nucleus of the SOC and innervates afferent terminals of type I afferent fibers under the IHCs. The MOC and LOC contain fibers from both sides of the brain, although most LOC fibers are ipsilateral and most MOC fibers are contralateral, as drawn. The exact ratios vary with the animal

first auditory structure in the brain; the type I fibers plus the neurons in the core of the cochlear nucleus make up the main pathway for auditory information entering the brain.

The remaining 5–10% of the spiral ganglion neurons, type II, innervate OHCs. The fibers cross the fluid spaces between the IHC and the OHC (shown by the asterisks in Fig. 12.4c) and spiral along the basilar membrane beneath the OHCs toward the base of the cochlea (toward the stapes, not shown) innervating a few OHCs along the way. Although the type II fibers are like type I fibers in that they connect hair cells (in this case OHC) to the cochlear nucleus, there are some important differences. The axons of type II SG neurons are unmyelinated, unlike the type I fibers; their central processes terminate in the granule-cell areas of the cochlear nucleus, where they contact interneurons, meaning neurons that participate in the internal circuitry of the nucleus but do not contribute their axons to the outputs of the nucleus. Finally, type II fibers do not seem to respond to sound [12.26–28], even though they do propagate action potentials [12.29, 30]. It is clear that the type I fibers are the main afferent auditory pathway, but the role of the type II fibers is unknown. In the remainder of this chapter, the term *AN fiber* refers to type I fibers only.

The efferent neurons have their cell bodies in an auditory structure in the brain called the superior olivary complex; for this reason, they are called the olivocochlear bundle (OCB). The anatomy and function of the OCB efferents are reviewed by *Warr* [12.25] and by *Guinan* [12.31]. Anatomically, there are two groups of OCB neurons. So-called lateral efferents (LOC) travel mainly to the ipsilateral cochlea and make synapses on the dendrites of type I afferents under the IHC. Thus they affect the afferent pathway directly. The second group, medial efferents (MOC), travel to both the ipsilateral and contralateral ears and make synapses on the OHCs. Their effect on the afferent pathway is thus indirect, in that they act through the OHC's effects on the transduction process in the cochlea.

12.2.2 Basilar-Membrane Vibration and Frequency Analysis in the Cochlea

An overview of the steps in cochlear transduction is shown in Fig. 12.6a. The sound pressure at the eardrum (p_T) is transduced into motion of the middle-ear ossicles, ultimately resulting in the stapes velocity (v_S), which couples sound energy into the SV. The transformations in this part of the system have mainly to do with acoustical mechanics and were described in the first section above. The stapes motion produces a sound pressure signal in the cochlea that results in vibration of the basilar membrane, which is the topic of this section. The vibration results in stimulation of hair cells, through opening and closing of transduction channels

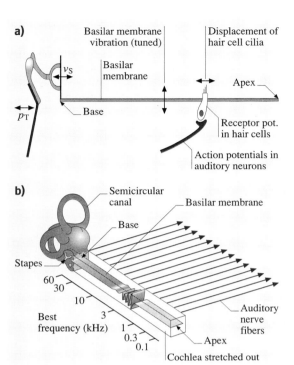

Fig. 12.6 (a) Summary of the steps in direct cochlear transduction. **(b)** Schematic diagram showing the basilar membrane in an unrolled cochlea and the nature of cochlear frequency analysis. A snapshot of basilar-membrane displacement (greatly magnified) by a tone with frequency near 3 kHz is shown. The frequency *left scale* shows the location of the maximum membrane displacement at various frequencies for a cat cochlea. For a human cochlea, the frequency scale runs from about 20 Hz to 15 kHz. The array of AN fibers is shown by the *parallel lines on the right*. Each fiber innervates a hair cell at one place on the basilar membrane, so the fiber's sensitivity is maximal at the frequency corresponding to that place, as shown by the *left scale*. In other words, the separation of frequencies done by the basilar membrane is preserved in the AN fiber array. (After [12.32] with permission)

built into the cilia that protrude from the top of the hair cell, described in a later section. The hair cells are synaptically coupled to AN fibers, so that basilar-membrane vibration is eventually coupled to activation of the nerve.

A key aspect of cochlear function is frequency analysis. The nature of cochlear frequency analysis is illustrated in Fig. 12.6b, which shows the cochlea uncoiled, with the basilar membrane stretching from the base, the end near the stapes, to the apex (the helicotrema). As described above, the basilar membrane lies between (separates) the ST and the SM/SV. The sound entering the cochlea through the vibration of the stapes is coupled into the fluids of the SV and SM at the base, above the basilar membrane in Fig. 12.6. The sound energy produces a pressure difference between the SV/SM and the ST, which causes vertical displacement of the basilar membrane. The displacement is tuned, in that the motion produced by sound of a particular frequency is maximum at a particular place along the cochlea. The resulting place map is shown by the frequency scale on the left in Fig. 12.6. Basilar-membrane displacement was first observed and described by *von Békésy* [12.33]; more recent data are reviewed by *Robles* and *Ruggero* [12.34].

Basilar-Membrane Motion

The displacement of any single point on the basilar membrane is an oscillation vertically (perpendicular to the surface of the membrane); the relative phase (or timing) of the oscillations of adjacent points is such that the overall displacement looks like a traveling wave, frequently described as similar to the waves propagating away from an object dropped into a pool of water. Essentially, the motion of the basilar membrane is delayed in time as the observation point moves from base to apex. Figure 12.7a shows an estimate of the cochlear traveling wave on the guinea-pig basilar membrane, based on electrical potentials recorded across the membrane [12.35]. The horizontal dimension in this figure is distance along the basilar membrane, with the base at left and the apex at right. The vertical dimension is the estimate of displacement, shown at a greatly expanded scale. The oscillations marked "2 kHz" show membrane deflections in response to a 2 kHz tone at five successive instants in time (labeled 1–5). The wave appears to travel rightward, from the base toward the apex; as it does so, its amplitude changes. When the stimulus is a tone, the displacement envelope (i.e., the maximum amplitude of the displacement at each point along the basilar membrane) has a single peak at a location that depends on the tone frequency (approximately at point 2 in this case). In Fig. 12.7a, displacement envelopes are shown by dashed lines for 0.5 kHz and 0.15 kHz tones; these envelopes peak at more apical locations, compared to 2 kHz. The locations of the envelope peaks are given by the cochlear frequency map for the cat cochlea, shown by the scale running along the basilar membrane in Fig. 12.6b. Notice that the scale is logarithmic: the position along the basilar membrane corresponds more

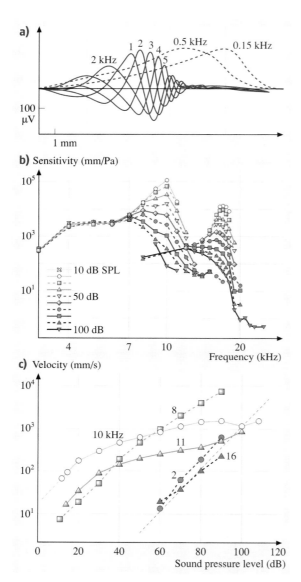

Fig. 12.7 (a) An estimate of the instantaneous displacement of the basilar membrane in the guinea-pig cochlea for a 2 kHz tone (*solid lines*) and estimates of the envelope of the displacement for two lower-frequency tones (*dashed lines*). Distance along the basilar membrane runs from left (*base*) to right (*apex*) and the displacement of the membrane is plotted vertically. The estimates were obtained by measuring the cochlear microphonic at three sites, indicated by the three vertical lines, and then interpolating or extrapolating to other locations using an informal method [12.35]; this method can be expected to produce only qualitatively correct results. The 2 kHz estimates are shown at 0.1 ms intervals. Waveforms at successive times are identified with the *numbers "1" through "5"*. The *dashed curves* were drawn through the maximum displacements of similar data for 0.5 and 0.15 kHz tones, which were scaled to have the same maximum displacement as the 2 kHz data. The cochlear microphonic is produced by hair cells, mainly OHCs, and is roughly proportional to basilar-membrane displacement. The ordinate scale has been left as microvolts of cochlear microphonic. The scale markers at lower left show distance from the origin. (b) Measurements of the gain of the basilar membrane (root mean square displacement divided by pressure at the eardrum) for two data sets. In each case, basilar-membrane displacement at one place was measured with an interferometer. Both frequency (abscissa) and sound level (symbols) were varied. The curves peaking near 9–10 kHz are from the chinchilla cochlea [12.36] and the curves peaking near 20 kHz are from the guinea pig cochlea [12.37] (After [12.34]). (c) Basilar-membrane input–output curves, showing the velocity of membrane displacement (*ordinate*) versus sound intensity (*abscissa*), at various frequencies, marked on the plots (kHz). The *dashed line at right* shows the linear growth of basilar-membrane motion (velocity proportional to sound pressure). The curve for 10 kHz is extrapolated to low sound levels (*dashed line at left*) under the assumption of linear growth at low sound levels (Fig. 12.7a after [12.35], Fig. 12.7b after [12.34], and Fig. 12.7c after [12.36])

closely to log than to linear frequency. The logarithmic frequency organization corresponds to a number of phenomena in the perception of sound, such as the fact that musical notes are spaced logarithmically in frequency.

The basilar-membrane responses in Fig. 12.7a illustrate the property of tuning, in that the displacement in response to a tone is confined to a region of the membrane centered on a place of maximal displacement. Tuning can be further understood by plotting basilar-membrane displacement as in Fig. 12.7b. This plot shows the displacement plotted against frequency at a fixed location, thus providing a direct measure of the frequency sensitivity of a place on the membrane. Data are shown at two locations from separate experiments (in different species). The ordinate actually shows basilar-membrane *gain*, defined as membrane displacement divided by sound pressure at the eardrum. At low sound levels (20 dB, plotted with empty circles), the gain peaks at a particular best frequency (BF) and decreases at adjacent frequencies. If such measurements could be repeated at multiple locations along the basilar membrane (which is difficult because of the restricted anatomical access to the basilar membrane), these gain functions would move toward higher frequencies at more basal locations in the cochlea, as predicted by the displacement functions in Fig. 12.7a.

Notice that the gain and tuning of the basilar-membrane response near the BF varies with sound level. At low sound levels (20 dB in Fig. 12.7b), gain is high and the tuning is sharp (i. e. the width of the gain functions is small), giving a clear BF; at high sound levels (100 dB), the gain is substantially reduced, especially in the vicinity of the low-sound-level BF, and the frequency at which the gain is maximum moves to a lower frequency, often about a half-octave lower. The tuning also becomes much broader.

Basilar Membrane and Compression

The fact that basilar-membrane gain changes with sound level was first described by *Rhode* [12.38]; this finding revolutionized our understanding of the auditory system, most importantly by introducing the idea of an active cochlea, meaning one in which the acoustic energy entering through the middle ear is amplified by an internal energy-utilizing process to produce cochlear responses [12.39–44]. The existence of amplification has been demonstrated by calculations of energy flow in the cochlea. However, a simpler evidence for cochlear amplification is the change in cochlear gain functions with death. The dependence of gain on sound level seen in Fig. 12.7b occurs only in the live, intact cochlea; after death, the cochlear gain functions resemble those at the highest sound levels in living animals (e.g. at 100 dB in Fig. 12.7b). Most important, the post-mortem gain functions are linear, meaning that the gain is constant, regardless of stimulus intensity, at all frequencies. In the live cochlea, as shown in Fig. 12.7b, the gain functions are linear only at low frequencies (below 6 kHz or 12 kHz for the two sets of data shown). Presumably, the post-mortem gain functions are the result of passive basilar-membrane mechanics, i. e., the result of the mechanical properties of the cochlea without any energy sources. The difference between the gain at the highest sound levels in Fig. 12.7b and those at low sound levels thus reflects an amplification process, often called the cochlear amplifier.

Additional evidence for a cochlear amplifier comes from the study of otoacoustic emissions, OAE [12.45, 46]. OAEs are sounds produced in the cochlea that can be measured at the eardrum, after propagating in the backwards direction through the middle ear. OAEs can be produced by reflection of the cochlear traveling wave from irregularities in the cochlea [12.47, 48] or from nonlinear distortion in elements of the cochlea [12.49]. In either case amplification of the sound is necessary to explain the characteristics of the emitted sound [12.46], but see also [12.50]. There are several different kinds of OAEs, varying in the mode of production. Most relevant for the present discussion are the spontaneous emissions, sounds that are present in the ear canal without any external sound source. Such sounds necessitate an acoustic energy source in the cochlea, of the type postulated for the cochlear amplifier.

A variety of evidence suggests that cochlear amplification depends on the OHCs. When these cells are destroyed, as by an ototoxic antibiotic, the sharply tuned high-sensitivity portion of the tuning of auditory-nerve fibers is lost [12.51]; presumably this change reflects a similar change at the level of the basilar membrane, i.e., loss of the high-gain sharply tuned portion of the basilar-membrane response. Electrical stimulation of the MOC (Fig. 12.5), which affects only the OHC, produces similar losses in both basilar-membrane motion [12.52] and neural responses [12.53, 54]. Finally, stimulation of the MOC reduces OAE [12.55, 56]. In each case, a neural input to a cellular element of the cochlea, the OHC, affects mechanical processes in the cochlea. These data suggest that the OHCs participate in producing basilar-membrane motion. The possible mechanisms for this effect are discussed in Sect. 12.2.4.

The sound-level dependence of gain observed in Fig. 12.7b shows that the sensitivity of the cochlea is regulated across sound levels. The gain is high at low sound levels and decreases at higher sound levels; as a result, the output varies over a narrower dynamic range than the input, which is compression. A different view of compression is shown in Fig. 12.7c, which plots basilar-membrane input/output functions at various frequencies for a 10 kHz site on a chinchilla basilar membrane [12.36]. Basilar-membrane response is measured as velocity, instead of displacement as in previous figures. However, velocity and displacement are proportional for a fixed frequency, so in terms of input/output functions it does not matter which variable is plotted. Figure 12.7c has logarithmic axes, so a linear growth of response, in which the output is proportional to the input, corresponds to a line with slope 1, as for the dashed line at right. At low sound levels (< 40 dB), the response is largest at 10 kHz, as it should be for the BF. The response to 10 kHz is roughly linear at very low sound levels (< 20 dB), but has a lower slope at higher levels. Over the range of input sound levels from 20–80 dB SPL, the output velocity increases only 10-

fold or 20 dB, a compression factor of about 0.3. At lower frequencies (8 and 2 kHz), the slope is closer to 1, as expected from the constant gain at low frequencies in Fig. 12.7b. At higher frequencies (11 kHz), the growth of response is also compressive, but with a lower gain. Finally, at very high frequencies (16 kHz), the growth is again approximately linear.

The behavior shown in Figs. 12.7b,c is typical of cochleae that are judged to be in the best condition during the measurements. As the condition of the cochlea deteriorates, the input/output functions become more linear and their gain decreases [12.57], reflecting a loss of cochlear amplification. Although such data are not available for the example in Fig. 12.7c, the typical post-mortem behavior of input/output functions at BF is something like the dashed line used to illustrate linear growth. Post-mortem functions have a slope of 1 at all sound levels and typically intercept the compressive BF input/output function at levels of 70–100 dB. Assuming that the cochlear amplifier is not functioning post-mortem, the gain of the amplifier can be defined as the horizontal distance between the linear-growth portion of the BF curve in the normal ear and the post-mortem curve. With this definition of gain, the compression region of the input/output function can be understood as resulting from a gradual decrease in amplification as the sound grows louder.

The compression of basilar-membrane response shown in Fig. 12.7c has a number of perceptual correlates [12.58]. The degree of compression can be measured in human observers using a masking technique, motivated by the behavior of basilar-membrane data like Fig. 12.7 [12.59, 60]; the compression measured in these studies is comparable to that measured in basilar-membrane data. Moreover, compression can be used to explain a number of perceptual phenomena, including aspects of temporal processing and loudness growth. In hearing-impaired persons, compression is lost to varying degrees, consistent with the effects of loss of OHCs on basilar-membrane responses discussed above (i. e., the slopes of basilar-membrane input/output functions steepen). Some of the effects of hearing impairment can be explained by a change in compression ratio from 0.2–0.3 to a value near 1. An important example is loudness recruitment. If loudness is somehow proportional to the overall degree of basilar-membrane motion, then a steepening of the growth of basilar-membrane motion with sound intensity should lead to a steepening of loudness growth with intensity. This is the major effect observed on loudness growth with hearing impairment.

Mechanisms Underlying Basilar-Membrane Tuning

The current understanding of how the properties of basilar-membrane motion arise is incomplete. Work on this subject has proceeded along two lines:

1. collection of evermore accurate data on the motion of the basilar membrane and the components of the organ of Corti; and
2. models of the observed motion of the basilar membrane based on physical principles of mechanics and fluid mechanics.

Progress in this field has been limited by the difficulty of obtaining data, however, and modelers have generally responded vigorously to each new advance in observing and measuring basilar-membrane motion. At present, the major impasse in this work seems to be the difficulty of observing the independent motion of the components of the organ of Corti. Such data are necessary to resolve questions about the mode of stimulation of hair cells by the motion of the basilar membrane and questions about the effects of OHC and stereociliary movements on the basilar membrane.

Modeling of basilar-membrane motion has been approached with a variety of methods in an attempt to account for the properties of the motion. Comprehensive reviews of this literature and its relationship to important aspects of data on basilar-membrane motion are available [12.61, 62]. The next paragraphs provide a rough summary of the current understanding of this topic.

Tuning or the separation of frequencies in basilar-membrane motion is usually attributed to the variation in the stiffness and (sometimes) the mass of the basilar membrane along its length. In most basilar-membrane models, the points on the basilar membrane are assumed to move independently; that is, the longitudinal stiffness coupling adjacent points on the membrane is assumed to be small. With this assumption, the resonant frequency of a point on the basilar membrane should vary approximately as the square root of its stiffness divided by its mass, using the usual formula for the resonant frequency of a mass–spring oscillator. In so-called long-wave models, it is this resonant frequency that determines the mapping of frequency into place in the cochlea. Experimentally, the stiffness of the basilar membrane varies from high near the base to low at the apex [12.33, 63]. Thus the stiffness gradient is qualitatively consistent with the observed place–frequency map; however, whether the stiffness gradient is sufficient to fully account for tuning in the cochlea is not settled, because of the difficulty of making measurements.

Recent efforts in cochlear modeling have been devoted to micromechanical models, i. e., models of the detailed mechanics of the organ of Corti and the tectorial membrane [12.49, 62]. Movement of the basilar membrane is coupled to the hair cells through the organ of Corti, particularly through the relative movement of the top surface of the organ (the reticular lamina) in which the hair cells are inserted and the overlying tectorial membrane (refer to Fig. 12.4c). Mechanical feedback from the OHCs is similarly coupled into the basilar membrane through the same structures. The structure of the organ of Corti is complex and heterogeneous, with considerable variation in the stiffness of its different parts [12.64]. Thus a full account of the properties of the basilar membrane awaits an adequate micromechanical model, based on accurate data on the mechanical properties of its components.

In both models and measurements, the sound pressure difference across the basilar membrane drops rapidly to zero past the point of resonance. This behavior is illustrated by the high-frequency part of the data in Fig. 12.7b and the lack of deflection beyond the peak of the responses in Fig. 12.7a. This fact means that there is no pressure difference across the basilar membrane at the apical end of the cochlea, so the helicotrema does not affect basilar-membrane mechanics (in particular it does not *short out* the pressure difference driving the basilar membrane) except at very low frequencies, where the displacement is large near the apex. The role of the helicotrema is presumably to prevent the generation of a constant pressure difference across the basilar membrane; such a difference would interfere with the delicate transducer system in the organ of Corti.

The frequency analysis done by the basilar membrane is preserved by the arrangement of hair cells and nerve fibers in the cochlea (Fig. 12.6b). A particular hair cell senses only the local motion of the basilar membrane, so that the cochlear place map is recapitulated in the population of hair cells spread out along the basilar membrane. Moreover, each AN fiber innervates only one IHC, so the cochlear place map is again recapitulated in the AN fiber array. As a result, AN fibers that innervate, say, the 3 kHz place on the basilar membrane are maximally activated by 3 kHz energy in the sound and it is the level of activation of those fibers that provides information to the brain about the 3 kHz energy in the sound. The representation of sound in an array of neurons tuned to different frequencies is the basic organizing principle of the whole auditory system and is called *tonotopic* organization.

12.2.3 Representation of Sound in the Auditory Nerve

Before discussing cochlear transduction in hair cells, it is useful to introduce some properties of the responses to sound of AN fibers. Of course, the properties of AN fibers derive directly from the properties of the basilar membrane and the transduction process in IHCs. AN fibers encode sound using trains of action potentials, which are pulses fired by a fiber when stimulated by a hair-cell synapse. Information is encoded in the rate at which the fiber discharges action potentials or by the temporal pattern of action potentials. Fibers are active spontaneously at rates that vary from near 0 to over 100 spikes/s. When stimulated by an appropriate sound, the fiber's discharge rate increases and the temporal patterning of action potentials often changes as well. Figure 12.8 shows some basic features of the encoding process. Reviews of the representation of sound by auditory neurons are provided by *Sachs* [12.65], *Eggermont* [12.66] and *Moore* [12.67]. Another view is provided by recent attempts to model comprehensively the responses of AN fibers [12.68–70] or to analyze the information encoded using theoretical methods [12.71].

The tuning of the basilar membrane is reflected in the tuning of AN fibers as is shown by the tuning curves in Fig. 12.8a. This figure shows tuning curves from 11 AN fibers with different BFs. Each curve shows how intense a tone has to be (ordinate) in order to produce a criterion change in discharge rate from a fiber, as a function of sound frequency (abscissa). The tuning curves qualitatively resemble the basilar-membrane response functions shown in Fig. 12.7b for the lowest sound levels (after being turned upside down). The comparison should not be exact because the data are from different species and the tuning curves in Fig. 12.8a are constant-response contours whereas the basilar-membrane functions in Fig. 12.7b are constant-input-level response functions. However, a quantitative comparison of threshold functions for the basilar membrane and AN fibers from the same species yields a good agreement, and it is generally considered that AN tuning is accounted for by basilar-membrane tuning [12.34]. The AN consists of an array of fibers tuned across the range of frequencies that the animal can hear. The frequency content of a sound is conveyed by which fibers are activated, the tonotopic representation discussed in the previous section.

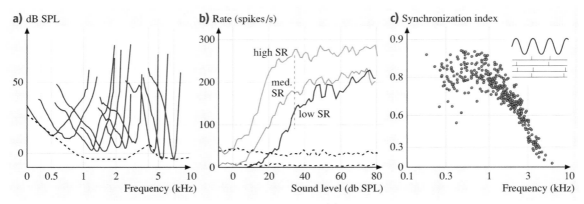

Fig. 12.8a–c Basic properties of responses to sound in AN fibers. (**a**) Tuning curves, showing the threshold sound level for a response plotted versus frequency. the *dashed line* shows the lowest thresholds across a population of animals. (**b**) Rate versus sound-level plots for responses to BF tones in three AN fibers. Rate in response to a 200 ms tone is shown by the *solid lines* and SR by the *dashed lines*, which actually plot the SR during the 400 ms immediately preceding each stimulus. The fibers had similar BFs (5.36, 6.18, and 5.74 kHz) but different spontaneous rates and were recorded successively in the same experimental preparation. The fluctuations in the curves derive from the randomness in the responses remaining after three-point smoothing of the curves. (**c**) Strength of phase-locking in a population of AN fibers to a BF tone plotted as a function of BF. Phase-locking is measured as the synchronization index, equal to the magnitude of the Fourier transform of the spike train at the stimulus frequency divided by the average rate. The *inset* illustrates phase-locked spike trains (Fig. 12.8a after [12.74], Fig. 12.8c after [12.75])

The perceptual sense of sound frequency, of the highness or lowness of a simple sound like a tone, is strongly correlated with stimulus frequency and, therefore, with the population of AN fibers activated by the sound. However, the pitch of a complex sound, as for musical sounds or speech, is a more complex attribute both in terms of the relationship of pitch to the physical qualities of the sound [12.72] and in terms of the representation of pitch in the AN. A discussion of this issue is provided by *Cariani* and *Delgutte* [12.73].

The intensity of a sound is conveyed by the rate at which fibers respond. AN fibers increase their discharge rates as sound intensity increases (Fig. 12.8b). In mammalian ears, fibers vary in their threshold sensitivity, where threshold means the sound level at which the rate increases from the spontaneous rate. The threshold variation is correlated with spontaneous discharge rate (SR), so that very sensitive fibers, those with the lowest thresholds, have relatively high SR and less-sensitive fibers have lower SR [12.76]. Fibers are broken into three classes, low, medium, and high, using the somewhat arbitrary criteria of 0.5–1 spikes/s to divide low from medium and 15–20 spikes/s to divide medium from high, depending on the experimenters. Figure 12.8b shows plots of discharge rate versus sound level for an example fiber in each SR group, labeled on the figure. Fibers of all SRs have similar rate-level functions; for tones, these increase monotonically with sound level (except for the effect of noisy fluctuations) from spontaneous rate to a saturation rate. In high-SR fibers, the dynamic range, the range over which sound intensity increases produce rate increases, is narrow and the saturation of discharge rate at high sound levels is clear. In low- and medium-SR fibers, the saturation is gradual (sloping) and the dynamic range is wider.

The sloping saturation in low- and medium-SR fibers is thought to reflect compression in basilar-membrane responses [12.21, 77]. High-SR fibers do not show sloping saturation because the fiber reaches its maximum discharge rate at sound levels where growth of the basilar-membrane response is linear (e.g. below 20 dB in Fig. 12.7). By contrast, the low- and medium-SR fibers sample both the linear and compressed portion of the basilar-membrane input/output function. The vertical dashed line in Fig. 12.8 shows the approximate threshold for basilar-membrane compression inferred for these fibers.

The relationship of AN responses to the perceptual sense of loudness is an apparently simple problem that has not been fully solved. While it is generally assumed that loudness is proportional to the overall increase in discharge rate across all BFs [12.78], this calculation does not predict important details of loudness growth [12.79–81], especially in ears with damage to the hair cells. Presumably there are additional transformations in the central auditory system that determine the final properties of loudness growth.

A property of AN discharge that is important for many sounds is the ability to represent the temporal waveform. Sounds such as speech have information encoded at multiple levels of temporal precision [12.82]; these include

1. syllables and other aspects of the envelope at frequencies below 50 Hz;
2. periodicity (pitch) in sounds such as vowels at frequencies from 50–500 Hz; and
3. the fine structure of the actual oscillations in the acoustic waveform at frequencies above 500 Hz.

AN fibers represent the waveform of sounds (the fine structure) by phase-locking to the stimulus. A schematic example is shown in the inset of Fig. 12.8c, which shows a sinusoidal acoustic waveform (a tone) and four examples of AN spike trains in response to the stimulus. The important point is that the spikes in the responses do not occur at random, but rather at a particular stimulus phase, near the positive peak of the stimulus waveform in this example. The phase-locking is not perfect, in that a spike does not occur on every cycle of the stimulus, and the alignment of spikes and the waveform varies somewhat. However, a histogram of spike-occurrence times shows a strong locking to the stimulus waveform and the Fourier transform of the spike train generally has its largest component at the frequency of the stimulus.

Phase-locking occurs at stimulus frequencies up to a few kHz, depending on the animal. The main part of Fig. 12.8c shows the strength of phase-locking in terms of the synchronization index (defined in the caption) plotted against the frequency of the tone. These data are from the cat ear [12.75] where phase-locking is strongest below 1 kHz and disappears by about 6 kHz. Phase-locking always shows this low-pass property of being strong at low frequencies and nonexistent at high frequencies. The highest frequency at which phase-locking is seen varies in different species, being lower in rodents than in cats [12.83].

AN phase-locking is essential for the perception of sound localization. The relative time of occurrence of spikes from the two ears is used to compute the delays in the stimulus waveform across the head that provide information about the azimuth of a sound source [12.84]. Phase-locking has also been suggested as a basis for processing of speech and other complex stimuli [12.71, 85]. Examples of the usefulness of phase-locking in analyzing responses to complex stimuli are given in Sect. 12.3.1.

12.2.4 Hair Cells

The transduction of basilar-membrane motion into electrical signals occurs in the IHCs. The OHCs participate in the generation of sensitive and sharply tuned basilar-membrane motion. In this section, the physiology of IHCs and OHCs are described with reference to these tasks. At the time of this writing, hair-cell research is one of the most active areas in auditory neuroscience and many of the details of hair-cell function are just being worked out. Useful recent reviews of the literature are available [12.86–89].

IHCs and Transduction

Both IHCs and OHCs transduce the motion of the basilar membrane. This section describes transduction with reference to the IHCs, but transduction works in essentially the same fashion in OHCs. The transduction apparatus depends on the arrangement of stereocilia on the apical surface of hair cells. Stereocilia are structures that protrude from the hair cells into the endolymphatic space of the SM (Fig. 12.9a). They consist of bundles of actin filaments anchored in an actin/myosin matrix just under the apical surface of the cell. The membrane of the cell wraps the actin rods so that they are intracellular and the membrane potential of the cell appears across the membranes of the stereocilia. The structure of stereocilia is intricate and involves a number of structural proteins that are important for organizing the development of the cilia and for maintaining them in proper position on the cell [12.90]. The stereocilia form a precisely organized bundle; in the cochlea, the bundle consists of several roughly V- (IHCs) or W-shaped (OHCs) rows of cilia (typically three, as in Fig. 12.9) of graded length, so that the cilia in the row nearest the lateral edge of the hair cell are the longest and those in the adjacent rows are successively shorter. Each row consists of 20–30 cilia precisely aligned with the cilia in adjacent rows. The tips of the cilia are connected by tip links [12.91], a string-like bundle of protein that connects the tip of a shorter cilium to the side of the taller adjacent cilium (Fig. 12.9b).

Transduction occurs when vertical motion of the basilar membrane is converted into a shearing motion of the stereocilia (the arrow in Fig. 12.9b). It has long been known that hair cells are functionally polarized, in the sense that they respond most strongly to displacement of the cilia in the direction of the arrow, are

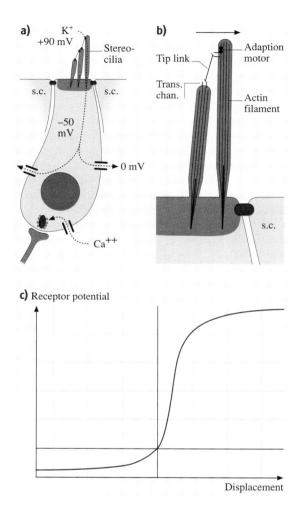

Fig. 12.9a–c Transduction in IHCs. (**a**) Schematic of an IHC showing the components important for transduction. The stereocilia protrude into the endolymphatic space (*top*), containing mainly K^+; the extracellular potential here is about +90 mV (the endolymphatic potential). The transduction apparatus consists of tip links and channels near the tips of the cilia. The *dashed lines* show the transduction current. It enters the cell through the transduction channel and exits through K^+ channels in the basolateral membrane of the cell into the perilymphatic space. The intracellular and extracellular potentials of the cell and the perilymphatic space are given. There is a mechanically stiff and electrically insulating boundary between the endolymphatic and perilymphatic spaces, formed by tight junctions between the apical surfaces of hair cells and supporting cells (s.c.), indicated by the *small black ovals*. The synapse, at the base of the cell, is a standard chemical synapse, in which voltage-gated Ca^{2+} channels open in response to the depolarization caused by the transduction current and admit Ca^{2+} to activate neurotransmitter release. (**b**) Detailed view of two stereocilia with a tip link connecting ion channels in the membrane of each. The upper ion channel is connected to an adaptation motor that moves up and down along the actin filaments to tension the tip link. (**c**) Transduction function showing the receptor potential (the depolarization or hyperpolarization) of a hair cell versus the displacement of the stereocilia tips. Typical axis scales would be 0.01–1 μm per tick on the abscissa and ≈2 mV per tick on the ordinate

inhibited by displacement in the opposite direction, and respond weakly to displacement of the cilia in lateral directions [12.92, 93]. In fact, hair cells are depolarized when the stereocilia are displaced in the direction that stretches the tip links. Further evidence to associate the tip links to the transduction process is that when the tip links are destroyed by lowering the extracellular calcium concentration, transduction disappears and reappears as the tip links are regenerated [12.94].

The mechanical connection between basilar-membrane motion and cilia motion is not fully understood. As shown in Fig. 12.4c, the stereocilia bundles are oriented so that lateral displacement of the cilia (i. e., displacement away from the central axis of the cochlea) should be excitatory. Furthermore the tips of the OHC cilia are embedded in the tectorial membrane, whereas those of the IHC are not. Based on the geometry of the organ, upward motion of the basilar membrane (i. e., away from the ST and toward the SM and SV) results in lateral displacement of the cilia through the relative motions of the organ of Corti and the tectorial membrane. Presumably the coupling to the OHC is a direct mechanical one, whereas the coupling to the IHC is via fluid movements, so the IHC stereocilia are bent through viscous drag of the endolymph in the space between the organ of Corti and the tectorial membrane. While this model is commonly offered and is probably basically correct, it is not consistent in detail with the phase relations between the stimulus (or basilar-membrane motion) and action potential phase locking in AN fibers [12.95]. Presumably the inconsistencies reflect complexities in the motions of the organ of Corti that connect basilar-membrane displacement to stereociliary displacement.

The current model for the tip link and transduction channel is shown in Fig. 12.9b. The tip link connects (probably) two transduction channels, one in each cilium. When the bundle moves in the excitatory direction (the direction of the arrow), the tension in the tip link increases, which opens the transduction channels and allows current flow into the cell. Movement in the opposite direction relaxes the tension in the tip link and allows the channels to close. The current through the transduction channels is shown by the dashed line in Fig. 12.9a. Outside of the stereociliary membrane is the endolymph of the SM. The transduction channels are nonspecific channels which allow small cations to pass; in the cochlea, these are mainly Na^+, K^+, and Ca^{2+}. However, the predominant ion in both the extracellular (endolymph) and intracellular spaces is K^+, so the transduction current is mostly K^+. The energy source producing the current is the potential difference between the SM (the endolymphatic potential, approximately $+90\,mV$) and the intracellular space in the hair cell (approximately $-50\,mV$). The endolymphatic potential is produced by the active transport of K^+ into the endolymph and Na^+ out of the endolymph in the stria vascularis (Fig. 12.4b). This transport process also produces the endolymphatic potential [12.22]. The negative potential inside the hair cell is produced by the usual mechanisms in which active transport of K^+ into the cell and Na^+ out of the cell produces a negative diffusion potential through the primarily K^+ conductances of the hair cell membrane. Because the transduction current is primarily K^+, it does not burden the hair cell with the necessity for additional active transport of ions brought into the hair cell by transduction.

When the transduction channels open, the hair cell is depolarized by the entry of positive charge into the cell. The transduction current flows out of the cell through the potassium channels in its membrane. Figure 12.9c shows the typical transduction curve for a cochlear hair cell, as the depolarization of the membrane (ordinate) produced by a certain displacement of the stereociliary bundle (abscissa). At rest (zero displacement), some transduction channels are open, so that both positive and negative displacements of the bundle can be signaled. However, hair cells generally have only a fraction of channels open at rest so that a much larger depolarization is possible than hyperpolarization, i.e., more channels are available for opening than for closing. The transduction curve saturates when all channels are closed or all channels are open.

The transduction process is completed by the opening of voltage-gated Ca^{2+} channels in the hair cell membrane, which allows Ca^{2+} ions to enter the cytoplasm (dotted line at the bottom of Fig. 12.9a) and activate the synapse. The hair-cell synapse has an unusual morphology, containing a synaptic ribbon surrounded by vesicles [12.96, 97]. This ribbon presumably reflects molecular specializations that allow the synapse to be activated steadily over a long period of time without losing its effectiveness and also to produce an action potential in the postsynaptic fiber on each release event, so that summation of postsynaptic events is not necessary, as is required for phase-locking at kilohertz frequencies. The synapse appears to be glutamatergic [12.98] with specializations for fast recovery from one synaptic release, also necessary for high-frequency phase-locking.

The final component of the transduction apparatus is the adaptation motor, shown as two black circles in Fig. 12.9b [12.99]. Presumably the transduction channel and the adaptation motor are mechanically linked together so that when the motor moves, the channel moves also. The sensitivity of the transduction process depends on having an appropriate tension in the tip link. If the tension is too high, the channels are always open; if the tip link is slack, then the threshold for transduction will by elevated because larger motions of the cilia will be required to open a transduction channel. The tip-link tension is adjusted by a myosin motor that moves upward along the actin filaments when the transduction channel is closed. When the channel opens, Ca^{2+} flows into the stereocilium through the transduction channel and causes the motor to slip downward. Thus the adaptation motor serves as a negative feedback system to set the resting tension in the tip link. The zero-displacement point on the transduction curve in Fig. 12.9c is set by an equilibrium between the motor's tendency to climb upward and the tension in the tip link, which causes the channel to open and admit Ca^{2+}, pulling it downward. Adaptation also regulates the sensitivity of the transduction process in the presence of a steady stimulus.

There is a second, fast, mechanism for adaptation in which Ca^{2+} entering through an open transduction channel acts directly on the channel to close it, thus moving the transduction curve in the direction of the applied stimulus [12.88]. This mechanism is discussed again in a later section, because it also can serve as a kind of cochlear amplifier.

IHC Transduction and the Properties of AN Fibers

The description of transduction in the previous section suggests that the stimulus to an AN fiber should be approximately the basilar-membrane motion at the point in the cochlea where the IHC innervated by the fiber is located. It follows that the tuning and rate responses of the fiber should be similar to the tuning and response amplitude of the basilar membrane, discussed in connection with Fig. 12.7.

The receptor potentials in an IHC responding to tones are shown in Fig. 12.10 [12.83]. The traces are the membrane potentials recorded in an IHC, each trace at a different stimulus frequency. These potentials are the driving force for the Ca^{2+} signal to the synapse, and thus indirectly for the AN fibers connected to the cell. The properties of these potentials can be predicted from Fig. 12.9c by the thought experiment of moving a stimulus sinusoidally back and forth on the abscissa and following the potential traced out on the ordinate. The potential should be a distorted sinusoid, because depolarizing responses are larger than hyperpolarizing ones, and it should have a steady (DC) offset, also because of the asymmetry favoring depolarization. At low frequencies, both components can be seen in Fig. 12.10; the response is a distorted sinusoid riding on a steady depolarization. At high frequencies (>1000 Hz), the sinusoidal component is reduced by low-pass filtering by the cell's membrane capacitance and only the steady component is seen. The transition from sinusoidal receptor potentials at low frequencies to rectified DC potentials at high frequencies corresponds to the loss of phase-locking at higher frequencies in AN fibers (Fig. 12.8c). At low frequencies, the IHC/AN synapse releases transmitter on the positive half cycles of the receptor potential and the AN fiber is phase-locked; at high frequencies, there is a steady release of neurotransmitter and the AN fiber responds with a rate increase but without phase-locking.

OHCs and the Cochlear Amplifier

In the discussion of the basilar membrane, the need for a source of energy in the cochlea, a cochlear amplifier, was discussed. At present, two possible sources of the amplification have been suggested. The first is OHC motility [12.86, 100] and the second is hair bundle motility [12.101]. Although hair bundle motility could function in both IHCs and OHCs, it seems likely that

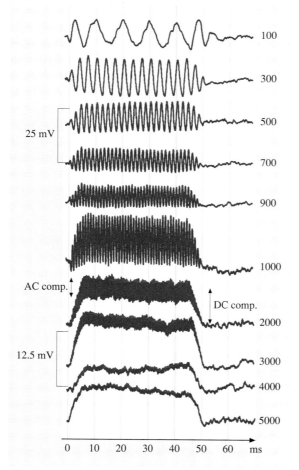

Fig. 12.10 Receptor potentials in an IHC in the basal turn of a guinea-pig cochlea. The stimulus level was set at 80 dB SPL so that the cell would respond across a wide range of frequencies. Stimulus frequency is given at right. Notice the change in ordinate scale between the 900 Hz and 1000 Hz traces. (After [12.83], with permission)

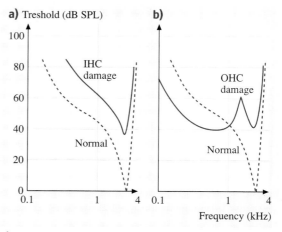

Fig. 12.11a,b Schematic summary of AN tuning curves in normal ears and in ears with damaged cochleae. (a) In regions of the cochlea with remaining OHCs but damaged IHCs tuning curves mainly show a threshold elevation. (b) In regions with missing OHCs but intact IHCs there is an elevation of threshold at frequencies near BF and a substantial broadening of tuning. (After [12.51])

the OHC is the principal element of the cochlear amplifier. The evidence for this is that the properties of the cochlea change in a way consistent with modification of cochlear mechanics when the OHC are damaged or otherwise manipulated, but not the IHC. For example, electrical stimulation of the MOC, which projects only to the OHCs (Fig. 12.5), modifies otoacoustic emissions [12.31] and reduces the amplitude of motion of the basilar membrane [12.102, 103].

Figure 12.11 shows tuning curves in AN fibers for intact ears ("normal") and for ears with IHC or OHC damage [12.51, 105]. These tuning curves are schematics that summarize the results of studies in ears with damage due to acoustic trauma or ototoxic antibiotics. Complete destruction of IHCs, of course, eliminates AN activity entirely. However, it is possible to find regions of the cochlea with surviving IHCs that have damaged stereocilia and OHCs that appear intact. The sensitivity of AN fibers in such regions is reduced, reflected in the elevated threshold in Fig. 12.11a, but the tuning is still sharp and appears to be minimally altered. By contrast, in regions with intact IHCs and OHCs that are damaged or missing, the tuning curves are dramatically altered, as in Fig. 12.11b. There is a threshold shift reflecting a loss of sensitivity but also a substantial broadening of tuning. These are the effects that are expected from the loss of the cochlear amplifier. Note that the thresholds well below BF (<1 kHz in Fig. 12.11b) can actually be lower than normal with OHC damage. This phenomenon is not fully understood and seems to reflect the existence of two modes of stimulation of IHCs [12.106].

Some properties of OHCs are shown in Fig. 12.12. These cells have a transduction apparatus similar to that in IHCs, but Fig. 12.12 focuses on unique aspects of OHC anatomy and physiology. The OHCs are located in an unusual system of structural or supporting cells. Unlike virtually all tissues, the OHC are surrounded by a fluid space containing perilymph for distances comparable to the cells' widths (asterisks in Fig. 12.4c). The OHC are held by supporting cells called Deiter's cells

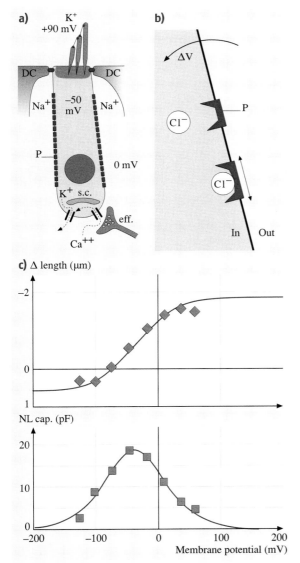

Fig. 12.12 (a) Schematic picture of an OHC showing elements important for OHC function. The diagram is labeled as for Fig. 12.9a except for the following: DC – Deiter's cell extensions that are linked to OHCs by tight junctions to form the top surface of the organ of Corti; Na^+ – the extracellular spaces around the lateral membranes of the OHCs are filled with perilymph, a high-Na^+, low-K^+ fluid; P – prestin molecules in the lateral membrane; eff. – efferent terminal of a MOC neuron on the OHC; s.c. – subsynaptic cistern associated with the efferent synapse. K^+ and Ca^{2+} currents activated by the synapse are shown by dashed and dotted lines, respectively. (b) Schematic of prestin molecules in the lateral membrane of the cell. When the membrane potential ΔV is negative, Cl^- ions are driven into the prestin molecule, increasing its membrane area (as indicated by the *double-headed arrow*). (c) The *top plot* shows the length of an isolated OHC as a function of membrane potential showing the amplitude of the electromotility. The *bottom plot* shows the nonlinear capacitance of the same cell. (Fig. 12.12c after [12.104] with permission)

that form a cup around the base of the cell and have a stiff extension up to the top surface of the organ of Corti. The extension terminates next to the apical surface of the OHC, where a system of tight junctions similar to that in the IHC region holds the OHCs and Deiter's cells to form a mechanically stiff and electrically insulating barrier between the endolymph and perilymph.

In response to depolarization of their membranes, say in response to a transduction current, OHCs shorten, a process called electromotility [12.107, 108]. Electromotility does not use a direct chemical energy source [such as adenosine triphosphate (ATP)] as do other cellular motility processes. It is also very fast, responding to frequencies of 10 kHz or above, limited mainly by the limits of the experimental observations. This unusual motile process is produced by a molecule called prestin ("P" in Fig. 12.12a) found in a dense array in the lateral wall of the OHCs [12.109–111], where it is associated with a complex mechanical matrix that forms the cell's skeleton [12.112]. Prestin causes the cell to shorten in response to electrical depolarization of its membrane, so the energy source for electromotility is the electrical membrane potential. The mechanism of electromotility is shown schematically in Fig. 12.12b. The prestin molecule is similar to anion transporters, molecules that normally move anions through the cell membrane. It is thought that prestin is a modified transporter that can bind an anion, but the ion moves only part of the way through the membrane. In OHCs, the relevant anion is Cl^- because electromotility is blocked by removing Cl^- from the cytoplasm [12.113]. When a Cl^- ion binds to a prestin molecule, the cross-sectional area of the molecule increases, thus increasing the membrane area and lengthening the cell (indicated by the double-headed arrow). Cl^- ions are driven into the membrane by negative membrane potentials, so depolarization pulls them out and decreases the membrane area, shortening the cell.

When a Cl^- ion binds to a prestin molecule, it moves part way through the cell's electrical membrane potential (ΔV in Fig. 12.12b). This movement behaves like charging a capacitance in the cell's membrane and appears in electrical recordings as an additional membrane capacitance. However the extent of charge movement depends on the membrane potential, making the capacitance nonlinear. At very negative membrane potentials, all the prestin molecules have a bound Cl^- and no charge movement can occur; similarly, at positive potentials, the Cl^- is unlikely to bind to prestin and again no charge movement can occur. Thus the nonlinear capacitance is significant only over the range of membrane potentials where the prestin molecules are partially bound to Cl^-. The bottom part of Fig. 12.12c shows the nonlinear capacitance as a function of membrane potential, showing the predicted behavior [12.104]. The top part of Fig. 12.12c shows the electrically produced change in cell length; as expected, changes in cell length are observed only over the range of membrane potentials where the nonlinear membrane capacitance is observed.

Prestin is found essentially only in OHCs and, in particular, is not seen in IHCs [12.111]. When expressed by genetic methods in cells that normally do not contain it, prestin confers electromotility on those cells. In addition, when the prestin gene was knocked out, the morphology of the cochlea and the hair cells was not affected (except that the OHCs were shorter), but electromotility was not observed and the sensitivity of auditory neurons was decreased; otoacoustic emissions were also decreased [12.114]. These are the effects expected if prestin is the energy source for the cochlear amplifier.

The means by which OHC motility amplifies basilar-membrane motion is somewhat uncertain. Models that use physiologically accurate OHC motility to produce cochlear amplification have been suggested and analyzed, e.g., [12.115]. However, such models require assumptions about the details of the mechanical interactions within the organ of Corti, assumptions that have not been tested experimentally. For example, the OHCs lie at an angle to the vertical extensions of the Deiter's cells, so that successive locations along the basilar membrane are coupled to one another [12.108]. The mechanical effects of structures like this are quite important in models of basilar-membrane movement that incorporate OHC electromotility [12.62]. An additional major uncertainty at present about a prestin-derived cochlear amplifier is the fact that prestin depends on the membrane potential as its driving signal [12.88]. Because the membrane potential is low-pass filtered by the cell's membrane capacitance, in the fashion seen in Fig. 12.10, it is not clear that a prestin-based mechanism can generate sufficient force at high frequencies.

The second possible mechanism for the cochlear amplifier is by stereocilia bundle movements that can amplify hair-cell stimulation either through nonlinearities in the compliance of the bundle or through active hair-bundle movements [12.89, 101]. Essentially these mechanisms work by moving the bundle further in the direction of its displacement, which would amplify the stimulation of the hair cell. Calculations suggest that the speed of bundle movement is sufficient and that the stiffness of the bundles is a significant fraction of the basilar-membrane stiffness, both necessary conditions for a stereociliary-bundle motor to be the cochlear amplifier. Of course both the prestin and stereociliary-bundle mechanisms may operate.

The final aspect of OHC physiology shown in Fig. 12.12a is the efferent synapse made by the MOC neuron on the OHC. This synapse activates an unusual cholinergic postsynaptic receptor which admits a significant Ca^{2+} current (dotted line), along with other cations, to the cytoplasm. The Ca^{2+} current activates a calcium-dependent potassium channel producing a larger potassium current (dashed line) [12.116]. The efferent synapse inhibits OHC function, decreasing the sensitivity of AN fibers and broadening their tuning, effects consistent with a decrease in cochlear amplification [12.31]. The mechanism is thought to be an increase in the membrane conductance of the OHC, from opening the calcium-dependent K^+ channel, which hyperpolarizes the cell and shorts the transducer current, producing a smaller receptor potential.

12.3 Auditory Nerve and Central Nervous System

In a previous section, the basic properties of AN fiber responses to acoustic stimuli were described. These properties can be related directly to the properties of the basilar membrane and hair cells. The auditory system does not usually deal with the kinds of simple stimuli that have been considered so far, i.e., tones of a fixed frequency. When multiple-tone complexes or other stimuli with multiple frequency components are presented to the ear, there are interactions among the frequency components that are important in shaping the responses. This section describes two such interactions and then provides an overview of the tasks performed by the remainder of the auditory system.

12.3.1 AN Responses to Complex Stimuli

Many of the features of responses of AN fibers to complex stimuli can be seen from the responses to a two-tone complex. An example is shown in Fig. 12.13 which shows responses of AN fibers to 2.17 and 2.79 kHz tones (called f_1 and f_2, respectively) presented simultaneously and separately [12.117]. In this experiment a large population of AN fibers were recorded in one animal and the same set of stimuli were presented to each fiber. Because fibers were recorded across a wide range of BFs, this approach allows the construction of an estimate of the response of the whole AN population. Responses are plotted in Fig. 12.13 as the strength of phase-locking to various frequency components of the stimulus. Phase-locking is used here to allow the complex responses of the fibers to be separated into responses to the various individual frequency components. The abscissa is plotted as BF, on a reversed frequency scale, and also in terms of the distance of the fibers' points of innervation from the stapes.

The distribution of responses to the f_1 frequency component (2.17 kHz) presented alone is shown in Fig. 12.13a by the solid line. The response peaks near the point of maximum response to f_1, indicated by the arrow labeled f_1 at the top of the plot. The dotted line shows the phase-locking to f_1 when the stimulus is a simultaneous presentation of f_1 and f_2. The response to f_1 is similar in both cases, except near the point of maximum response to f_2 (at the arrow marked f_2) where the response to f_1 is reduced by the addition of f_2 to the stimulus. This phenomenon is called two-tone suppression [12.118, 119]; it can be suppression of phase-locking as in this case or it can be a decrease of the discharge rate in response to an excitor tone when a suppressor tone is added. Suppression can be seen on the basilar membrane as a reduction in the membrane motion at one frequency caused by the addition of a second frequency [12.34], and is usually explained as resulting from compression in the basilar-membrane input/output relationship.

The importance of suppression for responses to complex stimuli is that it improves the tonotopic separation of the components of the stimulus. In Fig. 12.13a, for example, f_1 presented by itself (solid line) gives a broad response that spreads away from the f_1 place and encompasses the f_2 place. In the presence of f_2, the f_1 response is confined to points near the f_1 place, thus improving the effective *tuning* of the fibers for f_1. The bottom part of Fig. 12.13b shows that the responses at the f_2 place are dominated by phase-locking to f_2 when f_1 and f_2 are presented simultaneously (dashed curve). The suppression effect is symmetric and a dip in the response to f_2 is observed at the f_1 place (at the asterisk in the bottom plot of Fig. 12.13b) which represents suppression of f_2 by f_1.

The top part of Fig. 12.13b shows responses to combination tones, another phenomenon that can be important in responses to complex stimuli. When two tones are presented simultaneously, observers can hear

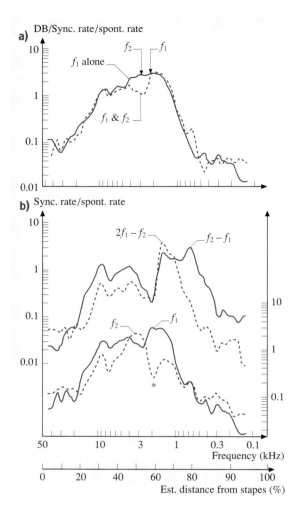

Fig. 12.13a,b Responses of a population of AN fibers to a two-tone complex consisting of $f_1=2.17$ kHz and $f_2=2.79$ kHz at 65 dB SPL. The *ordinates* show the strength of the phase-locked response as the synchronized rate divided by the spontaneous rate. Synchronized rate is the Fourier transform of the spike train at the appropriate frequency normalized to have units of spikes/s. Responses of a representative sample of fibers of different BFs were recorded. The lines are smoothed versions of the data computed as a moving-window average of the responses of individual fibers, for fibers with SR>15 /s. (**a**) Phase-locking to f_1 for responses to the f_1 stimulus tone alone (*solid line*) and to both f_1 and f_2 (*dashed line*). The arrows at top show the BF places for the two frequencies. (**b**) The *top plots* and *left ordinate* show phase-locking to the combination frequencies $2f_1 - f_2$ (*dashed*) and $f_2 - f_1$ (*solid*) when the stimulus was f_1 and f_2. The *bottom plots* and the *right ordinate* show phase-locking to f_1 (*solid*) and f_2 (*dashed*). For both plots, data were taken only from fibers to which all the stimuli shown were presented. Thus the f_1 phase-locking in (**a**) (*dashed line*) and in the bottom part of (**b**) (*solid line*) differ slightly, even though they estimate the same response to the same stimulus, because they were computed from somewhat different populations of fibers. (After [12.117] with permission)

distortion tones, most strongly at the cubic distortion frequency $2f_1 - f_2$ (1.55 kHz here). This distortion component is produced in the cochlea and is a prominent part of otoacoustic emissions for two-tone stimuli. It is used for both diagnostic and research purposes as a measure of OHC function [12.46]. In the cochlea, phase-locking at the frequency $2f_1 - f_2$ is seen, as shown by the dotted curve in the top part of Fig. 12.13b. It is important that the response to $2f_1 - f_2$ peaks at the place appropriate to the frequency $2f_1 - f_2$ (indicated by the arrow). The distribution of phase-locking to $2f_1 - f_2$ is similar to the distribution of phase-locking to a single tone at the frequency $2f_1 - f_2$, when account is taken of suppression effects and of sources of distortion in the region near the f_1 and f_2 places [12.117]. This behavior has been interpreted as showing that cochlear nonlinearities in the region of the primary tones (f_1 and f_2) lead to the creation of energy at the frequency $2f_1 - f_2$, which propagates on the basilar membrane in a fashion similar to an acoustic tone of that frequency presented at the ear. The phase of the responses (not shown) is consistent with this conclusion.

There is also a distortion tone at the frequency $f_2 - f_1$ (solid line in the top part of Fig. 12.13b), which appears to propagate on the basilar membrane and shows a peak of phase-locking at the appropriate place. This difference tone can also be heard by observers; however, there are differences in the rate of growth of the percept of $f_2 - f_1$ versus $2f_1 - f_2$, such that the former grows nonlinearly and the latter linearly as the level of f_1 and f_2 are increased. As a result, $f_2 - f_1$ is audible at high sound levels only.

Responses to a somewhat more complex stimulus are shown in Fig. 12.14, in this case an approximation to the vowel /eh/, as in *"met"* [12.65]. The magnitude spectrum of the stimulus is shown in Fig. 12.14a. It consists of the harmonics of a 125 Hz fundamental with their amplitudes adjusted to approximate the spectrum of /eh/. The actual vowel has peaks of energy at the resonant frequencies of the vocal tract, called formants; these are indicated by F1, F2, and F3 in Fig. 12.14a. Again the ex-

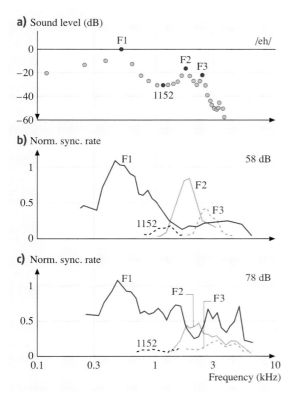

Fig. 12.14a–c Responses to a stimulus similar to the vowel /eh/ presented at two sound levels in a population of AN fibers. (**a**) The magnitude spectrum of the stimulus. It is periodic and consists of harmonics of a 125 Hz fundamental. There are peaks of energy at the formant frequencies of a typical /eh/, near 0.5 (F1), 1.75 (F2), and 2.275 kHz (F3). The 1.152 kHz component in the trough between F1 and F2 is used for comparison in parts (**b**) and (**c**). (**b**) Distribution of phase-locking to four frequency components of the vowel when presented at 58 dB SPL, as labeled. These are moving-window averages of the phase-locking of individual fibers. Phase-locking here is the magnitude of the Fourier transform of the spike train at the frequency of interest normalized by the maximum discharge rate of the fiber. Maximum rate is the rate in response to a BF tone 50 dB above threshold. (**c**) Same for responses to the vowel at a sound level of 78 dB SPL. (After [12.120, 121])

perimental approach is to record from a large population of AN fibers in one animal and present the same stimulus to each fiber. The response of the whole population is estimated by a moving average of the data points for individual fibers. This stimulus contains many frequency components, so the responses are quite complex. However, the data show that formant frequencies dominate the responses. Note that combination tones always occur at the frequency of a real acoustic stimulus tone for a harmonic series like this, so it is not possible to analyze combination tones without making assumptions that allow the responses to combination and real tones to be separated [12.120].

The distribution of phase-locking in response to the formants and to one non-formant frequency are shown in Figs. 12.14b and 12.14c; these show responses at two sound levels. At 58 dB (Fig. 12.14b), the formant responses are clearly separated into different populations of fibers, so that the response to a formant is largest among fibers with BFs near the formant frequency. There are significant responses to all three formants, but there is little response to frequency components between the formants, such as the response to 1.152 kHz (dotted line). These results demonstrate a clear tonotopic representation of the vowel.

At 78 dB (Fig. 12.14c), the responses to F2 and F3 decrease in amplitude and the response to F1 spreads to occupy almost the entire population. This behavior is typically observed at high sound levels, where the phase-locking to a large low-frequency component of the stimulus spreads to higher BFs in the population [12.122]. Suppression plays two roles in this process. First, the spread of F1 and the decrease in response to F2 and F3 behave quantitatively like suppression of a tone at F2 or F3 by a tone at F1 and therefore seem to represent suppression of F2 and F3 by F1 [12.121]. Second, the response to F1 is suppressed among fibers with BFs near F2, as shown by dips in the phase-locking to F1 near the F2 place. Suppression acts to improve the representation in the latter case, but has the opposite effect in the former case.

12.3.2 Tasks of the Central Auditory System

The discussion so far concerns only the most basic aspects of physiological acoustics. The representation of sound in the AN is the input to a complex of neural pathways in the central auditory system. Discussion of specifics of the central processing of auditory stimuli is beyond the scope of this chapter. However, several recent books provide comprehensive coverage of the anatomy and physiology of this system, e.g., [12.123–126].

The representation of the auditory stimulus provided to the brain by the AN is a spectrotemporal one. That is, the responses of AN fibers provide an accurate representation of the moment-by-moment distribution of energy across frequency for the sound entering the ear.

In terms of processing sound, the major steps taken in the cochlea are frequency analysis, compression, and suppression. Frequency analysis is essential to all of the processing that follows in the brain. Indeed central auditory centers are all organized by frequency and most properties of the perception of sound appear to include frequency analysis as a fundamental component. Compression is important for extending the dynamic range of hearing. Its importance is illustrated by the problems of loudness abnormality and oversensitivity to loud sounds in persons with damaged cochleae that are missing compression [12.127]. Suppression is important for maintaining the quality of the spectrotemporal representation, at least at moderate sound levels, and is the first example of interaction across frequencies in the auditory system. In the following paragraphs, some of the central nervous system's secondary analyses on the cochlear spectrotemporal representation are described.

The first function of the central auditory system is to stabilize the spectrotemporal representation provided in the AN. As shown in Fig. 12.14, the representation changes across sound level and the quality of the representation is generally lower at high sound levels and at low sound levels. In the cochlear nucleus the representation is more stable as sound level changes, e.g., for speech [12.128]. Presumably the stabilization occurs through optimal combination of information across SR groups in the AN and perhaps also through inhibitory interactions.

A second function of the lower central auditory system is binaural interaction. The location of sound sources is computed by the auditory system from small differences in the sounds at the two ears [12.84, 129]. Essentially, a sound is delayed in reaching the ear on the side of the head away from the source and is less intense there. Differences in interaural stimulus timing are small and require a specialized system of neurons in the first and second stages of the central auditory system to allow accurate representation of the interaural timing cues. A similar system is present for interaural differences in sound level. The representation of these cues is elaborated in higher levels, where neurons show response characteristics that might explain perceptual phenomena like binaural masking level differences [12.130] and the precedence effect [12.131].

A third aspect of central processing is that the representation of sound switches away from a spectrotemporal description of the stimulus, as in the AN, and moves to derived representations, such as one that represents auditory objects [12.132]. For example, the representation of speech in the auditory cortex does not take a recognizable tonotopic form, in that peaks of activity at BFs corresponding to the formant frequencies are not seen [12.133]. The nature of the representation used in the cortex is unknown; it is clearly based on a tonotopic axis, but neurons' responses are not determined in a straightforward way by the amount of energy in their tuning curves, as is observed in lower auditory nuclei [12.134]. In some cases, neurons specialized for particular auditory tasks have been found; neurons in one region of the marmoset cortex are tuned to pitch, for example [12.135]. Perhaps the best-studied cases are neurons that respond specifically to species-specific vocalizations in marmosets [12.136] and songbirds [12.137, 138].

12.4 Summary

Research on the auditory system has defined the means by which the system efficiently captures sound and couples it into the cochlea, how the frequency analysis of the cochlea is conducted, how transduction occurs, how nonlinear mechanisms involving the OHCs sharpen the frequency tuning, increase the sensitivity, and compress the stimulus intensity, and how suppression acting at the level of AN fibers maintains the sharpness of the frequency analysis for complex, multicomponent stimuli. Over the next decade, research on the auditory system will increasingly be focused on the organization and function of the central auditory system, with particular reference to the way in which central neurons derive information from the spectrotemporal display received from the AN.

References

12.1 G.A. Manley, A.N. Popper, R.R. Fay: *Evolution of the Vertebrate Auditory System* (Springer, Berlin, New York 2004)

12.2 J.J. Rosowski: Outer and middle ears. In: *Comparative Hearing: Mammals*, ed. by R.R. Fay, A.N. Popper (Springer, Berlin, New York 1994)

12.3 J.J. Rosowski: External- and middle-ear function. In: *Auditory Computation*, ed. by H.L. Hawkins, T.A. McMullen, A.N. Popper, R.R. Fay (Springer, Berlin, New York 1996)

12.4 E.A.G. Shaw: Transformation of sound pressure level from the free field to the eardrum in the horizontal plane, J. Acoust. Soc. Am. **56**, 1848–1861 (1974)

12.5 F.M. Wiener, R.R. Pfeiffer, A.S.N. Backus: On the sound pressure transformation by the head and auditory meatus of the cat, Acta Otolaryngol. **61**, 255–269 (1966)

12.6 J.J. Rice, B.J. May, G.A. Spiron, E.D. Young: Pinna-based spectral cues for sound localization in cat, Hearing Res. **58**, 132–152 (1992)

12.7 E.A.G. Shaw: External ear response and sound localization. In: *Localization of Sound: Theory and Applications*, ed. by R.W. Gatehouse (Amphora, Groton 1982)

12.8 M. Brodel: *Three Unpublished Drawings of the Anatomy of the Human Ear* (Saunders, Philadelphia 1946)

12.9 E.A.G. Shaw: 1979 Rayleigh medal lecture: The elusive connection. In: *Localization of Sound: Theory and Applications*, ed. by R.W. Gatehouse (Amphora, Groton 1982)

12.10 J.J. Rosowski, L.H. Carney, W.T. Peake: The radiation impedance of the external ear of cat: measurements and applications, J. Acoust. Soc. Am. **84**, 1695–1708 (1988)

12.11 J.C. Middlebrooks, D.M. Green: Sound localization by human listeners, Annu. Rev. Psychol. **42**, 135–159 (1991)

12.12 B.J. May, A.Y. Huang: Sound orientation behavior in cats. I. Localization of broadband noise, J. Acoust. Soc. Am. **100**, 1059–1069 (1996)

12.13 J. Blauert: *Spatial Hearing: The Psychophysics of Human Sound Localization* (MIT Press, Cambridge 1997)

12.14 G.F. Kuhn: Physical acoustics and measurements pertaining to directional hearing. In: *Directional Hearing*, ed. by W.A. Yost, G. Gourevitch (Springer, Berlin, New York 1987)

12.15 E.D. Young, J.J. Rice, S.C. Tong: Effects of pinna position on head-related transfer functions in the cat, J. Acoust. Soc. Am. **99**, 3064–3076 (1996)

12.16 A.D. Musicant, J.C.K. Chan, J.E. Hind: Direction-dependent spectral properties of cat external ear: New data and cross-species comparisons, J. Acoust. Soc. Am. **87**, 757–781 (1990)

12.17 A.Y. Huang, B.J. May: Spectral cues for sound localization in cats: Effects of frequency domain on minimum audible angles in the median and horizontal planes, J. Acoust. Soc. Am. **100**, 2341–2348 (1996)

12.18 P. Dallos: *The Auditory Periphery, Biophysics and Physiology* (Academic, New York 1973)

12.19 J. Zwislocki: Analysis of some auditory characteristics. In: *Handbook of Mathematical Psychology*, ed. by R.D. Luce, R.R. Bush, E. Galanter (Wiley, New York 1965)

12.20 W. Bialek: Physical limits to sensation and perception, Ann. Rev. Biophys. Biophys. Chem. **16**, 455–478 (1987)

12.21 M.B. Sachs, P.J. Abbas: Rate versus level functions for auditory-nerve fibers in cats: tone-burst stimuli, J. Acoust. Soc. Am. **56**, 1835–1847 (1974)

12.22 P. Wangemann, J. Schacht: Homeostatic mechanisms in the cochlea. In: *The Cochlea*, ed. by P. Dallos, A.N. Popper, R.R. Fay (Springer, Berlin, New York 1996)

12.23 W. Bloom, D.W. Fawcett: *A Textbook of Histology* (Saunders, Philadelphia 1975)

12.24 D.K. Ryugo: The auditory nerve: Peripheral innervation, cell body morphology, and central projections. In: *The Mammalian Auditory Pathway: Neuroanatomy*, ed. by D.B. Webster, A.N. Popper, R.R. Fay (Springer, Berlin, New York 1992)

12.25 W.B. Warr: Organization of olivocochlear efferent systems in mammals. In: *The Mammalian Auditory Pathway: Neuroanatomy*, ed. by D.B. Webster, A.N. Popper, R.R. Fay (Springer, Berlin, New York 1992)

12.26 M.C. Brown: Antidromic responses of single units from the spiral ganglion, J. Neurophysiol. **71**, 1835–1847 (1994)

12.27 D. Robertson: Horseradish peroxidase injection of physiologically characterized afferent and efferent neurons in the guinea pig spiral ganglion, Hearing Res. **15**, 113–121 (1984)

12.28 D. Robertson, P.M. Sellick, R. Patuzzi: The continuing search for outer hair cell afferents in the guinea pig spiral ganglion, Hearing Res. **136**, 151–158 (1999)

12.29 D.J. Jagger, G.D. Housley: Membrane properties of type II spiral ganglion neurones identified in a neonatal rat cochlear slice, J. Physiol. **552**, 525–533 (2003)

12.30 M.A. Reid, J. Flores-Otero, R.L. Davis: Firing patterns of type II spiral ganglion neurons in vitro, J. Neurosci. **24**, 733–742 (2004)

12.31 J.J. Guinan: Physiology of olivocochlear efferents. In: *The Cochlea*, ed. by P. Dallos, A.N. Popper, R.R. Fay (Springer, Berlin, New York 1996)

12.32 M.B. Sachs, I.C. Bruce, E.D. Young: The biological basis of hearing-aid design, Ann. BME **30**, 157–168 (2002)

12.33 G. von Bekesy: *Experiments in Hearing* (McGraw-Hill, New York 1960)

12.34 L. Robles, M.A. Ruggero: Mechanics of the mammalian cochlea, Physiol. Rev. **81**, 1305–1352 (2001)

12.35 D.H. Eldredge: Electrical equivalents of the Bekesy traveling wave in the mammalian cochlea. In: *Evoked Electrical Activity in the Auditory Nervous System*, ed. by R.F. Naunton, C. Fernandez (Academic, New York 1978)

12.36 M.A. Ruggero, N.C. Rich, A. Recio, S.S. Narayan: Basilar-membrane responses to tones at the base of the chinchilla cochlea, J. Acoust. Soc. Am. **101**, 2151–2163 (1997)

12.37 N.P. Cooper, W.S. Rhode: Mechanical responses to two-tone distortion products in the apical and basal turns of the mammalian cochlea, J. Neurophysiol. **78**, 261–270 (1997)

12.38 W.S. Rhode: Observations of the vibration of the basilar membrane in squirrel monkeys using the Mossbauer technique, J. Acoust. Soc. Am. **49**, 1218–1231 (1971)

12.39 D. Brass, D.T. Kemp: Analyses of Mossbauer mechanical measurements indicate that the cochlea is mechanically active, J. Acoust. Soc. Am. **93**, 1502–1515 (1993)

12.40 P. Dallos: The active cochlea, J. Neurosci. **12**, 4575–4585 (1992)

12.41 H. Davis: An active process in cochlear mechanics, Hear Res. **9**, 79–90 (1983)

12.42 R.J. Diependaal, E. de Boer, M.A. Viergever: Cochlear power flux as an indicator of mechanical activity, J. Acoust. Soc. Am. **82**, 917–926 (1987)

12.43 G.A. Manley: Evidence for an active process and a cochlear amplifier in nonmammals, J. Neurophysiol. **86**, 541–549 (2001)

12.44 E. Zwicker: A model describing nonlinearities in hearing by active processes with saturation at 40 dB, Biol. Cybern. **35**, 243–250 (1979)

12.45 D.T. Kemp: Stimulated acoustic emissions from within the human auditory system, J. Acoust. Soc. Am. **64**, 1386–1391 (1978)

12.46 C.A. Shera: Mechanisms of mammalian otoacoustic emission and their implications for the clinical utility of otoacoustic emissions, Ear Hearing **25**, 86–97 (2004)

12.47 C.L. Talmadge, A. Tubis, G.R. Long, C. Tong: Modeling the combined effects of basilar membrane nonlinearity and roughness on stimulus frequency otoacoustic emission fine structure, J. Acoust. Soc. Am. **108**, 2911–2932 (2000)

12.48 G. Zweig, C.A. Shera: The origin of periodicity in the spectrum of evoked otoacoustic emissions, J. Acoust. Soc. Am. **98**, 2018–2047 (1995)

12.49 R. Patuzzi: Cochlear micromechanics and macromechanics. In: *The Cochlea*, ed. by P. Dallos, A.N. Popper, R.R. Fay (Springer, Berlin, New York 1996)

12.50 E. de Boer, A.L. Nuttall, N. Hu, Y. Zou, J. Zheng: The Allen-Fahey experiment extended, J. Acoust. Soc. Am. **117**, 1260–1266 (2005)

12.51 M.C. Liberman, L.W. Dodds: Single-neuron labeling and chronic cochlear pathology. III. Stereocilia damage and alterations of threshold tuning curves, Hearing Res. **16**, 55–74 (1984)

12.52 E. Murugasu, I.J. Russell: The effect of efferent stimulation on basilar membrane displacement in the basal turn of the guinea pig cochlea, J. Neurosci. **16**, 325–332 (1996)

12.53 M.C. Brown, A.L. Nuttall: Efferent control of cochlear inner hair cell responses in the guinea-pig, J. Physiol. **354**, 625–646 (1984)

12.54 M.L. Wiederhold, N.Y.S. Kiang: Effects of electric stimulation of the crossed olivocochlear bundle on single auditory-nerve fibers in the cat, J. Acoust. Soc. Am. **48**, 950–965 (1970)

12.55 M.C. Liberman, S. Puria, J.J. Guinan: The ipsilaterally evoked olivocochlear reflex causes rapid adaptation of the $2f_1-f_2$ distortion product otoacoustic emission, J. Acoust. Soc. Am. **99**, 3572–3584 (1996)

12.56 D.C. Mountain: Changes in endolymphatic potential and crossed olivocochlear bundle stimulation alter cochlear mechanics, Science **210**, 71–72 (1980)

12.57 M.A. Ruggero, N.C. Rich, A. Recio: The effect of intense acoustic stimulation on basilar-membrane vibrations, Aud. Neurosci. **2**, 329–345 (1996)

12.58 A.J. Oxenham, S.P. Bacon: Cochlear compression: Perceptual measures and implications for normal and impaired hearing, Ear Hearing **24**, 352–366 (2003)

12.59 D.A. Nelson, A.C. Schroder, M. Wojtczak: A new procedure for measuring peripheral compression in normal-hearing and hearing-impaired listeners, J. Acoust. Soc. Am. **110**, 2045–2064 (2001)

12.60 A.J. Oxenham, C.J. Plack: A behavioral measure of basilar-membrane nonlinearity in listeners with normal and impaired hearing, J. Acoust. Soc. Am. **101**, 3666–3675 (1997)

12.61 E. de Boer: Mechanics of the cochlea: Modeling efforts. In: *The Cochlea*, ed. by P. Dallos, A.N. Popper, R.R. Fay (Springer, Berlin, New York 1996)

12.62 A.E. Hubbard, D.C. Mountain: Analysis and synthesis of cochlear mechanical function using models. In: *Auditory Computation*, ed. by H.L. Hawkins, T.A. McMullen, A.N. Popper, R.R. Fay (Springer, Berlin, New York 1996)

12.63 R.C. Naidu, D.C. Mountain: Measurements of the stiffness map challenge a basic tenet of cochlear theories, Hearing Res. **124**, 124–131 (1998)

12.64 M.P. Sherer, A.W. Gummer: Impedance analysis of the organ of corti with magnetically actuated probes, Biophys. J. **87**, 1378–1391 (2004)

12.65 M.B. Sachs: Speech encoding in the auditory nerve. In: *Hearing Science Recent Advances*, ed. by C.I. Berlin (College Hill, San Diego 1984)

12.66 J.J. Eggermont: Between sound and perception: reviewing the search for a neural code, Hear Res. **157**, 1–42 (2001)

12.67 B.C.J. Moore: Coding of sounds in the auditory system and its relevance to signal processing and coding in cochlear implants, Otol. Neurotol. **24**, 243–254 (2003)

12.68 I.C. Bruce, M.B. Sachs, E.D. Young: An auditory-periphery model of the effects of acoustic trauma on auditory nerve responses, J. Acoust. Soc. Am. **113**, 369–388 (2003)

12.69 S.D. Holmes, C. Sumner, L.P. O'Mard, R. Meddis: The temporal representation of speech in a nonlinear model of the guinea pig cochlea, J. Acoust. Soc. Am. **116**, 3534–3545 (2004)

12.70 X. Zhang, M.G. Heinz, I.C. Bruce, L.H. Carney: A phenomenological model for the responses of auditory-nerve fibers: I. Nonlinear tuning with compression and suppression, J. Acoust. Soc. Am. **109**, 648–670 (2001)

12.71 H.S. Colburn, L.H. Carney, M.G. Heinz: Quantifying the information in auditory-nerve responses for level discrimination, J. Assoc. Res. Otolaryngol. **4**, 294–311 (2003)

12.72 B.C.J. Moore: *An Introduction to the Psychology of Hearing* (Elsevier, London 2004)

12.73 P.A. Cariani, B. Delgutte: Neural correlates of the pitch of complex tones. II. Pitch shift, pitch ambiguity, phase invariance, pitch circularity, rate pitch, and the dominance region for pitch, J. Neurophysiol. **76**, 1717–1734 (1996)

12.74 R.L. Miller, J.R. Schilling, K.R. Franck, E.D. Young: Effects of acoustic trauma on the representation of the vowel /ɛ/ in cat auditory nerve fibers, J. Acoust. Soc. Am. **101**, 3602–3616 (1997)

12.75 D.H. Johnson: The relationship of post-stimulus time and interval histograms to the timing characteristics of spike trains, Biophys. J. **22**(3), 413–430 (1978)

12.76 M.C. Liberman: Auditory-nerve response from cats raised in a low-noise chamber, J. Acoust. Soc. Am. **63**, 442–455 (1978)

12.77 G.K. Yates, I.M. Winter, D. Robertson: Basilar membrane nonlinearity determines auditory nerve rate-intensity functions and cochlear dynamic range, Hearing Res. **45**, 203–220 (1990)

12.78 B.C.J. Moore: *Perceptual Consequences of Cochlear Damage* (Oxford Univ. Press, Oxford 1995)

12.79 M.G. Heinz, J.B. Issa, E.D. Young: Auditory-nerve rate responses are inconsistent with common hypotheses for the neural correlates of loudness recruitment, J. Assoc. Res. Otolaryngol. **6**, 91–105 (2005)

12.80 J.O. Pickles: Auditory-nerve correlates of loudness summation with stimulus bandwidth, in normal and pathological cochleae, Hearing Res. **12**, 239–250 (1983)

12.81 E.M. Relkin, J.R. Doucet: Is loudness simply proportional to the auditory nerve spike count?, J. Acoust. Soc. Am. **101**, 2735–2740 (1997)

12.82 S. Rosen: Temporal information in speech: acoustic, auditory and linguistic aspects, Philos. Trans. R. Soc. London B. **336**, 367–373 (1992)

12.83 A.R. Palmer, I.J. Russell: Phase-locking in the cochlear nerve of the guinea-pig and its relation to the receptor potential of inner hair-cells, Hear Res. **24**, 1–15 (1986)

12.84 T.C.T. Yin: Neural mechanisms of encoding binaural localization cues in the auditory brainstem. In: *Integrative Functions in the Mammalian Auditory Pathway*, ed. by D. Oertel, A.N. Popper, R.R. Fay (Springer, Berlin, New York 2002)

12.85 S.A. Shamma: Speech processing in the auditory system. II: Lateral inhibition and the central processing of speech evoked activity in the auditory nerve, J. Acoust. Soc. Am. **78**, 1622–1632 (1985)

12.86 P. Dallos, B. Fakler: Prestin, a new type of motor protein, Nature Rev. Molec. Bio. **3**, 104–111 (2002)

12.87 R.A. Eatock, K.M. Hurley: Functional development of hair cells, Curr. Top. Dev. Bio. **57**, 389–448 (2003)

12.88 R. Fettiplace, C.M. Hackney: The sensory and motor roles of auditory hair cells, Nature Rev. Neurosci. **7**, 19–29 (2006)

12.89 M. LeMasurier, P.G. Gillespie: Hair-cell mechanotransduction and cochlear amplification, Neuron. **48**, 403–415 (2005)

12.90 G.I. Frolenkov, I.A. Belyantseva, T.B. Friedmann, A.J. Griffith: Genetic insights into the morphogenesis of inner ear hair cells, Nature Rev. Genet. **5**, 489–498 (2004)

12.91 J.O. Pickles, S.D. Comis, M.P. Osborne: Cross-links between stereocilia in the guinea pig organ of Corti, and their possible relation to sensory transduction, Hear Res. **15**, 103–112 (1984)

12.92 A. Flock: Transducing mechanisms in the lateral line canal organ receptors, Cold Spring Harb. Symp. Quant. Biol. **30**, 133–145 (1965)

12.93 S.L. Shotwell, R. Jacobs, A.J. Hudspeth: Directional sensitivity of individual vertebrate hair cells to controlled deflection of their hair bundles, Ann. NY Acad. Sci. **374**, 1–10 (1981)

12.94 J.A. Assad, G.M.G. Shepherd, D.P. Corey: Tip-link integrity and mechanical transduction in vertebrate hair cells, Neuron **7**, 985–994 (1991)

12.95 M.A. Ruggero, N.C. Rich: Timing of spike initiation in cochlear afferents: dependence on site of innervation, J. Neurophysiol. **58**, 379–403 (1987)

12.96 E. Glowatzki, P.A. Fuchs: Transmitter release at the hair cell ribbon synapse, Nature Neurosci. **5**, 147–154 (2002)

12.97 L.O. Trussell: Transmission at the hair cell synapse, Nature Neurosci. **5**, 85–86 (2002)

12.98 Y. Raphael, R.A. Altschuler: Structure and innervation of the cochlea, Brain Res. Bull. **60**, 397–422 (2003)

12.99 P.G. Gillespie, J.L. Cyr: Myosin-1c, the hair cell's adaptation motor, Annu. Rev. Physiol. **66**, 521–545 (2004)

12.100 M.C. Holley: Outer hair cell motility. In: *The Cochlea*, ed. by P. Dallos, A.N. Popper, R.R. Fay (Springer, Berlin, New York 1996)

12.101 R. Fettiplace: Active hair bundle movements in auditory hair cells, J. Physiol. **576**, 29–36 (2006)

12.102 M.C. Brown, A.L. Nuttall, R.I. Masta: Intracellular recordings from cochlear inner hair cells: effects of stimulation of the crossed olivocochlear efferents, Science **222**, 69–72 (1983)

12.103 D.F. Dolan, M.H. Guo, A.L. Nuttall: Frequency-dependent enhancement of basilar membrane velocity during olivocochlear bundle stimulation, J. Acoust. Soc. Am. **102**, 3587–3596 (1997)

12.104 J. Santos-Sacchi: Reversible inhibition of voltage-dependent outer hair cell motility and capacitance, J. Neurosci. **11**, 3096–3110 (1991)

12.105 P. Dallos, D. Harris: Properties of auditory nerve responses in absence of outer hair cells, J. Neurophysiol. **41**, 365–383 (1978)

12.106 M.C. Liberman, N.Y. Kiang: Single-neuron labeling and chronic cochlear pathology. IV. Stereocilia damage and alterations in rate- and phase-level functions, Hearing Res. **16**, 75–90 (1984)

12.107 J.F. Ashmore: A fast motile response in guinea pig outer hair cells: the basis of the cochlear amplifier, J. Physiol. **388**, 323–347 (1987)

12.108 W.E. Brownell, C.R. Bader, D. Bertrand, Y. de Ribaupierre: Evoked mechanical responses of isolated cochlear hair cells, Science **227**, 194–196 (1985)

12.109 I.A. Belyantseva, H.J. Adler, R. Curi, G.I. Frolenkov, B. Kachar: Expression and localization of prestin and the sugar transporter GLUT-5 during development of electromotility in cochlear outer hair cell, J. Neurosci. **20**(RC116), 1–5 (2000)

12.110 F. Kalinec, M.C. Holley, K.H. Iwasa, D.J. Lim, B. Kachar: A membrane-based force generation mechanism in auditory sensory cells, Proc. Natl. Acad. Sci. USA **89**, 8671–8675 (1992)

12.111 J. Zheng, W. Shen, D.Z. He, K.B. Long, D. Madison, P. Dallos: Prestin is the motor protein of cochlear outer hair cells, Nature **405**, 149–155 (2000)

12.112 M.C. Holley, F. Kalinec, B. Kachar: Structure of the cortical cytoskeleton in mammalian outer hair cells, J. Cell Sci. **102**, 569–580 (1992)

12.113 D. Oliver, D.Z. He, N. Klocker, J. Ludwig, U. Schutte, S. Waldegger, J.P. Ruppersberg, P. Dallos, B. Fakler: Intracellular anions as the voltage sensor of prestin, the outer hair cell motor protein, Science **292**, 2340–2343 (2001)

12.114 M.C. Liberman, J. Galilo, D.Z.Z. He, X. Wu, S. Jia, J. Zuo: Prestin is required for electromotility of the outer hair cell and for the cochlear amplifier, Nature **419**, 300–304 (2002)

12.115 C.D. Geisler, C. Sang: A cochlear model using feed-forward outer-hair-cell forces, Hear Res. **86**, 132–146 (1995)

12.116 P.A. Fuchs: Synaptic transmission at vertebrate hair cells, Curr. Opin. Neurobiol. **6**, 514–519 (1996)

12.117 D.O. Kim, C.E. Molnar, J.W. Matthews: Cochlear mechanics: Nonlinear behavior in two-tone responses as reflected in cochlear-nerve-fiber responses and in ear-canal sound pressure, J. Acoust. Soc. Am. **67**, 1704–1721 (1980)

12.118 E. Javel, C.D. Geisler, A. Ravindran: Two-tone suppression in auditory nerve of the cat: rate-intensity and temporal analyses, J. Acoust. Soc. Am. **63**(4), 1093–1104 (1978)

12.119 M.B. Sachs, N.Y. Kiang: Two-tone inhibition in auditory-nerve fibers, J. Acoust. Soc. Am. **43**(5), 1120–1128 (1968)

12.120 E.D. Young, M.B. Sachs: Representation of steady-state vowels in the temporal aspects of the discharge patterns of populations of auditory-nerve fibers, J. Acoust. Soc. Am. **66**, 1381–1403 (1979)

12.121 M.B. Sachs, E.D. Young: Encoding of steady-state vowels in the auditory nerve: Representation in terms of discharge rate, J. Acoust. Soc. Am. **66**, 470–479 (1979)

12.122 J.C. Wong, R.L. Miller, B.M. Callhoun, M.B. Sachs, E.D. Young: Effects of high sound levels on responses to the vowel /ε/ in cat auditory nerve, Hear Res. **123**, 61–77 (1998)

12.123 G. Ehret, R. Romand: *The Central Auditory System* (Oxford Univ. Press, New York 1997)

12.124 M.S. Malmierca, D.R.F. Irvine: *International Review of Neurobiology – Auditory Spectral Processing*, Vol. 70 (Elsevier, Amsterdam 2005)

12.125 D. Oertel, A.N. Popper, R.R. Fay: *Integrative Functions in the Mammalian Auditory Pathway* (Springer, Berlin, New York 2001)

12.126 J.A. Winer, C.E. Schreiner: *The Inferior Colliculus* (Springer, Berlin, New York 2005)

12.127 S.P. Bacon, A.J. Oxenham: Psychophysical manifestations of compression: Hearing-impaired listeners. In: *Compression: From cochlea to cochlear implants*, ed. by S.P. Bacon, R.R. Fay, A.N. Popper (Springer, Berlin, New York 2004)

12.128 C.C. Blackburn, M.B. Sachs: The representations of the steady-state vowel sound /ε/ in the discharge patterns of cat anteroventral cochlear nucleus neurons, J. Neurophysiol. **63**, 1191–1212 (1990)

12.129 A.R. Palmer: Reassessing mechanisms of low-frequency sound localisation, Curr. Opin. Neurobiol. **14**, 457–460 (2004)

12.130 A.R. Palmer, D. Jiang, D. McAlpine: Neural responses in the inferior colliculus to binaural masking level differences created by inverting the noise in one ear, J. Neurophysiol. **84**, 844–852 (2000)

12.131 T.C. Yin: Physiological correlates of the precedence effect and summing localization in the inferior colliculus of the cat, J. Neurosci. **14**, 5170–5186 (1994)

12.132 I. Nelken: Processing of complex stimuli and natural scenes in the auditory cortex, Curr. Opin. Neurobiol. **14**, 474–480 (2004)

12.133 S.W. Wong, C.E. Schreiner: Representation of CV-sounds in cat primary auditory cortex: Intensity dependence, Speech Commun. **41**, 93–106 (2003)

12.134 C.K. Machens, M.S. Wehr, A.M. Zador: Linearity of cortical receptive fields measured with natural sounds, J. Neurosci. **24**, 1089–1100 (2004)

12.135 D. Bendor, X. Wang: The neuronal representation of pitch in primate auditory cortex, Nature **436**, 1161–1165 (2005)

12.136 X. Wang, S.C. Kadia: Differential representation of species-specific primate vocalizations in the auditory cortices of marmoset and cat, J. Neurophysiol. **86**, 2616–2620 (2001)

12.137 H. Cousillas, H.J. Leppelsack, E. Leppelsack, J.P. Richard, M. Mathelier, M. Hausberger: Functional organization of the forebrain auditory centres of the European starling: A study based on natural sounds, Hear Res. **207**, 10–21 (2005)

12.138 J.A. Grace, N. Amin, N.C. Singh, F.E. Theunissen: Selectivity for conspecific song in the zebra finch auditory forebrain, J. Neurophysiol. **89**, 472–487 (2003)

13. Psychoacoustics

Psychoacoustics is concerned with the relationships between the physical characteristics of sounds and their perceptual attributes. This chapter describes: the absolute sensitivity of the auditory system for detecting weak sounds and how that sensitivity varies with frequency; the frequency selectivity of the auditory system (the ability to resolve or *hear out* the sinusoidal components in a complex sound) and its characterization in terms of an array of auditory filters; the processes that influence the masking of one sound by another; the range of sound levels that can be processed by the auditory system; the perception and modeling of loudness; level discrimination; the temporal resolution of the auditory system (the ability to detect changes over time); the perception and modeling of pitch for pure and complex tones; the perception of timbre for steady and time-varying sounds; the perception of space and sound localization; and the mechanisms underlying *auditory scene analysis* that allow the construction of percepts corresponding to individual sounds sources when listening to complex mixtures of sounds.

13.1 **Absolute Thresholds** 460

13.2 **Frequency Selectivity and Masking** 461
    13.2.1 The Concept of the Auditory Filter 462
    13.2.2 Psychophysical Tuning Curves 462
    13.2.3 The Notched-Noise Method 463
    13.2.4 Masking Patterns and Excitation Patterns.............. 464
    13.2.5 Forward Masking...................... 465
    13.2.6 Hearing Out Partials in Complex Tones 467

13.3 **Loudness** 468
    13.3.1 Loudness Level and Equal-Loudness Contours 468
    13.3.2 The Scaling of Loudness 469
    13.3.3 Neural Coding and Modeling of Loudness 469
    13.3.4 The Effect of Bandwidth on Loudness............................. 470
    13.3.5 Intensity Discrimination............. 472

13.4 **Temporal Processing in the Auditory System**........................ 473
    13.4.1 Temporal Resolution Based on Within-Channel Processes 473
    13.4.2 Modeling Temporal Resolution.... 474
    13.4.3 A Modulation Filter Bank? 475
    13.4.4 Duration Discrimination............. 476
    13.4.5 Temporal Analysis Based on Across-Channel Processes 476

13.5 **Pitch Perception**.................................. 477
    13.5.1 Theories of Pitch Perception 477
    13.5.2 The Perception of the Pitch of Pure Tones............................ 478
    13.5.3 The Perception of the Pitch of Complex Tones 480

13.6 **Timbre Perception** 483
    13.6.1 Time-Invariant Patterns and Timbre............................... 483
    13.6.2 Time-Varying Patterns and Auditory Object Identification 483

13.7 **The Localization of Sounds** 484
    13.7.1 Binaural Cues 484
    13.7.2 The Role of the Pinna and Torso.. 485
    13.7.3 The Precedence Effect 485

13.8 **Auditory Scene Analysis**....................... 485
    13.8.1 Information Used to Separate Auditory Objects....................... 486
    13.8.2 The Perception of Sequences of Sounds............................... 489
    13.8.3 General Principles of Perceptual Organization 492

13.9 **Further Reading and Supplementary Materials** 494

References 495

13.1 Absolute Thresholds

The absolute threshold of a sound is the lowest detectable level of that sound in the absence of any other sounds. In practice, there is no distinct sound level at which a sound suddenly becomes detectable. Rather, the probability of detecting a sound increases progressively as the sound level is increased from a very low value. Hence, the absolute threshold is defined as the sound level at which an individual detects the sound with a certain probability, such as 75% (in a two-alternative forced-choice task, where guessing leads to 50% correct, on average). Typically, results are averaged across many listeners with normal hearing (i. e., with no known history of hearing disorders and no obvious signs of hearing problems) to obtain representative results.

The absolute threshold for detecting sinusoids is partly determined by the sound transmission through the outer and middle ear (see Chap. 12); to a first approximation, the inner ear (the cochlea) is equally sensitive to all frequencies, except perhaps at very low frequencies and very high frequencies [13.3, 4]. Figure 13.1 shows estimates of the absolute threshold, measured in two ways. For the curve labeled MAP, standing for minimum audible pressure, the sound level was measured at a point close to the eardrum [13.2]. For the curve labeled MAF, standing for minimum audible field, the measurement of sound level was made after the listener had been removed from the sound field, at the point which had been occupied by the center of the listener's head [13.1]. For the MAF curve, the sound was presented in free field (in an anechoic chamber) from a direction directly in front of the listener. Note that the MAP estimates are for monaural listening and the MAF estimates are for binaural listening. On average, thresholds are about 2 dB lower when two ears are used as opposed to one, although the exact value of the difference varies across studies from 0 to 3 dB, and it can depend on the interaural phase of the tone [13.5–7]. Both curves represent the average data from many young listeners with normal hearing. It should be noted, however, that individual people may have thresholds as much as 20 dB above or below the mean at a specific frequency and still be considered as normal.

The MAP and MAF curves are somewhat differently shaped, since the head, the pinna and the meatus have an influence on the sound field. The MAP curve shows only minor peaks and dips (± 5 dB) for frequencies between about 0.2 kHz and 13 kHz, whereas the MAF curve shows a distinct dip around 3–4 kHz and a peak around 8–9 kHz. The difference derives mainly from a broad resonance produced by the meatus and pinna. The sound level at the eardrum is enhanced markedly for frequencies in the range 1.5–6 kHz, with a maximum enhancement at 3 kHz of about 15 dB.

The highest audible frequency varies considerably with age. Young children can often hear tones as high as 20 kHz, but for most adults the threshold rises rapidly above about 15 kHz. The loss of sensitivity with increasing age (presbyacusis) is much greater at high frequencies than at low, and the variability between different people is also greater at high frequencies. There seems to be no well-defined low-frequency limit to hearing, although very intense low-frequency tones can sometimes be felt as vibration before they are heard. The low-frequency limit for the true hearing of pure tones probably lies at about 20 Hz. This is close to the lowest frequency which evokes a pitch sensation [13.8].

A third method of specifying absolute thresholds is commonly used when measuring hearing in clinical situations, for example, when a hearing impairment is

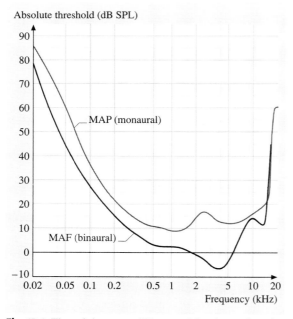

Fig. 13.1 The minimum audible sound level as a function of frequency. The *solid curve* shows the minimum audible field (MAF) for binaural listening published in an International Standards Organization (ISO) standard [13.1]. The *dashed curve* shows the minimum audible pressure (MAP) for monaural listening [13.2]

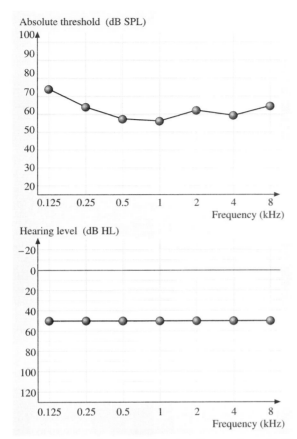

Fig. 13.2 Comparison of a clinical audiogram for a 50 dB hearing loss at all frequencies *(bottom)* and the absolute threshold curve for the same hearing loss plotted in terms of the MAP *(top)*

suspected; thresholds are specified relative to the average threshold at each frequency for young, healthy listeners with normal hearing. In this case, the sound level is usually specified relative to standardized values produced by a specific earphone in a specific coupler. A coupler is a device which contains a cavity or series of cavities and a microphone for measuring the sound produced by the earphone. The preferred earphone varies from one country to another. For example, the Telephonics TDH49 or TDH50 is often used in the UK and USA, while the Beyer DT48 is used in Germany. Thresholds specified in this way have units of dB HL (hearing level) in Europe or dB HTL (hearing threshold level) in the USA. For example, a threshold of 40 dB HL at 1 kHz would mean that the person had a threshold which was 40 dB higher than normal at that frequency. In psychoacoustic work, thresholds are normally plotted with threshold increasing upward, as in Fig. 13.1. However, in audiology, threshold elevations are shown as hearing losses, plotted downward. The average normal threshold is represented as a horizontal line at the top of the plot, and the degree of hearing loss is indicated by how much the threshold falls below this line. This type of plot is called an audiogram. Figure 13.2 compares an audiogram for a hypothetical hearing-impaired person with a flat hearing loss, with a plot of the same thresholds expressed as MAP values. Notice that, although the audiogram is flat, the corresponding MAP curve is not flat. Note also that thresholds in dB HL can be negative. For example, a threshold of −10 dB simply means that the individual is 10 dB more sensitive than the average.

The absolute thresholds described above were measured using tone bursts of relatively long duration. For durations exceeding about 500 ms, the sound level at threshold is roughly independent of duration. However, for durations less than about 200 ms, the sound level necessary for detection increases as duration decreases, by about 3 dB per halving of the duration [13.9]. Thus, the sound energy required for threshold is roughly constant.

13.2 Frequency Selectivity and Masking

Frequency selectivity refers to the ability to resolve or separate the sinusoidal components in a complex sound. It is a key aspect of the analysis of sounds by the auditory system, and it influences many aspects of auditory perception, including the perception of loudness, pitch and timbre. It is often demonstrated and measured by studying masking, which has been defined as 'The process by which the threshold of audibility for one sound is raised by the presence of another (masking) sound' [13.10]. It has been known for many years that a signal is most easily masked by a sound having frequency components close to, or the same as, those of the signal [13.11]. This led to the idea that our ability to separate the components of a complex sound depends, at least in part, on the frequency analysis that takes place on the basilar membrane (see Chap. 12).

13.2.1 The Concept of the Auditory Filter

Fletcher [13.13], following *Helmholtz* [13.14], suggested that the peripheral auditory system behaves as if it contains a bank of bandpass filters, with overlapping passbands. These filters are now called the auditory filters. Fletcher thought that the basilar membrane provided the basis for the auditory filters. Each location on the basilar membrane responds to a limited range of frequencies, so each different point corresponds to a filter with a different center frequency. When trying to detect a signal in a broadband noise background, the listener is assumed to make use of a filter with a center frequency close to that of the signal. This filter passes the signal but removes a great deal of the noise. Only the components in the noise which pass through the filter have any effect in masking the signal. It is usually assumed that the threshold for detecting the signal is determined by the amount of noise passing through the auditory filter; specifically, threshold is assumed to correspond to a certain signal-to-noise ratio at the output of the filter. This set of assumptions has come to be known as the *power spectrum model* of masking [13.15], since the stimuli are represented by their long-term power spectra, i.e., the short-term fluctuations in the masker are ignored.

The question considered next is: What is the shape of the auditory filter? In other words, how does its relative response change as a function of the input frequency? Most methods for estimating the shape of the auditory filter at a given center frequency are based on the assumptions of the power spectrum model of masking. The threshold of a signal whose frequency is fixed is measured in the presence of a masker whose spectral content is varied. It is assumed, as a first approximation, that the signal is detected using the single auditory filter which is centered on the frequency of the signal, and that threshold corresponds to a constant signal-to-masker ratio at the output of that filter. The methods described below both measure the shape of the filter using this technique.

13.2.2 Psychophysical Tuning Curves

One method of measuring the shape of the filter involves a procedure which is analogous in many ways to the determination of a neural tuning curve (see Chap. 12), and the resulting function is often called a psychophysical tuning curve (PTC). To determine a PTC, the signal is fixed in level, usually at a very low level, say, 10 dB above absolute threshold (called 10 dB sensation level,

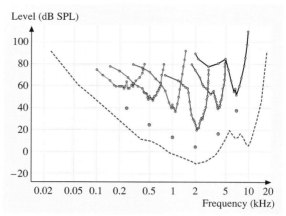

Fig. 13.3 Psychophysical tuning curves (PTCs) determined in simultaneous masking, using sinusoidal signals at 10 dB SL. For each curve, the *circle* below it indicates the frequency and level of the signal. The masker was a sinusoid which had a fixed starting phase relationship to the 50 ms signal. The masker level required for threshold is plotted as a function of masker frequency on a logarithmic scale. The *dashed line* shows the absolute threshold for the signal. (After *Vogten* [13.12])

SL). The masker can be either a sinusoid or a band of noise covering a small frequency range.

For each of several masker frequencies, the level of the masker needed just to mask the signal is determined. Because the signal is at a low level it is assumed that it produces activity primarily at the output of a single auditory filter. It is assumed further that at threshold the masker produces a constant output from that filter, in order to mask the fixed signal. Thus the PTC indicates the masker level required to produce a fixed output from the auditory filter as a function of frequency. Normally a filter characteristic is determined by plotting the output from the filter for an input varying in frequency and fixed in level. However, if the filter is linear the two methods give the same result. Thus, assuming linearity, the shape of the auditory filter can be obtained simply by inverting the PTC. Examples of some PTCs are given in Fig. 13.3.

One problem in interpreting PTCs is that, in practice, the listener may use the information from more than one auditory filter. When the masker frequency is above the signal frequency the listener might do better to use the information from a filter centered just below the signal frequency. If the filter has a relatively flat top, and sloping edges, this will considerably attenuate the masker at the filter output, while only slightly attenuating the signal. By using this off-center filter the listener can improve performance. This is known as *off-frequency listening*, and there is now good evidence that humans do indeed listen off-frequency when it is advantageous to do so [13.16, 17]. The result of off-frequency listening is that the PTC has a sharper tip than would be obtained if only one auditory filter were involved [13.18].

13.2.3 The Notched-Noise Method

Patterson [13.19] described a method of determining auditory filter shape which limits off-frequency listening. The method is illustrated in Fig. 13.4. The signal (indicated by the bold vertical line) is fixed in frequency, and the masker is a noise with a bandstop or notch centered at the signal frequency. The deviation of each edge of the noise from the center frequency is denoted by Δf. The width of the notch is varied, and the threshold of the signal is determined as a function of notch width. Since the notch is symmetrically placed around the signal frequency, the method cannot reveal asymmetries in the auditory filter, and the analysis assumes that the filter is symmetric on a linear frequency scale. This assumption appears not unreasonable, at least for the top part of the filter and at moderate sound levels, since PTCs are quite symmetric around the tips. For a signal symmetrically placed in a bandstop noise, the optimum signal-to-masker ratio at the output of the auditory filter is achieved with a filter centered at the signal frequency, as illustrated in Fig. 13.4.

As the width of the spectral notch is increased, less noise passes through the auditory filter. Thus the threshold of the signal drops. The amount of noise passing through the auditory filter is proportional to the area under the filter in the frequency range covered by the noise. This is shown as the dark shaded areas in Fig. 13.4. Assuming that threshold corresponds to a constant signal-to-masker ratio at the output of the filter, the change in signal threshold with notch width indicates how the area under the filter varies with Δf. The area under a function between certain limits is obtained by integrating the value of the function over those limits. Hence by differentiating the function relating threshold to Δf, the relative response of the filter at that value of Δf is obtained. In other words, the relative response of the filter for a given deviation, Δf, from the center frequency is equal to the slope of the function relating signal threshold to notch width, at that value of Δf.

A typical auditory filter derived using this method is shown in Fig. 13.5. It has a rounded top and quite steep skirts. The sharpness of the filter is often specified as the bandwidth of the filter at which the response has fallen by a factor of two in power, i.e., by 3 dB. The 3 dB bandwidths of the auditory filters derived using the notched-noise method are typically between 10% and

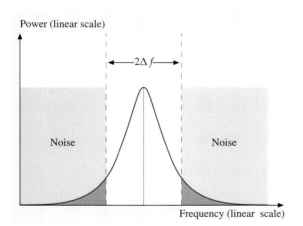

Fig. 13.4 Schematic illustration of the technique used by *Patterson*. [13.19] to determine the shape of the auditory filter. The threshold of the sinusoidal signal (indicated by the *bold vertical line*) is measured as a function of the width of a spectral notch in the noise masker. The amount of noise passing through the auditory filter centered at the signal frequency is proportional to the shaded areas

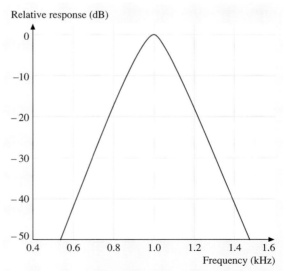

Fig. 13.5 A typical auditory filter shape determined using Patterson's method. The filter is centered at 1 kHz. The relative response of the filter (in dB) is plotted as a function of frequency

15% of the center frequency. An alternative measure is the equivalent rectangular bandwidth (ERB), which is the bandwidth of a rectangular filter which has the same peak transmission as the filter of interest and which passes the same total power for a white noise input. The ERB of the auditory filter is a little larger than the 3 dB bandwidth. In what follows, the mean ERB of the auditory filter determined using young listeners with normal hearing and using a moderate noise level is denoted ERB_N (where the subscript N denotes normal hearing). An equation describing the value of ERB_N as a function of center frequency, F (in Hz), is [13.20]:

$$ERB_N = 24.7(0.00437F + 1). \quad (13.1)$$

Sometimes it is useful to plot psychoacoustical data on a frequency scale related to ERB_N. Essentially, the value of ERB_N is used as the unit of frequency. For example, the value of ERB_N for a center frequency of 1 kHz is about 132 Hz, so an increase in frequency from 934 to 1066 Hz represents a step of one ERB_N. A formula relating ERB_N number to frequency is [13.20]:

$$ERB_N \text{ number} = 21.4 \log_{10}(0.00437F + 1), \quad (13.2)$$

where F is frequency in Hz. This scale is conceptually similar to the Bark scale proposed by *Zwicker* and coworkers [13.22], although it differs somewhat in numerical values.

The notched-noise method has been extended to include conditions where the spectral notch in the noise is placed asymmetrically about the signal frequency. This allows the measurement of any asymmetry in the auditory filter, but the analysis of the results is more difficult, and has to take off-frequency listening into account [13.23]. It is beyond the scope of this chapter to give details of the method of analysis; the interested reader is referred to [13.15, 20, 24, 25]. The results show that the auditory filter is reasonably symmetric at moderate sound levels, but becomes increasingly asymmetric at high levels, the low-frequency side becoming shallower than the high-frequency side. The filter shapes derived using the notched-noise method are quite similar to inverted PTCs [13.26], except that PTCs are slightly sharper around their tips, probably as a result of off-frequency listening.

13.2.4 Masking Patterns and Excitation Patterns

In the masking experiments described so far, the frequency of the signal was held constant, while the masker was varied. These experiments are most appropriate for estimating the shape of the auditory filter at a given center frequency. However, many of the early experiments on masking did the opposite; the masker was held constant in both level and frequency and the signal threshold was measured as a function of the signal frequency. The resulting functions are called masking patterns or masked audiograms.

Masking patterns show steep slopes on the low-frequency side (when the signal frequency is below that of the masker), of between 55 and 240 dB/octave. The slopes on the high-frequency side are less steep and depend on the level of the masker. A typical set of results is shown in Fig. 13.6. Notice that on the high-frequency side the curve is shallower at the highest level. Around the tip of the masking pattern, the growth of masking is approximately linear; a 10 dB increase in masker level leads to roughly a 10 dB increase in the signal threshold. However, for signal frequencies in the range from about 1300 to 2000 Hz, when the level of the masker is increased by 10 dB (e.g., from 70 to 80 dB SPL), the masked threshold increases by more than 10 dB; the amount of masking grows nonlinearly on the high-frequency side. This has been called the *upward spread of masking*.

The masking patterns do not reflect the use of a single auditory filter. Rather, for each signal frequency the listener uses a filter centered close to the signal frequency. Thus the auditory filter is shifted as the signal frequency is altered. One way of interpreting the masking pattern

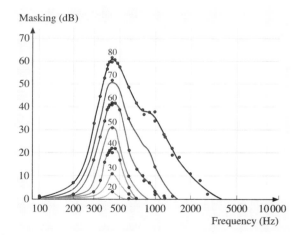

Fig. 13.6 Masking patterns for a narrow-band noise masker centered at 410 Hz. Each curve shows the elevation in threshold of a pure-tone signal as a function of signal frequency. The overall noise level for each curve is indicated in the figure (After *Egan* and *Hake* [13.21])

is as a crude indicator of the excitation pattern of the masker [13.27]. The excitation pattern is a representation of the effective amount of excitation produced by a stimulus as a function of characteristic frequency (CF) on the basilar membrane (BM; see Chap. 12), and is plotted as effective level (in dB) against CF. In the case of a masking sound, the excitation pattern can be thought of as representing the relative amount of vibration produced by the masker at different places along the basilar membrane. The signal is detected when the excitation it produces is some constant proportion of the excitation produced by the masker at places with CFs close to the signal frequency. Thus the threshold of the signal as a function of frequency is proportional to the masker excitation level. The masking pattern should be parallel to the excitation pattern of the masker, but shifted vertically by a small amount. In practice, the situation is not so straightforward, since the shape of the masking pattern is influenced by factors such as off-frequency listening, and the detection of beats and combination tones [13.28].

Moore and *Glasberg* [13.29] have described a way of deriving the shapes of excitation patterns using the concept of the auditory filter. They suggested that the excitation pattern of a given sound can be thought of as the output of the auditory filters plotted as a function of their center frequency. To calculate the excitation pattern of a given sound, it is necessary to calculate the output of each auditory filter in response to that sound, and to plot the output as a function of the filter center frequency. The characteristics of the auditory filters are determined using the notched-noise method described earlier. Figure 13.7 shows excitation patterns calculated in this way for 1000 Hz sinusoids with various levels. The patterns are similar in form to the masking patterns shown in Fig. 13.6. Software for calculating excitation patterns can be downloaded from http://hearing.psychol.cam.ac.uk/Demos/demos.html.

13.2.5 Forward Masking

Masking can occur when the signal is presented just before or after the masker. This is called non-simultaneous masking and it is studied using brief signals, often called *probes*. Two basic types of non-simultaneous masking can be distinguished: (1) backward masking, in which the probe precedes the masker; and (2) forward masking, in which the probe follows the masker. Although many studies of backward masking have been published, the phenomenon is poorly understood. The amount of backward masking obtained depends strongly on how much practice the subjects have received, and practiced subjects often show little or no backward masking [13.30, 31]. The larger masking effects found for unpracticed subjects may reflect some sort of confusion of the signal with the masker. In the following, I will focus on forward masking, which can be substantial even in highly practiced subjects. The main properties of forward masking are as follows:

1. Forward masking is greater the nearer in time to the masker that the signal occurs. This is illustrated in the left panel of Fig. 13.8. When the delay D of the signal after the end of the masker is plotted on a logarithmic scale, the data fall roughly on a straight line. In other words, the amount of forward masking, in dB, is a linear function of log D.
2. The rate of recovery from forward masking is greater for higher masker levels. Regardless of the initial amount of forward masking, the masking decays to zero after 100–200 ms.
3. A given increment in masker level does not produce an equal increment in amount of forward masking. For example, if the masker level is increased by 10 dB, the masked threshold may only increase by 3 dB. This contrasts with simultaneous mask-

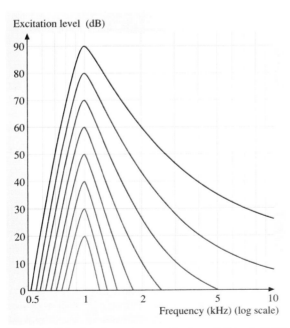

Fig. 13.7 Calculated psychoacoustical excitation patterns for a 1 kHz sinusoid at levels ranging from 20 to 90 dB SPL in 10 dB steps

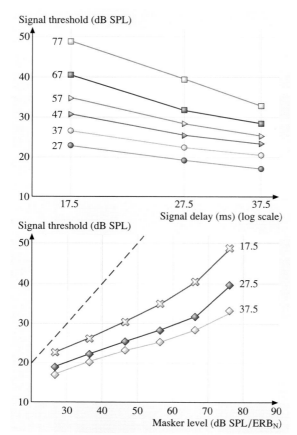

Fig. 13.8 The top panel shows the amount of forward masking of a brief 4 kHz signal, plotted as a function of the time delay of the signal after the end of the noise masker. Each curve shows results for a different noise level, specified as the level in a one-ERB$_N$-wide band centered at 4 kHz. The results for each level fall roughly on a straight line when the signal delay is plotted on a logarithmic scale, as here. The bottom panel shows the same thresholds plotted as a function of masker level. Each curve shows results for a different signal delay time (17.5, 27.5, or 37.5 ms). Note that the slopes of these growth of masking functions decrease with increasing signal delay. The *dashed line* indicates a slope of 1. (After *Moore* and *Glasberg* [13.32])

ing, where, at least for wide-band maskers, threshold corresponds approximately to a constant signal-to-masker ratio. This effect can be quantified by plotting the signal threshold as a function of the masker level. The resulting function is called a growth of masking function. Several such functions are shown in the right panel of Fig. 13.8. In simultaneous mask-

ing such functions would have slopes close to one, as indicated by the dashed line. In forward masking the slopes are usually less than one, and the slopes decrease as the value of D increases.

4. The amount of forward masking increases with increasing masker duration for durations up to at least 50 ms. The results for greater masker durations vary somewhat across studies. Some studies show an effect of masker duration for durations up to 200 ms [13.33, 34], while others show little effect for durations beyond 50 ms [13.35].
5. Forward masking is influenced by the relation between the frequencies of the signal and the masker (just as in the case of simultaneous masking).

The basis of forward masking is still not clear. Four factors may contribute:

1. The response of the BM to the masker continues for some time after the end of the masker, an effect known as *ringing*. If the ringing overlaps with the response to the signal, then this may contribute to the masking of the signal. The duration of the ringing is less at places tuned to high frequencies, whose bandwidth is larger than at low frequencies. Hence, ringing plays a significant role only at low frequencies [13.36–38]. For frequencies above 200–300 Hz, the amount of forward masking for a given masker level and signal delay time is roughly independent of frequency [13.39].
2. The masker produces short-term adaptation or fatigue in the auditory nerve or at higher centers in the auditory system, which reduces the response to a signal presented just after the end of the masker [13.40]. However, the effect in the auditory nerve appears to be too small to account for the forward masking observed behaviorally [13.41].
3. The neural activity evoked by the masker persists at some level in the auditory system higher than the auditory nerve, and this persisting activity masks the signal [13.42].
4. The masker may evoke a form of inhibition in the central auditory system, and this inhibition persists for some time after the end of the masker [13.43].

Oxenham and *Moore* [13.44] have suggested that the shallow slopes of the growth of masking functions, as shown in the bottom panel of Fig. 13.8, can be explained, at least qualitatively, in terms of the compressive input–output function of the BM (see Chap. 12). Such an input–output function is shown schematically

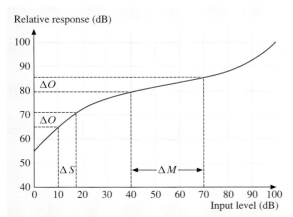

Fig. 13.9 Illustration of why growth of masking functions in forward masking usually have shallow slopes. The *solid curve* shows a schematic input–output function on the basilar membrane. The relative response is plotted on a dB scale with an arbitrary origin. When the masker is increased in level by ΔM, this produces an increase in response of ΔO. To restore signal threshold, the response to the signal also has to be increased by ΔO. This requires an increase in signal level, ΔS, which is markedly smaller than ΔM

in Fig. 13.9. It has a shallow slope for medium input levels, but a steeper slope at very low input levels. Assume that, for a given time delay of the signal relative to the masker, the response evoked by the signal at threshold is directly proportional to the response evoked by the masker. Assume, as an example, that a masker with a level of 40 dB produces a signal threshold of 10 dB. Consider now what happens when the masker level is increased by 30 dB. The increase in masker level, denoted by ΔM in Fig. 13.9, produces a relatively small increase in response, ΔO. To restore the signal to threshold, the signal has to be increased in level so that the response to it also increases by ΔO. However, this requires a relatively small increase in signal level, ΔS, as the signal level falls in the range where the input–output function is relatively steep. Thus, the growth of masking function has a shallow slope.

According to this explanation, the shallow slope of the growth of masking function arises from the fact that the signal level is lower than the masker level, so the masker is subject to more compression than the signal. The input–output function on the BM has a slope which decreases progressively with increasing level over the range 0 to about 50 dB. Hence the slope of the growth of masking function should decrease with increasing difference in level between the masker and signal. This can account for the progressive decrease in the slopes of the growth of masking functions with increasing delay between the signal and masker (see the right-hand panel of Fig. 13.8); longer delays are associated with greater differences in level between the signal and masker. Another prediction is that the growth of masking function for a given signal delay should increase in slope if the signal level is high enough to fall in the compressive region of the input–output function. Such an effect can be seen in the growth of masking function for the shortest delay time in Fig. 13.8; the function steepens for the highest signal level.

In summary, the processes underlying forward masking are not fully understood. Contributions from a number of different sources may be important. Temporal overlap of patterns of vibration on the BM may be important, especially for small delay times between the signal and masker. Short-term adaptation or fatigue in the auditory nerve may also play a role. At higher neural levels, a persistence of the excitation or inhibition evoked by the masker may occur. The form of the growth of masking functions can be explained, at least qualitatively, in terms of the nonlinear input–output functions observed on the BM.

13.2.6 Hearing Out Partials in Complex Tones

A complex tone, such as a tone produced by a musical instrument, usually evokes a single pitch; pitches corresponding to the frequencies of individual partials are not usually perceived. However, such pitches can be heard if attention is directed appropriately. In other words, individual partials can be *heard out*. Plomp [13.45] and *Plomp* and *Mimpen* [13.46] used complex tones with 12 equal-amplitude sinusoidal components to investigate the limits of this ability. The listener was presented with two comparison tones, one of which was of the same frequency as a partial in the complex; the other lay halfway between that frequency and the frequency of the adjacent higher or lower partial. The listener had to judge which of these two tones was a component of the complex. Plomp used two types of complex: a harmonic complex containing harmonics 1 to 12, where the frequencies of the components were integer multiples of that of the fundamental; and a nonharmonic complex, where the frequencies of the components were mistuned from simple frequency ratios. He found that for both kinds of complex, partials could only be heard out if they were sufficiently far in frequency from neighboring partials. The data, and other more recent data [13.47], are consistent with the hypothesis that a partial can be heard out (with 75% accuracy) when it is separated from neighboring (equal-amplitude) partials by 1.25 times the ERB_N of the auditory filter. For harmonic complex tones, only the first (lowest) five to eight harmonics can be heard out, as higher harmonics are separated by less than 1.25 ERB_N.

It seems likely that the analysis of partials from a complex sound depends in part on factors other than the frequency analysis that takes place on the basilar membrane. *Soderquist* [13.48] compared musicians and non-musicians in a task very similar to that of Plomp, and found that the musicians were markedly superior. This result could mean that musicians have smaller auditory filter bandwidths than non-musicians. However, *Fine* and *Moore* [13.49] showed that auditory filter bandwidths, as estimated in a notched-noise masking experiment, did not differ for musicians and non-musicians. It seems that some mechanism other than peripheral filtering is involved in hearing out partials from complex tones and that musicians, because of their greater experience, are able to make more efficient use of this mechanism.

13.3 Loudness

The human ear is remarkable both in terms of its absolute sensitivity and the range of sound intensities to which it can respond. The most intense sound that can be heard without damaging the ear has a level about 120 dB above the absolute threshold; this range is referred to as the dynamic range of the auditory system and it corresponds to a ratio of intensities of 1 000 000 000 000:1.

Loudness corresponds to the subjective impression of the magnitude of a sound. The formal definition of loudness is: that intensive attribute of auditory sensation in terms of which sounds can be ordered on a scale extending from soft to loud [13.10]. Because loudness is subjective, it is very difficult to measure in a quantitative way. Estimates of loudness can be strongly affected by bias and context effects of various kinds [13.50, 51]. For example the impression of loudness of a sound with a moderate level (say, 60 dB SPL) can be affected by presenting a high level sound (say, 100 dB SPL) before the moderate level sound.

13.3.1 Loudness Level and Equal-Loudness Contours

It is often useful to be able to compare the loudness of sounds with that of a standard, reference sound. The most common reference sound is a 1000 Hz sinusoid, presented binaurally in a free field, with the sound coming from directly in front of the listener. The loudness level of a sound is defined as the intensity level of a 1000 Hz sinusoid that is equal in loudness to the sound. The unit of loudness level is the phon. Thus, the loudness level of any sound in phons is the level (in dB SPL) of the 1000 Hz sinusoid to which it sounds equal in loudness. For example, if a sound appears to be as loud as a 1000 Hz sinusoid with a level of 45 dB SPL, then the sound has a loudness level of 45 phons. To determine the loudness level of a given sound, the subject is asked to adjust the level of a 1000 Hz sinusoid until it appears to have the same loudness as that sound. The 1000 Hz sinusoid and the test sound are presented alternately rather than simultaneously.

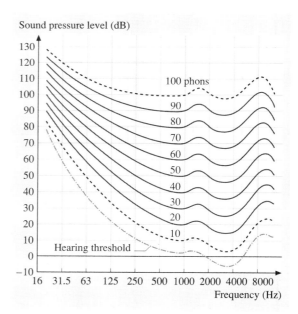

Fig. 13.10 Equal-loudness contours for loudness levels from 10 to 100 phons for sounds presented binaurally from the frontal direction. The absolute threshold curve (the MAF) is also shown. The curves for loudness levels of 10 and 100 phons are *dashed*, as they are based on interpolation and extrapolation, respectively

In a variation of this procedure, the 1000 Hz sinusoid is fixed in level, and the test sound is adjusted to give a loudness match, again with alternating presentation. If this is repeated for various different frequencies of a sinusoidal test sound, an equal-loudness contour is generated [13.52, 53]. For example, if the 1000 Hz sinusoid is fixed in level at 40 dB SPL, then the 40 phon equal-loudness contour is generated. Figure 13.10 shows equal-loudness contours as published in a recent standard [13.53]. The figure shows equal-loudness contours for binaural listening for loudness levels from 10–100 phons, and it also includes the absolute threshold (MAF) curve. The listening conditions were similar to those for determining the MAF curve; the sound came from a frontal direction in a free field. The equal-loudness contours are of similar shape to the MAF curve, but tend to become flatter at high loudness levels. Note that the MAF curve in Fig. 13.10 differs somewhat from that in Fig. 13.1, as the two curves are based on different standards.

Note that the subjective loudness of a sound is not directly proportional to its loudness level in phons. For example, a sound with a loudness level of 80 phons sounds much more than twice as loud as a sound with a loudness level of 40 phons. This is discussed in more detail in the next section.

13.3.2 The Scaling of Loudness

Several methods have been developed that attempt to measure directly the relationship between the physical magnitude of sound and perceived loudness [13.54]. In one, called magnitude estimation, sounds with various different levels are presented, and the subject is asked to assign a number to each one according to its perceived loudness. In a second method, called magnitude production, the subject is asked to adjust the level of a sound until it has a loudness corresponding to a specified number.

On the basis of results from these two methods, Stevens suggested that loudness, L, was a power function of physical intensity, I:

$$L = kI^{0.3}x,\tag{13.3}$$

where k is a constant depending on the subject and the units used. In other words, the loudness of a given sound is proportional to its intensity raised to the power 0.3. Note that this implies that loudness is *not* linearly related to intensity; rather, it is a *compressive* function of intensity. An approximation to this equation is that the loudness doubles when the intensity is increased

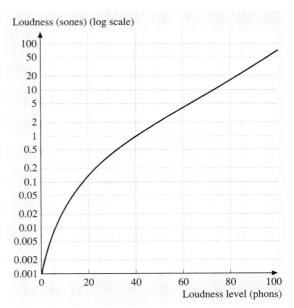

Fig. 13.11 The relationship between loudness in sones and loudness level in phons for a 1000 Hz sinusoid. The curve is based on the loudness model of *Moore* et al.[13.3]

by a factor of 10, or, equivalently, when the level is increased by 10 dB. In practice, this relationship only holds for sound levels above about 40 dB SPL. For lower levels than this, the loudness changes with intensity more rapidly than predicted by the power-law equation.

The unit of loudness is the *sone*. One sone is defined arbitrarily as the loudness of a 1000 Hz sinusoid at 40 dB SPL, presented binaurally from a frontal direction in a free field. Fig. 13.11 shows the relationship between loudness in sones and the physical level of a 1000 Hz sinusoid, presented binaurally from a frontal direction in a free-field; the level of the 1000 Hz tone is equal to its loudness level in phons. This figure is based on predictions of a loudness model [13.3], but it is consistent with empirical data obtained using scaling methods [13.55]. Since the loudness in sones is plotted on a logarithmic scale, and the decibel scale is itself logarithmic, the curve shown in Fig. 13.11 approximates a straight line for levels above 40 dB SPL. The slope corresponds to a doubling of loudness for each 10 dB increase in sound level.

13.3.3 Neural Coding and Modeling of Loudness

The mechanisms underlying the perception of loudness are not fully understood. A common assumption is that

Fig. 13.12 Block diagram of a typical loudness model

loudness is somehow related to the total neural activity evoked by a sound, although this concept has been questioned [13.56]. In any case, it is commonly assumed that loudness depends upon a summation of loudness contributions from different frequency channels (i.e., different auditory filters). Models incorporating this basic concept have been proposed by *Fletcher* and *Munson* [13.57], by *Zwicker* [13.58, 59] and by *Moore* and coworkers [13.3, 60]. The models attempt to calculate the average loudness that would be perceived by a large group of listeners with normal hearing under conditions where biases are minimized as far as possible. The models all have the basic form illustrated in Fig. 13.12. The first stage is a fixed filter to account for the transmission of sound through the outer and middle ear. The next stage is the calculation of an excitation pattern for the sound under consideration. In most of the models, the excitation pattern is calculated from psychoacoustical masking data, as described earlier. From the excitation-pattern stage onwards, the models should be considered as multichannel; the excitation pattern is sampled at regular intervals along the ERB_N number scale, each sample value corresponding to the amount of excitation at a specific center frequency.

The next stage is the transformation from excitation level (dB) to specific loudness, which is a kind of *loudness density*. It represents the loudness that would be evoked by the excitation within a small fixed distance along the basilar membrane if it were possible to present that excitation alone (without any excitation at adjacent regions on the basilar membrane). In the model of *Moore* et al. [13.3], the distance is 0.89 mm, which corresponds to one ERB_N, so the specific loudness represents the loudness per ERB_N. The specific loudness plotted as a function of ERB_N number is called the *specific loudness pattern*. The specific loudness cannot be measured either physically or subjectively. It is a theoretical construct used as an intermediate stage in the loudness models. The transformation from excitation level to specific loudness involves a compressive nonlinearity. Although the models are based on psychoacoustical data, this transformation can be thought of as representing the way that physical excitation is transformed into neural activity; the specific loudness is assumed to be related to the amount of neural activity at the corresponding CF. The overall loudness of a given sound, in sones, is assumed to be proportional to the total area under the specific loudness pattern. In other words, the overall loudness is calculated by summing the specific loudness values across all ERB_N numbers (corresponding to all center frequencies).

Loudness models of this type have been rather successful in accounting for experimental data on the loudness of both simple sounds and complex sounds [13.3]. They have also been incorporated into loudness meters, which can give an appropriate indication of the loudness of sounds even when they fluctuate over time [13.27, 60, 61]. Software for calculating loudness using the model of *Moore* et al. [13.3] can be downloaded from http://hearing.psychol.cam.ac.uk/Demos/demos.html. The new American National Standards Institute (ANSI) standard for calculating loudness [13.62] is based on this model.

13.3.4 The Effect of Bandwidth on Loudness

Consider a complex sound of fixed energy (or intensity) having a bandwidth W. If W is less than a certain bandwidth, called the critical bandwidth for loudness, CB_L, then the loudness of the sound is almost independent of W; the sound is judged to be about as loud as a pure tone or narrow band of noise of equal intensity lying at the center frequency of the band. However, if W is increased beyond CB_L, the loudness of the complex begins to increase. This has been found to be the case for bands of noise [13.27, 63] and for complex sounds consisting of several pure tones whose frequency separation is varied [13.64, 65]. An example for bands of noise is given in Fig. 13.13. The CB_L for the data in Fig. 13.13 is about 250–300 Hz for a center frequency of 1420 Hz, although the exact value of CB_L is hard to determine. The value of CB_L is similar to, but a little greater than, the ERB_N of the auditory filter. Thus, for a given amount of energy, a complex sound is louder if its bandwidth exceeds one ERB_N than if its bandwidth is less than one ERB_N.

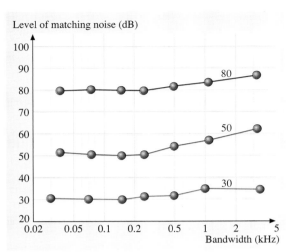

Fig. 13.13 The level of a 210 Hz-wide noise required to match the loudness of a noise of variable bandwidth is plotted as a function of the variable bandwidth; the bands of noise were geometrically centered at 1420 Hz. The number by each curve is the overall noise level in dB SPL. (After *Zwicker* et al.[13.63])

The pattern of results shown in Fig. 13.13 can be understood in terms of the loudness models described above. First, a qualitative account is given to illustrate the basic concept. When a sound has a bandwidth less than one ERB_N at a given center frequency, the excitation pattern and the specific loudness pattern are roughly independent of the bandwidth of the sound [13.66]. Further, the overall loudness is dominated by the specific loudness at the peak of the pattern. Consider now the effect of increasing the bandwidth of a sound from one ERB_N to N ERB_N, keeping the overall power, $P_{overall}$, constant. The power in each one-ERB_N-wide band is now $P_{overall}/N$. However, the peak specific loudness evoked by each band decreases by much less than a factor of N, because of the compressive relationship between excitation and specific loudness. For example, if $N = 10$, the peak specific loudness evoked by each band decreases by a factor of about two. Thus, in this example, there are ten bands each about half as loud as the original one-ERB_N-wide band, so the overall loudness is about a factor of five greater for the ten-ERB_N-wide than for the one-ERB_N-wide band. Generally, once the bandwidth is increased beyond one ERB_N, the loudness goes up because the decrease in loudness per ERB_N is more than offset by the increase in the number of channels contributing significantly to the overall loudness.

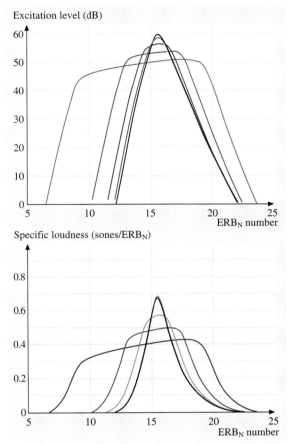

Fig. 13.14 The upper panel shows excitation patterns for a 1 kHz sinusoid with a level of 60 dB SPL (the narrowest pattern with the highest peak) and for noise bands of various widths, all centered at 1 kHz and with an overall level of 60 dB SPL. The frequency scale has been transformed to an ERB_N number scale. The noise bands have widths of 20, 60, 140, 300, 620 and 1260 Hz. As the bandwidth is increased, the patterns decrease in height but spread over a greater number of ERB_Ns. The lower panel shows specific loudness patterns corresponding to the excitation patterns in the upper panel. For bandwidths up to 140 Hz, the area under the specific loudness patterns is constant. For greater bandwidths, the total area increases

This argument is illustrated in a more quantitative way in Fig. 13.14, which shows excitation patterns (top) and specific loudness patterns (bottom) for a sinusoid and for bands of noise of various widths (20, 40, 60, 140, 300 and 620 Hz), all with a level of 60 dB SPL, calculated using the model of *Moore* et al. [13.3]. With increasing bandwidth, the specific loudness patterns become lower at their tips, but broader. For the first three bandwidths, the small decrease in area around the tip is almost exactly canceled by the increase on the skirts, so the total area remains almost constant. When the bandwidth is increased beyond 140 Hz, the increase on the skirts is greater than the decrease around the tip, and so the total area, and hence the predicted loudness, increases. The value of CB_L in this case is a little greater than 140 Hz. Since the increase in loudness depends on the summation of specific loudness at different center frequencies, the increase in loudness is often described as loudness summation.

At low sensation levels (around 10–20 dB SL), the loudness of a complex sound is roughly independent of bandwidth. This can also be explained in terms of the loudness models described above. At these low levels, specific loudness changes rapidly with excitation level, and so does loudness. As a result, the total area under the specific loudness pattern remains almost constant as the bandwidth is altered, even for bandwidths greater than CB_L. Thus, loudness is independent of bandwidth. When a narrow-band sound has a very low sensation level (below 10 dB), then if the bandwidth is increased keeping the total energy constant, the output of each auditory filter becomes insufficient to make the sound audible. Accordingly, near threshold, loudness must decrease as the bandwidth of a complex sound is increased from a small value. As a consequence, if the intensity of a complex sound is increased slowly from a subthreshold value, the rate of growth of loudness is greater for a wide-band sound than for a narrow-band sound.

13.3.5 Intensity Discrimination

The smallest detectable change in intensity of a sound has been measured for many different types of stimuli by a variety of methods. The three main methods are:

1. Modulation detection. The stimulus is amplitude modulated (i.e., made to vary in amplitude) at a slow regular rate and the listener is required to detect the modulation. Usually, the modulation is sinusoidal.
2. Increment detection. A continuous background stimulus is presented, and the subject is required to detect a brief increment in the level of the background. Often the increment is presented at one of two possible times, indicated by lights, and the listener is required to indicate whether the increment occurred synchronously with the first light or the second light.
3. Intensity discrimination of gated or pulsed stimuli. Two (or more) separate pulses of sound are presented successively, one being more intense than the other(s), and the subject is required to indicate which pulse was the most intense.

In all of these tasks, the subjective impression of the listener is of a change in loudness. For example, in method 1 the modulation is heard as a fluctuation in loudness. In method 2 the increment is heard as a brief increase in loudness of the background, or sometimes as an extra sound superimposed on the background. In method 3, the most intense pulse appears louder than the other(s). Although there are some minor discrepancies in the experimental results for the different methods, the general trend is similar. For wide-band noise, or for bandpass-filtered noise, the smallest detectable intensity change, ΔI, is approximately a constant fraction of the intensity of the stimulus, I. In other words, $\Delta I/I$ is roughly constant. This is an example of Weber's Law, which states that the smallest detectable change in a stimulus is proportional to the magnitude of that stimulus. The value of $\Delta I/I$ is called the Weber fraction. Thresholds for detecting intensity changes are often specified as the change in level at threshold, ΔL, in decibels. The value of ΔL is given by

$$\Delta L = 10 \log_{10}[(I + \Delta I)/I] . \tag{13.4}$$

As $\Delta I/I$ is constant, ΔL is also constant, regardless of the absolute level, and for wide-band noise has a value of about 0.5–1 dB. This holds from about 20 dB above threshold to 100 dB above threshold [13.67]. The value of ΔL increases for sounds which are close to the absolute threshold.

For sinusoidal stimuli, the situation is somewhat different. If ΔI (in dB) is plotted against I (also in dB), a straight line is obtained with a slope of about 0.9; Weber's law would give a slope of 1.0. Thus discrimination, as measured by the Weber fraction, improves at high levels. This has been termed the "near miss" to Weber's Law. The data of Riesz [13.68] for modulation detection show a value of ΔL of 1.5 dB at 20 dB SL, 0.7 dB at 40 dB SL, and 0.3 dB at 80 dB SL (all at 1000 Hz). The Weber fraction may increase somewhat at very high sound levels (above 100 dB SPL) [13.69]. In everyday life, a change in level of 1 dB would hardly be noticed, but a change in level of 3 dB (corresponding to a doubling or halving of intensity) would be fairly easily heard.

13.4 Temporal Processing in the Auditory System

This section is concerned mainly with temporal resolution (or acuity), which refers to the ability to detect changes in stimuli over time, for example, to detect a brief gap between two stimuli or to detect that a sound is modulated in some way. As pointed out by *Viemeister* and *Plack* [13.71], it is also important to distinguish the rapid pressure variations in a sound (the *fine structure*) from the slower overall changes in the amplitude of those fluctuations (the *envelope*). Temporal resolution normally refers to the resolution of changes in the envelope, not in the fine structure. In characterizing temporal resolution in the auditory system, it is important to take account of the filtering that takes place in the peripheral auditory system. Temporal resolution depends on two main processes: analysis of the time pattern occurring within each frequency channel and comparison of the time patterns across channels.

A major difficulty in measuring the temporal resolution of the auditory system is that changes in the time pattern of a sound are generally associated with changes in its magnitude spectrum, for example its power spectrum (see Chap. 14). Thus, the detection of a change in time pattern can sometimes depend not on temporal resolution per se, but on the detection of the change in magnitude spectrum. There have been two general approaches to getting around this problem. One is to use signals whose magnitude spectrum is not changed when the time pattern is altered. For example, the magnitude spectrum of white noise remains flat if the noise is interrupted, i.e., if a gap is introduced into the noise. The second approach uses stimuli whose magnitude spectra are altered by the change in time pattern, but extra background sounds are added to mask the spectral changes. Both approaches will be described.

13.4.1 Temporal Resolution Based on Within-Channel Processes

The threshold for detecting a gap in a broadband noise provides a simple and convenient measure of temporal resolution. Usually a two-alternative forced-choice (2AFC) procedure is used: the subject is presented with two successive bursts of noise and either the first or the second burst (at random) is interrupted to produce the gap. The task of the subject is to indicate which burst contained the gap. The gap threshold is typically 2–3 ms [13.72, 73]. The threshold increases at very low sound levels, when the level of the noise approaches the absolute threshold, but is relatively invariant with level for moderate to high levels.

The long-term magnitude spectrum of a sound is not changed when that sound is time reversed (played backward in time). Thus, if a time-reversed sound can be discriminated from the original, this must reflect a sensitivity to the difference in time pattern of the two sounds. This was exploited by *Ronken* [13.74], who used as stimuli pairs of clicks differing in amplitude. One click, labeled A, had an amplitude greater than that of the other click, labeled B. Typically the amplitude of A was twice that of B. Subjects were required to distinguish click pairs differing in the order of A and B: either AB or BA. The ability to do this was measured as a function of the time interval or gap between A and B. *Ronken* found that subjects could distinguish the click pairs for gaps down to 2–3 ms. Thus, the limit to temporal resolution found in this task is similar to that found for the detection of a gap in broadband noise. It should be noted that, in this task, subjects do not hear the individual clicks within a click pair. Rather, each click pair is heard as a single sound with its own characteristic quality.

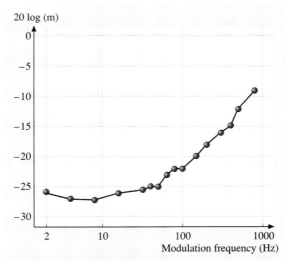

Fig. 13.15 A temporal modulation transfer function (TMTF). A broadband white noise was sinusoidally amplitude modulated, and the threshold amount of modulation required for detection is plotted as a function of modulation rate. The amount of modulation is specified as $20\log(m)$, where m is the modulation index (see Chap. 14). The higher the sensitivity to modulation, the more negative is this quantity. (After *Bacon* and *Viemeister* [13.70])

The experiments described above each give a single number to describe temporal resolution. A more comprehensive approach is to measure the threshold for detecting changes in the amplitude of a sound as a function of the rapidity of the changes. In the simplest case, white noise is sinusoidally amplitude modulated, and the threshold for detecting the modulation is determined as a function of modulation rate. The function relating threshold to modulation rate is known as a temporal modulation transfer function (TMTF). Modulation of white noise does not change its long-term magnitude spectrum. An example of the results is shown in Fig. 13.15; data are taken from *Bacon* and *Viemeister* [13.70]. For low modulation rates, performance is limited by the amplitude resolution of the ear, rather than by temporal resolution. Thus, the threshold is independent of modulation rate for rates up to about 50 Hz. As the rate increases beyond 50 Hz, temporal resolution starts to have an effect; the threshold increases, and for rates above about 1000 Hz the modulation cannot be detected at all. Thus, sensitivity to amplitude modulation decreases progressively as the rate of modulation increases. The shapes of TMTFs do not vary much with overall sound level, but the ability to detect the modulation does worsen at low sound levels.

To explore whether temporal resolution varies with center frequency, *Green* [13.75] used stimuli which consisted of a brief pulse of a sinusoid in which the level of the first half of the pulse was 10 dB different from that of the second half. Subjects were required to distinguish two signals, differing in whether the half with the high level was first or second. Green measured performance as a function of the total duration of the stimuli. The threshold, corresponding to 75% correct discrimination, was similar for center frequencies of 2 and 4 kHz and was between 1 and 2 ms. However, the threshold was slightly higher for a center frequency of 1 kHz, being between 2 and 4 ms.

Moore et al. [13.76] measured the threshold for detecting a gap in a sinusoid, for signal frequencies of 100, 200, 400, 800, 1000, and 2000 Hz. A background noise was used to mask the spectral *splatter* produced by turning the sound off and on to produce the gap. The gap thresholds were almost constant, at 6–8 ms over the frequency range 400–2000 Hz, but increased somewhat at 200 Hz and increased markedly, to about 18 ms, at 100 Hz. Individual variability also increased markedly at 100 Hz.

Overall, it seems that temporal resolution is roughly independent of frequency for medium to high frequencies, but worsens somewhat at very low center frequencies.

The measurement of TMTFs using sinusoidal carriers is complicated by the fact that the modulation introduces spectral sidebands, which may be detected as separate components if they are sufficiently far in frequency from the carrier frequency. When the carrier frequency is high, the effect of resolution of sidebands is likely to be small for modulation frequencies up to a few hundred Hertz, as the auditory filter bandwidth is large for high center frequencies. Consistent with this, TMTFs for high carrier frequencies generally show an initial flat portion (sensitivity independent of modulation frequency), then a portion where threshold increases with increasing modulation frequency, presumably reflecting the limits of temporal resolution, and then a portion where threshold decreases again, presumably reflecting the detection of spectral sidebands [13.77, 78].

The initial flat portion of the TMTF extends to about 100–120 Hz for sinusoidal carriers, but only to 50 or 60 Hz for a broadband noise carrier. It has been suggested that the discrepancy occurs because the inherent amplitude fluctuations in a noise carrier limit the detectability of the imposed modulation [13.79–82]; see below for further discussion of this point. The effect of the inherent fluctuations depends upon their similarity to the imposed modulation. When a narrow-band noise carrier is used, which has relatively slow inherent amplitude fluctuations, TMTFs show the poorest sensitivity for low modulation frequencies [13.79, 80]. In principle, then, TMTFs obtained using sinusoidal carriers provide a better measure of the inherent temporal resolution of the auditory system than TMTFs obtained using noise carriers, provided that the modulation frequency is within the range where spectral resolution does not play a major role.

13.4.2 Modeling Temporal Resolution

Most models of temporal resolution are based on the idea that there is a process at levels of the auditory system higher than the auditory nerve which is sluggish in some way, thereby limiting temporal resolution. The models assume that the internal representation of stimuli is smoothed over time, so that rapid temporal changes are reduced in magnitude but slower ones are preserved. Although this smoothing process almost certainly operates on neural activity, the most widely used models are based on smoothing a simple transformation of the stimulus, rather than its neural representation.

Most models include an initial stage of bandpass filtering, reflecting the action of the auditory filters. Each filter is followed by a nonlinear device. This nonlinear device is meant to reflect the operation of several processes that occur in the peripheral auditory system such as compression on the basilar membrane and neural transduction, whose effects resemble half-wave rectification (see Chap. 12). The output of the nonlinear device is fed to a smoothing device, which can be implemented either as a low-pass filter [13.83] or (equivalently) as a sliding temporal integrator [13.38, 84]. The device determines a kind of weighted average of the output of the compressive nonlinearity over a certain time interval or window. This weighting function is sometimes called the shape of the temporal window. The window itself is assumed to slide in time, so that the output of the temporal integrator is like a weighted running average of the input. This has the effect of smoothing rapid fluctuations while preserving slower ones. When a sound is turned on abruptly, the output of the temporal integrator takes some time to build up. Similarly, when a sound is turned off, the output of the integrator takes some time to decay. The shape of the window is assumed to be asymmetric in time, such that the build up of its output in response to the onset of a sound is more rapid than the decay of its output in response to the cessation of a sound. The output of the sliding temporal integrator is fed to a decision device. The decision device may use different rules depending on the task required. For example, if the task is to detect a brief temporal gap in a signal, the decision device might look for a dip in the output of the temporal integrator. If the task is to detect amplitude modulation of a sound, the device might assess the amount of modulation at the output of the sliding temporal integrator [13.83].

It is often assumed that backward and forward masking depend on the process of build up and decay. For example, if a brief signal is rapidly followed by a masker (backward masking), the response to the signal may still be building up when the masker occurs. If the masker is sufficiently intense, then its internal effects may swamp those of the signal. Similarly, if a brief signal follows soon after a masker (forward masking), the decaying response to the masker may swamp the response to the signal. The asymmetry in the shape of the window accounts for the fact that forward masking occurs over longer masker-signal intervals than backward masking.

13.4.3 A Modulation Filter Bank?

Some researchers have suggested that the analysis of sounds that are amplitude modulated depends on a specialized part of the brain that contains an array of neurons, each tuned to a different modulation rate [13.85]. Each neuron can be considered as a filter in the modulation domain, and the array of neurons is known collectively as a modulation filter bank. The modulation filter bank has been suggested as a possible explanation for certain perceptual phenomena, which are described below. It should be emphasized, however, that this is still a controversial concept.

The threshold for detecting amplitude modulation of a given carrier generally increases if additional amplitude modulation is superimposed on that carrier. This effect has been called modulation masking. *Houtgast* [13.86] studied the detection of sinusoidal amplitude modulation of a pink noise carrier. Thresholds for detecting the modulation were measured when no other modulation was present and when a masker modulator was present in addition. In one experiment, the masker modulation was a half-octave-wide band of noise, with a center frequency of 4, 8, or 16 Hz. For each masker, the masking pattern showed a peak at the masker frequency. This could be interpreted as indicating selectivity in the modulation-frequency domain, analogous to the frequency selectivity in the audio-frequency domain that was described earlier.

Bacon and *Grantham* [13.87] measured thresholds for detecting sinusoidal amplitude modulation of a broadband white noise in the presence of a sinusoidal masker modulator. When the masker modulation frequency was 16 or 64 Hz, most modulation masking occurred when the signal modulation frequency was near the masker frequency. In other words, the masking patterns were roughly bandpass, although they showed an increase for very low signal frequencies. For a 4 Hz masker, the masking patterns had a low-pass characteristic, i.e., there was a downward spread of modulation masking.

It should be noted that the sharpness of tuning of the hypothetical modulation filter bank is much less than the sharpness of tuning of the auditory filters in the audio-frequency domain. The bandwidths have been estimated as between 0.5 and 1 times the center frequency [13.80, 88–90]. The modulation filters, if they exist, are not highly selective.

13.4.4 Duration Discrimination

Duration discrimination has typically been studied by presenting two successive sounds which have the same power spectrum but differ in duration. The subject is required to indicate which sound had the longer duration. Both *Creelman* [13.91] and *Abel* [13.92] found that the smallest detectable increase in duration, ΔT, increased with the baseline duration T. The data of Abel showed that, for $T = 10$, 100, and 1000 ms, ΔT was about 4, 15, and 60 ms, respectively. Thus, the Weber fraction, $\Delta T/T$, decreased with increasing T. The results were relatively independent of the overall level of the stimuli and were similar for noise bursts of various bandwidths and for bursts of a 1000 Hz sine wave.

Abel [13.93] reported somewhat different results for the discrimination of the duration of the silent interval between two *markers*. For silent durations, T, less than 160 ms, the results showed that discrimination improved as the level of the markers increased. The function relating $\Delta T/T$ to T was non-monotonic, reaching a local minimum for $T = 2.5$ ms and a local maximum for $T = 10$ ms. The value of ΔT ranged from 6 to 19 ms for a base duration of 10 ms and from 61 to 96 ms for a base duration of 320 ms.

Divenyi and *Danner* [13.94] required subjects to discriminate the duration of the silent interval defined by two 20 ms sounds. When the markers were identical high-level (86 dB SPL) bursts of sinusoids or noise, performance was similar across markers varying in center frequency (500–4000 Hz) and bandwidth. In contrast to the results of *Abel* [13.93], $\Delta T/T$ was almost independent of T over the range of T from 25 to 320 ms. Thresholds were markedly lower than those reported by *Abel* [13.93], being about 1.7 ms at $T = 25$ ms and 15 ms at $T = 320$ ms. This may have been a result of the extensive training of the subjects of Divenyi and Danner. For bursts of a 1 kHz sinusoid, performance worsened markedly when the level of the markers was decreased below 25 dB SL. Performance also worsened markedly when the two markers on either side of a silent interval were made different in level or frequency.

In summary, all studies show that, for values of T exceeding 10 ms, ΔT increases with T and ΔT is roughly independent of the spectral characteristics of the sounds. This is true both for the duration discrimination of sounds and for the discrimination of silent intervals bounded by acoustic markers, provided that the markers are identical on either side of the interval. However, ΔT increases at low sound levels, and also increases when the markers differ in level or frequency on either side of the interval.

13.4.5 Temporal Analysis Based on Across-Channel Processes

Studies of the ability to compare timing across different frequency channels can give very different results depending on whether the different frequency components in the sound are perceived as part of a single sound or as part of more than one sound. Also, it should be realized that subjects may be able to *distinguish* different time patterns, for example, a change in the relative onset time of two different frequencies, without the subjective impression of a change in time pattern; some sort of change in the quality of the sound may be all that is heard. The studies described next indicate the limits of the ability to compare timing across channels, using highly trained subjects.

Patterson and *Green* [13.95] and *Green* [13.75] have studied the discrimination of a class of signals which have the same long-term magnitude spectrum, but which differ in their short-term spectra. These sounds, called Huffman sequences, are brief, broadband click-like sounds, except that the energy in a certain frequency region is delayed relative to that in other regions. The amount of the delay, the center frequency of the delayed frequency region, and the width of the delayed frequency region can all be varied. If subjects can distinguish a pair of Huffman sequences differing, for example, in the amount of delay in a given frequency region, this implies that they are sensitive to the difference in time pattern, i. e., they must be detecting that one frequency region is delayed relative to other regions. *Green* [13.75] measured the ability of subjects to detect differences in the amount of delay in three frequency regions: 650, 1900 and 4200 Hz. He found similar results for all three center frequencies: subjects could detect differences in delay time of about 2 ms regardless of the center frequency of the delayed region.

It should be noted that subjects did not report hearing one part of the sound after the rest of the sound. Rather, the differences in time pattern were perceived as subtle changes in sound quality. Further, some subjects required extensive training to achieve the fine acuity of 2 ms, and even after this training the task required considerable concentration.

Zera and *Green* [13.96] measured thresholds for detecting asynchrony in the onset or offset of complex signals composed of many sinusoidal components.

The components were either uniformly spaced on a logarithmic frequency scale or formed a harmonic series. In one stimulus, the standard, all components started and stopped synchronously. In the signal stimulus, one component was presented with an onset or offset asynchrony. The task of the subjects was to discriminate the standard stimulus from the signal stimulus. They found that onset asynchrony was easier to detect than offset asynchrony. For harmonic signals, onset asynchronies less than 1 ms could generally be detected, whether the asynchronous component was leading or lagging the other components. Thresholds for detecting offset asynchronies were larger, being about 3–10 ms when the asynchronous component ended after the other components and 10–30 ms when the asynchronous component ended before the other components. Thresholds for detecting asynchronies in logarithmically spaced complexes were generally 2–50 times larger than for harmonic complexes.

The difference between harmonically and logarithmically spaced complexes may be explicable in terms of perceptual grouping (see later for more details). The harmonic signal was perceived as a single sound source, i.e., all of the components appeared to belong together. The logarithmically spaced complex was perceived as a series of separate tones, like many notes being played at once on an organ. It seems that it is difficult to compare the timing of sound elements that are perceived as coming from different sources, a point that will be expanded later. The high sensitivity to onset asynchronies for harmonic complexes is consistent with the finding that the perceived timbres of musical instruments are partly dependent on the exact onset times and rates of rise of individual harmonics within each musical note [13.97]. This is described in more detail later on.

13.5 Pitch Perception

Pitch is an attribute of sound defined in terms of what is *heard*. It is defined formally as "that attribute of auditory sensation in terms of which sounds can be ordered on a scale extending from low to high" [13.10]. It is related to the physical repetition rate of the waveform of a sound; for a pure tone (a sinusoid) this corresponds to the frequency, and for a periodic complex tone to the fundamental frequency. Increasing the repetition rate gives a sensation of increasing pitch. Appropriate variations in repetition rate can give rise to a sense of melody. Variations in pitch are also associated with the intonation of voices, and they provide cues as to whether an utterance is a question or a statement and as to the emotion of the talker. Since pitch is a subjective attribute, it cannot be measured directly. Often, the pitch of a complex sound is assessed by adjusting the frequency of a sinusoid until the pitch of the sinusoid matches the pitch of the sound in question. The frequency of the sinusoid then gives a measure of the pitch of the sound. Sometimes a periodic complex sound, such as a pulse train, is used as a matching stimulus. In this case, the repetition rate of the pulse train gives a measure of pitch. Results are generally similar for the two methods, although it is easier to make a pitch match when the sounds to be matched do not differ very much in timbre (see Sect. 13.6.)

13.5.1 Theories of Pitch Perception

There are two traditional theories of pitch perception. One, the *place* theory, is based on the fact that different frequencies (or frequency components in a complex sound) excite different places along the basilar membrane, and hence neurons with different CFs. The place theory assumes that the pitch of a sound is related to the excitation pattern produced by that sound; for a pure tone the pitch is generally assumed to be determined by the position of maximum excitation [13.98].

An alternative theory, called the *temporal* theory, is based on the assumption that the pitch of a sound is related to the time pattern of the neural impulses evoked by that sound [13.99]. These impulses tend to occur at a particular phase of the waveform on the basilar membrane, a phenomenon called phase locking (see Chap. 18). The intervals between successive neural impulses approximate integer multiples of the period of the waveform and these intervals are assumed to determine the perceived pitch. The temporal theory cannot be applicable at very high frequencies, since phase locking does not occur for frequencies above about 5 kHz. However, the tones produced by most musical instruments, the human voice, and most everyday sound sources have fundamental frequencies well below this range.

Many researchers believe that the perception of pitch involves both place mechanisms and temporal mechanisms. However, one mechanism may be dominant for a specific task or aspect of pitch perception, and the relative role of the two mechanisms almost certainly varies with center frequency.

13.5.2 The Perception of the Pitch of Pure Tones

The Frequency Discrimination of Pure Tones

It is important to distinguish between frequency selectivity and frequency discrimination. The former refers to the ability to resolve the frequency components of a complex sound. The latter refers to the ability to detect changes in frequency over time. Usually, the changes in frequency are heard as changes in pitch. The smallest detectable change in frequency is called the frequency difference limen (DL).

Place models of frequency discrimination [13.100, 101] predict that frequency discrimination should be related to frequency selectivity; both should depend on the sharpness of tuning on the basilar membrane. *Zwicker* [13.101] described a model of frequency discrimination based on changes in the excitation pattern evoked by the sound when the frequency is altered, infer-

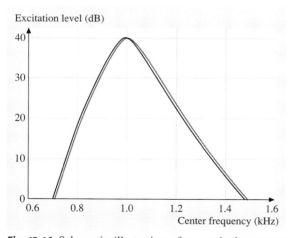

Fig. 13.16 Schematic illustration of an excitation-pattern model for frequency discrimination. Excitation patterns are shown for two sinusoidal tones differing slightly in frequency; the two tones have frequencies of 995 Hz and 1005 Hz. It is assumed that the difference in frequency, ΔF, can be detected if the excitation level changes anywhere by more than a criterion amount. The biggest change in excitation level is on the low-frequency side

ring the shapes of the excitation patterns from masking patterns such as those shown in Fig. 13.6. The model is illustrated in Fig. 13.16. The figure shows two excitation patterns, corresponding to two tones with slightly different frequencies. A change in frequency results in a sideways shift of the excitation pattern. The change is assumed to be detectable whenever the excitation level at some point on the excitation pattern changes by more than about 1 dB.

The change in excitation level is greatest on the steeply sloping low-frequency (low-CF) side of the excitation pattern. Thus, in this model, the detection of a change in frequency is functionally equivalent to the detection of a change in level on the low-frequency side of the excitation pattern. The steepness of the low-frequency side is roughly constant when the frequency scale is expressed in units of ERB_N, rather than in terms of linear frequency. The slope is about 18 dB per ERB_N. To achieve a change in excitation level of 1 dB, the frequency has to be changed by one eighteenth of an ERB_N. Thus, Zwicker's model predicts that the frequency DL at any given frequency should be about one eighteenth ($= 0.056$) of the value of ERB_N at that frequency.

To test Zwicker's model, frequency DLs have been measured as a function of center frequency. There have been two common ways of measuring frequency discrimination. One involves the discrimination of two successive steady tones with slightly different frequencies. On each trial, the tones are presented in a random order and the listener is required to indicate whether the first or second tone is higher in frequency. The frequency difference between the two tones is adjusted until the listener achieves a criterion percentage correct, for example 75%. This measure will be called the difference limen for frequency (DLF). A second measure, called the frequency modulation detection limen (FMDL), uses tones which are frequency modulated. In such tones, the frequency moves up and down in a regular periodic manner about the mean (carrier) frequency. The number of times per second that the frequency goes up and down is called the modulation rate. Typically, the modulation rate is rather low (2–20 Hz), and the change in frequency is heard as a fluctuation in pitch – a kind of *warble*. To determine a threshold for detecting frequency modulation, two tones are presented successively; one is modulated in frequency and the other has a steady frequency. The order of the tones on each trial is random. The listener is required to indicate whether the first or the second tone is modulated. The amount of modulation (also called the modulation depth) required to achieve a criterion response (e.g. 75% correct) is determined.

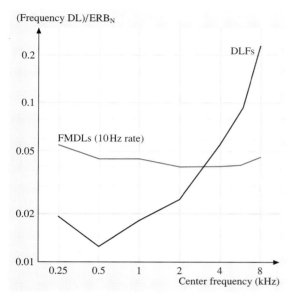

Fig. 13.17 Thresholds for detecting differences in frequency between steady pulsed tones (DLFs) and for detecting frequency modulation (FMDLs), plotted relative to the ERB_N of the auditory filter at each center frequency. (After *Sek* and *Moore* [13.103])

It turns out that the results obtained with these two methods are quite different [13.102], although the difference depends on the modulation rate used to measure the FMDLs. An example of results obtained with the two methods is given in Fig. 13.17 (data from *Sek* and *Moore* [13.103]). For the FMDLs the modulation rate was 10 Hz. To test Zwicker's model, the DLFs and FMDLs were plotted as a proportion of the value of ERB_N at the same center frequency. According to this model, the proportion should be independent of frequency. The proportion for FMDLs using a 10 Hz modulation rate, shown as the brighter line, is roughly constant, and its value is about 0.05, close to the value predicted by the model. However, the proportion for DLFs varies markedly with frequency [13.103, 104]. This is illustrated by the dark line in Fig. 13.17. The DLFs for frequencies of 2 kHz and below are smaller than predicted by Zwicker's model, while those for frequencies of 6 and 8 kHz are larger than predicted.

The results for the FMDLs are consistent with the place model, but the results for the DLFs are not. The reason for the deviation for DLFs is probably that DLFs at low frequencies depend on the use of temporal information from phase locking. Phase locking becomes less precise at frequencies above 1 kHz, and it is com-

pletely lost above 5 kHz. This can account for the marked increase in the DLFs at high frequencies [13.105].

The ratio $FMDL/ERB_N$ is not constant across center frequency when the modulation rate is very low (around 2 Hz), but increases with increasing frequency [13.103, 106]. For low center frequencies, FMDLs are smaller for a 2 Hz modulation rate than for a 10 Hz rate, while for high carrier frequencies (above 4 kHz) the reverse is true. Thus, for a 2 Hz modulation rate, the agreement between DLFs and FMDLs is better than for a 10 Hz rate, but discrepancies remain. For very low modulation rates, frequency modulation may be detected by virtue of the changes in phase locking to the carrier that occur over time. In other words, the frequency is determined over short intervals of time, using phase-locking information, and changes in the estimated frequency over time indicate the presence of frequency modulation. *Moore* and *Sek* [13.106] suggested that the mechanism for decoding the phase-locking information was sluggish; it had to sample the sound for a certain time in order to estimate its frequency. Hence, it could not follow rapid changes in frequency and it played little role for high modulation rates.

In summary, measures of frequency discrimination are consistent with the idea that DLFs, and FMDLs for very low modulation rates, are determined by temporal information (phase locking) for frequencies up to about 4–5 kHz. The precision of phase locking decreases with increasing frequency above 1–2 kHz, and it is almost absent above about 5 kHz. This can explain why DLFs increase markedly at high frequencies. FMDLs for medium to high modulation rates may be determined by a place mechanism, i.e., by the detection of changes in the excitation pattern. This mechanism may also account for DLFs and for FMDLs for low modulation rates, when the center frequency is above about 5 kHz.

The Perception of Musical Intervals

If temporal information plays a role in determining the pitch of pure tones, then we would expect changes in perception to occur for frequencies above 5 kHz, at which phase locking does not occur. Two aspects of perception do indeed change in the expected way, namely the perception of musical intervals, and the perception of melodies.

Two tones which are separated in frequency by an interval of one *octave* (i.e., one has twice the frequency of the other) sound similar. They are judged to have the same name on the musical scale (for example, C3 and C4). This has led several theorists to suggest that there are at least two dimensions to musical pitch. One aspect is related monotonically to frequency (for a pure tone) and is known as *tone height*. The other is related to pitch class (i.e., the name of the note) and is called *tone chroma* [13.107, 108]. For example, two sinusoids with frequencies of 220 and 440 Hz would have the same tone chroma (they would both be called A on the musical scale) but, as they are separated by an octave, they would have different tone heights.

If subjects are presented with a pure tone of a given frequency, f_1, and are asked to adjust the frequency, f_2 of a second tone (presented so as to alternate in time with the fixed tone) so that it appears to be an octave higher in pitch, they generally adjust f_2 to be roughly twice f_1. However, when f_1 lies above 2.5 kHz, so that f_2 would lie above 5 kHz, octave matches become very erratic [13.109]. It appears that the musical interval of an octave is only clearly perceived when both tones are below 5 kHz.

Other aspects of the perception of pitch also change above 5 kHz. A sequence of pure tones above 5 kHz does not produce a clear sense of melody [13.110]. It is possible to hear that the pitch changes when the frequency is changed, but the musical intervals are not heard clearly. Also, subjects with absolute pitch (the ability to assign names to notes without reference to other notes) are very poor at naming notes above 4–5 kHz [13.111].

These results are consistent with the idea that the pitch of pure tones is determined by different mechanisms above and below 5 kHz, specifically, by a temporal mechanism at low frequencies and a place mechanism at high frequencies. It appears that the perceptual dimension of tone height persists over the whole audible frequency range, but tone chroma only occurs in the frequency range below 5 kHz. Musical intervals are only clearly perceived when the frequencies of the tones lie in the range where temporal information is available.

The Effect of Level on Pitch

The pitch of a pure tone is primarily determined by its frequency. However, sound level also plays a small role. On average, the pitch of tones with frequencies below about 2 kHz decreases with increasing level, while the pitch of tones with frequencies above about 4 kHz increases with increasing sound level. The early data of *Stevens* [13.112] showed rather large effects of sound level on pitch, but other data generally show much smaller effects [13.113]. For tones with frequencies between 1 and 2 kHz, changes in pitch with level are generally less than 1%. For tones of lower and higher frequencies, the changes can be larger (up to 5%). There are also considerable individual differences both in the size of the pitch shifts with level, and in the direction of the shifts [13.113].

13.5.3 The Perception of the Pitch of Complex Tones

The Phenomenon of the Missing Fundamental

For complex tones the pitch does not, in general, correspond to the position of maximum excitation on the basilar membrane. Consider, as an example, a sound consisting of short impulses (clicks) occurring 200 times per second. This sound contains harmonics with frequencies at integer multiples of 200 Hz (200, 400, 600, 800 ... Hz). The harmonic at 200 Hz is called the fundamental frequency. The sound has a low pitch, which is very close to the pitch of its fundamental component (200 Hz), and a sharp timbre (a buzzy tone quality). However, if the sound is filtered so as to remove the fundamental component, the pitch does not alter; the only result is a slight change in timbre. This is called the phenomenon of the missing fundamental [13.99, 114]. Indeed, all except a small group of mid-frequency harmonics can be eliminated, and the low pitch remains the same, although the timbre becomes markedly different.

Schouten [13.99, 115] called the low pitch associated with a group of high harmonics the *residue*. Several other names have been used to describe this pitch, including *periodicity pitch*, *virtual pitch*, and *low pitch*. The term *low pitch* will be used here. Schouten pointed out that it is possible to hear the change produced by removing the fundamental component and then reintroducing it. Indeed, when the fundamental component is present, it is possible to hear it out as a separate sound. The pitch of that component is almost the same as the pitch of the whole sound. Therefore, the presence or absence of the fundamental component does not markedly affect the pitch of the whole sound.

The perception of a low pitch does not require activity at the point on the basilar membrane which would respond maximally to the fundamental component. *Licklider* [13.116] showed that the low pitch could be heard when low-frequency noise was present that would mask any component at the fundamental frequency. Even when the fundamental component of a complex tone is present, the pitch of the tone is usually determined by harmonics other than the fundamental.

The phenomenon of the missing fundamental is not consistent with a simple place model of pitch based on the idea that pitch is determined by the position of the peak excitation on the basilar membrane. However, more elaborate place models have been proposed, and these are discussed below.

Theories of Pitch Perception for Complex Tones

To understand theories of pitch perception for complex tones, it is helpful to consider how complex tones are represented in the peripheral auditory system. A simulation of the response of the basilar membrane to a complex tone is illustrated in Fig. 13.18. In this example, the complex tone is a regular series of brief pulses, whose spectrum contains many equal-amplitude harmonics. The number of pulses per second (also called the repetition rate) is 200, so the harmonics have frequencies that are integer multiples of 200 Hz. The lower harmonics are partly resolved on the basilar membrane, and give rise to distinct peaks in the pattern of activity along the basilar membrane. At a place tuned to the frequency of a low harmonic, the waveform on the basilar membrane is approximately a sinusoid at the harmonic frequency. For example, at the place with a characteristic frequency of 400 Hz the waveform is a 400 Hz sinusoid. At a place tuned between two low harmonics, e.g., the place tuned to 317 Hz, there is very little response. In contrast, the higher harmonics are not resolved, and do not give rise to distinct peaks on the basilar membrane. The waveforms at places on the basilar membrane responding to higher harmonics are complex, but they all have a repetition rate equal to the fundamental frequency of the sound.

There are two main (nonexclusive) ways in which the low pitch of a complex sound might be extracted. Firstly, it might be derived from the frequencies of the lower harmonics that are resolved on the basilar membrane. The frequencies of the harmonics might be determined either by place mechanisms (e.g., from the positions of local maxima on the basilar membrane) or by temporal mechanisms (from the inter-spike intervals in neurons with CFs close to the frequencies of individual harmonics). For example, for the complex tone whose analysis is illustrated in Fig. 13.18, the second harmonic, with a frequency of 400 Hz, would give rise to a local maximum at the place on the basilar membrane tuned to 400 Hz. The inter-spike intervals in neurons innervating that place would reflect the frequency of that harmonic; the intervals would cluster around integer multiples of 2.5 ms. Both of these forms of information may allow the auditory system to determine that there is a harmonic at 400 Hz.

The auditory system may contain a pattern recognizer which determines the low pitch of the complex sound from the frequencies of the resolved components [13.117–119]. In essence the pattern recognizer tries to find the harmonic series giving the best match to the resolved frequency components; the fundamental frequency of this harmonic series determines the perceived pitch. Say, for example, that the initial analysis establishes frequencies of 800, 1000 and 1200 Hz to be present. The fundamental frequency whose harmonics would match these frequencies is 200 Hz. The perceived pitch corresponds to this inferred fundamental frequency of 200 Hz. Note that the inferred fundamen-

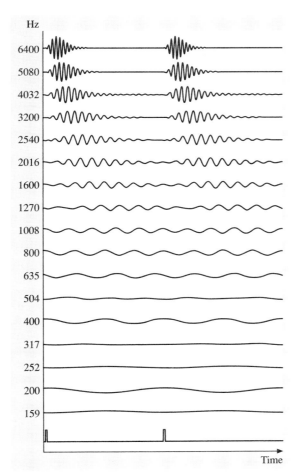

Fig. 13.18 A simulation of the responses on the basilar membrane to periodic impulses of rate 200 pulses per second. The input waveform is shown at the bottom; impulses occur every 5 ms. Each number on the left represents the frequency which would maximally excite a given point on the basilar membrane. The waveform which would be observed at that point, as a function of time, is plotted opposite that number

tal frequency is always the highest possible value that fits the frequencies determined in the initial analysis. For example, a fundamental frequency of 100 Hz would also have harmonics at 800, 1000 and 1200 Hz, but a pitch corresponding to 100 Hz is *not* perceived. It should also be noted that when a complex tone contains only two or three low harmonics, some people, called *analytic listeners*, do not hear the low pitch, but rather hear pitches corresponding to individual harmonics [13.120, 121]. Others, called *synthetic listeners*, usually hear only the low pitch. When many harmonics are present, the low pitch is usually the dominant percept.

Evidence supporting the idea that the low pitch of a complex tone is derived by combining information from several harmonics comes from studies of the ability to detect changes in repetition rate (equivalent to the number of periods per second). When the repetition rate of a complex tone changes, all of the components change in frequency by the same *ratio*, and a change in low pitch is heard. The ability to detect such changes is better than the ability to detect changes in a sinusoid at the fundamental frequency [13.122] and it can be better than the ability to detect changes in the frequency of any of the individual sinusoidal components in the complex tone [13.123]. This indicates that information from the different harmonics is combined or integrated in the determination of low pitch. This can lead to very fine discrimination; changes in repetition rate of about 0.2% can be detected for fundamental frequencies in the range 100–400 Hz provided that low harmonics are present (e.g., the third, fourth and fifth).

The pitch of a complex tone may also be extracted from the higher unresolved harmonics. As shown in Fig. 13.18, the waveforms at places on the basilar membrane responding to higher harmonics are complex, but they all have a repetition rate equal to the fundamental frequency of the sound, namely 200 Hz. For the neurons with CFs corresponding to the higher harmonics, nerve impulses tend to be evoked by the biggest peaks in the waveform, i.e., by the waveform peaks close to envelope maxima. Hence, the nerve impulses are separated by times corresponding to the period of the sound. For example, in Fig. 13.18 the input has a repetition rate of 200 periods per second, so the period is 5 ms. The time intervals between nerve spike would cluster around integer multiples of 5 ms, i.e., 5, 10 15, 20 ... ms. The pitch may be determined from these time intervals. In this example, the time intervals are integer multiples of 5 ms, so the pitch corresponds to 200 Hz.

Experimental evidence suggests that pitch can be extracted *both* from the lower harmonics and from the higher harmonics. Usually, the lower, resolved harmonics give a clearer low pitch, and are more important in determining low pitch, than the upper unresolved harmonics [13.124–126]. This idea is called the principle of dominance; when a complex tone contains many harmonics, including both low- and high-numbered harmonics, the pitch is mainly determined by a small group of lower harmonics. Also, the discrimination of changes in repetition rate of complex tones is better for tones containing only low harmonics than for tones containing only high harmonics [13.123, 127–129]. However, a low pitch can be heard when only high unresolvable harmonics are present. Although, this pitch is not as clear as when lower harmonics are present, it is clear enough to allow the recognition of musical intervals and of simple melodies [13.129, 130].

Several researchers have proposed theories in which both place (spectral) and temporal mechanisms play a role; these are referred to as spectro-temporal theories. The theories assume that information from both low harmonics and high harmonics contributes to the determination of pitch. The initial place/spectral analysis in the cochlea is followed by an analysis of the time pattern of the neural spikes evoked at each CF [13.131–135]. The temporal analysis is assumed to occur at a level of the auditory system higher than the auditory nerve, perhaps in the cochlear nucleus. In the model proposed by *Moore* [13.133], the sound is passed through an array of bandpass filters, each corresponding to a specific place on the basilar membrane. The time pattern of the neural impulses at each CF is determined by the waveform at the corresponding point on the basilar membrane. The inter-spike intervals at each CF are determined. Then, a device compares the time intervals present at different CFs, and searches for common time intervals. The device may also integrate information over time. In general the time interval which is found most often corresponds to the period of the fundamental component. The perceived pitch corresponds to the reciprocal of this interval. For example, if the most prominent time interval is 5 ms, the perceived pitch corresponds to a frequency of 200 Hz.

13.6 Timbre Perception

13.6.1 Time-Invariant Patterns and Timbre

Almost all of the sounds that we encounter in everyday life contain a multitude of frequency components with particular levels and relative phases. The distribution of energy over frequency is one of the major determinants of the quality of a sound or its timbre. Timbre is usually defined as "that attribute of auditory sensation in terms of which a listener can judge that two sounds similarly presented and having the same loudness and pitch are dissimilar" [13.10]. Differences in timbre enable us to distinguish between the same note played on, say, the piano, the violin or the flute.

Timbre depends upon more than just the frequency spectrum of the sound; fluctuations over time can play an important role (see the next section). For the purpose of this section we can adopt a more restricted definition suggested by *Plomp* [13.136] as applicable to steady tones: "Timbre is that attribute of sensation in terms of which a listener can judge that two steady complex tones having the same loudness, pitch and duration are dissimilar." Timbre defined in this way depends mainly on the relative magnitudes of the partials of the tones.

Timbre is multidimensional; there is no single scale along which the timbres of different sounds can be compared or ordered. Thus, a way is needed of describing the spectrum of a sound which takes into account this multidimensional aspect, and which can be related to the subjective timbre. A crude first approach is to look at the overall distribution of spectral energy. The *brightness* or *sharpness* [13.137] of sounds seems to be related to the spectral centroid. However, a much more quantitative approach has been described by *Plomp* and his colleagues [13.136, 138]. They showed that the perceptual differences between different sounds, such as vowels, or steady tones produced by musical instruments, were closely related to the differences in the spectra of the sounds, when the spectra were specified as the levels in 18 1/3-octave frequency bands. A bandwidth of 1/3 octave is slightly greater than the ERB_N of the auditory filter over most of the audible frequency range. Thus, timbre is related to the relative level produced at the output of each auditory filter. Put another way, the timbre of a sound is related to the excitation pattern of that sound.

It is likely that the number of dimensions required to characterize timbre is limited by the number of ERB_Ns required to cover the audible frequency range. This would give a maximum of about 37 dimensions. For a restricted class of sounds, however, a much smaller number of dimensions may be involved. It appears to be generally true, both for speech and non-speech sounds, that the timbres of steady tones are determined primarily by their magnitude spectra, although the relative phases of the components may also play a small role [13.139, 140].

13.6.2 Time-Varying Patterns and Auditory Object Identification

Differences in static timbre are not always sufficient to allow the absolute identification of an *auditory object*, such as a musical instrument. One reason for this is that the magnitude and phase spectrum of the sound may be markedly altered by the transmission path and room reflections [13.141]. In practice, the recognition of a particular timbre, and hence of an auditory object, may depend upon several other factors. *Schouten* [13.142] has suggested that these include: (1) whether the sound is periodic, having a tonal quality for repetition rates between about 20 and 20 000 per second, or irregular, and having a noise-like character; (2) whether the waveform envelope is constant, or fluctuates as a function of time, and in the latter case what the fluctuations are like; (3) whether any other aspect of the sound (e.g., spectrum or periodicity) is changing as a function of time; (4) what the preceding and following sounds are like.

The recognition of musical instruments, for example, depends quite strongly on onset transients and on the temporal structure of the sound envelope. The characteristic tone of a piano depends upon the fact that the notes have a rapid onset and a gradual decay. If a recording of a piano is reversed in time, the timbre is completely different. It now resembles that of a harmonium or accordion, in spite of the fact that the long-term magnitude spectrum is unchanged by time reversal. The perception of sounds with temporally asymmetric envelopes has been studied by *Patterson* [13.143, 144]. He used sinusoidal carriers that were amplitude modulated by a repeating exponential function. The envelope either increased abruptly and decayed gradually (damped sounds) or increased gradually and decayed abruptly (ramped sounds). The ramped sounds were time-reversed versions of the damped sounds and had the same long-term magnitude spectrum. The sounds were characterized by the repetition period of the envelope, which was 25 ms, and by the half life. For a damped sinusoid, the half life is the time taken for the amplitude to decrease by a factor of two.

Patterson reported that the ramped and damped sounds had different qualities. For a half life of 4 ms, the damped sound was perceived as a single source rather like a drum roll played on a hollow, resonant surface (like a drummer's wood block). The ramped sound was perceived as two sounds: a drum roll on a non-resonant surface (such as a leather table top) and a continuous tone corresponding to the carrier frequency. *Akeroyd* and *Patterson* [13.145] used sounds with similar envelopes, but the carrier was broadband noise rather than a sinusoid. They reported that the damped sound was heard as a drum struck by wire brushes. It did not have any hiss-like quality. In contrast, the ramped sound was heard as a noise, with a hiss-like quality, that was sharply cut off in time. These experiments clearly demonstrate the important role of temporal envelope in timbre perception

Pollard and *Jansson* [13.146] described a perceptually relevant way of characterizing the characteristics of time-varying sounds. The sound is filtered in bands of width 1/3 octave. The loudness of each band is calculated at 5 ms intervals. The loudness values are then converted into three coordinates, based on the loudness of:

1. the fundamental component,
2. a group containing partials 2–4, and
3. a group containing partials 5–n,

where n is the highest significant partial. This *tristimulus* representation appears to be related quite closely to perceptual judgements of musical sounds.

Many instruments have noise-like qualities which strongly influence their subjective quality. A flute, for example, has a relatively simple harmonic structure, but synthetic tones with the same harmonic structure do not sound flute-like unless each note is preceded by a small puff of noise. In general, tones of standard musical instruments are poorly simulated by the summation of steady component frequencies, since such a synthesis cannot produce the dynamic variation with time characteristic of these instruments. Thus traditional electronic organs (pre-1965), which produced only tones with a fixed envelope shape, could produce a good simulation of the bagpipes, but could not be made to sound like a piano. Modern synthesizers shape the envelopes of the sounds they produce, and hence are capable of more accurate and convincing imitations of musical instruments. For a simulation to be convincing, it is often necessary to give different time envelopes to different harmonics within a complex sound [13.97, 147].

13.7 The Localization of Sounds

13.7.1 Binaural Cues

It has long been recognized that slight differences in the sounds reaching the two ears can be used as cues in sound localization. The two major cues are differences in the time of arrival at the two ears and differences in intensity at the two ears. For example, a sound coming from the left arrives first at the left ear and is more intense in the left ear. For steady sinusoids, a difference in time of arrival is equivalent to a phase difference between the sounds at the two ears. However, phase differences are not usable over the whole audible frequency range. Experiments using sounds delivered by headphones have shown that a phase difference at the two ears can be detected and used to judge location only for frequencies below about 1500 Hz. This is reasonable because, at high frequencies, the wavelength of sound is small compared to the dimensions of the head, so the listener cannot determine which cycle in the left ear corresponds to a given cycle in the right; there may be many cycles of phase difference. Thus phase differences become ambiguous and unusable at high frequencies. On the other hand, at low frequencies our accuracy at detecting changes in relative time at the two ears is remarkably good; changes of 10–20 μs can be detected, which is equivalent to a movement of the sound source of 1–2° laterally when the sound comes from straight ahead [13.148].

Intensity differences are usually largest at high frequencies. This is because low frequencies bend or diffract around the head, so that there is little difference in intensity at the two ears whatever the location of the sound source, except when the source is very close to the head. At high frequencies, the head casts more of a shadow, and above 2–3 kHz the intensity differences provide useful cues. For complex sounds, containing a range of frequencies, the difference in spectral patterning at the two ears may also be important.

The idea that sound localization is based on interaural time differences at low frequencies and interaural intensity differences at high frequencies has been called the duplex theory of sound localization, proposed by Lord *Rayleigh* [13.149]. However, complex sounds, containing only high frequencies (above 1500 Hz), can be localized on the basis of interaural time delays, provided that they have an appropriate temporal structure. For example, a single click can be localized in this way no matter what its frequency content. Periodic sounds containing only high-frequency harmonics can also be localized on the basis of interaural time differences, provided that the envelope repetition rate (usually equal to the fundamental frequency) is below about 600 Hz [13.150, 151]. Since most of the sounds we encounter in everyday life are complex, and have repetition rates below 600 Hz, interaural time differences can be used for localization in most listening situations.

13.7.2 The Role of the Pinna and Torso

Binaural cues are not sufficient to account for all aspects of sound localization. For example, an interaural time or intensity difference will not indicate whether a sound is coming from in front or behind, or above or below, but such judgments can clearly be made. Also, under some conditions localization with one ear can be as accurate as with two. It has been shown that reflections of sounds from the pinnae and torso play an important role in sound localization [13.152, 153]. The spectra of sounds entering the ear are modified by these reflections in a way which depends upon the direction of the sound source. This direction-dependent filtering provides cues for sound source location. The spectral cues are important not just in providing information about the direction of sound sources, but also in enabling us to judge whether a sound comes from within the head or from the outside world. The pinnae alter the sound spectrum primarily at high frequencies. Only when the wavelength of the sound is comparable with or smaller than the dimensions of the pinnae is the spectrum significantly affected. This occurs mostly above about 6 kHz. Thus, cues provided by the pinnae are most effective for broadband high-frequency sounds. However, reflections from other structures, such as the shoulders, result in spectral changes at lower frequencies, and these may be important for front–back discrimination [13.154].

13.7.3 The Precedence Effect

In everyday conditions the sound from a given source reaches the ears by many different paths. Some of it arrives via a direct path, but a great deal may only reach the ears after reflections from one or more surfaces. However, listeners are not normally aware of these reflections, and the reflections do not markedly impair the ability to localize sound sources. The reason for this seems to lie in a phenomenon known as the precedence effect [13.155, 156]; for a review, see [13.157]. When several sounds reach the ears in close succession (i.e., the direct sound and its reflections) the sounds are perceptually fused into a single sound (an effect called echo suppression), and the location of the total sound is primarily determined by the location of the first (direct) sound (the precedence effect). Thus the reflections have little influence on the perception of direction. Furthermore, there is little awareness of the reflections, although they may influence the timbre and loudness of the sound.

The precedence effect only occurs for sounds of a discontinuous or transient character, such as speech or music, and it can break down if the reflections have a level 10 dB or more above that of the direct sound. However, in normal conditions the precedence effect plays an important role in the localization and identification of sounds in reverberant conditions.

13.8 Auditory Scene Analysis

It is hardly ever the case that the sound reaching our ears comes from a single source. Usually the sound arises from several different sources. However, usually we are able to decompose the mixture of sounds and to perceive each source separately. An auditory object can be defined as the percept of a group of successive and/or simultaneous sound elements as a coherent whole, appearing to emanate from a single source.

As discussed earlier, the peripheral auditory system acts as a frequency analyzer, separating the different frequency components in a complex sound. Somewhere in the brain, the internal representations of these frequency components have to be assigned to their appropriate sources. If the input comes from two sources, A and B, then the frequency components must be split into two groups; the components emanating from source A should be assigned to one source and the components emanating from source B should be assigned to another. The process of doing this is often called perceptual grouping. It is also given the name auditory scene analysis [13.158]. The process of separating the elements arising from two or more different sources is sometimes called segregation.

Many different physical cues may be used to derive separate perceptual objects corresponding to the individual sources which give rise to a complex acoustic input. There are two aspects to this process: the grouping together of all the simultaneous frequency components that emanate from a single source at a given moment, and the connecting over time of the changing frequencies that a single source produces from one moment to the next [13.159]. These two aspects are sometimes described as *simultaneous grouping* and *sequential grouping*, respectively.

Most experiments on perceptual grouping have studied the effect of grouping on one specific attribute of sounds, for example, their pitch, their subjective location, or their timbre. These experiments have shown that a cue which is effective for one attribute may be less effective or completely ineffective for another attribute [13.160]. Also, the effectiveness of the cues may differ for simultaneous and sequential grouping.

13.8.1 Information Used to Separate Auditory Objects

Fundamental Frequency and Spectral Regularity

When we listen to two steady complex tones together (e.g., two musical instruments or two vowel sounds), we do not generally confuse which harmonics belong to which tone. If the complex tones overlap spectrally, two sounds are heard only if the two tones have different values of fundamental frequency (F_0) [13.161]. *Scheffers* [13.162] has shown that, if two vowels are presented simultaneously, they can be identified better when they have F_0s that differ by more than 6% than when they have the same F_0. Other researchers have reported similar findings [13.163, 164].

F_0 may be important in several ways. The components in a periodic sound have frequencies which form a simple harmonic series; the frequencies are integer multiples of F_0. This property is referred to as harmonicity. The lower harmonics are resolved in the peripheral auditory system. The regular spacing of the lower harmonics may promote their perceptual fusion, causing them to be heard as a single sound. If a sinusoidal component does not form part of this harmonic series, it tends to be heard as a separate sound. This is illustrated by some experiments of *Moore* et al.[13.165]. They investigated the effect of mistuning a single low harmonic in a harmonic complex tone. When the harmonic was mistuned sufficiently, it was heard as a separate pure tone standing out from the complex as a whole. The degree of mistuning required varied somewhat with the duration of the sounds; for 400 ms tones, a mistuning of 3% was sufficient to make the harmonic stand out as a separate tone.

Roberts and *Brunstrom* [13.166, 167] have suggested that the important feature determining whether a group of frequency components sounds fused is not harmonicity, but spectral regularity; if a group of components form a regular spectral pattern, they tend to be heard as fused, while if a single component does not fit the pattern, it is heard to pop out. For example, a sequence of components with frequencies 623, 823, 1023, 1223, and 1423 Hz is heard as relatively fused. If the frequency of the middle component is shifted to, say, 923 or 1123 Hz, that component no longer forms part of the regular pattern, and it tends to be heard as a separate tone, standing out from the complex.

For the higher harmonics in a complex sound, F_0 may play a different role. The higher harmonics of a periodic complex sound are not resolved on the basilar membrane, but give rise to a complex waveform with a periodicity equal to F_0 (see Fig. 13.18). When two complex sounds with different F_0s are presented simultaneously, then each will give rise to a waveform on the basilar membrane with periodicity equal to its respective F_0. If the two sounds have different spectra, then each will dominate the response at certain points on the basilar membrane. The auditory system may group together regions with common F_0 and segregate them from regions with different F_0 [13.163]. It may also be the case that both resolved and unresolved components can be grouped on the basis of the detailed time pattern of the neural spikes [13.168].

This process can be explained in a qualitative way by extending the model of pitch perception presented earlier. Assume that the pitch of a complex tone results from a correlation or comparison of time intervals between successive nerve firings in neurons with different CFs. Only those channels which show a high correlation would be classified as belonging to the same sound. Such a mechanism would automatically group together components with a common F_0. However, *de Cheveigné* et al. [13.169] presented evidence against such a mechanism. They measured the identification of a target vowel in the presence of a background vowel; the nominal fundamental frequencies of the two vowels differed by 6.45%. Identification was better when the background was harmonic than when it was made inharmonic (by shifting the frequency of each harmonic by a random amount between 0 and 6.45% or by less than half of the spacing between harmonics, whichever was smaller). In contrast, identification of the target did not depend upon whether or not the target was harmonic. *De Cheveigné* and coworkers [13.169–171] proposed a mechanism based on the idea that a harmonic background sound can be canceled in the auditory system, thus enhancing the representation of a target vowel.

Onset and Offset Disparities

Another cue for the perceptual separation of (near-)simultaneous sounds is onset and offset disparity. *Rasch* [13.172] investigated the ability to hear one complex tone in the presence of another. One of the tones was treated as a masker and the level of the signal tone (the higher in F_0) was adjusted to find the point where it was just detectable. When the two tones started at the same time and had exactly the same temporal envelope, the threshold of the signal was between 0 and -20 dB relative to the level of the masker (Fig. 13.19a). Thus, when a difference in F_0 was the only cue, the signal could not be heard when its level was more than 20 dB below that of the masker.

Rasch also investigated the effect of starting the signal just before the masker (Fig. 13.19b). He found that threshold depended strongly on onset asynchrony, reaching a value of -60 dB for an asynchrony of 30 ms. Thus, when the signal started 30 ms before the masker, it could be heard much more easily and with much greater differences in level between the two tones. It should be emphasized that the lower threshold was a result of the signal occurring for a brief time on its own; essentially performance was limited by backward masking of the 30 ms asynchronous segment, rather than by simultaneous masking. However, the experiment does illustrate the large benefit that can be obtained from a relatively small asynchrony.

Although the percept of his subjects was that the signal continued throughout the masker, Rasch showed that this percept was not based upon sensory information received during the presentation time of the masker. He found that identical thresholds were obtained if the signal terminated immediately after the onset of the masker (Fig. 13.19c). It appears that the perceptual system "assumes" that the signal continues, since there is no evidence to the contrary; the part of the signal that occurs simultaneously with the masker would be completely masked.

Rasch [13.172] showed that, if the two tones have simultaneous onsets but different rise times, this also can give very low thresholds for the signal, provided it has the shorter rise time. Under these conditions and those of onset asynchronies up to 30 ms, the notes sound as though they start synchronously. Thus, we do not need to be consciously aware of the onset differences for the auditory system to be able to exploit them in the perceptual separation of complex tones. Rasch also pointed out

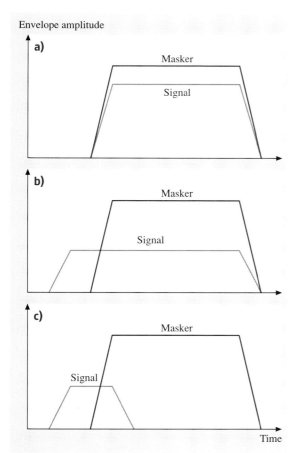

Fig. 13.19a–c Schematic illustration of the stimuli used by *Rasch* [13.172]. Both the signal and the masker were periodic complex tones, with the signal having the higher fundamental frequency. When the signal and masker were gated on and off synchronously (**a**), the threshold for the signal was relatively high. When the signal started slightly before the masker (**b**), the threshold was markedly reduced. When the signal was turned off as soon as the masker was turned on (**c**), the signal was perceived as continuing through the masker, and the threshold was the same as when the signal did continue through the masker

that, in ensemble music, different musicians do not play exactly in synchrony even if the score indicates that they should. The onset differences used in his experiments correspond roughly with the onset asynchronies of nominally simultaneous notes found in performed music. This supports the view that the asynchronies are an important factor in the perception of the separate parts or voices in polyphonic music.

Onset asynchronies can also play a role in determining the timbre of complex sounds. *Darwin* and *Sutherland* [13.173] showed that a tone that starts or stops at a different time from a vowel is less likely to be heard as part of that vowel than if it is simultaneous with it. For example, increasing the level of a single harmonic can produce a significant change in the quality (timbre) of a vowel. However, if the incremented harmonic starts before the vowel, the change in vowel quality is markedly reduced. Similarly, *Roberts* and *Moore* [13.174] showed that extraneous sinusoidal components added to a vowel could influence vowel quality, but the influence was markedly reduced when the extraneous components were turned on before the vowel or turned off after the vowel.

Contrast with Previous Sounds

The auditory system seems well suited to the analysis of changes in the sensory input, and particularly to changes in spectrum over time. The changed aspect stands out perceptually from the rest. It is possible that there are specialized central mechanisms for detecting changes in spectrum. Additionally, stimulation with a steady sound may result in some kind of adaptation. When some aspect of a stimulus is changed, that aspect is freed from the effects of adaptation and thus will be enhanced perceptually. While the underlying mechanism is a matter of debate, the perceptual effect certainly is not.

A powerful demonstration of this effect may be obtained by listening to a stimulus with a particular spectral structure and then switching rapidly to a stimulus with a flat spectrum, such as white noise. A white noise heard in isolation may be described as colorless; it has no pitch and has a neutral sort of timbre. However, when a white noise follows soon after a stimulus with spectral structure, the noise sounds colored [13.175]. A harmonic complex tone with a flat spectrum may be given a speech-like quality if it is preceded by a harmonic complex having a spectrum which is the inverse of that of a speech sound, such as a vowel [13.176].

Another demonstration of the effects of a change in a stimulus can be obtained by listening to a steady complex tone with many harmonics. Usually such a tone is heard with a single pitch corresponding to F_0, and the individual harmonics are not separately perceived. However, if one of the harmonics is changed in some way, by altering either its relative phase or its level, then that harmonic stands out perceptually from the complex as a whole. For a short time after the change is made, a pure-tone quality is perceived. The perception of the harmonic then gradually fades, until it merges with the complex once more.

Change detection is obviously of importance in assigning sound components to their appropriate sources. Normally we listen against a background of sounds which may be relatively unchanging, such as the humming of machinery, traffic noises, and so on. A sudden change in the sound is usually indicative that a new source has been activated, and the change detection mechanisms enable us to isolate the effects of the change and interpret them appropriately.

Correlated Changes in Amplitude or Frequency

Rasch [13.172] also showed that, when the two complex tones start and end synchronously, the detection of the tone with the higher F_0 could be enhanced by frequency modulating it. The modulation was similar to the vibrato which often occurs for musical tones, and it was applied so that all the components in the higher tone moved up and down in synchrony. Rasch found that the modulation could reduce the threshold for detecting the higher tone by 17 dB. A similar effect can be produced by amplitude modulation (AM) of one of the tones. The modulation seems to enhance the salience of the modulated sound, making it appear to stand out from the unmodulated sound.

It is less clear whether the perceptual segregation of simultaneous sounds is affected by the coherence of changes in amplitude or frequency when *both* sounds are modulated. Coherence here refers to whether the changes in amplitude or frequency of the two sounds have the same pattern over time or different patterns over time. Several experiments have been reported suggesting that coherence of amplitude changes plays a role; sounds with coherent changes tend to fuse perceptually, whereas sounds with incoherent changes tend to segregate [13.177–180]. However, *Summerfield* and *Culling* [13.181] found that the coherence of AM did not affect the identification of pairs of simultaneous vowels when the vowels were composed of components placed randomly in frequency (to avoid effects of harmonicity).

Evidence for a role of frequency modulation (FM) coherence in perceptual grouping has been more elusive. While some studies have been interpreted as indicating a weak role for frequency modulation coherence [13.182, 183], the majority of studies have failed to indicate such sensitivity [13.181, 184–188]. *Furukawa* and *Moore* [13.189] have shown that the detectability of FM imposed on two widely separated carrier frequencies is better when the modulation is coherent on the two carriers than when it is incoherent. However, this may occur because the overall pitch evoked by the two carriers fluctuates more when the carriers are modulated coherently than when they are modulated incoherently [13.190–192]. There is at present no clear evidence that the coherence of FM influences perceptual grouping when both sounds are modulated.

Sound Location

The cues used in sound localization may also help in the analysis of complex auditory inputs. A phenomenon that is related to this is called the binaural masking level difference (MLD). The phenomenon can be summarized as follows: whenever the phase or level differences of a signal at the two ears are not the same as those of a masker, the ability to detect the signal is improved relative to the case where the signal and masker have the same phase and level relationships at the two ears. The practical implication is that a signal is easier to detect when it is located in a different position in space from the masker. Although most studies of the MLD have been concerned with threshold measurements, it seems clear that similar advantages of binaural listening can be gained in the identification and discrimination of signals presented against a background of other sound.

An example of the use of binaural cues in separating an auditory object from its background comes from an experiment by *Kubovy* et al. [13.193]. They presented eight continuous sinusoids to each ear via earphones. The sinusoids had frequencies corresponding to the notes in a musical scale, the lowest having a frequency of 300 Hz. The input to one ear, say the left, was presented with a delay of 1 ms relative to the input to the other ear, so the sinusoids were all heard toward the right side of the head. Then, the phase of one of the sinusoids was advanced in the left ear, while its phase in the right ear was delayed, until the input to the left ear led the input to the right ear by 1 ms; this phase-shifting process occurred over a time of 45 ms. The phase remained at the shifted value for a certain time and was then smoothly returned to its original value, again over 45 ms. During the time that the phase shift was present, the phase-shifted sinusoid appeared toward the opposite (left) side of the head, making it stand out perceptually. A sequence of phase shifts in different components was clearly heard as a melody. This melody was completely undetectable when listening to the input to one ear alone. Kubovy et al. interpreted their results as indicating that differences in relative phase at the two ears can allow an auditory object to be isolated in the absence of any other cues.

Culling [13.194] performed a similar experiment to that of *Kubovy* et al. [13.193], but he examined the importance of the phase transitions. He found that, when one component of a complex sound changed rapidly but smoothly in interaural time difference (ITD), it perceptually segregated from the complex. When different components were changed in ITD in succession, a recognizable melody was heard, as reported by Kubovy et al. However, when the transitions were replaced by silent intervals, leaving only static ITDs as a cue, the melody was much less salient. Thus, transitions in ITD seem to be more important than static differences in ITD in producing segregation of one component from a background of other components. Nevertheless, static differences in ITD do seem to be sufficient to produce segregation under some conditions [13.195].

Other experiments suggest that binaural processing often plays a relatively minor role in simultaneous grouping. *Shackleton* and *Meddis* [13.196] investigated the ability to identify each vowel in pairs of concurrent vowels. They found that a difference in F_0 between the two vowels improved identification by about 22%. In contrast, a 400 μs interaural delay in one vowel (which corresponds to an azimuth of about 45 degrees) improved performance by only 7%. *Culling* and *Summerfield* [13.197] investigated the identification of concurrent whispered vowels, synthesized using bands of noise. They showed that listeners were able to identify the vowels accurately when each vowel was presented to a different ear. However, they were unable to identify the vowels when they were presented to both ears but with different ITDs. In other words, listeners could not group the noise bands in different frequency regions with the same ITD and thereby separate them from noise bands in other frequency regions with a different ITD.

In summary, when two simultaneous sounds differ in their interaural level or time, this can contribute to the perceptual segregation of the sounds and enhance their detection and discrimination. However, such binaural processing is not always effective, and in some situations it appears to play little role.

13.8.2 The Perception of Sequences of Sounds

Stream Segregation

When we listen to rapid sequences of sounds, the sounds may be grouped together (i.e., perceived as if they come from a single source, called fusion or coherence), or they may be perceived as different streams (i.e., as coming from more than one source, called fission or stream segregation) [13.158, 199–201]. The term streaming is used to denote the processes determining whether one stream or multiple streams are heard. *Van Noorden* [13.202] investigated this phenomenon using a sequence of pure tones where every second B was omitted from the regular sequence ABABAB ..., producing a sequence ABA ABA .... He found that this could be perceived in two ways, depending on the frequency separation of A and B. For small separations, a single rhythm, resembling a gallop, is heard (fusion). For larger separations, two separate tone sequences can be heard, one of which (A A A) is running twice as fast as the other (B B B) (fission). Components are more likely to be assigned to separate streams if they differ widely in frequency or if there are rapid jumps in frequency between them. The latter point is illustrated by a study of *Bregman* and *Dannenbring* [13.203]. They used tone sequences in which successive tones were connected by frequency glides. They found that these glides reduced the tendency for the sequences to split into high and low streams. Conditions using partial glides also showed decreased stream segregation, although the partial glides were not quite as effective as complete glides. Thus, complete continuity between tones is not required to reduce stream segregation; a frequency change pointing toward the next tone allows the listener to follow the pattern more easily.

The effects of frequency glides and other types of transitions in preventing stream segregation or fission are probably of considerable importance in the perception of speech. Speech sounds may follow one another in very rapid sequences, and the glides and partial glides observed in the acoustic components of speech may be a strong factor in maintaining the percept of speech as a unified stream.

Van Noorden found that, for intermediate frequency separations of the tones A and B in a rapid sequence, either fusion or fission could be heard, according to the instructions given and the attentional set of the subject. When the percept is ambiguous, the tendency for fission to occur increases with increasing exposure time to the tone sequence [13.204]. The auditory system seems to start with the assumption that there is a single sound source, and fission is only perceived when sufficient evidence has built up to contradict this assumption. Sudden changes in a sequence, or in the perception of a sequence, can cause the percept to revert to its initial, default condition, which is fusion [13.205, 206]; for a review, see [13.207].

For rapid sequences of complex tones, strong fission can be produced by differences in spectrum of successive tones, even when all tones have the same F_0 [13.201, 208–210]. However, when successive complex tones are filtered to have the same spectral envelope, stream segregation can also be produced by differences between successive tones in F_0 [13.210, 211], in temporal envelope [13.209, 212], in the relative phases of the components [13.213], or in apparent location [13.214]. *Moore* and *Gockel* [13.207] proposed that any salient perceptual difference between successive tones may lead to stream segregation. Consistent with this idea, *Dowling* [13.215, 216] has shown that stream segregation may also occur when successive pure tones differ in intensity or in spatial location. He presented a melody composed of equal-intensity notes and inserted between each note of the melody a tone of the same intensity, with a frequency randomly selected from the same range. He found that the resulting tone sequence produced a meaningless jumble. Making the interposed notes different from those of the melody, in either intensity, frequency range, or spatial location, caused them to be heard as a separate stream, enabling subjects to pick out the melody.

Darwin and *Hukin* [13.198] have shown that sequential grouping can be strongly influenced by ITD. In one experiment, they simultaneously presented two

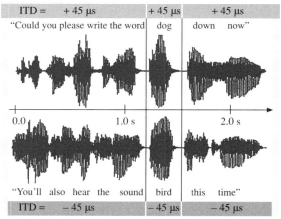

Fig. 13.20 Example of the stimuli used by *Darwin* and *Hukin* [13.198] (After Fig. 1 of [13.198])

sentences (see Fig. 13.20). They varied the ITDs of the two sentences in the range 0 to $\pm 181\,\mu$s. For example, one sentence might lead in the left ear by $45\,\mu$s, while the other sentence would lead in the right ear by $45\,\mu$s (as in Fig. 13.20). The sentences were based on natural speech but were processed so that each was spoken on a monotone, i.e., with constant F_0. The F_0 difference between the two sentences was varied from 0 to 4 semitones. Subjects were instructed to attend to one particular sentence. At a certain point, the two sentences contained two different target words aligned in starting time and duration ("dog" and "bird"). The F_0s and the ITDs of the two target words were varied independently from those of the two sentences. Subjects had to indicate which of the two target words they heard in the attended sentence. They reported the target word that had the same ITD as the attended sentence much more often than the target word with the opposite ITD. In other words, the target word with the same ITD as the attended sentence was grouped with that sentence. This was true even when the target word had the same ITD as the attended sentence but a different F_0. Thus, subjects grouped words across time according to their perceived location, independent of F_0 differences. *Darwin* and *Hukin* [13.198] concluded that listeners who try to track a particular sound source over time direct attention to auditory objects at a particular subjective location. The auditory objects themselves may be formed using cues other than ITD, for example, onset and offset asynchrony and harmonicity.

It should be noted that, for discrete sequences of musical tones, the auditory system does not necessarily form streams according to perceived location, especially when that cue competes with other cues. This is illustrated by an effect, called the scale illusion, reported by *Deutsch* [13.217]. She presented two sequences of tones via headphones, one sequence to each ear. The nth tone in the left ear was synchronous with the nth tone in the right ear. The sequences were created by repetitive presentation of the C major scale in both ascending and descending form, such that when a component of the ascending scale was in one ear, a component of the descending scale was in the other, and vice versa. However, the tones from each scale alternated between ears. Within each ear there were often large jumps in frequency between successive tones. Most subjects perceived the sounds as two streams, organized by the frequency proximity of successive tones. One stream (which was often heard towards one ear) was heard as a musical scale that started high, descended and then increased again, while the other stream (which was usually heard towards the opposite ear) was heard as a scale that started low, ascended, and then decreased again. Thus, the true location of the tones had little influence on the formation of the perceptual streams.

Another example come from the opening bars of the last movement of Tchaikovsky's sixth symphony. This contains interleaved notes played by the first and second violins, who according to 19th century custom sat on opposite sides of the stage. These notes are perceived as a single stream, despite the difference in location, presumably because of the frequency proximity between successive notes.

A number of composers have exploited the fact that stream segregation occurs for tones that are widely separated in frequency. By playing a sequence of tones in which alternate notes are chosen from separate frequency ranges, an instrument such as the flute, which is only capable of playing one note at a time, can appear to be playing two themes at once. Many fine examples of this are available in the works of Bach, Telemann and Vivaldi.

Judgment of Temporal Order

It is difficult to judge the temporal order of sounds that are perceived in different streams. An example of this comes from the work of *Broadbent* and *Ladefoged* [13.218]. They reported that extraneous sounds in sentences were grossly mislocated. For example, a click might be reported as occurring a word or two away from its actual position. Surprisingly poor performance was also reported by *Warren* et al. [13.219] for judgments of the temporal order of three or four unrelated items, such as a hiss, a tone, and a buzz. Most subjects could not identify the order when each successive item lasted as long as 200 ms. Naive subjects required that each item last at least 700 ms to identify the order of four sounds presented in an uninterrupted repeated sequence. These durations are well above those which are normally considered necessary for temporal resolution.

The poor order discrimination described by Warren et al. is probably a result of stream segregation. The sounds they used do not represent a coherent class. They have different temporal and spectral characteristics, and, as for tones widely differing in frequency, they do not form a single perceptual stream. Items in different streams appear to float about with respect to each other in subjective time. Thus, temporal order judgments are difficult. It should be emphasized that the relatively poor performance reported by *Warren* et al. [13.219] is found only in tasks requiring absolute identification of the order of sounds and not in tasks which simply require the discrimination of different sequences. Also, with extended training and feedback subjects can learn to distinguish between and identify orders within sequences of unrelated sounds lasting only 10 ms or less [13.220].

To explain these effects, *Divenyi* and *Hirsh* [13.221] suggested that two kinds of perceptual judgments are involved. At longer item durations the listener is able to hear a clear sequence of steady state sounds, whereas at shorter durations a change in the order of items introduces qualitative changes that can be discriminated by trained listeners. Similar explanations have been put forward by *Green* [13.75] and *Warren* [13.220].

Bregman and *Campbell* [13.200] investigated the factors that make temporal order judgments for tone sequences difficult. They used naive subjects, so performance presumably depended on the subjects actually perceiving the sounds as a sequence, rather than on their learning the overall sound pattern. They found that, in a repeating cycle of mixed high and low tones, subjects could discriminate the order of the high tones relative to one another or of the low tones among themselves, but they could not order the high tones relative to the low ones. The authors suggested that this was because the two groups of sounds split into separate perceptual streams and that judgments across streams are difficult. Several more recent studies have used tasks involving the discrimination of changes in timing or rhythm as a tool for studying stream segregation [13.210, 213, 222]. The rationale of these studies is that, if the ability to judge the relative timing of successive sound elements is good, this indicates that the elements are perceived as part of a single stream, while if the ability is poor, this indicates that the elements are perceived in different streams.

13.8.3 General Principles of Perceptual Organization

The Gestalt psychologists [13.223] described many of the factors which govern perceptual organization, and their descriptions and principles apply reasonably well to the way physical cues are used to achieve perceptual grouping of the acoustic input. It seems likely that the rules of perceptual organization have arisen because, on the whole, they tend to give the right answers. That is, use of the rules generally results in a grouping of those parts of the acoustic input that arose from the same source and a segregation of those that did not. No single rule will always work, but it appears that the rules can generally be used together, in a coordinated and probably quite complex way, in order to arrive at a correct interpretation of the input. In the following sections, I outline the major principles or rules of perceptual organization. Many, but not all, of the rules apply to both vision and hearing, and they were mostly described first in relation to vision.

Similarity
This principle is that elements will be grouped if they are similar. In hearing, similarity usually implies closeness of timbre, pitch, loudness, or subjective location. Examples of this principle have already been described. If we listen to a rapid sequence of pure tones, say 10 tones per second, then tones which are closely spaced in frequency, and are therefore similar, form a single perceptual stream, whereas tones which are widely spaced form separate streams.

For pure tones, frequency is the most important factor governing similarity, although differences in level and subjective location between successive tones can also lead to stream segregation. For complex tones, differences in timbre produced by spectral differences seem to be the most important factor. Again, however, other factors may play a role. These include differences in F_0, differences in timbre produced by temporal envelope differences, and differences in perceived location.

Good Continuation
This principle exploits a physical property of sound sources, that changes in frequency, intensity, location, or spectrum tend to be smooth and continuous, rather than abrupt. Hence, a smooth change in any of these aspects indicates a change within a single source, whereas a sudden change indicates that a new source has been activated. One example has already been described; *Bregman* and *Dannenbring* [13.203] showed that the tendency of a sequence of high and low tones to split into two streams was reduced when successive tones were connected by frequency glides.

A second example comes from studies using synthetic speech. In such speech, large fluctuations of an unexpected kind in F_0 (and correspondingly in the pitch) give the impression that a new speaker has stepped in to take over a few syllables from the primary speaker. *Darwin* and *Bethell-Fox* [13.224] synthesized spectral patterns which changed smoothly and repeatedly between two vowel sounds. When the F_0 of the sound patterns was constant they were heard as coming from a single source, and the speech sounds heard included glides ("l" as in let) and semivowels ("w" as in we). When a discontinuous, step-like F_0 contour was imposed on the patterns, they were perceived as two distinct speech streams, and the speech was perceived as containing predominantly stop consonants (e.g., "b" as in be, and "d" as in day). Apparently, a given group of components is usually only perceived as part of one stream. Thus, the perceptual segregation produces illusory silences in each stream during the portions of the signal attributed to the other stream, and these silences are interpreted, together with the gliding spectral patterns in the vowels, as indicating the presence of stop consonants. It is clear that the perception of speech sounds can be strongly influenced by stream organization.

Common Fate

The different frequency components arising from a single sound source usually vary in a highly coherent way. They tend to start and finish together, change in intensity together, and change in frequency together. This fact is exploited by the perceptual system and gives rise to the principle of common fate: if two or more components in a complex sound undergo the same kinds of changes at the same time, then they are grouped and perceived as part of the same source.

Two examples of common fate were described earlier. The first concerns the role of the onsets and offsets of sounds. Components will be grouped together if they start and stop synchronously; otherwise they will form separate streams. The onset asynchronies necessary to allow the separation of two complex tones are not large, about 30 ms being sufficient. The asynchronies which are observed in performed music are typically as large as or larger than this, so when we listen to polyphonic music we are easily able to hear separately the melodic line of each instrument. Secondly, components which are amplitude modulated in a synchronous way tend to be grouped together. There is at present little evidence that the coherence of modulation in frequency affects perceptual grouping, although frequency modulation of a group of components in a complex sound can promote the perceptual segregation of those components from an unchanging background.

Disjoint Allocation

Broadly speaking, this principle, also known as belongingness, is that a single component in a sound can only be assigned to one source at a time. In other words, once a component has been used in the formation of one stream, it cannot be used in the formation of a second stream. For certain types of stimuli, the perceptual organization may be ambiguous, there being more than one way to interpret the sensory input. When a given component might belong to one of a number of streams, the percept may alter depending on the stream within which that component is included.

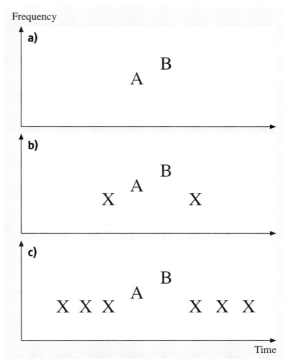

Fig. 13.21 Schematic illustration of the stimuli used by *Bregman* and *Rudnicky* [13.225]. When the tones A and B are presented alone *(panel a)*, it is easy to tell their order. When the tones A and B are presented as part of a four-tone complex XABX *(panel b)*, it is more difficult to tell their order. If the four-tone complex is embedded in a longer sequence of X tones *(panel c)*, the Xs form a separate perceptual stream, and it is easy to tell the order of A and B

An example is provided by the work of *Bregman* and *Rudnicky* [13.225]. They presented a sequence of four brief tones in rapid succession. Two of the tones, labeled X, had the same frequency, but the middle two, A and B, were different. The four-tone sequence was either XABX or XBAX. The listeners had to judge the order of A and B. This was harder than when the tones AB occurred in isolation (Fig. 13.21a) because A and B were perceived as part of a longer four-tone pattern, including the two "distracter" tones, labeled X (Fig. 13.21b). They then embedded the four-tone sequence into a longer sequence of tones, called captor tones (Fig. 13.21c). When the captor tones had frequencies close to those of the distracter tones, they captured the distracters into a separate perceptual stream, leaving the tones AB in a stream of their own. This made the order of A and B easy to judge. It seems that the tones X could not be perceived as part of both streams. When only one stream is the subject of judgment and hence of attention, the other one may serve to remove distracters from the domain of attention.

It should be noted that the principle of disjoint allocation does not always work, particularly in situations where there are two or more plausible perceptual organizations [13.226]. In such situations, a sound element may sometimes be heard as part of more than one stream.

Closure

In everyday life, the sound from a given source may be temporarily masked by other sounds. While the masking sound is present there may be no sensory evidence which can be used to determine whether the masked sound has continued or not. Under these conditions the masked sound tends to be perceived as continuous. The Gestalt psychologists called this process closure.

A laboratory example of this phenomenon is the continuity effect [13.227–229]. When a sound A is alternated with a sound B, and B is more intense than A, then A may be heard as continuous, even though it is interrupted. The sounds do not have to be steady. For example, if B is noise and A is a tone which is gliding upward in frequency, the glide is heard as continuous even though certain parts of the glide are missing [13.230].

Notice that, for this to be the case, the gaps in the tone must be filled with noise and the noise must be a potential masker of the tone (if they were presented simultaneously). In the absence of noise, discrete jumps in frequency are clearly heard.

The continuity effect also works for speech stimuli alternated with noise. In the absence of noise to fill in the gaps, interrupted speech sounds hoarse and raucous. When noise is presented in the gaps, the speech sounds more natural and continuous [13.231]. For connected speech at moderate interruption rates, the intervening noise actually leads to an improvement in intelligibility [13.232]. This may occur because the abrupt switching of the speech produces misleading cues as to which speech sounds were present. The noise serves to mask these misleading cues.

It is clear from these examples that the perceptual filling in of missing sounds does not take place solely on the basis of evidence in the acoustic waveform. Our past experience with speech, music, and other stimuli must play a role, and the context of surrounding sounds is important [13.233]. However, the filling in only occurs when one source is perceived as masking or occluding another. This percept must be based on acoustic evidence that the occluded sound has been masked. Thus, if a gap is not filled by a noise or other sound, the perceptual closure does not occur; a gap is heard.

13.9 Further Reading and Supplementary Materials

More information about the topics discussed in this chapter can be found in: A. S. Bregman: *Auditory Scene Analysis: The Perceptual Organization of Sound* (Bradford Books, MIT Press, Cambridge 1990); W. M. Hartmann: *Signals, Sound, and Sensation* (AIP Press, Woodbury 1997); B. C. J. Moore: *An Introduction to the Psychology of Hearing, 5th Ed.* (Academic, San Diego 2003); R. Plomp: *The Intelligent Ear* (Erlbaum, Mahwah 2002)

A compact disc (CD) of auditory demonstrations has been produced by A. J. M. Houtsma, T. D. Rossing, W. M. Wagenaars (1987). The disc can be obtained through the Acoustical Society of America; contact asapubs@abdintl.com for details.

The following CD has a large variety of demonstrations relevant to perceptual grouping: A. S. Bregman, P. Ahad (1995). *Demonstrations of Auditory Scene Analysis: The Perceptual Organization of Sound*, (Auditory Perception Laboratory, Department of Psychology, McGill University, distributed by MIT Press, Cambridge, MA). It can be ordered from The MIT Press, 55 Hayward St., Cambridge, MA 02142, USA. Further relevant demonstrations can be found at: http://www.kyushu-id.ac.jp/~ynhome/.

A CD simulating the effects of a hearing loss on the perception of speech and music is *Perceptual Consequences of Cochlear Damage*, which may be obtained by writing to B. C. J. Moore, Department of Experimental Psychology, University of Cambridge, Downing Street, Cambridge, CB2 3EB, England, and enclosing a check for 20 dollars or a cheque for 12 pounds sterling, made payable to B. C. J. Moore.

References

13.1 ISO 389-7: *Acoustics – Reference zero for the calibration of audiometric equipment. Part 7: Reference threshold of hearing under free-field and diffuse-field listening conditions* (International Organization for Standardization, Geneva 1996)

13.2 M.C. Killion: Revised estimate of minimal audible pressure: Where is the "missing 6 dB"?, J. Acoust. Soc. Am. **63**, 1501–1510 (1978)

13.3 B.C.J. Moore, B.R. Glasberg, T. Baer: A model for the prediction of thresholds, loudness and partial loudness, J. Audio Eng. Soc. **45**, 224–240 (1997)

13.4 M.A. Cheatham, P. Dallos: Inner hair cell response patterns: implications for low-frequency hearing, J. Acoust. Soc. Am. **110**, 2034–2044 (2001)

13.5 L.J. Sivian, S.D. White: On minimum audible sound fields, J. Acoust. Soc. Am. **4**, 288–321 (1933)

13.6 I. Pollack: Monaural and binaural threshold sensitivity for tones and for white noise, J. Acoust. Soc. Am. **20**, 52–57 (1948)

13.7 J.K. Dierks, L.A. Jeffress: Interaural phase and the absolute threshold for tone, J. Acoust. Soc. Am. **34**, 981–986 (1962)

13.8 K. Krumbholz, R.D. Patterson, D. Pressnitzer: The lower limit of pitch as determined by rate discrimination, J. Acoust. Soc. Am. **108**, 1170–1180 (2000)

13.9 R. Plomp, M.A. Bouman: Relation between hearing threshold and duration for tone pulses, J. Acoust. Soc. Am. **31**, 749–758 (1959)

13.10 ANSI: *ANSI S1.1-1994. American National Standard Acoustical Terminology* (American National Standards Institute, New York 1994)

13.11 R.L. Wegel, C.E. Lane: The auditory masking of one sound by another and its probable relation to the dynamics of the inner ear, Phys. Rev. **23**, 266–285 (1924)

13.12 L.L. Vogten: Low-level pure-tone masking: a comparison of 'tuning curves' obtained with simultaneous and forward masking, J. Acoust. Soc. Am. **63**, 1520–1527 (1978)

13.13 H. Fletcher: Auditory patterns, Rev. Mod. Phys. **12**, 47–65 (1940)

13.14 H.L.F. Helmholtz: *Die Lehre von den Tonempfindungen als physiologische Grundlage für die Theorie der Musik* (Vieweg, Braunschweig 1863)

13.15 R.D. Patterson, B.C.J. Moore: Auditory filters and excitation patterns as representations of frequency resolution. In: *Frequency Selectivity in Hearing*, ed. by B.C.J. Moore (Academic, London 1986)

13.16 D. Johnson-Davies, R.D. Patterson: Psychophysical tuning curves: restricting the listening band to the signal region, J. Acoust. Soc. Am. **65**, 765–770 (1979)

13.17 B.J. O'Loughlin, B.C.J. Moore: Off-frequency listening: effects on psychoacoustical tuning curves obtained in simultaneous and forward masking, J. Acoust. Soc. Am. **69**, 1119–1125 (1981)

13.18 B.J. O'Loughlin, B.C.J. Moore: Improving psychoacoustical tuning curves, Hear. Res. **5**, 343–346 (1981)

13.19 R.D. Patterson: Auditory filter shapes derived with noise stimuli, J. Acoust. Soc. Am. **59**, 640–654 (1976)

13.20 B.R. Glasberg, B.C.J. Moore: Derivation of auditory filter shapes from notched-noise data, Hear. Res. **47**, 103–138 (1990)

13.21 J.P. Egan, H.W. Hake: On the masking pattern of a simple auditory stimulus, J. Acoust. Soc. Am. **22**, 622–630 (1950)

13.22 E. Zwicker, E. Terhardt: Analytical expressions for critical band rate and critical bandwidth as a function of frequency, J. Acoust. Soc. Am. **68**, 1523–1525 (1980)

13.23 R.D. Patterson, I. Nimmo-Smith: Off-frequency listening and auditory filter asymmetry, J. Acoust. Soc. Am. **67**, 229–245 (1980)

13.24 B.C.J. Moore, B.R. Glasberg: Formulae describing frequency selectivity as a function of frequency and level and their use in calculating excitation patterns, Hear. Res. **28**, 209–225 (1987)

13.25 S. Rosen, R.J. Baker, A. Darling: Auditory filter non-linearity at 2 kHz in normal hearing listeners, J. Acoust. Soc. Am. **103**, 2539–2550 (1998)

13.26 B.R. Glasberg, B.C.J. Moore, R.D. Patterson, I. Nimmo-Smith: Dynamic range and asymmetry of the auditory filter, J. Acoust. Soc. Am. **76**, 419–427 (1984)

13.27 E. Zwicker, H. Fastl: *Psychoacoustics – Facts and Models*, 2nd edn. (Springer, Berlin 1999)

13.28 B.C.J. Moore, J.I. Alcántara, T. Dau: Masking patterns for sinusoidal and narrowband noise maskers, J. Acoust. Soc. Am. **104**, 1023–1038 (1998)

13.29 B.C.J. Moore, B.R. Glasberg: Suggested formulae for calculating auditory-filter bandwidths and excitation patterns, J. Acoust. Soc. Am. **74**, 750–753 (1983)

13.30 K. Miyazaki, T. Sasaki: Pure-tone masking patterns in nonsimultaneous masking conditions, Jap. Psychol. Res. **26**, 110–119 (1984)

13.31 A.J. Oxenham, B.C.J. Moore: Modeling the additivity of nonsimultaneous masking, Hear. Res. **80**, 105–118 (1994)

13.32 B.C.J. Moore, B.R. Glasberg: Growth of forward masking for sinusoidal and noise maskers as a function of signal delay: implications for suppression in noise, J. Acoust. Soc. Am. **73**, 1249–1259 (1983)

13.33 G. Kidd, L.L. Feth: Effects of masker duration in pure-tone forward masking, J. Acoust. Soc. Am. **72**, 1384–1386 (1982)

13.34 E. Zwicker: Dependence of post-masking on masker duration and its relation to temporal effects in loudness, J. Acoust. Soc. Am. **75**, 219–223 (1984)

13.35 H. Fastl: Temporal masking effects: I. Broad band noise masker, Acustica **35**, 287–302 (1976)

13.36 H. Duifhuis: Audibility of high harmonics in a periodic pulse II. Time effects, J. Acoust. Soc. Am. **49**, 1155–1162 (1971)

13.37 H. Duifhuis: Consequences of peripheral frequency selectivity for nonsimultaneous masking, J. Acoust. Soc. Am. **54**, 1471–1488 (1973)

13.38 C.J. Plack, B.C.J. Moore: Temporal window shape as a function of frequency and level, J. Acoust. Soc. Am. **87**, 2178–2187 (1990)

13.39 W. Jesteadt, S.P. Bacon, J.R. Lehman: Forward masking as a function of frequency, masker level, and signal delay, J. Acoust. Soc. Am. **71**, 950–962 (1982)

13.40 R.L. Smith: Short-term adaptation in single auditory-nerve fibers: Some poststimulatory effects, J. Neurophysiol. **49**, 1098–1112 (1977)

13.41 C.W. Turner, E.M. Relkin, J. Doucet: Psychophysical and physiological forward masking studies: Probe duration and rise-time effects, J. Acoust. Soc. Am. **96**, 795–800 (1994)

13.42 A.J. Oxenham: Forward masking: adaptation or integration?, J. Acoust. Soc. Am. **109**, 732–741 (2001)

13.43 M. Brosch, C.E. Schreiner: Time course of forward masking tuning curves in cat primary auditory cortex, J. Neurophysiol. **77**, 923–943 (1997)

13.44 A.J. Oxenham, B.C.J. Moore: Additivity of masking in normally hearing and hearing-impaired subjects, J. Acoust. Soc. Am. **98**, 1921–1934 (1995)

13.45 R. Plomp: The ear as a frequency analyzer, J. Acoust. Soc. Am. **36**, 1628–1636 (1964)

13.46 R. Plomp, A.M. Mimpen: The ear as a frequency analyzer II, J. Acoust. Soc. Am. **43**, 764–767 (1968)

13.47 B.C.J. Moore, K. Ohgushi: Audibility of partials in inharmonic complex tones, J. Acoust. Soc. Am. **93**, 452–461 (1993)

13.48 D.R. Soderquist: Frequency analysis and the critical band, Psychon. Sci. **21**, 117–119 (1970)

13.49 P.A. Fine, B.C.J. Moore: Frequency analysis and musical ability, Music Percept. **11**, 39–53 (1993)

13.50 B. Gabriel, B. Kollmeier, V. Mellert: Influence of individual listener, measurement room and choice of test-tone levels on the shape of equal-loudness level contours, Acustica Acta Acustica **83**, 670–683 (1997)

13.51 D. Laming: *The Measurement of Sensation* (Oxford University Press, Oxford 1997)

13.52 H. Fletcher, W.A. Munson: Loudness, its definition, measurement and calculation, J. Acoust. Soc. Am. **5**, 82–108 (1933)

13.53 ISO 226: *Acoustics – normal equal-loudness contours* (International Organization for Standardization, Geneva 2003)

13.54 S.S. Stevens: On the psychophysical law, Psych. Rev. **64**, 153–181 (1957)

13.55 R.P. Hellman, J.J. Zwislocki: Some factors affecting the estimation of loudness, J. Acoust. Soc. Am. **35**, 687–694 (1961)

13.56 E.M. Relkin, J.R. Doucet: Is loudness simply proportional to the auditory nerve spike count?, J. Acoust. Soc. Am. **191**, 2735–2740 (1997)

13.57 H. Fletcher, W.A. Munson: Relation between loudness and masking, J. Acoust. Soc. Am. **9**, 1–10 (1937)

13.58 E. Zwicker: Über psychologische und methodische Grundlagen der Lautheit, Acustica **8**, 237–258 (1958)

13.59 E. Zwicker, B. Scharf: A model of loudness summation, Psych. Rev. **72**, 3–26 (1965)

13.60 B.R. Glasberg, B.C.J. Moore: A model of loudness applicable to time-varying sounds, J. Audio Eng. Soc. **50**, 331–342 (2002)

13.61 H. Fastl: Loudness evaluation by subjects and by a loudness meter. In: *Sensory Research – Multimodal Perspectives*, ed. by R.T. Verrillo (Erlbaum, Hillsdale, New Jersey 1993)

13.62 ANSI: *ANSI S3.4-2005. Procedure for the Computation of Loudness of Steady Sounds* (American National Standards Institute, New York 2005)

13.63 E. Zwicker, G. Flottorp, S.S. Stevens: Critical bandwidth in loudness summation, J. Acoust. Soc. Am. **29**, 548–557 (1957)

13.64 B. Scharf: Complex sounds and critical bands, Psychol. Bull. **58**, 205–217 (1961)

13.65 B. Scharf: Critical bands. In: *Foundations of Modern Auditory Theory*, ed. by J.V. Tobias (Academic, New York 1970)

13.66 B.C.J. Moore, B.R. Glasberg: The role of frequency selectivity in the perception of loudness, pitch and time. In: *Frequency Selectivity in Hearing*, ed. by B.C.J. Moore (Academic, London 1986)

13.67 G.A. Miller: Sensitivity to changes in the intensity of white noise and its relation to masking and loudness, J. Acoust. Soc. Am. **191**, 609–619 (1947)

13.68 R.R. Riesz: Differential intensity sensitivity of the ear for pure tones, Phys. Rev. **31**, 867–875 (1928)

13.69 N.F. Viemeister, S.P. Bacon: Intensity discrimination, increment detection, and magnitude estimation for 1-kHz tones, J. Acoust. Soc. Am. **84**, 172–178 (1988)

13.70 S.P. Bacon, N.F. Viemeister: Temporal modulation transfer functions in normal-hearing and hearing-impaired subjects, Audiology **24**, 117–134 (1985)

13.71 N.F. Viemeister, C.J. Plack: Time analysis. In: *Human Psychophysics*, ed. by W.A. Yost, A.N. Popper, R.R. Fay (Springer, New York 1993)

13.72 R. Plomp: The rate of decay of auditory sensation, J. Acoust. Soc. Am. **36**, 277–282 (1964)

13.73 M.J. Penner: Detection of temporal gaps in noise as a measure of the decay of auditory sensation, J. Acoust. Soc. Am. **61**, 552–557 (1977)

13.74 D. Ronken: Monaural detection of a phase difference between clicks, J. Acoust. Soc. Am. **47**, 1091–1099 (1970)

13.75 D.M. Green: Temporal acuity as a function of frequency, J. Acoust. Soc. Am. **54**, 373–379 (1973)

13.76 B.C.J. Moore, R.W. Peters, B.R. Glasberg: Detection of temporal gaps in sinusoids: Effects of frequency and level, J. Acoust. Soc. Am. **93**, 1563–1570 (1993)

13.77 A. Kohlrausch, R. Fassel, T. Dau: The influence of carrier level and frequency on modulation and beat-detection thresholds for sinusoidal carriers, J. Acoust. Soc. Am. **108**, 723–734 (2000)

13.78 B.C.J. Moore, B.R. Glasberg: Temporal modulation transfer functions obtained using sinusoidal carriers with normally hearing and hearing-impaired listeners, J. Acoust. Soc. Am. **110**, 1067–1073 (2001)

13.79 H. Fleischer: Modulationsschwellen von Schmalbandrauschen, Acustica **51**, 154–161 (1982)

13.80 T. Dau, B. Kollmeier, A. Kohlrausch: Modeling auditory processing of amplitude modulation: I. Detection and masking with narrowband carriers, J. Acoust. Soc. Am. **102**, 2892–2905 (1997)

13.81 T. Dau, B. Kollmeier, A. Kohlrausch: Modeling auditory processing of amplitude modulation: II. Spectral and temporal integration, J. Acoust. Soc. Am. **102**, 2906–2919 (1997)

13.82 T. Dau, J.L. Verhey, A. Kohlrausch: Intrinsic envelope fluctuations and modulation-detection thresholds for narrow-band noise carriers, J. Acoust. Soc. Am. **106**, 2752–2760 (1999)

13.83 N.F. Viemeister: Temporal modulation transfer functions based on modulation thresholds, J. Acoust. Soc. Am. **66**, 1364–1380 (1979)

13.84 B.C.J. Moore, B.R. Glasberg, C.J. Plack, A.K. Biswas: The shape of the ear's temporal window, J. Acoust. Soc. Am. **83**, 1102–1116 (1988)

13.85 R.H. Kay: Hearing of modulation in sounds, Physiol. Rev. **62**, 894–975 (1982)

13.86 T. Houtgast: Frequency selectivity in amplitude-modulation detection, J. Acoust. Soc. Am. **85**, 1676–1680 (1989)

13.87 S.P. Bacon, D.W. Grantham: Modulation masking: effects of modulation frequency, depth and phase, J. Acoust. Soc. Am. **85**, 2575–2580 (1989)

13.88 S.D. Ewert, T. Dau: Characterizing frequency selectivity for envelope fluctuations, J. Acoust. Soc. Am. **108**, 1181–1196 (2000)

13.89 C. Lorenzi, C. Soares, T. Vonner: Second-order temporal modulation transfer functions, J. Acoust. Soc. Am. **110**, 1030–1038 (2001)

13.90 A. Sek, B.C.J. Moore: Testing the concept of a modulation filter bank: The audibility of component modulation and detection of phase change in three-component modulators, J. Acoust. Soc. Am. **113**, 2801–2811 (2003)

13.91 C.D. Creelman: Human discrimination of auditory duration, J. Acoust. Soc. Am. **34**, 582–593 (1962)

13.92 S.M. Abel: Duration discrimination of noise and tone bursts, J. Acoust. Soc. Am. **51**, 1219–1223 (1972)

13.93 S.M. Abel: Discrimination of temporal gaps, J. Acoust. Soc. Am. **52**, 519–524 (1972)

13.94 P.L. Divenyi, W.F. Danner: Discrimination of time intervals marked by brief acoustic pulses of various intensities and spectra, Percept. Psychophys. **21**, 125–142 (1977)

13.95 J.H. Patterson, D.M. Green: Discrimination of transient signals having identical energy spectra, J. Acoust. Soc. Am. **48**, 894–905 (1970)

13.96 J. Zera, D.M. Green: Detecting temporal onset and offset asynchrony in multicomponent complexes, J. Acoust. Soc. Am. **93**, 1038–1052 (1993)

13.97 J.C. Risset, D.L. Wessel: Exploration of timbre by analysis and synthesis. In: *The Psychology of Music*, 2nd edn., ed. by D. Deutsch (Academic, San Diego 1999)

13.98 G. von Békésy: *Experiments in Hearing* (McGraw-Hill, New York 1960)

13.99 J.F. Schouten: The residue and the mechanism of hearing, Proc. Kon. Ned. Akad. Wetenschap. **43**, 991–999 (1940)

13.100 W.M. Siebert: Frequency discrimination in the auditory system: place or periodicity mechanisms, Proc. IEEE **58**, 723–730 (1970)

13.101 E. Zwicker: Masking and psychological excitation as consequences of the ear's frequency analysis. In: *Frequency Analysis and Periodicity Detection in Hearing*, ed. by R. Plomp, G.F. Smoorenburg (Sijthoff, Leiden 1970)

13.102 C.C. Wier, W. Jesteadt, D.M. Green: Frequency discrimination as a function of frequency and sensation level, J. Acoust. Soc. Am. **61**, 178–184 (1977)

13.103 A. Sek, B.C.J. Moore: Frequency discrimination as a function of frequency, measured in several ways, J. Acoust. Soc. Am. **97**, 2479–2486 (1995)

13.104 B.C.J. Moore: Relation between the critical bandwidth and the frequency-difference limen, J. Acoust. Soc. Am. **55**, 359 (1974)

13.105 J.L. Goldstein, P. Srulovicz: Auditory-nerve spike intervals as an adequate basis for aural frequency measurement. In: *Psychophysics and Physiology of Hearing*, ed. by E.F. Evans, J.P. Wilson (Academic, London 1977)

13.106 B.C.J. Moore, A. Sek: Detection of frequency modulation at low modulation rates: Evidence for a mechanism based on phase locking, J. Acoust. Soc. Am. **100**, 2320–2331 (1996)

13.107 G. Revesz: *Zur Grundlegung der Tonpsychologie* (Veit, Leipzig 1913)

13.108 A. Bachem: Tone height and tone chroma as two different pitch qualities, Acta Psych. **7**, 80–88 (1950)

13.109 W.D. Ward: Subjective musical pitch, J. Acoust. Soc. Am. **26**, 369–380 (1954)

13.110 F. Attneave, R.K. Olson: Pitch as a medium: A new approach to psychophysical scaling, Am. J. Psychol. **84**, 147–166 (1971)

13.111 K. Ohgushi, T. Hatoh: Perception of the musical pitch of high frequency tones. In: *Ninth International Symposium on Hearing: Auditory Physiology and Perception*, ed. by Y. Cazals, L. Demany, K. Horner (Pergamon, Oxford 1991)

13.112 S.S. Stevens: The relation of pitch to intensity, J. Acoust. Soc. Am. **6**, 150–154 (1935)

13.113 J. Verschuure, A.A. van Meeteren: The effect of intensity on pitch, Acustica **32**, 33–44 (1975)

13.114 G.S. Ohm: Über die Definition des Tones, nebst daran geknüpfter Theorie der Sirene und ähnlicher tonbildender Vorrichtungen, Annalen der Physik und Chemie **59**, 513–565 (1843)

13.115 J.F. Schouten: The residue revisited. In: *Frequency Analysis and Periodicity Detection in Hearing*, ed. by R. Plomp, G.F. Smoorenburg (Sijthoff, Leiden, The Netherlands 1970)

13.116 J.C.R. Licklider: Auditory frequency analysis. In: *Information Theorie*, ed. by C. Cherry (Academic, New York 1956)

13.117 E. de Boer: On the 'residue' in hearing, Ph.D. Thesis (University of Amsterdam, Amsterdam 1956)

13.118 J.L. Goldstein: An optimum processor theory for the central formation of the pitch of complex tones, J. Acoust. Soc. Am. **54**, 1496–1516 (1973)

13.119 E. Terhardt: Pitch, consonance, and harmony, J. Acoust. Soc. Am. **55**, 1061–1069 (1974)

13.120 G.F. Smoorenburg: Pitch perception of two-frequency stimuli, J. Acoust. Soc. Am. **48**, 924–941 (1970)

13.121 A.J.M. Houtsma, J.F.M. Fleuren: Analytic and synthetic pitch of two-tone complexes, J. Acoust. Soc. Am. **90**, 1674–1676 (1991)

13.122 J.L. Flanagan, M.G. Saslow: Pitch discrimination for synthetic vowels, J. Acoust. Soc. Am. **30**, 435–442 (1958)

13.123 B.C.J. Moore, B.R. Glasberg, M.J. Shailer: Frequency and intensity difference limens for harmonics within complex tones, J. Acoust. Soc. Am. **75**, 550–561 (1984)

13.124 B.C.J. Moore, B.R. Glasberg, R.W. Peters: Relative dominance of individual partials in determining the pitch of complex tones, J. Acoust. Soc. Am. **77**, 1853–1860 (1985)

13.125 R. Plomp: Pitch of complex tones, J. Acoust. Soc. Am. **41**, 1526–1533 (1967)

13.126 R.J. Ritsma: Frequencies dominant in the perception of the pitch of complex sounds, J. Acoust. Soc. Am. **42**, 191–198 (1967)

13.127 R.P. Carlyon, T.M. Shackleton: Comparing the fundamental frequencies of resolved and unresolved harmonics: Evidence for two pitch mechanisms?, J. Acoust. Soc. Am. **95**, 3541–3554 (1994)

13.128 A. Hoekstra, R.J. Ritsma: Perceptive hearing loss and frequency selectivity. In: *Psychophysics and Physiology of Hearing*, ed. by E.F. Evans, J.P. Wilson (Academic, London, England 1977)

13.129 A.J.M. Houtsma, J. Smurzynski: Pitch identification and discrimination for complex tones with many harmonics, J. Acoust. Soc. Am. **87**, 304–310 (1990)

13.130 B.C.J. Moore, S.M. Rosen: Tune recognition with reduced pitch and interval information, Q. J. Exp. Psychol. **31**, 229–240 (1979)

13.131 R. Meddis, M. Hewitt: A computational model of low pitch judgement. In: *Basic Issues in Hearing*, ed. by H. Duifhuis, J.W. Horst, H.P. Wit (Academic, London 1988)

13.132 R. Meddis, M. Hewitt: Virtual pitch and phase sensitivity of a computer model of the auditory periphery. I: Pitch identification, J. Acoust. Soc. Am. **89**, 2866–2882 (1991)

13.133 B.C.J. Moore: *An Introduction to the Psychology of Hearing*, 2nd edn. (Academic, London 1982)

13.134 B.C.J. Moore: *An Introduction to the Psychology of Hearing*, 5th edn. (Academic, San Diego 2003)

13.135 P. Srulovicz, J.L. Goldstein: A central spectrum model: a synthesis of auditory-nerve timing and place cues in monaural communication of frequency spectrum, J. Acoust. Soc. Am. **73**, 1266–1276 (1983)

13.136 R. Plomp: Timbre as a multidimensional attribute of complex tones. In: *Frequency Analysis and Periodicity Detection in Hearing*, ed. by R. Plomp, G.F. Smoorenburg (Sijthoff, Leiden 1970)

13.137 G. von Bismarck: Sharpness as an attribute of the timbre of steady sounds, Acustica **30**, 159–172 (1974)

13.138 R. Plomp: *Aspects of Tone Sensation* (Academic, London 1976)

13.139 R.D. Patterson: A pulse ribbon model of monaural phase perception, J. Acoust. Soc. Am. **82**, 1560–1586 (1987)

13.140 R. Plomp, H.J.M. Steeneken: Effect of phase on the timbre of complex tones, J. Acoust. Soc. Am. **46**, 409–421 (1969)

13.141 A.J. Watkins: Central, auditory mechanisms of perceptual compensation for spectral-envelope distortion, J. Acoust. Soc. Am. **90**, 2942–2955 (1991)

13.142 J.F. Schouten: The perception of timbre, 6th International Conference on Acoustics 1, GP-6-2 (1968)

13.143 R.D. Patterson: The sound of a sinusoid: Spectral models, J. Acoust. Soc. Am. **96**, 1409–1418 (1994)

13.144 R.D. Patterson: The sound of a sinusoid: Time-interval models, J. Acoust. Soc. Am. **96**, 1419–1428 (1994)

13.145 M.A. Akeroyd, R.D. Patterson: Discrimination of wideband noises modulated by a temporally

13.145 asymmetric function, J. Acoust. Soc. Am. **98**, 2466–2474 (1995)
13.146 H.F. Pollard, E.V. Jansson: A tristimulus method for the specification of musical timbre, Acustica **51**, 162–171 (1982)
13.147 S. Handel: Timbre perception and auditory object identification. In: *Hearing*, ed. by B.C.J. Moore (Academic, San Diego 1995)
13.148 A.W. Mills: On the minimum audible angle, J. Acoust. Soc. Am. **30**, 237–246 (1958)
13.149 L. Rayleigh: On our perception of sound direction, Phil. Mag. **13**, 214–232 (1907)
13.150 E.R. Hafter: Spatial hearing and the duplex theory: How viable?. In: *Dynamic Aspects of Neocortical Function*, ed. by G.M. Edelman, W.E. Gall, W.M. Cowan (Wiley, New York 1984)
13.151 G.B. Henning: Detectability of interaural delay in high-frequency complex waveforms, J. Acoust. Soc. Am. **55**, 84–90 (1974)
13.152 D.W. Batteau: The role of the pinna in human localization, Proc. Roy. Soc. B. **168**, 158–180 (1967)
13.153 J. Blauert: *Spatial Hearing: The Psychophysics of Human Sound Localization* (MIT Press, Cambridge, Mass 1997)
13.154 W.M. Hartmann, A. Wittenberg: On the externalization of sound images, J. Acoust. Soc. Am. **99**, 3678–3688 (1996)
13.155 H. Haas: Über den Einfluss eines Einfachechos an die Hörsamkeit von Sprache, Acustica **1**, 49–58 (1951)
13.156 H. Wallach, E.B. Newman, M.R. Rosenzweig: The precedence effect in sound localization, Am. J. Psychol. **62**, 315–336 (1949)
13.157 R.Y. Litovsky, H.S. Colburn, W.A. Yost, S.J. Guzman: The precedence effect, J. Acoust. Soc. Am. **106**, 1633–1654 (1999)
13.158 A.S. Bregman: *Auditory Scene Analysis: The Perceptual Organization of Sound* (Bradford Books, MIT Press, Cambridge, Mass. 1990)
13.159 A.S. Bregman, S. Pinker: Auditory streaming and the building of timbre, Canad. J. Psychol. **32**, 19–31 (1978)
13.160 C.J. Darwin, R.P. Carlyon: Auditory grouping. In: *Hearing*, ed. by B.C.J. Moore (Academic, San Diego 1995)
13.161 D.E. Broadbent, P. Ladefoged: On the fusion of sounds reaching different sense organs, J. Acoust. Soc. Am. **29**, 708–710 (1957)
13.162 M.T.M. Scheffers: *Sifting vowels: auditory pitch analysis and sound segregation*, Ph.D. Thesis (Groningen University, The Netherlands 1983)
13.163 P.F. Assmann, A.Q. Summerfield: Modeling the perception of concurrent vowels: Vowels with different fundamental frequencies, J. Acoust. Soc. Am. **88**, 680–697 (1990)
13.164 J.D. McKeown, R.D. Patterson: The time course of auditory segregation: Concurrent vowels that vary in duration, J. Acoust. Soc. Am. **98**, 1866–1877 (1995)
13.165 B.C.J. Moore, B.R. Glasberg, R.W. Peters: Thresholds for hearing mistuned partials as separate tones in harmonic complexes, J. Acoust. Soc. Am. **80**, 479–483 (1986)
13.166 B. Roberts, J.M. Brunstrom: Perceptual segregation and pitch shifts of mistuned components in harmonic complexes and in regular inharmonic complexes, J. Acoust. Soc. Am. **104**, 2326–2338 (1998)
13.167 B. Roberts, J.M. Brunstrom: Perceptual fusion and fragmentation of complex tones made inharmonic by applying different degrees of frequency shift and spectral stretch, J. Acoust. Soc. Am. **110**, 2479–2490 (2001)
13.168 R. Meddis, M. Hewitt: Modeling the identification of concurrent vowels with different fundamental frequencies, J. Acoust. Soc. Am. **91**, 233–245 (1992)
13.169 A. de Cheveigné, S. McAdams, C.M.H. Marin: Concurrent vowel identification. II. Effects of phase, harmonicity and task, J. Acoust. Soc. Am. **101**, 2848–2856 (1997)
13.170 A. de Cheveigné, H. Kawahara, M. Tsuzaki, K. Aikawa: Concurrent vowel identification. I. Effects of relative amplitude and F0 difference, J. Acoust. Soc. Am. **101**, 2839–2847 (1997)
13.171 A. de Cheveigné: Concurrent vowel identification. III. A neural model of harmonic interference cancellation, J. Acoust. Soc. Am. **101**, 2857–2865 (1997)
13.172 R.A. Rasch: The perception of simultaneous notes such as in polyphonic music, Acustica **40**, 21–33 (1978)
13.173 C.J. Darwin, N.S. Sutherland: Grouping frequency components of vowels: when is a harmonic not a harmonic?, Q. J. Exp. Psychol. **36A**, 193–208 (1984)
13.174 B. Roberts, B.C.J. Moore: The influence of extraneous sounds on the perceptual estimation of first-formant frequency in vowels under conditions of asynchrony, J. Acoust. Soc. Am. **89**, 2922–2932 (1991)
13.175 E. Zwicker: 'Negative afterimage' in hearing, J. Acoust. Soc. Am. **36**, 2413–2415 (1964)
13.176 A.Q. Summerfield, A.S. Sidwell, T. Nelson: Auditory enhancement of changes in spectral amplitude, J. Acoust. Soc. Am. **81**, 700–708 (1987)
13.177 A.S. Bregman, J. Abramson, P. Doehring, C.J. Darwin: Spectral integration based on common amplitude modulation, Percept. Psychophys. **37**, 483–493 (1985)
13.178 J.W. Hall, J.H. Grose: Comodulation masking release and auditory grouping, J. Acoust. Soc. Am. **88**, 119–125 (1990)
13.179 B.C.J. Moore, M.J. Shailer: Comodulation masking release as a function of level, J. Acoust. Soc. Am. **90**, 829–835 (1991)

13.180 B.C.J. Moore, M.J. Shailer, M.J. Black: Dichotic interference effects in gap detection, J. Acoust. Soc. Am. **93**, 2130–2133 (1993)

13.181 Q. Summerfield, J.F. Culling: Auditory segregation of competing voices: absence of effects of FM or AM coherence, Phil. Trans. R. Soc. Lond. B **336**, 357–366 (1992)

13.182 M.F. Cohen, X. Chen: Dynamic frequency change among stimulus components: Effects of coherence on detectability, J. Acoust. Soc. Am. **92**, 766–772 (1992)

13.183 M.H. Chalikia, A.S. Bregman: The perceptual segregation of simultaneous vowels with harmonic, shifted, and random components, Percept. Psychophys. **53**, 125–133 (1993)

13.184 S. McAdams: Segregation of concurrent sounds. I.: Effects of frequency modulation coherence, J. Acoust. Soc. Am. **86**, 2148–2159 (1989)

13.185 R.P. Carlyon: Discriminating between coherent and incoherent frequency modulation of complex tones, J. Acoust. Soc. Am. **89**, 329–340 (1991)

13.186 R.P. Carlyon: Further evidence against an across-frequency mechanism specific to the detection of frequency modulation (FM) incoherence between resolved frequency components, J. Acoust. Soc. Am. **95**, 949–961 (1994)

13.187 J. Lyzenga, B.C.J. Moore: Effect of FM coherence for inharmonic stimuli: FM-phase discrimination and identification of artificial double vowels, J. Acoust. Soc. Am. **117**, 1314–1325 (2005)

13.188 C.M.H. Marin, S. McAdams: Segregation of concurrent sounds. II: Effects of spectral envelope tracing, frequency modulation coherence, and frequency modulation width, J. Acoust. Soc. Am. **89**, 341–351 (1991)

13.189 S. Furukawa, B.C.J. Moore: Across-frequency processes in frequency modulation detection, J. Acoust. Soc. Am. **100**, 2299–2312 (1996)

13.190 S. Furukawa, B.C.J. Moore: Dependence of frequency modulation detection on frequency modulation coherence across carriers: Effects of modulation rate, harmonicity and roving of the carrier frequencies, J. Acoust. Soc. Am. **101**, 1632–1643 (1997)

13.191 S. Furukawa, B.C.J. Moore: Effect of the relative phase of amplitude modulation on the detection of modulation on two carriers, J. Acoust. Soc. Am. **102**, 3657–3664 (1997)

13.192 R.P. Carlyon: Detecting coherent and incoherent frequency modulation, Hear. Res. **140**, 173–188 (2000)

13.193 M. Kubovy, J.E. Cutting, R.M. McGuire: Hearing with the third ear: dichotic perception of a melody without monaural familiarity cues, Science **186**, 272–274 (1974)

13.194 J.F. Culling: Auditory motion segregation: a limited analogy with vision, J. Exp. Psychol.: Human Percept. Perf. **26**, 1760–1769 (2000)

13.195 M.A. Akeroyd, B.C.J. Moore, G.A. Moore: Melody recognition using three types of dichotic-pitch stimulus, J. Acoust. Soc. Am. **110**, 1498–1504 (2001)

13.196 T.M. Shackleton, R. Meddis: The role of interaural time difference and fundamental frequency difference in the identification of concurrent vowel pairs, J. Acoust. Soc. Am. **91**, 3579–3581 (1992)

13.197 J.F. Culling, Q. Summerfield: Perceptual separation of concurrent speech sounds: Absence of across-frequency grouping by common interaural delay, J. Acoust. Soc. Am. **98**, 785–797 (1995)

13.198 C.J. Darwin, R.W. Hukin: Auditory objects of attention: the role of interaural time differences, J. Exp. Psychol.: Human Percept. Perf. **25**, 617–629 (1999)

13.199 G.A. Miller, G.A. Heise: The trill threshold, J. Acoust. Soc. Am. **22**, 637–638 (1950)

13.200 A.S. Bregman, J. Campbell: Primary auditory stream segregation and perception of order in rapid sequences of tones, J. Exp. Psychol. **89**, 244–249 (1971)

13.201 L.P.A.S. van Noorden: *Temporal coherence in the perception of tone sequences*, Ph.D. Thesis (Eindhoven University of Technology, Eindhoven 1975)

13.202 L.P.A.S. van Noorden: Rhythmic fission as a function of tone rate, IPO Annual Prog. Rep. **6**, 9–12 (1971)

13.203 A.S. Bregman, G. Dannenbring: The effect of continuity on auditory stream segregation, Percept. Psychophys. **13**, 308–312 (1973)

13.204 A.S. Bregman: Auditory streaming is cumulative, J. Exp. Psychol.: Human Percept. Perf. **4**, 380–387 (1978)

13.205 W.L. Rogers, A.S. Bregman: An experimental evaluation of three theories of auditory stream segregation, Percept. Psychophys. **53**, 179–189 (1993)

13.206 W.L. Rogers, A.S. Bregman: Cumulation of the tendency to segregate auditory streams: resetting by changes in location and loudness, Percept. Psychophys. **60**, 1216–1227 (1998)

13.207 B.C.J. Moore, H. Gockel: Factors influencing sequential stream segregation, Acta Acustica - Acustica **88**, 320–333 (2002)

13.208 W.M. Hartmann, D. Johnson: Stream segregation and peripheral channeling, Music Percept. **9**, 155–184 (1991)

13.209 P.G. Singh, A.S. Bregman: The influence of different timbre attributes on the perceptual segregation of complex-tone sequences, J. Acoust. Soc. Am. **102**, 1943–1952 (1997)

13.210 J. Vliegen, B.C.J. Moore, A.J. Oxenham: The role of spectral and periodicity cues in auditory stream segregation, measured using a temporal discrimination task, J. Acoust. Soc. Am. **106**, 938–945 (1999)

13.211 J. Vliegen, A.J. Oxenham: Sequential stream segregation in the absence of spectral cues, J. Acoust. Soc. Am. **105**, 339–346 (1999)

13.212 P. Iverson: Auditory stream segregation by musical timbre: effects of static and dynamic acoustic attributes, J. Exp. Psychol.: Human Percept. Perf. **21**, 751–763 (1995)

13.213 B. Roberts, B.R. Glasberg, B.C.J. Moore: Primitive stream segregation of tone sequences without differences in F0 or passband, J. Acoust. Soc. Am. **112**, 2074–2085 (2002)

13.214 H. Gockel, R.P. Carlyon, C. Micheyl: Context dependence of fundamental-frequency discrimination: Lateralized temporal fringes, J. Acoust. Soc. Am. **106**, 3553–3563 (1999)

13.215 W.J. Dowling: Rhythmic fission and perceptual organization, J. Acoust. Soc. Am. **44**, 369 (1968)

13.216 W.J. Dowling: The perception of interleaved melodies, Cognitive Psychol. **5**, 322–337 (1973)

13.217 D. Deutsch: Two-channel listening to musical scales, J. Acoust. Soc. Am. **57**, 1156–1160 (1975)

13.218 D.E. Broadbent, P. Ladefoged: Auditory perception of temporal order, J. Acoust. Soc. Am. **31**, 151–159 (1959)

13.219 R.M. Warren, C.J. Obusek, R.M. Farmer, R.P. Warren: Auditory sequence: confusion of patterns other than speech or music, Science N.Y. **164**, 586–587 (1969)

13.220 R.M. Warren: Auditory temporal discrimination by trained listeners, Cognitive Psychol. **6**, 237–256 (1974)

13.221 P.L. Divenyi, I.J. Hirsh: Identification of temporal order in three-tone sequences, J. Acoust. Soc. Am. **56**, 144–151 (1974)

13.222 R. Cusack, B. Roberts: Effects of differences in timbre on sequential grouping, Percept. Psychophys. **62**, 1112–1120 (2000)

13.223 K. Koffka: *Principles of Gestalt Psychology* (Harcourt and Brace, New York 1935)

13.224 C.J. Darwin, C.E. Bethell-Fox: Pitch continuity and speech source attribution, J. Exp. Psychol.: Hum. Perc. Perf. **3**, 665–672 (1977)

13.225 A.S. Bregman, A. Rudnicky: Auditory segregation: stream or streams?, J. Exp. Psychol.: Human Percept. Perf. **1**, 263–267 (1975)

13.226 A.S. Bregman: The meaning of duplex perception: sounds as transparent objects. In: *The Psychophysics of Speech Perception*, ed. by M.E.H. Schouten (Martinus Nijhoff, Dordrecht 1987)

13.227 T. Houtgast: Psychophysical evidence for lateral inhibition in hearing, J. Acoust. Soc. Am. **51**, 1885–1894 (1972)

13.228 W.R. Thurlow: An auditory figure-ground effect, Am. J. Psychol. **70**, 653–654 (1957)

13.229 R.M. Warren, C.J. Obusek, J.M. Ackroff: Auditory induction: perceptual synthesis of absent sounds, Science **176**, 1149–1151 (1972)

13.230 V. Ciocca, A.S. Bregman: Perceived continuity of gliding and steady-state tones through interrupting noise, Percept. Psychophys. **42**, 476–484 (1987)

13.231 G.A. Miller, J.C.R. Licklider: The intelligibility of interrupted speech, J. Acoust. Soc. Am. **22**, 167–173 (1950)

13.232 D. Dirks, D. Bower: Effects of forward and backward masking on speech intelligibility, J. Acoust. Soc. Am. **47**, 1003–1008 (1970)

13.233 R.M. Warren: Perceptual restoration of missing speech sounds, Science **167**, 392–393 (1970)

14. Acoustic Signal Processing

Signal processing refers to the acquisition, storage, display, and generation of signals – also to the extraction of information from signals and the re-encoding of information. As such, signal processing in some form is an essential element in the practice of all aspects of acoustics. Signal processing algorithms enable acousticians to separate signals from noise, to perform automatic speech recognition, or to compress information for more efficient storage or transmission. Signal processing concepts are the blocks used to build models of speech and hearing. As we enter the 21st century, all signal processing is effectively digital signal processing. Widespread access to high-speed processing, massive memory, and inexpensive software makes signal processing procedures of enormous sophistication and power available to anyone who wants to use them. Because advanced signal processing is now accessible to everybody, there is a need for primers that introduce basic mathematical concepts that underlie the digital algorithms. The present handbook chapter is intended to serve such a purpose.

The chapter emphasizes careful definition of essential terms used in the description of signals per international standards. It introduces the Fourier series for signals that are periodic and the Fourier transform for signals that are not. Both begin with analog, continuous signals, appropriate for the real acoustical world. Emphasis is placed on the consequences of signal symmetry and on formal relationships. The autocorrelation function is related to the energy and power spectra for finite-duration and infinite-duration signals. The chapter provides careful definitions of statistical terms, moments, and single- and multi-variate distributions. The Hilbert transform is introduced, again in terms of continuous functions. It is applied both to the development of the analytic signal – envelope and phase – and to the dispersion relations for linear, time-invariant systems. The bare essentials of filtering are presented, mostly to provide real-world examples of fundamental concepts – asymptotic responses, group delay, phase delay, etc. There is a brief introduction to cepstrology, with emphasis on acoustical applications. The treatment of the mathematical properties of noise emphasizes the generation of different kinds of noise. Digital signal processing with sampled data is specifically introduced with emphasis on digital-to-analog conversion and analog-to-digital conversion. It continues with the discrete Fourier transform and with the z-transform, applied to both signals and linear, time-invariant systems. Digital signal processing continues with an introduction to maximum length sequences as used in acoustical measurements, with an emphasis on formal properties. The chapter ends with a section on information theory including developments of Shannon entropy and mutual information.

14.1	**Definitions**	504
14.2	**Fourier Series**	505
	14.2.1 The Spectrum	506
	14.2.2 Symmetry	506
14.3	**Fourier Transform**	507
	14.3.1 Examples	508
	14.3.2 Time-Shifted Function	509
	14.3.3 Derivatives and Integrals	509
	14.3.4 Products and Convolution	509
14.4	**Power, Energy, and Power Spectrum**	510
	14.4.1 Autocorrelation	510
	14.4.2 Cross-Correlation	511
14.5	**Statistics**	511
	14.5.1 Signals and Processes	512
	14.5.2 Distributions	512
	14.5.3 Multivariate Distributions	513
	14.5.4 Moments	513
14.6	**Hilbert Transform and the Envelope**	514
	14.6.1 The Analytic Signal	514
14.7	**Filters**	515
	14.7.1 One-Pole Low-Pass Filter	515
	14.7.2 Phase Delay and Group Delay	516
	14.7.3 Resonant Filters	516
	14.7.4 Impulse Response	516
	14.7.5 Dispersion Relations	516

- 14.8 **The Cepstrum** 517
- 14.9 **Noise** 518
  - 14.9.1 Thermal Noise 518
  - 14.9.2 Gaussian Noise 519
  - 14.9.3 Band-Limited Noise 519
  - 14.9.4 Generating Noise 519
  - 14.9.5 Equal-Amplitude Random-Phase Noise 520
  - 14.9.6 Noise Color 520
- 14.10 **Sampled data** 520
  - 14.10.1 Quantization and Quantization Noise 520
  - 14.10.2 Binary Representation 520
  - 14.10.3 Sampling Operation 521
  - 14.10.4 Digital-to-Analog Conversion 521
  - 14.10.5 The Sampled Signal 522
  - 14.10.6 Interpolation 522
- 14.11 **Discrete Fourier Transform** 522
  - 14.11.1 Interpolation for the Spectrum ... 523
- 14.12 **The z-Transform** 524
  - 14.12.1 Transfer Function 525
- 14.13 **Maximum Length Sequences** 526
  - 14.13.1 The MLS as a Signal 527
  - 14.13.2 Application of the MLS 527
  - 14.13.3 Long Sequences 527
- 14.14 **Information Theory** 528
  - 14.14.1 Shannon Entropy 529
  - 14.14.2 Mutual Information 530
- **References** 530

14.1 Definitions

Signal processing begins with signals. The simplest signal is a sine wave with a single spectral component, i.e., with a single frequency, as shown in Fig. 14.1. It is sometimes called a pure tone. A sine wave function of time t with *amplitude* C, *angular frequency* ω, and *starting phase* φ, is given by

$$x(t) = C \sin(\omega t + \varphi). \tag{14.1}$$

The amplitude has the same units as the waveform x, the angular frequency has units of radians per second, and the phase has units of radians.

Because there are 2π radians in one cycle

$$\omega = 2\pi f, \tag{14.2}$$

and (14.1) can be written as

$$x(t) = C \sin(2\pi f t + \varphi) \tag{14.3}$$

or as

$$x(t) = C \sin(2\pi t/T + \varphi), \tag{14.4}$$

where f is the *frequency* in cycles per second (or Hertz) and T is the *period* in units of seconds per cycle, $T = 1/f$.

A *complex wave* is the sum of two or more sine waves, each with its own amplitude, frequency, and phase. For example,

$$x(t) = C_1 \sin(\omega_1 t + \varphi_1) + C_2 \sin(\omega_2 t + \varphi_2) \tag{14.5}$$

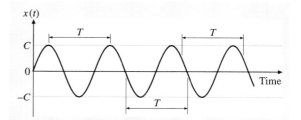

Fig. 14.1 A sine wave with amplitude C and period T. A little more than three and a half cycles are shown. The starting phase is $\varphi = 0$

is a complex wave with two spectral components having frequencies f_1 and f_2. The period of a complex wave is the reciprocal of the greatest common divisor of f_1 and f_2. For instance, if $f_1 = 400$ Hz and $f_2 = 600$ Hz, then the period is $1/(200\,\text{Hz})$ or 5 ms. The *fundamental frequency* is the reciprocal of the period.

A general waveform can be written as a sum of N components,

$$x(t) = \sum_{n=1}^{N} C_n \sin(\omega_n t + \varphi_n), \tag{14.6}$$

and the fundamental frequency is the greatest common divisor of the set of frequencies $\{f_n\}$.

An alternative description of the general waveform can be derived by using the trigonometric identity

$$\sin(\theta_1 + \theta_2) = \sin\theta_1 \cos\theta_2 + \sin\theta_2 \cos\theta_1 \tag{14.7}$$

so that

$$x(t) = \sum_{n=1}^{N} A_n \cos(\omega_n t) + B_n \sin(\omega_n t), \quad (14.8)$$

where $A_n = C_n \sin \varphi_n$, and $B_n = C_n \cos \varphi_n$, are the cosine and sine partial amplitudes respectively. Thus the two parameters C_n and φ_n are replaced by two other parameters A_n and B_n.

Because of the trigonometric identity

$$\sin^2 \theta + \cos^2 \theta = 1, \quad (14.9)$$

the amplitude C_n can be written in terms of the partial amplitudes,

$$C_n^2 = A_n^2 + B_n^2, \quad (14.10)$$

as can the component phase

$$\varphi_n = \text{Arg}(A_n, B_n). \quad (14.11)$$

The Arg function is essentially an inverse tangent, but because the principal value of the arctangent function only runs from $-\pi/2$ to $\pi/2$, an adjustment needs to be made when B_n is negative. In the end,

$$\text{Arg}(A_n, B_n) = \arctan(A_n/B_n) \quad (\text{for } B_n \geq 0) \quad (14.12)$$

and

$$\text{Arg}(A_n, B_n) = \arctan(A_n/B_n) + \pi \quad (\text{for } B_n < 0).$$

The remaining sections of this chapter provide a brief treatment of real signals $x(t)$ – first as continuous functions of time and then as sampled data. Readers who are less familiar with the continuous approach may wish to refer to the more extensive treatment in [14.1].

14.2 Fourier Series

The Fourier series applies to a function $x(t)$ that is periodic. Periodicity means that we can add any integral multiple m of T to the running time variable t and the function will have the same value as at time t, i.e.

$$x(t + mT) = x(t), \quad \text{for all integral } m. \quad (14.13)$$

Because m can be either positive or negative and as large as we like, it is clear that x is periodic into the infinite future and past. Then Fourier's theorem says that x can be represented as a Fourier series like

$$x(t) = A_0 + \sum_{n=1}^{\infty} [A_n \cos(\omega_n t) + B_n \sin(\omega_n t)]. \quad (14.14)$$

All the cosines and sines have angular frequencies ω_n that are harmonics, i.e., they are integral multiples of a fundamental angular frequency ω_o,

$$\omega_n = n\omega_o = 2\pi n/T, \quad (14.15)$$

where n is an integer.

The fundamental frequency f_0 is given by $f_0 = \omega_o/(2\pi)$. The fundamental frequency is the lowest frequency that a sine or cosine wave can have and still fit exactly into one period of the function $x(t)$ because $f_0 = 1/T$. In order to make a function $x(t)$ with period T, the only sines and cosines that are allowed to enter the sum are those that fit exactly into the same period T. These are those sines and cosines with frequencies that are integral multiples of the fundamental.

The factors A_n and B_n in (14.14) are the Fourier coefficients. They can be calculated by projecting the function $x(t)$ onto sine and cosine functions of the harmonic frequencies ω_n. Projecting means to integrate the product of $x(t)$ and a sine or cosine function over a duration of time equal to a period of $x(t)$. Sines and cosines with different harmonic frequencies are orthogonal over a period. Consequently, projecting $x(t)$ onto, for example $\cos(3\omega_o t)$, gives exactly the Fourier coefficient A_3.

It does not matter which time interval is used for integration, as long as it is exactly one period in duration. It is common to use the interval $-T/2$ to $T/2$.

The orthogonality and normality of the sine and cosine functions are described by the following equations:

$$\frac{2}{T} \int_{-T/2}^{T/2} dt \, \sin(\omega_n t) \cos(\omega_m t) = 0, \quad (14.16)$$

for all m and n;

$$\frac{2}{T} \int_{-T/2}^{T/2} dt \, \cos(\omega_n t) \cos(\omega_m t) = \delta_{n,m} \quad (14.17)$$

and

$$\frac{2}{T} \int_{-T/2}^{T/2} dt \, \sin(\omega_n t) \sin(\omega_m t) = \delta_{n,m}, \quad (14.18)$$

where $\delta_{n,m}$ is the Kronecker delta, equal to one if $m = n$ and equal to zero otherwise.

It follows that the equations for A_n and B_n are

$$A_n = \frac{2}{T} \int_{-T/2}^{T/2} dt\, x(t) \cos(\omega_n t) \quad \text{for } n > 0, \quad (14.19)$$

$$B_n = \frac{2}{T} \int_{-T/2}^{T/2} dt\, x(t) \sin(\omega_n t) \quad \text{for } n > 0. \quad (14.20)$$

The coefficient A_0 is simply a constant that shifts the function $x(t)$ up or down. The constant A_0 is the only term in the Fourier series (14.14) that could possibly have a nonzero value when averaged over a period. All the other terms are sines and cosines; they are negative as much as they are positive and average to zero. Therefore, A_0 is the average value of $x(t)$. It is the direct-current (DC) component of x. To find A_0 we project the function $x(t)$ onto a cosine of zero frequency, i.e. onto the number 1, which leads to the average value of x,

$$A_0 = \frac{1}{T} \int_{-T/2}^{T/2} dt\, x(t). \quad (14.21)$$

14.2.1 The Spectrum

The Fourier series is a function of time, where A_n and B_n are coefficients that weight the cosine and sine contributions to the series. The coefficients A_n and B_n are real numbers that may be positive or negative.

An alternative approach to the function $x(t)$ deemphasizes the time dependence and considers mainly the coefficients themselves. This is the spectral approach. The spectrum simply consists of the values of A_n and B_n, plotted against frequency, or equivalently, plotted against the harmonic number n. For example, if we have a signal given by

$$x(t) = 5\,\sin(2\pi\,150t) + 3\,\cos(2\pi\,300t)$$
$$- 2\cos(2\pi\,450t) + 4\,\sin(2\pi\,450t) \quad (14.22)$$

then the spectrum consists of only a few terms. The period of the signal is $1/150$ s, the fundamental frequency is 150 Hz, and there are two additional harmonics: a second harmonic at 300 Hz and a third at 450 Hz. The spectrum is shown in Fig. 14.2.

14.2.2 Symmetry

Many important periodic functions have symmetries that simplify the Fourier series. If the function $x(t)$ is an even function $[x(-t) = x(t)]$ then the Fourier series for x contains only cosine terms. All coefficients of the sine terms B_n are zero. If $x(t)$ is odd $[x(-t) = -x(t)]$, the the Fourier series contains only sine terms, and all the coefficients A_n are zero. Sometimes it is possible to shift the origin of time to obtain a symmetrical function. Such a time shift is allowed if the physical situation at hand does not require that $x(t)$ be synchronized with some other function of time or with some other time-referenced process. For example, the sawtooth function

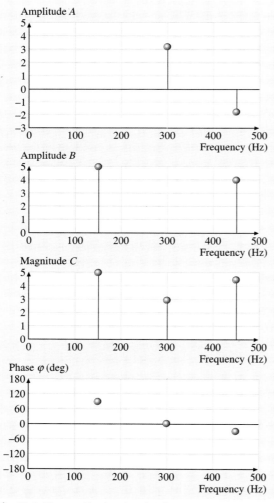

Fig. 14.2 The amplitudes A and B for the signal in (14.22) are shown in the *top* two plots. The corresponding magnitude and phases are shown in the *bottom* two

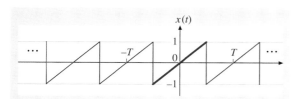

Fig. 14.3 The Fourier series of an odd function like this sawtooth consists of sine terms only. The Fourier coefficients can be computed by an integral over a single period from $-T/2$ to $T/2$

in Fig. 14.3 is an odd function. Therefore, only sine terms are present in the series.

The Fourier coefficients can be calculated by doing the integral over the interval shown by the heavy line. The integral is easy to do analytically because $x(t)$ is just a straight line. The answer is

$$B_n = \frac{2}{\pi} \frac{(-1)^{(n+1)}}{n} . \tag{14.23}$$

Consequently, the sawtooth function itself is given by

$$x(t) = \frac{2}{\pi} \sum_{n=1}^{\infty} \frac{(-1)^{(n+1)}}{n} \sin(2\pi n t/T) . \tag{14.24}$$

A bridge between the Fourier series and the Fourier transform is the complex form for the spectrum,

$$X_n = A_n + \mathrm{i} B_n . \tag{14.25}$$

Because of Euler's formula, namely

$$\mathrm{e}^{\mathrm{i}\theta} = \cos\theta + \mathrm{i}\sin\theta , \tag{14.26}$$

it follows that

$$X_n = \frac{2}{T} \int_{-T/2}^{T/2} \mathrm{d}t\, x(t)\, \mathrm{e}^{\mathrm{i}\omega_n t} . \tag{14.27}$$

14.3 Fourier Transform

The Fourier transform of a time-dependent signal is a frequency-dependent representation of the signal, whether or not the time dependence is periodic. Compared to the frequency representation in the Fourier series, the Fourier transform differs in several ways. In general the Fourier transform is a complex function with real and imaginary parts. Whereas the Fourier series representation consists of discrete frequencies, the Fourier transform is a continuous function of frequency. The Fourier transform also requires the concept of negative frequencies. The transformation tends to be symmetrical with respect to the appearance of positive and negative frequencies and so negative frequencies are just as important as positive frequencies. The treatment of the Fourier integral transform that follows mainly states results. For proof and further applications the reader may wish to consult [14.1, mostly Chap. 8].

The Fourier transform of signal $x(t)$ is given by the integral

$$X(\omega) = \mathcal{F}[x(t)] = \int \mathrm{d}t\, \mathrm{e}^{-\mathrm{i}\omega t}\, x(t) . \tag{14.28}$$

Here, and elsewhere unless otherwise noted, integrals range over all negative and positive values, i.e. $-\infty$ to $+\infty$.

The inverse Fourier transform expresses the signal as a function of time in terms of the Fourier transform,

$$x(t) = \frac{1}{2\pi} \int \mathrm{d}\omega\, \mathrm{e}^{\mathrm{i}\omega t}\, X(\omega) . \tag{14.29}$$

These expressions for the transform and inverse transform can be shown to be self-consistent. A key fact in the proof is that the Dirac delta function can be written as an integral over all time,

$$\delta(\omega) = \frac{1}{2\pi} \int \mathrm{d}t\, \mathrm{e}^{\pm \mathrm{i}\omega t} , \tag{14.30}$$

and similarly

$$\delta(t) = \frac{1}{2\pi} \int \mathrm{d}\omega\, \mathrm{e}^{\pm \mathrm{i}\omega t} . \tag{14.31}$$

Because a delta function is an even function of its argument, it does not matter if the $+$ or $-$ sign is used in these equations.

Reality and Symmetry

The Fourier transform $X(\omega)$ is generally complex. However, signals like $x(t)$ are real functions of time. In that connection (14.29) would seem to pose a problem, because it expresses the real function x as an integral involving the complex exponential multiplied by the complex Fourier transform. The requirement that x be real leads to a simple requirement on its Fourier transform X. The requirement is that $X(-\omega)$ must be the complex conjugate of $X(\omega)$, i.e., $X(-\omega) = X^*(\omega)$. That means that

$$\text{Re } X(-\omega) = \text{Re } X(\omega) \tag{14.32}$$

and

$$\text{Im } X(-\omega) = -\text{Im } X(\omega) \,.$$

Similar reasoning leads to special results for signals $x(t)$ that are even or odd functions of time t. If x is even $[x(-t) = x(t)]$ then the Fourier transform X is not complex but is entirely real. If x is odd $[x(-t) = -x(t)]$ then the Fourier transform X is not complex but is entirely imaginary.

The polar form of the Fourier transform is normally a more useful representation than the real and imaginary parts. It is the product of a magnitude, or absolute value, and an exponential phase factor,

$$X(\omega) = |X(\omega)| \exp[i\varphi(\omega)] \,. \tag{14.33}$$

The magnitude is a positive real number. Negative or complex values of X arise from the phase factor. For instance, if X is entirely real then $\varphi(\omega)$ can only be zero or 180°.

14.3.1 Examples

A few example Fourier transforms are insightful.

The Gaussian

The Fourier transform of a Gaussian is a Gaussian. The Gaussian function of time is

$$g(t) = \frac{1}{\sigma\sqrt{2\pi}} \, \mathrm{e}^{-t^2/(2\sigma^2)} \,. \tag{14.34}$$

The function is normalized to unit area, in the sense that the integral of $g(t)$ over all time is 1.0. The Fourier transform is

$$G(\omega) = \mathrm{e}^{-\omega^2\sigma^2/2} \,. \tag{14.35}$$

The Unit Rectangle Pulse

The unit rectangle pulse, $r(t)$, is a function of time that is zero except on the interval $-T_0/2$ to $T_0/2$. During that interval the function has the value $1/T_0$, so that the function has unit area. The Fourier transform of this pulse is

$$R(\omega) = [\sin(\omega T_0/2)]/(\omega T_0/2) \,, \tag{14.36}$$

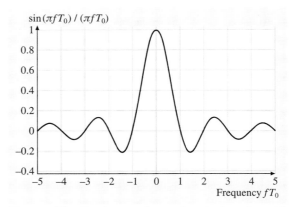

Fig. 14.4 The Fourier transform of a single pulse with duration T_0 as a function of frequency f expressed in dimensionless form fT_0

or, in terms of frequency

$$R(f) = [\sin(\pi f T_0)]/(\pi f T_0) \,,$$

as shown in Fig. 14.4.

The function of the form $(\sin x)/x$ is sometimes called the sinc function. However, $(\sin \pi x)/(\pi x)$ is also called the sinc function. Therefore, whenever the sinc function is used by name it must be defined.

Both the Gaussian and the unit rectangle illustrate a reciprocal effect sometimes called the *uncertainty principle*. The Gaussian function of time $g(t)$ is narrow if σ is small because σ appears in the denominator of the exponential in $g(t)$. Then the Fourier transform $G(\omega)$ is wide because σ appears in the numerator of the exponential in $G(\omega)$. Similarly, the unit rectangle is narrow if T_0 is small. Then the Fourier transform $R(\omega)$ is broad because $R(\omega)$ depends on the product ωT_0. The general statement of the principle is that, if a function of one variable (e.g. time) is compact, then the transform representation, that is the function of the conjugate variable (e.g. frequency), is broad, and vice versa. The extreme expression of the uncertainty principle appears in the Fourier transform of a function that is constant for all time. According to (14.30), that transform is a delta function of frequency. Conversely, the Fourier transform of a delta function is a constant for all frequency. That means that the spectrum of an ideal impulse contains all frequencies equally.

A contrast between the Fourier transforms of Gaussian and rectangle pulses is also revealing. Because the Gaussian is a smooth function of time, the transform has a single peak. Because the rectangle has sharp edges, there are oscillations in the transform. If the rectangle is given sloping or rounded edges, the amplitude of the oscillations is reduced.

14.3.2 Time-Shifted Function

If $y(t)$ is a time-shifted version of $x(t)$, i.e.

$$y(t) = x(t - t_0), \qquad (14.37)$$

then the Fourier transform of y is related to the Fourier transform of x by the equation

$$Y(\omega) = \exp(-i\omega t_0) X(\omega). \qquad (14.38)$$

The transform Y is the same as X except for a phase shift that increases linearly with frequency. There are two important implications of this equation. First, because the magnitude of the exponential with imaginary argument is 1.0, the magnitude of Y is the same as the magnitude of X for all values of ω. Second, reversing the logic of the equation shows that, if the phase of a signal is changed in such a way that the phase shift is a linear function of frequency, then the change corresponds only to a shift along the time axis for the function of time, and not to a distortion of the shape of the wave. A general phase-shift function of frequency can be separated into the best-fitting straight line and a residual. Only the residual distorts the shape of the signal as a function of time.

14.3.3 Derivatives and Integrals

If $v(t)$ is the derivative of $x(t)$, i.e., $v(t) = dx/dt$, then the Fourier transform of v is related to the transform of x by the equation

$$V(\omega) = i\omega X(\omega). \qquad (14.39)$$

Thus, differentiating a signal is equivalent to ideal high-pass filtering with a boost of 6 dB per octave, i.e., doubling the frequency doubles the ratio of the output to the input, as processed by the differentiator. Differentiating also leads to a simple phase shift of 90° ($\pi/2$ radians) in the sense that the new factor of i equals $\exp(i\pi/2)$. The differentiation equation can be iterated. The Fourier transform of the n-th derivative of $x(t)$ is $(i\omega)^n X(\omega)$.

Integration is the inverse of differentiation, and that fact becomes apparent in the Fourier transforms. If $w(t)$ is the integral of $x(t)$, i.e.,

$$w(t) = \int_{-\infty}^{t} dt' \, x(t'), \qquad (14.40)$$

then the Fourier transform of w is related to the Fourier transform of x by the equation

$$W(\omega) = X(\omega)/(i\omega) + X(0)\delta(\omega). \qquad (14.41)$$

The first term above could have been anticipated based on the transform of the derivative. The second term corresponds to the additive constant of integration that always appears in the context of an integral. The number $X(0)$ is the average (DC) value of the signal $x(t)$, and if this average value is zero then the second term can be neglected.

14.3.4 Products and Convolution

If the signal x is the product of two functions y and w, i.e. $x(t) = y(t)w(t)$ then, according to the convolution theorem, the Fourier transform of x is the convolution of the Fourier transforms of y and w, i.e.

$$X(\omega) = \frac{1}{2\pi} Y(\omega) * W(\omega). \qquad (14.42)$$

The convolution, indicated by the symbol *, is defined by the integral

$$X(\omega) = \frac{1}{2\pi} \int d\omega' \, Y(\omega') \, W(\omega - \omega'). \qquad (14.43)$$

The convolution theorem works in reverse as well. If X is the product of Y and W, i.e.

$$X(\omega) = Y(\omega) W(\omega) \qquad (14.44)$$

then the functions of time, x, y, and w are related by a convolution,

$$x(t) = \int_{-\infty}^{\infty} dt' \, y(t') \, w(t - t') \qquad (14.45)$$

or

$$x(t) = y(t) * w(t).$$

The symmetry of the convolution equations for multiplication of functions of frequency and multiplication of functions of time is misleading. Multiplication of frequency functions, e.g. $X(\omega) = Y(\omega)W(\omega)$, corresponds to a linear operation on signals generally known as filtering. Multiplication of signal functions of time, e.g. $y(t)w(t)$, is a nonlinear operation such as modulation.

14.4 Power, Energy, and Power Spectrum

The instantaneous power in a signal is defined as $P(t) = x^2(t)$. This definition corresponds to the power that would be transferred by a signal to a unit load that is purely resistive, or dissipative. Such a load is not at all reactive, wherein energy is stored for some fraction of a cycle.

The energy in a signal is the accumulation of power over time,

$$E = \int dt\, P(t) = \int dt\, x^2(t)\,. \tag{14.46}$$

At this point, a distinction must be made between finite-duration signals and infinite-duration signals. For a finite-duration signal, the above integral exits. By substituting the Fourier transform for $x(t)$, one finds that

$$E = \frac{1}{2\pi} \int d\omega\, X(\omega)\, X(-\omega) \quad \text{or} \quad \int d\omega\, E(\omega)\,. \tag{14.47}$$

Thus the energy in the signal is written as the accumulation of of the energy spectral density,

$$E(\omega) = \frac{1}{2\pi} X(\omega)\, X(-\omega) = \frac{1}{2\pi} |X(\omega)|^2\,. \tag{14.48}$$

The symmetry between (14.46) and (14.47) is known as Parseval's theorem. It says that one can compute the energy in a signal by either a time or a frequency integral.

The power spectral density is obtained by dividing the energy spectral density by the duration of the signal, T_D,

$$P(\omega) = E(\omega)/T_D\,. \tag{14.49}$$

For white noise, the power density is constant on average, $P(\omega) = P_0$. From (14.47) it is evident that a signal cannot be white over the entire range of frequencies out to infinite frequency without having infinite energy. One is therefore limited to noise that is white over a finite frequency band.

For pink noise the power density is inversely proportional to frequency, $P(\omega) = c/\omega$, where c is a constant. The energy integral in (14.47) for pink noise also diverges. Therefore, pink noise must be limited to a finite frequency band.

Turning now to infinite-duration signals, for an infinite-duration signal the energy is not well defined. It is likely that one would never even think about an infinite-duration signal if it were not for the useful concept of a *periodic* signal. Although the energy is undefined, the power P is well defined, and so is the power spectrum, or power spectral density $P(\omega)$. As expected, the power is the integral of the power spectral density,

$$P = \int d\omega\, P(\omega)\,, \tag{14.50}$$

where $P(\omega)$ is given in terms of X from (14.27),

$$P(\omega) = \frac{\pi}{2} \sum_{n=0}^{\infty} |X_n|^2 [\delta(\omega - \omega_n) + \delta(\omega + \omega_n)]\,. \tag{14.51}$$

It is not hard to convert densities to different units. For instance, the power spectral density can be written in terms of frequency f instead of ω ($\omega = 2\pi f$). By the definition of a density we must have that

$$P = \int df\, P(f)\,. \tag{14.52}$$

This definition is consistent with the fact that a delta function has dimensions that are the inverse of its argument dimensions. Therefore, $\delta(\omega) = \delta(2\pi f) = \delta(f)/(2\pi)$, and

$$P(f) = \frac{1}{4} \sum_{n=0}^{\infty} |X_n|^2 [\delta(f - f_n) + \delta(f + f_n)]\,. \tag{14.53}$$

14.4.1 Autocorrelation

The autocorrelation function a_f of a signal $x(t)$ provides a measure of the similarity between the signal at time t and the same signal at a different time, $t + \tau$. The variable τ is called the *lag*, and the autocorrelation function is given by

$$a_f(\tau) = \int_{-\infty}^{\infty} dt\, x(t)\, x(t + \tau)\,. \tag{14.54}$$

When τ is zero then the integral is just the square of $x(t)$, and this leads to the largest possible value for the autocorrelation, namely E. For a signal of finite duration, the autocorrelation must always be strictly less than its value at $\tau = 0$. Consequently, the normalized autocorrelation function $a(\tau)/a(0)$ is always less than 1.0 ($\tau \neq 0$).

By substituting (14.29) for $x(t)$ one finds a frequency integral for the autocorrelation function,

$$a_f(\tau) = \frac{1}{2\pi} \int_{-\infty}^{\infty} d\omega\, e^{i\omega\tau} |X(\omega)|^2\,, \tag{14.55}$$

or, from (14.47),

$$a_f(\tau) = \int_{-\infty}^{\infty} d\omega \, e^{i\omega\tau} E(\omega) \,. \tag{14.56}$$

Equation (14.56) says that the autocorrelation function is the Fourier transform of the energy spectral density. This relationship is known as the Wiener–Khintchine relation. Because $E(-\omega) = E(\omega)$, one can write a_f in a way that proves that it is a real function with no imaginary part,

$$a_f(\tau) = 2 \int_0^{\infty} d\omega \, \cos(\omega\tau) E(\omega) \,. \tag{14.57}$$

Furthermore, because the cosine is an even function of its argument $[a_f(-\tau) = a_f(\tau)]$, the autocorrelation function only needs to be given for positive values of the lag.

A signal does not have finite duration if it is periodic. Then the autocorrelation function is defined as

$$a(\tau) = \lim_{T_D \to \infty} \frac{1}{2T_D} \int_{-T_D}^{T_D} dt \, x(t) x(t+\tau) \,. \tag{14.58}$$

If the period is T then $a(\tau) = a(\tau + nT)$ for all integer n, and the maximum value occurs at $a(0)$ or $a(nT)$. Because of the factor of time in the denominator of (14.58), the function $a(\tau)$ is the Fourier transform of the power spectral density and not of the energy spectral density.

A critical point for both $a_f(\tau)$ and $a(\tau)$ is that autocorrelation functions are independent of the phases of spectral components. This point seems counterintuitive because waveforms depend on phases and it seems only natural that the correlation of a waveform with itself at some later time should reflect this phase dependence. However, the fact that autocorrelation is the Fourier transform of the energy or power spectrum proves that the autocorrelation function must be independent of phases because the spectra are independent of phases.

For example, if $x(t)$ is a periodic function with zero average value, it is defined by (14.6). Then it is not hard to show that the autocorrelation function is given by

$$a(\tau) = \frac{1}{2} \sum_{n=1}^{N} C_n^2 \, \cos(\omega_n \tau) \,. \tag{14.59}$$

The autocorrelation function is only a sum of cosines with none of the phase information. Only the harmonic frequencies and amplitudes play a role.

14.4.2 Cross-Correlation

Parallel to the autocorrelation function, the cross-correlation function is a measure of the similarity of the signal $x(t)$ to the signal $y(t)$ at a different time, i.e. the similarity to $y(t+\tau)$. The cross-correlation function is

$$\rho_o(\tau) = \int dt \, x(t) \, y(t+\tau) \,. \tag{14.60}$$

In practice, the cross-correlation is usually normalized,

$$\rho(\tau) = \frac{\int dt \, x(t) \, y(t+\tau)}{[\int dt_1 \, x^2(t_1) \, \int dt_2 \, y^2(t_2)]^{1/2}} \,, \tag{14.61}$$

so that the maximum value of $\rho(\tau)$ is equal to 1.0. Unlike the autocorrelation function, the maximum of $\rho(\tau)$ does not necessarily occur at $\tau = 0$. For example, if signal $y(t)$ is the same as signal $x(t)$ except that $y(t)$ has been delayed by T_{del} then $\rho(\tau)$ has its maximum value 1.0 when $\tau = T_{\text{del}}$.

14.5 Statistics

Measured signals are always finite in length. Definitions of statistical terms for measured signals, together with their continuum limits are given in this section.

The number of samples in a measurement is N. The *duration* of the measured signal is T_D, and $T_D = N\delta t$, where δt is the inverse of the sample rate.

The sampled signal has values x_i, $(1 \le i \le N)$, and the continuum analog is the signal $x(t)$, $(0 \le t \le T_D)$.

The *average* value, or mean, is

$$\bar{x} = \frac{1}{N} \sum_{i=1}^{N} x_i \quad \text{or} \quad \frac{1}{T_D} \int_0^{T_D} dt \, x(t) \,. \tag{14.62}$$

The *variance* is

$$\sigma^2 = \frac{1}{N-1} \sum_{i=1}^{N} (x_i - \bar{x})^2$$

or

$$\frac{1}{T_D} \int_0^{T_D} dt\, [x(t) - \bar{x}]^2 \,. \tag{14.63}$$

The *standard deviation* is the square root of the variance, $\sigma = \sqrt{\sigma^2}$.

The *energy* is

$$E = \delta t \sum_{i=1}^{N} x_i^2 \quad \text{or} \quad \int_0^{T_D} dt\, x^2(t) \,. \tag{14.64}$$

The *average power* is

$$\bar{P} = \frac{1}{N} \sum_{i=1}^{N} x_i^2 \quad \text{or} \quad \frac{E}{T_D} \,. \tag{14.65}$$

The *root-mean-square* (RMS) value is the square root of the average power, $x_{RMS} = \sqrt{\bar{P}}$.

14.5.1 Signals and Processes

Signals are the observed results of processes. A process is *stationary* if its stochastic properties, such as mean and standard deviation, do not change during the time for which a signal is observed. Signals provide incomplete glimpses into processes.

The best estimate of the mean of the underlying process is equal to the mean of an observed signal. The expected error in the estimate of the mean of the underlying process, the so-called *standard deviation of the mean*, is

$$s = \sigma/\sqrt{N} \,, \tag{14.66}$$

where N is the number of data points contributing to the mean of the observed signal.

14.5.2 Distributions

Digitized signals are often regarded as sampled data $\{x\}$. If the data are integers or are put into bins j then the probability that the signal has value x_j is the probability mass function $\text{PMF}(j) = N_j/N$, the ratio of the number of samples in bin j to the total number of samples. If data are continuous floating-point numbers, the analogous distribution is the probability density function $\text{PDF}(x)$. In terms of these distributions, the mean is given by

$$\bar{x} = \sum x_j\, \text{PMF}(j) \quad \text{or} \quad \int_{-\infty}^{\infty} dx\, x\, \text{PDF}(x) \,. \tag{14.67}$$

The most important PDF is the normal (Gaussian) density $G(x)$,

$$G(x) = \frac{1}{\sigma\sqrt{2\pi}} \exp[(x - \bar{x})^2/2\sigma^2] \,. \tag{14.68}$$

Like all PDFs, $G(x)$ is normalized to unit area, i.e.

$$\int_{-\infty}^{\infty} dx\, G(x) = 1 \,. \tag{14.69}$$

The probability that x lies between some value x_1 and $x_1 + dx$ is $\text{PDF}(x_1)\, dx$, and normalization reflects the simple fact that x must have *some* value.

The probability that variable x is less than some value x_1 is the cumulative distribution function (CDF),

$$\text{CDF}(x_1) = \int_{-\infty}^{x_1} dx'\, \text{PDF}(x') \,. \tag{14.70}$$

If the density is normal, the integral is the cumulative normal distribution (CND),

$$\text{CND}(x) = \frac{1}{\sigma\sqrt{2\pi}} \int_{-\infty}^{x} dx'\, \exp(x'^2/2\sigma^2) \,. \tag{14.71}$$

It is convenient to think of the CND as a function of x compared to the standard deviation, i.e., as a function of $y = (x - \bar{x})/\sigma$, as shown in Fig. 14.5.

$$C(y) = \frac{1}{\sqrt{2\pi}} \int_{-\infty}^{y} dy'\, \exp(y'^2/2) \,. \tag{14.72}$$

Because of the symmetry of the normal density,

$$C(-y) = 1 - C(y) \,. \tag{14.73}$$

Therefore, it is enough to know $C(y)$ for $y > 0$. A few important values follow.

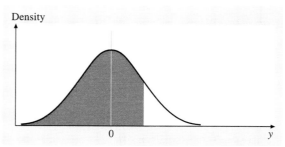

Fig. 14.5 The *area* under the normal density is the cumulative normal. Here the area is the function $C(y)$

Table 14.1 Selected values of the cumulative normal distribution

C(0)	0.5000
C(0.675)	0.7500
C(1)	0.8413
C(2)	0.9773
C(3)	0.9987
C(∞)	1.0000

Table 14.1 can be used to find probabilities. For example, the probability that a normally distributed variable lies between its mean and its mean plus a standard deviation, i.e., between $\bar{x}$ and $\bar{x} + \sigma$, is $C(1) - 0.5 = 0.3413$. The probability that it lies within plus or minus two standard deviations ($\pm 2\sigma$) of the mean is $2[C(2) - 0.5] = 0.9546$.

The importance of the normal density lies in the central limit theorem, which says that the distribution for a sum of random variables approaches a normal distribution as the number of variables becomes large. In other words, if the variable x is a sum

$$x = x_1 + x_2 + x_3 + \ldots x_N = \sum_{i=1}^{N} x_i, \quad (14.74)$$

then no matter how the individual x_i are distributed, x will be normally distributed in the limit of large N.

14.5.3 Multivariate Distributions

A multivariate distribution is described by a *joint probability density* $\mathrm{PDF}(x, y)$, where the probability that variable x has a value between x_1 and $x_1 + \mathrm{d}x$ and simultaneously variable y has a value between y_1 and $y_1 + \mathrm{d}y$ is

$$P(x_1, y_1) = \mathrm{PDF}(x_1, y_1) \, \mathrm{d}x \, \mathrm{d}y. \quad (14.75)$$

The normalization requirement is

$$\int \mathrm{d}x \int \mathrm{d}y \, \mathrm{PDF}(x, y) = 1. \quad (14.76)$$

The *marginal probability density* for x, $\mathrm{PDF}(x)$, is the probability density for x itself, regardless of the value of y. Hence,

$$\mathrm{PDF}(x) = \int \mathrm{d}y \, \mathrm{PDF}(x, y). \quad (14.77)$$

The y dependence has been integrated out.

The *conditional probability density* $\mathrm{PDF}(x|y)$ describes the probability of a value x, given a specific value of y, for instance, if $y = y_1$, then

$$\mathrm{PDF}(x|y_1) = \mathrm{PDF}(x, y_1) / \int \mathrm{d}x \, \mathrm{PDF}(x, y_1). \quad (14.78)$$

or

$$\mathrm{PDF}(x|y_1) = \mathrm{PDF}(x, y_1) / \mathrm{PDF}(y_1). \quad (14.79)$$

The probability that $x = x_1$ and $y = y_1$ is equal to the probability that $y = y_1$ multiplied by the conditional probability that if $y = y_1$ then $x = x_1$, i.e.,

$$P(x_1, y_1) = P(x_1|y_1) P(y_1). \quad (14.80)$$

Similarly, the probability that $x = x_1$ and $y = y_1$ is equal to the probability that $x = x_1$ multiplied by the conditional probability that if $x = x_1$ then $y = y_1$, i.e.

$$P(x_1, y_1) = P(y_1|x_1) P(x_1). \quad (14.81)$$

The two expressions for $P(x_1, y_1)$ must be the same, and that leads to Bayes's Theorem,

$$P(x_1|y_1) = P(y_1|x_1) P(x_1) / P(y_1). \quad (14.82)$$

14.5.4 Moments

The m-th moment of a signal is defined as

$$\overline{x^m} = \frac{1}{N} \sum_{i=1}^{N} x_i^m \quad \text{or} \quad \frac{1}{T_\mathrm{D}} \int_0^{T_\mathrm{D}} \mathrm{d}t \, x^m(t). \quad (14.83)$$

Hence the first moment is the mean (14.62) and the second moment is the average power (14.65).

The m-th central moment is

$$\mu_m = \frac{1}{N} \sum_{i=1}^{N} (x_i - \bar{x})^m \quad \text{or} \quad \frac{1}{T_\mathrm{D}} \int_0^{T_\mathrm{D}} \mathrm{d}t \, [x(t) - \bar{x}]^m. \quad (14.84)$$

The first central moment is zero by definition. The second central moment is the alternating-current (AC) power, which is equal to the average power (14.65) minus the time-independent (or DC) component of the power.

The third central moment is zero if the signal probability density function is symmetrical about the mean. Otherwise, the third moment is a simple way to measure how the PDF is skewed. The *skewness* is the normalized third moment,

$$\text{skewness} = \mu_3/\mu_2^{3/2}. \tag{14.85}$$

The fourth central moment leads to an impression about how much strength there is in the wings of a probability density compared to the standard deviation. The normalized fourth moment is the *kurtosis*,

$$\text{kurtosis} = \mu_4/\mu_2^2. \tag{14.86}$$

For instance, the kurtosis of a normal density, which has significant wings, is 3. But the kurtosis of a rectangular density, which is sharply cut off, is only 9/5.

14.6 Hilbert Transform and the Envelope

The Hilbert transform of a signal $x(t)$ is $\mathcal{H}[x(t)]$ or function $x_I(t)$, where

$$x_I(t) = \mathcal{H}[x(t)] = \frac{1}{\pi} \int_{-\infty}^{\infty} dt' \, \frac{x(t')}{t-t'}. \tag{14.87}$$

Some facts about the Hilbert transform are stated here without proof. Proofs and further applications may be found in appendices to [14.1].

First, the Hilbert transform is its own inverse, except for a minus sign,

$$x(t) = \mathcal{H}[x_I(t)] = -\frac{1}{\pi} \int_{-\infty}^{\infty} dt' \, \frac{x_I(t')}{t-t'}. \tag{14.88}$$

Second, a signal and its Hilbert transform are orthogonal in the sense that

$$\int dt \, x(t) \, x_I(t) = 0. \tag{14.89}$$

Third, the Hilbert transform of $\sin(\omega t + \varphi)$ is $-\cos(\omega t + \varphi)$, and the Hilbert transform of $\cos(\omega t + \varphi)$ is $\sin(\omega t + \varphi)$.

Further the Hilbert transform is linear. Consequently, for any function for which a Fourier transform exists,

$$\mathcal{H}\left[\sum_n A_n \cos(\omega_n t) + B_n \sin(\omega_n t)\right]$$
$$= \sum_n A_n \sin(\omega_n t) - B_n \cos(\omega_n t) \tag{14.90}$$

or

$$\mathcal{H}\left[\sum_n C_n \sin(\omega_n t + \varphi_n)\right]$$
$$= -\sum_n C_n \cos(\omega_n t + \varphi_n)$$
$$= \sum_n C_n \sin(\omega_n t + \varphi_n - \pi/2). \tag{14.91}$$

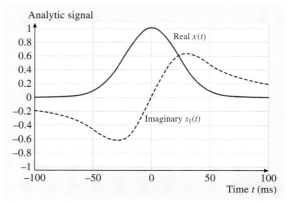

Fig. 14.6 A Gaussian pulse $x(t)$ and its Hilbert transform $x_I(t)$ are the real and imaginary parts of the analytic signal corresponding to the Gaussian pulse

Comparing the two sine functions above makes it clear why a Hilbert transform is sometimes called a 90-degree rotation of the signal.

Figure 14.6 shows a Gaussian pulse, $x(t)$, and its Hilbert transform, $x_I(t)$. The Gaussian pulse was made by adding up 100 cosine harmonics with amplitudes given by a Gaussian spectrum per (14.35). The Hilbert transform was computed by using the same amplitude spectrum and replacing all the cosine functions by sine functions.

Figure 14.6 illustrates the difficulty often encountered in computing the Hilbert transform using the time integrals that define the transform and its inverse. If we had to calculate $x(t)$ by transforming $x_I(t)$ using (14.88) we would be troubled by the fact that $x_I(t)$ goes to zero so slowly. An accurate calculation of $x(t)$ would require a longer time span than that shown in the figure.

14.6.1 The Analytic Signal

The *analytic signal* $\tilde{x}(t)$ for $x(t)$ is given by the complex sum of the original signal and an imaginary part equal to the Hilbert transform of $x(t)$,

$$\tilde{x}(t) = x(t) + \mathrm{i}\, x_\mathrm{I}(t) \,. \tag{14.92}$$

The analytic signal, in turn, can be used to calculate the envelope of signal $x(t)$. The envelope $e(t)$ is the absolute value – or magnitude – of the analytic signal

$$e(t) = |\tilde{x}(t)| \,. \tag{14.93}$$

For instance, if $x(t) = A\cos(\omega t + \varphi)$, then $x_\mathrm{I}(t) = A\sin(\omega t + \varphi)$ and

$$\tilde{x}(t) = A[\cos(\omega t + \varphi) + \mathrm{i}\sin(\omega t + \varphi)] \,. \tag{14.94}$$

By Euler's theorem

$$\tilde{x}(t) = A\exp[\mathrm{i}(\omega t + \varphi)] \,, \tag{14.95}$$

and the absolute value is

$$e(t) = \{A\exp[\mathrm{i}(\omega t + \varphi)]A\exp[-\mathrm{i}(\omega t + \varphi)]\}^{1/2}$$
$$= A \,. \tag{14.96}$$

14.7 Filters

Filtering is an operation on a signal that is typically defined in frequency space. If $x(t)$ is the input to a filter and $y(t)$ is the output then the Fourier transforms of x and y are related by

$$Y(\omega) = H(\omega)X(\omega) \,, \tag{14.97}$$

where $H(\omega)$ is the transfer function of the filter. The transfer function has a magnitude and a phase

$$H(\omega) = |H(\omega)|\exp[\mathrm{i}\Phi(\omega)] \,. \tag{14.98}$$

The frequency-dependent magnitude is the amplitude response, and it characterizes the filter type – low pass, high pass, bandpass, band-reject, etc. The phase $\Phi(\omega)$ is the phase shift for a spectral component with frequency ω. The amplitude and phase responses of a filter are explicitly separated by taking the natural logarithm of the transfer function

$$\ln H(\omega) = \ln[|H(\omega)|] + \mathrm{i}\Phi(\omega) \,. \tag{14.99}$$

Because $\ln|H| = \ln(10)\log|H|$,

$$\ln H(\omega) = 0.1151\,\Gamma(\omega) + \mathrm{i}\Phi(\omega) \,, \tag{14.100}$$

where G is the filter gain in decibels, and Φ is the phase shift in radians.

14.7.1 One-Pole Low-Pass Filter

The one-pole low-pass filter serves as an example to illustrate filter concepts. This filter can be made from a single resistor (R) and a single capacitor (C) with a time constant $\tau = RC$. The transfer function of this filter is

$$H(\omega) = \frac{1}{1+\mathrm{i}\omega\tau} = \frac{1-\mathrm{i}\omega\tau}{1+\omega^2\tau^2} \,. \tag{14.101}$$

The filter is called *one-pole* because there is a single value of ω for which the denominator of the transfer function is zero, namely $\omega = 1/(\mathrm{i}\tau) = -\mathrm{i}/\tau$.

The magnitude (or amplitude) response is

$$|H(\omega)| = \sqrt{\frac{1}{1+\omega^2\tau^2}} \,. \tag{14.102}$$

The filter cut-off frequency is the *half-power* point (or 3-dB-down point), where the magnitude of the transfer function is $1/\sqrt{2}$ compared to its maximum value. For the one-pole low-pass filter, the half-power point occurs when $\omega = 1/\tau$.

Filters are often described by their asymptotic frequency response. For a low-pass filter asymptotic behavior occurs at high frequency, where, for the one-pole filter $|H(\omega)| \propto 1/\omega$. The $1/\omega$ dependence is equivalent to a high-frequency slope of $-6\,\mathrm{dB/octave}$, i.e., for octave frequencies,

$$L_2 - L_1 = 20\log\left(\frac{\omega_1}{2\omega_1}\right) = 20\log\frac{1}{2} = -6 \,. \tag{14.103}$$

A filter with an asymptotic dependence of $1/\omega^2$ has a slope of $-12\,\mathrm{dB/octave}$, etc.

The phase shift of the low-pass filter is the arctangent of the ratio of the imaginary and real parts of the transfer function,

$$\Phi(\omega) = \tan^{-1}\left(\frac{\mathrm{Im}[H(\omega)]}{\mathrm{Re}[H(\omega)]}\right) \,, \tag{14.104}$$

which, for the one-pole filter, is $\Phi(\omega) = \tan^{-1}(-\omega\tau)$. The phase shift is zero at zero frequency, and approaches 90° at high frequency. This phase behavior is typical of simple filters in that important phase shifts occur in frequency regions where the magnitude shows large attenuation.

14.7.2 Phase Delay and Group Delay

The phase shifts introduced by filters can be interpreted as delays, whereby the output is delayed in time compared to the input. In general, the delay is different for different frequencies, and therefore, a complex signal composed of several frequencies is bent out of shape by the filtering process. Systems in which the delay is different for different frequencies are said to be *dispersive*.

Two kinds of delay are of interest. The *phase delay* simply reinterprets the phase shift as a delay. The phase delay T_φ is given by $T_\varphi = -\Phi(\omega)/\omega$. The *group delay* T_g is given by the derivative $T_g = -\mathrm{d}\Phi(\omega)/\mathrm{d}\omega$. Phase and group delays for a one-pole low-pass filter are shown in Fig. 14.7 together with the phase shift.

14.7.3 Resonant Filters

Resonant filters, or *tuned* systems, have an amplitude response that has a peak at some frequency where $\omega = \omega_o$. Such filters are characterized by the resonant frequency, ω_o, and by the bandwidth, $2\Delta\omega$. The bandwidth is specified by half-power points such that $|H(\omega_o + \Delta\omega)|^2 \approx |H(\omega_o)|^2/2$ and $|H(\omega_o - \Delta\omega)|^2 \approx |H(\omega_o)|^2/2$. The sharpness of a tuned system is often quoted as a Q value, where Q is a dimensionless number given by

$$Q = \omega_o/(2\Delta\omega). \qquad (14.105)$$

As an example, a two-pole low-pass filter with a resonant peak near the angular frequency ω_o is described by the transfer function

$$H(\omega) = \frac{\omega_o^2}{\omega_o^2 - \omega^2 + j\omega\omega_o/Q}. \qquad (14.106)$$

14.7.4 Impulse Response

Because filtering is described as a product of Fourier transforms, i.e., in frequency space, the temporal representation of filtering is a convolution

$$y(t) = \int \mathrm{d}t'\, h(t-t')x(t') = \int \mathrm{d}t'\, h(t')x(t-t'). \qquad (14.107)$$

The two integrals on the right are equivalent.

Equation (14.107) is a special form of linear processor. A more general linear processor is described by the equation

$$y(t) = \int \mathrm{d}t'\, h(t,t')x(t'), \qquad (14.108)$$

where $h(t,t')$ permits a perfectly general dependence on t and t'. The special system in which only the difference in time values is important, i.e. $h(t,t') = h(t-t')$, is a linear *time-invariant* system. Filters are time invariant.

A system that operates in real time obeys a further filter condition, namely *causality*. A system is causal if the output $y(t)$ depends on the input $x(t')$ only for $t' < t$. In words, this says that the present output cannot depend on the future input. Causality requires that $h(t) = 0$ for $t < 0$. For the one-pole corona, low-pass filter of (14.101) the impulse response is

$$h(t) = \frac{1}{\tau}\mathrm{e}^{-t/\tau} \quad \text{for } t > 0,$$
$$h(t) = 0 \quad \text{for } t < 0,$$
$$h(t) = \frac{1}{2\tau} \quad \text{for } t = 0. \qquad (14.109)$$

For the two-pole low-pass resonant filter of (14.106), the impulse response is

$$h(t) = \frac{\omega_o}{\sqrt{1-[1/(2Q)]^2}}\mathrm{e}^{-\frac{\omega_o}{2Q}t}$$
$$\times \sin\left\{\omega_o t\sqrt{1-[1/(2Q)]^2}\right\}, \quad t \geq 0,$$
$$h(t) = 0, \quad t < 0. \qquad (14.110)$$

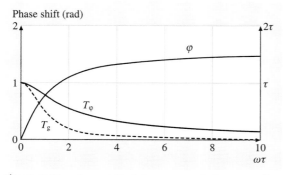

Fig. 14.7 The phase shift Φ for a one-pole low-pass filter can be read on the *left ordinate*. The phase and group delays can be read on the *right ordinate*

14.7.5 Dispersion Relations

The causality requirement on the impulse response, $h(t) = 0$ for $t < 0$, has implications for the transfer function. Causality means that the real and imaginary parts of the transfer function are Hilbert transforms of one another. Specifically, if the real and imaginary parts of H are defined as $H(\omega) = H_R(\omega) + iH_I(\omega)$ then

$$H_R(\omega) = \frac{1}{\pi} \mathcal{P} \int_{-\infty}^{\infty} d\omega' \frac{H_I(\omega')}{\omega - \omega'}, \qquad (14.111)$$

and

$$H_I(\omega) = \frac{-1}{\pi} \mathcal{P} \int_{-\infty}^{\infty} d\omega' \frac{H_R(\omega')}{\omega - \omega'}.$$

The symbol $\mathcal{P}$ signifies that the principal value of a divergent integral should be taken. In many cases, this requires no special steps, and definite integrals from integral tables give the correct answers.

These equations are known as *dispersion relations*. They arise from doing an integral in frequency space to calculate the impulse response for $t < 0$. The fact that this calculation must return zero means that $H(\omega)$ must have no singularities in the complex frequency plane for frequencies with a negative imaginary part. Similar dispersion relations apply to the natural log of the transfer function, relating the filter gain to the phase shift as in (14.100)

$$\Gamma(\omega) = \Gamma(0) - \frac{\omega^2}{0.1151\pi} \mathcal{P} \int_{-\infty}^{\infty} d\omega' \frac{\Phi(\omega')}{\omega'(\omega'^2 - \omega^2)} \qquad (14.112)$$

and

$$\Phi(\omega) = \frac{0.1151\,\omega}{\pi} \mathcal{P} \int_{-\infty}^{\infty} d\omega' \frac{\Gamma(\omega')}{\omega'^2 - \omega^2}.$$

Because $\Gamma(\omega)$ is even and $\Phi(\omega)$ is odd, both integrands are even in ω', and these integrals can be replaced by twice the integral from zero to infinity. The second equation above is particularly powerful. It says that, if we want to find the phase shift of a system, we only have to measure the gain of the system in decibels, multiply by 0.1151, and do the integral. Of course, it is in the nature of the integral that in order to find the phase shift at any given frequency we need to know the gain over a wide frequency range.

The dispersion relations for gain and phase shift also arise from a contour integral over frequencies with a negative imaginary part, but now the conditions on $H(\omega)$ are more stringent. Not only must $H(\omega)$ have no poles for $\text{Im}(\omega) < 0$, but $\ln H(\omega)$ must also have no poles. Consequently $H(\omega)$ must have no zeros for $\text{Im}(\omega) < 0$. A system that has neither poles nor zeros for $\text{Im}(\omega) < 0$ is said to be *minimum phase*. The dispersion relations in (14.112) only apply to a system that is minimum phase.

14.8 The Cepstrum

The cepstrum (pronounced *kepstrum*) is the inverse Fourier transform of the natural logarithm of the spectrum. Because it is the inverse transform of a function of frequency, the cepstrum is a function of a time-like variable. But just as the word *cepstrum* is an anagram of the word *spectrum*, the time-like coordinate is called the *quefrency*, an anagram of *frequency*. The field of cepstrology is full of word fun like this.

Fig. 14.8 The cepstrum of an original signal to which is added a delayed version of the same signal, with a delay of 2 ms ($a = 1$). The original signal is the sum of two female talkers

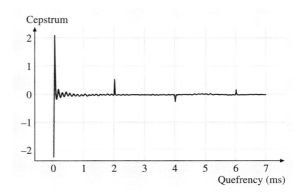

The complex cepstrum of complex spectrum $Y(\omega)$ is

$$q(\tau) = \frac{1}{2\pi} \int_{-\infty}^{\infty} d\omega \, e^{i\omega\tau} \ln[Y(\omega)], \quad (14.113)$$

where τ is the quefrency. Because $Y(\omega) = |Y(\omega)| e^{i\varphi(\omega)}$,

$$q(\tau) = \frac{1}{2\pi} \int_{0}^{\infty} d\omega \, e^{i\omega\tau} [\ln|Y(\omega)| + i\varphi(\omega)]$$

$$+ \frac{1}{2\pi} \int_{0}^{\infty} d\omega \, e^{-i\omega\tau} [\ln|Y(-\omega)| + i\varphi(-\omega)]. \quad (14.114)$$

For a real signal $y(t)$, the magnitude $|Y(\omega)|$ is an even function of ω, and $\varphi(\omega)$ is odd. Therefore,

$$q(\tau) = \frac{1}{\pi} \int_{0}^{\infty} d\omega \, [\ln|Y(\omega)|] \cos(\omega\tau)$$

$$+ \frac{i}{\pi} \int_{0}^{\infty} d\omega \, \varphi(\omega) \sin(\omega\tau). \quad (14.115)$$

The real part of q comes from the magnitude, the imaginary part from the phase. The phase must be unwrapped; it cannot be artificially restricted to a 2π range.

It is common to deal only with the real part of the cepstrum q_R. It is evident that the calculation will fail if $|Y(\omega)|$ is zero. The cepstrum is not applied to theoretical objects such as periodic functions of time that have delta function spectra – hence zeros. The cepstrum is applied to measured data, where it can lead to insight into features of the underlying processes.

The cepstrum is used in the acoustical and vibrational monitoring of machinery. Bearings and other rotating parts tend to produce sounds with interleaved periodic spectra. These periodicities lead to peaks at the corresponding quefrencies, revealing features that may not be apparent in the spectrum.

The cepstrum is particularly suited to the separation of source and filter functions. If Y is a filtered version of X, where the transfer function is H, then

$$|Y(\omega)| = |H(\omega)| \, |X(\omega)|. \quad (14.116)$$

The logarithm operation turns the product on the right-hand side into a sum, so that

$$q_R(\tau) = \frac{1}{\pi} \int_{0}^{\infty} d\omega \, [\ln|H(\omega)|] \cos(\omega\tau)$$

$$+ \frac{1}{\pi} \int_{0}^{\infty} d\omega \, [\ln|X(\omega)|] \cos(\omega\tau). \quad (14.117)$$

For instance, if $|Y|$ is the spectrum of a spoken vowel, then the term involving the formant filter $|H|$ leads to a low-quefrency structure, and the term involving source spectrum $|X|$ leads to a high-quefrency peak characteristic of the glottal pulse period.

The cepstrum can reveal reflections. As a simple example, we consider a direct sound X plus its reflection with relative amplitude a and delay T_D. The sum then has a spectrum Y,

$$|Y(\omega)| = [1 + a\cos(\omega T_D)]|X(\omega)| \quad (a < 1). \quad (14.118)$$

The logarithm of the factor in square brackets is periodic in ω with period $2\pi/T_D$. The corresponding term in the cepstrum leads to a peak at quefrency $\tau = T_D$, as shown in Fig. 14.8. The addition of more reflections with other delays will lead to additional peaks. Maintaining the anagram game, the separation of peaks along the quefrency axis is sometimes called *liftering*.

14.9 Noise

Noise has many definitions in acoustics. Commonly, noise is any unwanted signal. In the context of communications, it is an excitation that competes with the information that one wishes to transmit. In signal processing, noise is defined as a random signal that can only be defined in statistical terms with no long-term predictability.

14.9.1 Thermal Noise

Thermal noise, or Johnson noise, is generated in a resistor. An electrical circuit that describes this source of noise is a resistor R in series with a voltage source that depends on R, such that the RMS voltage is given by the equation

$$V = \sqrt{4Rk_B T \Delta f}, \quad (14.119)$$

where R is the resistance in ohms, k_B is Boltzmann's constant, T is the absolute temperature, and Δf is the bandwidth over which the noise is measured.

The corresponding noise power can be defined by measuring the maximum power that is transferred to a load resistor connected across the series circuit above. Maximum power occurs when the load resistor also has a resistance R and has zero temperature so that the load resistor produces no Johnson noise of its own. Then the thermal noise power is given by

$$P = k_B T \Delta f. \tag{14.120}$$

Because $k_B T$ has dimensions of Joules and Δf has dimensions of inverse seconds, the quantity P has dimensions of watts, as expected. Boltzmann's constant is 1.38×10^{-23} J/K, and room temperature is 293 K. Therefore, the noise power density is 4×10^{-21} W/Hz. Because the power is proportional to the first power of the bandwidth, the noise is white. Johnson noise is also Gaussian.

14.9.2 Gaussian Noise

A noise is Gaussian if its instantaneous values form a Gaussian (normal) distribution. A noise distribution is illustrated in an experiment wherein an observer makes hundreds of instantaneous measurements of a noise voltage and plots these instantaneous values as a histogram. Unless there is some form of bias, the measured values are equally often positive and negative, and so the mean of the distribution is zero. The noise is Gaussian if the histogram derived in this way is a Gaussian function. The more intense the noise, the larger is the standard deviation of the Gaussian function. Because of the central limit theorem, there is a tendency for noise to be Gaussian. However, non-Gaussian noises are easily generated. Random telegraph noise, where instantaneous values can only be $+1$ or -1, is an example.

14.9.3 Band-Limited Noise

Band-limited noise can be written in terms of Fourier components,

$$x(t) = \sum_{n=1}^{N} A_n \cos(\omega_n t) + B_n \sin(\omega_n t). \tag{14.121}$$

The amplitudes A_n and B_n are defined only statistically. According to a famous paper by *Einstein* and *Hopf* [14.2], these amplitudes are normally distributed with zero mean, and the distributions of A_n and B_n have the same variance σ_n^2. The distributions themselves can be thought of as representative of an ensemble of noises, all of which are intended by the creator to be the same: same duration and power, same frequency range and bandwidth.

Because the average power in a sine or cosine is 0.5, the average power in band-limited noise is

$$P = \sum_{n=1}^{N} \sigma_n^2. \tag{14.122}$$

An alternative description of band-limited noise is the amplitude and phase form

$$x(t) = \sum_{n=1}^{N} C_n \cos(\omega_n t + \varphi_n), \tag{14.123}$$

where φ_n are random variables with a rectangular distribution from 0 to 2π, and $C_n = \sqrt{A_n^2 + B_n^2}$.

Given that A_n and B_n follow a Gaussian distribution with variance σ_n, the amplitude C_n follows a Rayleigh distribution f_{Rayl}

$$f_{\text{Rayl}}(C_n) = \frac{C_n}{\sigma_n^2} e^{-C_n^2/(2\sigma_n^2)} \quad (C_n > 0). \tag{14.124}$$

The peak of the Rayleigh distribution occurs at $C_n = \sigma$. The zeroth moment is 1.0 because the distribution is normalized. The first moment, or $\overline{C_n}$, is $\sigma_n\sqrt{\pi/2}$. The second moment is $2\sigma_n^2$, and the fourth moment is $8\sigma_n^4$.

The cumulative Rayleigh distribution can be calculated in closed form,

$$F_{\text{Rayl}}(C_n) = \int_0^{C_n} dC_n' \, f_{\text{Rayl}}(C_n') = 1 - e^{-C_n^2/(2\sigma_n^2)}. \tag{14.125}$$

14.9.4 Generating Noise

To generate the amplitudes A_n and B_n with normal distributions using a computer random-number generator, one can add up twelve random numbers and subtract 6. On the average, the amplitudes will have a normal distribution, because of the central limit theorem, with a mean of zero and a variance of 1.0.

To generate the amplitudes C_n with a Rayleigh distribution, one can transform the random numbers r_n that come from a computer random-number generator, according to the formula

$$C_n = \sigma\sqrt{-2\ln(1-r_n)}. \tag{14.126}$$

14.9.5 Equal-Amplitude Random-Phase Noise

Equal-amplitude random-phase (EARP) noise is of the form

$$x(t) = C \sum_{n=1}^{N} \cos(\omega_n t + \varphi_n), \quad (14.127)$$

where φ_n is again a random variable over the range 0 to 2π.

The advantage of EARP noise is that every noise sample has the same power spectrum. A possible disadvantage is that the amplitudes A_n and B_n are no longer normally distributed. Instead, they are distributed like the probability density functions for the sine or cosine functions, with square-root singularities at $A_n = \pm C$ and $B_n = \pm C$. However, the actual values of noise are normally distributed as long as the number of noise components is more than about five.

14.9.6 Noise Color

White noise has a constant spectral density, which means that the power in white noise is proportional to the bandwidth. On the average, every band with given bandwidth Δf has the same amount of power. Pink noise has a spectral density that decreases inversely with the frequency. Consequently, pink noise decreases at a rate of 6 dB per octave. On the average, every octave band has the same amount of power.

14.10 Sampled data

Converting an analog signal, such as a time-dependent voltage, into a digital representation dices the signal in two dimensions, the dimension of the signal voltage and the dimension of time. Dicing the signal voltage is known as quantization, dicing with respect to time is known as sampling.

14.10.1 Quantization and Quantization Noise

It is common for an analog-to-digital converter (ADC) to represent the values of input voltages as integers. The range of the integers is determined by the number of bits per sample in the conversion process. A conversion into an M-bit sample (or word) allows the voltage value to be represented by 2^M bits. For instance, a 10-bit ADC that is restricted to converting positive voltages would represent 0 V by the number 0 and $+10$ V by $2^{10} - 1$ or 1023.

A 16-bit ADC would allow 2^{16} or 65 536 different values. A 16-bit ADC that converts voltages between -10 and $+10$ V would represent -10 V by $-32\,768$ and $+10$ V by $+32\,767$. Conversion is linear. Thus 0.3052 V would be converted to the sample value 1000 and 0.3055 V to the value 1001. A voltage of 0.3053 would also be converted to a value of 1000, no different from 0.3052. The discrepancy is an error known as the quantization error or *quantization noise*.

Quantization noise referenced to the signal is a signal-to-noise ratio. Standard practice makes this ratio as large as possible by assuming a signal with the maximum possible power. For the positive and negative ADC described above, maximum power occurs for a square wave between a sampled waveform value of $-2^{(M-1)}$ and $+2^{(M-1)}$. The power is the square of the waveform or $\frac{1}{4} \times 2^{2M}$.

For its part, the noise is a random variable that represents the difference between an accurately converted voltage and the actual converted value as limited by the number of bits in the sample word. This error is never more than 0.5 and never less than -0.5. The power in noise that fluctuates randomly over the range -0.5 to $+0.5$ is 1/12. Consequently the signal-to-noise (S/N) ratio is 3×2^{2M}. Expressed in decibels, this value is $10 \log(3 \times 2^{2M})$, or $20M \log(2) + 4.8$ dB, or $6M + 4.8$ dB. For a 16-bit word, this would be $96 + 4.8$ or about 101 dB. An alternative calculation would assume that the maximum power is the power for the largest sine wave that can be reproduced by such a system. This sine has half the power of the square, and the S/N ratio is then about $6M$ dB.

14.10.2 Binary Representation

Digitized data, like a sampled waveform are represented in binary form by numbers (or words) consisting of digits 0 and 1. For example, an eight-bit word consisting of two four-bit bytes and representing the decimal number 7, would be written as

0 0 0 0 0 1 1 1.

This number has 1 in the ones column, 1 in the twos column, 1 in the fours column, and nothing in any other column. One plus two plus four is equal to 7, which is what was desired.

An eight-bit word ($M = 8$) could represent decimal integers from 0 to 255. It cannot represent 2^M, which is decimal 256. If one starts with the decimal number 255 and adds 1, the binary representation becomes all zeros, i.e. $255 + 1 = 0$. It is like the 100 000-mile odometer on an automobile. If the odometer reads 99 999 and the car goes one more mile, the odometer reads 00 000.

Signals are generally negative as often as they are positive, and that leads to a need for a binary representation of negative numbers. The usual standard is a representation known as *twos-complement*. In twos-complement representation, any number that begins with a 1 is negative. Thus, the leading digit serves as a *sign bit*.

In order to represent the number $-x$ in an M-bit system one computes $2^M - x$. That way, if one adds x and $-x$ one ends up with 2^M, which is zero.

A convenient algorithm for calculating the twos-complement of a binary number is to reverse each bit, 0 for 1 and 1 for 0, and then add 1. Thus, in an eight-bit system the number -7 is given by

1 1 1 1   1 0 0 1 .

14.10.3 Sampling Operation

The sampling process replaces an analog signal, which is a continuous function of time, by a sequence of points. The operation is equivalent to the process shown in Fig. 14.9, where the analog signal $x(t)$ is multiplied by a train of evenly spaced delta functions to create a sequence of sampled values $y(t)$.

Intuitively, it seems evident that this operation is a sensible thing to do if the delta functions come along rapidly enough – rapid compared to the speed of the

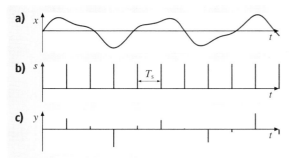

Fig. 14.9 An analog signal $x(t)$ is multiplied by a train of delta functions $s(t)$ to produce a sampled signal $y(t)$

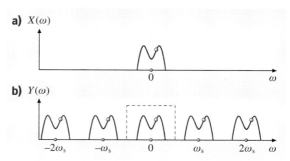

Fig. 14.10 (a) The spectrum of the analog signal $X(\omega)$ is bounded in frequency. (b) The spectrum of the sampled signal, $Y(\omega)$, is the convolution of $X(\omega)$ and the Fourier transform of the sampling train of delta functions. Consequently, multiple images of $X(\omega)$ appear. Frequencies that are allowed by the sampling theorem are included in the *dashed box*. A particular frequency (*circled*) is followed through the multiple imaging

temporal changes in the waveform. That concept is most clearly seen by studying the Fourier transforms of functions x, s and y.

The Fourier transform of the analog signal is $X(\omega)$, with a spectrum that is limited to some highest frequency $\omega_{\max}$. By contrast, the Fourier transform of the train of delta functions is, itself, a train of delta functions, $S(\omega)$ that extends over the entire frequency axis. Because the delta functions in time have period T_s, the delta functions in $S(\omega)$ are separated by ω_s, equal to $2\pi/T_s$. Because $y(t)$ is the product of the time-dependent analog signal and the train of delta functions, the Fourier transform $Y(\omega)$ is the convolution of $X(\omega)$ and $S(\omega)$, as shown in part (b) of Fig. 14.10. Because of the convolution operation, $Y(\omega)$ includes multiple images of the original spectrum.

It is evident from Fig. 14.10b that, if the $\omega_{\max}$ is less than half of ω_s, the multiple images will not overlap. That observation has the status of a theorem known as the *sampling theorem*, which says that the sampled signal is an adequate representation of an analog signal if the sample rate is more than twice the highest frequency in the analog signal, i.e., $\omega_s > 2\omega_{\max}$.

As an example of a failure to apply the sampling theorem, suppose that a 600 Hz sine tone is sampled at a rate of 1000 Hz. The spectrum of the sampled signal will contain 600 Hz as expected, and it will also contain a component at $1000 - 600 = 400$ Hz. The 400 Hz component was not present in the original spectrum; it is an alias, an unwanted image of the 600 Hz tone.

14.10.4 Digital-to-Analog Conversion

In converting a signal from digital to analog form, one can begin with the train of delta functions that is signal $y(t)$ as shown in Fig. 14.9c. An electronic device to do that is a digital-to-analog converter (DAC). However, as shown in Fig. 14.10b, this signal includes many high frequencies that are unwanted byproducts of the sampling process. Consequently, one needs to low-pass filter the signal so as to pass only the frequencies less than half the sample rate, i.e., the frequencies in the dashed box. Such a low-pass filter is called a *reconstruction filter*.

Practical DACs do not produce delta-function voltage spikes. Instead, they produce rectangular functions with durations pT_s, where p is a fraction of a sample period $0 < p \leq 1$. If $p = 1$, the output of the DAC resembles a staircase function. Mathematically, replacing the delta function train of Fig. 14.9c by the train of rectangles is equivalent to convolving the function $y(t)$ with a rectangular function. The consequence of this convolution is that the output is filtered, and the transfer function of the filter is the Fourier transform of the rectangle. The magnitude of the transfer function is

$$|H(\omega)| = \frac{\sin(\omega p T_s/2)}{\omega p T_s/2} . \tag{14.128}$$

The phase shift of the filter is a pure delay and consequently unimportant. The effective filtering that results from the rectangles, known as *sin(x)-over-x* filtering, can be corrected by the reconstruction filter.

14.10.5 The Sampled Signal

This brief section will introduce a notation that will be useful in later discussions of sampled signals. It is supposed at the outset that one begins with a total of N samples, equally spaced in time by the sampling period T_s. By convention, the first sample occurs at time $t = 0$ and the last sample occurs at time $t = (N-1)T_s$. Consequently, the signal duration is $T_D = (N-1)T_s$.

In dealing with sampled signals, it is common to replace the time variable with a discrete index k. Thus,

$$x(t) = x(kT_s) = x_k , \tag{14.129}$$

where the equation on the left indicates that the original data exist only at discrete time values.

14.10.6 Interpolation

The discrete-time values of a sampled waveform can be used to compute an approximate Fourier transform of the original signal. This Fourier transform is valid up to a frequency as high as half the sample rate, i.e., as high as $\omega_s/2$, or π/T_s. The Fourier transform can then be used to estimate the values of the original signal $x(t)$ at times other than the sample times. In this way, the Fourier transform computed from the samples serves to interpolate between the samples. Such an interpolation scheme proceeds as follows.

First, the Fourier transform is

$$X(\omega) = T_s \sum_k x_k \exp(-i\omega T_s k) , \tag{14.130}$$

where, as noted above, x_k is the signal $x(t)$ at the times $t = T_s k$, and the leading factor of T_s gets the dimensions right.

Then the inverse Fourier transform is

$$x(t) = \frac{T_s}{2\pi} \int_{-\omega_s/2}^{\omega_s/2} d\omega \, e^{i\omega t} \sum_k x_k \, e^{-i\omega T_s k} . \tag{14.131}$$

Reversing the order of sum and integral and using the fact that $T_s \omega_s/2 = \pi$, we find that

$$x(t) = \sum_k x_k \frac{\sin \pi(t/T_s - k)}{\pi(t/T_s - k)} . \tag{14.132}$$

The sinc function is 1.0 whenever $t = T_s k$, and is zero whenever t is some other integer multiple of T_s. Therefore, the sum on the right only interpolates; it does not change the values of $x(t)$ when t is equal to a sample time.

14.11 Discrete Fourier Transform

The Fourier transform of a signal with finite duration is well defined in principle. The finite signal itself can be regarded as some base function that is multiplied by a rectangular window to limit the duration. Then the Fourier transform proceeds by convolving with the transform of the window. For example, a truncated exponentially decaying sine function can be regarded as a decaying sine, with the usual infinite duration, multiplied by a rectangular window. Then the Fourier transform of the truncated function is the Fourier transform of the decaying sine convolved with a sinc function – the Fourier transform of the rectangular window. Such a Fourier transform is a function of a continuous frequency, and it shows the broad spectrum associated with the abrupt truncation.

In digital signal processing the frequency axis is not continuous. Instead, the Fourier transform of a signal is defined at discrete frequencies, just as the signal itself is defined at discrete time points. This kind of Fourier transform is known as the discrete Fourier transform (DFT).

To compute the DFT of a function, one begins by periodically repeating the function over the entire time axis. For example, the truncated decaying sine in Fig. 14.11a is repeated in Fig. 14.11b where it should be imagined that the repetition precedes indefinitely to the left and right.

Then the Fourier transform of the periodically repeated signal becomes a Fourier series. The fundamental frequency of the Fourier series is the reciprocal of the duration, $f_0 = 1/T_D$, and the spectrum becomes a set of discrete frequencies, which are the harmonics of f_0. For instance, if the signal is one second in duration, the spectrum consists of the harmonics of 1 Hz, and if the duration is two seconds then the spectrum has all the harmonics of 0.5 Hz. As expected, the highest harmonic is limited to half the sample rate. That Fourier series is the DFT. Using x_k to define the periodic repetition of the original discrete function, x_k, the DFT $\underline{X}(\omega)$ is defined for $\omega = 2\pi n/T_D$, where n indicates the n-th harmonic. In terms of the fundamental angular frequency $\omega_o = 2\pi/T_D$, the DFT is

$$\underline{X}(n\omega_o) = T_s \sum_{k=0}^{N-1} x_k \, \mathrm{e}^{-\mathrm{i}n\omega_o k T_s} , \qquad (14.133)$$

where the prefactor T_s keeps the dimensions right. The product $\omega_o T_s$ is equal to $\omega_o T_D/(N-1)$ or $2\pi/(N-1)$, and so

$$\underline{X}(n\omega_o) = T_s \sum_{k=0}^{N-1} x_k \, \mathrm{e}^{-\mathrm{i}2\pi nk/(N-1)} . \qquad (14.134)$$

Both positive and negative frequencies occur in the Fourier transform. Because the maximum frequency is equal to $[1/(2T_s)]/(1/T_D)$ times the fundamental frequency, the number of discrete positive frequencies is $(N-1)/2$, and the number of discrete negative frequencies is the same. Consequently the inverse DFT can be

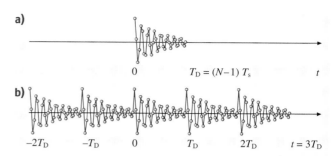

Fig. 14.11a,b A decaying function in part (**a**) is periodically repeated in part (**b**) to create a periodic signal with period T_D

written

$$x_k = \frac{1}{T_D} \sum_{n=-(N-1)/2}^{(N-1)/2} X(n\omega_o) \, \mathrm{e}^{\mathrm{i}2\pi nk/(N-1)} ,$$

$$(14.135)$$

or

$$x(t) = \frac{1}{T_D} \sum_{n=-(N-1)/2}^{(N-1)/2} X(n\omega_o) \, \mathrm{e}^{\mathrm{i}n\omega_o t} . \qquad (14.136)$$

A virtue of the DFT is that the information in the DFT is exactly what is needed to create the original truncated function $x(t)$ – no more and no less. The fact that the DFT spectrum actually creates the periodically repeated function $\underline{x}_k$ and not the original x_k is not a problem if we agree in advance to ignore $\underline{x}_k$ for k outside the range of the original time-limited signal. However, it should be noted that certain operations, such as time translations, products, and convolution, that have familiar characteristics in the context of the Fourier transform, retain those characteristics only for the periodically extended signal $\underline{x}_k$ and its Fourier transform $\underline{X}(n\omega_o)$ and not for the finite-duration signal.

14.11.1 Interpolation for the Spectrum

It is possible to estimate the Fourier transform at values of frequency between the harmonics of ω_o. The procedure begins with the definition of the Fourier transform of a finite function,

$$X(\omega) = \int_0^{T_D} \mathrm{d}t \, x(t) \mathrm{e}^{-\mathrm{i}\omega t} . \qquad (14.137)$$

Next, the function $x(t)$ is replaced by the inverse DFT from (14.136), and the variable of integration t is replaced by t', which has symmetrical upper and lower limits,

$$X(\omega) = \frac{1}{T_D} \int_{-T_D/2}^{T_D/2} dt' \, e^{-i\omega t'} \times \sum_{n=-(N-1)/2}^{(N-1)/2} X(n\omega_o) \, e^{in\omega_o t'} \, e^{-i\omega T_D/2} \, e^{in\omega_o T_D/2} ,$$
(14.138)

which reduces to

$$X(\omega) = \sum_{n=-(N-1)/2}^{(N-1)/2} X(n\omega_o) \frac{\sin[(\omega - n\omega_o)T_D/2]}{(\omega - n\omega_o)T_D/2} \times e^{-i\omega T_D/2} \, e^{i\pi n} .$$
(14.139)

14.12 The z-Transform

Like the discrete Fourier transform, the z-transform is well suited to describing sampled signals. We consider $x(t)$ to be a sampled signal so that it is defined at discrete time points $t = t_k = kT_s$, where T_s is the sampling period. Then the time dependence of x can be described by an index, $x_k = x(t_k)$. The z-transform of x is

$$X(z) = \sum_{k=-\infty}^{\infty} x_k z^{-k} .$$
(14.140)

The quantity z is complex, with amplitude A and phase φ,

$$z = A \, e^{i\varphi} = A \, e^{i\omega T_s} ,$$
(14.141)

where φ is the phase advance in radians per sample.

In the special case where $A = 1$, all values of z lie on a circle of radius 1 (the *unit circle*) in the complex z plane. In that case the z-transform is equivalent to the discrete Fourier transform. An often-overlooked alternative view is that the z-transform is an extension of the Fourier transform wherein the angular frequency ω becomes complex,

$$\omega = \omega_R + i\omega_I ,$$
(14.142)

so that

$$z = e^{-\omega_I T_s} \, e^{i\omega_R T_s} .$$
(14.143)

The extended Fourier transform will not be pursued further in this chapter.

A well-defined z-transform naturally includes a function of variable z, but the function itself is not enough. In order for the inverse transform to be unique, the definition also requires that the region of the complex plane in which the transform converges must also be specified.

Table 14.2 z-Transform pairs

x_k	$X(z)$	Radius of convergence
δ_{k,k_0}	z^{-k_0}	all z
$a^k u_k$	$z/(z-a)$	$|z| > a$
$k a^k u_k$	$az/(z-a)^2$	$|z| > a$
$a^k \cos(\omega_o T_s k) u_k$	$\frac{z^2 - az\cos(\omega_o T_s)}{z^2 - 2az\cos(\omega_o T_s) + a^2}$	$|z| > a$
$a^k \sin(\omega_o T_s k) u_k$	$\frac{az\sin(\omega_o T_s)}{z^2 - 2az\cos(\omega_o T_s) + a^2}$	$|z| > a$

To illustrate that point, one can consider two different functions x_k that have the same z-transform function, but different regions of convergence.

Consider first the function

$$x_k = 2^k \quad \text{for} \quad k \geq 0 ,$$
(14.144)
$$x_k = 0 \quad \text{for} \quad k < 0 .$$

This two-line function can be written as a single line by using the discrete Heaviside function u_k. The function u_k is defined as zero when k is a negative integer and as $+1$ when k is any other integer, including zero. Then x_k above becomes

$$x_k = 2^k u_k .$$
(14.145)

The z-transform of x_k is

$$X(z) = \sum_{k=0}^{\infty} (2/z)^k .$$
(14.146)

The sum is a geometric series, which converges to

$$X(z) = \frac{1}{1 - 2/z} = z/(z-2)$$
(14.147)

if $|z| > 2$. The region of convergence is therefore the entire complex plane except for the portion inside and on a circle of radius 2.

Next consider the function

$$x_k = -2^k u_{-k-1} . \quad (14.148)$$

The z-transform of x_k is

$$X(z) = -\sum_{k=-\infty}^{-1}(2/z)^k \quad \text{or} \quad -\sum_{k=1}^{\infty}(z/2)^k . \quad (14.149)$$

The sum converges to

$$X(z) = -\frac{(z/2)}{1-z/2} = \frac{z}{z-2} \quad (14.150)$$

if $|z| < 2$. The function is identical to the function in (14.147), but the region of convergence is now the portion of the complex plane entirely inside the circle of radius 2.

The inverse z-transform is given by a counterclockwise contour integral circling the origin

$$x_k = \frac{1}{2\pi \mathrm{i}} \oint_C \mathrm{d}z X(z) z^{k-1} . \quad (14.151)$$

The contour C must lie entirely within the region of convergence of x and must enclose all the poles of $X(z)$.

The regions of convergence when the functions x and y are combined in some way are at least the intersection of the regions of convergence for x and y separately. Scaling and time reversal lead to regions of convergence that are scaled and inverted, respectively. For instance, if $X(z)$ converges in the region between radii r_1 and r_2, them $X(1/z)$ converges in the region between $1/r_2$ and $1/r_1$.

14.12.1 Transfer Function

The output of a process at time point k, namely y_k, may depend on the inputs x at earlier times and also on the outputs at earlier times. In equation form,

$$y_k = \sum_{q=0}^{Nq} \beta_q x_{k-q} - \sum_{p=1}^{Np} \alpha_p y_{k-p} . \quad (14.152)$$

This equation can be z-transformed using the time-shift property in Table 14.3,

$$\sum_{q=0}^{Nq} \beta_q z^{-q} X(z) = \sum_{p=0}^{Np} \alpha_p z^{-p} Y(z) , \quad (14.153)$$

where $\alpha_0 = 1$. The transfer function is the ratio of the transformed output over the transformed input,

$$H(z) = Y(z)/X(z) , \quad (14.154)$$

which is

$$H(z) = \frac{\sum_{q=0}^{Nq} \beta_q z^{-q}}{\sum_{p=0}^{Np} \alpha_p z^{-p}} . \quad (14.155)$$

From the fundamental theorem of algebra, the numerator of the fraction above has Nq roots and the denominator has Np roots, so that $H(z)$ can be written as

$$H(z) = \frac{(1-q_1 z^{-1})(1-q_2 z^{-1})\ldots(1-q_{Nq}z^{-1})}{(1-p_1 z^{-1})(1-p_2 z^{-1})\ldots(1-p_{Np}z^{-1})} . \quad (14.156)$$

This equation and its development are of central importance to digital filters, also known as linear time-invariant systems. If the system is recursive, outputs from a previous point in time are sent back to the input. Therefore, some values of the coefficients α_p are finite for $p \geq 1$ and so are the values of some poles, such as p_2. Such filters are called *infinite impulse response* (IIR) filters because it is possible that the response of the system to an impulse put in at time zero will never entirely die out. Some of the output is always fed back into the input. A similar conclusion is reached by recognizing that the expansion of $1/(1-pz^{-1})$ in powers of z^{-1} goes on forever. Because the system has poles, there are concerns about stability.

If the system is nonrecursive, no values of the output are ever sent back to the input. Therefore, the denominator of $H(z)$ is simply the number 1. Such filters are called *finite impulse response* (FIR) filters because their response to a delta function input will always die out eventually as long as Nq is finite. The system is said to be an *all-zero* system. The order of the filter is estab-

Table 14.3 Properties of the z-transform

Property	Signal	z-transform
Definition	x_k	$X(z)$
Linearity	$ax_k + by_k$	$aX(z) + bY(z)$
Time shift	x_{k-k_o}	$z^{-k_o}X(z)$
Scaling z	$a^k x_k$	$X(z/a)$
Time reversal	x_{-k}	$X(1/z)$
Derivative w.r.t. z	kx_k	$-z \, \mathrm{d}X(z)/\mathrm{d}z$
Convolution	$x_k * y_k$	$X(z)Y(z)$
Multiplication	$x_k y_k$	$\frac{1}{2\pi\mathrm{i}} \oint_C \mathrm{d}z'/z' \; X(z')Y(z/z')$

lished by Nq or Np, the number of time points back to the earliest input or output that contribute to the current output value.

The formal z-transform,

$$H(z) = \sum_{k=-\infty}^{\infty} h_k z^{-k} \tag{14.157}$$

leads to conclusions about causality and stability.

A filter is causal if the current value of the output does not depend on future inputs. For a causal filter h_k is zero for $k < 0$. Then this sum has no terms with positive powers of z, and the region of convergence of $H(z)$ includes $|z| = \infty$.

A filter is stable if

$$S = \sum_{k=-\infty}^{\infty} |h_k| \tag{14.158}$$

is finite. It follows, that $H(z)$ is finite for $|z| = 1$, i. e., for z on the unit circle. Thus, if the region of convergence includes the unit circle, the filter is stable.

14.13 Maximum Length Sequences

A maximum length sequence (MLS) is a train of ones and zeros that makes a useful signal for measuring the impulse response of a linear system. An MLS can be generated by a bit-shift register, which resembles a bucket brigade. To make an N-bit MLS, one needs an N-stage shift register. Each stage can hold either a one or a zero. The register is imagined to have a clock which synchronizes the transfer of bits from each stage to the next. On every clock tick the content of each stage of the register is transferred to the next stage down the line. The content of the last stage is regarded as the output of the register, and it is also fed back into the first stage. In addition, the output can be fed back into one or more of the other stages, and when that occurs the stage receiving the output, in addition to the content of the previous stage, performs an exclusive OR (XOR) operation on those two inputs. The XOR operation obeys the truth table shown in Table 14.4. In words, the XOR of inputs A and B is zero if A and B are the same and is 1 if A and B are different.

A shift register with three stages is shown in Fig. 14.12. With three stages and feedback *taps* to stages 1 and 2, it is defined as [3: 1,2].

At the instant shown in the figure, the register holds the value 1,1,1. The subsequent development of the register values is given in Table 14.5. The sequence repeats after seven steps. The table shows that every possible pattern of *ones* and *zeros* occurs once, and only once, before the pattern repeats. There are $2^N - 1 = 2^3 - 1 = 7$ such patterns. There is one exception, namely the pattern 0,0,0. If this pattern should ever appear in the register then the process gets stuck forever. Therefore, this pattern is not allowed. The output sequence is the contents of the stage on the right, here, 1,1,0,0,1,0,1. Because all seven register patterns appear before repetition, this

Table 14.4 Truth table for the exclusive or (XOR) operation

A	B	A XOR B
0	0	0
0	1	1
1	0	1
1	1	0

Table 14.5 Successive values in the shift register of Fig. 14.12

Step			
0	1	1	1
1	1	0	1
2	1	0	0
3	0	1	0
4	0	0	1
5	1	1	0
6	0	1	1
7	1	1	1
8	1	0	1
9	1	0	0

Table 14.6 Successive values in the shift register of Fig. 14.13

Step			
0	1	1	1
1	1	0	0
2	0	1	0
3	0	0	1
4	1	1	1
5	1	0	0
6	0	1	0

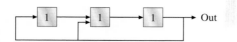

Fig. 14.12 A three-stage shift register [3: 1, 2] in which the output is fed back into the first and second stages

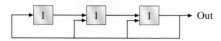

Fig. 14.13 A three-stage shift with feedback into all the stages does not produce an MLS

output is a maximum length sequence. There is nothing special about the starting register value, 1,1,1. Therefore, any cyclic permutation of the MLS is also an MLS. For instance, the sequence, 1,0,0,1,0,1,1 is the same sequence.

An example of a three-bit shift register that does not produce an MLS is [3: 1,2,3], shown in Fig. 14.13. The pattern for this shift register is shown in Table 14.6. The pattern of register values begins to repeat after only four steps. Therefore, the sequence of output values, namely 1,0,0,1,1,0,0,1,1,0,0,1, is not an MLS.

14.13.1 The MLS as a Signal

To make a signal from an MLS requires only one step: every 0 in the sequence is replaced by -1. Therefore, the MLS for the shift register in Fig. 14.12 becomes: $1, 1, -1, -1, 1, -1, 1$. For this three-stage register ($N = 3$) the MLS has a length of seven; there are four $+1$ values and three -1 values. These results can be generalized to an N-stage register which has $2^N - 1$ values; $2^{(N-1)}$ are $+1$ values and $2^{(N-1)} - 1$ are -1 values. The average value is therefore $1/(2^N - 1)$.

14.13.2 Application of the MLS

The key fact about an MLS is that its autocorrelation function is very nearly a delta function. To express that idea, one can write the autocorrelation function in the form appropriate for discrete samples,

$$c_k = \frac{1}{2^N - 1} \sum_{k1} x_{k1} \, x_{k1+k} \,. \tag{14.159}$$

This sum, and all sums to follow, are over the $2^N - 1$ values of the MLS sequence x. Because the sequence is cyclical, it does not matter where one starts the sum.

An MLS has the property that

$$c_k = \left(1 + \frac{1}{2^N - 1}\right) \delta_{k,0} - \frac{1}{2^N - 1} \,. \tag{14.160}$$

Therefore, c_k is approximately a Kronnecker delta function

$$c_k \approx \delta_{k,0} \,. \tag{14.161}$$

If we would like to know the impulse response h of a linear system, we can excite the system with the MLS x, and record the output y. As for filters, the linear response y is the convolution of x and h, i.e.,

$$y_k = x * h = \sum_{k1} x_{k1+k} \, h_{k1} \,. \tag{14.162}$$

To find the impulse response, one can form the quantity d, by convolving the recording y with the original MLS x, i.e.,

$$d_k = \frac{1}{2^N - 1} \sum_{k2} x_{k2+k} \, y_{k2} \tag{14.163}$$

or from (14.162)

$$d_k = \frac{1}{2^N - 1} \sum_{k1,k2} x_{k2+k} \, x_{k1+k2} \, h_{k1} \,. \tag{14.164}$$

Only $x * x$ involves the index $k2$, and doing the sum over $k2$ leads to

$$d_k = \sum_{k1} \delta_{k,k1} \, h_{k1} \tag{14.165}$$

so that $d_k = h_k$. In this way, we have found the desired impulse response.

As applied in architectural acoustics, the MLS is an alternative to recording the response to a popping balloon or gun shot. Because the MLS is continuous, it avoids the dynamic-range problem associated with an impulsive test signal, and by repeating the sequence one can achieve remarkable noise immunity.

Similarly, the MLS is an alternative to recording the response to white noise (the MLS is white). However, digital white noise, such as random telegraph noise, has an autocorrelation function that is zero only for a long-term or ensemble average. In practice, the white-noise response of a linear system is much noisier than the MLS response.

14.13.3 Long Sequences

Table 14.7 gives the taps for some MLSs generated by shift registers with 2–20 stages, i.e., orders 2–20. The

Table 14.7 Taps for maximum length sequences

Number of stages	Length (bits)	Number of taps	Number of sets	Set
2	3	2	1	[2: 1,2]
3	7	2	1	[3: 1,3]
4	15	2	1	[4: 1,4]
5	31	2	1	[5: 1,4]
6	63	2	1	[6: 1,6]
7	127	2	2	[7: 1,7], [7: 1,5]
8	255	4	6	[8: 1,2,7,8]
9	511	2	1	[9: 1,6]
10	1023	2	1	[10: 1,8]
11	2047	2	1	[11: 1,10]
12	4095	4	9	[12: 1,6,7,9]
13	8191	4	33	[13: 1,7,8,9]
14	16383	4	21	[14: 1,6,9,10]
15	32767	2	3	[15: 1,9], [15: 1,12], [15: 1,15]
16	65535	4	26	[16: 1,7,10,11]
17	131071	2	3	[17: 1,12], [17: 1,13], [17: 1,15]
18	262143	2	1	[18: 1,12]
19	524287	4	79	[19: 1,10,11,14]
20	1048575	2	1	[20: 1,18]

longest sequence has a length of more than one million bits,

For orders 2, 3, and 4, there is only one possible set of taps. These sets have two taps, including the feedback to stage 1. For orders 7, 15, and 17 there is more than one set with two taps, and all of them are shown in the table.

Beginning with order 5 there are four-tap sets as well as two-tap sets, except that for some orders, such as 8, there are no two-tap sets. For every order the table gives a set with the smallest possible number of taps.

Beginning with order 7 there are six-tap sets. As the order increases the number of sets also increases. For order 19, there are 79 four-tap sets.

14.14 Information Theory

Information theory provides a way to quantify information by computing information content. The information content of a message depends on the context, and the context determines the initial uncertainty about the message. Suppose, for example, that we receive one character, but we know in advance that the context is one in which the character must be a digit between 0 and 9. Our uncertainty before receiving that actual character is described by the number of possible outcomes, which is $\Omega = 10$ in this case. Suppose instead, that the context is one in which the character must be a letter of the alphabet. Then our initial uncertainty is greater because the number of possible outcomes is now $\Omega = 26$. The first step of information theory is to recognize that, when we actually receive and identify a character, the information content of that character is greater in the second context than in the first because in the second context the character has eliminated a greater number of a priori possibilities.

The second step in information theory is to consider a message with two characters. If the context of the message is decimal digits then the number of possibilities is the product of 10 for the first digit and 10 for the second, namely $\Omega = 100$ possibilities. Compared to a message with one character, the number of possibilities has been multiplied by 10. However, it is only logical to expect that two characters will give twice as much information as one, not 10 times as much. The logical problem can be solved by quantifying information in terms of entropy, which is the logarithm of the number of possibilities

$$H = \log \Omega . \qquad (14.166)$$

Because log 100 is just twice log 10, the logical problem is solved. The information measured in bits is obtained by using a base 2 logarithm.

A few simple features follow immediately. If the number of possible messages is $\Omega = 1$ then the message provides no information, which agrees with $\log 1 = 0$. If the context is binary, where a character can be only 1 or 0 ($\Omega = 2$), then receiving a character provides 1 bit of information, which agrees with $\log_2 2 = 1$.

If the context is an alphabet with M possible symbols, and all of the symbols are equally probable, then a message with N characters has $\Omega = M^N$ possible outcomes and the information entropy is

$$H = \log M^N = N \log M , \qquad (14.167)$$

illustrating the additivity of information over the characters of the message.

14.14.1 Shannon Entropy

Information theory becomes interesting when the probabilities of different symbols are different. Shannon [14.3, 4] showed that the information content per character is given by

$$H_c = -\sum_{i=1}^{M} p_i \log p_i , \qquad (14.168)$$

where p_i is the probability of symbol i in the given context.

The rest of this section proves Shannon's formula. The proof begins with the plausible assumption that, if the probability of symbol i is p_i, then in a very long message of N characters, the number of occurrences of character i, m_i will be exactly $m_i = N p_i$.

The number of possibilities for a message of N characters in which the set of $\{m_i\}$ is fixed by the corresponding $\{p_i\}$ is

$$\Omega = \frac{N!}{m_1! \, m_2! \, \ldots \, m_M!} . \qquad (14.169)$$

Therefore,

$$H = \log N! - \log m_1! - \log m_2! - \ldots \log m_M! . \qquad (14.170)$$

One can write $\log N!$ as a sum

$$\log N! = \sum_{k=1}^{N} \log k \qquad (14.171)$$

and similarly for $\log m_i!$.

For a long message one can replace the sum by an integral,

$$\log N! = \int_{1}^{N} dx \log x = N \log N - N + 1 \qquad (14.172)$$

and similarly for $\log m_i!$.

Therefore,

$$H = N \log N - N + 1 \\ - \sum_{i=1}^{M} m_i \log m_i + \sum_{i=1}^{M} m_i - \sum_{i=1}^{M} 1 . \qquad (14.173)$$

Because $\sum_{i=1}^{M} m_i = N$, this reduces to

$$H = N \log N + 1 - \sum_{i=1}^{M} m_i \log m_i - M . \qquad (14.174)$$

The information per character is obtained by dividing the message entropy by the number of characters in the message,

$$H_c = \log N - \sum_{i=1}^{M} p_i \log m_i + (1-M)/N , \qquad (14.175)$$

where we have used the fact that $m_i/N = p_i$.

In a long message, the last term can be ignored as small. Then because the sum of probabilities p_i is equal to 1,

$$H_c = -\sum_{i=1}^{M} p_i (\log m_i - \log N) , \qquad (14.176)$$

or

$$H_c = -\sum_{i=1}^{M} p_i \log p_i , \qquad (14.177)$$

which is (14.168) as advertised.

If the context of written English consists of 27 symbols (26 letters and a space), and if all symbols are equally probable, then the information content of a single character is

$$H_c = -1.443 \sum_{i=1}^{27} \frac{1}{27} \ln \frac{1}{27} = 4.75 \text{ (bits)} , \qquad (14.178)$$

where the factor $1/\ln(2) = 1.443$ converts the natural log to a base 2 log. However, in written English all symbols are not equally probable. For example, the most common letter, 'E', is more than 100 times more likely to occur than the letter 'J'. Because equal probability of symbols always leads to the highest entropy, the unequal probability in written English is bound to reduce the information content – to about 4 bits per character. An even greater reduction comes from the redundancy in larger units, such as words, so that the information content of written English is no more than 1.6 bits per character.

The concept of information entropy can be extended to continuous distributions defined by a probability density function

$$h = -\int_{-\infty}^{\infty} dx \, \text{PDF}(x) \, \log[\text{PDF}(x)] \,. \quad (14.179)$$

14.14.2 Mutual Information

The mutual information between sets of variables $\{i\}$ and $\{j\}$ is a measure of the amount of uncertainty about one of these variables that is eliminated by knowing the other variable. Mutual information H_m is given in terms of the joint probability mass function $p(i, j)$

$$H_m = \sum_{i=1}^{M} \sum_{j=1}^{M} p(i,j) \log \frac{p(i,j)}{p(i)p(j)} \,. \quad (14.180)$$

Using written English as an example again, $p(i)$ might describe the probability for the first letter of a word and $p(j)$ might describe the probability for the second. It is convenient to let the indices i and j be integers, e.g., $p(i = 1)$ is the probability that the first letter is an 'A', and $p(j = 2)$ is the probability that the second letter is a 'B'. Then $p(1, 2)$ is the probability that the word starts with the two letters 'AB'. It is evident that in a context where the first two letters are completely independent of one another so that $p(i, j) = p(i)p(j)$ then the amount of mutual information is zero because $\log(1) = 0$. In the opposite limit the context is one in which the second letter is completely determined by the first. For instance, if the second letter is always the letter of the alphabet that immediately follows the first letter then $p(i, j) = p(j) = p(i)\delta(j, i+1)$, and

$$H_m = \sum_{i=1}^{M} p(i) \log \frac{p(i)}{p(i)p(i)} \quad (14.181)$$

which simply reduces to (14.168) for H_c, the information content of the first letter of the word.

In the general case, the mutual information is a difference in information content. It is equal to the information provided by the second letter of the word given no prior knowledge at all, minus the information provided by the second letter of the word given knowledge of the first letter. Mathematically, $p(i, j) = p(i)p(j|i)$, where $p(j|i)$ is the probability that the second letter is j given that the first letter is i. Then

$$H_m = \sum_{j=1}^{M} p(j) \log \frac{1}{p(j)}$$
$$- \sum_{i=1}^{M} \sum_{j=1}^{M} p(i,j) \log \frac{1}{p(j|i)} \,. \quad (14.182)$$

The information transfer ratio T is the degree to which the information in the first letter predicts the information in the second. Equivalently, it describes the transfer of information from an input to an output

$$T = \frac{-H_m}{\sum_{i=1}^{M} p(i) \log p(i)} \,. \quad (14.183)$$

This ratio ranges between 0 and 1, where 1 indicates that the second letter, or output, can be predicted from the first letter, or input, with perfect reliability. The mutual information is the basis for the calculation of the information capacity of a noisy communications channel.

References

14.1 W.M. Hartmann: *Signals, Sound, and Sensation* (Springer, Berlin, New York 1998)
14.2 A. Einstein, L. Hopf: A principle of the calculus of probabilities and its application to radiation theory, Annal. Phys. **33**, 1096–1115 (1910)
14.3 C.E. Shannon: A mathematical theory of communication, Part I, Bell Syst. Tech. J. **27**, 379–423 (1948)
14.4 C.E. Shannon: A mathematical theory of communication, Part II, Bell Syst. Tech. J. **27**, 623–656 (1948)

Part E Music, Speech, Electroacoustics

15 Musical Acoustics
Colin Gough, Birmingham, UK

16 The Human Voice in Speech and Singing
Björn Lindblom, Stockholm, Sweden
Johan Sundberg, Stockholm, Sweden

17 Computer Music
Perry R. Cook, Princeton, USA

18 Audio and Electroacoustics
Mark F. Davis, San Francisco, USA

15. Musical Acoustics

This chapter provides an introduction to the physical and psycho-acoustic principles underlying the production and perception of the sounds of musical instruments. The first section introduces generic aspects of musical acoustics and the perception of musical sounds, followed by separate sections on string, wind and percussion instruments.

In all sections, we start by considering the vibrations of simple systems – like stretched strings, simple air columns, stretched membranes, thin plates and shells. We show that, for almost all musical instruments, the usual text-book description of such systems is strongly perturbed by material properties, geometrical factors and acoustical coupling between the drive mechanism, vibrating system and radiated sound.

For stringed, woodwind and brass instruments, we discuss excitation by the bow, reed and vibrating lips, which all involve strongly non-linear processes, even though the vibrations of the excited system usually remains well within the linear regime. However, the amplitudes of vibration of very strongly excited strings, air columns, thin plates and membranes can sometimes exceed the linear approximation limit, resulting in a number of interesting non-linear phenomena, often of significant musical importance.

Musical acoustics therefore provides an excellent introduction to the physics of both linear and non-linear acoustical systems, in a context of rather general interest to professional acousticians, teachers and students, at both school and college levels.

The subject continues its long tradition in advancing the frontiers of experimental, computational and theoretical acoustics, in an area of wide general appeal and contemporary relevance.

By discussing the theoretical models and experimental methods used to investigate the acoustics of many musical instruments, we have aimed to provide a useful background for professional acousticians, students and their teachers, for whom musical acoustics provides an exceedingly rich area for original research projects at all educational levels.

15.1	**Vibrational Modes of Instruments**	535
	15.1.1 Normal Modes	535
	15.1.2 Radiation from Instruments	537
	15.1.3 The Anatomy of Musical Sounds	540
	15.1.4 Perception and Psychoacoustics	551
15.2	**Stringed Instruments**	554
	15.2.1 String Vibrations	555
	15.2.2 Nonlinear String Vibrations	563
	15.2.3 The Bowed String	566
	15.2.4 Bridge and Soundpost	570
	15.2.5 String–Bridge–Body Coupling	575
	15.2.6 Body Modes	581
	15.2.7 Measurements	594
	15.2.8 Radiation and Sound Quality	598
15.3	**Wind Instruments**	601
	15.3.1 Resonances in Cylindrical Tubes	602
	15.3.2 Non-Cylindrical Tubes	606
	15.3.3 Reed Excitation	619
	15.3.4 Brass-Mouthpiece Excitation	628
	15.3.5 Air-Jet Excitation	633
	15.3.6 Woodwind and Brass Instruments	637
15.4	**Percussion Instruments**	641
	15.4.1 Membranes	642
	15.4.2 Bars	648
	15.4.3 Plates	652
	15.4.4 Shells	658
References		661

Because the subject is ultimately about the sounds produced by musical instruments, a large number of audio illustrations have been provided on a CD accompanying this volume, which can also be accessed by the electronic version of the Handbook on springerlink.com. The extensive list of references is intended as a useful starting point for entry to the current research literature, but makes no attempt to provide a comprehensive list of all important research.

This chapter highlights the acoustics of musical instruments. Other related topics, such as the human voice, the perception and psychology of sound, architectural acoustics, sound recording and reproduction, and many experimental, computational and analytic techniques are described in more detail elsewhere in this volume.

Musical acoustics is one the oldest of all the experimental Sciences (see *Levenson* [15.1] for an informative account of the interactions between Music and Science over the ages). The observation of the relationship between the notes produced by the exact fraction divisions of a stretched string and consonant musical intervals like the octave (2:1), perfect fifth (3:2) and fourth (4:3), resulted in the first physical law to be expressed in mathematical terms. It also led to the idea of a divinely created cosmos based on exact fractions, filled with the *music of the spheres* (see, for example, Kepler's account of the ellipticity of the planetary orbits described as notes on a musical scale, in *Harmonies of the World* (1618) [15.2]). Ultimately, such observations led to Newton's discovery of celestial dynamics and the laws of gravity leading to the modern view of the universe subject to physical laws rather than numerical relationships.

In the nineteenth century, musical acoustics continued to occupy a central scientific role. This culminated in Lord Rayleigh's monumental two volumes on the *Theory of Sound* [15.3], which still provide the foundations for almost every branch of modern acoustics. The 19th century advances in understanding waves in acoustics also laid the mathematical framework for quantum wave mechanics in the early part of the 20th century. More recently, the physics of vibrating strings can be said to have come full circle, with the suggestion that string-like vibrations of the quantum field equations account for the mass to the elementary particles (*Hawkins* [15.4]).

Musical acoustics still remains a challenging and exciting field of research and continues to advance mainstream acoustics in many ways. Examples include nonlinear physics and the use of laser holography and both modal and finite-element analysis to investigate complicated vibrating systems. Such developments are described in this chapter and in more detail in other chapters of this Handbook and in the *Physics of Musical Instruments* by *Fletcher* and *Rossing* [15.5], which will often be cited, as an authoritative text and source of additional references for most topics discussed. The *Science of Sound* by *Rossing* et al. [15.6] covers an even wider range of topics at a somewhat less mathematical level. An informative overview of the history, technology and performance of western musical instruments has recently been published by *Campbell*, *Greated* and *Myers* [15.7].

The first section of this chapter deals with the generic properties of the vibrations and sounds of musical instruments. A brief description is first given of the properties of both simple and coupled resonators, typifying the vibrational modes of stringed, wind and percussion instruments, where the sound is generated by vibrating strings, air columns, plates, membranes and shells. The radiation of sound by such structures is then described in terms of multipole sources. This is followed by a brief description of the envelopes, waveforms and spectra of the sounds that characterize the sound of individual instruments. The section ends with a consideration of the way the listener perceives such sounds.

The section on stringed instruments first considers the general properties of string vibrations and their excitation by plucking, bowing and striking. Large amplitude vibrations are shown to provide a particular interesting illustration of non-linearity of much wider applicability than to musical acoustics alone. The coupling of the vibrating string via the bridge to the acoustically radiating surfaces of the instrument is then discussed in some detail, followed by a more detailed discussion of excitation of a string via the bowed slip-stick mechanism. The vibrational modes of the main shell of the instrument and the importance of the bridge and soundpost in determining the efficiency of energy transfer to the radiating surfaces of the instrument are then discussed. The section ends with a description of some of the experiment and computational techniques used to describe the vibrational modes, followed by a brief description of the radiated sound and the subjective assessment of the quality of stringed instruments.

The section on woodwind and brass instruments starts with a consideration of oscillating air columns and sound radiation from cylindrical and conical tubes and the more complicated shapes used for woodwind and brass instruments. This is followed by sections on the highly non-linear processes involved in the excitation of such vibrations by reed and lip vibrations and air-jets. The section concludes with a brief description of the acoustical properties of various wind instruments.

The final section on percussion instruments describes the acoustical properties of a range of instruments based on the vibrations of stretched membranes, bars, plates and shells. Typical waveform envelopes and time-dependent spectra are used to illustrate the relationship between the vibrational modes of such instruments and the radiated sound. Non-linearity at large amplitude excitation is again shown to be important and accounts for the characteristic sounds of certain instruments like gongs and cymbals.

15.1 Vibrational Modes of Instruments

15.1.1 Normal Modes

All musical instruments produce sound via the excitation of a vibrating structure. Woodwind, brass and percussion instruments radiate sound directly. However, stringed instruments radiate sound indirectly, because the vibrating string itself radiates an insignificant amount of energy. Energy from the vibrating string therefore has to be transferred to the much larger area, acoustically efficient, radiating surfaces of the body of the instrument. The resultant modes of vibration are complex and involve the interactions and vibrations of all the component parts, such as the strings, bridge, front and back plates, soundpost, neck, and even the air inside the volume of the violin body.

Any vibrating structure, however complicated, will have a number of what are called normal modes of vibration (see Chap. 22). The important influence of damping on the nature of the normal modes will be described in the section on stringed instruments. The normal modes satisfy exactly the same equations of motion as a simple damped mass–spring resonator. The displacement ξ_n of a given excited mode measured at any chosen point p on the structure is given by

$$m_n \left(\frac{\partial^2 \xi_n}{\partial t^2} + \frac{\omega_n}{Q_n} \frac{\partial \xi_n}{\partial t} + \omega_n^2 \xi_n \right) = F(t), \qquad (15.1)$$

where the effective mass m_n at the point p is defined in terms of the kinetic energy of the excited mode, $\frac{1}{2} m_n (\partial \xi_n / \partial t)_p^2$, $\omega_n = 2\pi f_n$ is the eigenfrequency (the angular frequency) of free vibration of the excited mode in the absence of damping and Q_n is the quality factor describing its damping. Initially, we consider a local driving force $F(t)$ at the point p, though it could be applied at any chosen point on the structure or distributed over the whole surface.

The effective mass of a one-dimensional string, solid bar or air column, at the point of maximum displacement, is half the mass of the vibrating system, the factor half resulting from averaging the kinetic energy over the sinusoidal spatial displacement. Likewise, the effective mass of a two-dimensional vibrating object at maximum displacement, like a violin plate or drum skin, is of order 1/4 its mass. The effective mass is very large close to nodal positions, where the displacement is small, and is small at antinodes, where the displacement is large.

Typical driving forces are those acting on the bridge of a bowed or plucked string instrument and the pressure fluctuations at the input end of the air column of a blown woodwind or brass instrument. Such forces are generated by highly nonlinear excitation mechanisms. In contrast, the vibrations of the vibrating structure are generally linear with displacements proportional to the driving force. However, there are important exceptions for almost all types of instruments, when nonlinearity becomes significant at sufficiently strong excitation, as discussed later.

In any continuously bowed or blown musical instruments, feedback from the vibrating system results in a periodic driving force, which will not in general be sinusoidal. Nevertheless, by the Fourier theorem, any periodic force can always be represented as a superposition of sinusoidally varying, harmonic, partials with frequencies that are integer multiple of the periodic repetition frequency. We can therefore consider the induced vibrations of any musical instruments in terms of the induced response of its vibrational modes to a harmonic series of sinusoidal driving forces.

Resonance and Admittance

In the harmonic approach, the applied forces and induced motions are assumed to vary sinusoidally as $e^{i\omega t}$. We will generally use this complex notation for notational and algebraic simplicity, where $\text{Re}(e^{i\omega t}) = \cos\omega t$ and $\text{Im}(e^{i\omega t}) = \sin\omega t$. The resonant response, with displacement $\xi_n e^{i\omega t}$ and velocity $i\omega \xi_n e^{i\omega t}$ at the driving

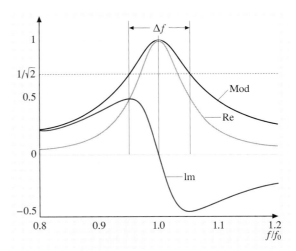

Fig. 15.1 Normalised real (Re) and imaginary (Im) components and the modulus (Mod) of the induced velocity of a simple harmonic resonator driven by a constant amplitude sinusoidal force for a Q-factor of 10

point p, for an applied sinusoidal force $Fe^{i\omega t}$, is then given by

$$\frac{\partial \xi_n}{\partial t} = i\omega \xi_n = \frac{F}{m_n} \frac{i\omega}{(\omega_n^2 - \omega^2 + i\omega\omega_n/Q_n)}. \quad (15.2)$$

The ratio of induced velocity to driving force is known as the local admittance at the driving point p, and is plotted in normalised form in Fig. 15.1 for a Q-value of 10. The admittance has both real and imaginary components. The real part describes the component of the induced velocity in phase with the driving force, while the imaginary part describes the component in phase quadrature (with phase leading that of the force by 90° degrees).

Well below resonance the induced displacement is in phase with the driving force, while at resonance the phase lags behind the driving force by 90°, and well above resonance lags by 180°. The velocity $v(t)$ leads the displacement by 90° degree and is thus in phase with the driving force at resonance, which corresponds to the maximum rate of energy transfer $i\omega F\xi_n$ to the excited mode. The 180° change in the phase of the response, as the excitation frequency passes from well below to well above resonance, is especially important in interpreting the multiple resonances of any musical instrument.

Provided the damping of an excited mode is not too strong (i.e. Q_n is significantly larger than unity), the peak in the modulus or real part of the admittance occurs at $\omega_n(1 - 1/8Q^2)$, which is very close to the natural resonant frequency ω_n. The width of the resonance is $\Delta f = f_n/Q$, where Δf is defined as the difference in frequency between the points on the resonance curve when the modulus of the induced displacement has fallen to $1/\sqrt{2}$ of its maximum value (i.e. the stored energy is half that at resonance). The displacement at resonance is $Q\times$ the static displacement.

Multi-Mode Systems

For any musical instrument having a number of vibrational modes, the admittance at the driving point p can be written as

$$A_{pp} = \sum_n \frac{1}{m_n} \frac{i\omega}{\omega_n^2 - \omega^2 + i\omega\omega_n/Q_n}, \quad (15.3)$$

with admittances A_{pp} of individual modes adding in series, equivalent to impedances in parallel. The vibrational response of a multi-resonant mode musical instrument can therefore be characterised by fitting the measured admittance to such a function giving the effective mass at the point of excitation, resonant frequency and Q-value for each of the excited modes. Using such a procedure, *Bissinger* [15.8] typically identifies up to around 40 vibrational modes for the violin below 4 kHz. However, at high frequencies, the width of individual resonances exceeds the spacing between them, making it increasingly difficult to identify individual modes.

It is important to recognise that damping is only important in a relatively narrow frequency range $\sim f_n/Q$ around the individual resonance peaks. Outside such regions, the reactive component associated with each vibrational mode continues to contribute significantly to the overall response. For example, well below resonance, each mode acts as a spring with effective spring constant $m_n(\omega_n^2 - \omega^2)$, while well above resonance it acts as a mass with effective mass $m_n(1 - \omega_n^2/\omega^2)$. The static displacement (at $\omega = 0$) is given by $\xi = F/K_o = \sum_n 1/(m_n\omega_n^2)$. Note that this involves contributions from all the vibrational modes of the structure, which is an important global property describing the low-frequency response of a multi-resonant structure such as the violin or guitar. If displacements are measured at a point p for an applied force at q, a *nonlocal admittance* can be expressed as

$$A_{pq} = \sum_n \frac{1}{m_{n,p}} \frac{i\omega}{(\omega_n^2 - \omega^2 + i\omega\omega_n/Q_n)} \frac{\xi_{n,p}\xi_{n,q}}{\xi_{n,p}^2}, \quad (15.4)$$

where $\xi_{n,p}$ and $\xi_{n,q}$ are the simultaneous displacements of the nth mode at the points p and q, with identical stored modal energy $1/2 m_{n,p}\omega^2\xi_{n,p}^2 = 1/2 m_{n,q}\omega^2\xi_{n,q}^2$.

Equation (15.4) illustrates the principle of reciprocity in acoustics, which states that the motion at a point p induced by a force at q is identical to the motion at q induced by the same force at p. Equation (15.4) also shows that, by applying a force at a particular position and measuring the induced motion (amplitude and phase) at a large number of different points $p(x, y, z)$ on the structure, it is possible to map out the amplitude of the modal vibrations $\xi_n(x, y, z)$ over the whole of any excited structure. Alternatively, the measurement point can be fixed and the excitation point moved across the structure. This is the basis of the powerful technique of modal analysis, which has been widely used to investigate the vibrational modes of many stringed and percussion instruments, as described by Rossing in Chap. 28 of this Handbook.

It also follows from (15.4) that a particular mode of vibration will never be excited if the driving force is located at a node of its vibrational state. This has important consequences for the spectrum of sound produced by bowed, plucked and struck stringed instruments and all percussion instruments.

Time-Domain Measurements

The vibrational characteristics of an instrument can also be investigated in the time domain. For example, by striking a stringed instrument with a light hammer or exciting the vibrational modes of a woodwind or brass instrument with a short puff of air, the frequencies of free vibration of the vibrational modes and their damping can be determined from their time-dependent decay. Provided the damping is not too strong ($Q \gg 1$), the modes will decay with time as

$$\xi_n(t) = \xi_0 \, \mathrm{e}^{-t/\tau_n} \, \mathrm{e}^{\mathrm{i}\omega_n t}, \qquad (15.5)$$

where $\tau_n = 2Q_n/\omega_n = Q_n/\pi f_n$. The frequency f_n of a given mode can be determined from its inverse period and Q_n from $\pi \times$ the number of periods for the amplitude to fall by the factor exponential e. The Q-value of strongly excited modes of a musical instrument can be estimated from $\tau_{60\mathrm{dB}} = 13.6\tau$, the Sabine decay time (Chap. 10 *Concert Hall Acoustics*). This is the perceptually significant time taken for the sound pressure to fall by a factor of 10^3 – from a very loud level to just being detectable. Hence, $Q_n = \pi f_n \tau_{60}/13.6 \sim 0.23 f_n \tau_{60}$. For example, the sound of a strongly plucked cello open A-string (220 Hz) can be heard for at least ~ 2 s, corresponding to a Q-value of ≈ 100 or more.

Damping results in a loss of stored energy given by

$$\frac{\mathrm{d}E_n}{\mathrm{d}t} = -\frac{\omega_n}{Q_n} E_n = -2\frac{E_n}{\tau_n}. \qquad (15.6)$$

Hence, the power P required to maintain a constant amplitude at resonance is $\frac{\omega_n}{Q_n} E_n$, where E_n is the energy stored. This tends to be the way that Q is defined and measured by physicists, whereas in acoustic spectroscopy it is more usual to define and measure Q-values from either the width of resonances in spectroscopic measurements or from decay times after transient excitation. As illustrated above, all such definitions are equivalent.

15.1.2 Radiation from Instruments

Although a large number of vibrational modes of a musical instrument may be excited simultaneously, they will not be equally important in radiating sound, which has important consequences for the quality of the sound. This section therefore provides a brief introduction to the radiation of sound from the vibrational modes of musical instruments.

Sound Waves in Air

In free space, the longitudinal displacement $\xi(x,t) = \xi_0 \, \mathrm{e}^{\mathrm{i}(\omega t - kx)}$ of plane sound waves satisfies the wave equation

$$\frac{\partial^2 \xi}{\partial x^2} = \frac{1}{c_0^2}\frac{\partial^2 \xi}{\partial t^2}. \qquad (15.7)$$

The dispersionless (independent of frequency) velocity of sound $c_0 = \sqrt{\gamma P_0/\rho}$, where $\gamma (\approx 1.4)$ is the ratio of specific heats at constant pressure and volume, P_0 ($\approx 10^5$ Pa or N/m^2) is the ambient pressure and ρ (≈ 1 kg/m^3) is the density (the brackets give the values for air at ambient pressure and temperature). The ratio of acoustic pressure $p = -\gamma P_0 \partial \xi/\partial x$ to the particle velocity $v = \partial \xi/\partial t$ is referred to as the specific impedance, $z_0 = p/v = \rho c_0$.

The appearance of γ in the expression for the velocity of sound reflects the adiabatic nature of acoustic waves. This arises because acoustic wavelengths are far too long to allow any significant equalisation of the longitudinal temperature fluctuations arising from the compressions and rarefactions of a sound wave. In free space longitudinal heat flow between the fluctuating regions is only important at very high ultrasonic frequencies (MHz), where it leads to significant attenuation. The major source of attenuation of freely propagating acoustic sound waves arises from the water vapour present. However, both viscous and transverse thermal losses to the side walls of woodwind and brass instruments can result in significant attenuation, as described later.

The above expressions neglect first-order, nonlinear, corrections to the compressibility, proportional to $\partial \xi/\partial x$, and other inertial correction terms in the nonlinear Navier–Stokes equation. This approximation breaks down at the very high intensities in the bores of the trumpet and trombone when played very loudly [15.9], which results in shockwave propagation, with a transition from a relatively smooth to a very brassy sound (*son cuivré* in French). For the present, such corrections will be neglected.

The speed of sound in air depends on the temperature θ (degrees centigrade) and, to a lesser extent, on the humidity. For 50% humidity,

$$c_0(\theta) = 332\,(1 + \theta/273)^{1/2} \approx 332(1 + 1.7\,10^{-3}\theta), \qquad (15.8)$$

giving a value of 343 m/s at 20 °C. Note that the air inside a woodwind or brass instrument, once the instrument is *warmed up*, will always be warm and humid, which will affect the playing pitch.

Pressure and Intensity

The intensity I of sound radiated by a musical instrument is given by the flow of acoustic energy ($1/2 \rho v^2$ per unit volume) crossing unit area per unit time,

$$I = \frac{1}{2}\rho c_0 v_{\max}^2 = \frac{1}{2}z_0 v_{\max}^2 = \frac{1}{2z_0}p_{\max}^2. \quad (15.9)$$

Sound pressure levels (SPL) are measured in dB relative to a reference sound pressure p_0 of 2×10^{-5} Pa or N/m^2, so that SPL(dB) $= 20\log_{10}(p/p_0)$. The reference pressure is approximately equal to the lowest level of sound that can be heard at around 1–3 kHz in a noise-free environment. Relative changes in sound pressure levels are given by $20\log_{10}(p_1/p_2)$ dB. A sound pressure of 2×10^{-5} Pa is very close to an intensity of $I_0 = 10^{-12}$ W/m^2, which is used to define the almost identical intensity level, IL(dB) $= 10\log_{10}(I/I_0)$. The difference between the factor 10 and 20 arises because sound intensity is proportional to the square of the sound pressure.

Spherical Waves

In free space, sound from a localised source will propagate as a spherical wave satisfying the three-dimensional wave equation (*Fletcher* and *Rossing* [15.5], Sect. 6.2), with pressure

$$p(r) = A\frac{e^{i(\omega t - kr)}}{r} \quad (15.10)$$

and particle velocity

$$v(r) = \frac{A}{z_0}\left(1 + \frac{1}{ikr}\right)\frac{e^{i(\omega t - kr)}}{r}. \quad (15.11)$$

Near and Far Fields

Note that, unlike plane-wave solutions, the velocity and pressure differ in phase by an amount that depends on the distance from the source and the wavelength. Close to the source, in the *near field* ($kr \ll 1$), the pressure and induced velocity are in phase quadrature. Such terms therefore involve no work being done (proportional to $\int pv \, dt$) and hence no radiation of sound. The near-field term describes the motion of the air that is forced to vibrate backwards and forwards with the vibrating surface of the source, which simply adds inertial mass to the vibrating mode. This term is responsible for the *end correction* ($\Delta L \sim a$, where a is the pipe radius), which extends the effective length of an open-ended vibrating air column. The additional inertial mass also lowers the vibrational frequency of the relatively light vibrating membranes of a stretched drum skin.

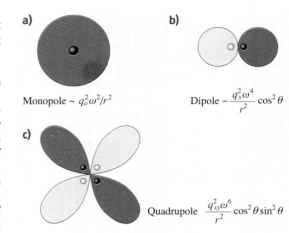

Fig. 15.2 Typical radiation patterns and intensities for monopole, dipole and quadrupole sources. The *two colours* represent monopole sources and sound pressures of opposite signs

In contrast, in the *far field* ($kr \gg 1$), the pressure is in phase with the velocity, so that work is done on the surrounding gas. This accounts for the fact that sound radiation varies in intensity as $1/r^2$.

The transition from the near- to far-field regions occurs when $r \sim \lambda/2\pi$, where λ is the acoustic wavelength of the radiated sound. At 340 Hz, this corresponds to a distance of only ≈ 15 cm. The difference in the frequency dependencies of the near- and far-field sound means that a violinist or piccolo player, with their ears relatively close to the instrument, experiences a rather different sound from that heard by the listener in the far field. However, for most musical instruments, the distance between the source of radiated sound and the player's head is already at least $\lambda/2$, so that even the player is in the *far field* ($kr > 1$), at least for the high-frequency partials of a musical tone.

Directionality and Multipole Sources

At very low frequencies, the acoustic wavelength λ is often considerably larger than the physical size of the radiating source (e.g. the open ends of woodwind and brass instrument bores and the body of most stringed instruments), which can then be considered as a point source radiating isotropically into space. However, as soon as the wavelength becomes comparable with the size of the radiating source, the radiated sound will acquire directional properties determined by the geometry of the instrument and the vibrational characteristics of the excited modes. The directional properties can then be described by treating the instrument as a superposition of monopole, dipole, quadrupole and higher-order multipole acoustic sources, with the directional radiating properties shown schematically in Fig. 15.2.

A monopole source can be considered as a pulsating sphere of radius a with surface velocity $v e^{i\omega t}$ resulting in a pulsating volume source $4\pi a^2 v e^{i\omega t} = Q e^{i\omega t}$. Equations (15.10) and (15.11) describe the sound field generated by such a source. Equating the velocities on the surface of the sphere to that of the induced air motion gives, at low frequencies such that $ka \ll 1$,

$$p(r, t) = \frac{i\omega \rho}{4\pi r} Q\, e^{i(\omega t - kr)}\;. \tag{15.12}$$

The radiated power P is then given by $\frac{1}{2} p^2/\rho c_0$ integrated over the surface of a sphere at radius r, so that

$$P(ka \ll 1) = \frac{\omega^2 \rho Q^2}{8\pi c_0}\;. \tag{15.13}$$

In the high-frequency limit ($ka \gg 1$), when the acoustic wavelength is much less than the size of the sphere,

$$P(ka \gg 1) = \frac{\rho c_0}{8\pi a^2} Q^2 = 4\pi a^2 \frac{1}{2} z_0 v^2\;. \tag{15.14}$$

Equation (15.12) is a special case of the general result that, at sufficiently high frequencies such that the size of the radiating object $\gg \lambda$, the radiated sound is simply $\frac{1}{2} z_0 v^2$ per unit area, though the sound at a distance also has to take into account the phase differences from different parts of the vibrating surface. Note that

$$\frac{P(ka \ll 1)}{P(ka \gg 1)} = (ka)^2\;. \tag{15.15}$$

The radiated sound intensity from a monopole source therefore initially increases with the square of the frequency but becomes independent of frequency above the crossover frequency when $ka > 1$. Hence members of the violin family and guitar families are rather poor acoustic radiators for the fundamental component of notes played on their lowest strings, as are wind and brass instruments, which radiate sound from the relatively small open ends and side holes. However, it is only because of such low radiation efficiencies, that strong resonances can be excited in the air columns of brass and woodwind instruments.

A dipole source can be formed by displacing two oppositely signed monopoles $\pm Q$ a short distance along the x-, y- or z-directions. For a dipole aligned along the x-axis of strength $q_x = Q \Delta x$. The sound pressure is simply the difference in pressure from monopoles of opposite sign a distance Δx apart, so that in the far field ($kr \gg 1$)

$$p(\theta)_{\text{dipole}} = p(\theta)_{\text{monopole}} \times (ik\Delta x)\cos\theta\;. \tag{15.16}$$

A polar plot of the sound pressure from a dipole is illustrated schematically in Fig. 15.2, with intensity and radiated power now proportional to ω^4 and q_x^2. In general, any radiating three-dimensional object will involve three dipole components (p_x, p_y and p_z), with radiation lobes along the three directions.

A quadrupole source is generated by two oppositely signed dipole sources displaced a small distance along the x-, y-, or z-directions (e.g. of the general form $q_{xy} = Q\Delta x \Delta y$). The pressure is now given by the differential of the dipole radiation in the newly displaced direction, so that, for example, the pressure from a quadrupole source q_{xy} in the xy-plane is given by

$$p_{\text{dipole}} = p_{\text{monopole}} \times \left(-k^2 \Delta x \Delta y\right) \cos\theta \sin\theta\;, \tag{15.17}$$

as illustrated in Fig. 15.2. Note that each time the order of the multipole source increases, the radiated pressure depends on one higher power of frequency, while the intensity increases by two powers of the frequency. The radiated power from multipole sources therefore decreases dramatically at low frequencies relative to that of a monopole source. At low frequencies, radiation from most musical instruments is dominated by monopole components.

In general, six quadrupole sources ($q_{xx}, \dots, q_{yz}$) would be required to describe radiation from a three-dimensional source. However, because the acoustic power radiated by a quadrupole source at low frequencies is proportional to ω^6, one need often only consider the monopole and three dipole components to describe the low-frequency radiation pattern of instruments like the violin and guitar family, as described in a recent study of the low-frequency radiativity of a number of quality guitars by *Hill* et al. [15.10]. However, at high frequencies, when λ is comparable with or less than the size of an instrument, the above simplifications break down. The directionality of the radiated sound then has to be computed from the known velocities over the whole surface, taking into account phase differences and baffling effects from the body of the instrument.

Radiation from Surfaces

Many musical instruments produce sound from the vibrations of two-dimensional surfaces – like the plates of a violin or the stretched membrane of a drum. Imagine first a standing wave set up in the two-dimensional

xy-plane with displacements in the z-direction varying as $\sin(k_x x)\,\mathrm{e}^{\mathrm{i}\omega t}$. We look for propagating sound waves solutions radiating from the surface of the form $\sin(k_x x)\,\mathrm{e}^{\mathrm{i}(\omega t - k_z z)}$, which must satisfy the wave equation and hence the relationship,

$$k_z^2 = \frac{\omega^2}{c_0^2} - k_x^2 = \omega^2 \left(\frac{1}{c_0^2} - \frac{1}{c_\mathrm{m}^2}\right), \qquad (15.18)$$

where c_m is the phase velocity of transverse waves on the membrane or plate in the xy-plane. Sound will therefore only propagate away from the surface ($k_z^2 > 0$) when $c_\mathrm{m} > c_0$. If the sound velocity is greater than the phase velocity in the plate or membrane, energy will flow from regions of positive to negative vertical displacements and vice versa, with an exponentially decaying sound field, varying as $\mathrm{e}^{-z/\delta}$ where $\delta = |k_z|^{-1}$.

Typical dispersionless wave velocities for the stretched drum heads of timpani are around 100 m/s (*Fletcher* and *Rossing* [15.5], Sect. 18.1.2), so that they are not very efficient radiators of sound. This is particular relevant for asymmetrical modes, when sound energy can flow from the regions of positive to negative displacement and vice versa. However, for even modes, the cancellation between adjacent regions moving out of phase with each other can never be complete, so that such modes will radiate more effectively.

A particularly interesting case occurs for stringed instruments, where the phase velocity of the transverse vibrations of the thin front and back plates increases with frequency as $\omega^{1/2}$ (Sect. 15.2.6). Hence, below a critical crossover or coincidence frequency, when the phase velocity in the plates is less than the speed of sound in air, standing waves on the vibrating plates are relatively inefficient radiators of sound, while above the crossover frequency the plates radiate sound rather efficiently. *Cremer* [15.11] estimates the critical frequency for a 4 mm-thick cello plate as 2.8 kHz; for a 2.5 mm violin plate the equivalent frequency would be ≈ 2 kHz.

Radiation from Wind Instruments

The holes at the ends or in the side walls of wind instruments can be considered as piston-like radiation sources. At high frequencies, such that $ka \gg 1$, where a is their radius, the holes will be very efficient radiators radiating acoustic energy $\sim 1/2 z_0 v^2$ per unit hole area. However, over most of the playing range $ka \ll 1$, so that the radiation efficiency drops off as $(ka)^2$, just like the spherical monopole source. Most of the sound impinging on the end of the instrument is therefore reflected, so that strong acoustic resonances can be excited, as discussed in the later section on woodwind and brass instruments.

15.1.3 The Anatomy of Musical Sounds

The singing voice, bowed string, and blown wind instruments produce continuous sounds with repetitive waveforms giving musical notes with a well-defined sense of musical pitch. In contrast, many percussion instruments produce sounds with non-repetitive waveforms composed of a large number of unrelated frequencies with no definite sense of pitch, such as the side drum, cymbal or rattle. There are also other stringed instruments and percussion instruments, such as the guitar, piano, harp, xylophone, bells and gongs, which produce relatively long sounds, where the slowly decaying vibrations produce a definite sense of pitch.

In all such cases, the complexity of the waveforms of real musical instruments distinguishes their sound from the highly predictable sounds of simple electronic synthesisers. This is why the sounds of computer-generated synthesised instruments lack realism and are musically unsatisfying. In this section, we introduce the way that sound waveforms are analysed and described.

Sinusoidal Waves

The most important, but musically least interesting, waveform is the pure sinusoid. This can be expressed in several alternative forms,

$$a\cos(2\pi f t + \phi) = a\cos(\omega t + \phi) = \mathrm{Re}(\boldsymbol{a}\,\mathrm{e}^{\mathrm{i}\omega t}), \qquad (15.19)$$

where $\boldsymbol{a}$ is in general complex to account for phase, f is the frequency measured in Hz and equal to the inverse of the period T, $\omega = 2\pi f$ is the angular frequency measured in radians per second, t is time, and ϕ is the phase, which depends on the origin taken for time.

Any sound, however complex, can be described in terms of a superposition of sinusoidal waveforms with a spectrum of frequencies. Figure 15.3 contrasts the envelopes, waveforms and spectra of a synthesised sawtooth waveform and the much more complex and musically interesting waveform of a note played on the oboe (🎵 CD-ROM provides an audio comparison). Note the much more complex fluctuating envelope and less predictable amplitudes of the frequency components in the spectrum of the oboe.

As we will show later, in defining the sound and quality of any musical instrument, the shape and fluctuations in amplitude of the overall envelope are just as important as the waveform and spectrum.

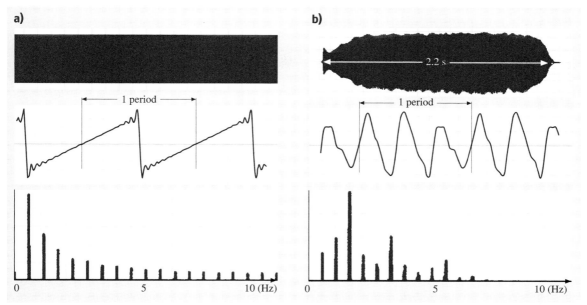

Fig. 15.3a,b Comparison of the envelope, repetitive waveform and spectrum of (**a**) a synthesised sawtooth and (**b**) a note played on an oboe

Range of Hearing and Musical Instruments

A young adult can usually hear musical sounds from around 20 Hz to 16 kHz, with the high-frequency response decreasing somewhat with age (typically down to between 10–12 kHz for 60-year olds). Audio ⟨?⟩CD-ROM provides a sequence of 1 s-long sine waves starting from 25 Hz to 12.8 kHz, doubling in frequency each time. Doubling the frequency of a sinusoidal wave is equivalent in musical terms of increasing the pitch of the note by an octave. Audio ⟨?⟩CD-ROM is a similar sequence of pure sine waves from 8 kHz to 18 kHz in 2 kHz steps. Any loss of sound at the low frequencies in ⟨?⟩CD-ROM will almost certainly be due to the limitations of the reproduction system used, which is particularly poor below ≈ 200 Hz on most PC laptops and notebooks, while the decrease in intensity at high frequencies in ⟨?⟩CD-ROM simply reflects the loss of high-frequency sensitivity of the ear (see Fig. 15.16 and Chap. 13 for more details on the amplitude and frequency response of the human ear).

The above sounds should be compared with the much smaller range of notes on a concert grand piano, typically from the lowest note A_0 at 27.5 Hz to the highest note C_9 at 4.18 kHz, as illustrated in Fig. 15.4. The nomenclature for musical notes is based on octave sequences of C-major scales with, for example, the note C_1 followed by the white keys D_1, E_1, F_1, G_1, A_1, B_1, C_2, D_2, ... Alternatively, the octave is indicated by a subscript (e.g. A_4 is concert A). Where the white notes are a tone apart, a black key is inserted to play the semitone between the adjacent white keys. This is indicated by the symbol # from the note below or ♭ from the note above. Figure 15.4 also illustrates the playing range of many of the instruments to be considered in this chapter.

Frequency and Pitch

It is important to distinguish between the terms frequency and pitch. The frequency of a waveform is strictly only defined in terms of a continuous sinusoidal waveform. In contrast, the waveforms of real musical instruments are in general complex, as illustrated by the oboe waveform in Fig. 15.3. However, despite such complexity, the repetition frequency and period T can still be defined provided the waveform does not vary too rapidly with time. The periodicity of a note (measured in Hz) can then be defined as the inverse of T. This is generally the note that the player reads from the written music. However, as described later, a repetitive waveform does not necessarily include a sinusoidal component at the repetition frequency, an effect referred to as the *missing fundamental*. Furthermore, depending on the strength of the various sinusoidal components present, there can often be an ambiguity in the pitch of a note perceived by the listener. The subjective pitch, when matched against

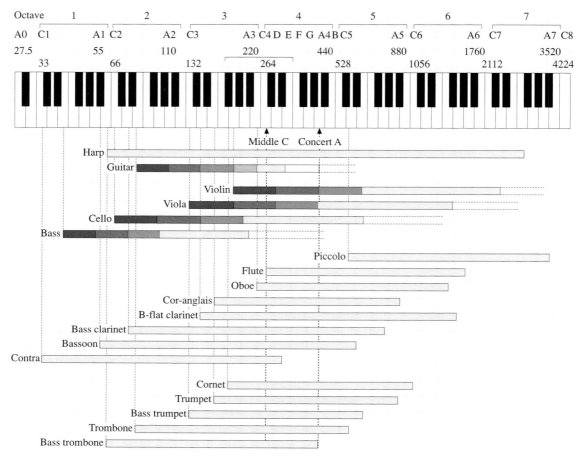

Fig. 15.4 Notation used for notes of the musical scale and the playing range of classical western musical instruments. Subdivisions for stringed instruments represent the tuning of the open strings

a pure sinusoidal wave, can often be an octave below or above the repetition frequency. The subjective pitch, as its name implies, can differ from person to person and within the musical context of the note being played.

Musical Intervals and Tuning

In western music the octave interval is divided into six tones (a whole-tone scale) and 12 semitones (the chromatic scale). Today, an equal temperament, logarithmic scale is used to tune a piano, with a fractional increase in frequency of $2^{1/12} = 1.059$ ($\approx 6\%$) between any two notes a semitone apart. The fractional increase between the frequencies of a given musical interval (a given number of semitones) is then always the same, whatever the starting note. Twelve successive semitones played in sequence therefore raises the frequency by an octave $[(2^{1/12})^{12} = 2]$. Any music played on the piano keyboard can therefore be transposed up or down by a given number of semitones, changing the pitch but leaving the relationship between the musical intervals unchanged. Such a scale was advocated as early as 1581, in a treatise by the lutenist Vincenzo Galileo (the father of Galileo Galilei). Although it is sometimes claimed that Bach exploited such a tuning in his 48 Preludes and Fugues, which uses all possible major and minor keys of the diatonic scale, historical research now suggests that Bach used a form of mean-tone tuning, which preserved some of the characteristic qualities of music written in particular keys [15.1].

To provide a greater discrimination in the measurements of frequency, the semitone is divided into 100 further logarithmic increments called cents. The octave is therefore equivalent to 1200 cents and a quarter-tone to ≈ 50 cents, with the exact relationship between frequencies given by

$$\text{interval} = \frac{1200}{\ln 2} \ln(f_2/f_1) \text{ cents} \quad (15.20)$$

corresponding to $\sim 1.73 \times 10^3 (\Delta f/f)$ cents for small fractional changes $\Delta f/f$.

Early musical scales were based on various variants of the natural harmonic series of frequencies, $f_n = nf_1$, where n is an integer (e.g. 200, 400, 600, ... 1600 Hz), illustrated by the audio ⟩CD-ROM. These notes correspond to the *harmonics* produced when lightly touching a bowed string at integral subdivisions of its length (1/2, 1/3, 1/4, etc.) ⟩CD-ROM. These simple divisions give successive musical intervals of the octave, perfect fifth, perfect fourth, major third and minor third, with frequency ratios 2/1, 3/2, 4/3, 5/4 and 6/5, respectively. The seventh member of the harmonic sequence has no counterpart in traditional western classical music, although it is sometimes used by modern composers for special effect [15.12].

Just temperament corresponds to musical scales based on these integer fraction intervals. The Pythagorean scale is based on the particularly consonant intervals of the octave (2/1) and perfect fifth (3/2), which can be used to generate individual intervals of the form $3^p/2^q$ or $2^p/3^q$, where p and q are positive integers. Although the Pythagorean and just-tempered scales coincide for the octave, perfect fifth and fourth, there are musically significant differences in the tuning for all other defined intervals, and all intervals other than the octave differ slightly from those of the equally tempered scale. A comparison between the musical intervals of just and equal temperament tuning is shown in Table 15.1, with the fractional mistuning indicated in cents. Because of the differences in tunings of the musical intervals, music transposed from one key to another will generally sound badly out of tune (particularly for commonly used intervals like the major and minor third) – unlike those played on a modern equal-tempered keyboard. Prior to the now almost universal practice of tuning keyboard instruments to a equal-tempered scale, many tuning schemes were devised which partially overcame the problems of tuning when playing in a succession of different keys (see *Fletcher* and *Rossing* [15.5], Sect. 17.6, and *Barbour* [15.13] for further discussion). Singers, stringed and wind instrument players can adjust the pitch of the notes they produce to optimise the tuning with other performers and for musical effect.

Figure 15.5 and audio ⟩CD-ROM illustrate the difference in the sounds of a major triad formed from the just intervals (1, 5/4, 3/2) and the equivalent equal-tempered scale intervals (1 : 1.260 : 1.498). The rational Pythagorean intervals give a repetitive waveform of constant amplitude, while the less-consonant, inharmonic, equal-tempered intervals have a non-repetitive waveform with an easily discernable periodic beat in amplitude resulting from the departures in harmonicity of its component frequencies, as illustrated in Fig. 15.5. Interestingly, the pitch of the equally tempered intervals also sounds slightly higher, though both share the same fundamental.

A sequence of upward fifths (frequency ratio 3/2) and downward octaves (ratio 1/2) can be used to fill in all the semitones of an octave scale on the piano keyboard. However, the resulting octave formed from a succession of 12 upward fifths and six downward

Table 15.1 Principal intervals and differences between just- and equal-temperament intervals

Interval	Just	Equal	$\Delta f/f$ Just-equal cents
Octave	2/1	2.00	0
Perfect fifth	3/2	$2^{7/12} = 1.498$	+2
Perfect fourth	4/3	$2^{5/12} = 1.334$	−2
Major third	5/4	$2^{4/12} = 1.260$	−13
Minor third	6/5	$2^{3/12} = 1.189$	+15
Tone	9/8	$2^{2/12} = 1.122$	−4
Semitone	16/15	$2^{1/12} = 1.066$	+1

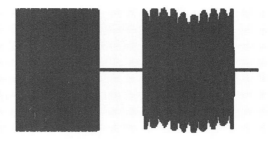

Fig. 15.5 Wave envelope of a major triad chord based on the Pythagorean scale followed by the same chord on the equal-tempered scale, with pronounced beats in the amplitude arising from the departures from harmonicity in the frequencies of the major third and perfect fifth

octaves gives a frequency ratio of $(3/2)^{12}/2^6 = 2.027$, which is significantly *sharper* (higher in frequency) than a true octave. In practice, a skilled piano tuner listens to the beats produced when playing the above intervals and tunes the upward fifth slightly flat, so that the sequence returns to the exact octave. However, there are striking psychoacoustic effects, in addition to physical shifts in the frequencies of upper partials arising from the finite rigidity of the strings, which result in pianos being tuned on a slightly stretched tuning scale with the octaves purposely tuned sharp at higher frequencies and flat at lower frequencies (*Fletcher* and *Rossing* [15.5], Sect. 12.8).

Repetitive Waveforms

Before considering the more complex waveforms of musical instruments, we consider the simple square, triangular, sawtooth and triangular repetitive waveforms (audio 19 CD-ROM) and the corresponding Fourier spectra illustrated in Fig. 15.6.

Fourier Theorem

Fourier showed that any repetitive waveform, $f(t+nT) = f(t)$, can be described as a linear combination of sinusoidal components with frequencies that are exact multiples of the inverse repetition period T or physical pitch of the wave. This is formally expressed by the Fourier theorem,

$$f(t) = \sum_{n=-\infty}^{\infty} c_n e^{in\omega_1 t}, \quad (15.21)$$

where $\omega_1 = 2\pi/T$ and

$$c_n = \frac{1}{T} \int_0^T f(t) e^{-i\omega t} dt, \quad (15.22)$$

where n takes on all positive and negative integer values. The Fourier coefficients c_n can be evaluated by performing the integral over any single period of the waveform. In general, the Fourier coefficients will have both real and imaginary components describing both the amplitude and phase.

For simple waveforms, such as the square, sawtooth and triangular waveforms, the origin of time can be chosen to make the waveforms symmetric or antisymmetric in time. The Fourier expansion can then be expressed in terms of the even cosine or odd sine functions,

$$f(t) = \sum_{n=0}^{\infty} \left\{ \begin{array}{l} a_n \sin(n\omega_1 t) \\ \text{or} \\ b_n \cos(n\omega_1 t) \end{array} \right\}, \quad (15.23)$$

with corresponding coefficients given by

$$\begin{pmatrix} a_n \\ b_n \end{pmatrix} = \frac{2}{T} \int_0^T f(t) \begin{pmatrix} \sin(n\omega_1 t) \\ \cos(n\omega_1 t) \end{pmatrix} dt, \quad (15.24)$$

where n is now restricted to positive integer values. The factor 2 is replaced by unity for the zero frequency average component b_0.

The first few terms of the square, sawtooth and triangular waveforms are listed in Table 15.2. The origin of time has been selected to make the waves odd-functions of time, as illustrated in Fig. 15.6, with the Fourier series only including sine terms. The Fourier components at integral multiples of the fundamental repetition frequency are referred to as partials, harmonics, or overtones. The nth partial has a frequency $f_n = nf_1$. This differs from the terminology used by musicians, who refer to f_2 as the first harmonic or overtone. Interestingly, a waveform depends critically on the sign (phase) of the individual Fourier components. In contrast, the ear is largely insensitive to the phase

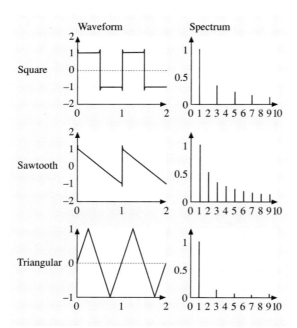

Fig. 15.6 Typical repetitive waveforms (synthesised from the first 100 components of a Fourier series) and the amplitudes of the first few partials normalised to the amplitude of the fundamental

of the individual partials, with little change in the perceived sound when the sign or phase of a component partial is changed, though the waveforms will be very different.

For an arbitrarily chosen origin of time, the Fourier expansion will include both sine and cosine terms. The energy or intensity is proportional to the resultant amplitudes squared, $a_n^2 + b_n^2$, which is independent of the origin of time. The phase ϕ_n is given by $\tan^{-1}(b_n/a_n)$. The spectrum of a waveform is often plotted in terms of the modulus of the amplitude as a function of frequency, without reference to phase, as in Fig. 15.6. However, measurements of both amplitude and phase are important in any detailed comparison with theoretical models and in analytic measurements, such as modal analysis.

The square and sawtooth waveforms are closely related to the waveforms excited on bowed and plucked strings and loudly played notes on wind and brass instruments. The discontinuities in waveform generate a very rich harmonic spectrum with Fourier components or partials that decrease relatively slowly (as $1/n$) with increasing n. The strong higher partials give a much harsher and more penetrating sound than simple sinusoids, which is why the oboe, which has a sound that is very rich in higher partials, is used to sound concert A when an orchestra tunes up. In contrast, the partials of the triangular wave, with discontinuities in slope instead of amplitude, decrease more rapidly as $1/n^2$, with a resultant sound little different from that of a simple sinusoidal wave.

Note the large difference between the sound of a sawtooth waveform, which is closely related to the sound of an oboe in having a complete set of harmonic components, and the "hollow" sound of the square waveform, which is more like the sound of the lowest notes on a clarinet, with rather weak even-integer harmonics or overtones on its lowest notes (Fig. 15.7).

Table 15.2 Fourier expansions of the square, sawtooth and triangular waveforms

Square	$\dfrac{4}{\pi}\left(\sin\omega_1 t + \dfrac{1}{3}\sin 3\omega_1 t + \dfrac{1}{5}\sin 5\omega_1 t \ldots\right)$
Sawtooth	$\dfrac{2}{\pi}\left(\sin\omega_1 t - \dfrac{1}{2}\sin 2\omega_1 t + \dfrac{1}{3}\sin 3\omega_1 t - \dfrac{1}{4}\sin 4\omega_1 t \ldots\right)$
Triangular	$\dfrac{2}{\pi}\left(\sin\omega_1 t - \dfrac{1}{3^2}\sin 3\omega_1 t + \dfrac{1}{5^2}\sin 5\omega_1 t \ldots\right)$

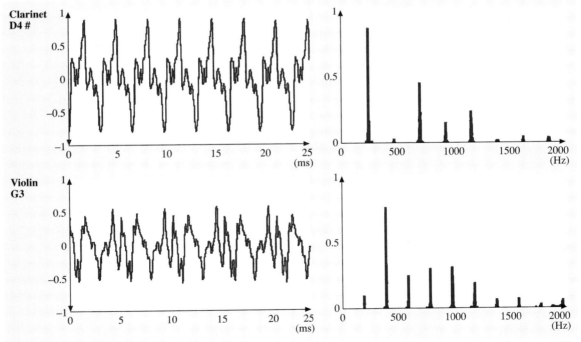

Fig. 15.7 Short-period samples of clarinet (D#4) and bowed violin (G3) tones and the corresponding Fourier spectra. The *vertical scales* are linear

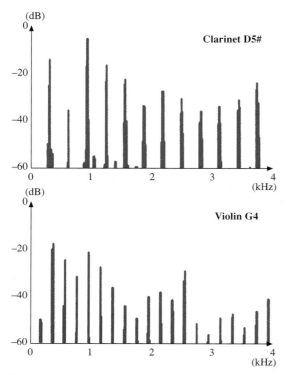

Fig. 15.8 Typical spectra for a clarinet and violin note plotted on a dB scale illustrating the large number of harmonics or overtones excited by bowed and blown instruments

Musical Waveforms and Spectra

The waveforms produced by musical instruments are generally far more complicated than the above simple examples, as already illustrated for an oboe note in Fig. 15.3. Waveforms and associated spectra of the clarinet note D#4 (309 Hz) and the violin bowed open-G-string G3 (195 Hz) are illustrated in Fig. 15.7. These are simply representative waveforms. Unlike the impression given in some elementary textbooks, there is no such thing as a defining violin or clarinet waveform or spectrum. Both the waveforms and the spectra change significantly from one note to the next – and even within a note when played with vibrato, particularly on stringed instruments. Despite the complexity of the waveforms, any repetitive waveform can be described as a linear superposition of sine waves, with frequencies that are integer multiples of the fundamental, as illustrated by the spectra.

Plotting the amplitudes of the Fourier coefficients on a linear scale often highlights the physical processes involved in the production of the sound. For example, the relatively small amplitudes of the second harmonic or partial in the sound of the clarinet reflects the absence of even-n modes of a cylindrical tube closed at one end, which approximates to that of the clarinet. Similarly, the small amplitude of the first partial in the sound of a violin reflects the absence of efficient radiating modes at low frequencies. However, because of the very wide dynamic range of hearing (a factor of $\approx 10^{10}$–10^{12} in intensity), it is often more appropriate to plot the Fourier coefficients in decibels on a logarithmic scale. Figure 15.8 shows the spectrum for clarinet and violin notes re-plotted on a dB scale, which illustrates the strength of all the partials over a very wide dynamical range. For bowed instruments such as the cello, well over 40 harmonic partials can be identified below 8 kHz. The sound of an instrument will be determined by all such components and not simply by the fundamental, which may make a relatively small contribution to the perceived sound. This is illustrated for a scale played on the violin with each note first played as recorded and then with the fundamental component removed by a digital filter (audio ⃝13 CD-ROM). The lowest notes of the scale, for which the fundamental component is already very weak, are little affected by the removal of the fundamental component; however, the sound gets progressively "thinner" in the second half of the scale for notes for which the fundamental partial makes an increasingly significant contribution to the "richness" or "warmth" of the sound.

Transient and Non-Repetitive Tones

No musical note lasts for ever, so that musical sounds are all, to some extent, transient. Moreover, the sound of many percussion instruments is composed of many strongly inharmonic partials, with no regime in which the waveforms can be considered even quasi-repetitive. Nevertheless, one can still use the Fourier theorem to extract the spectrum of such a note, by considering each transient signal as one of a sequence of such transients repeated, say, every second, minute or even year. The spectrum of such a repeated waveform will therefore involve frequency components at integer multiples of the inverse repetition period, which we can make as long as we choose. In the limit of infinitely long repetition times, the Fourier series of a non-repeating waveform can therefore be replaced by a continuous spectral distribution over all possible frequencies, known as the Fourier transform $F(\omega)$,

$$f(t) = \int_{-\infty}^{\infty} F(\omega) e^{i\omega t} \, dt \, , \tag{15.25}$$

where

$$F(\omega) = \frac{1}{2\pi} \int_{-\infty}^{\infty} f(t) e^{-i\omega t} \, dt \, . \tag{15.26}$$

The Fourier transform spectra of important non-repetitive waveforms are shown in Fig. 15.9. In all cases, the width Δf of the Fourier spectrum is inversely proportional to the length τ of the input waveform, with $\Delta f \tau \sim 1$. For a rectangular pulse, the spectrum extends over a rather wide frequency range with the first zero when $f\tau = 1$, but with many ripples of decreasing amplitude extending to higher frequencies. For an impulse of negligible width (the delta-function), the spectrum is flat out to very high frequencies. The spectrum of a Gaussian waveform varying as $\exp[-(t/\tau)^2]$ is also a Gaussian proportional to $\exp[-(\pi f \tau)^2]$, with a width $\Delta f = 1/\pi\tau$. Similarly, the spectrum of a sinusoidal waveform with a Gaussian envelope of width τ is broadened by $\Delta f = 1/\pi\tau$.

Any waveform that involves variations on a time scale τ will have Fourier components extending out to frequencies $\sim 1/\tau$. To reproduce such waveforms faithfully, the bandwidth of any recording or reproduction system must therefore extend to frequencies of at least $1/\tau$. Examples of non-repetitive waveforms and their associated spectra are illustrated in Fig. 15.10 for an orchestral rattle, a cymbal crash and timpani.

The ratchet sound consists of a sequence of short *clicks* illustrated by the selected short-section waveform. The spectrum is very broad with no individual frequencies particularly dominant. The crash of a cymbal generates a very large number of very closely spaced resonances, which appear as a fairly random set of peaks giving an overall broadband spectrum. The timpani spectrum shows a small number of large peaks corresponding to the prominent modes of vibration of the drum head, superimposed on a very wide-band spectrum largely

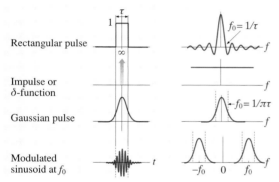

Fig. 15.9 Fourier transforms of transient waveforms

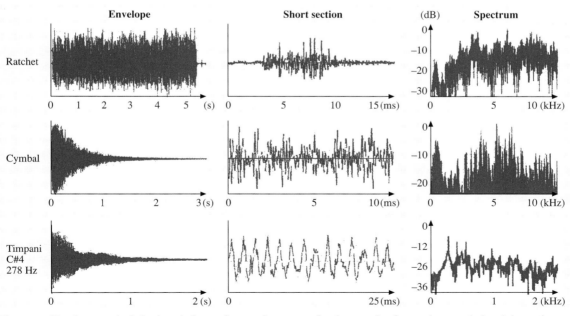

Fig. 15.10 Envelope, typical short-period waveform and spectrum for the sounds of a ratchet, cymbal and timpani

Table 15.3 Dynamic range of an analogue-to-digital converter (ADC)

N-bit ADC	Dynamic range $\pm 2^{n-1}$	Dynamic range (dB)
8 bit	128	42
16 bit	32.8 k	90
24 bit	8.39 M	138

associated with the initial transients involving sound from all parts of drum.

Digital Recording

Nowadays almost all sound is recorded digitally using an analogue-to-digital converter (ADC). This converts the continuously varying analogue input signal into a stream of numbers, which can be recorded digitally on a computer or compact disc. Audio signals are typically recorded at a sampling rate of 44.1 kHz with 16-bit resolution corresponding to 1 part in 2^{16}. This allows signals to be recorded in 65 536 equally divided levels between the maximum positive and negative input signals (i.e. between ± 32.8 k levels).

For the highest-quality digital recordings, even faster recording rates with higher-bit resolution are used (typically 24-bit sampling at 96 kHz). This allows for over-sampling of the recorded signal, so that signals can be averaged, any errors detected and eliminated, and filtered more easily. As already noted, the dynamic range of human hearing can be as large as 100 dB. To exploit such a large range and to capture the details of both loud and soft sounds from a large orchestra accurately requires the recording system to have a large dynamic range. Table 15.3 shows the dynamic range in terms of the number of bits used to record the sound. Audio 1?) CD-ROM illustrates the greatly enhanced signal-to-noise ratio and hence increased dynamic range when sound is recorded at 16-bit rather than 8-bit resolution.

Aliasing

When sound is recorded digitally, ambiguities can arise when any of the input frequencies is larger than half the sampling rate f_D. This is known as the Nyquist limit $f_{Nyquist} = f_D/2$. For example, if a 2 kHz sine wave were to be sampled at 2 kHz, the digital signal would be recorded at exactly the same point of the waveform each cycle. The recorded digital signal would then be indistinguishable from a DC signal. It can easily be shown that sinusoidal inputs at $f_{Nyquist} + \Delta f$ give the same digital output as at $f_{Nyquist} - \Delta f$ and that the recorded signal is the same for all frequencies differing in frequency by

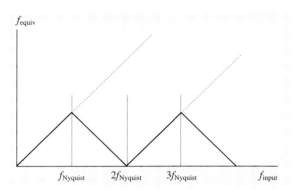

Fig. 15.11 Ambiguity of digital output for a steadily increasing frequency exceeding the Nyquist limit

the digitising frequency, $2f_{Nyquist}$. Thus for a steadily increasing input frequency the digital output is equivalent to that of a frequency which first increases up to $f_{Nyquist}$ then decreases to zero when $f_{in} = f_D = 2f_{Nyquist}$, with the process repeating for higher input frequencies, as illustrated in Fig. 15.11. In any replay system, an analogue output is generated that assumes a smooth curve between the sampled points. Hence, recorded frequencies above the Nyquist limit will be misinterpreted and will produce sounds below $f_{Nyquist}$ with no harmonic relevance to the original input.

This ambiguity is illustrated in audio 1?) CD-ROM, in which a sinusoidal input is swept in frequency from 200 Hz to 6 kHz. This is first recorded at 22.05 kHz, when aliasing is not a problem, and then at 6 kHz, when halfway through the increasing frequency signal, at 3 kHz, the replayed sound starts to descend to zero frequency at the end of the sweep, when the input signal has the same frequency as the sampling rate. The single-frequency sweep is then followed by an ascending major triad (with intervals in the ratios 1 : 5/4 : 3/2) recorded at 6 kHz, which illustrating the severe problems of aliasing in terms of musical harmonies, as soon as any of the higher-frequency components in a signal exceed the Nyquist limit, with the frequency of some partials ascending while others are descending.

To avoid such problems, a high-frequency cut-off input filter is generally used, with a cut-off frequency slightly below the Nyquist frequency (see Chap. 14 by *Hartmann* for further details).

Sound File Formats

A sound signal is frequently recorded and stored as an encoded WAVE file of the generic form *.wav, which includes additional information on data acqui-

Table 15.4 Decoded WAVE file information

Data provided	Typical example
Mono (1), stereo (2)	2
Data acquisition rate	2.205×10^4
Resolution in bits	16
Data per second	4.41×10^4
First measurement from left channel	237
Simultaneous measurement from right channel	-1356
Second measurement from left channel	456
Simultaneous measurement from right channel	-1972
Repeated sequence until end of recorded sound	...

sition rate, stereo or mono format and bit resolution. The decoded structure of a WAVE file is shown in Table 15.4.

Recording audio signals as WAVE files is very expensive in memory, with 1 hour of stereo music recorded at 22 kHz requiring ≈ 300 MB. Music files on CDs are encoded so that the input is redistributed over time and therefore over the surface of the disc. This enables the original signal to be reproduced even in the presence of dust, scratches and other small imperfections on the disc surface, eliminating the clicks that were a familiar feature of older vinyl records. More sophisticated, adaptive, encoding schemes can be used to significantly reduce the amount of memory used, such as the now widely used mp3 format. An algorithm is used, based on physical principles and on the way the ear responds to musical sounds, to continuously analyse and process the incoming data. The input data can then be recorded using a much reduced number of bits, in much the same sort of way that digital pictures are encoded more efficiently in ZIP files and compact image formats. The information used to encode the digital signal is also recorded, so that the processed data can be unscrambled on playback with relatively little loss in perceived quality.

Discrete Fourier Transform

The digital form of the recorded data allows certain computational efficiencies in calculating the Fourier spectrum. Consider a recorded sample of N measurements, corresponding to a sample of length $T_s = N/f_D$. To calculate the spectrum, this data set is assumed to repeat indefinitely, to form a continuous waveform with a repetition frequency. From the Fourier theorem, the resulting spectrum is composed of Fourier components that are exact multiples of the repetition frequency, so that $f_n = n/T_s$. Hence, a 1 s set of data points will give a discrete Fourier spectrum with frequencies at $1, 2, \ldots n \ldots$ Hz. In practice, the number of Fourier components is limited to $N/2$, because each component has both an amplitude and a phase, which requires at least two independent measurements to be made per Fourier component.

Windowing Functions

Using the sampled waveform to form a continuously repeating waveform will, in general, introduce a repeating discontinuity Δ at the beginning and end of each repeated data set, since the start and end values will not usually be the same. Any such discontinuity will generate spurious contributions to the spectrum, with additional Fourier coefficients with amplitudes proportional to Δ/n. To circumvent this problem, a *windowing function* is used. This applies a smooth envelope to the data set, which reduces the values at the start and end to zero, thus eliminating the discontinuities. However, as described above, the application of such an envelope will give an extra width $\Delta f \sim 1/T_s$ to the spectral features.

A typical windowing function is the Hanning function $\sin^2(2\pi t/T_s)$. A number of other windowing functions are illustrated in Fig. 15.12, each of which has advantages for specific applications [15.14]. The Hanning windowing is widely used for general-purpose measurements, while the Hamming function is used to separate closely spaced sine waves. In general there is a trade-off between the accuracy that can be achieved in determining the frequency of individual spectral components and the width of the low-amplitude side lobes generated by application of a windowing function. Various forms of the Blackman–Harris windowing function

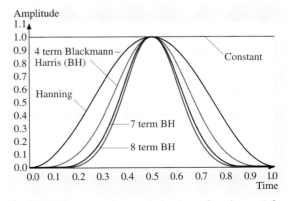

Fig. 15.12 Representative windowing functions (after [15.14])

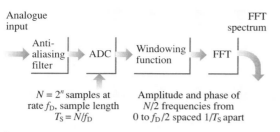

Fig. 15.13 A typical digital sampling and FFT analysis scheme

can be used to optimise the fast Fourier transform (FFT) for specific measurements. Windowing need not be used for the accurate measurements of widely separated sinusoidal waves with similar amplitudes, though one should be aware of the existence of the rather wide side lobes generated unless the sampling period is an exact integer multiple of the period of the waveform being measured.

Fast Fourier Transform (FFT)

To determining the amplitude and phase of the $N/2$ Fourier components from N input measurements requires the inversion of an $N \times N$ matrix, requiring a computation time proportional to N^2. However, if N is an integral power of 2 (e.g. $2^8 = 256$, $2^{16} = 65536$), the FFT computing algorithm can be used to reduce the computing time by many orders of magnitude (by a factor $\sim N/\log N$). The speed of modern computers is such that FFT spectra of the sound of musical instruments can be calculated and displayed with delays of only a few milliseconds, though any such delay will always be limited by the length and hence frequency resolution of the data set being analysed.

A typical implementation of the FFT method for spectral analysis is shown schematically in Fig. 15.13. An input anti-aliasing filter is first used to remove frequency components above the Nyquist limit $f_D/2$; an ADC then converts the incoming signal to a digital output to give a data set of $N = 2^n$ measurements over a time $T = N/f_D$. A windowing function removes problems from discontinuities at the start and end of the recorded set of data, and the computer evaluates the FFT giving the amplitudes and phases of the Fourier components at $N/2$ discrete frequencies spaced $1/T = f_D/N$ Hz apart.

A sequence of FFTs from data taken over successive short periods of time can be used to illustrate the decay of individual partials in transient and decaying waveforms, such as those of a plucked string, a piano note or struck bell, as illustrated for the sound of a plucked violin A-string in Fig. 15.14.

Envelopes of Sound Waveforms

The time dependence or envelope of the amplitude of a sound signal is just as important a factor in the recognition of any musical instrument as the spectrum of the sound produced. In general, the envelope has a starting

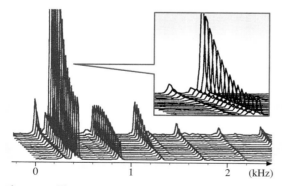

Fig. 15.14 Time sequence of delayed FFTs illustrating the decay of excited modes of a violin, when the A-string is plucked, with an expanded section of the frequency scale for the lowest resonances excited. The time between successive traces is 10 ms

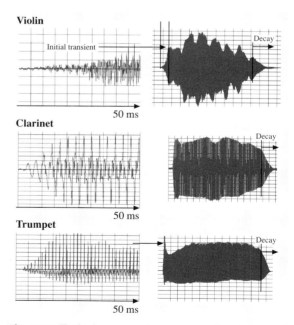

Fig. 15.15 Typical wave envelopes for a violin, clarinet and trumpet with a 50 ms expanded view of the initial starting transient

transient, a period with a quasi-constant amplitude for a continuously bowed or blown instrument, and a period of free decay, when the instrument is no longer being excited. The sound produced by musical instruments is also significantly affected by the acoustic environment in which the instrument is played, but this will be ignored for the moment. Typical initial transients and overall envelopes of single notes played on a violin, clarinet and trumpet are shown in Fig. 15.15.

The starting transient provides an immediate clue to the ear enabling the listener to recognise the instrument being played quickly. However, the characteristic fluctuations in frequency and amplitude within the overall envelope and noise associated with the method of excitation (e.g. bowing and blowing) are just as important in the recognition of specific instruments. This can easily be shown by removing the starting transient from a musical sound altogether, as illustrated in !?⟩CD-ROM. In this example comparisons are made between the sounds of a violin, flute, trombone and oboe played first with the initial 200 ms transient removed and then with the transient reinserted. In each case the instrument can immediately be identified even in the absence of the starting transient. The audio example ends with a constant amplitude sawtooth waveform having an unvarying sound quite unlike the sound of any real musical instrument.

Nevertheless, the starting transient and subsequent decay of sound are extremely important in the identification of the sounds of plucked or hammered strings and all percussion instruments, where the waveform and spectral content changes very rapidly with time after the start of the note. This is illustrated by the dramatic difference in the unrecognisable sound of a piano when played backwards and then replayed in the normal direction (!?⟩CD-ROM).

Noise

There are several potential sources of fluctuations in the envelope of musical instruments, which help to characterise their characteristic sounds, such as the breathiness induced by the noise of turbulent air passing over the sound hole in a recorder, flute or organ pipe (*Verge* and *Hirschberg* [15.15]) and irregularities in the sound of any bowed instrument due to inherent noise in the slip–stick bowing mechanism (*McIntyre* et al. [15.16]).

Amplitude and Frequency Modulation

Another important source of fluctuations is vibrato, which involves periodic changes in the amplitude, frequency, or spectral content of a note and often all three (*Meyer* [15.17], *Gough* [15.18]). Vibrato is produced on a stringed instrument by periodically changing the length of the bowed string by rocking the finger stopping the string backwards and forwards. In singing (*Prame* [15.19]) and wind instruments (*Gilbert* et al. [15.20]) vibrato is produced by periodic modulations of the pressure exerted by the lungs or mouth on the exciting reed or air passage.

Amplitude modulation of a sinusoidal frequency component can be expressed as

$$y(t) = (1 + a_m \cos \Omega t) \sin \omega t$$
$$= \sin \omega t + \frac{a_m}{2} [\sin(\omega + \Omega)t + \sin(\omega - \Omega)t] ,$$
(15.27)

where Ω is the modulation frequency and a the modulation parameter. Amplitude modulation introduces two "side-bands" with amplitude $a_m/2$ at frequencies Ω above and below that of the principal central component. The side-bands have a net resultant that remains in phase with the central component giving a fractional change in amplitude $[1 + a_m \cos \Omega t]$.

Frequency modulation should more strictly be referred to as phase modulation, with the phase varying as

$$\phi(t) = \omega t + a_f \cos \Omega t .$$
(15.28)

where a_f is frequency-modulation index. The time-varying frequency can then be defined by the rate of change of phase, such that

$$\frac{d\phi}{dt} = \omega - a_f \Omega \sin \Omega t ,$$
(15.29)

with a fractional shift in frequency varying as

$$\frac{\Delta \omega(t)}{\omega} = -a_f \frac{\Omega}{\omega} \sin \Omega t .$$
(15.30)

For small modulation index, a phase-modulated wave can be written as

$$y(t) = \sin \omega t + \frac{a_f}{2} [\cos(\omega + \Omega)t + \cos(\omega - \Omega)t] ,$$
(15.31)

which again results in equally spaced side-bands about the central frequency with amplitude $a_f/2$, but with a resultant now in phase-quadrature with that of the central frequency giving the above phase modulation.

Because of the multi-resonant frequency response of all musical instruments, changes in driving frequency also induce significant fluctuations in amplitude. Such fluctuations are particularly important for the strongly peaked multi-resonant instruments of the violin family, as illustrated in Fig. 15.15.

15.1.4 Perception and Psychoacoustics

In this section, we briefly highlight a number of psychoacoustic aspects of particularly importance in any discussion of musical acoustics. See also Chap. 13 on Psychoacoustics by *Brian Moore*.

Sensitivity of Hearing

We have already commented that the brain interprets both frequency and intensity on a logarithmic scale. The recognition of familiar intervals such as the octave, perfect fifth, irrespective of the absolute frequencies, provide an immediate example, as is the use of the dB scale in the measurement of sound levels.

In the 1930s, *Fletcher* and *Munson* [15.22] undertook a survey of a large population of subjects to investigate how the sensitivity of the ear varies with frequency and intensity (and age). These measurements were later refined by *Robinson* and *Dadson* [15.21]. Their published values for normal equal-loudness level contours, shown in Fig. 15.16, were adopted by the International Standards Organization, as the original ISO 226 standard for audio sensitivity, with data recently refined to define the new ISO 226:2003 standard.

The plotted curves show population-averaged equal loudness contours for sinusoidal sound waves measured in phons on a dB scale, which equate to sound pressure level (SPL) measurements in dB at 1 kHz. The SPL dB scale is based on a reference root-mean-square pressure of $20\,\mu$Pa (equivalent to $2\times 10^{-5}\,\mathrm{Nm}^{-2}$), which is very close to an qintensity of $10^{-12}\,\mathrm{Wm}^{-2}$. The threshold contour is the population-averaged minimum sound pressure that can be just detected under the quietest environmental conditions. Above sound pressures of $\approx 120\,\mathrm{dB}$, the ear experiences pain and potential permanent damage.

The equal subjective sound level contours reflect the dynamics of the ear's detection system. There is a rapid fall-off in sensitivity at low frequencies, where the efficiency of the outer ear drum considered as a piston detector falls off as ω^4. The fall-off at high frequencies is due to the increasing inertial impedance of the ear drum and bones in the inner ear. However, the fall-off is partially compensated by peaks in sensitivity from the resonances of the outer air channel between the ears and ear-drum. Older people experience a considerable loss in sensitivity at high frequencies, which is strongly correlated with age. Fortunately, the losses are at relatively high frequencies and are generally not too important for the appreciation of music.

The sensitivity of the ear is particularly strong in the frequency range 2–6 kHz, which is important for recognising the consonants in speech. One would therefore expect such frequencies to be equally important in the identification and assessment of sound quality of musical instruments. Below around 200–400 Hz there is an increasingly rapid fall-off in sensitivity, which will differentially affect the subjective loudness of the lower partials of any complex musical sound at these and lower frequencies. At low frequencies the contours of equal amplitude are more closely spaced, so that the effect of increasing the SLP by 20 dB increases the subjective loudness by considerably more. Turning up the volume on any reproduction system changes the perceived quality from a rather thin sound to a more exciting sound with a much stronger bass and a somewhat stronger high-frequency response.

From a musical acoustics viewpoint, it is often sensible to invert the equal contour plot, as the inverted plot essentially acts as a subjective, mid-frequency band filter, de-emphasising the perceived intensities of the lowest and highest-frequency partials in a complex waveform.

Loudness levels will clearly vary with distance from any source. Sounds levels exceed 120 dB close to an aeroplane on take off or close to a loudspeaker in a noisy rock concert, resulting in potential permanent damage to the ear. Sound levels close to a heavily used motorway are typically around 90 dB, around 70 dB inside a car, about 50 dB in an office, 30 dB inside a quiet house at night, 20 dB in a very quiet recording studio and 0 dB inside an anechoic chamber. At the quietest levels, one begins to hear the beating of the heart and

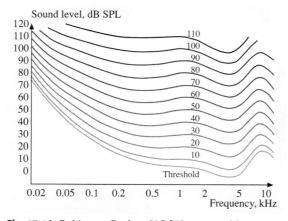

Fig. 15.16 *Robinson–Dadson* [15.21] curves with contours of perceived equal loudness measured in phons on a dB scale

workings of other internal organs, which can be a somewhat disquieting experience. There would clearly be no evolutionary reason to have developed a more sensitive hearing system.

Audio ▶ CD-ROM is a short orchestral excerpt played at successively decreasing 6 dB steps (half the amplitude or a quarter the intensity). Musicians indicate the loudness with which music should be played using the dynamic markings *pp*, *p*, *mp*, *mf*, *f* and *ff*, which roughly correspond to a subjective doubling of intensity between each level. Such levels are clearly only relative, since absolute values will vary strongly with the distance of the listener from the source with a 12 dB decrease in intensity on doubling the distance in free space. Although the dynamic range of an individual note on a musical instrument rarely exceeds 20 dB, with only about six distinguishable dynamic levels within this range, the total dynamic range of an instrument is more like 45 dB (*Patterson* [15.23]). However, there is a much larger range of sounds produced by different instruments (e.g. the trombone and violin. The carrying power, penetration and prominence of musical sounds is not simply a matter of absolute intensity, but also depends on the harmonic content and transient structures of the complex tones produced. This helps to explain how a solo violinist can still be heard over the massive sound of a large orchestra.

Subjective Assessment of Pitch

We have already noted that the perceived pitch of a note is determined by the inverse period of a waveform and does not necessarily require the presence of a Fourier component at that frequency. This is illustrated in audio ▶ CD-ROM, in which simple sinusoidal tones at 300 Hz and 200 Hz are first played in succession and then played together to produce a repeating waveform sounding an octave lower at 100 Hz, which is then followed by a pure 100 Hz tone of the same amplitude but sounding very much quieter. The final sinusoidal tone at 100 Hz may well not be heard on a typical PC laptop or notebook sound production systems, which radiate little sound below around 200 Hz. The absence of a Fourier component at the pitch of a complex tone is often referred to as the *missing fundamental* phenomenon. It is important in many stringed instruments, which produce very little sound at the actual frequency of their lowest open strings.

The missing fundamental phenomenon is a psychoacoustic rather than a nonlinear effect produced by a distortion of the waveform in the ear. It reflects the way that the ear processes sounds in the time domain at low frequencies (*Moore* [15.24] and Chap. 13).

The ability of the ear to recognise the pitch at which an instrument is playing, even though the lower partials of the sound of individual instruments may be missing is very important in sound reproduction systems. It enables the listener to recognise the distinctive sounds of all the instruments in an orchestra, even when the recording or reproduction system may have a very poor low-frequency response – as in early gramophones and the loudspeakers used in cheap radios and typical PC laptops and notebooks.

Combining tones to produce a lower tone is exploited on the *quint* combination stop on the organ to produce low-pitched sounds (e.g. a 16 ft pipe and a $16 \times 2/3 = 10.66$ ft pipe sounding a fifth above, when sounded together, reproduce the sound of a 32 ft pipe, as illustrated for the combination of 200 and 300 Hz sine waves in ▶ CD-ROM above). Interesting, the effect is nothing like so strong in playing two bowed strings a fifth apart (e.g. open A and open E on a violin), presumably because of the very rich spectrum of higher partials and independent fluctuations of the two sounds. However, such sounds can often be heard when two flutes play well-tuned intervals together. The early 18th century virtuoso violinist Tartini recognised the existence of such mysterious tones, whenever pairs of notes were played together in exact intonation [e.g. integer ratios such as 3/2 (perfect fifth), 5/4 (major third), 6/5 (major third)], and reputedly attributed them to the devil. The effect is small, but is still used by violinists when practising playing such intervals exactly in tune.

In general, complex tones are composed of a number of spectral components which have no particular harmonic relationship to each other. One then has to consider what determines the subjective pitch of the perceived sound. This involves the way the brain processes the signal and the relative emphasis given to the spectral components present, which will depend on their frequencies and intensities. It is important to recognise that the perceived pitch is not necessarily that of the lowest-frequency component present. This is illustrated in ▶ CD-ROM in which the fundamental and first octave are fixed in frequency, while an intermediate partial is swept upwards from the lower to the higher note. First the fundamental is sounded by itself and then with the octave added producing a note at the same pitch but with a different timbre. An intermediate partial is then added and swept upwards in steps from the lower to the upper note, giving the sense of a note of continuously rising pitch, though the fundamental and its octave remain fixed. Although the fundamental and octave remain fixed, the rising partial gives the sense of a note of increasing pitch. In this particularly simple example, it is relatively easy to identify and follow the pitch of each harmonic component separately. However, for a musical instrument like a gong or bell, with no preconceived knowledge of the likely pitch of the individual partials, this is far more difficult. The perceived pitch of the strike note of a bell and many other percussion instruments, with an inharmonic combination of excited modes, depends in a rather complex way on the relative weighting of the partials present and the musical context.

In assessing the subjective pitch of a note there can often be an ambiguity of an octave in the apparent pitch. This is illustrated by the famous example of the apparent, ever-rising pitch of a note generated by a continuously rising comb of logarithmically spaced frequencies passing through a fixed hearing band of frequencies, 🎵 CD-ROM [15.25], which appears to be increasing in pitch at all times though clearly repeating itself. This illustrates the *circularity* of pitch perception and is the audio equivalent of the visual illusion of continuously rising steps which return to the same point in space in an Escher drawing.

Precedence Effect

Another important time-domain phenomenon in the perception of musical sounds is the Haas precedence effect, which enables a listener to locate the source of a distant sound from the small difference in time that sound arriving at an angle to the head takes to reach the two ears. The brain gives precedence to the sound arriving first, even though later sounds from other directions may be significantly stronger. Any sound arriving within the first 20–40 ms (depending somewhat on frequency and intensity) of the first sound to arrive simply adds to the perceived intensity of the first sound. This is very important in musical performance, with reflections from close reflecting surfaces adding strongly to the intensity and definition of the music.

The precedence effect is illustrated in 🎵 CD-ROM. This is a stereo recording of identical clicks recorded on the left and right channels with a delay of 20 ms between them, which is then reversed. Although the clicks are too close together for the ear to distinguish them separately, when replayed through a pair of stereo loudspeakers (not earphones), the sound will appear to come from the speaker providing the earlier click.

The precedence effect is one of the ways in which one can locate the origin of a particular sound within an orchestra or the sound of a particular voice in a crowded room. Once located, the brain is able to focus on the subsequent source even against a highly confusing background of other sources. It is likely that fluctuations within the characteristic sound of an individual person or musical instrument enable the brain to focus continuously on a particular source. In musical acoustics one must always recognise the formidable power of the brain's auditory processing capabilities, which is far beyond what can be achieved using present-day computers. Consequently, even very small effects on a physical measurement scale can have a very significant effect on the listener's subjective response to the sound of a particular instrument.

15.2 Stringed Instruments

In this section we describe the production of sound by the great variety of musical instruments based on the plucking, bowing and striking of stretched strings. This will include an introduction to the different modes of string vibrations excited by the player, the transfer of energy from the vibrating string to the acoustically radiating structural vibrations of the body of the instrument via the bridge, and the modification of such sound by the environment in which the instrument is played. Although the production of sound is based on the vibrations of relatively simple structures, such as strings and plates, it is the interactions between these, extending the physics well beyond introductory text-book treatments, which results in the characteristic sounds of individual stringed instrument, as summarised in this section.

The Physics of Musical Instruments by *Fletcher* and *Rossing* ([15.5], Chaps. 9–11) provides an authoritative account of the acoustics of a wide range of string instruments, and a comprehensive set of references to the research literature prior to 1998. The four volumes of research papers on violin acoustics, collated and edited by *Hutchins* [15.26] and *Hutchins* and *Benade* [15.27], also includes excellent introductions to almost every aspect of the acoustics of instruments of the violin family, much of which is just as relevant to other stringed instruments. Carleen Hutchins has been an inspirational figure in the field of violin acoustics. The Catgut Acoustical Society, which she cofounded, published a journal and an earlier newsletter [15.28] containing many important papers on violin research of interest to both professional acousticians and violin makers. Her inspiration has encouraged a world-wide school of violin makers, who use scientific measurements and plate tuning in particular as an aide to making high-quality instruments. The comprehensive monograph on the *Physics of the Violin* by *Cremer* [15.29] provides an invaluable theoretical and experimental survey of research on instruments of the violin family, with particular emphasis on the bowed string, the action of the bridge, the vibrations of the body and the radiation of sound.

The production of sound by any stringed instrument is based on the same acoustic principles. The player excites the vibrations of a stretched string by bowing, plucking or striking. Energy from the vibrating string is then transferred via the supporting bridge to the acoustically radiating structural vibrations of the instrument. The radiated sound is then conditioned by the performing environment.

There are many different types of stringed instruments formally classified as chordophones. Harp-like lyres appear in Sumerian art from around 2800 BC. However, more primitive instruments, like a plucked string stretched over a bent stick and resonated across the mouth, probably date back to soon after the emergence of man the hunter [15.30, 31].

Stokes [15.32] was the first to recognise that the vibrating string was essentially a linear dipole, which radiated a negligible amount of sound at low frequencies (see also *Rayleigh* [15.3] Vol. 2, Sect. 341). To produce sound, the vibrating string has to excite the vibrations of a much larger area radiating surface. For bowed and plucked instruments, such as members of violin, lute and guitar families, almost all the sound is radiated by the shell of the instrument, with the acoustic output at low frequencies usually boosted by the Helmholtz resonance of the air inside the instrument vibrating in and out of the f- or rose-holes cut into the front plate. On larger instruments, such as the piano and harp, the sound is radiated by a large soundboard.

For any continuously bowed (or blown) instrument, the sound is conditioned by a complex feedback loop involving the instrument, player and surrounding acoustic, illustrated schematically for the violin in Fig. 15.17. The expert string player controls the intonation and quality of the sound produced using slight adjustments of the position of the left-hand fingers stopping the string, and the pressure, velocity and position of the bow on the string, in response to the sound heard from both the instrument and the surrounding acoustic. In addition, there is direct tactile feedback through the fingers of both the left hand controlling the pitch of the note and the right hand controlling the bow. A similar overall feedback system is also involved in playing woodwind or brass instrument. The perception of the sound by both player and listener is also strongly influenced by the performing acoustic and the way the brain processes the sound received by the sensory organs in the ears, as illustrated schematically in Fig. 15.17. All such factors are involved in determining the perceived quality of the sound produced by a musical instrument. However, for simplicity and physical insight into the various mechanisms involved, it is convenient to consider the acoustics of musical instruments in terms of their component parts, like the vibrating string, the supporting bridge and shell of the instrument. Nevertheless, it is important not to lose sight of the fact that the sound produced by any instrument will involve the interactions of all such subsystems and, even more importantly, the skill of the player in exciting and controlling the vibrations ultimately responsible for the sound produced.

15.2.1 String Vibrations

The transverse vibrations $\xi(x, t)$ of a perfectly flexible stretched string, of mass μ per unit length and tension T, satisfy the one-dimensional wave equation (d'Alembert, 1747)

$$\frac{\partial^2 \xi}{\partial x^2} = \frac{1}{c_T^2} \frac{\partial^2 \xi}{\partial t^2}, \qquad (15.32)$$

where the velocity of transverse waves $c_T = \sqrt{T/\mu}$. The tension $T = ES\Delta L/L$, where E is Young's modulus, S is the cross-sectional area of the string and $\Delta L/L$ is the fractional stretching of the string over its length L. For the relatively small transverse displacements of bowed and plucked strings on musical instruments, changes in tension can be ignored. However, at larger ampli-

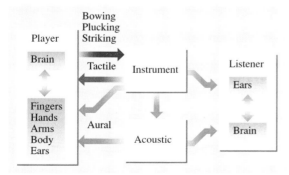

Fig. 15.17 A schematic representation of the complex feedback and sound radiation systems involved in the generation of sound by a bowed string instrument

tudes, a number of interesting nonlinear effects can be observed, which will be described in Sect. 15.2.2.

A string can also support longitudinal and torsional modes, with velocities $c_L = \sqrt{E/\rho}$ and $c_\theta = \sqrt{E/\rho(1+\nu)}$, where ρ is the density and ν is the Poisson ratio (≈ 0.35 for most materials). The Poisson ratio ν is the ratio of transverse to longitudinal strain when the material is stretched along a given direction. For strings on musical instruments, the longitudinal and torsional wave velocities are typically an order of magnitude larger than the transverse velocity, with $c_L/c_T \sim \sqrt{L/\Delta L}$.

Although both longitudinal and torsional waves play important roles in the detailed physics of the bowed, plucked and struck string, the musically important modes of string vibration are the transverse modes – apart from unwanted squeaks from longitudinal modes, which are often excited by the beginner on the violin. Unless otherwise stated, we only consider transverse waves and drop the defining subscript, unless a distinction needs to be made.

Waves on an ideal string are dispersionless (independent of frequency), so that any wave initially excited on the string will travel along the string without change in amplitude or shape. D'Alembert obtained a general solution of the wave equation of the form

$$f(x, t) = f_1(x + ct) + f_2(x - ct), \quad (15.33)$$

corresponding to two waves of unchanging shape travelling with wave velocity c in opposite directions along the string.

If the string is supported rigidly at its ends, the propagated waves are reflected with a change in sign giving zero displacement at the nodal end-points. Each propagating wave will continue to be reflected with change of sign on reflection at each end. For a string of length L, the string displacement will therefore return to its initial state in multiples of the transit time $2L/c$. The same is also true for the velocity and acceleration waveforms, since, if f satisfies the wave equation, then so must all its temporal and spatial derivatives, $\partial^n f/\partial x^n$ and $\partial^n f/\partial t^n$. It follows that the repetition frequency of any freely propagating wave on a given length of a stretched string will always be the same, however the string is excited (e.g. sinusoidally or by plucking, bowing or striking).

Excitation of Vibrations

First consider a string subject to a localised force F applied suddenly at a point along its length. This causes the string to move with velocity v at the point of contact exciting transverse waves travelling outwards in both

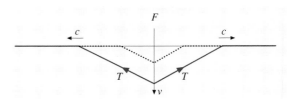

Fig. 15.18 Transverse motion of string induced by a localised force, with the *dotted lines* indicating the displacement at an earlier time

directions with velocity c, as illustrated schematically in Fig. 15.18.

In a short time δt, the transverse waves travel a distance $c\delta t$ along the string while the string at the point of contact is displaced by a transverse distance $v\delta t$. For $v \ll c$, one can make the usual small-angle approximations, so that equating the applied force to the transverse force from the deformed string, we obtain

$$v = \left(\frac{c}{2T}\right) F = \frac{1}{2R_0} F, \quad (15.34)$$

where $R_0 = \mu c$ is the characteristic impedance (force/induced velocity) of the string, which for an ideal string ignoring intrinsic losses is purely resistive. The factor of two in the above equation arises because the force acts on the two semi-infinite lengths of string in parallel. In practice, any discontinuity in slope will be rounded by the finite flexibility of real strings, as discussed later.

Force on End-Supports

The characteristic resistance R_0 of the string is an important parameter, because it determines the transfer of energy from the vibrating string to the acoustically radiating modes of the instrument via the supporting bridge at the end of the string. The transverse force exerted by the string on an end-support at the origin can be written as $F_B = T(\partial \xi/\partial x)_0$. This induces a transverse velocity at the point of string support given by

$$v_B = \frac{1}{Z_B} F_B = A_B F_B, \quad (15.35)$$

where Z_B and A_B are the frequency-dependent characteristic impedance and admittance at the end-support. In general, the induced velocity at the point of string support on the supporting bridge will differ in phase from that of the driving force, so that $Z(\omega)$ and $A(\omega)$ will be complex quantities.

The bridge on a musical instrument is never a perfect node otherwise no energy could be transferred to the radiating surfaces of the instrument. Waves on the string are reflected at the bridge with a frequency-dependent reflection coefficient r and a fractional loss of energy ε given by

$$r = \frac{R_0 - Z_B}{R_0 + Z_B} \text{ and } \varepsilon = \frac{2R_0 \left(Z_B + Z_B^*\right)}{|R_0 + Z_B|^2}, \quad (15.36)$$

where Z_B^* is the complex conjugate of the complex impedance at the terminating bridge.

For strings on musical instruments, $R_0 \ll |Z_B|$, so that to a first approximation we can consider the bridge as a node. If this were not so, the vibrational frequencies of strings would by strongly perturbed from their harmonic values. Nevertheless, first-order corrections are important, as they determine the energy transfer from the strings to the body of the instrument and hence the intensity of the radiated sound. The coupling via the bridge also affects the string vibrations themselves, with the resistive losses at the bridge causing damping and the reactive component of the admittance perturbing their vibrational frequencies, as described in Sect. 15.2.3. Such perturbations can sometimes be so large that it is no longer possible to sustain a stable bowed note, resulting in what is known as a *wolf-note* (for an illustration of a bad wolf-note on the cello listen to 🎵 CD-ROM).

Before considering the interaction of real strings with the supporting structure, we first consider the simplest cases of sinusoidal and simple Helmholtz modes of vibration on an ideal string with perfectly rigid end-supports.

Sine-Wave Modes

An ideally flexible string stretched between rigid end-supports a distance L apart can support standing waves, or eigenmodes, with transverse string displacements given by

$$\xi_n(x, t) = a_n \sin\left(\frac{n\pi x}{L}\right) \cos\left(\omega_n t + \phi_n\right), \quad (15.37)$$

where $\omega_n = 2\pi f_n$ and a_n is the amplitude of the nth mode with frequency $f_n = nc/2L$ and phase ϕ_n. Such modes can be considered as the sum of two sine waves of the d'Alembert form (15.33) travelling in opposite directions. For an ideal string, these solutions form a complete orthogonal set of *eigenmodes* with a *harmonic set* of *eigenfrequencies*, which are integer multiples of the fundamental frequency $c/2L$.

The resonant response of individual modes of a metal or metal-covered string can be investigated, for example, with a photosensitive device to detect the transverse string motion induced by a sinusoidal current passing through the string placed in a magnetic field to give a transverse Lorentz force (*Gough* [15.33]).

Because the wave equation is linear, any waveform, however excited, can be described as a Fourier sum of harmonic modes, such that

$$\xi_n(x, t) = \sum_{n=1}^{\infty} \sin\left(\frac{n\pi x}{L}\right)$$
$$[A_n \cos\left(\omega_n t\right) + B_n \sin\left(\omega_n t\right)], \quad (15.38)$$

where the Fourier coefficients A_n and B_n are determined by the initial transverse displacement and velocity along the length of the string, so that

$$A_n = \frac{2}{L} \int_0^L \xi(x, 0) \sin\left(\frac{n\pi x}{L}\right) dx, \quad (15.39)$$

and

$$B_n = \frac{2}{L\omega_n} \int_0^L \frac{d\xi(x, 0)}{dt} \sin\left(\frac{n\pi x}{L}\right) dx. \quad (15.40)$$

The transverse force on the end-support at $x = L$ is given by

$$F_{\text{end}} = -T \left(\frac{\partial \xi}{\partial x}\right)_L$$
$$= -T \sum_n \left(\frac{n\pi}{L}\right) (-1)^n$$
$$[A_n \cos\left(\omega_n t_n\right) + B_n \sin\left(\omega_n t\right)]. \quad (15.41)$$

Helmholtz Modes

Although many physicists and most musicians intuitively associate waves on strings with the sinusoidal waves of textbook physics, in practice, the vibrations of a bowed, plucked or struck string are very different. Nevertheless, because such waves are repetitive, it follows from the Fourier theorem that all such solutions can be described as a sum of sinusoidal wave components. However, the motions of plucked, bowed and struck strings are much more easily described by what are known as Helmholtz solutions to the wave equation [15.34]. These are illustrated for the plucked and bowed string in Fig. 15.19a,b.

The Helmholtz solutions are made up of straight-line sections of string. There is no net force acting on any small segment within any such section, because the transverse tension forces acting on its ends are equal and opposite. By Newton's laws, any such segment must therefore be either at rest or moving with constant velocity. Only where there is a *kink* or discontinuity in the

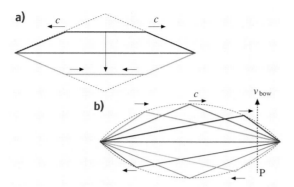

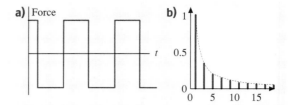

Fig. 15.20 (a) Square-wave time dependence of transverse force acting on the bridge from a string plucked at its centre and (b) the corresponding amplitudes of the odd Fourier components n varying as $1/n$ (*dotted curve*)

Fig. 15.19a,b Helmholtz waveforms for (**a**) a centrally plucked and (**b**) a bowed string. The *horizontal arrows* indicate the directions that the kinks are travelling in and the *vertical arrows* the directions of the moving string sections. The *different colours* represent string displacements at different times. P indicates a typical bowing position along the string

slope between adjacent straight-line sections (equivalent to a δ function in the spatial double derivative) can there be any acceleration. From our earlier discussion, any such kink must travel backwards and forwards along the string at the transverse string velocity c, reversing its sign on reflection at the ends. As the kink moves past a specific position along the string, the difference in the transverse components of the tension on either side of the kink results in a localised impulse, which changes the local velocity of the string from one moving or stationary straight-line section to the next. In general, there can be any number of Helmholtz kinks travelling along the string in either direction, each kink marking the boundary between straight-line sections either at rest or moving with constant velocity. Similar solutions also exist for torsional and longitudinal waves.

We now consider the Helmholtz wave solutions for the plucked, bowed and hammered string in a little more detail.

Plucked String

Consider an ideal string initially at rest with an initial transverse displacement a at its mid-point, as illustrated in Fig. 15.19a. On release, kinks will propagate away from the central point in both directions with velocity c, but points on the string beyond the kinks will remain at rest. When the kink arrives at a particular point along the string, the associated impulse will accelerate the string from rest to the uniform velocity of the central section of the string. After a time t, the solution therefore comprises a straight central section of the string of width $2ct$ moving downward with constant velocity c $(2a/L)$, with the outer regions remaining at rest until a kink arrives. After a time $L/2c$, the kinks separating the straight-line sections reach the ends and are reflected with change of sign. After half a single period L/c, the initial displacement will therefore be reversed and will return to the original displacement after one full period $2L/c$. In the absence of damping, the process would repeat indefinitely.

Now consider the transverse force acting on the end-support responsible for exciting sound through the induced motion of the supporting bridge and vibrational modes of the instrument. The initial transverse force on the bridge is $2Ta/L$, where we assume $a \ll L$. This force is unchanged until the first kink arrives. On reflection, the direction of the force is reversed and is reversed again when the second kink returns after reflection from the other end of the string. The two circulating kinks therefore cause a reversal in sign of the force on the end-supports every half-cycle, resulting in a square-wave waveform, as illustrated in Fig. 15.20. The spectrum of a square wave has Fourier components at odd multiples n of the fundamental frequency with amplitudes proportional to $1/n$, Fig. 15.20b.

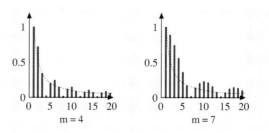

Fig. 15.21 Normalised Fourier amplitudes for the force on the bridge for a string plucked 1/4 and 1/7 of the string length from the bridge. The *dashed curves* show the $1/n$ envelope of the partials of a sawtooth waveform

Note that plucking a string at the mid-point excites only the odd-n modes. This is a consequence of the initial force being applied at a node of all the even-n modes. If a string is plucked at a fractional position $1/m$ along its length, any partial that is an integer multiples of m will be missing. This is illustrated in Fig. 15.21, showing the spectra of the force on the supporting bridge for a string plucked at points 1/4 and 1/7-th along its length. By selecting the plucking position along the string, the guitar or lute player can change the harmonic content of the sound produced. When plucked near the bridge, the sound of the plucked guitar string is rather bright, with nearly all the prominent partials almost equally strongly excited (audio CD-ROM).

In practice, the finite width of the plucking point, the finite rigidity of the string and the loss of energy at the bridge perturb the Helmholtz wave, removing the unphysical discontinuities of the idealised model. This results in a more rapid decrease in the intensities of the higher partials excited.

Bowed String

The motion of the bowed string can be described rather accurately by a simple Helmholtz wave with a single kink circulating backwards and forwards along the string. The kink now separates two straight sections moving with constant angular velocity about the nodal end-points, as illustrated in Fig. 15.19b. This is again a solution that satisfies Newton's laws of motion, with the only acceleration occurring as the kink arrives at a particular point along the string. Such a wave is just as valid a solution to the wave equation as a sine wave and once excited would continue indefinitely, if there were no damping or energy losses on reflection at the bow or supported ends.

The energy required to excite and maintain such a wave is provided by frictional forces between the moving bow hair and the string, involving what is known as the *slip–stick* excitation mechanism. For a typical bowing position, marked by the line at P in Fig. 15.19b, the friction between the bow and string forces the string to remain in contact with the bow hair moving with constant bow velocity. This is referred to as the *sticking regime* and occurs all the time the kink is travelling to the left of the bowing position. However, when the kink is between the bow and supporting bridge, the string moves in the opposite direction to the bow. This is the *slipping regime*. Such motion is possible because the sliding friction between the bow and string can be much smaller than the sticking friction, when the bow and string are in contact. In this highly idealised model, the frictional force is assumed to be infinite in the sticking regime and zero in the slipping regime.

A more detailed discussion of the slip–stick bowing mechanism will be given later (Sect. 15.2.3), taking into account more-realistic models for the frictional forces between the bow and string and the transfer of energy from the string to the vibrational modes of the structure via the bridge. However, the idealised Helmholtz motion provides a surprisingly good description of the vibrations of real strings, as confirmed in early measurements by *Raman* [15.35] and many more-recent publications to be cited later.

The amplitude of the Helmholtz bowed waveform is determined by the velocity of the bow v_{bow} and its distance L_B from the bridge. The transverse displacement of the kink maps out a parabolic path as it traverses the string (Fig. 15.19b). At the mid-point, the string displacement executes a triangular-wave motion with time, moving with velocity $\pm 2ca/L$ in alternate half-periods, where the maximum displacement $a = (L^2/4L_B)(v_B/c)$, for $a \ll L$. At the bowing position, the transverse string velocity alternates between v_{bow} in the sticking regime and $-v_{\text{bow}}(L - L_B)/L_B$ in the slipping regime, as illustrated in Fig. 15.22.

To increase the sound, the player can therefore either use a faster bow speed or play with the bow nearer the bridge. *Schelling* [15.36] has shown that more-realistic frictional models limit the playing range, as discussed later (Fig. 15.31).

The transverse force on the bridge produced by an idealised Helmholtz bowed wave has a sawtooth-

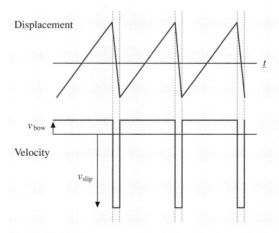

Fig. 15.22 Displacement and velocity of string at the bowing point. The mark-to-space ratio in the velocity is the same as the division of the string by the bow

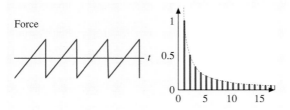

Fig. 15.23 Sawtooth time-dependence of force on the end supports from Helmholtz bowed waveform and corresponding amplitudes of the normalised Fourier spectrum with partials varying as $1/n$

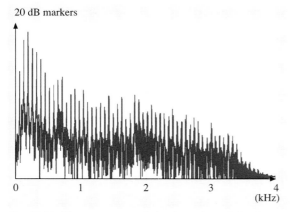

Fig. 15.24 The spectrum of the intensity of the lowest bowed note on a cello, illustrating the very large number of partials contributing to the sound of the instrument

waveform time dependence, as shown in Fig. 15.23. Each time the kink is reflected at the bridge, the transverse force acting on the bridge reverses in sign. It then increases monotonically with time until the process repeats again. The sense of the sawtooth motion reverses with bow direction. The spectrum of the force acting on the bridge includes both even and odd partials, with amplitudes varying as $1/n$.

The spectrum of the sound produced by the lowest plucked and bowed notes on stringed instruments can typically involve 40 or more significant harmonic partials, as illustrated in Fig. 15.24 by the spectrum of the sound produced by a bowed cello open C-string (C2 at ~64 Hz, audio ⓘ CD-ROM). The FFT spectrum is plotted on a dB scale to illustrate the large range of amplitudes of the partials (Fourier components) excited. The amplitudes of the individual partials depend not only on the force at the bridge exerted by the plucked or bowed strings, but also on the frequency dependent response

and radiative properties of the supporting structure, as discussed later.

Struck String

Many musical instruments are played by striking the string with a hammer. The hammer can be quite light and hard, as used for playing the dulcimer, Japanese koto and many other related Asian instruments, or relatively heavy and soft, like the felted hammers on a piano. Some time after the initial impact, the striking hammer bounces away from the string, leaving the string in a free state of vibration. There are a few instruments, such as the clavichord (*Thwaites* and *Fletcher* [15.37]), where the string is struck with a metal bar (the tangent), which remains in contact with the string, defining its vibrating length and hence the note produced.

Consider first a point mass m moving with velocity v striking an ideal stretched string of infinite extent. In any small increment of time, the moving mass will generate a wave moving outwards from the point of impact. This will result in a decelerating force on the mass equal to $2Tv/c = mcv$, as illustrated in Fig. 15.18. The displacement of the mass will then be described by the following equation of motion

$$m\frac{d^2\xi_m}{dt^2} = -\frac{2T}{c_T}\frac{d\xi_m}{dt} \ . \qquad (15.42)$$

The transverse velocity of the impacting mass therefore decays exponentially with time as

$$\frac{d\xi_m}{dt} = v_m \exp(-t/\tau) \ , \qquad (15.43)$$

with $\tau = mc/2T$. The is identical to the dynamics of a trapeze artist dropping onto a stretched wire, with waves of displacement and velocity travelling outwards in both directions away from the point of impact, as illustrated in Fig. 15.25.

In general, the string will be struck at a distance a from one of its end-supports. Hence, in a time $(a/2L)T_0$, a reflected wave will return to the mass and exert an additional force, which will tend to throw the mass back

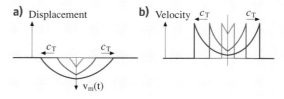

Fig. 15.25a,b Time sequences of (**a**) string displacement and (**b**) string velocity for a mass striking a string

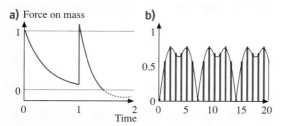

Fig. 15.26 (a) Time dependence of the upward force acting on a light hammer impacting a string in arbitrary units of time, and (b) amplitudes of Fourier coefficients of force acting on end supports for a hammer after hitting the string 1/7 of the length from an end-support. The *continuous line* shows the continuous spectrum for a string of semi-infinite length

off the string. However, because the mass cannot change its velocity instantaneously, any returning wave will be partially reflected, so that the mass acts as a source of secondary reflected waves travelling outwards in both directions. The total force acting on the hammer is then given by any residual force from the first impact plus the subsequent forces created by the succession of reflections from the end-supports. This problem was first correctly solved by Hall, in the first of four seminal papers on the string–piano hammer interaction [15.38–41].

Hall showed that the first reflected wave exerts an additional decelerating force $g(t') \sim (1 - t'/\tau)\mathrm{e}^{-t'/\tau}$ on the mass, where t' is the time after arrival of the first reflection. This is illustrated in Fig. 15.26 for a relatively light mass impacting the string at a position 1/7-th of the string length from an end. Provided the mass is sufficiently small, the force from the initial impact will have decayed significantly by the time the first reflection returns, so that the force acting on the mass will become negative (the dotted section in Fig. 15.26a), and the mass will detach itself from the string. The string will then move away from the mass and will vibrate freely, provided the hammer is prevented from falling back onto the string. An elaborate mechanism is used on the piano to prevent this from happening (see *Rossing*, *Fletcher* [15.5], Sect. 12.2), while the zither or dulcimer player quickly lifts the hammer well away from the string after the initial impact using much the same striking action as a percussionist playing a drum, where the same considerations apply.

The heavier the mass, the longer it will remain in contact with the string. Hall showed that it may then take several reflections from both ends of the string and sometimes several periods of attachment and detachment before the mass is finally thrown away from the string. A sufficiently heavy mass will never bounce back off the string.

In general, the waveforms excited on the string will therefore be rather complicated functions of the properties of the string, hammer and striking position. However, for a very light mass ($\ll$ mass of the string), which is thrown off the string by the first reflected wave, the Fourier coefficients of the induced velocity waveform, and hence the force on the end-supports, are given by $v_n \approx \left(1 + \mathrm{e}^{-1+in\pi\alpha}\right) \sin(n\pi\alpha)$, where $\alpha = a/L$, illustrated in Fig. 15.26b for an impact 1/7-th of the way along the string. Note that the seventh harmonic is missing, as again expected from general arguments, since no work can be transferred to a particular mode of string vibration for a force applied at a nodal position.

In practice, the spectrum is affected by the finite size of the hammer, multiple reflections occurring before the hammer is thrown from the string, and the elastic and often hysteretic properties of the hammer material [15.38].

Striking Tangent

On the clavichord (*Fletcher*, *Rossing* [15.5], Sect. 11.6), a string is struck by a rising end-support, or tangent, which remains in contact with the string, exciting transverse vibrations of the string on both sides of the tangent. If we assume a simplified model in which the rising tangent moves with constant velocity until its final displacement a is reached, there is again a simple Helmholtz wave solution. In practice, the length of string on one side of the tangent is damped, so that free vibrations are only excited on one side of the striking point. We therefore need only consider the length of string between the tangent and the end connected to the soundboard. The discontinuities $\pm v$ in the tangent velocity, occurring on initial impact and on reaching its final displacement after a time Δt, generate propagating kinks and discontinuities of velocity of opposite sign separated in time by Δt. The striking therefore excites waves with kinks, velocities and displacements along the string shown in Fig. 15.27a. The solutions are again Helmholtz waves, but now with two kinks of opposite signs travelling around the string in the same direction.

The Fourier coefficients of the velocity waveform shown in Fig. 15.27b can be written as

$$c_n \sim \frac{1}{n}\frac{a}{L'}\left(1 - \mathrm{e}^{in2\pi\beta}\right), \tag{15.44}$$

where $\beta = \Delta t/T_1$ is the fraction of the period $T_1 = L'/2c$ of the freely vibrating length of string during which the

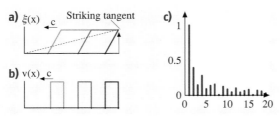

Fig. 15.27a–c Waveforms for a tangent hitting and sticking to string: (**a**) displacement and (**b**) velocity profiles along the string, at a succession of times (different colours) after the tangent hits the string, and (**c**) the spectrum of the resultant force on the end-supports for $\beta = 3/8$ (see text)

striking tangent moves from its initial to final position. Figure 15.27 also includes the spectrum of the force acting on the end-supports for $\beta = 3/8$.

Very similar modes to the above will be excited during the time a heavy hammer is initially in contact with the string on instruments like the dulcimer, zither and piano. Such modes therefore contribute to the initial transient sounds of such instruments. Another related example is the use of *col legno* on stringed instruments, when the strings are struck by the wooden part of the bow. By hitting the string at specific positions along the string, pitched initial transients can be produced, creating special sound effects, *nageln*, sometimes used in avant-garde contemporary music, audio ⟨♪ CD-ROM⟩). The above simplistic model for striking a clavichord string will, in practice, be modified by the way the player depresses the key, which is directly coupled to the rising tangent, both during and after the initial impact. The player can therefore influence the initial transient and the after-sound, including the use of a small amount of vibrato on the after-note, resulting in a particularly responsive and intimate but quiet instrument, which was particularly popular in the baroque period.

Real Strings
We now consider a number of departures from the above idealised models for real strings including:

1. the finite size of the plucking or striking point
2. the finite flexibility of the string
3. nonlinear effects

In a subsequent section, we consider the even larger perturbations resulting from coupling to the acoustically radiating modes of the body of the instrument via the bridge.

Finite Spatial Variation
Idealised models for the string, with infinitely sharp kinks produced by plucking, bowing or striking, involve waveforms with discontinuities in amplitude and slope and an infinite number of Fourier components are clearly unphysical. In practice, physics and geometrical limitations, like the finite size of the player's finger or plectrum, will always limit the maximum curvature of the string at the point of excitation. The kinks will therefore no longer be δ-functions (infinitely narrow) but will have a finite size. For illustration, travelling kinks can be modelled as Gaussian waveforms, $\xi_{\pm}(x,t) \sim \exp[-(x \pm ct)^2/2(\Delta x)^2]$, which approximate to δ-functions when $\Delta x \to 0$, where Δx characterises the width of the kink. The Fourier transform of such a function has a Gaussian distribution of Fourier coefficients varying as $c(k) \sim \exp[-(k/\Delta k)^2/2]$, where $\Delta k \Delta x = 1$. This is analogous to the uncertainty principle in position and momentum in quantum wave mechanics. For long bending lengths, the amplitudes of the higher-frequency Fourier components will be strongly attenuated.

The sound of a guitar string played with a sharp plectrum is therefore much brighter, with many more contributing higher partials, than when played with the fleshy part of a finger, which limits the bending radius to a few mm. This is illustrated by the sound of an acoustic guitar plucked first with a plectrum and then with the thumb, both at a distance of ≈ 10 cm from the bridge (audio ⟨♪ CD-ROM⟩).

Finite Rigidity
Even for an infinitely narrow plectrum, the bending at the plucking point will be limited by the finite flexibility of the string. The wave equation is then modified by an additional fourth-order bending stiffness term (*Morse* and *Ingard* [15.42, (5.1.25)]),

$$\rho S \frac{\partial^2 \xi}{\partial t^2} = T \frac{\partial^2 \xi}{\partial x^2} - E S \kappa^2 \frac{\partial^4 \xi}{\partial x^4}, \qquad (15.45)$$

where E is Young's Modulus, S is the cross-sectional area of the string (assumed homogeneous) and κ its radius of gyration. For a uniform circular wire of radius a, $S\kappa^2 = \pi a^4/4$. Using dimensional arguments, any changes in slope of the string will take place over a characteristic length $\delta \sim (ES\kappa^2/T)^{1/2} = (a^2L/2\Delta L)^{1/2}$, where ΔL is the extension of the string required to bring it to tension. This provides an intrinsic limit to the sharpness with which the string is bent and hence to the wavelength and frequency of the highest partials contributing significantly to the sound of a plucked, bowed or struck string.

The additional stiffness energy required to bend the string will also affect wave propagation on the string and the frequencies of the excited modes. Assuming sinusoidal wave solutions varying as $e^{i(\omega t \pm kx)}$, the modified wave equation (15.45) gives modes with resonant frequencies

$$\omega_n^2 = c^2 k_n^2 \left(1 + \delta^2 k_n^2\right). \qquad (15.46)$$

Waves on a real string are therefore no longer dispersionless, but travel with a phase and group velocity that depends on their frequency and wavelength. Any Helmholtz kink travelling around a real string will therefore decrease in amplitude and will broaden with time. To maintain the Helmholtz slip–stick bowed waveform, with a well-defined single kink circulating around the string, the bow has to transfer energy to the string to compensate for such broadening each time the kink moves past the bow (*Cremer* [15.29], Chapt. 7 and Sect. 15.2.2).

If a rigidly supported string is free to flex at its ends (known as a hinged boundary condition), solutions of the form $\sin(n\pi x/L)\sin(\omega t)$. However, the mode frequencies remain are no longer harmonic;

$$\frac{\omega_n^*}{\omega_n} = \left(1 + Bn^2\right)^{1/2} \sim 1 + \frac{1}{2}Bn^2, \qquad (15.47)$$

with $B = (\pi/L)^2 \delta^2$, where the expansion assumes $Bn^2 \ll 1$.

When a string is clamped (e.g. by a circular collet), it is forced to remain straight at its ends. *Fletcher* [15.43] showed that this raises all the modal frequencies by an additional factor $\sim [1 + 2/\pi B^{1/2} + (2/\pi)^2 B]$. For a real string supported on a bridge, connected to another length of tensioned string behind the bridge, the boundary conditions will be intermediate between hinged and clamped.

Kent [15.44] has demonstrated that finite-flexibility corrections raise the frequency of the fourth partial of the relatively short C5 (an octave above middle-C) string on an upright piano by 18 cents relative to the fundamental. The inharmonicity would be even larger for the very short, almost bar-like, strings at the very top of the piano. However, the higher partials of the highest notes on a piano rapidly exceed the limits of hearing, so that the resulting inharmonicity becomes somewhat less of a problem. The inharmonicity of the harmonics of a plucked or struck string results in dissonances and beats between partials, providing an *edge* to the sound, which helps the sound of an instrument to penetrate

more easily. This is particularly true for instruments like the harpsichord and the guitar when strung with metal strings.

Finite-rigidity effects are particularly pronounced for solid metal strings with a high Young's modulus. To circumvent this problem, modern strings for musical instruments are usually composite structures using a strong but relatively thin and flexible inner core, which is over-wound with one or more flexible layers of thin metal tape or wire to achieve the required mass (*Pickering* [15.45, 46]). The difference in sound of an acoustic guitar strung with metal strings and the same instrument strung with more flexible gut or over-wound strings is illustrated in ⟨?⟩ CD-ROM .

15.2.2 Nonlinear String Vibrations

Large-amplitude transverse string vibrations can result in significant stretching of the string giving a time-varying component in the tension proportional to the square of the periodically varying string displacement. This leads to a number of nonlinear effects of considerable scientific interest, though rarely of musical importance.

Morse and *Ingard* [15.42] and (*Fletcher* and *Rossing* [15.5], Chap. 5) provide theoretical introductions to the physics of nonlinear resonant systems and to nonlinear string vibrations in particular. *Vallette* [15.47] has recently reviewed the nonlinear physics of both driven and freely vibrating strings.

The Nonlinear Wave Equation

Transverse displacements of a string result in a fractional increase of its length L by an amount $1/L_0 \int_0^L 1/2(\partial \xi/\partial x)^2 \, \mathrm{d}x$ and hence to a similar fractional increase in tension and related frequency of excited modes. For a spatially varying sinusoidal wave, the induced strain and hence tension will vary with both position and time along the string. Any spatially localised changes in the tension will propagate along the string with the speed of longitudinal waves. As this is typically an order of magnitude larger than for transverse waves, $c_L/c_T \sim \sqrt{L/\Delta L}$, where ΔL is the amount that the string is stretched to bring it to tension, such perturbations will propagate backwards and forwards along the string many times during a single cycle of the transverse waves. Hence, as pointed out by *Morse* and *Ingard* [15.42], to a rather good approximation, transverse wave propagation is determined by the spatially averaged perturbation of the tension.

Consider a stretched string vibrating with large amplitude in its fundamental mode with transverse displacement $u = a \sin(\pi x/L) \cos \omega t$. The spatially averaged increase in tension is given by

$$\left(1 + \frac{\pi^2}{4} \frac{a^2}{\Delta L L} \cos^2 \omega_1 t\right)$$
$$= \left[1 + \beta \left(1 - \cos 2\omega_1 t\right) a^2\right], \qquad (15.48)$$

where $\beta = \frac{\pi^2}{8} \frac{1}{L \Delta L}$. Inserting this change in tension into the equation of motion for transverse string vibrations coplanar with a localised external driving force $f(t)$, we can write

$$\frac{\partial^2 u}{\partial t^2} + \frac{\omega_1}{Q} \frac{\partial u}{\partial t} + \omega_1^2 \left[1 + \beta \left(1 - \cos 2\omega_1 t\right) a^2\right] u$$
$$= \frac{2}{m} f(t). \qquad (15.49)$$

Mode Conversion

Nonlinearity results in an increase in the static tension by the factor $(1 + \beta a^2)$ and hence an increase in frequency of all the string modes. In addition, the term varying as $\cos 2\omega_1 t$, at double the frequency of the principal mode excited, will interact with any other modes present to excite additional with frequencies $f_n \pm 2f_1$. Of special note is the effect of this term on the principal mode of vibration itself, exciting a new mode at three times the fundamental frequency $3f_1$ and an additional parametric term (acting on itself) from the $f_1 - 2f_1 = -f_1$ contribution. The parametric term causes an additional increase in frequency of the principal mode excited, so that in total

$$\omega_{1*}^2 = \omega_1^2 \left(1 + \frac{3}{2} \beta a^2\right), \qquad (15.50)$$

where ω_1 is the small-amplitude resonant frequency.

Nonlinear effects depend on the square of the amplitude of the strongly excited mode and inversely on the amount by which the string has been stretched to bring it to tension. To investigate nonlinear effects, it is therefore advantageous to use weakly stretched strings at low initial tension. Conversely, because the tension of strings on musical instruments tends to be rather high, nonlinear effects are not in general important within a musical context.

Figure 15.28 shows measurements by *Legge* and *Fletcher* [15.48], which illustrate the nonlinear excitation and subsequent decay of the third partial of a guitar string plucked one third of the way along its length, so that the third partial was initially absent.

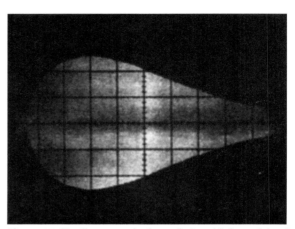

Fig. 15.28 Nonlinear excitation of the third partial of a stretched string plucked 1/3 of the way along its length; the graticule divisions are 50 ms apart (after *Legge* and *Fletcher* [15.48])

In general, bowed, plucked and struck waveforms have many Fourier components, each of which will contribute a term proportional to a_n^2 to the nonlinear increase in tension. However, in most cases, the fundamental will be the most strongly excited mode and will therefore dominate the nonlinearity.

The inharmonicity and changes in frequency associated with nonlinearity at large amplitudes can give a strongly plucked string an initial rather *twangy* sound. Nonlinear effects can also raise the frequency of a very strongly bowed open C-string of a cello by almost a semitone. However, under normal playing conditions, nonlinearity is rarely musically significant, at least in comparison with other more important perturbations of string vibrations, such as their interaction with the acoustically important structural resonances of an instrument, to be considered later.

Nonlinear Resonances

The nonlinear increase in frequency of modes with increasing amplitude leads to string resonances, which become increasingly skewed towards higher frequencies at large amplitudes, as illustrated in Fig. 15.29. For sufficiently large amplitudes and small damping, the resonance curves develop an overhang. On sweeping through resonance from the low-frequency side, the amplitude rises causing the resonance frequency to shift to higher frequencies, as indicated by the dashed line in Fig. 15.29a. Damping eventually leads to a sudden collapse, with the amplitude

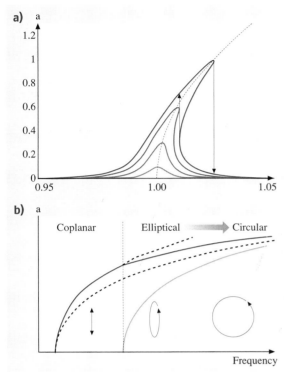

Orbital Motion

Nonlinearity results in another surprising effect on the driven resonant response. At sufficiently large amplitudes of vibration, a sinusoidally driven string suddenly develops motion in a direction orthogonal to and in phase-quadrature with the driving force, illustrated schematically in Fig. 15.29b. The transverse displacements then execute elliptical orbits about the central axis approaching circular motion at very large amplitudes (*Miles* [15.50]). In this limit, the string is under constant increased tension, producing an amplitude-dependent inward force balancing the centrifugal force of the orbiting string, resulting in an amplitude-dependent orbital frequency $\omega_{1*}^2 = \omega_1^2(1 + 2\beta a^2)$. For circular motion, the extension of the string and hence the increase in tension and resonant frequency are determined by the orbital radius of the whirling string, so there is now no variation in tension with time. The sense of clockwise or anticlockwise rotation is determined by chance or in practice by slight geometrical or material anisotropies of the string or supporting structure.

Such transitions have been investigated by *Hanson* and coworkers [15.51, 52] using a brass harpsichord string stretched to playing tension. The transition from linear to elliptically polarised motion was observed in addition to chaotic behaviour at very large amplitude. However, their measurements were complicated by the very long time constants predicted to reach equilibrium behaviour close to the transitional region and to rather strong and not well-understood splitting of the degeneracy of the transverse modes, even at low amplitudes when nonlinearity is unimportant.

A related effect occurs when a string is plucked so that it is given some orbital motion, as is invariably the case when plucking a string on a stringed instrument such as the guitar. Nonlinearity introduces

Fig. 15.29 (a) The effect of nonlinearity on the resonance curves of a stretched string with a Q of 100, for increasing drive excitation plotted against normalised resonant frequency. The *dashed curve* represents the nonlinear amplitude of the frequency for free decay. Note the hysteretic transitions at large amplitude; (b) the transition at large amplitudes from linearly polarised vibrations coplanar with the driving force to elliptical and finally circular orbital motion of the string at very large amplitudes. The two *continuous curves* represent the induced amplitudes in the directions parallel and perpendicular to the driving force

dropping to a much lower high-frequency value, illustrated by the downward arrow. On decreasing the frequency, the response initially remains on the low-amplitude curve before making a sudden hysteretic transition back to the large amplitude, strongly nonlinear, regime.

This behaviour is characteristic of any nonlinear oscillator with a restoring force that increases in strength on increasing amplitude. For a spring constant that softens with increases displacement, as we will discuss later in relation to Chinese gongs, the resonance curves are skewed in the opposite direction.

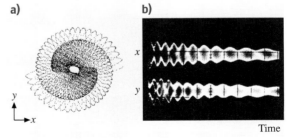

Fig. 15.30 (a) The computed precession of the damped elliptical orbits of a strongly plucked string, and (b) measurements of the orthogonal transverse components of such motion for a string plucked close to its mid-point (after [15.49])

coupling between motions in orthogonal transverse directions, causing the orbits to precess (*Elliot* [15.53], *Gough* [15.49], *Villagier* [15.47]), as illustrated by computational simulations and measurements in Fig. 15.30. The precessional frequency Ω is given by $\frac{\Omega}{\omega} = \frac{ab}{L\Delta L}$, where a and b are the major and minor semi-axes of the orbital motion and ΔL is the amount by which the string is stretched to bring it to tension [15.49].

Such precession can lead to the rattling of the string against the fingerboard on a strongly plucked instrument, as the major axis of the orbiting string precesses towards the fingerboard. The nonlinear origin of such effects can easily by distinguished from other linear effects causing degeneracy of the string modes and hence beats in the measured waveform, by the very strong dependence of the precession rate on amplitude, as illustrated in Fig. 15.30b.

15.2.3 The Bowed String

Realistic Models

Although the main features of the bowed string can be described by a simple Helmholtz wave, it is important to consider how such waves are excited and maintained by the frictional forces between the bow and string. The simple Helmholtz solution is clearly incomplete for a number of reasons including:

1. the unphysical nature of infinitely sharp kinks,
2. the insensitivity of the Helmholtz bowed waveform to the position and pressure of the bow on the string. In particular, the simple Helmholtz waveform involves partials with amplitudes proportional to $1/n$, whereas such partials must be absent if the string is bowed at any integer multiple of the fraction $1/n$ along its length, since energy cannot be transferred from the bow to the string at a nodal position of a partial,
3. the neglect of frictional forces in the slipping regime,
4. the neglect of losses and reaction from mechanical coupling to structural modes at the supporting bridge,
5. the excitation of the string via its surface, which must involve the excitation of additional torsional modes.

Understanding the detailed mechanics of the strongly nonlinear coupling between the bow and string has been a very active area of research over the last few decades, with major advances in our understanding made possible by the advent of the computer and the ability to simulate the problem using fast computational methods. *Cremer* ([15.29], Sects. 3–8) provides a detailed account of many of the important ideas and techniques used to investigate the dynamics of the bowed string. In addition, *Hutchins* and *Benade* ([15.27], Vol. 1), includes a useful introduction to both historical and recent research prefacing 20 reprinted research papers on the bowed string. *Woodhouse* and *Galluzzo* [15.54] have recently reviewed present understanding of the bowed string.

Pressure, Speed and Position Dependence

In the early part of the 20th century, *Raman* [15.35], later to be awarded the Nobel prize for his research on opto-acoustic spectroscopy, confirmed and extended many of Helmholtz's earlier measurements and theoretical models of the bowed string. Raman used an automated bowing machine to investigate systematically the effect of bow speed, position and pressure on bowed string waveforms. He also considered the attenuation of waves on the string and dissipation at the bridge. From both measurements and theoretical models, he showed that a minimum downward force was required to maintain the Helmholtz bowed waveforms on the string, which was proportional to bow speed and the square of bow distance from the bridge. He also measured and was able to explain the wolf-note phenomenon, which occurs when the pitch of a bowed note coincides with an over-strongly coupled mechanical resonance of the supporting structure. At such a coincidence, it is almost impossible for the player to maintain a steady bowed note, which tends to *stutter* and jump in a quasi-periodic way to the note an octave above, illustrated previously for a cello with a bad wolf note, audio ⓘ CD-ROM.

Saunders [15.55], well known for his work in atomic spectroscopy (Russel–Saunders spin-orbit coupling) was a keen violinist and a cofounder of the Catgut Acoustical Society. He showed that, for any given distance of the bow from the bridge, there was both a minimum and a maximum bow pressure required for the Helmholtz kink to trigger a clean transition from the sticking to slipping regimes and vice versa. Subsequently, *Schelling* [15.56] derived explicit formulae for these pressures in terms of the downward bow force F as a function of bow speed v_B, assuming a simple model for friction between bow hair and string in the slipping region of $\mu_d F$ and a maximum sticking force of $\mu_s F$,

$$F_{\min} = \frac{R_0^2 v_B}{2R\beta^2 (\mu_s - \mu_d)} \quad \text{and}$$
$$F_{\max} = \frac{2R_0 v_B}{\beta (\mu_s - \mu_d)} = 4\beta \frac{R}{R_0} F_{\min}, \quad (15.51)$$

where R_0 is the characteristic impedance of the string terminated by a purely resistive load R at the bridge, and

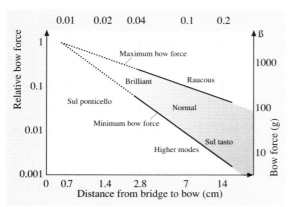

Fig. 15.31 The playing range for a bowed string as a function of bow force and distance from bridge, with the bottom and right-hand axis giving values for a cello open A-string with a constant bow velocity of $20\,\mathrm{cm\,s^{-1}}$ (after *Schelling* [15.56])

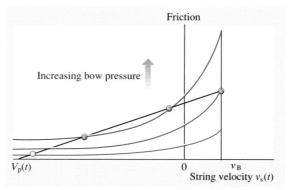

Fig. 15.32 Schematic representation of the dependence of the frictional force between bow and string on their relative velocity and downward pressure of the bow on the string. The *straight line* with slope $2R_0$ passes through the velocity v_p of the string determined by its past history and the intersection with the friction curves determines its current velocity. The *open circle* represents the single intersection in the slipping regime at low bow pressures, while the *closed circles* illustrate three intersections at higher pressures

β is the fractional bowing point along the string. If the downwards force is larger than $F_{\max}$ the string remains stuck to the string instead of springing free into the slipping regime, while for downward forces less than $F_{\min}$ an additional slip occurs leading to a double-slipping motion.

Figure 15.31 is taken from the article by *Schelling* on the bowed string in the Scientific American special issue on the *Physics of Musical Instruments* [15.57]. It shows how the sound produced by a bowed cello string changes with bow position and downward bow pressure for a typical bow speed of 20 cm/s. Note the logarithmic scales on both axes. In practice, a string can be bowed over a quite a large range of distances from the bridge, bow speeds and pressures with relatively little change in the frequency dependence of the spectrum and quality of the sound of an instrument, apart from regions very close and very distant from the bridge. Nevertheless, the ability to adjust the bow pressure, speed and distance from the bridge, to produce a good-quality steady tone, is one of the major factors that distinguish an experienced performer from the beginner.

Slip–Stick Friction
An important advance was the use of a more realistic frictional force, dependent on the relative velocity between bow and string, shown schematically for three downward bow pressures in Fig. 15.32. Such a dependence was subsequently observed by *Schumacher* [15.58] in measurements of steady-state sliding friction between a string and a uniformly moving bow.

The frictional force is proportional to the downward bow pressure.

Friedlander [15.59] showed that a simple graphical construction could be used to compute the instantaneous velocity v at the bowing point from the velocity $v_p(t)$ at the bowing point induced by the previous action of the bow. The new velocity is given by the intersection of a straight line with slope $2R_0$ drawn through v_p with the friction curve, where R_0 is the characteristic string impedance. This follows because the localised force between the bow and string generate secondary waves with velocity $F/2Z_0$ at the bowing point as previously described (15.34). In the slipping region well away from capture, there will be just a single point of intersection, so the problem is well defined. However, close to capture, as illustrated by the intersections marked by the black dots with the upper frictional curve, the straight line can intersect in three points (two in the slipping regime and one in the sticking regime) as first noted by Friedlander.

Computational Models
This model has been used in a number of detailed computational investigations of both the transient and steady-state dynamics of the bowed string, notably by the Cambridge group lead by *McIntyre* and *Woodhouse* [15.60–62], their close collaborator *Schumacher* [15.58, 63] from Carnegie-Mellon, and

Guettler [15.64], who is also a leading international double-bass virtuoso. Readers are directed to the original publications for details of the various computational schemes used, which are also discussed in some detail by *Cremer* ([15.29], Sect. 8.2).

Whenever a string is bowed at an integer interval along its length, the secondary waves excited by the frictional forces between bow and string can give rise to coherent reflections between the bow and bridge, giving rise to pronounced *Schelling ripples* on the Helmholtz waveform and hence significant changes in the spectrum of the radiated sound. However, because the bowing force tends to be distributed across the ≈1 cm width of the bow hairs, such effects tend to be smeared out and are not generally of significant musical importance. *McIntyre* et al. [15.62] have also shown that uncertainties in the sticking point from the finite-width strand of bow hairs leads to a certain amount of jitter or aperiodicity in the pitch of the bowed string amounting of a few cents, which is again of little musical significance, though the noise generated may be significant in contributing to the characteristic sound of bowed string instruments.

It is instructive to consider the kind of computational methods developed by Woodhouse and his collaborators to investigate both the initial transient and the steady-state dynamics of the bowed string. This is illustrated schematically in Fig. 15.33, where u and u' represent the velocity under the bow from waves travelling towards the bow from the bridge and from the stopped end of the string respectively, and v and v' are the velocities at the bowing point of the waves travelling away from the bow. In the absence of any bowing force $v = u'$ and $v' = u$. However, in the presence of a frictional force between the bow and string, the outgoing waves will acquire an additional velocity $f/2R_0$, where the frictional

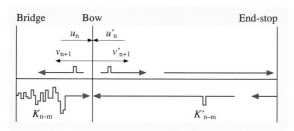

Fig. 15.33 Schematic representation of the model used by *McIntyre* and *Woodhouse* to compute bowed string dynamics. The velocities u and v represent incoming and outgoing waves from the two ends, with reflections of impulse functions from the bridge and end-stop represented in their digitised form

force is determined by the velocity from the incoming waves $u + u'$ excited by previous events. The outgoing wave travelling towards the stopped end or nut of the string will simply be reflected, while the outgoing wave reaching the bridge will not only be reflected, but will also excite continuing vibrations at the bridge from the excitation of the coupled structural modes.

Such problems can be solved using a Green's function approach *Cremer* ([15.29], Sect. 8.4), in which the outgoing waves can be considered in terms of the response to forces represented as a succession of short impulses. The problem is then reduced to understanding the response of the system for the reflection of a sequence of short impulses or δ functions. At the end-stop, an impulse will simply be reflected with reversed sign, but reduced amplitude in the case of a soft finger stopping the string. The incoming wave u' generated by the reflected impulse will therefore be an impulse function delayed in time by the transit time from the bow to the end-stop and back. Similarly, the impulse returning from the bridge will be an impulse delayed by the transit time between bow and bridge and back followed by a wave generated by the induced motions of the bridge on reflection. The time-delayed impulse responses from reflections at the bridge and end-stop can described by the functions $K(t)$ and $K'(t)$. The incoming waves $(u(t), u'(t))$ can then be described by the convolution of $K(t)$ and $K'(t)$ with the outgoing waves $(v(t), v'(t))$ considered as a succession of impulse functions at all previous times t', such that

$$u(t) = \int^t v(t')K(t-t')\,dt' \quad \text{and}$$
$$u'(t) = \int^t v'(t')K'(t-t')\,dt\,. \qquad (15.52)$$

To compute the resulting dynamics of string motion digitally, one simply computes the above velocities at a succession of short time intervals, with the outgoing waves determined from the incoming waves plus the secondary waves induced by the resulting frictional force, such that

$$v'_{n+1} = u_n + f_n/2R_0 \quad \text{and}$$
$$v_{n+1} = u'_n + f_n/2R_0 \qquad (15.53)$$

and

$$u_{n+1} = \sum^n v_m K_{n-m} \quad \text{and}$$
$$u'_{n+1} = \sum^n v'_m K'_{n-m}\,, \qquad (15.54)$$

where K_j and K'_j are now the digital equivalents of the time-delayed impulse responses, illustrated schematically in Fig. 15.33. The frictional force f_n entering (15.53) is evaluated from the pressure- and velocity-dependent frictional force using the Friedlander construction with the computed string velocity under the bow given by $u_n + u'_n$.

Pressure Broadening and Flattening

As an example, Fig. 15.34 illustrates the computed velocity of the string under the bow as a function of increasing bow pressure (*McIntyre* et al. [15.60]). In contrast to the rectangular waveform predicted by the simple Raman model, the waveform is considerably rounded, especially at low bow pressures. This results in a less strident, less intense sound, with the higher partials strongly attenuated. At higher pressures, but at the same position and with the same bow velocity, the rounding is less pronounced, so that higher partials become increasingly important. The increased intensity of the higher partials leads to an increased perceived intensity with bow pressure, in contrast to the Raman model, in which the waveform and hence intensity remains independent of bow pressure. This is referred to as the *pressure effect*. At even higher pressures, the ambiguity in intersections noted by Friedlander leads to a pronounced increase in the capture period and hence the pitch of the bowed note, known as the *flattening effect*. These features are discussed in considerable detail along with his own important research and that of his collaborators on such effects by *Cremer* ([15.29], Chaps. 7 and 8).

Initial Transients

Computational models can also describe the initial transients of the bowed string before the steady-state Helmholtz wave is established. Figure 15.35 compares the computed and measured initial transients of the string velocity under the bow for a string played with a sharp attack (a martelé stroke) (*McIntyre*, *Woodhouse* [15.60]). These computations also include the additional excitation of torsional waves, which are excited because the bowing force acts on the outer diameter of the wire, exerting a couple in addition to a transverse force. The excitation and loss of energy to the torsional waves appears to encourage the rapid stabilisation of the bowed Helmholtz waveform.

For low-pitched stringed instruments such as the double bass, it is very important that the Raman bowed waveform is established very quickly, otherwise there will be a significant delay in establishing the required pitch. Remarkably, *Guettler* [15.64] has shown that, by simultaneously controlling both bow speed and downward pressure, the player can establish a regular Raman waveform in a single period. The speed with which a steady-state bowed note can be established can be represented on a *Guettler diagram*, where the number of slips before a steady-state Helmholtz motion is established can be illustrated as a two-dimensional histogram as a function of bowing force and acceleration of the bow speed from zero.

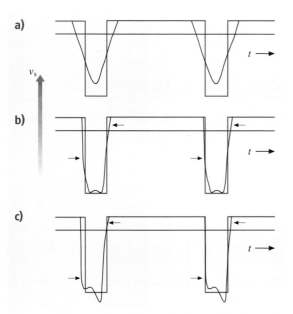

Fig. 15.34 Computed velocity of string at bowing point for increasing bow pressures in the ratios 0.4 : 3 : 5 (after *McIntyre* and *Woodhouse* [15.60]) illustrating both the broadened waveform and pitch dependence on bow pressure compared with the idealised rectangular Helmholtz bowed waveform

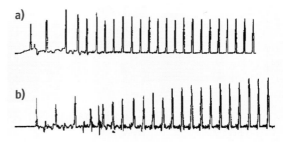

Fig. 15.35 (a) Computed transient string velocity at the bowing point for a strongly bowed string including coupling to both transverse and torsional modes and (b) the measured string velocity for a strongly played martelé bow stroke (after *McIntyre* and *Woodhouse* [15.60])

To investigate such effects experimentally, *Galluzzo* and *Woodhouse* [15.54, 65] have recently developed a dynamically controlled bowing machine with active feedback, providing programmable control of both downward bow pressure and bow speed. This enables reliable and reproducible results to be made over a very wide range of possible playing parameters, extending Guettler's original measurements.

Viscoelastic Friction

Recent measurements have shown that the frictional model assumed in these investigations is over-simplistic. The force between the bow hairs and the string is maintained by a thin layer of rosin which coats them both. Rosin is a rather soft, sticky substance, with a glass-to-liquid transition not far above room temperature, resulting in viscoelastic properties, which are very sensitive to temperature (*Smith, Woodhouse* [15.66]). As the bow slides past the bow hair, the frictional forces will heat the rosin and hence reduce its viscoelasticity frictional properties. During the sticking regime, with no work being done at the bow–string interface, the rosin will cool down and the friction will increase. The frictional forces are therefore hysteretic and will be strongly dependent on past history within a given period of string vibration. *Woodhouse* et al. [15.68] and *Smith* [15.69] have investigated this hysteretic behaviour in some detail using rosin-coated glass rods. The hysteretic properties shown in Fig. 15.36 were deduced from measurements at the two supported ends of the string. *Woodhouse* [15.70] subsequently extended

his computational models to incorporate the hysteretic frictional properties. Somewhat surprisingly, this more realistic model made little qualitative difference to the predicted behaviour. Such measurements contribute to our understanding of the physical processes underlying viscoelastic properties of various coatings and lubricants and have become an important tool in the field of tribology (studies of friction).

15.2.4 Bridge and Soundpost

We now consider the role of the bridge and soundpost in providing the coupling between the vibrating strings and the vibrational modes of the body of instruments of the violin family. We also consider the influence of such coupling on the modes of string vibration, which involves a discussion of the very important influence of damping on the normal modes of any coupled system.

Bridges

Many plucked and struck stringed instruments, such as the piano or guitar, use a rather low solid bridge to support the strings and transfer energy directly from the transverse string vibrations perpendicular to the supporting soundboard or front-plate of the instrument. The bridge needs only to be sufficiently high to prevent the strings from vibrating against the fingerboard or shell of the instrument. This is also true for the Chinese two-string violin, the *erhu*, which is held and played so that the bow excites string vibrations perpendicular rather than parallel to the stretched snake-skin membrane supporting the bridge and strings. The strings of a harp are attached to an angled sounding board, so that transverse string vibrations in the plane of the strings couple directly to the perpendicular vibrations of the supporting soundbox *Fletcher* and *Rossing* ([15.5], Sect. 11.2).

For such instruments, the bridge and other string terminations play a relatively insignificant acoustic role,

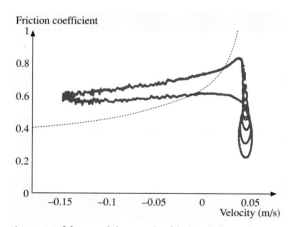

Fig. 15.36 Measured hysteretic frictional force between string and a glass bow coated with rosin, with the *dashed line* indicating previously assumed velocity dependence (after *Smith* and *Woodhouse* [15.66])

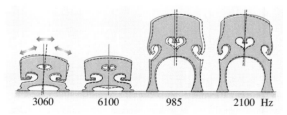

Fig. 15.37 The lowest in-plane resonant modes and frequencies of violin and cello bridges (after *Reinicke* [15.67]). The *arrows* represent the vibrational directions of the bowed outer and middle strings

apart from adding a small inertial mass and additional stiffness to the soundboard or top plate, which only slightly perturbs the frequencies of the structural modes of vibration.

In contrast, the rather high bridges on instruments of the violin and viol families have a profound influence on the acoustical properties, particularly at frequencies comparable with and above any mechanical resonances of such structures. Figure 15.37 illustrates the shape of modern violin (and viola) and cello bridges and indicates their principal vibrational modes, as measured by *Reinicke* and *Cremer* [15.67, 71] using laser interference holography. Bridges are cut from maple and taper in thickness from the two feet to the top surface supporting the four strings, which are set in small v-shaped locating grooves.

Reinicke [15.67, 71] showed that the lowest violin bridge resonance at typically around 3 kHz involves a rotational motion of the top half of the bridge about its waist. The rotational motion induced by the vibrating strings supported on the top of the bridge results in a couple acting on the top plate via the two feet. The next most important in-plane resonance is at ≈ 6 kHz and involves the top of the bridge bouncing up and down on its feet, resulting in forces via the legs perpendicular to the supporting surface. The cello bridge has rather longer legs, resulting in two low-frequency twisting modes with resonances at around 1 and 2 kHz, both of which exert a couple on the top plate. Longitudinal forces from the vibrating strings can also induce bridge motion perpendicular to its plane (at double the frequency of the vibrating strings), but such motion is generally rather small and will be ignored for the purposes of this chapter.

Any transverse string force at the top of the bridge, from bowing, plucking or striking the string, will be transferred to the supporting body via the two feet. This will induce a linear motion of the centre of mass of the instrument, rotation about its centre of mass and the excitation of both flexural and longitudinal waves in the plates of the instrument. Because bowing involves a static force which reverses with bow direction, a bowed instrument has to be held fairly firmly by the player, which introduces an extra channel for energy loss through the supporting chin and fingers. The induced linear and rotational motions of an instrument are relatively unimportant at audio frequencies as they involve the whole mass M of the instrument with admittances varying $\sim 1/\mathrm{i}M\omega$.

If a tall bridge is placed centrally on a symmetric shell structure, like the body of an early renaissance viol, the plucked or bowed motion of the strings parallel to the supporting top plate would excite only asymmetrical modes of the supporting structure, whereas perpendicular string vibrations would excite only symmetrical modes. For instruments of the violin family, an offset soundpost is wedged between the front and back plates, which destroys the symmetry. The coupled modes will then involve a linear combination of symmetrical and asymmetric body modes, as discussed later (Sect. 15.2.6).

The arching of the top of the bridge allows each of the supported four strings to be bowed separately or together (double stopping), with the bow direction making an angle of around $\pm 15-20°$ relative to the top plate for the outer two strings and almost parallel for the middle two strings. Bowing on the outer two strings therefore involves significant perpendicular in addition to parallel forces, but only slightly different sounds from a single type of string when supported in different positions on the bridge. Audio 13 CD-ROM compares the sound of a bowed covered-gut D-string mounted in the normal position and in the G-, A- and E-string positions on the same violin. On a guitar almost all the sound is produced by the vertical motion of the plucked string rather than by parallel vibrations, which primarily excite non-radiating longitudinal modes of the top plate.

Simplified Bridge Model

Cremer ([15.29], Chap. 9) gives a detailed historical and scientific introduction to research on violin and cello bridges and their coupling to the body of the instrument. Relatively complicated mechanical models are described composed of several masses and springs to account for the various possible vibrational modes of the bridge. However, the principal resonances of the violin bridge shown in Fig. 15.37 can be modelled very simply by a two-degree-of-freedom mechanical model, with effective masses representing the linear and rotational energy of the top of the bridge coupled to the supporting surface through the two supporting feet via a rotational or vertical spring, illustrated schematically in Fig. 15.38. The relatively light mass and added rigidity of the lower

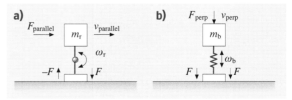

Fig. 15.38a,b Simplified mechanical models for the lowest (**a**) rotational and (**b**) bouncing motions of the bridge supported by its two feet on a rigid surface

half of the bridge will only slightly perturb the resonant frequencies of the more massive supporting plates and can therefore be ignored as a first approximation. The effective masses and strength of the coupling springs can be chosen to reproduce the vibrational characteristics of the first two vibrational modes of the violin (or cello) bridge, which dominate the acoustical properties of the instrument.

At low frequencies, well below any resonant frequency, the bridge will vibrate as a rigid body, adding a small amount of additional mass, moment of inertia and rigidity to the top plate, which will again only slightly perturb the vibrational frequencies of the supporting shell structure. The additional relative height of the cello bridge compared with that of the violin bridge enables a rather larger couple to be exerted by the bowed string on the more massive top plate. There is a delicate balance between increasing the coupling to enhance the intensity at low frequencies without making it so strong that troublesome wolf-note problems arise, as referred to earlier.

Bridge-Hill (BH) Feature

Reinicke [15.67, 71] and *Cremer* [15.29] highlighted the importance of the bridge resonance on both the sound of the violin and on admittance measurements, which are traditionally made by exciting the violin at the top of the bridge using an external force parallel to the top supporting plate. In recent year, this problem has attracted renewed interest, in an attempt to describe the rather broad peak and associated phase changes superimposed on the multi-resonant response of the instrument, which Jansson refers to as the *Bridge-Hill* (BH) feature [15.74, 75].

Figure 15.39 shows recent measurements by *Woodhouse* [15.72] of the modulus of the admittance at the bridge for a particular instrument using a series of bridges with different masses but the same resonant frequency at ≈ 2 kHz. A strong but rather wide overall BH peak is observed in the vicinity of the bridge resonance. Note the marked decrease in admittance with increasing bridge mass above the bridge resonance. There is also an associated overall 90° change in the phase of the admittance on passing through the peak.

Evidence for the BH feature can also be seen in *Dünnwald*'s [15.73] superimposed measurements of the sound output of a large number of high-quality Italian, modern master and factory violins as a function of sinusoidal input force at top of the bridge, shown in

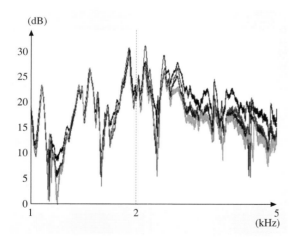

Fig. 15.39 The admittance at the top of the bridge on a single violin, plotted on the same but arbitrary dB scale, for a number of violin bridges having the same height and resonant frequency (≈ 2 kHz) but different masses. The upper curve corresponds to the lightest bridge that could be fabricated from a standard bridge blank and the lowest curve by the heaviest. Subtracting 45 dB from the results would give the approximate admittance in units of ms^{-1}N^{-1} (data kindly provided by *Woodhouse* [15.72])

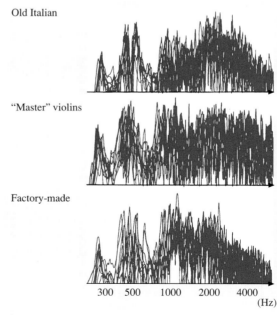

Fig. 15.40 Overlays of the sound output of 10 typical old Italian, modern master instruments, and 10 factory instruments for a constant sinusoidal force at the top of the bridge (after *Dünnwald* [15.73])

Fig. 15.40. A surprising aspect of these measurements is the apparent lack of any such feature for modern master violins, possibly because of a wider variation in bridge resonances and effective masses of bridge and plate resonances in the chosen instruments. From measurements of the radiated sound of over 700 violins, Dünnwald proposed that the presence of a number of strong acoustic resonances in the broad frequency band from 1.5 to 4 kHz was one of the distinguishing features of a really fine instrument. The influence of the bridge in accounting for such a peak and the reduced response at higher frequencies is clearly important.

Woodhouse [15.76] has recently revisited the problem of the coupling between bridge and body of the instrument and the origin of the BH peak. A simple theoretical model shows that the peak depends on many factors, such as the effective masses, Q-values and resonant frequencies of the major vibrational modes of the bridge and the multi-resonant properties of the instrument. To demonstrate the overall effect of the bridge without having to consider the detailed vibrational response of a particular instrument, Woodhouse first considered coupling to a simplified model for the vibrational modes of the coupled instrument. This assumed a set of coupled vibrational modes each having the same effective mass M and Q-value, with a constant spacing of resonances $\omega_0 = 2\pi \Delta f$. Different values for these parameters would need to be used to model the independent rotational or bouncing modes, though Woodhouse concentrates on the influence of the lowest frequency "rocking" bridge mode. The merit of such a model is that the multi-resonant response of such a system varies monotonically with frequency. The features introduced by the resonant properties of the bridge can then be easily identified and the input admittance expressed relative to the admittance A_V for a completely rigid bridge of the same mass, where

$$A_V(\omega) = \frac{1}{M} \sum_n \frac{i\omega}{(n\omega_0)^2 - \omega^2 + i\omega n\omega_0/Q} \,. \tag{15.55}$$

The corresponding input admittance for the one-degree-of-freedom model bridge is then given by

$$A_{BB}(\omega) = \frac{A_V + i\omega/m\omega_B^2}{1 - (\omega/\omega_B)^2 + i\omega m A_V} \,, \tag{15.56}$$

where m is the effective mass of the bridge and ω_B its resonant frequency and internal damping of the bridge has been neglected.

We can also define a nonlocal admittance or mobility A_{VB} to describe the induced body motion per unit force

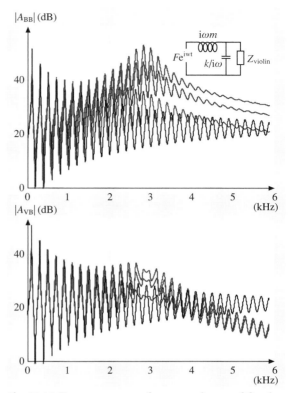

Fig. 15.41 Response curves for a one-degree-of-freedom bridge coupled to an artificial set of regularly spaced (200 Hz), constant effective mass (100 g) and constant Q (50) structural resonances. The *upper curves* illustrate the effect of bridge mass on the admittance A_{BB} measured at the point of excitation at the top of the bridge, while the *lower curves* illustrate the corresponding induced body mobility A_{VB}. The *coloured response curves* are for lossless bridges with effective masses 1, 1.5 and 3 g (highest to lowest response), having the same resonant frequency at 3 kH (after *Woodhouse* [15.76]) The *black curves* show the violin body response A_V that would be measured using a massless rigid bridge

at the foot of the bridge given by

$$A_{VB}(\omega) = \frac{A_V}{1 - (\omega/\omega_B)^2 + i\omega m A_V} \,. \tag{15.57}$$

The simulations in Fig. 15.41 illustrate the major effect of the bridge resonance on both the input response and induced body motion and hence radiated sound at, around and above the resonant frequency of the bridge (3 kHz in the above example). For a real instrument, the spacing and Q-values of the individ-

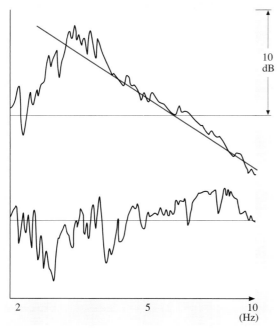

Fig. 15.42 Admittances of a violin measured at the top of the bridge (top trace) and at the left foot of the bridge (lower trace) illustrating a strong BH peak when measured at the top of the bridge but a relatively monotonic dependence of the body of the instrument (after *Cremer* [15.29], Fig. 12.9). The added *solid line* represents the $1/f$ reduction in the predicted BH response above the bridge resonance

ual modes will be very irregular and highly instrument dependent; nevertheless, the effect of the bridge resonance on the overall response will be very similar. In particular, the bridge resonance gives a broad peak in input admittance followed by a 6 dB/octave decrease in the admittance above resonance, where the response is largely dominated by the bridge dynamics rather than that of the instrument itself, with $A_{BB} \sim 1/im\omega$. Note that the height and width of the peak is largely determined by energy lost to the coupled structural vibrations (including, in practice, additional energy lost to all the supported strings) rather than from internal bridge losses, which have been neglected in this example.

The bridge resonance introduces a somewhat smaller peak in the induced body mobility and hence radiated sound. Well above the bridge resonance, the induced body velocity is given by $A_{VB}(\omega_B/\omega)^2$, with an intensity decreasing by 12 dB/octave. Unlike the input bridge admittance, the induced body motion and output sound retains the characteristic resonances of the instrument, though attenuated.

The predicted difference in admittance at the top of the bridge A_{BB} and top of the instrument A_V is illustrated in Fig. 15.42, in measurements by *Moral* and *Jansson* [15.77] reproduced by *Cremer* ([15.29], Fig. 15.9). Whereas the average admittance of the violin varies relatively little with frequency, the admittance at the bridge shows a pronounced BH peak with a relatively featureless and approximately $1/f$ (the added solid line) variation above the peak, as anticipated from the above model.

Woodhouse [15.76] has extended this idealised model to describe the coupling of the bridge to a more realistic, but still simplified, model for the vibrational modes of the violin with a soundpost. This changes the detailed response, but not the overall qualitative features. Because the response of a violin depends rather randomly at higher frequencies on the positions and Q-values of the structural modes, Woodhouse uses a logarithmic scale to average the peaks and troughs at the maxima and minima of the admittance (approximately proportional to Q and $1/Q$), to give a *skeleton curve* describing the global variation of the violin's complex admittance (more details are given in the later Sect. 15.2.3 on shell modes). This enables Woodhouse to illustrate the influence of various bridge parameters on the acoustical properties of the instrument, suggesting ways in which violin makers could vary bridge properties to optimise the sound quality of an instrument, though that will always be a matter of personal taste rather than being scientifically defined.

The important role of the bridge in controlling the sound of the violin or cello has often been overlooked, even by many skilled violin makers. Indeed one of the reasons why Cremonese violins generally produce such highly valued sounds is the experience and skill involved in adjusting the mass, size and fitting of the bridge (and the position of the soundpost) to optimize the sound quality, investigated experimentally by Hacklinger [15.78].

Added Mass and Muting
A familiar demonstration of the importance of the mass of the bridge on the sound of an instrument is to place a light mass or mute on the top of the bridge. This dramatically softens the tone of the instrument by decreasing the resonant frequency of the bridge and hence amplitude of the higher-frequency components in the spectrum of sound. The added mass Δm lowers the resonant frequency ω_B by a factor $[m/(m + \Delta m)]^{1/2}$.

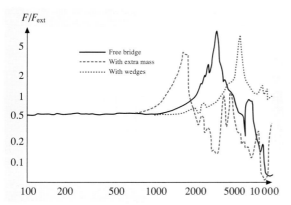

Fig. 15.43 Measurements of bridge resonances from measurements of the ratio of the force exerted by one bridge foot on a rigid surface to the applied force, for an added mass of 1.5 g and for wedges introduced between the *wings* of the bridge to increase its rotational stiffness (after *Reinicke* data reproduced in *Fletcher* and *Rossing* [15.5], Fig. 10.19)

Figure 15.43 illustrates changes in resonant frequency measured by Reinicke for a bridge mounted on a rigid support for an additional mass of 1.5 g and when wedges are inserted between the *wings* of the bridge to inhibit the rotational motion of the top of the bridge and hence the resonant frequency. Audio [?] CD-ROM illustrates the changes in sound of a violin before and after first placing a commonly used 1.8 g mute and then a much heavier practice mute on top of the bridge, and after wedges were inserted in the bridge to inhibit the rocking motion.

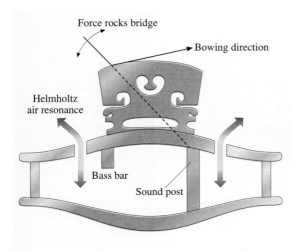

Fig. 15.44 Schematic cross section of the violin illustrating the position of the soundpost, bass-bar and f-hole openings

Soundpost and Bass Bar

In instruments of the violin family, a soundpost is wedged asymmetrically between the top and back plates, as illustrated schematically in Fig. 15.44. Additionally, a bass bar runs longitudinally along much of the length of the bass-side of the front plate. The soundpost and bass bar give added mechanical strength to the instrument, helping it to withstand the rather large downward force from the angled stretched strings passing over the bridge, which is typically ≈ 10 kg weight for the violin.

The influence of the soundpost on the quality of sound is so strong that the French refer to it as the *âme* (soul) of the instrument. Its acoustic function is to provide a rather direct coupling of the induced bridge vibrations to both the back and the front plates of the instrument and to provide an additional mechanical constraint, so that the bowed string vibrations excite normal modes, which are linear combinations of the asymmetric and symmetric modes of vibration of the front and back plates of the instrument.

Of modern stringed instruments, only the violin family makes use of a soundpost. However, soundposts were probably used in the medieval fiddle and other early instruments including the viol. The ancient Celtic *crwyth* effectively combined the functions of the bridge and soundpost by using a bridge with feet of unequal length, the first resting on the top plate and the second passing through a hole in the front face to rest on the back plate – a bridge design still used today in the folk-style Greek *rebec* (see *Gill* [15.79]).

15.2.5 String–Bridge–Body Coupling

We now consider the interaction of the strings with the vibrational modes of the body of the instrument via the bridge. Because we are dealing with the coupling of the vibrational modes of the strings, bridge and body of the instrument, the problem has to be considered in terms of the normal modes of the coupled system. An important aspect of this problem that is often not widely recognised, but is always important in dealing with musical instruments, is the profound influence of damping on the nature of the coupled modes. This is a generic phenomenon for any system of coupled oscillators. As we will see, the strength of the damping relative to the strength of the coupling determines whether a system can be considered as weakly or strongly coupled.

String–Body Mode Coupling

For simplicity, we only consider the perturbation of string resonances from the induced motion of the bridge

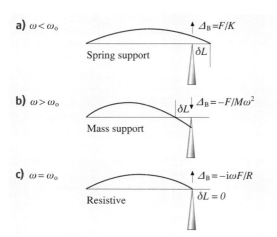

Fig. 15.45a–c Coupling of vibrating string to a weakly coupled normal mode via the bridge for the string resonance (**a**) below the resonant frequency of the coupled mode, (**b**) above the resonant frequency, and (**c**) at resonance

and ignore any damping introduced by, for example, a finger stopping the string at its opposite end. In general, as we have already seen, the admittance at the point of string support on the bridge will be a complicated, multi-resonant, function of frequency reflecting the normal modes of vibration of the coupled structure. The normal modes will include the combined motions of all parts of the violin body, including the body, bridge, neck, tailpiece, etc.

Each coupled normal mode will contribute a characteristic admittance, which will be spring-like below its resonant frequency, resistive at resonance and mass-like above resonance. The effect of such terminations on the vibrating string is therefore to shift its effective nodal position, as illustrated in Fig. 15.45a–c, for a spring-like string termination with spring constant K, an effective mass M and a lossy support with resistance R.

For a spring-like termination with spring constant K, the bridge will move in phase with the force acting on it. This will increase the effective length of the vibrating string between nodes by a distance $\Delta_B = T/K$, lowering the frequency of a string mode by a fraction T/KL. For a mass-like termination M, the end-support will move in anti-phase with the forces acting on it, so that the effective string length is shortened. The string frequencies are then increased by the fraction $T/M\omega_n^2 = (1/n\pi)^2 m/M$, where m is the mass of the string. It is less easy to visualise the effect of a resistive support because the induced displacement is in phase-quadrature with the driving force. Mathematically, however, a resistive termination

can be considered as an imaginary mass $m^* = R/i\omega$ leading to an imaginary fractional increase in frequency $i\omega_n 1/(n\pi)^2 m/R$. This imaginary frequency is equivalent to an exponential decay $e^{-t/\tau}$ for all modes with $\tau = \pi^2 R/m\omega_1^2$, where ω_1 is the frequency of the fundamental string mode. This result can also be derived using somewhat more physical arguments, by equating the loss of stored vibrational energy to the energy dissipated at the end-support.

The terminating admittance at the bridge for a single coupled vibrational mode can be written in the form

$$A_n(\omega) = \frac{1}{M_n} \frac{i\omega}{\omega_n^2 - \omega^2 + i\omega\omega_n/Q_n}, \quad (15.58)$$

with the real part of this function determining the decay time of the coupled string resonances and the imaginary part the perturbation in their resonant frequencies. The perturbations are proportional to the ratio of mass of the vibrating string to the effective mass of the coupled resonance at the point of string support on the bridge and vary with frequency with the familiar dispersion and dissipation curves of a simple harmonic oscillator. For a multi-resonant system like the body of any stringed instrument, the string perturbations from each of the coupled structural resonances are additive.

Normal Modes and Damping

Strictly speaking, whenever one considers the coupling between any two or more vibrating systems, one should always consider the normal modes or coupled vibrations rather than treat the systems separately, as we have done above. However, the inclusion of damping has a profound influence on the normal modes of any system of coupled oscillators (*Gough* [15.80]) and justifies the above *weak-coupling approximation*, provided that the coupling at the bridge is not over-strong. Although we consider the effect of damping in the specific context of a lightly damped string coupled to a more strongly damped structural resonance, the following discussion is completely general and is applicable to the normal modes of any coupled system of damped resonators.

Consider a string vibrating in its fundamental mode coupled via the bridge to a single damped structural resonance. The string has mass m, an unperturbed resonant frequency of ω_s a Q-value of Q_s and a displacement at its mid-point of v. The coupled structural resonance has an effective mass M at the point of string support, an unperturbed resonant frequency of ω_M a Q-value of Q_M and displacement of u.

The vibrating string exerts a force on the coupled body mode, such that

$$M\left(\frac{\partial^2 v}{\partial t^2} + \frac{\omega}{Q_m}\frac{\partial v}{\partial t} + \omega_M^2 v\right) = T\left(\frac{\pi}{L}\right)u. \quad (15.59)$$

Multiplying this expression through by $\partial v/\partial t$, one recovers the required result that the rate of increase in stored kinetic and potential energy of the coupled mode is simply the work done on it by the vibrating string less the energy lost from damping. Similar energy balance arguments enable us to write down an equivalent expression for the influence of the coupling on the string vibrations,

$$\frac{m}{2}\left(\frac{\partial^2 u}{\partial t^2} + \frac{\omega}{Q_m}\frac{\partial u}{\partial t} + \omega_M^2 u\right) = T\left(\frac{\pi}{L}\right)v, \quad (15.60)$$

where the effective mass of the vibrating string is $m/2$ (i.e. its energy is $1/4\, m\omega^2 u^2$). To determine the normal-mode frequencies, we look for solutions varying as $e^{i\omega t}$. Solving the resultant simultaneous equations we obtain

$$\left(\omega_M^2 - \omega^2(1 - i/Q_M)\right)\left(\omega_m^2 - \omega^2(1 - i/Q_m)\right)$$
$$= \left(T\frac{\pi}{l}\right)^2 \frac{2}{mM} = \alpha^4, \quad (15.61)$$

where α is a measure of the coupling strength.

Solving to first order in $1/Q$-values and α^2 we obtain the frequencies of the normal modes $\Omega_\pm$ of the coupled system,

$$\Omega_\pm^2 = \omega_+^2 \pm \left(\omega_-^4 + \alpha^4\right)^{1/2}, \quad (15.62)$$

where

$$\omega_\pm^2 = \frac{1}{2}\left[\omega_M^2 \pm \omega_m^2 + i\left(\frac{\omega_M^2}{Q_M} \pm \frac{\omega_m^2}{Q_m}\right)\right]. \quad (15.63)$$

If the damping terms are ignored, we recover the standard perturbation result with a splitting in the frequencies of the normal modes at the crossover frequency (when the uncoupled resonant frequencies of the two systems coincide) such that $\Omega_\pm^2 = \omega_M^2 \pm \alpha^2$.

In the absence of damping, the two normal modes at the crossover frequency are linear combinations of the coupled modes vibrating either in or out of phase with each other, with equal energy in each, so that $v/u = \pm\sqrt{m/2M}$. Well away from the crossover region, the mutual coupling only slightly perturbs the individual coupled modes, which therefore retain their separate identities. However, close to the crossover region, when $|\omega_M - \omega_m| \leqslant 2\alpha^2/(\omega_M + \omega_m)$, the coupled modes lose their separate identities, with the normal modes involving a significant admixture of both.

The inclusion of damping significantly changes the above result. If we focus on the crossover region, coupling between the modes will be significant when

$$\alpha^2 \sim \omega_M^2 - \omega_m^2. \quad (15.64)$$

At the crossing point, when the uncoupled resonances coincide, the frequencies of the coupled normal are given by

$$\Omega_\pm^2 = \omega_M^2\left(1 + i/2Q_+\right) \pm \left(\alpha^4 - \left(\frac{\omega_M^2}{2Q_-}\right)^2\right)^{1/2}, \quad (15.65)$$

where

$$\frac{1}{Q_\pm} = \frac{1}{Q_M} \pm \frac{1}{Q_m}. \quad (15.66)$$

The sign of the terms under the square root clearly depends on the relative strengths of the coupling and damping terms. When the damping is large and the coupling is weak, such that $(\omega_M^2/2Q_-)^2 > \alpha^4$, one is in the *weak-coupling* regime, with no splitting in frequency of the modes in the crossover region. In contrast, when the coupling is strong and the damping is weak, such that $(\omega_M^2/2Q_-)^2 < \alpha^4$, the normal modes are split, but by a somewhat smaller amount than had there been no damping.

Figure 15.46 illustrates the very different character of the normal modes in the crossover region in the weak- and strong-coupling regimes. The examples shown are for an undamped string interacting with a structural resonance with a Q of 25, evaluated for coupling factors,

$$K = \frac{2Q_M}{\omega_M^2}\alpha = \frac{2Q_M}{n\pi}\sqrt{\frac{2m}{M}}, \quad (15.67)$$

of 0.75 and $\sqrt{5}$, in the weak- and strong-coupling regimes, respectively.

In the weak-coupling limit, the frequency of the vibrating string exhibits the characteristic perturbation described in the previous section, with a shift in frequency proportional to the imaginary component of the terminating admittance and an increased damping proportional to the real part. Note that the coupling also weakly perturbs the frequency and damping of the coupled structural resonance. However, there is no splitting of modes at the crossover point and the normal modes retain their predominantly string-like or body-like character throughout the transition region.

higher frequencies, and vice versa for the other normal mode.

Our earlier discussion of the perturbation of string resonances by the terminating admittance is therefore justified in the weak-coupling regime ($K \ll 1$), which is the usual situation for most string resonances on musical instruments. However, if the fundamental mode of a string is over-strongly coupled at the bridge to a rather light, weakly damped body resonance, such that $K > 1$, the normal-mode resonant frequency of the vibrating string, when coincident in frequency with the coupled body mode, will be significantly shifted away from its position as the fundamental member of the harmonic set of partials. It is then impossible to maintain a steady Helmholtz bowed waveform on the string at the pitch of the now perturbed fundamental, which is the origin of the wolf-note problem frequently encountered on otherwise often very fine-stringed instruments, and cellos in particular.

To overcome such problems, it is sometimes possible to reduce K by using a lighter string, but more commonly the effective Q-value is reduced by extracting energy from the coupled system by fitting a resonating mass on one of the strings between the bridge and tailpiece. A lossy material can be placed between the added mass and the string to extract energy from the system, which might otherwise simply move the wolf note to a nearby frequency.

String Resonances

Figure 15.47 illustrates: (a) the frequency dependence of the in-phase and phase-quadrature resonant response of an A-string as its tension increased, so that its frequency passes through a relatively strongly coupled body resonance at ≈ 460 Hz; (b) the splitting in frequency of the normal modes of the second partial of the heavier G-string frequency tuned to coincide with the frequency of the coupled body resonance. Superimposed on these relatively broad resonances is a very sharp resonance arising from transverse string vibrations perpendicular to the strong coupling direction, to be explained in the next section. This very weakly perturbed string resonance provides a marker, which enables us to quantify the shifts and additional damping of string vibrations in the strong coupling direction.

When the frequency of the lighter A-string is tuned below that of the strongly coupled body resonance, the coupling lowers the frequency of the coupled string mode, as anticipated from our earlier discussion. In contrast, when tuned above the coupled resonance the frequency of the coupled string mode is increased, while

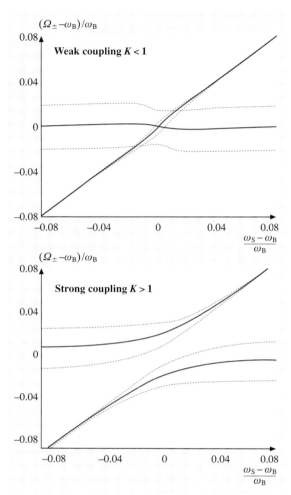

Fig. 15.46 Normal modes of coupled oscillators illustrating the profound effect of damping on the behaviour in the cross-over region illustrated for K-values of 0.75 and $\sqrt{5}$ for an undamped string resonance coupled to a body resonance with a typical $Q = 25$. The *solid line* shows the shifted frequencies of the normal modes as the string frequency is scanned through the body resonance, while the *dashed lines* show the 3 dB points on their damped resonant response (*Gough* [15.80])

In the strong-coupling limit, $K > 1$, the normal modes are split at the crossover point. The losses are also shared equally between the split modes. As the string frequency is varied across the body resonance, one mode changes smoothly from a normal mode with a predominantly string-like character, to a mixed mode at cross over, and to a body-like mode at

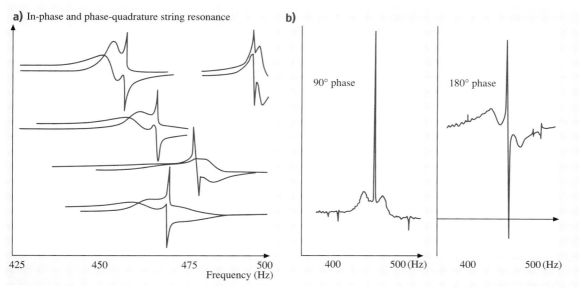

Fig. 15.47a,b Measurements of the in-phase and in-quadrature resonant response of violin strings coupled via the bridge to a strong body resonance (*Gough* [15.33]). The shift of the broader resonances relative to the unperturbed narrow resonance indicates the extent of the perturbative coupling. (**a**) tuning the A-string resonance through a coupled resonance at ≈ 460 Hz; (**b**) the splitting of the string–body normal modes for the more strongly coupled, second partial, of the heavier G-string

at coincidence there is a slight indication of split modes somewhat smaller than the widths. The splitting of modes is clearly seen for the second partial of the much heavier G-string (Fig. 15.47b), with symmetrically split broad string/body modes above and below the narrow *uncoupled mode*. Not surprisingly, this violin suffered from a pronounced wolf note when played at 460 Hz in a high position on the G-string, but not on the lighter D- or A-string. Such effects tend to be even more pronounced on cellos due to the very high bridge providing strong coupling between the vibrating strings and body of the instrument.

On plucked string instruments the inharmonicity of the partials of a plucked note induced by coupling at the bridge to prominent structural resonances causes beats in the sound of plucked string, which contribute to the characteristic sound of individual instruments. *Woodhouse* [15.81, 82] has recently made a detailed theoretical, computational and experimental study of such effects for plucked notes on a guitar taking account of the effect of damping on the coupled string–corpus normal modes. This is sometimes not taken into proper account in finite-element software, in which the normal modes of an interacting system are first calculated ignoring damping, with the damping of the modes then added.

As is clear from Fig. 15.46, such an approach will always break down whenever the width of resonances associated with damping becomes comparable with the splitting of the normal modes in the absence of damping, as is frequently the case in mechanical and acoustical systems.

Polarisation

We have already commented on the response of a bridge mounted centrally on a symmetrically constructed instrument, with string vibrations perpendicular to the front plate exciting only symmetric modes of the body of the instrument, while string vibrations parallel to the front plate induce a couple on the front plate exciting only asymmetric modes. The terminating admittance at the bridge end of the string will therefore be a strongly frequency dependent function of the polarisation direction of the transverse string modes. The angular dependence of the terminating admittance lifts the degeneracy of the string modes resulting in two independent orthogonal modes of transverse string vibration, with different perturbed frequencies and damping, polarised along the frequency-dependent principal directions of the admittance tensor. If a string is excited at an arbitrary angle, both modes will be excited, so that in free decay the directional polarisation will precess at the difference frequency. The resultant radiated sound from the excited body resonances will also exhibit beats, which unlike the nonlinear effects considered earlier will not vary with amplitude of string vibration.

In instruments of the violin family, the soundpost removes the symmetry of the instrument, with normal modes involving a mixture of symmetric and asymmetric modes. Measurements like those shown in Fig. 15.47 demonstrate that below ≈ 700 Hz, the effect of the soundpost is to cause the bridge to rock backwards and forwards about the treble foot closest to the soundpost, which acts as a rather rigid fulcrum. This accounts for the very narrow string resonances shown in Fig. 15.47, which correspond to string vibrations polarised parallel to the line between the point of string support and the rigidly constrained right-hand foot, as indicated in Fig. 15.44. In contrast, string vibrations polarised in the orthogonal direction result in a twisting couple acting on the bridge, with the left-hand foot strongly exciting the vibrational modes of the front plate giving the frequency-shifted and broadened string resonances of the strongly coupled string modes.

By varying the polarisation direction of an electromagnetically excited string, one can isolate the two modes and determine their polarisations (*Baker et al.* [15.84]). When such a string is bowed, it will in general be coupled to both orthogonal string modes. The unperturbed string mode may well help stabilise the repetitive Helmholtz bowed waveform.

String–String Coupling

A vibrating string on any multi-stringed instrument is coupled to all the other strings supported on a common supporting bridge. This is particularly important on the piano, where pairs and triplets of strings tuned to the same pitch are used to increase the intensity of the notes in the upper half of the keyboard. Such coupling is also important on instruments like the harp, where the strings and their partials are coupled via the soundboard. On many ancient bowed and plucked stringed instruments, a set of coupled sympathetic string were used to enhance the sonority and long-term decay of plucked and bowed notes. Even on modern instruments like the violin and cello, the coupling of the partials of a bowed or plucked string with those of the other freely vibrating open (unstopped) strings enhances the decaying after-sound of a bowed or plucked note. This may be one of the reasons why string players have a preference for playing in the bright key signatures of G, D and A major associated with the open strings, where both direct and sympathetic vibrations can easily be excited.

The musical importance of such coupling on the piano is easily demonstrated by first playing a single note and holding the key down so that the note remains undamped and then holding the sustaining pedal down, so that many other strings can also vibrate in sympathy and especially those with partials coincident with those of the played note. Composers, such as Debussy, exploit the additional sonorities produced by such coupling, as in La Cathédrale Engloutie 12) CD-ROM.

The influence of coupling at the bridge of the normal modes of string vibration on the piano has been discussed by *Weinreich* [15.83] and for sympathetic string in general by the present author [15.80]. Consider first two identically tuned string terminated by a common bridge with string vibrations perpendicular to the soundboard and relative phases represented by arrows. The normal modes can therefore be described by the combination ↑↑ and ↓↑ with the strings vibrating in phase or in anti-phase. When the strings vibrate in anti-phase ↓↑, they exert no net force on the bridge, which therefore remains a perfect node inducing no perturbation in frequency or additional damping or transfer of energy

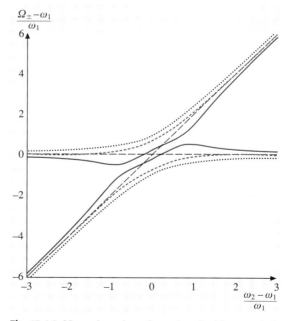

Fig. 15.48 Normal modes of a string doublet coupled at the bridge by a complex impedance. The *dashed* and *dotted curves* illustrate the effect of increasing the reactive component (after *Weinreich* [15.83])

to the soundboard. In contrast, when the strings vibrate in the same phase ↑↑, the force on the bridge and resultant amplitude of sound produced will be doubled, as will the perturbation in frequency and damping of the normal modes, and the amplitude of the resultant sound, relative to that of a single string.

Reactive terms in the common bridge admittance tend to split the frequencies of the normal modes in the vicinity of the crossover frequency region, while resistive coupling at the bridge tends to draw the modal frequencies together over an appreciable frequency range. This is illustrated by Weinreich's predictions for two strings coupled at the bridge by a complex admittance shown in Fig. 15.48, which shows the veering together of the normal modes induced by resistive coupling and the increase in splitting as the reactive component of the coupling is increased.

If the two strings are only slightly mistuned, the amplitudes of the string vibrations involved in the normal modes can still be represented as ↑↑ and ↓↑, but the amplitudes of the two string vibrations will no longer be identical. Hence, when the two strings of a doublet are struck with equal amplitude by a hammer, the mode can be represented by a combination of vibrations ↑↑ with equal amplitude with a small component with opposite amplitudes the ↓↑ dependent on the mistuning. This leads to a double decay in the sound intensity, with the strongly excited ↑↑ mode decaying relatively quickly, leaving the smaller amplitude but weakly damped ↓↑ mode persisting at longer times. Figure 15.49, from *Weinreich* [15.83], shows the rapid decay of a single C4 string, when all other members of the parent string triplet are damped, followed by

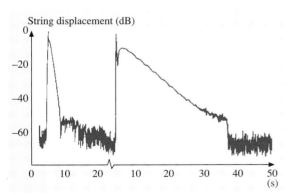

Fig. 15.49 Decay in string vibration of a struck C4 (262 Hz) piano string, first with other members of the string triplet damped and then with one other similarly tuned string allowed to vibrate also (after *Weinreich* [15.83])

the much longer long-term decay of the normal mode excited when one other member of the triplet is also allowed to vibrate freely. More-complicated decay patterns with superimposed slow beats are observed for various degrees of mistuning, from interference with small-amplitude orthogonally polarised string modes excited when the strong-coupling direction is not exactly perpendicular to the soundboard. Weinreich suggests that skilled piano tuners deliberately mistune the individual strings of string doublets and triplets to maximise their long-term ringing sound. Any weakly decaying component of a decaying sound is acoustically important because of the logarithmic response of the ear to sound intensity.

15.2.6 Body Modes

Stringed instruments come in a great variety of shapes and sizes, from strings mounted on simple boxes or on skins stretched over hollow gourds to the more complex renaissance shapes of the viols, guitars and members of the violin family. The vibrational modes of all such instruments, which are ultimately responsible for the radiated sound, involve the collective modes of vibration of all their many component parts. For example, when a violin or guitar is played, all parts of the instrument vibrate – the individual strings, the bridge, the front and back plates and ribs that make up the body of the instrument, the air inside its hollow cavity, the neck, the fingerboard, the tailpiece and, for members of the violin family, the internal soundpost also.

Because of the complexity of the dynamical structures, it would be well nigh impossible to work out the modal shapes and frequencies of even the simplest stringed instruments from first principles. However, as we will show later in this section, with the advent of powerful computers and finite-element analysis software, it is possible to compute the modal vibrations and frequencies of typically the first 20 or more normal modes for the violin and guitar below around 1 kHz. Such calculations do indeed show a remarkable variety of vibrational modes, with every part of the instrument involved in the vibrations to some extent. Such modes can be observed by direct experiment using Chladni plate vibrations, laser holography and modal analysis techniques, as briefly described in this section.

The frequencies of the vibrational modes can be obtained even more simply from the admittance measured at the position of string support or other selected position on the body of an instrument, when the instrument is excited at the bridge by a sinusoidal electromagnetic force or a simple tap. However, unless a large number of measurements over the whole body of the instrument (normal-mode analysis) are made, such measurements provide very little direct information about the nature of the normal modes and the parts of the violin which contribute most strongly to the excited vibrations.

Although a particular structural mode can be very strongly excited, it may contribute very little to the radiated sound and hence the quality of sound of an instrument. Examples of such resonances on the violin or guitar include the strong resonances of the neck and fingerboard. However, even if such resonances produce very little sound, their coupling to the strings via the body and bridge of the instrument can lead to considerable inharmonicity and damping of the string resonances, as discussed in the previous section. Such effects can have a significant effect on the sound of the plucked string of a guitar and the ease with which a repetitive waveform can be established on the bowed string.

To produce an appreciable volume of sound, the normal modes of instruments like the violin and guitar have to involve a net change in volume of the shell structure forming the main body of the instrument. This then acts as a monopole source radiating sound uniformly in all directions. However, when the acoustic wavelength becomes comparable with the size of the instrument, dipole and higher-order contributions also become important.

For the guitar and instruments of the violin family, there are several low-frequency modes of vibration which involve the flexing, twisting and bending of the whole body of the instrument, contributing very little sound to the lowest notes of the instruments. To boost the sound at low frequencies, use is often made of a Helmholtz resonance involving the resonant vibrations of the air inside the body cavity passing in and out of f-holes or rose-hole cut into the front plate of the instrument. This is similar to the way in which the low-frequency sound of a loudspeaker can be boosted by mounting it in a bass-reflex cabinet. The use of a resonant air cavity to boost the low-frequency response has been a common feature of almost every stringed instrument from ancient times.

Although finite-element analysis and modal analysis measurement techniques provide a great wealth of detailed information about the vibrational states of an instrument, considerable physical insight and a degree of simplification is necessary to interpret such measurements. This was recognised by *Savart* [15.85] in the early part of the 19th century, when he embarked on a number of ingenious experiments on the physics of the violin in collaboration with the great French violin maker Vuillaume. To understand the essential physics involved in the production of sound by a violin, he replaced the beautiful, ergonomically designed, renaissance shape of the violin body by a simple trapezoidal shell structure fabricated from flat plates with two central straight slits replacing the elegant f-holes cut into the front. As Savart appears to have recognised, the detailed shape is relatively unimportant in defining the essential acoustics involved in the production of sound by a stringed instrument.

We will adopt a similar philosophy in this section and will consider a stringed instrument made up of its many vibrating components – the strings and bridge, which we have already considered, the supporting shell structure, the vibrations of the individual plates of such a structure, the soundpost which couples the front and back plates, the fingerboard, neck and tailpiece, which vibrate like bars, and the air inside the cavity. Although we have already emphasised that it is never possible to consider the vibrations of any individual component of an instrument in isolation, as we have already shown for the string coupled to a structural resonance at the bridge, it is only when the resonant frequencies of the coupled resonators are close together that their mutual interactions are so important that they change the character of the vibrational modes. Otherwise, the mutual interactions between the various subsystems simply provide a first-order correction to modal frequencies without any very significant change in their modal shapes.

Flexural Thin-Plate Modes

To radiate an appreciable intensity of sound, energy has to be transferred via the bridge from the vibrating strings to the much larger surfaces of a soundboard or body of an instrument. The soundboards of the harp and keyboard instruments and the shell structures of stringed instruments like the violin and guitar can formally be considered as thin plates. Transverse or flexural waves on their surface satisfy the fourth-order thin-plate equation (*Morse* and *Ingard* [15.42] Sect. 5.3), which for an isotropic material can be written as

$$\frac{\partial^2 z}{\partial t^2} + \frac{Eh^2}{12\rho(1-\nu^2)}\left(\frac{\partial^4 z}{\partial x^4} + 2\frac{\partial^2 z}{\partial x^2}\frac{\partial^2 z}{\partial y^2} + \frac{\partial^4 z}{\partial y^4}\right) = 0,$$
(15.68)

where z is the displacement perpendicular to the xy-plane, h is the plate thickness, E is the Young's modulus, ν is the Poisson ratio, and ρ the density.

It is instructive first to consider solutions for a narrow quasi-one-dimensional thin plate, like a wooden ruler or the fingerboard on a violin. One-dimensional solutions can be written in the general form

$$z = (a\cos kx + b\sin kx + c\cosh kx + d\sinh kx)e^{i\omega t}, \quad (15.69)$$

where

$$\omega = \left(\frac{E}{12\rho(1-\nu^2)}\right)^{1/2} hk^2. \quad (15.70)$$

The hyperbolic functions correspond to displacements that decay exponentially away from the ends of the bar as $\exp(\pm kx)$. Well away from the ends, the solutions are therefore very similar to transverse waves on a string, except that the frequency now depends on k^2 rather than k with a phase velocity $c = \omega/k$ proportional to $k \sim \omega^{1/2}$. Flexural waves on thin plates are therefore dispersive and unlike waves travelling on strings any disturbance will be attenuated and broadened in shape as it propagates across the surface.

The k values are determined by the boundary conditions at the two ends of the bar, which can involve the displacement, couple $M = -ES\kappa^2 \partial^2 z/\partial x^2$ and shearing force $F = \partial M/\partial x = -ES\kappa^2 \partial^3 z/\partial x^3$ at the ends of the bar, where κ is the radius of gyration of the cross section (*Morse* and *Ingard* [15.42] Sect. 5.1).

For a flexible bar there are three important boundary conditions:

1. freely hinged, where the free hinge cannot exert a couple on the bar, so that

$$z = 0 \quad \text{and} \quad \frac{\partial^2 z}{\partial x^2} = 0, \quad (15.71)$$

2. clamped, where the geometrical constraints require

$$z = 0 \quad \text{and} \quad \frac{\partial z}{\partial x} = 0, \quad (15.72)$$

3. free, where both the couple and the shearing force at the ends are zero, so that

$$\frac{\partial^3 z}{\partial x^3} = \frac{\partial^2 z}{\partial x^2} = 0. \quad (15.73)$$

A bar of length L, freely hinged at both ends, supports simple sinusoidal spatial solutions with m half-wavelengths between the ends and modal frequencies

$$\omega_m = h\sqrt{\frac{E}{12\rho(1-\nu^2)}}\left(\frac{m\pi}{L}\right)^2. \quad (15.74)$$

For long bars with clamped or free ends, the nodes of the sinusoidal component are moved inwards by a quarter of a wavelength and an additional exponentially decaying solution has to be added to satisfy the boundary conditions, so that at the $x = 0$ end of the bar

$$z \sim A\left[\sin(k_m x - \pi/4) \pm \frac{1}{\sqrt{2}} e^{-k_m x}\right], \quad (15.75)$$

where the plus sign corresponds to a clamped end and the minus to a free end, and $k_m = (m+1/2)\pi/L$. The modal frequencies are given by

$$\omega_m = h\sqrt{\frac{E}{12\rho(1-\nu^2)}}\left(\frac{(m+1/2)\pi}{L}\right)^2 \quad (15.76)$$

which, for the same m value, are raised slightly above those of a bar with hinged ends. Corrections to these formulae from the leakage of the exponentially decaying function from the other end of the bar are only significant for the $m = 1$ mode and are then still less than 1%.

The solutions close to the end of a bar for hinged, clamped and free boundary conditions are illustrated in Fig. 15.50, with the phase-shifted sinusoidal component for the latter two indicated by the dotted line. The exponential contribution is only significant out to distances $\sim \lambda/2$.

The above formulae can be applied to the bending waves of quasi-one-dimensional bars of any cross section, by replacing the radius of gyration $\kappa = h/\sqrt{12}$ of the thin rectangular bar with $a/2$ for a bar of circular cross section and radius a, and $\sqrt{a^2+b^2}/2$ for a hollow cylinder with inner and outer radii a and b (*Fletcher* and *Rossing* [15.5], Fig. 2.19).

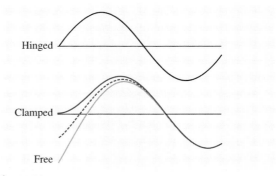

Fig. 15.50 Boundary conditions for flexural waves at the end of a one-dimensional bar. The *dashed line* represents the phase-shifted sinusoidal component, to which the exponentially decaying component has to be added to satisfy the boundary conditions

Another case of practical importance in musical acoustics is a bar clamped at one end and free at the other. This would, for example, describe the bars of a tuning fork or could be used to model the vibrations of the neck or finger board on a stringed instrument. In this case, there is an addition $m = 0$ vibrational mode, with exponential decay length comparable with the length of the bar. The modal frequencies are then given [*Fletcher* and *Rossing* [15.5], (2.64)] by

$$\omega_m = \frac{h}{4}\left(\frac{\pi}{L}\right)^2 \sqrt{\frac{E}{12\rho(1-\nu^2)}}$$
$$\left[1.194^2, 2.988^2, 5^2, \ldots, (2m+1)^2\right].$$
(15.77)

In the above discussion, we have described the modes in terms of the number m of half-wavelengths of the sinusoidal component of the wave solutions within the length of the bar. A different nomenclature is frequently used in the musical acoustics literature, with the mode number classified by the number of nodal lines (or points in one dimension) m in a given direction *not including the boundaries* rather than the number of half-wavelengths m between the boundaries, as in Fig. 15.51.

Twisting or Torsional Modes

In addition to flexural or bending modes, bars can also support twisting (torsional) modes, as illustrated in Fig. 15.51 for the $z = xy$ (1,1) mode.

The frequencies of the twisting modes are determined by the cross section and shear modulus G, equal to $E/2(1+\nu)$ for most materials (*Fletcher* and *Rossing* [15.5], Sect. 2.20). The wave velocity of torsional waves is dispersionless (independent of frequency) with $\omega_n = nc_T k$, where

$$c_T = \frac{\omega}{k} = \sqrt{\frac{GK_T}{\rho I}} = \alpha\sqrt{\frac{E}{2\rho(1+\nu)}},$$
(15.78)

where GK_T is the torsional stiffness given by the couple, $C = GK_T \partial\theta/\partial x$, required to maintain a twist of the bar through an angle θ and $I = \int \rho r^2 \, dS$ is the moment of inertia per unit length along the bar. For a bar of circular cross section $\alpha = 1$, for square cross section $\alpha = 0.92$, and for a thin plate with width $w > 6h$, $\alpha = (2h/w)$. For a bar that is fixed at both ends, $f_n = nc_T/2L$, while for a bar that is fixed at one end and free at the other, $f_n = (2n+1)c_T/4L$, where n is an integer including zero.

Thin bars also support longitudinal vibrational modes, but since they do not involve any motion perpendicular to the surface they are generally of little acoustic

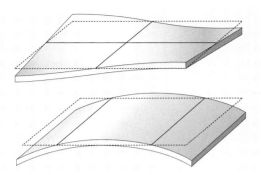

Fig. 15.51 Schematic illustration of the lowest-frequency twisting (1,1) and bending (2,0) modes of a thin bar with free ends

importance, other than possibly for the lowest-frequency soundpost modes for the larger instruments of the violin family.

Two-Dimensional Bending Modes

Solutions of the thin-plate bending wave solutions in two dimensions are generally less straightforward, largely because of the more-complicated boundary conditions, which couple the bending in the x- and y-directions. For a free edge parallel to the y-axis, the boundary conditions are (*Rayleigh* [15.3] Vol. 1, Sect. 216)

$$\frac{\partial^2 z}{\partial x^2} + \nu\frac{\partial^2 z}{\partial y^2} = 0$$

and

$$\frac{\partial}{\partial x}\left[\frac{\partial^2 z}{\partial x^2} + (2-\nu)\frac{\partial^2 z}{\partial y^2}\right] = 0.$$
(15.79)

Thus, when a rectangular plate is bent downwards along its length, it automatically bends upwards along its width and vice versa. This arises because downward bending causes the top surface of the plate to stretch and the bottom surface to contract along its length. But by Poisson coupling, this causes the top surface to contract and lower surface to stretch in the orthogonal direction, causing the plate to bend in the opposite direction across its width. This is referred to as anticlastic bending. The Poisson ratio ν can be determined from the ratio of the curvatures along the bending and perpendicular directions.

In addition, for orthotropic materials like wood, from which soundboards and the front plates of most stringed instruments are traditionally made, the elastic constants are very different parallel and perpendicular to the grain structure associated with the growth rings. *McIntyre* and *Woodhouse* [15.86] have published a detailed account of the theory and derivation of elastic constants from measurement of plate vibrations in both uniform and orthotropic thin plates, including the influence of damping.

For an isotropic rectangular thin plate, hinged along on all its edges, a simple two-dimensional (2-D) sine-wave solution satisfies both the wave equation (15.68) and the boundary conditions, with m and n half-wavelengths along the x- and y-directions, respectively, giving modal frequencies

$$\omega_{mn} = h\sqrt{\frac{E}{12\rho(1-\nu^2)}\left[\left(\frac{m\pi}{L_x}\right)^2 + \left(\frac{n\pi}{L_y}\right)^2\right]}. \tag{15.80}$$

By analogy with our discussion of flexural waves in one-dimensional bars, we would expect the modal frequencies of plates with clamped or free edges to be raised, with the nodes of the sinusoidal components of the wave solution moved inwards from the edges by approximately quarter of a wavelength. For the higher-order modes, the modal frequencies would therefore be given to a good approximation by

$$\omega_{mn} = h\sqrt{\frac{E}{12\rho(1-\nu^2)}} \\ \left[\left(\frac{(m+1/2)\pi}{L_x}\right)^2 + \left(\frac{(n+1/2)\pi}{L_y}\right)^2\right]. \tag{15.81}$$

As recognised by *Rayleigh* ([15.3] Vol. 1, Sect. 223), it is difficult to evaluate the modal shapes and modal frequencies of plates with free edges. The method used by Rayleigh was to make an intelligent guess of the wavefunctions which satisfied the boundary conditions and to determine the frequencies by equating the resulting potential and kinetic energies. *Leissa* [15.88] has reviewed various refinements of the original calculations. For a plate with free edges, the nodal lines are also no longer necessarily straight, as they were for plates with freely hinged edges.

Chladni Patterns

The modal shapes of vibrating plates can readily be visualised using Chladni patterns. These are obtained by supporting the plate at a node of a chosen mode excited electromagnetically, acoustically or with a rosined bow drawn across an edge. A light powder is sprinkled onto the surface. The plate vibrations cause the powder to bounce up and down and move towards the nodes

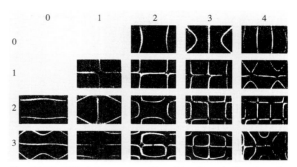

Fig. 15.52 Chladni pattern with *white lines* indicating the nodal lines of the first few modes of a rectangular plate (after *Waller* [15.87])

of the excited mode, allowing the nodal line patterns to be visualised. Figure 15.52 illustrates Chladni patterns measured by *Waller* [15.87] for a rectangular plate with dimensions $L_x/L_y = 1.5$, with the number of nodal lines between the boundary edges determining the nomenclature of the modes. Note the curvature of the nodal lines resulting from the boundary conditions at the free edges.

Figure 15.53 illustrates the nodal line shapes and relative frequencies of the first 10 modes of a square plate with free edges, where $f_{11} = hc_L/L^2\sqrt{1-\nu/2}$ (after *Fletcher* and *Rossing*, [1] Fig. 3.13).

Another important consequence of the anticlastic bending is the splitting in frequencies of combination modes that would otherwise be degenerate. This is illustrated in Fig. 15.54 by the combination $(2, 0) \pm (0, 2)$ normal modes of a square plate with free edges. The $(2, 0) \pm (0, 2)$ modes are referred to as the X- and ring-modes from their characteristic nodal line shapes. The $(2, 0)-(0, 2)$ X-mode exhibits anticlastic bending in the same sense as that induced naturally by the Poisson coupling. It therefore has a lower elastic energy and hence lower vibrational frequency than the $(0, 2)+(2, 0)$ ring-

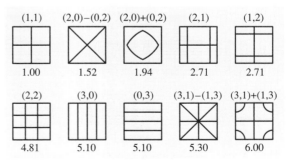

Fig. 15.53 Schematic representation of the lowest 10 vibrational modes of a square plate with free edge

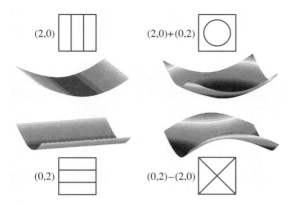

Fig. 15.54 Formation of the ring- and *X*-modes by the superposition of the (2, 0) and (0, 2) bending modes

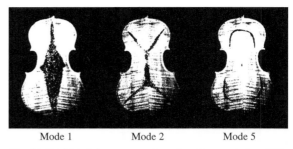

Fig. 15.55 Chladni patterns for the first twisting-(#1), X-(#2) and ring-(#5) modes of a viola back plate (after *Hutchins* [15.89])

mode, with curvatures in the same sense in both the *x*- and *y*-directions. The ring- and *X*-modes will therefore be split in frequency above and below the otherwise degenerate mode, as illustrated in *Fletcher* and *Rossing* ([15.5], Fig. 13.11).

Plate Tuning

The modal shapes and frequencies of the lowest-order (1,1) twisting mode and the *X*- and ring-modes are widely used for the *scientific tuning* of the front and back plates of violins following plate-tuning guidelines developed by *Hutchins* [15.89, 91]. These are referred to as violin plate modes 1, 2 and 5, as illustrated in Fig. 15.55 by Chladni patterns for a "well-tuned" back plate. The violin maker aims to adjust the thinning of the plates across the area of the plate to achieve these symmetrical nodal line shapes at specified modal frequencies.

The use of such methods undoubtedly results in a high degree of quality control and reproducibility of the acoustic properties of the individual plates before assembly and, presumably, of the assembled instrument also, especially for the lower-frequency structural resonances. Unfortunately, they do not necessarily result in instruments comparable with the finest Italian instruments, which were made without recourse to such sophisticated scientific methods. Traditional violin makers instinctively assess the elastic properties of the plates by their feel as they are twisted and bent, and also by listening to the sound of the plates as they are tapped or even rubbed around their edges, rather like a bowed plate. From our earlier discussion, it is clear that the mass of the plates is also important in governing the acoustical properties.

Geometrical Shape Dependence

The above examples demonstrate that the lower-frequency vibrational modes of quite complicated shaped plates can often be readily identified with those of simple rectangular plates, though the frequencies of such modes will clearly depend on the exact geometry involved. This is further illustrated in Fig. 15.56 by the modal shapes of a guitar front plate obtained from time-averaged holography measurements by *Richardson* and *Roberts* [15.90], where the contours indicate lines of constant displacement perpendicular to the surface. For the guitar, the edges of the top plate are rather good nodes, because of the rather heavy supporting ribs and general construction of the instrument. The boundary conditions along the edges of the plate are probably intermediate between hinged and clamped. The modes can be denoted by the number of half-wavelengths along the length and width of the instrument. Note that cir-

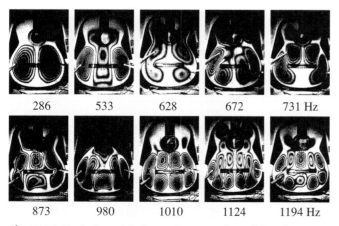

Fig. 15.56 Typical modal shapes for a number of low-frequency modes of a guitar top plate from time-averaged holographic measurements by *Richardson* and *Roberts* [15.90]

cular rose-hole opening in the front face, which plays an important role in determining the frequency of the Helmholtz air resonance boosting the low-frequency response of the instrument, tends to concentrate most of the vibrational activity to the lower half of the front plate.

Mode Spacing

Although the frequencies of the modes of complicated shapes such as the violin, guitar and piano soundboard are rather irregularly spaced, at sufficiently high frequencies, one can use a statistical approach to estimate the spacing of the modal frequencies (*Cremer* [15.29], Sect. 11.2). For large m and n values, the modal frequencies of an isotropic rectangular plate are given by

$$\omega_{mn} \sim h \sqrt{\frac{E}{12\rho(1-\nu^2)}} \left(k_m^2 + k_n^2\right), \quad (15.82)$$

where $k_m = m\pi/L_x$ and $k_n = n\pi/L_y$. The modes can be represented as points (m, n) on a rectangular grid in k-space, with a grid spacing of π/L_x and π/L_y along the k_x and k_y directions. Each mode therefore occupies an area in k-space of π^2/L_xL_y. For large m and n, the number of modes ΔN between k and $k + \Delta k$ is therefore on average just the number of modes in the k-space area between k and $k + \Delta k$, so that

$$\Delta N = \frac{\pi}{2} k \Delta k \frac{L_x L_y}{\pi^2} = \frac{\pi}{2} \frac{\Delta \omega}{2\beta} \frac{L_x L_y}{\pi^2}, \quad (15.83)$$

where we have made use of the dispersion relationship $\omega = \beta k^2$.

The density of modes per unit frequency is then constant and independent of frequency,

$$\frac{dN}{df} = \frac{1}{2\beta} L_x L_y = \frac{\sqrt{3(1-\nu^2)}}{c_L h} S \sim 1.5 \frac{S}{c_L h}, \quad (15.84)$$

where S is the area of the plate. The spacing of modes Δf will therefore on average be $\sim c_L h/1.5 S$, proportional to the plate thickness and inversely proportional to plate area. For large k values, this result becomes independent of the shape of the plate. For an orthotropic plate like the front plate of the violin or guitar, the spacing is determined by the geometric mean $(c_x c_y)^{1/2}$ of the longitudinal velocities parallel and along the grain. For the violin, *Cremer* ([15.29], p. 292) estimates an asymptotic average mode spacing of 73 Hz for the top plate and 108 Hz for the back plate. Above around 1.5 kHz the width of the resonances on violin and guitar plates becomes comparable with their spacing, so that experimentally it becomes increasingly difficult to excite or distinguish individual modes.

On many musical instruments such as the violin and guitar, the presence of the f- and rose-hole openings introduce additional free-edge internal boundary conditions, which largely confine the lower-frequency modes to the remaining larger areas of the plate. The effective area determining the density of lower-frequency modes for both instruments will therefore be significantly less that that of the whole plate. Any reduction in plate dimensions, such as the island region on the front plate of the violin between the f-holes, will limit the spatial variation of flexural waves in that direction. Such a region will therefore not contribute significantly to the normal modes of vibration of the plate until $\lambda(\omega)/2$ is less than the limiting dimension.

Anisotropy of Wood

Wood is a highly anisotropic material with different elastic properties perpendicular and parallel to the grain. Furthermore, the wood used for soundboards and plates of stringed instruments are cut from nonuniform circular logs (slab or quarter cut), so that their properties can vary significantly across their area.

McIntyre and *Woodhouse* [15.86] have described how the anisotropic properties affect the vibrational modes of rectangular thin plates and have shown how the most important elastic constants including their loss factors can be determined from the vibrational frequencies and damping of selected vibrational modes. For a rectangular plate with hinged edges

$$\omega_{mn}^2 = \frac{h^2}{\rho} \left[D_1 k_m^4 + D_3 k_n^4 + (D_2 + D_4) k_m^2 k_n^2 \right], \quad (15.85)$$

where D_1–D_4 are the four elastic constants required to describe the potential energy of a thin plate with orthotropic symmetry. These are related to the more familiar elastic constants by the following relationships

$$D_1 = E_x/12\mu, \; D_3 = E_y/12\mu, \; D_4 = G_{xy}/3,$$
$$D_2 = \nu_{xy} E_y/6\mu = \nu_{yx} E_x/6\mu, \quad (15.86)$$

where $\mu = 1 - \nu_{xy}\nu_{yx}$. G_{xy} gives the in-plane shear energy when a rectangular area on the surface is distorted into a parallelogram. This is the only energy term involved in a pure twisting mode (i.e. $z = xy$). For an isotropic plate, $D_4 = E/6(1+\nu)$.

Table 15.5 Typical densities and elastic properties of wood used for stringed instrument modelling (after *Woodhouse* [15.76]). (The values with asterisks are intelligent guesses in the absence of experimental data)

Property	Symbol	Units	Spruce	Maple
Density	ρ	kg/m³	420	650
	D_1	MPa	1100	860
	D_2	MPa	67	140*
	D_3	MPa	84	170
	D_4	MPa	230	230*
Relative scaling factors	$\sqrt[4]{D_1/D_3}$		1.9	1.4

For many materials, $(D_2 + D_4) \sim 2\sqrt{D_1 D_2}$, so that (15.82) can be rewritten as

$$\omega_{mn}^2 = \frac{h^2}{\rho}\sqrt{D_1 D_3}\left[\sqrt[4]{\frac{D_1}{D_3}}\left(\frac{m\pi}{L_x}\right)^2 + \sqrt[4]{\frac{D_3}{D_1}}\left(\frac{n\pi}{L_y}\right)^2\right]^2. \tag{15.87}$$

The vibrational frequencies are therefore equivalent to those of a shape of the same area with averaged elastic constant $\sqrt{D_1 D_3}$ and length scales L_x multiplied by the factor $\sqrt[8]{D_1/D_3}$ and L_y by its inverse. The relative change in scaled dimensions is therefore $\sqrt[4]{D_1/D_3}$. These scaling factors account for the elongation of the equal contour shapes along the stiffer bending direction in the holographic measurements of mode shapes for the front plate of the guitar (Fig. 15.56), where the higher elastic modulus along the grains is further increased by strengthening bars glued to the underside of the top plate at a shallow angle to the length of the top plate.

Typical values for the elastic constants of spruce and maple traditionally used for modelling the violin are listed in Table 15.5 from *Woodhouse* [15.76]. The anisotropy of the elastic constants along and perpendicular to the grain of a spruce plate cut with the growth rings running perpendicular to the surface would give a relative scaling factor for a violin front plate of almost double the relative width, if one wanted to consider the flexural vibrations in terms of an equivalent isotropic thin plate. The anisotropy is therefore very important in determining the vibrational modes of such instruments.

Plate Arching

The front and back plates of instruments of the violin family have arched profiles, which give the instrument a greatly enhanced structural rigidity to support the downward component of the string tension (≈ 10 kg weight). The arching also significantly increases the frequencies of the lowest flexural plate modes. In the case of lutes, guitars and keyboard instruments with a flat sounding board or front plate, the additional rigidity is achieved by additional cross-struts glued to the back of the sounding board. The bass bar in members of the violin family serves a similar purpose in providing addition strengthening to that of the arching.

The influence of arching on flexural vibration frequencies is easily understood by considering the transverse vibrations of a thin circular disc. For a flat disc, the modal frequencies are determined by the flexural energy associated with the transverse vibrations. The longitudinal strains associated with the transverse vibrations are only second order in displacement and can therefore be neglected. However, if the disc is *belled out* to raise the centre to a height H, the transverse vibrations now involve additional first-order longitudinal strains stretching the disc away from its edges. The energy involved in such stretching, which is resisted by the rigidity of the circumferential regions of the disc, introduces an additional potential energy proportional to H^2. By equating the kinetic to the increased potential energy, is follows that the frequency of the lowest-order symmetrical mode will be increased by a factor $[1+\alpha(H/h)^2]^{1/2}$, where $\alpha \approx 1$ has to be determined by detailed calculation. *Reissner* [15.92] showed that, when the arching is larger than the plate thickness, $H \gg h$, the frequency of the fundamental mode is raised by a factor $\omega/\omega_0 = 0.68 H/h$ for a circular disc with clamped edges, and $0.84 H/h$ with free edges. For a shallow shell with $H/a < 0.25$, where a is the radius of the disc, the asymptotic frequency can conveniently be expressed as $\omega_n \sim 2(E/\rho)^{1/2} H/a^2$. The arching dependence of the modal frequencies is greatest for the lowest-frequency modes. At high frequencies, the radius of curvature is large compared to the wavelength, so arching is much less important.

The combined effect of the arching and the f- and rose-holes cut into the front face of many stringed instruments is to raise the frequency of the acoustically important lower-frequency modes to well above the asymptotic spacing of modal frequencies predicted by (15.82). For example, the lowest-frequency plate modes of the violin front and back plates are typically in the range 400–500 Hz compared with Cremer's predictions for an asymptotic spacing of modes $\approx$73 Hz for the top plate and 108 Hz for the back plate ([15.29], p. 292).

The more highly arched the plates the stiffer they will be and, for a given mass, the higher will be their

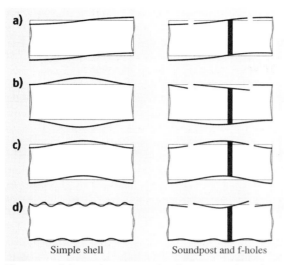

Fig. 15.57 Schematic cross-sectional representation of typical shell modes for a simple box and a "violin-type" structure with f-holes and a soundpost

associated vibrational frequencies. High arching may well contribute to the relatively soft and sweet sounds of the more highly arched early Amati and Stainer violins and the more powerful and brilliant sounds of the flatter later Stradivari and Guarneri models.

Shell Structures

Although it is interesting to investigate the free plates of violins and other instruments before they are assembled into the instrument as a whole, once assembled their vibrational properties will generally be very different, as they are subject to completely different boundary conditions at their supporting edges and by the soundpost constraint for instruments of the violin family. The supporting ribs tie the outer edges of the back and front plates together. The curvature of the outer edge shape gives the 1–2 mm-thick ribs of a violin considerable structural strength and rigidity, in much the same way as the bending of a serpentine brick wall. In many instruments there are extra strips and blocks attached to the ribs and plate edges to strengthen the joint, which still allow a certain amount of angular flexing, as indicated by the schematic normal-mode vibrations illustrated in Fig. 15.57. The supporting ribs add mass loading at the edges of the plates and impose a boundary condition for the flexing plates intermediate between freely hinged and clamped.

For a simple shell structure, Fig. 15.57a represents a low-frequency twisting mode in which the two ends of the instrument twist in opposite directions, just like the simple ($z = xy$) twisting mode of a rectangular plate. Figure 15.57b–d schematically represent normal modes involving flexural modes of the front and back plates. Mode (b) is the important *breathing mode*, which produces a strong monopole source of acoustic radiation at relatively low frequencies. In mode (c), the two plates vibrate in the same direction, resulting in a much weaker dipole radiation source along the vertical axis. The above examples assumed identical top and back plates, whereas in general they will have different thicknesses arching, and will be constructed from different types of wood with different anisotropic elastic properties: spruce for the front and maple for the back plate of the violin. Hence, for typical high-frequency normal modes (e.g. shown schematically in Fig. 15.57d), the wavelengths of flexural vibrations will be different in the top and back plates.

Note that, at low frequencies, several of the normal modes involve significant motion of the outer edges of the front and back plate, since the centre of mass of the freely supported structure cannot move. Hence, when an instrument is supported by the player, additional mode damping can occur by energy transfer to the chin, shoulder or fingers supporting the instrument at its edges, as indeed observed in modal analysis investigations on hand-held violins by *Marshall* [15.93] and *Bissinger* [15.94].

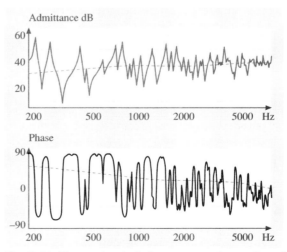

Fig. 15.58 The rotational admittance across the two feet of a bridge on an idealised rectangular violin structure with a soundpost under the treble foot but with no f-holes (after Woodhouse [15.76])

Skeleton Curves

In this idealised model, the normal modes of the structure at high frequencies will be similar to those of the individual plates. Such modes will only be significantly perturbed when the resonances of the separate plates are close together, apart from a general background interaction from the average weak but cumulative interaction with all other distant modes. *Woodhouse* [15.76] has recently shown that the averaged amplitude and phase of the admittance can be described by *skeleton curves*, indicated by the dashed lines in Fig. 15.58, on which peaks and troughs of height Q and $1/Q$ and phase changes from individual resonances are superimposed. These curves were evaluated analytically for the rotational admittance across the two feet of a bridge mounted on the front plate of an idealised rectangular box-like violin without f-holes but with a soundpost close to the treble foot of the bridge.

At low frequencies the averaged input impedance across the two feet of the bridge is largely reactive, with a value close to the static *springiness*, which can be identified with the low-frequency limit of the normal-mode admittance $\sum_n i\omega/m_n\omega_n^2$, where the effective mass of each mode will depend on where and how the instrument is excited. However, at high frequencies the admittance becomes largely resistive resulting from internal damping and energy loss to the closely overlapping modes. The use of skeleton curves enables Woodhouse to illustrate the effect of various different bridge designs on the overall frequency response of a violin, without having to consider the exact positions, spacing or damping of the individual resonances of the shell structure. Although the idealised model is clearly over-simplistic, the general trends predicted by such a model will clearly be relevant to any multi-resonant shell model.

Soundpost and f-Holes

The soundpost and f-holes cut into the front plate of the violin and related instruments have a profound effect on the frequencies and waveforms of the normal modes, illustrated schematically by the right-hand set of examples in Fig. 15.57. The f-holes create an island area on which the bridge sits, which separates the top and lower areas of the front plate. Like the rose-hole on a guitar illustrated in Fig. 15.56, the additional internal free edges introduced by the f-holes tend to localise the vibrations of the front plate to the regions above and below the *island area*. In addition, the soundpost acts as a rather rigid spring locking the vibrations of the top and back plates together at it its ends. At low frequencies, the soundpost introduces an approximate node of vibration on both the top and back plates, unless the frequencies of the uncoupled front and back plates modes are close together.

For some low-frequency modes, the soundpost and f-hole have a relatively small effect on the modes of the shell structure, such as the twisting mode (a) and mode (c), when the plates vibrate in the same direction. However, the breathing mode (c) will be strongly affected by the soundpost forcing the front and back plates to move together across its ends.

As indicated earlier, any string motion parallel to the plates will exert a couple on the top of the bridge. In the absence of the soundpost, only asymmetric modes of the top plate could then be excited. However, to satisfy the boundary conditions at the soundpost position, the rocking action now induces a combination of symmetric and antisymmetric plate modes (illustrated schematically in Fig. 15.57b), approximately doubling the number of modes that can contribute to the sound of an instrument including the very important lower-frequency symmetric *breathing modes*. Because of the f-holes, the central island can vibrate in the opposite direction to the wings on the outer edges of the instrument. The mixing of symmetric and antisymmetric modes is strongly dependent on the position of the soundpost relative to the nodes of the coupled waveforms. As a result, the sound of a violin instrument is very sensitive to the exact placing of the soundpost. The difference in the sound of a violin with the soundpost first in place and then removed is illustrated in ⏵ CD-ROM .

To a good approximation, in the audible frequency range, the violin soundpost can be considered as a rigid body, as its first longitudinal resonance is ≈ 100 kHz, though lower-frequency bending modes can also be excited, particularly if the upper and lower faces of the soundpost fail to make a flat contact with the top and back plates (*Fang* and *Rogers* [15.95]). At high frequencies, there is relatively little induced motion of the outer edges of top and back plates, so that the impedance $Z(\omega)$ (force/induced velocity) measured at the soundpost position is simply given by the sum of the impedances at the soundpost position, $Z(\omega)_{\text{top}} + Z(\omega)_{\text{back}}$, of the individual plates with fixed outer edges. If one knows the waveforms of the individual coupled modes, it is relatively straightforward to evaluate the admittance at any other point on the two surfaces, and hence to evaluate the rotational admittance across the two feet of the bridge (*Woodhouse* [15.76]).

We have already described the important role of the bridge dynamics in the coupling between the strings and the vibrational modes of the instrument. For instruments of the violin family, the island region between the f-holes probably plays a rather similar role to the bridge, as it is via the vibrations of this central region that the larger-area radiating surfaces of the front plate are excited. At low frequencies this will be mainly by the lowest-order twisting and flexing modes of the central island region. It therefore seems likely that the dynamics of the central island region also contributes significantly to the BH hill feature and the resulting acoustical properties of the violin in the perceptually important frequency range of $\approx 2\text{--}4\,\text{kHz}$, as recognised by Cremer and his colleagues [15.11].

Historically, the role of the soundpost and the coupling of plates through enclosed air resonances were first considered analytically using relatively simple mass–spring models with a limited number of degrees of freedom to mimic the first few resonances of the violin, as described in some detail by *Cremer* ([15.29], Chap. 10). Now that we can obtain detailed information about not only the frequencies, but also the shapes of the important structural modes of an instrument from finite-element calculations, holography and modal analysis, there is greater emphasis on analytic methods based on the observed set of coupled modes.

The Complete Instrument

Bowing, plucking or striking a string can excite every conceivable vibration of the supporting structure including, where appropriate, the neck, fingerboard, tailpiece and the partials of all strings both in front of and behind the bridge. Many of the whole-body lower-frequency modes can be visualised by considering all the possible ways in which a piece of soft foam, cut into the shape of the instrument with an attached foam neck and fingerboard, can be flexed and bent about its centre of mass.

Figure 15.59 illustrates the flexing, twisting and changes in volume of the shell of a freely supported violin for two prominent structural resonances computed by *Knott* [15.96] using finite-element analysis. However, not all modes involve a significant change in net volume of the shell, so that many of the lower-frequency modes are relatively inefficient acoustic radiators. Nevertheless, since almost all such modes involve significant bridge motion, they will be strongly excited by the player and will produce prominent resonant features in the input admittance at the point of string support on the bridge. They can therefore significantly perturb the vibrations of the string destroying the harmonicity of the string resonances and resulting playability of particular notes on the instrument, especially for bowed stringed instrument.

Helmholtz Resonance

Almost all hand-held stringed instruments and many larger ones such as the concert harp make use of a Helmholtz air resonance to boost the sound of their lowest notes, which are often well below the frequencies of the lowest strongly excited, acoustically efficient, structural resonances. For example, the lowest acoustically efficient body resonance on the violin is generally around 450 Hz, well above the bottom note G3 of the instrument at $\approx 196\,\text{Hz}$. Similarly, the first strong structural resonance on the classical acoustic guitar is $\approx 200\,\text{Hz}$, well above the lowest note of $\approx 82\,\text{Hz}$.

To boost the sound in the lower octave, a relatively large circular rose-hole is cut into the front plate of the guitar and two symmetrically facing f-holes are cut into the front plate of instruments of the violin family. The air inside the enclosed volume of the shell of such instruments vibrates in and out through these openings to form a Helmholtz resonator.

The frequency of an ideal Helmholtz cavity resonator of volume V, with a hole of area S in one of its rigid walls is given by

$$\omega_H = \sqrt{\frac{\gamma P}{\rho} \frac{S}{L'V}} = c_0 \sqrt{\frac{S}{L'V}}, \qquad (15.88)$$

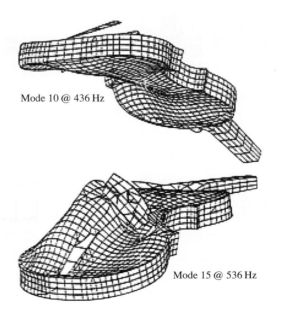

Fig. 15.59 Representative finite element simulations of the structural vibrations of a violin, with greatly exaggerated vibrational amplitudes (after *Knott* [15.96])

where L' is the effective length of the open hole. For a circular hole of radius a, *Rayleigh* ([15.3] Vol. 2, Sect. 306) showed that $L' = \frac{\pi}{2}a$, while for an ellipse $L' \sim \frac{\pi}{2}(ab)^{1/2}$, provided the eccentricity is not too large. Noting that the effective length depends largely on area, *Cremer* ([15.29], Fig. 10.6) modelled the f-hole as an ellipse having the same width and area as the f-hole. The two f-holes act in parallel to give an air resonance for the violin at ≈ 270 Hz, at an interval of just over a fifth above the lowest open string. For the acoustic guitar, the circular rose-hole produces an air resonance around 100 Hz, which, like for the violin, is close to the frequency of the second-lowest open string on the instrument.

Any induced motion of the top and bottom plates that involves a net change in volume results in coupling to the Helmholtz mode. Such coupling will perturb the Helmholtz and body-mode frequencies, in just the same way that string resonances are perturbed by coupling to the body resonances (see *Cremer* [15.29], Sect. 10.3 for a detailed discussion of such coupling). Since the acoustically important coupled modes are at considerably higher frequencies than the Helmholtz resonance, the mutual perturbation is not very large. Because of such coupling, purists often object to describing this resonance as a Helmholtz resonance. Similar objections could apply equally well to string resonances, since they too are perturbed by their coupling to body modes. But, as already discussed, in many situations the normal modes largely retain the character of the individually coupled modes other then when their frequencies are close together and, even then, when the damping of either of the coupled modes is large compared to the splitting in frequencies induced by the coupling in the absence of damping (Fig. 15.46).

Well below the Helmholtz resonance, any change in volume of the shell of the violin or guitar induced by the vibrating strings will be matched by an identical volume of air flowing out through the rose- or f-holes, with no net volume flow from the instrument as a whole. Since at low frequencies almost all the radiated sound is monopole radiation associated with the net flow of air across the whole surface of an instrument, little sound will be radiated. However, above the air resonance, the response of the air resonance will lag in phase by 180°, so that the flow from body and cavity will now be in phase, resulting in a net volume flow and strong acoustic radiation. The Helmholtz resonance serves the same purpose as mounting a loudspeaker in a bass-reflex cabinet, with the air cavity resonance boosting the intensity close to and above its resonant frequency.

A number of authors have considered the influence of the enclosed air on the lowest acoustically important modes of the violin (*Beldie* [15.97]) and guitar (*Meyer* [15.98], *Christensen* [15.99] and *Rossing* et al. [15.100]) using simple mechanical modes of interacting springs and masses with damping and their equivalent electric circuits. Figure 15.60 shows the mechanical and equivalent electrical circuits and resulting admittance curve for the top plate for the illustrative three-mass model used by Rossing et al., which accounts for the qualitative features of the first three most important resonances of a guitar body. To emphasise a number of important points, we have calculated the admittance for a cavity with identical front and back plates with uncoupled resonances at 300 Hz, coupled via a cavity Helmholtz resonance at 250 Hz. The closeness in frequencies of the coupled resonators has been chosen to emphasise the influence of the coupling on the modal frequencies.

Without concerning oneself with mathematical detail, one can immediately recognise an unshifted normal mode associated with the uncoupled body resonances

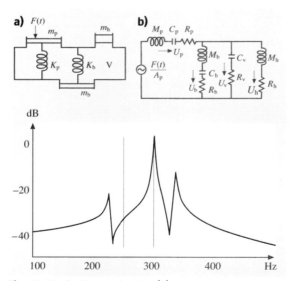

Fig. 15.60a,b The mechanical (**a**) and equivalent electrical (**b**) circuit for a three-mass model describing the vibrations of the front and back plates of a stringed instruments coupled via a Helmholtz resonance (after *Fletcher* and *Rossing* [15.5]). The modulus of the admittance at the top plate has been evaluated for identical front and back plates with uncoupled frequencies of 300 Hz, coupled via a Helmholtz air resonator at 250 Hz in the absence of coupling. The frequencies of the uncoupled air and body resonances are indicated by the *vertical lines*

at 300 Hz, corresponding to the two plates vibrating in the same phase, with no volume change and hence no coupling to the air resonance. However, the coupling via the Helmholtz resonance splits the degenerate plate modes, to give a normal mode at a raised frequency, with the plates vibrating in opposite directions in a breathing mode. The coupling also decreases the frequency of the Helmholtz cavity resonance. The unperturbed mode may dominate the measured admittance and affect the playability of the instrument via its perturbation of string resonances. But, because there is no associated volume change, it will be an inefficient acoustic radiator. One should note that, because of the changes in phase of the air resonance on passing through resonance, it appears as a dispersive curve superimposed on the low-frequency wings of the stronger higher-frequency body resonances. The frequency of the excited normal mode is not the peak in the admittance curve (i.e. its modulus), as often assumed but is more nearly mid-way between the maximum and minimum, where its phase lags 90° relative to the phase of the higher frequency normal modes. Similarly, the upper body mode results in a dispersive feature in the opposite sense, as its phase changes from almost 180° to 0° relative to the unshifted normal mode. Very similar, but narrower, dispersive features are also observed in admittance-curve measurements from string resonances, unless they are purposely damped.

Cavity Modes

In addition to the Helmholtz air resonance, there will be many other cavity resonances of the air enclosed within the shell of stringed instruments, all of which can in principle radiate sound through the f- or rose-holes. Alternatively, the internal air resonances can radiate sound via the vibrations they induce in the shell of the instrument, as discussed in some detail by *Cremer* ([15.29], Sect. 11.4). Because of the relatively small height of the violin ribs, below around 4 kHz the cavity air modes are effectively two dimensional in character. Simple statistical arguments based on the overall volume of the violin cavity show that there are typically $\approx$ 28 resonances below this frequency, as observed in measurements by *Jansson* [15.101]. Whether or not such modes play a significant role in determining the tonal quality of an instrument remains a somewhat contentious issue. However, at a given frequency, the wavelengths of the flexural modes of the individual plates and the internal sound modes will not, in general, coincide. The mutual coupling and consequent perturbation of modes will therefore tend to be rather weak. Even if such coupling were to be significant, it is likely to be far smaller than the major changes in modal frequencies and shapes introduced by the f-holes and soundpost.

Finite-Element Analysis

To progress further in our understanding of the complex vibrations of instruments like the violin and guitar, it is necessary to include the coupled motions of every single part of the instrument and to consider the higher-order front and back plate modes, which will be strongly modified by their mutual coupling via the connecting ribs and, for the violin, the soundpost as well.

Such a task can be performed by numerical simulations of the vibrations of the complete structure using finite-element analysis (FEA). This involves modelling any plate or shell structure in terms of a large number of interconnected smaller elements of known shape and elastic properties. This division into smaller segments is known as tessellation. Provided the scale of the tessellation is much smaller than the acoustic wavelengths at the frequencies being considered, the motion of the structure as a whole can be described by the three-dimensional translations and rotations of the tessellated elements. The motion of each element can be related to the forces and couples acting on the adjoining faces of each three-dimensional (3-D) element. The problem is then reduced to the solution of N simultaneous equations proportional to the number of tessellated elements. Deriving the frequencies and mode shapes of the resulting normal modes of the system involves the inversion of a $N \times N$ matrix. Such calculations can be performed very efficiently on modern computer systems, though the computation time, proportional to N^2, can still be considerable for complex structures, particularly if a fine tessellation is used to evaluate the higher-frequency, shorter-wavelength, modes.

Figure 15.59 has already illustrated the potential complexity of the vibrational modes of a violin. The displacements have been greatly exaggerated for graphical illustration. In practice, the displacements of the plates are typically only a few microns, but can easily be sensed by placing the pad of a finger lightly on the vibrating surfaces. The first example shows a typical low-frequency mode involving the flexing and bending of every part of the violin, but with little change in its volume, so that it will radiate very little sound. The second example illustrates a mode involving a very strong asymmetrical vibration of the front plate, excited by the rocking action of the bridge with the soundpost inhibiting motion on the treble side of the instrument. Such a mode involves an appreciable change in volume of the

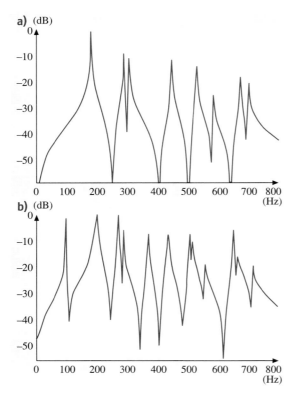

Fig. 15.61a,b FEA computations of admittance at bridge for a guitar with a 2.9 mm-thick front plate (**a**) in vacuo and (**b**) coupled to the air-cavity resonances (after *Derveaux* et al. [15.102])

shell-like structure, which will therefore radiate sound quite strongly.

One of the virtues of FEA is that the effects of changes in design of an instrument, or of the materials used in its construction, can be investigated systematically, without having to physically build a new instrument each time a change is made. For example, *Roberts* [15.103] has used FEA to investigate the changes in normal-mode frequencies of freely supported violin as a function of thickness and arching, the effects of cutting the f-holes and adding the bass bar, and the affect of the soundpost and mass of the ribs on the normal modes of the assembled body of the instrument, but without the neck and fingerboard. This enables a systematic correlation to be made between the modes and frequencies of the assembled instrument with the modes of the individual plates before assembly. Without the soundpost, the modes of the assembled violin were highly symmetric, with the bass-bar having only a marginal effect on the symmetry and frequency of modes. As expected, adding the soundpost removed the symmetry and changed the frequencies of almost all the modes, demonstrating the critical role of the soundpost and its exact position in determining the acoustic response of the violin.

Similar FEA investigations have been made of several other stringed instruments including the guitar (*Richardson* and *Roberts* [15.104]). Of special interest is the recent FEA simulation from first principles of the sound of a plucked guitar string by *Derveaux* et al. [15.102]. Their model includes interactions of the guitar plates, the plucked strings, the internal cavity air modes and the radiated sound. A DVD illustrating the methodology involved in such calculations [15.105] recently won an international prize, as an outstanding example of science communication. The effects of changing plate thickness, internal air resonances and radiation of sound on both admittance curves and the decay times and sound waveforms of plucked strings were investigated. Figure 15.61 compares the admittance curves at the guitar bridge computed for damped strings for a front-plate thickness of 2.9 mm first in vacuo and then in air. Note the addition of the low-frequency Helmholtz and higher-order cavity resonances in air and the perturbations of the structural resonances by coupling to the air modes.

15.2.7 Measurements

In this section we briefly consider the various methods used to measure the acoustical properties of stringed instruments, a number of which have already been referred to illustrate specific topics in the preceding section.

Admittance

The most common and easiest method used to investigate and characterise the acoustical properties of any stringed instrument is to measure the admittance $A(\omega)$ (velocity/force) at the point of string support on the bridge. *Fletcher* and *Rossing* [15.5] give examples of typical admittance curves for many stringed (and percussion) instruments including the violin, guitar, lute, clavichord and piano soundboard.

As described earlier, the admittance is in reality a complex tensor quantity, with the induced velocity dependent on and not necessarily parallel to the direction of the exciting force. In practice, most published admittance curves for the high-bridge instruments of the violin family show the amplitude and phase of the component of induced bridge velocity in the direction of an applied force parallel to the top plate. In contrast, for

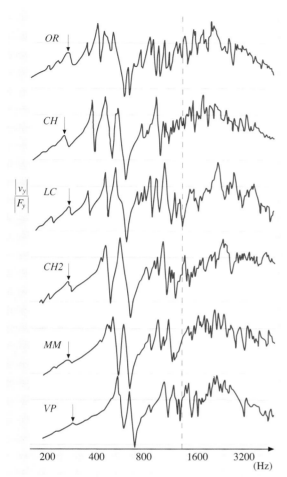

Fig. 15.62 Admittance measurements at bridge for six violins (after *Beldie* [15.97]). The *arrows* indicate the position of the Helmholtz air resonance. The horizontal lines are 20 dB markers

low-bridge instruments like the guitar, piano or harpsichord, the induced motion perpendicular to the top plate or soundboard is of primary interest.

The admittance at the bridge can be expressed in terms of the additive response of all the damped normal modes, which includes the mutual interactions of the plates of the instrument and all the component parts including, where appropriate, the neck, tailpiece, fingerboard, and all the strings. The admittance can then be written as

$$A(\omega) = \sum_n \frac{1}{m_n} \frac{i\omega}{\omega_n^2 - \omega^2 + i\omega\omega_n/Q_n} , \qquad (15.89)$$

where ω_n and Q_n are the frequency and Q-value of the nth normal mode and m_n is the associated effective mass at the point of measurement. The value of m_n depends on how well the normal mode is excited by a force at the point of measurement on the bridge. If, for example, the bridge on a guitar is at a position close to the nodes of particular normal modes, then the coupling will be weak and the corresponding effective mass will be rather large. Conversely, the low-frequency rocking action of the bridge on a bowed stringed instrument couples strongly into the breathing mode of the violin shell, so that the effective mass will be relatively low. The strength of this coupling plays an important role in determining the sound output from a particular instrument and also affects the playability of the bowed string and the sound of a plucked string.

In practice, by measuring the frequency response of the admittance, including both amplitude and phase, it is possible to decompose the admittance into the sum of the individual modal contributions and hence determine the effective mass, frequency and damping of the contributing normal modes. For the violin there are ≈ 100 identifiable modes below ≈ 4 kHz (*Bissinger* [15.106]), though not all of these are efficient acoustic radiators.

To avoid complications from the numerous sympathetic string resonances that can be excited, which includes all their higher-frequency partials, measurements are often made with all the strings damped by a piece of soft foam or a piece of card threaded between the strings. However, it should always be remembered that the damped strings still contribute significantly to the measured admittance. At low frequencies the strings still exert the same lateral restoring force on the bridge whether damped or not, while at high frequencies the damped strings present a resistive loading with their characteristic string impedances μc in parallel. When undamped, the strings present an additional impedance and transient response, which reflects the resonances of all the partials of the supported strings. This can make a significant difference to the sound of an instrument, notably when the sustaining pedal is depressed on a piano and in the ringing sound of any multi-stringed instrument, when a note is plucked or bowed and especially instruments like the theorbo and viola d'amore with freely vibrating sympathetic strings.

Figure 15.62 illustrates admittance measurements for six different violins by *Beldie* [15.97] reproduced from *Cremer* ([15.29], Fig. 10.1). The arrows indicate the position of the dispersive-shaped Helmholtz air resonance, which is the only predictable feature in such measurements, though its relative amplitude varies significantly from one instrument to the next. Such measurements provide a fingerprint for an individual instrument, highlighting the very large number of almost randomly positioned resonances that can be excited, which ultimately must be responsible for the distinctive sound of an instrument. As described earlier, above ≈ 1500 Hz the admittance often exhibits an underlying BH peak at ≈ 2 kHz followed by a characteristic decrease at higher frequencies, which can be attributed to a resonance of the bridge/central island region [15.75, 76].

Although most instruments have very different acoustic fingerprints, Woodhouse (private communication), in a collaboration with the English violin maker David Rubio, demonstrated that it is possible to construct instruments with almost identical admittance characteristics, provided one uses closely matching wood from the same log, with a nearly perfect match of plate thickness and arching. The German violin maker *Martin Schleske* [15.107] also claims considerable success in producing *tonal copies* with almost identical acoustical properties to those of distinguished Cremonese instruments by grading the thickness and arching of the plates to reproduce both the input admittance and radiated sound.

In contrast, slavishly copying the dimensions of a master violin rarely produces an instrument with anything like the same tonal quality. This is easily understood in terms of the differing elastic and damping properties of the wood used to carve the plates, which remains a problem of great interest, but beyond the scope of this article.

Traditionally, the admittance is usually measured using a swept sinusoidal frequency source, often generated by a small magnet waxed to the bridge and driven by a sinusoidally varying magnetic field. The admittance can equally well be determined from the transient response $f(t)$ following a very short impulse to the bridge, since it is simply the Fourier transform,

$$A(\omega) = \int_0^\infty f(t) e^{i\omega t} \, dt \, . \tag{15.90}$$

If signal-to-noise ratio from a single measurement is insufficient, one can use a sequence of impulses or a noise source, which is equivalent to a random succession of short pulses. In addition to many professional systems, relatively inexpensive PC-based versions using sound cards have been developed for researchers and instrument makers, such as the WinMLS system by *Morset* [15.108].

Laser Holography

Admittance measurements at the bridge provide detailed information on the frequencies, damping and effective masses of the normal modes of vibration of an instrument, but provide no information on the nature of the modes excited. Laser holography, which is essentially the modern-day equivalent of Chladni plate measurements, enables one to visualise the vibrational modes of stringed and percussion instruments. In such measurements, photographs or real-time closed-circuit television images of the interference patterns of laser light reflected from a stationary mirror and from the vibrating object are recorded. Using photographic or electronic/software reconstruction of the original image from the recorded holograms, a 3-D image of the vibrating surface is formed with superimposed contours indicating lines of equal vibrational amplitude, as already illustrated for a number of prominent guitar modes in Fig. 15.56.

Recent developments in laser and electronic data-acquisition technology allow one to record such interferograms electronically and to display them in real time on a video monitor (for example, *Saldner* et al. [15.109]). To record the shapes of individual vibrational modes of an instrument excited by a sinusoidal force, care has to be taken to avoid contamination from neighbouring resonances, by judicious placing of the force transducer (e.g. placing it at a node of an unwanted mode). *Cremer* ([15.29], Chap. 12) reproduces an interesting set of holograms by *Jansson* et al. [15.110] for the front plate of a violin at various stages of its construction, before and after the f-holes are cut, before and after the bass-bar is added and with a soundpost supported on a rigid back plate. These highlight the major effect of the f-holes and soundpost on the modal shapes and frequencies, but the relatively small influence of the bass-bar, consistent with the FEA computations by *Roberts* [15.103] referred to earlier. However, the bass bar strengthens the coupling between the island area between the f-holes and the larger radiating surfaces of the top plate and therefore has a strong influence on the intensity of radiated energy.

With modern intense pulsed laser sources, one can also investigate the transient response of instruments using single pulses. For example, *Fletcher* and *Rossing* ([15.5], Fig. 10.15) reproduce interferograms of the front and back plate of a violin by *Molin* et al. [15.111, 112], which illustrate flexural waves propagating out from the feet of the bridge on the front face and from the end of the soundpost on the back plate at intervals from 100–450 µs after the application of a sharp impulse at the bridge. Holograms can even be recorded while the instrument is being bowed [15.111, 112].

Modal Analysis

Modal analysis measurements have been extensively used to investigate the vibrational modes of the violin, guitar, the piano soundboards and many other stringed and percussion instruments (Chap. 28). Briefly, the method involves applying a known impulse at one point and measuring the response at a large number of other points on the surface of an instrument, which allows one to determine both the modal frequencies and the vibrations at all points on the surface. The Fourier transform of the impulse response is directly related to the nonlocal admittance, which in terms of the normal modes excited can be written as

$$A(r_1, r_2, \omega) = i\omega \sum_n \frac{1}{m_n} \frac{\psi_n(r_1)\psi_n(r_2)}{\omega_n^2 - \omega^2 + i\omega\omega_n/Q_n} \, , \quad (15.91)$$

where $\psi_n(r_1)\psi_n(r_2)$ is the product of the wavefunctions describing the displacements at the measurment and excitation points normalised to the product at the point of maximum displacement, and m_n is now the effective mass of the normal mode at its point of maximum amplitude of vibration (i.e. $KE_{max} = 1/2 m_n \omega^2 \psi_n^2|_{max}$). An FFT of the recorded transient response will give peaks in the frequency response, which can be decomposed into contributions from all the excited normal modes. By keeping the point of excitation fixed and moving the measurement point, one can record the amplitude and phase of the induced motion for a specific mode and, using the spatial dependence in (15.91), can map out the nodal waveform. Alternatively, one can keep the measurement point fixed and apply the impulse over the surface of the structure to derive similar information.

One of the first detailed modal analysis investigations of the violin was made by *Marshall* [15.93], who used a fixed measurement point on the top plate of the violin near the bass-side foot of the bridge with a force hammer providing calibrated impulses at a large number of points over the surface of the violin. From the FFT of the resultant transient responses, the amplitudes and phases of the excited normal modes of the violin involving all its component parts including the body shell, neck, fingerboard and tailpiece could be determined. Marshall identified and characterised around 30 normal modes below ≈ 1 kHz. Many of the modes involved the relatively low-frequency flexing and twisting of the instrument as a whole. However, because such modes involved little appreciable change in overall volume of the shell of the instrument structure, they resulted in little radiated sound. Nevertheless, it was suggested that such modes might well play an important role for the performer in determining the feel of the instrument and its playability.

In any physical measurement, the instrument has to be supported in some way. Rigid supports introduce additional boundary conditions, which can significantly perturb the normal modes of the instrument. Many measurements are made with the instrument supported by rubber bands, which provide static stability without significant perturbation of the higher-frequency structural modes. However, *Marshall* [15.93] showed that, when an instrument is held and supported by the player under the chin, the damping of many of the normal modes was significantly increased, which will clearly affect the sound of the instrument when played. This observation has also been confirmed in more recent modal analysis measurements by *Bissinger* [15.94] and by direct measurements of the decaying transient sound of a freely and conventionally supported violin by the present author [15.18].

Bissinger [15.106] has made extensive admittance, modal analysis and sound radiation measurements on a large number of instruments. Measurements were made using impulsive excitation at the bridge and a laser Doppler interferometer to record the induced velocities at over 550 points on the surface of the violin. Simultaneous measurements of the overall radiation and directivity were made using 266 microphone positions over a complete sphere. Figure 15.63 shows cross sections illustrating the displacements associated with four low frequency modes. The 159 Hz mode |⚙CD-ROM involves major vibrations of the neck, fingerboard, tailpiece and body of the instrument. The second example at 281 Hz |⚙CD-ROM illustrates the body displacements associated with the Helmholtz air resonance. The mode

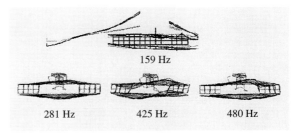

Fig. 15.63 Modal analysis measurements illustrating the displacements associated with four representative low-frequency modes of a violin (data provided by Bissinger)

at 425 Hz |⌾⃝ CD-ROM illustrates a mode with asymmetric in-phase vibrations of the front and back plates, with little net volume change and hence little radiated sound, while the mode at 480 Hz |⌾⃝ CD-ROM is a strong breathing mode.

By combining the modal analysis and radiativity measurements, Bissinger has shown that the radiation efficiency of the plate modes (i.e. the fraction of sound energy radiated by the violin relative to a baffled piston of the same surface area having the same root mean square surface velocity displacement) rises to nearly 100% at a *critical frequency* of $\approx 2-4$ kHz, when the wavelength of the flexural vibrations of the plates matches that of sound waves in air. Little apparent correlation was observed between the perceived quality of the measured violins and the frequencies and strengths of prominent structural resonances below ≈ 1 kHz or with the internal damping of the front and back plates. This runs contrary to the general view of violin makers that the front and back plates of a fine violin should be made of wood with a very long ringing time when tapped. Interestingly, the American violin maker Joseph Curtin has also observed that individual plates of old Italian violins often appear to be more heavily damped than their modern counterparts [15.113]. This is clearly an area that merits further research.

15.2.8 Radiation and Sound Quality

As already emphasised, at low frequencies, when the acoustic wavelength is smaller than the size of an instrument, the radiated sound is dominated by isotropic monopole radiation. As the frequency is increased, higher-order dipole and then quadrupole radiation become progressively important, while above the critical frequency, when the acoustic wavelength is shorter than the that of the flexural waves on the shell of an instrument, the radiation patterns become increasingly directional, so that it is no longer appropriate to consider the radiation in terms of a multipole expansions.

Fletcher and *Rossing* ([15.5], Fig. 10.30) reproduce measurements on both the violin and cello by *Meyer* [15.114], which highlight the increasing directionality of the sound produced with increasing frequency and the rather strong masking effect of the player at high frequencies. More recently, *Weinreich* and *Arnold* [15.115, 116] have made detailed theoretical and experimental studies of multipole radiation from the freely suspended violin at low frequencies (typically below 1 kHz). Interestingly, they made use of the principle

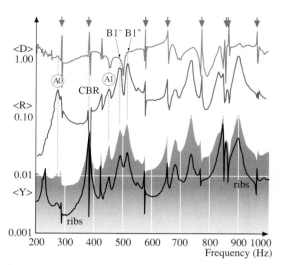

Fig. 15.64 Plots of the root-mean-square (rms) mobility $\langle Y \rangle$ (m/sN) averaged over the surface of the front and back plate (*solid shaded region*) and ribs (*white curve*) of the instrument, the radiativity $\langle R \rangle$ (Pa/N) averaged over a sphere, and directivity $\langle D \rangle$, the ratio of forward to backward radiation averaged over hemispheres. The *top arrows* represent the positions of the open-string resonances and their partials. A0 is the position of the Helmholtz air resonance and A1 the first internal cavity air resonance, CBR is a strong corpus bending mode and B1 and B2 are the two strong structural normal mode resonances of the coupled front and back plates (data kindly supplied by Bissinger)

of acoustic reciprocity, based on the fact that the amplitude of vibration at the top of the bridge produced by incoming sound waves is directly related to the sound radiated by a force applied to the violin at the same point. The violin was radiated by an array of loudspeakers to simulate incoming spherical or dipole sound fields and the induced velocity at the bridge recorded by a very light gramophone pick-up stylus.

Hill et al. [15.117] have used direct measurements to investigate the angular dependence of the sound radiated by a number of high-quality modern acoustic guitars with different cross-strutting, when excited by a sinusoidal force at the bridge. From such measurements, they were able to derive the fraction of sound radiated as the dominant monopole and dipole (with components in three directions) radiation, in addition to effective masses and Q-values, for a number of prominent modes up to ≈ 600 Hz. Significant differences were observed for the three different strutting schemes investigated.

Bissinger [15.118] has made an extensive investigation of radiation from both freely supported and hand-held violins, including measurements above 1 kHz, where the multimodal radiation expansion is no longer appropriate. Bissinger correlates the sound radiated over a large number of points on a sphere surrounding the violin with measurements of the input admittance at the bridge and the induced surface velocities over the whole violin structure. Figure 15.64 shows a typical set of simultaneous measurements up to 1 kHz. Although the low-frequency Helmholtz resonance contributes strongly to the radiated sound, it results in a relatively small feature on the mobility curves for the body of the instrument (or on the measured admittance at the bridge, not shown). Bissinger was unable to find any significant correlation between the frequencies and Q-values of the prominent *signature* modes excited (see, for example, *Jansson* [15.119]) below ≈ 1 kHz and the perceived quality of the instruments investigated. Above ≈ 1 kHz the modes strongly overlap, so that it becomes more appropriate to compare the frequency averaged global features. The measurements show that the fraction of mode energy radiated increases monotonically from close to zero at low frequencies up to around almost 100% efficient at 4 kHz and above, where almost all the energy is lost by radiation rather than internal damping. The ultimate aim of these detailed modal analysis studies is to correlate the measured acoustical properties with the results obtained from finite-element analysis and to produce sufficient information about the acoustical properties that might allow a more realistic comparison between physical properties and the properties of an instrument judged from their perceived sound quality and playability.

Directional Tone Colour

Although the acoustic power radiated averaged over a given frequency range is clearly an important signature of the sound of a particular instrument, the intensity at a particular frequency in the important auditory range above 1 kHz can vary wildly from note to note. This is illustrated in Fig. 15.65 by *Weinreich* [15.120], which compares the intensities of radiated sound in an anechoic chamber along the direction of the neck and perpendicular to the front plate for four violins of widely differing quality, with 0 dB representing an isotropic response. Above around 1 kHz, the wavelengths of flexural waves on the plates of the instrument become comparable with the wavelength of the radiated sound. This leads to strong diffraction effects in the radiated sound, which fluctuate wildly with direction as different modes are preferentially excited. At a particular point in the listening space, the spectral content of the bowed violin therefore varies markedly from note to note, as will the sound within a single note played with vibrato resulting in frequency modulation. The spectral content will also vary from position to position around the violin, especially if the player moves whilst playing. Weinreich has emphasised the importance of such effects in producing a greater sense of *presence* and *vibrancy* in the perceived sound from a violin than would be produced by a localised isotropic sound source, such as a small loudspeaker. Weinreich has coined the term *directional tone colour* to describe such effects. He has also designed a loudspeaker system based on the same principles, which gives a greater sense of realism to the sound of the recorded violin than a simple loudspeaker.

In addition to the intrinsic directionality of the violin, the time-delayed echoes from the surrounding walls

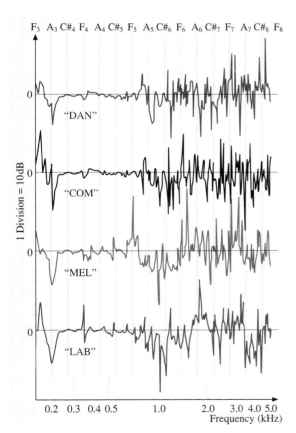

Fig. 15.65 Ratio of sound intensities along neck and perpendicular to front plate for four violins of widely different qualities (after *Weinreich* [15.120])

of a performing space also have a major influence on the complexity of the sound waveforms produced by a violin (or any other instrument) played with vibrato, as first noted by *Meyer* [15.121]. This arises from the interference between the different frequencies associated with the prompt sound reaching the listener (or microphone) and the sound generated at earlier times reflected from the surrounding surfaces. As discussed by the present author (*Gough* [15.18]), the additional complexity is largely a dynamic effect associated with the time-delayed interference between signals of different frequencies rather than caused by the amplitude modulation of individual partials associated with the multi-resonant response of the violin, first highlighted by *Fletcher* and *Sanders* [15.122] and *Matthews* and *Kohut* [15.123].

Perceived Quality

No problem in musical acoustics has attracted more attention or interest than the attempts made over the last 150 or so years to explain the apparent superiority of old Cremonese violins, such as those made by Stradivarius and Guarnerius, over their later counterparts, which have often been near exact copies. Many explanations have been proposed – a magic recipe for the varnish, chemical treatment of the wood and finish of the plates prior to varnishing [15.124], the special quality of wood resulting from micro-climate changes [15.125], etc. However, despite the committed advocacy for particular explanations by individuals, there is, as yet, little agreement between researchers, players or dealers on the acoustical attributes that distinguish a fine Italian violin worth $1M or more from that of a $100 student instrument.

From a physicist's point of view, given wood of the same quality as that used by the old Italian makers, there is no rational reason why a modern violin should not be just as good from an acoustic point of view as the very best Italian instrument. We have already commented on Martin Schleske's attempts to replicate the sounds of fine Italian instruments, by making *tonal copies* having as near as possible the same acoustical properties [15.107]. In addition, we have also highlighted Dünnewald's attempt to correlate the physical properties of well over 200 violins with their acoustical properties [15.73], including the comparison of selected student, modern and master violins reproduced in Fig. 15.62. Such studies appear to show a correlation between the amount of sound radiation in the acoustically important range around 3–4 kHz. As we have emphasized, this is just the region where the resonant properties of the bridge have a major influence on the spectrum.

It must also be remembered that the changed design of the bridge, the increase in string tension, higher pitch, increased size of the bass-bar, neck and soundpost, and the use of metal-covered rather than gut strings have resulted in a modern instrument sounding very different from the instruments heard by the 17th and early 18th century maker and performer. Even amongst the violins of the most famous Cremonese luthiers, individual instruments have very different distinctive tones and degrees of playability, particular as judged by the player. The gold standard itself is therefore very elusive. What is currently and may always be lacking is reliable measurements on the individual plates and shells of a large number of really fine instruments. We still largely rely on a small number of measurements performed by Savart in the nineteenth century and a few measurements by Saunders in the 1950s on which to base scientific guidelines for modern violin makers.

Performance Acoustic

It should also be recognised that, when a violin (or any other instrument) is played, the performer excites not only the vibrational modes of the instrument but also the multi-resonant normal modes of the performance space. Whereas the sound heard by a violinist is dominated by the sound of the violin, for the listener the acoustics of the performance space can dominate the timbre and quality of the perceived sound. To distinguish between the intrinsic sound qualities of violins, comparisons should presumably best be made in a rather dry acoustic, even though such an acoustic is generally disliked by the performer (and listener), who appreciates the improvement in sound quality provided by a resonant room acoustic.

One cannot review progress towards our understanding of what makes a good violin without recognising the inspiration and enthusiastic leadership of Carleen Hutchins, the doyenne and founder of the Catgut Society and scientific school of violin making, which has attracted many followers world-wide. By matching the frequencies and shapes of the first few modes of free plates before they are assembled into the completed instrument (Fig. 15.55), the scientific school of violin makers clearly achieve a high degree of quality control, which goes some way towards compensating for the inherent inhomogeneities and variable elastic properties of the wood used to carve the plates. However, in practice, there is probably just as much variability in the sound of such instruments as there is in instruments made by more traditional methods, where makers tap, flex and bend the plates until they feel and sound about right, this being part of the traditional skills handed down from master to apprentice even today. That is certainly the way that the Italian masters must have worked, without the aid of any scientific measurements beyond the feel and the sound of the individual plates as they are flexed and tapped.

Scientific Scaling

The other interesting development inspired by Carleen Hutchins and her scientific coworkers Schelling and Saunders has been the development of the modern violin octet [15.126], a set of eight instruments designed according to physical scaling laws based on the premise that the violin is the most successful of all the bowed stringed instruments. The aim is to produce a consort of instruments all having what are considered to be the optimised acoustical properties of the violin. Each member of the family is therefore designed to have the frequencies of the main body and Helmholtz resonances in the same relationship to the open strings as that on the violin, where the Helmholtz air resonance strongly supports the fundamental of notes around the open D-string, while the main structural resonances support notes around the open A-string and the second and generally strongest partial of the lowest notes played on the G-string. Several sets of such instruments have been constructed and admired in performance, though not all musicians would wish to sacrifice the diversity and richness of sounds produced by the different traditional violin, viola, cello and double bass in a string quartet or orchestra. Nevertheless, the scaling methods have led to rather successful intermediate and small-sized instruments.

15.3 Wind Instruments

In this section we consider the acoustics of wind instruments. These are traditionally divided into the woodwind, played with a vibrating reed or by blowing air across an open hole or against a wedge, and brass instruments, usually made of thin-walled brass tubing and played by buzzing the lips inside a metal mouthpiece attached to the input end of the instrument.

In general, the playing pitch of woodwind instruments is based on the first two modes of the resonating air column, with the pitch changed by varying the effective length by opening and closing holes along its length. In contrast, brass players pitch notes based on a wide range of modes up to and some times beyond the 10th. The effective length of brass instruments can be changed by sliding interpenetrating cylindrical sections of tubing (e.g. the trombone) or by a series of valves, which connect in additional length of tubing (e.g. trumpet and French horn). The pitch of many other instruments, such as the organ, piano-accordion and harmonium, is determined by the resonances of a set of separate pipes or reeds to excite the full chromatic range of notes, rather like the individual strings on a piano.

A detailed discussion of the physics and acoustical properties underlying the production of sound in all types of wind instruments is given by *Fletcher* and *Rossing* [15.5], which includes a comprehensive list of references to the most important research literature prior to 1998. As in many fields of acoustics, *Helmholtz* [15.127] and *Rayleigh* [15.3] laid the foundations of our present-day understanding of the acoustics of wind instruments. In the early part of the 20th century, *Bouasse* [15.128] significantly advanced our understanding of the generation of sound by the vibrating reed. More recently, *Campbell* and *Greated* [15.129] have written an authoritative textbook on musical acoustics with a particular emphasis on musical aspects, including extensive information on wind and brass instruments. Recent reviews by *Nederveen* [15.130] and *Hirschberg* et al. [15.131] provide valuable introductions to recent research on both wind and brass instruments. Earlier texts by *Backus* [15.132] and *Benade* [15.133], both leading pioneers in research on wind-instrument acoustics, provide illuminating insights into the physics involved and provide many practical details about the instruments themselves. A recent issue of *Acta Acustica* [15.134] includes a number of useful review articles, especially on problems related to the generation of sound by vibrating reeds and air jets and on modern methods used to visualise the associated air motions. For a mathematical treatment of the physics underlying the acoustics of wind instruments, *Morse* and *Ingard* [15.135] remains the authoritative modern text. Other important review papers will be cited in the appropriate sections, and selected publications will be used to illustrate the text, without attempting to provide a comprehensive list of references.

We first summarise the essential physics of sound propagation in air and simple acoustic structures before considering the more complicated column shapes used for woodwind and brass instruments. An introduction is then given to the excitation of sound by vibrating lips and reeds, and by air jets blown over a sharp edge.

The physical and acoustical properties of a number of woodwind and brass instruments will be included to illustrate the above topics.

A brief introduction to freely propagating sound in air was given in Sect. 15.1.3. In this section, we will be primarily concerned with the propagation of sound in the bores of wind and brass instruments, the excitation of standing-wave modes in such bores, the mechanics involved in the excitation of such modes and the resultant radiation of sound.

15.3.1 Resonances in Cylindrical Tubes

Standing waves in cylindrical tubes with closed or open ends provide the simplest introduction to the acoustics of wind instruments. For example, the flute can be considered as a first approximation as a cylindrical tube open at both ends, while the clarinet and trombone are closed at one end by the reed or the player's lips. For a cylindrical pipe open at both ends, wave solutions are of the general form

$$p_n(x, t) = A \sin(k_x x) \sin(\omega_n t), \qquad (15.92)$$

with the acoustic pressure zero at both ends. Neglecting end-corrections, open ends are therefore displacement antinodes and pressure nodes. These *boundary conditions* result in eigenmodes with $k_n = n\pi/L$ and $\omega_n = nc_0\pi/L$, where L is the length of the pipe and n is an integer.

Such modes are closely analogous to the transverse standing-wave solutions on a stretched string having n half-wavelengths along the length L and a *harmonic* set of frequencies $f_n = nc_0/L$, which are integral multiples of the *fundamental* (lowest) frequency $f_1 = c_0/2L$. When a cylindrical pipe open at both ends, such as a flute, is blown softly, the pitch is determined by the fundamental mode, but when it is *overblown* the frequency doubles, with the pitch stabilising on the second mode an octave above (audio ⟨?⟩ CD-ROM).

A cylindrical pipe played by a reed or vibrating lips has a pressure antinode and displacement node at the playing end. This results in standing-wave solutions with an odd number of 1/4-wavelengths between the two ends, such that $k_n = n\pi/4L$, where n is now limited to odd integer values. The corresponding modal frequencies, $\omega_n = n\pi/4L$, are therefore in the ratios 1:3:5:7: etc. The lowest note on the cylindrical bore clarinet, closed at one end by the mouthpiece, is therefore an octave below the lowest note on a flute of the same length. Furthermore, when overblown, the clarinet sounds a note three times higher than the fundamental, musical interval of an octave plus a perfect fifth (audio ⟨?⟩ CD-ROM). The weak intensity of the even-n-value modes in the spectrum accounts for the clarinet's characteristic hollow sound, particularly for the lowest notes on the instrument.

Real Instruments

For real wind and brass instruments, the idealised model of cylindrical tube resonators is strongly perturbed by a number of important factors. These include:

1. the shape of the internal bore of an instrument, which is often non-cylindrical including conical and often flared tubes with a flared bell on the radiating end,
2. the finite terminating impedance of the reed or mouthpiece used to excite the resonances, no longer providing a perfect displacement node,
3. radiation of sound from the end of the instrument, which is therefore no longer a perfect displacement antinode,
4. viscous and thermal losses to the walls of the instrument,
5. open and shut tone holes in the sides of wind instruments used to vary the pitch of the sounded note, and
6. bends and valves along the length of brass instruments, connecting additional lengths of tubing, which allow the player to play all the notes of a chromatic scale within the playing range of the instrument.

The skill of wind-instrument makers lies in their largely intuitive understanding of the way that changes in bore shape and similar factors affect the resonant modes of an instrument. This allows the design of instruments that retain, as closely as possible, a full set of harmonic resonances across the whole playing range of the instrument. This facilitates the stable production of continuous notes by the player, as the resulting harmonic set of Fourier components or partials coincide with the natural resonances of the instrument. For brass instruments with a flaring end this can often be achieved for all but the lowest natural resonance of the air column.

In discussing the acoustics of wind instruments with variable cross-sectional area S, the flow rate $U = Sv$ is a more useful parameter than the longitudinal particle velocity v. For example, the force acting on an element of air of length Δx along the bore length of an air column is then given by

$$-S\frac{\partial p}{\partial x}\Delta x = \rho S\frac{\partial v}{\partial t}\Delta x = \rho \frac{\partial U}{\partial t}\Delta x. \qquad (15.93)$$

For travelling waves, $e^{i(\omega t \pm kx)}$, this results in a ratio between the pressure and flow rate, defined as the tube impedance

$$Z = \frac{p}{U} = \pm \frac{\rho c_0}{S}, \qquad (15.94)$$

where the plus and minus signs refer to waves travelling in the positive and negative x-directions, respectively. There is a very close analogy with an electrical transmission line, with pressure and flow rate the analogue of voltage and current, as discussed later. Because the impedance is inversely proportional to area, it can be appreciably higher at the input end of a brass or wind instrument than at its flared output end. The flared bore of a brass instrument or the horn on an old wind-up gramophone can therefore be considered as an acoustic transformer, which improves the match between the high impedance of the vibrating source of sound to the much lower impedance presented by air at the end of the instrument. There is clearly an optimum matching, which enhances the radiated sound without serious degradation of the excited resonant modes.

Acoustic Radiation

In elementary textbook treatments, the pressure at the end of an open pipe is assumed to be zero and the flow rate a maximum, so that $Z_{\text{closed}} = p/U = 0$. However, in practice, the oscillatory motion of the air extends

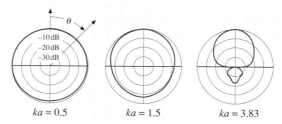

Fig. 15.67 Polar plots of the intensity radiation from the end of a cylindrical pipe of radius a for representative ka values, calculated by *Levine* and *Schwinger* [15.137]. The radial gradations are in units of 10 dB. The intensities in the forward direction ($\theta = 0$) relative to those of an isotropic source are 1.1, 4.8 and 11.8 dB (*Beranek* [15.136])

somewhat beyond the open end, providing a pulsating source that radiates sound, as described by *Rayleigh* ([15.3] Vol. 1, Sect. 313). Such effects can be described by a complex terminating load impedance, $Z_L = R + jx$. Figure 15.66 shows the real (*radiation resistance*) R and imaginary (inertial *end-correction*) x components of Z_L as a function of ka, where a is the radius of the open-ended pipe. The impedance is normalised to the tube impedance $\rho c_0/\pi a^2$.

When $ka \ll 1$, the reactive component is proportional to ka and corresponds to an effective increase in the tube length or end-correction of $0.61a$. At low frequencies, the real part of the impedance represents the radiation resistance $R_{\text{rad}} = \rho c/4S(ka)^2$. In this regime, the sound will be radiated isotropically as a monopole source of strength $U e^{i\omega t}$, illustrated in Fig. 15.67 by the polar plots of sound intensity as a function of angle and frequency.

When ka is of the order of and greater than unity, the real part of the impedance approaches that of a plane wave acting across the same area as that of the tube. Almost all the energy incident on the end of the tube is then radiated and little is reflected. For $ka \gg 1$, sound would be radiated from the end of the pipe as a beam of sound waves. The transition from isotropic to highly directional sound radiation is illustrated for a sequence of ka values in Fig. 15.67. The ripples in the impedance in Fig. 15.66 arise from diffraction effects, when the wavelength becomes comparable with the tube diameter.

For all woodwind and brass instruments, there is therefore a crossover frequency $f_c \sim c_0/2\pi a$, below which incident sound waves are reflected at the open end to form standing waves. Above f_c waves generated by the reed or vibrating lips will be radiated from the ends of the instrument strongly with very little reflec-

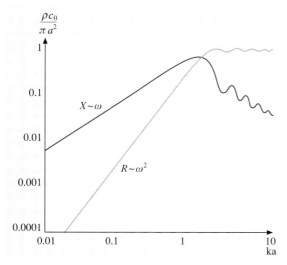

Fig. 15.66 Real and imaginary components of f, the impedance at the unbaffled open end of a cylindrical tube of radius a, in units of $\rho c_0/\pi a^2$, as a function of ka (after *Beranek* [15.136])

tion. Narrow-bore instruments can have a large number of resonant modes below the cut-off, while instruments with a large output bell, like many brass instruments, have far fewer.

For a narrow-bore cylindrical bore wind instrument with end radius $a \approx 1$ cm, the cut-off frequency ($ka \approx 1$) is ≈ 5.5 kHz. Below this frequency the instrument will support a number of relatively weakly damped resonant modes, which will radiate isotropically from the ends of the instrument or from open holes cut in its sides. In contrast, for brass instruments the detailed shape and size ot the flared end-bell determines the cut-off frequency. The large size of the bell leads to an increase in intensity of the higher partials and hence brilliance of tone-color, especially when the bell is pointed directly towards the listener. For French horns, much of the higher-frequency sound is therefore projected backwards relative to the player, unless there is a strongly reflecting surface behind.

For $ka \ll 1$, the open end of a musical instrument acts as an isotropic monopole source with radiated power P given by

$$P = U_{\text{rms}}^2 R_{\text{rad}} = \omega^2 \frac{\rho}{8\pi c} (S\omega\xi)^2 \ . \tag{15.95}$$

For a given vibrational displacement, the radiated power therefore increases with the fourth power of both frequency and radius. This very strong dependence on size explains why brass instruments tend to have rather large bells and why high-fidelity (HI-FI) *woofer* speakers and the horns of public address loudspeakers tend to be rather large. Conversely, it explains why the sound of small loudspeakers, such as those used in PC notebooks, fall off very rapidly below a few hundred Hz.

Acoustic radiation will lower the height and increase the width of resonances in a cylindrical tube. The resulting Q-values can be determined from

$$\begin{aligned} Q &= \omega \frac{\text{stored energy}}{\text{radiated energy}} \\ &= \omega \frac{\frac{1}{4}\rho SL\omega^2\xi^2}{\omega^4 (\rho/8\pi c) S^2\xi^2} = 2\pi cL/\omega S \ . \end{aligned} \tag{15.96}$$

Narrow-bore instruments will therefore have larger Q-values and narrower resonances than wide-bore instruments such as brass instruments, where the flared end-sections enhance the radiated energy at the expense of increasing the net losses.

The increased damping introduced by radiation from the end of an instrument is illustrated in Fig. 15.68, which compares the resonances of a length of 1 cm-diameter trumpet tubing, first with a normal open end and then with a bell attached (*Benade* [15.132], Fig. 20.4). Attaching a bell to such a tube dramatically increases the radiated sound from the higher partials and perceived intensity, but at the expense of a cut-off frequency at around ≈ 1.5 kHz and a significant broadening of the resonances at lower frequencies. Audio 13 CD-ROM demonstrates the sound of a mouthpiece-blown length of hose pipe with and without a conical chemical filter funnel attached to its end.

Viscous and Thermal Losses

In addition to radiation losses, there can be significant losses from viscous damping and heat transfer to the walls, as discussed in detail in *Fletcher* and *Rossing* ([15.5], Sect. 8.2). Although simple models for waves

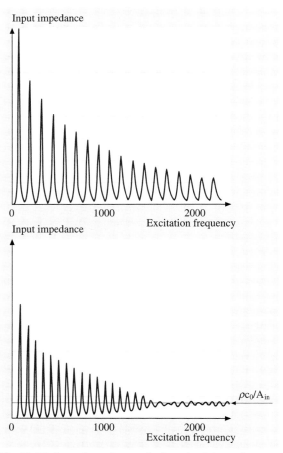

Fig. 15.68 Input impedance of a length of 1 cm diameter trumpet tubing with and without a bell attached to the output end (after *Benade* [15.133])

propagating along tubes assume a constant particle displacement across the whole cross-section, in reality the air on the surface of the walls remains at rest. The particle velocity only attains the assumed plane-wave constant value over a boundary-layer distance of δ_η from the walls. This is determined by the viscosity η, where $\delta_\eta = (\eta/\omega\rho)^{1/2}$, which can be expressed as $\sim 1.6/f^{1/2}$ mm for air at room temperature. At 100 Hz, $\delta_\eta \sim 0.16$ mm. which is relatively small in comparison with typical wind-instrument pipe diameters. Nevertheless, it introduces a significant amount of additional damping of the lower-frequency resonances of wind and brass instruments.

The viscous losses lead to an attenuation of sound waves, which can be described by adding an imaginary component to the k value such that $k' = k - i\alpha$. Waves therefore propagate as $e^{-\alpha x} e^{i(\omega t - kx)}$, with an attenuation coefficient

$$\alpha = \frac{1}{ac_0}\sqrt{\frac{\eta\omega}{2\rho}} = \frac{k\delta_\eta}{a}. \qquad (15.97)$$

In addition, heat can flow from the sinusoidally varying adiabatic temperature fluctuations of the vibrating column of air into the walls of the cylinder. At acoustic frequencies, this takes place over the thermal diffusion boundary length $\delta_\theta = (\kappa/\omega\rho C_p)^{1/2}$, where κ is the thermal conductivity and C_p is the heat capacity of the gas at constant pressure. In practice, $\delta_\theta \sim \delta_\eta$, as anticipated from simple kinetic theory (for air, the values differ by only 20%). Viscous and heating losses are therefore comparable in size giving an effective damping factor for air at room temperature, $\alpha = 2.2 10^4 k^{1/2}/a$ m^{-1} (*Fletcher* and *Rossing* [15.5], Sect. 8.2). The ratio of the real to imaginary components of k determines the damping and effective Q-value of the acoustic resonances from wall losses alone, with $Q_{\text{walls}} = k/2\alpha$.

The combination of radiation and wall losses leads to an effective Q_{total} of the resonant modes given by

$$\frac{1}{Q_{\text{total}}} = \frac{1}{Q_{\text{radiation}}} + \frac{1}{Q_{\text{wall-damping}}}. \qquad (15.98)$$

Because of the different frequency dependencies, wall damping tends to be the strongest cause of damping of the lowest-frequency resonances of an instrument. It can also be significant in the narrow-bore tubes and *crooks* used to attach reeds to wind instruments.

Input Impedance

The method used to characterize the acoustical properties of a wind or brass instrument is to measure the input impedance $Z_{\text{in}} = p_{\text{in}}/U_{\text{in}}$ at the mouthpiece or reed end of the instrument. Such measurements are frequently made using the capillary tube method. This involves connecting an oscillating source of pressure fluctuations to the input of the instrument through a narrow-bore tube. This maintains a constant oscillating flow of air into the instrument, which is largely independent of the frequency-dependent induced pressure fluctuations at the input of the instrument. Several examples of such measurements, similar to those for the length of trumpet tubing (Fig. 15.68), for woodwind and brass instruments are shown and discussed by *Backus* [15.132] and *Benade* [15.133], who pioneered such measurements, and in *Fletcher* and *Rossing* ([15.5], Chap. 15). Alternatively, a piezoelectric driver coupled to the end of the instrument can provide a known source of acoustic volume flow.

The input impedance of a cylindrical tube is a function of both the tube impedance $Z_0 = \rho c_0/S$ and the terminating impedance Z_L at its end. It can be calculated using standard transmission-line theory, which takes into account the amplitude and phases of the reflected waves from the terminating load. The reflection and transmission coefficients R and T for a sound wave impinging on a terminating load Z_L are given by

$$R = \frac{Z_L - Z_0}{Z_L + Z_0} \quad \text{and} \quad T = \frac{2Z_L}{Z_L + Z_0}. \qquad (15.99)$$

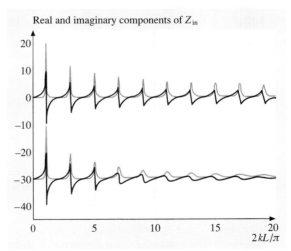

Fig. 15.69a,b Real and imaginary components of the input impedance in units of Z_0 as a function of $2kL/\pi$ for (**a**) an ideally open-ended, $Z_L = 0$, cylindrical tube with wall losses varying as $k^{1/2}$, and (**b**) the same components shifted downwards for a pipe with radiation resistance proportional to k^2 also included

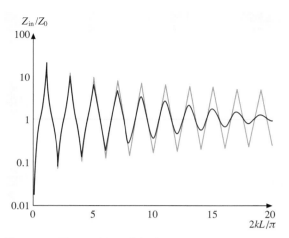

Fig. 15.70 The modulus of the input impedance plotted on a logarithmic scale of cylindrical pipe with wall damping alone and with additional radiative damping as a function of k in units of $\pi/2$

For a cylindrical tube of length L, the input impedance is given by

$$Z_{\text{in}} = Z_0 \frac{Z_L \cos kl + iZ_0 \sin kl}{iZ_L \sin kl + Z_0 \cos kl}, \quad (15.100)$$

with complex k values $k - i\alpha$, if wall losses need to be included.

In Fig. 15.69 we have plotted the kL dependence of the real and imaginary components of the input impedance of an open-ended cylindrical pipe. The upper plot includes wall losses alone proportional to $k^{1/2}$ while the lower plot includes losses from the end of the instrument with radiation resistance $\text{Re}(Z_L)$ varying as k^2. The input impedance is high when $k_n = nc_0/4L$, where n is an odd integer. The input impedance is a minimum when n is even.

It is instructive to consider the magnitude of the input impedances on a logarithmic scale, as shown in Fig. 15.70. When plotted in this way, the resonances and anti-resonances are symmetrically placed about the tube impedance Z_0. The magnitude of the impedance of the nth resonance is $Q_n Z_0$, where Q_n is the quality factor of the excited mode. In contrast the anti-resonances have values of Z_0/Q. The widths $\Delta\omega/\omega$ of the modes at half or double intensity are given by $1/Q_n$.

For efficient transfer of sound, the input impedance of a wind instrument has to match the impedance of the sound generator. For instruments like the flute, recorder and pan pipes, sound is excited by imposed fluctuations in air pressure across an open hole, so that the generator impedance is small. The resonances of such instruments are therefore located at the minima of the anti-resonance impedance dips, corresponding to the evenly spaced resonances, $f_n = nc_0/2L$, of a cylindrical tube with both ends open. In contrast, for all the brass instruments and woodwind instruments played by a reed, the playing end of the tube is closed by a relatively massive generator (the lips or reed). Resonances then occur at the peaks of the input impedance with frequencies $f_n = nc_0/4L$, where n is now an odd integer, corresponding to the resonances of a tube closed at one end. If we had plotted the magnitude of the input admittance, $A(\omega) = 1/Z(\omega)$, instead of the impedance, the positions of the resonances and anti-resonances would have been reversed. The resonant modes of a double-open-ended wind instrument therefore occur at the peaks of the input admittance, whereas the resonant modes of wind or brass instruments played with a reed or mouthpiece are at the peaks of the input impedance. This is a general property of wind instruments, whatever the size or shape of their internal bores.

15.3.2 Non-Cylindrical Tubes

Although there are simple wind instruments with cylindrical bores along their whole length, the vast majority of modern instruments and many ancient and ethnologically important instruments have internal bores that flare out towards their ends. One of the principle reasons for such flares is that they act as acoustic transformers, which help to match the high impedance at the mouthpiece to the much lower radiation impedance of the larger-area radiating output end. However, increasing the fraction of sound radiated decreases the amplitude of the reflected waves and hence the height and sharpness of the natural resonances of the air resonances. In addition, the shape of the bore can strongly influence the frequencies of the resonating air column, which destroys the harmonicity of the modes. This makes it more difficult for the player to produce a continuous note that is rich in partials, since any repetitive waveform requires the excitation of a harmonic set of frequencies.

Conical Tube

We first consider sound propagation in a conical tube, approximating to the internal bore of the oboe, saxophone, cornet, renaissance cornett and bugle. If side-wall interactions are neglected, the solutions for wave propagation in a conical tube are identical to those of spherical wave propagating from a central point. Such waves satisfy the wave equation, which may be written in spherical coordinates as

$$\nabla^2 (rp) = \frac{1}{c_0^2} \frac{\partial^2 (rp)}{\partial t^2} .\quad (15.101)$$

We therefore have standing-wave solutions for rp that are very similar to those of a cylindrical tube, with

$$p = C \frac{\sin kr}{r} e^{i\omega t} .\quad (15.102)$$

Note that the pressure remains finite at the apex of the cone, $r = 0$, where $\sin(kr)/r \to k$. For a conical tube with a pressure node $p = 0$ at the open end, we therefore have standing wave modes with $k_n L = n\pi$ and $f_n = nc_0/2L$, where n is any integer. The frequencies of the excited modes are therefore identical to the modes of a cylindrical tube of the same length that is open at both ends. The lowest note at $f_1 = c_0/2L$ for a conical tube instrument with a reed at one end (e.g. the oboe and saxophone) is therefore an octave above a reed instrument of the same length with a cylindrical bore (e.g. the clarinet) with a fundamental frequency of $c_0/4L$.

The flow velocity U is determined by the acceleration of the air resulting from the spatial variation of the pressure, so that

$$\rho \frac{\partial U}{\partial t} = \frac{\partial (r^2 p)}{\partial r} = C (\sin kr + kr \cos kr) e^{i\omega t} .\quad (15.103)$$

Figure 15.71 illustrates the pressure and flow velocity for the $n = 5$ mode of a conical tube. Unlike the modes of cylindrical tube, the nodes of U no longer coincide with the peaks in p, which is especially apparent for the first few cycles along the tube. Furthermore, the amplitude fluctuations increase with distance r from the apex ($\sim r$), whilst the fluctuations in pressure decrease $\approx 1/r$. A conical section therefore acts as an acoustic transformer helping to match the high impedance at the input mouthpiece end to the low impedance at the output radiating end.

Attaching a mouthpiece or reed to the end of a conical tube requires truncation of the cone, which will clearly perturb the frequencies of the harmonic modes. However, using a mouthpiece or reed unit having the same internal volume as the volume of the truncated section removed will leave the frequencies of the lowest modes unchanged. Only when the acoustic wavelength becomes comparable with the length of truncated section will the perturbation be large.

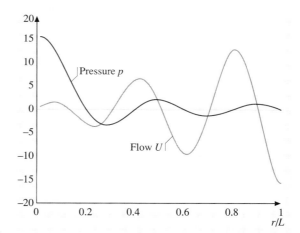

Fig. 15.71 Pressure and flow velocity of the $n = 5$ mode along the length of a conical tube

Fletcher and *Rossing* ([15.5], Sect. 8.7) consider the physics of the truncated conical tube and give the input impedance derived by *Olson* [15.139]

$$Z_{in} = \frac{\rho c_0}{S_1} \frac{i Z_L \left(\frac{\sin(kL-\theta_2)}{\sin \theta_2} \right) + \left(\frac{\rho c_0}{S_2} \right) \sin kL}{Z_L \frac{\sin(kL+\theta_1-\theta_2)}{\sin \theta_1 \sin \theta_2} + j \left(\frac{\rho c_0}{S_2} \right) \frac{\sin(kl+\theta_1)}{\sin \theta_1}} ,\quad (15.104)$$

where x_1 and x_2 are the distances of the two ends from the apex of the truncated conical section. The length $L = x_2 - x_1$, the end areas are S_1 and S_2, with $\theta_1 = \tan^{-1} kx_1$ and $\theta_2 = \tan^{-1} kx_2$.

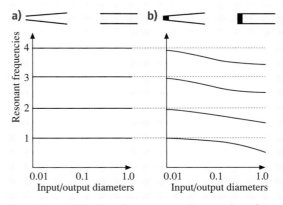

Fig. 15.72a,b The first four resonant frequencies of truncated cones with (**a**) both ends open, and (**b**) the input end closed, as a function of the ratio of their input to output diameters (after *Ayers* et al. [15.138])

For a cone with $Z_L = 0$ at the open end, (15.104) reduces to

$$Z_{\text{in}} = -j\frac{\rho c_0}{S_1}\frac{\sin kL \sin\theta_1}{\sin(kL+\theta_1)}, \qquad (15.105)$$

which is zero for $kL = n\pi$. The resonant frequencies of truncated cones with both ends open are therefore independent of cone angle and are the same as the equally spaced harmonic modes of a cylinder of the same length with both ends open, as shown in Fig. 15.72a. In contrast, the resonant frequencies of a truncated cone with one end closed (e.g. by the reed of an oboe or saxophone or mouthpiece of a bugle) are strongly dependent on the cone angle or ratio of input to output diameter, as shown in Fig. 15.72b, adapted from *Ayers* et al. [15.138]. As the ratio of input to output diameters of a truncated cone increases, the modes change from the evenly spaced harmonics of an open-ended cylinder of the same length, to the odd harmonics of a cylinder closed at one end. In the transitional regime, the frequencies of the modes are no longer harmonically related. This has a significant effect on the playability of the instrument, as the upper harmonics are no longer coincident with the Fourier components of a continuously sounded note. However, for an instrument such as the oboe, with a rather small truncated cone length, the perturbation of the upper modes is relatively small, as can be seen from Fig. 15.71b.

Cylindrical and non-truncated conical tubes are the only tubes that can produce a harmonically related set of resonant modes, independent of their length. Hence, when a hole is opened in the side walls of such a tube, to reduce the effective length and hence pitch of the note played, to first order, the harmonicity of the modes is retained. This assumes a node at the open hole, which will not be strictly correct, as discussed later in Sect. 15.3.3.

In reality, the bores of wind instruments are rarely exactly cylindrical or conical along their whole length. Moreover, many wind instruments have a small flare at the end to enhance the radiated sound, while others, like the cor anglais and oboe d'amore, have an egg-shaped cavity resonator towards their ends, which contributes to their characteristic timbre or tone colour. Table 15.6 lists representative wind instruments that are at least approximately based on cylindrical and conical bore shapes. The modern organ makes use of almost every conceivable combination of closed- and open-ended cylindrical and conical pipes.

Hybrid Tubes

Although many brass instruments include considerable lengths of cylindrical section, they generally have a fairly long, gently flared, end-section terminated by a very strongly flared output bell to enhance the radiated sound. The shape of such flares can be optimized to preserve the near harmonicity of the resonant modes, as described in the following section.

One can use (15.104) to model the input impedance of a flared tube of any shape, by approximating the shape by a number of short truncated conical sections joined together. Starting from the radiating end, one evaluates the input impedance of each cone in turn and uses it to provide the terminating impedance for the next, until one reached the mouthpiece end.

A weakness of all such models is the assumed plane wavefront across the air column, whereas it must always be perpendicular to the walls and belled outwards in any rapidly flaring region. We will return to this problem later.

Typical brass instruments, like the trumpet and trombone, have bores that are approximately cylindrical for around half their length followed by a gently flared section and end bell, while others have initial conical sections, like the bugle and horn. The affect on the resonant frequencies of the first six modes of adding a truncated conical section to a length of cylindrical tubing is shown as a function their relative lengths in Fig. 15.73, from *Fletcher* and *Rossing* ([15.5], Fig. 8.9). Note the major deviations from harmonicity of the resonant modes, apart from when the two sections are of nearly equal lengths. These results highlight the com-

Table 15.6 Instruments approximately based on cylindrical and conical air columns

$f_n = nc_0/2L$ n even and odd	$f_n = nc_0/4L$ n odd	$f_n = nc_0/2L$ n even and odd
Flute	Clarinet	Oboe
Recorders	Crumhorn	Bassoon
Shakuhachi	Pan pipes	Saxophone
Organ flue pipes (e.g. diapason)	Organ flue pipes (e.g. bourdon)	Cornett
	Organ reed pipes (e.g. clarinet)	Serpent
		Organ reed pipes (e.g. trumpet)

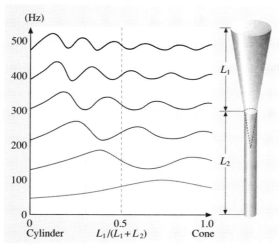

Fig. 15.73 The frequencies of the first six modes of a compound horn formed from different fractional lengths of cylindrical and conical section (after *Fletcher* and *Rossing* [15.5])

plexity involved, when adding flaring sections to brass instruments to increase the radiated sound.

Horn Equation

Physical insight into the influence of bore shape on the modes of typical brass instruments is given by the *horn equation* introduced by *Webster* [15.141], though similar models date back to the time of *Bernoulli* (*Rossing* and *Fletcher* [15.5], Sect. 8.6). In its simplest form, the horn equation can be written as

$$\frac{1}{S}\frac{\partial}{\partial x}\left(S\frac{\partial p}{\partial x}\right) = \frac{1}{c_0^2}\frac{\partial^2 p}{\partial t^2}, \qquad (15.106)$$

where $S(x)$ is the cross-sectional area of the horn at a distance x along its length. Provided the flare is not too

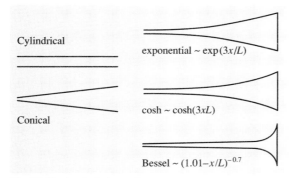

Fig. 15.74 Analytic horn shapes

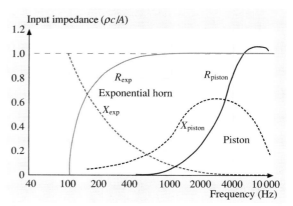

Fig. 15.75 Comparison of input impedance at the input throat of an infinitely long exponential horn and a piston of the same area set into an infinite baffle (after *Kinsler* et al. [15.140])

large, the above plane-wave approximation gives a good approximation to the exact solutions and preserves the essential physics involved.

If we make the substitution $p = \psi S^{1/2}$ and look for solutions varying as $\psi(x) e^{i\omega t}$, the horn equation can be expressed as

$$\frac{\partial^2 \psi}{\partial x^2} + \left[\left(\frac{\omega}{c_0}\right)^2 - \frac{1}{a}\frac{\partial^2 a}{\partial x^2}\right]\psi = 0, \qquad (15.107)$$

where the radius $a(x)$ is now a function of position along the length. The above equation is closely related to the Schrödinger wave equation in quantum mechanics, with $1/a\, \partial^2 a/\partial x^2$ the analogue of potential energy and $-\partial^2 \psi/\partial x^2$ the analogue of kinetic energy $-\hbar^2/2m\, \partial^2\psi/\partial x^2$, where m is the mass of the particle and $\hbar$ is Planck's constant. One can look for solutions of the form $e^{i(\omega t \pm kx)}$. At any point along the horn at radius x the radius of curvature of the horn walls, $R = (\partial^2 a/\partial x^2)^{-1}$, so that

$$k^2 = \left(\frac{\omega}{c_0}\right)^2 - \frac{1}{aR}. \qquad (15.108)$$

If $\omega > \omega_c = c_0/(aR)^{1/2}$, k is real, so that unattenuated travelling and standing-wave solutions exist. However, when $\omega < \omega_c$, k is imaginary and waves no longer propagate, but are exponentially damped as $e^{-x/\delta} e^{i\omega t}$ with a decay length of $c_0/\left(\omega_c^2 - \omega^2\right)^{1/2}$.

The propagation of sound waves in a horn is therefore directly analogous to the propagation of particle waves in a spatially varying potential. If the curvature is sufficiently large sound waves will be reflected before they reach the end of the instrument. However, just like particle waves in a potential well, sound waves can still tunnel through the potential barrier and radiate into free space at the end of the flared section. For a horn with a rapidly increasing flare, the reflection point occurs when the wavelength $\lambda^2 \sim (2\pi)Ra$. The effective length of an instrument with a flared horn on its end is therefore shorter for low-frequency modes than for the higher-frequency modes. This is illustrated schematically in Fig. 15.77 for resonant modes of a flared Bessel horn, which will be considered in more detail in the next section.

Exponential Horn

We now consider solutions of the horn equation, for a number of special shapes that closely describe sections of the internal bore of typical brass instrument. Cylindrical and conical section horns are special solutions with $(1/a)\partial^2 a/\partial x^2 = 0$, so that ψ satisfies the simple dispersionless wave equation. Figure 15.74 illustrates a number of other horn shapes described by analytic functions.

The radii of exponential and cosh function horns vary exponentially as $A\,e^{mx}$ and $A\cosh(mx)$, respectively, so that $(1/a)\partial^2 a/\partial x^2 = m^2$. The cosh mx function provides a smooth connection to a cylindrical tube at the input end. For both shapes, the horn equation can then be written as

$$\frac{\partial^2 \psi}{\partial x^2} + \left[\left(\frac{\omega}{c_0}\right)^2 - m^2\right]\psi = 0, \quad (15.109)$$

which has travelling solutions for the sound pressure $p = \psi/S^{1/2}$, where

$$p(x) = e^{-mx} e^{i\left(\omega t - \sqrt{k^2 - m^2}x^2\right)}, \quad (15.110)$$

and $k = \omega/c_0$. Above a critical cut-off frequency, $f_c = c_0 m/2$, waves can propagate freely along the air column with a dispersive phase velocity of $c_0/\sqrt{1-(\omega_c/\omega)^2}$, while below the cut-off frequency the waves are exponentially damped. The cut-off frequency occurs when the free-space wavelength is approximately six times the length for the radius to increase by the exponential factor e.

Figure 15.75 compares the input resistance and reactance of an infinite exponential horn with that of a baffled piston having the same input area (*Kinsler* et al. [15.140] Fig. 14.19). The plots are for an exponential horn with $m = 3.7\,\mathrm{m}^{-1}$, which corresponds to a cut-off frequency

of $\approx 100\,\mathrm{Hz}$, and a baffled piston having the same radius of 2 cm as the throat of the exponential horn. Above $\approx 400\,\mathrm{Hz}$, there is very little difference between the impedance of an infinitely long horn and a horn with a finite length of $\approx 1.5\,\mathrm{m}$ or longer, though below this frequency reflections cause additional fluctuations around the plotted values. Below the cut-off frequency, no acoustic energy can be radiated. Above the cut off the input resistance rises rather rapidly towards its limiting 100% radiating value. The exponential horn with a piston source at its throat is therefore a much more efficient radiator of sound than a baffled piston at all frequencies above the cut-off frequency.

The exponential horn illustrates how the flared cross section of brass instruments enhances the radiation of sound, though brass instruments are never based on exponential horns, otherwise no resonant modes could be set up. However, exponential horns were widely used in the early days of the gramophone. In the absence of any electronic amplification, they amplified the sound produced by the input diaphragm excited by the pick-up stylus on the recording cylinder or disc. They are still widely used in powerful public address systems. Such horns can also be used in reverse, as very efficient detectors of sound, with a microphone placed at the apex of the horn.

Bessel Horn

We now consider more realistic horns with a rapidly flaring end bell, which can often be modelled by what are known as Bessel horn shapes, with the radius varying

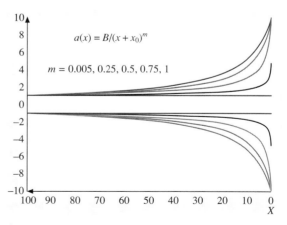

Fig. 15.76 Bessel horns representing the rapid outward flare of the bell on the end of a brass instrument, for a sequence of m values giving a ratio of input to output diameters of 10

as $1/x^m$ from their open end. Typical flared horn shapes are shown in Fig. 15.76 for various values of m, where the horn functions $A/(x+x_0)^m$ have been normalised by suitable choice of A and x_0, to model horns with input and output values 1 and 10. Increasing the value of m increases the rapidity with which the flare opens out at the end.

Again assuming the plane-wave approximation, the horn equation can be written as

$$\frac{\partial^2 \psi}{\partial x^2} + \left[\left(\frac{\omega}{c_0}\right)^2 - \frac{m(m+1)}{x^2}\right]\psi = 0, \quad (15.111)$$

with solutions

$$\psi(kx) = x^{1/2} J_{m+1/2}(kx) \quad (15.112)$$

and pressure varying from the end as

$$p(kx) = \frac{1}{x^{1/2}} J_{m+1/2}(kx), \quad (15.113)$$

where $J_{m+1/2}(kx)$ is a Bessel function of order $m+1/2$, giving the name to such horns.

In the plane-wave approximation, the sharpness and height of the barrier to wave propagation arising from the curvature could result in total reflection of the incident waves, so that no sound would be emitted from the end of the instrument. In reality, the curvature of the waveform will smear out any singularity in the horn function over a distance somewhat smaller than the output bell radius. Nevertheless, despite its limitations, the plane-wave model provides an instructive description of the influence of a rapidly flaring bell on a brass instrument. This is illustrated in Fig. 15.77 for the fundamental and fourth modes of a Bessel horn with $m = 1/2$, with the pressure $p(x)$ varying from the output end as $xJ_1(kx)$.

The most important point to note is the way that the flare pushes the effective node of the incident sine-wave solutions (extended into the flared section as dashed curves) away from the end of the instrument. The effective length is therefore shortened and resonant frequencies increased, the effect being largest for the lower frequency modes. The flare and general outward curvature of the horn cross section therefore destroys the harmonicity of the modal frequencies. This is a completely general result for any horn with a flared end. In practice, the nodal positions will also be affected by the curvature of the wavefront, which will further perturb the modal frequencies, but without changing the above qualitative behaviour.

Benade ([15.133], Sect. 20.5) notes that, from the early 17th century, trumpets and trombones have been

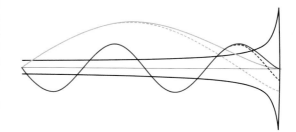

Fig. 15.77 Fundamental and fourth mode of an $m = 1/2$ Bessel horn, illustrating the increase in wavelength and resulting shift inwards of the effective nodal position. The *dashed lines* illustrate the extrapolated sine-wave solutions from well inside the bore. The plot is of $p(x)/x$

designed with strongly flaring bell corresponding to m values of 0.5–0.65, while French horns have bells with a less sudden flare with m values of 0.7–0.9.

From Fig. 15.77, it is easy to see how the player can significantly affect the pitch of a note on the French horn, by moving the supporting hand up into the bell of the instrument, which is referred to as hand-stopping. The pitch can be lowered by around a semitone, by placing the downwardly cupped hand and wrist against the top of the flared bell, effectively reducing the flare and increasing the effective length of the instrument. Alternatively, the pitch can be raised by a similar amount when the hand almost completely closes the inner bore. This leads to a major perturbation of the boundary conditions, effectively shortening the air column and reducing the output sound. The increase in frequency can be explained by the player using an almost unchanged embouchure to excite a higher frequency mode of the shortened tube. Because of the increased reflection of sound by the presence of the hand, the resultant sound although quieter is also much richer in higher partials (Fig. 15.113). Both effects are illustrated in Audio example 13 CD-ROM.

Flared Horns

Use is made of the dependence of modal frequencies on the curvature of horn shapes to design brass instruments with a set of resonances, which closely approximate to a harmonic series with frequencies $f_n = nf_1$. This should be contrasted with the modes of a cylindrical tube closed at one end by the mouthpiece, which would only involve the odd-n integer modes with frequencies of $f_n = c/2L$.

Historically, this has been achieved by empirical methods, with makers adjusting the shapes of brass instrument to give as near perfect a set of harmonic resonances as possible, as illustrated in Fig. 15.78 from *Backus* ([15.132] Chap. 12). A harmonic set of modes enhances the playability of the instrument, as the partials of any continuously blown note then coincide with the natural resonances of the instrument. However, the fundamental mode is always significantly flatter than required and is therefore not normally used in musical performance. Nevertheless, the brass player can still sound a *pedal note* corresponding to the *virtual fundamental* of the higher harmonics by exciting a repetitive waveform involving the higher harmonics, but with only a weak Fourier component at the pitch of the sounded note.

The way that this is achieved is shown schematically in Fig. 15.79 starting from the odd-n resonances

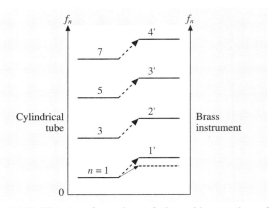

Fig. 15.79 The transformation of the odd-n modes of a cylindrical air column closed at one end to the near harmonic, all integer, n' modes of a flared brass instruments. The *lower dashed line* indicates schematically what can be achieved in practice for the lowest

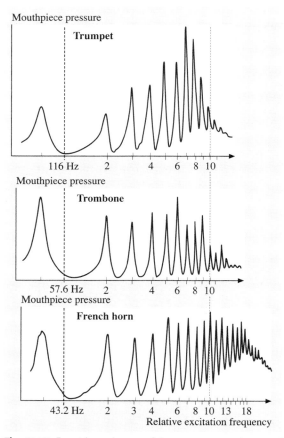

Fig. 15.78 Input impedances of the trumpet, trombone and French horn with the positions of the resonant peaks marked above the axis to show their position relative to a harmonic series based on the second harmonic of the instrument (after *Backus* [15.132])

of a cylindrical tube closed at one end by the mouthpiece to an appropriately flared horn of the same length. In practice, one can achieve a nearly perfect set of harmonic resonances, midway between the odd-integer modes of a cylindrical tube closed at one end, for all but the fundamental mode, which cannot be shifted upwards by a sufficient amount to form a harmonic fundamental, stopping and of the new set of modes.

Benade ([15.133], Sect. 20.5) has given an empirical expression for the frequencies of the partials of Bessel horns closed at the mouthpiece end, which closely describes these perturbations,

$$\frac{f'_n}{f_n} \sim \left(1 + 0.637 \frac{\sqrt{m(m+1)}}{2n-1}\right), \quad (15.114)$$

where the $(2n-1)$ in the denominator emphasising the preferential raising in frequency of the lower-frequency modes. This gives frequencies for the first six modes of a Bessel function horn with $m = 0.7$ are in the ratios 0.94, 2.00, 3.06, 4.12, 5.18 and 6.24, normalised to the $n = 2$ mode. These should be compared with the ideal 1, 2, 3, 4, 5, 6 ratios. Apart from the lowest note, which is a semitone flat, the higher modes are less than a semitone sharp compared with their ideal values.

Perturbation Models

Perturbation theory can be used to describe how changes in bore shape perturb the resonant modes of brass and woodwind instruments. *Fletcher* and *Rossing* ([15.5], Sect. 8.10) show that the change of frequency of a resonant mode $\Delta\omega$ resulting from small distributed changes $\Delta S(x)$ in bore area $S(x)$ is given by

$$\frac{\Delta\omega_n}{\omega_n} = -\frac{1}{2}\left(\frac{c_0}{\omega_n}\right) \int_0^L \left[\frac{\partial}{\partial x}\left(\frac{\Delta S(x)}{S(x)}\right) p_n \frac{\partial p_n}{\partial x} \mathrm{d}x\right] / \int_0^L \left[S(x) p_n^2 \mathrm{d}x\right]. \quad (15.115)$$

An alternative equivalent derivation uses Rayleigh's harmonic balance argument and equates the peak kinetic energy to the peak potential energy. To first order, the perturbation is assumed to leave the shape of the modal wavefunction unchanged. The kinetic and potential energy stored in a particular resonant mode can be expressed in terms of the local kinetic $\frac{1}{2}\rho\omega_n^2\xi_n^2$ and strain $\frac{1}{2}\gamma P_0(\partial\xi_n/\partial x)^2$ energy densities. For simplicity, we consider the perturbation of the nth resonant mode of a cylindrical air column open at one end, with particle displacement $\xi_n \approx \sin(n\pi x/L)\cos(\omega_n t)$, where n is an odd integer. Equating the peak kinetic and potential energy over the perturbed bore of the cylinder, we can then write

$$\omega_n'^2 \int_0^L \rho [S+\Delta S(x)] \sin^2(kx)\, \mathrm{d}x$$
$$= \gamma P_0 k_n^2 \int_0^L [S+\Delta S(x)] \cos^2(kx)\, \mathrm{d}x, \quad (15.116)$$

where ω_n' is the perturbed frequency. This can be rewritten as

$$\frac{\omega_n'^2}{\omega_n^2} = \int_0^L [S+\Delta S(x)] \cos^2(kx)\, \mathrm{d}x \Big/ \int_0^L [S+\Delta S(x)] \sin^2(kx)\, \mathrm{d}x. \quad (15.117)$$

Because the perturbations are assumed to be small, we can rearrange (15.117) to give the fractional change in frequency

$$\frac{\Delta\omega_n}{\omega_n} = \frac{1}{L}\int_0^L \frac{\Delta S(x)}{S}\left(\cos^2 kx - \sin^2 kx\right) \mathrm{d}x. \quad (15.118)$$

Hence, if the tube is increased in area close to a displacement antinode, where the particle flow is large (low pressure), the modal frequency will increase, whereas the frequency will decrease, if constricted close to a nodal position (large pressure) (*Benade* [15.133], Sect. 22.3). This result can be generalised to a tube of any shape. Hence, by changing the radius over an extended region close to a node or antinode, the frequencies of a particular mode can be either raised or lowered, but at the expense of similar perturbations to other modes. Considerable art and experience is therefore needed to correct for the inharmonicity of several modes simultaneously.

Electric Circuit Analogues

It is often instructive to consider acoustical systems in terms of equivalent electric circuit analogues, where voltage V and electrical current I can represent the acoustic pressure p and flow along a pipe U. For example, a volume of air with flow velocity U in a pipe of area S and length l has a pressure drop $(\rho l/S)\partial U/\partial t$ across its length, which is equivalent to the voltage $L\partial I/\partial t$ across an inductor in an electrical circuit. Likewise, the rate of pressure rise, $\partial p/\partial t = \gamma P_0 U/V$, as gas flows into a volume V, is equivalent to the rate of voltage rise, $\partial V/\partial t = I/C$ across a capacitance $C \equiv V/\gamma P_0$.

As a simple example, we re-derive the Helmholtz resonance frequency, previously considered in relation to the principal air resonance of the air inside a violin or guitar body (Sect. 15.2.4), but equally important, as we will show later, in describing the resonance of air within the mouthpiece of brass instruments.

In its simplest form, the Helmholtz resonator consists of a closed volume V with an attached cylindrical pipe of length l and area S attached, through which the air vibrates in and out of the volume. All dimensions are assumed small compared to the acoustic wavelength, so that the pressure p in the volume and the flow in the pipe U can be assumed to be spatially uniform. The volume acts as an acoustic capacitance $C = V/\gamma P_0$, which resonates with the acoustic inductance $L = \rho l/S$ of the air in the neck. The resonant frequency is therefore given by

$$\omega_{\text{Helmholtz}} = \frac{1}{\sqrt{LC}} = \sqrt{\frac{S}{\rho l}\frac{\gamma P_0}{V}} = c_0\sqrt{\frac{S}{lV}}, \quad (15.119)$$

as derived earlier.

Any enclosed air volume with holes in its containing walls acts as a Helmholtz resonator, with an effective *kinetic inductance* of the hole region equivalent to a tube of the same diameter with an effective length of wall thickness plus ≈ 0.61 hole radius (*Kinsler* et al. [15.140] Sect. 9.2). This is the familiar end-correction for an open-ended pipe (*Kinsler* et al. [15.140] Sect. 9.2). Open holes of different diameters will therefore give resonances corresponding to different musical tones. The ocarina is a very simple musical instrument based on such resonances, in which typically four or five holes with different areas can be opened and closed in combination, to give a full range of notes on a chosen musical scale (audio l?)CD-ROM). Because the sound is based on the single resonance of a Helmholtz resonator, there are no simply related higher-frequency modes that can be excited. Ocarinas appear in many ancient and ethnic cultures around the world and are often sold as ceramic toys.

Acoustic Transmission Line

There is also a close equivalence between acoustic waves in wind instruments and electrical waves on transmission lines, with an acoustic pipe having an equivalent inductance $L_0 = \rho/S$ and capacitance $C_0 = S/\gamma P_0$ per unit length. For a transmission line the wave velocity is therefore $c_0 = \sqrt{1/L_0 C_0} = \sqrt{\gamma P_0/\rho}$ and characteristic impedance $Z_0 = \sqrt{L_0/C_0} = \rho c_0/S$, as expected. The input impedance of a transmission line as a function of its characteristic impedance and terminating load is given by (15.100).

Valves and Bends

To enable brass instruments to play all the notes of the chromatic scale, short lengths of coiled-up tubing are connected in series with the main bore by a series of piston- or lever-operated air valves. The constriction of air flow through the air channels within the valve structures and the bends in the tubing, used to reduce the size of the instruments to a convenient size for the player to support, will clearly present discontinuities in the acoustic impedance of the air bore and will lead to reflections. Such reflections will influence the feel of the instrument for the player exciting the instrument via the mouthpiece and will also perturb the frequencies of the resonant modes of the instrument.

If the discontinuities are short in size relative to the acoustic wavelengths involved, the discontinuity can be considered as a discrete (localised) lumped circuit element. Using our electromechanical equivalent, a short, constricted channel through a valve can be represented as an inductance $\rho L_{\text{valve}}/S_{\text{valve}}$ in series with the acoustic transmission line, or an equivalent additional extra

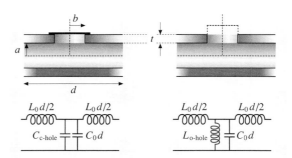

Fig. 15.80 Equivalent circuits for a short length d of cylindrical pipe containing a closed and an open tone hole, shunting the acoustic transmission line with a capacitance and inductance, respectively

length of bore tubing $L_{\text{valve}} S_{\text{tube}}/S_{\text{valve}}$ of cross section S_{tube}. For all frequencies such that $kL_{\text{valve}} \ll 1$, the valve simply increases the length of the acoustic air column slightly and the frequencies of all the lower modes by the same fractional amount. Only at very high frequencies, outside the normal playing range, will the constricted air channel significantly change the modal frequencies.

When a straight length of cylindrical tube of radius a is connected to the same size tubing but bent into a circle of radius R, there will a small change in the acoustic impedance and velocity of sound waves, which arises because the forces acting on each element induces rotational in additional to linear motion. The presence of bends will lead to reflections and slight perturbations of resonant frequencies, though these effects will again be relatively small. *Nederveen* [15.142] showed that fractional increase in phase velocity and decrease in wave impedance of a rectangular duct is given by the factor $F^{1/2}$, where

$$F = \frac{B^2}{2} \bigg/ \left[1 - \left(1 - B^2\right)^{1/2}\right]$$
$$= 1 - B^2/4 \quad \text{for} \quad B \ll 1, \quad (15.120)$$

$B = a/R$, a is the half-width of the duct and R its radius. *Keefe* and *Benade* [15.143] subsequently generalised this result to a bent circular tube, with its radius r replacing a.

Finger Holes

In many woodwind instruments, tone holes can be opened or closed to change the effective resonating length of an air column and hence pitch of the sounded note. The holes can be closed by the pad of the finger or a hinged felt-covered pad operated with levers.

To a first approximation, opening a side hole creates a pressure node at that position, shortening the effective length of the instrument and raising the modal frequencies. However, as described in *Fletcher* and *Rossing* ([15.5], Sect. 15.2) and in detail by *Benade* ([15.133], Chaps. 21 and 22), the influence of the side holes is in practice strongly dependent on the hole size, position and frequency, as summarised below.

At low frequencies, when the acoustic wavelength is considerably longer than the size and spacing of the tone holes, one can account for the effect of the tone holes by considering their equivalent capacitance when closed and their inductance when open, as illustrated schematically in Fig. 15.80.

Because the walls of wind instruments and particularly woodwind instruments have a significant thickness, the tone holes when shut introduce additional small volumes distributed along the length of the vibrating air column. Each closed hole will introduce an additional volume and equivalent capacitance $C_{\text{c-hole}} = \pi b^2/\gamma P_0$, which will perturb the frequencies of the individual partials upwards or downwards by a small amount that will depend on its position relative to the pressure and displacement nodes and the closed volume of the hole. In severe cases, the perturbations can be as large as a few per cent (one semitone is 6%), which requires compensating changes in bore diameter along the length of the instrument, to retain the harmonicity of the partials. However, this is essentially a problem that depends on geometrical factors involving the air column alone. Once solved, like all acoustic problems involving the shape and detailed design, instruments can be mass-produced with almost identical acoustic properties, quite unlike the problems that arise for stringed instruments.

The more interesting situation is when the holes are opened, introducing a pressure node at the exit of the tone hole and shortening the effective acoustical length of the instrument. An open hole can be considered as an inductance, $L \sim \rho(t+0.6b)/\pi b^2$, where the effective length of the hole is increased by the unflanged hole end-correction. Neglecting radiation losses from the hole [*Fletcher* and *Rossing* [15.5], (15.21, 22)], the effective impedance Z^* of an open-ended cylindrical pipe of length l and radius a shunted by the inductive impedance of a circular hole of radius b set into the wall of thickness t is given by

$$\frac{1}{Z^*} \sim \frac{\pi b^2}{i\omega\rho(t+0.6b)} + \frac{\pi a^2}{i\rho c_0 \tan kl}$$
$$= \frac{\pi a^2}{i\rho c_0 \tan kl'} \ . \tag{15.121}$$

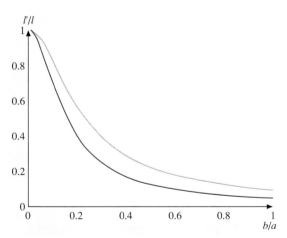

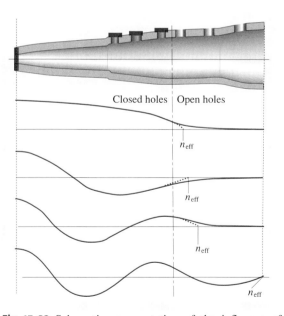

Fig. 15.81 Low-frequency ($kl \ll 1$) fractional reduction of effective length of a cylindrical end-pipe as a function of hole to cylinder radius, for additional lengths of 10 (lower curve) and 20 (upper curve) times the tube radius in length. The side wall thickness is 0.4 times the tube radius

Fig. 15.82 Schematic representation of the influence of open holes on the first four partials of a woodwind instrument, with the effective length indicated by the intercept n_{eff} on the axis of the extrapolated incident wave (after *Benade* [15.133])

Thus can be expressed in terms of an impedance of an effectively reduced length l'.

For $kl \ll 1$, $\quad \dfrac{l'}{l} = \left[1 + \dfrac{t + 0.6b}{l}\left(\dfrac{a}{b}\right)^2\right]^{-1}.$

(15.122)

The change in effective length introduced by the open hole depends strongly on its area relative to that of the cylinder, the thickness of the wall and its length from the end. This gives the instrument designer a large amount of flexibility in the positioning of individual holes on an instrument. Figure 15.81 illustrates the dependence of the effective pipe length on the ratio of hole to cylinder radii for two lengths of pipe between the hole and end of the instrument.

Not surprisingly, a very small hole with $b/a \ll 1$ has a relatively small effect on the effective length of an instrument. In contrast, a hole with the same diameter as that of the cylinder shortens the effective added length to about one hole diameter.

In practice, there will often be several holes open beyond the first open tone hole, all of which can affect the pitch of the higher partials.

Consider a regular array of open tone holes spaced a distance d apart. The shunting kinetic inductance of each open hole is in parallel with the capacitance associated with the volume of pipe between the holes. At low frequencies, such that $\omega \ll 1/\sqrt{L_{\text{hole}}C_0 d}$, the impedance is dominated by the hole inductance, so that each hole attenuates any incident wave by approximately the ratio

$\sim L_{\text{hole}}/(L_{\text{hole}} + L_0 d) = \left[1 + \dfrac{d}{t + 1.5b}\left(\dfrac{a}{b}\right)^2\right]^{-1},$

(15.123)

where L_0 is the inductance of the pipe per unit length. Incident waves are therefore attenuated with an effective node just beyond the actual hole as discussed above.

However, for frequencies such that $\omega \gg 1/\sqrt{L_{\text{hole}}C_0 d}$, the impedance of the shunting hole inductance is much larger than that of the capacitance of the air column, so that the propagating properties of the incident waves is little affected be the presence of the open hole. There is therefore a *crossover* or *cut-off* frequency

$\omega \approx 1/\sqrt{L_{\text{hole}}C_0 d} = c_0 \dfrac{a}{b}\left(\dfrac{1}{t_{\text{eff}}d}\right)^{1/2},$ (15.124)

below which the incident waves are reflected to give a pressure node just beyond the first hole of the array and above which waves propagate increasingly freely through the array to the open end of the instrument.

Figure 15.82 (*Benade* [15.133], Fig. 21.1) illustrates the effect of an array of open holes on the first few partials of a typical woodwind instrument, highlighting the increase in acoustic length of the instrument (indicated by the intercept of the extrapolated incident waveform)

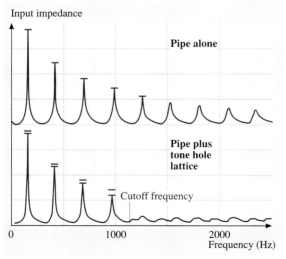

Fig. 15.83 Illustration of the cut-off-frequency effect, when adding an addition length of tubing with an array of open tone holes (after *Benade* [15.133])

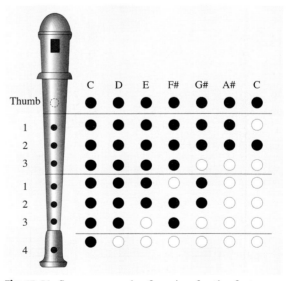

Fig. 15.84 Soprano recorder fingering for the first seven notes of a whole-tone scale (after *Fletcher* and *Rossing* [15.5])

with increasing frequency. The dependence of the effective length of the acoustic air column on frequency is therefore rather similar to the influence of the flare on the partials of a brass instrument.

A consequence of the greater penetration at high frequencies of the acoustic wave through the array of open tone holes is the greater attenuation of such waves by radiation and the consequent reduction in the amplitude of the higher resonant modes in measurements of the input impedance. This is illustrated in Fig. 15.83 for a length of clarinet tubing first without and then with an added section containing an array of equally spaced tone holes (*Benade* [15.133], Fig. 21.3).

Benade ([15.133], Sect. 21.1) states that "specifying the cut-off frequency for a woodwind instrument is tantamount to describing almost the whole of its musical personality" – assuming the proper tuning and correct alignment of resonances for good oscillation. His measured values of the cut-off frequency for the upper partials of classical and baroque instruments are 1200–2400 Hz for oboes, 400–500 Hz for bassoons, and 1500–1800 Hz for clarinets.

Cross-Fingering

The notes of an ascending scale can be played by successively opening tone holes starting from the far end of the instrument. In addition, by overblowing, the player can excite notes in the *second register* based on the second mode. As remarked earlier, instruments like the flute and oboe overblow at the octave, whereas the clarinet overblows at the twelfth (an octave plus a perfect fifth). To sound all the semitones of the western classical scale on the flute or oboe would therefore require 12 tone holes and the clarinet 20 – rather more than the fingers on the two hands! In practice, the player generally uses only three fingers on the left hand and four on the right to open and close the finger holes. The thumb on the left hand is frequently used to open a small *register* hole near the mouthpiece, which aids the excitation of the overblown notes in the higher register.

In practice, *cross-* or *fork-fingering* enables all the notes of the chromatic scale to be played using the seven available fingers and combinations of open and closed tone holes. This is illustrated in Fig. 15.84 for the baroque recorder (*Fletcher* and *Rossing* [15.5], Fig. 16.21). The bottom two notes can be sharpened by a semitone by half-covering the lower two holes and the overblown notes an octave above are played with the thumb hole either fully open or half closed. Cross-fingering makes use of the fact that the standing waves

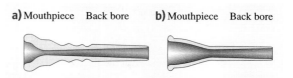

Fig. 15.85a,b Cross sections of (**a**) trumpet mouthpiece and (**b**) horn mouthpiece (after *Backus* [15.132])

set up in a pipe extend an appreciable distance into an array of open tone holes (Fig. 15.82), so that opening and closing holes beyond the first open hole can have an appreciable influence on the effective length of the resonating air column.

Modern woodwind instruments use a series of interconnected levers operated by individual keys, which facilitates the ease with which the various hole-opening combinations can be made.

Radiated Sound

Although the reactive loading of an open hole determines the effective length of the resonant air column, particularly at low frequencies, it does not follow that all the sound is radiated from the open tone holes. Indeed, since the intensity of the radiated sound depends on $(ka)^2$, very little sound will be radiated by a small hole relative to the much wider opening at the end of an instrument. The loss in intensity of sound passing an open side hole may therefore, in large part, be compensated by the much larger radiating area at end of the instrument. This also explains why the characteristic hollow sound quality of a cor anglais, derived in part from the egg-shaped resonating cavity near its end, is retained, even when the tone holes are opened on the mouthpiece side of the cavity.

In practice, the sound from the end and open tone holes of a woodwind instrument act as independent monopole sources. When the acoustic wavelength becomes comparable with the hole spacing, interesting interference effects in the output sound can occur contributing to strongly directional radiation patterns, as discussed by *Benade* ([15.133], Sect. 21.4). Similarly, reciprocity allows one to make use of such interference effects to produce a highly directional microphone by placing a microphone at the end of a cylindrical tube with an array of open side holes.

Brass Mouthpiece

Brass instruments are played using a mouthpiece insert, against which the lips are pressed and forced to vibrate by the passage of air between them. The mouthpiece not

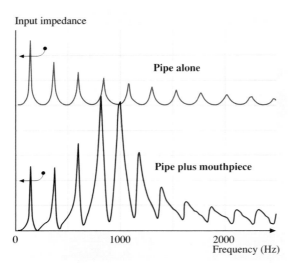

Fig. 15.86 Input impedance of a length of cylindrical trumpet pipe with and without a mouthpiece attached (after *Benade* [15.133])

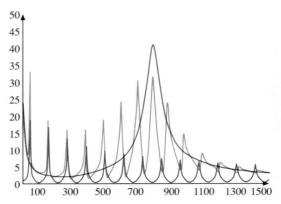

Fig. 15.88 The calculated impedance of an 800 Hz Helmholtz mouthpiece (*dark brown*), an attached pipe (*black*) and the combination of mouthpiece and pipe (*light brown*)

only enables the player to vibrate their lips over a wide range of frequency, but also provides a very important acoustic function in significantly boosting the amplitude

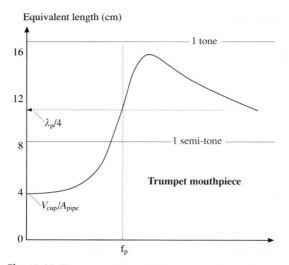

Fig. 15.87 The amount by which a trumpet tube of length 137 cm would have to be lengthened to compensate for the lowering in frequency of the instrument's resonant frequencies when a mouthpiece is attached to the input (after *Benade* [15.133]). The changes in length to give a semitone and a whole-tone change in frequency are indicated by the *horizontal lines*

of the higher partials, helping to give brass instruments their bright and powerful sounds.

Typical mouthpiece shapes are shown in Fig. 15.85. Mouthpieces can be characterized by the mouthpiece volume and the *popping* frequency characterizing the Helmholtz resonator comprising the mouthpiece volume and backbore. The popping frequency can easily be estimated from the sound produced when the mouthpiece is slapped against the open palm of the hand (audio).

By adjusting the tension in the lips, the shape of the lips within the mouthpiece (the embouchure), and the flow of air between the lips via the pressure in the mouth, the skilled brass player forces the lip to vibrate at the required frequency of the note to be played. This can easily be demonstrated by making a pitched buzzing sound with the lips compressed against the rim of the mouthpiece cup. The circular rim constrains the lateral motion of the lips making it far easier to produce stable high notes. A brass player can sound all the notes on an instrument by simply blowing into the mouthpiece alone, but the mouthpiece alone produces relatively little volume. The instrument both stabilises the playing frequencies and increases the coupling between the vibrating lips and radiated sound.

Figure 15.86 illustrates the enhancement in the input impedance around the popping frequency, when a mouthpiece is attached to the input of a cylindrical pipe, as measured by *Benade* [15.133]. Benade showed that the influence of the mouthpiece on the acoustical characteristics of a brass instrument is, to a first approximation, independent of the internal bore shape and can be characterized by just two parameters, the internal volume of the mouthpiece and the popping frequency.

Benade also measured the perturbation of the resonant frequencies of an instrument by the addition of a mouthpiece, as illustrated in Fig. 15.87. At low frequencies, the mouthpiece simply extends the effective input end of a terminated tube by an equivalent length of tubing having the same internal volume as the mouthpiece. In the measurements shown, Benade removed lengths of the attached tube to keep the resonant frequencies unchanged on adding the mouthpiece. However, since the fractional changes in frequency are small, the measurements are almost identical to the effective increase in length from the addition of the mouthpiece.

At the mouthpiece *popping* frequency (typically in the range 500 Hz to 1 kHz depending on the mouthpiece and instrument considered), the effective increase in length is $\lambda/4$. This can result in decreases in resonant frequencies by as much as a tone, which could have a significant influence on the harmonicity, and hence the playability, of an instrument. The effective length continues to increase above the popping frequency before decreasing at higher frequencies. In many brass instruments, such as the trumpet, there is also a longer transitional conical section (*the lead pipe*) between the narrow bore of the mouthpiece and the larger-diameter main tubing. This reduces the influence of the mouthpiece on the tuning of individual resonances and the overall formant structure of resonances.

It is straightforward to write down the input impedance inside the cup of a mouthpiece attached to an instrument using an equivalent electrical circuit. The volume within the cup is represented by a capacitance C in parallel with the inductance L and resistance R of air flowing through the backbore, which is in series with the input impedance of the instrument itself, so that

$$Z_{in} = \frac{1}{i\omega C} \frac{i\omega L + R + Z_{horn}}{(1/i\omega C) + i\omega L + R + Z_{horn}}. \quad (15.125)$$

Figure 15.88 shows the calculated input impedance of an 800 Hz Helmholtz mouthpiece resonator, of volume 5 cm³ with a narrow-backbore neck section resulting in a Q-value of 10, before and after attachment to a cylindrical pipe of length 1.5 m and radius 1 cm, radiating into free space at its open end. The input impedance of the pipe alone is also shown. Note the marked increase in heights and strong frequency shifts of the partials in the neighbourhood of the mouthpiece resonance. As anticipated from our previous treatment of coupled resonators in the section on stringed instruments, the addition of the mouthpiece introduces

an additional normal mode resonance in the vicinity of the Helmholtz resonance. In addition, it lowers the frequency of all the resonant modes below the popping frequency and increases the frequency of all the modes above.

Above the mouthpiece resonance, the input impedance is dominated by the inertial input impedance of the mouthpiece. The resonances of the air column are superimposed on this response and exhibit the familiar dispersive features already noted for narrow violin string resonances superimposed on the much broader body resonances. The calculated behaviour is very similar to the measured input admittance of typical brass instrument (Fig. 15.78) as extended to instruments with realistic bore shapes by *Caussé*, *Kergomard* and *Lurton* [15.144].

15.3.3 Reed Excitation

In the next sections, we consider the excitation of sound by: (a) the single and double reeds used for many woodwind instruments and selected organ pipes, (b) the vibrating lips in the mouthpiece of brass instrument, and (c) air jets used for the flute, certain organ stops and many ethnic instruments such as pan pipes.

Reed Types
Figure 15.89 shows a number of reed types used in woodwind and brass instruments (*Fletcher* and *Rossing* [15.5], Figs. 13.1, 7).

Helmholtz [15.127] classified two main types of reed: *inward-striking reeds*, which are forced shut by an overpressure within the mouth, and *outward-striking reeds*, forced open by an overpressure. Modern authors often prefer to call such reeds inward-closing and

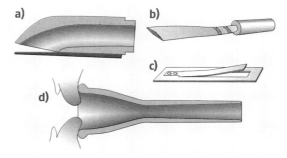

Fig. 15.89a–d Examples of wind and brass instrument reeds: (**a**) a single reed (clarinet), (**b**) a double reed (oboe), (**c**) a cantilever reed (harmonium) and (**d**) the mouthpiece lip-reed (horn) (after *Fletcher* and *Rossing* [15.5])

outward-opening or swinging-door reeds. In addition there are reeds that are pulled shut by the decreased Bernoulli pressure created by the flow of air between them. Such reeds are often referred to as sideways-striking or sliding-door reeds.

A more formal classification (*Fletcher* and *Rossing* [15.5], Sect. 13.3) characterises such reeds by a doublet symbol (σ_1, σ_2), where the values of $\sigma_{1,2} = \pm 1$ describe the action of over- and under-pressures at the input and output ends of the reed. When the valve is forced open by an overpressure at either end, $\sigma_{1,2} = +1$; if forced open by an under-pressure, $\sigma_{1,2} = -1$. The force tending to open the valve can then be written as $(\sigma_1 p_1 S_1 + \sigma_2 p_2 S_2)$, where $S_{1,2}$ and $p_{1,2}$ are the areas and pressures at the reed input and output. The operation of reeds can therefore be classified as $(+, -)$, $(-, +), (-, -)$ or $(+, +)$. Single and double woodwind reeds are inward-striking $(-, +)$ valves, while the vibrating lips in a mouthpiece and the vocal cords involve both outward-swinging $(+, -)$ and sideways-striking $(+, +)$ actions.

Figure 15.90 summarises the steady-state and dynamic flow characteristics of the above reeds for typical operating pressures across the valve, $\Delta p = p_m - p_{ins}$,

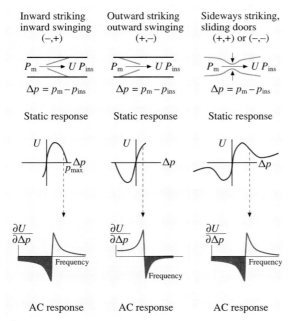

Fig. 15.90 Main classifications of vibrating reeds summarising reed operation, nomenclature and the associated static and *ac* conductance, with the negative resistance frequency regimes indicated by *solid shading*

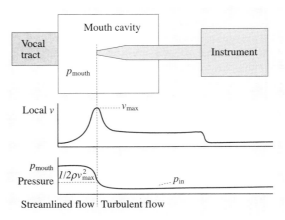

Fig. 15.91 Schematic representation of vocal tract, mouth cavity, reed and instrument, illustrating the variation of local velocity and pressure for air flowing into and along the reed and attached instrument

where p_m and p_{ins} are the input and output pressures in the mouth and instrument input, respectively. For the inward-swinging $(-, +)$ reed, the flow rate initially increases for a small pressure difference across the valve, but then decreases as the difference in pressures tends to close the valve, leading to complete closure above a certain pressure difference p_{max}. Before closure, there is an extended range of pressures where the flow rate decreases for increasing pressure difference across the reed. This is equivalent to an input with a negative resistance to flow. This results in positive feedback exciting resonances of any attached air column, provided the feedback is sufficient to overcome viscous, thermal and radiation losses.

It is less obvious why the outward-swinging $(-, +)$ reed can give positive feedback, because the steady-state flow velocity always increases with increasing pressure across the valve. However, this is only true at low frequencies below the mechanical resonance of the reed. Above its resonant frequency, the reed will move in anti-phase with any sinusoidally varying fluctuations in pressure. This will result in a regime of negative resistance and the resonant excitation of any attached air column, as discussed by *Fletcher* et al. [15.145].

Sideways-striking $(+, +)$ or $(-, -)$ reeds behave rather like inward-striking reeds, with an extended region of negative conductance. However, such reeds will never cut off the flow completely, so that for large pressure differences the dynamic conductance again becomes positive, as indicated in Fig. 15.90.

Bernoulli Pressures

Figure 15.91 schematically illustrates the variation of flow velocity and pressure as air flows from the mouth into the reed and attached air column. To solve the detailed dynamic response from first principles for a specific reed geometry would require massive computer modelling facilities. Fortunately, the physics involved is reasonably well understood, so that relatively simple models can be used to reproduce reed characteristics rather well, as illustrated for the clarinet reed in the next section.

The operation of all reed generators is controlled by the spatial variations in Bernoulli pressure exerted by the air flowing across the reed surfaces. Such variations in P arise because, within any region of streamlined flow with velocity v, $P + \frac{1}{2}\rho v^2$ remains constant. Hence the pressure will be lowered on any surface over which the air is flowing. The flow of air is determined by the specific reed assembly geometry and the nonlinear Navier–Stokes equation, which also includes the effects of viscous damping.

After passing through the narrow reed constriction, the air emerges as a jet, which breaks down into turbulent motion on the downstream side of the reed. The turbulence leads to a rapid lateral mixing of the air, so that the flow is no longer streamlined. As a result, the pressure on the downstream end of the reed opening remains low and fails to recover to the initial pressure inside the mouth. The double reeds used for playing the oboe, bassoon and bagpipe chanter are mounted on a relatively long, narrow tube connected to the wider bore of the instrument. Turbulent flow in this region could contribute significantly to the flow characteristics, though recent measurements by *Almeida* et al. [15.146] have shown that such effects are less important than initially envisaged, as discussed later.

Single Reed

We first consider the clarinet reed, which is one of the simplest and most extensively studied of all woodwind reeds (*Benade* [15.133], Sect. 21.2) *Fletcher* and *Rossing* ([15.5], Chap. 13 and recent investigations by *Dalmont* and collaborators [15.147, 148]). Figure 15.92 shows a cross section of a clarinet mouthpiece, defining the physical parameters of a highly simplified but surprisingly realistic model.

The lungs are assumed to supply a steady flow of air U, which maintains a steady pressure P_{mouth} within the mouth. Air flows towards the narrow entrance or lip of the reed through which it passes with velocity

Fig. 15.92 Cross section of air flow through a clarinet mouthpiece and reed assembly, illustrating the streamlined flow into the gap with jet formation and turbulence on exiting the reed entrance

v. Because the air flowing into the reed is streamlined, the pressure drops by $\frac{1}{2}\rho v^2$ on entering the reed, while the much slower-moving air on the outer surfaces of the reed leaves the pressure on the outer reed surfaces largely unchanged. The air is then assumed to stream through the narrow gap of the reed to form an outward-going jet, which breaks up into vortices and turbulent flow on the far side of the input constriction, with no further change in overall pressure p in the relatively wide channel on the downstream side of the reed entrance.

The resulting pressure difference $\frac{1}{2}\rho v^2$ across the reed forces the reed back towards its closing position on the curved *lay* of the mouthpiece, indicated by the dashed line in Fig. 15.92. The pressure difference Δp is assumed to reduce the area of reed opening from S_0 to $S_0(1 - \Delta p/p_{\mathrm{max}})$, where p_{max} is the pressure difference required to close the reed. The net flow of air through

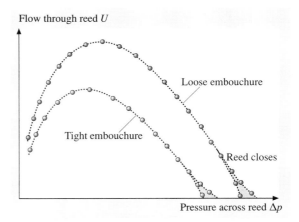

Fig. 15.93 Quasistatic flow through a clarinet single reed as a function of pressure across it illustrating the influence of the player's embouchure on the shape (after *Benade* [15.133])

the reed is therefore given by

$$U(\Delta p) = \alpha(\Delta p)^{1/2}(1 - \Delta p/p_{\max}). \quad (15.126)$$

The player can control these characteristic by varying the position and pressure of the lips on the reed, which is referred to as the *embouchure*. A lower pressure is required to close the reed, if the reed is already partially closed by pressing the lips against the reed to constrict the entrance gap.

The flow rate U as a function of static pressure across a clarinet reed is illustrated by the measurements of *Backus* and *Nederween* [15.149] redrawn in Fig. 15.93. Apart from a small region near closure, where the exact details of the closing geometry and viscous losses may also be important, the shape of these curves and later measurements by *Dalmont* et al. [15.148, 150], which exhibit a small amount of hysteresis from viscoelastic effects on increasing and decreasing pressure, are in excellent agreement with the above model. The measurements also illustrate how the player is able to control the flow characteristics by changing the pressure of the lips on the reed.

The *reed equation* can be written in the universal form

$$\frac{U\left(\frac{\Delta p}{p_{\max}}\right)}{U_{\max}} = \frac{3^{3/2}}{2}\left(\frac{\Delta p}{p_{\max}}\right)^{1/2}\left(1 - \frac{\Delta p}{p_{\max}}\right), \quad (15.127)$$

with just two adjustable parameters: $U_{\max}$ the maximum flow rate and $p_{\max}$, the static pressure required to force the reed completely shut. The maximum flow occurs when $\Delta p/p_{\max} = 1/3$.

Double Reeds

Instruments like the oboe, bassoon and bagpipe chanters use double reeds, which close against each other with a relatively long and narrow constricted air channel on the downstream side before entering the instrument. The turbulent air motion in the constricted air passage would result in an additional turbulent resistance proportional to the flow velocity squared, which would add to the pressure difference across the reed. This could result in strongly hysteretic re-entrant static velocity flow characteristics as a function of the total pressure across the reed and lead pipe (see, for example, *Wijnands* and *Hirschberg* [15.151]).

A recent comparison of the flow-pressure characteristics of oboe and bassoon double reeds and a clarinet single reed, Fig. 15.94, by *Almeida* [15.146]) shows no evidence for re-entrant double-reed features. Nevertheless, the measurements are strongly hysteretic, because of changes in the properties of the reeds (elasticity and mass), as they absorb and desorb moisture from the damp air passing through them. In the measurements the static pressure was slowly increased from zero to its maximum value and then back again. Under normal playing conditions, one might expect to play on a non-hysteretic operating characteristic somewhere between the two extremes of the hysteretic static measurements. Thus, although the shape of the flow-pressure curves for the double reeds differs significantly from those of the clarinet single reed, the general form is qualitatively similar, with a region of dynamic negative resistance above the peak flow. The strongly moisture dependent properties of reeds are very familiar to the player, who has to moisten and "play-in" a reed before it is ready for performance.

There therefore appears to be no fundamental difference between the way single and double reeds operate. Indeed, the sound of an oboe is apparently scarcely changed, when played with a miniature clarinet reed mouthpiece instead of a conventional double reed (Campbell and Gilbert, private communication).

Dynamic Characteristics

Fletcher [15.152] extended the quasistatic model by assuming the reed could be described as a simple mass–

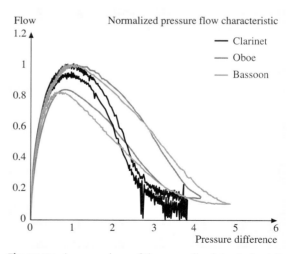

Fig. 15.94 A comparison of the normalised, hysteric static pressure/flow characteristics of single (clarinet) and double (oboe and bassoon) reeds measured on first increasing and then decreasing the flow rate of moist air through the reeds (after [15.146])

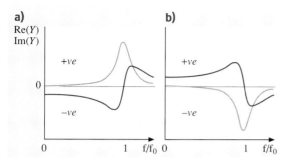

Fig. 15.95a,b Real (*dark brown*) and imaginary (*light brown*) components of the reed admittance $Y(\omega)$ for (**a**) an inward-closing reed in the negative dynamic conductance regime and for (**b**) an outward-opening reed, as a function of frequency normalised to the resonant frequency of the reed for reeds having Q-values of 5

spring resonator resulting in a dynamic conductance of

$$Y(\omega) = Y(0) \frac{1}{1 - (\omega/\omega_0)^2 + i\omega/\omega_0 Q} \quad (15.128)$$

where $Y(0) = \partial U/\partial (\Delta p)|_{\omega=0}$ is the quasistatic flow conductance and the denominator describes the dynamic resonant response of the reed. The Q-value is determined by viscous and mechanical losses in the reed.

The resistive and reactive components of $Y(\omega)_{\text{reed}}$ given by (15.128) are plotted in Fig. 15.95a,b for an inward-closing reed $(-, +)$, in the negative flow conductance regime above the velocity flow maximum, and for the outward-closing $(+, -)$ reed. As discussed qualitatively above, the negative input dynamic conductance of the inward-striking reed remains negative at all frequencies below its resonant frequency, whereas the conductance of the outward-opening reed only becomes negative above its resonant frequency.

For the oscillations of any attached air column to grow, feedback theory requires that

$$\begin{aligned} \text{Im}(Y_r + Y_p) &= 0 \\ \text{Re}(Y_r + Y_p) &< 0, \end{aligned} \quad (15.129)$$

where Y_p and Y_r are the admittances of the pipe and reed, respectively. The negative dynamic conductance of the reed must therefore be sufficiently small to overcome the losses in the instrument. Furthermore, the reactive components of the reed conductance will perturb the frequencies of the attached air column.

Fletcher and *Rossing* ([15.5], Chap. 13) give an extended discussion of the dynamics of reed generators including polar plots of admittance curves for typical outward and inward-striking reed generators as a function of blowing pressure.

For the inward-striking reeds of the clarinet, oboe and bassoon, the real part of the reed admittance is negative below the resonant frequency of the reed. For the oboe this is typically around 3 kHz, above the pitch of the reed attached to its staple (joining section) alone (CD-ROM). However, when attached to an instrument, the negative conductance will excite the lower-frequency natural resonances of the attached tube (CD-ROM). In this regime, the reactive load presented by the reed is relatively small and positive and equivalent to a capacitive or spring loading at the input end of the attached pipe. This results in a slight increase in the effective length of the pipe and a slight lowering of the frequencies of the resonating air column.

Free reeds, like the vibrating brass cantilevers used in the mouth organ, harmonium and certain organ reed pipes (Fig. 15.89c), are rather weakly damped inward-closing $(-, +)$ reeds (*Fletcher* and *Rossing* [15.5], Sect. 13.4). The reed is initially open. High pressure on one side or suction on the other (as in the harmonium or American organ) forces the reed back into the aperture, controlling the air flow. Like the clarinet reed, above a certain applied pressure the reed will close and restrict the flow resulting in a negative conductance regime. If the reed is forced right through the aperture, it becomes an outward-opening $(+, -)$ reed with a positive conductance. Because of its low damping, the blown-closed reed tends to vibrate at a frequency rather close to its resonant frequency. In practice, as soon as a threshold pressure is reached that is significantly below the maximum in the static characteristics, a harmonium reed starts to vibrate with a rather large sinusoidal amplitude (typically ≈ 4 mm) resulting in highly non-sinusoidal flow of air through the aperture (*Koopman* et al. [15.153]). For such high-Q-value mechanical resonators, the vibrational frequency is strongly controlled by the resonant frequency of the reed itself, so that a separate reed is needed for each note, as in the piano accordion, harmonium, mouth organ and reed organ pipes. The reeds in a mouth organ are arranged in pairs in line with the flow of air. They are individually excited by overpressure and suction.

In contrast, the dynamic conductance of an outward-opening reed $(+, -)$ is only negative above its resonant frequency. The conductance then decreases rather rapidly with increasing frequency, so that there may only be a relatively narrow range of frequencies above resonance over which oscillations can occur. The vibrating lips provide a possible example of such a reed, with an increase in steady-state pressure always increasing the static flow through them. Above their resonant frequency, the dynamic conductance becomes negative and could excite oscillations in an attached pipe. In such a regime, the reactive component of the reed admittance is negative. This corresponds to an inductive or inertial load, which will shorten the effective length of the air column and increase its resonant frequencies. The influence of the reed on the resonant frequencies of an attached instrument therefore provides a valuable clue to the way in which a valve is operating, as we will discuss later in relation to the vibrations of the lips in a brass-instrument mouthpiece.

Small-Amplitude Oscillations

We now consider the stability of the oscillations excited by the negative dynamic conductance of the reed. In particular, it is interesting to consider whether the oscillations, once initiated, are stabilised or grow quickly in amplitude into a highly nonlinear regime. Surprisingly, this depends on the bore shape of the attached air column, as discussed by *Dalmont* et al. [15.154]. Several authors have investigated such problems, including *Backus* [15.155] using simple theoretical models and measurements, *Fletcher* [15.156] using analytic models, *Schumacher* [15.157, 158] in the time rather than the frequency domain, and *Gilbert* et al. [15.154, 159] using a harmonic balance approach, which we will briefly outline in the following section. Recent overviews of the nonlinear dynamics of both wind and brass instruments have been published by *Campbell* [15.160] and by *Dalmont* et al. [15.154].

We first consider the excitation of small-amplitude oscillations based on the reed equation, which is replotted in Fig. 15.96 as a universal curve together with the negative dynamic admittance or conductance, $-\partial U/\partial p$ above the flow-rate maximum.

The onset of self-oscillations occurs when the sum of the real and the imaginary parts of the admittance of the reed and attached instrument are both zero (15.129). If losses in the reed and the attached instrument were negligible, resonances of the air column would be excited as soon as the mouthpiece pressure exceeded $\frac{1}{3}p_{\max}$. However, when losses are included, the negative conductance of the reed has to be sufficiently large to overcome the losses in the instrument. The onset then occurs at a higher pressure, as illustrated schematically in Fig. 15.96.

The onset of oscillations depends not only on the mouthpiece pressure but also on the properties of the reed, such as the initial air gap and its flexibility, which will depend on its thickness and elastic properties. The elastic properties also change on the take up of moisture during playing. It is not surprising that wind players take great care in selecting their reeds. Furthermore, notes are generally tongued. This involves pressing the tongue against the lip of the reed to stop air passing through it, so that the pressure builds up to a level well above that required to just excite the note. When the tongue is removed, the note sounds almost immediately giving a much more precise definition to the start of a note.

The transition to the oscillatory state can be considered using the method of small variations and harmonic balance (*Gilbert* et al. [15.159]). For a given mouth pressure defining the overall flow rate, oscillations of the flow rate u can be written as a Taylor expansion of the accompanying small pressure fluctuations p at the output of the reed, such that

$$u = Ap + Bp^2 + Cp^3 + \dots. \tag{15.130}$$

From the reed equation plotted in Fig. 15.96, the coefficient A is positive for mouthpiece pressures above $p_{\max}/3$, while B and C are positive and negative respectively. We look for a periodic solution with Fourier components that are integer multiples of the fundamental

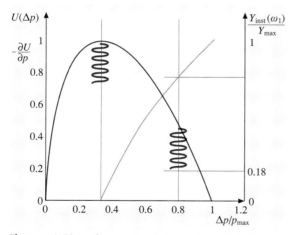

Fig. 15.96 Plot of normalised flow rate and differential negative conductance of an inward-striking reed valve as a function of pressure across the reed normalised to the pressure required for closure. The intersection of the negative conductance curve with the real part of the input admittance $Y_{\text{inst}}(\omega)$ of the instrument determines the pressure in the mouthpiece for the onset of oscillations, illustrated schematically for $Y_{\text{inst}}(\omega) \approx 0.78 Y_{\max}$

frequency ω, so that

$$p(t) = \sum p_n \mathrm{e}^{\mathrm{i}n\omega t}, \quad (15.131)$$

with a corresponding oscillatory flow $u(t)$ superimposed on the static flow U,

$$u(t) = \sum u_n \mathrm{e}^{\mathrm{i}n\omega t}. \quad (15.132)$$

At the input to the instrument, the oscillatory flow can be expressed in terms of the Fourier components of the input pressure and admittance, so that

$$u(t) = \sum p_n Y(n\omega) \mathrm{e}^{\mathrm{i}n\omega t}. \quad (15.133)$$

Using the method of harmonic balance, we equate the coefficients of the Fourier components in (15.130) with those in having substituted for the pressure. The first three Fourier components are then given by

$$p_1 = \pm \left(\frac{Y_1 - A}{2B^2/(Y_2 - A) + 3C} \right),$$

$$p_2 = \left(\frac{B}{Y_2 - A} \right) p_1^2,$$

$$p_3 = \left(\frac{C}{Y_3 - A} \right)^{1/2} p_1^3. \quad (15.134)$$

For a cylindrical-bore instrument like the clarinet at low frequencies, only the odd-n partials will be strongly excited, so that Y_2 is very large. The amplitude of the fundamental component is then given by $\pm[(Y_1 - A)/C]^{1/2}$, with a vanishingly small second harmonic p_2. The amplitude p_1 of small oscillations is then stabilised by the cubic coefficient C. Stable, small-amplitude oscillations can therefore be excited as soon as the negative conductance of the reed exceeds the combined admittance of the instrument and any additional losses in the reed itself.

Because the transition is continuous, p_1 rises smoothly from zero, taking either positive or negative values (simply solutions with opposite phases). The transition is therefore referred to as a *direct bifurcation*. The player can vary the pressure in the mouthpiece and the pressure of the lips on the reed to vary the coefficients A and C and hence can control the amplitude of the excited sound continuously from a very quiet to a loud sound, as often exploited by the skilled clarinet player.

In contrast, for a conical-bore instrument, the amplitude of the fundamental,

$$p_1 = \pm \left(\frac{(Y_1 - A)(Y_2 - A)}{2B^2 + 3C(Y_2 - A)} \right)^{1/2}, \quad (15.135)$$

involves the admittance of both the fundamental and second partial. On smoothly increasing A by increasing the pressure on the reed, *Grand* et al. [15.161] showed that there can again be a direct smooth bifurcation to small-amplitude oscillations, if $2B^2 > -3C(Y_2 - Y_1)$. However, if this condition is not met, there will be an indirect transition, with a sudden jump to a finite-amplitude oscillating state. This gives rise to the hysteresis in the amplitude as the mouth pressure is first increased and then decreased. This means that the player may have to exert a larger pressure to sound the note initially, but can then relax the pressure to produce a rather quieter sound. It may also explain why it is more difficult to initiate a very quiet note on the saxophone with a conical bore than it is on the clarinet with a cylindrical bore.

For a direct bifurcation transition, the small non-linearities in the dynamic conductance will result in a spectrum of partials with amplitudes varying as p_1^n, where p_1 is the amplitude of the fundamental component excited. The spectral content or *timbre* of wind and brass instruments, as discussed later, therefore changes with increasing amplitude. This is illustrated by measurements of the amplitude dependence of the partials of a trumpet, clarinet and oboe by *Benade* ([15.133], Fig. 21.6c), which are reproduced for the trumpet in Fig. 15.104. For the largely cylindrical-bore trumpet and clarinet, nonlinearity results in partials varying as p_1^n over quite a large range of amplitudes. However, for the oboe, with its conical bore, the relative increase in strength of the partials is rather more complicated (*Benade* [15.133], Sect. 21.3). Eventually, the small-amplitude approximation will always break down, with a transition to a strongly nonlinear regime. Benade associates this transition with a change in timbre and responsiveness of the instrument for the player.

Large-Amplitude Oscillations

For a *lossless* cylindrical-bore instrument with only odd integer partials, the large-amplitude solutions are particularly simple. The flow of air from the lungs and pressure in the mouth is assumed to remain constant resulting in an average flow rate through the instrument. The pressure at the exit of the reed then switches periodically from a high-pressure to a low-pressure state, with equal amplitudes above and below the mean mouth pressure, spending equal times in each. The net acoustic energy fed into the resonating air column per cycle is therefore zero, $U \int p(t) \, \mathrm{d}t = 0$, as required for a lossless system.

Such a solution can easily be understood in terms of the excess-pressure wave propagating to the open

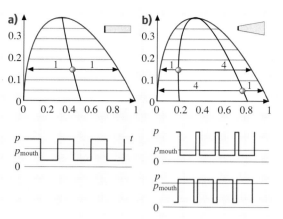

Fig. 15.97a,b Large-amplitude Helmholtz pressure fluctuation of (**a**) a cylinder and (**b**) a truncated cone with length to apex of 1/4 of its length, illustrating the dependence of fluctuation amplitudes as a function of mouthpiece pressure. For the cylinder, the mouthpiece pressure is single valued for a given flow rate, but for the truncated cone there are two possible solutions referred to as the standard and inverted Helmholtz solutions

end of the instrument, where it is reflected with change of sign. On return to the reed it reverses the pressure difference across the reed, which switches to the reduced pressure state. The subsequent reflection of the reduced pressure wave then switches the reed back to its original high-pressure state and the process re-peats indefinitely, with a periodic time of $4L/c_0$, as expected.

The dependence of the square-wave pressure fluctuations on the applied pressure can be obtained by the simple graphical construction illustrated in Fig. 15.97. The locus of the static pressure required to excite square-wave pressure fluctuations above and below the mouth pressure is shown by the solid line drawn from $p_{\text{mouth}} = 1/3$ to $1/2 p_{\text{max}}$, which bisects the high and low pressures for a given flow rate. If losses are taken into account, the horizontal lines are replaced by load lines with a downward slope given by the real part of the instrument's input admittance (*Fletcher* and *Rossing* [15.5], Fig. 15.9). At large amplitudes, the solutions can then involve periods during which the reed is completely closed. The transition from small-amplitude to large-amplitude solutions is clearly of musical importance, as it changes the sound of an instrument, and remains an active area of research [15.154].

Analogy with Bowed String

In recent years, an interesting analogy has been noted between the large-amplitude pressure fluctuations of a vibrating air column in a cylindrical or truncated conical tube and simple Helmholtz waves excited on a bowed string (*Dalmont* and *Kergomard* [15.162]). For example, the square-wave pressure fluctuations at the output of the reed attached to a cylindrical tube are analogous to the velocity of the bowed Helmholtz transverse waves of a string bowed at its centre, illustrated schematically in Fig. 15.98. Helmholtz waves could equally well be excited on a bowed string by a transducer with a square-wave velocity output placed halfway along the length of the string, in just the same way that the reed with a square-wave pressure output excites Helmholtz sound pressure waves into a cylinder, which acts like half the string length.

The analogy is particularly useful in discussing the large-amplitude pressure fluctuations in conical-bore instruments such as the oboe or saxophone. As described earlier, the conical tube has the same set of resonances as a cylindrical tube that is open at both ends. Therefore, in addition to having the same set of standing-wave sinusoidal solutions for the transverse oscillations of a stretched string, a conical tube can also support Helmholtz wave solutions. For the bowed string, the closer one gets to either end of the string the larger becomes the mark-to-space ratio between the regions of high to low transverse velocity. The same is also true for the switched fluctuations in pressure in a lossless conical tube, shown schematically in Fig. 15.97. Hence, if one

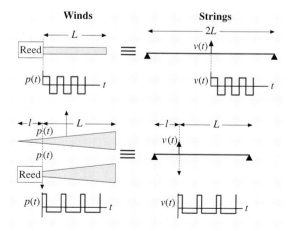

Fig. 15.98 Analogy between large-amplitude pressure waves in the bores of wind instruments and the transverse velocity of Helmholtz waves on a bowed stretched string, where the reed position is equivalent to the bowing position

truncates a conical tube with a vibrating reed system, the resonant modes of the remaining air column will be unchanged, provided the vibrating reed produces the same pressure fluctuations that would otherwise have been produced by the reflected Helmholtz waves returning from the removed apex end of the cone. Hence a conical tube, truncated by a vibrating reed at a distance l from the apex, can support Helmholtz wave solutions in the remaining length L. To produce such a wave the reed has to generate a rectangular pressure wave with a mark-to-space ratio and pressure fluctuations about the mean in the ratio $L:l$, as illustrated in Fig. 15.98, for a truncated cone with $L/l = 4$.

The period of the Helmholtz wave solutions of a conical bore instrument modelled as a truncated cone will therefore be $2(L+l)/c_0$, with a spectrum including all the harmonics $f_n = nc_0/2(L+l)$, other than those with integer values of $n = (L+l)$. To determine the amplitude of the rectangular pressure wave as a function of mouthpiece pressure, a graphical construction can be used similar to that used for the cylindrical tube, except that the pressures have now to be in the ratio L/l, as indicated in Fig. 15.97b. For the lossless large-amplitude modes of a truncated cone, there are two possible solutions involving high and low mouth pressures, which are known as the standard and inverted Helmholtz solutions, respectively.

Any complete model of a reed driven instrument must include losses and departures from harmonicity of an instrument's partials. This leads to a rounding of the rectangular edges of the Helmholtz waveforms and additional structure, in much the same way that bowed string waveforms are perturbed by frictional forces and non-ideal reflections at the end-supports (Fig. 15.34). Figure 15.99 shows a typical pressure waveform input for the conical-bore saxophone, which is compared with the Helmholtz waveform predicted for an ideal lossless system (*Dalmont* et al. [15.162]).

A completely realistic model for the excitation of sound in wind instruments must also include coupling to the vocal tracts (*Backus* [15.163] and *Scavone* [15.164]), since the assumption of a constant flow rate and constant mouth pressure is clearly over-simplistic.

Register Key

This analysis implicitly assumes that the reed excites the fundamental mode of the attached instrument. In practice, the reed will generally excite the partial with the lowest admittance corresponding to the highest peak in impedance measurements. For most instruments this is usually the fundamental resonance. However, the amplitude of the fundamental can be reduced relative to the higher resonances by opening a small hole, the register hole, positioned between the reed and first hole used in normal fingering of the instrument. Because of the difference in wavelengths and position of nodes, opening the register hole preferentially reduces the Q-value and shifts the frequency and amplitude of the fundamental relative to the higher partials. This allows the player to excite the upper register of notes based on the second mode, which in the case of the conical-bore saxophone, oboe and bassoon is an octave above the fundamental but is an octave and a perfect fifth above the fundamental for the cylindrical-bore clarinet. The lower and upper registers of the clarinet are sometimes referred to as the chalumeau and chanter registers, after the earlier instruments from which the clarinet was derived.

Figure 15.100 illustrates the lowering in amplitude and shift in resonant frequency of the fundamental mode on opening the register hole for the note E3 on a clarinet, which leaves the upper partials relatively unchanged (*Backus* [15.132]). The measurements also show the sig-

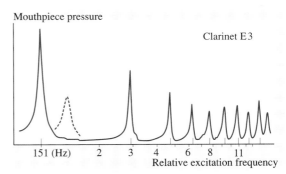

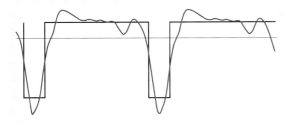

Fig. 15.99 Measured pressure waveform at input to a saxophone compared with the Helmholtz waveform expected for a truncated cone (after *Dalmont* et al. [15.162])

Fig. 15.100 Resonance curves for the note E3 on a clarinet, showing the shift of the lowest partial on opening the register hole (after *Backus* [15.132])

nificant departures from the 1, 3, 5, 7 harmonicity of the resonant modes of the instrument, which act as a warning not to take ideal models for the harmonicity of modes in real wind instruments too literally. Fletcher has developed a mode-locking model to account for the excitation of periodic waveforms on instruments with inharmonic partials [15.165].

15.3.4 Brass-Mouthpiece Excitation

The excitation of sound by the vibrating lips in the mouthpiece of a brass instrument cannot be described by any of the above simple models alone, which consider the reed as a simple mass–spring resonator. As we will show, the vibrations of the lips are three-dimensional and much more complicated. As a result, in some regimes the lips behave rather like outward-swinging-door valves, as first proposed by Helmholtz, and Bernoulli pressure operated sliding-door reeds in others. In addition the air flow is also affected by three-dimensional wave-like vibrations on the surface of the lips.

When playing brass instruments, the lips are firmly pressed against the rim of the mouthpiece with the lips pouted inwards. Pitched notes are produced by blowing air through the tightly clenched lips to produce a buzzing sound. The excitation mechanism can easily be demonstrated by buzzing the lips alone, though it is difficult to produce a very wide range of pitched sounds However, if the lips are buzzed when pressed against the rim of a mouthpiece, the input rim provides an additional constraint on the motion of the lips, which makes it much easier to produce pitched notes over a wide range of frequencies (I♪ CD-ROM). The audio demonstrates the "popping" sounds of trumpet and horn mouthpieces followed by the sound of the player buzzing the mouthpiece alone up to a pitch close to the popping frequency and back again. Attaching the mouthpiece to an instrument locks the oscillations to the various possible resonances of the instrument (I♪ CD-ROM).

Figure 15.101 shows a series of time-sequence plots of spectra of the sound produced by a player "buzzing" into a trumpet mouthpiece (*Ayers* [15.166]), which acts rather like a simple Helmholtz resonator. In the lower sequence, the player excites well-defined pitched notes from low frequencies up to slightly above the mouthpiece *popping frequency* (see Sect. 15.3.3). The middle set of traces shows the spectrum as the player starts at a high frequency and lowers the pitch. The upper traces shows the spectrum of a loudly sounded, low-frequency, note, illustrating the rich spectrum of harmonics produced by the strongly nonlinear sound-generation processes involved (see, for example, *Elliot* and *Bowsher* [15.167]).

These measurements on the mouthpiece alone strongly suggest that the lips can generate periodic fluctuations at frequencies up to, but not significantly beyond, the resonant frequency of any coupled acoustic resonator. This was confirmed in an investigation of lip-reed excitation using a simple single-resonant-frequency Helmholtz resonator by *Chen* and *Weinreich* [15.168], who used a microphone and loudspeaker in a feedback loop to vary the Q-value of the Helmholtz resonator played using a normal mouthpiece. They concluded that a player could adjust the way they vibrated their lips in the mouthpiece to produce notes that were slightly higher or lower than the Helmholtz frequency, though the most natural playing parameters generated frequencies in the range 20–350 Hz below the resonator frequency.

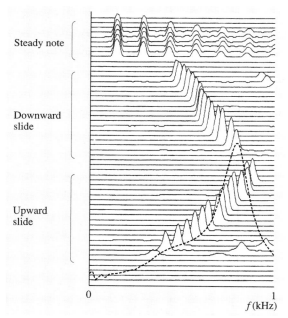

Fig. 15.101 Time sequence (*from bottom to top*) of spectra of the sound produced by a player "buzzing" into a trumpet mouthpiece, first for an upward slide in frequency, then for a downward slide and finally for a steady low note at high intensity showing the excitation of a note with many Fourier components. The *dashed line* is the spectrum of the *popping* note excited by slapping the open end of the mouthpiece against the palm of the hand (after *Ayers* [15.166])

Attached Mouthpiece

Ayers [15.166] also compared the frequencies produced in the mouthpiece before and immediately after attachment of the instrument. In these measurements the player first excited a pitched note in the mouthpiece with the instrument effectively decoupled by opening a large hole in its bore close to the mouthpiece. The hole was then closed and the immediate change in frequency measured before the player had time to make any adjustments to the embouchure. The results of such measurements are shown in Fig. 15.102, where the diagonal line represents the pitched notes before the instrument was connected and the discontinuous solid line through the triangular points are the modified frequencies for the same embouchure with the instrument connected.

At the higher frequencies, the jumps between successive branches are from just above the resonant frequency of one partial to just below the resonant frequency of the next, with a monotonic increase in frequency on tightening the embouchure in between. However, for the first two branches, the instrument resonances initially have a relatively small effect on the frequencies excited by the mouthpiece alone until such frequencies approach a particular partial frequency. The frequency then approaches the resonant frequency of the instrument before jumping to a frequency well below the next partial and the sequence repeats. The difference in behaviour of the lower and higher branches suggests that more than one type of lip-reed action is involved.

Comparison with computational models by *Adachi* and *Sato* [15.169] appear to rule out the outward-swinging-door model first proposed by Helmholtz, indicated by the squares in Fig. 15.101, as the predicted frequencies are always well above those of the instrument's partials. The computed predictions for the Bernoulli sliding door model, indicated by the circles, are in better agreement with measurements, but with predicted frequencies rather lower than those observed and never rising above the resonant frequencies of the instrument, in contrast to the observed behaviour for the higher modes excited.

Any model for the lip-reed sound generator has to explain all such measurements. Such measurements also highlight the way that a brass player can adjust the embouchure and pressure acting on the lips in the mouthpiece to change the frequency of the excited mode. On tightening the embouchure and pressure the player can progressively excite successive modes and can "lip" the pitch of the note up and down by surprisingly large amounts, used with great expressive effect by jazz trumpeters.

Vibrating Lips

In practice, the production of sound by the vibrating lips inside a mouthpiece is a highly complex three-dimensional problem closely analogous to the production of sound by the vocal folds – see, for example,

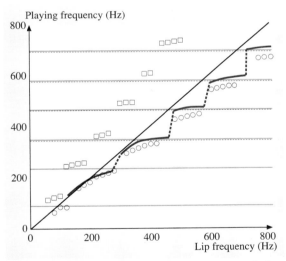

Fig. 15.102 Frequencies produced by a trumpet mouthpiece without (*diagonal line*) and with (*broken line*) the instrument strongly coupled using an unchanged embouchure under the same playing conditions. The *solid horizontal lines* are the resonant frequencies of the assembled trumpet. The *squares* and *circles* are predictions for the Helmholtz outwardly opening-door and sliding-door models computed by *Adachi* and *Sato* [15.169] for a trumpet with slightly lower-frequency resonant modes indicated by the *dashed horizontal lines* (after *Ayers* [15.166])

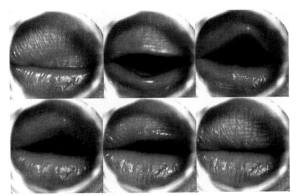

Fig. 15.103 High-speed photograph clips showing one cycle of lip vibration in a trombone mouthpiece

Titze [15.170]. A complete model would involve solving the coupled solutions of the Navier–Stokes equations describing the flow of air from the mouth, through the lips and into the mouthpiece, and the three-dimensional motions of the soft tissues of the lips induced by the Bernoulli pressures acting on their surfaces.

Stroboscopic and ultra-fast photography of the brass players lips while playing reveal highly complex three-dimensional motions (*Coppley* and *Strong* [15.171], *Yoshikawa* and *Muto* [15.172]) suggest that the upper lip is primarily involved in the generation of sound. Figure 15.103 and the video clip CD-ROM provided by Murray Campbell show high-speed photography clips of such motion. The lips inside the mouthpiece open and shut rather like the gasping opening and shutting of the mouth of a goldfish, but speeded up several hundred times. Points on the surface of the upper lip exhibit cyclic orbital motions involving the in-quadrature motions of the upper lip parallel and perpendicular to the flow. To model such motion clearly requires at least two independent mass–spring systems to account for the induced motions of the lips along and perpendicular to the flow (*Adachi* and *Sato* [15.169]). In addition, there is a pulsating wave-like motion along the surface of the lips in the direction of air flow, with the rear portion of lips moving in anti-phase with the front. *Yoshikawa* and *Muto* [15.172] identify such motion as strongly damped Rayleigh surface waves travelling through the mucous tissue of the upper lip.

The simplest possible model to describe such motion therefore requires at least three interacting mass–spring elements; one to describe the lip motion along the direction of flow, and two to describe the motions of the front and back surfaces of the lips. But even then, the model will still only be an approximation to the three-dimensional bulk tissue motions involved. Not surprisingly, research into the lip-reed sound-excitation mechanism remains a problem of considerable interest.

Artificial Lips

To achieve a better understanding of lip dynamics and its effect on the sound produced by brass instrument, several groups have developed artificial lips to excite brass instrument (e.g. *Gilbert* et al. [15.173] and *Cullen* et al. [15.174]). These can be used to investigate instruments under well-controlled and reproducible experimental conditions. Typically, the lips are simulated by two slightly separated thin-walled (0.3 mm thickness) latex tubes filled with water under a controlled pressure. The tubes are rigidly supported from behind so that the internal pressure forces the lips together. The tubes are placed across an opening in an otherwise hermetically sealed unit that represents the mouth and throat cavities. Air is fed into the mouth cavity at a constant flow rate. A fixed mouthpiece is then pushed against the artificial lips with a measured force. By varying this force, the applied pressure and the pressure within the artificial lips, the experimenter can simulate the various ways in which a player can control the dynamics of the lips (the *embouchure*) to produce different sounding notes. Despite the considerable simplification in comparison with the dynamics of real lips, the sound of brass instruments played by artificial lips is extremely close to that produced by a real player. Such systems enable acoustical studies to be made on brass instruments with a much greater degree of flexibility, reproducibility and stability than can be achieved by a player. Using a fixed mouthpiece, the playing characteristics of different attached instruments can easily be compared.

Nonlinear Sound Excitation

When played very quietly, brass instruments can produce sounds that are quasi-sinusoidal with relatively weak higher harmonics. However, as previously noted for vi-

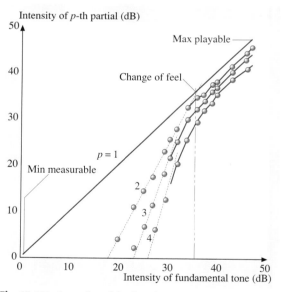

Fig. 15.104 Intensity of the first four partials of the note C4 on a trumpet as a function of the intensity of the first partial, measured from the minimum to the maximum playing intensity (after *Benade* [15.133] Fig. 21.8). On this logarithmic scale, the dashed lines through the measurements for the second, third and fourth Fourier components have slopes 2, 3 and 4, respectively

brating reeds, any nonlinearity will lead to the generation of harmonics of the fundamental frequency ω at frequencies 2ω, 3ω, etc., with initially amplitudes increasing as $|p_0(\omega)|^n$, where $p_0(\omega)$ is the amplitude of the fundamental. However, in the strongly nonlinear region at high amplitudes, all partials become important and increase in much the same way with increasing driving force (*Fletcher* [15.175] and *Fletcher* and *Rossing* [15.5], Sects. 14.6 and 14.7). Such effects are illustrated in Fig. 15.104 for measurements on a B-flat trumpet by *Benade* and *Worman* ([15.133], Sect. 21.3). The spectral content and resulting brilliance of the sound or timbre of a trumpet, or any other brass instrument, therefore depends on the intensity with which the instrument is played. Benade noted a change in the sound and feel of an instrument by the player in the transition region between the power-law dependence of the Fourier component and the high-amplitude regime, where the harmonic ratios remain almost constant. Similar characteristics were observed for the clarinet though rather different characteristics for the oboe.

Examples of the strongly non-sinusoidal periodic fluctuations of the pressure and flow velocity within the mouthpiece for two loudly played notes on a trombone are shown in Fig. 15.105 (*Elliot* and *Bowsher* [15.167]). *Fletcher* and *Rossing* ([15.5], Sect. 14.7) discuss such waveforms in terms of the lips operating slightly above

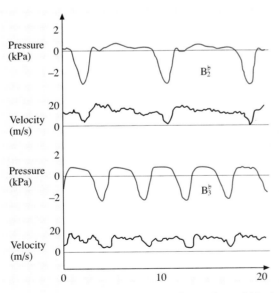

Fig. 15.105 Non-sinusoidal pressure fluctuations within the mouthpiece for two notes played at large amplitudes on the trombone (after *Elliot* and *Bowsher* [15.167])

the resonant frequency of their outward-swinging-door resonant frequencies.

The nonlinearity of the lip-reed excitation mechanism enables the player to vibrate the lips at the frequency of the missing fundamental of the quasi-harmonic series of modes of brass instruments, illustrated in Fig. 15.38. This is referred to as the *pedal note* and is an octave below the lowest mode normally excited on the instrument. The lips vibrate at the pedal-note frequency but only excite the quasi-harmonic $n = 2, 3, 4, \ldots$ modes.

The pressure fluctuations in Fig. 15.104 of ≈ 3 kPa correspond to a sound intensity of nearly 160 dB. As the static pressure in the mouthpiece is only a few percent above atmospheric pressure (10^5 Pa), such pressure excursions are a significant fraction of the excess static pressure. Even larger-amplitude pressure fluctuations can be excited on the trumpet and trombone when played really loudly, to produce a brassy sound. *Long* [15.176] has recorded pressure levels in a trumpet mouthpiece as high as 175 dB, corresponding to pressure fluctuations of ≈ 20 kPa.

At such high amplitudes, one can no longer neglect the change in density of a gas when considering its acceleration under the influence of the pressure gradient. To a first approximation, the wave equation then becomes

$$\frac{\partial^2 \xi}{\partial x^2} = \left(1 + \frac{\partial \xi}{\partial x}\right) \frac{1}{c_0^2} \frac{\partial^2 \xi}{\partial t^2} \, . \tag{15.136}$$

The speed of sound will now depend on both frequency and wave shape, with the velocity varying as

$$c' = c_0 \left\langle \left(1 + \frac{\partial \xi}{\partial x}\right)^{-1/2} \right\rangle_{x,t} \sim c_0 \left[1 + \alpha \, (k\xi)^2\right] \, , \tag{15.137}$$

where the averaging is taken over both time and wavelength giving a value for $\alpha \approx 1/8$. In practice, for waves propagating down a tube, other terms involving momentum transport, viscosity and heat transfer also have to be included in any exact solution. However, the essential physics remains unchanged. The net effect of the nonlinearity is to progressively increase the slope of the leading edge of any wave propagating along the tube. A propagating sine wave is then transformed into a sawtooth waveform, or shockwave, with an infinitely steep leading edge, at sufficiently large distances along the tube.

Such waves have indeed been observed in trumpet and trombone bores, which are long and relatively narrow. The effect is illustrated in Fig. 15.106 by the measurements of *Hirschberg* et al. [15.177], for waves

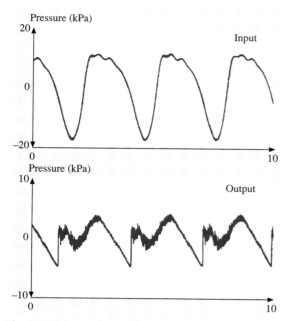

Fig. 15.106 Internal acoustic pressure in a trombone played at a dynamic level fortissimo (ff). *Upper curve*: pressure at the input to the air column in the mouthpiece. *Lower curve*: pressure at the output of the slide section, showing the characteristic profile of a shock wave (after *Hirschberg* et al. [15.177])

propagating along a trombone tube. The sound intensity of $\approx 175\,\mathrm{dB}$ is considerably higher than the intensities illustrated in Fig. 15.99. As predicted, the sharpness of the leading edge of the waveform progressively increased on propagating along the bore. The discontinuity in waveform of the fully developed shockwave dramatically increases the intensities of the higher harmonics of a continuously played note and gives the trumpet and trombone (and trompette organ pipes) their characteristic brassy sound at very high intensities (🔊 CD-ROM). The high-frequency components of all such sounds will make them highly directional. *Campbell* [15.160] reviews nonlinear effects in woodwind and brass instruments, with many references to recent research.

To achieve such brassy sounds, the instrument must have a sufficiently long length of relatively narrow pipe, like the trumpet and trombone, in which the pressure fluctuations remain high and have time to build up into a shock wave. Shockwaves are far more difficult to set up in instruments like the horn and cornet, with flaring conical bores, because the pressure drops rather rapidly with increasing diameter along the bore.

Time–Domain Analysis

When nonlinearity is important or when the initial transient response is of interest, it is more appropriate to consider the dynamics in the time- rather than frequency-domain, just as it was for analysing the transient dynamics of the bowed string. Time-domain analysis in wind and brass instruments was pioneered by *Schumacher* [15.158], *McIntyre* et al. [15.178] and *Ayers* [15.179], and is discussed in *Fletcher* and *Rossing* ([15.5], Sect. 8.14). Time-delayed reflectometry measurements are made by producing short pressure pulses inside the mouthpiece generated by a spark or by a sudden piezoelectric displacement of the mouthpiece end-wall. The pressure in the mouthpiece is then recorded as a function of time after the event.

In the linear response regime, measurement in the time domain gives exactly the same information about an instrument as measurements in the frequency domain, assuming both the magnitude and phase of the frequency response is recorded. This follows because the frequency response $Z(\omega)$ measured in the mouthpiece is simply the Fourier transform of the transient pressure response $p(t)$ for a δ- or impulse-function flow (Av) induced by the spark or wall motion. Knowing $p(t)$ or $Z(\omega)$ one can obtain the other by applying the appropriate Fourier transform

$$Z(\omega) = \int_0^\infty p(t)\,\mathrm{e}^{-\mathrm{i}\omega t}\,\mathrm{d}t \quad \text{or}$$

$$p(t) = \frac{1}{2\pi} \int_{-\infty}^\infty Z(\omega)\,\mathrm{e}^{\mathrm{i}\omega t}\,\mathrm{d}\omega\,. \tag{15.138}$$

Measurements of the impulse response are particularly useful in identifying large discontinuities in the acoustic impedance along the bore of instruments produced by tone holes, valves and bends, which can significantly perturb particular partials. The position of any such discontinuity can be determined by the time it takes for the reflected impulse to return to the mouthpiece.

Time-domain analysis is essential, if one wishes to investigate starting transients, where reflections from the end of the instrument are required to stabilise the pitch of a note. This problem is particularly pronounced for horn players pitching, for example, notes as high as the 12th resonant mode of the instrument. The player must buzz the lips a dozen or so cycles into the mouthpiece before the first reflection from the end of the instrument returns to stabilise the pitch. If the player gets the initial buzzing frequency slightly wrong, the instrument may lock on to the 11th or 13th harmonic rather than the intended 12th, leading to the familiar *cracked note* of the beginner and sometimes even professional horn players. Furthermore, *false* reflections from discontinuities along the length of the tube, may well confuse the initial feedback, making it more difficult to pitch particular notes. This leads to small pitch-dependent changes in the playing characteristics of instruments made to different designs adopted by different manufacturers.

15.3.5 Air-Jet Excitation

Many ancient and modern musical instruments are excited by blowing a jet of air across a hole in a hollow tube or some other acoustic resonator. Familiar examples include the flute, pan pipes, the ocarina and simple whistle in addition to many organ pipes. Sound excitation in flutes and organ pipes was first considered by *Helmholtz* [15.127] in terms of an interaction between the air jet produced by the lips or a flue channel in the mouthpiece of an instrument and the coupled air column resonator.

In practice, the dynamics of sound production is a very complex aerodynamic flow problem requiring the solution of the Navier–Stokes equations governing fluid flow in often complex geometries. Various simplified solutions have been considered by many authors since the time of *Helmholtz* and *Rayleigh* [15.3]. *Fletcher* and *Rossing* ([15.5], Sect. 16.1) provide references to both historic and more-recent research. *Fabre* and *Hirschberg* [15.180] have also written a recent review of simple models for what are sometimes referred to as flue instruments.

Rayleigh showed that the interface separating two fluids moving with different velocities was intrinsically unstable, resulting in an oscillating sinuous lateral disturbance of the interface that grows exponentially with time Fig. 15.107. This arises because, in the frame of reference in which the two fluids move with the same speed in opposite directions, any disturbance of the interface towards one of the fluids will increase the local surface velocity on that side. This will result in a decrease in Bernoulli pressure on that side of the interface and increase it on the other, creating a net force in the same sense as the disturbance, which will therefore grow exponentially with time. For a layer of air moving at velocity V without friction over a stationary layer, a sinusoidally perturbed deflection of the jet in the laboratory frame of reference at rest increases exponentially with distance as it travel along the interface with velocity $V/2$.

Similar arguments were used by Rayleigh to describe the instability of an air jet of finite width b and velocity V produced by blowing through an arrow constriction between the pouted lips when playing the flute, or by blowing air through an air channel towards the sharp lip of the recorder or an organ pipe. *Fletcher* [15.181] showed that the lateral displacement $h(x)$ of the jet induced by an acoustic velocity field $v e^{i\omega t}$ between the jet orifice and the lip varies with position and time as

$$h(x) = -j\left(\frac{v}{\omega}\right)\left[1 - \cosh \mu x \exp(-i\omega x/u)\right] e^{i\omega t},$$
(15.139)

illustrated schematically in Fig. 15.107. The first term simply corresponds to the jet moving with the impressed acoustic field, while the second describes the induced travelling-wave instability moving along the jet with velocity $u \approx V/2$, which dominates the jet displacement at the lip of the instrument. For long-wavelength instabilities on a narrow jet, such that the characteristic wavevector $k \ll 1/b$, Rayleigh showed that the phase velocity $u = \omega/k \sim kbV/2 = (\omega b V/2)^{1/2}$, while the exponential growth factor $\mu = (k/b)^{1/2}$.

In practice, the velocity profile of the jet is never exactly rectangular but in general will be bell-shaped. This results in a slightly different frequency, width b and velocity dependence of the phase velocity, with

$$u = 0.55(\omega b)^{1/3} V^{2/3} \qquad (15.140)$$

and corresponding changes in the exponential growth factor (*Savic* [15.182]). Typical propagation velocities of disturbances on the jet of flute and organ pipes measured by *Coltman* [15.183] range from 6.7 m/s to 3.7 m/s with velocity ratios u/V from 0.35 to 0.5 for blowing pressures from 1 to 0.15 inches of water.

Resonances of the attached resonator can be excited when the air-jet instability is in phase with oscillations within the resonator, as shown for the two positions indicated in Fig. 15.107. This corresponds to an odd number of half-wavelengths of the propagating instability, so that $\omega l/u = n\pi$, which is equivalent to $f \sim nV/4$,

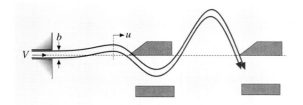

Fig. 15.107 Propagating sinuous instability of an air jet emerging from a flow channel with two possible positions of the angled labium or lip to excite resonances of an attached air column for a given jet velocity V

where n is an odd integer and we have assumed $u \approx V/2$. For a given length between jet orifice and lip, different frequencies can be excited by varying the jet velocity $V = \sqrt{2P_{\text{mouth}}/\rho}$, where P_{mouth} is the pressure in the mouth creating the jet. When coupled to an acoustic resonator with a number of possible resonances, such as an organ pipe or the pipe of a flute or recorder, the interaction between the jet and oscillating resonator causes the frequency to lock on to a particular resonance, with hysteretic jumps between the resonance excited as the blowing pressure is increased and decreased, as illustrated schematically in Fig. 15.108 (*Coltman* [15.183]). This explains why the pitch of an instrument like the recorder or flute doubles when it is blown more strongly. The line marked "edge tones" indicates the frequency of the excited jet mode instability for the same orifice–lip geometry without an attached pipe and the line marked "f = pipe resonance" indicates when the frequency of the coupled jet coincides with the free vibrations of the attached air column. In practice the instrument is played in close vicinity to the pipe resonances.

The above model is strictly only applicable to small perturbations in jet shape ($\ll b$) and to a nonviscous medium. In practice, measurements of jet displacement by Schlieren photography, hot-wire anemometry (*Nolle* [15.185]), smoke trails *Cremer* and *Ising* [15.186] and *Coltman* [15.187] and particle image velocimetry (PIV) (*Raffel* et al. [15.188]) show that within an

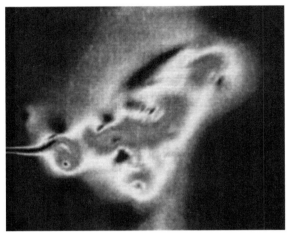

Fig. 15.109 A computed snapshot showing the breakup of a jet and generation of vortices (*Adachi* [15.184])

acoustic cycle the jet moves to either side of the lip of the instrument with large displacement amplitudes comparable with the physical dimensions of the distance between the orifice and lip. In addition, viscous forces lead to a change in profile of the jet as it moves through the liquid from rectangular to bell-shaped (*Ségoufin* et al. [15.189]). Furthermore, the associated shear forces eventually induce vorticity downstream, with individual vortices shearing away from the central axis of the jet on alternate sides, as observed by *Thwaites* and *Fletcher* [15.190, 191].

In recent years, major advances have also been made in studying such problems by computation of solutions of the Navier–Stokes equations describing the nonlinear aerodynamic flow. An example is illustrated by the simulated jet deflections by *Adachi* [15.184] shown in Fig. 15.109. Adachi's computational results are in reasonably good agreement with Nolle's flow measurements [15.185] made with a hot-wire anemometer. A sequence of such computations by Macherey for the jet in a mouthpiece with flute-like geometry is included in a recent review paper on the acoustics of woodwind instruments by *Fabre* and *Hirschberg* [15.192]. The computations show a relatively simple jet structure switching from one side of the lip of a flute to the other during each period of the oscillation.

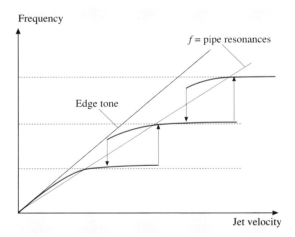

Fig. 15.108 Schematic representation of the frequency of a jet-edge oscillator before and after coupling to a multiresonant acoustic resonator. The *line* marked f = pipe resonator indicates the pressures when the frequencies are the same as the modes of the uncoupled acoustic resonator (after *Coltman* [15.183])

Air–Jet Resonator Interaction
Despite the obvious limitations of any small-amplitude linear approach to the jet–lip interaction and the excitation of resonator modes, it is instructive to consider the analytic model introduced by *Fletcher* [15.181],

which is discussed in some detail in *Fletcher* and *Rossing* ([15.5], Sect. 16.3), as it includes much of the essential physics in a way that is not always apparent from purely computational models. The assumed geometry is illustrated schematically in Fig. 15.110 and can easily be generalised to model sound excitation in air-jet-driven instruments with different mouthpiece geometries such as the recorder, organ pipe or flute. A uniform jet with a rectangular top-hat velocity profile of width b and velocity V is assumed to impinge on the lip or labium of the instrument, with the jet passing through a fraction A of the attached pipe area S_p. The oscillating travelling-wave jet instability will result in a periodic variation of the fractional area varying as $\alpha e^{i\omega t}$. On entering the pipe channel, the jet will couple to all possible modes of the attached pipe, which in addition to the principle acoustic modes excited, will include many other modes involving radial and azimuthal variations [15.193] that are very strongly attenuated. Much of the energy of the incident jet will therefore be lost by such coupling with typically only a few percent transferred to the important acoustic resonances of the instrument.

To evaluate the transfer of energy to the principle acoustic mode, we consider the pipe impedance across the plane M. In the absence of the jet, the impedances of the attached pipe and mouthpiece section are Z_p and Z_m, where the mouthpiece section has a certain volume and hole area, with the volume in the case of a flute involving an adjustable length of pipe used for fine-tuning. Fletcher simplified an earlier model introduced by *Elder* [15.194], in assuming the principle of linear superposition, with a jet flow superimposed on that of the instrument. The oscillating incident jet then has two principal effects: the oscillating fraction of area α of the jet entering the pipe introduces a flow into the attached pipe

$$U_1 = \alpha S_p Z_m / (Z_m + Z_p) \,, \quad (15.141)$$

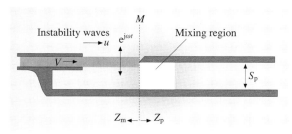

Fig. 15.110 Model to illustrate interaction between air jet and pipe modes (after *Fletcher* [15.193])

while the effective pressure acting on the plane M derived from momentum-balance arguments results in a pipe flow of

$$U_2 = \alpha \rho V^2 / (Z_m + Z_p) \,. \quad (15.142)$$

There is also an additional nonlinear term U_3 mathematically arising from the nonlinear effect of the fractional insertion area of the jet insertion at large amplitudes, which is negligible in comparison with the many other nonlinearities in any realistic model. This model integrates and simplifies earlier models introduced by *Cremer* and *Ising* [15.186], *Coltman* [15.187] and *Elder* [15.194].

It immediately follows from the above arguments that, provided the phase of the jet instabilities are appropriate, instabilities on the jet will excite strong resonances of the instrument when the series impedance $(Z_m + Z_p)$ is a minimum, which is just the impedance of the pipe loaded by the impedance of the mouthpiece assembly. Provided the lengths involved are much less than an acoustic wavelength, $Z_m = i\rho \Delta L / S_p$ where ΔL is the end correction introduced by the mouthpiece assembly. The net flow is then given by

$$U_p = \frac{(V + i\omega \Delta L)\rho V \alpha}{S_p (Z_m + Z_p)} \,. \quad (15.143)$$

The resonances are therefore those of a pipe that is open at both ends, but with a small end-correction for the effective volume of the mouthpiece and any additional closed tuning tube and flow in and out of the mouthpiece.

Because the mouthpiece impedance is reactive, the induced vibrations in flow from direct jet flow (U_1) and that induced by the jet pressure (U_2) are in phase quadrature. In addition, the two terms have a different dependence on jet velocity and frequency. In practice, $\omega \Delta L$ is often larger than V, so that the second term usually dominates, though this will not necessarily be true in more realistic models.

For small sinusoidally varying jet perturbations and a top-hat velocity profile, the driving force would only include single frequency components. However, in practice, viscous damping results in a spreading out of the velocity profile in the lateral direction with a bell-shaped profile that increases in width with distance along the jet. If such a profile is offset from the lip, any sinusoidal disturbance of the jet will introduced additional harmonics at frequencies, 2ω, 3ω, etc., with increasing amplitudes for increasing jet oscillations. In practice, the amplitudes of jet oscillation are so large that the jet undergoes a near switching action, alternating its position from one side to the other of the lip. The driving force is therefore strongly non-sinusoidal and provides a rich spectrum of harmonics to excite the upper partials of any sounded note. A flute player, for example, has considerable control over the quality of the sound produced, by variation of mouth pressure and jet velocity, its velocity profile on leaving the lips, and its direction in relation to the labium or lip of the instrument.

For most musical instruments excited by an air jet, the sinuous instability is the most important, though *Chanaud* [15.195] and *Wilson* et al. [15.196] have highlighted the importance of varicose instabilities (periodic variations in area of the jet), which were also investigated by Rayleigh, as in whistles and whistling, where the air passes through an aperture rather than striking an edge.

Regenerative Feedback

We now consider the effective acoustic impedance of the exciting air column as we did for the vibrating reed. Our emphasis here is to highlight the essential physics rather than provide a rigorous treatment. More details and references are given in *Fletcher* and *Rossing* ([15.5], Sect. 16.4) and *Fletcher* [15.193].

The flow U_m into the mouth of the resonator can be expressed as

$$U_m = v_m S_m = p_m Y_m \sim \frac{p_m S_p}{i\omega \rho \Delta L} \, . \quad (15.144)$$

From (15.139), the lateral displacement of the jet at the lip a distance l from the exit channel of the jet is then given by

$$hl \sim i\left(\frac{v_m}{\omega}\right) \cosh \mu l \exp(-\omega l/u) \, , \quad (15.145)$$

where u is the phase velocity for disturbances travelling along the jet and the implicit time variation has been omitted. From the ratio of the net flow into the mouthpiece-end and attached resonator to the pressure acting on the air jet, Fletcher and Rossing derive the effective admittance of the air-jet generator,

$$Y_j \sim \frac{VW}{\rho \omega^2 \Delta L} \left(\frac{S_p}{S_m}\right) \cosh \mu l \exp\left[-i\left(\frac{\omega l}{u}\right)\right] \, . \quad (15.146)$$

Apart from a small phase factor ($\phi \approx V/\omega L$), which we have omitted, the admittance is entirely real and *negative* when $\omega l/u = \pi$, which corresponds to the first half-wavelength of the instability just bridging the distance from channel exit to lip, as shown in Fig. 15.106. This is also true for $\omega l/u = n\pi$, where n is any odd integer, corresponding to any odd number of half-wavelengths

of the growing instability between the jet exit and lip of the instrument.

These are just the conditions for positive feedback and the growth of acoustic resonances in the pipe resonances will be excited when $\text{Im}(Y_p + Y_m) = 0$, which is equivalent to the condition that Z_s should be a minimum, as expected from (15.129). As *Coltman* [15.183] pointed out, the locus of $\text{Im}(Y_j)$ plotted as a function or $\text{Re}(Y_j)$ as a function of increasing frequency is a spiral about the origin in a clockwise direction (*Fletcher* and *Rossing* [15.5], Fig. 16.10). The jet admittance therefore has a negative real component at all frequencies when the locus point is in the negative half-plane. Resonances can therefore be set up over frequency ranges from $\approx (1/2 \text{ to } 3/2)\omega^*$, $(5/2 \text{ to } 7/2)\omega^*$, etc. where ω^*, $3\omega^*$, $5\omega^*$ are the frequencies when the admittance is purely conductive and negative. By varying the blowing pressure and associated phase velocity of the jet instability, the player can therefore excite instabilities of the jet with the appropriate frequencies to lock on to the resonances of the attached air column, as illustrated schematically in Fig. 15.108.

Measurements by *Thwaites* and *Fletcher* [15.190] are in moderately good agreement with the above model at low blowing pressures and reasonably high frequencies, but deviate somewhat at low frequencies and high blowing pressures. This is scarcely surprising in view of the approximations made in deriving the above result.

Edge Tones and Vortices

Edge tones are set up when jet of air impinges on a lip or thin wire without any coupling to an acoustic resonator. The high flow rates in the vicinity of the lip or wire can generate vortices on the downstream side, which spin off on alternating sides, setting up an alternating force on the lip or wire. If the object is itself part of a mechanical resonating structure, such as a stretched telegraph wire or the strings of an Aeolian harp, wind-blown resonances can be set up, with different resonant modes excited dependent on the strength of the wind. In extreme cases, the excitation of vortices can result in catastrophic build up of mechanical resonances, as in the Tacoma bridge disaster.

Before 1970, many treatments of wind instruments discussed air-jet sound generation in such terms. *Holger* [15.197], for example, proposed a nonlinear theory for edge-tone excitation of sound in wind instruments based on the formation of a vortex sheet, with a succession of vortices already created on alternate sides of the mid-plane of the emerging jet before it hit the lip or labium of the instrument. Indeed, measurements of the flow instabilities and phase velocity of instabilities in a recorder-like instrument by *Ségoufin* et al. [15.189], as a function of Strouhal number $\omega b/u$, fit the Holger theory rather better than models based on the Rayleigh instability and refined by later authors for both short and long jets, but the experimental errors are rather large. However, the vortex-sheet model does not include the growth of disturbances in the sound field with distance (as measured by *Yoshikawa* [15.198]), which is a crucial parameter for the prediction of the oscillation threshold observed for instruments such as the recorder.

It is also clear that vortex production is important in many wind instruments, especially where the acoustic amplitude is large, as in the vicinity of sharp edges or corners of both open and closed tone holes, as observed in direct measurements of the flow field. For example, *Fabre* et al. [15.199] have recently shown that vortex generation is a significant source of energy dissipation for the fundamental component of a flute note.

In view of the complexity of the fluid dynamics involves, it seems likely that future progress in our understanding of jet-driven wind instruments will largely come from computational simulations, though physical models still provide valuable insight into the basic physics involved.

15.3.6 Woodwind and Brass Instruments

In this last section on wind instruments, we briefly describe a number of woodwind and brass instruments of the modern classical orchestra. All such instruments were developed from much earlier instruments, many of which still exist in folk and ethnic cultures from all around the World. Illustrated guides to a very large number of such instruments are to be found in *Musical Instruments* by *Baines* [15.200] and the encyclopaedia *Musical Instruments of the World* [15.30]. The two text by *Backus* [15.132] and Benade [15.133], both leading researchers in the acoustics of wind instruments, provide many more technical details concerning the construction and acoustics of specific woodwind and brass instruments than space allows here, as does *Fletcher* and *Rossing* ([15.5], Chaps. 13–17) and *Campbell*, *Myers* and *Greated* [15.7].

Woodwind Instruments
The simplest instruments are those based on cylindrical pipes, such as bamboo pan pipes excited by a jet of air blown over one end, or hollow resonators, such as primitive ocarinas, which act as simple Helmholtz resonators with the pitch determined by the openings of the mouthpiece and fingered open holes. Woodwind instruments use approximately cylindrical or conical tubes excited by a reed or a jet of air blown over a hole in the wall of the tube. As we have seen, simple cylindrical and conical tubes retain a harmonic set of resonances independent of their length, which in principle allows a full set of harmonic partials to be sounded when the instrument is artificially shortened by opening the tone holes. In practice, as discussed in the previous section, the harmonicity of the modes is strongly perturbed by a large number of factors including the strongly frequency-dependent end-corrections from tone holes and significant departures from simple cylindrical and conical structures. Such perturbations can, to some extent, be controlled by the skilled instrument maker to preserve the harmonicity of the lowest modes responsible for establishing the playing pitch of an instrument. We will describe the various methods of exciting vibrations by single and double reeds and by air flow in the next section.

Figure 15.111 shows four typical modern orchestral woodwind instruments. All such instruments have distinctive tone qualities and come in various sizes, which cover a wide range of pitches and different tone-colours.

The flute, bass flute and piccolo are based on the resonances of a cylindrical tube, with the open end and mouthpiece hole giving pressure nodes at both ends.

Fig. 15.111 The modern flute, oboe, clarinet and bassoon (not to scale)

Like the recorder, all the chromatic notes of the musical scale can be played by selectively opening and shutting a number of tone holes in the walls of the instrument using the player's fingers (on ancient and baroque flutes) or felted hinged pads operated by a system of keys and levers (on modern instruments). Primitive flutes appear in most ancient cultures.

The clarinet is based on a cylindrical tube excited by a single reed at one end. The reed and mouthpiece close one end of the tube, so that the odd-integer partials are more strongly excited than the even partials, particularly for the lowest notes, when most sideholes are closed. In addition, when overblown, the clarinet sounds a note three times higher (an octave and a fifth). Like all real instruments, perturbations from the tone holes, variations in tube diameter and the nonlinear processes involved in the production of sound vibrations by the reed strongly influence the strength of the excited partials, all of which contribute to the characteristic sound of the instrument.

The single-reed clarinet is a relatively modern instrument developed around 1700 by the German instrument maker Denner. It evolved from the chalumeau, an earlier simple single-reed instrument with a recorder-like body, which still gives its name to the lower register of the clarinet's playing range. In the 1840s, the modern system of keys was introduced based on the Boehm system previously developed for the flute [15.200].

The oboe is based on a conical tube truncated and closed at the playing end by a double reed. As described earlier, a conical tube supports all the integer harmonic partials giving a full and penetrating sound quality that is very rich in upper partials. This is why an oboe note is used to define the playing pitch (usually A4 = 440 Hz) of the modern symphony orchestra. Like all modern instruments, today's oboe developed from much earlier instruments, in this case from the double-reed shawn and bagpipe chanters, which still exist in many ethnic cultures in Europe, Asia and parts of Africa. In addition to the bass oboe, the oboe d'amore and cor anglais, with their bulbous end cavity just before the output bell, have been used for their distinctive plaintive sounds by Bach and many later composers. Like the flute and clarinet, early oboes used mostly open-side holes closed by the fingers, with only one or two holes operated by a key, but developed an increasingly sophisticated key system over time to facilitate the playing of the instrument.

The bassoon is a much larger instrument, producing lower notes of the musical scale. Because of the length of the air column, the spacing of the tone holes would be far too wide to operate by the player's fingers alone. To circumvent this problem, the instrument is folded along its length and relatively long finger holes are cut diagonally through the large wall thickness, so that normally fingered holes can connect to the much wider separation of holes in the resonant air column. The bore of the instrument is based on a largely conical cross section, with the mouthpiece end terminated by a narrow bent tube or *crook* to which a large double reed is attached. Early bassoons included a single key to operate the most distant tone hole on the instrument. Modern instruments have an extended key system to facilitate playing all the notes of the chromatic scale. The contrabassoon includes an additional folded length of tube. Like the oboe, the sound of the bassoon is very rich in upper partials and has a very rich, mellow sound.

Related to the bassoon is the renaissance racket played with a crook and double reed. The instrument looks like a simple cylinder with a set of playing holes cut into its surface. However, in reality, it is a highly convoluted pipe with twelve parallel pipes arranged round the inner diameter of the cylinder and interconnected with short bends at their ends to produce a very long acoustic resonator with easily accessible tone holes. This provides a beautiful example of the centuries-old ingenuity of instrument makers in solving the many acoustic and ergonomic problems involved in the design of musically successful wind instruments.

Brass Instruments

Figure 15.112 illustrates the trumpet, trombone and horn, which like all brass instruments are based on lengths of cylindrical, conical and flared resonant air columns. They are excited by a mouthpiece at one end and a bell at the open end, as described in the previous

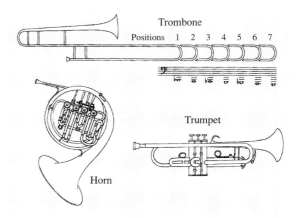

Fig. 15.112 The trombone, French horn and trumpet

section. The player selects the note to be sounded by buzzing the lips, usually at a frequency corresponding to one of the natural resonances of the coupled air column. The essential nonlinearity of this excitation process also excites multiples in frequency of the playing pitch. Ideally, for ease of playing, these harmonics should coincide with the higher modes of the excited air column. As already described, brass instruments are therefore designed to have a full harmonic set of modes. However, because of their shape and outward flare, it is impossible to achieve this for the fundamental mode (Fig. 15.77).

By adjusting the pressure and the tightness of the lips in the mouthpiece, the player can pitch notes based on the $n = 2, 3, 4 \ldots$ modes – the $n = 2$ mode, a fifth above, an octave above, an octave and a fourth above, etc. Trumpet players typically sound notes up to the 8–10 th mode, while skilled horn players can pitch notes up to and sometimes above the 15*th*. In the higher registers, the instruments can therefore play nearly all the notes of a major diatonic scale. A few of the notes can be rather badly out of tune, but a skilled player can usually correct for this by adjusting the lip pressure and flow of air through the mouthpiece. The low notes are based on simple intervals: the perfect fifth, octave, perfect fourth, etc. Trumpets and horns were therefore often used in early classical music to add a sense of military excitement and to emphasise the harmony of the key in which the music is written. However, in later classical music and music of the romantic period, all the notes of the chromatic scale were required. To achieve this, brass instruments such as the trumpet and horn were developed with a set of air valves, which enabled the player to switch in and out various combinations of different lengths of tube, to change the effective resonating length of the vibrating air column and hence playing pitch. Uniquely, the pitch of the trombone is changed by the use of interpenetrating cylindrical sliding tubes, which change the effective length. Modern instruments generally use a folded or coiled tube structure to keep the size of the instrument to manageable proportions.

The trombone can sound all the semitones of the chromatic scale, by the player sliding lengths of closely fitting cylindrical tubing inside each other. In the *first position*, with the shortest length tube (Fig. 15.112), the B-flat tenor trombone sounds the note B-flat at ≈ 115 Hz, corresponding to the $n = 2$ mode, an octave below the lowest note on the B-flat trumpet. To play notes at successive semitones lower, the total length has to be extended sequentially by fractional increases of 1.059. From the shortest to longest lengths there are seven such increasingly spaced *positions*. When fully extended, the trombone then plays a note six semitones lower (E) than the initial note sounded. One can then switch to the $n = 3$ mode to increase the pitch by a perfect fifth, to the note B a semitone higher than the initial note sounded. Using the closer positions enables the next six higher semitones to be played. Higher notes can be excited by suitable combinations of both position and mode excited. The trombone is one of the few musical instruments that can slide continuously over a large range of frequencies, simply by smoothly changing its length. This is widely used in jazz, where it also enables the player to use a very wide, frequency-modulated, vibrato effect and bending of the pitch of a note for expressive effect.

The fully extended length of the vibrating air column in the first position is ≈ 2.5 m. Two-thirds of the length is made up of 1.3 cm-diameter cylindrical tubing with the remaining gently flared end-section opening out to a bell diameter of 16–20 cm.

The trumpet achieves the full chromatic range by the use of three piston valves, which enable additional lengths of tubing to be switched in and out of the resonating air column. In the inactive *up* position, the sound travels directly through a hole passing directly through the valve. When the piston is depressed, the valve enables the tube on either side of the valve to be connected to an additional length of tubing, which includes a small, preset, sliding section for fine tuning. The pitch is decreased by a tone on depressing the first piston and a semitone by the second. Pressing them down together therefore lowers the pitch by three semitones (a minor third). Depressing the third valve also lowers the pitch by three semitones, so that when all three valves are depressed the pitch is lowered by six semitones. However, the tuning is not exact, because whenever any single valve is depressed the effective tube length is lengthened. Therefore, when a second (or third) valve is depressed, the fractional increase in effective length is less when the second valve alone is used. This is related to the need to *increase* the spacing of the semitone positions on the trombone as it is extended. Similar mistuning problems arise for all combinations of valves used.

To circumvent these difficulties, compromises have to be made, if the instrument is to play in tune (*Backus* [15.132], pp. 270–271). The added length of tubing to produce the semitone and tone intervals are therefore purposely made slightly too long, giving semitone and tone intervals that are slightly flat, but which in combination produce a three-semitone interval which is slightly sharp. Similarly, the third valve is tuned to give a pitch change of slightly more than three semitones. This allows the full of range of semitones to be played with only slight fluctuations above and below the correct tuning. The error is greatest when all three pistons are depressed. In the modern trumpet, the mistuning can be compensated by a small length of tubing operated by an additional small valve operated by the little finger of the playing hand. To regulate the overall tuning of the instrument, the playing length of the instrument can be varied by a sliding U-tube section at the first bend away from the mouthpiece. As we will show later, the skilled player can adjust the muscles controlling the dynamics of the lip-reed excitation to correct for any slight mistuning inherent in the design of the instrument.

The bends and valves incorporated into the structure of brass instruments will clearly result in sudden changes in acoustic impedance of the resonating air column, which will produce reflections and perturbations in the frequency of the frequency of the excited modes. Surprising, as shown earlier, such effects are acoustically relatively unimportant, though for the player they may affect the feel of the instrument and ease with which it can be played. In particular, when a brass player is pitching one of the higher modes of an instrument such as the horn several oscillations have to be produced before any feedback returns from the end of the instrument to stabilise the playing frequency. For example, when pitching the 12th mode on a horn, the player has to buzz the lips for about 12 cycles before the first reflection returns from the end of the instrument, which demonstrated the difficulty of exact pitching of notes in the higher registers. Note that, because of the dispersion of sound waves in a flared tube, the group velocity determining the transit time will not be the same as the phase velocity determining modal frequencies. Any strong reflections from sharp bends and discontinuities in acoustic impedance introduced by the valve structures can potentially confuse the pitching of notes and the playability of an instrument. Such transients can be investigated directly by time-domain acoustic reflectometry (see, for example, *Ayers* [15.179]). Different manufacturers choose different methods to deal with the various tuning and other acoustic problems involved in the design of brass instruments and players develop strong preferences for a particular type of instrument based on both the sound they produce and their ease of playing.

The trumpet bore is ≈ 137 cm long with largely cylindrical tubing with a diameter of ≈ 1.1, which tapers down to ≈ 0.9 cm at the mouthpiece end over a distance of ≈ 12–24 cm. It opens up over about the last third of its length with an end bell of diameter ≈ 11 cm. To reduce the overall length, its length is coiled with a single complete turn, as illustrated in Fig. 15.112.

The cornet is closely related to the trumpet but has a largely conical rather than cylindrical bore and is further shortened by having two coiled sections. This results in a somewhat lower cut-off frequency giving a slightly warmer but less brilliant sound quality. The bugle is an even simpler double-coiled valveless instrument of fixed length, so that it can only sound the notes of the natural harmonics. It was widely used by the military to send simple messages to armies and is still used today in sounding the *last post*, accompanying the burial of the military dead.

The modern French horn developed from long straight pipes with flared ends played with a mouthpiece. Technology then enabled horns to be made with coiled tubes, like hunting horns, greatly reducing their size. In early classical music, horns were only expected to play a few simple intervals in the key in which the music was written. For music written in different keys, the player had to add an additional section of coiled tube called a *crook*, to extend the length of the instrument accordingly.

To extend the range of notes that a horn could play, the player can place his hand into the end of the instrument. Depending how the hand is inserted, the pitch of individual harmonics can be lowered or raised by around a semitone, producing what is referred to as

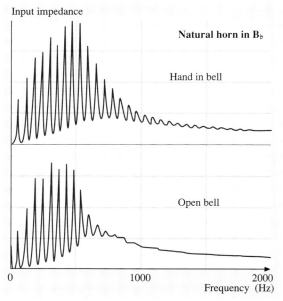

Fig. 15.113 Input impedance measurements of a natural horn, with and without a hand in the bell (after *Benade* [15.133])

a *hand-stopped* note. If the hand is partially inserted into the bell, it dramatically increases the cut-off frequency giving the player access to a much larger number of higher modes, as illustrated in measurements by *Benade* ([15.133], Fig. 20.17) reproduced in Fig. 15.113. Although the associated changes in pitch may be relatively small, inserting the hand into the end of a horn significantly changes the spectrum of the radiated sound, which allows the horn player some additional freedom in the quality of the sound produced.

The modern orchestral horn produces all the notes of the chromatic scale using rotary valves to switch in combinations of different length tube, rather like the trumpet. The modern instrument is a combination of a horn in F and a horn in B-flat, which can be interchanged by a rotary valve operated by the thumb of the left hand, while the first three fingers operate the three main valves, which are common to both horns.

The total length of the F horn is about 375 cm, a third longer than the B-flat trombone, enabling it to play down to the note F2. The F-horn is used for the lowest notes on the instrument, while the B-flat horn is used to give the player a much higher degree of security in pitching the higher notes. Like the primitive hunting horn, the modern horn is coiled to accommodate its great length and has a bore that opens up gently over its whole length from a diameter of ≈ 0.9 mm at the mouthpiece end to a rapidly flared output bell with a diameter of ≈ 30 cm.

There are many other instruments played with a mouthpiece in both ancient and ethnic cultures around the world. Many primitive instruments are simply hollowed-out tubes of wood, bamboo or animal bones. The notes that such instruments can produce are limited to the principal, quasi-harmonic, resonances of the instrument. There are also many hybrid instruments played with a mouthpiece, which use finger-stopped holes along their length, just like woodwind instruments. Important examples are the renaissance *cornett* and the now almost obsolete *serpent*, a spectacularly large, multiple bend, s-shaped, instrument. Many modern players of baroque-period *natural* trumpets have also added finger holes to the sides of their instruments, to facilitate the pitching of the highest notes.

15.4 Percussion Instruments

Compared with the extensive literature on stringed, woodwind and brass instruments, the number of publications on percussion instruments is somewhat smaller. This is largely because the acoustics of percussion instruments is almost entirely dominated by the well-understood, free vibrations of relatively simple structures, without complications from the highly nonlinear excitation processes involved in exciting string and wind instruments. However, as this section will highlight, the physics of percussion instruments involves a number of fascinating and often unexpected features of considerable acoustic interest.

The two most important references on the acoustics of percussion instruments are *The Physics of Musical Instruments*, *Fletcher* and *Rossing* ([15.5], Chaps. 18–21) and the *Science of Percussion Instruments* [15.201] by *Rossing*, the general editor of this Handbook, who has pioneered research on a very wide range of classical and ethnic percussion instruments. *James Blade*'s *Percussion Instruments and their History* [15.202] provides an authoritative survey of the development of percussion instruments from their primitive origins to their modern use.

Percussion instruments are amongst the oldest and most numerous of all musical instruments. Primitive instruments played by hitting sticks against each other or against hollowed-out tree stumps or gourds are likely to have evolved very soon after man discovered tools to make weapons and other simple artefacts essential for survival. The rhythmic beating of drums by Japanese Kodo drummers, the marching bands of soldiers down the centuries, and the massed percussion section of a classical orchestra still instil the same primeval feelings of power and excitement used to frighten away the beasts of the forest and to raise the fighting spirits of early groups of hunters. The beating of drums would also have provided a simple way of communicating messages over large distances, the rhythmic patterns providing the foundation of the earliest musical language – the organisation of sound to convey information or emotion. Martial music, relying heavily on the beating of drums continues to be used, and as often misused, to instil a sense of belonging to a particular group or nation and to instil fear in the enemy.

In China, bells made of clay and copper were already in use well before 2000 BC. The discovery of bronze quickly led to the development of some of the most sophisticated and remarkable sets of cast bells ever made, reaching its peak in the western Zhou (1122–771 BC) and eastern Zhou (770–249 BC) dynasties (*Fletcher* and *Rossing* ([15.5], Sect. 21.15). Inscriptions on the set of 65 tuned chime bells in the tomb of Zeng Hou Yi (433 BC), show that the Chinese had already established a 12-note scale, closely related to our present, but much later, western scale. The ceremonial use of bells and gongs is widespread in religious cultures all over the world and in western countries has the traditional use of summoning worshippers to church and accompanying the dead to their graves.

In the 18th century classical symphony orchestra of Haydn and Mozart's time, the timpani helped emphasise the beat and pitch of the music being played, particularly in loud sections, with the occasional use of cymbals and triangle to emphasise exotic and often Turkish influences. The percussion section of today's symphony orchestra may well be required to play up to 100 different instruments for a single piece, as in *Notations I–IV* by *Boulez* [15.203]. This typical modern score includes timpani, gongs, bells, metals and glass chimes, claves, wooden blocks, cowbells, tom-toms, marimbas, glockenspiels, xylophones, vibraphones, sizzle and suspended cymbals, tablas, timbales, metal blocks, log drums, boobams, bell plates in C and B flat, side drums, Chinese cymbals, triangles, maracas, a bell tree, etc. Modern composers can include almost anything that makes a noise – everything from typewriters to vacuum cleaners. The percussion section of the orchestra is required to play them all – often simultaneously.

We will necessarily have to be selective in the instruments that we choose to consider and will concentrate largely on the more familiar instruments of the modern classical symphony orchestra. We will also constrict our attention to instruments that are struck and will ignore instruments like whistles, rattles, scrapers, whips and other similar instruments that percussion players are also often required to play.

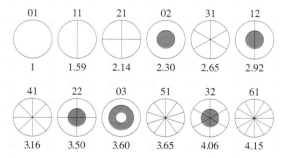

Fig. 15.114 The first 12 modes of a circular membrane illustrating the mode nomenclature, nodal lines and frequencies relative to the fundamental 01 mode

15.4.1 Membranes

Circular Membrane
A uniform membrane with areal density σ, stretched with uniform tension T over a rigid circular supporting frame, supports acoustically important transverse displacements z perpendicular to the surface described by the wave equation

$$T\left(\frac{\partial^2 z}{\partial r^2} + \frac{1}{r}\frac{\partial z}{\partial r} + \frac{1}{r^2}\frac{\partial^2 z}{\partial \phi^2}\right) = \sigma \frac{\partial^2 z}{\partial t^2}, \quad (15.147)$$

which has solutions of the form

$$z(r,\phi,t) = J_m(k_{mn}r) \begin{Bmatrix} A\cos m\phi \\ +B\sin m\phi \end{Bmatrix} e^{i\omega t}. \quad (15.148)$$

$J_m(k_{mn}r)$ are Bessel functions of order m, where n denotes the number of radial nodes and m the number of nodal diameters. The eigenfrequencies $\omega_{mn} = k_{mn}\sqrt{T/\sigma}$ are determined by the requirement that $J_m(k_{mn}a) = 0$ on the boundary at $r = a$. The frequency of the fundamental (01) mode is $(2.405/2\pi a)\sqrt{T/\sigma}$, where $J_0(k_{01}a) = 0$. The relative frequencies of the first 12 modes and associated nodal lines and circles are shown in Fig. 15.114.

For ideal circular symmetry, the independent azimuthal cosine and sine solutions result in degenerate modes having the same resonant frequencies. The degeneracy of such modes can be lifted by a nonuniform tension, variations in thickness when calfskin is used, and a lack of ideal circularity of the supporting rim. Any resulting splitting of the frequencies of such modes can result in beats between the sound of given partials, which the player can eliminate by selectively adjusting the tension around the perimeter of the membrane or by hitting the membrane at a nodal position of one of the contributing modes.

Unlike transverse waves on a stretched string, the modes of a circular membrane are inharmonic. As a consequence, the waveforms formed by the combination of such partials excited when the drumhead is struck are non-repetitive. Audio I2⟩ CD-ROM illustrates the frequencies of the first 12 modes played in sequence. I2⟩ CD-ROM illustrates their sound when played together as a damped chord, which already produces the realistic sound of a typical drum, having a reasonably well-defined sense of pitch, despite the inharmonicity of the modes.

Although percussion instruments may often lack a particularly well-defined sense of pitch, one can nevertheless describe the overall pitch as being high or low. For example, a side drum has a higher overall pitch than a bass drum and a triangle higher than a large gong. From an acoustical point of view, we will be particularly interested in the lower-frequency quasi-harmonic modes. However, one must never forget the importance of the higher-frequency inharmonic modes in defining the initial transient, which is very important in characterising the sound of an instrument.

Nonlinear effects arise in drums in much the same way as in strings (Sect. 15.2.2). The increase in tension, proportional to the square of the vibrational amplitude, leads to an increase in modal frequencies. In addition, nonlinearity can result in mode conversion (Sect. 15.2.2) and the transfer of energy from initially excited lower-frequency modes with large amplitudes to higher partials. Although the pitch of a drum is raised when strongly hit, this may to some extent be compensated by the psychoacoustic effect of a low-frequency note sounding flatter as its intensity increased. Changes in perceived pitch of a drum with time can often be emphasised by the use of digital processing, to increase the frequency of the recorded playback without changing the overall envelope in time (audio ❰?❱ CD-ROM).

Air Loading and Radiation

The above description of the vibrational states of a membrane neglects the induced motion of the air on either side of the drum skin. At low frequencies, this adds a mass $\approx \frac{8}{3}\rho a^3$ to the membrane (*Fletcher* and *Rossing* [15.5], Sect. 18.1.2), approximating to a cylinder of air with the same thickness as the radius a of the drum head. The added mass lowers the vibrational frequencies relative to those of an ideal membrane vibrating in vacuum. The effect is largest at low frequencies, when the wavelength in air is larger than or comparable with the size of the drumhead. For higher-frequency modes, with a number of wavelengths λ across the width of the drumhead, the induced air motion only extends a distance $\approx \lambda 2\pi (\ll a)$ from the membrane. Air loading therefore leaves the higher-frequency modes relatively unperturbed.

Drums can have a single drum skin stretched over a hollow body, such as the kettle drum of the timpani, or two drum heads on either side of a supporting cylinder or hollowed out block of wood, like the side drum and southern Indian mrdanga (*Fletcher* and *Rossing* [15.5], Sect. 18.5). By stretching the drum head over a hollow body, the sound radiated from the back surface is eliminated, just like mounting a loudspeaker cone in an enclosure. At low frequencies, the drum then acts as a monopole source with isotropic and significantly enhanced radiation efficiency. This is illustrated by the much reduced 60 dB decay time of the (11) dipole mode of a stretched timpani skin, when the drum skin was attached to the kettle –

Table 15.7 Ideal and measured frequencies of the modal frequencies of a timpani drum head normalised to the acoustically important (11) mode before and after mounting on the kettle, and the internal air resonances of the kettle. The *arrows* indicate the sense of the most significant perturbations of drum head frequencies and the *asterisks* indicate the resulting quasi-harmonic set of acoustically important modes (adapted from Fletcher and Rossing)

Mode	Ideal membrane	Drumhead in air		Coupled internal air resonances		Drumhead on kettle	
01	0.63	82 Hz	0.53	(0,1,0) (0,1,1)	385 Hz	127 Hz	0.85 ↑
11	1.00	160	1.0	(1,1,0) (1,1,1)	337 Hz 566 Hz	150	1.00 ↓ ***
21	1.34	229	1.48	(2,1,0) (2,1,1)	537 Hz 747 Hz	227	1.51 ↓ ***
02	1.44	241	1.55	(0,1,0)	(0,2,0)	252	1.68 ↑
31	1.66	297	1.92	(3,1,0)	(3,1,1)	298	1.99 ***
12	1.83	323	2.08	(1,2,0)	(1,2,1)	314	2.09 ↓
41	1.98	366	2.36			366	2.44 ***
22	2.20	402	2.59			401	2.67
03	2.26	407	2.63	(0,1,0)		418	2.79 ↑
51	2.29	431	2.78			434	2.89
32	2.55	479	3.09			448	2.99 ***
61	2.61	484	3.12			462	3.08

from 2.5 s to 0.5 s (*Fletcher* and *Rossing* [15.5], Table 18.4).

In addition, any net change in the volume of the enclosed air resulting from vibrations of the drum skin will increase the restoring forces acting on it and hence raise the modal frequencies, see Table 15.7. In this example, the coupling raises the frequency of the (01) drumhead mode from 82 to 127 Hz, the (02) mode from 241 to 252 Hz, and the (03) mode from 407 to 418 Hz. In contrast, the asymmetric, volume-conserving, (11) mode is lowered in frequency from 160 to 150 Hz, which probably results from coupling to the higher frequency 337 Hz internal air mode having the same aerial symmetry.

As in any coupled system in the absence of significant damping, the separation in frequency of the normal modes resulting from coupling will tend to be greater than the separation of the uncoupled resonators (Fig. 15.46b). In addition, any enclosed air will provide a coupling between the opposing heads of a double-sided drum. For example, the 227 and 299 Hz (01) uncoupled (01) modes of the opposing heads of a snare drum become the 182 and 330 Hz normal modes of the double-sided drum (*Fletcher* and *Rossing* [15.5], Fig. 18.7).

Excitation by Player

The quality of the sound produced by any percussion instrument depends as much on the player's skills as on the inherent qualities of the instrument itself. Drums are not simply hit. A player uses considerable manual dexterity in controlling the way the drumstick strikes and is allowed to bounce away from the stretched drum skin, xylophone bar or tubular bell. It is important that contact of the stick with the instrument is kept to a minimum; otherwise the stick will strongly inhibit the vibrational modes that the player intends to excite. The *lift off* is just as important as *the strike*, and it takes years of practice to perfect, for example, a continuous side drum or triangle roll.

It is also important to strike an instrument in the right place using the right kind of stick or beater to produce the required tone, resonance or loudness required for a specific musical performance. As discussed earlier in relation to the excitation of modes on a stretched string and modal analysis measurements, one can selectively emphasise or eliminate particular partials by striking an instrument at antinodes or nodes of particular modes. A skilled timpani player can therefore produce a large number of different sounds by hitting the drumhead at different positions from the rim. Striking timpani close to their outer rim preferentially excites the higher-frequency modes, while striking close to the centre results in a rather dull *thud*. This is due to the preferential excitation of the inefficient (00) mode and elimination of the acoustically important (0n) modes. The most sonorous sounding notes are generally struck about a quarter of the way in from the edge of the drum.

The sound is also strongly affected by the dynamics of the beater–drumhead interaction, which is rather soft and *soggy* near the centre and much harder and *springy* near the outer rim. The sound is also strongly affected by the type of drumstick used. Light, hard wooden sticks will make a more immediate impact with a bar or drum skin than heavily felted beaters. Such difference are similar to the effect of using heavy or light force hammers to preferentially excite lower- or higher-frequency modes in modal analysis investigations. A professional timpanist or percussion player will use completely different sticks for different styles of music and obtain effects on the same instrument ranging from the muffled drum beats of the *Death March* from Handel's *Saul* to the sound of cannons firing in Tchaikovsky's 1812 overture.

There has been much less research on the mechanics of the drumstick–skin interaction than on the excitation of piano strings by the piano hammer (Sect. 15.2.2), though much of the physics involved is very similar. In particular, the shortest time that the drumstick can remain in contact with the skin will be determined by the time taken for the first reflected wave to return from the rim of the drum to bounce it off. This is illustrated schematically in Fig. 15.115, which illustrates qualitatively the force applied to the drumhead as a function of time for a hard and a soft drumstick struck near the centre (*solid line*) and then played nearer the edge (*dashed*). The overall spectrum of sound of the drum will be controlled by the frequency content of such impulses. Short contact times will emphasise the higher partials and give

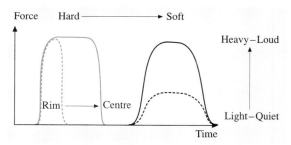

Fig. 15.115 Schematic impulses from a striking drumstick, illustrating the effect of exciting the drum head at different positions, with different strengths, and with a hard and soft drumstick

rise to a more percussive and brighter sound. Higher partials will also be emphasised by the use of metal beaters or drumsticks with hard-wooden striking heads rather than leather or soft felt-covered stick heads. This is illustrated by the second pulse, which would produce a softer, mellower sound, without such a strong initial attack. Clearly, the loudness of the drum note will be proportional to the mass m of the striking drumstick head and its impact velocity v, delivering an impulse of $\approx mv$.

Audio 13 CD-ROM illustrates the change in sound of a timpani note, as the player progressively strikes the drum with a hard felt stick, starting from the outside edge and moving towards the centre, in equal intervals of $\approx$ one eighth of the radius. Audio 13 CD-ROM illustrates the sound of a timpani when struck at one quarter of the radius from the edge, using a succession of drumsticks of increasing hardness, from a large softly felted beater to a wooden beater.

In modern performances of baroque and early classical music, the timpanist will use relatively light sticks, with leather-coated striking heads, while for music of the romantic period larger and softer felt-covered drumsticks will often be used.

Many drums of ethnic origin are played with the hands, hitting the drum head with fingers, clenched fists or open palms to create quite different kinds of sounds. In some cases, the player can also press down on the drum head to increase the tension and hence change pitch of the note. For a double-headed drum, the coupling of the air between the drum heads can even enable the player to change the pitch and sound of a given note by applying pressure to the drum head not being struck.

We now consider a number of well-known percussion instruments based on stretched membranes, which illustrate the above principles. These will include drums with a well defined pitch, such as kettle drums (timpani) and the Indian tabla and mrdanga, and drums with no defined pitch, such as the side and bass drum.

Kettle Drums (Timpani)

The kettle drum or timpani traditionally used a specially prepared calfskin stretched over a hollow, approximately hemispherical, copper kettle generally beaten out of copper sheet. Nowadays, thin (0.19 mm) mylar sheet is often used in preference to calfskin for the drum skin, because of its uniformity and reduced susceptibility to changes in tension from variations in temperature and humidity. The drum skin is stretched over a supporting ring attached to the kettle, with the tension of the skin typically adjusted using 6–8 tuning screws equally spaced around the circumference. The player adjusts these screws to tune the instrument and to optimise the quality of tone produced. In modern instruments, a mechanical pedal arrangement can be used to quickly change the tension and thereby the tuning, by pushing the supporting ring up against the drumhead. Typically, such an arrangement can increase the tension by up to a factor of two, raising the pitch by a perfect fifth. In the modern classical symphony orchestra, the timpanist will use two or three timpani of different sizes to cover the range of pitched notes required.

Figure 15.116 shows the waveform and spectrum of the immediate and delayed sound of a timpani note (the first drum note in audio 13 CD-ROM). The initial sound includes contributions from all the modes excited. This includes not only the vibrational modes of the drum head, but also the air inside the kettle, the kettle itself and even the supporting legs and vibrations induced in the floor. Many of these vibrations die away rather quickly, leaving a number of prominent, slowly decaying, drumskin modes. Note in particular, the strongly damped (01) mode at ≈ 140 Hz and the less strongly damped modes (02) and (03) modes at 210 Hz and 284 Hz, tuned ap-

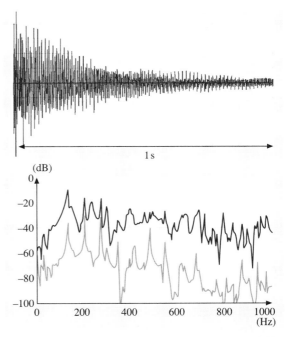

Fig. 15.116 Decaying waveform of a timpani note and FFT spectra at the start of a note (upper trace) and after 1 s (lower trace)

proximately to a perfect fifth and an octave above the fundamental. As noted by Rayleigh in relation to church bells ([15.3] Vol. 1, Sect. 394), the pitch of a note is often determined by the higher quasi-harmonically related partials rather than the lowest partial present. This is demonstrated by the second drum beat in |?) CD-ROM, which has all frequency components below 250 Hz removed. The perceived pitch at long times is unchanged, though there is a considerable loss in *depth* or *body* of the resulting sound.

Modal frequencies for a typical kettle drum have already been listed in Table 15.7, which includes a set of nearly harmonic modes indicated by asterisks achieved, in part, by empirical design of the coupled membrane and kettle air vibrations. To a first approximation, the modal frequencies are determined by the volume of the kettle rather than its shape. The smaller the enclosed volume, the larger its effect on the lowest-order drumhead modes. Nevertheless, there are distinct differences in the sounds of timpani used by orchestras in Vienna and those used elsewhere in Europe (*Bertsch* [15.204]). Such differences can be attributed to the Viennese preference for calfskin rather than mylar drum heads, a small shape dependence affecting the coupling to the internal air resonances, and a different tuning mechanism. The modal frequencies of the Viennese timpani measured by Bertsch were similar to those listed in Table 15.7, with the (11), (21), (31) and (41) modes again forming a quasi-harmonic set of partials, in the approximate ratios 1:1.5:2.0:2.4:2.9. Rather surprisingly, the relative frequencies of the lower two modes could be interchanged with tuning.

Indian Tabla and Mrdanga

Another way of achieving a near harmonic set of resonances of a vibrating drumhead is to add mass to the drum head and hence change the frequencies of its normal modes of vibration. For the single- and double-headed Indian tabla and mrdanga drums, this is achieved by selectively loading the drum skin with several coatings of a paste of starch, gum, iron oxide and other materials – see *Fletcher* and *Rossing* ([15.5], Sect. 18.5). The acoustics of the tabla was first investigated by *Raman* [15.205], who obtained Chladni patterns for many of the lower-frequency modes of the drum head. *Rossing* and *Sykes* [15.206] measured the incremental changes in frequency of the loaded membrane as each additional layer was added. A 100 layers lowered the fundamental mode by about an octave. The resulting five lowest modes were harmonically related and including several degenerate modes derived from the smoothly transformed modes of the original unloaded membrane. The results were very similar to those obtained earlier by Raman. Investigations by *Ramakrishna* and *Sondhi* [15.207] and by *De* [15.208] showed that, to achieve a quasi-harmonic set of low-frequency modes, the areal density at the centre of such drums should be approximately 10 times that of the unloaded sections.

Figure 15.117 illustrates the decaying waveform and spectra of a well-tuned tabla drum (audio |?) CD-ROM) from 200 ms FFTs of the initial sound and after 0.5 s. The spectra show three prominent partials at 549, 826 and 1107 Hz, in the near-harmonic ratios 1:1.51:2.02, which results in a well-defined sense of pitch. In contrast to the timpani, these partials dominate the sound and determine the pitch from the very beginning of the note. Note too the very wide spectrum of the rapidly decaying initial transient.

Side and Snare Drum

The side or snare drum is the classical two-headed drum of the modern symphony orchestra. It is usually played with either very short percussive beats or as a roll, with rapidly alternating notes from two alternating drumsticks. This results in a quasi-continuous

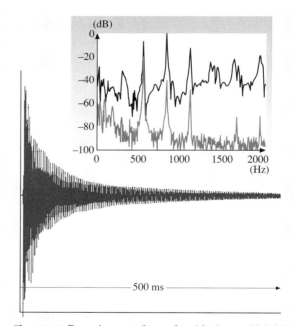

Fig. 15.117 Decaying waveform of a tabla drum with initial FFT spectrum (upper trace) and after 0.5 s (lower) illustrating the weakly damped, near-harmonic, resonances of the loaded drumhead

noise source, which can be played over a very wide range of intensities, from very soft to very loud, to support, for example, a Rossini crescendo. Because the side drum is designed to produce short percussive sounds or a wide-band source of noise, little effort is made to tune the partial frequencies of the two drum heads.

Like the timpani and Indian drums, the vibrational modes of the drumheads can be strongly perturbed in frequency by the air coupling. When used as a snare drum, the induced vibrations of the non-striking head can be sufficient for it to rattle against a number of metal cables tightly stretched a few mm above the surface of the non-striking head. The resulting interruption of the vibrations, on impact with the snares, leads to the generation of additional high-frequency noise and the *sizzle* effect of the sound excited. A not dissimilar effect is used on the Indian tambura, an Indian stringed instrument investigated by *Raman* [15.210], which has a bridge purposely designed to cause the strings to rattle *Fletcher* and *Rossing* ([15.5], Fig. 9.30).

Figure 15.118 shows the waveform and time-averaged FFT of a side-drum roll (audio ⒀ CD-ROM). The spectrum is lacking in spectral features other than a modest peak in noise at around 100–200 Hz, associated with the vibration of the lower head against the snares.

Although the exact placing of the vibrational modes of the strike and snare heads are of little acoustic importance, their coupling via the enclosed air illustrates the general properties of double-headed drums of all types. The first four coupled normal modes are shown in Fig. 15.119, which is based on data from *Rossing* ([15.211] Sect. 4.4). For a freely supported drum, momentum has to be conserved, so that normal modes with the two heads vibrating in the same phase will also involve motion of the supporting shell of the drum, as indicated by the arrows in Fig. 15.119.

As anticipated, the air coupling increases the separation of the (01) modes from 227 and 299 Hz to 182 and 330 Hz, and the (11) modes from 284 and 331 Hz to 278 and 341 Hz. The perturbations in modal frequencies will always be largest when the coupled modes have similar frequencies. Such perturbations becomes progressively weaker at higher frequencies, partly because the coupling from the enclosed air modes becomes weaker and partly because the frequencies of the two drum-head modes having the same symmetry become more widely separated. The higher modal frequencies are therefore little different from those of the individual drum heads in isolation.

Figure 15.119 also illustrates the anticipated polar radiation patterns for the normal modes measured by *Zhao* [15.209] and reproduced in *Rossing* ([15.201] Figs. 4.7 and 4.8). The coupled (10) normal modes act as a monopole radiation source, when the heads move in opposite directions, and a dipole source, when vibrating in anti-phase. In contrast, the (11) modes with heads vibrating in phase act as a quadrupole source, and a dipole source, when vibrating in anti-phase.

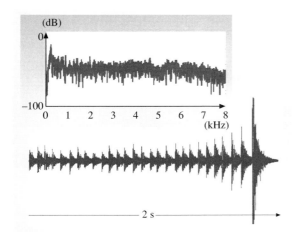

Fig. 15.118 Time-averaged FFT spectrum and waveform of the sound of a snared side-drum roll of increasing intensity

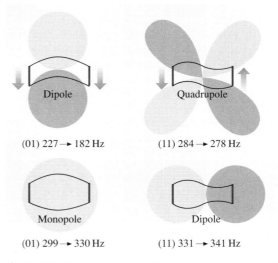

Fig. 15.119 Coupled motions of the two drumheads of a side drum, indicating the change in frequency of the drumhead modes from air coupling within the drum and the associated polar radiation patterns (after *Zhao* et al. [15.209])

Although any induced motion of the relatively heavy supporting structure will not significantly affect the frequencies of the normal modes, it can result in appreciable additional damping. As a consequence, the sound of a side drum can sound very different depending on how it is supported – freely suspended on rubber bands or rigidly supported by a heavy stand (*Rossing* [15.201] Sect. 4.4). Rossing has also made detailed vibrational and holographic studies of the free drum shell ([15.201] Fig. 4.5). As the induced motions are only a fraction of a percent of those of the drumhead, such vibrations will not radiate a very large amount of sound. Nevertheless, they may play an important role in defining the initial sound.

Bass Drum

The bass drum is large with a typical diameter of 80–100 cm. It can produce a peak sound output of 20 W, the largest of any orchestral instrument. Single-headed drums are used when a well-defined sense of pitch is required, but double-headed drums sound louder because they act as monopole rather than dipole sources. Modern bass-drum heads generally use 0.25 mm-thick mylar, though calfskin is also used.

The *batter* or beating drum head is normally tuned to about a fourth above the *carry* or resonating head (*Fletcher* and *Rossing* [15.5], Sect. 18.2). The change in modal frequencies induced by the enclosed air is illustrated in Table 15.8 (*Fletcher* and *Rossing* [15.5], Table 18.5). Note the strong splitting of the lowest frequency (01) and (11) normal modes, when the two heads are tuned to the same tension. In this example, the frequencies of the first five modes are almost harmonic, giving a sense of pitch to the sound (audio ⟨?⟩ CD-ROM illustrates the rather realistic synthesised sound of the first six modes of the batter head tuned to the carry head with equal amplitude and decay times). Drums with heads tuned to the same pitch have a distinctive timbre.

Table 15.8 Modal frequencies in Hz of the batter head of a 82 cm bass drum (after *Fletcher* and *Rossing* [15.5])

Mode	Batter head with carry head at lower tension	Batter head with heads at same tension (Hz)
(01)	39	44 , 104 split normal modes
(11)	80	76, 82 split normal modes
(21)	121	120
(31)	162	160
(41)	204	198
(51)	248	240

15.4.2 Bars

This section is devoted to percussion instruments based on the vibration of wooden and metallic bars, both in isolation and in combination with resonating air columns. Such instruments are referred to as idiophones – bars, plates and other structures that vibrate and produce sound without having to be tensioned, unlike the skins of a drum (membranophones). Representative instruments considered in this section include the glockenspiel, celeste, xylophone, marimbas, vibraphone and triangle.

The vibrations of thick and thin plates have already been considered in the context of the vibrational modes of the wooden plates of stringed instruments (Sect. 15.2.6). The most important acoustic modes of a rectangular plate are the torsional (twisting) and flexural (bending) modes, both of which involve acoustically radiating displacements perpendicular the surface of the plate.

The torsional vibrations of a bar are discussed by *Fletcher* and *Rossing* ([15.5], Sect. 2.20). For a bar of length L, the frequency of the twisting modes is given by $f_n = nc_\theta/2L$, where c_θ is the dispersionless velocity of torsional waves. For a rectangular bar with width w significantly larger than its thickness h, by $c_\theta \sim 2t/w\sqrt{E/2\rho(1+\nu)}$, where E is the Young's modulus and ν the Poisson ratio. For a bar with circular, cross-section, like the sides of a triangle, $c_\theta = \sqrt{E/2\rho(1+\nu)}$.

Musically, the most important modes of a thin bar are the flexural modes involving a displacement z perpendicular to their length, which for a rectangular bar satisfies the fourth-order wave equation

$$\frac{E}{12(1-\nu^2)}h^2\frac{\partial^4 z}{\partial x^4} + \rho\frac{\partial^2 z}{\partial t^2} = 0 , \quad (15.149)$$

with standing-wave solutions of the general form

$$z(x,t) = (A \sin kx + B \cos kx \\ + C \sinh kx + D \cosh kx)e^{i\omega t} , \quad (15.150)$$

where

$$\omega = \sqrt{\frac{E}{12\rho(1-\nu^2)}}hk^2 . \quad (15.151)$$

As discussed in the earlier section on the vibrational modes of soundboards and the plates of a violin or guitar, the sinh and cosh functions decay away from the ends of the bar or from any perturbation in the geometry, such

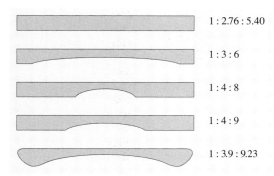

Fig. 15.120 Ratio of the frequencies of the first three partials of a simple rectangular bar for three selectively thinned xylophone bars and a typical marimba bar (after *Fletcher* and *Rossing* [15.5])

as local thinning or added mass, over a distance

$$k^{-1} \sim \left(\frac{E}{12\rho(1-\nu^2)}\right)^{1/4} \sqrt{\frac{h}{\omega}} \, .$$

Well away from the ends of a bar, the standing-wave solutions at high frequencies are therefore dominated by the sinusoidal wave components.

The lowest flexural modes of a freely supported thin rectangular bar are inharmonic, with frequencies in the ratios 1:2.76:5.40:8.93. However, by selectively thinning the central section, the frequency of the lowest mode can be lowered, to bring the first four harmonics more closely into a harmonic ratio, as illustrated schematically for a number of longitudinal cross-sections in Fig. 15.120, which also includes the measured frequencies of a more-complex-shaped marimba bar (*Fletcher* and *Rossing* [15.5], Figs. 19.2 and 19.7).

Pitch Perception

The audio ⟨?⟩ CD-ROM contrasts the synthesised sounds of the first four modes of a simple rectangular bar, followed by a note having the same fundamental but with partials in the ratio 1:3:6, while the final note has the inharmonic (1:8.96) fourth partial of the rectangular bar added. Despite the inharmonicity of the partials, the synthesised sound of a rectangular bar has a surprisingly well-defined sense of pitch. The main effect of replacing the second and third partials with partials in the ratio 1:3:6 is to raise the perceived pitch by around a tone, even though the first partial is unchanged at 400 Hz. This again emphasises that the perceived pitch is determined by a weighted average of the partials present and not by the fundamental tone alone. Adding the fourth inharmonic partial gives an increased edge or *metallic* characteristic to the perceived sound, without changing the perceived pitch.

The metal or wooden bars of tuned percussion instruments are usually suspended on light threads or rest on soft pads at the nodal positions of their fundamental mode, which reduces the damping to a minimum. The resulting 60 dB decay time for an aluminium vibraphone bar can be as long as 40 s (*Rossing* [15.201] Sect. 7.3) compared with a few seconds for the lower-frequency notes on the wooden bars of a marimba (*Rossing* [15.201], Sect. 6.4). The damping of vibrating bars is therefore highly material dependent and is largely determined by internal damping losses rather than radiation. This accounts for the very different sounds of wooden and metal bars on instruments like the glockenspiel and xylophone.

Glockenspiel and Celeste

The simplest of all idiophones are those instruments based on the vibrations of freely supported thin rectangular plates. Such instruments include the glockenspiel played with a variety of hard and soft round-headed hammers and the celeste played with strikers operated from a keyboard, with a sustaining pedal to control the damping. The playing range of the glockenspiel is typically

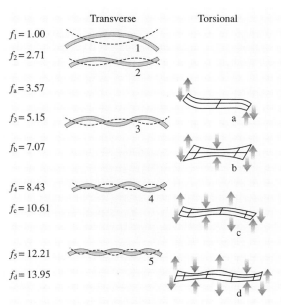

Fig. 15.121 Measured flexural and torsional modes of a glockenspiel bar (after *Fletcher* and *Rossing* [15.5])

two and a half octaves from G5 to C8, while the celeste has a range of 4–5 octaves, with a separate box-resonator used for each note.

Both instruments produce a bright, high-pitched, bell-like, sparkling sound, as in the *Dance of the Sugar Plum Fairy* in Tchaikovsky's *Nutcracker Suite*. No attempt is made to adjust the thickness of the plates to achieve a more nearly harmonic set of modes.

Figure 15.121 illustrates the lowest order flexural and torsional modes and measured ratios of frequencies for a C6 glockenspiel bar (*Fletcher* and *Rossing* [15.5], Fig. 19.1). A typical wave-envelope and spectrum of a glockenspiel note is shown in Fig. 15.122. FFT spectra are shown for 200 ms sections from the initial transient and after 200 ms. There are two strongly contributing weakly damped partials at 1045 Hz and 2840 Hz (in the ratio 1:2.72), which can be identified as the first two flexural modes of the bar. The lower of the two long sounding partials gives the sense of pitch, while the strong inharmonic upper partial gives the note its "glockenspiel" character. Audio 13 CD-ROM compares the recorded glockenspiel note with synthesised tones at 1045 and 2840 Hz, first played separately then together. In this case, the inharmonicity of the strongly sounded partials plays a distinctive role in defining the character of the sound. The spectrum is typical of all the notes on a glockenspiel, which demonstrates that only a few of the modes shown in Fig. 15.121 contribute significantly to the perceived sound.

Xylophone, Marimba and Vibraphone

We now consider a number of acoustically related idiophones, with bars that are selectively thinned to produce a more nearly harmonic set of resonances and well-defined sense of pitch.

The modern xylophone has a playing range of typically 3 to 4 octaves and uses wooden bars, which are

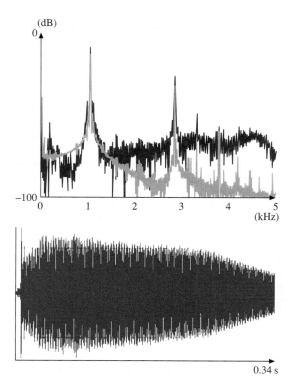

Fig. 15.122 The waveform envelope and FFT spectra of the prompt sound (upper) and the sound after ≈ 0.2 s (lower) of a typical glockenspiel note illustrating the long-time dominance of a few slowly decaying, inharmonic partials

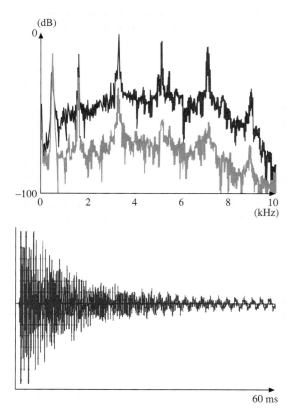

Fig. 15.123 The initial 60 ms of a xylophone waveform showing the rapid decay of high-frequency components and FFT spectra at the start of the note (upper trace) and after 0.2 s (lower trace), highlighting the persistence of the strong low-frequency air resonances excited

undercut on their back surface to improve the harmonicity of the lower frequency modes (Fig. 15.120). Each bar has an acoustic resonator immediately below it, consisting of a cylindrical tube, which is closed at the far end. Any of the flexural and torsional modes can contribute to the initial sound, when the bar is struck by a hammer; however, most modes die away rather quickly so that at longer times the sound is dominated by the resonances of the coupled pipe resonator.

Figure 15.123 shows the initial part of the waveform and spectrum of a typical xylophone note (🔊 CD-ROM), illustrating the initial large amplitudes and rapid decay of the higher frequency bar modes excited and the strongly excited but slowly decaying resonances of the first two modes of the air resonator. All modes contribute to the initial sound but the sound at longer times is dominated by the lowest-frequency bar modes and resonantly tuned air resonators.

The marimba is closely related to the xylophone, but differs largely in its playing range of typically two to four and a half octaves from A2 (110 Hz) to C7 (2093), though some instruments play down to C2 (65 Hz). In contrast to xylophone bars, which are undercut near their centre to raise the frequency of their second partial from 2.71 to 3.0 above the fundamental, marimba bars are often thinned still further to raise the frequency of the second partial to four times the fundamental frequency (Fig. 15.119).

Marimbas produce a rather mellow sound and are usually played with much softer sticks than traditionally used for the xylophone. Although the marimba is nowadays used mostly as a solo percussion instrument, in the 1930s ensembles with as many as 100 marimbas were played together. In many ways, such ensembles were the forerunners of today's Caribbean steelbands, to be described later in this section.

The vibraphone is similar to the marimba, but uses longer-sounding aluminium rather than wooden bars and typically plays over a range of three octaves from F3 to F6. Like the marimba, the bar thickness is varied to give a second partial two octaves above the fundamental. They are usually played with soft yarn-covered mallets, which produce a soft, mellow tone. In addition, the vibraphone incorporates electrically driven rotating discs at the top of each tuned air resonator, which periodically changes the coupling. This results in a strong amplitude-modulated vibrato effect. The wave envelope of audio 🔊 CD-ROM is shown in Fig. 15.124 for a succession of notes played on the vibraphone with vibrato, which are then allowed to decay freely. The vibrato rate can be adjusted by changing the speed of the electric motor.

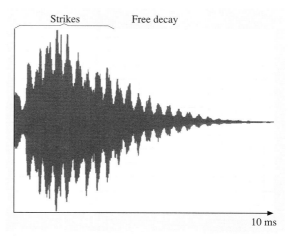

Fig. 15.124 Envelope of a succession of notes on the vibraphone, which freely decay with modulated coupling to tuned air resonators to produce an amplitude modulated vibrato effect

Note the very long decay of the sound, which can be controlled by a pedal-operated damper.

Triangle

The triangle is a very ancient musical instrument formed from a cylindrical metal bar bent into the shape of a triangle, with typical straight arm lengths of 15–25 cm. They are usually struck with a similar-diameter metal rod. Although the instrument is small and therefore a very inefficient acoustic radiator, it produces a characteristic high-frequency *ping* or repetitive high-pitched rattle, which is easily heard over the sound of a large symphony orchestra (audio 🔊 CD-ROM). The quality of the sound can be varied by beating at different positions along the straight arms. The triangle is usually supported be a thread around the top bend of the hanging instrument.

The flexural modes of a freely suspended bar of circular cross section are $f_n \sim (a/2)\sqrt{E/\rho}(2n+1)^2\pi/8L^2$, with frequencies in the ratios $3^2:5^2:7^2:9^2:11^2:13^2$ (see Sect. 15.2.4). Transverse flexural vibrations can be excited perpendicular or parallel to the plane of the instrument. For vibrations perpendicular to the plane, the bends are only small perturbations. Transverse modes in this polarisation are therefore almost identical to those of the straight bar from which the triangle are bent (*Rossing* [15.212] Sect. 7.6). However, for flexural vibrations in the plane, there is a major discontinuity in impedance at each bend, because the transverse vibrations within one arm couple strongly to longitudinal vibrations in

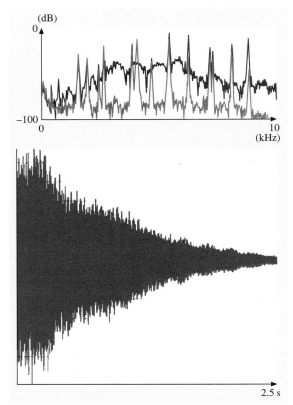

Fig. 15.125 The decaying waveform and FFT spectra at the start (upper trace) and after 1 s (lower trace) of a struck triangle note

the adjacent arms. Hence each arm will support its own vibrational modes, which will be coupled to form sets of normal-mode triplets, since each arm is of similar length.

Figure 15.125 illustrates the envelope and 50 ms FFTs of the initial waveform and after 1 s, illustrating the very well-defined and only weakly attenuated high-frequency modes of the triangle. Note the wide-band spectrum at the start of the note from the initial impact with the metal beater.

Chimes and Tubular Bells

We include orchestral chimes and bells in this section because their acoustically important vibrations are flexural modes, just like those of a xylophone or triangle. The radius of gyration of a thin-walled cylindrical tube of radius a is $\approx a/\sqrt{2}$. The frequency of the lowest flexural modes is then be given by $f_n \sim a\sqrt{E/2\rho}(2n+1)^2\pi/8L^2$.

Orchestral chimes are generally fabricated from lengths of 32–38 mm-diameter thin-walled tubing, with the striking end often plugged by a solid mass of brass with an overhanging lip, which provides a convenient striking point.

Fletcher and *Rossing* ([15.5], Sect. 19.8) note that the perceived pitch of tubular bells is determined by the frequencies of the higher modes, excited with frequencies proportional to 9^2, 11^2, and 13^2, in the approximate ratios 2:3:4. The pitch should therefore sound an octave below the lowest of these. Readers can make there own judgement from audio 12 CD-ROM, which compares the rather realistic sound of a tubular bell synthesised from the first six equal amplitude modes of an ideal bar sounded together, followed by the 9^2, 11^2, and 13^2 modes in combination, and then by a pure tone an octave below the 9^2 partial. Such comparisons highlight the problem of subjective pitch perception in any sound involving a combination of inharmonic partials.

15.4.3 Plates

Flexural Vibrations

This section describes the acoustics of plates, cymbals and gongs, which involve the two-dimensional flexural vibrations of thin plates described by the two-dimensional version of (15.149). Unlike stringed instruments, we can usually assume that the plates of percussion instruments are isotropic. Well away from any edges or other perturbing effects such as slots or added masses, the standing-wave solutions at high frequencies will be simple sinusoids. However, close to the free edges, and across the whole plate at low frequencies, contributions from the exponentially decaying solutions will be equally important over a distance $\sim (E/12\rho(1-\nu^2))^{1/4}(h/\omega)^{1/2}$. The nodes of the sinusoidal wave contributions will be displaced a distance $\sim 1/4\lambda$ from the edges. Hence, the higher frequency modes of a freely supported rectangular plate of length a, width b and thickness h will be given, to a first approximation, by

$$\omega_{mn} \sim h\left(\frac{E}{12\rho(1-\nu^2)}\right)^{1/4}\pi^2 \left[\left(\frac{m+1/2}{a}\right)^2 + \left(\frac{n+1/2}{b}\right)^2\right]. \quad (15.152)$$

A musical instrument based on the free vibrations of a thin rectangular metal plate is the thunder plate, which when shaken excites a very wide range of closely spaced modes, which can mimic the sharp clap followed by the rolling sound of thunder in the clouds.

Before the age of digital sound processing, such plates were often used in radio and recording companies to add artificial reverberation to the recorded sound. The plate was suspended in a chamber along with a loudspeaker and pick-up microphone. The sound to which reverberation was to be added was played through the loudspeaker, which excited the weakly damped vibrational modes of the plate, which were then re-recorded using the microphone to give the added reverberant sound. As described earlier (Sect. 15.2.4), the density of vibrational modes of a flat plate of area A and thickness h is given by $1.75A/c_L h$. This can be very high for a large-area thin metal sheet, giving a reverberant response with a fairly uniform frequency response.

Most percussion instruments which involve the flexural vibrations of thin sheets are axially symmetric, such as cymbals, gongs of many forms, and the vibrating plate regions of steeldrums or pans used in Caribbean steelbands. Such instruments have many interesting acoustical properties, many derived from highly nonlinear effects when the instrument is instrument is struck strongly.

The displacements of the flexural modes of an axially symmetric thin plate in polar coordinates are given by

$$z(r, \phi, t) = [A J_m (k_{mn}r) + B I_m (k_{mn}r)]$$
$$[C \cos(m\phi) + D \sin(m\phi)] \, e^{i\omega_{mn}t} ,$$
(15.153)

where $J_m(kr)$ and $I_m(kr)$ are ordinary and hyperbolic Bessel functions, the equivalent of the sinusoidally varying and exponentially damped sinh and cosh functions used to describe flexural waves in a rectangular geometry. The hyperbolic Bessel functions are important near the edges of a plate or at any perturbation, but decay over a length scale of $\sim k_{mn}^{-1}$. The values of k_{mn} are determined from the boundary conditions, in just the same way as considered earlier for rectangular plates.

The first six vibrational modes for circular plates with free, hinged and clamped outside edges are shown in Fig. 15.126, with frequencies expressed as a ratio relative to that of the fundamental mode (*Fletcher* and *Rossing* [15.5], Sect. 3.6). In each case, for large values of m and n, the frequency is given by the empirical Chladni's Law (1802),

$$\omega_{mn} \sim \sqrt{\frac{E}{12\rho(1-\sigma^2)}} \frac{\pi^2 h}{4a^2} (m+2n)^2 ,$$
(15.154)

justified much later by *Rayleigh* ([15.3] Vol. 1, Chap. 10). For arched plates, *Rossing* [15.213] showed

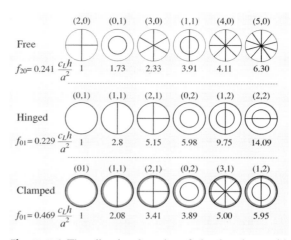

Fig. 15.126 The vibrational modes of circular plates with free, hinged and clamped outer edges, with the lowest frequencies and ratio of frequencies of higher modes indicated

that the frequencies are more closely proportional to $(m + bn)^p$, where p is somewhat less than 2 and b is in the the range 2–4.

All the axially symmetric modes involving nodal diameters are doubly degenerate, with a complementary solution with nodal diameters bisecting those drawn in Fig. 15.126. Any perturbation of the structure from cylindrical symmetry will split the degeneracy of such modes. Modes with a nodal diameter passing through the point at which the plate is struck will not be excited.

Instruments like the cymbal are slightly curved over their major surfaces but with a sudden break to a more strongly arched central section, to which the leather holding straps or support stand are attached. The outer edges can therefore be treated as free surfaces with the transition to the central cupped region providing an additional internal boundary condition, which will be intermediate between clamped and hinged. In contrast, gongs tend to have a relatively flat surface but with their edges turned though a right angle to form a cylindrical outer rim. The rim will add mass to the modes involving a significant radial displacement at the edge, but will also increase the rigidity of any mode having an azimuthal dependence. Thus, although the modes of an ideal circular plate provide a guide to modal frequencies and waveforms, we would expect significant deviations in modal frequencies for real percussion instruments. Any sudden change in plate profile on a length scale smaller than an acoustic wavelength will involve a strong coupling between the transverse flexural and longitudinal waves resulting in reflections from the discontinuity in acoustic impedance. This is why, for example, the indented region on the surface of a steeldrum pan can support localised standing waves on the indented regions. Indented areas of different sizes can then be used to produce a wide range of different notes on a single drum head with relatively little leakage in vibrations between them.

Nonlinear Effects

Figure 15.127 illustrates the cross section of some typical axially symmetric cymbals, gongs and a steelpan. Nonlinear effects in such instruments can be important when excited at large amplitudes. Such effects are particularly marked for gongs with relatively thin plates. For Chinese opera gongs, the nonlinearity can result in dramatic upward or downward changes in the pitch of the sound after striking. In addition, nonlinearity results in mode conversion, with the high-frequency content of cymbal and large gong sounds increasing with time, giving a characteristic *shimmer* to the sound quality.

The shape dependence of the nonlinearity arises from the arching of the plate. For the lowest mode of a flat plate, the potential energy initially increases quadratically with displacement, though the energy increases more rapidly at large-amplitude excursions from stretching, as indicated schematically in Fig. 15.128a. Although the energy of an arched plate initially also increases quadratically with distance about its displaced equilibrium position, the energy will first increase then decrease when the plate is pushed through the central plane, (Fig. 15.128b). If the plate were to be pushed downwards with increasing force, it would suddenly spring to a new equilibrium position displaced about the central plane by the same initial displacement, but in the opposite direction. In combination with a Helmholtz radiator, this is indeed how some insects such as cicadas generate such strong acoustic signals – as high as 1 mW at 3 kHz (see Chap. 19 by Neville Fletcher). The nonlinear bistable flexing of a belled-out plate can be disastrous, turning a cheap pair of thin cymbals inside out, when crashed together too strongly.

The central peak in potential energy of an arched plate, like the gently domed Chinese gong illustrated in Fig. 15.127, therefore leads to a reduced restoring force for large-amplitude vibrations and a lower pitch. In contrast, the restoring force of a flat plate, like the central playing region of the upper of the two Chinese gongs, increases with vibration amplitude. This arises from the additional potential energy involved in stretching the plate, in just the same way as we considered earlier for the large-amplitude vibrations of a stretched string.

For a flat plate of thickness h, Fletcher [15.214] has shown that the increase in frequency of the lowest axisymmetric mode increases with the amplitude of vibration a approximately as

$$\omega \sim \omega_0 \left[1 + 0.16(a/h)^2 \right] . \qquad (15.155)$$

The nonlinearity also generates additional components at three times the frequency of any initial modes present and at multiples of any new modes excited. In addition it produces cross-modulation products when more than one mode is present. For example, for modes with frequencies f_i and f_2 initially present, the nonlinearity will generate inter-modulation products at $2f_1 \pm f_2$ and $2f_2 \pm f_1$.

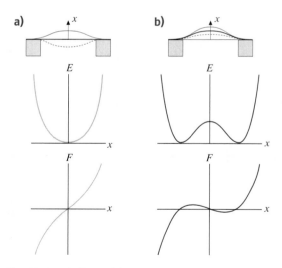

Fig. 15.128a,b Potential energy and restoring force of (**a**) a clamped flat and (**b**) arched circular plate as a function of displacement from the central plane

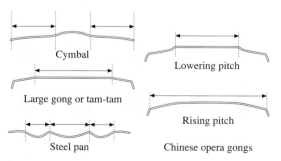

Fig. 15.127 Schematic cross sections of axially symmetric plate instruments and a steelpan, with *arrows* indicating the principal areas producing sound

Grossmann et al. [15.215] and *Fletcher* [15.214] have considered the vibrations of a spherical-cap shell of height H and thickness h. Interestingly, the change in frequency of the asymmetric vibrations about the equilibrium point depends only on the ratio of the amplitude to the arching height, a/H, as illustrated in Fig. 15.129. When the height of the dome is much less than the thickness, the frequency increases approximately as a^2, as expected from the induced increase in tension with amplitude. However, when the arching becomes comparable with and greater than the thickness, the asymmetry of the potential energy dominates the dynamics and results in an initial decrease in frequency, which increases strongly with the ratio of arching height to thickness. At very large amplitude, $a \gg H$, the frequency is dominated by the increase in tension and therefore again increases with amplitude like a flat plate. At large amplitudes, *Legge* and *Fletcher* [15.216] have shown that changes in the curvature of the plate profile result in a large transfer of energy to the higher-frequency plate modes.

We now show how many of the above properties relate to the sounds of cymbals, gongs of various types and steelpans.

Cymbals

Many types of cymbals are used in the classical symphony orchestra, marching bands and jazz groups. They are normally made of bronze and have a diameter of 20–70 cm. The low-frequency modes of a cymbal are very similar to those of a flat circular plate and can be described using the same (mn) mode nomenclature (*Fletcher* and *Rossing* [15.5], Fig. 20.2). However, small changes in curvature across a cymbal will results in modes that are linear combinations of ideal circular-plate modes.

Cymbals are usually played by striking with a wooden stick or soft beater or a pair can be crashed together, each method producing a distinctive sound. They can even be played by bowing with a heavy rosined bow against the outer rim. Rossing and Shepherd showed that the characteristic 60 dB decay time of the excited modes of a large cymbal varies approximately inversly proportional with frequency, with a typical decay time for the lowest (20) mode as long as 300 s (*Fletcher* and *Rossing* [15.5], Fig. 20.5)

Figure 15.130 shows the waveform envelope and spectra of the initial sound of a cymbal crash and after 1 s. Audio 12 illustrates a recorded cymbal crash followed by the same sound played first through a 0–1 kHz and then a 1–10 kHz band-pass filter, illustrating the decay of the low- and high-frequency wide-band noise.

When a cymbal is excited with a drumstick, waves travel out from the excitation point with a dispersive group velocity proportional to k, inversely proportional to the dimensions of the initial flexural indention of the surface made by the drumstick. The dispersive pulse strikes and is reflected from the edges of the cymbal

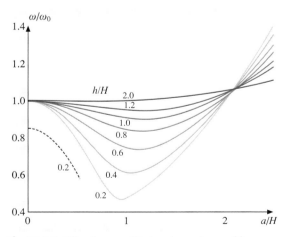

Fig. 15.129 Vibration amplitude a dependence of frequency of lowest axisymmetric mode of a spherical cap of dome height H as a function of thickness h to H ratio

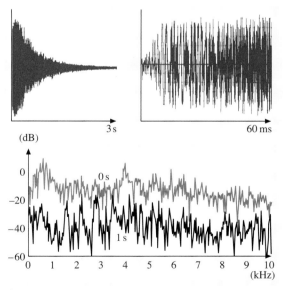

Fig. 15.130 Wave envelope and spectrum at start and after 1 s of a cymbal clash illustrating wide-band noise at all times

and the transitional region to the central curved cup, so that eventually the energy will be dispersed across the whole vibrating surface. This has been investigated using pulsed video holography by *Schedin* et al. [15.217]. On reflection there will also be considerable mixing of modes. In addition, because the plates are rather thin and are often hit extremely strongly, nonlinear effects are important. On large cylindrical plates this results in the continuous transfer of energy from strongly excited low-frequency modes to higher modes. Many of the nonlinear effects can be investigated in the laboratory using sinusoidal excitation. Measurements by *Legge* and *Fletcher* [15.216] have revealed a wide range of nonlinear effects including the generation of harmonic, bifurcations and even chaotic behaviours at large intensities.

When a plate is struck by a beater, the acoustic energy is distributed across a very wide spectrum of closely spaced resonances of the plate. To distinguish individual partials requires the sound to be sampled over a time of at least $\approx 1/\Delta f$, where Δf is the separation of the modes at the frequencies of interest. However, because the modes of a large cymbal are so closely spaced, the times involved can be rather long. Unlike the sound of pitched instruments such as the glockenspiel and xylophone, there are no particular resonances of the cymbal that dominate the sound, which is characterised instead by the *sizzle* produced by the very wide spectrum of very closely spaced resonances almost indistinguishable from wide-band high-frequency noise.

Large Gongs and Tam-Tams

Gongs are also very ancient instruments, which have a very characteristic sound when strongly struck by a soft beater, notably as a prelude to classic films by the Rank organisation. A typical tam-tam gong used in a symphony is a metre or even larger in diameter. It is made of bronze and, like cymbals, is sufficiently ductile not to shatter when strongly hit. The damping is very low, so the sound of large gongs can persist for very long times.

When initially struck strongly by a soft beater, the initial sound is largely associated with the lower-frequency partials that are strongly excited. However, on a time scale of a second or so, the sound can appear to grow in intensity, as nonlinear effects transfer energy from lower- to higher-frequency modes (audio I3⟩CD-ROM). This is illustrated in Fig. 15.131 (*Fletcher* and *Rossing* [15.5], Fig 20.8), which shows the build up and subsequent decay of acoustic energy in the higher-frequency bands at considerable times after the initial

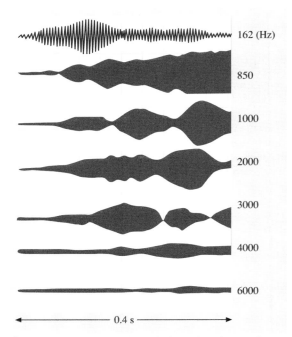

Fig. 15.131 Buildup and decay in intensity of a struck tam-tam sound in frequency bands centred on the indicated frequencies (after *Fletcher* and *Rossing* [15.5])

impact. The fluctuations in intensity within these bands were taken as evidence for chaotic behaviour. However, even in a linear system, interference between the large number of very closely spaced inharmonic partials would also result in apparently random fluctuations in amplitude.

Chinese Opera Gongs

Chinese gongs provide the most dramatic illustration of nonlinearity in percussion instruments, with upwards or downwards pitch glides of several semitones over a sizeable fraction of a second after being strongly struck. The direction of the pitch *glide* depends on the profile of the vibrating surface as previously described.

Figure 15.132 shows the decaying waveforms of the sound of three Chinese gongs with a downward pitch glide (audio I3⟩CD-ROM) played in succession. The initial spectrum of the third note played is followed by spectra at 0.75 and 1.5 s after striking, illustrating the transfer of energy to lower-frequency modes. The much broader width of the initial spectrum reflects the decrease in lifetime of the initial modes struck resulting from the nonlinear loss of energy to higher-frequency modes. The two well-defined peaks between the two major peaks are

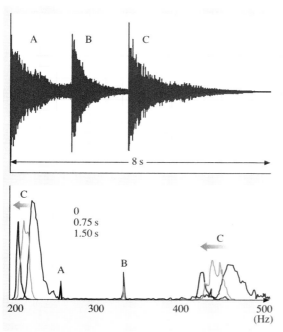

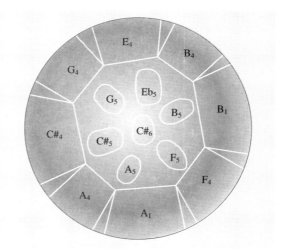

Fig. 15.133 Typical indentation areas in a tenor steelpan (after *Fletcher* and *Rossing* [15.5])

Fig. 15.132 The wave envelope of sounds from three downward-sliding Chinese gongs followed by the spectrum of the third gong at the start, after 0.75 and 1.5 s illustrating the nonlinear frequency shifts

from the long-ringing principal partials of the first two gongs.

Steelpans

Finally, in this section on percussion instruments based on vibrating plates, we consider steelpans originating from the Caribbean, which were initially fabricated by indenting the top of oil cans left on the beaches by the British navy after World War II. They have become a immensely popular instrument in that part of the world and are just as interesting from a musical acoustics viewpoint (see *Fletcher* and *Rossing* [15.5], Sect. 20.7 for further details).

Pitched notes on a given drum are produced by hammered indentations of different sizes on the top face of the drum. Different ranges of notes are produced by drums or pans of different sizes (e.g. lead tenor, double tenor, alto, cello and bass). Typical indented regions on a double-tenor steelpan are shown in Fig. 15.133, adapted from drawings for a full set of pans in *Fletcher* and *Rossing* ([15.5], Fig. 20.17). The various indented areas on the drum head can be considered as an array of relatively weakly coupled resonators. An individual in-

dented area on an infinite sheet would have very similar acoustic modes as those of a hemispherical cap indented in an infinite plate. The frequency of the modes would be determined by the size and arching of the indented areas. The effective cap size would be defined by the rate of change of the curvature of the plate and the associated change in the acoustic impedance at the edge of the indented region. However, all such regions on a steelpan will be coupled together by the relatively weak transfer of acoustic energy between them, to form a set of coupled modes. Hitting one particular region

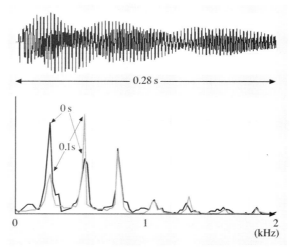

Fig. 15.134 Waveform and initial and time-delayed spectra of the note C#4 on a steeldrum

will therefore excite other regions, especially those that have closely matching partials. The coupling between such regions has been investigated holographically by *Rossing* ([15.201] Fig. 20.20).

Audio I⟩ CD-ROM illustrates a succession of notes played on a steeldrum. Figure 15.134 shows a typical decaying waveform with initial and time-delayed spectra of a single note. A relatively large number of well-defined modes can easily be identified. However, the subjective absolute pitch of the note is not particularly well defined and there is a strong sense of pitch circularity in the sound of an ascending scale (Sect. 15.1.3 and audio I⟩ CD-ROM), sometimes making it difficult to identify the octave to which a particular note should be attributed. Note the *increase* in the amplitude of the second partial with time, which could result from nonlinear effects in such thin-walled structures, or possibly from interference beats between degenerate modes split in frequency by the lack of axial symmetry of the hand-beaten indentations.

15.4.4 Shells

Blocks and Bells

Finally, we consider the acoustics of three-dimensional shell-like structures. This could include percussion instruments such as the wooden-block and hollow gourd instruments like the maracas. However, the physics of such instruments is essentially the same as that of the soundbox of stringed instruments and involves little of additional acoustic interest. In this section, we will therefore concentrate on the acoustics of bell-like structures, which are usually axially symmetric structures, closed at one end, and of variable thickness and radius along their length. We will also consider non-axisymmetric bells with a quasi-elliptical cross section, which produce two different pitched notes, depending on where the bell is struck. All such structures have a rich spectrum of modes, which are generally tuned to give long-ringing notes with a well-defined sense of pitch. The bronze used in their construction is typically an alloy of 80% copper and 20% tin and has to be sufficiently ductile not to crack under the impact of the beater or clapper. Metallurgical treatment is required to produce a grain structure producing little damping at acoustic frequencies.

Some of the oldest bells are to be found in China. Such bells are supreme examples of the art of bronze casting dating to the fifth century BC. Bells in church towers have traditionally marked the passage of time, while peals of bells with internal swinging clappers con-

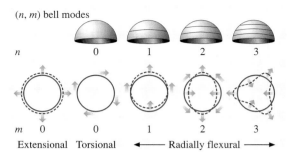

Fig. 15.135 Nomenclature of the (m, n) modes of a bell illustrating displacements of rim for given m-values

tinue to summon the faithful to worship. In more recent times, carillons with up to 77 tuned bells have been developed to play keyboard music from the top of specially constructed bell towers, notably in the centres of Dutch and American towns and college campuses. Bells come in all sorts of shapes and sizes ranging from small hand bells to the giant church bells on display inside the Kremlin walls. However, the acoustics of all bells is essentially the same, so no attempt will be made to provide a comprehensive coverage of every type, [for such information, see *Fletcher* and *Rossing* ([15.5], Chap. 21), and *Rossing* [15.201]].

Bell modes can be related to the longitudinal, torsional and flexural modes of a cylindrical disc that is axially deformed into a bell-shaped domed structure. Although the modal frequencies will clearly be strongly perturbed by such a transformation, the modal shapes will remain unchanged, as illustrated in Fig. 15.135, where m represents the number of radial nodal lines and n the number of nodal circles between the (fixed) centre and free edge.

The first example in Fig. 15.135 illustrates the rim displacements of the $(0, n)$ extensional modes. Although such "breathing" modes would be efficient sound radiators the energy involved in stretching the surfaces of the bell leads to very high modal frequencies, so that such modes are not strongly excited. Likewise, the torsional $(0, n)$–modes involve no motion perpendicular to the bell surfaces, and therefore generate a negligible amount of sound. With the bell rigidly supported at its top, the $m = 1$ swinging modes again involve large elastic strains and cannot be strongly excited. The first modes to produce a significant amount of sound are therefore the $m = 2$ and above flexural modes, which involve the transverse motions of the outer edges with negligible extension in circumferential length for small amplitude vibrations. When the wavelength of the flexural waves is much smaller than the overall curvature, the vibrational modes will be closely related to the flexural waves of a circular disc, with frequencies satisfying Chladni's generalised empirical law, $f_{mn} \sim c(m+2n)^p$, as confirmed by *Perrin* et al. [15.218].

The flexural modes involve radial displacements proportional to $\cos(m\phi)$. Continuity requires that there must also be a tangential displacement, such that $u + \partial v/\partial \phi = 0$, where u and v are the radial and tangential velocities respectively. Coupling to the tangential motion explains why it is possible to feed energy into a vibrating wine glass or the individual glass resonators of a glass harmonica (*Rossing* [15.201] Chap. 14), by rubbing a wetted finger around the rim. The excitation is very similar to the slip–stick mechanism used to excite the bowed string (Sect. 15.2.4).

Figure 15.136 illustrates a set of holographic measurements by *Rossing* ([15.201], Fig. 12.4), which is typical of most bell shapes. The $(m, 1)$ and $(m, 2)$ modes can immediatly be related to the (m, n) modes of a cupped disc. However, there is a distinct change in character for $n > 0$, with an additional node appearing close to the rim – referred to as a $(m, 1\#)$ mode. Some insight is provided by the finite element solutions for a typical large English church bell illustrated in Fig. 15.137 [15.221]. The three modes illustrated are very similar to the $(3, 0)$, $(3, 1)$ and $(3, 2)$ modes expected from the simple cupped disc model, except for the lowest frequency $(3, 0)$ mode,

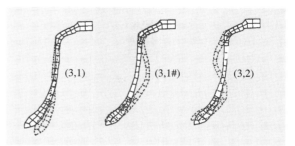

Fig. 15.137 Finite-element solutions for the lowest-order $m = 3$ modes of an English church bell (after *Perrin* et al. [15.219]) illustrating the (3,1), (3,1#) and (3,2) modes

in which the top surfaces of the bell move in antiphase with the rim, to give a nodal line about half-way along the length. Similarly, the anticipated $(4, 0)$, $(5, 0)$ and $(6, 0)$ modes of the handbell investigated by *Rossing* [15.201] acquire an additional nodal line close to the end rim denoted as $(4, 1\#)$, $(5, 1\#)$ and $(6, \#)$ modes. Such features can generally only be accounted for by detailed computational analysis.

Bell Tuning

Figure 15.138 shows the frequencies and associated modes of a well-tuned traditional church bell (*Rossing* and *Perrin* [15.220]), ordered into groups based on mode

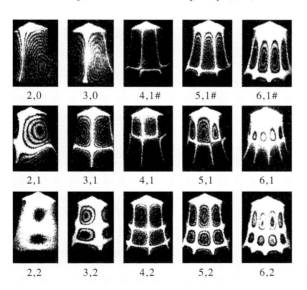

Fig. 15.136 Holographic interferograms and nomenclature for vibrational modes of a hand bell (after *Rossing* [15.201])

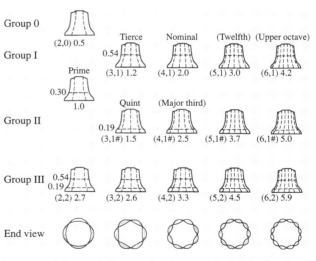

Fig. 15.138 Measured frequencies for a typical D5 church bell, indicating the relative frequencies of the observed mode, with the traditional names associated with such modes indicated (after *Rossing* and *Perrin* [15.220])

shapes, which enable correlations to be made between bells of different shapes and sizes.

Fine-quality bells are carefully tuned by the bell maker, so that the lowest modes have harmonically related frequencies, thereby achieving a well-defined sense of pitch. This is achieved by selective thinning of the thickness of the bell on a very large lathe after its initial casting. In an ideally tuned bell, the principal modes are designated as the hum (2,0) mode, an octave below the prime or fundamental (2,1#) mode. A minor third above the prime (ratio 5:4) is the tierce (3,1) mode, a perfect fifth above that (ratio 3:2) is the quint or fifth (3,1#) mode, and an octave above is the nominal (4,1) mode. *Fletcher* and *Rossing* ([15.5], Table 21.1) compare these and higher modes with measured values for a particular bell.

The art of tuning the partials of bells to achieve a well-sounding note was initially developed in the seventeenth century in the low countries (what is now Holland and the northern parts of Germany and France). Many fine bells from this period by François and Pieter Hemony are still in use in carillons today (*Rossing* [15.201]). However, the art of bell tuning then appears to have been lost until the end of the 19th century, when it was rediscovered by Canon Arthur Simpson in England, following Lord Rayleigh's pioneering research on the sound of bells ([15.3] Vol. 1, Sect. 235).

Acoustic Radiation

The strongest-sounding partials of most bells are the group I ($m,1$) modes, with a nodal circle approximately halfway up the bell (*Fletcher* and *Rossing* [15.5], Sect. 21.11). Such modes have $2m$ antinodal areas providing spatially alternating sound sources in antiphase. The radiation efficiency of such modes increases rapidly with size of bell and frequency. If the acoustic wavelength is much larger than the separation of such antinodes, the sound from such sources will tend to cancel out. However, above a crossover frequency, such that the velocity of sound is equal to that of the flexural waves on the surface of the bell, $v_{\text{flex}} = \sqrt{1.8 c_L h f}$, the spacing between the antinodes will exceed the wavelength in air and the modes will radiate more efficiently. For large church bells, this condition is satisfied for almost all but the very lowest partials, so almost all partials radiate sound rather efficiently. In contrast, hand bells with rather thin walls are significantly less efficient. There is also a small intensity of sound radiated axially at double the modal frequencies, from the induced fluctuations in the volume of air enclosed within the bell.

When a bell is struck, usually by a cast or wrought-iron, ball-shaped, clapper, the first sound heard, the strike note, contains contributions from the very large range of largely inharmonic partials of the bell. Nevertheless the listener can usually identify a pitch to the initial note, determined by the prominent partials excited. However, it is not always easy to attribute the pitch of a note to a particular octave, which is a common feature of many struck percussion instruments (e.g. notes on a xylophone, steeldrum and even timpani). The pitch of the strike note appears to be determined principally by the excited partials with frequencies in the ratios 2:3:4. The ear attributes the pitch to be that of the missing fundamental an octave below the lowest of the partials principally excited, which does not necessarily correspond to the pitch of the lowest partial excited, unless the bell is well tuned.

The majority of the higher partials decay rather quickly, with the long-term sound dominated by the hum note. For a 70 cm church bell, *Perrin* et al. [15.218] measured T_{60} decay times of 52 s for the (2,0) hum mode, 16 s for the (2,1#) and (3,1) prime and minor third modes, 6 s for the (4,1) octave and 3 s for the (4,1#) upper major third, with progressively shorter decay times

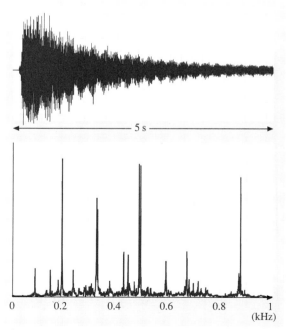

Fig. 15.139 Decaying vibrations and spectrum of Big Ben, London

for the higher modes. Audio ⟨♪⟩CD-ROM illustrates the synthesised sound of the above harmonic modes excited with equal amplitudes. This is followed by a synthesised bell with a major rather than a minor third partial. Such a bell has recently been realised (1999) based on FEA studies at the Technical University in Eindhoven in collaboration with the Eijsbouts Bell foundry. It produces a less mournful sound when played in the major scale music in a carillon. In both cases the synthesised sound reproduces the gentle sound of a bell rather well, though it lacks the initial clang that results from the higher-frequency inharmonic modes of a real bell.

As an example of the waveform and spectrum of a large bell, in Fig. 15.139 and audio signal ⟨♪⟩CD-ROM, we illustrate the sound and rich spectrum of partials of one of the most famous large bells in the world, Big Ben, hung high above the Houses of Parliament in London. This bell is broadcast each day following the Westminster Chimes, marking the hours and end of broadcasting on the BBC each night for UK listeners.

Non-Axial Symmetry

Bell modes with $m > 1$ are doubly degenerate with orthogonal modes varying as $\cos m\phi$ and $\sin m\phi$ and nodal lines in the azimuthal directions that bisect each other. Any departures from axial symmetry will lift the degeneracy and give rise to a split pair of orthogonal modes. If both modes are excited together, the two frequencies will beat against each other, as already evident in the sound of Big Ben and in the sound (⟨♪⟩CD-ROM) of a slightly asymmetrical glass beaker with a pouring spout, which lifts the degeneracy of the otherwise axially symmetric modes.

Bells with strongly distorted or elliptical cross sections can have two completely different pitches, which can be sounded independently by beating at the antinodal positions of one set of modes mode and nodes of the other. Drums with quasi-elliptical cross sections will therefore sound a single note when struck at the narrowest and widest cross-sectional radii, and a second note, when struck at appropriate positions in between.

A dramatic example of axially asymmetric bells is provided by the 65 ancient bells from the tomb of Zeng Hou Yi found at Sui Xiang from around 433 BC. These bells from the second millennium BC are masterpieces of Chinese art and bell casting. They have oval cross sections and range from small hand bells to well over 1.5 m in height. When struck at different positions along the flattened surfaces, two quite distinct tones can be excited, which were designed to be about a major or minor third apart. For further details of these and other Chinese and other eastern bells see *Rossing* ([15.201] Chap. 13) and [15.222].

References

15.1 T. Levenson: *Measure for Measure: How Music and Science together have explored the Universe* (Oxford Univ. Press, Oxford 1997)
15.2 S. Hawkins (Ed.): *On the Shoulders of Giants* (Running, Philadelphia 2002)
15.3 Lord Rayleigh: *The Theory of Sound: I and II (1896)*, 2nd edn. (Dover, New York 1945), Reprint
15.4 S. Hawkins: *The Universe in a Nutshell* (Bantam, London 2001)
15.5 N.H. Fletcher, T.D. Rossing: *The Physics of Musical Instruments*, 2nd edn. (Springer, New York, Berlin 1998)
15.6 T.D. Rossing, F.R. Moore, P.A. Wheeler: *The Science of Sound*, 3rd edn. (Addison Wesley, San Francisco 2002)
15.7 D.M. Campbell, A. Myers, C.A. Greated: *Musical Instruments: History, Technology and Performance of Instruments of Western Music* (Oxford Univ. Press, Oxford 2006)
15.8 G. Bissinger: Contemporary generalised normal mode violin acoustics, Acustica **90**, 590–599 (2004)
15.9 A. Hirschberg, J. Gilbert, R. Msallam, A.P.J. Wijnands: Shock waves in trombones, J. Acoust. Soc. Am. **99**, 1754–1758 (1996)
15.10 T.J.W. Hill, B.E. Richardson, S.J. Richardson: *Modal Radiation from Classical Guitars: Experimental Measurements and Theoretical Predictions.* Proc SMAC 03 (KTH, Stockholm 2003) pp. 129–132
15.11 L. Cremer: *The Physics of the Violin, section 11.2* (MIT Press, Mass 1984)
15.12 B. Britten: *Serenade for Tenor, Horn and Strings* (Hawkes & Son, London 1944)
15.13 J.M. Barbour: *Tuning and Temperament* (Michigan State College Press, East Lansing 1953)
15.14 National Instruments: *Windowing: Optimizing FFTs Using Windowing Functions* (National Instruments, Austin 2006), http://zone.ni.com/devzone/conceptd.nsf/
15.15 M.-P. Verge, A. Hirschberg: Turbulence noise in flue instruments. In: *Proc. Int. Symp. Mus. Acoust.*, SMAC 95 (IRCAM, Paris 1995) pp. 94–99
15.16 M.E. McIntyre, R.T. Schumacher, J. Woodhouse: Aperiodicity in bowed-string motion: On the dif-

ferential slipping mechanism, Acustica **49**, 13–32 (1981)

15.17 J. Meyer: Zur klanglichen wirkung des streichervibratos, Acustica **76**, 283–291 (1992)

15.18 C.E. Gough: Measurement, modelling and synthesis of violin vibrato sounds, Acta Acustica **91**, 229–240 (2005)

15.19 E.I. Prame: Measurement of the vibrato rate of ten singers, J. Acoust. Soc. Am. **96**, 1979–1984 (1994)

15.20 J. Gilbert, L. Simon, J. Terroir: *Vibrato in single reed instruments*, SMAC 03 (Royal Swedish Academy of Music, Stockholm 2003) pp. 271–273

15.21 D.W. Robinson, R.S. Dadson: A re-determination of the equal-loudness relations for pure tones, Brit. J. Appl. Phys. **7**, 166–181 (1956)

15.22 H.A. Fletcher, W.A. Munson: Loudness, its definition, measurement and calculation, J. Acoust. Soc. Am. **9**, 82–108 (1933)

15.23 B. Patterson: Musical dynamics, Sci. Am. **231**(5), 78–95 (1974)

15.24 B.C.J. Moore: *Psychology of Hearing* (Academic, London 1997)

15.25 A.J.M. Houtsma, T.D. Rossing, W.M. Wagenaars: *Auditory Demonstrations*, Phillips compact disc 1126-061 (Acoust. Soc. Am., Sewickley 1989)

15.26 C.M. Hutchins (Ed.): Musical Acoustics, Part I (Violin Family Components) and II (Violin Family Functions) Benchmark papers in Acoustics, Nos. 5 and 6 (Dowden Hutchinson, Ross, Stroudsburg 1975,1976)

15.27 C. Hutchins, V. Benade (Eds.): *Research Papers in Violin Acoustics 1975-1993*, Vols. 1 and 2, (Acoust. Soc. Am., Melville 1997)

15.28 CAS Newsletters (1994-1987), CAS Journals (1988-2004), www.catgutacoustical.org

15.29 L. Cremer: *The Physics of the Violin* (MIT Press, Cambridge 1984)

15.30 R. Midgley: *Musical Instruments of the World* (Paddington, New York 1978)

15.31 A. Baines: *European and American Musical Instruments* (Chancellor, London 1983)

15.32 G.G. Stokes: *On the communication of vibrations from a vibrating body to a surrounding gas* (Philos. Trans., London 1868)

15.33 C.E. Gough: The resonant response of a violin G-string and the excitation of the wolf-note, Acustica **44**(2), 673–684 (1980)

15.34 H.F. Helmholtz: *The Sensation of Tone (1877), translated by A.J. Ellis* (Dover, New York 1954)

15.35 C.V. Raman: On the mechanical theory of the vibrations of bowed strings and of musical instruments of then violin family, with experimental verification of the results: Part 1, Indian Assoc. Cultivation Sci. Bull. **15**, 1–158 (1918)

15.36 J.C. Schelling: The bowed string and player, J. Acoust. Soc. Am. **9**, 91–98 (1973)

15.37 S. Thwaites, N.H. Fletcher: Some notes on the clavichord, J. Acoust. Soc. Am. **69**, 1476–1483 (1981)

15.38 D.E. Hall: Piano string excitation in the case of a small hammer mass, J. Acoust. Soc. Am. **79**, 141–147 (1986)

15.39 D.E. Hall: Piano string excitation II: General solution for a hard narrow hammer, J. Acoust. Soc. Am. **81**, 535–546 (1987)

15.40 D.E. Hall: Piano string excitation III: General solution for a soft narrow hammer, J. Acoust. Soc. Am. **81**, 547–555 (1987)

15.41 D.E. Hall: Piano string excitation IV: Non-linear modelling, J. Acoust. Soc. Am. **92**, 95–105 (1992)

15.42 P.M. Morse, K.U. Ingard: *Theoretical Acoustics* (McGraw-Hill, New York 1968), Reprinted by Princeton Univ. Press, Princeton 1986

15.43 H. Fletcher: Normal vibration frequencies of a stiff piano string, J. Acoust. Soc. Am. **36**, 203–209 (1964)

15.44 E.L. Kent: Influence of irregular patterns in the inharmonicity of piano-tone partials upon tuning practice, Das Musikinstrument **31**, 1008–1013 (1982),

15.45 N.C. Pickering: *The Bowed String* (Bowed Instruments, Southampton 1991)

15.46 N.C. Pickering: Physical Properties of Violin Strings, J. Catgut Soc. **44**, 6–8 (1985), paper 21 in ref [15.42]

15.47 C. Vallette: The mechanics of vibrating strings. In: *Mechanics of musical instruments*, ed. by A. Hirschberg, J. Kergomard, G. Weinreich (Springer, Berlin, New York 1995) pp. 115–183

15.48 K. Legge, N.H. Fletcher: Nonlinear generation of missing modes on a vibrating string, J. Acoust. Soc. Am. **76**, 5–12 (1984)

15.49 C.E. Gough: The nonlinear free vibration of a damped elastic string, J. Acoust. Soc. Am. **75**, 1770–1776 (1984)

15.50 J. Miles: Resonant non-planar motion of a stretched string, J. Acoust. Soc. Am. **75**, 1505–1510 (1984)

15.51 R.J. Hanson, J.M. Anderson, H.K. Macomber: Measurement of nonlinear effects in a driven vibrating wire, J. Acoust. Soc. Am. **96**, 1549–1556 (1994)

15.52 R.J. Hanson, H.K. Macomber, A.C. Morrison, M.A. Boucher: Primarily nonlinear effects observed in a driven asymmetrical vibrating wire, J. Acoust. Soc. Am. **117**, 400–412 (2005)

15.53 J.A. Elliot: Intrinsic nonlinear effects in vibrating strings, Am. J. Phys. **48**, 478–480 (1980)

15.54 J. Woodhouse, P.M. Galluzzo: The bowed string as we know it today, Acustica **90**, 579–590 (2004)

15.55 F.A. Saunders: Recent work on violins, J. Acoust. Soc. Am. **25**, 491–498 (1953)

15.56 J.C. Schelling: The bowed string and the player, J. Acoust. Soc. Am. **53**, 26–41 (1973)

15.57 J.C. Schelling: The physics of the bowed string. In: *The Physics of Music*, ed. by C. Hutchins, Scientific American (W.H. Freeman & Co., San Fransisco 1978)

15.58 R.T. Schumacher: Measurements of some parameters of bowing, J. Acoust. Soc. Am. **96**, 1985–1998 (1994)

15.59 F.G. Friedlander: On the oscillation of the bowed string, Proc. Cambridge Phil. Soc. **49**, 516–530 (1953)

15.60 M.E. McIntyre, J. Woodhouse: On the fundamentals of bowed string dynamics, Acustica **43**, 93–108 (1979)

15.61 J. Woodhouse: On the playability of violins, Part II: Minimum bow force and transients, Acustica **78**, 137–153 (1993)

15.62 M.E. McIntyre, R.T. Schumacher, J. Woodhouse: Aperiodicity in bowed-string motion: On the differential slipping mechanism, Acustica **49**, 13–32 (1981)

15.63 R.T. Schumacher: Oscillations of bowed strings, J. Acoust. Soc. Am. **43**, 109–120 (1979)

15.64 K. Guettler: On the creation of the Helmholtz motion in bowed strings, Acustica **88**, 2002 (2002)

15.65 P.M. Galluzzo: *On the Playability of stringed instruments*, PhD Thesis (Cambridge University, Cambridge 2003)

15.66 J.H. Smith, J. Woodhouse: The tribology of rosin, J. Mech. Phys. Solids **48**, 1633–1681 (2000)

15.67 W. Reinicke: *Dissertation, Institute for Technical Acoustics* (Technical University of Berlin, Berlin 1973)

15.68 J. Woodhouse, R.T. Schumacher, S. Garoff: Reconstruction of bowing point friction force in a bowed string, J. Acoust. Soc. Am. **108**, 357–368 (2000)

15.69 J.H. Smith: *Stick-slip vibration and its constitutive laws*, PhD thesis (Cambridge University, Cambridge 1990)

15.70 J. Woodhouse: Bowed string simulation using a thermal friction model, Acustica **89**, 355–368 (2003)

15.71 W. Reinicke, L. Cremer: Application of holographic interferometry to the bodies of sting instruments, J. Acoust. Soc. Am. **48**, 988–992 (1970)

15.72 J. Woodhouse: data kindly provided for this chapter

15.73 H. Dünnewald: Ein erweitertes Verfahren zur objektiven Bestimmung der Klangqualität von Violinen, Acustica **71**, 269–276 (1990), reprinted in English as Deduction of objective quality parameters on old and new violins, J. Catgut Acoust. Soc. 2nd Ser. 1(7) 1-5 (1991), included in Hutchins [15.27, Vol 1]

15.74 E.V. Jansson: Admittance measurements of 25 high quality violins, Acta Acustica **83**, 337–341 (1997)

15.75 E.V. Jansson: Violin frequency response - bridge mobility and bridge feet distance, Appl. Acoust. **65**, 1197–1205 (2004)

15.76 J. Woodhouse: On the "bridge hill" of the violin, Acustica **91**, 155–165 (2005)

15.77 J.A. Moral, E.V. Jansson: Eigenmodes, input admittance and the function of the violin, Acustica **50**, 329–337 (1982)

15.78 M. Hacklinger: Violin timbre and bridge frequency response, Acustica **39**, 324–330 (1978)

15.79 D. Gill: *The Book of the Violin*, ed. by D. Gill (Phaedon, Oxford 1984) p. 11, Prologue

15.80 C.E. Gough: The theory of string resonances on musical instruments, Acustica **49**, 124–141 (1981)

15.81 J. Woodhouse: On the synthesis of guitar plucks, Acustica **89**, 928–944 (2003)

15.82 J. Woodhouse: Plucked guitar transients: comparison of measurements and synthesis, Acustica **89**, 945–965 (2003)

15.83 G. Weinreich: Coupled piano strings, J. Acoust. Soc. Am. **62**, 1474–1484 (1977)

15.84 C.G.B. Baker, C.M. Thair, C.E. Gough: A photodetector for measuring resonances of violin strings, Acustica **44**, 70 (1980)

15.85 F. Savart: *Mémoires sur le construction des Instruments Ã Cordes et Ã Archet* (Deterville, Paris 1819), See also the brief summary and references to modern translations of Savart's research in Hutchins 15.26 Part 1, page 8

15.86 M.E. McIntyre, J. Woodhouse: On measuring the elastic and damping of orthotropic sheet materials, Acta. Metall. **36**, 1397–1416 (1988)

15.87 M.D. Waller: Vibrations of free rectangular plates, Proc. Phys. Soc. London B **62**, 277–285 (1949)

15.88 A.W. Leissa: *Vibration of Plates* (NASA SP-160, Washington 1993), Reprinted by Acoust. Soc. Am., Woodbury 1993

15.89 C. Hutchins: The Acoustics of Violin Plates, Sci. Am. **1**, 71–186 (1981)

15.90 B.E. Richardson, G.W. Roberts: *The adjustment of mode frequencies in guitars: a study by means of holographic interferometry and finite element analysis*. SMAC 83 (Royal Swedish Academy of Music, Stockholm 1985) pp. 285–302

15.91 C. Hutchins: A rationale for BI-TRI octave plate tuning, Catgut Acoust. Soc. J. **1**(8), 36–39 (1991), Series II

15.92 E. Reissner: On axi-symmetrical vibrations of shallow spherical shells, Q. Appl. Math. **13**, 279–290 (1955)

15.93 K.D. Marshall: Modal analysis of a violin, J. Acoust. Soc. Am. **77**(2), 695–709 (1985)

15.94 G. Bissinger: Contemporary generalised normal mode violin acoustics, Acustica **90**, 590–599 (2004)

15.95 N.J.-J. Fang, O.E. Rodgers: Violin soundpost elastic vibration, Catgut Acoust. Soc. J. **2**(1), 39–40 (1992)

15.96 G.A. Knott: *A modal Analysis of the Violin using MSC.NASTRAN and PATRAN*. MSc Thesis (Naval Postgraduate School, Monterey 1987), Reproduced in Hutchins/Benade 15.27

15.97 J.P. Beldie: *Dissertation* (Technical University of Berlin, Berlin 1975), as described in Cremer 15.29 Sect. 10.4-5

15.98 J. Meyer: Die Abstimmung der Grundresonanzen von Guitarren, Das Musikinstrument **23**, 179–186

15.99 (1974), English translation in J. Guitar Acoustics No.5, 19 (1982)
15.99 Q. Christensen: Qualitative models for low frequency guitar function, J. Guitar Acoust. **6**, 10–25 (1982)
15.100 T.D. Rossing, J. Popp, D. Polstein: *Acoustical Response of Guitars. SMAC 83* (Royal Swedish Academy of Music, Stockholm 1985) pp. 311–332
15.101 E.V. Jansson: On higher air modes in the violin, Catgut Acoust. Soc. Newsletter **19**, 13–15 (1973), Reprinted in Musical Acoustics, Part 2, ed. C.M.Hutchins (Dowden, Hutchinson & Ross, Stroudsberg 1976)
15.102 G. Derveaux, A. Chaigne, P. Joly, J. Bécache: Time-domain simulation of guitar: Model and method, J. Acoust. Soc. Am. **114**, 3368–3383 (2003)
15.103 G. Roberts, G. Finite: *Element analysis of the violin, PhD Thesis* (Cardiff University, Cardiff 1986), extract reproduced in Hutchins and Benade 15.27, pp.575–590
15.104 B.E. Richardson, G.W. Roberts: *The adjustment of mode frequencies in guitars: A study by means of holographic and finite element analysis. Proc. SMAC. 83* (Royal Swedish Academy of Music, Stockholm 1985) pp. 285–302
15.105 G. Derveaux, A. Chaigne, P. Joly, J. Bécache: *Numerical simulation of the acoustic guitar* (INRIA, Rocquencourt 2003), from http://www.inria.fr/multimedia/Videotheque-fra.html
15.106 G. Bissinger: A unified materials-normal mode approach to violin acoustics, Acustica **91**, 214–228 (2005)
15.107 M. Schleske: On making "tonal copies" of a violin, J. Catgut Acoust. Soc. **3**(2), 18–28 (1996)
15.108 L.S. Morset: *A low-cost pc-based tool for violin acoustics measurements, ISMA 2001* (Fondazione Scuola Di San Giorgio, Venice 2001) pp. 627–630
15.109 H.O. Saldner, N.-E. Molin, E.V. Jansson: Vibrational modes of the violin forced via the bridge and action of the soundpost, J. Acoust. Soc. Am. **100**, 1168–1177 (1996)
15.110 E.V. Jansson, N.-E. Molin, H. Sundin: Resonances of a violin body studied by hologram interferometry and acoustical methods, Phys. Scipta **2**, 243–256 (1970)
15.111 N.-E. Molin, A.O. Wåhlin, E.V. Jansson: Transient response of the violin, J. Acoust. Soc. Am. **88**, 2479–2481 (1990)
15.112 N.-E. Molin, A.O. Wåhlin, E.V. Jansson: Transient response of the violin revisited, J. Acoust. Soc. Am. **90**, 2192–2195 (1991)
15.113 J. Curtin: Innovation in violin making, Proc. Int. Symp. Musical Acoustics, CSA-CAS **1**, 11–16 (1998)
15.114 J. Meyer: Directivity of the bowed string instruments and its effect on orchestral sound in concert halls, J. Acoust. Soc. Am. **51**, 1994–2009 (1972)
15.115 G. Weinreich, E.B. Arnold: Method for measuring acoustic fields, J. Acoust. Soc. Am. **68**, 404–411 (1982)
15.116 G. Weinreich: Sound hole sum rule and dipole moment of the violin, J. Acoust. Soc. Am. **77**, 710–718 (1985)
15.117 T.J.W. Hill, B.E. Richardson, S.J. Richardson: Acoustical parameters for the characterisation of the classical guitar, Acustica **89**, 335–348 (2003)
15.118 G. Bissinger, A. Gregorian: Relating normal mode properties of violins to overall quality signature modes, Catgut Acoust. Soc. J. **4**(8), 37–45 (2003)
15.119 E.V. Jansson: Admittance measurements of 25 high quality violins, Acustica **83**, 337–341 (1997)
15.120 G. Weinreich: Directional tone colour, J. Acoust. Soc. Am. **101**, 2338–2346 (1997)
15.121 J. Meyer: Zur klangichen Wirkung des Streicher-Vibratos, Acustica **76**, 283–291 (1992)
15.122 H. Fletcher, L.C. Sanders: Quality of violin vibrato tones, J. Acoust. Soc. Am. **41**, 1534–1544 (1967)
15.123 M.V. Matthews, K. Kohut: Electronic simulation of violin resonances, J. Acoust. Soc. Am. **53**, 1620–1626 (1973)
15.124 J. Nagyvary: Modern science and the classical violin, Chem. Intel. **2**, 24–31 (1996)
15.125 L. Burckle, H. Grissino-Mayer: Stradivari, violins, tree rings and the Maunder minimum, Dendrichronologia **21**, 41–45 (2003)
15.126 C.M. Hutchins: A 30-year experiment in the acoustical and musical development of violin family instruments, J. Acoust. Soc. Am. **92**, 639–650 (1992),
15.127 H.L.F. Helmholtz: *On the Sensations of Tone*, 4th edn. (Dover, New York 1954), trans. by A.J. Ellis
15.128 H. Bouasse: *Instruments á Vent* (Delagrave, Paris 1929)
15.129 M. Campbell, C. Greated: *The Musicians Guide to Acoustics* (Oxford Univ. Press, Oxford 1987)
15.130 C.J. Nederveen: *Acoustical Aspects of Woodwind Instruments* (Northern Illinois Univ. Press, DeKalb 1998)
15.131 A. Hirschberg, J. Kergomard, G. Weinreich: *Mechanics of Musical Instruments* (Springer, Berlin, New York 1995)
15.132 J. Backus: *The Acoustical Foundations of Music*, 2nd edn. (Norton, New York 1977)
15.133 A.H. Benade: *Fundamentals of Musical Acoustics* (Oxford Univ. Press, Oxford 1975)
15.134 D.M. Campbell (Ed.): Special Issue on Musical Wind Instruments Acoustics, Acustica **86**(4), 599–755 (2000)
15.135 P.M. Moorse, K.U. Ingard: *Theoretical Acoustics* (McGraw-Hill, New York 1968), reprinted by Princeton Univ. Press, Princeton (1986)
15.136 L.L. Beranek: *Acoustics* (McGraw Hill, New York 1954) pp. 91–115, reprinted by Acoust. Soc. Am. (1986)

15.137 H. Levine, L. Schwinger: On the radiation of sound from an unflanged pipe, Phys. Rev. **73**, 383–406 (1948)

15.138 R.D. Ayers, L.J. Eliason, D. Mahrgereth: The conical bore in musical acoustics, Am. J. Phys. **53**, 528–527 (1985)

15.139 H.F. Olson: *Acoustic Engineering* (Van Nostrand-Reinhold, Princeton 1957) pp. 88–123

15.140 L.E. Kinsler, A.R. Frey, A.B. Coppens, J.V. Sanders: *Fundamentals of Acoustics* (Wiley, New York 1982), Fig. 14.19

15.141 A.G. Webster: Acoustical impedance, and the theory of horns and the phonograph, Proc. Nat. Acad. Sci. (US) **5**, 275–282 (1919)

15.142 C.J. Nederveen: *Acoustical Aspects of Woodwind Instruments* (Knuf, Amsterdam 1969) p. 60

15.143 D.H. Keefe, A.H. Benade: Wave propagation in strongly curved ducts, J. Acoust. Soc. Am. **74**, 320–332 (1983)

15.144 R. Caussé, J. Kergomard, X. Luxton: Input impedance of brass instruments – Comparison between experiment and numerical models, J. Acoust. Soc. Am. **75**, 241–254 (1984)

15.145 N.H. Fletcher, R.K. Silk, L.M. Douglas: Acoustic admittance of air-driven reed generators, Acustica **50**, 155–159 (1982)

15.146 A. Almeida, C. Verges, R. Caussé: Quasistatic nonlinear characteristics of double-reed instruments, J. Acoust. Soc. Am. **121**, 536–546 (2007)

15.147 S. Ollivier, J.-P. Dalmont: *Experimental investigation of clarinet reed operation*, SMAC 03 (Royal Swedish Academy of Music, Stockholm 2003) pp. 283–289

15.148 J.-P. Dalmont, J. Gilbert, S. Ollivier: Nonlinear characteristics of single-reed instruments: Quasistatic flow and reed opening measurements, J. Acoust. Soc. Am. **114**, 2253–2262 (2003)

15.149 J. Backus, C.J. Nederveen: *Acoustical Aspects of Woodwind Instruments* (Knuf, Amsterdam 1969) pp. 28–37

15.150 S. Ollivier, J.-P. Dalmont: *Experimental investigation of clarinet reed operation*, Proc. SMAC 03 (Royal Swedish Academy of Music, Stockholm 2003) pp. 283–289

15.151 A.P.J. Wijnands, A. Hirschberg: Effect of pipeneck downstream of a double reed. In: *Proc. Int. Symp. Musical Acoustics* (IRCAM, Paris 1995) pp. 148–151

15.152 N.H. Fletcher: Excitation mechanisms in woodwind and brass instruments, Acustica **43**, 63–72 (1979), erratum Acustica 50, 155–159 (1982)

15.153 P.D. Koopman, C.D. Hanzelka, J.P. Cottingham: Frequency and amplitude of vibration of reeds from American reed organs as a function of pressure, J. Acoust. Soc. Am. **99**, 2506 (1996)

15.154 J.-P. Dalmont, J. Gilbert, J. Kergomard: Reed instruments, from small to large amplitude periodic oscillations and the Helmholtz motion analogy, Acustica **86**, 671–684 (2000)

15.155 J. Backus: Vibrations of the reed and air column of the clarinet, J. Acoust. Soc. Am. **33**, 806–809 (1961)

15.156 N.H. Fletcher: Nonlinear theory of musical wind instruments, Appl. Acoust. **30**, 85–115 (1990)

15.157 R.T. Schumacher: Self-sustained oscillations of a clarinet; an integral equation approach, Acustica **40**, 298–309 (1978)

15.158 R.T. Schumacher: Ab initio calculations of the oscillations of a clarinet, Acustica **48**, 71–85 (1981)

15.159 J. Gilbert, J. Kergomard, E. Ngoya: Calculation of steady state oscillations of a clarinet using the harmonic balance technique, J. Acoust. Soc. Am. **86**, 35–41 (1989)

15.160 D.M. Campbell: Nonlinear dynamics of musical reed and brass wind instruments, Contemp. Phys. **40**, 415–431 (1999)

15.161 N. Gand, J. Gilbert, F. Laloë: Oscillation threshold of wood-wind instruments, Acta Acustica **1**, 137–151 (1997)

15.162 J.-P. Dalmont, J. Kergomard: Elementary model and experiments for the Helmholtz motion of single reed wind instruments. In: *Proc. Int. Symp. Mus. Acoust.* (IRCAM, Paris 1995) pp. 115–120

15.163 A.H. Benade: Air column reed and player's windway interaction in musical instruments. In: *Vocal Fold Physiology, Biomechanics, Acoustics and Phonatory Control*, ed. by I.R. Titze, R.C. Scherer (Denver Centre for the Performing Arts, Denver 1985) pp. 425–452

15.164 G.P. Scavone: *Modelling vocal tract influence in reed wind instruments*, Proc. SMAC 03 (Royal Swedish Academy of Music, Stockholm 2003) pp. 291–294

15.165 N.H. Fletcher: Mode locking in non-linearly excited inharmonic musical oscillators, J. Acoust. Soc. Am. **64**, 1566–1569 (1978)

15.166 D. Ayers: *Basic tests for models of the lip reed*, ISMA 2001 (Fondazione Scuola Di San Giorgio, Venice 2001) pp. 83–86

15.167 S.J. Elliot, J.M. Bowsher: Regeneration in brass wind instruments, J. Sound Vibrat. **83**, 181–207 (1982)

15.168 F.C. Chen, G. Weinreich: Nature of the lip-reed, J. Acoust. Soc. Am. **99**, 1227–1233 (1996)

15.169 S. Adachi, M. Sato: Trumpet sound simulation using a two-dimensional lip vibration model, J. Acoust. Soc. Am. **99**, 1200–1209 (1996)

15.170 I.R. Titze: The physics of small-amplitude oscillation of the vocal folds, J. Acoust. Soc. Am. **83**, 1536–1552 (1988)

15.171 D.C. Copley, W.J. Strong: A stroboscopic study of lip vibrations in a trombone, J. Acoust. Soc. Am. **99**, 1219–1226 (1996)

15.172 S. Yoshikawa, Y. Muto: *Brass player's skill and the associated li-p wave propagation*, ISMA 2001 (Fondazione Scuola Di San Giorgio, Venice 2002) pp. 91–949

15.173 J. Gilbert, S. Ponthus, J.F. Petoit: Artificial buzzing lips and brass instruments: experimental results, J. Acoust. Soc. Am. **104**, 1627–1632m (1998)

15.174 J. Cullen, J.A. Gilbert, D.M. Campbell: Brass instruments: linear stability analysis and experiments with an artificial mouth, Acustica **86**, 704–724 (2000)

15.175 N.H. Fletcher: Nonlinear theory of musical wind instruments, Appl. Acoust. **30**, 85–115 (1990)

15.176 T.H. Long: The performance of cup-mouthpiece instruments, J. Acoust. Soc. Am. **19**, 892–901 (1947)

15.177 A. Hirschberg, J. Gilbert, R. Msallam, A.P.J. Wijnands: Shock waves in trombones, J. Acoust. Soc. Am. **99**, 1754–1758 (1996)

15.178 M.E. McIntyre, R.T. Schumacher, J. Woodhouse: On the oscillations of musical instruments, J. Acoust. Soc. Am. **74**, 1325–1345 (1983)

15.179 R.D. Ayers: Impulse responses for feedback to the driver of a musical wind instrument, J. Acoust. Soc. Am. **100**, 1190–1198 (1996)

15.180 B. Fabre, A. Hirschberg: Physical modelling of flue instruments: a review of lumped models, Acustica **86**, 599–610 (2000)

15.181 N.H. Fletcher: Sound production in organ pipes, J. Acoust. Soc. Am. **60**, 926–936 (1976)

15.182 P. Savic: On acoustically effective vortex motions in gaseous jets, Philos. Mag. **32**, 287–252 (1941)

15.183 J.W. Coltman: Jet drive mechanism in edge tones and organ pipes, J. Acoust. Soc. Am. **60**, 723–733 (1976)

15.184 S. Adachi: *CFD analysis of air jet deflection-comparison with Nolle's measurements*, SMAC 03 (Royal Swedish Academy of Music, Stockholm 2003) pp. 313–319

15.185 A.W. Nolle, Sinuous instability of a planar air jet: Propagation parameters and acoustic excitation, J. Acoust. Soc. Am. **103**, 3690–3705 (1998)

15.186 L. Cremer, H. Ising: Die selbsterregten Schwingungen von Orgelpfeifen, Acustica **19**, 143–153 (1967)

15.187 J.W. Coltman: Sounding mechanism of the flute and organ pipe, J. Acoust. Soc. Am. **44**, 983–992 (1968)

15.188 M. Raffel, C. Willert, J. Kompenhans: *Particle Image Velocimetry - A Practical Guide* (Springer, Berlin, Heidelberg 1998)

15.189 C. Ségoufin, B. Fabre, M.P. Verge, A. Hirschberg, A.P.J. Wijnands: Experimental study of the influence of mouth geometry on sound production in recorder-like instruments: windway length and chamfers, Acta Acustica **86**, 649–661 (2000)

15.190 S. Thwaites, N.H. Fletcher: Wave propagation on turbulent jets, Acustica **45**, 175–179 (1980)

15.191 S. Thwaites, N.H. Fletcher: Wave propagation of turbulent jets: II, Acustica **51**, 44–49 (1982)

15.192 B. Fabre, A. Hirschberg: *From sound synthesis to instrument making: an overview of recent researches on woodwinds*, SMAC 03 (Royal Swedish Academy of Music, Stockholm 2003) pp. 239–242

15.193 N.H. Fletcher: Jet-drive mechanism in organ pipes, J. Acoust. Soc. Am. **60**, 481–483 (1976)

15.194 S.A. Elder: On the mechanism of sound production in organ pipes, J. Acoust. Soc. Am. **54**, 1554–1564 (1973)

15.195 R.C. Chanaud: Aerodynamic whistles, Sci. Am. **222**(1), 40–46 (1970)

15.196 T.A. Wilson, G.S. Beavers, M.A. DeCoster, D.K. Holger, M.D. Regenfuss: Experiments on the fluid mechanics of whistling, J. Acoust. Soc. Am. **50**, 366–372 (1971)

15.197 D.K. Holger, T. Wilson, G. Beavers: Fluid mechanics of the edge-tone, J. Acoust. Soc. Am. **62**, 1116–1128 (1977)

15.198 S. Yoshikawa: Jet wave amplification in organ pipes, J. Acoust. Soc. Am. **103**, 2706–2717 (1998)

15.199 B. Fabre, A. Hirschberg, A.P.J. Wijnands: Vortex shedding in steady oscillations of a flue organ pipe, Acustica - Acta Acustica **82**, 877–883 (1996)

15.200 A. Baines: *European and American Musical Instruments* (Chancellor, London 1983)

15.201 T.D. Rossing: *Science of Percussion Instruments* (World Scientific, Singapore 2000)

15.202 J. Blades: *Percussion Instruments and Their History* (Faber, Faber, London 1974)

15.203 P. Boulez: *Notations I-IV* (Universal Edition, Vienna 1945, 78, 84)

15.204 M. Bertsch: *Vibration patterns and sound analysis of the Viennese timpani*. ISMA 2001 (Fondazione Scuola Di San Giorgio, Venice 2001)

15.205 C.V. Raman: The Indian Musical Drum, Proc. Indian Acad. Sci. A1 179 (1934). In: *Reprinted in Musical Acoustics Selected Reprints*, ed. by T.D. Rossing (Am. Assoc. Phys. Teach., College Park 1988)

15.206 T.D. Rossing, W.A. Sykes: Acoustics of indian drums, Percuss. Notes **19**(3), 58 (1982)

15.207 B.S. Ramakrishna, M.M. Sondhi: Vibrations of indian musical drums regarded as composite membranes, J. Acoust. Soc. Am. **26**, 523 (1954)

15.208 S. De: Experimental study of the vibrational characteristics of a loaded kettledrum, Acustica **40**, 206 (1978)

15.209 H. Zhao: *Acoustics of Snare Drums: An Experimental Study of the Modes of Vibration, Mode Coupling and Sound Radiation Pattern*, M.S. thesis (Northern Illinois Univ., DeKalb 1990)

15.210 C.V. Raman: On some Indian stringed instruments, Proc. Indian Assoc. Adv. Sci. **7**, 29–33 (1922)

15.211 T.D. Rossing, I. Bork, H. Zhao, D. Fystrom: Acoustics of snare drums, J. Acoust. Soc. Am. **92**, 84–94 (1992)

15.212 T.D. Rossing: Acoustics of percussion instruments: Part I, Phys. Teacher **14**, 546–556 (1976)

15.213 T.D. Rossing: Chladni's law for vibrating plates, Am. J. Phys. **50**, 271–274 (1982)

15.214 N.H. Fletcher: Nonlinear frequency shifts in quasi-spherical-cap shells: Pitch glide in Chinese Gongs, J. Acoust. Soc. Am. **78**, 2069 (1985)

15.215 P.L. Grossman, B. Koplik, Y.-Y. Yu: Nonlinear vibration of shallow spherical shells, J. Appl. Mech. **36**, 451–458 (1969)

15.216 K.A. Legge, N.H. Fletcher: Non-linear mode coupling in symmetrically kinked bars, J. Sound Vib. **118**, 23–34 (1987)

15.217 S. Schedin, P.O. Gren, T.D. Rossing: Transient wave response of a cymbal using double pulsed TV holography, J. Acoust. Soc. Am. **103**, 1217–1220 (1998)

15.218 R. Perrin, T. Charnley, H. Banu, T.D. Rossing: Chladni's law and the modern English church bell, J. Sound Vib. **102**, 11–19 (1985)

15.219 R. Perrin, T. Charnley, J. de Pont: Normal modes of the modern English church bell, J. Sound Vib. **102**, 28048 (1983)

15.220 T.D. Rossing, R. Perrin: Vibration of bells, Appl. Acoust. **20**, 41–70 (1988)

15.221 R. Perrin, T. Charnley, J. de Pont: Normal modes of the modern English church bell, J. Sound Vibr. **102**, 28048 (1983)

15.222 T.D. Rossing, D.S. Hampton, B.E. Richardson, H.J. Sathoff: Vibrational modes of Chinese two-tone bells, J. Acoust. Soc. Am. **83**, 369–373 (1988)

16. The Human Voice in Speech and Singing

This chapter describes various aspects of the human voice as a means of communication in speech and singing. From the point of view of function, vocal sounds can be regarded as the end result of a three stage process: (1) the compression of air in the respiratory system, which produces an exhalatory airstream, (2) the vibrating vocal folds' transformation of this air stream to an intermittent or pulsating air stream, which is a complex tone, referred to as the voice source, and (3) the filtering of this complex tone in the vocal tract resonator. The main function of the respiratory system is to generate an overpressure of air under the glottis, or a subglottal pressure. Section 16.1 describes different aspects of the respiratory system of significance to speech and singing, including lung volume ranges, subglottal pressures, and how this pressure is affected by the ever-varying recoil forces. The complex tone generated when the air stream from the lungs passes the vibrating vocal folds can be varied in at least three dimensions: fundamental frequency, amplitude and spectrum. Section 16.2 describes how these properties of the voice source are affected by the subglottal pressure, the length and stiffness of the vocal folds and how firmly the vocal folds are adducted. Section 16.3 gives an account of the vocal tract filter, how its form determines the frequencies of its resonances, and Sect. 16.4 gives an account for how these resonance frequencies

16.1	Breathing	669
16.2	The Glottal Sound Source	676
16.3	The Vocal Tract Filter	682
16.4	Articulatory Processes, Vowels and Consonants	687
16.5	The Syllable	695
16.6	Rhythm and Timing	699
16.7	Prosody and Speech Dynamics	701
16.8	Control of Sound in Speech and Singing	703
16.9	The Expressive Power of the Human Voice	706
References		706

or formants shape the vocal sounds by imposing spectrum peaks separated by spectrum valleys, and how the frequencies of these peaks determine vowel and voice qualities. The remaining sections of the chapter describe various aspects of the acoustic signals used for vocal communication in speech and singing. The syllable structure is discussed in Sect. 16.5, the closely related aspects of rhythmicity and timing in speech and singing is described in Sect. 16.6, and pitch and rhythm aspects in Sect. 16.7. The impressive control of all these acoustic characteristics of vocal signals is discussed in Sect. 16.8, while Sect. 16.9 considers expressive aspects of vocal communication.

16.1 Breathing

The process of breathing depends both on mechanical and muscular forces (Fig. 16.1).

During *inspiration* the volume of the chest cavity is expanded and air rushes into the lungs. This happens mainly because of the contraction of the *external intercostals* and the *diaphragm*. The external intercostal muscles raise the ribs. The diaphragm which is the dome-shaped muscle located below the lungs, flattens on contraction and thus lowers the floor of the thoracic cavity.

The respiratory structures form an elastic mechanical system that produces expiratory or inspiratory subglottal pressures, depending on the size of the lung volume, Fig. 16.2. Thus, exhalation and inhalation will always produce forces whose effect is to move the ribs and the lungs back to their resting state, often referred

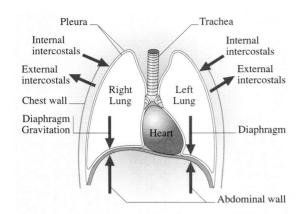

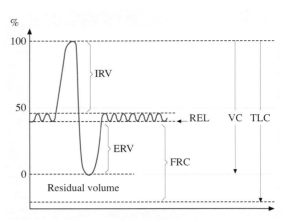

Fig. 16.1 Directions of the muscular and elastic forces. The direction of the gravitation force is applicable to the upright position

to as the resting expiratory level (REL). The deeper the breath, the greater this force of *elastic recoil*. This component plays a significant part in pushing air out of the lungs, both in speech and singing, and especially at large lung volumes. The elasticity forces originate both from the rib cage and the lungs. As illustrated in Fig. 16.2, the rib cage produces an exhalatory force at high lung volumes and an inhalatory force at low lung volumes, and the lungs always exert an exhalatory force. As a consequence, activation of inspiratory muscles is needed for producing a low subglottal pressure, e.g. for singing a soft (pianissimo) tone, at high lung volume. Conversely, activation of expiratory muscles is needed for producing a high subglottal pressure, e.g. for singing a loud (fortissimo) tone, at low lung volume.

In addition to mechanical factors, exhaling may involve the activity of the *internal intercostals* and the *abdominal muscles*. Contraction of the former has the effect of lowering the ribs and thus compressing the chest cavity. Activating the abdominal muscles generates upward forces that also contribute towards reducing the volume of the rib cage and the lungs. The function of these muscles is thus expiratory.

Fig. 16.2 Subglottal pressures produced at different lung volumes in a subject by the recoil forces of rib cage and lungs. The resting expiratory level (REL) is the lung volume at which the inhalatory and the exhalatory recoil forces are equal. The *thin* and *heavy chain-dashed lines* represent subglottal pressures typically needed for soft and very loud phonation

Fig. 16.3 Definition of the various terms for lung volumes. The graph illustrates the lung volume changes during quiet breathing interrupted by a maximal inhalation followed by a maximal exhalation. VC is the vital capacity, TLC is the total lung capacity, IRV and ERV are the inspiratory and expiratory reserve volume, REL is the resting expiratory level, FRC is the functional residual capacity

Another significant factor is gravity whose role depends on body posture. In an upright position, the diaphragm and adjacent structures tend to be pulled down, increasing the volume of the thoracic cavity. In this sit-

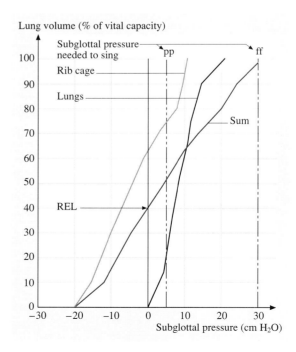

uation, the effect of gravity is inspiratory. In contrast, in the supine position the diaphragm tends to get pushed up into the rib cage, and expiration is promoted [16.1].

The total air volume that is contained in a maximally expanded rib cage is called the *total lung capacity* (TLC in Fig. 16.3). After maximum exhalation a small air volume, the *residual volume*, is still left in the airways. The greatest air volume that can be exhaled after a maximum inhalation is called the *vital capacity* (VC) and thus equals the difference between the TLC and the residual volume. The lung volume at which the exhalatory and inhalatory recoil forces are equal, or REL, is reached after a relaxed sigh, see Figs. 16.2 and 16.3. During tidal breathing inhalation is initiated from REL, so that inhalation is active resulting from an activation of inspiratory muscles, and exhalation is passive, produced by the recoil forces. In tidal breathing only some 10% of VC is inhaled, such that a great portion of the VC, the *inspiratory reserve volume*, is left. The air volume between REL and the residual volume is called the *expiratory reserve volume*.

VC varies depending on age, body height and gender. At the age of about 20 years, an adult female has a vital capacity of 3–3.6 l depending on body height, and for males the corresponding values are about 4–5.5 l.

Experimental data [16.4] show that, during tidal breathing, lung volume variations are characterized by

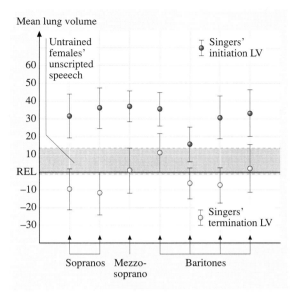

Fig. 16.5 Lung volume averages used in speech and operatic singing expressed as percentages of vital capacity relative to the resting expiratory level REL. The shaded band represents the mean lung volume range observed in untrained female voices' unscripted speech (after [16.2]). The filled and open symbols show the corresponding measures for professional opera singers of the indicated classifications when performing opera arias according to *Thomasson* [16.3]. The bars represent ± one SD

a regular quasi-sinusoidal pattern with alternating segments of inspiration and expiration of roughly equal duration (Fig. 16.4). In speech and singing, the pattern is transformed. Inspirations become more rapid and expiration occurs at a slow and relatively steady rate. Increasing vocal loudness raises the amplitude of the lung volume records but leaves its shape relatively unchanged.

Figure 16.5 shows mean lung volumes used by professional singers when singing well-rehearsed songs. The darker band represents the mean lung volume range observed in spontaneous speech [16.2]. The lung volumes of conversational speech are similar to those in tidal breathing. Loud speech shows greater air consumption and thus higher volumes (Fig. 16.4). Breath groups in speech typically last for 3–5 seconds and are terminated when lung volumes approach the relaxation expiratory level REL, as illustrated in Fig. 16.5. Thus, in phonation, lung volumes below REL are mostly avoided.

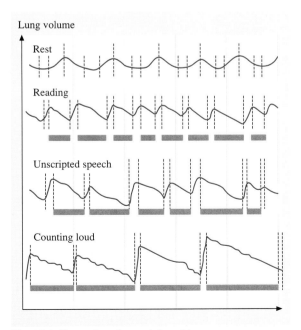

Fig. 16.4 Examples of speech breathing

Fig. 16.6 The records show lung volume (relative to the mid-respiratory level), subglottal (esophageal) pressure and stylized muscular activity for speaker counting from one to 32 at a conversational vocal effort. (After Draper et al. [16.5]) . To the *left* of the vertical line recoil forces are strongly expiratory, to the *right* they are inspiratory. *Arrows* have been added to the x-axis to summarize the original EMG measurements which indicate that the recoil forces are balanced by muscular activity (EMG = electromyography, measurement of the electrical activity of muscles). To the *left of the vertical line* (as indicating by *left-pointing arrow*) the net muscular force is inspiratory, to the *right* it is expiratory (*right-pointing arrow*). To keep loudness constant the talker maintains a stable subglottal pressure and recruits muscles according to the current value of the lung volume. This behavior exemplifies the phenomenon known as *motor equivalence*

In singing, breath groups tend to be about twice as long or more, and air consumption is typically much greater than in conversational speech. Mostly they are terminated close to the relaxation expiratory level, as in speech, but sometimes extend into the *expiratory reserve volume* as illustrated in Fig. 16.5. This implies that, in singing, breath groups typically start at much higher lung volumes than in speech. This use of high lung volumes implies that singers have to deal with much greater recoil forces than in speech.

Figure 16.6 replots, in stylized form, a diagram published by Ladefoged et al. [16.7]. It summarizes measurements of lung volume and subglottal pressure, henceforth referred to as Ps, recorded from a speaker asked to take a deep breath and then to start counting. The dashed line intersecting the Ps record represents the relaxation pressure. This line tells us that elastic recoil forces are strongly expiratory to the left of the vertical line. To the right they produce a pressure lower than the target pressure and eventually an inspiratory pressure. The Ps curve remains reasonably flat throughout the entire utterance.

The question arises: how can this relative constancy be achieved despite the continually changing contribution of the relaxation forces? In a recent replication of this classical work [16.8], various criticisms of the original study are addressed but basically the original answer to this question is restated: the motor system adapts to the external goal of keeping the Ps fairly constant (Fig. 16.6). Initially, when recoil forces are strong, muscle activity is predominantly in the *inspiratory* muscles such as the *diaphragm* and the *external intercostals*. Gradually, as recoil forces decline, the ex-

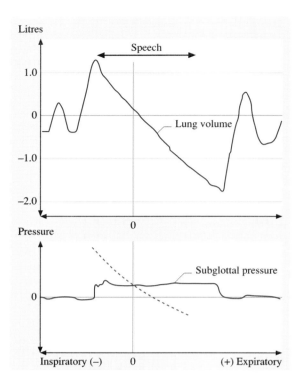

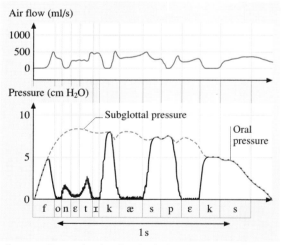

Fig. 16.7 Example of oral and subglottal pressure variations for the phrase "*phonetic aspects*" (after *Netsell* [16.6]). As the glottis opens and closes for voiceless and voiced sounds and the vocal tract opens or closes for vowels and consonants, the expired air is opposed by varying degrees of glottal and supraglottal impedance. Both the oral and the subglottal pressure records reflect the combined effect of these resistance variations

piratory muscles (the *internal intercostals*, the *rectus abdominis* among others) take over increasingly as the other group relaxes (cf. arrows, Fig. 16.6). According to our present understanding [16.8, 9], this adaptation of breathing for speech is achieved by constantly updating the balance between agonist and antagonist muscles in accordance with current conditions (lung volume, body posture, etc.) and in order to meet the goal of constant Ps. Muscle recruitment depends on lung volume.

Figure 16.7 presents a representative example of oral and Ps records. The phrase is *"phonetic aspects"* (After *Netsell* [16.6]). The top panel shows a record of oral air flow. Vertical lines indicate acoustic segment boundaries. The bottom diagram superimposes the curves for oral and subglottal pressure.

The Ps shows a falling pattern which becomes more pronounced towards the end of the phrase and which is reminiscent of the declination pattern of the fundamental frequency contour typical of declarative sentences [16.10, p. 127]. The highest values occur at the midpoints of [ɛ] and [æ], the stressed vowels of the utterance. For vowels, oral pressure is close to atmospheric pressure (near zero on the *y*-axis cm H$_2$O scale). The [k] of *phonetic* and the [p] of *aspects* show highly similar traces. As the tongue makes the closure for [k], the air flow is blocked and the trace is brought down to zero. This is paralleled by the oral pressure rising until it equals the Ps. As the [k] closure progresses, a slight increase builds up in both curves. The release of the [k] is signaled by a peak in the air flow and a rapid decrease in Ps. An almost identical pattern is seen for [p].

In analyzing Ps records, phoneticians aim at identifying variations based on an *active control* of the respiratory system and phenomena that can be attributed to the system's *passive response* to ongoing activity elsewhere, e.g., in the vocal tract and or at the level of the vocal folds [16.11].

To exemplify passive effects let us consider the events associated with [k] and [p] just discussed. As suggested by the data of Figs. 16.4 and 16.6, speech breathing proceeds at a relatively steady lung volume decrement. However, the open or closed state of the glottis, or the presence of a vocal tract constriction/closure, is capable of creating varying degrees of impedance to the expired air. The oral pressure record reflects the combined effect of glottal and articulatory resistance variations. Ps is also affected by such changing conditions. As is evident from Fig. 16.7, the Ps traces during the [k] and [p] segments first increase during the stop closures. Then they decrease rapidly during the release and the aspiration phases. These effects are passive responses to the segment-based changes in supraglottal resistance and are unlikely to be actively programmed [16.12, 13].

Does respiration play an active role in the production of stressed syllables? In *"phonetic aspects"* main stress occurs on [ɛ] and [æ]. In terms of Ps, these vowels exhibit the highest values. Are they due to an active participation of the respiratory system in signaling stress, or are they fortuitous by-products of other factors?

An early contribution to research on breathing and speech is the work by *Stetson* [16.14]. On the basis of aerodynamic, electromyographic and chest movement measurements, Stetson proposed the notion of the *chest pulse*, a chunk of expiratory activity corresponding to the production of an individual syllable.

In the late fifties, Ladefoged and colleagues published an electromyographic study [16.7] which cast doubt on Stetson's interpretations. It reported increased activity in expiratory muscles (*internal intercostals*) for English syllables. However, it failed to support Stetson's chest pulse idea in that the increases were found only on stressed syllables. *Ladefoged* [16.8] reports findings from a replication of the 1958 study in which greater activity in the internal intercostals for stressed syllables was confirmed. Moreover, reduced activity in inspiratory muscles (*external intercostals*) was observed to occur immediately before each stressed syllable.

Ps measurements provide further evidence for a positive role for respiration in the implementation of stress. *Ladefoged* [16.15, p. 143] states: "Accompanying every stressed syllable there is always an increase in the Ps". This claim is based on data on disyllabic English noun–verb pairs differing in the position of the main stress: TORment (noun)–torMENT (verb), INsult (noun)–inSULT (verb) etc. A clear distinction between the noun and verb forms was observed, the former having Ps peaks in the first syllable and the latter on the second syllable.

As for segment-related Ps variations (e.g., the stops in Fig. 16.7), there is wide agreement that such local perturbations are induced as automatic consequences of speech production aerodynamics. Explaining short-term ripple on Ps curves in terms of aerodynamics is consistent with the observation that the respiratory system is mechanically sluggish and therefore less suited to implement rapid Ps changes in individual phonetic segments. Basically, its primary task is to produce a Ps contour stable enough to maintain vocal intensity at a fairly constant level (Fig. 16.6).

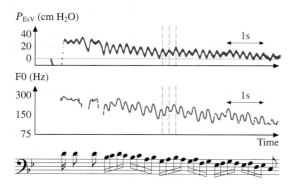

Fig. 16.8 Changes in esophageal pressure, corresponding to subglottic pressure changes (*upper curve*) and phonation frequency (*middle*) in a professional baritone singer performing the coloratura passage shown at the bottom. (After *Leanderson* et al. [16.17])

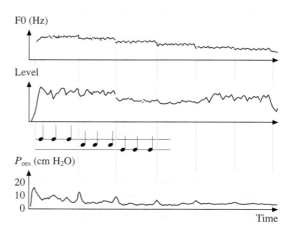

Fig. 16.9 The three *curves* show, from *top to bottom* F0 sound level and esophageal pressure in a professional baritone singer singing the music example shown in the *graph*, i. e. a descending scale with three tones on each scale tone. The singer marked the first beat in each bar by a pressure pulse. (After *Sundberg* et al. [16.18])

Singing provides experimental data on the behavior of the respiratory system that tend to reinforce the view that a time varying respiration activity can be an active participant in the sound generation process. Although there is ample justification for saying that the mechanical response of the respiratory structures is characterized by a long time constant, Ps in singing varies quickly and accurately. One example is presented in Fig. 16.8. It shows variations in pressure, captured as the pressure variations in the esophagus and thus mirroring the Ps variations [16.16], during a professional baritone singer's performance of the music example shown at the bottom. Each tone in the sequence of sixteenth notes is produced with a pressure pulse. Note also that in the fundamental frequency curve each note corresponds to a small rise and fall. The tempo of about six sixteenth notes per second implies that the duration of each rise–fall cycle is about 160 ms. It would take quite special skills to produce such carefully synchronized Ps and fundamental frequency patterns.

Synthesis experiments have demonstrated that this particular fundamental frequency pattern is what produces what is perceived as a sung legato coloratura [16.19, 20]. The voice organ seems incapable of producing a stepwise-changing F0 curve in legato, and such a performance therefore sounds as if it was produced by a music instrument rather than by a voice. In this sense this F0 variation pattern seems to be needed for eliciting the perception of a sung sequence of short legato tones.

The pressure variations are not likely to result from a modulation of glottal adduction. A weak glottal adduction should result in a voice source dominated by the fundamental and with very weak high spectrum partials, and mostly the amplitudes of the high overtones do not vary in coloratura sequences.

It is quite possible that the F0 variations are caused by the Ps variations. In the example shown in Fig. 16.8, the pressure variations amount to about 10 cm H_2O, which should cause a F0 modulation of about 30 Hz or so, corresponding to two semitones in the vicinity of 220 Hz. This means that the coloratura pattern may simply be the result of a ramp produced by the F0 regulating system which is modulated by the pressure pulses produced by the respiratory system.

As another example of the skilled use of Ps in singing, Fig. 16.9 shows the pressure variation in the esophagus in a professional baritone singer performing a descending scale pattern in 3/4 time, with three tones on each scale step as shown by the score fragment in the first three bars. The singer was instructed to mark the first tone in each bar. The pressure record demonstrates quite clearly that the first beat was produced with a marked pulse approximately doubling the pressure. Pulses of this magnitude must be produced by the respiratory apparatus. When the subject was instructed to avoid marking the first tone in each bar, no pressure pulses were observed [16.18].

The effect of Ps on fundamental frequency has been investigated in numerous studies. A commonly used

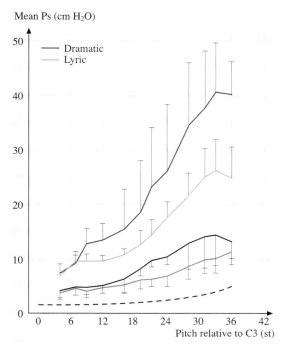

Fig. 16.10 Mean subglottal pressure, captured as the oral pressure during /p/-occlusion, in five lyric and six dramatic professional sopranos (*heavy* and *thin curves*, respectively) who sang as loudly and as softly as possible at different F0. The *dashed curve* shows Titze's prediction of threshold pressure, i.e., the lowest pressure that produces vocal fold vibration. The *bars* represent one SD (After [16.21])

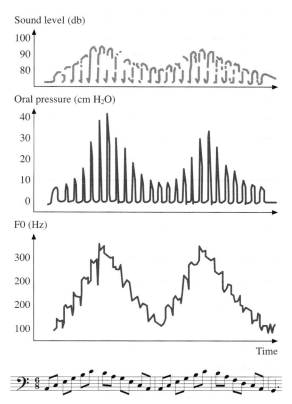

Fig. 16.11 Simultaneous recordings of sound level, subglottic pressure (captured as the oral pressure during /p/ occlusion) and F0 in a professional singer singing the note example shown at the bottom. (After *Leanderson* et al. [16.17])

method is to ask the subject to produce a sustained vowel at a given steady pitch and then, at unpredictable moments, change the Ps by applying a push to the subject's abdomen or chest. The assumption underlying these studies is that there will be an initial interval during which possible reflex action of laryngeal muscles will not yet come into play. Hence the data from this time segment should give a valid picture of the relationship between fundamental frequency and Ps.

In a study using the push method, *Baer* [16.22] established this relationship for a single male subject. Data on steady phonations at 94, 110, 220 Hz and a falsetto condition (240 Hz) were collected. From electromyograms from the *vocalis* and the *interarytenoid* it was possible to tell that the first 30 ms after the push onset were uncontaminated by reflex muscle responses. Fundamental frequency was plotted against Ps for the four F0 conditions. These plots showed linear data clusters with slopes of 4, 4, 3 and 9 Hz/cm H_2O respectively. Other studies which assumed a longer reflex response latency (100 ms) report that F0's dependence on Ps occurs between 2 to 7 Hz/cm H_2O.

From such observations phoneticians have concluded that, in producing the F0 variations of natural speech, Ps plays only a secondary role. F0 control is primarily based on laryngeal muscle activity. Nonetheless, the falling F0 contours of phrases with statement intonation tend to have Ps contours that are also falling. Moreover, in questions with final F0 increases, the Ps records are higher towards the end of the phrase than in the corresponding statements [16.15].

There is clear evidence that the Ps needs to be carefully adapted to the target fundamental frequency in singing, especially in the high range ([16.23], [16.24,

p. 36]). An example is presented in Fig. 16.10 which shows mean Ps for five dramatic and five lyric sopranos when they sang pp and ff tones throughout their range [16.21]. The bars represent one standard deviation. The dashed curve is the threshold pressure, i. e. the lowest pressure that produced vocal fold vibration, for a female voice according to *Titze* [16.25, 26].

The singers mostly used more than twice as high Ps values when they were singing their loudest as compared with their softest tones. The subjects reach up to 40 cm H_2O. This can be compared with values typical of normal adult conversational speech, which tend to occur in the range of 5–10 cm H_2O. However, loud speech and particularly stage speech may occasionally approach the higher values for singing.

Also interesting is the fact that lyric sopranos use significantly lower Ps values than dramatic sopranos both in soft and loud phonation. It seems likely that this difference reflects difference in the mechanical properties of the vocal folds.

Figure 16.11 presents another example of the close link between Ps and fundamental frequency. From top to bottom it plots the time functions for sound level (SPL in dB), pressure (cm H_2O), fundamental frequency (Hz) for a professional baritone singing an ascending triad followed by a descending dominant-seventh triad (according to the score at the bottom). The pressure record was obtained by recording the oral pressure as the subject repeated the syllable [pæ] on each note. During the occlusion of the stop the peak oral pressure becomes equal to the Ps (as discussed in connection with Fig. 16.7). That means that the peak values shown in the middle diagram are good estimates of the actual Ps. It is important to note that, when the trace repeatedly returns to a value near zero, what we see is not the Ps, but the oral pressure reading for the [æ] vowel, which is approximately zero.

The exercise in Fig. 16.11 is often sung staccato, i. e. with short pauses rather than with a /p/ consonant between the tones. During these pauses the singer has to get ready for the next fundamental frequency value and must therefore avoid exhaling the pressurized air being built up in the lungs. Singers do this by opening the glottis between the tones and simultaneously reducing their Ps to zero, so that no air will be exhaled during the silent intervals between the tones. A remarkable fact demonstrated here is that, particularly when sung loudly – so that high Ps values are used – this exercise requires nothing less than a virtuoso mastering of both the timing and tuning of the breathing apparatus and the pitch control process. A failure to reach a target pressure is likely to result in a failure to reach the target F0.

16.2 The Glottal Sound Source

In speech and singing the general method to generate sound is to make a constriction and to let a strong flow of air pass through it. The respiratory component serves as the power source providing the energy necessary for sound production. At the glottis the steady flow of air generated by the respiratory component is transformed into a quasiperiodic series of glottal pulses. In the vocal tract, the glottally modified breath stream undergoes further modifications by the resonance characteristics of the oral, pharyngeal and nasal cavities.

Constrictions are formed at the glottis - by adjusting the separation of the vocal folds – and above the glottis – by positioning the articulators of the vocal tract. As the folds are brought close together, they respond to the air rushing through by initiating an open–close vibration and thereby imposing a quasiperiodic modulation of airflow. Thus, in a manner of speaking, the glottal structures operate as a device that imposes an AC modulation on a DC flow. This is basically the way that *voicing*, the sound source of voiced vowels and consonants and the carrier of intonation and melody, gets made.

A second mechanism is found in the production of *noise*, the acoustic raw materials for voiceless sounds (e.g., [f], [s], [p], [k]). The term refers to irregular turbulent fluctuations in airflow which arise when air comes out from a constriction at a high speed. This process can occur at the glottis – e.g., in [h] sounds, whisper and breathy voice qualities – or at various places of articulation in the vocal tract.

The framework for describing both singing and speech is that of the *source-filter theory of speech production* [16.27, 28]. The goal of this section is to put speech and singing side by side within that framework and to describe how the speaker and the singer coordinate respiration, phonation and articulation to shape the final product: the acoustic wave to be perceived by the listener.

Figure 16.12 [16.29] is an attempt to capture a few key aspects of vocal fold vibrations. At the center a sin-

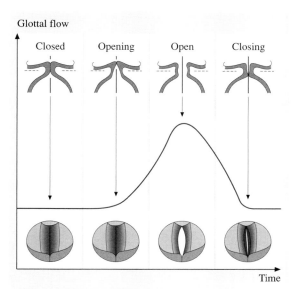

Fig. 16.12 Relationship between the flow glottogram, showing transglottal airflow versus time, and glottal configurations in a coronal plane (*upper series* of images) and from above, as through a laryngeal mirror (*lower series* of images). The airflow increases when the folds open the glottis and allow air to pass and decreases quickly when they close the glottis and arrest the airflow. (After [16.29])

gle cycle of a glottal waveform is seen. It plots airflow through the glottis as a function of time. Alternatively, the graph can be used to picture the time variations of the glottal area which present a pattern very similar to that for airflow. The top row shows stylized cross sections of the vocal folds at selected time points during the glottal cycle. From left to right they refer to the opening of the folds, the point of maximum area and the point of closure. Below is a view of the vocal folds from above corresponding to the profiles at the top of the diagram.

There are a number of different methods for visualizing vocal fold vibrations. By placing an electrode on each side of the thyroid cartilage, a minute current can be transferred across the glottis. This current increases substantially when the folds make contact. The resulting *electroglottogram*, also called a *laryngogram*, thus shows how the contact area varies with time. It is quite efficient in measurement of F0 and closed phase. *Optical glottograms* are obtained by illuminating the trachea from outside by means of a strong light source and capturing the light traveling through the glottis by means of an optical sensor in the pharynx. The signal therefore reflects the glottal area, but only as long as the light successfully finds it way to the sensor. A posterior tilting of the epiglottis may easily disturb or eliminate the signal.

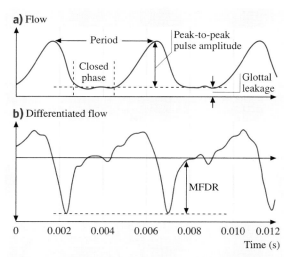

Fig. 16.13a,b Illustration of measures commonly used to characterize a flow glottogram (**a**) and its time derivative, the differentiated flow glottogram (**b**). (The *wiggles* in the latter are artifacts caused by an imperfect inverse filtering)

Flow glottograms show transglottal airflow versus time and are derived by inverse filtering the audio signal, often picked up as a flow signal by means of a pneumotachograph mask [16.30]. Inverse filtering implies that the signal is passed through a filter with a transfer function equalling the inverted transfer function of the vocal tract. Therefore correct inverse filtering requires that the inverted resonance peaks of the inverse filter are tuned to the formant frequencies of the vowel being filtered.

As transglottal airflow is zero when the glottis is closed and nonzero when it is open, the flow glottogram is physiologically relevant. At the same time it is a representation of the sound of the voice source.

A typical example of a flow glottogram is given in the upper graph of Fig. 16.13. The classical parameters derived from flow glottograms are the durations of the period and of the closed phase, pulse peak-to-peak amplitude, and glottal leakage. The lower graph shows the differentiated glottogram. The negative peak amplitude is often referred to as the *maximum flow declination rate* (MFDR). As we shall see it has special status in the process of voice production.

In the study of both speech and singing, the acoustic parameter of main relevance is the time variations in sound pressure produced by the vocal system and received by the listener's ears. Theoretically, this signal is roughly proportional to the derivative of the output airflow at the lips [16.28, 31] and it is related to the derivative of the glottal waveform via the transfer function of the vocal tract. Formally, the excitation signal for voiced sounds is defined in terms of this differentiated signal. Accordingly, in source-filter theory, it is the derivative of glottal flow that represents the source and is applied to the filter or resonance system of the vocal tract. The amplitude of the vocal tract excitation, generally referred to as the *excitation strength*, is quantified by the maximum velocity of flow decrease during vocal-fold closing movement (the MFDR, Fig. 16.13) which is a determinant of the level of the radiated sound. At the moment of glottal closure a drastic modification of the air flow takes place. This change is what generates voicing for both spoken and sung sounds and produces a sound with energy across a wide range of frequencies.

The Liljencrants–Fant (LF) model [16.32, 33] is an attempt to model glottal waveforms using parameters such as fundamental frequency, excitation strength, dynamic leakage, open quotient and glottal frequency (defined by the time period of glottal opening phase). Other proposals based on waveform parameters have been made by Klatt and Klatt [16.34], Ljungqvist and Fujisaki [16.35], Rosenberg [16.36] and Rothenberg et al. [16.37]. A second line of research starts out from assumptions about vocal fold mechanics and applies aerodynamics to simulate glottal vibrations [16.38, 39]. Insights from such work indicate the importance of parameters such as Ps, the adducted/abducted position of vocal folds and their stiffness [16.28].

During the early days of speech synthesis it became clear that the simplifying assumption of a constant voice source was not sufficient to produce high-quality natural-sounding copy synthesis. Experimental work on the voice source and on speech synthesis has shown that, in the course of an utterance, source parameters undergo a great deal of variation. The determinants of this dynamics are in part prosodic, in part segmental. Figure 16.14 [16.32] presents a plot of the time variations of the *excitation strength* parameter (i. e. MFDR) during the Swedish utterance: *Inte i DETta århundrade* [ɪntɪ ˣdɛtːaɔːrhəndradɛ]. The upper-case letters indicate that the greatest prominence was on the first syllable of *detta*. Vertical lines represent acoustic segment boundaries.

Gobl collected flow data using the mask developed by Rothenberg [16.30] and applied inverse filtering to obtain records of glottal flow which, after differentiation, enabled him to make measurements of excitation strength and other LF parameters.

Figure 16.14 makes clear that excitation strength is in no way constant. It varies depending on both prosodic and segmental factors. The effect of the segments is seen near the consonant boundaries. As the vocal tract is constricted, e.g., in [d] and [tː], and as transglottal pressure therefore decreases (cf. the pressure records of Fig. 16.7), excitation strength is reduced. In part these variations also occur to accommodate the voicing and the voicelessness of the consonant [16.28]. This influence of consonants on the voice source has been documented in greater detail by *Ni Chasaide* and *Gobl* [16.40] for German, English, Swedish, French, and Italian. Particularly striking effects were observed in the context of voiceless consonants.

Prosodically, we note in Fig. 16.14 that excitation strength exhibits a peak on the contrastively stressed syllable in *detta* and that the overall pattern of the phrase is similar to the falling declination contour earlier mentioned for declarative statements.

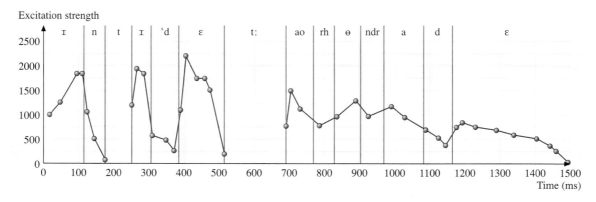

Fig. 16.14 Running speech data on the 'excitation strength parameter' of the LF model. (After [16.32])

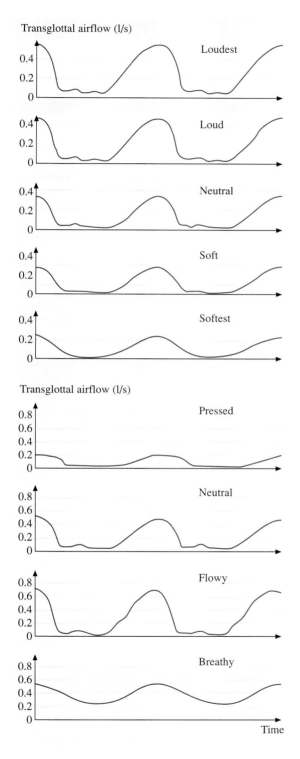

Table 16.1 Measurements of subglottal pressure (cm H_2O) and SPL at 0.3 m (dB) measured for the glottograms shown in Fig. 16.15

Table A	Ps (cm H_2O)	SPL at 0.3 m (dB)
loudest	14.3	84
loud	7.3	82
neutral	5.7	78
soft	3.9	75
softest	3	70
Table B	Ps (cm H_2O)	SPL at 0.3 m (dB)
pressed	11.4	83
neutral	5.1	79
flowy	8	88
breathy	6.6	84

Examples of the fact that the Ps has a strong influence on the flow glottogram are given in the upper set of graphs of Fig. 16.15, which shows a set of flow glottograms for phonations produced at the same pitch but with varying degrees of *vocal loudness*. As we examine the series of patterns from loudest to softest we note that both the peak flow and the maximum steepness of the trailing end of the pulse, i.e., MFDR, increase significantly with increasing Ps. These shape changes are lawfully related to the variations in Ps and are directly reflected in sound pressure levels as indicated by the numbers in Table 16.1.

Holmberg and colleagues [16.41] made acoustic and airflow recordings of 25 male and 20 female speakers producing repetitions of [pæ] at soft, normal and loud vocal efforts [16.42, p. 136]. Estimates of Ps and glottal airflow were made from recordings of oral pressure and oral airflow. Ps was derived by interpolating between peak oral pressures for successive [p] segments and then averaging across repetitions. A measure of average flow was obtained by low-pass filtering the airflow signal and averaging values sampled at the vowel midpoints.

Fig. 16.15 Typical flow glottogram changes associated with changes of loudness or mode of phonation (*top* and *bottom series*). As loudness of phonation is raised, the closing part of the curve becomes more steep. When phonation is pressed, the glottogram amplitude is low and the closed phase is long. As the adduction force is reduced, pulse amplitude grows and the closed phase becomes shorter (note that the flow scales are different for the *top* and *bottom series* of glottograms. Flowy phonation is the least adducted, and yet not leaky phonation

A copy of the airflow signal was low-pass filtered and inverse filtered to remove the effect of F1 and other formants. The output was then differentiated for the purpose of determining the MFDR (Fig. 16.13).

Figure 16.15 also illustrates how the voice source can be continuously varied between different *modes of phonation*. These modes range from hyperfunctional, or pressed, over neutral and flowy to hypofunctional, or breathy. The corresponding physiological control parameter can be postulated to be glottal adduction, i.e. the force by which the folds press against each other. It varies from minimum in hypofunctional to extreme in hyperfunctional. Flowy phonation is produced with the weakest degree of glottal adduction compatible with a full glottal closure. The physiologically relevant property that is affected is the vibration amplitude of the vocal folds, which is small in hyperfunctional/pressed phonation and wide in breathy phonation.

As illustrated in Fig. 16.15 the flow glottogram is strongly affected by these variations in phonation mode [16.43]. In pressed phonation, the pulse amplitude is small and the closed phase is long. It is larger in neutral and even more so in flowy. In breathy phonation, typically showing a waveform similar to a sine wave, airflow is considerable, mainly because of a large leakage, so there is no glottal closure.

Phonation mode affects the relation between Ps and the SPL of the sound produced. As shown in Table 16.1B pressed phonation is less economical from an acoustic point of view: a Ps of 11.4 cm H_2O produces an SPL at 0.3 m of only 83 dB, while in flowy phonation a lower Ps produces a higher SPL.

Pitch, loudness and phonation mode are voice qualities that we can vary continuously. By contrast, vocal registers, also controlled by glottal parameters, appear more like toggles, at least in untrained voices. The voice is operating either in one or another register. There are at least three vocal registers, *vocal fry*, *modal* and *falsetto*. When shifting between the modal and falsetto registers, F0 discontinuities are often observed [16.44].

The definition of vocal registers is quite vague, a set of tones along the F0 continuum that sound similar and are felt to be produced in a similar way. As registers depend on glottal function, they produce different flow glottogram characteristics. Figure 16.16 shows typical examples of flow glottograms for the falsetto and modal registers as produced by professional baritone, tenor and countertenor singers. The pulses are more rounded, the closed phase is shorter, and the glottal leakage is greater in the falsetto than in the modal register. However, the waveform of a given register often varies substantially between individuals. Classically trained sopranos, altos, and tenors learn to make continuous transitions between the modal and the falsetto registers, avoiding abrupt changes in voice timbre.

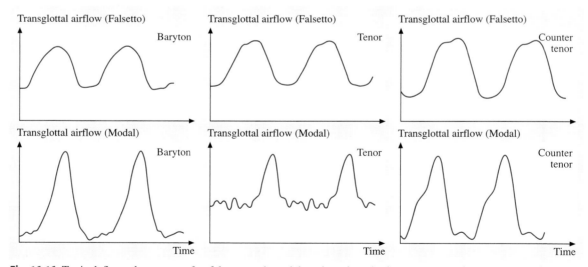

Fig. 16.16 Typical flow glottograms for falsetto and modal register in a baritone, tenor and countertenor singer, all professional. The flow scale is the same within subjects. The ripple during the closed phase in the lower middle glottogram is an artifact. In spite of the great inter-subject variability, it can be seen that the glottogram pulses are wider and more rounded in the falsetto register

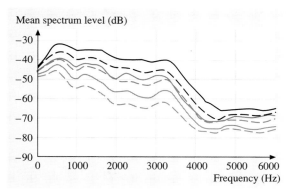

Fig. 16.17 Long-term-average spectra curves obtained from an untrained male speaker reading the same text at 6 different degrees of vocal loudness. From *top to bottom* the corresponding L_{eq} values at 0.3 m were 93 dB, 88 dB, 85 dB, 84 dB, 80 dB, and 76 dB

Variation of vocal loudness affects the spectrum slope as illustrated in Fig. 16.17, which shows long-term-average spectra (LTAS) from a male untrained voice. In the figure loudness is specified in terms of the so-called equivalent sound level L_{eq}. This is a commonly used time average of sound level, defined as

$$L_{eq} = 10 \log \frac{1}{T} \int_0^T \frac{p^2}{p_0^2} \, dt \, ,$$

where t is time and T the size of the time window. p and p_0 are the sound pressure and the reference pressure, respectively.

When vocal loudness is changed, the higher overtones change much more in sound level than the lower overtones. In the figure, a 14 dB change of the level near 600 Hz is associated with a 22 dB change near 3000 Hz, i. e., about 1.5 times the level change near 600 Hz. Similar relationships have been observed for professional singers [16.47]. In other words, the slope of the voice source spectrum decreases with increasing vocal loudness.

The physiological variable used for variation of vocal loudness is Ps. This is illustrated in the upper graph of Fig. 16.18, comparing averaged data observed in untrained female and male subjects and data obtained from professional operatic baritone singers [16.45, 46]. The relationship between the Ps and MFDR is approximately linear. It can be observed that the pressure range used by the singer is considerably wider than that used by the untrained voices. The MFDR produced with a given Ps by the untrained female and male subjects is mostly

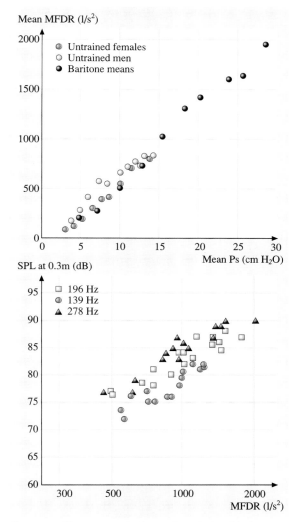

Fig. 16.18 The *top* graph shows the relationship between the mean subglottal pressure and the mean MFDR for the indicated subject groups. The *bottom* graph shows the relationship between MFDR and the SPL at 0.3 m for a professional baritone singing the vowels /a/ and /æ/ at different F0s. (After [16.45] p. 183, [16.46] p. 184)

higher than that produced by the baritones with the same pressure. This may depend on different mechanical characteristics of the vocal folds.

As we will see later, SPL depends on the strength of the excitation of the vocal tract, i. e. on MFDR. This variable, in turn, depends on Ps and F0; the higher the pressure, the greater the MFDR value and the higher the F0, the greater the MFDR. The top graph of Fig. 16.18

shows how accurately MFDR could be predicted from Ps and F0 for previously published data for untrained male and female singers and for professional baritone singers [16.45, 46]. Both Ps and F0 are linearly related to MFDR. However, the singers showed a much greater variation with F0 than the untrained voices. This difference reflected the fact that unlike the untrained subjects the singers could sing a high F0 much more softly than the untrained voices. The ability to sing high notes also softly would belong to the essential expressive skills of a singer. Recalling that an increase of Ps increases F0 by a few Hz/cm H_2O, we realize that singing high tones softly requires more forceful contraction of the pitch-raising laryngeal muscles than singing such tones loudly.

16.3 The Vocal Tract Filter

The source-filter theory, schematically illustrated in Fig. 16.19, describes vocal sound production as a three-step process: (1) generation of a steady flow of air from the lungs (DC component); (2) conversion of this airflow into a pseudo-periodically pulsating transglottal airflow (DC-to-AC conversion), referred to as the voice source; and (3) response of the vocal tract to this excitation signal (modulation of AC signal) which is characterized by the frequency curve or transfer function of the vocal tract. So far the first two stages, respiration and phonation, have been considered.

In this section we will discuss the third step, viz. how the vocal tract filter, i.e. the resonance characteristics of the vocal tract, modifies, and to some extent interacts with, the glottal source and shapes the final sound output radiated from the talker's/singer's lips.

Resonance is a key feature of the filter response. The oral, pharyngeal and nasal cavities of the vocal tract form a system of resonators. During each glottal cycle the air enclosed by these cavities is set in motion by the glottal pulse, the main moment of excitation occurring during the closing of the vocal folds, more precisely at the time of the MFDR, the maximum flow declination rate (cf. the previous section on source).

The behavior of a vocal tract resonance, or *formant*, is specified both in the time and the frequency domains. For any transient excitation, the time response is an exponentially decaying cosine [16.27, p. 46]. The frequency response is a continuous amplitude-frequency spectrum with a single peak. The shape of either function is uniquely determined by two numbers (in Hz): the formant frequency F and the bandwidth B. The bandwidth quantifies the degree of damping, i.e., how fast the formant oscillation decays. Expressed as sound pressure variations, the time response is

$$p(t) = A e^{-\pi B t} \cos(2\pi F t) \,. \tag{16.1}$$

For a single formant curve, the amplitude variations as a function of frequency f is given (in dB) by

$$L(f) = 20 \log \frac{[F^2 + \left(\frac{B}{2}\right)^2]}{\sqrt{(f-F)^2 + \left(\frac{B}{2}\right)^2}\sqrt{(f+F)^2 + \left(\frac{B}{2}\right)^2}} \,. \tag{16.2}$$

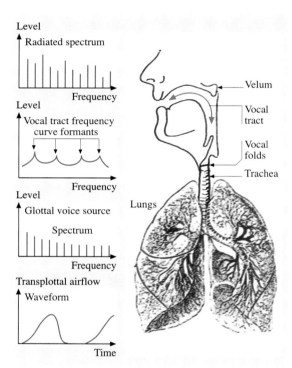

Fig. 16.19 Schematic illustration of the generation of voice sounds. The vocal fold vibrations result in a sequence of voice pulses (*bottom*) corresponding to a series of harmonic overtones, the amplitudes of which decrease monotonically with frequency (*second from bottom*). This spectrum is filtered according to the sound transfer characteristics of the vocal tract with its peaks, the formants, and the valleys between them. In the spectrum radiated from the lip opening, the formants are depicted in terms of peaks, because the partials closest to a formant frequency reach higher amplitudes than neighboring partials

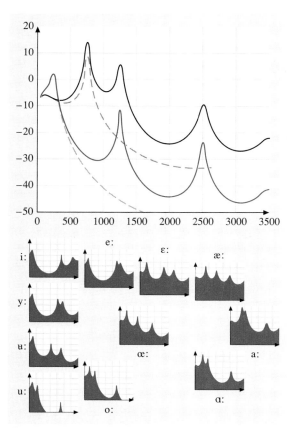

Fig. 16.20 Predictability of formant levels. Calculated spectral envelopes for a set of Swedish vowels. The *upper graph* illustrates how varying only the frequency of F1 affects the amplitudes of other formants

In the frequency domain the bandwidth is defined as the width of the formant 3 dB down from the peak, that is, at the half-power points. A large bandwidth produces a flat peak whereas a small value (less damping, reduced acoustic losses) makes the peak higher and sharper.

Figure 16.20 shows spectral envelopes for a set of Swedish vowels. They were calculated from formant frequency data [16.48, 49] in accordance with source-filter theory which assumes that a vowel spectrum can be decomposed into formants (all specified with respect to frequency and bandwidth), the spectrum of the voice source, the contribution of formants above F4 and radiation [16.27].

The individual panels are amplitude versus frequency plots. Vowels are portrayed in terms of their envelopes rather than as line spectra with harmonics.

The panels are arranged in a *formant space*. From top to bottom the F2 in the panels decreases. From left to right F1 increases. The frequency scale is limited to showing the first three formant peaks.

We note that all envelopes have falling overall slopes and that the amplitudes of the formant peaks vary a great deal from vowel to vowel and depending on the relative configuration of formant frequency positions.

In acoustic phonetic specifications of vowels, it is customary to report no more than the frequencies of the first two or three formants ([16.50, p. 59], [16.51]). Experiments in speech synthesis [16.52] have indicated that this compact description is sufficient to capture the quality of steady-state vowels reasonably well. Its relative success is explained by the fact that most of the building blocks of a vowel spectrum are either predictable (bandwidths and formant amplitudes), or show only limited spectral variations (source, radiation, higher-formant correction) [16.27].

Formant bandwidths [16.28, 53] reflect acoustic losses. They depend on factors such as radiation, sound transmission through the vocal tract walls, viscosity, heat conduction, constriction size as well as the state of the glottis. For example, a more open glottis, as in a breathy voice, will markedly increase the bandwidth of the first formant.

Despite the complex interrelations among these factors, bandwidths pattern in regular ways as a function of formant frequency. Empirical formulas have been proposed [16.54] that summarize bandwidth measurements made using transient and sweep-tone excitation of the vocal tract for closed-glottis conditions [16.55, 56].

To better understand how formant levels vary let us consider the top diagram of Fig. 16.20. It compares envelopes for two vowel spectra differing only in terms of F1. It is evident that the lowering of F1 (from 750 to 250 Hz) reduces the amplitudes of F2 and F3 by about 15 dB. This effect is predicted by acoustic theory which derives the spectral envelope of an arbitrary vowel as a summation of individual formant curves on a dB scale [16.27]. Figure 16.20 makes clear that, as F1 is shifted, its contribution to the envelope is drastically changed. In this case the shift moves the entire F1 curve down in frequency and, as a result, its upper skirt (dashed line) provides less of a *lift* to the upper formants. This interplay between formant frequency and formant levels is the major determinant of the various envelope shapes seen in Fig. 16.20.

One of the main lessons of Fig. 16.20 is accordingly that, under normal conditions of a stable voice source, formant amplitudes are predictable. Another important consequence of the source-filter theory is that, since knowing the formants will enable us to reconstruct the vowel's envelope, it should also make it possible to derive estimates about a vowel's overall intensity.

A vowel's intensity can be calculated from its power spectrum as

$$I = 10 \log \left[\sum (A_i)^2 \right], \qquad (16.3)$$

where A_i is the sound pressure of the i-th harmonic. This measure tends to be dominated by the strongest partial or partials. In very soft phonation the strongest partial is generally the fundamental while in neutral and louder phonation it is the partial that lies closest to the first formant. Thus, typically all partials which are more than a few dB weaker than the strongest partial in the spectrum do not contribute appreciably to the vowel's intensity.

As suggested by the envelopes in Fig. 16.20, the strongest harmonics in vowels tend to be found in the F1 region. This implies that a vowel's intensity is primarily determined by its F1. Accordingly, in the set shown in Fig. 16.20, we would expect [iːyːʉːuː] to be least intense and [æːɑɑː] the most intense vowels. This is in good agreement with experimental observations [16.57].

The intrinsic intensity of vowels and other speech sounds has been a topic of interest to phoneticians studying the acoustic correlates of stress [16.58]. It is related to sonority, a perceptual attribute of speech sounds that tends to vary in a regular way in syllables [16.59]. We will return below to that topic in Sect. 16.5.

Let us continue to illustrate how formant levels depend on formant frequencies with an example from singing: the singer's formant. This striking spectral phenomenon is characteristic of classically trained male singers. It is illustrated in Fig. 16.21. The figure shows a spectrogram of a commercial recording of an operatic tenor voice (Jussi Björling) performing a *recitativo* from Verdi's opera Aida. Apart from the vibrato undulation of the partials, the high levels of partials in the frequency range 2200–3200 Hz are apparent. They represent the singer's formant.

The singer's formant is present in all voiced sounds as sung by operatic male singers. It was first discovered by *Bartholomew* [16.60]. It manifests itself as a high, marked peak in the long-term-average spectrum (LTAS).

A second example is presented in Fig. 16.22 showing an LTAS of the same tenor voice as in Fig. 16.21, accompanied by an orchestra of the traditional western

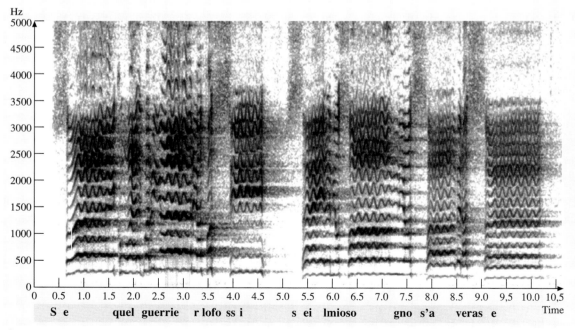

Fig. 16.21 Spectrogram of operatic tenor Jussi Björling's performance of an excerpt from Verdi's opera *Aida*. The lyrics are given below the graph

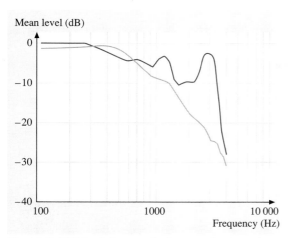

Fig. 16.22 LTAS of the sound of a symphonic orchestra with and without a singer soloist (*dark* and *light curves*). The singer's formant constitutes a major difference between the orchestra with and without the singer soloist. (After [16.61])

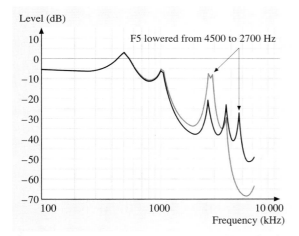

Fig. 16.23 Effect on the spectrum envelope of lowering formants F4 and F5 from 3500 Hz and 4500 Hz to 2700 Hz and 3500 Hz, respectively. The resulting gain in level at F3 is more than 10 dB

opera type. The LTAS peak caused by the singer's formant is just a few dB lower than the main peak in the low-frequency range.

In the same figure an LTAS of a classical symphony orchestra is also shown. On the average, the partials in the low-frequency range are quite strong but those above 600 Hz decrease by about 10 dB/octave with rising frequency. Even though this decrease varies depending on how loudly the orchestra is playing, this implies that the level of the orchestral accompaniment is much lower at 3000 Hz than at about 600 Hz. In other words, the orchestral sound offers the singer a rather reasonable competition in the frequency range of the singer's formant. The fact of the matter is that a voice possessing a singer's formant is much easier to hear when the orchestral accompaniment is loud than a voice lacking a singer's formant. Thus, it helps the singer's voice to be heard when the orchestral accompaniment is loud.

The singer's formant can be explained as a resonance phenomenon [16.62]. It is a product of the same rules that we invoked above to account for the formant amplitudes of vowels and for intrinsic vowel intensities. The strategy of a classically trained male singer is to shape his vocal tract so as to make F3, F4, and F5 form a tight cluster in frequency. As the frequency separations among these formants are decreased, their individual levels increase, and hence a high spectral peak is obtained between 2500 and 3000 Hz.

Figure 16.23 shows the effects on the spectrum envelope resulting from lowering F5 and F4 from 3500 Hz and 4500 Hz to 2700 Hz and 3500 Hz, respectively. The resulting increase of the level of F3 amounts to 12 dB, approximately. This means that male operatic singers produce a sound that can be heard more easily through a loud orchestral accompaniment by tuning vocal tract resonances rather than by means of producing an excessive Ps.

The acoustical situation producing the clustering of F3, F4, and F5 is obtained by acoustically mismatching the aperture of the larynx tube, also referred to as the epilaryngeal tube, with the pharynx [16.62]. This can be achieved by narrowing this aperture. Then, the larynx tube acts as a resonator with a resonance that is not much affected by the rest of the vocal tract but rather by the shape of the larynx tube. Apart from the size of the aperture, the size of the laryngeal ventricle would be influential: the larger the ventricle, the lower the larynx tube resonance. Presumably singers tune the larynx tube resonance to a frequency close to F3. The articulatory means used to establish this cavity condition seems mainly to be a lowering of the larynx, since this tends to widen both the pharynx and the laryngeal ventricle. Many singing teachers recommend students to sing with a comfortably low larynx position.

The level of the singer's formant is influenced also by the slope of the source spectrum which, in turn, depends on vocal loudness, i.e. on Ps, as mentioned. Thus, the singer's formant tends to increase by about 15 dB for a 10 dB change of the overall SPL [16.47].

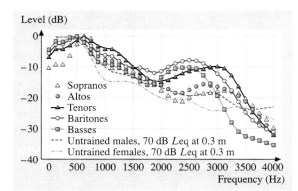

Fig. 16.24 Mean LTAS derived from commercial recordings of four representatives of each of the indicated voice classifications. The *dashed curves* show LTAS of untrained female (red) and male (blue) voices' speech with an L_{eq} of 70 dB at 0.3 m. Note that the singer's formant produces a prominent peak with a level well above the level of the untrained voices' LTAS only for the male singers

The center frequency of the singer's formant varies slightly between voice classifications, as illustrated in the mean LTAS in Fig. 16.24 which shows mean LTAS derived from commercial recordings of singers classified as soprano, alto, tenor, baritone and bass [16.63]. Each group has four singers. The center frequency of the singer's formant for basses, baritones and tenors are about 2.4, 2.6, and 2.8 kHz, respectively. These small differences are quite relevant to the typical voice timbres of these classifications [16.64]. Their origin is likely to be vocal tract length differences, basses tending to have longer vocal tracts than baritones who in turn have longer vocal tracts than tenors [16.65]. On the other hand, substantial variation in the center frequency of the singer's formant occurs also within classifications.

Also shown for comparison in Fig. 16.24 is a mean LTAS of untrained voices reading at an L_{eq} of 70 dB at 0.3 m distance. The singer's formant produces a marked peak some 15 dB above the LTAS of the untrained voices for the male singers. The female singers, on the other hand, do not show any comparable peak in this frequency range. This implies that female singers do not have a singer's formant [16.67–69].

Female operatic singers' lack of a singer's formant is not surprising, given the fact that: (1) they sing at high F0, i.e. have widely spaced spectrum partials, and (2) the F3, F4, F5 cluster that produces a singer's formant is rather narrow in frequency. The latter means that a narrow formant cluster will be hit by a partial only in

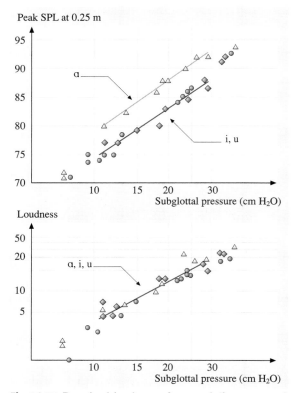

Fig. 16.25 Perceived loudness of a vowel (*bottom panel*) correlates better with subglottal pressure Ps than does SPL (*top panel*) which depends on the intrinsic intensity of the vowel. This, in turn, is determined by its formant pattern. (After [16.66])

some tones of a scale while in other scale tones there will be no partial in the formant cluster. This would lead to salient and quasi-random variation of voice timbre between scale tones.

The singer's formant is a characteristic of classically trained male singers. It is not found in nonclassical singing, e.g. in pop or musical theater singing, where audibility is the responsibility of the sound engineer rather than of the singer. Likewise, choir singers generally do not possess a singer's formant.

From the evidence in Fig. 16.18 we concluded that the logarithm of Ps does a good job of predicting the SPL over a large range of vocal intensities. This was shown for untrained male and female speakers as well as a professional baritone singer. In the next few paragraphs we will look at how vowel identity affects that prediction and how listeners perceive loudness in the presence of vowel-dependent variations in SPL at constant Ps.

This topic was addressed by Ladefoged [16.66] who measured the peak SPL and the peak Ps in 12 repetitions of *bee, bay, bar, bore* and *boo* spoken by a British talker at varying degrees of loudness. SPL values were plotted against log(Ps) for all tokens. Straight lines could readily be fitted to the vowels individually. The vowels of *bay* and *bore* tended to have SPL values in between those of *bar* and *bee/boo*. The left half of Fig. 16.25 replots the original data for [ɑ] and [i]/[u] pooled. The [ɑ] line is higher by 5–6 dB as it should be in view of F1 being higher in [ɑ] than in [i] and [u] (cf. preceding discussion).

In a second experiment listeners were asked to judge the loudness of the test words. Each item was presented after a carrier phrase: *Compare the words: bar and __*. The word bar served as reference. The subjects were instructed to compare the test syllable and the reference in terms of loudness, to give the value of 10 to the reference and another relative number to the test word. The analysis of the responses in terms of SPL indicated that, for a given SPL, it was consistently the case that *bee* and *boo* tended to be judged as louder than *bar*. On the other hand, there were instances of *bee* and *bar* with similar Ps that were judged to be equally loud. This effect stood out clearly when loudness judgements were plotted against log(Ps) as in the bottom half of Fig. 16.25.

Ladefoged concludes that "... in the case of speech sounds, loudness is directly related to the physiological effort" – in other words, Ps – rather than the SPL as for many other sounds.

Speech Sounds with Noise and Transient Excitation

A comprehensive quantitative treatment of the mechanisms of noise production in speech sounds is found in *Stevens* [16.28, pp. 100–121]. This work also provides detailed analyses of how the noise source and vocal tract filtering interact in shaping the bursts of stops and the spectral characteristics of voiced and voiceless fricatives. While the normal source mechanism for vowels always involves the glottis, noise generation may take place not only at the glottis but at a wide range of locations along the vocal tract. A useful rule of thumb for articulations excited by noise is that the output will spectrally be dominated by the cavity in front of the noise source. This also holds true for sounds with transient excitation such as stop releases and click sounds [16.28, Chapts. 7,8].

While most of the preceding remarks were devoted to the acoustics of vowels, we should stress that the source-filter theory applies with equal force to the production of consonants.

16.4 Articulatory Processes, Vowels and Consonants

X-ray films of normal speech movements reveal a highly dynamic process. Articulatory activity comes across as a complex flow of rapid, parallel lip, tongue and other movements which shows few, if any, steady states. Although the speaker may be saying no more than a few simple syllables, one nonetheless has the impression of a virtuoso performance of a *polyphonic* motor score. As these events unfold, they are instantly reflected in the acoustic output. The articulatory movements modify the geometry of the vocal tract. Accordingly, the filter (transfer function) undergoes continual change and, as a result, so do the output formant frequencies.

Quantitative modeling is a powerful tool for investigating the speech production process. It has been successfully applied to study the relations between formant and cavities. Significant insights into this mapping have been obtained by representing a given vocal tract configuration as an *area function* – i.e., a series of cylindrical cross sections of variable lengths and cross-sectional areas – and by simulating the effects of changes in the area function on the formant frequencies [16.27].

In the past, pursuing this approach, investigators have used lateral X-ray images – similar to the magnetic

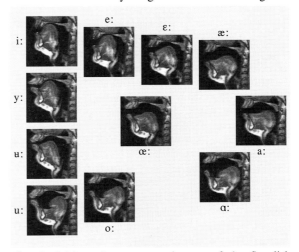

Fig. 16.26 Magnetic resonance images of the Swedish vowels of Fig. 16.20

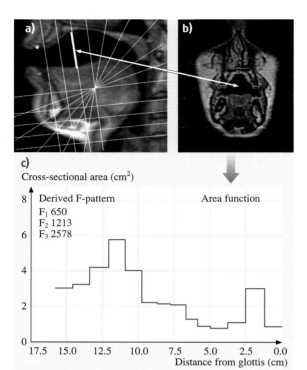

ticulatory movements, e.g., the lips, the tongue and the jaw [16.77–81].

Figure 16.27 highlights some of the steps involved in deriving the formant pattern from lateral articulatory profiles such those of Fig. 16.26. The top-left picture is the profile for the subject's [a]. The white lines indicate the coronal, coronal oblique and axial planes where ad-

Fig. 16.27a–c From articulatory profile to acoustic output. Deriving the formant patterns of a given vowel articulation involves (**a**) representing the vocal tract profile in terms of the cross-distances from the glottis to the lips; (**b**) converting this specification into a cross-sectional area function; and (**c**) calculating the formant pattern from this area function. To go from (**a**) to (**b**), profile data (*top left*) need to be supplemented by area measurements obtained from transversal (coronal and axial) images of the cross sections

resonance imaging (MRI) pictures in Fig. 16.26 – to trace the outlines of the acoustically relevant articulatory structures. To make estimates of cross-sectional areas along the vocal tract such lateral profiles need to be supplemented with information on the transverse geometry of the cross sections, e.g., from casts of the vocal tract [16.70] and tomographic pictures [16.27, 71].

More currently, magnetic resonance imaging methods have become available, making it possible to obtain three-dimensional (3-D) data on sustained steady articulations [16.72–76]. Figure 16.26 presents MRI images taken in the mid-sagittal plane of a male subject during steady-state productions of a set of Swedish vowels. These data were collected in connection with work on APEX, an articulatory model developed for studying the acoustic consequences of ar-

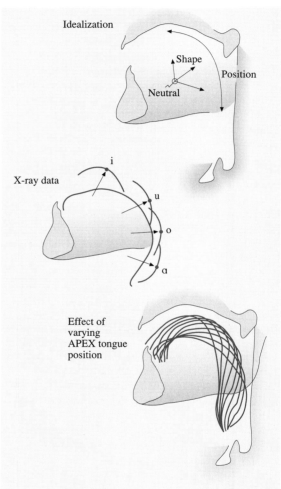

Fig. 16.28 The top drawing shows superimposed observed tongue shapes for [i], [u], [o] and [ɑ] differing in the place of the main constriction. Further analyses of similar data suggest that, for vowels, two main dimensions are used: The anterior–posterior location of the tongue body and its displacement from neutral which controls the degree of vocal tract constriction (*lower right*). These parameters are implemented numerically to produce vowels in the APEX articulatory model

Fig. 16.29 Acoustic consequences of tongue body, jaw and lip movements as examined in APEX simulations. The tongue moves continuously between palatal, velar and pharyngeal locations while maintaining a minimum constriction area $A_{\min}$ of $0.25\,\text{cm}$, a 7 mm jaw opening and a fixed larynx height. The *dashed* and *solid lines* refer to rounded and spread conditions. The *lower diagram* illustrates the effect of varying the jaw. The *thin lines* show tongue movement at various degrees of jaw opening (6, 7, 8, 9, 12, 15, 19 and 23 mm). The *bold lines* pertain to fixed palatal, neutral or pharyngeal tongue contours

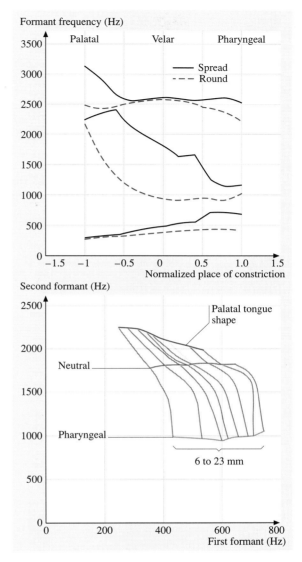

ditional MR images were taken to get information on the cross sections in the transverse dimension.

The location of the transverse cut is indicated by the bold white line segment in the lateral profile. In the coronal section to the right the airway of the vocal tract is the dark area at the center. Below it, the wavy surface of the tongue is evident. Since the teeth lack the density of hydrogen nuclei needed to show up on MR images, their intersection contours (green) were reconstructed from casts and added computationally to the image. The red lines indicate the outline of a dental plate custom-fitted to the subject's upper and lower teeth and designed to carry a contrast agent used to provide reference landmarks (tiny white dots) [16.82].

At any given point in the vocal tract, it would appear that changes in cross-sectional area depend primarily on the midsagittal distance from the tongue surface to the vocal tract wall. An empirically adequate approach to capturing distance-to-area relations is to use power functions of the form $A_x = \alpha d_x^\beta$, where d_x is the mid-sagittal cross distance at a location x in the vocal tract and α and β are constants whose values tend to be different in different regions and depend on the geometry of the vocal tract walls [16.24, 70, 72]. The cross-sectional area function can be obtained by applying a set of d-to-A rules of this type to cross distances measured along the vocal tract perpendicular to the midline. The final step consists in calculating the acoustic resonance frequencies of the derived area function.

In producing a vowel and the vocal tract shape appropriate for it, what are the main articulatory parameters that the speaker has to control? The APEX model [16.80] assumes that the significant information about the vowel articulations in Fig. 16.26 is the following. All the vowels exhibit tongue shapes with a single constriction. They differ with respect to how narrow this constriction is and where it is located. In other words, the vowels appear to be produced with control of two degrees of freedom: the palatal–pharyngeal dimension (*position*) and tongue height, or, in APEX terminology, *displacement* from neutral. This interpretation is presented in stylized form in Fig. 16.28 together with APEX tongue shapes sampled along the palatal–pharyngeal continuum. The choice of tongue parameters parallels the degree and place of constriction in the area-function models of *Stevens* and *House* [16.83] and *Fant* [16.27].

A useful application of articulatory models is that they can be set up to change a certain variable while keeping others constant [16.84]. Clearly, asking a human subject to follow such an instruction does not create an easily controlled task.

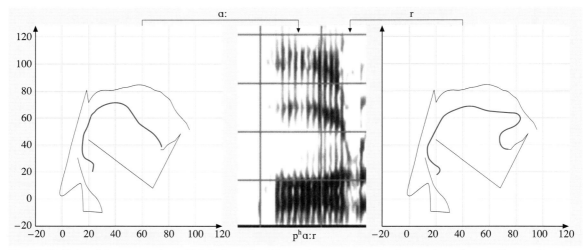

Fig. 16.30 Role of the tongue tip and blade in tuning F3. As the tongue is raised to produce the retroflex configuration required for the Swedish [r] sound an acoustically significant volume is created under the tongue blade. Simulations using APEX indicate that this cavity is responsible for the considerable lowering of the third formant associated with the [r]

Figure 16.29 plots results from APEX simulation experiments. It demonstrates how F1, F2 and F3 vary in response to the tongue moving continuously between palatal, velar and pharyngeal locations and maintaining a minimum constriction area $A_{\min}$ of $0.25\,\mathrm{cm}^2$, a 7 mm jaw opening and keeping the larynx height constant. The dashed and solid lines refer to rounded and spread conditions. The calculations included a correction for the impedance of the vocal tract walls ([16.54], [16.28, p. 158]).

It is seen that the tongue movement has strong effect on F2. As the constriction becomes more posterior F2 decreases and F1 rises. In general, rounding lowers formants by varying degrees that depends on the formant-cavity relations that apply to the articulation in question. However, for the most palatal position in Fig. 16.29 – an articulation similar to an [i] or an [y], it has little effect on F2 whereas F3 is affected strongly.

The F2 versus F1 plot of the lower diagram of Fig. 16.29 was drawn to illustrate the effect of varying the jaw. The thin lines show how, at various degrees of jaw opening (6, 7, 8, 9, 12, 15, 19 and 23 mm), the tongue moves between palatal and pharyngeal constrictions by way of the neutral tongue shape. The bold lines pertain to fixed palatal, neutral or pharyngeal tongue contours. We note that increasing the jaw opening while keeping the tongue configuration constant shifts F1 upward.

Figure 16.30 exemplifies the role of the tongue tip and blade in tuning F3. The data come from an X-ray study with synchronized sound [16.79, 85]. At the center a spectrogram of the syllable "*par*" [pʰɑːr]. Perhaps the most striking feature of the formant pattern is the extensive lowering of F3 and F4 into the [r].

The articulatory correlates are illustrated in the tracings. As we compare the profiles for [ɑː] and [r], we see that the major change is the raising of the tongue blade for [r] and the emergence of a significant cavity in front of, and below, the tongue blade. Simulations us-

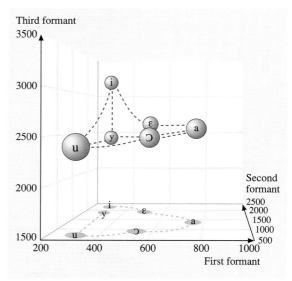

Fig. 16.31 The acoustic vowel space: *possible* vowel formant combinations according to the APEX model

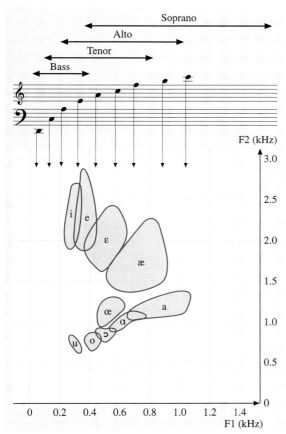

Fig. 16.32 Schematic illustration of the variation ranges of F1 and F2 for various vowels as pronounced by female and male adults. Above the graph F1 is given in musical score notation together with typical F0 ranges for the indicated voice classifications

ing APEX and other models [16.28, 86] indicate that this subapical cavity is responsible for the drastic lowering of F3.

The curves of Fig. 16.30 are qualitatively consistent with the nomograms published for three-parameter area-function models, e.g., Fant [16.27]. An advantage of more-realistic physiological models is that the relationships between articulatory parameters and formant patterns become more transparent. We can summarize observations about APEX and the articulation-to-formants mapping as:

- F1 is controlled by the jaw in a direct way;
- F2 is significantly changed by front–back movement of the tongue;

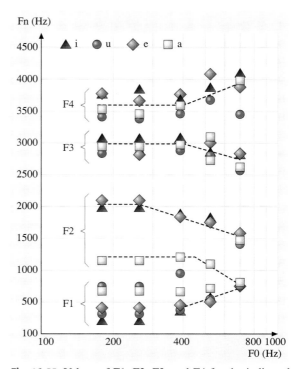

Fig. 16.33 Values of F1, F2, F3, and F4 for the indicated vowels in a professional soprano as function of F0. F1 and F3 are represented by open symbols and F2 and F4 by filled symbols. The values for F0 $\approx$ 180 Hz were obtained when the singer sustained the vowels in a speech mode. (After *Sundberg* [16.87])

- F3 is influenced by the action of the tongue blade.

Another way of summarizing the acoustic properties of a speech production model is to translate all of its articulatory capabilities into a formant space. By definition that contains all the formant patterns that the model is capable of producing (and specifying articulatorily). Suppose that a model uses n independent parameters and each parameter is quantized into a certain number of steps. By forming all *legal* combinations of those parametric values and deriving their vocal tract shapes and formant patterns, we obtain the data needed to represent that space graphically.

The APEX vowel space is shown in 3-D in Fig. 16.31. The x-axis is F1. The depth dimension is F2 and the vertical axis is F3. Smooth lines were drawn to enclose individual data points (omitted for clarity).

A cloud-like structure emerges. Its projection on the F2/F1 floor plane takes the form of the familiar trian-

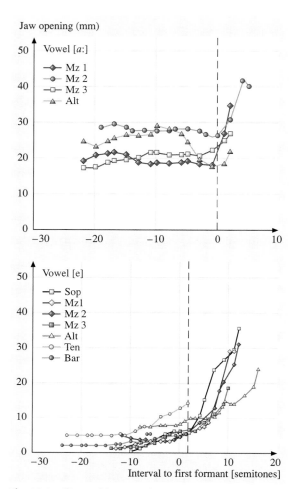

Fig. 16.34 Singers' jaw opening in the vowels [ɑː] and [e] plotted as functions of the distance in semitones between F0 and the individual singer's normal F1 value for these vowels. The singers belonged to different classifications: Sop = soprano, Mz = mezzosoprano, Alt = alto, Ten = tenor, Bar = baritone. Symbols refer to subjects. (After *Sundberg* and *Skoog* [16.88])

gular pattern appears with [i], [a] and [u] at the corners. Adding F3 along the vertical axis offers additional room for vowel timber variations especially in the [i/y] region.

Figure 16.32 shows typical values for F1 and F2 for various vowels. At the top of the diagram, F1 is given also on the musical staff. The graph thus demonstrates that the fundamental frequency in singing is often higher than the normal value of F1. For example, the first formant of [iː] and [uː] is about 250 Hz (close to the pitch of C4), which certainly is a quite low note for a soprano.

Theory predicts that the vocal fold vibration will be greatly disturbed when F0 = F1 [16.89], and singers seem to avoid allowing F0 to pass F1 [16.87, 90]. Instead they raise F1 to a frequency slightly higher than F0. Some formant frequency values for a soprano singer are shown in Fig. 16.33. The results can be idealized in terms of lines, also shown in the figure, relating F1, F2, F3, and F4 to F0. Also shown are the subject's formant frequencies in speech. The main principle seems to be as follows. As long as fundamental frequency is lower than the normal value of the vowel's first formant frequency, this formant frequency is used. At higher pitches, the first formant is raised to a value somewhat higher than the fundamental frequency. In this way, the singer avoids having the fundamental frequency exceed the first formant frequency. With rising fundamental frequency, F2 of front vowels is lowered, while F2 of back vowels is raised to a frequency just above the second spectrum partial; F3 is lowered, and F4 is raised.

A commonly used articulatory trick for achieving the increase of F1 with F0 is a widening of the jaw opening [16.85]. The graphs of Fig. 16.34 show the jaw opening of professional singers as a function of the frequency separation in semitones between F0 and the F1 value that the singer used at low pitches. The graph referring to the vowel [ɑː] shows that most singers began to widen their jaw opening when the F0 was about four semitones below the normal F1 value. The lower graph of Fig. 16.34 shows the corresponding data for the vowel [e]. For this vowel, most subjects started to widen the jaw opening when the fundamental was about four semitones above the normal value of F1. It is likely that below this pitch singers increase the first formant by other articulatory means than the jaw opening. A plausible candidate in front vowels is the degree of vocal tract constriction; a reduced constriction increases the first formant. Many singing teachers recommend their students to *release the jaw* or to *give space to the tone*; it seems likely that the acoustic target of these recommendations is to raise F1.

There are also other articulators that can be recruited for the purpose of raising F1. One is the lip opening. By retracting the mouth corners, the vocal tract is shortened; and hence the frequencies of the formants will increase. The vocal tract can also be shortened by raising the larynx, and some professional female singers take advantage of this tool when singing at high pitches.

This principle of tuning formant frequencies depending on F0 has been found in all singers who encounter the situation that the normal value of the first formant is lower than their highest pitches. In fact, all singers except basses encounter this situation at least for some

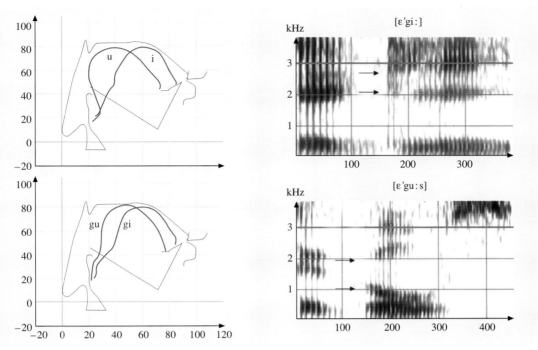

Fig. 16.35 The universal phenomenon of coarticulation illustrated with articulatory and acoustic data on Swedish [g]. The influence of the following vowel extends throughout the articulation of the stop closure. There is no point in time at which a *pure* (context-free) sample of the [g] could be obtained

vowels sung at high pitches. The benefit of these arrangements of the formant frequencies is an enormous increase of sound level, gained by sheer resonance.

Vowel quality is determined mainly by F1 and F2, as mentioned. Therefore, one would expect drastic consequences regarding vowel intelligibility in these cases. However, the vowel quality of sustained vowels seems to survive these pitch-dependent formant frequency changes surprisingly well, except when F0 exceeds the pitch of F5 (about 700 Hz). Above that frequency no formant frequency combination seems to help, and below it, the vowel quality would not be better if normal formant frequencies were chosen. The amount of text intelligibility which occurs at very high pitches relies almost exclusively on the consonants surrounding the vowel. Thus, facing the choice between inaudible tones with normal formant frequencies or audible tones with strange vowel quality, singers probably make a wise choice.

One of the major challenges both for applied and theoretical speech research is the great variability of the speech signal. Consider a given utterance as pronounced by speakers of the same dialect. A few moments' reflection will convince us that this utterance is certain to come in a large variety of physical shapes. Some variations reflect the speaker's age, gender, vocal anatomy as well as emotional and physiological state. Others are stylistic and situational as exemplified by speaking clearly in noisy environments, speaking formally in public, addressing large audiences, chatting with a close friend or talking to oneself while trying to solve a problem. Style and situation make significant contributions to the variety of acoustic waveforms that instantiate what we linguistically judge as the *same utterance*.

In phonetic experimentation investigators aim at keeping all of these stylistic and situational factors constant. This goal tends to limit the scope of the research to a style labeled *laboratory speech*, which consists of test items that are chosen by the experimenter and are typically read from a list. Despite the marked focus on this type of material, laboratory speech nonetheless presents several variability puzzles. We will mention two: segmental interaction, or *coarticulation*, and prosodic modulation. Both exemplify the ubiquitous context dependence of phonetic segments. First a few remarks on coarticulation.

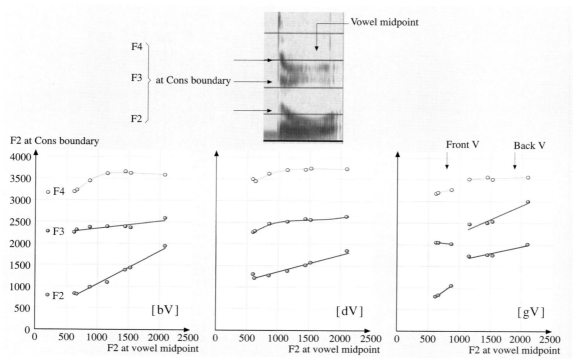

Fig. 16.36 Coarticulation in [bV], [dV] and [gV] as reflected by measurements of F2, F3 and F4 onsets plotted against an acoustic correlate of front back tongue movement, viz., F2 at the vowel midpoint

In Fig. 16.35 the phoneme [g] occurs in two words taken from an X-ray film [16.79, 85] of a Swedish speaker: [ɛˈgiː] and [ɛˈguːs]. In these words the first three phonemes correspond to the first three segments on the spectrogram: an initial [ɛ] segment, the [g] stop gap and then the final vowel. So far, so good. However, if we were to try to draw a vertical line on the spectrogram to mark the point in time where [ɛ] ends and [g] begins, or where [g] ends and the final vowel begins, we would soon realize that we have an impossible task. We would perhaps be able to detect formant movements in the [ɛ] segment indicating that articulatory activity towards the [g] had been initiated. However, that point in time occurs during the acoustic [ɛ] segment. Similarly, we could identify the endpoint of the formant transitions following [g] but that event occurs when the next segment, the final vowel, is already under way.

What we arrive at here is the classical conclusion that strings of phonemes are not organized as *beads on a necklace* [16.91–93]. The acoustic correlates of phonemes, the acoustic segments, are produced according to a motor schema that requires parallel activity in several articulatory channels and weaves the sequence of phonemes into a smooth fabric of overlapping movements. We are talking about *coarticulation*, the overlap of articulatory gestures in space and time.

Not only does this universal of motor organization give rise to the segmentation problem, i. e., make it impossible to chop up the time scale of the speech wave into phoneme-sized chunks, it creates another dilemma known as the *invariance issue*. We can exemplify it by referring to Fig. 16.35 again and the arrows indicating the frequencies of F2 and F3 at the moment of [g] release. With [iː] following they are high in frequency. Next to [uː] they are lower. What acoustic attributes define the [g] phoneme? How do we specify the [g] phoneme acoustically in a context-independent way?

The answer is that, because of coarticulation, we cannot provide an acoustic definition that is context independent. There is no such thing as an acoustic *pure sample* of a phoneme.

Articulatory observations confirm this account. There is no point in time where the shape of the tongue shows zero influence from the surrounding vowels. The tracings at the top left of Fig. 16.35 show the tongue contours for [iː] and [uː] sampled at the vowel midpoints. The bottom left profiles show the tongue shapes at the [g] release. The effect of the following vowel is readily apparent.

So where do we look for the invariant phonetic correlates for [g]? Work on articulatory modeling [16.49, 94] indicates that, if there is such a thing as a single context-free target underlying the surface variants of [g], it is likely to occur at a deeper level of speech production located before the motor commands for [g] closure and for the surrounding vowels blend.

This picture of speech production raises a number of questions about speech perception that have been addressed by a large body of experimental work [16.95, 96] but which will not be reviewed here.

It should be remarked though that, the segmentation and invariance issues notwithstanding, the context sensitivity of phonetic segments is systematic. As an illustration of that point Fig. 16.36 is presented. It shows average data on formant transitions that come from the Swedish speaker of Fig. 16.36 and Figs. 16.26–16.28. The measurements are from repetitions of CV test words in which the consonants were [b], [d] or [g] and were combined with [iː] [ɛː] [æː] [a] [ɑː] [oː] and [uː]. Formant transition onsets for F2, F3 and F4 are plotted against F2 midpoints for the vowels.

If the consonants are coarticulated with the vowels following, we would expect consonant onset patterns to co-vary with the vowel formant patterns. As shown by Fig. 16.36 that is also what we find. Recall that, in the section on articulatory modeling, we demonstrated that F2 correlates strongly with the front–back movement of the tongue. This implies that, in an indirect way, the x-axis labeled 'F2 at vowel midpoint' can be said to range from back to front. The same reasoning applies to F2 onsets.

Figure 16.36 shows that the relationship between F2 onsets and F2 at vowel midpoint is linear for bV and dV. For gV, the data points break up into back (low F2) and front (high F2) groups. These straight lines – known as *locus equations* [16.97] – have received considerable attention since they provide a compact way of quantifying coarticulation. Data are available for several languages showing robustly that slopes and intercepts vary in systematic ways with places of articulation.

Furthermore, we see from Fig. 16.36 that lawful patterns are obtained also for F3 and F4 onsets. This makes sense if we assume that vocal tract cavities are not completely uncoupled and that hence, all formants – not only F2 – are to some extent influenced by where along the front–back dimension the vowel is articulated.

16.5 The Syllable

A central unit in both speech and singing is the syllable. It resembles the phoneme in that it is hard to define but it can be described in a number of ways.

Linguists characterize it in terms of how vowels and consonants pattern within it. The central portion, the nucleus, is normally a vowel. One or more consonants can precede and/or follow forming the onset and the coda respectively. The vowel/nucleus is always there; the onset and coda are optional.

Languages vary with respect to the way they combine consonants and vowels into syllables. Most of them favor a frame with only two slots: the CV syllable. Others allow more-elaborated syllable structures with up to three consonants initially and the mirror image in syllable final position. If there is also a length distinction in the vowel and/or consonant system, syllables frames can become quite complex. A rich pattern with consonant clusters and phonological length usually implies that the language has a strong contrast between stressed and unstressed syllables.

In languages that allow consonant sequences, there is a universal tendency for the segments to be serially ordered on an articulatory continuum with the consonants compatible with the vowel's greater jaw opening occurring next to the vowel, e.g., [l] and [r], while those less compatible, e.g. [s], are recruited at the syllable margins [16.98, 99]. In keeping with this observation, English and other languages use [spr] as an initial, but not final, cluster. The reverse sequence [rps] occurs in final, but not initial, position, cf. *sprawl* and *harps*. Traditionally and currently, this trend is explained in terms of an auditory attribute of speech sounds, sonority. The *sonority principle* [16.59] states that, as the most *sonorous* segments, vowels take the central nucleus position of the syllable and that the *sonority* of the surrounding consonants must decrease to the left and to the right starting from the vowel. Recalling that the degree of articulatory opening affects F1 which in turn affects sound intensity, we realize that these articulatory and auditory accounts are not incompatible. However, the reason for the syllabic variations in sonority is articulatory: the tendency for syllables to alternate close and open articulations in a cyclical manner.

The syllable is also illuminated by a developmental perspective. An important milestone of normal speech acquisition is *canonical babbling*. This type of vocalization makes its appearance sometime between 6–10 months. It consists of sequences of CV-like events, e.g., [dædæ], [baba] [16.101–105]. The phonetic output of deaf infants differs from canonical babbling both quantitatively and quantitatively [16.106–108], suggesting that auditory input from the ambient language is a prerequisite for canonical babbling [16.109–112]. What babbling shares with adult speech is its "syllabic" organization, that is, the alternation of open and close articulations in which jaw movement is a major component [16.113].

As mentioned, the regular repetition of open-close vocal tract states gives rise to an amplitude modulation of the speech waveform. Vowels tend to show the highest amplitudes contrasting with the surrounding consonants which have various degrees of constriction and hence more reduced amplitudes. At the acoustic boundary between a consonant and a vowel, there is often an abrupt rise in the amplitude envelope of the waveform.

When a Fourier analysis is performed on the waveform envelope, a spectrum with primarily low, sub-audio frequency components is obtained. This is to be expected given the fact that amplitude envelopes vary slowly as a function of time. This representation is known as the modulation spectrum [16.114]. It reflects recurring events such as the amplitude changes at consonant-vowel boundaries. It provides an approximate record of the rhythmic pulsating stream of stressed and unstressed syllables.

The time envelope may at first appear to be a rather crude attribute of the signal. However, its perceptual importance should not be underestimated. Room acoustics and noise distort speech by modifying and destroying its modulation spectrum. The modulation transfer function was proposed by Houtgast and Steeneken as a measure of the effect of the auditorium on the speech signal and as a basis for an index, the speech transmission index (STI), used to predict speech intelligibility under different types of reverberation and noise. The success of this approach tells us that the modulation spectrum, and hence the waveform envelope, contains information that is crucial for robust speech perception [16.115]. Experimental manipulation of the temporal envelope has been performed by Drullman et al. [16.116, 117] whose work reinforces the conclusions reached by Houtgast and Steeneken.

There seems to be something special about the front ends of syllables. First, languages prefer CVs to VCs. Second, what children begin with is strings of CV-like pseudosyllables that emulate the syllable onsets of adult speech. Third there is perceptually significant information for the listener in the initial dynamics of the syllable. Let us add another phenomenon to this list: the *syllable beat*, or the syllable's P-center [16.118, 119].

In reading poetry, or in singing, we have a very strong sense that the syllables are spoken/sung in accordance with the rhythmic pattern of the meter. Native speakers agree more or less on how many syllables there are in a word or a phrase. In making such judgments, they seem to experience syllables as unitary events. Although it may take several hundred milliseconds to pronounce, subjectively a syllable appears to occur at a specific moment in time. It is this impression to which the phonetic term *syllable beat* refers and that has been studied experimentally in a sizable number of publications [16.120–126].

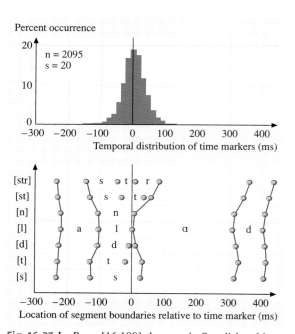

Fig. 16.37 In *Rapp* [16.100] three male Swedish subjects were asked to produce test words based on an [a_'ɑːd] frame with [s, t, d, l, n, st, str] as intervocalic segment(s). These items were pronounced in synchrony with metronome beats presented over headphones. The figure shows the temporal distribution of the metronome beats (*upper panel*) in relation to the average time location of acoustic segment boundaries (*lower panel*)

Rapp [16.100] asked three native speakers of Swedish to produce random sets of test words built from [aC'aːd] were the consonant C was selected from [s, t, d, l, n, st, str]. The instruction was to synchronize the stressed syllable with a metronome beat presented over earphones.

The results are summarized in Fig. 16.37. The *x*-axis represents distance in milliseconds from the point of reference, the metronome beat. The top diagram shows the total distribution of about 2000 time markers around the mean. The lower graph indicates the relative location of the major acoustic segment boundaries.

Several phonetic correlates have been proposed for the syllable beat: some acoustic/auditory, others articulatory. They all hover around the vowel onset, e.g., the amplitude envelope of a signal [16.127], rapid increase in energy in spectral bands [16.128, 129] or the onset of articulatory movement towards the vowel [16.130].

Rapp's data in Fig. 16.37 indicate that the mean beat time tends to fall close to the release or articulatory opening in [t, d, l, n] but that it significantly precedes the acoustic vowel onsets of [str-], [st-] and [s-]. However, when segment boundaries were arranged in relation to a fixed landmark on the F0 contour and vowel onsets were measured relative to that landmark, the range of vowel onsets was reduced. It is possible that the syllable beat may have its origin, not at the acoustic surface, nor at some kinematic level, but in a deeper motor control process that coordinates and imposes coherence on respiratory, phonatory and articulatory activity needed to produce a syllable.

Whatever the definitive explanation for the syllable's psychological *moment of occurrence* will be, the syllable beat provides a useful point of entry for attempts to understand how the control of rhythm and pitch works in speech and singing. Figure 16.38 compares spectrograms of the first few bars of *Over the rainbow*, spoken (left) and sung (right).

Vertical lines have been drawn at vowel onsets and points where articulators begin to move towards a more open configuration. The lines form an isochronous temporal pattern in the sung version which was performed at a regular rhythm. In the spoken example, they occur at intervals that seem more determined by the syllable's degree of prominence.

The subject reaches F0 targets at points near the beats (the vertical lines). From there target frequencies are maintained at stationary values until shortly before it is time to go to the next pitch. Thus the F0 curve resembles a step function with some smoothing applied to the steps.

On the other hand, the F0 contour for speech shows no such steady states. It makes few dramatic moves as it

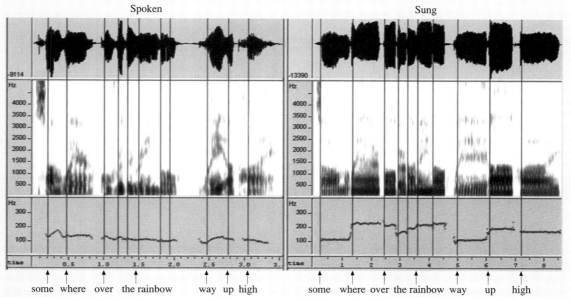

Fig. 16.38 Waveforms and spectrograms of the first few bars of Over the rainbow, spoken (*left*) and sung (*right*). Below: F0 traces in Hz. Vertical lines were drawn at time points corresponding to "vowel onsets" defined in terms of onset of voicing after a voiceless consonant, or the abrupt vocal tract opening following a consonant

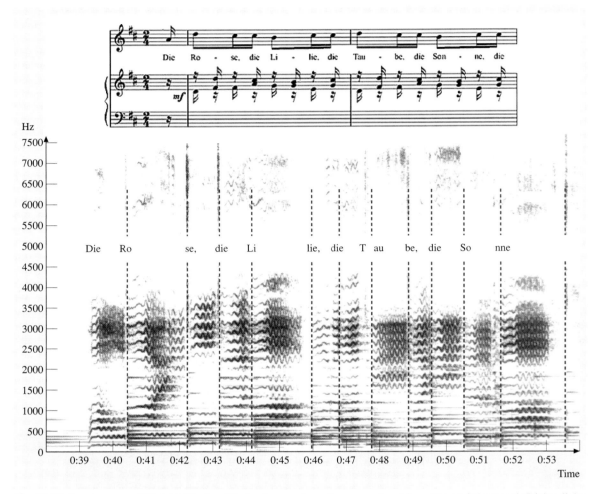

Fig. 16.39 Spectrogram of Dietrich Fischer-Dieskau's recording of song number 4 from Robert Schumann's Dichterliebe, op. 48. The *vertical bars* mark the onset of the tones of the piano accompaniment. Note the almost perfect synchrony between vowel onset and the piano. Note also the simultaneous appearance of high and low partials after the consonants also after the unvoiced /t/ in Taube

gradually drifts downward in frequency (the declination effect).

Figure 16.39 shows a typical example of classical singing, a spectrogram of a commercial recording of Dietrich Fischer-Dieskau's rendering of the song "*Die Rose, die Lilie*" from Robert Schumann's *Dichterliebe*, op. 48. The vertical dashed lines show the onsets of the piano accompaniment. The wavy patterns, often occurring somewhat after the vowel onset, reflect the vibrato. Apart from the vibrato undulation of the partials the gaps in the pattern of harmonics are quite apparent. At these points we see the effect of the more constricted artic- ulations for the consonants. Note the rapid and crisply synchronous amplitude rise in all partials at the end of the consonant segments, also after unvoiced consonants, e.g. the /t/ in *Taube*. These rises are synchronized with the onset of the piano accompaniment tones, thus demonstrating that in singing a beat is marked by the vowel. The simultaneous appearing of low and high partials seems to belong to the characteristics of classical singing as opposed to speech, where the higher partials often arrive with a slight delay after unvoiced consonants. As mentioned before, such consonants are produced with abduction of the vocal folds. To generate

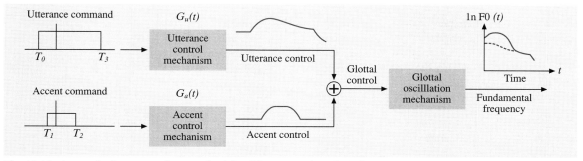

Fig. 16.40 A quantitative approach to synthesizing F0 contours by rule (After *Fujisaki* [16.131]). The framework shown here is an example of the so-called *superposition models* which generate F0 patterns by adding strongly damped responses to input step function target commands for phrase and accents

also high partials in this context, the vocal folds need to close the glottis at the very first vibratory cycle at the vowel onset. A potential benefit of this might be that this enhances the higher formants which are important to text intelligibility.

Several quantitative frameworks have been proposed for generating speech F0 contours by rule (for overview see Frid [16.132]). A subgroup of these has been called *superposition models* [16.131, 133]. A characteristic of such models is that they decompose the F0 curve into separate phrase and accent components. The accent commands are F0 step or impulse functions temporally linked to the syllables carrying stress. Accent pulses are superimposed on the phrase component. For a declarative sentence, the phrase component is a falling F0 contour produced from rectangular step function *hat patterns* [16.134]. The phrase and accent commands are passed through critically damped filters to convert their sum into the smoothly varying output F0 contour. An example of this approach is shown in Fig. 16.40 [16.131].

The last few figures suggest that the F0 patterns of singing and speech may be rather different if compared in terms of the raw F0 curves. However, the superposition models suggest that speech F0 contours have characteristics in part attributable to the passive response characteristics of the neuro-mechanical system that produces them, and in part due to active control signals. These control commands take the form of stepwise changes, some of short, others of longer duration. This representation is not unlike the sequence of F0 targets of a melody.

The implication of this analysis is that F0 is controlled in similar ways in speech and singing in the sense that both are based on sequences of underlying steady-state targets. On the other hand, a significant difference is that in singing high accuracy in attainment of acoustic target frequencies is required whereas in speech such demands are relaxed and smoothing is stronger.

16.6 Rhythm and Timing

A striking characteristic of a foreign language is its rhythm. Phoneticians distinguish between stress-timed and syllable-timed languages. English, Russian, Arabic and Thai are placed in the first group [16.135]. French, Spanish, Greek, Italian, Yoruba and Telugu are examples of the second.

Stress timing means that stressed syllables recur at approximately equal intervals. For instance, it is possible to say "in ENGlish STRESses reCUR at EQual INtervals", spacing the stressed syllables evenly in time and without sounding too unnatural. Stress increases syllable duration. In case several unstressed syllables occur in between stresses they are expected to undergo compensatory shortening. Syllable, or *machine-gun*, timing [16.136] implies that syllables recur at approximately equal intervals. In the first case it is the stresses that are isochronous, in the second the syllables.

To what extent are speech rhythms explicitly controlled by the speaker? To what extent are they fortuitous by-products of other factors? Does acquiring a new language involve learning new rhythmic target patterns?

The answer depends on how speech is defined. In the reading of poetry and nursery rhymes it is clear that an external target, viz. the metric pattern, is a key determinant of how syllables are produced. Similarly, the groupings and the pausing in narrative prose, as read by a professional actor, can exhibit extrinsic rhythmic stylization compared with unscripted speech [16.137].

Dauer [16.138] points the way towards addressing such issues presenting statistics on the most frequently occurring syllable structures in English, Spanish and French. In Spanish and French, syllables were found to be predominantly open, i.e. ending with a vowel, whereas English showed a preference for closed syllables, i.e. ending with a consonant, especially in stressed syllables. Duration measurements made for English and Thai (stress-timed) and Greek, Spanish and Italian (syllable-timed) indicated that the duration of the interstress intervals grew at the same constant rate as a function of the number of syllables between the interstress intervals. In other words, no durational evidence was found to support the distinction between stress timing and syllable timing. How do we square that finding with the widely shared impression that some languages do indeed sound *stress-timed* and others *syllable-timed*?

Dauer [16.138, p. 55] concludes her study by stating that: " ... *the rhythmic differences we feel to exist between languages such as English and Spanish are more a result of phonological, phonetic, lexical and syntactic facts about that language than any attempt on the part of the speaker to equalize interstress or intersyllable intervals.*"

It is clear that speakers are certainly capable of imposing a rhythmic template in the serial read-out of syllables. But do they put such templates in play also when they speak spontaneously? According to *Dauer* and others [16.139, 140] rhythm is not normally an explicitly controlled variable. It is better seen as an emergent product of interaction among the choices languages make (and do not make) in building their syllables: e.g., from open versus closed syllable structures, heavy versus weak clusters, length distinctions and strong stressed/unstressed contrast. We thus learn that the distinction between syllable timing and stress timing may be a useful descriptive term but should primarily be applied to the phonetic output, more seldom to its input control.

Do speech rhythms carry over into music? In other words, would music composed by speakers of syllable-timed or stress-timed languages also be syllable timed or stress timed? The question has been addressed [16.141, 142] and some preliminary evidence has been reported.

There is more to speech timing than what happens inside the syllable. Processes at the word and phrase levels also influence the time intervals between syllables. Consider the English words *digest*, *insult* and *pervert*. As verbs their stress pattern can be described as a sequence of *weak–strong* syllables. As nouns the order is reversed: *strong–weak*. The syllables are the same but changing the word's stress contour (the lexical stress) affects their timing.

The *word length effect* has been reported for several languages [16.58, 143]. It refers to the tendency for the stressed vowel of the word to shorten as more syllables are appended, cf. English *speed*, *speedy*, *speedily*. In Lehiste's formulation: "It appears that in some languages the word as a whole has a certain duration that tends to remain relatively constant, and if the word contains a greater number of segmental sounds, the duration of the segmental sounds decreases as their number in the word increases."

At the phrase level we find that rhythmic patterns can be used to signal differences in syntactic structure. Compare:

1. *The 2000-year-old skeletons*,
2. *The two 1000-year-old skeletons*.

The phrases contain syllables with identical phonetic content but are clearly timed differently. The difference is linked to that fact that in (1) *2000-year-old* forms a single unit, whereas in (2) *two 1000-year-old* consists of two constituents.

A further example is:

3. *Johan greeted the girl with the flowers*.

Hearing this utterance in a specific context a listener might not find it ambiguous. But it has two interpretations: (a) either Johan greeted the girl who was carrying flowers, or (b) he used the flowers to greet her. If spoken clearly to disambiguate, the speaker is likely to provide temporal cues in the form of shortening the syllables within each syntactic group to signal the coherence of the constituent syllables (cf. the word length effect above) and to introduce a short pause between the two groups. There would be a lengthening of the last syllable before the pause and of the utterance-final syllable.

Version (a) can be represented as

4. [*Johan greeted*] # [*the girl with the flowers*].

In (b) the grouping is

5. [*Johan greeted the girl*] # [*with the flowers*].

The # symbol indicates the possibility of a short juncture or pause and a lengthening of the segments preceding the pause. This boundary cue is known as *pre-pausal lengthening*, or more generally as *final lengthening* [16.143, 144], a process that has been observed for a great many languages. It is not known whether this process is a language universal. It is fair to say that is typologically widespread.

Curiously it resembles a phenomenon found in poetry, folk melodies and nursery tunes, called *catalexis*. It consists in omitting the last syllable(s) in a line or other metrical unit of poetry. Instead of four mechanically repeated trochaic feet as in:

|−⏑|−⏑|−⏑|−⏑|
|−⏑|−⏑|−⏑|−⏑|

we typically find catalectic lines with durationally implied but deleted final syllables as in:

|−⏑|−⏑|−⏑|− |

Old McDonald had a farm

|−⏑|−⏑|− | |

ee-i ee-i oh

Final lengthening is an essential feature of speech prosody. Speech synthesized without it sounds both highly unnatural and is harder to perceive. In music performance frameworks, instrumental as well as sung, final lengthening serves the purpose of grouping and constituent marking [16.145, 146].

16.7 Prosody and Speech Dynamics

Degree of prominence is an important determinant of segment and syllable duration in English. Figure 16.41 shows spectrograms of the word '*squeal*' spoken with four degrees of stress in sentences read in response to a list of questions (source: [16.147]). The idea behind this method was to elicit tokens having emphatic, strong (as in focus position), neutral and weak (unfocused) stress on the test syllable. The lower row compares the strong and the emphatic pronunciations (left and right respectively). The top row presents syllables with weaker degrees of stress.

The representations demonstrate that the differences in stress have a marked effect on the word's spectrographic pattern. Greater prominence makes it longer. Formants, notably F2, show more extensive and more rapid transitions.

A similar experimental protocol was used in two studies of the dynamics of vowel production [16.147, 148]. Both report formant data on the English front vowels [i], [ɪ], [ɛ] and [eɪ] occurring in syllables selected to maximize and minimize contextual assimilation to the place of articulation of the surrounding consonants. To obtain a maximally assimilatory frame, words containing the sequence [w_ɫ] were chosen, e.g., as in *wheel*, *will*, *well* and *wail*. (The [w] is a labio-velar and English [ɫ] is velarized. Both are thus made with the tongue in a retracted position). The minimally assimilatory syllable was [h_d].

Moon's primary concern was *speaking style* (clear versus casual speech). Brownlee investigated changes due to *stress variations*. In both studies measurements were made of vowel duration and extent of formant transitions. Vowel segment boundaries were defined in terms of fixed transition onset and endpoint values for [w] and [ɫ] respectively. The [h_d] frame served as a *null context* reference. The [w_ɫ] environment produced formant transitions large enough to provide a sensitive and robust index of articulatory movement (basically the front–back movement of the tongue).

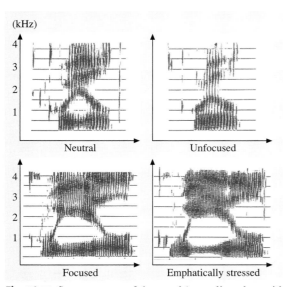

Fig. 16.41 Spectrograms of the word 'squeal' spoken with four degrees of stress in sentences read in response to a list of questions. (After *Brownlee* [16.147])

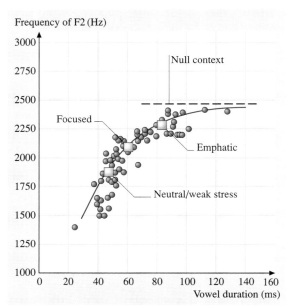

Fig. 16.42 Measurements of F2 as function of vowel duration from tokens of [i] in the word 'squeal' spoken by a male subject under four conditions of stress: emphatically stressed, focused, neutral stress and weak stress (out of focus). *Filled circles* pertain to the individual measurements from all the four stress conditions. The *dashed horizontal line*, is the average F2 value of citation form productions of [h_d]. The averages for the four stress conditions are indicated by the *open squares*. As duration increases – which is approximately correlated with increasing stress – the *square symbols* are seen to move up along the smooth curve implying decreasing undershoot. (After *Brownlee* [16.147])

Figure 16.42 presents measurements of F2 and vowel duration from tokens of [i] in the word '*squeal*' spoken by a male subject under four conditions of stress: emphatically stressed, focused, neutral stress and weak stress (out of focus). The subject was helped to produce the appropriate stress in the right place by reading a question before each test sentence.

Filled circles pertain to individual measurements pooled for all stress conditions. The points form a coherent cluster which is fairly well captured by an exponential curve. The asymptote is the dashed horizontal line, i. e., the average F2 value of the [h_d] data for [i]. The averages for the stress conditions are indicated by the unfilled squares. They are seen to move up along the curve with increasing stress (equivalently, vowel duration).

To understand the articulatory processes underlying the dynamics of vowel production it is useful to adopt a biomechanical perspective. In the [w_ɫ] test words the tongue body starts from a posterior position in [w] and moves towards a vowel target located in the front region of the vocal tract, e.g., [i], [ɪ], [ɛ] or [e]. Then it returns to a back configuration for the dark [ɫ]. At short vowel durations there is not enough time for the vowel gesture to be completed. As a result, the F2 movement misses the reference target by several hundred Hz. Note that in unstressed tokens the approach to target is off by almost an octave. As the syllable is spoken with more stress, and the vowel accordingly gets longer, the F2 transition falls short to a lesser degree. What is illustrated here is the phenomenon known as *formant undershoot* [16.149, 150].

Formant undershoot has received a fair amount of attention in the literature [16.151–155]. It is generally seen as an expression of the sluggish response characteristics of the speech production system. The term *sluggish* here describes both neural delays and mechanical time constants. The response to the neural commands for a given movement is not instantaneous. It takes time for neural impulses to be transformed into muscular contractions and it takes time for the tongue, the jaw and other articulatory structures to respond mechanically to the forces generated by those contractions. In other words, several stages of filtering take place between control signals and the unfolding of the movements. It is this filtering that makes the articulators sluggish. When commands for successive phonemes arrive faster than the articulatory systems are able to respond, the output is a set of incomplete movements. There is undershoot and failure to reach spatial, and thus also acoustic, targets.

However, biomechanics tells us that, in principle, it should be possible to compensate for system characteristics by applying greater force to the articulators thereby speeding up movements and improving target attainment [16.156]. Such considerations make us expect that, in speech movement dynamics, a given trajectory is shaped by

1. the distance between successive articulatory goals (the extent of movement);
2. articulatory effort (input force); and
3. duration (the time available to execute the movement).

The data of Brownlee are compatible with such a mechanism in that the stress-dependent undershoot observed is likely to be a joint consequence of: (1) the large distances between back and front targets; (2) the stress differences corresponding to variations in articulatory effort; and (3) durational limitations.

syllabic words to produce items such as *wheel*, *wheeling*, *Wheelingham*, *will*, *willing*, *willingly*, etc.

In the first part of his experiment, five male subjects were asked to produce these words in casual citation-form style. In the second section the instruction was to say them clearly in an *overarticulated* way. For the citation-form pronunciation the results replicated previously observed facts in that: (a) all subjects exhibited duration-dependent formant undershoot, and (b) the magnitude of this effect varied in proportion to the extent of the [w]–[vowel] transition. In the clear style, undershoot effects were reduced. The mechanism by which this was achieved varied somewhat from subject to subject. It involved combinations of increased vowel duration and more rapid and more extensive formant transitions. Figure 16.43 shows the response of one of the subjects to the two tasks [16.157].

Casual style is represented by the solid circles, clear speech by open circles. The frequency of F2 in [w] is indicated by the arrows at 600 Hz. The mean F2 for the [h_d] data is entered as dashed horizontal lines.

The solid points show duration-dependent undershoot effects in relation to the reference values for the [h_d] environment. The open circles, on the other hand, overshoot those values suggesting that this talker used more extreme F2 targets for the clear condition. The center of gravity of the open circles is shifted to the right for all four vowels showing that clear forms were longer than citation forms. Accordingly, this is a subject who used all three methods to decrease undershoot: clear pronunciations exhibited consistently more extreme targets (F2 higher), longer durations and more rapid formant movements.

These results demonstrate that duration and context are not the only determinants of vowel reduction since, for any given duration, the talker is free to vary the degree of undershoot by choosing to articulate more forcefully (as in overarticulated *hyperspeech*) or in a more relaxed manner (as in casual hypospeech). Taken together, the studies by Moon and Brownlee suggest that the dynamics of articulatory movement, and thus of formant transitions, are significantly constrained by three factors: extent of movement, articulatory effort and duration.

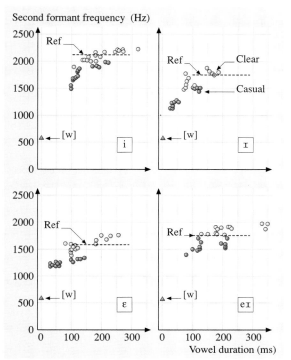

Fig. 16.43 Measurements F2 as function of vowel duration for five male subjects' casual citation-form style pronunciation of the words *wheel*, *wheeling*, *Wheelingham*, *will*, *willing*, *willingly* and for their pronunciation of the same words when asked to say them clearly in an overarticulated way. Casual style is represented by the *solid circles*, clear speech by *open circles*. The frequency of F2 in [w] is indicated by the arrows at 600 Hz. The mean F2 for the [h_d] data is entered as dashed horizontal lines. (After [16.157], Fig. 4)

The three factors – movement extent, effort and duration – are clearly demonstrated in Moon's research on clear and casual speech. As mentioned, Moon's test syllables were the same as Brownlee's. He varied speaking style while keeping stress constant. This was achieved by taking advantage of the word-length effect. The segment strings [wiɫ-], [wɪɫ-], [wɛɫ-] and [weɫ-] were used as the first main-stressed syllable in mono-, bi- and tri-

16.8 Control of Sound in Speech and Singing

When we perform an action – walk, run, or reach for and manipulate objects – our motor systems are faced with the fact that the contexts under which movements are made are never exactly the same. They change significantly from one situation to the next. Nonetheless, motor systems adapt effortlessly to the continually changing conditions presumably because, during evolution, they were shaped by the need to cope with unforeseen events and obstacles [16.158]. Their default mode of operation is compensatory [16.159].

Handwriting provides a good illustration of this ability. A familiar fact is that it does not matter if something is written on the blackboard or on a piece of paper. The characteristic features of someone's handwriting are nevertheless easily recognized. Different sets of muscles are recruited and the size of the letters is different but their shapes remain basically similar. What this observation tells us is that movements are not specified in terms of fixed set of muscles and constant contraction patterns. They are recruited in functionally defined groups, *coordinative structures* [16.160]. They are planned and executed in an external coordinate space, in other words in relation to the 3-D world in which they occur so as to attain goals defined outside the motor system itself. The literature on motor mechanisms teaches us that voluntary movements are prospectively organized or future-oriented.

Speech and singing provide numerous examples of this output-oriented mode of motor control [16.161–163]. Earlier in the chapter we pointed out that, in upright position, the diaphragm and adjacent structures are influenced by gravity and tend to be pulled down, thereby causing the volume of the thoracic cavity to increase. In this position, gravity contributes to the inspiratory forces. By contrast, in the supine position, the diaphragm tends to get pushed up into the rib cage, which promotes expiration [16.1, p. 40].

Sundberg et al. [16.164] investigated the effect of upright and supine positions in two baritone singers using synchronous records of esophageal and gastric pressure, EMG from inspiratory and expiratory muscles, lung volume and sound. Reorganization of respiratory activity was found and was interpreted as compensation for the different mechanical situations arising from the upright and supine conditions.

This finding is closely related to what we know about breathing during speech. As mentioned above (Fig. 16.6), the Ps tends to stay fairly constant for any given vocal effort. It is known that this result is achieved by tuning the balance between inspiratory and expiratory muscles. When the lungs are expanded so that the effect of elastic recoil creates a significant expiration force, inspiratory muscles predominate to put a brake on that force. For reduced lung volumes the situation is the opposite. The effect of elastic recoil is rather to increase lung volume. Accordingly, the muscle recruitment needed to maintain Ps is expected to be primarily expiratory. That is indeed what the data show [16.8, 66, 165].

The bite-block paradigm offers another speech example. In one set of experiments [16.166] subjects were instructed to pronounce syllables consisting only of a long vowel under two conditions: first normally, then with a bite block (BB) between the teeth. The subjects all non-phoneticians were told to try to sound as normal as possible despite the BB. No practice was allowed. The purpose of the BB was to create an abnormally large jaw opening for close vowels such as [i] and [u]. It was argued that, if no tongue compensation occurred, this large opening would drastically change the area function of the close vowels and disrupt their formant patterns. In other words, the question investigated was whether the subjects were able to sound normal despite the BB.

Acoustic recordings were made and formant pattern data were collected for comparisons between conditions, vowels and subjects. The analyses demonstrated clearly that subjects were indeed able to produce normal sounding vowels in spite of the BB. At the moment of the first glottal pulse formant patters were well within the normal ranges of variation.

In a follow-up X-ray investigation [16.167] it was found that compensatory productions of [i] and [u] were made with super-palatal and super-velar tongue shapes. In other words, tongue bodies were raised high above normal positions so as to approximate the normal cross-sectional area functions for the test vowels.

Speech and singing share a lot of features but significant differences are brought to light when we consider what the goals of the two behaviors are.

It is more than 50 years since the sound spectrograph became commercially available [16.91]. In this time we have learned from visible speech displays and other records that the relation between phonetic units and the acoustic signal is extremely complex [16.168]. The lesson taught is that invariant correlates of linguistic categories are not readily apparent in the speech wave [16.169]. In the preceding discussion we touched on some of the sources of this variability: coarticulation, reduction and elaboration, prosodic modulation, stylistic, situational and speaker-specific characteristics. From this vantage point it is astonishing to note that, even under noisy conditions, speech communication is a reliable and robust process. How is this remarkable fact to be accounted for?

In response to this problem a number of ideas and explanatory frameworks have been proposed. For instance, it has been suggested that acoustic invariants are relational rather than absolute (à la tone intervals defined as frequency ratios).

Speech is rather a set of movements made audible than a set of sounds produced by movements [16.14, p. 33].

In keeping with this classical statement, many investigators have argued that speech entities are to be found at some upstream speech production level and should be defined as *gestures* [16.170–173].

At the opposite end, we find researchers (e.g., Perkell [16.163]) who endorse Roman Jakobson's view which, in the search for units gives primacy to the perceptual representation of speech sounds. To Jakobson the stages of the speech chain form an " . . . *operational hierarchy of levels of decreasing pertinence: perceptual, aural, acoustical and articulatory (the latter carrying no direct information to the receiver)*." [16.174]

These viewpoints appear to conflict and do indeed divide the field into those who see speech as a motoric code (the *gesturalist* camp) and those who maintain that it is primarily shaped by perceptual processes (the *listener-oriented* school).

There is a great deal of experimental evidence for both sides, suggesting that this dilemma is not an either–or issue but that both parties offer valuable complementary perspectives.

A different approach is taken by the H & H (Hyper and Hypo) theory [16.175, 176]. This account is developed from the following key observations about speaker–listener interaction:

1. Speech perception is always a product of signal information and listener knowledge;
2. Speech production is adaptively organized.

Here is an experiment that illustrates the first claim about perceptual processing. Two groups of subjects listen to a sequence of two phrases: a question followed by an answer. The subject groups hear different questions but a single physically identical reply. The subjects' task is to say how many words the reply contains.

The point made here is that Group 1 subjects hear [lɛsn̩ faɪv] as "*less than five*". Those in Group 2 interpret it as "*lesson five*". The first group's response is "*three words*", and the answer of the second is "*two words*". This is *despite the fact* that physically the [lɛsn̩ faɪv] stimulus is exactly the same. The syllabic [n̩] signals the word *than* in one case and the syllable *–on* in the other. Looking for the invariant correlates of the initial consonant *than* is doomed to failure because of the severe degree of reduction.

To proponents of H & H theory this is not an isolated case. This is the way that perception in general works. The speech percepts can never be raw records of the signal because listener knowledge will inevitably interact with the stimulus and will contribute to shaping the percept.

Furthermore, H & H theory highlights the fact that spoken messages show a non-uniform distribution of information in that predictability varies from situation to situation, from word to word and from syllable to syllable. Compare (a) and (b) below. What word is likely to be represented by the gap?

1. "*The next word is __.*"
2. "*A bird in the hand is worth two in the __.*"

Any word can be expected in (1) whereas in (2) the predictability of the word "*bush*" is high.

H & H theory assumes that, while learning and using their native language, speakers develop a sense of this informational dynamics. Introducing the abstraction of an *ideal speaker*, it proposes that the talker estimates the running contribution that signal-complementary information (listener knowledge) will make during the course of the utterance and then tunes his articulatory performance to the presumed short-term listener needs. Statistically, this type of behavior has the long-term consequence of distributing phonetic output forms along a continuum with clear and elaborated forms (hyperspeech) at one end and casual and reduced pronunciations (hypospeech) at the other.

The implication for the invariance issue is that the task of the speaker is not to encode linguistic units as physical invariants, but to make sure that the signal attributes (of phonemes, syllables, words and phrases) carry discriminative power *sufficient* for successful lexical access. To make the same point in slightly different terms, the task of the signal is not to embody phonetic patterns of constancy but to provide *missing information.*

It is interesting to put this account of speech processes next to what we know about singing. Experimental observations indicate that singing in tune is not simply about invariant F0 targets. F0 is affected by the singer's expressive performance [16.177] and by the tonal context in which a given note is embedded [16.178]. Certain deviations from nominal target frequencies are not unlike the coarticulation and undershoot effects ubiquitously present in speech. Accordingly, with respect to frequency control, speaking and singing are qualitatively similar. However, quantitatively, they differ drastically. Recall that in describing the dynamics of vowel reduction we noted that formant frequencies can be displaced by as much as 50% from target values. Clearly, a musician or singer with a comparable under-/overshoot record would be recommended to embark on an alternative career, the margins of perceptual tolerance being much narrower for singing.

What accounts for this discrepancy in target attainment? Our short answer is that singing or playing *out of tune* is a bad thing. Where does this taboo come from? From consonance and harmony constraints. In simplified terms, an arbitrary sample of tonal music can be analyzed into a sequence of chords. Its melodic line is a rhythmic and tonal elaboration of this harmonic structure. Statistically, long prominent melody tones tend to attract chord notes. Notes of less prominence typically *interpolate* along scales in smaller intervals between the metrically heavier chord notes. The notion of *consonance* goes a long way towards explaining why singing or playing out of tune is tabooed in music: hitting the right pitch is required by the consonance and harmony constraints expected by the listener and historically presumably linked with a combination of the polyphony in our Western tradition of music composition and the fact that most of our music instruments produce tones with harmonic spectra. This implies that intervals departing too much from just intonation will generate beats between simultaneously sounding and only nearly coinciding partials, particularly if the fundamental frequencies are constant, lacking vibrato and flutter.

16.9 The Expressive Power of the Human Voice

The human voice is an extremely expressive instrument both when used for speech and for singing. By means of subtle variations of timing and pitch contour speakers and singers add a substantial amount of expressiveness to the linguistic or musical content and we are quite skilled in deciphering this information. Indeed a good deal of vocal artistry seems to lie in the artist's skill in making nothing but such changes of pitch, timbre, loudness and timing that a listener can perceive as carrying some meaning.

We perceive the extra-linguistic or expressive information in speech and singing in various shapes. For example, we can interpret certain combinations of acoustic characteristics in speech in terms of a smile or a particular forming of the lips on the face of the speaker [16.179]. For example *Fónagy* [16.180] found that listeners were able to replicate rather accurately the facial expression of speakers only by listening to their voices.

The emotive transforms in speech and singing seem partly similar and sometimes identical [16.181, 182]. Final lengthening, mentioned above, uses the same code for marking the end of a structural element, such as a sentence in speech or a phrase in sung and played music performance. Emphasis by delayed arrival is another example, i. e., delaying an emphasized stressed syllable or note by lengthening the unstressed syllable/note preceding it [16.183].

The expressive potential of the human voice is indeed enormous, and would transpire from the ubiquitous use of the voice for the purpose of communication. Correct interpretation of the extra-linguistic content of a spoken utterance is certainly important in our daily life, so we are skilled in deciphering vocal signals also along those dimensions. The importance of correct encoding of the extra-linguistic implies that speakers acquire a great skill in this respect. This skill would be the basic requirement for vocal art, in singing as well as in acting.

References

16.1 T.J. Hixon: *Respiratory Function in Speech and Song* (Singular, San Diego 1991) pp. 1–54
16.2 A. L. Winkworth, P. J. Davis, E. Ellis, R. D. Adams: Variability and consistency in speech breathing during reading: Lung volumes, speech intensity, and linguistic factors, JSHR **37**, 535–556 (1994)
16.3 M. Thomasson: *From Air to Aria*, Ph.D. Thesis (Music Acoustics, KTH 2003)
16.4 B. Conrad, P. Schönle: Speech and respiration, Arch. Psychiat. Nervenkr. **226**, 251–268 (1979)
16.5 M.H. Draper, P. Ladefoged, D. Whitteridge: Respiratory muscles in speech, J. Speech Hear. Disord. **2**, 16–27 (1959)
16.6 R. Netsell: Speech physiology. In: *Normal aspects of speech, hearing, and language*, ed. by P.D. Minifie, T.J. Hixon, P. Hixon, P. Williams (Prentice-Hall, Englewood Cliffs 1973) pp. 211–234

16.7 P. Ladefoged, M.H. Draper, D. Whitteridge: Syllables and stress, Misc. Phonetica **3**, 1–14 (1958)

16.8 P. Ladefoged: Speculations on the control of speech. In: *A Figure of Speech: A Festschrift for John Laver*, ed. by W.J. Hardcastle, J. Mackenzie Beck (Lawrence Erlbaum, Mahwah 2005) pp. 3–21

16.9 T.J. Hixon, G. Weismer: Perspectives on the Edinburgh study of speech breathing, J. Speech Hear. Disord. **38**, 42–60 (1995)

16.10 S. Nooteboom: The prosody of speech: melody and rhythm. In: *The Handbook of Phonetic Sciences*, ed. by W.J. Hardcastle, J. Laver (Blackwell, Oxford 1997) pp. 640–673

16.11 M. Rothenberg: *The breath-stream dynamics of simple-released-plosive production Bibliotheca Phonetica 6* (Karger, Basel 1968)

16.12 D.H. Klatt, K.N. Stevens, J. Mead: Studies of articulatory activity and airflow during speech, Ann. NY Acad. Sci. **155**, 42–55 (1968)

16.13 J.J. Ohala: Respiratory activity in speech. In: *Speech production and speech modeling*, ed. by W.J. Hardcastle, A. Marchal (Dordrecht, Kluwer 1990) pp. 23–53

16.14 R.H. Stetson: *Motor Phonetics: A Study of Movements in Action* (North Holland, Amsterdam 1951)

16.15 P. Ladefoged: Linguistic aspects of respiratory phenomena, Ann. NY Acad. Sci. **155**, 141–151 (1968)

16.16 L.H. Kunze: Evaluation of methods of estimating sub-glottal air pressure muscles, J. Speech Hear. Disord. **7**, 151–164 (1964)

16.17 R. Leanderson, J. Sundberg, C. von Euler: Role of the diaphragmatic activity during singing: a study of transdiaphragmatic pressures, J. Appl. Physiol. **62**, 259–270 (1987)

16.18 J. Sundberg, N. Elliot, P. Gramming, L. Nord: Short-term variation of subglottal pressure for expressive purposes in singing and stage speech. A preliminary investigation, J. Voice **7**, 227–234 (1993)

16.19 J. Sundberg: Synthesis of singing, in Musica e Technologia: Industria e Cultura per lo Sviluppo del Mezzagiorno. In: *Proceedings of a symposium in Venice*, ed. by R. Favaro (Unicopli, Milan 1987) pp. 145–162

16.20 J. Sundberg: Synthesis of singing by rule. In: *Current Directions in Computer Music Research, System Development Foundation Benchmark Series*, ed. by M. Mathews, J. Pierce (MIT, Cambridge 1989), 45–55 & 401–403

16.21 J. Molinder: *Akustiska och perceptuella skillnader mellan röstfacken lyrisk och dramatisk sopran, unpublished thesis work* (Lund Univ. Hospital, Dept of Logopedics, Lund 1997)

16.22 T. Baer: Reflex activation of laryngeal muscles by sudden induced subglottal pressure changes, J. Acoust. Soc. Am. **65**, 1271–1275 (1979)

16.23 T. Cleveland, J. Sundberg: Acoustic analyses of three male voices of different quality. In: *SMAC 83. Proceedings of the Stockholm Internat Music Acoustics Conf*, Vol. 1, ed. by A. Askenfelt, S. Felicetti, E. Jansson, J. Sundberg (Roy. Sw. Acad. Music, Stockholm 1985) pp. 143–156, No. 46:1

16.24 J. Sundberg, C. Johansson, H. Willbrand, C. Ytterbergh: From sagittal distance to area, Phonetica **44**, 76–90 (1987)

16.25 I.R. Titze: Phonation threshold pressure: A missing link in glottal aerodynamics, J. Acoust. Soc. Am. **91**, 2926–2935 (1992)

16.26 I.R. Titze: *Principles of Voice Production* (Prentice-Hall, Englewood Cliffs 1994)

16.27 G. Fant: *Acoustic theory of speech production* (Mouton, The Hague 1960)

16.28 K.N. Stevens: *Acoustic Phonetics* (MIT, Cambridge 1998)

16.29 M. Hirano: *Clinical Examination of Voice* (Springer, New York 1981)

16.30 M. Rothenberg: A new inversefiltering technique for deriving the glottal air flow waveform during voicing, J. Acoust. Soc. Am. **53**, 1632–1645 (1973)

16.31 G. Fant: The voice source – Acoustic modeling. In: *STL/Quart. Prog. Status Rep. 4* (Royal Inst. of Technology, Stockholm 1982) pp. 28–48

16.32 C. Gobl: *The voice source in speech communication production and perception experiments involving inverse filtering and synthesis. D.Sc. thesis* (Royal Inst. of Technology (KTH), Stockholm 2003)

16.33 G. Fant, J. Liljencrants, Q. Lin: A four-parameter model of glottal flow. In: *STL/Quart. Prog. Status Rep. 4, Speech, Music and Hearing* (Royal Inst. of Technology, Stockholm 1985) pp. 1–13

16.34 D.H. Klatt, L.C. Klatt: Analysis, synthesis and pereception of voice quality variations among female and male talkers, J. Acoust. Soc. Am. **87**(2), 820–857 (1990)

16.35 M. Ljungqvist, H. Fujisaki: A comparative study of glottal waveform models. In: *Technical Report of the Institute of Electronics and Communications Engineers*, Vol. EA85-58 (Institute of Electronics and Communications Engineers, Tokyo 1985) pp. 23–29

16.36 A.E. Rosenberg: Effect of glottal pulse shape on the quality of natural vowels, J. Acoust. Soc. Am. **49**, 583–598 (1971)

16.37 M. Rothenberg, R. Carlson, B. Granström, J. Lindqvist-Gauffin: A three- parameter voice source for speech synthesis. In: *Proceedings of the Speech Communication Seminar 2*, ed. by G. Fant (Almqvist & Wiksell, Stockholm 1975) pp. 235–243

16.38 K. Ishizaka, J.L. Flanagan: Synthesis of voiced sounds from a two-mass model of the vocal cords, The Bell Syst. Tech. J. **52**, 1233–1268 (1972)

16.39 Liljencrants: Chapter A translating and rotating mass model of the vocal folds. In: *STL/Quart. Prog. Status Rep. 1, Speech, Music and Hearing* (Royal Inst. of Technology, Stockholm 1991) pp. 1–18

16.40 A. Ní Chasaide, C. Gobl: Voice source variation. In: *The Handbook of Phonetic Sciences*, ed. by

W.J. Hardcastle, J. Laver (Blackwell, Oxford 1997) pp. 427–462

16.41 E.B. Holmberg, R.E. Hillman, J.S. Perkell: Glottal air flow and pressure measurements for loudness variation by male and female speakers, J. Acoust. Soc. Am. **84**, 511–529 (1988)

16.42 J.S. Perkell, R.E. Hillman, E.B. Holmberg: Group differences in measures of voice production and revised values of maximum airflow declination rate, J. Acoust. Soc. Am. **96**, 695–698 (1994)

16.43 J. Gauffin, J. Sundberg: Spectral correlates of glottal voice source waveform characteristics, J. Speech Hear. Res. **32**, 556–565 (1989)

16.44 J. Svec, H. Schutte, D. Miller: On pitch jumps between chest and falsetto registers in voice: Data on living and excised human larynges, J. Acoust. Soc. Am. **106**, 1523–1531 (1999)

16.45 J. Sundberg, M. Andersson, C. Hultqvist: Effects of subglottal pressure variation on professional baritone singers voice sources, J. Acoust. Soc. Am. **105**, 1965–1971 (1999)

16.46 J. Sundberg, E. Fahlstedt, A. Morell: Effects on the glottal voice source of vocal loudness variation in untrained female and male subjects, J. Acoust. Soc. Am. **117**, 879–885 (2005)

16.47 P. Sjölander, J. Sundberg: Spectrum effects of subglottal pressure variation in professional baritone singers, J. Acoust. Soc. Am. **115**, 1270–1273 (2004)

16.48 P. Branderud, H. Lundberg, J. Lander, H. Djamshidpey, I. Wäneland, D. Krull, B. Lindblom: X-ray analyses of speech: Methodological aspects, *Proc. of 11th Swedish Phonetics Conference* (Stockholm Univ., Stockholm 1996) pp. 168–171

16.49 B. Lindblom: A numerical model of coarticulation based on a Principal Components analysis of tongue shapes. In: *Proc. 15th Int. Congress of the Phonetic Sciences*, ed. by D. Recasens, M. Josep Solé, J. Romero (Universitat Autònoma de Barcelona, Barcelona 2003), CD-ROM

16.50 G.E. Peterson, H. Barney: Control methods used in a study of the vowels, J. Acoust. Soc. Am. **24**, 175–184 (1952)

16.51 Hillenbrand et al.: Acoustic characteristics of American English vowels, J. Acoust. Soc. Am. **97**(5), 3099–3111 (1995)

16.52 G. Fant: Analysis and synthesis of speech processes. In: *Manual of Phonetics*, ed. by B. Malmberg (North-Holland, Amsterdam 1968) pp. 173–277

16.53 G. Fant: Formant bandwidth data. In: *STL/Quart. Prog. Status Rep.* 7 (Royal Inst. of Technology, Stockholm 1962) pp. 1–3

16.54 G. Fant: Vocal tract wall effects, losses, and resonance bandwidths. In: *STL/Quart. Prog. Status Rep.* 2–3 (Royal Inst. of Technology, Stockholm 1972) pp. 173–277

16.55 A.S. House, K.N. Stevens: Estimation of formant bandwidths from measurements of transient response of the vocal tract, J. Speech Hear. Disord. **1**, 309–315 (1958)

16.56 O. Fujimura, J. Lindqvist: Sweep-tone measurements of vocal-tract characteristics, J. Acoust. Soc. Am. **49**, 541–558 (1971)

16.57 I. Lehiste, G.E. Peterson: Vowel amplitude and phonemic stress in American English, J. Acoust. Soc. Am. **3**, 428–435 (1959)

16.58 I. Lehiste: *Suprasegmentals* (MIT Press, Cambridge 1970)

16.59 O. Jespersen: *Lehrbuch der Phonetik* (Teubner, Leipzig 1926)

16.60 T. Bartholomew: A physical definition of good voice quality in the male voice, J. Acoust. Soc. Am. **6**, 25–33 (1934)

16.61 J. Sundberg: Production and function of the singing formant. In: *Report of the eleventh congress Copenhagen 1972 (Proceedings of the 11th international congress of musicology)*, ed. by H. Glahn, S. Sörensen, P. Ryom (Wilhelm Hansen, Copenhagen 1972) pp. 679–686

16.62 J. Sundberg: Articulatory interpretation of the 'singing formant', J. Acoust. Soc. Am. **55**, 838–844 (1974)

16.63 J. Sundberg: Level and center frequency of the singer's formant, J. Voice. **15**, 176–186 (2001)

16.64 G. Berndtsson, J. Sundberg: Perceptual significance of the center frequency of the singers formant, Scand. J. Logopedics Phoniatrics **20**, 35–41 (1995)

16.65 L. Dmitriev, A. Kiselev: Relationship between the formant structure of different types of singing voices and the dimension of supraglottal cavities, Fol. Phoniat. **31**, 238–41 (1979)

16.66 P. Ladefoged: *Three areas of experimental phonetics* (Oxford Univ. Press, London 1967)

16.67 J. Barnes, P. Davis, J. Oates, J. Chapman: The relationship between professional operatic soprano voice and high range spectral energy, J. Acoust. Soc. Am. **116**, 530–538 (2004)

16.68 M. Nordenberg, J. Sundberg: Effect on LTAS on vocal loudness variation, Logopedics Phoniatrics Vocology **29**, 183–191 (2004)

16.69 R. Weiss, W.S. Brown, J. Morris: Singer's formant in sopranos: Fact or fiction, J. Voice **15**, 457–468 (2001)

16.70 J.M. Heinz, K.N. Stevens: On the relations between lateral cineradiographs, area functions, and acoustics of speech. In: *Proc. Fourth Int. Congress on Acoustics*, Vol. 1a (1965), paper A44

16.71 C. Johansson, J. Sundberg, H. Willbrand: X-ray study of articulation and formant frequencies in two female singers. In: *Proc. of Stockholm Music Acoustics Conference 1983 (SMAC 83)*, Vol. 46(1), ed. by A. Askenfelt, S. Felicetti, E. Jansson, J Sundberg (Kgl. Musikaliska Akad., Stockholm 1985) pp. 203–218

16.72 T. Baer, J.C. Gore, L.C. Gracco, P. Nye: Analysis of vocal tract shape and dimensions using magnetic

16.73 D. Demolin, M. George, V. Lecuit, T. Metens, A. Soquet: Détermination par IRM de l'ouverture du velum des voyelles nasales du français. In: *Actes des XXièmes Journées d'Études sur la Parole* (1996)

16.74 A. Foldvik, K. Kristiansen, J. Kvaerness, A. Torp, H. Torp: *Three-dimensional ultrasound and magnetic resonance imaging: a new dimension in phonetic research* (Proc. Fut. Congress Phonetic Science Stockholm Univ., Stockholm 1995), Vol. 4, 46–49

16.75 B.H. Story, I.R. Titze, E.A. Hoffman: Vocal tract area functions from magnetic resonance imaging, J. Acoust. Soc. Am. **100**, 537–554 (1996)

16.76 O. Engwall: *Talking tongues, D.Sc. thesis* (Royal Institute of Technology (KTH), Stockholm 2002)

16.77 B. Lindblom, J. Sundberg: Acoustical consequences of lip, tongue, jaw and larynx movement, J. Acoust. Soc. Am. **50**, 1166–1179 (1971), also in Papers in Speech Communication: Speech Production, ed. by R.D. Kent, B.S. Atal, J.L. Miller (Acoust. Soc. Am., New York 1991) pp.329–342

16.78 J. Stark, B. Lindblom, J. Sundberg: APEX – an articulatory synthesis model for experimental and computational studies of speech production. In: *Fonetik 96: Papers presented at the Swedish Phonetics Conference TMH-QPSR 2/1996* (Royal Institute of Technology, Stockholm 1996) pp.45–48

16.79 J. Stark, C. Ericsdotter, B. Lindblom, J. Sundberg: Using X-ray data to calibrate the APEX the synthesis. In: *Fonetik 98: Papers presented at the Swedish Phonetics Conference* (Stockholm Univ., Stockholm 1998)

16.80 J. Stark, C. Ericsdotter, P. Branderud, J. Sundberg, H.-J. Lundberg, J. Lander: The APEX model as a tool in the specification of speaker-specific articulatory behavior. In: *Proc. 14th Int. Congress of the Phonetic Sciences*, ed. by J.J. Ohala (1999)

16.81 C. Ericsdotter: Articulatory copy synthesis: Acoustic performane of an MRI and X-ray-based framework. In: *Proc. 15th Int. Congress of the Phonetic Sciences*, ed. by D. Recasens, M. Josep Solé, J. Romero (Universitat Autònoma de Barcelona, Barcelona 2003), CD-ROM

16.82 C. Ericsdotter: *Articulatory-acoustic relationships in Swedish vowel sounds, PhD thesis* (Stockholm University, Stockholm 2005)

16.83 K.N. Stevens, A.S. House: Development of a quantitative description of vowel articulation, J. Acoust. Soc. Am. **27**, 484–493 (1955)

16.84 S. Maeda: Compensatory articulation during speech: Evidence from the analysis and synthesis of vocal-tract shapes using an articulatory model. In: *Speech Production and Speech Modeling*, ed. by W.J. Hardcastle, A. Marchal (Dordrecht, Kluwer 1990) pp.131–150

16.85 P. Branderud, H. Lundberg, J. Lander, H. Djamshidpey, I. Wäneland, D. Krull, B. Lindblom: X-ray analyses of speech: methodological aspects. In: *Proc. XIIIth Swedish Phonetics Conf. (FONETIK 1998)* (KTH, Stockholm 1998)

16.86 C.Y. Espy-Wilson: Articulatory strategies, speech acoustics and variability. In: *From sound to Sense: 50+ Years of Discoveries in Speech Communication*, ed. by J. Slifka, S. Manuel, M. Mathies (MIT, Cambridge 2004)

16.87 J. Sundberg: Formant technique in a professional female singer, Acustica **32**, 89–96 (1975)

16.88 J. Sundberg, J. Skoog: Dependence of jaw opening on pitch and vowel in singers, J. Voice **11**, 301–306 (1997)

16.89 G. Fant: Glottal flow, models and interaction, J. Phon. **14**, 393–399 (1986)

16.90 E. Joliveau, J. Smith, J. Wolfe: Vocal tract resonances in singing: The soprano voice, J. Acoust. Soc. Am. **116**, 2434–2439 (2004)

16.91 R.K. Potter, A.G. Kopp, H.C. Green: *Visible Speech* (Van Norstrand, New York 1947)

16.92 M. Joosg: Acoustic phonetics, Language **24**, 447–460 (2003), supplement 2

16.93 C.F. Hockett: *A Manual of Phonology* (Indiana Univ. Publ., Bloomington 1955)

16.94 F.H. Guenther: Speech sound acquisition, coarticulation, and rate effects in a neural network model of speech production, Psychol. Rev. **102**, 594–621 (1995)

16.95 R.D. Kent, B.S. Atal, J.L. Miller: *Papers in Speech Communication: Speech Perception* (Acoust. Soc. Am., New York 1991)

16.96 S.D. Goldinger, D.B. Pisoni, P. Luce: Speech perception and spoken word recognition. In: *Principles of experimental phonetics*, ed. by N.J. Lass (Mosby, St Louis 1996) pp.277–327

16.97 H.M. Sussman, D. Fruchter, J. Hilbert, J. Sirosh: Linear correlates in the speech signal: The orderly output constraint, Behav. Brain Sci. **21**, 241–299 (1998)

16.98 B. Lindblom: Economy of speech gestures. In: *The Production of Speech*, ed. by P.F. MacNeilage (Springer, New York 1983) pp.217–245

16.99 P.A. Keating, B. Lindblom, J. Lubker, J. Kreiman: Variability in jaw height for segments in English and Swedish VCVs, J. Phonetics **22**, 407–422 (1994)

16.100 K. Rapp: A study of syllable timing. In: *STL/Quart. Prog. Status Rep. 1* (Royal Inst. of Technology, Stockholm(1971) pp.14–19

16.101 F. Koopmans-van Beinum, J. Van der Stelt (Eds.): *Early stages in the development of speech movements* (Stockton, New York 1986)

16.102 K. Oller: Metaphonology and infant vocalizations. In: *Precursors of early speech*, ed. by B. Lindblom, R. Zetterström (Stockton, New York 1986) pp.21–36

16.103 L. Roug, L. Landberg, L. Lundberg: Phonetic development in early infancy, J. Child Language **16**, 19–40 (1989)

16.104 R. Stark: Stages of speech development in the first year of life. In: *Child Phonology: Volume 1: Production*, ed. by G. Yeni-Komshian, J. Kavanagh, C. Ferguson (Academic, New York 1980) pp. 73–90

16.105 R. Stark: Prespeech segmental feature development. In: *Language Acquisition*, ed. by P. Fletcher, M. Garman (Cambridge UP, New York 1986) pp. 149–173

16.106 D.K. Oller, R.E. Eilers: The role of audition in infant babbling, Child Devel. **59**(2), 441–449 (1988)

16.107 C. Stoel-Gammon, D. Otomo: Babbling development of hearing-impaired and normally hearing subjects, J. Speech Hear. Dis. **51**, 33–41 (1986)

16.108 R.E. Eilers, D.K. Oller: Infant vocalizations and the early diagnosis of severe hearing impairment, J. Pediatr. **124**(2), 99–203 (1994)

16.109 D. Ertmer, J. Mellon: Beginning to talk at 20 months: Early vocal development in a young cochlear implant recipient, J. Speech Lang. Hear. Res. **44**, 192–206 (2001)

16.110 R.D. Kent, M.J. Osberger, R. Netsell, C.G. Hustedde: Phonetic development in identical twins who differ in auditory function, J. Speech Hear. Dis. **52**, 64–75 (1991)

16.111 M. Lynch, D. Oller, M. Steffens: Development of speech-like vocalizations in a child with congenital absence of cochleas: The case of total deafness, Appl. Psychol. **10**, 315–333 (1989)

16.112 C. Stoel-Gammon: Prelinguistic vocalizations of hearing-impaired and normally hearing subjects: A comparison of consonantal inventories, J. Speech Hear. Dis. **53**, 302–315 (1988)

16.113 P.F. MacNeilage, B.L. Davis: Acquisition of speech production: The achievement of segmental independence. In: *Speech production and speech modeling*, ed. by W.J. Hardcastle, A. Marchal (Dordrecht, Kluwer 1990) pp. 55–68

16.114 T. Houtgast, H.J.M. Steeneken: A review of the MTF concept in room acoustics and its use for estimating speech intelligibility in auditoria, J. Acoust. Soc. Am. **77**, 1069–1077 (1985)

16.115 T. Houtgast, H.J.M. Steeneken: *Past, Present and Future of the Speech Transmission Index*, ed. by S.J. van Wijngaarden (NTO Human Factors, Soesterberg 2002)

16.116 R. Drullman, J.M. Festen, R. Plomp: Effect of temporal envelope smearing on speech reception, J. Acoust. Soc. Am. **95**, 1053–1064 (1994)

16.117 R. Drullman, J.M. Festen, R. Plomp: Effect of reducing slow temporal modulations on speech reception, J. Acoust. Soc. Am. **95**, 2670–2680 (1994)

16.118 J. Morton, S. Marcus, C. Frankish: Perceptual centers (P-centers), Psych. Rev. **83**, 405–408 (1976)

16.119 S. Marcus: Acoustic determinants of perceptual center (P-center) location, Perception & Psychophysics **30**, 247–256 (1981)

16.120 G. Allen: The place of rhythm in a theory of language, UCLA Working Papers **10**, 60–84 (1968)

16.121 G. Allen: The location of rhythmic stress beats in English: An experimental study, UCLA Working Papers **14**, 80–132 (1970)

16.122 J. Eggermont: Location of the syllable beat in routine scansion recitations of a Dutch poem, IPO Annu. Prog. Rep. **4**, 60–69 (1969)

16.123 V.A. Kozhevnikov, L.A. Chistovich: Speech Articulation and Perception, JPRS **30**, 543 (1965)

16.124 C.E. Hoequist: The perceptual center and rhythm categories, Lang. Speech **26**, 367–376 (1983)

16.125 K.J. deJong: The correlation of P-center adjustments with articulatory and acoustic events, Perception Psychophys. **56**, 447–460 (1994)

16.126 A.D. Patel, A. Löfqvist, W. Naito: The acoustics and kinematics of regularly timed speech: a database and method for the study of the P-center problem. In: *Proc. 14th Int. Congress of the Phonetic Sciences*, ed. by J.J. Ohala (1999)

16.127 P. Howell: Prediction of P-centre location from the distribution of energy in the amplitude envelope: I & II, Perception Psychophys. **43**, 90–93, 99 (1988)

16.128 B. Pompino-Marschall: On the psychoacoustic nature of the Pcenter phenomenon, J. Phonetics **17**, 175–192 (1989)

16.129 C.A. Harsin: Perceptual-center modeling is affected by including acoustic rate-of-change modulations, Perception Psychophys. **59**, 243–251 (1997)

16.130 C.A. Fowler: Converging sources of evidence on spoken and perceived rhythms of speech: Cyclic production of vowels in monosyllabic stress feet, J. Exp. Psychol. Gen. **112**, 386–412 (1983)

16.131 H. Fujisaki: Dynamic characteristics of voice fundamental frequency in speech and singing. In: *The Production of Speech*, ed. by P.F. MacNeilage (Springer, New York 1983) pp. 39–55

16.132 J. Frid: *Lexical and acoustic modelling of Swedish prosody, Dissertation* (Lund University, Lund 2003)

16.133 S. Öhman: Numerical model of coarticulation, J. Acoust. Soc. Am. **41**, 310–320 (1967)

16.134 J. t'Hart: F0 stylization in speech: Straight lines versus parabolas, J. Acoust. Soc. Am. **90**(6), 3368–3370 (1991)

16.135 D. Abercrombie: *Elements of General Phonetics* (Edinburgh Univ. Press, Edinburgh 1967)

16.136 K.L. Pike: *The intonation of America English* (Univ. of Michigan Press, Ann Arbor 1945)

16.137 G. Fant, A. Kruckenberg: Notes on stress and word accent in Swedish. In: *STL/Quart. Prog. Status Rep. 2-3* (Royal Inst. of Technology, Stockholm 1994) pp. 125–144

16.138 R. Dauer: Stress timing and syllable-timing reanalyzed, J. Phonetics **11**, 51–62 (1983)

16.139 A. Eriksson: *Aspects of Swedish rhythm, PhD thesis, Gothenburg Monographs in Linguistics* (Gothenburg University, Gothenburg 1991)

16.140 O. Engstrand, D. Krull: Duration of syllable-sized units in casual and elaborated speech: cross-language observations on Swedish and Spanish, TMH-QPSR **44**, 69–72 (2002)

16.141 A.D. Patel, J.R. Daniele: An empirical comparison of rhythm in language and music, Cognition **87**, B35–B45 (2003)

16.142 D. Huron, J. Ollen: Agogic contrast in French and English themes: Further support for Patel and Daniele (2003), Music Perception **21**, 267–271 (2003)

16.143 D.H. Klatt: Synthesis by rule of segmental durations in English sentences. In: *Frontiers of speech communication research*, ed. by B. Lindblom, S. Öhman (Academic, London 1979) pp. 287–299

16.144 B. Lindblom: Final lengthening in speech and music. In: *Nordic Prosody*, ed. by E. Gårding, R. Bannert (Department of Linguistics Lund University, Lund 1978) pp. 85–102

16.145 A. Friberg, U Battel: Structural communication. In: *The Science and Psychology of Music Performance*, ed. by R. Parncutt, GE McPherson (Oxford Univ., Oxford 2001) pp. 199–218

16.146 J. Sundberg: Emotive transforms, Phonetica **57**, 95–112 (2000)

16.147 Brownlee: *The role of sentence stress in vowel reduction and formant undershoot: A study of lab speech and informal spontaneous speech*, PhD thesis (University of Texas, Austin 1996)

16.148 S.-J. Moon: *An acoustic and perceptual study of undershoot in clear and citation-form speech*, PhD dissertation (Univ. of Texas, Austin 1991)

16.149 K.N. Stevens, A.S. House: Perturbation of vowel articulations by consonantal context. An acoustical study, JSHR **6**, 111–128 (1963)

16.150 B. Lindblom: Spectrographic study of vowel reduction, J. Acoust. Soc. Am. **35**, 1773–1781 (1963)

16.151 P. Delattre: An acoustic and articulatory study of vowel reduction in four languages, IRAL-Int. Ref. Appl. VII/ **4**, 295–325 (1969)

16.152 D.P. Kuehn, K.L. Moll: A cineradiographic study of VC and CV articulatory velocities, J. Phonetics **4**, 303–320 (1976)

16.153 J.E. Flege: Effects of speaking rate on tongue position and velocity of movement in vowel production, J. Acoust. Soc. Am. **84**(3), 901–916 (1988)

16.154 R.J.J.H. van Son, L.C.W. Pols: "Formant movements of Dutch vowels in a text, read at normal and fast rate, J. Acoust. Soc. Am. **92**(1), 121–127 (1992)

16.155 D. van Bergem: *Acoustic and Lexical Vowel Reduction, Doctoral Dissertation* (University of Amsterdam, Amsterdam 1995)

16.156 W.L. Nelson, J.S. Perkell, J.R. Westbury: Mandible movements during increasingly rapid articulations of single syllables: Preliminary observations, J. Acoust. Soc. Am. **75**(3), 945–951 (1984)

16.157 S.-J. Moon, B. Lindblom: Interaction between duration, context and speaking style in English stressed vowels, J. Acoust. Soc. Am. **96**(1), 40–55 (1994)

16.158 C.S. Sherrington: *Man on his nature* (MacMillan, London 1986)

16.159 R. Granit: *The Purposive Brain* (MIT, Cambridge 1979)

16.160 N. Bernstein: *The coordination and regulation of movements* (Pergamon, Oxford 1967)

16.161 P.F. MacNeilage: Motor control of serial ordering of speech, Psychol. Rev. **77**, 182–196 (1970)

16.162 A. Löfqvist: Theories and Models of Speech Production. In: *The Handbook of Phonetic Sciences*, ed. by W.J. Hardcastle, J. Laver (Blackwell, Oxford 1997) pp. 405–426

16.163 J.S. Perkell: Articulatory processes. In: *The Handbook of Phonetic Sciences*. 5, ed. by W.J. Hardcastle, J. Laver. (Blackwell, Oxford 1997) pp. 333–370

16.164 J. Sundberg, R. Leanderson, C. von Euler, E. Knutsson: Influence of body posture and lung volume on subglottal pressure control during singing, J. Voice **5**, 283–291 (1991)

16.165 T. Sears, J. Newsom Davis: The control of respiratory muscles during voluntary breathing. In: *Sound production in man*, ed. by A. Bouhuys et al. (Annals of the New York Academy of Science, New York 1968) pp. 183–190

16.166 B. Lindblom, J. Lubker, T. Gay: Formant frequencies of some fixed-mandible vowels and a model of motor programming by predictive simulation, J. Phonetics **7**, 147–161 (1979)

16.167 T. Gay, B. Lindblom, J. Lubker: Production of bite-block vowels: Acoustic equivalence by selective compensation, J. Acoust. Soc. Am. **69**(3), 802–810 (1981)

16.168 W.J. Hardcastle, J. Laver (Eds.): *The Handbook of Phonetic Sciences* (Blackwell, Oxford 1997)

16.169 J. S. Perkell, D. H. Klatt: *Invariance and variability in speech processes* (LEA, Hillsdale 1986)

16.170 A. Liberman, I. Mattingly: The motor theory of speech perception revised, Cognition **21**, 1–36 (1985)

16.171 C.A. Fowler: An event approach to the study of speech perception from a direct-realist perspective, J. Phon. **14**(1), 3–28 (1986)

16.172 E.L. Saltzman, K.G. Munhall: A dynamical approach to gestural patterning in speech production, Ecol. Psychol. **1**, 91–163 (1989)

16.173 M. Studdert-Kennedy: How did language go discrete?. In: *Evolutionary Prerequisites of Language*, ed. by M. Tallerman (Oxford Univ., Oxford 2005) pp. 47–68

16.174 R. Jakobson, G. Fant, M. Halle: *Preliminaries to Speech Analysis, Acoustics Laboratory, MIT Tech. Rep. No. 13* (MIT, Cambridge 1952)

16.175 B. Lindblom: Explaining phonetic variation: A sketch of the H&H theory. In: *Speech Produc-*

16.176 B. Lindblom: Role of articulation in speech perception: Clues from production, J. Acoust. Soc. Am. **99**(3), 1683–1692 (1996)
16.177 E. Rapoport: Emotional expression code in opera and lied singing, J. New Music Res. **25**, 109–149 (1996)
16.178 J. Sundberg, E. Prame, J. Iwarsson: Replicability and accuracy of pitch patterns in professional singers. In: *Vocal Fold Physiology, Controlling Complexity and Chaos*, ed. by P. Davis, N. Fletcher (Singular, San Diego 1996) pp. 291–306, Chap. 20

... tion and Speech Modeling*, ed. by W.J. Hardcastle, A. Marchal (Dordrecht, Kluwer 1990) pp. 403–439

16.179 J.J. Ohala: An ethological perspective on common cross-language utilization of F0 of voice, Phonetica **41**, 1–16 (1984)
16.180 I. Fónagy: Hörbare Mimik, Phonetica **1**, 25–35 (1967)
16.181 K. Scherer: Expression of emotion in voice and music, J. Voice **9**, 235–248 (1995)
16.182 P. Juslin, P. Laukka: Communication of emotions in vocal expression and music performance: Different channels, same code?, Psychol. Rev. **129**, 770–814 (2003)
16.183 J. Sundberg, J. Iwarsson, H. Hagegård: A singers expression of emotions in sung performance,. In: *Vocal Fold Physiology: Voice Quality Control*, ed. by O. Fujimura, M. Hirano (Singular, San Diego 1995) pp. 217–229

17. Computer Music

This chapter covers algorithms, technologies, computer languages, and systems for computer music. Computer music involves the application of computers and other digital/electronic technologies to music composition, performance, theory, history, and perception. The field combines digital signal processing, computational algorithms, computer languages, hardware and software systems, acoustics, psychoacoustics (low-level perception of sounds from the raw acoustic signal), and music cognition (higher-level perception of musical style, form, emotion, etc.). Although most people would think that analog synthesizers and electronic music substantially predate the use of computers in music, many experiments and complete computer music systems were being constructed and used as early as the 1950s.

Because of this rich legacy, and the large number of researchers working on digital audio (primarily in speech research laboratories), there are a large number of algorithms for synthesizing sound using computers. Thus, a significant emphasis in this chapter will be placed on digital sound synthesis and processing, first providing

17.1	Computer Audio Basics	714
17.2	Pulse Code Modulation Synthesis	717
17.3	Additive (Fourier, Sinusoidal) Synthesis	719
17.4	Modal (Damped Sinusoidal) Synthesis	722
17.5	Subtractive (Source-Filter) Synthesis	724
17.6	Frequency Modulation (FM) Synthesis	727
17.7	FOFs, Wavelets, and Grains	728
17.8	Physical Modeling (The Wave Equation)	730
17.9	Music Description and Control	735
17.10	Composition	737
17.11	Controllers and Performance Systems	737
17.12	Music Understanding and Modeling by Computer	738
17.13	Conclusions, and the Future	740
References		740

an overview of the representation of audio in digital systems, then covering most of the popular algorithms for digital analysis and synthesis of sound.

Pulse code modulation (PCM) is the means for sampling and retrieval of audio in computers, and PCM synthesis uses combinations of prerecorded waveforms to reconstruct speech, sound effects, and music instrument sounds, based largely on the work of Joseph Fourier, who gave us a technique for formulating any waveform from sine waves. Additive synthesis uses fundamental waveforms (often sine waves), to construct more complicated waves. PCM and sinusoidal (Fourier) additive synthesis are often called *nonparametric* techniques. The word parametric in this context means that an algorithm with a few (but hopefully expressive) parameters can be used to generate a large variety of waveforms and spectral properties by varying the parameters.

Since PCM involves recording and playing back waveforms, no actual expressive parameters are available in PCM systems. Similarly, Fourier's theorem says that we can represent any waveform with a sum of sine waves, but in fact it might take a huge number of sine waves to represent a particular waveform accurately. So, without further work to *parameterize* PCM samples or large groups of additive sine waves, these techniques are nonparametric.

Modal synthesis recognizes the fact that many vibrating systems exhibit a relatively few strong sinusoidal modes (natural frequencies) of vibration. Rigid bars and plates, plucked strings, and other structures are good candidates for modal synthesis, where after being excited (struck, plucked, etc.) the vibration/sound generation is restricted to a limited set of exponentially decaying sine waves. Modal synthesis is considered by some to be parametric, since the number of sine waves needed is often quite limited. Others claim that modal synthesis is nonparametric, since there is no small set of meaningful and expressive parameters that control variation in the sound.

Subtractive synthesis begins with spectrally complex waveforms (such as pulses or noise), and uses filters to shape the spectrum and waveform to the desired result. Subtractive synthesis works particularly well for certain classes of sounds, such as human speech, where a spectrally rich source (pulses of the vocal folds) is filtered by a time-varying filter (the acoustic tube of the vocal tract). Subtractive synthesis is parametric, because only a few descriptive numbers control such a model. For example, a subtractive speech model might be controlled by numbers expressing how pitched versus noisy the voice is, the voice pitch, the loudness, and 8–10 locations of the resonances of the vocal tract. Using only a dozen or so variable numbers (parameters), a rich variety of vowel and consonant sounds can be synthesized.

Frequency modulation (FM) synthesis exploits properties of nonlinearity to generate complex spectra and waveforms from much simpler ones. Modulating the frequency or phase of a sine wave (the carrier) by another sine wave (the modulator) causes multiple *sideband* sinusoids to appear at frequencies related to the ratio of the carrier and modulator frequencies. Thus an important parameter in FM is the carrier:modulator (C:M) ratio, and another is the amount (index) of modulation, which when increased causes the number of sidebands to increase.

Wavelets are waveforms that are usually compact in time and frequency. They are viewed by some as related to the fundamental unit of sound, by others as convenient building blocks for constructing sonic textures, and by others as related to important small events in nature and acoustics. *Formes d'onde formantiques* (FOFs, French for formant wave functions) are special wavelets that model the individual resonances of the vocal tract, and thus have been successfully used for modeling speech and singing.

Physical models solve the wave equation in space and time to generate waveforms in the same ways that physical acoustics does. Efficient techniques combining discrete simulation of physics with modern digital filtering techniques allow models of plucked strings, wind instruments, percussion, the vocal tract, and many other musical systems to be solved efficiently by computers.

After the presentation and examples of the main synthesis algorithms, some industry standards for representing and manipulating musical data in computerized systems will be discussed. The musical instrument digital interface (MIDI), downloadable sounds (DLS), open sound control (OSC), structured audio orchestra language (SAOL), and other score and control standards will be briefly covered.

Some languages and systems for computer music composition are discussed, including the historical MusicX languages, and many modern languages such as Cmusic and MAX/MSP, followed by a few real-time performance and interaction systems. Finally, some aspects of computer modeling of human listening and systems for machine-assisted and automatic composition are covered.

17.1 Computer Audio Basics

Typically, digital audio signals are formed by *sampling* analog (continuous in time and amplitude) signals at regular intervals in time, and then *quantizing* the amplitudes to discrete values. The process of sampling a waveform, holding the value, and quantizing the value to the nearest number that can be digitally represented (as a specific integer on a finite range of integers) is called analog-to-digital (A to D, or A/D) conversion [17.1]. A device that does A/D conversion is called an analog-to-digital converter (ADC). Coding and representing waveforms in sampled digital form is called pulse code modulation (PCM), and digital audio signals are often called PCM audio. The process of converting a sampled signal back into an analog signal is called digital-to-analog conversion (D to A, or D/A), and the device is called a digital-to-analog converter (DAC). Low-pass filtering (smoothing the samples to remove unwanted high frequencies) is necessary to reconstruct the sampled signal back into a smooth continuous-time analog signal. This filtering is usually contained in the DAC hardware.

The time between successive samples is usually denoted by T. Sampling an analog signal first requires filtering it to remove unwanted high frequencies (more on this shortly), holding the value steady for a period while a stable measurement can be made, then associating the analog value with a digital number (coding). Analog signals can have any of the infinity of real-numbered amplitude values. Since computers work with fixed word sizes (8 bit bytes, 16 bit words, etc.), digital signals can only have a finite number of amplitude val-

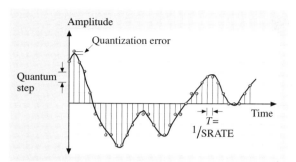

Fig. 17.1 Linear sampling and quantization

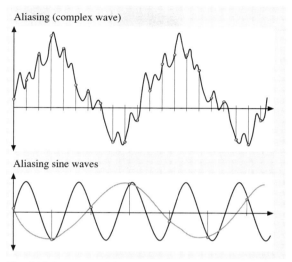

Fig. 17.2 Because of inadequate sampling rate, aliasing causes important features to be lost

ues. In converting from analog to digital, rounding takes place and a given analog value must be quantized to the nearest digital value. The difference between quantization steps is called the quantum (not as in quantum physics or leaps, but the Latin word for a fixed-sized jump in value or magnitude). Sampling and quantization is shown in Fig. 17.1; note the errors introduced in some sample values due to the quantization process.

A fundamental law of digital signal processing states that, if an analog signal is bandlimited with bandwidth B Hz, (Hz = Hertz = samples per second), the signal can be periodically sampled at a rate of $2B$ Hz or greater, and exactly reconstructed from the samples. Bandlimited with bandwidth B means that no frequencies above B exist in the signal. The rate $2B$ is called the sampling rate, and B is called the Nyquist frequency. Intuitively, a sine wave at the highest frequency B present in a bandlimited signal can be represented using two samples per period (one sample at each of the positive and negative peaks), corresponding to a sampling frequency of $2B$. All signal components at lower frequencies can be uniquely represented and distinguished from each other using this same sampling frequency of $2B$. If there are components present in a signal at frequencies greater than $1/2$ the sampling rate, these components will not be represented properly, and will *alias* (i. e.,, show up as frequencies different from their original values).

To avoid aliasing, ADC hardware often includes filters that limit the bandwidth of the incoming signal before sampling takes place, automatically changing as a function of the selected sampling rate. Figure 17.2 shows aliasing in complex and simple (sinusoidal) waveforms. Note the loss of detail in the complex waveform. Also note that samples at less than two times the frequency of a sine wave could also have arisen from a sine wave of much lower frequency. This is the fundamental nature of aliasing, because frequencies higher than $1/2$ sample rate *alias* as lower frequencies.

Humans can perceive frequencies from roughly 20 Hz to 20 kHz, thus requiring a minimum sampling rate of at least 40 kHz. Speech signals are often sampled at 8 kHz (*telephone quality*) or 11.025 kHz, while music is usually sampled at 22.05 kHz, 44.1 kHz (the sampling rate used on audio compact discs), or 48 kHz. Some new formats allow for sampling rates of 96 kHz, and even 192 kHz. This is because some engineers believe that we can actually hear things, or the effects of things, higher than 20 kHz. Dolphins, dogs, and some other animals can actually hear higher than us, so these new formats could be considered as catering to those potential markets.

In a digital system, a fixed number of binary digits (bits) are used to sample the analog waveform, by quantizing it to the closest number that can be represented. This quantization is accomplished either by rounding to the quantum value nearest the actual analog value, or by truncation to the nearest quantum value less than or equal to the actual analog value. With uniform sampling in time, a properly bandlimited signal can be exactly recovered provided that the sampling rate is twice the bandwidth or greater, but only if there is no quantization. When the signal values are rounded or truncated, the amplitude difference between the original signal and the quantized signal is lost forever. This can be viewed as an additive noise component upon reconstruction. Using the additive-noise assumption gives an approximate best-case signal-to-quantization-noise ratio (SNR) of approximately $6N$ dB, where N is the number of

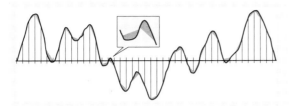

Fig. 17.3 Resampling by linear interpolation. Grey regions show the errors due to linear interpolation

bits. Using this approximation implies that a 16 bit linear quantization system will exhibit an SNR of approximately 96 dB. 8 bit quantization exhibits a signal-to-quantization-noise ratio of approximately 48 dB. Each extra bit improves the signal-to-noise ratio by about 6 dB.

Most computer audio systems use two or three types of audio data words. 16 bit (per channel) data is quite common, and this is the data format used in compact disc systems. Many recent (high-definition) formats allow for 24 bit samples. 8 bit data is common for speech data in personal computer (PC) and telephone systems, usually using methods of quantization that are nonlinear. In such systems the quantum is smaller for small amplitudes, and larger for large amplitudes.

Changing the playback sample rate on sampled sound results in a pitch shift. Many systems for recording, playback, processing, and synthesis of music, speech, or other sounds allow or require flexible control of pitch (sample rate). The most accurate pitch control is necessary for music synthesis. In sampling synthesis, this is accomplished by dynamic sample-rate conversion (interpolation). In order to convert a sampled data signal from one sampling rate to another, three steps are required: bandlimiting, interpolation, and resampling.

Bandlimiting: First, it must be assured that the original signal was properly bandlimited to less than half the new sampling rate. This is always true when *upsampling* (i. e., converting from a lower sampling rate to a higher one). When *downsampling*, however, the signal must be first low-pass filtered to exclude frequencies greater than half the new target rate.

Interpolation: When resampling by any but simple integer factors such as 2, 4, 1/2, etc., the task requires the computation of the values of the original bandlimited analog signal at the correct points between the existing sampled points. This is called interpolation. The most common form of interpolation used is linear interpolation, where the fractional time samples needed are computed as a linear combination of the two surrounding samples. Many assume that linear interpolation is the correct means, or at least an adequate means for accomplishing audio sample rate conversion. Figure 17.3 shows resampling by linear interpolation; note the errors shown by the shaded regions.

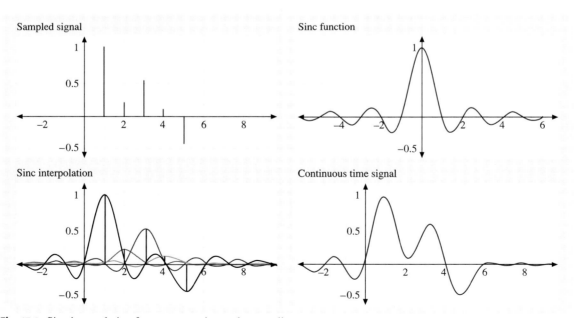

Fig. 17.4 Sinc interpolation for reconstruction and resampling

Some might notice that linear interpolation is not quite adequate, so they might assume that the correct solution must lie in more elaborate curve-fitting techniques using quadratics, cubics, or higher-order splines, and indeed these types of interpolation can be adequate for some applications. To arrive at the correct answer (provably correct from theory) the interpolation task should be viewed and accomplished as a filtering problem, with the filter designed to meet some appropriate error criterion. Linear time-invariant filtering is accomplished by convolution with a filter function. If the resampling filter is defined appropriately, we can exactly reconstruct the original analog waveform from the samples.

The correct (ideal in a provable digital signal processing sense) way to perform interpolation is convolution with the sinc function, defined as:

$$\text{sinc}(t/T) = \sin(\pi t/T)/(\pi t/T),$$

where

$$T = 1/\text{SRATE}.$$

The sinc function is the ideal low-pass filter with a cutoff of SRATE/2, where SRATE is the sampling rate. Figure 17.4 shows reconstruction of a continuous waveform by convolution of a digital signal with the sinc function. Each sample is multiplied by a corresponding continuous sinc function, and those are added up to arrive at the continuous reconstructed signal [17.2].

Resampling: This is usually accomplished at the same time as interpolation, because it is not necessary to reconstruct the entire continuous waveform in order to acquire new discrete samples. The resampling ratio can be time varying, making the problem a little more difficult. However, viewing the problem as a filter-design and implementation issue allows for guaranteed tradeoffs of quality and computational complexity.

17.2 Pulse Code Modulation Synthesis

The majority of digital sound and music synthesis today is accomplished via the playback of stored pulse code modulation (PCM) waveforms. Single-shot playback of entire segments of stored sounds is common for sound effects, narrations, prompts, segments of music, etc. Most high-quality modern electronic music synthesizers, speech synthesis systems, and PC software systems for sound synthesis use pre-stored PCM as the basic data. This data is sometimes manipulated to yield the final output sound(s).

There are a number of different ways to look at sound for computer music, with PCM being only one. We can look at the physics that produce the sound and try to model those. We could also look at the spectrum of the sound and other characteristics having to do with the perception of those sounds. Indeed, much of the legacy of computer music has revolved around parametric (using mathematical algorithms, controlled by a few well-chosen/-designed control parameters) analysis and synthesis algorithms. We will discuss most of the commonly used algorithms later, but first we should look at PCM in more depth.

For speech, the most common synthesis technique is *concatenative* synthesis [17.3]. Concatenative phoneme synthesis relies on the concatenation of roughly 40 pre-stored phonemes (for English). Examples of vowel phonemes are /i/ as in beet, /I/ as in bit, /a/ as in father, etc. Examples of nasals are /m/ as in mom, /n/ as in none, /ng/ as in sing, etc. Examples of fricative consonant phonemes are /s/ as in sit, /sh/ as in ship, /f/ as in fifty, etc. Examples of voiced fricative consonants are /z/, /v/ (visualize), etc. Examples of plosive consonants are /t/ as in tat, /p/ as in pop, /k/ as in kick, etc. Examples of voiced plosives include /d/, /b/, /g/ (dude, bob, gag) etc. Vowels and nasals are essentially periodic pitched sounds, so the minimal required stored waveform is only one single period of each. Consonants require more storage because of their noisy (non-pitched, aperiodic) nature.

The quality of concatenative phoneme synthesis is generally considered quite low, due to the simplistic assumption that all of the pitched sounds (vowels, etc.) are purely periodic. Also, simply *gluing* /s/ /I/ and /ng/ together does not make for a high-quality realistic synthesis of the word "sing". In actual speech, phonemes gradually blend into each other as the jaw, tongue, and other articulators move with time.

Accurately capturing the transitions between phonemes with PCM requires recording transitions from phoneme to phoneme, called *diphones*. A concatenative diphone synthesizer blends together stored diphones. Examples of diphones include see, she, thee, and most of the roughly 40x40 possible combinations of phonemes. Much more storage is necessary for a diphone synthesizer, but the resulting increase in quality is significant. PCM speech synthesis can be improved further by stor-

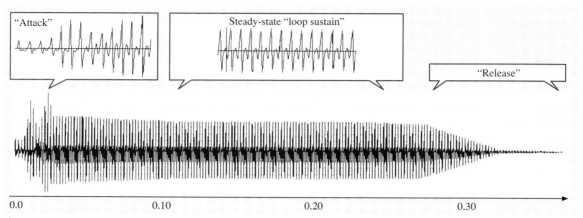

Fig. 17.5 Wave-table synthesis of a trumpet tone

ing multiple samples, for different pitch ranges, genders, voice qualities, etc.

For musical sounds, it is common to store only a loop, or table, of the periodic component of a recorded sound waveform and play that loop back repeatedly. This is called "wave-table synthesis" [17.4]. For more realism, the attack or beginning portion of the recorded sound can be stored in addition to the periodic steady-state part. Figure 17.5 shows the synthesis of a trumpet tone starting with an attack segment, followed by repetition of a periodic loop, ending with an enveloped decay (or release). The *envelope* is a synthesizer/computer music term for a time-varying change applied to a waveform amplitude, or other parameter. Envelopes are often described by four components: the *attack time*, the *decay time* (decay here means the initial decay down to the steady state segment), the *sustain level*, and the *release time* (final decay). Hence, envelopes are sometimes called ADSRs.

Originally called *sampling synthesis* in the music industry, any synthesis using stored PCM waveforms has now become commonly known as "wave-table synthesis". Filters are usually added to high-quality wave-table synthesis, allowing control of spectral brightness as a function of intensity, and to get more variety of sounds from a given set of samples.

Pitch shifting of a PCM sample or wave table is accomplished via interpolation as discussed in the previous section. A given sample can be pitch-shifted only so far in either direction before it begins to sound unnatural. This can be dealt with by storing multiple recordings of the sound at different pitches, and switching or interpolating between these upon resynthesis. This is called *multi-sampling*. Multi-sampling might also include the storage of separate samples for loud and soft sounds. Linear or other interpolation is used to blend the loudness multi-samples as a function of the desired synthesized volume. This adds realism, because loudness is not simply a matter of amplitude or power, and most sound sources exhibit spectral variations as a function of loudness. For example, there is usually more high-frequency energy (*brightness*) in loud sounds than in soft sounds. Filters can also be used to add spectral variation.

A common tool used to describe the various components and steps of signal processing in performing digital music synthesis is the *synthesizer patch* (historically named from hooking various electrical components together using patch cords). In a patch, a set of fairly commonly agreed building blocks, called *unit generators* (also called modules, plug-ins, operators, op-codes, and other terms) are hooked together in a signal flow diagram. This historical [17.5] graphical method of describing signal generation and processing affords a visual representation that is easily printed in pa-

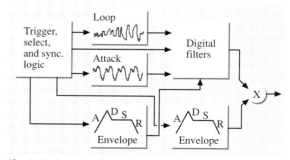

Fig. 17.6 A synthesis *patch* showing interconnection of *unit generators*

pers, textbooks, patents, etc. Further, graphical patching systems and languages have been important to the development and popularization of certain algorithms, and computer music in general. Figure 17.6 shows a PCM synthesizer patch with attack and loop wave tables whose amplitudes are controlled by an envelope generator, and a time-varying filter (also controlled by another envelope generator).

17.3 Additive (Fourier, Sinusoidal) Synthesis

Lots of sound-producing objects and systems exhibit sinusoidal modes. A plucked string might exhibit many modes, with the strength of each mode determined by the boundary conditions of the terminations, and the shape of the excitation pluck. Striking a metal plate with a hammer excites many of the vibrational modes of the plate, determined by the shape of the plate, and by where it is struck. A singing voice, struck drum head, bowed violin string, struck bell, or blown trumpet exhibit oscillations characterized by a sum of sinusoids. The recognition of the fundamental nature of the sinusoid gives rise to a powerful model of sound synthesis based on summing sinusoidal modes.

These modes have a very special relationship in the case of the plucked string, a singing voice, and some other limited systems, in that their frequencies are all integer multiples (at least approximately) of one basic sinusoid, called the *fundamental*. This special series of sinusoids is called a *harmonic series*, and lies at the basis of the Fourier-series representation of shapes, waveforms, oscillations, etc. The Fourier series [17.6] solves many types of problems, including physical problems with boundary constraints, but is also applicable to any shape or function. Any periodic waveform (repeating over and over again) F_{per} can be transformed into a Fourier series, written as:

$$F_{\text{per}}(t) = a_0 + \Sigma_m [b_m \cos(2\pi f_0 mt) + c_m \sin(2\pi f_0 mt)] \,. \quad (17.1)$$

The limits of the summation are technically infinite, but we know that we can cut off our frequencies at the Nyquist frequency for digital signals. The a_0 term is a constant offset, or the average of the waveform. The b_m and c_m coefficients are the weights of the mth harmonic cosine and sine terms. If the function $F_{\text{per}}(t)$ is purely even about $t = 0$ [$(F(-t) = F(t))$], only cosines are required to represent it, and only the b_m terms would be nonzero. Similarly, if the function $F_{\text{per}}(t)$ is odd [$(F(-t) = -F(t))$], only the c_m terms would be required. An arbitrary function $F_{\text{per}}(t)$ will require sinusoidal harmonics of arbitrary (but specific) amplitudes and phases. The magnitude and phase of the mth harmonic in the Fourier series can be found by:

$$A_m = \sqrt{b_m^2 + c_m^2} \,, \quad (17.2a)$$
$$\theta_m = \arctan(c_m/b_m) \,. \quad (17.2b)$$

Phase is defined relative to the cosine, so if c_m is zero, θ_m is zero. As a brief example, Fig. 17.7 shows the first few sinusoidal harmonics required to build up an approximation of a square wave. Note that due to symmetries (boundary conditions), only odd sine harmonics are required. Using more sines improves the approximation.

The process of solving for the sine and cosine components of a signal or waveform is called Fourier analysis, or the Fourier transform. If the frequency variable is sampled (as is the case in the Fourier series, represented by m), and the time variable t is sampled as well (as it is in PCM waveform data, represented by n), then the Fourier transform is called the *discrete Fourier transform* (DFT). The DFT is given by

$$F(m) = \sum_{n=0}^{N-1} f(n)[\cos(2\pi mn/N) - i\sin(2\pi mn/N)] \,. \quad (17.3)$$

Where N is the length of the signal being analyzed. The inverse DFT (IDFT) is very similar to the Fourier

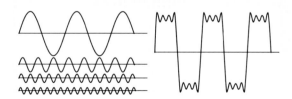

Fig. 17.7 A sum of odd harmonics approximates a square wave

series

$$f(n) = 1/N \sum_{m=0}^{N-1} F(m)[\cos(2\pi mn/N) + \mathrm{i}\sin(2\pi mn/N)]\,. \tag{17.4}$$

The imaginary number $\mathrm{i}(=\sqrt{-1})$ is used to place the cosine and sine components in a unique mathematical arrangement, where odd $[(x(-n) = -x(n))]$ terms of the waveform are represented as imaginary components, and even $[(x(-n) = x(n))]$ terms are represented as real components. This gives us a means to talk about the magnitude and phase in terms of the magnitude and phase of $F(m)$ (a complex number).

There is a near-mystical expression of equality in mathematics known as Euler's identity, which links trigonometry, exponential functions, and complex numbers in a single equation

$$\mathrm{e}^{\mathrm{i}\theta} = \cos(\theta) + \mathrm{i}\sin(\theta)\,.$$

We can use Euler's identity to write the DFT and IDFT in shorthand

$$F(m) = \sum_{n=0}^{N-1} f(n)\mathrm{e}^{-\mathrm{i}2\pi mn/N}\,, \tag{17.5}$$

$$f(n) = 1/N \sum_{m=0}^{N-1} F(m)\mathrm{e}^{\mathrm{i}2\pi mn/N}\,. \tag{17.6}$$

Converting the cosine/sine form to the complex exponential form allows lots of manipulations that would be difficult otherwise. But we can also write the DFT in real-number terms as a form of the Fourier series

$$f(n) = 1/N \sum_{m=0}^{N-1} F_b(n)\cos(2\pi mn/N) + F_c(n)\sin(2\pi mn/N)\,, \tag{17.7}$$

where

$$F_b(m) = \sum_{n=0}^{N-1} f(n)\cos(2\pi mn/N)\,, \tag{17.8}$$

$$F_c(m) = \sum_{n=0}^{N-1} -f(n)\sin(2\pi mn/N)\,. \tag{17.9}$$

The fast Fourier transform (FFT) is a cheap way of calculating the DFT. There are thousands of references on the FFT [17.6], and scores of implementations of it, so for our purposes we will just say that it is much more efficient than trying to compute the DFT directly from the definition. A well-crafted FFT algorithm for real input data takes on the order of $N\log_2(N)$ multiply–adds to compute. Comparing this to the N^2 multiplies of the DFT, N does not have to be very big before the FFT is a winner. There are some down sides, such as the fact that FFTs can only be computed for signals whose lengths are exactly powers of 2, but the advantages of using it often outweigh the annoying power-of-two problems. Practically speaking, users of FFTs usually carve up signals into small chunks (powers of two), or *zero pad* a signal out to the next biggest power of two.

The short-time Fourier transform (STFT) breaks up the signal and applies the Fourier transform to each segment individually [17.7]. By selecting the window size (length of the segments), and hop size (how far the window is advanced along the signal) to be perceptually relevant, the STFT can be used as an approximation of human audio perception. Figure 17.8 shows the waveform of the utterance of the word "synthesize", and some STFT spectra corresponding to windows at particular points in time.

If we inspect the various spectra in Fig. 17.8, we can note that the vowels exhibit harmonic spectra (clear, evenly spaced peaks corresponding to the harmonics of the pitched voice), while the consonants exhibit noisy spectra (no clear sinusoidal peaks). Recognizing that some sounds are well approximated/modeled by additive sine waves [17.8], while other sounds are essentially noisy, spectral modeling [17.7] breaks the sound into deterministic (sines) and stochastic (noise) components. Figure 17.9 shows a general sines+noise additive synthesis model, allowing us to control the amplitudes and frequencies of a number of sinusoidal oscillators, and model the noisy component with a noise source and a spectral shaping filter.

The beauty of this type of model is that it recognizes the dominant sinusoidal nature of many sounds, while still recognizing the noisy components that might be also present. More efficient and parametric representations, and many interesting modifications can be made to the signal on resynthesis. For example, removing the harmonics from voiced speech, followed by resynthesizing with a scaled version of the noise residual, can result in the synthesis of whispered speech.

One further improvement to spectral modeling is the recognition [17.9] that there are often brief (impulsive) moments in sounds that are really too short in time to be adequately analyzed by spectrum analysis. Further, such moments in the signal usually corrupt the sinusoidal/noise analysis process. Such

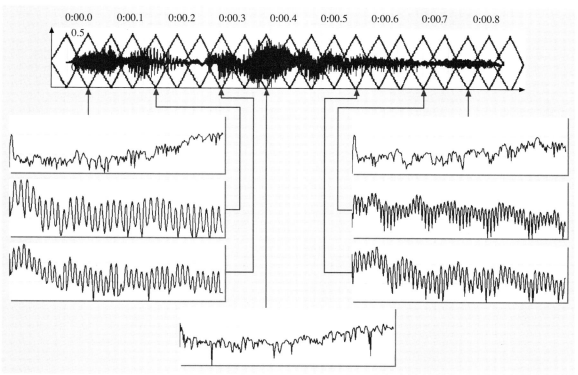

Fig. 17.8 Some STFT frames of the word 'synthesize'

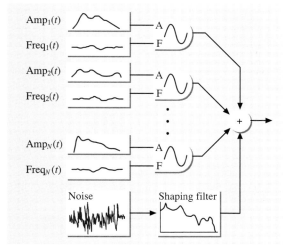

Fig. 17.9 Sinusoidal additive model with filtered noise added for spectral modeling synthesis

events, called transients, can be modeled in other ways (often by simply keeping the stored PCM for that segment).

Using the short-time Fourier transform, the *phase vocoder* (VoiceCoder) [17.10, 11] processes sound by calculating and maintaining both magnitude and phase. The frequency bins (basis sinusoids) of the DFT can be viewed as narrow-band filters, so the Fourier transform of an input signal can be viewed as passing it through a bank of narrow band-pass filters. This means that on the order of hundreds to thousands of sub-bands are used.

The phase vocoder has found extensive use in computer music composition. Many interesting practical and artistic transformations can be accomplished using the phase vocoder, including nearly artifact-free and independent time and pitch shifting. A technique called *cross synthesis* assumes that one signal is the analysis signal. The time-varying magnitude spectrum of the analysis signal (usually smoothed in the frequency domain) is multiplied by the spectral frames of another input (or filtered) signal (sometimes brightened first by high-frequency emphasis pre-filtering), yielding a composite signal that has the attributes of both. Cross-synthesis has produced the sounds of talking cows, *morphs* between people and cats, trumpet/flute hybrids, etc.

17.4 Modal (Damped Sinusoidal) Synthesis

The simplest physical system that does something acoustically (and musically) interesting is the mass/spring/damper (Fig. 17.10 [17.12]. The differential equation describing that system has a solution that is a single exponentially decaying cosine wave. The Helmholtz resonator (large, contained air cavity with a small long-necked opening, like a pop bottle, Fig. 17.11 behaves like a mass/spring/damper system, with the same exponentially damped cosine behavior. The equations describing the behavior of these systems is:

$$d^2y/dt^2 + (r/m)dy/dt + (k/m)y = 0 \,, \quad (17.10)$$

$$y(t) = y_0 e^{(-rt/2m)} \cos\left(t\sqrt{[k/m - (r/2m)^2]}\right) \,. \quad (17.11)$$

Of course, most systems that produce sound are more complex than the ideal mass/spring/damper system, or a pop bottle. And of course most sounds are more complex than a simple damped exponential sinusoid. Mathematical expressions of the physical forces (thus the accelerations) can be written for nearly any system, but solving such equations is often difficult or impossible. Some systems have simple enough properties and geometries to allow an exact solution to be written out for their vibrational behavior. An ideal string under tension is one such system.

Here we will resort to some graphical arguments and our prior discussion of the Fourier transform to motivate further the notion of sinusoids in real physical systems. The top of Fig. 17.12 shows a string, lifted from a point in the center (halfway along its length). Below that is shown a set of sinusoidal mode that the center-plucked

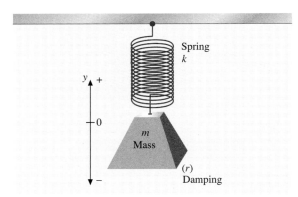

Fig. 17.10 Mass/spring/damper system

Fig. 17.11 Helmholtz resonator

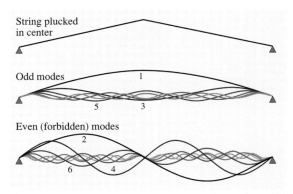

Fig. 17.12 Plucked string (*top*). The *center* shows sinusoidal *modes* of vibration of a center-plucked string. The *bottom* shows the even modes, which would not be excited by the center-plucked condition

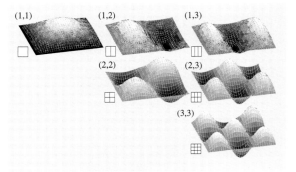

Fig. 17.13 Square-membrane modes

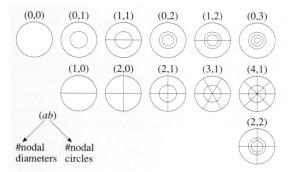

Fig. 17.14 Circular-membrane nodal lines

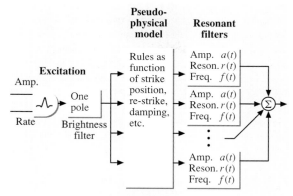

Fig. 17.16 Flexible parametric modal synthesis algorithm

string vibration would have. These are spatial functions (sine as function of position along the string), but they also correspond to natural frequencies of vibration of the string. At the bottom of Fig. 17.12 is another set of modes that would not be possible with the center-plucked condition, because all of these even modes are restricted to have no vibration in the center of the string, and thus they could not contribute to the triangular shape of the string. These conditions of no displacement, corresponding to the zero crossings of the sine functions, are called *nodes*. Note that the end points are forced nodes of the plucked string system for all possible conditions of excitation.

Physical constraints on a system, such as the pinned ends of a string, and the center plucked initial shape, are known as *boundary conditions*. Spatial sinusoidal solutions like those shown in Fig. 17.12 are called *boundary solutions* (the legal sinusoidal modes of displacement and vibration) [17.13].

Just as one can use Fourier boundary methods to solve the one-dimensional (1-D) string, we can also extend boundary methods to two dimensions. Figures 17.13 and 17.14 show the first few vibration modes of uniform square and circular membranes. The small boxes at the lower left corners of each square modal-shape diagram depict the modes in a purely two-dimensional way, showing lines corresponding to the spatial sinusoidal nodes (regions of no displacement vibration). This is how the circular modes are presented. The natural modes must obey the two-dimensional boundary conditions at the edges, but unlike the string, the square membrane modes are not integer-related harmonic frequencies. In fact they obey the relationship

$$f_{mn} = f_{11}\sqrt{[(m^2 + n^2)/2]}, \quad (17.12)$$

where m and n range from 1 to (potentially) infinity, and f_{11} is $c/2L$ (the speed of sound on the membrane divided by the square edge lengths). The circular modes are predictable from mathematics, but have a much more complex form than the square modes.

Unfortunately, circles, rectangles, and other simple geometries turn out to be the only ones for which the boundary conditions yield a closed-form solution in terms of spatial and temporal sinusoidal terms. However, we can measure and model the modes of any system by using the Fourier transform of the sound it produces, and looking for exponentially decaying sinusoidal components.

We can approximate the differential equation describing the mass/spring/damper system of (17.11) by replacing the derivatives (velocity as the derivative of position, and acceleration as the second derivative of position) with sampled time differences (normalized by the sampling interval T seconds). In doing so we would arrive at an equation that is a recursion in past values of $y(n)$, the position variable

$$[y(n) - 2y(n-1) + y(n-2)]/T^2$$
$$+ \frac{r/m[y(n) - y(n-1)]}{T + k/m\, y(n)} = 0. \quad (17.13)$$

Note that if the values of mass, damping, spring constant, and sampling rate are constant, then the coeffi-

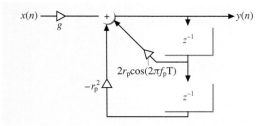

Fig. 17.15 Flexible parametric modal synthesis algorithm

cients $(2m + Tr)/(m + Tr + T^2k)$ for the single delay, and $m/(m + Tr + T^2k)$ for the twice-delayed signal applied to past y values are constant. Digital signal processing (DSP) engineers would note that a standard infinite impulse response (IIR) recursive filter as shown in Fig. 17.15 can be used to implement (17.13) (the Z^{-1} represents a single sample of delay). In fact, the second order two-pole feedback filter can be used to generate an exponentially decaying sinusoid, called a *phasor* in DSP literature [17.14]. The connection between the second order digital filter and the physical notion of a mode of vibration forms the basis for modal sound synthesis [17.15]. Figure 17.16 shows a general model for modal synthesis of struck/plucked objects, in which an impulsive excitation function is used to excite a number of filters that model the modes. Rules for controlling the modes as a function of strike position, striking object, changes in damping, and other physical constraints are included in the model.

17.5 Subtractive (Source–Filter) Synthesis

Subtractive synthesis uses a complex source wave, such as an impulse, a periodic train of impulses, or white noise, to excite spectral shaping filters. One of the earliest uses of electronic subtractive synthesis dates back to the 1920s and 1930s, with the invention of the *channel vocoder* (or VOiceCODER) [17.16]. In the this device, the spectrum is broken into sections called sub-bands, and the information in each sub-band is converted to a signal representing (generally slowly varying) power. The analyzed parameters are then stored or transmitted (potentially compressed) for reconstruction at another time or physical site. The parametric data representing the information in each sub-band can be manipulated in various ways, yielding transformations such as pitch or time shifting, spectral shaping, cross-synthesis, and other effects. Figure 17.17 shows a block diagram of a channel vocoder. The detected envelopes serve as control signals for a bank of bandpass synthesis filters (identical to the analysis filters used to extract the sub-band envelopes). The synthesis filters have gain inputs that are fed by the analysis control signals.

When used to encode and process speech, the channel vocoder explicitly makes an assumption that the signal being modeled is a single human voice. The *source analysis* block extracts parameters related to finer spectral details, such as whether the sound is pitched (vowel) or noisy (consonant or whispered). If the sound is pitched, the pitch is estimated. The overall energy in the signal is also estimated. These parameters become additional low-bandwidth control signals for the synthesizer. Intelligible speech can be synthesized using only a few hundred numbers per second. An example coding scheme might use eight channel gains + pitch + power, per frame, at 40 frames per second, yielding a total of only 400 numbers per second. The channel vocoder, as designed for speech coding, does not generalize to arbitrary sounds, and fails horribly when the source parameters deviate from expected harmonicity, reasonable pitch range, etc. But the ideas of sub-band decomposition, envelope detection, and driving a synthesis filter bank with control signals give rise to many other interesting applications and implementations of the vocoder concept.

A number of analog hardware devices were produced and sold as musical instrument processing devices in the 1970s and 1980s. These were called simply *vocoders*, but had a different purpose than speech coding. Figure 17.18 shows the block diagram of a cross-

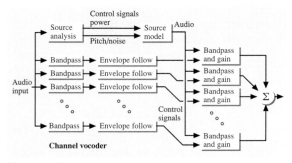

Fig. 17.17 Channel vocoder block diagram

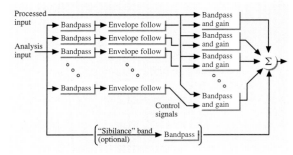

Fig. 17.18 Cross-synthesizing vocoder block diagram

synthesizing vocoder. The main difference is that the parametric source has been replaced by an arbitrary audio signal. The cross-synthesizer can be used to make non-speech sounds talk or sing. A typical example would be feeding a voice into the analysis filter bank, and an electric guitar sound into the synthesis bank audio inputs to make a talking guitar. To be effective, the synthesis audio input should have suitably rich spectral content (for instance, distorted electric guitar works better than undistorted). Some vocoders allowed an additional direct *sibilance band* (like the highest band shown in Fig. 17.18) pass-through of the analyzed sound. This would allow consonants to pass through directly, creating a more intelligible speech-like effect.

Modal synthesis, as discussed before, is a form of subtractive synthesis, but the spectral characteristics of modes are sinusoidal and thus exhibit very narrow spectral peaks. For modeling the gross peaks in a spectrum, which could correspond to weaker resonances, we can exploit the same two-pole resonance filters. This type of source-filter synthesis has been very popular for voice synthesis.

Having its origins and applications in many different disciplines, time-series prediction is the task of estimating future sample values from prior samples. Linear prediction is the task of estimating a future sample (usually the next in the time series) by forming a linear combination of some number of prior samples. Linear predictive coding (LPC) is a technique that can automatically extract the gross spectral features and design filters to match those, and give us a source that we can use to drive the filters [17.17, 18]. Figure 17.19 shows linear prediction in block diagram form. The difference equation for a linear predictor is

$$y(n) = \hat{x}(n+1) = \sum_{i=0}^{m} a_i x(n-i) \, . \tag{17.14}$$

The task of linear prediction is to select the vector of predictor coefficients

$$A = (a_0, a_1, a_2, a_2, \ldots, a_m)$$

such that $\hat{x}(n+1)$ (the estimate) is as close as possible to $x(n+1)$ (the real sample) over a set of samples (often called a frame) $x(0)$ to $x(N-1)$. Usually, "close as possible" is defined by minimizing the mean square error (MSE):

$$MSE = 1/N \sum_{i=0}^{N-1} [\hat{x}(n+1) - x(n+1)]^2 \, . \tag{17.15}$$

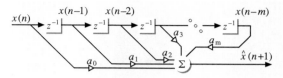

Fig. 17.19 A linear prediction filter

Many methods exist to arrive at the set of predictor coefficients a_i that yield a minimum MSE. The most common uses correlation or covariance data from each frame of samples to be predicted. The *autocorrelation* function, defined as

$$x \diamond x(n) = \sum_{i=0}^{N-1} x(i) x(i+n) \, . \tag{17.16}$$

Autocorrelation computes a time (lag) domain function that expresses the similarity of a signal to delayed versions of itself. It can also be viewed as the convolution of a signal with the time-reversed version of itself. Purely periodic signals exhibit periodic peaks in the autocorrelation function. Autocorrelations of white-noise sequences exhibit only one clear peak at the zero-lag position, and are small (zero for infinite-length sequences) for all other lags. Since the autocorrelation function contains useful information about a signal's similarity to itself at various delays, it can be used to compute the optimal set of predictor coefficients by forming the autocorrelation matrix

$$R = \begin{pmatrix} x \blacklozenge x(0) & x \blacklozenge x(1) & x \blacklozenge x(2) & \ldots & x \blacklozenge x(m) \\ x \blacklozenge x(1) & x \blacklozenge x(0) & x \blacklozenge x(1) & \ldots & x \blacklozenge x(m-1) \\ x \blacklozenge x(2) & x \blacklozenge x(1) & x \blacklozenge x(0) & \ldots & x \blacklozenge x(m-2) \\ x \blacklozenge x(3) & x \blacklozenge x(2) & x \blacklozenge x(1) & \ldots & x \blacklozenge x(m-3) \\ \vdots & \vdots & \vdots & & \vdots \\ x \blacklozenge x(m) & x \blacklozenge x(m-1) & x \blacklozenge x(m-2) & \ldots & x \blacklozenge x(0) \end{pmatrix} \tag{17.17}$$

We can get the least-squares predictor coefficients by forming

$$A = (a_0 a_1 a_2 a_3 \ldots a_m) = PR^{-1} \, , \tag{17.18}$$

where P is the vector of prediction correlation coefficients

$$P = (x \blacklozenge x(1) \ \ x \blacklozenge x(2) \ \ x \blacklozenge x(3) \ \ \ldots \ \ x \blacklozenge x(m+1)) \, . \tag{17.19}$$

For low-order LPC (delay order of 6–20 or so), the filter fits itself to the coarse spectral features, and the

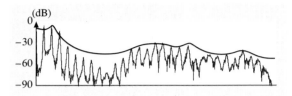

Fig. 17.20 Tenth order LPC filter fit to a voiced 'ooo' spectrum

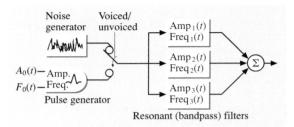

Fig. 17.22 Parallel factored formant subtractive synthesizer

residue contains the remaining part of the sound that cannot be linearly predicted. A common and popular use of LPC is for speech analysis, synthesis, and compression. The reason for this is that the voice can be viewed as a source-filter model, where a spectrally rich input (pulses from the vocal folds or noise from turbulence) excites a filter (the resonances of the vocal tract). LPC is another form of spectral vocoder (voice coder) as discussed previously, but since LPC filters are not fixed in frequency or shape, fewer bands are needed to model the changing speech spectral shape dynamically.

LPC speech analysis/coding involves processing the signal in blocks and computing a set of filter coefficients for each block. Based on the slowly varying nature of speech sounds (the speech articulators can only move so fast), the coefficients are relatively stable for milliseconds at a time (typically 5–20 ms is used in speech coders). If we store the coefficients and information about the residual signal for each block, we will have captured many of the essential aspects of the signal. Figure 17.20 shows an LPC fit to a speech spectrum. Note that the fit is better at the peak locations than at the valleys. This is due to the least-squares minimization criterion. Missing the mark on low-amplitude parts of the spectrum is not as important as missing it on high-amplitude parts. This is fortunate for audio signal modeling, in that the human auditory system is more sensitive to spectral peaks (formants, poles, resonances) than valleys (zeroes, anti-resonances).

Once we have performed LPC on speech, if we inspect the residual we might note that it is often a stream of pulses for voiced speech, or white noise for unvoiced speech. Thus, if we store parameters about the residual, such as whether it is periodic pulses or noise, the frequency of the pulses, and the energy in the residual, then we can get back a signal that is very close to the original. This is the basis of much of modern speech compression. If a signal is entirely predictable using a linear combination of prior samples, and if the predictor filter is doing its job perfectly, we should be able to hook the output back to the input and let the filter predict the rest of the signal automatically. This form of filter, with feedback from output to input, is called *recursive*. The recursive LPC reconstruction is sometimes called *all pole*, to reflect the high-gain poles corresponding to the primary resonances of the vocal tract. The poles do not capture all of the acoustic effects going on in speech, however, such as zeroes that are introduced in nasalization, aerodynamic effects, etc. However, as mentioned before, since our auditory systems are most sensitive to peaks (poles), the LPC representation does quite a good job of capturing the most important aspects of speech spectra.

In recursive resynthesis form, any deviation of the predicted signal from the actual original signal will show up in the error signal, so if we excite the recursive LPC reconstruction filter with the residual signal itself, we can get back the original signal exactly. This is a form of identity analysis/resynthesis, performing deconvolution to separate the source from the filter, and using the residue to excite the filter to arrive at the original signal, by convolution, or using the residue as the input to the filter.

For computer music composition, LPC can be used for cross-synthesis, by keeping the time-varying filter coefficients, and replacing the residual with some other sound source. Using the parametric source model also allows for flexible time and pitch shifting, without modifying the basic timbre. The voiced pulse period can be modified, or the frame rate update of the filter coefficients can be modified, independently. So it is easy to speed up a speech sound while making the pitch lower

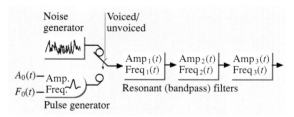

Fig. 17.21 Cascade factored formant subtractive synthesizer

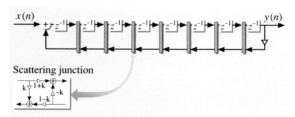

Fig. 17.23 Ladder filter implementation of an all-pole LPC filter

and still retaining the basic spectral shapes of all vowels and consonants.

In decomposing signals into a source and a filter, LPC can be a marvelous aid in analyzing and understanding some sound-producing systems. The recursive LPC reconstruction filter can be implemented in a variety of ways. Three different filter forms are commonly used to perform subtractive voice synthesis [17.19]. The filter can be implemented in series (cascade) as shown in Fig. 17.21, factoring each resonance into a separate filter block with control over center frequency, width, and amplitude. The filter can also be implemented in parallel (separate sub-band sections of the spectrum added together), as shown in Fig. 17.22.

One additional implementation of the resonant filter is the ladder filter structure, which carries with it a notion of one-dimensional spatial propagation as well [17.20]. Figure 17.23 shows a ladder filter realization of a 10th-order IIR filter. We will refer to this type of filter later when we investigate some 1-D waveguide physical models.

17.6 Frequency Modulation (FM) Synthesis

Wave-shaping synthesis involves warping a simple (usually a sine or sawtooth wave) waveform with a nonlinear function or lookup table. One popular form of wave-shaping synthesis, called frequency modulation (FM), uses sine waves for both input and warping waveforms [17.21–23]. Frequency modulation relies on modulating the frequency of a simple periodic waveform with another simple periodic waveform. When the frequency of a sine wave of average frequency f_c (called the carrier), is modulated by another sine wave of frequency f_m (called the modulator), sinusoidal sidebands are created at frequencies equal to the carrier frequency plus and minus integer multiples of the modulator frequency. Figure 17.24 shows a block diagram for simple FM synthesis (one sinusoidal carrier and one sinusoidal modulator). Mathematically, FM is expressed as

$$y(t) = \sin[2\pi t f_c + \Delta f_c \cos(2\pi t f_m)] \,. \quad (17.20)$$

The index of modulation I is defined as $\Delta f_c / f_c$. The equation as shown is actually phase modulation, but FM and PM can be related by a differentiation/integration of the modulation, and thus do not differ for cosine modulation.

Carson's rule (a rule of thumb) states that the number of significant bands on each side of the carrier frequency (sidebands) is roughly equal to $I + 2$. For example, a carrier sinusoid of frequency 600 Hz, a modulator sinusoid of frequency 100 Hz, and a modulation index of 3 would produce sinusoidal components of frequencies 600, 700, 500, 800, 400, 900, 300, 1000, 200, 1100, and 100 Hz. Inspecting these components reveals that a harmonic spectrum with 11 significant harmonics, based on a fundamental frequency of 100 Hz can be produced by using only two sinusoidal generating functions. Figure 17.25 shows the spectrum of this synthesis.

Selecting carrier and modulator frequencies that are not related by simple integer ratios yields an inharmonic spectrum. For example, a carrier of 500 Hz, modula-

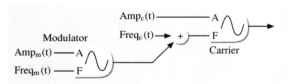

Fig. 17.24 Simple FM (one carrier and one modulator sine wave) synthesis

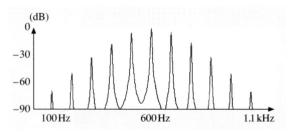

Fig. 17.25 Simple FM with 600 Hz carrier, 100 Hz modulator, and index of modulation of 3

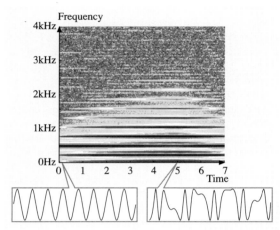

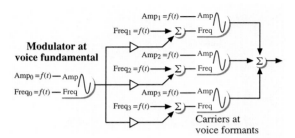

Fig. 17.27 Frequency-modulation voice synthesis diagram

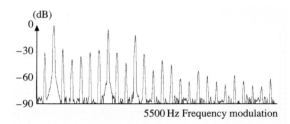

Fig. 17.26 Inharmonic simple FM with 500 Hz carrier and 273 Hz modulator. The index of modulation is ramped from zero to 5 then back to zero

Fig. 17.28 Frequency-modulation voice spectrum

tor of 273 Hz, and an index of 5 yields frequencies of 500 (carrier), 227, 46, 319, 592, 865, 1138, 1411 (negative sidebands), and 773, 1046, 1319, 1592, 1865, 2138, 2411 (positive sidebands). Figure 17.26 shows a spectrogram of this FM tone, as the index of modulation is ramped from zero to 5. The synthesized waveforms at $I = 0$ and $I = 5$ are shown as well.

By setting the modulation index high enough, huge numbers of sidebands are generated, and the aliasing and addition of these (in most cases) results in noise. By careful selection of the component frequencies and index of modulation, and combining multiple carrier/modulator pairs, many spectra can be approximated using FM. However, the amplitudes and phases (described by Bessel functions) of the individual components cannot be independently controlled, so FM is not a truly generic waveform or spectral synthesis method. Using multiple carriers and modulators, connection topologies (algorithms) have been designed for the synthesis of complex sounds such as human voices, violins, brass instruments, percussion, etc. Figures 17.27 and 17.28 show an FM connection topology for human voice synthesis, and a typical resulting spectrum [17.24]. Because of the extreme efficiency of FM, it became popular in the 1980s as a music synthesizer algorithm. Later FM found wide use in early PC sound cards.

17.7 FOFs, Wavelets, and Grains

In a source/filter vocal model such as LPC or parallel/cascade formant subtractive synthesis, periodic impulses are used to excite resonant filters to produce vowels. We could construct a simple alternative model using three, four, or more tables storing the impulse responses of the individual vowel formants. Note that it is not necessary for the tables to contain pure exponentially decaying sinusoids. We could include any aspects of the voice source, etc., as long as those effects are periodic. Formes d'onde formantiques (FOFs, French for formant wave functions) were created for voice synthesis using such tables, overlapped and added at the repetition period of the voice source [17.25]. Figure 17.29 depicts FOF synthesis of a vowel. FOFs are composed of a sinusoid at the formant center frequency, with an amplitude that rises rapidly upon excitation, then decays exponentially. The control parameters define the center frequency and bandwidth of the formant being modeled, and the rate at which the FOFs are generated and added determines the fundamental frequency of the voice.

Note that each individual FOF is a simple wavelet (local and compact wave both in frequency and time).

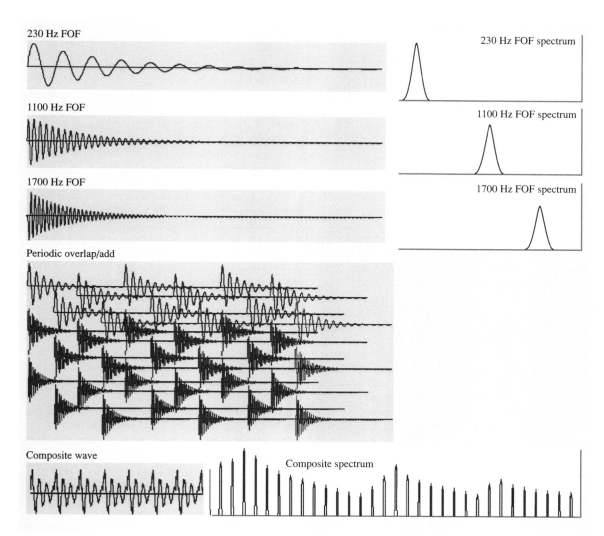

Fig. 17.29 FOF synthesis of a vowel

A family of filter-based frequency transforms known as *wavelet transforms* have been used for the analysis and synthesis of sound. Instead of being based on steady sinusoids like the Fourier transform, wavelet transforms are based on the decomposition of signals into fairly arbitrary wavelets [17.26]. These wavelets can be thought of as mathematical sound grains. Some benefits of wavelet transforms over Fourier transforms are that they can be implemented using fairly arbitrary filter criteria, on a logarithmic frequency scale rather than a linear scale as in the DFT, and that time resolution can be a function of the frequency range of interest. This latter point means that we can say accurate things about high frequencies as well as low. This contrasts to the Fourier transform, which requires the analysis window width to be the same for all frequencies. This means that we must either average out lots of the interesting high-frequency time information in favor of being able to resolve low-frequency information (large window), or opt for good time resolution (small window) at the expense of low-frequency resolution, or perform multiple transforms with different sized windows to catch both time and frequency details. There are a number of fast wavelet transform techniques that allow the sub-band decomposition to be accomplished in essentially $N \log_2(N)$ time just like the FFT.

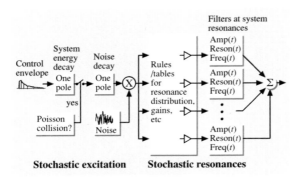

Fig. 17.30 Complete PhISEM model showing stochastic resonances

Most of classical physics can be modeled as objects interacting with each other. Lots of small objects are usually called particles. Granular synthesis involves cutting sound into grains and reassembling them by mixing [17.27]. The grains usually range in length from 5 to 100 ms. The reassembly can be systematic, but often granular synthesis involves randomized grain sizes, locations, and amplitudes. The transformed result usually bears some characteristics of the original sound, just as a mildly blended mixture of fruits still bears some attributes of the original fruits, as well as taking on new attributes due to the mixture. Granular synthesis is mostly used as a compositional and musical-effect type of signal processing, but some also take a physically motivated look at sound grains.

The physically informed stochastic event modeling (PhISEM) algorithm is based on pseudo-random overlapping and adding of parametrically synthesized sound grains [17.28]. At the heart of PhISEM algorithms are particle models, characterized by basic Newtonian equations governing the motion and collisions of point masses as can be found in any introductory physics textbook. By reducing the physical interactions of many particles to their statistical behavior, exhaustive calculation of the position, and velocity of each individual particle can be avoided. By factoring out the resonances of the system, the wavelets can be shortened to impulses or short bursts of exponentially decaying noise. The main PhISEM assumption is that the sound-producing particle collisions follow a Poisson process. Another assumption is that the system energy decays exponentially (the decay of the sound of a maraca after being shaken once, for example). Figure 17.30 shows the PhISEM algorithm block diagram.

The PhISEM maraca synthesis algorithm requires only two random number calculations, two exponential decays, and one resonant filter calculation per sample. Other musical instruments that are quite similar to the maraca include the sekere and cabasa (afuche). Outside the realm of multicultural musical instruments, there are many real-world particle systems that exhibit one or two fixed resonances like the maraca. A bag or box of candy or gum, a salt shaker, a box of wooden matches, and gravel or leaves under walking feet all fit well within this modeling technique.

In contrast to the maraca and guiro-like gourd resonator instruments, which exhibit one or two weak resonances, instruments such as the tamborine (timbrel) and sleigh bells use metal cymbals, coins, or bells suspended on a frame or stick. The interactions of the metal objects produce much more pronounced resonances than the maraca-type instruments, but the Poisson event and exponential system energy statistics are similar enough to justify the use of the PhISEM algorithm for synthesis. To implement these in PhISEM, more filters are used to model the individual partials, and at each collision, the resonant frequencies of the filters are randomly set to frequencies around the main resonances. Other sounds that can be modeled using stochastic filter resonances include bamboo wind chimes (related to a musical instrument as well in the Javanese anklung) [17.29].

17.8 Physical Modeling (The Wave Equation)

One way to model a vibrating string physically would be to build it up as a chain of coupled masses and springs, as shown in Fig. 17.31. The masses slide on frictionless guide rods to restrict their motion to transverse (at right angles to the length), as is the case in the ideal string. In fact, some synthesis projects have used mass/spring/damper systems to represent arbitrary objects [17.30].

If we force the masses to move only up and down, and restrict them to small displacements, we do not actually have to solve the acceleration of each mass as a function of the spring forces pushing them up and down. There is a simple differential equation that completely describes the motions of the ideal string:

$$\mathrm{d}^2 y / \mathrm{d}x^2 = (1/c^2)\mathrm{d}^2 y/\mathrm{d}t^2 \ . \tag{17.21}$$

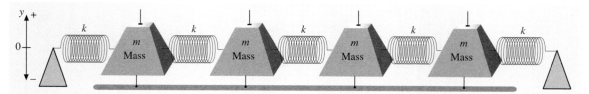

Fig. 17.31 Mass/spring string network, with frictionless *guide rods* to restrict motion to one dimension

This equation (called the wave equation) means that the acceleration (up and down) of any point on the string is equal to a constant times the curvature of the string at that point. The constant c is the speed of wave motion on the string, and is proportional to the square root of the string tension, and inversely proportional to the square root of the mass per unit length. This equation could be solved numerically, by sampling it in both time and space, and using the difference approximations for acceleration and curvature (much like we did with the mass/spring/damper system earlier). With boundary conditions (like rigid terminations at each end), we could express the solution of this equation as a Fourier series, as we did earlier in graphical form (Fig. 17.12). However, there is one more wonderfully simple solution to (17.21), given by

$$y(x,t) = y_L(t + x/c) + y_r(t - x/c) \,. \quad (17.22)$$

This equation says that any vibration of the string can be expressed as a combination of two separate traveling waves, one traveling left (y_L) and one traveling right (y_r). They move at the rate c, which is the speed of sound propagation on the string. For an ideal (no damping) string and ideally rigid boundaries at the ends, the wave reflects with an inversion at each end, and would travel back and forth indefinitely. This view of two traveling waves summing to make a displacement wave gives rise to the waveguide filter technique of modeling the vibrating string [17.31, 32]. Figure 17.32 shows a waveguide filter model of the ideal string. The two delay lines model the propagation of left- and right-going traveling waves. The conditions at the ends model the reflection of the traveling waves at the ends. The -1 in the left reflection models the reflection with inversion of a displacement wave when it hits an ideally rigid termination (like a fret on a guitar neck). The -0.99 on the right reflection models the slight loss that happens when the wave hits a termination that yields slightly (like the bridge of the guitar which couples the string motion to the body), and models all other losses the wave might experience (internal damping in the string, viscous losses as the string cuts the air, etc.) in making its round-trip path around the string.

Figure 17.33 shows the waveguide string as a digital filter block diagram. The $Z^{-P/2}$ blocks represent a delay equal to the time required for a wave to travel down the string. Thus a wave completes a round trip each P samples (down and back), which is the fundamental period of oscillation of the string, expressed in samples. Initial conditions can be injected into the string via the input $x(n)$. The output $y(n)$ would yield the right-going traveling wave component. Of course, neither of these conditions is actually physical in terms of the way a real

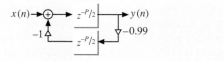

Fig. 17.33 Filter view of a waveguide string

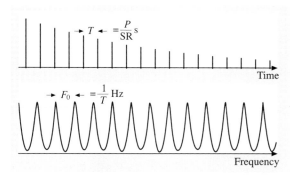

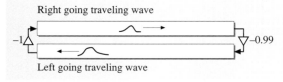

Fig. 17.32 Waveguide string modeled as two delay lines

Fig. 17.34 Impulse response and spectrum of a comb (string) filter

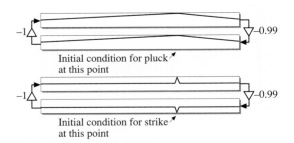

Fig. 17.35 Waveguide pluck and strike initial conditions

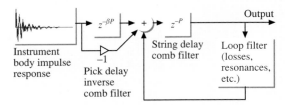

Fig. 17.37 Commuted plucked string model

string is plucked and listened to, but feeding the correct signal into x is identical to loading the delay lines with a predetermined shape.

The impulse response and spectrum of the filter shown in Fig. 17.32 is shown in Fig. 17.34. As would be expected, the impulse response is an exponentially decaying train of pulses spaced $T = P/\text{SRATE}$ seconds apart, and the spectrum is a set of harmonics spaced $F_0 = 1/T$ Hz apart. This type of filter response and spectrum is called a *comb filter*, so named because of the comb-like appearance of the time-domain impulse response, and of the frequency-domain harmonics.

The two delay lines taken together are called a *waveguide filter*. The sum of the contents of the two delay lines is the displacement of the string, and the difference of the contents of the two delay lines is the velocity of the string. If we wish to pluck the string, we simply need to load $1/2$ of the initial-shape magnitude into each of the upper and lower delay lines. If we wish to strike the string, we would load in an initial velocity by entering a positive pulse into one delay line and a negative pulse into the other (difference = initial velocity, sum = initial position = 0). These conditions are shown in Fig. 17.35.

Figure 17.36 shows a relatively complete model of a plucked string simulation system using digital filters. The inverse comb filters model the nodal effects of picking, and the output of an electrical pickup, emphasizing certain harmonics and forbidding others based on the pick (pickup) position [17.33]. Output channels for the pickup position and body radiation are provided separately. A solid-body electric guitar would have no direct radiation and only pickup output(s), while a purely acoustic guitar would have no pickup output, but possibly a family of directional filters to model body radiation in different directions [17.34].

We should note that everything in the block diagram of Fig. 17.35 is linear and time invariant (LTI). Linearity means that for input x and output y, if $x_1(t) \rightarrow y_1(t)$ and $x_1(t) \rightarrow y_2(t)$, then $x_2(t) + x_2(t) \rightarrow y_1(t) + y_2(t)$, for any functions $x_{1,2}$ (signals add and scale linearly). Time invariant means that if $x(t) \rightarrow y(t)$, then $x(t-T) \rightarrow y(t-T)$ for any $x(t)$ and any T (the system doesn't change with time). Given that the LTI condition is satisfied, we can then happily (legally) commute (swap the order) of any of the blocks. For example, we could put the pick position filter at the end, or move the body radiation filter to the beginning. Since the body radiation filter is just the impulse response of the instrument body, we can actually record the impulse response with the strings damped, and use that as the input to the string model. This is called *commuted synthesis* [17.35]. Figure 17.37 shows a commuted synthesis plucked string model.

Adding a model of bowing friction allows the string model to be used for the violin and other bowed strings. This focused nonlinearity is what is responsible for turning the steady linear motion of a bow into an oscillation of the string [17.36, 37]. The bow sticks to the string

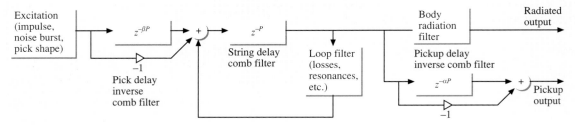

Fig. 17.36 Fairly complete digital filter simulation of the plucked string system

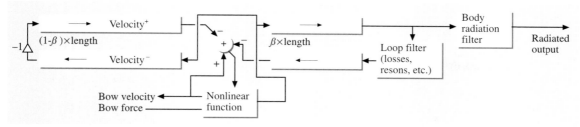

Fig. 17.38 Bowed string model

for a while, pulling it along, then the forces become too great and the string breaks away, flying back toward rest position. This process repeats, yielding a periodic oscillation. Figure 17.38 shows a simple bowed string model, in which string velocity is compared to bow velocity, then put through a nonlinear friction function controlled by bow force. The output of the nonlinear function is the velocity input back into the string.

In mathematically describing the air within a cylindrical acoustic tube (like a trombone slide, clarinet bore, or human vocal tract), the defining equation is:

$$d^2 P / dx^2 = (1/c^2)(d^2 P / dt^2) \qquad (17.23)$$

which we note has exactly the same form as (17.21), except that the displacement y is replaced by the pressure P. A very important paper in the history of physical modeling [17.36] noted that many musical instruments can be characterized as a linear resonator, modeled by filters such as all-pole resonators or waveguides, and a single nonlinear oscillator like the reed of the clarinet, the lips of the brass player, the jet of the flute, or the bow–string friction of the violin. Since the wave equation says that we can model a simple tube as a pair of bidirectional delay lines (waveguides), then we can build models using this simple structure. If we would like to do something interesting with a tube, we could use it to build a flute or clarinet. Our simple clarinet model might look like the block diagram shown in Fig. 17.39.

To model the reed, we assume that the mass of the reed is so small that the only thing that must be considered is the instantaneous force on the reed (spring). The pressure inside the bore P_b is the calculated pressure in our waveguide model, the mouth pressure P_m is an external control parameter representing the breath pressure inside the mouth of the player (Fig. 17.40). The net force acting on the reed/spring can be calculated as the difference between the internal and external pressures, multiplied by the area of the reed (pressure is force per unit area). This can be used to calculate a reed opening position from the spring constant of the reed. From the reed opening, we can compute the amount of pressure that is allowed to leak into the bore from the player's mouth. If bore pressure is much greater than mouth pressure, the reed opens far. If the mouth pressure is much greater than the bore pressure, the reed slams shut. These two extreme conditions represent an asym-

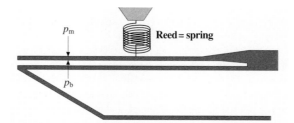

Fig. 17.40 Reed model

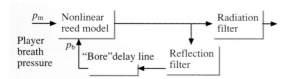

Fig. 17.39 Simple clarinet model

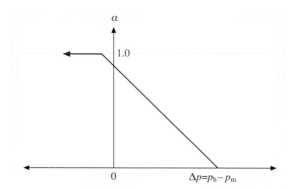

Fig. 17.41 Reed reflection table

metric nonlinearity in the reed response. Even a grossly simplified model of this nonlinear spring action results in a pretty good model of a clarinet [17.37]. Figure 17.41 shows a plot of a simple reed reflection function (as seen from within the bore) as a function of differential pressure. Once this nonlinear signal-dependent reflection coefficient is calculated (or looked up in a table), the right-going pressure injected into the bore can be calculated as $P_b^+ = \alpha P_b^- + (1-\alpha) P_m$.

The clarinet is open at the bell end, and essentially closed at the reed end. This results in a reflection with inversion at the bell and a reflection without inversion (plus any added pressure from the mouth through the reed opening) at the reed end. These odd boundary conditions cause odd harmonics to dominate in the clarinet spectrum.

We noted that the ideal string equation and the ideal acoustic tube equation are essentially identical. Just as there are many refinements possible to the plucked string model to make it more realistic, there are many possible improvements for the clarinet model. Replacing the simple reed model with a variable mass/spring/damper allows the modeling of a lip reed as is found in brass instruments. Replacing the reed model with an air-jet model allows the modeling of flute and recorder-like instruments. With all wind or friction (bowed) excited resonant systems, adding a little noise in the reed/jet/bow region adds greatly to the quality (and behavior) of the synthesized sound.

In an ideal string or membrane, the only restoring force is assumed to be the tension under which they are stretched. We can further refine solid systems such as strings and membranes to model more-rigid objects, such as bars and plates, by noting that the more-rigid objects exhibit internal restoring forces due to their stiffness. We know that if we bend a stiff string, it wants to return back to straightness even when there is no tension on the string. Cloth string or thread has almost no stiffness. Nylon and gut strings have some stiffness, but not as much as steel strings. Larger-diameter strings have more stiffness than thin ones. In the musical world, piano strings exhibit the most stiffness. Stiffness results in the restoring force being higher (thus the speed of sound propagation as well) for high frequencies than for low. So the traveling wave solution is still true in stiff systems, but a frequency-dependent propagation speed is needed

$$y(x, t) = y_l\left[t + x/c(f)\right] + y_r\left[t - x/c(f)\right]. \quad (17.24)$$

and the waveguide filter must be modified to simulate frequency-dependent delay, as shown in Fig. 17.42.

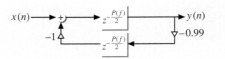

Fig. 17.42 Stiffness-modified waveguide string filter

For basic stiff strings, a function that predicts the frequencies of the partials has this form:

$$f_n = n\left[f_0\left(1 + Bn^2\right)\right], \quad (17.25)$$

where B is a number slightly greater than 0, equal to zero for perfect harmonicity (no stiffness), and increasing for increasing stiffness. Typical values of B are 0.00001 for guitar strings, and 0.004 or so for piano strings. This means that $P(f)$ should modify P by the inverse of the $(1 + Bn^2)^{1/2}$ factor.

Unfortunately, implementing the $Z^{-P(f)/2}$ frequency-dependent delay function is not simple, especially for arbitrary functions of frequency. One way to implement the $P(f)$ function is by replacing each of the Z^{-1} with a first-order all-pass (phase) filter, as shown in Fig. 17.43 [17.33]. The first-order all-pass filter has one pole and one zero, controlled by the same coefficient. The all-pass implements a frequency-dependent phase delay, but exhibits a gain of 1.0 for all frequencies. The coefficient α can take on values between -1.0 and 1.0. For $\alpha = 0$, the filter behaves as a standard unit delay. For $\alpha > 0.0$, the filter exhibits delays longer than one sample, increasingly long for higher frequencies. For $\alpha < 0.0$ the filter exhibits delays shorter than one sample, decreasingly so for high frequencies.

It is much less efficient to implement a chain of all-pass filters than a simple delay line. But for weak stiffness it is possible that only a few all-pass sections will provide a good frequency-dependent delay. Another option is to implement a higher-order all-pass filter, designed to give the correct stretching of the upper frequencies, added to simple delay lines to give the correct longest bulk delay required.

For very stiff systems such as rigid bars, a single waveguide with all-pass filters is not adequate to give enough delay, or far too inefficient to calculate. A technique called *banded waveguides* employs

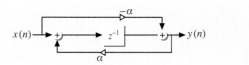

Fig. 17.43 First-order all-pass filter

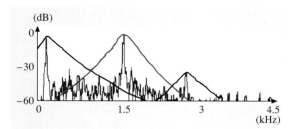

Fig. 17.44 Banded decomposition of a struck-bar spectrum

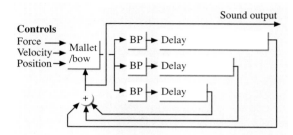

Fig. 17.45 Banded waveguide model

sampling in time, space, and frequency to model stiff one-dimensional systems [17.38]. This can be viewed as a hybrid of modal and waveguide synthesis, in that each waveguide models the speed of sound in the region around each significant mode of the system. As an example, Fig. 17.44 shows the spectrum of a struck marimba bar, with additional bandpass filters superimposed on the spectrum, centered at the three main modes. In the banded waveguide technique, each mode is modeled by a bandpass filter, plus a delay line to impose the correct round-trip delay, as shown in Fig. 17.45.

17.9 Music Description and Control

The musical instrument digital interface (MIDI) standard, adopted in 1984, revolutionized electronic music, and also profoundly affected the computer industry [17.39]. A simple two-wire serial electrical connection standard allows interconnection of musical devices and computers over cable distances of up to 15 m (longer over networks and extensions to the basic MIDI standard). The MIDI software protocol is best described as musical keyboard gestural, meaning that the messages carried over MIDI are the gestures that a pianist or organist uses to control a traditional keyboard instrument. There is no time information contained in the basic MIDI messages; they are intended to take effect as soon as they come over the wire. As such it is a real-time gestural protocol, and can be adapted for real-time non-musical sound synthesis applications. There are complaints about MIDI, however, mostly related to limited control bandwidth (approximately 1000–1500 continuous messages per second maximum), and the keyboard-centric bias of the messages.

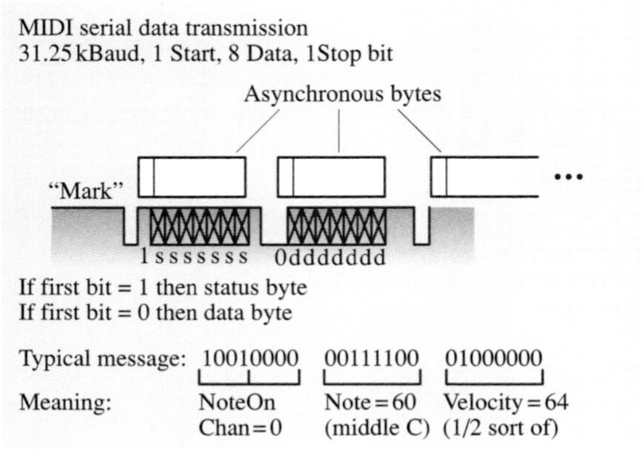

Basic MIDI message types include NoteOn and NoteOff, sustain pedal up and down, amount of modulation (vibrato), pitch bend, key pressure (also called AfterTouch), breath pressure, volume, pan, balance, reverberation amount, and others. NoteOn and NoteOff messages carry a note number corresponding to a particular piano key, and a velocity corresponding to how hard that key is hit. Figure 17.46 shows the software serial format of MIDI, and gives an example of a NoteOn message.

Another MIDI message is program change, which is used to select the particular sound being controlled in the synthesizer. MIDI provides for 16 channels, and the channel number is encoded into each message. Each channel is capable of one type of sound, but possibly many simultaneous *voices* (a voice is an individual sound) of that same sound. Channels allow instruments and devices to all listen on the same network, and choose to respond to the messages sent on particular channels.

General MIDI (1991), and the standard MIDI file specifications serve to extend MIDI and made it even more popular [17.40]. General MIDI helps to assure

Fig. 17.46 MIDI software transmission protocol

the performance of MIDI on different synthesizers, by specifying that a particular program (the algorithm for producing the sound of an instrument) number must call up a program the approximates the same instrument sound on all general MIDI-compliant synthesizers. There are 128 such defined instrument sounds available on MIDI channels 1–9 and 11–16. For example, MIDI program 1 is grand piano, and 57 is trumpet. Some of the basic instrument sounds are sound effects. Program #125 is a telephone ring, #126 is helicopter, #127 is applause, and #128 is a gunshot sound. On general MIDI channel 10, each note is mapped to a different percussion sound. For example, the bass drum is note number 35, and the cowbell is note number 56.

The MIDI file formats provide a means for the standardized exchange of musical/sonic information. A MIDI level 0 file carries the basic information that would be carried over a MIDI serial connection, which is the basic bytes in order, with added time stamps. Time stamps allow a very simple program to play back MIDI files. A MIDI level 1 file is more suited to manipulation by a notation program or MIDI sequencer (a form of multi-track recorder program that records and manipulates MIDI events). Data is arranged by *tracks*, (each track transmits on only one MIDI channel) which are the individual instruments in the virtual synthesized orchestra. Meta messages allow for information which is not actually required for a real-time MIDI playback, but might carry information related to score markings, lyrics, composer and copyright information, etc.

Recognizing the limitations of fixed PCM samples in music synthesizers, the downloadable sounds (DLS) specification was added to MIDI in 1999. This provides a means to load PCM into a synthesizer, then use MIDI commands to play it back. Essentially any instrument/percussion sound can be replaced with arbitrary PCM, making it possible to control large amounts of recorded sounds. The emergence of software synthesizers, on PC platforms has made the use of DLS more feasible than in earlier days when the samples were contained in a static ROM within a hardware synthesizer.

The simplicity and ubiquity of MIDI, with thousands of supporting devices and software systems, leads many to use it for all sorts of projects and systems. While not being directly compatible with popular music synthesizers, using MIDI for the control of synthesis still allows the use of the large variety of MIDI control signal generators and processors, all of which emit standard and user-programmable MIDI messages. Examples include MIDI sequencers and other programs for general-purpose computers, as well as breath controllers, drum controllers, and fader boxes with rows of sliders and buttons.

Sadly, one thing that is still missing from MIDI at the time of writing is the ability to download arbitrary synthesis algorithms. Beyond MIDI, there are a number of emerging standards for sound and multimedia control. As mentioned in the last section, one of the things MIDI lacks is high-bandwidth message capability for handling many streams of gestural/model control data. Another complaint about MIDI relates to the small word sizes, such as semitone quantized pitches (regular musical notes), and only 128 levels of velocity, volume, pan, and other controllers. MIDI does support some extended precision messages, allowing 14 bits rather than 7. Other specifications have proposed floating-point representations, or at least larger integer word sizes. A number of new specifications and systems have arisen recently.

One system for describing sounds themselves is called the sound description interchange format (SDIF 1999) [17.41], originally called the spectral description interchange file (SDIF) format. It is largely intended to encapsulate the parameters arising from spectral analysis, such as spectral modeling sinusoidal/noise tracks, or source/filter model parameters using LPC.

Related to SDIF, but on the control side, is open sound control (OSC, 2002) [17.41]. Open sound control allows for networked, extended-precision, object-oriented control of synthesis and processing algorithms over high-bandwidth transmission control protocol/internet protocol (TCP/IP), FireWire, and other high-bandwidth protocols.

As part of the moving-picture experts group 4 (MPEG4) standard, the structured audio orchestra language (SAOL, 1999) allows for parametric descriptions of sound algorithms, controls, effects processing, mixing, and essentially all layers of the audio synthesis and processing chain [17.42]. One fundamental notion of structured audio is that the more understanding we have about how a sound is made, the more flexible we are in compressing, manipulating, and controlling it.

17.10 Composition

The history of computer languages for music dates back well into the 1950s. Beginning with Max Mathew's Music I, many features of these languages have remained constant, such as the notion of unit generators as discussed before, and the idea of instruments within an orchestra, controlled by a score. Early popular languages include Music V (a direct descendent of Music I) [17.5], Csound, CMix, and CMusic (written in C) [17.43]. More recent languages developed in the last decade include SuperCollider [17.44], Jsyn (in Java) [17.45], STK in C^{++} [17.46], Aura [17.47], Nyquist [17.48], and ChucK [17.49].

One class of languages that has advanced education and participation in computer music are graphical systems such as MAX/MSP [17.50] and Pure Data (PD) [17.51]. These software systems allow novice (and expert) users to construct patches by dragging unit generator icons around on the desktop, connecting them with virtual patch cords (Fig. 17.47).

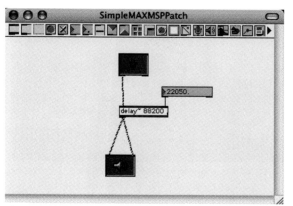

Fig. 17.47 A simple MAX/MSP patch connects a microphone through a 1/2 s (22050 samples at 44.1 k sample rate) delay to the speaker output

17.11 Controllers and Performance Systems

Music performance is structured real-time control of sound, to achieve an aesthetic goal (hopefully for both the performers and audience). So one obvious application area for real-time computer sound synthesis and processing is in creating new forms of musical expression. Exciting developments in the last few years have emerged in the area of new controllers and systems for using parametric sound in real-time music performance, or in interactive installation art.

Figure 17.48 shows three recent real-time music performance systems, all using parametric digital sound synthesis and a variety of sensors mounted on the instruments and players. The author's DigitalDoo is a sensor/speaker enhanced/augmented didgeridoo. Dan Trueman's bowed sensor speaker array (BoSSA) resulted from work in modeling, refining, and redefining the violin player's interface [17.52]. Curtis Bahn's 'r!g' consists of a sensor enhanced upright (stick) bass, pedal boards, and other controllers. All three of these new instruments use spherical speakers to allow flexible control of the sound radiation into the performance space [17.53]. These instruments have been used with computers to control percussion sounds, animal noises, string models, flute models, and a large va-

Fig. 17.48 Cook's DigitalDoo, Trueman's BoSSA, Bahn's 'r!g'

Fig. 17.49 Gigapop networked concert between Princeton (shown on screen to *left*) and Montreal (players on stage). The right screen shows real-time graphics generated from gestural data captured from sensors on the instruments

riety of other sample-based and parametric synthesis algorithms.

Various conducting systems, including Max Mathew's radio baton (Stanford's Center for Computer Research in Music and Acoustics) [17.54], and Theresa Marrin Nakra's conducting jacket (MIT Media Lab) [17.55] allow novices and experts to conduct music performed by the computer. Also, the growth and development of computer music has helped to advance greatly the technological capabilities of the familiar karaoke industry. Some modern karaoke systems now can deduce the song that is being sung and retrieve the appropriate accompaniment within just a few notes. Also, these systems can automatically correct the pitch and vocal qualities of a bad singer, making them sound more expert and professional. The next section will describe some of the algorithms being applied now in karaoke and other automatic accompaniment systems.

Some researchers and performers feel that there is significant potential in networked computer music performances and interactions. Projects range from MIT's Brain Opera [17.56], which allowed observers to participate and contribute content to the performance via the web, through the Auricle website, which is a kind of audio analysis/synthesis enhanced chat room [17.57]. Stanford's SoundWire Project [17.58] and Princeton's GigaPop Project [17.59] (Fig. 17.49) are aimed at synchronized multi-site live performance through high-bandwidth low-latency streaming of uncompressed audio and video.

17.12 Music Understanding and Modeling by Computer

Over the years, new research and applications have arisen in the areas of content-based audio analysis, retrieval, and modeling. Also called music information retrieval (MIR), machine listening, and other terms, the main tasks involve segmenting, separating, classifying, and modeling the textures and structures present in pieces of music or audio.

Audio segmentation is the process of breaking up an audio stream into sections that are perceptually different from adjacent sections. The audio texture within a given segment is relatively stable. Examples of segment boundaries could be a transition from background sound texture to a human beginning to speak over that background. Another segment boundary might occur when the scene changes, such as leaving an office building lobby and going outside onto a busy street. In movie production, the audio scene often changes just prior to the visual scene, allowing a noncausal mode of detecting a visual scene change.

Segmentation [17.60] can be accomplished in two primary ways: blind segmentation based on sudden changes in extracted audio features, and classification-based segmentation based on comparing audio features to a set of trained target feature values. The blind method works well and is preferred when the segment textures are varied and unpredictable, but requires the setting of thresholds to yield the best results. The classification-based method works well on a corpus of pre-labeled data, such as speech/music, musical genres, indoor/outdoor scenes, etc. databases. Both methods require the extraction of audio features.

Audio feature extraction is the process of computing a compact numerical representation of a sound segment. A variety of audio features exist, and have been used in systems for speech recognition, music/speech discrimination, musical genre (rock, pop, country, classical, jazz, etc.) labeling, and other audio classification tasks. Most features are extracted from short moving windows (5–100 ms in length, moving along at a rate of 5–20 windows per second) by using the short-time Fourier transform, wavelets, or compressed data coefficients (such as MP3). Each feature described below can be computed at different time resolutions, and the value of each feature, along with the mean and variance of the features can be used as features themselves.

- One common feature for segmentation is the gross power in each window. If the audio stream suddenly gets louder or softer, then there is a high likelihood that something different is occurring. In speech recognition and some other tasks, however, we would like the classification to be loudness invariant (over some threshold used to determine if anyone is speaking).
- Another feature is the spectral centroid, which is closely related to the brightness of the sound, or the relative amounts of high- and low-frequency energy. The spectral roll-off is another important feature that captures more information about the brightness of an audio signal.
- Spectral flux is the amount of frame-to-frame variance in the spectral shape. A steady sound or texture will exhibit little spectral flux, while a modulating sound will exhibit more flux.
- One popular set of features used for speech and speaker recognition are mel-frequency cepstral coefficients, which are a compact (between 4 and 10 numbers) representation of spectral shape.
- For multi-time-scale analysis systems, the low-energy feature is often used. This feature is defined as the percentage of small analysis windows that contain less power than the average over the larger window that includes the smaller windows. This is a coarser-time-scale version of flux, but computed only for energy.
- One other commonly used feature is zero-crossing rate, which is a simple measure of high-frequency energy.
- For music query/recognition, features such as the parametric pitch histogram, and beat/periodicity histogram can be calculated and used.

Selection of the correct feature set for a given task has proven to be an important part of building successful systems for machine audio understanding. For a fixed corpus, computing many features (40 dimensions or more), then using principal-components analysis (reduction of dimensionality through regression) has proven successful for reducing the dimensionality of the feature/search space. As an example, mapping the first three principal components onto color (red, green, blue) allows the automatic coloration of sound plots, as shown in Fig. 17.50. The three *timbregrams* on the left are speech, and the three on the right are classical music. The top two right timbregrams are orchestral recordings, while the lower right one is opera singing (hence the color similarity to speech).

One holy grail in computer music research is automatic transcription, where in the ideal case a computer could generate a perfect musical score by analyzing only a recording of the music. This problem has been around for many decades, and has generated significant funding for computer music researchers. While research still continues on this difficult and unsolved problem, the side effects of the research have given us a large number of developments in signal

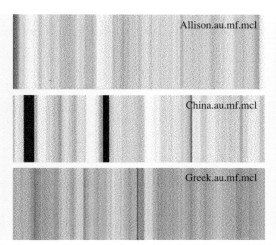
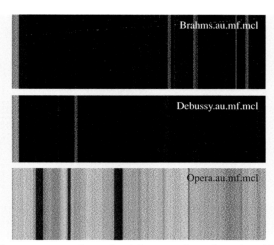

Fig. 17.50 *Timbregrams* of speech (*left*) and music (*right*). The top two sounds on the right are orchestral music, while the lower right is opera

processing, analysis/synthesis algorithms, and human perception and cognition.

One other cognition-related research area in computer music is musical style modeling, also called automatic composition. One of the leading researchers in this area is David Cope, whose experiments in musical intelligence (EMI) projects have been developed over many years now. Through musical rules and data entered by Cope in the LISP computer language, the simple analytic recombinant algorithm (SARA) program has generated imitative music in styles as varied as those of Palestrina, Bach, Brahms, Chopin, Mozart, Prokofiev, Stravinsky, Gershwin, and Scott Joplin [17.61].

Another interesting project in this area is The Brain, created by Tom Hadju and Andy Milburn of the production company "tomandandy". The Brain is partially responsible for a large amount of commercial and movie music created over the last few years, including music for *The Mothman Prophesies*, and *Mean Creek*.

17.13 Conclusions, and the Future

With the advent of cheap and powerful computers, especially laptops, the last few years have greatly democratized the tools and techniques of computer music. Many free or inexpensive sound-editing tools now include features and algorithms that were the sole domain of academic researchers only ten years ago. Further, the World Wide Web has made computer tools and music available to more people. Computer-augmented performers and performance spaces have become more common, and more commonly accepted by the public. It seems certain that the future of computer music is likely to include more developments in the areas of modeling and understanding of human perception, preference, and creativity.

References

17.1 K. Pohlmann: *Principles of Digital Audio* (McGraw Hill, New York 2000)

17.2 J.O. Smith, P. Gossett: A flexible sampling-rate conversion method. In: *Proc. Int. Conf. Acoustics Speech and Signal Processing*, Vol. 2 (IEEE, Piscataway 1984) pp. 19.4.1–19.4.2

17.3 T. Dutoit: *An Introduction to Text-To-Speech Synthesis* (Kluwer Academic, Dordrecht 1996)

17.4 R. Bristow-Johnson: *Wavetable synthesis 101: a fundamental perspective AES 101 Conference* (AES, New York 1996), Music-DSPSource Archive, (http://www.musicdsp.org)

17.5 M.V. Mathews: *The Technology of Computer Music* (MIT Press, Cambridge 1969)

17.6 R. Bracewell: *The Fourier Transform and It's Applications* (McGraw Hill, New York 1999)

17.7 X. Serra, J.O. Smith: Spectral modeling synthesis: Sound analysis/synthesis based on a deterministic plus stochastic decomposition, Comp. Music J. **14**(4), 1–2 (1990)

17.8 R.J. McAulay, T. Quatieri: Speech analysis/synthesis based on a sinusoidal representation, IEEE Trans. ASSP **34**, 744–754 (1986)

17.9 T. Verma, T. Meng: *An analysis/synthesis tool for transient signals that allows a flexible sines+transients+noise model for audio 1998 IEEE ICASSP-98* (IEEE, Piscataway 1998)

17.10 A. Moorer: The use of the phase vocoder in computer music applications, J. Audio Eng. Soc. **26**(1/2), 42–45 (1978)

17.11 M. Dolson: The phase vocoder: A tutorial, CMJ **10**(4), 14–27 (1986)

17.12 P. Cook: *Real Sound Synthesis for Interactive Applications* (AK Peters, Wellesley 2002)

17.13 P. Morse: *Vibration and Sound* (for the Acoustical Society of America, New York 1986)

17.14 K. Steiglitz: *A Digital Signal Processing Primer* (Addison Wesley, Boston 1996)

17.15 J.M. Adrien: The missing link: Modal synthesis Chap. 8. In: *Representations of Musical Signals*, ed. by De Poli, Picialli, Roads (MIT Press, Cambridge 1991) pp. 269–297,)

17.16 H. Dudley: *The Vocoder* (Bell Laboratories Record, Mirray Hill 1939), Reprinted in IEEE Trans. Acoust. Speech Signal Process. ASSP-29(3):347-351 (1981)

17.17 B. Atal: Speech analysis and synthesis by linear prediction of the speech wave, J. Acoust. Soc. Am. **A47**(65), 65 (1970)

17.18 A. Moorer: The use of linear prediction of speech in computer music applications, J. Audio Eng. Soc. **27**(3), 134–140 (1979)

17.19 D. Klatt: Software for a cascade/parallel formant synthesizer, J. Acoust. Soc. Am. **67**(3), 971–995 (1980)

17.20 J.L. Kelly Jr., C.C. Lochbaum: *Speech Synthesis, in Proceedings of the Fourth ICA* (Paper G42, Copenhagen 1962)

17.21 J.M. Chowning: The synthesis of complex audio spectra by means of frequency modulation, J. Audio Eng. Soc. **21**(7), 526–534 (1973)

17.22 J. Chowning: The synthesis of complex audio spectra by means of frequency modulation, Comp. Music J. **1**(2) (1977), reprinted

17.23 J. Chowning: The synthesis of complex audio spectra by means of frequency modulation. In: *Foundations of Computer Music*, ed. by C. Roads, J. Strawn (MIT Press, Cambridge 1985), reprinted

17.24 J. Chowning: Frequency modulation synthesis of the singing voice. In: *Current Directions in Computer Music Research*, ed. by M.V. Mathews, J.R. Pierce (MIT Press, Cambridge 1980)

17.25 X. Rodet: Time-domain formant-wave-function synthesis, Comp. Music J. **8**(3), 9–14 (1984)

17.26 I. Daubechies: Orthonormal bases of compactly supported wavelets, Commun. Pure Appl. Math. **41**, 909–996 (1988)

17.27 C. Roads: Asynchronous granular synthesis. In: *Representations of Musical Signals*, ed. by G. De Poli, A. Piccialli, C. Roads (MIT Press, Cambridge 1991) pp. 143–185

17.28 P. Cook: *Physically Informed Sonic Modeling (PhISM): Percussion Instruments, in Proceedings of the ICMC* (International Computer Music Association, San Francisco 1996)

17.29 P. Cook: Physically informed sonic modeling (PhISM): Synthesis of percussive sounds, Comp. Music J. **21**, 3 (1997)

17.30 C. Cadoz, A. Luciani, J.L. Florens: CORDIS-ANIMA: A modeling and simulation system for sound image synthesis. The general formalization, Comp. Music J. **17**(1), 23–29 (1993)

17.31 K. Karplus, Strong: Digital synthesis of plucked-string and drum timbres, Comp. Music J. **7**(2), 43–55 (1983)

17.32 J.O. Smith: Acoustic modeling using digital waveguides. In: *Musical Signal Processing*, ed. by Roads et.al. (Swets, Zeitlinger, Netherlands 1997)

17.33 D.A. Jaffe, Smith: Extensions of the Karplus-Strong plucked string algorithm, Comp. Music J. **7**(2), 56–69 (1983)

17.34 P.R. Cook, D. Trueman: Spherical radiation from stringed instruments: Measured, modeled, and reproduced, J. Catgut Acoust. Soc. **1**, 1 (1999)

17.35 J.O. Smith, S.A. Van Duyne: Overview of the commuted piano synthesis technique. In: *Proc. IEEE Workshop on Applications of Signal Processing to Audio and Acoustics*, ed. by E. Wenzel (IEEE, New York 1995)

17.36 M.E. McIntyre, R.T. Schumacher, J. Woodhouse: On the oscillations of musical instruments, J. Acoust. Soc. Am. **74**(5), 1325–1345 (1983)

17.37 J.O. Smith: Efficient simulation of the reed-bore and bow-string mechanisms. In: *Proceedings of the International Computer Music Conference* (International Computer Music Association, San Francisco 1986) pp. 275–280

17.38 G. Essl, P. Cook: *Banded waveguides: Towards physical modeling of bar percussion instruments, in Proc. 1999 Int. Computer Music Conf., Beijing, China* (International Computer Music Association, San Francisco 1999)

17.39 MMA: *The Complete MIDI 1.0 Detailed Specification* (MIDI Manufacturers Association, La Habra 1996)

17.40 S. Jungleib: *General MIDI, A-R Editions* (A-R Editions, Middleton 1995)

17.41 M. Wright, A. Chaudhary, A. Freed, D. Wessel, X. Rodet, D. Virolle, R. Woehrmann, X. Serra: *New Applications for the Sound Description Interchange Format, Proc. Of the ICMC 1998* (International Computer Music Association, San Francisco 1998)

17.42 M. Wright, A. Freed: Open sound control: A new protocol for communicating with sound synthesizers. In: *Proceedings of the International Computer Music Conference, 1997* (International Computer Music Association, San Francisco 1997),

17.43 S.T. Pope: Machine Tongues XV: Three packages for software sound synthesis, Comp. Music J. **17**(2), 23–54 (1993), Note: these are cmix, csound, cmusic

17.44 J. McCartney: SuperCollider: A new real-time synthesis language. In: *Proc. Int. Computer Music Conference* (International Computer Music Association, San Francisco 1996) pp. 257–258

17.45 P. Burk: JSyn - A real-time synthesis API for java. In: *Proc. Int. Computer Music Conference* (International Computer Music Association, San Francisco 1998) pp. 252–255

17.46 Cook, G. Scavone: The synthesis toolkit (STK). In: *Proceedings of the International Computer Music Conference* (International Computer Music Association, San Francisco 1999) pp. 164–166

17.47 R.B. Dannenberg, E. Brandt: A flexible realtime software synthesis system. In: *Proceedings of the International Computer Music Conference* (International Computer Music Association, San Francisco 1996) pp. 270–273

17.48 R.B Dannenberg: Machine Tongues XIX: Nyquist, a language for composition and sound synthesis, Comp. Music J. **21**(3), 50–60 (1997)

17.49 G. Wang, P.R. Cook: ChucK: A concurrent. In: *On-the-fly Audio Programming Language, in Proceedings of International Computer Music Conference* (International Computer Music Association, San Francisco 2003) pp. 219–226

17.50 M. Puckette: Combining event and signal processing in the MAX graphical programming environment, Comp. Music J. **15**(3), 68–77 (1991)

17.51 M. Puckett: Pure data. In: *Proceedings of International Computer Music Conference* (International Computer Music Association, San Francisco 1996) pp. 269–272

17.52 D. Trueman, P. Cook: BoSSA: The deconstructed violin reconstructed, J. New Music Res. **29**(2), 120–130 (2000)

17.53 D. Trueman, C. Bahn, P. Cook: Alternative voices for electronic sound. In: *Spherical speakers and sensor-speaker arrays (SenSAs), International Computer Music Conference, Berlin, Aug. 2000* (International Computer Music Association, San Francisco 2000)

17.54 M.V. Mathews: The radio baton and conductor program, or: Pitch, the most important and least expressive part of music, Comp. Music J. **15**(4), 37–46 (1991)

17.55 T. Marin Nakra: *Inside the Conductor's Jacket: Analysis, Interpretation and Musical Synthesis of Expressive Gesture*, Ph.D. Thesis (MIT Media Lab, Cambridge 2000)

17.56 J.A. Paradiso: The brain opera technology: New instruments and gestural sensors for musical interaction and performance, J. New Music Res. **28**, 2 (1999)

17.57 J. Freeman, S. Ramakrishnan, K. Varnik, M. Neuhaus, P. Burk, D. Birchfeld: *Adaptive High-level Classification of Vocal Gestures Within a Networked Sound Instrument, Proc. Of the International Computer Music Conference* (International Computer Music Association, San Francisco 2004)

17.58 C. Chafe, S. Wilson, R. Leistikow, D. Chisholm, G. Scavone: A simplified approach to high quality music and sound over IP. In: *Proceedings of the Conference on Digital Audio Effects (DAFX)*, ed. by D. Rocchesso (University of Verona, Italy 2000)

17.59 A. Kapur, G. Wang, P. Davidson, P.R. Cook, D. Trueman, T.H. Park, M. Bhargava: *The Gigapop Ritual: A Live Networked Performance Piece for Two Electronic Dholaks, Digital Spoon, DigitalDoo, 6 String Electric Violin, Rbow, Sitar, Tabla, and Bass Guitar, New Interfaces for Musical Expression (NIME)* (McGill University, Montreal 2003)

17.60 G. Tzanetakis, P. Cook: Multi-feature audio segmentation for browsing and annotation. In: *Proc. IEEE Workshop on Applications of Signal Processing to Audio and Acoustics (WASPAA)* (IEEE, Piscataway 1999)

17.61 D. Cope: *Experiments in Musical Intelligence*, Vol. 12 (A-R Editions, Madison 1996)

18. Audio and Electroacoustics

This chapter surveys devices and systems associated with audio and electroacoustics: the acquisition, transmission, storage, and reproduction of audio. The chapter provides an historical overview of the field since before the days of Edison and Bell to the present day, and analyzes performance of audio transducers, components and systems from basic psychoacoustic principles, to arrive at an assessment of the perceptual performance of such elements and an indication of possible directions for future progress.

The first, introductory section is an overall historical review of audio reproduction and spatial audio to establish the context of events. The next section surveys relevant psychoacoustic principles, including performance related to frequency response, amplitude, timing, and spatial acuity. Section 3 examines common audio specifications, with reference to the psychoacoustic limitations discussed in Sect. 2. The specifications include frequency and phase response, distortion, noise, dynamic range and speed accuracy. Section 4 examines some of the common audio components in light of the psychoacoustics and specifications established in the preceding sections. The components in question include microphones, loudspeakers, record players, amplifiers, magnetic recorders, radio, and optical media. Section 5 is concerned with digital audio, including the basics of sampling, digital signal processing, and audio coding. Section 6 is devoted to an examination of complete audio systems and their ability to reproduce an arbitrary acoustic environment. The specific systems include monaural, stereo, binaural, Ambisonics, and 5.1-channel surround sound. The final section provides an overall appraisal of the current state of audio and electroacoustics, and speculates on possible future directions for research and development.

18.1	**Historical Review**	744
	18.1.1 Spatial Audio History	746
18.2	**The Psychoacoustics of Audio and Electroacoustics**	747
	18.2.1 Frequency Response	747
	18.2.2 Amplitude (Loudness)	748
	18.2.3 Timing	749
	18.2.4 Spatial Acuity	750
18.3	**Audio Specifications**	751
	18.3.1 Bandwidth	752
	18.3.2 Amplitude Response Variation	753
	18.3.3 Phase Response	753
	18.3.4 Harmonic Distortion	754
	18.3.5 Intermodulation Distortion	755
	18.3.6 Speed Accuracy	755
	18.3.7 Noise	756
	18.3.8 Dynamic Range	756
18.4	**Audio Components**	757
	18.4.1 Microphones	757
	18.4.2 Records and Phonograph Cartridges	761
	18.4.3 Loudspeakers	763
	18.4.4 Amplifiers	766
	18.4.5 Magnetic and Optical Media	767
	18.4.6 Radio	768
18.5	**Digital Audio**	768
	18.5.1 Digital Signal Processing	770
	18.5.2 Audio Coding	771
18.6	**Complete Audio Systems**	775
	18.6.1 Monaural	776
	18.6.2 Stereo	776
	18.6.3 Binaural	777
	18.6.4 Ambisonics	777
	18.6.5 5.1-Channel Surround	777
18.7	**Appraisal and Speculation**	778
	References	778

Considering that human records extend back to cave drawings thousands of years old [18.1], it is notable that the technology to record, transmit, and reproduce sound has only existed for the last century or so. One can only wonder at the effect it might have had on the course of history, not to mention the effect it may yet have [18.2].

Sound is hardly the easiest quantity to work with. As a vibration in air, it is invisible, dynamic, and three-dimensional (3-D), not counting the dimensions of time and frequency. Human interaction with a sound field is often dynamic: natural movements of the head and body alter the perceived sounds in a way that is difficult to predict or reproduce. Despite over a century of progress, the transducers available for recording and reproducing sound, microphones and loudspeakers, are still limited to being point-in-space devices, and are not necessarily well suited to interacting with a 3-D physical phenomenon. And yet, human beings make do with just two ears; so how hard can it be to capture and reproduce all the characteristics and nuances of what the ear perceives in nature?

The intent of this chapter is to review and evaluate the performance of audio systems based on the known performance of the human auditory system. Starting with a brief historical overview, the salient aspects of human auditory performance will be described. These in turn will be used to derive a set of general audio specifications that may be expected to apply to a broad range of audio systems. Then, some of the more common audio components will be examined in light of these specifications. Finally, the performance of complete systems will be considered, in the hope of gaining some overall perspective of progress to date and possibly fertile directions for further investigation.

18.1 Historical Review

The first golden age of audio research arguably started in the late 1800s with the inventions of the telephone by A. G. Bell and the phonograph by T. A. Edison [18.3], and ended in 1982 with the introduction of digital audio as a practical reality as embodied in the audio compact disc (CD). The inventions that inaugurated this age established that an audio channel could indeed be transmitted and recorded, as a continuous mechanical, magnetic, optical, or electrical analog of the sound vibrations, but brought with them a raft of problems, including inadequate recording sensitivity and playback volume, noise, distortion, speed irregularities such as wow and flutter, nonlinear and limited frequency response, and media wear. Although steady incremental progress was made on these problems in the decades that followed, they were all largely resolved once and for all with digital audio. Efforts since then have focused on reduction of the required data rate (audio coding) and on extending the realm of audio from single isolated channels to full three-dimensional systems. Ongoing parallel efforts to understand the physiology and psychoacoustics of the human ear more fully have continued to extend the knowledge base, and in some cases have yet to be exploited by commercial audio systems.

Why wasn't sound recording invented sooner? It is hard to say. Bell's telephone may have required a steady electric current, which presumably was not available in ancient times, but Edison's phonograph was entirely mechanical. The key to the latter invention seems to have been the recognition that sound is a vibration and that the vibration is mechanically transferable, but this notion that is readily demonstrable by putting your fingers to your lips or throat, and humming. The inventions seem to have been part of a general flourishing of physical sciences and mathematics, one that has continued to the present day. Prior to that, however, there is certainly evidence of some understanding of the acoustic processes involved, as seen in the creation of musical instruments and the design of concert halls.

Actually, both Edison and Bell were driven in part by a desire to improve telegraph systems. The telegraph was popularized by Samuel Morse, starting around 1837, although the notion of an electrical telegraph harks back at least to a demonstration of sending a simple signal over a long wire by William Watson in 1747 [18.4]. The cause was greatly helped by the invention of the electric battery by Alessandro Volta in 1799, and the storage battery in 1859 by Gaston Plante [18.5].

Many of the theoretical and mathematical foundations of early audio technologies were established in the 1800s, with work in the electrical sciences by Ohm, Faraday, Henry and others, together with research in acoustics and hearing physiology by Helmholtz, Lissajous and others. Although Edison is credited with the invention of the first device to reproduce sound, the first audio recording device is generally credited to Leon Scott, who on March 25, 1857, patented the phonautograph [18.6]. This device combined a diaphragm and a stylus to inscribe a trace of a sound on a visual medium, such as a smoke-blackened glass or cylinder, but it had no way to reproduce the recording audibly. John Tyndall seems to have followed Scott's work on visualizing sound vibrations, using a variety of different devices.

Alexander Graham Bell filed his patent for the telephone on February 14, 1876, just two hours before Elisha Gray filed for a similar device. Aside from its intrinsic value, the invention was significant for employing acoustical to electrical and electrical to acoustical transduction, principles that are still almost universally employed in telephony, broadcast, and audio recording.

Transduction between acoustical and electrical energy was initially less important to Edison's invention of the phonograph on December 8, 1877, since a physical medium to store the recording was necessary in any case, and there neither was nor is a ready way to directly store a dynamic electrical signal directly on such a physical medium. So it is perhaps not too surprising that the original invention employed purely mechanical transduction for both recording and playback.

With these two inventions, the foundations for analog audio were established, and the race was on to refine and commercialize the systems. In 1887, Emile Berliner patented a recording format that used lateral groove modulation on a flat disk [18.8], in contrast to Edison's use of vertical modulation on a cylinder. The underlying principles may have been much the same, but Berliner's arrangement qualified as a separate patent, and the practical aspects of mass-producing, playing, and storing disks instead of cylinders led to the grooved disk becoming the commercial recording medium of choice for over half a century.

Lacking the means to record an electrical signal directly, but seeking to avoid the wear accompanying the playback of grooved recordings with a stylus, other media were soon being explored, such as the use of magnetic recording by Valdemar Poulson in 1898 and optical recording, by Leon Gaumont in 1901. Meanwhile, the eventual foundation of the broadcast industry was begun in 1894, with the invention of the wireless telegraph by Guglielmo Marconi.

These systems were initially handicapped by a lack of viable amplification means, and although there were some imaginative attempts at mechanical amplification, the key breakthrough occurred in 1906, with Lee de Forest's invention of the *audion* vacuum-tube triode amplifier. Vacuum tubes could and would be used to amplify all manner of electrical signals, but de Forest's choice of the name *audion* seems derived directly from the word audio. It is therefore a matter of some speculation as to whether the tube amplifier would have come about when it did had it not been for the formative invention of sound recording just 29 years earlier.

Aside from resolving the problems of providing better recording sensitivity and adequate playback volume,

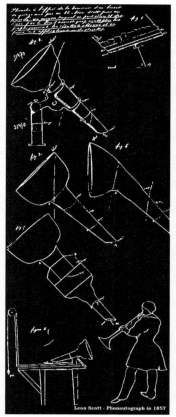

Fig. 18.1 Leon Scott and the phonautograph (After [18.7])

the vacuum tube also enabled the wireless transmission of audio, giving rise to the radio receiver and broadcast industries.

So by 1910, the basic principles, physics, and devices of audio and electroacoustics were in place, resulting in widespread continuing efforts to develop, patent, and refine components and systems, establish standards where required, and commercialize a growing variety of formats and services. Amidst competitive struggles and numerous patent fights, the following years saw a dramatic series of audio innovations, including [18.9, 10]:

- 1913: An early *talking* movie by Edison, using a cylinder synchronized with the picture.
- 1916: A superheterodyne radio circuit patented by Edwin F. Armstrong.
- 1917: A condenser microphone developed by E. C. Wente at Bell Laboratories.
- 1921: The first commercial amplitude-modulated (AM) broadcast, KDKA, Pittsburgh, PA.

- 1925: An electrical record cutter developed at Bell Laboratories, followed by the commercial release of 78 revolutions per minute (RPM) records.
- 1926: Iron-oxide-coated paper tape for magnetic recoding developed by O'Neill.
- 1927: *The Jazz Singer*, the first commercial sync-sound film, using audio on disk synchronized to the picture.
- 1929: The sampling theorem, the basis of digital signal processing, developed by Harry Nyquist.
- 1935: Plastic-based magnetic recording tape, from BASF.
- 1936: The first tape recording of a symphony concert, by Sir Thomas Beecham.
- 1936: A cardioid condenser microphone, by Von Braunmuhl and Weber.
- 1941: Commercial frequency-modulated (FM) broadcasting inaugurated in the US.
- 1947: The Williamson high-fidelity power amplifier developed.
- 1948: The 33 1/3-RPM long-play record, developed by CBS laboratories.
- 1951: The transistor developed at Bell Laboratories.
- 1954: The transistor radio, Sony.
- 1963: Tape cassette format developed by Philips.
- 1965: Dolby *A* noise-reduction system developed.
- 1969: Dr. Thomas Stockham experiments with digital recording.
- 1976: Stockham produces the first 16-bit digital recording, of the Santa Fe Opera.
- 1982: The digital compact disc is released commercially.

Much of the thrust of these and countless other developments in this period was to improve the quality of a given audio channel, by extending the frequency response, reducing noise, etc. As noted earlier, once digital recording came of age, these concerns were rendered somewhat moot, as adequate quality became a matter of using a sufficient number of bits per sample and adequately sampling rate to avoid audible degradation. Of course, certain parts of the audio chain, notably transducers such as microphones and loudspeakers, and associated interface electronics, have remained analog, as is likely to continue to be true at least until a direct link is established to the human nervous system, bypassing the ears altogether.

18.1.1 Spatial Audio History

Perhaps the one area of audio that was not immediately improved with the adoption of digital recording techniques was spatial reproduction. It is unlikely that the importance of the spatial aspect of audio was lost on early audio developers, nor the fact that most humans do in fact, possess two ears. It was probably just a case of priorities. Spatial audio systems would almost certainly require more channels and/or other information, which was hard to accommodate with early audio media, and the emphasis was, quite reasonably, on refining the quality of the single-channel monophonic format.

Still, there were some promising early experiments, notably a demonstration in 1881 by Clement Ader, who set up a series of microphones in front of the stage of the Paris Opera, and ran their outputs to earphones set up in a nearby room [18.9, 10]. Whether by accident or design, some of the listeners elected to use two headphones, one on each ear, and were rewarded with a crude but effective binaural spatial presentation.

Despite isolated experiments such as Ader's, there was little organized investigation of spatial audio until 1931, when Alan Blumlein single-handedly developed a range of devices and protocols to support stereo sound, including techniques to upgrade the entire audio broadcast and recording infrastructure of the time. Blumlein filed a far-ranging patent covering the bulk of his work; unfortunately, it seems to have been about a quarter century ahead of its time, as it would be the late 1950s before stereophonic sound became a commercial reality.

In the 1930s, Bell Laboratories also became interested in spatial audio, albeit with more of an emphasis on audience presentation. They concluded that stereo was ineffective for such uses, as the virtual imaging of just two channels would not work for listeners removed from the centerline. They further concluded that adding a center channel produced a far more satisfactory result, as demonstrated by Fletcher, Steinberg, and Snow in 1933.

With consumer media still limited to single-channel operation, it fell for a time to film studios to engage in occasional commercial forays into spatial audio. A notable experiment in the venue was the film *Fantasia* (1940), the road show productions of which used a synchronized three-channel soundtrack, plus control track, to feed as many as 10 loudspeakers [18.9, 10].

The use of multiple film channels accelerated somewhat in the 1950s, as film studios sought ways to differentiate their product from television. One popular approach was to use wider-than-standard 70 mm film stock, instead of the normal 35 mm, which provided enough room along the edge for several audio channels on magnetic stripes. A common arrangement was three front channels plus a surround channel, the latter usually fed to an array of loudspeakers arranged around the sides and rear of the theater.

In the 1970s, the blockbuster movie genre was kicked off with the releases of *Star Wars* and *Close Encounters of the Third Kind*, along with the synergistic Dolby matrix surround system, which encodes control signals in the audio to pan two optical film channels to four output channels (again, three in front, plus surround).

In the 1990s, multichannel digital soundtracks came into use, first in movie theaters and then in home systems, generally supporting three front channels, a pair of surround channels, and a low-frequency channel, otherwise referred to as a 5.1-channel system.

18.2 The Psychoacoustics of Audio and Electroacoustics

In order to evaluate the requirements and performance of audio systems, it is necessary to establish their relation to the capabilities and limitations of the human hearing. People perceive sound on the basis of frequency, amplitude, time and, somewhat indirectly, space. The perceptual resolution limits of these quantities are fairly well established, even if the underlying mechanisms are still being explored.

18.2.1 Frequency Response

1. Range: The commonly quoted frequency range of human hearing is 20–20 000 Hz, representing the range of frequencies that can be heard as sounds. In many cases, this is a somewhat optimistic specification, and in any case certainly varies with the individual. Only young people are likely to hear out to 20 kHz. With age generally comes a reduction in the upper limit (along with other reductions in hearing acuity, a process known as presbycusis [18.11]), with typical values of the upper limit of hearing in the range of 8–15 kHz common for middle-aged adults. There has been considerable debate about the ability to hear above 20 kHz, with some claiming some sort of perception out to frequencies of 40–50 kHz. Hearing limits above 20 kHz are known to exist in some small animals. At the low-frequency end of the spectrum, a frequency of perhaps 32 Hz is typically the minimum value at which a pitch can be discerned, below which the sensation becomes more of a series of discrete vibrations. Although it is difficult to construct loudspeakers with flat response below 20 Hz, humans can still perceive such vibrations if they are sufficiently large in amplitude.
2. Sensitivity: The ear is not uniformly sensitive to all frequencies, due in part to resonances and mechanical limitations of structures in the outer and middle ear, and in part to neural effects. The sensitivity also tends to be a function of absolute loudness (defined later). Contours of perceived equal loudness were first measured by Fletcher and Munson at Bell Laboratories in the 1930s [18.12]. The figure below shows typical equal loudness curves as a function of absolute level [dB sound pressure level (SPL)].
3. Resolution: The ear exhibits a curious duality with respect to the degree it can differentiate frequencies. On the one hand, certain frequency-dependent processes, such as masking, appear to indicate

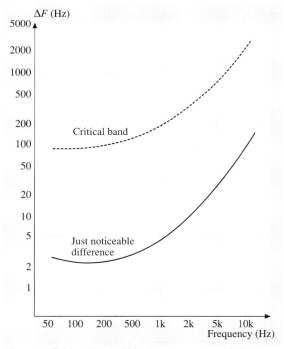

Fig. 18.2 Critical bandwidths (*upper curve*) and sine-wave pitch just noticeable difference (JND) as a function of frequency

a grouping of the frequency range into somewhat coarse sub-bands about 1/4 to 1/3 octave wide, called *critical bands*. On the other hand, frequency discrimination of individual sine waves on the basis of their perceived pitch has a typical resolution limit on the order of 1/100th of an octave, corresponding to a frequency change of about 0.7%. This is one of the cases where the underlying mechanisms are uncertain, and a mixture of temporal and spectral discrimination appears to be necessary to explain the various observed phenomena.

4. Pitch: In general terms, pitch is the perceptual response to differences in the frequency content of a source, but it is far more complex than a simple mapping of frequency to a perceptual scale, and can be affected by parameters such as the loudness of the signal, which ear it is played to, and the timing of the signal. The perception of pitch of spectrally complex signals is especially difficult to model, in part because a given signal may be perceived as having more than one pitch, and the signal itself may or may not actually contain spectral components corresponding to the perceived pitch [18.13].

18.2.2 Amplitude (Loudness)

Absolute Sensitivity

The total dynamic range of the human ear, that is, the range from loudest to softest sound one can perceive, is generally given to be about 120 dB a pressure range of about a million to one, or a power range of about a trillion to one. The term *sound pressure level* (SPL) is used to indicate where in this range a given sound falls, with 0 dB SPL taken as the approximate threshold of hearing at 1 kHz, corresponding to a level of about 20 μPa [18.14]. The mathematical relation between a sound pressure p and corresponding SPL value is

$$L_p = \text{SPL} = 20 \log(p/p_0) \,,$$

where p_0 corresponds to the 0 dB SPL quoted above.

The upper limit of loudness perception is not well defined, but corresponds in principle to the onset of non-audible phenomena such as tingling and outright pain. Depending on the listener, the upper limit is likely to be in the range of 115–140 dB SPL. (Note that exposure to such levels, especially for extended periods, can lead to hearing loss [18.11]).

As noted, sensitivity is a function of frequency and absolute level, with the greatest sensitivity observed in the range 3500–4000 Hz, mainly due to resonance of

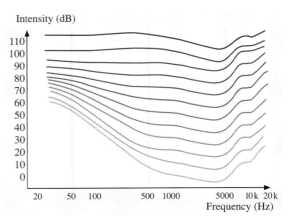

Fig. 18.3 Equal-loudness curves as a function of frequency and absolute level (After [18.9])

the ear canals, where the hearing threshold may actually be slightly lower in level than 0 dB SPL. The greatest variation as a function of frequency is at the lowest levels, where the threshold of hearing may rise to in excess of 20 dB SPL at high frequencies, and fully 60 dB SPL at low frequencies.

The variation of perceived loudness with actual sound pressure, while monotonic, is certainly nonlinear, as can be seen from the differences in spacing between the equal loudness curves at different frequencies. The actual perception of a difference in loudness will generally be less than the actual difference in SPL. According to the Sone scale of loudness perception, a difference in SPL of 10 dB at 1 kHz is roughly equivalent to a doubling of perceived loudness Sones $= 2^{[(\text{Phons}-40)/10]}$ [18.15].

Because of the variation in perceived loudness with the actual SPL and frequency of a signal, it is a nontrivial matter to calculate the perceived loudness of an arbitrary, spectrally complex signal. Recent efforts using multi-band analysis and power summation have improved the predictability to within close to a loudness just noticeable difference (JND) for most signals [18.16].

Resolution
The amplitude resolution of the ear is generally taken to be about 0.25 dB under best-case conditions, although for some situations it is considered to be slightly larger, on the order of 0.5–1.0 dB.

Masking
The audibility of a sine wave in isolation may be modified in the presence of other spectral components. In particular, the addition of spectral components is likely

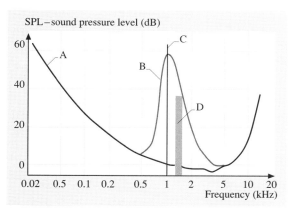

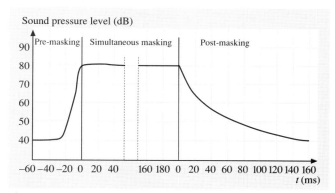

Fig. 18.4 Frequency Domain Masking: Curve A is hearing threshold with no signal present. Curve B is revised threshold in presence of 1 kHz tone C. Signal D, rising above curve A but below curve B, will be audible by itself but masked by signal C

Fig. 18.5 Temporal masking characteristic

to render an existing component less audible, a process referred to as masking.

The basic notion of masking is that a louder sound will tend to render as less audible or inaudible a softer sound that would otherwise be audible in isolation. The degree of masking will tend to depend in part on the frequency spacing of the spectral components in question. This gives rise to the notion of a prototype masking curve, delineating the threshold of audibility in the presence of a sine wave of some frequency and amplitude, as a function of frequency, as shown in Fig. 18.4.

To at least first-order approximation, the masking effect of two or more sine waves can be calculated by summing their respective individual masking curves. Since, from Fourier theory, any signal can be decomposed into a sum of sinusoids, a composite masking curve from an arbitrary signal can be calculated as the sum of the masking curves of each spectral component. This constitutes a convolution (filtering operation) of a prototype masking curve with the signal spectrum. In practice, the prototype masking curve used for the convolution is likely to vary as a function of frequency, amplitude, or signal type (noise versus sine wave). Further, nonlinearities in the ear may result in significant deviations from the predicted curve at times.

Although the masking effect of a signal is, quite logically, maximized while the signal is present, some residual of the masking effect will extend beyond the termination of the signal, and even slightly before its onset. This process is referred to as temporal masking, and typical behavior is illustrated in Fig. 18.5) [18.18].

18.2.3 Timing

Because the transduction structure of the ear – the basilar membrane and associated hair cells – functionally resembles a filter bank, questions of timing must be considered as applying to the filtered output signals of that

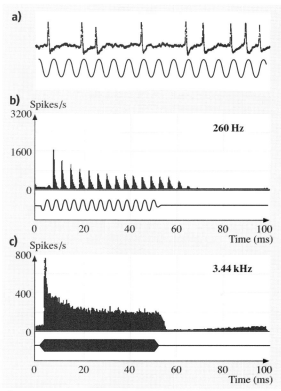

Fig. 18.6 Neural phase locking at low frequencies and lack of phase locking at higher frequencies [18.17]

filter bank. The brain does not *see* the wide-band audio signal arriving at the ear, as it might appear on an oscilloscope, but instead processes the outputs of the basilar membrane filter bank, subject to effective half-wave rectification and loss of high-frequency synchrony by the neurons (Fig. 18.6). Output neural signals from each ear are ultimately processed by multiple areas of the brain. So, it is not surprising that issues of timing are a little ill-defined, and that there are a number of time constants associated with different aspects of hearing.

For starters, there is the basic ability of the neurons in the auditory nerve to follow the individual cycles of audio impinging on the basilar membrane; i.e. to exhibit phase locking. Current research seems to put the upper frequency of this activity at about 5 kHz, although this seems somewhat at odds with lateralization of sine waves on the basic of interaural time difference, which only extends up to about 1500 Hz.

The ability to follow individual cycles of audio may aid in pitch perception, as the effective shapes of the basilar membrane filters appear too broad to account for the observed pitch resolution. However, the physiological mechanisms for how this might be accomplished are still the subject of some debate.

Other time constants appear to apply to the aggregate audio, regardless of the presence of a filter bank at the input. Already noted are the masking time constants, which substantially extend only 1–2 ms prior to the onset of a loud sound, but continue for several tens of milliseconds after its cessation.

Each of these may be related to a more fundamental time constant. The short time constant of about 1–2 ms may represent the shortest time one can perceive without relying on spectral cues, while the post-masking interval may be related to the fusion time. The fusion time in particular seems to represent a kind of acoustic integration time, or the limit of primary acoustic memory, typically on the order of 30–50 ms, and appears to be associated with the lower frequency limit of the audio band (20–30 Hz). It also seems to explain why echoes are only heard as such in large cathedral-like rooms, as they are integrated with the direct arrivals in normal-size rooms.

Table 18.1 Summary of approximate perceptual time/amplitude limits

Parameter	Range	JND
Amplitude	120 dB	0.25 dB
Premask	N/A	2 ms
Postmask	N/A	30–50 ms
Binaural timing	700 μs	10 μs

On the other hand, binaural timing clearly exhibits higher resolution, as differences of interaural timing on the order of 10 ms may be audible.

18.2.4 Spatial Acuity

Strictly speaking, the spatiality of a sound field is not perceived directly, but is instead derived from analysis of the physical attributes already described. However, given its importance to audio and electroacoustics, it is nonetheless useful to review the processes involved.

Specification of the position of a sound source relative to a listener in three-space requires three coordinates. From the mechanisms used by the human ear, the natural selection is to use a combination of azimuth, elevation, and distance. Of these, distance is generally considered to be an inferred metric, based on the amplitude, spectral balance, and reverberation content of a source relative to other sources, and is not especially accurate in any absolute sense.

This leaves directional estimation as the primary localization task. It appears that the ear uses separate processes to determine direction in a left/right sense and a front/top/back sense. Given the physical placement of the ears, it is not surprising that the left/right decision is the more direct and accurate, depending on the interaural amplitude difference (IAD) and interaural time difference (ITD) localization cues.

Although the ear is sensitive to IAD at all audible frequencies, as can be verified by headphone listening, the head provides little shadowing effect at low frequencies, since the wavelengths become much longer than the size of the head, so IAD is mostly useful at middle to high frequencies.

The JND for IAD is about 1 dB, corresponding to about 1 degree of an arc for a front/centered high-frequency source. As a sound moves around to the side of the head, the absolute IAD will tend to increase, and the differential IAD per degree will decrease, causing the angular JND to decrease accordingly.

The ITD lateralization cue has a rather astonishing JND of just 10 ms, for a source on the centerline between the ears, corresponding to an angle of about 1 degree. For signals off to one side, the ITD will typically reach a maximum value of just 750 ms – 3/4 of a millisecond. As with the IAD cue, the sensitivity of the ITD cue declines as a source moves around to the side of the head.

Also like IAD, this cue is usable at all frequencies, albeit with a key qualification. At frequencies below about 1500 Hz, the neural response from the basilar

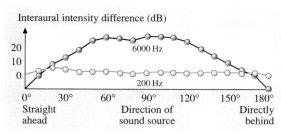

Fig. 18.7 Interaural amplitude difference as a function of direction [18.19]

membrane remains phase locked to the bandpass-filtered audio waveform. At such frequencies, even a modest amount of interaural phase shift of a sine wave can result in a shift in the position of the audio image. Above 1500 Hz, neural phase synchrony is effectively lost, at least as far as ITD perception is concerned, so the interaural phase of a high-frequency sine wave has little or no perceptual effect on its apparent position. However, the brain can still make fairly accurate judgments of the ITD of the *envelopes* of the high-frequency bandpass-filtered signals, assuming there is sufficient variation in the filtered envelopes. This will tend to make spectrally dense high-frequency signals easier to localize than sparse signals, since the latter may only exhibit constant-envelope sine waves after the critical band filtering of the basilar membrane.

While localization in the horizontal direction is firmly mediated by two strong, well-behaved cues, localization in the vertical dimension is decidedly less well behaved. The goal of vertical localization is to differentiate the positions of sources on locus of constant IAD/ITD. For sources equidistant from the two ears, the locus is a plane passing though the center of the head, referred to as the median plane. For sources off to one side, the locus is a cone, sometimes referred to as the *cone of confusion*. The best-case resolution, for front-centered sources, is typically on the order of 5°, which is quite a bit less precise than the horizontal resolution of about 1° under the same conditions. As with horizontal localization, vertical resolution appears to diminish for side sources, although since the cone of confusion becomes progressively narrower, the net spatial error resulting from a vertical localization error is likely to diminish. For sources at the side, the cone of confusion collapses to a line.

One demonstrated vertical localization cue [18.20] is generated by the interaction of the arriving sound and the folds of the pinnae. The pattern of reflections off the folds depends on the direction of arrival of a sound, although not in any simple, monotonic fashion. One consequence of these pinna reflections is the introduction of sharp peaks and dips in the frequency response, which the brain can apparently learn to associate with specific directions. Because of the size of the pinna, and resulting reflection delays, these direction-dependent response delays only appear at high frequencies, above about 7 kHz, and so are less effective for low-frequency or narrow-band signals. Vertical localization of signals below 7 kHz may be possible on the basis of pinna cues if the brain is performing some sort of autocorrelation of each ear's signal, and can detect the delays in the time domain, but so far there is little evidence to support this hypothesis.

A second vertical localization cue is dynamic in nature, namely the modulation of the horizontal localization cues – ITD and IAD – with movement of the head, which will vary with the vertical position of the source. Given a front source on the median plane, for example, a rightward movement of the head will result in a left ear arrival that is louder and sooner than the right ear arrival. For a rear source, the correspondence will be opposite, and for a top source, there will be essentially no change in the ITD/IAD values. By correlating head movement with these sonic changes, the brain can deduce the vertical position of the source.

Finally, the presence and pattern of room echoes, once learned, and in concert with the above cues, can reinforce the vertical position of a source, especially whether in front of or behind the listener.

Even so, the vertical localization cues tend to be somewhat individual, like fingerprints, and, for example, playing a recording made with microphones placed in the ear canals of one subject may not result in good vertical localization for another listener.

18.3 Audio Specifications

With the basic limits of auditory resolution now reviewed, consideration can be given to the requirements for high-quality audio systems in terms of those limits. While individual audio components can be expected to exhibit correspondingly individual performance requirements, as will overall systems, there are some specifications that can be expected to apply to all elements of the signal chain. Some of these are considered traditional audio specs that have been used as general metrics for many years. Happily, in at least some cases, modern audio systems are of such high quality as to render some specifications moot. In other cases, complete fulfillment of some of the requirements is something that may never be completely realized.

18.3.1 Bandwidth

It is reasonably evident that, given the nominal 20–20 kHz range of human hearing, it will be desirable for audio systems with aspirations to sonic accuracy to support at least that frequency range. Since an audio signal may pass through any number of processors on its way from acquisition to ultimate reproduction, it is generally advisable for each processor to have a wide-enough pass band to avoid progressive bandwidth limitation.

For most audio devices during the formative years of audio development, this requirement was daunting, and rarely achieved outside the laboratory, if then. There were too many mechanical vibrating structures in the chain, and it is very difficult to make them respond consistently over a 1000:1 frequency range. Grooved disc recordings were limited by large-diameter, low-compliance styli, and slow inner-groove velocities of constant-RPM discs. Optical media, like film, and magnetic media, were both limited by linear speed, and the size of the effective apertures used for recording and playback. The linear speed of such media was chosen as a compromise with playing time, and the smallness of the apertures was limited by high noise levels and diminishing signal level. Broadcast media were limited by bandwidth constraints and noise.

Not counting a straight piece of wire, amplifiers are the audio component that have probably had the easiest time achieving full audio bandwidth, particularly small-signal amplifiers, thanks in part to the relatively low mass of the electron. Power amplifiers have had a slightly harder time, partly because devices that handle high powers often come at the expense of compromises in other parameters, and partly because loudspeaker systems often exhibit ill-controlled, non-resistive loads. The response and linearity of amplifiers of all sorts was materially improved by the invention of negative feedback by Harold Black of Bell Laboratories, on August 2, 1927 [18.21].

An early recording system with fairly wide bandwidth was demonstrated on April 9–10, 1940, by Harvey Fletcher of Bell Laboratories, and conductor Leopold Stokowski. This employed three optical audio tracks on high-speed film, with a fourth track for a gain control signal, apparently implementing a wide-band compressor/expander for noise reduction, and possessing a frequency response of about 30–15 000 Hz.

It was probably not until the mid-1950s that such a bandwidth became commonplace in consumer media, with improvements such as the long-play record, commercial FM broadcasting, and refinements in magnetic tape, heads, and circuits.

As with a number of other basic audio specifications, the question of full-bandwidth frequency response was, to a great extent, rendered moot with the commercial introduction of digital audio in 1982. As discussed in a subsequent section, the bandwidth of a digitally sampled signal is largely dependent on the sampling rate chosen, which must be at least twice as high as the highest audio frequency of interest. Thus, conveying a bandwidth of 20 kHz requires a sampling frequency of at least 40 kHz. Current standard sampling frequencies in common use for full-bandwidth audio systems are 44 100 Hz and 48 000 Hz, depending upon the medium. Sometimes, a slight reduction in high-frequency response is tolerated, and a sampling frequency of 32 000 Hz is used to reduce the data rate slightly. Audio perfectionists, on the other hand, often prefer sampling rates of 96 000 Hz or more, to support an effective bandwidth in excess of 20 kHz and thereby provide additional safety margin to avoid alias distortion, preserve harmonics above 20 kHz, avoid possible intermodulation distortion, and mitigate possible filter distortion.

Once converted to digital representation, an audio signal tends to maintain its bandwidth unless explicitly limited by some digital signal processing (DSP) operation.

Despite the advances of digital audio, full-range audio systems are still the exception. The chief culprits for this situation are the transducers, especially loudspeakers. At low frequencies, naturally occurring combinations of mass and compliance lead to a fundamental resonance for most typical loudspeaker drivers that falls well inside the audio band, imparting a high-pass characteristic that usually rolls off at lower frequencies at a rate of at least 12 dB per octave. At high frequencies, a similar fundamental resonance leads to a similar roll-off, with a lowpass characteristic. Some of these roll-offs can be partially mitigated with electronic equalization, but the steepness of the characteristic makes it difficult to extend the response very far with equalization, without incurring unduly high-power requirements.

In general, most musical and speech energy is concentrated in a subset of the audio band, say from 100 Hz to 10 kHz, allowing the majority of audio signals to be satisfactorily reproduced with modestly compromised system bandwidth, to the benefit of size, cost, and bulk.

18.3.2 Amplitude Response Variation

From the basic amplitude acuity of the ear, it is logical to require that the pass-band deviation in frequency response be less than an amplitude JND, say within ± 0.25 dB of flat. Of course, this is the desired net response of an entire signal chain, and if deviations are known to exist in one part of the chain, it may be possible to apply compensating equalization elsewhere in the chain.

As with the quest for full audio bandwidth, early audio systems for many years had trouble meeting this requirement, in part from mechanical limitations. Amplifiers again have traditionally had the easiest time meeting this specification, and modern-day DSP is limited only by the precision of the analog/digital converters, which is generally quite good in this regard, plus any deliberate signal-processing spectral modifications.

The worst offender is many cases is the listening room, where echoes and resonances are likely to make nearly any sine-wave frequency response plot a jumble of peaks and dips covering many dB. The three-dimensionality of the room and the fact that echoes arrive from diverse directions work to reduce the audibility of many of these measured deviations.

After rooms, loudspeakers have the hardest time meeting this flat-response requirement, being mechanical devices and having to handle a lot of power and move a lot of air. In order to cover the bulk of the audio band, loudspeaker designs typically employ multiple drivers, each optimized for a portion of the audio band, but this in turn introduces potential response deviations from required crossover networks.

In considering reasonable values for most audio specifications based on psychophysical performance, it is prudent to consider both monotic and dichotic performance; that is, each ear considered in isolation, and then the binaural performance of the two ears together. In the case of amplitude response variation, similar JNDs apply, so there is no compelling reason to differentiate the situations in this case.

18.3.3 Phase Response

In the course of passing through an audio component, an audio signal may encounter delays, which may be frequency dependent, and may therefore be categorized as either frequency-dependent time delays, group delays, or phase shifts. It is therefore prudent to establish preferred perceptual limits on the allowable deviations of these quantities.

Of course, a pure wide-band time delay will be manifested as a linear change in phase shift proportional to frequency, so to separate the effects of time delay and phase shift, so the latter ("frequency-differential single-channel phase shift") may commonly be understood to represent the deviation from a linear-with-frequency phase characteristic.

It is a little difficult to be completely definitive about a phase-shift specification, in part because there are multiple neural timing mechanisms involved. For a sound channel considered in isolation, phase perception is rather weak. There does not seem to be an explicit means of comparing the fine-grain phase or timing of signals at significantly different frequencies (more than a critical bandwidth), at least for steady-state signals. This is sometimes used as the basis of a claim that the ear is virtually phase deaf, although that is probably an overstatement. Still, if one constructs a steady-state signal with widely space spectral components that are in slow precession, say a combination of 100 Hz and 501 Hz (360° per second relative phase precession), the resulting percept is almost entirely constant.

So under what conditions are phase shifts audible? There seem to be at least three such mechanisms: fusion time, transient fusion time, and critical band envelopes.

As noted earlier, fusion time refers to the audio event integration time of the ear, generally in the range of 20–50 ms. Separate audio events that occur farther apart than the fusion time will generally be heard as separate. If a spectrally complex signal is put through a system where some frequencies are phase- or time-shifted by more than the fusion time, the signal may temporally defuse, rendering the phase shift quite audible. (A 20 ms phase delay at 30 Hz corresponds to 216 degrees of phase shift, not an unrealistically large value.) This can occur, for example, if a sharp pulse is passed through a third-order, 1/3-octave filter bank tuned for flat magnitude response. What starts out as a click will emerge as a "boing", quite audibly distinct from the source.

Transient fusion time refers to the shortest interval the ear can differentiate temporally, as exemplified by the premasking interval of about 1–2 ms. If a sharp transient, shorter than the transient fusion time, is passed through an audio channel where the high-frequency phase shift exceeds the transient fusion time, the output signal may sound audibly time-smeared compared to the source. Although the effect can be subtle, it is often manifested as the difference, e.g., between a sharp strike with a hard mallet, and a wire-brush mallet hit.

Critical band envelopes come into play with regard to the audibility of phase shifts with steady-state spectrally dense signals, more specifically signals in which two or more significant signal components fall within a common critical band. Since the basilar membrane filter bank will be unable to segregate these signals into separate bands, they will beat with one another, and the pattern of beats may be quite audible, for example, a complex composed of equal parts 500 Hz and 501 Hz, which can be considered the 500th and 501st harmonic of a 1 Hz pulse train. These two spectral components are certainly within a critical band of one another, and the instantaneous phase relation between them will certainly be audible; indeed the signal will cancel altogether once per second. Compare that to the earlier example of 100 Hz and 501 Hz, the 100th and 501st harmonic of a 1 Hz pulse train, where the phase relation remained virtually inaudible because of the wider frequency spacing.

So, it can be safely hypothesized that keeping frequency-differential single-channel phase shift under 1 ms is probably pretty inaudible, assuming small changes over the width of a critical band, and in practice, with most non-transient music, speech or other natural sounds, it can be expected that phase shifts corresponding to frequency-differential time shifts much under the fusion time of 20–50 ms (at least at low frequencies) will be fairly benign, if not almost totally inaudible under most conditions.

Once again, amplifiers have the easiest time meeting this requirement, with multi-driver loudspeaker systems being one of the more problematic components. Although sharp resonances, the bane of mechanical acoustic components, may exhibit audible amounts of phase shift, they will often be rendered insignificant compared to the amplitude deviations introduced by such resonances. And, again, with DSP systems, audible phase shift is not likely to be an issue as long as the signal remains in the digital domain, unless such phase shift is purposely introduced with signal processing.

As can be readily gleaned from binaural psychophysics, dichotic phase shift – the phase shift between a pair of channels – can be more audible at lower delay levels than monotic phase perception. A timing difference of even a single sample at 48 000 Hz sampling, about 20 µs, is twice the JND of this parameter. Of course, the perceptual metric for binaural phase difference is a change in apparent image position or width, rather than the spectral or amplitude change associated with monotic phase shift. It is possible for audio systems to have audible amounts of monotic phase shift and still meet the dichotic (interchannel) phase requirement, as long as all channels have the same amount of phase shift.

As usual, amplifiers and digital signals have the easiest time with this specification. In the past, some multichannel analog recording media, such as magnetic tape recorders or optical film recorders, tended to exhibit small amounts of interchannel timing error, owing to misalignment of the recording and playback apertures. Real-world rooms and loudspeakers have the hardest time, since differential timing of 10 µs corresponds to a path-length difference of only about one eighth of an inch, and few listeners are willing or able to hold their heads that still. This effect is somewhat mitigated because the plethora of echoes in a typical room leads the brain to discount timing information in favor of first-arrival amplitudes, which are somewhat better behaved. Still, it can be readily appreciated that the virtual imaging of stereo reproduction is likely to be fairly sweet-spot dependent, since listener positions near a loudspeaker will tend to receive sound both sooner and louder from the near loudspeaker, resulting in spatial distortion of the sound stage.

18.3.4 Harmonic Distortion

The presence of nonlinearities in the transfer function of an audio device or system is likely to have the effect of introducing new, spurious spectral components. For a single isolated sine wave, the nonlinearity will result in a new waveform with the same periodicity as the original signal, which can then be expressed via Fourier analysis as a sum of the original sine wave plus harmonics thereof [and perhaps a direct-current (DC) component]. For more spectrally complex signals, each discrete spectral component will be subject to such spurious harmonic generation. This process is referred to as harmonic distortion, and the power sum of the generated harmonics, divided by the power of the original signal, is referred to as total harmonic distortion (THD), usually expressed as a percentage.

The limit of acceptable quantities of THD is the point at which the generated harmonics become audible. In part, this may be signal dependent, since distortion products of a low-level signal may themselves fall be-

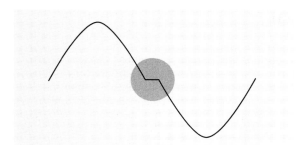

Fig. 18.8 Sine wave with crossover distortion (http://www.duncanamps.com/technical/ampclasses.html)

low the threshold of hearing, rendering them inaudible. For higher-level signals, the distortion products will be inaudible if they are masked, rendering the effect of harmonic distortion decidedly signal dependent. (The amplitude dependency may by further emphasized by nonlinearities that are themselves amplitude dependent, such as clipping.) For example, a wide-band noise signal may mask all harmonic distortion components of a moderate level. Harmonic distortion may therefore be most audible with spectrally sparse signals, with minimal masking components, such as a single sine wave.

For a sine-wave signal, the maximum acceptable amount of harmonic distortion can be deduced from the masking curve of that sine wave. From the masking curve shown in Fig. 18.4, it may be deduced that, since the masking level for a 1 kHz tone is around −35 dB at 2 kHz, the tolerable amount of second-harmonic distortion is about 2%. By 3 kHz, the masking curve is down around 50 dB, or about 0.3% third-harmonic distortion. Higher-order harmonics would benefit from even less masking, making them even more likely to be audible, until their amplitude falls below the threshold of hearing. Thus, the audibility of harmonic distortion will depend on the order of the distortion, with low-order distortion generally being less audible than high-order distortion.

Low-order harmonic distortion is generally associated with smooth, gentle nonlinearities, such as the typical onset of clipping of a vacuum-tube amplifier or a loudspeaker driver. High-order harmonic distortion, in contrast, is likely to result from sudden, sharp nonlinearities, such as the crossover distortion of an improperly designed solid-state power amplifier.

Among more traditional audio devices, playback of grooved media was probably fraught with the greatest amount of high-order THD, caused by sharp mistracking of the stylus. In the age of digital audio, low-level undithered digital signals will likely exhibit stair-step behavior which will also tend to be rich in high harmonic content, rendering them highly objectionable from the perspective of THD. Although this can lead to a desire for very high quantization accuracy (say 32-bits per sample), proper use of dither to spread the harmonic energy out as noise is generally the preferred solution.

18.3.5 Intermodulation Distortion

Intermodulation (IM) distortion results when two or more spectral components interact with a nonlinearity. The resulting distortion products will tend to be generated at sums and differences of the original spectral components. For even low-order IM distortion, these distortion products may be far in frequency from the source signal components, and may in fact be much lower in frequency. Since masking tends to be most effective when the signal to be masked is higher in frequency than the masker, having IM distortion products lower in frequency than the generating components renders them more likely to be audible, in turn making IM distortion potentially more bothersome than THD.

Serious IM distortion tends to be associated with sharp nonlinearities, and was a problem with analog grooved recordings when the stylus mistracked, as well as with poorly designed power amplifiers exhibiting crossover distortion. It can also show up with low-level signals and digital converters. Loudspeakers usually have less trouble with IM distortion, in part because any nonlinearities are usually more gentle (associated with maximum output limiting), and in part because IM distortion products located outside the range of a driver will tend not to be radiated.

Although the theoretical audible limit of IM distortion is a miniscule fraction of 1%, a value of 0.1% IM distortion or less is usually pretty safe.

18.3.6 Speed Accuracy

Considerations of speed accuracy encompass both absolute accuracy and dynamic variations, sometimes subdivided into slow variations called *wow* and faster variations called *flutter*. Except in extreme cases, timing error is usually perceived as an error in pitch, rather than time. Unless it is drastically in error, absolute speed accuracy is likely to be of concern mostly to folks with absolute pitch.

Speed variations such as wow and flutter are likely to be audible to anyone with normal pitch perception. Given a pitch JND of 0.7%, which would correspond to peak-to-peak just-noticeable speed variation, the corresponding root-mean-square (RMS) speed variation would be about 0.25%. Since speed variation may be cumulative, a preferred figure of 0.1% is likely to be specified [18.22]. There may also be cases in which cancellations from room echoes cause amplitude variations resulting from flutter that may require still lower limits of flutter to be rendered inaudible.

The analog mechanical and electromechanical recording devices of the past all have had to contend with speed accuracy. Turntables, tape drives of all stripe, and film projectors all fought off the effects of speed variation, usually with large flywheels and elaborate bearings. Getting a small lightweight tape drive like the Sony Walkman® to have low flutter was a good engineering trick. Properly designed and maintained equipment was generally able to meet the required speed accuracy specification.

In the age of digital audio, speed errors of all sorts are mostly a dead issue. Sampling clocks controlled by quartz oscillators provide speed accuracy and steadiness far in excess of what the ear can hope to detect.

18.3.7 Noise

Noise is a first cousin to distortion in its characteristics and considerations. It is a random signal, hopefully low in level, that gets tacked onto an audio signal in the course of almost any processing, analog or digital, and it is generally cumulative in nature. It often sounds like a hiss. Noise will be inaudible if it is either below the threshold of hearing, or masked by the audio signal.

Sources of Noise
Noise sources include:

Acoustical: Random fluctuations of air molecules against the eardrum or a microphone diaphragm

Electrical: Tiny fluctuations of electrical charge in analog electronic circuits from discrete electrons

Mechanical: Surface irregularities of grooved media

Magnetic: Irregularities of magnetic strength from discrete magnetic particles in recording tape

Arithmetic: Random errors from DSP quantization of numerical signals to discrete values.

There are other common spurious signals, such as hums and buzzes, that can corrupt an otherwise pristine audio signal, and although strictly speaking they may not be considered noise, they are generally at least as undesirable.

In any case, the level of noise can be considered as a function of frequency, giving rise to the notion of the spectrum or spectral density of the noise. If the value of an audio noise signal at any point in time is uncorrelated with the value of the noise at any other point in time, the signal is an example of what is called *white noise*. Spectrally, white noise is said to be flat in a linear frequency sense. That is, in essence, the output energy of a bandpass filter of bandwidth B Hz. supplied with white noise, will be the same regardless of the center frequency of the filter (providing, of course, that the pass-band of the filter is substantially within the audio band).

Because the filters on the basilar membrane are approximately logarithmic (not linear) in center frequency and bandwidth, and because they tend to integrate the signal energy within each pass-band, white noise will likely be perceived as having a rising high-frequency spectrum, even though it is flat on a linear frequency basis. In other words, as the center frequency of each effective basilar membrane bandpass filter increases, so too does its bandwidth, roughly in proportion, ergo the total filtered energy of a white-noise source will increase with frequency, leading to the bright, top-heavy perception. For each successive octave, or doubling of center frequency, the bandwidth will double, ergo so too will the bandpass white-noise signal energy, leading to a characteristic spectral rise of 3 dB per octave.

Note: dB = $10 \log$ (ratio of power) = $20 \log$ (ratio of amplitude).

If white noise is processed with a filter having a constant roll-off characteristic of -3 dB per octave, the output noise will have constant energy per octave, or fraction thereof, and will sound flatter to the ear. Such a signal is referred to as pink noise.

Although many forms of noise are independent of the presence of an actual signal, there are mechanisms whereby noise, or noise-like artifacts, are signal dependent. The most common such behavior is for the level of the noise to be roughly proportional to signal level. This type of artifact can be especially pernicious, as the dynamic behavior tends to call attention to the noise.

Although there are numerous ways of dynamically suppressing noise if it is unavoidable, it is by far preferable to avoid it in the first place as much as possible via proper low-noise design, proper shielding and grounding, etc.

18.3.8 Dynamic Range

Dynamic range is the amplitude range from the largest to the smallest signal an audio device or channel can handle. In cases where there is no minimum signal, other than zero level, the minimum useable level may be taken as the noise level of the channel.

If the noise level of the channel is substantially independent of signal level, the dynamic range of the channel may alternately be referred to as the signal-to-noise ratio (SNR). If the noise is signal dependent, the audibility of the noise will depend on the instantaneous signal-to-noise ratio, which is likely to be quite different from the dynamic range.

Since both the noise level and the overload point of a channel are likely to be functions of frequency, so too will the dynamic range. A common characteristic of the noise of traditional analog recording systems has been rising noise levels at the ends of the audio band, coupled with diminishing maximum signal levels, a double whammy that tends to limit the resulting dynamic range, in some cases rather severely.

Historically, unadorned analog channels have tended to have overall dynamic-range specifications on the order of 45–60 dB. This is true of such media as FM stereo, grooved records, optical film soundtracks, and magnetic tape. Perhaps the most limited hi-fi medium in this regard has been cassette tape, which manages about 55 dB at midband and on the order of 30 dB dynamic range at high frequencies.

Happily, DSP systems usually have flat overload characteristics and flat, or at least white, noise floors. The dynamic range of a DSP system is largely a question of the number of bits used to represent the signal. A rough rule of thumb is that each additional bit in the representation of sample values adds 6 dB to the dynamic range. The common 16-bit format used on digital compact discs yields a theoretical dynamic range of 96 dB. In practice, the combined effect of real-world imperfections and converter noise limits the practical attainable dynamic range with 16-bit audio to a little short of 96 dB, perhaps about 93 dB. A 24-bit converter can approach 144 dB dynamic range, which not only allows coverage of the full human auditory dynamic range, but allows margin for error in setting recording levels and subsequent mixing and processing operations.

18.4 Audio Components

With the common audio specifications now delineated, consideration can be given to some of the common audio components, their typical specifications, and specific characteristics and limitations.

18.4.1 Microphones

For any acoustical recording, the type, pickup pattern, position, and characteristics of the microphone are likely to have the greatest effect on the resulting quality of the recording. A microphone is, of course, an analog transducer, generally used to convert acoustic signals to proportional electrical signals. It is the functional analog of the eardrum in the human auditory system.

The term microphone was coined by Sir Charles Wheatstone in 1827, although at that point it referred to a purely acoustical device, such as a stethoscope. Its start as a transducer was pretty much coincident with Bell's invention of the telephone. Bell's initial *transmitter* used a vibrating needle connected to a diaphragm and immersed in an acid bath to implement a variable resistance responsive to acoustic waves [18.23]. This was followed in short order by more practical, rugged designs by Bell, Edison, Berliner, Hughes, and others.

Common to virtually all microphones ever made is a diaphragm. Most microphones do not directly sense the instantaneous pressure deviations of an airborne acoustic wave. Instead, the sound waves impinging on a diaphragm cause movement of the diaphragm, and it is this movement that is sensed and converted to an electrical audio signal.

The methods used for the mechanical-to-electrical transduction are many and varied. Bell's telephone patent seems to describe an odd type of electromagnetic microphone, in which the diaphragm was mechanically coupled to an armature in a magnetic circuit. Acoustic vibrations caused vibration of the armature, in turn inducing changes in the reluctance of the magnetic circuit, which induced a dynamic current in a coil of wire.

Later dynamic microphones simply attached a fine coil of wire to the diaphragm, and suspended it in a magnetic field. Movement of the diaphragm induces a voltage in the coil of wire, presumably proportional to the velocity of the diaphragm, and said voltage can then be amplified as necessary. Dynamic microphones do not require an external source of power. Their popularity was limited prior to World War II by a lack of powerful permanent magnets.

Many early telephone transmitters used carbon in one form or another, typically carbon granules coupled to the diaphragm such that vibration of the diaphragm would dynamically squeeze the granules, causing a variation in resistance that could be electrically sensed by an applied voltage. This arrangement could be scaled up in size and voltage enough to produce a sizeable signal, of particular importance prior to the advent of electrical amplification.

Another type of self-generating microphone is the crystal microphone, based on the piezoelectric effect discovered by the Curies in 1880. Here, the diaphragm constricts a thin crystal made of Rochelle salt, ceramic, or similar material, directly producing a voltage. The low compliance of the piezoelectric element typically limits the performance of such a microphone, especially the extent of its frequency response.

Indeed, the performance of most of the microphone types mentioned so far is somewhat limited by the need to attach some object or structure to the diaphragm to detect its motion, increasing the mass of the moving elements, which in turn tends to limit the high-frequency response.

Two types of microphones avoid any mechanical attachment to the diaphragm: the ribbon microphone and the condenser microphone.

A popular variant of the dynamic microphone, the ribbon microphone, uses a thin metal ribbon diaphragm suspended in a magnetic field. Movement of the diaphragm induces a current in the diaphragm, which can then be amplified. In effect, the diaphragm acts as a single-turn moving coil. The ribbon microphone was

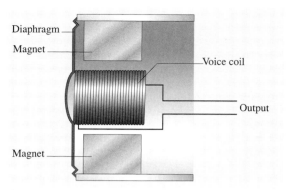

Fig. 18.10 A dynamic microphone uses a coil of wire attached to a diaphragm, suspended in a magnetic field (After [18.24])

developed by Schottky and Gerlach in 1923, and considerably refined by Harry Olsen of RCA in 1930 [18.26].

The condenser microphone is based on the notion of electrical capacitance. A metal, or other conducting, diaphragm is placed close to a second, fixed conducting plate, forming a capacitor, and a polarizing voltage is applied between the plates. Movement of the diaphragm causes alterations in the capacitance, in turn causing corresponding alteration of the charge on the plates, which is converted to a dynamic voltage by a resistor, and amplified as necessary. The invention of the condenser

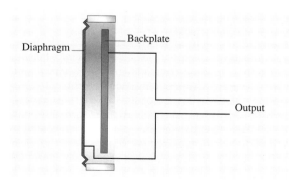

Fig. 18.9 A condenser microphone consists of a very thin diaphragm suspended parallel to a backplate (After [18.24])

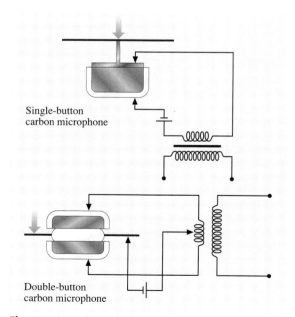

Fig. 18.11 A carbon microphone (After [18.25])

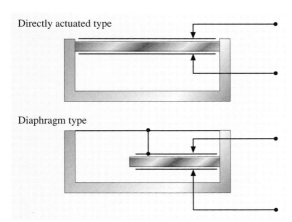

Fig. 18.12 Piezoelectric microphones (After [18.29])

microphone is generally credited to E. C. Wente of Bell Laboratories, in 1917 [18.27]. A popular modern variant is the electret condenser microphone, perfected in 1962 by Sessler and West of Bell Laboratories [18.28]. An electret is an object possessing a permanent electrostatic charge, making it rather like the electrical analog of the permanent magnet. Use of an electret to polarize the capacitor in a condenser microphone eliminates the need to supply a polarizing voltage for the capacitor. Since electret condenser microphones are used almost universally in cell phones, as well as other applications, they are probably the most popular type in current use. The diaphragm of electret condenser microphones is usually made of a thin sheet of flexible plastic which has had a metal coating deposited on it, a notably lightweight arrangement that permits good sensitivity and frequency response. Such microphones also usually exhibit good sample-to-sample consistency, making them practical for multi-microphone arrays.

A more recent variant of the condenser microphone is the silicon microphone. This is a microscopic condenser microphone constructed using micro electromechanical system (MEMS) technology. The required electrostatic polarizing field is provided by a charge pump, and the diaphragm can be free floating, being held in place by electrostatic attraction.

In recent years, the motion of a diaphragm has been measured using optical means, including laser interferometry, which may in time become a practical alternative to the capacitor microphone.

Microphone Dynamic Range
The noise level of a microphone is limited by Brownian motion of air molecules impinging on the diaphragm. Of course, it can be further limited by electrical noise of the amplification which follows it, but even discounting those effects, the so-called self-noise of a microphone is likely to be 10–20 dB higher than the threshold of hearing [18.30]. Lower self-noise is possible with larger diaphragms, say on the order of one inch, but this comes at a price of making the microphone more directional at high frequencies.

The loud signal limit of a microphone is often set by the maximum signal capacity of the associated preamplifier. Using large rail voltages can be of some aid in this regard. Eventually the diaphragm will reach the limits of its travel, and experience soft clipping. There has been some work on using force feedback to linearize the characteristics of the diaphragm [18.31, 32]. With a little effort and care, it is possible to make microphones that can handle high-level signals about as well as the human ear can.

Finally it should be noted that there has been some work on creating a microphone that does not require a diaphragm. Blondell and Chambers did some work on a flame microphone in 1902 and 1910 [18.23]. More recent efforts have used lasers to detect index of refraction changes due to acoustic pressure waves [18.33].

Microphone Pickup Pattern
Interacting with a real, three-dimensional sound field, a microphone is an audio device inextricably tied to spatial considerations. Immersed in the sound field, the microphone responds to the sound at essentially just one point, so its output represents a spatial sample of the total sonic event. This is not to say that it necessarily responds equally to sounds arriving from all directions; on the contrary, most microphones exhibit some directional bias, whether or not intended, and it is probably not possible to specify a single universally preferred pickup pattern.

The simplest arrangement is to put a sealed chamber behind the diaphragm, allowing only a slow leak for stabilization of quiescent atmospheric pressure. With short-term pressure behind the diaphragm fixed, the microphone responds to the sound pressure at the front of the diaphragm. Since pressure is a spatial scalar, it might be expected that a pressure microphone might respond equally to sound from all directions, and indeed this might be the case if the diaphragm had zero dimension. As it is, a finite-dimension diaphragm will spatially integrate the sound pressure across its surface. This will tend to increase output for normal-incident waves and waves with long wavelength, but will have less boosting effect for random incidence sounds, such as Brownian

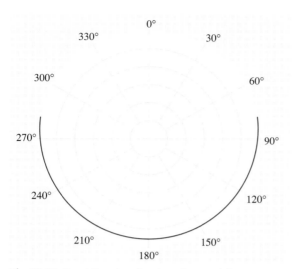

Fig. 18.13 *Omnidirectional* microphone pickup pattern. Note the slight attenuation of sounds arriving from the rear (After [18.35])

noise, and may reduce response to sounds with oblique incidence and wavelengths shorter than twice the diameter of the diaphragm. The resulting pickup pattern may therefore become directional at high frequencies, while remaining omnidirectional at lower frequencies. To remain omnidirectional up to 20 kHz, a diaphragm will have to have a maximum diameter of half the wavelength at the upper frequency limit, or about 8 mm. Whether by

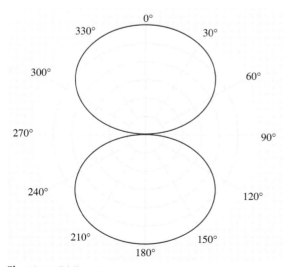

Fig. 18.14 Bidirectional microphone pickup pattern, typical of an open-back ribbon microphone (After [18.35])

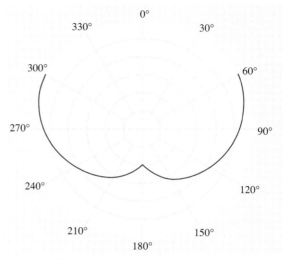

Fig. 18.15 Cardioid microphone pickup pattern, obtained by combining omnidirectional and bidirectional pickup patterns (After [18.35])

accident or design, this happens to be the approximate diameter of the human eardrum [18.34]. The fact that the eardrum is so small, and is coupled to the structures of the inner ear, and yet the ear still manages to exhibit a maximum sensitivity of 0 dB SPL or better has to be considered impressive performance, especially considering that quiet sounds just fade into perceived silence, rather than some inherent background perceptual noise.

More elaborate microphone pickup patterns (Figs. 18.13–18.15) can be achieved by increasing the number and complexity of acoustic paths to the diaphragm. For example, simply opening the back surface of the diaphragm to the atmosphere, the arrangement used in the classic open ribbon microphone, results in a microphone sensitive to the difference between front and back pressure, so sounds arriving from the side produce no response, and the pickup pattern usually resembles that of Fig. 18.8, albeit with the front arrival signal of opposite phase to the rear arrival signal.

If a pickup pattern as in Fig. 18.8 is summed appropriately with an omnidirectional pickup pattern, the rear arrival sound can be made to cancel, producing a cardioid front-directional pickup pattern.

With more elaborate channeling of the audio reading the diaphragm, more specialized or directional pickup patterns may be obtained.

Greater flexibility in the choice of pickup pattern can be obtained by using arrays of microphones, typically arranged in fixed physical orientations, where the effective pickup pattern of the array is dependent on how the individual microphone signals are electrically summed. By recording the individual signals, the array can be effectively focused and oriented after the fact by postprocessing the recorded signals. The large-volume low-cost manufacture of certain types of microphones has made this approach especially attractive.

18.4.2 Records and Phonograph Cartridges

Although by now semi-obsolete, the phonograph cartridge has historically played a key role in sound reproduction, and is fairly close to a microphone in many aspects of operation. As with a microphone, the intent is to act as a transducer for acoustical vibration, in this case the vibration originating from a stylus following the undulations of a groove, rather than a diaphragm responding to vibrations in air.

Although early phonographs employed direct mechanical-to-acoustical transduction, the advent of electronic amplification made the electromechanical phono cartridge the default device for record playing.

Most of the methods of transduction used for microphones have been tried for phono cartridges as well, with piezoelectric and magnetic being the two most popular. Whereas most magnetic microphones have been of the moving-coil type, the magnetic cartridge design has employed both moving magnet and moving coil designs. Perhaps the use of a cantilever design of the stylus to reduce the effective mass of the coil or magnet seen by the stylus tip, coupled with the lack of a diaphragm, led to the preference of magnetic transduction for phonograph cartridges, while the use of capacitive transduction in condenser microphones has tended to be preferred in that transducer.

Despite the popularity of the grooved phonograph record for well over half a century, it has always been a difficult medium with which to achieve full fidelity.

For one thing, the stylus assembly is fraught with high-frequency resonance problems, much like the microphone, which have tended to limit the high-frequency response. In many cases, the high-frequency resonance has been underdamped, resulting in a serious peak in the mid-high-frequency response. The material used to mount the stylus in a way to secure it but still allow it to vibrate, often little bits of rubber, can sometimes degrade with age, impairing the characteristics of the cartridge. Then there is the need for the cartridge to drag a tonearm across the record with it.

In theory, the tonearm should provide a rock-stable platform for the cartridge at audio frequencies, so that only the stylus vibrates. Below the audio band, the tonearm should move as freely as possible, to follow the slowly spiraling groove and miscellaneous movements induced by the rotating disc, including horizontal eccentricity and vertical warp. In practice, this behavior cannot be achieved perfectly (since a perfect *brick-wall* filter is impossible), but at least the cartridge and tonearm should be matched, and the low-frequency resonance peak resulting from the compliance of the stylus and the effective mass of the tonearm should be properly damped, to avoid a peak either in the low-frequency audio response, or in the subsonic region where a resonance could make the system overly sensitive to record warp. Even in the best arrangement of properly damped resonance centered an octave or so below the audio band, the system is only second order, so some low-frequency tonearm vibration and some subsonic stressing of the stylus from pulling the tonearm is inevitable [18.36], which can lead to distortion and uneven frequency response. To avoid the build up of acoustic traveling waves in the tonearm, it is desirable to have its construction and material chosen to damp out such vibrations; not all tonearms have done this well.

Another problem that has hampered record playback, although not strictly the province of the phono cartridge, is that the playback stylus usually does not exactly follow the same path and orientation as the cutting stylus. This is because discs are cut on a cutting lathe, in which the cutter slowly traverses a straight-line path from the outer grooves to the inner grooves. The playback cartridge, in turn, is usually mounted on a pivoted arm, so its path is an arc. (The longer the tonearm, the smaller the disparity between record and playback orientations.) In order to minimize the path mismatch, a right dogleg bend has often been incorporated into the shape of the tonearm. The cartridge has to be carefully mounted on the tonearm so that the stylus is a specified distance from the pivot if this feature is to be properly exploited. Unfortunately, the bent tonearm traversing an arc results in a steady-state force from friction between the moving record groove and stylus tip, one which does not pull straight down the length of the tonearm, but instead generates a sideways force, called the skating force, which pulls the stylus tip against one or the other of the groove walls, which is another potential source of distortion. The skating force varies with groove speed and stylus tip size and shape, among other things, making it challenging to fully compensate.

At various times straight-line tracking tonearms have been produced, usually with the back of the tonearm mounted via bearings to a platform intended to move in a straight line to follow the front of the arm (and the stylus). Since the stylus cannot be expected to drag the rear platform along with it, some sort of servo is used to insure that the platform follows the slow movements of the stylus/cartridge combination. Although this should mitigate skating-force problems, assuming the cartridge is mounted precisely, it still leaves the other difficulties of disc playback.

Another issue with phono cartridges is the shape of the stylus tip. The cutting stylus is chisel-shaped, with sharp corners, so it is capable of inscribing very fine undulations. But the traditional stylus tip shape is round, so the points of contact with the groove walls vary with the curvature of the groove. A round stylus will therefore follow a slightly different path than the cutting stylus, another source of distortion. In the 1960s, elliptically shaped styli became popular to reduce the tracing error, by more closely approximating the shape of the cutting stylus. The smaller stylus tip radius of curvature at the point of contact, compared to a spherical stylus tip, should allow the stylus to more accurately track the motion of the cutting stylus, but results in higher localized pressure, which can increase groove wear unless low tracking pressure is employed. Even so, the precision with which a given playback stylus follows the motions of the cutting stylus is unlikely to be sufficient to avoid some alteration in sound quality.

Around the time elliptical styli were becoming popular, RCA attempted to improve tracing performance by pre-distorting the cutter signal, anticipating playback with a round stylus. The system was not universally welcomed [18.37], in part because it arguably impaired performance with elliptical styli, and was eventually dropped.

Ultimately it has to be accepted that there will be some, hopefully slight, difference between the motion of the cutting stylus and the motion of the playback stylus, and that there are some signals that the playback stylus cannot hope to track properly, such as a constant-amplitude-cut square wave. The best one can hope for is that the deviation will be small and gentle, and that the resulting distortion will be mostly low-order THD. Even so, distortion figures for records were not uncommonly in excess of 10%, especially for the slower inner grooves. The technique of stereo recording on discs, pioneered by Blumlein in 1931 and introduced commercially in 1957, further complicated the task of the playback stylus, as it then had to track both horizontal and vertical undulations simultaneously.

Getting adequate dynamic range from records was a tricky process, fraught with compromises. For one thing, the maximum signal level used tended to be traded off against total playing time. Large-amplitude grooves took up more real estate than quiet grooves, which led to the use of variable-pitch, look-ahead record cutters that could dynamically adjust groove spacing in response to groove amplitude. Such cutters functioned by increasing the spacing one revolution before the arrival of a loud sound, and maintaining the spacing for the revolution that followed. Further complicating the amplitude/playing-time conundrum is the desire on the part of some recording engineers to minimize use of the inner groove area, to minimize inner groove distortion and bandwidth limitations.

Maximum amplitude was also limited by considerations of what the playback styli and cartridges of the consumers could accommodate, which could lead to lowest-common-denominator compromises in performance. To better match tracking ability and noise floor to the average statistics of music spectra, the Recording Industry Association of America (RIAA) standard equalization curve was adopted in 1954 that progressively reduced low frequencies and boosted high frequencies on recordings, applying inverse compensation during playback. This, along with the characteristics of the stylus, record surface and groove speed, made the overload characteristic of records strong functions of both frequency and playing time, so the precise specification of overload characteristic of records is problematical at best. To be sure, there are specified suggestions for maximum cutting amplitude and velocity, but given the commercial realities, at least with pop music, that louder all too often equates with better, the suggested limits have not always been observed.

The noise level of a record is dependent on the smoothness with which a groove can be cut, which in turn depends on the cutting technique and the material used. Although wax and shellac were used in the early years of record production, more modern practice has been to use lacquer deposited on an aluminum disc base. In conjunction with the hot-stylus cutting technique introduced in 1951 [18.9], a lacquer master disc is capable of a dynamic range approaching 70 dB, although in practice a somewhat lower figure is sometimes encountered for the end product on a production basis. Most of the steady-state noise encountered with records is concentrated at mid to low frequencies. In pursuit of the goal of fitting a performance to the dynamic-range characteristics and limitations of grooved discs, it has traditionally been common practice to employ a combination of compression and limiting to the signal in the course of cutting a master disc. There have been some attempts to apply complementary compression and expansion to record audio to improve the system dynamic range, notably discs produced by dbx Inc. in the 1970s.

Successful record playback can also run afoul of a problem known as acoustic feedback, if sound from the loudspeakers causes the disc itself to vibrate as a diaphragm. Solutions include the use of an inert turntable mat, remote positioning of the turntable, and limiting the playback level.

Of course, dust, dirt, scratches and wear will tend not only to degrade the recorded signal but also to raise the noise level, including the possible introduction of clicks and pops.

Given the litany of compromises and limitations in disc recording and playback, the level of performance that has been achieved with this medium is laudable. Although largely eclipsed by digital audio, some refinement of grooved record playback continues, including efforts to achieve the holy grail of disc playback, contactless playback via light or laser beam, notably by Fadeyev and Haber at Lawrence Berkeley National Laboratory [18.38]. The availability of advanced DSP has allowed after-the-fact improvements in noise reduction and distortion reduction that would have been impossible in an earlier age. It is tantalizing to consider the performance that might have been possible with this medium if such processing had been available and incorporated into the basic architecture of the disc recording system.

18.4.3 Loudspeakers

Loudspeakers and headphones provide complementary transduction to microphones, namely converting electrical audio energy to acoustical energy. Many of the principles and notions regarding microphones can be applied to loudspeakers either directly or by, in a sense, turning the ideas upside down.

In a manner reminiscent of microphones, few loudspeakers convert directly between electrical and acoustical energy. Instead, the electrical energy is converted to mechanical energy, specifically the movement of a diaphragm, often cone-shaped or dome-shaped, which then radiates acoustical energy, sometimes aided by an acoustical coupling horn or lens.

Indeed, many of the same transduction methods that have been used in microphones have at least been tried for loudspeakers including magnetic (also known as *dynamic*), piezoelectric, and condenser.

The genesis of loudspeaker-like transducer actually predates Bell's invention of the telephone by a couple of years when, in 1874, Ernst W. Siemens described a dynamic moving-coil transducer, although he does not seem to have initially done much with it [18.39].

Bell's telephone, patented in 1876, also used a dynamic receiver. With a magnetic armature mechanically coupled to a diaphragm secured at its edges, it was probably closer in spirit to a headphone than a loudspeaker.

There followed a series of refinements that progressively brought the output transducer closer to the modern concept of a loudspeaker. In 1898, Oliver Lodge patented a spacer to maintain the air gaps in the magnetic circuit. Three years later, John Stroh came up with the notion of attaching the cone to the frame via a flexible, corrugated surround that improved the linear travel of the cone. Proper, flexible centering of the rear of the cone in the magnetic gap was provided by the *spider*, developed by Anton Pollak in 1909.

Many of these early loudspeakers used an acoustic horn of some sort to project the loudspeaker output into the room. The horn improves efficiency, partly by improving the acoustic coupling from the loudspeaker to the room, and partly by collimating the sound into a beam directed at the listener. Credit for the first true direct radiator loudspeaker design to include all the elements of what can be regarded as the modern-day loudspeaker is generally accorded to Rice and Kellogg in 1925 [18.40]. Their design took explicit account of the baffle on which the driver was mounted in the overall acoustic design.

There were of course, countless follow-on refinements to the basic electrodynamic driver, a process that continues to present day. Other methods of transduction were also explored. In 1929, E. W. Kellogg filed a patent for an electrostatic loudspeaker, consisting of panels not unlike large condenser microphones. The design has not been as widely accepted as the dynamic driver, in part due to some limitations in maximum output. There have also been piezoelectric drivers, but the limited travel of the piezoelectric element has limited their use to mid-range drivers, tweeters, and earphones.

So, as the Rice/Kellogg baffle indicated, with the basic dynamic loudspeaker configuration established, much of the attention focused on refining the loudspeaker/cabinet combination as a system.

The output ultimately obtained from a raw loudspeaker driver is profoundly affected by the surrounding baffle or cabinetry. In isolation, a loudspeaker driver produces two acoustical outputs: one from the front of the diaphragm or cone, and the second from the back. Unfortunately, the two outputs are out of phase. Were they generated at the same place and time, they would cancel; the loudspeaker would produce nothing but heat. So part of the goal in designing a complete loudspeaker system is to keep those two outputs from annihilating each other.

At frequencies above the point where the wavelength is comparable to the size of the speaker cone, the radiation will tend to be a directional beam, and the front and back waves will tend to naturally go their own separate ways. At lower frequencies, the front and back radiation will tend to be more omnidirectional, and will therefore tend to cancel. This can be avoided by requiring that the size of the loudspeaker be at least as large as the wavelength of the lowest frequency of interest. For a low-frequency limit of 20 Hz, this would correspond to a diameter of at least 15 m, which at the very least would make for rather unwieldy boom boxes.

Happily, it is not necessary to rely on the loudspeaker itself to avoid low-frequency cancellation. Instead, a smaller driver can be mounted on a baffle, and the baffle can be used to prevent interaction of the front and back radiation. Indeed, a common concept in loudspeaker system theory is that of the *infinite baffle*, which precludes cancellation at any frequency, at the cost of being physically impossible. Even assuming a willingness to forego response below 20 Hz, a baffle of at least 15 m would still be required, which is not much more practical than an unadorned 15 m woofer. This leads to the notion of putting the loudspeaker in a box, or loudspeaker cabinet. (A speaker mounted in a car door or trunk lid is operating in an approximate infinite baffle, although the car interior can be considered a semi-sealed cabinet.)

The simplest arrangement is to make the box completely sealed, so that the rear radiation is trapped within the box, and cannot emerge to cancel the front radiation. Of course, this requires making the box quite rigid, else the rear speaker radiation will simply be conducted through the walls of the box. One problem with sealed-box systems is that the springiness of the air in the cabinet will decrease the natural compliance of the loudspeaker, so that a driver with a respectably low free-air resonance may have a much higher resonance mounted in a sealed box, with correspondingly reduced low-frequency response.

This issue was addressed in 1954 by Edgar Villchur, in a manner which in hindsight seems remarkably straightforward, namely increasing the compliance of the speaker to obtain the desired performance when it was mounted in a sealed box. This approach, referred to as an acoustic suspension design, made the cabinet an inherent part of the design, and is still in use today.

The alternative to the sealed-box loudspeaker enclosure is, logically enough, the vented box, in which at least a portion of the back radiation of the driver is allowed to emerge from the cabinet, often delayed or phase-shifted to better align it with the front radiation. One of the early examples of this approach was the bass reflex design by Albert Thuras of Bell Laboratories, in 1930. Other vented designs have used elaborate internal tubing, internal ductwork (such as the Klipsch folded horns), and or provided mass loading of the exit port via a diaphragm (the *passive radiator*). The goal is to improve efficiency, partially by relieving some of the back pressure behind the driver, but mostly by adding some of the back radiation to the front radiation in a nondestructive fashion, to achieve some increase in output, typically on the order of 3 dB.

More recently, hybrid enclosures have been developed, sometimes called bandpass enclosures, in which a vented or sealed box is used behind the driver and a vented box is used in front of the driver, to provide improved low-frequency efficiency over a somewhat narrow range of frequencies [18.41]. This design is popular in some smaller low-frequency speakers.

Regardless of its other characteristics, a bare loudspeaker cabinet is likely to impart a variety of spurious resonances to the response from sound waves bouncing helter-skelter around the inside of the cabinet. For this reason, loudspeaker cabinets are often lined or stuffed with sound-absorbing material. This alters the overall response as well as damping the resonances, and must be accounted for in the overall design.

Much of the modern modeling of loudspeaker systems is based on the work of Thiele and Small in the 1970s, who established the basic parameters governing the performance of loudspeaker systems, and developed accurate mathematical models incorporating those parameters, allowing rapid and precise mathematical prototyping [18.42].

Still, the implications of achievable, cost-effective drivers and practical cabinet size is that there is precious little that can be done to get usable bass response down to 20 Hz from a small loudspeaker system. It is possible to apply low-level equalization to the signal upstream of the loudspeaker, but this can increase the cost and the power requirements, and will not be effective unless the driver can handle increased signal level without serious distortion. The closed-box design is considered one of the better choices for the use of equalization, as its response roll-off below primary resonance is only 12 dB per octave, which is gentler than most vented systems.

Most of the issues relating to speaker cabinetry pertain to low frequencies. Midrange and high-frequency drivers may also have their backs sealed, but the wavelengths are shorter, and the required resonant frequencies are a lot easier to attain.

The notion of using multiple drivers in a loudspeaker system seems to have originated in 1931 with a two-way system designed by Frederick of Bell Laboratories [18.43]. A three-way system followed within two years. It is simply much easier to design a system to have flat frequency response with multiple drivers, optimized to handle a subset of the audio band, than to try to accomplish the task with a single driver.

Part of the motivation for using multiple drivers is that the loudspeaker, like the microphone, is an inherently spatial entity. Sound radiates from a loudspeaker in all directions, depending on angle and frequency. In an anechoic environment, with a listener situated on the main axis of the loudspeaker, the response on that axis is well defined, and the only one that really matters. In a real room, however, the multidirectional acoustical output of the loudspeaker will reflect off walls and other surfaces, generally producing a complicated multiplicity of arrivals at the listener's ears. There is no easy way to characterize this process, or of the resulting perception, although in the 1970s, Roy Allison made significant strides in characterizing the loudspeaker–room interaction at low frequencies [18.44]. However, it certainly establishes the importance of the loudspeaker spatial response at all angles, and suggests the desirability of a loudspeaker having a consistent radiated frequency response (radiation pattern) at any angle.

This is another area in which multiple drivers can be beneficial. Again, the output of a driver will begin to get directional at a frequency corresponding to a wavelength about twice the diameter of the driver. Assuming consistently wide dispersion is desired, a 30 cm woofer will be useful up to about 500 Hz, a 7.5 cm midrange continuing to 2 kHz and a 2.5 cm tweeter maintaining wide dispersion to at least 6 kHz. These figures may be conservative, in part because controlled decoupling of the speaker cone can result in a progressively smaller radiation area at successively higher frequencies, decreasing the effective diameter and broadening the resulting dispersion. Other approaches to maintaining a desired radiation pattern across frequency include the use of acoustic lenses and phased arrays of multiple drivers within a given frequency range. It is rare for a speaker to have truly constant directionality across its entire pass-band, in part from that fact that most are at least somewhat directional at mid and high frequencies, and, because of the long wavelengths involved, almost unavoidably omnidirectional at low frequencies. (Such long wavelengths may be difficult to localize, enabling the use of satellite systems with only one woofer to support two or more channels.)

The use of two or more frequency-selective drivers generally imposes the need for a crossover network to apportion the signal on the basis of frequency. Not only does this avoid wasting power, but it can be absolutely necessary to avoid overdriving or burning-out a driver. Some care is generally required in the transition frequency region to maintain flat response, and although there are well-regarded canonical topologies, such as the Linkwitz–Riley crossover developed in 1976 [18.45], real-world deviations from ideal driver behavior may engender the use of recursive approximation techniques to optimize system response fully.

The combination of crossover networks, drivers, and room loading will often result in a loudspeaker having an electrical input impedance characteristic that is strongly frequency dependent, and which poses a challenge for a power amplifier.

As is no doubt evident, a loudspeaker system and its associated radiation pattern is a fiendishly complicated device that does not readily lend itself to characterization by traditional audio metrics or specifications. In consideration of such specifications, it is difficult to make definitive statements that fully characterize the sound of a speaker.

The bandwidth of most speakers may approach the ear's 20–20 kHz bandwidth, but few speakers make it all the way, particularly at low frequencies, and bandwidths on the order of, say, 70–14 kHz are far more common, at least in consumer equipment.

The deviation of the frequency response is difficult to talk about meaningfully, but it seems desirable for speakers to be within 1 dB of flat response across their useful bandwidth, at least on axis, and better speakers can approach or achieve this, although at low frequencies the room exerts enough effect as to likely require tuning or experimentation with speaker position to maintain a flat response. The response at other angles should probably be smooth and well controlled, but the exact optimal specification has not been established. Perhaps the ideal speaker would adjust its radiation pattern to match the instruments being reproduced. (But then what is the preferred radiation pattern of a purely electronic instrument, such as a synthesizer?) Although it is necessary to measure the response of a loudspeaker in a room in order to accurately gauge the low-frequency response, a direct narrow-band or sine-wave measurement of the system response in a room at middle and high frequencies is likely to yield a useless, mad scribble of a response, owing to the complex interactions of the echoes at the point of measurement. Fortunately, the perceived response should be a lot smoother than what is measured, in part because the ear differentiates sound arrivals on the basis of direction. Measurement techniques that attempt to address this include the use of directional microphones, anechoic and semi-anechoic measurement, and time-gated response measurement.

Timing precision is another specification that is difficult to pin down with loudspeakers, again partly due to the effects of room echoes. In one respect, most speakers do pretty well in first-arrival timing, since the ear's temporal pre-masking resolution is on the order of 1–2 ms, corresponding to about 30–60 cm of path length, and speaker systems with all drivers mounted on a common front panel will generally be well within this tolerance for inter-driver path-length differences. Still, audible time disparities may result from primary resonances, particularly at low frequencies, and high-order crossovers.

In the area of distortion, loudspeakers also do pretty well, all things considered. At normal listening levels, a properly designed loudspeaker exhibits fairly low nonlinearity, with a fairly gentle limiting characteristic, leading to low IM distortion and fairly low THD, mostly third order, which is likely to be masked by most signals. Most properly operating loudspeakers do not produce drastic amounts of distortion until they reach the limits of their travel. To be sure, some loudspeakers can sound audibly harsh, but this is not necessarily a byproduct of distortion, it may be more related to an uneven response or radiation pattern. One bugaboo of many drivers is cone breakup, wherein the cone no longer acts as a piston. Although this can look unsightly in slow motion, it will not necessarily produce distortion unless there is nonlinearity present, and most speaker cones, being passive structures of paper, plastic, or metal in and of themselves, are not especially nonlinear. While such breakup is likely to affect the response and/or the radiation pattern of the speaker, it may not produce much distortion per se.

Although the basic dynamic loudspeaker has now been in use for over 70 years, and has been greatly refined in that time, it is still the subject of active research, in both design and evaluation. One of the holy grails yet to be achieved is, like the diaphragm-free microphone, the cone-less loudspeaker that more directly converts electrical energy to acoustical energy. Aside from eliminating some troubling mechanical elements and their associated resonances, this might improve the efficiency, which is currently typically around 1% for consumer loudspeakers.

Another long-sought goal is the projection loudspeaker, which can make a sound appear to emanate from a specified remote location. The have been some encouraging results in the use of ultrasonics [18.46] and large-scale speaker arrays [18.47] in this regard.

18.4.4 Amplifiers

Of all the common audio analog devices, electronic amplifiers of all types probably have the easiest time meeting the preferred specifications of a good-quality audio system. This in no small part derives from the fact that they generally contain no mechanical processing stages, and simply have to push a bunch of very lightweight electrons around. Properly designed, an amplifier should have little difficulty achieving a dynamic range approaching 120 dB, nor is it likely to deviate significantly from the preferred specifications for bandwidth, response deviation, phase/timing response, or distortion.

This is despite the fact that most amplification devices, mainly vacuum tubes, transistors, and field-effect transistors (FETs), are not themselves inherently very linear. In the years following de Forest's invention of the triode, considerable and rapid strides were made in not only the refinement of the devices themselves, but in circuit designs which optimized the linearity and response of the complete amplifiers. A major milestone was the invention in 1928 of the negative-feedback amplifier by *Black* of Bell Laboratories [18.9]. This arrangement compares the output signal with the input signal, and to the extent that they may tend to differ, generates an instantaneous correction signal. Successful application of negative feedback carries with it specific requirements on the performance of the raw amplifier, such as maximum phase shift, and some of these requirements were still being uncovered years after the initial development of negative feedback.

Several standard types of amplifiers are commonly found in audio use, each with specific requirements and design challenges, including preamplifiers, line amplifiers, power amplifiers, and radio frequency (RF) and intermediate frequency (IF) amplifiers for use in radio. Preamplifiers, for example, are required to handle the generally very small signal from transducers such as microphones, phonograph cartridges, tape heads, and optical photocells. Linearity is usually not a major problem, because of the small signals involved, but low noise and impedance matching to the transducer are often significant design considerations. At the other end of the chain, power amplifiers usually have a fairly easy time with dynamic range, but high required power-handling capability and the need to drive sometimes ill-behaved loudspeaker loads make power amplifier design an art unto itself, with the 1948 Williamson amplifier being just one early example of a notable design [18.48].

The advent of the transistor, by Brattain, Bardeen, and Shockley of Bell Laboratories in 1947, sparked a revolution in active devices, and stands as one of the foremost inventions of the 20th century [18.49, 50]. Operational differences between tubes and transistors were sufficiently great that, to an extent, the discipline of amplifier design had to be reinvented from the ground up, and it was some time before solid-state amplifiers were accepted in some high-end audio systems.

In time, the development of analog integrated circuits gave rise to the multi-transistor operational amplifier, an amazingly versatile device that facilitated cost-effective analog processors of great sophistication and complexity. These days, entire analog audio subsystems can be implemented on monolithic chips with many transistors and associated components.

Special mention should be made of one particularly challenging amplifier variant, the voltage-controlled variable-gain amplifier (VCA). Such a device, basically an analog multiplier, is essential for exercising automatic control of signal level, a common building block in many audio signal processors. Where fixed-gain amplifiers depend on fixed, highly linear passive components such as resistors to maintain operating point and linearity, VCAs require linear active elements, like FETs. In the late 1960s, Barry Blesser designed a novel variable pulse width multiplier, and a few years later, David Blackmer of dbx, Inc. designed a wide-range variable-transconductance VCA, elements of which are still used in commercial VCAs [18.51]. The complexity and cost of VCAs limited their use somewhat, which is perhaps ironic in these days of digital audio, where DSP chips typically perform millions of multiplications per second.

18.4.5 Magnetic and Optical Media

Part of the motivation for the use of magnetic and optical recording on linear media (tape, film) undoubtedly came from the desire to avoid the wear inherent in the playback grooved records. Both media involve the use of narrow apertures to record and play audio striations perpendicular to the motion of a linear medium.

Magnetic recording got its start in 1898, with Valdemar Poulsen recording on steel wire, an especially notable accomplishment considering that the invention of the triode amplifier tube was still several years away. The use of magnetic tape originated with O'Neill's patent of paper tape coated with iron oxide, in 1926, followed by the introduction of plastic tape by BASF in 1935. A serious nonlinearity problem with magnetic tape was largely resolved in 1939 with the development of alternating-current (AC) recording bias. Considerable refinement was achieved during and following World War II, and consumer reel-to-reel tape recording became a reality in the 1950s. There followed a series of formats which expanded the performance envelope and consumer friendliness, most notably the compact cassette and the venerable eight-track cartridge.

Optical recording on film was first investigated around 1901 [18.9], five years before the vacuum tube, and started to receive significant research and development attention starting around 1915, with the work of Arnold and others at Bell Laboratories, followed by a number of efforts by RCA and other groups. Lee De Forest may have come up with the first viable optical sound system, *Phonofilm*, in 1922 [18.52]. By 1928, synchronized optical sound on film was used commercially on Disney's *Steamboat Willie* cartoon.

Both of these formats rely on some sort of device with a narrow aperture to record and playback, and therefore both face a tradeoff between high-frequency response, corresponding to the smallest resolvable feature, and media speed/playback time.

Although there may be no vibrating mechanical elements in either the magnetic system or optical playback, both media exhibit difficulty reaching the full audible bandwidth, or in maintaining ruler-flat response, although, properly designed and adjusted, they can come fairly close. Beyond that, both media bear some similarities to modern record-based reproduction, including the use of fixed equalization, modest, gentle distortion at typical operating levels, a typical dynamic range on the order of 50–70 dB, and both absolute and dynamic speed error sensitivity.

The viability of these formats was significantly enhanced by the invention in 1965 of analog noise reduction by Ray Dolby. Noise-reduction systems usually consist of a compressor/encoder and an expander/decoder. The compressor boosts low-level signals to keep them above the noise floor of the channel, while the expander restores the signals to their proper level, reducing the noise in the process. The Dolby A-type noise-reduction system and follow-on systems exploited a range of solid-state circuit tools, including active filters, precision rectifiers, and linear FET-based VCAs, to implement a comprehensive solution to the problems faced by noise-reduction systems. The frequency-selective nature of masking by the human ear was addressed by splitting the signal processing into multiple bands, and the problem of potential overload associated with the sudden onset of loud signals was resolved by the use of a novel dual-path limiter. Other well-known NR systems used wide-range bandpass compressors (Telcom C-4), wide-band compressors with preemphasis (dbx I and II), sliding shelf filters (Dolby B), cascaded sliding shelf filters (Dolby C), cascaded wideband and variable slope compressors (dbx/MTS TV compressor), and cascaded combined sliding shelf and bandpass compressors (Dolby SR and S). Similar principles would subsequently be adapted for use in digital low-bitrate coders.

18.4.6 Radio

While the medium of analog radio has its own share of unique attributes, it shares with other analog media many of the same sorts of characteristics and limitations, including somewhat limited frequency response and dynamic range, which is partly associated with bandwidth conservation.

Commercial amplitude modulated (AM) radio broadcast kicked off in 1921 when KDKA went on air in Pittsburgh [18.9]. The use of dual-sideband modulation made for simpler receivers at the cost of channel spacing being twice the highest transmitted audio frequency. Although in theory AM radio was capable of supporting wide audio bandwidth, a channel spacing of just 10 kHz was chosen to conserve spectrum space, limiting audio bandwidth to something less than 5 kHz in practice.

Wide-band frequency modulated (FM) radio was developed by Edwin Armstrong in 1933, motivated in part by a desire to reduce the static that was common to AM radio, with commercial broadcast originating in 1941 [18.9, 53]. The dual sidebands of standard FM extend quite a bit farther than the maximum audio frequency supported, with channel spacing of 200 kHz for a typical audio bandwidth of 15 kHz. Dynamic range in excess of 60 dB is possible with mono FM, although this was compromised somewhat by the choice (in the US at least) made by the Federal Communications Commission (FCC) of the stereo multiplex system in 1961 [18.9].

18.5 Digital Audio

The advent of digital audio, perhaps best signified by the introduction of the audio compact disc in 1982 by Sony and Philips, completely revolutionized the world of audio engineering, in effect completely rewriting the rule book on what was readily achievable in mainstream audio systems, not to mention how signals were processed.

The basic notion of digital audio is discreteness. Instead of specifying a signal as a continuous function of time and amplitude, it is expressed as a rapid series of sample values at discrete time intervals, often with integer or finite-precision values. The signal becomes, in effect, a completely specified list of numbers, and as long as no explicit errors are made, can be copied an arbitrary number of generations without any loss or alteration whatsoever. (Error-correction schemes commonly help guard against the occasional error slipping through.)

The mathematical foundations of discrete sampling of continuous functions go back surprisingly far, with at least one claim on behalf of Archimedes, around 250 BC [18.54, 55]. Additional support for sampling theory was provided by *Fourier* in 1822 [18.56, 57] and *Cauchy* in 1841 [18.58, 59]. Modern sampling theory is usually credited to the combined work of *Whittaker* (1915) [18.60], *Nyquist* (1928), *Kotelnikov* (1933), and *Shannon* (1949) [18.61, 62], who provided a rigorous proof [18.63].

The key point of the sampling theorem is that a continuous band-limited signal can be discretely sampled without loss of information, and the original continuous signal subsequently recovered from the sampled version, as long as the sampling frequency is higher than twice the signal bandwidth.

In order to exploit the sampling theorem and digital audio practically, quite a few other pieces of the puzzle had to be supplied. George Boole contributed the binary number system in 1854, which would provide the numerical and logical basis of modern digital processors. Alec Reeves used Boole's binary numbers to represent audio samples, creating the pulse code modulation (PCM) format, in 1937. By the late 1960s, prototype digital audio recorders were operating at companies like Sony, NHK, and Philips [18.58, 59], leading to the 1982 debut of the audio CD.

Some appreciation of the performance of digital audio can be had by considering the traditional audio specifications that analog systems have for so long tried to fully embody.

Bandwidth. Traditionally an uphill battle fighting mechanical resonances and slit-loss effects, successfully meeting a bandwidth specification in the digital domain is simply a question of selecting a high enough sampling rate. Similar principles apply to maintaining the precision of a frequency response specification, usually a flat response: without parasitic or extraneous influences, the response will be ruler practically-flat by default, although it can otherwise be explicitly and predictably altered by application of appropriately designed DSP filters and equalizers.

Distortion. With signal processing implemented as a series of digital additions and multiplications, which are presumably accurate to the least significant bit, non-linearity and distortion in the digital domain is absent unless, like deviation from flat response, it is explicitly introduced into the processing, to which end it can usually be precisely applied and controlled.

Dynamic range and signal-to-noise ratio. The dynamic range is usually expressed as the logarithmic difference between the maximal signal level and the noise level. By default, the maximal signal capacity of a PCM DSP system as a function of frequency is flat. The noise level is once again a design choice, ergo so is the net dynamic range. In somewhat simplified terms, the noise level of a digital signal process drops by a factor of two (6 dB) for each additional digital bit of precision employed in the representation of the sample values. So, maintaining a desired dynamic range is fundamentally a design choice to employ a sufficient number of bits of precision.

Speed. Once a signal is in the digital domain, there is of course no error or variation in speed unless it is explicitly imposed with appropriate processing.

In short, the attainment of performance that in the analog domain might require considerable design skill, and even then might simply be impractical or impossible, becomes in the digital domain a matter of design choices.

Of course, since most audio signals begin and end as analog vibrations in air (and, so far, correspondingly analog electrical signals associated with input and output transducers), there must be means to convert between analog and digital domains. This is the province of analog-to-digital converters (ADCs) and digital-to-analog converters (DACs). With so much of the overall performance of a digital audio system riding on the performance of these devices, their design has been the subject of intense activity since early experimental digital audio systems. Those early systems were fortunate to muster a precision of just 10-bits per sample, but that has slowly evolved over time to the point that current converters can operate at 24-bits per sample or better in a practical, cost-effective manner. Similarly, the various sources of subtle distortions that might otherwise limit ultimate converter performance have progressively been identified and, on the whole, reduced to the point of insignificance. And, where once speed accuracy of an audio system depended on the skill with which one turned a crank, or perhaps the vagaries of a wind-up clock motor and mechanical speed regulator, the sampling rate of a digital/audio converter is usually controlled by a high-precision electronic timing circuit, such as a quartz oscillator, reducing speed anomalies of all sorts to a level far below that of human perception.

Lest the case for digital audio appear at this point a bit overly rosy, it must be allowed that there are limitations, restrictions and pitfalls; and that it is perfectly possible to produce miserable sound via digital means.

For one thing, there is the matter of the Nyquist frequency, defined as half the sampling frequency and otherwise the highest frequency that can be represented at a particular sampling frequency. In some respects, the Nyquist frequency of a digital audio system corresponds to infinite frequency of an analog system. This introduces certain response-warping effects, such that it may be difficult or impossible to replicate a given analog system response in the digital domain. (Example: the phase shift at the Nyquist frequency can only be $0°$ or $180°$.)

An attempt to represent a frequency greater than the Nyquist frequency in the digital domain will result in a different frequency being generated within the band, usually by reflecting the original frequency value back through the Nyquist frequency, a form of intermodulation distortion. Unless the goal is to produce weird spacey effects, this process is usually not highly desirable, and a low-pass filter is commonly placed ahead of an ADC to prevent that occurrence.

Then there is the issue of performance levels being matters of system architectural choices, rather than the result of possibly elaborate or laborious designs. Such design choices are not always free, in that they may be tradeoffs with either other system resources, or simply cost more monetarily, which may therefore still have the result of limiting overall performance to something less than idyllic near-perfection.

For example, the dynamic range of the human ear was cited earlier to be on the order of 120 dB or more. If the digital representation of an audio signal yields an approximate dynamic range of 6 dB for each bit used, it would follow that the logical word size to use for digital audio samples would be at least 120 dB/6 dB per bit = 20 bits. Yet the standard chosen for compact discs, as well as many other current generation digital devices, is just 16-bits per sample, corresponding to a best-case dynamic range of only 96 dB. This decision was predicated in part on the performance limitations of analog/digital converters at the time the CD specification was cast, and further from a desire to provide a playing time of 74 min on a 12.7 cm disc. Had a larger word size been chosen, either the disc would have had to be larger, or the playing time reduced.

Does this result in audible impairment of the sound? It can, but it will not necessarily be the case, so it will be evident some hopefully small percentage of the time. For example, there should be no audible alteration if the maximum playback level does not exceed 96 dB SPL. Nor should there be a problem if the playback level is higher, but the room noise is high enough to mask the digital noise. Since the ear is only maximally sensitive in the upper-middle part of the audio band, the elevated thresholds towards the band edges provide additional noise margin. And the presence of signal components will tend to mask digital noise that might otherwise have been heard. So, consistent with practical experience, 16 bit PCM digital audio is free of serious imperfections for most listeners and most source material, most of the time.

18.5.1 Digital Signal Processing

With the practice of representing audio signals in the digital domain as PCM samples comes the need to perform signal processing in the digital domain. To some extent, this required a reinvention of signal processing in terms of discrete numerical sample values, but since the foundations of signal processing were mathematical to begin with, in some respects there was a more direct connection between theory and practice.

The most common signal processing operations – amplification, filtering, and equalization – are mostly implemented with the basic four mathematical functions: addition, subtraction, multiplication, and, perhaps to a lesser extent, division.

More elaborate signal processing functions, such as rectification or level derivation, may involve geometric or transcendental functions such as square roots, cosine, or logarithms, virtually all of which are easier to implement, more direct, more wide-range, and more stable as digital entities than analog alternatives. Even some seemingly simple functions, such as a pure delay, can be challenging to implement well as analog devices, but become more-or-less trivial with DSP.

Clearly, the requirement of audio DSP to perform large numbers of numerical calculations and the ability to perform such calculations with modern digital hardware, including personal computers and specialized DSP chips, whether by accident or design, has been auspiciously symbiotic. Large, elaborate analog breadboards of spaghetti wiring have been replaced with even more elaborate programs of (sometimes) spaghetti C code.

The ability to create and combine a virtually limitless number of new processing functions by simply typing in the code for them, using a computer language with full mathematical support, is an understandably powerful tool to extend signal processing, a little like a bottomless Tinker Toy. The effect is amplified by the ability to nest functions to an almost arbitrary extent, allowing the creation of labyrinthine tree-structured processes. And whereas every block of an analog circuit must be physically implemented, DSP programs usually get by with just one physical copy of each routine, regardless of how many times the function may appear in the block diagram.

With all this processing power available, one may inquire as to what has been applied. Certainly, one function category that has *not* been strongly needed has been aids to correct deficiencies inherent in the recording system. If a clean recording is obtained to begin with, there is often little need to clean it up further. Digital signal processors have, however, been very effective in cleaning up and correcting the imperfections of older, analog recordings, and have been used for noise suppression, speed correction, and even distortion reduction.

The additional processing power afforded by DSP processors has enabled re-engineering of most traditional audio processors, such as equalizers, with greater complexity and precision. The availability of a simple, cheap delay mechanism with equalization has allowed extremely elaborate and faithful reverberators to be employed in both professional and consumer applications. This is one area for which there was arguably never a truly effective analog electronic solution. The best pre-digital reverberators were electromechanical, using transducers attached to a metal reverberation plate.

In general, the availability of a cheap, low-noise, high-precision multiplication operation has, along with other DSP elements, made it much easier to implement elaborate dynamic systems, once limited by the relative scarcity and high cost of analog VCAs. This has also made it much easier to include specific nonlinearities, where appropriate, usually in the signal analysis of a processor. And the ability to combine multiple processing algorithms into a single DSP program has facilitated multifunctional systems implemented on a single DSP chip.

One of the key building blocks to be brought to bear in a variety of digital signal processors has been a digital model of some or all of the human auditory system. If a processor can be made to hear the same way as humans do, it can make processing decisions that reflect the truly audible elements of a signal, resulting in far more effective processing. A key component of such an approach has been the digital filter bank, essentially a bank of bandpass filters that, in most cases, cover the entire audio band. This component provides a cost-effective analog of the filter-bank-like processing performed by the basilar membrane of the inner ear. A major element of such processing has been the so-called fast Fourier transform (FFT), first devised in 1805 by *Gauss* [18.64–66], then rediscovered in 1965 by *Cooley* and *Tukey* [18.67, 68]. This one algorithm allows full-range filter banks of great complexity, typically implementing hundreds or thousands of bandpass filters, with dramatically less processing power than would otherwise be needed by direct implementation of such filters. The use of this technique does, however, carry with it some compromise, as the bandpass filters of an FFT filter bank all have the same bandwidth, measured in Hertz, and are linearly spaced in center frequency, whereas the filters of the basilar membrane are logarithmically arranged in both bandwidth and center frequency. This discrepancy is sometimes compensated by processing the FFT outputs in groups, or bands, whose width and center frequency are arranged in an approximately logarithmic manner.

The transform-based filter bank is used both as an analysis tool, to present signals in a perceptually relevant domain for subsequent processing, and as a component of the actual signal path. In the latter instance, signals are processed in the transform domain, usually in overlapping blocks of samples, and ultimately restored to time-domain PCM values by performing running inverse transforms. This has elevated transform-based processing to the level of a new category of digital signal processing. In some situations, a variant transform of some sort may be used to provide, for example, logarithmic frequency selectivity, but in many cases such transforms are still based on a core FFT.

A final note on numerical precision: virtually any digital processor will employ finite-word-length symbols for numerical values, be they integers or floating-point numbers. This gives rise to the generation of digital noise, resulting from the round-off error of each arithmetic operation. Since there may be many thousands of such performed in the course of a DSP algorithm, the numerical precision of the processor must be great enough that the accumulated noise remains insignificant. For a 120 dB dynamic range, corresponding to 20 bits of precision, one might prefer at least 24 bit processing (that is, either 24 bit integers or 24 bit mantissas of float values). The actual precision required will depend some on the actual operations performed, with subtraction (or addition of opposite-signed values) especially troublesome, since it can leave a very small residual, potentially exposing the digital noise. As it happens, this is an issue in performing the FFT. As an additional hedge against the buildup of digital noise, many DSP processors employ double-length registers, so accumulated quantities are only rounded back to standard precision when stored back to memory, usually after multiple multiply–accumulate operations.

18.5.2 Audio Coding

The commercial introduction of digital audio in the form of the compact disc brought with it the reincarnation of a tradeoff that had long been a source of compromise in analog audio, namely that between quality and capacity, or equivalently playing time, or bandwidth. Mechanical audio recordings (i.e. records) had started with cylinders that played for a minute or two, and had progressed to long-play records with about 20 minutes per side. Magnetic tape had used media running at speeds like 76 cm per second, and had progressed to cassette tapes with typical playback times of 45 minutes per side, usually aided with noise reduction. Broadcast channels were less flexible, but techniques had at least progressed to the practice of piggy-backing low-energy subcarriers onto to the main signal, e.g. to advance to stereo operation, sometimes accompanied by noise reduction, at least in the case of US multichannel television sound (MTS) audio. If the use of digital audio for recording and broadcast was to be adopted for media beyond the CD, the need to reduce the number of bits required was apparent. This has given rise to one of the premier pursuits of digital development since that time: the use of audio coding to convey a signal faithfully using the smallest possible number of bits.

The data rate of an audio CD derives from the use of two channels, a sampling rate of 44 100 Hz, and 16 bit PCM samples, a total of 1.4 megabits per second. This is greater than can be accommodated on a standard 35 mm movie print, or a standard cassette tape, or an FM radio channel. (It is also more than will fit on a standard long-play record, but since that medium is not especially compatible with digital storage in the first place, it is a moot point.) Even the CD can only accommodate the two channels; a 5.1-channel PCM format would reduce the playing time to something under 25 min.

So the need to reduce the data rate without sacrificing quality was readily apparent, which motivated interest and activity in the development of low-bitrate audio coders, and made practical such applications as digital cinema soundtracks, portable music players, and satellite radio. It is perhaps not self-evident that the data rate can be reduced in the first place, but for starters, one could simply use smaller PCM words. This would, of course, raise the effective noise floor, which in turn could be addressed by adapting an analog noise-reduction system as an associated DSP algorithm. However, the resulting reduction in data rate would still be relatively modest, as the compression ratios of noise-reduction systems never exceeded a range of about 2–3:1, to avoid mistracking problems, and this is less data compression than is required by many coding applications.

Overall, two general classes of techniques have been exploited in digital audio coders, lossless and lossy, or perceptual, coding. They share in common the arrangement of an encoder and a decoder which, instead of just conveying PCM samples, employ a common vocabulary or protocol for describing and conveying the audio at an elevated level of abstraction. This protocol is sometimes referred to as a bit-stream syntax. Although the signal connecting encoder and decoder is indeed usually just a stream of bits, the meaning and significance of a given bit will depend on the context in which it appears. This represents an additional advantage of digital audio systems over analog systems: a bit stream can consist of interleaved sequences of diverse information packets. Some of these will generally be representations of some derived form of the audio, and the rest will be descriptive information about the signal, referred to as side-chain information.

Lossless Audio Coding

The goal of lossless coding is to convey digital information in a more compact form while reliably recovering the original data in the decoder with bit-for-bit accuracy. Obviously, no psychoacoustic or perceptual considerations need enter into the picture.

The key to the ability to losslessly compress digital data is redundancy. Any time there is a pattern or common element to a block of data, it may be possible to express it in a more compact form by conveying the essence of the pattern along with the deviations from the pattern, such that the original data may be reconstructed. Of course, in order for this to happen, the bit-stream syntax must support a compact means of specifying the necessary range of patterns to the decoder.

The number of potentially useful patterns is actually fairly large, and considerable ingenuity has gone into the development of lossless coding algorithms which can recognize and concisely encapsulate the details of redundant information patterns. Some of the more common data structures that can be losslessly compressed are listed below.

1. A repeating sample value: rather than send the same value multiple times, the sample can be sent just once, along with a replication factor. This technique is sometimes referred to as *run-length coding*.
2. Similar values: a variant of run-length coding, this type of pattern can be compressed by sending the average value, followed by concise representations of the difference between each sample value and the average. Alternatively, differential coding may be used, in which the difference between successive samples is coded.
3. Repeating sinusoids: for an ongoing sinusoidal signal, the PCM samples can be replaced with parameters indicating the amplitude, frequency, phase, and length of the sample sequence. In general form, this is sometimes referred to as *linear prediction*. If there are small deviations, they may be sent as concise difference values.
4. Small values: a sequence of samples, each with one or more leading zeros, may be sent with the zeros omitted, along with a factor indicating how many leading zeros were suppressed. (Note: negative values expressed in two's-complement form will have leading ones, but the same factor should otherwise apply.) This is known as *block floating point* representation.
5. Commonly used values: sample values which appear more often than the average – like the letters 'e'

and 's' in English text – may be assigned special, short transmission symbols to reduce the average data rate. Representative techniques which employ this include arithmetic coding and Huffman coding.
6. Interchannel difference coding: if two or more channels are used and are sufficiently similar, data-rate savings may be attained by sending sum and difference information, or a variant thereof. The expectation is that the difference values will be small, and can be coded in fewer bits using technique (4), above. In the case of stereo content, this approach is sometimes referred to as mid/side (M/S) coding.

In addition to the techniques described above (and others), a lossless coder may either directly code time-domain PCM samples, or may instead use running transforms and operate in the frequency domain. In the latter case, a specialized bit-exact transform may be used to guarantee that the final output is an exact clone of the input.

A fully realized lossless coder is likely to use a combination of techniques like those described above, requiring a flexible and perhaps elaborate bit stream syntax. The process of selecting which coding options will be brought to bear on a given block of samples is often not readily apparent from inspection or simple calculation, so the encoder may be required to try an exhaustive search of all possibilities for each block in order to select the best option. Needless to say, this can make for a rather slow-running encoder, but usually does not affect the speed of the decoder.

As may be evident, the effectiveness of a lossless coder is very much signal dependent: some signals can be compressed considerably more than others, and some, such as full-level white noise, basically cannot be losslessly compressed at all. Typical compression ratios run in the range of 2:1 to 3:1. Because lossless compression is inherently a variable bitrate (VBR) coding technology, it is not well suited to real-time streaming applications, such as broadcast or fixed-speed digital tape.

One weakness of lossless coding, and indeed of lossy coding as well, is that an error of even a single bit (in, say, one of the control codes specifying which method to use to unpack a block of data) can cause the loss of significant audio data. Some care must be used in the design of the bitstream to ensure that re-synchronization of the decoder in the face of data errors is reliably possible within an acceptably short period. More generally, the complexity and low error tolerance of almost any data-reduction algorithm raises the regrettable possibility that digital records, audio and otherwise, may be lost to future generations unless care is taken to avoid data errors, and to ensure that decoding algorithms are carefully documented and preserved in full detail. This is especially critical in the case of proprietary algorithms owned and maintained by commercial entities, whose long-term future operation may not be assured.

Perceptual/Lossy Audio Coding

While one might wish that lossless coding be used exclusively for digital audio compression, its limited performance and variable bitrate render it inappropriate for a large number of coding applications. For more-aggressive compression and acceptable performance within the constraint of constant bit rate, one must resort to the use of perceptual audio coding.

The goal of perceptual audio coding is to preserve and convey all of the perceptual aspects of audio content, without regard to preserving the signal literally, usually employing the fewest bits possible. Recall that the data rate for a standard audio CD was about 1.4 Mbit/s. As this is written, the lowest data rate in commercial use for conveying high-quality stereo audio is no higher than 48 Kbits/s, used by satellite radio and for Internet streaming, a data-compression ratio on the order of 30:1 or better.

How is this accomplished? For starters, the use of lossless coding is not retained and incorporated in overall lossy algorithms. Every packet of information generated and reduced to its ultimate compactness by any part of a perceptual coder is a candidate for further reduction in size if one or more lossless techniques can be brought to bear effectively.

Beyond that, the keys to high-efficiency perceptual coding are to:

1. suppress any audio components not contributing to the ultimate perception,
2. ensure that all the audible components are conveyed with no alteration in perceived quality, while using the minimum possible number of bits, and
3. employ, as much as possible, high-level abstractions to describe audio signals in purely perceptual terms, while retaining enough precision and detail to allow the decoder to accurately reconstitute a signal indistinguishable from the original encoder input signal.

One fundamental building block that has almost always been part of the foundation of perceptual coders is a digital filter bank. This, of course, mirrors the functionality of the basilar membrane; and the frequency-domain output signals are most commonly used for both the analysis sections of the coder and the signal path. Indeed, the fundamental data conveyed by most perceptual coders is not PCM time-domain data, but quantized frequency-domain information from which the decoder eventually produces PCM output by way of a synthesis filter bank. The most commonly used digital filter bank is a variant of the FFT called the time-domain alias cancellation (TDAC) transform, developed by Princen and Bradley in 1986 [18.69, 70]. This transform has the highly desirable property of being critically sampled, meaning that it produces exactly the same number of output frequency-domain samples as there are input PCM samples. There is also a companion inverse TDAC transform for producing output PCM from a set of decoded frequency-domain input samples.

In operation, PCM samples input to the encoder are divided into regular blocks of some predetermined length, and each block is transformed to the frequency domain, conveyed to the decoder, reconstituted as a block of PCM samples, and the successive blocks strung together, often in an overlapping manner, to recover the final outputs. This process is sometimes described as block-oriented processing using running transforms. The practice of coding an entire block of samples as a single entity facilitates data rates of less than one bit per original audio sample. One downside of block processing is that events which occupy a small fraction of a block, like a sharp transient, may become smeared across a range of frequencies, consuming a large number of bits. A number of techniques have evolved to deal with this situation, most commonly the use of block switching, wherein such short events are detected (by, for example, a transient detector routine), and the block size is temporarily shortened to more closely isolate the transient within the shortened block.

The filter-bank output is of direct use in pursuit of the first two perceptual coding techniques listed above, principally by deriving the magnitude of the frequency-domain signals as a function of frequency to obtain a discrete approximation to the power spectral density (PSD) of the signal block. This in turn is processed by a routine implementing a perceptual model of human hearing to derive a masking curve, which specifies the threshold of hearing as a function of frequency for that signal block. Any spectral component falling below the masking curve will be inaudible, so need not be coded, as per the first listed coding technique. The remaining spectral components must be preserved if audible alteration is to be avoided, but the quantization precision is only that which is required to render the level of the quantization noise below the masking curve. Instead of the 120 dB/20 bit range one might require for PCM audio, the instantaneous masking range is more often on the order of 20–30 dB, about 4–5 bits per sample, assuming 6 dB per bit SNR. Thus, between suppression of inaudible components and dynamic quantization of audible components, the data rate can be expected to be reduced to something less than 4 bits per sample. Of course, an effective bit-stream syntax protocol must be devised to signal these conditions efficiently to the decoder, and considerable ingenuity has been brought to bear on that issue to ensure the requirement is met.

Notable lossy audio coders based on these principles include AC-3, MP3, DTS, WMA, Ogg, and AAC. These have been instrumental enablers of such technologies as portable music players, DVD's, digital soundtracks on 35 mm film, and satellite radio.

It should be noted that perceptual coding is much more compatible with constant-bit-rate operation than is lossless coding, for as the complexity of a signal increases, which might otherwise increase the required data rate, the masking afforded by that signal also increases, reducing the average quantization accuracy required, thereby holding the required data rate to a more nearly constant rate. In effect, the human auditory system can only absorb so much information per unit time, so as long as the coder accurately models that behavior, the required data rate should be largely constant. Of course, very simple signals, like silence, are not likely to require the same data rate as more complex signals, in which case a constant-bitrate coder may simply use far more than the minimum required data rate in order to maintain a constant transmission rate. Some coders can defer use of bits in the presence of simple signals, saving them in a *bit bucket* until needed by a more complex signal.

The third listed technique used in perceptual coding, the application of higher-level abstractions to describe the signal, is much more general and open-ended; and is still an area of active investigation. It can be something as simple as using a single PSD curve to approximate the actual PSD curves of several successive transform blocks, or using decoder-generated noise in place of actually transmitting noise-like signals. More abstract approaches may fall under the heading of *parametric coding*.

One notable example of parametric coding is *bandwidth extension*, in which the entire high-frequency end of the spectrum is suppressed by the encoder, and is reconstituted by the decoder from analysis of the harmonic content of rest of the signal spectrum. This technique alone can reduce the data rate by 30% or more.

Another widely pursued approach to parametric coding is to segregate signal components into categories, such as sinusoidal, noise-like, or transient, then convey numerical descriptions of each to a synthesizing decoder. Although this is arguably far closer to a true perceptually grounded coder than a simple transform-based coder, the complexity of the descriptors needed to accurately portray the full range of possible signals has so far made it difficult to reduce the data rate required by more conventional coders dramatically.

Spatial Audio Coding

Although strictly speaking it may be a subset of perceptual audio coding, spatial audio coding (SAC) is a sufficiently independent approach to justify separate examination. SAC can be used to improve the transmission efficiency of multichannel content, typically either stereo or 5.1-channel surround material.

The approach involves coding the ensemble of channels as a single entity to achieve data savings that would be unavailable if each channel were to be coded individually. An element common of many SAC algorithms is to combine the source channels additively to a smaller number of channels, perhaps as few as a single channel, from which the decoded output channels are eventually reconstructed. The data rate is thereby reduced by the ratio of the number of input channels to the number of down-mixed channels, minus any side-chain information needed by the decoder to reconstruct the output channels.

Probably the simplest approach to SAC is to down-mix the input channels to mono in the encoder and then simply distribute the common signal equally to the output channels. If this is done to the full-bandwidth signal, the result will simply be monaural sound reproduced from multiple speakers, which is not likely to be a very satisfactory multichannel experience, but if limited to a subset of the audio range, say just high frequencies, the ear will tend to associate the spread mono content of the reproduced channels with the direction of the lower frequency, discrete content, and the alteration will be less noticeable.

A slightly more elaborate approach is to send channel-by-channel scale factors, representing relative signal amplitude in each input channel, which are subsequently used by the decoder to distribute signal components to each output channel. Sometimes referred to as either intensity stereo or channel coupling, the technique can be surprisingly effective, especially at high frequencies when individual scale factors are sent on something like a critical band frequency basis.

More recent spatial coding systems have augmented the use of amplitude coupling with transient flags and information about interchannel phase and coherence, allowing extension to progressively lower frequencies, with commensurately greater amounts of data rate reduction.

Although in most cases the spatial information about individual channels is sent as side-chain information comprising part of the total transmitted bit stream, there has been some success in encoding the information in the signal characteristics of the actual transmitted audio, which can reduce the net data rate.

Spatial audio coding remains at this time an area of very active investigation.

18.6 Complete Audio Systems

So far, this discussion of audio and electroacoustics has focused on the task of capturing, transmitting, and reproducing one or more individual channels of audio as accurately as possible. However, the goal of sound reproduction is not strictly to convey some designated number of channels, but to faithfully reproduce a full acoustic event, be it live, synthesized, or some combination thereof, to an audience that, at any given venue, may range in number from one to many hundreds or more.

The use of individual channels is mandated by the available technology. Microphones and loudspeakers are basically point-in-space devices, and most audio media and processors are designed to handle individual channels. But sound is, by nature, a three-dimensional entity, not counting time and frequency, and a few isolated channels can at best represent a sparse spatial sampling of a full acoustic sound field.

So the question is raised as to just how effective are complete audio systems, from recording to reproduction, at least in perceptual, psychoacoustic terms.

Any attempt to perform such an evaluation is immediately faced with the conundrum as to just what should be the ultimate goal of sound reproduction. Should we be trying to transport the listener to the performance site ("we are there"), or is the goal to transport the performers to the listener's living room ("they are here").

The answer perhaps depends on the nature of the source material, as well as listener preference. For situational acoustics, from walking down a busy street to riding in a submarine, it would seem logical to go with "we are there". This is most commonly the case for, say, cinema soundtracks, making a good case for surround-sound systems in commercial and home theaters. A similar case can be made for symphonic music, since the concert hall is such an integral part of the experience. However, a small ensemble or a solo artist might be more compelling if holographically projected into the listening room. Given unlimited wishes, one might hope to have both, with the ability choose which option according to one's predilections. In actual practice, we might be well served if either one is possible. But in any case, it is perhaps worth keeping in mind in evaluating the effectiveness of complete audio systems.

18.6.1 Monaural

For the first half century of recorded sound, monaural was all there was. In that time, considerable expertise was developed in microphone techniques for capturing sound with a viable balance of direct and ambient content. In this era of multichannel audio, it is easy to dismiss mono for failing to support any formal spatial audio cues. Still, there are times when a single channel should, at least in theory, be adequate for the task, such as a solo performer or other single, localized source, in a dry (relatively echo-free) ambient environment, or "here". Even in that case, however, the reproducing loudspeaker is unlikely to have the same spatial radiation pattern as the original source, a problem common to many sound formats to one degree or another, so that even with good frequency response and low distortion, the sound may be recognizable as emanating from a loudspeaker, rather than being a clone of the original performance. To its credit, mono does not exhibit the double-arrival problem of phantom images common to stereo and other multichannel formats (below), arguably resulting in a purer, less time-smeared sound. It is also gratifyingly free of sweet-spot dependence: one can sit pretty much anywhere in a listening room without worrying about impairment of imaging. And historically at least, mono was a big improvement over nothing at all.

18.6.2 Stereo

Two-channel stereophonic sound has been a commercial reality for nearly half a century, and shows little sign of being rendered obsolete anytime soon. The intuitive attraction of stereo is self-evident: we have two ears, and stereo provides two channels. On the surface, this is an apparently compelling argument. The result should perfect sound reproduction, yes? Well, no, although stereo did pretty much render mono obsolete for most media within a few years of its commercial deployment.

Stereo provides the smallest possible increment in spatial fidelity over mono: a horizontal line spread before the listener. But this is likely the most important single improvement one could hope to make, as the human ear is most sensitive to front-center horizontal position. In addition, performing ensembles are most often arrayed in such a manner. The arrangement also allows rendering, where appropriate, of frontal ambiance in a manner more natural than can be achieved with a single audio channel.

An important element of stereo reproduction is the creation of virtual images. These are sonic images that appear suspended in space between the two speakers, and they undoubtedly add considerably to the viability of the format. Virtual images depend for their existence on the primary horizontal localization cues: interaural time differences and interaural amplitude differences. The former cue is especially critical to the creation of virtual images, as the total range of the cue is only about $\pm 700\,\mu\text{s}$, corresponding to a horizontal displacement of the listener of only about a foot. This rather undermines the viability of stereo virtual imaging somewhat, and renders the format decidedly sweet-spot dependent. The situation is partially compensated by the flexibility of the human auditory system, and its apparent ability to use dynamically whatever cues appear to be providing reliable localization information. A listener a little off the center line may still experience some degree of virtual imaging by virtue of the interaural amplitude cue, which may not degrade as rapidly as the interaural time delay cue. Indeed, in the course of mixing a multichannel source to a stereo track, a recording engineer will typically position a virtual source between the speakers using simple amplitude panning, with zero interaural time delay, and yet the ear generally accepts this.

But sensitivity to listener position is not the only issue relating to virtual images. There is also that double-arrival problem. The direct sound from almost any real sound source will generally arrive at each ear once, and that is it. But a virtual image from a stereo speaker array will result in two direct arrivals at each ear, one from each speaker. Assuming a centered listener, the left ear will hear from the left speaker at about the same time that the right ear hears the first arrival from the right speaker. Then the sound from each speaker will cross the head, in opposite directions, and the left ear will hear the right speaker arrival, and vice versa, a process sometimes referred to as loudspeaker crosstalk. The time delay between arrivals is typically on the order of 250 μs, resulting in a small, subtle, but audible notch in the resulting response around 2 kHz. Some temporal smearing of sharp transients may also be evident.

A final difficulty with virtual imaging of stereo systems is the relative inability to position virtual images outside the range of the speakers. This is another consequence of loudspeaker crosstalk, limiting the maximum interaural time difference to that corresponding to the position of the speaker, which again is about 250 μs, and similarly limiting the maximum interaural amplitude difference. One way around this would be to put the speakers in different rooms and sit between them, one ear in each room, with a custom-fitted baffle doorway to block any crosstalk, which is a little impractical. Another approach is to employ a crosstalk cancellation (virtual-speaker) system, to cancel the second arrival at each ear, but such systems are themselves very sweet-spot dependent, and even then usually limited to creating images on the horizontal plane. The most direct way to provide sound from additional directions is to position additional speakers around the listener, including above and possibly below, assuming additional source channels are available to feed them, but that is no longer a stereo sound system.

18.6.3 Binaural

Like stereophonic sound, binaural makes use of just two channels, but it uses headphones instead of loudspeakers to avoid the crosstalk problem. On paper, this system seems like it should be even more perfect than stereo. However, the effect is often to place the sound images within the head, with little or no front/back differentiation.

Part of the problem appears to be that the presentation does not get altered when the listener's head moves, so that the image follows the head movement, which is rather unnatural. There has been some work exploring the use of tracking headphones, which alter the signal in response to head motion, but these have not yet been completely successful, and require more than two channels to be available.

Another issue with binaural reproduction seems to be the lack of pinna cues, those alterations in timing and spectra that are imparted by the pinnae to an arriving sound as a function of direction. Efforts to impart pinna cues electronically have had some success, but they tend to be as individual as fingerprints, making it difficult to have a single binaural track which works for everyone. There has also been some success in producing a universal binaural track by imparting a direction-dependent room reverberation characteristic to the signals, although some purists object to the addition, and the effect is still somewhat listener dependent.

One experimental system that mitigates some of the shortcomings of two-channel binaural is four-channel binaural. This employs two dummy heads or equivalently arranged directional microphones, with one head/microphone-pair in front of the other, preferably with an isolating baffle separating them. The resulting four-channel recording is played over four speakers, preferably placed close to the listener in fixed positions around the listener's head, corresponding to the microphone positions, possibly with a left–right isolating panel or crosstalk-canceling circuits to minimize crosstalk. Because the front–back pair each preserve ITD and IAD, the left–right space is properly preserved, while front–back differentiation and externalization are provided by using separate front and back external drivers. Dynamic IAD and ITD cues with head motion are also at least crudely preserved by the use of front and back channels. The resulting presentation can exhibit good spatial fidelity.

Although it may never be suitable for general audience presentation, the binaural format has compelling qualities for the single listener, and remains an area of active investigation.

18.6.4 Ambisonics

This system, devised by Michael Gerzon, approximates the sound field in the vicinity of a listener using an array of precisely positioned loudspeakers, It is sweet-spot dependent, so not amenable to audience presentation, but can provide compellingly good spatial reproduction.

18.6.5 5.1-Channel Surround

The 5.1-channel format is a well-thought-out response to the shortcomings of stereo. The format employs three front speakers, two surround speakers placed to the side and/or back of the listening room, and the 0.1 low-frequency channel that only occupies a small fraction of the total bandwidth.

The addition of the center-front speaker anchors the front left–right spread, significantly mitigating the problems stereo has with virtual images, even with audience presentation. The surround channels provide sound more to the side than can be achieved with the front speakers (or stereo), exercising the full range of the ITD/IAD sensitivity of the ear, and providing the option of a far more immersive experience than can be attained with stereo.

The 5.1 format with high-quality source material provides the most accurate overall audio reproduction of any widely used format to date, within either the "here" or "there" reproduction paradigms.

18.7 Appraisal and Speculation

We have been witness, from the crude but noble origins of sound reproduction, to an astonishing march of progress that has, by degrees, rolled down the problems and imperfections of sound reproduction, reaching a notably high level of performance, when considered in basic perceptual terms.

A cynic might note that, fifty years ago, children played little yellow plastic 78 RPM 17.78 cm discs on phonographs that had 12.7 cm loudspeakers, and now, after fifty years of dramatic, hard-won progress, many people play little 12.7 cm plastic (compact) discs on boom boxes with 12.7 cm loudspeakers. Yes, the fidelity is better, as is the playing time, but have we really come that far?

The apparent answer from all the above is that a high-quality system of modern vintage can now satisfy a major percentage of the electroacoustic and spatial requirements implied by the capabilities of the human auditory system.

This does not mean that audio engineers call all just pack up and go home just yet. Indeed, the quantities of audio research papers and patents published each year show no sign of abating, giving rise to the speculation that further meaningful improvements may yet be attained.

One can certainly wish for a few more channels, to fill in the remaining gaps and perhaps even provide explicit coverage above and below the listener. The issue in that regard at this point is more of practicality than of theory. There have been some experiments with the addition of height or ceiling (*voice of God*) channels, and the specifications for digital cinema systems allow for the use of over a dozen channels.

One can also hope for enhanced ability to project sound into a room, rather than having it just arrive from the walls. That is another area of active investigation. One approach, referred to as wave field synthesis [18.71] has been to approximate a complete sound field by reproducing the sound-pressure profile around a periphery, and relying on the Huygens principle [18.72] to recreate the complete interior sound field. This requires large number of loudspeakers (192 in one experimental setup) to avoid spatial aliasing effects.

The other well-known approach to generating interior sounds is to use an ultrasonic audio projector, which projects acoustic signals above the audio band which heterodyne to produce audible sound which appears to come from a remote location [18.73].

Another problem that has only partly been solved is that of sweet-spot sensitivity. This has long been an issue with cinema surround-sound systems, but is no less challenging for automotive surround systems, where there are often no centrally located seats and speaker locations may be less than optimal. The issues go beyond balanced sound to trying to provide a coherent, immersive experience at all listening positions.

Indeed, each new venue for audio seems to come with its own idiosyncratic problems, leading to active investigation of how to make compelling presentations for portable music players, cell phones, and interactive media such as game consoles.

It is also expected that there will be continuing refinement of binaural techniques for personal listening.

Ultimately there may come a wholesale revolution in audio techniques, perhaps with laser transducers replacing loudspeakers and microphones, along with holographic recording of entire sound fields. For personal listening, your music player might interface directly to your brain, bypassing the ear altogether, allowing otherwise deaf people to hear, and getting rid of that troublesome 3 kHz peak in the ear canal once and for all.

It should make for some interesting vibes.

References

18.1   N. Aujoulat: *The Cave of Lascaux* (Ministry of Culture and Communication, Paris August 2005), http://www.culture.gouv.fr/culture/arcnat/lascaux/en/

18.2 J. Sterne: *The Audible Past* (Duke Univ. Press, Durham 2003)

18.3 Thomas Edison (Wikipedia Aug. 2005) http://en.wikipedia.org/wiki/Thomas_Edison

18.4 J.H. Lienhard: *The Telegraph, Engines of Our Ingenuity* (Univ. Houston, Houston 2004), http://www.uh.edu/engines/epi15.htm

18.5 P. Ament: The Electric Battery (The Great Idea Finder, Mar. 2005) http://www.ideafinder.com/history/inventions/story066.htm

18.6 Phonograph (Wikipedia, Aug. 2005) http://en.wikipedia.org/wiki/Phonograph#History

18.7 D. Marty: The History of the Phonautograph (Jan. 2004) http://www.phonautograph.com/

18.8 Emile Berliner and the Birth of the Recording Industry, Library of Congress, Motion Picture, Broadcasting and Recorded Sound Division (April, 2002) http://memory.loc.gov/ammem/berlhtml/berlhome.html

18.9 J. Bruck, A. Grundy, I. Joel: *An Audio Timeline* (Audio Engineering Society, New York 1997), http://www.aes.org/aeshc/docs/audio.history.timeline.html

18.10 M.F. Davis: History of spatial coding, J. Audio Eng. Soc. **51**(6), 554–569 (2003)

18.11 NIH: *Hearing Loss and Older Adults, NIH Publication No. 01-4913, National Institute on Deafness and Other Communications Disorders* (National Inst. Health, Bethesda 2002), http://www.nidcd.nih.gov/health/hearing/older.asp

18.12 T. Weber: *Equal Loudness* (WeberVST, Kokomo 2005), http://www.webervst.com/fm.htm

18.13 E. Terhardt: *Pitch Perception* (Technical Univ. Munich, Munich 2000), http://www.mmk.e-technik.tu-muenchen.de/persons/ter/top/pitch.html

18.14 B. Truax: *Sound pressure level [SPL], Handbook for Acoustic Ecology, Simon Fraser University, Vancouver* (Cambridge Street Publ., Burnaby 1999), http://www2.sfu.ca/sonic-studio/handbook/Sound_Pressure_Level.html

18.15 B. Truax: *Phon, Handbook for Acoustic Ecology, Simon Fraser University, Vancouver* (Cambridge Street Publ., Burnaby 1999), http://www2.sfu.ca/sonic-studio/handbook/Phon.html

18.16 B.G. Crockett, A. Seefeldt, M. Smithers: *A new loudness measurement algorithm, AES 117th Convention Preprint 6236* (Audio Engineering Society, New York 2004)

18.17 L.H. Carney: *Studies of Information Coding in the Auditory Nerve* (Syracuse Univ., Syracuse 2005)

18.18 E.D. Haritaoglu: *Temporal masking* (University of Maryland, College Park 1997), http://www.umiacs.umd.edu/desin/Speech1/node11.html

18.19 K.M. Steele: *Binaural processing* (Appalachian State Univ., Boone April 2003), http://www.acs.appstate.edu/kms/classes/psy3203/SoundLocalize/intensity.jpg

18.20 J. Blauert: *Spatial Hearing* (MIT Press, Cambridge 1983)

18.21 R. Kline: Harold Black and the negative-feedback amplifier, IEEE Control Syst. Mag. **13**(4), 82–85 (1993), DOI: 10.1109/37.229565

18.22 N. Moffatt: *Wow and Flutter* (Vinyl Record Collectors, United Kingdom 2004), http://www.vinylrecordscollector.co.uk/wow.html

18.23 B. Paquette: *History of the microphone* (Belgium 2004), http://users.pandora.be/oldmicrophones/microphone_history.htm

18.24 J. Strong: *Understanding microphone types, Pro Tools for Dummies* (Wiley, New York 2005), http://www.dummies.com/WileyCDA/DummiesArticle/id-2509.html

18.25 P.V. Murphy: *Carbon Microphone* (Concord University, Athens 2003), http://students.concord.edu/murphypv/images/35107.gif

18.26 D.A. Bohn: *Ribbon Microphones* (Rane Corp., Mukilteo 2005), http://www.rane.com/par-r.html

18.27 H. Robjohns: *A brief history of microphones, Rycote Microphone Data Book* (Rycote Microphone Windshields, Stroud 2001), http://www.microphone-data.com/pdfs/History.pdf (Rycote Microphone Windshields, Stroud 2003)

18.28 G. M. Sessler, J. E. West: Electroacoustic Transducer Electret Microphone, US Patent Number 3118022 (1964) http://www.invent.org/hall_of_fame/132.html (National Inventors Hall of Fame, 2002)

18.29 P.V. Murphy: *Piezoelectric Microphone* (Concord University, Athens 2003), http://students.concord.edu/murphypv/images/35108.gif

18.30 D. Stewart: *Understanding microphone self noise* (ProSoundWeb, Niles 2004), http://www.prosoundweb.com/studio/sw/micnoise.php

18.31 A.G.H. van der Donk, P.R. Scheeper, P. Bergveld: Amplitude-modulated electro-mechanical feedback system for silicon condenser microphones, J. Micromech. Microeng. **2**, 211–214 (1992), http://www.iop.org/EJ/abstract/0960-1317/2/3/024

18.32 A. Oja: *Capacitive Sensors and their readout electronics, Helsinki Univ. Technology, Physics Laboratory* (Helsinki Univ. Technology, Espoo 2005), http://www.fyslab.hut.fi/kurssit/Tfy3.480/Spring 2004/Jan28_Oja/Microsensing_2004_Oja2.pdf

18.33 J.N. Caron: *Gas-coupled laser acoustic detection* (Quarktet, Lanham 2005), http://www.quarktet.com/GCLAD.html

18.34 H.M. Ladak: *Anatomy of the middle ear, Finite-Element Modelling Of Middle- Ear Prostheses in Cat, Master of Engineering Thesis* (McGill University, Montreal 1993), Chap. 2 http://audilab.bmed.mcgill.ca/funnell/AudiLab/ladak_master/chapter2.htm

18.35 Microphone Types, *The Way of the Classic CD* (Musikproduktion Dabringhaus und Grimm, Detmold 2005), http://www.mdg.de/mikroe.htm
18.36 G. Carol: *Tonearm/Cartridge Capability* (Galen Carol Audio, San Antonio 2005), http://www.gcaudio.com/resources/howtos/tonearmcartridge.html
18.37 J.G. Holt: *Down With Dynagroove, Stereophile, December 1964* (Primedia Magazines, New York 1964), http://www.stereophile.com/asweseeit/95/
18.38 D. Krotz: *From Top Quarks to the Blues, Berkeley Lab physicists develop way to digitally restore and preserve audio recordings, Berkeley Lab. Res. News 4/16/2004* (Lawrence Berkeley National Lab., Berkeley 2004), http://www.lbl.gov/Science-Articles/Archive/Phys-quarks-to-blues.html
18.39 S.E. Schoenherr: *Loudspeaker history, Recording Technology History* (Univ. San Diego, San Diego July 2005), http://history.sandiego.edu/gen/recording/loudspeaker.html
18.40 C.W. Rice, E.W. Kellogg: Notes on the development of a new type of hornless loudspeaker, Trans. Am. Inst. El. Eng. **44**, 461–475 (1925)
18.41 AudioVideo 101: *Speakers Enclosures* (AudioVideo 101, Frisco 2005), http://www.audiovideo101.com/learn/knowit/speakers/speakers16.asp
18.42 Eminence: *Understanding loudspeaker data* (Eminence Loudspeaker Corp., Eminence 2005), http://editweb.iglou.com/eminence/eminence/pages/params02/params.htm
18.43 M. Sleiman: *Speaker Engineering Analysis* (Northern Illinois University, DeKalb 2005), www.students.niu.edu/ z109728/UEET101third%20draft.doc
18.44 Allison Acoustics: *Company History* (Allison Acoustics LLC, Danville 2005), http://www.allisonacoustics.com/history.html
18.45 D. Bohn: A fourth-order state-variable filter for Linkwitz-Riley active crossover designs, J. Audio Eng. Soc. Abstr. **31**, 960 (1983), preprint 2011
18.46 W. Norris: *Overview of hypersonic technology* (American Technology Corp., San Diego 2005), http://www.atcsd.com/pdf/HSSdatasheet.pdf
18.47 T. Sporer: *Conveying sonic spaces* (Fraunhofer Elektronische Medientechnologie, Ilmenau 2005), http://www.emt.iis.fhg.de/projects/carrouso/
18.48 B. DePalma: Analog Audio Power Amplifier Design (January, 1997) http://depalma.pair.com/Analog/analog.html
18.49 E. Augenbraun: *The Transistor, Public Broadcasting System* (ScienCentral, The American Institute of Physics, New York 1999), http://www.pbs.org/transistor (July 2005)
18.50 S. Portz: *Who Invented The Transistor* (PhysLink, Long Beach 2005), http://www.physlink.com/Education/AskExperts/ae414.cfm
18.51 THAT Corp.: *A brief history of VCA development* (THAT Corp., Milford 2005), http://www.thatcorp.com/vcahist.html
18.52 M. Hart: *History of Sound Motion Pictures* (American WideScreen Museum, 2000), http://www.widescreenmuseum.com/sound/sound03.htm
18.53 L.P. Lessing: *Edwin H. Armstrong, Dictionary of American Biography, Supplement V* (Scribner, New York 1977) pp. 21–23, http://users.erols.com/oldradio/ (2005)
18.54 A. Antoniou: *Digital Filters: Analysis, Design, and Applications*, 2nd edn. (McGraw-Hill, New York 1993)
18.55 A. Antoniou: *On the Origins of Digital Signal Processing* (Univ. Victoria, Victoria 2004), http://www.ece.uvic.ca/ andreas/OriginsDSP.pdf
18.56 R.W.T. Preater, R. Swain: Fourier transform fringe analysis of electronic speckle pattern interferometry fringes from hinge-speed rotating components, Opt. Eng. **33**, 1271–1279 (1994)
18.57 E.O. Bringham: *The Fast Fourier Transform* (Prentice-Hall, Upper Saddle River 1974)
18.58 K.C. Pohlmann: *The Compact Disc Handbook*, Comp. Music Digital Audio Ser., Vol. 5, 2nd edn. (A-R Editions, Middleton 1992)
18.59 J. Despain: *History of CD Technology* (OneOff Media, Salt Lake City 1999), http://www.oneoffcd.com/info/historycd.cfm?CFID=435613&CFTOKEN=10955739
18.60 E.T. Whittaker: On the functions which are represented by the expansions of interpolation-theory, Proc. R. Soc. Edinburgh. **35**, 181–194 (1915)
18.61 C.E. Shannon: A mathematical theory of communication, Bell Syst. Tech. J. **27**, 379–423 (1948)
18.62 C.E. Shannon: A mathematical theory of communication, Bell Syst. Tech. J. **27**, 623–656 (1948)
18.63 E. Meijering: Chronology of interpolation, Proc. IEEE **90**(3), 319–342 (2002), http://imagescience.bigr.nl/meijering/research/chronology/
18.64 M.T. Heideman, D.H. Johnson, C.S. Burrus: Gauss and the history of the fast Fourier transform, IEEE Acoust. Speech Signal Process. Mag. **1**, 14–21 (1984),
18.65 M.T. Heideman, D.H. Johnson, C.S. Burrus: Gauss and the history of the fast Fourier transform, Arch. Hist. Exact Sci. **34**, 265–277 (1985)
18.66 C.S. Burrus: *Notes on the FFT* (Rice University, Houston 1997), http://www.fftw.org/burrus-notes.html
18.67 P. Duhamel, M. Vetterli: Fast Fourier transforms: A tutorial review and a state of the art, Signal Process. **19**, 259–299 (1990)
18.68 Fast Fourier Transform (Wikipedia, August 2005) http://en.wikipedia.org/wiki/Fast_Fourier_transform
18.69 M.F. Davis: *The AC-3 multichannel coder, Audio Engineering Society. 95th Convention 1993* (Audio Engineering Society, New York 1993), Preprint Number 3774
18.70 M.F. Davis: *The AC-3 multichannel coder* (Digital Audio Development, Newark 1993), http://www.dadev.com/ac3faq.asp

18.71 T. Sporer: *Creating, assessing and rendering in real time of high quality audio- visual environments in MPEG-4 context, Information Society Technologies* (CORDIS, Luxemburg 2002), http://www.cordis.lu/ist/ka3/iaf/projects/carrouso.htm

18.72 G. Theile, H. Wittek: Wave field synthesis: A promising spatial audio rendering concept, Acoust. Sci. Tech. **25**, 393–399 (2004), DOI: 10.1250/ast.25.393

18.73 E.I. Schwartz: *The sound war, Technol. Rev. May 2004* (MIT, Cambridge 2004), http://www.woodynorris.com/Articles/TechnologyReview.htm

Part F Biological and Medical Acoustics

19 Animal Bioacoustics
Neville H. Fletcher, Canberra, Australia

20 Cetacean Acoustics
Whitlow W. L. Au, Kailua, USA
Marc O. Lammers, Kailua, USA

21 Medical Acoustics
Kirk W. Beach, Seattle, USA
Barbrina Dunmire, Seattle, USA

19. Animal Bioacoustics

Animals rely upon their acoustic and vibrational senses and abilities to detect the presence of both predators and prey and to communicate with members of the same species. This chapter surveys the physical bases of these abilities and their evolutionary optimization in insects, birds, and other land animals, and in a variety of aquatic animals other than cetaceans, which are treated in Chap. 20. While there are many individual variations, and some animals devote an immense fraction of their time and energy to acoustic communication, there are also many common features in their sound production and in the detection of sounds and vibrations. Excellent treatments of these matters from a biological viewpoint are given in several notable books [19.1, 2] and collections of papers [19.3–8], together with other more specialized books to be mentioned in the following sections, but treatments from an acoustical viewpoint [19.9] are rare. The main difference between these two approaches is that biological books tend to concentrate on anatomical and physiological details and on behavioral outcomes, while acoustical books use simplified anatomical models and quantitative analysis to model whole-system behavior. This latter is the approach to be adopted here.

19.1	Optimized Communication	785
19.2	Hearing and Sound Production	787
19.3	Vibrational Communication	788
19.4	Insects	788
19.5	Land Vertebrates	790
19.6	Birds	795
19.7	Bats	796
19.8	Aquatic Animals	797
19.9	Generalities	799
19.10	Quantitative System Analysis	799
References		802

19.1 Optimized Communication

Since animals use their acoustic and vibrational senses both to monitor their environment and to communicate with other animals of the same species, we should expect that natural selection has optimized these sensing and sound production abilities. One particular obvious optimization is to maximize the range over which they can communicate with others of the same species. Simple observation shows that small animals generally use high frequencies for communication while large animals use low frequencies – what determines the best choice? The belief that there is likely to be some sort of universal scaling law involved goes back to the classic work of *D'Arcy Thompson* [19.10], while a modern overview of physical scaling laws (though not including auditory communication) is given by *West* and *Brown* [19.11].

The simplest assumption is that the frequency is determined simply by the physical properties of the sound-producing mechanism. For a category of animals differing only in size, the vibration frequency of the sound-producing organ depends upon the linear dimensions of the vibrating structure, which are all proportional to the linear size L of the animal, and upon the density ρ and elastic modulus E of the material from which it is made. We can thus write that the song frequency $f = A\rho^x E^y L^z$, where A, x, y, and z are constants. Since the dimensions of each side of the equation must agree, we must have $x = -1$, $y = 1$ and $z = -1$, which leads to the conclusion that song frequency should be inversely proportional to the linear size of the animal or, equivalently, that $f \propto M^{-1/3}$, where M is the mass of the animal. This simple result agrees quite well with

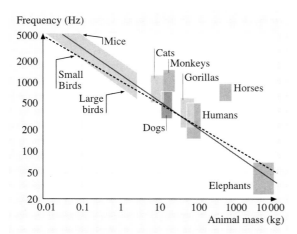

Fig. 19.1 The frequency ranges of the emphasized frequencies of vocalization in a large range of land-dwelling animals, plotted as a function of the mass of the animal. The *dashed line* shows the regression $f \propto M^{-0.33}$ while the *solid line* is the regression $f \propto M^{-0.4}$, as discussed in the text [19.12]

observation, as shown by the dashed line in Fig. 19.1, for a wide variety of animals, but a more detailed analysis is desirable.

Over reasonable distances in the air, sound spreads nearly hemispherically, so that its intensity decays like $1/R^2$, where R is the distance from the source. But sound is also attenuated by atmospheric absorption, with the attenuation coefficient varying as f^2. An optimally evolved animal should have maximized its communication distance under the influence of these two effects. An analysis [19.12] gives the result that $f \propto M^{-0.4}$ which, as shown in Fig. 19.1, fits the observations even better than the simpler result. There are, however, many outliers even among the animals considered, due to different anatomies and habitats. Insects, which must produce their sound in an entirely different way as is discussed later, are not shown on this graph, but there is a similar but not identical relative size relation for them too.

The total sound power produced by an animal is also a function of its size, typically scaling about as $M^{0.53}$ for air-breathing animals of a given category. When the variation of song frequency is included, this leads [19.12] to a conspecific communication distance proportional about to $M^{0.6}$. Again, while this trend agrees with general observations, there are many very notable outliers and great differences between different types of animals. Thus, while mammals comparable in size with humans typically produce sound power in the range 0.1–10 mW, and large birds may produce comparable power, some small insects such as cicadas of mass not much more than 1 g can also produce almost continuous calls with a power of 1 mW, as will be discussed in Sect. 19.4. At intermediate sizes, however, many animals, particularly reptiles, are almost mute.

Elephants represent an interesting extreme in the animal world because of their very large mass – as much as 10 tonnes. Their calls, which have a fundamental in the range 14–35 Hz, can have an acoustic power as large as 5 W, leading to a detection distance as large as 5 km, or even up to 10 km after sunset on a clear night [19.13] when very low frequency propagation is aided by atmospheric inversion layers (Chap. 4). Vegetation too, of course, can have a significant effect upon transmission distance. These elephant calls are often misleadingly referred to as "infrasonic" in the biological literature, despite the fact that only the fundamental and perhaps the second harmonic satisfy this criterion, and the calls have many harmonics well within the hearing range of humans. Indeed, even other elephants depend upon these upper harmonics to convey information, and usually cannot recognize the calls of members of the same family group above a distance of about 1.5 km because of the attenuation of harmonics above about 100 Hz through atmospheric absorption [19.14].

When it comes to sound detection, rather similar scaling principles operate. As will be discussed briefly in Sect. 19.2, the basic neuro-physiological mechanisms for the conversion of vibrations to neural impulses are remarkably similar in different animal classes, so that it is to be expected that auditory sensitivity should vary roughly as the area of the hearing receptor, and thus about as $M^{2/3}$, and this feature is built into the analysis referred to above. While a few animals have narrow-band hearing adapted to detecting particular predators – for example caterpillars detecting wing beats from wasps – or for conspecific communication, as in some insects, most higher animals require a wide frequency range so as to detect larger predators, which generally produce sounds of lower frequency, and smaller prey, which may produce higher frequencies. The auditory range for most higher animals therefore extends with reasonable sensitivity over a frequency range of about a factor 300, with the central frequency being higher for smaller animals. In the case of mammals comparable in size to humans, the range for mature adults is typically about 50 Hz to 15 kHz. This wide range, however, means that irrelevant background noise can become a problem, so that conspecific call frequencies may be adapted, even over a short time, to optimize communication. Small echolocating animals, such as bats, generally have a band of enhanced hearing sensitivity and resolution near their call frequency.

Aquatic animals, of course, operate in very different acoustic conditions to land animals, because the density of water nearly matches that of their body tissues. In addition, while the water medium may be considered to be essentially three-dimensional over small distances, it becomes effectively two-dimensional for the very long-distance communication of which some large aquatic animals are capable. Some of these matters will be discussed in Sect. 19.8.

19.2 Hearing and Sound Production

Despite the large differences in anatomy and the great separation in evolutionary time between different animals, there are many surprising similarities in their mechanisms of sound production and hearing. The basic mechanism by which sound or vibration is converted to neural sensation is usually one involving displacement of a set of tiny hairs (cilia) mounted on a cell in such a way that their displacement opens an ion channel and causes an electric potential change in the cell, ultimately leading to a nerve impulse. Such sensory hairs occur in the lateral line organs of some species of fish, supporting the otoliths in other aquatic species, and in the cochlea of land-dwelling mammals. Even the sensory cells of insects are only a little different. In the subsequent discussion there will not be space to consider these matters more fully, but they are treated in more detail in Chap. 12.

Somewhat surprisingly, the auditory sensitivities of animals differing widely in both size and anatomy vary less than one might expect. Human hearing, for example, has a threshold of about 20 μPa in its most sensitive frequency range between 500 Hz and 5 kHz, where most auditory information about conspecific communication and the cries of predators and prey is concentrated, and the sensitivity is within about 10 dB of this value over a frequency range from about 200 Hz to 7 kHz. Other mammals have very similar sensitivities with frequency ranges scaled to vary roughly as the inverse of their linear dimensions, as discussed in Sect. 19.1. Insects in general have narrower bandwidth hearing matched to song frequencies for conspecific communication. Of course, with all of these generalizations there are many outliers with very different hearing abilities.

Sound production mechanisms fall into two categories depending upon whether or not the animal is actively air-breathing. For air-breathing animals there is an internal air reservoir, the volume and pressure of which are under muscular control, so that air can be either inhaled or exhaled. When exhalation is done through some sort of valve with mechanically resonant flaps or membranes, it can set the valve into vibration, thus producing an oscillatory exhalation of air and so a sound source. In the case of aquatic mammals, which are also air-breathing, it would be wasteful to exhale the air, so that it is instead moved from one reservoir to another through the oscillating valve, the vibration of the thin walls of one of the reservoirs then radiating the sound, as is discussed in more detail in Chap. 19. Animals such as insects that do not breathe actively must generally make use of muscle-driven mechanical vibrations to produce their calls, though a few such as the cockroach *Blaberus* can actually make "hissing" noises through its respiratory system when alarmed.

The amount of muscular effort an animal is prepared to expend on vocalization varies very widely. Humans, apart from a few exceptions such as trained singers, can produce a maximum sound output of a few hundred milliwatts for a few seconds, and only about 10 mW for an extended time. In terms of body mass, this amounts to something around $0.1\,\mathrm{mW\,kg^{-1}}$. At the other end of the scale, some species of cicada with body mass of about 1 g can produce about 1 mW of sound output at a frequency of about 3 kHz, or about $1000\,\mathrm{mW\,kg^{-1}}$ on a nearly continuous basis. An interesting comparative study of birdsongs by *Brackenbury* [19.15] showed sound powers ranging from 0.15 to 200 mW and relative powers ranging from 10 to $870\,\mathrm{mW\,kg^{-1}}$. The clear winner on the relative power criterion was the small *Turdus philomelos* with a sound output of 60 mW and a body mass of only 69 g, equivalent to $870\,\mathrm{mW\,kg^{-1}}$, while the loudest bird measured was the common rooster *Gallus domesticus* with a sound output during crowing of 200 mW. Its body mass of 3500 g meant, however, that its relative power was only $57\,\mathrm{mW\,kg^{-1}}$.

Nearly all sound-production mechanisms, ranging from animals through musical instruments to electrically driven loudspeakers, are quite inefficient, with a ratio of radiated acoustic power to input power of not much more than 1%, and often a good deal less than this. In the case of air-breathing animals, the losses are largely due to viscosity-induced turbulence above the vocal valve, while in insects that rely upon vibrating body structures internal losses in body tissue are dominant. Wide-bandwidth systems such as the human vocal apparatus are generally much less efficient than narrow-band systems, the parameters of which can be optimized to give good results at a single song frequency. To all this must then be added the internal metabolic loss in muscles, which has not been considered in the 1% figure.

19.3 Vibrational Communication

Most animals are sensitive to vibrations in materials or structures with which they have physical contact. Even humans can sense vibrations over a reasonably large frequency range by simply pressing finger tips against the vibrating object, while vibrations of lower frequency can also be felt through the feet, without requiring any specialized vibration receptors. Many insects make use of this ability to track prey and also for conspecific communication, and some for which this mode of perception is very important have specialized receptors, called subgenual organs, just below a leg joint. A brief survey for the case of insects has been given by *Ewing* [19.16].

For water gliders and other insects that hunt insects that have become trapped by surface tension on the surface of relatively still ponds, the obvious vibratory signal is the surface waves spreading out circularly from the struggles of the trapped insect. The propagation of these surface waves on water that is deep compared with the wavelength has been the subject of analysis, and the calculated wave velocity c_s, which is influenced by both the surface tension T of water and the gravitational acceleration g, is found to depend upon the wavelength λ according to the relation

$$c_s = \left(\frac{2\pi T}{\rho_w \lambda} + \frac{g\lambda}{2\pi}\right)^{1/2}, \quad (19.1)$$

where ρ_w is the density of water. From the form of this relation it is clear that the wave speed has a minimum value when $\lambda = 2\pi(T/g\rho_w)^{1/2}$, and substituting numerical values then gives a wavelength of about 18 mm, a propagation speed of about $24\,\mathrm{cm\,s^{-1}}$, and a frequency of about 1.3 Hz. Waves of about this wavelength will continue to move slowly across the surface after other waves have dissipated. The attenuation of all these surface waves is, however, very great.

From (19.1) it is clear that, at low frequencies and long wavelengths, the wave speed c_s is controlled by gravitational forces and $c_s \approx g/\omega$. These are called gravity waves. At high frequencies and short wavelengths, surface tension effects are dominant and $c_s \approx (T\omega/\rho_w)^{1/3}$. These are called capillary waves. The frequencies of biological interest are typically in the range 10 to 100 Hz, and so are in the in the lower part of the capillary-wave regime, a little above the crossover between these two influences, and the propagation speeds are typically in the range 30 to $50\,\mathrm{cm\,s^{-1}}$.

For insects and other animals living aloft in trees and plants, the waves responsible for transmission of vibration are mostly bending waves in the leaves or branches. The wave speed is then proportional to the square root of the thickness of the structure and varies also as the square root of frequency. Transmission is very slow compared with the speed of sound in air, although generally much faster than the speed of surface waves on water. Once again, the attenuation with propagation distance is usually large.

Disturbances on heavy solid branches or on solid ground also generate surface waves that can be detected at a distance, but these waves propagate with nearly the speed of shear waves in the solid, and thus at speeds of order $2000\,\mathrm{m\,s^{-1}}$. On the ground they are closely related to seismic waves, which generally have much larger amplitude, and animals can also detect these. Underground, of course, as when an animal is seeking prey in a shallow burrow, the vibration is propagated by bulk shear and compressional waves.

19.4 Insects

The variety of sound production and detection mechanisms across insect species is immense [19.3, 16–18], but their acoustic abilities and senses possess many features in common. Many, particularly immature forms such as caterpillars but also many other insects, use external sensory hairs to detect air movement and these can sense both acoustic stimuli, particularly if tuned to resonance, or the larger near-field air flows produced by the wings of predators [19.19]. These hairs respond to the viscous drag created by acoustic air flow past them and, since this is directional, the insect may be able to gain some information about the location of the sound source.

Analysis of the behavior of sensory hairs is complex [19.19], but some general statements can be made. Since what is being sensed is an oscillatory flow of air across the surface of the insect, with the hair standing normal to this surface, it is important that the hair be long enough to protrude above the viscous boundary layer, the thickness of which varies inversely with the square root of frequency and is about 0.1 mm at 150 Hz. This means that the length of the hair should also be about inversely proportional to the frequency of the stimulus it is optimized to detect, and that it should typically be at least a few tenths of a millimeter in length. The thickness of the hair is not of immense importance, provided that it is less than about 10 μm for a typical hair, since much of the effective mass is contributed by co-moving fluid. At the mechanical resonance frequency of the hair, the sensitivity is a maximum, and the hair tip displacement is typically about twice the fluid displacement, provided the damping at the hair root is not large, but the response falls off above and below this frequency.

Mature insects of some species, however, have auditory systems that bear a superficial resemblance to those of modern humans in their overall structure. Thus for example, and referring to Fig. 19.2a, the cricket has thin sensory membranes (tympana or eardrums) on its forelegs, and these are connected by tubes (Eustachian tubes) that run to spiracles (nostrils) on its upper body, the left and right tubes interacting through a thin membrane (nasal septum) inside the body. Analysis of this system [19.9] shows that the tubes and membranes are configured so that each ear has almost cardioid response, directed ipsilaterally, at the song frequency of the insect, allowing the insect to detect the direction of sound arrival. Some simpler auditory systems showing similar directional response are discussed in Sect. 19.5.

Detailed analysis of the response of anatomically complex systems such as this is most easily carried out using electric network analogs, in which pressure is represented by electric potential and acoustic flow by electric current. Tubes are then represented as 2×2 matrices, diaphragms by L, R, C series resonant circuits, and so on, as shown in Fig. 19.2b. Brief details of this approach are given in Sect. 19.10 and fuller treatments can be found in the published literature [19.9, 20].

When it comes to sound production, insects are very different from other animals because they do not have the equivalent of lungs and a muscle-driven respiratory systems. They must therefore generally rely upon muscle-driven vibration of diaphragms somewhere on the body in order to produce sound. One significant exception is the Death's Head Hawk Moth *Acherontia atropos*, which takes in and then expels pulses of air from a cavity closed by a small flap-valve that is set into oscillation by the expelled air to produce short sound pulses [19.16].

Some insects, such as cicadas, have a large abdominal air cavity with two areas of flexible ribbed cartilage that can be made to vibrate in a pulsatory manner under muscle action, progressive buckling of the membrane as the ribs flip between different shapes effectively multiplying the frequency of excitation by as much as a factor 50 compared with the frequency of muscle contraction. The coupling between these tymbal membranes and the air cavity determines the oscillation frequency, and radiation is efficient because of the monopole nature of the source. The sound frequency is typically in the range 3 to 5 kHz, depending upon insect size, and the radiated acoustic power can be as large as 1 mW [19.21] on an almost continuous basis, though the

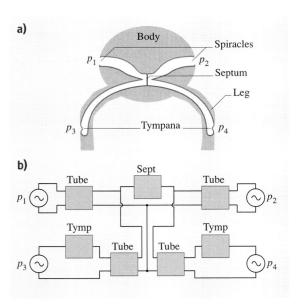

Fig. 19.2 (a) Schematic drawing of the auditory anatomy of a cricket. (b) Simplified electric network analog for this system. The impedances of the tympana and the septum are each represented by a set of components L, R, C connected in series; the tube impedances are represented by 2×2 matrices. The acoustic pressures p_i acting on the system are generally all nearly equal in magnitude, but differ in phase depending upon the direction of incidence of the sound

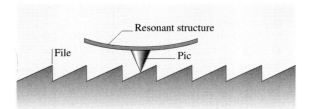

Fig. 19.3 Configuration of a typical file and pic as used by crickets and other animals to excite resonant structures on their wing cases

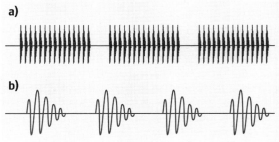

Fig. 19.4a,b Structure of the sound emitted by a typical insect: (**a**) three strokes of the file, the individual oscillations being the pic impacts on the file teeth; (**b**) expanded view of four of the pulses in (**a**), showing resonant oscillations of the structure produced by successive pic impacts on the file

actual sound signal may consist of a series of regularly spaced bursts corresponding to tymbal collapses and rebounds. Again, there are occasional insects that have developed extreme versions of this sound producing mechanism, such as the bladder cicada *Cystosoma Saundersii (Westwood)*, in which the whole greatly extended abdomen is a resonant air sac and the song frequency is only 800 Hz.

Most other insects do not have significant bodily air cavities and must produce sound by causing parts of their body shell or wings to vibrate by drawing a ribbed file on their legs, or in some cases on their wings, across the hard rim of the membrane, as shown in Fig. 19.3. The anatomical features making up the file and vibrating structure vary very greatly across species, as has been discussed and illustrated in detail by *Dumortier* [19.22]. In some cases the file is located on the vibrating structure and the sharp pic on the part that is moving. The passage of the pic across each tooth or rib on the file generates a sharp impulse that excites a heavily damped transient oscillation of the vibrator, and each repetitive motion of the leg or other structure passes the pic over many teeth. A typical insect call can therefore be subdivided into (a) the leg motion frequency, (b) the pic impact frequency on the file, and (c) the oscillation frequency of the vibrating structure. Since the file may have 10 or more teeth, the frequency of the pic impacts will be more than 10 times the frequency of leg motion, and the structural oscillation may be 10 times this frequency. The structure of a typical call is thus of the form shown in Fig. 19.4. The dominant frequency is usually that of the structural vibration, and this varies widely from about 2 to 80 kHz depending upon species and size [19.23]. Because of the structure of the call, the frequency spectrum appears as a broad band centered on the structural vibration frequency.

Such a vibrating membrane is a dipole source and so is a much less efficient radiator than is a monopole source, except at very high frequencies where the wing dimensions become comparable with the sound wavelength excited. Small insects therefore generally have songs of higher frequency than those that are larger, in much the same way as discussed in Sect. 19.1 for other animals. Some crickets, however, have evolved the strategy of digging a horn-shaped burrow in the earth and positioning themselves at an appropriate place in the horn so as to couple their wing vibrations efficiently to the fundamental mode of the horn resonator, thus taking advantage of the dipole nature of their wing source and greatly enhancing the radiated sound power [19.24].

19.5 Land Vertebrates

The class of vertebrates that live on land is in many ways the most largely studied, since they bear the closest relation to humans, but these animals vary widely in size and behavior. Detailed discussion of birds is deferred to Sect. 19.6, since their songs warrant special attention, while human auditory and vocal abilities are not considered specifically since they are the subject of several other chapters. Reptiles, too, are largely omitted from detailed discussion, since most of them are not notably vocal, and bats are given special attention in Sect. 19.7.

The feature possessed by all the land-dwelling vertebrates is that they vocalize using air expelled from their inflated lungs through some sort of oscillating vocal valve. The available internal air pressure depends upon the emphasis put by the species on vocalization, but is typically in the range 100–2000 Pa (or 1–20 cm water gauge) almost independent of animal size, which is what is to be expected if the breathing muscles and lung cavity diameter scale about proportionally with the linear size of the animal. There are, however, very wide variations, as discussed in Sect. 19.2. The auditory abilities of most animals are also rather similar, the 20 dB bandwidth extending over a factor of about 100 in frequency and with a center-frequency roughly inversely proportional to linear size, though again there are outliers, as shown in Fig. 19.1. Although the auditory and vocalization frequencies of different animals both follow the same trend with size, the auditory systems generally have a much larger frequency range than the vocalizations for several reasons. The first is that hearing serves many purposes in addition to conspecific communication, particularly the detection of prey and predators. The second is that, particularly with the more sophisticated animals, there are many nuances of communication, such as the vowels and consonants in human speech, for which information is coded in the upper parts of the vocalization spectrum.

For humans, for example, the primary vocalization frequency is typically in the range 100 to 300 Hz, but the frequency bands that are emphasized in vowel formants lie between 1 and 3 kHz. The same pattern is exhibited in the vocalizations of other air-breathing animals, as shown for the case of a raven in Fig. 19.5. The same formant structure is seen in the cries of elephants, with a fundamental in the range 25–30 Hz and in high-pitched bird songs, which may have a fundamental above 4 kHz.

In most such animals, sound is produced by exhaling the air stored in the lungs through a valve consisting of two taut tissue flaps or membranes that can be made to almost close the base of the vocal tract, as shown in Fig. 19.6a. The lungs in mammals are a complicated

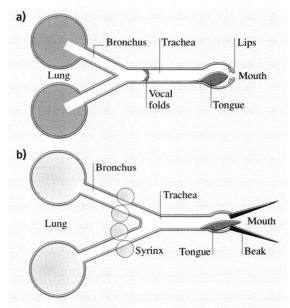

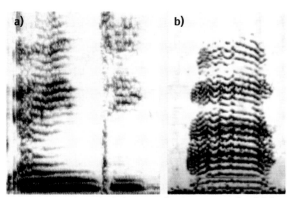

Fig. 19.5a,b A typical time-resolved spectrum for (**a**) human female speech, and (**b**) the cry of a raven, the level of each frequency component being indicated by its darkness. The frequency range is 0–6 kHz and the duration about 0.7 s in (**a**) and 0.5 s in (**b**). It can be seen that at any instant the spectrum consists of harmonics of the fundamental, with particular bands of frequencies being emphasized. These are the formant bands that distinguish one vowel from another in humans and enable similar distinctions to be made by other animals. Consonants in human speech consist of broadband noise, an example occurring at the beginning of the second syllable in (**a**), and other animals may have similar interpolations between the tonal sounds

Fig. 19.6 (**a**) Sketch of the vocal system of a typical land mammal. The lungs force air through the dual-flap vocal-fold valve, producing a pulsating flow of air that is rich in harmonics of the fundamental frequency. Resonances of the upper vocal tract, which can be modified by motion of the tongue and lips, produce formants in the sound spectrum which encode vocal information. (**b**) Sketch of the vocal system of a typical song-bird. There are two inflated-membrane valves in the syrinx just below the junction of the bronchi with the trachea, which may be operated either separately or together. Again a harmonic-rich airflow is produced that can either have formants imposed by resonances of the vocal tract and beak, or can be filtered to near pure-tone form by an inflated sac in the upper vocal tract

quasi-fractal dendritic network of tubules with as many as 16 stages of subdivision, the final stage being terminated by alveolar sacs that provide most of storage volume. The interaction of exhaled air pressure and air flow with the flaps of this valve when they have been brought together (or adducted) causes them to vibrate at very nearly their natural frequency, as determined by mass and tension, and this in turn leads to an oscillating air flow through the valve. The classic treatment of the human vocal valve is that of *Ishizaka* and *Flanagan* [19.25], but there have been many more recent treatments exploring modifications and refinements of this model. Very similar vocal valves are found in other mammals, while the major difference in the case of birds is that the flaps of the vocal folds are replaced by opposed taut membranes inflated by air pressure in cavities behind them, and there may be two such valves, as shown in Fig. 19.6b and discussed later in Sect. 19.6.

In most cases, the vibration frequency of the valve is very much lower than the frequency of the first maximum in the upper vocal tract impedance. The valve therefore operates in a nearly autonomous manner at a frequency determined by its geometry, mass and tension and, to a less extent, by the air pressure in the animal's lungs. Because any such valve is necessarily nonlinear in its flow properties, particularly if it actually closes once in each oscillation cycle, this mechanism generates an air flow, and thus a sound wave in the vocal tract, containing all harmonics of the fundamental oscillation. The radiated amplitudes of these harmonics can be modified by resonances in the air column of the vocal tract to produce variations in sound quality and peaks in the radiated spectrum known generally as vocal formants. Changing the tension of the valve flaps changes the fundamental frequency of the vocal sound, but the frequencies of the formants can be changed independently by moving the tongue, jaws, and lips or beak.

To be more quantitative, suppose that the pressure below the valve is p_0 and that above the valve p_1. If the width of the opening is W and its oscillating dimension is $x(t)$, then the volume flow U_1 through it is determined primarily by Bernoulli's equation and is approximately

$$U_1(t) = \left(\frac{2(p_0 - p_1)}{\rho}\right)^{1/2} Wx(t), \quad (19.2)$$

where ρ is the density of air. In order for the valve to be maintained in oscillation near its natural frequency f, the pressure difference $p_0 - p_1$ must vary with the flow and with a phase that is about 90° in advance of the valve opening $x(t)$. This can be achieved if the acoustic impedance of the lungs and bronchi is essentially compliant, and that of the upper vocal tract small, at the oscillation frequency. The fact that this pressure difference appears as a square root in (19.2) then introduces upper harmonic terms at frequencies $2f, 3f, \ldots$ into the flow $U(t)$. Allowance must then be made for the fact that the vocal valve normally closes for part of each oscillation cycle, so that $x(t)$ is no longer simply sinusoidal, and this introduces further upper-harmonic terms into the flow. This is, of course, a very condensed and simplified treatment of the vocal flow dynamics, and further details can be found in the literature for the case of mammalian, and specifically human, animals [19.25] and also for birds [19.26].

This is, however, only the beginning of the analysis, for it is the acoustic flow out of the mouth that determines the radiated sound, rather than the flow from the vocal valve into the upper vocal tract. Suppose that the acoustic behavior of the upper vocal tract is represented by a 2×2 matrix, as shown in Fig. 19.11 of Sect. 19.10, and that the mouth or beak is regarded as effectively open so that the acoustic pressure p_2 at this end of the tract is almost zero. Then the acoustic volume flow U_2 out of the mouth at a particular angular frequency ω is

$$U_2 = \frac{Z_{21}}{Z_{22}} U_1, \quad (19.3)$$

where U_1 is the flow through the larynx or syrinx at this frequency. For the case of a simple cylindrical tube of length L at angular frequency ω, this gives

$$U_2 = \frac{U_1}{\cos kL}, \quad (19.4)$$

where $k = (\omega/c) - i\alpha$, c is the speed of sound in air, and α is the wall damping in the vocal tract. If the spectrum of the valve flow U_1 has a simple envelope, as is normally the case, then (19.4) shows that the radiated sound power, which is proportional to $\omega^2 U_2^2$, has maxima when $\omega L/c = (2n - 1)\pi/2$, and thus in a sequence $1, 3, 5, \ldots$ These are the formant bands, the precise frequency relationship of which can be varied by changing the geometry of the upper vocal tract, which in turn changes the forms of Z_{21} and Z_{22}.

Some other air-breathing animals (and even human sopranos singing in their highest register) however, adjust their vocal system so that the frequency of the vocal valve closely matches a major resonance of the upper vocal tract, usually that of lowest frequency but not necessarily so. Some species of frogs and birds achieve this by the incorporation of an inflatable sac in the upper vocal tract. In some cases the sac serves simply to provide a resonance of appropriate frequency, and the sound is still radiated through the open mouth, but in others the walls of the sac are very thin so that they vibrate under the acoustic pressure and provide the primary radiation source. This will be discussed again in Sect. 19.6 in relation to bird song, but the analysis for frog calls is essentially the same.

The overall efficiency of sound production in vertebrates is generally only 0.1–1%, which is about the same as for musical instruments. The acoustic output power output is typically in the milliwatt range even for quite large animals, but there is a very large variation between different species in the amount of effort that is expended in vocalization.

The hearing mechanism of land-dwelling vertebrates shows considerable similarity across very many species. Sound is detected through the vibrations it induces in a light taut membrane or tympanum, and these vibrations are then communicated to a transduction organ where they are converted to nerve impulses, generally through the agency of hair cells. The mechanism of neural transduction and frequency discrimination is complex, and its explanation goes back to the days of *Helmholtz* [19.27] in the 19th century, with the generally accepted current interpretation being that of *von Békésy* [19.28] in the mid-20th century. Although these studies related specifically to human hearing, the models are generally applicable to other land-dwelling vertebrates as well, as surveyed in the volume edited by *Lewis* et al. [19.29]. All common animals have two auditory transducers, or ears, located on opposite sides of the head, so that they are able to gain information about the direction from which the sound is coming.

In the simplest such auditory system, found for example in frogs, the two ears open directly into the mouth cavity, as shown in Fig. 19.7a, one neural transducer being closely coupled to each of the two tympana. Such a system is very easily analyzed at low frequencies where the cavity presents just a simple compliance with no resonances of its own. Details of the approach are similar to those for the more complex system to be discussed next. For a typical case in which the ears are separated by 20 mm, the cavity volume is 1 cm^3, and the loaded tympanum resonance is at 500 Hz, the calculated response has the form shown in Fig. 19.7b. Directional discrimination is best at the tympanum resonance frequency and can be as much as 20 dB. In a more realistic model for the case of a frog, the nostrils must also be included, since they lead directly into the mouth and allow the entry of sound. The calculated results in this case show that the direction of maximum response for each ear is

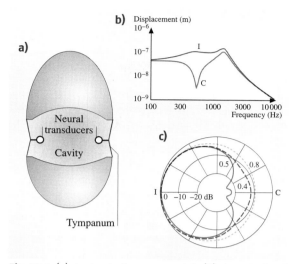

Fig. 19.7 (a) A simple frog-ear model; (b) the calculated frequency response for ipsilateral (I) and contralateral (C) sound signals for a particular optimized set of parameter values; (c) the calculated directivity at different frequencies, shown in kHz as a parameter (from [19.9])

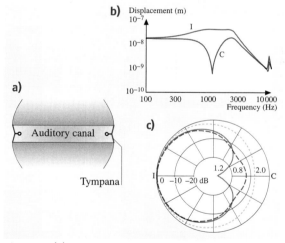

Fig. 19.8 (a) Idealized model of the auditory system of a lizard or bird. Each tympanum is connected to a neural transducer and the two are joined by a simple tube. (b) Response of a particular case of the system in (a) to ipsilateral (I) and contralateral (C) sound of amplitude 0.1 Pa (equivalent to 74 dB re 20 μPa). The tube diameter is assumed to be 5 mm and its length 20 mm, while the tympana have a thickness of 10 μm and are tensioned to a resonance frequency of 1000 Hz. (c) Directional response of this hearing system (from [19.9])

shifted towards the rear of the animal, typically by as much as 30° [19.9].

In another simple auditory system such as that of most reptiles, and surprisingly birds, the two tympana are located at the ends of a tube passing through the head and each is supplied with its own neural transducer, as shown in Fig. 19.8a. The tympana may be recessed at the end of short tubes to provide protection, these tubes only slightly modifying the frequency response. The behavior of such a system can be analyzed using the electric network analogs discussed in Sect. 19.10, with each tympanum involving a series L, R, C combination and the tube being represented by a 2×2 impedance matrix Z_{ij}. If the sound comes from straight in front, then it is the same in magnitude and phase at each ear, so that their responses are necessarily the same. When the sound comes from one side, however, there is a phase difference between the two ear pressures, along with a rather less significant amplitude difference. The motion of each tympanum is determined by the difference between the internal and external pressures acting upon it, and the internal pressure depends upon the signal transferred from the opposite tympanum, modified by the phase delay or resonance of the coupling tube. The analysis is straightforward, but the resulting behavior, which depends upon the mass and tension of the tympana and the length and diameter of the connecting tube, can only be determined by explicit calculation [19.9]. When these parameters are appropriately chosen, each ear has a strongly directional response near the resonance frequency of the tympana, as shown in Fig. 19.8b, c. The balance between the depth and sharpness of the contralateral minimum is affected by the acoustic absorption in the walls of the connecting canal. In some animals, the bone lining of this canal is actually porous, which reduces the effective sound-wave velocity inside it and so increases the phase shift between internal and external signals, a feature that is of assistance when the physical separation of the ears is small.

In most mammals and other large animals, the auditory system is modified in several significant ways. The first is that the auditory canal joining the two ears in birds and reptiles has generally degenerated in mammals to the extent that each ear functions nearly independently. The connecting canal in humans has become the pair of Eustachian tubes running from the middle ear cavity, which contains the bones linking the tympanum to the cochlea, down to the nasal cavities, and its main purpose is now simply to equalize internal and external static pressures and to drain excess fluids. The middle ear cavity itself is necessary in order that the enclosed air volume be large enough that it does not raise the resonance frequency of the tympanum by too much. The topology of the whole system is surprisingly similar to that of the cricket auditory system shown in Fig. 19.2a, but the functions of some of the elements are now rather different.

The other major change is that, instead of the tympana being located almost on the surface of the animal, they are buried below the surface at the end of short auditory canals (meati), which lead to external ears (pinnae) in the shape of obliquely truncated horns. As well as protecting the tympana from mechanical damage, the canals add a minor resonance of their own, generally in the upper part of the auditory range of the animal concerned. The pinnae both increase the level of the pressure signal, typically by about 10 dB and in some cases even more, and impart a directionality to the response [19.9, 30]. The convoluted form of some pinnae also imparts directionally-excited resonances that help distinguish direction from tonal cues, a feature that is necessary in order to distinguish up/down directions for sounds coming from immediately ahead or immediately behind. Some animals with particularly large pinnae, such as kangaroos, are able to rotate these independently to help locate a sound source. In all such cases, however, the neural system plays a large part in determining sound direction by comparing the phases and amplitudes, and sometimes the precise timing of transients, in the signals received by the two ears.

In a hearing system of any simple type it is clear that, if geometrical similarity is maintained and the system is simply scaled to the linear size of the animal concerned, then the frequency of maximum discrimination will vary as the inverse of the linear size of the animal, giving an approximate match to general trend of vocalization behavior. There is, however, one anomaly to be noted [19.31]. Vocalizations by some frogs have fundamental frequencies as high as 8 kHz and the vocal signal contains prominent formant bands in the case studied at 20 kHz and 80 kHz, so that they might almost be classed as ultrasonic. Despite this, behavioral evidence indicates that these frogs cannot hear signals much above 10 kHz, so that perhaps the formant bands are simply an epiphenomenon of the sound production system. The same is true of the calls of some birds.

19.6 Birds

Because of their generally melodious songs, birds have excited a great deal of attention among researchers, as described in several classic books [19.32, 33]. Hearing does not require further discussion because the auditory system is of the simple tube-coupled type discussed in Sect. 19.5 and illustrated in Fig. 19.8. Research interest has focused instead on the wide range of song types that are produced by different species. These range from almost pure-tone single-frequency calls produced by doves, and whistle-like sounds sweeping over more than a 2:1 frequency range produced by many other species, through wide-band calls with prominent formants, to the loud and apparently chaotic calls of the Australian sulfur-crested cockatoo. Some birds can even produce two-toned calls by using both syringeal valves simultaneously but with different tuning.

As noted in Sect. 19.5 and illustrated in Fig. 19.6b, song birds have a syrinx consisting of dual inflated-membrane valves, one in each bronchus just below their junction with the trachea. These valves can be operated simultaneously and sometimes at different frequencies, but more usually separately, and produce a pulsating air-flow rich in harmonics. In birds such as ravens, the impedance maxima of the vocal tract, as measured at the syringeal valve, produce emphasized spectral bands or formants much as in human speech, as is shown in Fig. 19.5. The vocal tract of a bird is much less flexible in shape than is that of a mammal, but the formant frequencies can still be altered by changes in tongue position and beak opening. Studies with laboratory models [19.34] show that the beak has several acoustic functions. When the beak is nearly closed, it presents an end-correction to the trachea that is about half the beak length at high frequencies, but only about half this amount at low frequencies, so that the formant resonances are brought closer together in frequency. As the beak gape is widened, the end-correction reduces towards the normal value for a simple open tube at all frequencies. The beak also improves radiation efficiency, particularly at higher frequencies, by about 10 dB for the typical beak modeled. Finally, the beak enhances the directionality of the radiated sound, particularly at high frequencies and for wide gapes, this enhanced directionality being as much as 10 dB compared with an isotropic radiator.

The role of the tongue has received little attention as yet, partly because it is rather stiff in most species of birds. It does, however, tend to constrict the passage between the top of the trachea and the root of the beak, and this constriction can be altered simply by raising the tongue, thereby exaggerating the end-correction provided to the trachea by the beak. These models of beak behavior can be combined with models of the syrinx and upper vocal tract [19.26] to produce a fairly good understanding of vocalization in birds such as ravens. The fundamental frequency of the song can be changed by changing the muscle tension and thus the natural resonance frequency of the syringeal valve, while the formants, and thus the tone of the song, can be varied by changing the beak opening and tongue position. The role of particular anatomical features and muscular activities in controlling the fundamental pitch of the song has since been verified by careful measurement and modeling [19.35, 36].

Some birds have developed the ability to mimic others around them, a notable example being Australian Lyrebirds of the family *Menuridae* which, as well as imitating other birds, have even learned to mimic human-generated sounds such as axe blows, train whistles, and even the sound of film being wound in cameras. There has also been considerable interest in the vocalization of certain parrots and cockatoos, which can produce, as well as melodious songs, quite intelligible imitations of human speech. The important thing here is to produce a sound with large harmonic content and then to tune the vocal tract resonances, particularly those that have frequencies in the range 1–2.5 kHz, to match those of the second and third resonances of the human vocal tract, which are responsible for differentiating vowel sounds. Consonants, of course, are wide-band noise-like transients. Careful studies of various parrots [19.37, 38] show that they can achieve this by careful positioning of the tongue, much in the way that humans achieve the same thing but with quantitative differences because of different vocal tract size. The match to human vowels in the second and third human formants can be quite good. The first formant, around 500 Hz in humans, is not reproduced but this is not important for vowel recognition.

These studies produce a good understanding of the vocalization of normal song-birds, the calls of which consist of a dominant fundamental accompanied by a range of upper harmonics. Less work has been done on understanding the nearly pure-tone cries of other birds. For many years there has been the supposition that the sound production mechanism in this case was quite different, perhaps involving some sort of aerodynamic whistle rather than a syringeal vibration. More recent experimental observations on actual singing birds, using modern technology [19.39], have established however that, at least in the cases studied, the syringeal valve is in fact normally active, though it does not close completely in any part of the cycle, thus avoiding the generation of large amplitudes of upper harmonics. Suppression of upper harmonics can also be aided by filtering in the upper vocal tract [19.40].

In the case of doves, which produce a brief pure-tone "coo" sound at a single frequency, the explanation appears to lie in the use of a thin-walled inflated esophageal sac and a closed beak, with fine tuning provided by an esophageal constriction [19.41]. The volume V of air enclosed in the sac provides an acoustic compliance $C = V/\rho c^2$ that is in parallel with an acoustic inertance $L = m/S^2$ provided by the mass m and surface area S of the sac walls. The resonance frequency f of this sac filter is given by

$$f = \frac{1}{2\pi} \left(\frac{\rho c^2 S^2}{mV} \right)^{1/2}, \quad (19.5)$$

where ρ is the density of air and c is the speed of sound in air. When excited at this frequency by an inflow of air from the trachea, the very thin expanded walls of the sac vibrate and radiate sound to the surrounding air. For a typical dove using this strategy, the wall-mass and dimensions of the inflated sac are about right to couple to the "coo" frequency of about 900 Hz. From (19.5), the resonance frequency varies only as the square root of the diameter of the sac, so that a moderate exhalation of breath can be accommodated without much disturbance to the resonance. The dove must, however, learn to adjust the glottal constriction to couple the tracheal resonance efficiently to that of the sac.

Some other birds, such as the Cardinal, that produce nearly pure-tone calls over a wide frequency range, appear to do so with the help of an inflated sac in the upper vocal tract that leads to the opened beak, with the tracheal length, sac volume, glottal constriction, tongue position, and beak opening all contributing to determine the variable resonance frequency of the filter. Because the sac is not exposed as in the dove, its walls are heavy and do not vibrate significantly. It thus acts as a Helmholtz resonator, excited by the input flow from the trachea and vented through the beak. The bird presumably learns to adjust the anatomical variables mentioned above to the syringeal vibration frequency, which is in turn controlled by syringeal membrane tension, and can then produce a nearly pure-tone song over quite a large frequency range. Despite this explanation of the mechanism of pure-tone generation in some cases, others have still to be understood, for the variety of birdsong is so great that variant anatomies and vocalization skills may well have evolved.

With most of the basic acoustic principles of birdsong understood, most of the interest in birdsong centers upon the individual differences between species. The variety found in nature is extremely great and, while sometimes correlated with large variations in bird anatomy, it also has many environmental influences. Some birds, surprisingly, have songs with prominent formant bands extending well into the ultrasonic region, despite the fact that behavioral studies and even auditory brainstem measurements show that they have no response above about 8 kHz [19.31, 42]. The formant bands in these cases are therefore presumably just an incidental product of the vocalization mechanism.

19.7 Bats

Since human hearing, even in young people, is limited to the range from about 20 Hz to about 20 kHz, frequencies lying outside this range are referred to as either infrasonic or ultrasonic. Very large animals such as elephants may produce sounds with appreciable infrasonic components, though the higher harmonics of these sounds are quite audible. Small animals such as bats, however, produce echo-location calls with dominant frequencies around 80 kHz that are inaudible to humans, though some associated components of lower frequency may be heard. In this section brief attention will be given to the sonar abilities of bats, though some other animals also make use of ultrasonic techniques [19.43].

Sound waves are scattered by any inhomogeneities in the medium through which they are propagating, the density of an animal body in air provides a large contrast in acoustic impedance and therefore a large scattered signal. The basic physics shows that, for a target in the form of a sphere of diameter d and a signal of wavelength $\lambda \gg d$, the echo strength is proportional to d^6/λ^4 times the incident intensity. This incident intensity is itself determined by the initial signal power and the angular width of the scanning signal, which varies about as λ/a, where a is the radius of the mouth of the probing animal. If R is the distance between the probing animal and the target, then the sound power incident on the target varies as $a^2/\lambda^2 R^2$ and the returned echo strength E thus varies as

$$E \propto \frac{d^6 a^2}{\lambda^6 R^4}, \qquad (19.6)$$

provided the wavelength λ is much greater than the diameter d of the scattering object and the radius a of the mouth. Another factor must, however, be added to this equation and that is the attenuation of the sound due to atmospheric absorption. This attenuation depends on relative humidity, but is about $0.1\,\mathrm{dB\,m^{-1}}$ at $10\,\mathrm{kHz}$, increasing to about $0.3\,\mathrm{dB\,m^{-1}}$ at $100\,\mathrm{kHz}$, and so is not a very significant factor at the short ranges involved.

Once the size of the target becomes comparable with the incident sound wavelength, the scattering tends towards becoming reflection and geometrical-optics techniques can be used. In this limit, the reflected intensity is proportional to the area of the target, but depends in a complicated way upon its shape and orientation. Finally, if a pure-tone sound is used, any Doppler shift in the frequency of the echo can give information about the relative motion of prey and predator. The reflected sound can therefore give information about the nature of the target and its motion, both of which are important to the pursuing bat.

The sonar-location cries of bats and other animals are, however, not pure-tone continuous signals but rather sound pulses with a typical frequency in the range $40-80\,\mathrm{kHz}$ and so a wavelength of about $4-8\,\mathrm{mm}$. This wavelength is comparable to the size of typical prey insects so that the echo can give information about size, shape and wing motion, while the echo delay reveals the distance. The bat's large ears are also quite directional at these frequencies, so that further information on location is obtained. Rather than using short bursts of pure tone, some predators use longer bursts in which the frequency is swept rapidly through a range of several kilohertz. This technique is known as "chirping", and it allows the returned echo to be reconstructed as a single pulse by frequency-dependent time delay, although animal neural systems may not actually do this. The advantage of chirping is that the total energy contained in the probing pulse, and thus the echo, can be made much larger without requiring a large peak power.

Because of the high frequencies involved in such sonar detection, bats have a hearing system that is specialized to cover this high-frequency range and to provide extended frequency resolution near the signal frequency, perhaps to aid in the neural equivalent of "de-chirping" the signal and detecting Doppler frequency shifts. The auditory sensitivity curve of bats is similar in shape to that of humans (threshold about $20\,\mu\mathrm{Pa}$), but shifted up in frequency by a factor of about 20 to give high sensitivity in the range $10-100\,\mathrm{kHz}$ [19.1]. The threshold curve generally shows a particular sensitivity maximum near the frequency of the echo-location call, as might be expected. Bats also generally have forward-facing ears that are large in comparison with their body size. The functions of these in terms of increased directionality and larger receiving area are clear [19.30].

Sound production in bats is generally similar to that in other animals, but with the system dimensions tailored to the higher frequencies involved, and often with the sound emitted through the nose rather than the mouth. The complicated geometry of the nasal tract, which has several spaced enlargements along its length, appears to act as an acoustic filter which suppresses the fundamental and third harmonic of the emitted sound while reinforcing the second harmonic, at least in some species [19.44]. The physics of this arrangement is very similar to that of matching stubs applied to high-frequency transmission lines.

19.8 Aquatic Animals

The generation and propagation of sound under water, as discussed in Chap. 19, have many features that are different from those applying to sound in air. From a physical point of view, a major difference is the fact that water is often quite shallow compared with the range of sound, so that propagation tends to be two-dimensional at long distances. From the present biological viewpoint the major difference is that the acoustic impedance of biological tissue is nearly equal to that of the surrounding water, while in air these differ by a factor of about 3500. This leads to some very significant differences in the auditory anatomy and behavior between aquatic and air-dwelling animals.

The variety of aquatic animal species is comparable to that of land-dwelling animals, and they can be divided into three main categories for the purposes of the present chapter. First come the crustaceans, such as shrimp, lobsters and crabs, which are analogous to insects, then the wide variety of fish with backbones, and finally the air-breathing aquatic mammals such as the cetaceans (whales, dolphins and porpoises, which are the subject of Chap. 19), though animals such as turtles also come into this category. There are, of course, many other categories of marine animal life, but they are not generally significant from an acoustic point of view. A good review has been given by *Hawkins* and *Myrberg* [19.45].

The crustaceans, like most insects, generally produce sound by rubbing a toothed leg-file against one of the plates covering their body, thus producing a pulse of rhythmic clicks. Some animals, such as shrimp, may produce a single loud click or pop by snapping their claws in much the same way that a human can do with fingers. Because this activity takes place under water, which has density ρ about 800 times greater and sound velocity c about 4.4 times greater than that of air, and thus a specific acoustic impedance ρc about 3500 times greater than air, the displacement amplitude of the vibration does not need to be large, and the vibrating structure is generally small and stiff. The peak radiated acoustic power in the "click" can, however, be very large – as much as several watts in the Pistol shrimp *Alpheidae* – probably from a mechanism involving acoustic cavitation. The click lasts for only a very short time, however, so that the total radiated energy is small.

Fish species that lack an air-filled "swim-bladder", the main purpose of which is to provide buoyancy, must produce sound in a similar manner. For fish that do have a swim-bladder, however, the sound production process is much more like that of the cicada, with some sort of plate or membrane over the bladder, or even the bladder wall itself, being set into oscillation by muscular effort and its displacement being aided by the compressibility of the air contained in the bladder. The surrounding tissue and water provide a very substantial mass loading which also contributes to determining the primary resonance frequency of the bladder. This is a much more efficient sound production process than that of the crustaceans and leads to much louder sounds, again generally consisting of a train of repetitive pulses at a frequency typically in the range 100–1000 Hz.

When it comes to sound detection, hair cells again come into play in a variety of ways. Because the density of the animal body is very similar to that of the surrounding water, a plane wave of sound tends to simply move the animal and the water in just the same way. The slight difference in density, however, does allow for some differential motion. The animal body is, however, relatively rigid, so that its response to diverging waves in the near field is rather different from that of the liquid. Any such relative motion between the body and the surrounding water can be detected by light structures such as hairs protruding from the body of the animal, and these can form the basis of hair-cell acoustic detectors. Even more than this, since such cells are generally sensitive to hair deflection in just one direction, arrays of hair cells can be structured so as to give information about the direction of origin of the sound. Although some such detector hairs may be mechanically strong and protrude from the body of the animal, others may be very small and delicate and may be protected from damage by being located in trench-like structures, these structures being called lateral-line organs. Their geometry also has an effect on sound localization sensitivity.

There is another type of hair detector that responds to acceleration rather than relative displacement and that therefore takes advantage of the fact that the body of the animal tends to follow the acoustic displacement of the surrounding water. This is the otolith, which consists of a small dense calcareous stone supported by one or more hair cells, as shown in Fig. 19.9. Since the motion of this stone tends to lag behind the motion of the liquid, and thus of its supports, the supporting hairs are deflected in synchrony with the acoustic stimulus, so that it can be detected by the cells to which they are attached. These sensors will normally have evolved so that combined mass of the otolith and stiffness of the hair give a resonance frequency in the range of interest to the animal, and detection sensitivity will be greatest at this frequency.

If the fish possesses an air-filled swim-bladder, then this can also be incorporated in the hearing sys-

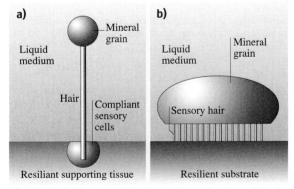

Fig. 19.9a,b Two possible forms for a simple otolith detector: (**a**) a single-cell detector, (**b**) a multi-cell detector. As the mineral grain moves sideways relative to the substrate, the sensory hairs are deflected and open ion channels in the supporting transducer cells (from [19.9])

tem [19.46]. An air bubble in water is made to move in a sound wave by the differential pressure across it. Because its mass is small and the pressure gradient is the same as that on an equivalent volume of water, the bubble tends to move more than the surrounding liquid but, in order to do so, it must displace a significant amount of this liquid. The final result is that the amplitude of free bubble motion is about three times the amplitude of the motion of the surrounding liquid. Since the body of the fish tends to follow the liquid motion because its density is nearly the same, there is relative motion between the bladder and the rest of the fish, provided it is not rigidly enclosed. This relative motion can be detected by neural cells pressed against the bladder wall and provides an efficient sensor of underwater sound. The radial compressive resonances of the swim bladder can also be important in the detection of sound and, since these are also involved in sound production in these fish, there is a good match between call frequency and the frequency of maximum hearing sensitivity, which typically lies in the range 100 to 1000 Hz. Fish with swim bladders have thus evolved to use these primarily buoyancy-regulating structures for both the production and detection of sound.

19.9 Generalities

This brief survey has shown the strong resemblance between the basic sound-producing and auditory mechanisms in a wide variety of animals. Certainly there is a distinction in the matter of sound production between those animals that store air within their bodies and those that do not, and a rather similar distinction in the matter of sound detection, but the similarities are more striking than the differences. In particular, the universal importance of hair-cells in effecting the transduction from acoustic to neural information is most striking.

Also notable in its generality is the scaling of the dominant frequency of produced sound and the associated frequency of greatest hearing sensitivity approximately inversely with the linear dimensions of the animal involved. Certainly there are notable outliers in this scaling, but that in itself is a clue to the significantly different acoustic behavior of the animal species concerned.

Fortunately, quite straightforward acoustic analysis provides a good semi-quantitative description of the operation of most of these animal acoustic systems, and in some cases such analysis can reveal evolutionarily optimized solutions to modern acoustical design problems.

19.10 Quantitative System Analysis

Biological systems are almost always anatomically complex, and this is certainly true of vocal and auditory systems, which typically comprise sets of interconnecting tubes, horns, cavities and membranes driven over a wide frequency range by some sort of internal or external excitation. While a qualitative description of the way in which such a system works can be helpful, this must be largely based upon conjectures unless one is able to produce a quantitative model for its behavior. Fortunately a method has been developed for the construction of detailed analytical models for such systems from which their acoustic behavior can be quantitatively predicted, thus allowing detailed comparison with experiment. This section will outline how this can be done, more detail being given in [19.9] and [19.20].

In a simple mechanical system consisting of masses and springs, the quantities of interest are the force F_i applied at point i and the resulting velocity of motion v_j at point j, and we can define the mechanical impedance Z^{mech} by the equation $Z_{jj}^{\text{mech}} = F_j/v_j$. Note that the subscripts on F and v are here both the same, so that Z_{jj}^{mech} is the ratio of these two quantities measured at a single point j. It is also possible to define a quantity $Z_{ij}^{\text{mech}} = F_i/v_j$ which is called the transfer impedance between the two points i and j. There is an important theorem called the reciprocity theorem, which shows that $Z_{ji} = Z_{ij}$.

In such a simple mechanical system, as well as the masses and springs, there is normally some sort of loss mechanism, usually associated with internal viscous-type losses in the springs, and the situation of interest involves the application of an oscillating force of magnitude F and angular frequency ω. We denote this force in complex notation as $F e^{i\omega t}$ where $i = \sqrt{-1}$. (Note that, in the interest of consistency throughout this Handbook, this notation differs from the standard electrical engineering notation in which i is replaced by j.) For the motion of a simple mass m under the influence of a force F we know that $F = m\,dv/dt$ or $F = i\omega m v$, so that the impedance provided by the mass is $i\omega m$. Similarly, for a spring of compliance C, $F = \int v\,dt/C = -i/(\omega C)$, giving an impedance $-i/(\omega C)$. For a viscous loss within the spring there is an equivalent resistance R proportional to the speed of motion, so that $F = Rv$. Examination of these equations shows that they are closely similar to those for electric circuits if we assume that voltage is the electrical analog of force and current the analog of velocity. The analog of a mass is thus an inductance of magnitude m, the analog of a compliant spring is a capacitance C, and the analog of a viscous resistance is an electrical resistance R. A frictional resistance is much more complicated to model since its magnitude is constant once sliding occurs but changes direction with the direction of motion, so that its whole behavior is nonlinear.

These mechanical elements can then be combined into a simple circuit, two examples being shown in Fig. 19.10. In example (a) it can be seen that the motion of the mass, the top of the spring, and the viscous drag are all the same, so that in the electrical analog the same current must flow through each, implying a simple series circuit as in (b). The total input impedance is therefore the sum of the three component impedances, so that

$$Z^{\text{mech}} = R + i\omega m - \frac{i}{\omega C} = R + \frac{im}{\omega}\left(\omega^2 - \frac{1}{mC}\right), \tag{19.7}$$

which has an impedance minimum, and thus maximum motion, at the resonance frequency $\omega^* = 1/(mC)^{1/2}$. In example (c), the displacement of the top of the spring is the sum of the compressive displacement of the spring and the motion of the suspended mass, and so the analog current is split between the spring on the one hand and the mass on the other, the viscous drag having the same motion as the spring, leading to the parallel resonant circuit shown in (d). There is an impedance maximum, and thus minimum motion of the top of the spring, at the resonance frequency, which is again $\omega^* = 1/(mC)^{1/2}$. In a simple variant of these models, the spring could be undamped and the mass immersed in a viscous fluid, in which case the resistance R would be in series with the inductance rather than the capacitance. These analogs can be extended to include things such as levers, which appear as electrical transformers with turns ratio equal to the length ratio of the lever arms.

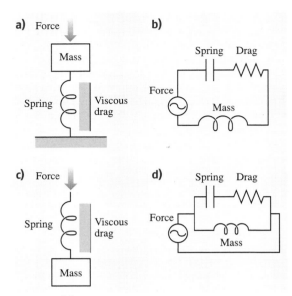

Fig. 19.10 (a) A simple resonant assembly with an oscillating force applied to a mass that is resting on a rigidly supported spring, the spring having internal viscous losses. (b) Analog electric network for the system shown in (a). (c) An alternative assembly with the mass being freely suspended from the spring and the oscillating force being applied to the free end of the spring. (d) Analog electric network for the system shown in (c). (Note the potential source of confusion in that an electrical inductance symbol looks like a physical spring but actually represents a mass)

While such combinations of mechanical elements are important in some aspects of animal acoustics, such as the motion of otolith detectors, most of the systems of interest involve the generation and detection of acoustic waves rather than of mechanical vibrations. The important physical variables are then the acoustic pressure p and the acoustic volume flow U, both being functions of the frequency ω. By analogy with mechanical systems, the acoustic impedance is now defined as $Z^{\text{acoust}} = p/U$. For a system with a well-defined measurement aperture of area S, it is clear that $p = F/S$ and $U = vS$, so that $Z^{\text{acoust}} = Z^{\text{mech}}/S^2$, and once again we can define both the impedance at an aperture and the transfer impedance between two apertures. In what follows, the superscript "acoust" will be omitted for simplicity.

Consider first the case of a circular diaphragm under elastic tension and excited by an acoustic pressure $p\mathrm{e}^{i\omega t}$ on one side. This problem can easily be solved using a simple differential equation [19.9], but here we seek a network analog. The pressure will deflect the diaphragm into a nearly spherical-cap shape, displacing an air volume as it does so. If the small mass of the deflected air is neglected, then deflection of the diaphragm will be opposed by its own mass inertia, by the tension force and by internal losses. The mass contributes an acoustic inertance (or inductance in the network analog) with magnitude about $2\rho_s d/S$, where ρ_s is the density of the membrane material, d is the membrane thickness, and S is its area, while the tension force T contributes an equivalent capacitance of magnitude about $2S^2/T$. This predicts a resonance frequency $\omega^* \approx 0.5(T/S\rho_s d)^{1/2}$, while the rigorously derived frequency for the first mode of the diaphragm replaces the 0.5 with 0.38. Because biological membranes are not uniform in thickness and are usually loaded by neural transducers, however, it is not appropriate to worry unduly about the exact values of numerical factors. For a membrane there are also higher modes with nodal diameters and nodal circles, but these are of almost negligible importance in biological systems.

The next thing to consider is the behavior of cavities, apertures, and short tubes. The air enclosed in a rigid container of volume V can be compressed by the addition of a further small volume dV of air and the pressure will then rise by an amount $p_0\gamma\,dV/V$ where p_0 is atmospheric pressure and $\gamma = 1.4$ is the ratio of specific heats of air at constant pressure and constant volume. The electric analog for this case is a capacitance of value $\gamma p_0/V = \rho c^2/V$ where c is the speed of sound in air and ρ is the density of air.

In the case of a tube of length l and cross-section S, both very much less than the wavelength of the sound, the air within the tube behaves like a mass $\rho l S$ that is set into motion by the pressure difference p across it. The acoustic inertance, or analog inductance, is then $Z = \rho l/S$. For a very short tube, such as a simple aperture, the motion induced in the air just outside the two ends must be considered as well, and this adds an "end-correction" of 0.6–0.8 times the tube radius [19.47] or about $0.4S^{1/2}$ to each end of the tube, thus increasing its effective length by twice this amount.

These elements can be combined simply by considering the fact that acoustic flow must be continuous and conserved. Thus a Helmholtz resonator, which consists of a rigid enclosure with a simple aperture, is modeled as a capacitor and an inductor in series, with some resistance added due to the viscosity of the air moving through the aperture and to a less extent to radiation resistance. The analog circuit is thus just as in Fig. 19.10b and leads to a simple resonance.

If any of the components in the system under analysis is not negligibly small compared with the wavelength of

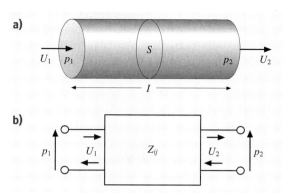

Fig. 19.11 (a) A simple uniform tube showing the acoustic pressures p_i and acoustic volume flows U_i. (b) The electrical analog impedance element Z_{ij}. Note that in both cases the current flow has been assumed to be antisymmetric rather than symmetric

the sound involved, then it is necessary to use a more complex analysis. For the case of a uniform tube of length l and cross section S, the pressure and flow at one end can be denoted by p_1 and U_1 respectively and those at the other end by p_2 and U_2. The analog impedance is then a four-terminal element, as shown in Fig. 19.11. The equations describing the behavior of this element are

$$p_1 = Z_{11}U_1 - Z_{12}U_2,$$
$$p_2 = Z_{21}U_1 - Z_{22}U_2. \qquad (19.8)$$

Note that the acoustic flow is taken to be inwards at the first aperture and outwards at the second, and this asymmetry gives rise to the minus signs in (19.8). Some texts take both currents to be inflowing, which removes the minus signs. For a simple uniform tube, symmetry demands that $Z_{21} = Z_{12}$ and that $Z_{22} = Z_{11}$. The first of these relations, known as reciprocity, is universal and does not depend upon symmetry, while the second is not true for asymmetric tubes or horns.

For a uniform tube of length l and cross-section S, consideration of wave propagation between the two ends shows that

$$Z_{11} = Z_{22} = -iZ_0 \cot kl,$$
$$Z_{21} = Z_{12} = -iZ_0 \csc kl, \qquad (19.9)$$

where $Z_0 = \rho c/S$ is the characteristic impedance of the tube. The propagation number k is given by

$$k = \frac{\omega}{c} - i\alpha \qquad (19.10)$$

where α represents the attenuation caused by viscous and thermal losses to the tube walls, and has the approximate value

$$\alpha \approx 10^{-5} \frac{\omega^{1/2}}{a} \tag{19.11}$$

where a is the tube radius. Since auditory systems often involve quite narrow tubes, this term can become very important.

Similar but much more complicated relations apply for horns of various shapes, the simplest being conical, exponential or parabolic in profile [19.9]. As discussed elsewhere [19.47], horns act as acoustic transformers, basically because $Z_{22} \neq Z_{11}$, and can raise the acoustic pressure in the narrow throat by about the ratio of the mouth diameter to the throat diameter. This amplification is limited, however, to frequencies high enough that the horn diameter does not change greatly over an axial distance of about one wavelength.

With the aid of these electric circuit element analogs it is now possible to construct an analog network to represent the whole acoustic system, as shown by the example in Fig. 19.2. The major guiding principle in constructing such a network is the conservation of acoustic flow, or equivalently of electric current. In the case of an auditory system, there may be several openings connecting to the environment, and the acoustic pressures at these openings will generally differ in both phase and amplitude, depending upon the direction from which the sound is coming. Phase differences are generally more important for directional discrimination than are amplitude differences, particularly for small animals, just as in directional microphones, [19.47] and the combination of tubes, cavities and membranes in the system provides additional phase shifts. The sensory input is normally generated by the motion of tympanic membranes on the two sides of the animal, with the combination of external and internal pressure determining the motion of each membrane, as shown for example in Figs. 19.7 and 19.8.

When sound production is considered, it is usual to make the approximation of separating the basic sound-generation mechanism, such as the air flow through vibrating vocal folds, from the acoustics of the subsequent tubes and cavities, thus leading to a "source/filter model". With this simplification, the source can be assumed to produce an oscillating acoustic flow at a particular frequency ω, generally accompanied by a set of harmonics at frequencies $n\omega$ generated by nonlinearity in the vibration of the flow valve. The sound production at each frequency is then determined by the final acoustic flow $U(n\omega)$ through an aperture into the surroundings. The square of this flow, multiplied the acoustic radiation resistance of the aperture, which depends upon its size, [19.47] then determines the radiated acoustic power at that frequency. In a much more complex model, the pressure generated by the acoustic input flow can be recognized to act back upon the acoustic source itself, thus modifying its behavior. Typically this results in a small shift in the source frequency to align it more closely with a neighboring strong resonance of the filter. Such interactions are well known in other systems such as musical wind instruments and probably assist in coupling the oscillations of the syrinx to the resonance of the vocal filter in "pure tone" birdsong.

Finally, it is worthwhile to emphasize that biological acoustic systems do not consist of straight tubes of uniform diameter with rigid walls, unloaded circular membranes, or other idealized elements. While it is possible in principle to include such refinements in a computational model, this will generally make that model so complex to analyze that it reveals little about the actual operation of the system. It is usually better, therefore, to construct the simplest possible model that includes all the relevant acoustic elements and to examine its predictions and the effect that changing the dimensions of some of those elements has on the overall behavior. In this way the behavior of the real biological system can be largely understood.

Remark

The number of books and papers published in this field is very large, so that the reference list below is a rather eclectic selection. Generally descriptive and behavioral references have not been included, and the list concentrates on those that apply physical and mathematical analysis to bioacoustic systems.

References

19.1 W.C. Stebbins: *The Acoustic Sense of Animals* (Harvard Univ. Press, Cambridge 1983)

19.2 J.W. Bradbury, S.L. Vehrenkamp: *Principles of Animal Communication* (Sinauer, Sunderland 1998)

19.3 R.-G. Busnel (Ed.): *Acoustic Behavior of Animals* (Elsevier, New York 1963)

19.4 A.N. Popper, R.R. Fay (Eds.): *Comparative Studies of Hearing in Vertebrates* (Springer, Berlin, New York 1980)

19.5 B. Lewis (Ed.): *Bioacoustics: A Comparative Approach* (Academic, London 1983)
19.6 D.B. Webster, R.R. Fay, A.N. Popper (Eds.): *The Evolutionary Biology of Hearing* (Springer, Berlin, New York 1992)
19.7 R.R. Fay, A.N. Popper (Eds.): *Comparative Hearing: Mammals* (Springer, Berlin, New York 1994)
19.8 R.R. Hoy, A.N. Popper, R.R. Fay (Eds.): *Comparative Hearing: Insects* (Springer, Berlin, New York 1998)
19.9 N.H. Fletcher: *Acoustic Systems in Biology* (Oxford Univ. Press, New York 1992)
19.10 D.A.W. Thompson: *On Growth and Form (abridged)*, 2nd edn., ed. by J.T. Bonner (Cambridge Univ. Press, Cambridge 1961)
19.11 G.B. West, J.H. Brown: Life's universal scaling laws, Phys. Today **57**(9), 36–42 (2004)
19.12 N.H. Fletcher: A simple frequency-scaling rule for animal communication, J. Acoust. Soc. Am. **115**, 2334–2338 (2004)
19.13 M. Garstang, D. Larom, R.Raspet, M. Lindeque: Atmospheric controls on elephant communication, J. Exper. Biol. **198**, 939–951 (1995)
19.14 K. McComb, D. Reby, L. Baker, C. Moss, S. Sayialel: Long-distance communication of acoustic cues to social identity in African elephants, Anim. Behav. **65**, 317–329 (2003)
19.15 J.H. Brackenbury: Power capabilities of the avian sound-producing system, J. Exp. Biol. **78**, 163–166 (1979)
19.16 A.W. Ewing: *Arthropod Bioacoustics: Neurobiology and Behavior* (Cornell Univ. Press, Ithaca 1989)
19.17 A. Michelsen, H. Nocke: Biophysical aspects of sound communication in insects, Adv. Insect Physiol. **10**, 247–296 (1974)
19.18 K. Kalmring, N. Elsner (Eds.): *Acoustic and Vibrational Communication in Insects* (Parey, Berlin, Hamburg 1985)
19.19 N.H. Fletcher: Acoustical response of hair receptors in insects, J. Comp. Physiol. **127**, 185–189 (1978)
19.20 N.H. Fletcher, S. Thwaites: Physical models for the analysis of acoustical systems in biology, Quart. Rev. Biophys. **12**(1), 25–65 (1979)
19.21 H.C. Bennet-Clark: Resonators in insect sound production: how insects produce loud pure-tone songs, J. Exp. Biol. **202**, 3347–3357 (1999)
19.22 B. Dumortier: Morphology of sound emission apparatus in arthropoda. In: *Acoustic Behaviour of Animals*, ed. by R.-G. Busnel (Elsevier, Amsterdam 1963) pp. 277–345
19.23 B. Dumortier: The physical characteristics of sound emission in arthropoda. In: *Acoustic Behaviour of Animals*, ed. by R.-G. Busnel (Elsevier, Amsterdam 1963) pp. 346–373
19.24 A.G. Daws, H.C. Bennet-Clark, N.H. Fletcher: The mechanism of tuning of the mole cricket singing burrow, Bioacoust. **7**, 81–117 (1996)
19.25 K. Ishizaka, J.L. Flanagan: Synthesis of voiced sounds from a two-mass model of the vocal cords, Bell Syst. Tech. J. **51**(6), 1233–1268 (1972)
19.26 N.H. Fletcher: Bird song – a quantitative acoustic model, J. Theor. Biol. **135**, 455–481 (1988)
19.27 H.L.F. Helmholtz: *On the Sensations of Tone as a Physiological Basis for the Theory of Music* (Dover, New York 1954), trans. A.J. Ellis
19.28 G. von Békésy: *Experiments in Hearing* (McGraw-Hill, New York 1960)
19.29 E.R. Lewis, E.L. Leverenz, W.S. Bialek: *The Vertebrate Inner Ear* (CRC, Boca Raton 1985)
19.30 N.H. Fletcher, S. Thwaites: Obliquely truncated simple horns: idealized models for vertebrate pinnae, Acustica **65**, 194–204 (1988)
19.31 P.M. Narins, A.S. Feng, W. Lin, H.-U. Schnitzler, A. Densinger, R.A. Suthers, C.-H. Xu: Old world frog and bird vocalizations contain prominent ultrasonic harmonics, J. Acoust. Soc. Am. **115**, 910–913 (2004)
19.32 C.H. Greenewalt: *Bird Song: Acoustics and Physiology* (Smithsonian Inst. Press, Washington 1968)
19.33 D.E. Kroodsma, E.H. Miller (Eds.): *Acoustic Communication in Birds*, Vol.1-2 (Academic, New York 1982)
19.34 N.H. Fletcher, A. Tarnopolsky: Acoustics of the avian vocal tract, J. Acoust. Soc. Am. **105**, 35–49 (1999)
19.35 F. Goller, O.N. Larsen: A new mechanism of sound generation in songbirds, Proc. Nat. Acad. Sci. U.S.A. **94**, 14787–14791 (1997)
19.36 G.B. Mindlin, T.J. Gardner, F. Goller, R.A. Suthers: Experimental support for a model of birdsong production, Phys. Rev. E **68**, 041908 (2003)
19.37 D.K. Patterson, I.M. Pepperberg: A comparative study of human and parrot phonation: Acoustic and articulatory correlates of vowels, J. Acoust. Soc. Am. **96**, 634–648 (1994)
19.38 G.J.L. Beckers, B.S. Nelson, R.A. Suthers: Vocal-tract filtering by lingual articulation in a parrot, Current Biol. **14**, 1592–1597 (2004)
19.39 O.N. Larsen, F. Goller: Role of syringeal vibrations in bird vocalizations, Proc. R. Soc. London B **266**, 1609–1615 (1999)
19.40 G.J.L. Beckers, R.A. Suthers, C. ten Cate: Pure-tone birdsong by resonant filtering of harmonic overtones, Proc. Nat. Acad. Sci. U.S.A. **100**, 7372–7376 (2003)
19.41 N.H. Fletcher, T. Riede, G.A. Beckers, R.A. Suthers: Vocal tract filtering and the "coo" of doves, J. Acoust. Soc. Am. **116**, 3750–3756 (2004)
19.42 C. Pytte, M.S. Ficken, A. Moiseff: Ultrasonic singing by the blue-throated hummingbird: A comparison between production and perception, J. Comp. Physiol. A **190**, 665–673 (2004)
19.43 G. Sales, D. Pye: *Ultrasonic Communication in Animals* (Chapman Hall, London 1974)

19.44 D.J. Hartley, R.A. Suthers: The acoustics of the vocal tract in the horseshoe bat, Rhinolophus hildebrandti, J. Acoust. Soc. Am. **84**, 1201–1213 (1988)

19.45 A.D. Hawkins, A.A. Myrberg: Hearing and sound communication under water. In: *Bioacoustics: A Comparative Approach*, ed. by B. Lewis (Academic, London 1983) pp. 347–405

19.46 N.A.M. Schellart, A.N. Popper: Functional aspects of the evolution of the auditory system in Actinopterygian fish. In: *The Evolutionary Biology of Hearing*, ed. by D.B. Webster, R.R. Fay, A.N. Popper (Springer, Berlin, New York 1992) pp. 295–322

19.47 T.D. Rossing, N.H. Fletcher: *Principles of Vibration and Sound* (Springer, New York 2004), Chapts. 8–10

20. Cetacean Acoustics

The mammalian order cetacea consist of dolphins and whales, animals that are found in all the oceans and seas of the world. A few species even inhabit fresh water lakes and rivers. A list of 80 species of cetaceans in a convenient table is presented by *Ridgway* [20.1]. These mammals vary considerably in size, from the largest living mammal, the large blue whale (*balaenoptera musculus*), to the very small harbor porpoise (*phocoena phocoena*) and Commerson's dolphin (*cephalorhynchus commersonnii*), which are typically slightly over a meter in length.

Cetaceans are subdivided into two suborders, odontoceti and mysticeti. Odontocetes are the toothed whales and dolphins, the largest being the sperm whale (*physeter catodon*), followed by the Baird's beaked whale (*berardius bairdii*) and the killer whale (*orcinus orca*). Within the suborder odontoceti there are four superfamilies: *platanistoidea*, *delphinoidea*, *ziphioidea*, and *physeteridea*. Over half of all cetaceans belong to the superfamily *delphinoidea*, consisting of seven species of medium whales and 35 species of small whales also known as dolphins and porpoises [20.1]. Dolphins generally have a sickle-shaped dorsal fin, conical teeth, and a long rostrum. Porpoises have a more triangular dorsal fin, more spade-shaped teeth, and a much shorter rostrum [20.1].

Mysticetes are toothless, and in the place of teeth they have rigid brush-like whalebone plate material called baleen hanging from their upper jaw. The baleen is used to strain shrimp, krill, micronekton, and zooplankton. All the great whales are mysticetes or baleen whales and all are quite large. The sperm and Baird's beaked whales are the only odontocetes that are larger than the smaller mysticetes such as Minke whales and pygmy right whales. Baleen whales are subdivided into four families, *balaenidae* (right and bowhead whales), *eschrichtiidae* (gray whales), *balaenopteridae* (Minke, sei, Bryde's, blue, fin, and humpback whales), and *neobalaenidae* (pygmy right whale).

Acoustics play a large role in the lives of cetaceans since sound travels underwater better than any other form of energy. Vision underwater is limited to tens of meters under the best conditions and less than a fraction of a meter in turbid and murky waters. Visibility is also limited by the lack of light at great depths during the day and at almost any depth on a moonless night. Sounds are used by marine mammals for myriad reasons such as group cohesion, group coordination, communications, mate selection, navigation and locating food. Sound is also used over different spatial scales from tens of km for some species and tens of meters for other species, emphasizing the fact that different species utilize sound in different ways. All odontocetes seem to have the capability to echolocate, while mysticetes do not echolocate except in a very broad sense, such as listening to their sound bouncing off the bottom, sea mounts, underwater canyon walls, and other large objects.

The general rule of thumb is that larger animals tend to emit lower-frequency sounds and the frequency range utilized by a specific species may be dictated more from anatomical constraints than any other factors. If resonance is involved with sound production, then anatomical dimensions become critical, that is, larger volumes resonate at lower frequencies than smaller volumes. The use of a particular frequency band will also have implications as to the distance other animals, including conspecifics, will be able to hear the sounds. Acoustic energy is lost in the propagation process by geometric spreading and absorption. Absorption losses are frequency dependent, increasing with frequency. Therefore, the sounds of baleen whales such as the blue whale that emit sounds with fundamental frequencies as low as 15 Hz can propagate to much longer distances than the whistles of dolphins that contain frequencies between 5 and 25 kHz.

20.1	**Hearing in Cetaceans**	806
	20.1.1 Hearing Sensitivity of Odontocetes	807

	20.1.2	Directional Hearing in Dolphins	808
	20.1.3	Hearing by Mysticetes	812
20.2	**Echolocation Signals**	813
	20.2.1	Echolocation Signals of Dolphins that also Whistle	813
	20.2.2	Echolocation Signals of Smaller Odontocetes that Do not Whistle	817
	20.2.3	Transmission Beam Pattern	819

20.3	**Odontocete Acoustic Communication**	821
	20.3.1	Social Acoustic Signals	821
	20.3.2	Signal Design Characteristics	823
20.4	**Acoustic Signals of Mysticetes**	827
	20.4.1	Songs of Mysticete Whales	827
20.5	**Discussion**	830
References		831

20.1 Hearing in Cetaceans

One of the obvious adaptations for life in the sea is the absence of a pinna in cetaceans. The pinna probably disappeared through a natural selection process because it would obstruct the flow of water of a swimming animal and therefore be a source of noise. In the place of a pinna, there is a pin-hole on the surface of a dolphin's head which leads to a ligament inside the head, essentially rendering the pinna nonfunctional in conducting sounds to the middle ear. So, how does sounds enter into the heads of cetacean? Several electrophysiological experiments have been performed in which a hydrophone is held at different locations on an animal's head and the electrophysiological thresholds are determined as a function of the position of the hydrophone [20.2–4]. All three studies revealed greatest sensitivity on the dolphin's lower jaw.

The experimental configuration of *Møhl* et al. [20.2] is shown in Fig. 20.1a, which shows a *suction-cup hydrophone* and attached a bottlenose dolphin's lower jaw and the surface contact electrodes embedded in suction cups used to measure the auditory brainstem potential signals for different levels of sound intensity. The important differences in the experiment of *Møhl* et al. [20.2] are that the subject was trained and the measurements were done in air so that the sounds were limited to the immediate area where they were applied. The results of the experiment are shown in Fig. 20.1b. The location of maximum sensitivity to sound is slightly forward of the pan-bone area of the lower jaw, a location where *Norris* [20.5] hypothesized that sounds enter the head of a dolphin. Numerical simulation work by *Aroyan* [20.6] suggest that sounds enter the dolphin's head forward of the pan bone, but a good portion of the sound propagates below the skin to the pan bone and enters the head through the pan bone. The fact that sounds probably propagate through the lower jaw of dolphins and other

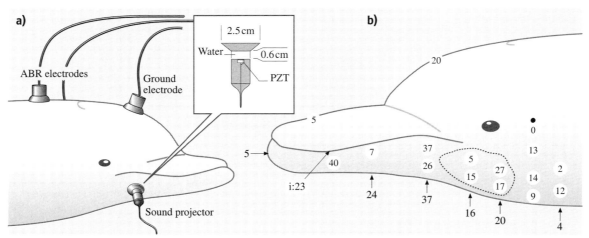

Fig. 20.1 (a) The experimental geometry used by *Møhl* et al. [20.2], (b) results of the auditory brainstem response (ABR) threshold measurements. The numerical values represent the amount of attenuation of the sound needed to obtain an ABR threshold. Therefore, the higher the number the more sensitive the location

odontocetes does not necessarily mean that the same or a similar propagation process is occurring with baleen whales.

20.1.1 Hearing Sensitivity of Odontocetes

Almost all our knowledge of hearing in cetaceans comes from studies performed with small odontocetes. The most studied species is the Atlantic bottlenose dolphin (*tursiops truncatus*). Despite the amount of research performed with the bottlenose dolphin, our understanding of auditory processes in these animals lags considerably behind that for humans and other terrestrial mammals. There are still many large gaps in our knowledge of various auditory processes occurring within the most studied odontocetes. The first audiogram for a cetacean was measured by *Johnson* [20.9] for a *tursiops truncatus*. Since then, audiograms have been determined for the harbor porpoise (*phocoena phocoena*) by *Andersen* [20.10] and *Kastelein* et al. [20.11], the killer whale (*orcinus orca*) by *Hall* and *Johnson* [20.12] and *Szymanski* et al. [20.13], the beluga whale (*delphinapterus leucas*) by *White* et al. [20.14], the Pacific bottlenose dolphin (*tursiops gilli*) by *Ljungblad* et al. [20.15], the false killer whale (*pseudorca crassidens*) by *Thomas* et al. [20.16], the Chinese river dolphin (*lipotes vexillifer*) by *Wang* et al. [20.17], Risso's dolphins (*grampus griseus*) by *Nachtigall* et al. [20.18], the tucuxi (*sotalia fluviatilis*) by *Sauerland* and *Dehnhardt* [20.19], and the striped dolphin (*stenella coeruleoalba*) by *Kastelein*

et al. [20.20]. The audiograms of these odontocetes are shown in Fig. 20.2. It is relatively striking to see how similar the audiograms are between species considering the vastly different habitats and areas of the world where some of these animals are found and the large differences in body size. All the audiograms suggest high-frequency hearing capabilities, with the smallest animal, *phocoena phocoena* having the highest hearing limit close to 180 kHz. However, the *orcinus orca*, which is over 95 times heavier and about six times longer can hear up to about 105 kHz. The actual threshold values shown in Fig. 20.2 should not be compared between species because the different methods of determining the threshold can lead to different results and because of the difficulties of obtaining good sound pressure level measurements in a reverberant environment. For example, *Kastelein* et al. [20.11] used a narrow-band frequency-modulated (FM) signal to avoid multi-path problems and the FM signals may provide additional cues not present in a pure-tone signal. Nevertheless, the audiograms shown in Fig. 20.2 suggest that all the animals had similar thresholds of 10–15 dB.

A summary of some important properties of the different audiograms depicted in Fig. 20.2 is given in Table 20.1. In the table, the frequency of best hearing is arbitrary defined as the frequency region in which the auditory sensitivity is within 10 dB of the maximum sensitivity depicted in each audiogram of Fig. 20.2. With the exception of the *Orcinus* and the *Lipotes*, the maximum sensitivity of the rest of the species represented

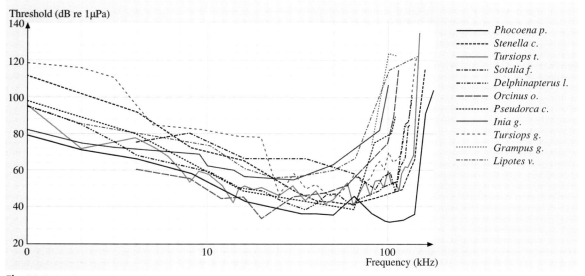

Fig. 20.2 Audiogram for different odontocetes species (after [20.7, 8])

Table 20.1 Some important properties of the audiograms plotted in Fig. 20.2

Species	Maximum sensitivity (dB re 1 : Pa)	Frequency of best hearing (kHZ)	Upper frequency limit (kHz)
phocoena phocoena	32	18 – 130	180
stenella coeruleoalba	42	30 – 125	160
tursiops truncatus	42	15 – 110	150
tursiops gilli	47	30 – 80	135
sotalia fluviatilis	50	35 – 50	135
delphinapterus leucas	40	11 – 105	120
pseudorca crassidens	39	17 – 74	115
orcinus orca	34	15 – 40	110
inia geoffrensis	51	12 – 64	100
lipotes vexillifer	55	15 – 60	100
grampus griseus	–	–	100

in Fig. 20.2 and Table 20.1 are very similar, and within experimental uncertainties, especially for audiograms obtained with the staircase procedure using relatively large step sizes of 5 dB or greater. At the frequency of best hearing, the threshold for *Orcinus* and *Phocoena* are much lower than for the other animals. It is not clear whether this keen sensitivity is a reflection of a real difference or a result of some experimental artifact. The data of Table 20.1 also indicate that *Phocoena phocoena*, *Stenella coeruleoalba* and *Tursiops truncatus* seem to have the widest auditory bandwidth.

20.1.2 Directional Hearing in Dolphins

Sound Localization

The ability to localize or determine the position of a sound source is important in order to navigate, detect prey and avoid predators and avoid hazards producing an acoustic signatures. The capability to localize sounds has been studied extensively in humans and in many vertebrates (see [20.21]). Lord Rayleigh in 1907 proposed that humans localized in the horizontal plane by using interaural time differences for low-frequency sounds and interaural intensity differences for high-frequency sounds. The sound localization acuity of a subject is normally defined in terms of a minimum audible angle (MAA), defined as the angle subtended at the subject by two sound sources, one being at a reference azimuth, at which the subject can just discriminate the sound sources as two discrete sources [20.22]. If the sound sources are separated symmetrically about a midline, the MAA is one half the threshold angular separation between the sound sources. If one of the sound sources is located at the midline, then the MAA is the same as the threshold angular separation between the two sources.

Renaud and *Popper* [20.23] examined the sound localization capabilities of a *tursiops truncatus* by measuring the MAA in both the horizontal and vertical planes. During a test trial the dolphin was required to station on a bite plate facing two transducers positioned at equal angles from an imaginary line running through the center of the bite plate. An acoustic signal was then transmitted from one of the transducers and the dolphin was required to swim and touch the paddle on the same side as the emitting transducer. The angle between the transducers was varied in a modified staircase fashion. If the dolphin was correct for two consecutive trials, the transducers were moved an incremental distance closer together, decreasing the angle between the transducer by 0.5°. After each incorrect trial, the transducers were moved an incremental distance apart, increasing the angle between the transducers by 0.5°. This modified staircase procedure allowed threshold determination at the 70% level. The threshold angle was determined by averaging a minimum of 16 reversals. A randomized schedule for sound presentation through the right or left transducer was used.

The localization threshold determined in the horizontal and vertical planes as a function of frequency for narrow-band pure-tone signals is shown in Fig. 20.3. The MAA had a U-shaped pattern, with a large value of 3.6° at 6 kHz, decreasing to a minimum of 2.1° at 20 kHz and then slowly increasing in an irregular fashion to 3.8° at 100 kHz. MAAs for humans vary between 1° and 3.7°, with the minimum at a frequency of approximately 700 Hz [20.24]. The region where the MAA decreased to a minimum in Figs. 20.19 and 20.20 (about 20 kHz) may be close to the frequency at which the dolphin switches from using interaural time difference cues to interaural intensity difference cues. The MAAs for the

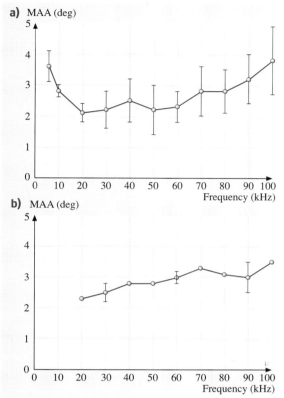

Fig. 20.3 (a) Localization threshold determined in the horizontal plane as a function of frequency. The mean ± one standard deviation is shown for seven determinations per frequency. The animal faced directly ahead at 0° azimuth. **(b)** Localization threshold determined in the vertical plane as a function of frequency. Standard deviation are indicated for 30, 60 and 90 kHz vertical data (seven sessions each). The dolphin's azimuth was 0° (after *Renauld* and *Popper*, [20.27]) Figs. 20.19 and 20.20

bottlenose dolphin were smaller than the MAAs (in the horizontal plane) of 3.5° at 3.5 kHz and 6° at 6 kHz for a harbor porpoise measured by *Dudok van Heel* [20.25], and 3° at 2 kHz, measured by *Andersen* [20.26] also for a harbor porpoise.

In order to measure the MAA in the vertical plane, the dolphin was trained to turn to its side (rotate its body 90° along its longitudinal axis) and to bite on a sheet of plexiglass which was used as the stationing device. The MAA in the vertical plane varied between 2.3° at 20 kHz to 3.5° at 100 kHz. These results indicate that the animal could localize in the vertical plane nearly as well as in the horizontal plane. These results were not expected since binaural affects, whether interaural time or intensity, should not be present in the vertical plane. However, the dolphin's ability to localize sounds in the vertical plane may be explained in part by the asymmetry in the receive beam patterns in the vertical plane discussed in the next section.

Renaud and *Popper* [20.23] also determined the MAA for a broadband transient signal or click signal, having a peak frequency of 64 kHz and presented to the dolphin at a repetition rate of 333 Hz. The MAA in the horizontal plane was found to be approximately 0.9° and 0.7° in the vertical plane. It is not surprising that a broadband signal should result in a lower MAA than a pure-tone signal of the same frequency as the peak frequency of the click. The short onset time and the broad frequency spectrum of a click signal should provide additional cues for localization.

Receiving Beam Patterns

Having narrow transmission and reception beams allows the dolphin to localize objects in a three-dimensional volume, spatially separate objects within a multi-object field, resolve features of extended or elongated objects, and to minimize the amount of interference received. The amount of ambient noise from an isotropic noise field and the amount of reverberation interference received is directly proportional to the width of the receiving beam. The effects of discrete or partially extended interfering noise or reverberant sources can be minimized by simply directing the beams away from the sources.

The receiving beam pattern of a dolphin was determined by measuring the masked hearing threshold as a function of the azimuth about the animal's head. The relative masked hearing threshold as a function of azimuth is equivalent to the received beam pattern since the receiving beam pattern is the spatial pattern of hearing sensitivity. *Au* and *Moore* [20.28] measured the dolphin's masked hearing threshold as the position of either the noise or signal sources varied in their angular position about the animal's head. The dolphin was required to voluntarily assume a stationary position on a bite plate constructed out of a polystyrene plastic material. The noise and signal transducers were positioned along an arc with the center of the arc located approximately at the pan bone of the animal's lower jaw. In order to measure the dolphin's receiving beam in the vertical plane, the animal was trained to turn onto its side before biting the specially designed vertical bite-plate stationing device. For the measurement in the vertical plane, the position of the signal transducer was fixed directly in

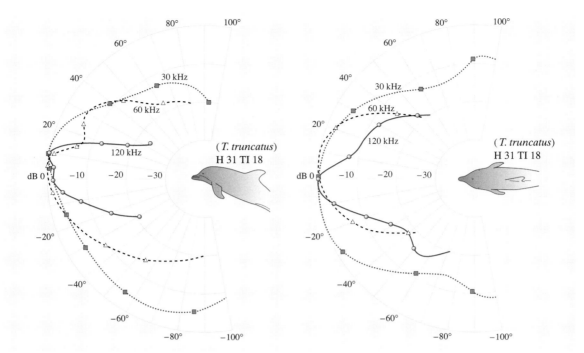

Fig. 20.4 Receive beam patterns in the vertical and horizontal planes for frequencies of 30, 60 and 120 kHz. The relative masked thresholds as a function of the elevation angle of the asking noise source are plotted for each signal frequency (after *Au* and *Moore*, [20.28])

line with the bite plate and its acoustic output was held constant. Masked thresholds were measured for different angular position of the noise transducer along the arc. The level of the noise was varied in order to obtain the masked threshold. A threshold estimate was considered complete when at least 20 reversals (10 per session) at a test angle had been obtained over at least two consecutive sessions, and if the average reversal values of the two sessions were within 3 dB. After a threshold estimate was achieved, the noise transducer was moved to a new azimuth over a set of randomly predesignated azimuths. As the azimuth about the dolphin's head increased, the hearing sensitivity of the dolphin tended to decrease, requiring higher levels of masking noise from a transducer located at that azimuth to mask the signal from a source located directly ahead of the animal.

The receiving beam patterns in both the vertical and horizontal plane are plotted for signal frequencies of 30, 60 and 120 kHz in Fig. 20.4. The radial axis of Fig. 20.4 represents the difference in dB between the noise level needed to mask the test signal at any azimuth and the minimum noise level needed to mask the test signal at the azimuth corresponding to the major axis of the vertical beam. The shape of the beams in Fig. 20.4 indicates that the patterns were dependent on frequency, becoming narrower, or more directional as the frequency increased. The beam of a planar hydrophone also becomes narrower as frequency increases. The 3 dB beam widths were approximately 30.4°, 22.7°, and 17.0° for frequencies of 30, 60, and 120 kHz, respectively. There was also an asymmetry between the portion of the beam above and below the dolphin's head. The shape of the beams dropped off more rapidly as the angle above the animal's head increased than for angles below the animal's head, indicating a more rapid decrease in the animal's hearing sensitivity for angles above the head than for angles below the head. If the dolphin receives sounds through the lower jaw, the more rapid reduction in hearing sensitivity for angles above the head may have been caused by shadowing of the received sound by the upper portion of the head structure including air in the nasal sacs [20.29]. There is a slight peculiarity in the 60 kHz beam which shows almost the same masked threshold values for 15° and 25° elevation angles.

The radial line passing through the angle of maximum sensitivity is commonly referred to as the major axis of the beam. The major axis of the vertical beams is elevated between 5° and 10° above the reference axis. The 30 and 120 kHz results show the major axis at 10° while the 60 kHz results showed the major axis at 5°. It will be shown in a later section that the major axis of the received beam in the vertical plane is elevated at approximately the same angle as the major axis of the transmitted beam in the vertical plane.

In the horizontal beam pattern measurement, two noise sources were fixed at azimuth angles of ±20°. The level of the noise sources was also fixed. The position of the signal transducer was varied from session to session. Masked thresholds were determined as a function of the azimuth of the signal transducer, by varying the signal level of the signal transducer in a staircase fashion. A threshold estimate was considered completed when at least 20 reversals at a test angle had been obtained over at least two consecutive sessions, if the average reversal values were within 3 dB of each other. After a threshold estimate was determined, the signal transducer was moved to a new azimuth over a set of randomly predesignated azimuths. Two noise sources were used in order to discourage the dolphin from internally steering its beam in the horizontal plane. If the animal could steer its beam, it would receive more noise from one of the two hydrophones, and therefore not experience any improvement in the signal-to-noise ratio. The masking noise from the two sources was uncorrelated but equal in amplitude. The radial axis represents the difference in dB between the signal level at the masked threshold for the various azimuths and the signal level of the masked threshold for 0° azimuth (along the major axis of the horizontal beam). The horizontal receiving beams were directed forward with the major axis being parallel to the longitudinal axis of the dolphin. The beams were nearly symmetrical about the major axis. Any asymmetry was within the margin of experimental error involved in estimating the relative thresholds. The horizontal beam patterns exhibited a similar frequency dependence as the vertical beam patterns, becoming narrower or more directional as the frequency increased. The 3 dB beam widths were 59.1°, 32.0°, and 13.7° for frequencies of 30, 60, and 120 kHz, respectively.

Zaytseva et al. [20.30] measured the horizontal beam pattern of a dolphin by measuring the masked hearing threshold as a function of azimuth. Their beam width of 8.2° for a frequency of 80 kHz was much narrower than the 13.7° for a frequency of 120 kHz. The difference in beam width is even larger if the results of *Au and Moore* [20.28] are linearly interpolated to 80 kHz.

We calculated an interpolated beam width of 25.9° at 80 kHz, which was considerably greater than the 8.2° obtained by *Zaytseva* et al. [20.30]. The difference in beam width measured by Zaytseva et al. and Au and Moore may be attributed to the use of only one noise source by Zaytseva et al. compared to the two noise sources used by Au and Moore. With a single masking noise source in the horizontal plane, there is the possibility of the animal performing a spatial filtering operation by internally steering the axis of its beam in order to maximize the signal-to-noise ratio. Another possibility is that Zaytseva et al. did not use a fixed stationing device. Rather, the dolphin approached the signal hydrophone from a start line, always oriented in the direction of the signal hydrophone. The animal responded to the presence or absence of a signal by either swimming or not swimming to the hydrophone. In such a procedure, it is impossible to control the orientation of the animal's head with respect to the noise masker so that the dolphin could move its head to minimize the effects of the noise.

Directivity Index

The directivity index is a measure of the sharpness of the beam or major lobe of either a receiving or transmitting beam pattern. For a spherical coordinate system, the directivity index of a transducer is given by the equation [20.31]

$$\mathrm{DI} = 10 \log \frac{4\pi}{\int_{0}^{2\pi} \int_{-\pi/2}^{\pi/2} \left(\frac{p(\theta,\phi)}{p_0}\right)^2 \sin\theta \, d\theta \, d\phi}. \quad (20.1)$$

Although the expression for directivity index is relatively simple, using it to obtain numerical values can be quite involved unless transducers of relatively simple shapes (cylinders, lines and circular apertures) with symmetry about one axis is involved. Otherwise, the beam pattern needs to be measured as a function of both θ, and ϕ. This can be done by choosing various discrete values of θ and measuring the beam pattern as a function of ϕ, a tedious process. Equation (20.1) can then be evaluated by numerically evaluating the double integral with a digital computer. The directivity indices associated with the dolphin's beam patterns in Fig. 20.4 were estimated by *Au and Moore* [20.28] using (20.1) and a two-dimensional Simpson's 1/3-rule algorithm [20.32]. The results of the numerical evaluation of are plotted as a function of frequency in Fig. 20.5. DIs of 10.4, 15.3 and 20.6 dB were obtained for frequencies of 30, 60,

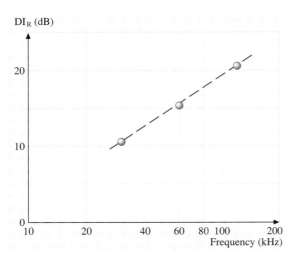

Fig. 20.5 Receiving directivity index as function of frequency for a *tursiops truncatus*

and 120 kHz, respectively. A linear curve fitted to the computed DIs in a least-square-error manner is also shown in the Fig. 20.5. The equation of the line is

$$\mathrm{DI(dB)} = 16.9 \log f(\mathrm{kHz}) - 14.5 \,. \qquad (20.2)$$

The results of Fig. 20.5 indicate that the dolphin's receive directivity index increased with frequency in a manner similar to that of a linear transducer (Bobber, 1970). The expression in (20.2) is only valid for frequencies at which DI(dB) is greater or equal to 0.

Although the directivity index expressed by (20.1) is for a *tursiops*, it can also be used to estimate the directivity index of other dolphins by applying an appropriate correction factor, so that (20.2) can be rewritten as

$$\mathrm{DI(dB)} = 16.9 \log f(\mathrm{kHz}) - 14.5 + \mathrm{CF(dB)} \,, \qquad (20.3)$$

where CF(dB) is a correction factor taking into account different head sizes. The directivity index of a planar circular plate is proportional to its diameter so if we let d_T be the diameter of the head of a *tursiops* at about the location of the blowhole and d_D be the diameter of the head of a different species of dolphin, then the correction factor can be expressed as

$$\mathrm{CF(dB)} = 20 \log(d_\mathrm{D}/d_\mathrm{T}) \,. \qquad (20.4)$$

The correction factor will be positive for a dolphin with a larger head and negative for a dolphin with a smaller head than *tursiops*.

20.1.3 Hearing by Mysticetes

Our knowledge of the hearing characteristics of baleen whales is extremely limited. We do not know how they receive sounds, the frequency range of hearing, and the sensitivity of hearing at any frequency. Much of our understanding of hearing in baleen whales comes from anatomical studies of the ears of different species. Baleen whales have occluded external auditory canals that are filled with a homogeneous wax [20.33]. The lower jaws of mysticetes are designed for sieving or gulp feeding and have no evident connection to the temporal bones [20.33] making it very difficult to understand how sounds enter into the ears of these whales.

Various types of whales have been observed by many investigators to react strongly and drastically change their behavior in the presence of boats and low-flying aircraft, however, the sound pressure levels at the whales' locations are often difficult to define and measure. In some situations, the sound levels of the aversive sound could be estimated and a data point obtained. Bowhead whales were observed fleeing from a 13 m diesel-powered boat having a noise level at the location of the whales of about 84 dB re 1 μPa in the dominant 1/3-octave band, or about 6 dB above the ambient noise in that band [20.34]. Playback experiments with bowhead whales indicated that they took evasive action when the noise was about 110 dB re 1 μPa or 30 dB above the ambient noise in the same 1/3-octave band [20.35]. Playback experiments with migrating gray whales indicated that about 10% of the whales made avoidance behavior when the noise was about 110 dB in a 1/3 octave band, 50% at about 117 dB and 90% at about 122 dB or greater. These playback signals consisted of anthropogenic noise associated with the oil and gas industry.

Frankel [20.36] played back natural humpback whale sounds and a synthetic sound to humpback whales wintering in the waters of the Hawaiian islands. Twenty seven of the 1433 trials produced rapid approach response. Most of the responses were to the feeding call. Feeding call and social sounds produced changes in both separation and whale speed, indicating that these sounds can alter a whale's behavior. The humpback whales responded to sounds as low as 102–105 dB but the strongest responses occurred when the sounds were 111 to 114 dB re 1 μPa.

All of the playback experiments suggest that sounds must be between 85 and 120 dB before whales will react to them. These levels are very high compared to those that dolphins can hear and may suggest that it is very difficult to relate reaction to hearing sensitivity. Whales may not be reacting strongly unless the sounds are much higher than their hearing threshold.

Determining the hearing sensitivity or audiogram of baleen whales represents an extremely difficult challenge and will probably require the use of some sort of electrophysiological technique as suggested by *Ridgway* et al. [20.37]. Perhaps a technique measuring auditory-evoked potentials with beached whales may provide a way to estimate hearing sensitivity. Even with evoked potential measurements, there are many issues that have to be considered. For example, if an airborne source is used, the results cannot be translated directly to the underwater situation. If a sound source is placed on a whale's head, relating that to a whale receiving a plane wave will also not be simple. Measurement of evoked potentials themselves may not be simple because of the amount of flesh, muscles and blubber that the brain waves would have to travel through in order to reach the measurement electrodes. *Ridgway* and *Carder* [20.38] reported on some attempts to use the evoked potential method with a young gray whale held at Sea World in California. They also showed that the pygmy whale auditory system was most sensitive to very high frequencies (120–130 kHz) in the same range as their narrow-band echolocation pulses. A young sperm whale was most sensitive in the 10–20 kHz region of the spectrum [20.38].

20.2 Echolocation Signals

Echolocation is the process in which an organism projects acoustic signals and obtains a sense of its surroundings from the echoes it receives. In a general sense, any animal with a capability to hear sounds can echolocate by emitting sounds and listening to the echoes. A person in an empty room can gain an idea of the size and shape of the room by emitting sounds and listening to the echoes from the different walls. However, in this chapter, echolocation is used in a more specific sense in which an animal has a very specialized capability to determine the presence of objects considerably smaller than itself, discriminate between various objects, recognize specific objects and localize objects in three-dimensional space (determine range and azimuth). Dolphins and bats have this specialized capability of echolocation.

The echolocation system of a dolphin can be broken down into three subsystems: the transmission, reception and signal processing/decision subsystems. The reception system has to do with hearing and localization. The transmission system consist of the sound production mechanism, acoustic propagation from within the head of the dolphin to into the water, and the characteristics of the signals traveling in the surrounding environment. The third subsystem has to do with processing of auditory information by the peripheral and central nervous system. The capability of a dolphin to detect objects in noise and in clutter, and to discriminate between various objects depends to a large extent on the information-carrying capabilities of the emitted signals.

Dolphins most likely produce sounds within their nasal system and the signals are projected out through the melon. Although there has been a long-standing controversy on whether sounds are produced in the larynx or in the nasal system of odontocetes, almost all experimental data with dolphins indicate that sounds are produced in the nasal system [20.39]. For example, *Ridgway* and *Carder* [20.40] used catheters accepted into the nasal cavity by trained belugas. The belugas preformed an echolocation task in open water, detecting targets and reported the presence of targets by whistling. Air pressure within the nasal cavity was shown to be essential for echolocation and for whistling [20.40].

The melon immediately in front of the nasal plug may play a role in channeling sounds into the water, a notion first introduced by *Wood* [20.41]. *Norris* and *Harvey* [20.42] found a low-velocity core extending from just below the anterior surface towards the right nasal plug, and a graded outer shell of high-velocity tissue. Such a velocity gradient could channel signals originating in the nasal region in both the vertical and horizontal planes. Using both a two-dimensional [20.43] and a three-dimensional model [20.6] to study sound propagation in a dolphin's head, Aroyan has shown that echolocation signals most likely are generated in the nasal system and are channeled into the water by the melon. *Cranford* [20.44] has also collected evidence from nasal endoscopy of trained echolocating dolphins that suggest that echolocation signals are most likely produced in the nasal system at the location of the monkey-lips, dorsal bursae complex just beneath the blow hole.

20.2.1 Echolocation Signals of Dolphins that also Whistle

Most dolphin species are able to produce whistle signals. Among some of the species in this category in which echolocation signals have been measured include the bottlenose dolphin (*tursiops sp*), beluga whale (*delphinapterus leucas*), killer whale (*orcinus orca*), false killer whale (*pseudorca crassidens*), Pacific white-sided dolphin (*lagenorhynchus obliquidens*), Amazon river dolphin (*inia geoffrensis*), Risso's dolphin (*grampus griseus*), tucuxi (*sotalia fluviatilis*), Atlantic spotted dolphin (*stenella frontalis*), Pacific spotted dolphin (*Stenella attenuata*), spinner dolphin (*Stenella longirostris*), pilot whale (*globicephala sp*), rough-toothed dolphin (*steno bredanesis*), Chinese river dolphin (*lipotes vexillifer*) and sperm whales (*physeter catodon*). However, most of the available data have been obtained for three species: the bottlenose dolphin, the beluga whale and the false killer whale.

Prior to 1973, most echolocation signals of *tursiops* were measured in relatively small tanks and the results provided a completely different picture of what we currently understand. It was not until the study of *Au* et al. [20.45] that certain features of biosonar signals used by *tursiops* and other dolphins in open waters were discovered. We discovered that the signals had peak frequencies between 120 and 130 kHz, over an octave higher than previously reported peak frequencies between 30 and 60 kHz [20.46]. We also measured an average peak-to-peak click source level on the order of 220 dB re 1 μPa at 1 m, which represents a level 30 to 50 dB higher than previously measured for *tursiops*. Examples of typical echolocation signals emitted by *tursiops truncatus* are shown in Fig. 20.6 for two situations. The top signal is typical of signals used in the open waters of Kaneohe Bay, Oahu, Hawaii, and the second signal represents typical signals for a *tursiops* in a tank. Signals measured in Kaneohe Bay regularly have duration between 40 and 70 μs, 4 to 10 positive excursion, peak frequencies between 110 and 130 kHz and peak-to-peak source levels between 210 and 228 dB re 1 μPa. The signals in Fig. 20.6 are not drawn to scale; if they were, the tank signal would resemble a flat line.

Au et al. [20.47] postulated that high-frequency echolocation signals were a byproduct of the animals producing high-intensity clicks to overcome snapping shrimp noise. In other words, dolphins can only emit high-level clicks (greater than 210 dB) if they use high frequencies. The effects of a noisy environment on the echolocation signals used by a beluga or white whale (*Delphinapterus leucas*) was vividly demonstrated by *Au* et al. [20.47]. The echolocation signal of a beluga was measured in San Diego Bay, California before the whale was moved to Kaneohe Bay. The ambient noise in Kaneohe Bay is between 15 to 20 dB greater than in San Diego Bay. The whale emitted echolocation signals with peak frequencies between 40 and 60 kHz and with a maximum averaged peak-to-peak source level of 202 dB re 1 μPa in San Diego Bay. In Kaneohe Bay, the whale shifted the peak frequency of its signals over an octave higher to 100 and 120 kHz. The source level also increased to over 210 dB re 1 μPa [20.47]. Examples of typical echolocation signals used by the whale in San Diego and in Kaneohe Bay are shown in Fig. 20.7a. Here, the signals are drawn to scale with respect to each other. Echolocation signals used by belugas in tanks also resemble the low-frequency signals shown in Fig. 20.2a [20.48, 49]. *Turl* et al. [20.50] measured the sonar signals of a beluga in a target-in-clutter detection task in San Diego Bay and found the animal used high-frequency (peak frequency above 100 kHz) and high-intensity (greater than 210 dB re 1 μPa) signals. Therefore low-amplitude clicks of the beluga had

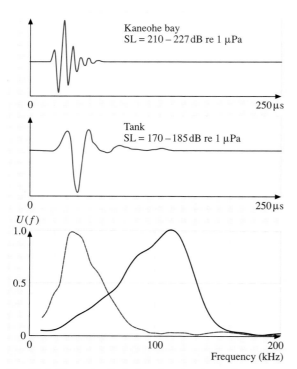

Fig. 20.6 Example of echolocation signals used by *tursiops truncatus* in Kaneohe Bay (after *Au* [20.45]), and in a tank

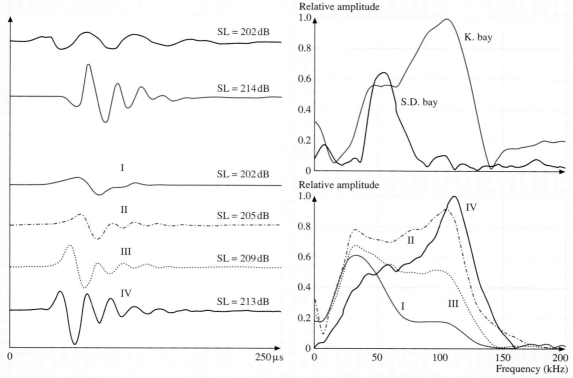

Fig. 20.7 (a) Example of beluga echolocation signals measured in San Diego Bay and in Kaneohe Bay (after *Au* et al. [20.47]); **(b)** Examples of *pseudorca* echolocation signals, Sl is the averaged peak-to-peak source level (after *Au* et al. [20.52])

low peak frequencies and the high-amplitude clicks had high peak frequencies. The data of *Moore* and *Pawloski* [20.51] for *Tursiops* also seem to support the notion that the shape of the signal in the frequency domain is related to the intensity of the signal.

Recent results with a false killer whale showed a clear relationship between the frequency content of echolocation signals and source level [20.52]. The *Pseudorca* emitted four basic types of signals, which are shown in Fig. 20.7b. The four signal types have spectra that are bimodal (having two peaks); the spectra in Fig. 20.2 are also bimodal. The type I signals were defined as those with the low-frequency peak (< 70 kHz) being the primary peak and the high-frequency peak being the secondary peak with its amplitude at least 3 dB below that of the primary peak. Type II signals were defined as those with a low-frequency primary peak and a high-frequency secondary peak having an amplitude within 3 dB of the primary peak. Type III signals were those with a high-frequency primary peak (> 70 kHz),

and a low-frequency secondary peak having an amplitude within 3 dB of the primary peak. Finally, type IV signals were those with a high-frequency primary peak having an amplitude that was at least 3 dB higher than that of the secondary low-frequency peak.

The data of *Thomas* et al. [20.53, 54] also indicated a similar relationship between intensity and the spectrum of the signal. The echolocation signals of a *pseudorca* measured in a tank had peak frequencies between 20 and 60 kHz and source levels of approximately 180 dB re 1 μPa [20.53]. Most of the sonar signals used by another *pseudorca* performing a detection task in the open waters of Kaneohe Bay had peak frequencies between 100 and 110 kHz and source levels between 220 and 225 dB re 1 μPa [20.54].

A bimodal spectrum is best described by its center frequency, which is defined as the centroid of the spectrum, and is the frequency which divides the spectrum into two parts of equal energy. From Fig. 20.7, we can see that, as the source level of the signal in-

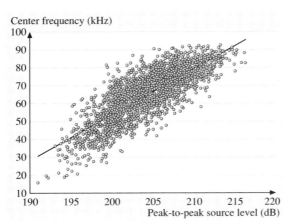

Fig. 20.8 Center frequency of echolocation signals emitted by a *pseudorca* as a function of the peak-to-peak source level (after *Au* et al. [20.52])

creased, the frequency components at higher frequencies also increased in amplitude, suggesting a relationship between source level and center frequency. This relationship can be examined by considering the scatter diagram of Fig. 20.8 showing center frequency plotted against source level. The solid line in the figure is a linear-regression curve fit of the data and has a correlation coefficient of 0.80.

The bimodal property of the echolocation signals of Fig. 20.7 seems to suggest that the response of the sound generator may be determined by the intensity of the driving force that eventually causes an echolocation signal to be produced. When the intensity of the driving force is low, only signals with low amplitudes and low-frequency peak are produced. Therefore, in small tanks, the signals resemble the tank signal of Fig. 20.1, and the bimodal feature is suppressed since the high-frequency portion of the source cannot be used for a low driving force. As the driving force increases to a moderate level, the low-frequency peak also increases in amplitude, and the high-frequency portion of the signal begins to come into use. As the driving force increases further the amplitude of the high-frequency peak becomes larger than that of the low-frequency peak, resulting in type III signals. As the driving force continues to increase to a high level, the amplitude of the high-frequency peak becomes much greater than the amplitude of the low-frequency peak and completely dominates the low-frequency peak causing the bimodal feature to be suppressed. Recent field measurements of free-ranging dolphins [20.55–57] suggest that the majority of echolocation clicks emitted by dolphins are bimodal.

The largest odontocete species is the sperm whale (*Physeter catodon*) and it too emits echolocation signals. Prior to the late 1990s the most prevalent understanding of sperm whale signals is that the clicks were broadband with peak frequencies between 4 and 8 kHz [20.58–61]. There were only two reports on source levels. *Dunn* [20.62] using sonobouys measured 148 sperm whale clicks from a solitary sperm whale and estimated an average peak-to-peak source level of 183 dB re 1 μPa. *Levenson* [20.59] estimated peak-to-peak source level of 180 dB re 1 μPa. The clicks were also thought to be essentially nondirectional and projected in codas [20.63]. Therefore, the notion of sperm whales echolocating was somewhat questionable. Part of the problem was the lack of measurements of sperm whale signaling in conjunction with foraging and the fact that click signals can have more than one function, such as communications and echolocation. In a review paper *Watkins* [20.63] spelled out his rational for not supporting the notion of a sonar function for sperm whale clicks, "Other features of their sounds however, do not so easily fit echolocation:" Watkin's rational included his observations that clicks do not appear to be very directional, the inter-click interval does not varied as if a prey or obstacle is being approached, solitary sperm whales are silent for long periods, the level of their clicks appears to be generally greater than that required for echolocating prey or obstacles and the individual clicks are usually too long for good range resolution. It is important to state that most of Watkins measurements were conducted at low-temperate latitudes where females and calves are found.

Sperm whale echolocation began to be more fully understood with data obtained in ground-breaking work by Bertel Møhl and his students from the University of Aarhus, along with other Danish colleagues. They began to perform large-array aperture measurements of large bull sperm whales foraging along the slope of the continental shelf off Andenes, Norway beginning in the summer of 1977 [20.64]. Up to seven multi-platforms spaced on the order of 1 km apart were used in their study with hydrophones placed at depths varying from 5 to 327 m [20.65]. They also came up with a unique but logistically simple scheme of obtaining global positioning system (GPS) information to localize the position of each platform. Each platform continuously logged its position and time stamps on one track of a digital audio tape (DAT) recorder, the other track being used for measuring sperm whale clicks [20.64]. The GPS signals were converted to an analog signal by frequency-shift keying (FSK) modulation. In this way, each platform can be operated essentially autonomously and yet its location and time stamps could be related to all the other platforms.

An important finding of *Møhl* et al. [20.64] is the monopulsed nature of on-axis clicks emitted by the sperm whales that were similar in shape and spectrum to dolphin echolocation signals but with peak frequencies between 15 and 25 kHz (this peak frequency range is consistent with the auditory sensitivity observed from evoked potential responses by a young sperm whale to clicks presented by *Ridgway* and *Carder* [20.38]). *Møhl* et al. [20.64] also estimated very high source levels as high as 223 dB re 1 μPa per RMS (root mean square). The per RMS measure is the level of a continuous sine wave having the same peak-to-peak amplitude as the click. This measure is clearly an overestimate of the RMS value of a sperm whale click. Nevertheless, the 223 dB reported by *Møhl* et al. [20.64] can easily be given a peak-to-peak value of 232 dB (adding 9 dB to the per-RMS value). In a follow-on study, *Møhl* et al. [20.65] measured clicks with RMS source levels as high as 236 dB re 1 μPa using the expression

$$p_{\text{RMS}} = \sqrt{\frac{1}{T}\int_0^T p^2(t)\,dt}\,. \tag{20.5}$$

For this measurement, they used a time interval corresponding to the 3 dB down points of the signal waveform. For this waveform, an RMS source level of 236 dB corresponds to a peak-to-peak source level of 243 dB re 1 μPa. Finally, they found that the clicks were very directional, as directional as dolphin clicks.

20.2.2 Echolocation Signals of Smaller Odontocetes that Do not Whistle

The second class of echolocation signals are produced by dolphins and porpoises that do not emit whistle signals. Not many odontocetes fall into this category and these non-whistling animals tend to be smaller than their whistling cousins. Included in this group of non-whistling odontocetes are the harbor porpoise (*phocoena phocoena*), finless porpoise (*neophocaena phocaenoides*), Dall's porpoise (*phocoenoides dalli*), Commerson's dolphin, *cephalorhynchus commersonii*), Hector's dolphin (*cephalorhynchus hectori*) and pygmy sperm whale (*kogia sp*).

Examples of harbor porpoise and Atlantic bottlenose dolphin echolocation signals presented in a manner for easy comparison are shown in Fig. 20.9. There are four fundamental differences in the two types of echoloca-

tion signals. The non-whistling animal emit a signal with longer duration, narrower band, lower amplitude and single mode. The length of the *Phocoena phocoena* signal vary from about 125–150 μs compared to 50–70 μs for the *tursiops truncatus* signal. The bandwidth of the *Phocoena* signal is almost 0.2–0.3 that of the *tursiops* signal. Since the non-whistling dolphins are usually much smaller in length and weight than the whistling dolphins, the smaller animals might be amplitude limited in terms of their echolocation signals and compensate by emitting longer signals to increase the energy output. This issue can be exemplified by characterizing an echolocation click as [20.39]

$$p(t) = As(t)\,, \tag{20.6}$$

where $A = |p|_{\max}$ is the absolute value of the peak amplitude of the signal and $s(t)$ is the normalized wave-

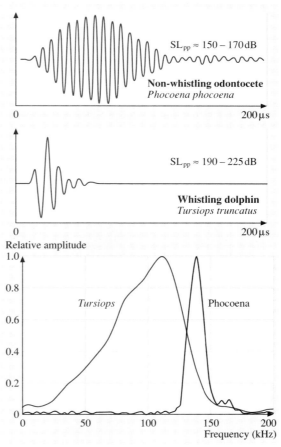

Fig. 20.9 Examples of typical echolocation signals of *phocoena phocoena* and *tursiops truncatus* (after [20.8])

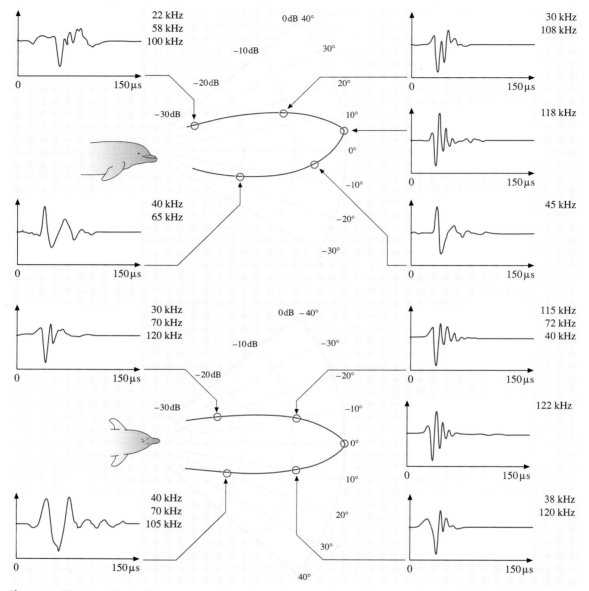

Fig. 20.10 The transmission beam pattern of a *tursiops truncatus* planes with the waveform of a click measured by 5–7 hydrophones (after Au [20.39]) in the vertical and horizontal planes

form having a maximum amplitude of unity. The source energy-flux density of the signal in dB can be expressed as

$$SE = 10 \log \left(\int_0^T p(t)^2 \, dt \right)$$

$$= 10 \log(A) + 10 \log \left(\int_0^T s(t)^2 \, dt \right), \quad (20.7)$$

where T is the duration of the signal. Letting $2A \approx$ be the peak-to-peak sound pressure level, (20.4) can be rewritten as

$$SE = SL - 6 + 10\log\left(\int_0^T s(t)^2 \, dt\right), \quad (20.8)$$

where SL is the the peak-to-peak source level and is approximately equal to $10\log(2A)$. From Fig. 20.9, the *Tursiops* can emit signals with peak-to-peak source levels in open water that are greater than 50 dB for *phocoena*. The portion of the source energy that can be attributed to the length of the signal is only about 2–4 dB greater for *phocoena* than *Tursiops* [20.39]. Therefore, one can conclude that the target detection range of *phocoena* is considerably shorter than for *Tursiops*. This has been demonstrated by *Kastelein* et al. [20.67] who measured a target detection threshold range of 26 m for a 7.62 cm water-filled sphere in a quiet environment. The target detection range in a noisy environment and a similar 7.62 cm water-filled sphere was 113 m [20.68].

The echolocation signals of other non-whistling odontocetes are very similar to the *Phocoena phocoena* signal shown in Fig. 20.9. Examples of the echolocation signals of other non-whistling odontocetes can be found in *Au* [20.7, 39, 69]. Unfortunately, reliable source-level data have been collected only for *Phocoena phocoena* [20.70] and *Kogia* [20.69].

20.2.3 Transmission Beam Pattern

The outgoing echolocation signals of dolphins and other odontocetes are projected in a directional beam that have been measured in the vertical and horizontal planes for the bottlenose dolphin [20.39], beluga whale [20.66] and the false killer whale [20.52]. The composite beam pattern from the three measurements on *Tursiops* along with the averaged waveform from a single trial measured by 5 or the 7 hydrophones are shown in Fig. 20.10. The 3 dB beam width for the bottlenose dolphin was approximately 10.2° in the vertical plane and 9.7° in the horizontal plane. The waveforms detected by the various hydrophones in Fig. 20.10 indicate that signals measured away from the beam axis will be distorted with respect to the signals measured on the beam axis. The further away from the beam axis the more distorted the signals will be. This distortion come from the broadband nature of the click signals emitted by whistling dolphins. This characteristics also make it difficult to get good measurements of echolocation signals in the field even with an array of hydrophones since it is extremely difficult to obtain on-axis echolocation signals and also to know the orientation of the animal with respect to the measuring hydrophones. The frequencies shown with each waveform are the fre-

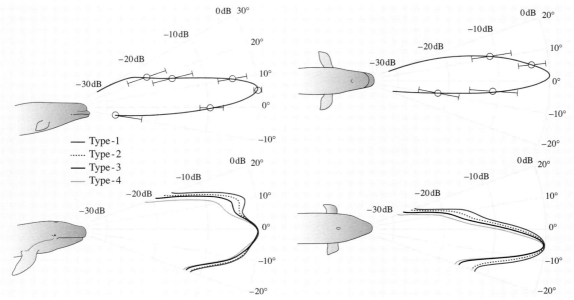

Fig. 20.11 The transmission beam pattern of a *delphinapterus leucas* (after *Au* et al. [20.66]) and a *pseudorca crassidens* (after *Au* et al. [20.52]) in the vertical and horizontal planes

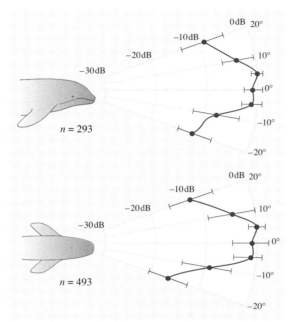

Fig. 20.12 Beam patterns in the vertical and horizontal planes for *phocoena phocoena* (*Au* et al. [20.70])

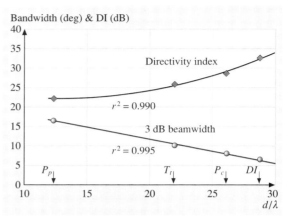

Fig. 20.13 Transmission directivity index and 3 dB beam width for four odontocetes. The directivity index and beam width for *Tursiops truncatus* and *delphinapterus leucas* came from *Au* [20.39] and for *pseudorca crassidens* after *Au* et al. [20.52]. The wavelength λ corresponds to the average peak frequency of the animals' echolocation signals. The directivity index is fitted with a second-order polynomial curve the the beam width is fitted with a linear curve (after *Au* et al. [20.70])

quencies of local maxima in the spectrum. At some angles the averaged signal had multiple peaks in the spectrum and these peaks are listed in order of prominence.

The beam patterns for the beluga and false killer whale are shown in Fig. 20.11. The beam width for the beluga whale was approximately 6.5° in both planes. Four beam patterns corresponding to the four signal types described in Fig. 20.7 are shown for the false killer whale. For the highest frequency (type IV signal) the 3 dB beam width in the vertical plane was approximately 10°, and 7° in the horizontal plane. The beam axis in the vertical plane for the bottlenose dolphin and the beluga whale was +5° above the horizontal plane. For the false killer whale, 49% of the beam axis was at 0° and 32% at −5°. The four beam patterns for the false killer whale indicate that, like a linear transducer, the lower the frequency the wider the beam pattern. The type I signal has a peak frequency of about 30 kHz and has the widest beam. However, even though the peak frequency of the type IV signal is about 3.5 times higher than the type I signal, the beam does not seem to be substantially larger. This property can be used when discussing lower-frequency whistle signals to suggest that whistle signals are also emitted in a beam.

The only transmission beam pattern for a non-whistling odontocete, a *phocoena phocoena*, was measured by *Au* et al. [20.70]. Their results are shown in Fig. 20.12 in both the horizontal and vertical planes. One of the obvious difference between the beam patterns in Fig. 20.12 and those in Figs. 20.10 and 20.11 is the width of the beam. Although the harbor porpoise emitted the highest-frequency signals, its small head size caused the beam to be wider than for the other animals. The beam patterns of Figs. 20.10–20.12 were inserted into (20.1) and numerically evaluated to estimate the corresponding directivity index for *tursiops* and *delphinapterus* [20.39] and for *pseudorca* by *Au* et al. [20.52] and for *phocoena* [20.70], and the results are shown in Fig. 20.13 plotted as a function of the ratio of the head diameter (at the blow hole) of the subject and the average peak frequency of the animals echolocation signal. Also shown in Fig. 20.13 are the 3 dB beam width, where the 3 dB beam width in the vertical and horizontal planes were averaged.

The second-order polynomial fit of the directivity index is given by the equation

$$\mathrm{DI} = 28.89 - 1.04 \left(\frac{d}{\lambda}\right) + 0.04 \left(\frac{d}{\lambda}\right)^2 . \quad (20.9)$$

The linear fit of the 3 dB beam width data is given by the equation

$$\text{BW} = 23.90 - 0.60 \left(\frac{d}{\lambda}\right). \quad (20.10)$$

The results of Fig. 20.13 provides a way of estimating the directivity index and 3 dB beam width of other odontocetes by knowing their head size and the peak frequency of their echolocation signals.

Another interesting characteristics of the beam pattern measurement on the *phocoena phocoena* is that the off-axis signals are not distorted in comparison to the on-axis signal. This is consistent with narrow-band signals whether for a linear technological transducer or a live animal. Therefore, measurements of echolocation signals of non-whistling odontocetes in the field can be performed much easier than for whistling odontocetes.

20.3 Odontocete Acoustic Communication

Members of the *Odontocete* suborder are an intriguing example of the adaptability of social mammalian life to the aquatic habitat. Most odontocetes, particularly marine dolphins, live in groups ranging from several to hundreds of animals in size, forage cooperatively, develop hierarchies, engage in alloparental care and form strong pair bonds and coalitions between both kin and non-kin alike (see [20.71–73] for a review of the literature on dolphin societies). These social traits are analogous to patterns found in many avian and terrestrial mammalian species. It is not surprising, therefore, that odontocetes mediate much social information via communication. Some river dolphins are usually found as solitary individuals or mother–calf pairs although they may occasionally congregate into larger groups.

Odontocetes communicate through a combination of sensory modalities that include the visual, tactile, acoustic and chemosensory channels [20.74]. Visual signals in the form of postural displays are thought to convey levels of aggression, changes in direction of movement, and affiliative states [20.72, 75]. Tactile interactions vary in purpose from sexual and affiliative signals to expressions of dominance and aggression [20.76, 77]. Although not well documented yet, it is thought that a derivative of chemosensory perception, termed *quasiolfaction* allows dolphins to relay chemical messages about stress and reproductive status [20.78, 79]. However, it is the acoustic communication channel in particular that is believed to be the main communication tool that enables odontocetes to function cohesively as groups in the vast, visually limited environment of the open sea [20.80]. Examining how odontocetes have adapted their use of sound is therefore an important step to understanding how social marine mammal life evolved in the sea and how it has found a way to thrive in a habitat so drastically different from our own.

When compared to all that has been learned about dolphin echolocation over the past several decades, surprisingly little is still known about how dolphins and other odontocetes use acoustic signals for communication. A major reason for this is that it is very difficult to observe the acoustic and behavioral interactions between the producer and the receiver(s) of social signals in the underwater environment. Sound propagation in the sea makes even a simple task like identifying the location of a sound source very challenging for air-adapted listeners. Therefore, matching signals with specific individuals and their behavior in the field is problematic without the use of sophisticated recording arrays [20.57, 76, 81, 82], or directional transducers. In addition, many questions remain unanswered about the nature of social signals themselves. Much still remains unknown about the functional design of dolphin social sounds. This is largely due to the fact that most species specialize in producing and hearing sounds with frequencies well beyond the limits of the human hearing range. Limitations in the technology available to study social signaling in the field have until recently restricted most analyses to the human auditory bandwidth (< 20 kHz). Yet, despite these significant challenges, a great deal of progress has been made in our understanding of odontocete acoustic communication. Rapidly advancing technologies are contributing greatly to our ability to study social signaling both in the field and in the laboratory. The emerging picture reveals that odontocetes have adapted their acoustic signaling to fit the aquatic world in a remarkably elegant way.

20.3.1 Social Acoustic Signals

Odontocetes have evolved in both marine and freshwater habitats for over 50 million years. Over such a long period of adaptive radiation one might expect that a variety of signaling strategies would have evolved among the approximately 65 species of small toothed whales and dolphins. Yet, remarkably, the vast majority of these species have either conserved or converged on the production of two types of sounds: whistles and short pulse trains.

Whistles

Whistles are frequency- and amplitude-modulated tones that typically last anywhere from 50 ms to 3 s or more, as illustrated in Fig. 20.14. They are arguably the most variable signals produced by dolphins. Both within and across species they range widely in duration, bandwidth and degree of frequency modulation. How whistles are used in communication is an ongoing topic of debate among researchers, but most agree that they play an important role in maintaining contact between dispersed individuals and are important in social interactions [20.85].

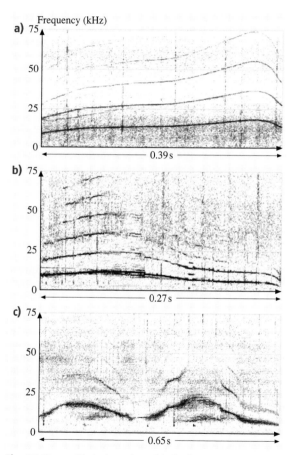

Fig. 20.14a–c Variations in whistle forms produced by Hawaiian spinner dolphins (**a**) and Atlantic spotted dolphins (**b**), (**c**) exhibiting multiple harmonics (after *Lammers* et al. [20.83, 84])

Dolphin whistles exhibit both species and geographic specificity [20.86–90]. Differences are greatest between distantly related taxa and between species of different size. As a general rule, larger species tend to produce whistles with lower fundamental frequencies than smaller ones [20.88]. Geographic variations within a species are usually smaller than interspecific differences [20.87], but some species do develop regional dialects between populations [20.89, 90]. These dialects tend to become more pronounced with increasing geographic separation. In addition, pelagic species tend to have whistles in a higher-frequency range and with more modulation than coastal and riverine species [20.86, 88]. Such differences have been proposed as species-specific reproductive-isolating characteristics [20.86], as ecological adaptations to different environmental conditions and/or resulting from restrictions imposed by physiology [20.88]. On the other hand, it must also be noted that numerous whistle forms are also shared by both sympatric and allopatric species [20.91]. The communicative function of shared whistle forms is unknown, but it is intriguing as it suggests a further tendency towards a convergence in signal design.

Dolphins produce whistles with fundamental frequencies usually in the human audible range (below 20 kHz). However, whistles typically also have harmonics (Fig. 20.14), which occur at integer multiples of the fundamental and extend well beyond the range of human hearing [20.84]. Harmonics are integral components of tonal signals produced by departures of the waveform from a sinusoidal pattern. Dolphin whistle harmonics have a *mixed directionality* property, which refers to the fact they become more directional with increasing frequency [20.83, 92]. It has been proposed that this signal feature functions as a cue, allowing listening animals to infer the orientation and direction of movement of a signaling dolphin [20.83, 92]. Harmonics may therefore be important in mediating group cohesion and coordination.

Burst Pulses

In addition to whistles, most odontocetes also produce pulsed social sounds known as *burst pulse* signals. Burst pulse signals are broadband click trains similar to those used in echolocation but with inter-click intervals of only 2–10 ms [20.93]. Because these intervals are considerably shorter than the processing period generally associated with echolocation and because they are often recorded during periods of high social activity, burst pulse click trains are thought instead to play an important role in communication.

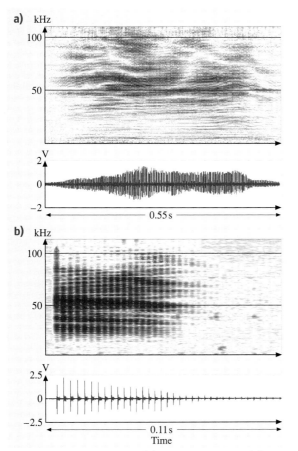

Fig. 20.15a,b Examples of (**a**) high-quantity and (**b**) low-quantity burst pulses produced by Atlantic spotted dolphins (*stenella frontalis*). Click train *a* has 255 clicks with mean ICI (interclick interval) of 1.7 ms. Click train *b* has 35 clicks with a mean ICI of 2.9 ms (after *Lammers* et al. [20.83, 84])

Burst pulses vary greatly in the inter-pulse interval and in the number of clicks that occur in a train, which can number anywhere from three to hundreds of pulses, as depicted in Fig. 20.15. This variation gives them distinctive aural qualities. Consequently, burst pulse sounds have been given many subjective labels, including *yelps* [20.94], *cracks* [20.95], *screams* [20.96] and *squawks* [20.97]. Their production has been reported in a variety of odontocete species including: the Atlantic bottlenose dolphin (*tursiops truncatus*) [20.95], the Hawaiian spinner dolphin (*stenella longirostris*) [20.72], the Atlantic spotted dolphin (*stenella frontalis*) [20.98], the narwhal (*monodon monoceros*) [20.99], Hector's dolphin (*cephalorhynchus hectori*) [20.100], the pilot whale (*globicephala melaena*) [20.101], and the harbor porpoise (*phocoena phocoena*) [20.101]. To date, much remains unknown about how burst pulses function as communication signals. It is generally believed that they play an important role in agonistic encounters because they are commonly observed during confrontational head-to-head behaviors between individuals [20.95, 102–104]. However, some authors have suggested they may represent emotive signals in a broader sense [20.97, 105, 106], possibly functioning as graded signals [20.107].

Given that dolphins have temporal discrimination abilities well within the range required to resolve individual clicks in a burst pulse [20.39, 108], it is possible that the quantity of clicks and their temporal spacing could form an important basis for communication. However, as with whistles, no data presently exist on the classification and discrimination tendencies of dolphins with respect to different burst pulses.

20.3.2 Signal Design Characteristics

Although much remains unknown about how odontocetes use acoustic signals for communication, the communicative potential of their signals can, in part, be inferred by considering the design characteristics that have been uncovered thus far. Signal features such as their detectable range, the production duty cycle, the identification level, the modulation potential and the form–content linkage provide useful clues about how odontocetes might use whistles and burst pulses.

The Active Space of Social Signals

The effective range of signals used for communication is generally termed the *active space*. Janik [20.109] investigated the active space of whistles produced by bottlenose dolphins in the Moray Firth, Scotland, using a dispersed hydrophone array to infer geometrically the location of signaling animals and establish the source level of their whistles. The mean source level was calculated to be 158 dB ± 0.6 re 1 μPa, with a maximum recorded level of 169 dB. By factoring in transmission loss, ambient noise levels, the critical ratios and auditory sensitivity of the species involved the active space of an unmodulated whistle between 3.5 and 10 kHz in frequency was estimated to be between 20 and 25 km in a habitat 10 meters deep at sea state 0. At sea state 4 the estimated active space ranged from 14 to 22 km, while for whistles at 12 kHz it dropped to between 1.5 and 4 km. These estimations were made for dolphins occurring in a relatively shallow, mostly mud-bottom, quiet environment. Presently, no data exist on the active space characteristics of delphinid whistles in pelagic waters and comparatively louder tropical near-shore environments. Similarly, no estimates have yet been made for the active space of burst pulses.

The Duty Cycle of Social Signals

Duty cycle refers to the proportion of time that any given signal or class of signals is *on* versus *off*. Signals can vary in the fine structure of their temporal patterning (e.g. duration), in their temporal spacing within a bout, and in their occurrence within a larger cyclical time frame, such as a 24 hour day. Each of these aspects of the temporal delivery of signals carries its own implication for communication. Fine-scale characteristics can define the nature of the signal itself and provide information about its relationship to other signals as well as constraints associated with its mechanism of production. The occurrence and timing of a signal within a bout can help convey information about the urgency of a situation, the level of arousal, the fitness of an individual and can assist the receiver in the task of localization. Finally, the periodicity of signals over hours or days can be an indicator of variables such as activity levels and reproductive state.

Murray et al. [20.110] investigated the fine-scale duty-cycle characteristics of delphinid social signals. Using a captive false killer whale (*pseudorca crassidens*) as their subject, they examined the temporal relationship between the units of a click train (individual clicks) and tonal signals. Their analysis revealed that false killer whales modulate the temporal occurrence of clicks to the point of grading them into a continuous wave (CW) signal such as a whistle. This finding was interpreted as evidence that *pseudorca* may employ graded signaling for communication (a topic discussed in more detail below), as well as to suggest that clicks and whistles are produced by the same anatomical mechanism.

The timing of signals within a bout has not been investigated with much success among delphinids. The primary obstacle towards this line of work has been the difficulty of identifying the signaler(s) involved for even short periods of time under field conditions. Early work by *Caldwell* and *Caldwell* [20.111] on a group or four naïve captive common dolphins (*delphinus delphis*) suggests that whistle exchanges do have temporal structure. In whistling bouts involving more than one signaler, the onset of a whistle specific to an individual was followed within two seconds by that of another. Furthermore, initiation of a whistle by two animals within 0.3 s of one another always resulted in the inhibition of one of them, with some individuals deferring more than others. Initiations separated by 0.4–0.5 s caused inhibition less frequently while those longer than 0.6 s resulted in almost no inhibition. Repeated whistles produced without an intervening response by another animal were usually delayed by less than one second. Thus, the duty cycle of whistling bouts and chorusing behavior among common dolphins does appear to follow certain temporal rules, but their significance is not clear.

Periodicity in social signaling has been investigated in captive bottlenose dolphins and common dolphins [20.112–114], as well as free-ranging spinner dolphins (*stenella longirostris*) [20.72] and common dolphins [20.115]. In the captive studies, signaling activity was linked to feeding schedules, nearby human activities and responses to different forms of introduced stress [20.113]. For spinner and common dolphins in the wild, the occurrence of social acoustic signals was highest at night, when both species were foraging, and lowest in the middle of the day.

The Identification Level of Social Signals

The role of delphinid whistles as individual-specific signals has been the focus of more scientific attention than any other aspect of their social acoustic signaling. *Caldwell* and *Caldwell* [20.116] were the first to propose that individual dolphins each possess their own distinct *signature* whistle. The idea was borne out of the observation that recently captured dolphins each produce a unique whistle contour that makes up over 90% of the individual's whistle output. Since being proposed, the so-called *signature whistle hypothesis* has emerged as the most widely accepted explanation for whistling behavior among dolphins. The idea has received support from numerous studies involving captive and restrained animals [20.85, 117–121] as well as from field studies of free-ranging animals [20.97, 122, 123]. Some, however, have argued that a simple signature function for dolphin whistling cannot account for the diversity of signals observed in socially interactive dolphin groups [20.124, 125]. While not denying the presence of signature whistles per se, these authors have argued that the large percentage of stereotyped signature signals observed in other studies may be an artifact of the unusual circumstances under which they were obtained (isolation, temporary restraint, separation, captivity). The debate over the prevalence of signature whistles among captive and free-ranging dolphins remains a contested topic in the literature and at scientific meetings. On the other hand, no evidence or formal discussions presently exist suggesting burst pulse signals carry any individual-specific information.

The Modulation Potential of Social Signals

The modulation potential describes the amount of variation present in any given signal as measured by its position along a scale from stereotyped to graded signaling [20.126]. Stereotyped signals are repeats of structurally identical forms and vary discretely between one another. They are used most often for communication when the prospect of signal distortion is high, such as in noisy or reverberant environments or for communicating over long ranges. Graded signals have more variants than stereotyped ones and are encoded by changing one or more signal dimensions. These can include intensity, repetition rate and/or frequency and amplitude modulation. Graded signals are usually employed when a continuous or relative condition must be communicated in a favorable propagation environment.

Many studies to date have made an a priori assumption that dolphin signaling is categorical in nature, with signals belonging to mutually exclusive classes on the basis of their shared similarities. Some evidence in support of this assumption comes from the signature-whistle hypothesis work, where restrained and isolated individuals are often observed producing highly stereotyped bouts of signaling. However, few data are presently available to suggest how dolphins perceive and distinguish social signals [20.127]. Therefore, it is unclear whether an assumption based on the occurrence of signature whistles is broadly applicable towards other forms of social acoustic signaling (i.e. non-signature whistles and burst pulses).

A few studies have explored the occurrence of graded signaling in the communication of delphinids and the larger whales. *Taruski* [20.128] created a graded model for the whistles produced by North Atlantic pilot whales (*globicephala melaena*). He concluded that these signals could be arranged as a "continuum or matrix (of signals) from simple to complex through a series of intermediates" (*Taruski* [20.128] p. 349). *Murray* et al. [20.110], examining the signals of a captive false killer whale (*Pseudorca crassidens*), came to a similar conclusion and proposed that *Pseudorca* signals were also best represented along a continuum of signal characteristics, rather than categorically. Finally, *Clark* [20.129], examining the "call" signals of southern right whales (a mysticete), made the observation that " ... the total repertoire of calls is best described as a sequence of intergraded types along a continuum" (*Taruski* [20.128] p. 1066). Therefore, graded signaling may well play a role in odontocete communication. Controlled perceptual experiments and a better understanding of the acoustic environment dolphins inhabit are needed to test how odontocetes discriminate and classify social sounds.

The Form–Content Linkage of Social Signals

The degree to which the form of a signal is linked to its content depends on how information is coded and on proximate factors associated with the signal's production. The relationship between a signal's structural form and its message can range from being rather arbitrary to tightly linked to a specific condition [20.126]. Among odontocetes, a clear relationship between signal form and content has only been demonstrated for signature whistles (discussed above). Dolphins have been experimentally shown to be capable of labeling objects acoustically (also known as vocal learning), as well as mimicking model sounds following only a single exposure [20.130]. Additional evidence of vocal learning exists from culturally transmitted signals in the wild [20.131] and in captivity [20.132], as well as from the spontaneous production of acoustic labels in captivity [20.127]. Context-specific variations in signature whistle form have also been demonstrated [20.121]. However, to date, no semantic rules have been identified for naturally occurring social signals among odontocetes.

Geographic Difference and Dialect

There is a distinct difference between geographical difference and dialect. Geographical differences are associated with widely separated populations that do not normally mix. Dialect is best reserved for sound emission differences on a local scale among neighboring populations which can potentially intermix [20.133]. Geographic variations are generally considered to result from acoustic adaptations to different environments, or a functionless byproduct of isolation and genetic divergence caused by isolation [20.134]. The functional significance of dialects is controversial, with some maintaining that dialects are epiphenomena of song learning and social adaptation, whereas others believe that they play a role in assorted mating and are of evolutionary significance [20.134]. Dialects are known to occur in many species of birds [20.135] but appear to be very rare in mammals.

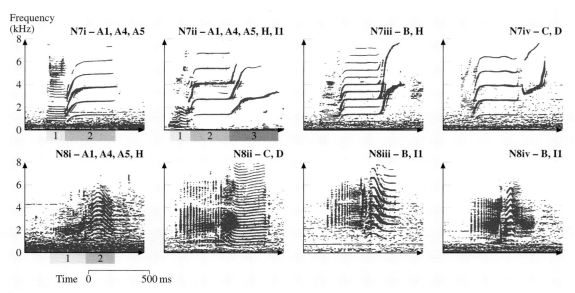

Fig. 20.16 Spectrograms of call types N7 and N8 for clan A. Above each spectrogram is the subtype identification and the pods that produce the variant, and below certain spectrograms are division marks separating calls into their component parts (after *Ford* [20.134]).

Killer Whale

Killer whales (*orcinus orca*) produce a specific type of burst pulse termed the *discrete call*. Discrete calls are thought to serve as contact calls between individuals, much like whistles in other odontocete species [20.136]. Discrete calls are population specific and even pod specific. The dialects of resident killer whales (*Orcinus orca*) in the coastal waters of British Columbia and Washington have been studied over a prolonged period by *Ford* and his colleagues [20.134, 137, 138]. They used photographic identification techniques, keying on unique natural markings on the whales to identify 16 pods or stable kin groups of 232 resident killer whales. Differences in acoustic behavior formed a system of related pod-specific dialects within the call tradition of each clan. *Ford* [20.134] has proposed that each clan is comprised of related pods that have descended from a common ancestral group and pod-specific repertoires probably serve to enhance the efficiency of acoustic communications within the group and act as behavioral indicators of pod affiliation.

Killer whale calls are typically made up of rapidly emitted pulses that to the human ear have a tonal quality [20.134]. Many calls have several abrupt shifts in pulse repetition rate allowing them to be divided into different segments or parts. Although all pods belonging to a clan share a number of calls, these calls were often rendered in consistently different form by different pods. Also, certain pods produced calls that were not used by the rest of the clan. Such variations produced a set of related group-specific dialects within the call tradition of each clan.

Spectrograms of call types N7 and N8 are shown in Fig. 20.16. In this example, we can see that different pods produce similar but different versions of call type N7 and N8. Pods A1, A4 and A5 produced two versions of call type N7 and pod B and I1 produced two versions of call type N8. All of the spectrograms have two parts except for call type N7ii, which had three parts.

Ford [20.134] contended that pod-specific repertories can be retained for periods in excess of 25 years. Discrete calls generally serve as signals for maintaining contact within the pod and that the use of repertoire of pod-specific calls enhances this function by conveying group identity and affiliation [20.134]. Killer whales, like most other small dolphins are able to learn and mimic a wide variety of sounds. Even from a young age, killer whale infants can selectively learn specific calls, especially the calls of their mothers. *Bowles* et al. [20.139] studied the development of calls of a captive-born killer whale calf and found that it learned and reproduced only the calls of its mother and ignored the calls of other killer whales in the same pool. *Ford* [20.134] also observed killer whales imitating the call types of different pods, and even those from other clans. These instances were rare, but it does show a capacity for learning and mimicry of acoustic signals.

Sperm Whales

Sperm whales live in a matrilineal family unit where there exist cooperative behaviors including communal care of the young in ways similar to killer whales. The family units are very stable and females may live as long as 60–70 y [20.140]. One type of signals that sperm whales emit are denoted as codas, which are sequences of click signals that may be the primary means of acoustic communications for these animals [20.58]. *Weilgart* and *Whitehead* [20.141] recorded, for over a year, codas from a number of sperm whales around the South Pacific and in the Caribbean Sea. Photographic identification also allowed them to assign recording sessions to particular groups. They found that the coda repertoire recorded from the same group on the same or different days were much more similar than those recorded from different groups in the same place. Groups recorded in the same place had more similar coda repertoire than those in the same broad area but different places. Groups from the same area were in turn marginally similar to those in the same ocean but different than those in different oceans. Coda class repertoires of groups in different oceans and in different areas within the same ocean were statistically significantly different. They concluded that strong group-specific dialects were apparently overlaid on weaker geographic variation. Sperm whales, killer whales and possibly bottlenose dolphins are the only cetaceans known to have dialects.

20.4 Acoustic Signals of Mysticetes

There are eleven species of mysticetes or baleen whales and sounds have been recorded from all but the pygmy right whale [20.142]. The vocalization of baleen whales can be divided into two general categories: (1) songs and (2) calls [20.142]. The calls can be further subdivided into three categories: (1) simple calls, (2) complex calls and (3) clicks, pulses, knocks and grunts [20.142]. Simple calls are often low-frequency, frequency-modulated signals with narrow instantaneous bandwidth that sound like moans if a recording is speeded up or slowed down, depending on the specific animal. Amplitude modulation and the presence of harmonics are usually part of a simple call, with most of the energy below 1 kHz. Complex calls are pulse-like broadband signals with a variable mixture of amplitude and/or frequency modulation. They sound like screams, roars, and growls, with most of the energy between 500–5000 Hz. Clicks, pulses, knocks and grunts are short-duration (< 0.1 s) signals with little or no frequency modulation. Clicks and pulses are very short (< 2 ms) signals with frequencies between 3–31 kHz, while grunts and knocks are longer (50–100 ms) signals in the 100–1000 Hz range [20.142].

20.4.1 Songs of Mysticete Whales

Songs are defined as "sequences of notes occurring in a regular sequence and patterned in time" [20.142]. Songs are easily discriminated from calls in most instances. Four mysticetes species have been reported to produce songs; the blue whale (*balaenoptera musculus*) [20.143], the fin whale (*balaenoptera physalus*) [20.144, 145], the bowhead whale (*balaena mysticetus*) [20.15], and the humpback whale [20.146]. Songs of humpback whales have without a doubt received the most attention from researchers. Part of the reason for this is the relative ease for investigators to travel and do research in the summer grounds of humpback whales, especially in Hawaii in the Pacific and Puerto Rico in the Atlantic.

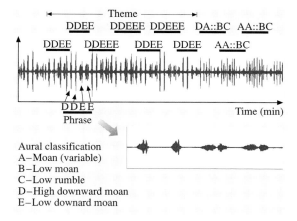

Fig. 20.17 Examples of two themes from a humpback whale song (after *Frankel* [20.36])

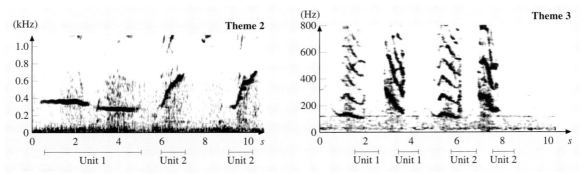

Fig. 20.18 An example of a portion of a humpback whale song. The individual pulse sounds are defined as units and units are aurally categorized based on how they sound. In this example, there are five units. Phrases are formed by a combination of units and themes are formed by a combination of phrases. The whole song consist of a combination of themes that are continuously repeated for the length of the song

Humpback Whale Songs

The list of studies involving humpback whale songs is long and extends from 1971 to the present time, *Helweg* et al. [20.147]. Although some songs may sound almost continuous to human listeners the basic units in a song are pulsed sounds. Songs are sung only by males and consist of distinct units that are produced in some sequence to form a phrase and a repeated set of phrases form a theme and repeated themes form a song. A song can last from minutes to hours depending on the disposition of the singer. An example of a portion of song is shown in Fig. 20.17. Example of two themes of a song in spectrogram format are shown in Fig. 20.18. The variation in frequency as a function of time is clearly shown in the figure.

Some general properties of songs and the whales that are singing are:

1. Songs from the North Pacific, South Pacific and Atlantic populations are different.
2. Singing peaks during the winter months when humpback whales migrate to warmer waters at lower latitudes.
3. Whales within a population sing the same basic song in any one year, although the song may undergo slight changes during a breeding season.
4. Changes in songs are not due to forgetting during the summer months, which are non-singing months, since songs recorded early in the winter breeding season are nearly the same as songs recorded late in the previous breeding season.
5. Songs from consecutive years are very similar but songs across nonconsecutive years will have fewer similarities.
6. Singers are most probably only males, since no females have been observed singing.
7. Some singing also occurs during the summer and fall.
8. Singing whales are often alone, although they have been occasionally observed singing in the presence of other humpback whales.
9. Singers tend to remain stationary. However, they have also been observed singing while swimming.

Other Mysticetes

Bowhead whales emit a variety of different types of simple and complex sounds that sound like moans to the human ear. They also emit sequential sounds that contain repeatable phrases that can be classified as songs during their spring migration [20.15] but not during the summer or autumn [20.148]. The bowhead song had just one theme with basically only two sounds repeated over and over. *Ljungblad* et al. [20.15] reported that songs were very tonal with clear pitch even though they were produced by pulsive moans, whereas *Cummings* and *Holliday* [20.149] described songs as sounding like raucous elephant roars and trumpeting in discrete repetitions or phases that were put together to form longer sequences. Differences in the songs recorded by *Ljungblad* et al. [20.15] and by *Cummings* and *Holliday* [20.149] may be due to bowhead whales changing their songs from year to year.

The sounds from finback whales include single 20 Hz pulses, irregular series of 20 Hz pulses, and stereotyped 20 Hz signal bouts of repetitive sequences of 20 Hz pulses [20.145]. The 20 Hz signals are emitted in bouts that can last for hours. The pulse intervals in a bout were very regular. In general, signals are produced in

Table 20.2 Characteristics of mysticete whales vocalizations

Species of whales	Signal type	Frequency limits(HZ)	Dominat frequency (Hz)	Source level (dB re 1 µPa) at 1 m	References
Blue	FM moans	12.5 – 200	16 – 25	188	*Cummings, Thompson* [20.150], *Edds* [20.151]
	Songs	16 – 60	16– 60	–	*McDonald* et al. [20.152]
Bowhead	Tonal moans	25 –900	100 – 400	129 – 178	*Cummings, Holliday* [20.149]; *Wursig, Clark* [20.148]
	Pulses	25 –3500	152 – 185		
	Songs	20 – 500	158 – 189		*Cummings, Holliday* [20.149]; *Ljungblad* et al. [20.15]
Bryde's	FM moans	70 – 245	124 – 132	152 – 174	*Cummings* et al. [20.153]; *Edds* et al. [20.154]
	Pulsed moans	100 - 930	165 – 900	–	*Edds* et al. [20.154]
	Discrete pulses	700- 950	700 – 950	–	*Edds* et al. [20.154]
Finback	FM moans	14 – 118	20	160 – 186	*Watkins* [20.145], *Edds* [20.155], *Cummings, Thompson* [20.156]
	Tonals	34 – 150	34 – 150	–	*Edds* [20.155]
	Songs	17 – 25	17 – 25	186	*Watkins* [20.145]
Gray	Pulses	100 – 2000	300 – 825	–	*Dalheim* et al. [20.157]; *Crane* et al. [20.158]
	FM moans	250 – 300	250 – 300	152	*Cummings* et al. [20.159]; *Dalheim* et al. [20.157]
	LF-FM-moans	125 – 1250	< 430	175	*Cummings* et al. [20.159]; *Dalheim* et al. [20.157]
	PM pulses	150 – 1570	225 – 600		*Cummings* et al. [20.159]; *Dalheim* et al. [20.157]
	Complex moans	35 – 360	35 – 360		*Cummings* et al. [20.159]
Humpback	Grunts (pulse & FM)	25 – 1900	25 – 1900	176	*Thompson* et al. [20.160]
	Pulses	25 – 89	25 – 80	144 – 174	*Thompson* et al. [20.160]
	Songs	30 – 8000	120 – 4000	–	*Payne, Payne* [20.161]
Minke	FM tones	60 – 130	60 – 130	165	*Schevill, Watkins* [20.162],
	Thumps	100 – 200	100 – 200	–	*Winn, Perkins* [20.163]
	Grunts	60 – 140	60 – 140	151 – 175	*Winn, Perkins* [20.163]
	Ratchets	850 – 6000	850	–	*Winn, Perkins* [20.163]
Right-N	Moans	< 400	–	–	*Schevill, Watkins* [20.162]
Right-S	Tonal	30 – 1250	160 – 500	–	*Cummings* et al. [20.164], *Clark* [20.129, 165]
	Pulses	30 – 2200	50 – 500	172 – 187	*Cummings* et al. [20.164], *Clark* [20.129, 165]
Sei	FM sweeps	1500 – 3500	1500 – 3500	–	*Knowlton* et al. [20.166]

a relatively regular sequence of repetitions at intervals ranging from about 7–26 s, with bouts that can last as long as 32.5 h. During a bout, periodic rests averaged about 115 s at roughly 15 min intervals and sometimes longer irregular gaps between 20 and 120 min were observed. There was also some variability in the 20 Hz signals in that they were never exactly replicated.

The songs of the blue whales (*balaenoptera musculus*) have been observed by a number of researchers [20.143, 152, 167]. A typical two-part blue whale song time series and corresponding spectrogram is shown in Fig. 20.18. The spectrogram of the first part of the two part song had six spectral lines separated by about 1.5 Hz. This type of spectrogram is typically generated by pulses. *Cummings* and *Thompson* [20.150] previously reported on the pulsive nature of some blue whale moans. The second part of the song was tonal in nature with a slight FM down sweep varying from 19 Hz to 18 Hz in the first 3–4 s. The 18 Hz tone is then carried until the last 5 s when there is an abrupt step down to

17 Hz. The amplitude modulation in the second part of the song was probably caused by multi-path propagation of the signal from the whale to the hydrophones. These two-part songs basically follow a pattern of a 19 s pulsive signal followed by a 24.5 s gap and a 19 s monotonic signal [20.152, 167].

Calls of Mysticete Whales

The calls of mysticete whales have been the subject of much research over the past three decades. As with dolphin sounds, there is also a lack of any standard nomenclature for describing emitted sounds. Similar calls are often given different names by different researchers. A summary of some of the acoustic properties of the different baleen whales is indicated in Table 20.2. Calls and songs are most likely used for some sort of communication, however, at this time the specific meanings of these sounds are not known. It is extremely difficult to study the context and functions of baleen whale vocalization. The sounds of mysticete whales have also been summarized nicely by *Richardson* et al. [20.34]. Instead of discussing the characteristics of various sounds, the properties of calls are summarized in Table 20.2. The calls of mysticete whales are mainly in the low-frequency range, from an infrasonic frequency of about 12.5 Hz for the blue whale to about 3.5 kHz for the sei whale. There are a variety of different types of calls, from FM tones to moans, grunts and discrete pulses. How these sounds are used by whales is still an open question since it is often very difficult to observe behavior associated with different calls.

20.5 Discussion

Cetaceans use a wide variety of sounds from brief echolocation signals used by dolphins having durations less than 100 μs to very long duration songs that can last for hours by are emitted by humpback whales. The range of frequency is also very large, from low-frequency (infrasonic) sounds between 10–20 Hz used by blue and finback whales to high-frequency echolocation signals that extend to 130–140 kHz or perhaps higher (see earlier mention of 180 kHz). It is quite clear that acoustics is important in the natural history of cetaceans since all species regularly emit sounds throughout their daily routine. Yet, it is not at all clear how sounds are used by these animals. The difficulty in observing the behavior of these animals has made it difficult to attach any function to a particular or specific sound. It is nearly impossible to determine a one-to-one relationship between the reception or transmission of a specific sound to specific behavioral response. There are many possible functions of sounds, some of which may include providing contact information in a population of animals, signaling alarm or warning of approaching predators, attracting potential mates, echolocation for prey detection and discrimination while foraging, a way of establishing a hierarchy, providing individual identification, and a method for disciplining juveniles. Although researchers have not been successful in uncovering the role or function of specific sounds in any cetacean species, researchers should not be discouraged but should be innovative and imaginative in designing experiments that can be conducted in the wild to delve deeper into this problem. Even with echolocation sounds, there are many unanswered question. It seems logical that echolocation sounds are used to detect and discriminate prey and to navigate, yet we still know very little about how often they are emitted during a daily routine for dolphins in the field.

In this chapter, we have seen that the characteristics of the sounds emitted by cetaceans depends a lot on the size of the animals. Therefore the active space of different species is indirectly related to the size of the animals since the amount of sound absorption by

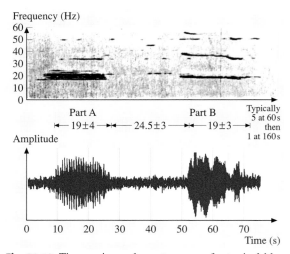

Fig. 20.19 Time series and spectrogram of a typical blue whale song (after [20.152])

sea water increases with frequency. Dolphins and small whales usually emit echolocation signals that can have peak frequencies up to 120–140 kHz and whistles between 10–20 kHz. These signals are not meant to travel large distances so that the active space is rather small, tens of meters for echolocations and burst pulse signals and several hundred meters for whistles. These signals will certainly propagate to much shorter distances than the low-frequency signals used by many of the baleen whales. The sound pressure level (SPL) of an emitted signal at any range from the animal is given by the equation

$$\text{SPL} = \text{SL} - \text{geometric spreading loss} - \alpha R , \quad (20.11)$$

where SL is the source level in dB, α is the sound absorption coefficient in dB/m and R is the distance the sound has traveled. The geometric spreading loss does not depend on frequency but is affected by the sound velocity profile, the depth of the source and the receiver. The most severe geometric spreading loss is associated with spherical spreading loss in which the amount of geometric spreading loss is equal to $20 \log R$. The least severe geometric spreading loss is associated with sound channels, such as the sound fixing and ranging (SOFAR) channel. In order to gain an appreciation of the effects of absorption losses for different types of cetacean signals, $-\alpha R$ in (20.9) was calculated as a function of range and the results are shown in Fig. 20.20. The results indicate that the low-frequency sounds used by baleen whales do not suffer as much absorption loss as the sounds of small odontocetes. Therefore, baleen whale signals can propagate long distances if some type of channel is present in the water column.

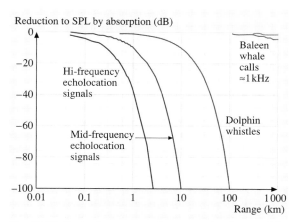

Fig. 20.20 Reduction to sound pressure level caused by sound absorption for different types of cetacean signals. The curve for the high-frequency echolocation signals was calculated at 120 kHz, and for the mid-frequency echolocation signals at 50 kHz. The curve for the dolphin whistle was calculated at 15 kHz. The baleen whale signal curve was determined at 1 kHz.

We have just scraped the top of the iceberg in regards to our understanding of cetacean acoustics. There is much to learn and understand about how these animals produce and receive sounds and how acoustics is used in their daily lives. The future looks extremely promising as new technology arrives on the scene that will allow researchers to delve into the many unanswered questions. As computer chips and satellite transmission tags get smaller and we apply state-of-the-art technologies to the study of cetaceans, our knowledge and understanding can but increase.

References

20.1 S.H. Ridgway: Who are the Whales, Bioacoust. **8**, 3–20 (1997)

20.2 B. Møhl, W.W.L. Au, J. Pawloski, P.E. Nachtigall: Dolphin hearing: Relative sensitivity as a function of point of application of a contact sound source in the jaw and head region, J. Acoust. Soc. Am. **105**, 3421–3424 (1999)

20.3 T.H. Bullock, A.D. Grinnell, E. Ikezono, K. Kameda, Y. Katsuki, M. Nomoto, O. Sato, N. Suga, K. Yanagisawa: Electrophysiological studies of central auditory mechanisms in Cetaceans, Z. Vergl. Phys. **59**, 117–316 (1968)

20.4 J.G. McCormick, E.G. Wever, J. Palin, S.J.H. Ridgway: Sound conduction in the dolphin ear, J. Acoust. Soc. Am. **48**, 1418–1428 (1970)

20.5 K.S. Norris: The evolution of acoustic mechanisms in odontocete cetaceans. In: *Evolution and Environment*, ed. by E.T. Drake (Yale Univ. Press, New Haven 1968) pp. 297–324

20.6 J.L. Aroyan: Three-dimensional modeling of hearing in Delphinus delphis, J. Acoust. Soc. Am. **110**, 3305–3318 (2001)

20.7 W.W.L. Au: Hearing in whales and dolphins: An overview. In: *Hearing by Whales and Dolphins*, ed. by W.W.L. Au, A.N. Popper, R.R. Fay (Springer, Berlin, New York. 2000) pp. 1–42

20.8 W.W.L. Au: Echolocation in dolphins. In: *Hearing by Whales and Dolphins*, ed. by W.W.L. Au, A.N. Popper, R.R. Fay (Springer, Berlin, New York 2000) pp. 364–408

20.9 C.S. Johnson: Sound detection thresholds in marine mammals. In: *Marine Bioacoustics*, ed. by W. Tavolga (Pergamon, New York 1967) pp. 247–260

20.10 S. Andersen: Auditory sensitivity of the harbor porpoise, Phocoena phocoena. Invest. Cetacea **2**, 255–259 (1970)

20.11 R.A. Kastelein, P. Bunskoek, M. Hagedoorn, W.W.L. Au, D. de Haan: Audiogram of a harbor porpoise (Phocoena phocoena) measured with narrow-band frequency modulated signals, J. Acoust. Soc. Am. **112**, 334–344 (2002)

20.12 J.D. Hall, C.S. Johnson: Auditory thresholds of a killer whale, J. Acoust.Soc. Am. **51**, 515–517 (1971)

20.13 M.D. Szymanski, D.E. Bain, K. Kiehi, S. Pennington, S. Wong, K.R. Henry: Killer whale (Orcinus orca) hearing: Auditory brainstem response and behavioral audiograms, J. Acoustic. Soc. Am. **106**, 1134–1141 (1999)

20.14 M.J. White Jr., J. Norris, D. Ljungblad, K. Baron, G. di Sciara: *Auditory thresholds of two beluga whales (Delphinapterus leucas). HSWRI Tech. Rep. No. 78-109* (Hubbs Marine Research Institute, San Diego 1979)

20.15 D.K. Ljungblad, P.D. Scoggins, W.G. Gilmartin: Auditory thresholds of a captive eastern Pacific bottlenose dolphin, Tursiops spp., J. Acoust. Soc. Am. **72**, 1726–1729 (1982)

20.16 J. Thomas, N. Chun, W. Au, K. Pugh: Underwater audiogram of a false killer whale (Pseudorca crassidens), J. Acoust. Soc. Am. **84**, 936–940 (1988)

20.17 D. Wang, K. Wang, Y. Xiao, G. Sheng: Auditory sensitivity of a Chinese River Dolphin, Lipotes vexillifer. In: *Marine Mammal Sensory Systems*, ed. by J.A. Thomas, R.A. Kastelein, Aya Supin (Plenum, New York 1992) pp. 213–221

20.18 P.E. Nachtigall, W.W.L. Au, J.L. Pawloski, P.W.B. Moore: Risso's Dolphin (Grampus griseus) Hearing Thresholds in Kaneohe Bay, Hawaii. In: *Sensory Systems of Aquatic Mammals*, ed. by R.A. Kastelein, J.A. Thomas, P.E. Nachtigall (DeSpiel, Woerden 1995) pp. 49–53

20.19 M. Sauerland, D. Dehnhardt: Underwater Audiogram of a Tucuxi (Sotalia fluviatilis guianensis), J. Acoust. Soc. Am. **103**, 1199–1204 (1998)

20.20 R.A. Kastelein, M. Hagedoorn, W.W.L. Au, D. de Haan: Audiogram of a striped dolphin (Stenella coeruleoalba), J. Acoust. Soc. Am. **113**, 1130–1144 (2003)

20.21 R.R. Fay: *Hearing in Vertebrates: A Psychophysics Databook* (Hill-Fay Associates, Winnetka 1988)

20.22 A.W. Mills: On the Minimum Audible Angle, J. Acoust. Soc. Am. **30**, 237–246 (1958)

20.23 D.L. Renaud, A.N. Popper: Sound localization by the bottlenose porpoise Tursiops truncatus, J. Exp. Biol. **63**, 569–585 (1978)

20.24 A.W. Mills: Auditory localization. In: *Foundations of Modern Auditory Theory*, Vol. 2, ed. by J.V. Tobias (Academic, New York 1972) pp. 303–348

20.25 W.H. van Dudok Heel: Sound and Cetacea, Netherlands J. Sea Res. **1**, 407–507 (1962)

20.26 S. Andersen: Directional hearing in the harbour porpoise, Phocoena phocoena. Invest. Cetacea **2**, 261–263 (1970)

20.27 D.L. Renaud, A.N. Popper: Sound localization by the bottlenose porpoise, tursiops truncatus, J. Exp. Biol. **63**, 569–585 (1975)

20.28 W.W.L. Au, P.W.B. Moore: Receiving Beam Patterns and Directivity Indices of the Atlantic Bottlenose Dolphin Tursiops truncates, J. Acoust. Soc. Am. **75**, 255–262 (1984)

20.29 D.S. Houser, J. Finneran, D. Carder, W. Van Bonn, C. Smith, C. Hoh, R. Mattrey, S. Ridgway: Structural and Functional Imaging of BottlenoseDolphin (Tursiops truncatus) Cranial Anatomy, J. Exp. Biol. **207**, 3657–3665 (2004)

20.30 K.A. Zaytseva, A.I. Akopian, V.P. Morozov: Noise resistance of the dolphin auditory analyzer as a function of noise direction, BioFizika **20**, 519–521 (1975)

20.31 R.J. Urick: *Principles of Underwater Sounds for Engineers* (McGraw-Hill, New York 1980)

20.32 J.M. McCormick, M.G. Salvadori: *Numerical Methods in FORTRAN* (Prentice-Hall, Englewood Cliffs 1964),

20.33 D.R. Ketten: The marine mammal ear: specializations for aquatic audition and echolocation. In: *The Evolutionary Biology of Hearing*, ed. by D. Webster, R. Fay, A. Popper (Springer, Berlin, New York 1992) pp. 717–754

20.34 W.J. Richardson, C.R. Greene Jr., C.I. Malme, D.H. Thomson: *Marine Mammals and Noise* (Academic, San Diego 1995)

20.35 W.J. Richardson, B. Würsig, C.R. Greene Jr: Reactions of bowhead whales, Balaena mysticetus, to drilling and dredging noise in the Canadian Beaufort Sea, Mar. Envr. Res. **29**, 135–160 (1990),

20.36 A.S. Frankel: *Acoustic and visual tracking reveals distribution, song variability and social roles of humpback whales in Hawai'ian waters. Ph.D. Dissertation* (Univ. Hawaii, Honolulu, HI 1994)

20.37 C.R. Ridgway: Electrophysiological experiments on hearing in odontocetes. In: *Animal Sonar*, ed. by P.E. Nachtigall, P.W.B. Moore (Plenum, New York 1980) pp. 483–493

20.38 S.H. Ridgway, D.A. Carder: Assessing hearing and sound production in cetaceans not available for behavioral audiograms: Experiences with sperm, pygmy sperm and gray whales, Aquatic Mammals. **27**, 267–276 (2001)

20.39 W.W.L. Au: *The Sonar of Dolphins* (Springer, Berlin, New York 1993)

20.40 S.H. Ridgway, D.A. Carder: Nasal pressure and sound production in an echolocating white whale, Delphinapterus leucas. In: *Animal Sonar: Processes and Performance*, ed. by P.E. Nachtigall,

20.41 F.G. Wood: Discussion. In: *Marine Bio-Acoustics Vol. II*, ed. by W. Tavolga (Pergamon, Oxford 1964) pp. 395–396

20.42 K.S. Norris, G.W. Harvey: Sound transmission in the porpoise head, J. Acoust. Soc. Am. **56**, 659–664 (1974)

20.43 J.L. Aroyan, T.W. Cranford, J. Kent, K.S. Norris: Computer modeling of acoustic beam formation in Delphinus delphis, J. Acoust. Soc. Am. **95**, 2539–2545 (1992)

20.44 T.W. Cranford: In Search of Impulse sound sources in odontocetes. In: *Hearing by Whales and Dolphins*, ed. by W.W.L. Au, A.N. Popper, R.R. Fay (Springer, New York 2000) pp. 109–155

20.45 W.W.L. Au, R.W. Floyd, R.H. Penner, A.E. Murchison: Measurement of echolocation signals of the Atlantic bottlenose dolphin Tursiops truncatus Montagu, in Open Waters, J. Acoust. Soc. Am. **56**, 280–1290 (1974)

20.46 W.E. Evans: Echolocation by marine delphinids and one species of fresh-water dolphin, J. Acoust. Soc. Am. **54**, 191–199 (1973)

20.47 W.W.L. Au, D.A. Carder, R.H. Penner, B.L. Scronce: Demonstration of adaptation in beluga whale echolocation signals, J. Acoust. Soc. Am. **77**, 726–730 (1985)

20.48 B.S. Gurevich, W.E. Evans: Echolocation discrimination of complex planar targets by the beluga whale (Delphinapterus leucas), J. Acoust. Soc. Am. **60**, S5 (1976)

20.49 C. Kamminga, H. Wiersma: Investigations on cetacean sonar V. The true nature of the sonar sound of Cephaloryncus Commersonii, Aqua. Mamm. **9**, 95–104 (1981)

20.50 C.W. Turl, D.J. Skaar, W.W.L. Au: The Echolocation Ability of the Beluga (Delphinapterus leucas) to detect Targets in Clutter, J. Acoust. Soc. Am. **89**, 896–901 (1991)

20.51 P.W.B. Moore, D.A. Pawloski: Investigations on the control of echolocation pulses in the dolphin (Tursiops truncatus). In: *Sensory Abilities of Cetaceans Laboratory and Field Evidence*, ed. by J.A. Thomas, R.A. Kastelein (Plenum, New York 1990) pp. 305–316

20.52 W.W.L. Au, J.L. Pawloski, P.E. Nachtigall: Echolocation signals and transmission beam pattern of a false killer whale (Pseudorc crassidens), J. Acoust. Soc. Am. **98**, 51–59 (1995)

20.53 J. Thomas, M. Stoermer, C. Bowers, L. Anderson, A. Garver: Detection abilities and signal characteristics of echolocating false killer whales (Pseudorca crassidens). In: *Animal Sonar Processing and Performance*, ed. by P.E. Nachtigall, P.W.B. Moore (Plenum, New York 1988) pp. 323–328

20.54 J.A. Thomas, C.W. Turl: Echolocation characteristics and range detection by a false killer whale Pseudorca crassidens. In: *Sensory Abilities of Cateceans:* P.W.B. Moore (Plenum, New York 1988) pp. 53–60,

Laboratory and Field Evidence, ed. by J.A. Thomas, R.A. Kastelein (Plenum, New York 1990) pp. 321–334

20.55 W.W.L. Au, J.K.B. Ford, J.K. Horne, K.A. Newman-Allman: Echolocation signals of free-ranging killer whales (Orcinus orca) and modeling of foraging for chinook salmon (Oncorhynchus tshawytscha), J. Acoust. Soc. Am. **115**, 901–909 (2004)

20.56 W.W.L. Au, B. Würsig: Echolocation signals of dusky dolphins (Lagenorhynchus obscurus) in Kaikoura, New Zealand, J. Acoust. Soc. Am. **115**, 2307–1313 (2004)

20.57 W.W.L. Au, D. Herzing: Echolocation signals of wild Atlantic spotted dolphin (Stenella frontalis), J. Acoust. Soc. Am. **113**, 598–604 (2003)

20.58 W.A. Watkins, W.E. Schevill: Sperm whale codas, J. Acoust. Soc. Am. **62**, 1485–1490 (1977)

20.59 C. Levenson: Source level and bistatic target strength of sperm whale (Physeter catodon) measured from an oceanographic aircraft, J. Acoust. Soc. Am. **55**, 1100–1103 (1974)

20.60 L. Weilgart, H. Whitehead: Distinctive vocalizations from mature male sperm whales (Physter macrocephalus), Can. J. Zool. **66**, 1931–1937 (1988)

20.61 B. Møhl, E. Larsen, M. Amundin: Sperm Whale Size Determination: Outlines of an Acoustic Approach. In: *Mammals in the Seas. Fisheries series No. 5, Mammals of the Seas*, Vol. 3 (FAO (Food and Agriculture Organization of the United Nations) Publications, Rome 1981) pp. 327–331

20.62 J.L. Dunn: Airborne Measurements of the Acoustic Characteristics of a Sperm Whale, J. Acoust. Soc. Am. **46**, 1052–1054 (1969)

20.63 W.A. Watkins: Acoustics and the behavior of sperm whales. In: *Animal Sonar Systems*, ed. by R.G. Busnel, J.F. Fish (Plenum, New York 1980) pp. 291–297

20.64 B. Møhl, M. Wahlberg, A. Heerfordt: A GPS-linked array of independent receiver for bioacoustics, J. Acoust. Am. **109**, 434–437 (2001)

20.65 B. Møhl, M. Wahlberg, P.T. Madsen, A. Heerfordt, A. Lund: The monpulsed natue of sperm whale clicks, J. Acoust. Soc. Am. **114**, 1143–1154 (2003)

20.66 W.W.L. Au, R.H. Penner, C.W. Turl: Propagation of beluga echolocation signal, J. Acoust. Soc. Am. **82**, 807–813 (1987)

20.67 R.A. Kastelein, W.W.L. Au, H.T. Rippe, N.M. Schooneman: Target detection by an echolocating harbor porpoise (Phocoena phocoena), J. Acoustic. Soc. Am. **105**, 2493–2498 (1999)

20.68 W.W.L. Au, K.J. Snyder: Long-range Target Detection in Open Waters by an Echolocating Atlantic Bottlenose Dolphin (Tursiops truncatus), J. Acoust. Soc. Am. **68**, 1077–1084 (1980)

20.69 P.T. Madsen, D.A. Carder, K. Bedholm, S.H. Ridgway: Porpoise clicks from a sperm whale nose –convergent evolution of 130 kHz pulses in toothed whale sonars?, Bioacoustics **15**, 192–206 (2005)

20.70 W.W.L. Au, R.A. Kastelein, T. Ripper, N.M. Schooneman: Transmission beam pattern and echolocation signals of a harbor porpoise (Phocoena phocoena), J. Acoustic. Soc. Am. **106**, 3699–3705 (1999)

20.71 K.W. Pryor, K.S. Norris: *Dolphin Societies: Discoveries and Puzzles* (Univ. California Press, Berkley 1991)

20.72 K.S. Norris, B. Würsig, R.S. Wells, M. Würsig: *The Hawai'ian Spinner Dolphin* (Univ. California Press, Berkeley 1994) p. 408

20.73 J. Mann, P.L. Tyack, H. Whitehead: *Cetacean societies: field studies of dolphins and whales* (Univ. Chicago Press, Chicago 2000)

20.74 L.M. Herman, W.N. Tavolga: The communications systems of cetaceans. In: *Cetacean Behavior: Mechanisms and Function*, ed. by L.M. Herman (Wiley-Interscience, New York 1980) pp. 149–209

20.75 K.W. Pryor: Non-acoustic communication in small cetaceans: glance touch position gesture and bubbles. In: *Sensory Abilities of Cetaceans*, ed. by J. Thomas, R. Kastelein (Plenum, New York 1990) pp. 537–544

20.76 K.M. Dudzinski: Contact behavior and signal exchange in Atlantic spotted dolphins (Stenella frontalis), Aqua. Mamm. **24**, 129–142 (1998)

20.77 C.M. Johnson, K. Moewe: Pectoral fin preference during contact in Commerson's dolphins (Cephalorhynchus commersonii), Aqua. Mamm. **25**, 73–77 (1999)

20.78 P.E. Nachtigall: Vision, audition and chemoreception in dolphins and other marine mammals. In: *Dolphin Cognition and Behavior: A Comparative Approach*, ed. by R.J. Schusterman, J.A. Thomas, F.G. Wood (Lawrence Erlbaum, Hillsdale 1986) pp. 79–113

20.79 V.B. Kuznetsov: Chemical sense of dolphins: quasi-olfaction. In: *Sensory Abilities of Cetaceans*, ed. by J. Thomas, R. Kastelein (Plenum, New York 1990) pp. 481–503

20.80 K.S. Norris, E.C. Evans III.: On the evolution of acoustic communication systems in vertebrates, Part I: Historical aspects. In: *Animal Sonar: Processes and Performance*, ed. by P.E. Nachtigall, P.W.B. Moore (Plenum, New York 1988) pp. 655–669

20.81 R.E. Thomas, K.M. Fristrup, P.L. Tyack: Linking the sounds of dolphins to their locations and behavior using video and multichannel acoustic recordings, J. Acoust. Soc. Am. **112**, 1692–1701 (2002)

20.82 K.R. Ball, J.R. Buck: A beamforming video recorder for integrated observations of dolphin behavior and vocalizations, J. Acoust. Soc. Am. **117**, 1005–1008 (2005)

20.83 M.O. Lammers, W.W.L. Au, D.L. Herzing: The broadband social acoustic signaling behavior of spinner and spotted dolphins, J. Acoust. Soc. Am. **114**, 1629–1639 (2003)

20.84 M.O. Lammers, W.W.L. Au: Directionality in the whistles of Hawai'ian spinner dolphins (Stenella longirostris): A signal feature to cue direction of movement?, Mar. Mamm. Sci. **19**, 249–264 (2003)

20.85 V.M. Janik, P.J.B. Slater: Context specific use suggests that bottlenose dolphin signature whistles are cohesion calls, Anim. Behav. **56**, 829–838 (1998)

20.86 W.W. Steiner: Species-specific differences in puretonal whistle vocalizations of five western north Atlantic dolphin species, Behav. Ecol. Sociobiol. **9**, 241–246 (1981)

20.87 L.E. Rendell, J.N. Matthews, A. Gill, J.C.D. Gordon, D.W. MacDonald: Quantitative analysis of tonal calls from five odontocete species, examining interspecific and intraspecific variation, J. Zool. **249**, 403–410 (1999)

20.88 W. Ding, B. Wursig, W.E. Evans: Comparisons of whistles among seven odontocete species. In: *Sensory Systems of Aquatic Mammals*, ed. by R.A. Kastelein, J.A. Thomas (De Spil, Woerden 1995) pp. 299–323

20.89 W. Ding, B. Wursig, W.E. Evans: Whistles of bottlenose dolphins: comparisons among populations, Aqua. Mamm. **21**, 65–77 (1995)

20.90 C. Bazua-Duran, W.W.L. Au: Geographic variations in the whistles of spinner dolphins (Stenella longirostris) of the Main Hawai'ian Islands, J. Acoust. Soc. Am. **116**, 3757–3769 (2004)

20.91 J.N. Oswald, J. Barlow, T.F. Norris: Acoustic identification of nine delphinid species in the eastern tropical Pacific Ocean, Mar. Mamm. Sci. **19**, 20–37 (2003)

20.92 P.J.O. Miller: Mixed-directionality of killer whale stereotyped calls: a direction-of-movement cue?, Behav. Ecol. Sociobiol. **52**, 262–270 (2002)

20.93 M.O. Lammers, W.W.L. Au, R. Aubauer: A comparative analysis of echolocation and burst-pulse click trains in Stenella longirostris. In: *Echolocation in Bats and Dolphins*, ed. by J. Thomas, C. Moss, M. Vater (Univ. Chicago Press, Chicago 2004) pp. 414–419

20.94 A.E. Puente, D.A. Dewsbury: Courtship and copulatory behavior of bottlenose dolphins (Tursiops truncatus), Cetology **21**, 1–9 (1976)

20.95 M.C. Caldwell, D.K. Caldwell: Intraspecific transfer of information via the pulsed sound in captive Odontocete Cetaceans. In: *Animal Sonar Systems: Biology and Bionics*, ed. by R.G. Busnel (Laboratoire de Physiologie Acoustic: Jouy-en-Josas, France 1967) pp. 879–936

20.96 S.M. Dawson: Clicks and communication: the behavioural and social contexts of Hector's dolphin vocalizations, Ethology **88**, 265–276 (1991)

20.97 D.L. Herzing: Vocalizations and associated underwater behavior of free-ranging Atlantic spotted dolphins, Stenella frontalis and bottlenose dolphin, Tursiops truncatus, Aqua. Mamm. **22**, 61–79 (1996)

20.98 M.C. Caldwell, D.K. Caldwell: Underwater pulsed sounds produced by captive spotted dolphins, Stenella plagiodon, Cetology **1**, 1–7 (1971)

20.99 J.K.B. Ford, H.D. Fisher: Underwater acoustic signals of the narwhal (Monodon monoceros), Can. J. Zool. **56**, 552–560 (1978)

20.100 S.M. Dawson: The high frequency sounds of free-ranging hector's dolphins, Cephalorhynchus hectori, Rep. Int. Whal. Comm. **Spec. Issue 9**, 339–344 (1988)

20.101 R.G. Busnel, A. Dziedzic: Acoustic signals of the Pilot whale Globicephala melaena, Delpinus delphis and Phocoena phocoena. In: *Whales, Dolphins and porpoises*, ed. by K.S. Norris (Univ.California Press, Berkley 1966) pp. 607–648

20.102 N.A. Overstrom: Association between burst-pulse sounds and aggressive behavior in captive Atlantic bottlenose dolphins (Tursiops truncatus), Zoo Biol. **2**, 93–103 (1983)

20.103 B. McCowan, D. Reiss: Maternal aggressive contact vocalizations in captive bottlenose dolphins (Tursiops truncatus): wide-band, low frequency signals during mother/aunt-infant interactions, Zoo Biol. **14**, 293–309 (1995)

20.104 C. Bloomqvist, M. Amundin: High frequency burst-pulse sounds in agonistic/ aggressive interactions in bottlenose dolphins (Tursiops truncatus). In: *Echolocation in Bats and Dolphins*, ed. by J. Thomas, C. Moss, M. Vater (Univ.Chicago Press, Chicago, IL 2004) pp. 425–431

20.105 J.C. Lilly, A.M. Miller: Sounds emitted by the bottlenose dolphin, Science **133**, 1689–1693 (1961),

20.106 D.L. Herzing: *A Quantitative description and behavioral association of a burst-pulsed sound, the squawk, in captive bottlenose dolphins, Tursiops truncatus, Masters Thesis* (San Francisco State University, San Francisco 1988) p. 87

20.107 S.M. Brownlee: *Correlations between sounds and behavior in wild Hawai'ian spinner dolphins (Stenella longirostris). Masters Thesis* (Univ. California, Santa Cruz 1983) p. 26

20.108 V.A. Vel'min, N.A. Dubrovskiy: The critical interval of active hearing in dolphins, Sov. Phys. Acoust. **2**, 351–352 (1976)

20.109 V.M. Janik: Source levels and the estimated active space of bottlenose dolphin (Tursiops truncatus) whistles in the Moray Firth, Scotland, J. Comp. Psych. **186**, 673–680 (2000)

20.110 S.O. Murray, E. Mercado, H.L. Roitblat: Characterizing the graded structure of false killer whale (Pseudorca crassidens) vocalizations, J. Acoust. Soc. Am. **104**, 1679–1688 (1998)

20.111 M.C. Caldwell, D.K. Caldwell: Vocalizations of naïve captive dolphins in small groups, Science **159**, 1121–1123 (1968)

20.112 W.A. Powell: Periodicity of vocal activity of captive Atlantic bottlenose dolphins: Tursiops truncatus, Bull. Soc. Ca. Acad. Sci. **65**, 237–244 (1966),

20.113 I.E. Sidorova, V.I. Markov, V.M. Ostrovskaya: Signalization of the bottlenose dolphin during the adaptation to different stressors. In: *Sensory Abilities of Cetaceans*, ed. by J. Thomas, R. Kastelein (Plenum, New York 1990) pp. 623–638

20.114 S.E. Moore, S.H. Ridgway: Whistles produced by common dolphins from the Southern California Bight, Aqua. Mamm. **21**, 55–63 (1995)

20.115 J.C. Goold: A diel pattern in vocal activity of short-beaked common dolphins, Delphinus delphis, Mar. Mamm. Sci. **16**, 240–244 (2000)

20.116 M.C. Caldwell, D.K. Caldwell: Individualized whistle contours in bottlenosed dolphins (Tursiops truncatus), Nature **207**, 434–435 (1965)

20.117 P.L. Tyack: Whistle repertoires of two bottlenose dolphins, Tursiops truncatus: mimicricy of signature whistles?, Behav. Ecol. Sociobiol. **18**, 251–257 (1986)

20.118 M.C. Caldwell, D.K. Caldwell, P.L. Tyack: A review of the signature whistle hypothesis for the Atlantic bottlenose dolphin. In: *The Bottlenose Dolphin*, ed. by S. Leatherwood, R.R. Reeves (Academic, San Diego 1990) pp. 199–234

20.119 L.S. Sayigh, P.L. Tyack, R.S. Wells, M.D. Scott, A.B. Irvine: Signature differences in signature whistles production of free-ranging bottlenose dolphins Tursiops truncatus, Behav. Ecol. Sociobiol. **36**, 171–177 (1990)

20.120 L.S. Sayigh, P.L. Tyack, R.W. Wells, M.D. Scott, A.B. Irvine: Sex difference in signature whistle production of free-ranging bottlenose dolphins, Tursiops truncatus, Behav. Ecol. Sociobiol. **36**, 171–177 (1995)

20.121 V.M. Janik, G. Denhardt, T. Dietmar: Signature whistle variation in a bottlenosed dolphin, Tursiops truncatus, Behav. Ecol. Sociobiol. **35**, 243–248 (1994)

20.122 R.A. Smolker, J. Mann, B.B. Smuts: Use of signature whistles during separations and reunions by wild bottlenose dolphin mothers and infants, Behav. Ecol. Sociobiol. **33**, 393–402 (1993)

20.123 V.M. Janik: Whistle matching in wild bottlenose dolphins (Tursiops truncatus), Science **289**, 1355–1357 (2000)

20.124 B. McCowan, D. Reiss: Quantitative comparison of whistle repertoires from captive adult bottlenose dolphins (Delphinidae, Tursiops truncatus): A re-evaluation of the signature whistle hypothesis, Ethology **100**, 194–209 (1995)

20.125 B. McCowan, D. Reiss: The fallacy of 'signature whistles' in bottlenose dolphins: a comparative perspective of 'signature information' in animal vocalizations, Anim. Behav. **62**, 1151–1162 (2001)

20.126 J.W. Bradbury, S.L. Vehrencamp: *Principles of Animal Communication* (Sinauer Associates, Sunderland 1998) p. 882

20.127 D. Reiss, B. McCowan: Spontaneous vocal mimicry and production by bottlenose dolphins (Tursiops truncatus): Evidence for vocal learning, J. Comp. Psychol. **101**, 301–312 (1993)

20.128 A.G. Taruski: The whistle repertoire of the North Atlantic pilot whale (Globicephala melaena) and its relationship to behavior and environment. In: *Behavior of Marine Animals: Current Perspectives on Research. Vol. 3: Cetaceans*, ed. by H.E. Winn, B.L. Olla (Plenum, New York 1979) pp. 345–368

20.129 C.W. Clark: The acoustic repertoire of the southern right whale: A quantitative analysis, Anim. Behav. **30**, 1060–1071 (1982)

20.130 D.G. Richards, J.P. Wolz, L.M. Herman: Vocal mimicry of computer-generated sounds and vocal labeling of objects by a bottlenose dolphin, Tursiops truncatus, J. Comp. Psych. **98**, 10–28 (1984)

20.131 V.B. Deecke, J.K.B. Ford, P. Spong: Dialect change in resident killer whales: Implications for vocal learning and cultural transmission, Anim. Behav. **60**, 629–638 (2000)

20.132 B. McCowan, D. Reiss, C. Gubbins: Social familiarity influences whistle acoustic structure in adult female bottlenose dolphins (Tursiops truncatus), Aqua. Mamm. **24**, 27–40 (1998)

20.133 L.G. Grimes: Dialects and geographical variation in the song on the splendid sunbird Nectarinia coccinigaster, Ibis **116**, 314–329 (1974)

20.134 J.K.B. Ford: Vocal traditions among resident killer whales (Orcinus orca) in coastal waters of British Columbia, Can. J. Zool. **69**, 1454–1483 (1991)

20.135 F. Nottebohm: The song of the chingolo, Zonotrichia capensis, in Agentina: description and evaluation of a system of dialect, Condor **71**, 299–315 (1969)

20.136 J.K.B. Ford: Acoustic behaviour of resident killer whales (Orcinus orca) off Vancouver Island, British.Columbia, Can. J. Zool. **67**, 727–745 (1989)

20.137 J.K.B. Ford, H.D. Fisher: Group-specific dialects of killer whales (Orcinus orca) in British Columbia. In: *Communication and Behavior of Whales. AAAS Sel Symp. 76. Boulder* (Westview Press, Boulder, CO 1983) pp. 129–161

20.138 J.K.B. Ford: *Call traditions and dialects of killer whales (Orcinus orca) in British Columbia, Ph.D. Dissertation* (Univ. British Columbia, Vancouver, BC 1984)

20.139 A.E. Bowles, W.G. Young, E.D. Asper: On ontogeny of stereotyped calling of a killer whale calf, orcinus orca, during her first year, Rit Fiskideildar **11**, 251–275 (1988)

20.140 H. Whitehead, S. Waters, L. Weilgart: Social organization in female sperm whales nd their offspring: Constant companions and casual acquaintance, Behav. Ecol. Sociobiol. **29**, 385–389 (1991)

20.141 L. Weilgart, H. Whitehead: Group-Specific Dialects and Geographical Variation in Coda Repertoire in South Pacific Sperm Whale, Behav. Ecol. Sociobiol. **40**, 277–285 (1997)

20.142 C.W. Clark: Acoustic behavior of mysticete whales. In: *Sensory Abilities of Cetaceans*, ed. by J. Thomas, R. Kastelein (Plenum, New York 1990) pp. 571–583

20.143 A.E. Alling, E.M. Dorsey, J.C. Gordon: Blue whales (Balaenoptera musculus) off the northeast coast of Sri Lanka: distribution, feeding, and individual identification. In: *Cetaceans and Cetacean Research in the Indian Ocean Sanctuary*, Mar. Mammal Tech. Rep., Vol. 3, ed. by S. Leaderwood, G.P. Donovan (UNEP, New York, NY 1991) pp. 247–258

20.144 W.A. Watkins, P.L. Tyack, K.E. Moore, J.E. Bird: The 20-Hz signals of finback whales (Balaenoptera physalus), J. Acoust. Soc. Am. **82**, 1901–1912 (1987)

20.145 W.A. Watkins: The activities and underwater sounds of fin whales, Sci. Rep. Whales Res. Inst. **33**, 83–117 (1981)

20.146 R.S. Payne, S. Mcvay: Songs of humpback whales. Humpbacks emit sounds in long, predictable patterns ranging over frequencies audible to humans, Science **173**, 585–597 (1971)

20.147 D.A. Helweg, A.S. Frankel, J.R. Mobley Jr, L.M. Herman: Humpback whale song: our current understanding. In: *Sensory Abilities of Aquatic Mammals*, ed. by J.A. Thomas, R.A. Kastelein, Aya Supin (Plenum, New York 1992) pp. 459–483

20.148 B. Wursig, C. Clark: Behavior. In: *The Bowhead Whale*, Lawrence Soc. Mar. Mammal Spec. Publ., Vol. 2, ed. by J.J. Burns, J.J. Montague, C.J. Cowles (Allen, Lawrence 1993) pp. 157–199

20.149 W.C. Cummings, D.V. Holliday: Sounds and source levels from bowhead whales off Pt. Barrow, Alaska. J. Acoust. Soc. Am. **82**, 814–821 (1987)

20.150 W.C. Cummings, P.O. Thompson: Underwater Sounds from the Blue Whale, Balaenoptera musculus, J. Acoust. Soc. Am. **50**, 1193–1198 (1971)

20.151 P.L. Edds: Characteristics of the blue whale, balaenoptera musculus, in St. Lawrence river, J. Mamm. **63**, 345–347 (1982)

20.152 M.A. McDonald, J.A. Hilderbrand, S.C. Webb: Blue and fin whales observed on a seafloor array in the northeast Pacific, J. Acoust. Soc. Am. **98**, 712–721 (1995)

20.153 W.C. Cummings, P.O. Thompson, S.J. Ha: Sounds from bryde's whale balaenoptera edeni, and finback, b. physalus, whales in the gulf of California: Fishery bulletin, Breviora Mus. Comp. Zool. **84**, 359–370 (1986)

20.154 P.L. Edds, D.K. Odell, B.R. Tershy: Vocalization of a captive juvenile and free-ranging adult-calf pairs of bryde's whales, balaenoptera edeni, Mar. Mamm. Sci. **9**, 269–284 (1993)

20.155 P.L. Edds: Characteristic of finback, balaenoptera physalus, vocalization in the St. Lawrence estuary, Bioacoustics **1**, 131–149 (1988)

20.156 W.C. Cummings, P.O. Thompson: Characteristics and seasons of blue and finback whale sounds along the U.S. west coast as recorded at SOSUS stations, J. Acoust. Soc. Am. **95**, 2853 (1994)
20.157 M.E. Dalheim, D.H. Fisher, J.D. Schempp: Sound Production by the Grey Whale and Ambient Noise Levels in Laguna San Ignacio, Baja California Sur, Mexico. In: *The Grey Whale*, ed. by M. Jones, S. Swartz, S. Leatherwood (Academic Press, S.D. 1984) pp. 511–541
20.158 N.L. Crane: *Sound Production of Gray Whales, Eschrichtius tobustus Along Their Migration Route* (MS Thesis, San Francisco St. Univ., San Francisco, CA 1992)
20.159 W.C. Cummings, P.O. Thompson, R. Cook: Underwater sounds of migrating gray whales, eschrichtius glaucus (cope), J. Acoust. Soc. Am. **44**, 278–1281 (1968)
20.160 P.O. Thompson, W.C. Cummings, S.J. Ha: Sounds, Source Levels, and Associated Behavior of Humpback Whales, Southeast Alaska, J. Acoust. Soc. Am. **80**, 735–740 (1986)
20.161 R.S. Payne, K. Payne: Large scale changes over 19 years in songs of humpback whales in bermuda, Z. Tierpsychol. **68**, 89–114 (1985)
20.162 W.E. Schevill, W.A. Watkins: Intense low-frequency sounds from an antarctic minke whale, balaenoptera acutorostrata, Breviora Mus. Comp. Zool., 1–8 (1972)
20.163 H.E. Winn, P.J. Perkins: Distribution and sounds of the minke whale, with a review of mysticete sounds, Cetology **19**, 1–11 (1976)
20.164 W.C. Cummings, J.F. Fish, P.O. Thompson: Sound production and other behavior of southern right wales, eubalena glacialis, Trans. San Diego Soc. Nat. Hist. **17**, 1–13 (1972)
20.165 C.W. Clark: Acoustic Communication and Behavior of the Southern Right Whale. In: *Behavior and Communication of Whales*, ed. by R.S. Payne (Westview Press, Boulder, CO 1983) pp. 643–653
20.166 A.R. Knowlton, W.W. Clark, S.D. Kraus: Sounds Recorded in the Presence of Sei Whales, Balaenoptera borealis, in Proc. 9th Bienn. Conf. Biol. Mar. Mamm., Chicago, IL, Dec. 1991, p. 40 (1991)
20.167 P.O. Thompson, L.T. Findlay, O. Vidal: Underwater sounds of blue whales, Balaenoptera musculus, in the Gulf of California, Mexico, Mar. Mamm. Sci. **12**, 288–293 (1987)

21. Medical Acoustics

Medical acoustics can be subdivided into diagnostics and therapy. Diagnostics are further separated into auditory and ultrasonic methods, and both employ low amplitudes. Therapy (excluding medical advice) uses ultrasound for heating, cooking, permeablizing, activating and fracturing tissues and structures within the body, usually at much higher amplitudes than in diagnostics. Because ultrasound is a wave, linear wave physics are generally applicable, but recently nonlinear effects have become more important, even in low-intensity diagnostic applications.

This document is designed to provide the nonmedical acoustic scientist or engineer with some insights into acoustic practices in medicine. Auscultation with a stethoscope is the most basic use of acoustics in medicine and is dependent on the fields of incompressible (circulation) and compressible (respiration) fluid mechanics and frictional mechanics. Detailed discussions of tribology, laminar and turbulent hemodynamics, subsonic and supersonic compressional flow, and surfactants and inflation dynamics are beyond the scope of this document. However, some of the basic concepts of auscultation are presented as a starting point for the study of natural body sounds. Ultrasonic engineers have dedicated over half a century of effort to the development of ultrasound beam patterns and beam scanning methods, stretching the current technical and economic limits of analog and digital electronics and signal processing at each stage. The depth of these efforts cannot be covered in these few pages. However, the basic progression of progress in the fields of transducers and signal processing will be covered. The study of the interaction of ultrasound with living tissues is complicated by complex anatomic structures, the high density of scatterers, and the constantly changing nature of the tissues with ongoing life processes including cardiac pulsations, the formation of edema and intrinsic noise sources. A great deal of work remains to be done on the ultrasonic characterization of tissues. Finally, the effect of ultrasound on tissues, both inadvertent and therapeutic will be discussed.

Much of the medical acoustic literature published since 1987 is searchable online, so this document has included key words that will be helpful in performing a search. However, much of the important basic work was done before 1987. In an attempt to help the reader to access that literature, Denis White and associates have compiled a complete bibliography of the medical ultrasound literature prior to 1987. Under Further Reading in this chapter, the reader will find a link to a complete compilation of 99 citations from *Ultrasound in Medicine and Biology* which list the thousands of articles on medical acoustics written prior to 1987.

The academically based authors develop, use and commercialize diagnostic ultrasonic Doppler systems for the benefit of patients with cardiovascular diseases. To translate ultrasonic and acoustic innovation into widespread clinical application requires as much knowledge about the economics of medicine, the training and practices of medical personnel, and the pathology and prevalence of diseases as about the diffraction patterns of ultrasound beams and signal-to-noise ratio of an echo. Although a discussion of these factors is beyond the scope of this chapter, a few comments will help to provide perspective on the likely future contribution of medical acoustics to improved public health.

21.1	**Introduction to Medical Acoustics**..........	841
21.2	**Medical Diagnosis; Physical Examination**	842
	21.2.1 Auscultation – Listening for Sounds...............................	842
	21.2.2 Phonation and Auscultation	847
	21.2.3 Percussion................................	847
21.3	**Basic Physics of Ultrasound Propagation in Tissue**	848
	21.3.1 Reflection of Normal-Angle-Incident Ultrasound	850

	21.3.2	Acute-Angle Reflection of Ultrasound	850	21.5.3	Agitated Saline and Patent Foramen Ovale (PFO)	884
	21.3.3	Diagnostic Ultrasound Propagation in Tissue	851	21.5.4	Ultrasound Contrast Agent Motivation	886
	21.3.4	Amplitude of Ultrasound Echoes	851	21.5.5	Ultrasound Contrast Agent Development	886
	21.3.5	Fresnel Zone (Near Field), Transition Zone, and Fraunhofer Zone (Far Field)	853	21.5.6	Interactions Between Ultrasound and Microbubbles	886
	21.3.6	Measurement of Ultrasound Wavelength	855	21.5.7	Bubble Destruction	887
	21.3.7	Attenuation of Ultrasound	855	21.6	**Ultrasound Hyperthermia in Physical Therapy**	889
21.4	**Methods of Medical Ultrasound Examination**		857	21.7	**High-Intensity Focused Ultrasound (HIFU) in Surgery**	890
	21.4.1	Continuous-Wave Doppler Systems	857	21.8	**Lithotripsy of Kidney Stones**	891
	21.4.2	Pulse-Echo Backscatter Systems	859	21.9	**Thrombolysis**	892
	21.4.3	B-mode Imaging Instruments	862	21.10	**Lower-Frequency Therapies**	892
21.5	**Medical Contrast Agents**		882	21.11	**Ultrasound Safety**	892
	21.5.1	Ultrasound Contrast Agents	883	**References**		895
	21.5.2	Stability of Large Bubbles	883			

Standard Symbols
c sound propagation velocity [km/s = mm/μs]
p acoustic pressure fluctuation [d/cm^2; Pa = N/m^2]
t time [s]
Z tissue impedance [Rayles = ρc = kg/(m^{2}s) = 0.1 g/(cm^{2}s)]
λ acoustic wavelength [m]
ρ tissue density [g/cm^3]
σ surface tension [d/cm = mN/m]

Special Symbols
d diameter
f frequency [Hz]
f_D Doppler frequency [Hz]
f_{us} ultrasound center frequency [MHz]
fd vibration frequency [Hz]
P total instantaneous pressure [d/cm^2, kPa]
q acoustic molecular displacement [cm]
Q volume flow [cm^3/s]
R reflection coefficient
T transmission coefficient
v_f fluid velocity [cm/s]
v molecular velocity [cm/s]
w transducer element width [cm]
x distance in propagation direction [cm]

κ tissue stiffness [d/cm^2]
θ angle of sound propagation from the normal at a refraction interface
Ψ distance to transition zone
θ_D Doppler angle of sound propagation to vessel axis or velocity vector axis
$\Delta\phi$ ultrasound echo phase shift
$\Delta\tau$ ultrasound echo time shift [ns]
PRF pulse repetition frequency [Hz]
D thermal dose

Subscripts
i incident
r reflected
s shear wave
t transmitted
1 incident material
2 second material

Dimensionless Numbers
St Strouhal number St = $(fD)/V$
Re Reynolds Number $(DV\rho)/\mu = (4Q\rho/\pi D)/\mu$

Units
Np nepers
dB decibels

21.1 Introduction to Medical Acoustics

Medical acoustics covers the use of sub-audio vibration, audible sound and ultrasound for medical diagnostics and treatment. Sound is a popular and powerful tool because of its low cost, noninvasive nature, and wide range of potential applications from passive listening to the application of high-energy pulses used for the destruction of kidney stones.

There are four frequency ranges of sound and three levels of intensity that will be presented in this chapter

Table 21.1 Medical frequencies

Title	Period	Frequency	Power/Intensity
Generation	25 years		
Ovulation	28 days		
Diurnal	1 day		
Respiration	6 s	0.15 Hz	
Heart Rate	1 s	1 Hz	
ECG surface bandwidth	10 ms	100 Hz	
ECG intramyocardial BW	250 µs	4 KHz	
Tremor	100 ms	3–18 Hz	
Electromyography surface	5 ms	200 Hz	
Electromyography intramuscle	100 µs	10 kHz	
Flicker fusion (visual)	60 ms	16 Hz	
Vibration sensation	10 ms	100 Hz	
Sound low	50 ms	20 Hz	
Telephone sound speech	3–0.3 ms	0.3–3 kHz	
Sound high	50 µs	20 kHz	
Electro-encephalography (EEG) surface waves			
Delta waves	300 ms	< 3.5 Hz	
Theta waves	200 ms	3.5–7.5 Hz	
Alpha waves	100 ms	7.5–13 Hz	
Beta waves	50 ms	13–30 Hz	
Intracerebral EEG	100 µs	10 kHz	
Ultrasound cleaners		42 kHz	100 W, 1 W/cm^2
Ultrasound bacteriocidal		22.5 kHz	
Ultrasound therapeutic	0.5 µs	1 3 MHz	15 W, 3 W/cm^2
Ultrasound diagnostic			
Transcranial		1–3 MHz	100 mW/cm^2
Heart, abdomen		2–5 MHz	100 mW/cm^2
Vascular, skeletal		4–10 MHz	100 mW/cm^2
Skin, eye		10–40 MHz	8 mW/cm^2
Ultrasound high intensity focused (HIFU)		2–6 MHz	2000 W/cm^2
Diathermy, shortwave, physiotherapy		13–40 MHz	2–15 W
Diathermy, shortwave, cell destruction		300 kHz	100 W
Diathermy, cancer ablation		500 kHz	500 W
Diathermy microwave		0.4–2.5 GHz	
Light		10^{14} Hz	
X-ray		1×10^{18} Hz	
		4×10^{18} Hz	

1. Infrasound, acoustic waves with frequencies below the level of human hearing. These vibrations, called *thrills*, are often caused by turbulent blood flow (which is abnormal). At the lowest extreme, the pulse can be considered infrasound with a frequency below 2 Hz. With fingertips, the physician can feel vibrations with amplitudes of 1 μm.
2. Audible sound, acoustic waves with frequencies in the range of human hearing. The audible frequency range for the average human is approximately 20 Hz–20 kHz, with a peak sensitivity around 3–4 kHz. At 1 kHz, the standard reference frequency, the ear is able to detect pressure variations on the order of one billionth of an atmosphere. This corresponds to intensities of about $0.1\,\text{fW/cm}^2$ ($10^{-16}\,\text{W/cm}^2 = 10^{-12}\,\text{W/m}^2$). The oscillatory velocity of air molecules carrying such sound intensities is $v = 50\,\text{nm/s}$ ($1\,\text{nm} = 10^{-9}\,\text{m}$) and the maximum pressure fluctuation is $20\,\mu\text{N/m}^2$ or 0.2 nanoatmospheres (nanobar). At 1 kHz, the air molecules oscillate back and forth about 0.05 nm. The intensity of conversation is $0.1\,\text{nW/cm}^2$ ($1\,\mu\text{W/m}^2$) causing molecules to oscillate with amplitudes of 50 nm, and oscillatory pressures of $20\,\text{mN/m}^2$.
3. Low-frequency ultrasound, which is used for cleaning and tissue disruption.
4. Radio-frequency ultrasound, acoustic waves with frequencies near 1 MHz and wavelengths in the body near 1 mm.

A list of interesting frequencies in medicine is provided for reference Table 21.1.

21.2 Medical Diagnosis; Physical Examination

The most valuable application of acoustics in medicine is the communication from a patient to their physician. In the modern world of *efficient medicine*, however, physicians are encouraged to avoid wasting time listening to the patient so that objective tests can be ordered without delay. This approach avoids physician frustration from sorting out the complaints and symptoms of the patient, even though the patient initiated the appointment with the physician.

Having gathered background information, the physician can proceed to the physical examination, which includes optical methods, olfaction, palpation, testing the mechanical response of tissues and systems to manipulation, percussion, remote listening and aided auscultation, and acoustic transmission and reflection techniques. Only those techniques related to medical acoustics will be discussed further here. Of future importance in medical practice though, are those methods that can be performed remotely, via the internet.

21.2.1 Auscultation – Listening for Sounds

Sounds in the body are generated through the presence of disturbed flow, such as eddies and turbulence, by frictional rubs, and/or by other sources with high accelerations. There are both normal and abnormal sounds within the body. Auscultation is the act of listening for sounds within the body for diagnostic purposes. This may be conducted on any part of the body from head to toe (Fig. 21.1). Anesthesiologists even use esophageal stethoscopes for convenient acquisition of chest sounds. Information gathered during auscultation includes the frequency, intensity, duration, timing, and quality of the sound.

Lung. Both normal and abnormal sounds can be present in the lungs. The source of normal lung sounds is turbulent flow associated with air movement through orifices along the airways. These are large, low-frequency oscillations that cause the surrounding tissues to vibrate within the audible frequency range. The respiratory system is naturally optimized to minimize the work needed to exchange air, so little energy is converted into acoustic power, resulting in low-amplitude sounds. With constriction of the airways, in such conditions as asthma, the air velocity through the airways increases (conservation of mass). Through conservation of energy and momentum, an increase in flow velocity results in an increase in pressure loss, raising the power needed to drive the system. The energy is of course not lost, but converted, in part, into acoustic power. The increase in flow velocity results in an increase in *turbulent diffusion*, or the transfer of energy from the flow to the surrounding tissues through a growing chain of turbulent eddies. When the passages are smaller, the eddies decrease in size, causing an increase in the frequency of the audible sound. The pressure drop caused by a small lumen is higher, causing an increase in eddy velocity and therefore an increase in acoustic power, and the flow rate is decreased,

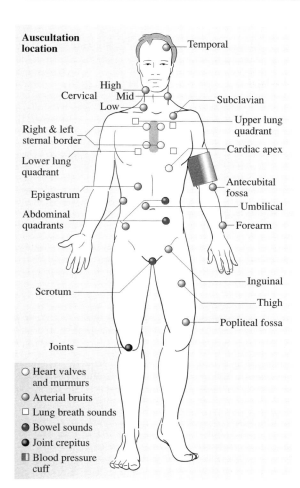

Fig. 21.1 Auscultation sites. Arterial bruits can be heard bilaterally indicating arterial stenoses. Heart valve sounds and heart murmurs are primarily heard in the peristernal intercostal spaces indicating cardiac timing, and valve stenosis and incompetence (insufficiency or regurgitation). Breath sounds in the lung indicate both airway sufficiency, lung inflation, alveolar inflation and pleural friction. Bowel sounds [21.1, 2] indicate normal peristalsis; bowel in the scrotum indicate intestinal hernia. Crepitus on motion indicates air in the tissue, bone fracture or joint [21.3–6] cartilage damage. Auscultation in the antecubital fossa during blood pressure cuff deflation yields Korotkoff sounds [21.7, 8], 1 = a sharp sound each systolic peak when the cuff pressure equals the systolic pressure indicating the momentary separation of the artery walls and the onset of blood flow, 2 = a loud blowing sound due to blood turbulence, 3 = a soft thud when the artery under the cuff is exhibiting maximum change in arterial diameter, 4 = a soft sound indicating slight arterial deformation near diastolic pressure, 5 = silence when the cuff pressure is below the diastolic arterial pressure

some medical conditions, the flow dynamics may be just right to produce coherent eddy structures, such as one would see in a smoke stack. These signals result in a narrow-band frequency tone.

Abnormal snapping or popping sounds may also be present in the lung. In the normal lung, the alveoli (terminal air sacks, 280 μm in diameter, 3×10^8 in lung) are filled with air. In pneumonia, regions of alveoli are filled with fluid, which prevents a rapid increase in size without the introduction of a gas phase. The rapid introduction of a gas phase when the lung is expanded during the inspiration phase of respiration causes a tiny snapping sound (broadband acoustic impulse) in each alveolus; many snaps together cause the sound of *rales*.

causing a longer duration of flow to fill and empty the lungs. Thus the breath sounds are louder, longer, higher-pitched sounds compared with normal lung sounds. In

Table 21.2 Diagnosing valve disease from heart murmurs

	Time	Side	Location	Position
Tricuspid stenosis	Diastolic	Parasternal	3rd ICS	Supine
Tricuspid regurgitation	Systolic	Peristernal	3rd ICS	Supine
Pulmonic stenosis	Systolic	Right	Superior	Supine
Pulmonic regurgitation	Diastolic	Right	Superior	Seated
Mitral stenosis	Diastolic	Left	Apex	Supine
Mitral regurgitation	Systolic	Left	Apex-Axilla	Supine
Aortic stenosis	Systolic	Parasternal	Superior	Supine
Aortic regurgitation	Diastolic	Parasternal	Superior	Seated

Side parasternal = in the intercostal spaces on either side of the sternum
Location 3rd ICS = the space between the 3rd and 4th rib, Erb's point
Position seated = Seated, leaning forward, breath in expiration

Because of the lubricant normally present in the pleura located between the lungs and the chest wall, there are normally no sounds generated through a friction rub due to the sliding of the lung along the chest wall. When the pleura are inflamed, coarse rubbing sounds can be heard during breathing.

Cardiac. The cardiovascular system also generates normal and abnormal sounds that can be heard in the chest with a stethoscope [21.9] (Table 21.2). Normal heart sounds [21.10–12] are caused by rapid blood decelerations resulting from the closure of the heart valves. Each of the four heart valves, the tricuspid, pulmonic, mitral and aortic (in order from the venous circulation to the arterial circulation), are flaccid sheets of tissue securely tethered in the closed position to form check (one-way) valves. Like a parachute opening, the valve leaflets move with the reversing blood flow until the tethers become taught, then the leaflets force a rapid deceleration of the blood which generates a sudden (broadband) *thump* with an acoustic power of several watts. No similar rapid acceleration is associated with valve opening. Because the sounds are primarily radiated along the axis of acceleration and conduct better through tissue than the air-filled lungs, the sound of each valve is most prominent at a particular location on the chest wall.

Though four valves are present, usually only two thumps are heard, and they occur at vary precise times in the cardiac cycle. 20 ms after the onset of ventricular contraction, the tricuspid and mitral valves close, causing the first heart sound. 50 ms after closure of the atrioventricular valves, the pulmonic and aortic valves open silently. 320 ms after the onset of ventricular contraction, when ventricular ejection is complete and the ventricular pressures drop below the pulmonary trunk and aortic pressures, the pulmonic and aortic valves snap shut, generating the second heart sound. These times do not vary with heart rate; each cardiac cycle acts like an independent impulse-response system. The systolic ejection volume does vary with the diastolic filling period from the prior cardiac cycle.

Thus, each of the two heart sounds is generated by two valve closures, one in the left heart and one in the right heart. These usually occur simultaneously, however, the sounds do split into two if the inflow valves (tricuspid and mitral) or outflow valves (pulmonic and aortic) close at different times. These *split heart sounds* [21.13] can be indicative of an abnormality. The closure of the tricuspid valve is dependent on contraction of the right ventricle, which in turn is dependent on the electrical conduction system from the pacemaker of the heart. As a result, a delay in conduction or a slow response of the right ventricular muscle (myocardium) will delay the tricuspid sound, causing splitting. Similarly, if there is a delay in the conduction system of the left heart, or the left ventricular myocardium contraction is slow, then the mitral sound is delayed. The closure of the outflow valves is dependent on contraction strength, pressure in the outflow vessels, and ejection volume. A mismatch of any of those factors will cause the pulmonic and aortic valves to close at different times, splitting the second sound. Even the effects of respiration on the filling and emptying of the heart chambers cause some normal heart sound splitting.

Abnormal sounds that occur in the cardiovascular system are called *bruits* and *murmurs* [21.14–17]. A murmur is a sound lasting longer than 10 ms heard by a cardiologist from the heart; a bruit is a similar sound, but heard by an angiologist from an artery. Veins rarely make sounds; venous sounds are called *hums* because they are continuous. A *thrill* is a similar vibration felt by the hand. Because these represent power dissipation from the cardiovascular system, all are considered abnormal. The absence of normal bruits and murmurs results from a biological feedback that forms the shape of the cardiovascular system. The feedback mechanism is not well understood. In fluid mechanics, the transition from laminar flow to turbulent flow occurs near a Reynolds number of 2000. Turbulent flow is highly dissipative while laminar flow is minimally dissipative. Coincidentally, peak flow through the cardiovascular system maintains a condition of Reynolds number less than 2000. If a heart valve or an artery is required to carry higher flow rates, then the conduit will dilate until the Reynolds number (proportional to the flow divided by the diameter) becomes less than 2000. Distal to a stenosis, where turbulence occurs, the conduit dilates, causing post-stenotic dilation. Thus, it appears that sustained turbulence, which dissipates power, is the driving force for remodeling to avoid turbulence and make the hemodynamic system more energy efficient.

The vibrations indicate the presence of disturbed blood flow, caused by deceleration of the blood beyond a narrow stenosis. The vibrations are associated with a pressure drop across the stenosis because of a transfer of hydraulic power to acoustic power through a cascade of eddies. In turbulent flow, the vibrations are broadband. When coherent eddies are present, the vibrations are narrow band, and the frequency of the eddies can be related to the flow velocity (v_f) and vessel diameter (d) through the Strouhal number (St $= fd/v_f$). In addition to the location and power of a bruit, the frequency of the bruit is also helpful in evaluating its clinical importance. The Strouhal number has been used for the prediction of the residual lumen diameters of carotid stenoses. If the frequency of a bruit is divided into the number 500 [mm Hz], the result is the minimum lumen diameter [21.18–25] in mm. In clinical testing, this method accurately predicted the stenosis in hundreds of cases. However, because cumbersome technology is involved, and because the method fails in tight stenoses, which are most clinically important, this clever method has been abandoned.

Heart murmurs are auscultated at a series of locations on the precordium (the front middle of the chest around the sternum). They are graded from grade 1 (barely audible) to grade 6 (no stethoscope needed). Murmurs may result from leaky (incompetent, regurgitant) valves, stenotic (narrow) valves, or other holes between heart chambers. The differential diagnosis of heart murmurs is based on the timing of the murmur and the locations on the chest wall [21.26, 27] where the murmur is best heard. A murmur can be early systolic, mid-systolic, late systolic, or early diastolic, or mid-diastolic. If a murmur is best heard in the left second intercostal space (the space between the second rib and the third rib along the left boarder of the sternum), it is most probably an aortic stenosis [21.28–30]. Each of the other valve pathologies has a corresponding location where the murmur is likely to be highest. Of course, there is a great deal of variability between patients, and the examiner must apply a full set of detective skills to sort out a proper diagnosis.

There are some common systolic heart murmurs. Athletes, with temporary increased blood flow due to exercise programs have physiologic systolic heart murmurs, as do pregnant women, who have a 30% increase in cardiac output to supply the placenta. Hemodialysis patients also often have systolic heart murmurs because of the increased cardiac output required to supply the dialysis access shunt [21.31] in addition to the supply to the body. In these cases, remodeling of the blood vessels to increase the flow diameter has not had time to occur.

Bruits are almost always associated with stenoses in arteries, which often require treatment. They are classified by their pitch and duration. The occurrence of bruits in the head, in the abdomen after eating and in the legs are all indications of arterial stenoses, or high-velocity blood entering vascular dilations. The presence of a bruit heard in the neck is usually indicative of a carotid stenosis. The carotid arteries are the major vessels located on the left and right side of the neck that supply blood to the brain. The development of an atherosclerotic plaque causes luminal narrowing and the corresponding onset of disturbed flow. This lesion is associated with a 20% risk of stroke in two years from material that can be released by the atherosclerotic plaque and travels to the brain to occlude a branch artery.

Carotid bruits are usually present when the residual lumen (stenotic) diameter is greater than 50% of the original lumen diameter (called a 50% diameter reduction), but less than 90% of the original lumen diameter (90% diameter reduction) In this range, the blood flow rate to the brain through the carotid artery is approximately $5\,\mathrm{cm}^3/\mathrm{s}$. If the pressure [21.32] drop across the stenotic region is about 20 mmHg, then the power dissipation at the stenosis is 13 mW, which can appear as sound energy in a bruit. Unfortunately, as the stenosis becomes more severe, the flow rate along the artery decreases, reducing the acoustic power for the bruit, so the bruit is no longer detectable, even though this more severe stenosis carries a higher risk for stroke. The decreased blood flow will not cause symptoms in most people because 95% of the population have collateral connections (the circle of Willis) that can provide the required blood flow that is not delivered by the stenotic artery. In the remaining 5% of the population, a severe carotid stenosis causes impaired mental function.

One of the barriers to new methods in the classification of carotid stenosis is that the first description of the relationship between carotid artery stenosis and apoplexy (stroke) was provided by Egaz Moniz in 1938. Moniz used percent diameter reduction and everyone has followed suit, through a confusing and contentious evolution, even when more-rational and useful alternatives have been developed. Thus, a new method that predicts residual lumen diameter will not be easily adopted by the medical community because of the tradition of using percent diameter reduction.

Bruits may be generated in the abdomen by stenoses in the renal arteries, which supply the kidneys. A renal artery stenosis will cause renovascular hypertension because the kidney is a transducer for detecting low blood pressure. The pressure drop across a renal artery stenosis causes the kidney to measure a low pressure, and deliver the signals to raise blood pressure in the arterial system, causing hypertension in all arteries except those in the affected kidney. The transduction system is called the renin-angiotensin-aldosterone system.

Some cardiovascular sounds are so loud that they cannot be missed. A patient in renal failure, refusing dialysis, had a cardiac friction rub (friction in the pericardial sack from inflammation). The 1 kHz *squeak* occurring with each systole could be heard from the doorway of the patient's room.

A stethoscope can also be used to detect bowel sounds from the abdomen. These sounds are generated by the liquid and gaseous contents of the bowel being forced from one intestinal segment to another through an orifice created by peristaltic action. The same fluid eddies described above are responsible for these sounds. The presence of bowel sounds is an important indicator of a normally functioning gut. It is common, the day after abdominal surgery, to check to be sure that normal bowel sounds are present.

Palpation, though not an acoustic technique, is used to feel, rather than hear, the same vibrations. Hemodialysis patients have a *vascular access* shunt created by tying an artery directly to a vein. This shunt may have a flow rate of 3 liters per minute, half of the cardiac output. The pressure drop is equal to the systolic blood pressure 120 mmHg. So the hemodynamic power dissipated in the shunt is 800 mW of power. This energy is converted to a vibration (thrill) which can be felt with the fingers.

Acoustic Stethoscope

Auscultation may be performed directly, but is most commonly conducted with a stethoscope. The device was invented by Dr. Rene Theophile Hyacinthe Laennec in 1816. While examining a patient, Dr. Laennec rolled up a sheet of paper and placed one end over the patient's heart and the other end over his ear. Dr. Laennec later replaced the rolled paper with a wooden tube (similar in appearance to a candlestick), which was called a stethoscope from the Greek words *stethos* (chest) and *skopein* (to look at). Two modern versions of the stethoscope are notable: the binaural stethoscope and the fetal stethoscope. The common binaural stethoscope consists of a bell connected by flexible tubing to a pair of earpieces. Each earpiece is sealed into the corresponding external auditory meatus. The bell is sometimes covered with a diaphragm, which acts as a high-pass filter. For example, normal heart sounds have frequencies near 50 Hz while heart murmurs have frequencies exceeding 100 Hz. Filtering out the normal heart sounds with the diaphragm of the stethoscope often allows pathologic murmurs to be detected. When using an open bell stethoscope, a diaphragm can be formed by pressing the bell tighter to the skin, stretching the skin and deeper tissue, which attenuates the lower frequencies [21.33–36].

The fetal stethoscope is designed for use in the auscultation of fetal heart sounds, which can be detected during the final half of a 40 week pregnancy. The fetal stethoscope uses a rod connecting the mother's abdominal wall to the examiner's forehead to transmit the sound by bone conduction in the examiner's head to the inner ear where it is converted into nerve impulses. This avoids the impedance changes between tissue and air at the skin surface and then again from air to tissue in the middle ear.

For the last 20 years, small ultrasonic Doppler ultrasound stethoscope systems have been available to detect the fetal heart beat [21.37, 38]. They operate like the ultrasound imaging *phono-angiograph* to easily detect solid tissue vibrations from deep in the body. These continuous-wave instruments allow the pregnant mother to perform easy self-examination. They are available on a rental basis without a medical prescription.

Electronic Stethoscope

Electronic stethoscopes [21.39–41] provide improved amplification and background noise cancelation over traditional stethoscopes. Some also have recording capabilities to provide quantitative, rather than just qualitative, assessment of the sounds heard. For example, by listening to the neck with a phono-angiograph (Fig. 21.2), the location and duration of the bruit can be documented. A phono-angiograph is just an electronic stethoscope with an attached oscilloscope. If applied to the heart, then it is called a phono-cardiograph. This bruit (Fig. 21.2), is a pan-systolic bruit (lasting all of systole) but not a pan-diastolic bruit (lasting all of systole and diastole). Since arterial blood velocities are higher

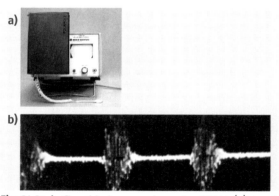

Fig. 21.2a,b Carotid phono-angiography (CPA) (**a**) A microphone fitted with an acoustic coupling to the neck gathers sounds, which are displayed on an oscilloscope and captured with an instant camera. (**b**) The sounds displayed are classified by the location, amplitude, and duration of the bruit. A similar method, phono-cardiography, was used for the heart. A method, quantitative phono-angiography (QCPA) provided an analysis of the frequency content of the bruit

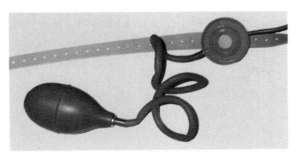

Fig. 21.3 Phono-cardiograph transducer. To avoid sounds and motions introduced by the examiner's hand on the chest, this phonocardiography transducer is fitted with a suction ring and elastic strap to secure the transducer element to the chest. Early echocardiographs had provision for amplifying and displaying the phono-cardiograph signals, which were filtered to display audible frequencies or infrasonic frequencies

during systole than diastole, all arterial bruits include systole. The exceptions, where arterial flow in a collateral pathway is reversed due to a *steal syndrome* are so rare that common names have not been adopted for this condition. To be sure that the bruit is systolic, the examiner can palpate the pulse at the wrist while listening, to associate the bruit timing with systolic arterial motion.

However, if such an image were gathered with a phono-cardiograph (Fig. 21.3) from the heart, it would be impossible to know whether the murmur was coming from aortic or pulmonic stenosis (during systole), aortic or pulmonic regurgitation (during diastole), from tricuspid or mitral regurgitation (during systole), or tricuspid or mitral stenosis (during diastole), unless there were some indication of the timing of the cardiac cycle. A phono-cardiogram displays an associated electro-cardiograph (ECG) tracing to provide that timing information.

Unfortunately, electronic stethoscopes, phono-angiographs and phono-cardiographs are considered too complicated for use in everyday medical practice. Although they do provide a graphical output for the patient chart (required documentation to receive payment for diagnostic procedures), a separate current procedural terminology (CPT) code is not available for phono-angiography, phonocardiography or for stethoscope examination. Like the stethoscope examination, these procedures, if performed, are considered part of a standard office visit. Physicians are reluctant to use technical examination methods if they cannot charge for the extra time involved. Although a few research facilities documented bruits and murmurs detected with these systems between 1965 and 1985, the systems have nearly disappeared from medical practice. Like the first half of the 20th century, physicians today are trained to use acoustic stethoscopes. Of course, as physicians become more experienced and more mature, their hearing deteriorates. Even the most respected cardiologist will fail to correctly diagnose cardiac valve disease in 50% of patients. Bruits with 35 μm, 300 Hz vibrations from the jugular vein (Fig. 21.4) documented by a new ultrasound imaging phono-angiograph were not detected by stethoscopic auscultation. So, in spite of the ubiquitous use of stethoscopes for bruit/murmur detection, their sensitivity may not be adequate for the task.

21.2.2 Phonation and Auscultation

In addition to listening for sounds with a stethoscope, an examiner may cause the generation of sounds to assess transmission or resonance. During auscultation of the lungs, an examiner may ask the patient to phonate (make voice sounds) at different frequencies "Say 'eeeeeee', now say 'aaaaaaaa'". During that period, the examiner will either palpate the chest wall for vibration or listen to the chest wall with a stethoscope. Poor sound transmission may indicate that the airways are obstructed, enhanced sound transmission (enhanced *fremitus*) indicates fluid has filled a space normally occupied by air.

21.2.3 Percussion

Percussion is another method of sound generation by striking the chest or abdomen to assess transmission or resonance. The skin may be struck with the hand or finger, or one hand may be placed on the skin and struck with the other hand. A higher-frequency shorter pulse can be generated by placing a coin on the skin and striking the coin with another coin. The resultant sound can be assessed by listening to the resonance through the air or by using a hand or stethoscope to assess transmission of the impulse. Percussion that results in a dull sound indicates a fluid-filled space below the skin; a resonant sound indicates an air-filled space below the skin. These methods are often used to find the lower boarder of the lungs, to identify fluid-filled consolidation in the lungs, and to find the edge of the liver (fluid-filled) in the abdomen (air-filled intestines).

These acoustic examination techniques have been developed in medical practice over centuries, and although efforts have been made to use electronic aids for

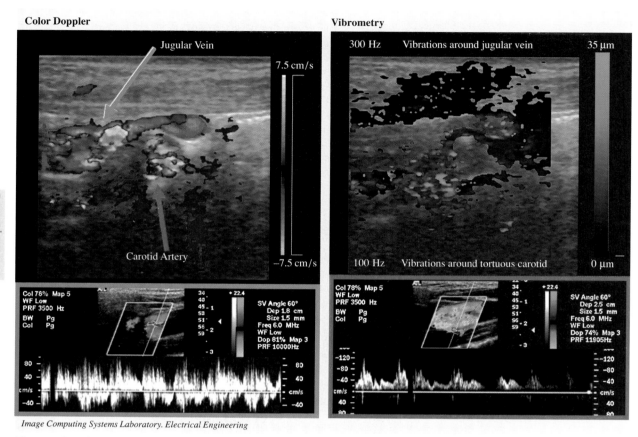

Fig. 21.4 Imaging vibrometry of the neck. Color Doppler image (*upper left*), vibration image (*upper right*), jugular vein Doppler (*lower left*) and carotid artery Doppler (*lower right*) of a patient complaining of "vibrations in the head". The 300 Hz vibrations with an amplitude of 35 μm could not be heard with a stethoscope by either of two experienced examiners (Courtesy M. Paun, S. Sikdar)

diagnosis and for training, still, the common practice is to learn these methods from a mentor. In the last quarter of the 20th century, ultrasound has become adopted in specialty medical practice, but has not made inroads in general practice. It is likely that, if and when ultrasound diagnostic methods become widely used in general medical practice, it will be used with the same level of skill as stethoscope auscultation.

21.3 Basic Physics of Ultrasound Propagation in Tissue

The vibrations of molecules at any location (or time) in a sound wave can be characterized by four parameters: the pressure fluctuation from the mean static pressure (p), the molecular velocity (v), the displacement of the molecules from the resting location (d), and the molecular acceleration (a). The two most basic parameters, pressure deviation and molecular velocity are dependent only on the intensity of the sound. Their ratio, p/v, is determined by the acoustic impedance ($Z = p/v$), a property of the tissue. At the sound level shown in Fig. 21.5, 300 mW/cm^2 (3 kW/m^2), the maximum acoustic pressure fluctuation equals 1 atm. If the sound intensity is increased, then the negative peaks represent a pressure below zero. Most ultrasound examinations are performed at pulse peak intensities 25 times this high, and therefore, peak amplitudes five times atmo-

Medical Acoustics | 21.3 Basic Physics of Ultrasound Propagation in Tissue

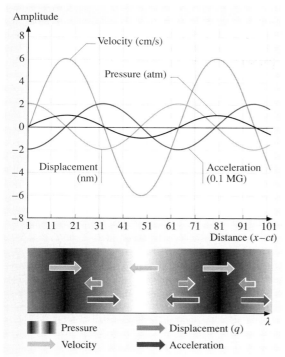

Fig. 21.5 Typical acoustic parameters of a 300 mW/cm², 5 MHz ultrasonic wave in tissue impedance. Input parameters: Tissue impedance = Z = 1.5 MRayles = 1.5 MN s/m³ = 1.5×10^6 kg/(m²s), pressure amplitude = p = 1 atm (bar) = 101 kPa (kN/m²) = 760 mmHg (Torr), frequency = 5 MHz; resultant wave parameters: intensity = 0.33 W/cm² = $p^2/2Z$, peak molecular velocity = 6 cm/s = p/Z, molecular displacement = 1.9 nm, molecular acceleration of 0.19 million times the acceleration of gravity

spheric pressure, making the compressions 6 atm and the decompressions −4 atm.

The linear acoustic relationships can be easily derived using a simple linear elastic model of tissue (Fig. 21.5) and Newton's first law of motion on the molecules. The molecules in the tissue will accelerate along the axis of the ultrasound propagation (x) and displace a distance (q) away from equilibrium because of a gradient in pressure (dp/dx), which is related to the tissue compression (dq/dx) and the tissue stiffness (κ).

$$p = -\frac{\kappa \, dq}{dx} = -\kappa q' , \quad (21.1)$$

$$\kappa \frac{d^2 q}{dx^2} = -\frac{dp}{dx} = \rho \frac{dq^2}{dt^2} . \quad (21.2)$$

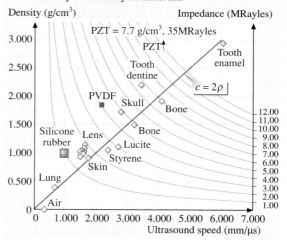

Fig. 21.6 Acoustic properties of materials of medical interest. Empirically, the acoustic properties of tissues fall on a line relating ultrasound speed (c) and tissue density (ρ); $c = 2\rho$. Using the equation for impedance $Z = \rho c$, then $Z = c^2/2 = 2\rho^2$. Polyvinylidene fluoride (PVDF) transducer material has acoustic properties near soft tissue, lead zirconate titanate (PZT) transducer material has much higher density and acoustic impedance

Solving this equation with the variable $q(x - Ct)$ produces

$$\kappa q'' = \rho C^2 q'' , \quad (21.3)$$

so

$$C = \sqrt{\frac{\kappa}{\rho}} . \quad (21.4)$$

The molecular velocity ($v = dq/dt = -Cq'$) and the tissue pressure ($p = -\kappa q'$) form an important ratio

$$Z = \frac{p}{v} = \frac{\kappa}{c} = \sqrt{\kappa \rho} = \rho C . \quad (21.5)$$

Although the tissue density (ρ) and stiffness (κ) determine the acoustic impedance (Z) and acoustic wave speed (C), it is more convenient to measure the density and wave speed of a material and compute the impedance.

A good deal of effort has been devoted to measuring the properties of tissues (Figs. 21.6–21.8). Measurement differences between laboratories and differences in tissue preparation contribute to the variability of tissue property values. While living mammalian body tissues

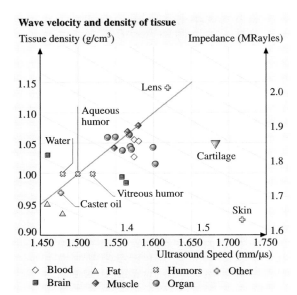

Fig. 21.7 Acoustic properties of soft tissues

are held to a temperature near 37 °C, and have blood flow that swells the tissue and changes the average composition with the heart rate and respiratory rate, it is convenient to measure properties of dead tissue at bench-top temperatures (25 °C). Fortunately, the exact

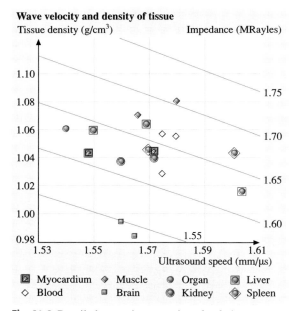

Fig. 21.8 Detailed acoustic properties of soft tissues

acoustic properties of tissues have not proven very useful in clinical diagnostic applications. Ultrasound imaging methods have not been able to differentiate tissue types in spite of a half century of efforts. The identification of disease with ultrasound imaging uses the shapes of large structures as the primary diagnostic tool.

21.3.1 Reflection of Normal-Angle-Incident Ultrasound

When ultrasound intersects an interface between two materials with differing acoustic impedances (Z_1 and Z_2), perpendicular to the interface, then the amplitude of the reflection of ultrasound at the interface can be computed from the requirement that the pressure change and molecular velocity of the incident wave (p_i, v_i) must equal sum of the pressure changes and molecular velocities of the transmitted and reflected waves (p_t, p_r, v_t, v_r). The impedances of the materials relate the pressures and molecular velocities ($Z_1 = p_i/v_i = -p_r/v_r$, $Z_2 = p_t/v_t$). The pressure reflection and transmission coefficients are

$$\frac{p_r}{p_i} = R = \frac{Z_2 - Z_1}{Z_2 + Z_1}, \qquad (21.6)$$

$$\frac{p_t}{p_i} = T = 2\frac{Z_2}{Z_2 + Z_1}. \qquad (21.7)$$

Note that if the acoustic impedance of the second material is lower than the acoustic impedance of the first material, then the reflected wave is inverted. The sum of the transmitted and reflected power equals the incident power.

21.3.2 Acute-Angle Reflection of Ultrasound

If the ultrasound is incident on the interface at an acute angle, the ultrasound that passes into the second tissue is tilted to a new direction according to Snell's law (Fig. 21.9).

$$\frac{C_1}{\sin\theta_1} = \frac{C_2}{\sin\theta_2}. \qquad (21.8)$$

Because the component of molecular velocity normal to the surface must be preserved:

$$\frac{p_r}{p_i} = R = \frac{\frac{Z_2}{\cos\theta_2} - \frac{Z_1}{\cos\theta_1}}{\frac{Z_2}{\cos\theta_2} + \frac{Z_1}{\cos\theta_1}}. \qquad (21.9)$$

In this case, not all of the ultrasound power is found in the sum of the reflected wave and the transmitted wave. The remaining portion of the power appears as a surface *capillary* wave propagating along the surface of

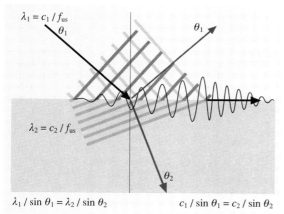

Fig. 21.9 Refraction of ultrasound as it passes from one material to another. *Dark brown*: incident compressional wavefronts and wave vector incident at angle θ_1. *Light brown*: reflected compressional wavefronts and wave vector reflecting at angle θ_1. *Gray*: transmitted compressional wave. The interfacial shear wave increases in intensity as the ultrasound beam cross section coherently contributes to the wave; attenuation of the shear wave causes a rapid decrease in amplitude outside the beam

the interface. This capillary wave is subject to high attenuation. Because tissue has an interface on every cell boundary and every tissue boundary, this mechanism is responsible for some of the conversion of incident ultrasound power into heat as ultrasound propagates into tissue. It is important to notice that the propagation speed of the capillary wave is not the same as that of the component of the compressional wave directed along the interface, so the capillary wave is out of phase with the ultrasound wave, contributing to the transfer of energy into the capillary wave. Outside the ultrasound beam, the capillary wave takes on its conventional wavelength (c_s/f), where f is the ultrasound frequency. The propagation speed of the capillary wave (c_s) is proportional to the interfacial tension.

21.3.3 Diagnostic Ultrasound Propagation in Tissue

Diagnostic ultrasound uses ultrasound frequencies of 1–30 MHz. (MHz = cycles per μs). The speed of ultrasound in most tissues is near 1.5 mm/μs (1500 m/s). There are some notable exceptions. In bone the speed is between 3.5 mm/μs and 4 mm/μs, and in air the speed is 0.3 mm/μs. In cartilage the speed is 1.75 mm/μs which of course varies with composition. In fat the speed is 1.45 mm/μs (Table 21.3). In muscle, the speed of ultrasound varies by about 1.5%, depending on the angle of propagation to fiber orientation.

Although most authors focus attention on ultrasound frequency, ultrasound wavelength may be more convenient for understanding the important issues. There are three issues of importance:

- Ultrasound cannot resolve structures smaller than the wavelength.
- Scattering and attenuation [21.42] both depend on ultrasound wavelength and its relationship to the size of the scatterers.
- The focusing properties of ultrasound beam patterns and the formation of side-lobes depend on the ratio between the wavelength and the width of the transducer.

21.3.4 Amplitude of Ultrasound Echoes

Red blood cells (erythrocytes) are 0.007 mm in diameter and about 0.001 mm thick. The largest cells in the body, nerve axons, may be 1 m long, but their diameter is about 0.001 mm. A liver cell is about 0.04 mm in diameter. The smallest structures imaged by ultrasound in clinical practice are layers less than 1 mm thick, such as the media lining major arteries (the intima-media thickness, IMT) [21.43], corneal layers (with 30 MHz ultrasound) and skin surfaces. However, these structures can only be observed in the range (depth) direction because resolution is best in that direction. Lateral resolution is dependent on the numeric aperture of the ultrasound beam pattern and the relationship between the object of interest and the focus as well as the ultrasound wavelength. Only the ultrasound wavelength affects depth resolution. The result is that ultrasound images are anisotropic, with much better resolution in the depth direction than in the lateral direction. The image is delivered to the examiner as a pictorial map in two dimensions. In our optical experience, resolution is equal in the two dimensions. Therefore, we expect that, if we can see detail in one dimension, we can also see the detail in the other, even when we cannot. This gives the impression that ultrasound has higher resolving power in the lateral direction than it in fact does. This is particularly important in measurements such as the IMT and measuring the thickness of the atherosclerotic cap. Such structures can only be visualized when the ultrasound beam is oriented perpendicular to the structure.

To identify tissues and organs, the lateral dimensions must by at least 10 wavelengths wide. This

Table 21.3 Ultrasound wavelength versus tissue thickness. To resolve the thickness of these structures, the ultrasound frequency must be much higher than listed in this table. Much lower-frequency ultrasound passes through with little interaction. For particles or cells, lower-frequency ultrasound is subject to Rayleigh scattering

Tissue	Property	Dimension (mm)	Ultrasound Speed (mm/µs)	Frequency (MHz)
Trachea	Lumen diameter	25	0.32	0.013
Aorta	Lumen diameter	30	1.57	0.063
Fat layer	Subcutaneous	20	1.45	0.080
Myocardium	Thickness	10	1.6	0.16
Skull	Thickness	10	4	0.4
Brain	Cortex	3	1.5	0.5
Artery	Wall	0.8	1.6	2
Trachea	Wall	0.6	1.75	2.9
Bone	Cortex	1	4	4
Eye	Cornea	0.6	1.64	2.73
Skin	Dermis	0.3	1.73	5.76
Fat	Cell	0.1	1.45	14.5
Brain	Cell, large	0.1	1.5	15
Skin	Epidermis	0.05	1.73	35
Erythrocyte	Cell	0.007	1.57	225
Bacteria	Cell, large	0.005	1.5	300
Brain	Cell, small	0.004	1.5	375

Notes:
1. Trachea lumen: for phonation below 13 kHz, the trachea diameter is smaller than the wavelength so that the trachea acts like a waveguide, bounded by cartilage with acoustic velocity and impedance higher than the luminal air.
2. Bone cortex and thin skull have thicknesses near the wavelength of transcranial ultrasound and a much higher ultrasound impedance, so the skull can act as an interference filter reflecting some wavelengths and passing others.
3. As the upper limit of most medical ultrasound systems is 15 MHz, small animal research instruments image at frequencies as high as 70 MHz, high enough to resolve the epidermis and large brain cells, but not high enough to image most cells.

number varies depending on the numeric aperture of the imaging system and on the acoustic contrast with the surrounding material. Small round structures such as tendons can be imaged using tricks, such as watching for motion, but most objects successfully imaged by ultrasound are larger. It is important to remember that it is possible to see an object that you cannot resolve. For instance, you are able see stars or lights at night, even though you cannot resolve closely spaced pairs of stars. When the signal from a target is strong, the target can be detected and located, even though a pair of targets cannot be resolved as two. If Doppler demodulation is used, rather than amplitude demodulation (used for B-mode), small vessels with

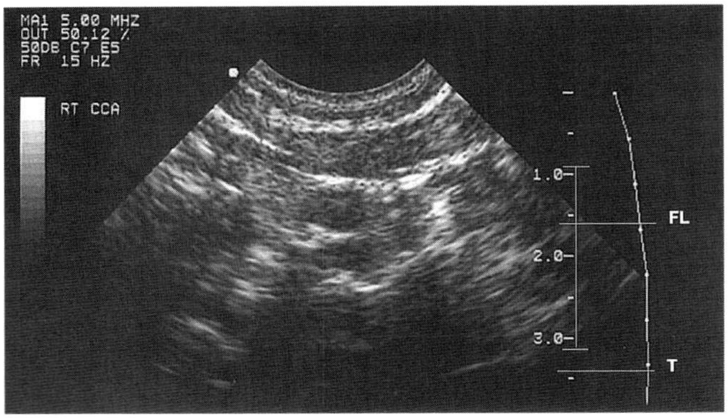

Fig. 21.10 Near-field versus far-field difference in ultrasound image. This is an image taken with a diagnostic scanner using a 5 MHz ultrasound transducer (wavelength = 0.3 mm) with a concavity radius of 16 mm and a transducer diameter of 6.4 mm. The computed *transition zone* is $d^2/4\lambda = 34$ mm. The *focus* must be less than the 34 mm distance to the transition zone to be effective. With a fixed-focus fixed-aperture ultrasound transducer, the transition from the near field to the far field is marked by lateral spreading of the speckle in the far field. There is no effect on the range (depth) speckle dimension

blood flow can be resolved if the signals are strong enough.

Ultrasound imaging is a coherent process that generates speckle in the image. This is often called *tissue texture*, but the spacing of the speckle is about equal to the burst length of the ultrasound pulse, which is up to several wavelengths, and this is much greater than the fine cellular structure of the tissues under observation. Our visual processing is so good that we integrate prior knowledge, and outline shape seamlessly into our recognition process, while believing that we are recognizing texture.

Ultrasound imaging and aeronautical radar imaging are electronically identical. However, in aeronautics, the space between aircraft is very large compared to the size of the aircraft and to the wavelength of the radar wave. In medical ultrasound imaging, the cells are tight packed together, and the cell size ($10\,\mu$m) is smaller than the wavelength of ultrasound ($500\,\mu$m) so that the scattering is complex. In addition, ultrasound medical imaging systems often generate images from the complex near field (Fresnel zone) of the ultrasound transducer (Fig. 21.10). This becomes particularly confusing when tissues are moving laterally through an array of ultrasound beams. Because of the Moire effect of this system, tissue motions from left to right can cause speckle patterns to move right to left.

21.3.5 Fresnel Zone (Near Field), Transition Zone, and Fraunhofer Zone (Far Field)

The transition between the Fresnel and Fraunhofer zones can be computed from geometric considerations (Fig. 21.11). As in the computation of the basic wave equations, we assume that ultrasound propagation is a linear process and that superposition holds (pressure fluctuations and molecular velocity fluctuations can be added). Although some authors use Huygens' principle [21.44], dividing the transducer into tiny segments and adding the contribution of each to the wave, the computation is simplified if reciprocity is used (the transmitted ultrasound beam pattern and the received ultrasound beam pattern are the same), and the wave is assumed to be emitted from a test point in the ultrasound beam and received by the transducer.

A wave propagating from a stationary point will expand as a series of concentric spherical surfaces, spaced by the ultrasound wavelength with the source at the center. If the source is close to a planar transducer face, then at every time point some compressional spherical regions and some rarefactional spherical regions will simultaneously be present on the transducer face plane. This provides a mixed signal to the transducer, the compressions and rarefactions destructively inter-

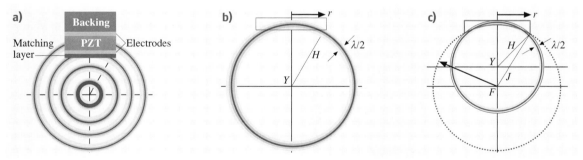

Fig. 21.11a–c Geometry of transition zone between near field (Fresnel zone) and far field (Fraunhofer zone) (**a**) A transducer consists of a piezoelectric element (PZT) with electrodes on the face with an impedance matching layer and on the back with a backing material. For computation, only the wave intersection with the matching layer face needs to be considered. (**b**) By geometry, the distance to the transition zone (Ψ) between the Fresnel zone and the Fraunhofer zone for a flat transducer can be estimated. The transition zone is located at the furthest distance from the transducer that the origin of a compression half-cycle of a wave can completely envelope the transducer face. r = transducer radius, Ψ = distance to transition zone, $\lambda/2$ = half wavelength, H = hypotenuse of transition radius. $H^2 = \Psi^2 + r^2$, $H - \Psi = \lambda/2$. If $\lambda/4 \ll \Psi$, then $\Psi = r^2/\lambda = d^2/4\lambda$, the accepted formula for the transition zone. A wave originating from a closer location (Fresnel zone), will always produce a combination of compressions and decompressions at the transducer surface. (**c**) Drawings can be made for focused transducers whether concave, lens or phased focus array, but the algebra is more complex. The equivalent concave curvature is shown by the *dotted circle*. F = transducer focal radius, J = hypotenuse of focal radius

fering. There are some locations of the source point that result in equal areas of compression and decompression of the transducer surface at all times. These are locations where the transducer has no sensitivity, and are called null points. By reciprocity, if the transducer emitted a wave, a null point will have no sound intensity from the transducer. Ultrasound transducer beam plots often show these null locations along the axis of the ultrasound beam. There may be several of these null points along the beam axis, and at other points in the beam pattern. The *near field* or *Fresnel zone* is the region of the ultrasound beam pattern where these null points are present. In the Fresnel zone, there is always some destructive interference of the ultrasound wave. The Fresnel zone is the region where beam focusing can be achieved.

When the source point is at a greater distance from the transducer plane, these regions of compression and rarefaction become nearly planar. There are times when the planar transducer surface can be completely contained in either a compression zone or a decompression zone. At this range, at least near the axis of the ultrasound beam pattern, there are no regions of destructive interference. This is the *far field* or *Fraunhofer zone*.

There is a transition between the Fresnel and Fraunhofer zones along the ultrasound beam axis, where it is first possible for the ultrasound transducer to be just barely completely contained in a compression half-wave region at some time within the ultrasound wave period. Half a period later, the transducer will be completely contained in a decompression half wave. This location is easily diagrammed geometrically (Fig. 21.11b) and the distance to the transition zone computed. Although the geometry for a similar computation for a concave transducer can be easily drawn (Fig. 21.11c), the algebraic solution for the focused transition, which is now retracted toward the transducer, is more difficult.

In addition to using a concave transducer, focusing of the ultrasound beam pattern can also be achieved by the use of a lens on the front of the transducer (Fig. 21.12a). The lens is often made of silicone rubber because of the low ultrasound propagation speed. Currently, most ultrasound scan heads consist of a linear array of 128 transducer elements; each element is 2 cm long and 0.5 mm wide arranged in a side-by-side row with a footprint of 2 cm by 6.4 cm. Focus is pro-

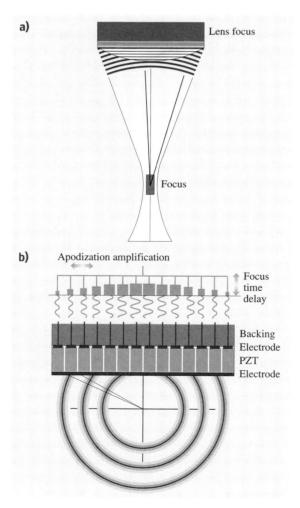

Fig. 21.12a,b Mechanical lens focusing and electronic delay focussing. (**a**) A low acoustic velocity in the lens allows the ultrasound traveling along the shorter central path to be delayed so that the ultrasound along the longer marginal paths can catch up, aligning the compressions and decompressions to provide the highest voltage excursions in the piezoelectric transducer element for ultrasound waves originating from the focal region. (**b**) Electronic delay lines in the signal path after the piezoelectric elements provide the same focusing function as the lens. Apodization *shading* of the sensitivity of the transducers at the edge of the aperture suppresses aperture diffraction side-lobes in the beam. The mechanical focus depth is determined at the time that the ultrasound transducer is manufactured; it cannot be altered by the examiner. The electronic focus depth can be adjusted to a different depth in a microsecond, the time that ultrasound travels 1.5 mm, so that the electronic focus can be **adj**usted to each depth during the time that the ultrasound is traversing the tissue

vided in the image thickness direction by a cylindrical silicone rubber lens and in the lateral image direction by adjusting the relative phase of the adjacent transducers (Fig. 21.12b).

21.3.6 Measurement of Ultrasound Wavelength

Modern diagnostic ultrasound systems use sophisticated transducers and filters to improve resolution by shifting and broadening the ultrasound frequency band used for imaging, Doppler and other methods. It is possible, however, to use basic wave physics to measure the effective ultrasound wavelength of the system. In addition to refraction, ultrasound waves are also subject to diffraction. By placing a diffraction grating in the ultrasound image path in water, and imaging a point reflector, the diffraction pattern of the wave can be imaged and the wavelengths computed (Figs. 21.13, 21.14).

21.3.7 Attenuation of Ultrasound

As an ultrasound passes through tissue as a pulse or as a *continuous wave* along the beam pattern, some of the ultrasound energy is converted to heat in the tissue and some is reflected or scattered out of the beam pattern. In addition, nonlinear propagation converts some of the energy to higher harmonics ($2f$, $3f$ ...) of the fundamental frequency (f) of the wave. Attenuation increases with frequency, and the harmonic frequencies also may be outside the bandwidth of the detection system. These factors contribute both to actual and apparent attenuation of the ultrasound energy in the pulse and the average power passing through a cross section of the ultrasound beam as a function of depth. Changes in intensity are related to these changes in power, but the intensity is power divided by the cross-sectional area of the beam, so changes in beam cross-sectional area cause intensity changes that are not related to attenuation. Focusing the beam pattern can decrease the area, increasing intensity; diffraction and refraction increase the area, decreasing intensity.

Ultrasound attenuation is usually expressed as a ratio of the output power divided by the input power. In linear systems, this ratio is constant for all ultrasound intensities. In nonlinear systems, the relationship is more complicated. The attenuation rate of ultrasound in tissue

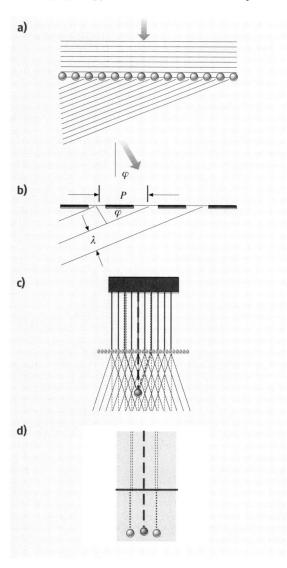

Fig. 21.13a–d Grating diffraction. Like any wave, an ultrasound beam can be tilted by a diffraction grating. The angle of tilt can be computed using Huygens' principle. If the pitch spacing of the grating is P, and the wavelength is λ, then the angle of deflection of the wave is expressed as $\sin\varphi = P/\lambda$. An ultrasound transducer array can act as a diffraction grating. This should be avoided. A diffraction grating in a water tank can be used to measure the wavelength of ultrasound. An image of a reflector deeper than the grating appears three times in the image, once at the true location and once for each of the diffraction orders. (**a**) Plane wave diffracted by grating. (**b**) Computaion of diffraction angle. (**c**) Ultrasound scanhead, diffraction grating and target. (**d**) Expected ultrasound image for a single frequency transducer

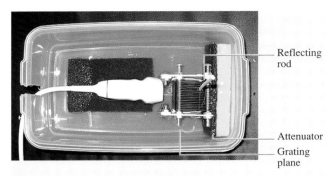

$\lambda = (1.30/1.74) \times (25.4\,\text{mm}/32) = 0.593\,\text{mm}$
$F = (1.5\,\text{mm}/\mu\text{s})/(0.593\,\text{mm}) = 2.53\,\text{MHz}$

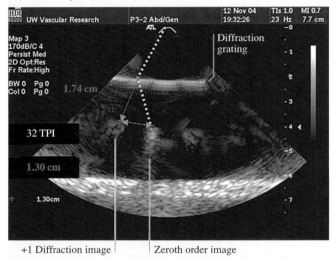

Fig. 21.14 Measuring ultrasound frequency and bandwidth by viewing a reflecting rod through a diffraction grating. The cross section of a rod forms the zeroth-order spot in the image. In addition to passing the ultrasound beam pattern directly (the zeroth image) the diffraction grating forms a +1 branch of the beam pattern (shown in *yellow*), which makes an image of the rod to the left (+1 diffraction image) and a symmetrical −1 image on the right. By measuring the geometric sine of the deflection angle, (1.30/1.74), the ultrasound wavelength can be computed using the spacing of the grating. The grating is made of monofilament fishing line wound on a frame made of 10–32 threaded screws (32 threads per inch, pitch spacing = 0.8 mm)

can be computed using base 10 (dB/cm, for engineers), base "e" (Np/cm, for scientists), compound percentage (% per cm, for physical therapists) or base 2 (half-value layers, for systems designers). The decibel, derived from the bel can be traced to Alexander Graham Bell, and the neper to John Napier. The bel is a factor of 10

in power, the neper is a factor of e (2.718) in amplitude. Ultrasound attenuation in soft tissue is between 3 dB/(MHz cm) and 0.3 dB/(MHz cm). Using intensity as the dependent variable in attenuation computations is complicated by changes in beam area. Some authors apply the bel to intensity. In acoustics the pressure amplitude of the wave, which is proportional to the square root of intensity (assuming constant impedance) is often used. The base 2 system is used to identify the thickness of tissue that attenuates the power by half, the intensity by half or the amplitude by half. The thickness is called the *half-value layer*, often without qualifying whether it is *half power*, *half intensity* or *half amplitude*. Intensity is proportional to the square of the amplitude, which is a multiplier of 2 in logarithms. If the cross-sectional area of the ultrasound beam is constant

$$\begin{aligned}0.7\,\text{nepers} &= 6\,\text{dB}\\ &= 1\,\text{half-amplitude layer}\\ &= 2\,\text{half-power layers}\,. \end{aligned} \qquad (21.10)$$

Often the attenuation rate is nearly proportional to frequency, and frequency is inversely proportional to wavelength, so the attenuation can be most easily expressed in wavelength (Table 21.4). The use of half-value layers has the advantage that it can be related directly to the number of bits in a digitizer, simplifying computation.

Because of the high instantaneous intensities used in pulse-echo diagnostic ultrasound, all of the ultrasound propagation is nonlinear. The commonly used linear equations do not apply. The beam patterns are

Table 21.4 Ultrasound attenuation

Tissue type	Half amplitude (quarter power) thickness in wavelengths	60 dB pulse-echo attenuation depth in wavelengths (10 bits dynamic range)	Attenuation dB/(cm MHz)
Bone	1	5	30
Cartilage	8	40	10
Tendon	8	40	10
Skin	25	125	3
Muscle	25	125	3
Fat	80	400	1
Liver	80	400	1
Blood	300	1500	0.25

not as predicted, because the wave propagation speed in the higher-pressure regions is faster than in the lower-pressure regions. The waves convert from sinusoidal to sawtooth, introducing a series of harmonic frequencies into the wave. The change in frequency (wavelength) affects the ultrasound beam pattern. The conversion of ultrasound energy into harmonic frequencies also increases the attenuation because the attenuation increases with frequency.

A significant development was the use of tissue-generated harmonic signals for diagnostic imaging. The utility of these harmonics was discovered by accident when investigators were searching for better methods of detecting ultrasound contrast agents. Ultrasound contrast agents consist of microbubbles that have natural oscillation frequencies near diagnostic frequencies. The bubbles therefore oscillate in response to incident ultrasound. The bubble oscillation stores some energy, which is reradiated at the original ultrasound frequency as well as at harmonics of that frequency. Older ultrasound systems were, by nature, insensitive to the second harmonic of the natural transducer frequency. By using a transducer with an increased bandwidth (3 MHz center frequency transducer manufactured with the quality factor reduced spreading the bandwidth to 1.9–4.1 MHz) then lowering the transmit frequency (2 MHz) and raising the receive frequency (4 MHz), a system sensitive to the harmonic emissions from the contrast bubbles can be achieved. Surprisingly, harmonic echoes are generated by tissues even when contrast bubbles are not present.

21.4 Methods of Medical Ultrasound Examination

Over the half century of ultrasonic medical examination, a variety of examination methods have been tried, discarded, with some to be resurrected and honed for use.

A passive method was evaluated to detect local ultrasound emissions above background due to local temperature elevations of cancerous tumors. The emissions of about 10^{-13} W/cm^2 (1 nW/m^2) are present in all warm tissues and provide a minimum intensity for the operation of active ultrasound systems. Transmission ultrasound systems have been developed using an ultrasound transmitter on one side of a body part and a receiver on the other side to detect differences in attenuation (like X-ray imaging). These systems determine ultrasound attenuation and ultrasound speed through the intervening tissues. One version of this system used an acousto-optical holographic system to form the image. Several systems have been explored to insonify tissue from one direction and gather the scattered ultrasound from another direction. Such systems usually require an array of receiving transducers.

The only ultrasound systems that have gained wide acceptance in clinical use are: (1) the continuous-wave (CW) backscatter Doppler system, and (2) common-axis pulse-echo backscatter imaging and Doppler systems. The reason for the failure of the other systems is due to two features of medical ultrasound examination: refraction of ultrasound in tissue and the cost of receiving and processing data from array receivers. The CW methods are used only for Doppler measurements. The first of these systems was constructed before 1960 [21.45, 46]; their application in medical diagnosis was well established by 1970 [21.47–52]. Pulse-echo methods have been under continuous development since 1950. The majority of medical diagnostic imaging examinations in the world are performed with pulse-echo instruments.

21.4.1 Continuous-Wave Doppler Systems

Continuous-wave Doppler was used in one of the earliest ultrasound systems to be commercialized. Developed

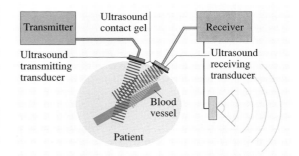

Fig. 21.15 Nondirectional continuous-wave Doppler. In a typical CW 5 MHz nondirectional Doppler, the incident ultrasound beam is scattered by stationary tissue forming a 5 MHz echo and by moving blood forming a 5.0004 MHz echo. The two echoes arrive at the receiving transducer simultaneously. By constructive and destructive interference between these waves, the received 5 MHz signal is amplitude modulated by the 5.0004 MHz echo. The frequency of the amplitude modulation is the difference frequency 5.0004 − 5 MHz or 400 Hz, which is amplitude demodulated, providing a 400 Hz audio signal to the speaker

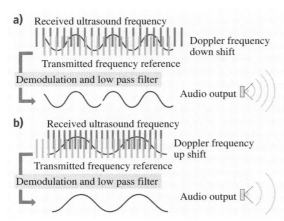

Fig. 21.16a,b Interference between echoes to produce audio frequencies. Constructive interference between the Doppler frequency-shifted echo and the original frequency wave increases the resultant amplitude; destructive interference reduces the amplitude. This occurs whether the Doppler shift is downward (**a**) due to reflections from blood cells receding from the transducer, or is upward (**b**) due to reflections from blood cells advancing toward the transducer. By amplitude demodulating the result, an audio frequency signal results. Amplitude demodulation can be done by multiplying the two ultrasound frequency signals together and low-pass filtering the result or by squaring (rectifying) the sum of the two ultrasound signals and low-pass filtering the result

independently in Japan in 1957 (Satamura and Koneko), and in the US in 1959 (Baker et al.), a system can be constructed from an RF transmitter, transmitting piezoelectric transducer, receiving piezoelectric transducer and amplitude demodulating receiver (Fig. 21.15). In the region of tissue where the transmitting beam pattern overlaps with the receiving beam pattern, the reflected wave from moving tissue will combine to interfere with the reflected wave from stationary tissue. The reflected wave from moving tissue has a Doppler shift. The Doppler-shifted frequency will interfere with the reflected wave from stationary tissue to cause an amplitude modulation in the combined wave. The frequency of that modulation is within the hearing range.

Ultrasonic Doppler examination produced an audible frequency by chance. The best ultrasound frequency to reflect by Rayleigh scattering from blood at a depth of 2 cm is 5 MHz. This frequency is shifted by the ratio of twice the blood speed (2×0.75 m/s $= 1.5$ m/s) to the ultrasound speed 1500 m/s ($1.5/1500 = 0.001$) to cause a downshift in the ultrasound frequency of 5 kHz. The interference between the unshifted and shifted waves causes a 5 kHz modulation in the amplitude of the 5 MHz wave. This modulation is converted to sound by common amplitude modulation (AM) radio circuitry.

The introduction of the transistor in 1960 allowed this simple circuit to be produced in a palm-size package so that convenient clinical examination could be done. Whether the blood is approaching the transducer or receding from the transducer (Fig. 21.16), the audio signal is the same, with audio frequency proportional to the blood velocity.

These CW Doppler systems allowed the detection of narrow arterial stenosis by the high velocities present in the stenotic lumen. Normal velocities are below 125 cm/s, causing a Doppler shift of 4 kHz if the Doppler angle is 60°. The blood velocity in a stenosis is limited by the blood pressure via the Bernoulli effect; the velocity can be no greater than 500 cm/s for a systolic blood pressure of 100 mmHg. A blood velocity of 500 cm/s causes a Doppler shift of 16 kHz. These CW Doppler systems also allow the detection of venous obstruction by the loss of variations in venous velocity synchronized with respiration, and the testing of venous valve incompetence by the detection of reverse flow resulting from distal compression release or sudden proximal compression. An examiner with one year of

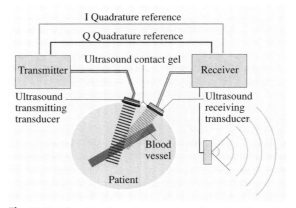

Fig. 21.17 Directional continuous-wave Doppler. By using a pair of reference waves at the transmitted ultrasound frequency, but phase shifted with respect to each other, the receiver Doppler demodulation can be performed against each of the reference waves. This allows the instrument to differentiate Doppler shifted echoes with increased frequency due to blood cells approaching the transducer from those with decreased frequency due to blood cells receding from the transducer

Fig. 21.18a,b Quadrature Doppler demodulation. (**a**) Demodulation of the of a Doppler down shifted echo against an-I reference and Q reference produces a quarter-wave advanced signal (shifted to the right) on the I product compared to the Q product. (**b**) Demodulation of the of a Doppler upshifted echo against an I reference and Q reference produces a quarter wave delayed signal (shifted to the left) on the I product compared to the Q product

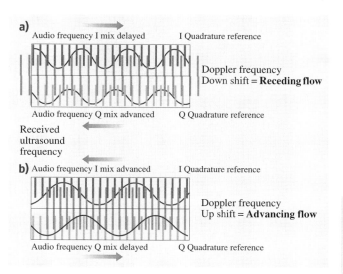

training and experience can diagnose all superficial arterial obstructions and venous obstructions and reflux with accuracies exceeding 95% against any gold standard.

The addition of directional detection by Fran McLeod in 1968 simplified the separation of mixed arterial and venous flow, and allowed more-objective evaluation of velocities (Fig. 21.17). Directional detection was accomplished by providing a pair of quadrature signals derived from the transmitted frequency directly to the demodulator in the receiver. Quadrature signals differ only in phase by a quarter cycle. When the Doppler-shifted ultrasound echo is demodulated against each of the quadrature reference signals, the pair of difference waves produced are phase shifted by 90° (Fig. 21.18). If the reflecting blood is advancing toward the transducer then audio frequency (AF) I wave is advanced compared to the AF Q wave, if the reflecting blood is receding, then the AF I wave is delayed with respect to the Q wave. By further delaying the AF Q wave by a quarter cycle (the shift time is dependent on frequency), and adding or subtracting the resultant AF I and AF Q delayed, the sound channels can be separated (Fig. 21.19). After separation, the Doppler signal representing advancing flow can be delivered to one ear and the Doppler signal representing receding flow can be delivered to the other ear. Although these continuous-wave Doppler systems are simple and have not changed substantially since 1970, they still have tremendous clinical utility and are used by angiologists and vascular surgeons in daily practice. A CW Doppler transducer is provided with modern echo-cardiology systems (Fig. 21.20) for the

Fig. 21.19a,b Audio direction separation. By retarding the I quadrature channel by 1/4 of the period of the audio frequency, and then adding or subtracting the resultant pairs, the directions of flow can be separated. (**a**) The added pair results in sound for only the downshifted Doppler signals representing receding flow. (**b**) The subtracted pair results in sound for only the upshifted Doppler signals representing advancing flow

measurement of cardiac output and for accurate velocity measurements of blood flow jets associated with valve disease.

During 1970–1985, several arterial mapping angiographs (Fig. 21.21) were developed to associate high-velocity Doppler signals, indicating stenoses with arterial landmarks such as bifurcations. These angiographs used mechanical or other methods to track the location of the ultrasound beam and display the results on an image that could be compared with an X-ray angiogram. Modern, large-field-of-view three-dimensional ultrasound imaging systems use extensions of this method.

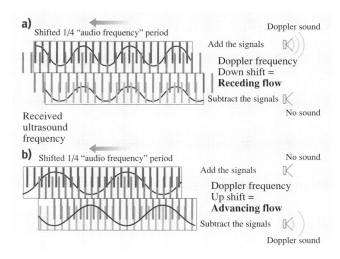

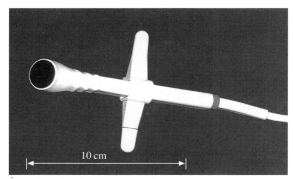

Fig. 21.20 Dual-element continuous-wave Doppler transducer for cardiac output measurements. This transducer is designed to fit in the supersternal notch between the clavicles at the base of the neck. From this location, the Doppler ultrasound beam patterns can be nearly aligned with the axes of the cardiac valves. The ultrasound avoids the lungs by passing directly down through the mediastinum

Table 21.5 Ultrasound examination parameters

Maximum Examination depth (cm)	Ultrasound wavelength (mm)	Ultrasound frequency (MHz)	Pulse repetition Period (μs)	Frequency (kHz)
3	0.075	20	40	25
5	0.125	12	67	15
6	0.15	10	80	12.5
10	0.250	6	133	7.5
12	0.3	5	160	6.25
15	0.375	4	200	5
20	0.5	3	266	3.75

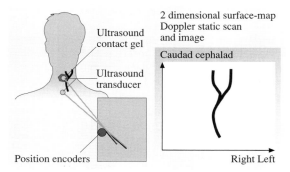

Fig. 21.21 Doppler arteriography. By monitoring the position of a Doppler ultrasound transducer, transferring the position on a storage image screen, and marking the display screen only when the Doppler detects blood flow, an image of the flow can be generated. Most of these systems used continuous-wave Doppler; the Hokanson system used pulsed Doppler

21.4.2 Pulse–Echo Backscatter Systems

Pulse-echo ultrasound consists of sending a pulse of ultrasound into tissue and receiving the echoes from various depths. The length of the transmitted pulse may be as short as 0.5 μs (for B-mode imaging) or as long as 5 μs (for trans-cranial Doppler). Usually the same transducer or transducer aperture is used for both transmitting and receiving. The depth of each reflector is determined by the time of flight between the transmitted pulse and the received echo: a reflector 3 cm deep returns an echo at 40 μs; a reflector 15 cm deep returns an echo at 200 μs. If the system is examining to a depth of 18 cm, then a new pulse can be transmitted after 240 μs. In that case, an echo returning at 40 μs after the later transmit pulse might come from 21 cm deep (280 μs) rather than 3 cm. Tissue attenuation will suppress the echo from deep tissue by about 54 dB (3 MHz ultrasound) compared to the echo from 3 cm.

To perform a pulse-echo examination, an ultrasound transducer (or array of transducers) is positioned to form a beam pattern along an axis, which is assumed to penetrate tissue in a single straight line. A short ultrasound burst is launched along that line. After allowing for the round-trip travel time to the depth of interest (Table 21.5), the ultrasound echo from a segment of tissue in the beam pattern is received, demodulated and prepared for display (Fig. 21.22). The ultrasound transmit pulse can be adjusted for frequency, duration (Fig. 21.23), energy, and aperture size and shape to fine-tune the examination if needed. The ultrasound echo can be demodulated to yield the amplitude (Fig. 21.24) and/or the phase. The results of multiple pulse-echo cycles along a single beam pattern can be combined. The echo is usually amplified with a fast-slew programmed amplifier to increase the gain of later echoes coming from deeper tissues, because they will be attenuated more than shallow echoes. If tissue visualization is desired, the echo strength is displayed in an image along a line representing the direction of the ultrasound beam, with the location along the line (representing depth) determined by the time after transmission that the echo returns. If a measurement of tissue motion is desired, then additional identical pulses are transmitted along the same beam pattern to obtain a series of nearly identical echoes from each depth of interest. These echoes are examined for changes in phase, which is the most sen-

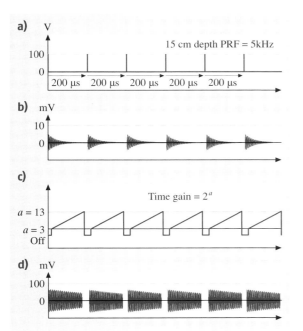

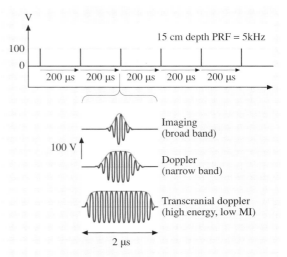

Fig. 21.22a–d Time gain compensation (TGC). (a) A new ultrasound pulse is transmitted each pulse repetition period (200 μs in this case). (b) Echoes from shallow and deep depths return at known times because of the speed of ultrasound and the expected round-trip travel time. The later echoes received from deeper tissues are weaker than the echoes received from shallow tissues because the echoes from deeper tissues have passed through more-attenuating tissue. (c) An amplifier programmed to increase the gain for deep echoes more than for shallow echoes is used to compensate for attenuation. This depth gain compensation (DGC) or time gain compensation (TGC) amplifier is under operator control. (d) The echoes, after TGC are ready for demodulation by analog or digital methods

sitive indicator of displacement of the tissue reflecting the ultrasound.

This simple model places constraints on the ultrasound examination. First, at any depth, the strongest echoes from a bone or air interface perpendicular to the beam are 10^6 (60 dB) stronger than the weakest echoes from cells in liquid (blood). So, a 10 bit digital echo-processing system is the minimum to handle the range of echo strengths expected in an image. Second, because the speed of ultrasound is about 1.5 mm/μs in most tissues, echoes returning from 15 cm deep arrive 200 μs after ultrasound pulse transmission. A new pulse cannot be sent into the tissue along the same line or along

Fig. 21.23 Different ultrasound transmit bursts for imaging and Doppler applications. Short broadband transmit bursts are used for B-mode imaging; long narrow-band transmit bursts are used for Doppler applications. A longer transmit burst contains more energy if the maximum acoustic negative pressure (mechanical index) are limiting factors. A short transmit burst allows better depth resolution and also spreads the diffraction side-lobes of the beam pattern, suppressing lateral ambiguity. A longer transmit burst is preferable for Doppler applications where echoes from blood are weak and coherence length is important for demodulation

another line until all of the echoes from the first pulse have returned. So, the maximum pulse repetition frequency (PRF) for an examination to a depth of 15 cm is 5 kHz (1/200 μs). Ultrasound energy is attenuated to 1/4 of the incident energy in most body tissues after passing through a depth of 80 wavelengths. So after passing through a depth of 400 wavelengths and returning through 400 wavelengths, the echo strength is only 0.25^{10} or 10^{-6} of the strength of a similar echo from the shallowest tissue. To compensate for this attenuation, an amplifier must amplify the deep echoes 60 dB more than the shallow echoes (Fig. 21.22). An alternative is to digitize the echoes with a 20 bit digitizer, 10 bits for the dynamic range of the echoes from different tissues and 10 bits to account for attenuation. In either case, the maximum depth of an examination is limited to 400 wavelengths of ultrasound. An ultrasound depth resolution cell (a pixel) is equal to about twice the wavelength of ultrasound. So, an ultrasound image can contain 200 pixels in the depth direction (Table 21.5). These con-

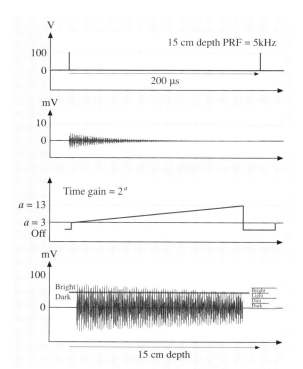

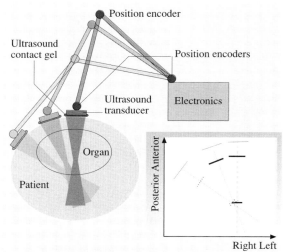

Fig. 21.24 Preparing the amplitude demodulated brightness mode echo for display. The amplitude of the amplified echo can be converted into brightness values by a digitizer. In early B-mode scanners, the display could show bright (echo) or dark (no echo) based on a one bit threshold detector (*dark line*). A two-bit digitizer allows four levels of gray scale (scale on *right*), making the result less sensitive to the setting of the TGC amplifier. Modern ultrasound B-mode imaging systems use an eight-bit digitizer to display 256 levels of gray. Depending on the method of presentation, the eye can distinguish no more than 64 stepped levels of gray or 16 isolated levels of gray

Fig. 21.25 B-mode static scanner. The traditional B-scanner restricted the transducer position and direction to movements in one plane that would form a cross-sectional image through the patient. The position of the transducer was located on the image display, and a line projected from the origin along the display in the direction of sound beam propagation. During the scan a marker was moved along the ultrasound beam direction at a speed of $0.75\,\text{mm}/\mu\text{s}$, representing the time that an echo would return after a round trip to each depth when traveling at $1.5\,\text{mm}/\mu\text{s}$. If a strong echo was detected, then a mark was stored on the screen to contribute to the image. The transducer was scanned across the skin to form an ultrasound scan cross-sectional image of the tissue

siderations provide a practical relationship between the ultrasound frequency and the PRF, based on the attenuation of tissue. The PRF is 1/800 of the ultrasound frequency in most applications.

21.4.3 B-mode Imaging Instruments

A series of B-mode (brightness mode) ultrasound imaging methods have been developed over the last 40 years. The methods have used the latest electronic technologies to improve the speed, convenience, capability and utility of the systems. The original imaging systems were based on radar technology and analog electronic parts. Today, ultrasound systems are personal computers using standard operating systems with specialized display software coupled to custom ultrasound circuitry on the front end. A review of the history is useful because the modern instruments keep incorporating older methods.

Two-Dimensional Static B-mode

The original B-scanners (Figs. 21.25, 21.26) consisted of a single ultrasound transducer element coupled to a mechanical system to track the origin and direction of the ultrasound beam. The systems had the advantage that large fields of view could be scanned, such as a plane through a full-term fetus, and the sonographer could return to portions of the image to *paint in* anatomical features by approaching each of the reflecting interfaces from a perpendicular direction. As the image display system showed the brightest echo for

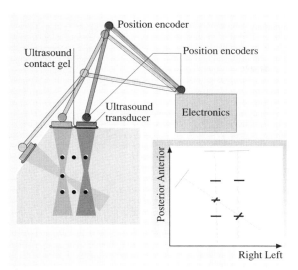

Fig. 21.26 B-mode calibration phantom. Analog electronics were not stable at the time that static B-scanners were constructed. Calibration phantoms were used to assure that a target would appear in the same location when viewed from different locations. Adjustments were provided to align the electronics

each anatomic location, *compound scanning* was used to obtain echoes from a variety of directions at each tissue location to render a useful image. Since a scan might take 5 min, movement of the patient during the scanning period was a problem, rendering the image useless. Because these systems used analog electronic

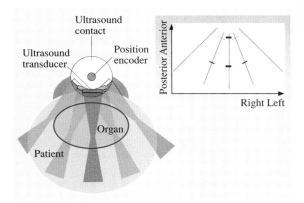

Fig. 21.27 Real-time two-dimensional ultrasound images. Motor-driven ultrasound transducers were able to rapidly tilt to a series of angles obtaining data from a series of lines in a plane of tissue to form a new two-dimensional image 10 times or 30 times per second

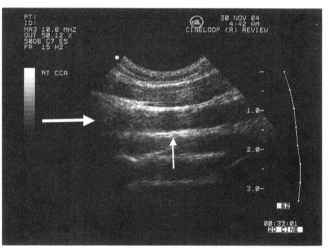

Fig. 21.28 Two-dimensional B-mode (brightness mode) common carotid artery image. The lumen of the common carotid artery in longitudinal section appears as a darker streak (*wide arrow*) bounded by the superficial and deep walls separated by 6 mm. The walls appear brighter where the ultrasound scan lines are perpendicular to the walls (*narrow arrow*) because of the specular backscatter.

devices, which were subject to deterioration over time, the stability and alignment of the instruments was important. In addition, since the same tissue was examined from multiple locations, the co-registered alignment of the tissue voxels (small volumes of tissue) on the image

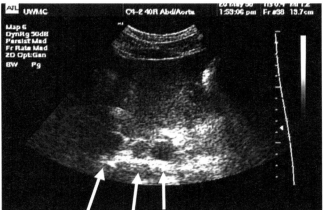

Fig. 21.29 Duplication of aorta [21.53] in the image. Refractive tissues composed of fat ($c = 1.45$ mm/μs) and muscle ($c = 1.58$ mm/μs) at the midline of the abdomen form two prisms. Each prism allows the aorta to be viewed resulting in two images of the aorta. This effect has also caused double viewing of a single pregnancy

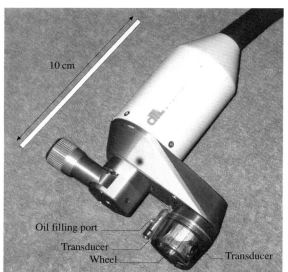

Fig. 21.30 Mechanical scan head. Three transducers are mounted around a rotating wheel encased in a cover filled with oil. Two of the 3 transducers can be seen in the image. Between the transducers is an air bubble in the oil; the air bubble must be removed through the oil filling port before scanning. Under the bubble is a rectangular shape that is the housing for a switch. Each of the transducers has a switch mounted opposite the transducer on the wheel which is activated magnetically to connect the transducer when the transducer is pointed toward the patient. The transducer wheel is driven by a motor in the white housing through a belt. A control knob on a joystick allows adjustment of the Doppler ultrasound beam, when needed; swinging the arm tilts the beam, rotating the knob advances the sample volume along the line. Electrical coupling to the transducer was accomplished by using a transformer aligned on the axle of the wheel with one coil stationary and one rotating. The magnet-activated switches mentioned above selected the proper transducer that was pointing at the patient

was critical. Regular testing of the instrument on alignment phantoms was required. The speed of ultrasound in the phantoms was controlled by the chemical composition of the phantom at the time of fabrication and the temperature of the phantom at the time of use. Of course, because the speed of ultrasound is not the same in different tissues, the success of the phantom in controlling image distortions in the depth direction during actual examination was limited. In addition, because the ultrasound beam can bend due to refraction as it passes from tissue to tissue, a factor ignored in the phantom, preventing lateral image distortions was impossible. However

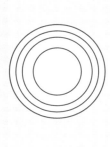

Fig. 21.31 Mechanically steered annular array scan head. The circular transducer is divided into concentric circles with each segment having the same area. By delaying by different amounts, the signals to the more central transducers, the depth of focus can be adjusted. A 10 ns delay between the transducer rings with the central transducer operating last will retract the focus a centimeter. The oil between the transducer and the dome shaped cap provides refractive focus in addition to the focus provided by the delays to the transducer. To point the ultrasound beam along different lines, the transducer wobbled back and forth rather than rotated. A rotating motion was not possible because the four coaxial cables could not cross the rotating axis. The arrow is pointing at the transducer along its axis

the phantoms did serve to prevent huge errors. Modern real-time ultrasound scanners that complete a scan in 30 ms (Fig. 21.27) form each image without overlapping beams so object co-registration is not tested. Thus, the value of phantoms has diminished. They still retain educational value.

Two-Dimensional Real-Time B-mode Mechanical Scanning

The earliest real-time systems scanned the ultrasound beam by tilting the ultrasound transducer, tracking the position and direction of the ultrasound beam with timing or an optical encoding system. The images had tremendous utility (Fig. 21.28) in spite of the poor lateral resolution compared to the depth resolution. Unfortunately, neither phantoms nor digital tracking of the ultrasound beam direction avoid image distortions due to refraction of the ultrasound beam in tissue (Fig. 21.29). These distortions, which are still a problem in the lateral direction, are ignored.

The ultrasound scan heads were convenient, ergonomic and compact (Fig. 21.30), located at the end of a cable attached to the ultrasound scanner. A variety

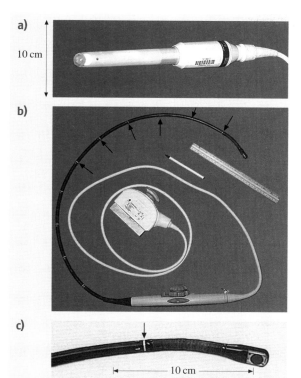

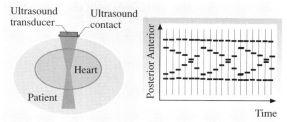

Fig. 21.33 M-mode (motion-mode) examination. For the study of cardiovascular wall motion, a single ultrasound beam is directed through the heart or other tissue. A *horizontal mark* showing the depth of each reflecting tissue in the vertical direction is shown on each vertical image line. By transmitting a new ultrasound pulse every millisecond, the echoes from tissue interfaces can be plotted for each pulse with time on the *horizontal axis*. By placing the lines side by side, the trajectories of tissue motion along the ultrasound beam are displayed. Speed of the reflecting tissue interface along the direction of the ultrasound beam is shown as a tilt or slope of an interface displayed on the image

Fig. 21.32a–c Intracavity scan heads. (**a**) Intracavity *wobbler* scan head for transvaginal or transrectal examination. To provide acoustic coupling between the transducer and the housing, the scan head is filled with liquid. The transducer, at the left end of the image, is pointing out of the page. (**b**) Trans-esophageal (TEE) ultrasound scan head. The patient swallows the end with the transducer array (*upper right* in the picture) mounted in the end of the flexible esophageal probe. Marks every 10 cm (*arrows*) allow the cardiologist to monitor the location of the transducer array from the patient's teeth. (**c**) The trans-esophageal ultrasound transducer 5 MHz linear phased array has a circular boarder so that the array can be rotated to rotate the scan plane once the transducer is in the esophagus. A pair of control handles ((**b**) at *bottom*) allows the cardiologist to rotate the transducer array and to tilt the tip housing the array to examine the heart in different planes. Some patients tolerate this procedure well with little anesthesia, others require more sedation

of scan heads were developed including some with electronic focusing (Fig. 21.31) and specialized shapes for placement in the esophagus, rectum, vagina (Fig. 21.32), and some 1 mm in diameter for insertion inside blood vessels.

M-mode Ultrasound

Among the early applications of ultrasound was the study of the motions of the heart using M-mode (motion mode) imaging [21.54, 55] (Fig. 21.33). By holding the ultrasound beam stationary and plotting the image width as a function of time, motions of the heart and other vascular structures could be documented. The velocity of a valve could be determined (Fig. 21.34) by measuring the slope of the trajectory on the M-mode tracing.

Two-Dimensional Real-Time B-mode Electronic Scanning

An alternative to moving the ultrasound transducer to point in different directions, is to have an array of transducers [21.56]. By switching to each transducer in turn to acquire echoes from the corresponding beam pattern, a two-dimensional (2-D) image is formed (Fig. 21.35). Several ultrasound scanners developed between 1970 and 1980 used low-density arrays like this. A scanhead with more transducer elements, spaced closely together (Fig. 21.36), could use groups of transducers to form an aperture. These high-density arrays permit phase focusing to a series of depths (Figs. 21.36–21.38), phase tilting of the ultrasound beam patterns (Fig. 21.39) and apodization of the aperture to smooth the ultrasound beam pattern (Fig. 21.40). The transducer array is usually formed from a single rectangular piezoelectric crystal, cut into a row of thin rectangles to form

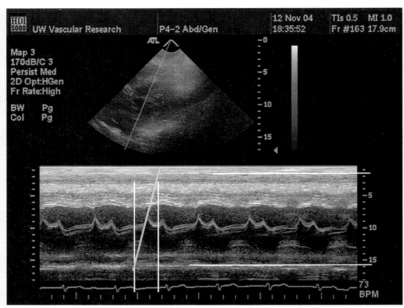

Fig. 21.34 M-mode (motion mode). Motion of a tissue interface along the ultrasound beam line is shown as depth (*vertical*) versus time (*horizontal*). The speed of the heart structure can be computed from the slope of the tracing. In this example, the speed is $158 - 16\,\text{mm}/0.4\,\text{s} = 355\,\text{mm/s}$. The green line across the bottom of the image is the Electrocardiograph (ECG, EKG) timing signal

an array in one direction. Because of the cuts, the array becomes flexible in the long dimension and can be curved (Fig. 21.41) to form a sector scan. Each of the 128 elements is connected via an electrical impedance-matching network to a coaxial cable, which forms part of the scan-head cable running back to the scanner. The plug connecting the 128 coaxial connectors to the ultrasound scanner has separate contacts for all 128 cables. Behind the plug in the scanner are separate ultrasound transmitters and receivers for each transducer element. Often each of the receivers will have a separate digitizer to convert the TGC-amplified signal into digital form.

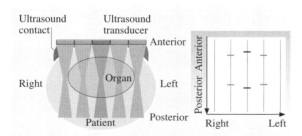

Fig. 21.35 Low-density linear-array two-dimensional B-mode. If the tissue does not move very much during the acquisition of an image, and the ultrasound beam pattern is moved across the tissue by successively activating different transducers and plotting the echoes with depth along the lines of the corresponding beam patterns, a two-dimensional image is formed. By convention in radiologic images, the images are usually displayed as if the patient is facing the physician. There is disagreement between clinics about how images should be oriented

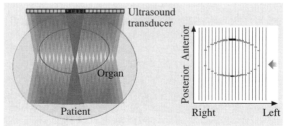

Fig. 21.36 High-density linear-array scan head. The transducers can be spaced more closely than the width of the ultrasound beam. Then each ultrasound aperture is formed by multiple transducers at the scan head face. The beam patterns overlap. The use of multiple transducers in the aperture allows the shaping of the aperture and electronic focusing of the ultrasound beam pattern

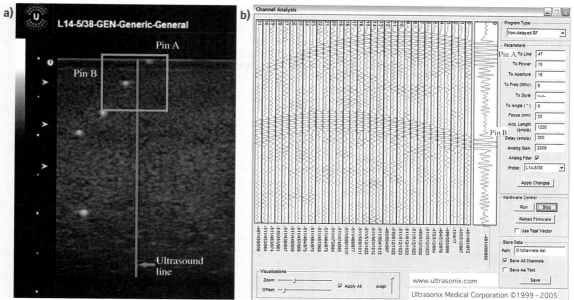

Fig. 21.37a,b Echoes received by each transducer in an array. By imaging highly reflective pins in a phantom, the effect of distance on travel time from a reflector to each transducer element in an array can be appreciated. By applying an appropriate delay on each of the transducer signals so that the echoes from the target align in time, and then adding the transducer signals together, a receiver beam pattern is formed, which is focused on the target. The time delays can be adjusted in 6 ns steps. (**a**) The transmit pulse was launched along the *red ultrasound line*. The echoes are displayed for data gathered from within the *green box*. (**b**) Radio-frequency echoes from each of 32 transducers are displayed for the echoes returning from a single transmit pulse (Courtesy Ultrasonix Medical Corp., Burnaby, www.ultrasonix.com)

The receivers and digitizers together form the ultrasound beam former for receive mode. Early array transducers had 32 elements, and the electronics were located in the scan head. Current instruments have 128 elements, a number limited by the costly cable containing 128 coaxial leads that connects the array to the receivers and transmitters in the console. The number of transducers will increase in the future as higher-capacity cables become available or, alternatively as it becomes possible to house the transmitting and receiving electronics in the ultrasound scan head.

A linear array transducer allows for adjustment of the transmit focus and aperture separate from the receive focus and aperture (Fig. 21.42). Because during the receiving period the relationship between depth and time is 0.75 mm/µs, a system can change receive focus

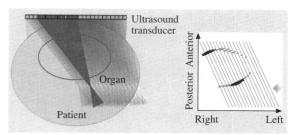

Fig. 21.38 Adjustable focus with large aperture. Using a high-density linear array to generate a large aperture and focus at a deep depth

Fig. 21.39 Steering of the ultrasound beam pattern. Multiple transducers in the aperture allows the ultrasound beam to be steered at an angle for either imaging or Doppler by introducing phase delays in the signals to and from the transducers progressively along the transducer array

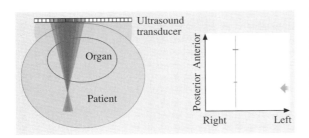

Fig. 21.40 Apodized aperture. Apodization is used to limit the side-lobes of the ultrasound beam pattern. Transducers near the edge of the aperture are excited with a low voltage during transmit compared to the central transducers and are amplified with a lower gain compared to the central transducers during receive. The drawing on the right shows the image line corresponding to the beam pattern with the superficial and deep echoes plus the mark on the right, indicating the focal depth

every 20 μs, advancing the focus 15 mm each time along with the corresponding aperture. This focal adjustment is called dynamic focusing and is similar in concept to time-gain compensation. It is also possible to acquire data along the ultrasound line using several pulse-echo cycles, each with a transmit burst that is focused at a different depth. Although dynamic receive focusing takes no more time than static focusing, using double transmit focusing to further improve the lateral resolution of the image at a greater range of depths does reduce the frame rate by half. The location of the transmit focus is indicated by a carrot on the right of the image.

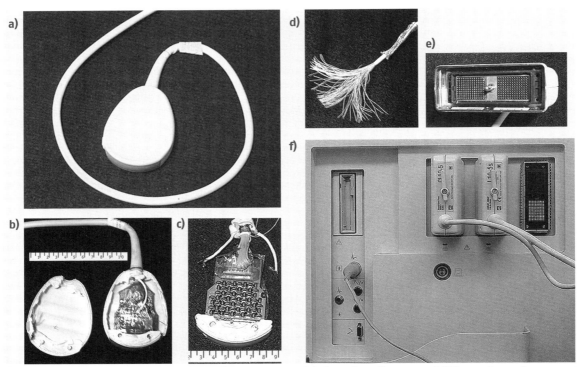

Fig. 21.41a–f Scan head details. This curved linear-array scan head consists of an array of 128 transducer elements cut from a single block of PZT piezoelectric ceramic. The array block is 64 mm long and 5 mm wide, cut into elements 0.4 mm long and 5 mm wide with a 0.1 mm kerf between. The block is curved to form a curved linear array. (**a**) Transducer in housing with cable; (**b**) Transducer housing removed showing the copper foil shielding over the internal electronics; (**c**) Shielding removed to show the 64 electronic matching circuits on this side of the circuit board; (**d**) 128 individual coaxial cables housed in the ultrasound scan head cable; (**e**) Scan head plug showing the connector electrodes for the 128 coaxial cables; (**f**) Front panel of the ultrasound scanner showing the sockets for the scan head connections. The front panel also shows connectors for other instruments including: electrocardiograph, plethysmograph, phonocardiograph microphone and computer disk drive for data exchange

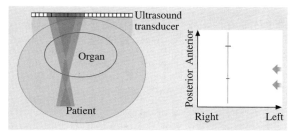

Fig. 21.42 Adjustable focus with large aperture. Using a high = density linear array to generate a large aperture allows electronic adjustment of the focal depth by adjusting the electronic delay of the transmit and receive signals so that the signal at the center transducers is delayed more than the signal at the edge transducers in both transmit and receive. Dual transmit focal depths, indicated by the pair of *arrows* on the right of the image, require that separate transmit pulses be sent into tissue for each of the two focal depths. On receive, the focus can be changed between the time that shallow echoes are received and the time that deep echoes are received

Because a linear array can steer the ultrasound beam, it is possible to form an image of each voxel in the tissue plane from different views, combining the results into a compound image (Fig. 21.43). The algorithm for combining the data from different angles is important in creating a pleasing image.

Phased Array Transducers. By applying a selected time delay to each transducer in an array, the ultrasound beam can be focused or tilted as desired (Fig. 21.39). Phased array transducers (Fig. 21.44) form the ultrasound image by tilting the ultrasound beam using selected delays on each transducer on transmit and receive to achieve the desired tilt. A scanhead with a transducer array of width

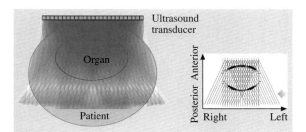

Fig. 21.43 Compound real-time imaging. By tilting the beam patterns to cross during acquisition of the image plane, system can take advantage of the specular backscatter of planar reflectors in the tissue

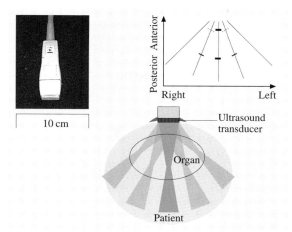

Fig. 21.44 Phased-array scan head. A phased-array scan head is able to use the same aperture for all ultrasound scan line beam patterns, directing the beam along different lines by adjusting the relative phase of the signals to the transducers

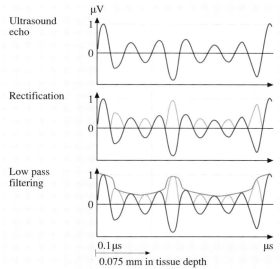

Fig. 21.45 Amplitude demodulation. To create a B-mode image of tissue anatomy, the strength or amplitude of the received ultrasound echo is determined as a function of depth. The RF echo is rectified (converting 5 MHz to 10 MHz) and low-pass-filtered with a 1 MHz cutoff to yield a depth resolution of 0.75 mm. That is effective if the ultrasound transmit burst is shorter than 1 μs. Amplitude demodulation requires one pulse-echo cycle along each ultrasound beam pattern line

D can tilt an ultrasound beam to an angle θ in tissue with an ultrasound velocity of c by applying a time delay gradient along the transducer of

$$\frac{tc}{D} = \sin\theta. \tag{21.11}$$

The time t is often represented as a function of the phase of the ultrasound ϕ for the ultrasound frequency f_{us}

$$t = \frac{\frac{\phi}{2\pi}}{f_{us}}. \tag{21.12}$$

To allow the ultrasound beam to tilt to θ degrees, the width of each transducer element must be less than $\lambda/2$ but the pitch (w = center to center spacing of the elements) must be $\lambda/4$ in order that, at the maximum deflection angle, a single beam pattern is defined when ultrasound intersects the array at angle θ

$$w < \frac{\lambda}{4\sin\theta} = \frac{c}{4f_{us}\sin\theta}. \tag{21.13}$$

A 3 MHz scan head tilting the ultrasound beam 30° to the left and to the right requires a transducer element pitch of 0.3 mm.

Pulsed Doppler [21.57]

The series of echoes from a single ultrasound transmit pulse provides information about the echogenicity of the tissue along the ultrasound beam pattern at various depths. The echogenicity of the tissue at each depth is determined by amplitude demodulation (Fig. 21.45). The measurement of the velocity of blood or tissue requires pairs of echoes from two (or more) ultrasound transmit pulses along the same beam pattern line, separated by a short time (0.1 ms). By comparing a second echo series from the same transducer beam pattern from a later time to the echo series from the first pulse, motion of the echogenic tissue along the beam pattern can be detected and measured. If the second echo series is identical to the first echo series, then none of the tissue along the ultrasound beam pattern has moved or changed during the period. However, if the second echo series is different, then the tissue has moved or changed. Motion in the direction of ultrasound propagation causes a simple time shift in the echo series, lateral motion or a change in tissue properties results in a loss of coherence between the echoes.

If the tissue has moved toward the ultrasound transducer, then the second echo pattern will arrive earlier after the transmit pulse compared to the first (Fig. 21.46). If the difference is detected by Doppler processing, then the echoes from each region of tissue are characterized by a vector in a complex plane (Fig. 21.46). One rotation of the vector is equivalent to the reflector moving toward the transducer to shorten the round-trip path by one wavelength of the ultrasound. If the ultrasound is 5 MHz, then the wavelength is 320 μm so one rotation represents a shortening of the round trip ultrasound travel distance of 320 μm, or 160 μm each way. A phase rotation of one quarter of a cycle ($\pi/2$ rad) represents 40 μm of motion. Doppler processing cannot tell the difference between 20 and 180 μm of motion because the vector is pointed in the same direction after the 160 μm displacement. This is called aliasing and is responsible for a waveform with the highest forward velocities incorrectly shown as reverse velocities (Fig. 21.47). Doppler

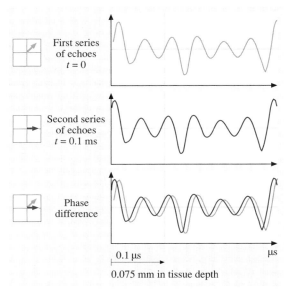

Fig. 21.46 Phase demodulation. To measure blood velocity, or tissue displacement or strain, the change in phase (or echo displacement) between successive pulse-echo cycles at a particular depth must be determined. In this case, a phase change of $\pi/4$ of 5 MHz ultrasound represents an echo time shift of 0.05 μs/8 = 0.006 25 μs, which is equivalent to a displacement of the scatterers toward the transducer of 0.006 25 μs × 0.75 mm/μs = 0.0047 mm in the time between pulse-echo cycles or 0.1 ms, yielding a velocity of 0.047 mm/ms or 0.047 m/s. The Doppler equation, time-domain velocimetry and strain imaging are all based on this computation. In Doppler demodulation, the phase and amplitude of the echo at any depth can be represented as a vector on a phase diagram. These vectors can be represented as complex numbers if the vertical direction is represented as the "imaginary" ($\sqrt{-1}$) direction and the horizontal direction represents the "real" direction

systems cannot tell the difference between displacements of 100 μm toward the transducer and 60 μm away from the transducer, because both are represented by the same phase shift. A motion of 20 μm will result in a rotation of the phase vector by 45° ($\pi/4$ rad). Often the interval between the first echo and the second echo is 0.1 ms. The velocity of the tissue in this case is $20/0.1$ μm/ms $= 200$ mm/s $= 20$ cm/s $= 0.2$ m/s.

The Doppler frequency is the phase change of the echo in cycles during the pulse interval divided by the pulse interval. In this case the phase change is 1/8 cycle, so the Doppler frequency is $0.125/0.1$ ms $= 1250$ cy/s $= 1.25$ kHz.

Usually, pulsed Doppler systems are used to measure blood velocity. Unfortunately, it is rarely convenient to align the ultrasound beam pattern along the axis of a blood vessel so typically the angle between the ultrasound beam pattern and the vessel axis is 60°. This is called the Doppler angle. The ultrasound system can only measure the projection of the blood velocity vector along the ultrasound beam. That projection is the cosine of the angle between the velocity vector and the ultrasound beam. The Doppler equation relating velocity to the ultrasound echo is:

$$f = \frac{2v_f f_{us} \cos\theta_D}{c} \quad \text{or} \quad v_f = c\frac{\frac{f}{f_{us}}}{2\cos\theta_D}. \quad (21.14)$$

where f is the Doppler frequency that you hear, the factor of 2 accounts for the round-trip of the ultrasound, v_f is the velocity of the blood, f_{us} is the ultrasound frequency, θ_D is the angle between the blood velocity and the ultrasound beam, and c is the speed of ultrasound in tissue. The form

$$\frac{v_f}{c} = \frac{\frac{f}{f_{us}}}{2\cos\theta_D} \quad (21.15)$$

emphasizes the relationship between the ratio of the blood velocity to the ultrasound velocity and the Doppler frequency to the ultrasound frequency. Since the ultrasound wavelength $\lambda = c/f_{us}$,

$$v_f = \frac{f\lambda}{2\cos\theta_D}. \quad (21.16)$$

To provide a measurement of blood velocity, ultrasound instruments have a cursor on the screen to align with the axis of the vessel under examination (Fig. 21.48). The associated Doppler data, here represented as a spectral waveform, is provided with a scale showing the angle-adjusted velocity based on the projection of the measured velocity onto the vessel axis. Although this adjustment is theoretically justified by the

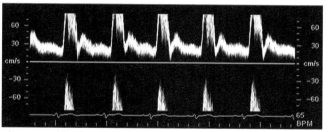

Fig. 21.47 Aliasing. Blood velocities that cause high phase shifts (greater than 180° or π radians) between pairs of pulsed Doppler echoes cause the Doppler system to mistake high-velocity forward flow for reverse flow. Continuity in blood flow allows the examiner to detect this error

geometry, unfortunately, it does not provide a useful velocity value. If a Doppler angle of zero degrees is used, then the value is correct, however, if other angles between the ultrasound beam and the presumed velocity vector are used, then the value of the velocity is not correct. The reason is that blood velocity vectors are rarely aligned with the axis of the artery. Instead, blood velocities form helical patterns along the artery [21.58–60]. Angle-adjusted Doppler velocity measurements are systematically higher as the Doppler angle increases from zero to 70 degrees or more. Although at a Doppler angle of zero, velocities are accurately used to determine both the volume flow rate and the Bernoulli pressure drop, neither result is correct when Doppler angles other than zero are used [21.61–63].

Pulsed Doppler Range of Measurement. The highest Doppler frequency shift that can be measured with pulsed Doppler occurs when the phase change of the echoes during the interval between echo acquisitions [pulse repetition interval (PRI)] is 180° (half a cycle). The Doppler frequency at that condition is half of the pulse repetition frequency (PRF, the Doppler sampling frequency). This is called the Nyquist limit, after Harry Nyquist. This establishes the highest velocity that the Doppler can detect, although since the signal is analyzed as a complex number with twice the data (half as a "real" number and half as an "imaginary" number) Doppler frequencies between $+0.5$ PRF and -0.5 PRF are reported. As long as there is no reverse flow the range of Doppler frequencies available is equal to the PRF. If the phase vector is characterized by a 2 bit complex number into four values, then the smallest phase change that can be classified is 1/4 cycle. So, a 5 MHz Doppler system with the data recorded as 2 bits, and a PRF of 10 kHz, can only report four Doppler frequency ranges,

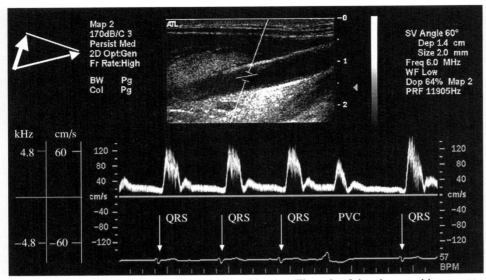

Fig. 21.48 Duplex Doppler image and waveform of a carotid artery. The axis of the ultrasound beam pattern used for Doppler is shown crossing the artery at an angle of 60° to the vessel axis. A cursor is aligned parallel to the vessel axis to provide θ_D for the Doppler equation. The Doppler frequency, which is produced by the speakers of the instrument, is shown by a scale drawn on the *left*. The scale showing the speed at which the blood is approaching the transducer is drawn adjacent to the frequency scale. The instrument computes the scale on the screen by dividing the cosine of the Doppler angle into the approach speed. Along the lower edge of the image is the electrocardiograph tracing (ECG) showing the "QRS complex" (small up and down) indicating the onset of ventricular contraction in each cardiac cycle. The lump in the trace 200 ms later is the T wave indication the end of ventricular contraction. The fourth beat is too early and the QRS has an odd shape. This is a premature ventricular contraction (PVC) originating from an abnormal spot on the ventricle. The PVC causes the ventricle to contract before filling is complete, producing a low-volume contraction, and low systolic blood velocities. The ventricle overfills before the next normal contraction, causing elevated systolic velocities

−1.25 to +1.25 kHz, +1.25 to +3.75 kHz, +3.75 to +6.25 kHz, and +6.25 to +8.75 kHz. −1.25 kHz is equal to +8.75 kHz. Such a Doppler system would probably classify −1.25 to +1.25 kHz as no motion. By increasing the number of bits in the digitizers representing the phase vector, the resolution of the phase shift is increased, and the number of possible measured velocities is increased. This does not increase the Nyquist limit.

Another way to increase the number of velocities that can be measured is to increase the number of pulse-echo cycles or Doppler phase samples that are gathered. The number of samples used to resolve the velocity is called the ensemble length or packet length. In a typical spectral waveform pulsed Doppler system, Doppler samples will be gathered from a voxel at a PRF of 12.8 kHz for 10 ms to provide 128 Doppler samples for frequency analysis. The Nyquist limit (upper frequency limit) is 6.4 kHz. The lowest frequency that can be detected is 100 Hz, one full Doppler cycle in the 10 ms acquisition period. If the signals are digitized into 2 bit complex numbers, then each of the 128 possible frequencies can be tested to see whether each is present or absent, but no information is available about the strength of each frequency signal. To determine the strength of each frequency, the signals must be digitized into numbers with more binary bits. Spectral waveform analysis assumes that there are many Doppler frequencies present in the Doppler signal from the voxel of interest.

Pulsed Doppler Pulse–Pair analysis. The Doppler frequency shift can be measured from just two echo samples separated by a known time. The assumption is that only a single Doppler frequency is present in the signal. The phase change divided by the sample interval time provides the Doppler frequency. The upper limit of the Doppler frequency is the Nyquist limit. To resolve the Doppler frequency, the phase angle change must be resolved. Representing the phase angle by a two bit complex number only resolves the phase angle into four divisions. For 5 MHz ultrasound, displacements of

40 μm can be resolved. For a 10 kHz PRF sample frequency, this allows a velocity of ±40 cm/s (±0.4 m/s) to be detected. However, if the phases are represented by 10 bit complex numbers, the phase angles can be resolved into 1000 divisions and the displacement and velocities can also be resolved. With 10 bit quadrature digitizers, velocities between plus and minus 80 cm/s can be measured with a resolution of 0.8 cm/s, equivalent to displacements of 80 nm during the pulse interval.

Pulsed Doppler Wall/Clutter Filtering. The measurement of blood velocity with ultrasound is complicated by the containment in vascular walls. The echogenicity of the blood is 60 dB less than the echogenicity of the walls at 5 MHz. The blood cells are Rayleigh scatterers because they are smaller (7 μm) than the wavelength of ultrasound (300 μm for 5 MHz). The strength of the echo from blood increases with the fourth power of the frequency. However, frequency-dependent attenuation of the ultrasound by overlying tissues limits most Doppler ultrasound studies to frequencies of 2–6 MHz. 5 MHz produces the strongest echoes for vessels under 2 cm of tissue, deeper vessels, and those behind the skull require 2 MHz ultrasound waves to penetrate these tissues [21.64–66].

In many examinations, Doppler measurements from blood adjacent to the wall are desired, so Doppler echoes from the voxel next to the wall will include strong wall echoes as well as weak echoes from the adjacent blood. Unfortunately, vascular walls move with velocities near 2 mm/10 ms or 0.2 m/s (20 cm/s). The strong echoes from these moving walls must be rejected to reveal the adjacent blood velocity, which may be only slightly higher than 20 cm/s. The strong echoes from the wall are called clutter. 10 bits of the system dynamic range are required to handle the clutter and the blood echoes must be resolved above that, so most ultrasound systems now utilize a 14–16 bit dynamic range, which is achieved by combining the outputs of separate digitizers (one for each transducer) and digitizing faster than the required four times the ultrasound center frequency, then filtering the result for processing.

Pulsed Doppler Spectral Analysis. To measure the Doppler frequency shift, any method of frequency analysis can be used. Zero-crossing rate meters and zero-crossing time interval histograms have been replaced with real-time spectrum analyzers. Although the time-compression analyzer produces identical results, the popular fast Fourier transform (FFT) analyzer is used to produce spectral waveforms of Doppler signals

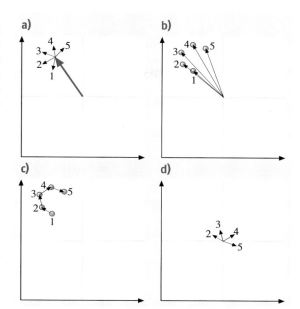

Fig. 21.49a–d Autocorrelation clutter (high-pass) filter. By representing each pulse-echo from a particular depth as a point on a phase diagram, the trajectory of the echo phase in time can be plotted. (a) The Doppler echo comes from two types of reflectors, strong reflecting stationary structures (*clutter*) and weak reflecting moving blood cells. In each of the five pulse-echo cycles (ensemble or packet of 5), the contribution from the strong reflectors is the same. The moving blood cells reflect the same amplitude (vector length) but a different phase during each pulse-echo cycle. (b) The ultrasound receiver cannot differentiate the clutter contributed by the stationary structures from the signal contributed by the moving blood cells. High-pass filter. (c) By taking the difference between successive results, the effect of the clutter is eliminated. (d) By centering the resultant vectors on a new phase diagram, the phase angle change between each pulse-echo cycle can be used to compute the blood velocity

(Figs. 21.47, 21.48). The FFT analyzer is coupled to the pulsed Doppler so that the FFT analyzer receives a quadrature pair of data from the depth of interest for each pulse-echo cycle. The advantage of the spectral analysis is that if the voxel contains several flow velocities, the FFT spectrum will display the Doppler frequency for each velocity. Unfortunately, in blood flow, eddy oscillation frequencies up to 500 Hz occur; the typical FFT length of 10 ms (100 spectra per second) is only capable of resolving eddies of 50 Hz or lower. When a voxel contains higher-frequency eddies, the result appears as spectral broadening or as harmonics on the spectral waveform.

Pulsed Doppler Autocorrelation. Autocorrelation is a frequency analysis method which is an extension of pulse-pair analysis. Autocorrelation analysis assumes that there is only one valid Doppler frequency in the blood flow. Autocorrelation analysis identifies that frequency. However, to reject the stationary and slow moving strong echoes of the wall, a clutter filter must be applied. Clutter is a stationary echo superimposed on the changing Doppler echo. The analysis is based on an ensemble of N echoes, usually eight echoes. The difference between adjacent pairs of echoes n and $n-1$ (for $n = 2, N$) creates a new group of $N - 1$ vectors (Fig. 21.49). By taking the average of the change in angles, a Doppler velocity can be obtained. Sometimes two filtering steps are required rather than one. Such filters are called finite impulse response (FIR) and infinite impulse response (IIR) filters.

Time–Domain Velocimetry. The Doppler equation contains the Doppler frequency and the phase angle. Doppler signals are subject to aliasing when the between-pulse phase angle change ($\Delta\phi$, a circular variable) has a value greater than π. In addition, echoes from deep tissues have a lower center frequency than echoes from shallow tissues, causing confusion about the correct value of the ultrasound frequency f_{us} to use in the equation.

$$\frac{v_f}{c} = \frac{\frac{\Delta\phi}{\Delta t}}{2 \cos \theta_D} \; . \qquad (21.17)$$

Of course, the center frequency of deep echoes is still within the bandwidth of the transmitted ultrasound; the center frequency is depressed because of the greater attenuation of the higher-frequency portions of the transmitted bandwidth.

Since bandwidth is inversely proportional to pulse duration and pulse duration is related to spatial resolution, it is desirable to have a short pulse duration so that spatial resolution is as small as possible. This means broadband pulses, permitting depression of the center frequency of the echo. The desire for improved resolution and the desire to avoid aliasing combine in the use of time-domain velocimetry as an alternative to Doppler.

$$\frac{v_f}{c} = \frac{\frac{\Delta\tau}{\Delta t}}{2 \cos \theta_D} \; . \qquad (21.18)$$

In time-domain velocimetry, two successive ultrasound echoes are compared to see if a time shift ($\Delta\tau$)

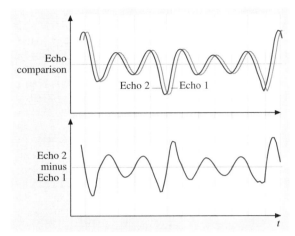

Fig. 21.50 Time-domain demodulation removing clutter. To remove clutter in time-domain demodulation, the RF echo signal from each pulse-echo cycle is subtracted from the preceding RF echo signal to yield the residual due to tissue motion

has occurred in the echo in the intervening time (Δt), which indicates tissue motion. By measuring the time shift, the velocity can be calculated. Of course, clutter from solid tissue in the voxel may provide a substantial stationary echo superimposed on the small shifted echo. To remove the clutter echo, subtraction of adjacent pairs

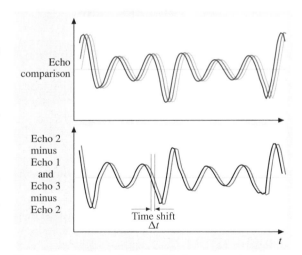

Fig. 21.51 Time-domain demodulation measuring the velocity component. By measuring the time shift between successive echo pairs, multiplying by half the ultrasound velocity and dividing by the pulse repetition interval time, the velocity of the tissue along the ultrasound beam results

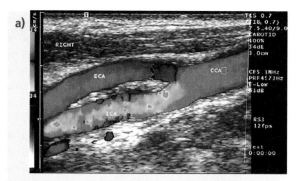
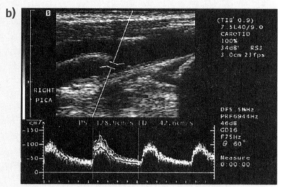
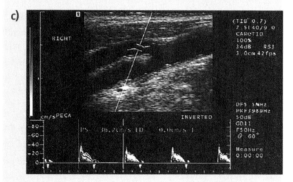
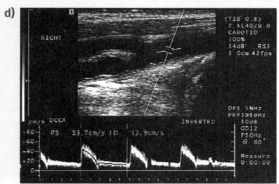

Fig. 21.52a–d Ultrasound images from a conventional carotid artery Doppler examination in the neck [21.67]. *Orientation*: a linear array transducer located at the top of each image acquires the 128 vertical scan lines in sequence across the 35 mm-wide image. Two of the monochrome B-mode images (**c**) and (**d**) were acquired in 24 ms (42 frames per second) so the scan lines progress across the image at a speed of (35/24) 1.46 m/s, a speed similar to blood velocity. The color flow image (**a**) was acquired at 12 fps; the other monochrome image (**b**) was acquired at 23 fps. *Anatomy*: (**a**) ECA external carotid artery supplying the face, ICA internal carotid artery supplying the brain, CCA common carotid artery from the heart via the Brachiocephalic trunk. (**a**) Color Doppler image of the carotid bifurcation. The color scale on the left of the image shows that forward flow is shown as *red* and reverse flow is shown as *blue*. The flow directions are with respect to the transducer array at the top of the image rather than with respect to the heart to the right of the image. The blue region at the origin of the external carotid artery shows that the vertical component of the velocity vector is pointing toward the ultrasound transducer rather than showing that flow is reversed toward the heart. Notice that this region is bounded by black bands indicating *zero* velocity toward the transducer. The *blue spots* in the lumen of the internal carotid artery are aliasing (Fig. 21.47). Notice that each of those regions is surrounded by an *orange* color indicating high blood velocity components with respect to the transducer. *Doppler waveform orientation*: each of the three Doppler waveforms is acquired from an ultrasound beam pattern that is tilted at an angle with respect to the transducer, selected by the examiner to form an angle of 60° with respect to the artery axis. (**b**) *lower* Normal Doppler spectral waveform from the internal carotid artery with high diastolic flow. Because the brain continuously requires 10% (10 ml/s) of the cardiac output (100 ml/s), the peripheral resistance is low so blood flows to the brain during systole and diastole. (**c**) *lower* Normal Doppler spectral waveform from the external carotid artery with low diastolic flow. Because the face at rest requires little blood flow but when active (chewing, blushing) requires more blood flow, at rest the arteriolar sphincters are constricted, only accepting blood flow during systole. When the arteriolar resistance decreases, the external carotid also has flow during diastole. (**d**) *lower* Normal Doppler spectral waveform from the common carotid artery which has the combined internal and external flow waveform. A pulsatility index showing the ratio between the diastolic and systolic velocities is sometimes used to classify waveforms

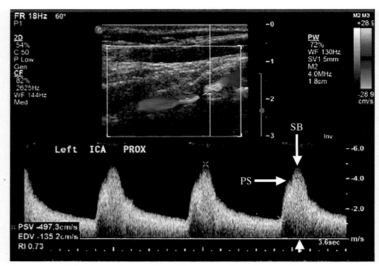

Fig. 21.53 Stenotic internal carotid artery. Even after 25 years of exploring the meaning of carotid artery spectral waveforms, there is no agreement on how the data should be interpreted. Although this color Doppler image shows an interruption in the color flow image of the lumen, the interruption is not due to an arterial occlusion, but to a local region of high ultrasound attenuation which causes the color Doppler signal to drop out. Thus, the diameter of the color Doppler streak cannot be used to assess the lumen. In this case, the vessel walls cannot be easily seen. A stenosis is present at the Doppler sample volume as indicated by the high velocity. The spectral waveform is confusing. The highest Doppler frequency shift which was measured by the examiner 220 ms after the onset of systole at 497.3 cm/s is probably not accurate because the measurement is at the time of greatest spectral broadening (SB), indicating eddies in the flow. Use of the Doppler equation is not appropriate for these eddy velocities because the heading of the vectors is not known. The true peak systolic (PS) velocity probably occurs 100 ms after the onset of systole where the measurement is near 4 m/s and the Doppler equation is probably accurate if the velocity vector is assumed to be parallel to the artery axis. An appreciating of the difference between the believed accuracy of 497.3 cm/s versus 'about 4 m/s' is important because of the clinical decisions that result. The PSV is used to decide whether a patient with symptoms of mini-stroke should have surgery. In this case, the end diastolic velocity (EDV) is measured at 135.2 cm/s. An EDV value exceeding 140 cm/s in a single measurement would cause a vascular surgeon to schedule the patient for carotid endarterectomy surgery even if the patient has no symptoms

of echoes (Fig. 21.50) provides the difference waveform containing the motion information. Then the time shift between difference waveforms can be measured (Fig. 21.51).

A comparison between the time-domain equation and the Doppler equation shows that $\Delta \tau = \Delta \phi / f_{us}$. This indicates that it is the demodulator frequency, not the frequency of the echo, that is relevant for the Doppler equation; the former must be close (and is usually equal) to the transmitted center frequency, i.e., the frequency required in the Doppler equation. There are substantial differences between the Doppler and time-domain systems. First the time-domain system usually transmits a broadband, short ultrasound pulse. The signal-to-noise ratio of the echo is determined by the transmit pulse energy. To achieve the same energy in a short pulse, the temporal peak power must be high, leading to a high acoustic pressure. When analyzing the time shift between the echoes, it is easiest to test for the time-lagged correlation by stepping one digitizing interval at a time. So, if the echo is digitized at a rate of 20 MHz, the digitizing interval is 50 ns ($\Delta \tau$), equivalent to a tissue displacement of 0.0375 mm. If the pulse repetition frequency is 10 kHz, so the inter-pulse interval is 100 μs (Δt), the velocity resolution is 38 cm/s (0.4 m/s). Time-domain systems usually digitize at a higher frequency and usually interpolate the wave between the measurements to improve velocity resolution. For large time shifts, when noise is present, the time domain system usually has trouble with peak hopping, associating the wrong cycles in the wave when determining the time shift. This results in a measurement error identical to

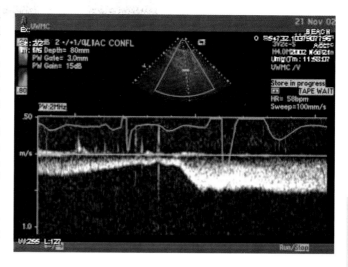

Fig. 21.54 Doppler spectral waveform signals from the inferior vena cava showing augmented flow. Doppler spectral waveforms show velocity (*vertical axis*) versus time (*horizontal axis*). In addition to the difference in flow direction, venous flow is differentiated from cardiac flow because the venous flow varies with respiration and with manual compressions of the tissues. Here the sudden increase in venous flow in the distal inferior vena cava was created by releasing an occlusion of the venous outflow from the legs. ECG tracing (*green*) shows cardiac timing. The heart beat to the right of center is due to a premature ventricular contraction, an abnormal heart rhythm which does not affect the venous flow

aliasing. So, the advantages of time-domain processing over Doppler processing, while promising, have not been fulfilled.

Clinical Applications of Ultrasound Velocimetry

The most common application of pulse Doppler ultrasound spectral waveform analysis is the examination of the carotid arteries [21.68] for stenoses that might lead to stroke (Figs. 21.52, 21.53). The measurement of high systolic (> 125 cm/s) carotid blood velocity is associated with a 20% chance of stroke in two years in a patient reporting stroke-like symptoms

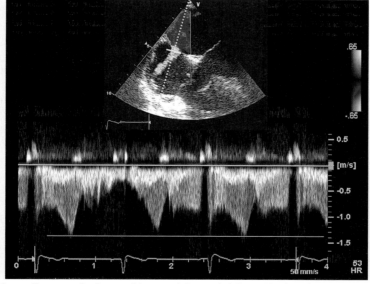

Fig. 21.55 Duplex echocardiogram of a heart with an atrial septal defect. The jet of blood from the left atrium to the right atrium can easily be seen on the color flow image. The Doppler spectral waveform taken at the center of the jet orifice at a Doppler angle of 25° to the jet shows velocities as high as 1.4 m/s when the inter-atrial pressure difference is the greatest. Correcting the velocity for the Doppler angle ($\cos 25 = 0.906$) increases the velocity value to 1.54 m/s. Cardiologists usually attempt to orient the Doppler transducer to align the jet with the Doppler ultrasound beam to achieve a *zero* Doppler angle. The pressure difference between the chambers can be computed using the Bernoulli equation for blood velocity $\Delta p [\text{mmHg}] = 4 [\text{mmHg}/(\text{m/s})^2] V^2$. The pressure difference here is near 9 mmHg (courtesy Frances DeRook and Keith Comess)

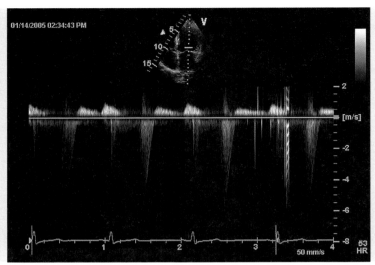

Fig. 21.56 Vibration from the mitral valve. Signals from blood flow are mixed with signals from solid tissue motion. Here, the signal of solid tissue motion is suppressed by the wall filter, which suppresses clutter at frequencies below 400 Hz. However, in this case with the Doppler sample volume in the mitral orifice, as the mitral valve closes for the fourth time, the valve vibrates at a frequency of 2 kHz for a period of 8 ms causing the harmonic (Bessel function) spectral pattern with seven harmonic terms, indicating that the amplitude of the vibration is high (courtesy Frances DeRook and Keith Comess)

(TIA = transient ischemic attack, RIND = reversible ischemic neurological deficit). The measurement of a high-end diastolic velocity (> 140 cm/s) in any patient is associated with a 20% chance of stroke in two years from emboli released from the carotid stenosis. Such findings, by themselves, can direct a patient to treatment that will prevent devastating disability.

Examination of the veins of the legs with Doppler for the detection of deep venous thrombosis (DVT) is another popular examination that can lead to life sav-

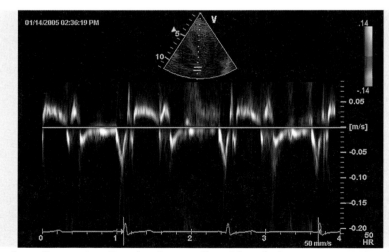

Fig. 21.57 Duplex echocardiogram tissue Doppler of myocardium. Like velocity measurements from blood flow, velocities of solid tissues can also be measured. Low velocities of the intra-ventricular septum of the heart indicates abnormal muscle contraction (courtesy Frances DeRook and Keith Comess)

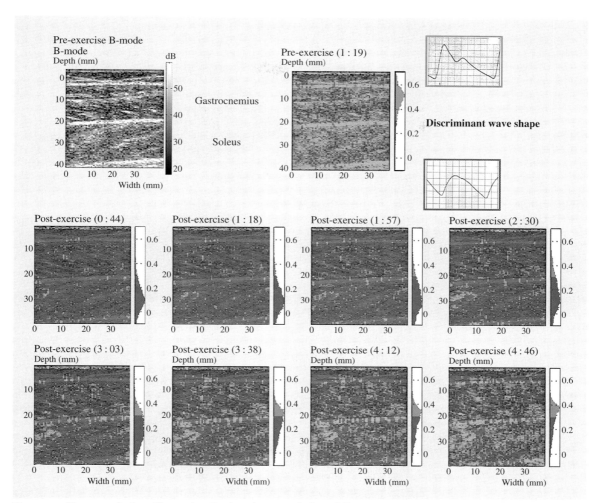

Fig. 21.58 Tissue strain in calf muscles. During tissue perfusion, different strain waveforms indicate resting perfusion (red) with high peripheral (arteriolar) resistance or hyperemic perfusion (blue) with low peripheral resistance. In this case, the calf muscle was imaged at rest, and then at intervals for a period of 5 min following a toe stand exercise. Immediately after exercise, the peripheral resistance is low to make up for the metabolic oxygen deficit after exercise. Five minutes after exercise, most of the muscle tissue has returned to the resting state (Courtesy John Kucewicz)

ing therapy. Because venous pressures are near 5 mmHg in the supine patient, normal venous flow is associated with respiration and minor local compression rather than pulsatile with the cardiac cycle (Fig. 21.54). Pulmonary embolus from venous emboli released from a DVT may account for one-third of sudden cardiac death cases [21.69]. A venous Doppler examination failing to show spontaneous venous flow with respiration indicates a venous obstruction that may be a DVT that could progress to threaten life if not treated with anticoagulation.

Cardiac Doppler examinations for blood flow are perfectly suited to characterize narrow holes between the cardiovascular structures in the chest (Fig. 21.55). By measuring the velocity of blood flow jets and applying the Bernoulli equation, the instantaneous pressure difference between the associated chambers of the heart can be measured. This works for all heart valves and wall defects so the method can be used to identify pulmonary artery hypertension as well as cardiac valve failure.

Although blood velocity through valves can be measured (Fig. 21.56), the dynamics of solid structures such

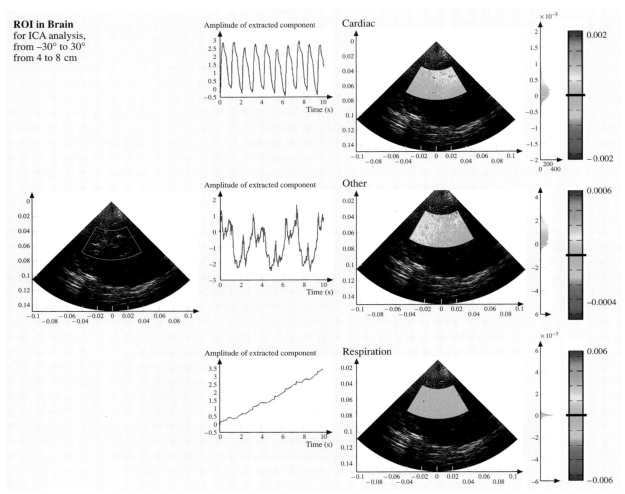

Fig. 21.59 Tissue pulsations in brain measured with ultrasonic phase gradient demodulation. Using ultrasound to measure tissue displacement over time with a resolution of 0.05 μm and differentiating motion with depth to measure strain, perfusion waveforms can be obtained. Independent component analysis of the brain perfusion signal reveals the cardiac and respiration contributions to the natural tissue strain (Courtesy Lingyun Huang)

as valve vibration, can also be characterized. By measuring the velocity of the cardiac wall, regions of poor myocardial function can be identified (Fig. 21.57).

Tissue Strain Imaging

A number of laboratories are exploring the ultrasound measurement of tissue strain [21.70] for diagnostic purposes. The most popular examinations involve the measurement of tissue response to externally applied vibration or pressure. These methods are called elastography or elastometry. During ultrasound examination, the ultrasound scan head or a nearby device is pressed on the tissue to induce a strain of several percent, either static or vibrating. By monitoring the relative motions of the tissues in response to the applied strain using Doppler or time-domain methods, the relative elastic properties of the tissues can be determined. The computation simply takes the derivative of the motion as a function of depth. To achieve 10% resolution on a 1% applied strain in 1 mm voxels requires that the ultrasound motion detection system resolve 1 μm tissue displacements. This requires a 1 GHz digitizing rate for a time-domain system without interpolation. A Doppler system with 6 bits

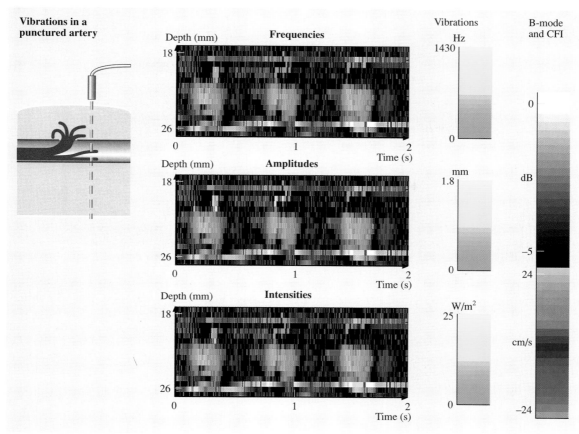

Fig. 21.60 Tissue vibrations of a punctured artery. Flow through any small orifice which results in a pressure drop can cause vibrations. Arterial bleeding can be detected by searching Doppler signals for such vibrations. (Courtesy M. Paun, M. Plett)

of dynamic range can achieve a 1 μm displacement resolution.

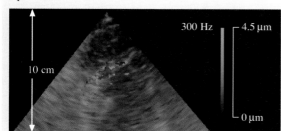

60 – 70% focal stenosis in top third of RCA
Apical view

Center frequency: 2 MHz; PRF = 1 kHz; ensemble = 10
Image Computing Systems Laboratory, Electrical Engineering

More challenging is to examine the natural tissue strain that occurs as the tissue fills and empties of blood each cardiac contraction and each respiratory cycle. Normal tissue strain for arteriolar filling is 0.1%. To resolve 0.1% strain with 10% resolution in 1 mm voxels requires the resolution of 0.1 μm displacements. This requires Doppler demodulation of a 5 MHz echo with at least 9 bits of dynamic range. Ultrasound pulsatil-

Fig. 21.61 Vibration of a coronary artery stenosis. Flow through any small orifice which results in a pressure drop can cause vibrations. Flow though a stenosis causes a bruit (Fig. 21.2), or murmur which can be detected and measured with ultrasound. These 4.5 μm-amplitude 300 Hz vibrations on the surface of the heart at the location of a stenosis in the right coronary artery occur only in diastole when the coronary flow rates are the highest. (Courtesy S. Sikdar, T. Zwink, S. Goldberg)

ity images can be made of most body tissues including muscle (Fig. 21.58) showing physiological responses to stresses, and brain [21.71] (Fig. 21.59) showing the separate effects of pulse and respiration.

Tissue Vibration Imaging

Several pathological conditions are associated with the presence of tissue vibrations. These were discussed at the beginning of the chapter where the detection of the vibrations with a stethoscope was described. Of particular interest are vibrations associated with arterial stenosis and vibrations and motions associated with internal bleeding [21.72]. Using the same Doppler demodulation methods, but filtering for the cyclic motions of the highly echogenic solid tissues rather than the continuous motions of the poorly echogenic blood, vibrations can be detected and rendered as an image showing pathologies such as arterial bleeding (Fig. 21.60). The amplitude and frequency of the vibrations are characteristic of the rate of bleeding. Arterial stenoses including stenoses of the coronary arteries (Fig. 21.61) can be demonstrated in vibration images.

Modern Ultrasound Scanners

Real-time diagnostic ultrasound scanners changed little on the outside between 1975 and 2000; they are narrow enough to fit through a doorway (Fig. 21.62) and draw less than 1.5 kW of electrical power from a standard power outlet in the hospital wall. However, inside the systems have evolved from analogue devices to digital devices (Fig. 21.63), with the conversion to digital processing determined by cutting-edge computer technology. Now an ultrasound scanner with a full array of functions is constructed from software, using a Windows laptop computer (Fig. 21.64) or a custom box. Ultrasound systems can be produced inexpensively and can obtain the diagnostic results automatically, so that a patient can use the system for self-management of some conditions (Fig. 21.64).

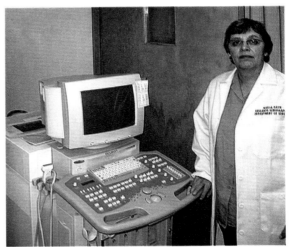

Fig. 21.62 Modern ultrasound imaging system. During the quarter century between 1975 and 2000, ultrasound systems retained the same exterior design, mounted on wheels, narrow enough to fit through a doorway, light enough that a sonographer could push the system, short enough so that a sonographer could see over the top and powered by a 1.5 kW outlet. To the right is the key to good ultrasound diagnosis, the well-trained, certified, experienced ultrasound examiner

21.5 Medical Contrast Agents

Contrast agents are used in medical imaging to make invisible structures visible. Contrast agents can be classified by anatomic/application: intraluminal/intracavity, excretory, diffusional, and chemotactic. Contrast mechanisms include reflection, absorption, fluorescence, emission and nuclear resonance frequency shift.

Intraluminal/intracavity agents are used to visualize the cardiovascular system, gastrointestinal system, reproductive system, body cavities, and other spaces. The contrast agent is injected into the space, where it spreads throughout the structure of interest. Excretory systems are used to visualize the renal system [excretory urogram or intravenous pyelogram (IVP)] or gall bladder and bile duct (cholangiopancreaticogram). The contrast agent is injected into the venous system and is excreted in concentrated form by the target organ. Because of the concentrating action of the kidney, a contrast agent in the urine will be 100 times more concentrated than in the blood. Diffusional agents are injected into the venous system and diffuse into the interstitial space through leaky capillaries. Newly formed capillaries are leaky, and such diffusion may mark local angiogenic factors secreted by tumors or retinopathy. Contrast magnetic resonance imaging (MRI) for breast tumor detection and fluorescene angiography of the retina are examples of diffusional agents. Chemotactic agents are injected into the veins and diffuse into the interstitial space through normal capillaries, where they bind to tissues of interest such as tumor cells. The time between the injection of a chemotactic contrast agent and the imaging study is several hours. This method is common in nuclear medicine imaging.

Ultrasound contrast agents cause increased backscatter of ultrasound and often cause the generation of harmonics. X-ray contrast agents contain heavy nuclei which absorb X-rays to provide contrast. Fluorescene, indocyanine green and other dyes are activated by high-energy (short-wavelength) photons and emit longer-wavelength photons, which are detected by optical camera systems. Chemotactic chemicals prepared with short-half-life (hours) nuclear isotopes are prepared and injected within a few hours. After allowing time for the agents to accumulate in the tissues of interest, the sources of radiation are imaged with pinhole or collimation nuclear radiation cameras. Gadolinium is used as a contrast agent in nuclear magnetic resonance imaging. The paramagnetism of gadolinium decreases the T_1 and T_2 relaxation times, which are displayed in the MR image.

21.5.1 Ultrasound Contrast Agents

The first intravenous ultrasound contrast agents [21.73] were large bubbles generated by intravenous injection of indocyanine green [21.74], other injectable liquids including saline and blood [21.75–77], and finally agitated saline [21.78]. These bubbles were so large that they would not pass through capillaries, so appearance in the left heart or arteries was proof of a shunt bypassing the lungs. These bubbles were generated inadvertently or deliberately by flow-induced cavitation. The bubbles contained air that was dissolved in the blood. These bubbles generated strong ultrasound reflections because of the large difference in ultrasound impedance between the body (water) and the air in the bubbles. Small bubbles were not stable, so only large bubbles survived.

Currently, most commercial ultrasound contrast agents are small bubbles or particles, which are able to pass through the capillaries of the lungs but are confined to the vascular system. Some ultrasound contrast agents are solid particles smaller than a micron in diameter. These are Rayleigh scatters of ultrasound at the fundamental frequency. Most ultrasound contrast agents are bubbles about 3 μm in diameter (Fig. 21.65) so that they will pass through pulmonary capillaries and recirculate in the cardiovascular system. They are

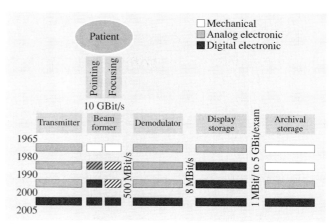

Fig. 21.63 Diagnostic ultrasound signal processing history. The amount of data transferred between the stages of the ultrasound signal-processing chain is a minimum around the display, the display systems were converted to digital processing during the early years of digital technology development. Beam forming occurs in two directions: in the plane of the image, where the evolution from mechanical to analog to digital proceeded with time, and perpendicular to the plane of the image, where mechanical focusing was replaced by digital electronic focusing and then steering with the advent of transducers segmented into two-dimensional arrays. Pointing of the ultrasound beam between 1980 and 1990 was done digitally in linear array raster scans by switching on selected transducers in the array, but in phased-array sector scans the direction of the beam was determined by adjusting the phase delay with nanosecond resolution on each transducer. Because analog signals have a continuum of possible signal levels, small differences in analog signals can be extracted as long as the differences are greater than the noise in the signal. Once a signal is digitized, no noise is added to the signal by processing. The 10 GBit/s data rate on the receiving path is based on the assumption that the transducer is divided into an array

also Rayleigh scatters, but because they oscillate, they convert the fundamental ultrasound frequency into harmonics because of the nonlinear properties of their oscillation.

21.5.2 Stability of Large Bubbles

The size of stable bubbles can be estimated from physical principles. The gas inside a stable bubble of radius r (Fig. 21.65) is under a pressure equal to the sum of the ambient atmospheric pressure plus the fluid pressure (1 atm = 10 m of water), and the pressure generated by the surface tension of the interface between the contained gas and the external fluid (water). In the body and in carbonated beverages, the depth of the bubbles is

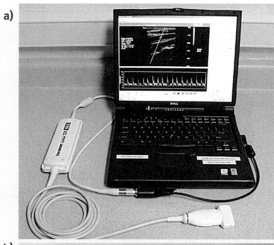

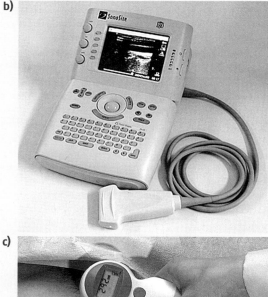

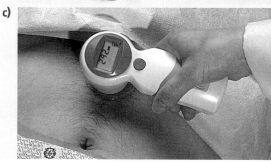

Fig. 21.64a–c Two small ultrasound imaging systems, based on the request of the Defense Advanced Research Project Administration for general-purpose handheld systems with full function (**a**) and (**b**) and an application-specific imaging system. (**a**) Terason scanner using a laptop computer; (**b**) SonoSite scanner using custom electronics; (**c**) Application-specific ultrasound scanners (BladderScan) are palm-sized and fully automatic. This instrument shows no image, but displays the volume of urine in the bladder as a number or an alarm, depending on the application

small compared to 10 m of water, so the pressure added because of the depth of the fluid can be ignored compared to atmospheric pressure. A force balance on the equator of the bubble equates the pressure force $P\pi r^2$ to the surface tension force $\sigma 2\pi r$. Thus the pressure in the bubble above the ambient pressure is $P = 2\sigma/r$. In the absence of a surfactant, the surface tension of water is 70 d/cm. When $r = 1.4\,\mu\text{m}$, $P = 1$ atm; this is a huge chemical potential to drive the contained gas into solution. If a bubble is 0.14 mm in diameter, too large to pass through a capillary (red blood cells are 8 μm in diameter), then the pressure elevation in the bubble caused by surface tension is 0.01 atm and the gas contained in the bubble has a low driving force to diffuse into solution. So, large bubbles are more stable than small bubbles. If the surface tension is reduced by applying a surfactant (shell) to the bubble, then small bubbles become stable. If the gas contained in the bubble is less soluble in blood and/or has a low diffusion constant, then the bubble has a longer life.

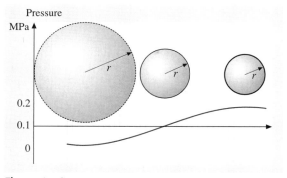

Fig. 21.65 Contrast agent bubble. An ultrasound contrast bubble is a system consisting of a gas inside the bubble, a surfactant, a surrounding liquid, and solutes including gases and surfactants in the liquid solvent. Thermodynamics and the transport of heat and of the dissolved gasses and dissolved surfactant materials all contribute to the response of a bubble to ultrasound exposure. Interactions between bubbles are also important factors in contrast agent dynamics. Here are the expected sizes for a bubble based on an isothermal ideal gas model without considering surface tension for an ultrasound intensity of 87.5 kPa

21.5.3 Agitated Saline and Patent Foramen Ovale (PFO)

Agitated saline contrast bubbles are generated by forcing fluid rapidly through a small orifice, such as an injection needle. They have an important diagnostic application: to determine whether there is a shunt from the right heart to the left heart, bypassing the lungs. The most common of these shunts is the *patent foramen ovale* (PFO). PFO may be a cardiovascular defect of considerable medical importance. If a person under the age of 50 has a stroke due to occluded vessels in the brain (ischemic stroke), the most likely cause is a PFO. One affected group is divers, who return from great depths with large amounts of gas dissolved in their body tissues, especially in fat. Bubbles which form in the low-pressure venous system as the gas is transported toward the lungs, might pass through a PFO during a cough, releasing these large bubbles into the arterial system where they cause occlusion of arteries and ischemia of the tissues that were supplied by the arteries. Ischemia of some tissues might be unnoticed. Ischemia of other tissues might cause pain that is ignored. Ischemia of some portions of the brain might result in undetected changes in personality or memory. Only ischemia of the motor and sensory cortices of the brain causes loss of sensory and motor function that are recognized as stroke symptoms. In divers, these symptoms are called neurological symptoms of the bends, in others they are called transient ischemic attacks (TIA) or stroke, depending on whether the symptoms last less than 24 h or longer than 24 h, respectively.

The foramen ovale is a normal structure before birth, along with the ductus arteriosus, allowing the blood to circulate in parallel through the fetal body, lungs and placenta. At birth, with the muscular constriction of the ductus arteriosus to obliterate the lumen, pressure shifts cause the flap valve of the foramen ovale to close and over time it will seal in most people. However, in some people the foramen ovale does not seal, opening occasionally when the person coughs or strains. The prevalence of PFO in the general population is 30% at age 20 and 10% at age 70. The reason for the decline in prevalence is either because in some people the foramen seals after age 20 or because these people suffer repeated embolic events from the venous to the arterial circulation, which progressively damage body organs leading to failure and death. In the absence of PFO, such venous emboli pass to the lungs where they temporarily occlude portions of the lung, but resolve over time.

The importance of the diagnosis of PFO is currently disputed, but two factors may make testing for PFO a part of every physical examination:

1. the cost of the contrast agent is only the 2 min that it takes to handle off-the-shelf saline and
2. recently catheter methods of sealing a PFO have been released for testing, avoiding the need for open-chest heart surgery to seal the defect. If 2/3 of the

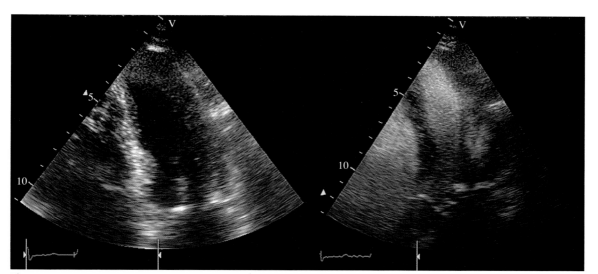

Fig. 21.66 Two-dimensional B-mode real-time echocardiogram of the left ventricle. *Left*: left ventricle without ultrasound contrast agent. *Right*: left ventricle with intravenously infused Optison ultrasound contrast agent (courtesy Frances DeRook and Keith Comess)

people with PFO at age 20 die before the age of 70, this diagnosis would be an issue of considerable importance.

21.5.4 Ultrasound Contrast Agent Motivation

The case for the development of stable ultrasound contrast bubbles small enough to pass through the capillaries of the lungs and appear in the systemic arteries is based on the wide use of ultrasound in medical imaging. Over half of all medical imaging is done with ultrasound, so there are many potential applications for contrast agents. Because these mechanical agents are confined to the vascular space, they cannot provide contrast in excretory, diffusional or extravascular chemotactic applications. Blood is a poor reflector of ultrasound, generating echoes 60 dB below the echoes from vascular walls. Contrast agents increase the backscatter. In most cases, the detection of blood flow is easy by Doppler methods which can reject the high-intensity stationary echoes from vascular walls and detect the moving echoes of blood using Doppler methods. However, in cases where the blood is moving slowly, these methods are more difficult and there is an advantage from the increased echo signals produced by contrast agents. Thus, in identifying the exact location of endocardium (Fig. 21.66), in regions deep within strong attenuators (such as skull) and for blood flow in capillaries, contrast agents promise to provide important additional diagnostic information.

21.5.5 Ultrasound Contrast Agent Development

Ultrasound microbubble contrast agents have been under development for a quarter century [21.79]. The first task in the development of microbubble contrast agents was to make stable bubbles small enough so that they could pass through the capillaries of the lungs (about 3 μm in diameter) after injection into the venous system, and stable enough to survive long enough to be useful (more than 1 min, preferably 30 min). Three strategies are available to stabilize the microbubbles:

1. lower the surface tension,
2. form bubbles of a gas with low solubility or diffusivity in water/blood and
3. produce a gas-impermeable barrier at the gas–liquid interface.

Surfactants lower surface tension; fluorocarbons have low solubility in water/blood; polymeric shells can be impermeable to gas diffusion. A series of contrast agents have been developed based on various combinations of surfactant, contained gas and method of formation.

The behavior of ultrasound contrast agent microbubbles is complicated and can easily consume an entire career. Only a few factors will be mentioned here. A microbubble consists of three phases: the contained gas, the surfactant, and the surrounding fluid. The size distribution of the bubbles is determined by the method of formation. Air-filled albumin-coated microspheres can be formed by whipping a mixture of albumin and water in a blender. This forms a broad distribution of microspheres. The desired size can be selected by centrifugation or flotation. A more nearly uniform microbubble size can be achieved by metering the relative volume of surfactant and gas through a calibrated nozzle orifice. The relative amounts of gas and surfactant are determined by the desired surface to volume ratio. The speed of bubble extrusion is crucial to the formation of bubbles of stable size. If prepared for medical use, in addition to a sterile preparation, formed using Food and Drug Administration (FDA)-approved good manufacturing practice, the agent must have sufficient stability and shelf life to permit distribution, storage and timely preparation for use.

When released into blood, the contained gas is usually not at equilibrium with the dissolved gases in the blood. A microbubble filled with fluorocarbon will take up nitrogen, oxygen and carbon dioxide without releasing much fluorocarbon, so the microbubble will swell after injection. The amount of swelling will depend, in part, on the amount and properties of the surfactant. If the bubble swells so that the surface area begins to exceed the available surfactant, the surface tension rises. This may favor bubble coalescence. As the microbubble passes through the lungs, some of the contained fluorocarbon will be released. Gas exchange between the microbubble and the dissolved gas in the blood is continuous during the life of the bubble. Some surfactants and some contained gases have chemical or physical affinities for the interior surfaces of the cardiovascular system of the lungs, arteries, capillaries or veins. The bubbles may become specifically adsorbed to the endoluminal surfaces of these vessels. This can be accidental or deliberate, and can become part of the functionality of the contrast agent. When specifically adsorbed, the surfactant may migrate to the vascular surface or may take on molecules from the vascular surface or surroundings. Of course, the contrast agent might be taken up by cells, transported by those cells and/or retained in particular tissues or organs.

21.5.6 Interactions Between Ultrasound and Microbubbles

When exposed to ultrasound at diagnostic frequencies (1–20 MHz), the contrast bubbles will oscillate at the driving frequency of the ultrasound, emitting the fundamental frequency as well as harmonics of the fundamental frequency. Microbubbles have a natural resonance frequency, primarily determined by their size, surface tension, shell rigidity (if any) and the density of surrounding material. If stimulated at a frequency near the resonant frequency, the amplitude of oscillations will be greater. The pressure of the interior gas varies with volume according to the rules of adiabatic compression and expansion, at least to the extent that the process is too fast for heat transfer. Although the waveform of the oscillation is primarily determined by the waveform of the driving ultrasound, the surfactant forces also play a great role in this process. When the ultrasound pressure is high (for $0.3\,\mu s$) the gas compresses, the temperature raises, the water vapor pressure increases as does the pressure of other gases. The gas pressures rise above the surrounding partial pressures of gas, providing a driving force to move gas out of the bubble into solution. Because the bubble is small, the surface area available for heat and mass transfer is small, and the amount of loss of heat and gas is small. Alternatively, when the ultrasound pressure is low (the subsequent $0.3\,\mu s$), the pressures and temperature drops, providing driving forces for heat and gas to enter the bubble. Now the surface area is large and it is easy for transport into the bubble, leading to rectified diffusion of gas and heat into the bubble.

At the body temperature of $37\,°C$, the vapor pressure of water is 6.2 kPa, 6% of atmospheric pressure. An acoustic wave of $285\,mW/cm^2$ SPTP (spatial peak temporal peak) intensity drops the pressure in a bubble below the vapor pressure of water (ignoring surface tension) which will cause the water to vaporize, filling the bubble with water vapor in addition to the other contained gasses. The latent heat of evaporation lowers the temperature of the bubble. On compression, the recondensing of the water adds heat to the bubble. When the ultrasound is compressing the bubble, the surfactant acts as a stable skin. The skin may even exhibit negative surface tension. Like the hull of a submarine deep in the ocean, the surfactant may hold the bubble gas pressure lower than the surrounding pressure. Similarly, when the ultrasound pressure has dropped below ambient, and the bubble expands to form a larger surface area, the surfactant, which is often a monolayer of molecules on the surface, will change in configuration, become patchy and result in increased surface tension, holding the internal gas pressure higher than expected from the surrounding pressure. As there is a driving force for the gaseous contents of the microbubble to experience rectified diffusion, there is also a driving force for the surface phase to experience rectified diffusion, changing the composition of the surface. A stable molecular monolayer on the surface of the bubble at one bubble volume will become compressed when the bubble becomes small, providing a driving chemical potential to both change the phase of the monolayer to a multilayer and to drive some of the surfactant into solution. The monolayer will be expanded when the bubble becomes large, providing a driving chemical potential to draw surfactant from solution. The solution my provide surfactant materials that are similar to or different from the original surfactant on the microbubble. Unless the concentration of potential surfactants in solution is high, this process is too slow to affect bubble behavior. In experimental solutions, such concentrations may be low, but the complex mixture of complex molecules in blood may favor such a process.

21.5.7 Bubble Destruction

The first experiments with microbubble contrast agents were disappointing. Contrast agents did not produce reflections as bright as expected. Increasing the transmit intensity did not improve the strength of the echoes coming from the agents. In an effort to improve the signal strength, broadband ultrasound transducers were used to form harmonic images. An ultrasound transducer oscillates at a resonant fundamental frequency determined by the thickness. The transducer has no efficiency or sensitivity at double the frequency, although it does have one-third the efficiency and sensitivity at triple the frequency. By damping the transducer, a transducer with a center resonant frequency of 3 MHz can be made to operate at frequenciesof 1.9–4.1 MHz. In harmonic imaging to detect contrast bubbles, a 2 MHz transmit burst is applied to this broadband transducer and the receiver is tuned to 4 MHz. Echoes from tissue are expected at 2 MHz, echoes from contrast microbubbles are expected at 2 MHz and 4 MHz. By selecting only echoes at 4 MHz, clear visualization of the contrast bubbles was expected. Unfortunately, for this application, because ultrasound transmit intensities are so high, ultrasound transmission in tissue is nonlinear, generating

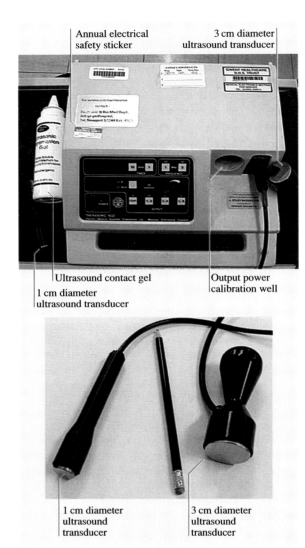

Fig. 21.67 Ultrasound therapy system

Fig. 21.68 Combined ultrasound diathermy and electrotherapy system. Deep heating of tissue can be achieved by using ultrasound or by using radio-frequency electric currents. This system is capable of delivering both kinds of heating. The two transducers shown are for ultrasound, but since the metallic surface is in contact with the patient, either can also apply RF diathermy when used with a counter-electrode

harmonics in the transmit beam on the way to the reflectors. So even without contrast microbubbles, tissue echoes contain 4 MHz, 6 MHz and 8 MHz components when exposed to 2 MHz transmit pulses. Even more disturbing, real-time B-mode images still did not show brighter echoes from contrast-filled blood spaces than from the same spaces before the introduction of the contrast agent. The reason for this puzzling result is that the transmitted ultrasound pulse destroys the echogenic bubbles, bleaching the contrast effect. Only bubbles of a particular size are bleached by the particular incident ultrasound frequency used, some becoming larger by coalescence and some dividing into smaller sizes, so they are not strongly echogenic at the incident frequency. The process of bubble destruction takes several milliseconds after exposure to a single ultrasound transmit burst. This provided the basis for two new methods of imaging with ultrasound contrast microbubbles: interrupted imaging and color Doppler imaging. If 2-D real-time B-mode imaging is used at a typical frame rate of 30 images per second, the interval between one pulse-echo cycle along an image line and the next is 33 ms, so the echogenic bubbles are gone after the first frame, and any new echogenic bubbles that enter the image field are destroyed as the imaging continues. This led to interrupted imaging of the heart muscle. In interrupted imaging, a series of ECG-triggered images is recorded for later analysis. Each image is acquired at the same time during the cardiac cycle so that successive images are separated by at least one cycle. During the sequence of ECG-triggered acquisition of images from a single plane, an initial image is obtained (i0), after an interval of five cardiac cycles a second image is obtained (i5), after the next interval of four cardiac cycles the next image is obtained (i4), then after three cycles (i3) and after two cycles (i2) and finally after an interval of one cycle the final image is obtained (i1). In theory, the prior image bleaches all of the contrast while acquiring the image. For contrast to appear in the next image, perfusion must carry the contrast bubbles into the image. By subtracting i5 from i0, all regions with significant

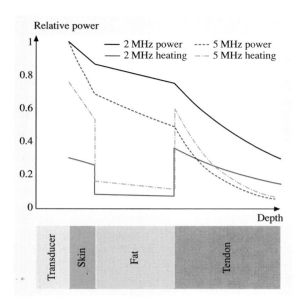

Fig. 21.69 5 MHz ultrasound is attenuated more in a short distance than 2 MHz ultrasound (*upper lines*). At shallow depths 5 MHz ultrasound delivers more heat because of greater absorption (*lower interrupted line*), but at the deepest depths, 2 MHz ultrasound provides greater heating (*lower solid line*). Because of the low absorption of ultrasound by fat, little heating occurs in fat

1.6 ms period, the contrast bubbles will begin and complete the process of vanishing. Thus there will be a large change in the echoes received from each depth during the period of the ensemble. This will be interpreted by the ultrasound system as a Doppler signal and displayed as color in the image.

The exact role of ultrasound contrast agents in medical diagnosis remains to be developed. However, there are a number of efforts to develop nanoparticles for useful applications. These nanoparticles are the same size as ultrasound contrast agents and may contain gas as well as liquid. The National Cancer Institute Unconventional Innovations Program supports some of this work for the diagnosis, treatment and monitoring of cancer. Nanoparticles can act as machines, using ultrasound energy as the supply to do useful work. This can be achieved using protein mechanics. Protein folding confirmations can change in nanoseconds, the molecular volume is different for each conformation and the energy states of folding change under high and low pressure, so ultrasound can be used to drive proteins from one conformational state to another. Nanoparticles can be used to deliver drugs [21.80] to target tissues, with ultrasound serving to trigger the release of the drug at desired times.

perfusion are dark. By subtracting i1 from i2, only the regions that are filled with contrast within a second are dark, those that fill during the time between 1 and 2 s are bright. By subtracting i4 from i5, the bright regions showing contrast bubbles in the image i5 that are not present in i4 show the regions that are filling slowly with blood.

During a color Doppler or color power angiography ultrasound examination, the ultrasound system sends an ensemble of eight transmit bursts at intervals of 0.2 ms along the ultrasound beam pattern to measure the changes in phase of the ultrasound echoes. In that

21.6 Ultrasound Hyperthermia in Physical Therapy

One of the earliest medical applications of ultrasound was the therapeutic heating of deep tissue because absorption of ultrasound by tissue results in the conversion of ultrasound energy into heat. In common practice, an ultrasound transducer 3 cm in diameter transmitting 20 W at a frequency of 3 MHz is applied to the skin (Figs. 21.67, 21.68).

As a rule, the size of the treatment area should be less than twice the surface area of the transducer. Because of diffraction effects, the ultrasound field has a complex pattern of peaks and valleys. The best transducers have a beam nonuniformity ratio (BNR) of 2 : 1 in intensity. A BNR of 6 : 1 is clinically acceptable. Some instruments have a BNR of 8 : 1.

The ultrasound passes through skin and subcutaneous fat to the muscles and connective tissues below. Because the attenuation of ultrasound in tendon is 10 times the attenuation in fat, the majority of the heating is in the tendon (Fig. 21.69). Heating is more superficial when a higher ultrasound frequency is used.

The attenuation in muscle and tendon is higher than in fat, favoring the heating of these tissues even when they are covered by fat. However, the attenuation of skin is quite variable; some reports show skin attenuation higher than tendon. The variability of the attenuation in skin results from two factors, the thin and nonuniform character of skin, and the variable air content of

21.7 High-Intensity Focused Ultrasound (HIFU) in Surgery

Starting in 1950, William J. Fry and colleagues at the University of Illinois, Urbana, developed an ultrasound method to heat small volumes of deep tissue to high temperatures. Using a 10 cm-diameter transducer operating at 2 MHz focused at a depth of 10 cm, tissue the size of a grain of rice can be heated at 25 °C per second, causing the tissue to boil in 2.5 s. This method allowed Fry to create thermal lesions in the brain for the treatment of brain tumors and other neurological conditions.

The technology is still in evolution, but holds promise of allowing the cautery of deep tissue without damaging the overlying superficial tissue. A 35 mm-diameter 5 MHz concave ultrasound transducer focused at a depth of 35 mm (Fig. 21.70) delivering 30 W of acoustic power to a treatment volume 1 mm in diameter and 10 mm long can deliver an intensity of $3\,\mathrm{kW/cm^2}$ ($30\,\mathrm{MW/m^2}$) to form a lesion (Fig. 21.71). At such intensities the nonlinearity of the tissue causes much of the ultrasound to be converted to higher harmonics (Fig. 21.72), increasing the fraction of the incident power absorbed in the treatment volume. During HIFU treatment, ultrasound images of the treatment volume show a spot of increased echogenicity caused by bubbles filled with water vapor (boiling) or with gas coming out of solution. When the HIFU treatment stops, the hyper-echogenic zone quickly becomes anechoic. During treatment, the bubbles may act as a shield, restricting the delivery of ultrasound to deeper tissues (Figs. 21.73, 21.74).

HIFU treatment can coagulate blood to stop bleeding or to render tissue ischemic, causing necrosis. The method holds promise for:

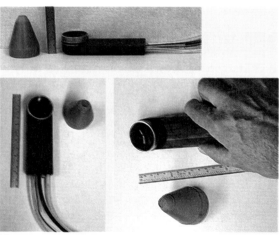

Fig. 21.70 3.5 MHz HIFU transducer used in abdominal surgery research. Notice that the black transducer face is concave (the reflection of the fluorescent light fixture is curved). The transducer is fitted with a signal cable (*black*), a pair of tubes to deliver and return cooling water (*clear*), and a thermocouple cable (*thin*) to monitor the temperature of the transducer. A plastic cone filled with water fits on the front of the transducer to conduct the sound to the tissue

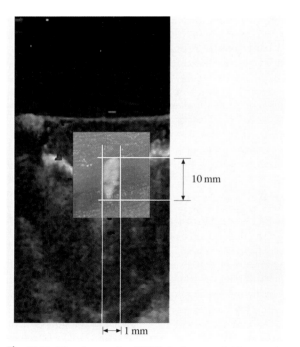

Fig. 21.71 Histology of a HIFU lesion overlaid onto the ultrasound image. 5 s high-intensity focused ultrasound cautry at a depth of 5 cm under tissue in porcine spleen. Typical lesions are 10 mm long (in depth from the transducer) and 1 mm in diameter

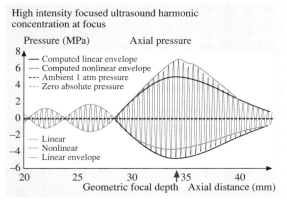

Fig. 21.72 Computed central beam instantaneous pressures from a HIFU transducer. The high acoustic pressures in the focal zone create nonlinear effects that generate harmonics in the transmitted ultrasound beam (courtesy F. Curra)

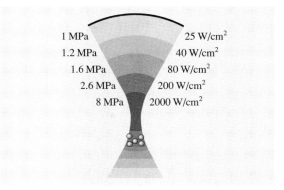

Fig. 21.73 Typical HIFU ultrasound beam profile. High-intensity focused ultrasound beam profile showing increase in beam intensity, not accounting for attenuation. The small circles post focus represent bubbles formed by a combination of mechanical-index-induced cavitation boiling. They are pushed into the post-focus zone by radiation pressure. As the exposure continues, more bubbles will be formed at shallower depths. Post-focal bubbles may provide a safety shield, preventing HIFU exposure of deeper, healthy tissue

1. stopping internal bleeding of the liver or spleen,
2. stopping bleeding due to catheter procedures,
3. reducing the size of tissues such as uterine fibroids and prostate hypertrophy,
4. destroying cancer cells and tumors,
5. birth-control procedures such as vasectomy and tubal ligation,
6. tissue welding, and
7. pushing tissues using radiation force to stretch or compress them.

Thermal dose is one way to summarize hyperthermia treatment conditions. Thermal dose implies that tissue changes occur above $43\,°\mathrm{C}$, and that the effect is increased with increasing time and increasing temperature. One empirical formula for thermal dose has been given [21.81] as:

$$D = \int_{t_1}^{t_2} n(T)^{(T-43)}\,\mathrm{d}t\,,$$

where t_1 and t_2 are the initial and final times of the heating profile, respectively, and $n(T)$ is 4 for $T < 43\,°\mathrm{C}$ and 2 for $T > 43\,°\mathrm{C}$. The temperature is determined by the rate of heat delivery per tissue volume, which is the product of the intensity times the attenuation minus the rate at which heat is removed from the tissue by conduction through tissues, and convection via the blood.

The most difficult issue with HIFU is the small treatment volume versus the relatively large access window. Under ideal conditions, the treatment of a single volume of tissue $10\,\mathrm{mm}^3$ takes about 2 s. A cubic region 50 mm on a side has a volume of $125\,000\,\mathrm{mm}^3$ and requires nearly 4 h to treat. These long treatment times present a considerable economic problem for the adoption of the technology.

21.8 Lithotripsy of Kidney Stones

Explosive ultrasound can be used to fracture kidney stones in the renal pelvis where the urine is collected from the kidney and directed down the ureter to the bladder. Stones which form in the renal pelvis from solutes in the urine are very painful. The pain occurs because urine is moved down the ureter by peristaltic action, and the peristalsis cannot handle a bolus the size of a stone. One type of ultrasound lithotripter uses a 10 cm-diameter concave reflector with a spark gap at the source focus. The elliptical curvature forms a target focus 12 cm from the face of the reflector. Equally intense ultrasound pulses can be formed using focused piezoelectric transducers or electromagnetically driven membranes with the ultrasound focused using an acoustic lens.

A water path is used to couple the reflector to the skin nearest the affected kidney so that the target focus can be positioned on the kidney stone. By firing the spark gap in the water at the source focus, pressures of $+100\,\text{MPa}$ followed by $-10\,\text{MPa}$ are achieved at the target focus. Treatment of a kidney stone may require 3000 shocks delivered over an hour. The object of the treatment is to break a large stone ($\geq 1\,\text{cm}$ in diameter) into smaller pieces ($\leq 2\,\text{mm}$ diameter) that the ureter can carry to the bladder. The mechanism of stone destruction includes erosion, spallation, fatigue, shear and circumferential compression associated with the primary shock wave or resulting cavitation. Erosion is caused by cavitation. Spallation is the ejection of the distal portion of the stone, resulting from a decompression wave reflected from the distal portion of the stone. Unfortunately, lithotripsy is often associated with injury to the kidney parenchyma [21.82].

21.9 Thrombolysis

A new therapeutic application of ultrasound is the destruction of blood clots occluding vessels. Two mechanical methods may contribute to clot destruction [21.83–85]: the formation of microbubbles in the clot (cavitation) and differential radiation force on the clot squeezing and expanding the clot like a sponge to take up thrombolytic drugs. There are research reports of successful thrombolysis in coronary arteries [21.86], intracranial arteries [21.87, 88] and vascular access shunts for dialysis. The ultrasound intensities required for thrombolysis are similar to those used in diagnostic ultrasound. The mechanisms include streaming and disaggregation of the clot [21.89].

21.10 Lower-Frequency Therapies

Lower-frequency ultrasound is used for tissue disruption and cleaning purposes. To remove an opacified lens (cataract) from the eye through a small incision, a needle oscillating at 25–60 kHz is used to emulsify the lens (phacoemulsification), which is then removed by aspiration (suction).

Dental scaling is performed with a similar instrument operating at frequencies of 18–30 kHz and applied power of 30 W. Cavitation is an important mechanism in the cleaning process. Chemical additives such as toothpaste can suppress cavitation and degrade the cleaning efficiency.

21.11 Ultrasound Safety

Undoubtedly, ultrasound is safe. In over a half century of widespread use of diagnostic ultrasound, there are no currently accepted reports of harm to patients or examiners from the use of diagnostic ultrasound. However, in 1950, unamplified X-ray fluoroscopic imaging was considered so safe that it was used for shoe fitting in nearly all shoe stores, and for the entertainment of children waiting to be served. In 1950, smoking was considered to be so beneficial to health that nearly all doctors smoked. Although there are reports that ultrasound examination in pregnancy is associated with increased birth complications in retrospective studies where ultrasound was applied to high-risk pregnancies, a causative effect is unlikely. Fewer complications were found in pregnancies randomized to ultrasound examination compared to those with conventional care [21.90] because of the increased knowledge of physicians about the expected date of delivery. In one discussion of the safety of Doppler ultrasound examination of the ophthalmic artery through the eye, an experienced investigator stated "If the patient doesn't see flashes of light, I figure that the exam is safe". Still today, diagnostic ultrasound is considered so safe that, although there are occasional disclaimers such as "to be used only under the supervision of a physician", ultrasound examinations are performed for educational purposes on students, demonstrated on hired actors at trade shows, available for home use, and advertised

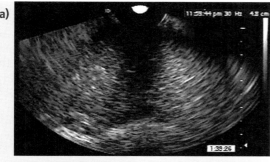
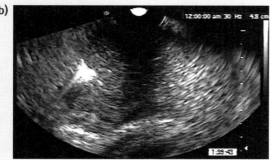
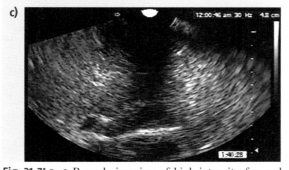

Fig. 21.74a–c B-mode imaging of high-intensity focused ultrasound lesion formation. Focal intensity 1000 W/cm2. (**a**) B-mode image of tissue before HIFU therapy. (**b**) Echogenic region at HIFU treatment zone after 15 s of exposure. (**c**) Echogenicity vanishes with time after cessation of HIFU exposure

for entertainment purposes. One prominent ultrasound safety expert warned that driving to the examination carried more risk than the risk associated with ultrasound examination.

In 1976, the United States congress passed the Medical Device Amendments providing that the Food and Drug Administration should have authority over medical devices [21.91]. At that time, any medical device in clinical use, in the absence of evidence of hazard, was considered safe. Any new medical device that was substantially equivalent to a pre-enactment device could be approved for distribution in interstate commerce and use by filing a 510k application showing equivalence. Devices that were not substantially equivalent required pre-market approval (PMA) as an investigational device. This started a search for existing ultrasound systems that had been sold across state lines so that the ultrasound output levels could be measured. These old systems are called pre-enactment systems. New ultrasound devices that had similar or lower acoustic outputs were considered equivalent and could be approved by the FDA under a 510k application, which was simpler and far less costly than the PMA process.

In 1976, continuous-wave (CW) Doppler systems were in use, pulsed Doppler systems were in use, M-mode, 2-D static B-mode and 2-D real-time B-mode systems were also in use. One CW Doppler system at the time caused red spots on the skin that were called *Doppler hickies*, but were still considered safe because there were no complaints.

The task was to characterize the equivalent acoustic output. Over the quarter century since enactment, various methods have been developed to describe the effect of ultrasound on tissue, to establish equivalence. Because of the common use of therapeutic ultrasound to create tissue hyperthermia, the first consideration was the heating effect of the ultrasound; later mechanical effects were included in the analysis.

The heating of tissue is governed by the bioheat equation. This equation relates the increase in tissue temperature to the rate of heat deposit minus the rate of heat removal divided by the heat capacity of the tissue. The rate of heat deposit is proportional to the incoming ultrasound power minus the outgoing ultrasound power, per unit mass. In a simple computation, the heat capacity of tissue is the same as water, 1 cal/gm or $4.2\,\mathrm{J/gm\,°C} = 4.2\,\mathrm{J/cm^3\,°C} = 4.2\,\mathrm{W\,s/cm^3\,°C}$. So, if ultrasound at an intensity of $4.2\,\mathrm{W/cm^2}$ is completely absorbed by a layer of tissue (water) that is 1 cm thick, the temperature will rise 1 °C per second, if no heat is carried away. $42\,\mathrm{mW/cm^2}$ ($420\,\mathrm{W/m^2}$) will raise the temperature by only 0.01 °C per second, allowing 100 s of exposure to raise the temperature from the normal body temperature of 37 °C to 38 °C, well below fever temperature considered safe by most people. Since at 4 MHz, only half of the ultrasound power is converted to heat in one cm, the heating rate is half of the values above.

Until 1985, ultrasound exposure levels were quoted in terms of $\mathrm{W/cm^2}$ related to tissue heating. Based on pre-enactment levels, diagnostic ultrasound systems had a temporal average (TA) acoustic intensity at the transducer focal zone measured in a water-filled test tank of $100\,\text{mW/cm}^2$ ($1\,\text{kW/m}^2$) or less. The major exception was ultrasound systems for the examination of the eye, which had pre-enactment intensities of $7\,\text{mW/cm}^2$ ($70\,\text{W/m}^2$). So, the generally accepted temporal average allowable intensities became: $7\,\text{mW/cm}^2$ ($70\,\text{W/m}^2$) for the eye and $100\,\text{mW/cm}^2$ ($1\,\text{kW/m}^2$) for all other diagnostic applications. The maximum intensity of sunlight is about $50\,\text{mW/cm}^2$ ($500\,\text{W/m}^2$), but the attenuation of sunlight at the skin is near 100%. Acceptable physical therapy ultrasound intensities were about 30 times higher: $3\,\text{W/cm}^2$ ($30\,\text{kW/m}^2$).

Ultrasound beam patterns are not uniform, especially in the Fresnel zone (near field). The intensities at the focus in diagnostic ultrasound beams are often three times higher than the average intensities. Because of this, ultrasound intensities were reported as spatial peak (SP) intensities, to indicate the variations, or spatial average (SA) intensities.

With pulsed-echo ultrasound imaging, the transmitted ultrasound is on for less than $0.5\,\mu\text{s}$ every millisecond along the same ultrasound beam pattern in M-mode examination; thus the ultrasound was on only 1/2000 of the time (duty factor or duty cycle of 0.05%). For a spatial peak temporal average intensity of $100\,\text{mW/cm}^2$, ($1\,\text{kW/m}^2$) the pulse average (PA) was 2000 times as great or $200\,\text{W/cm}^2$ ($2\,\text{kW/m}^2$) (SP PA) and the temporal peak (TP) intensity was twice that value (SPTP, Fig. 21.23, broadband pulse). A conventional pulsed Doppler examination with a $1\,\mu\text{s}$ transmit burst every $50\,\mu\text{s}$ has a duty factor of 2% and an SPTP intensity of $5\,\text{W/cm}^2$ ($50\,\text{kW/m}^2$). In both M-mode and Doppler examinations, the ultrasound beam pattern is held stationary for several seconds. But in real-time 2-D B-mode imaging, each successive beam axis is along a different line. Typically, the ultrasound is directed along the same line at the frame rate, about 30 times per second.

Over the next decade, the guidelines for ultrasound intensities in different applications became more sophisticated, separating application-specific output levels into four categories: general, cardiac, vascular, and ophthalmic. The motivation was to allow higher intensity levels to be used for Doppler blood velocity measurements because the echoes returned by blood were 0.000 001 (60 dB) lower than those from tissue. Rather than accepting the intensities measured at the focal zone in a water tank, the intensities were derated by computing the expected intensity if the water were replaced by attenuating tissue with an attenuation rate of $0.3\,\text{dB/(MHz cm)}$. Derating allows a 3 MHz cardiac Doppler system focused at a depth of 11 cm to derate the intensity by 10 dB, thus allowing a tenfold increase in transmit power. The derated SPTA limits were: fetal, neonatal and other general-purpose applications $-94\,\text{mW/cm}^2$; cardiac $-430\,\text{mW/cm}^2$; peripheral vascular $-720\,\text{mW/cm}^2$; ophthalmic $-17\,\text{mW/cm}^2$; the SPPA limits were $190\,\text{W/cm}^2$ for the first three categories and $28\,\text{W/cm}^2$ for ophthalmic exams.

Working together, the US Food and Drug Administration (FDA), National Electronic Manufacturers Association (NEMA) and the American Institute of Ultrasound in Medicine (AIUM) began to agreed on output display standards (ODS) for ultrasound instruments, including the thermal index (TI, the intensity required to raise the temperature of tissue by $1\,°\text{C}$) predictive of tissue heating, and the mechanical index (MI) predictive of cavitation and interactions with ultrasound contrast agents. The TI was further divided into subcategories for soft tissue (TIS), bone near the focus (TIB) and cranial bone near the skin (TIC) based on the expected absorption of ultrasound in the tissue. These values also include the fact that the ultrasound beam is scanning across the tissue in 2-D imaging. The computation of MI is based on the SPTP intensity values, includes attenuation derating but not the scanning of the ultrasound beam.

To better understand the MI values, refer to Fig. 21.5 which indicates that the amplitude of the pressure fluctuation for a $0.3\,\text{W/cm}^2$ wave is 1 atm, or $100\,\text{kPa} = 100\,\text{kN/m}^2$. The pressure increases by the square root of intensity, for a linear system (which this cannot be). So, for the M-mode SPTP intensity of $200\,\text{W/cm}^2$, ($2\,\text{MW/m}^2$) the pressure amplitude is 26 atm, making the peak positive pressure 27 atm (2.7 MPa) and the peak negative pressure (below zero) -25 atm ($-2.5\,\text{MPa}$). The MI is defined as the negative pressure amplitude [MPa] divided by the square root of the ultrasound frequency [MHz]. This provides a basis for the estimation of the chance of an adverse nonthermal event. At 3 MHz, the MI for this case is 1.5. An MI less than 1.9 is acceptable for obstetrical examinations in the absence of ultrasound contrast agents our other sources of gas bubbles.

In 1991, the FDA released its 510k guidance [21.92], establishing a new track for approval of non-PMA devices, regulatory track 3. This permitted increased outputs for devices implementing the output display standard developed jointly by the AIUM and NEMA [21.93]. This standard recognized two general classes of mechanisms by ultrasound could adversely affect biological tissue, thermal and mechanical, and provided simple metrics for the likelihood of such effects, the thermal index (TI) and mechanical index (MI).

Similar standards have been addressed in other countries by organizations and agencies including Health Canada, the European Committee for Medical Ultrasound Safety (ECMUS), the British Medical Ultrasound Society (BMUS), the European Federation of Societies for Ultrasound in Medicine and Biology (EFSUMB), the Australasian Society for Ultrasound in Medicine (ASUM), the Asian Federation for Societies of Ultrasound in Medicine and Biology (AFSUMB), the Latin American Federation of Ultrasound in Medicine and Biology (FLAUS), the Mediterranean and African Society of Ultrasound (MASU), and the World Federation for Ultrasound in Medicine and Biology (WFUMB).

In spite of the confidence that ultrasound is absolutely safe, all ultrasound agencies subscribe to the policy of using ultrasound exposure levels that are as low as reasonably achievable (ALARA) for diagnostic ultrasound examinations.

In the presence of ultrasound contrast agents or gas in the lungs or gut, there is a risk of damaging capillaries and creating interstitial hemorrhaging in tissues when high-MI examinations are performed.

However, there are some other considerations. In addition to heating tissue and the compression/decompression mechanical factors related to mechanical index, an ultrasound wave exerts radiation pressure on tissue as it passes through. The radiation pressure is proportional to intensity. Radiation force is used in the measurement of acoustic output. The body force is equal to the [power entering a region minus the power transmitted through and exiting the region] divided by the speed of sound. At a pulsed Doppler focus in the examination of the fetus, a temporal peak radiation pressure of 1700 Pa can be generated at a frequency equal to the pulse repetition frequency (PRF) of 5 kHz, which is well within the hearing range. If the ultrasound beam pattern is directed to intersect the ear, the radiation pressure oscillation can be heard by the fetus and cause increased fetal activity [21.94].

Of more serious concern is the increase in permeability in cell membranes and other biologic barriers to small and large molecules that can be induced by ultrasound [21.95]. This may prove useful for the delivery of therapeutic drugs through the skin (sonoporation) and genes across membranes including the placental barrier and the blood–brain barrier, but it might also allow the inadvertent transfer of pathogens. The effect is enhanced in the presence of gas and ultrasound contrast agents. It is possible that such effects have escaped detection by current monitoring methods.

The conceptual simplicity of the pulse-echo system has allowed instrument engineers to devote all efforts toward lowering the cost and improving the diagnostic utility of these instruments. Unlike most other medical imaging methods, ultrasound is low-cost, portable, real-time, and considered safe in most applications.

Further Reading

Please find a complete compilation of citations from Ultrasound in Medicine and Biology which list the thousands of articles on medical acoustics written prior to 1987: http://www.ncbi.nlm.nih.gov/entrez/query.fcgi?CMD=Display&DB=pubmed

References

21.1 G.E. Horn, J.M. Mynors: Recording the bowel sounds, Med. Biol. Eng. **4**(2), 205–208 (1966)

21.2 T. Tomomasa, A. Morikawa, R.H. Sandler, H.A. Mansy, H. Koneko, T. Masahiko, P.E. Hyman, Z. Itoh: Gastrointestinal sounds and migrating motor complex in fasted humans, Am. J. Gastroenterol. **94**(2), 374–381 (1999)

21.3 C. Leknius, B.J. Kenyon: Simple instrument for auscultation of temporomandibular joint sounds, J. Prosthet. Dent. **92**(6), 604 (2004)

21.4 S.E. Widmalm, W.J. Williams, D. Djurdjanovic, D.C. McKay: The frequency range of TMJ sounds, J. Oral. Rehabil. **30**(4), 335–346 (2003)

21.5 C. Oster, R.W. Katzberg, R.H. Tallents, T.W. Morris, J. Bartholomew, T.L. Miller, K. Hayakawa: Characterization of temporomandibular joint sounds. A preliminary investigation with arthrographic correlation, Oral Surg. Oral Med. Oral Pathol. **58**(1), 10–16 (1984)

21.6 A. Peylan: Direct auscultation of the joints; Preliminary clinical observations, Rheumatism **9**(4), 77–81 (1953)

21.7 E.P. McCutcheon, R.F. Rushmer: Korotkoff sounds. An experimental critique, Circ. Res. **20**(2), 149–161 (1967)

21.8 J.E. Meisner, R.F. Rushmer: Production of sounds in distensible tubes, Circ. Res. **12**, 651–658 (1963)

21.9 S. Jarcho: The young stethoscopist (H. I. Bowditch, 1846), Am. J. Cardiol. **13**, 808–819 (1964)

21.10 R.F. Rushmer, R.S. Bark, R.M. Ellis: Direct-writing heart-sound recorder (The sonvelograph), Ama. Am. J. Dis. Child **83**(6), 733–739 (1952)

21.11 M.A. Chizner: The diagnosis of heart disease by clinical assessment alone, Curr. Probl. Cardiol. **26**(5), 285–379 (2001)

21.12 M.A. Chizner: The diagnosis of heart disease by clinical assessment alone, Dis. Mon. **48**(1), 7–98 (2002)

21.13 A.E. Aubert, B.G. Denys, F. Meno, P.S. Reddy: Investigation of genesis of gallop sounds in dogs by quantitative phonocardiography and digital frequency analysis, Circulation **71**, 987–993 (1985)

21.14 R.F. Rushmer, C. Morgan: Meaning of murmurs, Am. J. Cardiol. **21**(5), 722–730 (1968)

21.15 J.E. Foreman, K.J. Hutchison: Arterial wall vibration distal to stenoses in isolated arteries, J. Physiol. **205**(2), 30–31 (1969)

21.16 J.E. Foreman, K.J. Hutchison: Arterial wall vibration distal to stenoses in isolated arteries of dog and man, Circ. Res. **26**(5), 583–590 (1970)

21.17 A.M. Khalifa, D.P. Giddens: Characterization and evolution poststenotic flow disturbances, J. Biomech. **14**(5), 279–296 (1981)

21.18 G.W. Duncan, J.O. Gruber, C.F. Dewey, G.S. Myers, R.S. Lees: Evaluation of carotid stenosis by phonoangiography, N. Engl. J. Med. **293**(22), 1124–1128 (1975)

21.19 J.P. Kistler, R.S. Lees, J. Friedman, M. Pressin, J.P. Mohr, G.S. Roberson, R.G. Ojemann: The bruit of carotid stenosis versus radiated basal heart murmurs. Differentiation by phonoangiography, Circulation **57**(5), 975–981 (1978)

21.20 J.P. Kistler, R.S. Lees, A. Miller, R.M. Crowell, G. Roberson: Correlation of spectral phonoangiography and carotid angiography with gross pathology in carotid stenosis, N. Engl. J. Med. **305**(8), 417–419 (1981)

21.21 R. Knox, P. Breslau, D.E. Strandness Jr.: Quantitative carotid phonoangiography, Stroke **12**, 798–803 (1981)

21.22 R.S. Lees: Phonoangiography: qualitative and quantitative, Ann. Biomed. Eng. **12**(1), 55–62 (1984)

21.23 R.S. Lees, C.F. Dewey Jr.: Phonoangiography: a new noninvasive diagnostic method for studying arterial disease, Proc. Natl. Acad. Sci. USA **67**, 935–942 (1970)

21.24 A. Miller, R.S. Lees, J.P. Kistler, W.M. Abbott: Spectral analysis of arterial bruits (phonoangiography): Experimental validation, Circulation **61**(3), 515–520 (1980)

21.25 A. Miller, R.S. Lees, J.P. Kistler, W.M. Abbott: Effects of surrounding tissue on the sound spectrum of arterial bruits in vivo, Stroke **11**(4), 394–398 (1980)

21.26 D. Smith, T. Ishimitsu, E. Craige: Mechanical vibration transmission characteristics of the left ventricle: implications with regard to auscultation and phonocardiography, J. Am. Coll. Cardiol. **4**, 517–521 (1984)

21.27 D. Smith, T. Ishimitsu, E. Craige: Abnormal diastolic mechanical vibration transmission characteristics of the left ventricle, J. Cardiogr. **15**, 507–512 (1985)

21.28 J.P. Murgo: Systolic ejection murmurs in the era of modern cardiology: What do we really know?, J. Am. Coll. Cardiol. **32**, 1596–1602 (1998)

21.29 H. Nygaard, L. Thuesen, J.M. Hasenkam, E.M. Pedersen, P.K. Paulsen: Assessing the severity of aortic valve stenosis by spectral analysis of cardiac murmurs (spectral vibrocardiography). Part I: Technical aspects, J. Heart Valve Dis. **2**, 454–467 (1993)

21.30 Z. Sun, K.K. Poh, L.H. Ling, G.S. Hong, C.H. Chew: Acoustic diagnosis of aortic stenosis, J. Heart Valve Dis. **14**, 186–194 (2005)

21.31 D.J. Doyle, D.M. Mandell, R.M. Richardson: Monitoring hemodialysis vascular access by digital phonoangiography, Ann. Biomed. Eng. **30**(7), 982 (2002)

21.32 C. Clark: Turbulent wall pressure measurements in a model of aortic stenosis, J. Biomech. **10**(8), 461–472 (1977)

21.33 M.B. Rappoport, H.B. Sprague: Physiologic and physical laws that govern auscultation and their clinical application, Am. Heart J. **21**, 257 (1941)

21.34 R. Tidwell, R. Rushmer, R. Polley: The office diagnosis of operable congenital heart lesions, Northwest Med. **52**(7), 546 (1953)

21.35 R. Tidwell, R. Rushmer, R. Polley: The office diagnosis of operable congenital heart lesions, Northwest Med. **52**(8), 635 (1953)

21.36 R. Tidwell, R. Rushmer, R. Polley: The office diagnosis of operable congenital heart lesions; coarctation of the aorta, Northwest Med. **52**(11), 927 (1953)

21.37 F.D. Fielder, P. Pocock: Foetal blood flow detector, Ultrasonics **6**(4), 240–241 (1968)

21.38 W.L. Johnson, H.F. Stegall, J.N. Lein, R.F. Rushmer: Detection of fetal life in early pregnancy with an ultrasonic doppler flowmeter, Obstet. Gynecol. **26**, 305–307 (1965)

21.39 R.F. Rushmer, C.L. Morgan, D.C. Harding: Electronic aids to auscultation, Med. Res. Eng. **7**(4), 28–36 (1968)

21.40 R.F. Rushmer, D.R. Sparkman, R.F. Polley, E.E. Bryan, R.R. Bruce, G.B. Welch, W.C. Bridges: Variability in detection and interpretation of heart murmurs; A comparison of auscultation and stethography, AMA Am. J. Dis. Child **83**(6), 740–754 (1952)

21.41 R.F. Rushmer, R.A. Tidwell, R.M. Ellis: Sonvelographic recording of murmurs during acute myocarditis, Am. Heart J. **48**(6), 835–846 (1954)

21.42 P.N.T. Wells: Review, absorption and dispersion of ultrasound in biological tissue, Ultrasound Med. Biol. **1**, 369–376 (1975)

21.43 M.A. Allison, J. Tiefenbrun, R.D. Langer, C.M. Wright: Atherosclerotic calcification and intimal medial

21.44 J.A. Jensen: FIELD (1997) http://www.es.oersted.dtu.dk/staff/jaj/field/index.html; http://www.es.oersted.dtu.dk/staff/jaj/old_field/

21.45 R.F. Rushmer, D.W. Baker, H.F. Stegall: Transcutaneous doppler flow detection as a nondestructive technique, J. Appl. Physiol. **21**(2), 554–566 (1966)

21.46 D.L. Franklin, W. Schlegel, R.F. Rushmer: Blood flow measured by doppler frequency shift of back-scattered ultrasound, Science **134**, 564–565 (1961)

21.47 H.F. Stegall, R.F. Rushmer, D.W. Baker: A transcutaneous ultrasonic blood-velocity meter, J. Appl. Physiol. **21**(2), 707–711 (1966)

21.48 D.E. Strandness Jr., E.P. McCutcheon, R.F. Rushmer: Application of a transcutaneous doppler flowmeter in evaluation of occlusive arterial disease, Surg. Gynecol. Obstet. **122**(5), 1039–1045 (1966)

21.49 D.E. Strandness Jr., R.D. Schultz, D.S. Sumner, R.F. Rushmer: Ultrasonic flow detection. A useful technic in the evaluation of peripheral vascular disease, Am. J. Surg. **113**(3), 311–320 (1967)

21.50 R.F. Rushmer, D.W. Baker, W.L. Johnson, D.E. Strandness: Clinical applications of a transcutaneous ultrasonic flow detector, JAMA **199**(5), 326–328 (1967)

21.51 C.R. Hill: Medical ultrasonics: an historical review, Br. J. Radiol. **46**(550), 899–905 (1973)

21.52 A.H. Crawford: Ultrasonic medical equipment, Ultrasonics **8**(2), 105–111 (1970)

21.53 B. Buttery, G. Davison: The ghost artifact, J. Ultras. Med. **3**, 49–52 (1984)

21.54 R.M. Ellis, D.L. Franklin, R.F. Rushmer: Left ventricular dimensions recorded by sonocardiometry, Circ. Res. **4**(6), 684–688 (1956)

21.55 R.F. Rushmer: Continuous measurements of left ventricular dimensions in intact, unanesthetized dogs, Circ. Res. **2**(1), 14–21 (1954)

21.56 J.C. Somer: Electronic Sector Scanning fo Ultrasonic Diagnosis, Ultrasonics **6**(3), 153–159 (1968)

21.57 F.E. Barber, D.W. Baker, A.W. Nation, D.E. Strandness Jr., J.M. Reid: Ultrasonic duplex-echo doppler scanner, IEEE Trans. Biomed. Eng. vBME **21**(2), 109–113 (1974)

21.58 D.N. Ku, D.P. Giddens: Pulsatile flow in a model carotid bifurcation, Arteriosclerosis **3**(1), 31–39 (1983)

21.59 T.L. Yearwood, K.B. Chandran: Experimental investigation of steady flow through a model of the human aortic arch, J. Biomech. **13**(12), 1075–1088 (1980)

21.60 T.L. Yearwood, K.B. Chandran: Physiological pulsatile flow experiments in a model of the human aortic arch, J. Biomech. **15**(9), 683–704 (1982)

21.61 D.E. Strandness Jr.: *Duplex Scanning in Vascular Disorders*, 3rd edn. (Lippincott Williams Wilkins, Philadelphia 2002)

21.62 G. Papanicolaou, K.W. Beach, R.E. Zierler, D.E. Strandness Jr.: The relationship between arm-ankle pressure difference and peak systolic velocity in patients with stenotic lower extremity vein grafts, Ann. Vasc. Surg. **9**(6), 554–560 (1995)

21.63 T.R. Kohler, S.C. Nicholls, R.E. Zierler, K.W. Beach, P.J. Schubert, D.E. Strandness Jr.: Assessment of pressure gradient by Doppler ultrasound: experimental and clinical observations, J. Vasc. Surg. **6**(5), 460–469 (1987)

21.64 D.N. White, J.M. Clark, M.N. White: Studies in ultrasonic echoencephalography. 7. General principles of recording information in ultrasonic B- and C-scanning and the effects of scatter, reflection and refraction by cadaver skull on this information, Med. Biol. Eng. **5**(1), 3–14 (1967)

21.65 D.N. White, J.M. Clark, M.N. White: Studies in ultrasonic echoencephalography. 8. The effects on resolution of irregularities in attenuation of an ultrasonic beam traversing cadaver skull, Med. Biol. Eng. **5**(1), 15–23 (1967)

21.66 D.N. White: Ultrasonic encephalography, Acta Radiol. Diagn.(Stockh.) **9**, 671–674 (1969)

21.67 D.N. Ku, D.P. Giddens, D.J. Phillips, D.E. Strandness Jr.: Hemodynamics of the normal human carotid bifurcation: in vitro and in vivo studies, Ultrasound Med. Biol. **11**(1), 13–26 (1985)

21.68 W.M. Blackshear Jr., D.J. Phillips, B.L. Thiele, J.H. Hirsch, P.M. Chikos, M.R. Marinelli, K.J. Ward, D.E. Strandness Jr.: Detection of carotid occlusive disease by ultrasonic imaging and pulsed doppler spectrum analysis, Surgery **86**(5), 698–706 (1979)

21.69 K.A. Comess, F.A. DeRook, M.L. Russell, T.A. Tognazzi-Evans, K.W. Beach: The incidence of pulmonary embolism in unexplained sudden cardiac arrest with pulseless electrical activity, Am. J. Med. **109**(5), 351–356 (2000)

21.70 R.C. Bahler, T. Mohyuddin, R.S. Finkelhor, I.B. Jacobs: Contribution of doppler tissue imaging and myocardial performance index to assessment of left ventricular function in patients with duchenne's muscular dystrophy, J. Am. Soc. Echocardiogr. **18**(6), 666–673 (2005)

21.71 J.K. Campbell, J.M. Clark, C.O. Jenkins, D.N. White: Ultrasonic studies of brain movements, Neurology **20**(4), 418 (1970)

21.72 S. Sikdar, K.W. Beach, S. Vaezy, Y. Kim: Ultrasonic technique for imaging tissue vibrations: preliminary results, Ultrasound Med. Biol. **31**, 221–232 (2005)

21.73 A. Evans, D.N. Walder: Detection of circulating bubbles in the intact mammal, Ultrasonics **8**(4), 216–217 (1970)

21.74 C. Shub, A.J. Tajik, J.B. Seward, D.E. Dines: Detecting intrapulmonary right-to-left shunt with contrast echocardiography. Observations in a pa-

21.75 E. Grube, H. Simon, M. Zywietz, G. Bodem, G. Neumann, A. Schaede: Echocardiographic contrast studies for the delineation of the heart cavities using peripheral venous sodium chloride injections. A contribution to shunt diagnosis, Verh. Dtsch. Ges. Inn. Med. **83**, 366–367 (1977)

tient with diffuse pulmonary arteriovenous fistulas, Mayo Clin. Proc. **51**(2), 81–84 (1976)

21.76 P.W. Serruys, W.B. Vletter, F. Hagemeijer, C.M. Ligtvoet: Bidimensional real-time echocardiological visualization of a ventricular right-to-left shunt following peripheral vein injection, Eur. J. Cardiol. **6**(2), 99–107 (1977)

21.77 D. Danilowicz, I. Kronzon: Use of contrast echocardiography in the diagnosis of partial anomalous pulmonary venous connection, Am. J. Cardiol. **43**(2), 248–252 (1979)

21.78 C. Tei, T. Sakamaki, P.M. Shah, S. Meerbaum, K. Shimoura, S. Kondo, E. Corday: Myocardial contrast echocardiography: a reproducible technique of myocardial opacification for identifying regional perfusion deficits, Circulation **67**(3), 585–593 (1983)

21.79 J.S. Rasor, E.G. Tickner: Microbubble precursors and methods for their production and use. US Patent 4 442 843 (1984)

21.80 J.L. Nelson, B.L. Roeder, J.C. Carinen, F. Roloff, W.G. Pitt: Ultrasonically activated chemotherapeutic drug delivery in a rat model, Cancer Res. **62**, 7280–7283 (2002)

21.81 S.A. Sapareto, W.C. Dewey: Thermal dose determination in cancer therapy, Int. J. Radiat. Oncol. Biol. Phys. **10**, 787–800 (1984)

21.82 M.R. Bailey, L.A. Crum, A.P. Evan, J.A. McAteer, J.C. Williams, O.A. Sapozhnikov, R.O. Cleveland, T. Colonius: *Cavitation in Shock Wave Lithotripsy (Cav03-0s-2-1-006)* (Fifth Int. Symp. Cavitation (CAV2003), Osaka 2003), http://cav2003.me.es.osaka-u.ac.jp/Cav2003/Papers/Cav03-OS-2-1-006.pdf

21.83 E.A. Noser, H.M. Shaltoni, C.E. Hall, A.V. Alexandrov, Z. Garami, E.D. Cacayorin, J.K. Song, J.C. Grotta, M.S. Campbell: Aggressive mechanical clot disruption. A safe adjunct to thrombolytic therapy in acute stroke?, Stroke **36**(2), 292–296 (2005)

21.84 S. Pfaffenberger, B. Devcic-Kuhar, C. Kollmann, S.P. Kastl, C. Kaun, W.S. Speidl, T.W. Weiss, S. Demyanets, R. Ullrich, H. Sochor, C. Wober, J. Zeitlhofer, K. Huber, M. Groschl, E. Benes, G. Maurer, J. Wojta, M. Gottsauner-Wolf: Can a commercial diagnostic ultrasound device accelerate thrombolysis? An in vitro skull model, Stroke **36**(1), 124–128 (2005)

21.85 S. Schafer, S. Kliner, L. Klinghammer, H. Kaarmann, I. Lucic, U. Nixdorff, U. Rosenschein, W.G. Daniel, F.A. Flachskampf: Influence of ultrasound operating parameters on ultrasound-induced thrombolysis in vitro, Ultrasound Med. Biol. **31**(6), 841–847 (2005)

21.86 D.S. Jeon, H. Luo, M.C. Fishbein, T. Miyamoto, M. Horzewski, T. Iwami, J.M. Mirocha, F. Ikeno, Y. Honda, R.J. Siegel: Noninvasive transcutaneous ultrasound augments thrombolysis in the left circumflex coronary artery – An in vivo canine study, Thromb. Res. **110**(2–3), 149–158 (2003)

21.87 A.V. Alexandrov, C.A. Molina, J.C. Grotta, Z. Garami, S.R. Ford, J. Alvarez-Sabin, J. Montaner, M. Saqqur, A.M. Demchuk, L.A. Moye, M.D. Hill, A.W. Wojner, CLOTBUST Investigators: Ultrasound-enhanced systemic thrombolysis for acute ischemic stroke, N. Engl. J. Med. **351**(21), 2170–2178 (2004)

21.88 J. Eggers, G. Seidel, B. Koch, I.R. Konig: Sonothrombolysis in acute ischemic stroke for patients ineligible for rt-PA, Neurology **64**(6), 1052–1054 (2005)

21.89 D. Harpaz: Ultrasound enhancement of thrombolytic therapy: observations and mechanisms, Int. J. Cardiovasc. Intervent. **3**(2), 81–89 (2000)

21.90 S.H. Eik-Nes, O. Okland, J.C. Aure, M. Ulstein: Ultrasound screening in pregnancy: A randomised controlled trial, Lancet **1**(8390), 1347 (1984)

21.91 FDA, U.S. Food and Drug Administration: *Information for Manufacturers Seeking Marketing Clearance of Diagnostic Ultrasound Systems and Transducers* (Center for Devices and Radiological Health U.S. Food and Drug Administration, Rockville 1997)

21.92 FDA: *510(k) Guide for Measuring and Reporting Output of Diagnostic Ultrasound Medical Devices* (Center for Devices and Radiological Health, U.S. Food and Drug Administration, Rockville 1991)

21.93 AIUM/NEMA: *Standard for Real-Time Display of Thermal and Mechanical Acoustic Output Indices on Diagnostic Ultrasound Equipment* (Am. Inst. Ultrasound Medicine, Laurel 1992)

21.94 M. Fatemi, P.L. Ogburn, J.F. Greenleaf: Fetal stimulation by pulsed diagnostic ultrasound, J. Ultrasound Med. **20**, 883–889 (2001)

21.95 M.A. Dinno, M. Dyson, S.R. Young, A.J. Mortimer, J. Hart, L.A. Crum: The significance of membrane changes in the safe and effective use of therapeutic and diagnostic ultrasound, Phys. Med. Biol. **34**(11), 1543–1552 (1989)

Part G Structural Acoustics and Noise

22 Structural Acoustics and Vibrations
Antoine Chaigne, Palaiseau, France

23 Noise
George C. Maling, Jr., Harpswell, USA

22. Structural Acoustics and Vibrations

This chapter is devoted to vibrations of structures and to their coupling with the acoustic field. Depending on the context, the radiated sound can be judged as desirable, as is mostly the case for musical instruments, or undesirable, like noise generated by machinery. In architectural acoustics, one main goal is to limit the transmission of sound through walls. In the automobile industry, the engineers have to control the noise generated inside and outside the passenger compartment. This can be achieved by means of passive or active damping. In general, there is a strong need for quieter products and better sound quality generated by the structures in our daily environment.

Structural acoustics and vibration is an interdisciplinary area, with many different potential applications. Depending on the specific problem under investigation, one has to deal with material properties, structural modifications, signal processing and measurements, active control, modal analysis, identification and localization of sources or nonlinear vibrations, among other hot topics.

In this chapter, the fundamental methods for the analysis of vibrations and sound radiation of structures are presented. It mainly focuses on general physical concepts rather than on specific applications such as those encountered in ships, planes, automobiles or buildings. The fluid–structure coupling is restricted to the case of light compressible fluids (such as air). Practical examples are given at the end of each section.

After a brief presentation of the properties of the basic linear single-degree-of-freedom oscillator, the linear vibrations of strings, beams, membranes, plates and shells are reviewed. Then, the structural–acoustic coupling of some elementary systems is presented, followed by a presentation of the main dissipation mechanisms in structures. The last section is devoted to nonlinear vibrations. In conclusion, a brief overview of some advanced topics in structural acoustics and vibrations is given.

22.1 Dynamics of the Linear Single-Degree-of-Freedom (1-DOF) Oscillator 903
  22.1.1 General Solution 903
  22.1.2 Free Vibrations 903
  22.1.3 Impulse Response and Green's Function 904
  22.1.4 Harmonic Excitation 904
  22.1.5 Energetic Approach 905
  22.1.6 Mechanical Power 905
  22.1.7 Single-DOF Structural–Acoustic System 906
  22.1.8 Application: Accelerometer 907

22.2 Discrete Systems 907
  22.2.1 Lagrange Equations 907
  22.2.2 Eigenmodes and Eigenfrequencies 909
  22.2.3 Admittances 909
  22.2.4 Example: 2-DOF Plate–Cavity Coupling 911
  22.2.5 Statistical Energy Analysis 912

22.3 Strings and Membranes 913
  22.3.1 Equations of Motion 913
  22.3.2 Heterogeneous String. Modal Approach 914
  22.3.3 Ideal String 916
  22.3.4 Circular Membrane in Vacuo 919

22.4 Bars, Plates and Shells 920
  22.4.1 Longitudinal Vibrations of Bars ... 920
  22.4.2 Flexural Vibrations of Beams 920
  22.4.3 Flexural Vibrations of Thin Plates 923
  22.4.4 Vibrations of Thin Shallow Spherical Shells 925
  22.4.5 Combinations of Elementary Structures 926

22.5 Structural–Acoustic Coupling 926
  22.5.1 Longitudinally Vibrating Bar Coupled to an External Fluid 927
  22.5.2 Energetic Approach to Structural–Acoustic Systems 932
  22.5.3 Oscillator Coupled to a Tube of Finite Length 934
  22.5.4 Two-Dimensional Elasto–Acoustic Coupling 936

22.6	**Damping**............ 940		22.7.2	Duffing Equation......... 949
	22.6.1 Modal Projection in Damped Systems 940		22.7.3	Coupled Nonlinear Oscillators 951
	22.6.2 Damping Mechanisms in Plates... 943		22.7.4	Nonlinear Vibrations of Strings ... 955
	22.6.3 Friction 945		22.7.5	Review of Nonlinear Equations for Other Continuous Systems 956
	22.6.4 Hysteretic Damping 947			
22.7	**Nonlinear Vibrations**............ 947	22.8	**Conclusion. Advanced Topics**............ 957	
	22.7.1 Example of a Nonlinear Oscillator 947	**References** 958		

In this chapter, the fundamental methods for the analysis of vibrations and sound radiation of structures are presented. It mainly focuses on general physical concepts rather than on specific applications such as those encountered in ships, planes, automobiles or buildings. The fluid–structure coupling is restricted to the case of light compressible fluid (such as air). The case of strong coupling with heavy fluid (water) is not treated.

If the magnitude of the vibratory and acoustical quantities of interest can be considered as sufficiently small, then the equations that govern the structural–acoustics phenomena can be linearized. The linear theory of vibrations applied to simple continuous elastic structures, such as bars, membranes, shells and plates, forms a reference framework for numerous studies involving more-complex geometries. In some practical situations, it is not necessary to know the exact vibratory shape at each point of a given system: to take advantage of this, a continuous structure can often be represented as a discrete system built with rigid masses connected to springs and dampers. The main results of the linear theory for the vibration of discrete and continuous systems are presented in Sects. 22.1–22.4. Section 22.1 is devoted to the linear single-degree-of-freedom (DOF) oscillator for which the fundamental results are derived with the help of the Laplace transform. In Sect. 22.2, the case of discrete systems with multiple DOFs is used to introduce the concepts of eigenmodes, eigenfrequencies, the admittance and modal analysis. As an illustration, one example of coupled oscillators is solved in detail. Also the fundamental energy equation of statistical energy analysis (SEA) that governs the mean power flow between coupled oscillators is summarized. Section 22.3 deals with the transverse vibrations of strings and membranes, which can be viewed as limiting cases of prestressed structures with negligible stiffness. In Sect. 22.4, longitudinal and transverse vibrations of bars are presented. The case of flexural vibrations is extended to two-dimensional (2-D) structures such as plates and shallow shells. In this latter case, the influence of curvature is emphasized using the example of a spherical cap for which analytical results can be obtained.

The three remaining sections focus on specific problems of great significance in structural acoustics. The case of vibroacoustic coupling is treated in Sect. 22.5 through examples of increasing complexity. Starting from the simple case of a longitudinally vibrating bar coupled to a semi-infinite tube, it is shown that modes of the structure are coupled to the radiated field, and that eigenfrequencies and modal shapes of the in vacuo structure are modified by this coupling. An energetic approach to the structural–acoustic coupling allows the introduction of radiation filters and radiation efficiencies. A state-space formulation of the phenomena is presented, which is of particular interest for the control of radiated sound. The example of a single-DOF oscillator coupled to a tube illustrates the coupling between a structure and a cavity. Finally, attention is paid to fluid–plate coupling, with the introduction of the concept of the critical frequency. Section 22.6 is devoted to the presentation of important causes of damping in structures. Also some indications are given with regard to the validity of the modal approach for damped discrete and continuous systems.

For structures subjected to motion with large amplitude, most of the concepts and methods developed for linear vibrations no longer apply. Specific methods must be used to account for the observed phenomena of distortion, amplitude-dependent resonance, jump, hysteresis, instability and chaos. Section 22.7 introduces the basic concepts of nonlinear vibrations with some examples. The Duffing equation and the example of two nonlinearly coupled oscillators are analyzed in detail. These examples illustrate the use of perturbation methods for solving nonlinear equations of motion for vibrating systems. Finally, some equations for the vibrations of nonlinear continuous systems (strings, beams, shells and plates) are briefly reviewed.

22.1 Dynamics of the Linear Single-Degree-of-Freedom (1-DOF) Oscillator

In the low-frequency range, the linear motion of an electrodynamical loudspeaker, for example, can be described by a single-degree-of-freedom oscillator. This is also the case for any complex structure considered as a rigid body for which the equivalent stiffness is represented by a unique spring, and where all the dissipation mechanisms are approximated by a unique dashpot or *fluid damping* constant; one can think of the rigid-body mode of vibration of an automobile. Also the 1-DOF oscillator has fundamental mathematical properties of great interest in the field of vibrations, since linear vibrations of discrete and continuous structures can be represented by a set of oscillators, as will be shown in the following sections.

The motion of a standard 1-DOF oscillator is governed by a second-order differential equation involving an excitation force $F(t)$ and the response of the structure. Using a formulation with the velocity $v(t)$, we get

$$F = M \frac{\mathrm{d}v}{\mathrm{d}t} + Rv + K \int v \, \mathrm{d}t . \tag{22.1}$$

This equation accounts for the motion of a mass M subjected to a restoring force due to a linear spring of stiffness K and a dashpot R which represents a linear viscous fluid damping (Fig. 22.1).

Equation (22.1) can be written alternatively, using the mass displacement $\xi(t)$ and the usual reduced parameters:

$$F = M \left(\frac{\mathrm{d}^2 \xi}{\mathrm{d}t^2} + 2\zeta_0 \omega_0 \frac{\mathrm{d}\xi}{\mathrm{d}t} + \omega_0^2 \xi \right) , \tag{22.2}$$

where $\omega_0 = \sqrt{M/K}$ is the eigenfrequency of the lossless system and $\zeta_0 = R/2M\omega_0$ is the nondimensional damping coefficient.

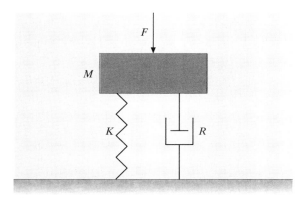

Fig. 22.1 Single-DOF oscillator

22.1.1 General Solution

One strategy for solving a linear equation such as (22.2) is to make use of the Laplace transform [22.1]. Using this transform, (22.2) becomes

$$s^2 \mathcal{X}(s) - s\xi(0) - \dot{\xi}(0) + 2\zeta_0 \omega_0 \left[s\mathcal{X}(s) - \xi(0) \right] + \omega_0^2 \mathcal{X}(s) = \frac{\mathcal{F}(s)}{M} , \tag{22.3}$$

where $\mathcal{X}(s)$ and $\mathcal{F}(s)$ are the Laplace transforms of $\xi(t)$ and $F(t)$, respectively. Using the tables and properties of the Laplace transforms, we obtain the general solution

$$\xi(t) = \frac{1}{M} \int_0^t F(\theta) G(t - \theta) \, \mathrm{d}\theta + \left[2\zeta_0 \omega_0 \xi(0) + \dot{\xi}(0) \right] G(t) + \xi(0) \dot{G}(t) , \tag{22.4}$$

where $G(t)$ can take different forms, depending on the value of the damping coefficient ζ_0:

- if $\zeta_0 < 1$, which corresponds to the so-called *underdamped* case, then

$$G(t) = \mathrm{e}^{-\zeta_0 \omega_0 t} \frac{\sin \omega_0 \sqrt{1 - \zeta_0^2}\, t}{\omega_0 \sqrt{1 - \zeta_0^2}} ; \tag{22.5}$$

- if $\zeta_0 = 1$, which corresponds to the *critical* case, we obtain

$$G(t) = t \mathrm{e}^{-\omega_0 t} ; \tag{22.6}$$

- finally, if $\zeta_0 > 1$, which corresponds to the *overdamped* case, we get

$$G(t) = \mathrm{e}^{-\zeta_0 \omega_0 t} \frac{\sinh \omega_0 \sqrt{\zeta_0^2 - 1}\, t}{\omega_0 \sqrt{\zeta_0^2 - 1}} . \tag{22.7}$$

22.1.2 Free Vibrations

Free vibrations are characterized by the absence of an external force. With $F(t) = 0$ in (22.4), we get

$$\xi(t) = \left[2\zeta_0 \omega_0 \xi(0) + \dot{\xi}(0) \right] G(t) + \xi(0) \dot{G}(t) , \tag{22.8}$$

- if $\zeta_0 < 1$, the displacement $\xi(t)$ is given by

$$\xi(t) = e^{-\zeta_0 \omega_0 t} \left\{ \xi(0) \cos \omega_0 \sqrt{1-\zeta_0^2}\, t \right.$$

$$+ \left[\xi(0) \frac{\zeta_0}{\sqrt{1-\zeta_0^2}} + \dot{\xi}(0) \frac{1}{\omega_0 \sqrt{1-\zeta_0^2}} \right]$$

$$\left. \times \sin \omega_0 \sqrt{1-\zeta_0^2}\, t \right\} . \qquad (22.9)$$

In this case, the free regime is characterized by a damped sinusoid with oscillation frequency $\omega_0 \sqrt{1-\zeta_0^2}$ and decay time $\tau = 1/\zeta_0 \omega_0$. The total energy of the system decreases with time.

- if $\zeta_0 = 1$, we obtain

$$\xi(t) = e^{-\omega_0 t} \left[\xi(0)(1+\omega_0 t) + \dot{\xi}(0) t \right]. \qquad (22.10)$$

The motion exhibits no oscillation. This critical case corresponds to the situation where we get the fastest non-oscillatory motion toward equilibrium;

- if $\zeta_0 > 1$ the displacement $\xi(t)$ becomes

$$\xi(t) = e^{-\zeta_0 \omega_0 t} \left\{ \xi(0) \cosh \omega_0 \sqrt{\zeta_0^2 - 1}\, t \right.$$

$$+ \left[\xi(0) \frac{\zeta_0}{\sqrt{\zeta_0^2 - 1}} + \dot{\xi}(0) \frac{1}{\omega_0 \sqrt{\zeta_0^2 - 1}} \right]$$

$$\left. \times \sinh \omega_0 \sqrt{\zeta_0^2 - 1}\, t \right\} . \qquad (22.11)$$

Here again, no oscillation exists. The decay is governed by the combination of two exponentials with decay time τ such that $1/\tau = \omega_0 (\zeta_0 \pm \sqrt{\zeta_0^2 - 1})$.

22.1.3 Impulse Response and Green's Function

Consider now the case where $\xi(0) = \dot{\xi}(0) = 0$ and where $F(t)$ is the Dirac delta function $\delta(t)$ whose fundamental property is

$$\int_{-\infty}^{\infty} u(t) \delta(t-a)\, dt = u(a), \qquad (22.12)$$

where $u(t)$ is a test function. The response $\xi(t)$ can therefore be written as

$$\xi(t) = \frac{1}{M} \int_0^t \delta(\theta) G(t-\theta)\, d\theta$$

$$= \frac{1}{M} \int_0^t \delta(t-\theta) G(\theta)\, d\theta = \frac{G(t)}{M}. \qquad (22.13)$$

The function $G(t)$ defined in Sect. 22.1.1 can therefore be defined as the impulse response of the oscillator with mass equal to unity. This function is also referred to as the *Green's function* of the oscillator. For an oscillator initially at rest and excited by an arbitrary time function $F(t)$, the response is given by the convolution integral of $F(t)$ and the Green's function $G(t)$. Equation (22.4) shows, in addition, that for $F(t) = 0$ and $\xi(0) = 0$, the displacement is given by

$$\xi(t) = \dot{\xi}(0) G(t). \qquad (22.14)$$

Therefore, we can see that the solution obtained with a Dirac delta force is proportional to the one obtained with an initial velocity $\dot{\xi}(0)$.

22.1.4 Harmonic Excitation

For harmonic excitation, $F(t) = F_M \sin \Omega t$, of the oscillator in the underdamped case, (22.4) gives

$$\xi(t) = \frac{F_M}{M} \frac{(\omega_0^2 - \Omega^2) \sin \Omega t - 2\zeta_0 \omega_0 \Omega \cos \Omega t}{(\omega_0^2 - \Omega^2)^2 + (2\zeta_0 \omega_0 \Omega)^2}$$

$$+ \alpha G(t) + \beta \dot{G}(t), \qquad (22.15)$$

where α and β are constants that depend on the initial conditions. The Green's function $G(t)$ and its time derivative are proportional to $e^{-\zeta_0 \omega_0 t}$ and become negligible for $\zeta_0 \omega_0 t \gg 1$. With this assumption, the response to the forcing term at frequency Ω is given by the first term on the right-hand side of (22.15). This term can be written alternatively as

$$\xi(t) = A(\Omega) \sin[\Omega t - \Phi(\Omega)] \qquad (22.16)$$

with

$$A(\Omega) = \frac{F_M}{M} \frac{1}{\sqrt{(\omega_0^2 - \Omega^2)^2 + (2\zeta_0 \omega_0 \Omega)^2}}$$

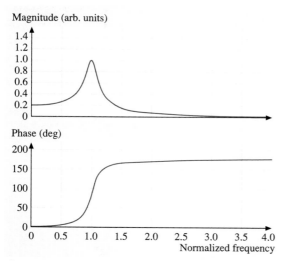

Fig. 22.2 Magnitude and phase of the mass displacement for the forced underdamped oscillator as a function of the normalized frequency Ω/ω_0

and

$$\tan \Phi(\Omega) = \frac{2\zeta_0 \omega_0 \Omega}{\omega_0^2 - \Omega^2} \,. \tag{22.17}$$

Figure 22.2 shows the variations of A and Φ with Ω in the underdamped case. The amplitude A reaches its maximum for frequency $\Omega = \omega_0\sqrt{1 - 2\zeta_0^2}$, which is very close to ω_0 for $\zeta_0 \ll 1$. The maximum is equal to $F_M/2M\zeta_0\omega_0\Omega$.

22.1.5 Energetic Approach

The instantaneous mechanical power $p_m(t)$ into the system is given by the scalar product of F and v, which yields

$$p_m = \frac{d}{dt}\left(\frac{1}{2}Mv^2 + \frac{1}{2}K\xi^2\right) + Rv^2 \,. \tag{22.18}$$

The terms on the right-hand side of (22.18) represent the kinetic energy of the mass M, the elastic energy of the spring K and the energy dissipated in the mechanical resistance R, respectively. In a number of applications, we are mostly interested in the time-averaged value of $p_m(t)$ rather than in details of its time evolution. In audio acoustics, for example, the human ear is sensitive to the sound level, which is correlated to the average value of the instantaneous acoustic power, after integration over a duration of nearly 50 ms. Therefore, after defining an integration duration T, whose appropriate selection is discussed later in this section, we can define the average mechanical power $\mathcal{P}_m(T)$ as

$$\begin{aligned}\mathcal{P}_m(T) &= \frac{1}{2T}\left[Mv^2(T) + K\xi^2(T) - Mv^2(0) - K\xi^2(0)\right] \\ &\quad + \frac{1}{T}\int_0^T Rv^2(t)\,dt \,.\end{aligned} \tag{22.19}$$

In a conservative system, the sum of the kinetic and elastic energy remains constant over time, so that the term between the square brackets in (22.19) is equal to zero. As a consequence, the average input power becomes:

$$\mathcal{P}_m(T) = \frac{1}{T}\int_0^T Rv^2(t)\,dt = \mathcal{P}_s(T) \,, \tag{22.20}$$

where $\mathcal{P}_s(T)$ represents the mean *structural* power dissipated in the mechanical resistance R.

22.1.6 Mechanical Power

A particular case of importance is the steady-state harmonic motion of the mechanical oscillator with angular frequency ω. Given an excitation force $F(t) = F_M \sin \omega t$, then, due to the assumed linearity of the system, the mass velocity is written $v(t) = V_M \sin(\omega t + \phi)$. Therefore, $\mathcal{P}_m(T)$ is given by

$$\mathcal{P}_m(T) = \frac{1}{T}\int_0^T F_M V_M \sin \omega t \sin(\omega t + \phi)\,dt \,. \tag{22.21}$$

Denoting the period of motion by $\tau = 2\pi/\omega$, we can write $T = n\tau + \tau_0$, where n is a positive integer. In this case, the mean power can be rewritten

$$\begin{aligned}\mathcal{P}_m(T) &= \frac{1}{2}F_M V_M \cos \phi + \frac{F_M V_M}{4(2\pi n + \tau_0 \omega)} \\ &\quad \times [\sin \phi - \sin(2\omega\tau_0 + \phi)] \,.\end{aligned} \tag{22.22}$$

Comment. This equation (22.22) shows that the mean (or average) power $\mathcal{P}_m(T)$ is nearly equal to $1/2 F_M V_M \cos \phi$ only if the average duration T contains a sufficiently large number n of periods. If T is equal to $n\tau$, then the previous result is strict. In what follows, it will be assumed that this condition is fulfilled, so that the dependence versus integration time T of the mean power terms will be suppressed. For a given force, the velocity amplitude V_M and phase angle ϕ are given by

$$V_M = \frac{F_M}{M} \frac{\omega}{\sqrt{(\omega^2 - \omega_0^2)^2 + 4\zeta_0^2 \omega^2 \omega_0^2}}, \quad (22.23)$$

$$\cos\phi = \frac{2\zeta_0 \omega_0 \omega}{\sqrt{(\omega^2 - \omega_0^2)^2 + 4\zeta_0^2 \omega^2 \omega_0^2}}, \quad (22.24)$$

and the mean power becomes:

$$\mathcal{P}_m = \frac{F_M^2}{2R} \frac{4\zeta_0^2 \omega_0^2 \omega^2}{(\omega^2 - \omega_0^2)^2 + 4\zeta_0^2 \omega^2 \omega_0^2}. \quad (22.25)$$

Equation (22.25) shows that the maximum of the dissipated power, and thus the maximum of the input power, is obtained for $\omega = \omega_0$, i.e. when the excitation frequency is equal to the eigenfrequency of the oscillator. In this case, we obtain:

$$\text{Max}(\mathcal{P}_m) = \frac{F_M^2}{2R}. \quad (22.26)$$

Since F_M is known, in general, (22.26) can be used for estimating the mechanical resistance R.

Remark. Recall that (22.25) is only valid in the dynamic equilibrium, which does not correspond to many experimental (or numerical) situations where, for obvious causality reasons, the force is applied at a given instant of time, usually taken as origin. In this more realistic case, the force signal should be written as $F(t) = F_M H(t) \sin \omega t$, where $H(t)$ is the Heaviside function. In this case, the Laplace transform of the velocity is

$$V(s) = \frac{F_M}{M} \frac{s\omega}{(s^2 + \omega^2)(s^2 + 2\zeta_0 \omega_0 s + \omega_0^2)}. \quad (22.27)$$

From which the velocity is derived

$$v(t) = \frac{F_M \omega}{M\sqrt{D(\omega)}} + A(\omega) \exp(-\zeta_0 \omega_0 t) \times \sin\left[\omega_0 \sqrt{1-\zeta_0^2}\, t + \psi(\omega)\right], \quad (22.28)$$

where $D(\omega) = (\omega^2 - \omega_0^2)^2 + 4\zeta_0^2 \omega^2 \omega_0^2$. $A(\omega)$ and $\psi(\omega)$ are also functions of the excitation frequency whose exact expressions do not add significant matter to the present discussion. The first term in the expression of $v(t)$ corresponds to the steady-state regime. The important features of the second term are the following:

- It is non-negligible as long as the time is small compared to the decay time $(\zeta_0 \omega_0)^{-1}$. For lightly damped structural modes, the decay time can be of the order of 0.1 ms or more. The second term cannot then be neglected during the first 0.5–1.0 s of the sound, if one wants to estimate the mean sound power correctly.
- When multiplying $v(t)$ by $F(t)$ and integrating over time, terms with frequencies $\omega + \omega_0 \sqrt{1-\zeta_0^2}$ and $\left|\omega - \omega_0\sqrt{1-\zeta_0^2}\right|$ appear in the expression of $\mathcal{P}_m$. As a consequence, the mean power fluctuates at low frequency, which is another cause of difficulty for estimating its value properly.

22.1.7 Single-DOF Structural–Acoustic System

We now investigate the simple example of a mechanical oscillator acting as a piston at one end of an air-filled semi-infinite tube presenting an additional acoustical resistance $R_a = \rho c S$. The oscillator motion is now governed by the equation

$$F = M\frac{\mathrm{d}v}{\mathrm{d}t} + Rv + \int v\,\mathrm{d}t + R_a v. \quad (22.29)$$

The instantaneous input power is given by

$$p_m(t) = \frac{\mathrm{d}}{\mathrm{d}t}\left(\frac{1}{2}Mv^2 + \frac{1}{2}K\xi^2\right) + (R + R_a)v^2. \quad (22.30)$$

The average input power becomes

$$\mathcal{P}_m(T) = \frac{1}{T}\int_0^T Rv^2(t)\,\mathrm{d}t + \frac{1}{T}\int_0^T R_a v^2(t)\,\mathrm{d}t$$
$$= \mathcal{P}_s(T) + \mathcal{P}_a(T), \quad (22.31)$$

where $\mathcal{P}_a(T)$ is the acoustic mean power radiated in the tube. The acoustical efficiency of the system is given by .

$$\eta = \frac{\mathcal{P}_a(T)}{\mathcal{P}_m(T)} = \frac{\mathcal{P}_a(T)}{\mathcal{P}_s(T) + \mathcal{P}_a(T)} = \frac{R_a}{R + R_a} \quad (22.32)$$

Some remarks can be formulated with regards to (22.32).

- The expression for the acoustical efficiency is independent of the integration time T.
- Though (22.32) appears simple in form, the experimental (or numerical) determination of the efficiency is generally not straightforward. R can be estimated through calculation of the mechanical power in vacuo $\mathcal{P}_{mo}(T)$.
- In the simple example discussed here, the acoustical resistance, and hence the acoustic power, is known.

This does not include the case where the acoustic power is not known, and has to be estimated by measurements of the acoustic intensity in the fluid, for example.

Link Between Sound Power and Free-Vibration Decay Times

For a single-DOF system, an alternative method for estimating the sound power consists of estimating the decay times of free vibrations through experiments or numerical calculations. Let us take the example of the previously described oscillator loaded by the semi-infinite tube. We consider the case $F(t) = 0$, with the mass initially moved from equilibrium by a quantity $\xi(0) = \xi_0$ and released with zero velocity at $t = 0$. The equation of motion is written as

$$\frac{d^2\xi}{dt^2} + 2\zeta\omega_0 \frac{d\xi}{dt} + \omega_0^2 \xi = 0 \tag{22.33}$$

with

$$\zeta = \frac{R + R_a}{2M\omega_0}.$$

The Laplace transform of the displacement is given by

$$\xi(s) = \xi_0 \frac{s + 2\zeta\omega_0}{s^2 + 2\zeta\omega_0 s + \omega_0^2} \tag{22.34}$$

from which the time evolution of the mass displacement is obtained (assuming $\zeta < 1$)

$$\xi(t) = \exp(-\zeta\omega_0 t)\left[\cos\left(\omega_0\sqrt{1-\zeta^2}\,t\right) + \frac{\zeta}{\sqrt{1-\zeta^2}}\sin\left(\omega_0\sqrt{1-\zeta^2}\,t\right)\right]. \tag{22.35}$$

Equation (22.35) shows that the *decay factor* α (equal to the inverse of the decay time) is equal to $\zeta\omega_0 = (R + R_a)/2M$. The same mathematical derivations applied to the oscillator in vacuo yields $\alpha_0 = \zeta_0\omega_0 = R/2M$. In conclusion, this shows that, for the 1-DOF system studied here, the acoustical efficiency can be estimated in the time domain by the expression

$$\eta = \frac{\alpha - \alpha_0}{\alpha}. \tag{22.36}$$

22.1.8 Application: Accelerometer

Accelerometers are piezoelectric transducers which are widely used for vibration measurements. An accelerometer has a base, a piezoelectric crystal and a seismic mass (Fig. 22.3). It delivers an electric signal which is proportional to the compression force applied to the crystal. Such a device can be modeled by a 1-DOF oscillator equation. Let us denote by k the equivalent stiffness of the crystal, by m the seismic mass, and by $\omega_0 = \sqrt{k/m}$ the resonance frequency of the accelerometer. Internal damping is neglected here, for simplicity. For a sinusoidal excitation $A\sin\Omega t$ of the base, the compression force is [22.2]:

$$F \simeq -\frac{k\Omega^2}{\omega_0^2 - \Omega^2} A \sin \Omega t. \tag{22.37}$$

Therefore, we can see that if $\Omega \ll \omega_0$, then the signal delivered by the accelerometer is proportional to the acceleration of the base. The frequency response curve of an accelerometer is flat below its resonance frequency, which gives the upper limit of its valid frequency range.

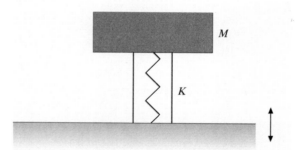

Fig. 22.3 Accelerometer

22.2 Discrete Systems

In a number of applications, real structures can be approximated by an assembly of discrete rigid substructures, which leads to enormous computational simplifications. In fact numerical modeling (finite elements, finite differences) is such an example where continuous structures are replaced by an equivalent set of discretely connected elements. In experimental techniques, such as modal analysis, measurements are taken at discrete positions on the structure under study, and thus the measurements are treated by methods applicable to discrete systems.

22.2.1 Lagrange Equations

For a discrete system with n degrees of freedom, the kinetic energy E_k can be formulated as a function of the generalized coordinates q_n and velocities $\dot{q}_n$

$$E_k = E_k(q_1, q_2, \ldots, q_n, \dot{q}_1, \dot{q}_2, \ldots \dot{q}_n) = E_k(q, \dot{q}) \tag{22.38}$$

Thus, given virtual vectors of displacements δq and velocities $\delta \dot{q}$, the variation of the kinetic energy is written

$$\delta E_k = \sum_{k=1}^{n} \frac{\partial E_k}{\partial q_k} \delta q_k + \frac{\partial E_k}{\partial \dot{q}_k} \delta \dot{q}_k . \tag{22.39}$$

Remark. Recall that δq has to be kinematically compatible with the system, though not necessarily equal to a real displacement, so that the partial derivatives of T versus δq and $\delta \dot{q}$ are independent.

Applying Hamilton's principle, with $\delta q = 0$ at t_1 and t_2, we can write the minimization integral as

$$\int_{t_1}^{t_2} (\delta E_k - \delta V + \delta W_{nc}) \, \mathrm{d}t = 0 \tag{22.40}$$

where V is the potential energy of the system and W_{nc} is the energy of the nonconservative applied external forces. Through integration by parts, we can write

$$\int_{t_1}^{t_2} \frac{\partial E_k}{\partial \dot{q}_k} \delta \dot{q}_k \, \mathrm{d}t$$

$$= \left[\frac{\partial E_k}{\partial \dot{q}_k} \delta q_k \right]_{t_1}^{t_2} - \int_{t_1}^{t_2} \frac{\mathrm{d}}{\mathrm{d}t}\left(\frac{\partial T}{\partial \dot{q}_k}\right) \delta q_k \, \mathrm{d}t$$

$$= -\int_{t_1}^{t_2} \frac{\mathrm{d}}{\mathrm{d}t}\left(\frac{\partial E_k}{\partial \dot{q}_k}\right) \delta q_k \, \mathrm{d}t . \tag{22.41}$$

Assuming that $V = V(q)$, which corresponds to a large class of problems in dynamics and vibrations,

$$\delta V = \sum_{k=1}^{n} \frac{\partial V}{\partial q_k} \delta q_k . \tag{22.42}$$

Finally, the virtual work δW_{nc} can be written as the scalar product of the vectors F of generalized forces and virtual displacement δq in the form $\delta W_{nc} = \sum_{k=1}^{n} F_k \delta q_k$, so that (22.40) becomes

$$\sum_{k=1}^{n} \int_{t_1}^{t_2} \left[\frac{\partial E_k}{\partial q_k} - \frac{\mathrm{d}}{\mathrm{d}t}\left(\frac{\partial E_k}{\partial \dot{q}_k}\right) - \frac{\partial V}{\partial q_k} + F_k \right] \delta q_k \, \mathrm{d}t = 0 . \tag{22.43}$$

Since the integral in (22.43) must be equal to zero for any δq this implies, for each k

$$\frac{\mathrm{d}}{\mathrm{d}t}\left(\frac{\partial E_k}{\partial \dot{q}_k}\right) - \frac{\partial E_k}{\partial q_k} + \frac{\partial V}{\partial q_k} = F_k . \tag{22.44}$$

The set of n differential equations (22.44) are the Lagrange equations of the discrete system. This set yields the equations of motion in a very elegant and practical manner.

Small Displacements. Linearization

For small perturbations from equilibrium, we can write $q_k = Q_k + \varepsilon X_k$, where $\varepsilon \ll 1$. Assuming that Q is given by the Lagrange equations and that the generalized forces F are independent of q, a first-order approximation of (22.44), for each k, gives

$$\frac{\partial^2 E_k}{\partial \dot{q}_k \partial \dot{q}_j} \ddot{X}_k +$$
$$\left[\frac{\mathrm{d}}{\mathrm{d}t}\left(\frac{\partial^2 E_k}{\partial \dot{q}_k \partial \dot{q}_j}\right) + \frac{\partial^2 E_k}{\partial q_k \partial \dot{q}_j} - \frac{\partial^2 E_k}{\partial \dot{q}_k \partial q_j} \right] \dot{X}_k +$$
$$\left[\frac{\mathrm{d}}{\mathrm{d}t}\left(\frac{\partial^2 E_k}{\partial q_k \partial \dot{q}_j}\right) + \frac{\partial^2 V}{\partial q_k \partial q_j} - \frac{\partial^2 E_k}{\partial q_k \partial q_j} \right] X_k = 0 ,$$
$$\tag{22.45}$$

which becomes in matrix form

$$\mathbf{M}\ddot{X} + \mathbf{C}\dot{X} + \mathbf{K}X = 0 . \tag{22.46}$$

1. Since the kinetic energy E_k is a positive quadratic form of the velocity, the mass matrix $\mathbf{M}$ is a symmetric positive operator.
2. The matrix $\mathbf{C}$ is a generalized damping matrix made up of three terms:
   - The first, with generic element $C_{kj1} = \frac{\mathrm{d}}{\mathrm{d}t}\left(\frac{\partial^2 E_k}{\partial \dot{q}_k \partial \dot{q}_j}\right)$ is associated with the temporal variation of the mass matrix;
   - The two other terms $\frac{\partial^2 E_k}{\partial q_k \partial \dot{q}_j} - \frac{\partial^2 E_k}{\partial \dot{q}_k \partial q_j}$ relate to gyroscopic forces. This part of the matrix $\mathbf{C}$ is antisymmetric and does not lead to dissipation.
3. The stiffness matrix $\mathbf{K}$ also is a sum of three terms:
   - The first is associated with the time variation of the first gyroscopic term in $\mathbf{C}$;
   - The second is governed by the q dependence of V;
   - The third is governed by the q dependence of E_k.

In what follows, we restrict our attention to conservative systems where E_k is independent of q and where the mass matrix is taken to be constant with time. In this case, (22.46) becomes

$$\mathbf{M}\ddot{X} + \mathbf{K}X = 0 \qquad (22.47)$$

with

$$\begin{cases} M_{kj} = M_{jk} = \dfrac{\partial^2 E_k}{\partial \dot{q}_k \partial \dot{q}_j} \\ K_{kj} = K_{jk} = \dfrac{\partial^2 V}{\partial q_k \partial q_j} \end{cases} \qquad (22.48)$$

22.2.2 Eigenmodes and Eigenfrequencies

Natural modes (or eigenmodes) are the sinusoidal (or harmonic) solutions of (22.47) with frequency ω in the absence of a driving term. As a consequence, the natural modes are solutions of the equation

$$\left(-\omega^2 \mathbf{M} + \mathbf{K}\right) X = 0 \qquad (22.49)$$

with roots (or eigenvalue, or eigenfrequency) ω_n solutions of the characteristic equation

$$\det\left(-\omega^2 \mathbf{M} + \mathbf{K}\right) = 0. \qquad (22.50)$$

The eigenvector $\boldsymbol{\Phi}_n$ associated to each eigenfrequency ω_n is given by

$$\left(-\omega_n^2 \mathbf{M} + \mathbf{K}\right) \boldsymbol{\Phi}_n = 0. \qquad (22.51)$$

$\boldsymbol{\Phi}_n$ is defined with an arbitrary multiplicative constant. The natural modes are defined by the set of eigenvalues ω_n and associated eigenvectors $\boldsymbol{\Phi}_n$. Spectral analysis theory shows that the $\boldsymbol{\Phi}_n$ form an $\mathbf{M}$ and $\mathbf{K}$ orthogonal basis set, so that

$${}^t\boldsymbol{\Phi}_m \mathbf{M} \boldsymbol{\Phi}_n = 0$$

and

$${}^t\boldsymbol{\Phi}_m \mathbf{K} \boldsymbol{\Phi}_n = 0 \quad \text{for} \quad m \neq n \qquad (22.52)$$

The orthogonality properties of the eigenmodes mean that the inertial (stiffness) forces developed in a given mode do not affect the motion of the other modes. The modes are mechanically independent. As a consequence, it is possible to expand any motion onto the eigenmodes. Given a force distribution F, the motion of the system is governed by the equation

$$\mathbf{M}\ddot{X} + \mathbf{K}X = F. \qquad (22.53)$$

The modal projection is written in the form:

$$X = \sum_n \boldsymbol{\Phi}_n q_n(t). \qquad (22.54)$$

The functions $q_n(t)$ in (22.54) are the modal participation factors. Inserting (22.54) into (22.53) and, after taking the scalar product of both sides of the equation with an eigenfunction $\boldsymbol{\Phi}_m$, we find

$$\langle \boldsymbol{\Phi}_n, \mathbf{M}\boldsymbol{\Phi}_n \rangle \ddot{q}_n + \langle \boldsymbol{\Phi}_n, \mathbf{K}\boldsymbol{\Phi}_n \rangle q_n = \langle \boldsymbol{\Phi}_n, F \rangle \qquad (22.55)$$

where the notation $\langle A, B \rangle$ denotes the scalar product ${}^t A \cdot B$ between vectors the A and B. Equation (22.55) shows that the generalized displacements are uncoupled. Each q_n is governed by a single-DOF oscillator differential equation. The quantity

$$m_n = \langle \boldsymbol{\Phi}_n, \mathbf{M}\boldsymbol{\Phi}_n \rangle \qquad (22.56)$$

is the modal mass associated with the mode n. These coefficients are defined with an arbitrary multiplicative constant. Similarly,

$$\kappa_n = \langle \boldsymbol{\Phi}_n, \mathbf{K}\boldsymbol{\Phi}_n \rangle \qquad (22.57)$$

is the modal stiffness, related to the modal mass through the relationship

$$\kappa_n = m_n \omega_n^2. \qquad (22.58)$$

Finally, the quantity

$$f_n = \langle \boldsymbol{\Phi}_n, F \rangle \qquad (22.59)$$

is the projection of the nonconservative forces onto mode n. Each independent oscillator equation can then be rewritten as

$$\ddot{q}_n + \omega_n^2 q_n = \frac{f_n}{m_n}. \qquad (22.60)$$

Remark. Since the eigenvectors are defined with arbitrary multiplicative constants C_n, (22.56) shows that the modal mass is proportional to C_n^2. In addition, (22.58) shows that f_n is also proportional to C_n. Therefore, through (22.60), q_n is proportional to C_n^{-1} and from (22.54), X is independent of C_n.

Normal Modes. The normal modes Ψ_n correspond to the case where the arbitrary constant is such that the modal masses become unity. In this case, we can write

$$\Psi_n = \frac{\boldsymbol{\Phi}_n}{\sqrt{m_n}}. \qquad (22.61)$$

Consequently, (22.57) becomes

$$\langle \Psi_n, \mathbf{K}\Psi_n \rangle = \omega_n^2 \,. \tag{22.62}$$

22.2.3 Admittances

In general, a given structure vibrates as a result of the action of localized and distributed forces and moments. For both numerical and experimental reasons, one often has to work on a discrete representation of the structure. In practice, the geometry of the structure is represented by a mesh consisting of a number N of areas with dimensions smaller than the wavelengths under consideration. This results in having to consider the structure as an N-DOF system. At each point of the mesh, the motion is defined by three translation components plus three rotation components. Similarly, the action of the external medium on the structure can be reduced to three force components plus three moment components, here denoted by F_l. The translational and rotational velocity components V_k at each point are thus related to the forces and moments through a 6×6 admittance matrix, such that

$$\mathbf{V} = \mathbf{Y}\mathbf{F} \tag{22.63}$$

For each force component at a given point j on the structure, denoted $F_{j|l}$, a motion at another point i, denoted by $V_{i|k}$, can be generated. As a consequence one can define the transfer admittance

$$Y_{ij|kl} = \frac{V_{i|k}}{F_{j|l}} \quad \text{with} \quad 1 \leq i, j \leq N$$
$$\text{and} \quad 1 \leq k, l \leq 6 \,.$$

In summary, the transfer admittance matrix is defined by $6N \times 6N$ coefficients such as $Y_{ij|kl}$.

Notation. In what follows, the indices (k, l) are omitted, for the sake of simplicity. The transfer admittances are written as Y_{ij}, which reduces to Y_{ii} (or, even to Y_i) in the case of a driving-point admittance. Attention here is mostly focused on translation, though the formalism remains valid for rotation.

The previous results obtained on modal decomposition are now used to investigate the frequency dependence of the admittances. Here, $X(x_i) \equiv X_i$ is one component of displacement at point x_i and $F(x_j) \equiv F_j$ is one force component at point x_j. The structure is subjected to a forced motion at frequency ω. We have

$$(\mathbf{K} - \omega^2 \mathbf{M}) X_i = F_j \,, \tag{22.64}$$

where $X_i = \sum \Phi_n(x_i) q_n(t)$ and

$$(\omega_n^2 - \omega^2) q_n = \frac{f_n}{m_n} = \frac{{}^t\Phi_n(x_j) F_j(\omega)}{m_n} \,. \tag{22.65}$$

For a structure discretized on N points, the displacement at point x_i is given by

$$X_i(\omega) = \sum_{n=1}^{N} \frac{\Phi_n(x_i) \, {}^t\Phi_n(x_j)}{m_n (\omega_n^2 - \omega^2)} F_j(\omega) \,. \tag{22.66}$$

The complete set of functions $X_i(\omega)$ given in (22.66) is the operating deflexion shape (ODS) at frequency ω. The transfer admittance between points x_i and x_j is written

$$Y_{ij}(\omega) = \mathrm{i}\omega \sum_{n=1}^{N} \frac{\Phi_n(x_i) \, {}^t\Phi_n(x_j)}{m_n (\omega_n^2 - \omega^2)} \,. \tag{22.67}$$

The driving-point admittance at point x_i becomes

$$Y_i(\omega) = \mathrm{i}\omega \sum_{n=1}^{N} \frac{[\Phi_n(x_i)]^2}{m_n (\omega_n^2 - \omega^2)} \tag{22.68}$$

Remark. To include damping, (22.67) can be written

$$Y_{ij}(\omega) = \mathrm{i}\omega \sum_{n=1}^{N} \frac{\Phi_n(x_i) \, {}^t\Phi_n(x_j)}{m_n (\omega_n^2 + 2\mathrm{i}\zeta_n \omega_n \omega - \omega^2)} \,, \tag{22.69}$$

where ζ_n is a nondimensional damping factor. The physical origin of the damping terms will be presented in more detail in Sect. 22.6.1.

Frequency Analysis and Approximations
For weak damping, and low modal density, the modulus of $Y_{ij}(\omega)$ passes through maxima at frequencies close to the eigenfrequencies $\omega = \omega_n$ (Fig. 22.4).

- For $\omega \approx \omega_n$, the main term in $Y_{ij}(\omega)$ is equal to

$$\mathrm{i}\omega \frac{\Phi_n(x_i) \, {}^t\Phi_n(x_j)}{m_n (\omega_n^2 - \omega^2)} \,. \tag{22.70}$$

- For $\omega \gg \omega_n$, the admittance becomes

$$Y_{ij} \approx \mathrm{i}\omega \sum_{l>n} \frac{\Phi_l(x_i) \, {}^t\Phi_l(x_j)}{-m_l \omega^2} \approx \frac{1}{\mathrm{i}M\omega}$$

with

$$\frac{1}{M} = \sum_{l>n} \frac{\Phi_l(x_i) \, {}^t\Phi_l(x_j)}{m_l} \tag{22.71}$$

It can be seen that the modes with a rank larger than a given mode n play the role of a mass M.

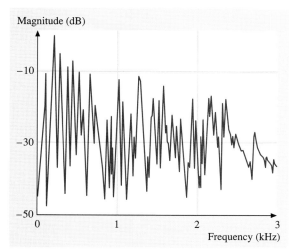

Fig. 22.4 Typical frequency dependence of an admittance modulus for a lightly damped structure. (After *Derveaux* et al. [22.3])

- Similarly, for modes with a rank smaller than n, we find
$$Y_{ij} \approx i\omega \sum_{l<n} \frac{\Phi_l(x_i)\,{}^t\Phi_l(x_j)}{m_l \omega_l^2} \approx \frac{i\omega}{K}$$
with
$$\frac{1}{K} = \sum_{l<n} \frac{\Phi_l(x_i)\,{}^t\Phi_l(x_j)}{m_l \omega_l^2}\,. \tag{22.72}$$

Here, the contribution of the modes is equivalent to a stiffness K.

In summary, in the vicinity of a resonance of mode n, the transfer admittance can be written
$$Y_{ij}(\omega) \approx i\omega \frac{\Phi_n(x_i)\,{}^t\Phi_n(x_j)}{m_n(\omega_n^2 - \omega^2)} + \frac{i\omega}{K} + \frac{1}{iM\omega}\,. \tag{22.73}$$

22.2.4 Example: 2-DOF Plate–Cavity Coupling

To illustrate the previous concepts, a simple example of structural–acoustic coupling represented by a moving plate over a cavity with a hole is considered. This corresponds to a crude low-frequency model for stringed instruments or vented boxes [22.4]. V is the volume of the cavity, A_h is the section of the air piston of mass m_h and displacement x_h, and m_p is the mass of the plate with area A_p (Fig. 22.4). It is assumed that all points of the plate move in phase with the displacement x_p;

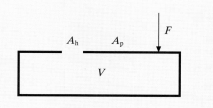

Fig. 22.5 Two-DOF plate–cavity coupling

k_p is the equivalent stiffness of the elastic plate and $\omega_{p,0} = \sqrt{k_p/m_p}$ denotes its eigenfrequency in the absence of coupling to the cavity. As the plate and piston are set in motion, the change of volume in the cavity is equal to $\Delta V = A_p x_p + A_h x_h$. Consequently, the pressure change in the cavity is $\Delta p = -\frac{\rho c^2 \Delta V}{V}$. F is the vertical external force applied to the plate, R_p and R_h are fluid damping coefficients. The equations that govern the motion of the coupled oscillators can be written
$$\begin{cases} m_p \ddot{x}_p = F - k_p x_p - R_p \dot{x}_p + A_p \Delta p\,, \\ m_h \ddot{x}_h = A_h \Delta p - R_h \dot{x}_h \end{cases} \tag{22.74}$$

Let us write $\mu = c^2 \rho / V$. If the cavity is closed ($A_h = 0$), then the eigenfrequency of the plate is given by $\omega_p = \sqrt{(k_p + \mu A_p^2)/m_p}$. If the stiffness k_p of the plate tends to zero, then the eigenfrequency of the plate coupled to the cavity becomes $\omega_a = \sqrt{\mu A_p^2/m_p}$. Finally, if the plate is assumed to be completely rigid ($x_p = 0$), then the eigenfrequency of the cavity–hole system (Helmholtz resonance) is $\omega_h = \sqrt{\mu A_h^2/m_h}$. Solving (22.74) for a sinusoidal excitation $F e^{i\omega t}$, with $\gamma_p = R_p/m_p$ and $\gamma_h = R_h/m_h$, yields for the plate velocity u_p and air velocity u_h in the piston:
$$u_p = i\omega \frac{F}{m_p} \frac{(\omega_h^2 - \omega^2 + j\omega\gamma_h)}{D} \tag{22.75}$$
and
$$u_h = -i\omega \frac{F}{m_p} \frac{A_p}{A_h} \frac{\omega_h^2}{D} \tag{22.76}$$
with
$$D = (\omega_p^2 - \omega^2 + i\omega\gamma_p)(\omega_h^2 - \omega^2 + i\omega\gamma_h) - \omega_{ph}^4\,, \tag{22.77}$$
where
$$\omega_{ph}^4 = \omega_h^2 \omega_a^2\,.$$

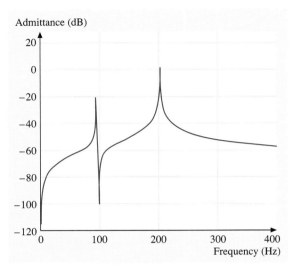

Fig. 22.6 Driving-point admittance of a plate loaded by a cavity. $A_p = 0.01\,\mathrm{m}^2$; $V = 0.1\,\mathrm{m}^3$; $f_h = 100\,\mathrm{Hz}$; $f_p = 200\,\mathrm{Hz}$; $\gamma_h = 1.0\,\mathrm{s}^{-1}$; $\gamma_p = 2.0\,\mathrm{s}^{-1}$; $m_p = 0.4\,\mathrm{kg}$; Admittance Y (in dB) $= 20\log|u_p/F|$

The eigenfrequencies of the coupled system correspond to the roots of the denominator D. Neglecting the influence of losses, these eigenfrequencies are solutions of the equation

$$(\omega_p^2 - \omega^2)(\omega_h^2 - \omega^2) - \omega_{ph}^4 = 0. \tag{22.78}$$

The solutions ω_+ and ω_- satisfy the relation

$$\omega_+^2 + \omega_-^2 = \omega_p^2 + \omega_h^2, \tag{22.79}$$

which shows that the difference between the eigenfrequencies increases due to the coupling between the plate and the Helmholtz resonator, compared to the uncoupled case. Figure 22.6 shows the modulus of the driving-point admittance $|u_p/F|$ of the plate, showing two maxima at ω_+ and ω_- and a zero at ω_h.

In stringed instruments and vented boxes, the main effect of the open cavity is to enhance the radiated sound in the low-frequency range, below the lowest resonance of the moving structure (the top plate or loudspeaker diaphragm).

22.2.5 Statistical Energy Analysis

The aim of statistical energy analysis (SEA) is to use power flows as a means for estimating the responses of complex systems. It is of particular interest for the analysis of coupled structures such as those encountered in the transportation industry. It is often used for the preliminary design of structures, or for identifying energy transmission paths [22.5].

One main advantage of SEA is that it usually leads to a substantial reduction of the number of degrees of freedom. The price to pay is that results are only given in terms of average and variance over frequency and space. It is particularly suitable when structural details are not known, or when measurements yield uncertainty in the modal parameters. SEA techniques focus on the analysis of energy levels in resonant modes of dynamical systems. Its main principle is based on the property that the average power flow between two coupled systems (or subsystems) is proportional to the difference in the average modal energies of each system. In order to illustrate this basic concept of SEA, the case of two coupled 1-DOF oscillators is summarized below. Detailed developments can be found in [22.6].

To consider a general case, it is assumed that both oscillators are coupled by means of a coupling spring with stiffness coefficient k_{12} and a gyroscopic constant B. The equations of motion of this system are given by

$$\begin{cases} m_1 \dfrac{\mathrm{d}^2 \xi_1}{\mathrm{d}t^2} + r_1 \dfrac{\mathrm{d}\xi_1}{\mathrm{d}t} + s_1 \xi_1 + B \dfrac{\mathrm{d}\xi_2}{\mathrm{d}t} + k_{12}\xi_2 = F_1, \\ m_2 \dfrac{\mathrm{d}^2 \xi_2}{\mathrm{d}t^2} + r_2 \dfrac{\mathrm{d}\xi_2}{\mathrm{d}t} + s_2 \xi_2 - B \dfrac{\mathrm{d}\xi_1}{\mathrm{d}t} + k_{12}\xi_1 = F_2, \end{cases} \tag{22.80}$$

where m_1 and m_2 are the masses of the oscillators, $r_1 = 2\delta_1 m_1$ and $r_2 = 2\delta_2 m_2$ are the mechanical resistances, $s_1 = k_1 - k_{12}$ and $s_2 = k_2 - k_{12}$ are stiffness coefficients, and F_1 and F_2 are the magnitudes of the forces applied to each mass. For a harmonic motion with frequency ω, the power flow between the oscillators is given by

$$P_{12} = \frac{1}{2}\operatorname{Re}\left(F_{12} v_2^*\right), \tag{22.81}$$

where v_2^* is the complex conjugate of the velocity for the second oscillator and F_{12} is the force applied by oscillator 1 to oscillator 2. Using (22.80), the power flow at frequency ω is written [22.6]

$$P_{12} = \frac{\omega^2 s_{12}^2 + \omega^4 B^2}{m_1^2 m_2^2 |D(\omega)|^2}\left(m_2 \delta_2 |F_1|^2 - m_1 \delta_1 |F_2|^2\right), \tag{22.82}$$

where

$$D(\omega) = \omega^4 - 2i\omega^3(\delta_1 + \delta_2)$$
$$- \omega^2\left(\omega_1^2 + \omega_2^2 + 4\delta_1\delta_2 + \frac{B^2}{m_1 m_2}\right)$$
$$+ 2i\omega\left(\delta_1\omega_2^2 + \delta_2\omega_1^2\right) - \frac{k_{12}}{m_1 m_2} + \omega_1^2\omega_2^2 \,.$$
(22.83)

In (22.83), ω_1 and ω_2 are the natural frequencies of the undamped uncoupled oscillators. It can be shown that (22.82) can be generalized to a broadband excitation with frequency bandwidth $\Delta\omega$. One basic property of SEA follows from the fact that P_{12} is proportional to the energy difference $W_1 - W_2$ of the oscillators. For a broadband excitation, these energies are given by

$$W_i = \frac{m_i}{2\Delta\omega} \int_{-\infty}^{+\infty} |v_i(\omega)|^2 \, d\omega \qquad (22.84)$$

and the basic SEA equation becomes

$$P_{12} = \beta(W_1 - W_2)$$

give the rate at which energy is transferred from one subsystem to the other. The SEA theory can be used, for example, for predicting the vibration level of coupled

with

$$\beta = \frac{2}{m_1 m_2} \frac{(\delta_1\omega_2^2 + \delta_2\omega_1^2)B^2 + (\delta_1 + \delta_2)k_{12}^2}{(\omega_1^2 - \omega_2^2)^2 + 2(\delta_1 + \delta_2)(\delta_1\omega_2^2 + \delta_2\omega_1^2)} \,.$$
(22.85)

Equation (22.85) shows that β is positive and depends on the parameters of the oscillators and on the coupling parameters B and k_{12}. This equation can be generalized to the coupling between any mode m_1 of a given subsystem with N_1 modes and a mode m_2 of another subsystem with N_2 modes. Considering a frequency band with central frequency ω_c, the generalization leads to the following expression for the power flow from subsystem 1 to subsystem 2

$$P_{12} = \omega_c(\eta_{12}W_1 - \eta_{21}W_2) \,, \qquad (22.86)$$

where η_{12} and η_{21} are the coupling loss factors obtained through averaging of the parameters $\beta_{m_1 m_2}$ analogous to that defined in (22.85)

$$\eta_{12} = \frac{N_2}{\omega_c} \langle \beta_{m_1 m_2} \rangle_{N_1 N_2} \quad \text{and} \quad \eta_{21} = \frac{N_1}{N_2}\eta_{12} \,. \quad (22.87)$$

$W_1 = N_1 W_{m_1}$ and $W_2 = N_2 W_{m_2}$ are the total modal energies in both subsystems, averaged over a frequency band, with the assumption that both energies are equally divided among the modes. The coupling loss factors plates over a large frequency range, where standard numerical methods, such as the finite-element method are too costly [22.6].

22.3 Strings and Membranes

Strings and membranes belong to that category of structures where the stiffness is due to external tension. This is sometimes called *geometrical* stiffness, in contrast with the *elastic* stiffness of most vibrating solids. In practice, strings and membranes also are made of elastic materials and, consequently, show more or less elastic stiffness due to their Young's moduli. This elastic stiffness will be neglected in the present section. The relative contributions of both stiffnesses will be discussed in Sect. 22.4 in the case of prestressed beams.

22.3.1 Equations of Motion

Consider a small element at a given point of Cartesian coordinates x on a stretched membrane of density $\rho(x)$. The membrane is assumed to be in equilibrium in the plane e_x, e_y and subjected to a tension field $\tau(x)$ so that the resulting forces per unit length are (Fig. 22.7):

$$\begin{cases} \text{on both sides oriented in the } e_x \text{ direction:} \\ \tau_x = \tau_{11}e_x + \tau_{21}e_y \,, \\ \text{on both sides oriented in the } e_y \text{ direction:} \\ \tau_y = \tau_{12}e_x + \tau_{22}e_y \,, \end{cases}$$
(22.88)

where $\tau_{12} = \tau_{21}$ to ensure the moment equilibrium (reciprocity principle). In the general case, τ_{ij} are functions of the coordinates x. The membrane is subjected to a small displacement in the direction e_z and released, so that its vertical motion $u(x, y, t)$ is governed by the balance between inertial forces and elastic forces due to tension. Gravity is neglected. After balance of forces on the four sides of the membrane element and projection on the vertical axis, with the assumption of small displace-

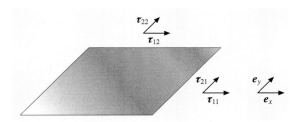

Fig. 22.7 Balance of forces on a membrane element

ments, one obtains the equation that governs the flexural motion of a nonuniformly stretched membrane:

$$\rho(\mathbf{x})h\ddot{u} = \frac{\partial}{\partial x}\left(\tau_{11}\frac{\partial u}{\partial x} + \tau_{12}\frac{\partial u}{\partial y}\right) + \frac{\partial}{\partial y}\left(\tau_{12}\frac{\partial u}{\partial x} + \tau_{22}\frac{\partial u}{\partial y}\right), \quad (22.89)$$

where h is the thickness. The tension field

$$\boldsymbol{\tau} = \begin{bmatrix} \tau_{11} & \tau_{12} \\ \tau_{12} & \tau_{22} \end{bmatrix} \quad (22.90)$$

is a symmetric tensor of order two, and

$$\mathbf{grad}\, u = \frac{\partial u}{\partial x}\mathbf{e}_x + \frac{\partial u}{\partial y}\mathbf{e}_y. \quad (22.91)$$

The previous equation can be written in the more compact form

$$\mathrm{div}(\boldsymbol{\tau}\cdot\mathbf{grad}\, u) = \rho(\mathbf{x})h\ddot{u}, \quad (22.92)$$

where this last formulation has the advantage of being independent of the selected system of coordinates.

One-Dimensional (1-D) Approximation. Transverse Motion of Strings

Strings can be viewed as membranes where one dimension (the length) is much larger than the other two, so that a 1-D approximation is justified. Thus, rewriting (22.89) after integrating mass and tension along the axis $\mathbf{e}_y$, one obtains the equation of transverse motion for a string with nonuniform tension

$$\mu(x)\ddot{u} = \frac{\partial}{\partial x}\left[T(x)\frac{\partial u}{\partial x}\right], \quad (22.93)$$

where $\mu = \rho S$ is the linear mass density, S is the cross-sectional area and T is the tension.

Remark.

1. A similar equation can be obtained for the transverse motion $v(x, t)$ of the string in the direction $\mathbf{e}_y$. Therefore, two transverse motions polarized in orthogonal directions can exist on a string.
2. In the absence of coupling terms, u and v are independent. However, induced motion at the ends and/or nonlinear terms usually lead to coupling between these motions (Sect. 22.7).
3. Strings and membranes are also subjected to longitudinal motion. This motion is due to the variation of tension resulting from the variation in strain associated with the motion.

Homogeneous and Uniformly Stretched Strings and Membranes

For a uniformly stretched and homogeneous membrane, (22.92) reduces to

$$\tau\,\mathrm{div}(\mathbf{grad}\, u) = \tau\Delta u = \rho h\ddot{u}, \quad (22.94)$$

where the Laplacian is written in Cartesian coordinates as

$$\Delta u = \frac{\partial^2 u}{\partial x^2} + \frac{\partial^2 u}{\partial y^2} \quad (22.95)$$

For a string, we get the well-known wave equation

$$\mu\ddot{u} = T\frac{\partial^2 u}{\partial x^2}. \quad (22.96)$$

22.3.2 Heterogeneous String. Modal Approach

Extending the previous results to continuous structures shows that, in the linear regime, these structures have an infinite number of eigenmodes with similar properties of orthogonality with respect to mass and stiffness to those presented in Sect. 22.2 for discrete systems. In this case, the previously defined eigenvectors $\boldsymbol{\Phi}_n$ become continuous functions of the space coordinates $\Phi(\mathbf{x})$. To illustrate the modal approach for continuous systems, consider the case of an heterogeneous string (i.e., a nonuniformly stretched string with variable density), rigidly fixed at both ends and subjected to a force density $f(x, t)$:

$$\rho(x)S(x)\frac{\partial^2 y}{\partial t^2} - \frac{\partial}{\partial x}\left[T(x)\frac{\partial y}{\partial x}\right] = f(x, t), \quad (22.97)$$

$\Phi_n(x)$ are the eigenfunctions of the string, which account for the boundary conditions. We seek solutions of the form

$$y(x,t) = \sum_n \Phi_n(x) q_n(t) \,. \qquad (22.98)$$

Multiplying both sides of (22.97) by the eigenfunction $\Phi_m(x)$ and integrating over the length of the string yields

$$\sum_n \ddot{q}_n(t) \int_0^L \Phi_n(x) \Phi_m(x) \rho(x) S(x) \, dx$$

$$- \sum_n q_n(t) \int_0^L \Phi_m(x) \frac{d}{dx}\left[T(x) \frac{d\Phi_n(x)}{dx}\right] dx$$

$$= \int_0^L \Phi_m(x) f(x,t) \, dx \qquad (22.99)$$

which can be written in symbolic form as

$$\mathcal{M}(\ddot{y}, \Phi_m) + \mathcal{K}(y, \Phi_m) = \langle f, \Phi_m \rangle \,. \qquad (22.100)$$

Equation (22.100) is general and can be applied to all conservative systems. $\mathcal{M}$ denotes the mass operator and $\mathcal{K}$ the stiffness operator.

Orthogonality Properties of the Eigenmodes
Each eigenmode $\Phi_n(x)$ is a solution of the equation

$$-\omega_n^2 \rho(x) S(x) \Phi_n(x) = \frac{d}{dx}\left[T(x) \frac{d\Phi_n(x)}{dx}\right]. \qquad (22.101)$$

These eigenfunctions are orthogonal with respect to the mass

$$\int_0^L \Phi_n(x) \Phi_m(x) \rho(x) S(x) \, dx = 0 \qquad (22.102)$$

and orthogonal with respect to the stiffness

$$\int_0^L T(x) \frac{d\Phi_m(x)}{dx} \frac{d\Phi_n(x)}{dx} \, dx = 0 \,. \qquad (22.103)$$

Generalized Coordinates
Taking the orthogonality properties of the eigenmodes into account, we can write

$$-\frac{d}{dx}\left[T(x) \frac{d\Phi}{dx}\right] = \omega^2 \rho(x) S(x) \Phi(x) \qquad (22.104)$$

so that (22.100) becomes:

$$\ddot{q}_n(t) + \omega_n^2 q_n(t) = \frac{f_n(t)}{m_n} \,, \qquad (22.105)$$

where the modal mass is given by:

$$m_n = \int_0^L \Phi_n^2(x) \rho(x) S(x) \, dx \qquad (22.106)$$

with

$$f_n(t) = \int_0^L f(x,t) \Phi_n(x) \, dx \,. \qquad (22.107)$$

The equations that govern the generalized displacements are therefore formally identical to those obtained for discrete systems. For continuous systems, truncation of the theoretically infinite number of differential equations has to be performed. The truncation criteria are often determined by the frequency domain under consideration. For a lossless string of length L with perfect boundary conditions excited by a source term $f(x,t)$, the general solution can be written in the modal domain as

$$y(x,t) = \sum_n \Phi_n(x) q_n(t) \qquad (22.108)$$

where

$$\ddot{q}_n(t) + \omega_n^2 q_n(t) = \frac{f_n(t)}{m_n}$$

with

$$f_n(t) = \langle f(x,t), \Phi_n(x) \rangle = \int_0^L \Phi_n(x) f(x,t) \, dx \,.$$

Let us assume an initial shape $y(0,t)$ and an initial velocity $\dot{y}(0,t)$. The initial values of the $q_n(t)$ are thus given by

$$\begin{cases} q_n(0) = \dfrac{1}{m_n} \int_0^L \rho(x) S(x) y(0,t) \Phi_n(x) \, dx \\ \text{and} \\ \dot{q}_n(0) = \dfrac{1}{m_n} \int_0^L \rho(x) S(x) \dot{y}(0,t) \Phi_n(x) \, dx \end{cases}$$

$$(22.109)$$

Denoting the Laplace transform of $q_n(t)$ by $Q_n(s)$, we have

$$Q_n(s) = \frac{1}{s^2 + \omega_n^2}\left[\frac{F_n(s)}{m_n} + sq_n(0) + \dot{q}_n(0)\right], \quad (22.110)$$

where $F_n(s)$ is the Laplace transform of $f_n(t)$. The first term of $Q_n(s)$ is a product of two transforms which, in the time domain, corresponds to a convolution. Using transform tables, we obtain

$$q_n(t) = \frac{1}{m_n \omega_n}\int_0^t f_n(\theta)\sin\omega_n(t-\theta)\,d\theta + q_n(0)\cos\omega_n t + \dot{q}_n(0)\frac{\sin\omega_n t}{\omega_n}. \quad (22.111)$$

If $f_n(t)$ is a Dirac delta function of the form $f_{n0}\delta(t)$, the Laplace transform reduces to $F_n(s) = f_{n0}$. Equation (22.111) can then be written

$$q_n(t) = \frac{f_{n0}}{m_n}\frac{\sin\omega_n t}{\omega_n} + q_n(0)\cos\omega_n t + \dot{q}_n(0)\frac{\sin\omega_n t}{\omega_n}. \quad (22.112)$$

From the above impulse response for mode n of the string, we note the following:

1. There is an equivalence between $q_n(t)$ from the impulse source term and the initial velocity term. In other words, the motion of the string induced by an initial velocity profile is equivalent if the string is excited by a spatial distribution of Dirac delta function forces.
2. A source term localized at a given point x_0 can be written $f(x,t) = A\delta(x-x_0)\delta(t)$. In this case, $f_n(t) = \langle f(x,t), \Phi_n(x)\rangle = A\Phi_n(x_0)\delta(t)$. Therefore, the mode n will not be excited if the force is applied on a node of vibration.
3. The first term of the function $g_n(t) = \sin\omega_n t/m_n\omega_n$ is the Green's function of mode n.

String with a Finite Mass at One End

To present general results, the string is still assumed to be heterogeneous. The purpose of this section is to give the orthogonality properties of the modes for a heterogeneous string terminated by a finite mass M. The boundary conditions are

$$y(0,t) = 0 \quad \text{and} \quad -T(x)\frac{\partial y}{\partial x} = M\frac{\partial^2 y}{\partial t^2},$$
$$\text{at} \quad x = L. \quad (22.113)$$

The eigenfunctions $\Phi(x)$ therefore have to satisfy

$$\rho(x)S(x)\omega^2\Phi(x) = -\frac{d}{dx}\left[T(x)\frac{d\Phi(x)}{dx}\right] = 0,$$
for $\quad 0 < x < L$
with $\quad \Phi(0) = 0$
and $\quad T(x)d\Phi/dx = \omega^2 M\Phi(x),$
at $\quad x = L.$

For $n \neq m$, the orthogonality condition with respect to mass becomes

$$\int_0^L \rho(x)S(x)\Phi(x)\Phi_m(x)\,dx + M\Phi_n(L)\Phi_m(L) = 0 \quad (22.114)$$

and the orthogonality condition with respect to stiffness remains

$$\int_0^L T(x)\frac{d\Phi_n(x)}{dx}\frac{d\Phi_m(x)}{dx}\,dx = 0. \quad (22.115)$$

For a homogeneous string with uniform tension and constant diameter, these conditions reduce to

$$\int_0^L \Phi_n(x)\Phi_m(x)\,dx = -\frac{M}{\rho S}\Phi_n(L)\Phi_m(L) \quad (22.116)$$

and

$$\int_0^L \frac{d\Phi_n(x)}{dx}\frac{d\Phi_m(x)}{dx}\,dx = 0. \quad (22.117)$$

22.3.3 Ideal String

To solve initial-boundary problems in simple cases, we assume that the lossless string is homogeneous with density ρ, constant tension T and constant cross section S. In this case, the vertical displacement is given by the wave equation

$$\frac{1}{c^2}\frac{\partial^2 u}{\partial t^2} = \frac{\partial^2 u}{\partial x^2}, \quad (22.118)$$

where $c = \sqrt{T/\rho S}$ is the transverse wave velocity.

D'Alembert's Solution in the Time Domain

In order to show the correspondence between the wave and modal approaches, a time-domain method is used first to solve the wave equation (22.118) with initial conditions [22.7]

$$u(x, 0) = f(x) \quad \text{and} \quad \frac{\partial u}{\partial t}(x, 0) = g(x). \quad (22.119)$$

Using the change of variables $\xi = x - ct$ and $\eta = x + ct$, we have

$$u(x, t) = U(\xi, \eta) \quad (22.120)$$

from (22.120), we obtain:

$$\begin{cases} \dfrac{\partial^2 u}{\partial t^2} = c^2 \left(\dfrac{\partial^2 U}{\partial \xi^2} - 2 \dfrac{\partial^2 U}{\partial \xi \partial \eta} + \dfrac{\partial^2 U}{\partial \eta^2} \right) \\ \dfrac{\partial^2 u}{\partial x^2} = \dfrac{\partial^2 U}{\partial \xi^2} + 2 \dfrac{\partial^2 U}{\partial \xi \partial \eta} + \dfrac{\partial^2 U}{\partial \eta^2} \end{cases}. \quad (22.121)$$

Inserting (22.121) into (22.118) yields $(\partial^2 U)/(\partial \xi \partial \eta) = 0$ which implies

$$u(x, t) = F(x - ct) + G(x + ct), \quad (22.122)$$

where F and G are are two twice-differentiable functions. $F(x - ct)$ represents a travelling wave moving to the right (direction of increasing values for x), while $G(x + ct)$ is a travelling wave moving to the left. Using (22.119) implies that F and G must fulfill the conditions

$$\begin{cases} F(x) + G(x) = f(x) \\ -cF'(x) + cG'(x) = g(x) \end{cases}. \quad (22.123)$$

Solving (22.123) yields

$$\begin{cases} F(x) = \dfrac{1}{2} f(x) - \dfrac{1}{2} [-F(0) + G(0)] \\ \quad - \dfrac{1}{2c} \displaystyle\int_0^x g(s)\,ds \\ G(x) = \dfrac{1}{2} f(x) + \dfrac{1}{2} [-F(0) + G(0)] \\ \quad + \dfrac{1}{2c} \displaystyle\int_0^x g(s)\,ds \end{cases}. \quad (22.124)$$

Finally, the solution of the wave equation is written explicitly

$$u(x, t) = \frac{1}{2} [f(x - ct) + f(x + ct)] + \frac{1}{2c} \int_{x-ct}^{x+ct} g(s)\,ds. \quad (22.125)$$

Semi-infinite String. For a semi-infinite string rigidly fixed at $x = 0$, we have the boundary condition $u(0, t) = 0$. This requires:

$$F(-ct) + G(ct) = 0 \quad (22.126)$$

and, finally, with the appropriate change of variables

$$u(x, t) = -G(x - ct) + G(x + ct)$$
$$= F(x - ct) - F(-x - ct). \quad (22.127)$$

Equation (22.127) expresses the fact that, due to the fixed end at $x = 0$, the left-traveling wave is reflected with a change of sign and becomes a right-traveling wave. The validity domain for (22.127) is $0 \leq x < +\infty$ and $t > 0$.

String of Finite Length. In the case of an ideal string fixed at both ends, the wave approach can still be used, with the additional boundary condition $u(L, t) = 0$, which yields

$$u(L, t) = -G(L - ct) + G(L + ct)$$
$$= F(L - ct) - F(-L - ct) = 0. \quad (22.128)$$

Equation (22.128) can be rewritten $F(z) = F(z - 2L)$, which shows that F (and G) are now periodic functions with spatial period $2L$ or, equivalently, temporal period $2L/c$. The validity domain of (22.128) is now $0 \leq x \leq L$ and $t > 0$. Equations (22.126)–(22.128) can be used for step-by-step constructions of the string shape at successive instants of time.

String Fixed at Both Ends. Eigenmodes

Injecting a harmonic wave of the form $u(x, t) = e^{i(\omega t - kx)}$ in (22.118), we find the dispersion equation $D(\omega, k)$ that governs the relationship between frequency ω and wavenumber k. Here, we obtain

$$D(\omega, k) = \omega^2 - c^2 k^2 = 0. \quad (22.129)$$

This equation shows that the ratio between the frequency and wavenumber is constant, which is a characteristic property of a nondispersive medium. If we assume further that the string is rigidly fixed at both ends, the eigenmodes must satisfy the equation

$$\frac{d^2 \Phi(x)}{dx^2} + \frac{\omega^2}{c^2} \Phi(x) = 0 \quad (22.130)$$

with the boundary conditions $\Phi(0) = \Phi(L) = 0$ from which we obtain

$$\Phi_n(x) = \sin k_n x. \quad (22.131)$$

The only possible values for the wavenumber are given by the discrete series

$$k_n = n\pi/L \quad \text{so that} \quad \omega_n = n\pi c/L. \tag{22.132}$$

Using (22.106), the modal mass is $m_n = \rho S L/2 = M_s/2$, where M_s is the total mass of the string. Recall, however, that the ratio $m_n/M_s = 1/2$ is purely arbitrary since the modal masses are defined with an arbitrary multiplicative constant. The important result here is that all modal masses are equal and do not depend on the rank n of the mode.

Application: Plucked String. Modal Approach

The particular case where the string is released from an initial triangular shape at the origin of time without initial velocity is now examined. With the assumption of no stiffness, the initial profile of the string is given by

$$u(0,t) = \begin{cases} \dfrac{hx}{x_0} & \text{for } 0 \leq x \leq x_0 \\ \dfrac{h(L-x)}{L-x_0} & \text{for } x_0 \leq x \leq L \end{cases}. \tag{22.133}$$

The modal method consists of looking for solutions of the form

$$u(x,t) = \sum_n \Phi_n(x) q_n(t). \tag{22.134}$$

At the origin of time, we can write

$$u(0,t) = \sum_n \Phi_n(x) q_n(0). \tag{22.135}$$

The unknowns of the problem are the functions $q_n(0)$. Exploiting again the orthogonality properties of the eigenmodes, we find

$$q_n(0) = \frac{2hL^2}{n^2\pi^2 x_0(L-x_0)} \sin k_n x_0. \tag{22.136}$$

The functions $q_n(t)$ are given by the oscillator equations

$$\ddot{q}_n + \omega_n^2 q_n = 0 \tag{22.137}$$

which, in the case of zero initial velocity, leads to $q_n(t) = q_n(0)\cos \omega_n t$. In summary, the transverse displacement of the string is given by

$$u(x,t) = \sum_n \frac{2hL^2}{n^2\pi^2 x_0(L-x_0)} \times \sin k_n x_0 \sin k_n x \cos \omega_n t. \tag{22.138}$$

Despite the simplicity of this example, a number of important issues can be derived from this result:

1. The eigenfrequencies are integer multiples of the fundamental $f_1 = c/2L$. The solution is periodic and the fundamental frequency is the inverse of the period of vibration.
2. The amplitudes of the modal components decrease as $1/n^2$ with the rank n of the modes.
3. Because the magnitude is proportional to $\sin k_n x_0$, a modal component n can be suppressed by selecting the plucking point $x_0 = pL/n$, where p is an integer $< n$.
4. The excitation point x_0 and observation point x can be interchanged in (22.138). This is an illustration of the reciprocity principle.

Moving End

To a first approximation, a vibrating string radiates as a dipole. Because of its small diameter compared to the acoustic wavelength, it is a poor radiator. If significant sound is to be radiated, one end of the string must be coupled to a component with a considerable vibrating area (plate, shell, etc.). In Sect. 22.3.2, the general properties of a heterogeneous string attached to a mass were obtained. In this section, the simple example of an ideal string coupled to a spring will be considered to illustrate the influence of such coupling on the eigenmodes, eigenfrequencies and modal masses. We consider a string fixed at point $x = 0$ to a spring of stiffness K_0. The boundary condition at this end is then

$$T\left(\frac{\partial u}{\partial x}\right)_{x=0} = K_0 u(0,t). \tag{22.139}$$

The string is rigidly fixed at the other end, so that $u(L,t) = 0$. We look for solutions of the form $u(x,t) = \Phi(x)\cos \omega t$. Following (22.118), the eigenfunctions $\Phi(x)$ must satisfy the equation

$$\frac{d^2\Phi}{dx^2} + k^2\Phi = 0 \tag{22.140}$$

and the boundary conditions. Thus, we derive the equation for the eigenvalues k_n

$$\tan(k_n L) = -\frac{k_n T}{K_0} \quad \text{with } k_n > 0. \tag{22.141}$$

The graphical representation of (22.141) shows that the k_n are no longer integer multiples of k_1 (Fig. 22.8). As a consequence, the free vibration of the string is no longer periodic. Note that the spring decreases the eigenfrequencies relative to those of a perfectly rigid end.

The eigenfunctions become

$$\Phi_n(x) = \sin k_n x + \frac{k_n T}{K_0}\cos k_n x = \frac{\sin k_n(L-x)}{\cos k_n L} \tag{22.142}$$

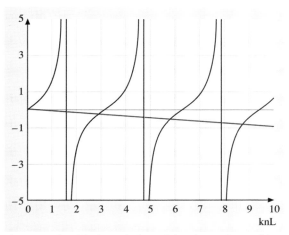

Fig. 22.8 Graphical resolution of the eigenvalue problem for a string attached to a spring (22.141)

from which the initial values $q_n(0)$ and the modal masses can be derived. We obtain:

$$m_n = \int_0^L \rho S \left(\sin k_n x + \frac{k_n T}{K_0} \cos k_n x \right)^2 dx$$

$$= \frac{M_s}{2} \left[1 + \left(\frac{k_n T}{K_0} \right)^2 \right]. \quad (22.143)$$

Equation (22.143) shows here that the modal masses increase with the rank n of the mode. In the case of a mass termination, a similar procedure would lead to the eigenvalue equation

$$\tan(k_n L) = \frac{\rho S}{M_0 k_n}. \quad (22.144)$$

Here, the inertial load at the end leads to an increase of the eigenfrequencies.

Application to Stringed Instruments. In plucked and struck strings instruments (guitar, piano, harp, etc.), the main part of the sound is due to the free regime of vibration, and the spectrum is composed of the eigenfrequencies of the vibrating system. Previous considerations show that the coupling between the strings and the radiating body (top plate, soundboard) alters the frequencies, compared to the perfect harmonic case. Though this alteration is usually relatively small, it has the consequence that the waveform is not perfectly periodic anymore. This might have in turn some perceptual effects on the produced sound. See the chapter on musical acoustics (Chap. 15) in this Handbook for more details.

22.3.4 Circular Membrane in Vacuo

Membranes are used in percussion instruments, microphones, loudspeakers and wall resonators. Because of their small thickness, the influence of air loading cannot be neglected in most applications. However, we limit ourselves here by solving the equation for transverse vibrations of a homogeneous membrane in vacuo. Fluid-loading effects on structures are discussed in later sections. From (22.94) the membrane equation, in circular geometry, is

$$\sigma \frac{\partial^2 u}{\partial t^2} = \tau \Delta u = \tau \left(\frac{\partial^2 u}{\partial r^2} + \frac{1}{r} \frac{\partial u}{\partial r} + \frac{1}{r^2} \frac{\partial^2 u}{\partial \theta^2} \right), \quad (22.145)$$

where $\sigma = \rho h$. Equation (22.145) can be solved by using the method of separation of variables [22.8]. For zero displacement at the edge, $z(r = a, \theta, t) = 0$, and neglecting losses, the expansion of the displacement on the eigenmodes can be expressed as

$$u(r, \theta, t) = \sum_{m=1}^{\infty} \left[\sum_{n=0}^{\infty} U_{nm}(r, \theta) \times \right.$$
$$\left(A_{nm} \cos \omega t + B_{nm} \sin \omega t \right) +$$
$$\sum_{n=1}^{\infty} \tilde{U}_{nm}(r, \theta) \times$$
$$\left. \left(\tilde{A}_{nm} \cos \omega t + \tilde{B}_{nm} \sin \omega t \right) \right]. \quad (22.146)$$

In (22.146), the eigenfunctions are

$$U_{nm}(r, \theta) = J_n(\beta_{nm} r) \cos n\theta$$

and

$$\tilde{U}_{nm}(r, \theta) = J_n(\beta_{nm} r) \sin n\theta, \quad (22.147)$$

where J_n is the order-n Bessel function of the first kind [22.9]. The indices n and m refer to the number of nodal lines and to the number of nodal circles, respectively (Fig. 22.9).

The fixed edge corresponds to one nodal circle. The symmetrical modes are those where there are no nodal diameters. In this case, the eigenfunctions are of the form $J_n(\beta_{0m} r)$. All other modes can be viewed as *pairs of modes* with the same eigenfrequency and the same dependence versus r. They only differ by an angle of $\pi/2n$. The discrete values β_{mn} of the wavenumbers are derived from the boundary condition at the edge, such that

$$J_n(\beta a) = 0. \quad (22.148)$$

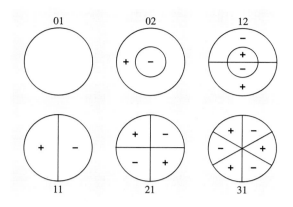

Fig. 22.9 Eigenmodes (n, m) of a circular membrane invacuo

For each value of n, we obtain an infinite series of roots β_{nm}. For $n=0$, for example, $J_0(\beta a) = 0$ yields $\beta_{0m} a = 2.405, 5.520, 8.654, 11,792, 14,931, \ldots$. Similarly, the roots of $J_1(\beta a) = 0$ are given by $\beta_{1m} a = 3.832, 7.016, 10.173, 13.324, 16.471, \ldots$.

Finally, the eigenfrequencies ω_{nm} are obtained from the 2-D wave equation (22.145):

$$\omega_{nm} = c\beta_{nm}, \quad \text{where} \quad c = \sqrt{\frac{\tau}{\sigma}}. \tag{22.149}$$

Unlike the case of ideal strings, the eigenfrequencies of a circular membrane in vacuo are not harmonically related. It has been shown by different authors that the fluid and cavity loading both contribute to restore the harmonic character of timpani sound, although the fundamental is missing [22.10, 11].

22.4 Bars, Plates and Shells

In this section, we consider the flexural vibrations of bars, plates and shells, since these structures are of the most practical use for the radiation of sound. First, the longitudinal vibrations of bars are presented, to show the analogies with the string wave equation, and also because this example will be used in Sect. 22.5.1 to illustrate the basic concepts of fluid–structure interactions.

22.4.1 Longitudinal Vibrations of Bars

The longitudinal vibrations of a 1-D bar in vacuo, clamped at one end and free at the other, are governed by the equations

$$\begin{cases} \rho_s S \dfrac{\partial^2 \xi}{\partial t^2} = ES \dfrac{\partial^2 \xi}{\partial x^2}, \\ \xi(0, t) = 0; \quad \dfrac{\partial \xi}{\partial x}(L, t) = 0. \end{cases} \tag{22.150}$$

The eigenfunctions are

$$\phi_n(x) = \sin(2n-1)\frac{\pi x}{2L} = \sin k_n x \tag{22.151}$$

with $k_n = \omega_n/c_L$ and thus the linear solution can be written as

$$\xi(x, t) = \sum_n \phi_n(x) q_n(t), \tag{22.152}$$

where the generalized displacements obey

$$\ddot{q}_n + \omega_n^2 q_n = 0. \tag{22.153}$$

Equation (22.153) yields a harmonic solution with eigenfrequency ω_n.

22.4.2 Flexural Vibrations of Beams

The flexural vibrations of a slender isotropic beam are presented, within the framework of the Euler–Bernoulli assumptions [22.8]. The beam is oriented along the x-axis, and the cross section $S(x)$ is in the (y, z) plane. The flexural displacement $v(x, t)$ is in the direction of the y-axis. $E(x)$ is the Young's modulus, $\rho(x)$ is the density, and $I_z(x)$ is the moment of inertia of a cross section with respect to its neutral midplane. $f(x, t)$ represents a source term. Applying Hamilton's principle, we obtain the well-known equation

$$\rho(x)S(x)\ddot{v} + \frac{\partial^2}{\partial x^2}\left[E(x)I_z(x)\frac{\partial^2 v}{\partial x^2}\right] = f(x, t) \tag{22.154}$$

with the additional conditions to be satisfied

$$\left[E(x)I_z(x)\frac{\partial^2 v}{\partial x^2}\frac{\partial v}{\partial x}\right]_0^L = 0$$

and

$$\left[\frac{\partial}{\partial x}\left(E(x)I_z(x)\frac{\partial^2 v}{\partial x^2}\right)v\right]_0^L = 0. \tag{22.155}$$

Equation (22.154) is fourth order in the spatial coordinate. Therefore, four boundary conditions (two conditions at each end) are necessary to solve the boundary-value problem. Following (22.155), four different situations are compatible at each end:

Supported $v = 0$

and $\mathcal{M}(x) = E(x)I_z(x)\dfrac{\partial^2 v}{\partial x^2} = 0$;

Clamped $\dfrac{\partial v}{\partial x} = 0$

and $v = 0$;

Free $\mathcal{T}(x) = \dfrac{\partial}{\partial x}\left[E(x)I_z(x)\dfrac{\partial^2 v}{\partial x^2}\right] = 0$

and $\mathcal{M}(x) = E(x)I_z(x)\dfrac{\partial^2 v}{\partial x^2} = 0$;

Guided $\mathcal{T}(x) = \dfrac{\partial}{\partial x}\left[E(x)I_z(x)\dfrac{\partial^2 v}{\partial x^2}\right] = 0$

and $\dfrac{\partial v}{\partial x} = 0$,

where $\mathcal{M}(x)$ is the bending moment and $\mathcal{T}(x)$ is the shear force.

Free–Free Beam with a Constant Section

As a specific example of (22.154), we consider a homogeneous and isotropic beam of length L, thickness h, width b, with a constant rectangular section $S = bh$ whose moment of inertia with respect to the z-axis is $I_z = I = bh^3/12$ (Fig. 22.10).

In this case, (22.154) reduces to

$$EI\dfrac{\partial^4 v}{\partial x^4} + \rho S\dfrac{\partial^2 v}{\partial t^2} = 0. \qquad (22.156)$$

Looking for solutions for (22.156) of the form $v(x,t) = \Phi(x)\cos \omega t$ yields the general solution

$$\Phi(x) = A\cosh\dfrac{\omega x}{c} + B\sinh\dfrac{\omega x}{c} + C\cos\dfrac{\omega x}{c} + D\sin\dfrac{\omega x}{c} \qquad (22.157)$$

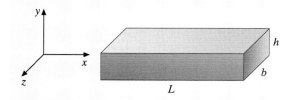

Fig. 22.10 Geometry of a beam with constant section

with

$$c = \sqrt{\omega c_L}\sqrt[4]{\dfrac{I}{S}} \quad \text{and} \quad c_L = \dfrac{E}{\rho}. \qquad (22.158)$$

This expression shows that the phase velocity c of the flexural waves is proportional to $\sqrt{\omega}$. The dispersion relationship is given by

$$EIk^4 + \rho S\omega^2 = 0. \qquad (22.159)$$

From (22.159), we can see that the group velocity $c_g = d\omega/dk$ is also proportional to $\sqrt{\omega}$. The group velocity refers to the propagation velocity of wave energy and envelope of slowly varying amplitude. Flexural waves in the beam are thus dispersive with the higher frequencies propagating faster than the lower ones.

Remark. There is a paradox in (22.157) in the sense that the group velocity tends to infinity as the frequency tends to infinity. This results from the Euler–Bernoulli assumptions, where shear and rotation inertia of the cross sections are neglected. Introducing these two effects in the model (the Timoshenko model) leads to a bounded asymptotic value of the group velocity as frequency increases.

For a beam free at both ends, the boundary conditions are

$$\dfrac{\partial^2 y}{\partial x^2}(0,t) = \dfrac{\partial^2 y}{\partial x^2}(L,t) = 0 \qquad (22.160)$$

and

$$\dfrac{\partial^3 y}{\partial x^3}(0,t) = \dfrac{\partial^3 y}{\partial x^3}(L,t) = 0, \qquad (22.161)$$

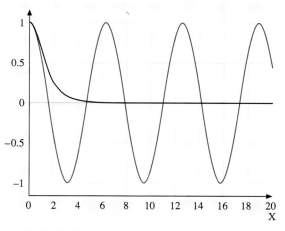

Fig. 22.11 Graphical resolution of the eigenfrequency equation for a free–free beam, with $X = \omega L/c$

which yields the equation of the eigenfrequencies (Fig. 22.11)

$$\cos \frac{\omega L}{c} \cosh \frac{\omega L}{c} = 1 . \quad (22.162)$$

The numerical resolution of (22.162) shows that the eigenfrequencies f_n (in Hz) are

$$f_n \approx \sqrt{\frac{EI}{\rho S}} \frac{\pi}{8L^2} \left[3.011^2, 5^2, 7^2, \ldots, (2n+1)^2 \right] . \quad (22.163)$$

The eigenfrequencies are not harmonically related so that the impulse response of the beam is not periodic.

Beam with a Variable Cross Section

As another example, we consider the flexural vibrations for a beam with variable cross section. Here, we describe the Galerkin method [22.12]. It has the advantage that it remains valid even in the case of nonconservative systems. In this method, the eigenfunctions $\Phi(x)$ of (22.154) are approximated by a finite sum of p terms

$$\Phi^{(p)}(x) = \sum_{j=1}^{p} a_j \phi_j(x) , \quad (22.164)$$

where $\phi_j(x)$ are arbitrary functions that satisfy the boundary conditions at both ends of the beam. Inserting (22.164) in (22.154), and defining $\lambda^{(p)} = (\omega^2)^{(p)}$ as the approximate eigenvalues of order p, one can define the Galerkin residue

$$\mathcal{R}\left[\Phi^{(p)}(x)\right] = \frac{d^2}{dx^2} \left(EI(x) \frac{d^2 \Phi^{(p)}}{dx^2} \right) \\ - \lambda^{(p)} \rho(x) S(x) \Phi^{(p)} \quad (22.165)$$

following (22.164), this can be written as

$$\mathcal{R}\left[\Phi^{(p)}(x)\right] = \sum_{j=1}^{p} a_j \times \\ \left\{ \frac{d^2}{dx^2} \left[EI(x) \frac{d^2 \phi_j(x)}{dx^2} \right] - \lambda^{(p)} \rho(x) S(x) \phi_j(x) \right\} . \quad (22.166)$$

The weak formulation of the problem is given by

$$\int_0^L \phi_i(x) \sum_{j=1}^{p} a_j \left\{ \frac{d^2}{dx^2} \left[EI(x) \frac{d^2 \phi_j(x)}{dx^2} \right] \\ - \lambda^{(p)} \rho(x) S(x) \phi_j(x) \right\} dx = 0 . \quad (22.167)$$

The goal of the method is to find the coefficients a_j for which the residue $\mathcal{R}\left[\Phi^{(p)}(x)\right]$ is equal to zero. Thus (22.167) becomes

$$\sum_{j=1}^{p} k_{ij} a_j - \lambda^{(p)} \sum_{j=1}^{p} m_{ij} a_j = 0 \\ \text{for} \quad i = 1, 2, \ldots p , \quad (22.168)$$

where the mass and stiffness coefficients are

$$k_{ij} = \int_0^L \phi_i(x) \frac{d^2}{dx^2} \left[EI(x) \frac{d^2 \phi_j(x)}{dx^2} \right] dx \quad (22.169)$$

and

$$m_{ij} = \int_0^L \phi_i(x) \rho(x) S(x) \phi_j(x) \, dx . \quad (22.170)$$

Equation (22.168) can be written equivalently in matrix form:

$$[\mathbf{K} - \lambda \mathbf{M}] \mathbf{a} = 0 , \quad (22.171)$$

where $\mathbf{a}$ is a vector with dimension p and $\mathbf{K}$ and $\mathbf{M}$ are matrices with dimension $p \times p$. The problem to be solved is therefore equivalent in form to the eigenvalue problem for a p-component discrete system (22.49).

Prestressed Beam or Stiff String

The flexural motion for a beam subjected to tension (or prestressed beam) is governed by the same equation as a string with stiffness. For a homogeneous and isotropic beam of constant section subjected to a uniform tension T, we have

$$\rho S \frac{\partial^2 v}{\partial t^2} = T \frac{\partial^2 v}{\partial x^2} - EI \frac{\partial^4 v}{\partial x^4} . \quad (22.172)$$

For a propagating wave of the form $y(x, t) = e^{i(\omega t - kx)}$, we obtain the dispersion relationship

$$\omega^2 = k^2 c^2 \left(1 + \frac{EI}{T} k^2 \right) . \quad (22.173)$$

For a stiff string of length L, the usual order of magnitude generally implies $\varepsilon = \frac{EI}{TL^2} \ll 1$ so that (22.173) can be written:

$$\omega^2 = k^2 c^2 \left(1 + \varepsilon k^2 L^2 \right) . \quad (22.174)$$

For a stiff string with hinged ends the boundary conditions impose

$$\sin kL = 0 \quad \text{so that} \quad k_n L = n\pi \quad (22.175)$$

from which we can derive through (22.174) the eigenfrequencies

$$\omega_n \approx \frac{n\pi c}{L}\left(1 + \varepsilon \frac{n^2\pi^2}{2}\right). \quad (22.176)$$

The eigenfrequencies are raised by the stiffness. The difference increases as n^2. The inharmonicity coefficient can be defined as

$$i = \varepsilon \frac{n^2\pi^2}{2}. \quad (22.177)$$

22.4.3 Flexural Vibrations of Thin Plates

The *thin plate*, or Kirchhoff–Love plate, described below is a generalization of the Euler–Bernoulli beam in 2-D [22.13]. The general case of orthotropic plates is treated here. The problem is solved in Cartesian coordinates, where $w(x, y, t)$ denotes the transverse displacement. It is assumed that the coordinates coincide with the symmetry axes of the material. The assumption of orthotropy leads to the following relations between plane stresses σ_{ij} and plane strains ε_{kl}

$$\begin{pmatrix} \sigma_{xx} \\ \sigma_{yy} \\ \sigma_{xy} \end{pmatrix} = \begin{pmatrix} \frac{E_x}{1-\nu_{xy}\nu_{yx}} & \frac{\nu_{yx}E_x}{1-\nu_{xy}\nu_{yx}} & 0 \\ \frac{\nu_{yx}E_x}{1-\nu_{xy}\nu_{yx}} & \frac{E_y}{1-\nu_{xy}\nu_{yx}} & 0 \\ 0 & 0 & 2G_{xy} \end{pmatrix}$$

$$\times \begin{pmatrix} \varepsilon_{xx} \\ \varepsilon_{yy} \\ \varepsilon_{xy} \end{pmatrix}. \quad (22.178)$$

These relations involve two elastic moduli E_x and E_y, two Poisson's ratios ν_{xy} and one torsional modulus G_{xy}. In addition, we have the property $E_x\nu_{yx} = E_y\nu_{xy}$ [22.14]. The bending and twisting moments are obtained by integration of the elementary moments over the thickness h of the plate:

$$\mathcal{M}_x = \int_{-h/2}^{h/2} z\sigma_{xx}\,dz\,;$$

$$\mathcal{M}_y = \int_{-h/2}^{h/2} z\sigma_{yy}\,dz\,;$$

$$\mathcal{M}_{xy} = \mathcal{M}_{yx} = \int_{-h/2}^{h/2} z\sigma_{xy}\,dz \quad (22.179)$$

which leads to the matrix relation between the moments and curvatures

$$\begin{pmatrix} \mathcal{M}_x \\ \mathcal{M}_y \\ \mathcal{M}_{xy} \end{pmatrix} = -\begin{pmatrix} D_1 & D_2/2 & 0 \\ D_2/2 & D_3 & 0 \\ 0 & 0 & D_4/2 \end{pmatrix} \times \begin{pmatrix} \frac{\partial^2 w}{\partial x^2} \\ \frac{\partial^2 w}{\partial y^2} \\ \frac{\partial^2 w}{\partial x \partial y} \end{pmatrix}, \quad (22.180)$$

where

$$D_1 = \frac{E_x h^3}{12(1-\nu_{xy}\nu_{yx})};$$

$$D_2 = \frac{E_x \nu_{yx} h^3}{6(1-\nu_{xy}\nu_{yx})} = \frac{E_y \nu_{xy} h^3}{6(1-\nu_{xy}\nu_{yx})}, \quad (22.181)$$

$$D_3 = \frac{E_y h^3}{12(1-\nu_{xy}\nu_{yx})};$$

$$D_4 = \frac{G_{xy} h^3}{3}. \quad (22.182)$$

Again, the equation of motion can be obtained using Hamilton's principle [22.12]. Finally, given an external force density term $f(x, y, t)$, we obtain the equation of motion for the plate:

$$\rho_p h \frac{\partial^2 w}{\partial t^2} = \frac{\partial^2 \mathcal{M}_x}{\partial x^2} + \frac{\partial^2 \mathcal{M}_y}{\partial y^2} + 2\frac{\partial^2 \mathcal{M}_{xy}}{\partial x \partial y} + f(x, y, t). \quad (22.183)$$

Boundary Conditions. As for the beam, the boundary conditions follow from integration by parts of the elastic energy. The number of possible conditions depends on the selected geometry. For rectangular plates, for example, Leissa lists 21 possible boundary conditions [22.15]. Along the principal directions, for example, the conditions of greatest practical interest are:

1. Clamped edge: displacement $w = 0$ and rotation $\frac{\partial w}{\partial x} = 0$;

2. Simply supported edge: displacement $w = 0$ and bending moment $\mathcal{M}_x = 0$;

3. Free edge: bending moment $\mathcal{M}_x = 0$ and shear force $\frac{\partial \mathcal{M}_x}{\partial x} + 2\frac{\partial \mathcal{M}_{xy}}{\partial y} = 0$.

Example: Homogeneous Rectangular Simply Supported Plate. From (22.180) and (22.183), we obtain for a homogeneous orthotropic plate

$$\rho_p h \frac{\partial^2 w}{\partial t^2} + D_1 \frac{\partial^4 w}{\partial x^4} + (D_2 + D_4) \times \frac{\partial^4 w}{\partial x^2 \partial y^2} + D_3 \frac{\partial^2 w}{\partial y^4} = 0. \quad (22.184)$$

For a rectangular plate of length a and width b, simply supported at its edges, the boundary conditions are

$$\begin{cases} W(0, y, t) = W(a, y, t) = W(x, 0, t) \\ \qquad = W(x, b, t) = 0, \\ \mathcal{M}_x(0, y, t) = \mathcal{M}_x(a, y, t) = \mathcal{M}_y(x, 0, t) \\ \qquad = \mathcal{M}_y(x, b, t) = 0. \end{cases} \quad (22.185)$$

The eigenfunctions satisfying (22.185) are of the form

$$\Phi_{mn}(x, y) = \sin \frac{m\pi x}{a} \sin \frac{n\pi y}{b} \quad (22.186)$$

and the associated eigenfrequencies are given by:

$$\omega_{mn} = \pi^2 \sqrt{\frac{1}{\rho_p h}} \\ \times \sqrt{D_1 \frac{m^4}{a^4} + D_3 \frac{n^4}{b^4} + (D_2 + D_4) \frac{m^2 n^2}{a^2 b^2}}. \quad (22.187)$$

In both expressions, m and n are positive integers. As for strings and beams, the eigenfunctions Φ_{mn} are orthogonal with respect to mass and stiffness, so that

$$\int_0^a \int_0^b \rho_p h \Phi_{mn}(x, y) \Phi_{m'n'}(x, y) \, dx \, dy = \\ \begin{cases} 0 & \text{if } m \neq m' \text{ or } n \neq n', \\ M_{mn} & \text{if } m = m' \text{ and } n = n'; \end{cases} \quad (22.188)$$

where M_{mn} is the modal mass for the mode (m, n). The eigenfrequencies for the orthotropic plate are distributed between two limiting curves in the (k, f)-plane (Fig. 22.12).

Particular Case: Isotropic Plate

In the isotropic case, the rigidity coefficients become

$$D_1 = D_3 = D = \frac{Eh^3}{12(1-\nu^2)} \quad (22.189)$$

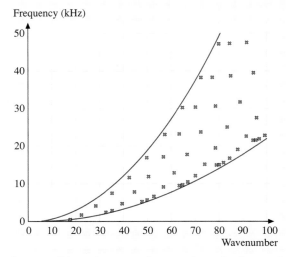

Fig. 22.12 Dispersion curves for an orthotropic plate made of carbon fibers with $D_1 = 8437$ MPa; $D_2 = 463$ MPa; $D_3 = 852$ MPa; $D_4 = 2267$ MPa; $h = 2$ mm; $\rho_p = 1540$ kg/m³; $a = 0.4$ m; $b = 0.2$ m

and

$$D_4 = \frac{Eh^3}{6(1+\nu)},$$

$$D_2 = 2D_1 - D_4 = \frac{E\nu h^3}{6(1-\nu^2)}. \quad (22.190)$$

Here, the elastic behavior of the material is fully determined by two constants: the Young's modulus E and the Poisson's ratio ν. The eigenfunctions $\Phi_{mn}(x, y)$ are the same as in (22.186). However, the eigenfrequencies now reduce to

$$\omega_{mn} = \pi^2 \sqrt{\frac{D}{\rho_p h}} \left(\frac{m^2}{a^2} + \frac{n^2}{b^2} \right). \quad (22.191)$$

Prestressed Isotropic Plate. If a tension (or compression) T_x in the x-direction and, simultaneously, a tension (or compression) T_y in the y-direction are applied in the plate plane, then the flexural equation is modified as follows [22.1]:

$$\rho_p h \frac{\partial^2 w}{\partial t^2} + D \left(\frac{\partial^4 w}{\partial x^4} + 2 \frac{\partial^4 w}{\partial x^2 \partial y^2} + \frac{\partial^4 w}{\partial y^4} \right) \\ - T_x \frac{\partial^2 w}{\partial x^2} - T_y \frac{\partial^2 w}{\partial y^2} = 0. \quad (22.192)$$

In this case, the eigenfrequencies become

$$\omega_{mn} = \sqrt{\frac{1}{\rho_p h}}$$

$$\times \sqrt{D\left(\frac{\pi^2 m^2}{a^2} + \frac{\pi^2 n^2}{b^2}\right)^2 + T_x \frac{m^2 \pi^2}{a^2} + T_y \frac{n^2 \pi^2}{b^2}}\,.$$

(22.193)

22.4.4 Vibrations of Thin Shallow Spherical Shells

One interest of presenting the vibrations of shallow spherical shells is to show the influence of curvature, compared to the case of flat plates. The present section focuses on the particular case of free edge, which is scarcely treated in the literature [22.16]. Linear vibrations of cymbals and gongs, for example, are conveniently described by such an idealized structure.

An isotropic spherical cap is considered, with thickness h, radius of curvature R and whose projection on a plane is a circle of radius a (Fig. 22.13). Using the Donnell–Mushtari–Vlasov assumptions for thin ($h \ll a$) and shallow ($a \ll R$) shells, the equations of motion for the free transverse displacement $w(r, \theta, t)$ of the cap is given by [22.17]

$$D\nabla^6 w + \frac{Eh}{R^2}\nabla^2 w + \rho h \nabla^2 \ddot{w} = 0\,,$$

(22.194)

where E is the Young's modulus, ρ is the density, ν is the Poisson's ratio and $D = Eh^3/12(1-\nu^2)$ is the flexural rigidity of the shell. Equation (22.194) can be conveniently put into nondimensional form using the reduced variables $\overline{w} = w/w_0$, $\overline{r} = r/a$ and $\overline{t} = t/t_0$ with $t_0 = a^2\sqrt{\rho h/D}$. In this case, (22.194) becomes

$$\nabla^6 \overline{w} + \chi \nabla^2 \overline{w} + \nabla^2 \ddot{\overline{w}} = 0\,,$$

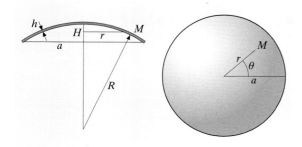

Fig. 22.13 Spherical cap

where

$$\chi = 12(1-\nu^2)\frac{a^4}{R^2 h^2}\,.$$

(22.195)

Notice that w_0 can be selected arbitrarily in the linear case. For nonlinear vibrations, this parameter should be selected in accordance with the order of magnitude of the nonlinear perturbation terms (Sect. 22.7).

Remark. In what follows, the nondimensional equation is solved but the overlines are removed from the variables, for convenience.

Eigenmodes of the Spherical Cap with Free Edge. We look for harmonic solutions of (22.195) of the form $w(r, \theta, t) = \Phi(r, \theta)q(t)$. The solutions must fulfill one of the two following cases [22.18]:

$$\begin{cases} \text{if} & \omega^2 - \chi = \zeta^4 > 0 \\ \text{then} & \nabla^2\left(\nabla^4 - \zeta^4\right)\Phi = 0 \text{ case 1}\,, \\ \text{if} & \omega^2 - \chi = -\zeta^4 < 0 \\ \text{then} & \nabla^2\left(\nabla^4 + \zeta^4\right)\Phi = 0 \text{ case 2}\,. \end{cases}$$

(22.196)

In case 1, the eigenfunctions are given by

$$\Phi_{nm}(r, \theta) =$$

$$\left[A_n r^n + B_n J_n(\zeta_{nm} r) + C_n I_n(\zeta_{nm} r)\right] \begin{vmatrix} \cos m\theta \\ \sin m\theta \end{vmatrix},$$

(22.197)

where A_n, B_n and C_n are constants, ζ_{nm} are determined by the boundary conditions, J_n are the Bessel functions of the first kind and I_n are the modified Bessel functions of the first kind. In case 2, the eigenfunctions are given by

$$\Phi_{nm}(r, \theta) =$$

$$\left[A_n r^n + B_n \operatorname{ber}_n(\zeta_{nm} r) + C_n \operatorname{bei}_n(\zeta_{nm} r)\right] \begin{vmatrix} \cos m\theta \\ \sin m\theta \end{vmatrix},$$

(22.198)

where ber_n and bei_n are the Kelvin functions defined by [22.9]

$$\operatorname{ber}_n(x) + i\operatorname{bei}_n(x) = J_n[x \exp(3i\pi/4)]\,.$$

(22.199)

For a free edge, the eigenvalues ζ_{mn} are determined from the equation obtained by expressing that forces and moments are equal to zero at the edge $r = a$. The four possible situations for the eigenfrequencies are summarized below [22.19]:

$$\begin{cases} \text{for} \quad n \in 0, 1, \quad \forall m \geq 1 \\ \qquad \omega_{nm} = \sqrt{\chi + \omega_{nm}^{(0)2}}, \\ \text{for} \quad n \geq 2, \quad m = 0 \quad \text{and} \quad \chi < \chi_n^{\lim} \\ \qquad \omega_{n0} = \sqrt{\chi + \zeta_{n0}^4}, \\ \text{for} \quad n \geq 2, \quad m = 0 \quad \text{and} \quad \chi > \chi_n^{\lim} \\ \qquad \omega_{n0} = \sqrt{\chi - \zeta_{n0}^4}, \\ \text{for} n \quad \geq 2, \quad m \geq 1, \\ \qquad \omega_{nm} = \sqrt{\chi + \zeta_{nm}^4}, \end{cases}$$
(22.200)

where $\omega_{nm}^{(0)}$ are the eigenfrequencies obtained in the case of the circular plate with free edge, corresponding to the limiting case of the spherical shell with no curvature. The limiting value $\chi_n^{\lim}$ that separates the Bessel from the Kelvin axisymmetric modes is given by [22.18]:

$$\chi_n^{\lim} = \frac{(1-v)(3+v)n^2(n^2-1)}{1 + \frac{1}{4}(1-v)(n-2) - \frac{n^2(n-1)(1-v)(4n-v+9)}{16(n+2)^2(n+3)}}.$$
(22.201)

22.4.5 Combinations of Elementary Structures

Interesting and powerful models of complex structures can be obtained by combining together elementary components. One-dimensional models of intersecting walls, for example, can be modeled by means of connected bars, with appropriate boundary conditions at the junctions: equality of displacements, angles, forces and moments. This allows calculations of reflection and transmission coefficients in the different parts of a wall or a frame. Other examples of interest are sandwich plates. A sandwich is usually made of two plates with a thin viscoelastic (absorbing) layer between them. Again, the model contains the equation of motion for each layer plus additional boundary conditions. As a result, one can calculate, for example, the equivalent complex stiffness of the sandwich. One main advantage of such a configuration is that one can obtain a substantial increase of loss factor over a relatively wide frequency range, compared to a sytem of equal thickness composed by only one plate with a simple viscoelastic layer. However, the main drawback is that each application needs careful design of a specific sandwich in order to be effective in the frequency range of interest [22.13, 20].

Periodic structures can be found in many applications, such as ships, planes or roofs where a relatively light plate (or shell) is reinforced by stiffeners and ribs at uniform intervals. Such problems have been studied extensively in recent decades. A simple example of such a structure is the periodically supported infinite beam subjected to bending motion. By generalizing methods previously applied to atomic chains and crystals by Brillouin, different authors represent the spatial part of the wave propagation in such a beam by an equation of the form

$$w(x + L) = \exp(\lambda) w(x), \qquad (22.202)$$

where w is the transverse displacement, L is the spatial period of the structure and λ is the propagation constant to be determined. The problem is solved by expressing the continuity and equilibrium equations at the junctions. Two situations may occur: if λ is purely imaginary, the bending wave can propagate freely in the periodic structure. The corresponding frequency range is called the *pass band*. If λ contains a real part, then the wave becomes evanescent, and the corresponding frequency range is called the *stop band*. Recently, some authors have shown that SEA can by useful for tackling such problems [22.21].

22.5 Structural–Acoustic Coupling

If the sound is generated by a light and flexible structure, then the effect of the acoustic loading back on the structure cannot be neglected, and the system has to be considered as a whole [22.22]. In the general case, the action of a fluid on a structure has several effects: radiation damping and modification of the eigenfrequencies. Approximations can be made in the case of a light fluid and weak coupling. These points are examined in Sect. 22.5.1. The simple example of a 1-D bar coupled to a semi-infinite tube is selected in order to facilitate the presentation of the various concepts. In a number of applications, the goal is to control the amount of energy radiated by a vibrating structure. This forms the heart of Sect. 22.5.2. In particular, the state-space formulation of structural–acoustic coupling facilitates the development of active noise-reduction systems [22.23]. The concept of radiation modes and their link with sound power and efficiency is also presented in this section [22.24, 25].

In the case of structure–cavity coupling, the effects of the cavity field are different from those of the free field. Strong structural–acoustic modes can appear, which may contribute to a substantial modification of the radiated spectrum. These features are illustrated in Sect. 22.5.3 for the example of a 1-DOF oscillator coupled to a finite tube.

The concept of coincidence (or critical) frequency is essential in the radiation of sound by structures. By comparing the dispersion equations for the structure and fluid, respectively, one can make a distinction between frequency domains where the radiation efficiency of the structure is strong and where it is weak. This is treated in Sect. 22.5.4 for an isotropic plate.

22.5.1 Longitudinally Vibrating Bar Coupled to an External Fluid

Model and Modal Projection

We consider here the simple case of a longitudinally vibrating bar of length L coupled at one end to a 1-D infinite tube filled with air, which presents a resistive loading $R_a = \rho c S$ at the end of the bar. It is assumed throughout this section that the bar has a constant cross-sectional area S and that it is clamped at one end ($x = 0$) and free at the other ($x = L$). ρ_s is the density of the bar, E its Young's modulus, $c_L = \sqrt{E/\rho_s}$ is the longitudinal wave speed and $\xi(x, t)$ the longitudinal displacement at a point M at position x along the bar ($0 \leq x \leq L$). Similarly, ρ denotes the air density, c the speed of sound and $p(x, t)$ the sound pressure in the tube ($L < x < \infty$) (Fig. 22.14). In the absence of an exciting force (free vibration), the equations of the problem are:

$$\begin{cases} \rho_s S \dfrac{\partial^2 \xi}{\partial t^2} = ES \dfrac{\partial^2 \xi}{\partial x^2} - S p(L,t) \delta(x-L) \\ \qquad\qquad \text{for } 0 \leq x \leq L, \\ p(L, t) = \rho c \dot{\xi}(L, t), \\ \xi(0, t) = 0, \\ p(x, t) = \rho c \dot{\xi}\left(L, t - \dfrac{x-L}{c}\right) \\ \qquad\qquad \text{for } L < x < \infty \end{cases} \tag{22.203}$$

We look for the solution $\xi(x, t)$ expanded in terms of the in vacuo modes $\phi_n(x)$ of the bar. Both sides of the first equation in (22.203) are multiplied by any eigenfunction and integrated over the length of the bar. This gives

$$\int_0^L \rho_s S \left[\sum_m \phi_m(x) \ddot{q}_m(t)\right] \phi_n(x) \, dx$$

$$- \int_0^L ES \left[\sum_m \phi_m''(x) q_m(t)\right] \phi_n(x) \, dx$$

$$= \int_0^L S \rho c \left[\sum_m \phi_m(L) \dot{q}_m(t)\right] \phi_n(x) \delta(x-L) \, dx. \tag{22.204}$$

This equation can be rewritten in a simpler form, by using the definition of the modal mass

$$m_n = \int_0^L \rho_s S \phi_n^2(x) \, dx. \tag{22.205}$$

Because of the mass orthogonality property of the eigenfunctions, the terms of the series in the first integral on the left-hand side of (22.204) are zero for $m \neq n$, so that the integral reduces to $m_n \ddot{q}_n(t)$. The second integral can be rewritten:

$$- \int_0^L ES \frac{\omega_m^2}{c_L^2} \left[\sum_m \phi_m(x) q_m(t)\right] \phi_n(x) \, dx \tag{22.206}$$

which, due to stiffness orthogonality properties of the eigenfunctions, reduces to $-\omega_n^2 m_n q_n(t)$. Finally, due to the properties of the delta function, the third integral can be rewritten:

$$- R_a \phi_n(L) \sum_m \phi_m(L) \dot{q}_m(t), \tag{22.207}$$

where $R_a = \rho c S$ is the acoustic radiation resistance.

Remark. In the case of the clamped–free bar, we would get $\phi_n(L) = (-1)^{n+1}$ and, similarly, $\phi_m(L) = (-1)^{m+1}$. However, in what follows, we choose to retain the more general formulation of (22.207) so that the equations remain general.

For a 1-D bar of length L coupled to a semi-infinite tube, the displacement is given by

$$\xi(x, t) = \sum_n \phi_n(x) q_n(t), \tag{22.208}$$

where the functions of times $q_n(t)$ obey the set of coupled equations

$$m_n \ddot{q}_n(t) + m_n \omega_n^2 q_n(t)$$
$$= -R_a \phi_n(L) \sum_m \phi_m(L) \dot{q}_m(t). \tag{22.209}$$

Once the displacement of the bar $\xi(x, t)$ is known, we can calculate the velocity $\dot{\xi}(x, t)$ at each point of the bar and, in particular, at the point $x = L$ and thus derive the radiated sound pressure inside the tube through (22.203). We now examine the consequences of the coupling on systems of small dimensions, to better understand its physical meaning.

Single-DOF System. Suppose that, for various reasons, we can reduce the previous system to one single mode. In this case, (22.209) reduces to

$$\ddot{q}_1(t) + \frac{R_a \phi_1^2(L)}{m_1} \dot{q}_1(t) + \omega_1^2 q_1(t) = 0 . \quad (22.210)$$

This is simply the equation of a damped oscillator, where the dimensionless damping factor ζ_1 is given by

$$2\zeta_1 \omega_1 = \frac{R_a \phi_1^2(L)}{m_1} . \quad (22.211)$$

Thus, for a single-DOF system, the acoustic coupling adds a radiation damping to the structure. This damping represents the amount of acoustic energy radiated by the vibrating bar.

Two-DOF System. Now, we truncate the system to the two lowest modes of the structure. In this case, (22.209) reduces to:

$$\begin{cases} \ddot{q}_1 + 2\zeta_1 \omega_1 \dot{q}_1 + \omega_1^2 q_1 = -\frac{R_a \phi_1(L) \phi_2(L)}{m_1} \dot{q}_2 \\ \qquad\qquad = C_{12} \dot{q}_2 , \\ \ddot{q}_2 + 2\zeta_2 \omega_2 \dot{q}_2 + \omega_2^2 q_2 = -\frac{R_a \phi_2(L) \phi_1(L)}{m_2} \dot{q}_1 \\ \qquad\qquad = C_{21} \dot{q}_1 . \end{cases} \quad (22.212)$$

Several conclusions can be drawn from this last result:

- The acoustic radiation introduces damping terms $2\zeta_i \omega_i \dot{q}_i$ in each equation.
- The time functions are coupled by the radiation. In general, the coupling coefficients C_{12} and C_{21} are not equal.

Taking the Laplace transform of the system in (22.212) leads to the characteristic equation

$$(s^2 + 2\zeta_1 \omega_1 s + \omega_1^2)(s^2 + 2\zeta_2 \omega_2 s + \omega_2^2) - 4\zeta_1 \zeta_2 \omega_1 \omega_2 s^2 = 0 . \quad (22.213)$$

Equation (22.213) shows that the structural–acoustic coupling modifies the eigenfrequencies and damping factors of the system, compared to the in vacuo case.

Light-Fluid Approximation. For $\zeta_1 \omega_1 \ll 1$ and $\zeta_2 \omega_2 \ll 1$ one can find first-order approximations for both the eigenfrequencies and decay times, due to the air–structure coupling. Denoting $s = \sigma + i\omega$ and replacing it in (22.213) yields the new damping factors:

$$\sigma_1 \approx -\left(\zeta_1 \omega_1 - \frac{\omega_2}{2\omega_1} C_{12}\right) ,$$

$$\sigma_2 \approx -\left(\zeta_2 \omega_2 - \frac{\omega_1}{2\omega_2} C_{21}\right) . \quad (22.214)$$

Similarly, the new eigenfrequencies become:

$$\begin{cases} \omega_1'^2 \approx \omega_1^2 - \left[2\zeta_1^2 \omega_1^2 - (\zeta_1 + \zeta_2)\omega_2 C_{12} \right. \\ \qquad\qquad \left. + \frac{\omega_1}{2\omega_2} C_{21} C_{12}\right] , \\ \omega_2'^2 \approx \omega_2^2 - \left[2\zeta_2^2 \omega_2^2 - (\zeta_1 + \zeta_2)\omega_1 C_{21} \right. \\ \qquad\qquad \left. + \frac{\omega_2}{2\omega_1} C_{21} C_{12}\right] . \end{cases} \quad (22.215)$$

Forced Vibrations

The transfer function between the excitation force and the bar displacement is now examined. Equation (22.203) is modified through the introduction of a force term $F(x_0, t)$ at point x_0:

$$\rho_s S \frac{\partial^2 \xi}{\partial t^2} = ES \frac{\partial^2 \xi}{\partial x^2} - Sp(L, t)\delta(x - L) + F(x_0, t)\delta(x - x_0)$$

$$\text{for } 0 \leq x \leq L$$
$$\text{and } 0 \leq x_0 \leq L . \quad (22.216)$$

Using the same modal projection as in the previous section, we find

$$\ddot{q}_n + 2\zeta_n \omega_n \dot{q}_n + \omega_n^2 q_n = \sum_{m \neq n} C_{nm} \dot{q}_m + F(x_0, t) \frac{\phi_n(x_0)}{m_n} \quad (22.217)$$

or, equivalently, using Laplace transforms, as

$$\tilde{q}_n(s) = H_n(s) \tilde{F}(x_0, s) + \sum_{m \neq n} K_{nm}(s) \tilde{q}_m(s) \quad (22.218)$$

with

$$H_n(s) = \frac{\phi_n(x_0)}{m_n (s^2 + 2\zeta_n \omega_n s + \omega_n^2)}$$

and

$$K_{nm}(s) = \frac{s C_{nm}}{s^2 + 2\zeta_n \omega_n s + \omega_n^2} . \quad (22.219)$$

Finally, the displacement is

$$\tilde{\xi}(x,s) = \sum_n \tilde{q}_n(s)\phi_n(x)$$
$$= \tilde{F}(x_0,s)\sum_n \phi_n(x)H_n(s)$$
$$+ \sum_n \phi_n(x)\sum_{m\neq n} K_{nm}(s)\tilde{q}_m(s). \quad (22.220)$$

Matrix Formulation. Equation (22.218) can then be rewritten (removing the tilde from the Laplace transforms, for convenience) as

$$\begin{pmatrix} 1 & -K_{12} & \cdots & -K_{1n} \\ -K_{21} & 1 & \cdots & -K_{2n} \\ \cdots & \cdots & \cdots & \cdots \\ -K_{n1} & -K_{n2} & \cdots & 1 \end{pmatrix} \begin{pmatrix} q_1 \\ q_2 \\ \cdots \\ q_n \end{pmatrix} = F \begin{pmatrix} H_1 \\ H_2 \\ \cdots \\ H_n \end{pmatrix} \quad (22.221)$$

which can be formulated in the more compact form:

$$\mathbf{K}\mathbf{Q} = F\mathbf{H}. \quad (22.222)$$

The displacement of the bar is

$$\xi = \mathbf{Q}^{\mathrm{t}}.\boldsymbol{\phi} = \mathbf{Q}.\boldsymbol{\phi}^{\mathrm{t}} = (\mathbf{K}^{-1}\mathbf{H})^{\mathrm{t}}.\boldsymbol{\phi}F = (\mathbf{K}^{-1}\mathbf{H}).\boldsymbol{\phi}^{\mathrm{t}}F \quad (22.223)$$

In some applications, it is interesting to write (22.218) in a form that allows the characteristic equation to be determined immediately. In this case, (22.221) becomes:

$$\begin{pmatrix} s^2+\omega_1^2 & & & \\ +2\zeta_1\omega_1 s & -sC_{12} & \cdots & -sC_{1n} \\ -sC_{21} & s^2+\omega_2^2 & \cdots & -sC_{2n} \\ & +2\zeta_2\omega_2 s & & \\ \cdots & \cdots & \cdots & \cdots \\ -sC_{n1} & -sC_{n2} & \cdots & s^2+\omega_n^2 \\ & & & +2\zeta_n\omega_n s \end{pmatrix} \begin{pmatrix} q_1 \\ q_2 \\ \cdots \\ q_n \end{pmatrix}$$

$$= F \begin{pmatrix} \beta_1 \\ \beta_2 \\ \cdots \\ \beta_n \end{pmatrix}, \quad (22.224)$$

where $\beta_n = \dfrac{\phi_n(x_0)}{m_n}$. The equivalent matrix notation is

$$\mathbf{C}\mathbf{Q} = F\boldsymbol{\beta}. \quad (22.225)$$

The displacement of the bar is given by

$$\xi = (\mathbf{C}^{-1}\boldsymbol{\beta})^{\mathrm{t}}.\boldsymbol{\phi}F = (\mathbf{C}^{-1}\boldsymbol{\beta}).\boldsymbol{\phi}^{\mathrm{t}}F. \quad (22.226)$$

Inversion of the Matrix **C**. Finding explicit solutions for the generalized displacements q_i in (22.224) requires inversion of the matrix **C**. An example is given below for a subsystem of order two, where we write for convenience $D_n = s^2 + 2\zeta_n\omega_n s + \omega_n^2$. We obtain easily

$$q_1 = \frac{\beta_1 D_2 + sC_{12}\beta_2}{D_1 D_2 - s^2 C_{12}C_{21}} F$$

and

$$q_2 = \frac{\beta_2 D_1 + sC_{21}\beta_1}{D_1 D_2 - s^2 C_{12}C_{21}} F. \quad (22.227)$$

Or, using the $(\mathbf{H},\mathbf{K})$ formulation

$$q_1 = \frac{H_1 + K_{12}H_2}{1 - K_{12}K_{21}} F = L_1 F$$

and

$$q_2 = \frac{H_2 + K_{21}H_1}{1 - K_{12}K_{21}} F = L_2 F \quad (22.228)$$

from which the displacement ξ can be derived. The coupling appears in (22.224) through the fact that q_1 is not zero, even if β_1 vanishes. In other words, one can excite vibrations at ω_n even if the structure is excited at a node of the associated in vacuo mode. The new eigenfrequencies and damping factors modified by the coupling are now given by the roots of the denominator $D_1 D_2 - s^2 C_{12}C_{21}$.

Orthogonalization of the Matrix **C**. The question is now to study whether one can find an appropriate basis for decoupling the system. Approximate formulations in the case of weak coupling are derived below. We start with a 2-DOF system before generalization. The matrix **C** is

$$\begin{pmatrix} D_1 & -sC_{12} \\ -sC_{21} & D_2 \end{pmatrix}$$

with

$$D_i = s^2 + 2\zeta_i\omega_i s + \omega_i^2. \quad (22.229)$$

The corresponding diagonal matrix $\boldsymbol{\Lambda}$ is written

$$\begin{pmatrix} \lambda_1 & 0 \\ 0 & \lambda_2 \end{pmatrix}, \quad (22.230)$$

where the λ_i are solutions of the equation

$$\begin{vmatrix} D_1 - \lambda & -sC_{12} \\ -sC_{21} & D_2 - \lambda \end{vmatrix}$$
$$= (D_2 - \lambda)(D_1 - \lambda) - s^2 C_{12}C_{21} = 0. \quad (22.231)$$

Denoting by e_i the corresponding eigenvectors and $\mathbf{T} = (e_1\ e_2)$, then we have the following matrix equations:
$$\mathbf{T\Lambda = CT} \Leftrightarrow \mathbf{\Lambda = T^{-1}CT}. \tag{22.232}$$

In the general case, the λ_i are given by
$$\lambda_{1,2} = \frac{1}{2}\left[D_1 + D_2 \pm \sqrt{(D_1 - D_2)^2 + 4s^2 C_{12}C_{21}}\right]. \tag{22.233}$$

Weak-Coupling Approximation. It is convenient to define the nondimensional coupling parameter
$$\varepsilon = \frac{C_{12}C_{21}}{D_2 - D_1}$$
$$= \frac{C_{12}C_{21}}{\omega_2^2 - \omega_1^2 + 2s(\zeta_2\omega_2 - \zeta_1\omega_1)} \tag{22.234}$$

so that, for $\varepsilon \ll 1$, the eigenvalues of $\mathbf{C}$ can be written to first-order approximation as
$$\lambda_1 = D_1 - \varepsilon s^2; \quad \lambda_2 = D_2 + \varepsilon s^2. \tag{22.235}$$

This approximation is justified if the C_{ij} are small, as in the case of a light fluid, or if the in vacuo eigenfrequencies ω_i and ω_j are not too close to each other (Sect. 22.6.1). In this case, the previously defined matrices become

$$\mathbf{T} = \begin{pmatrix} 1 & -\dfrac{\varepsilon s}{C_{21}} \\ \dfrac{\varepsilon s}{C_{12}} & 1 \end{pmatrix} \Rightarrow$$

$$\mathbf{T}^{-1} = \frac{C_{12}C_{21}}{C_{12}C_{21} + \varepsilon^2 s^2} \times \begin{pmatrix} 1 & \dfrac{\varepsilon s}{C_{21}} \\ -\dfrac{\varepsilon s}{C_{12}} & 1 \end{pmatrix}. \tag{22.236}$$

Defining further the vectors $\boldsymbol{\gamma} = \mathbf{T}^{-1}\boldsymbol{\beta}$ and $\boldsymbol{R} = \mathbf{T}^{-1}\boldsymbol{Q}$, (22.225) and (22.226) can be rewritten as
$$\mathbf{\Lambda R} = F\boldsymbol{\gamma} \Rightarrow$$
$$\xi = \mathbf{T}\mathbf{\Lambda}^{-1}\boldsymbol{\gamma}\,\boldsymbol{\phi}^t F. \tag{22.237}$$

Here, we have:
$$\mathbf{\Lambda}^{-1} = \frac{1}{D_1 D_2 - s^2 C_{12}C_{21}} \times \begin{bmatrix} D_2 + \varepsilon s^2 & 0 \\ 0 & D_1 - \varepsilon s^2 \end{bmatrix}, \tag{22.238}$$

$$\boldsymbol{\gamma} = \frac{C_{12}C_{21}}{C_{12}C_{21} + \varepsilon^2 s^2}\begin{pmatrix} \beta_1 + \dfrac{\varepsilon s}{C_{21}}\beta_2 \\ \beta_2 - \dfrac{\varepsilon s}{C_{12}}\beta_1 \end{pmatrix}$$

and
$$\boldsymbol{R} = \frac{C_{12}C_{21}}{C_{12}C_{21} + \varepsilon^2 s^2}\begin{pmatrix} q_1 + \dfrac{\varepsilon s}{C_{21}}q_2 \\ q_2 - \dfrac{\varepsilon s}{C_{12}}q_1 \end{pmatrix}. \tag{22.239}$$

To a first-order approximation, it can be shown that
$$\mathbf{\Lambda}^{-1} \approx \mathbf{\Lambda}_0^{-1} + \varepsilon \mathbf{\Lambda}_c^{-1}, \tag{22.240}$$

where $\mathbf{\Lambda}_0^{-1}$ is the in vacuo diagonal matrix given by
$$\mathbf{\Lambda}_0^{-1} = \begin{pmatrix} \dfrac{1}{D_1} & 0 \\ 0 & \dfrac{1}{D_2} \end{pmatrix} \tag{22.241}$$

and $\mathbf{\Lambda}_c^{-1}$ is the coupling diagonal matrix given by
$$\mathbf{\Lambda}_c^{-1} = \begin{pmatrix} \left(\dfrac{s}{D_1}\right)^2 & 0 \\ 0 & -\left(\dfrac{1}{D_2}\right)^2 \end{pmatrix}. \tag{22.242}$$

This result is of particular interest for computing the perturbation due to air loading of all significant variables of the system.

Generalization. Equation (22.235) can be generalized to n coupled modes. In this case, the eigenvalues become:
$$\lambda_i = D_i + \varepsilon_i$$
$$\text{with} \quad \varepsilon_i = -s^2 \sum_j \frac{C_{ij}C_{ji}}{D_j - D_i}$$
$$\text{and} \quad 1 \leq j \leq n$$
$$\text{and} \quad j \neq i. \tag{22.243}$$

Approximate Expressions for Displacement and Mode Shapes. From the above results, we can now find a first-order approximate expression for the bar displacement ξ. As in the previous sections, we start with the simple example of a 2-DOF system. Let us write first the in vacuo displacement
$$\xi_0 = \phi_{10}q_{10} + \phi_{20}q_{20}. \tag{22.244}$$

N.B. The notation ϕ_{i0}, denotes the eigenmode in vacuo and should not be confused with $\phi_i(x_0)$, which denotes the value of this eigenmode at one particular point of the structure. For the structure vibrating in air, the displacement becomes

$$\xi = \phi_{10} q_1 + \phi_{20} q_2 \,. \tag{22.245}$$

From (22.226) one can derive a first-order approximation for the bar displacement

$$\xi = \phi_{10}\left(q_{10} + \frac{sC_{12}}{D_1} q_{20}\right) + \phi_{20}\left(q_{20} + q_{10} \frac{sC_{21}}{D_2}\right). \tag{22.246}$$

Operating Deflexion Shapes. Equation (22.246) can be rewritten

$$\xi = q_{10}\left(\phi_{10} + \frac{sC_{21}}{D_2}\phi_{20}\right)$$
$$+ q_{20}\left(\phi_{20} + \phi_{10}\frac{sC_{12}}{D_1}\right)$$

with $\quad q_{i0} = \dfrac{\phi_{i0}}{m_i D_i} F \,. \tag{22.247}$

In experiments on structures, sinusoidal excitation is often used. Imagine that we apply a sudden harmonic force $F(t) = H(t) \sin \omega t$ at time $t = 0$ to the structure, with excitation location and frequency such that q_{20} is negligible compared to q_{10}. $H(t)$ is the Heaviside function. In this case, the spatial pattern of the structure is given by

$$\phi_1 = \phi_{10} + \phi_{20}\frac{sC_{21}}{D_2} \,. \tag{22.248}$$

Because of the time dependence of the second term (through the Laplace variable s), the spatial shape evolves with time. Here, we can use the Laplace limit theorem, which states that the value of $\phi(t)$ as time tends to infinity is given by the product of $s\phi(s)$ as s tends to zero. Since $s^2 C_{12}/D_2$ tends to zero as s tends to zero, the second term on the right-hand side of (22.248) vanishes after some time. Calculating the inverse Laplace transform shows that this decay time is of the order of the magnitude of the decay time of the second structural mode. After this transient regime, the spatial shape is nearly equal to the spatial shape in vacuo ϕ_{10}. In the more general case, (22.247) shows that F excites both q_{10} and q_{20}. After a transient regime, the bar displacement then finally converges to:

$$\xi(\omega, x) = \left[\frac{\phi_{10}\beta_1}{D_1(i\omega)} + \frac{\phi_{20}\beta_2}{D_2(i\omega)}\right]. \tag{22.249}$$

The quantity between in square brackets is called the operating deflexion shape (ODS) of the structure at frequency ω. Since it is very difficult in practice to excite a single q_{i0}, the ODS describes the multi-mode shapes that are observed for sinusoidal excitation of structures.

State-Space Formulation

The transfer function formulation is convenient if the system is initially at rest, and for time-invariant systems. For other applications, such as sound control, it is useful to express the results in terms of state-space variables [22.26–29]. In most cases, the mechanical state of the system is given by the position and the velocity of the DOFs. Since the unloaded structure is described by its eigenmodes ϕ_n, all useful information about the state of the system is contained in the modal participation factors q_n and in their first derivatives with time $\dot{q}_n$. This allows us to rewrite the equations for a 2-DOF coupled system as follows

$$\frac{d}{dt}\begin{pmatrix} X_1 \\ X_2 \\ X_3 \\ X_4 \end{pmatrix} = \begin{pmatrix} 0 & 0 & 1 & 0 \\ 0 & 0 & 0 & 1 \\ -\omega_1^2 & 0 & -2\zeta_1\omega_1 & C_{12} \\ 0 & -\omega_2^2 & C_{21} & -2\zeta_2\omega_2 \end{pmatrix}$$
$$\times \begin{pmatrix} X_1 \\ X_2 \\ X_3 \\ X_4 \end{pmatrix} + \begin{pmatrix} 0 \\ 0 \\ \beta_1 \\ \beta_2 \end{pmatrix} F \,, \tag{22.250}$$

where

$$X_1 = q_1; \ X_2 = q_2; \ X_3 = \dot{q}_1; \ X_4 = \dot{q}_2 \,. \tag{22.251}$$

Equation (22.250) can then be formulated equivalently as

$$\dot{X} = AX + BF \,, \tag{22.252}$$

where F is the *input* and X is the *state vector*. The output Y depends on the investigated mechanical problem. If we decide, for example, to investigate the displacement, then we can write for the *output*

$$Y = \begin{pmatrix} \phi_1 & \phi_2 & 0 & 0 \end{pmatrix}^t X = \Gamma X \,. \tag{22.253}$$

Equations (22.252) and (22.253) are general expressions for a linear system expressed in terms of state-space variables.

Remark 1. The representation presented in (22.250) is not unique. Selecting, for example

$$X_1 = q_1; \ X_2 = \dot{q}_1; \ X_3 = q_2; \ X_4 = \dot{q}_2 \tag{22.254}$$

leads to different values for A and B.

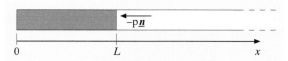

Fig. 22.14 Longitudinally vibrating bar coupled to a semi-infinite tube

Remark 2. Denoting by $\mathbf{M}$, $\mathbf{R}$, and $\mathbf{K}$, the mass, resistance and stiffness matrices, respectively, of the 2-DOF system, the matrix $\mathbf{A}$ can be rewritten more generally using submatrices as

$$\mathbf{A} = \begin{pmatrix} \mathbf{0} & \mathbf{I} \\ -\mathbf{M}^{-1}\mathbf{K} & -\mathbf{M}^{-1}\mathbf{R} \end{pmatrix}, \quad (22.255)$$

where $\mathbf{I}$ is the identity matrix.

22.5.2 Energetic Approach to Structural–Acoustic Systems

Two-DOF System

We now consider a 2-DOF structure coupled to a radiating wave, but now include damping terms in the structure r_1 and r_2, to compare the power dissipated in vacuo and air, respectively. The damping terms r_{a1} and r_{a2} are due to radiation. Modal masses and stiffnesses are written explicitly, so that comparison with the case of a single oscillator becomes easier. We can then write

$$\begin{cases} m_1 \ddot{q}_1 + (r_1 + r_{a1})\dot{q}_1 + k_1 q_1 + \gamma \dot{q}_2 = \phi_1(x_0) F, \\ m_2 \ddot{q}_2 + (r_2 + r_{a2})\dot{q}_2 + k_1 q_2 + \gamma \dot{q}_1 = \phi_2(x_0) F, \end{cases} \quad (22.256)$$

where $\gamma = -C_{12} m_1 = -C_{21} m_2$. In what follows, we use the notations:

$$\begin{cases} 2\zeta_1 \omega_1 = \frac{r_1 + r_{a1}}{m_1}; \\ 2\zeta_{10} \omega_1 = \frac{r_1}{m_1}; \\ \omega_1^2 = \frac{k_1}{m_1}; \\ 2\zeta_2 \omega_2 = \frac{r_2 + r_{a2}}{m_1}; \\ 2\zeta_{20} \omega_2 = \frac{r_2}{m_2}; \\ \omega_2^2 = \frac{k_2}{m_2} \end{cases} \quad (22.257)$$

to make a distinction between the damping factors in air and in vacuo. The force F is applied at point $x = x_0$ so that the mechanical input power is given by

$$p_m(t) = F \frac{d\xi}{dt}(x_0, t) = F [\phi_1(x_0)\dot{q}_1 + \phi_2(x_0)\dot{q}_2]. \quad (22.258)$$

Using (22.256) allows $p_m(t)$ to be written as

$$\begin{aligned} p_m(t) = & m_1 \ddot{q}_1 \dot{q}_1 + (r_1 + r_{a1})\dot{q}_1^2 + k_1 q_1 \dot{q}_1 \\ & + 2\gamma \dot{q}_1 \dot{q}_2 \\ & + m_2 \ddot{q}_2 \dot{q}_2 + (r_2 + r_{a2})\dot{q}_2^2 + k_2 q_2 \dot{q}_2. \end{aligned} \quad (22.259)$$

Integrating $p_m(t)$ over a duration T and removing the conservative energy terms yields the mean input power

$$\begin{aligned} \mathcal{P}_m(T) = & \frac{1}{T} \int_0^T (r_1 + r_{a1})\dot{q}_1^2 + (r_2 + r_{a2})\dot{q}_2^2 \\ & + 2\gamma \dot{q}_2 \dot{q}_1 \, dt. \end{aligned} \quad (22.260)$$

The input power is now seen as the sum of three terms:

- The mean power

$$\mathcal{P}_s(T) = \frac{1}{T} \int_0^T r_1 \dot{q}_1^2 + r_2 \dot{q}_2^2 \, dt$$

dissipated in the structure;

- The mean acoustic power

$$\mathcal{P}_a(T) = \frac{1}{T} \int_0^T r_{a1} \dot{q}_1^2 + r_{a2} \dot{q}_2^2 \, dt$$

radiated in the air;

- The mean coupling power

$$\mathcal{P}_c(T) = \frac{2}{T} \int_0^T \gamma \dot{q}_2 \dot{q}_1 \, dt$$

due to the exchange of energy between the two oscillators via the fluid.

Let us denote the steady-state excitation frequency by ω. The transient regime is neglected, and the integration time T is supposed to be taken equal to an integer multiple of the period $T = n\tau = n2\pi/\omega$. The mean input power can then be written as

$$\begin{aligned} \mathcal{P}_m = & \frac{1}{2} \Big[(r_1 + r_{a1})|\dot{q}_1|^2 \\ & + (r_2 + r_{a2})|\dot{q}_2|^2 + 2\gamma |\dot{q}_2||\dot{q}_1| \Big]. \end{aligned} \quad (22.261)$$

$\mathcal{P}_m$ can be written in matrix form. Writing $\dot{\mathbf{Q}} = \begin{pmatrix} \dot{q}_1 \\ \dot{q}_2 \end{pmatrix}$,

$$\mathbf{R}_s = \begin{pmatrix} r_1 & 0 \\ 0 & r_2 \end{pmatrix} \text{ and } \mathbf{R}_a = \begin{pmatrix} r_{a1} & \gamma \\ \gamma & r_{a2} \end{pmatrix}, \text{ we have}$$

$$\mathcal{P}_{\mathrm{m}} = \dot{Q}^{\mathrm{H}} [\mathbf{R}_{\mathrm{s}} + \mathbf{R}_{\mathrm{a}}] \dot{Q} , \qquad (22.262)$$

where $\dot{Q}^{\mathrm{H}}$ is the Hermitian conjugate (conjugate transpose) of $\dot{Q}$.

Generalization

The mean power for a multiple-DOF system is written:

$$\mathcal{P}_{\mathrm{m}}(T) = \frac{1}{T} \int_0^T \left[\sum_{i=1}^n (r_i + r_{ai}) \dot{q}_i^2 + \sum_{i=1}^n \sum_{\substack{j=1 \\ j \neq i}}^n \gamma_{ij} \dot{q}_i \dot{q}_j \right] \mathrm{d}t \qquad (22.263)$$

with $\gamma_{ij} = -m_i C_{ij}$.

This mean power can be written in the same form as (22.262), where the resistance matrix is defined as

$$\begin{pmatrix} r_1 + r_{a1} & \cdots & \gamma_{1i} & \cdots & \gamma_{1j} & \cdots & \gamma_{1n} \\ \cdots & \cdots & \cdots & \cdots & \cdots & \cdots & \cdots \\ \gamma_{i1} & \cdots & r_i + r_{ai} & \cdots & \gamma_{ij} & \cdots & \gamma_{in} \\ \cdots & \cdots & \cdots & \cdots & \cdots & \cdots & \cdots \\ \gamma_{j1} & \cdots & \gamma_{ji} & \cdots & r_j + r_{aj} & \cdots & \gamma_{jn} \\ \cdots & \cdots & \cdots & \cdots & \cdots & \cdots & \cdots \\ \gamma_{n1} & \cdots & \gamma_{ni} & \cdots & \gamma_{nj} & \cdots & r_n + r_{an} \end{pmatrix} . \qquad (22.264)$$

Remark. The resistance matrix can again be viewed as the sum of a structural resistance matrix $\mathbf{R}_{\mathrm{s}}$ and an acoustical resistance matrix $\mathbf{R}_{\mathrm{a}}$. This leads to the expression for the mean acoustic power

$$\mathcal{P}_{\mathrm{a}} = \dot{Q}^{\mathrm{H}} \mathbf{R}_{\mathrm{a}} \dot{Q} \qquad (22.265)$$

and acoustical efficiency

$$\eta_{\mathrm{m}} = \frac{\dot{Q}^{\mathrm{H}} [\mathbf{R}_{\mathrm{a}}] \dot{Q}}{\dot{Q}^{\mathrm{H}} [\mathbf{R}_{\mathrm{s}} + \mathbf{R}_{\mathrm{a}}] \dot{Q}} \qquad (22.266)$$

which generalizes (22.32). Note that we have assumed that the structural resistance matrix $\mathbf{R}_{\mathrm{s}}$ is diagonal, which is usually a reasonable assumption for lightly damped structures. However, strong structural damping can also be the source of intermodal coupling. In this case, $\mathbf{R}_{\mathrm{s}}$ is no longer diagonal, but the general results expressed in (22.262) and (22.266) remain valid. A comparison of efficiencies between the cases of light and heavy fluids, respectively has been conducted by *Rumerman* [22.30].

Radiation Filter

Because $\mathbf{R}_{\mathrm{a}}$ is real, symmetric and positive definite, we can write this matrix in the form

$$\mathbf{R}_{\mathrm{a}} = \mathbf{P}^{\mathrm{t}} \mathbf{\Omega} \mathbf{P} , \qquad (22.267)$$

where $\mathbf{\Omega}$ is a diagonal matrix [22.25]. As a consequence, the acoustic power becomes, removing for simplicity the integration time T,

$$P_{\mathrm{a}} = b^{\mathrm{H}} \mathbf{\Omega} b$$

where

$$b = \mathbf{P} \dot{Q} . \qquad (22.268)$$

This can be written explicitly as

$$P_{\mathrm{a}} = \sum_n \Omega_n |b_n|^2 . \qquad (22.269)$$

Equation (22.269) shows that, defining the appropriate basis, the acoustic power can be expressed as a sum of quadratic terms, thus removing the cross products between the q_i in the previous subsections. Another interesting consequence of the properties of $\mathbf{R}_{\mathrm{a}}$ is that the acoustic power can be decomposed using the Cholesky method. This leads to the expression:

$$P_{\mathrm{a}} = \dot{Q}^{\mathrm{H}} \mathbf{R}_{\mathrm{a}} \dot{Q}$$
$$= \dot{Q}^{\mathrm{H}} \mathbf{G}^{\mathrm{H}} \mathbf{G} \dot{Q} = z^{\mathrm{H}} z = \sum_n |z_n|^2 , \qquad (22.270)$$

where, by comparison with (22.269), the vector $z(\omega)$ can be viewed as the output of a set of radiation filters whose transfer functions $\mathbf{G}(\omega)$ are given by

$$\mathbf{G}(\omega) = \sqrt{\mathbf{\Omega}(\omega)} \mathbf{P}(\omega) \qquad (22.271)$$

with input vector $\dot{Q}$, so that

$$z = \mathbf{G} \dot{Q} . \qquad (22.272)$$

Impulsively Excited Structure: Total Radiated Energy

For an impulsively excited structure, the total radiated energy is given by:

$$E_T = \int_0^\infty \dot{Q}^{\mathrm{H}} \mathbf{R}_{\mathrm{a}} \dot{Q} \, \mathrm{d}\omega = \int_0^\infty z^{\mathrm{H}}(\omega) z(\omega) \, \mathrm{d}\omega \qquad (22.273)$$

which, by Parseval's theorem, is equivalent to [22.31]

$$E_T = \int_0^\infty z^{\mathrm{t}}(t) z(t) \, \mathrm{d}t . \qquad (22.274)$$

State-Space Analysis

The interest of formulating structural acoustic coupling in terms of state-space variables is now emphasized using an example. Denoting by r the internal state of filter $\mathbf{G}$ with input $\mathbf{X}$ (or $\dot{\mathbf{Q}}$) and output z, we can use a state-space realization of the form:

$$\begin{cases} \dot{r} = \mathbf{A}_G r + \mathbf{B}_G X, \\ z = \mathbf{C}_G r + \mathbf{D}_G X. \end{cases} \quad (22.275)$$

Combining these equations with the equations of motion of the structure gives

$$\begin{pmatrix} \dot{X} \\ \dot{r} \end{pmatrix} = \begin{pmatrix} \mathbf{A} & \mathbf{0} \\ \mathbf{B}_G & \mathbf{A}_G \end{pmatrix} \begin{pmatrix} X \\ r \end{pmatrix} + \begin{pmatrix} \mathbf{B} \\ 0 \end{pmatrix} F \quad (22.276)$$

with the output

$$z = \begin{pmatrix} \mathbf{D}_G & \mathbf{C}_G \end{pmatrix} \begin{pmatrix} X \\ r \end{pmatrix}. \quad (22.277)$$

If, for example, the purpose is to minimize the total energy radiated by a source, then the cost function can be defined as

$$\mathcal{C}_f = \frac{\int_0^\infty z^t(t) z(t)\,\mathrm{d}t}{\max\{E_T\}}. \quad (22.278)$$

22.5.3 Oscillator Coupled to a Tube of Finite Length

Presentation of the Model

The purpose of this section is to present the fundamental concepts for a vibrating structure coupled to an acoustic cavity. As a simple example, we consider a single-DOF oscillator coupled to a 1-D tube of cross-sectional area S and finite length L loaded at its end $x = L$ by an impedance Z_L, defined here as the ratio between the pressure and acoustic velocity. The selected structure is a mechanical oscillator of mass M, stiffness $K = M\omega_0^2$ and dashpot $R = 2M\zeta_0\omega_0$ driven by a force $T(0, t)$ at position $x = 0$ (Fig. 22.15). The amplitude of the motion of the mass is assumed to be small, so that the acoustic velocity $v(0, t)$ is equal to the mechanical velocity $\dot{\xi}(t)$. We assume lossless wave propagation in the tube itself. The set of equations for the model is written:

$$\begin{cases} \dfrac{1}{c^2}\dfrac{\partial^2 p}{\partial t^2} = \dfrac{\partial^2 p}{\partial x^2} \quad \text{for } 0 < x < L; \\ \rho\dfrac{\partial v}{\partial t} = -\dfrac{\partial p}{\partial x} \\ p(L, s) = Z_L v(L, s);\quad v(0, t) = \dot{\xi}(t) \\ M(\ddot{\xi} + 2\zeta_0\omega_0\dot{\xi} + \omega_0^2\xi) = -Sp(0, t) + T(t) \end{cases}$$

(22.279)

Mass Displacement

Using the Laplace transforms, we obtain

$$\begin{cases} p(x, s) = \exp\left(-\dfrac{sx}{c}\right)F(s) + \exp\left(\dfrac{sx}{c}\right)G(s), \\ v(x, s) = \dfrac{1}{\rho c}\left[\exp\left(-\dfrac{sx}{c}\right)F(s) - \exp\left(\dfrac{sx}{c}\right)G(s)\right], \end{cases}$$

(22.280)

where $F(s)$ and $G(s)$ are two functions to be determined. The boundary condition at $x = L$ gives

$$G(s) = F(s)\frac{Z_L - \rho c}{Z_L + \rho c}\exp\left(-\frac{2Ls}{c}\right). \quad (22.281)$$

The continuity of the displacement at position $x = 0$ yields:

$$v(0, s) = \frac{1}{\rho c}F(s)\left[1 - \frac{Z_L - \rho c}{Z_L + \rho c}\exp\left(-\frac{2Ls}{c}\right)\right]$$
$$= s\xi(s). \quad (22.282)$$

Finally, the equation governing the oscillator motion becomes

$$\left(s^2 + 2\zeta_0\omega_0 s + \omega_0^2\right)\xi(s)$$
$$= \frac{T(s)}{M} - s\frac{R_a}{M}\frac{z_L + \tanh\left(\frac{sL}{c}\right)}{1 + z_L \tanh\left(\frac{sL}{c}\right)}\xi(s) \quad (22.283)$$

with

$$z_L = \frac{Z_L}{\rho c} \quad \text{and} \quad R_a = \rho c S. \quad (22.284)$$

Equation (22.283) shows that, due to the loading of the oscillator by the finite tube, the damping term $2\zeta_0\omega_0$ becomes:

$$2\zeta_0\omega_0 + 2\zeta_a\omega_0 Z(s), \quad (22.285)$$

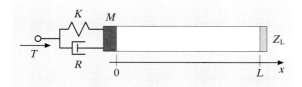

Fig. 22.15 Single-DOF mechanical oscillator coupled to a finite loaded tube

where
$$\frac{R_a}{M} = 2\zeta_a \omega_0$$
and
$$Z(s) = \frac{z_L + \tanh\left(\frac{sL}{c}\right)}{1 + z_L \tanh\left(\frac{sL}{c}\right)} \quad (22.286)$$

so that we can write:
$$\xi(s) = \frac{T(s)}{M} \frac{1}{s^2 + 2\omega_0 [\zeta_0 + \zeta_a Z(s)] s + \omega_0^2} . \quad (22.287)$$

$F(s)$ is derived from (22.282):
$$F(s) = \frac{\rho c s \xi(s)}{\left[1 - \frac{z_L - 1}{z_L + 1} \exp\left(-\frac{2sL}{c}\right)\right]}$$
$$= \frac{\rho c s \xi(s) \exp\left(\frac{sL}{c}\right)(z_L + 1)}{2\left[z_L \sinh\left(\frac{sL}{c}\right) + \cosh\left(\frac{sL}{c}\right)\right]} . \quad (22.288)$$

Equations (22.280) and (22.281) yield the pressure in the tube
$$p(x, s) = \rho c s \xi(s)$$
$$\times \frac{z_L \cosh\left(\frac{s(x-L)}{c}\right) - \sinh\left(\frac{s(x-L)}{c}\right)}{\cosh\left(\frac{sL}{c}\right) + z_L \sinh\left(\frac{sL}{c}\right)} . \quad (22.289)$$

The pressure $p(0, s)$ acting on the oscillator is
$$p(0, s) = \rho c s \xi(s) \frac{z_L \cosh\left(\frac{sL}{c}\right) + \sinh\left(\frac{sL}{c}\right)}{\cosh\left(\frac{sL}{c}\right) + z_L \sinh\left(\frac{sL}{c}\right)}$$
$$= \rho c s \xi(s) Z(s) . \quad (22.290)$$

Taking the inverse Laplace transform of (22.289) and (22.287) yields the pressure in the tube and the mass displacement in the time domain.

Discussion of $Z(s)$. Several interesting limiting cases can be examined:

- If the load at the end of the tube at $x = L$ is such that $Z_L = \rho c$, then $G(s) = 0$ and $F(s) = \rho c s \xi(s)$. This means that when the tube is loaded by its characteristic impedance there is no returning wave. In this case $Z(s) = 1$, and the only effect of the tube is to increase the damping of the oscillator, which becomes equal to $\xi_0 + \xi_a$. The increase of damping is entirely due to radiation and is, of course, identical to that obtained for the 1-DOF approximation of the vibrating bar loaded by the semi-infinite tube presented in Sect. 22.5.1.
- If the tube is closed at $x = L$, then z_L tends to ∞ so that $Z(s) = 1/\tanh(sL/c)$. If, in addition, the length L of the tube is sufficiently small so that we can make the approximation $\tanh(sL/c) \approx sL/c$, then the displacement can be written
$$\xi(s) = \frac{T(s)}{M} \frac{1}{s^2 + 2\omega_0 \zeta_0 s + \omega_0^2 + \frac{2\omega_0 \zeta_a c}{L}} . \quad (22.291)$$

This is equivalent to the tube acting as an added stiffness $K_a = 2\omega_0 M \zeta_a c / L$, which increases the eigenfrequency of the mechanical oscillator.
- If the tube is open at $x = L$ and if we neglect radiation, then $z_L = 0$ and $Z(s) = \tanh(sL/c)$. For a small tube, or more generally for $sL/c \ll 1$, we obtain
$$\xi(s) = \frac{T(s)}{M} \frac{1}{s^2 + 2\omega_0 \zeta_0 s + \omega_0^2 + \frac{2\omega_0 \zeta_a s^2 L}{c}} . \quad (22.292)$$

so that the tube acts as an added mass $M_a = 2M\omega_0 \zeta_a L / c$, which decreases the eigenfrequency of the oscillator.

In the most general case, the coupling between a structure and a cavity yields new vibroacoustic (or structural–acoustic) modes that differ from the in vacuo modes of the structure and from the cavity modes with rigid walls [22.32]. For the present example, assuming light damping, these modes are the poles of $\xi(\omega)$, which are obtained from the equation
$$-\omega^2 + \omega_0^2 - 2\zeta_a \omega_0 \omega \, \text{Im}[Z(\omega)] = 0 . \quad (22.293)$$

For a lossless tube closed at $x = L$, the eigenmodes are the solutions of
$$(\omega_0^2 - \omega^2) \tan \frac{\omega L}{c} + 2\zeta_a \omega_0 \omega = 0 . \quad (22.294)$$

For high frequencies, i.e., for $\omega \gg \omega_0$, the displacement of the oscillator tends to zero and (22.294) reduces to
$$\tan \frac{\omega L}{c} \approx 0 . \quad (22.295)$$

In this regime, the eigenfrequencies tend to those of the tube closed at both ends.

22.5.4 Two-Dimensional Elasto–Acoustic Coupling

We now turn to the case of 2-D systems. First, the case of an infinite isotropic plate is examined to introduce the concept of critical frequency. The presentation used the wave formalism and is inspired from *Filippi* et al., [22.33] and *Lesueur* [22.34]. We then consider the case of a finite plate using the wavenumber Fourier-transform method [22.35, 36].

Infinite Isotropic Plate. Critical Frequency

The case of thin plates submerged in compressible fluids and subjected to a transverse flexural displacement $\xi(x, y, t)$ is considered. Kirchhoff–Love approximations are assumed, so that shear stresses and rotary inertia are neglected. The infinite plate is located in the plane $z = 0$. In what follows, h is the plate thickness, E the Young's modulus, ν the Poisson's ratio and ρ_s the density of the plate. In the half-space $z < 0$, the pressure field is $p_1(x, y, z, t)$ in fluid 1 of density ρ_1 where the speed of sound is c_1. For $z > 0$, the pressure field is $p_2(x, y, z, t)$ in fluid 2 of density ρ_2 with speed of sound is c_2. The motion of the plate is harmonic with frequency ω. The equations of the problem are written in Cartesian coordinates:

- For $z < 0$
$$\nabla^2 p_1 + \frac{\omega^2}{c_1^2} p_1 = 0 \qquad (22.296)$$
with the continuity condition
$$\frac{\partial p_1}{\partial z}(x, y, 0-) = \omega^2 \rho_1 \xi(x, y). \qquad (22.297)$$

- For $z > 0$
$$\nabla^2 p_2 + \frac{\omega^2}{c_2^2} p_2 = 0 \qquad (22.298)$$
with the continuity condition
$$\frac{\partial p_2}{\partial z}(x, y, 0+) = \omega^2 \rho_2 \xi(x, y). \qquad (22.299)$$

- For $z = 0$, we have the loaded-plate equation:
$$p_1(x, y, 0-) - p_2(x, y, 0+)$$
$$= -\omega^2 \rho_s h \xi(x, y) + D \nabla^4 \xi(x, y), \qquad (22.300)$$

where $D = \dfrac{Eh^3}{12(1-\nu^2)}$.

General Solution. The continuity conditions yields the following equalities for the wavenumber components
$$k_x^1 = k_x^2 = k_x \quad \text{and} \quad k_y^1 = k_y^2 = k_y. \qquad (22.301)$$

Therefore the general solution of the problem can be written in the form:
$$\begin{cases} p_1(x, y, z) = \left(A_x^1 e^{-ik_x x} + B_x^1 e^{ik_x x}\right) \\ \qquad\qquad \left(A_y^1 e^{-ik_y y} + B_y^1 e^{ik_y y}\right) B_z^1 e^{ik_z^1 z}, \\ p_2(x, y, z) = \left(A_x^2 e^{-ik_x x} + B_x^2 e^{ik_x x}\right) \\ \qquad\qquad \left(A_y^2 e^{-ik_y y} + B_y^2 e^{ik_y y}\right) A_z^2 e^{-ik_z^2 z}, \\ \xi(x, y) = \left(C_x e^{-ik_x x} + D_x e^{ik_x x}\right) \\ \qquad\qquad \left(C_y e^{-ik_y y} + D_y e^{ik_y y}\right), \end{cases}$$
$$(22.302)$$

where the constants A, B, C, and D depend on the initial conditions. The wavenumbers in both fluids are written
$$k_1^2 = \frac{\omega^2}{c_1^2} = k_x^2 + k_y^2 + k_z^{1\,2}$$
and
$$k_2^2 = \frac{\omega^2}{c_2^2} = k_x^2 + k_y^2 + k_z^{2\,2}. \qquad (22.303)$$

From (22.296–22.303) we derive the dispersion equation for the fluid-loaded plate
$$i\omega^2 \left(\frac{\rho_1}{\sqrt{\frac{\omega^2}{c_1^2} - k_x^2 - k_y^2}} + \frac{\rho_2}{\sqrt{\frac{\omega^2}{c_2^2} - k_x^2 - k_y^2}} \right)$$
$$+ D\left(k_x^2 + k_y^2\right)^2 - \omega^2 \rho_s h = 0 \qquad (22.304)$$

and the expressions for the pressure fields
$$\begin{cases} p_1 = -\dfrac{i\omega \rho_1 c_1 k_1}{\sqrt{k_1^2 - k_x^2 - k_y^2}} \xi(x, y) e^{+i\sqrt{k_1^2 - k_x^2 - k_y^2}\, z} \\ \qquad \text{with} \quad k_1 = \dfrac{\omega}{c_1}, \\ p_2 = +\dfrac{i\omega \rho_2 c_2 k_2}{\sqrt{k_2^2 - k_x^2 - k_y^2}} \xi(x, y) e^{-i\sqrt{k_2^2 - k_x^2 - k_y^2}\, z} \\ \qquad \text{with} \quad k_2 = \dfrac{\omega}{c_2}. \end{cases}$$
$$(22.305)$$

Equation (22.305) shows that the acoustic waves can be either progressive or evanescent, depending on the values of the wavenumber components k_x and k_y.

Radiated Power. The mean acoustic power in the fluids are given by

$$\begin{cases} \langle \mathcal{P}_1 \rangle = \int_{(S)} \frac{1}{2} \operatorname{Re}\left(p_1 [i\omega\xi]^*\right) dS \\ \qquad = -\frac{1}{2}\omega^2 \rho_1 c_1 \int_{(S)} \operatorname{Re}\left(\frac{k_1}{\sqrt{k_1^2 - k_x^2 - k_y^2}}\right) \\ \qquad \quad \times |\xi(x,y)|^2 \, dS \, , \\ \langle \mathcal{P}_2 \rangle = \int_{(S)} \frac{1}{2} \operatorname{Re}\left(p_2 [i\omega\xi]^*\right) dS \\ \qquad = +\frac{1}{2}\omega^2 \rho_2 c_2 \int_{(S)} \operatorname{Re}\left(\frac{k_1}{\sqrt{k_2^2 - k_x^2 - k_y^2}}\right) \\ \qquad \quad \times |\xi(x,y)|^2 \, dS \, . \end{cases}$$

(22.306)

The minus sign in $\langle \mathcal{P}_1 \rangle$ follows from the fact that the acoustic intensity vector in fluid 1 is oriented towards the negative direction of the z-axis. The radiation efficiencies σ_i of the plate in both fluids (with $i \in [1, 2]$) are defined as

$$\sigma_i = \left| \frac{\langle \mathcal{P}_i \rangle}{\rho_i c_i \langle V^2 \rangle} \right|$$

with

$$\langle V^2 \rangle = \frac{1}{2} \int_{(S)} \omega^2 |\xi(x,y)|^2 \, dS \qquad (22.307)$$

which gives

$$\sigma_i = \operatorname{Re}\left(\frac{k_i}{k_z^i}\right). \qquad (22.308)$$

These quantities represent the ratio between the amount of acoustic energy radiated in fluid and the total kinetic energy of the plate. Equation (22.308) shows that, if k_i^z is real, we have a propagating acoustic wave and the mean radiated power is different from zero. In the opposite case, the acoustic wave is evanescent and the mean radiated power is equal to zero.

Particular Case of Two Identical Fluids. One particular case of interest corresponds to the practical situation where the plate is coupled to the same fluid on both sides. In this case, the dispersion equation becomes:

$$2i\omega^2 \frac{\rho}{\sqrt{k^2 - k_F^2}} - \omega^2 \rho_s h + D k_F^4 = 0 \, , \qquad (22.309)$$

where $k_x = k_F \cos\theta$ and $k_y = k_F \sin\theta$. k_F is the flexural wavenumber coupled to the fluid, θ is the angle of propagation and $k = \omega/c$ is the acoustic wavenumber.

For a plate vibrating in vacuo, the flexural wavenumber is written

$$k_{F0} = \left(\frac{\omega^2 \rho_s h}{D}\right)^{1/4} \qquad (22.310)$$

so that the dispersion equation can be rewritten

$$\frac{k_F^4}{k_{F0}^4} = 1 - \frac{2i\rho}{\rho_s h k_z} \qquad (22.311)$$

Light-Fluid Approximation. The light-fluid approximation corresponds to

$$\left| \frac{2i\rho}{\rho_s h k_z} \right| \ll 1 \, . \qquad (22.312)$$

In this case, the wavenumbers are given by:

$$k_F = k_{F0} \quad \text{or} \quad k_F = i k_{F0} \, . \qquad (22.313)$$

The case $k_F = k_{F0}$ corresponds to propagating waves. In this case, the radiation efficiency becomes

$$\begin{cases} \sigma = \operatorname{Re}\left(\frac{k}{k_z}\right) = \frac{1}{\sqrt{1 - \frac{k_{F0}^2}{k^2}}} \\ \qquad = \frac{1}{\sqrt{1 - \frac{\omega_c}{\omega}}} \quad \text{if} \quad \omega > \omega_c \\ \text{with} \quad \omega_c = \frac{c^2}{h}\sqrt{\frac{12(1-\nu^2)\mu}{E}} \\ \qquad = 0 \quad \text{if} \quad \omega < \omega_c \, ; \end{cases} \qquad (22.314)$$

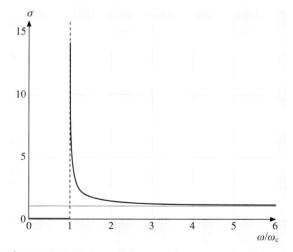

Fig. 22.16 Radiation efficiency σ for progressive flexural waves in an infinite isotropic plate submerged in a light fluid, ω_c is the critical frequency

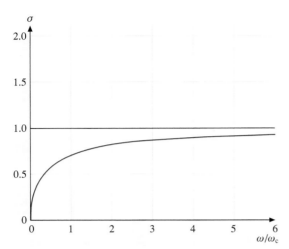

Fig. 22.17 Radiation efficiency for evanescent flexural waves in an infinite isotropic plate submerged in a light fluid

ω_c is the critical (or coincidence) frequency of the plate coupled to the light fluid. For this frequency, the wavelength in the plate is equal to the wavelength in the fluid. For $\omega > \omega_c$ ($k_{F0} < k$), the flexural waves are *supersonic*. The radiation efficiency tends to infinity as ω tends to ω_c and tends to unity as ω tends to infinity (Fig. 22.16). σ tends to infinity for $\omega = \omega_c$ because we are dealing with an infinite plate.

For $\omega < \omega_c$, the flexural waves are *subsonic*. The radiation efficiency is equal to zero which means that the plate does not radiate sound. This result follows from the fact that acoustic pressure and velocity are in quadrature.

The solution $k_F = ik_{F0}$ corresponds to evanescent waves in the plate. In this case, the radiation efficiency becomes

$$\sigma = \operatorname{Re}\left(\frac{k}{k_z}\right) = \frac{1}{\sqrt{1+\frac{k_{F0}^2}{k^2}}} = \frac{1}{\sqrt{1+\frac{\omega_c}{\omega}}}. \quad (22.315)$$

Equation (22.315) shows that the radiation efficiency of a flexural evanescent wave increases with frequency and tends to unity as the frequency tends to infinity (Fig. 22.17).

Simply Supported Radiating Isotropic Plate

Equations of Motion. We address the problem of structural–acoustic interaction for a 2-D, rectangular, baffled, simply supported isotropic thin plate radiating in air.

The system is described by the equations

$$\begin{cases} D\nabla^4 \xi(x,y,t) + \rho_s h \frac{\partial^2 \xi}{\partial t^2}(x,y,t) \\ \quad = p(x,y,0-,t) - p(x,y,0+) \\ \quad \text{for } 0 < x < L_x \text{ and } 0 < y < L_y, \\ \xi(0,y,t) = \xi(L_x,y,t) = \xi(x,0,t) \\ \quad = \xi(x,L_y,t) = 0, \\ \xi''(0,y,t) = \xi''(L_x,y,t) = \xi''(x,0,t) \\ \quad = \xi''(x,L_y,t) = 0, \\ v_z(x,y,0+,t) = -v_z(x,y,0-,t) = \dot{\xi}(x,y,t) \\ \quad \text{for } 0 < x < L_x \text{ and } 0 < y < L_y, \\ v(x,y,0) = 0 \\ \quad \text{for } x \notin]0, L_x[\text{ or } y \notin]0, L_y[, \\ \rho \frac{\partial v}{\partial t} + \nabla p = 0, \\ \nabla^2 p - \frac{1}{c^2} \frac{\partial^2 p}{\partial t^2} = 0, \end{cases} \quad (22.316)$$

where h is the thickness of the plate, $D = EI = \frac{Eh^3}{12(1-\nu^2)}$ is the flexural rigidity factor, E is the Young's modulus, ν is the Poisson's ratio and ρ_s is the plate's density. The transverse displacement of the plate is denoted by $\xi(x,y,t)$. The acoustic variables are the pressure $p(x,y,z,t)$ and the velocity $v(x,y,z,t)$. ρ is the air density and c is the speed of sound in air.

Eigenmodes in Vacuo. Solving the plate equations in vacuo with simply supported boundary conditions yields the eigenmodes

$$\phi_{mn}(x,y) = \phi_m(x)\phi_n(y) = \sin\frac{\pi m x}{L_x} \sin\frac{\pi n y}{L_y} \quad (22.317)$$

allowing the displacement to be expanded as follows

$$\xi(x,y,t) = \sum_{m,n} \phi_{mn}(x,y) q_{mn}(t)$$
$$= \phi_m(x)\phi_n(y) q_{mn}(t). \quad (22.318)$$

Defining the vectors

$$\begin{cases} \boldsymbol{Q}^t = (q_{01}\ q_{10}\ q_{11}\ \ldots\ q_{MN}), \\ \text{and} \\ \boldsymbol{\Phi}^t = (\phi_{01}\ \phi_{10}\ \phi_{11}\ \ldots\ \phi_{MN}), \end{cases} \quad (22.319)$$

(22.318) can be written

$$\xi(x, y, t) = \boldsymbol{Q}^{\mathrm{t}} \boldsymbol{\Phi} \quad (22.320)$$

which is similar in form to the 1-D systems studied in Sect. 22.5.1. In (22.318), the number of considered modes (with indices M and N in both directions, respectively) depends on the nature and frequency content of the practical problem to be solved.

Calculation of the Radiated Sound Field Using Wavenumber Fourier Transform. Using the spatial Fourier transform enables the pressure on the plate surface to be easily determined [22.36]. For any 2-D spatial function $f(x, y)$, the transform is defined by:

$$\begin{cases} F(k_x, k_y) = \int_{-\infty}^{+\infty}\int_{-\infty}^{+\infty} f(x, y) \mathrm{e}^{\mathrm{i}(k_x x + k_y y)} \, \mathrm{d}x \, \mathrm{d}y, \\ f(x, y) = \frac{1}{(2\pi)^2} \times \\ \quad \int_{-\infty}^{+\infty}\int_{-\infty}^{+\infty} F(k_x, k_y) \mathrm{e}^{-\mathrm{i}(k_x x + k_y y)} \, \mathrm{d}k_x \, \mathrm{d}k_y. \end{cases}$$
(22.321)

This allows us to transform the wave equation for a harmonic sound pressure in space as follows

$$\int_{-\infty}^{+\infty}\int_{-\infty}^{+\infty} (\nabla^2 + k^2) p(x, y, z) \mathrm{e}^{\mathrm{i}(k_x x + k_y y)} \, \mathrm{d}x \, \mathrm{d}y = 0.$$
(22.322)

This leads to:

$$\left(k^2 - k_x^2 - k_y^2 + \frac{\partial^2}{\partial z^2}\right) P(k_x, k_y, z) = 0 \quad (22.323)$$

with a solution given by

$$P(k_x, k_y, z) = \frac{\omega \rho \dot{\Xi}(k_x, k_y)}{k_z} \mathrm{e}^{-\mathrm{i}k_z z}$$

with

$$k_z = \sqrt{k^2 - k_x^2 - k_y^2}, \quad (22.324)$$

where $\dot{\Xi}(k_x, k_y)$ is the wavenumber transform of the transverse plate velocity $\dot{\xi}(x, y)$. The complex sound pressure in space is obtained by using the inverse transform, so that

$$p(x, y, z) = \frac{\omega \rho}{(2\pi)^2} \int_{-\infty}^{+\infty}\int_{-\infty}^{+\infty} \frac{\dot{\Xi}(k_x, k_y)}{k_z} \\ \times \mathrm{e}^{-\mathrm{i}(k_x x + k_y y + k_z z)} \, \mathrm{d}k_x \, \mathrm{d}k_y. \quad (22.325)$$

This equation can be solved by using the method of stationary phase or the fast Fourier transform algorithm [22.37].

Radiated Sound Power and the Radiation Impedance Matrix. In the harmonic case, the total sound power radiated by the plate is given by

$$\mathcal{P}_{\mathrm{a}} = \frac{1}{2} \operatorname{Re}\left[\int_{-\infty}^{+\infty}\int_{-\infty}^{+\infty} p(x, y, 0) \dot{\xi}^*(x, y) \, \mathrm{d}x \, \mathrm{d}y\right],$$
(22.326)

where $(^*)$ denotes the complex conjugates. Using Parseval's theorem and (22.324) yields

$$\mathcal{P}_{\mathrm{a}} = \frac{\omega \rho}{8\pi^2} \operatorname{Re}\left[\int_{-\infty}^{+\infty}\int_{-\infty}^{+\infty} \frac{|\dot{\Xi}(k_x, k_y)|^2}{k_z} \, \mathrm{d}k_x \, \mathrm{d}k_y\right].$$
(22.327)

Notice that this last result is only valid for wavenumbers such that $\sqrt{k_x^2 + k_y^2} \leq k$. Alternatively, the sound power can be written in terms of the wavenumber transform of the acoustic pressure, such that

$$\mathcal{P}_{\mathrm{a}} = \frac{1}{8\omega\rho\pi^2} \iint_{k_x^2 + k_y^2 \leq k^2} |P(k_x, k_y)|^2 \, k_z \, \mathrm{d}k_x \, \mathrm{d}k_y.$$
(22.328)

Using (22.320) for the plate velocity and $\boldsymbol{\Phi}(k_x, k_y)$ as the wavenumber transform of $\boldsymbol{\Phi}(x, y)$ yields the squared modulus of the velocity field

$$|\dot{\Xi}(k_x, k_y)|^2 = |\dot{\boldsymbol{Q}}^{\mathrm{t}} \boldsymbol{\Phi}(k_x, k_y)|^2 \\ = \dot{\boldsymbol{Q}}^{\mathrm{H}} \boldsymbol{\Phi}^*(k_x, k_y) \boldsymbol{\Phi}^{\mathrm{t}}(k_x, k_y) \dot{\boldsymbol{Q}}.$$
(22.329)

Finally, substituting this result into (22.327), the acoustic power can be written as

$$\mathcal{P}_{\mathrm{a}} = \dot{\boldsymbol{Q}}^{\mathrm{H}} \mathbf{R}_{\mathrm{a}} \dot{\boldsymbol{Q}}. \quad (22.330)$$

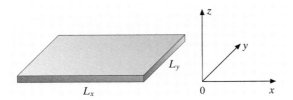

Fig. 22.18 Geometry of the thin isotropic baffled radiating plate

The radiation impedance matrix $\mathbf{R}_a$ is defined as

$$\mathbf{R}_a = \frac{\omega\rho}{8\pi^2}\,\mathrm{Re}\left(\iint \frac{\boldsymbol{\Phi}^*(k_x,k_y)\boldsymbol{\Phi}^{\mathrm{t}}(k_x,k_y)}{\sqrt{k^2-k_x^2-k_y^2}}\,\mathrm{d}k_x\,\mathrm{d}k_y\right) \quad (22.331)$$

which generalizes the results obtained in (22.265) to 2-D systems. Each element $(\mathbf{R}_a)_{ij}$ of the matrix $\mathbf{R}_a$ quantifies the mutual radiation resistance resulting from the interference between the fields of modes (m,n) and (m',n'), respectively. If $(m,n) = (m',n')$, we obtain the self-radiation resistances, which are the diagonal terms of the matrix $\mathbf{R}_a$. These terms can be written explicitly as

$$(\mathbf{R}_a)_{ij} = (\mathbf{R}_a)_{mn,m'n'}$$
$$= \frac{\omega\rho}{8\pi^2}\,\mathrm{Re}\iint \frac{\Phi_m^*(k_x)\,\Phi_n^*(k_y)\Phi_{m'}(k_x)\,\Phi_{n'}(k_y)}{\sqrt{k^2-k_x^2-k_y^2}}\,\mathrm{d}k_x\,\mathrm{d}k_y\,. \quad (22.332)$$

For a baffled simply supported plate, the radiation resistances become

$$(\mathbf{R}_a)_{mn,m'n'} = \frac{mm'nn'\omega\rho\pi^2}{8L_x^2 L_y^2}$$
$$\times \mathrm{Re}\left(\iint (f_{mm'}(k_x L_x)\,f_{nn'}(k_y L_y)\,\mathrm{d}k_x\,\mathrm{d}k_y\right.$$
$$\left./\left([k_x^2-(m\pi/L_x)^2][k_x^2-(m'\pi/L_x)^2]\right.\right.$$
$$\left.\left.[k_y^2-(n\pi/L_y)^2][k_y^2-(n'\pi/L_y)^2]\right]\right)\,, \quad (22.333)$$

where the functions of the form $f_{mm'}(k_x L_x)$ are given by

$$f_{mm'}(k_x L_x) = \begin{cases} 2(1-\cos k_x L_x) & \text{for } m \text{ even, } m' \text{ even} \\ 2(1+\cos k_x L_x) & \text{for } m \text{ odd, } m' \text{ odd} \\ 2\mathrm{i}\sin k_x L_x & \text{for } m \text{ odd, } m' \text{ even} \\ -2\mathrm{i}\sin k_x L_x & \text{for } m \text{ even, } m' \text{ odd} \end{cases}. \quad (22.334)$$

For the application of the radiation modal expansion to active structural control of sound, see, for example, the work by *Gibbs* et al.[22.38]. In a recent work, *Kim* investigated structural–acoustic coupling for active control purpose, using an impedance/mobility approach [22.39]. Alternative techniques make use of multipole expansion of the radiated sound pressure [22.40].

22.6 Damping

In this section, we start by summarizing briefly the conditions for modal decoupling in discrete damped systems, and the concept of proportional damping is introduced. The modal approach is convenient for treating the case of localized damping (Sect. 22.6.1). The following example of a string with a dissipative end illustrates the limit of the modal approach, and shows that such a system cannot exhibit stationary solutions. A physical interpretation is presented in terms of damped propagating waves. Other authors use a state-space modal approach to address the analysis of damped structures [22.41, 42].

The section continues with the presentation of three damping mechanisms in plates: thermoelasticity, viscoelasticity and radiation, with emphasis on the time-domain formulation. The section ends with a brief presentation of friction, stick–slip vibrations and hysteretic damping, which are often encountered in structural damping models.

22.6.1 Modal Projection in Damped Systems

Discrete Systems
For a linear system with multiple degrees of freedom, with a dissipation energy of the form

$$\mathcal{E}_D = \frac{1}{2}\,{}^{\mathrm{t}}\dot{X}\mathbf{C}\dot{X}\,, \quad (22.335)$$

where $\mathbf{C}$ is a symmetric damping matrix with positive elements, the equations of motion can be written as

$$\mathbf{M}\ddot{X} + \mathbf{C}\dot{X} + \mathbf{K}X = F\,. \quad (22.336)$$

Denoting the eigenfrequencies and eigenfunctions of the associated conservative system by ω_n and Φ_n, respectively, we use the expansion $X = \sum_m \Phi_m q_m(t)$. Denoting by ζ_{nm} the intermodal coefficients, the generalized coordinates are governed by the system of coupled differential equations

$$\ddot{q}_n + 2\omega_n \zeta_n \dot{q}_n + \omega_n^2 q_n = \frac{f_n}{m_n} - 2\omega_n \sum_{m \neq n} \zeta_{nm} \dot{q}_m . \quad (22.337)$$

Weakly Damped Systems. In the case of weak damping, a first-order development of both the eigenfrequencies and eigenvectors of the form

$$\omega'_n = \omega_n + \Delta\omega_n \; ; \quad \Phi'_n = \Phi_n + \Delta\Phi_n \quad (22.338)$$

in (22.336), with $F = 0$, leads to the matrix equation

$$(\mathbf{K} - \omega_n^2 \mathbf{M}) \Delta\Phi_n + \omega_n(-2\Delta\omega_n \mathbf{M} + \mathbf{C})\Phi_n \simeq 0 \quad (22.339)$$

from which we get:

$$\Delta\omega_n \simeq \mathrm{i}\zeta_n \omega_n . \quad (22.340)$$

Equation (22.340) yields two significant results:

1. The correction frequency is purely imaginary. As a consequence, the eigenmode is transformed into a damped sinusoid of the form $\mathrm{e}^{\mathrm{i}\omega_n t} \mathrm{e}^{-\zeta_n \omega_n t}$.
2. To first order, the correction frequency is independent of the intermodal coefficients, thus the damping matrix is diagonal.

The perturbation of the eigenvectors due to damping is given by

$$\Delta\Phi_n = \sum_{m \neq n} \alpha_m \Phi_m$$

with

$$\alpha_m = 2\mathrm{i}\zeta_{mn} \frac{m_n \omega_n^2}{m_m (\omega_n^2 - \omega_m^2)} . \quad (22.341)$$

Consequently, in the presence of damping, the eigenvectors become

$$\Phi'_n = \Phi_n + 2\mathrm{i}m_n \omega_n^2 \sum_{m \neq n} \Phi_m \frac{\zeta_{mn}}{m_m(\omega_n^2 - \omega_m^2)} . \quad (22.342)$$

Equation (22.342) shows that

1. If the eigenfrequencies of the associated conservative system are sufficiently different from each other then, to first order, the perturbation of the eigenvectors are of the same order of magnitude as the intermodal damping ζ_{mn}. If $\omega_n \approx \omega_m$, terms of the form $\omega_n^2 - \omega_m^2$ in the denominator can lead to significant corrections.
2. The correction terms for the eigenvectors are also purely imaginary. Thus, the eigenmodes are no longer in phase (or in antiphase) as for the conservative case.

Proportional Damping. The system can be decoupled without any approximations if the damping matrix can be written in the form of a linear combination of mass and stiffness matrices $\mathbf{C} = \alpha\mathbf{M} + \beta\mathbf{K}$ where α and β are real constants. This corresponds to the so-called proportional damping case. In this case, we obtain

$$\ddot{q}_n + 2\omega_n \zeta_n \dot{q}_n + \omega_n^2 q_n = \frac{f_n}{m_n}$$

with

$$\zeta_n = \frac{\alpha}{2\omega_n} + \frac{\beta\omega_n}{2} . \quad (22.343)$$

We therefore obtain a frequency-dependent modal damping that decreases with frequency below $\omega = \sqrt{\alpha/\beta}$, and increases with frequency above this value. This damping law does not correspond to particular physical phenomena, but is rather used for practical mathematical considerations. It can be useful in some cases for fitting experimental curves over a restricted frequency range.

Localized Damping in Continuous Systems

In some applications, the damping of the vibrations of continuous systems is localized at discrete points of the structure. In this section, this problem is illustrated by the case of a string of finite length L with eigenfrequencies ω_n and eigenfunctions $\Phi_n(x)$. A *fluid damping* term is introduced at position $x = x_0$, with $0 < x_0 < L$, so that this term is of the form $-R\dot{y}(x,t)\delta(x-x_0)$. The equation of motion can therefore be written

$$\rho(x)S(x)\frac{\partial^2 y}{\partial t^2} + R\frac{\partial y}{\partial t}\delta(x-x_0) - \frac{\partial}{\partial x}\left[T(x)\frac{\partial y}{\partial x}\right] = f(x,t) . \quad (22.344)$$

Applying the usual scalar products, we can write

$$\sum_m \ddot{q}_m(t) \int_0^L \Phi_m(x)\Phi_n(x)\rho(x)S(x)\,\mathrm{d}x$$

$$+ \sum_m \dot{q}_m(t) \int_0^L R\delta(x-x_0)\Phi_m(x)\Phi_n(x)\,\mathrm{d}x$$

$$- \sum_m q_m(t) \int_0^L \Phi_n(x) \frac{\mathrm{d}}{\mathrm{d}x}\left(T(x)\frac{\mathrm{d}\Phi_m(x)}{\mathrm{d}x}\right)\mathrm{d}x$$

$$= \int_0^L \Phi_n(x)f(x,t)\,\mathrm{d}x \, . \tag{22.345}$$

The generalized displacements are governed by the equations

$$\ddot{q}_n + 2\omega_n \sum_m \zeta_{nm}\dot{q}_m + \omega_n^2 q_n = \frac{f_n}{m_n} \, , \tag{22.346}$$

where the intermodal coefficients are written here

$$\zeta_{nm} = \frac{R\Phi_n(x_0)\Phi_m(x_0)}{2m_n\omega_n} \, . \tag{22.347}$$

For weak damping, we have

$$\ddot{q}_n + 2\omega_n \zeta_n \dot{q}_n + \omega_n^2 q_n \simeq \frac{f_n}{m_n} \, , \tag{22.348}$$

where the modal damping coefficient are given by

$$\zeta_n = \frac{R\Phi_n^2(x_0)}{2m_n\omega_n} \, . \tag{22.349}$$

Example. For a homogeneous string rigidly fixed at both ends, we have: $\Phi_n(x) = \sin k_n x$ and $\omega_n = ck_n = n\pi c/L$. With a damper fixed at position $x_0 = L/2$, we get $\zeta_n = \frac{R \sin^2(n\pi/2)}{2m_n\omega_n}$. Two cases can be differentiated:

1. If n is odd, $\sin^2(n\pi/2) = 1$ and $\zeta_n = \frac{R}{2m_n\omega_n}$;
2. If n is even, $\zeta_n = 0$.

In this case, only the odd modes are damped.

String With a Dissipative End. The case of a dissipative end is slightly more delicate to handle since the dissipation is localized at a boundary, and thus is directly involved in the definition of the eigenmodes. The equations of the problem are

$$\rho S \frac{\partial^2 y}{\partial t^2} - T\frac{\partial^2 y}{\partial x^2} = 0 \tag{22.350}$$

with the boundary conditions

$$y(0,t) = 0 \quad \text{and} \quad T\frac{\partial y}{\partial x}(L,t) = -R\frac{\partial y}{\partial t}(L,t) \, . \tag{22.351}$$

One strategy consists of looking for complex solutions of the separate variables $y(x,t) = f(x)g(t)$. The time dependence is assumed to be of the form

$$g(t) = A\,\mathrm{e}^{-st} + B\mathrm{e}^{st} \tag{22.352}$$

with A, B and $s \in \mathbb{C}$. Inserting (22.352) into (22.350), we find that $f(x)$ is of the form

$$f(x) = C\mathrm{e}^{-sx/c} + D\mathrm{e}^{sx/c}$$

$$\text{with} \quad c = \sqrt{\frac{T}{\rho S}} \, , \tag{22.353}$$

where C and $D \in \mathbb{C}$. The boundary solution at $x = 0$ yields $C = -D$. The boundary condition at $x = L$ yields two possible cases:

$$\begin{cases} \cosh \dfrac{sL}{c} = r \sinh \dfrac{sL}{c} \text{ and } B = 0 \, , \\ \cosh \dfrac{sL}{c} = -r \sinh \dfrac{sL}{c} \text{ and } A = 0 \, , \end{cases} \tag{22.354}$$

where $r = R/Z_c$ and $Z_c = T/c$ is the characteristic impedance of the string. Defining $s = \mathrm{i}\omega + \alpha$, we find two sets of solutions depending on whether R is larger or smaller than Z_c.

1. For $r < 1$,

$$\begin{cases} \cos \dfrac{\omega L}{c} = 0 \, , \quad \tanh \dfrac{\alpha L}{c} = r \text{ and } B = 0 \, , \\ \cos \dfrac{\omega L}{c} = 0 \, , \quad \tanh \dfrac{\alpha L}{c} = -r \text{ and } A = 0 \, . \end{cases} \tag{22.355}$$

2. For $r > 1$,

$$\begin{cases} \sin \dfrac{\omega L}{c} = 0 \, , \quad \tanh \dfrac{\alpha L}{c} = \dfrac{1}{r} \text{ and } B = 0 \, , \\ \sin \dfrac{\omega L}{c} = 0 \, , \quad \tanh \dfrac{\alpha L}{c} = -\dfrac{1}{r} \text{ and } A = 0 \, . \end{cases} \tag{22.356}$$

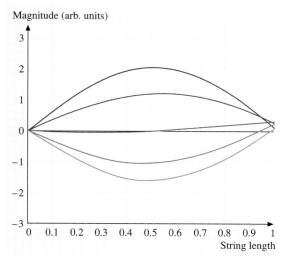

Fig. 22.19 Temporal evolution of a string with one dissipative end

As a consequence, we cannot get a steady-state solution.

In conclusion, as shown in Fig. 22.19, the string with a dissipative end cannot exhibit stationary wave solutions. We cannot make reference to nodes and antinodes, since the points of maximum (and minimum) amplitude are moving with time, though first-order approximations can be made for weak dissipation.

22.6.2 Damping Mechanisms in Plates

In vibrating structures, the dissipation of energy is partly due to radiation, and partly due to other internal damping mechanisms such as thermoelasticity and viscoelasticity. The relative prominence of these mechanisms is dependent on the material. In this section, models for these three mechanisms are briefly reviewed in the particular case of free rectangular plates, for which the losses at the boundaries are neglected.

We focus on the case $r > 1$. Defining $a = AC$ and $b = BC$, and taking (22.355) into account, we obtain

$$y(x,t) = e^{-\alpha t} \sum_n a_n e^{-i\omega_n t}$$
$$\times \left(e^{-(i\omega_n + \alpha)x/c} - e^{(i\omega_n + \alpha)x/c} \right)$$
$$+ e^{-\alpha t} \sum_n b_n e^{i\omega_n t}$$
$$\times \left(e^{-(i\omega_n - \alpha)x/c} - e^{(i\omega_n - \alpha)x/c} \right) \quad (22.357)$$

which can be rearranged, in condensed form, as:

$$y(x,t) = e^{-\alpha t} \left[e^{-\alpha x/c} F\left(t + \frac{x}{c}\right) - e^{\alpha x/c} F\left(t - \frac{x}{c}\right) \right] \quad (22.358)$$

with

$$F(t) = \sum_n \left(a_n e^{-i\omega_n t} - b_n e^{i\omega_n t} \right). \quad (22.359)$$

The complex constants a_n and b_n are determined by the initial conditions of the problem. It can be shown that (22.359) yields a real function $F(t)$. Several conclusions can be drawn from this example:

1. The mechanical resistance R yields a decay time $1/\alpha$ which is independent of frequency.
2. The term in square brackets in (22.358) corresponds to two traveling waves in opposite directions with different amplitudes, except at $x = 0$ where the rigidly fixed boundary condition must be fulfilled.

Thermoelasticity

Thermoelastic damping affects the vibrations of plates with significant thermal conductivity, such as metallic plates. A model for such damping can be derived from the equations that govern the coupling between plate vibrations and the diffusion of heat [22.43, 44]. Here, the method previously used by *Cremer* for isotropic bars is extended to orthotropic plates to derive the expressions for the complex rigidity constants [22.20]. The temperature change θ is assumed to be sufficiently small, so that the stress–strain relationships can be linearized. The plate is located in the (x, y) plane and is subjected to transverse flexural vibrations. $w(x, y, t)$ is the transverse displacement. The stress components are given by:

$$\sigma_{xx} = -12z \left(D_1 w_{,xx} + \frac{D_2}{2} w_{,yy} \right) - \phi_x \theta \,,$$
$$\sigma_{yy} = -12z \left(D_3 w_{,yy} + \frac{D_2}{2} w_{,xx} \right) - \phi_y \theta \,,$$
$$\sigma_{xy} = -6z D_4 w_{,xy} \,, \quad (22.360)$$

where ϕ_x and ϕ_y are the thermal coefficients of the material. In the particular case of an isotropic material, we have $\phi_x = \phi_y = \phi$. Equation (22.360) has to be complemented by the heat diffusion equation. Assuming that θ only depends on z, we have

$$\kappa \theta_{,zz} - \rho C s \theta = -z T_0 s \left(\phi_x w_{,xx} + \phi_y w_{,yy} \right), \quad (22.361)$$

where T_0 is the absolute temperature, C is the specific heat at constant strain, and κ is the thermal conductivity.

It is assumed that $\theta(z)$ is given by an equation of the form [22.20]

$$\theta(z) = \theta_0 \sin\frac{\pi z}{h} \quad \text{for } z \in \left[-\frac{h}{2}, \frac{h}{2}\right], \quad (22.362)$$

which takes into account the fact that there is no heat transmission between plate and air. Through integration of $z\sigma_{ij}$ over the thickness h of the plate, one obtains, in the Laplace domain, the relationship between the bending and twisting moments M_{ij} and the partial derivatives of the transverse displacement w of the plate. The rigidity factors are now complex functions of the form [22.45]:

$$\tilde{D}_1(s) = D_1 + \phi_x^2 \frac{s\zeta}{1+\tau s},$$

$$\tilde{D}_2(s) = D_2 + 2\phi_x\phi_y \frac{s\zeta}{1+\tau s},$$

$$\tilde{D}_3(s) = D_3 + \phi_y^2 \frac{s\zeta}{1+\tau s},$$

$$\tilde{D}_4(s) = D_4, \quad (22.363)$$

where the thermoelastic relaxation time τ and the thermal constant ζ are given by

$$\tau = \frac{\rho C h^2}{\kappa \pi^2} \quad \text{and} \quad \zeta = \frac{8 T_0 h^2}{\kappa \pi^6}. \quad (22.364)$$

As a consequence of (22.364), the thermoelastic losses decrease as the thickness h of the plate increases. Notice also that D_4 is real. As a consequence, the particular flexural modes of the plate, such as the torsional modes, which involve this rigidity factor will be relatively less affected by thermoelastic damping than the other modes. In other words, the thermoelastic damping factors depend on the modal shapes [22.46]. The complex rigidities can be rewritten in the form

$$\tilde{D}_i(s) = D_i \left[1 + \tilde{d}_{it}(s)\right]$$
$$= D_i \left[1 + \frac{sR_i}{s+c_1/h^2}\right], \quad i = [1,2,3],$$

$$\tilde{D}_4(s) = D_4. \quad (22.365)$$

Because the thermoelastic damping is small for most materials, the $\tilde{d}_{it}(s)$ can often be considered as perturbation terms of the complex rigidities. For convenience, these terms are written in (22.365) using the following nondimensional numbers

$$R_1 = \frac{8 T_0 \phi_x^2}{\pi^4 D_1 \rho C};$$

$$R_2 = \frac{16 T_0 \phi_x \phi_y}{\pi^4 D_3 \rho C};$$

$$R_3 = \frac{8 T_0 \phi_y^2}{\pi^4 D_3 \rho C}. \quad (22.366)$$

The decay factor $1/\tau$ is written here in the form c_1/h^2 in order to highlight the fact that this factor is proportional to the inverse squared thickness of the plate.

Viscoelasticity

A large class of materials is subject to viscoelastic damping [22.47]. A convenient method for representing viscoelastic phenomena is to use a differential formulation between the stress σ and strain ε tensors of the form

$$\sigma + \sum_{w=1}^{N} q_w \frac{\partial^w \sigma}{\partial t^w} = E\left(\varepsilon + \sum_{v=1}^{N} q_v \frac{\partial^v \varepsilon}{\partial t^v}\right). \quad (22.367)$$

As a consequence, the differential equations involving the flexural displacements and moments in thin plates contain time derivatives up to order N. By taking the Laplace transform of both sides in (22.367) and inserting it into the plate equation, one obtains the complex rigidities due to viscoelasticity:

$$\tilde{D}_i(s) = D_i\left[1 + \tilde{d}_{iv}(s)\right] = D_i \frac{1+\sum_{v=1}^{N} s^v p_{iv}}{1+\sum_{w=1}^{N} s^w q_w}. \quad (22.368)$$

Equation (22.368) is a particular class of representation for the complex rigidities. The operator is bounded by the condition [22.48]

$$q_N \neq 0. \quad (22.369)$$

Another restrictive condition on the coefficients in (22.368) follows from the fact that the energy for deforming the material from its initial undisturbed state must be positive. For a generalized viscoelastic strain–stress relationship, the Laplace transform is given by

$$\tilde{\sigma}_{ij} = \sum_{k,l} \tilde{a}_{ijkl}(s)\tilde{\varepsilon}_{kl}. \quad (22.370)$$

Gurtin and *Sternberg* [22.48] have proven that one necessary condition for the model to be dissipative is

$$\text{Im}[X_{ij}\tilde{a}_{ijkl}(i\omega)X_{kl}] \geq 0 \quad \text{for} \quad \omega \geq 0 \quad (22.371)$$

for any real symmetric tensor X_{ij}. For the viscoelastic orthotropic plate model, (22.371) together with (22.368) yield the following conditions

$$\begin{cases} p_{1N} > 0, \quad p_{3N} > 0?; , \quad p_{4N} > 0, \\ p_{1N} p_{3N} - \dfrac{D_2\, p_{2N}^2}{4 D_1 D_3} > 0. \end{cases} \quad (22.372)$$

As for thermoelastic damping, if the viscoelastic losses in the materials are sufficiently small, the corresponding terms can be viewed as first-order corrections to the rigidity.

Time–Domain Model of Radiation Damping in Isotropic Plates

In this section, an approximate time-domain model for isotropic plates is presented that takes the effect of radiation into account. The leading idea is to use a differential operator for the radiation losses (i. e., a polynomial formulation in the Laplace domain), similar to that presented in the previous paragraphs for thermoelastic and viscoelastic damping. Experiments performed on freely suspended plates show that in the *low-frequency* domain (i. e., below the critical frequency), viscoelastic and/or thermoelastic losses are the main causes of damping [22.45]. Above the critical frequency, damping is mainly governed by radiation.

The equations that govern the flexural motion of an isotropic plate surrounded by air are given in (22.316). For a travelling wave of the form $w(x, y, t) = W(x, y) \exp[i(\tilde{\omega} t - \boldsymbol{k}\cdot\boldsymbol{x})]$, where $\boldsymbol{k}$ is the wavenumber and $\tilde{\omega} = \omega + i\alpha_r$ is the complex frequency, one obtains the dispersion equation

$$-\tilde{\omega}^2 \left(1 + \frac{2\rho_a}{\rho h} \frac{1}{\sqrt{k^2 - \frac{\tilde{\omega}^2}{c^2}}}\right) + \frac{D h^2}{\rho} k^4 = 0. \quad (22.373)$$

With the assumption of a light fluid, one can derive the radiation decay factor α_r from (22.373) by reformulating this equation through the introduction of a rigidity modulus $\tilde{D}$ of the form

$$\tilde{D} \simeq D \left(1 - \frac{2\rho_a c}{\omega_c \rho h} \frac{1}{\sqrt{\Omega - \Omega^2}}\right)$$

with

$$\omega_c = c^2 \sqrt{\frac{\rho}{h^2 D}}$$

and

$$\Omega = \frac{\omega}{\omega_c}. \quad (22.374)$$

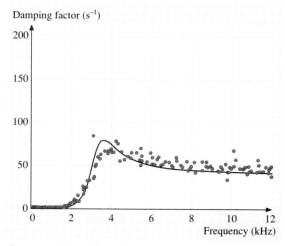

Fig. 22.20 Comparison between damping model (*solid line*) and measurements (o) for a free rectangular aluminum plate. Damping factors (in s^{-1}) as a function of frequency (in kHz). (After *Chaigne* et al. [22.45])

The complex rigidity $\tilde{D}$ can be rewritten in the form of a third-order Padé development:

$$\begin{aligned} \tilde{D}(s) &\hat{=} D\left(1 + \frac{2\rho_a c}{\omega_c \rho h} \frac{\sum_{m=1}^{3} b_m \left(\frac{s}{\omega_c}\right)^m}{\sum_{n=0}^{3} a_n \left(\frac{s}{\omega_c}\right)^n}\right) \\ &= D[1 + \tilde{d}_r(s)] \end{aligned} \quad (22.375)$$

with

$a_0 = 1.1669, a_1 = 1.6574,$
$a_2 = 1.5528, a_3 = 1,$
$b_1 = 0.0620, b_2 = 0.5950,$
$b_3 = 1.0272.$

Figure 22.20 shows a comparison between this asymptotic model and experimental data in the case of an aluminum plate. The present formulation of radiation damping cannot be easily generalized to anisotropic plates where the dispersion equation depends on the direction of wave propagation [22.14].

22.6.3 Friction

Elementary 1-DOF Oscillator with Coulomb Damping

The Coulomb damping corresponds to the simple modeling of a friction force applied to a solid sticking and/or sliding on dry surfaces. Figure 22.21 shows a 1-DOF oscillator subjected to Coulomb damping. For an initial

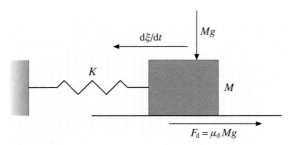

Fig. 22.21 One-DOF oscillator subjected to Coulomb damping

displacement $\xi(0) = \xi_0$, the motion of the mass M can occur under the condition that the restoring force $K\xi_0$ due to the spring is larger than the static friction force $F_s = \mu_s Mg$, where μ_s is the static friction coefficient with $0 < \mu_s < 1$. After the motion has started, the mass is subjected to a dynamic frictional force $F_d = \mu_d Mg$ where $\mu_d < \mu_s$. This force is in the opposite direction to the velocity, so that the equation of motion can be written [22.12]

$$M \frac{d^2 \xi}{dt^2} + F_d \, \text{sgn}\left(\frac{d\xi}{dt}\right) + K\xi = 0$$

$$\text{where} \quad \text{sgn}\left(\frac{d\xi}{dt}\right) = \frac{\frac{d\xi}{dt}}{\left|\frac{d\xi}{dt}\right|}. \tag{22.376}$$

During the initial part of the motion, we have

$$\frac{d^2 \xi}{dt^2} + \omega_0^2 \xi = \omega_0^2 \frac{F_d}{K} \tag{22.377}$$

whose solution is

$$\xi(t) = \left(\xi_0 - \frac{F_d}{K}\right) \cos \omega_0 t + \frac{F_d}{K}$$

$$\text{for} \quad 0 \le t \le t_1 = \frac{\pi}{\omega_0}. \tag{22.378}$$

The solution (22.378) remains valid until the time $t_1 = \pi/\omega_0$ where the velocity reduces to zero. At that time, the displacement is $\xi(t_1) = \xi_1 = 2F_d/K - \xi_0$. If $K\xi_1 > \mu_s Mg$, then the motion is governed by

$$\frac{d^2 \xi}{dt^2} + \omega_0^2 \xi = -\omega_0^2 \frac{F_d}{K} \tag{22.379}$$

whose solution is

$$\xi(t) = \left(\xi_0 - 3\frac{F_d}{K}\right) \cos \omega_0 t - \frac{F_d}{K}$$

$$\text{for} \quad t_1 \le t \le t_2 = \frac{2\pi}{\omega_0}. \tag{22.380}$$

At the time $t_2 = 2\pi/\omega_0$, the displacement is $\xi(t_2) = \xi_2 = \xi_0 - 4F_d/K$. The motion decreases linearly with time and the equation of the envelope is $\xi_0 - (2F_d)/(\pi\sqrt{KM})t$. The motion stops abruptly at time t_n where $\xi(t_n) < \mu_s F_d / \mu_d K$.

Stick–Slip Vibrations

A number of acoustic sources are due to frictional effects [22.49]. The sounds produced can be perceived as noise (such as squeal) or music (like for violin bowed strings). In all cases, the vibrations are the result of self-sustained oscillations. The simplest example of such systems is a 1-DOF oscillator whose mass is resting on a belt moving at constant velocity V_0 (Fig. 22.22).

The equation of motion of such an oscillator can be written in a general form

$$M\ddot{\xi} + R\dot{\xi} + K\xi = F(v_r, \xi)$$

$$\text{with} \quad v_r = V_0 - \dot{\xi}. \tag{22.381}$$

The previous Coulomb model presented in Sect. 22.6.3 does not account for the energy exchange between oscillator and belt. One widely used semi-empirical model assumes a friction force $F(t) \le F_s = \mu_s Mg$ during the sticking phase, while the sliding frictional force decreases with relative mass-belt velocity (Fig. 22.23). Various mathematical formulations have been proposed for such frictional forces. *Leine* et al. [22.50], for example, make use of the following model

$$F(v_r, \xi) = \begin{cases} F(\xi) = \min(|K\xi|, F_s) \, \text{sgn}(K\xi), \\ \qquad v_r = 0 \quad \text{stick}, \\ F(v_r) = \frac{F_s \, \text{sgn} v_r}{1 + \delta |v_r|} \\ \qquad v_r \ne 0 \quad \text{slip}, \end{cases} \tag{22.382}$$

where δ is an arbitrary constant. In fact, the energy exchange depends on many factors: material properties,

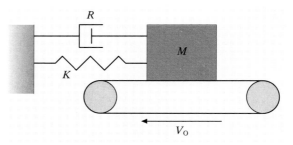

Fig. 22.22 One-DOF oscillator in stick–slip vibrations

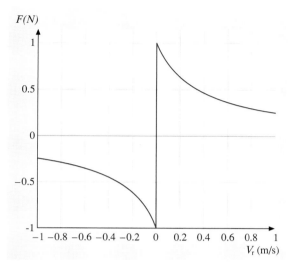

Fig. 22.23 Typical frictional force versus relative mass/belt velocity

equation of motion is written in the frequency domain as

$$[-M\omega^2 + i\omega R(\omega) + K]\Xi(\omega) = F(\omega) \quad (22.383)$$

or, equivalently,

$$\left\{-M\omega^2 + K\left[1 + i\eta\ \text{sgn}(\omega)\right]\right\}\Xi(\omega) = F(\omega). \quad (22.384)$$

This formulation accounts for steady-state oscillations of the system. However, its major drawback follows from the fact that the corresponding time-domain description is not causal. In the frequency domain, the dashpot force is $F_d(\omega) = i\omega R(\omega)\Xi(\omega) = iK\eta\ \text{sgn}(\omega)\Xi(\omega)$. Taking the inverse Fourier transform of $F_d(\omega)$ yields [22.54]

$$f_d(t) = \frac{iK\eta}{2\pi}\int_{-\infty}^{+\infty}\text{sgn}(\omega)e^{i\omega t}\,d\omega\int_{-\infty}^{+\infty}\xi(\tau)e^{-i\omega\tau}\,d\tau. \quad (22.385)$$

Inverting (22.385) yields the displacement of the mass

$$\xi(t) = \frac{1}{2\pi K\eta}\int_{-\infty}^{+\infty}\frac{e^{i\omega t}}{i\,\text{sgn}(\omega)}\,d\omega\int_{-\infty}^{+\infty}f_d(\tau)e^{-i\omega\tau}\,d\tau. \quad (22.386)$$

Selecting, for example, $f_d(t) = \delta(t)$ where $\delta(t)$ is the Dirac delta function leads to

$$\xi(t) = \frac{1}{\pi K\eta}\frac{1}{t}\quad -\infty < t < +\infty \quad (22.387)$$

which is clearly not causal, since the response anticipates the excitation [22.54]. Thus, hysteretic damping is generally not appropriate in the time domain, and one has to find other strategies for modeling frequency-dependent damping while preserving the causality.

surface roughness, thermal properties of the materials in contact [22.51], etc. See, for example, [22.52] and [22.49] for a review of friction models.

A periodic oscillation can be obtained for specific ranges of belt velocity and friction parameters. In many cases, such a complex system leads to bifurcations and chaos [22.53].

22.6.4 Hysteretic Damping

Hysteretic damping is a widely used model of structural damping. In the frequency domain, it corresponds to the case of a frequency-dependent dashpot that leads to a loss factor independent of frequency. Recall that the loss factor η is defined as the ratio between the energy lost in the system during a period and the maximum of the potential energy [22.54]. For a 1-DOF system, the

22.7 Nonlinear Vibrations

Nonlinearities in structures can be either due to nonlinear properties of the material or to large amplitude of vibration affecting the equations of motion. We limit our attention to this latter case, which is called *geometrical nonlinearity*. The most frequently encountered nonlinear terms are cubic and quadratic. These nonlinearities are responsible for many important phenomena: existence of harmonics of the excitation frequency in the response, combination of modes, jumps and hysteresis in response curves, instability and chaos. The purpose of this section is to introduce most of these properties for simple systems: single and coupled nonlinear oscillators. Finally, a number of nonlinear equations for continuous structures are briefly reviewed.

22.7.1 Example of a Nonlinear Oscillator

Equation of Motion

To introduce the fundamental concepts of nonlinear vibrations, we start with a simple example of a nonsymmetrical oscillator: the interrupted pendulum (Fig. 22.24). The oscillator consists of a point mass m suspended by a massless thread of length L and subjected to gravity g. The length of the thread varies with time, as it moves around the pulley of radius R [22.55].

The equation of motion can be derived from the Lagrange equations (Sect. 22.2). Selecting the angle θ between the vertical axis and the actual direction of the thread as the parameter of the system, the kinetic energy is written $E_k = 1/2 m(L - R\theta)^2 \dot{\theta}^2$ and the equation of motion becomes

$$\ddot{\theta} + \omega_0^2 \sin\theta = \rho \frac{d}{dt}(\theta \dot{\theta}), \qquad (22.388)$$

where $\omega_0^2 = g/L$ and $\rho = R/L$. As R tends to zero in (22.388), we find the well-known equation for a pendulum where the only nonlinear term is due to $\sin\theta$. The nonlinear right-hand side member of the equation is due to the rigid support. The validity domain of this equation is $\theta \in [0, \theta_{\max}]$ where $\theta_{\max} = \min(1/\rho, \pi/2)$. We assume that the pendulum is released at time $t = 0$ from its initial position θ_0 with zero initial velocity.

Resolution with the Harmonic Balance Method

It is observed experimentally that the solution is periodic for small amplitudes with $\rho < 1$, with a period of oscillation that depends on amplitude (Fig. 22.25). With increasing amplitude, the solution departs more and more from the linear harmonic reference motion. This encourages us to look for solutions of the form

$$\begin{cases} \theta &= \varepsilon A \cos\omega t + \varepsilon^2 B \cos 2\omega t \\ &\quad + \varepsilon^3 C \cos 3\omega t + \ldots \\ &= \varepsilon \theta_1 + \varepsilon^2 \theta_2 + \varepsilon^3 \theta_3 + \ldots \\ \omega^2 &= \omega_0^2 + \varepsilon \omega_1^2 + \varepsilon^2 \omega_2^2 + \varepsilon^3 \omega_3^2 + \ldots \end{cases} \qquad (22.389)$$

In (22.389), the solution is expanded in the form of a Fourier series, where the amplitudes of the expansion terms are proportional to increasing powers of a nondimensional parameter $\varepsilon \ll 1$. Similarly, the frequency ω is expanded in terms of increasing power of ε. The basic principle of the harmonic balance method used here for solving this nonlinear equation consists of inserting (22.389) in (22.388) and matching sinusoidal terms to calculate the unknowns $(A, B, C, \omega_1, \omega_2, \omega_3)$ from the equations, obtained separately for each power of ε [22.55]. An example up to the third order in ε is given below. We start by defining a nondimensional time variable $\tau = \omega t$. Limiting the development of $\sin\theta$ to the third order, we obtain

$$\omega^2 \frac{d^2\theta}{d\tau^2} + \omega_0^2(\theta - \theta^3/6) = \rho \omega^2 \frac{d}{d\tau}\left(\theta \frac{d\theta}{d\tau}\right). \qquad (22.390)$$

Inserting (22.389) in (22.390) and equating terms of equal power in ε, we find

$$\frac{d^2\theta_1}{d\tau^2} + \theta_1 = 0 \qquad (22.391)$$

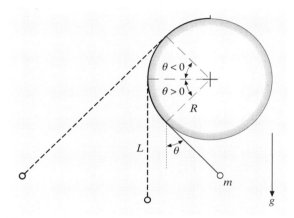

Fig. 22.24 Interrupted pendulum (After *Denardo* [22.55])

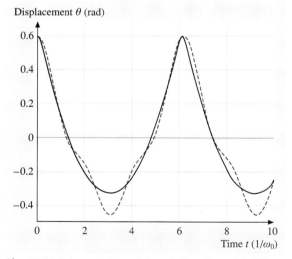

Fig. 22.25 Interrupted pendulum (after *Denardo*). Waveform: exact solution (*solid line*). Perturbative solution to the third order (*dashed line*)

whose solution is $\theta_1 = A \cos \omega \tau$. With the same procedure applied to the term in ε^2, one obtains

$$\frac{d^2 \theta_2}{d\tau^2} + \theta_2 = \frac{\omega_1^2}{\omega_0^2} \theta_1 + \rho \frac{d}{d\tau} \left(\theta_1 \frac{d\theta_1}{d\tau} \right). \qquad (22.392)$$

Equation (22.392) yields the following relation

$$-3B \cos 2\tau = \frac{\omega_1^2}{\omega_0^2} A \cos \tau - \rho A^2 \cos 2\tau. \qquad (22.393)$$

Using the harmonic balance method for each component $n\tau$ leads to

$$\omega_1 = 0 \quad \text{and} \quad B = \frac{\rho A^2}{3}. \qquad (22.394)$$

At this stage, (22.394) shows that the pulley is responsible for a nonlinear quadratic perturbation. To first order, the quadratic nonlinearity has no effect on the frequency. However, extending the calculations to the third order in ε, we get

$$\frac{d^2 \theta_3}{d\tau^2} + \theta_3 = \frac{\omega_2^2}{\omega_0^2} \theta_1 + \frac{\theta_1^3}{6} + \rho \frac{d^2}{d\tau^2}(\theta_1 \theta_2). \qquad (22.395)$$

Using the harmonic balance method in (22.395) now yields

$$C = \frac{A^3}{192}(36\rho^2 - 1)$$

and

$$\frac{\omega_2^2}{\omega_0^2} = \frac{A^2}{24}(4\rho^2 - 3). \qquad (22.396)$$

In summary, by defining $\theta_{10} = \varepsilon A$ as the amplitude of the fundamental, we find the third-order perturbation solution

$$\begin{cases} \theta &= \theta_{10} \cos \omega t + \frac{\rho \theta_{10}^2}{3} \cos 2\omega t \\ &\quad + (36\rho^2 - 1)\frac{\theta_{10}^3}{192} \cos 3\omega t \\ \text{and} \\ \omega^2 &= \omega_0^2 \left[1 + \frac{\theta_{10}^2}{24}(4\rho^2 - 3) \right] \end{cases} \qquad (22.397)$$

The following comments can be made:

- To third order, θ_{10} is linked to the initial angle θ_0 by the relation:

$$\theta_0 \simeq \theta_{10} + \frac{\rho \theta_{10}^2}{3} + (36\rho^2 - 1)\frac{\theta_{10}^3}{192}. \qquad (22.398)$$

- Figure 22.25 shows that the third-order approximation yields a good estimation of the period. In addition, the waveform is also correctly predicted for $\theta > 0$. However, the estimation is less good for $\theta < 0$.
- For $\rho = 0$, we have $\omega \simeq \omega_0(1 - \theta_{10}^2/16)$, and the term $\cos 2\omega t$ is equal to zero. The system exhibits a cubic nonlinearity due to gravity.
- For $0 < \rho < \sqrt{3}/2$, the oscillation frequency decreases as the amplitude increases. A softening effect is obtained.
- For $\rho > \sqrt{3}/2$, the oscillation frequency increases with amplitude we have an hardening effect.
- For $\rho = \sqrt{3}/2$, it is observed that the frequency is independent of the magnitude. The softening effect due to gravity is compensated by the hardening effect due to the rigid pulley. This result remains valid up to fourth order.

Many of the nonlinear effects derived for this simple oscillator, such as amplitude-dependent frequency and softening/hardening effects, are found in more-complex systems.

22.7.2 Duffing Equation

A number of nonlinear vibrations are governed by the Duffing equation. This presentation starts with the example of an elastic pendulum. General results for this cubic nonlinearity are then discussed.

Elastic Pendulum

To illustrate the Duffing equation, the elastic pendulum shown in Fig. 22.26 is considered [22.56]. The mass M moves horizontally due to the harmonic force $F \cos \Omega t$. Two springs with stiffness k, and length L_0 at rest, are

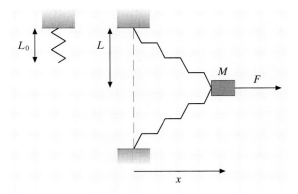

Fig. 22.26 Elastic pendulum

stretched to the length L when attached to the mass. The displacement of the mass from its equilibrium position is x. We introduce the dimensionless parameters $\lambda = L_0/L < 1$ and $y = x/L$. During the motion, the actual length of the spring is $l(y) = L\sqrt{1+y^2}$ and the elastic potential energy is $V = 1/2k(l-L_0)^2$. By differentiating this expression with respect to y, the elastic force is obtained from which the equation of motion for the mass is derived

$$M\frac{d^2y}{dt^2} + 2ky\left(1 - \frac{\lambda}{\sqrt{1+y^2}}\right) = \frac{F}{L}\cos\Omega t. \tag{22.399}$$

With the additional assumption of small amplitude ($y_{max} \ll 1$), and defining further $\omega_0^2 = 2k/M$ we obtain the first-order approximation

$$\frac{d^2y}{dt^2} + \omega_0^2(1-\lambda)y + \frac{\omega_0^2\lambda}{2}y^3 = \frac{F}{ML}\cos\Omega t \tag{22.400}$$

which can be written in nondimensional form, after defining $\tau = \omega t$ with $\omega^2 = \omega_0^2(1-\lambda)$ and $\gamma = \Omega/\omega$, as

$$\frac{d^2y}{d\tau^2} + y + \eta y^3 = \alpha \cos\gamma\tau, \tag{22.401}$$

where $\eta = \lambda/(2(1-\lambda))$ and $\alpha = F/ML\omega^2$. Equation (22.401) is a forced Duffing equation with a hardening cubic coefficient ($\eta > 0$).

Forced Vibrations

We consider the case where $\eta > 0$ and α defined in (22.401) are small compared to unity. By analogy with the linear case, a first approximation of the solution is $y_1(t) = A\cos\gamma\tau$, where A is the unknown. Inserting y_1 into (22.401), we get the differential equation that governs the second-order approximation $y_2(t)$

$$\frac{d^2y_2}{d\tau^2} = -A\cos\gamma\tau - \eta A^3\cos^3\gamma\tau + \alpha\cos\gamma\tau. \tag{22.402}$$

Using the equality $4\cos^3\gamma\tau = 3\cos\gamma\tau + \cos 3\gamma\tau$ and selecting the appropriate origin of time to avoid constant terms, we get

$$y_2(t) = \frac{A + 3\eta A^3/4 - \alpha}{\gamma^2}\cos\gamma\tau + \frac{\eta A^3}{36\gamma^2}\cos 3\gamma\tau. \tag{22.403}$$

Using the method originally used by Duffing, we look for the value of A that is equal to the lowest terms of the

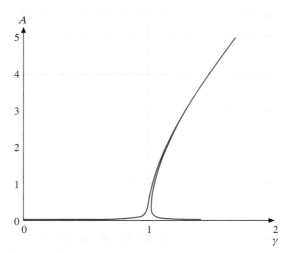

Fig. 22.27 Frequency response curve of the undamped Duffing equation with $\eta > 0$. $\gamma = \Omega/\omega$ is the reduced frequency

series in (22.403), which yields:

$$(1-\gamma^2)A + \frac{3\eta A^3}{4} = \alpha. \tag{22.404}$$

In order to plot the resonance curve, this can be written as

$$(\gamma^2 - 1) = \frac{3\eta A^2}{4} \pm \frac{\alpha}{|A|}. \tag{22.405}$$

With $\alpha = 0$, the free regime is obtained. For $\alpha \neq 0$, two branches of solution are obtained, depending on whether A is positive or negative. A nonlinear oscillator of the hardening type is obtained here, since the resonance frequency increases with the amplitude (Fig. 22.27).

Duffing Equation with a Viscous Damping Term. With the introduction of a viscous damping term, we obtain the nondimensional equation

$$\frac{d^2y}{d\tau^2} + \beta\frac{dy}{d\tau} + y + \eta y^3 = \alpha\cos\gamma\tau. \tag{22.406}$$

In this case, the response curves are governed by the equation

$$\left[(1-\gamma^2)A + \frac{3\eta A^3}{4}\right]^2 + \beta^2 A^2 = \alpha^2 \tag{22.407}$$

which corresponds to the curves shown in Fig. 22.28.

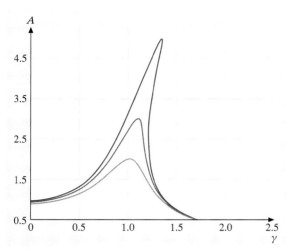

Fig. 22.28 Response curves of the damped Duffing equation

Jump and Hysteresis. It can be seen from Fig. 22.28 that a frequency range exists where three different values for $|A|$ correspond to one single value of the reduced frequency γ. Only the two extreme solutions are stable, while the intermediate value is unstable. As a consequence, on decreasing or increasing the frequency, a jump phenomenon is observed at points corresponding to the vertical tangents. The hysteresis cycle include these two jumps in addition to the stable portions of the upper and lower resonance curves.

Generation of Subharmonics. An important property of nonlinear oscillators is that they exhibit spectral components that are not present in the forcing terms, and which are also distinct from their eigenfrequencies. The generation of subharmonics is examined here. The Duffing equation with a forcing term at reduced frequency 3γ is considered

$$\frac{d^2 y}{d\tau^2} + y + \eta y^3 = \alpha \cos 3\gamma\tau \quad (22.408)$$

with $\eta \ll 1$. We look for solutions of the form

$$y(\tau) = y_0(\tau) + \eta y_1(\tau) \quad \text{with} \quad \gamma^2 = 1 + \eta \gamma_1^2 . \quad (22.409)$$

By inserting (22.409) in (22.408) and eliminating the higher-order terms in η, we get

$$\frac{d^2 y_0}{d\tau^2} + \gamma^2 y_0 - \eta \gamma_1^2 y_0 + \eta \frac{d^2 y_1}{d\tau^2} + \eta \gamma^2 y_1 + \eta y_0^3$$
$$= \alpha \cos 3\gamma\tau . \quad (22.410)$$

The linear solution (zero-order term) is obtained by imposing $\eta = 0$ in (22.410), which yields:

$$\frac{d^2 y_0}{d\tau^2} + \gamma^2 y_0 = \alpha \cos 3\gamma\tau . \quad (22.411)$$

With the appropriate initial conditions, we obtain the zero-order solution $y_0(\tau) = A \cos \gamma\tau + C \cos 3\gamma\tau$ with $C = -\alpha/8\gamma^2$. By pursuing the mathematical derivations iteratively, we obtain the first-order approximation y_1

$$\frac{d^2 y_1}{d\tau^2} + \gamma^2 y_1 = \gamma_1^2 y_0 - y_0^3 . \quad (22.412)$$

Inserting the expression for $y_0(\tau)$ in (22.412) it is found that $y_1(\tau)$ must be a solution of the equation

$$\frac{d^2 y_1}{d\tau^2} + \gamma^2 y_1$$
$$= A \left[\gamma_1^2 - \frac{3}{4}\left(A^2 + \frac{3\alpha^2}{64\gamma^4} - \frac{A\alpha}{8\gamma^2}\right) \right] \cos \gamma\tau$$
$$+ \text{terms in } 3\gamma, 5\gamma, 7\gamma, 9\gamma . \quad (22.413)$$

The first term on the right-hand side of (22.413) is a resonant excitation term or *secular term*, which leads to an infinitely growing solution. As such a solution is not physically possible, this secular terms must be eliminated, which leads to the condition:

$$\gamma_1^2 = \frac{3}{4}\left(A^2 + \frac{3\alpha^2}{64\gamma^4} - \frac{A\alpha}{8\gamma^2}\right) . \quad (22.414)$$

Equation (22.414) yields the condition of existence, in terms of amplitude A and nonlinear strength η, for subharmonics of order three

$$\gamma^6 - \gamma^4 - \frac{3\eta}{256}\left(64A^2\gamma^4 - 8A\alpha\gamma^2 + 2\alpha^2\right) = 0 . \quad (22.415)$$

Similarly, one can show that the Duffing oscillator can yield superharmonics of order three. The method consists of imposing a forcing term with reduced frequency γ and looking for the condition of existence of solutions with frequency 3γ.

22.7.3 Coupled Nonlinear Oscillators

Continuous systems and discrete systems with several degrees of freedom subjected to large amplitude of motion show multiple nonlinear coupling effects. In this section, we limit ourselves to nonlinear coupling between two oscillators, which represents a significant step forward in terms of complexity compared to the single-DOF nonlinear oscillator.

Example of a Quadratic Nonlinear Coupling

Here, the case of a quadratic nonlinear coupling between two oscillators is investigated using the multiple-scales method [22.57]. The equations are:

$$\begin{cases} \ddot{x}_1 + \omega_1^2 x_1 = \varepsilon(-\beta_{12} x_1 x_2 - 2\mu_1 \dot{x}_1), \\ \ddot{x}_2 + \omega_2^2 x_2 = \varepsilon(-\beta_{21} x_1^2 - 2\mu_2 \dot{x}_2 + P \cos \Omega t), \end{cases} \quad (22.416)$$

where x_1 and x_2 are the displacements, ω_1 and ω_2 are the eigenfrequencies of the oscillators in the linear range. The perturbation terms are grouped on the right-hand side of (22.416). The nondimensional parameter $\varepsilon \ll 1$ indicates that these terms are small. The quadratic nonlinear coupling is due to the presence of the terms $\beta_{12} x_1 x_2$ and $\beta_{21} x_1^2$. It is assumed that the system has a so-called *internal resonance* in the sense that $\omega_2 = 2\omega_1 + \varepsilon \sigma_1$, where σ_1 is the internal detuning parameter. The forcing frequency Ω is close to ω_2 so that we can write $\Omega = \omega_2 + \varepsilon \sigma_2$ where σ_2 is the external detuning parameter, and μ_1 and μ_2 are viscous damping parameters.

Resolution with the Method of Multiple Scales

Solving the above equations involves calculating the amplitudes a_1 and a_2 of both oscillators as a function of the external detuning parameter σ_2. Another goal is to determine the values of the threshold in terms of amplitude and frequency, at which the nonlinear set of oscillators exhibit bifurcations and unstable behavior. The present example contains most of the concepts and methods used in the theory of multiple scales applied to nonlinear oscillators. The main steps of the calculations are the following:

1. Definition of time scales and general form of the solution
2. Solvability conditions. Elimination of secular terms
3. Autonomous system and fixed points
4. Stability of the system
5. Amplitudes and phases of the solution

Definition of Time Scales and General Form of the Solution. The time scales are defined as

$$T_j = \varepsilon^j t \quad \text{with} \quad j \geq 0 \quad (22.417)$$

and the solutions are expanded into

$$\begin{cases} x_1(t) = x_{10}(T_0, T_1) + \varepsilon x_{11}(T_0, T_1) + O(\varepsilon^2), \\ x_2(t) = x_{20}(T_0, T_1) + \varepsilon x_{21}(T_0, T_1) + O(\varepsilon^2), \end{cases} \quad (22.418)$$

where the expansion is limited here to the first order in ε. The differentiation operators can be written

$$\begin{cases} \dfrac{\partial}{\partial t} = \dfrac{\partial}{\partial T_0} + \varepsilon \dfrac{\partial}{\partial T_1}, \\ \dfrac{\partial^2}{\partial t^2} = \dfrac{\partial^2}{\partial T_0^2} + 2\varepsilon \dfrac{\partial}{\partial T_0}\dfrac{\partial}{\partial T_1}. \end{cases} \quad (22.419)$$

In what follows, we write $D_j = \partial/\partial T_j$. Inserting (22.419) into (22.416) and matching the coefficients of terms with the same power in ε yields:

- for the zero-order term $\varepsilon^0 = 1$

$$D_0^2 x_{10} + \omega_1^2 x_{10} = 0; \quad D_0^2 x_{20} + \omega_2^2 x_{20} = 0; \quad (22.420)$$

- for the first-order term ε:

$$\begin{cases} D_0^2 x_{11} + \omega_1^2 x_{11} \\ = -2 D_0 D_1 x_{10} - \beta_{12} x_{10} x_{20} - 2\mu_1 D_0 x_{10}, \\ D_0^2 x_{21} + \omega_2^2 x_{21} \\ = -2 D_0 D_1 x_{20} - \beta_{21} x_{10}^2 \\ \quad - 2\mu_2 D_0 x_{20} + P \cos \Omega t. \end{cases} \quad (22.421)$$

The solutions of (22.420) can be written

$$\begin{aligned} x_{10}(t) &= A_1(T_1) e^{i\omega_1 t} + A_1^*(T_1) e^{-i\omega_1 t}; \\ x_{20}(t) &= A_2(T_1) e^{i\omega_2 t} + A_2^*(T_1) e^{-i\omega_2 t}, \end{aligned} \quad (22.422)$$

where the (*) indicates the complex conjugate.

Solvability Conditions. In order to calculate the complex terms $A_1(T_1)$ and $A_2(T_1)$, the expressions (22.422) are inserted into (22.421). Then, conditions are determined so that no secular terms, such as $t \cos \omega t$, exist in the solution. This leads to the so-called solvability conditions:

$$\begin{cases} -2i\omega_1 \left(\dfrac{\partial A_1}{\partial T_1} + \mu_1 A_1 \right) - \beta_{12} A_1^* A_2 e^{i\sigma_1 T_1} = 0, \\ -2i\omega_2 \left(\dfrac{\partial A_2}{\partial T_1} + \mu_2 A_2 \right) - \beta_{21} A_1^2 e^{-i\sigma_1 T_1} \\ + \dfrac{P}{2} e^{i\sigma_2 T_1} = 0. \end{cases} \quad (22.423)$$

Equations (22.423) are usually written in polar form

$$A_1(T_1) = \frac{a_1}{2} e^{i\theta_1} ;$$
$$A_2(T_1) = \frac{a_2}{2} e^{i\theta_2} , \qquad (22.424)$$

where the amplitudes a_i and phases θ_i are functions of T_1. Substituting (22.424) in (22.423) yields the system that governs the slow temporal evolution of the response, with scale $T_1 = \varepsilon t$:

$$\begin{cases} \dfrac{\partial a_1}{\partial T_1} = -\mu_1 a_1 - \dfrac{\beta_{12} a_1 a_2}{4\omega_1} \\ \qquad\qquad \sin(\sigma_1 T_1 + \theta_2 - 2\theta_1) , \\[4pt] a_1 \dfrac{\partial \theta_1}{\partial T_1} = \dfrac{\beta_{12} a_1 a_2}{4\omega_1} \cos(\sigma_1 T_1 + \theta_2 - 2\theta_1) , \\[4pt] \dfrac{\partial a_2}{\partial T_2} = -\mu_2 a_2 + \dfrac{\beta_{21} a_1^2}{4\omega_2} \sin(\sigma_1 T_1 + \theta_2 - 2\theta_1) \\ \qquad\qquad + \dfrac{P}{2\omega_2} \sin(\sigma_2 T_1 - \theta_2) , \\[4pt] a_2 \dfrac{\partial \theta_2}{\partial T_2} = \dfrac{\beta_{21} a_1^2}{4\omega_2} \cos(\sigma_1 T_1 + \theta_2 - 2\theta_1) \\ \qquad\qquad - \dfrac{P}{2\omega_2} \cos(\sigma_2 T_1 - \theta_2) . \end{cases}$$
(22.425)

Autonomous System and Fixed Points. Now, the system is written in autonomous form, i.e. in the form $\dot{X} = F(X)$. In practice, this is achieved with the changes of variables $\gamma_1 = \sigma_2 T_1 - \theta_2$ and $\gamma_2 = \sigma_1 T_1 + \theta_2 - 2\theta_1$. As a consequence (22.425) becomes:

$$\begin{cases} \dfrac{\partial a_1}{\partial T_1} = -\mu_1 a_1 - \dfrac{\beta_{12} a_1 a_2}{4\omega_1} \sin \gamma_2 , \\[4pt] \dfrac{\partial \gamma_1}{\partial T_1} = \sigma_2 - \dfrac{\beta_{21} a_1^2}{4\omega_2 a_2} \cos \gamma_2 + \dfrac{P}{2\omega_2 a_2} \cos \gamma_1 , \\[4pt] \dfrac{\partial a_2}{\partial T_2} = -\mu_2 a_2 + \dfrac{\beta_{21} a_1^2}{4\omega_2} \sin \gamma_2 + \dfrac{P}{2\omega_2} \sin \gamma_1 , \\[4pt] \dfrac{\partial \gamma_2}{\partial T_2} = \sigma_1 - \dfrac{\beta_{12} a_2}{2\omega_1} \cos \gamma_2 + \dfrac{\beta_{21} a_1^2}{4\omega_2 a_2} \cos \gamma_2 \\ \qquad\qquad - \dfrac{P}{2\omega_2 a_2} \cos \gamma_1 . \end{cases}$$
(22.426)

The fixed points correspond to the steady-state solutions, which are those of interest in the case of forced vibrations. These solutions are obtained through cancelation of the time derivatives in (22.426), which gives:

$$\begin{cases} a_1 \left(\mu_1 + \dfrac{\beta_{12} a_2}{4\omega_1} \sin \gamma_2 \right) = 0 , \\[4pt] \sigma_2 - \dfrac{\beta_{21} a_1^2}{4\omega_2 a_2} \cos \gamma_2 + \dfrac{P}{2\omega_2 a_2} \cos \gamma_1 = 0 , \\[4pt] -\mu_2 a_2 + \dfrac{\beta_{21} a_1^2}{4\omega_2} \sin \gamma_2 + \dfrac{P}{2\omega_2} \sin \gamma_1 = 0 , \\[4pt] \sigma_1 + \dfrac{\beta_{21} a_1^2}{4\omega_2 a_2} \cos \gamma_2 - \dfrac{P}{2\omega_2 a_2} \cos \gamma_1 \\ \qquad - \dfrac{\beta_{12} a_2}{2\omega_1} \cos \gamma_2 = 0 . \end{cases}$$
(22.427)

Imposing $a_1 = 0$ in (22.427) yields the response curve a_2 of the second oscillator in the linear case (Fig. 22.29)

$$a_2 = \frac{P}{2\omega_2 \sqrt{\sigma_2^2 + \mu_2^2}} . \qquad (22.428)$$

Stability of the Nonlinear Coupled System. Intuitively, a physically unstable system behaves in such a way that, subjected to a small departure from equilibrium, its motion will never bring it back to its initial position but will rather continue its deviation. From a mathematical point of view, the deviations from an equilibrium position for a system with multiple variables such as (22.427) are calculated from the partial derivatives of each equation with regard to all variables of the system. This leads to

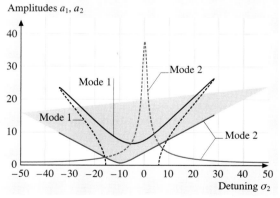

Fig. 22.29 Frequency response curves of the nonlinearly coupled oscillators as a function of the external detuning parameter

the definition of the Jacobian matrix $\mathcal{J}$ of the system:

$$\mathcal{J} = \begin{vmatrix} \frac{\partial f_1}{\partial a_1} & \frac{\partial f_1}{\partial \gamma_1} & \frac{\partial f_1}{\partial a_2} & \frac{\partial f_1}{\partial \gamma_2} \\ \frac{\partial f_2}{\partial a_1} & \frac{\partial f_2}{\partial \gamma_1} & \frac{\partial f_2}{\partial a_2} & \frac{\partial f_2}{\partial \gamma_2} \\ \frac{\partial f_3}{\partial a_1} & \frac{\partial f_3}{\partial \gamma_1} & \frac{\partial f_3}{\partial a_2} & \frac{\partial f_3}{\partial \gamma_2} \\ \frac{\partial f_4}{\partial a_1} & \frac{\partial f_4}{\partial \gamma_1} & \frac{\partial f_4}{\partial a_2} & \frac{\partial f_4}{\partial \gamma_2} \end{vmatrix}. \tag{22.429}$$

In our case, the eigenvalues of this Jacobian are

$$\begin{cases} \lambda_1 = -\mu_2 + i\sigma_2 \; ; \\ \lambda_2 = -\mu_2 - i\sigma_2 \; ; \\ \lambda_3 = -\dfrac{\beta_{12}a_2}{4\omega_1} \sin\gamma_2 - \mu_1 \; ; \\ \lambda_4 = \dfrac{\beta_{12}a_2}{2\omega_1} \sin\gamma_2 \; . \end{cases} \tag{22.430}$$

The system is unstable if the real part of any one of these eigenvalues is positive. In (22.430), the real parts of λ_1 and λ_2 are negative, because of the damping term μ_2. However, calculating the product of the two other eigenvalues yields

$$\lambda_3\lambda_4 = -\frac{\mu_1\beta_{12}a_2}{2\omega_1}\sin\gamma_2 - \frac{\beta_{12}^2 a_2^2}{8\omega_1^2}\sin^2\gamma_2 \;. \tag{22.431}$$

This product can be negative, leading to an instability, if

$$a_2 > \frac{2\omega_1}{|\beta_{12}|}\sqrt{4\mu_1^2 + (\sigma_1+\sigma_2)^2} \;. \tag{22.432}$$

The instability domain corresponding to (22.432) is the shaded area in Fig. 22.29.

Amplitudes and Phases of the Solution. Solutions for the nonlinear coupled system are obtained by combining (22.422) with (22.424), taking the definitions of T_1, σ_1, σ_2 γ_1 and γ_2 into account. We then obtain

$$\begin{cases} x_{10} = a_1 \cos(\omega_1 t + \theta_1) \\ \qquad = a_1 \cos\left(\frac{\Omega}{2}t - \frac{\gamma_1+\gamma_2}{2}\right), \\ x_{20} = a_2 \cos(\omega_2 t + \theta_2) = a_2 \cos\left(\Omega t - \gamma_1\right), \end{cases} \tag{22.433}$$

where, by solving the system (22.427), the amplitudes of the oscillators are given by

$$\begin{cases} a_2 = \dfrac{2\omega_1}{|\beta_{12}|}\sqrt{(\sigma_1+\sigma_2)^2 + 4\mu_1^2} \;, \\ a_1 = 2\left[-\Gamma_1 \pm \sqrt{P^2/(4\beta_{21}^2) - \Gamma_2^2}\right]^{1/2} \;, \\ \text{with} \quad \Gamma_1 = \dfrac{2\omega_1\omega_2}{\beta_{12}\beta_{21}}[2\mu_1\mu_2 - \sigma_2(\sigma_1+\sigma_2)] \;, \\ \text{and} \quad \Gamma_2 = \dfrac{2\omega_1\omega_2}{\beta_{12}\beta_{21}}[2\mu_1\sigma_2 - \mu_2(\sigma_1+\sigma_2)] \;. \end{cases} \tag{22.434}$$

Equation (22.433) shows that, for steady-state oscillations, the forcing frequency Ω is the same as the oscillation frequency of the second oscillator whereas, for appropriate conditions of instability, the first one oscillates with frequency $\Omega/2$, as a result of the nonlinear coupling. Figure 22.29 shows the resonance curves of the coupled nonlinear oscillators as a function of σ_2. By increasing this parameter progressively, successive phenomena occur:

1. First the resonance curve is that of the second oscillator. No subharmonics are observed as long as the amplitude remains below the instability region.
2. As the amplitude of the second oscillator reaches the instability limit given in (22.432), the first oscillator starts to oscillate. In our example, its magnitude a_1 is larger than a_2.

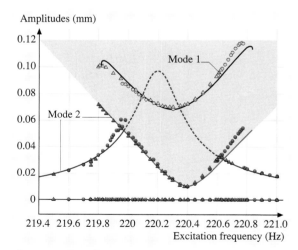

Fig. 22.30 Experimental response curves for two nonlinearly coupled degrees of freedom of a harmonically forced spherical shell (after *Thomas* et al. [22.19])

3. As the amplitude reaches the top of the resonance curve 1, it suddenly jumps back to the resonance curve 2, and oscillation at $\Omega/2$ disappears.
4. Following the frequency axis in the reverse direction, similar phenomena are observed, although the jumps do not occur at the same frequencies. Also the values of the instability thresholds are different.

To illustrate this theoretical part, Fig. 22.30 shows an example of experimental results for nonlinear coupling between two particular degrees of freedom of a spherical shell with free edge subjected to large-amplitude motion [22.19].

22.7.4 Nonlinear Vibrations of Strings

As an example of nonlinear coupling in a continuous system, the forced vibrations of a homogeneous elastic string is examined. Following the presentation by *Murthy* and *Ramakrishna*, only the coupling between the two polarizations $y(x,t)$ and $z(x,t)$ is presented [22.58, 59]. Other complicating effects such as torsional and longitudinal motions are neglected. See, for example, [22.60] for a more complete description of the nonlinear vibrating string.

Equations of Motion

As a result of the initial tension T_0 applied to the string, the length of a small element dx increases by a relative quantity λ, so that the actual length ds is given by:

$$ds - dx = \lambda \, dx . \tag{22.435}$$

As a result, for a string with a Young's modulus of E and cross-sectional area of A, the actual tension becomes

$$T = T_0 + EA\lambda . \tag{22.436}$$

In the general case, we have $ds = [dx^2 + dy^2 + dz^2]^{1/2}$, which, by means of a second-order development is written

$$ds = dx \left\{ 1 + \frac{1}{2} \left[\left(\frac{\partial y}{\partial x}\right)^2 + \left(\frac{\partial z}{\partial x}\right)^2 \right] \right.$$
$$\left. - \frac{1}{8} \left[\left(\frac{\partial y}{\partial x}\right)^2 + \left(\frac{\partial z}{\partial x}\right)^2 \right]^2 + \ldots \right\} . \tag{22.437}$$

With the additional assumption that $EA \gg T_0$, the potential energy of the string of length L rigidly fixed at both ends is written [22.58]

$$V = \int_0^L \left\{ \frac{T_0}{2} \left[\left(\frac{\partial y}{\partial x}\right)^2 + \left(\frac{\partial z}{\partial x}\right)^2 \right] \right.$$
$$\left. + \frac{EA}{8} \left[\left(\frac{\partial y}{\partial x}\right)^2 + \left(\frac{\partial z}{\partial x}\right)^2 \right]^2 \right\} dx . \tag{22.438}$$

The kinetic energy of the string is given by

$$E_k = \frac{\mu}{2} \int_0^L \left[\left(\frac{\partial y}{\partial t}\right)^2 + \left(\frac{\partial z}{\partial t}\right)^2 \right] dx . \tag{22.439}$$

Using Hamilton's principle, and considering further that a sinusoidal force $f(x)\cos\omega t$ is applied to the string in the y-direction, we get the coupled equations of motion, where the derivatives are written with subscripts for convenience

$$\begin{cases} y_{tt} - c_0^2 y_{xx} = \frac{c_1^2}{2} \left[3 y_{xx} y_x^2 + \frac{\partial}{\partial x}(y_x z_x^2) \right] \\ \qquad\qquad + \frac{f(x)}{\mu} \cos\omega t , \\ z_{tt} - c_0^2 z_{xx} = \frac{c_1^2}{2} \left[3 z_{xx} z_x^2 + \frac{\partial}{\partial x}(z_x y_x^2) \right] , \end{cases}$$
$$\tag{22.440}$$

where $c_0 = \sqrt{T_0/\mu}$ and $c_1 = \sqrt{EA/\mu}$ are the transverse and longitudinal velocities in the linear case. It can be seen that the nonlinear terms on the right-hand side of (22.440) are negligible, as long as the slopes y_x and z_x are small. The terms on the left-hand side of (22.440) correspond to the linear case. The coupling between the two polarizations is due to the presence of the nonlinear terms. As a consequence, a force applied in the y-direction can give rise to a motion in the z-direction.

Forced Vibrations

Neglecting the nonlinear terms in (22.440) is generally not justified near resonance. We assume that the excitation frequency ω is close to one particular (linear) eigenfrequency $\omega_n = n\pi c_0/L$ of the string. If the coupling between the modes is neglected (which might not always be justified, see [22.59]) then the motion of the string can be described, to a first approximation by

$$y(x,t) = a_{ny} \sin\frac{n\pi x}{L} \cos\omega t$$

and

$$z(x,t) = a_{nz} \sin\frac{n\pi x}{L} \sin\omega t . \tag{22.441}$$

This leads to the following relations between the amplitudes a_{ny} and a_{nz}:

$$\begin{cases} (\omega_n^2 - \omega^2)a_{ny} + \dfrac{9c_1^2}{32}\left(\dfrac{n\pi}{L}\right)^4 a_{ny}^3 \\ + \dfrac{3c_1^2}{32}\left(\dfrac{n\pi}{L}\right)^4 a_{ny} a_{nz}^2 = \dfrac{\alpha_n}{\mu}, \\ (\omega_n^2 - \omega^2)a_{nz} + \dfrac{9c_1^2}{32}\left(\dfrac{n\pi}{L}\right)^4 a_{nz}^3 \\ + \dfrac{3c_1^2}{32}\left(\dfrac{n\pi}{L}\right)^4 a_{nz} a_{ny}^2 = 0, \end{cases} \quad (22.442)$$

where α_n is the projection of the excitation force onto the nth mode of the string.

Planar Motion. The case of planar motion for the string is obtained by setting $a_{nz} = 0$ in (22.442). In this case, the amplitude in the y-direction is governed by the nonlinear equation

$$(\omega_n^2 - \omega^2)a_{ny} + \dfrac{9c_1^2}{32}\left(\dfrac{n\pi}{L}\right)^4 a_{ny}^3 = \dfrac{\alpha_n}{\mu} \quad (22.443)$$

which is similar in form to (22.404), obtained in the case of a discrete Duffing oscillator. Therefore, all properties of Duffing oscillators of the hardening type presented in Sect. 22.7.2 can be transposed here.

Nonplanar Motion. Eliminating a_{nz} between the two equations in (22.442) yields first for a_{ny}

$$(\omega_n^2 - \omega^2)a_{ny} + \dfrac{3c_1^2}{8}\left(\dfrac{n\pi}{L}\right)^4 a_{ny}^3 = \dfrac{\alpha_n}{\mu}, \quad (22.444)$$

which is similar in form to (22.443) except that the nonlinear term has increased by a factor of $4/3$ compared to the planar case. The second equation can be written in the form:

$$a_{nz}^2 = a_{ny}^2 - \dfrac{16}{3}\dfrac{\alpha_n}{\mu c_1^2 \left(\dfrac{n\pi}{L}\right)^4 a_{ny}}. \quad (22.445)$$

Equation (22.445) shows that the string can exhibit nonplanar motion under the condition

$$a_{ny} > a_{ny}^{\text{crit}} = \left(\dfrac{16\alpha_n}{3\mu c_1^2\left(\dfrac{n\pi}{L}\right)^4}\right)^{1/3}. \quad (22.446)$$

This condition can be obtained either by increasing the excitation force or by varying the frequency to come closer to resonance.

22.7.5 Review of Nonlinear Equations for Other Continuous Systems

Beams

In this paragraph, the influence of amplitude on the flexural equations of beams is summarized. For more-extensive developments, one can refer to the review by Nayfeh [22.57]. The same notations as in Sect. 22.4.2 are used. The case of a homogeneous beam with constant cross section is considered. The plane cross sections remain plane during the motion. The radius of gyration $r = \sqrt{I_z/S}$ is assumed to be small compared to the wavelength so that transverse shear and rotary inertia are neglected. We assume planar motion, and the losses are simply modeled by a linear viscous term. Given an arbitrary characteristic length w_0, we can define the set of dimensionless variables [22.57]

$$x^* = \dfrac{x}{w_0}, \quad w^* = \dfrac{w}{w_0}, \quad L^* = \dfrac{L}{w_0},$$

$$r^* = \dfrac{r}{w_0}, \quad t^* = t\dfrac{r}{w_0^2}\sqrt{\dfrac{E}{\rho}}. \quad (22.447)$$

In this case, the nonlinear flexural equation of a beam hinged at both ends is written in nondimensional form (removing the asterisk for convenience):

$$r^2\left(w_{tt} + \mu w_t + w_{xxxx}\right)$$
$$= \left(\dfrac{1}{2L}\int_0^L w_x^2\, dx\right)w_{xx} + f(x,t), \quad (22.448)$$

where μ is the dimensionless viscous damping parameter, and where the indices refer to the derivatives. Equation (22.448) shows that the nonlinear term is due to the increase in axial tension consecutive to large-amplitude motion. Expanding the solution on the linear eigenmodes $w(x,t) = \sum_n \Phi_n(x)q_n(t)$, we obtain, for the special case of the hinged–hinged beam, the following set of nonlinearly coupled differential equations:

$$\ddot{q}_n + \omega_n^2 q_n = -\varepsilon\left(\mu \dot{q}_n + \dfrac{\omega_n^2}{2n^2}q_n \sum_{m=1}^{\infty} m^2 q_m^2\right) + f_n(t), \quad (22.449)$$

where ε is a small dimensionless parameter indicating that the slenderness ratio of the beam is small. Equation (22.449) generalizes (22.416) obtained for two nonlinearly coupled oscillators, though an important difference is that we have to deal here with cubic nonlinear terms, instead of the quadratic terms in the previously mentioned 2-DOF nonlinear system.

Shallow Spherical Shells and Plates

Similarly, the case of large deflections for transverse vibrations of shallow spherical shells is now examined. The same notations and assumptions as in Sect. 22.4.4 are used. The only difference is that we assume here large displacements and moderate rotations. Moderate rotations mean that it is justified to linearize the angular functions $\sin\theta$ and $\cos\theta$ in the equations. The definition of large displacement depends on the order of magnitude of the characteristic length used for obtaining the dimensionless equations of motions (see below). These standard assumptions lead to the well-known von Kármán equations, also called Marguerre's or Koiter's equations in the literature [22.19, 61]. *Hamdouni* and *Millet* have shown recently that these equations can also be obtained through an asymptotic method applied to the general equations of elasticity [22.62]. Like in the linear case (Sect. 22.4.4), the assumption of thin shallow shell allows one to write the equations of motion as a function of the circular coordinates (r, θ) that refer to the projection of a current point of the shell on the horizontal plane. Finally, we obtain [22.19]

$$\begin{cases} \nabla^4 w + \dfrac{\nabla^2 F}{R} + \rho h \ddot{w} = L(w, F) - \mu \dot{w} + p, \\ \nabla^4 F - \dfrac{Eh}{R}\nabla^2 w = -\dfrac{Eh}{2} L(w, w), \end{cases}$$
(22.450)

where F is the Airy stress function, and where L is a bilinear quadratic operator which is written in circular coordinates as

$$L(w, F) = w_{rr}\left(\dfrac{F_r}{r} + \dfrac{F_{\theta\theta}}{r^2}\right) + F_{rr}\left(\dfrac{w_r}{r} + \dfrac{w_{\theta\theta}}{r^2}\right) \\ - 2\left(\dfrac{w_{r\theta}}{r} - \dfrac{w_\theta}{r^2}\right)\left(\dfrac{F_{r\theta}}{r} - \dfrac{F_\theta}{r^2}\right).$$
(22.451)

In (22.450), F contains linear and quadratic terms in w. Therefore, $L(w, F)$ contains quadratic and cubic terms in w. In total, the nonlinear equations of the shell contain linear, quadratic and cubic terms. With $R \to \infty$, (22.450) reduces to the nonlinear plate equations. In this case, only cubic terms are present.

Equation (22.450) can be put into nondimensional form with the change of variables [22.19]

$$r = a\bar{r},$$
$$t = a^2\sqrt{\rho h/D}\bar{t},$$
$$w = h^3/a^2\bar{w},$$
$$F = Eh^7/a^4\bar{F},$$
$$\mu = (2Eh^4/Ra^2)\sqrt{\rho h/D}\bar{\mu},$$
$$p = Eh^7/Ra^6\bar{p}.$$
(22.452)

Equation (22.450) becomes, with the overbars dropped for convenience,

$$\begin{cases} \nabla^4 w + \varepsilon_q \nabla^2 F + \ddot{w} \\ \quad = \varepsilon_c L(w, F) + \varepsilon_q(-\mu\dot{w} + p), \\ \nabla^4 F - \dfrac{a^4}{Rh^3}\nabla^2 w = -\dfrac{1}{2}L(w, w). \end{cases}$$
(22.453)

Equation (22.453) has the advantage of highlighting a quadratic perturbation coefficient $\varepsilon_q = 12(1-\nu^2)h/R$ and a cubic perturbation coefficient $\varepsilon_c = 12(1-\nu^2)h^4/a^4$. In view of the assumptions of a thin shallow shell, the cubic nonlinear terms are much smaller than the quadratic ones. The quadratic terms are due to the curvature of the shell.

Expanding the solution on the linear eigenmodes

$$w(r, \theta, t) = \sum_n \Phi_n(r, \theta) q_n(t)$$
(22.454)

yields the set of coupled differential equations

$$\ddot{q}_n + \omega_n^2 q_n = \varepsilon_q\left(-\sum_p\sum_q \alpha_{npq} q_p q_q - \mu\dot{q} + p_n\right) \\ + \varepsilon_c \sum_q\sum_q\sum_r \beta_{npqr} q_p q_q q_r,$$
(22.455)

which shows multiple cubic and quadratic coupling. For a plate, the coefficients α_{npq} are equal to zero. For shells, one can see an analogy with the interrupted pendulum presented in Sect. 22.7.1 in the sense that the quadratic nonlinearity is due to the asymmetry arising from the curvature.

22.8 Conclusion. Advanced Topics

A number of applications of interest in the field of structural acoustics and vibrations were not described in this chapter, since they are exhaustively presented in other chapters; this is, for example, the case of near-field acoustical holography and modal analysis. Also the area of active control of sound was only briefly mentioned, since the presentation of this technique would need a significant part devoted to signal processing.

In recent years, new theories have been developed in order to allow better prediction and modeling of structures and the related acoustic field in the medium frequency range, i.e., when modes overlap due to both damping and increasing modal density. In this context, a new model and the associated numerical method based on the theory of structural fuzzy has been given by *Soize* [22.63]. Fuzzy structures are systems composed by a well-defined *master* structure with other smaller subsystems attached on it, whose location, geometry and material properties are not known with certainty. One can think of the main body of a ship (or of a plane), with some attached equipment and/or ribs. The general idea is to describe the behavior of the subsystems with a probabilistic law; the input data are given in terms of mean value and dispersion. The advantage of such a technique is to greatly reduce the number of degrees of freedom. A recent paper by *Lyon* shows the analogy between SEA and structural fuzzy framework [22.64].

Several analytical models of structural acoustics and vibrations have been presented in this chapter. However, one should be aware of the fact that analytical solutions are generally hard to obtain in engineering applications. Therefore, numerical methods are required, and considerable work has been done in the field of computational structural dynamics over the last decades. Notice that analytical results are nonetheless very useful for testing the validity of numerical models. One can refer, for example, to the review by *Wachulec* et al. in order to get a comprehensive overview of the advantages and drawbacks of the main numerical methods applied to structure-borne sound, depending on the frequency range and structural complexity [22.65].

Finally, the design of smart materials for the active control of vibrations and sound is an active field and a good example of interdisciplinary work in structural acoustics. The goal in this area is often to use appropriate actuators and sensors in order to reduce the noise generated by structures. However, other applications such as enhancing or modifying intentionally some modal parameters, for example in musical instruments, are also potential applications. One difficulty is finding the appropriate locations for the actuators. Another problem is to find transducers that are insensitive to external electromagnetic fields. The interested reader can refer to a recent work by *Shih* et al. for more information [22.66].

References

22.1 F. Axisa: *Modelling of Mechanical Systems* (Kogan Page Science, London 2003)
22.2 K.G. McConnell: *Vibration Testing, Theory and Practice* (Wiley, New York 1995)
22.3 G. Derveaux, A. Chaigne, E. Becache, P. Joly: Time-domain simulation of a guitar: Model and method, J. Acoust. Soc. Am. **114**(6), 3368–3383 (2003)
22.4 O. Christensen, B.B. Vistisen: Simple model for low-frequency guitar function, J. Acoust. Soc. Am. **68**(3), 758–766 (1980)
22.5 R.H. Lyon, R.G. Dejong: *Theory and Applications of Statistical Energy Analysis* (Butterworth Heinemann, Newton 1995)
22.6 C.B. Burroughs, R.W. Fisher, F.R. Kern: An introduction to statistical energy analysis, J. Acoust. Soc. Am. **101**(4), 1779–1789 (1997)
22.7 R. Knobel: *An Introduction to the Mathematical Theory of Waves* (Am. Math. Soc., Providence 2000)
22.8 K.F. Graff: *Wave Motion in Elastics Solids* (Dover, Mineola 1991)
22.9 M. Abramowitz, I.A. Stegun: *Handbook of Mathematical Functions, with Formulas, Graphs, and Mathematical Tables* (Dover, Mineola 1972)
22.10 L. Rhaouti, A. Chaigne, P. Joly: Time-domain simulation and numerical modeling of the kettledrum, J. Acoust. Soc. Am. **105**(6), 3545–3562 (1999)
22.11 R.S. Christian, R.E. Davis, A. Tubis, C.A. Anderson, R.I. Mills, T.D. Rossing: Effect of air loading on timpani membrane vibrations, J. Acoust. Soc. Am. **76**(5), 1336–1345 (1984)
22.12 L. Meirovitch: *Fundamentals of Vibrations* (McGrawHill, New York 2001)
22.13 Y.Y. Yu: *Vibrations of Elastic Plates* (Springer, Berlin, Heidelberg 1996)
22.14 R.F.S. Hearmon: *An Introduction to Applied Anisotropic Elasticity* (Oxford Univ. Press, Oxford 1961)
22.15 A. Leissa: *Vibrations of Plates* (Acoust. Soc. Am., Melville 1993)
22.16 K.M. Liew, C.W. Lim, S. Kitipornchai: Vibrations of shallow shells: A review with bibliography, Trans. ASME Appl. Mech. Rev. **8**, 431–444 (1997)
22.17 W. Soedel: *Vibrations of Shells and Plates* (Marcel Dekker, New York 1981)
22.18 M.W. Johnson, E. Reissner: On transverse vibrations of spherical shells, Quart. Appl. Math. **15**(4), 367–380 (1956)
22.19 O. Thomas, C. Touzé, A. Chaigne: Non-linear vibrations of free-edge thin spherical shells: modal interaction rules and 1:1:2 internal resonance, Int. J. Solid Struct. **42**(1), 3339–3373 (2005),

22.20 L. Cremer, M. Heckl: *Structure-Borne Sound. Structural Vibration and Sound Radiation at Audio Frequencies* (Springer, Berlin, Heidelberg 1988)

22.21 Y.K. Tso, C.H. Hansen: The transmission of vibration through a complex periodic structure, J. Sound Vib. **215**(1), 63–79 (1998)

22.22 M.C. Junger, D. Feit: *Sound, Structures and Their Interaction* (Acoust. Soc. Am., Melville 1993)

22.23 S.D. Snyder, N. Tanaka: Calculating total acoustic power output using modal radiation efficiencies, J. Acoust. Soc. Am. **97**(3), 1702–1709 (1995)

22.24 P.T. Chen, J.H. Ginsberg: Complex power, reciprocity and radiation modes for submerged bodies, J. Acoust. Soc. Am. **98**(6), 3343–3351 (1995)

22.25 S.J. Elliott, M.E. Johnson: Radiation modes and the active control of sound power, J. Acoust. Soc. Am. **94**(4), 2194–2204 (1993)

22.26 L. Meirovitch: *Dynamics and Control of Structures* (Wiley, New York 1990)

22.27 J.N. Juang, M.Q. Phan: *Identification and control of mechanical systems* (Cambridge Univ. Press, New York 2001)

22.28 C.R. Fuller, S.J. Elliott, P.A. Nelson: *Active Control of Vibration* (Academic, London 1996)

22.29 A. Preumont: *Vibration Control of Active Structures* (Kluwer Academic, Dordrecht 2002)

22.30 M.L. Rumerman: The effect of fluid loading on radiation efficiency, J. Acoust. Soc. Am **111**(1), 75–79 (2002)

22.31 W.T. Baumann: Active suppression of acoustic radiation from impulsively excited structures, J. Acoust. Soc. Am. **90**(6), 3202–3208 (1991)

22.32 R. Ohayon, C. Soize: *Structural Acoustics and Vibration. Mechanical models, Variational Formulations and Discretization* (Academic, London 1998)

22.33 P. Filippi, D. Habault, J.P. Lefebvre, A. Bergassoli: *Acoustics: Basic physics, Theory and Methods* (Academic, London 1999)

22.34 C. Lesueur: *Rayonnement Acoustique des Structures, Collection de la Direction des Etudes et Recherches d'EDF* (Eyrolles, Paris 1988)

22.35 F. Fahy: *Foundations of Engineering Acoustics* (Academic, London 2001)

22.36 E.G. Williams: *Fourier Acoustics: Sound Radiation and NearField Acoustical Holography* (Academic, New York 1999)

22.37 A. Le Pichon, S. Berge, A. Chaigne: Comparison between experimental and predicted radiation of a guitar, Acustica united with Acta Acustica **84**, 136–145 (1998)

22.38 G.P. Gibbs, R.L. Clark, D.E. Cox, J.S. Vipperman: Radiation modal expansion: Application to active structural acoustic control, J. Acoust. Soc. Am. **107**(1), 332–339 (2000)

22.39 S.M. Kim, M.J. Brennan: Active control of harmonic sound transmission into an acoustic enclosure using both structural and acoustic actuators, J. Acoust. Soc. Am. **107**(5), 2523–2534 (2000)

22.40 S.D. Snyder, N.C. Burgan, N. Tanaka: An acoustic-based modal filtering approach to sensing system design for active control of structural acoustic radiation: theoretical development, Mech. Syst. Sig. Process. **16**(1), 123–139 (2002)

22.41 J.H. Ginsberg: *Mechanical and Structural Vibrations. Theory and Applications* (Wiley, New York 2001)

22.42 J. Woodhouse: Linear damping models for structural vibrations, J. Sound Vib. **215**(3), 547–569 (1998)

22.43 M.A. Biot: Thermoelasticity and irreversible thermodynamics, J. Appl. Phys. **27**(3), 240–253 (1956)

22.44 C. Zener: Internal friction in solids – I theory of internal friction in reeds, Phys. Rev. **215**, 230–235 (1937)

22.45 A. Chaigne, C. Lambourg: Time-domain simulation of damped impacted plates. Part I. Theory and experiments, J. Acoust. Soc. Am. **109**(4), 1422–1432 (2001)

22.46 M.E. McIntyre, J. Woodhouse: The influence of geometry on linear damping, Acustica **39**(4), 209–224 (1978)

22.47 R.M. Christensen: *Theory of Viscoelasticity. An Introduction* (Academic, New York 1982)

22.48 M.E. Gurtin, E. Sternberg: On the linear theory of viscoelasticity, Arch. Ration. Mech. Anal. **11**, 291–356 (1962)

22.49 A. Akay: Acoustics of friction, J. Acoust. Soc. Am **111**(4), 1525–1548 (2002)

22.50 R.I. Leine, D.H. Van Campen, A. De Kraker, I. Van den Steen: Stick-Slip Vibrations Induced by Alternate Friction Models, Nonlin. Dyn. **16**, 41–54 (1998)

22.51 J.H. Smith, J. Woodhouse: The tribology of rosin, J. Mech. Phys. Solids **48**, 1633–1681 (2000)

22.52 A.J. McMillan: A non-linear friction model for self-excited vibrations, J. Sound Vib. **205**(3), 323–335 (1997)

22.53 Y. Yoshitake, A. Sueoka: Forced self-excited vibration with dry friction. In: *Applied Nonlinear Dynamics and Chaos of Mechanical Systems with Discontinuities, number 28 in World Scientific Series on Nonlinear Science*, ed. by M. Wiercigroch, B. de Kraker (World Scientific, Singapore 2000) pp. 237–257

22.54 S.H. Crandall: The role of damping in vibration theory, J. Sound Vib. **11**(1), 3–18 (1970)

22.55 B. Denardo: Nonanalytic nonlinear oscillations: Christiaan Huygens, quadratic Schroedinger equations, and solitary waves, J. Acoust. Soc. Am. **104**(3), 1289–1300 (1998)

22.56 L.N. Virgin: *Introduction to Experimental Nonlinear Dynamics. A Case Study in Mechanical Vibrations* (Cambridge Univ. Press, New York 2000)

22.57 A.H. Nayfeh, D.T. Mook: *Nonlinear Oscillations* (Wiley, New York 1979)

22.58 G.S.S. Murthy, B.S. Ramakrishna: Nonlinear Character of Resonance in Stretched Strings, J. Acoust. Soc. Am. **38**, 461–471 (1965)
22.59 C. Valette: The mechanics of vibrating strings. In: *Mechanics of Musical Instruments, CISM Courses Lect.*, Vol. 355, ed. by A. Hirschberg, J. Kergomard, G. Weinreich (Springer, Wien 1995) pp. 115–183,
22.60 A. Watzky: Non-linear three-dimensional large-amplitude damped free vibration of a stiff elastic string, J. Sound Vib. **153**(1), 125–142 (1992)
22.61 M. Amabili, M.P. Paidoussis: Review of studies on geometrically nonlinear vibrations and dynamics of circular cylindrical shells and panels, with and without fluid-structure interaction, ASME Appl. Mech. Rev. **56**(4), 349–381 (2003)
22.62 A. Hamdouni, O. Millet: Classification of thin shell models deduced from the nonlinear three-dimensional elasticity. Part I: The shallow shells, Arch. Mech. **55**(2), 135–175 (2003)
22.63 C. Soize: A model and numerical method in the medium frequency range for vibroacoustic predictions using the theory of structural fuzzy, J. Acoust. Soc. Am. **94**(2), 849–865 (1993)
22.64 R.H. Lyon: Statistical energy analysis and structural fuzzy, J. Acoust. Soc. Am. **97**(5), 2878–2881 (1995)
22.65 M. Wachulec, P.H. Kirkegaard, S.R.K. Nielsen: *Methods of estimation of structure borne noise in structures – Review*, Techn. Rep., Vol. 20 (Aalborg Univ., Aalborg 2000)
22.66 H.R. Shih, H.S. Tzou, M. Saypuri: Structural vibration control using spatially configured opto-electromechanical actuators, J. Sound Vib. **284**, 361–378 (2005)

23. Noise

Noise is discussed in terms of a source–path–receiver model. After an introduction to sound propagation and radiation efficiency, the quantities measured for noise control are defined, and the instruments used for noise measurement and control are described.

The noise emission of sources is discussed with emphasis on the determination of the sound power level of a variety of sources. The properties of two very significant sources of environmental noise, aircraft and motor vehicles, are presented. Tire noise is identified as a major noise source for motor vehicles. Criteria for the noise emission of sources are given, and the basic principles of noise control are presented. A section on active control of noise is included.

The path from the source to the receiver includes propagation in the atmosphere, noise barriers, the use of sound-absorptive materials, and silencers. Guidance is given on the determination of sound pressure level in a room when the sound power output of the source is known.

At the receiver, the effects of noise are presented, including both hearing damage and annoyance. A brief section is devoted to sound quality.

Finally, noise regulations and policies are discussed. Many activities of the US government are discussed, and information on both state and local noise policies and regulations are presented. The activities of the European Union are included, as are the noise policies in many countries.

	23.0.1 The Source–Path–Receiver Model	961
	23.0.2 Properties of Sound Waves	962
	23.0.3 Radiation Efficiency	963
	23.0.4 Sound Pressure Level of Common Sounds	965
23.1	**Instruments for Noise Measurements**	965
	23.1.1 Introduction	965
	23.1.2 Sound Level	966
	23.1.3 Sound Exposure and Sound Exposure Level	967
	23.1.4 Frequency Weightings	967
	23.1.5 Octave and One-Third-Octave Bands	967
	23.1.6 Sound Level Meters	968
	23.1.7 Multichannel Instruments	969
	23.1.8 Sound Intensity Analyzers	969
	23.1.9 FFT Analyzers	969
23.2	**Noise Sources**	970
	23.2.1 Measures of Noise Emission	970
	23.2.2 International Standards for the Determination of Sound Power	973
	23.2.3 Emission Sound Pressure Level	977
	23.2.4 Other Noise Emission Standards	978
	23.2.5 Criteria for Noise Emissions	979
	23.2.6 Principles of Noise Control	981
	23.2.7 Noise From Stationary Sources	984
	23.2.8 Noise from Moving Sources	987
23.3	**Propagation Paths**	991
	23.3.1 Sound Propagation Outdoors	991
	23.3.2 Sound Propagation Indoors	993
	23.3.3 Sound-Absorptive Materials	995
	23.3.4 Ducts and Silencers	998
23.4	**Noise and the Receiver**	999
	23.4.1 Soundscapes	999
	23.4.2 Noise Metrics	1000
	23.4.3 Measurement of Immission Sound Pressure Level	1000
	23.4.4 Criteria for Noise Immission	1000
	23.4.5 Sound Quality	1004
23.5	**Regulations and Policy for Noise Control**	1006
	23.5.1 United States Noise Policies and Regulations	1006
	23.5.2 European Noise Policy and Regulations	1009
23.6	**Other Information Resources**	1010
References		1010

23.0.1 The Source–Path–Receiver Model

The standard definition of noise is *unwanted sound*. This definition implies that there is an observer who hears the noise and makes a judgment based on a number of factors that what is heard is not wanted. This judgment may be made because the sound is too loud, is annoying, or has an unpleasant quality. In all cases, the receiver applies some metric to what is heard in order to characterize sound as noise. In other cases, the level of the sound is high enough to cause hearing damage, and the listener may accept the noise levels, as in an industrial situation, or may actually want the sound, as at a loud concert or in an automobile with a very loud audio system. The concept of *unwanted sound* also implies that there is a person who wants the sound level reduced; hence, the field of noise control has been developed to satisfy these needs. In some cases, control of noise is an administrative matter such as setting hours of operation for noisy outdoor activities or otherwise managing noise sources to make the levels acceptable. In most cases, however, a technical analysis of the source, an understanding of the propagation of sound from the source to the receiver, and application of one or more psychoacoustic metrics is required to find a solution to a noise problem. This process may be called *noise control engineering*. This source–path–receiver model was first proposed by *Bolt* and *Ingard* in 1956 [23.1], and has proved to be very useful as a systematic way to approach noise control problems. There are many situations where the distinction between a source and a path is not clear; for example when sound travels from a source through a structure and is radiated by the structure Chap. 22. Nevertheless, the above model has proved to be useful in many practical situations.

Another concept that has proved to be useful is the distinction between *emission* and *immission*. The verb emit is common in the English language, and means to send out. The verb *immit* is currently much less common, but has a long history, and means to receive [23.2]. Noise control then deals with noise emission from sources – using a variety of metrics – and noise immission is the reception of sound by an observer – expressed as some metric based on sound pressure level.

23.0.2 Properties of Sound Waves

Noise propagation, for the purposes of this chapter, can be described in terms of linear acoustics Chap. 3. Exceptions include propagation of sonic booms and very-high-amplitude sound waves such as in gas turbines and other devices. For such problems, nonlinear acoustical theory is needed Chap. 8, which will not be covered here. The physical quantities most often used in noise control are sound pressure, particle velocity, sound intensity and sound power. A one-dimensional treatment serves to illustrate the principles involved. The linear one-dimensional wave equation may be expressed in terms of sound pressure as

$$\frac{\partial^2 p}{\partial x^2} - \frac{1}{c^2}\frac{\partial^2 p}{\partial t^2} = Q, \quad (23.1)$$

where p is the sound pressure, x is distance, and t is time. The speed of sound c is given by

$$c = 20.05\sqrt{T}, \quad (23.2)$$

where T is the temperature in Kelvin. At $20\,°C$, $c = 343.3\,m/s$, rounded to $344\,m/s$ for most noise control problems.

The *source term* Q in (23.2) represents the various sources of sound, and may be in the form of fluctuations in mass, force, or heat introduced into the medium. Or, it may be nonlinear terms placed on the right-hand side of the equation as in the *acoustic analogy* proposed by *Lighthill* [23.3] which explains the generation of sound by turbulence Chap. 3.

For $Q = 0$ in an unbounded medium at rest, any function, $f(x - ct)$, a propagating wave, satisfies the wave equation. The exact solution is determined by the nature of the sources and the boundary conditions. The particle velocity (for one-dimensional propagation) is found from

$$\rho \frac{\partial u}{\partial t} = -\frac{\partial p}{\partial x}. \quad (23.3)$$

Where u is the particle velocity. As will be shown below, the sound pressure and particle velocity in almost all cases are small compared with atmospheric pressure and the speed of sound, respectively. Therefore, air can be considered a linear medium for the propagation of sound waves, and the sound pressures as a function of time from two (or more) sources can be added to find the total pressure p

$$p(t) = p_1(t) + p_2(t). \quad (23.4)$$

However, the quantity measured by sound level meters and other analyzers is the root-mean-square (RMS) pressure over some time interval, which may be very short. In this case, the way two sound pressures are combined depends on the correlation coefficient between them. The total mean-square pressure is

$$\overline{p^2} = \overline{p_1^2} + \overline{p_2^2} + \overline{2p_1 p_2}, \quad (23.5)$$

where the overbars represent time averages. When p_1 and p_2 are uncorrelated, the cross-term in (23.5) is zero, and it is the mean square pressures that add. This is the case in most noise control situations where the noise comes from one or more different sources. When $p_1 = p_2$, it is the pressures that add. When $p_1 = -p_2$, generally the objective in active noise control, the resulting mean-square pressure is zero. For sinusoidal signals, the mean-square pressure depends on the amplitudes of the waves and the phase difference between them.

Other quantities of importance are the intensity $\mathbf{I}$ of the sound wave, a vector in the direction of propagation that is the sound energy per unit of area perpendicular to an element of surface area dS, Chap. 6. For a source of sound, the total sound power W radiated by the source is given by

$$W = \int_S \mathbf{I} \cdot d\mathbf{S}, \quad (23.6)$$

where the integral is over a surface that surrounds the source. Sound power is a widely used measure of the noise emission of the source. Its determination using sound intensity is advantageous in the presence of background noise because, in the absence of sound-absorptive materials inside the surface, any sound energy that enters the surface from the outside also leaves the surface from the inside, and thus does not affect the radiated power. In practice, the sound intensity on the surface is determined by scanning or by measurements at a finite number of measurement positions, and only an approximation to the ideal situation is obtained Chap. 6. Other techniques for the determination of the sound power of sources depend on an approximation of the intensity in terms of mean-square pressure, and are discussed later in this chapter.

The Decibel as a Unit of Level

As will be shown below, the magnitude of the quantities associated with a sound wave described above vary over a very wide range, and it is convenient to use logarithms to compress the scale. Logarithms are also useful in many noise transmission problems because quantities that are multiplicative in terms of the quantities themselves are additive using logarithms. The decibel is a unit of level, and is defined as

$$L_x = 10 \log \frac{X}{X_0}, \quad (23.7)$$

where X is a quantity related to energy, X_0 is a reference quantity, and log is the logarithm to the base 10. In the case of sound pressure, $X = p^2$, the mean square pressure and $X_0 = p_0^2$. The reference pressure p_0 is standardized at $20\,\mu\text{Pa}$ ($2 \times 10^{-5}\,\text{N/m}^2$). The sound pressure level can then be written as

$$L_p = 10 \log \frac{p^2}{(2 \times 10^{-5})^2} \text{dB}. \quad (23.8)$$

For sound power and sound intensity, the reference levels are $10^{-12}\,\text{W}$ and $10^{-12}\,\text{W/m}^2$, respectively, and the corresponding sound power and sound intensity levels are

$$L_W = 10 \log \frac{W}{10^{-12}} \text{dB}, \quad (23.9)$$

$$L_I = 10 \log \frac{I}{10^{-12}} \text{dB}. \quad (23.10)$$

Note that all three quantities above are expressed in decibels; the decibel is a *unit of level*, and may be used to express a variety of quantities related to energy relative to a reference level. This fact often causes confusion because the quantity being expressed in decibels is often not stated ("The level is 75 dB."), and the meaning of the statement is not clear. The wide use of sound power level as a measure of noise emission can easily cause confusion between sound power level and sound pressure level. In this situation, it is convenient, and common in the information technology industry, to omit the 10 before the logarithm in (23.9), and express the sound power level in bels, where one bel = 10 dB.

Relative Magnitudes

As mentioned above, the quantities associated with a sound wave are small. For example, a sound pressure level of 90 dB is relatively high and corresponds to an RMS sound pressure of

$$p_A = \sqrt{\left[(4 \times 10^{-10}) \times 10^{90/10}\right]} = 0.63\,\text{Pa}. \quad (23.11)$$

Since atmospheric pressure p_{at} is nominally 1.01×10^5 Pa, the ratio p_0/p_{at} is very small: 0.62×10^{-5}.

The particle velocity can be determined from (23.3) assuming plane wave propagation of a sinusoidal wave having a radian frequency ω ($\omega = 2\pi f$). In this case, (23.3) reduces to $p = \rho c u$ where ρ is the density of air, nominally $1.18\,\text{kg/m}^2$. The particle velocity is then

$$u = 0.63/(1.18 \times 344) = 1.6 \times 10^{-3}\,\text{m/s} \quad (23.12)$$

and the ratio u/c is 4.7×10^{-6}. The particle velocity is very small compared with the speed of sound, and the model of a wave traveling with speed c and having a particle velocity u is justified. When this is not the case, the compressional portion of the wave travels with

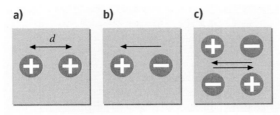

Fig. 23.1a–c Schematic of loudspeakers in an enclosure. (**a**) Two loudspeakers operating in phase to form a monopole. (**b**) Two loudspeakers operating out of phase to form a dipole. (**c**) Four speakers that form a quadrupole

speed $c + u$ and the rarefaction travels with speed $c - u$. This phenomenon is important when considering the propagation of sonic booms and other finite-amplitude waves.

23.0.3 Radiation Efficiency

The radiation efficiency of ideal sources is best described in terms of the sound power radiated by higher-order

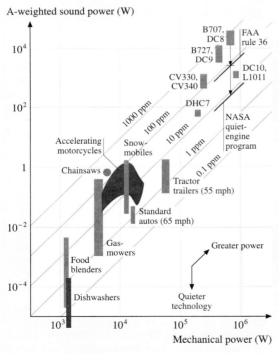

Fig. 23.2 A-weighted sound power versus mechanical power. (After Shaw [23.4] with permission, Phys. Today 28(1), 46 (1975) ©1975, AIP)

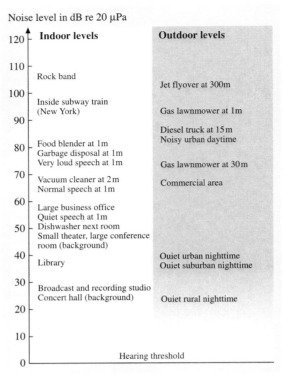

Fig. 23.3 Typical levels for indoor and outdoor environments. (After Burgé [23.5], with permission)

sources relative to monopole radiation, as discussed below. For actual sources, the radiation efficiency can be expressed in terms of the sound power output of the source relative to the mechanical power input. As shown below, the acoustic power radiated by an actual source is a very small fraction of the mechanical power of the source.

Radiation Efficiency of Ideal Sources

One way of classifying sources and determining the efficiency of the radiation of sound is in terms of a monopole and higher-order sources. If a sinusoidal source, such as two small baffled loudspeakers operating in phase having a radian frequency $\omega = 2\pi f$ injects mass per second with amplitude Q_o into the air, the power W_m radiated by the combination of the two sources can be shown to be [23.6]

$$W_m = \frac{(\omega^2 Q_o^2)}{8\pi \rho c} . \qquad (23.13)$$

The situation is illustrated in Fig. 23.1.

dipole moments are combined, the result is a quadrupole – as illustrated in (c). In this case, the total power radiated relative to monopole radiation can be shown [23.6] to be:

$$W_q = \frac{1}{120}(kd)^4 W_m . \quad (23.15)$$

It can be seen that at small values of kd (low frequencies), dipole and quadrupole sources of sound are much less efficient in radiating sound than monopoles. The dipole case is especially important because one way to reduce the source noise level is to create a secondary source of opposite phase near the primary sources. The sound radiated by the secondary source partially cancels the sound radiated by the primary source, and the radiation efficiency is reduced.

Radiation Efficiency of Machines

The sound power radiated by a source is also a very small fraction of the mechanical power driving the source. *Shaw* [23.4] studied the radiation efficiency of a number of practical sources, and found that only a small fraction of the mechanical power was converted to sound. Fig. 23.2 shows the relationship between A-weighted sound power and mechanical power for a wide variety of sources.

23.0.4 Sound Pressure Level of Common Sounds

Many authors have shown the relationship between sound pressure level and many common sounds. Data adapted from *Burgé* [23.5] are shown in Fig. 23.3. A-weighted sound pressure levels range from about 25 to 110 dB, which is a range in sound pressure of approximately 1:18 000. Similar data, but at lower sound pressure levels, have been adapted from Miller and are shown in Fig. 23.4. The vertical scales are labeled differently to conform to the practice of each author. As can be seen, there is a considerable difference for certain common terms; environmental levels may vary greatly in urban and suburban areas.

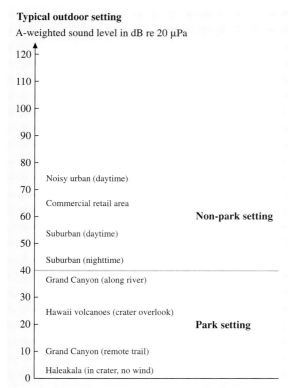

Fig. 23.4 Comparison of outdoor A-weighted levels in different environments. (After Miller [23.7], with permission)

The plus signs indicate that the two loudspeakers are operating in phase. The distance between all loudspeakers is d. In the second case (b) the two loudspeakers are operated out of phase and form a dipole with a dipole moment in the direction shown. In this case, the total power radiated relative to monopole radiation can be shown [23.6] to be

$$W_d = \frac{1}{24}(kd)^2 W_m , \quad (23.14)$$

where k is the wave number (ω/c) and d is the distance between the two sources. If two dipoles with opposite

23.1 Instruments for Noise Measurements

23.1.1 Introduction

The most common quantity for control and assessment of noise is the frequency-weighted sound level, as measured by a sound level meter. Exponential time weighting may also be employed for measurements of frequency-weighted sound level. Sound pressure levels are usually measured through constant-percentage-bandwidth filters. Sound intensity is commonly used for localization of noise sources and for direct determination of sound power level. As discussed in Sect. 23.2.1, measurements of time-average sound pressure levels are also widely used for the determination of sound power levels. The time-average sound level is also known as the equivalent-continuous sound level.

23.1.2 Sound Level

Sound pressure almost always varies with the position of the receiver in space and with time. When measured by a sound level meter, the instantaneous sound pressure signal is frequency weighted, squared, time integrated, and time averaged before the logarithm is taken to display the result in decibels relative to the standard reference pressure. The process is illustrated by the following expression for a time-averaged frequency-weighted sound level L_{wT} in decibels:

$$L_{wT} = 10 \log \left(\frac{\left[(1/T) \int_{-T}^{t} p_w^2(\xi) \, d\xi \right]^{1/2}}{p_0} \right)^2 ,$$

(23.16)

where T is the duration of the averaging time interval, ξ is a dummy variable for time integration over the averaging time interval ending at the time of observation t, $p_w(\xi)$ is the instantaneous frequency-weighted sound pressure signal in Pascals, and p_0 is the standard reference pressure of $20 \, \mu\text{Pa}$.

The numerator of the argument of the logarithm in (23.16) represents the root-mean-square frequency-weighted sound pressure over the duration of the averaging time interval. In principle, there is no time weighting involved in a determination of a root-mean-square sound pressure.

The frequency weighting applied to the instantaneous sound-pressure signal could be either the standard A or C weighting described below, or the weighting of a bandpass filter, in which case the result is a band sound pressure level. When reporting the result of a measurement of sound level, the frequency weighting should always be stated.

In general, for measurements of time-average sound level, it is important to record both the duration of the averaging time interval and the time of observation at the end of the averaging time. Sound level meters that display time-average sound levels are integrating–averaging sound level meters.

For applications such as determining estimates of the sound level that would have been indicated if the contaminating influence of background noise had not been present, for combining narrow-band sound pressure levels into wider-bandwidth sound pressure level, or for calculating sound power levels from measurements of corresponding sound pressure levels, it is necessary to work with mean-square sound pressures, not with root-mean-square sound pressures.

Before the development of digital technology, sound level meters utilized a dial-and-pointer system to display the level of the sound pressure signal. These instruments employ exponential time weighting as illustrated by the following expression for exponential-time-weighted frequency-weighted sound level $L_{w\tau}(t)$ in decibels:

$$L_{w\tau}(t) = 10 \log \left(\frac{\left[(1/\tau) \int_{-\infty}^{t} p_w^2(\xi) e^{-(t-\xi)/\tau} \, d\xi \right]^{1/2}}{p_0} \right)^2 ,$$

(23.17)

where τ is one of the standardized time constants, ξ is a dummy variable of time integration from some time in the past, as indicated by the $-\infty$ for the lower limit of the integral, to the time t when the level is observed, and the other terms are as described above for the time-average sound level. The numerator in the argument of the logarithm in (23.17) represents an exponentially-time-weighted quasi-RMS sound pressure. The exponential time weighting de-emphasizes the contributions to the integral from sound-pressure signals at earlier times, with the degree of de-emphasis increasing as the time constant increases.

By international agreement, the two standardized exponential time constants are 125 ms for *fast* or F time weighting and 1000 ms for *slow* or S time weighting. The time weighting is applied to the square of the instantaneous frequency-weighted sound-pressure signal. Instruments that measure time-weighted sound levels are conventional sound level meters.

Note that in contrast to a time-average sound level, indications of exponential-time-weighted sound level vary continuously on the display device of the sound level meter.

A maximum sound level is the highest time-weighted sound level occurring within a stated time interval. A *hold* capability is needed to display an indication of a maximum sound level on a conventional sound level meter. Maximum sound levels are not the same as peak sound levels.

Although a proper description of a measurement of a sound level is, for example, "the A-weighted sound level was 90 dB", the statement "the sound level was

90 dB(A)" is widely used, and implies that the decibel is A-weighted. The instantaneous sound pressure signal is frequency weighted, not the decibel.

23.1.3 Sound Exposure and Sound Exposure Level

For transient sounds (including the sounds produced by many different kinds of machines, the sound of the pass by of an automobile or motorcycle or the overflight of an aircraft, the sound from blasting or explosions, and sonic booms), the most appropriate quantity to measure is sound exposure or sound exposure level. Instruments that measure sound exposure level are integrating sound level meters. Sound exposure may be measured with one of the standardized frequency weightings.

Sound exposure encompasses the magnitude or level of the sound pressure and its duration in a single quantity. The expression for a frequency-weighted sound exposure E_w, in Pascal-squared seconds, is as follows

$$E_w = \int_{t_1}^{t_2} p_w^2(t)\, dt \qquad (23.18)$$

with running time in seconds. In principle, no time weighting is involved in a determination of sound exposure according to (23.18).

The frequency-weighted sound exposure level L_{wE} in decibels, is given by

$$L_{wE} = 10 \log E_w/E_0 , \qquad (23.19)$$

where E_0 is the reference sound exposure equal to $(p_0^2 T_0) = (20 \mu Pa)^2 \times (1s)$ or the product of the square of the reference pressure and the reference time for sound exposure of $T_0 = 1s$.

It is often necessary to make use of the link between a measurement of the sound exposure level for a transient sound occurring in a given time interval and the corresponding time-average sound level. For a given frequency weighting, the relation is given by the following

$$L_{wT} = L_{wE} - 10 \log (T/T_0) , \qquad (23.20)$$

where T is the averaging time for the determination of time-average sound level, in seconds, and T_0 is the reference time of 1 s. The relationship of (23.20) may be extended to the determination of the time-average sound level corresponding to the total sound exposure of a series of transient sounds occurring within a given time interval such as 8 h or 24 h.

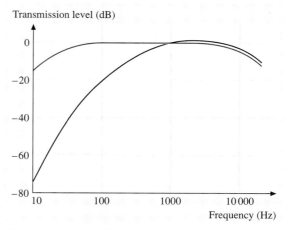

Fig. 23.5 Design goals for A- and C-frequency weightings

23.1.4 Frequency Weightings

Specifications for A- and C-weighting as well as a new Z-weighting are given in IEC 61672-1:2002 [23.8]. Design goals for the A- and C-weightings are shown in Fig. 23.5. The design goal for the Z-weighting is flat (transmission level = 0 dB) over the frequency range 10–20 kHz.

23.1.5 Octave and One-Third-Octave Bands

In general, frequency-weighted sound levels are not adequate for the diagnosis of noise problems and the design and implementation of engineering controls. Sound levels do not contain enough information to calculate measures of human response such as loudness level

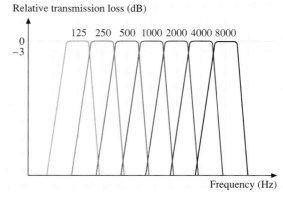

Fig. 23.6 Idealized octave-band filters for nominal midband frequencies from 125 Hz to 8 kHz

Table 23.1 Nominal octave and one-third-octave midband frequencies in Hz

ob	(1/3)ob	ob	(1/3)ob	ob	(1/3)ob
31.5	16	250	200	2000	1600
	31.5		250		2000
	40		315		2500
63	50	500	400	4000	3150
	63		500		4000
	80		630		5000
125	100	1000	800	8000	6300
	125		1000		8000
	160		1250		10000

ob = octave band

or ratings of room noise such as those obtained from noise criterion (NC) and other criteria by means of a curve-tangent method (Chap. 11). In these cases, a set of bandpass filters covering octave- or one-third-octave frequency bands should be used. The filters may be implemented by analog or digital techniques. Idealized octave-band filter shapes are shown in Fig. 23.6 for the 125 Hz to 8 kHz range of nominal midband frequencies. This range is adequate for many noise problems. Sound pressure levels in the octave bands with nominal midband frequencies of 31.5 Hz and 63 Hz are important for many applications, e.g., noise from heating, ventilating, and air-conditioning equipment in rooms.

There are three one-third-octave bands in each octave band. Nominal midband frequencies are shown in Table 23.1. The standardized midband frequencies increase by a factor of ten in each decade. Specifications for octave-band and fractional-octave-band filters are given in IEC 61260:1995.

23.1.6 Sound Level Meters

A sound level meter consists of a microphone, an amplifier and means to process the waveform of the sound-pressure signal from the microphone according to the equations above. There may be an analog or a digital readout or other device to indicate the measured sound levels. Extensive analog, or digital, or a combination of analog and digital signal processing may be utilized. Storage devices may include digital memory, computers, and printers.

A sound level meter should conform to the requirements of IEC 61672-1:2002 [23.8]. This standard provides design goals for various electroacoustical requirements along with appropriate tolerance limits around the design goals for class 1 and class 2 performance categories. The two performance classes differ mainly in the tolerance limits and the ranges of environmental conditions over which an instrument conforms to the requirements within the tolerance limits.

Sound level meters are intended for measurement of sounds that are audible to humans, generally assumed to be in the frequency range from 20 Hz to 20 kHz.

Microphones are covered in detail in Chap. 24. Sound level meters generally use a condenser microphone, either with a direct-current (DC) polarizing voltage or pre-polarization to provide the electrostatic field between the diaphragm and back plate.

The acoustical sensitivity of a measuring system should be checked using a class 1 or class 2 sound calibrator that produces a known sound pressure level at a specified frequency in the cavity of a coupler into which the microphone is inserted [23.9]. More details on means to establish the sensitivity of acoustical instruments may be found in Chap. 15.

In the presence of turbulent airflow, pressure fluctuations at the microphone diaphragm will occur that are not related to the sound pressure in the sound field. In this case, a windscreen – which may be made from an open-cell material, a metal, or a nonmetallic fabric – may be used. The insertion loss of a windscreen is the difference between the sound pressure level measured without and with the windscreen (in the absence of airflow), and may be determined in a free or a diffuse sound field. A standard for measurement of windscreen performance is available [23.10]. The insertion loss of windscreens that do not have spherical shapes may vary with sound incidence angle.

Microphones are generally designed to have their best frequency response from a specified reference direction of sound incidence in a free sound field or in response to sounds incident on the microphone from random directions, i.e., so-called free-field or random-incidence microphones. The specifications in IEC 61672-1:2002 apply equally to either microphone type when installed on a sound level meter.

A free-field microphone should be used when there is a principal direction from which the sound is incident on the microphone. A random-incidence microphone is preferred for applications where the direction of sound incidence is unknown, unpredictable, or varies with time, a situation commonly encountered when measuring noise levels in a factory or community.

In addition to the time-average, time-weighted, and peak sound levels described above, many sound level meters can record maximum and minimum levels in a given time interval, as well as statistical measures

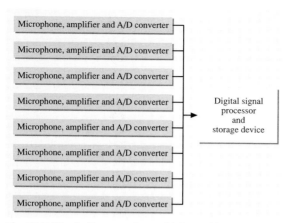

Fig. 23.7 Block diagram of a multichannel instrument using digital signal processing

(A/D) converter. High sampling rates (e.g., greater than 40 kHz) are required. The data are sent to a digital signal processor and storage device, for example, a digital computer. A block diagram of such a system is shown in Fig. 23.7.

The number of samples of sound pressure level data that can be stored in a multichannel measurement system depends on the available storage capacity, which can be quite large in the case of systems used for monitoring noise in the community and around airports.

23.1.8 Sound Intensity Analyzers

Sound intensity is covered in detail in Chap. 25. Sound intensity analyzers often use two closely spaced microphones to approximate the pressure gradient in a sound field. Particle velocity is found by integration of the time derivative of the sound-pressure gradient (23.3). Sound pressure is the average of the pressures at the two microphones. A block diagram of a sound intensity meter is shown in Fig. 23.8. Processing of the sampled signals may be accomplished either with analog circuits or digitally. Standards for sound intensity analyzers are available [23.11, 12].

such as the sound level exceeded at various percentages of a time interval, e.g., 10% of the time in an hour of observations. Performance specifications for the measurement of percentile sound level are not standardized, and the results may vary depending on the manufacturer of the instrument and the percentile of interest, with the greatest variations at the extremes of the percentiles.

23.1.7 Multichannel Instruments

Many applications for noise control are best satisfied by use of a microphone array. These include measurement of sound pressure level at multiple positions on a measurement surface for determination of sound power, and measurements at several locations around a time-varying sound source. In the first case, several microphones may be connected to a multiplexer, and the outputs sampled sequentially. In the second case, the microphones may each be connected to an amplifier and analog-to-digital

23.1.9 FFT Analyzers

Unlike octave-band and one-third-octave band filters, which are proportional (or constant-percentage) bandwidth filters, a fast Fourier transform (FFT) analyzer is essentially a set of fixed-bandwidth filters. The FFT is a computational implementation of the discrete Fourier transform that produces a set of discrete-frequency spectral lines. The number of spectral lines in a given frequency interval depends on the number of samples of the time series obtained from digital sampling of the sound-pressure signal. If a periodic wave is sampled

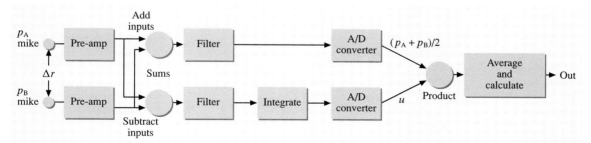

Fig. 23.8 Block diagram of an analog-to-digital sound intensity analyzer with octave-band filters. Alternatively, the preamplifier outputs of the microphones could be suitably amplified and converted to digital format; all further processing would then be done digitally

with a sampling period that is not an integral number of wavelengths of the wave, spectral *leakage* occurs. FFT analyzers contain *windows* to improve spectral estimates – the Hanning window is the most common. There are no national or international standards for the performance of FFT analyzers. The results of measurements by FFT analyzers may vary depending on the design implementation by the manufacturer or computer programmer.

23.2 Noise Sources

In this section, methods for the specification of noise emissions are given, noise emission criteria are described, and some basic principles of noise control are presented. A general description of noise control for stationary sources is presented, and some information on vehicle noise and aircraft noise is given. A short section on the principles of active noise control is included.

23.2.1 Measures of Noise Emission

This section is an edited version of a textbook chapter of [23.13].

Two quantities are needed to describe the strength of a noise source, its *sound power level* and its *directivity*.

The sound power level is a measure of the total sound power radiated by the source in all directions and is usually stated as a function of frequency, for example, in one-third octave bands. The sound power level is then the preferred descriptor for the emission of sound energy by noise sources.

The directivity of a source is a measure of the variation in its sound radiation with direction. Directivity is usually stated as a function of angular position around the acoustical center of the source and also as a function of frequency.

From the sound power level and directivity, it is possible to calculate the sound pressure levels produced by the source in the acoustical environment in which it operates. In Sect. 23.3.2, a classical method for this calculation is presented – as is an alternative method for long and flat rooms.

A source may set a nearby surface into vibration if it is rigidly attached to that surface, causing more sound power to be radiated than if the source were vibration isolated. Both the operating and mounting conditions of the source therefore influence the amount of sound power radiated as well as the directivity of the source. Nonetheless, the sound power level alone is useful for: comparing the noise radiated by machines of the same type and size as well as by machines of different types and sizes; determining whether a machine complies with a specified upper limit of noise emission; planning in order to determine the amount of transmission loss or noise control required; and engineering work to assist in developing quiet machinery and equipment.

Expanding the dot product in (23.6) allows the surface integral to be written in terms of scalar quantities

$$W = \int_S I_n \, \mathrm{d}S \, , \tag{23.21}$$

where $I_n = I \cos(\theta)$ is the component of sound intensity *normal* to the surface at the location of $\mathrm{d}S$; $\mathrm{d}S$ is the magnitude of the elemental surface area vector.

The integral may be carried out over a spherical or hemispherical surface that surrounds the source. Other regular surfaces, such as a parallelepiped or a cylinder, are also used in practice, and, in principle, any closed surface can be used. If the source is nondirectional and the integration is carried out over a spherical surface having a radius r and centered on the source, sound intensity and sound power are related by

$$I(\text{at } r) = I_n(\text{at } r) = \frac{W}{S} = \frac{W}{4\pi r^2} \, , \tag{23.22}$$

where I is the magnitude of intensity on the surface (at radius r), I_n is the normal component of intensity on the surface (at radius r), W is the sound power, S is the area of spherical surface ($4\pi r^2$), and r is the radius of the sphere.

In general, a source is directional, and the sound intensity is not the same at all points on the surface. Consequently, an approximation must be made to evaluate the integral of (23.21). It is customary to divide the measurement surface into a number of subsegments, each having an area S_i, and to approximate the normal component of the sound intensity on each surface subsegment. The sound power of the source may then be calculated by a summation over all of the surface subsegments:

$$W = \sum_i I_{ni} S_i \, , \tag{23.23}$$

here I_{ni} is the normal component of sound intensity averaged over the i-th area segment, S_i is the i-th area segment, and i is the number of segments.

When each segment of the measurement surface has the same area,

$$W = S \times \sum_i I_{ni} = S \times \langle I_n \rangle, \qquad (23.24)$$

where $\langle I_n \rangle$ is the average value of the normal sound intensity over the measurement surface, and S is the total area of the measurement surface.

The sound power level is then

$$L_W = 10 \log \frac{W}{W_0}, \qquad (23.25)$$

where W_0 is the standard reference power, 10^{-12} W.

Equation (23.24) is usually used to determine the sound power of a source from the sound intensity level except when the source is highly directional. For directional sources, the subareas of the measurement surface may be selected to be unequal and (23.23) should be used.

Table 23.2 A-frequency weightings for octave- and one-third-octave bands

Mid band center frequency (Hz)	One-third octave band weightings (dB)	Octave-band weightings (dB)
50	-30.2	-26.2
63	-26.2	
80	-22.5	
100	-19.1	-16.1
125	-16.1	
160	-13.4	
200	-10.9	-8.6
250	-8.6	
315	-6.6	
400	-4.8	-3.2
500	-3.2	
630	-1.9	
800	-0.8	0
1000	0	
1250	0.6	
1600	1.0	1.2
2000	1.2	
2500	1.3	
3150	1.2	1.0
4000	1.0	
5000	0.5	
6300	-0.1	-1.1
8000	-1.1	
10000	-2.5	

A-Weighted Sound Power

All procedures described in this section apply to the determination of sound power levels in octave or one-third-octave bands. The techniques are independent of bandwidth. The A-weighted sound power level is obtained by summing (on a mean-square basis) the octave-band or one-third octave-band data after applying the appropriate A-weighting corrections. A-weighting values are listed Table 23.2.

Measurement Environments

Three different types of laboratory environments in which noise sources are measured are found in modern acoustics laboratories: anechoic rooms (free field), hemi-anechoic rooms (free field over a reflecting plane), and reverberation rooms (diffuse field). In an anechoic room, all of the boundaries are highly absorbent, and the free-field region extends very nearly to the boundaries of the room. Because the floor itself is absorptive, anechoic rooms usually require a suspended wire grid or other mechanism to support the sound source, test personnel, and measurement instruments. A hemi-anechoic room has a hard, reflective floor, but all other boundaries are highly absorbent. Both anechoic and hemi-anechoic environments are used to determine the sound power level of a source, but the hemi-anechoic room is clearly more practical for testing large, heavy sources.

In a reverberation room, where all boundaries are acoustically hard and reflective, the reverberant field extends throughout the volume of the room except for a small region in the vicinity of the source. The sound power level of a source may be determined from an estimate of the average sound pressure level in the diffuse field region of the room coupled with a knowledge of the absorptive properties of the boundaries, or by comparison with a reference sound source of known sound power output. A standard for the calibration of reference sound sources is available [23.14].

Free-Field Approximation for the Determination of Sound Power

For the far field of a source radiating into free space, the magnitude of the intensity at a point a distance r from the source is:

$$I(\text{at } r) = \frac{p_{\text{RMS}}^2(\text{at } r)}{\rho c}, \qquad (23.26)$$

where ρc is the characteristic resistance of air equal to about 406 MKS Rayl at normal room conditions, and p_{RMS} is the root-mean-square sound pressure, at r.

Strictly speaking, this relationship is only correct in the far field of a source radiating into free space. Good approximations to free-space, or free-field, conditions can be achieved in properly designed anechoic or hemi-anechoic rooms, or outdoors. Hence, (23.26) is approximately correct in the far field of a source over a reflecting plane, provided that the space above the reflecting plane remains essentially a free field at distance r. Even if the free field is not perfect, and a small fraction of the sound is reflected from the walls and ceiling of the room, an environmental correction may be introduced to allow valid measurements to be taken in the room. The relations below are widely used in standards that determine the sound power of a source from a measurement of sound pressure level.

If a closed measurement surface is placed around a source so that all points on the surface are in the far field, and if the intensity vector is assumed to be essentially normal to the surface so that $I = I_n$ at all points on the surface, then (23.23) and (23.26) can be combined to yield

$$W = \frac{1}{\rho c} \sum_i p_i^2 S_i , \qquad (23.27)$$

here p_i is the average RMS sound pressure over area segment S_i.

The sound power level is as expressed in (23.25). If all segments are of equal area,

$$W = \frac{S}{\rho c} \frac{1}{N} \sum_i p_i^2 = \frac{S}{\rho c} \langle p^2 \rangle , \qquad (23.28)$$

where S is the total area of the measurement surface, N is the number of area segments, and $\langle p^2 \rangle$ is the average mean square pressure over the measurement surface.

In logarithmic form with reference power 10^{-12} W and reference pressure $20\,\mu\text{Pa}$, the sound power level is

$$L_W = L_p + 10 \log \frac{S}{S_0} + 10 \log \frac{400}{\rho c} , \qquad (23.29)$$

where $S_0 = 1\,\text{m}^2$.

For $\rho c = 406$ MKS Rayl, the last term in (23.29) is -0.064, and can usually be neglected.

Hence, the sound power level of a source can be computed from sound pressure level measurements made in a free field. Equation (23.29) is widely used in standardized methods for the determination of sound power levels in a free field or in a free field over a reflecting plane.

Sound Power Determination in a Diffuse Field

The sound power level of a source can also be computed from sound pressure level measurements made in an enclosure with a diffuse sound field because in such a field the sound energy density is constant; it is directly related to the mean-square sound pressure and, therefore, to the sound power radiated by the source. The sound pressure level in the reverberant room builds up until the total sound power absorbed by the walls of the room is equal to the sound power generated by the source. The sound power is determined by measuring the mean-square sound pressure in the reverberant field. This value is either compared with the mean-square pressure of a source of known sound power output (*comparison method*) or calculated directly from the mean-square pressure produced by the source and a knowledge of the sound-absorptive properties of the reverberant room (*direct method*).

The procedure for determining the sound power level of a noise source by the comparison method requires the use of a reference sound source of known sound power output. The procedure is essentially as follows.

With the equipment being evaluated at a suitable location in the room, determine, in each frequency band, the average sound pressure level (on a mean-square basis) in the reverberant field using the microphone array or traverse described above.

Replace the source under test with the reference sound source and repeat the measurement to obtain the average level for the reference sound source.

The sound power level of the source under test L_W for a given frequency band is calculated from

$$L_W = L_{Wr} + (\langle L_p \rangle - \langle L_p \rangle_r) \qquad (23.30)$$

where L_W is the one-third-octave band sound power level for the source being evaluated, $\langle L_p \rangle$ is the space-averaged one-third octave band sound pressure level of source being evaluated, L_{Wr} is the calibrated one-third-octave band sound power level of the reference source, and $\langle L_p \rangle_r$ is the space-averaged one-third-octave band sound pressure level of the reference sound source.

The direct method does not use a reference sound source. Instead, this method requires that the sound-absorptive properties of the room be determined by measuring the reverberation time in the room for each frequency band. Measurement of T_{60} is described in Chap. 11.

With this method, the space-averaged sound pressure level for each frequency band of the source being evaluated is determined as described above for the comparison method. The sound power level of the source is found from

$$L_W = \overline{L_p} + \left[10 \log \frac{A}{A_0} + 4.34 \frac{A}{S} \right.$$
$$+ 10 \log \left(1 + \frac{Sc}{8Vf} \right)$$
$$\left. - 25 \log \left(\frac{427}{400} \sqrt{\frac{273}{273+\theta}} \frac{B}{B_0} \right) - 6 \right]$$
$$\text{dB re } 10^{-12} \text{ W}, \qquad (23.31)$$

where L_W is the band sound power level of the sound source under test, $\overline{L_p}$ is the band space-averaged sound pressure level of the sound source under test, A is the equivalent absorption area of the room, given by:

$$A = \frac{55.26}{c} \left(\frac{V}{T_{\text{rev}}} \right),$$

V is the room volume, T_{rev} is the reverberation time for the particular band, A_0 is the reference absorption area, S is the total surface area of the room, f is the midband frequency of the measurement, c is the speed of sound at temperature θ in $°C$, B is the atmospheric pressure, with $B_0 = 1.013 \times 10^5$ Pa, $V_0 = 1$ m^3, $T_0 = 1$ s.

Diffuse sound fields can be obtained in laboratory reverberation rooms. Sufficiently close engineering approximations to diffuse-field conditions can be obtained in rooms that are fairly reverberant and irregularly shaped. When these environments are not available or when it is not possible to move the noise source under test, other techniques valid for the in situ determination of sound power level may be used and are described later in this section.

Non-steady and impulsive noises are difficult to measure under reverberant-field conditions. Measurements on such noise sources should be made under free-field conditions.

Sound Power Determination in an Ordinary Room

The sound pressure field in an ordinary room such as an office or laboratory space that has not been designed for acoustical measurements is neither a free-field nor a diffuse field. Here the relationship between the sound intensity and the mean-square pressure is more complicated. Instead of measuring the mean-square pressure, it is usually more advantageous to use a sound intensity analyzer that measures the sound intensity directly (Sect. 23.1.8). By sampling the sound intensity at defined locations in the vicinity of the source, the sound power level of the source can be determined. A standard for the in situ determination of sound power is also available, and is described below.

Source Directivity

Most sources of sound [23.15, 16] of practical interest are directional to some degree. If one measures the sound pressure level in a given frequency band a fixed distance away from the source, different levels will generally be found for different directions. A plot of these levels in polar fashion at the angles for which they were obtained is called the *directivity pattern* of the source.

The directivity factor Q_θ is defined as the ratio of the mean-square sound pressure, p_θ^2, at angle θ and distance r from an actual source radiating W and the mean-square sound pressure p_S^2 at the same distance from a nondirectional source radiating the same acoustic power W. Alternatively, Q_θ is defined as the ratio of the intensity in the direction of propagation W/m^2 at angle θ and distance r from an actual source to the intensity at the same distance from a nondirectional source, both sources radiating the same sound power W. It must be assumed that the directivity factor is independent of distance from the source.

The directivity index DI_θ of a sound source on a rigid plane (radiating into hemispherical space) at angle θ and for a given frequency band is computed from

$$DI_\theta = L_{p\theta} - \langle L_p \rangle_H + 3 \text{ dB}, \qquad (23.32)$$

where $L_{p\theta}$ is the sound pressure level measured a distance r and angle θ from the source, and $\langle L_p \rangle_H$ is the sound pressure level of the space-averaged mean-square pressure averaged over a test hemisphere of radius r (and area $2\pi r^2$) centered on and surrounding the source.

The 3 dB in this equation is added to $\langle L_p \rangle_H$ because the measurement was made over a hemisphere instead of a full sphere. The reason for this is that the intensity at radius r is twice as large if a source radiates into a hemisphere as compared to a sphere. That is, if a nondirectional source were to radiate uniformly into hemispherical space, $DI_\theta = DI = 3$ dB. Equations are available for sound radiation into a sphere and into a quarter sphere [23.15, 16].

23.2.2 International Standards for the Determination of Sound Power

The International Organization for Standardization (ISO) has published a series of international standards, the ISO 3740 series [23.17], which describes several methods for determining the sound power levels of noise sources. Table 23.3 summarizes the applicability of each of the basic standards of the ISO 3740 series. The most important factor in selecting an appropriate noise mea-

Table 23.3 Description of the ISO 3740 series of standards for determination of sound power

ISO standard	Test environment	Criteria for suitability of test environment	Volume of source	Character of noise	Limitation on background noise	Sound power levels obtainable	Optional information available
ISO 3741 precision (grade 1)	Reverberation room meeting specified requirements	Room volume and reverberation time to be qualified	Preferably less than 2% of room volume	Steady, broadband, narrow-band, or discrete-frequency	$\Delta L \geq 10$ dB $K_1 \leq 0.5$ dB	A-weighted and in one-third-octave or octave bands	Other frequency-weighted sound power levels
ISO 3743-1 Engineering (grade 2)	Hard-walled room	Volume ≥ 40 m^3 $\alpha \leq 0.20$	Preferably less than 1% of room volume	Any, but no isolated bursts	$\Delta L \geq 6$ dB $K_1 \leq 1.3$ dB	A-weighted and in octave bands	Other frequency-weighted sound power levels
ISO 3743-2 Engineering (grade 2)	Special reverberation room	$70\,\text{m}^3 \leq V \leq 300\,\text{m}^3$ $0.5\,\text{s} \leq T_{nom} \leq 1.0\,\text{s}$	Preferably less than 1% of room volume	Any, but no isolated bursts	$\Delta L \geq 4$ dB $K_1 \leq 2.0$ dB	A-weighted and in octave bands	Other frequency-weighted sound power levels
ISO 3744 Engineering (Grade 2)	Essentially free field over a reflecting plane	$K_2 \leq 2$ dB	No restrictions; limited only by available test environment	Any	$\Delta L \geq 6$ dB $K_1 \leq 1.3$ dB	A-weighted and in one-third-octave or octave bands	Source directivity; SPL as a function of time; single-event SPL; Other frequency-weighted sound power levels
ISO 3745 Precision (Grade 1)	Anechoic or hemi-anechoic room	Specified requirements; measurement surface must lie wholly inside qualified region	Characteristic dimension not greater than 1/2 measurement surface radius	Any	$\Delta L \geq 10$ dB $K_1 \leq 0.5$ dB	A-weighted and in one-third-octave or octave bands	Same as above plus sound energy level
ISO 3746 Survey (Grade 3)	No special test environment	$K_2 \leq 7$ dB	No restrictions; limited only by available test environment	Any	$\Delta L \geq 3$ dB $K_1 \leq 3$ dB	A-weighted	Sound pressure levels as a function of time
ISO 3747 Engineering or Survey (Grade 2 or 3)	Essentially reverberant field in situ, subject to stated qualification requirements	Specified requirements	No restrictions; limited only by available test environment	Steady, broadband, narrow-band, or discrete-frequency	$\Delta L \geq 6$ dB $K_1 \leq 1.3$ dB	A-weighted from octave bands	Sound pressure levels as a function of time

Notes:
1. ΔL is the difference between the sound pressure levels of the source-plus-background noise and the background noise alone
2. K_1 is the correction for background noise, defined in the associated standard
3. V is the test room volume
4. α is the sound absorption coefficient
5. T_{nom} is the nominal reverberation time for the test room, defined in the associated standard
6. K_2 is the environmental correction, defined in the associated standard
7. SPL is an abbreviation for *sound pressure level*

surement method is the ultimate use of the sound power level data that are to be obtained.

In making a decision on the appropriate measurement method to be used, several factors should be considered: (a) the size of the noise source, (b) the moveability of the noise source, (c) the test environments available for the measurements, (d) the character of the noise emitted by the noise source, and (e) the grade (classification) of accuracy required for the measurements. The methods described in this chapter are consistent with those of the ISO 3740 series. A set of standards with the same objectives is available from the American National Standards Institute (ANSI) or the Acoustical Society of America (ASA).

Determination of Sound Power in a Free Field

The basis for sound power level in a free field using sound pressure is the approximate relationships in (23.28) and (23.29) above. Essentially, a measurement surface is chosen and microphone positions are defined over this surface. Sound pressure level measurements are taken at each microphone position for each frequency band, and from these, the sound power levels are computed. The relevant ISO standards are ISO 3744, ISO 3745, and ISO 3746.

Selection of a Measurement Surface. The international standards allow a variety of measurement surfaces to be used; some are discussed here. In selecting the shape of the measurement surface to be used for a particular source, an attempt should be made to choose one where the direction of sound propagation is approximately normal to the surface at the various measurement points. For example, for small sources that approximate a point source, the selection of a spherical or hemispherical surface may be the most appropriate; the direction of sound propagation will be essentially normal to this surface. For machines in a hemi-anechoic environment that are large and in the shape of a box, the parallelepiped measurement surface may be preferable. For tall machines in a hemi-anechoic environment having a height much greater than the length and depth, a cylindrical measurement surface [23.18, 19] may be the most appropriate.

Measurement in Hemi-anechoic Space. The sound power determination in a hemi-anechoic space may be performed according to ISO 3744 for engineering-grade accuracy or according to ISO 3745 for precision-grade accuracy. ISO 3744 is strictly for hemi-anechoic en-

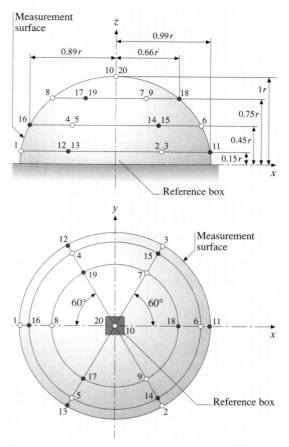

Fig. 23.9 Microphone positions for a hemispherical measurement surface according to ISO 3744. [ISO CD 3744 (N1497)]. (Courtesy of the International Organization for Standardization, Geneva, Switzerland. ©ISO, www.iso.org)

vironments, while ISO 3745 includes requirements for both hemi-anechoic and fully anechoic environments. These standards specify requirements for the measurement surfaces and locations of microphones, procedures for measuring the sound pressure levels and applying certain corrections, and the method for computing the sound power levels from the surface-average sound pressure levels. In addition, they provide detailed information and requirements on criteria for the adequacy of the test environment and background noise, calibration of instrumentation, installation and operation of the source, and information to be reported. Several annexes in each standard include information on measurement uncertainty and the qual-

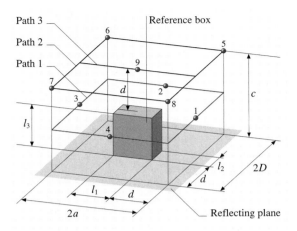

Fig. 23.10 Microphone positions for a rectangular measurement surface according to ISO 3744. [ISO CD 3744 (N1497)]. (Courtesy of the International Organization for Standardization, Geneva, Switzerland. ©ISO, www.iso.org)

ification of the test rooms. ISO 3746 is a survey method.

As examples, microphone positions for a hemispherical measurement surface, a rectangular measurement surface, and a cylindrical measurement surface are shown in Figs. 23.9–23.11, respectively.

Determination of Sound Power in a Diffuse Field

ISO 3741 is a precision method for the determination of sound power level in a reverberant test environment. The standard includes both the direct method and the comparison method described above. Requirements on the volume of the reverberation room are specified. Qualification procedures for both broadband and discrete frequency sound, are included, the minimum distance between the source and the microphone(s) are specified, and other measurement details are given.

Determination of Sound Power in Situ

In ISO 3747, engineering or survey methods are given for determination of the sound power of a source in situ, when the source cannot be moved.

Determination of Sound Power Using Sound Intensity

Sound intensity is discussed in detail in Chap. 25. The fundamental procedures for the determination of sound power from sound intensity are discussed above.

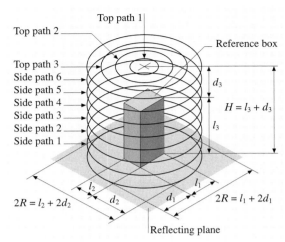

Fig. 23.11 Microphone positions for a cylindrical measurement surface

The sound intensity is measured over a selected surface enclosing the source. In principle, the integral over any surface totally enclosing the source of the scalar product (dot product) of the sound intensity vector and the associated elemental area vector provides a measure of the sound power radiated directly into the air by all sources located within the enclosing surface.

The use of sound intensity to determine sound power has several advantages over the free-field methods described above using sound pressure.

1. It is not necessary to make the approximation $I = p^2/\rho c$. Therefore, it is not necessary to make any assumption about the direction of sound intensity relative to the measurement surface. What is required is proper orientation of the intensity probe.
2. If there is no sound absorption within the measurement surface, the sound energy that enters the measurement surface from the outside also exits. Therefore, the sound power measurement is less sensitive to background noise levels and reflections from room surfaces than the pressure methods described above. The pressure methods require corrections for background noise and, in some cases, environmental corrections for reflections from room surfaces.
3. Measurements can often be made in an ordinary room.

There are, however, disadvantages to the intensity method.

An intensity analyzer is more complex than a sound level meter in the sense that the two channels required for the measurements (Sect. 23.1.8) must be carefully matched in phase. When using discrete microphone positions, a very large number of microphone positions may be required to obtain a specified accuracy in the measurements. Mean square sound pressures on a measurement surface are always positive, whereas sound intensity may be outward from the measurement surface (positive) or inward (negative). Thus, the summation of sound intensities over the measurement surface may produce a small number, and the accuracy of the measurement may be difficult to determine.

Standards have been developed to overcome these disadvantages. Instrument standards have been developed [23.11, 12] and there are standards available for the determination of sound power via sound intensity both for fixed intensity-probe positions [23.20] and for scanning [23.21–23] with an intensity probe over the measurement surface. The standards also describe *field indicators* that can be used to estimate the accuracy of the sound power determination, and technical information on the use of field indicators is available [23.24, 25].

After substantial experience with the determination of sound power via sound intensity has been achieved, it is expected that these standards will be revised.

Determination of Sound Power in a Duct

The most common application of in-duct measurements is to determine the sound power radiated by air-moving devices. The sound power level of a source in a duct can be determined according to ISO 5136 [23.26] from sound pressure level measurements, provided that the sound field in the duct is essentially a plane progressive wave, using the equation

$$L_W = L_p + 10 \log \frac{S}{S_0}, \quad (23.33)$$

where L_W is the level of total sound power traveling down duct, L_p is the sound pressure level measured just off centerline of duct, S is the cross-sectional area of duct, and $S_0 = 1 \, \text{m}^2$.

The above relation assumes not only a nonreflecting termination for the end of the duct opposite the source but also a uniform sound intensity across the duct. At frequencies near and above the first cross resonance of the duct the latter assumption is no longer satisfied. Also, when following the measurement procedures of ISO 5136, several correction factors are incorporated into (23.33) to account for microphone response and atmospheric conditions.

Equation (23.33) can still be used provided L_p, is replaced by a suitable space average $\langle L_p \rangle$ obtained by averaging the mean square sound pressures obtained at selected radial and circumferential positions in the duct, or by using a traversing circumferential microphone. The number of measurement positions across the cross section used to determine $\langle L_p \rangle$ will depend on the accuracy desired and the frequency [23.27].

In practical situations, reflections occur at the open end of the duct, especially at low frequencies. The effect of branches and bends must be considered [23.28]. When there is flow in the duct, it is also necessary to surround the microphone by a suitable windscreen. This is necessary to reduce turbulent pressure fluctuations at the microphone, which can cause an error in the measured sound pressure level.

23.2.3 Emission Sound Pressure Level

Another measure of the noise emission of a source is the *emission sound pressure level*. The sound pressure level measured in the vicinity of a noise source is dependent not only on its operating condition, but also the distance from the source and the acoustical environment – primarily reflections from the room surfaces – but also on the presence of other nearby sources. There are, however, cases where the noise emissions of a source can be described using sound pressure levels. These levels may be specified at an operator's position and at selected bystander positions, and in a controlled acoustical environment. Peak C-weighted sound levels (Sect. 23.1.4) may also be important when the levels are high enough to cause hearing damage, and these levels are not determined when using sound power level as a measure of noise emissions.

A series of international standards has been developed to ensure consistency in the measurement process and to estimate emission sound pressure levels when only sound power level information is available. ISO 11200 [23.29] provides an introduction and overview of the series. ISO 11201 [23.30] provides guidance for measurements at the operating position and bystander position when the acoustical environment is a free field over a reflecting plane. If the measurements cannot be made in a free field over a reflecting plane, it may be necessary to use a semi-reverberant environment. Using this method, A-weighted or C-weighted peak sound levels may be measured using ISO 11202 [23.31]. When only the sound power level of the source is available, it may be necessary to make estimates of the emission sound pressure level, and this procedure is covered in ISO 11203 [23.32]. As in the case of engineering and survey methods for determination of sound power level, environmental corrections may be needed in the measurement of emission sound pressure level, and this subject is covered in ISO 11204 [23.33]. Sound intensity level may be used to determine sound power level of a source, as described in the previous section. Sound intensity measurement may also be used in the determination of emission sound pressure level. The procedures are standardized in ISO 11205 [23.34].

23.2.4 Other Noise Emission Standards

Many standards organizations, trade associations, and other organizations have developed standards to determine the noise emissions of specific sources. This section is devoted to a brief description of such standards.

ISO Standards
The ISO has many standards for specific noise sources.

- Noise emitted by road vehicles [23.35–41]
- Information technology equipment [23.42–44]
- Noise emitted by rotating electrical machinery [23.45]
- Shipboard noise [23.46, 47]
- Aircraft noise [23.48, 49]
- Industrial plants [23.50, 51]
- Construction machinery [23.52]
- Agricultural machinery [23.53, 54]
- Earth-moving machinery [23.55–58]
- Lawn care equipment [23.59]
- Air terminal devices [23.60]
- Pipes and valves [23.61]
- Brush saws [23.62].

For other equipment, the ISO has published standards related to the writing of noise test codes [23.63].

American National Standards
The determination of noise emissions from a wide variety of mechanical equipment is the subject of two American national standards. ANSI S12.15-1992 [23.64] defines microphone positions and other information for determining noise emissions from a wide variety of tools. It defines measurement surfaces for portable tools and gives information on determination of sound power level for stationary tools. The tools covered include the following.

Portable electric tools:

- Circular hand saws
- Drills
- Grinders
- Polishers
- Reciprocating saws
- Nibblers
- Screwdrivers
- Nut settlers and tappers
- Hedge trimmers
- Shears
- Grass shears
- Routers
- Planers
- Edge trimmers
- Rotary cultivators

Stationary and fixed electric tools:

- Drill presses
- Scroll saws
- Scroll saws
- Disk sanders
- Joiner/planers
- Bench grinders
- Belt sanders
- Radial saws
- Hacksaws
- Band saws

A second American national standard, ANSI S12.16-1992(R2002) [23.65], gives guidance to users on how to request noise emission data for machinery, and refers to industry standards for a wide variety of equipment. The applicable standard is given and the data to be reported. Examples of equipment included in the standard are:

- Fans
- Hydraulic fluid power pumps
- Hydraulic fluid power motors
- Pneumatic tools
- Gear drives and gear motors
- Mechanical equipment (general)
- Liquid immersed transformers
- Dry type ventilated and non ventilated and sealed transformers
- Control valves and associated piping
- Air-conditioning equipment; refrigeration equipment, chillers, etc.
- Stationary heavy duty internal combustion engines
- Electric motors, turbines, and generators
- Machine tools

23.2.5 Criteria for Noise Emissions

Criteria for noise emissions can be divided into standards for making a noise declaration, specific criteria for both indoor and outdoor sources, and criteria for obtaining an environmental label for a product.

General Noise Declaration Standards

After the sound power level of a product has been determined, the results are usually communicated to users either in the form of a product noise declaration published in product literature, on the manufacturer's web site, or a physical label attached to the product. The emission sound pressure level may also be declared (Sect. 23.2.3). Noise declarations are generally in terms of a statistical upper limit determined according to national or international standards. The basic statistical procedures are contained in ISO 7574, a four-part series of standards [23.66].

The preferred descriptor of product noise emissions is the *declared A-weighted sound power level*, L_{WAd}. A widely used international standard is ISO 4871:1996 [23.67] for machinery in general. This standard is an implementation of ISO 7574 and must be followed by noise test codes. ISO 9296:1998 [23.68] is an example of such a code specifically for computer and business equipment (information technology equipment).

Noise Emission Limits

Below, two examples of upper limits for noise emissions are given, one for office equipment widely used for specification of noise emissions of information technology equipment, and the other for specification of outdoor equipment.

Indoor Equipment. Since the middle of the 1980s, recommended limits for equipment in the information technology industry have been published by the Swedish Agency for Public Management, Statskontoret [23.69]. The limits are in terms of the statistical upper-limit A-weighted sound power level of the equipment in bels. Since a wide variety of products are produced in this industry, it is necessary, in addition to upper limits, to specify the environment in which the equipment will be operated. A category 1 product is intended for use in a large data-processing installation, either generally attended or generally unattended, where the installation noise level is determined by a large number of machines. Category 2 products are intended for use in a business office where many persons operate workstations. A category 3 product is intended for use in a private office where, for example, personal computers or low-noise copiers are operated. Users are encouraged to obtain the latest version of the standard; it contains a great deal of information in addition to the recommended upper limits in the table below. Information on the presence of impulsive noise and prominent discrete tones is also required in the version below, and there is information on determination of the emission sound pressure level determined according to the ISO 11200 series of international standards [23.29–34]. While the limits in the table below are recommended, they can become a requirement when written into a purchase contract or other document.

Note that in the table below, the sound power level is expressed in bels (1 bel = 10 decibels).

Outdoor Equipment. A source of noise emission requirements for outdoor equipment may be found in European Union directive 2000/14/EC [23.70]. The directive specifies demanding noise limits and noise markings for about 55 different types of equipment. The intent is that the state of the art in noise reduction permits manufacturers to meet these limits without excessive additional costs. The limits are stated in terms of the A-weighted sound power level L_{WA} and the measurements are to be made according to European standards (EN) that correspond to ISO standards 3744 and 3746 [23.17].

The permissible sound power level shall be rounded to the nearest whole number (less than 0.5 use the lower number, greater than or equal to 0.5 use the higher number).

Environmental Labels

An alternative to the declaration of a specific noise level is the submission of product noise emission data to an independent body, which makes a judgment of the data

Fig. 23.12 German *Blue Angel* logo

relative to other similar products and allows an environmental label to be used. Such labels are not limited to noise levels, but are issued to products that display a variety of favorable environmental characteristics. The general objective of environmental labels (eco-labels) is to provide a voluntary system to both allow manufacturers to identify environmentally friendly products and to provide the public with a recognizable label that designates environmental quality.

The first eco-label was the German *Blue Angel* label [23.71]. Noise as a requirement to obtain the *Blue Angel* label, illustrated in Fig. 23.12, is covered in specific documents for a number of products. The verification criteria are prepared by the German Institute for Quality Assurance and Certification (RAL).

Fig. 23.13 European Union (EU) *Flower* logo

Table 23.6 shows product categories and document numbers for products with noise emission requirements.

Table 23.4 Recommended upper limits of declared A-weighted sound power level (From Statskontoret standard 26.5, 2002-10-01, Table 1)

Product description	Recommended upper limit A-weighted sound power level in bels	
	L_{WAd} Operating	L_{WAd} Idle
Category I: Equipment for use in data processing (DP) areas		
A. Products in unattended DP areas	$8.0 + L$	$8.0 + L$
B. Products in generally attended DP areas	$7.5 + L$	$7.5 + L$
Category II: Equipment for use in general business areas		
A. Fully-formed character typewriters and printers	7.2	5.5
B. Printers and copiers more than 4 m distance from workstations	7.2	6.5
C. Tabletop printers and tabletop copiers	7.0	5.5
D. Processors, controllers, disk and tape drives, etc. (more than 4 m distant from workstations)	$7.0 + L$	$7.0 + L$
E. Processors, controllers, disk and tape drives, etc. (less than 4 m distant from workstations)	6.8	6.6
Category III: Floor-standing use in quiet office areas		
A. Printers, typewriters, plotters	6.5	5.0
B. Keyboards	6.2	N/A
C. Processors	6.0	5.5
D. Tabletop processors, controllers, system units including built-in disk drives and/or tapes, display units with fans [4]	5.8	5.0
E. Display units (no moving parts)	4.5	4.5

Notes:
1. $L = \log(S/S_0)$. S is the footprint area of the product, in square meters. S_0 is the reference footprint area, equal to $1\,\mathrm{m}^2$. If $S < 1\,\mathrm{m}^2$, S shall be set equal to 1.0. (Thus the specification will not get lower than 8.0 and 7.5 and 7.0 bels for Category IA and IB and IID, respectively)
2. For products comprised of multiple racks or frames of equal footprint areas linked together, the reference footprint area S_0 shall be set equal to the footprint area of the individual racks or frames, or to $1.0\,\mathrm{m}^2$, whichever is smaller
3. The calculated value of the recommended upper limit (for Category IA, IB, and IID) may be rounded to the nearest upper 0.1 bel
4. For Category IIID products, especially for personal computers to be used in the home, lower values are often recommended (e.g., $L_{WAd} = 5.5$ B operating and $L_{WAd} = 4.8$ B idle; see, for instance, European Commission decision 2001/686/EC and Swedish TCO 99)

Table 23.5 Table of limit values from EU directive 2000/14/EC

Type of equipment	Net installed power P (in kW); Electric Electric power P_{el} in kW; Mass of appliance m in kg; cutting width L in cm	Permissible sound power level (dB/1pW)	
		Stage I as from 3 January 2002	Stage II as from 3 January 2006
Compaction machines (vibrating rollers, vibrating plates, vibratory hammers	$P \leq 8$	108	105
	$8 < P \leq 70$	109	106
	$P > 70$	$89 + 11 \log P$	$86 + 11 \log P$
Tracked dozers, tracked loaders, tracked excavators–loaders	$P \leq 55$	106	103
	$P > 55$	$87 + 11 \log P$	$84 + 11 \log P$
Wheeled dozers, wheeled loaders, dumpers, graders, loader-type landfill compactors, combustion-engine driven counterbalanced lift trucks, mobile cranes, compaction machines (non-vibrating rollers) paver finishers, hydraulic power packs	$P \leq 55$	104	101
	$P > 55$	$85 + 11 \log P$	$82 + 11 \log P$
Excavators, builder's hoists for the transport of goods, construction winches motor hoes	$P \leq 15$	96	93
	$P > 15$	$83 + 11 \log P$	$80 + 11 \log P$
Hand-held concrete breakers and picks	$m \leq 15$	107	105
	$15 < m < 30$	$94 + 11 \log P$	$92 + 11 \log m$
	$m \geq 30$	$96 + 11 \log P$	$94 + 11 \log m$
Tower cranes		$98 + \log P$	$96 + \log P$
Welding and power generators	$P_{el} \leq 2$	$97 + \log P_{el}$	$95 + \log P_{el}$
	$2 < P_{el} \leq 10$	$98 + \log P_{el}$	$96 + \log P_{el}$
	$P_{el} > 10$	$97 + \log P_{el}$	$95 + \log P_{el}$
Compressors	$P \leq 15$	99	97
	$P > 15$	$97 + 2 \log P$	$95 + 2 \log P$
Lawnmowers, lawn trimmers/ lawn edge trimmers	$L \leq 50$	96	$94(^2)$
	$50 < L \leq 70$	100	98
	$70 < L \leq 120$	100	$98(^2)$
	$L > 120$	105	$103(^2)$

Notes:
1. P_{el} for welding generators: conventional welding current multiplied by the conventional load voltage for the lowest value of the duty factor given by the manufacturer
2. P_{el} for power generators: prime power according to ISO 8528-1:1993, point 13.3.2
3. Indicative figures only. Definitive figures will depend on amendment of the directive following the report required in article 20(3). In the absence of any such amendment, the figures for stage I will continue to apply for stage II

A review of *Blue Angel* requirements specifically for noise emissions of construction equipment is available [23.72].

A more recent eco-label is the European Union *EU Flower* designation [23.73]. Currently, only personal computers [23.74] and portable computers [23.75] have noise emissions included with other environmental requirements. The *EU Flower* label is illustrated in Fig. 23.13.

Other European countries with eco-label programs include Austria, France, and the Nordic countries.

23.2.6 Principles of Noise Control

As illustrated in the previous section, the noise emissions of a wide variety of equipment are of importance, and it is not possible to present noise reduction on each type here. It is however, possible to present general principles

Table 23.6 German *Blue Angel* product categories with noise criteria

Product category	Relevant document
Automobile tires, low noise	RAL-UZ 89
Commercial vehicles and buses	RAL-UZ 59a and RAL-UZ 59b
Workstation computers	RAL-UZ 78
Construction machinery	RAL-UZ 53
Garden shredders	RAL-UZ 54
Portable computers	RAL-UZ 93
Printers	RAL-UZ 85
Copiers	RAL-UZ 62
Low-noise and low-pollutant chain saws	RAL-UZ 83
Low-noise waste-glass containers	RAL-UZ 21

of noise reduction. A widely distributed set of principles was prepared by *Ingemansson* [23.76]. One principle and its application is shown in Fig. 23.14.

The text below is partly organized according to those principles. In the following section, examples of the control of stationary noise sources are given.

As discussed in Sect. 23.0.2, sources of sound in the wave equation include mass introduced into a region, the application of a fluctuating force, transfer of heat, or stresses induced by turbulence. A fluctuating force or excess force produced by the speed of moving mechanical parts produce noise, and reduction of this force will reduce noise levels. In particular, high-frequency noise is generated by abrupt changes in force. Lower-frequency noise is produced by more gradual changes in force. Heavy objects impacting at high speeds generally produce less noise than lighter objects at low speeds.

Forces transmitted to large radiating surfaces also cause sound to be radiated, and these forces can be reduced by vibration isolation. Practical vibration isolators include rubber–plastic materials, cork, and various types of springs.

A reduction in the area of vibrating surfaces also reduces noise levels. When a plate, such as a cover over a portion of a rotating machine is not used to provide transmission loss, a perforated plate will reduce the vibrating area and reduce the radiation from the plate. Sound cancelation takes place between the front and back of a vibrating device such as a loudspeaker or a plate. Cancelation is more effective for a long narrow plate than for a square plate having the same area. This may have applications, for example, in the design of belt drives. Cancelation is also more effective for plates with free edges rather than edges which are clamped. Plates may also be covered with materials with a high degree of internal damping, thus producing absorption of sound that might otherwise be radiated as sound. This may apply, for example to objects that resonate such as saw blades. In general, damping is most effective at high frequencies.

In solid structures, structure-borne sound may travel over long distances, and its effect on radiated sound can be reduced by vibration isolation. Flexible connections are useful for prevention of noise transmission from a machine into a pipeline. Separate isolation pads may be used for mounting machinery, and heavy rigid foundations are generally most effective.

When the wavelength of sound is small (high-frequency sound), sound-absorptive materials are most effective, and enclosures may be an effective means of noise control. The sound attenuation in air is also greater for high-frequency sound than low-frequency sound, which means that there can be additional attenuation due to the distance between a source and an observer (Sect. 23.3.1 and Chap. 4). However, a high-frequency sound pressure level is perceived to be louder than a low-frequency sound of the same level. This difference is expressed in the A-weighting curves in Sect. 23.1.4. Thus, conversion of a high-frequency sound to a low-frequency sound may result in a reduction in perceived noise level, and a reduction measured using A-weighting on a sound level meter.

Tones can be produced when air flows past an object. An example is wind passing over high tension wires. Vortices are shed from the wire (Karman vortex street), and this produces a fluctuating force on the wire, which results in radiation of sound. Adding an irregular structure to the object may reduce the strength of such forces and reduce the noise level. Tones can also be generated by wind passing over resonators, and noise levels may be reduced by altering the geometry of the resonator.

In pipes carrying fluid, constrictions, as may be produced by a valve, bends, and other obstructions generate forces in the system which lead to sound radiation. Noise can be reduced by gradual changes in bends, cross sections, and pipe lengths that are long enough to allow the levels of turbulence to be reduced. Lower exhaust air flows also result in reduced noise levels, especially in high-Mach-number regions where the noise is generated aerodynamically, and the radiated noise is proportional to the eighth power of the Mach number. Another way to reduce noise from exhaust air flows is to add secondary flow of a lower speed around a high-speed exhaust jet.

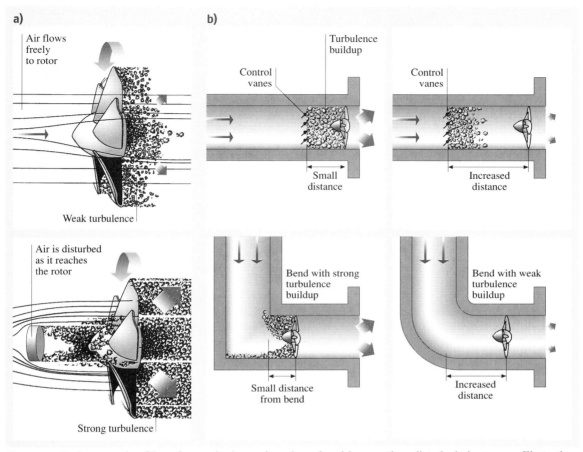

Fig. 23.14a,b An example of how fans make less noise when placed in smooth undisturbed air streams. Illustrations courtesy of Ingemansson Technology, Gothenburg, Sweden. (**a**) principle, (**b**) application with ventilation

This is the principle of the high-bypass-ratio jet engine, and may also apply to compressed-air nozzles when used as part of a cleaning system in industrial applications. In air-moving devices, particularly axial flow fans, turbulence in the inlet air stream causes the angle of attack of the flow incident on the fan blades (airfoils) to vary, and generates fluctuating lift forces on the blades. Thus, obstructions in inlet air streams should be avoided.

The presence of solid surfaces affects the radiation of sound. For idealized sources such as monopoles, dipoles, and quadrupoles (Sect. 23.0.3), the reflected pressure from one or more surfaces may be in phase with the particle velocity of the source on a small surface surrounding the source and thus produce a radiated sound power that is greater than the sound power that would be radiated by the source in a free field [23.77]. For actual machines, placement in, for example, a corner results in reflected sound waves that increase the measured sound pressure level. Thus, machines should be spaced a distance from walls and corners to reduce noise level. This effect is included in the prediction algorithm described in Sect. 23.3.3.

Acoustical materials for noise reduction are described in Sect. 23.3.3. Because good sound absorption requires a particle velocity in the material, and the particle velocity in a sound wave is small near a rigid surface, thick materials tend to be better sound absorbers than thin materials when placed on a rigid surface. Also, an air gap between a rigid wall and sound-absorptive material tends to increase sound absorption (Sect. 23.3.3). A layer of perforated metal with relatively large perforations is effective in protecting the surface of the material from abrasion, and has relatively little effect on the sound-absorptive performance of the material. The shape of the perforations may take many different forms. Sound-absorptive material is also effective when used on ceilings in conjunction with screens in an office environment, since sound transmitted over the screen may be reflected before reaching an observer on the other side of the screen.

When sound travels in a tube, an expansion chamber as part of the tube may reduce noise levels. This principle may be used in reactive silencers. An example of a generic reactive silencer is given in Sect. 23.3.4. Such silencers with added sound-absorptive material may provide additional sound absorption. Dissipative silencers contain sound-absorptive materials, as discussed in the same section. Sound cancelation at selected frequencies may be achieved by using a tube with a side path which is long enough to produce sound cancelation between the wave that travels down the duct and the wave that travels over the side path.

The effect of wall constructions on noise reduction is covered in Chap. 11.

23.2.7 Noise From Stationary Sources

One of the most pervasive sources of noise is noise generated by air-moving devices. The general principles of control of noise from these devices is described below. Noise from other stationary sources is the subject of current research.

Air-Moving Device Noise

Air moving devices (AMDs) (e.g., fans and blowers) are widely used in a variety of equipment, and are a pervasive source of noise. These devices (AMDs) may have a very large diameter, and used in ventilating systems for buildings as well as in industrial processes. Large fans are usually axial devices and large blowers are centrifugal devices that may have either forward-curved or backward-curved blades. Smaller units may be used for home heating and ventilating as well as in industrial equipment such as that produced for the information technology industry. Axial fans are most common, but centrifugal blowers are also used when the system resistance (see below) is relatively high. Other air-moving devices, such as cross-flow fans, motorized impellers, or mixed-flow devices may also be used.

The primary source of noise is fluctuating lift forces on the blades as turbulent air passes through the unit. For high values of inlet turbulence, there is a fluctuating angle of attack on the blades, and this leads to fluctuating lift forces. Forces are also produced by flow separation from the blade, and by vortices being shed from the blades. The radiated noise is broadband in character with discrete frequency components at the blade passage frequency and its harmonics. The blade passage frequency is

$$f = B \times \frac{N}{60}, \qquad (23.34)$$

where f is the frequency in Hz, B is the number of blades and N is the device speed in revolutions per minute (rpm).

A performance curve is common to all types of air-moving devices, illustrated in Fig. 23.15 for a small centrifugal device. All systems have a resistance to air flow, and the system resistance can generally be defined as a quadratic function of air flow, as illustrated in the figure. The intersection of these two curves is the operating point for the device. If the flow rate is zero, the condition is known as *shutoff*, and when the static pressure is zero, the condition is known as *free delivery*. The air flow through the blades under either condition is poor, the noise level tends to rise as discussed below.

The static efficiency of an air-moving device is generally defined as

$$\eta_\mathrm{s} = \frac{PQ}{W_\mathrm{in}}, \qquad (23.35)$$

where P is the static pressure rise in Pa (N/m^2), Q is the volume flow rate in m^3/s, and W_in is the input power to the device in W. The static efficiency is a maximum at some point on the performance curve, as illustrated in Fig. 23.16 for a constant input power. An operating point near the point of maximum static efficiency is desirable for low noise radiation, and is usually a design goal for

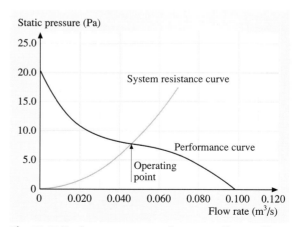

Fig. 23.15 Performance curve for a small centrifugal blower

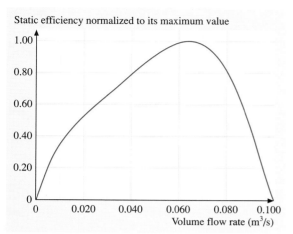

Fig. 23.16 Illustration of air-moving device static efficiency for a constant input power

large devices. In the design of small ventilating systems, the system resistance is often not known accurately, and optimal design may be difficult to achieve. However, an operating point to the left of the point of maximum static efficiency is usually undesirable, both from the viewpoint of flow instability and increased noise.

The choice of a type of air-moving device is usually dictated by the desired *specific speed* N_s defined as

$$N_s = \frac{NQ^{1/2}}{P^{3/4}}, \qquad (23.36)$$

where N is the rotational speed of the air moving device, and P and Q are defined above. Axial flow devices are generally used at low pressures and high-volume flow rates (high specific speed), and centrifugal devices are usually used at relatively high pressures and low-volume flow rates (low specific speed).

For noise radiation from large air-moving devices, it is often useful to use a sound law first proposed by *Madison* [23.78]. Its validity depends on the point of rating for a homologous series of air-moving devices. For such a series, it is useful to define dimensionless parameters, the *pressure coefficient* ψ and the *flow coefficient* ϕ as:

$$\psi = \frac{2P}{\rho(\pi DN)^2}, \qquad (23.37)$$

$$\phi = \frac{Q}{\pi D^3 N}, \qquad (23.38)$$

where D is a characteristic dimension (diameter), and N is the rotational speed of the device. Then, the performance curve for any device in the series can be expressed as a single curve of ψ versus ϕ and the operating point discussed above becomes the point of rating for the series.

The sound law is then

$$W = W_s \times \frac{P^2 Q}{P_0^2 Q_0}, \qquad (23.39)$$

where W_s is the specific sound power, and W is the radiated sound power of the device P_0 and Q_0 are 1 Pa and 1 m^3/s, respectively.

No frequency dependence was given in (23.39), but values of W_s are often published in octave bands. It must be emphasized that (23.39) is valid only at a given point of rating. If the operating point goes from shutoff ($P = 0$) to free discharge ($Q = 0$), the point of rating is changing, and the radiated sound power does not go to zero at these extremes; in fact, it generally increases, as illustrated in Fig. 23.17. The above sound law is most useful in the design of large systems when the system resistance is known and the air-moving device operates near its point of maximum static efficiency. Using (23.37), (23.38), and (23.39), it can be shown that at a given point-of-rating, the sound power is proportional to the fifth power of the speed of the device.

For small air-moving devices, it is most meaningful to obtain sound power level data on the device itself, usually as a function of speed, operating point, or both. An apparatus for determination of sound power of small devices has been standardized both nationally and internationally [23.79, 80] and is shown in Fig. 23.18 [23.81]. The AMD is mounted on a flexible membrane for airborne noise tests. Small fans mounted on lightweight structures are known to transmit energy into the structure, which can then be radiated as sound. The same test apparatus can also be used for evaluation of the structure-borne vibration of small AMDs if the membrane is replaced by a specially designed plate [23.82].

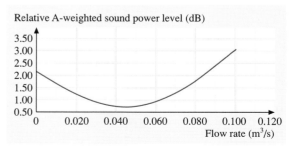

Fig. 23.17 An illustration of relative sound power level as a function of flow for a small air-moving device

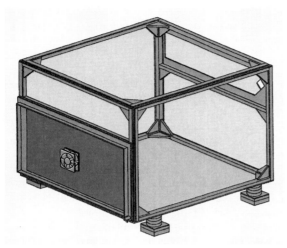

Fig. 23.18 Test plenum for determination of the sound power level of small air-moving devices (In 1962, an early version of this apparatus was first used to determine the sound power output of centrifugal blowers [23.81])

A history of noise control in small air-moving devices is available [23.83]. The following guidelines have proved to be useful for the design of cooling systems for electronic equipment [23.84].

1. Design the system to be cooled to have the lowest possible static pressure rise for the required air flow. A low static pressure rise indicates that the AMD can operate at a low tip speed, resulting in a low noise level. The static pressure rise across a system is caused by several sources of resistance such as the devices being ventilated and finger guards which may be required for safety. If unnecessary sources of resistance can be eliminated, the air flow will increase. It should then be possible to reduce the tip speed of the device (speed of the outer edge of the AMD blade) to obtain the desired air flow at a lower noise level.
2. Select an AMD so that it operates near its point of maximum static efficiency, considering the required air flow rate and the pressure drop through the system. Operation away from the point of maximum static efficiency should be in the direction of lower static pressure rise and higher air flow (see point 3).
3. Select a point of operation of a fan that is away from the best efficiency point in the direction of higher air flow and lower static pressure rise. Small fans are often unstable when operated at air flow rates less than the air flow rate at the best efficiency point. They are often very noisy under conditions of high static pressure rise and low air flow rate.
4. Select a fan or blower with a low sound power level and avoid devices that have high level peaks in their one-third-octave-band sound power spectrum. Such peaks usually indicate the presence of discrete frequency tones in the spectrum. Such tones can be difficult to eliminate and are generally a source of annoyance.
5. Select a fan or blower having the lowest speed and largest diameter consistent with the other requirements.
6. Minimize system noise levels by designing the system so that obstructions are not present within one fan diameter of the inlet to axial-flow fans so that the airflow into the inlet of axial-flow fans [23.85] is as spatially uniform as possible. Avoid the direct attachment of the AMD to lightweight sheet-metal parts.
7. Axial-flow fans should generally be mounted so that the air-flow direction is *toward* the equipment being cooled. *Pulling* air over equipment being cooled usually causes undesirable turbulence at the fan inlet, and produces an increase in noise level. See also item 6 above.
8. When possible, consider operating the AMD at a lower rotational speed. Then noise generally varies as the fifth power of the rotational speed, speed reductions produce significant reductions in noise level.

Active Noise Control

As shown in Sect. 23.0.2, in a linear medium sound pressures at a given point in space add. Therefore, in theory, if one sound pressure is created by a source and another pressure equal in amplitude and of opposite sign is created by a second source, the resulting sound pressure is zero. This is the principle of active control of noise. In practice, active control is difficult to achieve, and there are many technical issues involved.

The electronic sound absorber of *Olson* and *May* [23.86] was a very early attempt to produce active noise control. At the time, only analog circuits were available, it was necessary to control the low-frequency phase shift of both the loudspeaker and microphone used in the experiments – a subject of interest to both of the authors. With the introduction of signal processing techniques in the late 1960s and 1970s, and its rapid development in the 1980s and 1990s, active control became more practical, in spite of many limitations which still exist. Nevertheless, there have been successful applications. Good results are obtained only at low frequencies (long wavelengths) and for sources that are small compared with the wavelength of sound. In general, discrete-frequency noise is easier to cancel than broadband noise.

There are several configurations of a source and one or more loudspeakers that can produce active noise reduction:

1. Noise control in a duct. This generally requires a microphone to determine the sound coming along the duct, a loudspeaker mounted at the side of the duct, an error microphone, and signal processing to determine the signal to the loudspeaker which will reduce the sound pressure level at the error microphone (ideally) to zero. This configuration is most successful for low-frequency sound and for plane-wave propagation in the duct.
2. Reduction of radiation efficiency. Placement of one or more canceling sources very close to the noise source. In the case of a single canceling source, the effect of the canceling source is to create a dipole source and to reduce the radiation efficiency of the combined noise source – canceling source. In principle, one could use three canceling sources which, with proper phasing, reduces the radiation efficiency even further by creating a quadrupole source.
3. Creation of an outgoing wave field. Using several loudspeakers around a source, and processing the driving signals to the loudspeakers in such a way as to produce an outgoing wave equal in amplitude and opposite in phase to the wave field of the source.
4. Creation of a *zone of silence*. Using one or more loudspeakers to cancel the noise generated by a source at a point in space, creating a *zone of silence* in the vicinity of the point. Such zones are small compared with the wavelength of sound, and in any practical application, many loudspeakers are needed.
5. Global reduction inside an enclosed space. Investigation of the modal structure of an enclosed space and placement of a large number of microphones and cancelation sources within the space. A complex signal-processing system is then used to create a net noise reduction in the space. This technique has been used in the cabins of small turboprop aircraft.
6. Active headsets. A cancellation field is created in the ear canal when a headphone is worn, thereby creating a noise-canceling headset. Such headsets are commercially available.

The literature on active noise control includes books [23.87, 88], articles [23.89, 90], conference papers [23.91–95], and collected works [23.96].

Other Stationary Sources

Noise from other stationary sources covers a very wide variety of machinery and equipment. Many examples of noise control may be found in the *Noise Control Engineering Journal* and in the INTER-NOISE and NOISE-CON series of international congresses and conferences on noise control engineering(Sect. 23.6).

23.2.8 Noise from Moving Sources

Vehicle Noise

Noise emissions from motor vehicles are an important source of environmental noise in almost every industrialized country in the world, and many steps have been taken to reduce these levels. Unlike aircraft noise, vehicle noise affects populations in a very wide geographical area, and is pervasive in any area with a well-defined roadway network. In recreational areas, noise from snowmobiles and other recreational vehicles is a major source of annoyance. However, in this section, emphasis is on road vehicles – passenger cars, buses, and trucks. Aside from reduction of the noise emissions of the vehicles themselves, noise barriers and buffer zones are the primary methods of shielding persons from motor vehicle noise. The properties of noise barriers are discussed briefly in Sect. 23.3.1, and in more detail in Chap. 4.

The major sources of vehicle noise can be divided into power-train noise (cooling fan, engine, drive train, exhaust), tire–road interaction noise, and wind noise. The latter is not a major source of noise. Fan noise has been described in Sect. 23.2.7. Engine noise is produced by pressure fluctuations within the engine due to combustion, their interactions with mechanical components, transmission into the engine block and subsequent radiation. Gear noise is produced by the drive train, and exhaust noise is attenuated by the muffler. The relative importance of these sources is highly dependent on the type of vehicle and its design. Considerable progress has been made in the reduction of all of these sources, partly because of the development of regulatory limits on pass-by noise in nearly every industrialized country in the world, but also because of the desire of manufacturers to reduce the interior noise in vehicles. As a result, tire–road interaction noise has become a major source of exterior vehicle noise, especially at high speeds.

As part of a study on the effect of noise regulations on traffic noise [23.97], the International Institute of Noise Control Engineering summarized some of the noise reduction measures that have been taken to reduce the noise emissions of motor vehicles. Examples are given in Table 23.7 below.

The international standard for the determination of pass-by noise, a common descriptor of vehicle noise emissions, is ISO 362 [23.98], which specifies measurement methods, including microphone position, the characteristics of the test area in the immediate vicinity of the vehicle, driving mode, and the area of the site that must be free from obstructions. In summary, this document specifies that the test vehicle is driven over a designated test area with a standardized road surface. Fixed microphones are located 7.5 m from the vehicle path center line (In the United States, the microphone distance is 15 m and the relevant standards are SAE J986 and SAE J366.). Vehicles approach the test area at a constant speed, usually 20–50 km/h, but the throttle is fully open when the vehicle is within ±10 m of the microphone positions. The maximum A-weighted fast sound level during the acceleration is measured.

The relative importance of power train and tire/road interaction noise has been studied by *Sandberg* and *Ejsmont* [23.99]. The relative importance of these sources depends on the year of vehicle manufacture, the type of vehicle (cars, heavy trucks), and engine speed. The characteristics of the tire itself (tread, sidewall stiffness, etc.) are also important – as is the character of the road surface. The relative importance of these sources may be described in terms of a crossover speed – the speed at which tire–road noise becomes more important than power train (power unit) noise.

It can be seen from Table 23.8 that at normal speeds on major highways, tire–road interaction noise is the dominant source of vehicle noise emissions. In the table caption, "SMA" refers to a stone mastic asphalt surface and "chippings" refers to stones in an asphalt concrete surface that also consists of sand, filler, and an asphalt binder.

A qualitative description of the sources of noise radiation from tires includes radial and tangential vibration

Table 23.7 Examples of noise reduction measures in motor vehicles. (Table 1 from [23.97])

Engine in general	Other vehicle components
• Switchover to turbo-charged engines • Optimization of the engine combustion process, e.g., by using electronics or by improving the shape of the combustion chamber • Encapsulation or shielding of entire engines or especially noisy parts of them • Use of hood blankets or laminated covers • Sound-absorptive material in the engine compartment • Optimization of the stiffness of the cylinder block • Use of structure-borne noise-reducing material	• Improvement of gearboxes, damped propeller shafts • Improved rear-axle transmission • Shielding of transmission components • Regulation of the fan by thermostat • Decreased speed of fan by using a larger fan or optimization of fan shape • Silencers for air compression outlet noise • Improvements of brakes for reduction of brake squeal • Improved aerodynamics • Selection of suitable tires (low noise at acceleration)
Exhaust system	**Induction system**
• Minimization of outlet and mantle emission of exhaust silencers, e.g., by increased volume • Introduction of more than one silencer • Optimization of pipes to/from the silencer, e.g., by equal length pipes or air gap pipes • Dual-mode mufflers • Use of absorptive materials • Active noise control	• 1/4-wave tuners or other resonators • Thicker duct walls, and/or lined ducts • Increased volume of air cleaner • Intake covers or shields • Active noise control

Table 23.8 Crossover speeds for various cases. According to Sandberg [23.99] "The table assumes that 'normal' tires are used, and that the road surface is a dense asphalt concrete or an SMA with max 10-14 mm chippings. 'Cruising' is constant speed, and 'Accelerating' means an 'average' way of accelerating after a stop, but not as much as in the ISO 362 driving mode." (After [23.99] Table 5.1)

Vehicle type	Cruising	Accelerating
Cars made 1985-95	30–35 km/h	45–50 km/h
Cars made 1996-	15–25 km/h	30–45 km/h
Heavy trucks made 1985-1995	40–50 km/h	50–55 km/h
Heavy trucks made 1996-	30–35 km/h	45–50 km/h

of the tire treads, axial vibrations of the sidewalls, *air pumping* due to the deformation of the tread by irregularities in the road surface, and possibly vibration of the rim of the tire. At low frequencies, there is evidence that the tire treadband can be considered to be a cylindrical beam elastically supported by the sidewall [23.100].

A quantative description of tire–road noise is best obtained by development of a model of the sound radiation. A history of past modeling attempts has been given by *Kuijpers* and *van Blokland* [23.101], and more generalized models are also available [23.102, 103].

The road surface itself has a significant effect on the generation of tire noise. The roughness of the surface affects the excitation of the tire and subsequent generation of noise; the porosity of the surface effectively changes the compressibility of trapped air, and if the acoustic impedance is in the range to absorb sound, energy can be absorbed by the surface. Information on the characteristics of porous surfaces is given in [23.99].

Tire–road interaction noise is currently the subject of intensive research. *Bernhard* et al. have given a summary of research [23.104], *Sandberg* et al. [23.105] have discussed poroelastic road surfaces, and *Thornton* et al. [23.106] have studied the variability of existing pavements. Although porous surfaces have been shown to provide noticeable reductions in traffic noise levels, questions remain about the durability of these surfaces, especially in cold climates. The durability of such surfaces in a warm climate has been studied by *Donavan* et al. [23.107].

Traffic Noise Prediction Models. The US Federal Highway Administration has released version 2.5 of the *traffic noise model*, which may be used for a wide variety of calculations related to traffic noise. The characteristics of the model are described in Sect. 23.3.1 on *outdoor noise barriers*.

Aircraft Noise
Exterior noise emissions of aircraft are of major importance to persons on the ground, and interior noise emissions are a concern of both airline passengers and crews.

Aircraft Noise Emissions. Noise emissions from civil aircraft have been regulated for more than 35 years, first by the Federal Aviation Administration in the United States, and later, internationally, by the International Civil Aviation Organization (ICAO). Nevertheless, although the noise from individual airplanes has been significantly reduced, the total exposure to aircraft noise for residents in communities near many airports has increased because of the growth in the number of aircraft operations at airports throughout the world. Consequently, operators of the airplanes and the airports continue to face limitations because of aircraft noise. This section discusses the noise emissions of aircraft; noise at the receiver is discussed in Sect. 23.4.4. Requirements for the noise levels that may be produced by aircraft are given in title 14 of the US Code of Federal Regulations, 14CFR part 36 [23.108] and in annex 16 to the Convention on International Civil Aviation (ICAO annex 16) [23.109].

The methodology of determining the noise emission of aircraft for certification purposes is well developed. Certification noise levels are not intended to, and do not, represent noise levels that may be measured in communities around airports.

Certification noise limits, in terms of effective perceived noise level (EPNL), are specified at three points – known as *lateral*, *flyover*, and *landing approach*. The limits are a function of the maximum design takeoff gross mass, and apply to both jet-powered and propeller-driven transport-category airplanes, business/executive airplanes, and helicopters. The limits apply to new-design airplanes and to retrofit modifications of old-design aircraft.

Perceived noise levels are determined from 500 ms-average one-third-octave-band sound pressure levels, at 500 ms intervals, and are adjusted for the additional noisiness caused by prominent tones, if present, or other spectral irregularities. For each of the three noise-measurement points, effective perceived noise levels are determined from the time integral of tone-corrected perceived noisiness and adjusted to reference atmospheric conditions, reference flight paths, reference airspeeds, and reference engine power settings.

The first issue of 14CFR part 36 became effective in 1969; ICAO annex 16 was first issued in 1971. The main objective of those regulations was to ensure that noise produced by new aircraft designs would be less than the noise produced by earlier designs. In the years since the initial issue, the certification noise level limits have been gradually reduced in stages.

Stage 1 airplanes are the noisiest, and include the as-manufactured versions of the Boeing 707 and McDonnell Douglas DC8 and various business/executive jets. Stage-2-compliant airplanes have lower noise levels than stage 1 airplanes. Noise levels from airplanes that comply with stage 3 requirements are even lower. An airplane that complies with the Stage 1, 2, or 3 requirements of 14CFR Part 36 also complies with the corresponding Chap. 1, 2 or 3 requirements of ICAO Annex 16.

ICAO Annex 16 recently introduced the Chapter 4 requirements for jet-propelled airplanes and for propeller-driven airplanes having a takeoff gross mass of more than 8618 kg. The FAA is considering adoption of similar stage 4 requirements. Noise-level limits for chapter 4 compliance are the same as for chapter 3 compliance but compliance was made more stringent by eliminating the *tradeoff provisions* that had been part of the chapter 2 and 3 (or stage 2 and 3) requirements.

As an example of the changes to the certification noise-level limits, the effective perceived noise level limit at the lateral position for a takeoff gross mass of 200 000 kg (441 000 lbf) is 107.1 dB for a chapter 2/stage 2 airplane at 650 m to the side of the takeoff flight path, and 100.4 dB for a chapter 3/stage 3 airplane (or a chapter 4 airplane) at 450 m to the side of the takeoff flight path. Changes in lateral noise levels are representative of improvements in the technology of aircraft noise reduction.

Figure 23.19 illustrates trends in the progress in reducing the noise emission from commercial jet aircraft. There are no chapter 1/stage 1 airplanes operating in most of the industrialized countries of the world and only a few chapter 2/stage 2 airplanes.

For jet-powered and propeller-driven airplanes, the engines are the principal sources of noise during ground roll, climb-out, and landing. Sound produced by vortices shed by the extended landing gear and by wing leading-and trailing-edge devices (flaps) contributes to the noise level under landing-approach flight paths, and is an important consideration as engine noise is reduced.

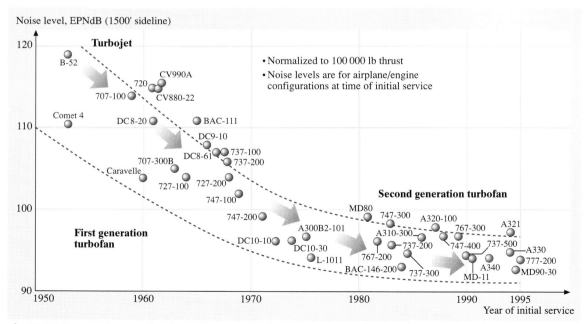

Fig. 23.19 Progress in aircraft noise reduction. Effective perceived noise level at the 450 m lateral noise-measurement point normalized to a total static thrust of 444 800 N for the noted airplane/engine configurations at the time of initial service. (After *Willshire* [23.110]; adopted from Condit [23.111])

Reduction of airplane engine noise during departure operations (as illustrated in Fig. 23.19) was achieved primarily by increasing the bypass ratio, and hence the diameter of the fan stage at the front of the turbofan engines. Bypass ratio is the ratio of the mass flow rate of the air that passes through the fan-discharge ducts to the mass flow rate through the turbine stages in the core of the engine. Design thrust is provided by moving a large amount of air at a lower exhaust-gas velocity and hence with lower levels of jet-mixing noise at high engine-power settings. Higher bypass ratios also result in better fuel efficiency for the same thrust, although at an increase in engine weight and diameter.

Discrete-frequency noise from the fan, compressor, and turbine stages is reduced by means of sound-absorbing linings in the engine inlet and discharge ducts. Tonal components in the sound from the fan stages are minimized or eliminated by careful selection of the number of blades and vanes and by increasing the spacing between the fan stage and the fan-outlet guide vanes. Inlet guide vanes ahead of the fan blades are no longer used in modern turbofan engines.

A summary of technology for engine noise-control designs is available [23.112]. A summary of retrofit applications intended to allow chapter 2/stage 2 airplanes to meet a phase-out deadline of the year 2000 for stage 3 (ICAO chapter 3) compliance has been published [23.113].

Interior Aircraft Noise. The engines and pressure fluctuations in the turbulent boundary layer outside the fuselage of an aircraft during flight are sources of noise in the interior of an airplane. Vibration from jet engines or propellers can cause the fuselage to vibrate and be radiated as sound into the cabin of an aircraft or helicopter. The air-conditioning system is often a source of noise in the interior of an aircraft.

Noise from external sound sources can be partially controlled by the design and construction of the fuselage. Sound-absorbing material is installed between the skin of the fuselage and the interior trim panels; this material also acts as a thermal insulation barrier to the cold outside air. Some propeller-driven airplanes have successfully used active noise and vibration control systems.

Control of noise in the interior of an aircraft requires a balance between the desire to achieve low cabin interior noise levels for the comfort of the passengers and the flight and cabin crews, while maintaining a degree of privacy, and an increase in airplane empty weight and maintenance costs.

Cabin noise levels vary greatly with seat location and operation of the aircraft (takeoff, cruise, and landing). Standardized test methods are now available for the specification of procedures to measure aircraft interior noise under specified cruise conditions [23.114, 115].

23.3 Propagation Paths

23.3.1 Sound Propagation Outdoors

Sound propagation in the atmosphere is discussed in Chap. 4, and only general information will be presented here. Geometrical spreading is the most important effect that reduces the sound pressure level as distance from a source is increased. There are, however, several other factors which influence outdoor sound propagation.

1. Atmospheric absorption. At long distances and at high frequencies, the effects of atmospheric absorption are significant. These effects are usually described by an attenuation coefficient in decibels per meter Chap. 4.
2. Ground effects. When the sound source is located above a ground surface, sound waves that reflect from the ground will constructively and destructively interfere with those propagating directly from the source. In general, the ground is partially reflecting; the reflected wave is modified in amplitude and phase by its interaction with the ground surface. The amount of attenuation attributable to this ground interaction, and its variation with frequency, depend on the surface characteristics, the source and receiver heights, and their separation. The effects of the ground are largest for intermediate frequencies (≈ 500 Hz) when the source is above the ground (1 m or more). If the source is very close to the ground, all frequencies above about 500 Hz are highly attenuated.
3. Temperature gradients and wind speed gradients. The speed of sound is proportional to the square root of the absolute temperature of the medium. The normal temperature lapse with height above the ground means that a sound wavefront moves more rapidly near the ground surface. This causes the wavefront

to bend upwards, creating a *shadow zone* of low sound pressure level near the surface. The opposite effect occurs when the temperature lapse is abnormal. Wind velocity gradients have a similar effect since the speed of the wave is the wind speed plus the speed of sound. Since the wind speed is usually lower close to the ground, propagation upwind tends to bend the wavefront upwards, creating a shadow zone similar to that produced by a normal temperature lapse. For sound propagation downwind, the opposite effect occurs, and the wavefront is bent toward the ground; there is no significant attenuation of the sound wave relative to the attenuation in the absence of wind.

4. Turbulence. Since turbulence always accompanies wind outdoors, the effects of atmospheric turbulence must be included in any analysis of sound propagation outdoors. There are three major effects of turbulence. First, at single frequencies, atmospheric turbulence affects the coherence between waves that propagate directly from the source and waves that are reflected from the ground surface. This results in sound pressure levels that are higher than would be expected in the absence of turbulence at those points where destructive interference between the waves occurs. Second, atmospheric turbulence produces scattering into the shadow zones caused by air-temperature and wind-speed gradients that tends to raise the sound pressure level in these areas. Third, in a highly directive sound beam, the presence of turbulence results in an excess attenuation produced by sound scattering out of the beam.

A review of sound propagation outdoors has been given by *Embleton* [23.116].

Outdoor Noise Barriers

Information on the design of noise barriers may be found in Chap. 4. In this section, the current use of outdoor noise barriers is described, and a summary of the effectiveness of these barriers is given. Barriers are widely used for the control of highway traffic noise, and are also used to control noise from rail vehicles and airport ground operations. Two common alternatives to barriers exist: sound insulation of buildings when noise indoors is the problem, and the use of quiet (porous) road surfaces. Information on tire–road interaction noise is presented in Sect. 23.2.8. However, noise barriers are currently the preferred solution for the reduction of traffic noise, and they have been constructed along many highways, as described below.

Barrier Characteristics. A typical noise barrier is shown in Fig. 23.20.

The most important parameter to describe the effectiveness of the noise barrier is the Fresnel number N

$$N = \frac{(d_1 + d_2 - d_3)}{\lambda/2} \, . \tag{23.40}$$

The distances d_1, d_2, and d_3 are defined in Fig. 23.20; λ is the wavelength of sound. The noise reduction performance of the barrier for single source and receiver points may be described in terms of its insertion loss L_i; $L_i = L_{\text{pb}} - L_{\text{pa}}$.

L_{pb} is the sound pressure level before installation of the barrier, and L_{pa} is the sound pressure level after installation. These levels may be described in octave- or one-third-octave bands, or for an overall sound level with frequency weightings such as A or C. Since the insertion loss depends on the location of both the sound source and the receiver, standardization of the measurement procedures must be available to allow specification of barrier performance (see below). Large Fresnel numbers lead to high insertion loss, and therefore performance is best at high frequencies. A technical assessment of the performance of many noise barriers [23.117] found insertion losses (A-weighted sound levels) to be in the range 5–25 dB. Barrier heights are typically 3–8 m. The performance of a barrier depends on its height, thickness, shape, edge conditions at the top of the barrier, and ground impedance on both sides of the barrier (because ground reflections contribute to the insertion loss). The transmission loss of the barrier material itself should exceed about 25 dB, but it is often of secondary importance because structural considerations usually require massive walls.

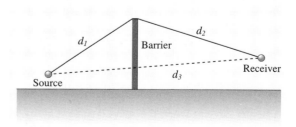

Fig. 23.20 A basic schematic showing the source of noise, barrier, and receiver

A variety of materials are used to construct barriers – precast concrete, block, earth berms, wood, metal, and plastics (including transparent plastics). Barriers with sound-absorptive surfaces have two advantages: the sound at the receiver point may be lowered, and reflections from the barrier are reduced. The latter may be of importance where barriers are installed on both sides of a highway and multiple reflections between the barriers can compromise insertion loss. Also, absorptive surfaces are useful where a barrier is installed on only one side of a highway, and neighbors on the opposite side may perceive increased noise due to reflections.

Determination of Barrier Insertion Loss. One method to determine the insertion loss of barriers is described in ANSI S12.8 [23.118]. The most reliable method is the direct method that uses data measured at the same site before and after construction of the barrier. Two "indirect" methods are described – one using *before* data at an equivalent site, and a second using a prediction method to obtain *before* data.

Steps must be taken to ensure that the source level is the same for the *before* and *after* measurements by using a reference microphone position not influenced by the barrier. The atmospheric conditions must be constant, the microphones must be calibrated, and corrections must be applied for calibration differences before and after a series of measurements. The standard specifies microphone heights and distances at which measurement must be made to determine the insertion loss.

Noise Barrier Construction. Data on noise barrier performance in many countries is available [23.117]. In the United States, construction of noise barriers is the responsibility of State Departments of Transportation guided by criteria specified by the US Federal Highway Administration (FHWA). According to the FHWA [23.119], by the end of 2001, the 10 states with the most noise barriers had constructed 8.53×10^6 m^2 of noise barriers (1993 linear km) at a cost of 1.89 billion US dollars (2001 dollars). Many of the barriers have been constructed using a combination of materials. However, the materials most used in single-material barriers are:

- Concrete/precast,
- block,
- wood/post and plank,
- concrete,
- berm,
- wood/glue laminated

- metal, and
- other wood construction.

A more detailed analysis of barrier construction by state, including area, length, and cost has been made by *Polcak* [23.120]. He concluded that cost figures are reliable only in general terms, and that many costs are variable from project to project. These include drainage, excavation, guard rails, utility relocation, landscaping, and the cost of the barrier system itself.

Noise Barrier Prediction Methods. Prediction of the performance of noise barriers is included in a more general noise model used to predict traffic noise, the *Traffic Noise Model* (TNM). According to the sponsor of the model, the US FHWA [23.121], the model contains the following components:

- Modeling of five standard vehicle types, including automobiles, medium trucks, heavy trucks, buses, and motorcycles, as well as user-defined vehicles,
- Modeling of both constant-flow and interrupted-flow traffic using a 1994/1995 field-measured database,
- Modeling of the effects of different pavement types, as well as the effects of graded roadways,
- Sound level computations based on a one-third-octave-band database and algorithms,
- Graphically interactive noise barrier design and optimization,
- Attenuation over/through rows of buildings and dense vegetation,
- Variable ground impedance,
- Multiple diffraction analysis,
- Parallel barrier analysis,
- Contour analysis, including sound level contours, barrier insertion loss contours, and sound level difference contours.

The TNM replaced the STAMINA/OPTIMA program, and several early versions were produced. Experience with the TNM and elements of its design have been described by *Menge* et al. [23.122], and comparisons of model predictions with experimental data have been made by *Rochat* [23.123].

23.3.2 Sound Propagation Indoors

In this section, the primary objective is to relate the sound pressure level measured in a room having specified acoustical characteristics to the sound power radiated by one or more sources in the room. Sound transmission through walls, window and other building elements is discussed in Chap. 11.

Classical Theory of Sound Propagation in Rooms

It is important to relate the sound pressure level in rooms to the sound power output of a source, and therefore the path indoors between the source and receiver should be discussed. In the classical theory of room acoustics, reflections from all of the surfaces in the room are equally important, and the difference between the sound pressure level L_p, and sound power level L_W at a distance r from a source emitting sound power L_W may be expressed as

$$L_p - L_W = 10 \log \left(\frac{Q}{4\pi r^2} + \frac{4}{R} \right), \quad (23.41)$$

where Q is the directivity of the source in the direction of the receiver point (Sect. 23.2.1), and R is the room constant (Chap. 9). The first term on the right of the equation is the direct sound, and the second term is the reverberant sound – the sound reflected from all of the room surfaces. For an omnidirectional source ($Q = 1$), the difference between sound pressure level and sound power level is as shown in Fig. 23.21.

In many rooms of practical interest, reflections from the room surfaces are not of equal importance, and the noise emission of one or more sources in the room has been determined according to the methods described earlier in this chapter. Rooms almost always contain objects that scatter the sound waves, and this scattering is not taken into account in (23.41). Therefore, a more detailed theory must be used to characterize the path between the source and the receiver.

Prediction of Installation Noise Levels

One procedure for determining the path attenuation in real rooms has been published by the ECMA [23.124]. This method allows for the prediction of sound levels in rooms, and, while it was developed for use with information technology equipment, it is applicable to other noise sources when the A-weighted sound power level of the source is known. The method predicts A-weighted sound pressure levels, and the sound-absorptive properties are described in terms of a sound absorption coefficient α, which is the average of the absorption coefficients in the 250, 500, 1000, and 2000 Hz octave bands.

Three different shapes of rooms are defined in the procedure:

An *ordinary room* has a length L and a width W, smaller than $3H$, where H is the height of the room. For this room, classical theory is used with the modifications described below.

A *flat room* has $L > 3H$ and $W > 3H$. For such a room, it is assumed that the sound absorption of only the floor and ceiling is important.

A *long room* has $L > 3H$ and $W < L/2$. For such rooms, reflections from the sidewalls are assumed to be important.

For an ordinary room, the classical theory is used with the following modifications.

In a free field, the radiation from the source is through a hemispherical surface, since widely used methods for the determination of sound power level depend on sound pressure level measurements on the surface of a hemisphere.

The measured directivity of the source itself is assumed to be unity because the source directivity is not normally reported with the sound power level. However, a directivity Q is included in the calculations if the source is near a wall ($Q = 2$) or in a corner ($Q = 4$).

With these modifications, the difference $L_{pA} - L_{WA}$ as a function of distance, r, from the source in an ordinary room is

$$L_{pA} - L_{WA} = 10 \log \left(\frac{Q}{2\pi r^2} + \frac{4}{S\alpha} \right), \quad (23.42)$$

where $S\alpha$ is the total absorption in the room weighted according to the area and absorption of each room surface and piece of furniture. Curves similar to those shown in Fig. 23.21 above may be plotted for this case.

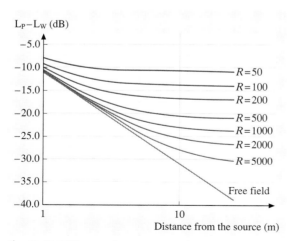

Fig. 23.21 Difference between sound pressure level and sound power level as a function of distance from an omnidirectional source

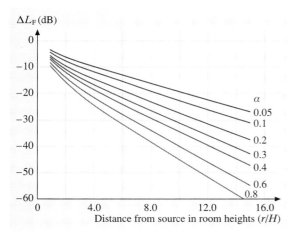

Fig. 23.22 Correction term for a furnished flat room having a density factor $q = 2$ as a function of distance from the source for several values of α

Similar, but more complex equations are given in [23.124] for flat rooms and long rooms. As an example, in a flat room with scattering objects, the absorption of the scatterers is assumed to be $\alpha_s = 0.1$ and the density factor of scattering objects q is given by

$$q = \frac{S_E}{4 S_F},$$

where S_E is the total area of scatterers as viewed from the source in m², and S_F is the surface area of the room floor in m².

The difference $L_{pA} - L_{WA}$ in a flat furnished room is then given by

$$L_{pA} - L_{WA} = -20 \log H + \Delta L_F, \quad (23.43)$$

where ΔL_F is a complicated function of distance from the source, density factor, and average absorption of the ceiling and floor. As an example, ΔL_F is given in Fig. 23.22 as a function of absorption coefficient for a density factor $q = 2$.

Other equations are given in [23.124], and correction terms are defined, both as equations and in graphical form.

Calculations for Multiple Sources

In general, there are several sources in a room at different distances from the receiver. In this case, the A-weighted sound pressure level is calculated for each source and distance, and converted to mean-square pressure. These mean-square pressures are added and converted to a total A-weighted sound pressure level at a given receiver position.

23.3.3 Sound-Absorptive Materials

Sound-absorptive materials are widely used for the control of noise in a variety of different situations. These include addition of absorptive materials to room surfaces to control reverberant sound generated by machinery and other equipment, installation of duct liners to increase the sound attenuation in air-conditioning ducts, liners for machine enclosures to reduce the sound radiated from the machine, hanging baffles for reduction of reverberant sound, and other applications.

Sound-absorptive materials exist in many different forms. These include:

- Glass-fiber materials,
- open-cell acoustical foams (urethane and similar materials, reticulated and partially reticulated),
- fiber board as frequently used for acoustical ceilings,
- hanging baffles used to reduce reverberation in factories, and other enclosed spaces
- felt materials,
- curtains and drapes,
- thin porous sheets, often mounted on a honeycomb structure,
- hollow concrete blocks with a small opening to the outside – to create a Helmholtz resonator,
- head liners for automobiles – of various materials, and
- carpets.

One characteristic common to nearly all sound-absorptive materials is that they are porous. That is, there is air flow through the material as a result of a pressure difference between the two sides of the material. Porous materials are frequently fragile, and, as a result, it is necessary to protect the exposed surface of the material. Typical protective surfaces include:

- Thin impervious membranes of plastic or other material,
- perforated facings of metal, plastic, or other material,
- fine wire-mesh screens,
- sprayed-on materials such as neoprene, and
- thin porous surfaces.

A full description of propagation in porous materials, the relationship between the propagation parameters and the physical properties of materials, and the matching of boundary conditions when composite structures are created is beyond the scope of this chapter. However, some of the properties of sound-absorptive materials will be discussed, methods of measurement of normal impedance will be presented, and some results of calculations on composite structures will be shown.

A key parameter is the flow resistance r which can be defined as the ratio of the pressure drop across a sample of a material and the velocity through it

$$r = \frac{\Delta p}{u}. \tag{23.44}$$

The flow resistance is usually measured for steady flow through the material, the assumption being made that this value is equivalent, at least at low frequencies, for the particle velocity in a sound wave.

The unit of of flow resistance is $\mathrm{N\,s/m^3}$, the MKS Rayl. Actual values are often specified as a dimensionless quantity, the specific flow resistance r_s, where $r_s = r/\rho c = r/406$ for a density of air ρ of $1.18\,\mathrm{kg/m^3}$ and a speed of sound c of $344\,\mathrm{m/s}$. It is also common to specify the flow resistance per unit thickness.

Another property of sound-absorptive materials is that they can often considered to be *locally reacting*. When a sound wave is incident on a locally reacting material, the local pressure at the surface produces a particle velocity normal to the surface. When the two are sinusoidal and expressed as complex numbers the ratio is the normal impedance of the surface

$$z_N = \frac{p}{u}. \tag{23.45}$$

The normal impedance is independent of the angle of incidence of the sound wave.

Two mechanisms are mainly responsible for the absorption of sound. First, friction in the boundary layer between the air and the internal structure of the material absorbs energy, and a large particle velocity is needed for effective absorption. Second, the temperature rise during the compression phase of the sound wave results in conduction of heat into the material, further absorbing energy from the sound wave. The former is generally most important.

The sound absorption coefficient α is the ratio of the sound energy absorbed by the surface and the incident sound energy. In general, it is a function of the angle of incidence of the sound wave on the material. It varies from 0 to 1.0, although some measurement methods – described later in the section – result in values of $\alpha > 1$. The unit of sound absorption is the metric Sabine, S. Material with an area of $1\,\mathrm{m^2}$ and $\alpha = 1$ has 1 Sabine of sound absorption (10.8 Sabins in English units). The sound-absorptive properties of, for example, hanging baffles are frequently expressed in Sabins.

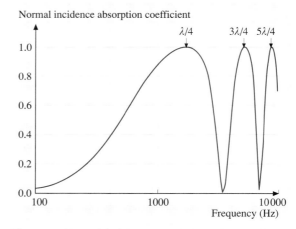

Fig. 23.23 Normal incidence sound absorption coefficient for a thin, rigid sound absorber having a flow resistance ρc placed 50 mm from a rigid termination. Calculated using a program by *Ingard* [23.125]

Sound absorption coefficients are frequently measured in octave bands, and the noise reduction coefficient (NRC) is the average absorption in the 250, 500, 1000, and 2000 Hz octave bands.

The importance of particle velocity can be illustrated for the case where a thin sound absorber having a flow resistance ρc is placed in a tube at a distance 50 mm from a rigid termination. The normal incidence absorption coefficient calculated as a function of frequency is shown in Fig. 23.23. It can be seen that the absorption coefficient is highest when the distance from the rigid termination is an odd multiple of one-quarter wavelength of the sound. It is at these distances that the particle velocity in the absorber is highest.

Galaitsis [23.126] showed that multiple resistive sheets can be used to improve the sound absorption at normal incidence, and *Ingard* [23.125] extended the analysis to diffuse fields for both locally and non-locally reacting materials. As an example, Fig. 23.24 shows the absorption coefficients for two resistive sheets having a flow resistance $1.5\,\rho c$ and a total absorber thickness of 50 mm. The first sheet is 33 mm from a rigid surface, and the spacing between the two sheets is 17 mm. The structure can be made approximately locally reacting by, for example, using a honeycomb material between the resistive layers.

Most sound-absorptive materials are relatively thick (e.g., 25 mm), and in this case, it is beneficial to have an air gap behind the material when it is mounted near a rigid surface. In the figure below, the sound-

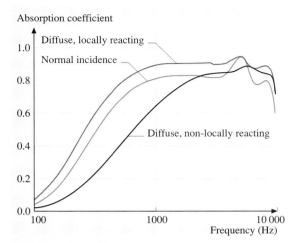

Fig. 23.24 Sound absorption coefficient for a two-screen sound absorber for normal incidence, a locally reacting structure, and a nonlocally reacting structure (After [23.127])

absorptive material is 25 mm thick, and has a total flow resistance of ρc. It is assumed that the material itself does not move. The normal incidence sound absorption coefficient as a function of spacing is shown in Fig. 23.25.

The sound-absorptive properties of flexible materials can be calculated [23.128], but this requires a knowledge

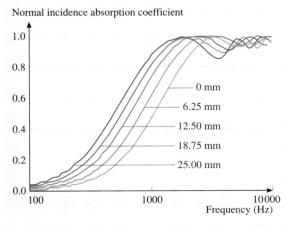

Fig. 23.25 Normal incidence sound absorption coefficient calculated for a porous material 25 mm-thick with a total flow resistance of ρc. The spacing between the material and the rigid backing is shown in the figure. Calculated using a program by *Ingard* [23.127]

of properties of the material other than flow resistance, and is beyond the scope of this chapter.

Measurement of Sound Absorption Coefficients

Two methods are commonly used to measure the performance of sound-absorptive materials, the impedance tube method and the reverberation room method.

The classic impedance tube method involves two tubes to cover two frequency ranges, a 100 mm-diameter tube and a 30 mm-diameter tube. A sample of the material is mounted at one end of the tube, and a loudspeaker at the other end is used to create a plane-wave sound field in the tube. A moving probe is used to detect the sound pressure an any point. The absorption coefficient is determined from

$$\alpha = 1 - \left| R^2 \right|, \qquad (23.46)$$

where $|R|$ is the magnitude of the pressure reflection coefficient in the tube, the ratio of the incident and reflected pressure. R is determined from the maximum and minimum pressures detected in the tube. The distance to the first minimum of sound pressure can also be measured, and together with R, can be used to determine the normal impedance of the material at its surface [23.129].

This method is useful to demonstrate plane-wave propagation and the interference of sound waves upon reflection, but it has several disadvantages:

1. The absorption coefficient is determined at single frequencies; it is therefore time consuming to determine its value as a function of frequency.
2. The sample size is small; 100 mm in diameter or 30 mm in diameter depending on the frequency of interest. The properties of a small sample may not be representative of a large sample of the same material.
3. The sample must be cut very accurately and very carefully mounted to achieve consistent results.

The first disadvantage may be eliminated by using the two-microphone method [23.130, 131] for probing the sound field. This method uses broadband excitation of the tube and two closely spaced phase-matched microphones; a digital signal processor is used to determine the transfer function between them. The absorption coefficient can then be calculated.

The reverberation room method [23.132] uses a large sample of material in a reverberation room, usually with rotating diffusers. The sound absorption coefficient is determined from the difference in reverberation time in the room with and without the sample present. The results approximate the diffuse-field absorption coefficient.

One additional method for measuring the spherical-wave absorption coefficient in a free field over a reflecting plane has been proposed by *Nobile* [23.133]. The method uses the two-microphone technique referenced above, and makes it possible to measure the absorption coefficient as a function of the angle of incidence of the sound wave for samples of absorptive materials, which can be very large. The method therefore overcomes the disadvantages of the impedance tube method and provides more information than the reverberation room method.

23.3.4 Ducts and Silencers

Attenuation of sound in heating, ventilating, and air-conditioning ducts is covered in Chap. 11. There are, however, many other applications for devices to reduce noise between the source and the receiver. These include:

- Pumps,
- compressors,
- fans,
- cooling towers,
- engines (diesel and gas),
- power plants, and
- gas turbines.

A general schematic of a *silencer* is shown in Fig. 23.26.

It is not always appropriate to describe input and output parameters in terms of impedances, but is is convenient for the purposes of illustration. Sources of sound have an internal impedance which is Z_{int} in the figure. This can be used to quantify the effect that an acoustic load has on the power output of a source [23.6]. For example, if a monopole source is characterized by a volume velocity that is independent of load, then a reflected pressure at the source which is in phase with the particle velocity will increase the power output of a source. In the figure below, the conditions at the inlet to the silencer are represented by Z_{in}. The conditions at the output of the silencer looking down the duct are characterized by Z_d. The radiation of sound into a reverberation room,

hemi-anechoic environment, or an environment in which reflections are present is characterized by Z_{rad}.

When the silencer does not contain any sound-absorptive materials (a *packless* silencer), the only mechanism that can produce an insertion loss (the ratio of the sound power radiated with the silencer in place and the sound power radiated with the silencer removed) is to create a Z_{in} which, in combination with Z_{int}, reduces the sound power input to the silencer. Such a reactive silencer works best at low frequencies, and clearly depends on the internal impedance of the source, a topic that has been studied for internal combustion engines. At low frequencies, the silencer itself can often be represented by transmission matrices [23.126]. Transmission matrices for a variety of duct elements have been defined by *Ingard* [23.134]. Characterization of the silencer in terms of lumped parameters (mass, compliance, resistance) is not generally effective. A tube, for example, is a mass element only if its end is open (low-impedance termination) whereas it is a compliance when the end is closed (high-impedance termination). Thus, a description using transmission lines in the form of T networks is an alternative. Characterization of the silencer for complex geometries and high frequencies requires a computer program.

In most practical applications, the silencer contains sound-absorptive materials and usually protective facings. The input impedance of the muffler is usually such that reflections back to the source are not important. At high frequencies, the radiation impedance is

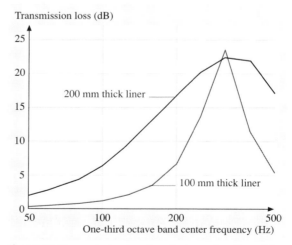

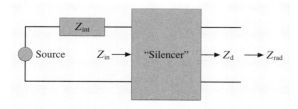

Fig. 23.26 Schematic diagram of a *silencer*

Fig. 23.27 Transmission loss in a duct as a function of frequency for two thicknesses of duct liner. Calculated using the program DuctMltl by *Ingard* [23.134]

high enough that there is little sound power reflected back down the outlet duct, and the transmission loss of the silencer (the ratio of the sound power at the outlet and the sound power at the inlet) is a good measure of silencer performance. This may not be true at low frequencies.

Duct silencers are manufactured in a wide variety of sizes, with linings, with splitters, and with aerodynamic shapes on the inlet and outlet. Standards are available for the measurement of silencer performance and guidance on their use [23.135–139]. The insertion loss of the silencer is usually measured, and its value may depend on the direction of air flow relative to the direction of propagation of sound. When the two are in the same direction (outlet duct), a sound wave is bent toward the lining whereas when the two are in opposite directions (inlet duct), sound is bent toward the center of the duct [23.140].

Calculation of Duct Transmission Loss

Computer programs for the determination of the transmission loss of ducts are available [23.134]. The duct lining may consist of air gaps, one or more layers of porous material, resistive sheets and perforated facings. As one example, Fig. 23.27 shows the transmission loss for a duct 2.5 m long having a cross section of 0.6 m × 0.6 m. The sound-absorptive material has a steel perforated facing 1.6 mm thick with perforations 3 mm in diameter and a 30% open area. The duct liner is 200 mm thick in one case, and 100 mm thick in the second. The transmission loss shown in the figure is from 50 Hz to 500 Hz because the calculation is for plane-wave transmission only. It can be seen that the peak attenuation is about the same for both cases, but the thicker liner provides a much higher transmission loss at low frequencies.

23.4 Noise and the Receiver

Section 23.2 was devoted to the characterization of sources in terms of their noise emission. The third part of the source–path–receiver model involves immission of sound at the receiver. The sound may not be *unwanted*, and is therefore not technically noise, but this section is generally devoted to the effects of noise on people. Section 23.4.5 is devoted to *sound quality*, a subject of increasing importance in the design of products.

23.4.1 Soundscapes

As discussed in Sect. 23.0.2, a sound field may be described in terms of a sound pressure, $p(r, t)$, that varies both in space and time. In practice, it is the RMS pressure that is measured since the time average of the pressure itself is zero, and several quantities measured by modern sound level meters are discussed in Sect. 23.1. The sound field can be described in the time domain or the frequency domain, or as a short-time spectrum that varies with time. Other descriptions are also possible.

The sound field $p(r, t)$ may, for the purposes of this section, be called a *soundscape*, an overall acoustical environment – both indoors and outdoors – that includes all sound, both wanted and unwanted. The interaction between an observer and this *physical* soundscape can then be described in terms of *immission* of sound. In some cases, it has been found that this interaction depends not only on the properties of the soundscape itself, but on the visual environment of the observer (the landscape). A discussion of this effect is beyond the scope of this chapter.

The soundscape includes indoor environments such as living space and industrial plants, urban and suburban areas, parks, and wilderness areas. The effects of the sound field on observers in different portions of the soundscape are varied, and in many cases difficult to quantify. These effects include hearing damage, annoyance in various forms – which can range from mild dissatisfaction to frustration and anger – as well as interference with speech communication in many settings, including meeting rooms and classrooms. Other effects include interference with sleep, and loss of productivity. Since the soundscape includes both wanted and unwanted sound, the soundscape, when properly managed can induce a sense of well-being in the observer which can have a positive effect on the quality of life. Examples include listing to natural sounds in remote areas where unwanted sound is either nonexistent, or has been reduced to an acceptable level. It is a fact that "acceptable" means different things to different persons.

A further complication is that using conventional measures of noise immission, the acceptability of a sound may depend on the source. For example, it has been found that the same level of noise from aircraft, road traffic, and rail traffic produces different human reactions. In recent years, the quality of the sound has become an important factor in the design of machinery, particularly in automobiles and other motor vehicles. The soundscape created by a snowmobile may, for example, be quite acceptable to the operator, but unacceptable to an observer some distance from the machine.

In this section, various measures of noise level will be discussed, and generally accepted criteria for the magnitude of the level will be discussed. The management of the soundscape can be thought of as a policy issue that is partly the responsibility of individuals and partly the responsibility of federal, state, and local governments.

23.4.2 Noise Metrics

The effects of noise on people is a complicated subject, and extensive research is still in progress. Several metrics are commonly used to describe noise immission at the receiver, and most of these are in terms of the A-weighted sound pressure level. These metrics work well, both indoors and outdoors provided that the spectrum of the noise, measured in octave or one-third-octave bands is somewhat neutral in character; that is, it is not perceived as *rumbly* or *hissy*, does not contain discrete frequency components (tones), is broadband, and steady without impulsive or time-varying characteristics. Other, more complex, metrics are required when any of the above characteristics are present.

The A-weighted sound level as a function of time can be expressed as L_{Af} or L_{As}, where the subscripts f and s refer to the dynamic characteristics of a sound level meter: *fast* or *slow*. These dynamic characteristics are defined in Sect. 23.1.2.

In the early 1970s, the day–night sound level (DNL, L_{dn}) was selected as a long-term measure of noise exposure. The day–night level is the time-weighted average level over a 24 hour period, but with 10 dB added to the level during the nighttime hours from 10:00 to 07:00.

Another time-weighted average level used in Europe is the day–evening–night level (LDEN, L_{den}) where the evening hours are normally between 19:00 and 23:00, the nighttime hours are from 23:00 to 07:00, and the daytime hours are from 07:00 to 19:00. In this case, 5 dB is added to the level in the evening hours, and 10 dB is added to the level in the nighttime hours. According to the European environmental noise directive, 2002/49/EC, the evening interval may be shortened by one or two hours, with corresponding lengthening of the day and/or night periods.

Other metrics for the sound level include, L_{10}, L_{50}, and L_{90}, the levels exceed 10%, 50%, and 90% of the time, respectively. Single events may be described in terms of the sound exposure level (Sect. 23.1.3).

Indoors, other metrics are commonly used to describe noise immission. In most of the world, the equivalent sound level over an eight hour period is the metric used to relate noise level to hearing loss. This level is usually described as having a 3 dB exchange rate because an equivalent level over a time interval T_1 is the same as for a level 3 dB higher over a time interval of $T_1/2$. In the United States, the Occupational Safety and Health Administration (OSHA) has adopted a 5 dB exchange rate. In this case, the time-weighted average level over a time interval T_1 is equivalent in noise dose to a level 5 dB higher measured over a time interval of $T_1/2$.

23.4.3 Measurement of Immission Sound Pressure Level

Before presenting criteria for noise immission, it is necessary to have procedures for the measurement of sound pressure level, both indoors and outdoors. For occupational noise exposure, ANSI S12.19 [23.141] defines terms used in occupational noise exposure, including noise dose, exchange rates, and the criterion sound level used in some cases. The use of sound level meters and dosimeters are also described.

ANSI S1.13 [23.142] is a very general standard for description of noise levels. The quantities discussed in Sect. 23.1 are defined as are various temporal characteristics of noise – continuous and steady, continuous sound, fluctuating sound, impulsive and intermittent sound, and other characteristics. Characteristics in the frequency domain are also described – broadband and narrow-band noise, and noise containing discrete-frequency sound. The relationship between discrete-frequency components in a spectrum and the critical bands (Chap. 14) is discussed, and the *prominence ratio* and *tone-to-noise ratio*, measures of the prominence of discrete tones in noise, are defined.

For the measurement and description of environmental noise, there is a complete series of standards [23.143–148] available. This series covers such topics as basic quantities to be measured, measurement positions, definition of noise zones, categories of land use, land-use criteria, treatment of prominent discrete tones, and measuring positions for determination of the effect of aircraft noise exposure on sleep.

23.4.4 Criteria for Noise Immission

In this section, criteria are divided into criteria for hearing damage, criteria for annoyance, and other currently used criteria.

Hearing Damage Criteria

In the United States, permissible levels for industrial noise exposure are set by the US Department of Labor, and are described in 29 CFR 1910.95. In summary, an employer shall administer a hearing conservation program when the eight hour time-weighted average level exceeds 85 dB. The requirements of such a program are described in detail in the above document – as is the requirements on audiometers for the determination of hearing loss, personnel conducting the tests, and hearing protectors when used to satisfy the requirements below. More details may be found in the *Hearing Conservation Manual* [23.149].

Permissible noise exposures are given in the regulation in Table G-16, reproduced below (Table 23.9) with an explanation of how the permissible levels are to be determined.

Currently, engineering controls are required only if the eight hour noise exposure exceeds 100 dB. For lower levels, hearing protection is an alternative.

The problem of specifying upper limits for noise in the workplace has been studied by a technical study group of the International Institute of Noise Control Engineering, and a report has been prepared and approved by the member societies of that organization [23.150]. Most countries in the world allow a time-weighted average sound level over eight hours of 85–90 dB as the upper limit, and an exchange rate of 3 dB. Selection of this exchange rate is widely believed to be the best alternative, and it greatly simplifies the measurement of levels that fluctuate during the working day. Table 23.10, from [23.150], shows hearing damage criteria in a number of countries as of 1997.

I-INCE publication 97-1 [23.150] makes specific recommendations with respect to allowable levels, exchange rate, and hearing conservation programs:

1. It is desirable for jurisdictions without regulations, or with currently higher limits, to set a limit on the level of exposure over a workshift, A-weighted and normalized to eight hours, of 85 dB as soon as may be possible given the particular economic and sociological factors that are pertinent.
2. This exposure level should include the contribution from all sounds that are present including short-term, high-intensity sounds. If such sounds are further limited in regulations to a maximum sound pressure level, then regulations should set a limit of 140 dB for C-weighted peak sound pressure level.
3. An exchange rate of 3 dB per doubling or halving of exposure time should be used. This exchange rate is implicit when the exposure level is stated in terms of eight-hour-average sound pressure level;
4. Efforts should be made to reduce levels of noise in the workplace to the lowest economically and technologically reasonable values, even when there may be no risk of long-term damage to hearing. Such action can reduce other negative effects of noise such as reduced productivity, stress and disturbed speech communication.
5. At the design stage of any new installation, consideration should be given to sound and vibration isolation between noisier and quieter areas of activity. Rooms normally occupied by people should have a significant amount of acoustical absorption in order to reduce the increase of sound due to excessive reverberation.
6. The purchase specifications for all new and replacement machinery should contain clauses specifying the maximum emission sound power level and emission sound pressure level at the operator's position when the machinery is operating.

Table 23.9 Criteria from Table G-16 of 29 CFR 1910.95. Permissible noise exposures[1]

Duration in hours per day	Sound level (dBA) slow response
8	90
6	92
4	95
3	97
2	100
1 1/2	102
1	105
1/2	110
1/4 or less	115

Notes:
1. When the daily noise exposure is composed of two or more periods of noise exposure of different levels, their combined effect should be considered, rather than the individual effect of each. If the sum of the following fractions: $C_1/T_1 + C_2/T_2 + \ldots C_n/T_n$ exceeds unity, then, the mixed exposure should be considered to exceed the limit value. C_n indicates the total time of exposure at a specified noise level, and T_n indicates the total time of exposure permitted at each level. Exposure to impulsive noise should not exceed 140 dB peak sound pressure level

7. A long-term noise control program should be established and implemented at each workplace where the level of the daily exposure, normalized to eight hours, exceeds 85 dB. This program should be reassessed periodically in order to exploit advances in noise control technology.
8. The use of personal hearing protection, either earplugs or other hearing protection devices, should

Table 23.10 Some features of legislation in other countries in 1997. After *Embleton* [23.150]. See the notes following the table

Some features of legislation in various countries					
Country (Jurisdiction)	Eight hour average A-weighted sound pressure level (dB)	Exchange rate(dB)	8h average A-wtd limit for engineering or administrative controls (dB)	Eight hour average A- wtd limit for monitoring hearing (dB)	Upper limit for peak sound pressure level (dB)
Argentina	90	3			110 A slow
Australia (varies by State)	85	3	85	85	140 unwgtd peak
Austria[a,c]	85		90		
Brazil	85	5	90, no exposure> 115 if no protection, no time limit	85	130 unwgtd peak or 115 A slow
Canada (Federal)	87	3	87	84	
(ON, PQ, NB)	90	5	90	85[b]	140 C peak
(Alta, NS, NF)	85	5	85		
(BC)	90	3	90		
Chile	85	5			140 unwgtd peak or 115 A slow
China	70–90	3			115 A slow
Finland[c]	85	3	90		
France[c]	85	3	90	85	135 C peak
Germany[c,d]	85	3	90	85	140 C peak
Hungary	85	3	90		140 C peak or 125 A slow
India	90				140 A peak
Israel	85	5			140 C peak or 115 A slow
Italy[c]	85	3	90	85	140 C peak
Japan	90		85 hearing protection mandatory at 90	85	
Netherlands[c]	85	3	90	80	140 C peak
New Zealand	85	3	85	85	140 unwgtd peak
Norway	85	3		80	110 A slow
Poland	85	3			135 C peak or 115 A slow
Spain[c]	85	3	90	80	140 C peak
Sweden[c]	85	3	90	80	140 C peak or 115 A fast
Switzerland	85 or 87	3	85	85	140 C peak or 125 ASEL

Table 23.10 (cont.)

Some features of legislation in various countries

Country (Jurisdiction)	Eight hour average A-weighted sound pressure level (dB)	Exchange rate(dB)	8h average A-wtd limit for engineering or administrative controls (dB)	Eight hour average A- wtd limit for monitoring hearing (dB)	Upper limit for peak sound pressure level (dB)
United Kingdom	85	3	90	85	140 C peak
USA[e]	90 (TWA)	5	90	85	140 C peak or
USA (army and air force)	85	3		85	115 A slow 140 C peak
Uruguay	90	3			110 A slow
This Report Recommends	85 for 8 hour normalized exposure level limit	3	85, see also text under recommended engineering controls	on hiring, and at intervals thereafter, see text under audiometric programs	140 C peak

Notes:

* Information for Austria, Japan, Poland and Switzerland was provided directly by these member societies of I-INCE. For other countries not represented by member societies participating in the working party the information is taken with permission from [23.14]

[a] Austria also proposes 85 dB AU-weighted according to IEC 1012) as a limit for high-frequency noise, and a separate limit for low-frequency noise varying inversely as the logarithm of frequency

[b] A more complex situation is simplified to fit this tabulation

[c] All countries of the European Union require the declaration of emission sound power levels of machinery, the use of the quietest machinery where reasonably possible, and reduced reflection of noise in the building, regardless of sound pressure or exposure levels. In column 4, the limit for other engineering or administrative controls is 90 dB or 140 dB C-weighted peak (or lower) or 130 dB A-weighted impulse

[d] The rating level consists of time-average, A-weighted sound pressure level plus adjustments for tonal character and impulsiveness.

[e] TWA is the time-weighted average. The regulations in the USA are unusually complicated. Only A-weighted sound pressure levels of 80 dB or greater are included in the computation of TWA to determine whether or not audiometric testing and noise exposure monitoring are required. A-weighted sound pressure levels less than 90 dB are not included in the computation of TWA when determining the need for engineering controls

be encouraged when engineering and other noise control measures are unable to reduce the daily A-weighted exposure level of workers normalized to eight hours to 85 dB. The use of hearing protection devices should be mandatory when the exposure level is over 90 dB.

9. All employers should conduct audiometric testing of workers exposed to more than 85 dB at least every three years, or at shorter intervals depending on current exposure levels and past history of the individual worker. Records of the results of the audiometric tests should be preserved in the employee's permanent file.

Effect of High Noise Levels on Hearing

The result of exposure to high levels of noise is first a temporary shift in the hearing threshold (TTS), and eventually permanent hearing damage called noise-induced permanent threshold shift (NIPTS). Quantitative data have been published in ISO 1999 [23.151], and in ANSI S3.44-1996 [23.152]. The two standards contain the same information except that the ANSI standard allows exchange rates other than 3 dB through the calculation of an equivalent level. As one example (informative), exposure to an A-weighted sound level of 95 dB for 20 years is said to result in a NIPTS greater than 23 dB at 4 kHz in 50% of the population.

Noise Levels and Annoyance

It is difficult to specify criteria for annoyance, even when the descriptor for the noise is selected – usually the day–night sound level L_{dn}. There is a high variability in the response of individuals to noise, and it is difficult to describe a physical soundscape in terms of a single number. Reporting guidelines for community noise were published in 1998 [23.153] – after many surveys were performed. However, the classic work of

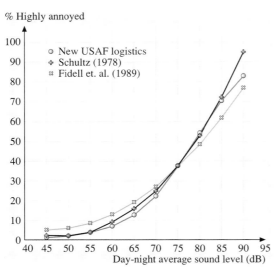

Fig. 23.28 New USAF logistic curve – together with data from *Schultz* [23.154], *Fidell* et. al. [23.156]. (After *Finegold* et al. [23.155], reprinted with permission)

Schultz [23.154] resulted in the *Schultz curve* that related the percent of persons highly annoyed and the day–night sound level. More recent data are also available, and a new logistic curve has been developed by the US Air Force [23.155]. Figure 23.28 shows this curve together with data from *Schultz* [23.154] and *Fidell* et al. [23.156].

Fidell et al. [23.156] showed that annoyance also depends on the source of sound (e.g., road traffic noise, railway noise, aircraft noise), and more recently studies by *Miedma* [23.157] and *Midema* and *Oudshoorn* [23.158] have quantified this effect with 95% confidence limits on the data. Their results are shown in Fig. 23.29.

In spite of the progress that has been made, there are doubts about the current approach to developing relationships between noise exposure and annoyance [23.159], and there may be improved methods for prediction of community response [23.160].

Criteria for Annoyance from Noise

While the figures above relate day–night sound level and the percentage of a population annoyed, there is still the question of criteria that can be used for land-use planning and other activities for which noise control is needed.

ANSI standard S12.9 [23.143–148] contains information on noise levels appropriate for various categories of land use. The data are a combination of recommendations from various sources. Table 2 from the standard is reproduced in Table 23.11.

Schomer [23.161] has studied the recommendations of several federal agencies in the United States, the World Health Organization, and the National Research Council. His recommendation is:

"For residential areas and other similarly noise sensitive land uses, noise impact becomes significant in urban areas when the DNL exceeds 55 dB. In suburban areas where the population density is between 1250 and 5000 inhabitants per square mile, noise impact becomes significant when the DNL exceeds 50 dB. And in rural areas where the population density is less than 1250 inhabitants per square mile, noise impact becomes significant when the DNL exceeds 45 dB."

Noise in Recreational Areas

Noise immission in recreational areas presents special problems. Many individuals expect a very quiet environment in parks and other recreational areas whereas others – particularly those who enjoy the use of motorized vehicles such as snowmobiles and off-road vehicles – enjoy sound levels, particularly if they are of good quality. Others enjoy recreational areas from a distance – such as in helicopters and airplanes, and the noise of these sources of concern to those on the ground. Other problems include the measurement of sounds which are often of very low level (Fig. 23.4), and may be intermittent. Measurement location is also a problem because the areas of interest are frequently very large.

A collection of 16 papers covering a variety of topics related to recreational noise is available. The papers were presented at a symposium on recreational noise held in New Zealand in 1998. *Miller* [23.7] has given a summary of the state of the art in recreational noise as of 2003.

Other Noise Criteria

A wide variety of noise criteria are discussed in a document on community noise prepared for the World Health Organization.

23.4.5 Sound Quality

In Sect. 23.0.1, noise was defined as *unwanted sound*. The field of sound quality or sound quality engineering has become important for two reasons. Noise often has characteristics that may make the sound more annoying than sounds of equivalent level using conventional measures. These include the presence of discrete-frequency

Table 23.11 A-weighted day, night, and day–night average sound levels in decibels and corresponding approximate population densities in people per square kilometer or square mile. Reprinted by permission of the Acoustical Society of America from ANSI S12.9/part 3, annex D [23.145]. This annex is for information, and is not part of the standard

	Land-use category	DNL range dB	Typical DNL dB	Day level dB	Night level dB	Approximate population density per km^2	per sq.mi.
1	Very noisy urban	> 67	70	69	61	24650	63840
2	Noisy urban residential	62–67	65	64	57	7722	20000
3	Urban and noisy suburban residential	57–62	60	58	52	2465	6384
4	Quiet urban and normal suburban residential	52–57	55	53	47	772	2000
5	Quiet suburban residential	47–52	50	48	42	247	638
6	Very quiet suburban and rural residential	< 47	45	43	37	77	200

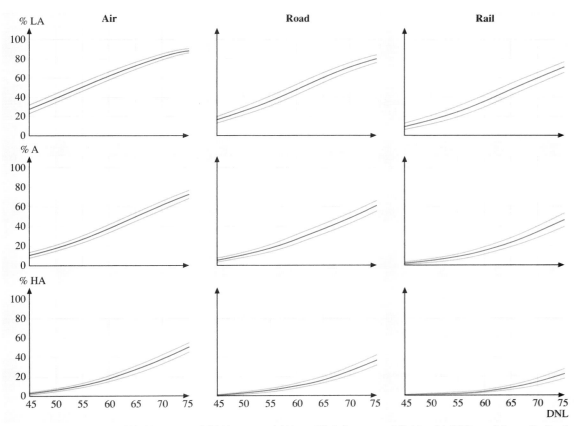

Fig. 23.29 Percentages of highly annoyed (HA), annoyed (A), and lightly annoyed (LA), with 95% confidence limits, for aircraft noise, road noise, and railway noise. (After *Miedema* [23.157] and *Miedema* and *Oudshoorn* [23.158]. Reprinted with permission)

sound, sound which is impulsive in character, and sounds that create a negative image – such as a noisy bearing which may indicate that a machine is failing. Reduction or elimination of such sounds improves the quality of the noise.

More recent work, particularly in the automobile industry has focused on the sound of engines and other automotive components to create a desirable sound for the consumer. Similarly, consumer products such as sewing machines and vacuum cleaners have been studied in an effort to eliminate sounds that the consumer finds objectionable and emphasize sounds that give a good impression of the quality of the product. There have been two sound-quality symposia that cover a wide range of topics in this field. In addition, Lyon has described the sound-quality design process [23.162], and has described a methodology by which various attributes of sound can be used to define the quality of that sound [23.163].

23.5 Regulations and Policy for Noise Control

In this section, an overview of noise policies and regulations is given. First, an overview of noise policy in the United States is given, and then information on noise policies in other countries and in the European Union is presented.

23.5.1 United States Noise Policies and Regulations

This material has been adapted from paper 302 titled "A Review of Untied States Noise Policy" by G. C. Maling, Jr. and L. S. Finegold. The paper was presented an INTER-NOISE 04, Czech Republic (2004).

The policies of the United States with respect to noise emission and noise immission have a long history. Many of the historical environmental noise documents produced in the US in the past are well known throughout the world, and have been reviewed elsewhere [23.164–170].

Funding for the US Office of Noise Abatement and Control of the US Environmental Protection Agency was discontinued in 1982, and since then there has been a lack of focus in the implementation of US noise policy. Since then, noise policies and regulations have been promulgated by a number of different federal agencies. There is a variety of state and local noise regulations; information on selected policies is given below.

Federal Government Noise Policies

The noise policies of the US federal government have been reviewed in detail elsewhere [23.169, 170]. A database of US noise policy documents has been published on CD-ROM [23.171].

Federal Aviation Administration. A number of policies related to aviation noise were in existence before 1969, but the first FAR 36 regulation [23.172] issued in 1969 by the Federal Aviation Administration, an agency of the US Department of Transportation. This regulation implemented a certification procedure for the noise emissions of aircraft, and has been amended many times to reduce noise emissions as the technology of aircraft engine design improved. More information may be found in Sect. 23.2.8.

US Environmental Protection Agency. The National Environmental Policy Act of 1969 established the Office of Noise Abatement and Control (ONAC) of the US Environmental Protection Agency (EPA), and in the early 1970s, ONAC held a series of public hearings to discuss the noise problem. These hearings led to the passage of the Noise Control Act of 1972 (NCA 72), which was landmark legislation in which the US Congress declared that it is *"the policy of the United States to promote an environment for all Americans free from noise that jeopardizes their health and welfare."* The program included:

- development of standards and criteria,
- identification of major sources of noise,
- regulation of the noise emissions of certain vehicles, machinery, and equipment,
- product noise labeling,
- interactions with other federal agencies,
- the establishment of regional noise technical centers,
- support of research in noise control, and
- assistance to state and local governments.

Funding for ONAC was discontinued in 1982. *Finegold*, *Finegold* and *Maling* [23.169, 170] discussed the successes and failures of the program, and a history of the program has been prepared by *Maling* [23.173]. A review of administrative procedures related to NCA 72 and ONAC, and recommendations for future noise abatement were prepared by the Administrative Conference of the United States in 1991 [23.174].

US Department of Housing and Urban Development. It is the policy of the US Department of Housing and Urban Development (HUD) (24 CFR part 51) to provide minimum national standards applicable to HUD programs to protect citizens against excessive noise in their communities and places of residence. HUD has defined noise exposure standards for new construction, interior noise goals, and acoustical privacy in multifamily buildings. These standards must be met if federal financing assistance is to be provided for new construction or residential rehabilitation. It is also HUD's policy to apply standards to prevent incompatible development around civil airports and military airfields. HUD requires that grantees give adequate consideration to noise exposures and sources of noise as an integral part of the urban environment when HUD assistance is provided for planning purposes. Particular emphasis is placed on the importance of compatible land-use planning in relation to airports, highways and other sources of high levels of noise exposure. HUD assistance for the construction of new noise sensitive uses is prohibited generally for projects with unacceptable noise exposures and is discouraged for projects with normally unacceptable noise exposure. Noise exposure by itself will not result in the denial of HUD support for the resale and purchase of otherwise acceptable existing buildings. However, environmental noise is a marketability factor, which HUD will consider in determining the amount of insurance or other assistance that may be given.

The Federal Energy Regulatory Commission. The Federal Energy Regulatory Commission (FERC) is an agency of the US Department of Energy, and has issued a regulation that controls noise emitted by compressors and other sources of noise related to the power industry. The regulation 18 CFR 157.206(d)(5) states that:

5. The noise attributable to any compressor facility installed pursuant to the blanket certificate shall not exceed a day–night sound level (L_{dn}) of 55 dB(A) at any noise-sensitive area unless the noise-sensitive areas (such as schools, hospitals, or residences) are established after facility construction.

US Department of Labor, Occupational Safety and Health Administration. Within the US Department of Labor, OSHA has regulations on occupational noise exposure which are detailed in the US code of federal regulations, 29 CFR 1910.95. Additional information may be found in the hearing conservation manual [23.175]. Noise policy for exposure to industrial noise is generally covered by the text in Sect. 23.4.4.

US Department of Labor, Mine Safety and Health Administration. The Mine Safety and Health Administration is part of the US Department of Labor. Regulations for the control of the noise exposure of miners is covered in 30 CFR 62. The basic requirements are:

- The eight hour time-weighted, A-frequency-weighted sound pressure level shall not exceed 90 dB. Levels below 90 dB and above 140 dB are not included in the integration.
- A 5 dB exchange rate is used.
- Above an *action level* of 85 dB, hearing protection must be used, and the individual must be enrolled in a hearing conservation program.

Additional information on the policies of other government agencies vis à vis occupational noise may be found in a paper by *Bruce* and *Wood* [23.176].

Federal Highway Administration. The primary activity of the Federal Highway Administration is to coordinate state efforts with regard to the construction of highway noise barriers. Noise barrier design in discussed in Chap. 4. Additional information on noise barriers is in Sect. 23.3.1, in the section on state policies below, and in [23.120]. More recently, there have been efforts to design highways so that tire–road noise is reduced.

State Government Policies

Policies of the state governments can generally be divided into policies with respect to exposure to occupational noise, transportation noise, and noise emissions from industrial plants. In the first case, states may establish occupational noise policies that are at least as stringent as federal regulations (see the section on OSHA).

Highway Noise Barriers. Many State governments have established noise policies with regard to the erection of noise barriers along highways [23.120]. Noise barrier policy is based on a cooperative arrangement between the federal government and state governments. Federal policy is set by the Federal Highway Administration, an agency of the US Department of Transportation, and is detailed in an FHWA policy document [23.177]. It is, however, the responsibility of each State Department of Transportation (SDoT) to determine the extent of noise abatement measures and to balance costs and environmental benefits in determining where noise barriers are to be erected.

Industrial Noise Emissions. The policy situation with regard to industrial noise immissions varies from state-to-state. The Internet site maintained by the Noise Pollution Clearing House [23.178] lists 12 states (with links to the requirements) with regulations on noise. One of the states not listed is the state of Maine – whose noise requirements have been described as the most complex in the United States [23.179]. These requirements show evidence of being carefully written. *Brooks* [23.180, 181] has identified two other states, Illinois [23.182] and Connecticut [23.183] as two other states with carefully crafted environmental noise regulations.

The Maine text is in the Department of Environmental Protection regulations 375.10, and is approximately 6000 words in length. The requirements are written in terms of hourly A-weighed sound levels. Daytime levels are between 7 a.m. and 7 p.m., and nighttime levels are between 7 p.m. and 7 a.m. The key elements are:

- In protected locations that are not predominantly commercial, transportation, or industrial, 60 dB daytime and 50 dB nighttime.
- In other areas, the required levels are 10 dB higher.
- Where pre-development levels are less than 45 dB in the daytime and 35 dB at night, the required levels are 5 dB lower.
- There are penalties for repetitive sounds and requirements on blast noise.
- Maximum levels are specified for construction activities.

Local Noise Regulations and Policies. A large number of cities and towns in the United States have ordinances and building codes to deal with local noise issues. For example, noise ordinances from cities and towns in 31 states have been posted on the Internet by the Noise Pollution Clearing House [23.177]. A model noise ordinance published by the EPA and dating from 1975 September has also been posted [23.184]. However, there is no consistency in how local noise ordinances are structured. Some use subjective noise exposure criteria, while others use objective, quantitative criteria. As described by *Finegold* and *Brooks* [23.185], an American National Standards Institute (ANSI) working group is currently developing a new ANSI standard with an updated model community noise ordinance for use by local jurisdictions to provide the needed guidance. There is a strong preference for objective noise standards as opposed to subjective standards. For example, a requirement for the prohibition of *excessive and unreasonable noises* without objective limits is difficult to enforce.

As one example of the purpose of local noise ordinances, the noise policy statement in the March 1998 New York City noise code [23.186] is reproduced below. This noise code is being revised.

24-202 Declaration of policy. It is hereby declared to be the public policy of the city to reduce the ambient noise level in the city, so as to preserve, protect and promote the public health, safety and welfare, and the peace and quiet of the inhabitants of the city, prevent injury to human, plant and animal life and property, foster the convenience and comfort of its inhabitants, and facilitate the enjoyment of the natural attractions of the city. It is the public policy of the city that every person is entitled to ambient noise levels that are not detrimental to life, health and enjoyment of his or her property. It is hereby declared that the making, creation or maintenance of excessive and unreasonable noises within the city affects and is a menace to public health, comfort, convenience, safety, welfare and the prosperity of the people of the city. For the purpose of controlling and reducing such noises, it is hereby declared to be the policy of the city to set the unreasonable noise standards and decibel levels contained herein and to consolidate certain of its noise control legislation into this code. The necessity for legislation by enactment of the provisions of this chapter is hereby declared as matter of legislative determination.

Noise Policies in Other Countries. Most industrialized countries have published noise regulations for a variety of sources. *Gottlob* [23.187] as well as *Flindell* and *McKenzie* [23.188] have presented information on community noise regulations in a large number of countries, and have listed the noise descriptors in use. More recently, technical study group 3 of the International Institute of Noise Control Engineering (I-INCE) prepared a comprehensive draft report on noise policies and regulations worldwide [23.189]. This report documents the results of an international survey on current environmental noise policies and regulations, and describes some similarities and differences in noise policies that exist at the global level. It also provides recommendations for a follow-on project to estimate national noise exposures in the I-INCE participating member countries as a next step towards the long-term goal of assessing the effectiveness of noise policies.

23.5.2 European Noise Policy and Regulations

Early European noise policy emphasized *old approach* directives that covered specific noise sources such as motor vehicles, construction equipment, lawnmowers, aircraft, and other equipment. The original safety directive, 89/392/EC, contained requirements on emission sound pressure level (Sect. 23.2.3). This document was superseded by directive 98/37/EC [23.190], but with the same acoustical requirements. This directive has been implemented throughout the European Union – the details being in European standards (EN). There are now several hundred of these standards covering a wide variety of machine types [23.191].

More recently, *new approach* directives give requirements, but are much more general in their approach to regulation of noise. There are still specific requirements on machines – as discussed in Sect. 23.2.5. These approaches have been summarized by *Higginson* et al. [23.192] in 1996, the European Union published a green paper that recognized the fact that in spite of limits on the noise emissions of individual machines, the noise to which persons are exposed has been increasing, and that more emphasis should be placed on noise in communities. A summary of the future noise policy as presented in the green paper is available [23.193], as is the green paper itself [23.194] and an edited version [23.195]. This activity led to directive 2002/49EC [23.196] designed to understand noise problems in communities through noise mapping and other activities, and to prepare action plans. The directive instructs the member states to develop a long-range plan to establish common assessment methods, to develop noise maps, to develop action plans to combat noise, to keep the public informed about noise issues. It also requires that the European Commission make regular evaluations of the implementation of the directive.

Some key action items and implementation dates in the directive are:

1. Common noise assessment methods will be established by the European Commission.
2. By 2005 July 18, the European Commission will be informed of limit values in force by the member states.
3. By 2005 June 30 and every five years thereafter, the member states will inform the European Commission of the major roads that have more than six million vehicle passages per year, major railways that have more than 60 000 train passages per year, and airports and the agglomerations with more than 250 000 inhabitants within their territories.
4. By 2007 June 30, the member states will ensure that strategic noise maps showing the situation in the previous year have been made and approved by competent authorities. By 2008 June 30, all agglomerations, major roads, and major railways shall be identified and by 2012 June 30, noise maps for these areas will be prepared and updated every five years.
5. By 2008 July 18, competent authorities shall draw up action plans to manage noise issues in areas defined by item 3 above.
6. By 2013 July 18, action plans for those areas defined by item 4 above shall be developed.
7. Action plans shall be reviewed whenever the existing noise situation changes, and at least every five years.
8. The public will be kept informed about noise maps and action plans.
9. By 2004 January 18, the European Commission will submit a report to the European parliament which reviews existing measures related to sources of environmental noise. A summary report of strategic noise maps and action plans will be prepared by 2009 July 18 and every five years thereafter. By this date, a report on the implementation of the directive will also be submitted.

Six annexes to the directive cover noise indicators, assessment methods for the noise indicators, assessment methods for harmful effects, minimum requirements for strategic noise mapping, minimum requirements for action plans, and definitions of data to be sent to the European Commission.

There are a number of ongoing projects related to this directive. Further information on EU noise policies can be found at http://europa.eu.int/comm/environment/noise/home.htm.

23.6 Other Information Resources

The following information will be helpful in locating many of the references below.

American National Standards are available from the Acoustical Society of America, asa.aip.org.

International Standards on electroacoustics (sound level meters, etc.) are published by the International Electrotechnical Commission (IEC), www.iec.ch.

International Standards on noise are published by the International Organization for Standardization (ISO), www.iso.ch.

The sponsor of the INTER-NOISE series of international congresses on noise control engineering is the International Institute of Noise Control Engineering, www.i-ince.org.

A technical journal, Noise Control Engineering Journal, has been published by INCE/USA since 1973, and all technical papers from that journal are now available on CD-ROM.

The organizer of the INTER-NOISE series when the congresses are held in North America is the Institute of Noise Control Engineering of the USA (www.inceusa.org). The institute also sponsors the NOISE-CON series of national conferences on noise control engineering. Technical papers on CD-ROM for both of these series are available from the institute (www.atlasbooks.com/marktplc/00726.htm).

References

23.1 R.H. Bolt, K.U. Ingard: System considerations in noise-control problems (Chap. 22). In: *Handbook of Noise Control*, ed. by C.M. Harris (McGraw-Hill, New York 1957)

23.2 Oxford English Dictionary, 2nd ed., Volume VII, Immission: The action of immitting; insertion, injection, admission, introduction. 1578 BANNISTER Hist.Man VIII 102, ...under judgement, as touchying emission and immission...

23.3 M.J. Lighthill: On sound generated aerodynamically, Proc. R. Soc. London **A 211**, 564–587 (1952)

23.4 E.A.G. Shaw: Noise pollution – what can be done?, Phys. Today **28**(1), 46 (1975)

23.5 P.L.Burgé: The power of public participation. In: *Proc. INTER-NOISE 02*, ed. by A. R. G.C. SelametSinghMaling (Int. Inst. Noise Control Eng., West Lafayette 2002), paper in02_501

23.6 K.U. Ingard, G.C. Maling: Physical principles of noise reduction: Properties of sound sources and their fields, Noise Control Eng. J. **2**, 37–48 (1974)

23.7 N.P. Miller: Transportation noise and recreational lands, Noise News Int. **11**, 9–21 (2003)

23.8 IEC: *IEC 61672-1:2002, Electroacoustics – Sound Level Meters – Part 1: Specifications* (IEC, Geneva 2002)

23.9 IEC: *IEC 60942:2003 Electroacoustics – Sound calibrators* (IEC, Geneva 2003)

23.10 ANSI: *ANSI S1.17-2004–Part 1. American National Standard microphone windscreens – Part 1, Measurement and Specification of Insertion Loss in Still or Slightly Moving Air* (Acoust. Soc. Am., Melville 2004)

23.11 IEC: *IEC 61043:1993 Electroacoustics – Instruments for measurement of sound intensity-measurements with pairs of pressure sensing microphones* (IEC, Geneva 1993)

23.12 ANSI: *ANSI S1.9-1996 (R2001) American National Standard instruments for the measurement of sound intensity* (Acoust. Soc. Am., Melville 2001)

23.13 I.L. Vér, L.L. Beranek: *Noise and Vibration Control Engineering*, 2nd edn. (Wiley, New York 2005), Chap. 4

23.14 ISO: *ISO 6926:1999 Acoustics – Requirements for the performance and calibration of reference sound sources used for the determination of sound power levels* (ISO, Geneva 1999)

23.15 L.L. Beranek: *Acoustical Measurements* (Acoust. Soc. Am., Woodbury 1988)

23.16 W.W. Lang, G.C. Maling, M.A. Nobile, J. Tichy: In: *Noise and Vibration Control Engineering*, 2nd edn., ed. by I.L. Vér, L.L. Beranek (Wiley, New York to be published), Chap. 4

23.17 ISO: *ISO 3740:2000 Acoustics – Determination of sound power levels of noise sources – Guidelines for the use of basic standards* (ISO, Geneva 2000)
ISO: *ISO 3741:1999 Acoustics – Determination of sound power levels of noise sources using sound pressure – Precision methods for reverberation rooms* (ISO, Geneva 1999)
ISO: *ISO 3741:1999/Cor 1:2001* (ISO, Geneva 2001)
ISO: *ISO 3743-1:1994 Acoustics – Determination of sound power levels of noise sources – Engineering methods for small, movable sources in reverberant fields – Part 1: Comparison method for hard-walled test rooms* (ISO, Geneva 1994)
ISO: *ISO 3743-2:1994 Acoustics – Determination of sound power levels of noise sources using sound pressure – Engineering methods for small, movable sources in reverberant fields – Part 2: Methods for special reverberation test rooms* (ISO, Geneva 1994)
ISO: *ISO 3744:1994 Acoustics – Determination of*

sound power levels of noise sources using sound pressure – Engineering method in an essentially free field over a reflecting plane (ISO, Geneva 1994)
ISO: *ISO 3745:2003 Acoustics – Determination of sound power levels of noise sources using sound pressure – Precision methods for anechoic and hemi-anechoic rooms* (ISO, Geneva 2003)
ISO: *ISO 3746:1995 Acoustics – Determination of sound power levels of noise sources using sound pressure – Survey method using an enveloping measurement surface over a reflecting plane* (ISO, Geneva 1995)
ISO: *ISO 3746:1995/Cor 1:1995* (ISO, Geneva 1995)
ISO: *ISO 3747:2000 Acoustics – Determination of sound power levels of noise sources using sound pressure – Comparison method in situ* (ISO, Geneva 2000)

23.18 M.A. Nobile, B. Donald, J.A. Shaw: *The cylindrical microphone array: A proposal for use in international standards for sound power measurements*, Proc. NOISE-CON 2000 (CD-ROM) (Int. Inst. Noise Control Eng., West Lafayette 2000), paper 1pNSc2

23.19 M.A. Nobile, J.A. Shaw, R.A. Boyes: *The cylindrical microphone array for the measurement of sound power level: Number and arrangement of microphones*, Proc. INTER-NOISE 2002 (CD-ROM) (Bookmaster, Mansfield 2002), paper N318

23.20 ISO: *ISO 9614: Determination of sound power levels of noise sources using sound intensity Part 1: Measurement at discrete points* (ISO, Geneva 1993)

23.21 ISO: *ISO 9614: Determination of sound power levels of noise sources using sound intensity Part 2: Measurement by scanning* (ISO, Geneva 1994)

23.22 ISO: *ISO 9614: Determination of sound power levels of noise sources using sound intensity Part 3: Precision method for measurement by scanning* (ISO, Geneva 2000)

23.23 ECMA: *ECMA-160: Determination of Sound Power Levels of Computer and Business Equipment using Sound Intensity Measurements; Scanning Method in Controlled Rooms*, 2nd edn. (ECMA International, Geneva 1992), A free download is available from www.ecma-international.org

23.24 F. Jacobsen: Sound field indicators, useful tools, Noise Control Eng. J. **35**, 37–46 (1990)

23.25 A.C. Balant, G.C. Maling, D.M. Yeager: Measurement of blower and fan noise using sound intensity techniques, Noise Control Eng. J. **33**, 77–88 (1989)

23.26 ISO: *ISO 5136:2003, Acoustics – Determination of sound power radiated into a duct by fans and other air moving devices – Induct method* (ISO, Geneva 2003)

23.27 ISO: *ISO 5136:2003, Acoustics – Determination of sound power radiated into a duct by fans and other air moving devices – Induct method* (ISO, Geneva 2003), Section 6.2

23.28 P.K. Baade: Effects of acoustic loading on axial flow fan noise generation, Noise Control Eng. J. **8**, 5–15 (1977)

23.29 ISO: *ISO 11200:1995 Acoustics – Noise emitted by machinery and equipment – Guidelines for the use of basic standards for the determination of emission sound pressure levels at a work station and at other specified positions. ISO 11200:1995/Cor 1:1997* (ISO, Geneva 1995)

23.30 ISO: *ISO 11201:1995 Acoustics – Noise emitted by machinery and equipment – Measurement of emission sound pressure levels at a work station and at other specified positions – Engineering method in an essentially free field over a reflecting plane. ISO 11201:1995/Cor 1:1997* (ISO, Geneva 1995)

23.31 ISO: *ISO 11202:1995 Acoustics – Noise emitted by machinery and equipment – Measurement of emission sound pressure levels at a work station and at other specified positions – Survey method in situ. ISO 11202:1995/Cor 1:1997* (ISO, Geneva 1995), Applies to French version only

23.32 ISO: *ISO 11203:1995 Acoustics – Noise emitted by machinery and equipment – Determination of emission sound pressure levels at a work station and at other specified positions from the sound power level* (ISO, Geneva 1995)

23.33 ISO: *ISO 11204:1995 Acoustics – Noise emitted by machinery and equipment – Measurement of emission sound pressure levels at a work station and at other specified positions – Method requiring environmental corrections. ISO 11204:1995/Cor 1:1997* (ISO, Geneva 1995)

23.34 ISO: *ISO 11205:2003 Acoustics – Noise emitted by machinery and equipment – Engineering method for the determination of emission sound pressure levels in situ at the work station and at other specified positions using sound intensity* (ISO, Geneva 2003)

23.35 ISO: *ISO 362:1998 Acoustics – Measurement of noise emitted by accelerating road vehicles – Engineering method* (ISO, Geneva 1998)

23.36 ISO: *ISO 5128:1980 Acoustics – Measurement of noise inside motor vehicles* (ISO, Geneva 1980)

23.37 ISO: *ISO 5130:1982 Acoustics – Measurement of noise emitted by stationary road vehicles – Survey method* (ISO, Geneva 1982)

23.38 ISO: *ISO 7188:1994 Acoustics – Measurement of noise emitted by passenger cars under conditions representative of urban driving* (ISO, Geneva 1994)

23.39 ISO: *ISO 9645:1990 Acoustics – Measurement of noise emitted by two wheeled mopeds in motion – Engineering method* (ISO, Geneva 1990)

23.40 ISO: *ISO 10844:1994 Acoustics – Specification of test tracks for the purpose of measuring noise emitted by road vehicles* (ISO, Geneva 1994)

23.41 ISO: *ISO 11819-1:1997 Acoustics – Measurement of the influence of road surfaces on traffic noise – Part 1: Statistical Pass-By method* (ISO, Geneva 1997)

23.42 ISO: *ISO 7779:1999 Acoustics – Measurement of airborne noise emitted by information technology and telecommunications equipment ISO 7779:1999/Amd 1:2003 Noise measurement specification for CD/DVD-ROM drives* (ISO, Geneva 1999)

23.43 ISO 9295: *1988 Acoustics – Measurement of high-frequency noise emitted by computer and business equipment* (ISO, Geneva 1988)

23.44 ISO: *ISO 9296:1988 Acoustics – Declared noise emission values of computer and business equipment* (ISO, Geneva 1988)

23.45 ISO: *ISO 1680:1999 Acoustics – Test code for the measurement of airborne noise emitted by rotating electrical machines* (ISO, Geneva 1999)

23.46 ISO: *ISO 2922:2000 Acoustics – Measurement of airborne sound emitted by vessels on inland waterways and harbours* (ISO, Geneva 2000)

23.47 ISO: *ISO 2923:1996 Acoustics – Measurement of noise on board vessels* (ISO, Geneva 1996), ISO 2923:1996/Cor 1:1997

23.48 ISO: *ISO 3891:1978 Acoustics – Procedure for describing aircraft noise heard on the ground* (ISO, Geneva 1978)

23.49 ISO: *ISO 5129:2001 Acoustics – Measurement of sound pressure levels in the interior of aircraft during flight* (ISO, Geneva 2001)

23.50 ISO: *ISO 8297:1994 Acoustics – Determination of sound power levels of multisource industrial plants for evaluation of sound pressure levels in the environment – Engineering method* (ISO, Geneva 1994)

23.51 ISO: *ISO 15664:2001 Acoustics – Noise control design procedures for open plant* (ISO, Geneva 2001)

23.52 ISO: *ISO 4872:1978 Acoustics – Measurement of airborne noise emitted by construction equipment intended for outdoor use – Method for determining compliance with noise limits* (ISO, Geneva 1978)

23.53 ISO: *ISO 5131:1996 Acoustics – Tractors and machinery for agriculture and forestry – Measurement of noise at the operator's position – Survey method* (ISO, Geneva 1996)

23.54 ISO: *ISO 7216:1992 Acoustics – Agricultural and forestry wheeled tractors and self-propelled machines – Measurement of noise emitted when in motion* (ISO, Geneva 1992)

23.55 ISO: *ISO 6393:1998 Acoustics – Measurement of exterior noise emitted by earth-moving machinery – Stationary test conditions* (ISO, Geneva 1998)

23.56 ISO: *ISO 6394:1998 Acoustics – Measurement at the operator's position of noise emitted by earth-moving machinery – Stationary test conditions* (ISO, Geneva 1998)

23.57 ISO: *ISO 6395:1988 Acoustics – Measurement of exterior noise emitted by earth-moving machinery – Dynamic test conditions ISO 6395:1988/Amd 1:1996* (ISO, Geneva 1988)

23.58 ISO: *ISO 6396:1992 Acoustics – Measurement at the operator's position of noise emitted by earth-moving machinery – Dynamic test conditions* (ISO, Geneva 1992)

23.59 ISO: *ISO 11094:1991 Acoustics – Test code for the measurement of airborne noise emitted by power lawn mowers, lawn tractors, lawn and garden tractors, professional mowers, and lawn and garden tractors with mowing attachments* (ISO, Geneva 1991)

23.60 ISO: *ISO 5135:1997 Acoustics – Determination of sound power levels of noise from air-terminal devices, air-terminal units, dampers and valves by measurement in a reverberation room* (ISO, Geneva 1997)

23.61 ISO: *ISO 15665:2003 Acoustics – Acoustic insulation for pipes, valves and flanges. ISO 15665:2003/Cor 1:2004* (ISO, Geneva 2003)

23.62 ISO: *ISO 7917:1987 Acoustics – Measurement at the operator's position of airborne noise emitted by brush saws* (ISO, Geneva 1987)

23.63 ISO: *ISO 12001:1996 Acoustics – Noise emitted by machinery and equipment – Rules for the drafting and presentation of a noise test code* (ISO, Geneva 1996)

23.64 ANSI: *ANSI S12.15-1992 (R2002), American National Standard for Acoustics – Portable electric power tools, stationary and fixed electric power tools, and gardening appliances – measurement of sound emitted* (Acoust. Soc. Am., Melville 1992)

23.65 ANSI: *ANSI S12.16-1992 (R2002), American National Standard guidelines for the specification of noise of new machinery* (Acoust. Soc. Am., Melville 1992)

23.66 ISO: *ISO 7574-1:1985 Acoustics – Statistical methods for determining and verifying stated noise emission values of machinery and equipment – Part 1: General considerations and definitions* (ISO, Geneva 1985)
ISO:ISO 7574-2:1985 *Acoustics – Statistical methods for determining and verifying stated noise emission values of machinery and equipment – Part 2: Methods for stated values for individual machines.* (ISO, Geneva 1985)
ISO:ISO 7574-3:1985 *Acoustics – Statistical methods for determining and verifying stated noise emission values of machinery and equipment – Part 3: Simple transition method for stated values for batches of machines* (ISO, Geneva 1985)
ISO:ISO 7574-4:1985 *Acoustics – Statistical methods for determining and verifying stated noise emission values of machinery and equipment – Part 4: Methods for stated values for batches of machines.*(ISO, Geneva 1985)

23.67 ISO: *ISO 4871:1996 Acoustics – Declaration and verification of noise emission values of machinery and equipment* (ISO, Geneva 1996)

23.68 ISO: *ISO 9296:1988 Acoustics – Declared noise emission values of computer and business equipment* (ISO, Geneva 1988)

23.69 Swedish Agency for Public Management: *Acoustical Noise Emission of Information Technology Equipment, Statskontoret*, Statskontoret Technical Standard, Vol. 26.5 (Swedish Agency for Public Management, Stockholm 2002)

23.70 EU Official: Directive 2000/14/EC of the European Parliament and of the Council of 8 May, 2000 on the approximation of the laws of the Member States relating to the noise emission in the environment by equipment for use outdoors, Official J. Eur. Communities **L162/1** (2000)

23.71 Blauer Engel: Umwelt Bundesamt, Dessau Germany, www.blauer-engel.de

23.72 V. Irmer, E.F.-S. Ali: Reduction of noise emission of construction machines due to the "Blue Angel" award, Noise News Int. **7**, 73–80 (1999)

23.73 www.europe.eu.int/comm/environment/ecolabel

23.74 EU Official: Commission decision of 22 August 2001 establishing the ecological criteria for the award of the Community eco-label to personal computers, Official J. Eur. Communities **L242/4** (2001)

23.75 EU Official: Commission decision of 28 August 2001 establishing the ecological criteria for the award of the Community eco-label to portable computers, Official J. Eur. Communities **L242/11** (2001)

23.76 S. Ingemansson: *Noise and Vibration Control – Principles and Applications* (Ingemansson Technology AB, Gothenburg 2000), The URL is www.ingemansson.com. The booklet is a set of principles of noise control accompanied by illustrations and an example of the application of each principle. The booklet was jointly published by Ingemansson AB and the INCE Foundation

23.77 R.V. Waterhouse: Output of a sound source in a reverberation chamber and other reflecting environments, J. Acoust. Soc. Am. **30**, 4–12 (1958)

23.78 R.D. Madison: *Fan Engineering*, 5th edn. (Buffalo Forge Company, Buffalo 1949)

23.79 ISO: *ISO 10302 :1996. Acoustics – Measurement of airborne noise emitted by small air moving devices* (ISO, Geneva 1996)

23.80 ANSI: *ANSI S12.11-2003/Part 1 ; ISO 10302 :1996 (MOD). American National Standard/Acoustics – Measurement of noise and vibration of small air moving devices. Part 1. Airborne noise emission* (Acoust. Soc. Am., Melville 2003)

23.81 G.C. Maling Jr.: Measurements of the noise generated by centrifugal blowers , J.Acoust. Soc. Am. **35**, 1913(A) (1963)

23.82 ANSI: *ANSI S12.11-2003/Part 2. Acoustics – Measurement of noise and vibration of small air moving devices. Part 2: Structure-borne vibration* (Acoust. Soc. Am., Melville 2003)

23.83 G.C. Maling: Historical developments in the control of noise generated by small air-moving devices, Noise Control Eng. J. **42**, 159–169 (1994)

23.84 G.C. Maling, D.M. Yeager: Acoustical noise measurement and control in electronic systems. In: *Thermal Measurements in Electronics Cooling*, ed. by K. Azar (CRC, Boca Raton 1997) pp. 425–467

23.85 K.B. Washburn, G.C. Lauchle: Inlet flow conditions and tonal sound radiation from a subsonic fan, Noise Control Eng. J. **31**, 101–110 (1988)

23.86 H.F. Olson, E.G. May: Electronic sound absorber, J. Acoust. Soc. Am. **25**, 1130–1136 (1953)

23.87 P.A. Nelson, S.J. Elliott: *Active Control of Sound* (Academic, San Diego 1992)

23.88 C.H. Hansen, S.D. Snyder: *Active Control of Noise and Vibration* (E &FN Spon, Florence 1997)

23.89 P.A. Nelson, S.J. Elliott: Active noise control, Noise News Int. **2**, 75–98 (1994)

23.90 J. Tichy: Applications for active control of sound and vibration, Noise News Int. **4**, 73–86 (1996)

23.91 C.R. Fuller: *Active control of sound radiation from structures, progress and future directions*, Proc. ACTIVE 02 (Institute of Sound and Vibration Research, Southampton 2002), paper P236

23.92 C.H. Hansen: *Current and future applications of active noise control*, Proc. ACTIVE 04 (Institute of Noise Control Engineering USA, Williamsburg 2004), paper a04_053

23.93 J. Scheuren: *Engineering applications of active sound and vibration control*, Proc. ACTIVE 04 (Institute of Noise Control Engineering USA, Williamsburg 2004), paper A04_100

23.94 B.B. Monson, S.D. Sommerfeldt: *Active control of tonal noise from small cooling fans*, Proc ACTIVE 04 (Institute of Noise Control Engineering USA, Williamsburg 2004), paper A04_040

23.95 A.J. Brammer, R.B. Crabtree, D.R. Peterson, M.G. Cherniack, S. Gullapappi: *Active headsets: Influence of control structure on communication signals and noise reduction*, Proc. ACTIVE 04 (Institute of Noise Control Engineering USA, Williamsburg 2004), paper A094_095

23.96 Institute of Noise Control Engineering of the USA: *Technical papers from the ACTIVE symposia in 1995, 1997, 1999, 2002, and 2004* (Institute of Noise Control Engineering of the USA, Ames 2004), Collected papers from the ACTIVE series on CD-ROM, see the URL www.atlasbooks.com/marktplc/00726.htm

23.97 U. Sandberg, Convenor: International INCE Final Report 01-1, Noise Emissions of Road Vehicles: Effects of regulations, Noise News Int. **9**, 147–206 (2001)

23.98 ISO: *ISO 362:1998 Acoustics – Measurement of noise emitted by accelerating road vehicles* (ISO, Geneva 1998)

23.99 U. Sandberg, J.A. Ejsmont: *Tyre/Road Reference Book* (Informex, Kisa 2002)

23.100 Y.-J. Kim, J.S. Bolton: *Modeling tire treadband vibration*, Proc. INTER-NOISE 01 (Int. Inst. Noise Control Eng., The Hague 2001), paper in01_716

23.101 A. Kuijpers, G. van Blokland: *Tyre/road noise models in the last two decades: a critical evaluation*, Proc. INTER-NOISE 01 (Int. Inst. Noise Control Eng., The Hague 2001), paper in01_706

23.102 W. Kropp, K. Larsson, F. Wullens, P. Andersson, F.-X. Bécot, T. Beckenbauer: *The modeling of tyre/road noise – A quasi three-dimensional model*, Proc. INTER-NOISE 01 (Int. Inst. Noise Control Eng., The Hague 2001), paper in01_657

23.103 F. de Roo, E. Gerretsen, E.H. Mulder: *Predictive performance of the tyre-road noise model TRIAS*, Proc. INTER-NOISE 01 (Int. Inst. Noise Control Eng., The Hague 2001), paper in01_706

23.104 R. Bernhard, N. Franchek, V. Drnevich: *Tire/pavement interaction noise research activities of the Institute for Safe, Quiet, and Durable Highways*, Proc. INTER-NOISE 00 (Int. Inst. Noise Control Eng., Nice 2000), paper in2000/543

23.105 U. Sandberg, H. Ohnishi, N. Kondo, S. Meiarashi: *Poroelastic road surfaces – state-of-the-art review*, Proc INTER-NOISE 00 (French Acoustical Society, Nice 2000), paper in2002/772

23.106 W.D. Thornton, R.J. Bernhard, D.I. Hanson, L. Scofield: *Acoustical variability of existing pavements*, Proc. NOISE-CON 04 (Int. Inst. Noise Control Eng., Baltimore 2004), paper nc040275

23.107 P.R. Donavan, L. Scofield: *Proc. NOISE-CON 04* (Int. Inst. Noise Control Eng., Baltimore 2004), paper nc040448

23.108 U.S. Code of Federal Regulations, 14CFR Part 36, Noise Standards, Aircraft Type and Airworthiness Certification, as amended (Federal Register, Vol. 67, No. 130, 45193-45237, July 08, 2002)

23.109 International Standards and Recommended Practices, Environmental Protection, Annex 16 to the Convention on International Civil Aviation, Volume 1: Aircraft Noise, 3rd edn., (July 1993), International Civil Aviation Organization, Montreal, Canada, including Amendments 1 through 7 as of 2002 March 21

23.110 W.L. Willshire Jr., D.G. Stephens: *Aircraft noise technology for the 21st century*, Proc. NOISE-CON 98 (Int. Inst. Noise Control Eng., Ypsilanti 1998), paper NC980007.pdf

23.111 P. Condit: Performance, Process and Value: Commercial aircraft design in the 21st century. 1996 Wright Brothers Lectureshipin Aeronautics, World Aviation Congress and Exposition, Los Angeles,California, October 22, 1996

23.112 H.H. Hubbard (Ed.): *NASA Reference Publication 1258, Vols. 1 and 2* (National Aeronautics and Space Administration, Hampton 1991)

23.113 A.H. Marsh: *Noise control for in-service jet transports*, Proc. INTER-NOISE 94, **1**, 199–204 (Int. Noise Control Eng. Japan, Yokohama 1994)

23.114 A.H. Marsh: *ISO 5129:2001 – The international standard for measurement of noise in the interior of aircraft*, Proc. INTER-NOISE 2004 (Czech Acoustical Society, Prague 2004), paper 218

23.115 ISO 5129: 2001, *Acoustics – Measurement of sound pressure levels in the interior of aircraft during flight* (ISO, Geneva 2001)

23.116 T.F.W. Embleton: Tutorial on sound propagation outdoors, J. Acoust. Soc. Am. **100**, 31–48 (1996)

23.117 G.A. Daigle, Convenor: International INCE Technical Report 99-1, Technical assessment of the effectiveness of noise walls, Noise News Int. **7**, 137–161 (1999)

23.118 ANSI: *American National Standard S12.8-1988, Methods for determining the insertion loss of outdoor noise barriers* (Acoust. Soc. Am., Melville 1988)

23.119 FHWA Internet Home Page, http://www.fhwa.gov. Search for "noise barriers"

23.120 K.D. Polcak: Highway traffic noise barriers in the U.S. – construction trends and cost analysis, Noise News Int. **11**, 96–108 (2003)

23.121 U.S. Federal Highway Administration: *FHWA Traffic Noise Model*, Version 2.5 (U.S. Department of Transportation, Washington 2004)

23.122 C.W. Menge, G.S. Anderson, D.E. Barrett: *Experiences with USDOTs Traffic Noise Model (TNM)*, Proc. NOISE-CON 01 (Int. Inst. Noise Control Eng., Portland 2001), paper nc01_083

23.123 J.L. Rochat: *Highway traffic noise measurements at acoustically hard ground sites compared to predictions from FHWA's Traffic Noise Model*, Proc. NOISE-CON 01 (Int. Inst.Noise Control Eng., Portland 2001), paper nc01_082

23.124 ECMA: *Technical Report TR/27, Method for the prediction of installation noise levels* (ECMA, Geneva 1995), Available for download at www.ecma.ch. ECMA was formerly the European Computer Manufacturers Association

23.125 K.U. Ingard: www.ingard.com. The program used was AbsMtl, Version 2.0

23.126 A.G. Galaitsis: Predicted sound interaction with a cascade of porous resistive layers in a duct, Noise Control Eng. J. **26**, 62–67 (1986)

23.127 K.U. Ingard: Notes on Sound Absorbers-N3 1999. (www.ingard.com)

23.128 K.U. Ingard: Notes on Sound Absorbers-N3 1999. (www.ingard.com, 1999) CD-ROM. See program AbsFlxSP

23.129 ISO: *ISO 10534-1: 1996. Acoustics – Determination of sound absorption coefficient and impedance in impedance tubes – Part 1: Method using standing wave ratio* (ISO, Geneva 1996)

23.130 ISO: *ISO 10534-2: 1998. Acoustics – Determination of sound absorption coefficient and impedance

23.131 ASTM E 1050-98:2003: *Standard test method for impedance and absorption of acoustical materials using a tube, two microphones, and a digital analyzer system* (Am. Soc. Testing and Materials, Philadelphia 2003), continuing from *in impedance tubes – Part 2: Transfer function method* (ISO, Geneva 1998)

23.132 ISO: *ISO 354:2003. Acoustics – Measurement of sound absorption in a reverberation room* (ISO, Geneva 2003)

23.133 M.A. Nobile: *Measurement of the spherical wave absorption coefficient at oblique incidence using the two-microphone transfer function method*, Proc. INTER-NOISE 89 (Int. Inst. Noise Control Eng., Newport Beach 1989) pp. 1067–1072

23.134 K.U. Ingard: Notes on duct attenuators-N4 (www.ingard.com, 1999) CD-ROM

23.135 ISO: *ISO 7235:2003 Acoustics – Laboratory measurement procedures, for ducted silencers and air-terminal units – Insertion loss, flow noise and total pressure loss* (ISO, Geneva 2003)

23.136 ISO: *ISO 11691:1995 Acoustics – Measurement of insertion loss of ducted silencers without flow – Laboratory survey method* (ISO, Geneva 1995)

23.137 ISO: *ISO 11820:1996 Acoustics – Measurements on silencers in situ* (ISO, Geneva 1996)

23.138 ISO: *ISO 14163:1998 Acoustics – Guidelines for noise control by silencers* (ISO, Geneva 1998)

23.139 ASTM E477-99: *Standard test method for measuring acoustical and airflow performance of duct liner materials and prefabricated silencers* (Am. Soc. Testing and Materials, Philadelphia 1999)

23.140 M. Hirschorn: The aeroacoustic rating of silencers for "forward" and "reverse" flow of air and sound, Noise Control Eng. **2**, 25–29 (1974)

23.141 ANSI: *ANSI S12.19-1996 (R2001). American National Standard measurement of occupational noise exposure* (Acoust. Soc. Am., Melville 1996)

23.142 ANSI: *ANSI S1.13-1995 (R1998). American National Standard measurement of sound pressure levels in air* (Acoust. Soc. Am., Melville 1996)

23.143 ANSI: *ANSI S12.9-1998 (R2003). Quantities and procedures for description and measurement of environmental sound, Part 1* (Acoust. Soc. Am., Melville 1998)

23.144 ANSI: *ANSI S12.9-1992 (R2003). Quantities and procedures for description and measurement of environmental sound. Part 2, measurement of long-term, wide area sound* (Acoust. Soc. Am., Melville 1992)

23.145 ANSI: *ANSI S12.9-1993 (R2003). Quantities and procedures for description and measurement of environmental sound. Part 3, short term measurements with an observer present* (Acoust. Soc. Am., Melville 1993)

23.146 ANSI: *ANSI S12.9-1996 (R2001), Quantities and procedures for description and measurement of environmental sound. Part 4, noise assessment and prediction of long term community response* (Acoust. Soc. Am., Melville 1996)

23.147 ANSI: *ANSI S12.9-1998, Quantities and procedures for description and measurement of environmental sound. Part 5, sound level descriptors for determination of compatible land use* (Acoust. Soc. Am., Melville 1998)

23.148 ANSI: *ANSI S12.9-1998 (R2003), Quantities and procedures for description and measurement of environmental sound. Part 6, methods for estimation of awakenings associated with aircraft noise events heard in homes* (Acoust. Soc. Am., Melville 1998)

23.149 A.H. Suter: *Hearing Conservation Manual*, 4th edn. (Council for Accreditation in Occupational Hearing Conservation, Milwaukee 1993)

23.150 T.F.W. Embleton, Convenor: I-INCE Publication 97-1, Technical Assessment of Upper Limits of Noise in the Workplace, Noise News Int. **5**, 203–216 (1997)

23.151 ISO: *ISO 1999:1990(E). Acoustics – Determination of occupational noise exposure and estimation of noise induced hearing impairment* (ISO, Geneva 1990)

23.152 ANSI: *ANSI S3.44-1996 (R2001) American National Standard determination of occupational noise exposure and estimation of noise induced hearing impairment* (Acoust. Soc. Am., Melville 1998)

23.153 J.M. Fields: Reporting guidelines for community noise reaction surveys, Noise News Int. **6**, 139–144 (1998)

23.154 T.J. Schultz: Synthesis of social surveys on noise annoyance, J. Acoust. Soc. Am. **64**, 377–405 (1978)

23.155 L.S. Finegold, C.S. Harris, H.E. von Gierke: Community annoyance and sleep disturbance: Updated criteria for assessing the impacts of general transportation noise on people, Noise Control Eng. J. **42**, 25–30 (1994)

23.156 S. Fidell, D.S. Barber, T.J. Schultz: Updating a dosage-effect relationship for the prevalence of annoyance due to general transportation noise, J. Acoust. Soc. Am. **89**, 221–233 (1991)

23.157 H.M.E. Miedema: *Noise & Health: How does noise affect us?* Proc. INTER-NOISE 01 (Int. Inst. Noise Control Eng., The Hague 2001), paper in01_740

23.158 H.M.E. Miedema, C.G.M.Oudshoorn: Annoyance from transportation noise: Relationships with exposure metrics DNL and DENL and their confidence intervals, Environmental Health **109**, 409–416 (2001)

23.159 S. Fidell: *Reliable prediction of community response to noise: Why you can't get there from here*, Proc. INTER-NOISE 02 (Int. Inst. Noise Control Eng., Dearborn 2002), paper N500

23.160 S. Fidell: The Schultz curve 25 years later: A research perspective, J. Acoust. Soc. Am. **114**, 3007–3015 (2003)

23.161 P.D. Schomer: *Criteria for assessment of noise annoyance*, Proc. NOISE-CON 01 (Int. Inst. Noise Control Eng., Portland 2001), paper nc01_018

23.162 R.H. Lyon: *Designing for product sound quality* (Marcel Dekker, New York 2000)

23.163 R.H. Lyon: *Product sound quality – from design to perception*, Proc. INTER-NOISE 04 (Int. Inst. Noise Control Eng., Prague 2004), keynote paper

23.164 Special issue on Global Noise Policy, Noise Control Engineering Journal **49**(4) (2001), The papers were reprinted on the CD-ROM prepared for INTER-NOISE 2002, Dearborn, Michigan, USA, 2002 August 19–21

23.165 L.L. Beranek, W.W. Lang: America needs a new national noise policy, Noise Control Eng. J. **51**(3), 123–130 (2003)

23.166 L.L. Beranek, W.W. Lang: *The need for a unified national noise policy*, Proc. NOISE-CON 01 (Int. Inst. Noise Control Eng., Portland 2001), paper NC01_002

23.167 L.L. Beranek, W.W. Lang: *U.S. noise policy on the 30th anniversary of INCE/USA*, Proc. NOISE-CON 01 (Int. Inst. Noise Control Eng., Portland 2001), paper NC01_001

23.168 L.S. Finegold, M.S. Finegold, G.C. Maling: An overview of U.S. national noise policy, Noise News Int. **10**(2), 51–63 (2002)

23.169 L.S. Finegold, M.S. Finegold, G.C. Maling: An overview of U.S. national noise policy, Noise Control Eng. J. **51**(3), 162–165 (2003)

23.170 M.S. Finegold, L.S. Finegold: Proc. INTER-NOISE 2002 on CD-ROM, Dearborn, Michigan, USA, 2002 August 19–21. The Excel database may be found at x:\policy\database\policy.xls, where x: is the designation of the CD-ROM drive.

23.171 U.S. Code of Federal Regulations, 14 CFR 36, 1969 (FAR 36)

23.172 G.C. Maling: An editor's view of the EPA noise program, Noise Control Eng. J. **51**(3), 143–150 (2003)

23.173 S.A. Shapiro: The Dormant Noise Control Act and Options to Abate Noise Pollution. In: *Noise and its Effects, Administrative Conference of the United States*, ed. by A.H. Suter (Washington 1991)

23.174 Council for Accreditation in Occupational Hearing Conservation: *Hearing Conservation Manual*, 4th edn. (Council for Accreditation in Occupational Hearing Conservation, Milwaukee 1993)

23.175 R.D. Bruce, E.W. Wood: The USA needs a new national policy for occupational noise, Noise Control Eng. J. **51**(3), 162–165 (2003)

23.176 U.S. Department of Transportation, Office of Environment and Planning, Noise and Air Quality Branch: *Highway traffic noise analysis and abatement policy and guidance* (U.S. Department of Transportation, Washington 1995), see also 23 CFR 772

23.177 Noise Pollution Clearing House: http://www.nonoise.org/lawlib/cities/cities.htm

23.178 T. Doyle: Twelve years with the most complex noise regulations in the United States. In: *Proc. NOISE-CON 01* (Int. Inst. Noise Control Eng., Portland 2001), paper NC01_067

23.179 B.M. Brooks: The need for a unified community noise policy. In: *Proc. NOISE-CON 01* (Int. Inst. Noise Control Eng., Portland 2001), paper NC01_016

23.180 B.M. Brooks: The need for a unified community noise policy, Noise Control Eng. J. **51**(3), 160–161 (2003)

23.181 State of Illinois Rules and Regulations, Title 35: Environmental Protection, Subtitle H: Noise, Chapter 1: Pollution Control Board, Parts 900 and 901 (1991)

23.182 State of Connecticut Department of Environmental Protection (DEP) regulations Title 22a, Environmental Protection, "Control of Noise", Connecticut General Statues Sections 22a-69-1 to 22a-69-7.4 (1978)

23.183 Noise Pollution Clearing House: www.nonoise.org/epa/roll2/roll2doc7blp7.pdf

23.184 L.S. Finegold, B.M. Brooks: Progress on Developing a Model Community Noise Ordinance as a National Standard in the US. In: *Proceedings of INTER-NOISE 2003*, ed. by S. Lee (Seogwipo, Jeju Island, Korea 2003), CD-ROM, paper N589

23.185 Noise Pollution Clearing House: www.nonoise.org/lawlib/cities/ny/newyork.htm

23.186 D. Gottlob: Regulations for Community Noise, Noise News Int. **3**, 223–236 (1995)

23.187 I.H. Flindell, A.R. McKenzie: *Technical Report: An inventory of current European methodologies and procedures for environmental noise management* (European Environmental Agency, Copenhagen 2000)

23.188 H. Tachibana, Convenor: International INCE Technical Study Group 3, Assessing the effectiveness of noise policies and regulations: Phase I, Noise Policies and Regulations, Draft Report dated 31 July (2004)

23.189 Directive 98/37/EC of the European Parliament and of the Council of 22 June 1998 on the approximation of the laws of the Member States relating to machinery

23.190 R. F. Higginson, personal communication, 2004 November 22

23.191 R.F. Higginson, J. Jacques, W.W. Lang: Directives, standards, and European noise requirements, Noise News Int. **2**, 156–184 (1994)

23.192 C. Grimwood, M. Ling: The EC Green Paper on future noise policy – A need for change, Acoustics Bulletin **1**, 5–8 (1997)

23.193 Commission of the European Communities: *Future Noise Policy*, European Commission Green Paper, Vol. COM(96) 540 (Commission of the European Communities, Brussels 1996)

23.194 Commission of the European Communities: Future noise policy – European Commission Green Paper, Noise News Int. **5**, 77–98 (1997)

23.195 EU Official: Directive 2002/49/EC of the European Parliament and of the Council of 25 June, 2002 relative to the assessment and management of environmental noise, Official J. Eur. Communities **L189**(12) (2002)

23.196 International Electrotechnical Commission, Geneva, Switzerland: www.iec.ch

Part H Engineering Acoustics

24 Microphones and Their Calibration
George S. K. Wong, Ottawa, Canada

25 Sound Intensity
Finn Jacobsen, Lyngby, Denmark

26 Acoustic Holography
Yang-Hann Kim, Daejeon, Korea

27 Optical Methods for Acoustics and Vibration Measurements
Nils-Erik Molin, Luleå, Sweden

28 Modal Analysis
Thomas D. Rossing, Stanford, USA

24. Microphones and Their Calibration

The condenser microphone continues to be the standard against which other microphones are calibrated. A brief discussion of the theory of the condenser microphone, including its open-circuit voltage, electrical transfer impedance, and mechanical response, is given. The most precise method of calibration, the reciprocity pressure calibration method for laboratory standard microphones is discussed in detail, beginning with the principles of the reciprocity method. Corrections for heat conduction, equivalent volume, capillary tube, wave motion, barometric pressure and temperature are necessary to achieve the most accurate open-circuit sensitivity of condenser microphones.

Free-field calibration is discussed briefly, and in view of the difficulties in obtaining more accurate results than those provided by the reciprocity method, references are given for more detailed consideration. Secondary microphone calibration methods by comparison are described. These methods include interchange microphone comparison, comparison with a calibrator, comparison pressure and free-field, and comparison with a precision attenuator. These secondary calibration methods, which are adequate for most industrial applications, are economically attractive and less time consuming.

The electrostatic actuator method for frequency response measurement of working standard microphones is discussed with some pros and cons presented. An example to demonstrate the stability of laboratory standard microphones and the stability of a laboratory calibration system is described.

Appendix A discusses acoustic transfer impedance evaluation, while appendix B contains physical properties of air, which are necessary for microphone calibration.

24.1 Historic References on Condenser Microphones and Calibration 1024
24.2 Theory 1024
    24.2.1 Diaphragm Deflection 1024
    24.2.2 Open-Circuit Voltage and Electrical Transfer Impedance 1024
    24.2.3 Mechanical Response 1025
24.3 Reciprocity Pressure Calibration 1026
    24.3.1 Introduction 1026
    24.3.2 Theoretical Considerations.......... 1026
    24.3.3 Practical Considerations 1027
24.4 Corrections... 1029
    24.4.1 Heat Conduction Correction 1029
    24.4.2 Equivalent Volume 1031
    24.4.3 Capillary Tube Correction 1032
    24.4.4 Cylindrical Couplers and Wave-Motion Correction 1034
    24.4.5 Barometric Pressure Correction ... 1035
    24.4.6 Temperature Correction.............. 1037
    24.4.7 Microphone Sensitivity Equations 1038
    24.4.8 Uncertainty on Pressure Sensitivity Level 1038
24.5 Free-Field Microphone Calibration 1039
24.6 Comparison Methods for Microphone Calibration.................... 1039
    24.6.1 Interchange Microphone Method of Comparison 1039
    24.6.2 Comparison Method with a Calibrator 1040
    24.6.3 Comparison Pressure and Free-Field Calibrations 1042
    24.6.4 Comparison Method with a Precision Attenuator 1042
24.7 Frequency Response Measurement with Electrostatic Actuators 1043
24.8 Overall View on Microphone Calibration. 1043
24.A Acoustic Transfer Impedance Evaluation 1045
24.B Physical Properties of Air....................... 1045
    24.B.1 Density of Humid Air.................. 1045
    24.B.2 Computation of the Speed of Sound in Air.......................... 1046
    24.B.3 Ratio of Specific Heats of Air 1047
    24.B.4 Viscosity and Thermal Diffusivity of Air for Capillary Correction 1047
References 1048

Nomenclature

A'_1	numerical coefficient
$\tilde{A}_1, \tilde{A}_2$	complex quantities
a	the radius of the capillary tube (m)
a_0 to a_{12}	coefficients
$B = (\eta/\rho\alpha_t)^{\frac{1}{2}}$	coefficient
B'_1	numerical coefficient
C	fixed capacitance (F)
C'_1	shunt capacitance (F)
C_p	specific heat capacity at constant pressure (kJ kg^{-1} K^{-1})
C_r	reference capacitance (F)
C_v	specific heat capacity at constant volume (kJ kg^{-1} K^{-1})
C_t	shunt capacitance (F)
c_c	large shunt capacitance (F)
c	sound speed (m/s)
c_0	reference dry air sound speed at $0\,°C$, 101.325 kPa and 400 ppm CO_2 content (m/s)
c_i	input capacitance of the preamplifier (F)
c_{stat}	static capacitance (F)
c_s	stray capacitance (F)
$c_t(t)$	variable capacitance (F)
Δ_c	capillary correction ($\Delta_{c(AB)}$, $\Delta_{c(BC)}$, and $\Delta_{c(CA)}$ are the capillary corrections for the microphone pairs)
D	complex function
D_1	function for computation of R'
D'_1	numerical coefficient
D_2	function for computation of R'
d	diameter of microphone cavity (m)
E	voltage across the microphone (V)
E_0	polarizing voltage (V)
E_v	complex temperature transfer function
e_A, e_B, e_C	open-circuit voltage (V)
e_o	output voltage (V)
e_{oc}	output voltage as function of the varying capacitor c_t (V)
$e_{i(B)}$	sinusoidal insert voltage (V)
$e(t)$	time-varying voltage (V)
e'_B	microphone output voltage with microphone shunting capacitance
e_{Bref}, e_{Aref}	open-circuit voltage and the source driving voltage when the reference microphone is in position
e_{Bx}, e_{Ax}	open-circuit voltage and the source driving voltage when the test microphone is in position
e_1	driving voltage across Z_x (V)
e_2	preamplifier output voltage (V)
f	frequency (Hz)
f_e	$= 1.00062 + p_s(3.14 \times 10^{-8}) + t^2(5.6 \times 10^{-7})$, enhancement factor
f_0	resonance frequency (Hz)
g_1	gain of the preamplifier
H	relative humidity, %
h_0	static deflection of diaphragm (m)
h_a	air gap between the diaphragm and the backplate when the polarizing voltage is applied (m)
$h_s = \frac{p_{sv}}{p_0}\frac{H}{100}$	fractional molar concentration of moisture
h_c	CO_2 content (%)
$h = H/100$	humidity (dimensionless)
ΔH	humidity correction (dB/100%)
Δ_H	complex correction factor
I	input current (A)
i_t	time varying current generated by the microphone (A)
K	wave number of sound in the membrane (m^{-1})
k	$= i\omega Z_x/(\gamma P_s)$
k_0	constant
k_c	$= (-i\omega\rho/\eta)^{1/2}$ complex wave number (m^{-1})
L_r	level reading of the reference microphone (dB)
L_t	level reading of the test microphone (dB)
ℓ	volume to surface ratio of the coupler (m)
ℓ_c	length of the capillary tube (m)
l_0	length of the cavity, i. e. distance between the two diaphragms (m)
M	microphone sensitivity
M_A, M_B, M_C	open-circuit sensitivity of microphone A,B,C (V/Pa)
M_m	mechanical response (m/Pa)
M_e	electrical transfer impedance
M'_A, M'_B, M'_C	modified microphone A, B, C sensitivity (V/Pa)
M_t	open-circuit sensitivity of the test microphone (dB re 1 V/Pa)
M_r	open-circuit sensitivity of the reference microphone (dB re 1 V/Pa)
M_{dB}	microphone sensitivity (dB re 1 V/Pa)
M_{corr}	microphone sensitivity with correction (V/Pa)
M_{com}	combined level sensitivity of microphone and preamplifier (dB)
n	number of capillary tubes

P_s	barometric pressure (Pa)	x_c	mole fraction of carbon dioxide in air
p	alternating pressure inside the cavity (Pa)	Δx	variation in depth of the cavity (m)
p_i	incident sound pressure (Pa)	y_0	initial deflection of diaphragm (m)
p_0	reference pressure = 101.325 kPa	y_r	diaphragm deflection (m)
p_{sv}	saturation water vapor pressure (Pa)	Z_{AB}	acoustic impedance of a cavity with a pair of microphones A and B (N s m^{-5})
p_s	static pressure (during calibration) (Pa)		
ΔP_c	capacitance correction (dB)	Z_{AC}	acoustic impedance of a cavity with a pair of microphones A and C (N s m^{-5})
ΔP_v	microphone sensitivity level pressure correction (dB)		
		Z_{CA}	acoustic impedance of a cavity with a pair of microphones C and A (N s m^{-5})
ΔP	$= (p_s - p_0)$ pressure difference (kPa)		
R	Universal gas constant, 8.314 51 J mol^{-1} K^{-1}. For dry air, 0 °C, 400 ppm CO$_2$ content and 101 325 Pa	Z'_{AB}	acoustic impedance of a cavity with a pair of microphones A and B with assumptions (N s m^{-5})
R'	parameter for computation of heat conduction correction	Z'_{AC}	acoustic impedance of a cavity with a pair of microphones A and C with assumptions (N s m^{-5})
R_p	impedance parallel with R_c and R_i (Ω)		
R_c	high resistance (Ω)	Z'_{CA}	acoustic impedance of a cavity with a pair of microphones C and A with assumptions (N s m^{-5})
R_i	input resistance of the preamplifier (Ω)		
r_a	radius of microphone diaphragm (m)		
r_b	radius of microphone backplate (Ω)	Z_x	electrical impedance (Ω)
r_c	capillary tube radius (m)	Z	$= \rho c S_0^{-1}$
S	parameter for computation of heat conduction correction	$\underline{Z}$	compressibility factor for humid air
		$\underline{Z}_{a,c}$	impedance of an open capillary tube (N s m^{-5})
S_0	cross-sectional area of the cavity (m^2)		
S_w	switch	$\underline{Z}_{a,t}$	complex acoustic wave impedance of an infinite tube (Pa s m^{-3})
T	$= (273.15 + t)$, thermodynamic temperature (K)		
		$\underline{Z}'_{a,12}$	acoustic transfer impedance of a pair of microphones
T_m	membrane tension (N/m)		
T_0	$= 273.15$ K	α_t	thermal diffusivity of the gas media (m^2 s^{-1})
t	temperature (°C)		
ΔT	microphone sensitivity temperature correction (db)	α_1, α_2	attenuator reading
		β_{AB}	voltage ratio with microphone pair A and B
Δt	$(T - 273.15)$, temperature difference (K)		
u	volume velocity (m^3/s)	β_{BC}	voltage ratio with microphone pair B and C
V	equivalent volumes of the cavity (m^3)		
V_{eA}, V_{eB}, V_{eC}	equivalent volumes of the microphones A, B, C (m^3)	β_{CA}	voltage ratio with microphone pair C and A
V_{AB}	$= (V + V_{eA} + V_{eB})$ (m^3)	$\beta_0, \beta_1, \beta_2$	attenuator reading
V_{BC}	$= (V + V_{eB} + V_{eC})$ (m^3)	β	voltage ratio reading
V_{CA}	$= (V + V_{eC} + V_{eA})$ (m^3)	γ	ratio of specific heats
ΔV_0	small change of volume (m^3)	γ_0	ratio of specific heats of dry air at 0 °C, 101.325 kPa and 314 ppm CO$_2$ content (dimensionless)
$V'_{AB}, V'_{BC}, V'_{CA}$	equivalent volume including capillary tubes (m^3)		
W	wave-motion correction (dB)	η	viscosity of air (Pa s)
X'	parameter for computation of heat conduction correction	θ	phase angle (deg)
		λ	wavelength (m)
$x = f/f_0$	frequency normalized by the resonance frequency f_0 of the individual microphone	ξ	complex propagation coefficient (m^{-1})
		ρ	density of the gas enclosed (kg/m^3)
$x_w = \frac{p_{sv}}{p_s} f_e h$	mole fraction of water vapor in air	ρ_0	$= 1.29295$ (dry air, 0 °C, 101.325 kPa and 400 ppm CO$_2$ content) (kg/m^3)

σ_M	membrane surface mass density (kg/m^2)	$\Delta_{c(CA)}$	capillary tube correction factor for microphones C and A
Δ	compensation for dispersion (dB)	ω	$= 2\pi f$, angular frequency (rad/s)
$\Delta_{c(AB)}$	capillary tube correction factor for microphones A and B	$J_0()$ and $J_1()$	zero- and first-order cylindrical Bessel functions of the first kind for complex argument
$\Delta_{c(BC)}$	capillary tube correction factor for microphones B and C	i	$\sqrt{-1}$

24.1 Historic References on Condenser Microphones and Calibration

In 1917, *Wente* [24.1] described a microphone close to what we find today as the modern condenser microphone. The theory of absolute pressure calibration for condenser microphones and some of the modern analyses and implementations including the early implementation of the Western Electric 640AA condenser microphone have been described [24.2–30]. In 1995, a comprehensive discussion on the history of condenser microphone development and calibration was published [24.29].

24.2 Theory

The theory of condenser microphone operation has been investigated [24.1–30], and a brief summary is discussed in the following.

24.2.1 Diaphragm Deflection

The basic condenser microphone consists of a stretched diaphragm (Fig. 24.1) over a backplate that is polarized with a voltage E_0, usually 200 V. Due to the small air gap h_a (approximately 20 μm for a 25 mm diameter microphone) between the backplate and the thin (approximately 1–5 μm) metallic diaphragm, the latter is deflected by electrostatic charge of the backplate. With the diameter of the planar (flat) backplate ($2r_b$) smaller than that of the diameter ($2r_a$) of the diaphragm, the relatively complex initial deflection (y_0) of the diaphragm shape was analyzed by *Hawley* et al. ([24.29], Chap. 2). It has been shown by *Fletcher* and *Thwaites* [24.30] that it is an advantage to modify the backplate to have a parabolic profile to accommodate the curvature of the diaphragm and, according to their analysis, this eliminates distortion and increases the microphone sensitivity. Until this idea is further developed, the following will concentrate on condenser microphones with planar backplates.

24.2.2 Open-Circuit Voltage and Electrical Transfer Impedance

For functional implementation of a condenser microphone, *Zuckerwar* gave a detailed analysis ([24.29], Chap. 3), and the electrical circuit is shown in Fig. 24.2. A microphone is represented by a variable capacitance c_t with a static value of c_{stat}, and a stray capacitance of c_s. The polarizing voltage E_0 is applied to the microphone backplate via a high resistance R_c. The preamplifier is represented by an input impedance consisting of a capacitance c_i and a resistance R_i; c_c is a large blocking capacitor to prevent the polarizing voltage from overloading the preamplifier.

When a time-varying sound pressure is applied to the diaphragm, the diaphragm vibrates and the microphone capacitance c_t and the voltage E across the microphone

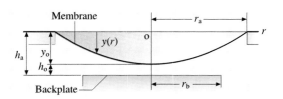

Fig. 24.1 Schematic diagram of a condenser microphone

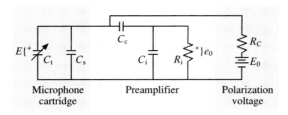

Fig. 24.2 Electrical circuit of a condenser microphone

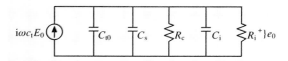

Fig. 24.3 Equivalent small-signal circuit

are represented by their corresponding static and time-varying components

$$c_t = c_{stat} + c_t(t), \quad (24.1)$$
$$E = E_0 + e(t). \quad (24.2)$$

The time-varying current i_t generated by the microphone is

$$i_t = E_0 \frac{dc_t}{dt} + c_{stat} \frac{de}{dt}. \quad (24.3)$$

For small electrical signals, Fig. 24.2 can be represented by an equivalent circuit shown in Fig. 24.3 and for small time-varying (ω) pressure at the diaphragm, the microphone can be represented by a fixed capacitance C and a current source i_t, and the output voltage e_o is:

$$e_o = E_0 \frac{c_t}{C} \frac{i\omega C R_p}{1 + i\omega C R_p}, \quad (24.4)$$

where R_p is the impedance of R_c in parallel with R_i, and C is the sum of c_{stat}, c_i and c_s. The microphone open-circuit voltage e_{oc} as a function of the varying capacitance c_t is

$$e_{oc} = E_0 \frac{c_t}{c_{stat}} \left(1 + \frac{c_i + c_s}{c_{stat}}\right)^{-1}. \quad (24.5)$$

Based on an approximate equation derived by *Hawley* ([24.29], Chap. 2, Eq. 5.8) for the capacitance of a microphone under the influence of the polarizing voltage, and assuming that the diaphragm can be modeled as a piston-like displacement, *Zuckerwar* ([24.29], Chap. 3, Eq. 2.23) has shown that the electrical transfer impedance M_e is

$$M_e = \frac{E_0}{h_0} \left(1 + \frac{r_b^2}{2r_a^2}\right) \left(1 + \frac{c_i + c_s}{c_{stat}}\right)^{-1}, \quad (24.6)$$

where r_a and r_b are the radii of the diaphragm and the backplate, respectively.

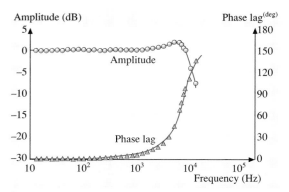

Fig. 24.4 Sample amplitude and phase response of a one-inch microphone (Brüel Kjær 4146) *Solid lines*: theoretical. *Symbols*: experimental (after *Zuckerwar* [24.29], Chap. 3)

Both *Hawley* et al. ([24.29], Chap. 2) and *Zuckerwar* ([24.29], Chap. 3) arrived at the conclusion that the optimum backplate size is

$$r_a/r_b = (2/3)^{1/2} = 0.8165. \quad (24.7)$$

24.2.3 Mechanical Response

The membrane motion is coupled to the air between the diaphragm and the backplate. The system exhibits damping due to the damping holes in the backplate. *Zuckerwar* ([24.29], Chap. 3, Eq. 3.51) arrived at a solution for the mechanical response as

$$M_m = \frac{[y(r)]}{p_i} = \frac{1}{TK^2} \frac{J_2(K_a)}{J_0(K_a) + D}, \quad (24.8)$$

where K is the wave number of sound in the membrane, given by

$$K = \omega \left(\frac{\sigma_M}{T_m}\right)^{1/2}, \quad (24.9)$$

ω is angular frequency, σ_M is the membrane surface mass density, and T_m is the membrane tension. D is a complex function ([24.29], Chap. 3, Eq. 3.50). With (24.8) the amplitude and phase response can be realized; some examples ([24.29], Chap. 3, Fig. 3.8) are shown in Fig. 24.4.

24.3 Reciprocity Pressure Calibration

24.3.1 Introduction

High precision in acoustical measurements is needed even though the human ear cannot discern a change in sound level much smaller than 1 dB. For example, to certify aircraft for regulations, it is necessary to obtain measurements with a known uncertainty, usually 0.1–0.2 dB (Unless otherwise stated, the uncertainties referred to in this chapter, may be taken as better than two standard deviations. However, in most cases, the quoted statements on uncertainty are unclear in the source documents. Further complications are introduced when the original data are converted to decibels. For a more rigorous treatment of uncertainties, one should consult [24.31]). It is reasonable to assume that laboratory standard microphones (working standards) should be approximately an order of magnitude better than this, say with an uncertainty ranging from 0.01 to better than 0.1 dB. For research and development purposes, including the study of microphones and monitoring their stability, the primary calibration of condenser microphones should have an uncertainty no larger than several thousandths of a decibel.

One of the most accurate techniques to calibrate a primary standard condenser microphone is the absolute method of reciprocity pressure calibration (sometimes called the coupler method), which currently has a best uncertainty of less than 0.01 dB at 250 Hz, under a fully controlled environment under reference conditions [24.20]. If the calibration were to be performed on a laboratory bench without any control on the environment, such as barometric pressure, temperature and humidity, the estimated uncertainty of the coupler method is approximately 0.05 dB at lower and middle frequencies.

The theory of absolute pressure calibration methods for condenser microphones and some of the relatively modern implementations have been described in [24.1–29].

24.3.2 Theoretical Considerations

The reciprocity method measures the product of the sensitivities of each pair of a set of three microphones in terms of related electrical and mechanical quantities, from which the absolute sensitivity of each microphone can be deduced.

The acoustic impedance Z_{AB}, of a cavity with a pair of microphones A and B, as shown in Fig. 24.5 is

$$Z_{AB} = \gamma P_s [i\omega(V + V_{eA} + V_{eB})]^{-1}, \quad (24.10)$$

where γ is the ratio of specific heats of the gas in the coupler, P_s is the barometric pressure, ω equals $2\pi f$, where f is the frequency of the driving sinusoidal signal, and V, V_{eA} and V_{eB} are the equivalent volumes of the cavity and of microphone A and microphone B, respectively. Equivalent volumes, as the name implies, are volumetric measures that represent the physical volumes of the cavity and of the coupled microphones modified by thermal and finite impedance factors.

The relationship between the alternating pressure p inside the cavity and the volume velocity u is

$$p = uZ_{AB} = u\gamma P_s (i\omega V_{AB})^{-1}, \quad (24.11)$$

where the equivalent volume of the cavity with microphones A and B is

$$V_{AB} = (V + V_{eA} + V_{eB}). \quad (24.12)$$

The reciprocity theorem states that, in a passive linear reversible (reciprocal) transducer, the ratio of the volume velocity u in the cavity to the input current I when used as a sound source is equal to the ratio of the open-circuit voltage e_A across the electrical terminals to the sound pressure p acting on the diaphragm when used as a receiver. The theorem enables one to write the

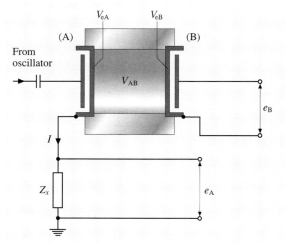

Fig. 24.5 Calibrating condenser microphones by the reciprocity method (after *Wong*, [24.29], Chap. 4)

following equations:

$$M_A = u/I = e_A/p, \quad (24.13)$$
$$M_B = e_B/p, \quad (24.14)$$
$$M_C = e_C/p, \quad (24.15)$$

where M_A, M_B and M_C are the open-circuit sensitivities of the microphones.

Multiplying (24.13) and (24.14), and substituting p from (24.11) gives

$$M_A M_B = e_B i\omega V_{AB}/(I\gamma P_s). \quad (24.16)$$

Since

$$I = e_A/Z_x \quad (24.17)$$

and Z_x is an electrical impedance, substitution of (24.17) into (24.16) gives

$$M_A M_B = V_{AB} k \beta_{AB}, \quad (24.18)$$

where

$$k = i\omega Z_x/(\gamma P_s) \quad (24.19)$$

and

$$\beta_{AB} = e_B/e_A. \quad (24.20)$$

By using the microphones A, B and C, configured in pairs to give equivalent volumes V_{AB}, V_{BC} and V_{CA}, and by measuring the corresponding voltage ratios β_{AB}, β_{BC}, and β_{CA}, respectively, the sensitivity product of three pairs of microphones can be obtained:

$$M_A M_B = k\beta_{AB} V_{AB},$$
$$M_B M_C = k\beta_{BC} V_{BC},$$
$$M_C M_A = k\beta_{CA} V_{CA}. \quad (24.21)$$

From (24.21), the open-circuit sensitivities of the individual microphones can be derived

$$M_A = [(V_{CA} V_{AB}/V_{BC})(\beta_{CA}\beta_{AB}/\beta_{BC})k]^{1/2},$$
$$M_B = [(V_{AB} V_{BC}/V_{CA})(\beta_{AB}\beta_{BC}/\beta_{CA})k]^{1/2},$$
$$M_C = [(V_{BC} V_{CA}/V_{AB})(\beta_{BC}\beta_{CA}/\beta_{AB})k]^{1/2}. \quad (24.22)$$

It can be seen from (24.22) that the sensitivity of each microphone can be deduced by measuring the voltage ratios if the equivalent volumes for each of the three pairs of microphones and the numerical value of the constant k are known.

Similarly, the open-circuit sensitivity shown in (24.22) can be expressed in terms of the acoustic impedance of the cavity (24.10):

$$M_A = \{[Z_{BC}/(Z_{CA} Z_{AB})](\beta_{CA}\beta_{AB}/\beta_{BC})Z_x\}^{1/2},$$
$$M_B = \{[Z_{CA}/(Z_{AB} Z_{BC})](\beta_{AB}\beta_{BC}/\beta_{CA})Z_x\}^{1/2},$$
$$M_C = \{[Z_{AB}/(Z_{BC} Z_{CA})](\beta_{BC}\beta_{CA}/\beta_{AB})Z_x\}^{1/2}, \quad (24.23)$$

where Z_{AB}, Z_{BC} and Z_{CA} are the acoustic impedances of the cavities.

An equation for calculating the acoustic impedance of a coupler is given in Appendix 24.A.

24.3.3 Practical Considerations

The general arrangement for implementation of pressure reciprocity calibration of condenser microphones is shown in Fig. 24.6. To simplify electrical measurements, it is necessary that the driving microphone A should be isolated electrically from the receiving microphone B. The preamplifier has a gain of g_1 and an input capacitance of C_i. Since the receiving microphone is shunted by C_i the microphone output voltage is modified to e'_B. When the switch S_w is at position 2, an insert voltage can be applied in order to obtain the open-circuit voltage. For both microphones, a polarization voltage (usually 200 V) is applied via a high resistance. A small shunt capacitor C_t, which closely approximates C_i, modifies the driving voltage e_A to e_1.

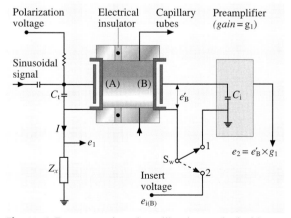

Fig. 24.6 Pressure reciprocity calibration method with provision for insert voltage (after *Wong* [24.29], Chap. 4)

The cavity usually has two capillary tubes for atmospheric pressure equalization and for the insertion of gases other than air. Hence the sensitivities of the microphones shown in (24.22) are modified to

$$M'_A = [(V'_{CA}V'_{AB}/V'_{BC})(\beta_{CA}\beta_{AB}/\beta_{BC})k]^{1/2},$$
$$M'_B = [(V'_{AB}V'_{BC}/V'_{CA})(\beta_{AB}\beta_{BC}/\beta_{CA})k]^{1/2},$$
$$M'_C = [(V'_{BC}V'_{CA}/V'_{AB})(\beta_{BC}\beta_{CA}/\beta_{AB})k]^{1/2},$$
(24.24)

where the cavity equivalent volume is marked with a prime to indicate the inclusion of the equivalent volume of the capillary tubes. The voltage ratio e_B/e_A becomes e_2/e_1 and it is represented by β_{AB} in (24.24). The next section discusses the parameters shown in (24.24).

Open-Circuit Sensitivity

The open-circuit sensitivity of the microphones is obtained with the insert voltage method [24.8]. From Fig. 24.6, the measurement procedure is as follows.

With the switch S_w at position 1, measure the output voltage e_2 from the preamplifier. Next, remove the drive sinusoidal signal from microphone A, and with the switch at position 2, apply a sinusoidal insert voltage $e_{i(B)}$ to drive the diaphragm of microphone B. The magnitude of $e_{i(B)}$ is adjusted such that the voltage e_2 is repeated at the output of the preamplifier. As the circuit shows, the insert voltage is identical to the desired open-circuit voltage of microphone B if the loading of the preamplifier's input impedance were to be removed. The open-circuit sensitivity of the microphones can be obtained from (24.24) with the voltage ratio $e_{i(B)}/e_1$, for each pair of microphones, substituted into (24.20). Alternatively, the insert voltage can be computed from an injected voltage when the switch is at position 2, and then measuring the corresponding output voltage e_2 to obtain the overall gain that enables the computation of the proper insert voltage.

Electrical Circuit for Measurement

The basic electrical measurement for pressure reciprocity calibration is that of the amplitude ratio of two sinusoidal signals (e_1 and e_2 in Fig. 24.6) that may not be in phase. The uncertainty of direct measurement of each of the two signals is limited by the uncertainty of the alternating-current (AC) voltmeter. Several practical circuits for this purpose have been described [24.27]. A higher accuracy can be achieved with voltage ratio measurements. Two examples are described below to give some insight into the diversity of the methods of implementation.

Rectified Signal Null Method

In this method, used in commercial reciprocity calibration apparatus, one of the AC signals is applied to both inputs (Fig. 24.7). With the attenuator removed but with the bypass connection in place, the signal is balanced at the amplifier–rectifier circuit, such that a null reading is indicated by the null-meter. With the two sinusoidal signals connected to the inputs, the attenuator is adjusted for a null reading. The attenuator reading gives the voltage ratio. The uncertainty of the voltage ratio is theoretically dependent on the uncertainty of the precision attenuator. However, the linearity and matching of the amplifiers, amplifier–rectifiers (with RC time constants, not shown) together with the amplitude difference of the two signals, also affect the uncertainty of the circuit. Even with good design, the uncertainty of the voltage ratio measured is approximately 0.003 dB. With some modifications, the user of the apparatus may reduce the uncertainty of measurements by exchanging the signals at the input, and by shifting the attenuator from positions A and B to positions C and D, then taking the mean of the ratio readings for the two attenuator positions.

Variations on this method have been used; for example, the insert voltage (Fig. 24.6) can be derived from a precision attenuator driven by the same source as the driving microphone [24.25].

Interchange Reference Method

The amplitude ratio R and the phase θ of two sinusoidal signals can be measured very precisely with the interchange reference method [24.32] (Fig. 24.8). Signals e_1 and e_2 are presented to the differential inputs of a lock-in amplifier via the switch S_w and the attenuators. When the switch is in position 1, e_1 is selected to be the reference signal for the lock-in amplifier. The attenuator readings β_1 and α_1 are adjusted for a null condition. Using the vector diagram in Fig. 24.9,

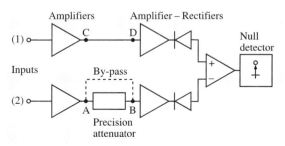

Fig. 24.7 Rectified signal null circuit (after *Wong* [24.29], Chap. 4)

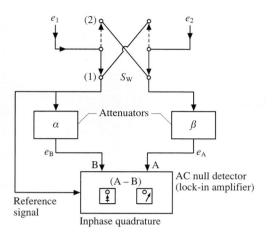

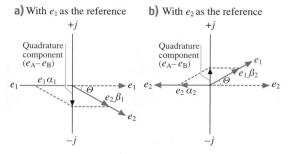

Fig. 24.9 Vector diagram for the in-phase null condition (after *Wong* [24.29], Chap. 4)

Fig. 24.8 Interchange reference method (after *Wong* [24.29], Chap. 4)

we have

$$e_2\beta_1\cos\theta - e_1\alpha_1 = 0. \quad (24.25)$$

Similarly, when e_2 is selected as the reference by the switch at position 2, the null condition with attenuator readings β_2 and α_2 provides the following:

$$e_1\beta_2\cos\theta - e_2\alpha_2 = 0. \quad (24.26)$$

From (24.5) and (24.6), the general equations for the amplitude ratio $R = e_2/e_1$ and $\cos\theta$ are found

$$R = [(\alpha_1/\alpha_2)(\beta_2/\beta_1)]^{1/2}, \quad (24.27)$$

$$\cos\theta = [(\alpha_1\alpha_2)/(\beta_1\beta_2)]^{1/2}. \quad (24.28)$$

Although for symmetry, Fig. 24.8 shows two attenuators, in practice only one attenuator is required. For the condition in Fig. 24.9, where $e_2 > e_1$, $\alpha_1 = \beta_2 = 1$.

The uncertainty of the attenuator that can be a seven-decade ratio transformer (inductive voltage dividers) is of the order of 0.1 ppm. The uncertainty of the measured amplitude ratio is of the order of 1 ppm. Other circuits employing inductive voltage dividers for voltage ratio measurements are given in [24.17, 23].

Reference Impedance Selection

The reference impedance Z_x shown in Fig. 24.6 needs to be carefully selected. The nature of the impedance, either purely resistive or purely capacitive, or a combination of electrical impedances, dictates the mode of evaluation of the factor k, which is required for the computation of microphone sensitivity in (24.24). If Z_x is a resistance, the angular frequency ω, shown in (24.19), has to be measured. When the impedance is a capacitance C_r the term $Z_x = 1/(i\omega C_r)$, and the frequency terms are canceled. The numerical value of Z_x is chosen such that the magnitudes of e_1 and e_2 are nearly equal in order to enhance the measurement of voltage ratios using null measurement techniques. For precise measurement of Z_x, the capacitance of the connecting cables to the impedance should be taken into consideration.

24.4 Corrections

24.4.1 Heat Conduction Correction

The alternating sound pressure induces compression and expansion in the gas medium inside the closed cavity. At sufficiently low frequencies, the induced temperature changes occur isothermally, and the heat exchange at the walls of the cavity has to be included when computing the acoustic impedance of the cavity. As the frequency increases, heat conduction between the gas and the walls decreases. Gradually, isothermal conditions change to adiabatic conditions, at which virtually no heat is exchanged with the walls.

Classical analytical approaches to calculate heat conduction in cavities are given by *Ballantine* [24.2] and *Daniels* [24.33]. Correction factors for heat conduction were obtained by *Biagi* and *Cook* [24.34]. Theoretical and experimental data, based on thermal diffusion, on the acoustic impedance of the cavity, are given by *Ger-*

Table 24.1 Tabulated values of E_v for the heat conduction correction (ANSI S1.15: 2005, and IEC 61094-2:1992-03)

Real part of E_v			X'	Imaginary part of E_v		
$R' = 0.2$	$R' = 0.5$	$R' = 1$		$R' = 0.2$	$R' = 0.5$	$R' = 1$
0.72127	0.71996	0.72003	1.0	0.24038	0.22323	0.22146
0.80092	0.80122	0.80128	2.0	0.17722	0.16986	0.16885
0.83727	0.83751	0.83754	3.0	0.14818	0.14304	0.14236
0.85907	0.85920	0.85922	4.0	0.13003	0.12614	0.12563
0.87393	0.87402	0.87403	5.0	0.11732	0.11421	0.11380
0.89343	0.89348	0.89349	7.0	0.10030	0.09807	0.09777
0.91082	0.91086	0.91086	10.0	0.08477	0.08321	0.08300
0.93693	0.93694	0.93694	20.0	0.06086	0.06007	0.05997
0.94850	0.94851	0.94851	30.0	0.05002	0.04950	0.04942
0.95540	0.95541	0.95541	40.0	0.04349	0.04310	0.04304
0.96358	0.96359	0.96359	60.0	0.03568	0.03541	0.03538
0.96846	0.96846	0.96846	80.0	0.03098	0.03078	0.03076
0.97179	0.97179	0.97179	100.0	0.02776	0.02761	0.02758
0.98005	0.98005	0.98005	200.0	0.01972	0.01964	0.01963
0.98590	0.98590	0.98590	400.0	0.01399	0.01395	0.01395
0.99003	0.99003	0.99003	800.0	0.00992	0.00990	0.00989

The numerical values given are considered accurate to 0.00001

ber [24.35]. *Ballagh* [24.36] took into consideration heat loss effects due to the microphone diaphragms at the ends of the cavity. *Jarvis* [24.37] pointed out the need to include the impedance of the driver microphone during computation of the heat conduction correction. It is difficult to decide accurately at which middle frequencies the isothermal–adiabatic transition occurs, and which portions of the cavity to include in the impedance of the microphone. The exact nature of this transition depends upon the frequency of the calibration and the dimensions of the coupler. The sound pressure generated by the transmitter microphone, i.e. a constant-volume displacement source, will change accordingly. The effect can be considered as an apparent increase in the coupler volume expressed by a complex correction factor Δ_H to the geometrical volume V in Sect. 24.A or to the cross-sectional area S_0 defined in $Z = \rho c S_0^{-1}$ in Sect. 24.B.

The heat conduction correction factor is given by:

$$\Delta_H = \frac{\gamma}{1 + (\gamma - 1)E_v}, \qquad (24.29)$$

where E_v is the complex temperature transfer function defined as the ratio of the space average of the sinusoidal temperature variation associated with the sound pressure to the sinusoidal temperature variation that would be generated if the walls of the coupler were perfectly nonconducting (24.30), and γ is the ratio of specific heats of the gas inside the coupler. Tabulated values for E_v can be found in Table 24.1 [24.35] as a function of the parameters R' and X', where R' is the length-to-diameter ratio of the coupler, $X' = f\ell^2/(\gamma\alpha_t)$, f is the frequency, ℓ is the volume-to-surface ratio of the coupler and α_t is the thermal diffusivity of the gas enclosed.

For finite cylindrical couplers as described in [24.38], annex C, the approximation described below for the complex quantity E_v will be satisfactory

$$E_v = 1 - S + D_1 S^2 + (3/4)\sqrt{\pi} D_2 S^3, \qquad (24.30)$$

where

$$S = \left(-i\frac{1}{2\pi X'}\right)^{1/2} = \frac{1-i}{2\sqrt{\pi X'}},$$

$$D_1 = \frac{\pi R'^2 + 8R'}{\pi(2R'+1)^2},$$

$$D_2 = \frac{R'^3 - 6R'^2}{3\sqrt{\pi}(2R'+1)^3}.$$

The modulus of E_v, as calculated from (24.30), is accurate to 0.01% within the range $0.125 < R' < 8$ and for $X' > 5$. The first two terms in (24.30) constitute an approximation that may be used for couplers that are not right circular cylinders.

24.4.2 Equivalent Volume

The equivalent volume without heat conduction correction, as defined in (24.12), consists of the volume of the cavity plus the equivalent volumes of the microphones. The volume of the cavity is usually measured with traditional methods offered by mechanical metrology. The equivalent volume of a standard microphone, which includes a small volume occupied by a screw thread and an effective volume caused by the non-rigidity of the microphone diaphragm, is very difficult to measure precisely with conventional mechanical means. Various methods for the measurement of the equivalent volume of microphones are available [24.39]. Optical techniques, such as the traveling microscope, have been used to measure the cavity length [24.25]. For higher precision, interferometric procedures can be applied.

An acoustical resonance method can be used to determine the front cavity volume of a standard microphone. A three-port coupler accommodates two smaller microphones (e.g. 1/4 inch microphones), and the standard microphone. The smaller microphones are used as a driver–receiver combination to give the electrical transfer impedance including the cavity volume with the standard microphone in place. The standard microphone is replaced with a movable plunger; the displacement of the plunger is calibrated to indicate volumetric changes covering the range of actual microphone volumes. The equivalent front volume of the test microphone can be deduced from the calibrated position of the plunger at which the two smaller microphones indicate the same electrical transfer impedance as when the test microphone is in place. See [24.12] for a similar three-port procedure and [24.16] for precise measurement of volumetric changes.

Wong and *Embleton* [24.20] developed a very precise method that assesses the equivalent volume with the heat conduction correction and includes the impedance of the capillary tubes for the coupler and two microphones in one procedure. The length of the cavity can be varied with a removable spacer to replace the electrical insulation, and the voltage ratio $\beta = e_2/e_1$, as shown in Fig. 24.6, is assessed at the frequency of interest with the interchange reference method [24.32]. From (24.10) and (24.11), the equivalent volume of the cavity can be expressed in terms of the volume velocity u and the alternating pressure p

$$V = (\gamma P_s/i\omega)(u/p) \,. \qquad (24.31)$$

Since the voltage ratio β is inversely proportional to the pressure p inside the cavity (an increase of sig-

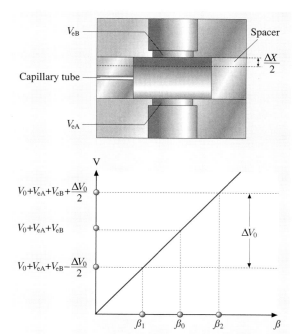

Fig. 24.10 Arrangement for determining the equivalent volume, including corrections for capillary tubes, and heat conduction (after [24.20])

nal magnitude requires a decrease in ratio reading to maintain a null condition), (24.31) can be modified to

$$V = k_0 \beta \,, \qquad (24.32)$$

where k_0 is a constant.

Three readings, β_0, β_1 and β_2 corresponding to cavity volume $(V_0 + V_{eA} + V_{eB})$, volume $(V_0 + V_{eA} + V_{eB} - \Delta V_0/2)$ and volume $(V_0 + V_{eA} + V_{eB} + \Delta V_0/2)$, are obtained. Here ΔV_0 is a small variation in volume corresponding to a small variation Δx in spacer thickness. Figure 24.10 illustrates the relationship between changes in volume and the ratio readings. The equivalent volume of the cavity is

$$V = (V_0 + V_{eA} + V_{eB}) = \Delta V_0 [\beta_0/(\beta_2 - \beta_1)] \,. \qquad (24.33)$$

From Fig. 24.10,

$$V = [\beta_0/(\beta_2 - \beta_1)]d^2 \Delta x \pi/4 \,, \qquad (24.34)$$

where d and Δx are the diameter and the variation in depth of the cavity.

For highly accurate measurements, the spacers are implemented with optical flats whose thicknesses are measured by interferometry. The cavity is manufactured with precision optical fabrication. The uncertainty in the equivalent volume of such a cavity is estimated to be approximately 20 ppm. For less stringent applications the three optical flats can be replaced with adaptor rings [24.25] or with three cavities of known volumes. The equivalent volume of the cavity can then be determined by interpolation. This method has been commercialized with a similar procedure that relied on two couplers (long and short couplers) in the Brüel and Kjær 5998 reciprocity apparatus. With the sensitivities of the microphone pairs obtained with the two couplers, the microphone cavity depths are then adjusted numerically until the final microphone sensitivities obtained with the two couplers are nearly identical or very close.

24.4.3 Capillary Tube Correction

For microphone calibration, the preferred expression [24.38, 40] adopted for the acoustic input impedance of an open capillary tube is:

$$Z_c = Z_t \tanh(\zeta L_c), \tag{24.35}$$

where Z_t is the complex characteristic impedance of a capillary tube with infinite length, L_c is the length of the capillary tube, ζ is the complex propagation coefficient for plane waves in a tube, and

$$Z_t = (\tilde{A}_1/\tilde{A}_2)^{1/2}, \tag{24.36}$$

$$\zeta = (\tilde{A}_1 \tilde{A}_2)^{1/2}. \tag{24.37}$$

$\tilde{A}_1$ and $\tilde{A}_2$ are complex quantities; $\tilde{A}_1$ represents viscosity loss at the walls and inertia of the gas medium, and $\tilde{A}_2$ represents heat conduction loss at the walls and the compliance of the gas medium, where

$$\tilde{A}_1 = i\frac{\omega \rho}{\pi r_c^2}\left(1 - \frac{2J_1(k_c r_c)}{k_c r_c J_0(k_c r_c)}\right)^{-1}, \tag{24.38}$$

$$\tilde{A}_2 = i\omega \frac{\pi r_c^2}{\rho c^2}\left(1 + \frac{2}{Bk_c r_c}(\gamma - 1)\frac{j_1(Bk_c r_c)}{j_0(Bk_c r_c)}\right), \tag{24.39}$$

J_0 and J_1 are, respectively, the zeroth- and first-order cylindrical Bessel functions of the first kind for complex arguments, r_c is tube radius (m), $k_c = (-i\omega\rho/\eta)^{1/2}$ is the complex wave number (m^{-1}), $B = [\eta/(\rho\alpha_t)]^{1/2}$, α_t is the thermal diffusivity of the gas media (m^2/s).

Table 24.2 Real part of Z_c in GPa s/m^3 (ANSI S1.15: 2005, and IEC 61094-2:1992-03), tube dimensions in millimeters

$\ell_c = 50$			Frequency (Hz)	$\ell_c = 100$		
$r_c = 0.1667$	$r_c = 0.20$	$r_c = 0.25$		$r_c = 0.1667$	$r_c = 0.20$	$r_c = 0.25$
3.018	1.457	0.597	20	6.041	2.916	1.195
3.019	1.457	0.597	25	6.044	2.919	1.196
3.020	1.458	0.597	31.5	6.049	2.922	1.198
3.022	1.459	0.598	40	6.059	2.928	1.201
3.025	1.460	0.599	50	6.072	2.937	1.205
3.029	1.463	0.600	63	6.094	2.951	1.212
3.036	1.467	0.602	80	6.130	2.975	1.225
3.047	1.473	0.605	100	6.185	3.011	1.243
3.063	1.482	0.610	125	6.270	3.069	1.272
3.093	1.499	0.620	160	6.422	3.173	1.326
3.137	1.524	0.633	200	6.643	3.331	1.408
3.207	1.564	0.654	250	6.989	3.595	1.550
3.326	1.631	0.689	315	7.542	4.066	1.817
3.534	1.750	0.750	400	8.353	4.944	2.381
3.871	1.943	0.849	500	9.068	6.288	3.535
4.504	2.314	1.034	630	8.670	7.336	5.631
5.807	3.113	1.435	800	6.375	5.311	4.375
8.332	4.890	2.378	1000	4.353	3.005	1.925
12.120	9.008	5.385	1250	3.545	2.127	1.146

The data used for this table are: $c = 345.7$ m/s, $\gamma = 1.40$, $\rho = 1.186$ kg/m^3, $\eta = 18.3 \times 10^{-6}$ Pa s, $\alpha_t = 21 \times 10^{-6}$ m^2/s
The values given in this table are valid at the reference environmental conditions: 23 °C, 101.325 kPa and 50% RH

Table 24.2 (continued)

$\ell_C = 50$			Frequency	$\ell_C = 100$		
$r_c = 0.1667$	$r_c = 0.20$	$r_c = 0.25$	(Hz)	$r_c = 0.1667$	$r_c = 0.20$	$r_c = 0.25$
9.191	7.926	6.740	1600	4.171	2.410	1.196
4.326	3.021	1.951	2000	6.325	4.409	2.527
2.694	1.637	0.893	2500	4.979	3.717	2.768
2.807	1.580	0.783	3150	4.411	2.661	1.392
5.923	3.536	1.749	4000	5.238	4.019	3.075
5.946	4.825	3.903	5000	5.059	3.262	1.770
3.306	1.939	1.011	6300	4.578	2.920	1.672
6.571	5.375	4.137	8000	4.695	3.034	1.749
4.184	2.465	1.258	10 000	4.977	3.363	1.952
3.902	2.539	1.540	12 500	4.760	3.331	2.271
4.043	2.590	1.534	16 000	4.753	3.263	2.137
4.535	2.813	1.517	20 000	4.844	3.324	2.023
0.097	0.074	0.049	20	0.096	0.114	0.090
0.122	0.092	0.061	25	0.120	0.143	0.112
0.153	0.116	0.077	31.5	0.151	0.180	0.141
0.195	0.147	0.098	40	0.191	0.228	0.180
0.244	0.184	0.123	50	0.238	0.285	0.225
0.307	0.232	0.155	63	0.299	0.359	0.283
0.390	0.295	0.197	80	0.376	0.455	0.360
0.488	0.369	0.246	100	0.465	0.569	0.452
0.610	0.461	0.308	125	0.569	0.710	0.567
0.782	0.592	0.396	160	0.701	0.905	0.731
0.980	0.743	0.496	200	0.824	1.123	0.923
1.228	0.933	0.623	250	0.916	1.380	1.170
1.556	1.186	0.792	315	0.888	1.664	1.500
1.990	1.527	1.021	400	0.479	1.842	1.922
2.511	1.948	1.306	500	−0.684	1.11	2.200
3.189	2.532	1.711	630	−2.739	−0.777	0.926
3.987	3.353	2.325	800	−3.890	−3.152	−2.510
4.280	4.213	3.186	1000	−3.031	−2.595	−2.130
1.338	3.162	3.730	1250	−1.382	−1.157	−0.944
−5.333	−4.384	−3.281	1600	0.429	0.456	0.282
−4.500	−3.768	−2.956	2000	0.260	0.971	1.221
−1.996	−1.663	−1.280	2500	−1.702	−1.552	−1.344
0.491	0.244	0.051	3150	0.205	0.199	0.053
2.428	2.283	1.692	4000	−1.074	−0.864	−0.524
−2.803	−2.434	−1.954	5000	0.208	0.438	0.406
0.186	−0.037	−0.190	6300	−0.070	−0.095	−0.219
−1.245	−0.607	0.209	8000	−0.041	−0.027	−0.138
0.872	0.643	0.336	10 000	−0.056	0.152	0.212
−0.542	−0.699	−0.764	12 500	−0.281	−0.295	−0.281
−0.210	−0.399	−0.532	16 000	−0.174	−0.187	−0.228
0.430	0.349	0.142	20 000	−0.109	−0.001	0.035

The data used for this table are: $c = 345.7\,\text{m/s}$, $\gamma = 1.40$, $\rho = 1.186\,\text{kg/m}^3$, $\eta = 18.3 \times 10^{-6}\,\text{Pa s}$, $\alpha_t = 21 \times 10^{-6}\,\text{m}^2/\text{s}$
The values given in this table are valid at the reference environmental conditions: 23 °C, 101.325 kPa and 50% RH

For n capillary tubes the correction factors $\Delta_{c(AB)}$, $\Delta_{c(BC)}$ and $\Delta_{c(CA)}$ for the respective cavity impedances in (24.23) are

$$\Delta_{c(AB)} = 1 + n(Z_{AB}/Z_c),$$
$$\Delta_{c(BC)} = 1 + n(Z_{BC}/Z_c),$$
$$\Delta_{c(CA)} = 1 + n(Z_{CA}/Z_c). \quad (24.40)$$

Equation (24.40) assumes that the heat conduction correction (Sect. 24.4.1) has been applied.

At a temperature of 23 °C, a relative humidity of 50% and a static pressure of 101.325 kPa, for a typical range of parameters and frequencies, the real and imaginary parts of Z_c are as shown in Table 24.2, respectively. The computation of the air density ρ, sound speed c in humid air and other physical quantities are discussed in Sect. 24.B.3.

It should be noted that the numerical values given in these tables are relatively sensitive to the change in the viscosity of air η. For example, when η deviates by 0.5%, the numerical values changes by approximately 1% in the real part of Z_c and approximately 8% in the corresponding imaginary part. Since the original data for the computation of η (Sect. 24.B.4) may have an uncertainty of several percent, the Table 24.2a,b should be used with caution.

24.4.4 Cylindrical Couplers and Wave-Motion Correction

Two types of couplers (cavities) are used for microphone calibration: plane-wave couplers, where the diameter of the cavity equals the diameter of the microphone diaphragm; and large-volume couplers, where the cavity diameter is larger than that of the diameter of the microphone diaphragm and the cavity volume is much larger than the sum of the equivalent volume and front volumes of the microphones. In reciprocity calibration, the dimensions of the large-volume coupler are chosen such that the pressure increase at the diaphragm of the receiving microphone, due to the first longitudinal mode, is partially canceled by the pressure decrease due to the lowest radial mode, whose maximum pressures occur at the side walls of the cavity and hence are removed from the vicinity of the microphone diaphragm. A generally accepted rule to ensure that the pressure is sufficiently uniform over the microphone diaphragm is that the maximum dimension of the cavity should not exceed $\lambda/20$, where λ is the wavelength in the gas. With a large-volume coupler there is less influence from the equivalent volumes of the capillary tubes and the microphones. However, with larger dimensions, the coupler is restricted to lower-frequency applications. In a plane-wave coupler with relatively dimensions there are smaller wave-motion effects, and the coupler can be used at higher frequencies. To extend the range of calibration frequency, a gas with a higher sound speed should be used. Theoretical and experimental data [24.38, 42–44] on wave motion in couplers have been published. In practice, it is difficult to obtain a precise theoretical expression for wave-motion correction. However, for experimental determination, the following method [24.44] has been used.

Assuming that the sensitivity of the microphones remains constant, the microphone sensitivity is measured with air as the gas at relatively high frequencies for which the pressure distribution in the cavity is nonuniform. The measurements are repeated with hydrogen or helium, which have higher sound speeds and for which the wave motion correction is correspondingly smaller.

Table 24.3 Experimental data on wave-motion corrections for the air-filled large-volume coupler used with type LS1P microphones (ANSI S1.15: 2005, IEC 61094-2:1992-03). LS1P: type designation [24.41] for laboratory standard (LS) microphones, where the last letter P or F represents pressure or free-field microphones, and the number 1 or 2 represent mechanical configuration for one-inch and half-inch microphones, respectively

Frequency (Hz)	Correction (dB)
800 and below	0.000
1000	–0.002
1250	–0.013
1600	–0.034
2000	–0.060
2500	–0.087

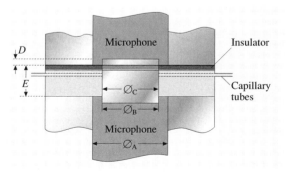

Fig. 24.11 Mechanical configuration of plane-wave couplers (ANSI S1.15: 2005, and IEC 61094-2:1992-03)

Table 24.4 Nominal dimensions for plane-wave couplers (ANSI S1.15: 2005, and IEC 61094-2:1992-03)

Dimensions (mm) Symbol	Laboratory standard microphones Type LS1P	Type LS2aP	Type LS2bP
⌀A	23.77	13.2	12.15
⌀B	18.6	9.3	9.8
⌀C	18.6	9.3	9.8
D	1.95	0.5	0.7
E	6.5–8.5	3.5–6	3.5–6

Table 24.5 Nominal dimensions and tolerances for large-volume couplers (ANSI S1.15: 2005, and IEC 61094-2:1992-03)

Dimensions (mm) Symbol	Laboratory standard microphones Type LS1P	Type LS2aP	Type LS2bP
⌀A	23.77	13.2	12.15
⌀B	18.6	9.3	9.8
⌀C	42.88 ± 0.03	18.30 ± 0.03	18.30 ± 0.03
D	1.95	0.5	0.7
E	12.55 ± 0.03	3.50 ± 0.03	3.50 ± 0.03
F	0.80 ± 0.03	0.40 ± 0.03	0.40 ± 0.03

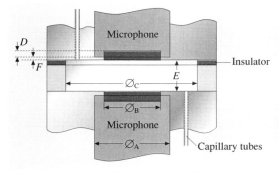

Fig. 24.12 Mechanical configuration of large-volume couplers (ANSI S1.15: 2005, and IEC 61094-2:1992-03)

Changes in the apparent sensitivity of the microphones give the wave-motion correction W for microphones calibrated in air.

Recommended [24.38] dimensions for the plane-wave and large-volume couplers are shown in Fig. 24.11, Table 24.4 and Fig. 24.12, Table 24.5, respectively. Representative wave-motion corrections [24.38] for both couplers used with LS1P (one-inch) microphones are shown in Table 24.3. These corrections should be added to the pressure sensitivity level measured with the coupler filled with air. When the coupler is filled with a gas other than air, the same correction can be used when the frequency scale is multiplied by a factor equal to the ratio of the velocity of sound in the gas to the corresponding velocity in air.

For LS1P microphones, the following empirical equation [24.25] for the wave-motion correction W, in decibels, may be used

$$W = -0.2242 F^2 + 1.2145 F - 1.6473 , \qquad (24.41)$$

where F equals $\log f$, and W should be added to the pressure sensitivity level of the microphone.

24.4.5 Barometric Pressure Correction

The sensitivity of a condenser microphone ([24.38], annex D) is inversely proportional to the acoustic impedance of the microphone. In a lumped parameter representation, the impedance is given by a serial connection of the impedance of the diaphragm (due primarily to its mass and compliance) and the impedance of the air behind the diaphragm is mainly determined by the following: (a) the thin air film between the diaphragm and the backplate, introducing loss and mass; (b) the air in slots or holes in the backplate, introducing loss and mass; and (c) the air in the cavity behind the backplate, acting at low frequencies as a compliance but at high frequencies as a mass due to wave motion in the cavity.

The density and the viscosity of air are considered linear functions of temperature and/or static pressure. Consequently the acoustic impedance of the microphone also depends upon the static pressure and the temperature. The resulting static pressure and temperature coefficients of the microphone are then considered to be determined by the ratio of the acoustic impedance

Table 24.6 Coefficients of the polynomial for calculating the microphone sensitivity pressure corrections using (24.42) or (24.43), and temperature corrections using (24.44) for specific LS1P and LS2P microphones (ANSI S1.15: 2005)

Coefficients	Pressure correction (24.42) [24.45, 46]			Pressure correction (24.43) [24.47]		Temperature correction (24.44) [24.47]	
	Brüel & Kjær Type 4160 LS1P	Brüel & Kjær Type 4180 LS2P	G.R.A.S. Type 40AG LS2P	Brüel & Kjær Type 4160 LS1P	Brüel & Kjær Type 4180 LS2P	Brüel & Kjær Type 4160 LS1P	Brüel & Kjær Type 4180 LS2P
a_0	-1.625×10^{-2}	-4.44613×10^{-3}	-3.93458×10^{-3}	-0.0152	-0.00519	-0.0020	-0.0012
a_1	-1.152×10^{-6}	-8.34314×10^{-7}	-5.68714×10^{-7}	-0.00584	-0.0304	0.00913	0.00633
a_2	1.816×10^{-9}	-3.51608×10^{-10}	-3.00641×10^{-10}	0.132	0.5976	-0.245	-0.242
a_3	-7.111×10^{-13}	-1.60278×10^{-14}	-1.43914×10^{-14}	-0.596	-3.912	1.673	1.656
a_4	2.056×10^{-16}			1.763	14.139	-6.058	-6.1833
a_5	-2.721×10^{-20}			-2.491	-27.561	11.766	11.81
a_6	1.188×10^{-24}			1.581	29.574	-13.11	-12.1366
a_7				-0.358	-17.6325	8.5138	6.875
a_8				-0.0364	5.4997	-3.0016	-2.0324
a_9				0.01894	-0.7017	0.4426	0.2457

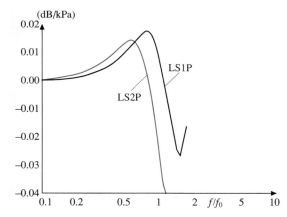

Fig. 24.13 Examples of static pressure coefficient of LS1P and LS2P microphones relative to the low-frequency value as a function of the relative frequency f/f_0 (ANSI S1.15: 2005, and IEC 61094-2:1992-03)

at reference conditions to the acoustic impedance at the relevant static pressure and temperature, respectively.

Both the mass and the compliance of the enclosed air depend on the static pressure, while the resistance can be considered independent of static pressure. The static pressure coefficient generally varies with frequency as shown in Fig. 24.13. For frequencies higher than about $0.5 f_0$ (f_0 being the resonance frequency of the microphone), the frequency variation depends strongly upon the wave motion in the cavity behind the backplate. In general, the pressure coefficient depends on constructional details in the shape of the backplate and back volume, and the actual values may differ considerably for two microphones of different manufacture although the microphones may belong to the same type, say LS1P. Consequently the pressure coefficients shown in Fig. 24.13 should not be applied to individual microphones.

The low-frequency values of the static pressure coefficient generally lie between -0.01 and -0.02 dB/kPa for LS1P microphones, and between -0.003 and -0.008 dB/kPa for LS2P microphones.

For the Brüel and Kjær type 4160 LS1P microphones, an empirical equation based on precise measurement of the microphone sensitivity [24.45] at various static pressures to derive the pressure coefficient, may be used to correct for microphone sensitivity variation with static pressure:

$$\Delta P = \left(a_0 + a_1 f + a_2 f^2 + a_3 f^3 + a_4 f^4 + a_5 f^5 + a_6 f^6\right) \Delta_p, \quad (24.42)$$

where ΔP is the microphone sensitivity level pressure correction (dB), (to be added to the sensitivity level of a microphone), a_0–a_6 are constant coefficients listed in Table 24.6, f is the frequency (Hz), $\Delta_p = (p_s - p_0)$ is the pressure difference (kPa), p_s is the static pressure during calibration (kPa), and p_0 is 101.325 kPa.

The variation of the pressure sensitivity correction with frequency is shown in Fig. 24.14. At a particular frequency, if the difference between the standard pressure and the barometric pressure during calibration is known, the sensitivity pressure correction can be obtained from the above equation. Over a pressure range 94–106 kPa, if (24.42) is used for sensitivity level pressure correction for a type 4160 microphone, the maximum deviation between the corrected microphone sensitivity level and the corresponding measured microphone sensitivities level is within 0.0085 dB over the frequency range 250–8000 Hz. For a wider frequency range of 63–10 000 Hz, the corresponding deviation is within 0.013 dB. Similar corrections can be obtained for LS2P (half-inch) microphones by applying the coefficients [24.46] given in Table 24.6.

Users of condenser microphones, particularly for very precise measurements such as during aircraft certification that involves huge resources, should be fully aware of the fact that the sensitivities of condenser microphones change with frequencies (frequency response) and these changes are functions of barometric pressure ([24.45], Fig. 3). Therefore for two sets of measurements at different barometric pressures, pressure corrections, such as with (24.43), can be applied at each frequency to arrive at more-precise measurements.

For LS2P microphones Brüel and Kjær type 4180 and GRAS-type 40AG, the pressure correction curves similar to that shown in Fig. 24.14 have been published [24.46]. With (24.42) their corrections can be computed from the coefficients shown in Table 24.6.

Alternatively, a similar equation for pressure correction developed at a later date also based on measurement [24.47], may be used

$$\delta_p = a_0 + a_1 x + a_2 x^2 + \ldots + a_9 x^9 , \quad (24.43)$$

where δ_p is the microphone sensitivity pressure correction (dB/kPa), a_0–a_9 are the constants listed in Table 24.6, $x = f/f_0$ is the frequency normalized to the resonance frequency f_0 of the individual microphone, and f is the frequency (Hz).

With (24.43), it is necessary to determine the resonant frequency f_0 of the microphones. With the assumption of a mean value for f_0, the numerical values obtained with (24.43) and (24.42) differ by a few thousandths of a decibel per kPa.

24.4.6 Temperature Correction

Both the mass and the resistance of the enclosed air depend on the temperature, while the compliance can

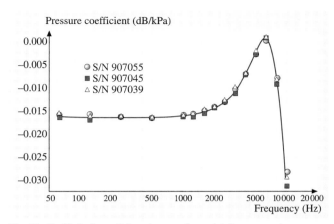

Fig. 24.14 Variation of the slopes of sensitivity correction curves with frequency for three Brüel and Kjær type 4160 microphones. The curve is obtained with an empirical equation for the computation of microphone sensitivity pressure correction. See (24.42). Similar corrections can be obtained for Brüel and Kjær Type 4180 and GRAS 40AG microphones with the coefficients shown in Table 24.6 (after [24.48])

be considered independent of temperature. The dependence on temperature is of secondary when compared with the pressure dependence. The resulting frequency dependence of the temperature coefficient is shown in Fig. 24.15.

In addition to the influence on the enclosed air, temperature variations also affect the mechanical parts of the microphone. The main effect will generally be a change in the tension of the diaphragm and thus a change in its

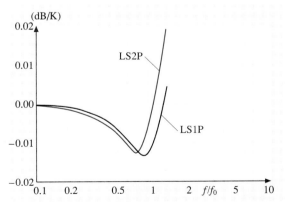

Fig. 24.15 General frequency dependence of the part of the temperature coefficient for LS1P and LS2P microphones caused by the variation in the impedance of the enclosed air (ANSI S1.15: 2005, and IEC 61094-2:1992-03)

Table 24.7 A numerical example for the calculation of estimated expanded uncertainty of microphone sensitivity level with a particular microphone calibration arrangement at 250 Hz (ANSI S1.15: 2005)

Source	Type	Estimated contribution (dB)	Distribution	Sensitivity coefficient	Standard uncertainty (dB)
Acoustic transfer impedance	B	0.025	Rectangular	1	0.0144
Voltage ratio	B	0.01	Rectangular	1	0.0058
Reference impedance	B	0.0042	Rectangular	1	0.0024
Specific heat ratio	B	0.005	Rectangular	1	0.0029
Density of air	B	0.001	Rectangular	1	0.0006
Barometric pressure	B	0.004	Rectangular	1	0.0023
Heat conduction correction	B	0.02	Rectangular	1	0.00115
Wave motion etc.	B	0.006	Rectangular	1	0.0035
Insert voltage	B	0.001	Rectangular	1	0.0006
Polarizing voltage	B	0.001	Rectangular	1	0.0006
Pressure coefficient	B	0.002	Rectangular	1	0.0012
Temperature coefficient	B	0.001	Rectangular	1	0.0006
Humidity coefficient	B	0.001	Rectangular	1	0.0006
Repeatability	A	0.004	Normal	1	0.0040
Combined uncertainty					0.0206
Expanded uncertainty ($k=2$)					0.0412
Rounded expanded uncertainty					**0.042**

compliance. This results in a constant change in sensitivity in the stiffness-controlled range and a slight change in the resonance frequency.

The low-frequency values of temperature coefficient generally lie close to $+0.005$ dB/K for both LS1P and LS2P microphones.

An equation developed for the normalized temperature coefficient based on measurements of LS1P and LS2P microphones [24.47, 49], may be used

$$\Delta T = (a_0 + a_1 x + a_2 x^2 + \ldots + a_9 x^9)\Delta t, \quad (24.44)$$

where ΔT is the microphone sensitivity temperature correction (dB), a_0–a_9 are the constants listed in Table 24.6, $x = f/f_0$ is the frequency normalized by the resonance frequency f_0 of the individual microphone, f is the frequency (Hz), $\Delta t = (T - 273.15)$, the temperature difference from the reference condition (K).

24.4.7 Microphone Sensitivity Equations

In general, based on (24.22) or (24.23), the open-circuit sensitivity of a microphone expressed in decibels relative to 1 V/Pa is

$$M_{\text{dB}} = 20 \log M + 10 \log \Delta_c + 10 \log \Delta_H \\ + 10 \log \Delta P_v + 10 \log \Delta T + 10 \log \Delta h + W, \quad (24.45)$$

where M is the microphone sensitivity, and Δ_c, Δ_H, ΔP, ΔT, Δh and W are the corrections for capillary, heat conduction, pressure, temperature, humidity and wave motion, respectively. For LS1P and LS2P microphones, the humidity corrections of the order of 0.0025 dB/100% RH may be ignored.

When expressed in the format of volts per pascal, and including corrections, the microphone sensitivity is:

$$M_{\text{corr}} = 10^{M_{\text{dB}}/20}. \quad (24.46)$$

24.4.8 Uncertainty on Pressure Sensitivity Level

The uncertainty of the pressure sensitivity level, expressed as a function of frequency, should be stated as the expanded uncertainty of measurement obtained by multiplying the derived standard uncertainty by a coverage factor of two.

It should be noted that not all of the contributing uncertainty components are known. Some of the uncertainty components that may be necessary for consideration for a given calibration arrangement are: (a) the electrical transfer impedance: series impedance, voltage ratio, cross-talk and noise, distortion etc., (b) the acoustic transfer impedance related to the coupler: coupler dimensions and surface area, capillary tube parameters and environmental conditions, and (c) the microphone parameters, such as loss factor, diaphragm: mass, compliance, resistance and heat conduction related to front cavity etc. Details of some of these uncertainties have been published [24.50].

For the determination of the pressure sensitivity level, it is estimated [24.33, 39] that a reciprocity calibration carried out under laboratory conditions can achieve an uncertainty of approximately 0.05 dB at low and middle frequencies. The uncertainty increases to about 0.1 dB at 10 kHz and 20 kHz for LS1P and LS2P laboratory standard microphones, respectively.

In order to clarify the computation process, a numerical example is given in Table 24.7 for the calculation of estimated expanded uncertainty of the microphone sensitivity level with a particular microphone calibration arrangement at 250 Hz.

24.5 Free-Field Microphone Calibration

The theory and procedure for free-field calibration have been discussed in detail [24.51, 52]. Similarly to pressure reciprocity calibration, one microphone is driven as a sound source and the microphone pair faces each other in a free field at a fixed distance. One of the major uncertainties in free-field calibration is to measure the positions of the acoustic centers of the microphones accurately so that the distance between them can be deduced. Procedures for measuring and estimating values of the positions of the acoustic centers for laboratory standard microphones have been published ([24.52], annex A). For free-field calibration of half-inch microphones, over a frequency range from about 1.25 kHz to more than 20 kHz the uncertainty (three standard deviations) [24.51] is approximately 0.1–0.2 dB.

The manufacturers of the microphones usually supply free-field corrections for their microphones, to be added to the pressure reciprocity measurements, for free-field applications.

24.6 Comparison Methods for Microphone Calibration

Reciprocity microphone calibration discussed in Sect. 24.3 is a primary method that provides high accuracy but requires stringent procedures and is relatively time consuming. A more economical approach to microphone calibration is by means of comparison methods with which the performance of the test microphone is compared with that of a standard microphone that has been calibrated with a primary method. Secondary methods, such as using comparison, have higher uncertainties but offer several advantages, such as simplicity in procedure, are less time consuming and, depending on the method, require little or no correction, and are very economically attractive.

24.6.1 Interchange Microphone Method of Comparison

This economically attractive method was developed to compare microphones in a cavity or in a small anechoic box (*Wong* and *Embleton* [24.53] and *Wong* and *Wu* [24.54]). The technique can be applied to microphone phase comparison calibration (*Wong* [24.55]). The method has been standardized in an international standard [24.56].

When two microphones with their diaphragms facing in close proximity to each other are excited in a sound field, the difference of level readings, L_{C12} between the two channels, where microphone 1 is connected to preamplifier 1, and microphone 2 is connected to preamplifier 2, is

$$L_{C12} = (L_1 + L_{m1} + L_d + L_{WA}) \\ - (L_2 + L_{m2} + L_d + L_{WB}), \quad (24.47)$$

where L_1 and L_2 are the sensitivity levels of the microphones 1 and 2, respectively, L_{m1} and L_{m2} are the gain sensitivity levels of the measuring systems 1 and 2, respectively, L_d is the level of the sound excitation, and L_{WA} and L_{WB} are the wave pattern effects at the microphone positions A and B, respectively.

When the microphones are interchanged so that microphone 1 is now connected to preamplifier 2 and microphone 2 is now connected to preamplifier 1, the difference of level readings between the two channels becomes

$$L_{C21} = (L_2 + L_{m1} + L_d + L_{WA})$$
$$- (L_1 + L_{m2} + L_d + L_{WB}) \,. \quad (24.48)$$

Subtracting (24.48) from (24.47), the difference in level between the two microphones is

$$(L_1 - L_2) = 1/2(L_{C12} - L_{C21}) \,. \quad (24.49)$$

If either L_1 or L_2 is the sensitivity level of a calibrated reference microphone, then the sensitivity level of the other microphone (the test microphone) can be deduced. It is interesting to point out that with this method, the level L_d of the driving sound need not have a flat response or known levels.

The above relatively simple theoretical consideration has assumed that the input capacitances of the two microphones are identical, i. e. the same model of microphones, and the same model of preamplifiers used for both channels such that their input capacitances have the same value.

If the input capacitances C_m of the preamplifiers are the same but the two models of microphones are different, then a correction L_{corr} should be added to the right-hand side of (24.49), where

$$L_{corr} = 20 \log[C_2/(C_m + C_2)]$$
$$- 20 \log[C_1/(C_m + C_1)] \,. \quad (24.50)$$

The two terms in (24.50) are the capacitance corrections for the two microphones; and C_1 and C_2 are the capacitances of microphone 1 and microphone 2, respectively.

It is important to realize that the physical positions of the reference microphone should be repeated accurately before and after the microphones are interchanged such that the sound field at the reference microphone is unchanged. The difference in level readings shown in (24.47) and (24.48) can be measured with the ratio provision of a precision AC voltmeter.

Uncertainties

The uncertainty for the above method depends on the reference microphone uncertainty that may be of the order of 0.05 dB. With very sophisticated measuring methods such as phase-lock amplifiers the uncertainty introduced by this comparison method to be added to the uncertainty of the reference microphone [24.57] is estimated to be less than 0.1 dB.

24.6.2 Comparison Method with a Calibrator

In general, there are five components and three basic steps in this comparison method [24.57]. The components are the reference microphone, the test microphone, a stable sound source, which is usually an acoustical calibrator that generates a constant sound pressure level, a microphone preamplifier and an acoustical level indicator such as a measuring amplifier or a voltmeter. Since the calibrator is used as a transfer standard, its sound pressure level need not be known.

The three basic steps are as follows

1. The reference microphone which is connected to the preamplifier is inserted into the acoustical calibrator and a level reading L_r is noted.
2. The reference microphone is replaced with the test microphone. With the same acoustical calibrator, a second level reading L_t is noted.
3. The test microphone is removed, and again with the reference microphone, step 1 is repeated to give a third level reading.

The agreement between the first and third level readings with the reference microphone is a confirmation of the validity of the second level reading obtained with the test microphone, that is, to assure that nothing has changed after the first level reading is taken. The difference between the level readings, $(L_t - L_r)$ is the difference in open-circuit sensitivities (expressed in decibels) between the two microphones if the following conditions are satisfied:

1. The equivalent volumes of the two microphones are identical, or the equivalent volume of the sound calibrator is much larger than that of the microphones, such that any difference in the front volumes of the microphones does not produce a significant change in the sound pressure levels measured in steps 1 and 2 above.
2. The capacitances of the two microphones are identical. Since the preamplifier has a finite input impedance (a very high resistance in parallel with a small known capacitance), the loading on the microphone will be different if the microphone capacitances were to be different, and this will produce a small change in the overall combined sensitivity of the microphone and preamplifier.
3. The environmental conditions remain constant during the measurements.

Corrections

When it is not possible to satisfy all of the above conditions during the implementation of the comparison procedure, the following corrections can be applied.

Equivalent Volume Correction

Depending on the type of microphones, the equivalent volume of the test microphone, as specified by the manufacturer, can differ substantially from that of the reference microphone. The following correction can be applied.

$$\Delta P_{\text{v}} = 20 \log[V/(V+\Delta V)] \text{dB}, \quad (24.51)$$

where ΔP_{v} is the volume correction to be added to the calibrated sensitivity of the microphone, V is the effective coupler volume of the acoustical calibrator, ΔV is the difference between the equivalent volumes of the microphones, i.e., the equivalent volume of the reference microphone minus the equivalent volume of the test microphone.

When the combined volume of the calibrator and the equivalent volume of the microphone increases, the sound pressure level inside the acoustical calibrator decreases. For example, if ΔV is negative, i.e., the equivalent volume of the test microphone is larger than that of the reference microphone the volume correction ΔP_{v} has a positive sign in order to compensate for the decrease in sound pressure level when the test microphone is measured. When the correction is added to the microphone sensitivity (24.45), the apparent sensitivity of the test microphone increases.

Depending on the frequency of the acoustical calibrator, the effective coupler volume V should include the additional volume due to heat conduction (Sect. 24.4.1). For example at a frequency of 250 Hz, the additional volume to account for heat conduction for the Brüel and Kjær type 4220 pistonphone is 0.16 cm^3. There are other acoustical calibrators that have relatively large cavity volumes, such as the Brüel and Kjær model 4230 (94 dB at 1 kHz, with an equivalent volume of $> 100 \text{ cm}^3$ between 10 and 40 °C); and the Larson–Davis model CA250. In these cases, from (24.51), the correction is negligible since V is much larger than ΔV. Most calibrators are supplied with adaptors to accommodate microphones of various diameters, and these adaptors are designed to include the volume correction for the particular types of microphones when used with the calibrator. If the adaptors and the calibrator are from different manufacturers, the instruction manual of the calibrator should be consulted to ensure proper usage.

Capacitance Correction

The combined level sensitivity of the microphone and preamplifier is:

$$M_{\text{com}} = M_{\text{o}} + g_1 + 20 \log[C/(C+C_{\text{i}})], \quad (24.52)$$

where M_{o} is the open-circuit sensitivity of the microphone (dB re 1 V/PA), g_1 is the gain of the preamplifier (dB), C is the capacitance of the microphone, C_{i} is the input capacitance of the preamplifier. The input capacitance of the preamplifier would load the microphone and the overall sensitivity decreases. With this comparison method, if the capacitances of the two microphones are different, the capacitance correction is

$$\Delta P_{\text{c}} = 20 \log[C_{\text{r}}/(C_{\text{r}}+c_{\text{i}})] - 20 \log[C_{\text{t}}/(C_{\text{t}}+c_{\text{i}})], \quad (24.53)$$

where ΔP_{c} is the capacitance correction to be added to the calibrated sensitivity of the microphone (dB), C_{r} is the capacitance of the reference microphone, c_{i} is the input capacitance of the preamplifier, and C_{t} is the capacitance of the test microphone.

The first and the second term on the right-hand side of (24.53) are the capacitance corrections for the reference microphone and the test microphone, respectively.

Environmental Conditions

One of the advantages of a comparison method is that environmental effects usually remain constant during measurements outlined in the above three basic steps. Changes in temperature and humidity affect both microphones. A rapid variation in barometric pressure may affect some acoustical calibrators, and precaution should be taken to avoid opening or closing the doors of the calibration laboratory during measurements.

Uncertainties

From the above discussion, the open-circuit sensitivity of the test microphone is

$$M_{\text{t}} = M_{\text{r}} + (L_{\text{t}} - L_{\text{r}}) + \Delta P_{\text{v}} + \Delta P_{\text{c}}, \quad (24.54)$$

where M_{r} is the open circuit sensitivity of the reference microphone (dB re 1 V/Pa), L_{t} is the level reading obtained with the test microphone (dB), and L_{r} is the level reading obtained with the reference microphone (dB).

The uncertainty of M_{r} depends on the primary reciprocity procedure that calibrates the reference microphone (Sect. 24.3). If the environmental conditions are controlled [24.20], the uncertainty of the open-circuit sensitivity can be less than 0.005 dB. The uncertainty expectation for reciprocity calibrations based on international standard [24.38, 40] is approximately 0.05 dB.

If the magnitudes of the levels L_t and L_r are nearly identical, the uncertainty can be much less than 0.05 dB. However, if it is necessary for the level-measuring device, such as a measuring amplifier, to engage different instrument ranges during the measurement of the above levels, the uncertainties can be up to ± 0.1 dB. The uncertainties of ΔP_v and ΔP_c are estimated to be approximately 0.01 dB each, and the overall uncertainty of calibrator comparison method is estimated to be 0.05–0.2 dB.

For this comparison method, the preamplifier should have very high input impedance, usually in the order of gigaohms with a parallel capacitance of a fraction of a picofarad.

24.6.3 Comparison Pressure and Free-Field Calibrations

The coupler method described in Sect. 24.3 for primary reciprocity microphone calibration can be utilized to implement comparison pressure calibration. A laboratory standard microphone of unknown pressure response can be calibrated against a reference microphone with a known pressure response. One example of the coupler arrangement shown in Fig. 24.6 with which microphone A is the driver that serves as a sound source. The reference microphone and the test microphone replace microphone B in succession on the right-hand side of the coupler. The ratio of the driving voltage e_A applied to the driver microphone to the open-circuit output voltage of the receiving microphones e_B, is measured as described in Sect. 24.3.3. The sensitivity of the test microphone can be calculated as follows

$$M_t = M_r - 20\log(e_{Bref}/e_{Aref}) + 20\log(e_{Bx}/e_{Ax}),\tag{24.55}$$

where M_t and M_r are the sensitivities of the test and the reference microphones, respectively, e_{Bref} and e_{Aref} are the open-circuit voltage and the source driving voltage when the reference microphone is in position, e_{Bx} and e_{Ax} are the open-circuit voltage and the source driving voltage when the test microphone is in position.

With (24.55), unless the equivalent volume of the coupler is large compared to those of the reference microphone and the test microphone, and the equivalent volumes of the reference microphone and the test microphone are similar, suitable volume correction should be applied such that the equivalent sound pressure levels are identical for both the reference and the test microphones.

Uncertainties

The uncertainties of the coupler pressure comparison method are very similar to those described in Sect. 24.6.2 with the exception that the open-circuit voltage measurements with the insert voltage method, capacitance corrections are unnecessary. The uncertainty of the coupler comparison method may approach approximately two times the uncertainty of the reference microphone plus the uncertainty involved with the equivalent volume corrections; assuming that the reference microphone has been calibrated with a similar method for open-circuit voltage measurement.

The uncertainty for free-field comparison method is higher and is due to electrical and mechanical air-transmitted interferences. Even with narrow-band filtering, the overall uncertainty for free-field comparison is of the order of 0.5 dB.

24.6.4 Comparison Method with a Precision Attenuator

A more precise method than that described in Sect. 24.6.2, comparison with a calibrator, is comparison with an attenuator. The acoustical level indicator is replaced with a calibrated attenuator [24.58]. Over the usable range of 5–100 mV/Pa of the open-circuit sensitivity of a microphone, the attenuator, i.e., a precision ten-turn potentiometer, is calibrated with a resolution of better than 0.05 mV/Pa. The comparison procedure is as follows:

1. The reference microphone is inserted into a calibrator. The sensitivity of the reference microphone is entered into the calibrated attenuator, say 50 mV/Pa. An indicator reading is taken.
2. The reference microphone is replaced with the test microphone, and the calibrated attenuator is adjusted until the indicator gives the same reading. The new attenuator reading gives the sensitivity of the test microphone.

With this method, it is still necessary to correct for capacitances and equivalent volumes of the microphones. However, the uncertainty is reduced by the precision of the attenuator, which may have an uncertainty of a few thousandths of a decibel. If the test microphone is of the same type and model as the reference microphones, the uncertainty of this method is the uncertainty of the reference microphone plus a few thousandths of a decibel, if the indicator has a resolution of a thousandth of a decibel.

24.7 Frequency Response Measurement with Electrostatic Actuators

Electrostatic actuators were used by *Ballantine* [24.2] to investigate microphone calibration and *Koidan* [24.59] gave a comprehensive discussion of the uncertainties between measurements obtained with electrostatic actuators compared with those measured with the coupler pressure reciprocity method. The description and analyses of some commercially available electrostatic actuators for microphone calibration have been published [24.60–62] and standardized [24.63] for frequency response measurement for working standard (WS) microphones. The electrostatic actuator method, Fig. 24.16, consists of a metal grid polarized with a high voltage direct-current (DC) supply (usually 800 V). An AC time-varying sweep signal provides the electrostatic pressure on the microphone diaphragm. With the usual amplification and a recorder, the frequency response is displayed graphically.

Madella [24.64] and *Nedzelnitsky* [24.65] pointed out that with electrostatic actuator measurements the effective mechanical radiation impedance loading of the microphone diaphragm in the presence of the electrostatic actuator is different from that of those used in a coupler or in a free field. Consequently, at higher frequencies the absolute values of the difference between the actuator-determined response of type L microphones and pressure calibrations determined in couplers by reciprocity can be as large as about 1.5 dB [24.65]. In practice, the design of the metal grid can be optimized for a particular microphone type. According to the IEC standard [24.63], the expanded uncertainty of the method is 0.1–0.2 dB for frequencies up to 10 kHz for WS1 (1-inch) and WS2 (half-inch) type microphones. Since the air gap between the metal grid and the microphone diaphragm is very small (nominally approximately 0.5 mm) the use of electrostatic actuators for models of microphones other than those specified is not recommended. Since the motion of the microphone diaphragm, caused by electrostatic pressure produce by the actuator, creates a sound pressure on the outer surface

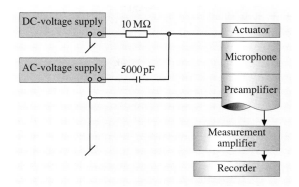

Fig. 24.16 Frequency response measurement with an electrostatic actuator (after *Frederiksen* [24.29], Chap. 15)

of the diaphragm that adds to the electrostatic pressure and influences the measure response, precautions should be taken to avoid blocking the actuator space that opens to the atmosphere.

Apart from being economically attractive, there are some salient features to the electrostatic actuator method.

1. The method is ideal for comparing the frequency responses of microphones, such as those in a production line, at which only changes from a nominal value is important.
2. Since the actuator method is unaffected by atmospheric pressure, humidity and temperature (assuming the air gap remains constant) [24.63], the method may be used to measure the change in microphone responses with respect to variations in the environment.

Given the fact that the international standard [24.63] only discusses type WS1 and WS2 working standard microphones and not the laboratory types LS1 and LS2 microphones, it is obvious that the electrostatic actuator method still requires further development to reduce uncertainties.

24.8 Overall View on Microphone Calibration

Primary microphone calibrations, such as those performed with the reciprocity method, provide uncertainty estimations. One may wonder what is the contribution to the calibration uncertainty from two major sources: (a) the stability of the microphones and (b) the stability of the calibration system that performed the microphone calibration. It is very difficult to separate these two sources of uncertainties.

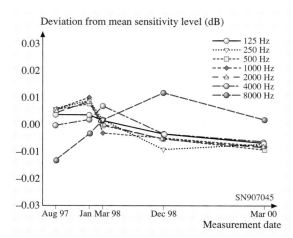

Fig. 24.17 Deviation from the mean sensitivity level of a transfer standard microphone s/n 907045 (after [24.67])

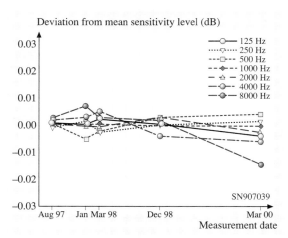

Fig. 24.19 Deviation from the mean sensitivity level of a transfer standard microphone s/n 907039 (after [24.67])

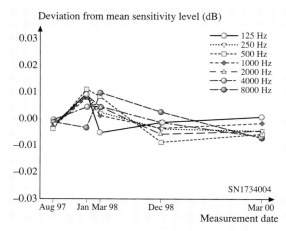

Fig. 24.18 Deviation from the mean sensitivity level of a transfer standard microphone s/n 1734004 (after [24.67])

In connection with an international microphone calibration comparison [24.66] that involved five National Metrology Institutes (NMIs), the pilot laboratory (Canada) had the opportunity to calibrate the same set of three laboratory standard LS1P microphones five times over a period of 31 months. The calibrations were made in a chamber with a controlled environment [24.45] at the reference condition of 23 °C, 101.325 kPa and 50% RH. The microphone sensitivity levels were measured at seven frequencies from 125 Hz to 8 kHz. Two of the microphones (the transfer standards) were *hand carried* to and from the pilot laboratory in a *star* configuration, i.e., the microphones were calibration by the pilot laboratory before and after delivery to each of the participating NMIs. The average of five sets of readings obtained by the pilot laboratory was taken as the mean sensitivity level. The deviation from the mean sensitivity level of the transfer standard microphones (LS1P Brüel and Kjær type 4160 microphones, s/n 907045 and s/n 1734004) are shown in Fig. 24.17 and Fig. 24.18, and are within approximately 0.011 dB. It is interesting to show that, for the above reciprocity calibration, the third type 4160 LS1P microphone s/n 907039 which remained at the pilot laboratory without the need to endure air transportation to and from the participating laboratories over the test period, the deviations shown in Fig. 24.19 were significantly smaller for frequencies below 8000 Hz, and are within +0.007 and −0.006 dB. It should be noted that the deviations shown for the three microphones are the combined contributions from the sensitivity level stability of the microphones and the stability of the primary calibration system. Since the specifications for laboratory standard microphones [24.41] states that the type 4160 LS1P microphone has a long-term stability coefficient of < 0.02 dB per year from 250 Hz to 1000 Hz, and the manufacturer Brüel and Kjær indicated the stability of their LS1P microphones may have a stability coefficient of approximately 0.01 dB per year, one may conclude that the primary microphones and the calibration system of the pilot laboratory have an excellent stability of approximately 0.01 dB over the period of 31 months.

24.A Acoustic Transfer Impedance Evaluation

Assume the sound pressure to be the same at any point inside the coupler (this will take place when the physical dimensions of the coupler are very small compared to the wavelength) the acoustic transfer impedance can be evaluated theoretically [24.38]. The air in the coupler then behaves as a pure compliance and assuming adiabatic compression and expansion of the gas:

$$\frac{1}{Z'_{AB}} = i\omega \left(\frac{V}{\gamma p_s} + \frac{V_{eA}}{\gamma_0 p_0} + \frac{V_{eB}}{\gamma_0 p_0} \right), \quad (24.A1)$$

where values for γ and γ_0 can be derived from the equations given in Sect. 24.B.

The computation of the acoustic transfer impedance becomes complicated at high frequencies when the dimensions of the coupler are not small compared with the wavelength. When the shape of the coupler is cylindrical and the cavity diameter is the same as that of the microphone diaphragm [24.38], and at frequencies where plane-wave transmission and adiabatic conditions can be assumed, the acoustic transfer impedance is

$$\frac{1}{Z'_{AB}} = \frac{1}{Z} \left[\left(\frac{Z}{Z_A} + \frac{Z}{Z_B} \right) \cosh(\xi \ell_0) \right.$$
$$\left. + \left(1 + \frac{Z}{Z_A} \frac{Z}{Z_B} \right) \sinh(\xi \ell_0) \right], \quad (24.A2)$$

where ξ is the complex propagation coefficient (m^{-1}), and $\xi = i(\omega/c)$, approximately.

Allowance should be made for the air volume associated with the microphones that is not enclosed by the circumference of the coupler and the two diaphragms.

24.B Physical Properties of Air

Certain physical properties, characterizing the enclosed gas in the coupler, enter into the expressions for calculating the sensitivity of the microphones, see (24.A1) and Sects. 24.4.1 and 24.4.3. These properties are: the speed of sound in the gas; the density of the gas; the ratio of the specific heats of the gas; the viscosity of the gas; and the thermal diffusivity of the gas. These properties depend on one or more of the variables: temperature, static pressure and humidity.

A large number of investigations have been published in the literature where reference values for the physical properties can be found for specified environmental conditions, i.e. for standard dry air [24.68] at 0 °C and at a static pressure of 101.325 kPa. In the following, unless otherwise stated, the recommended calculation procedures with the corresponding estimated uncertainties are valid over the temperature range 0–30 °C; barometric pressure of 60–110 kPa, and relative humidity from 10% to 90%.

24.B.1 Density of Humid Air

An equation for the determination of the density of moist air was first published by *Giacomo* [24.69] and some constants were modified by *Davies* [24.70]. The following equation may be used to compute the density of humid air ρ based on a CO_2 content of 400 ppm, which is slightly higher than the 314 ppm specified for the ISO standard air [24.68] constituents.

$$\rho = [3.48349 + 1.44 (x_c - 0.0004)]$$
$$\times 10^{-3} \frac{p_s}{ZT} (1 - 0.3780 x_w), \quad (24.B1)$$

where x_c and x_w are the mole fractions of carbon dioxide and water vapor in air, respectively; p_s is the static pressure, Z is the compressibility factor for humid air, T is the thermodynamic temperature, and

$$Z = 1 - \frac{p_s}{T} \left[a_0 + a_1 t + a_2 t^2 + (a_3 + a_4 t) x_w \right.$$
$$\left. + (a_5 + a_6 t) x_w^2 \right] + \frac{p_s^2}{T^2} \left(a_7 + a_8 x_w^2 \right),$$
$$x_w = \frac{p_{sv}}{p_s} f_e h,$$
$$p_{sv} = \exp \left(A'_1 T^2 + B'_1 T + C'_1 + D'_1 T^{-1} \right),$$
$$f_e = 1.00062 + p_s(3.14 \times 10^{-8}) + t^2(5.6 \times 10^{-7}).$$

For Z the numerical coefficients are:

$a_0 = 1.58123 \times 10^{-6}$, $a_1 = -2.9331 \times 10^{-8}$,
$a_2 = 1.1043 \times 10^{-10}$, $a_3 = 5.707 \times 10^{-6}$,
$a_4 = -2.051 \times 10^{-8}$, $a_5 = 1.9898 \times 10^{-4}$,
$a_6 = -2.376 \times 10^{-6}$, $a_7 = 1.83 \times 10^{-11}$,
$a_8 = -0.765 \times 10^{-8}$.

For p_{sv} the numerical coefficients are:

$A'_1 = 1.2378847 \times 10^{-5}$,
$B'_1 = -1.9121316 \times 10^{-2}$,
$C'_1 = 33.93711047$,
$D'_1 = -6.3431645 \times 10^3$.

Within the normal range of environmental conditions during calibrations, the following simplified equation for the density of humid air may be used

$$\rho = \rho_0 \frac{p_s}{p_0} \frac{T_0}{T}(1 - 0.378 h_s), \quad (24.B2)$$

where ρ and ρ_0 are the densities of air, p_0 and p_s are the reference and static pressure, respectively; T_0 and T are the temperatures in Kelvin and h_s is the fractional molar concentration of moisture.

Uncertainties in ρ

The uncertainty in ρ obtained with (24.B1) has been estimated [24.69, 70] at approximated 100 ppm. At the reference conditions of 101.325 kPa and 50% relative humidity, ρ obtained with (24.B2) are 239 ppm and 293 ppm larger than those obtained with (24.B1) at 23 °C and 30 °C, respectively.

24.B.2 Computation of the Speed of Sound in Air

Method 1

The speed of sound in air varies with temperature, carbon dioxide content, relative humidity and barometric pressure. The sequence of the above parameters is listed roughly in decreasing order of their influence on the speed of sound with respect to the day-to-day encounter of environmental conditions. A large number of investigations related to the speed of sound [24.50, 71–80] had been published, detailed references have been given [24.75], and an updated and comprehensive bibliography ([24.80], Chap. 17, pp. 265–284 and references therein) on publications relating to sound speed in air is included at the end of this appendix. It can be shown that the speed of sound remains relatively constant with barometric pressure [24.72]: at 23 °C and 50% RH, over the barometric pressure range 60–110 kPa, the sound speed varies by less than 76 ppm. In view of this, the following equation has a relatively small overall uncertainty, is based on a steady barometric pressure of 101.325 kPa and may be used to compute the variation of the speed of sound with temperature, carbon dioxide content and relative humidity.

A general empirical equation has been obtained [24.71] for calculation of the variation of c/c_0 with relative humidity h, temperature t and carbon dioxide content h_c:

$$\begin{aligned} c/c_0 = &\, a_0 + a_1 t + a_2 t^2 + a_3 h_c + a_4 h_c t + a_5 h_c t^2 \\ &+ a_6 h + a_7 h t + a_8 h t^2 + a_9 h t^3 + a_{10}(h_c)^2 \\ &+ a_{11} h^2 + a_{12} h t h_c, \end{aligned} \quad (24.B3)$$

where c and c_0 are the sound speed and the reference dry-air sound speed, respectively; a_0–a_{12} are coefficient constants listed in Table 24.8. With Table 24.8, the sound speed can be deduced by multiplying c/c_0 with the corresponding reference dry-air sound speed c_0.

For h_c values from 0% to 1%, and for t from 0 °C to 30 °C, for h from 0 to 1 (relative humidity 0% to 100%), and at a barometric pressure of 101.325 kPa, the sound speed computed using the numerical coefficients in Table 24.8 fits the theoretical data with a standard uncertainty of ± 48 ppm.

Uncertainties in c_0

Over the temperature range 0–30 °C, (24.B3) is fitted to a computation [24.71] for a real gas at 101.325 kPa at which the value $C_p - C_v$ is not greatly different from the universal constant R [24.75]. Based on this approximate assumption, the dry-air sound speed c_0 is 331.29 m/s, with an uncertainty of approximately 200 ppm [24.75], which encompasses sound speeds from 331.224 to 331.356 m/s [24.50].

Table 24.8 Coefficient constants for the computation of c/c_0 and γ/γ_0 (after [24.71])

Coefficient constants	c/c_0 (24.B3)	γ/γ_0 (24.B5)
a_0	1.000100	1.000034
a_1	1.8286×10^{-3}	-2.8100×10^{-6}
a_2	-1.6925×10^{-6}	-2.1210×10^{-7}
a_3	-3.1066×10^{-3}	-1.01223×10^{-3}
a_4	-7.9762×10^{-6}	-5.2500×10^{-6}
a_5	3.4000×10^{-9}	1.1290×10^{-8}
a_6	8.9180×10^{-4}	-3.4920×10^{-4}
a_7	7.7893×10^{-5}	-2.8560×10^{-5}
a_8	1.3795×10^{-6}	-5.9000×10^{-7}
a_9	9.5330×10^{-8}	-2.9710×10^{-8}
a_{10}	1.2990×10^{-5}	4.23427×10^{-6}
a_{11}	4.8016×10^{-5}	8.0000×10^{-7}
a_{12}	-1.4660×10^{-6}	5.1000×10^{-7}

Method 2

Alternatively, over the temperature range 0–30 °C, with a computation [24.72–74] that is based on the derivation of real-gas sound speed from virial coefficients, the dry-air sound speed c_0 calculated with 16 coefficients similar to those shown in Table 24.8, is 331.46 m/s. The uncertainty is approximately 545 ppm [24.73, 74], which encompasses sound speeds c_0 from 331.279 to 331.641 m/s. It should be noted that the sound speed of 331.29 m/s obtained with the first method is within the uncertainty range of the second method [24.50].

Within the normal range of environmental conditions during calibration, the following simplified equation [24.38, 40, 50] for the sound speed in humid air may be used:

$$c = c_0 \left(\frac{T}{T_0}\right)^{1/2} (1 + 0.165 h_s) \Delta \,, \qquad (24.B4)$$

where h_s is the fractional molar concentration of moisture, and Δ is a factor to compensate for dispersion. Values of $\Delta = 0.99935$ and $\Delta = 0.99965$ are found in [24.81] and [24.82], respectively. A value of 1.0001 has also been obtained [24.72].

Over a frequency range of 10 Hz to 10 kHz, by ignoring the dispersion factor Δ, the uncertainty in the sound speeds obtained with methods 1 and 2 increases by approximately 300 ppm.

By ignoring dispersion, at the reference conditions of 23 °C, 101.325 kPa and 50% relative humidity, the sound speed obtained with (24.B4) is 196 ppm higher than that obtained with (24.B3).

24.B.3 Ratio of Specific Heats of Air

There have been several experimental [24.83, 84] and theoretical [24.71–74, 85, 86] investigations into the variation of the ratio of specific heats of air with temperature, humidity, pressure and carbon dioxide content [24.75] ([24.86], pp. 36–39). A general empirical equation has been obtained [24.71] for the calculation of the variation of γ/γ_0 with relative humidity h, temperature t and carbon dioxide content h_c

$$\begin{aligned}\gamma/\gamma_0 = &\, a_0 + a_1 t + a_2 t^2 + a_3 h_c + a_4 h_c t + a_5 h_c t^2 \\ &+ a_6 h + a_7 h t + a_8 h t^2 + a_9 h t^3 + a_{10}(h_c)^2 \\ &+ a_{11} h^2 + a_{12} h t h_c \,, \end{aligned} \qquad (24.B5)$$

where a_0–a_{12} are coefficient constants listed in Table 24.8. With (24.B5), the above normalized ratio of specific heats can be deduced by multiplying γ/γ_0 by the corresponding reference dry-air ratio of specific heats γ_0. For CO_2 content h_c values from 0% to 1%, and for temperatures of 0–30 °C, humidity from 0 to 1 (relative humidity 0% to 100%), and at a barometric pressure of 101.325 kPa, the ratio of specific heats computed using the numerical coefficients in Table 24.8, fit the theoretical data with a standard uncertainty of ± 17 ppm.

Uncertainties in γ_0

Based on method 1 above [24.71], the dry-air specific heat ratio γ_0 is 1.3998. The standard uncertainty in γ_0 is approximately 400 ppm.

Similarly, based on method 2 [24.72–74], the dry-air specific heat ratio γ_0 is 1.4029; the standard uncertainty in γ_0 is over 760 ppm.

24.B.4 Viscosity and Thermal Diffusivity of Air for Capillary Correction

The viscosity η of air, in the parameters in (24.39) of Sect. 24.4.3, is a function of temperature t. An empirical equation which is based on a least-squares fit to published data [24.87] is

$$\begin{aligned}\eta = &\, [17.26797 + (5.0756 \times 10^{-2}) t \\ &- (4.4028 \times 10^{-5}) t^2 + (5.0000 \times 10^{-8}) t^3] \\ &\times 10^{-6} \, \text{Pa s} \end{aligned} \qquad (24.B6)$$

The equation for the thermal diffusivity α_t of air [24.38, 40] is

$$\alpha_t = \eta(9\kappa - 5)/(4\kappa\rho) \, \text{m}^2/\text{s} \,. \qquad (24.B7)$$

Uncertainties in η

Over the temperature range from -80 °C to 100 °C, when compared with the published data [24.87], the maximum deviation in η obtained with (24.B6) is ± 1 ppm from the data points. However, the data point uncertainty may be several percent [24.87].

For standard dry air at a pressure of 101.325 kPa, the numerical values calculated with the above equations for the viscosity are 1.7268×10^{-5} Pa s and 1.8413×10^{-5} Pa s, at 0 and 23 °C, respectively; and the corresponding values for the thermal diffusivity are $1.81234 \times 10^{-5} \, \text{m}^2/\text{s}$ and $2.09549 \times 10^{-5} \, \text{m}^2/\text{s}$, respectively.

Examples. Aiming at the highest accuracies (method 1), Table 24.9 gives the recommended values of the quantities given in (24.B2)–(24.B4) for the reference

Table 24.9 Recommended values for some physical quantities in (24.A1)–(24.B5) (ANSI S1.15: 2005)

Environmental conditions	Density of air ρ (kg/m^3)	Speed of sound c (m/s)	Ratio of specific heats γ	Viscosity of air η (Pa s)	Thermal diffusivity of air α_t (m^2/s)
$T = 23\,°C$ $p_s = 101325$ Pa $H = 50\%$	1.1859997	345.677519	1.39836198	1.841268×10^{-5}	2.105344×10^{-5}

Table 24.10 Recommended reference values applicable to dry air at 0 °C and 101.325 kPa. (ANSI S1.15: 2005)

Environmental conditions	Density of air ρ (kg/m^3)	Speed of sound c (m/s)	Ratio of specific heats γ	Viscosity of air η (Pa s)	Thermal diffusivity of air α_t (m^2/s)
$T = 0\,°C$ $p_0 = 101325$ Pa	1.29296	331.28	1.3998	1.7268×10^{-5}	1.8123×10^{-5}

conditions. The values in Table 24.9 are shown with more decimals than necessary and are intended for testing computation programs that are used to calculate these quantities.

Again, aiming at the highest accuracies (method 1) and in view of the need to have reference values c_0 and γ_0 in (24.B3), (24.B4) and (24.B5), Table 24.10 gives the recommended values for dry air.

References

24.1 E.C. Wente: A condenser transmitter as a uniformly sensitive instrument for the absolute measurement of sound intensity, Phys. Rev. **10**(1), 39–63 (1917)

24.2 S. Ballantine: Technique of microphone calibration, J. Acoust. Soc. Am. **3**, 319–360 (1932)

24.3 W.R. MacLean: Absolute measurement of sound without a primary standard, J. Acoust. Soc. Am. **12**, 140–146 (1940)

24.4 R.K. Cook: Absolute pressure calibration of microphones, Res. Nat. Bur. Standard. **25**, 489–555 (1940)

24.5 R.K. Cook: Absolute pressure calibration of microphones, J. Acoust. Soc. Am. **12**, 415–420 (1941), published in abbreviated form

24.6 M.S. Hawley: The condenser microphone as an acoustical standard, Bell Lab. Rec. **22**, 6–10 (1943)

24.7 A.L. DiMattia, F.M. Wiener: On the absolute pressure calibration of condenser microphones by the reciprocity method, J. Acoust. Soc Am. **18**(2), 341–344 (1946)

24.8 M.S. Hawley: The substitution method of measuring the open circuit voltage generated by a microphone, J. Acoust. Soc. Am. **21**(3), 183–189 (1949)

24.9 B.D. Simmons, F. Biagi: Pressure calibration of condenser microphones above 10 000 cps, J. Acoust. Soc. Am. **26**(5), 693–695 (1954)

24.10 T. Hayasaka, M. Suzuki, T. Akatsuka: New method of absolute pressure calibration of microphone sensitivity, Inst. Electr. Commun. Eng. Jpn. **33**, 674–680 (1955)

24.11 H.G. Diestel: Zur Bestimmung der Druckempfindlichkeit von Mikrophonen, Acustica **9**, 398–402 (1959), (in German)

24.12 T.F.W. Embleton, I.R. Dagg: Accurate coupler pressure calibration of condenser microphones at middle frequencies, J. Acoust. Soc. Am. **32**, 320–326 (1960)

24.13 W. Koidan: Method for measurement of $[E'/I']$ in the reciprocity calibration of microphones, J. Acoust. Soc. Am. **32**, 611 (1960)

24.14 A.N. Rivin, V.A. Cherpak: Pressure calibration of measuring microphones by the reciprocity method, Sov. Phys. Acoust. **6**, 246–253 (1960)

24.15 P.V. Brüel: *Accuracy of Condenser Microphone Calibration Methods, Part II*, Tech. Rev. 1 (Brüel Kjær, Nœrum 1965)

24.16 P. Riety, M. Lecollinet: Le dispositif d'etalonnage primaire des microphones de laboratoire de l'Institut National de Metrologie, Metrologia **10**, 17–34 (1974)

24.17 A.C. Corney: Capacitor microphone reciprocity calibration, Metrologia **11**, 25–32 (1975)

24.18 H. Miura, T. Takahashi, S. Sato: Correction for 1/2 inch laboratory standard condenser microphones, Trans. IECE Japan **EA77-41**, 15–22 (1977)

24.19 A. Suzuki, S. Yoshikawa: Simplified method for pressure calibration of condenser microphones using active couplers, Trans. IECE Japan **EA77-40**, 9–14 (1977)

24.20 G.S.K. Wong, T.F.W. Embleton: Arrangement for precision reciprocity calibration of condenser microphones, J. Acoust. Soc. Am. **66**(5), 1275–1280 (1979)

24.21 V. Nedzelnitsky, E. D. Burnett, W. B. Penzes: Calibration of laboratory condenser microphones. In: *Proc. 10th Transducer Workshop, Transducer Committee, Telemetry Group, Range Commanders Council* (DTIC, Colorado Springs 1979)

24.22 A. Suzuki, S. Yoshikawa: On the estimated error for pressure calibration of standard condenser microphones, Electr. Commun. Lab. J. (Jpn) **29**(7), 1251–1262 (1980)

24.23 D.L.H. Gibbings, A.V. Gibson: Contributions to the reciprocity calibration of microphones, Metrologia **17**, 7–15 (1981)

24.24 T. Takahashi, H. Miura: Corrections for the reciprocity calibration of laboratory standard condenser microphones, Bull. Electrotech. Lab. (Jpn) **46**(3-4), 78–81 (1982)

24.25 M.E. Delany, E.N. Bazley: *Uncertainties in Realizing the Standard of Sound Pressure by Closed-Coupler Reciprocity Technique*, Acoust. Rep. **99** (National Physical Laboratory, Teddington 1980), (see also 2nd ed. 1982)

24.26 D.L.H. Gibbings, A.V. Gibson: Wide-band calibration of capacitor microphones, Metrologia **20**, 95–99 (1984)

24.27 L. L. Beranek: *Acoustical Measurements* (Acoust. Soc. Am./AIP, Melville 1988), (**4**, 133–134)

24.28 K.O. Ballagh, A.C. Corney: Some developments in automated microphone reciprocity calibration, Acustica **71**, 200–209 (1990)

24.29 G. S. K. Wong, T. F. W. Embleton: *AIP Handbook of Condenser Microphones. Theory, Calibration, and Measurements* (American Institute of Physics, New York 1995)

24.30 N.H. Fletcher, S. Thwaites: Electrode surface profile and the performance of condenser microphones, J. Acoust. Soc. Am. **112**(6), 2779–2785 (2002)

24.31 BIPM: *Guide to the Expression of Uncertainties in Measurement* (BIPM, Paris 1995)

24.32 G.S.K. Wong: Precise measurement of phase difference and amplitude ratio of two coherent sinusoidal signals, J. Acoust. Soc. Am. **75**(3), 967–972 (1984)

24.33 B. Daniels: Acoustic impedance of enclosures, J. Acoust. Soc. Am. **19**, 569–571 (1947)

24.34 F. Biagi, R.K. Cook: Acoustic impedance of a right circular cylindrical enclosure, J. Acoust. Soc. Am. **26**, 506–509 (1954)

24.35 H. Gerber: Acoustic properties of fluid-filled chambers at infrasonic frequencies in the absence of convection, J. Acoust. Soc. Am. **36**, 1427–1434 (1964)

24.36 K.O. Ballagh: Acoustical admittance of cylindrical cavities, J. Sound Vibrat. **112**(3), 567–569 (1987)

24.37 D.R. Jarvis: Acoustical admittance of cylindrical cavities, J. Sound Vibrat. **117**(2), 390–392 (1987)

24.38 ANSI: ANSI S1.15 – 2005. American National Standard Measurement Microphones – Part 2: Primary Method for Pressure Calibration of Laboratory Standard Microphones by Reciprocity Technique, (NIST, Gaithersburg 2005)

24.39 T. Salava: *Measurement of the Acoustic Impedance of Standard Laboratory Microphones*, The Acoustics Laboratory, Report, Vol.18 (Technical University of Denmark, Lyngby 1976)

24.40 IEC: IEC 61094-2:1992-03 Measurement Microphones – Part 2: Primary Method for Pressure Calibration of Laboratory Standard Microphones by the Reciprocity Technique (IEC, Geneva 1992)

24.41 ANSI S1.15-1997/Part 1 (R2001) American National Standard Measurement Microphones – Part 1: Specifications for Laboratory Standard Microphones. See also: IEC International Standard 61094-1 1992-05: Measurement Microphones Part 1: Specifications for Laboratory Standard Microphones (NIST, Gaithersburg 1997)

24.42 K. Rasmussen: *Acoustical behaviour of cylindrical couplers*, The Acoustics Laboratory Report, Vol. 1 (Technical University of Denmark, Lyngby 1969)

24.43 H. Miura, E. Matsui: On the analysis of the wave motion in a coupler for the pressure calibration of laboratory standard microphones, J. Acoust. Soc. Jpn. **30**, 639–646 (1974)

24.44 F. Jacobsen: An improvement of the 20 cm³ coupler for reciprocity calibration of microphones, Acustica **38**, 151–153 (1977)

24.45 G.S.K. Wong, L. Wu: Controlled environment for reciprocity calibration of laboratory standard microphone and measurement of sensitivity pressure correction, Metrologia **36**, 275–280 (1999)

24.46 L. Wu, G.S.K. Wong, P. Hanes, W. Ohm: Measurement of sensitivity level pressure corrections for LS2P laboratory standard microphones, Metrologia **42**, 45–48 (2005)

24.47 K. Rasmussen: *The Influence of Environmental Conditions on the Pressure Sensitivity of Measurement Microphones*, Brüel Kjær Tech. Rev. **1** (Brüel Kjær, Nœrum 2001)

24.48 G.S.K. Wong, L. Wu: Controlled environment for reciprocity callibration of laboratory standard miccrophones and measurement of sensitivity pressure correction, Metrologica **36**, 275–280 (1999)

24.49 K. Rasmussen: The static pressure and temperature coefficients of laboratory standard microphones, Metrologia **36**(4), 265–273 (1999)

24.50 G.S.K. Wong: Primary pressure calibration by reciprocity. In: *AIP Handbook of Condenser Microphones. Theory, Calibration, and Measurements*, Chap.4, ed. by G.S.K. Wong, T.F.W. Embleton. (American Institute of Physics, New York 1995) p. 99

24.51 V. Nedzelnitsky: Primary method for calibrating free-field response. In: *AIP Handbook of Condenser*

24.52 IEC: IEC 61094-3 Ed. 1.0 b: 1995 Measurement Microphones – Part 3: Primary Method for Free-Field Calibration of Laboratory Standard Microphones by the Reciprocity Technique. (IEC, Geneva 1995)

Microphones. Theory, Calibration, and Measurements, ed. by G.S.K. Wong, T.F.W. Embleton (American Institute of Physics, New York 1995) pp. 103–119, Chap. 5

24.53 G.S.K. Wong, T.F.W. Embleton: Three-port two-microphone cavity for acoustical calibrations, J. Acoust. Soc. Am. **71**(5), 1276–1277 (1982)

24.54 G.S.K. Wong, L. Wu: *Interchange microphone method for calibration by comparison*, Congress on Noise Control Engineering, Christchurch (Institute of Noise Control Engineering, Ames 1998), paper #15

24.55 G.S.K. Wong: Precision method for phase match of microphones, J. Acoust. Soc. Am. **90**(3), 1253–1255 (1991)

24.56 IEC: IEC 61094-5: 2001 Measurement Microphones Part 5: Method for Pressure Calibration of Working Standard Microphones by Comparison (IEC, Geneva 2001)

24.57 G.S.K. Wong: Comparison methods for microphone calibration. In: *AIP Handbook of Condenser Microphones. Theory, Calibration, and Measurements*, ed. by G.S.K. Wong, T.F.W. Embleton (American Institute of Physics, New York 1995) pp. 215–222, Chap. 13

24.58 G.S.K. Wong: Precision A.C. voltage-level measuring system for acoustics, J. Acoust. Soc. Am. **65**(3), 830–837 (1979)

24.59 W. Koidan: Calibration of standard condenser microphones: Coupler versus electrostatic actuator, J. Acoust. Soc. Am. **44**(5), 1451–1453 (1968)

24.60 E. Frederiksen: Electrostatic Actuator. In: *AIP Handbook of Condenser Microphones. Theory, Calibration, and Measurements*, ed. by G.S.K. Wong, T.F.W. Embleton. (American Institute of Physics, New York 1995) pp. 231–246, Chap. 15

24.61 P.V. Brüel: *The accuracy of microphone calibration methods Part II*, Brüel and Kjær Tech. Rev. **1**, 3–26 (Brüel and Kjær, Nœrum 1965)

24.62 G. Rasmussen: *Free-Field and Pressure Calibration of Condenser Microphones Using Electrostatic Actuator*, Proc. 6th Int. Congress on Acoustics (Elsevier, New York 1968) pp. 25–28

24.63 IEC: *IEC 61094-6: 2004 Measurement Microphones – Part 6: Electrostatic Actuators for Determination of Frequency Response* (IEC, Geneva 2004)

24.64 G.B. Madella: Substitution method for calibrating a microphone, J. Acoust. Soc. Am. **20**(4), 550–551 (1948)

24.65 V. Nedzelnitsky: Laboratory microphone calibration methods at the National Institute of Standards and Technology U.S.A.. In: *AIP Handbook of Condenser Microphones. Theory, Calibration, and Measurements*, ed. by G.S.K. Wong, T.F.W. Embleton (American Institute of Physics, New York 1995) pp. 145–161, Chap. 8

24.66 G.S.K. Wong, L. Wu: Interlaboratory comparison of microphone calibration, J. Acoust. Soc. Am. **115**(2), 680–682 (2004)

24.67 G.S.K. Wong, L. Wu: Primary microphone calibration system stability, J. Acoust. Soc. Am. **114**(2), 577–579 (2003)

24.68 ISO: ISO 2533-1975(E), 1975 Standard Atmosphere, International Organization for Standardization (ISO, Geneva 1975)

24.69 P. Giacomo: Equation for the determination of the density of moist air (1981), Metrologia **18**, 33–40 (1982)

24.70 R.S. Davis: Equation for the determination of the density of moist air (1981/91), Metrologia **29**, 67–70 (1992)

24.71 G.S.K. Wong: Approximate equations for some acoustical and thermodynamic properties of standard air, J. Acoust. Soc. Jpn. E **11**, 145–155 (1990)

24.72 O. Cramer: The variation of the specific heat ratio and the speed of sound in air with temperature, pressure, humidity, and CO_2 concentration, J. Acoust. Soc. Am. **93**, 2510–2516 (1993)

24.73 G.S.K. Wong: Comments on The variation of the specific heat ratio and the speed of sound in air with temperature, pressure, humidity, and CO_2 concentration, J. Acoust. Soc. Am. **93**, 2510–2516 (1993)

24.74 G.S.K. Wong: Comments on The variation of the specific heat ratio and the speed of sound in air with temperature, pressure, humidity, and CO_2 concentration, J. Acoust. Soc. Am. **97**(5), 3177–3179 (1995)

24.75 G.S.K. Wong: Speed of sound in standard air, J. Acoust. Soc. Am. **79**, 1359–1366 (1986)

24.76 M. Greenspan: Comments on Speed of sound in standard air, J. Acoust. Soc. Am. **79**, 1359–1366 (1986)

24.77 M. Greenspan: Comments on Speed of sound in standard air, J. Acoust. Soc. Am. **82**, 370–372 (1987)

24.78 G.S.K. Wong: Response to Comments on 'Speed of sound in standard air', J. Acoust. Soc. Am. **82**, 370–372 (1987)

24.79 G.S.K. Wong: Response to Comments on 'Speed of sound in standard air', J. Acoust. Soc. Am. **83**, 373–374 (1987)

24.80 G.S.K. Wong: Air sound speed measurement and computation, A historical review. In: *Handbook of the Speed of Sound in Real Gases: Speed of Sound in Air*, Vol. III, ed. by A.J. Zuckerwar (Academic, New York 2002) pp. 265–284, Chap. 17, and references therein

24.81 C.M. Harris: Effects of humidity on the velocity of sound in air, J. Acoust. Soc. Am. **49**, 890–893 (1971)

24.82 G.P. Howell, C.L. Morfey: Frequency dependence of the speed of sound in air, J. Acoust. Soc. Am. **82**, 375–376 (1987)

24.83 G.S.K. Wong, T.F.W. Embleton: Experimental determination of the variation of specific heat ratio in air with humidity, J. Acoust. Soc. Am. **77**, 402–407 (1985)

24.84 J.R. Partington, W.G. Shilling: *The Specific Heats of Gases* (Ernest Benn, London 1924)

24.85 G.S.K. Wong, T.F.W. Embleton: Variation of specific heats and of specific heat ratios in air with humidity, J. Acoust. Soc. Am. **76**, 555–559 (1984)

24.86 A.J. Zuckerwar (Ed.): *Speed of Sound in Air*, Handbook of the Speed of Sound in real Gases, Vol. III (Academic, New York 2002)

24.87 E. Lide (Ed.): *CRC Handbook of Chemistry and Physics*, 83rd edn. (CRC, Boca Raton 2002-2003) pp. 6–182

25. Sound Intensity

Sound intensity is a vector that describes the flow of acoustic energy in a sound field. The idea of measuring this quantity directly, instead of deducing it from the sound pressure on the assumption of some idealized conditions, goes back to the early 1930s, but it took about 50 years before sound intensity probes and analyzers came on the market. The introduction of such instruments has had a significant influence on noise control engineering.

This chapter presents the *energy corollary*, which is the basis for sound power determination using sound intensity. The concept of reactive intensity is introduced, and relations between fundamental sound field characteristics and active and reactive intensity are presented and discussed.

Measurement of sound intensity involves the determination of the sound pressure and the particle velocity at the same position simultaneously. The established method of measuring sound intensity employs two closely spaced pressure microphones. An alternative method is based on the combination of a pressure microphone and a particle velocity transducer.

25.1	Conservation of Sound Energy	1054
25.2	Active and Reactive Sound Fields	1055
25.3	Measurement of Sound Intensity	1058
	25.3.1 The p–p Measurement Principle	1058
	25.3.2 The p–u Measurement Principle	1066
	25.3.3 Sound Field Indicators	1068
25.4	Applications of Sound Intensity	1068
	25.4.1 Noise Source Identification	1068
	25.4.2 Sound Power Determination	1070
	25.4.3 Radiation Efficiency of Structures	1071
	25.4.4 Transmission Loss of Structures and Partitions	1071
	25.4.5 Other Applications	1072
References		1072

Both methods are described, and their limitations are analyzed. Methods of calibrating and testing the two different measurement systems are also described. Finally the state of the art in the various areas of practical application of sound intensity measurement is summarized. These applications include the determination of the sound power of sources, identification and rank ordering of sources, and the measurement of the transmission of sound energy through partitions.

Sound waves are compressional oscillatory disturbances that propagate in a fluid. Moving fluid elements have kinetic energy, and changes in the pressure imply potential energy. Thus, there is a flow of energy involved in the phenomenon of sound; sources of sound emit sound power, and sound waves carry energy. In most cases the oscillatory changes undergone by the fluid are very small compared with the equilibrium values, from which it follows that typical values of the sound power emitted by sources of sound, which is an acoustic second-order quantity, are extremely small. The radiated sound power is a negligible part of the energy conversion of almost any source. However, energy considerations are nevertheless of great practical importance in acoustics. Their usefulness is mainly due to the fact that a statistical approach where the energy of the sound field is considered turns out to give extremely useful approximations in room acoustics and in noise control. In fact determining the sound power of sources is a central point in noise control engineering. The value and relevance of knowing the sound power radiated by a source is due to the fact that this quantity is largely independent of the surroundings of the source in the audible frequency range.

Sound intensity is a measure of the flow of acoustic energy in a sound field. More precisely, the sound intensity is a vector that describes the time average of the net flow of sound energy per unit area. The units of this quantity are power per unit area (W/m^2).

The advent of sound intensity measurement systems in the early 1980s had a significant influence on noise control engineering. One of the most important advantages of this measurement technique is that sound intensity measurements make it possible to determine the sound power of sources in situ without the use of costly special facilities such as anechoic and reverberation rooms, and sound intensity measurements are now routinely used in the determination of the sound power of machinery and other sources of noise in situ. Other important applications of sound intensity include the identification and rank ordering of partial noise sources, the determination of the transmission losses of partitions, and the determination of the radiation efficiencies of vibrating surfaces. Because the intensity is a vector it is also more suitable for visualization of sound fields than, for instance, the sound pressure.

25.1 Conservation of Sound Energy

In the absence of mean flow the instantaneous sound intensity is the product of the sound pressure $p(t)$ and the particle velocity $\boldsymbol{u}(t)$

$$\boldsymbol{I}(t) = p(t)\boldsymbol{u}(t) . \tag{25.1}$$

By combining the fundamental equations that govern a sound field, the equation of conservation of mass, the adiabatic relation between changes in the sound pressure and in the density, and Euler's equation of motion, one can show that the divergence of the instantaneous intensity equals the (negative) rate of change of the sum of the potential and kinetic energy density $w(t)$,

$$\nabla \cdot \boldsymbol{I}(t) = -\frac{\partial w(t)}{\partial t} . \tag{25.2}$$

This is the equation of conservation of sound energy [25.1, 2], expressing the simple fact that if there is net flow of energy away from a point in a sound field then the sound energy density at that point is reduced at a corresponding rate. Integrating this equation over a volume V enclosed by the surface S gives, with Gauss's divergence theorem,

$$\int_S \boldsymbol{I}(t) \cdot d\boldsymbol{S} = -\frac{\partial}{\partial t}\left(\int_V w(t) dV\right)$$
$$= -\frac{\partial E(t)}{\partial t} , \tag{25.3}$$

in which $E(t)$ is the total sound energy in the volume as a function of time. The left-hand term is the total net outflow of sound energy through the surface, and the right-hand term is the rate of change of the total sound energy in the volume. In other words, the net flow of sound energy out of a closed surface equals the (negative) rate of change of the sound energy in the volume enclosed by the surface because energy is conserved.

In practice we are often concerned with stationary sound fields and the time-averaged sound intensity rather than the instantaneous intensity, that is,

$$\langle \boldsymbol{I}(t) \rangle_t = \langle p(t)\boldsymbol{u}(t) \rangle_t . \tag{25.4}$$

For simplicity the symbol $\boldsymbol{I}$ is used for this quantity in what follows rather than the more precise notation $\langle \boldsymbol{I}(t) \rangle_t$. If the sound field is harmonic with angular frequency $\omega = 2\pi f$ the complex representation of the sound pressure and the particle velocity can be used, which leads to the expression

$$\boldsymbol{I} = 1/2 \operatorname{Re}(p\boldsymbol{u}^*) , \tag{25.5}$$

where $\boldsymbol{u}^*$ denotes the complex conjugate of $\boldsymbol{u}$.

A consequence of (25.3) is that the integral of the normal component of the time-averaged sound intensity over a closed surface is zero,

$$\int_S \boldsymbol{I} \cdot d\boldsymbol{S} = 0 , \tag{25.6}$$

when there is neither generation nor dissipation of sound energy in the volume enclosed by the surface, irrespective of the presence of steady sources *outside* the surface. If the surface encloses a steady source then the surface integral of the time-averaged intensity equals the sound power emitted by the source P_a, that is,

$$\int_S \boldsymbol{I} \cdot d\boldsymbol{S} = P_a, \tag{25.7}$$

irrespective of the presence of other steady sources outside the surface. This important equation is the basis for sound power determination using sound intensity.

In a plane wave propagating in the r-direction the sound pressure p and the particle velocity u_r are in phase and related by the characteristic impedance of air ρc, where ρ is the density of air and c is the speed of sound,

$$u_r(t) = \frac{p(t)}{\rho c} . \tag{25.8}$$

Under such conditions the intensity is

$$I_r = \langle p(t)u_r(t)\rangle_t = \langle p^2(t)\rangle_t / \rho c$$
$$= p_{rms}^2 / \rho c = \frac{|p|^2}{2\rho c}, \qquad (25.9)$$

where p_{rms}^2 is the mean square pressure (the square of the root-mean-square pressure) and p in the rightmost expression is the complex amplitude of the pressure in a harmonic plane wave. In the particular case of a plane propagating wave the sound intensity is seen to be simply related to the rms sound pressure, which can be measured with a single microphone. Under free-field conditions the sound field generated by any source of finite extent is *locally plane* sufficiently far from the source. This is the basis for the free-field method of sound power determination (which requires an anechoic room); the sound pressure is measured at a number of points on a sphere that encloses the source [25.3].

A practical consequence of (25.9) is the following extremely simple relation between the sound intensity level ($I_{ref} = 1\,\text{pW/m}^2$) and the sound pressure level ($p_{ref} = 20\,\mu\text{Pa}$),

$$L_I \simeq L_p. \qquad (25.10)$$

This is due to the fortuitous fact that

$$\rho c \simeq \frac{p_{ref}^2}{I_{ref}} = 400\,\text{kg/m}^2\text{s} \qquad (25.11)$$

in air under normal ambient conditions. At a static pressure of 101.3 kPa and a temperature of 23 °C the error of (25.10) is about 0.1 dB.

However, it should be emphasized that in the general case there is *no* simple relation between the sound intensity and the sound pressure, and both the sound pressure and the particle velocity must be measured simultaneously and their instantaneous product time-averaged as indicated by (25.4). This requires the use of a more complicated device than a single microphone.

25.2 Active and Reactive Sound Fields

Some typical sound field characteristics can be identified. For example, the sound field far from a source under free-field conditions has certain well-known properties (dealt with above); the sound field near a source has other characteristics, and some characteristics are typical of a reverberant sound field. One of the characteristics of the sound field near a source is that the sound pressure and the particle velocity are partly out of phase. To describe such a phenomenon one may introduce the concepts of *active* and *reactive* sound fields.

It takes four second-order quantities to describe the distributions and fluxes of sound energy in a stationary sound field completely [25.4–6]: potential energy density, kinetic energy density, active intensity (the quantity given by (25.4) and (25.5), usually simply referred to as the intensity), and the reactive intensity. The last of these quantities represents the non-propagating, oscillatory sound energy flux that is characteristic of a sound field in which the sound pressure and the particle velocity are in quadrature (90° out of phase), as for instance in the near field of a small source. The reactive intensity is a vector defined as the imaginary part of the product of the complex pressure and the complex conjugate of the particle velocity,

$$\boldsymbol{J} = 1/2\,\text{Im}\,(p\boldsymbol{u}^*) \qquad (25.12)$$

with the $e^{i\omega t}$ sign convention [cf. (25.5)]. Units: power per unit area (W/m^2). More-general time-domain formulations based on the Hilbert transform are also available [25.7]. Unlike the usual active intensity, the reactive intensity remains a somewhat controversial issue although the quantity was introduced more than half a century ago [25.8], perhaps because the vector $\boldsymbol{J}$ has no obvious physical meaning [25.9], or perhaps because describing an oscillatory flux by a time-averaged vector seems peculiar to some. However, even though the reactive intensity is of no obvious direct practical use it is nevertheless quite convenient that we have a quantity that makes it possible to describe and quantify the particular sound field conditions in the near field of sources in a precise manner. This will become apparent in Sect. 25.3.2.

It can be shown from (25.5) that the active intensity is proportional to the gradient of the phase of the sound pressure [25.10],

$$\boldsymbol{I} = -\frac{|p|^2}{2\rho c}\frac{\nabla\varphi}{k}, \qquad (25.13)$$

where $k = \omega/c$ is the wavenumber. Thus the active intensity is orthogonal to surfaces of equal phase, that is, the wavefronts [25.5]. Likewise it can be shown from (25.12) that the reactive intensity is proportional to the

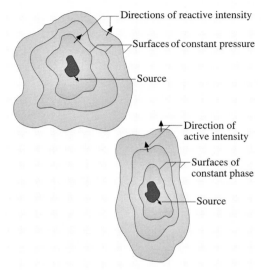

Fig. 25.1 Surfaces of constant phase and surfaces of constant pressure (After [25.5])

gradient of the mean square pressure [25.11],

$$J = -\frac{\nabla(|p|^2)}{4\rho c k}, \qquad (25.14)$$

and thus is orthogonal to surfaces of equal pressure [25.5]. See Fig. 25.1

Very near a sound source the reactive field is often stronger than the active field at low frequencies. However, the reactive field dies out rapidly with increasing distance to the source. Therefore, even at a fairly moderate distance from the source, the sound field is dominated by the active field. The extent of the reactive field depends on the frequency and the radiation characteristics of the sound source. In practice, the reactive field may usually be assumed to be negligible at a distance greater than, say, half a wavelength from the source.

The fact that I is the real part and J is the imaginary part of the product of the pressure and the complex conjugate of the particle velocity has led to the concept of complex sound intensity [25.4, 10],

$$\boldsymbol{I} + \mathrm{i}\boldsymbol{J} = 1/2 \, p\boldsymbol{u}^*. \qquad (25.15)$$

Note that

$$\frac{I_\mathrm{r} + \mathrm{i}J_\mathrm{r}}{1/2\,|u_\mathrm{r}|^2} = \frac{1/2\,|p|^2}{I_\mathrm{r} - \mathrm{i}J_\mathrm{r}} = \frac{p}{u_\mathrm{r}} = Z_\mathrm{s}, \qquad (25.16)$$

which shows that there is a simple relation between the complex intensity and the wave impedance of the sound

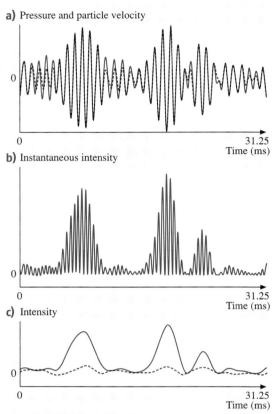

Fig. 25.2a–c Measurement 30 cm from a loudspeaker driven with one-third octave noise with a center frequency of 1 kHz. (**a**) Instantaneous sound pressure (*solid line*); instantaneous particle velocity multiplied by ρc (*dashed line*); (**b**) instantaneous sound intensity; (**c**) the real part of the complex instantaneous intensity (*solid line*); the imaginary part of the complex instantaneous intensity (*dashed line*) (After [25.7])

field Z_s [25.6, 10]. It is also worth noting that

$$\frac{I_\mathrm{r}^2 + J_\mathrm{r}^2}{1/4\,|p|^2\,|u_\mathrm{r}|^2} = 1. \qquad (25.17)$$

Figures 25.1–25.4 demonstrate the physical significance of the active and reactive intensities. The figures show the sound pressure and a component of the particle velocity as functions of time, the corresponding instantaneous sound intensity, and the real and imaginary part of the *complex instantaneous intensity*, that is, the quantity

$$\boldsymbol{I}_\mathrm{c}(t) = 1/2 \, (p(t) + \mathrm{i}\tilde{p}(t))(\boldsymbol{u}(t) - \mathrm{i}\tilde{\boldsymbol{u}}(t)), \qquad (25.18)$$

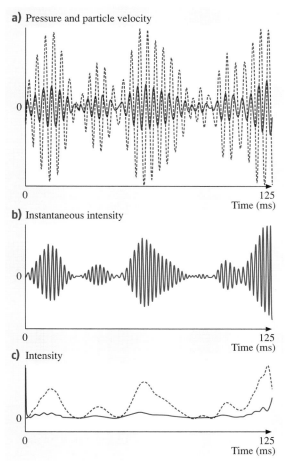

Fig. 25.3 Measurement in the near field of a loudspeaker driven with one-third octave noise with a center frequency of 250 Hz. *Key* as in Fig. 25.2 (After [25.7])

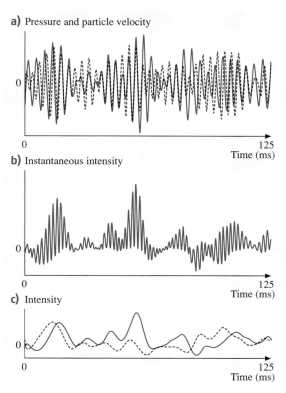

Fig. 25.4 Measurement 30 cm from a vibrating steel box driven with one-third octave noise with a center frequency of 250 Hz. *Key* as in Fig. 25.2 (After [25.7])

where $\tilde{p}(t)$ and $\tilde{u}(t)$ are the Hilbert transforms of the pressure and the particle velocity [25.12]. The time average of this quantity in a steady sound field is the complex intensity given by (25.15).

In a narrow-band sound field the real and imaginary parts of the complex instantaneous intensity are low-pass signals that represent short-time average values of the two components of (25.15) [25.7].

Figure 25.2 shows the result of a measurement at a position 30 cm (about one wavelength) from a small loudspeaker driven with a band of one-third octave noise. The sound pressure and the particle velocity (multiplied by ρc) are almost identical; therefore the instantaneous intensity is always positive, and the real part of the complex instantaneous intensity is much larger than the imaginary part. This is an active sound field. Figure 25.3 shows the result of a similar measurement at a distance of a few centimeters (less than one tenth of a wavelength) from the loudspeaker cone. In this case the sound pressure and the particle velocity are almost in quadrature, and as a result the instantaneous intensity fluctuates about zero, that is, sound energy flows back and forth, out of and into the loudspeaker, and the imaginary part of the complex instantaneous intensity is much larger than the real part. This is an example of a strongly reactive sound field. Figure 25.4 shows data measured about 30 cm from a vibrating box made of 3 mm steel plates. It is apparent that the vibrating structure generates a much more complicated sound field than a loudspeaker does. And finally Fig. 25.5 shows the result of a measurement in a reverberant room several meters from the loudspeaker generating the sound field. Here the sound pressure and the particle velocity appear to be uncorrelated signals; this is neither an active nor a reactive sound field; this is a diffuse sound field.

25.3 Measurement of Sound Intensity

Acousticians have attempted to measure sound intensity since the early 1930s, but measurement of sound intensity is more difficult than measurement of the sound pressure, and it took almost 50 years before sound intensity measurement systems came on the market. The first international standards for measurements using sound intensity and for instruments for such measurements were issued in the middle of the 1990s. A description of the history of the development of sound intensity measurement up to the middle of the 1990s is given in *Fahy*'s monograph *Sound Intensity* [25.2].

Measurement of sound intensity involves determination of the sound pressure and the particle velocity at the same position simultaneously. In the general case at least two transducers are required. There are three possible measurement principles: (i) one can determine the particle velocity from a finite-difference approximation of the pressure gradient using two closely spaced pressure microphones and use the average of the two microphone signals as the pressure [25.2, 13] (the *two-microphone* or *p–p* method); (ii) one can combine a pressure microphone with a particle velocity transducer [25.2] (the *p–u* method); and (iii) one can determine the pressure from a finite-difference approximation of the divergence of the particle velocity [25.14] (the *u–u* method). The first of these methods is well established. The second method has been hampered by the absence of reliable particle velocity transducers, but with the recent advent of the Microflown particle velocity sensor [25.15, 16] it seems to have potential [25.17]. The third method, which involves three matched pairs of particle velocity transducers, has never been used in air and is mentioned here for the sake of completeness.

25.3.1 The *p–p* Measurement Principle

For more than 25 years the *p–p* method based on two closely spaced pressure microphones (or hydrophones) has dominated sound intensity measurement. This method relies on a finite-difference approximation to the sound pressure gradient. Both the International Electrotechnical Commission (IEC) standard on instruments for the measurement of sound intensity and the corresponding American National Standards Institute (ANSI) standard deal exclusively with *p–p* measurement systems [25.18, 19]. The success of this method is related to the fact that condenser microphones are more stable and reliable than any other acoustic transducer.

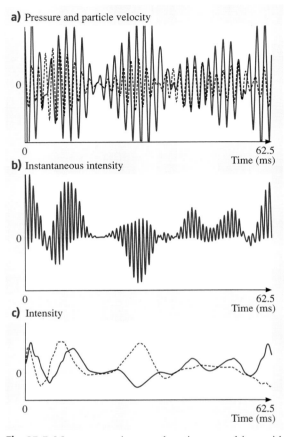

Fig. 25.5 Measurement in a reverberation room driven with one-third octave noise with a center frequency of 500 Hz. *Key* as in Fig. 25.2 (After [25.7])

Two pressure microphones are placed close together. The particle velocity component in the direction of the axis through the two microphones r is obtained from Euler's equation of motion

$$\frac{\partial p(t)}{\partial r} + \rho \frac{\partial u_r(t)}{\partial t} = 0 , \quad (25.19)$$

where the gradient of the pressure is approximated by a finite difference. Thus the particle velocity is determined as

$$\hat{u}_r(t) = -\frac{1}{\rho} \int_{-\infty}^{t} \frac{p_2(\tau) - p_1(\tau)}{\Delta r} \, d\tau , \quad (25.20)$$

where p_1 and p_2 are the signals from the two microphones, Δr is the microphone separation distance, and

τ is a dummy time variable. The caret indicates that the result is an estimate, which of course is an approximation to the true particle velocity. The sound pressure at the center of the probe is estimated as

$$\hat{p}(t) = \frac{p_1(t) + p_2(t)}{2}, \quad (25.21)$$

and the time-averaged intensity component in the r-direction is

$$\hat{I}_r = \langle \hat{p}(t)\hat{u}_r(t) \rangle_t$$
$$= \left\langle \frac{p_1(t) + p_2(t)}{2} \int_{-\infty}^{t} \frac{p_1(\tau) - p_2(\tau)}{\rho \Delta r} \, d\tau \right\rangle_t. \quad (25.22)$$

Some sound intensity analyzers use (25.22) to measure the intensity in frequency bands (usually one-third octave bands). Another type calculates the intensity from the imaginary part of the cross spectrum of the two microphone signals $S_{12}(\omega)$,

$$\hat{I}_r(\omega) = -\frac{1}{\omega \rho \Delta r} \text{Im}[S_{12}(\omega)]. \quad (25.23)$$

The frequency-domain formulation is equivalent to the time-domain formulation, and (25.23) gives exactly the same result as (25.22) when the intensity spectrum is integrated over the frequency band of concern [25.20, 21]. The frequency-domain formulation makes it possible to determine sound intensity with a dual-channel fast Fourier transform (FFT) analyzer.

The most common microphone arrangement is known as *face-to-face*, but another called *side-by-side* is occasionally used. The latter arrangement has the advantage that the diaphragms of the microphones can be placed very near a radiating surface, but the disadvantage that the microphones disturb each other acoustically more than they do in the other configuration. Figure 25.6 shows a three-dimensional (3-D) intensity probe produced by Ono-Sokki with yet another configuration. At high frequencies the face-to-face configuration with a solid plug between the microphones is superior [25.22]. Such a sound intensity probe produced by Brüel & Kjær is shown in Fig. 25.7. The *spacer* between the microphones tends to stabilize the *acoustic distance* between them.

Sources of Error in Measurement of Sound Intensity with p–p Measurement Systems

It is far more difficult to measure sound intensity than to measure sound pressure, and a surprisingly large part of

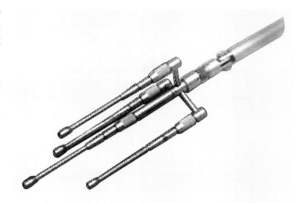

Fig. 25.6 Three-dimensional sound intensity probe for vector measurements (Ono Sokki, Japan)

Fig. 25.7 Sound intensity probe with the microphones in the *face-to-face* configuration (Brüel & Kjær, Denmark)

the literature on sound intensity is concerned with identifying and studying the sources of error. One confusing problem is that the accuracy of any sound intensity measurement system depends strongly on the sound field under study; sometimes such measurements are not very difficult, but under certain conditions even very small imperfections of the measuring equipment can have a significant influence on the results. Another complication is that small local errors are sometimes amplified into large global errors when the intensity is integrated over a closed surface [25.23]. The opposite may also happen. Yet another problem is that the distribution of the sound intensity in the near field of a source of sound often is far more complicated than the distribution of the sound pressure, indicating that sound fields can be much more complicated than earlier realized [25.2]. The problems are reflected in the fairly complicated international and national standards for sound power determination using sound intensity, ISO 9614-1, ISO 9614-2, ISO 9614-3, and ANSI S12.12 [25.24–27].

The most important limitations of sound intensity measurement systems based on the p–p approach are caused by the finite-difference approximation, scattering and diffraction, and instrumentation phase mismatch.

The Finite-Difference Error

The accuracy of the finite-difference approximation obviously depends on the separation distance and the wavelength. For a plane wave of axial incidence the finite-difference error can be shown to be [25.22]

$$\hat{I}_r / I_r = \frac{\sin k\Delta r}{k\Delta r}, \qquad (25.24)$$

This expression is shown in Fig. 25.8 for various values of the microphone separation distance. More complicated expressions for other sound field conditions are available in the literature [25.30]. The upper frequency limit of p–p intensity probes has generally been considered to be the frequency at which the finite-difference error for axial plane-wave incidence is acceptably small. With 12 mm between the microphones (a typical value) this gives an upper limiting frequency of about 5 kHz.

Scattering and Diffraction

It is evident that the effect of scattering and diffraction depends on the geometry of the microphone arrangement. Several configurations are possible, but in the early 1980s it was shown experimentally that the face-to-face configuration with a solid spacer between the two microphones is particularly favorable [25.22]. About 15 years later it was discovered that the effect of scattering and diffraction in combination with the resonance of the small cavity between the spacer and the diaphragm of each microphone not only tends to counterbalance the finite-difference error but in fact for a certain length of the spacer cancels it almost perfectly over a wide frequency range under fairly general sound field conditions [25.28]. (This finding had been anticipated much earlier [25.31].) 25.9 shows the increase of the pressure on two 1/2 inch microphones in the face-to-face configuration for axial sound incidence, and Fig. 25.10, which corresponds to Fig. 25.8, shows the error of the resulting sound intensity estimate. A practical consequence is that the upper frequency limit of a sound intensity probe based on two 1/2 inch microphones separated by a 12 mm spacer in the face-to-face arrangement is about 10 kHz, which is an octave higher than the frequency limit determined by the finite-difference

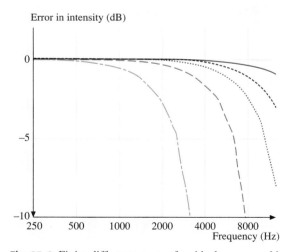

Fig. 25.8 Finite-difference error of an ideal p–p sound intensity probe in a plane wave of axial incidence for different values of the separation distance: 5 mm (*solid line*); 8.5 mm (*dashed line*); 12 mm (*dotted line*); 20 mm (*long dashes*); 50 mm (*dash-dotted line*) (After [25.28])

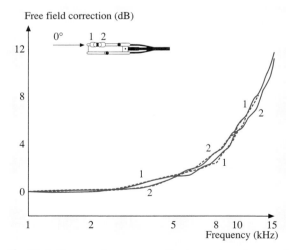

Fig. 25.9 Pressure increase on the two microphones of a sound intensity probe with 1/2 inch microphones separated by a 12 mm spacer for axial sound incidence. Experimental results (*solid line*); numerical results (*dashed line*) (After [25.29])

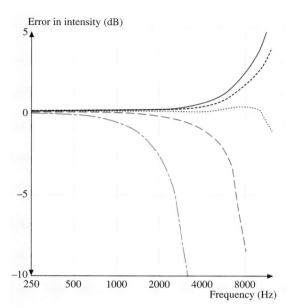

Fig. 25.10 Error of a sound intensity probe with 1/2 inch microphones in the face-to-face configuration in a plane wave of axial incidence for different spacer lengths: 5 mm (*solid line*); 8.5 mm (*dashed line*); 12 mm (*dotted line*); 20 mm (*long dashes*); 50 mm (*dash-dotted line*) (After [25.28])

Phase mismatch between the two measurement channels is the most serious source of error in measurement of sound intensity with p–p measurement systems, even with the best equipment that is available today. It can be shown that a small (positive or negative) phase mismatch error φ_{pe} gives rise to a bias error that can be approximated by the following expression,

$$\hat{I}_{\text{r}} \simeq I_{\text{r}} - \frac{\varphi_{\text{pe}}}{k\Delta r} \frac{p_{\text{rms}}^2}{\rho c} = I_{\text{r}} \left(1 - \frac{\varphi_{\text{pe}}}{k\Delta r} \frac{p_{\text{rms}}^2/\rho c}{I_{\text{r}}}\right), \tag{25.25}$$

where I_{r} is the true intensity (unaffected by phase mismatch) [25.35]. This expression shows that the effect of a given phase error is inversely proportional to the frequency and the microphone separation distance and is proportional to the ratio of the mean square sound pressure to the sound intensity. If this ratio (which in logarithmic form is known as the *pressure-intensity index*) is large, say, more than 10 dB, then the true phase difference between the two pressure signals in the sound field is small and even the small phase errors mentioned above will give rise to significant bias errors. Because of phase mismatch it will rarely be possible to make reliable measurements below, say, 80 Hz unless a longer spacer than the usual 12 mm spacer is used.

Figure 25.11 shows the effect of a positive phase error of 0.3° in a plane wave of axial incidence, calcu-

approximation. The combination of 1/2 inch microphones and a 12 mm spacer is now regarded as optimal, and longer spacers are only used when the focus is exclusively on low frequencies. With a longer spacer between 1/2 inch microphones the resonance occurs at too high a frequency to be of any help and the finite-difference error will dominate; thus an intensity probe with a 50 mm spacer has an upper frequency limit of about 1.2 kHz.

Phase Mismatch

Unless the measurement is compensated for phase mismatch the microphones for measurement of sound intensity with the p–p method have to be phase-matched extremely well, and state-of-the-art sound intensity microphones are matched to a maximum phase response difference of 0.05° below 250 Hz and a phase difference proportional to the frequency above 250 Hz (say, 0.2° at 1 kHz) [25.33] The proportionality to the frequency is a consequence of the fact that phase mismatch in this frequency range is caused by differences between the resonance frequencies and the damping of the two microphones [25.34].

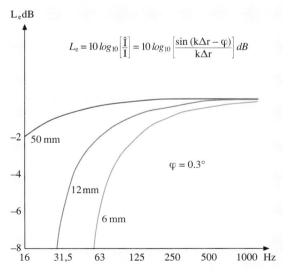

Fig. 25.11 Error due to a phase error of 0.3° in a plane propagating wave (After [25.32])

lated for different values of the microphone separation distance Δr. It is apparent that large underestimation errors occur at low frequencies. Note, however, that still larger errors will occur when the ratio of the mean square pressure to the sound intensity takes a larger value than unity, which is likely to happen unless the measurement takes place in an anechoic room. It should also be noted that a frequency-independent value of the phase error φ_{pe} is very unlikely [25.34]. Finally it is worth mentioning again that state-of-the-art microphone pairs for sound intensity measurements are much better matched now than within $0.3°$.

Inspection of (25.25) shows that the simple expedient of reversing a p–p probe makes it possible to eliminate the influence of p–p phase mismatch; the intensity changes sign but the error does not [25.20, 36]. Unfortunately, most p–p intensity probes are not symmetrical and are therefore not suitable for real measurements with the probe reversed.

The ratio of the phase error to the product of the frequency and the microphone separation distance can be measured (usually in the form of the so-called *pressure-residual intensity index*) by exposing the two pressure microphones to the same pressure. The residual intensity is the *false* sound intensity indicated by the instrument when the two microphones are exposed to the same pressure p_0, for instance in a small cavity driven by a wide-band source. A commercial example of such a device is shown in Fig. 25.12. When the pressure on the two microphones is the same (in amplitude as well as in phase) the true intensity is zero, and the magnitude of the indicated, residual intensity,

$$I_0 = -\frac{\varphi_{\text{pe}}}{k\Delta r} \frac{p_0^2}{\rho c}, \qquad (25.26)$$

should obviously be as small as possible. Expressed in terms of this quantity (25.25) takes the form

$$\hat{I}_r = I_r + \left(\frac{I_0}{p_0^2}\right) p_{\text{rms}}^2$$

$$= I_r \left(1 + \frac{I_0}{p_0^2/\rho c} \frac{p_{\text{rms}}^2/\rho c}{I_r}\right). \qquad (25.27)$$

An alternative form that follows directly from (25.27) has the advantage of expressing the error in terms of the available, biased intensity estimate [25.37],

$$\hat{I}_r = I_r \left(1 - \frac{I_0}{p_0^2/\rho c} \frac{p_{\text{rms}}^2/\rho c}{\hat{I}_r}\right)^{-1}. \qquad (25.28)$$

The pressure-residual intensity index of the measurement system can be measured once (and should be checked occasionally). Combined with the pressure-intensity index of the actual measurement it makes it possible to estimate the error. Some analyzers can give warnings when the error due to phase mismatch as predicted by (25.28) exceeds a specified level. The bias error predicted by (25.28) is less than $\pm 10\log(1\pm 10^{-K/10})$ dB if the pressure-intensity index of the measurement, δ_{pI}, is less than the pressure-residual intensity index, δ_{pIo}, minus the *bias error index K*, that is

$$\delta_{\text{pI}} = 10\log\left(\frac{p_{\text{rms}}^2/\rho c}{|I_r|}\right) < \delta_{\text{pIo}} - K$$

$$= 10\log\left(\frac{p_0^2/\rho c}{|I_0|}\right) - K. \qquad (25.29)$$

The larger the value of K the smaller the maximum error that can occur and the stronger and more restrictive on the range of measurement is the requirement, as demonstrated by Fig. 25.13. Note that a negative residual intensity (which leads to underestimation of a positive intensity component) gives larger errors (in decibels) than a corresponding positive residual intensity (which leads to overestimation of a positive intensity component) unless the error is small. A bias error index of 7 dB corresponds to the error due to phase mismatch being

Fig. 25.12 Coupler for measurement of the pressure-residual intensity index of p–p sound intensity probes (Brüel & Kjær, Denmark)

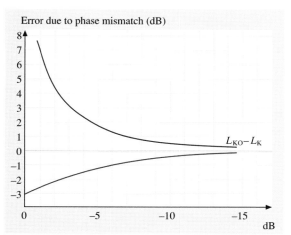

Fig. 25.13 Maximum error due to phase mismatch as a function of the bias error index K for negative residual intensity (*upper curve*) and for positive residual intensity (*lower curve*) (After [25.38])

less than 1 dB (*survey accuracy* [25.25]), and a bias error index of 10 dB corresponds to the error being less than 0.5 dB (*engineering accuracy* [25.25]). These values correspond to the phase error of the equipment being five and ten times less than the actual phase angle in the sound field, respectively. The quantity $\delta_{\text{pIo}} - K$ is known as the *dynamic capability* of the measurement system.

Many applications of sound intensity measurements involve integrating the normal component of the intensity over a surface. The global versions of (25.27) and (25.28) are found by integrating the normal component over a surface that encloses a source. The result is the expressions

$$\hat{P}_a = \int_S \hat{\boldsymbol{I}} \cdot d\boldsymbol{S}$$

$$\simeq P_a \left(1 - \frac{\varphi_{\text{pe}}}{k \Delta r} \frac{\int_S \left(p_{\text{rms}}^2/\rho c\right) dS}{\int_S \boldsymbol{I} \cdot d\boldsymbol{S}} \right)$$

$$= P_a \left(1 + \frac{I_0 \rho c}{p_0^2} \frac{\int_S \left(p_{\text{rms}}^2/\rho c\right) dS}{\int_S \boldsymbol{I} \cdot d\boldsymbol{S}} \right)$$

$$= P_a \left(1 - \frac{I_0 \rho c}{p_0^2} \frac{\int_S \left(p_{\text{rms}}^2/\rho c\right) dS}{\int_S \hat{\boldsymbol{I}} \cdot d\boldsymbol{S}} \right)^{-1}, \quad (25.30)$$

where P_a is the true sound power of the source within the surface [25.35]. The condition expressed by (25.29) still applies, although the pressure-intensity index now involves averaging over the measurement surface.

Sources outside the measurement surface do not contribute to the surface integral of the true intensity [the denominator of the second term on the right-hand side of (25.30)], but they invariably increase the surface integral of the mean square pressure (the numerator of the second term), as demonstrated by the results shown in Fig. 25.14. It follows that even a very small phase error imposes restrictions on the amount of extraneous noise that can be tolerated in sound power measurement with a *p–p* sound intensity measurement system.

Other Sources of Error

It can be difficult to avoid that the intensity probe is exposed to airflow, for example in measurements near air-cooled machinery. Strictly speaking the *p–p* measurement principle is simply not valid in such circumstances [25.40]. However, practice shows that the resulting fundamental error is insignificant in airflows of moderate velocities (say, up to 10 m/s), and that the false, low-frequency intensity signals produced by turbulence are a more serious problem under such conditions. Turbulence generates pressure fluctuations (flow noise, unrelated to the sound field) that contaminate the signals from the two microphones, and at low frequencies these signals are correlated and thus interpreted by the measurement system as intensity. The resulting false intensity is unpredictable and can be positive or negative. This is mainly a problem below 200 Hz [25.41]. A longer spacer helps reducing the problem, but the most

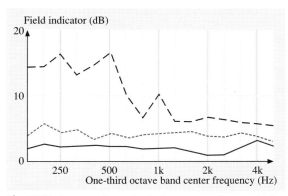

Fig. 25.14 Pressure-intensity index on a surface enclosing a noise source determined under three different conditions: measurement using a reasonable surface (*solid line*); measurement using an eccentric surface (*dashed line*); measurement with strong background noise at low frequencies (*long dashes*) (After [25.39])

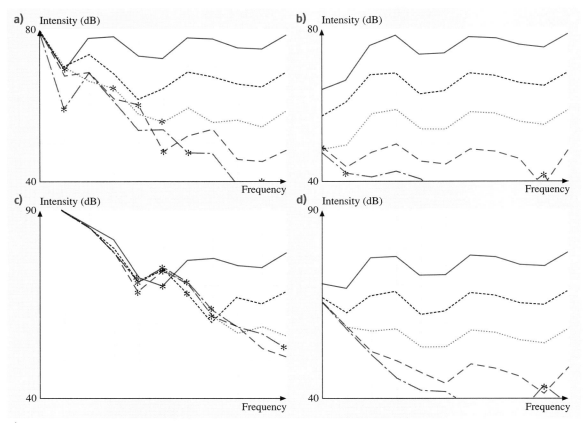

Fig. 25.15 Measurement 60 cm from a loudspeaker and a fan producing airflow of about 4 m/s at the measurement position. *Top figures*: 50 mm spacer; *bottom figures*: 12 mm spacer; *left figures*: no windscreen; *right figures*: windscreen. Asterisk: negative intensity estimate. Nominal sound power level of source: 100 dB (*solid line*); 90 dB (*dashed line*); 80 dB (*dotted line*); 70 dB (*long dashes*); 60 dB (*dash-dotted line*) (After [25.41])

efficient expedient is to use a windscreen of porous foam, as demonstrated by the results shown in Fig. 25.15. However, windscreens give rise to errors at low frequencies in highly reactive sound fields, because the losses of the foam modify Euler's equation of motion (25.19) [25.42, 43]. Thus measurements with windscreened probes very near sources should be avoided.

The finite averaging time used in any measurement results in a random error, and this random error is usually larger in sound intensity measurements than in sound pressure measurements – sometimes much larger [25.44–46]. Thus to maintain the same random error as in measurement of the sound pressure one will usually have to average over a longer time. However, most applications of sound intensity involve integrating over a measurement surface, and since it can be shown that it is the total averaging time that matters [25.47],

the problem is less serious in practice than one might expect from observations at discrete positions.

Yet another source of error is the electrical self-noise from the microphones and preamplifier circuits. Although the level of such noise is very low in modern 1/2 inch condenser microphones it can have a serious influence on sound intensity measurement at low levels. The noise increases the random error associated with a finite averaging time at low frequencies, in particular if the pressure-intensity index takes a large value. The problem reveals itself by poor or nonexistent repeatability [25.48].

Calibration of p–p Sound Intensity Measurement Systems

Calibration of p–p sound intensity measurement systems is fairly straightforward: the two pressure microphones are calibrated with a pistonphone in the usual manner. However, because of the serious influence of phase mismatch the pressure-residual intensity index should also be determined. The IEC standard for sound intensity instruments and its North American counterpart [25.18, 19] specify minimum values of the acceptable pressure-residual intensity index for the probe as well as for the processor; according to the results of a test the instruments are classified as being of class 1 or class 2. The test involves subjecting the two microphones of the probe to identical pressures in a small cavity driven with wide-band noise. A similar test of the processor involves feeding the same signal to the two channels. The pressure and intensity response of the probe should also be tested in a plane propagating wave as a function of the frequency, and the directional response of the probe is required to follow the ideal cosine law within a specified tolerance.

According to the two standards a special test is required in the frequency range below 400 Hz: the intensity probe should be exposed to the sound field in a standing-wave tube with a specified standing wave ratio (24 dB for probes of class 1). When the sound intensity probe is drawn through this interference field the sound intensity indicated by the measurement system should be within a certain tolerance.

Figure 25.16a illustrates how the sound pressure, the particle velocity and the sound intensity vary with

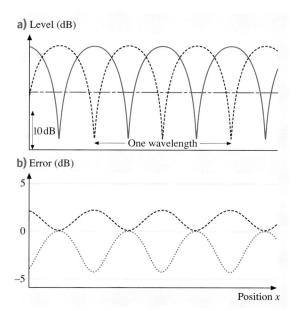

Fig. 25.16 (a) Sound pressure level (*solid line*), particle velocity level (*dashed line*), and sound intensity level (*dash-dotted line*) in a standing wave with a standing-wave ratio of 24 dB (b) Estimation error of a sound intensity measurement system with a residual pressure-intensity index of 14 dB (positive and negative residual intensity) (After [25.49])

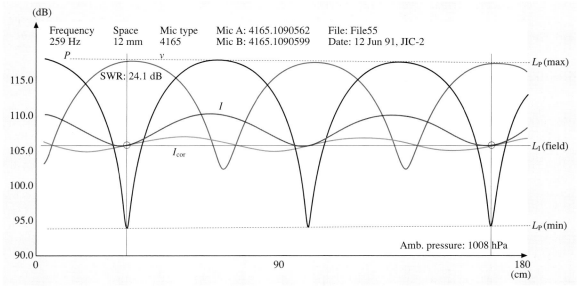

Fig. 25.17 Response of a sound intensity probe exposed to a standing wave: sound pressure, particle velocity, intensity, and phase-corrected intensity (After [25.50])

position in a one-dimensional interference field with a standing-wave ratio of 24 dB. It is apparent that the pressure-intensity index varies strongly with the position in such a sound field. Accordingly, the influence of a given phase error depends on the position, as shown in the calculations presented in Fig. 25.16b. Figure 25.17 shows an example of measurements with an intensity probe drawn through a standing-wave tube. The standing-wave test will also reveal other sources of error than phase mismatch, for example, the influence of an unacceptably high vent sensitivity of the microphones [25.51].

25.3.2 The p–u Measurement Principle

A p–u sound intensity measurement system combines two fundamentally different transducers, a pressure microphone and a particle velocity transducer. The sound intensity is simply the time average of the instantaneous product of the pressure and particle velocity signal,

$$I_r = \langle p(t)u_r(t)\rangle_t = 1/2 \operatorname{Re}\left(pu_r^*\right) , \quad (25.31)$$

where the latter expression is based on the complex exponential representation. In the frequency domain the expression takes the form

$$I_r(\omega) = S_{pu}(\omega). \quad (25.32)$$

Equation (25.32) gives the same result as (25.31) when the intensity spectrum is integrated over the frequency band of concern.

A p–u sound intensity probe that combined a pressure microphone with a transducer based on the convection of an ultrasonic beam by the particle velocity flow was produced by Norwegian Electronics for some years [25.2, 52], but the device was somewhat bulky, very sensitive to airflow, and difficult to calibrate, and production was stopped in the middle of the 1990s. More recently, a micromachined transducer called the *Microflown* has become available for measurement of the particle velocity [25.15], and an intensity probe based on this device in combination with a small pressure microphone is now in commercial production [25.16]; see Fig. 25.18. The Microflown particle velocity transducer consists of two short, thin, closely spaced wires, heated to about 300 °C [25.15]. Their resistance depends on the temperature. A particle velocity signal in the perpendicular direction changes the temperature distribution instantaneously, because one of the wires will be cooled more than the other by the airflow. The frequency response of this device is relatively flat up to a corner

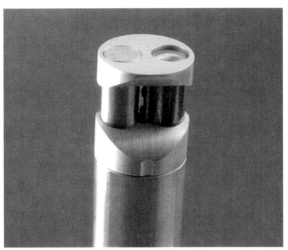

Fig. 25.18 A p–u sound intensity probe (by Microflown Technologies, The Netherlands)

frequency of the order of 1 kHz caused by diffusion effects related to the distance between the two wires. A second corner frequency at about 10 kHz is caused by the thermal heat capacity of the wires. Between 1 and 10 kHz there is a roll-off of 6 dB per octave. The particle velocity transducer is combined with a small electret condenser microphone in the 1/2 inch sound intensity probe shown in Fig. 25.18. The velocity transducer is mounted on a small, solid cylinder, and the condenser microphone is mounted inside another, hollow cylinder. The geometry of this arrangement increases the sensitivity of the velocity transducer. Unlike Norwegian Electronics' intensity probe the Microflown probe is very small – in fact much smaller than a standard p–p probe. Thus it is possible to measure very close to a vibrating surface with this device. At the time of writing there is still relatively little experimental evidence of the practical utility of the Microflown intensity probe, but it seems to be promising [25.17]. Figure 25.19 shows the results of sound power measurements with a Microflown intensity probe in comparison with similar results made with a Brüel & Kjær p–p sound intensity probe.

A variant of the p–u method, the *surface intensity method*, combines a pressure microphone with a transducer that measures the vibrational displacement, velocity or acceleration of a solid surface, for example with an accelerometer or a laser vibrometer [25.53–55]; see Fig. 25.20. Obviously, this method can only give the normal component of the sound intensity near vibrating surfaces. A disadvantage of the surface intensity method is that sound fields near complex sources are often very

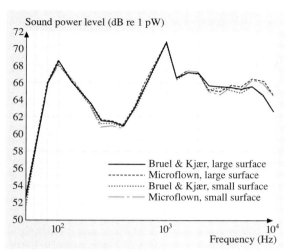

Fig. 25.19 Sound power of a source measured using a p–u intensity probe produced by Microflown and p–p intensity probe produced by Brüel & Kjær. Brüel & Kjær probe on large measurement surface (*solid line*); Microflown probe on large measurement surface (*dashed line*); Microflown probe on small measurement surface (*dotted line*) (After [25.17])

complicated, which makes it necessary to measure at many points.

Sources of Error in Measurement of Sound Intensity with p–u Measurement Systems

Evidently, some of the limitations of any p–u intensity probe must depend on the particulars of the particle velocity transducer. However, some general problems are described in the following.

Phase Mismatch

Irrespective of the measurement principle used in determining the particle velocity there is one fundamental problem: the pressure and the particle velocity transducer will invariably have different phase responses. One must compensate for this p–u phase mismatch, otherwise the result may well be meaningless. In fact even a small residual p–u mismatch error can have serious consequences under certain conditions. This can be seen by introducing such a small phase error, φ_{ue}, in (25.31). The result is

$$\hat{I}_r = 1/2 \, \mathrm{Re} \left(p u_r^* e^{-i\varphi_{ue}} \right)$$
$$= \mathrm{Re} \left[(I_r + i J_r)(\cos \varphi_{ue} - i \sin \varphi_{ue}) \right]$$
$$\simeq I_r + \varphi_{ue} J_r \,, \quad (25.33)$$

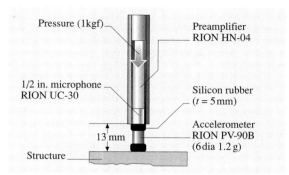

Fig. 25.20 Hand-held probe for surface intensity measurement (After [25.53])

where J_r is the reactive intensity, cf. (25.12). Equation (25.33) demonstrates that even a small uncompensated p–u phase mismatch error will give rise to a significant bias error when $J_r \gg I_r$. On the other hand it also shows that substantial p–u phase errors can be tolerated if $J_r \ll I_r$. For example, even a phase mismatch of 35° gives a bias error of less than 1 dB under such conditions. In other words, phase calibration is critical when measurements are carried out under near-field conditions, but not at all critical if the measurements are carried out in the far field of a source. The *reactivity* (the ratio of the reactive to the active intensity) indicates whether this source of error is of concern or not.

Whereas phase-mismatching of a p–p sound intensity measurement system can in principle be eliminated simply by reversing the probe and measuring again, it can be seen from (25.33) that reversing a p–u probe simply changes the sign of the result, including the bias error. In other words, probe reversal does not provide any new information.

The global version of (25.33) is found by integrating over a surface that encloses a source,

$$\hat{P}_a = \int_S \mathrm{Re}\left[(\boldsymbol{I}+i\boldsymbol{J})(\cos\varphi_{ue} - i\sin\varphi_{ue})\right] \cdot \mathrm{d}\boldsymbol{S}$$

$$\simeq P_a + \varphi_{ue} \int_S \boldsymbol{J} \cdot \mathrm{d}\boldsymbol{S}$$

$$= P_a \left(1 + \varphi_{ue} \frac{\int_S \boldsymbol{J} \cdot \mathrm{d}\boldsymbol{S}}{\int_S \boldsymbol{I} \cdot \mathrm{d}\boldsymbol{S}} \right)$$

$$= P_a \left(1 - \varphi_{ue} \frac{\int_S \boldsymbol{J} \cdot \mathrm{d}\boldsymbol{S}}{\int_S \hat{\boldsymbol{I}} \cdot \mathrm{d}\boldsymbol{S}} \right)^{-1}, \quad (25.34)$$

and this shows that uncompensated p–u phase mismatch is a potential source of error when the reactivity (which in the global case is the ratio of the surface integral of the reactive intensity to the surface integral of the active intensity) is large. This will typically occur at low frequencies when the measurement surface is close to the source. Thus the reactivity is an important error indicator for p–u probes [25.17]. In contrast the pressure-intensity index is not relevant for p–u probes.

Sources outside the measurement surface do not in general increase the reactivity, and thus they do not in general increase the error due to p–u phase mismatch.

Other Sources of Error

Airflow, and in particular unsteady flow, is a more serious problem for p–u than for p–p sound intensity measurement systems, irrespective of the particulars of the particle velocity transducer, since the resulting velocity cannot be distinguished from the velocity associated with the sound waves. Windscreens of porous foam reduce the problem.

Calibration of p–u Sound Intensity Measurement Systems

Calibration of p–u sound intensity measurement systems involves exposing the probe to a sound field with a known relation between the pressure and the particle velocity, for example a plane propagating wave, a simple spherical wave or a standing wave [25.15, 16]. One cannot calibrate a particle velocity transducer in a coupler. There are no standardized calibration procedures and no standards for testing such instruments. However, recent experimental results seem to indicate the possibility of calibrating the particle velocity channel of the Microflown intensity probe in the far field of a loudspeaker in an anechoic room, in the near field of a very small source (sound emitted from a small hole in a plane or spherical baffle), and in a rigidly terminated standing wave tube [25.17, 56].

25.3.3 Sound Field Indicators

A sound field indicator may be defined as a normalized, energy-related quantity that describes a local or global property of the sound field [25.39]. The pressure-intensity index defined by (25.29) is an example. The concept of sound field indicators is closely related to sound intensity. The idea is that useful information about the nature of the sound field might be derived from the signals from an intensity probe. The purpose can be to derive information that may be helpful in interpreting experimental data, with a view, for example, to improving measurement accuracy in intensity-based sound power estimation by choosing another measurement surface, by shielding the surface from extraneous sources, etc. The international standards on sound power determination using sound intensity [25.24–26] prescribe initial measurements of a number of indicators and specify corrective actions on the basis of the values of these quantities. The corresponding ANSI standard [25.27] is more pragmatic; no less than 26 indicators are described, but their use is optional and it is left to the user to interpret the data and decide what to do.

It is clear from the considerations in Sect. 25.3.1 that the (global) pressure-intensity index reflects the amount of extraneous noise in sound power measurements and, combined with the pressure-residual intensity index of a p–p sound intensity measurement system, makes it possible to predict the bias error due to p–p phase mismatch. Likewise, the considerations in Sect. 25.3.2 show that the reactivity reflects whether the measurement takes place in the near field of the source under test or not and, combined with knowledge about the residual phase mismatch of a p–u sound intensity measurement system, makes it possible to predict the bias error due to the mismatch. In comparison, most of the many quantities suggested in [25.27] are, as shown in [25.57], difficult to interpret and only vaguely related to the measurement accuracy.

25.4 Applications of Sound Intensity

Some of the most important practical applications of sound intensity measurements are now briefly discussed. In view of the relative lack of experimental evidence of the performance of p–u sound intensity measurement systems it is assumed in the following that p–p measurement systems are used unless otherwise mentioned.

25.4.1 Noise Source Identification

A noise reduction project usually starts with the identification and ranking of noise sources and transmission paths. Before sound intensity measurement systems became available this was often a difficult task, but now it is relatively straightforward since intensity measurements make it possible to determine the sound power contribution of the various components separately. Plots of the sound intensity normal to a measurement surface can be used in locating noise sources. Figure 25.21 shows the results of 3-D sound intensity measurements near a chain saw, and Fig. 25.22 shows an example of intensity measurements near a ship window. It should be mentioned, though, that sound intensity measurement can give misleading results; for example, the intensity vector will point to a position between two uncorrelated sources where no source actually exists.

Visualization of sound fields contributes to our understanding of the sound radiation of complicated sources, and sound intensity is more suitable for visualizing sound fields than sound pressure, which is a scalar. Figure 25.23 shows a map of the sound intensity in the sound field generated by a rectangular plate driven in a certain mode, measured with a microphone array using near-field acoustic holography. It is apparent that the generated sound field is fairly complicated. Note in particular the recirculation of sound energy; a region of the plate acts as a sink. This phe-

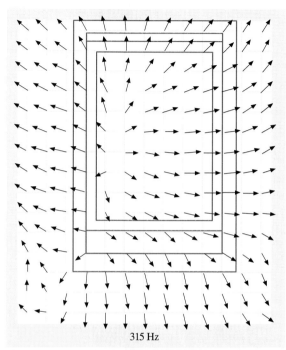

Fig. 25.22 Sound intensity distribution (tangential component) near a single cabin window (After [25.59])

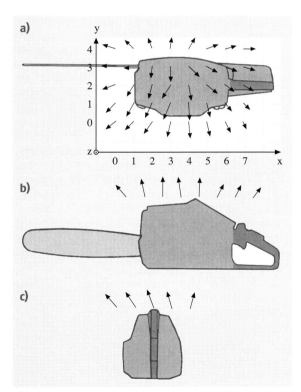

Fig. 25.21a–c Noise radiation from a chain saw in the 1600 Hz one-third octave band. (a) Sound intensity vector in the x–y plane; (b) sound intensity vector in the x–z plane; (c) sound intensity vector in the y–z plane (After [25.58])

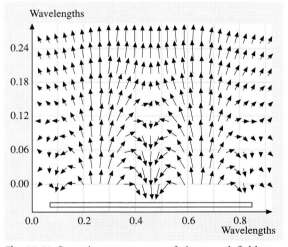

Fig. 25.23 Intensity vector map of the sound field generated by a rectangular plate excited in its (4, 2) mode (After [25.60])

nomenon is typical of vibrating panels of low radiation efficiency and demonstrates that spatial averaging of the sound intensity very near such sources is problematic. The circulation of sound energy, which implies that some regions act as sources without actually radiating to the far field, has led Williams to introduce the concept of *supersonic intensity*, which is the part of the sound intensity associated with wavenumber components within the *radiation circle*, in other words, the part of the intensity associated with radiation to the far field [25.61]. One cannot determine the supersonic intensity with an ordinary sound intensity probe, though; wavenumber processing of near-field holographic data is require.

25.4.2 Sound Power Determination

One of the most important applications of sound intensity measurement is the determination of the sound power of operating machinery in situ. Sound power determination using intensity measurements is based on (25.7), which shows that the sound power of a source is given by the integral of the normal component of the intensity over a surface that encloses the source, also in the presence of other sources outside the measurement surface. The analysis of errors and limitations presented in Sect. 25.3.1 leads to the conclusion that the sound intensity method is suitable for determining the sound power of stationary sources in stationary background noise provided that the pressure-intensity index is within the dynamic capability of the measurement system. On the other hand, the method is *not* suitable in nonstationary background noise (because the sound field will change during the measurement); it cannot be used for determining the sound power of very weak sources of low-frequency noise (because of large random errors caused by electrical noise in the microphone signals); and the absorption of the source under test should be negligible compared with the total absorption in the room where the measurement takes place (otherwise the sound power will be underestimated because the measurement gives the *net* sound power).

The surface integral can be approximated either by sampling at discrete points or by scanning manually or with a robot over the surface. With the scanning approach, the intensity probe is moved continuously over the measurement surface. A typical scanning path is shown in Fig. 25.24. The scanning procedure, which was introduced by Chung in the late 1970s on a purely empirical basis, was regarded with much skepticism for more than a decade [25.63], but is now generally regarded as more accurate and very much faster and more convenient than the procedure based on fixed points [25.62, 64, 65]. A moderate scanning rate, say 0.5 m/s, and a reasonable scan line density should be used, say 5 cm between adjacent lines if the surface is very close to the source, 20 cm if it is further away. However, whereas it may be possible to use the method based on discrete positions if the source is operating in cycles (simply by measuring over a full cycle at each position) one cannot use the scanning method under such conditions; both the source under test and possible extraneous noise sources must be perfectly stationary.

Usually the measurement surface is divided into a number of segments that are convenient to scan. The pressure-intensity index of each segment and the accuracy of each partial sound power estimate depends on whether (25.29) is satisfied or not, but it follows from (25.30) that it is the global pressure-intensity index associated with the entire measurement surface that determines the accuracy of the estimate of the (total) radiated sound power. It may be impossible to satisfy (25.29) on a certain segment, for example because the net sound power passing through the segment takes a very small value because of extraneous noise, but if the global criterion is satisfied then the total sound power estimate will nevertheless be accurate.

Theoretical considerations seem to indicate the existence of an optimum measurement surface that minimizes measurement errors [25.66]. In practice one uses a surface with a simple shape at some distance, say 25–50 cm, from the source. If there is a strong reverberant field or significant ambient noise from other sources, the measurement surface should be chosen to be somewhat closer to the source under study. One particular problem is that one might be tempted to forget to close a measurement surface that is very close to, say, a panel or a window by measuring only the component of the

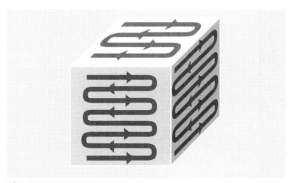

Fig. 25.24 Typical scanning path on a measurement surface (After [25.62])

intensity normal to the source. This can lead to serious errors in cases where the panel radiates very weakly, because the radiation will be nearly parallel to the source plane.

The three ISO standards for sound power determination using intensity measurement have been designed for sources of noise in their normal operating conditions, which may be very unfavorable [25.24–26]. In order to ensure accurate results under such general conditions the user must determine a number of field indicators and check whether various conditions are satisfied, as mentioned in Sect. 25.3.3. Fahy, who was the convener of the working group that developed ISO 9614-1 and 9614-2, has described the rationale, background, and principles of the procedures specified in these standards [25.65].

25.4.3 Radiation Efficiency of Structures

The radiation efficiency of a structure is a measure of how effectively it radiates sound. This dimensionless quantity is defined as

$$\sigma = \frac{P_a}{\rho c \langle v_n^2 \rangle S}, \qquad (25.35)$$

where P_a is the sound power radiated by the structure, S is its surface area, v_n is normal velocity of the surface, and the angular brackets indicate averaging over time as well as space. Comparing with (25.16) shows the close relation between this quantity and the real part of the specific impedance, which can be measured directly with a sound intensity probe if it can be placed sufficiently close to the structure [25.68]. The Microflown p–u probe has an advantage in this respect. However, measurements very near weak radiators of sound are difficult because of the circulating energy flow mentioned in Sect. 25.4.1.

25.4.4 Transmission Loss of Structures and Partitions

The transmission loss of a partition is the ratio of incident to transmitted sound power in logarithmic form. The traditional method of measuring this quantity requires a transmission suite consisting of two vibration-isolated reverberation rooms. The sound power incident on the partition under test in the source room is deduced from the spatial average of the mean square sound pressure in the room on the assumption that the sound field is diffuse, and the transmitted sound power is determined from a similar measurement in the receiving room where, in addition, the reverberation time must be deter-

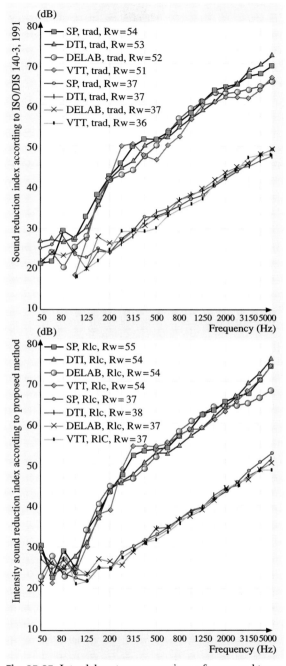

Fig. 25.25 Inter-laboratory comparison of measured transmission loss of a single metal leaf window (*lower curves*) and a double metal leaf window (*upper curves*). *Top panel*: Conventional method; *bottom panel*: intensity method (After [25.67])

mined. The sound intensity method makes it possible to measure the transmitted sound power directly. In contrast one cannot measure the incident sound power in the source room using sound intensity, since the method gives the *net* sound intensity. If the intensity method is used for determining the transmitted sound power it is not necessary that the sound field in the receiving room is diffuse, from which it follows that only one reverberation room is necessary. Thus sound intensity is suitable for field measurements of transmission loss. There are international standards both for laboratory and field measurements of transmission loss based on sound intensity [25.69, 70].

Figure 25.25 shows the results of a *round robin* investigation in which a single-leaf and a double-leaf construction were tested by four different laboratories using the conventional method and the intensity-based method. Apart from the fact that only one reverberation room is needed the main advantage of the intensity method is that it makes it possible to evaluate the transmission loss of individual parts of the partition. However, to be reliable each sound power measurement must obviously satisfy the condition expressed by (25.29). There are other sources of error than phase mismatch. If a significant part of the absorption in the receiving room is due to the partition under test then the net power is less than the transmitted power because a part of the transmitted sound energy is absorbed or retransmitted by the partition itself [25.71]. Under such conditions one must increase the absorption of the receiving room; otherwise the intensity method will overestimate the transmission loss because the transmitted sound power is underestimated. In contrast, the conventional method measures the transmitted sound power irrespective of the distribution of absorption in the receiving room.

Deviations between results determined using the traditional method and the intensity method led several authors to reanalyze the traditional method in the 1980s [25.72] and point out that the Waterhouse correction [25.73], well established in sound power determination using the reverberation room method [25.74], had been overlooked in the standards for conventional measurement of transmission loss. Recent results imply that the Waterhouse correction should be used not only for the receiving room but also for the source room [25.75].

25.4.5 Other Applications

The fact that the sound intensity level is considerably lower than the sound pressure level in a diffuse, reverberant sound field has led to the idea of replacing a measurement of the *emission sound pressure level* generated by machinery at the operator's position by a measurement of the sound intensity level, because the latter is less affected by diffuse background noise [25.76]. This method, which involves measuring three components of the intensity at a specified position near the source, has recently been standardized [25.77].

In principle, sound intensity may be used for measuring sound absorption in situ. As in measurement of transmission losses the incident sound power must be deduced from a spatial average of the mean square pressure in the room on the assumption that the sound field is diffuse, and the absorbed sound power is measured by integrating the normal component of the intensity over a surface that encloses the specimen under test [25.2]. In practice, however, this is one of the least successful applications of sound intensity, partly because of the assumption of diffuse sound incidence and partly because estimation errors in the absorbed power will be translated to relatively large fractional errors in the resulting absorption coefficients.

References

25.1 A.D. Pierce: *Acoustics: An Introduction to Its Physical Principles and Applications*, 2nd edn. (Acoustical Society of America, New York 1989)

25.2 F.J. Fahy: *Sound Intensity*, 2nd edn. (E & FN Spon, London 1995)

25.3 ISO: *ISO 3745 Acoustics – Determination of sound power levels of noise sources using sound pressure – Precision methods for anechoic and hemianechoic rooms* (ISO, Geneva 2003)

25.4 J.-C. Pascal: *Mesure de l'intensité active et réactive dans differents champs acoustiques*, Proc. Rec. Devel. Acoust. Intens (CETIM, Senlis 1981) pp. 11–19

25.5 J.A. Mann III, J. Tichy, A.J. Romano: Instantaneous and time-averaged energy transfer in acoustic fields, J. Acoust. Soc. Am. **82**, 17–30 (1987)

25.6 F. Jacobsen: Active and reactive, coherent and incoherent sound fields, J. Sound Vib. **130**, 493–507 (1989)

25.7 F. Jacobsen: A note on instantaneous and time-averaged active and reactive sound intensity, J. Sound Vib. **147**, 489–496 (1991)

25.8 P.J. Westervelt: Acoustical impedance in terms of energy functions, J. Acoust. Soc. Am. **23**, 347–348 (1951)

25.9 W. Maysenhölder: *The reactive intensity of general time-harmonic structure-borne sound fields*. Proc. Fourth Int. Congr. Intens. Tech. (CETIM, Senlis 1993) pp. 63–70

25.10 U. Kurze: Zur Entwicklung eines Gerätes für komplexe Schallfeldmessungen, Acustica **20**, 308–310 (1968)

25.11 J.C. Pascal, C. Carles: Systematic measurement errors with two microphone sound intensity meters, J. Sound Vib. **83**, 53–65 (1982)

25.12 R.C. Heyser: Instantaneous intensity. Preprint 2399, 81st Conv. Audio Eng. Soc. (1986)

25.13 M.P. Waser, M.J. Crocker: Introduction to the two-microphone cross-spectral method of determining sound intensity, Noise Control Eng. J. **22**, 76–85 (1984)

25.14 K.J. Bastyr, G.C. Lauchle, J.A. McConnell: Development of a velocity gradient underwater acoustic intensity sensor, J. Acoust. Soc. Am. **106**, 3178–3188 (1999)

25.15 H.-E. de Bree: The Microflown: An acoustic particle velocity sensor, Acoust. Australia **31**, 91–94 (2003)

25.16 R. Raangs, W.F. Druyvesteyn, H.E. de Bree: A low-cost intensity probe, J. Audio Eng. Soc. **51**, 344–357 (2003)

25.17 F. Jacobsen, H.-E. de Bree: A comparison of two different sound intensity measurement principles, J. Acoust. Soc. Am. **118**, 1510–1517 (2005)

25.18 IEC: *IEC (International Electrotechnical Commission) 1043 Electroacoustics – Instruments for the Measurement of Sound Intensity – Measurements with Pairs of Pressure Sensing Microphones* (IEC, Geneva 1993)

25.19 ANSI: *ANSI (American National Standards Institute) S1.9-1996 Instruments for the Measurement of Sound Intensity* (ANSI, Washington 1996)

25.20 J.Y. Chung: Cross-spectral method of measuring acoustic intensity without error caused by instrument phase mismatch, J. Acoust. Soc. Am. **64**, 1613–1616 (1978)

25.21 F.J. Fahy: Measurement of acoustic intensity using the cross-spectral density of two microphone signals, J. Acoust. Soc. Am. **62**, 1057–1059 (1977)

25.22 G. Rasmussen, M. Brock: *Acoustic intensity measurement probe*, Proc. Rec. Devel. Acoust. Intens. (CETIM, Senlis 1981) pp. 81–88

25.23 J. Pope: *Qualifying intensity measurements for sound power determination*, Proc. Inter-Noise 89 (Int. Inst. Noise Control Eng., West Lafayette 1989) pp. 1041–1046

25.24 ISO: *International Organization for Standardization) 9614-1 Acoustics – Determination of Sound Power Levels of Noise Sources Using Sound Intensity – Part 1: Measurement at Discrete Points* (ISO, Geneva 1993)

25.25 ISO: *International Organization for Standardization) 9614-2 Acoustics – Determination of Sound Power Levels of Noise Sources Using Sound Intensity – Part 2: Measurement by Scanning* (ISO, Geneva 1996)

25.26 ISO: *International Organization for Standardization) 9614-3 Acoustics – Determination of Sound Power Levels of Noise Sources Using Sound Intensity – Part 3: Precision Method for Measurement by Scanning* (ISO, Geneva 2002)

25.27 ANSI: *ANSI (American National Standards Institute) S12.12-1992 Engineering Method for the Determination of Sound Power Levels of Noise Sources Using Sound Intensity* (ANSI, Washington 1992)

25.28 F. Jacobsen, V. Cutanda, P.M. Juhl: A numerical and experimental investigation of the performance of sound intensity probes at high frequencies, J. Acoust. Soc. Am. **103**, 953–961 (1998)

25.29 P. Juhl, F. Jacobsen: A note on measurement of sound pressure with intensity probes, J. Acoust. Soc. Am. **116**, 1614–1620 (2004)

25.30 S. Shirahatti, M.J. Crocker: Two-microphone finite difference approximation errors in the interference fields of point dipole sources, J. Acoust. Soc. Am. **92**, 258–267 (1992)

25.31 P.S. Watkinson, F.J. Fahy: Characteristics of microphone arrangements for sound intensity measurement, J. Sound Vib. **94**, 299–306 (1984)

25.32 S. Gade: Sound intensity (Theory), Brüel, Kjær Tech. Rev. **3**, 3–39 (1982)

25.33 Anonymous: *Product data, sound intensity pair – type 4197* (Brüel, Kjær, Nærum 2000)

25.34 E. Frederiksen, O. Schultz: Pressure microphones for intensity measurements with significantly improved phase properties, Brüel, Kjær Tech. Rev. **4**, 11–12 (1986)

25.35 F. Jacobsen: A simple and effective correction for phase mismatch in intensity probes, Appl. Acoust. **33**, 165–180 (1991)

25.36 S.J. Elliott: Errors in acoustic intensity measurements, J. Sound Vib. **78**, 439–445 (1981)

25.37 M. Ren, F. Jacobsen: Phase mismatch errors and related indicators in sound intensity measurement, J. Sound Vib. **149**, 341–347 (1991)

25.38 S. Gade: Validity of intensity measurements in partially diffuse sound field, Brüel, Kjær Tech. Rev. **4**, 3–31 (1985)

25.39 F. Jacobsen: Sound field indicators: Useful tools, Noise Control Eng. J. **35**, 37–46 (1990)

25.40 D.H. Munro, K.U. Ingard: On acoustic intensity measurements in the presence of mean flow, J. Acoust. Soc. Am. **65**, 1402–1406 (1979)

25.41 F. Jacobsen: *Intensity measurements in the presence of moderate airflow*, Proc. Inter-Noise 94 (Int. Inst. Noise Control Eng., Yokohama 1994) pp. 1737–1742

25.42 F. Jacobsen: A note on measurement of sound intensity with windscreened probes, Appl. Acoust. **42**, 41–53 (1994)

25.43 P. Juhl, F. Jacobsen: A numerical investigation of the influence of windscreens on measurement of

25.44 A.F. Seybert: Statistical errors in acoustic intensity measurements, J. Sound Vib. **75**, 519–526 (1981)

sound intensity, J. Acoust. Soc. Am. **119**, 937–942 (2006)

25.45 O. Dyrlund: A note on statistical errors in acoustic intensity measurements, J. Sound Vib. **90**, 585–589 (1983)

25.46 F. Jacobsen: Random errors in sound intensity estimation, J. Sound Vib. **128**, 247–257 (1989)

25.47 F. Jacobsen: Random errors in sound power determination based on intensity measurement, J. Sound Vib. **131**, 475–487 (1989)

25.48 F. Jacobsen: Sound intensity measurement at low levels, J. Sound Vib. **166**, 195–207 (1993)

25.49 F. Jacobsen, E.S. Olsen: Testing sound intensity probes in interference fields, Acustica **80**, 115–126 (1994)

25.50 E. Frederiksen: *BCR-report, Sound intensity measurement instruments. Free-field intensity sensitivity calibration and standing wave testing* (Brüel, Kjær, Nærum 1992)

25.51 F. Jacobsen, E.S. Olsen: The influence of microphone vents on the performance of sound intensity probes, Appl. Acoust. **41**, 25–45 (1993)

25.52 S.A. Nordby, O.H. Bjor: *Measurement of sound intensity by use of a dual channel real-time analyzer and a special sound intensity microphone*, Proc. Inter-Noise 84 (Int. Inst. Noise Control Eng., Honolulu 1984) pp. 1107–1110

25.53 J.A. Macadam: The measurement of sound radiation from room surfaces in lightweight buildings, Appl. Acoust. **9**, 103–118 (1976)

25.54 M.C. McGary, M.J. Crocker: Surface intensity measurements on a diesel engine, Noise Control Eng. J. **16**, 27–36 (1981)

25.55 Y. Hirao, K. Yamamoto, K. Nakamura, S. Ueha: Development of a hand-held sensor probe for detection of sound components radiated from a specific device using surface intensity measurements, Appl. Acoust. **65**, 719–735 (2004)

25.56 F. Jacobsen, V. Jaud: A note on the calibration of pressure-velocity sound intensity probes, J. Acoust. Soc. Am. **120**, 830–837 (2006)

25.57 F. Jacobsen: *A critical examination of some of the field indicators proposed in connection with sound power determination using the intensity technique*, Proc. Fourth Internat. Congr. Sound Vib. (IIAV, St. Petersburg 1996) pp. 1889–1896

25.58 T. Astrup: Measurement of sound power using the acoustic intensity method – A consultant's viewpoint, Appl. Acoust. **50**, 111–123 (1997)

25.59 S. Weyna: The application of sound intensity technique in research on noise abatement in ships, Appl. Acoust. **44**, 341–351 (1995)

25.60 E.G. Williams, J.D. Maynard: *Intensity vector field mapping with nearfield holography*, Proc. Rec. Devel. Acoust. Intens (CETIM, Senlis 1981) pp. 31–36

25.61 E.G. Williams: Supersonic acoustic intensity on planar sources, J. Acoust. Soc. Am. **104**, 2845–2850 (1998)

25.62 U.S. Shirahatti, M.J. Crocker: Studies of the sound power estimation of a noise source using the two-microphone sound intensity technique, Acustica **80**, 378–387 (1994)

25.63 M.J. Crocker: Sound power determination from sound intensity – To scan or not to scan, Noise Contr. Eng. J. **27**, 67 (1986)

25.64 O.K.Ø. Pettersen, H. Olsen: On spatial sampling using the scanning intensity technique, Appl. Acoust. **50**, 141–153 (1997)

25.65 F.J. Fahy: International standards for the determination of sound power levels of sources using sound intensity measurement: An exposition, Appl. Acoust. **50**, 97–109 (1997)

25.66 F. Jacobsen: Sound power determination using the intensity technique in the presence of diffuse background noise, J. Sound Vib. **159**, 353–371 (1992)

25.67 H.G. Jonasson: Sound intensity and sound reduction index, Appl. Acoust. **40**, 281–293 (1993)

25.68 B. Forssen, M.J. Crocker: Estimation of acoustic velocity, surface velocity, and radiation efficiency by use of the two-microphone technique, J. Acoust. Soc. Am. **73**, 1047–1053 (1983)

25.69 ISO: *ISO 15186-1 Acoustics – Measurement of sound insulation in buildings and of building elements using sound intensity – Part 1: Laboratory measurements* (ISO, Geneva 2000)

25.70 ISO: *ISO 15186-2 Acoustics – Measurement of sound insulation in buildings and of building elements using sound intensity – Part 2: Field measurements* (ISO, Geneva 2003)

25.71 J. Roland, C. Martin, M. Villot: *Room to room transmission: What is really measured by intensity?* Proc. 2nd Intern. Congr. Acoust. Intens. (CETIM, Senlis 1985) pp. 539–546

25.72 E. Halliwell, A.C.C. Warnock: Sound transmission loss: Comparison of conventional techniques with sound intensity techniques, J. Acoust. Soc. Am. **77**, 2094–2103 (1985)

25.73 R.V. Waterhouse: Interference patters in reverberant sound fields, J. Acoust. Soc. Am. **27**, 247–258 (1955)

25.74 ISO: *ISO 3741:1999 Acoustics – Determination of sound power levels of noise sources using sound pressure – Precision methods for reverberation rooms* (ISO, Geneva 1999)

25.75 A. Ismai: *On the application of the Waterhouse correction in sound insulation measurement*. In: *Proc. Inter-Noise 2006* (Int. Inst. Noise Control. Eng., Honolulu 2006)

25.76 H.G. Jonasson: *Determination of emission sound pressure level and sound power level in situ*, SP Rep. 18 (Swedish National Testing and Research Institute, Borås 1999)

25.77 ISO: *ISO 11205 Acoustics – Noise emitted by machinery and equipment – Engineering method for the determination of emission sound pressure levels in situ at the work station and at other specified positions using sound intensity* (ISO, Geneva 2003)

26. Acoustic Holography

One of the subtle problems that make noise control difficult for engineers is the invisibility of noise or sound. A visual image of noise often helps to determine an appropriate means for noise control. There have been many attempts to fulfill this rather challenging objective. Theoretical (or numerical) means for visualizing the sound field have been attempted, and as a result, a great deal of progress has been made. However, most of these numerical methods are not quite ready for practical applications to noise control problems. In the meantime, rapid progress with instrumentation has made it possible to use multiple microphones and fast signal-processing systems. Although these systems are not perfect, they are useful. A state-of-the-art system has recently become available, but it still has many problematic issues; for example, how can one implement the visualized noise field. The constructed noise or sound picture always consists of bias and random errors, and consequently, it is often difficult to determine the origin of the noise and the spatial distribution of the noise field. Section 26.2 of this chapter introduces a brief history, which is associated with "sound visualization," acoustic source identification methods and what has been accomplished with a line or surface array. Section 26.2.3 introduces difficulties and recent studies, including de-Dopplerization and de-reverberation methods, both essential

26.1 The Methodology of Acoustic Source Identification 1077
26.2 Acoustic Holography: Measurement, Prediction and Analysis 1079
    26.2.1 Introduction and Problem Definitions 1079
    26.2.2 Prediction Process 1080
    26.2.3 Measurement 1083
    26.2.4 Analysis of Acoustic Holography .. 1089
26.3 Summary 1092
26.A Mathematical Derivations of Three Acoustic Holography Methods and Their Discrete Forms 1092
    26.A.1 Planar Acoustic Holography 1092
    26.A.2 Cylindrical Acoustic Holography... 1094
    26.A.3 Spherical Acoustic Holography 1095
References 1095

for visualizing a moving noise source, such as occurs for cars or trains. This section also addresses what produces ambiguity in realizing real sound sources in a room or closed space. Another major issue associated with sound/noise visualization is whether or not we can distinguish between mutual dependencies of noise in space (Sect. 26.2.4); for example, we are asked to answer the question, "Can we see two birds singing or one bird with two beaks?"

26.1 The Methodology of Acoustic Source Identification

The famous article written by *Kac* [26.1], "Can one hear the shape of the drum?" clearly addresses the essence of the inverse problem. It can be regarded as an attempt to obtain what is *not* available, using what *is* available, using the example of the relationship between sound generation by membrane vibration and its reception in space. One can find many other examples of inverse problems [26.2–8]. Often, in the inverse problem, it is hard to predict or describe data that are not measured because the available data are insufficient. This circumstance is commonly referred to as an *ill-posed problem* in the literature

(for examples, see [26.1–19]). Figure 26.1 demonstrates what might happen in practice; the prediction depends on how well the basis function (the elephants or dogs in Fig. 26.1) mimics what happens in reality. When we try to see the shape of noise/sound sources, how well we see the shape of a noise source completely depends on this basis function, because we predict what is not available by using this selected basis function.

One of the common methods of classifying methods used in noise/sound source identification, is by the type of basis function. According to this classification,

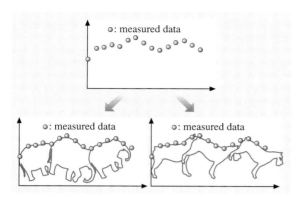

Fig. 26.1 The inverse problem and basis function. Are the measured data the parts of elephants or dogs?

one approach is the nonparametric method, which uses basis functions that do not model the signal. In other words, the basis functions do not map the unmeasured sound field; all orthogonal functions fall into this category. One of the typical methods of this kind uses Fourier transforms. Acoustic holography uses this type of basis function, mapping the sound field of interest with regard to every measured frequency; it therefore sees the sound field in the frequency domain. In fact, the ideas of acoustic holography originated from optics [26.20–30]. Acoustic holography was simply extended or modified from the basic idea of optical holography. Near-field acoustic holography [26.31, 32] has been recognized as a very useful means of predicting the true appearance of the source Fig. 26.2. (The near-field effect on resolution was first introduced in the field of microwaves [26.33].) The basis of this method is to include or measure exponentially decaying waves as they propagate from the sound source so that the sources can be completely reconstructed.

Another class of approaches are based on the parametric method, which derives its name from the fact that the signal is modeled using certain parameters. In other words, the basis function is chosen depending upon the prediction of the sound source. A typical method of this kind is the so-called *beam-forming method*. Different types of basis functions can be chosen for this method, entirely depending on the sound field that the basis function is trying to map [26.34, 35]. In Fig. 26.1, we can select either the elephants or dogs (or another choice), depending on what we want to predict. This type of mapping gives information about the source location. As illustrated in Fig. 26.1, the basis function maps the signal by changing its parameter; in the case of forming a plane-wave beam, the incident angle of the plane wave can be regarded as a parameter. The main issues that have been discussed for this kind of mapping method are directly related to the structure of the correlation matrix that comes from the measured acoustic pressure vector and its complex conjugate (see Fig. 26.3 for the details). In this method, each scan vector has a multiplicative parameter; for the plane wave in Fig. 26.3 it is the angle of arrival. The correlation matrix is given as illustrated in Fig. 26.3. The scan vector is a basis function in this case. As one can see immediately, this problem is directly related

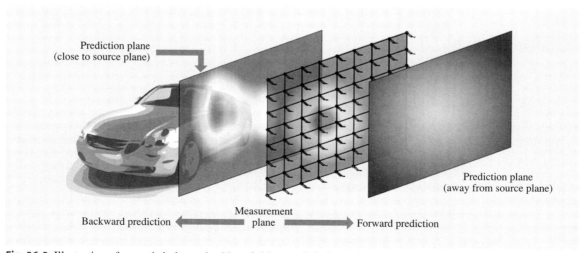

Fig. 26.2 Illustration of acoustic holography. Near-field acoustic holography measures evanescent waves on the measurement plane. The measurement plane is always finite, i.e. there is always a finite aperture. Therefore, we only get limited data)

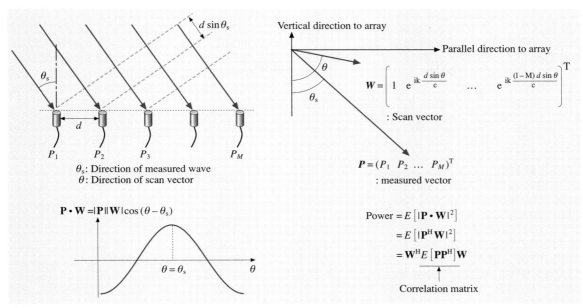

Fig. 26.3 The beam-forming method

to the structure of the correlation matrix and the basis function used. The signal-to-noise (S/N) ratio of the measured correlation matrix determines the effectiveness of the estimation. There have been many attempts to improve the estimator's performance with regard to the signal-to-noise ratio [26.35, 36]. These methods have mainly been developed for applications in the radar and underwater communities [26.37]. This technique has also been applied to a noise source location finding problem; high-speed-train noise source estimation [26.38–40] is one such example. Various shapes of arrays have been tried to improve the spatial resolution [26.41–43]. However, it is obvious that these methods cannot sense the shape of the sound or noise source; they only provide its location. Therefore, we will not discuss the beam-forming method in this chapter. In the next section, the problems that we have discussed will be defined.

26.2 Acoustic Holography: Measurement, Prediction and Analysis

26.2.1 Introduction and Problem Definitions

Acoustic holography consists of three components: measurement, which consists of measuring the sound pressure on the hologram plane, prediction of the acoustic variables, including the velocity distribution, on the plane of interest, and analysis of the holographic reconstruction. This last component was not recognized as important as the others in the past. However, it yields the real meaning of the sound picture: visualization.

The issues associated with measurement are all related to the hologram measurement configuration; we measure the sound pressure at discrete measurement points over a finite measurement area (finite aperture), as illustrated in Fig. 26.2. References [26.44–52] explain the necessary steps to avoid spatial aliasing, wrap-around errors, and the effect of including evanescent waves on the resolution (near-field acoustic holography). If sensors are incorrectly located on the hologram surface, errors result in the prediction results. Similar errors can be produced when there is a magnitude and phase mismatch between sensors. This is well summarized in [26.53]. There have been many attempts to reduce the aperture effect. One method is to extrapolate the pressure data based on the measurements taken [26.50, 52]. Another method allows the measurement of sound pressure in a sequence and interprets the measured sound pressures with respect to reference signals, assuming that the measured sound pressure field is stationary during the measurement and the number of independent sources is smaller than the number of reference microphones [26.54–61]. Another method allows scanning or moving of the microphone array, thereby extending the aperture size as much as possible [26.62–65]. This also allows one to measure the sound pressure generated by moving sound sources, such as a vehicle's exterior noise.

The prediction problem is rather well defined and relatively straightforward. Basically, the solution of the acoustic wave equation usually results in the sound pressure distribution on the measurement plane. Prediction can be attempted using a Green's function, an example of which may be found in the Kirchhoff–Helmholtz integral equation. It is noteworthy, however, that the prediction depends on the shape of the measurement and prediction surfaces, and also on the presence of sound reflections [26.54, 66–87].

The acoustic holography analysis problem was introduced rather recently. As mentioned earlier in this section, this is one of the essential issues connected to the general inverse problem. One basic question is whether what we see and imagine is related to what happens in reality. There are two different sound/noise sources, one of which is really radiating the sound, and the another that is reflecting the sound. The former is often called *active sound/noise*, while the latter is called *passive sound/noise*. This is an important practical concept for establishing noise control strategies; we want to eliminate the active noise source. Another concern is whether the sources are independently or dependently correlated (Fig. 26.23). The concept of an independent and dependent source has to be addressed properly to understand the issues.

26.2.2 Prediction Process

The prediction process is related to how we predict the unmeasured sound pressure or other acoustic variables based on the measured sound pressure information. The following equation relates the unmeasured and measured pressure:

$$P(\bm{x}; f) = \int_{S_h} \left[G(\bm{x}|\bm{x}_h; f) \left. \frac{\partial P}{\partial n} \right|_{(\bm{x} = \bm{x}_h; f)} - P(\bm{x}_h; f) \frac{\partial G(\bm{x}|\bm{x}_h; f)}{\partial n} \right] dS_h . \quad (26.1)$$

Equation (26.1) is the well-known Kirchhoff–Helmholtz integral equation, where $G(\bm{x}|\bm{x}_h; f)$ is the free-space Green's function. This equation essentially says that we can predict the sound pressure anywhere if we know the sound pressures and velocities on the boundary Fig. 26.4. However, it is noteworthy that measuring the velocity on the boundary is more difficult than measuring the sound pressure. This rather practical difficulty can be solved by introducing a Green's function that satisfies the Dirichlet boundary condition: $G_D(\bm{x}|\bm{x}_h; f)$. Then, (26.1) becomes

$$P(\bm{x}; f) = \int_{S_h} \left[-P(\bm{x}_h; f) \frac{\partial G_D(\bm{x}|\bm{x}_h; f)}{\partial n} \right] dS_h . \quad (26.2)$$

This equation allows us to predict the sound pressure on any surface of interest. It is noteworthy that we can choose a Green's function as long as it satisfies the

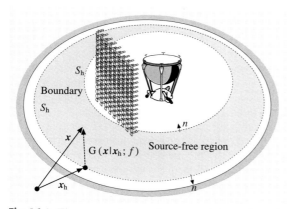

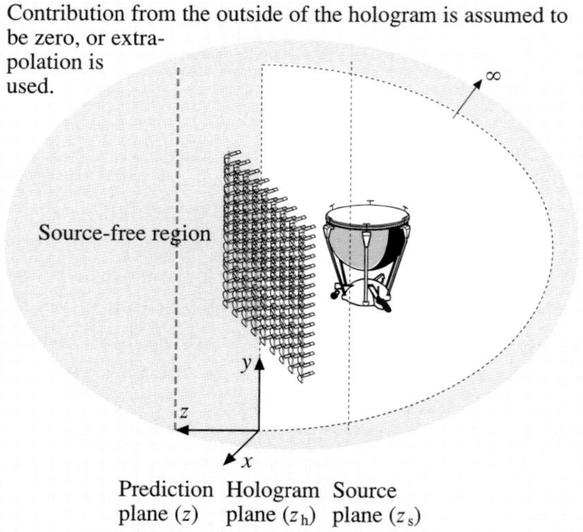

Fig. 26.4 The geometry and nomenclature for the Kirchhoff–Helmholtz integral (26.1)

Fig. 26.5 Illustration of the planar acoustic holography

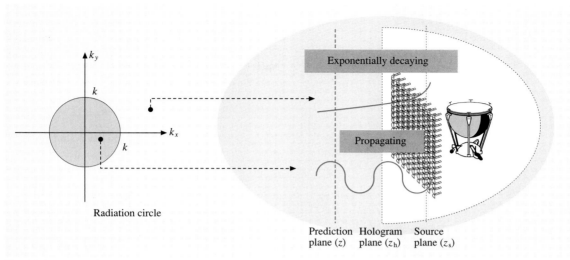

Fig. 26.6 Propagating and exponentially waves in acoustic holography

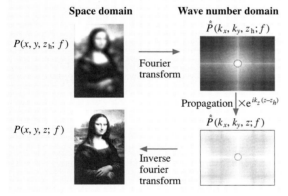

Fig. 26.7 The data-processing procedure for acoustic holography

linear inhomogeneous wave equation, or the inhomogeneous Helmholtz equation in the frequency domain. That is,

$$\nabla^2 G(\bm{x}|\bm{x}_h; f) + k^2 G(\bm{x}|\bm{x}_h; f) = -\delta(\bm{x} - \bm{x}_h) \,. \tag{26.3}$$

Therefore, we can select a Green's function in such a way that we can eliminate one of the terms on the right-hand side of (26.1); (26.2) is one such case.

To see what essentially happens in the prediction process, let us consider (26.2) when the measurement and prediction plane are both planar. Planar acoustic holography assumes that the sound field is free from reflection (Fig. 26.5); then we can write (26.2) as

$$P(x, y, z; f) = \int_{S_h} P(x_h, y_h, z_h; f) \\ \times K_{PP}(x - x_h, y - y_h, z - z_h; f) \mathrm{d}S_h \,, \tag{26.4}$$

$$K_{PP}(x, y, z; f) = \frac{1}{2\pi} \frac{z}{r^3} (1 - ikr) \exp(ikr) \,, \tag{26.5}$$

where $r = \sqrt{x^2 + y^2 + z^2}$,

$$k = \frac{2\pi f}{c} \,,$$
$$\bm{x} = (x, y, z) \,,$$
$$\bm{x}_h = (x_h, y_h, z_h) \,.$$

K_{PP} can be readily obtained by using two free-field Green's functions that are located at z_h and $-z_h$, so that it satisfies the Dirichlet boundary condition.

This is a convolution integral, and therefore we can write this in the wave-number domain as

$$\hat{P}(k_x, k_y, z; f) = \\ \hat{P}(k_x, k_y, z_h; f) \exp[ik_z(z - z_h)] \,, \tag{26.6}$$

where

$$\hat{P}(k_x, k_y, z; f) \\ = \int_{-\infty}^{\infty} \int_{-\infty}^{\infty} P(x, y, z; f) e^{-i(k_x x + k_y y)} \mathrm{d}x \, \mathrm{d}y \,, \tag{26.7}$$

$$k_z = \sqrt{k^2 - k_x^2 - k_y^2} \,.$$

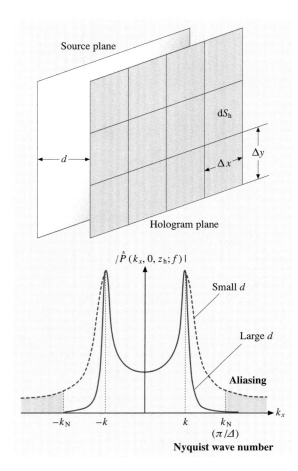

Fig. 26.8 The error due to discrete measurement: the spatial aliasing problem. The microphone spacing determines the Nyquist wave number. This wave number has to be smaller than the maximum wave number of the acoustic pressure distribution on the hologram plane. The microphone spacing, therefore, has to get smaller as the distance between the hologram (the measurement plane) and source decrease. The rule of thumb is $\delta < d$

This equation essentially predicts the sound pressure with respect to the wave number (k_x, k_y). If $k^2 \geq k_x^2 + k_y^2$, the wave in z-direction (k_z) is propagating in space. Otherwise, the z-direction wave decays exponentially, i.e. it is an evanescent wave (Fig. 26.6).

We have derived the formulas that can predict what we did not measure based on what we measured, by using a Green's function. It is noteworthy that we can get the same results if we use the characteristic solutions of the Helmholtz equation; the Appendix describes the details. The Appendix also includes discrete expressions for the formula, which are normally used in the computation.

Equation (26.6) also allows us to predict the sound pressure on the source plane, when $z = z_s$. This is an inverse problem because it predicts the pressure distribution on the source plane based on the hologram pressure (see Figs. 26.3 and 26.6).

Figure 26.7 essentially illustrates how we can process the data for predicting what we did not measure based on what we measure. There are four major areas that cause errors in acoustic holography prediction. One is related to the integration of (26.4). Equation (26.4) has to be implemented on the discretized surface Fig. 26.8. This surface, therefore, has to be spatially sampled according to the selected surface. This spatial sampling can produce spatial aliasing, depending on the spatial distribution of the sound source: the sampling wave number must be larger than twice the maximum wave number of interest. It is noteworthy that, as illustrated in Fig. 26.8, the distance between the hologram and source planes is usually related to the sampling distance d. The closer one is to the source, the smaller the sampling distance needs to be. We must also note that the size of the aperture determines the wave-number resolution of acoustic holography. The finite aperture inevitably produces very sharp data truncation, as illustrated in Fig. 26.9. This produces unrealistic high-wave-number noise (see "without window" in Fig. 26.9). Therefore, it is often required that

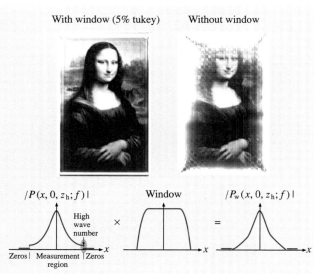

Fig. 26.9 The effect of a finite aperture: the rapid change at the aperture edges produces high-wave-number noise

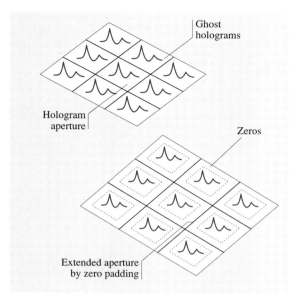

Fig. 26.10 The effect of the finite spatial Fourier transform on acoustic holography: the ghost image is due to the finite Fourier transform; circular convolution can be eliminated by adding zeros

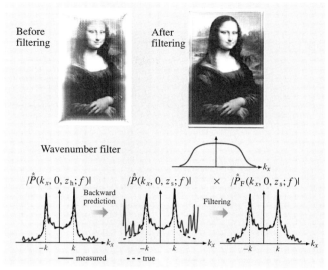

Fig. 26.11 Wave-number filtering in backward prediction. Evanescent wave components are magnified without filtering [26.47]

we use a window, which can result in a smoother data transition from what is on the measurement plane to what is not measured (Fig. 26.9).

The spatial Fourier transform that has to be done in the prediction process (26.6) has to be carried out in the domain for which data is available, i.e. a finite Fourier transform. It therefore produces a ghost hologram, as illustrated in Fig. 26.10 [26.46, 50]. This effect can be effectively removed by adding zeros to the hologram data (Fig. 26.11). The last thing to note is what can happen when we do backward propagation. As we can see in (26.6), when we predict the sound pressure distribution on a plane close to the sound source [$z < z_h$ (Fig. 26.5)] and k_z has an imaginary value (evanescent wave), then the sound pressure distribution of the exponentially decaying part will be unrealistically magnified (Fig. 26.11) [26.47]. Figure 26.12 graphically summarizes the entire processing steps of acoustic holography.

The issues related with the evanescent wave and its measurement are well addressed in the literature (for example, see [26.88]). The measurement of evanescent waves essentially allows us to achieve higher resolution than conventional acoustic holography [26.89–94]. However, it is noteworthy that the evanescent-wave component is substantially smaller than the other prop-

agating components. Therefore, it is easy to produce errors that are associated with sensor or position mismatch [26.53]; in other words, it is very sensitive to the signal-to-noise ratio. Errors due to position and sensor mismatch are bias errors and random errors, respectively. It has been shown [26.53] that the bias error due to the mismatches is negligible, but the random error is significant in backward prediction. This is related to the measurement spacing on the hologram plane (Δ_h), prediction plane (Δ_z), and the distance between the hologram plane and the prediction plane (d). It is approximately proportional to $24.9(d/\Delta_z) + 20\log_{10}(\Delta_h/\Delta_z)$ in a dB scale. The signal-to-noise ratio can be amplified when we try to reconstruct the source field: a typical ill-posed phenomena. There have been many attempts to reduce this effect by using a spatial filter [26.47, 95–99], which is often called the *regularization* of acoustic holography [26.100–115].

Depending on the separable coordinates that we use for acoustic holography, we can construct cylindrical or spherical coordinates [26.54, 67, 69] (Figs. 26.13 and 26.14). These methods predict the sound field in exactly the same manner as in planar holography but with respect to different coordinates. As expected, however, these methods have some advantages. For example, the wrap-around error is negligible – in fact, there is no such error in spherical acoustic holography – and there is no aperture-related error [26.54]. Recently, the advantage

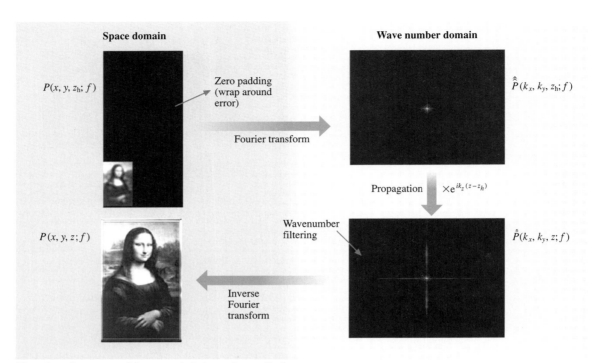

Fig. 26.12 Summary of the acoustic holography prediction process

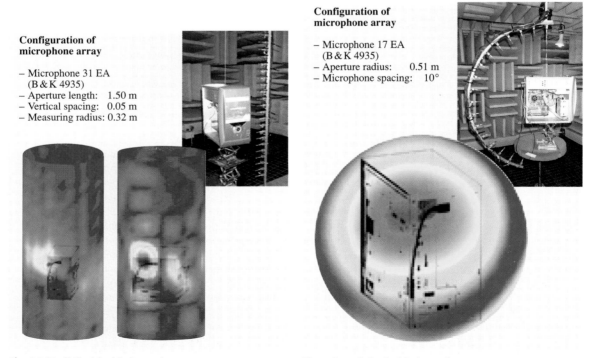

Fig. 26.13 Cylindrical holography

Fig. 26.14 Spherical holography

of using spherical functions has also been noted [26.79, 81, 86, 87, 113].

26.2.3 Measurement

To construct a hologram, we commonly measure the sound pressure at discrete positions, as illustrated in Fig. 26.2. However, if the sound generated by the source, and therefore the sound field, can be assumed to be stationary, then we do not have to measure them at the same time.

Figure 26.15 illustrates one way to accomplish this measurement. This method normally measures the sound pressure field using a stepped line array (Fig. 26.15a). To understand the issues associated with this measurement system for the sake of its simplicity, let us see how we process a signal of frequency f when there is a single source. The relationship between the sound source and sound pressure in the field, or measurement position (x_h), can be written as

$$P(x_h; f) = H(x_h; f)Q(f), \qquad (26.8)$$

where $Q(f)$ is the source input signal and $H(x_h; f)$ is the transfer function between the source input and the measured pressure. This means that, if we know the transfer function and the input, we can find the magnitude and phase between the measured positions. Because it is usually not practical to measure the input, we normally use reference signals (Fig. 26.15a). By using a reference signal, the pressure can be written as

$$P(x_h; f) = H'(x_h; f)R(f), \qquad (26.9)$$

where $R(f)$ is the reference signal. We can obtain $H'(x_h; f)$ by

$$H'(x_h; f) = \frac{P(x_h; f)}{R(f)}. \qquad (26.10)$$

The input and reference are related through:

$$R(f) = H_R(f)Q(f), \qquad (26.11)$$

where $H_R(f)$ is the transfer function between the input and the reference. As a result, we can see that (26.9) has the same form as (26.8).

It is noteworthy that (26.8) holds for the case that we have only one sound source and the sound field is stationary and random. However, if there are two sound sources, then (26.8) becomes

$$P(x_h; f) = H_1(x_h; f)Q_1(f) + H_2(x_h; f)Q_2(f), \qquad (26.12)$$

where $Q_i(f)$ is the i-th input and $H_i(x_h; f)$ is its transfer function. There are now two independent sound fields. This requires, of course, two independent reference signals. It has been well accepted that the number of reference microphones has to be greater than the number of independent sources [26.57]. However, if this is strictly true, then it means that we have to somehow know the number of sources, and this, in some degree, contradicts the the acoustic holography approach.

A recent study [26.61] demonstrated that the measured information, the location of the sources, and the number of independent sources converge to their true values as the number of reference microphones increases. This study also showed that high-power sources

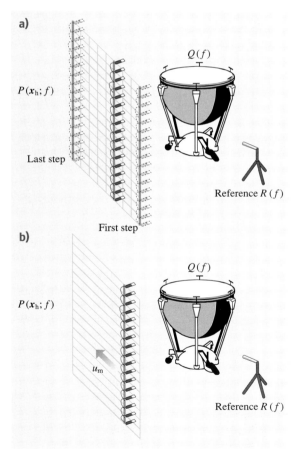

Fig. 26.15a,b Two measurement methods for the pressure on the hologram plane: (**a**) step-by-step scanning (**b**) continuous scanning

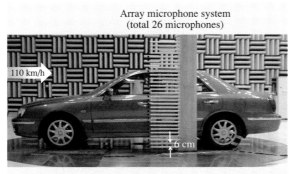

Fig. 26.16 Application result of the step-by-step scanning method to the wind noise of a car. This figure is the pressure distribution at $710 \sim 900\,\text{Hz}$ in a source plane when the flow velocity is $110\,\text{km/h}$. In this experiment, 17 reference microphones are randomly located in the car, to see the coherence between interior noise and what are measured by the array microphone system. The array microphone system was initially located at 3 m forward from the middle point of a car, and moved 6 cm in step until it reached at 3 m backward from the middle point

are likely to be identified even if the number of reference microphones is less than the number of sources. Figure 26.16 shows an example of this method when there are many independent sound fields. On the other hand, one study showed that we can even continuously scan the sound field by using a line array of microphones (Fig. 26.15b) [26.62–65]. This method essentially al-

lows us to extend the aperture size without any limit as long as the sound field is stationary. In fact, [26.65] also showed that this method can be used for a slowly varying (quasi-stationary) sound field.

This method has to deal with the Doppler shift. For example, let us consider a plane wave in the $(k_{x_0}, k_{y_0}, k_{z_0})$ direction and a pure tone of frequency f_{h_0}. Then the pressure on the hologram plane can be written as

$$p(x_h, y_h, z_h; t) = P_0 \exp\left[i(k_{x_0}x_h + k_{y_0}y_h + k_{z_0}z_h)\right]\exp(-i, f_{h_0}t)\,, \quad (26.13)$$

where P_0 denotes the complex magnitude of the plane wave. Spatial information about the plane wave with respect to the x-direction can be represented by a wavenumber spectrum, and can be described as

$$\hat{P}(k_x, y_h, z_h; t) = \int_{-\infty}^{\infty} p(x_h, y_h, z_h; t)\,\text{e}^{-ik_x x_h}\,\text{d}x_h$$
$$= P_0 \exp\left[i(k_{y_0}y_h + k_{z_0}z_h)\right]$$
$$\times \delta(k_x - k_{x_0})\exp(-i2\pi f_{h_0}t)$$
$$= P(k_{x_0}, y_h, z_h)\delta(k_x - k_{x_0})$$
$$\times \exp(-i2\pi f_{h_0}t)\,, \quad (26.14)$$

where $P(k_{x_0}, y_h, z_h) = P_0 \exp[i(k_{y_0}y_h + k_{z_0}z_h)]$ is the wave-number spectrum of the plane wave at $k_x = k_{x_0}$.

If a microphone is moving at an x-velocity u_m, the measured signal $p_m(x_h, y_h, z_h; t)$ is

$$p_m(x_h, y_h, z_h; t) = p(u_m t, y_h, z_h; t)\,. \quad (26.15)$$

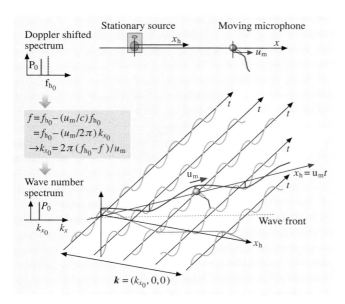

Fig. 26.17 The continuous scanning method for a plane wave and a pure tone (one-dimensional illustration). f_{h_0} is the source frequency, f is the measured frequency, u_m is the microphone velocity, c is the wave speed, k_{k_0} is the x-direction wave number, and P_0 is the complex amplitude of a plane wave

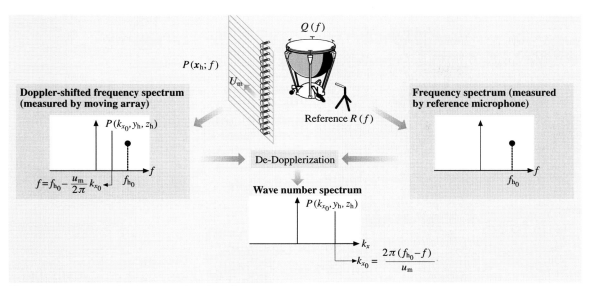

Fig. 26.18 De-Dopplerization procedure for a line spectrum

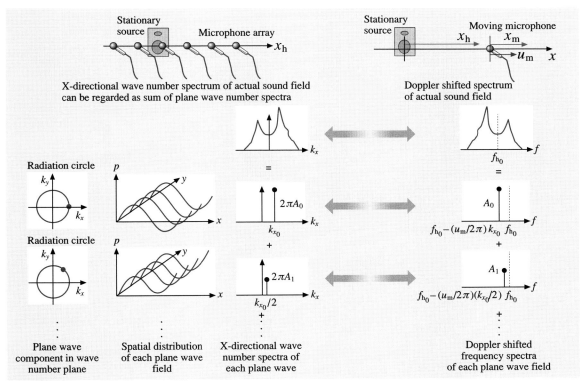

Fig. 26.19 The continuous scanning method for a more general case (one-dimensional illustration)

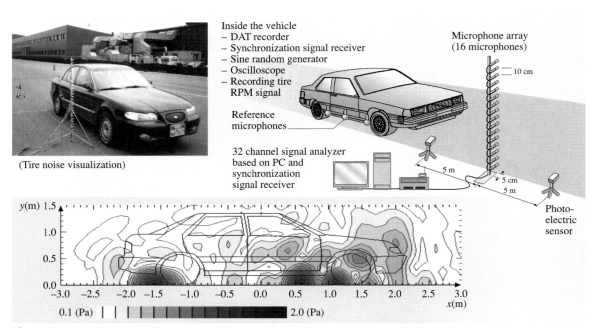

Fig. 26.20 Experimental configuration and result of the continuous scanning method to vehicle pass-by noise. The tire pattern noise distribution (pressure) on the source plane is shown when the car passed the microphone array with constant speed of 50 km/h

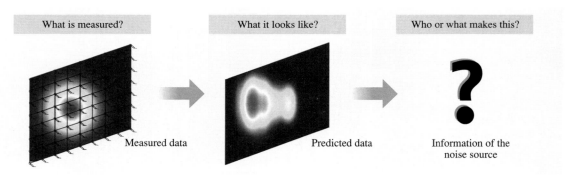

Fig. 26.21 Illustration of analysis problem in acoustic holography

The Fourier transform of (26.15) with respect to time F_T, using (26.13), can be expressed as

$$F_T[p_m(u_m t, y_h, z_h; t)]$$
$$= \int_{-\infty}^{\infty} p_m(u_m t, y_m, z_h; t) e^{i2\pi ft} dt$$
$$= P_0 \exp\left[i(k_{y_0} y_h + k_{z_0} z_h)\right] \delta\left(\frac{u_m}{2\pi} k_{x_0} - f_{h_0} + f\right)$$
$$= P(k_{x_0}, y_h, z_h) \delta\left(\frac{u_m}{2\pi} k_{x_0} - f_{h_0} + f\right). \quad (26.16)$$

Equation (26.16) means that the complex amplitude of the plane wave is located at the shifted frequency $f_{h_0} - u_m k_{x_0}/2\pi$, as shown in Fig. 26.17. In general, the relation between the shifted frequency f and x-direction wave number k_x is expressed as Fig. 26.18

$$k_x = \frac{2\pi(f_{h_0} - f)}{u_m}. \quad (26.17)$$

We can measure the original frequency f_{h_0} by a fixed reference microphone. Using the Doppler shift, we can therefore obtain the wave-number components from the frequency components of the moving microphone signal. Figure 26.19 illustrates how we obtain the wave-number spectrum.

This method essentially uses the relative coordinate between the hologram and microphone. Therefore, it can be used for measuring a hologram of moving noise sources (Fig. 26.20), which is one of the major contributions of this method [26.62–65].

26.2.4 Analysis of Acoustic Holography

Once we have a picture of the sound (acoustic holography), w questions about its meaning are the next topic of interest. What we have is usually a contour plot of the sound pressure distribution or a vector plot of the sound intensity on a plane of interest. This plot may help us to imagine where the sound source is and how it radiates into space with respect to a frequency of interest. However, in the strict sense, the only thing we can do from the two-dimensional expression of sound pressure or intensity distribution is to guess what was really there. We do not know, precisely, where the sound sources are (Fig. 26.21).

As mentioned earlier, there are two types of sound sources: active and passive sound source. The former is the source that radiates sound itself, while the latter only radiates reflected sound. These two different types of sound sources can be distinguished by eliminating reflected sound [26.116]. This is directly related to the way boundary conditions are treated in the analysis.

The boundary condition for a locally reacting surface can be written as [26.116–118]

$$V(\boldsymbol{x}_s; f) = A(\boldsymbol{x}_s; f)P(\boldsymbol{x}_s; f) + S(\boldsymbol{x}_s; f), \quad (26.18)$$

where $V(\boldsymbol{x}_s; f)$ and $P(\boldsymbol{x}_s; f)$ are the velocity and pressure on the wall. $A(\boldsymbol{x}_s; f)$ is the wall admittance and $S(\boldsymbol{x}_s; f)$ is the source strength on the wall. The active sound source is located at a position such that the source strength is not zero. This equation says that we can estimate the source strength if we measure the wall admittance. To do this, it is necessary to first turn off the source or sources, and then measure the wall admittance by putting a known source in the desired position (Fig. 26.22a). The next step is to turn on the sources and obtain the sound pressure and velocity distribution on the wall, using the admittance information (Fig. 26.22b). This provides us with the location

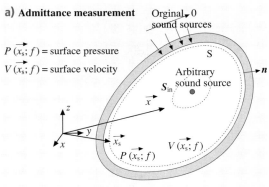

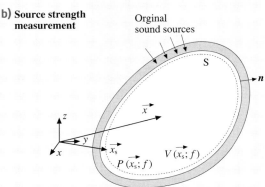

Fig. 26.22 Two steps to separate the active and passive source

Fig. 26.23 Spatially independent or dependent sources

and strength of the source (i.e. the source power; for example, see Fig. 26.24).

Another very important problem is whether or not we can distinguish between independent or dependent sources, i.e. two birds singing versus one bird with two beaks (Fig. 26.23). This has a rather significant practical application. For example, to control noise sources effectively, we only need to control independent noise sources. This can be achieved by using the statistical

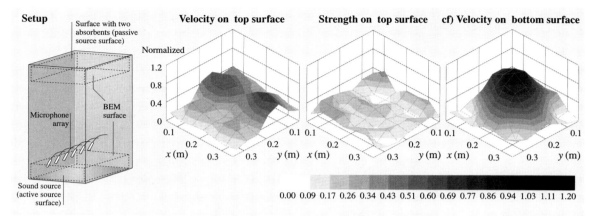

Fig. 26.24 Experiment results that separate the active and passive sources. The top surface is made by the sound absorption material. The speaker on the bottom surface, which is reflecting sound, is eliminated by this separation

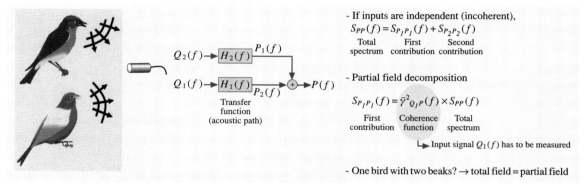

Fig. 26.25 A two-input single-output system and its partial field

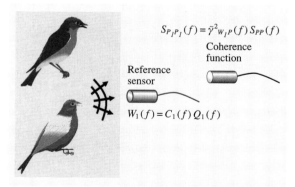

Fig. 26.26 The conventional method to obtain the partial field

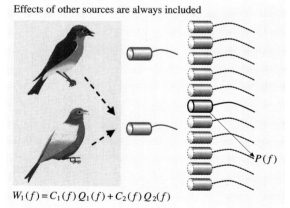

Fig. 26.27 Acoustic holography and partial field decomposition

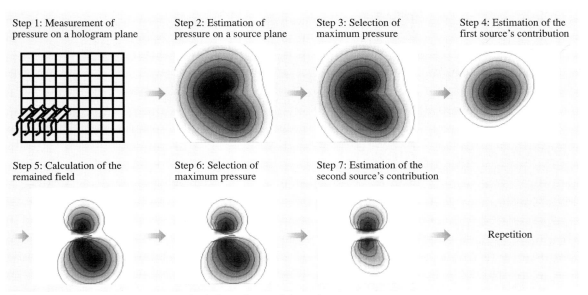

Fig. 26.28 The procedure to separate the independent and dependent sources

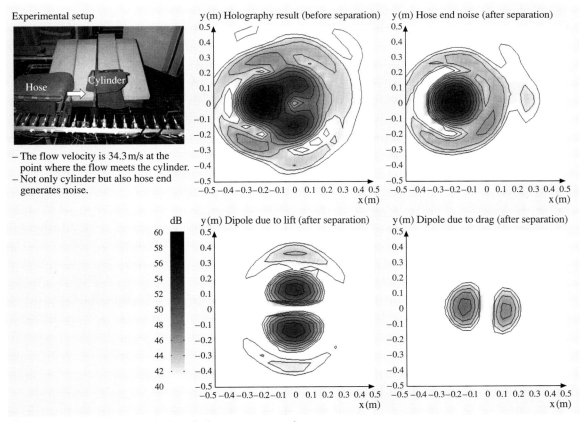

Fig. 26.29 The separation method applied to a vortex experiment

differences between signals that are induced by independent and dependent sound sources.

For example, let us consider a two-input single-output system (Fig. 26.25). If the two inputs are independent, the spectrum $S_{PP}(f)$ of the output $P(f)$ can be expressed as

$$S_{PP}(f) = |H_1(f)|^2 S_{Q_1 Q_1}(f) + |H_2(f)|^2 S_{Q_2 Q_2}(f), \tag{26.19}$$

where $S_{Q_i Q_i}(f)$ is the spectrum of the i-th input $Q_i(f)$, and $H_i(f)$ is its transfer function. The first and second terms represent the contributions of the first and second input to the output spectrum, respectively. If we can obtain a signal as

$$W_1(f) = C(f) Q_1(f), \tag{26.20}$$

then we can estimate the contribution of the first source as [26.119]

$$S_{P_1 P_1}(f) = |H_1(f)|^2 S_{Q_1 Q_1}(f) = \gamma_{W_1 P}^2(f) S_{PP}(f), \tag{26.21}$$

where $\gamma_{W_1 P}^2(f)$ is the coherence function between $W_1(f)$ and $Q_1(f)$ (Fig. 26.26).

We can simply extend (26.21) to the case of multiple outputs, as in the case of acoustic holography (Fig. 26.27). The main problem is how to obtain a signal that satisfies (26.20). We can generally say that, by putting sensors closer to the source or sources [26.57, 58, 120–123], we may have a better signal that can be used to distinguish between independent or dependent sources. However, this is neither well proven nor practical, as it is not always easy to put the sensors close to the sources. Very recently, a method that does not require this [26.124, 125] was developed. Figure 26.28 explains the method's procedures. The first and second steps are the same as in acoustic holography: measurement and prediction. The third step is to search for the maximum pressure on the source plane. This method assumes that the maximum pressure satisfies (26.20). The fourth step is to estimate the contribution of the first source by using the coherence functions between the maximum pressure and other points, as in (26.21). The fifth step is to calculate the remaining spectrum by subtracting the first contribution from the output spectrum. These steps are repeated until the contributions of the other sources are estimated (for example, see Fig. 26.29).

26.3 Summary

As expected, it is not simple to answer the question of whether we can see the sound field. However, it is now understood that the analysis of what we obtained, acoustic holography, needs to be properly addressed, although little attention was given to this problem in the past. We now understand better how to obtain information from the sound picture. Making a picture is the job of acoustic holography, but the interpretation of this picture is the responsibility of the observer. This paper has reviewed some useful guidelines for better interpretation of the sound field to deduce the right impression or information from the picture.

26.A Mathematical Derivations of Three Acoustic Holography Methods and Their Discrete Forms

We often use the Kirchhoff–Helmholtz integral equation to explain how we predict what we do not measure based on what we do measure. It is noteworthy, however, that the same result can be obtained by using the characteristic solutions of the Helmholtz equation. The following sections address how these can be obtained. Planar, cylindrical, and spherical acoustic holography are derived using characteristic equations in terms of a corresponding coordinate system. The equations for holography are also expressed in a discrete form.

26.A.1 Planar Acoustic Holography

If we see solutions of the Helmholtz equation

$$\nabla^2 P + k^2 P = 0, \tag{26.A1}$$

in terms of Cartesian coordinate, then we can write them as

$$P(x, y, z; f) = X(x) Y(y) Z(z), \tag{26.A2}$$

where $k = \frac{\omega}{c} = \frac{2\pi f}{c}$. We assume then P is separable with respect to X, Y and Z. Equations (26.A1) and (26.A2) yield the characteristic equation

$$\psi(x, y, z; k_x, k_y, k_z)$$
$$= \begin{pmatrix} e^{ik_x x} \\ e^{-ik_x x} \end{pmatrix} \begin{pmatrix} e^{ik_y y} \\ e^{-ik_y y} \end{pmatrix} \begin{pmatrix} e^{ik_z z} \\ e^{-ik_z z} \end{pmatrix}, \quad (26.A3)$$

where

$$k^2 = k_x^2 + k_y^2 + k_z^2. \quad (26.A4)$$

Now we can write

$$P(x, y, z; f) = \int \hat{P}(\boldsymbol{k}) \psi(x, y, z; k_x, k_y, k_z) \, d\boldsymbol{k}, \quad (26.A5)$$

where

$$\boldsymbol{k} = (k_x, k_y, k_z). \quad (26.A6)$$

Let us assume that the sound sources are all located at $z < z_s$ and we measure at $z = z_h > z_s$. Then we can write (26.A5) as

$$P(x, y, z; f)$$
$$= \int_{-\infty}^{\infty} \int_{-\infty}^{\infty} \hat{\hat{P}}(k_x, k_y) e^{i(k_x x + k_y y + k_z z)} \, dk_x \, dk_y. \quad (26.A7)$$

It is noteworthy that we selected only $+ik_z z$. This is because of the assumptions we made ($z < z_s$ and $z = z_h > z_s$). The k_x and k_y can be either positive or negative. Therefore it is not necessary to include $-ik_x x$ or $-ik_y y$ in (26.A3).

In (26.7),

$$k_z = \begin{cases} \sqrt{k^2 - k_x^2 - k_y^2}, & \text{when} \quad k^2 > k_x^2 + k_y^2 \\ i\sqrt{k_x^2 + k_y^2 - k^2}, & \text{when} \quad k^2 < k_x^2 + k_y^2. \end{cases}$$
$$(26.A8)$$

Figure 26.6 illustrates what these two different k_z values essentially mean. We measure $P(x, y, z = z_h; f)$, therefore we have data of the sound pressure data on $z = z_h$. A Fourier transform of (26.A7) leads to

$$\hat{\hat{P}}(k_x, k_y)$$
$$= \int_{-\infty}^{\infty} \int_{-\infty}^{\infty} P(x, y, z; f) e^{-i(k_x x + k_y y + k_z z)} \, dx \, dy. \quad (26.A9)$$

Using (26.A9) and (26.A7), we can always estimate the sound pressure on z, which is away from the source.

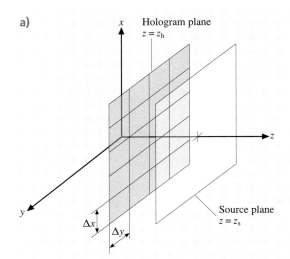

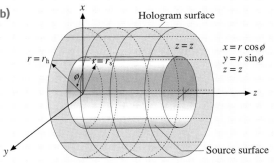

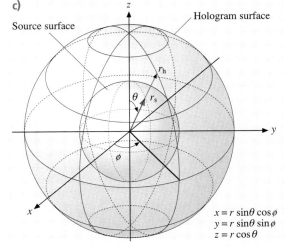

Fig. 26.30a–c Coordinates system for acoustic holography: (**a**) planar acoustic holography (**b**) cylindrical Acoustic holography (**c**) spherical acoustic holography

It is noteworthy that (26.A9) has to be preformed in the discrete domain. In other words, we have to use a finite rectangular aperture, which is spatially sampled (26.8 and 26.30a). If the number of measurement points along the x- and y-directions are M and N, respectively, and the corresponding sampling distances are Δx and Δy, then (26.A9) can be rewritten as

$$\hat{P}(k_x, k_y) = \frac{1}{(2\pi)^2} e^{-ik_z z_h} \Delta x \Delta y$$
$$\sum_{m=0}^{M-1} \sum_{n=0}^{N-1} P(x_m, y_n, z_h) e^{-ik_x x_m} e^{-ik_y y_n} . \quad (26.A10)$$

where

$$x_m = \left(m + \frac{1-M}{2}\right) \Delta x,$$
$$y_n = \left(n + \frac{1-N}{2}\right) \Delta y . \quad (26.A11)$$

M and N are the number of data points in the x- and y-directions, respectively.

26.A.2 Cylindrical Acoustic Holography

A solution can also found in cylindrical coordinate, that is

$$P(r, \phi, z) = R(r)\Phi(\phi)Z(z) . \quad (26.A12)$$

Figure 26.1 shows the coordinate systems. Then, its characteristic solutions are

$$\psi(r, \phi, z; k_r, k_z)$$
$$= \begin{pmatrix} H_m^{(1)}(k_r r) \\ H_m^{(2)}(k_r r) \end{pmatrix} \begin{bmatrix} e^{im\phi} \\ e^{-im\phi} \end{bmatrix} \begin{bmatrix} e^{ik_z z} \\ e^{-ik_z z} \end{bmatrix} , \quad (26.A13)$$

where

$$k^2 = k_r^2 + k_z^2 . \quad (26.A14)$$

It is noteworthy that m is a nonnegative integer. $H_m^{(1)}$ and $H_m^{(2)}$ are first and second cylindrical Hankel functions, respectively. $e^{im\phi}$ and $e^{-im\phi}$ express the mode shapes in the ϕ-direction.

Using the characteristic function (26.A6), we can write a solution of the Helmholtz equation with respect to cylindrical coordinate as

$$P(r, \phi, z; f) = \int \hat{P}_m(\mathbf{k}) \psi_m(r, \phi, z; k_r, k_z) d\mathbf{k} , \quad (26.A15)$$

where

$$\mathbf{k} = (k_r, k_z) . \quad (26.A16)$$

Assuming that the sound sources are all located at $r < r_s$ and that the hologram surface is situated on the surface $r = r_h$, and that $r_h > r_s$, then no waves propagate into the negative r-direction, in other words, toward the sources. Then (26.A15) can be rewritten as

$$P(r, \phi, z; f)$$
$$= \sum_{m=-\infty}^{\infty} \int_{-\infty}^{\infty} \hat{P}_m(k_z) e^{im\phi} e^{ik_z z} H_m^{(1)}(k_r r) dk_z , \quad (26.A17)$$

and k_r has to be

$$k_r = \begin{cases} \sqrt{k^2 - k_z^2}, & \text{when } k^2 > k_z^2 \\ i\sqrt{k_z^2 - k^2}, & \text{when } k^2 < k_z^2 . \end{cases} \quad (26.A18)$$

We measure the acoustic pressure at $r = r_h$, therefore $P(r_h, \phi, z)$ is available. $\hat{P}_m(k_z)$ can then be readily obtained.

That is

$$\hat{P}_m(k_z) = \frac{1}{(2\pi)^2} \int_0^{2\pi} \int_{-\infty}^{\infty} P(r_h, \phi, z) e^{-m\phi} e^{-ik_z z}$$
$$\left\{ H_m^{(1)}(k_r r_h) \right\}^{-1} dz d\phi . \quad (26.A19)$$

Inserting (26.A19) into (26.A17) provides us with the acoustic pressure at the unmeasured surface at r.

Discretization of (26.A19) leads to a formula that can be used in practical calculations:

$$\hat{P}_m(k_z) = \frac{1}{(2\pi)^2 H_m^{(1)}(k_r r_h)} \frac{2\pi}{L} \Delta z$$
$$\sum_{l=0}^{L-1} \sum_{n=0}^{N-1} P(r_h, \phi_l, z_n) e^{-im\phi_l} e^{-ik_z z_n} , \quad (26.A20)$$

where

$$\phi_l = \frac{(2l+1)\pi}{L} ,$$
$$z_n = \left(n + \frac{1-N}{2}\right) \Delta z . \quad (26.A21)$$

L and N are the number of data points in the ϕ- and z-directions, respectively.

26.A.3 Spherical Acoustic Holography

The Helmholtz equation can also be expressed in spherical coordinate (Fig. 26.30c). Assuming again that the separation of variable also holds in this case, we can write:

$$P(r, \theta, \phi) = R(r)\Theta(\theta)\Phi(\phi) . \quad (26.A22)$$

Substituting this into (26.A1) gives the characteristic equation

$$\psi_{mn}(r, \theta, \phi; k) = \begin{pmatrix} h_m^{(1)}(kr) \\ h_m^{(2)}(kr) \end{pmatrix} \begin{pmatrix} P_m^n \cos\theta \\ Q_m^n \cos\theta \end{pmatrix} \begin{pmatrix} e^{in\phi} \\ e^{-in\phi} \end{pmatrix} . \quad (26.A23)$$

m is a nonnegative integer and n can be any integer between 0 and m. $h_m^{(1)}$ and $h_m^{(2)}$ are first and second spherical Hankel functions. It is also noteworthy that P_m^n and Q_m^n are first and second Legendre polynomials.

Then we can write the solution of the Helmholtz equation as

$$P(r, \theta, \phi) = \sum_{m=0}^{\infty} \sum_{n=-m}^{m} \hat{P}_{mn} \psi_{mn}(r, \theta, \phi; k) . \quad (26.A24)$$

Suppose that we have sound sources at $r < r_s$ and the hologram is on the surface $r = r_h > r_s$; then (26.A24) can be simplified to:

$$P(r, \theta, \phi) = \sum_{m=0}^{\infty} \sum_{n=-m}^{m} \hat{P}_{mn} Y_{mn}(\theta, \phi) h_m^{(1)}(kr) , \quad (26.A25)$$

where

$$Y_{mn}(\theta, \phi) = P_m^{|n|}(\cos\theta) e^{in\phi} . \quad (26.A26)$$

This is a spherical harmonic function. It is noteworthy that we only have first spherical harmonic functions because all waves propagate away from the sources. The second Legendre function was discarded because it would have finite acoustic pressure at $\theta = 0$ or π.

Similarly, as previously stated, the sound pressure data on the hologram is available, therefore we can obtain $\hat{P}_{mn}$ in (26.A25) by

$$\hat{P}_{mn} = \frac{2m+1}{4\pi h_m^{(1)}(kr_h)} \frac{(m-|n|)!}{(m+|n|)!} \int_0^\pi \int_0^{2\pi} P(r_h, \theta, \phi) Y_{mn}^*(\theta, \phi) \sin\theta \, d\phi \, d\theta , \quad (26.A27)$$

where we have used the orthogonality property of Y_{mn}. The $*$ represents the complex conjugate. Using (26.A27) and (26.A25), we can estimate the acoustic pressure anywhere away from the sources.

The discrete form of (26.A27) can be written as

$$\hat{P}_{mn} = A_{mn} \frac{2\pi^2}{LQ} \sum_{l=0}^{L-1} \sum_{q=0}^{Q-1} P(r_h, \theta_l, \phi_q) P_m^{|n|}(\cos\theta_l)(\sin\theta_l) e^{-in\phi_q} , \quad (26.A28)$$

where

$$\theta_l = \frac{(2l+1)\pi}{2L} ,$$
$$\phi_q = \frac{(2q+1)\pi}{Q} . \quad (26.A29)$$

where L is the number of data points in θ and Q is what is in ϕ-direction.

References

26.1　M. Kac: Can one hear the shape of the drum?, Am. Math. Mon. **73**, 1–23 (1966)
26.2　G.T. Herman, H.K. Tuy, K.J. Langenberg: *Basic methods of tomography and inverse problems* (Malvern, Philadelphia 1987)
26.3　H.D. Bui: *Inverse Problems in the Mechanics of Materials: An Introduction* (CRC, Boca Raton 1994)
26.4　K. Kurpisz, A.J. Nowak: *Inverse Thermal Problems* (Computational Mechanics Publ., Southampton 1995)
26.5　A. Kirsch: *An Introduction to the Mathematical Theory of Inverse Problems* (Springer, Berlin Heidelberg New York 1996)
26.6　M. Bertero, P. Boccacci: *Introduction to Inverse Problems in Imaging* (IOP, Bristol 1998)
26.7　V. Isakov: *Inverse Problems for Partial Differential Equations* (Springer, New York 1998)
26.8　D.N. Ghosh Roy, L.S. Couchman: *Inverse Problems and Inverse Scattering of Plane Waves* (Academic, New York 2002)
26.9　J. Hardamard: *Lectures on Cauchy's Problem in Linear Partial Differential Equations* (Yale Univ. Press, New Haven 1923)
26.10　L. Landweber: An iteration formula for Fredholm integral equations of the first kind, Am. J. Math. **73**, 615–624 (1951)

26.11 A.M. Cormack: Representation of a function by its line integrals, with some radiological applications, J. App. Phys. **34**, 2722–2727 (1963)

26.12 A.M. Cormack: Representation of a function by its line integrals, with some radiological applications II, J. App. Phys. **35**, 2908–2913 (1964)

26.13 A.N. Tikhonov, V.Y. Arsenin: *Solutions of Ill-Posed Problems* (Winston, Washington 1977)

26.14 S.R. Deans: *The Radon Transform and Some of its Applications* (Wiley, New York 1983)

26.15 A.K. Louis: Mathematical problems of computerized tomography, Proc. IEEE **71**, 379–389 (1983)

26.16 M.H. Protter: Can one hear the shape of a drum? Revisited, SIAM Rev. **29**, 185–197 (1987)

26.17 K. Chadan, P.C. Sabatier: *Inverse Problems in Quantum Scattering Theory*, 2nd edn. (Springer, New York 1989)

26.18 A.K. Louis: Medical imaging: state of the art and future development, Inv. Probl. **8**, 709–738 (1992)

26.19 D. Colton, R. Kress: *Inverse Acoustic and Electromagnetic Scattering Theory*, 2nd edn. (Springer, New York 1998)

26.20 D. Gabor: A new microscopic principle, Nature **161**, 777 (1948)

26.21 B.P. Hilderbrand, B.B. Brenden: *An Introduction to Acoustical Holography* (Plenum, New York 1972)

26.22 E.N. Leith, J. Upatnieks: Reconstructed wavefronts and communication theory, J. Opt. Soc. Am. **52**, 1123–1130 (1962)

26.23 W.E. Kock: Hologram television, Proc. IEEE **54**, 331 (1966)

26.24 G. Tricoles, E.L. Rope: Reconstructions of visible images from reduced-scale replicas of microwave holograms, J. Opt. Soc. Am. **57**, 97–99 (1967)

26.25 G.C. Sherman: Reconstructed wave forms with large diffraction angles, J. Opt. Soc. Am. **57**, 1160–1161 (1967)

26.26 J.W. Goodman, R.W. Lawrence: Digital image formation from electronically detected holograms, Appl. Phys. Lett. **11**, 77–79 (1967)

26.27 Y. Aoki: Microwave holograms and optical reconstruction, Appl. Opt. **6**, 1943–1946 (1967)

26.28 R.P. Porter: Diffraction-limited, scalar image formation with holograms of arbitrary shape, J. Opt. Soc. Am. **60**, 1051–1059 (1970)

26.29 E. Wolf: Determination of the amplitude and the phase of scattered fields by holography, J. Opt. Soc. Am. **60**, 18–20 (1970)

26.30 W.H. Carter: Computational reconstruction of scattering objects from holograms, J. Opt. Soc. Am. **60**, 306–314 (1970)

26.31 E.G. Williams, J.D. Maynard, E. Skudrzyk: Sound source reconstruction using a microphone array, J. Acoust. Soc. Am. **68**, 340–344 (1980)

26.32 E.G. Williams, J.D. Maynard: Holographic imaging without the wavelength resolution limit, Phys. Rev. Lett. **45**, 554–557 (1980)

26.33 E.A. Ash, G. Nichols: Super-resolution aperture scanning microscope, Nature **237**, 510–512 (1972)

26.34 S.U. Pillai: *Array Signal Processing* (Springer, New York 1989)

26.35 D.H. Johnson, D.E. Dudgeon: *Array Signal Processing Concepts and Techniques* (Prentice-Hall, Upper Saddle River 1993)

26.36 M. Kaveh, A.J. Barabell: The statistical performance of the MUSIC and the minimum-Norm algorithms in resolving plane waves in noise, IEEE Trans. Acoust. Speech Signal Process. **34**, 331–341 (1986)

26.37 M. Lasky: Review of undersea acoustics to 1950, J. Acoust. Soc. Am. **61**, 283–297 (1977)

26.38 B. Barskow, W.F. King, E. Pfizenmaier: Wheel/rail noise generated by a high-speed train investigated with a line array of microphones, J. Sound Vib. **118**, 99–122 (1987)

26.39 J. Hald, J.J. Christensen: *A class of optimal broadband phased array geometries designed for easy construction*, Proc. Inter-Noise 2002 (Int. Inst. Noise Control Eng., West Lafayette 2002)

26.40 Y. Takano: *Development of visualization system for high-speed noise sources with a microphone array and a visual sensor*, Proc. Inter-Noise 2003 (Int. Inst. Noise Control Eng., West Lafayette 2003)

26.41 G. Elias: *Source localization with a two-dimensional focused array: Optimal Signal processing for a cross-shaped array*, Proc. Inter-Noise 95 (Int. Inst. Noise Control Eng., West Lafayette 1995) pp. 1175–1178

26.42 A. Nordborg, A. Martens, J. Wedemann, L. Wellenbrink: *Wheel/rail noise separation with microphone array measurements*, Proc. Inter-Noise 2001 (Int. Inst. Noise Control Eng., West Lafayette 2001) pp. 2083–2088

26.43 A. Nordborg, J. Wedemann, L. Wellenbrink: *Optimum array microphone configuration*, Proc. Inter-Noise 2000 (Int. Inst. Noise Control Eng., West Lafayette 2000)

26.44 P.R. Stepanishen, K.C. Benjamin: Forward and backward prediction of acoustic fields using FFT methods, J. Acoust. Soc. Am. **71**, 803–812 (1982)

26.45 E.G. Williams, J.D. Maynard: Numerical evaluation of the Rayleigh integral for planar radiators using the FFT, J. Acoust. Soc. Am. **72**, 2020–2030 (1982)

26.46 J.D. Maynard, E.G. Williams, Y. Lee: Nearfield acoustic holography (NAH): I. Theory of generalized holography and the development of NAH, J. Acoust. Soc. Am. **78**, 1395–1413 (1985)

26.47 W.A. Veronesi, J.D. Maynard: Nearfield acoustic holography (NAH) II. Holographic reconstruction algorithms and computer implementation, J. Acoust. Soc. Am. **81**, 1307–1322 (1987)

26.48 S.I. Hayek, T.W. Luce: Aperture effects in planar nearfield acoustical imaging, ASME J. Vib. Acoust. Stress Reliab. Des. **110**, 91–96 (1988)

26.49 A. Sarkissian, C.F. Gaumond, E.G. Williams, B.H. Houston: Reconstruction of the acoustic field

26.49 over a limited surface area on a vibrating cylinder, J. Acoust. Soc. Am. **93**, 48–54 (1993)
26.50 J. Hald: *Reduction of spatial windowing effects in acoustical holography*, Proc. Inter-Noise 94 (Int. Inst. Noise Control Eng., West Lafayette 1994) pp. 1887–1890
26.51 H.-S. Kwon, Y.-H. Kim: Minimization of bias error due to windows in planar acoustic holography using a minimum error window, J. Acoust. Soc. Am. **98**, 2104–2111 (1995)
26.52 K. Saijou, S. Yoshikawa: Reduction methods of the reconstruction error for large-scale implementation of near-field acoustical holography, J. Acoust. Soc. Am. **110**, 2007–2023 (2001)
26.53 K.-U. Nam, Y.-H. Kim: Errors due to sensor and position mismatch in planar acoustic holography, J. Acoust. Soc. Am. **106**, 1655–1665 (1999)
26.54 G. Weinreich, E.B. Arnold: Method for measuring acoustic radiation fields, J. Acoust. Soc. Am. **68**, 404–411 (1980)
26.55 E.G. Williams, H.D. Dardy, R.G. Fink: Nearfield acoustical holography using an underwater, automated scanner, J. Acoust. Soc. Am. **78**, 789–798 (1985)
26.56 D. Blacodon, S.M. Candel, G. Elias: Radial extrapolation of wave fields from synthetic measurements of the nearfield, J. Acoust. Soc. Am. **82**, 1060–1072 (1987)
26.57 J. Hald: *STSF – a unique technique for scan-based near-field acoustic holography without restrictions on coherence*, Bruel Kjær Tech. Rev. 1 (Bruel Kjær, Noerum 1989)
26.58 K.B. Ginn, J. Hald: *STSF – practical instrumentation and applications*, Bruel Kjær Tech. Rev. 2 (Bruel Kjær, Noerum 1989)
26.59 S.H. Yoon, P.A. Nelson: A method for the efficient construction of acoustic pressure cross-spectral matrices, J. Sound Vib. **233**, 897–920 (2000)
26.60 H.-S. Kwon, Y.-J. Kim, J.S. Bolton: Compensation for source nonstationarity in multireference, scan-based near-field acoustical holography, J. Acoust. Soc. Am. **113**, 360–368 (2003)
26.61 K.-U. Nam, Y.-H. Kim: *Low coherence acoustic holography*, Proc. Inter-Noise 2003 (Int. Inst. Noise Control Eng., West Lafayette 2003)
26.62 H.-S. Kwon, Y.-H. Kim: Moving frame technique for planar acoustic holography, J. Acoust. Soc. Am. **103**, 1734–1741 (1998)
26.63 S.-H. Park, Y.-H. Kim: An improved moving frame acoustic holography for coherent band-limited noise, J. Acoust. Soc. Am. **104**, 3179–3189 (1998)
26.64 S.-H. Park, Y.-H. Kim: Effects of the speed of moving noise sources on the sound visualization by means of moving frame acoustic holography, J. Acoust. Soc. Am. **108**, 2719–2728 (2000)
26.65 S.-H. Park, Y.-H. Kim: Visualization of pass-by noise by means of moving frame acoustic holography, J. Acoust. Soc. Am. **110**, 2326–2339 (2001)
26.66 S.M. Candel, C. Chassaignon: Radial extrapolation of wave fields by spectral methods, J. Acoust. Soc. Am. **76**, 1823–1828 (1984)
26.67 E.G. Williams, H.D. Dardy, K.B. Washburn: Generalized nearfield acoustical holography for cylindrical geometry: Theory and experiment, J. Acoust. Soc. Am. **81**, 389–407 (1987)
26.68 B.K. Gardner, R.J. Bernhard: A noise source identification technique using an inverse Helmholtz integral equation method, ASME J. Vib. Acoust. Stress Reliab. Des. **110**, 84–90 (1988)
26.69 E.G. Williams, B.H. Houston, J.A. Bucaro: Broadband nearfield acoustical holography for vibrating cylinders, J. Acoust. Soc. Am. **86**, 674–679 (1989)
26.70 G.H. Koopmann, L. Song, J.B. Fahnline: A method for computing acoustic fields based on the principle of wave superposition, J. Acoust. Soc. Am. **86**, 2433–2438 (1989)
26.71 A. Sarkissan: Near-field acoustic holography for an axisymmetric geometry: A new formulation, J. Acoust. Soc. Am. **88**, 961–966 (1990)
26.72 M. Tamura: Spatial Fourier transform method of measuring reflection coefficients at oblique incidence. I: Theory and numerical examples, J. Acoust. Soc. Am. **88**, 2259–2264 (1990)
26.73 L. Song, G.H. Koopmann, J.B. Fahnline: Numerical errors associated with the method of superposition for computing acoustic fields, J. Acoust. Soc. Am. **89**, 2625–2633 (1991)
26.74 M. Villot, G. Chaveriat, J. Ronald: Phonoscopy: An acoustical holography technique for plane structures radiating in enclosed spaces, J. Acoust. Soc. Am. **91**, 187–195 (1992)
26.75 D.L. Hallman, J.S. Bolton: *Multi-reference nearfield acoustical holography in reflective environments*, Proc. Inter-Noise 93 (Int. Inst. Noise Control Eng., West Lafayette 1993) pp. 1307–1310
26.76 M.-T. Cheng, J.A. Mann III, A. Pate: Wave-number domain separation of the incident and scattered sound field in Cartesian and cylindrical coordinates, J. Acoust. Soc. Am. **97**, 2293–2303 (1995)
26.77 M. Tamura, J.F. Allard, D. Lafarge: Spatial Fourier-transform method for measuring reflection coefficients at oblique incidence. II. Experimental results, J. Acoust. Soc. Am. **97**, 2255–2262 (1995)
26.78 Z. Wang, S.F. Wu: Helmholtz equation – least-squares method for reconstructing the acoustic pressure field, J. Acoust. Soc. Am. **102**, 2020–2032 (1997)
26.79 S.F. Wu, J. Yu: Reconstructing interior acoustic pressure fields via Helmholtz equation least-squares method, J. Acoust. Soc. Am. **104**, 2054–2060 (1998)
26.80 S.-C. Kang, J.-G. Ih: The use of partially measured source data in near-field acoustical holography based on the BEM, J. Acoust. Soc. Am. **107**, 2472–2479 (2000)

26.81 S.F. Wu: On reconstruction of acoustic pressure fields using the Helmholtz equation least squares method, J. Acoust. Soc. Am. **107**, 2511–2522 (2000)

26.82 N. Rayess, S.F. Wu: Experimental validation of the HELS method for reconstructing acoustic radiation from a complex vibrating structure, J. Acoust. Soc. Am. **107**, 2955–2964 (2000)

26.83 Z. Zhang, N. Vlahopoulos, S.T. Raveendra, T. Allen, K.Y. Zhang: A computational acoustic field reconstruction process based on an indirect boundary element formulation, J. Acoust. Soc. Am. **108**, 2167–2178 (2000)

26.84 S.-C. Kang, J.-G. Ih: On the accuracy of nearfield pressure predicted by the acoustic boundary element method, J. Sound Vib. **233**, 353–358 (2000)

26.85 S.-C. Kang, J.-G. Ih: Use of nonsingular boundary integral formulation for reducing errors due to near-field measurements in the boundary element method based near-field acoustic holography, J. Acoust. Soc. Am. **109**, 1320–1328 (2001)

26.86 S.F. Wu, N. Rayess, X. Zhao: Visualization of acoustic radiation from a vibrating bowling ball, J. Acoust. Soc. Am. **109**, 2771–2779 (2001)

26.87 J.D. Maynard: *A new technique combining eigenfunction expansions and boundary elements to solve acoustic radiation problems*, Proc. Inter-Noise 2003 (Int. Inst. Noise Control Eng., West Lafayette 2003)

26.88 E.G. Williams: *Fourier Acoustics: Sound Radiation and Near Field Acoustical Holography* (Academic, London 1999)

26.89 F.L. Thurstone: Ultrasound holography and visual reconstruction, Proc. Symp. Biomed. Eng. **1**, 12–15 (1966)

26.90 A.L. Boyer, J.A. Jordan Jr., D.L. van Rooy, P.M. Hirsch, L.B. Lesem: *Computer reconstruction of images from ultrasonic holograms*, Proc. Second International Symposium on Acoustical Holography (Plenum, New York 1969)

26.91 A.F. Metilerell, H.M.A. El-Sum, J.J. Dreher, L. Larmore: Introduction to acoustical holography, J. Acoust. Soc. Am. **42**, 733–742 (1967)

26.92 L.A. Cram, K.O. Rossiter: *Long-wavelength holography and visual reproduction methods*, Proc. Second International Symposium on Acoustical Holography (Plenum, New York 1969)

26.93 T.S. Graham: *A new method for studying acoustic radiation using long-wavelength acoustical holography*, Proc. Second International Symposium on Acoustical Holography (Plenum, New York 1969)

26.94 E.E. Watson: Detection of acoustic sources using long-wavelength acoustical holography, J. Acoust. Soc. Am. **54**, 685–691 (1973)

26.95 H. Fleischer, V. Axelrad: Restoring an acoustic source from pressure data using Wiener filtering, Acustica **60**, 172–175 (1986)

26.96 E.G. Williams: Supersonic acoustic intensity, J. Acoust. Soc. Am. **97**, 121–127 (1995)

26.97 M.R. Bai: Acoustical source characterization by using recursive Wiener filtering, J. Acoust. Soc. Am. **97**, 2657–2663 (1995)

26.98 J.C. Lee: Spherical acoustical holography of low-frequency noise sources, Appl. Acoust. **48**, 85–95 (1996)

26.99 G.P. Carroll: The effect of sensor placement errors on cylindrical near-field acoustic holography, J. Acoust. Soc. Am. **105**, 2269–2276 (1999)

26.100 W.A. Veronesi, J.D. Maynard: Digital holographic reconstruction of sources with arbitrarily shaped surfaces, J. Acoust. Soc. Am. **85**, 588–598 (1989)

26.101 G.V. Borgiotti, A. Sarkissan, E.G. Williams, L. Schuetz: Conformal generalized near-field acoustic holography for axisymmetric geometries, J. Acoust. Soc. Am. **88**, 199–209 (1990)

26.102 D.M. Photiadis: The relationship of singular value decomposition to wave-vector filtering in sound radiation problems, J. Acoust. Soc. Am. **88**, 1152–1159 (1990)

26.103 G.-T. Kim, B.-H. Lee: 3-D sound source reconstruction and field repredicion using the Helmholtz integral equation, J. Sound Vib. **136**, 245–261 (1990)

26.104 A. Sarkissan: Acoustic radiation from finite structures, J. Acoust. Soc. Am. **90**, 574–578 (1991)

26.105 J.B. Fahnline, G.H. Koopmann: A numerical solution for the general radiation problem based on the combined methods of superposition and singular value decomposition, J. Acoust. Soc. Am. **90**, 2808–2819 (1991)

26.106 M.R. Bai: Application of BEM (boundary element method)-based acoustic holography to radiation analysis of sound sources with arbitrarily shaped geometries, J. Acoust. Soc. Am. **92**, 533–549 (1992)

26.107 G.V. Borgiotti, E.M. Rosen: The determination of the far field of an acoustic radiator from sparse measurement samples in the near field, J. Acoust. Soc. Am. **92**, 807–818 (1992)

26.108 B.-K. Kim, J.-G. Ih: On the reconstruction of the vibro-acoustic field over the surface enclosing an interior space using the boundary element method, J. Acoust. Soc. Am. **100**, 3003–3016 (1996)

26.109 B.-K. Kim, J.-G. Ih: Design of an optimal wave-vector filter for enhancing the resolution of reconstructed source field by near-field acoustical holography (NAH), J. Acoust. Soc. Am. **107**, 3289–3297 (2000)

26.110 P.A. Nelson, S.H. Yoon: Estimation of acoustic source strength by inverse methods: Part I, conditioning of the inverse problem, J. Sound Vib. **233**, 643–668 (2000)

26.111 S.H. Yoon, P.A. Nelson: Estimation of acoustic source strength by inverse methods: Part II, experimental investigation of methods for choosing regularization parameters, J. Sound Vib. **233**, 669–705 (2000)

26.112 E.G. Williams: Regularization methods for near-field acoustical holography, J. Acoust. Soc. Am. **110**, 1976–1988 (2001)

26.113 S.F. Wu, X. Zhao: Combined Helmholtz equation – least squares method for reconstructing acoustic radiation from arbitrarily shaped objects, J. Acoust. Soc. Am. **112**, 179–188 (2002)

26.114 A. Schuhmacher, J. Hald, K.B. Rasmussen, P.C. Hansen: Sound source reconstruction using inverse boundary element calculations, J. Acoust. Soc. Am. **113**, 114–127 (2003)

26.115 Y. Kim, P.A. Nelson: Spatial resolution limits for the reconstruction of acoustic source strength by inverse methods, J. Sound Vib. **265**, 583–608 (2003)

26.116 Y.-K. Kim, Y.-H. Kim: Holographic reconstruction of active sources and surface admittance in an enclosure, J. Acoust. Soc. Am. **105**, 2377–2383 (1999)

26.117 P.M. Morse, H. Feshbach: *Methods of Theoretical Physics, Part I* (McGraw–Hill, New York 1992)

26.118 L.L. Beranek, I.L. Ver: *Noise and Vibration Control Engineering* (Wiley, New York 1992)

26.119 J.S. Bendat, A.G. Piersol: *Random Data: Analysis and Measurement Procedures*, 2nd edn. (Wiley, New York 1986)

26.120 D. Hallman, J.S. Bolton: *Multi-reference nearfield acoustical holography*, Proc. Inter-Noise 92 (Int. Inst. Noise Control Eng., West Lafayette 1992) pp. 1165–1170

26.121 R.J. Ruhala, C.B. Burroughs: *Separation of leading edge, trailing edge, and sidewall noise sources from rolling tires*, Proc. Noise-Con 98 (Noise Control Foundation, Poughkeepsie 1998) pp. 109–114

26.122 H.-S. Kwon, J.S. Bolton: *Partial field decomposition in nearfield acoustical holography by the use of singular value decomposition and partial coherence procedures*, Proc. Noise-Con 98 (Noise Control Foundation, Poughkeepsie 1998) pp. 649–654

26.123 M.A. Tomlinson: Partial source discrimination in near field acoustic holography, Appl. Acoust. **57**, 243–261 (1999)

26.124 K.-U. Nam, Y.-H. Kim: Visualization of multiple incoherent sources by the backward prediction of near-field acoustic holography, J. Acoust. Soc. Am. **109**, 1808–1816 (2001)

26.125 K.-U. Nam, Y.-H. Kim, Y.-C. Choi, D.-W. Kim, O.-J. Kwon, K.-T. Kang, S.-G. Jung: *Visualization of speaker, vortex shedding, engine, and wind noise of a car by partial field decomposition*, Proc. Inter-Noise 2002 (Int. Inst. Noise Control Eng., West Lafayette 2002)

27. Optical Methods for Acoustics and Vibration Measurements

Modern optical methods applicable to vibration analysis, monitoring bending-wave propagation in plates and shells as well as propagating acoustic waves in transparent media such as air and water are described. Field methods, which capture the whole object field in one recording, and point measuring (scanning) methods, which measure at one point (small area) at a time (but in that point as a function of time), will be addressed. Temporally, harmonic vibrations, multi-frequency repetitive motions and transient or dynamic motions are included.

Interferometric methods, such as time-average and real-time holographic interferometry, speckle interferometry methods such as television (TV) holography, pulsed TV holography and laser vibrometry, are addressed. Intensity methods such as speckle photography or speckle correlation methods and particle image velocimetry (PIV) will also be treated.

27.1	Introduction	1101
	27.1.1 Chladni Patterns, Phase-Contrast Methods, Schlieren, Shadowgraph	1101
	27.1.2 Holographic Interferometry, Acoustical Holography	1102
	27.1.3 Speckle Metrology: Speckle Interferometry and Speckle Photography	1102
	27.1.4 Moiré Techniques	1104
	27.1.5 Some Pros and Cons of Optical Metrology	1104
27.2	**Measurement Principles and Some Applications**	1105
	27.2.1 Holographic Interferometry for the Study of Vibrations	1105
	27.2.2 Speckle Interferometry – TV Holography, DSPI and ESPI for Vibration Analysis and for Studies of Acoustic Waves	1108
	27.2.3 Reciprocity and TV Holography	1113
	27.2.4 Pulsed TV Holography – Pulsed Lasers Freeze Propagating Bending Waves, Sound Fields and Other Transient Events	1114
	27.2.5 Scanning Vibrometry – for Vibration Analysis and for the Study of Acoustic Waves	1116
	27.2.6 Digital Speckle Photography (DSP), Correlation Methods and Particle Image Velocimetry (PIV)	1119
27.3	**Summary**	1122
References		1123

27.1 Introduction

27.1.1 Chladni Patterns, Phase-Contrast Methods, Schlieren, Shadowgraph

Visualization of vibration patterns in strings, plates and shells and propagating acoustic waves in transparent objects such as air and water have been of great importance for the understanding and description of different phenomena in acoustics. *Chladni* [27.1] (1756–1824) sprinkled sand on vibrating plates to show the nodal lines. His beautiful figures challenged the research community for a theoretical description and Napoleon provided a prize of 3000 francs for someone that could give a satisfactory explanation; historical introduction by Lindsay in Lord Rayleigh's, *Theory of Sound* [27.2]. Chladni patterns are in some applications a simple, cost-effective way to visualize vibration patterns, often competitive to and an alternative to modern optical methods. In certain cases, such as for curved surfaces, the Chladni patterns may however, give completely misleading results.

Transparent or semitransparent objects such as cells and bacteria in microscopy, air flow and pressure fields in fluid mechanics, sound fields in acoustics etc. are objects that was not readily observed with the unaided eye. *Zernike* [27.3] developed phase-contrast microscopy, one of many optical, spatial-filtering techniques that operate in the Fourier-transform plane of imaging systems, to obtain images of such normally invisible objects. These techniques became very important in medicine and biology and are today used both in routine investigations and in research. In *Atlas de phénomènes d'optique* (Atlas of optical phenomena) [27.4] a collection of beautiful photographs of interference, diffraction, polarization and phase-contrast experiments are found. In *An Album of Fluid Motion* [27.5] different kinds of fluid flows are visualized. Classical optical filtering methods such as shadowgraph and Schlieren [27.6] are, together with interferometric methods, used to illustrate phenomena in turbulence, convection, subsonic and supersonic flow, shock waves, etc. Some of them, such as shock waves in air and water, certainly also generate sound fields. Still, most of those more classical methods only find limited application in acoustics since they mostly give qualitative pictures only. Practical optical methods with higher sensitivity, higher spatial resolution and that also give quantitative data have therefore been developed.

27.1.2 Holographic Interferometry, Acoustical Holography

In 1965 *Horman* [27.7], *Burch* [27.8], *Powell* and *Stetson* [27.9, 10] proposed to use holography [27.11] for optical interferometry. With holography a method was invented by which a wavefront, including phase and amplitude information, could be recorded and stored in a hologram. Subsequently a copy of the recorded wavefront could be reconstructed from the hologram for later use. At that time most holograms were recorded on special, high-resolution photographic plates. In ordinary photography it is the intensity (the irradiance distribution) that is recorded and displayed. This quantity is proportional to the amplitude squared. In that operation the phase is lost. And it is the phase that carries the information of the optical path: that is the distance to the object, or the direction, for light passing through a transparent medium. Now, assume for instance that a hologram was exposed twice in the same setup (a double-exposed hologram) of an object in two different states, for instance with a small deformation introduced between the exposures. Then in the reconstruction both these two fields were reconstructed and displayed simultaneously. The total reconstruction then showed the object covered with a set of interference fringes (dark and bright bands) that display the difference in phase. A measure of the deformation field was thus obtained by letting an object interfere with itself. The techniques were called holographic interferometry or just hologram interferometry. Three sub-techniques evolved: two-exposure (or double exposure), time-average and real-time holographic interferometry. The research area evolved rapidly. A thorough, self-contained description of holographic interferometry of that time, including pulsed (two-exposure) holographic interferometry, is found in *Vest* [27.12]. For the first time diffusely scattering objects and transparent objects could be subjected to interferometric analysis. Holograms were, however, mostly recorded on photographic plates or film. The reconstruction of the recorded field was done optically and to evaluate data often a new set of photographs had to be made. All together, this was a quite time-consuming and complicated task and this fact stopped or hindered many attempts for industrial use of the technique. Holographic interferometry methods were also applied in acoustics quite early on [27.13], in what was called acoustical holography (see also Chap. 26).

27.1.3 Speckle Metrology: Speckle Interferometry and Speckle Photography

Speckle Interferometry and Shearography
The use of video systems to record holograms and speckle patterns was proposed by several groups in the early 1970s [27.14–17]. Several different names for such systems were proposed, among them, electronic speckle-pattern interferometry (ESPI). But these new holographic techniques also had quite limited commercial success. One drawback was the rather poor quality of the interferograms (poor resolution, noisy images); furthermore quantitative evaluation of deformation and vibration fields was still difficult to obtain. Any diffuse object that is illuminated by coherent laser light will normally appear very grainy (speckled) to an observer. These randomly distributed spots are called speckles. They carry information about the fine structure of the objects micro-surface and of the optical system (the illumination and the observation system) used. The smaller aperture of the imaging system is set, the larger the speckles are. Speckles are of

course also present in ordinary holographic interferometry but, as the apertures used here normally are much larger, the effects of speckles are not too severe. However, sometimes the speckles lowered the quality of the interferograms significantly, especially in early speckle interferometry. Vibration analysis using speckle interferometry was, however, developed quite far by a group in Trondheim [27.18–21]. With the improved capabilities during the 1980s of computers, sophisticated high-resolution charge-coupled device (CCD) cameras, fiber optics and ingenious piezoelectric devices, quantitative fringe interpretation became feasible. ESPI became digital and was renamed digital speckle-pattern interferometry (DSPI). Other names are electro-optic holography (EOH) and television (TV) holography. These methods are *all electronic* and they capture images at TV rate (25 pictures/s). Interferograms are also updated at such a rate that, for the human eye, they seem to be displayed in real time even if they are really time-average interferograms displayed at high rate, all in one system. Such systems work well both for sinusoidal vibrations and two-step motions. The quality of the interferograms was improved with the introduction of phase stepping [27.22–25] and speckle averaging methods [27.18, 19], which largely eliminated the noisy, speckled background.

Pulsed TV holography [27.12, 26–28] has a pulsed laser as a light source instead of continuous laser as in the ordinary TV holography system. It was used mainly for the study of transient and dynamic events both for solids and transparent objects.

Shearography or speckle shearing interferometry [27.29–31] is another branch of speckle metrology that can directly measure derivatives of surface displacements (without numerical differentiation of deformation fields) of surfaces undergoing a deformation. It is, for instance, used for nondestructive testing (NDT) of parts for the airplane industry and for tire testing. It is a much more robust technique than DSPI or TV holography. The laser-illuminated object is imaged twice, usually before and after some load or deformation is introduced, in such a way that two nearby object points in the image plane will overlap and interfere. This shearing effect can, for instance, be accomplished by letting the camera look at the object through a Michelson interferometer setup with the two mirrors slightly tilted with respect to each other, thus forming two overlapping sheared images. A more traditional method is to cover one half of the imaging aperture (the camera lens) with a thin prism so that two sheared images will overlap each other in the image plane. Since both these sheared fields have traveled almost along the same path from the object to the image plane, they will both be submitted to about equal disturbances caused by the surroundings. In the sheared interferogram, these two fields are subtracted and therefore such equal disturbances will disappear and not affect the final interferogram. If, however, local disturbances exist, for instance an opening crack causing large in-plane, local deformations, such events will show up in the interferogram. The setup is thus quite robust and can be used in the workshop. Another advantage is that the measuring sensitivity can easily be changed by changing the amount of shear (tilt of mirrors in the Michelson setup) that is introduced. Even very large deformations or vibration amplitudes can be measured but since the instrument measures slope change (surface derivatives) the measured result has to be integrated to give deformation or vibration amplitudes. In [27.31–33] shearography is compared to TV holography and applied to measure vibration fields.

Several different kinds of speckle interferometry systems manufactured in Germany, US, Norway etc. are available on the market.

Laser Doppler Anemometry (LDA) Velocimetry and Vibrometry (LDV)

Parallel to full-field methods, point-measuring, interferometric methods were developed quite early. Laser Doppler anemometry (LDA, also named LDV where V stands for velocimetry) is a technique that measures flow velocities in seeded fluids such as air or water. It can also be used to measure the in-plane motion of solid surfaces. It was proposed as early as 1964 [27.34, 35]. A single laser beam is split into two beams. The beams are focused at an angle to each other into a common small volume in the flow, where they interfere so an interference pattern is formed in this small volume. Particles moving through this striped interference pattern scatter light with varying intensities when moving through it. The frequency of this intensity variation is measured. The distance between the interference stripes is given by the angle between the interfering beams and the laser wavelength. A measure of the velocity component of the flow at a right angle to the striped pattern is therefore obtained. With several crossing beams at different wavelengths in one point, it is possible to obtain several velocity components. Another explanation of the same phenomenon states that laser-illuminated, moving particles will scatter light with a shifted frequency, a Doppler shift, relative to particles that have zero velocity. The frequency of this intensity variation is detected despite the very large difference in frequency between this Doppler signal and the frequency of undisturbed light by combining interferometry with heterodyne detection. It can be shown that this frequency shift is proportional to the flow-velocity component at a right angle to the interference pattern.

Laser Doppler vibrometry or just vibrometry [27.36] is a related interferometric point-measuring method that also uses a heterodyne technique. It is used to measure vibration amplitudes, or rather vibration velocities of vibrating surfaces instead of flow velocities. Here it is the Doppler shift that laser light experiences when it is reflected from a moving surface that is detected. It gives a measure of the out-of-plane velocity component of the moving object point if it is illuminated also at a right angle. Both LDA and vibrometry systems can be equipped with scanning facilities that allow whole-field measurements. To preserve relative phase information in the signals when moving from one measuring point to the next, some kind of reference signal from a microphone, accelerometer or other transducer is necessary. Hardware and software of point-measuring techniques are sold by several manufacturers and the equipment is today quite user-friendly but also rather expensive.

Digital Speckle Photography (DSP) Speckle Correlation, Particle Image Velocimetry (PIV)

Another branch of speckle metrology [27.37–40] in addition to speckle interferometry, is speckle photography or speckle correlation methods. Speckle photography measures the bulk motion of the speckle pattern, the random noisy pattern that is present. This random pattern is either created artificially (i.e. sprayed dots of paint), present naturally (i.e. as fibers in a wood sample) or caused by laser-light illumination of diffuse objects. In speckle interferometry it is the phase coded in intensity of the speckles (assumed more or less stationary in space) and in speckle photography it is the position of the speckles that is detected. Speckle photography is based upon the fact that, if the laser-illuminated object is moved or deformed, then the speckles in the image plane will also move. So speckle photography in its simplest form gives a measure of the in-plane motion of the object surface. In particle image velocimetry (PIV) a sheet of light illuminates a flow seeded with suitable scattering particles [27.41]. These particles follow the flow. They are imaged twice with a CCD camera at a known time interval. The distance the particles have moved is determined with a correlation technique. In this way a measure of the flow velocity field is obtained [27.42].

27.1.4 Moiré Techniques

The Moiré effect, sometimes also named mechanical interference, shows up in many practical viewing situations. It is called mechanical interference since it is not optical interference between light fields that is studied; instead two or more striped, meshed, physical intensity patterns are overlaid and imaged most often in white light illumination. Folded fine-meshed curtains, suits of people appearing on TV, netting along roads etc. sometimes become modulated by a coarse pattern overlapping the finer mesh itself; in many cases this finer mesh is not resolved at all only the coarse, so-called Moiré patterns remain. Moiré as a measuring technique is quite old [27.43, 44]. It is related to the more modern speckle photography technique with the great difference that, instead of a random pattern, a regular pattern, a grating or a mesh, is used. If two identical gratings are overlapped and imaged a regular pattern of so called Moiré fringes will show up. If one of the gratings is glued onto an object and this object is deformed a modulation of the Moiré fringes will appear when this grating is compared to the stationary grating. This exemplifies an in-plane measuring technique using the Moiré effect. Combined with high-speed photography this technique has been used for the study of very fast events in engineering mechanics [27.45].

Techniques to measure out-of-plane deformations, vibrations, shape (contouring) are also in use as well as shadow Moiré, projected fringe techniques etc. These techniques are also named triangulation techniques since they rely on the fact that, if a regular striped pattern is projected onto a curved object surface, initially straight fringes will appear curved. A viewer will record different curved fringe patterns from different positions in the room. Knowing the geometrical parameters it is then possible to determine, e.g., the shape of a body illuminated by the projected fringes. One great advantage of Moiré methods is that the sensitivity can be chosen at will by proper selection of the line separation of the grating lines. This distance can be compared to the wavelength in optical interferometry. Quite small up to very large deformation fields can therefore be studied. A good overview of Moiré techniques can be found in [27.46], a book in which many other different phenomena and methods in optical metrology are described.

27.1.5 Some Pros and Cons of Optical Metrology

In the following, measuring principles for holographic interferometry, speckle metrology, vibrometry, pulsed TV holography, Moiré methods and PIV are discussed together with some applications. The main advantages of these methods are:

- There are no contact probes. They are non-disturbing.
- Most of them are whole-field, all-electronic methods. They give pictures.
- They give not only qualitative but also quantitative measures.
- The (digital) processing is often very fast, sometimes as if in real time.
- Pulsed lasers give very short light pulses (exposure times ≈ 10 ns), that *freezes* propagating sound fields and vibrations as if stationary.

Some disadvantages are:

- Optical equipment and lasers are, in many cases, still quite expensive.
- Trained personnel are needed to get full use of high-technology equipment.
- Speckle interferometry methods such as TV holography (DSPI, ESPI) most often need auxiliary equipment such as vibration-isolated optical tables placed in rooms that are not too dusty or noisy. Others, such as pulsed TV holography and vibrometry instruments, work well in more hostile environments without vibration isolation.
- Lasers are used and must be handled in a safe way.

27.2 Measurement Principles and Some Applications

27.2.1 Holographic Interferometry for the Study of Vibrations

With the introduction of holographic interferometry in 1965, amplitude fields of vibrating, diffusely reflecting surfaces could be mapped. Earlier only mirror-like, polished objects were used in interferometry but now suddenly all kinds of objects could be studied. The interest in holographic methods increased dramatically. The most often used technique for vibration studies was time-average holographic interferometry [27.9, 10]. A hologram of the vibrating object was recorded while the object was vibrating. The exposure time of the photographic plates was quite long, often about one second. This time was long compared to the period time of the object vibration so light fields from many cycles contributed to the total exposure of the photographic plate. After development this plate is called a hologram. In the optical reconstruction of the hologram, these object fields were reconstructed simultaneously. Now, a sinusoidally vibrating object spends most of its time in each cycle at the turning points, where the velocity is momentarily zero. These parts (at a distance of twice the amplitude of vibration) therefore contribute the most (spend the longest time) to the total exposure and therefore these parts will also dominate the reconstructed fields. A time-average field is therefore reconstructed, and essentially light beams from the turning points interfere with each other. Iso-amplitude fringes that cover the object surface are mapped using this technique. The intensity of the reconstructed image of a sinusoidally vibrating surface can be described by

$$I(x, y) = I_0(x, y) J_0^2(\Omega) , \qquad (27.1)$$

where $I_0(x, y)$ is the intensity of the object at rest and J_0 is the zero-order Bessel function of the first kind. The exposure does not have to span many cycles to get the J_0^2 function in time-average techniques. It is also possible to record time-average interferograms of low-frequency vibrations.

The argument

$$\Omega = \boldsymbol{K} \cdot \boldsymbol{L} , \qquad (27.2)$$

where $\boldsymbol{K}$ is a sensitivity vector defined as

$$\boldsymbol{K} = \boldsymbol{k}_1 - \boldsymbol{k}_2 , \qquad (27.3)$$

and $\boldsymbol{L}(x, y)$ is the vibration amplitude field. $\boldsymbol{k}_1$ and $\boldsymbol{k}_2$ are the illumination and observation directions, respectively, measured relative to the normal of the object surface. With normal illumination and observation directions the maximum sensitivity becomes

$$|\boldsymbol{K}| = 4\pi/\lambda , \qquad (27.4)$$

where λ is the laser wavelength used.

A deformation field or a vibration field $\boldsymbol{L}(x, y)$ is however vectorial, that is, it has three components.

Equation (27.2) implies that it is the projected component of $L(x, y)$ along the sensitivity vector K that is measured. With normal illumination and observation directions only the out-of-plane component is therefore measured. To obtain the in-plane components the measurements can be performed with two or more inclined illumination and observation directions. Several reference beams may also be used; see for instance [27.12, 47, 48]. With two symmetrically placed and inclined illumination directions and with observation along the object normal, the sensitivity will be zero for the out-of-plane component and optimal for one in-plane component [27.49]. A device for planning and evaluation of holographic interferometry measurements is called a holo-diagram [27.50], proposed by Abramson and in [27.23], Kreis presents computer methods to optimize holographic arrangements for different purposes.

For a two-step (double-exposure or double-pulsed) motion the squared zero-order Bessel function in (27.1) is replaced by a squared-cosine function with half the argument, that is, $\cos^2(\Omega/2)$. $L(x, y)$ is now the displacement of the object field that has taken place between the two recordings (pulses).

Most studies of vibrating objects, however, involve a combination of the time-average and real-time holographic interferometry technique [27.10]. Real-time observation of the vibration field gives very useful information to the experimentalist. A search for resonant modes can be made in real time to find the settings of frequency, amplitude and position of exciters etc. The vibration pattern is then recorded by the time-average technique, which often gives very nice interferograms that allow a more detailed analysis of the vibration field. Vibration modes of a violin recorded at different steps in making the violin [27.51–53] used this combined technique.

To illustrate these combined traditional techniques, the following experiment is sketched. The setup used is pictured in Fig. 27.1.

The laser, optical and mechanical components and the object are placed on a vibration isolated optical table in an off-axes holographic setup. In the off-axes technique there is an angle at the hologram plate between the object-observation direction and the reference-beam direction so that the object appears against a dark background. This way an observer does not have to look into the bright laser light from the reference beam to see the object. Beautiful reconstructions are thus obtained [27.54]. First a hologram is recorded of the object at rest, in this case the bottom of a herring can. The hologram is recorded on a high-resolution photographic plate in a special holder, a liquid gate hologram plate

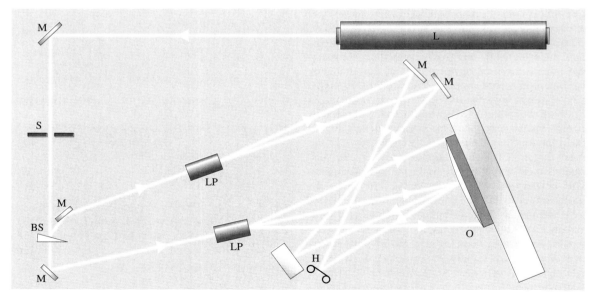

Fig. 27.1 A combined real-time and time-average hologram interferometry setup. Abbreviations: laser (L), front surface mirrors (M), glass wedge beam splitter (BS), beam expander lens–pinhole system (LP), shutter (S), a liquid gate hologram plate holder and a holder for 35 mm film (H) and an object (O), for instance a herring can or a guitar clamped to a rigid, heavy jig (J)

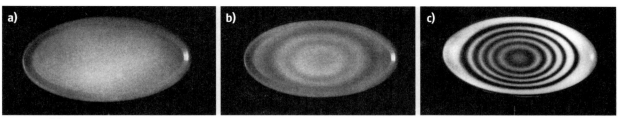

Fig. 27.2 (a) The test object, the bottom of a white-painted herring can, observed in its rest position in the real-time hologram interferometer using the liquid gate hologram holder (H) in Fig. 27.1, [27.55]. The object field emanating from the herring can is transmitted through the hologram and interferes with the field reconstructed by the hologram. These fields are so similar that no interference fringes are seen (actually it is the same broad white fringe that is covering the whole object). **(b)** Real-time interferogram obtained by taking a photo through the liquid gate when the herring can is vibrating in its fundamental mode. **(c)** Time-average interferogram obtained from a time-average hologram recording (using the film holder at H in Fig. 27.1) of the same vibration pattern as in Fig. 27.2b. (Permission to reproduce the Figs. 26.2a–c is given by IOP Publishing Ltd. 21/10/04)

holder [27.55]. It is exposed to the interfering light from the reference and object beam. The object field at rest is reconstructed by illuminating the hologram with the former reference wave or a laser wave proportional to it, in the same setup as before. The reconstructed stationary object field is then observed by looking through the hologram by eye or with a camera. Compare Fig. 27.2a which shows the bottom of the elliptical herring can at rest.

Now, if the original object is also illuminated by laser light in the same setup as before and observed through the hologram, the observer will see two images of the same object through the hologram, one as it is in real time and one as it was when the hologram was recorded. As both fields are coherent (the same laser reconstructs and illuminates the object), they will interfere. If the two fields are identical, no fringes will appear. To be more correct, since a negative hologram plate [a high irradiation gives a high exposure of a photographic plate, which is dark (high absorbing) and not bright] is used for the real-time holographic reconstruction and this field is added to the direct field observed through the hologram, the total object field should appear dark since the two fields are identical but out of phase (one field is negative and the other positive). If however the hologram plate holder or the object itself is moved slightly so that the first white cosine fringe covers the whole object, then the whole field will appear bright. Actually Fig. 27.2a is such a recording. The object is at rest but slightly translated from its initial position. When the object is vibrating it will become covered with interference fringes – compare Fig. 27.2b. The contrast of such real-time fringes are, however, quite low. It is also difficult, without special techniques, to avoid spurious fringes caused by rigid-body motions, emulsion shrinkage etc. of the real-time interferogram. Therefore a time-average hologram of interesting patterns is recorded; compare the time-average interferogram of the same vibrating herring can in Fig. 27.2c. This time-average hologram is recorded on 35 mm holographic film in a hologram film holder (a camera house without lenses), at H in Fig. 27.1. It is placed close to the real-time hologram liquid gate on the hologram table. The time-average interferogram contains twice as many fringes with higher contrast than the corresponding real-time one. The distance from peak to peak in a sinusoidal vibration is also twice as great as the distance from the peaks to the average (the amplitude) value. This is why the number of fringes is doubled. This combined real-time and time-average technique has been applied to many different structures, such as musical instruments, turbine blades, aircraft and car structures etc. Figure 27.3 shows an example of a time-average recording of a vibration mode of a guitar front plate identified using the real-time technique

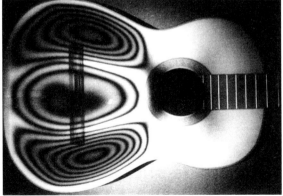

Fig. 27.3 Time-average interferogram of a guitar top plate vibrating at 505 Hz (Bolin guitar from 1969).

and recorded by time-average hologram interferometry. This picture was used as a Christmas card interferogram and it appeared in a number of physics textbooks, see also [27.56, 57]. It may be pointed out that the quality of interferograms as well as the spatial resolution, obtained in this way was high but it was a time-consuming task to obtain them compared to today's methods.

Practical resolution of vibration modes by visual observation in real time is often better than 100 nm ($\lambda/10$) – compare Fig. 27.2b. Practical maximum measurement range for time-average recordings as in Fig. 27.2c is about 4 µm, otherwise the fringe density might be too high to be resolved. The maximum possible object size in width and depth depends very much upon the laser used, both its output energy and its coherence length. The coherence length of the laser determines how big a difference in path length between the reference wave and the object beam is allowed if recording high-quality holograms. The difference in path length between the reference beam, from the laser via mirrors to the hologram, and the object beam, from the laser via the object to the hologram, should therefore not exceed the coherence length. This is easily determined with a soft string fastened at the laser and at the hologram. The power of the laser and the object reflectivity determines the width of the object and the coherence length, and thus the depth of the object volume. Objects that absorb light might be dusted with chalk or painted white. In difficult cases retro-reflective paint or tape might be used. The guitar interferogram shown in Fig. 27.3 was recorded using a 20 mW HeNe laser having a wavelength of 633 nm and a coherence length of about 25 cm, the guitar shown in Fig. 27.3 was recorded. Remember that a holographic setup is sensitive to disturbances; an optical vibration-isolated table is recommended. Powerful lasers shorten the exposure time and lower the requirements for vibration isolation.

Holographic interferometry has been further developed using pulsed or stroboscopic techniques, temporal phase modulation as well as for studies of transparent objects such as flames, pressure waves in air and water. The books by *Vest* [27.12], *Kreis* [27.23] and *Hariharan* [27.48] are recommended. In practice today, all electronic, digital speckle interferometry methods have more or less replaced wet-processing photographic holographic interferometry, at least for industrial use. The fundamental ideas are, however, still the same.

27.2.2 Speckle Interferometry – TV Holography, DSPI and ESPI for Vibration Analysis and for Studies of Acoustic Waves

Speckles [27.32–35, 46, 47] are the granular, speckled appearance that a laser-illuminated diffusely reflecting object gets, when imaged by an optical system such as the eye or a camera. The speckles can be viewed as a fingerprint that is as a unique, random pattern generated by

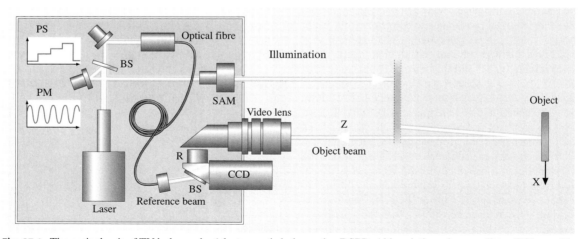

Fig. 27.4 The optical unit of TV holography (electro-optic holography, DSPI). Abbreviations: beam splitter (BS), speckle average mechanism (SAM), mirrors on piezoelectric devices used for phase stepping (PS) and sinusoidal phase modulation (PM), relay lenses (R). The distance between the optical head to the left and the object to the right can vary considerably depending of object size, magnification etc.

the fine structure of the object surface. One technically important property of the speckles is that the speckles seem to move as if glued to the object surface when the object is moved or deformed. A change in illumination or observation directions of the object will however also change the speckle pattern. Two different observers both see speckles but with different distributions. The speckles were earlier looked upon as unwanted noise but as they also carry information; they also form the basis for techniques such as speckle photography (SP) and speckle interferometry (SI). They both belong to a group of methods named speckle metrology [27.32–34]. In SP the in-plane motion of the speckles themselves is measured, and in SI the change in intensity of the speckles (i. e. the phase change of interfering beams) are used. Hologram interferometry is also based on recording the information of the same speckles.

The optical arrangement of a speckle interferometer setup called TV holography or electro-optic holography [27.25] (also named DSPI or ESPI) is shown in Fig. 27.4. A special computer board is developed that allows real-time presentations of phase-stepped interferograms.

Laser light is divided by a beam splitter (BS) into an object illumination part and a reference wave part. The object space, situated to the right, is illuminated and imaged almost along the z-axis of the object. The measuring sensitivity of this configuration therefore also becomes highest along the same direction and consequently zero along the x–y-axes. The object light amplitude field, E_{obj}, is imaged by a video lens and some transferring optics onto the photosensitive surface of the CCD chip, where the smooth reference beam, E_{ref}, also emanating from the end of an optical reference fiber is added. Each pixel of the CCD camera pictures a small area in the object space, i. e. this can be described as an in-line, image-plane holographic setup. Via a special personal computer (PC) image-processor board, results are presented at the rate allowed by the digital camera (25 or 30 Hz) on a monitor, which is a highly valuable feature in many experimental situations. In the setup there are mirrors mounted on piezoelectric crystals, allowing phase modulation (PM) and phase stepping (PS). These options are used to determine the relative phase of vibration fields and to extract quantitative data from the interferograms.

The optical detectors are so-called quadratic detectors. They measure intensity or irradiance which in turn is proportional to the square of the sum of the two interfering electromagnetic amplitude fields. The instantaneous intensity $I(x, y)$ at a pixel of the CCD detector can therefore be written:

$$I(x, y) = |(E_{\text{obj}} + E_{\text{ref}})|^2 = I_0 + I_A \cos(\Delta\Phi),$$
(27.5)

where I_0 is the background intensity and I_A the modulating intensity. The desired information however, is found in $\Delta\Phi$, the relative optical phase difference between the (constant) reference and the object field in each pixel. A difficulty exists since there are three unknowns in (27.5); $\Delta\Phi$, I_0 and I_A but only one equation. This can be solved by Fourier-filtering methods or phase-stepping (shifting) techniques [27.22–24]. In the phase-stepping technique, PS in Fig. 27.3 is used to shift the phase with, for instance, $\pi/2$ in steps between consecutive recorded frames to obtain four equations. With four equations and three unknowns, $\Delta\Phi$ can be calculated with good accuracy. In double-exposure and pulsed TV holography two series of recordings are made, one before and one after some deformation or change is introduced to the object. The corresponding phase changes $\Delta\Phi(x, y)$ for each of the two states are determined as above and then subtracted. Since the same reference beam is used in the recording of both states, ideally only the difference in object phase will remain after subtraction. The change in object phase, $\Omega = \Delta\Phi_1 - \Delta\Phi_2$, between the two objects fields are thus obtained. This quantity, Ω in turn is defined as the change (Δ) of the product of refractive index n, the geometrical path length l and the laser wave number k. That is, the measured change in optical phase between two object states can be written as:

$$\Omega = \Delta(knl) = k(n\Delta l + l\Delta n).$$
(27.6)

The wave number, $k = 2\pi/\lambda$, is (most often) a constant since λ, the laser wavelength, is the same in all recordings. Equation (27.6) illustrates two main application areas: one where the optical phase change is proportional to the change in path length (i. e. vibration amplitudes, deformation fields etc.) and one where it is proportional to the change in refractive index (caused by, for instance, wave propagation in fluids, flames etc.).

The TV holography system, shown in Fig. 27.4, illuminating and observing the object along its normal, is highly sensitive to out-of-plane vibrations or object motions and insensitive to in-plane motions. A frequency doubled, continuous-wave Nd:YAG laser with a wavelength of 532 nm, is often used. Such lasers often have a higher output power and a longer coherence length than most comparable He–Ne lasers, say 50 mW and a few meters in coherence length. The object size can now vary from a few mm up to meter-sized objects both

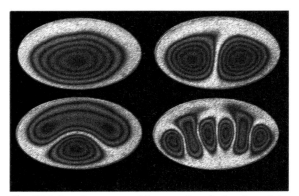

Fig. 27.5 Interferograms of four modes of vibration of the bottom of the same herring can as in Fig. 27.2, now recorded using TV holography. Both real-time observation to find the settings and time-average recordings are made at once

in width and depth. With visual observation of the monitor in real time, the resolution of the vibration amplitude is often better than 100 nm and, if the phase-stepping technique is used, the resolution limit is of the order of 20 nm, compare Fig. 27.5. The practical maximum amplitude range for vibration modes is about 2.5 μm. To be able to unwrap (numerically evaluate) fringe patterns automatically without problems, fringe densities should be kept low; not less than 15–20 pixels/fringe is recommendable. For example, a CCD camera may have 512×512 pixels or more. If the imaged object covers all 512 pixels this would allow about 25 fringes or more over the whole object field to be analyzed automatically. So it is preferable that as much as possible of the CCD chip is used to image the object. This is favorable both for the unwrapping and also for the spatial resolution of the imaged object. It is also a great advantage if a vibration-isolated optical table is used for the experiments.

The TV holography system, like many other commercial systems can also be arranged for maximum sensitivity for motion in the plane or in some other direction. Dual-beam, symmetric illumination directions about the object normal and normal object observation direction are used to achieve maximum in-plane motion sensitivity and no sensitivity to out-of-plane motions [27.47–49]. Practical resolution (adjusted for noise) is now about 30 nm with phase-stepped interferograms. The maximum measurement range for the vibration mode is about 2.5 μm. The object size in this case is often limited by the size of the illuminating lenses if equal sensitivity (with collimated light) over the whole object area is important. So, object diameters smaller than 5–10 cm are

often the case, but much larger objects can be studied if varying sensitivity over the object field is allowed. Other 2-D and 3-D arrangements that measure two or three vibration components are also commercially available.

Interferograms of Vibration Modes Using TV Holography

In normal use the TV holography setup is used for sinusoidally vibrating, diffusely reflecting objects, such as the bottom of the elliptical herring can in Fig. 27.5. This is the same can that was used in Fig. 27.2, 35 years earlier. Since then the ravage of time has changed the can somewhat. It has been dropped from the shelf etc. and despite its rigid construction the mechanical shape and behavior has changed somewhat during the years. It is not as symmetrical as before. This experiment is also an example of nondestructive testing of components, which is a large application area of TV holography where vibration analysis is used.

In Fig. 27.5 the setup is arranged so that the change in index of refraction can be neglected; only the first part, $\Omega = kn\Delta l$, of (27.6) remains. In this way we get a measure of the change in geometrical path length, Δl. A complication with sinusoidally vibrating objects such as the herring can is that the modulation of the interference fringes is, as in time-average holographic interferometry, not cosinusoidal as with two-step motions (i.e. double exposure, double-pulsed holographic and pulsed speckle interferometry) but instead modulated by a zero-order Bessel function squared $J_0^2(\Omega)$, (27.1). As before, this has to do with the integrated exposure during several cycles that the detector gets when the period time of the sinusoidally vibrating object is much shorter than the exposure time of the detector (about 1/25 s at a video rate of 25 frames per second). It is, however, also possible to record time-average interferograms of low-frequency vibrations. It is only at the start that the fringe function is somewhat dependent on which part of the vibration the integration spans — after 2–3 periods the difference between the recorded fringe function and the theoretical is negligible. The white closed line (fringe) along the rim of the can, in some higher modes connected to vertical white fringes, indicates zero amplitude, i.e. a nodal line. The bright and the dark closed loop fringes connect points with equal vibration amplitude, increasing towards the center towards a hill or a valley. Measurements such as this are quite simple to perform since the equipment works well in real time. Quantitative values of the vibration amplitude and phase in each pixel is also possible using the modulation options in the setup. This is exempli-

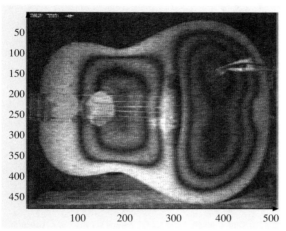

Fig. 27.6 TV holography interferogram of a baritone guitar showing a top-plate mode at 262 Hz [27.58]

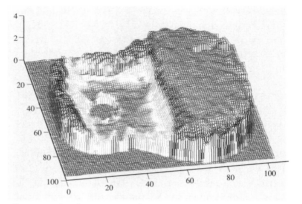

Fig. 27.7 A phase map of the same guitar mode as in Fig. 27.6 measured by phase-modulated TV holography. The units for the phase values at the z-axis are radians [27.58]

fied in Figs. 27.6 and 27.7 [27.58]. Figure 27.6 shows an ordinary TV holography recording of a mode shape of a baritone guitar mode at 262 Hz. Since this is a normal vibration mode of the front plate, the phase on each side of the vertical nodal line at the bridge should be in anti-phase.

It must, however, be stressed that, in many practical vibration situations, normal modes are often replaced by combinations or superpositions of a number of different modes. A loudspeaker, for instance, often only shows uniform, normal-mode behavior at the lowest modes. At higher frequencies the phase may vary quite a lot over the field of view; not only object areas vibrating in anti-phase are present but there is a continuous change of phase over the loudspeaker membrane. A simple test to see if mode combinations are present is to observe if nodal lines are stationary when the driving amplitude is increased from zero upwards or when the frequency is varied slightly about an amplitude resonance peak. If the nodal lines start to move sideways or twist, or so that nodal points or other peculiar phenomena occur, then mode combinations can be the source. If the experimentalist is not sure that it is a real normal mode that he has detected, i.e. one that does not have neighboring vibrating areas vibrating in anti-phase, he often calls it an operation deflection shape (ODS). So, there is a need for the experimentalist to measure not only the amplitude distribution but also the phase variation over the field of view to be sure. One technique is to set the mirror (PM in Fig. 27.4) in vibration at the same frequency and in phase with one of the vibrating areas of the object. If the amplitude of the mirror is then increased from zero, the white nodal line will start to move to new object points where the difference between the object amplitude and the mirror amplitude is zero. If, as in Fig. 27.6, a normal mode is studied, one vibrating area gets smaller (that is, there are fewer fringes in the area which is in phase with the mirror) and one area gets larger as the nodal line moves. Another technique is to use a small frequency difference, a *beat*, between the object frequency and the mirror frequency. By visual inspection, moving nodal lines are seen. A review of a number of different measurement techniques of mechanical vibrations is given by *Vest* in [27.12], by *Kreis* in [27.47] and by *Stetson* in [27.59].

The vibration phase can also be determined quantitatively using phase modulated TV holography (again using PM in Fig. 27.4). The phase difference between the upper and lower bouts, as shown in Fig. 27.7, was measured as π with such a method. Phase modulated TV holography is a technique used to measure phase and amplitude of small vibrations [27.21]. The method uses Bessel-slope evaluation of small amplitudes. Here the vibrating mirror and the object itself are given such small amplitudes such that the total maximum amplitude is kept within the linear part of the J_0^2 fringe function at low arguments. The vibrating reference wave is given such amplitude that it forms a working point, where the slope of the fringe function is the largest and around which the object amplitude may vary. By changing the phase in steps of the reference wave relative to the object wave a number of recordings are made with a known phase difference. From this the phase and amplitude distribution can be estimated. A review of such recent developments in speckle interferometry is given by *Lökberg* as one chapter in [27.37]. Visual use of the phase-modulated techniques works in near real time, but calculations based on phase-modulation take a short time.

Visualization of Aerial Waves Using TV Holography

To get a measure of the sound field, the geometrical path length l in (27.6), is kept constant so that the first part $kn\Delta l$, equals zero. Then only the $kl\Delta n$ part remains. Since an increased air density or air pressure gives an increase to the refractive index Δn the way to see the sound is simply to let the index of refraction vary along the probing length l in a measuring volume. This distance l has to be kept constant to be able to quantify the results. Figure 27.8 illustrates the measuring principle.

Inside the box a standing aerial wave is generated by a loudspeaker. The incident plane light wave is modulated by the aerial pressure and density field in the box. A curved wavefront is generated to the right of the box. This phase change may also be measured by the TV holography system in Fig. 27.4.

Projections of the box are shown in Figs. 27.9 and 27.10. Laser light travels through the measuring volume, in the simplest case a rather flat, two-dimensional (2-D), transparent box. The light field is phase modulated by a standing aerial wave inside the box. To increase the modulation the field is reflected back again by a rigid dif-

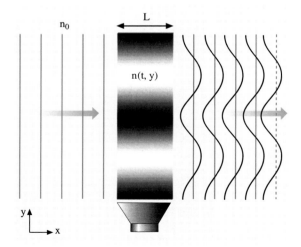

Fig. 27.8 A plane laser wave is traveling through a transparent box with dimensions $10\,\text{cm} \times 30\,\text{cm} \times 50\,\text{cm}$; the geometrical path length L is $10\,\text{cm}$ [27.60]. (Reprint from S. Hirzel Verlag, Stuttgart 2004)

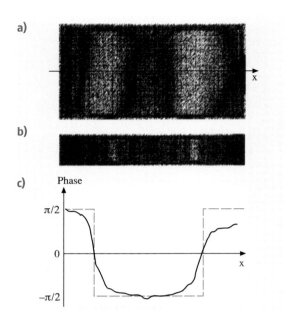

Fig. 27.9a–c Air mode $(2, 0, 0)$ in the rectangular transparent box in Fig. 27.8 [27.60]. (a) view along the shortest side (1 dm); (b) View along the next shortest side (3 dm); (c) phase along the x-axis in (a) measured by phase-modulated TV holography. (Reprint by S. Hirzel Verlag, Stuttgart)

fuse wall (not seen in Fig. 27.8) to pass through the box twice and then pass on into the interferometer (Fig. 27.4) as the object beam. The instantaneous phase shift compared to the undisturbed air (no standing wave) with index of refraction n_0 can then be written as

$$\Omega(t, y) = \frac{4\pi}{\lambda} \int_L [n(t, x, y) - n_0]\,\mathrm{d}x$$
$$\cong 2k[n(t, y) - n_0]L \,. \qquad (27.7)$$

Since L can be measured, the integrated product of L and $\Delta n = n - n_0$ in (27.3) is known. The refractive index n is related to the air density ρ by the Gladstone–Dale equation, $n - 1 = K\rho$, where K is the Gladstone–Dale constant [27.12]. If, for instance, adiabatic conditions are assumed $(p/p_0) = (\rho/\rho_0)^\gamma$, then the relations between air density ρ and pressure p are also known. By combining these equations, a quantitative measure of the sound pressure $p(x, y, t)$ at all points in the projected field of view can be obtained. Observe that it is the integrated field that is measured; thus when we have a 2-D distribution of the sound field like the one in Fig. 27.8, it can be assumed that the integral in (27.3) is reduced to the sim-

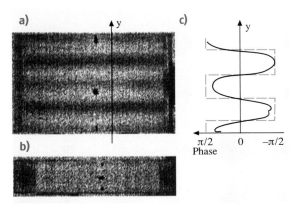

Fig. 27.10a–c Air mode (0, 4, 0) in the rectangular transparent box in Fig. 27.8 [27.60]. (**a**) View along the shortest side; (**b**) view along the next shortest side; (**c**) phase along the x–axis in (**a**) measured by phase-modulated TV holography. (Reprint by S. Hirzel Verlag, Stuttgart)

ple product $2k(n - n_0)L$. If not, the integral equation in (27.7) remains and tomography has to be used to reveal the distribution, see for instance Appendix B in [27.23]. Figs. 27.9 and 27.10 show measured standing waves inside the transparent box in Fig. 27.8 for two different air modes [27.60]. Fig. 27.11 shows standing waves inside a guitar body with a transparent top and back plate measured by phase-modulated TV holography [27.61].

If instead a probing laser ray is passing through the box in Fig. 27.8 in the x-direction, it will experience a time-varying phase change. A small frequency shift is thus generated by the standing aerial waves inside the box. This shift can be detected by a laser vibrometer; compare Sect. 27.2.5.

27.2.3 Reciprocity and TV Holography

An indirect pointwise method to measure the sound distribution from an instrument is to use reciprocity [27.62] combined with, for instance, TV holography [27.63]. An advantage of this method is that, even in a quite large spatial volume, radiativity [27.64] can be recorded. This is valuable since audible acoustical waves at low frequencies can have very long wavelengths.

An experiment using reciprocity and TV holography can be performed as follows. A movable loudspeaker at constant (unit) driving conditions excites, say a stringed instrument, from different positions outside it. The corresponding out-of-plane velocity component at the bridge foot is measured by TV holography for the specific excited vibration pattern. Then, it can be argued using reciprocity that, if a constant (unit) driving force now instead is acting at the bridge foot, it will create a sound pressure of the same magnitude as the measured velocity component at the same place as the loudspeaker was placed. By moving the loudspeaker around when

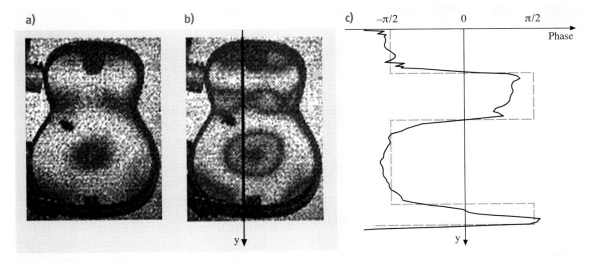

Fig. 27.11a–c Standing aerial waves inside a transparent guitar cavity measured by phase-modulated TV holography [27.61]; (**a**) covered sound hole (1195 Hz); (**b**) open sound hole (1182 Hz); (**c**) phase distribution along the y-axis in (**b**). (Reprint from S Hirzel Verlag, Stuttgart).

measuring the vibration amplitude at the bridge foot, the radiativity is estimated.

Experiences from the violin experiments in [27.63] were:

- Radiativity is measured in an ordinary laboratory both close to and at quite large distances from the instrument;
- the mode shape is simultaneously recorded,
- the violin radiativity is measured as being almost spherical up to the 500 Hz,
- complicated radiativity patterns result at higher modes.

Important conditions for dynamic reciprocity measurements are that the system is linear, passive and that the vibration system has stable equilibrium positions.

27.2.4 Pulsed TV Holography – Pulsed Lasers Freeze Propagating Bending Waves, Sound Fields and Other Transient Events

Pulsed lasers, like the traditional Q-switched ruby laser, emit one, two or more very short pulses (each pulse only some tenths of a ns long) of coherent laser light with a very high intensity. Most mechanical or acoustical events are frozen at such short exposure times. The time between the pulses can be set from about 10 μs to 800 μs with a ruby laser (during the burning time of the laser exciting flash lamps) and over a much broader range with a twin-cavity Nd:YAG laser. By double exposing a photographic plate of a transient event with two such pulses, one pulse of the object in the start position and another some time after the start, a double-exposed hologram is obtained. Both fields are reconstructed simultaneously with a continuous He–Ne laser. The technique is called pulsed holographic interferometry [27.12] and compares two states of an object. The optical setups used with pulsed lasers are in principle the same as with continuous ones with an object and a reference beam, compare Figs. 27.1 and 27.4. Both traditional recording on photographic plates (double-exposed hologram interferometry) and all-electronic methods (pulsed TV holography using a CCD chip for the recording as in TV holography) are available, although there are some differences. First, positive lenses are avoided and usually replaced by negative ones, since gas ions may otherwise occur at the focal points of positive lenses with the very strong light irradiances. With negative lenses light rays can be arranged so they seem to emanate from

Fig. 27.12 Transient bending wave propagation in a violin body, 125 μs after impact start. The top of the bridge is impacted horizontally by a 5 mm steel ball at the end of a 30 cm-long pendulum (seen in the figure as a thin line). In the top plate a *hill and a valley* centered on the two bridge feet can be seen. In the back plate the fringes are centered at the soundpost, which rapidly transfers the impact from the bridge to the back of the violin [27.65]. (Reprint American Institute of Physics, 10/22/04)

a virtual focal point. Secondly, to use the temporal phase-stepping technique with a moving mirror, such as PM in Fig. 27.4 is not applicable; there is simply not enough time to move them to record a sequence. Instead quantitative measuring data are obtained with a Fourier method where a small angle between the reference light and the object light is introduced to produce a modulated carrier wave at the CCD detector [27.66]. By Fourier filtering it is then possible to extract quantitative data.

In Fig. 27.12 [27.65], the transient bending wave response of a violin body to an impact is visualized with pulsed hologram interferometry. The instrument is impacted at the top of the bridge with a small pendulum. This is a crude and simple model of one of the pulses from the pulse train that is produced by the stick–slip bow–string interaction when the instrument is played. It is obvious that energy is effectively transferred to the back plate by the soundpost of the instrument and that the back is acting more like a monopole source. The top plate, on the other hand, is excited by the rocking motion of the bridge with a *hill* at one foot and a *valley* at the other, more like a dipole source. Interesting information about the violin is pictured in this way.

The pulsed TV holography setup is quite similar to ordinary TV holography (Fig. 27.4) in that it is all electronic, it uses fast CCD cameras and is computer op-

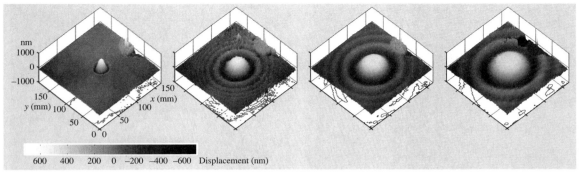

Fig. 27.13 Fig. 27.13. Measured traveling bending waves on an impacted steel plate surface 7 μs, 32 μs, 57 μs and 82 μs after impact start. These were recorded with a four-pulse TV holography technique [27.67]. The pattern at the upper right side is a shadow in the illumination path and can be ignored

erated. In an experiment [27.67], see Fig. 27.13, a pulsed TV holography setup using a four-pulse Nd:YAG laser technique is used that presents four interferograms of the same transient event. This is in contrast to double-pulse experiments with, for instance, the double-pulsed ruby laser, where the experiment must be repeated at increasing time delays. With the four pulse technique a short sequence from one single event is obtained. (Another way to get a time sequence of the motion but now only in one point at a time is to use a vibrometer, see the next section.)

What kind of sound field does an impacted plate or played violin create? It is difficult to measure that in one projection since it is a 3-D sound distribution. In principle however it is possible to create a tomographic picture from many projections [27.68, 69] but such an experiment is quite complicated to perform. A much simpler and quite informative way is to use a 2-D structure instead. A simple cantilever beam or cantilever plate, much longer than it is wide and thick is used, as shown from the thinnest side in Fig. 27.14. The transient sound field outside the cantilever beam [27.70], which is impacted at the top end by an air gun bullet at about 100 m/s is shown. It is obtained by pulsed hologram interferometry. It is interesting to note that the impact causes a spherical sound wave to propagate (at about 340 m/s) around the impact point and also fast bending waves that move down the cantilever to act as secondary supersonic sound sources. They create a pattern that looks like a bow wave behind a ship. It differs from that however, since the waves have opposite phases on each side of the cantilever, as could be expected. Bending waves are dispersive, that is, fast waves preceding the slow ones have shorter wavelength than the slower ones. The figure also shows that the signal detected by a microphone or the ear from the impacted plate depends strongly upon the position of the detector. At some positions the sound from the supersonic bending waves will reach the ear earlier than the direct impact sound.

Figure 27.15 shows the quite complicated density fields that are produced when an air gun bullet at 100 m/s

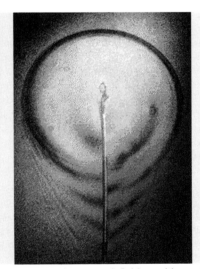

Fig. 27.14 The transient sound field outside an impacted cantilever steel beam 1 mm × 30 mm × 190 mm, clamped at the bottom. The 1 mm side is seen in the figure. The laser light passes outside and parallel to the cantilever surface along the 30 mm side. It is impacted at the *top right* by an air gun bullet. The interferogram is obtained 230 μs after impact start by pulsed hologram interferometry. Supersonic bending waves travel down the beam and create sound waves, *bow waves*, moving faster than the spherical sound wave, centered at the impact point

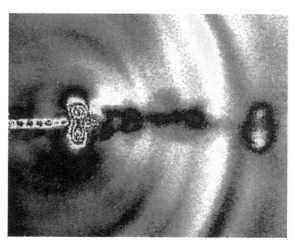

Fig. 27.15 The figure shows the density fields that are produced by an air-gun bullet traveling to the *right*. It emerged from the muzzle, which is situated at the left border. It has traveled in open air for 305 μs. Both the air that is pushed out by the bullet in front of the bullet in the barrel and the compressed air column behind it are seen, as well as the almost spherical and faster sound fields produced when the bullet emerged from the muzzle. Behind the bullet patterns reminiscent of vortex streets may be seen. The bullet speed is about 1/3 the speed of sound. Recorded by pulsed TV holography [27.71] (Reprint SPIE 21/10/04)

leaves the muzzle [27.39, 71]. The air column pressed out from the barrel preceding the bullet as well as the air column behind it is seen. The almost spherical sound wave produced when the bullet is leaving the muzzle is also seen. Behind the bullet another pattern, reminiscent of vortex streets, is seen. A number of fluid mechanics and acoustical phenomena are thus visualized in the figure.

27.2.5 Scanning Vibrometry – for Vibration Analysis and for the Study of Acoustic Waves

Laser vibrometers are a family of accurate, versatile, non-contact, optical vibration transducers. They are used in applications where it is impractical or impossible to use transducers mounted on the object and they quickly map the structural response at many measurement points in sequence. They are easy-to-use, turnkey systems that include optical and electronic hardware. With optional software many possibilities exist for advanced modal analysis, animations etc. Compared to TV holography scanning laser Doppler vibrometry is more of an off-

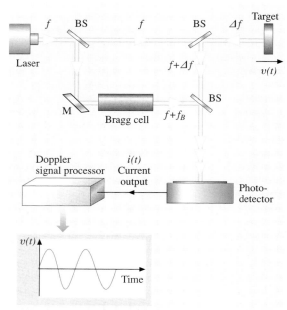

Fig. 27.16 Principle of laser vibrometer in a Mach–Zehnder-type arrangement. BS is the beam splitter, while M is the mirror. A laser emits light at a frequency f onto an object and into a reference beam. The target/object to the top right, is moving with velocity $v(t)$, causing a Doppler shift of the reflected light of $\Delta f(t)$. The photodetector receives a time-varying signal, the interference between the object and the reference light, which is processed to give a measure of the target velocity. The Bragg cell in the reference beam introduces a known and constant frequency shift f_b to the reference light using which the sign of the target velocity is determined.

the-shelf instrument, developed for industrial use. One drawback in that comparison is that it is not a real-time instrument. It is not as simple as with TV holography to find settings for unknown, interesting vibration patterns in a frequency scan. On the other hand it is highly sensitive and operates well not only with harmonic object motions but also with multi-frequency, repeatable object motions. Interesting vibration modes can be found by exciting the object with, for instance, a band-limited random signal and then performing modal analysis; compare Fig. 27.17. With scanning it is a full-field method but it has to be remembered that the measurements of different object points are made sequentially. In one measuring point, very fast events can be recorded as a function of time. One manufacturer however, makes a special unit that takes measurements at a limited

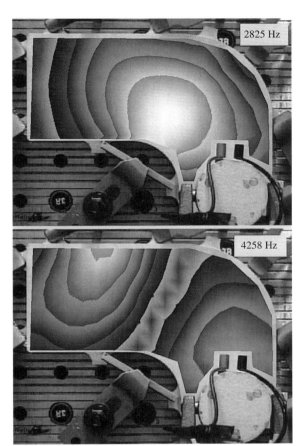

Fig. 27.17 Two vibration modes of a lightweight aluminium structure measured by scanning vibrometry, with vibration amplitude coded in grayscale. The structure was excited by a pseudo-random signal feeding a mechanical exciter. The vibration frequencies are 2825 Hz and 4258 Hz, respectively

number of points simultaneously. Informative, animated visualizations of vibrations and sound fields can be obtained. For more technical information the reader may visit the home pages of the manufacturers, since a wide variety of options are available.

Laser light of wavelength λ that is scattered back from a moving surface with velocity $v(t)$ undergoes a shift in frequency, a Doppler shift Δf

$$\Delta f(t) = 2v(t)/\lambda \, . \tag{27.8}$$

In an interferometer, often of the Mach–Zehnder type, the scattered light from a moving object is mixed with a reference light from the same laser. In Fig. 27.16 the optical unit of such a heterodyne vibrometer is pictured with the Bragg cell mounted in the reference arm of the interferometer. This instrument is commonly used for measuring displacement and velocity fields of vibrations of solid objects. The first beam splitter (BS) splits the laser beam into measuring and reference beams. The reference beam reaches the photodetector via a mirror (M) and a second beam splitter. The measuring beam hits the moving target/object, gets reflected, frequency shifted and guided by two beam splitters to the photodetector. There, the measuring and the reference beams interfere to give a signal that varies with time.

A fast photodetector measures the time-dependent intensity, the beat frequency, of the mixed light,

$$i(t) = I_{av} + 2I_{mod} \cos(2\pi \Delta f t + \Phi) \, . \tag{27.9}$$

I_{av} and I_{mod} are the average and modulating intensities, respectively, and Φ is the difference in optical phase between the diffuse object surface and the smooth reference beam. The Doppler frequency shift $\Delta f(t)$ is measured with high accuracy. Using (27.8) the target velocity is calculated. However, a sign ambiguity is present, that is, it is not obvious from to the sinusoidal nature of the signal whether the object is mowing towards or away from the detector. This sign problem is solved by adding a known optical frequency shift f_B (using the Bragg cell in Fig. 27.16) in the reference arm of the interferometer (the heterodyne technique is described in [27.46]) to obtain a virtual frequency offset. Another way is to add polarization components and one more detector (a homodyne technique) to get signals in quadrature.

Modern vibrometers are often equipped with a scanner. These scanning vibrometers scan the surface of a vibrating object pointwise and can present the results as animated video movies. In Fig. 27.17 two modes of vibration of a lightweight aluminium structure are shown. The structure is excited by a pseudo-random signal feeding a mechanical exciter and the different modes of vibration are estimated by Fourier analysis of the complex measuring signal. A reference signal from an accelerometer is necessary to ensure that phase at all measuring points are measured relative to the same reference.

Not only vibration modes but also sound fields are measured using vibrometry. In Fig. 27.18 the vibration behavior of the body of a guitar is pictured together with the emitted sound field from the instrument [27.72]. Vibration patterns of a guitar front plate at harmonic excitation are shown in the left part of the figure. A guitar string was continuously excited by a shaker at 600 Hz and 1200 Hz, respectively. To measure and visualize

Fig. 27.18a–d Measured vibration patterns of a guitar top plate (*left*) and the projected radiated sound fields from the guitar (*right*): (**a**) and (**b**) at 600 Hz; (**c**) and (**d**) at 1200 Hz. Scanning vibrometry was used for both recordings. Observe that in (**b**) and (**d**) it is the projected sound field across the guitar body that is measured [27.72]. (Reprint S. Hirzel Verlag, Stuttgart)

Fig. 27.19 The sound field at 1303 Hz inside a saxophone cavity model measured with scanning vibrometry. The model is a conical transparent model about 54 cm long with a rectangular cross section, excited at the mouthpiece end by a normal saxophone mouthpiece and it has an artificial mouth

the acoustic waves emitted from the guitar by this vibration, the guitar is rotated 90, so that the probing laser light rays is passing in front of and in parallel to the top plate. The probing laser then hits to a heavy, rigid reflector and is reflected back again into the measuring unit of the vibrometer instrument. The reflector must be absolutely rigid, as in TV holography and pulsed TV holography, to record sound fields, compare Fig. 27.8.

The temporal and spatial pressure fluctuations $\Delta p(x, y, z, t)$, which are connected with acoustic or fluidic phenomena, cause changes of the optical refractive index $\Delta n(x, y, z, t) = n(x, y, z, t) - n_0$ and, consequently, the rate of change of the measured optical phase. In the simplest case a linear acoustic wave travels in the x-direction in the measuring volume; compare Fig. 27.8. Then the optical phase varies from point to point but at each point it also varies sinusoidally with time. This signal therefore behaves as a *virtual* displacement of the rigid reflector. In real measurements the acoustic wave is not ideally shaped in the measuring volume (i.e. it is usually not two-dimensional) and a vibrometer senses, by phase demodulation, a projected virtual reflector displacement,

$$s(x, z, t) = \int_L \Delta n(x, y, z, t)\, \mathrm{d}y \qquad (27.10)$$

or by frequency demodulation, a virtual reflector velocity

$$v(x, z, t) = \int_L \dot{n}(x, y, z, t)\, \mathrm{d}y\,, \qquad (27.11)$$

i.e. the reflector seems to vibrate although it is immovable. These virtual vibrations represent the acoustic wave in the measuring volume. But, according to (27.10) and (27.11), the integrated sum of all fluctuations of the refractive index $\Delta n(x, y, z, t)$ along the laser beam contributes to the measured result. An ideal measurement situation is therefore a 2-D acoustic field as in Fig. 27.8. If not, several projections may be used or, if possible, enough that a tomographic reconstruction can be obtained.

To study the sound generation in, and the sound propagation from, a saxophone or a clarinet cavity, rectangular model pipes (two-dimensional structures) with one transparent side wall were fabricated. Figure 27.19 shows a standing acoustic wave at 1303 Hz of a model pipe when it is excited by an artificial mouth at the reed in a normal saxophone mouthpiece (to the left in the figure). Different air-column modes inside and at the bell of the pipe can be observed.

A transient pressure pulse in a water-filled cavity, $25 \times 25 \times 25$ cm^3, is shown in Fig. 27.20a. The pressure pulse is recorded through two opposite transparent walls from one side by pulsed TV holography. The impacted left side wall is made of steel and is hit in the middle by a pendulum (not seen in the interferogram) that creates a propagating, almost hemispherical, pressure wave in the water moving to the right at a speed of about 1500 m/s. The recording is taken 25.4 μs after the start of the impact. The pendulum is still in contact with the wall so energy and momentum are still being transferred to the tank. Figure 27.20b shows a line diagram of the same wave but now measured along the center horizontal line. The circles in the graph are measured using the vibrometer at different distances from the wall but at the same time instants, as in pulsed TV holography recording. The different points are recorded from different experiments (the scanning facility cannot be used since the event is not repetitive and is far too rapid). As seen in the figure, the optical path differences at coincident time and space values agree quite well between the methods. The pulsed TV holography experiment measures and compares the disturbed field with the undisturbed one at one time instant. To obtain a sequence the experiment has to be repeated or a multiple-pulsed laser has to be used to record several time instants of one event. The vibrometer on the other hand can take measurements of a single transient event one point at a time as a function of time, and at several points as a function of time if the event can be repeated. The vibrometer thus records a time sequence of the optical path (or velocity) difference at each point.

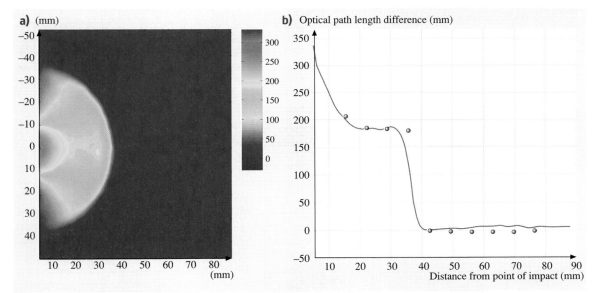

Fig. 27.20 (a) A pulsed TV holography recording of the wave propagation 25.4 μs after impact start in a water-filled transparent channel impacted at the *left wall* by a pendulum. The pendulum is hitting the *left wall* from the outside, not seen in the picture, and is still in contact with the wall when this interferogram was recorded, so energy and momentum are still being transferred to the water from the impact. The *bar* to the *right* shows the coding in optical path difference. (b) The *full line* shows the optical path difference along the *center horizontal line* measured in Fig. 27.20a, using pulsed TV holography. The *circles* in the graph are measured points using a vibrometer. The measurements are performed at different distances from the wall but evaluated at 25.4 μs. The different *points* are, however, obtained from different experiments (the scanning facility cannot be used since the event is far too rapid). The optical path differences at coincident time and space point values agree quite well between the two methods

27.2.6 Digital Speckle Photography (DSP), Correlation Methods and Particle Image Velocimetry (PIV)

The equipment needed in the interferometry-based methods described above is often rather expensive and some of the techniques can unfortunately be quite difficult for an untrained user to apply. Simpler and cheaper methods have therefore been sought. Less expensive, less sensitive methods are the speckle photography (SP) and digital speckle photography (DSP) methods [27.73], which are also called speckle correlation [27.32–34]. Here the motion of the speckles themselves is measured, rather than the change in intensity of the stationary speckles as with interferometric methods. As mentioned before, the speckled random pattern *acts as if glued* to the surface of the laser-illuminated object. This pattern is imaged twice, usually with a CCD camera on individual frames, before and after some external or internal load is applied to the object, which changes the speckle pattern field. The illumination and the camera are left stationary, so it is the motion of the microstructure of the object surface that is responsible for the motion of the speckles. The deformation field of the object may consist of the sum of rigid-body motions (translation and rotation) and some strain-induced deformation field (at crack tips, elastic and plastic deformation fields etc.). So far this is reminiscent of TV holography [(or digital speckle pattern interferometry (DSPI)] but no reference wave is added here. The experiments are not as sensitive to disturbances as interferometric measurements and can usually be performed in a workshop without vibration isolation of the quite simple optical setup. It is the position of each speckle in the speckled object wave that is recorded and the motion of small sub-images containing this pattern that is determined.

Two main types of random patterns are in use: (1) laser speckle, which is the random pattern generated by the surface microstructure of a laser-illuminated diffuse object (laser speckles in DSP) together with the specific illumination and recording geometry; and (2) white-light speckle in DSP, where there may already exist a random pattern at the object surface, such as random fibers in a paper or wood sample, which is imaged using an ordinary, broadband incandescent lamp. A random dot pattern may also be painted or sprayed onto the surface and imaged. White-light illumination can be used so that lasers can be avoided in these cases.

Two main imaging techniques are in use, one where the object surface is focused by the recording CCD camera and one where the camera is defocused behind or in front of the object surface. The focused case, which is the simplest and most commonly used, gives a measure of the in-plane motion field of the surface. However, it is important that the object is in focus, otherwise, using laser speckles, strain fields and rotations may disturb the result. It must be understood that laser speckles exist in all space, not only when focused at the object surface. Therefore, if the camera is defocused, other components of the strain tensor that affect the speckle motion will eventually become important. Using this defocused technique it has been possible to measure the strain field directly, i. e. without first measuring the deformation field and then differentiating the result to obtain the strain field [27.74].

A digital correlation technique is then used to determine how far small sub-areas in the first recording have to be moved to correlate as exactly as possible with those in the second recording. Performing this for all sub-areas, an in-plane deformation map (for the focused case) is formed. As well as yielding information

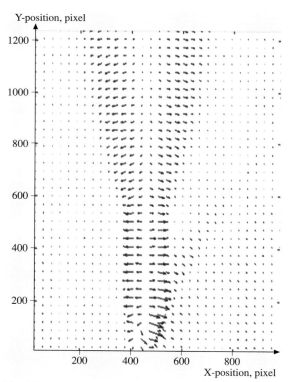

Fig. 27.21 Defocused digital speckle photography (DSP) recording, of a transparent helium jet entering air, recorded using a continuous HeNe laser

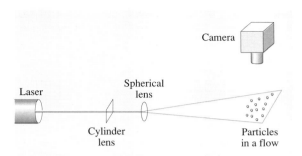

Fig. 27.22 The principle of particle image velocimetry (PIV). Pulsed laser light is sent through a cylinder lens system to form a sheet of light in the measuring volume. A CCD camera captures the seeded flow in this light sheet from above at two time instances

on the in-plane motion, the value of the correlation peak height is measured for each sub-area. This peak value gives a measure of how similar the sub-images are, even if they have not moved. If the correlation is high (almost 1) then the speckles in the two sub-surfaces are identical; if not, the correlation value will drop. Reasons for this drop include different kinds of noise and limitations in the recording system (limited and varying optical resolution and the digitalization of the intensity etc.); but it may also be that the microstructure itself has changed between the two exposures, for instance due to plastic deformation of parts of the object surface. Figure 27.21 shows an example of a defocused laser DSP recording of a transparent helium jet entering air. From similar recordings [27.75, 76] where pulsed lasers were used, it is possible to calculate where the gas jet is situated and how strong the refractive-index field of the gas is.

Particle image velocimetry (PIV) [27.77, 78] is highly reminiscent of the focused version of the speckle photography technique. However, the measurements are now made of a flow in a transparent medium, usually water or air. It is a nonintrusive optical measuring method. Tracer particles are illuminated with pulsed laser light in a light sheet, for instance smoke in air that follows the fluid motion. The displacements of the particles between two exposures are determined with a cross-correlation technique, as in digital speckle photography or speckle correlation. Figure 27.22 shows a PIV setup. Since the time between the exposures is known (often the time between repetitive pulses from a pulsed laser) and the motion of the small sub-areas between exposures of the imaged flow can be calculated, the velocity fields in the flow are determined.

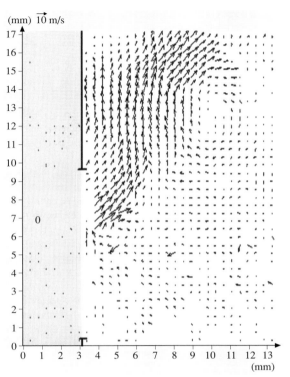

Fig. 27.23 Particle image velocimetry (PIV) measurements of the flow field at the labium (the opening to the *left*) of an organ pipe. A pulsed laser was used as an illumination source as the flow is quite fast at the labium; see the *reference arrow* in the *top left*

With two cameras viewing the illuminated plane from two directions and/or illuminating two different and crossing planes, not only the in-plane components but also out-of-plane velocity components can be estimated, so both 2-D and 3-D stereoscopic PIV techniques are available. With fast repetitive pulsed lasers, so called time-resolved PIV (TR-PIV), even fast turbulent flows can be measured. Microflow systems have been developed for flow studies at the microscopic scale [27.79]. By adding fluorescent particles to the flow and illuminating in a plane with a laser at a shorter wavelength and observing it at a longer (fluorescent) wavelength, concentration and temperatures can also be measured. This is called planar laser-induced fluorescent (PLIF). Sophisticated PIV systems, including hardware and software for industrial or research use, are sold by several manufacturers.

Figure 27.23 shows the flow field at the labium of a blown organ pipe, measured using PIV with a pulsed laser as the illuminating source [27.80]. The number of papers using PIV and LDA presented at acoustical conferences is rapidly increasing [27.81, 82].

27.3 Summary

Several new optical, all-electronic measuring techniques and systems have entered the market in recent decades. The rapid technical development of new lasers, new optical sensors such as the CCD camera, optical fibers and powerful and cheap PCs has made this possible. The demands for optical techniques have increased because validation of large numerical calculations is becoming more important. Optical methods have the advantage over traditional vibration detectors that they are fast and do not disturb what is being measured, and they also often yield an image field – for instance a flow velocity field – rather than just measuring values at one single point.

Equipment can be purchased from many manufacturers and in many cases you do not have to be a specialist in optics to use them. It is also possible to make, for instance, a speckle interferometer or a shearing interferometer in most laboratories [27.83]. Methods exist today that have the sensitivity and spatial resolution needed to visualize and measure sound fields in musical acoustics.

In Table 27.1 some modern optical measuring methods and their temporal application areas are indicated.

Optical metrology on the whole is growing in importance and applicability to scientists and engineers in solid mechanics, fluid mechanics and in acoustics. Techniques such as phase-modulated TV holography, pulsed TV holography and scanning vibrometry are commonly used for modal analysis and increasingly to measure phase shifts in transparent phase objects such as sound fields. Flow fields in air and water can be measured with speckle photography or correlation methods such as digital speckle photography (DSP) and particle image velocimetry (PIV); these methods provide both illustrative images as well as quantitative measures. Reciprocity methods can also be of great help in such experiments. Optical measuring methods have the advantage of being contact-less, non-disturbing and whole-field methods. The future for optical metrology in acoustics looks bright.

Table 27.1 Some modern optical measuring methods and their temporal application areas

Optical measuring methods and their temporal application areas. Int.=Interferometry	Static or (slow) quasistatic events	Harmonic motions, single frequency	Repetitive motions, multi-frequencies	Transients, fast, dynamic events
Real-time holographic int. with continuous lasers	×	×		
Double-exposure holographic int. with continuous lasers	×			
Time-average holographic int. with continuous lasers		×		
TV holography, DSPI, ESPI with continuous lasers	×	×		
Pulsed TV holography and pulsed DSPI and ESPI with pulsed lasers	(×)	(×)	×	×
Scanning laser vibrometry		×	×	× (one point, no scanning)
Speckle correlation/photography (DSP)	×			× (with pulsed lasers)
Particle image velocimetry, PIV	×	(×)	(×)	× (with pulsed lasers)

References

27.1 E.F.F. Chladni: *Die Akustik* (Breitkopf and Härtel, Leipzig 1802)

27.2 Lord Rayleigh: *Theory of Sound, Vol. I and II* (Dover, New York 1945)

27.3 F. Zernike: Das Phasenkontrastverfahren bei der Mikroskopischen Beobachtung, Tech. Phys. **16**, 454 (1935)

27.4 M. Cagnet, M. Francon, J.C. Thrierr: *Atlas de Phénomènes d'Optique (Atlas of Optical Phenomena)* (Springer, Berlin Heidelberg New York 1962)

27.5 M. Van Dyke: *An Album of Fluid Motion* (Parabolic, Stanford 1982)

27.6 G.S. Settles: *Schlieren and Shadowgraph Techniques* (Springer, Berlin Heidelberg New York 2001)

27.7 M.H. Horman: An application of wavefront reconstruction to Interferometry, Appl. Opt. **4**, 333–336 (1965)

27.8 J. Burch: The application of lasers in production engineering, Prod. Eng. **211**, 282 (1965)

27.9 R.L. Powell, K.A. Stetson: Interferometric analysis by wavefront reconstruction, J. Opt. Soc. Am. **55**, 1593–1598 (1965)

27.10 K.A. Stetson, R.L. Powell: Interferometric hologram evaluation and real-time vibration analysis of diffuse objects, J. Opt. Soc. Am. **55**, 1694–1695 (1965)

27.11 D. Gabor: Microscopy by reconstructed wavefronts, Proc. R. Soc. London A **197**, 454–487 (1949)

27.12 C.M. Vest: *Holographic Interferometry* (Wiley, New York 1979)

27.13 K. Suzuki, B.P. Hildebrandt: Holographic Interferometry with Acoustic Waves. In: *Acoustical Holography*, ed. by N. Booth (Plenum, New York 1975) pp. 577–595

27.14 R.K. Erf (Ed.): *Speckle Metrology* (Academic, New York 1978)

27.15 J.N. Butters, J.A. Leendertz: Holographic and videotechniques applied to engineering measurements, J. Meas. Control **4**, 349–354 (1971)

27.16 O. Schwomma: Austrian patent no. 298830 (1972)

27.17 A. Mocovski, D. Ramsey, L.F. Schaefer: Time lapse interferometry and contouring using television systems, Appl. Opt. **10**, 2722–2727 (1971)

27.18 G.Å. Slettemoen: Electronic speckle pattern interferometric systems based on a speckle reference beam, Appl. Opt. **19**, 616–623 (1980)

27.19 O. J Lökberg, G.Å. Slettemoen: *Improved fringe definition by speckle averaging in ESPI*, ICO-13 Proceedings (Sapporo 1984) pp. 116–117

27.20 O.J. Lökberg: Sound in flight: measurement of sound fields by use of TV holography, Appl. Opt. **33**(13), 2574–2584 (1994)

27.21 K. Högmoen, O.J. Lökberg: Detection and measurement of small vibrations using electronic speckle pattern interferometry, Appl. Opt. **16**, 1869–1875 (1976)

27.22 K. Creath: Phase-shifting speckle interferometry, Appl. Opt **13**, 2693–2703 (1985)

27.23 T.M. Kreis: Computer-aided evaluation of fringe patterns, Opt. Lasers Eng. **19**, 221–240 (1993)

27.24 J.M. Huntley: Automated analysis of speckle interferograms. In: *Digital Speckle Interferometry and Related Techniques*, ed. by K. Rastogi (Wiley, New York 2001), Chap. 2

27.25 K.A. Stetson, W.R. Brohinsky: Electro-optic holography and its application to hologram interferometry, Appl. Opt. **24**, 3631–3637 (1985)

27.26 R. Spooren: Double-pulse subtraction TV holography, Opt. Eng. **32**, 1000–1007 (1992)

27.27 G. Pedrini, B. Pfister, H.J. Tiziani: Double-pulse electronic speckle interferometry, J. Mod. Opt. **40**, 89–96 (1993)

27.28 A. Davila, D. Kerr, G.H. Kaufmann: Fast electro-optical system for pulsed ESPI carrier fringe generation, Opt. Commun. **123**, 457–464 (1996)

27.29 Y. Y. Hung, C.E. Taylor: Speckle-shearing interferometric camera – a tool for measurement of derivatives of surface displacements, Soc. Photo-Opt. Instrum. Eng. **41**, 169–175 (1973)

27.30 Y.Y. Hung, C.Y. Liang: Image-shearing camera for direct measurement of surface strain, Appl. Opt. **18**, 1046–1051 (1979)

27.31 Y. Y Hung: Electronic Shearography versus ESPI for Nondestructive Evaluation, Moire Techniques, Holographic Interferometry, Optical NDT and Applications to Fluid Dynamics proc. Soc. Photo-Opt. Instr. Eng. **1554B**, 692–700 (1991)

27.32 N.K. Mohan, H. Saldner, N.E. Molin: Electronic speckle pattern interferometry for simultaneous measurement of out-of-plane displacement and slope, Opt. Lett. **18**(21), 1861–1863 (1993)

27.33 N.K. Mohan, H.O. Saldner, N.-E. Molin: Electronic shearography applied to static and vibrating objects, Opt. commun. **108**, 197–202 (1994)

27.34 Y. Yeh, H.Z. Cummings: Localized fluid flow measurements with a He-Ne laser spectrometer, Appl. Phys. Lett. **4**, 176–178 (1964)

27.35 L.E. Drain: *Laser Doppler Technique* (Wiley Interscience, New York 1980)

27.36 P. Castellini, G.M. Revel, E.P. Tomasini: Laser doppler vibrometry: a review of advances and applications, Shock Vib. Dig. **30**(6), 443–456 (1998)

27.37 R.S. Sirohi (Ed.): *Speckle Metrology* (Marcel Decker, New York 1993)

27.38 R.K. Erf. (Ed.): *Speckle Metrology* (Academic, New York 1978)

27.39 P.M. Rastogi (Ed.): *Digital Speckle Pattern Interferometry and Related Techniques* (Wiley, Chichester 2001)

27.40 R. Jones, C. Wykes: *Holographic and Speckle Interferometry*, 2nd ed. (Cambridge Univ. Press, Cambridge 1989)

27.41 N.A. Fomin: *Speckle Photography for Fluid Mechanics Measurements* (Springer, Berlin Heidelberg New York 1998)

27.42 M. Raffel, C. Willert, J. Kompenhans: *Particle Image Velocimetry* (Springer, Berlin Heidelberg New York 1998)

27.43 Lord Rayleigh: On the Manufacture and Theory of Diffraction Gratings, Philos. Mag. **XLVII**, 193–205 (1874)

27.44 G. Indebetouw, R. Czarnek. (Eds.): *Selected Papers on Optical Moiré and Applications* (SPIE, Bellingham 1992)

27.45 C. Forno: Deformation measurement using high resolution Moiré photography, Opt. Lasers Eng. **7**, 189–212 (1988)

27.46 K.J. Gåsvik: *Optical metrology*, 3rd edn. (Wiley, New York 2002)

27.47 T. Kreis: *Holographic Interferometry Principles and Methods* (Academie Verlag, Berlin 1996)

27.48 P. Hariharan: *Optical Holography. Principles, Techniques and Applications* (Cambridge Univ. Press, Cambridge 1996)

27.49 J.A. Lendertz: Interferometric displacement measurement on scattered surface utilizing speckle effects, J. Phys. E **3**, 214–218 (1970)

27.50 N. Abramson: *The Making and Evaluation of Holograms* (Academic, New York 1981)

27.51 E. Jansson, N.-E. Molin, H. Sundin: Resonances of a Violin Body Studied by Hologram Interferometry and Acoustical Methods, Phys. Scripta **2**, 243–256 (1970)

27.52 L. Cremer: *The Physics of the Violin* (MIT Press, Cambridge 1984), (Physik der Geige (1981))

27.53 W. Reinecke, L. Cremer: J. Acoust. Soc. Am. **48**, 988 (1970)

27.54 E.N. Leith, J. Upatnieks: Reconstructed wave fronts and communication theory, J. Opt. Soc. Am. **52**, 1123–1130 (1962)

27.55 K. Biedermann, N.-E. Molin: Combining hypersensitization and in situ processing for time-average observation in real time hologram interferometry, J. Phys. E **3**, 669–680 (1970)

27.56 N.H. Fletcher, T.D. Rossing: *The Physics of Musical Instruments*, 2nd edn. (Springer, New York 1991)

27.57 E.V. Jansson: A study of acoustical and hologram interferometric measurements on the top plate vibrations of a guitar, Acustica **25**, 95–100 (1971)

27.58 T.D. Rossing, F. Engström: Using TV holography with phase modulation to determine the deflection phase in a baritone guitar. In: *Proceedings of the International Symposium on Musical Acoustics 1998 (ISMA-98), Leavenworth, Washington, USA, June 26-July*, ed. by D. Keefe, T. Rossing, C. Schmidt (Acoust. Soc. Am., Sunnyside 1998)

27.59 K.A. Stetson.: Fringe-shifting technique for numerical analysis of time-average holograms of vibrating objects, J. Opt. Soc. Am. **5**, 1472–1476 (1988)

27.60 A. Runnemalm: Standing waves in a rectangular sound box recorded by TV holography, J Sound Vib. **224**(4), 689–707 (1999)

27.61 A. Runnemalm, N.-E. Molin: Operating deflection shapes of the plates and standing aerial waves in a violin and a guitar model, Acustica Acta Acust. **86**(5), 883–890 (2000)

27.62 T. ten Volde: On the validity and application of reciprocity in acoustical mechano-acoustical and other systems, Acustica **28**, 23–32 (1973)

27.63 H.O. Saldner, N.-E. Molin, E.V. Jansson: Sound distribution from forced vibration modes of a violin measured by reciprocity and TV holography, CAS J. **3**(4), 10–16 (1997), (Ser. II)

27.64 G. Weinreich: Directional tone color, J. Acoust. Soc. Am. **101**, 4 (1997)

27.65 N.-E. Molin, A.O. Wåhlin, E.V. Jansson: Transient wave response of the violin body, J. Acoust. Soc. Am. **88**(5), 2479–2481 (1990)

27.66 H. O. Saldner, N-E Molin, K. A. Stetson: Fourier-transform evaluation of phase data in spatially phase-biased TV holograms, Appl. Opt. **35**(2), 332–336 (1996)

27.67 P. Gren: Four-pulse interferometric recordings of transient events by pulsed TV holography, Opt. Lasers Eng. **40**, 517–528 (2003)

27.68 O.J. Lökberg, M. Espeland, H.M. Pedersen: Tomographic reconstruction of sound fields using TV holography, Appl. Opt. **34**(10), 1640–1645 (1995)

27.69 P. Gren, S. Schedin, X. Li: Tomographic reconstruction of transient acoustic fields recorded by pulsed TV holography, Appl. Opt. **37**(5), 834–840 (1998)

27.70 A.O. Wåhlin, P.O. Gren, N.-E. Molin: On structure-borne sound: Experiments showing the initial transient acoustic wave field generated by an impacted plate, J. Acoust. Soc. Am. **96**(5), 2791–2797 (1994)

27.71 S. Schedin, P. Gren, M. Finnström: Measurement of the density field around an airgun muzzle by pulsed TV holography, Proc. EOS/SPIE **3823**, 13–19 (1999)

27.72 N.-E. Molin, L. Zipser: Optical methods of today for visualize sound fields in musical acoustics, Acta Acust./Acustica **90**, 618–628 (2004)

27.73 M. Sjödahl: Digital speckle photography. In: *Digital Speckle Pattern Interferometry and Related Techniques*, ed. by P.M. Rastogi (Wiley, Chichester 2001), (Chap. 5)

27.74 M. Sjödahl: Whole-field speckle strain sensor, ISCON99 Yokohama. SPIE **3740**(84-87), 16–18 (1999)

27.75 E.-L. Johansson, L. Benckert, M. Sjödahl: Phase object data obtained from defocused laser speckle displacement, Appl. Opt. **43**(16), 3229–3234 (2004)

27.76 E.-L. Johansson, L. Benckert, M. Sjödahl: Phase object data obtained by pulsed TV holography and defocused laser speckle displacement, Appl. Opt. **43**(16), 3235–3240 (2004)

27.77 M. Raffel, C. Willert, J. Kompengans: *Particle Image Velocimetry – A practical Guide* (Springer, Berlin Heidelberg ew York 1998)

27.78 K.D. Hinsch: *Particle image velocimetry*, Speckle Metrology, ed. by R.J. Sirohi (Marcel Decker, New York 1993)

27.79 P. Synnergren, L. Larsson, T.S. Lundström: Digital speckle photography: visualization of mesoflow through clustered fiber networks, Appl. Opt. **41**(7), 1368–1373 (2002)

27.80 E.-L. Johansson, L. Benckert, P. Gren: Particle image velocimetry (PIV) measurements of velocity fields at an organ pipe labium, TRITA-TMH **2003:8** (2003)

27.81 D.J. Skulina, D.M Campbell, C.A. Greated: Measurement of the Termination Impedance of a tube Using Particle Image Velocimetry, TRITA-TMH **2003:8** (2003)

27.82 E. Espositi, M. Marassi: Quantitative assessment of air flow from professional bass reflex systems ports by particle image velocimetry and laser doppler anemometry, TRITA-TMH **2003:8** (2003)

27.83 T.R. Moore: A simple design for an electronic speckle pattern interferometer, Am. J. Phys. **72**, 11 (2004)

28. Modal Analysis

Modal analysis is widely used to describe the dynamic properties of a structure in terms of the modal parameters: natural frequency, damping factor, modal mass and mode shape. The analysis may be done either experimentally or mathematically. In mathematical modal analysis, one attempts to uncouple the structural equations of motion so that each uncoupled equation can be solved separately. When exact solutions are not possible, numerical approximations such as finite-element and boundary-element methods are used.

In experimental modal testing, a measured force at one or more points excites the structure and the response is measured at one or more points to construct frequency response functions. The modal parameters can be determined from these functions by curve fitting with a computer. Various curve-fitting methods are used. Several convenient ways have developed for representing these modes graphically, either statically or dynamically. By substituting microphones or intensity probes for the accelerometers, modal analysis methods can be used to explore sound fields. In this chapter we mention some theoretical methods but we emphasize experimental modal testing applied to structural vibrations and also to acoustic fields.

28.1	**Modes of Vibration**	1127
28.2	**Experimental Modal Testing**	1128
	28.2.1 Frequency Response Function	1128
	28.2.2 Impact Testing	1130
	28.2.3 Shaker Testing	1131
	28.2.4 Obtaining Modal Parameters	1132
	28.2.5 Real and Complex Modes	1133
	28.2.6 Graphical Representation	1133
28.3	**Mathematical Modal Analysis**	1133
	28.3.1 Finite-Element Analysis	1134
	28.3.2 Boundary-Element Methods	1134
	28.3.3 Finite-Element Correlation	1135
28.4	**Sound-Field Analysis**	1136
28.5	**Holographic Modal Analysis**	1137
References		1138

28.1 Modes of Vibration

The complex vibrations of a complex structure can be described in terms of normal modes of vibration.

A normal mode of vibration represents the motion of a linear system at a normal frequency (eigenfrequency). Each mode is characterized by a natural frequency, a damping factor, and a mode shape. *Normal* implies that each shape is independent of, and orthogonal to, all other mode shapes of vibration for the system. Any deformation pattern the structure can exhibit can thus be expressed as a linear combination of the mode shapes.

In a stricter mathematical sense, *normal* modes provide a solution for an undamped system. Each mode shape is a list of displacements at various places and in various directions. These normal-mode vectors contain one real number for each motional degree of freedom (DOF) studied. Since real structures are invariably damped, normal modes represent an approximation that may prove imprecise when the damping level is significant. In instances where the energy absorbing damping mechanisms are distributed in a manner proportional to the structural stiffness, normal modes provide an exact solution. In this case, the damping factor of each mode is proportional to its associated natural frequency. The normal modes also provide an exact solution when the damping is proportional to the mass distribution, but in this case each damping factor is inversely proportional to its associated natural frequency.

A structure is said to be proportionally damped when the matrix describing its damping can be written as a linear combination of the corresponding mass and stiffness matrices. Normal-mode shapes provide the exact solution to such equations, and the damping factors may be expressed as a linear combination of the natural frequencies and their reciprocals. When these restrictive proportionality conditions do not exist, the structure may be better modeled by complex mode shapes.

Complex modes result when the instantaneous velocity at each DOF is treated as being independent of the displacement. This doubles the number of system differential equations, but simplifies them from second order to first order. The solution vectors are most commonly written in terms of complex displacements, one for each DOF. Expressing the modal displacements in this manner eliminates any constraints between the damping factors and the natural frequencies. All but the simplest experimental curve-fitting algorithms can identify complex modes. Most finite-element codes are restricted to the undamped analysis of structures and can therefore only identify the approximating real or normal modes.

Each modal vector represents a displacement pattern. A zero in the vector denotes a node, a point and direction on the structure that does not move in that mode. The element in a modal vector that is largest in displacement value is termed an anti-node. A given mode shape can be readily excited by a sinusoidal force at or near its natural frequency applied to an anti-node. The same force applied at a node will impart no motion in that shape.

It should be noted that the real or complex displacement values in each modal vector are relative numbers. The mode shape is inferred by the ratio between the vector elements, not their specific values. The elements within each modal vector may be scaled by any of various methods, the simplest of which is division by the anti-node value. This is useful for display or plotting so that all modes have a common maximum value. However, the inner product or length of each vector is related to the modal mass of the mode. For many analytical purposes, including structural dynamic modification, it is necessary to choose the length of each vector so that the corresponding modal mass is equal to 1. When the shape vectors are scaled in this manner, they are said to be orthonormal or unit modal mass (UMM) modal vectors.

It is normally possible to excite a mode of vibration from any point on a structure that is not a node and to observe motion at all points that are not nodes. A mode shape is a characteristic only of the structure itself, independent of the way it is excited or observed. In practical terms, however, instrumentation associated with both excitation and observation may modify the structure slightly by adding mass or stiffness (or both). This results in small shifts in frequency and mode shape, which in most cases are negligible. Acoustic excitation and optical monitoring are the least intrusive, but mechanical means of driving and observing are frequently more convenient and less costly.

Mode shapes are unique for a structure, whereas the deflection of a structure at a particular frequency, called an operating deflection shape (ODS), may result from the excitation of more than one normal mode. When exciting a structure at a resonance frequency, the ODS will be determined mainly by one mode, although if several modes have nearly the same frequency, special techniques may be required to determine their contributions to the observed ODS. Modes of a structure are functions of the entire structure. A mode shape describes how every point on the structure moves when it is excited at any point.

28.2 Experimental Modal Testing

Modal testing is a systematic method for identification of the modal parameters of a structure. Generally these include natural frequencies, modal damping, and UMM mode shapes. In experimental modal testing the structure is excited with a measured force at one or more points and the response is determined at one or more points. From these sets of data, the modal parameters are determined, often by the use of multidimensional curve-fitting routines on a digital computer.

Modal testing may use sinusoidal, random, pseudorandom, or impulsive excitation. The response may be measured mechanically, optically, or indirectly (by observing the radiated sound field, for example). The first step in experimental modal testing is generally to obtain a set of frequency response functions.

There is a vast amount of good literature describing experimental modal testing. Ewins [28.1] provides a good overall introduction. The proceedings of the annual International Modal Analysis Conference (IMAC), held every year since 1982, is a gold mine of papers on the subject [28.2]. The Structural Dynamics Research Laboratory at the University of Cincinnati has published many scientific papers and reports on the subject, and has included several tutorials on their website [28.3].

28.2.1 Frequency Response Function

The frequency response function (FRF) is a fundamental measurement that isolates the inherent dynamic properties of a mechanical structure. The FRF describes the motion-per-force input–output relationship between two points on a structure as a function of frequency. Since both force and motion are vector quantities, they have directions associated with them. An FRF is actually defined between a single-input DOF (point and direction) and a single-output DOF.

In practice, the force and response are usually measured as functions of time, and transformed into the frequency domain using a fast Fourier transform (FFT) analyzer. Due to this transformation, the functions end up as complex numbers; the functions contain real and imaginary components (or magnitude and phase components). Depending on whether the response motion is measured as displacement, velocity, or acceleration, the FRF can be expressed as *compliance* (displacement/force), *mobility* (velocity/force), *accelerance* or *inertance* (acceleration/force), *dynamic stiffness* (1/compliance), *impedance* (1/mobility), or *dynamic mass* (1/accelerance). Force can be measured with a piezoelectric force transducer or load cell; acceleration can be measured with an accelerometer; velocity can be measured with a laser velocimeter or obtained by integrating acceleration; displacement can be determined by holographic interferometry or by integrating velocity.

Because it is a complex quantity, the FRF cannot be fully displayed on a single two-dimensional plot. One way to present an FRF is to plot the magnitude and the phase as in Fig. 28.1a. At resonance, the magnitude is a maximum and is limited only by the damping in the system. The phase ranges from 0 to 180° and the response lags the input by 90° at resonance. The 3 dB width of each resonance peak (bounding locations of 0.707 resonance amplitude) is determined by the damping. At these half-power points the phase angle is ±45°.

Another way of presenting the FRF is to plot the real and imaginary parts as in Fig. 28.1b. The imaginary part has a peak at resonance, while the real part goes through zero at resonance. The real component exhibits peaks of opposite sign at the extremes of the half-power bandwidth.

A third method of presenting the FRF is to plot the real part versus the imaginary part as shown in Fig. 28.1c. This is called a Nyquist or vector response plot. Each mode produces a circular pattern. The diameter of the circle is proportional to the product of the UMM vector elements for the force and response DOFs. Maximum change in arc length with frequency occurs at the natural frequency.

If measurements are made at n points, the FRFs can be arranged in an $n \times n$ matrix with elements designated

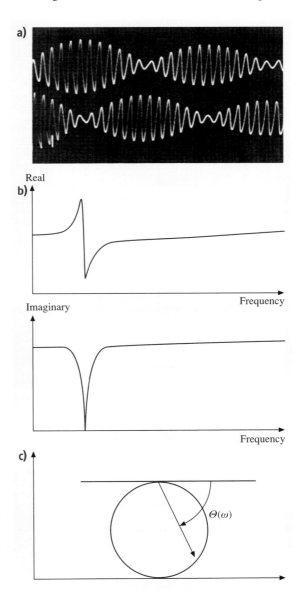

Fig. 28.1a–c Three methods for presenting the frequency response function (FRF) of a vibrating system graphically. (**a**) Magnitude and phase versus frequency; (**b**) Real and imaginary parts versus frequency; (**c**) Real part versus imaginary part (a Nyquist plot)

h_{ij}. The diagonal elements denote that the excitation and observation occurred at the same point. Such measurements are termed driving-point FRFs and they are of particular importance as at least one driving-point measurement is required to scale the mode shapes to UMM form. In a linear system the FRF matrix is symmetric: $h_{ij} = h_{ji}$. We do not have to measure all the terms of the FRF matrix. In practice, it is (generally) possible to obtain mode shapes from only one row or column of the FRF matrix.

Theoretically, it does not matter whether the measured frequency response comes from a shaker test or an impact test. If the structure is impacted at n points while the response is measured at one point, one row of the FRF is obtained. If a shaker is used to excite it at one point while an accelerometer (or other sensor) is moved from one point to another, one column of the FRF is obtained. In practice, there may be differences between the results of the *roving hammer* and the *roving accelerometer* methods, but these are the result of experimental compromises, not structural properties. Roving impact and roving response techniques provide essentially the same answers because a linear structure exhibits reciprocity. The motion produced at DOF a by a force applied at DOF b is identical to the motion at b due to a force at a.

The analog signals obtained from the measuring devices are generally digitized in an FFT analyzer. Sampling and quantization errors can occur in the process, but the most worrisome of signal-processing errors is probably leakage, which will be discussed later.

Serious consideration must be given to the way the structure is mounted and supported. Boundary conditions can be specified exactly in a completely free or completely constrained situation. In practice it is generally not possible to fully achieve these conditions. In order to approximate a free system, the structure can be suspended from very soft elastic cords or placed on a very soft cushion. The structure will be constrained to a degree and the rigid-body modes will not have zero frequency. However, they will generally be much lower than the frequencies of the flexible modes and will therefore have negligible effect. Some support fixtures affect damping as well.

28.2.2 Impact Testing

A roving hammer test is the most common type of impact test. An accelerometer is fixed at a single DOF, and the structure is impacted at as many DOFs as desired to define the mode shapes of the structure. Using a two-channel FFT analyzer, FRFs are computed, one at a time, between each impact DOF and the fixed response DOF. A suitable grid is usually marked on the structure to define the impact points.

Piezoelectric transducers are generally used to sense both force and acceleration. They generate an electrical charge when mechanically strained. Because of their high impedance, charge amplifiers are generally used to amplify the signals. More-modern sensors incorporate an amplifier within the transducer and power it from a constant-current source using the same two-conductor cable that transmits the signal (at low impedance). The resonant frequencies of the transducers must be well above the highest frequency that will be measured. The mass of the accelerometer should be small enough that the effect of mass loading is small. One way to determine if mass loading is significant is to measure an FRF with the accelerometer and compare it to a second measurement with an additional accelerometer (or equivalent mass) attached to it. An obvious advantage of this testing setup is that the response transducer is never moved; the structure is subjected to a consistent mass loading.

In some circumstances, the symmetry of the test object may produce repeated roots. These are modes with differen*t* mode shapes that have the same natural frequency. When such modes are encountered, they may be separated by using two or more fixed reference accelerometers and multi-reference curve-fitting techniques. Of course, an analyzer with three or more channels is required if the data is to be acquired in a single measurement set.

It is generally impossible to impact a structure in all three directions at all points, so three-dimensional (3-D) motion cannot be measured at all points. When 3-D motion at all points is desired, a roving triaxial accelerometer may be used and the structure is impacted at a fixed DOF with the hammer. Triaxial accelerometers are usually more massive, however. Since the triaxial accelerometer must be simultaneously sampled together with the force data, a four-channel FFT analyzer is required. [28.4] Note, however, that the roving accelerometer presents a different mass-load to the structure at each response site. This can lead to inconsistencies in the data that make an accurate curve fit more difficult.

Because the impulse signal exists for such a short period, it is important to capture all of it in the sampling window of the FFT analyzer. To insure that the entire signal is captured, the analyzer must be able to capture the impulse and response signals prior to the occurrence of the impulse. The analyzer must begin sampling data before the trigger point occurs. This is called a pre-trigger delay.

Modern FFT analyzers provide very high-resolution (24 bit) analog-to-digital converters (ADCs) and large capture block sizes (64 kpoint typical). These characteristics nullify the need to use weighting functions or *windows* when performing an impact test. The preferred analysis is conducted without weighting, also termed using a *rectangular* window.

Two common time-domain windows that are used in impact testing with older equipment are the force and exponential windows. These windows are applied to the signals after they are sampled but before the FFT is applied to them in the analyzer. The *force* window is used to remove noise from the force signal. Any nonzero data following the impulse signal in the sampling window is assumed to be measurement noise. The force window preserves the samples in the vicinity of the impulse but removes the noise from all other samples in the force signal.

The *exponential* window is used to reduce leakage in the spectrum of the response. The FFT analyzer assumes that the signal is periodic in the transform window. This is true of signals that are completely contained within the transform window or cyclic signals that complete an integer number of cycles within the transform window. If a time signal is not periodic in the transform window a smearing of its spectrum will occur when it is transformed to the frequency domain. This is called *leakage*. Leakage distorts the spectrum and makes it inaccurate. If the response does not decay to zero before the end of the sampling window, an exponential window can add artificial damping to all modes of the structure. This artificial damping must be removed by the subsequent curve-fitting algorithm to obtain proper damping factors.

It is important that the impact hammer provides a pulse that is well matched to the frequency span of the analysis. This is accomplished by fitting a striking tip of appropriate stiffness to the hammer's force gauge. A soft tip produces a broad pulse time-history with a narrow spectrum. A hard tip increases the force spectrum bandwidth by applying a narrow pulse. The spectrum of the force pulse has a *lobed* structure and all tests are done using the spectral content of the first lobe. Trial measurements are made to select a tip that provides a force spectrum that falls off no more than 25 dB from the direct-current (DC) point to the selected analysis bandwidth.

It is often difficult to strike at exactly the same place and angle multiple times. For this reason, averaging is less useful in an impact test than in other types of experimental modal analysis. Many experienced practitioners favor conducting tests with a single strike at each target DOF. This precludes calculating a coherence function, but that very useful causality measurement often tells you more about your ability to hit the same place twice than it does about structural nonlinearities or noise in measurement. For this reason, it is imperative to inspect every measurement set in the time domain before accepting it. Most analyzers automate this type of acquisition, giving you an accept/reject control.

28.2.3 Shaker Testing

Not all structures can be impact tested. Sometimes the surface is too delicate. Sometimes the impact force has too low an energy density over the entire frequency range of interest. In this case, FRF measurements must be made by attaching one or more shakers to the structure.

Since the FRF is a single input function, the shaker should transmit only one component of force in line with the main axis of the load cell. Often a structure tends to rotate slightly when it is displaced along an axis. To minimize the problem of forces being applied in other directions, the shaker is generally connected to the load cell through a slender rod called a stinger to allow the structure to move freely in the other directions. The stinger should have a strong axial stiffness but weak bending and shear stiffnesses.

A variety of broadband excitation signals have been developed for making shaker measurements with FFT analyzers: transient, random, pseudo-random, burst random, sine sweep (chirp). A true random signal is synthesized with a random number generator. Since it is nonperiodic in the sampling window, a Hanning window must always be used to minimize leakage.

A pseudo-random signal is specially synthesized by an FFT analyzer to coincide with the window parameters. It is synthesized over the desired frequency range and then passed through an inverse FFT algorithm to obtain a random time-domain signal, converted to an analog signal and used as the shaker excitation signal. Since the excitation signal is periodic in the sampling window, the acquired signals are leakage-free. However pseudo-random excitation does not excite nonlinearities differently between spectrum averages and will not, therefore, remove nonlinearities from FRF measurements.

Burst random excitation combines some advantages of both random and pseudo random testing. Its signals are leakage free and when used with spectrum averaging, will remove nonlinearities from the FRFs. In burst random testing, either a true random or time-varying pseudo-random signal can be used, but it is turned off prior to the end of the sampling window time period so that the response decays within the sampling window. Hence, they are periodic in the window and leakage-free [28.1].

In large structures more than one shaker may be needed to obtain sufficient excitation. These are driven simultaneously so that the structure is subjected to multiple inputs. The FRFs between each of the inputs and each of the multiple outputs are calculated by using a matrix inversion process. This type of measurement is called multiple-input multiple-output (MIMO) testing. It provides FRFs from multiple columns of the FRF matrix and two special types of coherence functions. One multiple coherence function is generated for each shaker used. This function may be thought of as a generalization of ordinary coherence; it asserts what fraction of all the response output power spectra may be attributed to the single measured force power spectrum and the FRFs in one column of the FRF matrix. One partial coherence function is generated for each FRF; it asserts what fraction of a single response power spectrum can be attributed to one force power spectrum and a single FRF. MIMO data is analyzed using a multi-reference curve fitter.

28.2.4 Obtaining Modal Parameters

Most experimental modal analysis relies on a modal parameter estimation (curve-fitting) technique to obtain modal parameters from the FRFs. Curve fitting is a process of matching a mathematical expression to a set of experimental points by minimizing the squared error between the analytical function and the measured data. Curve-fitting methods used for modal analysis fall into one of the following categories: local single degree of freedom, local multiple degree of freedom, global, or multi-reference.

Single-degree-of-freedom (SDOF) methods estimate modal parameters one mode at a time. Multiple-degree-of-freedom (MDOF), global, and multi-reference methods can estimate modal parameters for two or more modes at a time. Local methods are applied to one FRF at a time. Global and multi-reference methods are applied to an entire set of FRFs at once.

SDOF methods can be applied to most FRF data sets with low modal density, but MDOF methods must be used in cases of high modal density. Global methods work better than MDOF methods for cases with local modes. Multi-reference methods can find repeated roots (very closely coupled modes) where the other methods cannot [28.4].

Perhaps the simplest of the SDOF methods is the quadrature or peak picking method. Modal coefficients are estimated from the real and imaginary parts of the frequency response, so the method is really not a curve fit in the strict sense of the term. First the resonance peaks are detected on the imaginary plot and the frequencies of maximum response taken as the natural frequencies of these modes. Then the line width between the half-power points is determined from the real plot, from which damping can be estimated. The magnitude of the modal coefficient is taken as the value of the imaginary part at resonance. This method is adequate for structures whose modes are well separated and have modest damping [28.5].

Another SDOF method is the circle-fit method. This is based on the fact that a Nyquist plot of frequency response of a SDOF system near a resonance is a circle. Thus the circle that best fits the data points is determined. Half-power points are those frequencies for which $\theta = \pm 90$. The modal coefficient is determined from the diameter of the circle.

Two of the most popular MDOF methods are the complex exponential and rational fraction polynomial methods. The complex exponential method (or Prony algorithm) fits the impulse response function (IRF), which is the inverse Fourier transform of the FRF rather than the FRF itself. A potentially serious error, called wraparound error or time-domain leakage, can occur but can be mitigated by using an exponential window on the IRF This error is caused by the limited frequency range of the FRF measurement, and distorts the impulse response. The complex exponential algorithm works very well for FRFs with high modal density. In most applications, the algorithm is allowed to fit far more modes than the FRF form suggests are present. The resulting extraneous *computational modes* are then discarded, normally by using a stability diagram.

The rational fraction polynomial method fits a rational fraction of polynomials expression directly to an FRF measurement. Its advantage is that it can be applied over a user-selected frequency range of data, focusing its findings upon that frequency interval. It does not generate *computational modes* and therefore does not require the use of a stability diagram or other methods to filter its outputs.

Global curve fitting divides the curve-fitting process into two steps: estimating the frequency and damping parameters, and using these to obtain mode shape es-

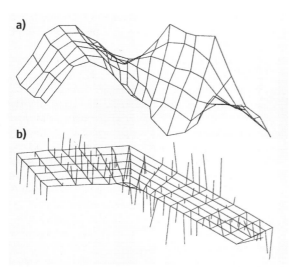

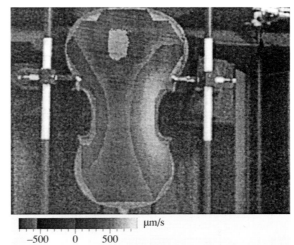

Fig. 28.2a,b Two ways of representing a vibrational mode in a Korean *pyeongyeong*: (**a**) The shape at maximum bending; (**b**) The neutral shape with vectors

Fig. 28.3 Gray-scale representation of a vibration mode in a violin (courtesy of George Bissinger)

timates by a second estimation process. The advantage of global curve fitting is that more-accurate frequency and damping estimates can potentially be obtained by processing all of the measurements rather than relying on a single measurement. Another advantage is that, because damping is already known and fixed as a result of the first step, the modal coefficients are more accurately estimated during the second step. Both the complex exponential and rational fraction polynomial methods can be formulated to obtain global estimates from a set of measurements [28.6].

28.2.5 Real and Complex Modes

The assumption of proportional viscous damping implies the existence of real, or normal modes. Mathematically, this implies that the damping matrix can be defined as a linear combination of the physical mass and stiffness matrices. Physically, all the points in a structure reach their maximum excursion, in one or the other direction, at the same time. The imaginary part of the FRF reaches a maximum at resonance, and the Nyquist circle lies along the imaginary axis.

Some structures exhibit a more complicated form of damping, and the mode shapes are complex, meaning the phase angles can have values other than 0 or 180°. Different points on the structure reach their maxima at various times as in a traveling wave pattern. The imaginary part of the FRF no longer reaches a maximum at resonance nor is the real part zero. The Nyquist circle is rotated at an angle in the complex plane. When damping is light, the proportional damping assumption is generally an accurate approximation, [28.5] although it can be argued that a complex-mode formulation is essential to preserve accuracy in damping evaluation [28.7].

28.2.6 Graphical Representation

One of the nice features of experimental modal testing is the way that the modes can be represented graphically. Animations of the vibration can be viewed from any angle to comprehend complex mode shapes. Static representations include the shape at maximum bending, and the neutral shape with vectors, as shown in Fig. 28.2. Another representation is shown in Fig. 28.3.

28.3 Mathematical Modal Analysis

In mathematical modal analysis, one attempts to uncouple the structural equation of motion by means of some suitable transformation, so that the uncoupled equations can be solved. The frequency response of the structure can then be found by summing the respective modal responses in accordance with their degree of participation in the structural motion.

Sometimes it is not possible to obtain exact solutions to the equation of motion. Computers have popularized the use of numerical approximations such as finite-element and boundary-element methods.

28.3.1 Finite-Element Analysis

The finite-element method of analysis (FEA or FEM) started in the late 1940s as a structural analysis tool useful in helping aerospace engineers design better aircraft structures. Aided by the rapid increase in computer power since then, it has become a very sophisticated tool for a wide array of engineering tasks. It can be used to predict how a system will react to environmental factors such as forces, heat, and vibration.

The technique is based on the premise that an approximate solution to any complex engineering problem can be reached by subdividing the problem into smaller more manageable (finite) elements. Using finite elements, solving complex equations that describe the behavior of a structure can often be reduced to a set of linear equations that can be solved using the standard techniques of matrix algebra.

The finite-element method works by breaking a real object down into a large number of elements, which are regular in shape and whose behavior can be predicted by a set of mathematical equations. The summation of the behavior of each individual element produces the expected behavior of the actual object. Finite-element analysis (FEA) uses a complex system of points called *nodes*, which make a grid called a *mesh*. The mesh is programmed to include the material and structural properties that define how the structure will react to certain loading conditions.

General-purpose finite-element codes, such as NASTRAN and ANSYS, are programmed to solve the matrix equation of motion for the structure of the form:

$$[M]\{\ddot{u}\} + [C]\{\dot{u}\} + [K]\{u\} = \{F\cos(\omega t + \phi)\}, \quad (28.1)$$

where $[M]$ is the mass matrix, $[C]$ is the damping matrix, $[K]$ is the stiffness matrix, $\{u\}$ is displacement, and $\{F\}$ is force.

The model details are entered in a standard format, and the computer assembles the matrix equation of the structure. The first step is to solve the matrix equation $[M]\{\ddot{u}\} + [K]\{u\} = \{0\}$ for free vibrations of the structure. The solution to this equation gives the natural frequencies (eigenvalues) and the undamped mode shapes (eigenvectors). These parameters are the basic dynamical properties of the structure, and they are used in subsequent analysis for dynamic displacements and stresses. For harmonic motions, $\{\ddot{u}\} = -\omega^2\{u\}$, so that $[K]^{-1}[M]\{u\} = [I]\{u\}$, where $[I]$ is the unity matrix. Typically the matrix equations of motion for the structure contain off-diagonal terms, but they may be decoupled by introducing a suitable transformation. The damping matrix may also be uncoupled on the condition that the damping terms are proportional to either the corresponding stiffness matrix terms or the corresponding mass matrix terms [28.8].

Finite-element analysis in conjunction with high-speed digital computers permits the efficient solution of large, complex structural dynamics problems. Structural dynamics problems that are linear can generally be solved in the frequency domain. There are other applications of finite-element analysis in acoustics. The sound field produced in an enclosure or by a radiating surface vibrating at a single frequency can be described by the Helmholtz equation

$$\nabla^2 p + k^2 p = 0, \quad (28.2)$$

where p is the acoustic pressure, k is the wave number ($k = \omega/c$) and ∇ is the Hamiltonian nabla operator.

The most common boundary conditions are fixed pressure and fixed velocity on the surface of the enclosed or radiating body. Other types of boundary conditions are normal and transfer impedances [28.9].

Exact analytical solutions of the Helmholtz equation exist for a very limited number of idealized geometries corresponding to situations where the geometry of the radiating surface can be described by orthogonal coordinate systems that are separable. In order to solve the Helmholtz equation for more general applications, numerical schemes are necessary.

28.3.2 Boundary-Element Methods

The boundary-element method (BEM) is also a powerful computational technique, providing numerical solutions to a wide range of scientific and engineering problems. Since only a mesh of the boundary of the domain is required, the method is generally easier to apply than the finite-element method. The boundary-element method has found application in stress analysis, structural vibrations, acoustic fields, heat transfer, and potential flow. The advantage in the boundary-element method is that only the boundaries of the domain of the partial differential equation require subdivision, compared to FEA in which the whole domain of the equation requires discretization.

The boundary-element method is especially effective for calculating the sound radiated by a vibrating body or for predicting the sound field inside a cavity. A typical BEM input file consists of a surface mesh, a normal velocity profile on the surface, the fluid density, speed of sound, and frequency. The output of the BEM includes the sound pressure distribution on the surface of the body and at other points in the field, the sound intensity, and the sound power. BEM may be formulated in either the time domain or the frequency domain [28.10].

In cases where the domain is exterior to the boundary, as in acoustic radiation, the extent of the domain may be infinite and hence the advantages of BEM are even more striking. [28.11] The boundary-element mesh, which consists of a series of points called *nodes* connected together to form triangular or quadrilateral elements, covers the entire surface of the radiating body. The magnitude and phase of the vibration velocity must be known at every point. BEM calculates the sound pressure distribution on the surface of the body from which the sound pressure and sound intensity at field points in the acoustic domain can be calculated. BEM will even calculate the sound pressure in the *shadow zone* behind a body due to radiation from the other side [28.10].

Another class of problems to which the BEM may be applied is the so-called interior problem, such as the prediction of the sound field inside an enclosure. The procedure is similar to the solution of exterior problems.

There are two different formulations for BEM: *direct* and *indirect* methods. The primary variables in the direct BEM are the acoustic pressure and acoustic velocity, while the indirect method uses the difference in acoustic pressure and the difference in the normal gradient of acoustic pressure across the boundary [28.12]. There is no distinction between an interior and an exterior problem when using indirect BEM, since it considers both sides of the boundary simultaneously.

Yet another class of problems is that in which an interior acoustic space is connected to an exterior space through one or more openings. The radiation of sound from a source within a partial enclosure is one example of such a problem. In one approach, the integral equations for the interior and exterior domains are coupled using continuity conditions at the interface surface between the two domains. The integral equations are then reduced to numerical form using second-order boundary elements [28.13].

To apply BEM to the Helmholtz equation, it must essentially be reduced to a two-dimensional integral equation by use of Green's theorem. This means that only a description of the radiating surface is necessary rather than a complete discretization of the surrounding medium. The technique is suited to both interior- and exterior-domain problems.

Some methods couple a finite element of the structure with a boundary-element model of the surrounding fluid. The surface fluid pressures and normal velocities are first calculated by coupling the finite-element model of the structure with a discretized form of the Helmholtz surface integral equation for the exterior fluid [28.14].

Both BEM and FEM can be computationally intensive if the model has a large number of nodes. The solution time is roughly proportional to the number of nodes squared for an FEM analysis and to the number of nodes cubed for a BEM analysis. The accuracy of both methods depends on having a sufficient number of nodes in the model, so engineers try to straddle the line between having a mesh that is accurate yet can be solved in a reasonable time. The general rule of thumb is that six linear or three parabolic elements are needed per wavelength [28.10].

28.3.3 Finite-Element Correlation

It is often desirable to validate the results of a finite-element analysis with measured data from a modal test. This correlation is generally an iterative process incorporating two major steps. First the modal frequencies and mode shapes are compared and the differences quantified. Then, adjustments and modifications are made to the finite-element model to achieve more-comparable results.

It is useful to graphically compare the sets of frequencies by plotting predicted versus measured frequencies. This shows the global trends and suggests possible causes of these differences. If the random scatter is too great, then the finite-element model may not be an accurate representation of the structure. If the points lie on a straight line, but with a slope other than 1, then the problem may be a mass loading problem in the modal test or an incorrect material property such as elastic modulus or material density in the finite-element model.

Numerical techniques have been developed to perform statistical comparisons between mode shapes. The modal assurance criterion (MAC) is a correlation coefficient between the measured and calculated mode shapes. Another technique, called direct system parameter identification, is the derivation of a physical model of a structure from measured force and response data [28.5]. Various forms of cross-orthogonality tests have also been employed. In general these methods see how well the experimental shapes can uncouple the mathematical model. As with many comparison methods, difficulty is often encountered in reducing the FEA degrees of freedom to a set of DOFs consistent with the experiment.

28.4 Sound-Field Analysis

Although modal analysis developed primarily as a way of analyzing mechanical vibrations, by substituting microphones or acoustic intensity probes for accelerometers, experimental modal testing techniques can be used to explore sound fields. Modal testing has been used to explore standing acoustic waves inside air columns and to explore radiated sound fields from a vibrating structure. It could also be used to explore acoustical modes in rooms.

Figure 28.4 shows an arrangement for measuring the sound field inside a flute. The exciter is a loudspeaker coupled to the flute by a capillary tube. A pressure microphone probes the sound field at successive points, and the input and output signals are fed to a 2-channel FFT analyzer in the usual manner. The sound pressure amplitudes and phases for four modes in the flute (D5 fingering) are shown in Fig. 28.5 [28.15].

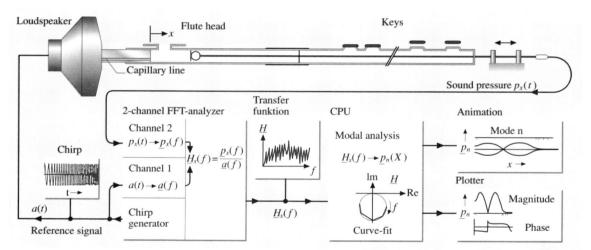

Fig. 28.4 Experimental arrangement for measuring the sound field inside a flute using modal analysis. (After [28.15])

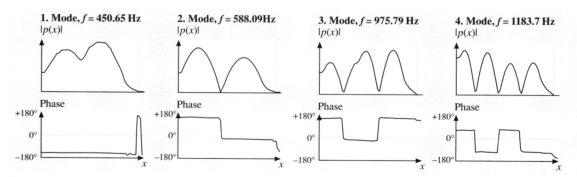

Fig. 28.5 The sound pressure amplitudes and phases for four modes in the flute (D5 fingering) (After [28.15])

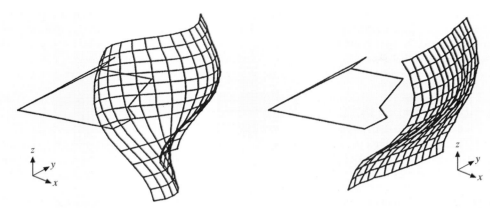

Fig. 28.6 Sound pressure in the radiated sound field of a piano at 162 Hz (*left*) and 107 Hz (*right*). Note the 180° phase difference above and below the plane of the soundboard (After [28.15])

The sound field radiated by a piano soundboard has been similarly explored by driving the soundboard with a shaker and probing the radiated field at 200 points with a pressure microphone. Figure 28.6 shows the sound pressure in the radiated sound field at two frequencies. The sound field shows a 180° phase difference above and below the plane of the soundboard as expected [28.15].

28.5 Holographic Modal Analysis

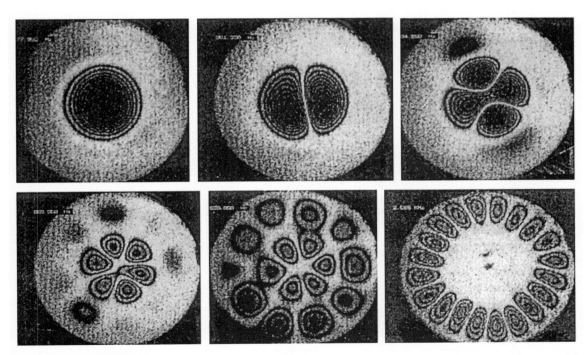

Fig. 28.7 Holographic interferograms of several vibrational modes in a Chinese gong. Note that some modes are mainly in the center of the gong, some in the outer portion, and some are global. (After [28.16])

Holographic interferometry offers the best spatial resolution of operating deflection shapes. In cases where the damping is small and the modes are well separated in frequency, the operating deflection shapes correspond closely to the normal mode shapes. Modal damping can be estimated with a fair degree of accuracy from half-power points determined by counting fringes [28.17]. Phase modulation allows analysis to be done at exceedingly small amplitudes and also offers a means to separate modes that overlap in frequency [28.18]. TV holography allows the observation of vibrational motion in real time, and it is a fast, convenient way to record deflection shapes. Holographic interferograms of several vibrational modes of a Chinese opera gong appear in Fig. 28.7.

An excellent discussion of holographic and other optical techniques for acoustic and vibration measurements appears in Chap. 27.

References

28.1 D.J. Ewins: *Modal Testing: Theory, Practice and Application* (Research Studies Press, Baldock 2001)

28.2 Proceedings of International Conference on Modal Analysis, 1982-2006 (Society for Experimental Mechanics, Bethel 1982-2006)

28.3 R. J. Allemang: *Academic Course Info, Dynamics Research Laboratory* (Univ. Cincinnati, Cincinnati 2003)

28.4 B.J. Schwarz, M.H. Richardson: *Experimental Modal Analysis, unpublished report* (Vibrant. Technology, Scotts Valley 1999), http://www.vibetech.com/papers/paper28.pdf

28.5 Agilent Technologies: *The Fundamentals of Modal Testing*, Applic. Rep. 243-2 (Agilent Technologies, Palo Alto 2000)

28.6 M.H. Richardson, D.L. Formenti: *Global curve fitting of frequency response measurements using the rational fraction polynomial method*, Proc., 3rd IMAC Conf. (Society for Experimental Mechanics, Bethel 1985)

28.7 G.F. Lang: *Demystifying complex modes*, Sound Vibration Magazine (Acoustical Publ., Bay Village 1989) pp. 36–40

28.8 N.F. Rieger: *The relationship between finite element analysis and modal analysis*, Sound Vibration Magazine (Acoustical Publ., Bay Village 1986) pp. 16–31

28.9 K.R. Fyfe, J.-P.G. Coyette, P.A. van Vooren: *Acoustic and elasto-acoustic analysis using finite element and boundary element methods*, Sound Vibration Magazine (Acoustical Publ., Bay Village 1991) pp. 16–22

28.10 A.F. Seybert, T.W. Wu: Acoustic modeling: Boundary element methods. In: *Encyclopedia of Acoustics*, ed. by M.J. Crocker (Wiley, New York 1997)

28.11 S. Kirkup: *The Boundary Element Method in Acoustics* (Integrated Sound Software, Heptonstall 1998)

28.12 D.W. Herrin, T.W. Wu, A.F. Seybert: Practical issues regarding the use of the finite and boundary element methods for acoustics, Building Acoustics **10**, 257–279 (2003)

28.13 A.F. Seybert, C.Y.R. Cheng, T.W. Wu: The solution of coupled interior/exterior acoustic problems using the boundary element method, J. Acoust. Soc. Am. **88**, 1612–1618 (1990)

28.14 U. J. Hansen, I. Bork: Adapting techniques of modal analysis to sound field representation. In: *Proc. Stockholm Music Acoustics Conference (SMAC 93)* (KTH Voice Res. Centre, Stockholm 1994) pp. 533–538

28.15 A. Peekna, T.D. Rossing: The acoustics of Baltic psaltery, Acta Acustica/Acustica **91**, 269–276 (2005)

28.16 T.D. Rossing, J. Yoo, A. Morrison: Acoustics of percussion instruments: An update, Acoust. Sci. Tech. **25**, 1–7 (2004)

28.17 T.D. Rossing, F. Engström: *Using TV holography with phase modulation to determine the deflection phase in a baritone guitar*, Proc. Int. Symp. Musical Acoustics (ISMA 98) (Acoustical Society of America, Woodbury 1998)

28.18 G.C. Everstine, F.M. Henderson: Coupled finite element/boundary element approach for fluid-structure interaction, J. Acoust. Soc. Am. **87**, 1938–1947 (1990)

Acknowledgements

B.8 Nonlinear Acoustics in Fluids
by Werner Lauterborn, Thomas Kurz, Iskander Akhatov

The authors would like to thank U. Parlitz, R. Geisler, D. Kröninger, K. Köhler, and D. Schanz for stimulating discussions and help with the manuscript, either compiling tables or designing figures.

C.9 Acoustics in Halls for Speech and Music
by Anders Christian Gade

The author of this chapter owes deep thanks to Jerald R. Hyde for his constant support throughout the preparation of this chapter and for detailed comments and suggestions for improvements in content and language. Comments from Leo Beranek and Thomas Rossing have also been very valuable.

D.13 Psychoacoustics
by Brian C. J. Moore

I thank Brian Glasberg, Aleksander Sek and Chris Plack for assistance in producing figures. I also thank William Hartmann and Adrian Houtsma for helpful and detailed comments on an earlier version of this chapter.

E.15 Musical Acoustics
by Colin Gough

Many musical acoustics colleagues and musicians have contributed directly or indirectly to the material in this chapter, though responsibility for the accuracy of the content and its interpretation remains my own. In particular, I am extremely grateful to Tom Rossing, Neville Fletcher and Murray Campbell for their critical reading of earlier drafts and their invaluable suggestions for improvements. I also gratefully acknowledge an Emeritus Fellowship from the Leverhulme Foundation, which supported research contributing to the writing of this manuscript.

F.21 Medical Acoustics
by Kirk W. Beach, Barbrina Dunmire

This work was supported by the generosity of the taxpayers of the United States trough the National Institutes of Health, National Cancer Institute NCI-N01-CO-07118, and the National Institute for Biomedical Imaging and Bioengineering 1 RO1 EB002198-01. Images and data were provided by Keith Comess, Larry Crum, Frances deRook, Lingyun Huang, John Kucewicz, Marla Paun, Siddhartha Sikdar, Shahram Vaezy, and Todd Zwink.

G.23 Noise
by George C. Maling, Jr.

The author would like to thank the following individuals who read portions of the manuscript and made many helpful suggestions for changes and inclusion of additional material: Douglas Barrett, Elliott Berger, Lawrence Finegold, Robert Hellweg, Uno Ingard, Alan Marsh, Christopher Menge, and Matthew Nobile, and Paul Schomer.

H.24 Microphones and Their Calibration
by George S. K. Wong

The author would like to thank the Acoustical Society of America, and the International Electrotechnical Commission (IEC) for permission to use data give in their publications. Special acknowledgement is given to the Journal of the Acoustical Society of Japan (JASJ) for duplicating data from their publication.

H.26 Acoustic Holography
by Yang-Hann Kim

This study was partly supported by the National Research Laboratory (NRL) project of the Korea Institute of Science and Technology Evaluation and Planning (KISTEP) and the Brain Korea 21 (BK21) project initiated by the Ministry of Education and Human Resources Development of Korea. We especially acknowledge Dr. K.-U. Nam's comments and contribution to the completion of the manuscript.

We also appreciate Dr. S.-M. Kim's contribution to the Appendix. This is mainly based on his M.S. thesis in 1994.

H.27 Optical Methods for Acoustics and Vibration Measurements
by Nils-Erik Molin

These projects were supported by the Swedish research council (VR and former TFR) with equipment from the Wallenberg and the Kempe foundations. I am also very grateful to present and former PhD students, coworkers, guest researchers etc. at our Division of Experimental Mechanics, LTU for their kind help with figures in this paper and other contributions over the years.

About the Authors

Iskander Akhatov

North Dakota State University
Center for Nanoscale Science and
Engineering, Department of Mechanical
Engineering
Fargo, ND, USA
iskander.akhatov@ndsu.edu

Chapter B.8

Iskander Akhatov earned his B.S. and M.S in Physics and Ph.D. in Mechanical Engineering from Lomonosov University of Moscow. He has extensive experience in multiphase fluid dynamics, nonlinear dynamics and acoustics of bubbles and bubbly liquids. Prior to joining faculty at NDSU, Professor Akhatov worked at the Russian Academy of Sciences, State University of Ufa (Russia), Göttingen University (Germany), Boston University and RPI (USA). His current research interests include fluid dynamics in micro and nano scales, nanotechnology.

Yoichi Ando

Makizono, Kirishima, Japan
andoy@cameo.plala.or.jp

Chapter C.10

Professor Yoichi Ando received his Ph.D. from Waseda University in 1975. He was an Alexander-von-Humboldt Fellow from 1975–1977 at the Drittes Physikalisches Institut of the Universität Göttingen. In 2001 he established the Journal of Temporal Design. In 2002 he received the Dottore *AD Honorem* from the University of Ferrara, Italy. Since 2003 he is Professor Emeritus of Kobe University, Japan. He is the Author of the books *Concert Hall Acoustics*, (1985) and *Architectural Acoustics* (1998) both published by Springer. His research works was on auditory and visual sensations and brain activies.

Keith Attenborough

The University of Hull
Department of Engineering
Hull, UK
k.attenborough@hull.ac.uk

Chapter A.4

Keith Attenborough is Research Professor in Engineering, Director of the University of Hull Acoustic Research Centre and a Chartered Engineer. In 1996 he received the Institute of Acoustics Rayleigh medal for distinguished contributions to acoustics. His research has included pioneering studies of acoustic-to-seismic coupling and blast noise reduction using granular materials. Current research uses laboratory simulations of blast noise propagation.

Whitlow W. L. Au

Hawaii Institute of Marine Biology
Kailua, HI, USA
wau@hawaii.edu

Chapter F.20

Dr. Au is the Chief Scientist of the Marine Mammal Research Program, Hawaii Institute of Marine Biology. He received his Ph.D. degree in Electrical Engineering from Washington State University in 1970. His research has focused on the sonar system of dolphins and on marine bioacoustics. He is a fellow of the Acoustical Society of America and a recipient of the society's Silver Medal in Animal Bioacoustics.

Kirk W. Beach

University of Washington
Department of Surgery
Seattle, WA, USA
kwbeach@u.washington.edu

Chapter F.21

Kirk Beach received his B.S. in Electrical Engineering from the University of Washington, his Ph.D. in Chemical Engineering from the University of California at Berkeley and his M.D. from the University of Washington. He develops noninvasive devices to explore arterial, venous and microvascular diseases using ultrasonic, optical and electronic methods and uses those devices to characterize disease.

Mack A. Breazeale

University of Mississippi
National Center for Physical Acoustics
University, MS, USA
breazeal@olemiss.edu

Chapter B.6

Mack A. Breazeale earned his Ph.D. in Physics at Michigan State University and is co-editor of a book. He has more than 150 publications on Physical Acoustics. He is Fellow of the Acoustical Society, the Institute of Acoustics, and Life Fellow of IEEE. He has been President's Lecturer and Distinguished Lecturer of IEEE and was awarded the Silver Medal in Physical Acoustics by the Acoustical Society of America.

Antoine Chaigne

Unité de Mécanique (UME)
Ecole Nationale Supérieure de
Techniques Avancées (ENSTA)
Palaiseau, France
antoine.chaigne@ensta.fr

Chapter G.22

Antoine Chaigne received his Ph.D. in Acoustics from the University of Strasbourg. He is currently the head of the Mechanical Engineering Department at The Ecole Nationale Supérieure de Technique Avancées (ENSTA), one of the top ten Institutes of higher education for Engineering in France, which belongs to the Paris Institute of Technology (ParisTech). Professor Chaigne's main areas of research are musical acoustics, transportation acoustics and modeling of sound sources. He is a Fellow member of the Acoustical Society of America and a member of the French Acoustical Society (SFA).

Perry R. Cook

Princeton University
Department of Computer Science
Princeton, NJ, USA
prc@cs.princeton.edu

Chapter E.17

Professor Cook received bachelor's degrees in music and EE at the University of Missouri-Kansas City, worked as a sound engineer, and received an EE Ph.D. from Stanford. He was Technical Director of Stanford's CCRMA, and is now a Princeton Computer Science Professor (jointly in Music). He is a 2003 Guggenheim Fellowship recipient for a new book on *Technology and the Voice*, he is co-founder of the Princeton Laptop Orchestra.

James Cowan

Resource Systems Group Inc.
White River Junction, VT, USA
jcowan@rsginc.com

Chapter C.11

James Cowan has consulted on hundreds of projects over the past 25 years in the areas of building acoustics and noise control. He has taught university courses in building acoustics for the past 20 years both in live classes and on the internet. He has authored 2 books and several interactive CD sets and book chapters on architectural and environmental acoustics.

Mark F. Davis

Dolby Laboratories
San Francisco, CA, USA
mfd@dolby.com

Chapter E.18

Dr. Davis obtained his Ph.D. in Electrical Engineering from M.I.T. in 1980. He is a Principal Member of the Technical Staff at Dolby Laboratories, where he has worked since 1985. He has been involved with the development of the AC-3 multichannel coder, MTS analog noise reduction system, DSP implementations of Pro Logic upmixer, virtual loudspeaker, and SR noise reduction. His current work is principally involved with investigation of advanced surround sound systems.

Barbrina Dunmire

University of Washington
Applied Physics Laboratory
Seattle, WA, USA
mrbean@u.washington.edu

Chapter F.21

Barbrina Dunmire is an engineer within the Center for Industrial and Medical Ultrasound (CIMU) division of the Applied Physics Laboratory at the University of Washington. She holds an M.S. in Aero/Astronautical Engineering and BioEngineering. Her current areas of research include ultrasonic tissue strain imaging and its application to functional brain imaging and breast cancer, and high intensity focused ultrasound (HIFU).

About the Authors

Neville H. Fletcher

Australian National University
Research School of Physical Sciences
and Engineering
Canberra, ACT, Australia
neville.fletcher@anu.edu.au

Chapter F.19

Professor Neville Fletcher has a Ph.D. from Harvard and a D.Sc. from Sydney University and is a Fellow of the Australian Academy of Science. He has also been an Institute Director in CSIRO, Australia's national research organisation. His research interests include solid-state physics, cloud physics, and both musical and biological acoustics. He has published six books on these topics.

Anders Christian Gade

Technical University of Denmark
Acoustic Technology, Oersted.DTU
Lyngby, Denmark
acg@oersted.dtu.dk,
anders.gade@get2net.dk

Chapter C.9

Anders Gade is an expert in architectural acoustics and shares his time between the Technical University of Denmark and private consultancy. His research areas are acoustic conditions for musicians on orchestra stages, the relationships between concert hall geometry and their acoustic properties and electro acoustic enhancement systems for auditoria. Anders Gade is a fellow of the Acoustical Society of America.

Colin Gough

University of Birmingham
School of Physics and Astronomy
Birmingham, UK
c.gough@bham.ac.uk

Chapter E.15

Colin Gough is an Emeritus Professor of Physics at the University of Birmingham, where he supervised research projects and taught courses in musical acoustics, in addition to leading a large interdisciplinary research group in superconductivity. He also led the University string quartet and many other local chamber and orchestral ensembles, providing a strong musical focus to his academic research.

William M. Hartmann

Michigan State University
East Lansing, MI, USA
hartmann@pa.msu.edu

Chapter D.14

William Hartmann studied electrical engineering (B.S., Iowa State) and theoretical physics (Dr. Phil., Oxford, England). He is currently a professor of physics at Michigan State University, where he studies psychoacoustics and signal processing. His work in human pitch perception and auditory organization was summarized in the book *Signals, Sound, and Sensation* (Springer, 1998). His current work concerns binaural hearing and sound localization. He was formerly president of the Acoustical Society of America and received the Society's Helmholtz–Rayleigh Award.

Finn Jacobsen

Ørsted DTU, Technical University
of Denmark
Acoustic Technology
Lyngby, Denmark
fja@oersted.dtu.dk

Chapter H.25

Finn Jacobsen received a Ph.D. in acoustics from the Technical University of Denmark in 1981. His research interests include general linear acoustics, numerical acoustics, statistical methods in acoustics, and acoustic measurement techniques. He has recently turned his research interests towards methods based on transducer arrays, e.g. acoustic holography.

Yang-Hann Kim

Korea Advanced Institute of Science
and Technology (KAIST)
Department of Mechanical Engineering
Center for Noise and Vibration Control
(NOVIC)
Acoustics and Vibration Laboratory
Daejeon, Korea
yanghannkim@kaist.ac.kr

Chapter H.26

Dr. Yang-Hann Kim is a Professor of Mechanical Engineering Department at the Korea Advanced Institute of Science and Technology (KAIST) and a Director of the Center for Noise and Vibration Control (NOVIC). He received his Ph.D. degree in the field of acoustics and vibration from M.I.T. in 1985. He has been working in the field of acoustics and noise/vibration control, especially sound field visualization (acoustic holography) and sound source identification. He has been extending his research area to acoustic holography analysis, sound manipulation (3D sound coloring), and MEMS-sensors/actuators. He is a member of Sigma Xi, KSME, ASME, ASA, INCE, the Acoustical Society of Korea, and the Korea Society for Noise and Vibration (KSNVE) and has been on the editorial board of Mechanical Systems and Signal Processing (MSSP) and the Journal of Sound and Vibration (JSV).

William A. Kuperman

University of California at San Diego
Scripps Institution of Oceanography
La Jolla, CA, USA
wkuperman@ucsd.edu

Chapter A.5

William Kuperman is an ocean acoustician who has spent a total of about three years at sea. He is a co-author of the textbook *Computational Ocean Acoustics*. His most recent ocean research interests have been in time reversal acoustics and in ambient noise imaging. Presently he is a Professor at the Scripps Institution of Oceanography and Director of its Marine Physical Laboratory.

Thomas Kurz

Universität Göttingen
Göttingen, Germany
t.kurz@dpi.physik.uni-goettingen.de

Chapter B.8

Thomas Kurz obtained his Ph.D.. in physics from Göttingen university in 1988. He is now a staff scientist at the Third Physics Institute of this university working on problems in nonlinear physics and ultrashort phenomena. His current research is focused on cavitation collapse, sonoluminescence and the propagation of ultrashort optical pulses.

Werner Lauterborn

Universität Göttingen
Drittes Physikalisches Institut
Göttingen, Germany
w.lauterborn@dpi.physik.uni-goettingen.de

Chapter B.8

Dr. Lauterborn is Professor of Physics at the University of Göttingen and Head of the Drittes Physikalisches Institut directing the research on vibration and waves in acoustics and optics. He is co-author of a book on *Coherent Optics* and Editor of Proceedings on cavitation and nonlinear acoustics. His main research interest is on nonlinear physics, with special interest in acoustic and optic cavitation, acoustic chaos, bubble dynamics and sonoluminescence, nonlinear time series analysis and investigation of chaotic systems.

Björn Lindblom

Stockholm University
Department of Linguistics
Stockholm, Sweden
lindblom@ling.su.se,
blindblom@mail.utexas.edu

Chapter E.16

After obtaining his Ph.D. in 1968, Lindblom set up a laboratory to do research on phonetics at Stockholm University. His academic experience includes teaching and research at various laboratories in Sweden and the US. Recently he became Fellow of the AAAS. Currently he is Professor Emeritus and he continues his research on phonetics at Stockholm University and the University of Texas at Austin.

George C. Maling, Jr.

Institute of Noise Control Engineering of the USA
Harpswell, ME, USA
maling@alum.mit.edu

Chapter G.23

George Maling received a Ph.D. degree in physics from the Massachusetts Institute of Technology. He is the author of 76 papers and 8 handbook articles related to acoustics and noise control. He has edited or co-edited 10 conference proceedings. He is the recipient of the Rayleigh Medal from the Institute of Acoustics in the UK, and is a member of the National Academy of Engineering.

Nils-Erik Molin

Luleå University of Technology
Experimental Mechanics
Luleå, Sweden
nem@ltu.se

Chapter H.27

Nils-Erik Molin received his Ph.D. in 1970 at the Royal Institute of Technology, Stockholm, Sweden, with ""On fringe formation in hologram interferometry and vibration analysis of stringed musical instruments. Thereafter he has worked as lecturer and professor in experimental mechanics at Luleå University of Technology, Sweden. His main research field is optical metrology to measure mechanical and acoustical quantities.

Brian C. J. Moore

University of Cambridge
Department of Experimental Psychology
Cambridge, UK
bcjm@cam.ac.uk

Chapter D.13

Brian Moore's research field is psychoacoustics. He is a Fellow of: the Royal Society, the Academy of Medical Sciences, and the Acoustical Society of America. He has written or edited 12 books and over 400 scientific papers and book chapters. He has received the Acoustical Society of America Silver Medal in physiological and psychological acoustics, the *International Award in Hearing* from the American Academy of Audiology and the Littler Prize of the British Society of Audiology (twice).

Alan D. Pierce

Boston University
College of Engineering
Boston, MA, USA
adp@bu.edu

Chapter A.3

Allan D. Pierce received his doctorate from the Massachusetts Institute of Technology (MIT) and has held research positions at the Rand Corporation, the Avco Corporation, the Max Planck Institute for Fluids Research, in Göttingen, Germany, resulting from a Humboldt award, and the U. S. Department of Transportation; he has held professorial positions at MIT, Georgia Institute of Technology (Regents Professor), and the Pennsylvania State University (Leonhard Chair in Engineering). He is currently Professor of Aerospace and Mechanical Engineering (formerly the Department Chair) at Boston University and also serves as the Editor-in-Chief for the Acoustical Society of America (ASA). His research and teaching in acoustics have been recognized with his being awarded the ASA's Silver Medal in Physical Acoustics, the ASA's Gold Medal, and the Per Bruel Gold Medal in Acoustics and Noise Control from the American Society of Mechanical Engineers. He was the first recipient of the ASA's Rossing Prize in Acoustics Education, and was a founding Co-Editor-in-Chief of the Journal of Computational Acoustics.

Thomas D. Rossing

Stanford University
Center for Computer Research
in Music and Acoustics (CCRMA)
Department of Music
Stanford, CA, USA
rossing@ccrma.stanford.edu

Chapters 1, A.2, H.28

Thomas Rossing received a B.A. from Luther College, and MS and PhD degrees in physics from Iowa State University. After three years as a research physicist with the UNIVAC Division of Sperry Rand, he joined the faculty of St. Olaf College (Minnesota), where he was professor of physics for 14 years and chaired the department for 6 years. Since 1971 he has been a professor of physics at Northern Illinois University. He was named distinguished Research Professor in 1987, and Professor Emeritus in 2002. He is presently a Visiting Professor of Music at Stanford University. He is a Fellow of the American Physical Society, the Acoustical Society of America, IEEE, and AAAS. He was awarded the Silver Medal in Musical Acoustics by ASA and the Robert A. Millikan Medal by the American Association of Physics Teachers. He was a Sigma Xi National Lecturer 1984-87 and a Visiting Exchange Scholar in China in 1988. He is the author of more than 350 publications (including 15 books, 9 U.S. and 11 foreign patents), mainly in acoustics, magnetism, environmental noise control, and physics education. His areas of research have included musical acoustics, psychoacoustics, speech and singing, vibration analysis, magnetic levitation, surface effects in fusion reactors, spin waves in metals, and physics education.

Philippe Roux

Université Joseph Fourier
Laboratoire de Géophysique Interne
et Tectonophysique
Grenoble, France
philippe.roux@obs.ujf-grenoble.fr

Chapter A.5

Philippe Roux is a physicist with a strong background in ultrasonic and underwater acoustics. He obtained his Ph.D. in 1997 from the University of Paris on the application of time-reversal to ultrasounds. He is now a full-time CNRS researcher in Grenoble where he develops small-scale laboratory experiments in geophysics. Since 2004, he is also an Associate Researcher at the Marine Physical Laboratory of the Scripps Institute of Oceanography (San Diego). He is a Fellow of the Acoustical Society of America.

Johan Sundberg

KTH–Royal Institute of Technology
Department of Speech, Music, and Hearing
Stockholm, Sweden
pjohan@speech.kth.se

Chapter E.16

Johan Sundberg (Ph.D. musicology, doctor honoris causae 1996 University of York, UK) had a personal chair in Music Acoustics at the department of Speech Music Hearing, KTH and founded and was head of its music acoustics research group until his retirement in 2001. His research concerns particularly the singing voice and music performance. Written *The Science of the Singing Voice* (1987) and *The Science of Musical Sounds* (1991), he edited or co-edited many proceedings of music acoustic meetings. He has practical experience of performing music (choir and solo singing). He is Member of the Royal Swedish Academy of Music (President of its Music Acoustics Committee 1974–1982), the Swedish Acoustical Society (President 1976–1981) and fellow of the Acoustical Society of America, receiving its Silver Medal in Musical Acoustics 2003.

Gregory W. Swift

Los Alamos National Laboratory
Condensed Matter
and Thermal Physics Group
Los Alamos, NM, USA
swift@lanl.gov

Chapter B.7

Greg Swift invents, studies, and develops novel energy-conversion technologies in the Condensed Matter and Thermal Physics Group at Los Alamos National Laboratory. He is a Fellow of the American Physical Society, of the Acoustical Society of America, and of Los Alamos. He is a co-author of thermoacoustics software used worldwide, and the author of a graduate-level textbook on thermoacoustics.

George S. K. Wong

Institute for National Measurement Standards (INMS)
National Research Council Canada (NRC)
Ottawa, ON, Canada
George.Wong@nrc-cnrc.gc.ca

Chapter H.24

Dr. George Wong is an expert in acoustical metrology and works in the development of measuring techniques for the calibration of microphones, sound calibrators, sound level meters and National and international acoustical standards including ultrasound and vibration. His research area includes velocity of sound in gases and water, shock and vibration, microphone calibration and acoustical measuring techniques. He is a Distinguished International member of the Institute of Noise Control Engineering.

Eric D. Young

Johns Hopkins University
Baltimore, MD, USA
eyoung@jhu.edu

Chapter D.12

Eric Young is a Professor of Biomedical Engineering at the Johns Hopkins University. His research concerns the representation of complex stimuli in the auditory parts of the brain. This includes normal function and impaired function following acoustic trauma.

Detailed Contents

List of Abbreviations XXI

1 Introduction to Acoustics
Thomas D. Rossing 1
- 1.1 Acoustics: The Science of Sound 1
- 1.2 Sounds We Hear 1
- 1.3 Sounds We Cannot Hear: Ultrasound and Infrasound 2
- 1.4 Sounds We Would Rather Not Hear: Environmental Noise Control 2
- 1.5 Aesthetic Sound: Music 3
- 1.6 Sound of the Human Voice: Speech and Singing 3
- 1.7 How We Hear: Physiological and Psychological Acoustics 4
- 1.8 Architectural Acoustics 4
- 1.9 Harnessing Sound: Physical and Engineering Acoustics 5
- 1.10 Medical Acoustics 5
- 1.11 Sounds of the Sea 6
- **References** 6

Part A Propagation of Sound

2 A Brief History of Acoustics
Thomas D. Rossing 9
- 2.1 Acoustics in Ancient Times 9
- 2.2 Early Experiments on Vibrating Strings, Membranes and Plates 10
- 2.3 Speed of Sound in Air 10
- 2.4 Speed of Sound in Liquids and Solids 11
- 2.5 Determining Frequency 11
- 2.6 Acoustics in the 19th Century 12
  - 2.6.1 Tyndall 12
  - 2.6.2 Helmholtz 12
  - 2.6.3 Rayleigh 13
  - 2.6.4 George Stokes 13
  - 2.6.5 Alexander Graham Bell 14
  - 2.6.6 Thomas Edison 14
  - 2.6.7 Rudolph Koenig 14
- 2.7 The 20th Century 15
  - 2.7.1 Architectural Acoustics 15
  - 2.7.2 Physical Acoustics 16
  - 2.7.3 Engineering Acoustics 18
  - 2.7.4 Structural Acoustics 19
  - 2.7.5 Underwater Acoustics 19
  - 2.7.6 Physiological and Psychological Acoustics 20
  - 2.7.7 Speech 21

		2.7.8	Musical Acoustics	21
	2.8		Conclusion	23
	References			23

3 Basic Linear Acoustics
Alan D. Pierce 25

	3.1	Introduction		27
	3.2	Equations of Continuum Mechanics		28
		3.2.1	Mass, Momentum, and Energy Equations	28
		3.2.2	Newtonian Fluids and the Shear Viscosity	30
		3.2.3	Equilibrium Thermodynamics	30
		3.2.4	Bulk Viscosity and Thermal Conductivity	31
		3.2.5	Navier–Stokes–Fourier Equations	31
		3.2.6	Thermodynamic Coefficients	31
		3.2.7	Ideal Compressible Fluids	32
		3.2.8	Suspensions and Bubbly Liquids	33
		3.2.9	Elastic Solids	33
	3.3	Equations of Linear Acoustics		35
		3.3.1	The Linearization Process	35
		3.3.2	Linearized Equations for an Ideal Fluid	36
		3.3.3	The Wave Equation	36
		3.3.4	Wave Equations for Isotropic Elastic Solids	36
		3.3.5	Linearized Equations for a Viscous Fluid	37
		3.3.6	Acoustic, Entropy, and Vorticity Modes	37
		3.3.7	Boundary Conditions at Interfaces	39
	3.4	Variational Formulations		40
		3.4.1	Hamilton's Principle	40
		3.4.2	Biot's Formulation for Porous Media	42
		3.4.3	Disturbance Modes in a Biot Medium	43
	3.5	Waves of Constant Frequency		45
		3.5.1	Spectral Density	45
		3.5.2	Fourier Transforms	45
		3.5.3	Complex Number Representation	46
		3.5.4	Time Averages of Products	47
	3.6	Plane Waves		47
		3.6.1	Plane Waves in Fluids	47
		3.6.2	Plane Waves in Solids	48
	3.7	Attenuation of Sound		49
		3.7.1	Classical Absorption	49
		3.7.2	Relaxation Processes	50
		3.7.3	Continuously Distributed Relaxations	52
		3.7.4	Kramers–Krönig Relations	52
		3.7.5	Attenuation of Sound in Air	55
		3.7.6	Attenuation of Sound in Sea Water	57
	3.8	Acoustic Intensity and Power		58
		3.8.1	Energy Conservation Interpretation	58
		3.8.2	Acoustic Energy Density and Intensity	58

		3.8.3	Acoustic Power	59
		3.8.4	Rate of Energy Dissipation	59
		3.8.5	Energy Corollary for Elastic Waves	60
	3.9	Impedance		60
		3.9.1	Mechanical Impedance	60
		3.9.2	Specific Acoustic Impedance	60
		3.9.3	Characteristic Impedance	60
		3.9.4	Radiation Impedance	61
		3.9.5	Acoustic Impedance	61
	3.10	Reflection and Transmission		61
		3.10.1	Reflection at a Plane Surface	61
		3.10.2	Reflection at an Interface	62
		3.10.3	Theory of the Impedance Tube	62
		3.10.4	Transmission through Walls and Slabs	63
		3.10.5	Transmission through Limp Plates	64
		3.10.6	Transmission through Porous Blankets	64
		3.10.7	Transmission through Elastic Plates	64
	3.11	Spherical Waves		65
		3.11.1	Spherically Symmetric Outgoing Waves	65
		3.11.2	Radially Oscillating Sphere	66
		3.11.3	Transversely Oscillating Sphere	67
		3.11.4	Axially Symmetric Solutions	68
		3.11.5	Scattering by a Rigid Sphere	73
	3.12	Cylindrical Waves		75
		3.12.1	Cylindrically Symmetric Outgoing Waves	75
		3.12.2	Bessel and Hankel Functions	77
		3.12.3	Radially Oscillating Cylinder	81
		3.12.4	Transversely Oscillating Cylinder	81
	3.13	Simple Sources of Sound		82
		3.13.1	Volume Sources	82
		3.13.2	Small Piston in a Rigid Baffle	82
		3.13.3	Multiple and Distributed Sources	82
		3.13.4	Piston of Finite Size in a Rigid Baffle	83
		3.13.5	Thermoacoustic Sources	84
		3.13.6	Green's Functions	85
		3.13.7	Multipole Series	85
		3.13.8	Acoustically Compact Sources	86
		3.13.9	Spherical Harmonics	86
	3.14	Integral Equations in Acoustics		87
		3.14.1	The Helmholtz–Kirchhoff Integral	87
		3.14.2	Integral Equations for Surface Fields	88
	3.15	Waveguides, Ducts, and Resonators		89
		3.15.1	Guided Modes in a Duct	89
		3.15.2	Cylindrical Ducts	90
		3.15.3	Low-Frequency Model for Ducts	90
		3.15.4	Sound Attenuation in Ducts	91
		3.15.5	Mufflers and Acoustic Filters	92

	3.15.6	Non-Reflecting Dissipative Mufflers	93
	3.15.7	Expansion Chamber Muffler	93
	3.15.8	Helmholtz Resonators	93
3.16	Ray Acoustics		94
	3.16.1	Wavefront Propagation	94
	3.16.2	Reflected and Diffracted Rays	95
	3.16.3	Inhomogeneous Moving Media	96
	3.16.4	The Eikonal Approximation	96
	3.16.5	Rectilinear Propagation of Amplitudes	97
3.17	Diffraction		98
	3.17.1	Posing of the Diffraction Problem	98
	3.17.2	Rays and Spatial Regions	98
	3.17.3	Residual Diffracted Wave	99
	3.17.4	Solution for Diffracted Waves	102
	3.17.5	Impulse Solution	102
	3.17.6	Constant-Frequency Diffraction	103
	3.17.7	Uniform Asymptotic Solution	103
	3.17.8	Special Functions for Diffraction	104
	3.17.9	Plane Wave Diffraction	105
	3.17.10	Small-Angle Diffraction	106
	3.17.11	Thin-Screen Diffraction	107
3.18	Parabolic Equation Methods		107
References			108

4 Sound Propagation in the Atmosphere
Keith Attenborough 113

4.1	A Short History of Outdoor Acoustics		113
4.2	Applications of Outdoor Acoustics		114
4.3	Spreading Losses		115
4.4	Atmospheric Absorption		116
4.5	Diffraction and Barriers		116
	4.5.1	Single-Edge Diffraction	116
	4.5.2	Effects of the Ground on Barrier Performance	118
	4.5.3	Diffraction by Finite-Length Barriers and Buildings	119
4.6	Ground Effects		120
	4.6.1	Boundary Conditions at the Ground	120
	4.6.2	Attenuation of Spherical Acoustic Waves over the Ground	120
	4.6.3	Surface Waves	122
	4.6.4	Acoustic Impedance of Ground Surfaces	122
	4.6.5	Effects of Small-Scale Roughness	123
	4.6.6	Examples of Ground Attenuation under Weakly Refracting Conditions	124
	4.6.7	Effects of Ground Elasticity	125
4.7	Attenuation Through Trees and Foliage		129
4.8	Wind and Temperature Gradient Effects on Outdoor Sound		131
	4.8.1	Inversions and Shadow Zones	131
	4.8.2	Meteorological Classes for Outdoor Sound Propagation	133

	4.8.3	Typical Speed of Sound Profiles	135
	4.8.4	Atmospheric Turbulence Effects	138
4.9	Concluding Remarks		142
	4.9.1	Modeling Meteorological and Topographical Effects	142
	4.9.2	Effects of Trees and Tall Vegetation	142
	4.9.3	Low-Frequency Interaction with the Ground	143
	4.9.4	Rough-Sea Effects	143
	4.9.5	Predicting Outdoor Noise	143
References			143

5 Underwater Acoustics
William A. Kuperman, Philippe Roux 149

5.1	Ocean Acoustic Environment		151
	5.1.1	Ocean Environment	151
	5.1.2	Basic Acoustic Propagation Paths	152
	5.1.3	Geometric Spreading Loss	154
5.2	Physical Mechanisms		155
	5.2.1	Transducers	155
	5.2.2	Volume Attenuation	157
	5.2.3	Bottom Loss	158
	5.2.4	Scattering and Reverberation	159
	5.2.5	Ambient Noise	160
	5.2.6	Bubbles and Bubbly Media	162
5.3	SONAR and the SONAR Equation		165
	5.3.1	Detection Threshold and Receiver Operating Characteristics Curves	165
	5.3.2	Passive SONAR Equation	166
	5.3.3	Active SONAR Equation	167
5.4	Sound Propagation Models		167
	5.4.1	The Wave Equation and Boundary Conditions	168
	5.4.2	Ray Theory	168
	5.4.3	Wavenumber Representation or Spectral Solution	169
	5.4.4	Normal-Mode Model	169
	5.4.5	Parabolic Equation (PE) Model	172
	5.4.6	Propagation and Transmission Loss	174
	5.4.7	Fourier Synthesis of Frequency-Domain Solutions	175
5.5	Quantitative Description of Propagation		177
5.6	SONAR Array Processing		179
	5.6.1	Linear Plane-Wave Beam-Forming and Spatio-Temporal Sampling	179
	5.6.2	Some Beam-Former Properties	181
	5.6.3	Adaptive Processing	182
	5.6.4	Matched Field Processing, Phase Conjugation and Time Reversal	182
5.7	Active SONAR Processing		185
	5.7.1	Active SONAR Signal Processing	185
	5.7.2	Underwater Acoustic Imaging	187

		5.7.3	Acoustic Telemetry	191
		5.7.4	Travel-Time Tomography	192
	5.8	Acoustics and Marine Animals		195
		5.8.1	Fisheries Acoustics	195
		5.8.2	Marine Mammal Acoustics	198
	5.A	Appendix: Units		201
	References			201

Part B Physical and Nonlinear Acoustics

6 Physical Acoustics
Mack A. Breazeale, Michael McPherson 207

	6.1	Theoretical Overview		209
		6.1.1	Basic Wave Concepts	209
		6.1.2	Properties of Waves	210
		6.1.3	Wave Propagation in Fluids	215
		6.1.4	Wave Propagation in Solids	217
		6.1.5	Attenuation	218
	6.2	Applications of Physical Acoustics		219
		6.2.1	Crystalline Elastic Constants	219
		6.2.2	Resonant Ultrasound Spectroscopy (RUS)	220
		6.2.3	Measurement Of Attenuation (Classical Approach)	221
		6.2.4	Acoustic Levitation	222
		6.2.5	Sonoluminescence	222
		6.2.6	Thermoacoustic Engines (Refrigerators and Prime Movers)	223
		6.2.7	Acoustic Detection of Land Mines	224
		6.2.8	Medical Ultrasonography	224
	6.3	Apparatus		226
		6.3.1	Examples of Apparatus	226
		6.3.2	Piezoelectricity and Transduction	226
		6.3.3	Schlieren Imaging	228
		6.3.4	Goniometer System	230
		6.3.5	Capacitive Receiver	231
	6.4	Surface Acoustic Waves		231
	6.5	Nonlinear Acoustics		234
		6.5.1	Nonlinearity of Fluids	234
		6.5.2	Nonlinearity of Solids	235
		6.5.3	Comparison of Fluids and Solids	236
	References			237

7 Thermoacoustics
Gregory W. Swift 239

7.1	History		239
7.2	Shared Concepts		240
	7.2.1	Pressure and Velocity	240
	7.2.2	Power	243

	7.3	Engines	244
		7.3.1 Standing-Wave Engines	244
		7.3.2 Traveling-Wave Engines	246
		7.3.3 Combustion	248
	7.4	Dissipation	249
	7.5	Refrigeration	250
		7.5.1 Standing-Wave Refrigeration	250
		7.5.2 Traveling-Wave Refrigeration	251
	7.6	Mixture Separation	253
	References		254

8 Nonlinear Acoustics in Fluids
Werner Lauterborn, Thomas Kurz, Iskander Akhatov 257

	8.1	Origin of Nonlinearity	258
	8.2	Equation of State	259
	8.3	The Nonlinearity Parameter B/A	260
	8.4	The Coefficient of Nonlinearity β	262
	8.5	Simple Nonlinear Waves	263
	8.6	Lossless Finite-Amplitude Acoustic Waves	264
	8.7	Thermoviscous Finite-Amplitude Acoustic Waves	268
	8.8	Shock Waves	271
	8.9	Interaction of Nonlinear Waves	273
	8.10	Bubbly Liquids	275
		8.10.1 Incompressible Liquids	276
		8.10.2 Compressible Liquids	278
		8.10.3 Low-Frequency Waves: The Korteweg–de Vries Equation	279
		8.10.4 Envelopes of Wave Trains: The Nonlinear Schrödinger Equation	282
		8.10.5 Interaction of Nonlinear Waves. Sound–Ultrasound Interaction	284
	8.11	Sonoluminescence	286
	8.12	Acoustic Chaos	289
		8.12.1 Methods of Chaos Physics	289
		8.12.2 Chaotic Sound Waves	291
	References		293

Part C Architectural Acoustics

9 Acoustics in Halls for Speech and Music
Anders Christian Gade 301

	9.1	Room Acoustic Concepts	302
	9.2	Subjective Room Acoustics	303
		9.2.1 The Impulse Response	303
		9.2.2 Subjective Room Acoustic Experiment Techniques	303
		9.2.3 Subjective Effects of Audible Reflections	305

9.3	Subjective and Objective Room Acoustic Parameters	306
	9.3.1 Reverberation Time	306
	9.3.2 Clarity	308
	9.3.3 Sound Strength	308
	9.3.4 Measures of Spaciousness	309
	9.3.5 Parameters Relating to Timbre or Tonal Color	310
	9.3.6 Measures of Conditions for Performers	310
	9.3.7 Speech Intelligibility	311
	9.3.8 Isn't One Objective Parameter Enough?	312
	9.3.9 Recommended Values of Objective Parameters	313
9.4	Measurement of Objective Parameters	314
	9.4.1 The Schroeder Method for the Measurement of Decay Curves	314
	9.4.2 Frequency Range of Measurements	314
	9.4.3 Sound Sources	315
	9.4.4 Microphones	315
	9.4.5 Signal Storage and Processing	315
9.5	Prediction of Room Acoustic Parameters	316
	9.5.1 Prediction of Reverberation Time by Means of Classical Reverberation Theory	316
	9.5.2 Prediction of Reverberation in Coupled Rooms	318
	9.5.3 Absorption Data for Seats and Audiences	319
	9.5.4 Prediction by Computer Simulations	320
	9.5.5 Scale Model Predictions	321
	9.5.6 Prediction from Empirical Data	322
9.6	Geometric Design Considerations	323
	9.6.1 General Room Shape and Seating Layout	323
	9.6.2 Seating Arrangement in Section	326
	9.6.3 Balcony Design	327
	9.6.4 Volume and Ceiling Height	328
	9.6.5 Main Dimensions and Risks of Echoes	329
	9.6.6 Room Shape Details Causing Risks of Focusing and Flutter	329
	9.6.7 Cultivating Early Reflections	330
	9.6.8 Suspended Reflectors	331
	9.6.9 Sound-Diffusing Surfaces	333
9.7	Room Acoustic Design of Auditoria for Specific Purposes	334
	9.7.1 Speech Auditoria, Drama Theaters and Lecture Halls	334
	9.7.2 Opera Halls	335
	9.7.3 Concert Halls for Classical Music	338
	9.7.4 Multipurpose Halls	342
	9.7.5 Halls for Rhythmic Music	344
	9.7.6 Worship Spaces/Churches	346
9.8	Sound Systems for Auditoria	346
	9.8.1 PA Systems	346
	9.8.2 Reverberation-Enhancement Systems	348
References		349

10 Concert Hall Acoustics Based on Subjective Preference Theory
Yoichi Ando 351
- 10.1 Theory of Subjective Preference for the Sound Field 353
  - 10.1.1 Sound Fields with a Single Reflection 353
  - 10.1.2 Optimal Conditions Maximizing Subjective Preference 356
  - 10.1.3 Theory of Subjective Preference for the Sound Field 357
  - 10.1.4 Auditory Temporal Window for ACF and IACF Processing ... 360
  - 10.1.5 Specialization of Cerebral Hemispheres for Temporal and Spatial Factors of the Sound Field 360
- 10.2 Design Studies 361
  - 10.2.1 Study of a Space-Form Design by Genetic Algorithms (GA) . 361
  - 10.2.2 Actual Design Studies 365
- 10.3 Individual Preferences of a Listener and a Performer 370
  - 10.3.1 Individual Subjective Preference of Each Listener 370
  - 10.3.2 Individual Subjective Preference of Each Cellist 374
- 10.4 Acoustical Measurements of the Sound Fields in Rooms 377
  - 10.4.1 Acoustic Test Techniques 377
  - 10.4.2 Subjective Preference Test in an Existing Hall 380
  - 10.4.3 Conclusions 383
- **References** 384

11 Building Acoustics
James Cowan 387
- 11.1 Room Acoustics 387
  - 11.1.1 Room Modes 388
  - 11.1.2 Sound Fields in Rooms 389
  - 11.1.3 Sound Absorption 390
  - 11.1.4 Reverberation 394
  - 11.1.5 Effects of Room Shapes 394
  - 11.1.6 Sound Insulation 395
- 11.2 General Noise Reduction Methods 400
  - 11.2.1 Space Planning 400
  - 11.2.2 Enclosures 400
  - 11.2.3 Barriers 402
  - 11.2.4 Mufflers 402
  - 11.2.5 Absorptive Treatment 402
  - 11.2.6 Direct Impact and Vibration Isolation 402
  - 11.2.7 Active Noise Control 402
  - 11.2.8 Masking 403
- 11.3 Noise Ratings for Steady Background Sound Levels 403
- 11.4 Noise Sources in Buildings 405
  - 11.4.1 HVAC Systems 405
  - 11.4.2 Plumbing Systems 406
  - 11.4.3 Electrical Systems 406
  - 11.4.4 Exterior Sources 406
- 11.5 Noise Control Methods for Building Systems 407
  - 11.5.1 Walls, Floor/Ceilings, Window and Door Assemblies 407

		11.5.2	HVAC Systems	412
		11.5.3	Plumbing Systems	415
		11.5.4	Electrical Systems	416
		11.5.5	Exterior Sources	417
	11.6	Acoustical Privacy in Buildings		419
		11.6.1	Office Acoustics Concerns	419
		11.6.2	Metrics for Speech Privacy	419
		11.6.3	Fully Enclosed Offices	422
		11.6.4	Open-Plan Offices	422
	11.7	Relevant Standards		424
	References			425

Part D Hearing and Signal Processing

12 Physiological Acoustics

Eric D. Young 429

	12.1	The External and Middle Ear		429
		12.1.1	External Ear	429
		12.1.2	Middle Ear	432
	12.2	Cochlea		434
		12.2.1	Anatomy of the Cochlea	434
		12.2.2	Basilar-Membrane Vibration and Frequency Analysis in the Cochlea	436
		12.2.3	Representation of Sound in the Auditory Nerve	441
		12.2.4	Hair Cells	443
	12.3	Auditory Nerve and Central Nervous System		449
		12.3.1	AN Responses to Complex Stimuli	449
		12.3.2	Tasks of the Central Auditory System	451
	12.4	Summary		452
	References			453

13 Psychoacoustics

Brian C. J. Moore 459

	13.1	Absolute Thresholds		460
	13.2	Frequency Selectivity and Masking		461
		13.2.1	The Concept of the Auditory Filter	462
		13.2.2	Psychophysical Tuning Curves	462
		13.2.3	The Notched-Noise Method	463
		13.2.4	Masking Patterns and Excitation Patterns	464
		13.2.5	Forward Masking	465
		13.2.6	Hearing Out Partials in Complex Tones	467
	13.3	Loudness		468
		13.3.1	Loudness Level and Equal-Loudness Contours	468
		13.3.2	The Scaling of Loudness	469
		13.3.3	Neural Coding and Modeling of Loudness	469
		13.3.4	The Effect of Bandwidth on Loudness	470
		13.3.5	Intensity Discrimination	472

13.4		Temporal Processing in the Auditory System	473
	13.4.1	Temporal Resolution Based on Within-Channel Processes	473
	13.4.2	Modeling Temporal Resolution	474
	13.4.3	A Modulation Filter Bank?	475
	13.4.4	Duration Discrimination	476
	13.4.5	Temporal Analysis Based on Across-Channel Processes	476
13.5		Pitch Perception	477
	13.5.1	Theories of Pitch Perception	477
	13.5.2	The Perception of the Pitch of Pure Tones	478
	13.5.3	The Perception of the Pitch of Complex Tones	480
13.6		Timbre Perception	483
	13.6.1	Time-Invariant Patterns and Timbre	483
	13.6.2	Time-Varying Patterns and Auditory Object Identification	483
13.7		The Localization of Sounds	484
	13.7.1	Binaural Cues	484
	13.7.2	The Role of the Pinna and Torso	485
	13.7.3	The Precedence Effect	485
13.8		Auditory Scene Analysis	485
	13.8.1	Information Used to Separate Auditory Objects	486
	13.8.2	The Perception of Sequences of Sounds	489
	13.8.3	General Principles of Perceptual Organization	492
13.9		Further Reading and Supplementary Materials	494
References			495

14 Acoustic Signal Processing

William M. Hartmann 503

14.1		Definitions	504
14.2		Fourier Series	505
	14.2.1	The Spectrum	506
	14.2.2	Symmetry	506
14.3		Fourier Transform	507
	14.3.1	Examples	508
	14.3.2	Time-Shifted Function	509
	14.3.3	Derivatives and Integrals	509
	14.3.4	Products and Convolution	509
14.4		Power, Energy, and Power Spectrum	510
	14.4.1	Autocorrelation	510
	14.4.2	Cross-Correlation	511
14.5		Statistics	511
	14.5.1	Signals and Processes	512
	14.5.2	Distributions	512
	14.5.3	Multivariate Distributions	513
	14.5.4	Moments	513
14.6		Hilbert Transform and the Envelope	514
	14.6.1	The Analytic Signal	514
14.7		Filters	515
	14.7.1	One-Pole Low-Pass Filter	515

		14.7.2 Phase Delay and Group Delay	516
		14.7.3 Resonant Filters	516
		14.7.4 Impulse Response	516
		14.7.5 Dispersion Relations	516
	14.8	The Cepstrum	517
	14.9	Noise	518
		14.9.1 Thermal Noise	518
		14.9.2 Gaussian Noise	519
		14.9.3 Band-Limited Noise	519
		14.9.4 Generating Noise	519
		14.9.5 Equal-Amplitude Random-Phase Noise	520
		14.9.6 Noise Color	520
	14.10	Sampled data	520
		14.10.1 Quantization and Quantization Noise	520
		14.10.2 Binary Representation	520
		14.10.3 Sampling Operation	521
		14.10.4 Digital-to-Analog Conversion	521
		14.10.5 The Sampled Signal	522
		14.10.6 Interpolation	522
	14.11	Discrete Fourier Transform	522
		14.11.1 Interpolation for the Spectrum	523
	14.12	The z-Transform	524
		14.12.1 Transfer Function	525
	14.13	Maximum Length Sequences	526
		14.13.1 The MLS as a Signal	527
		14.13.2 Application of the MLS	527
		14.13.3 Long Sequences	527
	14.14	Information Theory	528
		14.14.1 Shannon Entropy	529
		14.14.2 Mutual Information	530
	References		530

Part E Music, Speech, Electroacoustics

15 Musical Acoustics
Colin Gough 533

15.1	Vibrational Modes of Instruments		535
	15.1.1	Normal Modes	535
	15.1.2	Radiation from Instruments	537
	15.1.3	The Anatomy of Musical Sounds	540
	15.1.4	Perception and Psychoacoustics	551
15.2	Stringed Instruments		554
	15.2.1	String Vibrations	555
	15.2.2	Nonlinear String Vibrations	563
	15.2.3	The Bowed String	566
	15.2.4	Bridge and Soundpost	570

		15.2.5	String–Bridge–Body Coupling	575
		15.2.6	Body Modes	581
		15.2.7	Measurements	594
		15.2.8	Radiation and Sound Quality	598
	15.3	Wind Instruments		601
		15.3.1	Resonances in Cylindrical Tubes	602
		15.3.2	Non-Cylindrical Tubes	606
		15.3.3	Reed Excitation	619
		15.3.4	Brass-Mouthpiece Excitation	628
		15.3.5	Air-Jet Excitation	633
		15.3.6	Woodwind and Brass Instruments	637
	15.4	Percussion Instruments		641
		15.4.1	Membranes	642
		15.4.2	Bars	648
		15.4.3	Plates	652
		15.4.4	Shells	658
	References			661

16 The Human Voice in Speech and Singing
Björn Lindblom, Johan Sundberg 669

16.1	Breathing	669
16.2	The Glottal Sound Source	676
16.3	The Vocal Tract Filter	682
16.4	Articulatory Processes, Vowels and Consonants	687
16.5	The Syllable	695
16.6	Rhythm and Timing	699
16.7	Prosody and Speech Dynamics	701
16.8	Control of Sound in Speech and Singing	703
16.9	The Expressive Power of the Human Voice	706
References		706

17 Computer Music
Perry R. Cook 713

17.1	Computer Audio Basics	714
17.2	Pulse Code Modulation Synthesis	717
17.3	Additive (Fourier, Sinusoidal) Synthesis	719
17.4	Modal (Damped Sinusoidal) Synthesis	722
17.5	Subtractive (Source-Filter) Synthesis	724
17.6	Frequency Modulation (FM) Synthesis	727
17.7	FOFs, Wavelets, and Grains	728
17.8	Physical Modeling (The Wave Equation)	730
17.9	Music Description and Control	735
17.10	Composition	737
17.11	Controllers and Performance Systems	737
17.12	Music Understanding and Modeling by Computer	738
17.13	Conclusions, and the Future	740
References		740

18 Audio and Electroacoustics
Mark F. Davis 743
- 18.1 Historical Review 744
  - 18.1.1 Spatial Audio History 746
- 18.2 The Psychoacoustics of Audio and Electroacoustics 747
  - 18.2.1 Frequency Response 747
  - 18.2.2 Amplitude (Loudness) 748
  - 18.2.3 Timing 749
  - 18.2.4 Spatial Acuity 750
- 18.3 Audio Specifications 751
  - 18.3.1 Bandwidth 752
  - 18.3.2 Amplitude Response Variation 753
  - 18.3.3 Phase Response 753
  - 18.3.4 Harmonic Distortion 754
  - 18.3.5 Intermodulation Distortion 755
  - 18.3.6 Speed Accuracy 755
  - 18.3.7 Noise 756
  - 18.3.8 Dynamic Range 756
- 18.4 Audio Components 757
  - 18.4.1 Microphones 757
  - 18.4.2 Records and Phonograph Cartridges 761
  - 18.4.3 Loudspeakers 763
  - 18.4.4 Amplifiers 766
  - 18.4.5 Magnetic and Optical Media 767
  - 18.4.6 Radio 768
- 18.5 Digital Audio 768
  - 18.5.1 Digital Signal Processing 770
  - 18.5.2 Audio Coding 771
- 18.6 Complete Audio Systems 775
  - 18.6.1 Monaural 776
  - 18.6.2 Stereo 776
  - 18.6.3 Binaural 777
  - 18.6.4 Ambisonics 777
  - 18.6.5 5.1-Channel Surround 777
- 18.7 Appraisal and Speculation 778
- **References** 778

Part F Biological and Medical Acoustics

19 Animal Bioacoustics
Neville H. Fletcher 785
- 19.1 Optimized Communication 785
- 19.2 Hearing and Sound Production 787
- 19.3 Vibrational Communication 788
- 19.4 Insects 788
- 19.5 Land Vertebrates 790

19.6	Birds		795
19.7	Bats		796
19.8	Aquatic Animals		797
19.9	Generalities		799
19.10	Quantitative System Analysis		799
References			802

20 Cetacean Acoustics
Whitlow W. L. Au, Marc O. Lammers 805

20.1	Hearing in Cetaceans		806
	20.1.1	Hearing Sensitivity of Odontocetes	807
	20.1.2	Directional Hearing in Dolphins	808
	20.1.3	Hearing by Mysticetes	812
20.2	Echolocation Signals		813
	20.2.1	Echolocation Signals of Dolphins that also Whistle	813
	20.2.2	Echolocation Signals of Smaller Odontocetes that Do not Whistle	817
	20.2.3	Transmission Beam Pattern	819
20.3	Odontocete Acoustic Communication		821
	20.3.1	Social Acoustic Signals	821
	20.3.2	Signal Design Characteristics	823
20.4	Acoustic Signals of Mysticetes		827
	20.4.1	Songs of Mysticete Whales	827
20.5	Discussion		830
References			831

21 Medical Acoustics
Kirk W. Beach, Barbrina Dunmire 839

21.1	Introduction to Medical Acoustics		841
21.2	Medical Diagnosis; Physical Examination		842
	21.2.1	Auscultation – Listening for Sounds	842
	21.2.2	Phonation and Auscultation	847
	21.2.3	Percussion	847
21.3	Basic Physics of Ultrasound Propagation in Tissue		848
	21.3.1	Reflection of Normal-Angle-Incident Ultrasound	850
	21.3.2	Acute-Angle Reflection of Ultrasound	850
	21.3.3	Diagnostic Ultrasound Propagation in Tissue	851
	21.3.4	Amplitude of Ultrasound Echoes	851
	21.3.5	Fresnel Zone (Near Field), Transition Zone, and Fraunhofer Zone (Far Field)	853
	21.3.6	Measurement of Ultrasound Wavelength	855
	21.3.7	Attenuation of Ultrasound	855
21.4	Methods of Medical Ultrasound Examination		857
	21.4.1	Continuous-Wave Doppler Systems	857
	21.4.2	Pulse-Echo Backscatter Systems	859
	21.4.3	B-mode Imaging Instruments	862

21.5		Medical Contrast Agents	882
	21.5.1	Ultrasound Contrast Agents	883
	21.5.2	Stability of Large Bubbles	883
	21.5.3	Agitated Saline and Patent Foramen Ovale (PFO)	884
	21.5.4	Ultrasound Contrast Agent Motivation	886
	21.5.5	Ultrasound Contrast Agent Development	886
	21.5.6	Interactions Between Ultrasound and Microbubbles	886
	21.5.7	Bubble Destruction	887
21.6		Ultrasound Hyperthermia in Physical Therapy	889
21.7		High-Intensity Focused Ultrasound (HIFU) in Surgery	890
21.8		Lithotripsy of Kidney Stones	891
21.9		Thrombolysis	892
21.10		Lower-Frequency Therapies	892
21.11		Ultrasound Safety	892
References			**895**

Part G Structural Acoustics and Noise

22 Structural Acoustics and Vibrations
Antoine Chaigne 901

22.1		Dynamics of the Linear Single-Degree-of-Freedom (1-DOF) Oscillator	903
	22.1.1	General Solution	903
	22.1.2	Free Vibrations	903
	22.1.3	Impulse Response and Green's Function	904
	22.1.4	Harmonic Excitation	904
	22.1.5	Energetic Approach	905
	22.1.6	Mechanical Power	905
	22.1.7	Single-DOF Structural–Acoustic System	906
	22.1.8	Application: Accelerometer	907
22.2		Discrete Systems	907
	22.2.1	Lagrange Equations	907
	22.2.2	Eigenmodes and Eigenfrequencies	909
	22.2.3	Admittances	909
	22.2.4	Example: 2-DOF Plate–Cavity Coupling	911
	22.2.5	Statistical Energy Analysis	912
22.3		Strings and Membranes	913
	22.3.1	Equations of Motion	913
	22.3.2	Heterogeneous String. Modal Approach	914
	22.3.3	Ideal String	916
	22.3.4	Circular Membrane in Vacuo	919
22.4		Bars, Plates and Shells	920
	22.4.1	Longitudinal Vibrations of Bars	920
	22.4.2	Flexural Vibrations of Beams	920
	22.4.3	Flexural Vibrations of Thin Plates	923
	22.4.4	Vibrations of Thin Shallow Spherical Shells	925

	22.4.5	Combinations of Elementary Structures	926
22.5	Structural–Acoustic Coupling		926
	22.5.1	Longitudinally Vibrating Bar Coupled to an External Fluid	927
	22.5.2	Energetic Approach to Structural–Acoustic Systems	932
	22.5.3	Oscillator Coupled to a Tube of Finite Length	934
	22.5.4	Two-Dimensional Elasto–Acoustic Coupling	936
22.6	Damping		940
	22.6.1	Modal Projection in Damped Systems	940
	22.6.2	Damping Mechanisms in Plates	943
	22.6.3	Friction	945
	22.6.4	Hysteretic Damping	947
22.7	Nonlinear Vibrations		947
	22.7.1	Example of a Nonlinear Oscillator	947
	22.7.2	Duffing Equation	949
	22.7.3	Coupled Nonlinear Oscillators	951
	22.7.4	Nonlinear Vibrations of Strings	955
	22.7.5	Review of Nonlinear Equations for Other Continuous Systems	956
22.8	Conclusion. Advanced Topics		957
References			958

23 Noise

George C. Maling, Jr. 961

	23.0.1	The Source–Path–Receiver Model	961
	23.0.2	Properties of Sound Waves	962
	23.0.3	Radiation Efficiency	963
	23.0.4	Sound Pressure Level of Common Sounds	965
23.1	Instruments for Noise Measurements		965
	23.1.1	Introduction	965
	23.1.2	Sound Level	966
	23.1.3	Sound Exposure and Sound Exposure Level	967
	23.1.4	Frequency Weightings	967
	23.1.5	Octave and One-Third-Octave Bands	967
	23.1.6	Sound Level Meters	968
	23.1.7	Multichannel Instruments	969
	23.1.8	Sound Intensity Analyzers	969
	23.1.9	FFT Analyzers	969
23.2	Noise Sources		970
	23.2.1	Measures of Noise Emission	970
	23.2.2	International Standards for the Determination of Sound Power	973
	23.2.3	Emission Sound Pressure Level	977
	23.2.4	Other Noise Emission Standards	978
	23.2.5	Criteria for Noise Emissions	979
	23.2.6	Principles of Noise Control	981
	23.2.7	Noise From Stationary Sources	984
	23.2.8	Noise from Moving Sources	987

23.3	Propagation Paths		991
	23.3.1	Sound Propagation Outdoors	991
	23.3.2	Sound Propagation Indoors	993
	23.3.3	Sound-Absorptive Materials	995
	23.3.4	Ducts and Silencers	998
23.4	Noise and the Receiver		999
	23.4.1	Soundscapes	999
	23.4.2	Noise Metrics	1000
	23.4.3	Measurement of Immission Sound Pressure Level	1000
	23.4.4	Criteria for Noise Immission	1000
	23.4.5	Sound Quality	1004
23.5	Regulations and Policy for Noise Control		1006
	23.5.1	United States Noise Policies and Regulations	1006
	23.5.2	European Noise Policy and Regulations	1009
23.6	Other Information Resources		1010
References			1010

Part H Engineering Acoustics

24 Microphones and Their Calibration
George S. K. Wong 1021

24.1	Historic References on Condenser Microphones and Calibration		1024
24.2	Theory		1024
	24.2.1	Diaphragm Deflection	1024
	24.2.2	Open-Circuit Voltage and Electrical Transfer Impedance	1024
	24.2.3	Mechanical Response	1025
24.3	Reciprocity Pressure Calibration		1026
	24.3.1	Introduction	1026
	24.3.2	Theoretical Considerations	1026
	24.3.3	Practical Considerations	1027
24.4	Corrections		1029
	24.4.1	Heat Conduction Correction	1029
	24.4.2	Equivalent Volume	1031
	24.4.3	Capillary Tube Correction	1032
	24.4.4	Cylindrical Couplers and Wave-Motion Correction	1034
	24.4.5	Barometric Pressure Correction	1035
	24.4.6	Temperature Correction	1037
	24.4.7	Microphone Sensitivity Equations	1038
	24.4.8	Uncertainty on Pressure Sensitivity Level	1038
24.5	Free-Field Microphone Calibration		1039
24.6	Comparison Methods for Microphone Calibration		1039
	24.6.1	Interchange Microphone Method of Comparison	1039
	24.6.2	Comparison Method with a Calibrator	1040
	24.6.3	Comparison Pressure and Free-Field Calibrations	1042
	24.6.4	Comparison Method with a Precision Attenuator	1042
24.7	Frequency Response Measurement with Electrostatic Actuators		1043

	24.8	Overall View on Microphone Calibration	1043
	24.A	Acoustic Transfer Impedance Evaluation	1045
	24.B	Physical Properties of Air	1045
		24.B.1 Density of Humid Air	1045
		24.B.2 Computation of the Speed of Sound in Air	1046
		24.B.3 Ratio of Specific Heats of Air	1047
		24.B.4 Viscosity and Thermal Diffusivity of Air for Capillary Correction	1047
	References		1048

25 Sound Intensity
Finn Jacobsen 1053
- 25.1 Conservation of Sound Energy 1054
- 25.2 Active and Reactive Sound Fields 1055
- 25.3 Measurement of Sound Intensity 1058
  - 25.3.1 The $p-p$ Measurement Principle 1058
  - 25.3.2 The $p-u$ Measurement Principle 1066
  - 25.3.3 Sound Field Indicators 1068
- 25.4 Applications of Sound Intensity 1068
  - 25.4.1 Noise Source Identification 1068
  - 25.4.2 Sound Power Determination 1070
  - 25.4.3 Radiation Efficiency of Structures 1071
  - 25.4.4 Transmission Loss of Structures and Partitions 1071
  - 25.4.5 Other Applications 1072
- **References** 1072

26 Acoustic Holography
Yang-Hann Kim 1077
- 26.1 The Methodology of Acoustic Source Identification 1077
- 26.2 Acoustic Holography: Measurement, Prediction and Analysis ... 1079
  - 26.2.1 Introduction and Problem Definitions 1079
  - 26.2.2 Prediction Process 1080
  - 26.2.3 Measurement 1083
  - 26.2.4 Analysis of Acoustic Holography 1089
- 26.3 Summary 1092
- 26.A Mathematical Derivations of Three Acoustic Holography Methods and Their Discrete Forms 1092
  - 26.A.1 Planar Acoustic Holography 1092
  - 26.A.2 Cylindrical Acoustic Holography 1094
  - 26.A.3 Spherical Acoustic Holography 1095
- **References** 1095

27 Optical Methods for Acoustics and Vibration Measurements
Nils-Erik Molin 1101
- 27.1 Introduction 1101
  - 27.1.1 Chladni Patterns, Phase-Contrast Methods, Schlieren, Shadowgraph 1101

	27.1.2	Holographic Interferometry, Acoustical Holography	1102
	27.1.3	Speckle Metrology: Speckle Interferometry and Speckle Photography	1102
	27.1.4	Moiré Techniques	1104
	27.1.5	Some Pros and Cons of Optical Metrology	1104
27.2	Measurement Principles and Some Applications		1105
	27.2.1	Holographic Interferometry for the Study of Vibrations	1105
	27.2.2	Speckle Interferometry – TV Holography, DSPI and ESPI for Vibration Analysis and for Studies of Acoustic Waves	1108
	27.2.3	Reciprocity and TV Holography	1113
	27.2.4	Pulsed TV Holography – Pulsed Lasers Freeze Propagating Bending Waves, Sound Fields and Other Transient Events	1114
	27.2.5	Scanning Vibrometry – for Vibration Analysis and for the Study of Acoustic Waves	1116
	27.2.6	Digital Speckle Photography (DSP), Correlation Methods and Particle Image Velocimetry (PIV)	1119
27.3	Summary		1122
References			1123

28 Modal Analysis
Thomas D. Rossing 1127

28.1	Modes of Vibration		1127
28.2	Experimental Modal Testing		1128
	28.2.1	Frequency Response Function	1128
	28.2.2	Impact Testing	1130
	28.2.3	Shaker Testing	1131
	28.2.4	Obtaining Modal Parameters	1132
	28.2.5	Real and Complex Modes	1133
	28.2.6	Graphical Representation	1133
28.3	Mathematical Modal Analysis		1133
	28.3.1	Finite-Element Analysis	1134
	28.3.2	Boundary-Element Methods	1134
	28.3.3	Finite-Element Correlation	1135
28.4	Sound-Field Analysis		1136
28.5	Holographic Modal Analysis		1137
References			1138

Acknowledgements 1139
About the Authors 1141
Detailed Contents 1147
Subject Index 1167

Subject Index

3 dB bandwidth   463

A

abdominal muscles   670
ABR (auditory brainstem responses)   353
absolute threshold   460
absorption   390
– coefficient   390
– low frequency   392
AC (articulation class)   421
accelerometer   907
ACF (autocorrelation function)   351, 352
acoustic
– cavitation   292
– chaos   289
– distance   1059
– Doppler current profiler (ADCP)   187
– fingerprints   596
– holography   1079
– impedance   61
– impedance of ground surfaces   123
– intensity   208
– source identification   1077
– telemetry   191
– transfer impedance   1045
– trauma   447
– tube   733
acoustic system
– biological   785, 802
acoustic waves
– types of   208
acoustical
– efficiency   906
– holography   1102
– horn   431
– resistance   906, 927
acoustically hard/soft   120
acoustically neutral   134, 135
acoustics
– 19th century   12
– 20th century   15
– architectural   15, 299
– biological   785, 802
– engineering   18, 1019
– history   9
– musical   21, 22, 531

– physical   16, 205
– physiological   20, 429
– psychological   20, 459
– speech   21, 669
– structural   19, 901
– underwater   19, 149
actin   443
action potentials   441
active cochlea   439
active intensity   1055
active processing   185
adaptation motor   444, 445
ADC (analog-to-digital converter)   520, 714, 769, 1131
ADCP (acoustic Doppler current profiler)   187
additive synthesis   719
adiabatic mode theory   170
admittance
– local   536
– matrix   910
– non-local   536
– simple harmonic oscillator   535
– tensor   579
– violin   594
admittance measurement
– violin   595
ADP (ammonium dihydrogen phosphate)   18
AF (audio frequency)   859
afferent   435, 436
AFSUMB (Asian Federation for Societies of Ultrasound in Medicine and Biology)   895
AI (articulation index)   419
air absorption   116
air coupled Rayleigh waves   129
airflow   1063, 1068
air-jet
– coupling to air column   633
– modelling   633
– Rayleigh instability   633
– vorticity   634
air-jet resonator
– modelling   634
– mouthpiece impedance   635
air-moving devices (AMDs)   984
AIUM (American Institute of Ultrasound in Medicine)   894
ALARA (as low as reasonably achievable)   895

alias   521
aliasing   715
all-pass filter   734
AM (amplitude modulation)   768
American Institute of Ultrasound in Medicine (AIUM)   894
American National Standards Institute (ANSI)   1008
ammonium dihydrogen phosphate (ADP)   18
amplitude   504
– modulation (AM)   768
– wave   210
AN (auditory nerve)   434
AN fiber   441
AN population   449
analog
– electric network   789
– network   800–802
analog signal   521
– spectrum   521
analog-to-digital converter (ADC)   520, 714, 769, 1131
analytic listening   482
analytic signal   514
anatomy
– of sounds   540
– vocal   795
animal   785, 793, 802
– air breathing   787
– aquatic   797
– auditory system   794
– bioacoustics   785, 802
– hearing   785, 793, 802
– hearing range   787
– sound power   787
– sound production   785, 787, 802
– vocalization frequency   785, 786
anion transporters   448
ANSI (American National Standards Institute)   1008
ANSI standard   1058
anti-resonances   726
apparent source width (ASW)   309
AR (assisted resonance)   348
architectural acoustics   4
area function   687
articulation class (AC)   421
articulation index (AI)   419
articulatory processes   687

as low as reasonably achievable (ALARA) 895
assisted resonance (AR) 348
ASUM (Australasian Society for Ultrasound in Medicine) 895
ASW (apparent source width) 309
asynchrony detection 476
atmospheric instability
– shear and bouyancy 138
atmospheric stability 133
atmospheric turbulence
– Bragg reflection 139
– von Karman spectrum 140
attenuation 219
– atmospheric 786
– measurement of 221
– through trees 129
audibility of partials 467
audio classification 738
audio feature extraction 738
audio frequency (AF) 859
audiogram 461
auditory brainstem responses (ABR) 353
auditory filter 462
auditory grouping 485
auditory nerve (AN) 434
auditory scene analysis 485
auditory system
– animal 794
auditory temporal window 360
autocorrelation 725
autocorrelation function (ACF) 351, 352, 510, 527
– discrete samples 527
AUV (automated underwater vehicle) 191

B

backward masking 465, 474
backward propagation 1083
balanced noise criterion (NCB) curves 404
banded waveguides 734
bandlimiting 716
bandwidth 715
bandwidth and loudness 470
bandwidth-time response 547
bar vibrations 648
– celeste 649
– marimba 649, 650
– triangle 651
– vibraphone 649, 650
– xylophone 649, 650
Bark scale 464

barrier attenuation 117
barrier edge diffraction
– Maekawa's formula 117
barrier insertion loss 118
barriers 402
barriers of finite length 119
bars 920
basilar membrane 436–438, 440, 447
basis function 1077
bass bar 575
bass drum 648
bass ratio (BR) 310
bassoon 622, 637
bats 796
– echo-location 796
Bayes's theorem 513
BB (bite block) 704
beam nonuniformity ratio (BNR) 889
beam-forming method 1078
beam-forming properties 181
beams
– flexural vibrations 920
– free–free 921
– nonlinear vibrations 956
– prestressed 922
– with variable cross section 922
beats 214
Bell, Alexander Graham 14
bells 658
– FEA 659
– holograms 659
– mode degeneracy 661
– mode nomenclature 659
– modes 659
– non-axial symmetry 661
– tuning 659
BEM (boundary-element method) 1134
BER (bit error rate) 193
Bernoulli pressure 621
Bessel functions 919, 925
– modified 925
best frequency (BF) 438
BF (best frequency) 438
bias error index 1062
binaural cues for localization 484
binaural impulse response 378
binaural interaction 452
binaural listening level (LL) 351
bioacoustics 785
Biot theory for dynamic poroelasticity 126
bird
– auditory system 793

– sound power 787
– vocal anatomy 795
birdsong 795, 796
– formant 795, 796
– pure tone 796
bit error rate (BER) 193
bite block (BB) 704
BMUS (British Medical Ultrasound Society) 895
body modes 581
– admittance measurement 581
– collective motions 581
Boehm key system 638
BoSSA (bowed sensor speaker array) 737
bottom loss 157
boundary conditions 723
boundary-element method (BEM) 1134
boundary-layer limit 243, 250
bowed sensor speaker array (BoSSA) 737
bowed string 732
– bow position 566
– bow pressure 566
– bow speed 566
– bowing machine 570
– computer modelling 568
– flattening 569
– Green's function 568
– Helmholtz waves 566, 567
– playing regimes 567
– pressure broadening 569
– slip–stick friction 567
– transients 569
– viscoelastic friction 570
Brain Opera 738
breathing
– diaphragm 672
– in speech and singing 669, 671
– internal intercostals 670
– mode 589, 598, 658
bridge
– admittance 572, 573, 577
– bouncing mode 571
– bridge-hill (BH) feature 572
– coupling to body 571
– design 574
– dynamics 573
– mechanical models 571
– muting 574
– resonant frequency 572
– rocking motion 573
– rotational mode 571
– timbre 574
– vibrational modes 570

British Medical Ultrasound Society
  (BMUS)   895
bubble collapse   287
bubble oscillation   287
bubbles
– compressible liquid   278
– incompressible liquid   276
– Minnaert frequency   277
– Rayleigh–Plesset equation   276
bubbles and bubbly media   162
bubbly liquid   275
– Burgers–Korteweg–de Vries
  equation   281
– Korteweg–de Vries equation   279
Burgers equation   268

C

CAATI (computed angle-of-arrival
  transient imaging)   17
CAC (ceiling attenuation class)   422
calibration   1064, 1068
– condenser microphone   1028
caterpillar
– sound detection   786
Catgut Acoustical Society   554, 600
causality requirement   517
cavitation   6
cavity modes   593
CCD (charge-coupled device)   1103
CDF (cumulative distribution
  function)   512
ceiling attenuation class (CAC)   422
celeste   649
cello   560
central limit theorem   513
central moment   513
– kurtosis   513
– skewness   513
cepstrum   517
cerebral hemispheres   361
CFD (computational fluid dynamics)
  142
change sensitivity   488
channel vocoder   724
chaotic sound waves   292
– in musical instruments   293
– in speech production   293
charge-coupled device (CCD)   1103
Chinese gongs   654
Chinese Opera Gongs   656
chirping
– animal sonar   797
Chladni pattern   1101
– guitar   586
– holography   586

– rectangular plate   585
– violin plates   586
Chladni's Law   10, 653
cicada sound production   789
circular membranes   919
circulation of sound energy   1070
clarinet   550, 602, 622, 637
clarinet model   733
clavichord   560, 561
closure, perceptual   494
CMU (concrete masonry unit)   407
CN (cochlear nucleus)   435, 436,
  452
CND (cumulative normal
  distribution)   512
cochlea   4, 429, 430, 434, 437
cochlear
– amplification   439, 440, 448
– amplifier   439, 446
– frequency map   437
– nucleus (CN)   435, 436, 452
– transduction   436, 437
coherence   215
coherence of changes, role in
  perceptual grouping   488
col legno   562
coloratura singing   674
comb filter   732
combination of structures   926
combination tones   449, 553
combustion
– pulsed   248
common fate, principle of   493
communication
– vibrational   788
communication with sound   2
commuted synthesis   732
complex exponential representation
  1066
complex instantaneous intensity
  1056
complex notation   240
complex representation   1054
complex sound intensity   1056
complex wave   504
compliance   241
compression   439, 442, 449, 452
computational fluid dynamics (CFD)
  142
computed angle-of-arrival transient
  imaging (CAATI)   17
computer music language   737
computer speech recognition   3
concatenative synthesis   717
CONCAWE (CONservation of Clean
  Air and Water in Europe)   134

concert halls   4
concrete masonry unit (CMU)   407
condenser microphone
– calibration   1027
– functional implementation   1024
– mechanical response   1025
– open-circuit sensitivity   1028
– practical considerations   1027
– reciprocity pressure calibration
  1026
– theoretical considerations   1026
– theory   1024
conditional probability density   513
conducting jacket   738
conical tube   606
– input impedance   607
– Q-values   606
– truncation   607
CONservation of Clean Air and Water
  in Europe (CONCAWE)   134
conservation of sound energy   1054
conservative system   905
consonants   687
context dependence of acoustic
  properties of vowels and
  consonants
– vowel quality   693
continuity equation
– lossless   240
– thermoacoustic   242
control of intonation (speech melody)
  699
convolution   509, 521
– of the Fourier transforms   509
cornet   640
correlation matrix   1078
cost function   934
coupling
– air-jet to air column   633
– air-membrane   643
– body-air cavity   592
– instrument-room acoustic
  600
– string–body   575
– string–string   580
– vocal tract   627
CPT (current procedural
  terminology)   847
cricket   789
– sound production   790
critical angle of shadow zone
  formation   133
critical distance-reverberation
  distance   347
critical frequency   396, 938
cross synthesis   721

cross-correlation 511
cross-fingering 617
cross-synthesizing vocoder 724
crustacean 797
crystalline elastic constant 219
CSDM (cross-spectral-density matrix) 182
cubic nonlinearity 949
cumulative distribution function (CDF) 512
cumulative normal distribution (CND) 512
current procedural terminology (CPT) 847
cut-off-frequency 616
cylindric spreading 115
cylindrical coupler 1034
cylindrical pipe
– closed resonance 602
– tube impedance 602
cymbals 655

D

D'Alembert's solution 916
DAC (digital-to-analog converter) 714, 769
damped sounds 483
damping
– Coulomb 945
– hysteretic 947
– localized 941
– matrix 908
– modal projection 940
– proportional 941
– weak 941
deep scattering layer (DSL) 160
deep sound channel (DSC) 149
deep venous thrombosis (DVT) 878
degree of freedom (DOF) 902
Deiter's cell 447, 448
density
– Gaussian 512
deterministic (sines) components 720
detuning parameter 952
DF (directivity factor) 115
DFT 729
DFT (discrete Fourier transform) 719
DI (directivity index) 115, 347
diaphragm
– impedance 800
difference limen 478
diffraction 215

diffuse sound field 1057
digital recording 548
– aliasing 548
– dynamic range 548
– files 549
– Nyquist frequency 548
– resolution 548
– sampling rate 548
– sampling system 550
– sound 549
– sound file 548
– windowing functions 549
digital signal processing (DSP) 724, 752
digital speckle photography (DSP) 1104
digital speckle-pattern interferometry (DSPI) 1103
DigitalDoo 737
digital-to-analog converter (DAC) 714, 769
digitized data 520
diphones 717
Dirac delta function 507, 904, 916
directed reflection sequence (DRS) 340
directional sensitivity 429, 431
directional tone colour 599
directionality
– violin 598
directivity factor (DF) 115
directivity index (DI) 115, 347
discharge rate 441, 442
discrete Fourier transform (DFT) 719
discrete systems 907
disjoint allocation, principle of 493
dispersion equation 921, 922, 936
dispersion relation 279, 282, 517
dispersive system 516
dissipation 249
distortion tones 449
DLS (downloadable sounds) 714, 736
DOF (degree of freedom) 902
DOF (motional degree of freedom) 1127
dominance, pitch 482
downsampling 716
downloadable sounds (DLS) 714, 736
dramatic and lyric soprano 676
driving-point admittance 910, 912
DRS (directed reflection sequence) 340
drum sticks 644

drums
– air loading 643
– circcular membrane modes 642
– excitation stick 644
– radiation 647
dry air standard 1047
DSC (deep sound channel) 149
DSL (deep scattering layer) 160
DSP (digital signal processing) 724, 752
DSP (digital speckle photography) 1104
DSPI (digital speckle-pattern interferometry) 1103
Duffing equation 949, 950
dulcimer 561
duration discrimination 476
DVT (deep venous thrombosis) 878
dynamic capability 1063
dynamic range 434, 442

E

eardrum 429, 430, 432, 433
early decay time (EDT) 307, 379
EARP (equal-amplitude random-phase) 520
earthquakes: P-waves and N-waves 2
echoe suppression 485
echo-location
– bats 796
ECMUS (European Committee for Medical Ultrasound Safety) 895
edge tones 636
Edison, Thomas 14
EDT (early decay time) 307, 379
EDV (end diastolic velocity) 876
EEG (electroencephalography) 352
effect of loudness on glottal source 679
effect of subglottal pressure on fundamental frequency 674
effective area 433
effective cross sectional area 430
efferent 436
efferent synapse 449
EFSUMB (European Federation of Societies for Ultrasound in Medicine and Biology) 895
eigenfrequency 909
eigenmodes 909, 914
– string vibrations 557
elastic energy 905, 950
elasticity effects on ground impedance 125

electric circuit analogues
– acoustic transmission line   614
– lumped components   613
electrical self-noise   1064
electrical transfer impedance   1025
electroencephalography (EEG)   352
electroglottogram   677
electromotility   448
electronic music   22
electronic speckle-pattern interferometry (ESPI)   1102
electrooptic holography (EOH)   1103
elephant   786
embouchure
– brass instrument   618
emission sound pressure level   1072
emphasis   706
empirical orthogonal functions (EOF)   194
enclosures   400
end diastolic velocity (EDV)   876
endolymph   434, 445, 448
endolymphatic potential   444, 445
energy spectral density   510, 511
– Fourier transform   511
engine   244
– standing-wave   244
– Stirling   240
– thermoacoustic   240
– traveling-wave   246
engineering accuracy   1063
engineering acoustics   5, 1019
entropy   528
envelope   438, 718
– generator   718
– of signal   515
EOF (empirical orthogonal functions)   194
EOH (electro-optic holography)   1103
epilaryngeal tube   685
equal-amplitude random-phase (EARP)   520
equal-loudness contour   468
equation of state (pressure, density, entropy)   259
equivalent rectangular bandwidth (ERB)   464
equivalent sound level   681
errors   1082
ESPI (electronic speckle-pattern interferometry)   1102
Euler equation
– see continuity equation   240
Euler formula   507

Euler's equation of motion   1058, 1064
Euler's identity   720
European Committee for Medical Ultrasound Safety (ECMUS)   895
European Federation of Societies for Ultrasound in Medicine and Biology (EFSUMB)   895
evanescent wave   1082
excess attenuation (EA)   118
excitation pattern   464
– model   478
excitation strength   678
experiments in musical intelligence (EMI)   740
expressive power   706
external and middle ears   429
external ear   429
extraneous noise   1063

F

face-to-face   1059
fast field program (FFP)   114, 168, 169
fast field program for air–ground systems (FFLAGS)   128
fast Fourier transform (FFT)   169, 720, 771, 1129
FDA (Food and Drug Administration)   894
FEA (finite element analysis)   593, 1134
– bells   659
– finite-element correlation   1135
– guitar   594
– violin   591, 593
Federal Energy Regulatory Commission (FERC)   1007
FEM (finite element method of analysis)   1134
FERC (Federal Energy Regulatory Commission)   1007
FFLAGS (fast field program for air–ground systems)   128
FFP (fast field program)   114, 168, 169
FFT (fast Fourier transform)   169, 316, 720, 729, 771, 1129
figure of merit (FOM)   167
filter
– causality   526
– recursive   526
– stability   526
filter gain   515
filtering   717

final lengthening
– catalexis   701
finite averaging time   1064
finite element analysis (FEA)   593, 1134
– bells   659
– finite-element correlation   1135
– guitar   594
– violin   591, 593
finite element method of analysis (FEM)   1134
finite impulse response (FIR)   525
finite-difference error   1060
finite-difference time-domain (FDTD)   142
fish   798, 799
– hearing   797
fisheries   195
fission, sequences of sounds   490
FLAUS (Latin American Federation of Ultrasound in Medicine and Biology)   895
flow glottogram   677
flow resistivity   123
flute   602, 637
flute model   733
FM (frequency modulation)   714, 727, 768
FOF (Formes d'onde formantiques)   714, 728
FOFs   728
foliage attenuation   130
Food and Drug Administration (FDA)   894
formant   450, 682, 726, 728
– birdsong   795
– level   683
– undershoot   702
– vocal   795
Formes d'onde formantiques (FOFs)   714, 728
forward masking   465, 474
Fourier series   505, 523, 544
– fundamental frequency   523
Fourier synthesis   719
Fourier theorem   535, 544
– harmonic partials   535
Fourier transform   507, 508, 522, 524, 546, 722, 969
– delta-function   547
– derivative   509
– discrete   522, 549
– fast (FFT)   550
– Gaussian   547
– integral   509
– interpolation   522

– modulated sinewave 547
– rectangular pulse 547
– z-transform 524
free field 390
– conditions 1055
free vibrations 903
frequency analysis 434, 437, 441, 452
frequency difference limen 478
frequency discrimination 478
frequency modulation (FM) 714, 727, 768
frequency modulation detection limen 478
frequency response function (FRF)
– Nyquist plot 1129
frequency scaling
– in animals 785, 786
frequency selectivity 461
frequency sensitivity 438
frequency shift keying (FSK) 191, 816
frequency-domain formulation 1059
Fresnel number 117
FRF (frequency response function)
– Nyquist plot 1129
friction 946
frog
– auditory system 793
FSK (frequency shift keying) 191, 816
Fubini solution 266
fundamental frequency 504, 682
– role in perceptual grouping 486
fusion, sequences of sounds 490
fuzzystructures 958

G

GA (genetic algorithm) 362
gain
– thermoacoustic 242
gain of amplifier 440
Galerkin method 922
gap detection 473
gases
– properties of 243
Gaussian 508
– density 512
– function 508
– noise 519
– pulse 514
general MIDI 735
generalized coordinates 915
genetic algorithm (GA) 362

geometric spreading loss 154
Gestalt psychology 492
gesturalist 705
gestures 705
GigaPop Project 738
glottal waveform 678
Goldberg number 269
gongs 656
– pitch glides 656
goniometer system 231
good continuation, principle of 492
gradient of the mean square pressure 1056
gradient of the phase 1055
granular synthesis 730
Green's function 904, 1080
ground attenuation
– weakly refracting conditions 124
ground effect 120
ground impedance 123
ground wave 121
group delay 516
group velocity 921
growth of masking 466
guitar
– plate-cavity coupling 592
– rose-hole 590, 591
gyroscopic term 908

H

hair bundle motility 446
hair cells 436, 438, 441
hair, sensory 787, 788
hair-bundle movements 448
Hamilton's principle 908, 920, 923
hand stopping 611, 640, 641
harmonic balance 613, 625
– method 948
harmonic modes 557
harmonic motion 905
harmonic series 719
harmonic signal 1054
harmonic spectrum 727
harmonics 543
head shadow 484
hearing 552
– animal 785, 793, 802
– directional 793, 794
– frequency response 794
– ISO standards 552
– phons 552
– sensitivity 552
– vertebrates 793
hearing level 461
hearing out partials 467

hearing threshold level 461
heat conduction correction 1030
heating, ventilating and air conditioning (HVAC) 403
helicotrema 435, 441
Helmholtz modes
– struck 560
– woodwind 626
Helmholtz resonance 911
– brass mouthpieces 618
Helmholtz resonator 392, 722
– bass reflex cabinet 582
– coupling to walls 592
– guitar 592
– resonant frequency 591
– stringed instruments 582
– violin 599
Helmholtz waves
– bowed 558, 559
– kinks 558
– plucked 558
– spectra 558
– strings 557
hemispheric specialization in listening 359
Hilbert transform 514
hologram 1079
holographic interferometry 1102
holography 571
– cymbals 655
– guitar 586
– violin 596
Hopf–Cole transformation 269
horn 638
horn equation 609
horn shapes 609
– Bessel 610
– conical 606
– cylindrical 602
– exponential 609, 610
– flared 611
– hybrid 607
– perturbation models 612
horns
– impedance matrix 801
Huffman sequence 476
human voice 3
HVAC (heating, ventilating and air conditioning) 403
hybrid tubes 608
hydraulic radius 242
hydrophones 155
Hyper and Hypo (H & H) theory
– adaptive organization of speech 705

hyperspeech   703
hypospeech   703

I

IACC (interaural cross-correlation coefficient)   310
IACC (magnitude of the IACF)   351
IACF (interaural cross-correlation function)   351, 352, 357
IAD (interaural amplitude difference)   750
ICAO (International Civil Aircraft Organization)   989
identification and ranking of noise sources   1068
identity analysis/resynthesis   726
IDFT (inverse discrete Fourier transform)   719
IEC standard   1058, 1065
IFFT (inverse fast Fourier transform)   177
IHC   441, 443, 444
IHC (inner hair cells)   434–436
IIC (impact insulation class)   396
IL (insertion loss)   118
IM (intermodulation)   755
impact insulation class (IIC)   396
impedance
– acoustic   211, 799, 801
– cavity   801
– cylindrical pipe   605
– mechanical   799, 801
impedance matrix   794, 802
– horns   801
– tube   801, 802
impedance transformation   432
impedance transformer   432
impulse response   516, 904, 916
– function (IRF)   1132
IMT (intima-media thickness)   851
increment detection   472
incus   429, 433
inertance   241
infinite impulse response (IIR)   525, 724
infinite-duration signals   510
information content   529
information theory   528
information transfer ratio   530
infrasound   2
inharmonic spectrum   727, 728
inharmonicity   563, 923
initial time delay gap (ITDG)   310

initial time delay gap between the direct sound and the first reflection   351
inner hair cells (IHC)   434–436
input impedance   1024
– brass instrument   612
– cylindrical pipe   604
insects   788, 789
– sound production   788
insertion loss (IL)   118
instantaneous sound intensity   1054, 1056
intensity discrimination   472
interaural amplitude difference (IAD)   750
interaural cross-correlation coefficient (IACC)   310
interaural cross-correlation function (IACF)   351, 352, 357
interaural differences   484
interaural time difference (ITD)   750
interaural timing cues   452
interference   212, 213
– pattern   432
intermodulation (IM)   755
internal resonance   952
International Civil Aircraft Organization (ICAO)   989
interpolation   716
intersecting walls   926
intersymbol interference (ISI)   191
intima-media thickness (IMT)   851
intravenous pyelogram (IVP)   882
invariance issue   694
inverse discrete Fourier transform (IDFT)   719
inverse fast Fourier transform (IFFT)   177
inverse Fourier transform   517
inverse problems   1077
IRF (impulse response function)   1132
ISI (intersymbol interference)   191
ISO standards for sound power determination   1071
ITD (interaural time difference)   750
IVP (intravenous pyelogram)   882

J

JND (just noticeable difference)   747
Johnson noise   518
joint probability density   513

jump phenomenon   951
just noticeable difference (JND)   747

K

KDP (potassium dihydrogen phosphate)   18
Kelvin functions   925
Kettle drums   645
kinetic energy   905
Kirchhoff–Helmholtz integral equation   1080
Koenig, Rudolph   14
Kronecker delta   506
kurtosis   514

L

laboratory speech   693
laboratory standard microphone   1026
labyrinth   430
Lagrange equations   908
land mine detection
– acoustical methods   224
Laplace transform   903, 907, 916
large amplitude effects
– brass   630
– Helmholtz motion (wind)   626
– shock waves   632
– woodwind   626, 630
large-volume coupler   1035
laser Doppler anemometry (LDA)   1103
laser Doppler vibrometry (LDV)   1103
lateral energy fraction (LEF)   309
lateral olivocochlear system (LOC)   436
Latin American Federation of Ultrasound in Medicine and Biology (FLAUS)   895
LDA (laser Doppler anemometry)   1103
LDV (laser Doppler vibrometry)   1103
lead pipe   619
lead zirconate titanate (PZT)   849
leaf-shape concert hall   364
LEF (lateral energy fraction)   309
LEV (listener envelopment)   309
level, effect on pitch   480
Liljencrants–Fant (LF) model   678
line source
– finite line source   115

linear interpolation 716
linear predictive coding (LPC) 725
linear processor 516
lip vibrations
– artificial lips 630
– modelling 630
listener envelopment (LEV) 309
listener-oriented school 705
listening level (LL) 352, 380
LL (binaural listening level) 351
LL (listening level) 352, 380
LOC (lateral olivocochlear system) 436
localization 429
localization of sound 484
location
– bats 797
location, role in perceptual grouping 489
locus equations 695
longitudinal vibrations of bars 920
long-play vinyl record (LP) 18
long-term-average spectra (LTAS) 681
loudness 442, 468
– growth 440
– meter 470
– model 470
– perceptual correlates 686
– recruitment 440
– scaling 469
low pitch 480
low-pass filter 515
low-pass resonant filter 516
LP (long-play vinyl record) 18
LPC (linear predictive coding) 725
LPC vocoder 726
LTAS (long-term-average spectra) 681
lung
– reserve volume 671
– residual volume 671
– total capacity 671
– vital capacity 671
Lyapunov exponent 291

M

MAA (minimum audible angle) 808
machine-gun timing 699
MAF (minimum audible field) 460
magnetic resonance imaging (MRI) 688
magnetoencephalogram (MEG) 361
magnetoencephalography (MEG) 352

magnitude estimation 469
magnitude of the IACF (IACC) 351
magnitude production 469
main response axis (MRA) 181
malleus 429, 433
MAP (minimum audible pressure) 460
marginal probability density 513
marimba 649, 650, 735
marine animals 195
marine mammals 198
masking 403, 461
– pattern 464
mass
– law 398
– matrix 908
– modal 924
MASU (Mediterranean and African Society of Ultrasound) 895
matched field processing (MFP) 182
maximum flow declination rate (MFDR) 677
maximum length sequence (MLS) 526
maximum-likelihood method (MLM) 182
MCR (multi channel reverberation) 349
MDOF (multiple degree of freedom) 1132
mean square error (MSE) 725
measurement 1085
measurement principles 1058
mechanical index (MI) 894
medial olivocochlear system (MOC) 436
medical acoustics 5
medical ultrasonography 225
medical ultrasound 6
Mediterranean and African Society of Ultrasound (MASU) 895
MEG (magnetoencephalogram) 361
MEG (magnetoencephalography) 352
membrane capacitance 448
membranes 913
meteorologically-neutral 134
MFDR (maximum flow declination rate) 677–682
MFP (matched field processing) 182
MI (mechanical index) 894
micromechanical models 441

microphone
– acoustic transfer impedance 1044
– calibration 1044
– coupler 1045
– frequency dependence 1037
– frequency response measurement 1043
microphone calibration
– barometric pressure correction 1035
– capillary tube correction 1032
– comparison method 1039
– comparison method with a calibrator 1040
– cylindrical coupler 1034
– equivalent volume 1031
– free-field calibration 1039
– heat conduction correction 1029
– interchange microphone method 1039
– temperature correction 1037
– wave-motion correction 1034
microphone sensitivity
– correction 1038
– level 1036
– temperature correction 1036
middle ear 432, 433
middle-ear bones 429
middle-ear ossicles 436
MIDI (musical instrument digital interface) 714, 735
MIMO (multiple-input multiple-output) 191
– configuration 193
– mode 191
minimum audible angle (MAA) 808
minimum audible field (MAF) 460
minimum audible pressure (MAP) 460
minimum phase 517
minimum-variance distortionless processor (MV) 182
MIR (music information retrieval) 738
missing fundamental 480, 541, 553
mixture
– separation of 253
MLM (maximum-likelihood method) 182
MLS (maximum length sequence) 526
MOC (medial olivocochlear system) 436, 439, 447, 449
modal
– mass 909

– participation factors   909
– stiffness   909
modal analysis   536, 597
– holding instrument   598
– holographic   1137
– mathematical   1133
– sound-field analysis   1136
modal synthesis   713, 722
modal testing   1128
– complex modes   1133
– impact excitation   1130
– multiple-input multiple-output (MIMO)   1132
– obtaining modal parameters   1132
– pseudo-random signal   1131
– shaker excitation   1131
mode   722
models of temporal resolution   474
modulation
– amplitude   551
– detection   472
– filter bank   475
– frequency/phase   551
– masking   475
– timbre   551
– transfer function MTF   312
Moiré techniques   1104
momentum equation
– lossless   240
– thermoacoustic   242
motional degree of freedom (DOF)   1127
motor equivalence   672
mouthpiece
– brass instruments   617
– Helmholtz resonance   618
– input end-correction   619
– input impedance   619
– lip vibration   628
– popping frequency   618
MRA (main response axis)   181
Mrdanga   646
MRI (magnetic resonance imaging)   688
MSE (mean square error)   725
mufflers   402
multichannel reverberation (MCR)   349
multilayered partitions   399
multimode systems   536
multiple degree of freedom (MDOF)   1132
multiple-input multiple-output (MIMO)   191
– configuration   193

– mode   191
multiple-scales method   952
multiplication of frequency functions   509
multisampling   718
multivariate distribution   513
music information retrieval (MIR)   738
musical acoustics   3
musical instrument digital interface (MIDI)   714, 735
musical interval perception   479
musical intervals   542, 543
musical nomenclature   553
mutual information   530
MV (minimum-variance distortionless processor)   182

N

nageln   562
National Electronic Manufacturers Association (NEMA)   894
National Metrology Institutes (NMIs)   1044
NC (noise criterion) curves   404
NCB (balanced noise criterion curves)   404
NDT (nondestructive testing)   1103
near field   1057
near miss to Weber's law   472
near-field acoustic holography   1078, 1079
negative conductance   623
NEMA (National Electronic Manufacturers Association)   894
network
– analog   800–802
neurotransmitter   444
Newton, Isaac   11
NIC (noise isolation class)   396
NMI (National Metrology Institutes)   1044
noise   2, 518, 551
– band-limited   519
– barrier   116, 118
– components   520
– electrical systems   406
– Gaussian   519
– HVAC systems   405
– plumbing systems   406
– random telegraph   519
– thermal   518
noise control
– door designs   412
– electrical systems   417

– engineering   1053
– floor design   409
– HVAC systems   412
– plumbing systems   415
– wall designs   407
– windows design   410
noise criterion (NC) curves   404
noise isolation class (NIC)   396
noise reduction   400
– coefficient (NRC)   390, 996
– reverberant field   393
nondestructive testing (NDT)   1103
nonlinear
– capacitance   447, 448
– coupled oscillators   951
– oscillator   948
– vibrations   947
nonlinear acoustics
– in fluids   257
– of fluids   234
– of solids   235
nonlinear time-series analysis   290
nonlinear waves
– combination frequencies   275
– difference-frequency   275
– interaction   273
– sound–ultrasound interaction   284
nonlinearity
– amplitude dependence   564
– coefficient   262
– hard spring   259
– inharmonicity   564
– mode conversion   564, 654
– orbital motion   565
– origin   258
– parameter   260
– parametric excitation   564
– plate modes   654
– reed excitation   625
– shewed resonances   564
– soft spring   259
– spherical cap   655
nonparametric technique   713
nonsimultaneous masking   465
NORD2000   115
normal modes   535, 909
– coupled strings   580
– coupling factor   577
– damping   576
– effective mass   535
– string-body   576
– veering   581
– weak/strong-coupling   577

normal modes of vibration
- damping factor   1127
- eigenfrequency   1127
- mode shape   1127, 1128
normal-mode model   169
notch   432
notched-noise method   463
NRC (noise reduction coefficient)   390, 996
Nyquist wave number   1082

O

oboe   622, 637
Obukhov length   137
ocarina   614
Occupational Safety and Health Administration (OSHA)   1000
ocean acoustic environment   151
ocean coustic noise   161
octave   479
ODS (operating deflexion shape)   910, 931, 1128
ODS (output display standards)   894
off-frequency listening   463
OHC   439, 446–449
OHC (outer hair cells)   434–436
OHC motility   446
OITC (outdoor–indoor transmission class)   397
olivocochlear bundle   436
one-pole low-pass filter   515
onset asynchrony, role in perceptual grouping   487
open sound control (OSC)   736
operating deflexion shape (ODS)   910, 931, 1128
operation deflection shape (ODS)   1111
operation on
- OR   526
- XOR   526
optical glottogram   677
organ of Corti   434, 436, 441, 444
orthogonal components   504
orthogonality   915
- with respect to mass   916, 924
- with respect to stiffness   916, 924
orthotropy   923
OSC (open sound control)   736
OSHA (Occupational Safety and Health Administration)   1000
ossicles   430, 432

otoacoustic emissions   439, 447, 450
otolith   798
ototoxic antibiotics   447
outdoor–indoor transmission class (OITC)   397
outer hair cells (OHC)   434–436
output display standards (ODS)   894
oval window   432, 433
overblowing   602

P

PA (pulse average)   894
parabolic equation (PE)   114, 168
- model   172
parametric   713
parametric synthesis   717
Parseval's theorem   510
particle image velocimetry (PIV)   1104
particle models   730
particle velocity   207
- transducer   1066
Pasquill categories   134
patch synthesizer   718
patent foramen ovale (PFO)   885
PC (phase conjugation)   182, 183
PCM (pulse code modulation)   713, 717, 768
PD (probability of detection)   165
PDF (probability density function)   165, 512
PE (parabolic equation)   114, 168
- model   172
peak systolic (PS)   876
pedal note   612
pendulum
- elastic   949
- interrupted   948
penetration depth   242
perception   552
- violin quality   600
perceptual grouping   485
perceptual organization   492
percussion   641
- bars   648
- membrane   642
- plates   652
- shells   658
perilymph   434, 447
period-doubling cascade   293
periodic functions   506
periodic signal   510
periodic structures   926

perturbation
- bends   614
- bore profiles   613
- finger holes   614
- string vibrations   576
- valves   614
PFA (probability of false alarm)   165
PFO (patent foramen ovale)   885
phase   504
- conjugation (PC)   182, 183
- delay   516
- filter   734
- mismatch   1061
- modulation (PM)   1109
- shift   210
- shift keying (PSK)   191
- stepping (PS)   1109
- velocity   921
- vocoder   721
phase-contrast methods   1101
phase-locking   442–451
PhISEM   730
phon   468
phonation
- modes of   680
phonation types
- hyperfunctional   680
- hypofunctional   680
phoneme   693, 717
physical acoustics   5, 205
physical mechanisms   155
physical models   714
physical properties of air   1044
physiological acoustics   4, 459
PI (privacy index)   420
piano   560
- double decay   581
- string doublets/triplets   580
piezoelectric transducers   226
pink noise   510, 520
pinna   430, 431
- animal   794
pinna, role in localization   485
pipe
- end-correction   603
- input impedance   604, 605
- Q-valve   604
- radiation impedance   603
- reflection/transmission coefficients   605
- thermal and viscous losses   604
pitch   477, 540
- ambiguity   541
- circularity   554
- glides   656
- hearing range   541

– musical instruments   541
– musical notation   541
– shift   716
– subjective   553
pitch theory, complex tones   481
PIV (particle image velocimetry)   1104
PL (propagation loss)   175
place theory   477
planar acoustic holography   1081
planar laser-induced fluorescent   1121
plane propagating wave   1055
plane-wave coupler   1035
plate modes
– 1-D solutions   583
– 2-D solutions   584
– anisotropy   584, 587
– antielastic bending   584
– arching   588, 653
– boundary conditions   583
– Chladni pattern   585
– circular plate   653
– density of modes   587
– elastic constants   588
– flexural vibrations   582
– longitudinal modes   584
– measurement   585
– mode conversion   656
– mode spacing   587
– non-linearity   654
– rectangular plate   585
– shape dependence   586
– symmetry   571
– torsional modes   584
plates
– flexural vibrations   923
– isotropic   924, 938
– prestressed   924
– rectangular   924
player-instrument feedback   555
plucked string   731, 732
PM (phase modulation)   1109
PMA (pre-market approval)   893
point source   115
Poisson process   730
Poisson ratio   556
polar plot
– open pipe   603
poles   726
polyvinylidene fluoride (PVDF)   227
popping frequency   618, 628
position and sensor   1083
positive feedback
– air-jet interactions   635

potassium channels   445
potassium dihydrogen phosphate (KDP)   18
potential and kinetic energy density   1054
power
– acoustic   243, 249, 932, 933, 937
– mechanical   905
– spectral density (PSD)   510, 774
– spectrum model   462
– time-averaged thermal   243
– total   243
p–p method   1058
Prandtl number   243
precedence effect   485, 554
preferred delay time of single reflection   353
preferred horizontal direction of single reflection   355
pre-market approval (PMA)   893
pressure level band   214
pressure-intensity index   1061
pressure-residual intensity index   1062
prestin   447, 448
PRF (pulse repetition frequency)   861
PRI (pulse repetition interval)   871
primary microphone   1044
principal-components analysis   739
privacy index (PI)   420
probability   529
probability density function (PDF)   165, 512
probability mass function (PMF)   512
probability of detection (PD)   165
probe reversal   1067
propagation and transmission loss   175
propagation loss (PL)   175
PS (phase stepping)   1109
PSD (power spectral density)   510, 774
PSK (phase shift keying)   191
psychoacoustics   20, 552
psychological acoustics (psychoacoustics)   4
psychophysical tuning curve   462
p–u phase mismatch   1067
p–u sound intensity measurement system   1066
pulse average (PA)   894

pulse code modulation (PCM)   713, 717, 768
pulse repetition frequency (PRF)   861
pulse repetition interval (PRI)   871
pulsed combustion   248
pulsed TV holography   1103
pulse-tube
– refrigerator   240
PVDF (polyvinylidene fluoride)   227
PZT (lead zirconate titanate)   849

Q

Q factor   347
Q value   516
quadratic nonlinearity   949, 957
quality factor   535
quantitative description   177
quantization   715
– noise   520
quefrency   517
Q-values   537

R

r!g (sensor speader bass)   737
racket   638
radiation
– control   934
– critical frequency   598
– damping   928
   in plates   945
– efficiency   598, 937, 938, 1071
– energy   933
– filter   933
– impedance   603
– impedance matrix   939
– polar plot   603
– tone holes   616
– violin   598
– wavenumber Fourier transform   939
radio baton   738
ramped sounds   483
random error   1064
rapid speech transmission index (RASTI)   421
RASTI (rapid speech transmission index)   421
rate-level functions   442
rational wave in elastic medium   125
Rayleigh distribution   519
Rayleigh, Lord   13
RC (room criterion) curves   404

reactive intensity 1055
reactivity 1067, 1068
receiving operating characteristic (ROC) 186
receptor potentials 446
reciprocity 536, 1113
reconstruction filter 522
recursive filter 726
reed model 733
reeds
– bifurcation 625
– classification 619
– double reed 622
– dynamic characteristics 622, 623
– embouchure 622
– feedback 623
– hysteresis 622
– large-amplitude oscillations 625
– negative resistance 623
– positive feedback 623
– reed equation 622
– single reed 621
– small-amplitude oscillations 624
– static characteristics 619
– streamlined flow 620
– turbulent flow 620
– wind/brass instruments 619
reference microphone
– acoustical calibrator 1040
– uncertainty 1040
reflection 432
– coefficient of 212
– wave 210
refraction 131, 212
refrigerator 250
– pulse tube 240
– standing-wave 250
– Stirling 240
– thermoacoustic 240
– traveling-wave 251
regenerator 242, 246, 251
register key 627
regularization 1083
Reissner's membrane 434
REL (resting expiratory level) 670
relationship F0 and formant frequencies 692
relationship F0 and jaw opening 692
relationship first formant and F0 692
repeatability 1064
resampling 716
residual intensity 1062
residue pitch 480

resistance
– matrix 933
    acoustical 933
    structural 933
– thermal-relaxation 242
– viscous 242
resonance 213, 535, 726
– air column 602
– conical pipe 606
– cylindrical pipe 604
– dispersion 535
– loss 535
– phase 536
– strings 579
– width 536
resonant filter 516
resonant frequency 440
resonant ultrasound spectroscopy (RUS) 220
respiratory system
– active control 673
– passive control 673
resting expiratory level (REL) 670
reticular lamina 441
reverberation time 378
reversible ischemic neurological deficit (RIND) 878
reversing a p–p probe 1062
Riemann characteristics 263
Rijke oscillations 246
RIND (reversible ischemic neurological deficit) 878
RMS (root-mean-square) signal 512
ROC (receiving operating characteristic) 186
role of biomechanics in speech production 703
room acoustic 600
room criterion (RC) curves 404
room modes 388
room shapes 394
roughness effects on ground impedance 124
rough-sea effects 143
RUS (resonant ultrasound spectroscopy) 220

S

S/N (signal-to-noise) 520
SA (spatial average) 894
SAA (sound absorption average) 391
Sabine decay time 537
Sabine equation 394
Sabine reverberation formula 16

Sabine, Wallace Clement 15
SAC (spatial audio coding) 775
sampled data 520
sampling 521, 715
– rate 715
– synthesis 718
– theorem 521
sandwich plates 926
SAOL (structured audio orchestra language) 714, 736
SARA (simple analytic recombinant algorithm) 740
saturation rate 442
SAW (surface acoustic wave) 13
SAW (surface acoustic waves) 231
scala media (SM) 434
scala tympani (ST) 434
scala vestibuli (SV) 429, 433, 434
scan vector 1078
scanning 1070
scattering and diffraction 1060
scattering and reverberation 158
scattering by turbulence 139
Schlieren 1101
Schlieren imaging 228
Schroeder diffuser 368
scientific scaling 601
SDIF (sound description interchange format) 736
SDOF (single degree of freedom) 1132
– oscillator 928
SE (signal excess) 167
SEA (statistical energy analysis) 19, 902, 912
secular terms 952
segmentation of audio 738
segmentation problem 694
sensation level 462
sensory hair 789, 798
sequences of sounds, perception of 490
serpent 641
SG (spiral ganglion) 435
shadow zone boundary 132
shadowgraph 1101
shallow water 153
Shannon entropy 529
shearography 1103
shells
– bells 658
– blocks 658
– body modes 589
– breathing mode 589
– eigenmodes 925
– external constraints 589

– nonlinear vibrations 957
– plate modes 589
– spherical 925, 955
– vibrational modes 589
– violin body 591
shift register 526
– tap 527
shock distance 266
shock formation time 266
shock wave velocity 272
shock waves 271, 632
shoe-box concert hall 363
short-time Fourier transform (STFT) 720
SI (speckle interferometry) 1102, 1109
sibilance 725
side drum
– air loading 647
– directionality 647
– snare 646
side-by-side 1059
signal 514
– analog 520
– autocorrelation function 510
– average power 512
– average value 511
– conversion 522
– cross-correlation 511
– delays 516
– digitized 512
– envelope 515
– filter 515
– Gaussian pulse 514
– Hilbert transform 514
– moment of 513
– periodically repeated 523
– root-mean-square (RMS) 512
– sampling 521
– standard deviation 512
– variance 511
signal excess (SE) 167
signal functions 509
signal to noise ratio (SNR) 520, 715, 757
signal-based and signal-independent knowledge in speech perception 705
SIL (sound intensity level) 208
similarity, principle of 492
simple analytic recombinant algorithm (SARA) 740
sinc interpolation 716
singer's formant 684, 686
– larynx tube 685
singers' subglottal pressure 676

singing
– coordinative structures 704
single degree of freedom (SDOF) 1132
– oscillator 903, 928
single-bubble sonoluminescence 286
single-input single-output mode (SISO) 191
sinusoidal synthesis 719
sinusoidal waves
– complex numbers 540
SISO (single-input single-output mode) 191
skeleton curves 590
skewness 513
slip–stick model 559
sloping saturation 442
slow vertex response (SVR) 361
SM (scala media) 434
smart materials 958
SNR (signal to noise ratio) 715, 757
SOC (superior olivary complex) 436
SOFAR 6
soliton 281
SONAR 6, 165, 181, 185
sonar animals 797
SONAR array processing 179
Sondhauss oscillations 239, 246
sone 469
sonoluminescence 5, 17, 18, 286
sonority principle 695
SOSUS (sound ocean surveillance system) 150
sound
– dB sound level 538
– end corrections 538
– insulation 395
– intensity 538, 1053
– intensity level (SIL) 208
– intensity Robinson–Dadson hearing plots 552
– levels (SPL) 538
– localization 431, 443
– near and far fields 538
– ocean surveillance system (SOSUS) 150
– power 1053, 1054
– power determination 1054, 1070
– pressure 538
– radiation 537, 538
– specific impedance 537
– spherical waves 538, 606
– waves 537

sound absorption 1072
– average (SAA) 391
sound attenuation through trees and foliage 129
sound description interchange format (SDIF) 736
sound field 389
– indicator 1068
– spatial factors 352
– temporal factors 352
sound pressure level (SPL) 209, 358, 748, 831
sound production
– animal 785, 802
– birds 795
– insects 788
– vertebrates 793
sound propagation
– atmospheric turbulence effects 138
– effects of ground elasticity 125
– ground effect 114, 120
– meteorological classes 133
– rough-sea effects 143
– shadow inversions 130
– spherical acoustic waves 120
– surface wave 122
– wind and temperature gradient effects 130
sound source
– changing airflow 1
– crossover frequency 540
– dipole 539
– monopole 539
– pipe 603
– polar plots 539
– quadrupole 539
– size dependence 539
– supersonic flow 1
– surfaces 539
– time-dependent heat sources 1
– vibrating bodies 1
– wind instruments 540
sound speed profiles 136
sound transmission class (STC) 396
sound velocity in solids 220
soundpost 575, 590
SoundWire 738
source directivity 115
source-filter model 725
source-filter theory 676
SP (spatial peak) 894
SP (speckle photography) 1102, 1109
spatial aliasing 1082
spatial audio coding (SAC) 775

spatial average (SA) 894
spatial peak (SP) 894
speaking style 701
specific acoustic impedance 432
specific impedance 1071
specific loudness 470
– pattern 470
speckle correlation 1104
speckle interferometry (SI) 1102, 1109
speckle metrology 1102
speckle photography (SP) 1102, 1109
spectra 545
– Big Ben bell 660
– bowed string 560
– cello 560
– clarinet 545, 627
– cymbal 547
– glockenspiel 650
– gongs 656
– guitar (modelled) 594
– marimba 650
– plucked string 558
– ratchet 547
– steeldrum 657
– struck string 562
– tambla 647
– tam-tam 656
– timpani 547
– triangle 652
– vibraphone 650
– violin 545
– violins 595
– xylophone 650
spectral cues 431
spectral description interchange file format (SDIF) 736
spectral modeling 720
spectral regularity, role in perceptual grouping 486
spectrogram
– speech and singing 697
spectrotemporal 451
spectrum 506
speech 21
– coarticulation 694
– control of sound 703
– dynamics 701
– intelligibility index (SII) 420
– interference level (SIL) 421
– perceptual processing 705
– production 676
– prosodic modulation 693
– prosody 701
– rhythm and timing 699
– superposition model 699
– transmission index (STI) 311, 421, 696
speech privacy 419
– office design 422
speech synthesis 717
speed of sound
– computation 1046
speed of sound/in air 10
speed of sound/in liquids 11
speed of sound/in solids 11
spherical spreading 115
spherical waves
– standing waves 606
spiral ganglion (SG) 435
SPL (sound levels) 538
SPL (sound pressure level) 209, 358, 748, 831
spontaneous discharge rate (SR) 442
spontaneous emissions 439
SR (spontaneous discharge rate) 442
ST (scala tympani) 434
staccato 676
stack 242, 244, 250
standards
– building acoustics related 424
standing waves 388
standing-wave tube 1065
stapes 429, 433
– velocity 436
starting transient 551
state space variables 931, 934
stationary process 512
statistical energy analysis (SEA) 19, 902, 912
STC (sound transmission class) 396
steelpans 657
stereocilia 443, 444
stereocilium 445
STFT (short-time Fourier transform) 720
STI (speech transmission index) 311, 421, 696
stiffness 440, 448, 734
– matrix 908
Stirling
– engine 240
– refrigerator 240
stochastic (noise) components 720
Stokes, George 13
strange attractor 291
stream segregation 490
stress timing
– syllable timing 700
stria vascularis 434, 435
string vibrations
– bending stiffness 562
– characteristic impedance 556
– D'Alembert solution 556
– dipole source 555
– directional coupling 578
– force on bridge 556
– Helmholtz waves 557
– measurements 579
– non-linearity 563
– perturbation 576
– polarisation 579
– reflection coefficient 557
– sinusoidal 557
– transverse, longitudinal and torsional 556
– wave equation 555
stringed instruments 919
strings 913
– eigenmodes 917
– heterogeneous 914
– manufacture 563
– nonlinear vibrations 955
– nonplanar motion 956
– plucked 918
– semi-infinite 917
– tension 575
– transverse motion 914
– with dissipative end 942
– with moving end 918
structural resonance
– skeleton curve 574
structural–acoustic coupling 911, 926
– bar 927
– cavity 934
– energy approach 932
– light fluid 928
– plate 936
– weak-coupling 930
structured audio orchestra language (SAOL) 714, 736
sub-bands 724
subglottal and oral pressure 672
subglottal pressure 679
– elastic recoil 670
subharmonic 292, 951
subjective difference limen 308
subjective preference
– individual listeners 370
– measured and calculated values 383
– performers 374
– seat selection 371
– tests in existing halls 381

subjective preference theory 353
subjective preference, conditions for maximizing 356
subsequent reverberation time 351
subtractive synthesis 714, 724
superharmonics 951
superior olivary complex (SOC) 436
supersonic intensity 1070
supporting cells (s.c.) 434, 444, 447
suppression 450–452
surface acoustic wave (SAW) 13, 231
surface intensity method 1066
surface wave 122
surfaces of equal phase 1055
surfaces of equal pressure 1056
survey accuracy 1063
SV (scala vestibuli) 429, 433, 434
SVR (slow vertex response) 361
swim bladder 798, 799
syllable beat
– canonical babbling 696
syllable timing 699
syllables in speech and singing 695
sympathetic strings 580
synapse 435, 436, 444, 446
synaptic ribbon 445
synchronization index 443
synchronization sung syllable with piano accompaniment 698
synthesizer patch 718
synthetic listening 482
syrinx 795
system biological 799

T

TA (temporal average) 894
Tabla 646
Taconis oscillations 239, 246
Tait equation 259
tam-tam 656
TDAC (time domain alias cancellation) 774
TDGF (time-domain Green's function) 183
tectorial membrane 435, 441, 444
temperature gradient, critical 245
temporal average (TA) 894
temporal modulation transfer function 473, 474
temporal order judgment 491
temporal peak (TP) 894
temporal processing 473
temporal resolution 473

temporal theory 477
temporal waveform 443
THD (total harmonic distortion) 754
The Brain (composition system) 740
thermal index (TI) 894
thermoacoustic
– engine 223, 240
– refrigerator 240
thermoacoustics 5, 239
– history 239
thermoelasticity 943
three wave interaction 285
three-stage shift register 527
threshold 442
thunder plate 652
TI (thermal index) 894
TIA (transient ischemic attack) 878
timbre perception 483
timbre, effect of envelope 483
timbre, effect of spectrum 483
timbregrams 739
time domain alias cancellation (TDAC) 774
time reversal (TR) 183
time shift 506
time-/frequency-response equivalence 596
time-averaged sound intensity 1054
time-domain analysis
– brass 632
– FFTs 550
time-domain Green's function (TDGF) 183
time-domain response
– Big Ben bell 660
– chinese gongs 657
– glockenspiel 650
– gongs 656
– marimba 650
– non-linear string 564
– simple harmonic resonator 537
– steeldrum 657
– tabla 646
– tam-tam 656
– timpani 645
– triangle 652
– vibraphone 650
– violin 597
– violin string 569
– xylophone 650
time-reversal acoustics 183
time-varied gain (TVG) 196
timpani 645
– head-air cavity coupling 644

tip link 443–445
TL (transmission loss) 175, 395, 1071
TLC (total lung capacity) 671
TMTF 473
TNM (Traffic noise model) 993
tone holes
– array 616
– mode pertubation 614
– radiation 616
tonotopic 449, 451
tonotopic organization 441
total harmonic distortion (THD) 754
total lung capacity (TLC) 671
TP (temporal peak) 894
TR (time reversal) 183
TR (treble ratio) 310
Traffic noise model (TNM) 993
transducer
– coaxial 228
transduction 434, 443, 444, 447
– channel 444
– current 444, 445
transfer admittance 910
transfer function 430, 516, 525
transfer standard microphone
– mean sensitivity level 1044
transglottal airflow 680
transient ischemic attack (TIA) 878
transients 546, 721
transmission loss (TL) 175, 395, 1071
traveling wave 437, 731
treble ratio (TR) 310
triangle 651
tristimulus representation 484
trombone 638
trumpet 550, 638
tube
– impedance 603
– impedance matrix 801
tuning 438, 440, 441, 542
– by sliding tubes 639
– by valves 639
– cents 543
– curves 441, 442, 446
– equal temperament 542
– forks 12
– mean-tone 542
– measurement 543
– Pythagorean 543
– stretched 544
– temperament 543

turbulence 131
– effects 138
– spectra 140
TV holography 1103
TVG (time-varied gain) 196
two-pole feedback filter 724
two-pole filter 516
twos-complement 521
two-tone suppression 449
tympanum animal 794
Tyndall, John 12

U

ultrasound 2
UMM (unit modal mass) vectors 1128
uncertainty principle 508
underwater acoustic imaging 187
underwater propagation 152–155, 157, 158, 168–170, 172, 175, 177, 182, 183
– models 167
underwater travel-time topography 192
unit generator 718
unit modal mass (UMM) vectors 1128
unit rectangle pulse 508
upper frequency limit 1060
upsampling 716
upward spread of masking 464

V

valves and bends 614
variable bitrate (VBR) 773
VBR (variable bitrate) 773
VC (vital capacity) 671
velocity of sound
– fluids 219
vent sensitivity 1066
vertebrates 790
– hearing 793
very short-range paths 152
vibraphone 650
vibration isolation 414
vibrato 551
violin 550
– admittance measurements 595
– cross-section 575
– directionality 600

– FEA 591
– Helmholtz resonator 591
– octet 601
– quality 572
– signature modes 599
– tonal copies 600
– tone quality 600
virtual pitch 480
viscoelasticity 944
visualization of sound fields 1069
vital capacity (VC) 671
vocal folds 629
vocal formant 791, 792
vocal loudness 681
vocal registers 680
vocal sac 792
– birds 796
vocal tract 792
– filter 682
vocal valve 792
vocoder 721, 724, 726
voice 706
volume attenuation 157
von Helmholtz, Hermann 12, 13
von Kármán equations 957
vortex-sheet model 636
vortices 636
vorticity 634
vowel 450
vowel articulation 691
– APEX model 688–690
vowels 687

W

Waterhouse correction 1072
wave
– capillary 788
– equation 168, 731
– impedance 1056
– number 210
– surface 789
– train envelope 282
– velocity 207
wave propagation
– in fluids 215
– in solids 217
– nonlinear Schrödinger equation 283
waveform 520, 544
– binary form 520
– envelope 550

– non-repetitive 546
– periodic 544
– sawtooth 544, 545
– square 544, 545
– symmetry 506
– triangular 544, 545
wavefront steepening 264
wavefronts 1055
waveguide filter 731
wavelength 210
wavelet 714, 728
– transform 729
wave-motion correction 1034
waves
– finite-amplitude 264
– thermoviscous 268
wave-table synthesis 718
weak refraction 133
Weber's law 472
WFUMB (World Federation for Ultrasound in Medicine and Biology) 895
white noise 527
Wiener–Khintchine relation 511
wind instruments 601, 637
window 1083
windscreen 1064
wolf-note 557
wood
– elastic constants 588
working standard (WS) 1043
World Federation for Ultrasound in Medicine and Biology (WFUMB) 895
WS (working standard) 1043

X

XOR 526
– operation on 526
xylophone 649, 650

Z

zeroes 726
zither 562
z-transform
– convergence 524
– inverse 525
z-transform pairs 524